ISCAS '98

Proceedings of the 1998 IEEE International Symposium on Circuits and Systems

May 31 - June 3, 1998
Monterey Conference Center
Monterey, CA

Volume 6 of 6

Proceedings Of The IEEE 1998 International Symposium on Circuits and Systems

Copyright and Reprint Permission: Abstracting is permitted with credit to the source. Libraries are permitted to photocopy beyond the limit of U. S. Copyright law for private use of patrons those articles in this volume the carry a code at the bottom of the first page, provided the per-copy fee indicated in the code is paid through the Copyright Clearance Center, 27 Congress Street, Salem, MA 01970. Instructors are permitted to photocopy isolated articles for non-commercial classroom use without fee. For other copying, reprint, or republication permission, write to IEEE Copyright Manager, IEEE Service Center, 445 Hoes Lane, P.O. Box 1331, Pisataway, NJ 08855-1331. All rights reserved. Copyright 1998 for the Institute of Electrical and Electronics Engineers. Inc.

IEEE Catalog Number: 98CH36187

Library of Congress: 80-646530

ISBN Softbound: 0-7803-4455-3
      Casebound: 0-7803-4455-3
      Microfiche: 0-7803-4455-3
      CD-ROM: 0-7803-4455-3

Additional Copies of this Publication are Available from:

IEEE Service Center
445 Hoes Lane
P.O. Box 1331
Picatway, NJ 08855-1331
1-800-678-IEEE

Welcome to ISCAS' 98 and Monterey

On behalf of the Organizing Committee, it is our pleasure to invite and welcome you to the 1998 IEEE International Symposium on Circuit and Systems (ISCAS'98), to be held in the beautiful and historic city of Monterey, California. The 1998 ISCAS, sponsored by the IEEE Circuits and Systems Society and hosted by the Naval Postgraduate School of Monterey, CA, will be held at the Monterey Conference Center in conjunction with the Monterey Marriott and Doubletree Hotels, from May 31 through June 3, 1998.

The technical program this year consists of 97 sessions that cover a broad range of technical subjects. Among these are 16 Special Invited Sessions that have been organized and selected to bring you the most current thinking and research results in the field. Two panel discussions on Education and Government sponsored research, along with three plenary presentations are also planned. In addition to the regular technical program, 13 specially organized short courses are scheduled on Sunday, May 31, preceding the start of the regular program.

A number of social events are planned throughout the conference, including a Welcoming Reception on Sunday evening, a conference reception and concert at the Naval Postgraduate School on Monday evening, and a Banquet at the Monterey Bay Aquarium on Tuesday evening. Additionally, spouse and dependent activities are also organized which include a bus tour of Monterey, Pebble Beach, Carmel, Big Sur and a wine tasting tour to Carmel Valley.

The conference is an excellent opportunity for researchers to meet in a relaxing and stimulating environment. Apart from the beauty of its coastline, the Monterey Peninsula, situated in central California (2 hours south of San Francisco), is the habitat of a rich wildlife. Sea otters, sea lions and migrating whales can be seen in their natural settings.

The Monterey Conference Center with the Monterey Marriott and Doubletree Hotels, the venue of ISCAS'98, are located in downtown Monterey a few minutes walk from the Fisherman's Wharf and Cannery Row with their numerous restaurants, the beach and the main attractions of Monterey. Other close attractions like the 17 Mile Drive with its world famous Pebble Beach golf courses and Carmel-by-the-Sea with its European style boutiques, are within a short drive. They will conspire to pull you away from the Symposium,.but you will heroically resist.....most of the time.

We sincerely hope you enjoy your visit to Monterey, and you will remember both the technical and social aspects of ISCAS' 98 as a pleasant and worthwhile experience.

Sherif Michael
General Chairman
ISCAS '98

Stanley A. White
General Co-Chairman
ISCAS '98

Message from Technical Program Co-Chairs

On behalf of the Technical Program Committee, it is our pleasure to introduce the Technical Program for ISCAS'98. This program represents the integrated efforts of many individuals, namely, the authors, special session organizers, reviewers, and the Technical Program Committee. The entire review process was carried out on-line and a significant fraction of the papers were provided in publish ready Adobe Acrobat Portable Document Files (pdf).

We received over 1200 papers from various parts of the globe. In selecting papers, the Technical Program Committee had the excruciatingly difficult task of selecting among many papers of near equivalent quality. It is tempting to draw the conclusion that if a paper was not accepted, it must have been judged a poor or unqualified paper. Although there were such papers submitted, many of the papers that we could not fit into the ISCAS'98 technical program were fine papers.

The Technical Program is comprised of 779 contributed and 140 special session papers. There are three Plenary Talks, two Panel Discussions, 18 Special Sessions and 79 Regular Sessions. With the exception of the plenary session, there will be over 15 parallel sessions each morning or afternoon. About 43% of the papers will be presented in Poster Sessions which have the advantage of allowing attendees to meet the authors personally and to discuss their papers in depth. The Technical Program Committee made no quality differentiation in selecting papers for poster and oral sessions. Papers were assigned with the sole purpose of forming coherent sessions.

We would like to take this opportunity to thank all authors who submitted papers, the reviewers, the Track Chairs, the Members of Technical Program Committee, the Special Sessions Chair, and the special session organizers; they all have contributed mightily to the success of the Technical Program for ISCAS'98.

Kenneth R. Laker and Murali Tummala
Technical Program Co-Chairs
ISCAS '98

ISCAS '98 COMMITTEE CHAIRS

General Chair
Professor Sherif Michael
Naval Postgraduate School
gchair@iscas.nps.navy.mil

General Co-Chair
Dr. Stanley White
SPACE Corporation
cochair@iscas.nps.navy.mil

Technical Program Co-Chair
Professor Murali Tummala
Naval Postgraduate School
tchair@iscas.nps.navy.mil

Technical Program Co-Chair
Professor Kenneth Laker
University of Pennsylvania
k.laker@ieee.org

Special Sessions Chair
Professor W. Kenneth Jenkins
University of Illinois,
Urbana-Champaign
special@iscas.nps.navy.mil

Plenary Sessions Chair
Professor Herschel Loomis, Jr.
Naval Postgraduate School
plenary@iscas.nap.navy.mil

Publications Chair
Dr. Philip Lopresti
Independent Counsel
publish@iscas.nps.navy.mil

Publicity Chair
Professor Lawrence P. Huelsman
University of Arizona
publicity@iscas.nps.navy.mil

Electronic Media Chair
Professor John McEachen
Naval Postgraduate School
pubmedia@iscas.nps.navy.mil

Exhibits Chair
Professor Russ Duren
Naval Postgraduate School
exhibits@iscas.iscas.nps.navy.mil

Short Courses Co-Chair
Professor Wasfy Mikhael
University of Central Florida
wbm@ece.engr.ucf.edu

Short Courses Co-Chair
Professor Philip Pace
Naval Postgraduate School
shcource@iscas.nps.navy.mil

Finance Chair
Professor David Jenn
Naval postgraduate School
finance@iscas.nps.navy.mil

Registration Co-Chairs
Professor John Ciezki and
Professor Robert Ashton
Naval Postgraduate School
register@iscas.nps.navy.mil

Local Arrangements Chair
Professor Todd Weatherford
Naval Postgraduate School
arrange@iscas.nps.navy.mil

ISCAS Steering Committee Chair
Professor Hari C. Reddy
California State University, Long Beach
hreddy@engr.csulb.edu

International Coordinators

Europe Chair
Professor George Moschytz
Swiss Federal Institute of Technology
moschytz@isi.eee.ethz.ch

South and Central America Chair
Professor Paulo Diniz
University Fed do Rio de Janeiro
diniz@coe.ufrj.br

Far East Chair
Professor Yong Ching Lim
National University of Singapore
elelimyc@leonis.nus.sg

Technical Program Committee
Co-Chairs

Murali Tummala
Naval Postgraduate School
tchair@iscas.nps.navy.mil

Kenneth Laker
University of Pennsylvania
laker@iscas.nps.navy.mil

Track 1: Analog Circuits and Signal Processing
Randall L. Geiger
Iowa State University (Chair)
rlgeiger@iastate.edu

Track 2: Circuits and Power Systems
Wai-Kai Chen
University of Illinois at Chicago (Chair)
wkchen@eecs.uic.edu

Track 3: Computer aided Design
Ibrahim Hajj
University of California, Berkeley
(on leave from University of Illinois at Urbana) (Chair)
hajj@ic.eecs.berkeley.edu

Track 4: VLSI
Gordon Roberts
McGill University, Canada (Chair)
roberts@macs.ee.mcgill.ca

Track 5: Neural Systems
Jan Van der Spiegel
University of Pennsylvania (Chair)
an@ee.upenn.edu

Track 6: Digital Signal Processing I
P.P. Vaidyanathan
California Institute of Technology (Chair)
ppvnath@systems.caltech.edu

Track 7: Digital Signal Processing II
M.N.S. Swamy
Concordia University, Canada (Chair)
swamy@ece.concordia.ca

Track 8: Multimedia and Video Technology
Bing Sheu
University of Southern California (Chair)
sheu@pacific.usc.edu

Track 9: Communication Circuits and Systems
Donald F. Gingras
SPAWAR Systems Center, San Diego (Chair)
gingras@spawar.navy.mil

Track 10: Computer Communications
Chung-Sheng Li
IBM T. J. Watson Research Center (Chair)
csli@watson.ibm.com

Track 11: Applications
Xiaoping Yun
Naval Postgraduate School
yun@ece.nps.navy.mil

Technical Program Committee Members

M.O. Ahmad (member, Track 7)
Concordia University, Canada
A. Antoniou (member, Track 7)
University of Victoria Canada
Lex. A Akers (member, Track 5)
University of Texas at San Antonio
Jacob Baker (member, Track 1)
University of Idaho
Magdy A. Bayoumi (member, Track 10)
University of Southwestern Louisiana
Kwabena A. Boahen (member, Track 5)
University of Pennsylvania
Robert Caverly (member, Track 9)
Villanova University
Tsuhan Chen (member, Track 8)
Carnegie Mellon University
Paulo S. R. Diniz (member, Track 7)
Federal University of Rio de Janeiro, Brazil
Ramesh Harjani (member, Track 4)
University of Minnesota
Srinath Hosur (member, Track 9)
Texas Instruments
Yih-Fang Huang (member, Track 6)
University of Notre Dame
Joe Kahn (Member, Track 10)
Unversity of California, Berkeley
Alex C. Kot (member, Track 9)
Nanyang Technological University, Singapore
John Lazzaro (member, Track 5)
UC Berkeley

Edward Lee (member, Track 1)
Iowa State University
Yong Ching Lim (member, Track 6)
National University of Singapore
Paul Mueller (member, Track 5)
Corticon, Inc.
Michel S. Nakhla (member, Track 3)
Carleton University, Canada
Truong Nguyen (member, Track 6)
University of Wisconsin, Madison
Keshab K Parhi (member, Track 6)
University of Minnesota
Alison Payne (member, Track 1)
Imperial College UK,
E. I. Plotkin(member, Track 7)
Concordia University, Canada
Jaime Ramirez-Angulo (member, Track 1)
New Mexico State University
Majid Sarrafzadeh (member, Track 3)
Northwestern
Yvon Savaria (member, Track 4)
Ecole Polytechnique, Canada
Rolf Schaumann (member, Track 2)
Portland State University
Martin Snelgrove (member, Track 4)
Carleton University, Canada
Ming-Ting Sun (member, Track 8)
University of Washington, Seattle
Ken Suyama (member, Track 4)
Columbia University
K. Thulasiraman (member, Track 2)
University of Oklahoma
Chung-Yu Wu (member, Track 8)
National Chiao Tung University, Taiwan

Reviewers List

We would like to acknowledge the following reviewers for their assistance in reviewing papers for ISCAS '98. Over 200 reviewers were invited to review about 1200 papers submitted for possible presentation in regular sessions of the conference. (Every effort has been made to accurately list reviewers' names. However, if you find any omissions or mistakes, please contact the Technical Program Committee).

Abcarius, John
Ahmad, M. Omair
Akbari-Dilmaghani, R.
Akers, L.
Allen, Phillip E.
Allstot, David J.
Alves, Vladimir C.
Antonio, Jose
Antoniou, Andreas
Aronhime, P.
Assi, Ali
Au, O.
Baumgarte, F.
Bayoumi, Madgi
Bazargan, Kiarash
Bechman, Gary
Beh, Kian Teik
Belabbes, Nacer E.
Belhaouane, Adel
Bellaouar, Abdellatif
Benzler, U.
Berekovic, M.
Bishop, Andrew
Black, William
Boahen, Kwabena
Bobba, Sudhakar
Brannen, Robert
Burns, Steve
Carlosena, Alfonso
Caverly, Robert
Chai, Douglas
Champac, Victor H.
Chan, Brian Lum Shue
Chao, Kwong S.
Chen, Chang Wen
Chen, Huiting
Chen, Liang-Gee
Chen, Michael
Chen, Oscal T.-C.
Chen, Tsuhan
Chen, Wai-Kai
Chen, Yiqin
Chiang, David
Chik, Raymond
Chiprout, Eli
Choma, John
Chow, Francis
Chow, Martin
Chowdhury, Nasirul
Ciezki, John G.
Cijvat, Ellie
Cong, Jason
Cong, Lin

Cotter, Martin
Cozzie, James C.
Crenshaw, Jim
da Silva, Eduardo
Dabak, Anand G.
Dai, Liang
Davies, Anthony C.
de Figueiredo, Rui
De Veirman, Geert
Dempsey, Dennis A.
Diaz-Sanchez, Alejandro
Dietrich, G. Wayne
Diniz, Paulo S. R.
Djahani, Pouyan
Djemouai, Abdelouahab
Drakakis, E. M.
Dufort, Benoit
El-Gamal, Mourad
Elwan, Hassan O.
Eskiyerli, M.
Fakotakis, Nikos
Falkowski, Bogdan
Farrahi, Amir H.
Fiez, Terri
Filanovsky, I.
Fiori, S.
Fong, Ed
Freimann, A.
Frey, Doug
Furth, Paul M.
Gadiri, Abdelkarim
Garavan, P .J.
Geiger, Randy
Genzer, David
Gharpurey, Ranjit
Giesselmann, Michael
Ginesta, Xavier
Glover, Mark
Goel, Manish
Goh, Chee-Kiang
Goknar, Izzet Cem
Gondi, Srikanth
Gonzalez-Altamirano, G.
Gordon, F. V.
Gosti, Wilsin
Granger, Eric
Grung, Bernard
Guillermo Espinosa,
Gupta, Subodh
Hajj, I.
Hajjar, Ara
Hamilton, Alister
Hang, H.-M.

Hanzinger, T.
Harb, Adnan
Hariton, Dan
Harjani, Ramesh
Harvey, Jean-Francois
Hasler, Martin
Hayatleh, Khaled
He, Lei
He, Y.
Hegde, Raj
Helfenstein, Markus
Hella, Mona
Hematy, Arman
Herrmann, K.
Hiser, Doug
Hosur, Srinath
Huang, Chung-Lin
Huang, Y. F.
Huang, Yih-Fang
Ingino Jr., Joseph
Ioinovici, Adrian
Ismail, Mohammed
Iyer, Arathi
Jen, Chein-Wei
Jenkins, W. Kenneth
Jia-lin, Shen,
Jiang, Danchi
Johns, David
Johnson, Bruce
Jonsson, Bengt E.
Jove, Xavier
Karsilayan, Aydin
Kazimierczuk, Marian K.
Kennedy, Michael P
Khali, Hakim
Khoo, KeiYong
Khoury, J.
Kim, Jonghae
Klein, Hans
Knight, John
Koneru, Satyaki
Kot, Alex
Kouwenhoven, Michiel
Kozhaya, Joseph
Kraljic, Ivan
Kropp, H.
Kuhn, William B.
Kuo, C.-C. Jay
Laker, Ken
Lan, Mao-Feng
Laurin, Jean-Jacques
Lazzaro, John
Le, Vuong Kim

Lee, David C.
Lee, Edward
Lee, Michael
Leme, Carlos Azeredo
Leong, Choon H. C.
Leung, Vincent
Li, Chung-Sheng
Li, Harry
Li, Weiping
Lidgey, John
Lim, Drahoslav
Lim, Y. C.
Liou, M. L.
Loai, Louis
Loloee, Arash
Lopez, David Baez
Loui, Alex
Low, Seo-How
Lu , Yilong
Lu, W.-S.
Luong, Howard
Lustenberger, Felix
Mactaggart, I. Ross
Magdy, Mayoumi
Mahattanakul, Jirayuth
Mahmoud, Hanan
Malik, Saqib Q.
Manetakis, K.
Manku, Tajinder
Marston, Neil
Mayaram, Karti
McCartney, Damien
Mech, R.
Meier, Thomas
Mendonca, Gelson V.
Mirzai, Bahram
Mok, Philip K. T.
Mokhtari, Mehran
Monteiro, Fabrice
Moon, Gyu
Moore, P. A.
Moreno, Moran
Moschytz, G. S.
Mow, Wai Ho
Mueller, Paul
Mulder, Jan
Murata, Tad
Naiknaware, Ravi
Nair, Kavita S.
Nakhla, Michel
Narayana, Amit
Nekili, Mohamed
Ng, A. E. J.
Ng, Wai Tung
Nguyen, Truong
Nielsen, Asbeck
Noren, K.

Nowrouzian, B.
Ohmacht, M.
Ong, Adrian
Opal, Ajoy
Papathanasiou, K.
Parhi, Keshab K.
París, Jordi
Patel, R.
Payne, Alison
Perkins, Stephen J.
Piazza, F.
Plett, Calvin
Plotkin, E.
Raje, Salil
Ramprasad, Sumant
Reuter, C.
Ribas, J.
Roberts, Gordon
Rosenbaum, Elyse
Rost, U.
Roytman, L. M.
Rumin, Nicholas
Sánchez-Sinencio, Edgar
Sansen, W.
Sarkar, Nilanjan
Sarmiento-Reyes, A.
Sarraj, Maher
Savaria, Yvon
Sawan, Mohamad
Schaumann, Rolf
Schlarmann, Mark
Schmid, Hanspeter
Schuelke, Robert
Schuppener, Gerd
Sculley, Terry
Seevinck, E.
Serdijn, Wouter A.
Sewell, J.I.
Shanbhag, Naresh
Shen, G.B.
Shenai, Krishna
Shilman, Michael
Shiu, Da-shan
Shpak, Dale
Silva Martinez, Jose
Smy, Tom
Snelgrove, Martin
Soma, Mani
Song, B.
Spalding, George R
Sriram, S.
Stouraitis, Thanos
Suder, Ed
Sun, Ming-Ting
Suyama, K.
Swamy, M. S. S.
Tam, Derek

Tarr, Garry
Thanachayanont, Apinunt
Thanos, Stouraitis
Thorp, James S.
Thulasiraman, K.
Tong, Wen
Toumazou, C.
Tretz, C.
Tsai, Ching-Han
Tschanz, Jim
Tse, Michael C. K.
Tzou, Kou-Hu
Ubiergo, Gabriel F.
Vaidyanathan, P. P.
van der Woerd, A.
van Staveren, Arie
Veillette, Benoit R.
Vital, Joao
Vlach, Jiri
Wad, Paul E.
Walkey, David J.
Wang, J.
Wang, Janet Meiling
Wang, Jhing-fa
Wang, Maogang
Wang, X.-F.
Wang, Yao
Weisbin, Amy
Whiteside, Frank
Wing, Omar
Wittenburg, J.-P.
Wollborn, M.
Worapishet, Apisak
Wu, K.
Wu, Lin
Yamamoto, Yoshio
Yan, Jie
Yang, R.
Yazdanpanah, M.
Yeap, Gary
Yoh, Gilbert
Younis, Ahmed
Yu, Baiying
Yu, Chong-Gun
Yu, Qingjian
Yun, Xiaoping
Zaghloul, Mona
Zefran, Milos
Zeng, B.
Zeng, Fan-Gang
Zhang, Chengjin
Zhang, Q. J.
Zhou, Joe P
Zhu, Weiping
Zohios, Jerasimos
Zukowski, C.

	Sun	Monday, June 1		Tuesday, June 2			Wednesday, June 3	
		MA	MP	TA/TB	TP		WA/WB	WP
1	Short Courses	Parameter Estimation	Multidimensional Signal Processing	Filter Banks & Wavelets	Adaptive SP II	Panel Session I - Government Funded Research; Panel Session II – Teaching Circuits & Systems in the 21st Century	Adaptive Signal Processing III	Wavelets: Impl & Applications
2		Single-rate & Multirate Filters	**Opti of Subband Coders based on the Input**	**Model., Anal. & Design Switching Converters**	**Coding of Arbit.-ShapObjects**		**Steerable Filters & Applications**	**3D Data Modeling & Imaging**
3		**NN for Intelligent Signal Processing**	**Memory, Adaptation & Learning**	NNI: Algorithms & Computation	NNII: Implement Issues		Nets Biological Computing & Fuzzy Logic	Cellular NN
4		Image Processing & Coding	Multimedia Systems & Processing	Speech &Video Processing	Image & Video Processing		Image & Video Processing III	Hi-Level Synthesis / Gate Arrays
5		Signal Processing For Communications I	Equalization/ Modulation/ Decoding	**Communicating with Chaos I**	SP for Comms II		Wireless/Mobile Communications	**Arch, Alg & Imp Wireless Comm**
6		Low-Power IC Techniques	Circuit Techniques For Wireless Applications	/ Prog. Logic Device	C&S for Comm Nets I		Deep-Sub Dig Circuit Issues / DSP Arch	Topics in Analog & Digital Test
7		Chaos & Application	Linear Circuits	Nonlinear Networks & Systems	Analog VLSI		Power Electronics / **Impr ADCs**	**Controlling Bifurcation & Chaos**
8		Data Converters	Continuous-Time Filters	Amplifiers I			Amplifiers II	Current Mode Techniques
9		Sym. Anal. Meths &Appl. to Anal. Circuit Design	Hi-Speed Communications Circuits	Logdomain Filters	**Comm with Chaos II**		Switched-Capacitor Techniques	Amplifier Building Blocks
10		Low Power Digital Circuit Design	Interconnect Modeling & Design	Oversampled Data Converters	Circuit Simulation		**Communicating with Chaos III**	Oversampled & SD Tech II
11		VLSI Circuits for MM Signal Processing	**Multi-Sensor Data Fusion**	**DSP for Hearing Aids**	Robotics		VLSI Layout & Timing	Device Modeling
12		VLSI Digital Circuits	Power Distribution Systems	Analog Circuits Design	Feedback Systems		**Sys & Appls Next Generation Internet**	Filters & Electronics Circuits
13		**Communications Circuits**	**Image & Video Processing I**	**Adapt. SP I / DSP Implt**	**MM Processing**		**Digital Filter Des & Implementation**	**MM/ Comm / C&S Comm II**
14		**Circuits & Power Systems I**	**VLSI I**	**VLSI II / Circuit & PS**	**Oversampl & S-D I**		**VLSI Arch, Algorithms & CAD**	**Neural Networks**
15		**Analog Filters**	**Analog & Signal VLSI Design**	**CAD I**	**Sensors & Circuits**		**CAD II**	**Analog Circuits & Systems**

Special **Poster**

TABLE OF CONTENTS

VOLUME 1

MAA8 – DATA CONVERTERS
Chair: William Black
Iowa State University

Fast Pipelined A/D Converter in CMOS Technology — 1
Park, Sangbeom, *Texas Instruments*

A 200 MHz 6-bit folding and interpolating ADC in 0.5-um CMOS — 5
Jiang, Xicheng; Wang, Yunti; and Willson, Jr., Alan N., *UCLA*,

A CMOS Current-Mode Pipeline ADC Using Zero-Voltage Sampling Technique — 9
Luong, Howard Cam and Hui, Ronny C. C., *Hong Kong University of Science and Technology*

A Comparison of Monolithic Background Calibration In Two Time-Interleaved Analog-to-Digital Converters — 13
Dyer, Kenneth, *University of California at Davis*

Improving the linearity in high-speed analog-to-digital converters — 17
Gatti, Umberto, *Italtel S.p.A.*

On the Dynamic Performance of High-Speed ADC Architectures — 21
Gustavsson, Mikael, *Linkoping University*, and Tan, Nianxiong, *Microelectronics Research Center*

Modelling of CMOS Digital-to-Analog Convertors for Telecommunication — 25
Wikner, J. Jacob, *Linkoping University*, and Tan, Nianxiong, *MERC, Ericsson Components AB*

Distortion Analysis of Switched-Current Circuits — 29
Helfenstein, Markus and Moschytz, George S., *Swiss Federal Institute of Technology*

MAA15 – Analog Filters
Chair: Philip E. Allen
Georgia Institute of Technology

Distortion analysis of MOSFETs for application in MOSFET-C circuits — 33
Schneider, Márcio Cherem, and Galup-Montoro, Carlos, *Universidade Federal de Santa Catarina*

Design and implementation of an algorithmic S2I switched-current multiplier — 37
Pineda de Gyvez, Jose, *Texas A&M University*; and Manganaro, Gabriele, *Texas Instruments, Inc.*

Simulation of Coupled Tuned Circuits Using CFOAs — 41
Aronhime, Peter, and Deng, Jie, *University of Louisville*, and Maundy, Brent, *University of Calgary*

Improved Fully Differential Circuits Using Hybrid Structures — 45
Walker, Paul D., *Silicon Systems*, and Green, Michael M., *University of California, Irvine*

Feasible Designs for High Order Switched-Current Filters — 49
Sewell, John Isaac, and Ng, Andrew, *University of Glasgow*, Lo, Chun-Keung, Lo, Chun-Keung, and Lo, Chun-Keung, *The Hong Kong University of Science and Technology*, and Li, Dongju, *Tokyo Institute of Technology*

Accurate CMOS Switched-Current Divider Circuits — 53
Wey, Chin-Long, *Michgan State University*, and Wang, Jin-Sheng, *Michigan State University*

Fundamental Frequency Limitations in Current-Mode Sallen-Key Filters — 57
Schmid, Hanspeter, and Moschytz, George S., *Swiss Federal Institute of Technology*

BiCMOS OTA for high Q very high frequency continuous-time bandpass filters — 61
Minot, Sophie, and Degrugillier, Dominique, *Enst de Bretagne*

Automatic Tuning of Frequency and Q-Factor of Bandpass Filters Based on Envelope Detection — 65
Karsilayan, Aydin Ilker, *Portland State University*

A CMOS Multiplier/Divider based on Current Conveyors — 69
Cattet, Stephane, *CPE Lyon*

Reducing Spread Resistance in High Q State Variable Filters — 72
Báez-López, David J.M., Guillermo, Espinosa F. V., and Silva-Martinez, Jose, *Instituto Nacional de Astrofísica, Optica y Electrónica*

Low-Voltage S2I and S3I Cells for Sigma-Delta Signal Processing — 76
Musil, Vladislav, and Simek, Petr, *Technical University of Brno*,

Two-Step Current-Memory Cells with Optimal Dynamic Range for Advanced CMOS Technologies — 80
Kaiser, Andreas, *IEMN-ISEN*

A 4-Transistor Euclidean Distance Cell for Analog Classifiers — 84
Cilingiroglu, Ugur, Aksin, and Devrim Yilmaz, *Istanbul Technical University & ETA ASIC Design Center*

Multiple-Input Translinear Element Networks — 88
Minch, Bradley A., and Hasler, Paul, *Georgia Institute of Technology*, and Diorio, Chris. *University of Washington*

UCM - Universal Current-mode Structures — 92
Galvez-Durand, Federico, *Universidade Federal do Rio de Janeiro*

Generation of canonic multiple current output OTA sinusoidal oscillators with non-interacting controls — 96
Tao, Yufei, and Fidler, J. Kel, *University of York*

Very Low-Distortion Fully Differential Switched-Current Memory Cell — 100
Martins, Jorge Manuel, *INESC/IST*

Reliable Analog Bandpass Signal Generation — 104
Veillette, Benoit R., and Roberts, Gordon W., *McGill University*

Very Low-Distortion Fully Differential Switched-Current Memory Cell — 108
Martins, Jorge Manuel, *INESC/IST*

Phase-Tunable CMOS Triode Transconductor — 112
Jun, Sibum, *LG Semicon Co., Ltd.* and Ahn, Su Jin, *POSTECH*

TABLE OF CONTENTS

Efficient Design of Switched-Current Lowpass Elliptic Wave Filters Using Bruton Transformation — 115
Lancaster, Jason David, Al-Hashimi, Bashir, and Moniri, Mansour, *Staffordshire University*

Ladder Decompositions for Wideband Switched-Current Filter Applications — 119
Sewell, John Isaac, and NG, Andrew *University of Glasgow*

MPA8 – Continuous-Time Filters
Chair: Jamie Ramirez-Angulo
New Mexico State University

Design of a CMOS Fully-differential Continuous-Time Tenth-Order Filter Based on IFLF Topology — 123
Chiang, David, H., *Portland State University*

A Sixth-Order UHF Bandpass Filter Using Silicon Bipolar Active Inductors — 127
Roberts, Gordon W., and Leong, Choon Haw, *McGill University*

An Autozeroing Floating-Gate Bandpass Filter — 131
Hasler, Paul *Georgia Institute of Technology*, and Minch, Bradley A., *Cornell University*, and Diorio, Chris, *University of Washington*

A Novel Loss Control Feedback Loop for VCO Indirect Tuning of RF Integrated Filters 135
Li, Dandan, *Columbia University*

A 2V Low-Distortion CMOS Biquadratic cell — 139
Python, Dominique, *Swiss Federal Institute of Technology (EPFL)*

Analysis of Noise and Interference in Companding Signal Processors — 143
Krishnapura, Nagendra, and Tsividis, Yannis, *Columbia University*, and Toth, Laszlo, *TKU, Budapest*

Fundamental Limits to the Dynamic Range of Integrated Continuous-Time Integrators — 147
Moreira, Jose Pedro, *INESC*, and Verhoeven, Chris J. M., *Delft University of Technology*

Impedance Scalers for IC Active Filters — 151
Silva-Martínez, José, and Vázquez-González, Alejandro, *Instituto Nacional de Astrofísica, Optica y Electrónica*

MPA9 – High Speed Communication Circuits
Chair: R. Jacob Baker
University of Idaho

A Phase Detector with No Dead Zone And a Very Wide Output Voltage Range Chargepump — 155
Ahola, Rami, Halonen, Kari, and Routama, Jarkko, and Lindfors, Saska, *Helsinki University of Technology*

A 150mbit/s cmos clock recovery pll including a new improved phase detector and a fully integrated fll — 159
Routama, Jarkko, Antero, Koli, Kimmo, and Halonen, Kari, *Helsinki University of Technology*

Low Noise Clock Synthesizer Design Using Optimal Bandwidth — 163
Kim, Beomsup, Lim, Kyoohyun, and Park, Chan-Hong, *KAIST*

A Low Jitter 1.25GHz CMOS Analog PLL for Clock Recovery — 167
Wu, Lin, and Black, William C., *Iowa State University*

A Modified Costats Loop for Clock Recovery and Frequency Synthesis — 171
Amourah, Mezyad M., *Iowa State Univ.*

Effects of Random Jitter on High-Speed CMOS Oscillators — 176
Chen, Yiqin, and Geiger, Randall, *Iowa State University*

A novel ring-oscillator with a very small process and temperature variation — 181
Routama, Jarkko, Antero, Koli, and Kimmo, Kari, *Helsinki University of Technology*

Reduction of the 1/f noise induced phase noise in a CMOS ring oscillator by increasing the amplitude of oscillation. — 185
Gierkink, Sander L.J., *MESA Research Institute, University of Twente,*

MPA15 – Analog and Mixed-Signal VLSI Design
Chair: Sherif Embabi
Texas A & M University

Design of a Delta-Sigma Modulated Switching Power Supply — 189
Dunlap, Steven K., and Fiez, Terri, *Washington State University*

Theory and Implementation of a Gaussian Decay Low-pass Filter — 193
Harris, John G., and Pu, Chiang-Jung, *U. of Florida*

Circuit Tolerances and Word Lengths in Overlap Resolution — 197
Saed, Aryan, Jullien, Graham A., Ahamdi, Majid, and Miller, William C., *VLSI Research Group, University of Windsor*

A Design-for-Testability Technique for Detecting Delay faults in Logic Circuits — 201
Raahemifar, Kaamran, and Ahmadi, Majid, *University of Windsor*

About the Demodulation of PWM-Signals with Applications to Audio — 205
Bresch, Helmut, Streitenberger, Martin, and Mathis, Wolfgang, *Otto-von-Guericke Universitaet Magdeburg*

BiCMOS add-compare-select units for Viterbi decoders — 209
Demosthenous, Andreas, *University College London*

Deterministic Phase Jitter in Multi-Phase CMOS Ring Oscillators due to Transistor Mismatches — 213
Koneru, S., Chen, Y., Geiger, R., and Lee, E. *Iowa State*

TABLE OF CONTENTS

A winner-take-all network for large-scale analogue vector quantisers — 217
Demosthenous, Andreas, *University College London*

Noise Analysis of an Oscillator with an Mth-order Filter and Comparator-type Nonlinearity — 225
Dec, Aleksander M, *Columbia University*

A 4 GHz Differential Transimpedance Amplifier Channel for a Pulsed Time-of-Flight Laser Radar — 229
Pennala, Riku-Matti, *University of Oulu, Electronics Laboratory*

Design of Low Jitter PLL for Clock Generator with Supply Noise Insensitive VCO — 233
Lee, Chang-Hyeon, *USC,* and Cornish, Jack, and McClellan, Kelly, *ADM-Tek* and Choma, Jr,John, *USC*

Novel Palmo Techniques for Electronically Programmable Mixed Signal — 237
Papathanasiou, Konstandinos, and Hamilton, Alister, *University of Edinburgh*

A Constant Input Transconductance and Rail-to-Rail Input/Output Swing SiC CMOS OPAMP — 241
Chen, Jian-Song, *Purdue University* and Kornegay, Kevin T., *Cornell University*

CMOS analog multipliers based on a class-B squaring circuit. — 245
Gnudi, Antonio, Pellegrini, Aurelio, Baccarani, Giorgio, *DEIS, Universita' di Bologna*

A Baseband Pulse Shaping Filter for Gaussian Minimum Shift Keying — 249
Pavan, Shanthi, *Texas Instruments*, and Krishnapura, Nagendra, *Columbia University,* and Mathiazhagan, Chakravarthy and Ramamurthi, Bhaskar, *Indian Institute of Technology*

A 10-bit 220-MSample/s CMOS Sample-and-Hold Circuit — 253
Waltari, Mikko Eljas and Halonen, Kari, *Electronic Circuit Design Laboratory, Helsinki University of Technology*

A Novel Technique for Noise Reduction in CMOS Subsamplers — 257
Lindfors, Saska J., Parssinen, Aarno T., Ryynanen, Jussi H., and Halonen, Kari A, *Helsinki University of Technology*

A Novel Self-Error Correction Pulse Width Modulator for a Class D Amplifier for Hearing Instruments — 261
Tan, Meng Tong, *Nanyang Technological University*

An Approach to the Design of Low-Voltage SC Filters — 265
Giustolisi, Gianluca, *DEES - Università di Catania*

Design of a Micropower Signal Conditioning Circuit for a Piezoresistive Acceleration Sensor — 269
Silveira, Fernando, Baru, Marcelo Daniel, Picuñ, Gonzalo, and Arnaud, Alfredo, *Instituto de Ingenieria. Electrica*

A BiCMOS Current-Mode Track-and-Hold — 273
Vasseaux, Tony; Loumeau, Patrick; and Oliaei, Omid, *Ecole Nationale Superieure des Telecommunications*

Analog Implementation of Ratio Spectrum Computation — 277
Harris, John G., Harris, John G. and Lim, Shao-Jen, *University of Florida*

Analysis and Two Proposed Design Methodologies for Optimizing Power Efficiency of a Class D Amplifier Output Stage — 281
Tan, Meng Tong, *Nanyang Technological University*

TAA8 – Amplifiers I
Chair: Edward Lee
Iowa State University

A 3.3-V CMOS Wideband Exponential Control Variable-Gain-Amplifier — 285
Huang, Po-Chiun, Chiou, Li-Yu, and Wang and Chorng-Kuang, *National Central University*

Systematic Generation Of Transconductance Based Variable Gain Amplifier Topologies — 289
Klumperink, Eric A. M., *MESA Research Institute, University of Twente*

Low Noise Current-Mode CMOS Transimpedance Amplifier for Giga-bit Optical Communication — 293
PARK, SUNG MIN, *Imperial College of Science, Technology and Medicine*

A CMOS Automatic Gain Control for Hearing Aid Devices — 297
Silva-Martinez, Jose, Salcedo-Suner, and Jorge, *Instituto Nacional de Astrofisica Optica y Electronica*

An Amplifier Design Methodology Derived from a MOSFET Current-Based Model — 301
Schneider, Márcio Cherem, Pinto, Rodrigo and Luiz de Oliveira, *Universidade Federal de Santa Catarina*

A 3-V CMOS Optical Preamplifier with dc Photocurrent Rejection — 305
Phang, Khoman, Johns, David A., *University of Toronto*

TAA9 – Logdomain Filters
Chair: Alison Payne
Imperial College, London, England

A Fully-Programmable Analog Log-Domain Filter Circuit — 309
Roberts, Gordon W., and Hematy, Arman, *MACS Laboratory, McGill University*

Fully differential class-AB log-domain integrators — 313
WU, J. and El-Masry, Ezz I., *DalTech, Dalhousie University*

Mulitple Feedback Log-Domain Filters — 317
Drakakis, E.M.; Payne, A.J., and Toumazou, C., *Imperial College*

TABLE OF CONTENTS

Synthesis of Distortion Compensated Log-Domain Filters Using State Space Techniques — 321
Frey, Douglas, *Lehigh University*

An Auto-Biased 0.5um CMOS Transconductor for Very High Frequency Applications — 325
Garrido, Nuno Miguel de, Franca, and Jose Epifaneo da, *Instituto Superior Tecnico*

Noise in High-Order Log-Domain Filters — 329
Punzenberger, Manfred, *Swiss Federal Institute of Technology,(EPFL)* and Enz, Christian C., *Rockwell Semiconductor Systems,*

Low-Voltage Current-Mode CMOS IC Continuous-Time Filters with Orthogonal W0-Q Tuning — 333
Shana'a, Osama, *Stanford University* and Schaumann, Rolf, *Portland State University*

Analysis of noise in translinear filters — 337
Mulder, Jan, *Delft University of Technology*

TAA10 – Oversampled Data Converters
Chair: Raymond Chik

Micro-power 10-bit sigma-delta A/D-converter — 341
Rapakko, Harri Antero, *Nokia Mobile Phones*

An Area-Efficient Sigma-Delta DAC with a Current-Mode Semidigital IFIR Reconstruction Filter — 344
Kim, Jae-Wan, *Korea University,* Min, and Byung-Moo, *LG semicon Co., Ltd. ,and* Yoo, Jang-Sik, and Kim, Soo-Won, *Korea University*

A 1V Second-Order Sigma-Delta Modulator — 348
Ma, Stanley Jeh-Chun, and Salama, C. Andre T., *MOSAID Technologies Inc./University of Toronto*

Mismatch-Shaping DAC for Lowpass and Bandpass Multi-Bit Delta-Sigma Modulators — 352
Shui, Tao, *Oregon State University,* and Schreier, Richard, *Analog Devices Inc. and* Hudson, Forrest, *Veris Industries, Inc.*

Mismatch Cancellation for Double-Sampling Sigma-Delta Modulators — 356
Yu, Li, *Carleton University*

Power Optimization of Delta-Sigma Analog-to-Digital Converters Based on Slewing and Partial Settling Considerations — 360
Fiez, Terri, and Naiknaware, Ravindranath, *Washington State University*

Nonuniform-to-Uniform Decimation for Delta-Sigma Frequency-to-Digital Conversion — 365
Galton, Ian, *University of California, San Diego*

A Single-path Multi-bit DAC for LP Delta-Sigma A/D Converters — 369
Louis, Loai, and Roberts, Gordon W., *MACS Laboratory, McGill University*

TAA10 – Oversamplified and Sigma-Delta Techniques I
Chair: Terri Fiez
Washington State University

Harmonic Distortion in Switched-Current Sigma-Delta Modulators due to Clock Feedthrough — 373
Martins, J. M., and Dias, V. F., *INESC*

Analysis of Non-uniform Sampling Effects in Sigma-Delta Modulated Signals — 377
Lee, Eel-Wan, and Chae, Soo-Ik, *Seoul National University*

New Analytical Model of Interpolation Waveforms in Time-averaging Interpolative Digital to Analogue Convereters — 381
Moniri, Mansour, *Staffordshire University*

Encoding Hidden Data Channels in Sigma Delta Bitstreams — 385
Sandler, Mark B., and Magrath, A.J., *King's College London*

An Architecture of Delta Sigma A-to-D Converters Using A Voltage Controlled Oscillator as a Multi-Bit Quantizer — 389
Iwata, Atsushi, Sakimura, Noboru, Nagata, Makoto, and Morie, Takashi, *Hiroshima University*

Current Mode Approach To Sigma-Delta Modulators — 393
Wawryn, Krzysztof, and Suszynski, Robert, *Technical University of Koszalin*

A Folding ADC Employing a Robust Symmetrical Number System with Gray-Code Properties — 397
Pace, P.E., *Naval Postgraduate School,* and Styer, D., *University of Cincinnati,* and Akin, I.A., *Naval Postgraduate School*

Performance Analysis of Low Oversampling Ratio Ratio Sigma-Delta Noise Shapers for RF Applications — 401
Gothenberg, Andreas, and Tenhunen, Hannu, *Royal Inst. of Technology,*

Adaptive compensation of analog circuit imperfections for cascaded delta-sigma ADCs — 405
Sun, Tao, *Oregon State University*

A Two-Loop Third-Order Multistage Delta-Sigma Frequency To Digital Converter — 408
Filiol, Norman M.,*Carleton University,*

An Oversampled A/D Converter Using Cascaded Fourth Order Sigma-Delta Modulation and Current Steering Logic — 412
Miao, Guoqing, *Crosslink Semiconductor, Inc.,* and Yang, Howard C., *Newave Semiconductor Corp.,* and Tang, Pushan, *Fudan University*

A 12-bit, 100ns/b, 1.9mW CMOS Switched- — 416

TABLE OF CONTENTS

Current Cyclic A/D Converter
Wey, Chin-Long, and Wang, Jin-Sheng, *Michigan State University*

TPB2 – Panel Discussion II Teaching of Circuits and Systems in the 21st Century

Analog Signal Processing: A Better Way to Teach   420
Munson Jr., David C., and Jones, Douglas L., *University of Illinois*

Teaching Circuits and Electronics to First-Year Students   424
Tsividis, Yannis, *Columbia University*

Teaching, Circuits, Systems, and Signal Processing   428
Diniz, Paulo S. R., *Federal University of Rio de Janeiro*

Introductory Circuits: A Course in Crisis   432
Davis, Artice M., *San Jose State University*

WAA8 – Amplifiers II
Chair: Willy Sansen
Katholic University of Leuven

A 1.2 V CMOS Op-Amp with High Driving Capability   436
Ferri, Giuseppe, *Universita' di L'Aquila*

A 3-V RF CMOS Bandpass Amplifier Using An Active Inductor   440
Thanachayanont, Apinunt, *Imperial College of Science, Technology and Medicine*

A low voltage Op-Amp with Constant-Gm Rail-to-Rail Input and High Swing Self-Biasing Super Cascode Output Stage   444
Asmanis, Georgios, *Rockwell Semiconductor Systems*

A Low-Voltage CMOS Rail-to-Rail Class-AB Input/Output OpAmp with Slew-Rate and Settling Enhancement   448
Lin, Chi-Hung, and Ismail Mohammed, *Ohio State University*

A 1.6V 80uW Rail-to-Rail Constant-Gm Bipolar Adaptive Biased Op-Amp Input Stage   452
Cardarilli, G.C., and Re, M., *University of Rome "Tor Vergata"*, and Ferri, G., *University of L'Aquila*

A High-Drive High-Gain CMOS Currrent Operational Amplifier   456
Pennisi, Salvatore, *DEES - University of Catania*

High Accuracy, High Speed Voltage-Follower   460
Lidgey, John, *Oxford Brooks University*, and Su, Wenjun, *Qualcomm Inc.*

Feedforward Compensation Techniques in the Design of Low Voltage Opamps and OTAs   464
Setty, Suma, *Ericsson Components*

WAA9 – Switched Capacitor Techniques
Chair: Marcus Helfenstein
Swiss Federal Institute of Technology

Switched-Capacitor Decimation Filter for 0.8 mm CMOS   468
Petraglia, Antonio, *Federal University of Rio de Janeiro*

A Switched-Capacitor N-Path Decimating Filter   472
Neves, Rui Ferreira, and Franca, Jose E., *Instituto Superior Tecnico*

CMOS Switched-Opamp Based Sample-and-Hold Circuit   476
Harjani, Ramesh, and Dai, Liang, *University of Minnesota*

Companding Switched Capacitor Filters   480
Krishnapura, Nagendra, and Tsividis, Yannis, *Columbia University*, and Nagaraj, Krishnaswamy, *Texas Instruments, New Jersey*; and Suyama, Ken, *Columbia University*

A New Look at Analogue Computing Using Switched Capacitor Circuits   485
Sobhy, Mohamed Ibrahim, *The University of Kent at Canterbury*, and Makkey, Mostafa, *The University of Assiut*

Clocking Scheme for Switched-Capacitor Circuits   488
Steensgaard-Madsen, Jesper, *Technical University of Denmark*

Settling Time Design Considerations for SC Integrators   492
Chilakapati, Uma, and Fiez, Terri, *Washington State University*

Recursive Switched-Capacitor Hilbert Transformer   496
Petraglia, Antonio, *Federal University of Rio de Janeiro*

WAB7 – Digital Techniques for Improving Delta-Sigma ADCS
Chair: Gabor C. Temes
Oregon State University

Some Observations on Tone Behavior in Data Weighted Averaging   500
Chen, Feng, *Texas Instruments*

A Reduced-Complexity Mismatch-Shaping DAC For Delta-Sigma Data Converters   504
Jensen, Henrik, and Galton, Ian, *University of California, San Diego*

Blind On-Line Digital Calibration of Multi-Stage Nyquist-Rate and Oversampled A/D Converters   508
Cauwenberghs, Gert, *Johns Hopkins University*

WPA8 – Current Mode Techniques
Chair: Arash Loloee
Texas Instruments

TABLE OF CONTENTS

A 50th Order Elliptic LP-Filter Using Current Mode Gm-C Topology   512
Kosunen, Marko, Koli, Kimmo, and Halonen, Kari, *Helsinki University of Technology*

A Low Mismatch Sensitivity Fully-Balanced Current-mode Integrator   518
Sanchez-Sinencio, Edgar, *Texas A&M Univ*

A CMOS current-mode multiplier/divider circuit   520
Baturone, Iluminada, *Instituto de Microelectronica deSevilla (IMSE-CNM*

Sampling Jitter in High-Speed SI Circuits   524
Jonsson, Bengt E., *Ericsson Radio Systems AB*

The Multiple-Input Translinear Element: A Versatile Circuit Element   527
Minch, Bradley A., *Cornell University*, and Hasler, Paul, *Georgia Institute of Technology* and Diorio, Chris, *University of Washington*

Very low charge injection switched-current memory cell   531
Leelavattananon, Kritsapon, *Imperial College of Science, Technology and Medicine*

An improved CMOS offset-compensated current comparator for high-speed applications   535
Worapishet, Apisak, *Imperial College of Science, Technology and Medicine*

Wideband Current-Mode Absolute Value Circuits   539
Lidgey, John, and Hayatleh, K., *Oxford Brookes University* and Porta, S. *Universidad Publica de Navarra*

WPA9 – Amplifier Building Blocks
Chair: Jose Silva-Martinez
National Institute for Astrophysics, Optics, and Eng.

A Current Driven, Programmable Gain Differential Pair Using MOS Translinear Circuits   543
Conti, Massimo, Crippa, Paolo, Guaitini, Giovanni, Orcioni, Simone, and Turchetti, Claudio, *University of Ancona*

New High-Precision Circuits For On-Chip Capacitor Ratio Testing and Sensor Readout   500
Wang, Bo, *Oregon State University*

Voltage Clamping Current Mirrors with 13-decades Gain adjustment Range Suitable for Low Power MOS/Bipolar Current Mode Signal Processing   551
Linares-Barranco, Bernabe and Serrano-Gotarredona, Teresa, *National Microelectronics Center* and Andreou, Andreas G., *The Johns Hopkins University*

Matching Performance of Current Mirrors with Arbitrary Parameter Gradients Through the Active   555
Lan, Mao-Feng, Geiger, Randall, *Iowa State Univeristy*

An Active Tuning and Impedance Matching Element   559
Lapinoja, Mikko J., and Rahkonen, Timo E., *University of Oulu*

Voltage Controlled Resistor for Mismatch Adjustment in Analog CMOS   563
Yu, Baiying, *Iowa State University*

Harmonic Distortion in CMOS Current Mirrors   567
Bruun, Erik, *Technical University of Denmark*

WPA10 – Oversampled and Sigma-Delta Techniques II
Chair: Phillip E. Pace
Naval Postgraduate School

A Bandpass Sigma-Delta Demodulator   571
Keady, Aidan G., and Lyden, Colin, *National Microelectronics Research Centre*

A 5 GHZ continuous time Sigma-Delta modulator implemented in 0.4um InGaP/InGaAs HEMT technology   575
Olmos, Alfredo, *Laboratorio de Sistemas Integraveis Univ. Sao Paulo*

A 50 MHz Continuous-Time Switched-Current Sigma-Delta Modulator   579
Luh, Louis, Choma, John, and Draper, Jeffrey, *University of Southern California*

A Multi-Bit Sigma-Delta Modulator with Interstage Feedback   583
Fang, L., and Chao, K.S., *Texas Tech University*

Approaches to Simulating Continuous-Time Delta Sigma Modulators   587
Cherry, James A., Snelgrove, W. Martin, *Carleton University*

Architectural Coefficient Synthesis for the Implementation of Optimal Higher-Order Delta-Sigma Analog-to-Digital Converters   591
Fiez, Terri, and Naiknaware, Ravindranath, *Washington State University*

Loop Delay and Jitter in Continuous-Time Delta Sigma Modulators   596
Cherry, James A., and Snelgrove, W. Martin, *Carleton University*

Digital Correction of Non-Ideal Amplifier Effects in the MASH Modulator   600
Davis, Alan J., and Fischer, Godi, *University of Rhode Island*

VOLUME 2

MAA6 – Low-Power IC Techniques
Chair: Eby. G. Freidman
University of Rochester

Signal Coding for Low Power: Fundamental Limits and Practical Realizations   1

TABLE OF CONTENTS

Ramprasad, Sumant; Shanbhag, Naresh R.; and Hajj, Ibrahim N. *University of Illinois at Urbana-Champaign*

Finite-State Machine Partitioning of Low Power — 5
Benini, Luca and Giovanni, De Micheli, *Stanford University*; and Vermeulen, Frederik, *IMEC*

Use of Charge Sharing to Reduce Energy Consumption in Wide Fan-in Gates — 9
Khellah, Muhammad M., and Elmasry, Mohamed I., *University of Waterloo, Canada*

Low Power/Low Swing Domino CMOS Logic — 13
Rjoub, Abdoul and Koufopavlou, Odysseas, *University of Patras*; Nikolaides, S., *University of Thessaloniki*

Power Optimization of Combinational Modules Using Self-Timed — 17
Mota, Antonio S. and Monteiro, Jose C., *IST-INESC*; Oliveira, Arlindo L., *Cadence Europ0ean Labs/IST-INESC*

A Mathematical Approach to a Low Power FFT Architecture — 21
Stevens, Kenneth S., *Intel* and Suter, Bruce, *Rome Labs, US Air Force*

A Configurable 32nd Order Low Voltage Low Power Digital Filter for Portable Communications Systems — 25
Suvakovic, Dusan and Salama, C. Andre T., *University of Toronto*

Optimal Design of Low Power Nested Gm-C Compensation Amplifiers Using a Current-based MOS Transistor Model — 29
Xie, X.; Schneider, M.C.; Embabi, S.H.K.; and Sanchez-Sinencio, Edgar, *Texas A&M University*

MAA12 – VLSI Digital Circuits
Chair: Yvon Savaria
Ecole Polytechnique

Design of Low Power Differential Logic Using Adiabatic Switching Technique — 33
Lo, Chun-Keung and Chan, Philip C.H., *The Hong Kong University of Science and Technology*

Ultra Low-Voltage Digital Floating-Gate UVMOS (FGUVMOS) Circuits — 37
Berg, Yngvar; Wisland, Dag T.; Lande, Tor Sverre; and Mikkelsen, Sindre, *University of Olso*

Single Ended Swing Restoring Pass Transistor Cells for Logic Synthesis and Optimization — 41
Pihl, Johnny, *Royal Military Institute of Technology*

Edge Reversal-Based Asynchronous Timing Synthesis — 45
Franca, Felipe Maia Galvao; Alves, Vladimir Castro; and Granja, Edson do Prado, *COPPE/UFRJ*

A New True-Single-Phase-Clocking (TSPC) BiCMOS Dynamic Pipelined Logic — 49
Tseng, Yuh-Kuang and Wu, Chung-Yu, *National Chiao-Tung University*

Low Voltage BiCMOS TSPC Latch for High Performance Digital Systems — 53
Nikolic, Borivoje, *University of California* and Oklobdzija, Vojin G., *Integration/UC-Davis*

Low Ringing I/O Buffer Design — 57
Carro, Luigi and Bego, Lauro Jardim, *Universidade Federal do Rio Grande do Sul*

CMOS Circuit Design of Threshold Gates with Hysteresis — 61
Sobelman, Gerald E. and Fant, Karl, *Theseus Logic, Inc.*

MPA6 – Circuit Techniques for Wireless Applications
Chair: Martin Snelgraove
Carlton University

A 1.8 GHz Subsampling CMOS Downconversion Circuit for Integrated Radio Circuits — 65
Cijvat, Ellie and Eriksson, Patrik, *Royal Institute of Technology*; Tan Nianxiong, *Ericsson Components AB, (MERC)* and Tenhunen, Hannu, *Royal Institute of Technology*

A CMOS sampled-data system for IF-to-baseband demodulation and filtering — 69
Baschirotto, Andrea, *University of Pavia*; and Denti, G. and Smori, C., *Politecnico di Milano*

1.8 GHz CMOS LNA with On-Chip DC-Coupling for a Subsampling Direct Conversion Front-End — 73
Parssinen, Aarno T.; Lindfors, Saska J; Ryynanen; *Helsinki University of Technology*, Long, Stephen I., *University of California Santa Barbara* and Halonen, Kari A., *Helsinki University of Technology*

ASIC for 1-Ghz Wide Band Monobit Receiver — 77
Pok, David and Chen, Henry Chien-In, *Wright State University*; Montgomery, C.; Tsui, B.Y.; and Schamus, John; *Wright-Patterson, AFB*

RF Low-Noise Amplifiers — 81
Martinez, Jose Silva, *INAOE* and Carreto-Castro, M.F., *National Institute of Astrophysics, Optics, and Electronics*

A Low Power, Wide Linear-Range CMOS Voltage-Controlled Oscillator — 85
Rhee, Woogeun, *Rockwell Semiconductor Systems, Inc.*

A New GHz CMOS Cellular Oscillator Network — 89
Moon, Gyu; Kim, Hong-Sun; Ismail, Mohammed; and Hwang, Changku *The Ohio State University*

MPA14 - VLSI
Chair: Igor Filanovsky
University of Alberta

A Pulse-Triggered TSPC Flip-Flop for High-Speed Low-Power VLSI Design Applications — 93
Wang, Jinn Shyan and Yang, Po-Hui, *National Chung Cheng University*

A Programmable Interpolation Filter for Digital Communications Applications — 97
Kuo, Tzu-Chieh; Kwentus, Alan, and Willson, Jr., Alan N., *UCLA*

Validation of an Accurate and Simple Delay Model and its Application to Voltage Scaling — 101
Njoelstad, Tormod and Aas, Einar Johan, *Norwegian University of*

TABLE OF CONTENTS

Science and Technology (NTNU)

A Compact 31-input Programmable Majority Gate based on Capacitive Threshold Logic   105
Leblebici, Yusuf; Gurkaynak, Frank Kagan and Mlynek, Daniel, *Swiss Federal Institute of Technology*

A Scalable Shared Buffer ATM Switch Embedded SPRAMS   109
Jeong, Gab Joong, Shim, Jae Wook, and Lee, Moon Key, *Yonsei University* and Ahn, Seung Han, *Hyundai Electronics Industries Co., Ltd.*

Optimum Design for a Two-Stage CMOS I/O ESD Protection Circuit   113
Li, Tong; Bendix, P., *LSI Logic*; Suh, D.; Huh, Y.J.; and Rosenbaum, E., *Universtiy of Illinois at Urbana-Champaign*; Kapoor, A., *LSI Logic*; Kang, S.M., *University of Illinois at Urbana-Champaign*

Low-Swing Charge Recycle Bus Drivers   117
Karlsson, Magnus; Vesterbacka, Mark; and Wanhammar, Lars, *Linkoping University*

A Pipelined Architecture of Fast Modular Multiplication For RSA Cryptography   121
Sheu, Jia-Lin; Shieh, Ming-Der; Wu, Chien-Hsing; and Sheu, Ming-Hwa, *National Yunlin University of Science & Technology*

Design Methodology of Multiple-Valued Logic Voltage-Mode Storage Circuits   125
Thoidis, I.; Soudris, Dimitrios J.; Karafyllidis, I.; Thanailakis, A., *Democritus University of Thrace*; and Stouraitis, T., *University of Patras*

Current Sensing Differential Logic (CSDL) for Low-Power and High-Speed Systems   129
Park, Joonbae; Lee, Jeongho; and Kim, Wonchan, *Seoul National University*

A Divide-by-4 Circuit Implemented in Low Voltage, High Speed Silicon Bipolar Topology   133
Schuppener, Gerd; Mokhtari, Mehran; and Tenhunen, Hannu, *Royal Institute of Technology*

Automated Implementation Of RNS-To-Binary Converters   137
Henkelmann, Heiko, *University of Bremen*, Drolshagen, A., *Siemens AG*; Bagherinia, H.; Ahrens, H.; and Anheier, W., *University of Bremen*

A Programmable Image Processing Chip   141
LeRiguer, E. and Woods, R., *The Queen's University of Belfast*; Ridge, D., *Integrated Silicon Systems* and McCanny, J., *The Queen's University of Belfast*

An Implementation Technique of Dynamic CMOS Circuit Applicable Asynchronous/Synchronous Logic   145
Yoshizawa, Hiroyasu; Taniguchi, Kenji; and Nakashi, Kenichi, *Kyushu University*

Design Issues In Cross-Coupled Inverter Sense Amplifier   149
Hajimiri, Ali, *Stanford University*; Heald, Raymond, *SUN Microelectronics*

A Novel Low-Power Building Block CMOS Cell for Adders   153
Shams, Ahmed M. and Bayoumi, Magdy A., *University of Southwestern Louisiana*

Modified Half Rail Differential Logic for Reduced Internal Logic Swing   157
Won, Jae-Hee and Choi, Kiyoung, *Seoul National University*

A Reconfigurable Integrated Circuit for High Performance Computer Arithmetic   161
Miller, Neil Linton and Quigley, Steven Francis *University of Birmingham*

Data-driven Self-Timed Differential Cascode Voltage Switch Logic   165
Mathew, Sanu and Sridhar, Ramalingam, *State University of New York at Buffalo*

A Novel Asynchronous Control Unit and the Application to a Pipelined Multiplier   169
Chiang, Jen-Shiun, *Tamkang University* and Liao, Jun-Yao, *Holtek Corporation*

The Design and Implementation of an Asynchronous Radix-2 Non-Restoring 32-B/32-B Ring Divider   173
Chiang, Jen-Shiun, *Tamkang University* and Liao, Jun-Yao, *Holtek Corporation*

A Novel Digit-Serial Systolic Array For Modular Multiplication   177
Guo, Jyh-Huei and Wang, Chin-Liang, *National Tsing Hua University*

Dual Signal Configuration for Low Power Low Voltage High Performance Pipeline Multiplier   181
Wu, Angus and Ng, C.K., *City University of Hong Kong*

Circuit Design for Current-Sensing Completion Detection   185
Lampinen, Harri and Vainio, Olli, *Tampere University of Technology*

TAA14- VLSI II
Chair: Sudhakar Muddu
Silicon Graphics

Synthesis of Critical ASICs with Embedded Fully Concurrent Fault Resilience   189
Hamilton, Samuel Norman and Orailoglu, Alex, *University of California, San Diego*

A Low-Power GaAs MESFET Dual-Modulus Prescaler   193
Kanan, Riad; Hochet, B.; Kaess, F.; and Declercq, M., *Swiss Federal Institute of Technology*

A Noise-Based Random Bit Generator IC for Applications in Cryptography   197
Petrie, Craig Steven, and Connelly, J. Alvin, *Georgia Institute of Technology*

44 Gbit/s 4:1 Multiplexer and 50 Gbit/s 2:1 Multiplexer in pseudomorphic AlGaAs/GaAs-HEMT Technology   201
Nowotny, Ulrich; Lao, Z.; Thiede, A.; Lienhart, H.; Hornung, J.; Kanfel, G.; Hohler, K.; and Glorer, K., *Fraunhofer-Institut fuer Angewandte Festkoerperphysik*

TABLE OF CONTENTS

Floating-Gate CMOS Analog Memory Cell Array — 204
Harrison, Reid R., *California Institute of Technology*; Hasler, Paul, *Georgia Institute of Technology*; and Minch, Bradley A., *Cornell University*

The Most Resistive Model for The MOS Resistive Circuit — 208
Osa, Juan I.; Porta, Sonia; and Carlosena, Alfonso, *Universidad Publica de Navarra*

Novel Imput ESD Protection Circuit with Substrate-Triggering Technique In a 0.25-µm Shallow-Trench-Isolation CMOS Technology — 212
Ker, Ming-Dou; Chen, Tung-Yang, and Wu, Chung-Yu, *National Chiao-Tung University*; Tang, Howard; Su, Kuan-Cheng; and Sun, S.-W., *United Microelectronics Corp.*

Dynamic-Floating-Gate Design For Output ESD Protection In A 0.35-µm CMOS Cell Library — 216
Ker, Ming-Dou, *Industrial Technology Research, Institute*; Chang, Hun-Hsien; Wang; Chen-Chia; Yeng, Horng-Ru; and Tsao, Y.-F., *Taiwan Semiconductor Manufacturing Company*

Fully Integrated Readout Channel with Amplitude and Time Measurement for AMS Experiment on ISSA — 220
Baschirotto, Andrea, *University of Pavia/I.N.F.N.*, Boella, G., *I.N.F.N.*, Castello, R., *University of Pavia/I.N.F.N.*; Frattini, G., *University of Pavia*; Pessina, F., *I.N.F.N.*, Rancoita, P.G., *I.N.F.N.*

Optimum SNS to Binary Conversion Algorithm and FPGA Realization — 224
Pace, P.E., *Naval Postgraduate School*; Styer, D., *University of Cincinnati* and Ringer, W.P., *Naval Postgraduate School*

Switched-Capacitor Interpolator for Direct Digital Frequency Synthesizers — 228
Santos, Paulo Jorge and Franca, José E., *Instituto Superior Técnico Center for MicroSystems*

TAB6 - Programmable Logic Devices
Chair: John Sewell
University of Glasgow

A Three-Dimensional FPGA with an Integrated Memory for In-Application Reconfiguration Data — 232
Chiricescu, Silviu M.S.A. and Vai, M. Michael, *Northeastern University*

VLSI Design of A 1.0 GHz 0.6-µM 8-Bit CLA Using PLA-Styled All-N-Transistor Logic — 236
Wang, Chua-Chin and Tsai, Kun-Chu, *National Sun Tat-Sen University*

Thermal Testing on Programmable Logic Devices — 240
Lopez-Buedo, Sergio; Garrido, Gavier; and Boemo, Eduardo, *Universidad Autonoma de Madrid*

WAA6 – Deep-Submicron Digital Circuit Issues
Chair: Nicholas Rumin
McGill University

Performance Criteria for Evaluating the Importance of On-Chip Inductance — 244
Ismail, Yehea I. and Friedman, Eby G., *University of Rochester;* and Neves, Jose L., *IBM Microelectronics*

Interconnect Inductance Effects on Delay and Crosstalks for Long On-Chip Nets with Fast Input Slew Rates — 248
Lee, Mankoo; Hill, Anthony; and Darley, Merrick H., *Texas Instruments*

Linearized Sub-Optimum Method of Long Wire Interconnections with Uniform Wire Driver — 252
Mu, Fenghao; Alvandpour, Atila; and Svensson, Christer, *Linköping University*

An Interconnect Transient Coupling Induced Noise Susceptibility for Dynamic Circuits in Deep Submicron CMOS Technology — 256
Lee, Mankoo and Darley, Merrick H., *Texas Instruments*

WAB6 – DSP Architectures
Chair: Herschel H. Loomis, Jr.
Naval Postgraduate School

General Data-Path Organization of a MAC Unit for VLSI Implementation of DSP Processors — 260
Farooqui, Aamir Alam, *University of California, Davis,* and Oklobdzija, Vojin G., *Integration Berkeley*

Nonlinear DSP Coprocessor Cells -- One and Two Cycle Chips — 264
Jain, V.K. and Lin, L., *University of Southern Florida*

GAA: A VLSI Genetic Algorithm Accelerator with On-the Fly Adaptation of Crossover Operators — 268
Wakabayashi, Shin'ichi; Koide, Tetsushi; Hatta, Koichi; Nakayama, Yoshikatsu; Goto, Mutsuaki; and Toshine, Naoyoshi, *Hiroshima University*

VLSI Implementation of a DWT Architecture — 272
Acharya, Tinku, *Intel Corporation*, and Chen, Po-Yueh, *University of Maryland*

WPA6 – Topics in Analog and Digital Test
Chair: Ramesh Harjani
University of Minnesota

Low Expense Architectures for a Dynamic Spectrum Analyzer Based on SC-Filters — 276
Marschner, Uwe; Fischer, Wolf-Joachim; and Kranz, Ernst-Georg, *Dresden University of Technology*

A Multi-Pass A/D Conversion Technique For Extracting On-Chip Analog Signals — 280
Hajjar, Ara and Roberts, Gordon W., *McGill University*

Arbitrary Band-Limited Pulse Generation for Built-In Self-Test Applications — 284
Dufort, Benoit and Roberts, Gordon W., *McGill University*

An Effective BIST Scheme for Delay Testing — 288
Li, Xiaowei and Cheung, Paul Y.S., *The University of Hong Kong*

Design of Single-Ended SRAM with High Test Coverage and Short Test Time — 292

TABLE OF CONTENTS

Wu, Chi-Feng; Wang, Chua-Chin; Hwang, Rain-Ted; and Kao, Chia-Hsiung, *National Sun Yat-Sen University*

Reducing Power Consumption During Test Application by Test Vector Ordering — 296
Girard, Patrick; Landrault, C.; Pravossoudovitch, S.; and Severac, P.; *Université Montpellier II/CNPS*

A Simplicial Method for the Simulation of Transistor Shorts in CMOS Logic Gates — 300
Lin, Hung-Jen and Milor, Linda, *University of Maryland at College Park*

Artificial Neural Network Based Multiple Fault Diagnosis In Digital Circuits — 304
Al-Jumah, Abdullah A. and Arslan, T., *Cardiff University of Wales*

WPA15 – Analog Circuits and Systems
Chair: Todd R. Weatherford
Naval Postgraduate School

7 Gbit/s Measurements on a 0.8 μm CMOS Line-Receiver — 308
Johansson, Henrik O., *Linköping University*

AlGaAs/GaAs HEMT 5-12 GHz Integrated System for an Optical Receiver — 312
Reina, Rodrigo and Olmos, Alfredo, *Universidade de Sao Paulo* and Charry, Edgar, *CPqD - Telebras*

Multiple 1:N Interpolation FIR Filter Design Based on a Single Architecture — 316
Kang, In and Yeon, Kwang-Il, *Electronics and Telecommunications Research Institute (ETRI)*, Jo, Han-Cheol, *Doowon Technical College*, Chong, Jong-Hwa, *HanYang University* and Kim, Kyungsoo, *Electronics and Telecommunications Research Institute (ETRI)*

VLSI Architectures for Weighted Order Statistic (WOS) Filters — 320
Chakraharti, Chaitali, *Arizona State University*; and Lucke, Lori E., *Minnetronix, Inc.*

Analog CMOS Design of the Incremental Credit Assignment (ICRA) Scheme for Time Series Classification — 324
Vlassis, S.; Siskos, S.; Hatzopoulos, Alkiviades A., Petrides, Kehapious V.; and Kehagias, A., *Aristotle University of Thessaloniki*

MOSFET Stair-Shaped I-V Circuit and Applications — 328
Jun, Sibum, *LG Semicon Co., Ltd.* and Ahn, Su Jin, *POSTECH*

Statistical Design Techniques for Yield Enhancement of Low Voltage CMOS VLSI — 331
Tarim, Tuna B. and Kuntman, H. Hakan, *Istanbul Technical University*; and Ismail, Mohammed, *Ohio State University*

A Novel Digitally Controlled CMOS Current Follower for Low Voltage Low Power Applications — 335
Elwan, Hassan O. and Ismail, Mohammed, *Ohio State University*

An Analytical Solution for a Class of Oscillators, and Its Application to Filter Tuning — 339
Pavan, Shanthi, *Texas Instruments, New Jersey* and Tsividis, Yannis, *Columbia University*

Active Capacitance Multipliers Using Current Conveyors — 343
Di Cataldo, G., *Universita di Catania*; Ferri, G., *Universita Di L'Aquila*; and Pennisi, Salvatore, *Università di Catania*

A ±1.5V CMOS Four-Quadrant Analogue Multiplier Using 3GHz Analogue Squaring Circuits — 347
Li, Simon Cimon and Lin, Kaung-Long, *National Yunlin University of Science and Technology*

An Autozeroing Floating-Gate Second-Order Section — 351
Hasler, Paul, *Georgia Institute of Technology*; Stanford, Theron, *California Institute of Technology*; Minch, Bradley A., *Cornell University*; and Diorio, Chris, *University of Washington*

Design of Phase Equalizer using Phase Delay Characteristic — 355
Carvalho, Delmar B., *Federal University of Santa Catarina/Caikelic University of Peletas*; Filho, Sidnei Noceti, *LINSE/EEL/CTC/UFSC*; Seara, Rui, *Federal University of Santa Catarina*

New Switched-Current Circuits for Nonlinear Signal Processing — 359
Zeng, X. and Tang, P.S., *Fudan University*; and Tse, C. K., *Hong Kong Polytechnic University*

A Replica Biasing for Constant-gain CMOS Open-Loop Amplifiers — 363
Palmisano, Giuseppe and Salerno, R., *Università di Catania*

Switched-Capacitor Impedance Simulation Circuits Realized with Current Conveyor — 367
Ono, Toshio, *Saitama Institute of Technology*

A Simple CMOS Digital Controlled Oscillator with High Resolution and Linearity — 371
To, Cheuk-Him; Chan, Cheong-Fat; and Choy, Oliver Chiu-Sing, *The Chinese University of Hong Kong*

On Optimizing Micropower MOS Regulated Cascode Circuits on Switched Current Techniques — 374
De Lima, Jader Alves, *Universidade Estadual Paulista*

Elimination of Nonlinear Clock Feedthrough in Component-Simulation Switched-Current Circuits — 378
de Queiroz, Antonio Carlos M. and Schechtman, Jones, *Universidade Federal do Rio de Janeiro*

Analog Building Blocks for a Sampled Data Fast Wavelet Transform CMOS VLSI Implementation — 382
Gonzalez-Altamir, Gerardo and Ramirez-Angulo, Jaime, *New Mexico State University*

An Area Efficient Time-Interleaved Parallel Delta-Sigma A/D Converter — 386
Eshraghi, Aria and Fiez, Terri, *Washington State University*

Program Delivery Control with On-Screen Display — 390
Peng, Dennis Lee Chew; Cheung, Chu Ming; Liang, Joseph; and Yong, Teo Tee, *Siemens Components Pte Ltd.*

TABLE OF CONTENTS

WAA14- VLSI Architectures
Chair: Ken Suyama
Columbia University

Accuracy Analysis of Layout Parasitic Extraction Based on Boolean Methods — 394
Brambilla, Angelo, *Politecnico di Milano* and Mancini, Paolo, *Accent Srl*

Hardware Efficient Transform Designs with Cyclic Formulation and Subexpression Sharing — 398
Chang, Tian-Sheuan and Jen, Chein-Wei, *National Chiao-Tung University*

A Field Programmable Gate Array Chip with Hierarchical Interconnection Structure — 402
Lai, Yen-Tai, Kao, Chi-Chou, Chang, Tsun-Chen, and Chen, Kun-Nern, *National Cheng-Kung University*

Low-Energy Programmable Finite Field Data Path Architectures — 406
Song, Leilei and Parhi, Keshab K., *University of Minnesota;* Kuroda, Ichiro and Nishitani, Takao, *NEC Corporation*

The LEMMA Developer's Toolbox: Semi-Automated Test Development for Analog and Mixed-Signal Circuits — 410
Kennedy, Michael Peter, *University College Dublin,* Grogan, Paul; O'Donnell, John, and O'Dwyer, Tom, *Analog Devices;* Wrixon, Adrian, *University of California, Berekeley*

An High Speed VLSI Architecture for Scaled Residue to Binary Conversion — 414
Cardarilli, Gian Carlo; Re, M.; Lojacono, R., *University of Rome "Tor Vergataa",* Ferri, G., *University of L'Aquila*

VLSI Implementation of Phong Shader in 3D Graphics — 417
Shin, Yun Chul; Lee, Jin-Aeon; and Kim, Lee-Sup, *Korea Advanced Institute of Science and Technology*

State Encoding for Low Power Embedded Controllers — 421
Daldoss, L., *Universita di Brescia;* Sciuto, Donatella, *Politecnico di Milano;* Silvano, C., *Politecnico di Milano*

Design of a Scan Line Image Processor Chip — 425
Chang, Hyunman; Ong, Soahwan; Lee, Changhee; and Sunwoo, Myung H., *Ajou University*

An Efficient Programmable 2-D Convolver Chip — 429
Chang, Hyun Man and Sunwoo, Myung H., *Ajou University*

Synthesis of Folded Multidimensional DSP Systems — 433
Sundararajan, Vijay and Parhi, Keshab K., *University of Minnesota*

Low Power Scheduling with Resources Operating at Multiple voltages — 437
Shiue, Wen-Tsong and Chakrabarti, Chaitali, *Arizona State University*

A CPLD Design of a Self-Organizing System for Data Clustering — 441
Ohkubo, Jun'ya; Miyanaga, Yoshikazu; and Tochinai, Koji, *Hokkaido University*

High Performance Cell for Solving Real Time Field Problems Using the Resistive Grid Method — 445
Carneiro, Noel Carlos F. and Ramírez-Angulo, Jaime, *New Mexico State University*

VLSI Design of A CORDIC-based Derotator — 449
Ahn, Youngho; Nahm, Seunghyeon, and Sung, Wonyong *Seoul National University*

Theoretical Estimation of Power Consumption in Binary Adders — 453
Freking, Robert A. and Parhi, Keshab K., *University of Minnesota*

Implementation of the Fuzzy Art Neural Network for Fast Clustering of Radar Pulses — 458
Cantin, Marc-Andre, *Universite du Quebec a Montreal;* Blaquiere, Yves, *Universite du Quebec a Montreal;* Savaria, Yvon and Granger, Eric, *Ecole Polytechnique de Montreal;* Lavioe, Pierre, *Defense Research Establishment Ottawa*

Neural Core Module for Embedded Intelligence — 462
Diepenhorst, Marco; Ter Haseborg, Henrickus; Nijhuis, J.A.G; and Spaanenburg, L., *University of Groningen*

Cheap and Easy Systematic CMOS Transistor Mismatch Characterization — 466
Serrano-Gotarredona, Teresa and Linares-Barranco, Bernabe, *National Microelectronics Center*

Self-Calibrating Clock Distribution with Scheduled Skews — 470
Hsieh, Hong-Yean, *Next Wave Technology Inc.,* Liu, Wentai; Clements, Mark; and Franzon, Paul, *North Carolina State University*

Array Architecture and Design For Image Window Operation Processing Asics — 474
Li, Dongju; Jiang, Li; Isshiki, Tsuyoshi; and Kunieda, Hiroaki, *Tokyo Institute of Technology*

Novel Digit-Serial Systolic Array Implementation of Euclid's Algorithm for Division in $GF(2^m)$ — 478
Guo, Jyh-Huei and Wang, Chin-Liang, *National Tsing Hsu University*

VLSI Architectures of Divider for Finite Field $GF(2^m)$ — 482
Wei, Shyue-Win, *Chung-Hua University*

A High Throuput Variable Length Decoder with Modified Memory Based Architecture — 486
Shieh, Bai-Jue; Lee, Yew-San; and Lee, Chen-Yi, *National Chiao Tung University*

VOLUME 3

MAA3- Neural Networks for Intelligent Signal Processing
Chair: Chung-Yu Wu
National Chiao Tung University
Co-organizer: Fathi Salam
Michigan State University
Co-Organizer: Bing Sheu
University of Southern California

TABLE OF CONTENTS

Dynamical Functional Artificial Neural Networks Networks (D-FANNs) for Intelligent Signal Processing — 1
de Figueiredo, Rui J.P., *University of California, Irvine*

Voice Output Extraction by Signal Separation — 5
Erten, G., *IC Tech, Inc.* and Salam, Fathi M., *Michigan State University*

Compact Neural Network Detector for Hard-Disk Drive Using Zero-Forcing Preprocessing — 9
Wang, Michelle Y. and Sheu, Bing J., *University of Southern California*

A Multi-Resolution Image Registration for Multimedia Application — 13
Huang, Chung-Lin, and Chang, Pen-Yiing, *National Tsing-Hua University*

Blind Separation of Linear Convolutive Mixtures Through Parallel Stochastic Optimization — 17
Cohen, Marc, and Cauwenberghs, Gert, *Johns Hopkins University*

VLSI Chaotic Pulse Coded Modulator Using Neural Type Cells — 21
Sellami, Louiza, *US Naval Academy/ University of Maryland*; Zaghloul, Mona E., *The George Washington University*; and Newcomb, Robert W., *University of Maryland*

Recognition of Handwritten Chinese Postal Address Using Neural Networks — 25
Su, Yih-Ming, *I-Sheu University*; and Wang, Jhing-Fa, *National Cheng Kung University*

MPA3-Memory, Adaptation and Learning
Chair: Gert Cauwenberghs
The John Hopkins University
Co-Organizer: Fathi Salam
Michigan State University
Co-Organizer: Paul Hasler
Georgia Institute of Technology

A Four-Quadrant Floating-Gate Synapse — 29
Hasler, Paul; Diorio, Chris; and Minch, Bradley A., *Georgia Institute of Technology*

Programmable Current Mode Hebbian Learning Neural Network using Programmable Metallization Cell — 33
Swaroop, B.; West, W.C.; Martinez, G.; Kozidu, M.N.; and Akers, Lex A., *University of Texas at San Antonio*

A Programmable Neural Processor for Pulse-Coded Hippocampal Signal — 37
Tsai, Richard H.; Sheu, Bing J.; and Berger, Theodore W., *University of Southern California*

A Robust Hybrid Neural Architecture for An Industrial Sensor Application — 41
Djahanshahi, Hormoz; Ahmadi, Majid; Jullien, Graham A.; and Miller, William C., *University of Windsor*

Design of an Analog CMOS Self-Learning MLP Chip — 46
Bo, G.M.; Caviglia, D.D.; Chibl`e, H.; and Valle, M, *University of Genoa*

Mixed-signal CMOS high precision circuits for on chip learning — 50
Vidal-Verdú, Fernando; Navas, Rafael; and Rodriquez-Vasquez, Angel, *University of Málaga.*

A Neuro-Chip for Real-time Learning, Processing and Control — 54
Salam, Fathi M., *Michigan State University*

A Micropower Learning Vector Quantizer for Parallel Analog-to-Digital Data Compression — 58
Lubkin, Jeremy and Cauwenberghs, Gert, *Johns Hopkins University*

TAA3-Neural Networks I: Algorithms and Computation
Chair: Robert Newcomb
University of Maryland

On-Line Tracking Abilities of Neural Networks with Graded Responses — 62
Kuh, Anthony, *University of Hawaii*

A New Class of Apex-Like PCA Algorithms — 66
Fiori, Simone; Uncini, Aurelio; and Piazza, Francesco, *University of Ancona*

Multiresolution Neural Networks for Recursive Signal Decomposition — 70
Kan, Kai-chiu, and Wong, Kwok-wo, *City University of Hong Kong*

Characteristics of Gradient Descent Learning with Neuronal Gain Control — 74
Ho, Murphy, *City University of Hong Kong* and Kurokawa, Hiroaki, *Keio University*

Training of a Class of Recurrent Neural Networks — 78
Shaaban, Khaled M., *Assiut University*

Dynamical Systems Learning by a Circuit Theoretic Approach — 82
Campolucci, Paolo; Uncin, Aurelio; and Piazza, Francesco, *University of Ancona*

Training RBF Networks with the Kalman Filter — 86
Ciocoiu, Iulian B., *Technical University of Iasi*

TPA3 –Neural Networks II: Implementation Issues
Chair: G. Cauwenbreghs
John Hopkins University

Continuous-Time Feedback in Floating-Gate MOS Circuits — 90
Hasler, Paul; Diorio, Chris; and Minch, Bradley A., *Georgia Institute of Technology*

An Analog Neural Network Circuit with Simultaneous Perturbation Learning Rule — 94
Maeda, Yutaka and Kanata, Yakichi, *Kansai University*

A Self-Organizing Map with Resistive Fuse — 99
Kousuke, Katayama; Singo, Kawahara; and Saito, Toshimichi, *Hosei*

TABLE OF CONTENTS

University

Accuracy vs. Precision in General Purpose Neural Digital VLSI Architectures — 103
Alippi, Cesare, *CNR-CESTIA;* and Briozzo, Luciano, *SGS-Thomson Microelectronics*

WAA3 – Networks for Biological Computing & Fuzzy Logic
Chair: Lex A. Akers
University of Texas at San Antonio

An Adaptive Front End for Olfaction — 107
Apsel, Alyssa, *The John Hopkins University;* Stanford, Theron, *California Institute of Technology* and Hasler, Paul, *Georgia Institute of Technology*

Adaptation in an AVLSI Model of a Neuron — 111
Simoni, Mario; and DeWeerth, Stephen P., *Georgia Institute of Technology*

Biologically-Motivated Neural Learning in Situated Systems — 115
Damper, Robert Ivan, *University of Southampton;* and Scutti, T.W., *University of Nottingham*

FWNN for Interval Estimation with Interval Learning Algorithm — 119
Jiao, Licheng; Liu, F.; and Wang, L., *Xidian University*

FDSP: A VLSI Core for Adaptive Fuzzy and Digital Signal Processing Applications — 123
Sultan, Labib, *Microfuzz Technologies Inc*

A Current-Mode Piecewise-Linear Function Approximation Circuit Based on Fuzzy-Logic — 127
Manaresi, Nicolo'; Rovatti, Riccardo; Franchi, Eleonora; and Baccarani, Giorgio, *D.E.I.S. - University of Bologna*

Embedded Fuzzy Control on Monolithic DC/DC Converters — 131
Criscione, M., *SGS-THOMSON Microelectronics;* Lionetto, A., and Nunnari, G., *University of Cantania;* and Occhipinti, L., *SGS-THOMSON Microelectronic*

A Wavelet-Based Fuzzy Neural Network for Interpolation of Fuzzy If-Then Rules — 135
Jiao, Licheng, *Xidian University*

WPA3 – Cellular Neural Networks
Chair: Jacek M. Zurada
University of Louisville

A Modular g^{mC} Programmable CNN Implementation — 139
Lím, Drahoslav; and Moschytz, George S., *Swiss Federal Institute of Technology, Zurich*

An Improved Architecture for the Interconnections in a Multi-Chip CNN System — 143
Sargeni, Fausto; Sorgeni, Fausto; and Bonaiuto, Vincenzo, *University of Rome "Tor Vergata"*

VLSI Delta-Sigma Cellular Neural Network for Analog Random Vector Generation — 147
Cauwenberghs, Gert, *Johns Hopkins University*

On Evolving Hardware: On-Line Evolution by Cellular Programming — 151
Nicoletti, Guy M., *University of Pittsburgh at Greensburg*

An Analysis of CNN Settling Time — 155
Haenggi, Martin; and Moschytz, George S., *Swiss Federal Institute of Technology*

Learning Algorithms for Cellular Neural Networks — 159
Mirzai, Bahram; Cheng, Zhenlan; and Moschytz, George S., *Swiss Federal Institute of Technology*

Autowaves for Motion Control: A CNN Approach — 163
Arena, Paolo; Fortuna, Luigi; Branciforte, Marco; and Di Grazia, Pietro *Università degli Studi di Catania;*

A Time-Multiplexing Simulator for Cellular Neural Network (CNN) Using Simulink — 167
El-Shafei, Ahmed Abdel-Rahman, and Sobhy, Mohamed Ibrahim, *The University of Kent at Canterbury*

WPA14 – Neural Networks
Chair: Mona Zaghloul
George Washington University

Low Complexity CMOS Competitive Array for Approaching Assignments — 171
Gomez-Castaneda, Felipe; Flores-Nava, Luis M.; and Moreno-Cadenas, Jose A., *Center for Research and Advanced Studies of the NPL*

An Artificial Model for Biological Computation and Control for a Locomotion System — 175
Curran, K. and Wooten, E., *U. S. Naval Academy;* and Newcomb, Robert W., *University of Maryland*

An Efficient Method of Automatical Feature Extraction and Target Classification — 179
Zhang, Yanning and Jiao, Licheng, *Xidian University*

Pseudorandom Generator Based on Clipped Hopfield Neural Network — 183
Chan, Chi-Kwong and Cheng, L.M., *City University of Hong Kong*

Global Stability of a Larger Class of Dynamical Neural Networks — 187
Arik, Sabri, *Istanbul University*

Hardware Realization of a Hamming Neural Network with On-Chip Learning — 191
Schmid, Alexandre; Leblebici, Yusuf; and Mlynek, Daniel, *Swiss Federal Institute of Technology Lausanne*

Car Plate Recognition by Neural Networks and Image Processing — 195
Parisi, Raffaele; Lucarelli, G.; and Orlandi, G., *University of Rome "La Sapienza"*

TABLE OF CONTENTS

Architecture and Design Methodology of the RBF-DDA Neural Network — 199
Aberbour, Mourad and Mehvez, Habib, *University de Pierre et Marie Lovie*

Hardware Implementation of Post-Retinal Processing using Analog VLSI — 203
Satakopan, S.; James, S.; and Akers, Lex, *Arizona State University*

High Performance Programmable Bi-Phasic Pulse Generator Design for a Cochlear Speech Processor — 207
Ay Suat; Sheu, Bing; and Zeng, Fan-Gang, *University of Southern California*

Pulse Stream based CNN Hardware Implementation — 211
Colodro, Francisco; Torralba, A.; Carvajal, R.G.; and Franguelo, L.G., *Escuela Superior de Ingenieros*

Design of Cellular Neural Networks with Space-Invariant Cloning Template — 215
Lu, Zanjun and Liu, Derong, *Stevens Institute of Technology*

Synthesis of a Recurrent Double-Layer Transistor Network for Early-Vision Tasks — 219
Barbaro, Massimo; Nazzaro, Antonio; and Raffo, Luigi, *University of Cagliari*

Harmonic Retrieval Using Higher-Order Statistics and Hilbert Transform — 223
Li, Shenghong and Liu, Zemin, *Beijing University of Posts and Telecommunications*

Segmentation Coding for Object-Based Attentive Selection Systems — 227
Wilson, Charles S., *Georgia Institute of Technology*; Morris, Tonia G., *Intel Corporation*; and DeWeerth, Stephen P., *Georgia Institute of Technology*

Current-Mode Truth Value Evaluation Circuits for Complementary Fuzzy Logic Systems — 231
Chen, Chien-Yau; Yu, Gwo-Jeng; and Liu, Bin-Da, *National Cheng Kung University*

A Hydrid Fuzzy Neural Decoder for Convolutional Codes — 235
Wu, Meng; Zhu, Wei-Ping, *Nanjing University of Posts and Telecommunications*; and Nakamura, Shogo, *Tokyo Donki University*

A New Edge-Preserving Smoothing Filter Based on Fuzzy Control Laws And Local Features — 239
Muneyasu, Mitusji; Wada, Yuji; and Hinamoto, Takao, *Hiroshima University*

Current-Mode Circuit to Realize Fuzzy Classifier with Maximum Membership Value Decision — 243
Chen, Chuen-Yau; Tsao, Ju-Ying; and Liu, Bin-Da, *National Cheng Kung University*

A Novel CMOS Analogue Fuzzy Inference Processor — 247
Song, C.T. Peter; Quigley, Steven Francis; and Pammu, Sridhar, *University of Birmingham*

Mixed-Mode VLSI Implementation of Fuzzy ART — 251
Cohen, Marc H.; Abshire, Pamela; and Cauwenberghs, Gert, *Johns Hopkins University*

A Cellular Nonlinear Network for Digital Error Correction — 255
Kananen, Asko Tapio; Paasio, Ari Juhani; Lindfors, Saska; and Halonen, Kari, *Helsinki University of Technology*

A Fuzzy Reasoning Based Approach for ARMA Order Selection — 259
Haseyama, Miki; Emura, Masafumi; and Kitajma, Hideo, *Hokkaido University*

MAA7 – Chaos and Applications
Chair: Martin Hasler
Swiss Federal Institute of Technology Lausanne

Synchronous Phenomena from Chaotic Circuits with Intermittently Coupled Capacitors — 263
Takanori, Matsushita; Saito, T.; and Torikai, H., *Hosei University*

BER Performance of a Chaos Communication System Including Modulation-Demodulation Circuits — 267
Wada, Masahiro, *Tokushima University*; Kawata, Junji, *Tokushima Bunri University*; Nishio, Yoshifumi; and Ushida, Akio, *Tokushima University*

Chaos Shift Keying in the Presence of Noise: A Simple Discrete Time Example — 271
Hasler, Martin, Swiss Federal Institute of Technology Lausanne

Chaotic Signals for CW-Ranging Systems - a Baseband System Model for Distance and Bearing Estimation — 275
Bauer, Andreas, *Technical University of Dresden*

Design of Infinite Chaotic Polyphase Sequences with Perfect Correlation Properties — 279
Abel, Andreas; and Goetz, Marco, *Technical University of Dresden*

Design of Nonlinear Observers for Hyperchaos Synchronization Using a Scaler Signal — 283
Grassi, Giuseppe, *Universita di Lecce*; and Mascolo, Saverio, *Politecnico di Bari*

Synchronization in Arrays of Chaotic Circuits Coupled via Hypergraphs: Static and Dynamic Coupling — 287
Wu, Chai Wah, *IBM*

Chaotic and Bifurcation Behavior in an Autonomous Flip-Flop Circuit Used by Piecewise Linear Diodes — 291
Okazaki, Hideaki, *Gifu National College of Technology*; Nakano, Hideo, *Shonan Institute of Technology*; and Kawase, Takehiko, *Waseda University*

MAA14 – Circuits and Power Systems I
Chair: Graham R. Hellestrand
University of New South Wales

TABLE OF CONTENTS

Amplitude Bounds on Oscillations from a Sigma-Delta Modulator Structure — 298
Davies, Anthony C., *King's College London*

Global Synchronization in Coupled Map Lattices — 302
Wu, Chai Wah, *IBM*

On-off Intermittency from a Ring of Four Coupled PLLs — 306
Hasegawa, Akio; Igarashi, Ryo; Endo, Tetsuro, *Meiji University*; and Komuro, Motomasa, *Teikyo University of Science & Technology*

Performance Comparison of Communication Systems Using Chaos Synchronization — 310
Junji, Kawata, *Tokushima Bunri University*; Nishio, Yoshifumi, *Tokushima University*; Dedieu, Herve, *Swiss Federal Institute of Technology*; and Ushida, Akio, *Tokushima University*

Synchronization in Chaotic Oscillators Based on Classical Oscillator Coupled by One Resistor — 314
Sekiya, Hiroo; and Moro, Seiichiro, *Keio University*; Mori, Shinsaku, *Nippon Institute of Technology*; and Sasase, Iwao, *Keio University*

Neural-Network Based Adaptive Control of Uncertain Chaotic Systems — 318
Qin, Huashu and Zhang, Huaizhou, *Chinese Academy of Science*; and Chen, Guanrong, *University of Houston*

Time Domain Analysis of Modulated Carriers in (Non)-Linear Systems — 322
Leenaerts, Domine, *Technical University Eindhoven*

Singularities of Implicit Ordinary Differential Equations — 326
Reißig, Gunther, *Technische University Dresden*; Boche, Holger, *Heinrich-Hertz-Institut fuer Nachrichtentechnik Berlin*

Useful Necessary and Sufficient Condition for Reachability of Extended Marked Graphs — 330
Tsuji, Kohkichi, *Aichi Prefectural University*

Calculation of the Homoclinic Bifurcation Sets of PLL Equation with Five-Segment Piecewise-Linear Phase Detector Characteristic — 334
Ohno, Wataru; and Endo, Tetsuro, *Meiji University*

Algorithm for Non-Intrusive Identification of Residential Appliances — 338
Cole, Agnim I. and Albicki, Alexander, *University of Rochester*

The Resolution of Algebraic Loops in the Simulation of Finite-Inertia Power Systems — 342
Ciezki, John G.; and Ashton, R.W., *Naval Postgraduate School*

On Stability Robustness of Discrete-Time Systems: The Complex-Variable Approach of Mastorakis — 346
Lu, Wu-Sheng, *University of Victoria*

Phase-Jitter Dynamics of Second-Order Digital Phase-Locked Loops — 350
Rogers, Alan; Teplinsky, Alexey; and Feely, Orla, *University College Dublin*

Analysis of the DC Link Current Spectrum in Force Commutated Inverters — 354
Mariscotti, A., *University of Genoa*

On the Modeling of a Chaotic Circuit Containing a Bent Hysteresis Resistor — 358
Xia, Cheng-Quan, *Xian Jiaotong University*; and Du, Min, *Trident Microsystems, Inc.*

Time-Delay Neural Networks, Volterra Series, and Rates of Approximation — 362
Sandberg, Irwin W., *The University of Texas at Austin*

Topological Dimensionality Determination and 366Dimensionality Reduction Based on Minimum Spanning Trees — 366
de Figueiredo, Rui J.P. and Oten, Resnzi, *University of California, Irvine*

Investigations of Periodic Orbits in Electronic Circuits with Interval Newton Method — 370
Galias, Zbigniew, *University of Mining and Metal*

Stability Analysis and Robust Stabilization of a Class of Nonlinear Systems Based on Stability Radii — 374
Jannesari, S., *Wichita State University*

The Design and Fabrication of a Reconfigurable Hardware Testbed for the Interaction Analysis of Power Converters in a Reduced-Scale Navy DC Distribution System — 379
Ashton, R.W.; and Ciezki, John G., *Naval Postgraduate School*

A Computer Program for Accurate Time-Domain Analysis of 1D-Arrays of Chua's Oscillators — 383
Bicy, Mario; Gill, Marco; Maio, Ivano; and Premoli, Amedeo, *Politecnico di Torino*

The Analysis of Tradeoffs Between Power Section Hardware and Feedback Gains for a DC Distribution System DC-to-DC Converter — 387
Ashton, R.W.; and Ciezki, John G., *Naval Postgraduate School*

MPA7 – Linear Circuits
Chair: Isao Shirakawa
Osaka University

Flow Problems on Information Network — 391
Takatama, Hirokazu; and Watanabe, Hitoshi, *Soka University*

Hybrid Matrix Minors from Tableau Applied to a Multiport Generalization of NDR Related to Stability — 395
Chaiken, Seth, *State University of New York, Albany*

A Method for Automatic Design of Analog Circuits Based on a Behavioral Model — 399
Shojaei, M.; and Sharif-Bakhtiar, M., *Sharif University of Technology*

Modified Nodal Formulation Method Applied to Piecewise-Linear DC Analysis — 403
Roos, Janne Wilhelm; and Valtonen, Martti Erik, *Helsinki University*

TABLE OF CONTENTS

of Technology

N-Port Reciprocity and Irreversible   407
Thermodynamics
Weiss, Laurens and Mathis, Wolfgang, *Otto-von-Guericke-University Magdeburg*

NARX Approach to Black-Box Modeling of   411
Circuit Elements
Maio, Ivano; Adolfo,Stievano, I.S.; Canavaro, F.G., *Politecnico di Torino*

A General Method of Feedback Amplifier   415
Analysis
Nikolic, Borivoje, *University of California, Davis*; and Marjanovic, Slavoljub, *University of Belgrade*

The Index of the Standard Circuit Equations of   419
Passive RLCTG-Networks Does Not Exceed 2
Reißig, Gunther, *Technische University Dresden*

MPA12 –Power Distribution Systems
Chair: David A. Johns,
University of Toronto

A Power Distributor with Winner-Take-All   423
Function
Mokunaka, Naoki; and Saito, Toshimichi, *Hosei University*

A DSP Controlled Variable Frequency Resonant-   427
Commutated Converter
Chickamenahalli, Shamala A.; Liu, Jon; Nallapervimal V.; and Barker, Clashow M., *Wayne State University*

New Parallel Tabu Search for Voltage and   431
Reactive Power Control in Power Systems
Mori, Hiroyuki; and Hayashi, Takanori, *Meiji University*

Power Energy Metering Based on Random   435
Signal Processing (EC-RPS)
Toral, Sergio L.; Quero, Jose Manuel; and Franquelo, Leopoldo, *INGENIEROS*

Sensitivity Analysis of Power System   439
Trajectories: Recent Results
Hiskens, Ian A. and Pai, M.A., *University of Illinois at Urbana-Champaign*

Contingency Screening Using Interval Analysis   444
in Power Systems
Mori, Hiroyuki; and Yuihara, Atsushi, *Meiji University*

Estimation of Nonsinusoidal Bus Voltage   448
Waveforms in Power Systems
Abur, Ali, *Texas A&M University*

TAA7 –Nonlinear Networks and Systems
Chair: K.S. Chao,
Texas Tech University

A Simple Bracketing Algorithm for Determining   452
Transition Time Instants in PWL Circuits
Pastore, Stefano, *Universita di Trieste*; and Premoli, Amedeo, *Politecnico*

di Milano

Global Asymptotic Stability of a Class of   456
Nonlinear Dynamical Systems
Xiong, Kaiqi, *North Carolina State University*

A Discrete–Time Approach to the Steady State   460
Analysis of Distributed Nonlinear Autonomous
Circuits
Bonet, Jordi; Palà, Pere; and Miró, Joan Maria, *UPC-Department of Signal Theory and Communications*

Bifurcation of Switched Nonlinear Dynamical   465
Systems
Ueta, Tetsushi; Kousaka, Takuji; and Kawakami,Hiroshi, *Tokushima University*

Chaos Generators with Piecewise Linear   469
Trajectories
Tsubone, Tsdashi Sailo, T.; and Schwarz, W., *Hosei University*

Spatiotemporal Pattern from a Simple Hysteresis   473
Network
Jinno, Kenya; and Tanaka, Mamoru, *Sophia University*

Analysis of a Simple Hysteresis Neural Network   477
and Its Application
Nakaguchi, Toshiya; Jinno, Kenya; and Tanaka, Mamoru, *Sophia University*

Effects of the Deviation of Element Values in a   481
Ring of Three and Four Coupled Van Der Pol
Oscillators
Oakawara, Tsoyoshi and Endo, Tetsuro, *Meiji University*

TAB14 –Circuits and Power Systems II
Chair: Krishnaiyan Thulasiraman
University of Oklahoma

A Class of Systems with Symmetric Impulse   485
Response
Vucic, Mladen and Bahie, Hrvoje, *Faculty of Electrical Engineering and Computing*

The Formulation and Implementation of an   489
Analog/Digital Control System for a 100kW DC-to-DC
Buck Chopper
Ashton, R.W.; and Ciezki, John G., *Naval Postgraduate School*; and Mak, C., *Power Paragon, Inc.*

Synchronization of Subthreshold-CMOS Chaotic   493
Oscillators
Neeley, John E.; Overman, Charles H.; and Harris, John G., *University of Florida*

New Mode-Domain Representation of Transmis-   497
sion Line - Clarke Transformation Analysis
Tavares, Maria Cristina, *State University of Campinas*; Pissolato, J.; and Dortela, C.M., *Federal University of Rio De Janeiro*

New Mode-Domain Representation of Transmis-   501
sion Line for Power Systems and Studies
Tavares, Maria Cristina, *State University of Campinas*; Pissolato, J. and

TABLE OF CONTENTS

Portela, C.M., *Federal University of Rio De Janeiro*

Optimal Power Flow in Distribution Networks   505
by Newton's Optimization Methods
de Medeiros, Manoel Firmino, *Universidade Federal do Rio Grande do Norte – UFRN*; and Pimentel, Max Chianca, *Programa de Pós-graduação em Engenharia*

Placement of Variable Impedance Devices for   510
Enhancement Of Small Signal Stability in Power Systems
Messina, A.Roman; Begovich, O.; and Sanchez, Edgar N., *IPN-Unidad Guadalajara*

PLD Implementation of Control Algorithms: 514
Design and Validation
Carmeli, S.; Lazzaroni, M.; and Monti, Antonello, *Politecnico di Milano*

Time-Domain Analysis for Reflection   518
Characteristic of Tapered and Stepped Nonuniform Transmission Lines
Murakami, Kazuhito; and Ishii, Junya, *Kinki University*

A Unified Method for the Small-Signal Modeling   522
of Multi-Resonant and Quasi-Resonant Converters
Szabo, Adrian; Kansara, M.; and Ward, E.S., *The Nottingham Trent University*

The Application of Feedback Linearization   526
Techniques to the Stabilization of DC-to-DC Converters with Constant Power Loads
Ciezki, John G.; and Ashton, R.W., *Naval Postgraduate School*

Highly Efficient CMOS Class E Power   530
Amplifier for Wireless
Tu, Steve Hung-Lung; and Toumazou, Chris, *Imperial College of Science, Technology, and Medicine*

TPA7 – Analog VLSI
Chair: Yoji Kajitani
Tokyo Institute of Technology

Mutual Synchronization in 4 Coupled Oscillators   534
with Different Natural Frequencies
Moro, Seiichiro; Mori, Shinsaku; and Sasae, Iuao, *Fukui University*

Short Period Oscillations from a Sigma-Delta   538
Modulator
Davies, Anthony Christopher, *King's College London*

Spatiotemporal Dynamics of a Stochastic   542
VLSI Array
Neff, Joseph Daniel; Patel, Girish N.; Menders, Brian; DeWeerth, Stephen P.; and Ditto, William L. *Geogia Institute of Technology*

IC Implementation of a Current-Mode   546
Chaotic Neuron
Herrera, Ruben D.;Suyano, Ken; and Horio, Yoshihiko, *Columbia University*

TPA12 – Feedback Systems and Stability
Chair: Bell A. Shenoi
Wright State University

A General Operating-Point Instability Test   550
Based on Feedback Analysis
Fox, Robert M., *University of Florida*

A 3.3V All Digital Phase-Locked Loop with   554
Small DCO Hardware and Fast Phase Lock
Chiang, Jen-Shiun, *Tamkang University*; and Chen, Kuang-Yuan, *Key Tech Corporation*

A Novel Algorithm that Finds Multiple Operating   558
Points of Nonlinear Circuits Automatically
Goldgeisser, Leonid; and Green, Michael M.,*University of California, Irvine*

An Extension of the Classical Feedback Theory
Neag, Marius; and McCarthy, Oliver, *University of Limerick*

WAA7 – Power Electronics
Chair Krishna Shenai,
University of Illinois at Chicago

An Adaptive Stepwise Quadratic State-Space   566
Modeling Technique for Analysis of Power Electronics Circuits
Tse, K.K.; Chung, Henry S.H.; and Hui, S.Y.R., *City University of Hong Kong*

An Efficient Method for Calculating Power Flow   570
Solutions and the Closest Bifurcation Point Using Mathematical Programming
Mori, Hiroyuki; and Iizuka, Fumitaka, *Meiji University*

Computation of State Variable Sensitivities of   574
PWM DC/DC Regulators and Its Applications
Wong, Billy K.H. and Chung, Henry S.H., *City University of Hong Kong*

Single-Ended Compact MOS-FET Power Inverter   578
with Automatic Frequency Control for Maximizing RF Output Power Megasonic Transducer at 3 MHz
Mizutani, Yoko; Suzuki, Taiju;Ishikawa, Jun-ich; Ikeda, Hiroaki; Yoshida, Hirofumi; and Shinohara, Shigenobu, *Shizuoka University*

WAA12 – Systems and Appls. For Next Generation Internet
Chair: Chung-Shen Li
IBM T.J. Watson Research Center
Co-organizer: Tsu-han Chen
CMU
Co-Organizer: Horng-Dar Lin
NeoParaDigm, Inc.

System Resource Management for Network   582
Servers
Kandlur, Dilip D. *I.B.M. Thomas J. Watson Research Center*

TABLE OF CONTENTS

Wireless Systems and Portable Multimedia 586
Lin, Horng-Dar, *NeoParadigm Labs, Inc.*

Very High-Speed Digital Subscriber Lines (VDSL) 590
Cioffi, John M., *Information Systems Laboratory*

Issues for Image/Video Digital Libraries 595
Manjunath, Bangalore S. and Deng, Yining, *University of California at Santa Barbara*

Transcoding Internet Content for Heterogeneous Client Devices 599
Smith, John R.; Mohan, Rakesh; and Li, Chung-Sheng, *IBM T. J. Watson Research Center*

Real-Time Distributed and Parallel Processing for MPEG-4 603
He, Yong; Ahmad, Ishfaq; and Liou, Ming L., *The Hong Kong University of Science and Technology*

Video-Coding and Multimedia Communications Standards for Internet 607
Chen, Tsuhan, *Carnegie Mellon University*

A Virtual Classroom for Real-Time Interactive Distance Learning 611
Hwang, Jenq-Neng; Deshpande, Sachin G.; and Sun, Ming-Ting, *University of Washington*

Fast Browsing of Speech/Audio Material for Digital Library and Distance Learning 615
Wong, Peter H.W. and Au, Oscar C., *Hong Kong University of Science and Technology*

Predicting Period-Doubling Bifurcations in Nonlinear Time-Delayed Feedback Systems 619
Berns, Daniel W., *Universidad Nacional de la Patagonia San Juan Bosco* Moiola, Jorge L., *Universidad Nacional del Sur*; Chen, Guanrong *University of Houston*

Rotating Stall Control Via Bifurcation Stabilization 623
Chen, Xiang; Gu Guoxiang,; Martin, Phillip B.; and Zhou Kemin, *Louisiana State Universit*

Bifurcation Analysis and Control of Nonlinear Systems with a Nonsemisimple Zero at Criticality and Application 627
Fu, Jyun-Horng (Alex), *Systems Planning and Analysis Inc.*

Stabilization of a Class of Bifurcations via State Feedback 631
Fitch, Osa *NAVAIR*
Kang, Wei, *Naval Postgraduate School*

Suppressing Cardiac Alternans: Analysis and Control of a Border-Collision Bifurcation in a Cardiac Conduction Model 635
Chen, Dong and Wang, Hua O., *Duke University*; and Chin, Wai, *Colorado State University*

Feedback Control of Hopf Bifurcations 639
Chen, Guanrong; Yap, Keng C.; and Lu, Jialiang, *University of Houston*

Stability of a Continuous-Time State Variable Filter with ORA-L and Current Amplifier Integrators 643
Bakken, Tim W. and Choma, Jr., John, *University of Southern California*

A New Direct Digital Frequency Synthesizer Architecture for Mobile Transceivers 647
Hegazi, Emad Mahmoud; Ragaie, Hani Fikry; Haddara, H.; and Ghali, H., *In Shams University*

A Second-Order Log-Domain Bandpass Filter for Audio Frequency 651
Edwards, Robert Timothy and Cauwenberghs, Bert, *Johns Hopkins University*

A Theory of Information Network Analyzer PPN 655
Shinomiya, Norihiko and Wantanabe, Hitoshi, *Soka University*

CMOS Precision Half-Wave Rectifying Transconductor 659
Jun, Sibum, *LG Semicon Co., Ltd.* and Ahn, Su Jin, *POSTECH*

Analysis of Limit-Cycle Oscillations in a Log-Domain Filter 663
Fox, Robert M., *University of Florida*, and Ferrer, Enrique, *Motorola Inc.*

Large Signal Models for Oscillator Design 667
Kukk, Vello, *Tallinn Technical University*

An Approximate Analytical Approach for Predicting Period-Doubling in the Colpitts Oscillator 671
Maggio, Gian Mario and Kennedy, Michael Peter, *University College Dublin,* and Gilli, Marco, *Politecnico di Torino*

VOLUME 4

MAA4 – Image Processing and Coding
Chair: Charles Creusere
Naval Air Warfare Center

A Lapped Transform Progressive Image Coder 1
Tran, Trac D. and Nguyen, Truong, *Boston University*

Joint Channel and Source Decoding for Vector Quantized Images Using Turbo Codes 5
Peng, Zhishi; Huang, Yih-Fang; Costello, Daniel J.; and Stevenson, Robert L, *University of Notre Dame*

An Enhanced Trellis Coded Quantization Scheme for Robust Image 9
Chen, Chang Wen and Li, Hongzhi, *University of Missouri-Columbia*

Dimensional Adaptive Arithmetic Coding 13
Li, Weiping and Ling, Fan, *Lehigh University*

An Efficient Weight Optimization Algorithm for Image Representation using Nonorthogonal Basis Vectors 17
Chan, Yuk-Hee and Siu, Wan-Chi, *Hong Kong Polytechnic University*

TABLE OF CONTENTS

Morphological Signal Adaptive Median Filter for Still Image and Image Sequence Filtering — 21
Tsekeridou, Sofia; Kotropoulos, Constantine; and Pitas, Ioannis, *Aristotle University of Thessaloniki*

Vector Set Partitioning with Classified Successive Refinement VQ for Embedded Wavelet Image Coding — 25
Mukherjee, Debargha and Mitra, Sanjit K., *University of California, Santa Barbara.*

Perceptual Image Compression with Wavelet Transform — 29
Lai, Yung-Kai and Kuo, C.C.J., *University of Southern California*

MAA11 – VLSI Circuits for Multimedia Signal Processing
Chair: Magdy Bayoumi
University of Southwestern Louisiana

A Paradigm for Collaboration Across a Globally Networked Environment: Implementation of ISCAS '98 Internet Services — 33
Coffman, James W. and McEachen, John, *Naval Postgraduate School*

A Flexible MPEG Audio Decoder Layer III Chip Architecture — 37
Singh, P., Moreno, W., Ranganathan, N. and Neinhaus, H., *University of South Florida*

Low Power 2D DCT Chip Design For Wireless Multimedia Terminals — 41
Chen, Liang-Gee, Jiu, Juing-Ying, Chang, Hao-Chieh, Lee, Yung-Pin and Ku, Chung-Wei, *National Taiwan University*

Influences of Object Based Segmentation onto Multimedia Hardware Architectures — 45
Ohmacht, Martin, Wittenburg, Jens-Peter and Pirsch, Peter, *University of Hannover*

New Video And Multimedia Standards And Their Impact To Implementation — 49
Chen, Tsuhan, *Carnegie Mellon University*

Providing Multicast Video on Demand using Native-mode Asynchronous Transfer Mode — 53
Lockwood, John W.; Kang, Sung Mo; Hossain, Ashfaq; and Hiltenbrant, John, *University of Illinois*

An Introductin to Multi-Sensor Data Fusion — 57
Llinas, James, *State University of New York at Buffalo*, and Hall, David L., *Pennsylvania State University*

MPA4 – Multimedia Systems and Processing
Co-Chair: Bing Sheu
University of Southern California
Co-Chair: Ramalingam Sridhar
State University of New York, Buffalo

Bandwidth Planning in Near Video-on-Demand — 61
Chan, Shueng-Han Gary; Tobagi, Fouad; and Ko, T.M., *Stanford University*

A Low-Cost Architecture Design with Efficient Data Arrangement and Memory Configuration for MPEG-2 Audio Decoder — 65
Tsai, Tsung-Han, Chen, Liang-Gee, Huang, Sheng-Chieh, and Chang, Hao-Chieh, *National Taiwan University*

Real-time Digital Video Stabilization for Multimedia Applications — 69
Ratakonda, Krishna, *University of Illinois at Urbana-Champaign*

Scalable Image Sensor/Processor Architecture with Frame Memory Buffer and 2-D Cellular Neural Network — 73
Cho, Kwang-Bo; Sheu, Bing J.; and, Young, Wayne C. *University of Southern California*

The BJT-Based Silicon-Retina Sensory System for Direction- and Velocity-Selective Sensing — 77
Jiang, Hsin-Chin and Wu, Chung-Yu,, *National Chiao Tung University*

New View Generation from A Video Sequence — 81
Chen, Szo Sheng and Hang, Hsueh-Ming, *National Chiao Tung University*

A Fast Approach For Detecting Human Faces In A Complex Background — 86
Lam, Kin-Man, *The Hong Kong Polytechnic University*

A Memory Based Architecture for Real-time Convolution with Variable Kernels — 89
Moshnyaga, Vasily, *Fukuoka University*, Suzuki, Kazuhiro and Tamaru, Keikichi, *Kyoto University*

MPA13 – Image and Video Processing I
Co-Chair: Ya-Qin Zhang
Sarnoff Corporation
Co-Chair: Oscar Au
Hong Kong University of Science and Technology

A Multi-Transform Approach to Reversible Embedded Image Compression — 93
Adams, Michael David and Antoniou, Andreas, *University of Victoria*

Dynamic Load Balancing for Distributed Movie Based Browser Systems — 97
Hiraiwa, Atsunobu; Komatsu, Naohisa; Komiya, Kazumi; and Ikeda, Hiroaki, *Telecommunications Advancement Organization of Japan*

The Impact of Encoding Algorithms on MPEG VLSI Implementation — 102
Cheng, Sheu-Chih and Hang, Hsueh-Ming, *NCTU*

Efficient Subtree Splitting Algorithm for Wavelet-based Fractal Image Coding — 106
Po, Lai-Man, *City University of Hong Kong*; Zhang, Ying, *Guangdong Posts & Telecommunications*; Cheung, Kwok-Wai and Cheung, Chun-Ho, *City University of Hong Kong*

Error Resilient Image Coding with Rate-Compatible Punctured Convolutional Codes — 110
Cai, Jianfei and Chen, Chang Wen, *University of Missouri-Columbia*; and Sun, Zhaohui, *University of Rochester*

TABLE OF CONTENTS

Fast Motion Estimation Based on Total Least   114
Squares for Video Encoding
Deshpande, Sachin G. and Hwang, Jenq-Neng, *University of Washington*

Error Control for H.263 Video Transmission   118
over Wireless Channels
Chen, Yen-Lin and Lin, David W., *National Chiao Tung University*

Comparison between Block-based and Pixel-   122
based Temporal Interpolation for Video Coding
Tang, Chi Wah and Au, Oscar C., *HK University of Science & Technology*

A Scalable Hierarchical Motion Estimation   126
Algorithm for MPEG 2
Song, Xudong; Chiang, Tihao, and Zhang, Ya-Qin, *Sarnoff Corporation*

Transform Domain Motion Estimation without   130
Macroblock-based Repetitive Padding for MPEG-4
Video
Chen, Jie and Liu, K.J. Ray, *University of Maryland*

Using a Region-Based Blurring Method and Bits   134
Reallocation to Enhance Quality on Face Region in
Very Low Bitrate Video Coding
Chen, Chang-Hung; Chen, Liang-Gee; and Chang, Hao-Chieh, *DSP/IC Design Lab.*

A Novel and Fast Feature Based Motion   138
Estimation Algorithm Through Extraction of
Background and Object
Mok, Wai Hung and Yung, Hon Ching Nelson, *The University of Hong Kong*

An Adaptive Arithmetic Coding Method Using   142
Fuzzy Logic and Gray Theory
Jou, Jer Min and Chen, Pei-Yin, *National Cheng Kung University*

An Integrated Classifier in Classified Coding   146
Huang, Jiwu, *New Jersey Institute of Technology;* Chen, Li, *Shantou University;* and Shi, Yun Q., *New Jersey Institute of Technology*

Corner Detection using Gabor-Type Filtering   150
Quddus, Azhar and Fahmy, Moustafu, *King Fahd University of Pet. and Minerals*

Detection of Vehicle Occlusion Using a   154
Generalized Deformable Model
Yung, H. C. Nelson and Lai, Hon Seng, *The University of Hong Kong*

Error Resilient Coding for JPEG Image   158
Transmission over Wireless Fading Channels
Chandramouli, Rajarathnam, *University of South Florida,*

Noise Sensitivity Analysis for a Novel Error   162
Recovery Technique
Hasan, Moh'd Abdel Majid, *AMIEE* and Marvasti, Farokh I., *King's College, MIEEE*

On the Perceptual Interband Correlation for   166
Octave Subband Coding
Liu, Chi-Min and Wang, Chung-Neng, *National Chiao Tugn University*

Analyzing Memory Bandwidth Requirements of   170
Video Algorithms
Kapoor, Bhanu, *Texas Instruments Incorporated*

An Adaptive Network Control Scheme for   174
Region-Based Hybrid Coding Algorithm
Chen, Hsu-Tung, Chen, Liang-Gee, Huang, Sheng-Chieh, Tsai, Tsung-Han and Chang, Hao-Chieh, *National Taiwan University*

On Piecewise-Quadratic Filter for Gaussian   178
Noisy Image Filtering
Li, Wenzhe; and Lin, Ji-Nan; and Unbehauen, Rolf, *University Erlangen-Nuernberg*

MPEG-4 Accelerator for PC Based Codec   182
Implementation
Lim, Young-Kwon; Kwak, Jinsok; and Park, Sanggyu, *Electronics and Telecommunications Research Institute*

Rate Control in Video Coding by Adaptive Mode   186
Selection
Ryu, Chul, *Polytechnic University;* and Kim, Seung, *InterDigital Telecommunication*

TAA4 – Speech and Video Processing
Co-Chair: Chung-Yu Wu
National Chiao Tung University
Co-Chair: Thanos Stouraitis
University of Patras, Greece

Effcient Coding of Linear Predictive Coeffcients   190
for Wideband Speech
Magrath, A.J and Sandler, Mark B., *King's College London*

Regressive Linear Prediction with Triplets - An   194
Effective All-pole Modeling Technique for Speech
Processing
Varho, Susanna and Alku, Paavo, *University of Turku*

A Novel Algorithm to Estimate the Line Spectral   198
Frequencies from LPC Coefficients
Nakhai, Mohammad Reza and Marvasti, Farokh Alim, *King's College, University of London.*

Wideband Speech Recovery from Bandlimited   202
Speech In Telephone Communications
Yasukawa, Hiroshi, *Aichi Prefectural University*

An Efficient Method for the Removal of Impulse   206
Noise from Speech and Audio Signals
Chandra, Charu; Moore, Michael S., and Mitra, Sanjit K., *University of California, Santa Barbara*

Finite Wordlength Effects Analysis and Word-   209
length Optimization of Dolby Digital Audio Decoder
Lee, Seokjun and Sung, Wonyong, *Seoul National University*

The NLMS Algorithm Using a Quasi-Ortho-   213
normal Initialization Scheme for Acoustic Cancellation
Chen, Heng-Chou and Chen, Oscal T.-C., *National Chung Cheng University*

TABLE OF CONTENTS

Motion Estimation Using An Efficient Four-Step Search Method — 217
Chen, Oscal T.-C., *National Chung Cheng University*

TPA4 – Image and Video Processing
Co-Chair: Chung-Lin Huang
National Tsing Hua University
Co-Chair: Mohammed Ismail
Ohio State University

VLSI Implementation of Decoder for Decompressing Fractal-based Compressed Image — 221
Kim, Kyung-Hoon, Hong, Chang-Yu and Kim, Lee-Sup, *Korea Advanced Institute of Science and Technology*

Fast Integrated Algorithm and Implementations for the Interpolation and Color Correction of CCD-Sensed Color Signals — 225
Kuo, Kuo-Tang, *ITRI* and Chen, Sau-Gee, *National Chiao Tung University;*

Genetic Algorithms for Active Contour Optimization — 229
MacEachern, Leonard A. and Manku, Tajinder, *University of Waterloo*

TPA13 – Multimedia Processing
Co-Chair: Che-Ho Wei
National Chiao Tung University
Co-Chair: Yeong Ho Ha
Kyungpook National University

An LPC Cepstrum Processor for Speech Recognition — 233
Hwang, In-chul; Kim, Sung-Nam; Kim, Young-Woo; and Kim, Soo-Won, *Korea University*

A Visual Model for Subband Image Coding — 237
Fong, W.C., Chan, S.C., and Ho, K.L., *The University of Hong Kong*

A Fast and Accurate Scoreboard Algorithm for Estimating Stationary Backgrounds in an Image Sequence — 241
Lai, Hon Seng and Yung, Nelson, *The University of Hong Kong*

Subimage Error Concealment Techniques — 245
Hasan, Moh'd Abdel Majid, *AMIEE,* Sharaf, Atif Ibrahim and Marvasti, Farokh I. *King's College, MIEEE*

Wipe Scene Change Detector for Segmenting Uncompressed Video Sequences — 249
Alattar, Adnan Mohammed, *King Fahd University of Petroleum & Minerals*

Hierarchical Scene Change Detection in an MPEG-2 Compressed Video Sequence — 253
Shin, T, *Kwangju Institute of Science & Technology*; Kim, J.G. Lee, H. and Kim, J., *Electronic and Telecommunications Research Institute*

A Robust Linear Prediction Method for Noisy Speech — 257
Shimamura, Tetsuya, Kunieda, Nobuyuki and Suzuki, Jouji. *Saitama University*

Object Tracking and Hypermedia Links Creation in MPEG-2 Digital Video — 261
Favalli, Lorenzo; Mecocci, Alessandro; and Fulvio, Moschetti, *University of Pavia*

Syntax Based Error Concealment — 265
Papadakis, Vasilios, Lynch, William E. and Le-Ngoc, Tho, *Concordia University*

Digital Restoration of Painting Cracks — 269
Giakoumis, Ioannis and Pitas, Ioannis, *Aristotle University of Thessaloniki,*

WAA4 – Image and Video Processing III
Co-Chair: Peter Pirsch
University of Hannover,
Co-Chair: James Brailean
Motorola, Inc.

A Comparative Study of DCT and Wavelet Based Coding — 273
Xiong, Zixiang; *University of Hawaii*; Ramchandran, Kannan, *Princeton University*; Orchard, Michael T., *University of Illinios at Urbana-Champagne*; and Zhang, Ya-Qin, *Sarnoff Corporation*

A Two-Layer MPEG2-Compatible Video Coding Technique Using Wavelets — 277
Zan, Jinwen; Zhu, Wei-Ping; Ahmad, M. Omair, and Swamy, M.N.S., *Concordia University*

An Adaptive Video Sub-sampling Technique for the Conversion between High and Low Resolution — 281
Wong, Peter H.W.; Au, Oscar C.; Wong, Justy W.C.; and Tourapis, A., *Hong Kong University of Science and Technology*

Computation Reduction for Discrete Cosine Transform — 285
Pao, I-Ming and Sun, Ming-Ting, *University of Washington*

Scalable Coding of Video Objects — 289
Haridasan, Radhakrishnan and Barns, John S., *University of Maryland*

Hybrid DCT/Wavelet I-frame Coding for Efficient H.263+ Rate control at Low Bit Rates — 293
Song, Hwangjun; Kim, Jongwon; and Kuo, C.-C. Jay, *University of Southern California*

Hybrid Search Algorithm for Block Motion Estimation — 297
Cheung Chok-Kwan and Po, Lai-Man, *City University of Hong Kong*

Low-power MPEG2 Encoder Architecture for Digital CMOS Camera — 301
Hsieh, Jeff Yeu-Farn and Meng, Teresa H., *Stanford University*

MAA5 – Signal Processing for Communcations I
Chair: Alex Kot
Nanyang Technological University

FFT-Based Clipper Receiver for Fast Frequency-Hopping Spread-Spectrum System — 305
Teh, K. C.; Kot, Alex C.; and Li, K.H., *Nanyang Technological University*

TABLE OF CONTENTS

University, Singapore

Harmonic and Intermodulation due to   309
Requantization of Fixed-Point Numbers
Hentschel, Tim and Fettweis, Gerhard, *Dresden University of Technology*

A Comparison of CAP/QAM Architectures   316
Abdolhamid, Amir and Johns, David A., *University of Toronto*

Quantization for Robust Sequential M-ary Signal   317
Detection
Chandramouli, Rajarathnam and Ranganathan, Nagarajan, *University of South Florida,*

The Optimal RLS Parameter Tracking   321
Algorithm for a Power Amplifier Feed-Forward
Linearizer
Chen, Jiunn-Tsair; Tsai, Huan-Shong; and Chen, Young-Kai, *Stanford University*

Selectivity and Sensitivity Performances of   325
Superregenerative Receivers
Vouilloz, Alexandre and Declercq, M., *Swiss Federal Institute of Technology*

A New Model for the DOA Estimation of the   329
Coherent Signals
Jin, Liang; Yao, Minli; and Yin, Qinye, *Xi'an Jiaotong University*

A System Scheme for Downlink Selective   333
Beamforming In Smart Antenna
Jin, Liang, Li, Jiang, and Yin, Qinye, *Xi'an Jiaotong University*

MAA13 – Communications Circuits
Chair: Robert H. Caverly
Villanova University

Nonlinear Properties of Gallium Arsenide and   337
Silicon FET-based RF and Microwave Switches
Caverly, Robert H, *Villanova University*

A 3-V 45-mW CMOS Differential Bandpass   341
Amplifier for GSM Receivers
Leung, David Lap Chi and Luong, Howard Cam, *Hong Kong University of Science & Technology*

Gb/s Encoder/Decoder Circuits for Fiber Optical   345
links in Si-Bipolar Technology
Mokhtari, Mehran; Ellervee, Peter; Schuppar, Gerd; Juhola, Tarja; and Tenhunen, Hannu, *Royal Institute of Technology*

A Comparative Analysis of CMOS Low Noise   349
Amplifiers for RF Applications
Mayaram, Kartikeya and Ge, Yongmin, *Washington State University*

Characterization of Micromachined CMOS   353
Transmission Lines for RF Communications
Ozgur, Mehmet, Milanovic, Veljko, Zincke, Christian and Zaghloul, Mona E., *The George Washington University*

A 2.0-GHz Submicron CMOS LNA and a   357
Downconversion Mixer
Litmanen, Petteri Matti, *Texas Instruments*; Ikalainen, P.and Halonen, Kari A., *Helsinki University of Technology*

Easy Simulation and Design of On-Chip   360
Inductors in Standard CMOS Processes
Christensen, Kåre Tais and Jergensen, Allan, T*echnical University of Denmark*

Programmable Low Noise Amplifier with Active-   365
Inductor Load
Zhuo, Wei, Pineda de Gyvez, Jose and Sanchez-Sinencio, Edgar, *Texas A&M University*

Electro-Mechanical Properties of a Micro-   369
machined Varactor with a Wide Tuning Range
Dec, Aleksander M.and Suyama, Ken, *Columbia University*

A Low Voltage Design Technique for Low Noise   373
RF Integrated Circuits
Abou-Allam, Eyad and Manku, Tajinder, *University of Waterloo*

A 1.8GHz CMOS Quadrature Voltage-Controlled   378
Oscillator (VCO) using the Constant-Current LC
Ring Oscillator Structure
Wu, Chung-Yu and Kao, Hong-Sing, *National Chiao Tung University*

Low Voltage, 2 X 2, 25 Gb/s Crosspoint Switch in   382
InP-HBT Technology
Mokhtari, Mehran, *Royal Institute of Technology*

A Cordic-based Digital Quadrature Mixer :   385
Comparison with a ROM-based Architecture
Nahm, Seunghyeon, Han, Kyungtae and Sung, Wonyong, *Seoul National University*

Reconfigurable Signal Processing ASIC   389
Architecture for High Speed Data Communications
Grayver, Eugene, *UCLA*

Dual Loop DSP-PLL with Wide Frequency   393
Acquisition Range and Fast Frequency Acquisition
Obote, Shigeki, *Tottori University*; Sumi, Yasuaki, *Tottori Sanyo Electric Co. Ltd.*; Syoubu, Kouichi, Fukui, Yutaka and Itoh, Yoshio, *Tottori University*

Pipelined Arrays for Modular Multiplication   397
Ciminiera, Luigi, *Politecnico di Torino*

Distortion and Noise Performance of Bottom-   401
Plate Sampling Mixers
Yu, Wei , and Leung, Bosco, *University of Waterloo*

A Digital Frequency Modulator Circuit for   405
a Dual-Mode Cellular Telephone
Lahti, Jukka A. and Niemistö, Matti, *University of Oulu*

Design of a 2.4 GHz CMOS Frequency-Hopped   409
RF Transmitter IC
Kosunen, Marko, Vankka, Jouko, Waltari, Mikko, Sumanen, Lauri, Koli, Kimmo, and Halonen, Kari, *Helsinki University of Technology*

A Bipolar Semi-Custom PLL-based Synthesizer   413
for GSM and DCS Systems
Hakkinen, Juha Tapio; Rahkonen, Timo; and Kostamovaara, Juha,

TABLE OF CONTENTS

University of Oulu

Adaptive FEC on a Reconfigurable Processor for Wireless Multimedia Communications   417
Arrigo, Jeanette F., Page, Kevin J., Wang, Yuhe and Chau, Paul M., *University of California San Diego*

Low-Power, Low-Phase-Noise CMOS Voltage-Controlled-Oscillator with Integrated LC Resonator   421
Park, Byeong-Ha and Allen, Phillip E., *Georgia Institute of Technology*

PLL Frequency Synthesizer with Multi-Programmable Divider   608
Sumi, Y., Syoubu, K., Obote, S., and Fukui, Y., *Tottori University*

A 3.3V 600Mhz - 1.30Ghz CMOS Phase-Locked Loop for Clock Synchronization of Optical Chip-to-Chip Interconnects   425
Sheen, Robin R.-B. and Chen, Oscal T.-C., *National Chung Cheng University*

MPA5 – Equalization/Modulation/Decoding
Co-Chairs: Simone Fiori and Francesco Piazza
University of Ancona, Italy

A Novel Reinitialization Method for Successive Blind Equalization of MIMO Communication Channel   429
Ma, Chor Tin and Yau, Sze Fong, *The Hong Kong University of Science and Technology*

Narrow-Band Interference Rejection in OFDM-CDMA Transmission System   433
Hsiao, Hsu-Feng; Hseih, Meng-Han; and Wei, Che-Ho, *National Chiao Tung University*

BLADE: A New On-Line Blind Equalization Method based on the Burelian Distortion Measure   437
Fiori, Simone and Piazza, Francesco, *University of Ancona*

A Parallel Decoding Scheme for Turbo Codes   441
Hsu, Jah-Ming and Wang, Chin-Liang, *National Tsing Hua University*

Efficient Management of In-Place Path Metric Update and Its Implementation For Viterbi Decoders   445
Shieh, Ming-Der, Sheu, Ming-Hwa, Wu, Chien-Ming, and Ju, Wann-Shyang, *National Yunlin University of Science & Technology*

Three-Dimensional Equalization for the 3-D QAM System with Strength Reduction   449
Shalash, Ahmed F., and Parhi, Keshab K., *University of Minnesota*

A New Polynomial Structure For Channel Equalization and ACI Suppression in 64-QAM Reception   453
Srivastava, M.C. and Saini, J.P., *Kamla Nehru Institute of Technology*

TAA5 – Communicating with Chaos I
Chair: Michael Peter Kennedy
University College, Dublin
Organizer: Geza Kolumban
Technical University of Budapest

Recent Advances in Communicating with Chaos   457
Kennedy, Michael Peter, *University College Dublin;* Kolumban, Geza, Kis, Gabor and Jako, Zoltan, *Technical University of Budapest*

Statistical Analysis of Chaotic Communication Schemes   461
Abel, Andreas, Gotz, Marco and Schwarz, Wolfgang, *TU Dresden*

The Performance of Chaos Shift Keying: Synchronization Versus Symbolic Backtracking   465
Schweizer, Jorg, *Swiss Federal Institute of Technology*

Integrated Circuit Blocks for a DCSK Chaos Radio   469
Delgado-Restituto, Manuel and Rodriguez-Vazquez, Angel, *Instituto de Microelectronica de Sevilla;* and Porra, Veikko, *Helsinki University of Technology*

FM-DCSK: A Novel Method for Chaotic Communications   473
Kolumban, G., *Technical University of Budapest*; Kennedy, M.P., *University College of Dublin*; Kis, G, and Jako, Z., *Technical University of Budapest*

High Rate Data Communication using Dynamical Chaos   477
Dmitriev, Alexander Sergeevich; Yemetz, Sergei; and Starkov, Sergei Olegovich, *IRE RAS*

Sequence Synchronization in Chaos-Based DS-CDMA Systems   481
Mazzini, Gianluca, *CSITE – CNR*; Rovatti, Riccardo and Setti, Gianluca, *University of Bologna*

Implementing RF Broadband Chaotic Oscillators: Design Issues and Results   485
Silva, Christopher Patrick and Young, Albert M., *The Aerospace Corporation*

TPA5 – Signal Processing for Communications II
Chair: R. Clark Robertson
Naval Postgraduate School

A New Processor Architecture Dedicated to Digital Modem Applications   490
Monteiro, Fabrice; Philip Serge; Danddache, Abbas; and Lepley, Bernard, *University of Metz*

A GSM Modulator Using a Delta-Sigma Frequency Discriminator Based Synthesizer   494
Bax, Walt T. and Copeland, Miles A., *Carleton University*

Error Resilient Transmission of H.263 Coded Video over Mobile Networks   498
Lu, Jianhua, Liou, Ming L. and Letaief, K.B., *Hong Kong University of Science & Technology*; and Chuang, Justin C-I, *AT&T Labsh*

Discrete Fractional Hilbert Transform   502
Pei, Soo-Chang, and Yeh, Min-Hung, *National Taiwan University*

TPA9 – Communications with Chaos II
Chair: Michael Peter Kennedy
University College Dublin

TABLE OF CONTENTS

Organizer: Geza Kolumban
Technical University of Budapest

Communicating via Chaos Synchronization Generated by Noninvertible Maps — 506
Millerioux, Gilles and Mira, Christian, *Institut National des Sciences Appliquées*

From Chaotic Maps to Encryption Schemes — 510
Kocarev, Ljupco and Jakimoski, Goce, *"Sv. Kiril i Metodij" University*; Stojanovski, Toni, *RMIT University*; and Parlitz, Ulrich, *Universitat Gottingen*

Chaotic versus Classical Stream Ciphers - A Comparative Study — 514
Dachselt, Frank, *TU Dresden*; Kelber, Kristina; Schwarz, Wolfgang; Vandewalle, Joos, *KU Leuven*

Some Tools for Attacking Secure Communication Systems employing Chaotic Carriers — 518
Ogorzalek, Maciej J., *University of Mining and Metallurgy*; and Dedieu, Herve, *Swiss Federal Institute of Technology*

WAA5 – Wireless/Mobile Communications
Chair: Donald Gingras
SPAWAR Systems Center

A Priority-Based Random Access Spread-Spectrum Protocol for Integrated Voice/Data Packet Networks — 522
Gingras, Donald F. and Lapic, Stephan K., *SPAWAR Systems Center*

A Chip-Interleaving DS SS System and Its Performance under On-Off Wide-Band Jamming — 526
Gui, Xiang and Ng, Tung Sang, *The University of Hong Kong*

A Genetic-Algorithm-Based Multiuser Detector for Multiple-Access Communications — 530
Wang, Xiao-Feng; Lu, Wu-Sheng; and Antoniou, Andreas, *University of Victoria*

A New Multiple Access Protocol for Multimedia Wireless Networks — 534
Salles, Ronaldo Moreira and Gondim, Paulo Roberto de Lira, *Military Institute of Engineering*

A DS-CDMA Receiver Using Exponentially Weighted Despreading Waveforms — 538
Huang, Yuejin and Ng, Tung Sang, *University of Hong Kong*

Blind Joint Equalization and Multiuser Detection for DS-CDMA in Unknown Correlated Noise — 542
Wang, Xiaodong and Poor, H. Vincent, *Princeton University*

A Multiple-Access Interference Suppression Technique Employing Orthogonal/Random Spreading Sequences and a Novel Decentralized Receiver for B-CDMA Forward Link Systems in Multipath Channels — 546
Shin, Sung-Hyuk and Voltz, Peter J., *Polytechnic University*

WAA10 - Communicating with Chaos III
Chair: Michael Peter Kennedy
University College Dublin
Organizer: Geza Kolumban
Technical University of Budapest

Multiplex Communication Scheme based on Synchronization via Multiplex Pulse-Trains — 550
Torikai, Hiroyuki, *HOSEI University*; Saito, Toshimicho and Schwarz, Wolfgang, *Dresden University of Technology*

Synchronizing Autonomous Chaotic Circuits using Bandpass Filtered Signals — 554
Carroll, Thomas L. and Johnson, Gregg A., *US Naval Research Lab*

Master Stability Functions for Synchronized Chaos in Arrays of Oscillators — 558
Pecora, Louis M. and Carroll, Thomas L., *Naval Research Laboratory*

Exploiting the Concept of Conditional Transversal Lyapunov Exponents for Study of Synchronization of Chaotic Circuits — 564
Galias, Zbigniew, *University of Mining and Metallurgy*

Nonlinear Hinfinity Synchronization: Case Study for a Hyperchaotic System — 568
Suykens, Johan A.K. and Vandewalle, J., *K.U. Leuven*; Chua, L.O. *University of California at Berkeley*

Estimation Via Synchronization: FM Demodulation Example — 572
Hahs, Daniel W. and Corron, Ned J., *Dynetics, Inc.*

Chaos Synchronization in Coupled Phase System — 576
Shalfeev, Vladimir Dmitrievich, *University of Nizhny Novgorod*; Matsrov, V.V. and Korzinova, M.V., *University of California, San Diego*

WPA5 – Archs., Algors., and Impl. for Wireless Communications Systems
Chair: H. V. Poor
Princeton University
Organizer: G. W. Wornell
Massachusetts Institute of Technology

Creating and Exploiting Diversity in Wireless Systems: A Signal Processing Perspective — 579
Wornell, Gregory W., *Massachusetts Institute of Technology*

Blind Demodulation of High-Order QAM Signals in the Presence of Cross-Pole Interference — 581
Treichler, John and Bohanon, Jon, *Applied Signal Technology, Inc.*

Signal Processing Algorithms for Adaptive Interference Suppression — 585
Poor, H. Vincent, *Princeton University*

Design of a Wideband Spread Spectrum Radio Using Adaptive Multiuser Detection — 589
Teuscher, Craig, Yee, Dennis, Zhang, Ning and Brodersen, Robert, *UC Berkeley*

TABLE OF CONTENTS

Distributed Network Protocols for Wireless Communication — 596
Meng, Teresa H. and Rodoplu, Volkan, *Stanford University*

Trends in Low Power Digital Signal Processing — 600
Chandrakasan, Anantha, Amirtharajah, Rajeevan, Goodman, James and Rabiner, Wendi, *Massachusetts Institute of Technology*

Low-Power Radio Frequency Circuit Architectures for Portable Wireless Communications — 604
Larson, Lawrence E., *University of California - San Diego*

Volume 5

MAA1-Perimeter Estimation
Chair: Eugene Plotkin
Concordia University

Parameter-Free Structural Modeling: A Contribution to the Solution of the Separation of Highly Correlated AR-Signals — 1
Plotkin, Eugene I. and Swamy, M.N.S., *Concordia University*

On the Harmonic Analysis of Speech — 5
Stylianou, Yannis, *AT&T Labs Research*

A New Approach For Coherent Direction-Of-Arrival Estimation — 9
Lai, W.K. and Ching, Pak-Chung, *The Chinese University of Hong Kong*

Non-Minimum Phase FIR System Identification Using Cumulants with Selected Orders — 13
Li, Wei and Siu, W.C., *The Hong Kong Polytechnic University*

A Subspace Method For Blind Single Channel Identification Using Redundancy-Transform in Transmitters — 17
Choi, Jinho, *LGIC*

On Implementation of a Least-Squares Based Algorithm for Noisy Autoregressive Signals — 21
Zheng, Wei Xing, *University of Western Sydney*

Parallel Computation of SVD for High Resolution DOA Estimation — 25
Feng, Gang and Liu, Zemin, *Beijing University of Posts & Telecom*

Performance Analysis of a Class of Cyclic Weighted Subspace Fitting Method of Direction Estimation for Cyclostationary Signals — 29
Yu, Hongyi and Zheng, Bao, *Xidian University*

MAA2-Single-Rate and Multirate Filters
Chair: Charles Creusere
Naval Air Warfare Center

The Design of Optimum Filters for Quantizing a Class of Non Bandlimited Signals — 33
Tuqan, Jamal, and Vaidyanathan, P.P., *California Institute of Technology*

M-th Band Filter Design Based on Cosine Modulation — 37
Oraintara, Soontorn, and Nguyen, Truong Q., *Boston University*

An Iterative Quadratic Programming Method for Multirate Filter Design — 41
Mo, Y.-S ; Lu, Wu-Sheng; and Antoniou, Andreas, *University of Victoria*

Filter Structures Composed of Allpass and FIR Filters for Interpolation and Decimation with Factors of Two — 45
Johsnsson, Haken, and WanHammer, Lars, *Linkoping University*

Realization of General 2-D Linear-Phase FIR Filters Using Singular-Value Decomposition — 49
Zhu, Wei-Ping, Ahmad, M. Omair, and Swamy, M.N.S., *Concordia University*

New Insights Into Multirate Systems with Stochastic Inputs Using Bifrequency Analysis — 53
Akkarakaran, Sony and Vaidyanathan P.P., *California Institute of Technology*

Synthesis of 2-D Half-Band Filters using the Frequency Response Masking Technique — 57
Low, Seo-How, and Lim, Yong-Ching, *National University of Singapore*

Continuous-Time Signal Processing Based on Polynomial Approximation — 61
Vesma, Jussi; Hamila, Ridha; Renfors, Markku; and Saramaki, *Tampere University of Technology*

MPA1-Multidimensional Signal Processing
Chair: M.Omair Ahmed
Concordia University

Symmetry in the Frequency Response of Two-Dimensional {gamma1, gamma2} Complex Plane Discrete-Time Systems — 66
Reddy, Hari C., *California State University, Long Beach/UC Irvine*, Khoo, I-Hung, *University of California, Irvine;* Rajan. P. K.; *Tennessee Tech University,* and Stubberud, Allen R., *University of California, Irvine*

FPGA Implementation of Hierarchical Clustering Algorithms — 70
Niamat, Y., Bitter, D., and Jamali, M.M., *The University of Toledo*

Multidimensional Digital Filter Approach for Numerical Solution of a Class of PDEs of the Propagating Wave Type — 74
Basu, Sankar, *IBM T. J. Watson Research Center;* and Zerzghi, Amanuel, *Lucent Technologies*

On (P,Q)-Markov Covers for 2-D Separable Denominator Systems — 78
Sreeram, V. and Liu, W.Q., *University of Western Australia;* Agathoklis, Pan, *University of Victoria*

Weighted L2 Sensitivity Minimization of 2-D Discrete Systems — 82
Hinamoto, Takao and Yokoyama, Shuichi, *Hiroshima University;* Lu,

TABLE OF CONTENTS

Wu-Sheng, *University of Victoria*

The Two Dimensional Lapped Hadamard Transform — 86
Muramatsu, Shogo, Yamada, Akihiko, and Kiya, Hitoshi, *Tokyo Metropolitan University*

A New 2-D Adaptive Filter Using Affine Projection Algorithm — 90
Muneyasu, Mitsuji, and Hinamoto, Takao, *Hiroshima Univeristy*

A Stability Test of Reduced Complexity for 2-D Digital System Polynomials — 94
Bistritz, Yuval, *Tel Aviv University*

MPA2-Optimization of Subband Coders Based on the Input
Chair: P. P. Vaidyanathan, *California Institute of Technology*

Enhancing the Performance of Subband Audio Coders for Speech Signals — 98
Malvar, Henrique S., *Microsoft Research*

Optimized Orthogonal and Biorthogonal Wavelets Using Linear Parameterization of Halfband Filters — 102
Lu, Wu-Sheng, and Antoniou, Andreas, *University of Victoria*

Optimization of High-Energy-Compaction, Nearly-Orthonormal, Linear-Phase Filter Banks — 106
Yang, Xuguang; Ramchandran, Kannan; and Moulin, Pierre, *University of Illinois at Urbana-Champaign*

Post-Processing of Compressed Images with Side Information1 — 110
Nosratinia, Aria, *Rice University*

A Survey of the State-of-the Art and Utilization of Embedded, Tree-Based Coding — 114
Pearlman, William A., *Rensselaer Polytechnic Institute*; and Said, Amir, *Iterated Systems, Inc.*

A Performance Study of DCT and Subband Image Codecs with Zero-Zone Quantizers — 118
Ramkumar, Mahalingam and Akansu, Ali N., *New Jersey Institute of Technology*

The Role of the Discrete-Time Kalman-Yakubovitch-Popov Lemma in Designing Statistically Optimum Fir Orthonormal Filter Banks — 122
Tuqan, Jamal, and Vaidyanathan, P.P., *California Institute of Technology*

Design of Linear Phase Paraunitary Filter Banks with Suboptimal Coding Gain Without Nonlinear Optimization — 126
Ikehara, Masaaki, *Keio University* and Nguyen, Truong Q., *Boston University*

TAA1-Filter Banks and Wavelets
Chair: Yih-Fang Huang
University of Notre Dame

A Filter Bank - Mother Wavelet Relationship in the Context of the Discrete Time Wavelet Transform — 130
Hanna, Magdy Tawfik, *Cairo University* and Mansoori, Sana Ahmed, *University of Bahrain*

Design of Signal-Adapted Linear Phase Paraunitary Filter Banks — 134
Takeuchi, Tomoaki; Nagai, Takayuki; and Ikehara, Masaaki, *Keio University*

Mutual Relations between Arithmetic and Haar Functions — 138
Falkowski, Bogdan J. *Nanyang Technological University*

A New Approach to the Design of QMF Banks — 142
Kao, Min-Chi and Chen, Sau-Gee, *National Chiao Tung University*

Rationalizing the Coefficients of Popular Biorthogonal Wavelet Filters — 146
Tay, David Ban Hock, *Nanyang Technological University*

Analytical Design for a Family of Cosine Modulated Filter Banks — 150
Siohan, Pierre and Roche, Christian, *CNET/DMR*

Results on Optimal Biorthogonal Subband Coders — 154
Vaidyanathan, P.P. and Kirac, Ahmet, *California Institute of Technology*

An Efficient Algorithm To Design Perfect Reconstruction Regular Quadrature Mirror Filters Using Weighted Lp Error Criteria — 158
Goh, Chee Kiang, and Lim, Yong Ching, *National University of Singapore*

TAA13-Adaptive Signal Processing I
Chair: Paulo S. Diniz
Federal University of Rio De Janiero

A New Delayless Subband Adaptive Filter Structure — 162
Diniz, Paulo Sergio Ramirez, *Federal University of Rio de Janeiro*

On the Design of the Target-Signal Filter in Adaptive Beamforming — 166
Joho, Marcel, *and* Moschytz, George S., *Swiss Federal Institut of Technology*

A New Modified Variable Step Size for the LMS Algorithm — 170
Okello, James, Itoh, Yoshio, Fukui, Yutaka, and Nakanishi, Isao, *Tottori University* and Kobayashi, Masaki; *Ibaraki University*

Adaptive Prediction of Sample Values for Digital Transducers — 174
Hölling, Matthias; Thaler, Markus; and Troster, Gerhard; *SwissFederal Institute of Technology*

A DSP-Based Modular Architecture for Noise Cancellation and Speech Recognition — 178
Gómez, Pedro; Álvarez, Agustín; Martínez, Rafael; Rodellar Victoria; Pérez-Castellanos; and Nieto, Victor; *DATSI, Universidad Politecnica de Madrid*

TABLE OF CONTENTS

An Efficient Approach to Noise Suppression — 182
in Adaptive Filtering Subject to Output Envelope
Constraints
Zheng, Wei Xing, *University of Western Sydney,*

A Feedback ANC System Using Adaptive — 186
Lattice Filters
Yeung, Tak Keung and Yau, Sze Fong, *The Hong Kong University of Science and Technology,*

Pipelining of 2-Dimensional Adaptive Filters — 190
Based on the LDLMS Algorithm
Kimijima, Tadaaki; Nishikawa, Kiyoshi; and Kiya, Hitoshi, *Tokyo Metropolitan University*

Transform-Domain Delayed LMS — 194
Algorithm and Architecture
Wu, An-Yeu and Wu, Cheng-Shing, *National Central University*

Generalization of Exponentially Weighted — 198
RLS Algorithm Based on a State-Space Model
Lee, Yong Hoon; Kim, Beomsup; and Chun, Byungjin, *Korea Advanced Institute of Science and Technology*

Adaptive Spectral Estimation Based on EXP — 202
Model
Sanubari, Junibakti, and Tokuda, Keiichi, *Satya Wacana University*

LMS/LMF and RLS Volterra System — 206
Identification based on Nonlinear Wiener Model
Chang, Shue-Lee, and Ogunfunmi, Tokunbo, *Santa Clara University*

TAB13-DSP Implementations
Chair: Keshab Parhi
University of Minnesota

16-Point High Speed (I)FFT for OFDM — 210
Modulation
Bertazzoni, Stefano; Cardarilli, Gian Carlo; Iannuccelli, Manuele; Salmeri, Marcello; Salsano, Adelio; and Simonelli, Osvaldo, *University. of Rome "Tor Vergata"*

Use of the Chinese Abacus Method for — 213
Digital Arithmetic Functions
Maloberti, Franco, *University of Pavia; and* Gang, Chen, *Hunan Normal University*

Residue to Binary Number Converters for Three — 217
Moduli Set
Wang, Yuke, Swamy, M.N.S., Ahmad, M. O., *Concordia University*

A New Hybrid Low-Latency Serial-Parallel — 221
Multiplier
Al-Besher, B.; Bouridane, Ahmed; Ashur, A.S.; and Crookes, D., *The Queen's University of Belfast*

Efficient Prime Factor Decomposition — 225
Algorithm and Address Generation Techniques for the Computation of Discrete Cosine Transform
Chau, Lap-Pui, *Nanyang Technological University,* Lun, Daniel Pak-Kong, and Siu, Wan-Chi, *Hong Kong Polytechnic University*

A CORDIC Algorithm with Fast Rotation — 229
Prediction and Small Iteration
Lin, Chun-Fu *Vanguard International Semiconductor Corporation*; Chen, Sau-Gee, *National Chiao Tung University*

Efficient Algorithms for Binary Logarithmic — 233
Conversion and Addition
Wan, Yi and Wey, Chin-Long *Michigan State University*

High Level Performance Estimation For — 237
A Primitive Operator Filter FPGA
Arslan, Tughrul; Eskikurt; Halil Ibrahim; and D.H.Hoerocks, *Cardiff University of Wales*

Direct Digital Frequency Synthesis Using a — 241
Modified Cordic
Grayver, Eugene and Daneshrad, Babak *University of California-Los Angeles*

High - Speed Cordic Based Parallel Weight — 245
Extraction For QRD-RLS Adaptive Filtering
Ma, Jun, and Parhi, Keshab K., *University of Minnesota;* and Deprettere, Ed F., *Delft University of Technology,*

Design of Optimum Power Estimator Based on — 249
Wiener Model Applied to Mobile Transmitter
Power Control
Huang, A. *Helsinki University of Technology/ Zhejiang University*; Tanskanen, J.M.A.; and Hartimo, I.O., *IRC, Helsinki University of Technology*

An Implementation Of A Normalized ARMA — 253
Lattice Filter With A Cordic Algorithm
Shiraishi, Shin-ichi; Haseyama, Miki; and Kitajima, Hideo, *Hokkaido University*

TPA1-Adaptive Signal Processing II
Chair: M.N.S. Swany
Concordia University

A New Adaptive Algorithm for Reducing the — 257
Hardware Complexity
Lee, Haeng-Woo, *Byuksung College; and* Cha, Jin-Jong, and Kim, Kyung-Soo, *ETRI*

A New Approach To Least-Squares Adaptive — 261
Filtering
Kocal, Osman Hilmi, *Istanbul Technical University*

Effective Algorithms For Regressor Based — 265
Adaptive Infinite Impulse Response Filtering
Acar, Emrah, *Carnegie Mellon University;* Arikan, Orhan, *Bilkent University*

Simplified Realization of Cascaded Adaptive — 269
Notch Filters Using Complex Coefficients
Nishimura, Shotaro, and Jiang, Hai-Yun, *Shimane University*

TPA2-Coding of Arbitrarily Shaped Objects
Chair: Weiping, Li
Lehigh University

Coding of Arbitrarily Shaped Objects with — 273
Binary and Greyscale Alpha-Maps.
What can MPEG-4 Do for You?

TABLE OF CONTENTS

Ostermann, Joern, *AT&T Labs*

Predictive Shape Coding using Generic   277
Polygon Approximation
Kim, Jong-il, and Evans, Brian L., *The University of Texas at Austin*

Shape Adaptive Wavelet Coding   281
Li, Shipeng, *Sarnoff Corporation;* Li, Weiping, *Lehigh University;* Sun, Hongqiao, *Vector Vision, Inc.;* and Wu, Zhixiong, *Oki Corporation*

Joint Shape and Texture Rate Control for   285
MPEG-4 Encoders
Vetro, Anthony, and Sun, Huifang, *Mitsubishi Electric ITA;* and Wang, Yao, *Polytechnic University*

Rate-Distortion Optimal Boundary   289
Encoding Using an Area Distortion Measure
Melnikov, Gerry, *Northwestern University*

WAA1-Adaptive Signal Processing III
Chair: Adreas Antoniou
University of Victoria

DOA Estimation of Speech Source with   293
Microphone Arrays
Jian, Ming, *Nanyang Technological University*

Two New Model Order Selection Approaches   297
for ARMA System Modeling Using The Two-
Dimensional Frequency Domain Least Square
Algorithm
Mikhael, Wasfy B., *University of Central Florida;* Roman, Jaime R., *Scientific Studies Corporation;* Zhang, Qingwen, *University of Central Florida;* and Davis, Dennis, *Scientific Studies Corporation*

An Algorithm-Based Fault Tolerant Method   301
for the 2-D LMS Algorithm
Schmitz, Christopher Dale, and Jenkins, W. Kenneth, *University of Illinois at Urbana-Champaign*

An Adaptive Kalman Filter for the   305
Enhancement of Noisy AR Signals
Doblinger, Gerhard, *Vienna University of Technology*

Acceleration of Normalized Adaptive   309
Filtering Data-Reusing Methods using the
Tchebyshev and Conjugate Gradient Methods
Soni, Robert A., and Jenkins, W. Kenneth *University of Illinois at Urbana-Champaign;* and Gallivan, Kyle A., *Florida State University*

An Adaptive Beamspace Algorithm for   313
Mobile Satellite Communications Systems using
Orthogonal Waveforms and Convolutional Codes
Terry John D. and Williams, Douglas B., *Georgia Institute of Technology*

On Unbiased Adaptive IIR Filtering Algorithms   317
Cousseau, Juan E., and Don~ate, Pedro D., *Universidad Nacional del Sur;* and Diniz, Paulo S. *Universidade Federal do Rio de Janeiro*

A Context Based Lossless Compression   321
Algorithm for Ionogram Data
Ye, Hua, *La Trobe University*

WAA2-Steerable Filters and Applications
Chair: Vedat Tavsanoglu
Southbank University

Steerable Pyramid Filters for Selective   325
Image Enhancement Applications
Wu, Qiang, and Castleman, Kenneth R., *Perceptive Scientific Instruments, Inc.*

Simplified Design of Steerable Pyramid Filters   329
Castleman, Kenneth R., *Perceptive Scientific Instruments, Inc.*

Multirate Separable Implementation of   333
Steerable Filter Banks
Manduchi, Roberto, *Caltech*

A Common Framework for Steerability,   337
Motion Estimation, and Invariant Feature Detection
Teo, Patrick C., *Stanford University;* and Hel-Or, Yacov, *Hewlett-Packard Labs, Israel*

Handwritten Character Recognition using   341
Steerable Filters and Neural
Tufan, Emir *Istanbul University*

Motion Analysis Using Steerable Filters For   345
The Application To Low Quality Images
Tavsanoglu, Vedat, *Southbank University;* Buhmann, Sitta, *Hochschule Bremen*

The SVD Approach For Steerable Filter Design   349
Herpers, Rainer, and Sommer, Gerald, *University of Kiel; and* Michaelis, Markus, *Plettac Electronics*

WAA13-Digital Filter Designs and Implementation
Chair: Y.C. Lim
National University of Singapore

Efficient Parallel FIR Filter Implementations   354
Using Frequency Spectrum Characteristics
Parhi, Keshab K., *University of Minnesota*

Signed Power-Of-Two(SPT) Term Allocation   359
Scheme For The Design Of Digital Filters
Lim, Yong-Chang, and Yang, Rui, National University of Singapore;
Li, Dongning, Systems Technology; and Sing, Jianjian, National Superconducting Research Center

Peak-Constrained Design of Nonrecursive   363
Digital Filters with High Passband/Stopband
Energy Ratio
Netto, Sergio Lima, and Diniz, Paulo Sergio Ramirez, *COPPE - Universidade Federal do Rio de Janeiro*

An Iterative Reweighted Least Squares   367
Algorithm for Constrained Design of Nonlinear Phase
FIR Filters
Lang, Mathias C., *Vienna University of Technology*

Design of Linear Phase FIR Filters Using the   371

TABLE OF CONTENTS

Nonuniform DCT
Okuda, Maahiro; Kageyuki, Kiyose; Okuda, Masahiro; Kiyose, Kageyuki; Ikehara, Masaaki; and Takahashi, Shin-ichi, *Keio University*

Optimal Fixed-Point VLSI Structure of a   375
Floating-Point Based Digital Filter
Wu, An-Yeu and Hwang, Kuo-Fuo, *National Central University*

Improved Tuning Accuracy Design of   379
Parallel-Allpass- Structures-Based Variable Digital Filters
Stoyanov, Georgi Kostov, *Tohoku University*

Computationally Fast Lattice Bilinear Digital   383
Ladder Filters with Comparison to Circulator WDF's
Harnefors, Lennart, *Malardalen University*

Design of General-Order Bode-Type   387
Variable-Amplitude Digital Equalizers
Nowrouzian, Behrooz, *University of Alberta*

A Novel Modified Branch-and-Bound   391
Technique for Discrete Optimization over Canonical Signed-Digit Number Space
Nowrouzian, Behrooz, *University of Alberta*

Analytical Guess of Error for Nonlinear   395
FIR Filters to Approximate Linear Phase Response
YAGYU, Mitsuhiko, Nishihara, Akinori, and Fujii, Nobuo, *Tokyo Institute of Technology*

A Systematic Technique for Designing   399
Approximately Linear Phase recursive Digital Filters
Saramaki, Tapio Antero, *Tampere University of Technology*

Design of Very Low-Sensitivity and   404
Low-Noise Recursive Digital Filters using a Cascade of Low-Order Wave Lattice Filters
Saramaki, Tapio Antero, *Tampere University of Technology*

Analytical Design of Almost Equiripple FIR   409
Half-Band Filters
Zahradnik, Pavel, Vlcek, Miroslav. *Czech Technical University;* and Unbehauen, Rolf, *University Erlangen-Nurnberg*

Design and VLSI Implementation of   413
Multirate Filter Banks Based on Approximately Linear Phase Allpass Sections
Summerfield, Stephen, *Warwick University*

A Highly-Flexible FIR Processor with   417
Scaleable Dynamic Data Ranges
Chen, Oscal T.-C., *National Chung Cheng University*

A Computationally Efficient Design of   421
Two-band QMF Banks Based on the Frequency Sampling Approach.
Gandhi, Rajeev, and Mitra, Sanjit, *University of California, Santa Barbara.*

Low Power Implementation of Linear   425
Phase FIR Filters for Single Multiplier CMOS Based DSPs
Erdogan, Ahmet Teyfik *Cardiff University of Wales,*

Automated Design of Low Complexity FIR Filters   429
Redmill, David Wallace, and Bull, David R., *University of Bristol*

Architecture of a Programmable FIR   433
Filter Co-Processor
Gay-Bellile, Olivier, Dujardin, Eric, *Laboratoires d'Electronique Philips S.A.S.*

Digital Hilbert Transformers Composed   437
of Identical Allpass Subfilters
Johansson, Hakan, and Wanhammar, Lars, *Linkoping University*

Data Block Processing for Low Power   441
Implementation of Direct Form FIR Filters on Single Multiplier CMOS DSPs
Erdogan, Ahmet Teyfik, *Cardiff University of Wales,*

Constrained Genetic Algorithm Design   445
of Finite Precision FIR Linear Phase Raised Cosine Filters
Al-Hashimi, B. M., Somerset, W.P. and Moniri, M. *Staffordshire University*

WPA1-Wavelets: Implementation and Application
Chair: Truong Nguyen
Boston University

Nonstationary Signal Classification Using   449
Pseudo Power Signatures
Venkatachalam, Vidya, and Aravena, Jorge L., *Louisiana State University*

A Simple Scheme of Decomposition and   453
Reconstruction of Continuous-Time Signals by B-splines
Ichige, Koichi. *University of Tsukuba;* Kamada, Masaru, *Ibaraki University;* and Ishii, Rokuya, *Yokohama National University*

A VLSI Architechture Design with Lower   457
Hardware Cost and Less Memory for Separable 2-D Discrete Wavelet Transform
Sheu, Ming-Hwa; Shieh, Ming-Der; and Liu, Sheng-Wel, *National Yunlin University of Science and Technology*

Synthesis Filter Bank With Low Memory   461
Requirements For Image Subband Coding
Sundsbo, Ingil, *Nordic VLSI ASA;* and Ramstad, Tor A., *Norwegian University of Science and Technology*

Optimal Design of Interpolating Wavelet   465
Transform
Shui, PengLang, *XiDian University*

Semi-recursive VLSI Architecture   469
for Two-dimensional Discrete Wavelet Transform
Paek, Seung-Kwon, *Korea Advanced Institute of Science and Technology*

Implementation of 2-D Wavelet Transform   473
on TESH Connected Parallel
Maziarz, Bogdan M., and Jain, Vijay K., *University of South Florida*

TABLE OF CONTENTS

Polyphase Adaptive Filter Banks for Subband Decomposition — 480
Gerek, Omer Nezih, and Cetin, A. Enis, *Bilkent University*

WPA2-3D Data Modeling and Imaging
Chair: Guido Cortelazz
University of Padovao

Hybrid Modeling for Manufacturing using NURBS, Polygons, and 3D Scanner — 484
Besl, Paul J., *Alias Wavefront, Inc.*

Portable Digital 3-D Imaging System for Remote Sites — 488
Beraldin, J-Angelo, *National Research Council Canada*

3D photography Using Shadows — 494
Bouguet. Jean-Yves and Perona, Pietro, *California Institute of Technology*

View-dependent texture coding using the MPEG-4 video coding scheme — 498
Jordan, Fred D.; Ebrahimi, Touradj; and Kunt, Murat, *EPFL - LTS*

Systems for Disparity-Based Multiple-View Interpolation — 502
Ohm, Jens-Rainer, Izquierdo, Ebroul, and Muller, Karsten, *Heinrich-Hertz-Institut*

2-D Patterns for 3-D Surface Matching — 506
Johnson, Andrew Edie, *Jet Propulsion Laboratory*

Review of Methods for Object-Based Coding of Stereoscopic and 3D Image Sequences — 510
Strintzis, Michael G., *University of Thessaloniki*

Image-based Modeling and Rendering of Architecture with Interactive Photogrammetry and View-Dependent Texture Mapping — 514
Debevec, Paul Ernest, *University of California at Berkeley*

A Frequency Domain Method for Registration of Range Data — 518
Cortelazzo, G.M., Doretto, G., Lucchese, L., and Totaro S. *University of Padova*

WPA13-Multimedia Communications
Chair: Tsuhan Chen
Carnegie Mellon University

Low Complexity Equalization for Cable Modems — 525
Wolf, Tod D. *Texas Instruments Inc.*

A Consideration on the Computational Requirements of Blind Equalization Using the Orthogonal Projection — 526
Kitaoka, Yoshihiro, *Fukuoka Institute of Technology*; Matsumoto, Hiroki, *Maebashi City College of Technology* and Furukawa, Toshihiro, *Fukuoka Institute of Technology*

A Novel MPEG Audio Degrouping Algorithm — 530
Tsai, Tsung-Han, Chen, Liang-Gee, Chang, Hao-Chieh, and Huang, Sheng-Chieh *National Taiwan University*

A Robust Algorithm for Formant Frequency Extraction of Noisy Speech — 534
Zhao, Qifang, *Saitama University*

Realization of Multiwavelet-Based Transform Kernels for Image Coding — 538
Rieder, Peter, Schimpfle, Christian V., and Nossek, Josef A., *Munich University of Technology*

Asynchronous VLSI Architectures for Huffman Codecs — 542
Hauck, Oliver Friedrich, *Darmstadt University of Technology*; Sauerwein, Helmut, *BetaResearch;* and Huss, Sorin Alexander *Darmstadt University of Technology*

A Perceptual Based Rate Control Scheme for MPEG-2 — 546
Chan, Shing Chow, *The University of Hong Kong*

Fast time Scale Modification Using Envelope-Matching (EM-TSM) — 550
Au, Oscar C., and Wong, W.C., *The Hong Kong University of Science and Technology*

Performance Study of Time Delay Estimation in a Room Environment — 554
Jian, Ming, *Nanyang Technological University*

VOLUME 6

MAA9 – Symbolic Analysis Methods & Applications to Analog Circuit Design
Chair: Marwan Hassoun
Iowa State University

Applications of Symbolic Methods to Circuit Design: An Overview — 1
Konczylowska, Agnieska, *Laboratoratoire de Bagneux*; Hassoun, Marwan M., *Iowa State University*; Huelsman, Lawrence, *University of Arizona*

Exploring Data Conversion Architectures by Symbolic Computation — 5
Horta, Nuno Cavaco Gomes, *Universidade Nova de Lisboa*; and Franca, Jose E., *Instituto Superior Tecnico*

A symbolic approach for testability evaluation in fault diagnosis of nonlinear analog circuits — 9
Fedi, Giulio; Giomi, Riccardo; Manetti, Stefano, and Piccirilli; Maria Cristina, *University of Florence*

Symbolic Analysis of Microwave Circuits — 13
Benboudjema, Kamel, *COM DEV Space Group*; Boukadoum, M. *Universite du Quebec a Montreal*; Vasilescu, G. and Alquie, G., *LEAM, Universite Pierre et Marie Curie*

Behavioral modeling of PWL analog circuits using symbolic analysis — 17
Fernandez, Francisco V.; Perez-Verdu, Belen; and Rodriguez-Vazquez, Angel, *Instituto de Microelectronica de Sevilla*

TABLE OF CONTENTS

Efficient Statistical Analog IC Design Using Symbolic Methods — 21
Debyser, Geert; Leyn, F.; Gielen, Georges; and Sansen, W., *Katholieke Universiteit Leuven ESAT-MICAS*; and Styblinski, M., *Texas A&M University*

Approximate Symbolic Pole/Zero Extraction Using Equation-Based Simplification Driven by Eigenvalue Shift Prediction — 25
Henning, Eckhard; Sommer, Ralf; and Wiese, Michael, *Institut fuer Techno- und Wirtschaftsmathematik*

Efficient symbolic analysis of large analog circuits using sensitivity-driven ranking of matroid intersections — 29
Dobrovolny, Petr, *Technical University Brno*; Wambacq, Piet; Gielen, Georges; and Sansen, Willy, *Katholieke Universiteit Leuvan*

MAA10 – Low Power Digital Circuit Design
Chair: Naresh Shanbhag
University of Illinois

Analytical expressions for average bit statistics of signal lines in DSP Architectures — 33
Bobba, Sudhakar; Hajj, Ibrahim N.; and Shanbhag, Naresh R., *Univ. of Illinois at Urbana-Champaign*

Architecture selection of a flexible DSP core using reconfigurable system software — 37
Lee, Jong-Yeol; Lee, Dae-Hyun; Kim, Jong-Sun; Yoon, Hyun-Dhong; Kyung, Chong-Min; Park, Kyu-Ho; Lee, Yong-Hoon; and Hwan, Seung H., *Korea Advanced Institute of Science and Technology*

Analyzing Effects of Cache Parameters on Memory Power Consumption of Video Applications — 41
Kapoor, Bhanu, *Texas Instruments Incorporated*

Transformational-Based Synthesis Of VLSI Based DSP Systems For Low Power Using A Genetic Algorithm — 45
Bright, Marc Stephen and Arslan, T., *Cardiff University Of Wales*

Power Estimation Using Input/Output Transition Analysis(IOTA) — 49
Lucke, Lori E.; Lee, Junsoo; and Vinnakota, Bapi, *University of Minnesota*

Fast Delay-Dependent Power Estimation of Large Combinational Circuits — 53
Jou, Jer Min, *National Cheng Kung University*; Chen, Shung-Chih, *Nan-Tai Institute of Technology*; Wang, Chih-Liang, *National Cheng Kung University*

Resynthesis of sequential circuits for low power — 57
Roy, Sumit, *Ambit Design Systems;* and Banerjee, Prithviraj, *Northwestern University*

STG Optimization for Power and Area Reduction — 62
Panagiotaras, George, S. and Koufopavlou, Odysseas G. *University of Patras*

MPA10 – Interconnect Modeling and Design
Chair: Cem Goknar
University of Illinois

A Universal Closed-Loop High-Speed Interconnect Model for General Purpose Circuit Simulators — 66
Li, X.; Nakhla, M.; and Achar, R., *Carleton University*

A Novel Technique for Minimum-Order Macromodel Synthesis of High-Speed Interconnect Subnetworks — 70
Achar, R. and Nakhla, M., *Carleton University*

Multipoint multiport algoritnm for passive reduced-order model of interconnect networks — 74
Yu, Qingjian; Wang, Janet M.L., and Kuh, Ernest S., *U.C. Berkeley*

Applications of Complex Frequency Hopping Method in PCB Signal Integrity Simulation — 78
Mu, Zhen, *Cadence Design Systems, Canada*

Time Domain Method for Reduced Order Synthesis of Large RC Circuits — 82
Batterywala, Shabbir Hussain and Narayanan, H., *Indian Institute of Technology, Powai.*

CMOS Inverter Current and Delay Model Incorporating Interconnect Effects — 86
Hafed, Mohamed and Rumin, Nicholas, *McGill University*

Path Resizing Based on Incrememtal Technique — 90
Cremoux, Severine; Azemard-Crestani, Nadine; and Auvergne, Daniel; *LIRMM*

Substrate Coupling Analysis and Simulation for an Industrial Phase-Locked Loop — 94
Welch, Ryan Joseph and Young, Andrew T., *University of Washington*

TAA12 – Analog Circuit Design
Chair: Chorng-Kuang Wang
National Central University, Taiwan

Statistical Behavioral Modeling of Integrated Circuits — 98
Swidzinski, Jan Feliks; Styblinski, M.A.; and Xu, Gonggui, *Texas A&M University*

Robust Recursive Inverse Approximation and its Application to Parameter Extraction of Behavioral Models — 102
Xu, Gonggui and Styblinski, M.A., *Texas A&M University*

Study of Optimal Importance Sampling in Monte Carlo Estimation of Average Quality Index — 106
Keramat, Mansour and Kielbasa, Richard, *SUPELEC*

Worst-Case Analysis of Linear Analog Circuits Using Sensitivity Bands — 110
Tian, Michael W. and Shi, C.-J. Richard, *University of Iowa*

Fast time domain noise simulation of sigma-delta converters and periodically switched linear circuits — 114
Dong, Yikui and Opal, Ajoy, *University of Waterloo*

TABLE OF CONTENTS

Efficient Utilization Of On-chip Inductors in Silicon RF IC Design Using a Novel CAD Tool; The LNA Paradigm — 118
Papananos, Yannis, and Koutsoyannopoulos, Yorgos, *National Technical University of Athens*

OPTOMEGA: An Environment for Analog Circuit Optimization — 122
Keramat, Mansour, and Kielbasa, Richard, *Ecole Superieure d'Electricite (SUPELEC)*

AC Constraint Transformation for Top-Down Analog Design — 126
Arsintescu, Bogdan G., *Delft University of Technology*; Charbon, Edoardo; Malavasi, Enrico; and Kao, William, *Cadence Design Systems*

TAA15 – CAD I
Chair: Douglas J. Fouts
Naval Postgraduate School

Maximally Routable Switch Matrices for FPD Design — 131
Chang, Yao-Wen and Wu, Guang-Min, *National Chiao Tung University*

Fault Emulation with Optimized Assignment of Circuit Nodes to Fault Injectors — 135
Sedaghat-Maman, Reza, *Institute of Microelectronic Systems University of Hanover*

State-space Technique for Minimal Realisation of Analogue Circuits and Systems — 139
Kadim, H.J., *University of Hull*; and Arslan, Tughrul, *Cardiff University of Wales*

Precise-MD: A Software Tool for Resources Constrained Scheduling of Multi-Dimensional Applications — 143
Hua, Jia; Rashid, Obaidur; Passos, Nelson L.; Halverson, Ranette H.; and Simpson, Richard P., *Midwestern State University*

Redesignability Analysis of Digital VLSI Circuits with Incomplete Implementation Information — 147
Wey, Chin-Long and Khalil, Mohammad Athar, *Michigan State University*

Fuzzy Multiobjective Decision Making On Modeled VLSI Architecture Concepts — 151
Jeschke, Hartwig, *Universitaet Hannover*

Parallel Coprocessor Architectures for Molecular Dynamics Simulation: A Case Study in Design Space Exploration — 155
Gerber, Martin and Goessi, Thomas, *Swiss Federal Institute of Technology*

Dual Edge Operations in Reduced Ordered Binary Decision Diagrams — 159
Miller, D. Michael, *University of Victoria*; Drechsler, Rolf, *Albert-Ludwigs-University*

ROBDD as a recursively defined periodic bit-string — 163
Lee, Seong-Bong; Yeon, Kwang-Il; and Park, In-Hak, *Semiconductor Technology Division, ETRI*

Generation of Quasi-Optimal FBDDs through Paired Haar Spectra — 167
Chang, Chip-Hong, *Nanyang Polytechnic* and Falkowski, Bogdan J., *Nanyang Technological University*

Calculation of Paired Haar Spectra for Systems of Incompletely Specified Boolean Functions — 171
Falkowski, Bogdan J., *Nanyang Technological University*; and Chang, Chip-Hong, *Nanyang Polytechnic*

Pseudo-Symmetric Functional Decision Diagrams — 175
Chrzanowska-Jeske, Malgorzata; Ma, Xiang Ying; and Wang, Wei, *Portland State University*

A New Lock Based State Coding Methodology for Signal Transition Graphs — 179
Nagalla, Radhakrishna, *University of New South Wales*

Multi-input/multi-output Block Diagram Grammar — 183
Adachi, Yoshihiro; Kobayashi, Suguru and Tsuchida, Kensei, *Toyo University*

Improved Minimization Methods of Pseudo Kronecker Expressions for Multiple Output Functions — 187
Lindgren, Per, *Lulea University of Technology, Sweden* and Drechsler, Rolf and Becker, Bernd, *Albert-Ludwigs-University*

Computational experience with a primal-dual interior point method for smooth convex placement problems — 191
Kennings, Andrew A., *Ryerson Polytechnic University* and Frazer, Mark J. and Vannelli, Anthony, *University of Waterloo*

An Initial Placement Algorithm for 3-D VLSI — 195
Michiroh, Ohmura, *Hiroshima Institute of Technology*

A Novel Methodology for Power Consumption Reduction in a Class of DSP Algorithms — 199
Masselos, Konstantinos; Merakos, P.; Stouraitis, T.; and Goutis, C.E., *University of Patras*

Performance Modeling for System Design: an MPEG A/V Decoder Example — 203
Hocevar, Dale E.; Sriram, Sundararajan and Hung, Ching-Yu, *Texas Instruments Inc.*

Graph Transformation for Communication Minimization Using Retiming — 207
Sheliga, Michael; Yu, Zhihong, Chen; Fei; and Sha, Edwin H.-M., *University of Notre Dame*

Gate to Channel Shorts in PMOS Devices: Effects on Logic Gate Failures — 211
Sayeed, M. Shahen and Mourad, Samiha, *Santa Clara University*

Realistic Delay Modeling in Satisfiability-Based Timing Analysis — 215
Silva, Luis Guerra; Silva, Joao P. Marques; and Silveira, Luis Miguel, *Cadence European Labs/INESC Instituto Superior Tecnico*; Sakallah, Karem A, *University of Michigan*

Enhancing Circuit Performance under a Multiple-Phase Clocking Scheme — 219
Hsu, Yaun-chung, *IBM*; Sun, Shangzhi, *Synopsys, Inc.*; Du, David H. C.,

xl

University of Minnesota; and Chu, Xuedao, *Qufu Normal University*

TPA10 – Circuit Simulation
Chair: Michel Nakhla
Carleton University, Canada

Applications of the Variable Dimension Newton Method to Large Scale Circuits — 223
Ng, Shek-Wai, *Hong Kong Polytechnic University*

HomSPICE: Simulator with homotopy algorithms for finding dc and steady-state solutions of nonlinear circuits — 227
Trajkovic, Ljiljana, *Simon Fraser University*; Fung, Eula; and Sander, Seth, *University of California - Berkeley*

Convergence Conditions of Waveform Relaxation Methods for Circuit Simulation — 232
Jiang, Yao-Lin, *Xian Jiaotong University*; and Wing, Omar, *The Chinese University of Hong Kong*

A time-frequency algorithm for the simulation of the initial transient response of oscillators — 236
Brachtendorf, H.G., *Bell Labs*; and Laur, Rainer A., *University of Bremen*

WAA11 – VLSI Layout and Timing
Chair: Majid Sarrafzadeh
Northwestern University

On Orientation Metric and Euclidean Steiner Tree Constructions — 240
Li, Y. Y.; Leung, K.S.; and Wong, C. K., *The Chinese University of Hong Kong*

Computational Complexity Analysis of Set-Bin-Packing Problem — 244
Izumi, Tomonorim; Yokomaru, Toshihiko; Takahashi, Atsushi, and Kajitani, Yoji, *Tokyo Institute of Technology*

A New Approach to Over-the-Cell Channel Routing — 248
Wang, Ting-Chi, *Chung Yuan Christian University*; Wen, Shui-An, *Industrial Technology Research Institute*; Wong, D. F., *University of Texas at Austin*; and Wong, C. K., *Chinese University of Hong Kong*

A Heuristic Algorithm to Solve Constrained Via Minimization for Three-Layer Routing Problems — 254
Takahashi, Kazahiro and Watanabe, Toshimasa, *Hiroshima University*

Utility Function Based Hybrid Algorithm for Channel Routing — 258
Etawil, Hussein A. and Vannelli, Anthony, *University of Waterloo*

An Age-Controlled Evolutionary Algorithm for Optimization Problems in Physical Layout — 262
Huber, Andreas and Mlynski, Dieter A., *Universitat Karlsruhe*

Timing Optimization of Mixed Static and Domino Logic — 266
Zhao, Min and Sapatnekar, Sachin, *Univerity of Minnesota*

Optimizing Circuits with Confidence Probability using Probabilistic Retiming — 270
Tongsima, Sissades; Chantapornchai, Chantana; and Sha, Edwin H.-M., *University of Notre Dame*; Passos, Nelson L., *Midwestern State University*

WAA15 – CAD II
Chair: Ibrahim Hajj
University of Illinois, Urbana-Champaign

A Matlab based tool for bandpass continuous-time sigma-delta modulators design — 274
Benabes, Philippe; Aldebert, Patrick; and Kielbasa, Richard, *SUPELEC*

Bicmos versus Cmos Technology in Fully Differential OTA Designs — 278
Recoules, Hector; Bouchakour, R.; and Loumeau, P., *Ecole Nationale Superieure des Telecommunications*

Analysis of Modulator Circuits Based on Multi-Dimensional Fourier Transformation — 282
Ushida, Akido; Yamagami, Yoshihiro, and Nishio, Yoshifumi, *Tokushima University*

Joint Optimization of Multiple Behavioral and Implementation Properties of Analog Filter Designs — 286
Damera-Venkata, Niranjan and Evans, Brian L., *The Univerity of Texas at Austin*; and Lutovac, Miroslav D. and Tosic, Dejan V., *University of Belgrade*

New Description Language and Graphical User Interface for Module Generation in Analog Layouts — 290
Wolf, Markus; Kleine, Ulrich; and Schulze, Jens, *University of Magdeburg*

Analysis and Compensation of OTA Non-Ideal Effects in Video-Frequency CMOS sinc(x) Equalizers — 294
Dudek, Frank; Al-Hashimi, Bashir M.; and Moniri, M., *Staffordshire University*

Adjoint network of periodically switched linear Circuits — 298
Yuan, Fei and Opal, Ajoy, *University of Waterloo*

Layout Driven Macromodel of an Operational Amplifier — 302
Chung-Yuk, Or, *The Chinese University of Hong Kong*
Franca, Jose E., *Instituto Superior Tecnico*

On the algebraic reuse of hardware design — 306
de Melo, Ana Cristina Vieira, *State University of Sao Paulo (USP)*

Assessing the uniqueness of the dc solutions by tearing of cactus graphs via detection of hinging structures — 310
Sarmiento-Reyes, Arturo and Bernal Rafael Vargas, *INAOE*

On the High Level Canonical Representation of Piecewise Linear Functions — 314
Julián, Pedro; Desages, Alfredo; and Agamennoni, Osvaldo, *Universidad Nacional del Sur*

Hierarchical Symbolic Analysis of Large Analog Circuits with Determinant Decision Diagrams — 318

TABLE OF CONTENTS

Tan, Xiangdong and Shi, C.-J. Richard, *University of Iowa*

Reducing Operation Complexity in Symbolic   322
Techniques through Partitioning
Cabodi, Gianpiero; Camurari, Paolo; and Quer, Stefano, *Politecnico di Torino*

Waveform Approximation Technique in The   326
Switch-Level Timing Simulator BTS
Chang, Mao-Lin (Molin); Chen, Wang-Jin, *National Taiwan University*; Wand, Jyh-Herng, *National Center for High-Performance Computer*; Feng, Wu-Shiung, *National Taiwan University*

New and Efficient Method for the   330
Multitone Steady-State Circuit Simulation
Larcheveque, Remi and Bolcato, P., *ANACAD EES / MGC*; Ngoya, E., *IRCOM*

Generalized Centers of Gravity Algorithm for   334
Yield Optimization of Integrated Circuits
Keramat, Mansour and Kielbasa, Richard, *SUPELEC*

Statistical Design of Integrated Circuits Using   338
Maximum Likelihood Estimation of the Covariance Matrix
Seifi, Abbas; Vlach, J.; and Ponnambalam, K., *University of Waterloo*

Modeling the Dynamic Behavior of Series-   342
Connected MOSFETs for Delay Analysis of Multiple-Input CMOS Gates
Bisdounis, Labros and Koufopavlou, Odysseas, *University of Patras*

A New Curve Fitting Technique For Analysis of   346
Frequency-Dependent Lossy Transmission Lines
Tanji, Yuichi; Nishio, Yoshifumi, and Ushida, Akio, *Tokushima University*

Two-Pole Approximation for High Speed   350
Interconnect Design
Shao, Jianhua and Chen, Richard M. M., *City University of Hong Kong*

Analysis of Interconnected Lumped Distributed   354
Multi-Branch Multi-Stage Networks
Sakagami, Iwata, *Muroran Institute of Technology*

Precorrected-DCT Techniques for Modeling and   358
Simulation of Substrate Coupling in Mixed-Signal IC's
Costa, Joao Paulo, *INESC / Technical University of Lisbon*; Chou, Mike *Massachusetts Institute of Technology*; Silveira, Luis Miguel, *INESC / Technical University of Lisbon*

Analysis of the Transistor Chain Operation   363
in CMOS Gates for Short Channel Devices
Chatzigeorgiou, Alexander and Nikolaidis, Spyridon, *Aristotle University of Thessaloniki*

Delay and Power Estimation for a CMOS Inverter   368
Driving RC Interconnect Loads
Chatzigeorgiou, Alexander and Nikolaidis, Spyridon, *Aristotle University of Thessaloniki*; and Kyriakis-Bitzaros, Eystathios D., *NCSR "Demokritos"*

WPA4 – High Lever Synthesis
Chair: Forrest D. Brewer
University of California at Santa Barbara

Parallel Algorithms for Simultaneous Scheduling,   372
Binding and Floorplanning in High-level Synthesis
Banerjee, Prithviraj, *Northwestern University* and Prabhakaran, Pradeep Kumar, *University of Illinois, Urbana Champaign*

A Simple Alternative for Storage Allocation in   377
High-level Synthesis
Aloqeely, Mohammed A., *King Saud University*

A New Partitioning Framework for Uniform   381
Clock Distribution During High-Level Synthesis
Krishnamurthy, Harsha, *Intel Corp.*; and Maaz, Mohamad and Bayoumi, Magdy, *The University of Southwestern Louisiana*

A Binding Algorithm for Retargetable   385
Compilation to Non-Orthogonal Datapath Architectures
Masayuki, Yamaguchi, *SHARP Corporation*; Nagisa, Ishiura, *Osaka University*; Takashi, Kambe, *SHARP Corporation*

WPA11 – Device Modeling
Chair: Kartikeya Mayaram
Washington State University

Modeling, Extraction and Simulation of CMOS I/O Circuits under ESD Stress   389
Li, Tong, *Univ. of Illinois at Urbana-Champaign*

Single-Event Effects in Micromachined   393
PMOSFETs
Osman, Ashraf A.; Mojarradi, Mohammad and Mayaram, Kartikeya, *Washington State University*

A Generalized HSPICE Macro-Model for Pseudo-   397
Spin-Valve GMR Memory Bits
Black, William C. and Das, Bodhisattva, *Iowa State University*

Compact SPICE Modeling and Design   401
Optimization of Low Leakage a-Si:H TFTs for Large-Area Imaging Systems
Arokia, Nathan, *University of Waterloo*

SPICE Model for Mechanically Stressed   405
Device/Circuit Simulation
Maier, Christoph; Steiner, Ralph; Mayer, Michael; Vogt, Rolf, and Baltes, Henry, *ETH Zuerich*

Rapid Extraction of Capacitance in a-Si Imaging   409
Arrays
Nathan, Arokia and Pham, Hoan H., *University of Waterloo*

An Efficient MOS Transistor Charge/Capacitance   413
Model with Continuous Expressions for VLSI
Sheu, Bing J., *University of Southern California,*

Wavelet-Based Galerkin Method For   417
Semiconductor Devices Simulation
Chan, Chung-Kei Thomas and Chang, Fung-Yuel, *The Chinese*

TABLE OF CONTENTS

University of Hong Kong

WPB4 – Field Programmable Gate Arrays
Chair: Sachin Sapatnekar
University of Minnesota

Using PLAs to design Universal Logic Modules in FPGAs — 421
Lee, Kok Kiong and Wong, Martin D. F., *The University of Texas at Austin*

FPGA Maping of Sequential Circuits with Retiming — 426
Lee, Jun-yong, *Hongik University*

RAISE: A detailed routing algorithm for SRAM based Field-Programmable Gate Arrays using multiplexed switches — 430
Baena, Vicente; Aguirre, Miguel Angel, and Torralba, Antonio, *ESI- (Grupo de tecnologia electronica)* Faura, Julio, *Sidsa*

Rapid Prototype of a Fast Data Encryption Standard With Integrity Processing for Cryptographic Applications — 434
Guendouz, Hassina, *Ecole Centrale Electronique*; and Bouaziz, Samir, *Universite Paris*

TAA2 – Modeling Analysis and Design of Switching Mode Converters
Chair: Henry Chung
University of Hong Kong
Organizer: Adrian Ioinovici
Holon Institute for Technological Education

Design and Analysis of Quasi Switched-Capacitor Step-Up DC/DC Converters — 438
Chung, Henry S H., *City University of Hong Kong*

Compact Neural Network Detector for Hard-Disk Drive Using Zero-Forcing Preprocessing — 442
Sheu, Bing J. and Wang, Michelle Y., *University of Southern California*; and Lin, Horng-Dar, *NeoParadigm Labs, Inc.*

Bidirectional Buck-Boost Converter with Variable Output Voltage — 446
Czarkowski, Dariusz, *Polytechnic University* and Krishnamachari, Bhaskar, *The Cooper Union*

Issues for Image/Video Digital Libraries — 450
Deng, Yining and Manjunath, Bangalore S., *University of California at Santa Barbara*

Novel PWM Control Method of Switched Capacitor DC-DC Converter — 454
Suetsugu, Tadashi, *Fukuoka University*

True-Worst-Case Evaluation in Circuit Tolerance & Sensitivity Analysis using Genetic Algorithms and Affine Mathematics — 458
Femia, Nicola, *University of Salerno DIIIE*

Analytical Solution to Harmonic Characteristics of PWM H-Bridge Converters with Dead Time — 462
Lau, W.H., *City University of Hong Kong*

A Virtual Classroom for Real-Time Interactive Distance Learning — 466
Hwang, Jenq-Neng; Deshpande, Sachin G., and Sun, Ming-Ting, *University of Washington*

Fast Browsing of Speech/Audio Material for Digital Library and Distance — 470
Au, Oscar C., *Hong Kong University of Science and Technology*

TPA6 – Circuits and Systems for Communication Networks I
Chair: Magdi Bayoumi
Tokyo Institute of Technology

A Low-Power VLSI Design Methodology for High Bit-Rate Data Communications over UTP Channel — 474
Goel, Manish and Shanbhag, Naresh R., *University of Illinois at Urbana-Champaign*

VLSI Design of an ATM Switch with Automatic Fault Detection — 478
Tsui, Chi-Ying, *Hong Kong University of Science and Technology*

A Signaling Protocol Architecture for an ATM Mobile Simulator — 482
Yoo, Jeong-Ju; Nah, Jae Hoon; Yoo, Jea Hoon; Lee, Yoon Ju, *Electronics and Telecommunications Research Institute*; and Hutchison, David, *Lancaster University*

WPB13 – Circuits and Systems for Communication Networks II
Co-Chair: Joseph Kahn
University of California, Berkeley
Co-Chair: Chung-Sheng Li
IBM T.J. Watson Research Center

Compile-time Priority Assignment and Re-routing for Communication Minimization in Parallel Systems — 486
Sha, Edwin; Surma, David R., and Kogge, Peter M., *University of Notre Dame*

A VLSI Design of Dual-Loop Automatic Gain Control for Dual-Mode QAM/VSB — 490
Wang, Chorng-kuang; Shiue, Muh-Tian; Huang, Kuang-Hu, and Lu, Cheng-Chang, *National Central University*; and Way, Winston I., *National Chiao Tung University*

Routing Multipoint Connections in Computer Networks — 494
Wensheng, Sun, *Beijing University of Posts and Telecommunications*

Dynamic Routing Algorithms in ATM Networks — 498
Feng, Gang, *Beijing University. of Posts & Telecom.*

A New Memory Controller for the Shared Multibuffer ATM Switch with Multicast Functions — 502
Chang, Robert Chen-hao, *National Chung-Hsing Univ.*

TABLE OF CONTENTS

Scheduling Algorithm for Real-time Burst Traffic using Dynamic Weighted Round Robin — 506
Kwon, Taeck-Geun; Lee, Sook-Hyang; Rho, June-Kyung, and Kwon, Taeck-Geun, *LG Information & Communications, Ltd.*

Design and Implementation of an ATM Segmentation Engine with PCI Interface — 510
Kim, Chan; Jun, Jong Arm; Lee, Kyou Ho, and Kim, Hyup Jong, *Electronics and Telecommuncations Research Institute*

A Novel Neural Estimator for Call Admission Control and Buffer Design in ATM Network — 514
Zhang, Liang and Liu, Zemin, *Beijing University of Posts & Telecommunications*

DCT Based Error Concealment for RTSP Video over a Modem Internet — 518
Chung, Yon Jun, *University of Southern California*

A transmitting, and receiving method for cdma communications over indoor electrical power lines — 522
Okazaki, Hideaki, *Gifu National College of Tech.*; and Kawashima, Mitsusato, *Ibiden Industries Co.,LTD*

Network Design and Control for Multipoint-to-Multipoint Communications — 529
Kinoshita, Kazuhiko, *Osaka University*; Soeda, Junichiro, *Matsushita Electric Industrial Co.,LTD*; Yamai, Nariyoshi; Takine, Tetsuya; and Murakami, Koso, *Osaka University*

A Multi-rate Channelized Wireless LAN System with Fixed Channel Assignment — 533
Ko, Tsz-Mei and Lam, Chi-Wai, *Hong Kong Univ of Sci and Tech*

An Introduction To Multi-Sensor Data Fusion — 537
Llinas, James, *State University of New York at Buffalo* and Hall, David L., *Pennsylvania State University*

From GI Joe To Starship Trooper: The Evolution of Information Support for Individual Soldiers — 541
Hall, David L., *The Pennsylvania State University* and Llinas, James, *The State University of New York at Buffalo*

MPA11 – Multi Sensor Data Fusion: Application and Issues
Chair: Sean Midwood
Canadian Navy
Organizer: Ian Glenn
Canadian Army

A Multisensor Data Fusion Algorithm for the USCG's Vessel Traffic Services — 545
Glenn, Ian N., *Department of National Defence, Canada*

Microsimulation as a Tool for Traget Tracking and State Estimation — 549
Brown, Donald E. and Pittard, C. Louis, *University of Virginia*

Information Understanding: Integrating Data Fusion and Data Mining Processes — 553
Waltz, Edward, *ERIM International*

Strategically-Controlled Information Fusion — 557
Flank, Steven M., *Defense Advanced Research Projects Agency (DARPA)*

The CANADA-NETHERLANDS Collaboration on MultiSensor Data Fusion and Other Canada-NATO MSDF Activities — 561
Bosse, Eloi; Roy, Jean; and Bosse, Eloi, *Canada National Defence*

TAA11- Digital Signal Processing for Hearing Aids
Chair: Neeraj Magotra
University of New Mexico

Recruitment compensation as a hearing aid signal processing strategy — 565
Allen, Jont B., *AT&T Labs*

A Flexible Filterbank Structure for Extensive Signal Manipulations in Digital Hearing Aids — 569
Schneider, Todd and Brennan, Robert, *dspFactory*

Multiband Compression Hearing Aids: Developing a Performance Metric — 573
Rutledge, Janet C., *University of Maryland at Baltimore*; and Schmidt, Jon C., *Starkey Laboratories*

Multichannel Compression in the Normal Ear and as a Signal Processing Algorithm for the Hearing Impaired — 578
Yund, E. William, *Veterans Affairs Northern California Health Care System*

Multichannel Adaptive Noise Reduction in Digital Hearing Aids — 582
Magotra, N.; Kasthuri, P.; Yang, Y.; and Whitman, R., *University of New Mexico*; and Livingston, F., *Texas Instruments Incorporated*

Signal Processing Techniques for a DSP Hearing Aid — 586
Edwards, Brent W., *ReSound Corporation*

TPA11 – Robotics
Chair: David C. Jenn
Naval Postgraduate School

Wireless Power Transfer for a Micro Remotely Piloted Vehicle — 590
Jenn, David, *Naval Postgraduate School*

Sliding Mode Control For Elastic Multi-Link Manipulators Based On The Dynamic Compensation Method — 594
Utkin, Victor A., *Institute of Control Sciences*

New Broadband 100-MBPS Switch System Using Broadband Pin-Board Switch and High-Precision Pin-Handling Mechanism — 598
Inagaki, Shuichiro, *NTT Opto-electronics Laboratories*

An Agent-Based Structure for Mobile Robots Using Vision and Ultrasonic — 602
Bastos-Filho, Teodiano Freire; Freitas, Roger Alex de Castro; Sarcinelli-

TABLE OF CONTENTS

Filho, Mario, and Schneebeli, Hansjorg Andreas, *Federal University of Espirito Santo - UFES*

TPA15 – Sensors and Related Circuits
Co-Chair: Richard Colbeth
Varian Imaging Products
Co-Chair: Igor M. Filanovsky
University of Alberta

CMOS 1D and 2D N-well Tetra-lateral Position Sensitive Detectors 606
Chowdhury, Mohamed Foysol, *University of Kent at Canterbury*

High Performance CMOS Position-Sensitive Photodetectors (PSDs) 610
Mäkynen, Anssi Jaakko, *University of Oulu*

Measuring Temperature Calibration Free With Bipolar Transistors 617
Kanoun, Olfa, *Universität der Bundeswehr München*

A CMOS Integrated Infrared Radiation Detector for Flame Monitoring 625
Malcovati, Piero; Maloberti, Franco; and Poletti, Matteo, *University of Pavia* and Francesconi, Fabrizio, *Micronova Sistemi S.r.l.* Bendiscioli, Paolo and Valacca, Roberto, *University of Pavia*

Two Temperature Sensors Realized in BiCMOS Technology 621
Filanovsky, Igor M., *University of Alberta*

A Multi-mode X-ray Imager for Medical and Industrial Applications 629
Colbeth, Richard E., *Varian Imaging Products*

Novel Low Power Class-B Output Buffer 633
Yu, Pang-Cheng and Wu, Jiin-Chuan, *National Chiao-Tung University*

A Novel Image Sensor with Flexible Sampling Control 637
Ohtsuka, Yasuhiro; Hamamoto, Takayuki; Aizawa, Kiyoharu, and Hatori, Mitsutoshi, *University of Tokyo*

A 128x128 Imaging Array Using Lateral Bipolar Phototransistors in a Standard CMOS Process 641
Sandage, Robert, W., *Georgia Institute of Technology*

Single Chip CMOS Image Sensors for a Retina Implant System 645
Schwarz, M.; Hauschild, R.; Hosticka, B.J.; Huppertz, J.; Kneip, T.; Kolnsberh, S.; Mokwa, W., and Trieu, H.K., *Fraunhofer Institute of Microeletronic Circuits and Systems*

An Analog VLSI Velocity Sensor Using the Gradient Method 649
Deutschmann, Rainer Alexander, *Walter Schottky Institut*; and Koch, Christof, *California Institute of Technology*

A Single Fourier Series Technique For The Simulation And Analysis Of Asynchronous Pulse Width Modulation In Motor Drive Systems 653
Guinee, Richard Anthony, *Cork Institute Of Technology*; and Lyden, C., *NMRC University College*

APPLICATIONS OF SYMBOLIC METHODS TO CIRCUIT DESIGN: AN OVERVIEW

Agnieszka Konczykowska[1], Marwan Hassoun[2], Lawrence Huelsman[3]

[1] Laboratoire de Bagneux, Centre National d'Etude des Telecommunications, 92220 Bagneux, FRANCE
[2] Dept. of Electrical and Computer Engineering, Iowa State University, Ames, Iowa 50011, USA
[3] Dept. of Electrical and Computer Engineering, University of Arizona, Tucson, Arizona 85721, USA

ABSTRACT

This paper presents an overview of the popular applications of symbolic methods nowadays as they pertain to circuit design. The specific applications that are discussed are: insight into circuit behavior, device modeling and extraction, behavioral and functional modeling and structural synthesis. The application of symbolic analysis in each of these areas is illustrated via a description of a specific methodology.

1. INTRODUCTION

This paper is the introductory paper for the ISCAS 98 special session entitled "Symbolic Analysis Methods and their Applications to Analog Circuit Design". The goal here is to provide an overview of the applications of symbolic analysis to complement the other papers presented in the session. The authors must note that other excellent references have been published discussing, in part, the general topic of symbolic analysis applications [1], [2], [3]. This paper, in combination with the above mentioned publications, provide a survey of the most popular applications of symbolic analysis to-date. The paper will not discuss specific symbolic analysis engines, but survey publications on the topic can be found in [2], [3], [4], [5], [6], [7].

Classical applications of symbolic analysis include statistical analysis, design centering and optimization where the main advantage of using a symbolic approach is linked to the reduced computation effort. Recently, new application domains, such as structures generation, topological optimization, design automation and behavioral and functional modeling are emerging. The emergence of these new application domains is driven by the increased complexities and the automation needs in the analog and mixed-signal design fields. These fields are experiencing a rapid growth in interest from the electronics industry. Analog and mixed-signal applications are characterized by long design times and the need for highly-skilled designers due to the lack of design automation. This need for new and innovative approaches to design automation has resulted in the new applications of symbolic analysis.

Traditionally attractive aspects of symbolic analysis, consisting mainly of reduction in calculation time, are complemented by new enhancements in the design methodology, design automation and architecture optimization. Symbolic approaches can thus be used in conjunction with numerical or qualitative approaches to produce new solutions.

For calculation speedup applications, symbolic formulae can be an efficient solution in cases where multiple calculations on the same topology are required. Depending on the necessary flexibility, symbolic functions can be either compiled and linked to the software (for fixed topology), or generated and evaluated in an "on-line" way. Since calculation speedup is a very well-known application of symbolic analysis, it will not be discussed further in this paper except as it pertains to the rest of the applications described in the following sections.

2. INSIGHT INTO CIRCUIT BEHAVIOR

Insight into circuit behavior is gained by making available to the designer a compact and legible symbolic expression that represents the behavior of the circuit. The insight can only be gained by examining results with few symbolic expressions in them. This is achieved by advancements in the approximation methods in symbolic analysis. These approximation techniques identify the most significant parameters in a circuit and their impact on its performance. This allows a designer to efficiently evaluate several topologies and produce newer ones.

One of the traditional applications of symbolic analysis here is the support that is offered to the education arena in the areas of circuit theory and circuit design (e.g. [8]). As such, it can play an important companion role to the numerical analysis methods normally introduced as simulation tools. The early introduction of symbolic analysis methods in engineering studies provides avenues of insight to system behavior which are difficult, if not impossible, to obtain by other methods. For example, the system describing function obtained by symbolic methods includes literal symbols for each of the parameters of the system. If it is desired to study the effects of variations in any specific parameter on the overall system behavior, the symbolic result can be modified by including the nominal numerical values for the other parameters, resulting in a system description which is a function only of the single chosen parameter, and (usually) the complex frequency variable. It is now relatively simple to insert the desired values of the parameter directly into the describing function to determine the effects. This is, in general, a far simpler procedure, and one which is computationally more efficient, than the repeated re-determination of the function through a series of numerical analyses. In addition to such parameter variation studies, the symbolic system describing function provides the necessary starting point for a wide range of system studies which require the determination of a derivative with respect to one or more parameters. An example of such a study is the determination of various sensitivity measures, especially the function sensitivity which displays the variation of the sinusoidal steady-state behavior of the function with respect to frequency for each (or all) of the system parameters. Such sensitivity information is vital in determining the tolerance and other characteristics required of the individual elements which interact to form the system. It also provides a measure of the overall quality of a given design, and permits a comparison between competing designs. Thus, symbolic analysis provides a valuable design tool. The derivative information is also of prime importance when optimization tools are to be applied, both as initial synthesis methods, and as methods for improving a design to minimize (or maximize) some specified criteria. Most optimization or search techniques require such derivative information as part of their algorithmic approach to the optimization process. In summary of the above, we see that early introduction to symbolic analysis procedures provides the engineering student with a powerful tool which can be used to improve his or her insight into the behavior of systems, and provide ways in which his or her designs of these systems can be evaluated and improved.

3. DEVICE MODELING AND CHARACTERIZATION

Device modeling and characterization is an integral part of the circuit design process. It is a computationally intensive and complex process that benefits greatly from the use of symbolic analysis. The methodology of device characterization will be presented here as an example showing some attractive possibilities of symbolic analysis.

Rapid developments in semiconductor technologies creates new needs for device characterization and modeling methods. Increased performance, higher integration levels and circuit complexities impose a need for precise models not only for active devices, but also for passive, functional and parasitic, elements. Characterization of semiconductor devices associated with extraction methods is an established procedure to obtain accurate electrical models. Iterative methods which consist of adjusting parameter values during the optimization process are time consuming and can be sensitive to the starting point values and to the definition of the optimization goals. A different approach is used in direct extraction methods. Model parameters are determined from linear approximation of experimentally measured data. Direct extraction is rapid and simple in execution, but specific extraction procedures have to be established for each new type of measurement and for each new or modified technology. It is also difficult to extract some parameters of nonlinear models and precision may be a problem. Most often a combination of both methods provides the most attractive solution. Both direct and iterative extraction methods can be enhanced with the symbolic approach.

<u>Direct extraction methods</u>

Although direct approach can be used for different types of measurements, we will discuss problems related to the AC regime only. Measured data are most often acquired in the form of S parameters. The well known formulae allow the transformation of these S parameters in Z or Y matrices. On the other hand, analytic expressions for Z or Y can be obtained from the analysis of the small signal electrical scheme. Approximation of these characteristics (real and imaginary parts can be used) gives necessary formulae for direct extraction. These two operations (generation and approximation of analytic expressions) when done manually, are cumbersome, error prone, and must be repeated for each modification of the model topology and even when the range of parameter values changes.

Analytic formulas necessary in the case of direct extraction can be established with an aid of symbolic simulators. The most useful are symbolic simulators producing an approximated form of results. A practical example of direct extraction using symbolic approach in the case of InP HBT was presented in [9].

<u>Simulation-based extraction</u>

In the simulation-based iterative approach, a circuit simulator is used to provide circuit responses. The advantages of this method lay in its great flexibility. It allows for the elimination of inconsistencies between the extracted model and the one used in the simulation tools and the facility to include all packaging and parasitics elements in the extraction procedure. On the other hand, repeated simulations necessary for the extraction through optimization, can easily become time-consuming, especially when numerous parameters are extracted from large sets of measured data. For linear extraction, instead of repeating the analysis for each iteration, it is possible to derive circuit response in symbolic form. For each iteration, objective and error functions can be calculated using this formula and different variable values. An example of such an approach was presented in [10]. An interface providing a convenient access to a symbolic simulator was developed. To gain in flexibility, instead of using a compiled version of circuit functions, an intermediate representation of symbolic results is used, which is efficiently evaluated for different combinations of the variables' values. An important reduction of computation time was obtained.

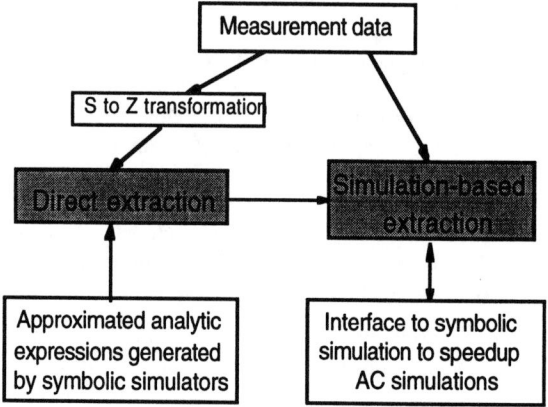

Figure 1. Direct and simulation-based extraction enhanced by symbolic simulation

In Figure 1 the organization of the extraction procedure combining direct and simulation-based extraction is presented. Direct extraction is used to rapidly obtain an initial set of parameters. These values are then used as a starting point for iterative simulation-based extraction.

4. BEHAVIORAL AND FUNCTIONAL MODELING

Behavioral and functional modeling in the digital design world is the corner stone of all successful large circuit design projects. The main, and well documented advantage, is the orders of magnitude savings in design time, and in turn time-to-market, due to the ability to understand the behavioral of a large circuit well before the details of the circuit components are realized. This is done with the use of hardware description languages such as Verilog and VHDL. Behavioral and functional modeling of analog and mixed-signal circuits is a fairly new field that has yet to find wide usage amongst circuit designers. While hardware description languages and simulators for analog and mixed-signal circuits do exist (e.g. AHDL, MAST-HDL, SPECTRE-HDL, etc..), one of the main problems is the lack of behavioral and functional macros for common building blocks of analog circuits. Symbolic analysis algorithms that have been presented so far, provide a very convenient way of generating such functional macros and thus standard libraries for common analog circuit building blocks. During initial design investigations, a high level of abstraction is desired in an analog system. The macros used here can be generated using symbolic simulators that utilize approximation techniques (e.g. [11], [12]) that result in very compact analog expressions that can be used to estimate the behavior of the block. These descriptions for the various blocks are generated only once, symbolically, and stored in a library in a format that is consistent with the specific behavioral simulator that is to be used. If more complex (accurate) descriptions are desired, hierarchical symbolic analyzers provide a compact way to represent the circuit behavior accurately (e.g. [13]).

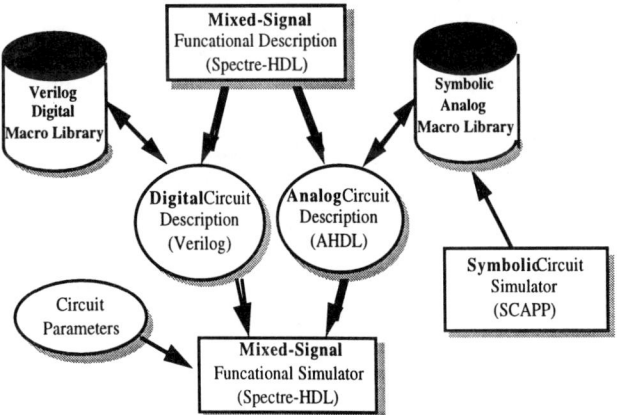

Figure 2. A symbolic, mixed-signal functional verification design flow

Figure 2 provides a general flow diagram of current work at Iowa State University that uses symbolic analysis for behavioral characterization of analog and mixed-signal circuits. The methodology uses a combination of the symbolic simulator SCAPP [13] to model the analog building blocks symbolically, the analog hardware description language AHDL to characterize the analog circuits that are the interconnection of the building blocks, the digital hardware description language Verilog to represent the digital blocks and circuits, and SPECTRE-HDL to simulate the behavior of the entire system. The results of this work has not yet appeared in the literature and will be published in the near future.

5. STRUCTURAL SYNTHESIS AND OPTIMIZATION

Circuit optimization is a term that describes traditionally the technique used to determine appropriate circuit parameter values. This process is performed for a given circuit structure in order to obtain the desired circuit performance. Structural synthesis and optimization consist of determining the most advantageous circuit topology in a specified class of structures. This procedure may be combined with numerical optimization for each of examined topologies. A general scheme of the synthesis/optimization organization is presented in Figure 3.

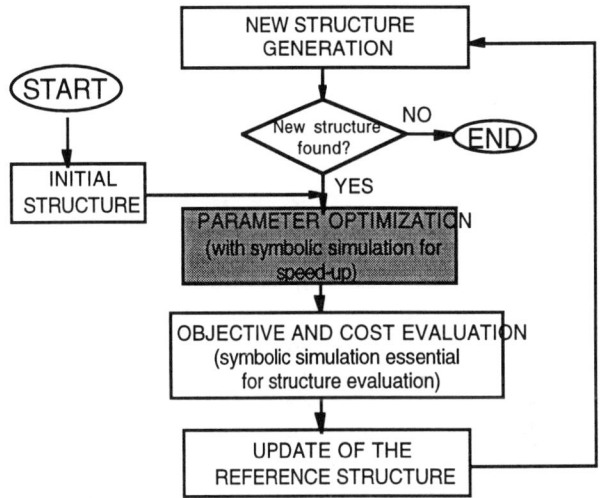

Figure 3. General organization of the synthesis/optimization process

Different important aspects are to be analyzed in the above presented scheme. The definition of the class of possible structures has to be done in correlation with the proper determination of the objective function. Various situations should be examined.

The <u>objective function</u> (technical performance) can depend either on the topology only, and in this case the circuit analytical function is examined, or can be determined in a classical form calculated for a set of parameters' values associated with the topology under consideration. In the second case, the parameter optimization step should be included the procedure as presented in Figure 3.

The second important aspect is the notion of the *cost* for each structure. The *cost* can be considered as an economic performance of the circuit. It can be determined for individual elements taking under consideration their type and if necessary value. The total cost is expressed as a sum of individual costs.

The class of structures can be composed of elements with equal or different costs. In the second case the structure cost has to be taken into account in the synthesis/optimization process. In [3] different formulations of the structural synthesis/optimization problem are presented. Most general formulation is the global optimization with combined objective/cost goal. In this case a trade-off between structural complexity (expressed as a cost function) and performance criteria has to be found. The strategy of the new structure generation is another important aspect of the procedure.

One specific class of circuit structure optimization is the systematic exploration of a given class of topologies in order to find the best suited one. An example of such a problem was presented in [14]. The problem was approached by looking for new structures for switched-capacitor integrators (integrators are widely used as basic building blocks in SC circuits). The class of structures taken into account was a family of circuits composed of one amplifier and three capacitors interconnected by an arbitrary number of switches. The method considered all elements from this class of equal cost. An additional condition was that each considered structure had to be stray-capacitance insensitive. The goal was to find structures least affected by finite amplifier gain. This was done with the goal of selecting architectures well-suited to high-frequency operation. The above problem was solved using the symbolic simulator of SC networks SCYMBAL and structure generating and selecting engine developed in PROLOG. In order to avoid a combinatorial explosion, new structures were built in successive steps with elimination of all unneeded configurations at each step. Thanks to this strategy, instead of impossible examination of 5^{12} potential structures, only about 2000 topologies were considered. About 200 structures were of the integrator type. Until this point all circuits were considered as ideal ones. For all generated integrator type structures an influence of a finite amplifier gain was evaluated. Symbol μ was introduced, equal to 1/A, where A was the finite amplifier gain. For each integrator a gain and phase error due to a finite amplifier gain were calculated in symbolic form (as a function of μ). The algorithm searched for structures with error functions depending only on higher orders of μ. As a result one new SC-integrator was found superior to all previously known structures. In this structure both gain and phase errors had only terms in μ^2 (linear term in μ was canceled) while previously proposed structures had at least one linear term in μ.

Symbolic simulation in structural synthesis is useful in two different phases. Obviously, it is essential when an optimization criterion is formulated using analytical characteristic of the circuit. Secondly, it can speed-up the classical parameter optimization stage with the use of symbolic formulas rather then performing repetitive numerical simulation at each iteration.

6. CONCLUSIONS

This paper presented several applications of Symbolic Analysis. The main observation here is that Symbolic Analysis has not replaced numerical methods, but rather provided additional functionality that facilitates better design automation in the circuit design arena.

7. REFERENCES

[1] P. M. Lin, "A survey of Applications of Symbolic Network Functions," IEEE Transactions on Circuit Theory, Vol. CT-20, No. 6, pp.732-737, Nov 1973.

[2] G. Gielen, P. Wambacq and W. Sansen, "Symbolic Analysis Methods and Applications for Analog Circuits: A Tutorial Overview," *Proceedings of the IEEE*, Vol 82, No. 2, pp. 287-304, February 1994.

[3] F.Fernandez, A.Rodriguez-Vazquez, J.Huertas, G.Gielen, Symbolic analysis techniques. Applications to Analog Design Automation, IEEE Press, 1997.

[4] P. M. Lin, "Computer Generation of Symbolic Network Functions - An Overview," CAD (Proc of IFIP working conf on principles of CAD (Oct 1982), edited by Vlietstra and Wielinga), North Holland, pp. 261-282, 1973.

[5] P. M. Lin, Symbolic Network Analysis. Elsevier Science, Amsterdam, 1991.

[6] G. Gielen, and W. Sansen, Symbolic Analysis for Automated Design of Analog Integrated Circuits. Kluwer Academic, MA, 1991.

[7] M. Hassoun and L. Huelsman, "Symbolic Circuit Analysis: An Overview," IEEE Midwest Symposium on Circuits and Systems, Rio de Janeiro, Brazil, June 1995.

[8] L. P. Huelsman, "Personal Computer Symbolic Analysis Programs for Undergraduate Engineering Courses," IEEE ISCAS, Portland, OR, pp. 798-801, May 1989.

[9] N. Zérounian, F.Aniel, N. Kauffmann, R. Adde, A. Konczykowska, Extraction strategies of semiconductor device parameters using symbolic approach and optimization methods, Proc of ECCTD'97, September 97, Budapest, Hungary, pp.1298-1303.

[10] W. Zuberek, A. Konczykowska, D. Martin, An approach to integrated numerical and symbolic circuit analysis, Proc. of ISCAS'94, May 1994, London, pp. 1.33-36 (vol.1).

[11] P. Wambacq, F. V. Fernández, G. Gielen, and W. Sansen, "Approximation During Expression Generation in Symbolic Analysis of Analog Integrated Circuits," Alta Frequenza Rivista Di Elettronica, Vol 5, No 6, pp. 48-55, November 1993.

[12] Q. Yu and C. Sechen, "Efficient Approximation of Symbolic Network Function Using Matroid Intersection Algorithms," IEEE International Symp. on Circuits and Systems, Seattle, Wa, May 1995.

[13] M. M. Hassoun and P.-M. Lin, "A Hierarchical Network Approach To Symbolic Analysis Of Large-Scale Networks," IEEE Trans on Circuits and Systems I, March 1995.

[14] A. Konczykowska, M. Bon, Automated Design Software for Switched-Capacitor IC's with Symbolic Simulator SCYMBAL, Proc of 25th Design Automation Conference, Anaheim, USA, 1988, pp.363-368

EXPLORING DATA CONVERSION ARCHITECTURES BY SYMBOLIC COMPUTATION

N. C. Horta[1] and J. E. Franca[2]

[1]Faculdade de Ciências e Tecnologia, DEE-UNL,
2825 Monte da Caparica, Portugal
Phone/Fax: (351)-1-2948545/32, E-mail: n.horta@ieee.org

[2]Instituto Superior Técnico, IST Center for Microsystems
Av. Rovisco Pais, 1, 1096 Lisboa Codex, Portugal
Phone/Fax: (351)-1-8417675

ABSTRACT

The implementation of an algorithm-driven methodology for the synthesis and characterization of data converter systems, using extensively symbolic methods, is described. The synthesis process is, here, started at the algorithm-level allowing an unconstrained specification of a broad range of conversion systems, when compared to the traditional architecture oriented design tools. More, the exploration of new architectures and their characterization at the sub-block level, is performed fully automatically.

1. INTRODUCTION

Today, the relentless trend of electronics integration and miniaturization is rapidly changing the way A/D converters are used, designed and even produced. Because such converters are being increasingly used as macromodels embedded into VLSI mixed-signal systems, their traditional label of "general purpose" components is quickly paving the way to "tailor-made" components that can optimally meet target specifications for performance, cost and energy consumption [1]. Dedicated computer-aided design tools have been largely instrumental in such popularization of data converters design [2].

In the later 80's, the first tools to approach the synthesis of data converters made it by implementing an architecture-constrained methodology, together with the use of standard cells to generate a semi-custom layout. This approach applies only to very specific cases due to its highly dependence on lower level circuitry. In the early 90's, the evolution of lower level tools together with the more systematic application of hierarchical concepts led to an increased flexibility of design automation by specifying the converter at the topological level and using lower level primitives / tools to generate the appropriate sub-blocks. Although a variety of tools were developed covering a larger range of specifications their expansion to other classes of converters and integration in higher level CAD systems is still limited by the lack of formal synthesis techniques. More recently, a new generation of synthesis tools has been proposed whereby formal symbolic methods are employed to achieve highly increased flexibility and levels of abstraction similar to those currently experienced in the digital domain. This paper describes the use of such a tool, TAGUS, which allows the exploration of data conversion algorithms by using extensively symbolic methods for the automatic generation and characterization of their architectures.

2. TAGUS ARCHITECTURE

The implementation of the synthesis methodology originated the system TAGUS whose architecture is shown in Fig. 1. The methodology consists of three different phases, along three hierarchical levels, with specific interfaces for simulation with ELDO [3]. The first phase is decomposed into two parts. First, the acquisition of the synthesis specifications, i.e., the desired algorithm description in FIDEL HDL [3] and the performance specs. The validation of the algorithm description is carried out by functional simulation using FMODEL models. Then, an equivalent description is achieved through the interpretation, in PROLOG, of the algorithm description in FIDEL HDL which is translated sequentially into a signal flow graph, generating a file with extension *gra*.

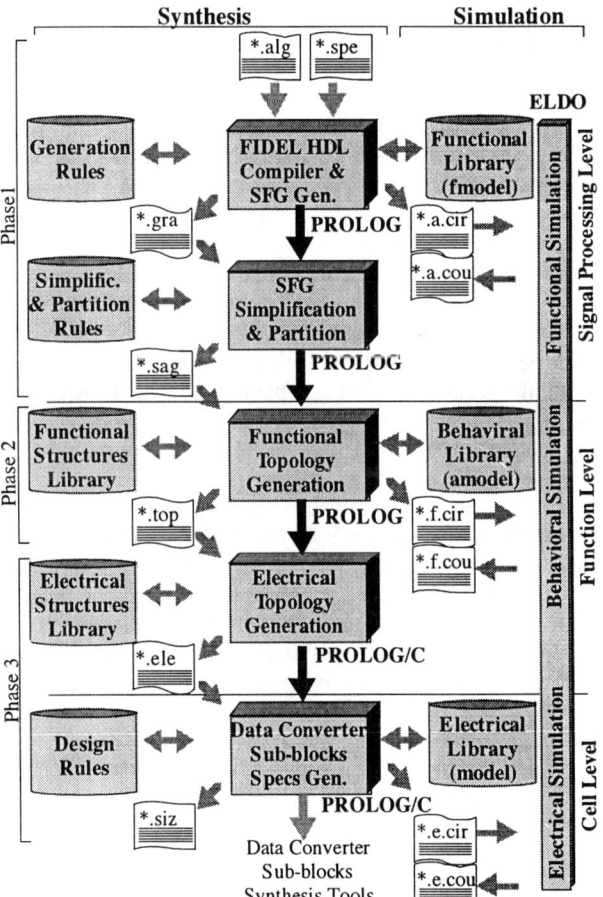

Figure 1. Organization of the TAGUS system architecture.

After that, a symbolic task is performed, in PROLOG, for the simplification and partitioning of the previously achieved graph, originating a file with extension *sag* including the graph in the canonical form. The second phase consists of the functional topology generation. In this phase, a library of functional blocks is described in terms of signal flow graphs, in order, to allow the topology generation process to act as a pattern recognition process, in PROLOG, over the graph, and originating the file with extension *top* containing the topology description. The topology validation is performed through behavioral simulation using AMODELs. The third phase consists of the electrical topology and sub-block specs generation. In this phase, although the topology generation follows a similar process to the previous phase, the generation of the specs for the sub-blocks was implemented using both PROLOG and C. Here, another library was developed to aid the task of identifying electrical structures. In this case, the generated topologies are described in a file with extension *ele* which is used as an entry for the sub-blocks specs generation task. These specs are obtained through the application of design rules and originate a file with extension *siz*. Finally, validation is carried out by electrical simulation. The achieved sub-block requirements are then submitted to lower level tools.

3. ARCHITECTURE GENERATION

The highly flexible input specification led to the methodology decomposition in three different phases, in order, to define a systematic and general synthesis process. First, the specification at the algorithm level, sampled in Fig. 2 (a) for a two-step flash A/D, allows the free use of the FIDEL HDL primitives and consequently permits the introduction of non-optimized and/or redundant algorithm descriptions which are the price to pay for flexibility. Naturally, the direct identification of functional structures from such an algorithm description [4] would lead to either a more complex process at lower levels of synthesis or a non-optimized solution, which in both cases would probably result in higher computational cost at lower levels of synthesis. Therefore, a simplification process allowing the generation of the analog and digital partitions and, also, a definition of a canonical form is here required. For this purpose a non-conventional signal flow graph representation allowing both the description of analog and digital signals, as well as their interfaces were developed [4]. In Fig. 2 (b) the resulting graph for the analog partition, is illustrated. Once the algorithm is represented in a canonical form the identification of functional structures and, so, the generation of the functional topology, illustrated in Fig. 2 (c), is based on the implementation of pattern recognition techniques applied to the SFG representation described above. Finally, the functional building blocks are instantiated by electrical sub-blocks, selected from the working library, yielding the architecture circuits shown in Fig. 2 (d). The results summarized in Table I show the applicability of the synthesis process to a broad range of data converters systems.

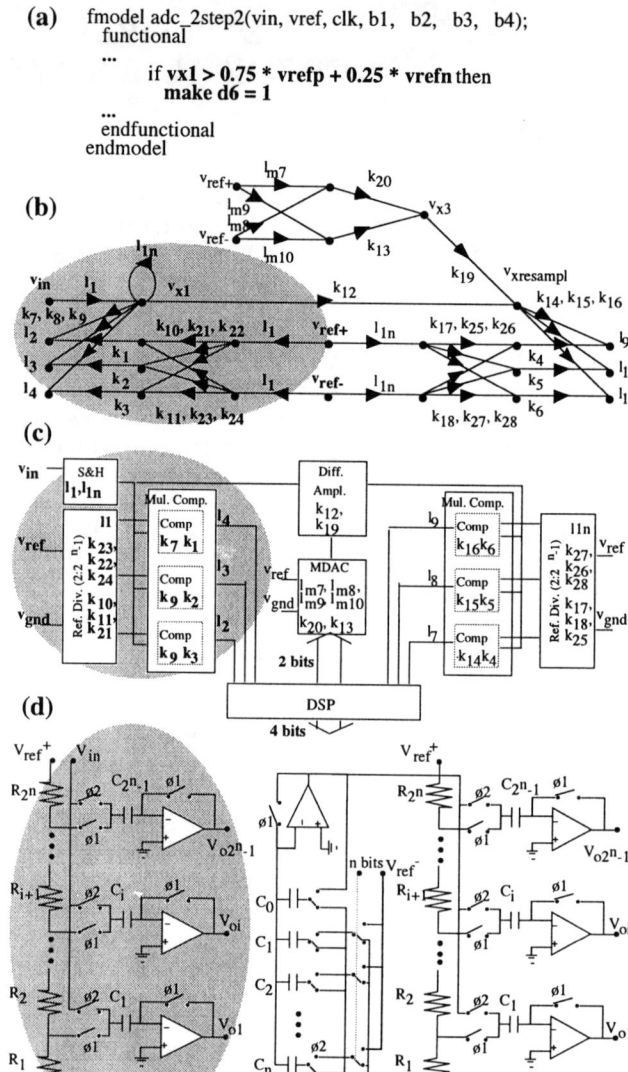

Figure 2. Architecture Generation Process: (a) Algorithm description sample. (b) Analog partition SFG. (c) Functional topology. (d) Data converter architecture.

Table I: TAGUS Performance for Architecture Generation

Algorithm/Graph/Topology/Architecture Generation								
Algor.	n bits	No. B.	CPU	No. B.	CPU	F. B.	CPU	E. B.	CPU
Succ.	8	112	2.3	28	1.2	3	0.30	1	0.50
Approx.	12	168	5.4	40	2.9	3	0.58	1	0.72
	16	224	10.9	52	5.3	3	0.80	1	0.90
Flash	4	169	1.3	64	1.3	3	0.23	1	0.63
	6	697	46.2	256	32.4	3	1.38	1	2.77
	8	2809	3742	1024	1405	3	9.28	1	16.23
Two-Step Flash	2+2	96	0.9	38	0.8	7	0.18	3	0.73
	4+4	388	13.2	140	13.1	7	0.80	3	1.83
	5+5	754	96.9	271	57.3	7	2.18	3	3.18
	6+6	1472	704	530	327	7	6.08	3	6.05
Combined Succ. Approx. and Flash	2+2	127	1.3	38	1.4	5	0.28	3	0.67
	2+2+2+2	239	4.4	62	5.7	5	0.65	3	0.82
	4+4	347	10.6	110	12.6	5	0.60	3	1.38
	4+4+4+4	571	46.0	158	39.9	5	1.68	3	1.62

4. ARCHITECTURE CHARACTERIZATION

Upon synthesis, the architecture is analyzed in order to automatically translate the high-level specifications, i.e., the data converter performance specs, shown in table II, into the first guess specifications for the electrical sub-blocks [4] and, therefore, allow the interface with lower level tools, e.g., OPCADSYS [5]. This process consist of four sub-tasks. First, determining the constraints imposed by the architecture. Then, estimating the required settling time and accuracy for each sub-block. Next, generating the sub-blocks specs and, finally, validating the specs through electrical simulation.

4.1 Constraints at the Architecture Level

The settling time and accuracy constraints are obtained based on the signal flow paths across the determined conversion architecture, due to their high dependency on each block operation phase. By contrast, power and area constraints are imposed by the static analysis of the existing blocks since the characterization can only be achieved with the interfacing with lower level tools. In Fig. 3, the signal flow paths for the case of a two-step conversion are illustrated. For a complete conversion cycle, at most, these paths are determined depending on the different operating phases for each block, starting with an analog input and ending with a digital output. In the present example the architecture is composed by only two types of electrical blocks and so the operation phases considered are the sampling and the comparison phases for the flash quantizer, and the sampling, the hold and the amplification phases for the MDAC. Table III shows the various propagation paths as a sum of time slots of each block. From the algorithm description the different operation phases are associated to a specific conversion cycle. The conjugation of both the signal propagation paths and the time slots associated with each conversion step result in a set of equations corresponding to the settling time constraints. For the generation of accuracy constraints the same propagation paths are used and an uniform error distribution of the maximum acceptable error on each path is considered, as usual in design practice [6-8]. Regarding power and area estimations [9], no restrictions are made at this level.

4.2. Settling Time and Accuracy Estimation

The previously obtained settling time and accuracy constraints, at the architecture level, allow the estimation of the requirements for each operation phase of each electrical block, and from which we can, then, generate a first set of subspecifications. The requirements for each operation phase are determined through the application of both knowledge-based rules, derived from the sub-block models, and heuristic rules obtained from design experience [6-8]. Fig. 4 shows two solutions obtained using two different approaches, respectively, an uniform and an heuristic requirements distribution.

4.3. Sub-Block Specs Generation

The specs generation for the electrical blocks are based on the use of behavioral models developed for each library element, e.g., flash quantizer, MDAC, etc. These specs are general, and so independent from any lower level architecture, e.g., OpAmp and comparators, which should be defined by lower level tools. In the present case, a dominant pole approach is considered for the OpAmps and the unit capacitor is imposed to allow the required linearity. Table IV shows the set of specifications obtained for each block of the two-step flash A/D converter. In general, the knowledge-based results are more favorable than results that could have been achieved by uniform distribution of time slots during conversion.

4.4. Validation

The fully synthesized conversion architecture is validated by electrical simulation using the same multi-level simulator, and where the various converter sub-blocks are described using electrical parametrized macromodels. Fig. 12 illustrates the results obtained for the five most significant bits and the final result obtained for the full 10-bit resolution of the converter. Upon validation of both the architecture and the generated specifications, the synthesis process can be continued with conventional lower level tools for further optimization. The traditional data conversion systems characteristics, e.g., INL, DNL, FFT etc., can only be achieved, in the present and in a reasonable time, by dedicated behavioral simulators and, therefore, are not discussed for this generic approach.

5. CONCLUSIONS

The algorithm-driven methodology developed for the synthesis and characterization of data converter systems, using extensively symbolic methods, allow the automatic exploration of a broad range of conversion algorithms without implying a previous description of the system architecture as in the traditional approaches. Although still limited at lower level by a required interface with other tools, the generality of this approach opens future opportunities to other classes of mixed signal systems.

REFERENCES

[1] R. Van de Plassche, *Integrated Analog-to-Digital and Digital-to-Analog Converters*, Kluwer Academic Publishers, 1994.
[2] G. G. E. Gielen, J. E. Franca, "CAD Tools for Data Converter Design: An Overview", *IEEE Transactions on Circuits and Systems - Part II*, Vol. 43, No. 2, pp. 77-89, 1996.
[3] ANACAD Computer Systems, ELDO V4.4.1 - Users Manual, 1994.
[4] N. Horta, J. Franca, "Algorithm-Driven Synthesis of Data Conversion Architectures", to appear in *IEEE Transactions on Computer-Aided Design*, Vol. 14, No. 10, 1997.
[5] N. E. Franca, M. A. Lança, J. E. Franca, "OpCadsys: An Open Tool for Automatic Synthesis of Circuits Components for Data Converters", *Proc. IEEE Midwest Symposium on Circuits and Systems*, pp. 343-346, 1994.
[6] B. S. Song, S. H. Lee, M. F. Tompsett, "A 10-b 15-MHz CMOS Recycling Two-Step A/D Converter", *IEEE Journal of Solid-State Circuits*, Vol. 25, No. 6, pp. 1328-1337, 1990.

[7] J. C. Vital, "Analog-Digital Data Conversion Integrated System with Functional Reconfiguration and Digital Testability", Ph.D. Thesis, Universidade Técnica de Lisboa - Instituto Superior Técnico, 1994.

[8] J. Goes, J. Vital, J. Franca, "A CMOS 4-bit MDAC with Self-Calibrated 14-bit Linearity for High-Resolution Pipelined A/D Converters", *Proc. IEEE Custom Integrated Circuits Conference*, pp. 6.6.1-6.6.4, 1996.

[9] G. Van der Plas, J. Vandenbussche, G. Gielen, W. Sansen, "EsteMate: a Tool for Automated Power and Area Estimation in Analog Top-Down Design and Synthesis", *Proc. IEEE Custom Integrated Circuits Conference*, pp. 7.7.1-7.7.4, 1997.

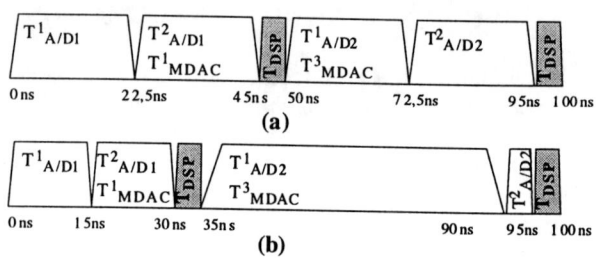

Figure 4. Time slot estimation considering (a) a uniform distribution, (b) a heuristic distribution.

Table II: Target Specifications for a Two-Step Flash A/D Converter

Parameter	Value	Parameter	Value
Resolution	10 bit	Power Supply	5 V
Conv. Rate	10 MS/s	Power	P1
Linearity	11 bit	Area	P2
Ref. Voltage	±1 V	Technology	0.8μm CMOS

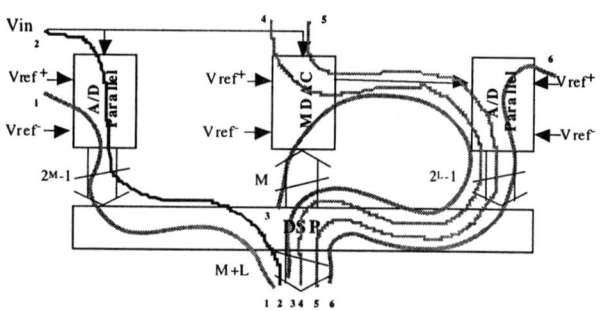

Figure 3. Signal Flow Paths Identification.

Table III: Propagation Paths, Functional Conditions and Temporal Constraints.

Paths	Conditions
$T_1 = T^1_{A/D1} + T_{DSP}$	Step 1:
$T_2 = T^2_{A/D1} + T_{DSP}$	$T^1_{A/D1}, T^2_{A/D1}, T^1_{MDAC}, T_{DSP}$
$T_3 = T^1_{MDAC} + T^2_{A/D2} + T_{DSP}$	Step 2:
$T_4 = T^2_{MDAC} + T^2_{A/D2} + T_{DSP}$	$T^2_{MDAC}, T^3_{MDAC}, T^2_{A/D2},$
$T_5 = T^3_{MDAC} + T^2_{A/D2} + T_{DSP}$	$T^1_{A/D2}, T_{DSP}$
$T_6 = T^1_{A/D2} + T_{DSP}$	
Constraints	
$T^1_{A/D1} + T^2_{A/D1} + T_{DSP} < T_{Step1}$	
$T^1_{MDAC} < T_{Step1}$	
$T^2_{MDAC} + T^3_{MDAC} + T^1_{A/D2} + T_{DSP} < T_{Step2}$	
$T^1_{A/D2} + T^2_{A/D2} + T_{DSP} < T_{Step2}$	
$T^1_{A/D1}, T^2_{A/D1}, T^1_{A/D2}, T^2_{A/D2}$ - Time Slots (TS) for Sampling and Comparison Phases on the first and Second flash,	
$T^1_{MDAC}, T^2_{MDAC}, T^3_{MDAC}$ - TS for Sampling and Hold Phases on the MDAC,	
T_{DSP} - TS for Digital Control,	
T_{Step1}, T_{Step2} - TS for the nth conversion step.	

Table IV: Data Converter Sub-Blocks Specs.

Elect. Block	Substructure	Specs	Uniform	Heuristic
MDAC	Cap. Array $C_i = 2^{i-1} C_u$	Cunit	0.25 pF	0.25 pF
		Cmax	4.00 pF	4.00 pF
		Ron_unit	< 1.16 KΩ	< 0.73 KΩ
		Ron_max	< 18.55 KΩ	< 11.67 KΩ
	OpAmp	Go	> 72 dB	> 72 dB
		GBW	> 1.47 GHz	> 583 MHz
		Cl	3.10 pF	3.10 pF
		Settling Time	< 22.50 ns	< 56.6 6ns
A/D Parallel (1)	Res. String $R_n = R_{unit}$	Runit	< 154 Ω	< 388 Ω
		Cunit	0.1 pF	0.1 pF
		Ron	< 15.40 KΩ	< 38.80 KΩ
	Comparator	Input Resolution	< 31.25 mV	< 31.25 mV
		Settling Time	< 22.5 ns	< 22.5 ns
A/D Parallel (2)	Res. String $R_n = R_{unit}$	Runit	< 154 Ω	< 97 Ω
		Cunit	0.1 pF	0.1 pF
		Ron	< 15.40 KΩ	< 9.67 KΩ
	Comparator	Input Resolution	< 15.63 mV	< 15.63 mV
		Settling Time	< 22.5 ns	< 5 ns

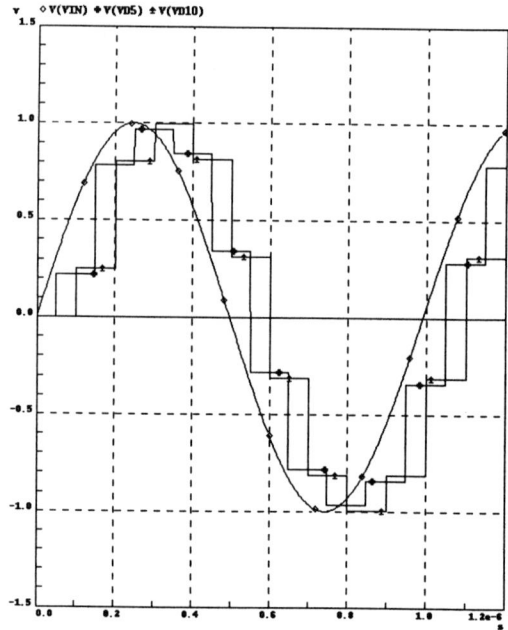

Figure 5. Simulation of the synthesized architecture for the two-step flash A/D conversion algorithm

A SYMBOLIC APPROACH FOR TESTABILITY EVALUATION IN FAULT DIAGNOSIS OF NONLINEAR ANALOG CIRCUITS

G. Fedi, R. Giomi, S. Manetti, M. C. Piccirilli

Department of Electronic Engineering, University of Florence,
Via S. Marta, 3 - 50139 Florence – Italy
E-mail: manetti@ingfi1.ing.unifi.it

ABSTRACT

A symbolic approach for testability evaluation in fault diagnosis of nonlinear analog circuits is presented. The new approach extends the methodologies developed for the linear case to circuits where nonlinear components, such as diodes or transistors, are present. The testability evaluation is a fundamental information for the fault diagnosis process, whatever method will be used, also in the nonlinear case. An example of circuit verifying this consideration and the validity of the proposed approach is briefly presented.

1. INTRODUCTION

Testing and fault diagnosis of electronic circuits are an essential part of the overall design and fabrication process. While for digital devices fully automated testing methodologies have been developed and are commonly used, for analog devices there is a lack of efficient and systematic methods of testing and fault diagnosis. Furthermore this problem is going to be even more critical with the increased complexity of electronic circuits and the increasing number of applications where mixed systems, characterized by the coexistence of both digital and analog parts, are present. As a consequence the analog fault diagnosis problem still has the attention of many researchers (see, for example, [1-3]).

In the analog fault diagnosis field an essential point is constituted by the concept of testability, which, independently of the method that will be effectively used in the fault location, gives theoretical and rigorous upper limits to the degree of solvability of the problem once the test point set has been chosen. A well-defined quantitative measure of testability can be deduced by referring to fault diagnosis techniques of parametric kind. These techniques, starting from a series of measurements carried out on previously selected test points, are aimed at determining the effective values of the circuit parameters by solving a set of equations nonlinear with respect to the component values. The solvability degree of these nonlinear equations constitutes one of the most used definitions of testability measurement, which indicates the ambiguity resulting from an attempt to solve such equations in a neighbourhood of almost any failure [4-6]. In other words, the testability measure provides information about the number of testable components with the selected test point set and then it is essential either to designers that must know which test points to make accessible for testing or to test engineers that must plan tests and know how many and what parameters can be uniquely isolated by these tests.

For the testability evaluation problem the symbolic approach is a natural choice, because, on the basis of the previous testability definition, a circuit description made by means of equations in which the component values are the unknowns is properly represented by symbolic relations. Algorithms based on a symbolic approach for evaluating this kind of testability measure in the case of analog linear circuits have been developed by the authors in the past [7-11]. The aim of this paper is to present an extension of the results obtained for linear circuits to the case of circuits containing highly nonlinear components such as diode and transistor. The theoretical basis of the developed procedure is the independence of testability with respect to the value of the circuit components. With the presented algorithms it is possible to determine also the ambiguity groups [12], which can be defined as sets of components that, if considered as potentially faulty, do not give unique solution in fault location phase and whose determination is of great importance, particularly in the case of low testability [13, 14]. Finally the proposed approach for testability and ambiguity group determination can constitute the first step in the development of whatever procedure for the fault location of nonlinear analog circuits.

The paper is organized as follows. In Section 2 the theoretical bases for the testability and ambiguity group determination of analog linear circuits are summarized. In Section 3 the extension of the results obtained in the linear case to the nonlinear one is presented and an example of circuit verifying the validity of the proposed approach is briefly presented. Finally in Section 4 the obtained results are summarized and the future work is outlined.

2. TESTABILITY EVALUATION FOR ANALOG LINEAR CIRCUITS

To better understand how the methodologies of testability evaluation obtained for analog linear circuits can be extended to the nonlinear case, firstly the fundamental concepts relevant to the linear case are summarized.

Referring to parametric fault diagnosis techniques, in the case of analog, linear, time-invariant circuits the fault diagnosis equations can be constituted by the network functions relevant to the selected test points [7, 9], which are nonlinear with respect to the potentially faulty circuit parameters. By assuming that the faults can be expressed as parameter variations, without influencing the circuit topology (i.e. faults as short and open are not considered), the testability measure T is given by the maximum number of linearly independent columns of the Jacobian matrix associated with the fault diagnosis equations and it represents a measure of the solvability degree of the nonlinear fault diagnosis equations [4-6]. The entries of the Jacobian matrix are rational functions depending on the complex frequency s and the potentially faulty parameters. So, in order to evaluate the testability, it is necessary to fix the potentially faulty parameter values and the complex frequency s. In [5] it has been shown that, once fixed the frequency values (generally a multifrequency approach is considered in order to use a reasonable number of test points), the rank of the obtained Jacobian matrix is constant almost everywhere, that is for all the potentially faulty parameter values except those lying in an algebraic variety. Following this way, the testability value is independent of component values. In order to make the testability evaluation independent also on the complex frequency s, it has been demonstrated that, starting from the network functions expressed in the following way:

$$h_l(\mathbf{p},s) = \frac{N_l(\mathbf{p},s)}{D(\mathbf{p},s)} = \frac{\sum_{i=0}^{n_l} \frac{a_i^{(l)}(\mathbf{p})}{b_m(\mathbf{p})} \cdot s^i}{s^m + \sum_{j=0}^{m-1} \frac{b_j(\mathbf{p})}{b_m(\mathbf{p})} \cdot s^j} \qquad l=1,\ldots K$$

where $\mathbf{p}=[p_1, p_2, \ldots p_R]^t$ is the vector of the potentially faulty parameters and K is the total number of equations, the testability is equal to the rank of a matrix $\mathbf{B}$, independent on the complex frequency s, whose entries are constituted by the derivatives of the coefficients of the fault diagnosis equations with respect to the potentially faulty circuit parameters [9]. So the testability matrix can be considered in the following form:

$$\mathbf{B} = \begin{bmatrix} \partial\frac{a_0^{(1)}}{b_m}/\partial p_1 & \partial\frac{a_0^{(1)}}{b_m}/\partial p_2 & \cdots & \partial\frac{a_0^{(1)}}{b_m}/\partial p_R \\ \vdots & \vdots & & \vdots \\ \partial\frac{a_{n_1}^{(1)}}{b_m}/\partial p_1 & \partial\frac{a_{n_1}^{(1)}}{b_m}/\partial p_2 & \cdots & \partial\frac{a_{n_1}^{(1)}}{b_m}/\partial p_R \\ \vdots & \vdots & & \vdots \\ \partial\frac{a_0^{(K)}}{b_m}/\partial p_1 & \partial\frac{a_0^{(K)}}{b_m}/\partial p_2 & \cdots & \partial\frac{a_0^{(K)}}{b_m}/\partial p_R \\ \vdots & \vdots & & \vdots \\ \partial\frac{a_{n_K}^{(K)}}{b_m}/\partial p_1 & \partial\frac{a_{n_K}^{(K)}}{b_m}/\partial p_2 & \cdots & \partial\frac{a_{n_K}^{(K)}}{b_m}/\partial p_R \\ \partial\frac{b_0}{b_m}/\partial p_1 & \partial\frac{b_0}{b_m}/\partial p_2 & \cdots & \partial\frac{b_0}{b_m}/\partial p_R \\ \vdots & \vdots & & \vdots \\ \partial\frac{b_{m-1}}{b_m}/\partial p_1 & \partial\frac{b_{m-1}}{b_m}/\partial p_2 & \cdots & \partial\frac{b_{m-1}}{b_m}/\partial p_R \end{bmatrix}$$

If the fault diagnosis equations are generated in completely symbolic form, the testability evaluation becomes easy to perform. In this case, the entries of the matrix $\mathbf{B}$ can be simply led back to derivatives of sums of products and testability evaluation can be performed by triangularizing $\mathbf{B}$ and assigning arbitrary values to the components, because, as already mentioned, testability does not depend on component values. In particular, if rank$\mathbf{B}$ is equal to the number of unknown parameters, the component values can be uniquely determined by solving the fault diagnosis equations through the consideration of a set of measurements carried out on the test points. If the testability T (T=rank$\mathbf{B}$) is less than the number of unknown parameters R, a locally unique solution can be determined only if R-T components are considered not faulty.

The matrix $\mathbf{B}$ does not give only information about the global solvability degree of the fault diagnosis problem. In fact, by noticing that each column is relevant to a specific component of the circuit and by considering the linearly dependent columns of $\mathbf{B}$, other information can be obtained. For example, if a column is linearly dependent with respect to another one, this means that a variation of the corresponding component provides a variation on the fault equation coefficients indistinguishable with respect to that produced by the variation of the component corresponding to the other column. This means that the two components are not testable and they constitute an ambiguity group of the second order. By extending this reasoning to groups of linearly dependent columns of $\mathbf{B}$, ambiguity groups of higher order can be found. The knowledge of the ambiguity groups allows to have information about the solvability of the fault diagnosis problem with respect to each component and in case of bounded number of faults (k-fault hypothesis) [14].

3. TESTABILITY EVALUATION FOR ANALOG NONLINEAR CIRCUITS

In order to extend all the previous results to the case of analog nonlinear circuits, it is possible to proceed in the following way. By taking into account the important result relevant to the independence of testability with respect to component values, each nonlinear component of the circuit under consideration is replaced with its approximated piece wise linear model (PWL). Using this technique the voltage-current characteristic of any nonlinear electronic device is replaced by piece wise linear segments obtained by means of the individuation of one or more corner points on the characteristic. A piece wise linear characteristic describes approximately the element behavior in the different operating regions in which it can work. It is evident that the increase of the corner point number and, consequently, of the linearity region number allows to obtain a higher precision in the simulation of the real component; obviously, in this way, the corresponding model becomes more complex. But, by facing the problem through symbolic techniques, the number of the PWL characteristic corner points is not of any importance: in fact, from a symbolic analysis point of view, each component of a PWL model is represented by a single symbol. So, by considering the PWL characteristics of each nonlinear device as succession of several very little linear regions, the parameters describing the piece wise linear models can be considered changing in a continuous way their numerical values. Being the testability value independent of the circuit parameter values, the results obtained for the linear case are still valid if there is not a change in the circuit topology. Let us consider, for example, the case of a diode. The diode characteristic can be piece wise linearized by considering a DC current generator plus a conductance R_i, assuming so much different values as the linear regions are. Hence, the testability value, obtained using the model with I_i and R_i, is valid under the condition that R_i is different from zero or infinite and the device is not working in a range of frequencies that can introduce parasitic components as C_p, L_p, etc.

In practice, the procedure for determining the testability value and the ambiguity groups in the case of analog circuits containing also nonlinear devices consists in the following steps:

- substitution of each nonlinear component with its piece wise linear model;
- determination of the circuit network functions in symbolic form by means of the MNA (Modified Nodal Analysis);
- determination of the testability matrix **B**;
- evaluation of the testability value through the calculation of rank **B** obtained by assigning arbitrary values to both the parameters relevant to linear components and the ones corresponding to piece wise linear models relevant to nonlinear components;
- determination of the ambiguity groups by means of the individuation of the linearly dependent columns of the matrix **B**.

It is worth pointing out that in the nonlinear case the network functions determined in the s domain have not any meaning and they cannot be used in phase of fault location in order to determine the effective value of the circuit components. They can give information only about the testability and ambiguity group determination. In phase of fault location it is necessary to use procedures which do not refer to the network functions or procedures based on the discretization of the network functions and, in any case, on time domain measurements. In this last case, the discretization is equivalent to consider, for the reactive elements of the circuit, suitable models derived from the discretization of the constitutive equations. In this way it is still possible to determine the faulty components by solving the discretized network functions by means of algorithms relevant to piece wise linearized circuits, as the Katznelson algorithm [15]. On the other hand, referring to fault location procedures that are not based on the solution of the network functions starting from the measurement results, a possible procedure could be given by a neural network approach.

In order to verify the validity of the proposed method for testability and ambiguity group determination in the case of circuits containing nonlinear devices, a fault location procedure based on a feedforward neural network has been considered. In fact, when this kind of network is used for fault location, the knowledge of both testability and ambiguity groups is of fundamental importance in order to train the network, because, for example, it is useless to train the network with data relevant to indistinguishable components (i.e. components belonging to an ambiguity group). Let us consider the simple analog nonlinear circuit shown in Fig. 1. By considering as test point the amplifier output voltage Vo and a simple piece wise linear model [15] for the transistor, a testability value equal to four and a set of testable components not belonging to any ambiguity group containing Rc, R1, Re and a parameter relevant to the transistor PWL model have been determined. Then a three layer feedforward neural network with a number of output neurons equal to the obtained testability value has been considered. In the learning phase, samples in the time domain of the output signal, corresponding to variations in the value of the testable parameters, have been used as inputs for the first layer neurons. After the learning phase, the neural network has been able to locate a single fault in the circuit simulated with the program SPICE, by using as inputs of the neural network the samples of the output, obtained with SPICE, corresponding to variation of the circuit parameters with respect to the nominal value. It has been also verified that, considering a number of output neurons different with respect to the testability value or a set of parameters different with respect to the testable ones, the learning process of the network does not converge. Finally it is worth pointing out that, if in the circuit of Fig. 1 the transistor works in a region of smooth nonlinearity, it is possible to use the small signal model for the transistor and consider, then, the circuit as a linear network for the testability evaluation.

4. SUMMARY

A symbolic approach for testability evaluation of analog nonlinear circuits has been presented. The proposed approach permits to extend the methodologies developed for linear circuits to the case of circuits containing nonlinear devices by exploiting the independence of testability with respect to the component value and a piece wise linear representation for the nonlinear components.

A simple example of application of the results obtained with the presented procedure has been considered. It will be matter of future work the realization of a fully automated fault location system, based on the use of feedforward neural networks, in which the measurement data are acquired directly on the circuit under test through an acquisition board.

5. REFERENCES

[1] M. Slamani, B. Kaminska, «Fault observability analysis of analog circuits in frequency domain», *IEEE Trans. Circ. Syst.*, vol. CAS-43, pp. 134-139, February 1996.

[2] H.T. Sheu, Y. H. Chang, «Robust fault diagnosis for large-scale analog circuits with measurement noises», *IEEE Trans. Circ. Syst.-I*, vol. CAS-44, pp. 198-209, March 1997.

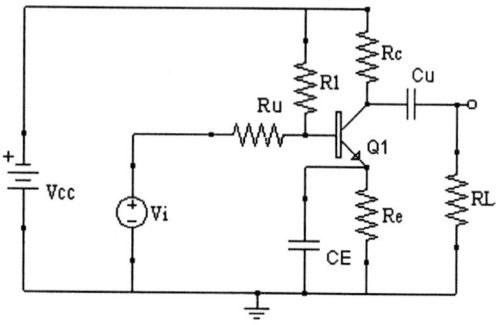

Fig.1. *Single transistor amplifier.*

[3] R. Spina, S. Upadhyaya, «Linear circuit fault diagnosis using neuromorphic analyzers», *IEEE Trans. Circ. Syst.-II*, vol. CAS-44, pp. 188-196, March 1997.

[4] R. Saeks, «A measure of testability and its application to test points selection theory», *20th Midwest Symp. Circuits and Syst.*, Texas Tech. Univ. Lubbock, August 1977.

[5] N. Sen and R. Saeks, «Fault diagnosis for linear systems via multifrequency measurement», *IEEE Trans. Circ. Syst.*, vol. CAS-26, pp. 457-465, July 1979.

[6] H. M. S. Chen and R. Saeks, «A search algorithm for the solution of multifrequency fault diagnosis equations», *IEEE Trans. Circ. Syst.*, vol. CAS-26, pp. 589-594, July 1979.

[7] R. Carmassi, M. Catelani, G. Iuculano, A. Liberatore, S. Manetti, M. Marini, «Analog network testability measurement: a symbolic formulation approach», *IEEE Trans. Instr. Meas.*, vol. IM-40, pp. 930-935, December 1991.

[8] S. Manetti, M. C. Piccirilli, «Symbolic method for circuit testability and fault diagnosis in time domain», *11th European Conference on Circuit Theory and Design, ECCTD'93*, Davos, Switzerland, August 1993, pp. 1681-1686.

[9] A. Liberatore, S. Manetti, M. C. Piccirilli, «A new efficient method for analog circuit testability measurement», *IEEE Instr. and Meas. Tech. Conf., IMTC'94*, Hamamatsu, Japan, May 1994, pp. 193-196.

[10] A. Liberatore, A. Luchetta, S. Manetti, M. C. Piccirilli, «Analog network testability measurement using symbolic techniques», *Third Int. Workshop on Symbolic Methods and Application to Circuit Design, SMACD'94*, Sevilla, Spain, October 1994, pp. 167-176.

[11] M. Catelani, G. Fedi, A. Luchetta, S. Manetti, M. Marini, M. C. Piccirilli, «A new symbolic approach for testability measurement of analog networks», *MELECON'96*, Bari, Italy, May 1996, pp. 517-520.

[12] G. N. Stenbakken, T. M. Souders and G. W. Stewart, «Ambiguity groups and testability», *IEEE Trans. Instr. Meas.*, vol. IM-38, pp. 941-947, October 1989.

[13] G. Fedi, R. Giomi, A. Luchetta, S. Manetti, M. C. Piccirilli, «The use of symbolic methods in the full automation of analog circuit fault location», *4th Int. Workshop on Symbolic Methods and Applications to Circuit Design, SMACD'96*, Heverlee, Belgium, October 1996.

[14] G. Fedi, R. Giomi, A. Luchetta, S. Manetti, M. C. Piccirilli, «Symbolic algorithm for ambiguity group determination in analog fault diagnosis», *European Conference on Circuit Theory and Design, ECCTD'97*, Budapest, Hungary, September 1997, pp. 1286-1291.

[15] S. Manetti, M. C. Piccirilli, «Symbolic simulators for the fault diagnosis of nonlinear analog circuits», *Analog Integrated Circuits and Signal Processing*, vol. 3, pp. 59-72, January 1993.

SYMBOLIC ANALYSIS OF MICROWAVE CIRCUITS

K. Benboudjema, M. Boukadoum*, G. Vasilescu** and G. Alquié**

COM DEV Space Group,
155 Sheldon Dr., Cambridge, Ontario,
N1R 7H6, Canada

* Université du Québec à Montréal, Montréal, Canada
** LEAM, Université Pierre et Marie Curie, Paris, France

ABSTRACT

The polynomial interpolation (PI) method is one of the most suitable symbolic analysis techniques for the computation of linear transfer functions when the only circuit variable is the complex frequency p. In this article, we present improvements to the PI method so that it may be used for the computation of network functions in a fully symbolic form, using only numerical algorithms. We provide examples of how to use the new approach to directly determine the microwave performances, such as the scattering parameters, of given networks.

1. INTRODUCTION

Symbolic Analysis refers to techniques which allow the determination of transfer function for an arbitrary linear electrical network, knowing the topology of one such network. The transfer function is obtained as the ratio of two polynomials:

$$H(p, x_1, x_2, \cdots, x_m) = \frac{N(p, x_1, x_2, \cdots, x_m)}{D(p, x_1, x_2, \cdots, x_m)} \quad (1)$$

Where f_i and g_j are functions of type sum of products of the network elements $x_{k=1,\ldots,m}$, and where p is the complex frequency.

In the past decades, several symbolic methods have been developed for the study of linear analogue circuits: they are summarized in [1], [2]. These methods have led to the development of numerous symbolic simulators: SCYMBAL, SAPEC, ISAAC, SCAAP, ASAP, SCCNAP... A comparative study of these simulators can be found in reference [3].

As the applications of microwaves in various domains, such as the mobile communications in the L-Band and satellite communications in the K-Band..., are growing, it has become important and useful to be able to specify the performances of any microwave circuit in a fully symbolic form to better understand, model and design that circuit. This paper presents a technique to achieve this goal.

Section 2 describes the basic principles of the polynomial interpolation method. Then, in section 3, improvements to this method are presented, so that the transfer function of an arbitrary linear circuit may be obtained in a fully symbolic form. We also show how to directly have access to the microwave performances. Section 4 provides practical examples of using the proposed method.

2. THE POLYNOMIAL INTERPOLATION METHOD

This method has been presented in the literature as the most suitable for the determination of transfer functions when the only circuit variable is the complex frequency p. The principle of this method consists of finding the coefficients of a polynomial Q(p):

$$Q(p) = \sum_{i=0}^{n} a_i \, p^i \quad (2)$$

given that Q(p) can be evaluated for any value of p. It has been shown [4-9] that for large value of n, the best choice for the (n+1) values of p is to distribute them along the complex unit circle:

$$p_k = e^{j\frac{k\Pi}{n+1}}, \quad k = 0, \cdots, n \quad (3)$$

In this case, the coefficients a_i of Q(p) can be found using the following formula (DFT):

$$a_i = \frac{1}{n+1} \sum_{k=0}^{n} Q(p_k) e^{-j\frac{2k\Pi}{n+1}}, \quad i = 0, \cdots, n \quad (4)$$

To obtain the transfer function in a rational form, the above technique is applied to both the numerator and the denominator, which are evaluated for each value of p_k (eq. 3) by applying Cramer's rule to the network equation, obtained by using the modified nodal analysis (MNA) method:

$$T x = w \quad (5)$$

In this equation, T is the modified nodal matrix, w the independent sources vector and x the unknown vector which contains the nodal voltages and some branches currents.

In reference [7], the authors have shown that by applying perturbations of amplitude δ_i to the network elements values $x_{i=1,\ldots,m}$, so that they become:

$$x_i = x_{i0} + \delta_i \qquad (6)$$

where x_{i0} is the nominal value of the element x_i, then, it is possible to generate the circuit's transfer function as the ratio of two polynomials which depend on the complex frequency p and on the deviations δ_i which affect the network elements:

$$H(p,\delta_1,\cdots,\delta_m) = \frac{N(p,\delta_1,\cdots,\delta_m)}{D(p,\delta_1,\cdots,\delta_m)}$$

$$= \frac{R_0(p) + \sum_{i=1}^{m} R_i(p)\delta_i + \sum_{i=1}^{m}\sum_{j>i}^{m} R_{ij}(p)\delta_i\delta_j + \ldots}{Q_0(p) + \sum_{i=1}^{m} Q_i(p)\delta_i + \sum_{i=1}^{m}\sum_{j>i}^{m} Q_{ij}(p)\delta_i\delta_j + \ldots}$$

$$(7)$$

where $R_{ij\ldots}(p)$ and $Q_{ij\ldots}(p)$ are polynomials in p. The coefficients of these polynomials are determined by applying the above technique (eq. 3 and 4). Then, the evaluation of the obtained polynomials for each value of $p_{k=0,\ldots,n}$ (n being the number of reactive elements) consists of finding the determinant and various co-factors of a matrix $\hat{F}$ of dimension (m+1) x (m+1), where m is the number of perturbed elements, defined by:

$$\hat{F} = \hat{Q}^t\, T^{-1}\, \hat{P} \qquad (8)$$

and where T^{-1} is the inverse of the modified nodal matrix. $\hat{P}$ and $\hat{Q}$ are matrices that allow the localization of the perturbed branches and the perturbed elements control branches [7].

The $R_{ij\ldots}(p)$ and $Q_{ij\ldots}(p)$ polynomials are evaluated for each value of p_k using the following formula:

$$R_{j_1 j_2 \cdots j_l}(p_k) = \det T \det(\hat{F}[j_1, j_2, \cdots, j_l, m+1]) \quad (9)$$

and $Q_{j_1 j_2 \cdots j_l}(p_k) = \det T \det(\hat{F}[j_1, j_2, \cdots, j_l]) \quad (10)$

where $\hat{F}[j_1, j_2, \ldots, j_l, m+1]$ is a sub-matrix of $\hat{F}$ created with rows and columns $j_1, j_2, \ldots, j_l, m+1$.

3. MOVING TO THE FULLY SYMBOLIC FORM

3.1 Method Description

It is possible to go from the symbolic form in terms of deviations (eq.7) to a fully symbolic form if all network elements values are greater than one, so that the product of two or more of these elements values is also greater than 1. When this condition is met, the fully symbolic form of the transfer function is derived from the symbolic form in terms of deviations by keeping only the coefficients of $R_{ij}(p)$ and $Q_{ij}(p)$ polynomials with values equal to ±1 (there exists at most one such coefficient per polynomial), and by replacing the δx_i by x_i in the resulting equation. It is usually not necessary to calculate all the coefficients α_i and β_i of, the $R_{ij\ldots}(p)$ and $Q_{ij\ldots}(p)$ polynomials in order to obtain the weight factors of the symbolic terms ($\Pi\, x_i$). Since each symbolic term is a product of circuit elements, it follows that the weight of any symbolic term, in the fully symbolic form, is of the form: $\alpha_i\, p^i$ for the numerator and $\beta_i\, p^i$ for the denominator, and may be extracted from the $R_{ij\ldots}(p)$ and $Q_{ij\ldots}(p)$ polynomials, respectively. Thus, it should be obvious by inspecting each symbolic term to deduce which coefficient of $R_{ij\ldots}(p)$ and $Q_{ij\ldots}(p)$ is to be calculated. This can be done by introducing an auxiliary variable pow which keeps track of the number of reactive elements in each symbolic term and which is precisely the power of p that is sought [9].

This technique of computation of transfer functions in a fully symbolic form can be summarized in the following pseudo-code:

Read the circuit description (Netlist form)
Build the network's matrix T and the $\hat{F}$ matrix
For each (2^m-1) symbolic term Πx_i do:
   for each value of $p_{k=0,\ldots,n}$, computed according to equation (3) do:
      Apply equation (9) to find the value of $R_{ij\ldots}(p_k)$
      Apply equation (10) to find the value of $Q_{ij\ldots}(p_k)$
   End for
   Compute the value of pow
   Compute the coefficient α_{pow} from the set $R_{ij\ldots}(p_k)$, according to equation (4)
   Compute the coefficient β_{pow} from the set $Q_{ij\ldots}(p_k)$, according to equation (4)
   Remove the symbolic term from the numerator of eq. (1) if α_{pow} is different from ±1
   Remove the symbolic term from the denominator of eq. (1) if β_{pow} is different from ±1
End for

3.2 Fully Symbolic Analysis of Microwave Two-port networks

The method developed in the previous section can easily be extended to the determination of the microwaves performances of linear circuits (figure. 1), specially the input and output reflection coefficients and the scattering parameters.

These parameters are usually expressed as nonlinear function of the output or input impedances and it is difficult, if not impossible, to find them in a fully symbolic form via others symbolic methods, particularly when the rational equations that express the impedances contain a large number of terms. This is due to the fact that the numerator and denominator multivariate polynomials are determined separately and independently, which is not the case of the method described in the previous section in which we select first a symbolic term and we calculate its contribution for the numerator and the denominator (i.e. α_i and β_i coefficients), then, we select an other symbolic term and do again the same process...etc.

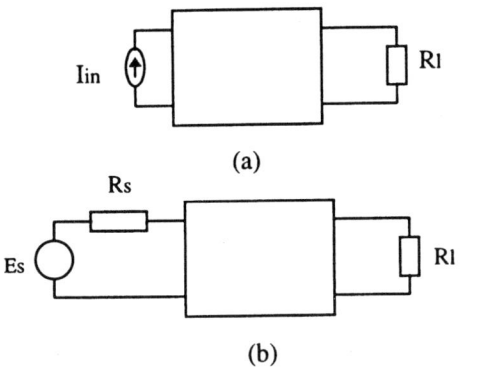

Figure 1: Determination of (a): Zin, and (b): the gain of a two-port.

It follows, that if the input impedance is obtained in the following form:

$$Z_{in} = \frac{\sum_i \alpha_i \, p^i \prod_k x_k}{\sum_i \beta_i \, p^i \prod_k x_k} \qquad (11)$$

then the input reflection coefficient and the S_{11} parameter of a two-port, can also be determined in a fully symbolic form, as followed:

$$\Gamma_{in} = \frac{\sum_i (\alpha_i - Z_0 \, \beta_i) \, p^i \prod_k x_k}{\sum_i (\alpha_i + Z_0 \, \beta_i) \, p^i \prod_k x_k} \doteq S_{11}\big|_{R_l = 50\,\Omega} \qquad (12)$$

where Z_0 is the characteristic impedance of the network (typically, $Z_0 = 50\ \Omega$). The same argument applies to the output reflection coefficient and the S_{22} parameter.

When the input voltage E_s is equal 1. volt and when the source R_s and load R_l resistances are equal to 50. Ω (figure 1-b), then, the S_{21} parameter of a two-port is obtained from the voltage gain:

$$S_{21} \doteq 2\, G_V\big|_{E_s=1.\,volt,\ R_s=R_l=50.\,\Omega} \qquad (13)$$

A similar reasoning applied to S_{12} parameter, with the input and output of the two-port reversed.

4. EXAMPLES

The previous method was implemented in a software called SYSMIC (Symbolic Simulator of Microwaves Circuits) which is available for both the DOS and Unix environments. This section presents two examples that were analyzed using SYSMIC. The simulation time of each parameter and for each example was of the order of milliseconds on a IBM R6000 workstation.

Example 1: FET small signal equivalent circuit

In this example we considered a small signal equivalent circuit of a Field Effect Transistor used in many microwaves and millimeter-waves applications (figure 2). The s_{21} parameter has been obtained in the following form:

$S_{21}=$ { [-2 G0 Gg Gd gm Gi Gs] **p**0** + 2 G0 Gg Gd Gi [Cg (Gs+gm+Gl)+Cs Gl] **p**1** + [Cg Cs G0 Gg Gd (Gi+Gs+Gl)+ G0 Gg Gd Gi Cd (Cg+Cs)+G0 Gg Gd Gi Gs (Cg Ls (gm+Gl)+Gl Cs Ls)] **p**2** + 2 G0 Gg Gd [Cg Cs Cd+Gs Cg Cs Ls (Gi+Gl)+Gi Gs Cd Ls (Cg+Cs)] **p**3** + 2 G0 Gg Gd Gs Cg Cs Cd Ls **p**4** } / { G0 Gg Gi (Gd GL (gm + Gl + Gs)+ Gl Gs (Gd + GL)) **p**0** + [Cg Gi (G0 (Gg Gd gm + Gg GL gm + Gd GL gm + Gg Gd Gl + Gg GL Gl + Gd GL Gl + Gg Gd Gs+ Gg GL Gs+ Gd GL Gs + Gd gm Gs + GL gm Gs + Gd Gl Gs + GL Gl Gs) + Gg (Gd GL gm + Gd GL Gl + Gd GL Gs + Gd gm Gs + GL gm Gs +Gd Gl Gs + GL Gl Gs)) + Cs (G0 (GL (Gg Gd Gl+ Gg Gd Gi + Gg Gl Gi + Gd Gl Gi + Gg Gd Gs + Gg Gl Gs + Gd Gi Gs + Gl Gi Gs) + Gg Gd Gl Gi + Gg Gd Gl Gs + Gd Gl Gi Gs) + Gg Gi (Gd GL Gl + Gd GL Gs + Gd Gl Gs + GL Gl Gs)) Cd G0 Gg Gi (Gd GL + Gd Gs+ GL Gs)+ (Ld Gl+ Ls (Gl + gm)) G0 Gg Gd GL Gi Gs] **p**1** + [Cg Cs (G0 (GL (Gg Gl+Gd Gl+Gg Gi+Gd Gi+Gg Gs+Gd Gs+Gl Gs+Gi Gs) + Gg Gd Gl+Gg Gd Gi+Gg Gl Gi+Gd Gl Gi + Gg Gd Gs+Gd Gl Gs+Gd Gi Gs) +Gg (Gd GL Gl+Gd GL Gs+Gd Gl Gs+GL Gl Gs+Gd Gi Gs+GL Gi Gs)) + Cg Cd Gi (G0 (Gg Gd +Gg GL + Gd GL + Gd Gs + GL Gs) + Gg (Gd GL + Gd Gs + GL Gs)) + Cs Cd (G0 (Gg Gd GL+Gg Gd Gi+Gg GL Gi + Gd GL Gi + Gg Gd Gs + Gg GL Gs + Gd Gi Gs + GL Gi Gs)) + Gg (Gd GL + Gd Gs + GL Gs)) + Cg Lg G0 Gg Gi (Gd GL gm + Gd GL Gl + Gd GL Gs + Gd gm Gs + GL gm Gs + GL Gl Gs + Gd Gl Gs) + Cs Lg G0 Gg Gi (Gd GL Gl + Gd GL Gs + Gd Gl Gs + GL Gl Gs)+ Cg Ld Gd Gi (G0 (Gg GL gm + Gg GL Gl + Gg GL Gs + GL gm Gs + GL Gl Gs) + Gg GL Gs (gm + Gl))+ Cs Ld Gd Gl GL (G0 Gg Gi + G0 Gg Gs + G0 Gi Gs + Gg Gi Gs) + Cd (Ld + Ls) G0 Gg Gd GL Gi Gs + Cs Ls Gs (G0 (Gg Gd (GL Gl + GL Gi + Gl Gi) + GL Gl Gi (Gg + Gd)) + Gg Gd GL Gl Gi)+ Cg Ls Gi Gs (G0 (Gg Gd gm + Gg GL gm + Gd GL gm + Gg Gd Gl + Gg GL Gl + Gd GL Gl) + Gg Gd GL gm + Gg Gd GL Gl)] **p**2** + [Cg Cs Cd (G0 Gg Gd + G0 (Gg GL + Gd GL + Gd Gs + GL Gs) + Gg (GL + Gs) + GL Gs)) + Cg Cs Lg G0 Gg (Gd (GL Gl + GL Gi + GL Gs + Gl Gs + Gi Gs) + GL Gl Gs + GL Gi Gs)+ Cg Cd Lg G0 Gg Gi (Gd (GL + Gs) + GL Gs) + Cs Cd Lg G0 Gg Gi (Gd (GL + Gs) + GL Gs)+ Cg Lg Ld G0 Gg Gd GL Gi Gs (gm + Gl)+ Cg Lg Ls G0 Gg Gd GL Gi Gs (gm + Gl) + Cs Lg (Ld + Ls) G0 Gg Gd GL Gl Gi Gs + Cs Ls Ld GL (G0 Gd (Gg Gl + Gg Gi + Gl Gi + Gs Gs) +Gg Gd Gs (Gl + Gi))+Cs Cd Ld Gd GL (G0 Gg + G0 Gs + Gg Gs)+Cs Cd Ld Gd GL (G0 Gg Gi + G0 Gg Gs + G0 Gi Gs + Gg Gi Gs) + Cg Ld Ls G0 Gg Gd GL Gi Gs (gm + Gl) + Cs Ld Ls G0 Gg Gd GL Gl Gi Gs + Cg Cs Ls Gs (G0 (Gg Gd Gl+Gg GL Gl+Gd GL Gl+Gg Gd Gi+Gg GL Gi+Gd GL Gi)+Gg Gd GL (Gl + Gi)) + Cg Cd Ls Gi Gs (G0 (Gg Gd +Gg GL+Gd GL) +Gg Gd GL) + Cs Cd Ls Gs (G0 (Gg Gd GL +Gg Gd Gi +Gg GL Gi +Gd GL Gi) +Gg Gd GL Gi)] **p**3** + [Cg Cs Cd Ld Gd GL Gs (G0 + Gg) + Cg Cs Cd Ls Gs (G0 Gg Gd +G0 Gg GL +G0 Gd GL + Gg Gd GL)+ Cg Cd Lg G0 Gg GL Gs (Cs + Ld Gd Gi)+ Cg Cs Lg Ld G0 Gg Gd GL Gs (Gl + Gi) + Cs Cd Lg (Ld+Ls) G0 Gg Gd GL Gi Gs + Cg Cs Ls Lg Ls G0 Gg Gd GL Gi Gs (Gl + Gi) + Cg Cd Lg Ls G0 Gg Gd GL Gi Gs + Cg Cs Ld Ls G0 Gg Gd GL Gl Gi Gs + (Cg + Cs)Cd Ld Ls G0 Gg Gd GL Gi Gs] **p**4** + [Cg Cs Cd Lg (Ld + Ls) G0 Gg Gd GL Gs + Cg Cs Cd Ld Ls G0 Gg Gd GL Gs] **p**5** }

Where: $G_0 = G_L = 1/50.\ \Omega$,

In this expression the common term factorization was manually accomplished. The others s-parameters have also been simulated with SYSMIC and numerically evaluated for several frequency points. The results have been compared to those obtained by others numerical microwaves simulators such as HP MDS or HP-Eesof Libra. The comparison showed an excellent agreement between the different results.

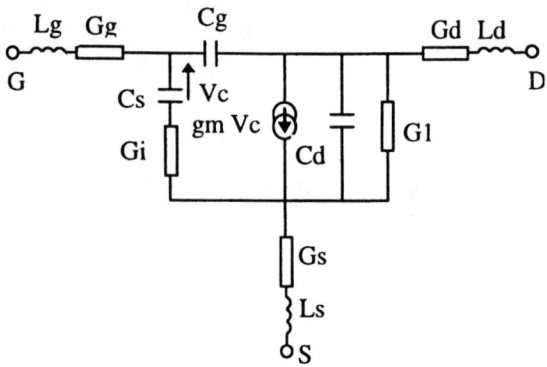

Figure 2: Small signal equivalent circuit of the FET.

Example 2: FET Amplifier with parallel feedback
In this second example, we present one microwaves application of a FET for wide band amplification, figure 3 [10].

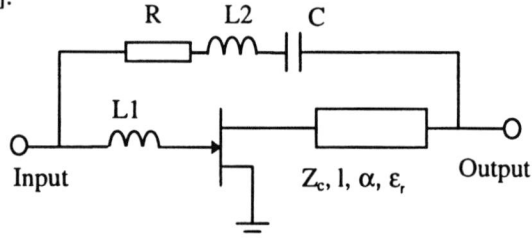

Figure3: Parallel Feedback FET Amplifier

where L_1=0.096 nH, L_2=0.63 nH, C=5 pF, R=200 Ω, Z_c=90. Ω, l=4.695 mm, $\alpha=10^{-5}$ and ε_r=1.

Z_c, l, α and ε_r are the parameters of the transmission line used in the amplifier.

Tables 1 and 2 compare the results for the input reflection coefficient and for the transducer power gain, obtained via SUPER-COMPACT and SYSMIC. The transducer power gain is calculated by using the s_{21} parameter. The obtained results are nearly identical.

5. CONCLUSION

A method for the computation of microwaves performances in a fully symbolic form was presented. The method offers the advantage, over other techniques, of being solely based on numerical algorithms such as the determinants and cofactors of scalar matrices and such as the DFT. This allows for its easy implementation in software. The method was successfully applied to determine the s-parameters expressions of microwaves devices, which are important and useful for many applications such as modeling [11] and design.

	Input Reflection Coefficient			
	SUPER-COMPACT		**SYSMIC**	
F (GHz)	**Mag**	**Ang (deg)**	**Mag**	**Ang (deg)**
2	0.209	-39.7	0.209	-39.7
4	0.245	-63.5	0.245	-63.5
6	0.354	-91.1	0.356	-91.2
8	0.522	-132.	0.525	-132.2

Table 1: Input Reflection Coefficient of the Amplifier.

	Transducer Power Gain			
	SUPER-COMPACT		**SYSMIC**	
F (GHz)	**Mag**	**Ang (deg)**	**Mag**	**Ang (deg)**
2	4.465	110.2	4.469	110.18
4	6.111	43.4	6.128	43.38
6	9.217	138.4	9.279	137.84
8	9.09	28.4	9.11	27.46

Table 2: The Amplifier Transducer Power Gain.

6. REFERENCES

[1] P. M. Lin, " Symbolic Network Analysis ", Elsievier, Amsterdam, 1991.
[2] G. Gielen and W. Sansen, " Symbolic Analysis for Automated Design of Analog Integrated circuits ", Kluwer Academic Publishers, Boston, 1991.
[3] G. Gielen, " Symbolic Analysis Methods - An Overview ", Proc. of IEEE ISCAS, pp. 1141-1144, San-Diego, May 1992.
[4] K. Singhal and J. Vlach, " Interpolation Using the Fast Fourier Transform ", IEEE Trans. on Circuit Theory, vol. 60, pp. 1558, 1972.
[5] K. Singhal and J. Vlach, " Generation of Immitance Functions in Symbolic Form for Lumped Distributed Elements ", IEEE Trans. on CAS, vol. CAS-21, no 1, pp. 57-67, 1974.
[6] K. Singhal and J. Vlach, " Symbolic Analysis of Analog and Digital Circuits ", IEEE Trans. on CAS, vol. CAS-24, no 11, pp. 598-609, 1977.
[7] K. Singhal and J. Vlach, " Symbolic Circuit Analysis ", Proc. of IEE, vol. 128, part G, pp. 81-86, April 1981.
[8] J. Vlach and K. Singhal, " Computer Methods for Circuits Analysis and Design ", Van Nostrand, New York, 1983.
[9] K. Benboudjema, " Analyse Symbolique des Circuits Microondes ", Ph.D. thesis, Université de Paris 6, France, 1996.
[10] K. B. Niclas, " Noise in Broad-Band GaAs MESFET Amplifiers with Parallel Feedback ", Trans. on MTT, vol. 30, no. 1, pp. 63-70, jan. 1982.
[11] K. Benboudjema, G. Vasilescu and G. Alquié, "A Novel, Bias-Dependant, Microwave Characterization of MESFET", Proc. of MIOP'95, pp. 418-422, Stuttgart, may 1995.

Behavioral Modeling of PWL Analog Circuits using Symbolic Analysis

Francisco V. Fernández, Belén Pérez-Verdú and Angel Rodríguez-Vázquez

Instituto de Microelectrónica de Sevilla
Centro Nacional de Microelectrónica-C.S.I.C.
Edificio CICA-CNM, Avda. Reina Mercedes s/n, 41012-Sevilla, SPAIN
FAX:: 34 5 4231832 Phone: 34 5 4239923 email: angel@imse.cnm.es

ABSTRACT[1]

Behavioral models are used both for top-down design and for bottom-up verification. During top-down design, models are created that reflect the nominal behavior of the different analog functions, as well as the constraints imposed by the parasitics. In this scenario, the availability of symbolic modeling expressions enable designers to get insight on the circuits, and reduces the computational cost of design space exploration. During bottom-up verification, models are created that capture the topological and constitutive equations of the underlying devices into behavioral descriptions. In this scenario symbolic analysis is useful because it enables to automatically obtain these descriptions in the form of equations. This paper includes an example to illustrate the use of symbolic analysis for top-down design.

1. INTRODUCTION

Circuit analysis is the cornerstone for electronic circuit engineering. On the one hand, it provides the keys to understanding the intricate mechanisms underneath the circuit operation. On the other hand, designers use analysis to obtain models of the circuit behavior – the basis on which this behavior can be predicted. However, manual analysis of even the simplest circuits encountered in practical applications is a complicated, time-consuming, and error-prone task. Symbolic analyzers are intended to relieve designers of systematic manual analysis tasks, thus letting them concentrate on creative issues.

The last generation symbolic analyzers are able to handle up to around 400 different symbols [1]. This permits, for example, to analyze circuits as complex as the rail-to-rail CMOS opamp of Fig.1(a) [2] with the small-signal transistor model of Fig.1(b) [3]-[5]. The analyzers capabilities include the calculation of s-domain expressions for all types of driving-point and transfer characteristics, the simplification of these expressions to retain only the dominant terms, the extraction of their poles and zeroes, etc.

Recently, different authors have focused also on the symbolic analysis of nonlinear circuits [6][7]. The most important advances have been obtained with those weak nonlinearities which can be represented through a few terms (typically

---

1. This work has been partially supported by the Spanish C.I.C.Y.T. under contract TIC97-0580 and the EEC in the project ESPRIT AMADEUS.

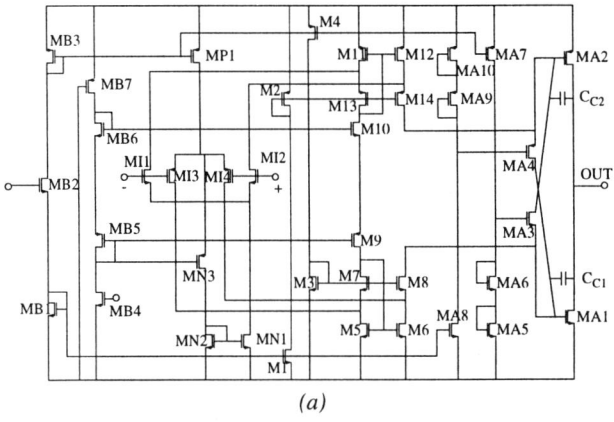

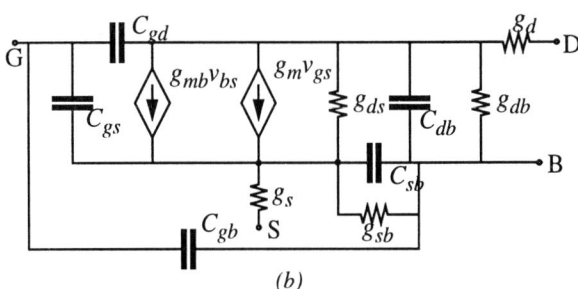

Figure 1. (a) Rail-to-rail opamp; (b) small-signal MOS transistor model.

3) of a power series expansion. In [6] systematic techniques are proposed for the automatic calculation of the Volterra kernels [8] that characterize the behavior of these weakly-nonlinear circuits in the frequency domain. However, the symbolic analysis of hardly-nonlinear circuits is yet in exploratory phase [7]. To date, no systematic technique is available to automate the calculation of a set of symbolic equations governing the large signal nonlinear behavior, either in the static or in the dynamic case.

This lack of nonlinear analysis limits the modeling capabilities of state-of-the-art symbolic analyzers. However, in many practical applications these limitations can be overcome by resorting to piecewise-linear (PWL) representations. For instance, these representations suffice to study the qualitative behavior of many practical dynamical circuits such as oscillators and comparators, as well as to approximate the transient response of opamps and comparators.

2. BEHAVIORAL MODELING

A behavioral model is a set of equations that capture the operation of a circuit from its terminals. Behavioral models can also be given as circuital representations of these equations. Depending on the nature on the excitations and responses, the set of equations and its associated circuital representation can be static or dynamic, linear or nonlinear. For example, Fig.2 shows different behavioral models for an opamp. The first one is static and the others dynamic; the first two are linear and the others nonlinear. Circuital representations are included for each model.

The construction of a behavioral model does not strictly require to know the detailed circuit schematic. Neither the model circuital representation must reflect the circuit topology. Models can be built by, first, applying a suitable set of excitation-response pairs and, then, finding and tuning a math structure that reproduce these pairs within some prescribed error margin. Consider for example the OTA shown in Fig.3(a) and assume that we are interested only in its linear AC behavior. This can be modeled by h-parameters, which can be experimentally calculated in the lab by using a network analyzer in the configurations of Fig.3(c).

Alternatively, behavioral models can be constructed through the electrical analysis of the internal circuit schematic. For instance, assuming that the OTA schematic of

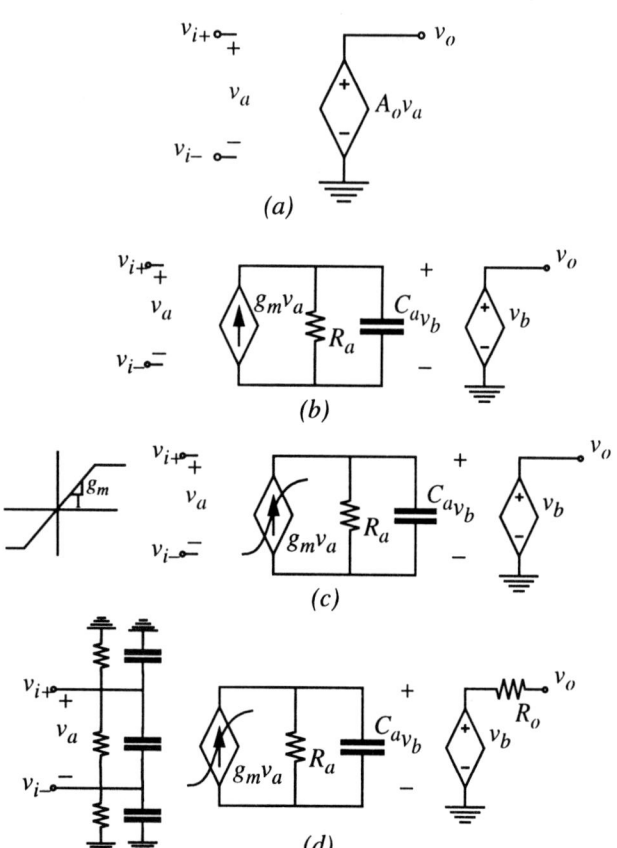

Figure 2. Opamp behavioral models.

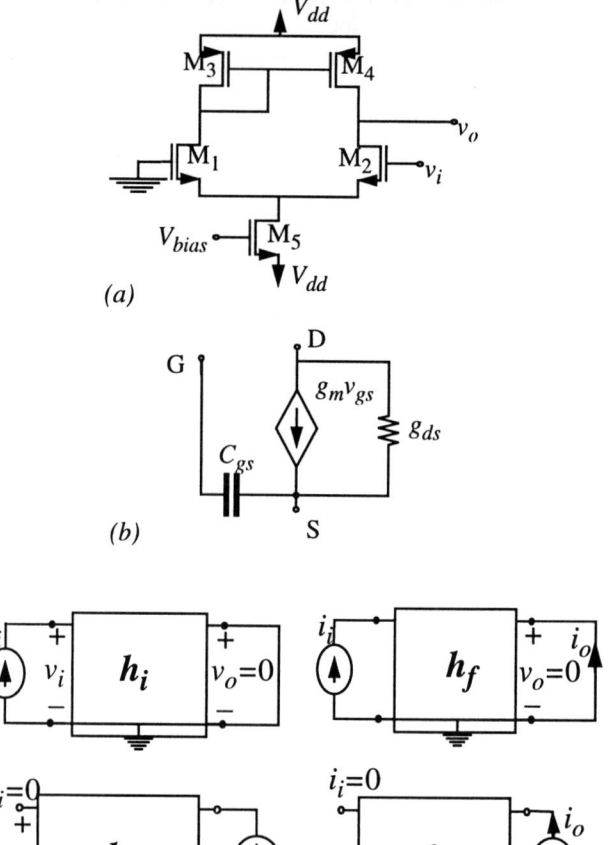

Figure 3. (a) Simple OTA; (b) simplified small-signal MOS transistor model; (c) black box modeling procedure for a two-port.

Fig.3(b) and the transistor model of Fig.3(c), a last generation symbolic analyzer such as SYMBA [9] would return the following h-parameter approximated expressions,

$$h_i(s) = \frac{g_{m3}(g_{m1}+g_{m2})}{sC_{gs2}g_{m1}g_{m3} + s^2 C_{gs2}[g_{m1}(C_{gs3}+C_{gs4}) + g_{m3}C_{gs1}]} +$$

$$+ \frac{s[g_{m3}(C_{gs1}+C_{gs2}) + (g_{m1}+g_{m2})(C_{gs3}+C_{gs4})]}{sC_{gs2}g_{m1}g_{m3} + s^2 C_{gs2}[g_{m1}(C_{gs3}+C_{gs4}) + g_{m3}C_{gs1}]}$$

$$h_f(s) = \frac{g_{ds1}g_{m3} + sg_{ds1}(C_{gs1}+C_{gs3}+C_{gs4})}{g_{m1}g_{m3} + s[g_{m1}(C_{gs3}+C_{gs4}) + g_{m3}C_{gs1}]}$$

$$h_r(s) = \frac{g_{ds2}}{g_{m1}+sC_{gs1}}$$

$$h_o(s) = \frac{g_{ds1}g_{m3} + sg_{ds1}(C_{gs1}+C_{gs3}+C_{gs4})}{g_{m1}g_{m3} + s[g_{m1}(C_{gs3}+C_{gs4}) + g_{m3}C_{gs1}]} \quad (1)$$

In practice, designers use their experience to build modular models, where functional substructures are used to represent the nominal circuit behavior as well as the parasitics. Thus, for example, Fig.2(a) captures the nominal opamp behavior; that of Fig.2(b) includes a first-order approximation to the dynamics associated to any voltage gain mechanism; and Fig.2(c) models the first stage transconductor nonlinearity. These three models focus on the transfer behavior. On the other hand, Fig.2(d) captures also to a first approach the driving-point behaviors at the circuit terminals.

The different functional substructures pertaining to a behavioral model can be tuned either through blackbox measurements or through analysis of the underlying circuit. In the case of PWL models, measurements are used to tune the nonlinear substructures while those linear ones are represented through their symbolic transfer functions. The example presented in the section below illustrates about the practical interest of these PWL models.

3. A PWL MODEL FOR TOP-DOWN TRANSIENT OPAMP OOPTIMIZATION

Consider the SC integrator of Fig.4(a) and assume the opamp represented by the PWL model of Fig.4(b) [10]. The transient observed in the transferring of the input voltage to the integrating capacitor contains a linear part an a nonlinear part. The linear part can be calculated symbolically from,

$$V_o(s) = \frac{C_i C_o}{\gamma_C} \frac{s^2 + \frac{g_1}{C_1}s - \frac{g_{m1}g_{m2}}{C_1 C_o}}{s^2 + 2\alpha s + \alpha^2 + \beta^2} \frac{V_I}{s}$$

$$V_1(s) = \frac{g_{m1}C_i(C_o + C_2)}{C_1 \gamma_C} \frac{s + \frac{g_{m1}}{C_o + C_2}}{s^2 + 2\alpha s + \alpha^2 + \beta^2} \frac{V_I}{s} \quad (2)$$

$$V_a(s) = \frac{C_i(C_o + C_2)}{\gamma_C} \frac{\left(s + \frac{g_2}{C_o + C_2}\right)\left(s + \frac{g_1}{C_1}\right)}{s^2 + 2\alpha s + \alpha^2 + \beta^2} \frac{V_I}{s}$$

where,

$$\gamma_C = C_i C_2 + C_i C_o + C_2 C_o + C_2 C_p + C_o C_p$$
$$C_a = C_i + C_p + C_o$$
$$\alpha = \frac{g_2 C_a}{2\gamma_C} + \frac{g_1}{2C_1} \quad (3)$$

$$\beta = \sqrt{\frac{g_{m1}g_{m2}C_o}{C_1\gamma_C} + \frac{g_1 g_2 C_a}{2C_1\gamma_C} - \frac{g_1^2}{4C_1^2} - \frac{g_2^2 C_a^2}{4\gamma_C^2}}$$

yielding,

$$v_o(t) = A_{o1} + B_{o1}\exp(-\alpha t)\cos\beta t + C_{o1}\exp(-\alpha t)\sin\beta t$$
$$v_1(t) = A_{11} + B_{11}\exp(-\alpha t)\cos\beta t + C_{11}\exp(-\alpha t)\sin\beta t \quad (4)$$
$$v_a(t) = A_{a1} + B_{a1}\exp(-\alpha t)\cos\beta t + C_{a1}\exp(-\alpha t)\sin\beta t$$

where the coefficients are given as functions of the model parameters [10]. These equations remain valid while the following condition is fulfilled,

$$g_{m2}|v_1(t)| \leq I_o \quad (5)$$

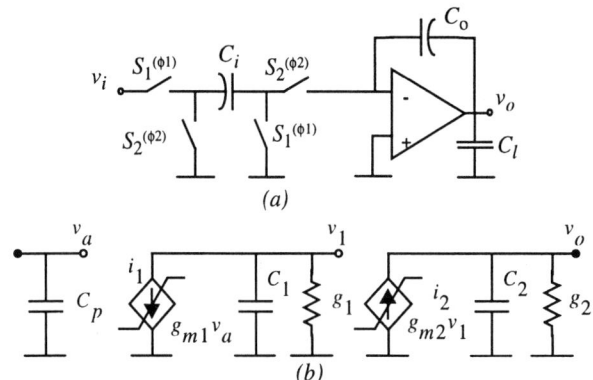

Figure 4. (a) SC integrator and, (b) opamp model.

Afterwards, the second model stage enters into saturation and the transient is described by the following equations,

$$V_o(s) = V_{o1}\frac{s + I_o C_a/(V_{o1}\gamma_C)}{s(s + g_2 C_a/\gamma_C)}$$

$$V_1(s) = \frac{I_o}{g_{m2}}\left(\frac{s^2 + \left(\frac{g_2 C_a}{\gamma_C} - \frac{g_{m1}g_{m2}V_{a1}}{C_1 I_o}\right)s - }{s(s + g_1/C_1)(s + g_2 C_a/\gamma_C)}\right.$$

$$\left.\frac{-\frac{g_{m1}g_{m2}}{C_1\gamma_C I_o}(C_o I_o + g_2 C_a V_{a1} - g_2 C_o V_{o1})}{s(s + g_1/C_1)(s + g_2 C_a/\gamma_C)}\right) \quad (6)$$

$$V_a(s) = V_{a1}\frac{s + (C_o I_o + g_2 C_a V_a - g_2 C_o V_o)/(\gamma_C V_{a1})}{s(s + g_2 C_a/\gamma_C)}$$

yielding,

$$v_o(t) = A_{o2} + B_{o2}\exp[-p_2(t - t_1)]$$
$$v_1(t) = A_{12} + B_{12}\exp[-p_1(t - t_1)] + C_{12}\exp[-p_2(t - t_1)] \quad (7)$$
$$v_a(t) = A_{a2} + B_{a2}\exp[-p_2(t - t_1)]$$

where,

$$p_1 = \frac{g_1}{C_1} \qquad p_2 = \frac{g_2 C_a}{\gamma_C} \quad (8)$$

and the remaining coefficients are given as functions of the model parameters [10]. This second transient persists while the condition (5) holds. Afterwards, the transient is given by,

$$V_o(s) = \frac{V_{o2}}{s}\frac{s^2 + \left(\frac{C_a I_o}{\gamma_C V_{o2}} + \frac{g_1}{C_1}\right)s + \frac{g_{m1}g_{m2}}{\gamma_C C_1 V_{o2}}(C_o V_{o2} - C_a V_{a2})}{s^2 + 2\alpha s + \alpha^2 + \beta^2} \quad (9)$$

yielding,

$$v_o(t) = A_{o3} + B_{o3}\exp(-\alpha t)\cos\beta t + C_{o3}\exp(-\alpha t)\sin\beta t \quad (10)$$

where the coefficients are, as in the previous expressions, given as functions of the model parameters.

The model has been validated by comparing its output waveform to that of obtained through detailed electrical simulation. The amplifier consisted of a fully-differential folded-cascode OTA whose core schematic and summarized performance are given in Fig.5. Said amplifier was designed to have small phase margin (around 45deg) to reduce power dissipation. Fig.6(a) shows the integrator output voltage during the integration phase obtained through electrical simulation (HSPICE) and that obtained using the model. Model parameters are also enclosed in Fig.5. A good concordance between both approximations is observed. A more general result is given in Fig.6(b) where the difference between the final value of the integrator output voltage and its ideal value is shown as a function of the input level.

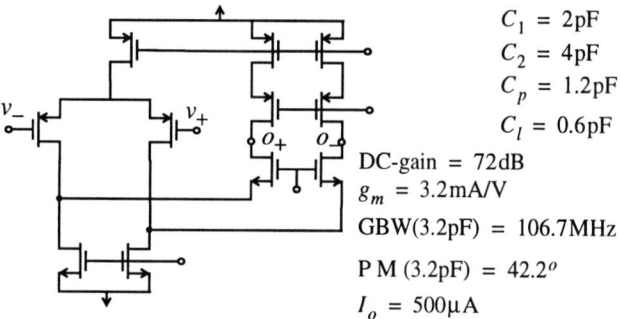

Figure 5. Fully differential OTA.

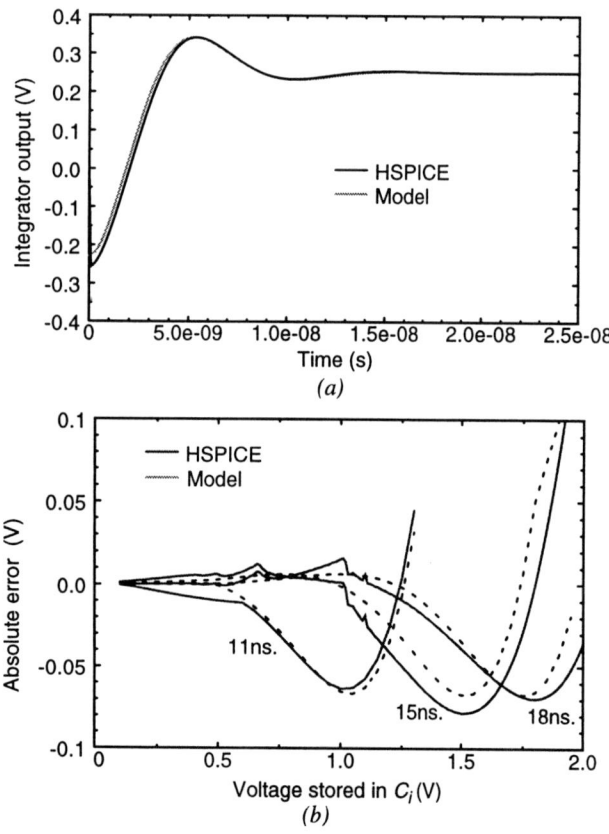

Figure 6. Simulated and calculated responses.

4. REFERENCES

[1] F.V. Fernández, A. Rodríguez-Vázquez, J.L. Huertas and G. Gielen, eds., *Symbolic Analysis Techniques. Applications to Analog Design Automation.* IEEE Press, 1998.

[2] W.S. Wu, W. Helms, J.A. Kuhn and B.E. Byrkett, "Digital-compatible high-performance operational amplifier with rail-to-rail input and output ranges," *IEEE J. Solid-State Circuits*, Vol. 29, pp. 63-66, Jan. 1994.

[3] P. Wambacq, F.V. Fernández, G. Gielen, W. Sansen and A. Rodríguez-Vázquez, "Efficient symbolic computation of approximated small-signal characteristics of analog integrated circuits," *IEEE J. Solid-State Circuits*, Vol. 30, pp. 327-330, March 1995.

[4] Q. Yu and C. Sechen, "A unified approach to the approximate symbolic analysis of large analog integrated circuits," *IEEE Trans. Circuits and Systems I*, Vol. 43, No. 8, pp. 656-669, Aug. 1996.

[5] F.V. Fernández, O. Guerra, J.D. García and A. Rodríguez-Vázquez, "Symbolic analysis of analog integrated circuits: the numerical reference generation problem," *IEEE Trans. Circuits and Systems*, 1998, to appear.

[6] P. Wambacq, *Symbolic Analysis of Large and Weakly Nonlinear Analog Integrated Circuits.* Ph.D. Dissertation, Katholieke Universiteit Leuven, Belgium, 1996.

[7] C. Borchers, L. Hedrich and E. Barke, "Equation-based model generation for nonlinear analog circuits," *Proc. 33rd Design Automation Conf.*, pp. 236-239, 1996.

[8] S. A. Maas, *Nonlinear Microwave Circuits,* Artech House 1988.

[9] IMSE-CNM, *Methods for SBG and SDG.* ESPRIT Project 21812 AMADEUS, Internal Report, March 1997.

[10] F. Medeiro, *ΣΔ-modulator Interfaces for Mixed-Signal CMOS IC's: Modeling, Simulation and Automated Design.* Ph.D. dissertation, University of Sevilla, Spain, 1997.

EFFICIENT STATISTICAL ANALOG IC DESIGN USING SYMBOLIC METHODS

G. Debyser, F. Leyn, G. Gielen* and W. Sansen

M. Styblinski[†]

Katholieke Universiteit Leuven

Texas A&M University

ABSTRACT

A new statistical design methodology is presented that uses symbolic methods to increase the efficiency of statistical IC design. The methodology is implemented in an environment that combines the symbolic design equation manipulation engine DONALD with the generic statistical design system GOSSIP. DONALD is used to generate the initial design and to create the symbolic computational plan that is used by GOSSIP for the actual statistical yield, Cp and Cpk optimization. The strengths and weaknesses of the proposed approach are discussed, and its accuracy and speed are compared with the traditional method of using a SPICE-type circuit simulator in the inner loop of the statistical optimization. Overall, a significant speed-up of the statistical IC design efficiency is observed.

1. INTRODUCTION

Designing an analog circuit is one thing, producing it is another. Real-life technology parameter variations make the circuits fail for some or all of the specifications if no precautions are taken. The ratio of the number of successful circuits over the number of produced circuits is the total yield. The total yield consists of yield due to production faults and yield due to soft faults. In this paper we concentrate on the yield based on the soft faults, generated by the technology parameter variations: the parametric yield.

The hard way to make the design more robust is to run multiple batch jobs of Monte Carlo simulations in the inner loop of a design centering procedure. A Monte Carlo simulation consists of tens or hundreds of SPICE simulations, so the whole design centering procedure can easily take a night, even some days.

In the following sections we demonstrate that we gain a lot of computation time using symbolic techniques without sacrificing too much accuracy. The paper is organized as follows. Section 2 describes how we used symbolic techniques to construct a yield estimation model. Section 3 explains the yield optimization strategy. Section 4 shows the experimental results of a comparison between GOSSIP-SPICE and GOSSIP-DONALD. Section 5 draws some conclusions.

2. YIELD ESTIMATION

2.1. Introduction

Our approach replaces the Monte Carlo simulations with a direct technique for yield estimation. We start with a statistical transistor model, which gives us a reduced set of quasi-independent technology parameters θ. Then we calculate the nominal design point and then we calculate the variances of all performance parameters y with respect to the reduced set of technology parameters θ. Using these variances we construct an efficient representation of "yield" based on Cp/Cpk indices, which we then use in the inner loop of the statistical optimization routine to replace the Monte Carlo simulations. The flow diagram in Fig. 1 is explained in the following subsections (x stands for designable parameters and e stands for simulator variables).

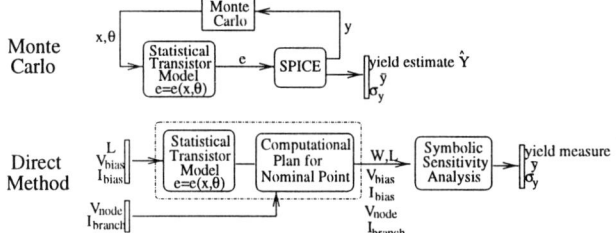

Figure 1. Yield estimation methods. Monte Carlo and SPICE versus direct method.

2.2. Statistical transistor model

The default transistor models have to be replaced by statistical models in order to take the correlations of the technology parameters into account. This is necessary to estimate the yield in a statistically correct way. The statistical model describes all technological variables as a function of only 7 quasi-uncorrelated input variables, which are TOX, $NSUB_n$, $NSUB_p$, CJ_n, CJ_p, LD_n and RSH_n. A method for deriving such a statistical transistor model has been discussed in [1]. The Monte Carlo routine perturbes only these 7 parameters and extracts performances from the SPICE output to construct the yield estimation figures based on a pass/fail mechanism. Also our symbolic yield estimator starts from these 7 parameters.

---

*research associate of the Belgian Fund of Scientific Research.
[†]this paper is in honor of the late Dr. Styblinski, who co-initiated this work during his sabattical stay in Leuven.

2.3. Using Monte Carlo & simulator

Parametric yield is estimated by perturbing the independent technology variables θ and measuring the perturbations of the performances variables y. The Monte Carlo (MC) algorithm takes samples $\theta_s = \theta + \Delta\theta_s$ from the distribution $f_\theta(\theta)$ and repeatedly performs the SPICE simulations to extract the performances as a function of the perturbed technology variables. So, per step the MC algorithm finds $y + \Delta y = f(\theta + \Delta\theta_s)$. These samples are then used to construct the complex distribution function of the perturbations of the performances $f_y(\theta)$. Therefore the variances of the performances are estimated.

The yield estimator $\hat{Y}$ gives the average number of "good" circuits:

$$\hat{Y} = \frac{1}{N}\sum_{i=1}^{N=7}\Phi(\theta^i) = \frac{N_s}{N} \quad (1)$$

where N_s is the number of "successful" circuits, for which $\theta^i \in A_\theta(x)$, i.e. $\Phi(e(x,\theta^i)) = 1$ and $A_\theta(x) = \{\theta \in R^t \mid \text{Spec}_j^L \leq y_j(e(x,\theta)) \leq \text{Spec}_j^U; j = 1\ldots m\}$ is the acceptability region in the e-space. Very important to note here is that $A_\theta(x)$ is very operating point dependent. We have taken this into account in our approach by representing the circuit behavior in symbolic form.

During a Monte Carlo Yield estimation a random number generator is used to generate a sequence of θ^i vectors with a $f_\theta(\theta)$ distribution.

This way of calculating the yield is very costly: a large number (e.g. 300) of SPICE simulations have to be executed.

2.4. Using symbolic approach

Let us now investigate how we accelerate the above brute force method. First let's analyze what exactly are the various parts of one SPICE simulation.

A SPICE simulation can be divided into three parts:

- operating point calculation: solving a set of nonlinear equations
- ac analysis (small signal): solving set of linear equations (MNA)
- transient analysis (large signal): consecutive operating point simulations over a certain period.

Our equation based method shortens the time needed to calculate the performances as follows:

- transient analysis: use of approximate formulas
- ac analysis: symbolic expressions for the transfer function and for the related performance variables (A_{v0}, GBW, f_d, ...), as shown in [2].

- operating point calculation: here we make the difference. The DONALD tool [3], a workbench for design equation manipulations, is the off-line tool used to construct all the computational paths. DONALD allows us to specify different sets of input variables. It then orders the design equations and builds an executable plan. We avoid solving the set of nonlinear (device) equations by reformulating the problem. Instead of solving for the node voltages and branch currents starting from the device sizes (W, L) we calculate the W for each transistor starting from an applied operating point. Instead of solving a set of N nonlinear equations we only solve N one-dimensional nonlinear equations, which are much cheaper to handle. This systematic approach has been formally elaborated in [4] and is illustrated in Fig. 2.

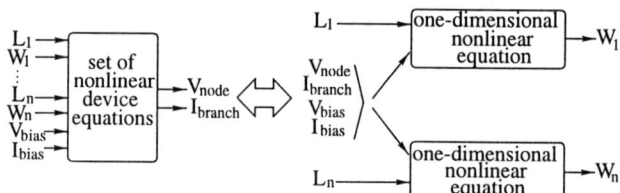

Figure 2. Comparison between SPICE-like operating point calculation and the DONALD way

This approach has already been used for optimization based nominal sizing [5] and a comparison has been made between the use of an equation based and a simulation based system for sizing [6]. In this paper we extend this to yield optimization.

2.5. Direct method of calculating variances

2.5.1. Yield measure

As the pass/fail measure used in Monte Carlo simulation is very rough and gives little or no information about which performance does not meet its specification, another quality measure is preferred. The Taguchi quality measure gives a much better idea of the quality of a circuit, because it takes absolute variability and bias to the target spec into account:

$$\begin{aligned} M_{TAG} &= E\{[y(x) - \text{Spec}^T]^2\} \\ &= var\{y(x)\} + (\overline{y} - \text{Spec}^T)^2 \end{aligned} \quad (2)$$

To ease the calculation of M_{TAG}, two capability indices were introduced in [7]: the capability *potential* index C_p:

$$C_p = \frac{\text{Spec}^U - \text{Spec}^L}{6\sigma_y} \quad (3)$$

and the capability *performance* index C_{pk}:

$$C_{pk} = \min\{\frac{\text{Spec}^U - \overline{y}}{3\sigma_y}, \frac{\overline{y} - \text{Spec}^L}{3\sigma_y}\} \quad (4)$$

The first index represents the variability, the index second the bias. As can be noticed, these indices strongly depend on the variances of the performances.

2.5.2. Propagation of Variances

The variances of the 7 technological parameters are propagated through the computational path by means of sensitivities. An extension has been made to DONALD (see Fig. 3), so that it can build also a computational plan to calculate all sensitivities from the output variables w.r.t. the input variables.

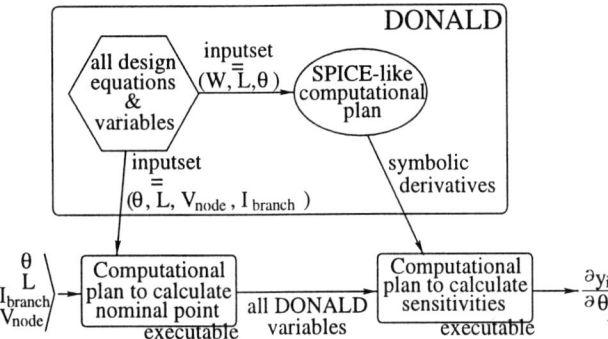

Figure 3. *Creation and use of direct yield estimation model*

This computational path is derived from the same set of equations used to determine the nominal point, but W and L have been chosen as input variables. This time however, we do *not* solve the set of nonlinear equations, but Symbolic derivatives from all equations have automatically been derived. Out of this a computational path results, which is a chain of one-dimensional and more-dimensional subsystems of equations. The local sensitivities are calculated using Jacobians in symbolic form. The global sensitivity matrix S_θ^y with elements $S_{\theta_j}^{y_i} = \partial y_i / \partial \theta_j$, is calculated by applying the chain rule according to the computational plan:

$$S_\theta^y \approx S_{z_n}^y S_{z_{n-1}}^{z_n} \ldots S_\theta^{z_1} \quad (5)$$

where z_i are the internal variables along the path. In case $\theta_1, \theta_2, \ldots \theta_n$ are not correlated, the variability of the performances can be written as follows:

$$var\{y_i\} = \sigma_{y_i}^2 \approx \sum_{j=1}^{n}(S_{\theta_j}^{y_i})^2 \sigma_{\theta_j}^2 \quad (6)$$

(assuming that $\sigma_\theta / \bar{\theta}$ is sufficiently small). The sensitivities have to be known for *every* x. Our Jacobians are in symbolic form, so the Jacobian updating for every x is relatively cheap.

2.5.3. Sensitivity Analysis

First, at each square subsystem with equations $f_1, \ldots, f_n$, a distinction is made between those variables of the subsystem that have been solved by the subsystem, and those variables that are solved by previous subsystems. The former are called the *output variables* $z^{out} = (z_1^{out}, \ldots, z_q^{out})^T$ of the subsystem, the others are called the *input variables* $z^{in} = (z_1^{in}, \ldots, z_p^{in})^T$. For each square subsystem the change δz^{out} is then calculated of the output variables with respect to a unit change δz^{in} of the input variables. This is done by solving the following system of linear equations for δz^{out}:

$$\nabla \mathbf{F}^{out} \delta z^{out} = -\nabla \mathbf{F}^{in} \delta z^{in} \quad (7)$$

where $\delta z_j^{in} = 1$ for $j = 0 \ldots p$ and

$$\nabla \mathbf{F}^{out} = \begin{bmatrix} \frac{\partial f_1}{\partial z_1^{out}} & \cdots & \frac{\partial f_1}{\partial z_q^{out}} \\ \vdots & \ddots & \vdots \\ \frac{\partial f_n}{\partial z_1^{out}} & \cdots & \frac{\partial f_n}{\partial z_q^{out}} \end{bmatrix} \quad (8)$$

and

$$\nabla \mathbf{F}^{in} = \begin{bmatrix} \frac{\partial f_1}{\partial z_1^{in}} & \cdots & \frac{\partial f_1}{\partial z_p^{in}} \\ \vdots & \ddots & \vdots \\ \frac{\partial f_n}{\partial z_1^{in}} & \cdots & \frac{\partial f_n}{\partial z_p^{in}} \end{bmatrix} \quad (9)$$

where all partial derivatives are in symbolic form. The δz^{out} value of an output variable is called a *local sensitivity* value. By multiplying local sensitivity values along the computational path between two variables r and s, a *global sensitivity* value $\partial r / \partial s$ can be calculated.

3. YIELD OPTIMIZATION

3.1. Optimization using propagation of variances

The GOSSIP environment [8] has been chosen as optimization framework. GOSSIP is a generic framework for statistical and deterministic circuit optimization. It is currently composed of several tools/packages, supporting design evaluation, optimization, yield estimation, deterministic and statistical gradient estimation, etc. The flow shown in Fig. 3 explains that in the inner loop of the optimization two computational plans are evaluated. The first calculates the nominal point starting from the node voltages and branch currents, the second updates the symbolic Jacobians, calculates the global sensitivities of the performances towards the quasi-uncorrelated technological parameters and gives a value for each σ_{y_i}, according to (7). The evaluation of the circuit and the yield estimation are done simultaneously.

The quality measure used in the cost function is the same as the one internally used in the *povopt* routine in GOSSIP

$$\Phi_i^{\pm}(x) = C_{p,i} \pm \lambda \frac{\frac{\text{Spec}_i^U + \text{Spec}_i^L}{2} - \bar{y}_i}{3\sigma_{y_i}} \quad (10)$$

$$\Phi_i(x) = \min\{\Phi_i^+(x), \Phi_i^-(x)\} \quad (11)$$

The weight factor λ acts as a penalty term on the bias, with respect to target $Spec_i^T$. If $\lambda = 0$ then $\Phi_i = C_{p,i}$, if $\lambda = 1$ then $\Phi_i = C_{pk,i}$. A value of $\lambda = 0.8$ has been chosen experimentally. The optimization problem to be solved is then

$$\Phi(x) = \max_{x \in R_x} \min_i \{\Phi_i(x); i = 1, .., m\} \quad (12)$$

where R_x is (usually) a hyperbox of constraints.

4. EXPERIMENTAL RESULTS

The circuit is a CMOS high-speed OTA (see Fig. 4).

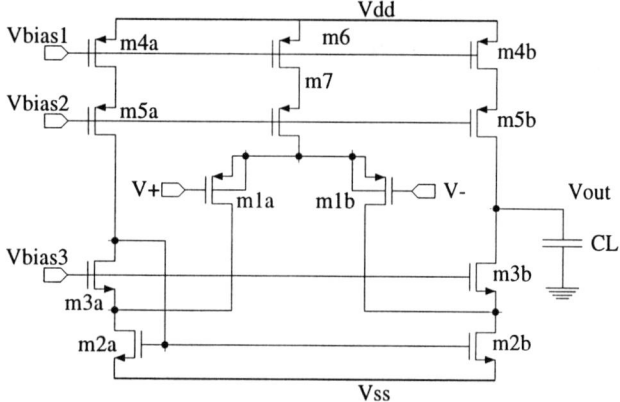

Figure 4. High-speed OTA

Specs	Initial					After povopt	
	SPICE	MC		Yield Model		MC	
	y	$\bar{y}$	σ_y	$\bar{y}$	σ_y	$\bar{y}$	σ_y
GBW>100MHz	307	304	36.7	301	37.1	172	21.4
A_{v0}>60dB	86.6	75.8	13.9	86.5	1.68	78.1	17.4
PM>60°	62	62.6	0.85	60.2	0.21	63.6	0.4
OR>2.75V	3.0	2.99	0.063	3.0	0.05	3.27	0.058
SR>120V/µs	150	151	31	150	32.7	149	28.5
V_{off}<5mV	4.9	5.15	0.4	5.0	0.14	3.5	0.7
I_{tot}<3.0mA	3.0	3.1	0.75	3.0	0.66	2.6	0.56
		yield = 5 %				yield = 59 %	

Table 1. Comparison between SPICE based and DONALD based yield optimization

The nominal starting point is the result of an interactive design with DONALD. In this nominal point a Monte Carlo simulation is performed (300 SPICE simulations). The obtained yield is 5% due to the too critical nominal design. More important are the variances, which are also estimated and can be found in Table 1. In the same point the symbolic approach is used to calculate the variances of the performances, which can also be found in Table 1. This yield model is not as accurate as Monte Carlo (especially for A_{v0}, due to its non-Gaussian behavior), but it is extremely fast (10s) compared to MC (300 samples: 2h20' on a Sparc I). Then a Cp/Cpk based optimization is performed (in 1h16') with the two computational plans (see Fig. 3) in the inner loop of *povopt*. In the resulting optimized point a MC is run, which gives a yield of 59% (see last two columns in Table 1). A better result is expected when the yield model for A_{v0} will be more accurate.

5. CONCLUSION

A new statistical design method using symbolic techniques has been presented. A direct yield estimation model has automatically been derived from the whole set of symbolic design equations by constructing a computational plan to calculate symbolically the sensitivities of the performances w.r.t. the technology parameters. By propagating the variances of 7 quasi-independent technology parameters the quality measures Cp and Cpk for the performances were obtained. This yield model was directly used in the inner loop of a yield optimization routine. The experimental results show that the model is accurate enough to steer the optimization in the direction of a "better" design in a much faster way than using a simulator. Research is still to be continued to improve the yield model and to take advantage of the very fast circuit and yield evaluations.

Acknowledgments

The authors acknowledge the financial support for parts of this work from Philips Research (NL), Robert Bosch GmbH (D) and MEDEA SADE.

6. REFERENCES

[1] J. Chen, M. A. Styblinski, "A systematic approach of statistical modeling and its applications to CMOS circuits". Proc. IEEE ISCAS, pp. 1805-1808, May 1993.

[2] F. Leyn, W. Daems, G. Gielen, W. Sansen, "A behavioral signal path modeling methodology for qualitative insight in and efficient sizing of CMOS opamps". Proc. IEEE ICCAD, pp. 374-381, Nov. 1997.

[3] K. Swings, W. Sansen, G. Gielen, "DONALD: An intelligent analog IC design system based on manipulation of design equations". Proc. IEEE CICC, pp. 8.6.1-4, May 1990.

[4] F. Leyn, W. Daems, G. Gielen, W. Sansen, "Analog circuit sizing with constraint programming modeling and minimax optimization". Proc. IEEE ISCAS, pp. 1500-1503, June 1997.

[5] G. Gielen, G. Debyser, et al., "Use of symbolic analysis in analog circuit synthesis". Proc. IEEE ISCAS, pp. 2205-2208, May 1995.

[6] G. Gielen, G. Debyser, et al., "Comparison of analog synthesis using symbolic equations and simulation". Proc. ECCTD, pp. 79-82, August 1995.

[7] M. A. Styblinski, S. A. Aftab, "IC variability minimization using a new Cp and Cpk based variability/performance measure". Proc. IEEE ISCAS, May-June 1994.

[8] L.J. Opalski, M.A. Styblinski, "GOSSIP. A Generic System for Statistical Improvement of Performance". User's Guide V1.2, 166pp, April 1995.

APPROXIMATE SYMBOLIC POLE/ZERO EXTRACTION USING EQUATION-BASED SIMPLIFICATION DRIVEN BY EIGENVALUE SHIFT PREDICTION

Eckhard Hennig, Ralf Sommer, Michael Wiese

ITWM – Institute of Industrial Mathematics, Kaiserslautern, Germany

ABSTRACT

This paper presents a new approach to symbolic pole/zero analysis of linear electronic circuits using equation-based simplification techniques driven by novel term ranking and error control strategies. Term rankings are determined by using a linear prediction formula to estimate the shifts of the eigenvalues of interest caused by removing terms from a matrix equation. The shift estimations are based on one initial solution of a numerical generalized eigenvalue problem. Error control is achieved by tracking the true eigenvalue shifts numerically. The algorithm allows for the computation of approximated symbolic expressions for selected poles and zeros, including non-dominant ones.

1. INTRODUCTION

To overcome the classical complexity problem of symbolic circuit analysis much research effort has been spent on the development of methods for generating approximate symbolic descriptions of circuit characteristics (e.g. [1, 2, 3]). Most of these simplification techniques are directed towards the extraction of rational approximations of transfer functions in selected frequency intervals or the entire frequency domain. However, circuit designers are not only interested in computing transfer functions but also in obtaining symbolic expressions for poles and zeros.

Provided that poles and zeros are spaced widely enough in the complex plane they can be extracted from symbolic transfer functions by applying Simplification-After or During-Generation algorithms in combination with root splitting or clustering techniques. By exploiting the order reduction effect of Simplification Before Generation methods, poles and zeros can be computed efficiently through equation-based approximation techniques which use the influence of matrix terms on the magnitude of a transfer function as ranking criterion [3]. A suitable choice of reference points on the frequency axis, or in the complex plane [4], allows for extracting a first or second-order approximation of a corner in the magnitude response of a transfer function, which can then be solved for the poles and zeros analytically [5].

However, the above symbolic approximation approaches have not been originally designed for such computations and, therefore, do not use the error on poles and zeros as primary term approximation or generation criterion. In this paper a novel equation-based approach is presented which is directly driven by the error on poles and zeros. A given system of Sparse Tableau or MNA equations is transformed into a generalized eigenvalue problem whose numerical solution is computed as a reference. Then, the symbolic system is simplified with respect to a selected eigenvalue, or set of eigenvalues. The simplification follows a term ranking computed via a fast linear prediction formula which estimates eigenvalue shifts due to removing matrix entries. The approximation algorithm stops as soon as a user-defined error bound for the eigenvalues in the complex plane is reached. Finally, poles or zeros are extracted from the symbolic determinant of the non-singular part of the approximated system.

2. THE GENERALIZED EIGENVALUE PROBLEM

From a theoretical point of view a very suitable starting point for approximate P/Z analysis would be the state-space representation of a system in the form

$$(s\mathbf{I} - \mathbf{A})\mathbf{x} = \mathbf{0}. \qquad (1)$$

In fact, there exist approaches to simplify symbolic state equations of electrical circuits [6]. However, since state equations cannot be set up directly from a network graph they have to be computed from other circuit analysis formulations, e.g. MNA, by elimination of variables. For most practical circuits, it is therefore impossible to obtain a state-space representation *symbolically* due to the complexity of the symbolic operations involved.

To compute poles[1] symbolically through equation-based approximation it is better to rewrite a system of linear circuit equations

$$\mathbf{Tx} = \mathbf{b} \qquad (2)$$

into a generalized eigenvalue problem [7] of the form

$$\mathbf{Tx} = (\mathbf{P} + s\mathbf{Q})\mathbf{x} = \mathbf{0}. \qquad (3)$$

and approximate the two matrices $\mathbf{P}$ and $\mathbf{Q}$ prior to carrying out any symbolic eliminations by discarding all terms which have negligible influence on the eigenvalue(s) of interest.

3. EIGENVALUE SENSITIVITY AND TERM RANKING

Approximating the symbolic matrix pencil $(\mathbf{P}, \mathbf{Q})$ w.r.t. a given set of eigenvalues $S = \{\lambda_1, ..., \lambda_n\}$ requires a ranking strategy by which the terms of the pencil can be sorted according to their influences on the elements of S. A ranking can be determined by estimating the eigenvalue shifts $\Delta\lambda_i$ due to the removal of a term by a linear, or quadratic, Taylor series approximation of (3) [8]: Consider the generalized eigenvalue problem

$$\begin{aligned}\mathbf{Pu} - \lambda\mathbf{Qu} &= 0 \\ \mathbf{v}^T\mathbf{P} - \lambda\mathbf{v}^T\mathbf{Q} &= 0\end{aligned} \qquad (4)$$

with $\mathbf{P}, \mathbf{Q} \in \mathbb{R}^{n \times n}$, $\mathbf{u}, \mathbf{v} \in \mathbb{C}^n$ and $\lambda \in \mathbb{C}^n$, where $\mathbf{u}$ and $\mathbf{v}$ denote the right and left eigenvectors, respectively. Each matrix entry of $\mathbf{P}$ and $\mathbf{Q}$ consists of a sum of terms, which shall be denoted as parameters. Thus, the MNA fill-in pattern of an admittance contributes four parameters.

Let there be a total of m parameters $p_1, ..., p_m$ in $\mathbf{P}$ and $\mathbf{Q}$. Then, $\mathbf{P}$ and $\mathbf{Q}$ depend linearly on the parameter vector $\mathbf{p} = (p_1, ..., p_m)$:

$$\begin{aligned}\mathbf{P} &= \mathbf{P}(\mathbf{p}) = p_1\mathbf{A}_1 + ... + p_m\mathbf{A}_m \\ \mathbf{Q} &= \mathbf{Q}(\mathbf{p}) = p_1\mathbf{B}_1 + ... + p_m\mathbf{B}_m\end{aligned} \qquad (5)$$

Hence, the eigenvalues and eigenvectors are functions of $\mathbf{p}$:

$$\lambda = \lambda(\mathbf{p}), \mathbf{u} = \mathbf{u}(\mathbf{p}), \mathbf{v} = \mathbf{v}(\mathbf{p}) \qquad (6)$$

According to the above definition of a parameter each $\mathbf{A}_i$ is either the zero matrix or can be expressed as

$$\mathbf{A}_i = \mathbf{e}_j \cdot \mathbf{e}_k^T \qquad (7)$$

where $\mathbf{e}_k = (0, .., \overset{k}{1} ..., 0)^T$. The same holds for all $\mathbf{B}_i$. Obviously, the first derivatives of $\mathbf{P}$ and $\mathbf{Q}$ exist w.r.t. all p_k. Differentiating (4) w.r.t. p_k thus yields

$$\begin{aligned}\frac{\partial(\mathbf{Pu} - \lambda\mathbf{Qu})}{\partial p_k} &= \left(\frac{\partial\mathbf{P}}{\partial p_k} - \lambda\frac{\partial\mathbf{Q}}{\partial p_k}\right)\mathbf{u} \\ -\frac{\partial\lambda}{\partial p_k}\mathbf{Qu} &+ (\mathbf{P} - \lambda\mathbf{Q})\frac{\partial\mathbf{u}}{\partial p_k} = 0\end{aligned} \qquad (8)$$

Premultiplying this equation by the corresponding left eigenvector $\mathbf{v}^T$ yields an expression for the partial derivative of an eigenvalue λ w.r.t. p_k. Note that $\mathbf{v}^T\mathbf{Qu} \in \mathbb{C}$ can never be zero.

$$0 = -\frac{\partial\lambda}{\partial p_k}(\mathbf{v}^T\mathbf{Qu}) + \mathbf{v}^T\left(\frac{\partial\mathbf{P}}{\partial p_k} - \lambda\frac{\partial\mathbf{Q}}{\partial p_k}\right)\mathbf{u}$$
$$+ \underbrace{\mathbf{v}^T(\mathbf{P} - \lambda\mathbf{Q})}_{= 0}\frac{\partial\mathbf{U}}{\partial p_k}$$

$$\Rightarrow \frac{\partial\lambda}{\partial p_k} = (\mathbf{v}^T\mathbf{Qu})^{-1}\mathbf{v}^T\left(\frac{\partial\mathbf{P}}{\partial p_k} - \lambda\frac{\partial\mathbf{Q}}{\partial p_k}\right)\mathbf{u} \qquad (9)$$

Now let $\mathbf{p}^0 = (p_1^0, ..., p_m^0)$ denote the design-point values of the parameters, and let $\mathbf{P}^0 = \mathbf{P}(\mathbf{p}^0)$, $\lambda^0 = \lambda(\mathbf{p}^0)$ etc. Then, consider the Taylor series expansion for $\lambda(\mathbf{p})$ w.r.t. $\mathbf{p}$ about $\mathbf{p}^0$, truncated after the linear term:

$$\lambda(\mathbf{p}) \approx \lambda(\mathbf{p}^0) + \sum_{k=1}^{m}\left(\frac{\partial\lambda}{\partial p_k}\bigg|_{\mathbf{p}=\mathbf{p}^0} \cdot \underbrace{(p_k - p_k^0)}_{= \Delta p_k}\right) \qquad (10)$$

Substituting (9) into (10) yields

$$\Delta\lambda \approx \mathbf{v}^{0T} \cdot \frac{\Delta\mathbf{P} - \lambda^0\Delta\mathbf{Q}}{\mathbf{v}^{0T}\mathbf{Q}^0\mathbf{u}^0} \cdot \mathbf{u}^0, \qquad (11)$$

where

$$\Delta\lambda = \lambda(\mathbf{p}) - \lambda(\mathbf{p}^0), \Delta\mathbf{P} = \mathbf{P}(\mathbf{p}) - \mathbf{P}^0, \Delta\mathbf{Q} = \mathbf{Q}(\mathbf{p}) - \mathbf{Q}^0.$$

The influence of a parameter p_i in $(\mathbf{P}, \mathbf{Q})$ on an eigenvalue λ_j can now be estimated by computing the approximate eigenvalue deviation $\Delta\lambda_j$ for $p_i \to 0$ according to (11).

The special structure of the matrix pencil allows for a further simplification of the deviation formula. During matrix approximation only one term is discarded at a time, therefore $\Delta\mathbf{T}(\mathbf{p}) = \Delta\mathbf{P}(\mathbf{p}) - s\Delta\mathbf{Q}(\mathbf{p})$ has only one non-

---

[1] Since zeros can be determined in a similar fashion all further considerations are, without loss of generality, restricted to the computation of poles.

zero entry t_{jk}. Thus, (11) is reduced to

$$\Delta\lambda \approx \frac{v_j \Delta t_{jk} u_k}{\mathbf{v}^{0T} \mathbf{Q}^0 \mathbf{u}^0} \quad (12)$$

Moreover, the sets of parameters in **P** and **Q** are disjoint, so for each parameter p_k either $\Delta\mathbf{P}(p_k)$ or $\Delta\mathbf{Q}(p_k)$ is zero. This simplifies the numerical calculation of the eigenvalue shift estimates.

With the above formula a term ranking can be determined easily and efficiently by computing the estimated eigenvalue shift for each matrix entry and sorting the resulting list by least influence on the eigenvalue(s) of interest. This requires only one initial numerical computation of the eigenvalues and eigenvectors of the matrix equation in the design point.

4. ERROR CONTROL

As opposed to the Sherman-Morrison formula which predicts magnitude errors accurately [3] equation (12) does not yield mathematically exact results for eigenvalue deviations. Since the formula has been derived from a first-order Taylor series approximation of the eigenvalue problem accurate results can only be guaranteed for parameter deviations $\Delta p_k \ll p_k$, therefore some eigenvalue deviations may be underestimated. Moreover, the cumulative influence of several parameters is not taken into account.

Nevertheless, the algorithm generally succeeds in finding the dominant elements for a given eigenvalue even for larger deviations of the non-dominant parameters. To ensure that occasional incorrect rankings are discovered and that the cumulative errors do not exceed the specified bounds the true eigenvalue shifts must be tracked after each approximation step. This can be done efficiently by an iterative eigenvalue solver, using e.g. the Jacobi-Davidson-QZ algorithm [9]. Bad convergence behavior after one or two iterations indicate that an eigenvalue shift has been underestimated so that the last approximation step must be undone.

5. THE ALGORITHM

- **Input:**
  1) symbolic matrix equation $\mathbf{Tx} = \mathbf{b}$, e.g. MNA
  2) set of design point values D

- **Step 1: initialization**
  decompose $\mathbf{T}$ into $\mathbf{P} + s\mathbf{Q}$, set up parameter list $\mathbf{p}$,
  compute $\mathbf{p}^0 = \mathbf{p}(D)$, $\mathbf{P}^0 = \mathbf{P}(\mathbf{p}^0)$, $\mathbf{Q}^0 = \mathbf{Q}(\mathbf{p}^0)$

- **Step 2: compute numerical reference solution**
  compute eigenvalues λ_i^0 and eigenvectors $\mathbf{u}_i^0, \mathbf{v}_i^0$ of $(\mathbf{P}^0, \mathbf{Q}^0)$ by QZ algorithm, select eigenvalue(s) λ_j of interest and specify error bound ε.

- **Step 3: compute a parameter ranking**
  for all $p_k \in \mathbf{p}$ compute $\Delta\lambda(p_k = 0)$ using (12),
  sort influence list by least $\|\Delta\lambda\|$.

- **Step 4: simplify the matrix pencil (P, Q)**
  remove parameter with smallest influence from $(\mathbf{P}, \mathbf{Q})$,
  calculate cumulative eigenvalue deviation $\Delta\lambda$,
  if $\Delta\lambda > \varepsilon$ undo last cancellation and stop
  else repeat Step 4

- **Step 5: compute pole from simplified pencil**
  let $\mathbf{T}' = \mathbf{P} + s\mathbf{Q}$,
  delete linearly dependent rows and columns from $\mathbf{T}'$
  compute zeros of $\det(\mathbf{T}')$ symbolically,
  select zero numerically closest to λ_j

- **Output:** pole = selected zero from Step 5.

6. RESULTS

The approximation algorithm has been tested on several circuits, including the bipolar buffer circuit displayed in Figure 1. A transient simulation shows that the circuit responds with a damped oscillation at the output node when a voltage step is applied to the input. Figure 2 shows a hardcopy of the simulation results.

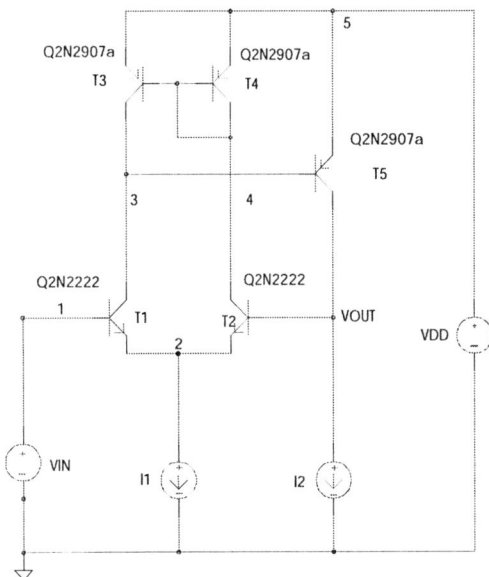

Figure 1: Buffer circuit

Using the AC transistor model from Figure 3 a numerical pole/zero analysis reveals that a complex conjugate pair of poles near the imaginary axis is responsible for the oscillation. The task is now to determine a symbolic expression for this pole pair in terms of those elements in the small-signal equivalent circuit which have dominant influence on these poles.

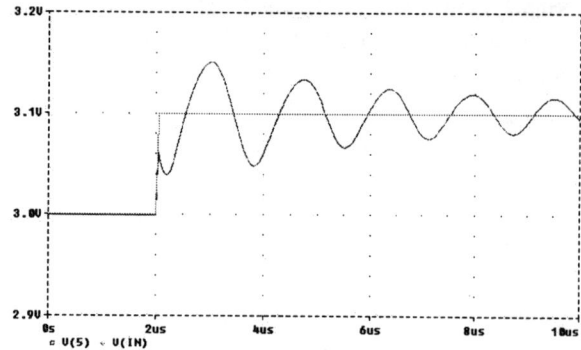

Figure 2: Step response of buffer circuit

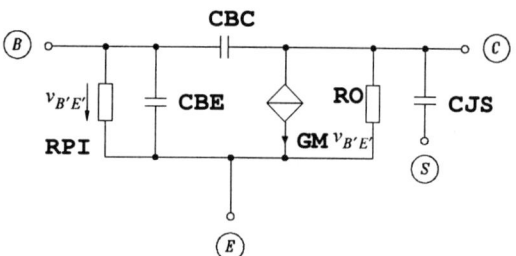

Figure 3: AC equivalent circuit for BJTs

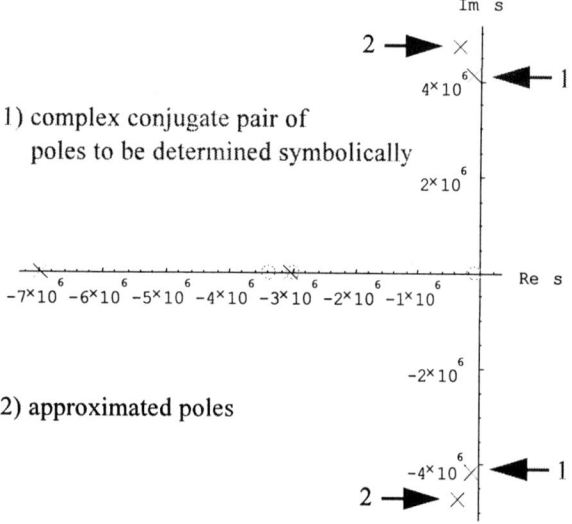

1) complex conjugate pair of poles to be determined symbolically

2) approximated poles

Figure 4: Pole/zero locations of buffer circuit

A full symbolic analysis would yield a characteristic polynomial of the order 4 with more than 250 million terms. Therefore, the approximation algorithm is applied to simplify the system of MNA equations with respect to the pole pair. This results in a compact second-order approximation of the characteristic polynomial from which the poles can be derived easily:

$$(CBE\_Q10 \; CBE\_Q12 \; GM\_Q11 - CBC\_Q13 \; CBE\_Q11 \; GM\_Q10 + CBC\_Q13 \; CBE\_Q10 \; GM\_Q10) \cdot s^2$$
$$+ GM\_Q10 \; ((CBE\_Q11 - CBE\_Q10) GM\_Q13 + (CBE\_Q12 - CBC\_Q13) GM\_Q11) \cdot s$$
$$+ GM\_Q10 \; GM\_Q11 \; GM\_Q13$$

For comparison with the original location of the poles the zeros of the approximated expression have been added to the plot in Figure 4.

7. CONCLUSIONS

An new algorithm for symbolic pole/zero extraction was presented, which is based on numerical eigenvalue shift estimations in a generalized eigenvalue problem. Future enhancements include improved accuracy of the ranking algorithm, e.g. by extending the eigenvalue shift prediction formula to the quadratic term.

ACKNOWLEDGMENTS

This work has been carried out within the MEDEA project A409 "Systematic Analog Design Environment", and with support of the "Stiftung Rheinland-Pfalz für Innovation" and the Deutsche Forschungsgemeinschaft within the "Schwerpunktprogramm Effiziente Algorithmen für diskrete Probleme und ihre Anwendungen".

REFERENCES

[1] G. Gielen, W. Sansen, *Symbolic Analysis for Automated Design of Analog Integrated Circuits*, Boston: Kluwer Academic Publishers, 1991

[2] F. V. Fernández, A. Rodríguez-Vázquez, J. L. Huertas, "A tool for symbolic analysis of analog integrated circuits including pole/zero extraction", in *Proc. ECCTD91*

[3] R. Sommer, E. Hennig, G. Dröge, E.-H. Horneber, "Equation-Based Symbolic Approximation by Matrix Reduction with Quantitative Error Prediction", Alta Frequenza–Rivista di Elettronica, No. 6, Dec. 1993, pp. 29–37

[4] G. Dröge, E.-H. Horneber, "Symbolic Calculation of Poles and Zeros", in *Proc. SMACD96*

[5] E. Hennig, *Analog Insydes Tutorial*, ITWM, Kaiserslautern, 1997

[6] F. Constantinescu, M. Nitescu, "A New Approach to Symbolic Pole Computation", in *Proc. ECCTD95*

[7] J. Vlach, K. Singhal, *Computer Methods for Circuit Analysis and Design*, 2nd Edition, New York: Van Nostrand Reinhold, 1994

[8] F. Kuhnert, "Über die Sensibilität von Eigenwerten und Eigenvektoren", ZAMM 71 (1991) 7/8, pp. 233–239

[9] Gerard L. G. Sleijpen, Albert G. L. Booten, Henk A. v. d. Vorst, Diederik R. Fokkema, "Jacobi-Davidson Type Methods for Generalized Eigenproblems and Polynomial Eigenproblems: Part I", Preprint no. 923, Univ. Utrecht, Dept. of Mathematics, Sept. 1995

EFFICIENT SYMBOLIC ANALYSIS OF LARGE ANALOG CIRCUITS USING SENSITIVITY-DRIVEN RANKING OF MATROID INTERSECTIONS

Petr Dobrovolny[1], Piet Wambacq[2], Georges Gielen[2], Willy Sansen[2]

[1] Technical University Brno, Dep. Microelectronics, Udolni 53, 60200 Brno, Czech Republic, Tel.: +420/5/43167127 - Fax: +420/5/43167298

[2] Katholieke Universiteit Leuven, Dep. Elektrotechniek, ESAT-MICAS, Kardinaal Mercierlaan 94, B-3001 Heverlee, Belgium, Tel.: +32/16/321077 - Fax:+32/16/321986

ABSTRACT

A new program for the generation of approximate symbolic network functions is presented. The approximation technique used in this tool is the simplification during expression generation technique, which in the general case suffers from a nonpolynomial CPU complexity. The technique makes use of the two-graph method. Using matroid intersection theory, spanning trees common to the voltage and the current graph are directly generated at selected sample frequencies. The generation of the approximate symbolic expression of a network function is driven by the sensitivity of the magnitude of the network function with respect to the different coefficients. In this way, very little terms are generated more than once. The total algorithm runs in $O(Kmn^3)$ time. Here, K is the average number of matroid intersections that must be generated for the approximation of each coefficient of the network function, n is the number of nodes in the linearized network and m is the number of circuit elements. Experimental results are presented and a comparison to previous methods based on matroid intersection theory is given.

1. INTRODUCTION

Symbolic simulation techniques are useful for the analysis and synthesis of analog integrated circuits [1,2]. Unlike numerical analysis, symbolic analysis maintains the input frequency and all or part of the circuit parameters as symbols. Symbolic analysis can accelerate the design process of analog integrated circuits. The symbolic techniques can be used for the fast automatic generation of equations that express the circuit performance for repeated evaluation or for the automatic generation of interpretable expressions that describe the behavior of a circuit. This information is complementary to information supplied by numerical simulation results. A lot of papers have been published where various tools for symbolic network analysis are presented. These tools are able to compute exact symbolic expressions or directly compute approximate expressions that are usually required in the case of the analysis of large analog circuits.

In this paper a new efficient algorithm is presented for the direct computation of approximate symbolic equations. The paper is organized as follows. Section 2 presents the approximation during generation technique. Section 3 describes the matroid intersection algorithm. The new sensitivity-driven ranking of matroid intersections is presented in section 4. Experimental results are provided in section 5, and conclusions in section 6.

2. SIMPLIFICATION DURING GENERATION

The generated symbolic expressions are in the general case network functions, which can be expressed in the cancellation-free expanded format as the ratio of two polynomials in the Laplace variable s :

$$T(s) = \frac{N(s)}{D(s)} = \frac{f_0 + f_1 s + \cdots + f_p s^p}{g_0 + g_1 s + \cdots g_q s^q} \quad (1)$$

in which the coefficients f_i ($i = 0, ..., p$) and g_i ($i = 0, ..., q$) are sums of products of the symbolic circuit elements. The alternative form of the representation of the network function is the nested representation, which is typical format of results evolved from a hierarchical analysis. This paper, however, addresses the symbolic analysis of analog building blocks like complex opamps that cannot be efficiently partitioned into loosely coupled subblocks.

Due to the exponential growth of the number of product terms in the symbolic network function when the number of nodes and circuit elements grows, it is virtually impossible to generate the exact expression. Some kind of approximation or simplification of the symbolic expression is required. Various symbolic approximation techniques have been proposed [3,4,5,6]. In our program a very efficient technique, so-called *simplification during generation,* was used. With this technique, the dominant terms of every coefficient f_i and g_i in the network function (1) are generated in decreasing order of magnitude until the sum of generated terms is within a user-defined error bound of the exact numerical value of the coefficient under consideration. This technique has forced a breakthrough in the symbolic analysis of analog circuits of practical size. The approach described in this paper addresses a new, more efficient technique for simplification during generation.

For the generation of terms different existing symbolic network analysis methods can be used. Our approach is based on the well-known two-graph enumeration method that is the most efficient one for the simplification during generation approach [6]. In the two-graph method, a valid term is found as a spanning tree that is common to a weighted voltage graph and a weighted current graph. These two graphs have the same nodes and edges, but the edges are connected in a different way. The two graphs are constructed from an inspection of the circuit. The weight of each edge in these graphs is equal to the admittance of the corresponding element in the circuit. The value of a term is equal to the product of the weights of the edges of the spanning tree. Hence, the two-graph method reduces to an enumeration of common spanning trees.

3. MATROID INTERSECTION ALGORITHM

Matroid theory [7] can be very helpful in the development of algorithms for the enumeration of terms of a network function using some symbolic network analysis method, such as the two-graph method. The enumeration of common spanning trees that correspond to a given coefficient f_i or g_i in a network function can then be formulated as the enumeration of bases that are common to three matroids. This problem however is *NP-complete* for general matroids [7].

The three matroids that are involved in our problem are the following ones. First there are two graphic matroids, one on the voltage graph, the other one on the current graph. Each tree of the voltage or current graph is an *independent set* of this matroid and each spanning tree is a *base* of this matroid. The third matroid is a partition matroid induced by the fact that the edges in the two-graphs either correspond to capacitors or to (trans)conductances. Terms (or common spanning trees) that correspond to a given coefficient f_i or g_i contain exactly i symbols of capacitors and for the rest conductance symbols. Hence, a common spanning tree must also be a base of a partition matroid in which a set is independent if it contains at most i capacitors and at most $n-i-1$ conductances, in which n is the number of nodes in the voltage or current graph.

Since the enumeration problem of bases common to three matroids is *NP-complete*, the best intersections of two matroids will be enumerated and each intersection will be examined afterwards whether or not it is also an independent set in the third matroid.

There are three options to choose two matroids whose intersections are generated before independence is checked in the third one:
1. Enumeration of the best intersections of the graphic matroid on the voltage graph with the partition matroid. This option has been suggested in [8]. Due to the simple structure of the partition matroid the problem can be solved more efficiently than it follows from the general matroid intersection theory. The most efficient algorithm reported up till now [8] runs in $O(K_1 m)$ time with $O(K_1 m)$ space requirements. K_1 is the number of generated spanning trees in the voltage graph and m is the number of edges in the voltage or current graph. This procedure is repeated for each coefficient of each power of s in the numerator or denominator of the network function. Because the number of powers of s is $O(n)$, the total procedure runs in $O(K_1 mn)$ time in which K_1 then corresponds to the mean value of spanning trees in the voltage graph that must be generated in order to obtain sufficient terms for the approximation of coefficient f_i or g_i. The disadvantage of this approach is that for very large transistor circuits many spanning trees of the voltage graph may have to be generated for a very small number of common spanning trees. Indeed, it is seen that the ratio of common spanning trees divided by the number of generated spanning trees in the voltage graph decreases as the circuit size increases. Hence K_1 can be much larger than the number of terms that need to be generated for f_i or g_i. Nevertheless, this approach gives good results for the simplification during generation of two-graphs that correspond to circuits of twenty transistors, but for larger two-graphs the CPU time and memory requirements increase sharply.

2. The same as option 1 but with the role of the voltage and the current graph interchanged. This option is less advantageous as the previous option, since the number of spanning trees in the current graph is usually higher than the number of spanning trees in the voltage graph [8].

3. Enumeration of intersections of the two graphic matroids. This means that common spanning trees are enumerated. Hereby, capacitive edges are considered in the same way as conductive edges. This is possible by choosing a value for the frequency which makes that the absolute value of the admittance of a capacitor ωC can be compared to the admittance g of a conductance. In this way, it is not known in advance how many capacitors a term contains, and hence it is not known in advance to which coefficient in the network function the generated terms will belong. Hence, the terms must be classified afterwards. This procedure has to be repeated for different frequencies and in this way an approximation can be constructed for each coefficient f_i or g_i of interest. For the enumeration of common spanning trees the algorithm of Camerini and Hamacher [9] can be used. This is an algorithm that enumerates in decreasing order the independent sets, which are common to two general matroids. The running time limit of this algorithm applied to two graphic matroids is $O(Kmn^2)$, in which K is the number of generated intersections, m is the number of edges in each graph and n is the number of nodes in each graph. However, this approach also has some disadvantages. First, when frequency points are close to each other, then it is very likely that a term that has been generated at the previous frequency will also be generated at the next frequency. However, in order to obtain a reasonable accuracy in the frequency range of interest, it is required to choose the frequency points close to each other. For example, if a symbolic expression for the response of an operational amplifier is to be generated, then in the vicinity of the unity-gain frequency a very fine grid of frequency points is usually required. This implies that many terms will be generated more than once. In this way, the efficiency of the approach goes down. The overall running time limit of the approach is $O(n_g K_2 mn^2)$, in which n_g is the number of chosen frequency points and K_2 is the mean value of the total number of enumerated intersections at every frequency point.

4. SIMPLIFICATION DURING GENERATION USING SENSITIVITY-DRIVEN RANKING OF MATROID INTERSECTIONS

The approach that is proposed in this paper starts from the third option mentioned in the previous section. The efficiency of that approach is enhanced by an intelligent choice of the frequency points. This choice is based on the sensitivities of the magnitude of the network function with respect to a coefficient f_i or g_i. The sensitivity with respect to coefficient h, which can be f_i or g_i, is defined as:

$$S_h^{|T|} = \frac{\partial |T|}{\partial h} \frac{h}{|T|} = \frac{\partial \ln(|T|)}{\partial h} \quad (2)$$

For every coefficient h there is a frequency at which the absolute value of the sensitivity of the magnitude with respect to that coefficient is maximal. If this frequency is chosen to generate common spanning trees, then it is very likely that the dominant terms that will be generated, will belong to coefficient h. In this way, the need for a fine frequency grid in frequency ranges where the network function varies rapidly, is not required anymore.

The occurrence of frequencies where the sensitivity with respect to the different coefficients is maximal, is illustrated for the fully differential BiCMOS amplifier of Fig. 1. The corresponding sensitivities of the transfer function with respect to the coefficients of the denominator are displayed in Fig. 2.

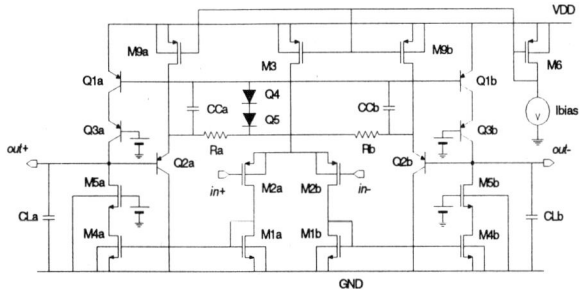

Figure 1. A BiCMOS fully-differential operational transconductance amplifier.

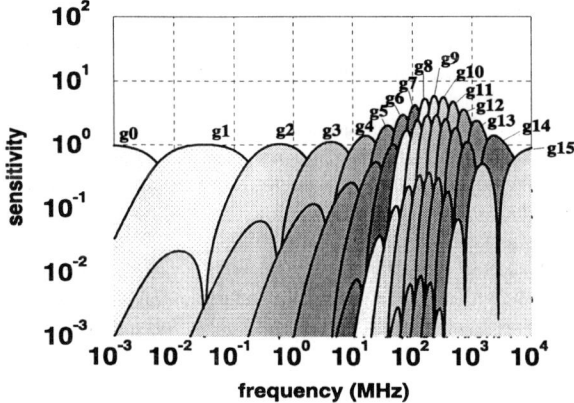

Figure 2. The sensitivities of the transfer function with respect to the coefficients of the denominator.

The overall algorithm now runs as follows. First, the numerical value of each coefficient is computed with the polynomial interpolation method [1]. Next, the sensitivity of the magnitude of the network function with respect to each coefficient is computed and the frequencies where the maxima occur are determined. Then the algorithm of Camerini and Hamacher is applied at each of these frequencies to enumerate terms (common spanning trees).

The running time limit of this approach can be determined as follows. Assume that the order of the polynomial in the numerator and the denominator of a network function is p and q, respectively. Then the algorithm of Camerini and Hamacher needs to be applied $p + q + 2 = O(n)$ times. Hence the running time limit is $O(Kmn^3)$, where K is the average number of intersections (common spanning trees) that is generated for the approximation of each coefficient f_i or g_i. This running time limit is smaller than the running time limit $O(n_g K_2 mn^2)$ of the approach reported in [5] (see also above), since usually $n<<n_g$, and, due to the intelligent choice of the frequencies, $K<<K_2$. This has also been tested experimentally.

5. EXPERIMENTAL RESULTS

A symbolic expression is generated for the network function that corresponds to the differential-mode gain of the fully-differential BiCMOS amplifier of Fig. 1. For this example we can restrict to the analysis of the denominator, without loss of generality. The denominator is a polynomial of order fifteen. The voltage and the current graph that are set up for the computation of this denominator contain 16 vertices and 70 edges, 25 of which corresponds to capacitors, while others correspond to (trans)conductances. The statistics of this example for three values of the approximation error are shown in Table 1.

	Relative error		
	25%	10%	5%
#AI	1459	4541	8395
#T	569	1678	3010
#AI / #T	2.56	2.7	2.79
CPU time [s]	2.4	7.2	13.2

Table 1. Statistics of the symbolic approximation.

The meaning of the abbreviations in Table 1 is as follows: #AI represents the number of all generated instersections, #T represents the number of all generated common bases that corresponds to the number of valid terms. It is seen that the ratio between the number of all generated intersections and the number of valid terms is small and grows very slowly with the circuit size. This is an indication of the high efficiency of the approach. Finally, the CPU time to generate an approximate symbolic expression of the denominator of the differential-mode gain is shown as well. These results were generated on a SG server PowerChallange L with MIPS®R10000(TM)-195 Mhz processor.

Fig. 3 shows the differences in magnitude between the exact and the approximate expression for this example for the same three user-prescribed relative approximation errors.

The CPU time as a function of the number of generated valid terms is shown in Fig. 4. From this graph it is clear that our new approach is superior to the old conventional approaches that are represented for instance by the symbolic analyzer ISAAC [3]. Moreover, for larger anolog circuits our program is also more effective and faster than the program described in [4] that also uses the matroid intersection theory for the implementation of the simplification during generation approach, but with option 1 for the matroid intersection (see above).

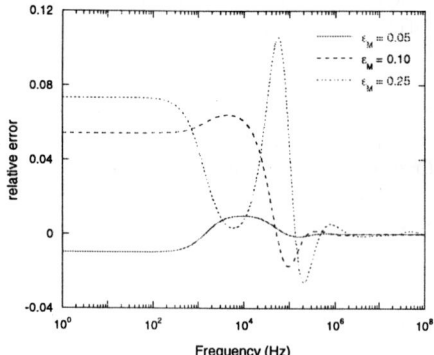

Figure 3. Differences in the magnitude of the exact and the approximate network function (with relative error ε_M) as a function of frequency.

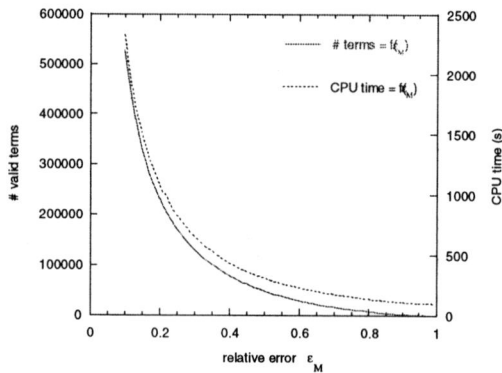

Figure 4. CPU time and number of generated valid terms as a function of the required relative error ε_M of the network function's magnitude.

The number of terms to obtain a required accuracy would be drastically reduced if a simplification before generation would be performed before this routine is used [2]. The method presented in this paper can be combined with any such technique, as it starts from a given two-graph, whether it has been simplified in advance or not. Finally, it should be remarked as well that the number of terms would be much smaller if corresponding circuit elements are assumed to match, such that they are represented by one symbol. In this case identical terms arise, which either cancel or can be combined to a symbolic term with an integer coefficient different from zero.

6. CONCLUSIONS

In this paper, a new approach has been presented for the simplification during generation technique, which is known to be a very efficient technique for the generation of approximate symbolic network functions for large analog circuits. The symbolic analysis method to generate the terms is the two-graph method. The underlying algorithm is based on matroid theory. By an intelligent choice of the frequencies at which common spanning trees are generated, namely at the frequencies where the sensitivities of the transfer function magnitude to the individual coefficients are maximal, the running time limit of the overall approach is $O(Kmn^3)$, where m is the number of edges in the voltage or current graph, n is the number of vertices in each graph and K is the total number of generated matroid intersections. K is higher than the total number of different valid terms, such that the approach - like all other approaches - yields some overhead. However, experimental results presented on practical circuits have shown that the overhead, both in terms of memory usage and CPU time, is slowly growing with the circuit size and for large circuits this overhead is much lower than for previously reported approaches. As a result, the approach presented here can - in conjunction with reliable simplification before generation techniques - be the core of a powerful symbolic analyzer.

REFERENCES

[1] P. Lin, "Symbolic network analysis," Elsevier, 1991.
[2] F. Fernandez, A. Rodriguez-Vazquez, J. Huertas, G. Gielen, "Symbolic analysis techniques," IEEE Press, 1998.
[3] G. Gielen, H. Walscharts, W. Sansen, "ISAAC: a symbolic simulator for analog integrated circuits," IEEE Journalf of Solid-State Circuits, Vol. 24, No. 6, pp. 1587-1597, December 1989.
[4] P. Wambacq et al., "Efficient symbolic computation of approximate small-signal characteristics," IEEE Journal of Solid-State Circuits, Vol. 30, No. 3, pp. 327-330, March 1995.
[5] Q. Yu, C. Sechen, "A unified approach to the approximate symbolic analysis of large analog integrated circuits," IEEE Transactions on Circuits and Systems, Part I, Vol. 43, No. 8, pp. 656-669, August 1996.
[6] P. Wambacq et al., "A family of matroid intersection algorithms for the computation of approximated symbolic network functions," proceedings ISCAS, Vol. IV, pp. 806-809, 1996.
[7] E. Lawler, "Combinatorial optimization: networks and matroids," Holt, Rinehart and Winston, 1976.
[8] P. Wambacq, "Symbolic analysis of large and weakly nonlinear analog integrated circuits," Ph.D. dissertation Katholieke Universiteit Leuven, 1996.
[9] P. Camerini, H. Hamacher, "Intersection of two matroids: (condensed) border graph and ranking", SIAM Journal of Discrete Mathematics, Vol. 2, pp. 16-27, February 1989.

ANALYTICAL EXPRESSIONS FOR AVERAGE BIT STATISTICS OF SIGNAL LINES IN DSP ARCHITECTURES*

S. Bobba, I. N. Hajj, and N. R. Shanbhag

Coordinated Science Lab & ECE Dept.
University of Illinois at Urbana-Champaign
Urbana, Illinois 61801

Abstract

Accurate high-level power estimation methods are required for exploring the design space to obtain an optimal low-power circuit. DSP architectures are regular and they consist of interconnected macro-blocks such as adders and multipliers. In [1], the power dissipation of macro-blocks was related to the average bit statistics. Given the input word-level statistics for a DSP architecture, the word-level statistics at all the internal signal lines can be computed analytically using transfer function evaluation or by propagating the statistics. In this paper, we present simple analytical expressions for computing the average bit statistics using the word-level statistics of the signal lines in a DSP architecture.

1 Introduction

The decreasing feature size and other developments in device technology have led to the design of complex circuits operating at high clock rates. These circuits have a higher power dissipation per unit area. High power densities lead to higher operating temperatures which decrease the lifetime of IC due to reliability problems. For reliable long-term operation, sophisticated and expensive packaging is required to ensure proper heat dissipation. The integrated circuits used in portable products, such as laptop computers and cellular phones need to consume less power in order to extend the battery life and reduce the overall weight of the final product. Hence, power minimization is an important concern in present day IC design.

An accurate estimate of average power dissipation at a high-level (RTL level or architectural level) is necessary for the exploration of the design space to obtain a power optimal design. DSP architectures are modular, regular and they consist of instantiations of DSP macro-blocks such as adders, multipliers, multiplexers. The word-level statistics of all internal signals can be computed from the input signal statistics using transfer function evaluation or propagating the input statistics. The power dissipation of a macro-block can be shown to be a function of the average bit statistics of the output and the input signal lines [1]. The total transition activity and the average bit transition activity of the bits can be computed by using the methods proposed in [2,3]. Both these methods do not present closed-form analytical expressions for bit statistics. In this work, we *derive* analytical expressions for bit statistics and develop simple and accurate closed-form analytical expressions for average bit statistics in terms of the word-level statistics.

Signals in DSP systems are represented in finite precision using a certain signal encoding. There exist a number of different representations such as the sign magnitude, one's-complement and two's-complement, but the two's-complement representation is the most commonly used representation [4]. The word values of a signal have a certain probability distribution. The bit statistics depend on both the signal probability density function and the type of encoding. Many DSP inputs can be closely approximated by a Gaussian process. In this work, we assume that the input is a zero-mean Gaussian signal with a two's-complement number representation. Many signals in real-life applications are zero-mean Gaussian signals. For audio signals, such as speech or music, the distribution of the amplitude is concentrated about zero and falls off rapidly with increasing amplitude [4]. Hence, these signals can be considered as zero-mean Gaussian signals. Given two signals with the same variance, for the same quantization error the signal with non-zero mean would require more bits than the signal with zero mean. Signals with a non-zero mean do not utilize the entire dynamic range. Even if the input signal is not a zero-mean signal, the mean value can be subtracted from the input and an appropriately scaled value of the mean can be added to the output of the DSP circuit.

If the input signal is a zero-mean Gaussian signal, then all the internal signals in a DSP architecture are also zero-mean Gaussian signals. Since, we consider zero-mean signals, the word-level statistics of a signal consist of the variance (σ^2) and the correlation coefficient (ρ). The bit statistics are the transition activity (t_i), probability (p_i) and the temporal correlation (ρ_i) respectively, where t_i denotes the probability of a transition of the bit, p_i denotes the probability that the bit is 1 (logic HIGH), and ρ_i denotes the lag-one temporal correlation of the bit. In this work, we present analytical expressions for the average bit statistics, *average bit probability*, *average bit temporal correlation* and *average bit transition activity* in terms of the signal word-level statistics (σ, ρ).

This paper is organized as follows. In the next section, we present expressions for the bit probability and the average bit probability. In section 3, we give the analytical expression for average bit transition activity. In section 4, we derive a relationship between the average bit transition activity and average bit temporal correlation. Finally, in section 5 we give the conclusions.

2 Average Bit Probability

In this section, we first present a theorem which relates the average bit probability of signals to certain properties of the signal. We then show that the theorem is valid for a zero-mean Gaussian signal in two's complement representation.

Consider a signal represented using a word-length of B bits. The signal values are represented using 2^B distinct binary representations. Let $N = 2^{B-1}$ and $\{a_1, a_2, \cdots, a_N\}$, $\{b_1, b_2, \cdots, b_N\}$ denote two sets of N symbols, where each of the symbols corresponds to a particular bit representation.

---

*This work was supported by NSF grant MIP-9710235 and NSF CAREER award MIP-9625737

Let $P(a_i), P(b_i)$ denote the probability values of the symbols a_i, b_i, respectively. The following theorem relates the bit probability to certain properties of the symbols $a_1, a_2, \cdots, a_N$, $b_1, b_2, \cdots, b_N$.

Theorem 1 *If the symbols $a_1, a_2, \cdots, a_N, b_1, b_2, \cdots, b_N$ of a signal are such that, for each a_i there is a b_i with $P(a_i) = P(b_i)$ and the bit representations of a_i, b_i are complements of each other then, bit probability is 0.5 for each of the B bits.*

Proof: To prove the theorem, we use the conditions that the symbols $a_1, a_2, \cdots, a_N, b_1, b_2, \cdots, b_N$ satisfy and the fact that the cumulative sum of the probability values associated with the symbols a_i's and b_i's is 1, i.e.,

$$\sum_{i=1}^{N} \{P(a_i) + P(b_i)\} = 1. \quad (1)$$

We will prove the above theorem for an arbitrary bit j. The bit j is 1 in the bit representation of exactly N symbols and 0 for the bit representation of the other N symbols. This is due to the fact that we have 2^B (2N) distinct binary representations for the signal values. Let S_j denote the set of symbols for which the bit j is 1 and $\overline{S_j}$ denote the set of symbols for which the bit j is 0. Since the symbols have distinct bit representations the sets $S_j, \overline{S_j}$ are disjoint sets. The probability of a bit j, denoted by p_j is given by the following expression,

$$p_j = \sum_{a_i \in S_j} P(a_i) + \sum_{b_i \in S_j} P(b_i). \quad (2)$$

The set S_j cannot contain symbols that have bit representations that are complements of each other. It would violate the condition that S_j contains symbols for which the bit j is 1. Hence for every symbol $a_i \in S_j$, there is a symbol $b_i \in \overline{S_j}$ that has the complement bit representation of symbol a_i. Since, the probability values of the symbols a_i and b_i are the same, we can relate the sum of the probability values of the symbols $a_i \in S_j$ and $b_i \in \overline{S_j}$ by the following equation,

$$\sum_{a_i \in S_j} P(a_i) = \sum_{b_i \in \overline{S_j}} P(b_i). \quad (3)$$

Similarly, the sum of the probability values of the symbols $b_i \in S_j$ and $a_i \in \overline{S_j}$ are related by the following equation,

$$\sum_{b_i \in S_j} P(b_i) = \sum_{a_i \in \overline{S_j}} P(a_i). \quad (4)$$

Since S_j and $\overline{S_j}$ are disjoint sets that contain all the symbols, Eqn. (1) which is an expression for the cumulative sum of the probability values associated with all the symbols can be rewritten as,

$$\sum_{a_i \in S_j} P(a_i) + \sum_{b_i \in S_j} P(b_i) + \sum_{a_i \in \overline{S_j}} P(a_i) + \sum_{b_i \in \overline{S_j}} P(b_i) = 1 \quad (5)$$

Using (3) and (4) in (5),

$$2 \sum_{a_i \in S_j} P(a_i) + 2 \sum_{b_i \in S_j} P(b_i) = 1 \quad (6)$$

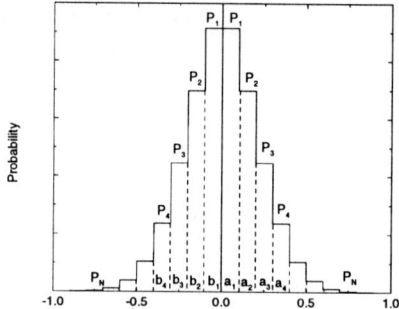

Figure 1: Probability distribution of the word values

Using (2) and (6), the expression for bit probability (p_j) can be written as,

$$p_j = \sum_{a_i \in S_j} P(a_i) + \sum_{b_i \in S_j} P(b_i) = 0.5 \quad (7)$$

∎

Theorem 1 relates the probability of the bits to the properties of the symbols which correspond to a binary representation of the signal word value. The properties of the symbols depend on the probability density function and the type of encoding. In general, a signal with any probability density function that is symmetric about zero amplitude and uses either two's-complement or one's-complement signal encoding with sufficient number of bits satisfies the conditions on the symbols given in *Theorem 1*. Hence, by *Theorem 1* the probability of each of the bits is 0.5 for all such signals. We will show this result for the two's-complement representation of a zero-mean Gaussian signal.

Corollary 1 *For a zero-mean Gaussian signal, the bit probability (p_i) in the two's-complement representation is 0.5*

Proof: Let $a_1, a_2, \cdots, a_N$ denote the two's-complement representation of the positive signal word values in the increasing order. Let $b_1, b_2, \cdots, b_N$ denote the two's-complement representation of the negative signal word values in the decreasing order. Figure 1 shows the notations described. Note that the bit representations of $(a_i, b_i) \forall i$, are complements of each other. Let $P_1, P_2, \cdots, P_N$ denote the probability values $P(a_i)$ corresponding to each of the a_i's. A Gaussian signal is symmetric about the mean and hence, we can relate $P(a_i)$ and $P(b_i)$. This relationship is shown below.

$$P(a_i) = P(b_i) = P_i \quad \forall i \quad (8)$$

Since the conditions on the symbols specified in theorem 1 are satisfied, the probability of each of the bits is 0.5.

∎

By *Corollary 1*, the probability of each of the bits in the two's-complement representation of a zero-mean Gaussian signal is 0.5. Hence the average value of the bit probability is also 0.5.

3 Average Bit Transition Activity

In this section, we first present a theorem that relates the probability of a sign change to the word-statistics of the signal. We then obtain the expression for the most significant bit (MSB) transition activity. The MSB bit transition activity and the break-points presented in [2] are used to obtain the expression for average bit transition activity.

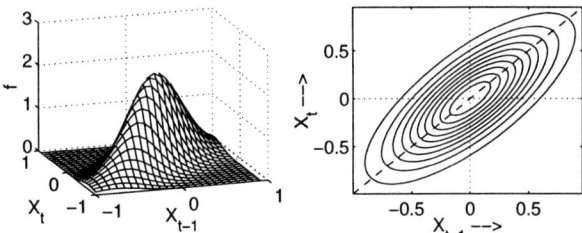

Figure 2: Probability density function and contour plot of jointly Gaussian random variables (X_t, X_{t-1})

The average bit transition activity is related to the word-statistics of the signal. Let $P(t^{-\to+})$ denote probability of a sign change from a negative signal value to a positive signal value. Let $P(t^{+\to-})$ denote probability of a sign change from a positive signal value to a negative signal value. The value of the sign change probability is dependent on the word-statistics. The following theorem relates the sign change probability and the word-statistics.

Theorem 2 *Given random variables (X_t, X_{t-1}) that are jointly Gaussian with zero-mean, variance σ^2 and correlation coefficient ρ, the probability of sign change is given by,*

$$P(t^{-\to+}) = P(X_{t-1} < 0, X_t \geq 0) = \frac{1}{2\pi}\cos^{-1}(\rho) \quad (9)$$

$$P(t^{+\to-}) = P(X_{t-1} \geq 0, X_t < 0) = \frac{1}{2\pi}\cos^{-1}(\rho) \quad (10)$$

Proof: The joint density function f of the random variables (X_t, X_{t-1}) is given by,

$$f(X_t, X_{t-1}) = \frac{1}{2\pi\sigma^2\sqrt{1-\rho^2}} e^{\left(-\frac{X_t^2+X_{t-1}^2-2\rho X_t X_{t-1}}{2\sigma^2(1-\rho^2)}\right)} \quad (11)$$

Fig. 2 shows the plot of the joint density function and the contour plot. Since the joint probability density function is symmetric, $P(t^{+\to-})$ is equal to $P(t^{-\to+})$. Hence, we will prove the probability of sign change from a negative to positive signal value $(P(t^{-\to+}))$. The probability of sign change $P(t^{-\to+})$ is given by,

$$P(t^{-\to+}) = P(X_{t-1} < 0, X_t \geq 0) \quad (12)$$

This can be obtained by finding the volume under the surface in the region $(X_{t-1} < 0, X_t \geq 0)$ and dividing by the total volume under the surface. In this proof, we consider a particular contour and find the ratio of the area in the desired region to the total area of the contour. Furthermore, we show that this ratio is independent of the contour chosen. Hence, this ratio is valid for all contours and it also denotes the ratio of the volume in the desired region to the total volume. The equation for a contour is given by,

$$X_t^2 + X_{t-1}^2 - 2\rho X_t X_{t-1} = k^2 \quad (13)$$

where k is a constant. The ratio of the shaded area to the total area in Fig. 3 gives the probability of sign change for this contour. Put $Z_1 = (X_t + X_{t-1}/(\sqrt{2}k)$ and $Z_2 = (X_t - X_{t-1}/(\sqrt{2}k)$. Using (Z_1, Z_2) and performing the change of variable, Eqn. (13) becomes,

$$(1-\rho)Z_1^2 + (1+\rho)Z_2^2 = 1 \quad (14)$$

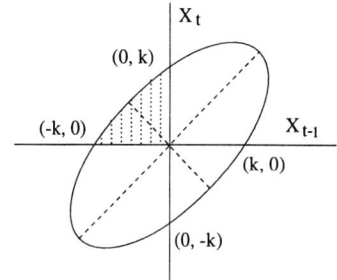

Figure 3: Contour plot for a particular value of k

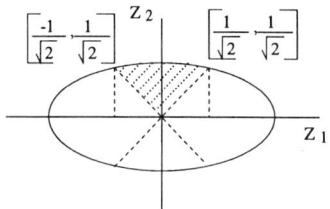

Figure 4: Transformed contour plot

Fig. 4 shows the plot of the above equation and the corresponding end points. Eqn. (14) can be rewritten as,

$$\frac{Z_1^2}{a^2} + \frac{Z_2^2}{b^2} = 1 \quad (15)$$

a, b denote $1/\sqrt{1-\rho}$ and $1/\sqrt{1+\rho}$ respectively. Let A_s denote the area of the shaded region. The value of A_s can be obtained by integrating the area under the curve $Z_2 = \frac{b}{a}\sqrt{a^2 - Z_1^2}$, and deleting the area of the two right triangles. The following expression relates the area under the curve and the area under the shaded region denoted by A_s.

$$A_s = -\frac{1}{2} + \frac{b}{a}\int_{-1/\sqrt{2}}^{1/\sqrt{2}} \sqrt{a^2 - Z_1^2}\, dZ_1 \quad (16)$$

Solving the integral yields,

$$A_s = \frac{\cos^{-1}(\rho)}{2\sqrt{1-\rho^2}} \quad (17)$$

The total area of the contour can be computed as the area of the ellipse and it is given by,

$$\frac{\pi}{ab} = \frac{\pi}{\sqrt{1-\rho^2}} \quad (18)$$

The ratio of the area in the shaded region to the total area of the contour is given by,

$$\frac{1}{2\pi}\cos^{-1}(\rho) \quad (19)$$

Observe that the ratio is independent of k, the constant for the contour. Hence, the probability of sign change $P(t^{-\to+})$ is given by the following expression.

$$P(t^{-\to+}) = \frac{1}{2\pi}\cos^{-1}(\rho) \quad (20)$$

∎

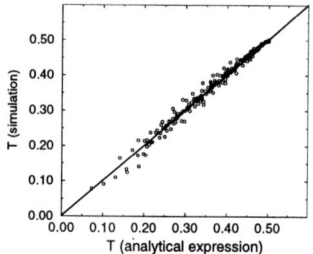

Figure 5: Validation of analytical expression for average bit transition activity

Theorem 2 relates the word-statistics to the probability of a sign change between consecutive word-values. The MSB bit is the sign-bit in one's-complement, two's-complement or the sign magnitude encoding of the signal. Hence, the MSB bit transition activity (t_{msb}) in the one's-complement, two's-complement or the sign magnitude encoding of the signal is given by,

$$t_{msb} = P(t^{-\to+}) + P(t^{+\to-}) = \frac{1}{\pi}cos^{-1}(\rho) \quad (21)$$

The dual bit-type (DBT) model for word-level signals was proposed in [2]. The DBT model has two break points, BP_0 and BP_1, that are computed using the word-level statistics of the signal. The expressions for the break points BP_0, BP_1 presented in [2] are given by Eqns. (22) and (23) below,

$$BP_0 = log_2(\sigma) + log_2(\sqrt{1-\rho^2} + |\rho|/8) \quad (22)$$

$$BP_1 = log_2(3\sigma) \quad (23)$$

In this work, we use the above break-points and Eqn. (21) to compute the average bit transition activity. The uniform white noise model is valid for the least significant bits up to the break point BP_0. The transition activity of these bits is 0.5. The transition activity of the sign bits, bits from the most significant bit to BP_1 is given by (21). For the bits between BP_0 and BP_1 a linear approximation between 0.5 and t_{msb} is used to obtain the bit transition activity. The average bit transition activity (T) is the sum of the bit transition activity values divided by the word-length. It is given by the following expression,

$$T = 0.5 \; r \; + \; (1-r) \; \frac{1}{\pi}cos^{-1}(\rho) \quad (24)$$

r is the ratio of the average value of the break points BP_0, BP_1 and the signal word-length B. To experimentally validate the analytical expression for average bit transition activity, a first order auto-regressive model [3] was used to generate the signal word values for the given word-statistics. Fig. 5 shows a scatter plot of the total bit transition activity T obtained using the analytical expression and simulation for different values of word-length B (4, 8, 12, 16, 20, 24, 28, 32), ρ (0.0, 0.25, 0.5, 0.75, 0.9, 0.95, 0.99) and different σ^2 values. It can be seen that there is a good match between the results obtained by simulations and the analytical expression. The absolute average error was found to be 2.86%.

In the next section, we derive a relationship between the average bit transition activity and average bit temporal correlation.

4 Average Bit Temporal Correlation

In [3], an exact relation between the transition activity, probability and temporal correlation for a single bit signal was presented. This relation is given by (25):

$$t_i = 2p_i(1-p_i)(1-\rho_i), \quad (25)$$

where t_i is the transition activity, p_i is the probability and ρ_i is the temporal correlation of bit i. By corollary 1, the bit probability (p_i) of a zero-mean Gaussian signal with a two's complement encoding is 0.5 for all the bits. Substituting the values of p_i's in (25),

$$t_i = 0.5(1-\rho_i) \quad (26)$$

The average bit transition activity T is given by,

$$T = \frac{1}{B}\sum_{i=1}^{B} t_i \quad (27)$$

Substituting (26) in (27) we obtain,

$$T = 0.5 - \frac{0.5}{B}\sum_{i=1}^{B}\rho_i = 0.5 - 0.5\rho_{avg}. \quad (28)$$

Rearranging (28) we obtain the following relationship between the average bit transition activity (T) and the average bit temporal correlation (ρ_{avg}).

$$\rho_{avg} = 1 - 2T \quad (29)$$

Thus, ρ_{avg} can be computed from the word-statistics using (24) and (29).

5 Conclusion

In this paper, we presented analytical expressions for the bit statistics. We derived the result for the bit probability and the average bit probability. This was used to relate the average bit transition activity to the average bit temporal correlation. We developed a simple analytical expression for the average bit transition activity. We compared the average bit transition activity obtained by our model with a long simulation to verify the accuracy of the analytical expression. Hence, given the word-statistics of a signal, the bit statistics *average bit probability*, *average bit temporal correlation*, and *average bit transition activity* can be computed analytically.

References

[1] S. Gupta and F. Najm, "Power macromodeling for high level power estimation," in *Proc. of 34th ACM/IEEE Design Automation Conference*, Anaheim, CA, June 6-10, 1997, pp. 365–370.

[2] P. Landman and J. M. Rabaey, "Architectural power analysis: the dual bit type method," *IEEE Transactions on VLSI systems*, vol. 3, pp. 173–187, June 1995.

[3] S. Ramprasad, N. R. Shanbhag and I. N. Hajj, "Analytical estimation of signal transition activity from word-level statistics," *IEEE Transactions on Computer-Aided design*, vol. 16, no. 7, pp. 718–733, July 1997.

[4] A. V. Oppenheim and R. V. Schafer, *Discrete-time signal processing* NJ: Prentice-Hall, 1989.

ARCHITECTURE SELECTION OF A FLEXIBLE DSP CORE USING RECONFIGURABLE SYSTEM SOFTWARE

Jong-Yeol Lee, Dea-Hyun Lee, Jong-Sun Kim, Hyun-Dhong Yoon, Chong-Min Kyung, Kyu-Ho Park, Yong-Hoon Lee, and Seung Ho Hwang

Department of Electrical Engineering, Korea Advanced Institute of Science and Technology
373-1, Kusung-dong, Yusung-gu, Taejon, Korea 305-701

ABSTRACT

MetaCore is a flexible DSP core in that the architecture of MetaCore can be modified easily by changing the hardware parameters. To fully exploit the merits of a flexible core, the system software must be re-configurable when the target architecture changes. In this paper, we present a re-configurable system software for a flexible DSP core and the architecture selection procedure called "compile-simulate-refine" cycle using the re-configurable system software. The "compile-simulate-refine" cycle can make it possible to select the best architecture for a given application by exploring the possible candidate architectures in short time.

1. INTRODUCTION

As the tendency towards more complex electronic systems continues, many of these systems come to be equipped with embedded processors. Essential advantages of these processors include their high flexibility and short design time. This contrasts with the low flexibility and high cost of application specific integrated circuits. Furthermore, their low flexibility makes the short time-to-market more difficult to achieve.

Embedded processors come in different types and can be classified according to three different criteria : flexibility of the architecture, architectural features for certain application domains, and the form in which the processor is available. By the criteria MetaCore is classified as a flexible DSP core. MetaCore is a core that has special features for DSP applications(e.g. MAC unit and AGU block, etc.). By flexibility, we mean that a user can easily modify the architecture by just changing hardware parameters. These parameters include the number of registers, the number of functional blocks, bus width, and so on. If new instructions are needed, they are included in the core by writing HDL descriptions of the functional units to be used for the execution of those new instructions.

To support the flexible core, the system software must be re-configurable i.e. the changes in the architecture must be reflected in the system software immediately. If it is not the case, each time the architecture changes the software must be re-implemented and the short time-to-market, one of the main advantages of the flexible core, can not be achieved. Our system software (MetaCore C Compiler, MetaCore Assembler, MetaCore Instruction Set Simulator) can be reconfigured fast and easily when the architecture changes.

For re-configurability, each software has its own form of machine description file. At run time each software reads its machine description file and reconfigure those parts that need changes. In this way, the system software need not be rewritten entirely but only the machine description has to be changed, when the architecture changes. This makes it possible that the designer can test many architecture candidates at the stage of architecture design in a short time.

2. METACORE C COMPILER

Our re-configurable C compiler(MCC) is based upon GNU C Compiler(GCC). GCC is a portable compiler based on the compiler-compiler method. If we want to port GCC to a new target machine, we have only to describe Target Description Macros and Machine Description File for specific target machine. But in case of GCC, if the target machine architecture is changed then Target Description Macros and Machine Description File must be re-described and re-compiled to generate new compiler for the new target machine architecture.

To implement our re-configurable C compiler(MCC), we make the best use of the portability of GCC by modifying the parts of GCC and adding new parts. We describe Macros and Machine Description File in such a way that all possible architecture changes are included and MCC is generated by a single compilation. The internal variables related to the target architecture are set according to the parameters in the hardware parameter file when an application program is compiled. The code generation is guided by the internal variables and the code is generated, which is composed of the instructions that use available functional blocks and registers in current target hardware structure.

An example of MCC reconfiguration is shown in **Figure 1**. In this example, MCC is reconfigured automatically according to the presence or the absence of minmaxALU block. minmaxALU block provides min or max instruction which is used to find minimum or maximum between 2 operands. If the target architecture has a minmaxALU block, m = min(m,t) statement in C code is compiled into min instruction. Otherwise, this statement is compiled into a group of instructions of which the operation is to find minimum between two operands.

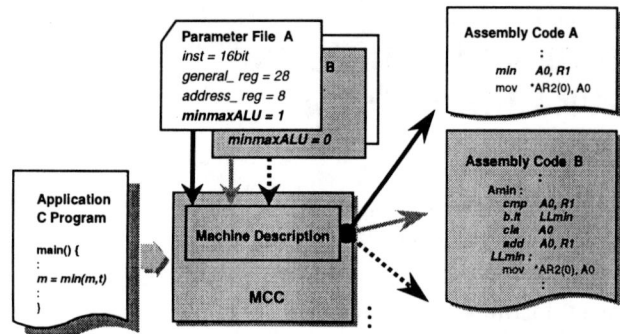

Figure 1. MCC reconfiguration when a hardware parameter changes

3. METACORE INSTRUCTION SET SIMULATOR

Instruction set simulators are programs that run on a host computer and simulate the execution of a processor. An instruction-set simulator simulates the target processor by interpreting the effects of instructions on that processor, one instruction at a time. An instruction-set simulator provides a view into the internal operation of the processor, showing the contents of registers and memory locations and allowing the programmer to set breakpoints, single step through the execution of a program, and collect performance information. MetaCore Instruction Set Simulator(MISS) is capable of tracing the usage of each functional unit. MISS knows what functional units are used when each instruction is executed. MISS also provides the designer with various performance information and statistics.

MISS gets the information about the architecture from the machine description file(IDF file). At each run-time, MISS reads IDF file and reconfigures the units that have to be changed according to the architecture such as the input/output unit and target memory management unit. If a new instruction is defined, the encoding information(this is needed because the input of MISS is object code) and the description of the behavior of that instruction are added to IDF file. The behavior of the instruction is described using the primitive functions defined in MISS. The group of primitive functions consists of data movement functions, arithmetic functions, and logical functions.

Figure 2 shows how a new instruction added to the instruction set can be simulated in MISS. If a new u_add instruction is defined, the encoding(B0000,2,B0010,6) and the behavior are described in IDF file. The tree builder reads the IDF file and constructs a decoding tree. The decoding tree has the information needed to decode the binary input stream. The simulation core uses the decoding tree to decode the input(object code) and find out the instructions in the object code and primitive functions to be called for each instruction. And then the core calls the appropriate primitive functions according to the decoding results.

4. METACORE ASSEMBLER

MetaCore Assembler(MASM) has also its own machine description format. In this format, the encoding of each instruction and the locations on the data-path or memory that can be sources or destinations of each instruction are described. MASM refers to this description during each run-time and translates the source code into an object code in common object file format(COFF).

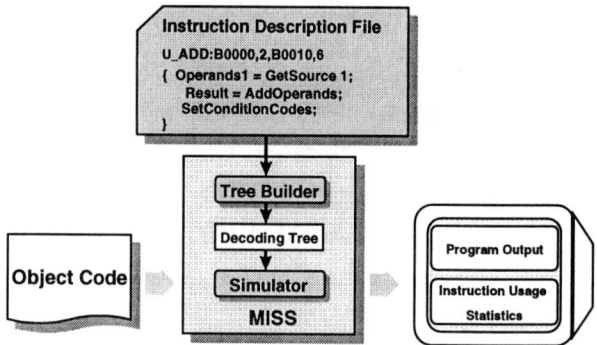

Figure 2. Addition of new instruction(U_ADD) into MISS

5. ARCHITECTURE SELECTION PROCEDURE

Using the re-configurable system software, it is possible to select a set of hardware parameters most suitable for a given application. The architecture selection procedure is called "compile-simulate-refine" cycle. The "compile-simulate-refine" cycle of the parameter selection is as follows : A given algorithm is coded in C language. The C code is compiled by the compiler(MCC) and the assembler(MASM). The generated object code is simulated by the instruction-set simulator(MISS). The simulation results (number of instructions, resource usage etc.) and the size of the object code can be used to determine a new architecture for the next iteration. This procedure is illustrated in **Figure 3**.

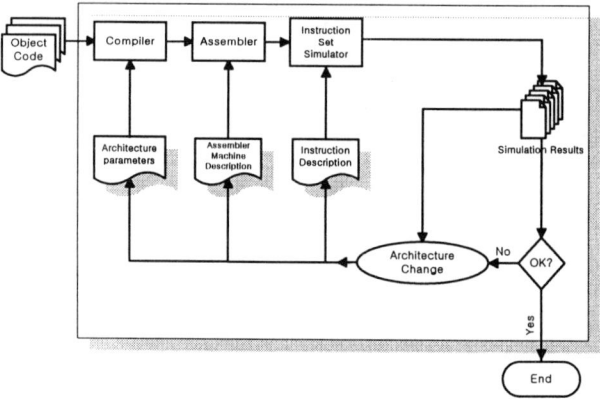

Figure 3. The Architecture Selection Procedure

In "refine" step, the architecture is tuned according to the simulation results. The size of the general register file can be reduced if the simulation results indicate that some of the general registers are rarely used or not used, for example.

Simulators of various abstraction levels can be used in the architecture selection procedure. For example, an architectural simulator that models the target architecture on lower abstraction level than an instruction set simulator can be used. But in that case the simulation time will be much longer. It is the general tendency that the simulation time becomes longer as the abstraction level goes lower. An instruction-set simulator is more appropriate than an architectural simulator because both the simulation time is very critical for a short design time and an instruction-set simulator can provide the designer with enough information to select suitable hardware parameters.

In "compile" and "simulate" step, the changes of the architecture made in "refine" step must be reflected in the compiler(MCC), the assembler(MASM), and the instruction-set simulator(MISS); that is, they have to be reconfigured according to the new architecture. As in the previous example if an instruction is removed from the instruction set(by excluding the functional block to be used for the execution of the instruction), the compiler should output a group of instructions instead of that instruction.

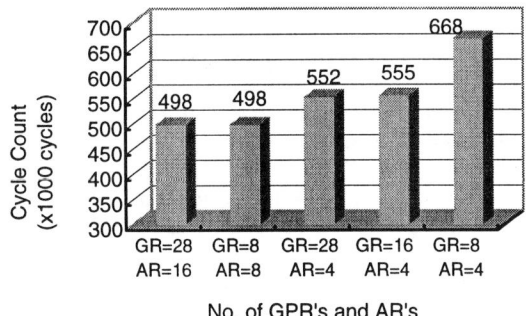

(a) Cycle Counts

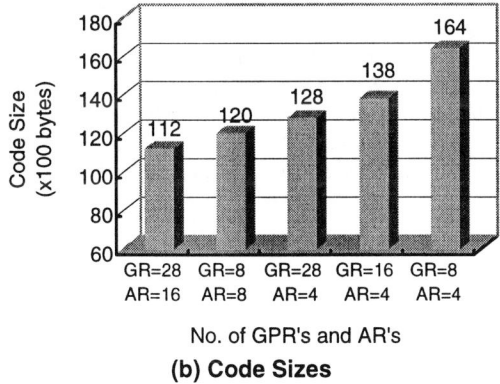

(b) Code Sizes

Figure 4. Cycle Count and code size of RPE-LTP block

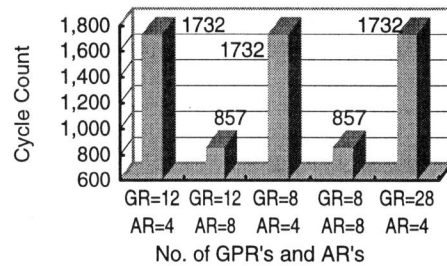

(a) Cycle Counts

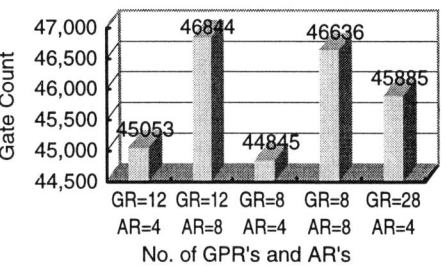

(b) Gate Counts

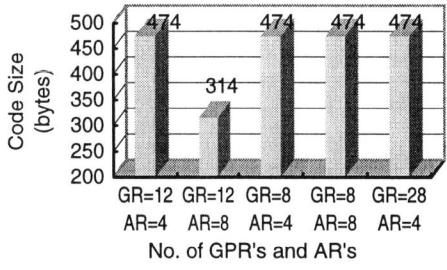

(c) Code Sizes

Figure 5. Cycle Counts and generated code size of the FIR block and the gate count of the processor

6. EXPERIMETAL RESULTS

In this section, we present the experimental results in which we determine the minimum number of registers that satisfies the code size and cycle count constraints by observing the code size and cycle time variation with the number of registers.

In the first experiment, we use the code of the RPE-LTP block of the GSM system in C language and repeat the "compile-simulate-refine" cycle. As can be seen in **Figure 4**, a designer can test various architectures with different number of GPRs(General Purpose Registers) and ARs(Address Registers) and can select an architecture considering the code size and cycle count. If the code size should be smaller than 12KBytes, 16 GPRs and 8 ARs may be enough.

FIR code is used for the second experiment. In this experiment, the gate count of the processor is also compared. The gate count is estimated by Architecture Simulator (this simulator is developed by another group in our project team). **Figure 5** shows that the number of GPRs has little effect on the code size and the cycle count but the number of ARs is very important to control the code size and the cycle count. If the gate count is not important and the cycle count should be as small as possible, 8 GPRs and 8 ARs can be used. However if the gate count is very important, 12 GPRs and 4 ARs will be more appropriate.

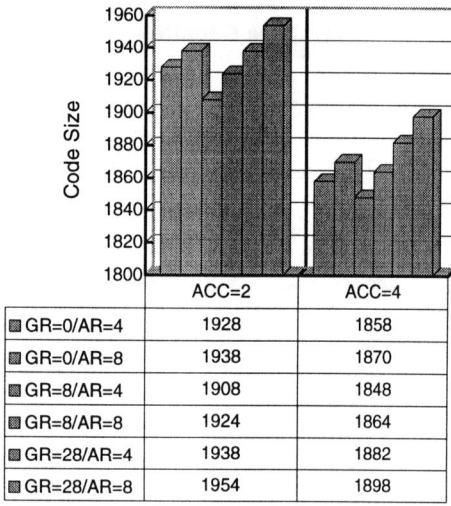

(a) Code Sizes

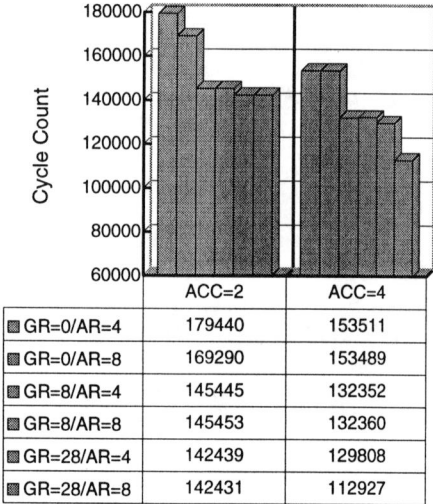

(b) Cycle Counts

Figure 6. Code sizes and cycle counts of ADPCM code with 2 accumulators and 4 accumulators

In the third experiment, we repeat the "compile-simulate-refine" cycle with 2 accumulators(ACC=2) and 4 accumulators(ACC=4). The result is shown in **Figure 6**. We know in this experiment that the code size and cycle count can be reduced significantly by varying the number of accumulators. If the current architecture has 28 general registers, 8 address registers and 2 accumulators, the cycle count reduction of 20% can achieved by increasing the number of accumulators to 4.

7. CONCLUSION

In this paper, we present a re-configurable system software for a flexible DSP core and show that the re-configurable system software can be used to select the most suitable architecture for a given application. Some design examples are presented, which illustrate how we select the architecture by determining the number of registers after the iteration of "compile-simulate-refine" cycle using the re-configurable system software.

Currently the overhead of the compiler-generated code is about 500% in code size and 200% in cycle count compared to the hand-optimized code. So the code optimization will be the major future work. To reduce the overhead, we are now developing an optimizing compiler.

8. REFERENCES

[1] P. Marwedel and G. Goossens(ed.), *Code Generation for Embedded Processors*, Kluwer Academic Publishers, 1995

[2] A. V. Aho, R. Sethi, J. D. Ullman, *Compilers -principles, techniques, and tools*, Addison-Wesley, Reading(MASS. U.S.A), 1996

[3] Richard M. Stallman, *Using and Porting GNU CC*, Free Software Foundation, 1993

[4] Christopher W. Fraser, David R. Hanson, *A Retargetable C Compiler: Design and Implementation*, The Benjamin/Cummings Publishing Company, Inc., 1995

[5] A. Fauth, A. Koll, "Automated generation of DSP program development tools using a machine description formalism," in *Proceedings of IEEE Int. Conf. Acoustics, Speech, Signal Processing.*, 1993, pages 457-460

[6] Mazen A. R Sahgir, Paul Chow, and Corrina G. Lee, "Application-driven design of DSP architecture and compilers," *Proceedings of IEEE Int. Conf. Acoustics, Speech, Signal Processing.*, 1994, pages. 437-440

Analyzing Effects of Cache Parameters on Memory Power Consumption of Video Applications

Bhanu Kapoor [1]

DSPS R&D Center
Texas Instruments Incorporated
P. O. Box 655303, MS 8344
8330, LBJ Freeway, Dallas, TX, 75243

ABSTRACT

Energy efficient computing is growing in demand as portable systems require energy efficiency in order to maximize the battery life. We provide data and insight into how the choice of cache parameters affects memory power consumption of video algorithms. We make use of memory traces generated as a result of running typical video algorithms to simulate a large number of cache configurations. The cache simulation data is then combined with on-chip and off-chip memory power models to compute memory power consumption. The configurations of particular interest are the ones that optimize power under certain constraints. We also study the role of process technology in these experiments. In particular, we look at how moving to a more advanced process technology for the on-chip cache affects optimal points of operation with respect to memory power consumption.

1. INTRODUCTION

A growing number of computer systems are incorporating multimedia capabilities for displaying and manipulating video data. At the same time, due to the remarkable success and growth of the portable consumer electronics, power consumption has become a critical constraint especially in the design of portable [1] systems. The interest in multimedia combined with the great popularity of portable devices provides the impetus for creating portable video-on-demand system. Even for the desktop units and large computing machines, the cost of removing the generated heat as well the reliability concerns are making power reduction a priority.

Spurred by the high computation and memory bandwidth requirements for popular signal processing applications such as digital wireless communications and multimedia processing, the need for high performance as well as low power processors has never been greater. In the arena of high performance digital signal processor and microprocessor design, a large number of on-chip transistors are being devoted to memory. For example, Digital's Alpha 21064 [2] has approximately 70% of its transistors devoted to the cache.

The growing inability of memory systems to keep up with processor requests has significant ramifications for the design of microprocessors in future. Much of the research so far has been focused on reducing memory access latencies [3]. Power consumption implications of CMOS microprocessor design decisions have been studied by Bunda, Athas, and Fussel [4]. They have studied the effects of instruction code densities, cache block buffering, and cache sub-blocks. They make use of bit-wise switching statistics to model power consumption and conclude that block buffering and sub-blocks are beneficial in reducing cache power consumption and off-chip memory traffic. Su and Despain [5] have studied the power-performance trade-offs in cache design. They have examined block buffering, banking, associativity, and gray code addressing. Their study shows that Gray coding

---

[1] Further author information: Email:kapoor@ti.com, Telephone:972-997-5295, Fax: 972-997-5598

and banking further reduce power consumption. They have also concluded that direct-mapped instruction and set-associative data caches result in better memory access times. Recently, some work on the integration of a microprocessor and DRAM memory on the same die, called Intelligent RAM (IRAM) [6], has been shown to have the potential for dramatic improvements in the energy consumption of the memory system. An IRAM will have far fewer external memory accesses, which consume a great deal of energy to drive high-capacitance off-chip buses.

A high-level analysis of energy optimization through the use of multiple-divided module (MDM) cache architecture was performed by Ko, Balsara, and Nanda [7]. Their model takes miss rates, power consumption of the cache and external memory, and the latencies of cache and external memory. It is concluded that MDMs reduce power consumption by a factor proportional to the number of modules.

Nachtergaele et al [8] describe a power exploration methodology for video applications using the low bit-rate H.263 video decoding algorithm. They show that memory consumes a large fraction of the system power consumption and describe a methodology for reducing power consumption by up to an order of magnitude. Liu and Svensson [9] have developed generalized models for memory power consumption in integrated circuits. These models were designed to model any VLSI system but do not consider factors such as miss rates, cache structure, on-chip as well as external bandwidth requirements.

In our study, we use a hierarchical motion estimation algorithm typically found in the implementations of MPEG-2 video compression [10] standard. The cache simulation data is then combined with on-chip and off-chip memory power models to compute memory power consumption. We provide a detailed study of how varying cache size, block size, and associativity affects memory power consumption. We also study the role of process technology in these experiments. In particular, we look at how moving to a more advanced process technology for the on-chip cache affects optimal points of operation with respect to memory power consumption.

The motion estimation algorithm used in the experiments described in this paper is a hierarchical [10] algorithm which uses four levels of sub-sampled images to come up with a motion vector for a given block. The algorithm is designed to work with the standard frame sizes used in MPEG-2 video codecs.

Section 2 describes the simulation methodology and the modeling aspects for our study. In Section 3, we discuss the experimental results and provide some guidelines for power-efficient memory system design. This is followed by conclusions and some suggestions for future work in Section 4.

2. SIMULATION METHODOLOGY

The study of cache behavior of video algorithms is a daunting task in itself. First of all, just encoding a few frames of MPEG-2 video sequences generate a huge amount of memory trace. While approximate miss rates can be found using much smaller traces, our study found that Gbytes of trace data is necessary to accurately determine on-chip and off-chip memory bandwidth requirements which are essential for computing memory power consumption.

The traces containing a sequence of memory references were generated using the Quick Profiler and Tracer (QPT) [11] program, an exact and efficient program profiling and tracing system. The cache simulation using the trace generated by QPT was carried out using a trace-driven cache simulator called DineroIII [11] which supports sub-block placement. We have generated the trace data for the hierarchical motion estimation algorithm compiled on a Sun Ultrasparc 1 workstation.

The on-chip memory bandwidth calculations use the total number of demand fetches to cache, including the instruction and data fetches. The read and write bandwidths for the data cache and the bandwidth requirement on the instruction cache is then combined with the SRAM read and write power consumption data to compute the on-chip memory power consumption. The bandwidth requirements along with the read and write widths of the SRAM determine the necessary clock rate of operation, which is then used for computing memory power consumption.

For the on-chip power modeling, we make use of spice-simulated data for caches designed in 0.25- and 0.18-micron technologies, operating at up to

750 MHz. This clock-rate is only sufficient to support smaller frame sizes. For CIF and larger frame sizes, the required clock rate to support the on-chip bandwidth requirements is higher than 750 MHz for the SRAMs used in our sudy. The on-chip power model is then extended to account for bandwidth requirements of an application, instruction and data cache miss rates, and bandwidth utilization factor for a given application. For off-chip power model, we make use of the data sheet numbers for the 64Mb Rambus DRAM [12] with the specified pin capacitance values for the memory controller and the DRAM.

3. SIMULATION RESULTS

The primary cache memory organizational characteristics that determine the external memory bandwidth (traffic) requirement are a cache's size, C, its associativity, A, and the block size B. In addition, the choice of cache replacement policy such as the least recently used, FIFO, and random as well as the choice unified cache versus a cache divided into instruction-only and data-only caches can also have an impact on the traffic.

Memory power consumption as a function of cache size attains a minimum value for a cache size depending on block size and associativity. The external memory bandwidth requirement decreases with the increasing cache size. The larger caches have lower miss-rates which results in less data movement between the cache and the external memory and this results in lower external power consumption. However, the power consumption of on-chip cache increases with the increasing size of the cache. The caches used in these experiments are direct-mapped and have separate instruction and data caches. The replacement policy for all the results presented in this paper is the least recently used (LRU). The memory power consumption is normalized with respect to the power consumption corresponding to the configuration using 4 Kbytes each of direct-mapped instruction and data cache with a block size of 8 Bytes. The data points for four block sizes, ranging from 8 Bytes to 32 Bytes, are shown in Figure 1. Each plot has a minimum which shifts to the right as the block size is increased. The on-chip power consumption numbers are for the cache designed 0.25-micron technology.

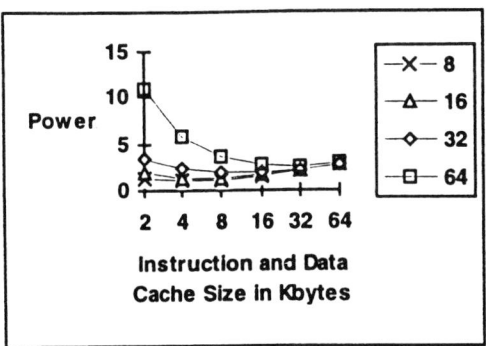

Figure 1: Memory power consumption versus cache size for various block sizes.

The system memory power consumption as a function of associativity typically decreases with the increasing value of associativity. The plots for four cache sizes, ranging from 2Kbytes to 32 Kbytes, are shown in Figure 2. For smaller caches, there is a big reduction in power consumption as we go from a direct-mapped cache to a two-way set associative cache. This is mainly due to the improved hit rates as a result of increasing associativity, which reduces the number of off-chip accesses to the external memory.

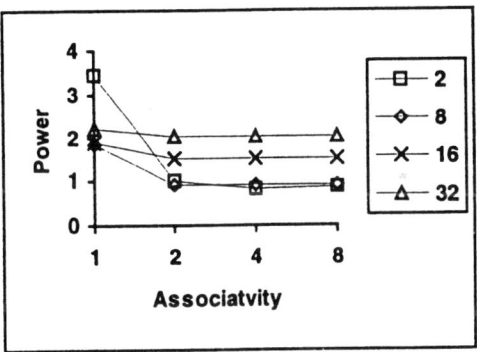

Figure 2: Memory power consumption versus associativity for various cache sizes.

The memory power consumption reduces, as expected, as we move to a more advanced technology with smaller device sizes. The plots for the 0.25- and 0.18-micron technologies are shown in Figure 3. Each plot has a point of minimum power consumption and this point shifts to the right as we move to a more advanced technology. There is an optimal power utilization point at a larger cache size in more advanced technologies.

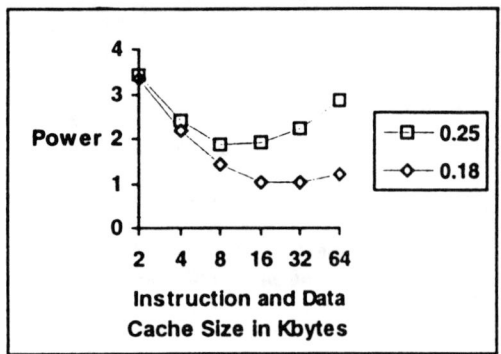

Figure 3: Memory power consumption versus cache size for two technologies.

In this case, the least power-consuming configuration is a cache which is 4X larger in the more advanced technology. This is a good news from system power consumption point of view as it is supporting the natural evolution of processor design with larger on-chip memories.

4. CONCLUSIONS

We have provided a study of how varying cache size, block size, and associativity affects memory power consumption. As it is clear from the experimental results, there is a point of diminishing return with respect to all the cache parameters in achieving low power memory architectures. Multi-level memory hierarchies may present a way out of this problem and are being used these days. It is evident from our experiments that power consumption can be reduced significantly by appropriately configuring the memory architecture of the system.

5. REFERENCES

[1] Anantha Chandrakasan, "Low Power Digital CMOS Design", PhD Thesis, University of California, Berkeley, 1994.

[2] D. Doberpuhl et al, "A 200 MHz 64b Dual-issue CMOS Microprocessor", *IEEE Journal of Solid-State Circuits*, vol. 27, no. 11, Nov., 1992.

[3] Mark D. Hill and Alan Jay Smith, "Experimental Evaluation of On-chip Microprocessor Cache Memories", *Proc. of Eleventh International Symposium on Computer Architecture*, June 1984, Ann Arbor, MI, pp. 158-174.

[4] J. Bunda, W. Athas, and D. Fussel, "Evaluating Power Implications of CMOS Microprocessor Design Decisions", *Proceedings of the 1994 International Workshop on Low Power Design*, April 1994, pp. 147-152.

[5] C. Su and A. Despain, "Cache Design Trade-offs for Power and Peformance Optimization: A Case Study", *Proceedings of the 1995 International Symposium on Low Power Design*, April 1995, pp. 63-68.

[6] Richard Fromm, Styllianos Perissakis, Neal Cardwell, Christoforos Kozyrakis, Bruce McGaughy, David Patterson, Tom Anderson, and Katherine Yelick, "The Energy Efficiency of IRAM Architectures", *Proc. of 24th Annual International Symposium on Computer Architecture*, June 1997.

[7] U. Ko, P. Balsara, and A. Nanda, "Energy Optimization of Multi-level Processor Cache Architecture", *Proceedings of the 1995 International Symposium on Low Power Design*, April 1995, pp. 45-49.

[8] L. Nachtergaele, F. Catthoor, B. Kapoor, D. Moolenaar, and S. Janssens, "Low-power Storage Exploration for H.263 Video Decoder System", *1996 IEEE Workshop on VLSI Signal Processing*, pp. 115-126, Nov., 1996.

[9] D. Liu and C. Svensson, "Power Consumption Estimation in CMOS VLSI Chips", *IEEE Journal of Solid-State Circuits*, pp. 663-670, Jun, 1994.

[10] V. Bhaskaran and K. Konstantinides, "Image and Video Compression Standards," pp. 105-128, Kluwer Academic Publishers, 1995.

[11] Mark D. Hill, James R. Larus, Alvin R. Lebeck, Madhusudhan Talluri, and David A. Wood, "Wisconsin Architectural Research Tool Set", Computer Architecture News, 21(4):8-10, August 1993.
http://www.cs.wisc.edu/ ~larus/warts.html

[12] Rambus Website: Products, http://www.rambus.com/html/products.html

TRANSFORMATIONAL-BASED SYNTHESIS OF VLSI BASED DSP SYSTEMS FOR LOW POWER USING A GENETIC ALGORITHM

M. S. Bright and T. Arslan

Circuits and Systems Research Group
Cardiff University Of Wales, Newport Road, Cardiff, UK, CF2 3TF
brightms1@cf.ac.uk

ABSTRACT

This paper describes a technique for the synthesis of CMOS based DSP systems under multiple design constraints. The primary target of the technique is to reduce operating power by applying high level transformations to designs. During the search for a low power solution the technique considers issues at circuit and layout levels, using appropriate capacitive models, together with tracking speed and area design constraints. In exploring the complex search space of the synthesis problem the technique uses a Genetic Algorithm which utilises a library of high level transformation based techniques within its operators. The paper describes the technique, the capacitive models used for power estimation and presents results for DSP systems of varying complexity. The results demonstrate the significant power savings achieved with the technique.

1. INTRODUCTION

The power consumption of VLSI systems has become an important design parameter alongside the traditional constraints of area and speed [1]. This is largely due to the increase in demand for portable computing systems with longer operating times. The operating time of these portable systems is constrained by both the battery capacity and system power requirements. Battery technology has peaked in recent years so attention has focused on the reduction of system power requirements, including the power consumption of the VLSI devices [1].

In addition to portable operation considerations, the increased power consumption of VLSI devices can lead to overheating, which reduces device speed and time to failure. This heat dissipation problem requires heat management systems that significantly add to device cost [2].

The issues described above have led to the development of a number of low power design techniques and methodologies that tackle various levels of the VLSI design process. The consideration of power as a high-level design parameter, to be optimised alongside speed and area, will have the greatest impact on power and will not require expensive modifications to the VLSI fabrication process [1].

A number of techniques have been developed for power reduction of digital CMOS based VLSI devices [3]. The authors in [4] proposed the use of high-level algorithmic transformations, traditionally used to optimise for speed and area, for power reduction. The transformations operate on a high-level description of the design, modifying elements within the design to produce lower power VLSI implementations.

However, the use of high-level transformations compounds the already complex nature of the low power design problem. Even with a restricted set of transformations no time efficient algorithm can be developed to determine the optimum low power solution [5]. An efficient algorithm is required to search the low power solution space while obeying constraints on design speed and area.

A Genetic Algorithm (GA) [6] is a heuristic search algorithm that has been applied to various VLSI design problems such as test pattern generation and bus size minimisation. Previous research has demonstrated the effectiveness of GAs when applied to the problem of structural synthesis [7], where there are multiple design constraints to satisfy. The authors have previously demonstrated a limited application of GAs to low power synthesis [8].

This paper presents a GA framework for the application of high-level algorithmic transformations to Digital Signal Processing (DSP) designs. The transformations modify the design characteristics to produce designs with lower power implementations. The GA is modified to suit the specific nature of the synthesis problem; unique genetic operators are developed to apply the high-level transformations. The developed GA is capable of reducing the power consumption of a wide range of signal processing circuits while obeying speed and area constraints.

2. PROCEDURE

The most significant factor affecting power consumption in a CMOS VLSI device is the switching power, which is expressed by the product [(supply voltage)^2 × switched capacitance] [9]. This equation identifies that reduction of supply voltage will yield a quadratic decrease in power. However, reduction of supply voltage decreases the speed of a CMOS device, as illustrated in figure 1 [4]. The application of high-level transformations is used to compensate for the speed decrease, producing a design with no loss in speed but a lower supply voltage.

High-level transformations have been well documented in the VLSI design literature [10-13]. They operate on a high-level description of the design represented as a Data Flow Graph (DFG). In order to be processed by a GA the DFGs are encoded into the chromosome representation shown in figure 2. A GA contains a number of chromosomes, collectively known as a population; each chromosome represents a possible solution to the design problem.

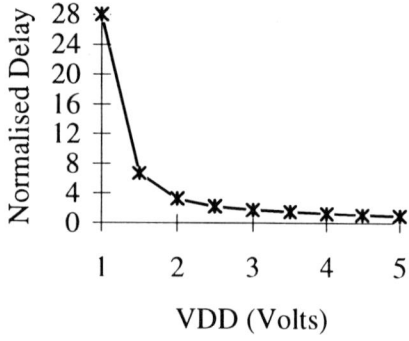

Figure 1. Relationship Of Supply Voltage And Delay [4]

The high-level transformations operate on the elements of the DFG (adders, delays, etc.) to modify its characteristics. The high-level transformations used within the GA synthesis system are; *Retiming* [10], which is the process of moving delays around the DFG to reduce the length of the critical path. The critical path is the longest computational path within the DFG so it places a bound on the maximum operating speed. A shorter critical path results in a faster design. *Pipelining* [11] attempts to minimise the critical path by inserting delay elements within the DFG. *Automatic Pipelining* [12] is a specialised form of retiming; delay elements are inserted on the inputs of the DFG and retimed through in an attempt to reduce the critical path length. *Loop Unfolding* [13] creates a parallel implementation of the DFG, which increases throughput at the cost of an increase in area.

The GA is used to apply the transformations to a population of candidate design chromosomes. The application rates were determined through a combination of design heuristics and experimental analyses. For example, both pipelining and retiming are powerful transformations for the reduction of critical path (increase in speed) but pipelining has the added overhead of increasing the number of delay elements within the system. Therefore retiming is applied at a greater rate than pipelining.

Genetic evolution proceeds with the creation of an initial population of candidate designs. The GA synthesis system creates a population of designs from an initial DFG specification. Random application of transformations is used to create a population of DFGs with different characteristics but the same function as the specified design. The power consumption of each design is assessed to calculate the fitness of each chromosome; lower power consumption corresponds to a higher fitness. Power estimation is a complex process at the high-level. A number of different high-level techniques have been reported in the literature [14-17].

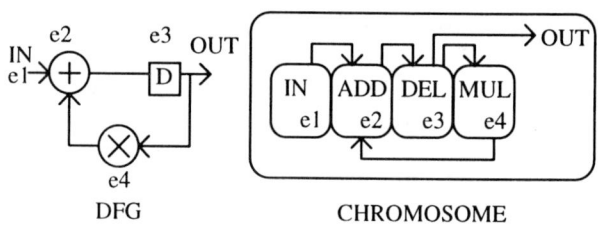

Figure 2. Data Flow Graph (DFG) Representation

The GA uses a capacitive model that combines data from practical VLSI designs with statistically derived relationships. A "good" model should be able to convey critical design information to the GA. In VLSI synthesis this usually requires circuit layout information for the optimum design synthesis. To provide this information to the GA a number of capacitive models were considered. The most simple model assigned unit size and delay to functional elements. This option requires very simple fitness calculations but is very inaccurate as it ignores the substantial size difference between multipliers, registers, adders, etc. The second model characterises each functional element with a gate count, providing imprecise comparative areas. This model ignores the effect of intra-connect capacitance within an element. The third model uses functional elements constructed in VLSI design CAD tools. The CAD tools are used to extrapolate accurate area and capacitance information for each element.

In addition to functional element capacitance, information on interconnect capacitance between elements is provided by a statistical model. This capacitive data is used to compute the capacitance contribution to power consumption. This third model was chosen as producing the best accuracy and speed of calculation trade-off.

The graph of figure 1 is used to estimate a supply voltage that will enable the design to run at the same speed as the initial design. For example, a design with a critical path half that of the original design will run at twice the speed. The graph is used to determine that a supply voltage of

2.9V will double the delays in the device, negating the speed increase, producing a design of the same speed but lower supply voltage than the original. Equation (1) is then used to compute the power consumption of the design.

The GA selects members of the current population for modification using the standard Roulette Wheel selection method [6]. Designs with a greater fitness (i.e. those best satisfying the design constraints) have a greater probability of being selected for reproduction; the GA attempts to find the optimum design by using the higher fitness members of the current population to create the next population.

Standard GAs use mutation and crossover operators to modify the characteristics of chromosomes [6]. In the case of DSP synthesis these standard genetic operators would corrupt the functionality of the DFG, therefore they are modified to suit the nature of the synthesis problem. The *mutation* process is modified to apply the high-level transformations, accessed from a transformation library. *Crossover* is a complex genetic operator that combines the characteristics of two parent chromosomes to produce child chromosomes. The GA uses a modified form of crossover that identifies which transformations have been applied to each parent, then produces child chromosomes with both sets of transformations applied.

The repeated application of the genetic operators to the current population breeds a new population of designs, which then becomes the current population. This process of genetic evolution is repeated until the design with the required specifications is produced. The design with the highest fitness over all generations is selected as the *best low power design* for the initial specified DFG.

3. RESULTS

A number of DSP designs, comprising a set of benchmarks, are used to illustrate the performance of the GA synthesis tool. The designs were selected to cover a range of recursive and non-recursive signal processing operations of varying complexity. FIR3 and FIR8 are non-recursive Finite Impulse Response filters, 3rd order and 8th order respectively. LAT is a 2nd order recursive filter. AV6D is a direct form representation of a 6th order Avenhaus filter. AV8P is an 8th order parallel implementation of the Avenhaus filter [5]. The Avenhaus filters contain complex recursive structures. ELLIP is the 5th Order Elliptic Wavelet filter presented in [18].

The results for each of the benchmark circuits are presented in graphical form in figure 3. As an example the GA has produced an FIR3 design with an 8 times increase in area, but the overall power consumption has been reduced by a factor of 19. The supply voltage for each optimised design, to produce the reported power reduction, is also presented.

The FIR filters have a significant estimated power reduction but with an associated increase in area. This increase is due to the amount of unfolding applied to these filters, producing parallel designs that are capable of operating at very low voltages with the same speed as non-parallel designs.

The Avenhaus filters have a very small increase in area as the unfolding transformation could not improve on the effect of the other transformations. The large power reductions obtained are possible because of the large critical paths of these filters, which offers plenty of scope for minimisation through retiming and pipelining.

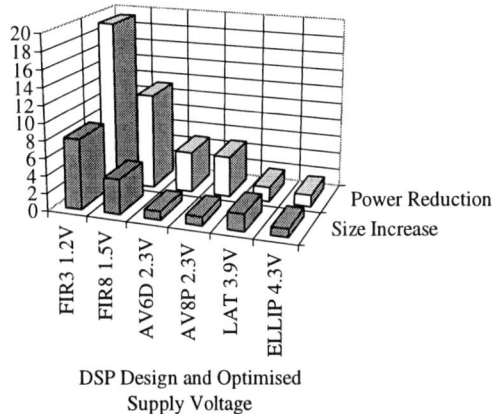

Figure 3 Power Reduction And Area Increase For Benchmark Designs

The relatively smaller reduction of power obtained with the elliptical filter is due to the fact that pipelining and retiming have limited effect on reducing the length of its feedback paths.

The GA produces a number of designs with the same power specifications, giving greater flexibility to the designer. Typically a population will converge on an optimum design within 500 generations. Subsequent generations will increase the number of designs that meet the design criteria, 3-8% of the final population will consist of unique *best low power designs*.

The population size of the GA was set to 500 for all designs to enable a comparison between the speed of each synthesis operation. The synthesis tool was run on a Pentium Pro Windows NT workstation with 64 Megabytes of RAM; the synthesis tool runs in 16 Megabytes of RAM. The execution time for each DSP algorithm to be synthesised is presented in table 1.

The relatively longer running time of the FIR8 filter is due to its complexity (22 elements) and the application of unfolding using the postponing principle [13], which postpones the application of the unfolding transformation until the other transformations are incapable of increasing the quality of the design any further. If unfolding is successful the unfolded design is fed back into the synthesis cycle for further optimisation, resulting in longer execution times for synthesising unfolded designs.

DSP Design	Execution Time (Seconds)
FIR3	83.32
FIR8	260.06
AV6D	40.51
AV8P	129.58
LAT	28.24
ELLIP	3.24

Table 1. Synthesis Tool Execution Times

4. CONCLUSION

A technique has been developed for the synthesis of low power DSP systems. At the core of the system is a Genetic Algorithm which accesses a library of high level transformations to modify designs, and a capacitive model which feeds device level information to the GA for power estimation. The technique shows flexibility with a wide range of DSP systems of varying complexity. Results have been provided for a number of benchmark DSP algorithms which show significant power reduction obtained in all cases.

5. REFERENCES

[1] D. Singh, J. M. Rabaey, M. Pedram, F. Catthoor, S. Rajgopal, N. Sehgal, and T. J. Mozdzen, "Power conscious CAD tools and methodologies : A perspective", *Proc. IEEE*, vol. 83, pp. 570-594, Apr. 1995.

[2] A. Raghunathan and N. K. Jha, "An ILP formulation for low power based on minimizing switched capacitance during data path allocation", *Proc. IEEE Int. Symp. Circuits And Systems '95*, 1995, vol. 2, pp. 1069-1073.

[3] A. P. Chandrakasan and R. W. Broderson, "Minimizing power consumption in digital CMOS circuits", *Proc. IEEE*, vol. 83, pp. 498-523, Apr. 1995.

[4] A. P. Chandrakasan, M. Potkonjak, R. Mehra, and R. W. Broderson, "Optimizing power using transformations", *IEEE Trans. CAD of Integrated Circuits and Systems*, vol. 14, pp. 12-31, Jan. 1995.

[5] A. P. Chandrakasan, M. Potkonjak, J. Rabaey, and R. W. Broderson, "HYPER-LP: A system for power minimization using architectural transformations", *Proc. IEEE/ACM Int. Conf. Computer Aided Design '92*, 1992.

[6] D. E. Goldberg, *Genetic Algorithms In Search, Optimization and Machine Learning*. Addison-Wesley, Reading, MA, 1988.

[7] T. Arslan, D. H. Horrocks, and E. Ozdemir, "Structural cell-based VLSI circuit design using a genetic algorithm", *Proc. IEEE Int. Symp. Circuits And Systems '96*, 1996, vol. 4, pp. 308-311.

[8] M. S. Bright and T. Arslan, "A genetic framework for the high-level optimisation of low power VLSI DSP systems", *IEE Electronics Letters*, vol. 32, pp. 1150-1151, June 1996.

[9] T. Arslan, D. H. Horrocks, and E. T. Erdogan, "Overview and design directions for low-power circuits and architectures for digital signal processing", *IEE Colloquium (Digest)*, No.122, 1995, pp 6/1-6/5.

[10] K. K. Parhi, "High-level algorithm and architecture transformations for DSP synthesis", *IEEE Journal Of VLSI Signal Processing*, vol. 9, pp. 121-143, Jan. 1995.

[11] M. Potkonjak, J. Rabaey, "Retiming for scheduling", *VLSI, Signal Processing IV*, H. S. Moscovitz, K. Yao, and R. Jain, (Ed.)., IEEE Press, New Jersey, 1991, pp. 23-32.

[12] K. K. Parhi, "Static rate-optimal scheduling of iterative data-flow programs via optimum unfolding", *IEEE Trans. Computers*, vol. 40, pp. 178-195, Feb. 1991.

[13] S. Huang and J. M. Rabaey, "Maximizing the throughput of high performance DSP applications using behavioural transformations", *Proc. European Design and Test Conference '94*, 1994, pp. 25-30.

[14] A. Raghunathan and N. K. Jha, "Behavioural synthesis for low power", *Proc. IEEE/ACM Int. Conf. Computer Aided Design '94*, 1994, pp. 318-322.

[15] R, Mehra and J. M. Rabaey, "Behavioural level power estimation and exploration", *1994 International Workshop On Low Power Design*, California, USA, 1994.

[16] P. E. Landman and J. M. Rabaey, "Power estimation for high level synthesis", *Proc. EDAC-EUROASIC '93, Paris, France*, 1993, pp. 361-366.

[17] P. M. Chau and S. R. Powell, "Power dissipation of VLSI array processing systems", in *Journal of VLSI Signal Processing*, vol. 4, pp. 199-212, 1992.

[18] S. Y. Kung, H. J. Whitehouse and T. Kailath, *VLSI And Modern Signal Processing*, Prentice-Hall, New Jersey, 1985

POWER ESTIMATION USING INPUT/OUTPUT TRANSITION ANAYLSIS(IOTA)

Junsoo Lee   *Bapiraju Vinnakota*   *Lori Lucke*

Department of Electrical and Computer Engineering
University of Minnesota
Minneapolis, Minnesota 55455, USA

ABSTRACT - Optimizing power dissipation is now a major concern in IC design. Accurate power estimation at the circuit level is too expensive. Tools which estimate power dissipation at higher levels of abstraction without sacrificing accuracy are of interest. We discuss a new cell-library based technique for power estimation. An IO transition model is developed to represent energy consumption in library elements. Power estimation based on the IO model requires less computational effort than with previous models. The IO model also addresses a accuracy limitation of previous models. A power estimation tool IOTA, based on this model, has been implemented in C and Matlab. For reasonably large combinational and sequential arithmetic circuits, IOTA has an average error of 5% and low simulation time compared to SPICE simulations.

I INTRODUCTION

Power estimation methods can be divided into three major approaches [5]: simulation based methods [4, 8], probabilistic methods [1, 6, 7], and statistical approaches [9]. In this paper, we present a new cell-library method for simulation-based power estimation. We model power dissipation in library elements using Input/Output Transition Analysis (IOTA). Power dissipation in large circuits is estimated by simulating them at the logic level, or at higher levels of abstraction. Information from the simulation is combined with library models to estimate power dissipation for the circuit. We will demonstrate that IOTA is both very fast and reasonably accurate. This estimation technique can be applied to both combinational and sequential circuits. We will compare IOTA with previous approaches to simulation-based power estimation. The analysis will show that our modeling technique addresses a significant accuracy limitation in previous modeling techniques, Without substantially increasing the information to be stored. We also demonstrate that it is possible to quantify the accuracy of IOTA, with respect to a circuit level simulator such as SPICE. At the circuit level, it is known that glitches can cause power dissipation to increase significantly. We develop inexpensive techniques to estimate the impact of glitches on power dissipation at the circuit level. At the circuit level, we demonstrate that IOTA is capable of estimating power reasonably accurately even for large circuits, but far more quickly than SPICE.

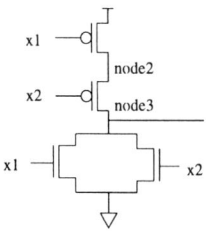

Figure 1: NOR gate

II INPUT OUTPUT TRANSITION ANALYSIS

Typically library elements are represented by their inputs and outputs. Lin [2] suggested a method to model internal nodes so as to account for power consumption at those nodes for small gates such as 2-input NAND and NOR gates. Janardhan [3] improved this method so that it could be used to estimate power dissipation in larger circuits. This method can be explained briefly as follows. In a state transition graph, nodes model the status of internal circuit nodes. State update equations are used to form the edges in the state transition graph(STG). For example, a NOR gate and its corresponding STG are shown in Figures 1 and 2. In Figure 2, the state variables are node2 and node3. The state update equations for these nodes are as follows:

$$node_2(n+1) = (1 - x_1(n)) + x_1(n) * x_2(n) * node_2(n) \quad (1)$$

$$node_3(n+1) = (1 - x_1(n)) * (1 - x_2(n)) \quad (2)$$

Since energy consumption occurs on transitions, each edge in the STG is associated with a value that represents its energy consumption. The models assume that the energy consumed on a transition is dependent on the inputs exciting and the state sourcing the transition.

A Complete FSM Model Consider the state transition graph model for a library element. An edge (excited by input) I_t in the state transition graph is sourced from state S. In addition to the values of I_t and S, the energy consumed by the edge I_t also depends on the input I_{t-1} used to enter state S. This dependence on the previous input is

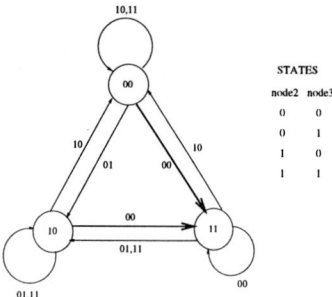

Figure 2: State transition diagram of NOR gate

neglected in [2, 3]. Consider the edge excited from state 10, on an input of 10 in Figure 2. If the circuit arrives in state 10 with an input of 01, the energy consumption on this transition is 3.1×10^{-12} Joules. If the circuit arrives in state 10 (from state 11) with an input 11, the energy consumption is 3.4×10^{-14} Joules, *almost two orders of magnitude less*. Previous models neglect this difference. For an edge I_t, sourced from state S, let Z be the set of inputs on the transitions whose destination state is S. Instead of a single energy, *in a complete STG model* a set of Z energy consumptions, one for each input that can be used to transition into state S, will have to be associated with edge I_t. By representing all possible energy consumptions on each transition, a *complete STG model* can potentially increase the accuracy of estimation. Each entry in a complete STG model is a triplet {Previous Input,Source State,Current Input}. For example, the triplet {01,10,10} refers to the energy consumed on a transition from state 10, with an input of 10, and a previous input of 01. For an element with m inputs and n nodes, the size of a complete FSM model is 2^{2m+n}. Such a model is too cumbersome.

Cell Library Power Estimation In this section we develop a new technique to model energy consumption by tracking vector transitions at element inputs. In many cases, in STGs for cell library elements, a single input is often a *synchronizing sequence*. That is, the input sets the values of all internal nodes, independent of their previous values. For example, in the STG for the NOR gate, shown in Figure 2, three single inputs 00, 01, 10 are synchronizing sequences. Though it is not a synchronizing sequence, the input 11, synchronizes the value of one of the internal nodes. In general, for many STG models of library elements, a single input usually synchronizes several internal nodes. If an input I_{t-1} is a synchronizing sequence, then the internal state need not be tracked to determine the energy consumption on a transition excited by any subsequent vector I_t. We model a library element by storing the energy consumption for every possible pair of input transitions I_{t-1}, I_t, where I_{t-1} and I_t are input vectors, into an Input/Output Transition energy Table(IOTT). This table lists all possible input/output transitions and energy consumption data corresponding to each transition. For comparison, a state transition table from the stochastic approach and IOTT(without energy consumption data) from our new approach are given in Tables 1 and 2 for the NOR gate. Clearly the size is similar. Not modeling internal nodes does not impact model quality. Consider the state and IO table representations of a NOR gate in Tables 1 and 2 respectively. Consider the individual vectors 00, 01, and 10. Each of these vectors is a synchronizing sequence. For example, the vector 00 sets the state to 11. The input transition pair $\{00,i_1i_2\}$ (where i_1i_2 is a vector at the NOR gate inputs) models the edge energy in the complete FSM model for the corresponding triplet $\{00,11,i_1i_2\}$. Though the state of the element is not tracked, all the energies on the transitions from the state 11 are represented in this model. For the states 00 and 10, all energies corresponding to triplets with previous inputs of 10 and 01 respectively, are represented in the IO model. The input 11 is not a synchronizing sequence. The destination state of an input of 11 may either be 00 or 10. Note that one of the two state elements is synchronized. Thus, a single input transition $\{11,i_1i_2\}$ represents two triplets $\{11,00,i_1i_2\}$ and $\{11,10,i_1i_2\}$ in the complete FSM model.

NOR STATE TRANSITION					
Pre_state		Inputs		Cur_state	
0	0	0	0	1	1
1	0	0	0	1	1
1	1	0	0	1	1
0	0	1	0	0	0
1	0	1	0	0	0
1	1	1	0	0	0
0	0	0	1	1	0
1	0	0	1	1	0
1	1	0	1	1	0
0	0	1	1	0	0
1	0	1	1	1	0
1	1	1	1	1	0

Table 1: State Transition Table

NOR INPUT/OUTPUT TRANSITION					
Prev_Inputs		Curr_inputs		Prev_out	Cur_out
0	0	0	0	1	1
0	0	0	1	1	0
0	0	1	0	1	0
0	0	1	1	1	0
0	1	0	0	0	1
0	1	0	1	0	0
0	1	1	0	0	0
0	1	1	1	0	0
1	0	0	0	0	1
1	0	0	1	0	0
1	0	1	0	0	0
1	0	1	1	0	0
1	1	0	0	0	1
1	1	0	1	0	0
1	1	1	0	0	0
1	1	1	1	0	0

Table 2: Input/Output Transition Table

Computing Transition Energy Consider an input transition vector pair I_1,I_2. Let I_1 be a synchronizing sequence. Since all internal nodes in the library element are synchronized, the energy associated with this transition is unique. This energy can be computed using SPICE.

If I_1 is not a synchronizing sequence, one may easily *derive bounds for energy consumption* for a specific pair. The state of unsynchronized internal nodes depends on the vectors applied prior to I_1. Let the set of nodes not set by I_1 be N. Let the initial value of a node $n \epsilon N$ be $v \epsilon B$. The value of v will depend on the inputs prior to I_1. When I_2 is applied after I_1, n may continue to float, or it may be set by I_2. If n is not synchronized for both I_1 and I_2 it does not consume any energy on the transition. If n is synchronized to value v by I_2 it will not consume energy. If n is synchronized to a value $\bar{v} = 1$ by I_2, then it will consume energy on the transition. The energy consumed is not unique, but depends on the inputs applied prior to I_1. In other words, a range of energies is actually associated with the transition. Let the set of nodes in N synchronized to 1 by I_2 be N_1. If every node $n \epsilon N_1$ has an initial value of 0, then all of them consume energy, and the *maximum energy consumption* occurs on the transition. If every node $n \epsilon N_1$ has an initial value of 1, then none of them consume energy, and the *minimum energy consumption* occurs on the transition. Both of these can be computed using SPICE. In general the cardinality of N_1 is very small. Consequently, the range of energies consumed is small. For three elements, the NAND, NOR, and full adder we constructed two tables, one with maximum energies for each transition, the second with minimum energies. For one set of vectors, we computed energy consumption using each of the tables and compared the results. The difference between the maximum and minimum estimates is 4.02% for the full adder, 1.82% for the NOR gate and 1.06% for the NAND gate. In our method, we used average of the minimum and maximum energies. This does not lead to significant errors.

Circuit Level Power Estimation Once library elements have been characterized, given a sequence of N inputs, logic simulation can be used to estimate the energy consumed by the circuit. For a logic gate G_i with m inputs, if we know the energy consumption and the number of occurrences of each input transition pair, we can estimate the energy consumption of G_i as follows.

$$Energy(G_i) = \sum_j TE_j \times W_j \quad (3)$$

where TE_j = Number of occurrences of transition j
   from the IOTT
W_j = Energy consumption of jth transition
i = gate number
j = corresponding in/out transition from IOTT

Assume that a combinational network CN has M logic gates. Assume the N patterns in the input sequence are input with clock cycle time T_{cycle}. Then, the average power dissipation of CN is:

$$P_{avg}(CN) = \frac{\sum_{i=1}^{M} Energy(G_i)}{N \times T_{cycle}} \quad (4)$$

Our power estimation technique has been implemented in a tool IOTA(Input Output Transition Analysis). IOTA con-

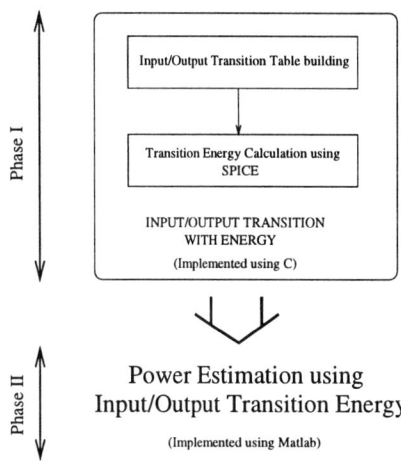

Figure 3: Power Estimation based on Input/Output Transition Analysis

sists of two phases, as shown in Fig. 3. At the first phase, an IOTT is built for each gate. Power estimation for the combinational/sequential circuit is done in the second phase using the IOTT built in the first phase.

Logic Simulation Logic simulation is used to identify the transitions that occur at the inputs to every component in a circuit. The logic simulation algorithm used has a significant impact on the accuracy and performance of the power estimation tool. The simplest method is to ignore circuit delays and use zero delay logic simulation. Though this is fast, this approach ignores glitches in the circuit that can consume significant amounts of energy. Consequently, *zero-delay simulation underestimates power dissipation*. A second method is to use transport delay models to accurately model the delays of circuit components. If the models are accurate, this method identifies all circuit hazards. The problems with this approach are two-fold. Not all hazards consume energy. Inertial delay models can be used to delete some spurious glitches. However, it is far more expensive than zero-delay simulation. *Consequently, transport delay simulation overestimates power dissipation even with inertial delay models*. Clearly, we only need to identify those hazards which may consume energy. To a first order, the energy consumed by a glitch depends on its duration. The accuracy with which glitches are identified depends on the quality of the delay model. A coarse model will only identify long duration glitches. A fine model will identify all glitches. To estimate power dissipation we are interested in the former. Thus, a coarse delay model, while not being very accurate, will provide sufficient detail to identify power consuming glitches. We use a simple *unit delay* model that provides sufficient accuracy while offering performance far superior to that of a transport delay model. The quality of this model was validated experimentally.

Circuit	SPICE mW	IOTA mW	C_S sec	C_I sec	C_H sec	Error
INV	0.08319	0.08327	24	< 0.1	N/A	0.09%
NOR	0.07247	0.07161	25.2	< 0.1	0.5	1.19%
NAND	0.08132	0.08133	34.9	< 0.1	0.5	0.01%
AND	0.1295	0.13063	45.5	< 0.1	N/A	0.81%
XOR	0.1241	0.12799	45.2	< 0.1	N/A	0.46%
MUX	0.02350	0.02357	39.3	< 0.1	N/A	0.31%
*F-A	0.2865	0.2812	360	< 0.1	2.5	1.85%
F-A	0.4362	0.4585	398	1	N/A	2.83%
D-FF	0.2120	0.2220	195	< 0.1	0.6	2.79%

C_S:CPU time of SPICE
C_I:CPU time of IOTA
C_H:CPU time of HEAT

Table 3: Basic Gate Power Estimation

Circuit	SPICE (mW)	IOTA0 (mW)	IOTAVI (mW)	IOTAU (mW)	Error
NOR_2	0.1440	0.1324	0.1399	0.1399	2.84%
NOR_3	0.2443	0.2354	0.2367	0.2367	3.11%
RCA_4	1.8350	1.4978	1.8583	1.8493	0.77%
RCA_8	4.2600	3.2847	4.0760	4.0560	4.79%
CSA_8	7.2370	5.6898	8.0333	7.8696	8.74%
$4Mult$	6.5790	5.1783	7.4355	6.9198	5.18%
$8Mult$	42.070	25.874	52.388	48.388	15.01%
$8Mult_P$	39.960	26.957	45.721	39.280	1.70%

Table 4: Power consumption for several circuits

III RESULTS AND DISCUSSION

We present results for IOTA in two sections. The first set presents results for cell library elements individually. The second set presents results for larger circuits. Table 3 summarizes the results of power estimation for library elements. All CPU times are seconds. It is clear that our method is significantly faster than SPICE and estimates power dissipation with less than 2% error on the average.

Table 4 summarizes simulation results for several large circuits under a variety of delay models. In the table, RCA refers to ripple carry adders, CSA refers to carry select adders, Mult refers to a multiplier, and $Mult_P$ refers to a pipelined multiplier. For the circuits used in Table 4, T_{clk} = 25ns and the number of input vectors was 100 for each circuit. As mentioned above, the columns titled IOTA0, IOTAVI, and IOTAU correspond to simulation with zero, variable-inertial and unit-inertial delay models respectively.

Circuit	CPU time SPICE(sec)	CPU time IOTA(sec)
NOR_2	54.1	< 0.1
NOR_3	74.4	< 0.1
RCA_4	618	4
RCA_8	2054	31
CSA_8	2721	56
$4Mult$	12973	52
$8Mult$	39856	314
$8Mult_P$	42827	209

Table 5: Simulation Time Comparison

In all cases, the zero delay model underestimated power consumption because it neglected glitches and the unit delay model consistently offers the estimate closest to that computed by SPICE. With this model, the average error for the six largest circuits in Table 4 is less than 5%. The simulation times shown in Table 5 are those for the most accurate estimator, unit-inertial delay simulation. In every instance, IOTA is at least two orders of magnitude faster than SPICE.

IV CONCLUSION

In this paper, we have developed a new method, input output transition analysis(IOTA) for simulation-based power estimation for digital circuits. In this approach, a cell is modeled by an input output transition table(IOTT) which lists all possible input transition pairs and energy consumptions. We discussed methods to form the IOTT model, methods to compute transition energies, and methods to utilize the model in circuit level power estimation. Previous methods modeled cells using state transition graphs. Though power consumption in internal nodes is represented, the models suffer from significant accuracy limitations. Though internal nodes are not modeled explicitly, we demonstrated that the IOTA does not ignore energy consumption in internal nodes, and that modeling inaccuracy is limited. For cell library elements, IOTA is able to estimate power dissipation with an error of 2%, far more quickly than SPICE. We also developed techniques to accurately estimate the impact of glitches on power dissipation in large circuits. Experimental results show the average error of our method is within 5% of SPICE simulations, while the computational time is two orders of magnitude faster than HSPICE.

References

[1] Farid N. Najm, "Transition density: a new measure of activity in digital circuits," *IEEE Trans. on CAD*, vol.12, pp. 310-323, Feb. 1993

[2] J. Lin, T. Liu, and W Shen, "A cell-based power estimation in CMOS combinational circuits," *Proc. International Conference on CAD* (1994), pp. 304-309

[3] J. Satyanarayana, K. Parhi, "HEAT: Hierarchical Energy Analysis Tool," *Proc. IEEE/ACM Design Automation Conference(DAC)*, June 1996, pp. 9-14

[4] T. Krodel, "PowerPlay - fast dynamic power estimation based on logic simulation," *IEEE International Conference on Computer Design*, pp. 96-100, Oct. 1991

[5] F. Najm, "A survey of power estimation techniques in VLSI circuits," *IEEE Trans. on Very Large Scale Integration*, vol.2, No.4, pp. 446-455, Dec. 1994

[6] F. Najm, "Low-pass filter for computing the transition density in digital circuits," *IEEE Trans. on Computer Aided Design*, vol.13, no.9, pp. 1123-1131, Sept. 1994

[7] A. Ghosh, S. Devadas, K. Keutzer, J. White, "Estimation of average switching activity in combination and sequential circuits," *ACM/IEEE 29th Design Automation Conference*, pp. 253-259, 1992

[8] C. Huzier, "Power dissipation analysis of CMOS VLSI circuits by means of switch-level simulation," *IEEE European Solid State Circuits Conference*, pp.61-64, 1990

[9] R. Burch, F. Najm, P. Yang, and T. Trick, "A Monte Carlo approach for power estimation," *IEEE Transactions on VLSI Systems*, vol.1, no.1, pp.63-71, Mar. 1993

FAST DELAY-DEPENDENT POWER ESTIMATION OF LARGE COMBINATIONAL CIRCUITS

Jer Min Jou[+], Shung-Chih Chen[*], and Chih-Liang Wang[+]

[+]: Department of Electrical Engineering
National Cheng Kung University
1 University Road
Tainan, Taiwan, R.O.C.

[*]: Department of Electrics Engineering
Nan-Tai Institute of Technology
1 Nan-Tai Street, Yung Kang City
Tainan County, Taiwan, R.O.C.

ABSTRACT

In this paper, we propose a fast and memory-efficient algorithm to estimate the glitch effects of the circuit under a general delay model without constructing global BDDs and without calculating Boolean difference. A new concept, where the circuit's signal activities with and without glitching effects are separately calculated by two newly developed calculation modules, is developed. The combined Markov chain and BAM method is used, and approximates the transient signals behavior with the steady state behavior, then the glitching effects as well as the temporal and spatial correlations among signals are all considered and processed efficiently. The analysis of our method indicates that it is applicable to large size circuits with acceptable errors.

1. INTRODUCTION

Due to the growing popularity of portable systems, power dissipation is rapidly becoming one of the major considering factors in the design of VLSI circuits. In CMOS circuits, a majority of power dissipation is due to charging and discharging of load capacitance. The average number of transitions per time unit is referred to as signal activity. Computing it fast and accurately is critical.

Two main approaches are now widely used for estimating signal activity: the simulation-based and the probability-based approaches [11]. The former can get accurate estimation, but very time-consuming. Extensions [11,16] of such techniques have been proposed which tradeoff accuracy for speed; however, the processing time is still long. To overcome the speed problem of this approach, the latter approach, with user-provided parameters in a probabilistic form that characterize the typical behavior at the primary inputs (PIs), propagates these parameters through the whole circuit and thereby fast obtains the average transition probabilities at internal nodes of the circuit. While this approach can provide very quick results, a number of problems about its accuracy is needed to pay attention to. The first issue is the modeling of the temporal and/or spatial correlations among signals[14]. A number of researchers have addressed this issue[3,5,7-9,13,15] based on zero delay model, which ignores the glitch effects on power consumption [1][4]. Though this important issue does not effect the operation of the circuit ([2], p.493), it has been shown that glitch consumes about 20-40% of the total power, even as high as 70% in some cases [11]. Computing the glitch power is one main challenge in power estimation and is the focus of the paper.

In this paper, we propose a new fast incremental probabilistic analysis and propagation technique to estimate the inertial-delay dependent switching activities in combinational circuits. As compared to other related approaches [5, 6, 10, 18], it can fast estimate the glitch effects under a general delay model without constructing global BDDs and without calculating Boolean difference for each possible transition time. Formerly unpredictable signal activities caused by gate and interconnection delays are now tractable without exhausting simulation runs.

2. TRANSITION RATIO FORMULATION

The average power dissipation in a CMOS circuit under the general delay model is given by:

$$P_{circuit} = \sum_{signal\ i} \frac{1}{2} C_i V_{dd}^2 fT(i) \quad (1)$$

where C_i is the sum of the input capacitances of all transistors driven by signal i, V_{dd} denotes the supply voltage, and $T(i)$ is the transition ratio, which needs to be estimated.

In the general delay model, there are two types of transitions at an output node: (i) *steady state transition* which is propagated to it from previous stages (Fig. 1(a)); (ii) *transient state transition* (*glitch*) which is generated due to different arrival time of its input signals (Fig. 1(b)). Since these two transitions are independent, they can be estimated separately, and because they constitute all the transitions, the transition ratio can be computed as $T(y) = A(y) + P_{glitch}(y)$, where $A(y)$ is the *propagation activity* of y and is defined as $A(y) = P(y(\tau) \oplus y(0))$, and $P_{glitch}(y)$ is the probability of the generation of glitch, called *glitch generation*, at y, and. Computating the $P_{glitch}(y)$ is the major work in this paper and will describe in next section; While for computing $A(y)$, the combined Markov chain [12-13] and BAM [15] methods are used. For example, the $A(Z)$ for an AND gate with two inputs i and j, which may have some correlations, is computed as follows;

$$A(Z) = 2 \cdot P(\overline{Z(\tau)}Z(0)) = 2 \cdot P((\overline{i(\tau)} + \overline{j(\tau)})i(0)j(0)) \quad (2)$$
$$= 2 \cdot \{P(\overline{i(\tau)}j(0)j(\tau)j(0)) + P(\overline{i(\tau)}i(0)\overline{j(\tau)}j(0)) + P(i(\tau)i(0)\overline{j(\tau)}j(0))\},$$

each term above can be computed directly by using the fundamental equations in BAM efficiently.

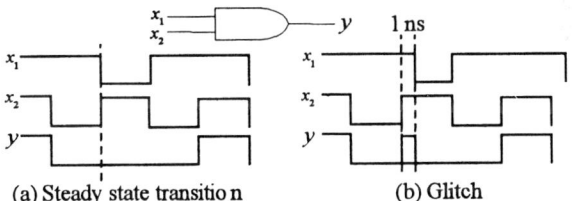

(a) Steady state transition (b) Glitch
Fig. 1. Transition types.

3. GENERAL DELAY ACTIVITY ESTIMATION

In this section, some symbols are first described, the formula for glitching probability is made, and then an efficient algorithm for calculating the transition ratio is presented.

If $y = f(x_1, x_2, ..., x_n)$ is a Boolean function (gate), the *k*-th *arrival time*, t_j^k, for x_j is the total delay value from some PI I_i along some viable path propagates to the input node x_j of the gate y, the *j*-th *arrival vecto* is defined as: $\alpha_y^j = (t_1^u, t_2^v, ..., t_n^w)$, $(j = 1, 2, ...)$; $V_y = \{\alpha_y^1, \alpha_y^2, ...\}$ as the set of all arrival vectors. For the example shown in Fig. 2, $V_{y_1} = \{(0,3)\}$ and $V_{y_3} = \{(1,1),(1,4),(4,1),(4,4)\}$.

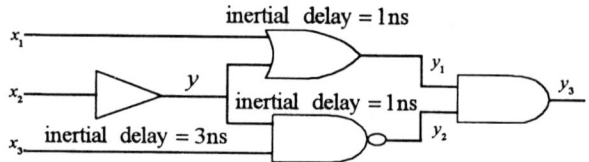

Fig. 2. An example of set of arrival vectors

The *transition mode of an input pattern* for gate y is defined as:

$$\omega_{y,j} = (x_1^{\alpha\beta}, x_2^{\alpha\beta}, ..., x_n^{\alpha\beta}) \quad (j = 1, 2, ...; \alpha, \beta = 0, 1), \quad (3)$$

where j is the index number of the input pattern, $x_i^{\alpha\beta}$ is the state transition property of input signal x_i switching from α to β. We further define $\Omega_y^j = \{\omega_{y,1}, \omega_{y,2}, ...\}$ and $P(\Omega_y^j)$, the *probability of the set of glitch-generating patterns* appearing at the inputs of y, for α_y^i with intertransition time $\zeta_y^j >$ inertial delay τ_y, is defined as: $P(\Omega_y^j) = \sum_{k=1}^{v_j} P(\omega_{y,k})$, (4)

where v_j is the number of glitch-generating patterns for α_y^i.

For the example shown in Fig. 3, $V_y = \{(t_1, t_2), (t_2, t_1)\}$, $\Omega_y^2 = \{(x_1^{10}, x_2^{01})\}$, and $P(\Omega_y^2) = \sum_{j=1}^{1} P(\omega_{y,j}) = P(x_1^{10} x_2^{01})$.

With the above definitions, we have the *glitching probability*, $P_{glitch}(y)$, is:

$$P_{glitch}(y) = 2 \cdot \sum_{i=1}^{u} \left(P(\alpha_y^i) P(\Omega_y^i) \right) = 2 \cdot \sum_{i=1}^{u} \left(P(\alpha_y^i) \sum_{j=1}^{v_i} P(\omega_{y,j}) \right), \quad (5)$$

where u is the total number of arrival vectors in V_y. The factor 2 is multiplied because one glitch causes two transitions at the output. In the case of Fig. 3, $P_{glitch}(y)$ is equal to

$$P_{glitch}(y) = 2 \cdot \sum_{i=1}^{2} \left(P(\alpha_y^i) P(\Omega_y^i) \right) = 2 \cdot \sum_{i=1}^{2} \left(P(\alpha_y^i) \sum_{j=1}^{1} P(\omega_{y,j}) \right)$$

$$= 2 \cdot \left(P(\alpha_y^1) P(x_1^{01} x_2^{10}) + P(\alpha_y^2) P(x_1^{10} x_2^{01}) \right) \cdot P(\alpha_y^j = (t_1^u, t_2^v, \cdots, t_n^w))$$

$$= P(t_1^u) P(t_2^v) \cdots P(t_n^w)$$

$$= \prod_{k=1}^{n} P(t_k^l) \quad (l \in \{u, v, ..., w\}), \quad (6)$$

based on the assumption that all of t_k^l are independent.

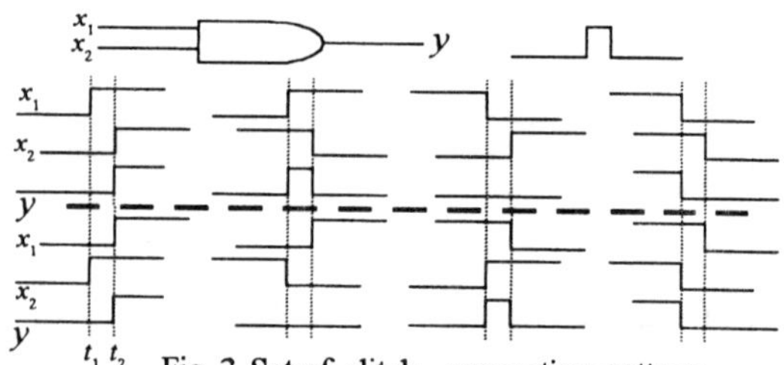

Fig. 3. Set of glitch-generating patterns for a 2-input AND gate.

To calculate the probability of each arrival time, we need two new definitions. The *path probability* of a signal path $p \equiv x_i \to \cdots \to y$, $P_{path}(x_i \to \cdots \to y)$, is the probability that the signal of PI x_i can propagate to y along p; The *path delay*

$D_{path}(x_i \to \cdots \to y)$ is the total delay along p. Based on these two definitions, we have $P(t_k^l) =$

$$\frac{\sum P_{path}(x_i \to \cdots \to y_k) \text{ whose } Delay_{path}(x_i \to \cdots \to y_k) = t_k^l}{\sum_{i=1}^{n} P_{path}(x_i \to \cdots \to y_k)}, \quad (7)$$

where n is the number of PIs. *Path delay* is easily obtained by summing all the gate inertial delays in the path. While for obtaining the path probability, we need to know the probability of each signal in the path whose value can be transferred to its output node, and then multiplying them. Therefore, we have the *transfer probability* $P_{tf}(x_i; y)$ which means the probability that the input signal x_i can propagate to its output node y.

We use the AND gate to illustrate how to obtain each $P_{tf}(x_i; y)$. Other gates can be obtained similarly. For a two-input AND gate, $y = x_1 x_2$, $P_{tf}(x_1; y) = P(x_2)$ and $P_{tf}(x_2; y) = P(x_1)$. As for the gate with more than 2 fanins, when considering the correlation among input signals, the n-input gate has been decomposed into the recursive representation, as shown in Fig. 4. The $P(x_{tmp})$, which is equal to $P(x_1)$, is calculated based on the combined Markov chain and BAM method. Table 1 lists the actual and approximated calculation for all the logic gates, in which the approximated calculation neglects the correlation between x_i and x_{tmp}.

If signals $x_1, x_2, ..., x_n$ are dependent, then we calculate $P_{prop}(x_i)$ by decomposition.

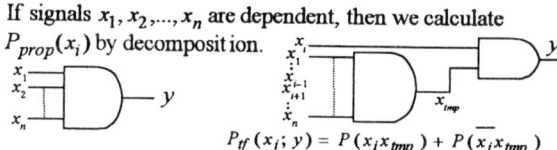

$P_{tf}(x_i; y) = P(x_i x_{tmp}) + P(\overline{x_i} x_{tmp})$

Fig. 4. Calculation of n-input AND gate transfer probability.

After knowing how to calculate the *path probability*, we then have $P_{path}(x_i \to x_{i+1} \to \cdots \to x_j) = \prod_{k=i}^{j} P_{prop}(x_k; x_{k+1})$

$= P_{path}(x_i \to x_{i+1} \to \cdots \to x_{j-1}) \cdot P_{tf}(x_{j-1}; x_j). \quad (8)$

Based on equation (8) and Table 1, we can derive the path probability incrementally.

Table 1 The calculation of logic gate transfer probabilities.

Gate	Transfer Probability	
	General	Approximated
BUFFER	1	1
NOT	1	1
AND	$P_{tf}(x_i; y) = P(x_i x_{tmp}) + P(\overline{x_i} x_{tmp})$	$P_{tf}(x_i; y) = P(x_{tmp})$
OR	$P_{tf}(x_i; y) = P(x_i \overline{x_{tmp}}) + P(\overline{x_i} \overline{x_{tmp}})$	$P_{tf}(x_i; y) = P(\overline{x_{tmp}})$
NAND	$P_{tf}(x_i; y) = P(x_i x_{tmp}) + P(\overline{x_i} x_{tmp})$	$P_{tf}(x_i; y) = P(x_{tmp})$
NOR	$P_{tf}(x_i; y) = P(x_i \overline{x_{tmp}}) + P(\overline{x_i} \overline{x_{tmp}})$	$P_{tf}(x_i; y) = P(\overline{x_{tmp}})$

For large size circuits, the number of signal paths for each gate grows exponentially with the number of levels in the circuit, and hence, the derivation of *path probability* for each gate will cost very much. Moreover, the number of arrival vectors of each gate is dependent on all the possible permutations of signal paths of the gate, the calculation time may take several times more.

Therefore, a heuristic, which limits the number of signal paths in the calculation, is introduced. During the level-by-level calculating process, only the paths with top Δ path probabilities are considered. The inaccurate effect caused by this limit may be ignored if Δ is large enough.

After knowing how to calculate $P(\alpha_y^i)$, we still need to calculate $P(\alpha_y^i)$ in order to obtain the glitching probability shown in (5). Suppose that $y = f(x_1, x_2, ..., x_n)$, and $\omega_{y,j} = x_1^{00} x_2^{00} \cdots x_i^{00} \cdot x_{i+1}^{01} \cdots x_j^{01} \cdot x_{k+1}^{11} \cdots x_n^{11}$, then $P(\omega_{y,j}) = P(x_1^{00} x_2^{00} \cdots x_i^{00} \cdot x_{i+1}^{01} \cdots x_j^{01} \cdot x_k^{10} \cdot x_{k+1}^{11} \cdots x_n^{11})$. The temporal and spatial correlations among the switching signals at the inputs of a gate depend on each gate delay, the circuit topology, and the signals applied to the PIs. It is expensive and sometimes even impossible to get the exact answer. We instead use the values of steady signals to approximate that of corresponding transient signals.

By using the signal probabilities and switching activity to approximate the probability, we have

$$P(\omega_{y,j}) = P(\overline{x_1(\tau)x_1(0)} \overline{x_2(\tau)x_2(0)} \cdots \overline{x_i(\tau)x_i(0)} \cdot x_{i+1}(\tau)\overline{x_{i+1}(0)} \cdots x_j(\tau)\overline{x_j(0)} \cdot \overline{x_{j+1}(\tau)}x_{j+1}(0) \cdots \overline{x_k(\tau)}x_k(0) \cdot x_{k+1}(\tau)x_{k+1}(0) \cdots x_n(\tau)x_n(0)). \quad (9)$$

When the input signals $x_1, x_2, ..., x_n$ are mutually independent, then (9) is equal to

$$P(\omega_{y,j}) = \left(P(\overline{x_1(\tau)x_1(0)})P(\overline{x_2(\tau)x_2(0)}) \cdots P(\overline{x_i(\tau)x_i(0)})\right)$$
$$\cdot \left(P(x_{i+1}(\tau)\overline{x_{i+1}(0)}) \cdots P(x_j(\tau)\overline{x_j(0)})\right)$$
$$\cdot \left(P(\overline{x_{j+1}(\tau)}x_{j+1}(0)) \cdots P(\overline{x_k(\tau)}x_k(0))\right)$$
$$\cdot \left(P(x_{k+1}(\tau)x_{k+1}(0)) \cdots P(x_n(\tau)x_n(0))\right)$$
$$= \left(\prod_{l=1}^{i}\left(P(\overline{x_l}) - \frac{1}{2}T(x_l)\right)\right) \cdot \left(\prod_{l=i+1}^{j}\frac{1}{2}T(x_l)\right) \cdot \left(\prod_{l=j+1}^{k}\frac{1}{2}T(x_l)\right) \cdot$$
$$\left(\prod_{l=k+1}^{n}\left(P(x_l) - \frac{1}{2}T(x_l)\right)\right). \quad (10)$$

If we neglect the signal correlations by using equations (5)-(7), and (10), the $P_{glitch}(y_3)$ is:

$$P_{glitch}(y_3) = 2 \cdot \sum_{i=1}^{2}\left(P(\alpha_y^i)\sum_{j=1}^{1}P(\omega_{y,j})\right)$$
$$= 2 \cdot \left(P(x_1^{01} x_2^{01} x_3^{10}) + P(x_1^{01} x_2^{10} x_3^{01}) + P(x_1^{01} x_2^{11} x_3^{11}) + P(x_1^{01} x_2^{11} x_3^{10}) + P(x_1^{11} x_2^{01} x_3^{10})\right) = 2 \cdot 0.25 \cdot \left(\frac{1}{2} \cdot T(y_1) \cdot \frac{1}{2} \cdot T(y_2)\right)$$
$$+ 2 \cdot 0.25 \cdot \left(\frac{1}{2} \cdot T(y_1) \cdot \frac{1}{2} \cdot T(y_2)\right) = 0.0625. \quad (11)$$

As for input signals with correlations, the calculation of $P(\omega_{y,j})$ becomes more complicated. Let us consider the example in Fig. 4 with n =10 and assumed that there is one glitch-generating pattern whose transition mode is: $\omega_{y,1} = x_1^{01} x_2^{01} x_3^{01} x_4^{01} x_5^{01} x_6^{01} x_7^{01} x_8^{01} x_9^{01} x_{10}^{10}$ for the arrival vector $\alpha_y^1 = (2, 4, 6, 8, 10, 12, 14, 16, 18, 20)$. By using equation (9), we get

$$P(x_1(\tau)\overline{x_1(0)} x_2(\tau)\overline{x_2(0)} x_3(\tau)\overline{x_3(0)} x_4(\tau)\overline{x_4(0)} x_5(\tau)\overline{x_5(0)}$$
$$x_6(\tau)\overline{x_6(0)} x_7(\tau)\overline{x_7(0)} x_8(\tau)\overline{x_8(0)} x_9(\tau)\overline{x_9(0)} x_{10}(\tau)\overline{x_{10}(0)}).$$

The calculation above is not feasible as there will have 4097 glitch-generating patterns for one arrival vector. To reduce the calculation complexity, a decomposition similar to Fig. 4 is done to approximate the glitch effects. Of course, some accuracy will be lost.

Let us consider the example shown in Fig. 5 again with signal correlations for the inputs of y_3:

$$P_{glitch}(y_3) = 2 \cdot \left(0.25 \cdot P(\overline{y_1(\tau)y_1(0)}y_2(\tau)\overline{y_2(0)}) + 0.25 \cdot P(y_1(\tau)\overline{y_1(0)}y_2(\tau)\overline{y_2(0)})\right)$$
$$= 2 \cdot (0.25 \cdot 0.10546875 + 0.25 \cdot 0.10546875) = 0.10546875.$$

Note that the error caused by neglecting the signals' correlations is about 40%. Now, the transition ratio for node y_3 of Fig. 5 is
$$T(y_3) = A(y_3) + P_{glitch}(y_3)$$
$$= 0.2265625 + 0.10546875 = 0.3323125,$$
which is near the result obtained by Verilog-XL simulation, which is $T(y_3) = 0.3125$.

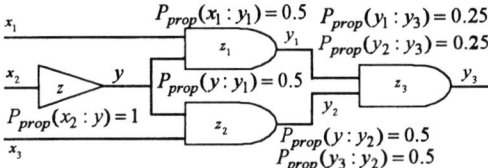

The inertial delays for z, z_1, z_2, z_3 are $2, 1, 1, 1$ ns.
Fig. 5. Example for path probability derivation.

4. EXPERIMENTAL RESULTS

To demonstrate the effectiveness, we made several experiments using the ISCAS-85 benchmark circuits [20] on SUN SPARC 10 workstation. Tables 2 and 3 show the CPU time and the average error of switching activity for the zero delay model, respectively. To obtain the reference switching activities, the circuits were simulated Cadence Verilog-XL [17] with randomly generated test vectors according to the specified signal probabilities and switching activities at PIs, which are all 0.5. For the general delay model, circuits were simulated with extra gate inertial delay. The stop criterion for simulation is the absolute value of (currentvalue-lastvalue)/currentvalue of each node's transition ratio converged to 0.0001. From the tables, we find that the three methods Schneider [13], Chou [3], and ours can maintain acceptable accuracy; but ours has the shortest estimation time.

Table 2. CPU time of switching activity estimation.

Ckt	CPU Time (sec.) for zero delay model					
	Simu.*	Xakellis**	NSS[3]	Schneider	Chou	Ours
c880	306.6	729	3	34	3	1.18
c1355	362.6	609	34	11	22	1.10
c1908	666.1	1978	26	8	27	1.25
c2670	1876.5	2911	22	N/A	18	11.72
c3540	2465.9	4579	454	65	498	3.68
c5315	4535.4	7314	92	N/A	68	20.37
c6288	6356.0	3448	2357	N/A	590	6.40
c7552	8250.9	8855	237	N/A	147	32.53

* Ran on SUN ULTRA 1 workstation.
** Ran on SUN SPARC-ELC with $\eta_{min} = 0.05$.

Table 3 Average error of switching activity estimation.

Circuit	Average error (%) for zero delay model			
	NSS	Schneider	Chou	Ours
c880	85	0.6	0.1	-1.04
c1355	>300	0.5	0.2	-2.17
c1908	221	1.0	-0.14	-0.24
c2670	98	N/A	0.8	-0.61
c3540	180	1.4	-1.4	1.49
c5315	93	N/A	-0.49	-0.32
c6288	>300	N/A	6.7	-15.54
c7552	116	N/A	-4.05	-0.79

Table 4 shows the results of CPU time and average errors for the general delay model, which reveals that introducing the general delay model only causes little CPU overhead, while still obtains acceptable average error rate. Therefore, our approach is applicable to large size circuits.

The effects caused by the value of path limit Δ have also experimented and the results show that the inaccuracy caused by $\Delta \geq 10$ can be neglected.

Table 4. Transition ratio estimation for general delay model.

Circuit	CPU Time (sec.)		Average error (%)	
	Simulation	Ours	Choi[19] *	Ours
c880	380.6	1.35	4.9	3.83
c1355	860.9	1.42	21.9	24.98
c1908	1482.9	1.56	25.7	23.58
c2670	2992.2	12.43	10.6	9.73
c3540	3112.4	4.53	20.6	16.50
c5315	8643.8	22.97	25.8	20.89
c6288	4437.7	8.02	N/A	26.91
c7552	8394.9	33.67	27.9	17.06

*CPU Time information is not available.

5. CONCLUSION

We have proposed a fast algorithm to estimate the glitch effects of a logic circuit under a general delay model without constructing global BDDs and without calculating Boolean difference. By taking the probabilistic analysis on transition effects; we can estimate the transition information of internal signals in about 1000 times shorter than simulation and maintain the acceptable errors.

6. ACKNOWLEDGEMENT

This work was supported in part by NSC of ROC under Contract NSC-85-2215-E-006-012.

7. REFERENCES

[1] L. Benini, M. Favalli, and B. Ricco, "Analysis of Hazard Contributions to Power Dissipation in CMOS ICs," *Int'l. Workshop on Low Power Design*, 1994, pp. 27-32.

[2] A. Bellaouar and M.I. Elmasry, Low-Power Digital VLSI Design: Circuits and Systems, Kluwer Academic Publishers, 1995.

[3] T.-L. Chou, K. Roy, and S. Prasad, "Estimation of Circuit Activity for Static and Domino CMOS Circuits Considering Signal Correlations and Simultaneous Switching," *IEEE Trans. Computer-Aided Design*, Vol.15, No.10, pp.1257-1265, 1996.

[4] M. Favalli, and L. Benini, "Analysis of Glitch Power Dissipation in CMOS ICs," *Int'l. Symp. on Low Power Design*, 1995, pp. 123-128.

[5] A. Ghosh, S. Devadas, K. Keutzer, and J. White, "Estimation of Average Switching Activity in Combinational and Sequential Circuits," *29th Design Automation Conf.*, 1992, pp. 253-259.

[6] Y. J. Lim, K.-I. Son, H.-J. Park, and M. Soma, "A Statistical Approach to the Estimation of Delay-Dependent Switching Activities in CMOS Combinational Circuits," *33rd Design Automation Conf.*, 1996.

[7] R. Marculescu, D. Marculescu, and M. Pedram, "Switching Activity Analysis Considering Spatiotemporal Correlations," *ICCAD*, 1994, pp. 294-299.

[8] R. Marculescu, D. Marculescu, and M. Pedram, "Efficient Power Estimation for Highly Correlated Input Streams," *32nd Design Automation Conf.*, 1995, pp. 628-634.

[9] H. Mehta, M. Borah, R. M. Qwens, and M. J. Irwin, "Accurate Estimation of Combinational Circuit Activity," *32nd Design Automation Conf.*, 1995, pp. 618-622.

[10] F. N. Najm, "Transition Density: A Stochastic Measure of Activity in Digital Circuits," *28th Design Automation Conf.*, 1991, pp.644-649.

[11] F. N. Najm, "A Survey of Power Estimation Techniques in VLSI Circuits," *IEEE Trans. VLSI Systems*, Vol. 2, No. 4, pp. 446-455, Dec. 1994.

[12] A. Papoulis, *Probability, Random Variables and Stochastic Processes*. McGraw-HILL, 1991.

[13] P. H. Schneider, U. Schlichtmann, and B. Wurth, "Fast Power Estimation of Large Circuits," *IEEE Design & Test of Computers*, pp. 70-78, 1996.

[14] P. H. Schneider and S. Krishnamoorthy, "Effects of Correlations on Accuracy of Power Analysis - An Experimental Study," *Int'l. Symp. on Low Power Design*, 1996, pp. 113-116.

[15] T. Uchino, F. Minami, T. Mitsuhashi, and N. Goto, "Switching Activity Analysis using Boolean Approximation Method," *ICCAD*, 1995.

[16] M. G. Xakellis and F. N. Najm, "Statistical Estimation of the Switching Activity in Digital Circuits," *31st Design Automation Conf.*, 1994, pp. 728-733.

[17] Verilog-XL, Cadence Design Systems, Inc. Training Manual, 1996.

[18] C.-Y. Tsui, M. Pedram, and A.M. Despain, "Efficient Estimation of Dynamic Power Consumption Under a Real Delay Model," *ICCAD*, 1993, pp.224-228.

[19] H. Choi, J. H. Yi, and S. H. Hwang, "Estimation of Inertial-Delay Dependent Switching Activities by Using Time-Stamped Transition Density," *ISCAS*, 1997, pp. 1528-1531.

[20] F. Brglez and H. Fujiwara, "A Neural Netlist of 10 Combinational Benchmark Circuits and a Target Translator in FORTRAN," *ISCAS*, 1985, pp. 663-698.

RESYNTHESIS OF SEQUENTIAL CIRCUITS FOR LOW POWER

Sumit Roy

Ambit Design Systems,
Santa Clara, USA.
sroy@ambit.com

Prithviraj Banerjee[†]

Center for Parallel & Distributed Computing,
Northwestern University, USA.
banerjee@ece.nwu.edu

ABSTRACT

At the logic level, a popular approach [1, 2] is to power down the sequential machine during the *self-loops* of the underlying finite state machine(FSM). In this work, we extend this idea to resynthesize existing sequential circuits to reduce power. We report a novel technique based on symbolic simulation of a sequential circuit to extract its self-loops without extracting the corresponding state transition diagram(STG). Since self loops may not be inherently present in the corresponding FSM, we partition the circuit heuristically and identify *partial-self-loops* for each partition to bring down the corresponding sub-circuit by gating the clock sub-tree feeding that partition. By using this approach, we could save upto **45%** of the total power on **a controller circuit of a microprocessor design**, where traditional techniques could not save any power.

1. INTRODUCTION

As portable devices proliferate and device sizes continue to shrink, allowing more devices to fit on a chip, power consumption takes on increased importance. Low power has thus emerged as a principal theme in today's electronic industry. Recent work has focused on accurate estimates of power consumption and on its reduction at all levels of abstraction, from high-level synthesis down to physical layout. Most power reduction techniques [3], and [4] emphasize reducing the level of activity in some portion of the circuit or bringing down the circuit by gating the system clock [1, 2] with a suitable function. These techniques focus on synthesizing a circuit for low power consumption from scratch. As a result they may not be applicable to the vast collection of synthesized circuits already present in the industry. Instead of starting from the scratch, we propose novel techniques for resynthesizing these sequential circuits for low power.

A recent work reported in [1, 2] tries to reduce the power of a sequential circuit by gating the clock during the self-loops of

---

[†] This research was supported in part by the National Science Foundation under grant MIP-9320854 and the Semiconductor Research Corporation under contract SRC 96-DP-109.

the underlying finite state machine (FSM). It saves the unnecessary power consumption on clocking latches and computing the next-state function during the self-loops. Unfortunately it is not directly applicable for re-engineering existing sequential circuits. In [5], a symbolic technique was used to identify the self-loops of a sequential machine. Further, the circuit may not have any self-loops making the approach futile even if the extraction of the STG is feasible.

In this paper, we propose techniques for minimizing the power of sequential circuits which inherently lack self-loops. We report a novel technique based on symbolic simulation of a sequential circuit to extract its self-loops without extracting the corresponding STG. We try to identify sets of latches which do not change in the next cycle and gate the clock sub-tree for this set of latches to bring down parts of the circuit when they are inactive. We partition the circuit heuristically and identify *partial-self-loops* for each partition to bring down the corresponding sub-circuit.

The rest of the paper is organized as follows; We provide definitions in Section 2. Section 3 describes the algorithms for extracting the self loops from the sequential machine. In Section 4, we formulate the partitioning problem, MPSL and describe a heuristic algorithm to solve it. In Section 4.3, we provide some experimental results. Finally, we conclude the paper in Section 5.

2. DEFINITIONS

Let B represent the Boolean set $\{0, 1\}$. The sequential machine $\mathcal{M}$ for a Moore machine is defined by the 5-tuple $(n, m, \omega, \delta, S)$ where ω and δ are the output and transition functions of the machine respectively. S is the initial state. The set of valid states, V_k, of a Moore machine are the states which are reachable from S. The self-loop function for a given state $s \in (V_k)$ is a function, $Self_s$, which represents the conditions under which the next state remains s. The self-loop function is defined as $Self_s(\vec{i}) = 1 \quad if \quad (\delta(\vec{i}, \vec{s}) = \vec{s})$ Overall, we have a set of functions $Self_s(i), s \in (V_k)$ that define the set

Table 1: Number of inputs and state variables(FFs) in some ISCAS 89 benchmarks

ISCAS89 circuits	PI	FFs	FFs/PI
s5378	35	179	5.1
s9234	19	228	12.0
s15850	14	597	42.6
s35932	35	1728	49.4
s38417	28	1636	58.4
s38584	12	1452	121.0
average	-	-	48.1

of self-loops for the entire sequential machine. We then define the self-loop function $Self$ as the union of the self-loops in the FSM:

$$Self(i,s) = \bigcup_{s \in (V_k)} s.Self_s(i), \quad i \in B^m \quad (1)$$

In the next section, we will describe efficient symbolic simulation techniques to find $Self(\vec{i},\vec{s})$ for a given $\mathcal{M}$ which avoids building the transition BDD for the circuit altogether.

3. EXTRACTION OF SELF LOOPS

In this section, we will describe an algorithm to extract self-loop from a given netlist of an existing sequential circuit without extracting the STG or even building the transition BDD of the circuit making them applicable to larger circuits.

3.1. Motivation

In [5], the authors reported a symbolic technique for identifying self-loops from the BDDs of the next state functions. Since these BDDs have the present state and input as variables, for large circuits with many state bits, the above approach is computationally expensive. In most FSMs, the number of state variables is much larger than the number of input variables. Table 1 validates the above point. This indicates that for a circuit with 20 primary inputs, on an average the bdds for the next state function will have (48.1 + 1)*20 = 982 variables. Since bdd sizes grows rapidly with the height of the bdd, we provide an approach which restricts the bdd size to the primary input variables.

3.2. Symbolic Simulation

In this section, we describe an algorithm to extract self-loops of large circuits based on symbolic simulation. We symbolically simulate each valid state s to find the $Self_s$ based on Equation 1. Note that the $Self_s$ function which detects all the self loop transitions for a given state s is a function of i only. Based on our observations that the number of primary inputs are a small percentage of number of state variables in a FSM, we can extract self-loops for very large circuits. The following algorithm in Figure 1 illustrates the mechanism to generate the $Self$ function from the netlist description using BDDs. We illustrate the above algorithm with Example 3.1.

Example 3.1 Let N be the given netlist. The network $N(x_0, x_1, s_0, s_1)$ is given below with the valid states being S_v = { 00, 01, 10 }.

$$Z = x_1 s_0 s_1' \quad S_0 = x_0' Z + x_0 x_1' s_0 s_1' + x_0' x_1' s_0' s_1$$

$$S_1 = x_0' x_1' s_0' S_0' + s_0' s_1 S_0' + S_0' Z$$

We invoke the $Symbolic\_Simulate$ routine with the network N. It starts with state 00. It assigns $s_0 = 0$, $s_1 = 0$ and then creates the BDD for S_0 and S_1. $Bdd_{S_0} = \mathbf{0}.Bdd_{S_1} = x_0' x_1'$. Since we want the next state to be 00, we find the complement BDD for S_0 and S_1. $Self_{00} = Bdd_{\overline{S_0}} \& Bdd_{\overline{S_1}} = x_0 + x_1$ Similarly, we get $Self_{01} = x_0 + x_1$ $Self_{10} = x_0 x_1' + x_0' x_1$

3.3. Results

Table 2 compares the symbolic simulation based extraction algorithm with that of [5]. *Previous* and *Ours* refer to the results obtained using [5] and our approach. The *time* is the execution time on Sun-Sparc10 in seconds and *size* is the maximum number of BDD node used during the extraction. We see that for large circuits, like s3330, the approach in [5], could not finish after **12 hours**. The dash indicates that it could not complete and the *star* entries are the maximum BDD nodes allocated till then. Hence our approach can work for large circuits. As the outer loop in Figure 1 iterates over all the valid states explicitly, it might fail when the set of valid states is huge. Fortunately most of the circuits have valid states much smaller

```
SYMBOLIC_SIMULATE()
1    Self ← 0;
2    for i ← 1 to |valid_states|
3        s ← valid_state[i];
4        substitute_present_state(s);
5        Self_s = 1;
6        for j ← 1 to n
7            bdd_ns_j ← construct_bdd_next_state(j);
8            if (s[j] = 1)
9                Self_s ← Self_i ∧ bdd_ns_j;
10           else Self_i ← Self_i ∧ NOT(bdd_ns_j);
11       Self ← Self ∨ Self_i;
```

Figure 1: Algorithm for generating the $Self_s$ function for a state s

Table 2: Comparison of our extraction algorithm with [5]

ISCAS89 circuits	PI	FFs	Previous		Ours	
			Size	Time	Size	Time
s27	4	3	40	0.1	35	0.1
s344	9	15	106	0.2	312	3.8
s349	9	15	105	0.2	195	2.4
s382	3	21	292	0.2	71	2.5
s386	7	6	215	0.2	56	0.2
s3271	26	116	760014*	-	890	116.2
s3330	40	132	1625480*	-	7964	81.3
s3384	43	183	1711053*	-	460	167.0

than the state space due to the presence of pipelined and data latches. In the next section, we will introduce the concept of *partial self-loops* and which enables us to save power on circuits without self-loops.

4. PARTIAL SELF LOOPS

In general, users design modules of a circuit and then synthesize these FSMs together to generate the sequential circuit, so as to share the logic. If we try to find self loops of these circuits, we would find very few of them, since the circuit has independent modules which behave in their own way. But if we could look at the modules individually, we could find plenty of self loops. We could then gate the clock of the latches of a particular module to bring down that module during the self loop. Also there exists certain circuits, like the counters, which do not have any self loops but contains portion of the circuit which can be powered down since it does not perform any useful computation. This is the underlying philosophy of our partitioned approach. Let us illustrate the idea with the following example.

Example 4.1 *Consider a 4-bit counter.*
From the state transition diagram given in Figure 2, it is clear that there are no self-loops. Hence the conventional technique of gating the clock of all the latches will fail. If we implement it using a synchronous circuit, the average power consumed with a 20MHz clock by the 4-bit counter is $0.807\mu W$. But if we look at the two most significant bits (MSB), we notice

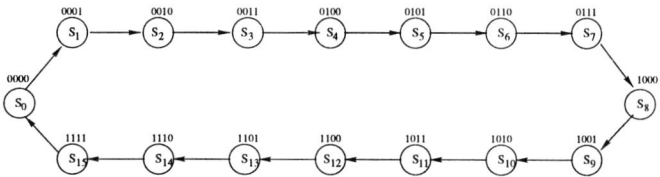

Figure 2: State transition graph of a 4 bit counter

that of the 16 transitions, it does not change for 12 of them. Hence during this time, we could save the power dissipated by the 2 MSB latches by gating their clock input. The gated counter only consumes $0.606\mu W$ with a 20MHz clock. Hence we observe that although there were no self loops, we could save 25% power by gating the latches selectively.

In [6], we have proposed the idea of decomposing an FSM into interacting FSMs such that the latter have plenty of self-loops although the original FSM lacked it. We can power down these smaller FSMs independently to save power consumed by the subcircuit. In [6] the STG was used to decomposing the FSM. Since extraction of the STG from the sequential machine is computationally expensive, in this work, we propose a novel partitioning algorithm for decomposing a sequential machine. In the following section, we formulate the partitioning problem.

4.1. Problem Formulation

In this section, we define the concept of partial self-loops. Let $\mathcal{M} = (n, m, \omega, \delta, Init)$ define a sequential machine corresponding to a Moore machine. Given p, the number of partitions, let $\Theta_n = \{\theta_1, \ldots, \theta_p\}$ define a disjoint partition of Z_n (set of natural numbers from 1 to n) into p parts. If $\vec{s}$ represents the state variables, $\vec{s_{\theta_i}}$ corresponds to those state variables which belongs to partition θ_i. Given a partition Θ_n, we define the *partial-self-loop*, $PSelf(\theta_k, \vec{i}, \vec{s}) = 1$ if the next state bits of $\vec{s}$ in the partition θ_k does not change on input $\vec{i}$. $PSelf(\theta_k, \vec{i}, \vec{s})$ represents the set of state and input combination for which the values of the next state variables in the θ_k partition remains unchanged. Hence we can gate the clocks of the latches corresponding to the θ_k partition without changing the circuit behavior.

Figure 3 shows the implementation of the gated clocks on latch partitions. Since we are interested in bringing down the total power consumed by the entire circuit, we would like the partition Θ_n not only to increase the partial-self-loops but also to increase the part of the circuit brought down by them. Hence we would like to maximize the product of the partial-self-loops times the portion of circuit it affects. To estimate the fraction of the circuit affected by a gating function $PSelf(\theta_i)$, we will use a function $est(\|\theta_i\|)$. Since the next state function is usually exponential in size of the number of next state variable, we will assume, $est(\|\theta_i\|) = \|\theta_i\|$

Definition 1 *Given a partition Θ_n, we define the **Cost** of the partition to be the following:*

$$Cost(\Theta_n) = \sum_{\theta_k \in \Theta_n} est(\|\theta_k\|).\|PSelf(\theta_k, \vec{i}, \vec{s})\| \quad (2)$$

By maximizing the cost function, we increase the set of partial-self-loops as well as the fraction of the circuit it can bring down.

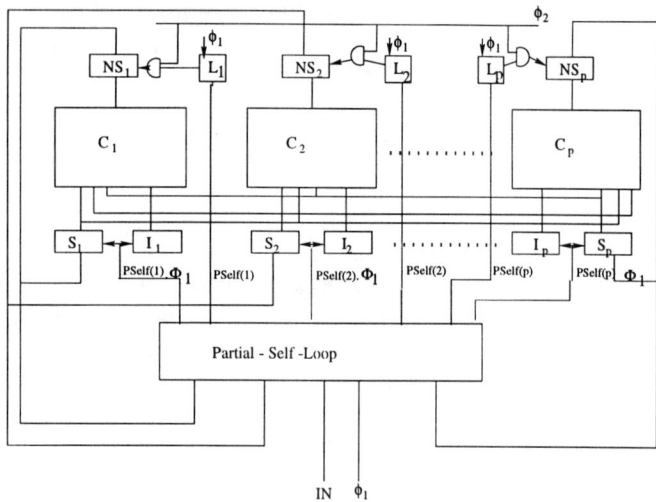

Figure 3: p partitioned gated implementation of a sequential circuit

Problem Formulation: Given a netlist of a sequential machine, we need to partition the latches of the circuit, Θ_n, in such a way so as to maximize the $Cost(\Theta_n)$ and and minimize the duplication among various parts of the circuits. We call this problem *maximal partial-self-loop extraction(MPSL)*. By solving the MPSL problem efficiently, we hope to achieve a low power resynthesized equivalence of the given sequential machine.

Theorem 1 *The problem of maximal partial-self-loop extraction is NP-hard.*
Proof omitted due to lack of space.

Since MPSL is NP-hard, we describe a technique to find a maximal partitioning of the latches of the sequential circuit.

4.2. S-graph based partitioning algorithm

In this section we provide an algorithm to obtain a maximal partition. In a sequential circuit, we define an affinity relation between a latch and a set of latches. A latch is said to have an affinity to a set of latches, if a transition in that latch causes transitions in the set of latches with a given probability. It can be easily seen that for any given partial-self-loop of this set of latches, the latch under consideration will not switch with the same probability. Hence, if the given probability for the affinity relation was high, a partial self loop for the set leads to no transition on the latch with high probability. Hence this latch should belong to the same partition as the set of latches to increase the size of the partial-self-loop.

We will abstract a graph from the given sequential circuit to establish the affinity relation. Given a sequential circuit $\mathcal{M} = (n, m, \omega, \delta, Init)$, we create a Sequential-graph (**S-graph**) $G = (V, E)$ in the following manner. Create a vertex v in G for every state variable and primary input of $\mathcal{M}$. A directed edge exists from u to v in G if there is a combinational path from node corresponding to u to that of v. Thus the edge from u to v signifies that a transition can propagate from node corresponding to u to that of v. Hence the affinity relation is given by the fanout relation of a vertex, v. Since the change in the value of a node in a circuit can cause a transition on the latches it fans out to, we include these latches in the affinity list of the vertex, v. Thus the affinity relation defines the hyper-edges on the graph. We notice that the affinity relation may not create a disjoint partition. Hence we need to come up with a partition which satisfies as many affinity relation as possible. We use a variant of the linear time network partitioning algorithm developed by Sanchis [7] to decompose the S-graph into p partitions. Since we have used an hyper-edge to indicate the affinity relation, minimizing the cut results in maximum satisfaction of the affinity relation between the latches. Hence this leads to maximizing the partial-self-loops. Also, since the edges in the S-graph indicate logic, minimizing the cut directly leads to minimizing the duplication of logic required for creating disjoint sub-circuits.

We applied an extension of the algorithm described in Section 3.2 to extract $PSelf$ for each of the latch set. We used a subset of the $PSelf$ to gate the clock of each of the latch set in order to reduce the power consumption of the gating function. First, all primes of $PSelf$ are generated using symbolic methods [8], then we keep choosing the primes with the largest sizes till the new function crosses the user defined threshold. We use this subset of $PSelf$ to gate the clock of latches in a partition. The combinational logic of each of the partitioned circuit is optimized in SIS [9] using the additional do not care set given by the subset of $PSelf$. This step is repeated for all the partitions of the original circuit. Next, all the gating functions are synthesized simultaneously using SIS to produce a multi-output function. Synthesizing the subsets of all the $PSelf$ helps in sharing of logic between the different gating functions. The gated-clock circuit is synthesized by combining all the partitioned circuits, the set of latches accompanying them and the multi-output gating function. The original circuit and the gated clock implementation are mapped on the library provided in the SIS package using the technology mapper in SIS. Finally, the mapped circuits are run through a power estimation tool based on the algorithms reported in [10].

The efficacy of the algorithm is shown in the following experimental section.

4.3. Results

Table 3 show the power consumed with a 20MHz clock by the original circuit, our method and the method described in [1]. The table summarizes the preliminary results based on our technique. We report results on some sequential circuits.

Table 3: Comparison of average power(in μW) consumption of gated circuits based on different approaches.

Circuit	PS	PI	Org	Gat	Part	Sav
Counter8	8	0	1.55	1.55	1.04	33%
Counter12	12	0	2.13	2.13	1.04	50%
Gray8	8	0	1.34	1.34	0.93	37%
UpDown8	8	1	2.35	2.35	2.14	10%
Ctrl1	11	7	2.81	2.92	1.71	45%
Ctrl2	19	7	5.16	5.02	4.86	6%
s349	15	9	2.31	1.99	2.41	-5%
average			2.52	2.49	2.01	25%

Counter8, *Counter12*, *Gray8* and *UpDown8* are 8 bit, 12 bit ripple carry counters, 8 bit gray and up-down counter respectively. *Ctrl1* and *Ctrl2* are parts of the controller of a latest microprocessor. We are trying to gather results on a wider variety of circuits. In the Table 3, **PS** and **PI** represent primary states and primary inputs respectively. **Org**, **Gat** and **Part** reports the power consumed by the original circuit, the gated-circuit [1] and the gated-circuit based on the partitioned approach respectively. **Sav** summarize the percentage of power saved by our partitioned approach. Based on the size of the circuits, we explored with different number of partitions. Running our tool with one partition results in finding self-loops of the entire circuit and hence corresponds to the technique described in [1]. The results show that our technique was able to save upto **45%** on **Ctrl1** where the conventional technique failed due to absence of self-loops. Ctrl1 is a queue controller and our tool was able to partition the read and the write logic of the queue controller and thereby obtain plenty of partial-self-loops. Also, the single partition gated clock performs better in circuits when there are plenty of self loops in the underlying FSM, like in s349. Hence partitioning is good for circuits with not many self-loops, such as counters and modular circuits. From the above table, we see that the partitioned approach for increasing the self loops have paid off as we are able to save 25% of the power consumed by the original circuit.

5. CONCLUSIONS

In this paper we have presented techniques for resynthesizing sequential machines targeting for low power. We describe a new symbolic simulation technique for extracting the self-loops from the netlist description of the sequential machine. Instead of extracting the entire **STG**, we use symbolic techniques to identify the self-loops of the underlying FSM. Since there are very few primary inputs compared to the state variables, our technique can be applied to much larger circuits than [5]. Our second contribution is in the novel partitioning algorithm which we developed to save power in circuits without self loops where conventional techniques [2] will fail to work. Instead of searching for self-loops, we try to find partial-self-loops, power down part of the circuit and thereby manage to save considerable power. We formulated the maximal partial-self-loop extraction problem and showed that it is NP-hard. We also provided efficient algorithms to partition the latches which results in maximizing the partial-self-loops. We achieve 45% power savings by applying our algorithm on industrial circuits. Hence we feel that our algorithm can be applied in conjunction with the conventional gated-clock algorithm. Finally, the timing analysis of our resynthesized circuit is exactly same as that reported in [2], since our techniques are similar but more general than the above work.

5.1. Acknowledgements

We will like to thank Dr. Gagan Hasteer for his constructive suggestions on certain parts of this paper.

6. REFERENCES

[1] L. Benini, P. Siegel, and G. DeMicheli, "Saving power by synthesizing gated clocks for sequential circuits," *IEEE Design & Test of Computers*, pp. 32–41, Dec. 1994.

[2] L. Benini and G. D. Micheli, "Automatic synthesis of low-power gated-clock finite-state machine," *IEEE Trans. Computer-Aided Design*, vol. 15, pp. 630–643, June 1996.

[3] A. Chandrakasan, *Low-Power digital CMOS design*. Boston, MA: Kluwer Academic Publishers, 1995.

[4] S. Devadas and S. Malik, "A survey of optimization techniques targeting low power VLSI circuits," in *Proceedings of the Design Automation Conference*, (San Francisco, CA), pp. 242–247, June 1995.

[5] L. Benini, G. D. Micheli, E. Macii, D. Sciuto, and C. Silvano, "Symbolic synthesis of clock-gating logic for power optimization of control-oriented synchronous networks," in *European Design and Test Conference*, (Paris, France), pp. 514–520, Mar. 1997.

[6] S. Roy and P. Banerjee, "Partitioning sequential circuits for low power," in 11th *International Conference on VLSI Design*, (Chennai, India), Jan. 1998.

[7] L. A. Sanchis, "Multiple-way network partitioning with different cost functions," *IEEE Trans. Computers*, vol. 42, pp. 1500–1504, 1993.

[8] O. Coudert and J. C. Madre, "Implicit and incremental computation of primes and essential primes of boolean functions," in *Proceedings of the Design Automation Conference*, (Anaheim, CA), pp. 36–39, June 1992.

[9] E. M. Sentovich, K. J. Singh, L. Lavagno, C. Moon, R. Murgai, A. Saldanha, H. Savoj, P. R. Stephan, R. K. Brayton, and A. Sangiovanni-Vincentelli, "SIS: A system for sequential circuit synthesis," Tech. Rep. UCB/ERL M92/41, Department of Electrical Engineering and Computer Science, University of California, Berkeley, CA, May 1992.

[10] V. Saxena, F. N. Najm, and I. N. Hajj, "Monte-carlo approach for power estimation in sequential circuits," in *European Design & Test Conference*, (Paris, France), Mar. 1997.

STG OPTIMIZATION FOR POWER AND AREA REDUCTION

G.S. Panagiotaras and O.G. Koufopavlou

VLSI Design Laboratory, Department of Electrical & Computer Engineering,
University of Patras, 26500 Patras, Greece

ABSTRACT

Graph transformations for deterministic Signal Transition Graphs (STGs) are presented. The proposed transformations can result in more efficient circuits with respect to power consumption and area, by reducing the STG signals concurrency, provided that the timing restrictions of the circuit are not violated. The circuits' function and speed are preserved. The proposed graph modifications steps are presented in a procedural, clearly defined algorithm. So the transformations can be very easily integrated in existed STG synthesis tools.

1. INTRODUCTION

Asynchronous circuits have recently returned to the foreground, because of their potential for lower power consumption, as compared to synchronous circuits. The main cause for this is the fact that there is no need to drive long clock lines at every cycle.

A design entry of many asynchronous circuit synthesis tools is the Signal Transition Graph (STG) description. In most of these methodologies dealing with STGs synthesis, the goal was the optimization of the resulting circuit in speed and area. A general transformation framework for asynchronous synthesis using concurrency reduction at the State Graph has been presented in [1].

In this paper we consider the important parameter of low-power dissipation by reducing the concurrency at the STG level rather than the State Graph. The number of nodes (states) in a State Graph increases exponentially as the number of signals increases. On the other hand, the STG specification is simpler and can be handled easier.

The rest of the paper is organized as follows: In the next section the basic concepts of the STGs are given. The proposed graph transformation is explained in section 3. STGs transformation example results are presented in section 4. Finally, we conclude the paper in section 5.

2. STG CHARACTERISTICS

Many asynchronous circuit synthesis methodologies use the Signal Transition Graph (STG) as the initial circuit specification [2] [3] [5-10]. The STGs were introduced by Chu [5] and can be considered as a subset of Petri nets. A Petri net is defined as triple <P,T,F>, where P is a set of places, T is a set of transitions, and F is the flow relation $F \subseteq (P \times T) \cup (T \times P)$ [2].

A *Token Marking* of a PN is a non-negative integer labeling of its places. If $(p, t) \in F$, a place $p \in P$ is called a predecessor of a transition $t \in T$, and t is called a successor of p. Conversely, if $(t, p) \in F$, a transition t is a predecessor of a place p, and p is a successor of t. A *Free-Choice Petri Net* (FCPN) is a PN where, if a place p has more than one transition as its successors, then p must be the only predecessor of its successor transitions [4].

A *Signal Transition Graph (STG)* is an interpreted FCPN where transitions are interpreted as value changes on input, output or internal signals of the specified circuit [2]. *Positive* transitions represent the rising $(0 \to 1)$ signal transitions and are labeled with "+". *Negative* transitions represent the falling $(1 \to 0)$ signal transitions and are labeled with "-". The conventional graphical representation of an STG is a directed graph, where the corresponding labels denote transitions and circles denote places. Places with one predecessor and one successor are usually omitted [2].

The set T of signal transitions can be partitioned in input, internal, and output signal transitions. *Input signal transitions* are those that occur on the input signals of the circuit, and in the STG representation are underlined. *Output signal transitions* are those that occur on its outputs. *Internal signal transitions* are the transitions that can happen on internal signals. The fundamental difference between transitions of input and non-input (internal and output) signals is that the former are caused by the external environment while the latter by the system [5],[2].

An *arc* in an STG represents a causal relation: for example $x^+ \to y^+$ means that x must go high before y can go high. When we say that a transition is *enabled* we mean that the corresponding event can happen in the circuit. A transition is enabled whenever all its predecessor places are marked with at least one token. An enabled transition may *fire*. This means that the corresponding signal changes value in the circuit. The firing of an enabled transition consists of removing one token from every predecessor place and adding one token to every successor place.

An arc is *redundant* if it never determines the firing of the signal transition to which it points. Two transitions of a marked PN are *concurrent* if there exists a reachable marking where both transitions are enabled and none of them disables the other one in any reachable marking. The *State Graph (SG)* of an STG, is a directed graph, where each node (called *state*) is in one-to-one correspondence with a marking of the STG reachable from its initial marking [2]. An STG has the

Unique State Coding (USC) property if two distinct states in the SG do not have the same labeling (coding)

In the subsequent paragraphs we refer to deterministic STGs which are defined as interpreted marked graphs, where transitions are interpreted as value changes on the input, output or internal signals of the circuit. A *marked graph* is defined as a PN in which every place has only one predecessor and one successor [3].

An STG is *live* if: a) it is strongly connected, b) every single loop contains exactly one token, and c) transitions t^+ and t^- are ordered. An STG is *safe* if no arc in it is ever assigned more than one token and if no sequence of firings can bring more than one token to one arc.

3. PROPOSED TRANSFORMATIONS

In this section the proposed transformations are described. It is shown that the reduction of the concurrency in a deterministic STG can result in simpler, functionally correct and thus lower power consumption circuits. As it is referred in [5], in order for the STG specification to provide correct logic implementation of the circuit, the STG must be live, safe and has the USC property. Then the State Graph (SG) is derived. In the SG, states there are binary vectors that represent the values of the signals in the circuit, while transitions are the transitions of these signals [2]. Next, the Karnaugh maps (K-maps) for each output and internal signals are constructed [6]. Finally, the Boolean expressions for the signals are derived.

The transformations are explained by using the general STG structure of Fig.1(a). Assume a part of a deterministic STG (that satisfies all the above mentioned properties) with a concurrency of the general form that is illustrated in Fig.1(a), and that the timing restrictions of the circuit are preserved in all changes. Signal transition a^* will be referred as the *start* transition of the concurrency, and b^* as the *end* of it. Signals c_i, i=1,2,...n, d_j, j=1,2,...m, (n, m >1) can be input, output or internal signals. (The arcs and the states in all figures that are illustrated in dashed lines, correspond to the original graphs; i.e. before the application of the transformation. In the transformed graphs they are removed).

If c_n^* is not an input signal transition then an arc from d_m^* to c_n^* is added. The arc from d_m^* to b^* becomes redundant and can be removed. This reduces the concurrency since the signals transitions d_j^*, j=1,2,...,m and c_n^* are not concurrent anymore.

The decreased concurrency can be identified in the corresponding SG as well. The states that correspond to the removed concurrency are deleted. So the resultant SG contains m less states than the initial SG, since there is a direct correspondence between these states and the signal transitions on the right side of the concurrency in the STG. Thus, m more "don't care" states exist in the K-map. These m new "don't care" states are used to achieve a more effective logic circuit optimization. It can be noticed that the SG keeps the properties of the initial SG since essentially it is the same SG with some deducted states.

If c_{n-1}^* is not an input signal transition the above procedure can be applied again, to obtain a simpler logic implementation. The procedure can be repeated until the first input signal transition is met or before the concurrency is fully eliminated. The first reason is because the environment cannot impose any restrictions on the inputs of the circuit and the second because the function of the circuit may change. If c_n is an input signal, the procedure cannot be applied on the left side of the concurrency. But if d_m^* is not an input signal transition, the transformation can be symmetrically performed on the right side.

The previous procedure is summarized in the following algorithm:

/* *The concurrency is determined by preprocessing. Let a^* be the start and b^* be the end signal transitions of the concurrency. Let L={ c_i }, i=1,2,...,n the signal transitions of the left side of the concurrency. Let R={ d_j }, j=1,2,...,m the signal transitions of the right side of the concurrency* */
    if ((n > 1) and (m > 1)) then
        { if (n ≥ m) then left_side
        else if (m > n) then right_side }
    else if (n > 1) then left_side
    else if (m > 1) then right_side
/* *application of the transformation on the left side* */
procedure left_side
    { i = n
    do while ((i > 1) and (c_i is a non-input signal)
      { add an arc from d_m^* to c_i^*
      if (i = n) then remove the arc from d_m^* to b^*
        else remove the arc from d_m^* to c_{i+1}^*
    i = i - 1 }}
/* *application of the transformation on the right side* */
procedure right_side
    { j = m
    do while ((j > 1) and (d_j is a non-input signal))
      { add an arc from c_n^* to d_j^*
      if (j = m) then remove the arc from c_n^* to b^*
        else remove the arc from c_n^* to d_{j+1}^*
    j = j - 1 }}

4. EXAMPLES AND RESULTS

The operation of the algorithm is illustrated by using an example STG, taken from [9],[10], that is shown in Fig. 2(a). The input signals are R_i and A_i, the output signal is R_0, and X is an internal signal. The signal transitions R_0^+ and X^- are the start and the end transition of the concurrency, respectively. The left side of the concurrency includes the signal transitions X^+, A_i^+, and R_0^- while only R_i^- belongs to the right side. This means that $L = \{X^+, A_i^+, R_0^-\}$ and $R = \{R_i^-\}$ Thus, the algorithm is applied on the left side.

The first signal transition that is being checked is R_0^-. It is not an input signal transition and so an arc from R_i^- to R_0^- can be added, as illustrated in Fig.2(a). The remaining arc from R_i^- to X^- now becomes redundant and can be removed. It appears in dashed lines in Fig.2(a). The signal transitions R_i^- and R_0^- are not concurrent anymore since there exists a simple path containing both of them ($R_i^- \to R_0^- \to X^- \to R_i^+ \to R_0^+ \to R_i^-$) [5], [9].

The next signal transition of the left side of the concurrency is A_i^+. This is an input signal. A circuit cannot interfere with its input signals, so the algorithm is terminated.

In Fig.2(b) the State Graphs of the STGs before and after the application of the algorithm are illustrated. The representation of the initially concurrent signal transitions R_i^- and R_0^- includes the SG states with binary coding: 1111, 0111, 1110, and 0110. In the transformed SG, the state 1110 is removed due to fact that the two signal transitions are now ordered. The state 1110 appears in dashed lines in Fig.2(b). The dashed arrows labeled R_0^-, R_i^- are also removed. The resultant K-maps for signal R_0 before and after the application of the algorithm are presented in Fig.2(c),(d).

For the hazard-free synthesis of the circuits, the algorithms presented in [8] have been used, as they are implemented within the Berkeley Sequential Interactive System SIS 1.1. In the derived circuits, the SR latches have been replaced with C-elements (with the additional logic where needed) in order to avoid the SR latches limitations [8]. The synthesized circuits have been implemented using the Mentor Graphics design environment (Design Architect and IC Station) The circuits have been simulated with HSPICE. Van Berkel's CMOS C-element implementation has been used, which has been proven to be better than other implementations with respect to energy consumption and area for the same delay [11] [12].

The logic circuit implementations are given in Fig.2(e) and Fig.2(f), before and after transformations respectively. The circuit after the transformations is smaller in area (5 gates less).

In Table 1 measurements for two derived circuits before (b) and after (a) the application of the proposed algorithm are shown. The area is measured in number of logic gates. The delay (in nsec) is the average rise-fall input to output delay for a primary output. The power consumption(in mW) measurements have been got using the method that is presented in [13]. Ref. [9] is the STG that is already analyzed. In Fig.3 the example of Ref. [7] is illustrated. In all the examples there is a significant reduction in area and power consumption without any significant cost in the delay. Similar results have been obtained for a number of STGs that have been tested.

Circuit	No. of Gates		nsec		mW	
	b	a	b	a	b	a
Ref.[9]	17	12	0.66	0.73	4.86	3.2
Ref. [7]	44	34	4.2	5.2	27.7	21.8

Table 1: Measurements for area, delay and power consumption which produced by applying the proposed algorithm in a number of STGs.

5. CONCLUSIONS

Emerging applications demand design methodologies for low power circuits. In this paper transformations for STGs are described. The transformations can be easily integrated in existed synthesis tools. The proposed method is given in a pseudocode algorithmical form. Results for typical examples are presented which prove the significant improvement in power consumption and area which is achieved without any cost in the circuit delay.

6. REFERENCES

[1] B. Lin, C. Ykman-Couvreur, and P. Vanbekbergen, "A General State Graph Transformation Framework for Asynchronous Synthesis", in Proc. of the European Design Automation Conference (EURO - DAC), 1994.

[2] L. Lavagno, K. Keutzer and L. Sangiovanni-Vincentelli, "Synthesis of Hazard -Free Asynchronous Circuits with Bounded Wire Delays", IEEE Trans. CAD of Integrated Circuits and Systems, vol. 14, no. 1, pp. 61-86, January 1995.

[3] S.T. Jung, and C. S. Jhon, "Direct Synthesis of Efficient Speed-Independent Circuits from Deterministic Signal Transition Graphs", in Proc. of the International Symposium on Circuits And Systems (ISCAS), pp. 307-310, 1994.

[4] P.A. Beerel, C. T. Hsieh and S. Wadekar, "Estimation of Energy Consumption in Speed-Independent Control Circuits", IEEE Trans. on CAD of Integrated Circuits and Systems, vol. 15, no. 6, pp. 672-680, June 1996.

[5] T. Chu, "Synthesis of Self-timed VLSI Circuits from Graph-theoretic Specifications", in Proc. of the International Conference on Computer Design (ICCD), pp. 220-223, 1987.

[6] T. H.-Y Meng, R. W. Brodersen and D.G.Messerschmitt, "Automatic Synthesis of Asynchronous Circuits from High-Level Specifications", IEEE Trans. On CAD, vol. 8, no.11,pp. 1185-1205, November 1989.

[7] P. Vanbekberken, G. Goossens, F. Catthoor and H.J. De Man, "Optimized Synthesis of Control Circuits from Graph-Theoretic Specifications", IEEE Trans. on CAD, vol. 11, no. 11, pp. 1426-1438, November 1992

[8] C.W. Moon, P. R. Stephan and R.K. Brayton, "Synthesis of Hazard-free Asynchronous Circuits from Graphical Specifications", in Proc. of the International Conference on Computer Aided Design (ICCAD), pp. 322-325, 1991.

[9] M.-L. Yu and P.A. Subrahmanyam, "A New Approach for Checking the Unique State Coding Property of Signal Transition Graphs", in Proc. of the European Design Automation Conference (EURO-DAC), pp. 312-321, 1992.

[10] L. Lavagno, C.W. Moon, R.K. Brayton and A. Sangiovanni-Vincentelli, "Solving the State Assignment Problem for Signal Transition Graphs", in Proc. of the Design Automation Conference, pp.568-572, 1992.

[11] M. Shams, J.C. Ebergen and M. I. Elmarsy, "A Comparison of CMOS Implementations of an Asynchronous Circuits Primitive: the C-Element", in Proc. of the International Symposium on Low Power Electronics and Design (ISLPED), pp.93-96, August 1996.

[12] K. van Berkel, "Beware the isochronic fork", Integration, the VLSI journal, vol. 13, pp. 103-128, June 1992.

[13] S. M. Kang, "Accurate Simulation of Power Dissipation in VLSI Circuits", IEEE Journal of Solid-State Circuits, vol. sc-21, no. 5, October 1986.

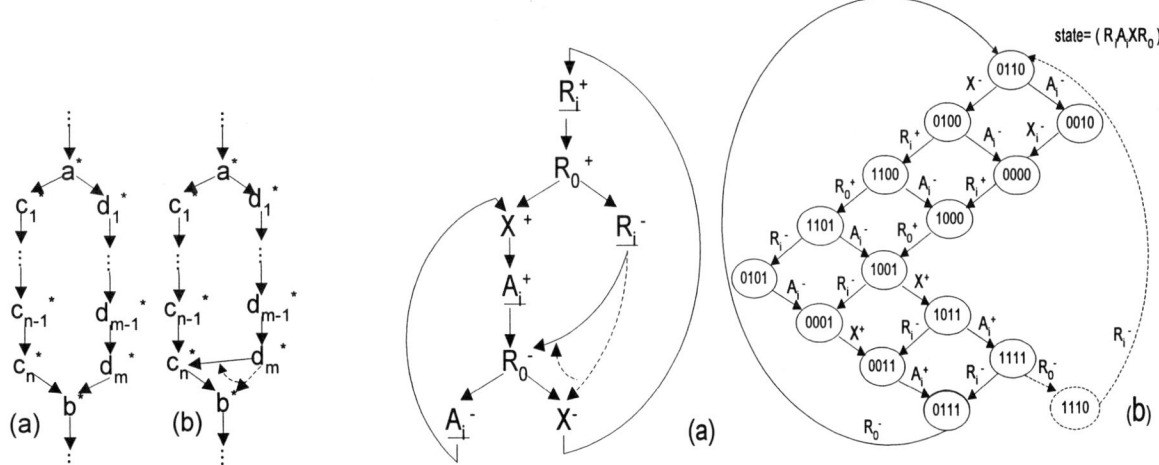

Fig.1: General STG structure (a) before (b) after transformation

Fig.2: (a) STG example (from ref. [9]) (b) the corresponding SGs

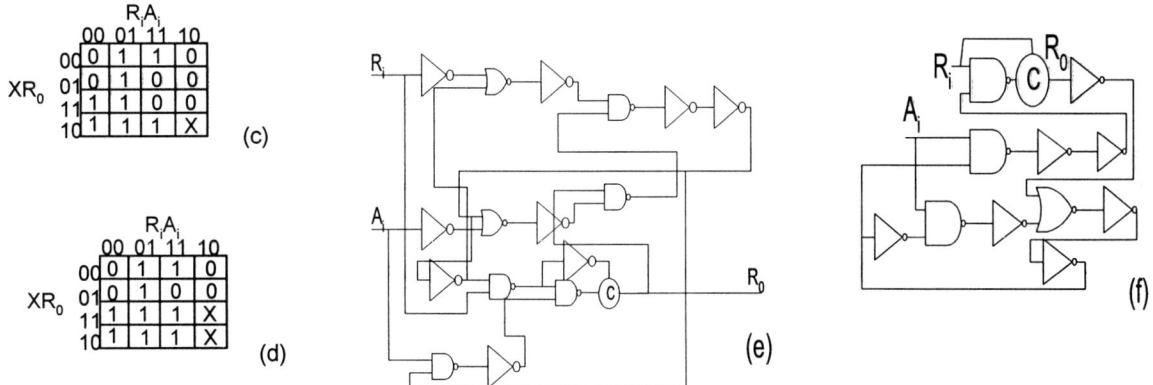

Fig.2: (c) K-map of R_0 before SG before transformation (d) K-map of R_0 after (e) logic diagram before (f) logic diagram after

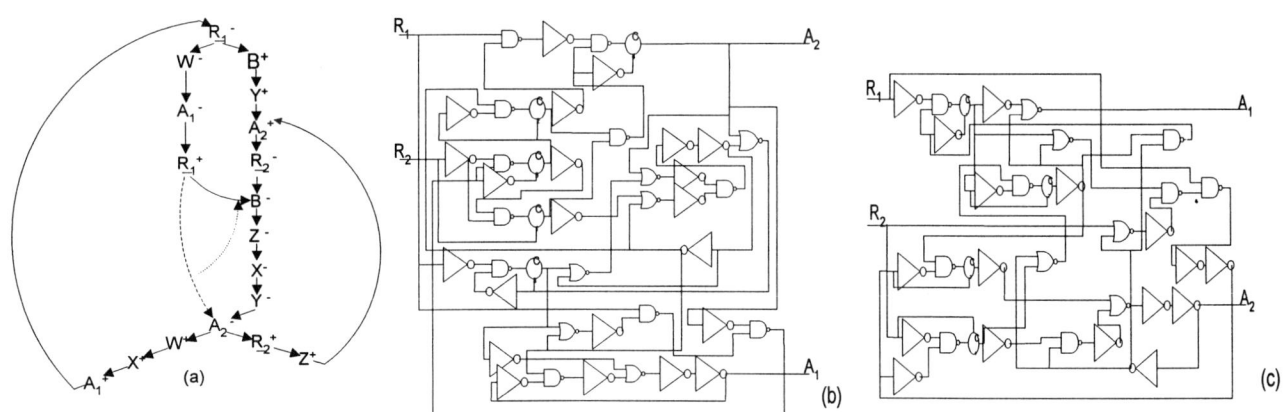

Fig.3: (a) STG example (from Ref. [7]) (b) Logic diagram before transformations (c) Logic diagram after transformations

A UNIVERSAL CLOSED-LOOP HIGH-SPEED INTERCONNECT MODEL FOR GENERAL PURPOSE CIRCUIT SIMULATORS

X. Li,   M. Nakhla   and   R. Achar

Department of Electronics, Carleton University, Ottawa, Canada, ON- K1S5B6
Ph: (613) 520-5780   Fax: (613) 520-5708   e-mail: msn@doe.carleton.ca

Abstract - In this paper, a novel technique is presented for the computation of multiconductor transmission line stencils suitable for inclusion in general purpose circuit simulators such as SPICE. Transmission lines can be lossy, coupled and distributed. The method offers an efficient means to discretize transmission lines compared to the conventional lumped discretization. Coefficients of the approximation are computed using closed-loop Padé approximants of exponential matrices. The algorithm also provides an error criterion for automatically selecting the order of the approximation for the frequency-bandwidth of interest. Numerical examples are presented to demonstrate the validity of the proposed model and to illustrate its application to variety of interconnect structures.

1. INTRODUCTION

The phenomenal growth in density, operating speeds and complexity of modern integrated circuits has made the interconnect analysis a requirement for all state-of-the-art circuit simulators. Interconnect effects such as ringing, signal delay, distortion, attenuation as well as crosstalk between adjacent lines can severely degrade the signal integrity [1], [2]. Interconnections can be from various levels of design hierarchy, such as on-chip, packaging structures, MCMs, PCBs and backplanes. As the frequency of operation increases the length of an interconnect becomes a significant fraction of the operating wavelength. In such cases conventional lumped models become inadequate in describing the interconnect performance and transmission line models become necessary. In addition, due to the presence of large number of interconnects in the circuit, besides the accuracy, the efficiency of simulation of these models will also become very important.

The major difficulty usually encountered while linking the distributed transmission line models and the nonlinear simulators is the problem of mixed frequency/time. This is because distributed elements are usually characterized in the frequency domain whereas the nonlinear components such as drivers and receivers are represented only in time-domain. There exists several publications to address this issue [2] - [10]. Approaches based on conventional lumped discretization of transmission lines provide easy solution to the problem of mixed frequency/time simulation [4]. However, the drawback of this approach is, not only it causes spurious oscillations but also leads to large size circuit matrix, rendering the simulation inefficient. Approaches based on method of characteristics are ideal for modeling lossless transmission lines [5]. However, its extension to lossy interconnects becomes computationally inefficient, especially in the presence of electrically long lines. Hence it becomes important to develop an accurate modeling scheme for transmission lines which can be efficiently simulated using general purpose nonlinear simulators such as SPICE.

In this paper, a novel method for discretization of distributed transmission lines is presented. The method utilizes closed-loop Padé expressions of exponential matrices to efficiently approximate Telegrapher's equations. The proposed technique overcomes the problem of spurious oscillations encountered by conventional discretization schemes. Also it keeps the order of the approximation very small compared to the conventional discretization. The algorithm also provides an error criterion for automatically selecting the order of the approximation needed for capturing the interconnect behavior in the frequency-bandwidth of interest. Resulting model can be directly linked to nonlinear simulators such as SPICE for the purpose of transient simulation. In addition, the resulting model can be easily incorporated into the recently developed model-reduction techniques [6] - [10].

2. TELEGRAPHER'S EQUATIONS

Consider the lossy, distributed transmission line described by a set of Telegrapher's equations

$$\frac{\partial}{\partial x}v(x,t) = -Ri(x,t) - L\frac{\partial}{\partial t}i(x,t) \quad (1)$$
$$\frac{\partial}{\partial x}i(x,t) = -Gv(x,t) - C\frac{\partial}{\partial t}v(x,t)$$

where R, L, C and G are the per-unit-length parameters. (1) can be written in the Laplace-domain using the exponential function as

$$\begin{bmatrix} V(d,s) \\ -I(d,s) \end{bmatrix} = e^Z \begin{bmatrix} V(0,s) \\ I(0,s) \end{bmatrix} \quad (2)$$

where

$$Z = (D + sE)d$$
$$D = \begin{bmatrix} 0 & -R \\ -G & 0 \end{bmatrix}; \quad E = \begin{bmatrix} 0 & -L \\ -C & 0 \end{bmatrix} \quad (3)$$

Here V, I represent the Laplace-domain terminal voltage and current vectors of the multiconductor transmission line and d is the length of the line. As is evident, (2) doesn't have a direct representation in time-domain, which makes it difficult to include it with nonlinear simulators.

3. DEVELOPMENT OF THE UNIVERSAL TRANSMISSION LINE MODEL

In this section we derive a universally compatible form for (2) so as to link it with general purpose simulators. Approximating the exponential matrix e^Z using matrix rational function, we have

$$P_M(Z)e^Z \approx Q_N(Z) \quad (4)$$

where $P_M(Z)$ and $Q_N(Z)$ are polynomial matrices of order M and N, respectively. However, it is not necessary to iteratively solve the system of equations arising from (4) in order to obtain the coefficients of $P_M(Z)$ and $Q_N(Z)$. This is possible as a closed-form relation can be derived to generate these matrices. (4) can be re-written as

$$\begin{bmatrix} P_{11} & P_{12} \\ P_{21} & P_{22} \end{bmatrix}_M e^Z \approx \begin{bmatrix} Q_{11} & Q_{12} \\ Q_{21} & Q_{22} \end{bmatrix}_N \quad (5)$$

where the polynomials P_{ij} and Q_{ij} are computed in a closed-form using the following relations

$$Q_{11} = \sum_{i=0}^{N} (M+N-i)! \binom{N}{i} \left[\frac{1}{2}(1+(-1)^i)(ab)^{i/2}\right] \quad (6)$$

$$Q_{22} = \sum_{i=0}^{N} (M+N-i)! \binom{N}{i} \left[\frac{1}{2}(1+(-1)^i)(ba)^{i/2}\right] \quad (7)$$

$$Q_{12} = \sum_{i=0}^{N} (M+N-i)! \binom{N}{i} \left[\frac{1}{2}(1-(-1)^i)(ab)^{\frac{i-1}{2}} a\right] \quad (8)$$

$$Q_{21} = \sum_{i=0}^{N} (M+N-i)! \binom{N}{i} \left[\frac{1}{2}(1-(-1)^i)(ba)^{\frac{i-1}{2}} b\right] \quad (9)$$

$$P_{11} = \sum_{i=0}^{M} (M+N-i)! \binom{M}{i} (-1)^i \left[\frac{1}{2}(1+(-1)^i)(ab)^{i/2}\right] \quad (10)$$

$$P_{22} = \sum_{i=0}^{M} (M+N-i)! \binom{M}{i} (-1)^i \left[\frac{1}{2}(1+(-1)^i)(ba)^{i/2}\right] \quad (11)$$

$$P_{12} = \sum_{i=0}^{M} (M+N-i)! \binom{M}{i} (-1)^i \left[\frac{1}{2}(1-(-1)^i)(ab)^{\frac{i-1}{2}} a\right] \quad (12)$$

$$P_{21} = \sum_{i=0}^{M} (M+N-i)! \binom{M}{i} (-1)^i \left[\frac{1}{2}(1-(-1)^i)(ba)^{\frac{i-1}{2}} b\right] \quad (13)$$

where a and b are matrices defined as

$$a = -(R+sL)d; \qquad b = -(G+sC)d \quad (14)$$

Using (6)-(14), (2) can be written in a closed-form as

$$\begin{bmatrix} \hat{P}_{11}(s) & \hat{P}_{12}(s) \\ \hat{P}_{21}(s) & \hat{P}_{22}(s) \end{bmatrix}_M \begin{bmatrix} V(d,s) \\ -I(d,s) \end{bmatrix} = \begin{bmatrix} \hat{Q}_{11}(s) & \hat{Q}_{12}(s) \\ \hat{Q}_{21}(s) & \hat{Q}_{22}(s) \end{bmatrix}_N \begin{bmatrix} V(0,s) \\ I(0,s) \end{bmatrix} \quad (15)$$

where coefficients of $\hat{P}_{ij}$ and $\hat{Q}_{ij}$ can be recursively obtained from the coefficients in (6) - (14). Since all the coefficients in (15) are known, it can be easily stenciled into a circuit simulator. In other words the transmission line can now be viewed as another "element" having a easy time-domain representation. Also, (15) can be directly converted to a set of ordinary differential equations and can be combined with rest of the circuitry containing nonlinear elements as

$$\frac{d}{dt}z_\pi(t) - A_\pi z_\pi(t) - B_\pi i_\pi(t) = 0$$

$$(L_\pi)^t v_\phi(t) - C_\pi z_\pi(t) - D_\pi i_\pi(t) = 0 \quad (16)$$

$$W_\phi \frac{d}{dt} v_\phi(t) + G_\phi v_\phi(t) + L_\pi i_\pi(t) + F(v_\phi(t)) - b_\phi(t) = 0$$

where i_π is the vector of terminal currents entering the transmission line subnetwork π. z_π, A_π, B_π, C_π and D_π are the matrices describing the state-space equations obtained from (15). Parameters W_ϕ, G_ϕ, F, b_ϕ, v_ϕ and L_π correspond to the original network excluding the transmission line, description of which can be found in [6], [7].

4. CRITERIA FOR SELECTING THE ORDER OF THE APPROXIMATION

Accuracy of the proposed model depends on the order of the approximation (M, N). Following error criterion is developed to efficiently decide the order of the approximation

$$\|e^Z - P_M^{-1}(Z)Q_N(Z)\| < \varepsilon_1 \quad (17)$$

where ε_1 is a pre-defined error tolerance in the frequency range of interest. Since all the elements in (15) are computed in a recursive and in a closed-loop form, the required order (M,N) can be easily estimated. Alternatively,

$$\|P_{M+1}^{-1}(Z)Q_{N+1}(Z) - P_M^{-1}(Z)Q_N(Z)\| < \varepsilon_2 \quad (18)$$

can also be used to efficiently estimate the order of the approximation.

5. COMPUTATIONAL RESULTS

An example is presented here to demonstrate the efficiency and accuracy of the proposed approximation of Telegrapher's equations compared to the conventional lumped discretization. In this experiment a circuit containing five lossy distributed transmission lines are simulated using the proposed method. Interconnects used had the following parameters: $L = 60 nH/m$, $R = 3\Omega/m$, $C = 100 pF/m$ and $d = 0.1 m$ and the circuit is shown in Fig. 1. Fig. 2 shows the comparison of frequency responses from both the proposed method and the conventional discretization with the exact solutions, up to 8GHz. Proposed method required an approximation of order (8, 8) to match the exact solutions accurately. However, the conventional lumped approximation using 8 sections deviates significantly from the exact solutions. As seen from Fig. 3, the conventional lumped approximation requires 50 sections to reasonably match the original solutions. A transformation of both these approximations into time-domain reveals the computational merits of the proposed model: conventional lumped approximation of 50 sections for five interconnects results in approximately 500 differential equations, whereas the proposed method with order (8, 8) results in only 40 differential equations.

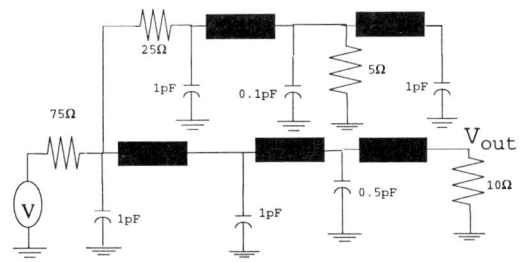

Fig. 1. An example circuit with 5 transmission lines

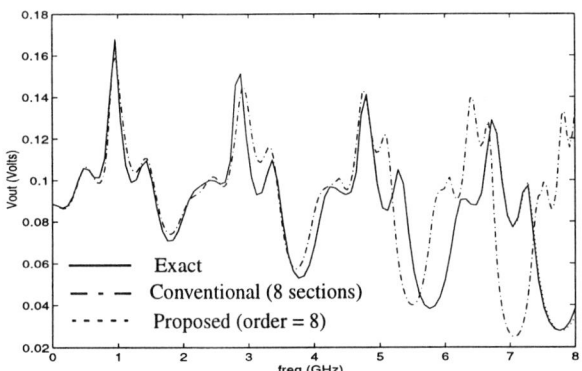

Fig. 2. Frequency response

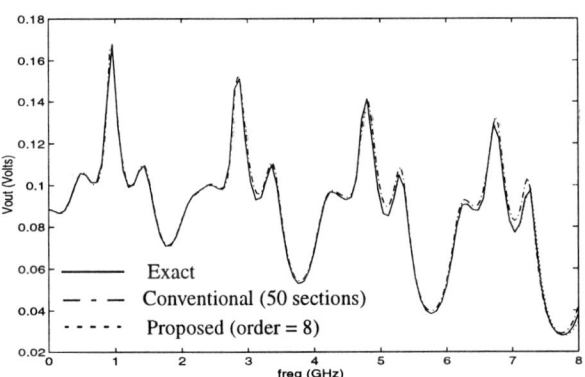

Fig. 3. Frequency response

6. CONCLUSIONS

In this paper, a new algorithm is presented for the computation of multiconductor transmission line stencils suitable for inclusion in general purpose circuit simulators such as SPICE. Transmission lines can be lossy, coupled and distributed. The algorithm also provides an error criterion for automatically selecting the order of the approximation. In addition, the resulting model can be easily incorporated into the recently developed model-reduction techniques.

REFERENCES

[1] H. B. Bakoglu, "*Circuits, Interconnections and packaging for VLSI*", Addison-Wesley, Reading MA, 1990

[2] W. W. M. Dai (Guest Editor) "Special issue on simulation, modeling, and electrical design of high-speed and high-density interconnects," *IEEE Trans. Circuits Syst.* vol. 39 no. 11, pp. 857-982, Nov. 1992.

[3] C. R. Paul, *Analysis of Multiconductor Transmission Lines*, New York: John Wiely and Sons, Inc., 1994.

[4] T. Dhane and D. Zutter, "Selection of lumped element models for coupled lossy transmission lines," *IEEE Trans. Computer-Aided Design*, vol. 11, pp.959-067, July 1992.

[5] F. Y. Chang, "The generalized method of characteristics for waveform relaxation analysis of lossy coupled transmission lines," *IEEE Trans. Microwave Theory Tech.*, vol. 37, pp. 2028-2038, Dec. 1989.

[6] T. Tang and M. S. Nakhla, "Analysis of high-speed VLSI interconnect using the asymptotic waveform evaluation technique," *IEEE Trans. Computer-Aided Design*, vol 11, pp. 2107-2116 Mar. 1992.

[7] D. Xie and M. S. Nakhla "Delay and crosstalk simulation of high speed VLSI interconnects with nonlinear terminations," *IEEE Trans. Computer-Aided Design*, vol. 12, no. 11, pp. 1798-1811, Nov. 1993.

[8] E. Chiprout, M. S. Nakhla, "Analysis of interconnect networks using complex frequency hopping," *IEEE Trans. Computer-Aided Design*, vol. 14, No. 2, pp.186-199, Feb. 1995.

[9] M. Celik and A. C. Cangellaris, "Simulation of dispersive multiconductor transmission lines by Padé via Lanczos process", *IEEE Trans. Microwave Theory Tech.*, vol. 44, pp. 2525-2535, Dec. 1996.

[10] M. Silveria, M. Kamon, I. Elfadel and J. White, "A coordinate-transformed Arnoldi algorithm for generating guaranteed stable reduced-order models of arbitrary RLC circuits", in *Proc. ICCAD*, Nov. 96.

A NOVEL TECHNIQUE FOR MINIMUM-ORDER MACROMODEL SYNTHESIS OF HIGH-SPEED INTERCONNECT SUBNETWORKS

R. Achar and *M. Nakhla*

Department of Electronics, Carleton University, Ottawa, Canada, ON- K1S5B6
Ph: (613) 520-5780   Fax: (613) 520-5708   e-mail: msn@doe.carleton.ca

Abstract - A new algorithm is presented to efficiently combine the process of model-reduction of high-speed interconnect subnetworks with the nonlinear simulation. The proposed algorithm guarantees the number of states required during the macromodel synthesis from reduced-order frequency-domain matrix-transfer functions to be minimum. This greatly speeds up the nonlinear simulation as the number of algebraic differential equations involved is reduced. An important advantage of the new technique is that it can be used in conjunction with any of the existing model-reduction techniques.

1. INTRODUCTION

During the recent years the intense drive for signal integrity has been at the forefront of rapid and new developments in CAD algorithms. Trend towards miniature design on an unprecedented scale, coupled with higher clock frequencies has highlighted the previously negligible effects of interconnects, such as ringing, signal delay, distortion, reflections and crosstalk, which are not always handled appropriately by current simulators [1]. Interconnects can be from various levels of design hierarchy, such as on-chip, packaging, multichip modules, printed circuit boards and backplanes. Depending on the operating frequency and nature of the structure, these interconnects can be modeled as lumped (RC/RLC), distributed (frequency independent/dependent RLCG parameters, lossy, coupled) and full-wave linear subnetworks, and these models are best described int the frequency-domain. However, since rest of the circuitry is comprising of nonlinear models that are analyzed in the time-domain, it becomes important to develop efficient techniques for time-domain characterization of interconnects in order to perform transient analysis on the entire circuitry [2].

Computational effort involved in circuit-simulation increases superlinearly with the number of nodes in a circuit. However, interconnect models generally tend to introduce large number of nodes which makes the nonlinear simulation practically difficult under the design-time and memory constraints. Recently developed model-reduction techniques [2] - [14] can be used to reduce the order of the network. AWE [3] is based on the Padé approximation of Laplace-domain transfer function of a linear network by a reduced-order model, containing only a relatively small number of dominant poles and zeros. Since Padé approximation is accurate only near the point of expansion [4] multi-point moment-matching techniques such as CFH [7] can be used to improve the accuracy and stability of AWE algorithm. Extension of AWE and CFH to various levels of interconnect modeling has been well addressed in the literature [4], [5] - [7], [11] - [14]. The other alternative approaches, PVL [8], Arnoldi [9], PRIMA [10], are more suitable for network-reduction of linear lumped subnetworks, *but they can't efficiently handle distributed/full-wave or measured subnetworks*. Depending on the type of circuitry under consideration, using any of the above model-reduction techniques, a reduced-order frequency-domain *matrix-transfer function* containing poles and residues can be obtained.

However, transient analysis of the entire circuitry requires coupling the frequency-domain reduced-order models with the nonlinear elements, in time-domain. An easy way to achieve this is to convert the frequency-domain matrix-transfer function into a set of ordinary differential equations and then linking them with the differential/algebraic equations describing the nonlinear components. Derivation of differential equations from reduced-order interconnect models is known as *macromodel synthesis*. In [2] a controller-canonical form of realization is described. In this method a common set of poles is used for each entry in the reduced-order model and the total number of state variables in this case is given by the product of total number of poles and total number of ports. State-space realization described in [11] also assumes a common denominator and it suggests a macromodel synthesis with total number of state variables equal to the product of total number of poles and total number of ports. *The above realizations become inefficient when the number of ports involved are large since the number of states introduced increase linearly with the number of ports.* None of the previously published techniques [8]-[10] guarantee a minimum realization for a given matrix-transfer function.

In this paper, a new algorithm has been developed for macromodel synthesis. The algorithm is based on the Gilbert's diagonal approach [15] and guarantees a *minimal-order realization* (requiring least number of state variables) for a given matrix-transfer function. The proposed technique can be used in conjunction with any of the existing network-reduction algorithms. Total number of states in the synthesized macromodel would be in a range having *a lower limit of total number of distinct poles present in the matrix-transfer function and an upper limit given by the product of total number of distinct poles and total number of ports*. It is to be noted that the macromodels from previously published realizations results in number of states represented by the upper limit. The resulting macromodel can easily be integrated in a nonlinear simulation.

2. REDUCED-ORDER MATRIX-TRANSFER FUNCTIONS OF LINEAR SUBNETWORKS

An interconnect linear subnetwork can be considered to be a multiport network. Such a network is usually characterized in terms of y- (admittance), z- (impedance), h- (hybrid) or s- (scattering) parameters. Using model-reduction techniques, a q- pole lower-order model can be obtained for any of these parameters (represented by a general notation $H(s)$)

$$H(s) = \begin{bmatrix} H_{11} & \cdots & H_{1m} \\ \cdots & \cdots & \cdots \\ H_{m1} & \cdots & H_{mm} \end{bmatrix}; \quad H_{jk}(s) = d^{j,k} + \sum_{i=1}^{q^{j,k}} \frac{r_i^{j,k}}{s - p_i^{j,k}} \quad (1)$$

where $1 \leq (j, k) \leq m$. $p_i^{j,k}$ is the ith dominant pole at a port k due to an input excitation at port j and the corresponding residue is $r_i^{j,k}$ and $d^{j,k}$ is the direct coupling constant. $q^{j,k}$ is the number of dominant poles used for approximating H_{jk}. Each entry H_{jk} may be obtained using one of the following two approaches: in the first approach different sets of poles are used for the computation of residues of each entry [6] and in the second approach a common set of poles are used for computation of residues [2], [11], [14].

Let P be a vector of distinct poles in $H(s)$ and q be the total number of distinct poles. Let p_i represent the ith pole in P. Next, $H(s)$ can be written as

$$H(s) = \sum_{i=1}^{q} \frac{R_i}{s - p_i}; \quad R_i = \begin{bmatrix} r_i^{1,1} & \cdots & r_i^{1,m} \\ \cdots & \cdots & \cdots \\ r_i^{m,1} & \cdots & r_i^{m,m} \end{bmatrix} \quad (2)$$

where R_i is the residue matrix and is computed as $R_i = \lim_{s \to p_i} (s - p_i) H(s)$.

Equation (2) represents the residue matrices for the general case of matrix transfer functions: with or without common poles among the entries h_{jk}. In the case of using a common set of poles for all the entries each $r_i^{j,k}$ in (2) will have a non-zero entry. In case of using different sets of poles for each h_{jk}, R_i will have only one non-zero entry. In the following sections a new algorithm for minimal-order macromodel synthesis is described.

3. PROPOSED MINIMUM-REALIZATION

Given a matrix-transfer function described by (2), several forms of time-domain realizations can be obtained. The realizations discussed in the literature requires the number of states equal to the product given by *(total number of distinct poles from all the ports)* x *(total number of ports)*, which makes the transient analysis inefficient in the presence of large number of ports. It may be noted from (2) that in the case of using different sets of poles for each h_{jk}, the rank of each R_i will be unity. Also, in the case of using a single set of dominant poles as a common denominator for the entire transfer-function matrix, not all the poles are seen distinctly at every input-output port combination. This would leave some poles having very small residues for few ports or sometimes zero for certain ports. It is observed that the singular values of residue matrices differ by a large ratio (example 1). Exploiting these features of residue matrices with little CPU effort as described in the rest of this section can significantly reduce the number of states.

3.1 Minimum realization through Gilbert's diagonal construction

A minimal realization is defined as the *one that has the smallest size 'A' matrix (requiring least number of state variables) for all triples {A, B, C} satisfying* $C(sI-A)^{-1}B = H(s)$, *a given transfer function.* Next, we present the following result [15], which is useful in minimum realization. *If the denominator polynomial of H(s) has q distinct roots and if ρ_i is the rank of each residue matrix R_i, then the minimum number of state variables needed for realizing such a system is given by*

$$n_{min} = \sum_{i=1}^{q} \rho_i \quad (3)$$

Next, in order to obtain a state-space representation, the residue matrix R_i is decomposed as product of two matrices as $R_i = C_i B_i$, where C_i and B_i are matrices of dimensions $m \times \rho_i$ and $\rho_i \times m$, respectively. The procedure for obtaining C_i and B_i from R_i is described in section 3.2. The state-space realization for matrix-transfer function $H(s)$ can be derived using Gilbert's diagonal approach as

$$A = Block\ diag\{p_i I_{\rho_i};\quad i = 1, 2, ..., q\}$$
$$B^t = \begin{bmatrix} B^t_1 & B^t_2 & ... & B^t_q \end{bmatrix} \quad (4)$$
$$C = \begin{bmatrix} C_1 & C_2 & ... & C_q \end{bmatrix}$$

where I_{ρ_i} is an identity matrix of dimension ρ_i. The above realization has the order n_{min} and is a minimal-order realization for the given $H(s)$.

3.2 Evaluating C_i and B_i from residue matrix R_i

The method discussed here is based on the singular value decomposition (SVD). Given a $m \times m$ matrix R (subscript i is dropped for simplicity) with first ρ rows of the matrix linearly independent, then R can be written using singular value decomposition as

$$R_{(m \times m)} = U_{(m \times m)} S_{(m \times m)} V^t_{(m \times m)} \quad (5)$$

where S is a diagonal matrix with dimensions same as R and with nonnegative diagonal elements in decreasing order. U and V are unitary matrices. Due to the fact that only ρ rows of matrix R are linearly independent, it can be approximated using subblocks of U, S and V as [15], [16]

$$\hat{R}_{(m \times m)} = C_{(m \times \rho)} B_{(\rho \times m)}$$
$$C_{(m \times \rho)} = \hat{U}_{(m \times \rho)} \hat{S}_{(\rho \times \rho)} \quad (6)$$
$$B_{(\rho \times m)} = \left(\hat{V}^t\right)_{(\rho \times m)}$$

Since the diagonal elements representing singular values (λ_i) of matrix S are automatically arranged in a descending order, if the rank of $S(\rho)$ is less than m, then the singular values beyond ρ^{th} diagonal element would be of value zero. However, due to round-off errors introduced by network-reduction techniques, singular values beyond ρ^{th} diagonal element will not be exactly equal to zero but they will have very small value. Hence in the proposed algorithm a threshold ratio ε is defined such that if $|\lambda_i/\lambda_{max}| \leq \varepsilon$, then λ_i is considered to be zero. Threshold ratio ε is defined such that $\|\hat{R} - R\| \leq \eta$, where η is a predefined tolerance value. Practically it has been observed that a value of ε below 10^{-3} is sufficient to reconstruct the residue matrix with a percentage norm error less than 0.1. Next, the time-domain state-space equations (4) obtained through minimum realization can be combined with the rest of the circuitry containing nonlinear elements very easily [11] to obtain unified transient solutions for the entire nonlinear circuit consisting of high-speed interconnect subnetworks.

4. COMPUTATIONAL RESULTS

A two port circuit was macromodeled using the proposed minimum realization algorithm (Fig. 1). Performing a model-reduction [14] a set consisting of five dominant poles was derived. These poles were then used to compute the residues corresponding to all the entries in the Z matrix. Next, the proposed minimum-realization algorithm was applied on each of the residue matrices and the corresponding singular values of each of the residue matrices is given Table 1. After applying the new method on each residue matrix, a macromodel (4) is obtained with $q = 5$.

The macromodel derived using the proposed technique contains a total of 6 states (only pole P_4 needs two states). However, a realization based on previous techniques would have resulted in two (number of ports) states for each pole or a total of 10 states. Savings in the number of states in this case is 40%. Next, a nonlinear simulation is performed using an input excitation having a rise time of 0.1ns and a pulse width of 3ns. Fig. 2 shows the transient responses at node V_2 obtained using both the proposed algorithm and SPICE, and they are indistinguishable.

Table 1: Singular values of residue matrices

Poles	P1	P2	P3	P4	P4
Singular values	4.655e+08	2.539e+09	9.380e+09	5.250e+08	1.377e+09
	1.074e+01	1.410e+01	4.066e+03	3.511e+07	9.766e+06

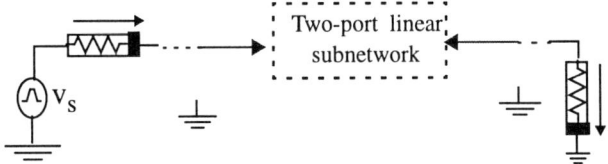

Fig. 1. Circuit for example 1.

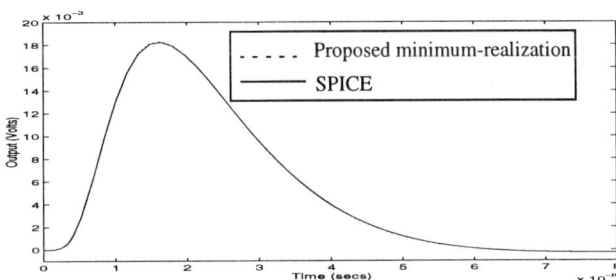

Fig. 2. Transient responses at node V_2

5. CONCLUSIONS

In this paper, an efficient technique is presented to combine model-reduction with nonlinear simulation. The proposed algorithm guarantees the number of states required during the macromodel synthesis from a frequency-domain reduced-order matrix-transfer function to be minimum. An important advantage of the new technique is that it can be used in conjunction with any of the existing model-reduction techniques.

REFERENCES

[1] H. B. Bakoglu, "*Circuits, Interconnections and packaging for VLSI*", Addison-Wesley, Reading MA, 1990

[2] S. Y. Kim, N. Gopal, and L. T. Pillage, "Time-domain macromodels for VLSI interconnect analysis," *IEEE T-CAD*, pp. 1257-1270, Oct. 94.

[3] L. T. Pillage and R. Rohrer, "Asymptotic waveform evaluation for timing analysis," *IEEE T-CAD*, vol. 9, pp. 352-366, Apr. 1990.

[4] T. Tang and M. Nakhla, "Analysis of high-speed VLSI interconnect using the AWE technique," *IEEE T-CAD*, vol 11, pp. 2107-2116, Mar. 92.

[5] J. Bracken, V. Raghavan, and R. Rohrer, "Interconnect simulation with asymptotic waveform evaluation," *IEEE T-CAS*, pp. 869-878, Nov. 92.

[6] V. Raghavan, J. Bracken and R. Rohrer, "AWESpice: A general tool for efficient simulation of interconnects," *Proc.DAC*, pp. 87-92, Jun 92.

[7] E. Chiprout and M. Nakhla, *Asymptotic Waveform Evaluation and Moment Matching for Interconnect Analysis*, Boston: Kluwer Publ., 93.

[8] P. Feldmann and R. Freund, "Reduced order modeling of large linear circuits via a block Lanczos algorithm", *Proc. DAC*, pp.474-479, Jun 95.

[9] M. Silveria, M. Kamon, I. Elfadel and J. White, "A coordinate-transformed Arnoldi algorithm for generating guaranteed stable reduced-order models of arbitrary RLC circuits", in *Proc. ICCAD*, Nov. 96.

[10] A. Odabasioglu, M. Celik and L. T. Pillage, "PRIMA: Passive Reduced-Order Interconnect Macromodeling Algorithm", *Proc. IEEE ICCAD*, pp. 58-65, Nov. 1997..

[11] D. Xie and M. Nakhla, "Delay and crosstalk simulation of VLSI interconnects with nonlinear terminations," *IEEE T-CAD*, pp 1798-1811, Nov. 92.

[12] R. Achar, M. Nakhla and Q. J. Zhang, "Addressing high frequency issues in VLSI interconnects with full wave model and Complex Frequency Hopping", *Proc. ICCAD*, pp. 53-56, Nov. 1995.

[13] G. Zheng, Q. J. Zhang, M. Nakhla and R. Achar, "An efficient approach for moment-matching simulation of linear subnetworks with measured or tabulated data", *Proc. ICCAD*, pp. 20-23, Nov. 96.

[14] R. Achar, M. Nakhla and E. Chiprout, "Block CFH: A model-reduction technique for multiport distributed interconnect networks", in *Proc. ECCTD*, pp. 396-401, Sept. 97.

[15] E. G. Gilbert, "Conrollability and observability in multivariable control systems", SIAM J. Control, pp. 128-151, 1963.

[16] T. Kailath, *Linear Systems*. Toronto: Printce-Hall Inc., 1980

Multipoint multiport algorithm for passive reduced-order model of interconnect networks

Qingjian Yu, Janet M.L. Wang and Ernest S. Kuh
Electronics Research Laboratory
University of California at Berkeley
Berkeley CA 94720

Abstract

In this paper, we prove that any real full-rank congruence transformation applied to a system of RLC interconnect described by MNA or state equations will result in a passive network, and provide an algorithm for multipoint multiport passive model-order reduction of RLC interconnect networks. The algorithm extends the matching points from real s-value to $j\omega$ and ∞ frequencies, and treats the moment matching for each port individually so that it is more flexible than the block-wise algorithms.

1 Introduction

With the rapid increase of signal frequency and decrease of feature sizes of IC's, interconnects play more and more important roles. Very often, interconnects are modeled as lumped or distributed RLC networks. The size of such networks is usually so large that simulation cost is not affordable, and to form a reduced order model for such networks is a key to the fast simulation of interconnects.

There are two basic requirements for the reduced order model. The first one is that it should be passive as in its original network so that when it is connected to other passive networks the whole circuit will remain stable. The second one is that the model should keep the characteristics of the original network to some extent. It has been proven in practice that if moment matching of the elements of the port matrix is kept to some degree the model will have good accuracy.

In recent years, a number of papers have been published for the formation of reduced order model, especially for passive model order reduction with multipoint moment matching of multiport matrix [1, 2, 3]. These algorithms are based on the MNA equations of circuits, where moment matching at finite real value in the s-domain only is concerned and block-wise implementation is used. In this paper, we prove that any real full-rank congruence transformation applied to a system of RLC interconnect described by MNA or state equations will result in a passive network, and provide an algorithm for multipoint multiport passive model-order reduction of RLC interconnect networks. The algorithm extends the matching points from real s-value to $j\omega$ and ∞ frequencies, and treats the moment matching for each port individually so that it is more flexible than the block-wise algorithms.

2 Passivity preservation by congruence transformation

2.1 MNA equations

For an RLC m-port driven by voltage sources, its MNA equations can be written in the form of

$$H_a x_a = b_a u \tag{1}$$

where $u = V_p$ is the port voltage vector, x_a consists of node voltage vector V_n, inductor current vector I_L and port current vector I_p, $b_a = [0\ 0\ I]^T$ and

$$H_a = \begin{bmatrix} sC+G & A_L & -A_p \\ -A_L^T & sL+R & \\ A_p^T & & \end{bmatrix} \tag{2}$$

where C and G are nodal capacitance and conductance matrices, L and R are branch inductance and resistance matrices of inductor-resistor branches, and A_L and A_p are the node-branch incidence matrices of the inductor-resistor branches and voltage source branches, respectively. Then,

$$I_p = b^T x_a \tag{3}$$

and the port admittance matrix

$$Y(s) = b^T H_a^{-1} b \tag{4}$$

2.2 State equations

Let $H_a = sM_a + N_a$. As M_a is singular, the MNA equations cannot be directly used to deal with moment matching at $s = \infty$ and we will rely on the state equations of the circuit. Let $V_n = [V_p^T V_x^T]^T$, where V_x is the unknown node voltages. For interconnects, each node is connected to a grounded capacitance, and $x = [V_x^T, I_L^T]^T$ is the state vector. The state equations can be derived from the MNA equations and be described in the following form

$$Hx = bu \qquad (5)$$

where

$$H = \begin{bmatrix} sC_{xx} + G_{xx} & A_{Lx} \\ -A_{Lx}^T & sL + R \end{bmatrix} \qquad (6)$$

$$b = \begin{bmatrix} -(G_{xp} + sC_{xp}) \\ A_{Lp}^T \end{bmatrix} \qquad (7)$$

and the output equations can be written in the form of

$$I_p = c^T x + d^T u \qquad (8)$$

where $c^T = [G_{px} + sC_{px} \; A_{Lp}]$ and $d^T = G_{pp} + sC_{pp}$. In the above expressions, matrices G_{xp}, C_{xp} and A_{Lp} etc. are submatrices of matrices G, C and A_L respectively, where the subscripts p and x refer to the port nodes and internal nodes, respectively. Then, the admittance matrix of the network can be expressed as

$$Y(s) = d^T + c^T H^{-1} b \qquad (9)$$

Note that $c \neq b$ in the general case.

2.3 Passivity preservation

It is easy to verify that $H_a(s)$ and $Y(s)$ are positive real matrices [4]. Now suppose that $x_a \in R^l$ and a congruence transform $P \in R^{l \times p}$ with $p \leq l$ is applied to the original system and results in a transformed system

$$\hat{H}_a \hat{x}_a = \hat{b}_a u \qquad (10)$$

and

$$y = \hat{b}_a^T \hat{x}_a \qquad (11)$$

where $\hat{x}_a = Px_a$, $\hat{H}_a = P^T H_a P$ and $\hat{b}_a = P^T b_a$. Then, the admittance matrix of the transformed system is

$$\hat{Y} = \hat{b}_a^T \hat{H}_a^{-1} \hat{b}_a \qquad (12)$$

and we have the following conclusion.
Proposition 1
If P is real and of full rank, then $\hat{H}_a$ is positive-real, and the transformed system is passive.
Proof.

$$\hat{H}_a(s) + \hat{H}_a(s^*)^T = P^T H_a(s) P + (P^T H_a(s^*) P)^T$$
$$= P^T (H_a(s) + H_a(s^*)^T) P$$

As matrix H_a is positive real, when $Re(s) > 0$ $H_a(s) + H_a(s^*)^T$ is nonnegative-definite, so is $P^T(H_a(s) + H_a(s^*)^T)P$ and $\hat{H}_a$ is positive-real. □

Now we consider the case that a congruence transform U is applied to the state equations, we will have the admittance matrix of the transformed system

$$\hat{Y}(s) = \hat{d}^T + \hat{c}^T \hat{H}^{-1} \hat{b} \qquad (13)$$

where $\hat{H} = U^T H U$, $\hat{b} = U^T b$ and $\hat{c}^T = c^T U$.
Def.1 Equivalent congruence transform
A congruence transform P applied to the MNA description of an RLC multi-port is said to be equivalent to a congruence transform U applied to the state description of the same port if they generate the same port matrix of the transformed system.

In fact, for any transform U applied to the state equations, we can form an equivalent transform P for its MNA equations as $P = diag(I, U, I)$ where I is an $m \times m$ identity matrix (See [5] for the proof). From Proposition 1, we have the following conclusion:
Proposition 2
Let n be the number of state variables of an RLC network. Any real full-rank transform $U \in R^{n \times p}$ ($p \leq n$) applied to the state equations of an RLC multi-port will result in a passive reduced-order network.

3 Multipoint moment matching

Let $H = sM + N$ and assume that d is null since it has no effects on the moment matching algorithms. Let $m_{ij}^k(s_q)$ be the k-th order moment of y_{ij} at matching point s_q, where s_q may be 0, ∞, or $j\omega_q$. Let $b = [b_1, \ldots, b_m]$ with $b_j = b_{j0} + sb_{j1}$. For finite s_q, let $H(s)^{-1}b_j$ be expanded as $H(s)^{-1}b_j = \sum_{k=0}^{\infty} r_{jk}(s_q)(s - s_q)^k$. Let $N(s_q) = N + s_q M$ and $b_{j0}(s_q) = b_{j0} + s_q b_{j1}$. Then,

$$r_{j0}(s_q) = N(s_q)^{-1} b_{j0}(s_q) \qquad (14)$$

$$r_{j1}(s_q) = A(s_q) r_{j0}(s_q) + N(s_q)^{-1} b_{j1} \qquad (15)$$

where $A(s_q) = -N(s_q)^{-1}M$, and

$$r_{jk}(s_q) = A(s_q) r_{j,k-1}(s_q) \quad k > 1 \qquad (16)$$

For $s_q = \infty$, let $H(s)^{-1}b_j$ be expanded as $H(s)^{-1}b_j = \sum_{k=0}^{\infty} r_{jk}(\infty)s^{-k}$. Let $B = -M^{-1}N$. Then,

$$r_{j0}(\infty) = M^{-1} b_{j1} \qquad (17)$$

$$r_{j1}(\infty) = Br_{j0}(\infty) + M^{-1}b_{j0} \quad (18)$$

and

$$r_{jk}(\infty) = Br_{j,k-1}(\infty) \quad k > 1 \quad (19)$$

For any j, let $n_j(s_q)$ be the maximum moment matching order for all y_{ij} at $s = s_q$ and denote the Krylov subspace $K^j(s_q, n_j(s_q)) = \{r_{j0}(s_q), r_{j1}(s_q), \ldots, r_{j,n_j(s_q)}(s_q)\}$. Then, we have the following conclusion:

Proposition 3

Let $\hat{m}_{ij}^k(s_q)$ be the k-th order moment of $\hat{y}_{ij}$ at $s = s_q$. If U is an orthonormal congruence transform and $K^j(s_q, n_j(s_q)) \in span(U)$, then for all $1 \leq i \leq m$, when s_q is finite or $s_q = \infty$ and $C_{px} = 0$,

$$\hat{m}_{ij}^k(s_q) = m_{ij}^k(s_q) \quad 0 \leq k \leq n_j(s_q) \quad 1 \leq i \leq m \quad (20)$$

and when $s_q = \infty$ and $C_{px} \neq 0$,

$$\hat{m}_{ij}^k(s_q) = m_{ij}^k(s_q) \quad -1 \leq k < n_j(s_q) \quad 1 \leq i \leq m \quad (21)$$

(See the proof in [6]).

4 Modified multipoint multiport Arnoldi algorithm

Let m be the port number of an RLC network, and $M_j = \{n_j(0), n_j(\infty), (n_{j1}, s_1), \ldots, (n_{j,q_j}, s_{q_j})\}$ be the matching set of the j-th input, where $n_j(0)$ and $n_j(\infty)$ are the maximum matching order at $s = 0$ and $s = \infty$, respectively; $s_i = j\omega_i$ and n_{ji} are the maximum matching order at $s = s_i$. Let $MA = \{M_1, M_2, \ldots, M_m\}$ be the matching set of all inputs. The Input-oriented Modified Multipoint Multiport Arnoldi algorithm will result in an orthonormal matrix U which will meet with the moment matching requirements specified to each input. This algorithm is described as follows:

IMMMA - *Input-oriented Modified Multipoint Multiport Arnoldi algorithm*
{Input: Matrices M, N and b, Matching set MA.
  Output: Orthonomal matrix U.
  $U = \phi$;
  $j = 0$; /* j – number of column vectors in U */
  for $i = 1 : m$
  {if($k_{i0} \geq 0$) moment_matching_at_s = $0(i, k_{i0}, U, j)$;
   if($k_{i\infty} \geq 0$)
     moment_matching_at_s = $\infty(i, k_{i\infty}, U, j)$;
   for $k = 1 : q_i$
     moment_matching_at_s = $j\omega(i, k_{ik}, \omega_{ik}, U, j)$;
  }
}

The three moment matching functions are as follows:
moment_matching_at_s = $0(i, k, U, j)$;
{ Solve $Nr = b_{i0}$ for r; orthonormal(r, U, j);
  if ($k < 1$) return;
  solve $Nr_1 = -Mr$ for r_1; solve $Nr_2 = b_{i1}$ for r_2;
  $r = r_1 + r_2$; orthonormal(r, U, j);
  for $kk = 2 : k$
  { solve $Nr_1 = -Mr$ for r_1; $r = r_1$;
    orthonormal(r, U, j); }}

The fuction moment_matching_at_s = $\infty(i, k, U, j)$ is similar to the above function with the interchange of M and N and b_{i0} and b_{i1}, respectively.

moment_matching_at_s = $j\omega(i, k, \omega, U, j)$;
{$\bar{N} = N + j\omega M$;
  $b_0 = b_{i0} + j\omega b_{i1}$; Solve $\bar{N}r = b_0$ for r;
  $r_a = real(r)$; orthonormal(r_a, U, j);
  $r_b = imag(r)$; orthonormal(r_b, U, j);
  if ($k < 1$) return;
  solve $\bar{N}r_1 = -Mr$ for r_1; solve $\bar{N}r_2 = b_{i1}$ for r_2;
  $r = r_1 + r_2$;
  $r_a = real(r)$; orthonormal(r, U, j);
  $r_b = imag(r)$; orthonormal(r_b, U, j);
  for $kk = 2 : k$
  { solve $\bar{N}r_1 = -Mr$ for r_1; $r = r_1$;
    $r_a = real(r)$; orthonormal(r_a, U, j);
    $r_b = imag(r)$; orthonormal(r_b, U, j); }}

In these algorithms, function orthonormal(r, U, j) is as follows:
orthonormal(r, U, j)
{ if ($U == \phi$) $u_1 = normalize(r)$;
  else
  { for $k = 1 : |U|$ $r = r - u_k^T r u_k$;
    $j = j + 1$; $u_j = normalize(r)$; } }

Correspondingly, we can form an output-oriented algorithm, where the i-th moment matching set is specified to output i, i.e., $\hat{y}_{ij}$ and y_{ij} will meet the moment matching requirement set by M_i for all j. Note that

$$Y(s)^T = d + b^T(H^T)^{-1}c \quad (22)$$

and $H^T = (sM + N)^T = sM + N^T$. Therefore, if in IMMMA, b is replaced by c and N is replaced by N^T, we will get OMMMA - an Output-oriented Modified Multipoint Multiport Arnoldi algorithm.

Note that if the matching sets for all inputs are the same, then IMMMA will result in an orthonormal matrix with the same size generated by the block Arnoldi algorithm. Therefore, IMMMA is generally more flexible and at least as good as the block Arnoldi algorithm.

5 Examples and Conclusions

We have successfully tested several examples, and we show one borrowed from [9]. The circuit is shown in Fig.1, where each set of coupled lines is modeled by a number of coupled RLGC sections. The transfer function $V_{out}(s)/V_{in}(s)$ is shown in Fig.2, and the time-domain response $V_{out}(t)$ is shown in Fig.3 when $V_{in}(t)$ is a pulse. The solid and dashed lines represent the exact solution and the solution from our model, respectively, and they are indistinguishable. Compared with [9] where moment matching at $s = 0$ up to the order of 40 is used, we use moment matching at $s = 0$ with order 4, and at 1.5, 3, 4, and 5 GHz with order 0, and the order of the our reduced model is about 1/3 of theirs.
We have proved that any real full-rank congruence transform applied to the MNA or state equations of an RLC multiport will result in a passive network, and provide a multipoint multiport algorithm for passive reduced-order vmodel of RLC networks. The algorithm includes ∞ and points on the $j\omega$ axis as matching points and can treat the moment matching regarding each input or output independently.

References

[1] K.J.Kerns and A.T.Yang,"Preservation of passivity during RLC network reduction via split congruence transformations," Proc. of 34-th DAC, pp.34-39, 1997.

[2] A.Odabasioglu el al, "PRIMA: Passive reduced- order interconnect macromodeling algorithm", Proc. of ICCAD'97, pp.58-65, Nov. 1997.

[3] I.M.Elfadel et al, "A block rational Arnoldi algorithm for multipoint passive model-order reduction of multiport RLC networks," Proc. of ICCAD'97, pp.66-71, Nov. 1997.

[4] E.S.Kuh and R.A.Rohrer, Theory of linear active networks. Holden-Day Inc., 1967.

[5] Q.Yu, "Equivalent congruence transforms", Technical Report M97/69, ERL U.C.Berkeley, July, 1997.

[6] Q.Yu, "Moment matching in congruence transform," Technical Report M97/71, ERL U.C.Berkeley, Sept., 1997.

[7] Q.Yu and E.S.Kuh,"Reduced order model of transmission lines with preservation of passivity and moment matching at multiple points", Technical Report M97/70, ERL U.C.Berkeley, July 1977 and to be published in Proc. Nolta'97, Nov. 1997.

[8] A.Ruhe, "The rational Krylov algorithm for nonsymmetric eigenvalue problem III: complex shifts for real matrices," BIT 34, pp.165-176, 1994.

[9] M.Celik et al, "Simulation of dispersive multiconductor transmission lines by *Padé* approximation via the Lanczos process," IEEE Trans. on MTT, vol.44, No.12, pp.2525-2535, Dec. 1966.

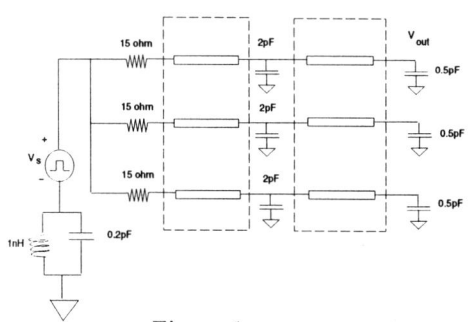

Figure 1.

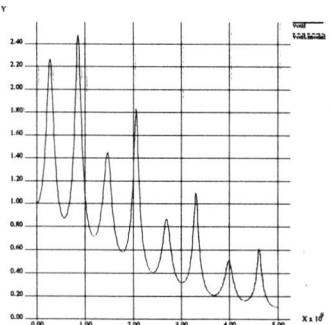

Figure 2.

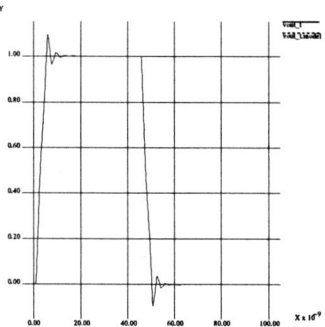

Figure 3.

APPLICATIONS OF COMPLEX FREQUENCY HOPPING METHOD IN PCB SIGNAL INTEGRITY SIMULATION

Zhen Mu

Cadence Design Systems, Canada Inc.
2745 Iris Street, Ottawa, Ontario, Canada K2C 3V5

ABSTRACT

This paper presents a fast Signal Integrity(SI) simulator using a newly published moment matching technique, Complex Frequency Hopping(CFH), which can efficiently simulate boards containing a large number of interconnects with lossy parameters. For the first level examination of an entire board, this simulator provides an excellent solution to delay, reflection, and cross-talk by using linear component models. For the detailed analysis of an interconnection network with non-linear components, this simulator creates equivalent time domain macro-models to replace the linear subnetwork containing lossy interconnects. Such models can be easily used in any SPICE like simulator to efficiently perform accurate time domain analysis of the network.

1. INTRODUCTION

In high speed PCB designs, there is a critical need for fast and accurate SI tools to provide reliable information on design validations. As interconnects play a dominant role due to the increased clockrate and the decreased rise/fall time, lumped element representation of interconnects is not accurate for delay and crosstalk analysis. Distributed transmission line models become necessary. This makes analysis of boards containing interconnects and non-linear components a difficult task. The distributed transmission line models are described in frequency domain, while non-linear components require time domain analysis. Currently used SI tools try to solve this problem in two ways. One is to generate distributed models with RLC ladder networks in time domain. The other is to solve transmission lines in frequency domain and then perform convolution with the response of non-linear components. In general, these two methods produce accurate simulations but the time consumed restrict their applications to large scale interconnection networks. In the past several years, the Asymptotic Waveform Evaluation(AWE) method has attracted great attention due to its efficient analysis of interconnection networks with more accurate results and less computation cost. More recent research shows that the errors in the waveform approximation using the AWE method can be reduced by applying the AWE technique at multiple complex frequency points. The improved method is called Complex Frequency Hopping(CFH). Compared with AWE, this method promises a means to guarantee a more accurate frequency response of interconnection networks at a little CPU expense. This paper presents a CFH based SI simulator, and gives examples of analyzing interconnection networks with linear and non-linear component models.

2. THEORY

The AWE method evaluates the transfer function of a linear network at one particular frequency point on the Laplace plane s = 0 using Pade approximation. The frequency and time domain responses can then be represented by the dominant poles extracted from the transfer function. It is well known that Pade approximation can accurately reflect the actual system transfer function near the expansion point, but inaccurately far from it[4][7]. This makes it difficult to estimate the simulation error of the AWE method, and causes inaccurate results, especially at the higher frequencies. The CFH method was therefore introduced to overcome the problem addressed with the single point expansion. It generates a complete set of dominant poles from the transfer functions evaluated at multiple complex frequency points. The accuracy of the impulse response in the frequency domain is guaranteed within the frequency region determined by the multiple complex frequency expansion points. The approximation errors can be estimated between the adjacent transfer functions, and controlled with the number of expansion points.

Assuming a multi-port network with linear components and distributed transmission line models, the system matrix in the complex frequency domain can be generated in the format of the Modified Nodal Admittance(MNA) matrix $Y(s)$. The system equation is

$$Y(s)X(s) = E \qquad (1)$$

where the vector $X(s)$ includes the unknown node voltages and some branch currents, and the vector E has all the excitations.

For an arbitrary s on the Laplace plane, $X(s)$ can be expanded in Taylor series as

$$X(s) = \sum_{n=0}^{\infty} M_n (s-\alpha)^n \qquad (2)$$

A recursive equation to evaluate the system moments at a particular frequency point $s = \alpha$ can be described as

$$[Y_\alpha] M_n = -\sum_{r=1}^{n} \frac{[Y]^{(r)} M_{n-r}}{r!} \qquad (3)$$

where $Y^{(r)} = \dfrac{\partial^r Y}{\partial s^r}\bigg|_{s=\alpha}$, $Y_\alpha = Y|_{s=\alpha}$,

with $[Y_\alpha]M_0 = E$.

Then, the transfer function $H(s)$ at a port $i = 1, 2, \ldots$ at each expansion point can be estimated from the moments using Pade approximation,

$$H(s) = X_{[i]}(s) = \frac{P_L(s)}{Q_M(s)} = \frac{\sum_{j=0}^{L} a_j s^j}{1 + \sum_{j=1}^{M} b_j s^j} \quad (4)$$

The transfer functions at different expansion points can be generated using binary search strategy. The poles can be extracted from each transfer function according to the Pade approximation. The existence of the common poles to be "seen" from two adjacent transfer functions is used as the accuracy criteria to ensure the reliable region of approximations. Finally, a complete set of the dominant poles within the frequency range of interest can be constructed from the poles extracted at each expansion point. This is the original CFH method. In general, there are about 2-10 hops needed to generate a complete dominant pole set. Such poles and the associated residues can be used for linear analysis, or to construct the equivalent time domain macromodel of interconnection subnetworks for non-linear analysis. Compared with the AWE method, CFH costs more CPU time when multiple point expansions are searched. However, this is a small price to pay for the significant increase in the reliability of the results obtained.

If only a linear analysis is required, a fast CFH method was proposed by using the two adjacent transfer functions as the accuracy criteria to ensure the reliable region of the Pade approximations[3]. It does not extract any poles, but calculates the frequency response directly from the multiple transfer functions in a binary fashion. The time domain response is then obtained with the inverse FFT. Compared with the original CFH method, a fewer number of hops are needed to obtain the same accuracy, and the CPU time to extract the poles is avoided. Tests show that most of the valid nets on a general board need only 2 to 4 hops.

When non-linear components are involved in board simulation, CFH is first applied to the subnetwork containing interconnects and some lumped elements[8][9]. Assume there are n input/output buffers on the board, the subnetwork becomes an n-port linear network, the port currents follow the convention of flowing into the port, whose Z-parameter matrix is described as

$$Z(s) = \begin{bmatrix} Z_{11}(s) & Z_{12}(s) & \ldots & Z_{1n}(s) \\ Z_{21}(s) & Z_{22}(s) & \ldots & Z_{2n}(s) \\ \vdots & \vdots & & \vdots \\ Z_{n1}(s) & Z_{n2}(s) & \ldots & Z_{nn}(s) \end{bmatrix} \quad (5)$$

where each entry of the matrix is the transfer function between a pair of ports. If two dominant poles are available for the whole network, the port voltages can be derived as follows for the case n = 2,

$$\begin{bmatrix} V_1 \\ V_2 \end{bmatrix} = \begin{bmatrix} d^{1,1} + \sum_{k=1}^{2} \dfrac{r_k^{1,1}}{s - p_k} & d^{1,2} + \sum_{k=1}^{2} \dfrac{r_k^{1,2}}{s - p_k} \\ d^{2,1} + \sum_{k=1}^{2} \dfrac{r_k^{2,1}}{s - p_k} & d^{2,2} + \sum_{k=1}^{2} \dfrac{r_k^{2,2}}{s - p_k} \end{bmatrix} \begin{bmatrix} I_1 \\ I_2 \end{bmatrix} \quad (6)$$

where d_{ij} denotes direct coupling between ports i and j, p_1 and p_2 are dominant poles, $r_k^{i,j}$ are the residues corresponding to the ports and poles $i, j, k = 1, 2$. From (6), a set of ordinary differential equations with state variables can be obtained as

$$\begin{bmatrix} \dot{x}_1 \\ \dot{x}_2 \\ \dot{x}_3 \\ \dot{x}_4 \end{bmatrix} = \begin{bmatrix} p_1 & 0 & 0 & 0 \\ 0 & p_1 & 0 & 0 \\ 0 & 0 & p_2 & 0 \\ 0 & 0 & 0 & p_2 \end{bmatrix} \begin{bmatrix} x_1 \\ x_2 \\ x_3 \\ x_4 \end{bmatrix} + \begin{bmatrix} 1 & 0 \\ 0 & 1 \\ 1 & 0 \\ 0 & 1 \end{bmatrix} \begin{bmatrix} I_1 \\ I_2 \end{bmatrix} \quad (7)$$

and

$$y = \begin{bmatrix} V_1 \\ V_2 \end{bmatrix} = \begin{bmatrix} r_1^{1,1} & r_1^{1,2} & r_2^{1,1} & r_2^{1,2} \\ r_1^{2,1} & r_1^{2,2} & r_2^{2,1} & r_2^{2,2} \end{bmatrix} \begin{bmatrix} x_1 \\ x_2 \\ x_3 \\ x_4 \end{bmatrix} + \begin{bmatrix} d^{1,1} & d^{1,2} \\ d^{2,1} & d^{2,2} \end{bmatrix} \begin{bmatrix} I_1 \\ I_2 \end{bmatrix} \quad (8)$$

which, in general, is in the format of

$$\begin{aligned} \dot{X} &= AX + Bu \\ y &= CX + Du \end{aligned} \quad (9)$$

According to (9), a macromodel containing only R, C, and dependent sources can be synthesized.

If complex poles exist, complex numbers appear in the matrices A, B, and C. Note that the complex conjugate λ^* of a complex number λ is also a root of $Z_{ij}(s)$ if the complex number λ is a root of $Z_{ij}(s)$, provided all coefficients of the denominator polynomials of $Z_{ij}(s)$ are real. Assume the matrices A, B, and C include only real poles and their corresponding residues, the matrices A_1, B_1, and C_1 include only complex poles and their corresponding residues, (9) becomes

$$\begin{bmatrix} \dot{x}_1 \\ \dot{x}_2 \\ \dot{x}_3 \end{bmatrix} = \begin{bmatrix} A & 0 & 0 \\ 0 & A_1 & 0 \\ 0 & 0 & A_1^* \end{bmatrix} \begin{bmatrix} x_1 \\ x_2 \\ x_3 \end{bmatrix} + \begin{bmatrix} b \\ b_1 \\ b_1^* \end{bmatrix} u \qquad (10)$$

$$y = \begin{bmatrix} C & C_1 & C_1^* \end{bmatrix} \begin{bmatrix} x_1 \\ x_2 \\ x_3 \end{bmatrix} + Du$$

Introduce a linear transform as

$$[\hat{x}] = [T][x] \qquad (11)$$

where

$$T = \begin{bmatrix} I & 0 & 0 \\ 0 & I & I \\ 0 & jI & -jI \end{bmatrix} \qquad (12)$$

then, (10) can be solved by

$$\begin{bmatrix} \dot{\hat{x}}_1 \\ \dot{\hat{x}}_2 \\ \dot{\hat{x}}_3 \end{bmatrix} = \begin{bmatrix} A & 0 & 0 \\ 0 & Re(A_1) & Im(A_1) \\ 0 & -Im(A_1) & Re(A_1) \end{bmatrix} \begin{bmatrix} \hat{x}_1 \\ \hat{x}_2 \\ \hat{x}_3 \end{bmatrix} + \begin{bmatrix} b \\ 2Re(b_1) \\ -2Im(b_1) \end{bmatrix}$$

$$y = \begin{bmatrix} C & Re(C_1) & Im(C_1) \end{bmatrix} \begin{bmatrix} \hat{x}_1 \\ \hat{x}_2 \\ \hat{x}_3 \end{bmatrix} + Du \qquad (13)$$

It should be noticed that the Z-parameter matrix may not exist for many nets on a practical board when SI simulation is performed. In this case, an artificial resistor is placed at the port where there is no connection between a port and the ground. A negative value resistor of the original artificial one is then put at the same port inside the macromodel for compensation. Tests show that the resistor with values close to interconnect's characteristic impedance produces more stable results. A 50 Ohm resistor is therefore set up in this simulator as the default.

3. ALGORITHMS

3.1 Fast algorithm for linear analysis of interconnection networks

Step 1 Generate the transmission line moments for each interconnect, and create the MNA matrix[6] of a network;

Step 2 Estimate the maximum frequency point s_{max} based on the input function;

Step 3 Calculate the system moments using equation (3) at the frequency points of $s_{low} = 0$ and $s_{high} = s_{max}$;

Step 4 Evaluate the coefficients of the transfer functions at s_{low} and s_{high} at each pair of input/output (driver/receiver) nodes by the system moments using Pade approximation;

Step 5 For each pair of input/output nodes, calculate the frequency response at $s = s_{middle} = (s_{low} + s_{high})/2$ using the two evaluated transfer functions. If the difference between the two frequency responses is smaller than a pre-defined error criteria, the algorithm stops; otherwise, replace s_{high} with s_{middle} and repeat this step.

Once the frequency responses at every pair of driver/receiver nodes are obtained, the time domain responses can be calculated with the inverse FFT.

3.2 Macromodel realization for interconnection networks

Step 1 Generate the linear subnetwork which includes only transmission line models for interconnection and linear elements of the net;

Step 2 Check the existence of the Z-parameter matrix of the linear subnetwork; if not, place artificial resistors as described in section 2;

Step 3 Generate the MNA matrix of the linear subnetwork;

Step 4 Extract dominant poles of the subnetwork and solve the residues using CFH;

Step 5 Generate the ordinary differential equations of (13) using the dominant poles and residues;

Step 6 Create macromodel subcircuit, in the syntax of a time domain simulator, using the matrices A, B, C, and D in (13); negative artificial resistors are connected at the ports if compensation is required.

4. SIMULATION EXAMPLES

A board with 360 nets is chosen to demonstrate the algorithms. The component models on the board are of TTL and CMOS technologies. The rise/fall times of the components vary from 1ns to 5ns. Figure 1 shows a net extracted from the board. There are 6 pieces of lossless transmission lines in the net. The characteristic impedances of the transmission lines vary from 58 Ohms to 91 Ohms. The output resistance of the driver is 37 Ohms. The receivers have input resistance 10,000 Ohms and input capacitance 5 pF. The rise/fall time of the input pulse is 1ns. The estimated maximal frequency is about 2 GHz. There are only 2 hops needed to obtain the frequency and time domain responses at all eight nodes with the fast CFH method using linear buffer models. Compared with HSPICE, the voltages at the driver and receiver ends are almost identical. When the entire board is analyzed on a SPRAC 10 machine, the linear analysis needs only approximately 2 minutes. When non-linear buffer models are required, macromodel of the tree-like transmission line network is generated first. Since the Z-parameter matrix does not exist, resisters of 50 Ohms are placed at port 1(node 1), port 2(node 6), and port 3(node 8). Apply the algorithm 3.2, 2 real poles and 7 complex poles are extracted. 27 state variables are needed to create the macromodel. Three negative resistors with the value of 50 Ohms are connected at ports of the macromodel. The macromodel and the non-linear buffer models are submitted to HSPICE. The comparison of the results is showed in Figure 2. The CPU time cost to analyze this network instead of the one containing transmission line models is reduced to 20% of the direct non-linear analysis.

5. CONCLUSIONS

There are promising applications of the CFH method in fast and reliable analysis for PCB SI simulations. The simulator presented in this paper guarantees accurate results for large scaled interconnection networks with linear terminations, and bridges the frequency domain and time domain analyses for networks with interconnects and non-linear components by replacing the interconnection subnetworks with time domain equivalent macromodels.

6. REFERENCES

[1] T. K. Tang and M. Nakhla, "Analysis of high-speed VLSI interconnects using the asymptotic waveform evaluation technique", IEEE Trans. Computer-Aided Design, vol.11, pp. 341-352, Mar. 1992

[2] E. Chiprout and M. Nakhla, "Analysis of interconnect networks using complex frequency hopping(CFH)", IEEE Trans. Computer-Aided Design of Integrated Circuits and Systems, vol.14, pp. 186-200, Feb. 1995

[3] R. Sanaie, E. Chiprout, M. Nakhla, and Q. J. Zhang, "A fast method for frequency and time domain simulation of high -speed VLSI interconnects", IEEE Trans. Microwave Theory Tech., vol.42, pp. 2562-2571, Dec. 1994

[4] E. Chiprout and M. Nakhla, Asymptotic Waveform Evaluation and Moment Matching for Interconnect Analysis. Boston: Kluwer, 1993

[5] J. Vlach and K. Singhal, Computer Methods for Circuit Analysis and Design, New York: VNR, 1993

[6] D. H. Xie and M. Nakhla, "Delay and crosstalk simulation of high-speed VLSI interconnects with nonlinear terminations", IEEE Trans. Computer-Aided Design of Integrated Circuits and Systems, vol.12, No.11, pp. 1798-1811, Nov. 1993

[7] R. Achar and M. Nakhla, "Minimum realization of reduced-order high-speed interconnect macromodels", Submitted for publishing, 1996

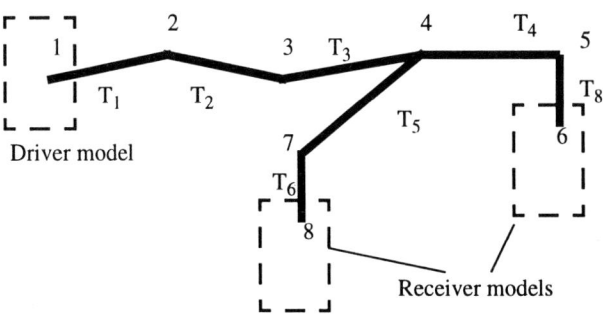

Figure 1

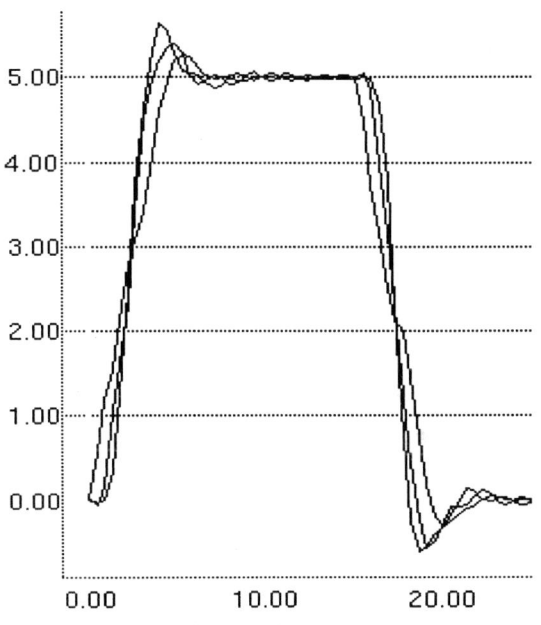

Figure 2

Time Domain Method for Reduced Order Network Synthesis of Large RC Circuits.

Shabbir H. Batterywala* H. Narayanan[†]

Abstract - In this article we present a time domain method to approximate linear RC multiport networks, accurate up to user specified frequency. We use Lanczos algorithm to compute eigenvalues, an electrical network based method to compute eigenvectors and use these to synthesize a reduced order RC network. A similar method for RC multiport approximation has been proposed by Yang and Kerns[4], which works with admittance matrices ($\mathbf{G}$ and $s\mathbf{C}$) of the multiport and also uses Lanczos algorithm to compute dominant eigenvalues and eigenvectors. However their method forbids capacitor cutsets in the multiport, which is a common occurrence in multilayered interconnect modeling, signal integrity analyses etc. We propose an algorithm which works **implicitly** on the state matrix description of the network and allows capacitor cutsets in the graph of the network. We use an efficient DC analysis algorithm[1] to make implicit computations of eigenvalues, eigenvectors and port admittance matrices.

1 Introduction

With increasing clock speed and higher levels of integration in VLSI fabrication technology, there is now a trend to consider effects of interconnect structures while doing circuit simulations to verify functional and timing specifications. Towards this the multilayered interconnects are replaced by their equivalent RC (or RLC) circuits in the SPICE netlist file of overall circuit. This increases the simulation time substantially. The overall circuit can be thought of as composed of various nonlinear multiports connected together by these linear multiports (containing only R, C and L). To perform signal integrity analyses one excites an interconnect with a spike signal and observes its effect on signals present in adjacent interconnects. This can only be done with equivalent linear multiport containing capacitor cutsets.

Among the methods used for approximate analysis of linear multiport networks, Asymptotic Waveform Evaluation (AWE) [5] figures out first. It uses Padé approximation to compute reduced order transfer functions between each pair of input and output port. In [3] a port reduction method based on block Lanczos algorithm is discussed. The method uses Padé approximation implicitly. A new method based on Congruence Transformation to reduce multiport networks is discusses in [4]. It also uses Lanczos algorithm, but is not general enough to handle capacitor cutsets in original network.

In this paper, we present a method of Spectral Approximation for Reducing Networks (SARN), to generate lower order approximations of linear RC multiports. Our method allows capacitor cutsets in the multiport to be approximated. We specify the multiport by SPICE like circuit file, with ports specified as boundary nodes, looking into which we give a smaller equivalent circuit accurate up to a user specified frequency. We compute eigenvalues and eigenvectors of state matrix of given RC multiport and use that to synthesize a lower order model. By retaining all eigenvalues with absolute value up to λ_{max} we get an approximation which (informally) is accurate up to frequency $f_{max} = \lambda_{max}/\alpha(2\pi)$, where α can be taken to lie between 5 and 10. If the original multiport is stable (i.e., having all eigenvalues as negative real numbers), the approximation is also stable, since it has eigenvalues which are also present in the original multiport and hence are negative real numbers.

2 Basic Formulation of SARN

We now give a brief sketch of our method. Let $\mathcal{N}$ be the given RC multiport, with p ports specified in terms of $p+1$ boundary nodes. We will refer to these ports as P ports. Let $\mathbf{I}_P(s)$ ($\mathbf{V}_P(s)$) be the Laplace transform of P port current (voltage) vector. Let $\mathbf{I}_P(s) = \mathbf{Y}(s)\mathbf{V}_P(s)$. Then the port admittance matrix $\mathbf{Y}(s)$ can be written as

$$\mathbf{Y}(s) = \mathbf{G} + \mathbf{C}s + \sum_{r=1}^{n} (\mathbf{K}^r)\frac{1}{s-\lambda_r} \quad (1)$$

where n is the total number of states (in this case total number of independent capacitor voltages), λ_rs are eigenvalues of a state matrix of $\mathcal{N}$ and $\mathbf{K}^r$s are rank one matrices, which can be obtained from the corresponding eigenvectors for λ_rs (we assume that all the eigenvalues are distinct, which is a very practical assumption). $\mathbf{G}_0 = \mathbf{G} - \sum_{r=1}^{n}(1/\lambda_r \mathbf{K}^r)$ represents the behavior of the multiport at $s = 0$. Hence $\mathbf{G}_0$ is the admittance matrix of the multiport obtained by open circuiting and deleting all capacitors in $\mathcal{N}$ (capacitor C has admittance sC which is zero at $s = 0$). Next $s\mathbf{C}$ represents the behavior of the multiport at $s = \infty$ and could be obtained as port admittance matrix by open circuiting and deleting all resistors in $\mathcal{N}$ (as $s \to \infty$, observe that admittance due to a capacitor would dominate over admittance

---

*(battery@ee.iitb.ernet.in) EE Dept. IIT Bombay
[†](hn@ee.iitb.ernet.in) EE Dept. IIT Bombay, Mumbai-400076, INDIA, Fax +91-22-5783480

G_0^i, the ith column of G_0, could be obtained by computing currents in P port branches when ith port branch is replaced by a voltage source of unit value, all other port branches replaced by short circuits (voltage source of zero value) and all capacitors inside the multiport $\mathcal{N}$ replaced by open circuits. This requires p solutions of a DC network containing voltage sources and positive resistors. With the method specified in [1] (where for a circuit containing voltage sources, current sources and positive resistors, KCE are written such that it results in a positive definite matrix), these p solutions of DC networks can be done very efficiently. Note that the same Cholesky factors can be used as only the voltage source values are changed in p network solutions. Similarly $\mathbf{C}^i$, the ith column of $\mathbf{C}$ could be obtained by open circuiting all resistors inside $\mathcal{N}$, replacing each capacitor of value C Farads by a conductance of value $C\mho$ and replacing port branches by voltage sources as above and then solving for current through them. Once again p solutions of a DC network are required and same Cholesky factors can be used.

We now talk about the computation of eigenvalues λ_r and matrices $\mathbf{K}^r$. First we sketch the implicit state variable description of $\mathcal{N}_0$, the network obtained by short circuiting the input P ports of $\mathcal{N}$. Let f_c be a forest of capacitors in $\mathcal{N}_0$. Let X be a new forest built up of star trees cospanning each connected component of f_c. We will refer to these X branches as X ports. Let $\mathcal{N}_C$ be the multiport in $\mathcal{N}_0$ containing only capacitors of $\mathcal{N}$ and port branches X in parallel with the capacitors in f_c. Let $\mathcal{N}_R$ be the multiport in $\mathcal{N}_0$ containing only resistors of $\mathcal{N}$ and port branches as above. Thus $\mathcal{N}_0$ can be thought of as being obtained by 'parallel' connection of $\mathcal{N}_C$ and $\mathcal{N}_R$. Observe that the rank of the subgraph of capacitors in $\mathcal{N}_0$ is the number of elements in f_c and *no condition is imposed to forbid the capacitor cutsets*. The branch voltages in ports X can be taken as state variables. Let $\mathbf{i}_C$ be the current entering $\mathcal{N}_C$ (and hence coming out of $\mathcal{N}_R$). Let $\mathbf{v}_C$ be the voltage across the port branches X in $\mathcal{N}_C$ (and hence in $\mathcal{N}_R$). Then $\mathbf{i}_C(t) = \mathcal{C}\dot{\mathbf{v}}_C(t)$, where $\mathcal{C} = A_{Cr}\hat{\mathcal{C}}A_{Cr}^T$ is the coupled port capacitance matrix of $\mathcal{N}_C$, $\hat{\mathcal{C}}$ is the diagonal matrix containing values of all capacitors in $\mathcal{N}_C$ and A_{Cr} is the reduced incidence matrix of $\mathcal{N}_C$ with port branches X open circuited.

We now derive the state matrix from port admittance matrix relating $\mathbf{i}_P(t), \mathbf{i}_C(t), \mathbf{v}_P(t), \dot{\mathbf{v}}_P(t)$ and $\mathbf{v}_C(t)$. It is easy to see that the current vectors $\mathbf{i}_P(t)$ and $\mathbf{i}_C(t)$, when $\mathbf{v}_P^j(t) = \delta(t), \mathbf{v}_P^i(.) = 0$ for $i \neq j$ would have a $\dot{\delta}(t)$ term, $\delta(t)$ term and $e^{\lambda_i t}$ terms. Let us therefore begin by taking

$$\begin{bmatrix} \mathbf{i}_P(t) \\ -\mathbf{i}_C(t) \end{bmatrix} = \begin{bmatrix} \hat{G}_{PP} & G_{PP} & G_{PC} \\ \hat{G}_{CP} & G_{CP} & G_{CC} \end{bmatrix} \begin{bmatrix} \dot{\mathbf{v}}_P(t) \\ \mathbf{v}_P(t) \\ \mathbf{v}_C(t) \end{bmatrix} \quad (2)$$

Then the state equations for $\mathcal{N}$, taking $\mathbf{v}_C$ as state variables is

$$\begin{aligned} \dot{\mathbf{v}}_C(t) = & -\mathcal{C}^{-1}G_{CC}\mathbf{v}_C(t) - \mathcal{C}^{-1}G_{CP}\mathbf{v}_P(t) \\ & -\mathcal{C}^{-1}\hat{G}_{CP}\dot{\mathbf{v}}_P(t) \end{aligned} \quad (3)$$

3 Transformation for Symmetric State Matrix

We need to compute eigenvalues of the matrix $-\mathcal{C}^{-1}G_{CC}$, which (though G_{CC} is symmetric) is asymmetric and hence difficult to handle with the Lanczos algorithm. So we transform the Equation 3 to another one, in terms of variables $\mathbf{z}$ for which the state matrix is symmetrical. Let $\mathcal{C} = \mathrm{LL}^T$, where L is a lower triangular matrix. This Cholesky decomposition is possible since $\mathcal{C}$ is a positive definite matrix. Let $\mathbf{z} = \mathrm{L}^T \mathbf{v}_C$. Then Equation 3 transforms to

$$\begin{aligned} \dot{\mathbf{z}}(t) = & -\mathrm{L}^{-1}G_{CC}\mathrm{L}^{-T}\mathbf{z}(t) - \mathrm{L}^{-1}G_{CP}\mathbf{v}_P(t) \\ & -\mathrm{L}^{-1}\hat{G}_{CP}\dot{\mathbf{v}}_P(t) \end{aligned} \quad (4)$$

The state matrix $\hat{A} = -\mathrm{L}^{-1}G_{CC}\mathrm{L}^{-T}$ is symmetric and hence the eigenvectors are mutually orthogonal.

4 Implicit Computations

Note that the matrices in Equation 2, 3 and 4 are never computed explicitly. We only compute $\mathcal{C}$ and its factors L explicitly. Computation of $\hat{A}\mathbf{y}, (\hat{A} - kI)^{-1}\mathbf{y}, G_{CC}\mathbf{y}, G_{CP}\mathbf{y}, G_{PC}\mathbf{y}$ and $\hat{G}_{CP}\mathbf{y}$ for a given $\mathbf{y}$ are all interpreted as DC analyses of appropriately constructed networks.

We compute the eigenvalues λ_r of $\hat{A}$ by Lanczos method [2], which requires the computation of $\hat{A}\mathbf{y}$ for a given vector $\mathbf{y}$. Having computed L explicitly, $\hat{A}\mathbf{y} = -\mathrm{L}^{-1}G_{CC}\mathrm{L}^{-T}\mathbf{y}$ can be computed implicitly, provided we can compute $G_{CC}\mathbf{u}$ for a given vector $\mathbf{u}$. For computing $G_{CC}\mathbf{u}$, we replace port branches X in $\mathcal{N}_R$ by voltage sources of value $\mathbf{u}$ and then find the current through these voltage sources.

In order to synthesize a lower order approximation we need to compute eigenvalues from the lower end of spectrum of $\hat{A}$. This can often be done with much lesser Lanczos iterations if we compute the eigenvalues of $\hat{A}^{-1}$. However, due to the presence of capacitor cutsets, $\hat{A}$ has zero eigenvalues and hence is not invertible. To make $\hat{A}$ invertible, we shift its eigenvalues λ_i to $\lambda_i - \mathbf{k}$ by adding in parallel with each capacitor of value C Farads a conductance of value $kC\mho$. This is equivalent to adding a diagonal matrix $-\mathbf{k}I$ to $\hat{A}$. Then we compute eigenvalues of $(\hat{A} - \mathbf{k}I)^{-1}$ using the Lanczos algorithm which requires computation of $(\hat{A} - \mathbf{k}I)^{-1}\mathbf{y}$ for a given $\mathbf{y}$. Equivalently, we need to compute $(\hat{A} - \mathbf{k}I)^{-1}\mathbf{y} = \mathrm{L}^T(G_{CC} + \mathbf{k}\mathcal{C})^{-1}\mathrm{L}\mathbf{y}$, for a given $\mathbf{y}$. This requires computation of $(G_{CC} + \mathbf{k}\mathcal{C})^{-1}\mathbf{u}$ for a given $\mathbf{u}$. For this, first we add the parallel conductance branches, then we replace all port X branches by current sources of value $\mathbf{u}$ and then find the voltage across these current sources. Having computed top r eigenvalues (in sense of absolute value) λ_{ir} of $(\hat{A} - \mathbf{k}I)^{-1}$, the corresponding eigenvalues λ_r of $\hat{A}$ can be computed as, $\lambda_r = (1/\lambda_{ir}) + \mathbf{k}$. Note that if $\mathbf{k}$ is negative then the electrical networks to be solved contains negative resistors. In such cases the method discussed in [1] has to factor an indefinite matrix using LU factors as opposed to Cholesky factors for positive definite matrices.

Having computed r eigenvalues of $\hat{A}$ with least absolute values, we refine it by shifted inverse power iteration [6], wherein we compute the largest eigenvalue of $(\hat{A}-\lambda_r)^{-1}$ to refine the eigenvalue λ_r.

To compute eigenvector $\mathbf{p}_r$ corresponding to λ_r we first compute eigenvector $\mathbf{q}_r$ of asymmetric state matrix $-\mathcal{C}^{-1}G_{CC}$ and then compute $\mathbf{p}_r = \mathbf{L}^T\mathbf{q}_r$. To compute $\mathbf{q}_r$ we replace all the capacitors (with value C_i) in $\mathcal{N}_0$ by conductances of value $\lambda_r C_i$, P ports by shorts and X (state variable) ports by open circuits. We put a voltage source of unit value at one of these state variable ports. Then the voltage vector that exists across ports X is the required vector $\mathbf{q}_r$.

Having computed the eigenvalues up to a user specified frequency we compute $\mathbf{K}^r$ as

$$\mathbf{K}^r = G_{PC}\mathbf{L}^{-T}\mathbf{p}_r\mathbf{p}_r^T\mathbf{L}^{-1}G_{CC}\mathcal{C}^{-1}\hat{G}_{CP}$$
$$-G_{PC}\mathbf{L}^{-T}\mathbf{p}_r\mathbf{p}_r^T\mathbf{L}^{-1}G_{CP} \quad (5)$$

Note that $\mathbf{K}^r$ is a symmetric matrix. It has rank one due to the presence of $\mathbf{p}_r\mathbf{p}_r^T$ in its expression. Hence it suffices to compute any linear combination of its rows or columns. We compute sum of all its columns. This can be computed by post multiplying Equation 5 by $\hat{1}$, vector with all entries of value 1. For this, we need $G_{CP}\hat{1}$ and $\hat{G}_{CP}\hat{1}$. $G_{CP}\hat{1}$ is computed by replacing the ports X by shorts in $\mathcal{N}$, then replacing all capacitors by open circuits, setting $\mathbf{v}_P^j(t) = 1$, for all j and solving for current flowing into the ports X. $\hat{G}_{CP}\hat{1}$ is computed by replacing the ports X by shorts, setting $\mathbf{v}_P^j(t) = 1$, for all j, replacing all resistors by open circuits, all capacitors by conductances of value C and solving for current flowing in the ports X. To compute $\hat{G}_{PC}\mathbf{y}$ for a given vector y we replace ports P by shorts, capacitors in $\mathcal{N}$ by open circuits and ports X by voltage vector $\mathbf{y}$. Then let the current vector leaving ports P be i_P^1 and the current vector entering ports X be i_C. Next we replace ports P by shorts, resistors in $\mathcal{N}$ by open circuits, capacitors (of value C_i) in $\mathcal{N}$ by conductances of value C_i, make the current vector i_C enter the ports X and compute the current vector i_P^2 leaving ports P. Then $\hat{G}_{PC}\mathbf{y} = i_P^1 + i_P^2$. Having computed $G_{CP}\hat{1}$ and $\hat{G}_{CP}\hat{1}$, $\mathbf{K}^r\hat{1}$ can be obtained by successive premultiplications by matrices (vectors) present in Equation 5.

5 Synthesis

Now finally, we briefly sketch the synthesis of an m-th order RC multiport as an approximation for $\mathcal{N}$. Note that each row (column) index of $\mathbf{K}^r$ corresponds to a boundary node (or a port) of $\mathcal{N}$. We can synthesize star like RC circuit connecting all the boundary nodes to an internal node via resistors (of positive or negative) value. The internal node is connected to ground via a capacitor of appropriate value. Then we subtract the zero frequency and infinite frequency admittance matrices of this star tree like RC circuit from $\mathbf{G}_0$ and $\mathbf{C}$ respectively (see Equation 1). We synthesize one such star like RC network for each distinct eigenvalue. Finally whatever remains in $\mathbf{G}_0$ ($\mathbf{C}$) can be synthesized as positive and negative resistor (capacitor) connecting boundary nodes.

6 Experimental Results

We conducted experiments on randomly generated RC grid circuits, which contained floating capacitors, grounded capacitors and capacitor cutsets. Table 1 gives the specifications of test circuits used in this paper. Table 2 gives the

Network	n	R	C	X	P
g5h	500	768	272	261	5
g1k	1000	1657	478	463	10
g2k	2000	3310	1070	1007	10
g5k	5000	8385	2265	2189	10
g10k	10000	19750	7805	7796	10
g20k	20000	29906	14644	13436	10
g30k	30000	44647	22203	20316	10

Table 1: Specifications of test circuits: Number of nodes, resistors, capacitors, state variables and ports

SPICE simulation time for original circuit and time taken to generate lower order models. All these lower order models took less than 0.5 seconds for their SPICE simulations. All experiments were performed on Pentium Pro machines with 64 MB RAM and 200 MHz clock speed. To ascertain

Network	2 pole	5 pole	10 pole	SPICE
g5h	1.28	1.65	2.37	12.44
g1k	4.03	5.07	7.34	45.66
g2k	9.84	13.41	17.88	115.08
g5k	53.51	70.5	98.83	644.34
g10k	170.46	215.81	298.81	1594.64
g20k	365.62	448.37	614.68	—
g30k	558.75	708.04	966.76	—

Table 2: Comparison of model reduction time with SPICE simulation time

the correctness and numerical stability of our port reduction method we compare plots of voltage magnitude vs frequency for g5h circuit and its resynthesised version, which was generated using all the eigenvalues and eigenvectors of g5h. An input source was applied to one of the ports. In Figure 1 we plot the voltages at remaining ports for original and resynthesised circuit. It may be noted that the plots for the original and resynthesised circuit are indistinguishable. In Figure 2 we compare a port voltage for g10k circuit and its 2nd, 5th and 10th order models. Note that as we take more eigenvalues for synthesising a lower order model the approximation gets more accurate. In this example a lower 10th order approximation gives a very accurate model. In Figure 3 we compare real and imaginary part for a port voltage of g5k and its 10th order model. Finally in Figure 4 we compare a port voltage for g10k circuit with its corresponding counterpart in 10th order models, one generated by taking the eigenvalues as computed through Lanczos algorithm and other generated after refining these eigenvalues. Note that unrefined eigenvalues give a poorer approximation.

Due to lack of space we are not giving complexity analysis

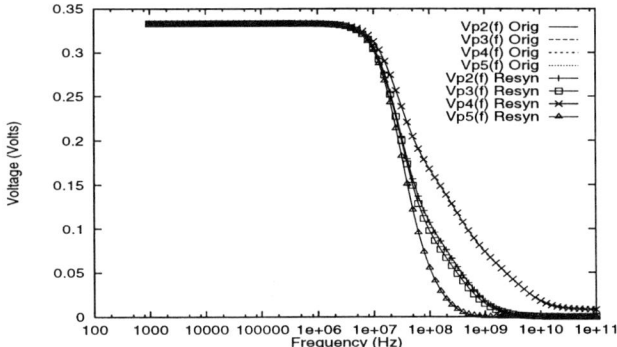

Figure 1: Comparison of port voltages for original and resynthesised circuit

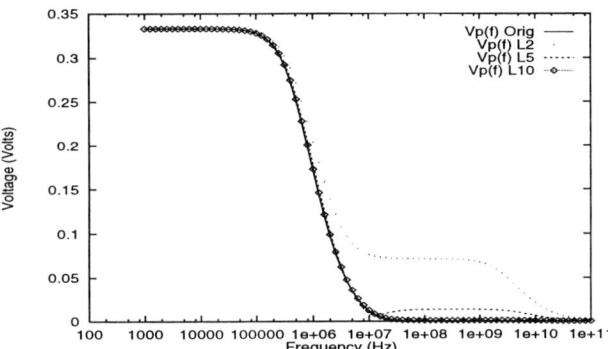

Figure 2: Comparison of a port voltage for original and resynthesised circuits of different order

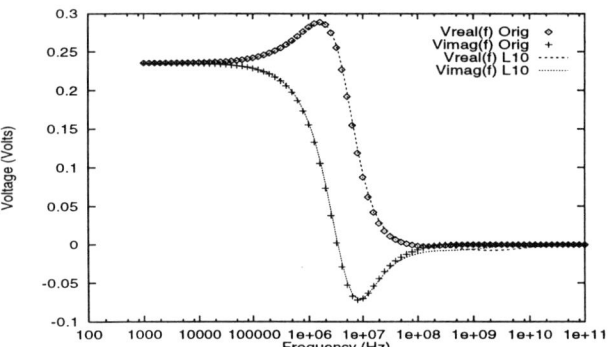

Figure 3: Comparison of real and imaginary part for a port voltage of g5k and its 10th order model

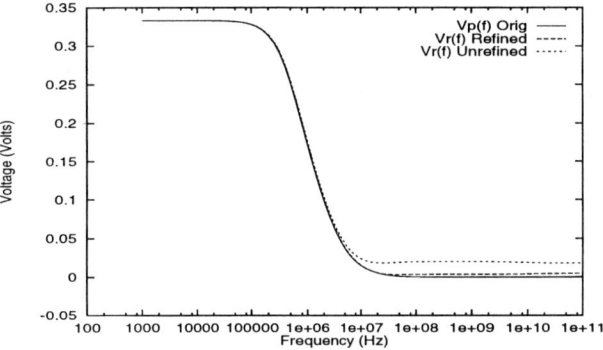

Figure 4: Comparison of a port voltage for g10k circuit with its 10th order models

of our algorithms. However from Table 2 one may note that the time required to synthesize a lower order approximation grows almost linearly with the size of original multiport and sublinearly with the number of poles in the approximation.

7 Conclusions

Although the work reported in this paper should be regarded as preliminary, our results encourage us to conclude that, for network reduction, the time domain method based on spectral values is the most promising. The essential ideas of the method are simple and the key subroutine is that of DC analysis (performed very efficiently). The 'nearness' of the reduced network to the original can be controlled to the extent we need. The time for building the reduced network can be seen to grow nearly linearly with the size of the network and sub linearly with the number of eigenvalues used in the approximation. Further, with a greater reliance on the Lanczos method for eigenvalue and eigenvector computation, the overall reduction time is likely to improve substantially.

References

[1] Shabbir H. Batterywala and H. Narayanan. Efficient analysis of RVJ circuits. *Technical Report VLSI-TR-97-1, VLSI Design Center, IIT Bombay, INDIA*, 1997.

[2] Jane K. Cullum and Ralph A. Willoughby. *Lanczos Algorithms for Large Symmetric Eigenvalue Computations Vol-1 Theory*. Birkhauser, Boston.Basel.Stuttgart, 1985.

[3] P. Feldmann and R. W. Freund. Reduced-order modeling of large linear subcircuits via a block lanczos algorithm. *32nd Design Automation Conference*, pages 474–479, 1995.

[4] Kevin J. Kerns and Andrew T. Yang. Stable and efficient reduction of large, multiport RC network by pole analysis via congruence transformations. *33rd Design Automation Conference*, pages 280–285, 1996.

[5] L. W. Pillage and R. A. Rohrer. Asymptotic waveform evaluation for timing analysis. *IEEE Trans. CAD*, 9(4):352–366, April 1990.

[6] David S. Watkins. *Fundamentals of Matrix Computations*. John Wiley, New York, USA, 1991.

CMOS INVERTER CURRENT AND DELAY MODEL INCORPORATING INTERCONNECT EFFECTS

M. Hafed and N. Rumin

Microelectronics and Computer Systems Laboratory
Department of Electrical and Computer Engineering, McGill University
3480 University Street, Montreal, Canada   H3A 2A7

ABSTRACT

We present a new model for predicting the switching current and delay in a CMOS inverter with an RC load. The model exploits the ability of an inverter model such as the one described in [1] to predict accurately the current peak time, t_m, as a function of inverter size, input slope and capacitive load. An iterative procedure computes the effective capacitance presented by the RC load, using an empirical model for the output voltage of the RC load driven by a reference inverter. Not only is the resulting model accurate but computationally efficient as well, so that the two to three order speed up over HSPICE achieved in [1] is preserved.

1. INTRODUCTION

The increasingly important role that interconnection layers play in determining the performance of deep sub-micron technologies makes it essential to develop analysis tools that take these RC effects into account. In recent years, CMOS gate delay models have improved dramatically to the extent that the effects of RC interconnection trees can now be efficiently incorporated into fast and accurate analysis tools [2][3][4]. For example, in [2], an "effective capacitance" is derived analytically and used in an existing delay and output slope model for CMOS gates driving capacitive loads. On the other hand, in [3], the authors choose to use a π-model along with a nonlinear macro-model of the inverter to find the delay and the output voltage waveform of the driving inverter.

During the same period of time, significant headway was made in the area of power estimation and supply current evaluation [1][5]. However, in contrast to gate delay modeling, there has been very little work on power models that incorporate the resistive effects of interconnects. Other than the works in [4][6], most models assume pure capacitive loads at the outputs of the CMOS gates. Moreover, there seems to be no model that can compute the supply current waveform in the presence of an RC load. This was the objective of the work reported below.

The organization of this paper is as follows. Section 2 gives a brief background of the modeling scheme that is assumed in this paper. Then, in Section 3, we discuss the proposed model in detail. Results are included in the subsequent section, and, finally, conclusions and comments on future work are presented.

2. BACKGROUND

The effect of the resistance of interconnection wires can be thought of as a "shielding" of the overall capacitive load seen by the inverter. As a result, the delay through the inverter is reduced, and the output voltage and supply current waveforms no longer resemble those of an inverter with a lumped capacitive load. These two effects, combined, create the need to model CMOS gates with RC loads.

Figure 1 shows the general modeling technique that is assumed in this paper. As can be seen, the actual RC load that is seen by an inverter is replaced by a reduced-order RC model. This simplified RC model is chosen in a way that would approximate the driving point admittance of the original load [7]. As a result, if the driving point admittance of the load of Figure 1(a) is the same as that of the load of Figure 1(b), then, the inverter's output voltage and supply current should be the same in both cases. Once the inverter output is evaluated using the circuit of Figure 1(b), then, it can be used as a voltage source in place of the inverter in Figure 1(a).

Concerning the supply current pulse of the CMOS inverter, it should be noted that this current tracks the current through the PMOS device. As a result, the behavior of the supply current depends on the voltage transition at the input of the inverter. Figure 2 shows the charging supply current and output waveforms of an inverter that is loaded by a lumped RC load. Also shown in the same figure are the corresponding waveforms when the inverter is loaded by an effective capacitance that captures the 50% point delay accurately. As can be seen, the effective capacitance voltage waveform only matches the actual waveform up to t_m, the time of occurrence of the supply current peak. Beyond that time, the actual output exhibits a longer rise-

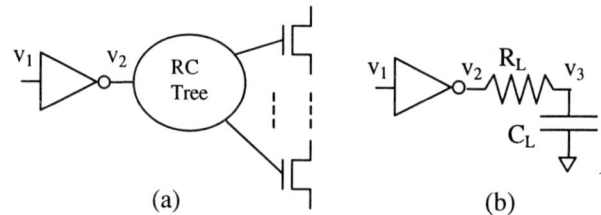

Figure 1   (a) Circuit under consideration, (b) Reduced-order model of RC tree

time, which can be explained by the fact that, during this time interval, the PMOS device is deep in the triode region, and its effective resistance is smaller than that of the RC load. Also, as should be expected, the shape of the current pulse that is obtained using an effective capacitance is not an accurate representation of the actual supply current waveform.

3. THE PROPOSED MODEL

The proposed model is divided into two main parts. The first deals with estimating an effective capacitance that captures the output voltage and the supply current up to t_m, the time of occurrence of the current peak. The second deals with obtaining the rest of the voltage and current waveforms. In the interest of brevity, only a falling input transition will be considered.

3.1 The Effective Capacitance

An effective capacitance for a given RC load can be obtained by equating, over a certain time interval, the average of the current that flows into a lumped capacitive load to the average current that flows into the actual RC load. Since, as can be seen from Figure 2, the use of an effective lumped capacitive load, C_{eff}, in place of the actual one yields good agreement in the inverter output voltage waveforms only up to t_m, the time when the supply current peak occurs, we define the time interval for the current averaging as $[0,t_m]$. In [1], it was shown that t_{st}, the time when the PMOS device leaves the saturation region, can be used as a very accurate estimate of t_m, and an accurate and efficient method for determining t_{st} was presented.

So, the value of C_{eff} is given by the solution of

$$\frac{1}{t_m}\int_0^{t_m} C_{eff}\frac{dv_c}{dt}dt = \frac{1}{t_m}\int_0^{t_m} C_L \frac{dv_3}{dt}dt \quad (1)$$

where v_c is the inverter output across C_{eff}, and v_3 is as defined in Figure 1(b). Usually [2], an analytical expression for v_c (which has to be equal to v_2 of Figure 1(b)) is assumed. Then, v_3, the voltage at the output of the RC lump, is expressed in terms of this v_c. Finally, these expressions for v_c and v_3 are substituted into (1) to get an expression for C_{eff}. However, in this paper, we note that while v_2, in Figure 1(b), assumes a shape that is different from the normal shape of CMOS transitions, the voltage at the output of the RC load, v_3, does have the more normal shape. If this waveform, or a good approximation to it, was known, then the inverter output voltage, v_2, could be determined, at least up to t = t_m, by replacing C_L in Figure 1(b) with a voltage source producing this approximation to v_3. This is demonstrated in Figure 3 where the actual waveforms of v_2 and v_3 are compared with the result of replacing the voltage across C_L by a piecewise-linear approximation to the true v_3, and determining the resulting v_2. Such an approach leads to a very simple expression for C_{eff}.

The piecewise-linear approximation to v_3 is obtained by passing a straight line through the 20% and 80% points of the actual waveform, respectively, at times t_{20} and t_{80}. It is therefore given by

$$v_3(t) = \begin{cases} 0 & t < t_{start} \\ K(t - t_{start}) & t_{start} \leq t < t_{stop} \\ V_{DD} & t \geq t_{stop} \end{cases} \quad (2)$$

where

$$K = \frac{0.6 V_{DD}}{t_{80} - t_{20}}, \quad (3)$$

$$t_{start} = t_{20} - \frac{0.2 V_{DD}}{K}, \quad (4)$$

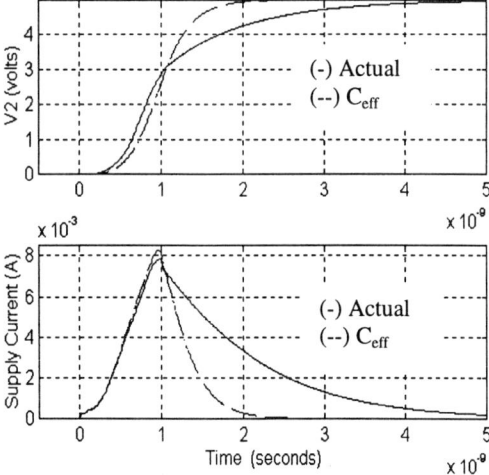

Figure 2 Output voltage & supply current waveforms for an inverter with an RC load. $W_p=60\mu m$, $W_n=20\mu m$, $R_L=250\Omega$, $C_L=2.2pF$.

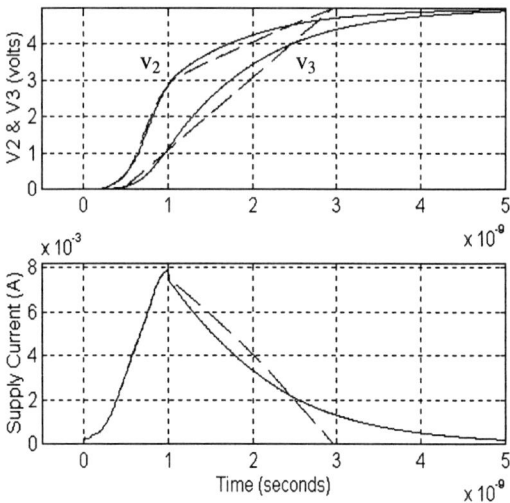

Figure 3 Results for the inverter of Figure 2. (-) Actual outputs using Spice, (--) simulation with C_L replaced by a PWL voltage source approximating v_3.

and
$$t_{stop} = t_{80} + \frac{0.2 V_{DD}}{K} \quad (5)$$

t_{20} and t_{80} are obtained from empirical equations with coefficients that are determined by pre-simulating the circuit in Figure 1(b) with various load sizes and input voltage slopes. Details on these empirical models are presented in the next section. Here, suffice it to say that it is easy to obtain the values of t_{20} and t_{80} using a very small set of coefficients that are determined only by the technology. So, having obtained an analytical description of v_3 as a function of time, the right hand side of (1) is now easy to evaluate.

In order to evaluate the left hand side of (1), we note that the shape of v_c (or v_2) has the normal CMOS form up to t_m. Hence, we approximate v_c by a straight line passing through t_{start} and t_m. That is,

$$v_c(t) = \begin{cases} 0 & t \leq t_{start} \\ \frac{v_c(t_m)}{(t_m - t_{start})}(t - t_{start}) & t_{start} < t \leq t_m \end{cases} \quad (6)$$

Finally, substituting (6) into (1), we get the following expression for C_{eff}

$$C_{eff} = \frac{v_3(t_m)}{v_c(t_m)} C_L \quad (7)$$

As can be seen, the piecewise-linear approximations for v_c and v_3 result in a very simple expression for C_{eff}. Also, notice that, for an inverter with a capacitive load, t_m and $v_c(t_m)$ can be computed using the model in [1]. Thus, the following iterative algorithm is used to solve for C_{eff}:

1. Set $C_{eff} = C_L$.

2. Use model in [1] and the current value of C_{eff} to compute t_m. Combine this t_m with *any* delay model to get $v_c(t_m)$. Compute $v_3(t_m)$ using (2). Finally, compute a new value of C_{eff} using (7).

3. If the new C_{eff} is different from the old one by more than some threshold (say 5%), then, set C_{eff} to the new value and go back to 2. Otherwise, terminate the iteration.

At the end of this procedure, which usually converges in three or four steps, the model in [1] or any other suitable inverter model can be used with C_{eff} to get quantities like delay and supply current peak as will be detailed in Section 3.3.

3.2 Empirical Parameter Extraction

In this section, we present the method for determining t_{20} and t_{80}. It should be noted that the same set of equations, with different coefficients, applies to both t_{20} and t_{80} and also for both charging and discharging transitions. Now, for a given reference inverter and RC load, it can be shown that t_{20} (t_{80}) depends on T_i, the input transition time, in a linear manner. Thus, we can write

$$t_{20} = a(R_L, C_L) + b(R_L, C_L) \cdot T_i \quad (8)$$

In a similar fashion, it can be shown experimentally that, for a given R_L, $a(R_L, C_L)$ depends linearly on C_L. In other words,

$$a(R_L, C_L) = c(R_L) + d(R_L) \cdot C_L \quad (9)$$

where $c(R_L)$ can be sufficiently approximated by a constant that depends only on the technology and the reference inverter of choice, and $d(R_L)$ can be very well approximated by another linear function of the form

$$d(R_L) = e + f \cdot R_L \quad (10)$$

As can be seen, the model, so far, requires three empirical coefficients: c, e, and f. As for the slope term in (8), a similar analysis leads to the following expression for b in terms of R_L and C_L

$$b(R_L, C_L) = \min[g + h(R_L) \cdot C_L, k + l \cdot R_L] \quad (11)$$

where

$$h(R_L) = i + j \cdot R_L \quad (12)$$

As can be seen, for a given inverter, equations (8)-(12) completely specify t_{20} (t_{80}) for any load and any input slope. So, in principle, for a general model, a lookup table of the empirical coefficients should be produced in order to account for the dependence of t_{20} (t_{80}) on the inverter size. However, since the effective capacitance that we are trying to compute is really a model of the admittance seen looking into the RC load, its value should not be affected by the characteristics of the driver. Consequently, for the purposes of computing C_{eff}, we only need to look at a single reference inverter. Specifically, with the reference inverter, we can apply the results of Section 3.1 and 3.2 in order to get the appropriate "effective" capacitive load that is seen by our original inverter (or CMOS gate). Once this C_{eff} is determined, and as will be described in the next section, obtaining the final output voltage waveform and supply current pulse of the actual circuit under test is a trivial matter.

3.3 The Complete Output Voltage and Supply Current Waveforms

If a lumped capacitance having the value of C_{eff} is used to load the inverter under consideration, then the output voltage can be captured accurately from the start of the input transition until t_m, defined in Section 2. In order to obtain the output voltage *waveform* for times later than t_m, we compute the effective resistance of the PMOS device at $t = t_m$ and use that, together with the load resistance, R_L, to obtain a single time constant solution to the voltage tail beyond t_m. Since the PMOS voltages and currents at $t = t_m$ are known (e.g. [1]), computing this device's effective resistance is quite straightforward.

Similarly, the rise time of the current pulse is captured very accurately using this method. If we use a triangular approximation for the current pulse, then, the only remaining unknown to completely specify the current waveform is the end of that pulse. In other words, the input transition and the turn-on conditions of the pertinent transistors determine the start, and the peak and its time of occurrence are found using the results of the previous section. Namely, we perform supply current analysis for the inverter (CMOS gate) under consideration when loaded by the C_{eff} that was obtained using the reference inverter. Experimentally, it was found that the end of the current pulse can be defined as the point at which the output of the inverter reaches 97% of V_{DD}.

4. RESULTS

Figure 4 shows typical waveforms of the inverter model presented in this paper. As can be seen, the model of Section 3.2 yields a very accurate piecewise-linear approximation to the output of the RC lump (for the purposes of the C_{eff} analysis). Also, the C_{eff} approximation as well as the single time constant approximation to the output voltage waveform result in very good agreement with HSPICE LEVEL 3 results. Finally, the current pulse obtained using this model is much more accurate than what would have been obtained using traditional lumped-capacitance approaches (Figure 2). Table 1 shows some sample results demonstrating the accuracy of the model. In all of these cases, an input transition time of 1nsec is assumed.

5. CONCLUSIONS AND COMMENTS

This paper presented an efficient and accurate method for extending existing inverter current and delay models to deal with RC loads. The semi-empirical approach developed in this paper accounts for the size of the inverter, the input transition slope, and the size of the RC load. To be more specific, a new iterative method for obtaining an effective capacitance that approximates the admittance presented by the RC load was developed. This effective capacitance was then used with the conventional CMOS current and delay models in order to obtain quantities like delay, total power dissipation, and supply current and output voltage waveforms. Also, the technique being developed in this project is

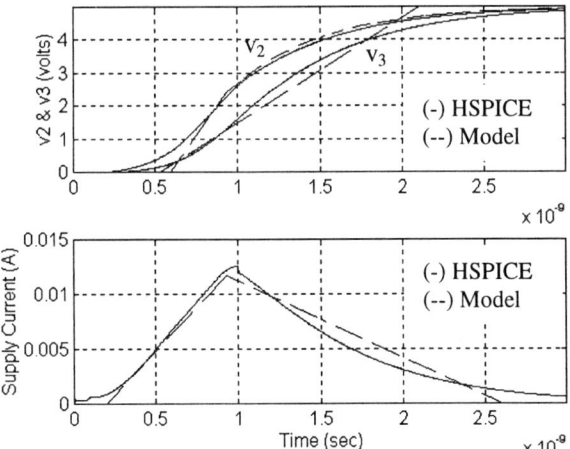

Figure 4 Output voltages & supply current for an inverter with W_p=90μm, W_n=30μm, R_L=90Ω, C_L=2.5pF, and T_i=1ns.

Table 1 Inverter output and supply current accuracy examples.

Wp/Wn	R_L	C_L	Current Peak (mA)		80% Delay (nsec)	
(μm/μm)	(Ω)	(pF)	Spice	Model	Spice	Model
120/40	90	1.5	12.90	12.50	1.087	1.080
120/40	230	3.8	12.49	13.30	1.393	1.950
150/50	74	2.2	16.78	15.84	1.145	1.137
90/50	70	2.0	12.52	11.94	1.407	1.280

computationally efficient, and it can yield a 2-3 order of magnitude speedup over Spice LEVEL 3 simulations. Current and future work involves extending this proposed model to deal with complex CMOS gates and involves using this technique to model the interaction between fully complementary logic and pass-transistor logic.

6. Acknowledgements

This work was supported by the Natural Sciences and Engineering Research Council of Canada. Also, gratitude has to be given to the Canadian Microelectronics Corporation for making available to us the necessary resources. Finally, the authors are grateful to M. Nakhla of Carleton University for his discussions about IC interconnects and for providing assistance with interconnect analysis and model reduction.

7. REFERENCES

[1] A. Nabavi-Lishi, N. Rumin, "Delay and Bus Current Evaluation in CMOS Logic Circuits," *Proc. IEEE Int. Conf. Computer-Aided Design*, pp. 189-203, 1992.

[2] J. Qian, S. Pullela, & L. Pillage, "Modeling the 'Effective Capacitance' for the RC Interconnect of CMOS gates," *IEEE Trans. Computer-Aided Design*, Vol. 13, pp. 1526-1535, Dec. 1994.

[3] J. Kong, D. Overhauser, "Combining RC-interconnect Effects with Nonlinear MOS Macromodels," *Proc. IEEE Int. Symp. Circuits & Systems*, pp. 570-573, 1995.

[4] F. Dartu, N. Menezes, & L. Pilleggi, "Performance Computation for Precharacterized CMOS Gates with RC Loads," *IEEE Trans. Computer-Aided Design*, Vol. 15, pp. 544-553, May 1996.

[5] F. Rouatbi, B. Haroun, & A. Al-Khalili, "Power Estimation Tool for Sub-Micron CMOS VLSI Circuits," *Proc. IEEE Int. Conf. Computer-Aided Design*, pp. 204-209, 1992.

[6] V. Adler, E. Friedman, "Delay and Power Expressions for a CMOS Inverter Driving a Resistive-Capacitive Load," *Proc. IEEE Int. Symp. Circuits & Systems*, pp. 101-104, 1996.

[7] T. Tang, M. Nakhla, "Analysis of High-Speed VLSI Interconnect using the Asymptotic Waveform Evaluation Technique," *IEEE. Trans. Computer-Aided Design*, Vol. 11, pp. 341-352, March 1992

PATH RESIZING BASED ON INCREMENTAL TECHNIQUE

S. Crémoux, N. Azémard and D. Auvergne

LIRMM : Laboratoire d'Informatique ,de Robotique et de Microélectronique de Montpellier,
UMR 5506 Université Montpellier II //CNRS,
161 rue ADA , 34392 Montpellier cedex 5, France
Phone: (33) 04 67 41 85 21, Fax: (33) 4 67 41 85 00, E-mail: azemard@lirmm.fr

ABSTRACT

Based on an incremental path search algorithm, this paper addresses the problem of longest combinational paths selection for performance optimization at physical level. A realistic evaluation of gate delay and controlled sizing techniques are used to manage the circuit path sizing alternatives, such as delay or power/area constraints. The efficiency of this technique is demonstrated and also illustrated on several ISCAS'85 benchmark circuits. A comparison is given between regular sizing alternatives to local optimization steps controlled by specific indicators.

1. INTRODUCTION

Path circuit classification constitutes the necessary step in identifying candidates for circuit performance optimization. The resulting path hierarchy allows the direct implementation of delay optimization and power saving design techniques such as gate up sizing on longest paths, respectively [1][2].
Identifying the performance bottleneck on a circuit is one of the most difficult task to be realized at the last step of the circuit design flow. The satisfaction of imposed delay constraints implies a full path enumeration with complete account of real (post layout) evaluation of delays on the different switching blocks. This imposes to develop fast and accurate path identification techniques [3]. If the circuit path exploration can be obtained from timing verifiers [4][5], the resulting topological delays are easily obtained using graph exploring techniques such as BFS or DFS [6] algorithms. In these enumeration techniques all the circuit paths must be considered and stored. The price to be paid appears very heavy with respect to the CPU time and memory allocation when considering usually large circuits. As a result no complete path enumeration, allowing speed-power trade-off on the different branches can be performed.
In this paper, we address the problem of path selection and optimization based on a new incremental technique [7] for path classification. This technique supplies the enumeration of the longest paths in a non increasing delay order. Validation of the observed longest path is obtained from standard static sensitization technique [8].
Using a realistic evaluation of the gate delays [9][10], the key idea is to determine in a preprocessing step of the circuit, the specific parameters allowing fast path ordering. For that, we considered the maximum propagation delay evaluated between the output of a gate and the output of its successors, and the delay difference between two edges of the divergence branches.
These parameters are evaluated here from a post layout circuit extraction (Cadence tool) on a 0.7µm submicronic CMOS process.
One advantage of this technique is to explore an user specified limited number of paths [11], allowing the easy application of different path optimization criteria. For that, we developed a sizing algorithm to satisfy delay (power) constraints. This algorithm sizes the gates belonging to the paths classified by the incremental technique. The selected criterion for gate resizing is defined through the gate load to drive ratio, evaluated on each node and which has been shown [12] to constitute a robust metric for the gate strength and the gate loading evaluation. This parameter associated to the physical performance description of the gate load (considering active, diffusion and routing capacitances), gives direct indication of its loading level.
Using different sizing strategies such as regular transistor sizing and selective gate [13] sizing, we applied this search algorithm on ISCAS'85 circuits

The incremental technique with the different parameters necessary for the path preprocessing step is detailed in the next part. The following part is devoted to the optimization technique used to size the gates on the paths consideration. In the last part, we give examples of path enumeration and optimization on several benchmark circuits.

2. INCREMENTAL SEARCH PATH ALGORITHM

We develop an incremental algorithm to apply optimization criteria for delay and power on combinational circuit paths. This technique supplies the enumeration in a non increasing delay order of the longest paths. The circuit is represented as an acyclic directed graph, where the nodes represent the gates and the edges the connections between components.

The path enumeration is obtained from this representation at which we associated two parameters :
- **MDS** : Maximum Delay to Sink.
- **BS** : Branch Slack.

The **MDS** represents the maximum propagation delay time evaluated from the output of a gate to the output of its successors.
The **BS** is the delay difference between two edges of a divergence branch.
For simplicity, the calculation of the *MDS* and *BS* will be given on the dotted line circle shown in figure 1.

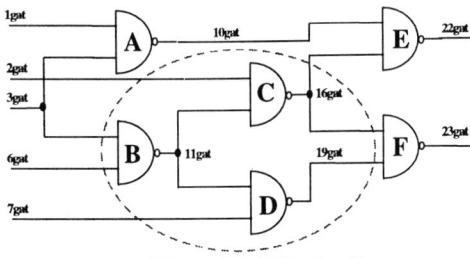

Figure 1. C17 circuit.

2.1 Data structure

For each node the *MDS* is calculated from the primary outputs to the primary inputs. All the primary inputs are represented by a source node and all the primary outputs have the MDS equal to 0. The MDS equation is calculated for each primary output node as (1):

$MDS_{gate} = Max_{\text{down-stream gates}}(MDS_i + Max(T_{HLi}, T_{LHi}))$ (1)

where T_{HLi}, T_{LHi} represent respectively the falling and rising edges.

With this parameter we obtain the maximum delay of the circuit at the primary input. Figure 2 illustrates the calculation of the MDS on the reduced set of the C17 corresponding to the BCD divergence branch.

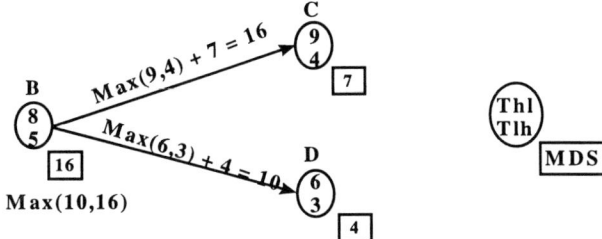

Figure 2. MDS calculation on the BCD.

To obtain a path enumeration in the non increasing delay order, we associate to each edge a value allowing to control the path search. This value is the *Branch Slack BS*, it represents the delay difference between two edges or equivalently between two possible paths.

$$BS = MDS_{ref} - MDS_{edge}$$ (2)

Figure 3 illustrates the calculation of the BS parameter on the reduced set of the C17 circuit (fig.1) corresponding to the **BCD** divergence branch.

The BS is set to 0 on the edge with the greater MDS value.

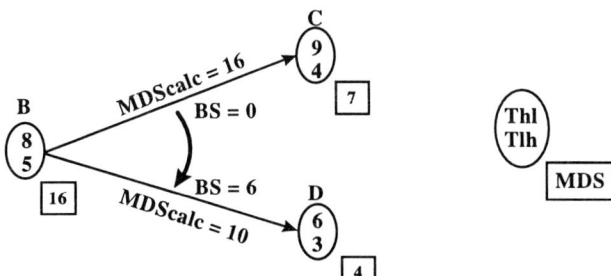

Figure 3. BS calculation on the BCD.

MDS and BS values calculated for all the graph are given in Figure 4. The nodes are characterized with their respective propagation delays. They will allow by direct inspection of the graph to determine the classification of paths with respect to the constraint defined by the user (delay, power.....). Then, we obtain easily on the figure 4 the longest path (BCE) where all the BS have a 0 value.

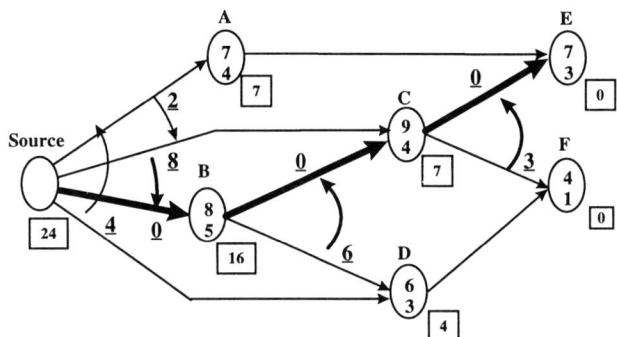

Figure 4. MDS and BS calculation on the C17 where The propagation delays (Thl,Tlh) are given for each node ⟦xx⟧ stands for MDS and x for BS.

2.2 Path enumeration technique

The path enumeration can be obtained using the MDS and BS parameters. Once the longest path is found, the second one is deduced from the divergence branches of the longest path, considering the edges with minimum BS. The corresponding path is processed until all the edges of the branches have been explored. All the paths are then registered in a list of paths. This general method is named **incremental technique** for path enumeration. This technique sorts directly the paths in the selected order of delays (or power). Processing the circuit graph representation by considering increasing order of branch slacks, it is then possible to sort directly the path with the greatest delay (paths with BS parameter to 0 value). The paths with immediately inferior delay values are then found by increasing order of branch slack.

Table 1 summarizes the path enumeration obtained on the C17 (circuit given in Figure 1, where the longest path is shown with bold lines in figure 4 (B,C,E path of Table 1).

Iterations	Paths	Delay (ns)	Available branch points (*Branch Slack*)
1	B, C, E	24	F(3), D(6), C(8)
2	B, C, F	21	
3	B, D, F	18	
3	C, E	16	A(2), F(3)
5	A, E	14	D(4)
6	C, F	13	
7	D, F	10	

Table 1. Complete path enumeration for the C17.

Circuits	Gate number	Regular sizing			Selective sizing		
		Delay (ns)			Delay (ns)	% sized gates	CPU Time
		W = 2µm	W = 5µm	W = 8µm			
C17	6	1.08	0.78	0.7	0.69	83.33	-
fpd	17	3.34	2.30	2.28	2.73	88.24	16 ms
C880	529	23.1	12.70	10.2	8.67	41.02	1,12 s
Adder16x2	328	43.5	26.00	20.8	21	33.23	117 ms
C432	249	27.1	15.30	11.3	10.3	60.64	350 ms
C499	700	21.8	11.78	9.42	9.38	66.29	2,33 s
C1908	1075	29.3	15.98	12	13.1	41.02	1,67 s
C1355	628	27.2	15.00	11	21.5	27.07	500 ms

Table 2. Evolution of the longest delays (in ns) on ISCAS'85, versus the transistor sizing alternatives.

3. TRANSISTORS SIZING METHODOLOGY

The delay path evaluation is based on a realistic MOS macro model. This model includes the carrier velocity saturation effect of submicronic MOSFET's and is developed in [9]. It takes into consideration the input-to-output coupling capacitances. It also considers the short circuit current effects allowing a realistic delay evaluation on post layout capacitances extraction of the circuit. We obtain a good accuracy compared to SPICE evaluations (Level 6).

The sizing algorithm modifies the sizing of the gates belonging to the longest paths. The path is identified with the incremental search algorithm presented in previous part. The selected criterion for gate resizing is defined through the gate loading factor noted Fc. This factor has been shown in [12] to be a robust criterion for gate strength and gate loading evaluation, and is defined as:

$$Fc = \frac{Cload}{Cin} \quad (3)$$

This parameter represents the ratio of the output load to the input capacitance of the gate.

The output load is constituted by the active, diffusion and routing capacitances. This factor associated to the physical performance description of the gate, gives direct indication of its loading level.

The sizing algorithm is based on a local sizing strategy [13] and proceeds as follows :
- initial uniform sizing of all the gates at a value which can be the technology minimum width or any value used as a reference,
- evaluation of the Fc loading factor of each gate on the identified longest path, and comparison to the loading limit value (noted Flimt) defined on each gate [10],
- sizing of the gates with a loading factor value Fc greater than the Flimt value in order to satisfy the constraint Fc < Flimt,
- evaluation of the path delay from the gate delay and the MDS and BS parameter values.

Then, a new longest path is considered and this procedure is iterated until the circuit delay constraint is satisfied.

4. EXPERIMENTAL RESULTS & CONCLUSION

For different circuits, Table 2 reports the delay of the longest path resulting from different sizing alternatives :
- regularly sized transistors (columns 3-5),
- optimally sized transistors, considering the heavy loaded nodes on the longest path (columns 6-8).

As shown, the incremental technique offers low CPU time opportunities to easily investigate various sizing alternatives with facilities in trading delay against gate area (power indicator). One example of path delay profiling is given in figure 5 where we compare for the C1908 (1075 Gates and about 730 000 Paths), the evolution of the number of paths with specific delay values, varying the transistor width from 2µm to 8µm.

The evolution of the delay distribution clearly shows the shortening of the distribution which can be used to trade speed for power. This is illustrated in figure 6 where starting from the initial sizing solution (A curve) we give the new delay profile (B curve) obtained by controlling the gate loading factor from selective optimization techniques [13]. The optimization has been applied to only the 25 longest paths among the 730 000 stored paths resulting in a 58% speed improvement with respect to the 2µm initial solution, at the expense of only a 30% increase in active area (used as a good indicator of dynamic power).

Number of Paths

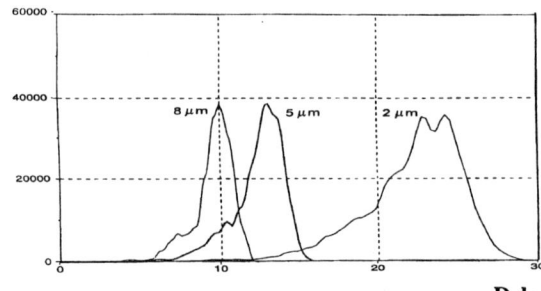

Figure 5. Comparison of the delay profile for regular sizing alternatives of the C1908.

Note here the reduced dispersion of delays obtained with the selective sizing technique.

Number of Paths

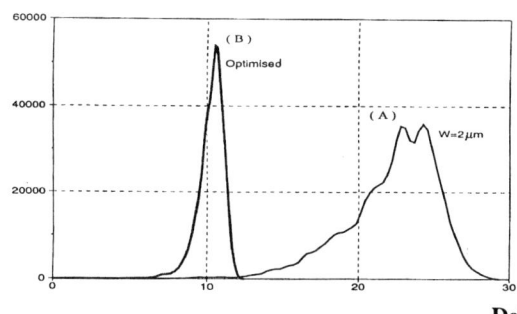

Figure 6: Comparison of the path delay distribution for the C1908 for two sizing alternatives: 2µm (curve A) regular sizing, load driven selective sizing (curve B).

Finally in the figure 7, we compare the delay distribution resulting from the application of selective and regular sizing techniques.

Number of Paths

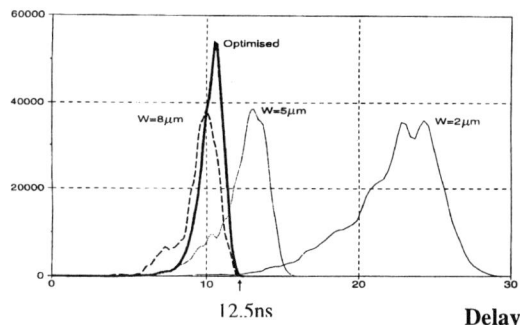

Figure 7: Comparison of the delay distribution on the C1908 resulting from selective and regular transistor sizing alternatives.

As shown, a delay constraint of 12.5ns can be satisfied using optimized and 8µm regular sizing methods. However regular sizing (blind sizing) involves all the transistors and results in a power increase of 49% with respect to the 2µm implementation, to be compared to the optimized solution (controlled sizing) applied to only 25 paths with a power penalty reduced to 30%.

In conclusion, these results show clearly the interest in using incremental technique for circuit path analysis. The control of the path number gives to this technique a preponderant advantage with respect to standard ones. The resulting reduction in required CPU time and memory allocation allows to treat large circuits (the limitation in this work has been given by the layout). Using realistic delay values for path evaluation and ordering gives interesting facilities in circuit verification and optimization. This can be obtained for path classification in increasing or non increasing order of delays as well than to implement power saving techniques or in trading speed for power. The application to the C1908 shows the importance in controlling the path number to be considered for circuit optimization. In this way different optimization techniques can be quickly evaluated.

It has been demonstrated that if delay constraints can be satisfied with regular transistor sizing, local optimization results in power minimized implementation. Considering the longest paths it has been shown, on a specific example, that circuit optimization can be obtained by resizing few paths of the circuit, only 25 paths over 730,000 for C1908. This gives evidence of the importance, for path optimization, of the definition and the control of specific indicators such as the loading factor and its associated limit for gates, as well as the degree of path degeneracy. The application to power and to delay-power path optimization is under progress.

5. REFERENCES

[1] D.H.C. Du, S.H.C. Yen and S. Ghanta, *"On the General False Path Problem in Timing Analysis"*, ACM / IEEE Design Automation Conference, pp. 555-560, June 1989.

[2] H.C. Chen, D. DU, *"Critical path selection for performance optimization"*, ACM / IEEE Design Automation Conference, Los Alamitos, California, pp. 547-550, 1991.

[3] J. Benkoski, E.V. Meersch, L. Claesen and H. De Man, *"Efficient Algorithms for Solving the False Path Problem in Timing Verification"*, IEEE International Conference on Computer-Aided Design, pp. 44-47, June 1987.

[4] N.P Jouppi , *"Timing analysis and performance improvements of MOS VLSI designs"*, IEEE trans. on CAD, vol CAD 6, n° 4, 1987.

[5] J.K. Ousterhout, *"CRYSTAL : a timing analyser for NMOS VLSI circuits"*, Proc. 3rd Caltech VLSI conf. R.BRYANT, ed. 1983.

[6] P.C. McGeer and R.K. Brayton, *"Efficient Algorithms for Computing the Longest Viable Path in a Combinational Network"*, Design Automation Conference, pp. 561-567, June 1989.

[7] Yun-Chen Ju, R.A. Saleh, *"Incremental techniques for the identification of statically sensitizable critical paths"*, Proc. 28th Design Automation Conf., pp.541-546, 1991.

[8] A. Dargelas, C. Gauthron, Y. BeMDSrand, *"MOSAIC: multiple strategy oriented sequential ATPG for integrated circuits"*, ED&TC, pp.29-36, Paris 1997.

[9] D. Auvergne, D. Deschacht, M. Robert, *"Input waveform slope effects in CMOS delays"*, IEEE Solid State circuits, vol. 25, n° 6, dec. 1990.

[10] J.M. Daga, S.Turgis, D. Auvergne,*"Inverter delay modeling for submicrometre CMOS process"*, IEE Electronic Letters, vol.32, n°22, pp.2070-2071, October 96

[11] S. Yen, D. Du and S. Ghanta,, *"Efficient Algorithms for Extracting the k Most Critical Paths in Timing Analysis"*, Design Automation Conference, pp. 649-654, June 1989.

[12] S. Turgis, N. Azemard, D. Auvergne, *"Design and selection of buffers for minimum power-delay product"*, ED&TC, pp.224-228, Paris, March 1996.

[13] D. Auvergne, N. Azemard, V. Bonzom, D. Deschacht, M. Robert, *"Formal Sizing Rules of CMOS Circuits"*, EDAC, pp.96-100, Amsterdam, The Netherlands, February 1991.

SUBSTRATE COUPLING ANALYSIS AND SIMULATION FOR AN INDUSTRIAL PHASE-LOCKED LOOP

Ryan J. Welch* and Andrew T. Yang

Department of Electrical Engineering, FT-10, University of Washington, Seattle, WA 98195

Abstract

Current injected into the common chip substrate from fast-switching digital devices can affect the operation of sensitive analog circuits in mixed signal designs. An industrial Phase-Locked Loop(PLL) is analyzed with the non-ideal substrate modeled to show the effects of substrate coupling. Detailed simulation results strongly correlate to the measured circuit jitter. Additional results show that well guard structures should not be used and the effectiveness of ohmic guarding structures depends on number, location in the layout, and the bias scheme.

Introduction

Switching devices in integrated circuits inject current into the chip substrate across p-n junctions, gate capacitance, and interconnect capacitance. Injected current leaves through ohmic contacts. In strictly digital systems, substrate current is not a problem as long as latch-up is does not occur. In monolithic mixed-signal designs, substrate current causes substrate voltage fluctuations which alters the quiet environment required for analog circuit precision. The substrate coupling problem has been identified as a significant design consideration in advanced silicon mixed-signal designs[1]. The problem worsens as chip integration increases, device sizes shrink, supply voltages reduce, and required analog signal precision increases. Strategies exist to reduce the amount of substrate crosstalk from digital to analog devices, but they tend to be heuristic and costly in chip area.

Designers use three general guidelines to prevent substrate coupling in mixed-signal designs: separate power supplies, physical separation, and guard structures. Power supply separation reduces the problem of ground bounce for analog circuits. Physical separation and guarding structures are both technology and layout dependent techniques that are used to prevent substrate coupling. Physical separation is self-explanatory. The first guarding technique is ohmic guarding, which is a low-impedance contact to the substrate. The effect is to collect substrate current before it reaches analog circuits. The second technique is well guarding. A diffused well cuts a hole in any low resistance channel-stop region forcing substrate current to flow below it. Generally, both of these techniques are used together.

Industry trends and the substrate coupling problem have inspired the CAD community to develop tools to model the non-ideal substrate for circuit simulation. Such tools must be able to incorporate both layout and technology information in the substrate model generation. Many different substrate features such as wells, channel-stops, field implants, ohmic contacts, epitaxial layers, and BICMOS devices add an additional dimension to the task of creating a general tool. Due to the size of resulting substrate models, previous techniques have been constrained to small problems or potentially inaccurate assumptions have been made to analyze large circuits. This paper will show the simulation results of an industrial circuit based on an efficient and accurate substrate analysis system that can model much larger substrates than other systems.

Substrate Modeling System

The effects of substrate coupling presented in this paper were produced on a standard SPICE-based circuit simulator. The system creates a netlist where the "designed" circuit components are attached to the "parasitic" substrate model (mesh). The mesh generation scheme[2] is based on a non-uniform gridding technique that localizes substrate node density to layout feature density which maintains accuracy while reducing complexity. The created mesh is then simplified to allow simulation to take place. This reduction is based on a well-conditioned algorithm that creates an approximation of the mesh within a specified error tolerance accurate to a given frequency[3]. Both the mesh generation and reduction have proven to be accurate and efficient. These tools allow a relatively large circuit (up to 1000 devices) to be simulated allowing an accurate evaluation of substrate coupling on present day computing systems.

Phase-Locked Loop

Phase-locked loops are commonly used in advanced microprocessors to more accurately control the clock signals, so they are placed on the same die as many other digital circuits. PLLs are good candidates to study substrate coupling effects since they consist of both analog and digital components operating at a relatively high frequency oscillation. Cycle-to-cycle jitter (referred to as jitter) is a measure of PLL performance referring to the output period deviation from the ideal. A clocked system's frequency is limited by the jitter of the clock generator in order to eliminate timing errors. Substrate coupling has been identified, through simulations and silicon testing, as a major source of PLL jitter.

An industrial source provided a frequency multiplier PLL for clock generation as a test circuit. A large noise generator consisting of 5 output buffers in series was placed next to the PLL. It was built in a 0.72 micron epitaxial process with

---

*Ryan Welch is currently working in the Electron Device Division of the Avionics Directorate of the Air Force Research Lab in Dayton, Ohio

approximately 1800 transistors. The sensitive circuitry has both ohmic and well guard rings around the analog devices. It was apparent that this entire design could not be evaluated on the current system without simplifications. The large transistor count and small minimum drawing dimension created a mesh that could not be reduced nor simulated. In order to correlate simulation jitter to silicon jitter, the simplifications had to be accurate for an epitaxial substrate process.

Simplifying Assumptions

Since the sensitive analog circuitry is the PLL's Voltage Controlled Oscillator(VCO) and the majority of the noise comes from the Noise Generator, it was decided to only simulate those two pieces of the layout (see Figure 1) to get jitter results. Despite cutting the transistor number in half, further reductions had to be made to make the simulations feasible. The first assumption was that the bulk substrate could be modeled as a single node which greatly reduced the size of the substrate model. This assumption has been validated in independent studies[4] and stems from the bulk substrate usually being three orders of magnitude less resistive than the epitaxial layer.

The second assumption was to omit diffused well regions used for guarding structures and zener diodes from the mesh formulation. The zener diode well was modeled as a single diode to the bulk substrate since the junction depth is near the epi-bulk boundary, the well is very large, and it is not near many ohmic contacts. The resulting well diodes were responsible for nearly 50 percent of the injected noise current so they could not be ignored.

In order to validate removing well guarding structures, a simple test circuit was simulated consisting of one NMOS device near a scaled noise generator. The voltage underneath the sensitive NMOS device was monitored for four different guarding structures: A (no guard), B (well guard), C (ohmic and well guard), and D (ohmic guard). Figure 2 shows these results. The difference between A,B and C,D was the inclusion of a p+ guard ring around the sensitive device. This graph shows that no guarding(A) and well guarding(B) were very similar, and well with p+ guarding(C) was very similar to just p+ guarding(D). Thus, well guard structures had little effect in this epitaxial process and could be omitted from the substrate model. The omission of guard wells in the entire PLL circuit reduced the draw area by 3.2 percent. Guard well

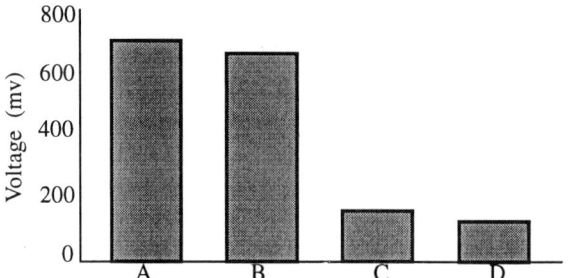

Figure 2. Voltage fluctuations under a sensitive device with guarding strategies A-D as described in the text.

ineffectiveness can be explained by examining how substrate current flows in an epitaxial process.

Substrate Coupling

The substrate coupling mechanism has two stages: injection and reception. Figure 3 shows a simplified resistive model for the injected current paths. As R1 gets much larger than R2+Reff, the majority of injected current goes directly to the less resistive bulk material (p+) and thus can spread more easily to other portions of the chip. R1 is a function of physical separation while R2 is a function of the epitaxial depth. Reff is the equivalent resistance from the bulk to the substrate bias and is inversely proportional to the number of ohmic contacts. This schematic shows the importance of placing ohmic guard structures as close as possible to noisy digital circuits to keep current from reaching the bulk substrate. Well guard structures block the current flow through the epitaxial layer, effectively increasing R1. They push current from the epitaxial layer to the bulk substrate. However, substrate current is already drawn to the bulk due to its relatively low resistivity, thus well guard structures do not significantly help in reducing substrate coupling.

Once the noise current is in the bulk substrate, there are two mechanisms that affect how much of this noise is received into the analog circuitry (Figure 4). First, the voltage level in the bulk substrate (p+) depends on the value of Reff since Reff is much less than R3. Second, the proximity of ohmic contacts to the sensitive device determines the substrate voltage for that individual device. If R1 is greater than R2, node A is more strongly coupled to the bulk node, but if R1 is less than R2, node A is more strongly coupled to the ground node. Again, the proximity of ohmic guard structures determine their effectiveness.

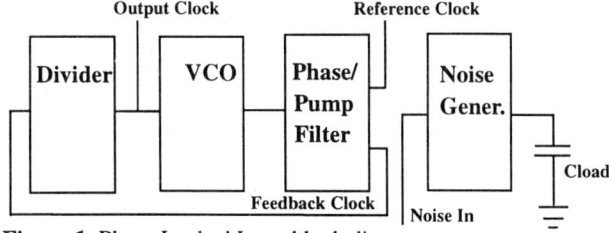

Figure 1. Phase-Locked Loop block diagram

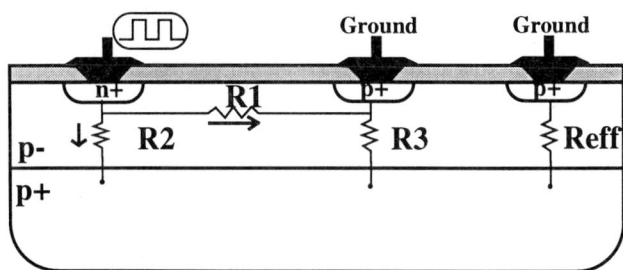

Figure 3. Cross-sectional diagram showing current injection into the bulk substrate.

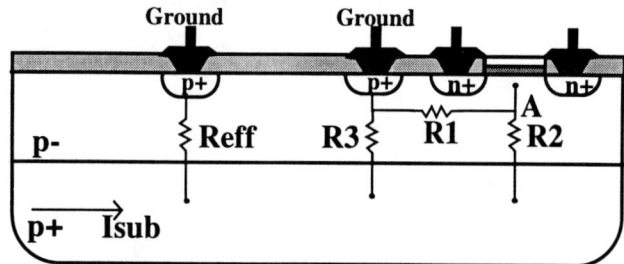

Figure 4. Cross-sectional diagram showing noise reception from the bulk substrate.

An important noise reduction strategy arises from analyzing the substrate in an epitaxial process: ohmic guard structures should be placed as close as possible to noisy and sensitive circuits. To demonstrate this strategy, a simple test circuit was simulated. A NMOS device was used as the sensitive circuit and a fast switching inverter with no load was used as the noise injector. The input voltage to the NMOS device was 1.2 volts, the input voltage to the inverter was a square wave with 2ns rise and fall times, and Vdd was 5 volts. An ohmic guard band was swept from the analog device (D=0um) to the digital device(D=110um) and the voltage under the sensitive device was monitored.

Figure 5 shows how the proximity of the guard structure to either the analog or digital device affected the amount of substrate fluctuation at the sensitive device. When the guard structure was close to the noisy device, the amount of current reaching the bulk was reduced. When the structure was near the sensitive device, the device was decoupled from the bulk. Some techniques such as [4] model epitaxial processes with semi-empirically determined resistances from the bulk layer to devices and ohmic contacts since the bulk can be modeled as one node. If the technique from [4] were used on this experiment, there would be no change in the sensitive device fluctuation in Figure 5 since no lateral resistances are modeled. Although the bulk node assumption is valid due to the relative doping levels in the epitaxial and bulk substrate, lateral resistance can only be ignored if there is a large physical separation between features which is rarely the case in mixed A/D designs.

Hardware Results Comparison

One of the goals in the VCO simulations was to correlate the previously measured jitter to the simulated jitter. Since the circuit was simulated in blocks, careful steps were taken to maintain accuracy. The measured results from the industrial source, Figure 6, show that the clean (noise generator off) case has a jitter of under 100ps while the dirty (noise generator on) case has a jitter of approximately 500ps.

To simulate this modified circuit as accurately as possible, several steps had to be taken beyond the simplifications previously discussed. The Phase_Pump and Filter block (Figure 1) was simulated with the Noise Generator to model the non-ideal input to the VCO. Ideally this value should be DC, but with substrate coupling there will be some peak-to-peak noise superimposed. All ohmic contacts were left in the layout to keep the value of Reff as accurate as possible. The simulations included pin parasitics provided by the industrial source. The VCO and the Noise Generator were then simulated with all ohmic contacts and the non-ideal input to the VCO. An additional output pad attached to the output of the VCO was included to model noise injected as the VCO signal leaves the chip.

Each simulation lasted for 100 nanoseconds. A 50 MHz square wave with 2ns rise and fall times drove the input to the noise generator. When the noise generator was turned off, the VCO input had a 150 mV peak-to-peak fluctuation and the VCO jitter was 27ps. With the noise generator turned on, the VCO input had a 1 V peak-to-peak fluctuation and the VCO jitter was 383ps. The jitters do not match exactly due to the assumptions made, no interconnect parasitics modeled, and possible variations in the doping profiles. Despite the possible variations from simulation to measurement, the simulation results with just the VCO and the noise source correlate to the measured PLL jitter values.

Guard Structure Effectiveness

The next step in analyzing the VCO and the Noise Generator consisted of varying the types of guarding strategies used. Simplifications were made to the simulations, such as reducing the size of the noise generator, reducing the number of ohmic contacts in the layout, and giving a DC input to the VCO, to show the relative effects of different guard structures. Since well guarding structures have been seen to be ineffective, only ohmic guard structures and a backplane contact were evaluated.

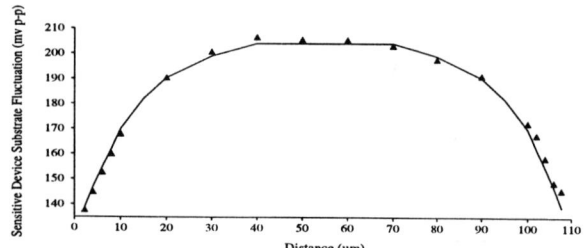

Figure 5. Simple test results for analog device swept away from digital device

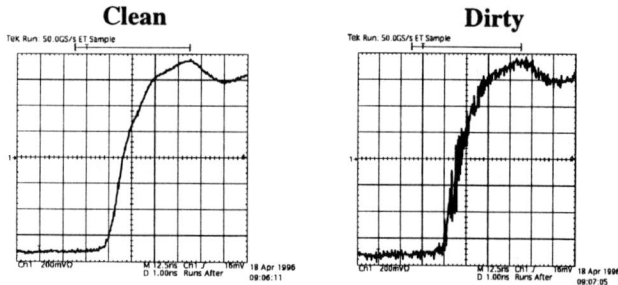

Figure 6. Clean and dirty PLL jitter measurement results

Six different guarding structures were chosen to evaluate guarding techniques. These include: A (no guard), B (no guard with a backplane contact), C (p+ ring), D (p+ ring with a backplane contact), E (ohmic guarding in the VCO), and F (ohmic guarding in the VCO with a p+ ring). The p+ ring and ohmic guarding are both ohmic guard structures but the p+ ring is a single ring placed around the entire section of analog devices while the ohmic guarding is many ohmic guard bands placed as close as possible to all sensitive devices not in well regions. The jitter results for these six test cases without and with pin parasitics are shown in Figure 7.

By first comparing the test cases with no pin parasitics, several conclusions can be drawn. First, the backplane contact(B) gave the best jitter performance. Second, the ohmic guarding of the VCO(E) was better than the p+ guard ring(C) in two aspects, the jitter and the drawn area. Additional area was not required to do the ohmic guarding but was necessary for the p+ guarding. These results are important because they validate the theory behind substrate coupling reduction techniques. These simulations do no tell the entire story.

The simulation results would not be complete without the parasitic pin values, especially the pin inductance. Switching current flowing through the pin inductance deteriorates power supply values. Not only did the pin parasitics increase the amount of jitter for each case drastically, but the effectiveness of each guarding configuration changed. Several guarding structures showed to worsen the jitter performance(F). With no pin parasitics, this case was the best performer without a backplane contact. With the pin parasitics, it was the worst performer.

The reason the jitter performance deteriorated with the pin parasitics is that the p+ guard ring was biased through the analog ground line. When the guard ring is biased by a certain node, the bulk substrate is strongly coupled to that node and a majority of the substrate current leaves through that node. When the ring was biased by the analog ground node, the bulk fluctuation remained low but the substrate current leaving through the analog ground node caused ground bounce to the sensitive devices. Switching the p+ guard ring

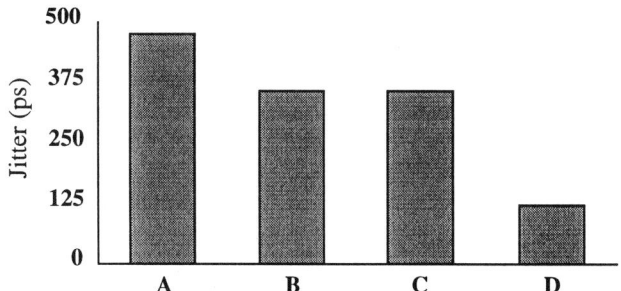

Figure 8. Results with 4 different p+ biases as listed in the text.

bias to the dirty ground node caused less analog ground bounce but increased the fluctuation on the bulk substrate since the ground bounce on the digital ground line is greater due to transient current in switching devices. A solution to this problem is to hook the p+ ring to a separate power supply. Figure 8 shows the results from the VCO simulations with 4 different biases on the p+ ring including: A (no bias), B (clean bias), C (dirty bias), and D (separate bias). Clearly, case D is the best performer since now the p+ ring bias is not hooked to other devices but collects a majority of the substrate current.

Conclusion

Since the PLL in question was too large to simulate as one piece when the non-ideal substrate is modeled, accurate simplifications were made resulting in a strong correlation between the VCO jitter simulation results and the PLL silicon jitter measured results. Additionally, heuristic guarding strategies have shown to have little worth in an epitaxial process. Guard wells were shown to be ineffectual. Ohmic guard structures are effective in reducing substrate coupling by either reducing the bulk substrate voltage fluctuation, which is a function of the number of substrate contacts, or locally decoupling devices from the noisy bulk substrate, which is a function of the proximity of the substrate contacts. It is important to bias a majority of the ohmic contacts by a separate bias supply to reduce analog ground bounce while keeping the bulk substrate fluctuation low.

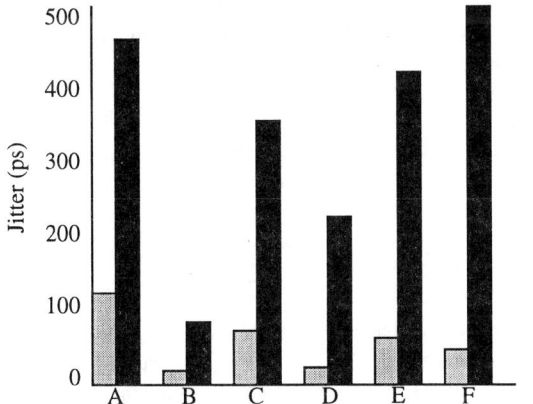

Figure 7. VCO jitter for cases A-F as described in the text.
▢ No pin parasitics, ■ pin parasitics.

References

[1] G.H. Warren and C. Jungo, "Noise, Crosstalk, and Distortion in Mixed Analog/Digital Integrated Circuits", *Prcdngs of the IEEE Custom Integrated Circuits Conference*, pp. 12.1.1-12.1.4, 1988.
[2] I.L. Wemple and A.T. Yang, "Integrated Circuit Substrate Coupling Models Based on Voronoi Tessellation", *IEEE Transactions on Computer-Aided Design*, pp. 1459-1469, Dec.1995.
[3] K.J. Kerns, I.L. Wemple, and A.T. Yang, "Stable and Efficient Reduction of Substrate Model Networks Using Congruence Transforms", *International Conference on Computer-Aided Design*, pp.207-214, Nov. 1995.
[4] D. K. Su, M.J. Lionaz, S. Masui, and B. A. Wooley, "Experimental Results and Modeling Techniques for Substrate Noise in Mixed-Signal Integrated Circuits," *IEEE Journal of Solid-State Circuits*, vol. 28, no. 4, pp. 420-430, Apr. 1993.

STATISTICAL BEHAVIORAL MODELING OF INTEGRATED CIRCUITS

J. F. Swidzinski, M. A. Styblinski, G. Xu

Department of Electrical Engineering
Texas A&M University, College Station, TX 77843
Phone: (409)845-5257   Email: jan@ee.tamu.edu

ABSTRACT

A full statistical model for the behavioral parameters of an analog cell is presented. Behavioral parameter variations with respect to manufacturing process disturbances are approximated utilizing multivariate modeling techniques which allow the means, standard deviations, parameter correlations and the actual distributions to be reproduced with reasonable accuracy. The modeling procedure is demonstrated in statistical behavioral modeling of a MOSFET-C notch filter at the cell level and a phase-locked loop (PLL) tunable filter at the system level. The accuracy of the results obtained utilizing the characterized behavioral MOSFET-C filter and the PLL models, relative to the circuit-level simulation is considered.

1. INTRODUCTION

Recent increase of activity in behavioral modeling and simulation has been focused on evaluating the nominal performance of the circuit only in terms of the nominal values of certain physical quantities. If fluctuations of the circuit due to process parameter variations must be considered (e.g., to determine manufacturing yield), such models are insufficient and cannot be applied to IC design. Statistical characterization has to be incorporated into the model to make the behavioral simulation an attractive alternative to the transistor level simulation.

The objective of this work is to develop methods for creating accurate statistical models of circuits at the cell level so that efficient statistical analysis and design at the system level become feasible. A practical examples are presented.

2. STATISTICAL BEHAVIORAL MODELING TECHNIQUES FOR IC'S

Statistical IC modeling approaches proposed here are designed to determine the mapping of the process disturbances from the circuit parameter space to the IC behavioral model space such that the statistical characteristics of the integrated circuit are well represented. Graphical interpretation of the complexity of the problem to be solved is shown in Fig. 1. Formally, our problem can be defined as follow: (1) Characterize the distribution of $\theta = (\theta_G, \theta_{GL}, \theta_L)$ for a given $\mathbf{x}$; (2) Create a statistical model for the j-th block

$$b^j = b^j(\theta, \mathbf{x}) = b^j(\theta_G, \theta_{GL}^j, \theta_L^j, \mathbf{x}) \quad (1)$$

Supported in part by Semiconductor Research Corporation and Texas Advanced Technology program.

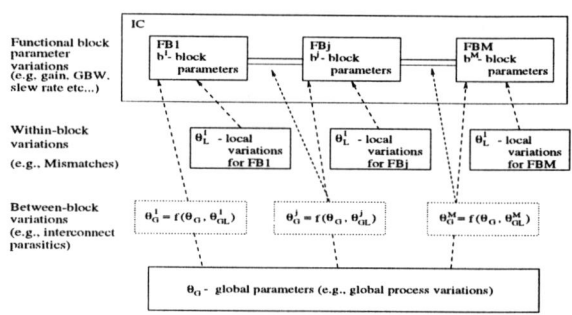

Figure 1: Representation of variability in integrated circuits

where, θ_G are the global random parameters (TOX, $NSUB$, etc.), θ_{GL} are between-block variations of global parameters, θ_L are within-block variations (mismatches), $\mathbf{x}$ is the vector of block designable parameters (e.g., transistor widths and lengths) - assumed fixed in what follows. Additionally, other mismatching and disturbing factors, such as interconnect parasitics have to be taken into account (not considered here).

2.1. Basic Additive Model - Model (A)

The modeling technique to be described in this section was originally proposed in [3]. Three levels of models which vary from the simple linear regression model to the complex hierarchical one are demonstrated. The basic principle is to statistically model the IC parameters as simply as possible. If a given behavioral model parameter, b_k, depends on the global noise parameters θ_G's only, the simplest linear or quadratic regression model results w.r.t. the global parameters only, called **Model(A1)**.

$$\hat{b}_k = b_k^I(\theta_G) = \beta_{k,o} + \sum_{i=1}^{N_G} \beta_{k,i} \theta_{Gi} \quad (2)$$

$$\hat{b}_k = b_k^{II}(\theta_G) = b_k^I(\theta_G) + \sum_{i=1}^{N_G} \sum_{j=1}^{N_G} \beta_{k,ij} \theta_{Gi} \theta_{Gj} \quad (3)$$

where, θ_{Gi}'s are the N_G global random parameters, $\beta_{k,o}, \beta_{k,i}, \beta_{k,ij}$ are regression coefficients, and $b_k^i(\theta_G)$ is the parameter b_k modeled by model Level i. This model can be often sufficient for digital blocks. If the mismatch effects are important for some model parameters, then Model (A1) become inaccurate. Linear additive model w.r.t. mismatches is created, called **Model(A2)**.

$$\hat{b}_k = b_k^{II}(\theta_G) + \sum_{r=1}^{N_L} \alpha_{k,r} \theta_{L,r} \qquad (4)$$

where, $\theta_{L,r}$ are the N_L local (mismatch) random parameters, $\alpha_{k,r}$ are regression coefficients. In our work we found that this approach is valid for some classes of analog circuits. However, to satisfy the parsimony principle and to avoid difficulties in screening the important mismatch parameters needed for regression analysis, the principal component or common factor model is adopted for some cases instead of Model(A2) to model the residual variations (mismatches). The residual variations are formed by substracting the results obtained using Model (A1) from the parameter data. These residual variations are due to mismatches and the nonlinear effects that cannot be sufficiently well modeled by the quadratic regression model using the global parameters only. Such newly created model, called **Model(A3)**, combines the regression analysis and Principal Component Analysis (PCA) or Factor Analysis (FA) to model the parameters and is given by

$$\hat{b}_k = b_k^{III}(\theta, \theta^h) = b_k^{II}(\theta) + b^H(\theta^h) \qquad (5)$$

where, $b_k^{III}(\theta)$ denotes a parameter modeled by Model(A3), and $b^H(\theta^h)$ is the hypothetical model which could be a reduced PCA model or FA model. Since we perform the PCA and FA on the residual correlation matrix, the resulting PCA model is

$$b^H(\theta^h) = \hat{\mu}_k^R + \hat{\sigma}_k^R \sum_{j=1}^{K'} \alpha_{kj} c_j \qquad (6)$$

and the FA model is

$$b^H(\theta^h) = \hat{\mu}_k^R + \hat{\sigma}_k^R (\sum_{j=1}^{K} a_{kj} F_j + u_k) \qquad (7)$$

where, $\hat{\mu}_k^R$'s and $\hat{\sigma}_k^R$'s are, respectively, the sample means and standard deviations of the residual parameter data. This approach is: (1) efficient when large number of θ's parameters is involved, but (2) lacking physical interpretation.

2.2. Direct Modeling (B)

Parameter matching is the basic principle in designing analog integrated circuits, and the mismatches due to process imperfections most often cannot be neglected. To enable the circuit designers get insight into the circuit behavior due to second order effects (mismatches) and at the same time maintain parsimony and clear physical interpretation of the model, the first order sensitivity matrix and regression analysis are used. In general, the IC performance functions y are the functions of the designable parameters (x) and the random or noise parameters (θ), and can be described by the vector function $y = y(x, \theta)$. If the mapping function $y = y(\theta)$ is known for a particular value of x, the performance variances can be computed approximately from the knowledge of K^θ, the variance-covariance matrix of the noise parameters. Let us postulate a linear model of the circuit performances with respect to the noise parameters at a particular setting of the designable parameters (x). At any particular point x in the space of designable parameters, $y(x, \theta) = y(\theta)$. Let y be the vector of performance functions evaluated at the nominal (mean) values of the noise parameters θ. If we assume that for small perturbations the linear approximation to $y = y(\theta)$ is sufficiently accurate to represent $\triangle y$, then the change of the performances y, can be expressed by the following relationship

$$\triangle y = y(\theta + \triangle \theta) - y(\theta) \approx J \triangle \theta, \qquad (8)$$

where J is a Jacobian matrix whose elements are given by $J[i, j] = \partial y_i / \partial \theta_j$. Thus, we get a linear system of y with respect to θ for a specific x. Since some or most of the derivatives in the Jacobian are often very small or near zero, the J is sparse and a good screening of θ's parameters is accomplished, which is important for both better circuit understanding by the IC designers and for more physical and accurate statistical circuit modeling. Analysis of the Jacobian matrix in conjunction with the regression analysis leads to the reduced quadratic model, (since only some linear, quadratic and mixed terms are used):

$$\hat{b}_k = \beta_{k,o} + \sum_{some\ i's} \beta_{k,i} \theta_{Gi} + \sum\sum_{some\ ij's} \beta_{k,ij} \theta_i \theta_j \qquad \theta = (\theta_G, \theta_L) \qquad (9)$$

Interpretation of Direct Modeling(B) is shown in Fig. 2. This ap-

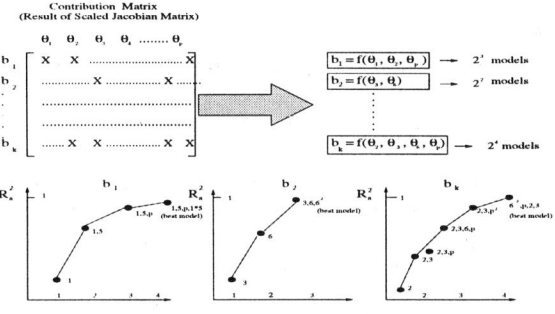

Figure 2: Representation of the Direct Modeling(B) method

proach models performance dependencies of a design accurately in terms of both random and designable parameter values and is well suited to design investigations when functional dependence on global process variations and mismatches between critical devices must be accurately accounted for.

2.3. Reduced Space Modeling Approach - Model (C)

The direct modeling methodology discussed above offers highly accurate statistical models, but when the number of performances of interest in the $b-space$ is significant, the methodology becomes computationally expensive. Savings in computational cost are proposed in Model(C) which we call "Reduced Space Modeling Approach", which interpretation is shown in Fig. 3. We begin investigation and modeling in the $b-space$. Modeling procedure is as follow: (1) **Principal Component Analysis in** $b-space$: determine the minimum number of dimensions needed to account for a high percentage of variance explained (2) **Factor Analysis in** $b-space$: develops a statistical model able to extract the underlying factors affecting the original variables. (2a) Identify, basic random variables as a subset of b_B of b parameters through the rotation of factor

loading matrix. (3) **Regression Analysis in** $b-space$: Remaining $b-space$ parameters can be generated from basic random variables using linear and quadratic regressional dependencies; $b_k = f(b_B)$.
(4) **Model Basic Random Variables in Terms of** θ: model b_B's as functions of $\theta = (\theta_G, \theta_L)$ using the Direct Modeling (B) approach. A hierarchical model is created:

$$\hat{b}_k = b_k(b_B(\theta_G, \theta_L)) \qquad (10)$$

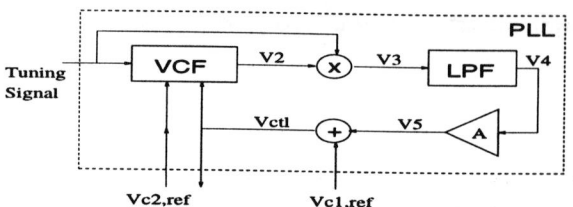

Figure 4: PLL-block diagram

voltage Vctl causes the phase shift at the internal node V2 to be 90 degrees. The characteristics of the MOSFET-C notch filter shown in Fig. 5 were mathematically defined by two transfer functions:

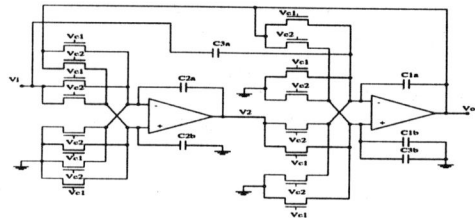

Figure 5: MOSFET-C notch filter

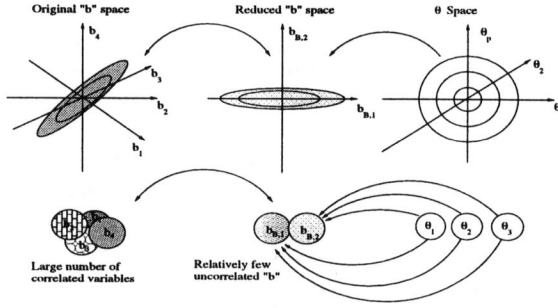

Figure 3: Interpretation of reduced space modeling method

3. STATISTICAL MODELING OF CMOS TRANSISTORS

In order to demonstrate the statistical modeling approaches, the statistical CMOS transistors had to be characterized. Based on our previous work in this area, we used PCA, FA and quadratic regression models in order to create suitable statistical models. The total number of parameters investigated for the n-channel and the p-channel was 34. Trough the PCA/FA analysis, the dimension of the transistor model was reduced to seven global parameters TOX, NSUBn, NSUBp, LDn, RSHn, CJp, CJn. These parameters accounted for 97 percent of the drain current variations of the CMOS transistor. Additionally, two mismatches: MMVTO mismatch in the threshold voltage, and the MMUO mismatch in the surface mobility were added to the model, in order to model statistically such critical circuits as differential pairs and current mirrors.

4. STATISTICAL BEHAVIORAL MODELING OF IC'S

Two approaches were undertaken for behavioral modeling: 1) A top-down hierarchical decomposition methodology, and 2) A bottom-up hierarchical verification methodology. The top-down phase involves circuit functional partitioning, mathematical definition of the characteristics of interest, and verification of the model equations. The bottom-up phase is used for statistical characterization, and simulation. The example chosen to demonstrate the modeling process is that of a phase locked loop (PLL) of the type commonly used to tune voltage controlled filters in master-slave structures, in order to correct the effects of process and other variations. As it can be seen in Fig. 4 the PLL can be modeled decomposing it into four major functional blocks - a master voltage controlled filter (MOSFET - C notch filter), an analog multiplier (phase detector), a loop filter and a loop amplifier. Tuning of the notch frequency of the filter to the input reference frequency is achieved when the applied control

one for the node $V2$ and the other one for node V_{out}, respectively. However, both transfer functions can be defined by the following general single equation. Correctness of the equation was verified.

$$H(s) = H_o \frac{s^2 + s(\omega_z/Q_z) + \omega_z^2}{s^2 + s(\omega_p/Q_p) + \omega_p^2} \left(\frac{s/\omega_{z_{p1}} + 1}{s/\omega_{p_{p1}} + 1} \right) \left(\frac{s/\omega_{z_{p2}} + 1}{s/\omega_{p_{p2}} + 1} \right) \qquad (11)$$

where H_o is the DC gain, $\omega_p, Q_p, \omega_z, Q_z$ are the locations and quality factors of the dominant poles and zeros, and $\omega_{p_{p1,2}}$ and $\omega_{z_{p1,2}}$ represent two parasitic poles and zeros. Statistical data regarding the modeled poles and zeros was obtained using GOSSIP (a statistical system developed at Texas A&M [4]) in conjunction with Hspice pole-zero analyses at node V_2 and the filter output, respectively. The values of H_o were extracted from the low frequency AC analysis. Dependencies of $\omega_p, Q_p, \omega_z, Q_z$ on the control voltage were also modeled. Upon verification of the modeling equations, Monte Carlo simulations with 1000 samples were performed using HSPICE in conjunction with GOSSIP. The statistical CMOS transistor model was used in Monte Carlo simulations. Statistical data extracted from that simulation was entered into the S-PLUS environment and three types of statistical models: Model(A3), Model(B), and Model(C) were created. Effectiveness of the modeling methodologies was judged based on the ability of the given model to recreate the distribution and correlations of the behavioral model parameters of the original device-level design. Probability density functions for some behavioral model parameters are shown in Fig. 6. Some of the parameters modeled by the Basic Additive Model(A3) had significant errors. Large errors in particular, in estimating the standard deviations for model(A3), resulted for those parameters whose regressional dependency required interaction terms (θ_G, θ_L). This discovery led us to the investigation of the correlations/relationships between parameters for all established models. Fig. 7 illustrates scatter plots of some performance variables against each other performance for all models. It is clearly visible

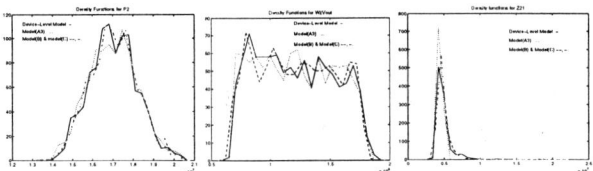

Figure 6: Density functions for $\omega_{p_{p2}}, \omega_{z_o}, \omega_{z_{21}}$: Device Level v. Models

that Basic Additive Model(A3) is unable to reproduce the relationships, the influence of the global parameters θ_G and mismatches θ_L are not additive as originally assumed! This important observation

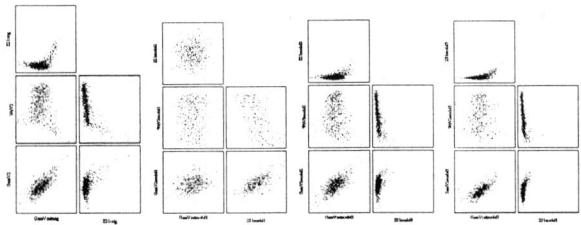

Figure 7: Relationships between performances: Device-level, Model(A3), Model(B), Model(C)

goes against the common belief that interactions terms between the global and local parameters are not important, and needs to be further investigated in the future. The remaining functional blocks of the PLL were also statistically behaviorally modeled. Modeling details (e.g., for the analog multiplier) can be found in previous publications [1,2].

5. SYSTEM LEVEL STATISTICAL BEHAVIORAL SIMULATION

Two complete statistical PLL system models (Model(A3) and Model(B)) were implemented using CADENCE's SpectreHDL behavioral modeling language. In order to determine the validity of the modeling results, a device-level system model was simulated using the original device-level statistical models. All three systems, were simulated in the CADENCE Analog Artist environment. For the comparison purpose, 200 Monte Carlo transient analyses were performed on all three systems. The control voltage $Vctl$ is locked when the notch frequency is tracked to the input signal at 200KHz. For the purpose of automatic tuning, we are interested in the steady state response of the tuning signal $V5$, since tuning can be successful only within a specified range of the Vctl voltage. Fig. 8 shows the variations of the steady state response of $V5$ due to process disturbances for the PLL model(A3), model(B), and the device-level design. The means and standard deviations differ by 8.5% and 9.75% for Model (B), respectively. Large errors were registered for Model(A3), which is to be expected after our discovery that the global effects and mismatch parameters are not additive. High speed of the model allowed us to capture other important characteristics of the PLL such as $Vctl$ control signal, notch frequency $Notch\_fo$, and gain at the notch frequency $Notch\_dB$, which pdf's are shown

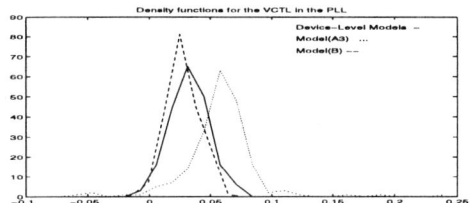

Figure 8: PLL tunable filter statistical simulation: $V5$ tuning signal Device-level vs. Models

in figure 9 and the table below summarize their statistic.

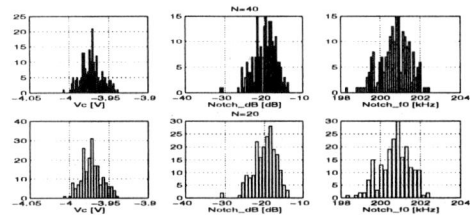

Figure 9: Histograms of the control voltage $Vctl$, $Notch\_dB$, $Notch\_fo$

Variable	Mean	Std. dev.	Min.	Max
$Vctl$ [V]	-3.972	1.294e-2	-4.008	-3.938
$Notch\_dB$ [dB]	-19.87	2.996	-31.33	-13.54
$Notch\_f0$ [kHz]	200.7	0.7593	198.2	202.5

6. CONCLUSIONS

Methods for efficient mapping from circuit parameter space to the behavioral space that is both computationally efficient and accurate have been developed. The models constructed based on developed techniques are able to relate the statistical performance variations of the entire circuit to the variations of a few global statistical transistor parameters (closely related to significant process variations) and most important mismatches between different devices on the chip. Demonstrated examples show that the proposed techniques can serve as a valuable tool in reduction of design cycle in today's competitive market.

7. REFERENCES

[1] J.F. Swidzinski, D.D. Alexander, M. Qu, and M.A. Styblinski, "A Systematic Approach to Statistical Simulation of Complex Analog Integrated Circuits," 2nd Int. Workshop on Statistical Metrology, Kyoto, Japan, June 8, 1997.

[2] J.F. Swidzinski, M.A. Styblinski, Statistical Behavioral Modeling and Simulation of an Analog Phase Locked Loop," 1st Int. Workshop on Design of Mixed-Mode Integrated Circuits and Applications, Cancun, Mexico, July 28-30, 1997.

[3] M. Qu, Statistical Modeling of Large Integrated Circuits and its Application to Design for Quality and Manufacturibility, PhD Diss., Electr. Eng. Dept., Texas A&M Univ., College Station,TX, Dec. 1995.

[4] L.J. Opalski and M.A. Styblinski, GOSSIP: A Generic System for Statistical Improvement of Performance, Electr. Eng. Dept., Texas A&M Univ., College Station, TX, April 1995.

ROBUST RECURSIVE INVERSE APPROXIMATION AND ITS APPLICATION TO PARAMETER EXTRACTION OF BEHAVIORAL MODELS

G. Xu, M. A. Styblinski

Department of Electrical Engineering
Texas A&M University, College Station, TX 77843
Phone: (409)845-5257   Email: gxu@ee.tamu.edu

ABSTRACT

An algorithm, Robust Recursive Inverse Approximation (RRIA), for parameter extraction in the statistical IC modeling is presented. RRIA gives very accurate modeling parameter formulas based on observation data after the behavioral model is set up. It can extract the parameters simultaneously for multiple circuits which have same structure but with different parameter values. RRIA combines Newton-Raphson, least-squares and Singular Value Decomposition techniques. The theory of this algorithm is presented in details. Its efficiency for different modeling levels is verified by two examples: the MOSFET device and the bandpass filter.

1. INTRODUCTION

In statistical integrated circuit behavioral modeling, the distributions of parameters in various modeling levels play an important role and they determine the various modeling techniques. Therefore, the statistical model parameter extraction is one of the basic issues in behavioral modeling. Recursive Inverse Approximation (RIA) is a new method to extract *statistical* device model parameters and behavioral model parameters proposed in [4,1]. Compared to the standard extraction process, for example, utilizing optimization-based algorithms, the new RIA-based approach avoids the uncertainties, allows direct analytical formula approximation. Once these formulas are constructed, parameter extraction is very fast and accurate. However, in some practical cases (especially for device-level modeling) the construction of RIA can be difficult or impossible using the original linear regression approach proposed in [4,1], due to the ill-conditioning of the RIA modeling problem and/or algorithm divergence during the model creation. In this paper, a Singular Value Decomposition (SVD) approach is proposed to implement a *Robust RIA* (RRIA) algorithm.

The objective of statistical parameter extraction at the device modeling level is to find the model parameters from the measured performances such as the I-V curves. The typical problem is formulated as follows:

Let $\mathbf{p} = [p_1, p_2, ..., p_n]^T \in R^n$ be the vector of IC model parameters to be extracted, $\mathbf{y} = [y_1, y_2, ..., y_m]^T \in R^m$, the vector of performances, $\mathbf{z}^k = [z_1^k, z_2^k, ..., z_m^k]^T \in R^m$, $k = 1, 2, ..., K$, the vector of the measured performances of the kth device, where

---
SUPPORTED IN PART BY SEMICONDUCTOR RESEARCH CORPORATION AND TEXAS ADVANCED TECHNOLOGY PROGRAM

K is the number of devices. A general IC model can be written as $\mathbf{y} = \mathbf{y}(\mathbf{p})$.

Statistical parameter extraction is defined as (an "ideal" solution): *Given the known IC model* $\mathbf{y} = \mathbf{y}(\mathbf{p})$, *find the parameter vectors* $\mathbf{p}_{(*)}^k \in R^n$, $k = 0, 1, 2, ..., K$ *such that*

$$\mathbf{p}_{(*)}^k = arg \left\{ \min_{\mathbf{p}^k \in R^n} \varepsilon(\mathbf{p}^k) = \frac{1}{2}[\mathbf{y}(\mathbf{p}^k) - \mathbf{z}^k]^T [\mathbf{y}(\mathbf{p}^k) - \mathbf{z}^k] \right\} \quad (1)$$

The practical problem is that *exact* solution of Eq.(1) is, in general, not possible, so various approximation, mostly iterative schemes, were proposed in practice (see [4,1] for references).

2. RECURSIVE INVERSE APPROXIMATION

The objective of *Recursive Inverse Approximation* is to find an approximation $\mathbf{p} \cong \tilde{\psi}(\mathbf{y})$ to the *inverse mapping* $\mathbf{p} = \psi(\mathbf{y})$ from the model performance space to the model parameter space, under the least-squares criterion. If we approximate $\mathbf{y} = \mathbf{y}(\mathbf{p})$ about the nominal parameter vector by a linear function $\mathbf{y} \approx \mathbf{y}_L(\mathbf{p})$, then under the least-squares criterion, there exists a linearized inverse approximation $\hat{\mathbf{p}} \cong \tilde{\psi}_L(\mathbf{y})$[1]. The extracted (approximate) model parameter vector for the kth device is then found from $\hat{\mathbf{p}}_{(0)}^k = \tilde{\psi}_L(\mathbf{z}^k)$, where the subscript "(0)" indicates the initial estimate (i.e., **prediction**) of the parameter vector. In general, $\hat{\mathbf{p}}_{(i)}^k = \tilde{\psi}_i(\mathbf{z}^k)$, where $\tilde{\psi}_i(\cdot)$ is the ith approximation to the inverse mapping and $\tilde{\psi}_0(\cdot) \equiv \tilde{\psi}_L(\cdot)$. Recursive refinement of the RIA model for the kth device is performed using the following procedure

$$\hat{\mathbf{p}}_{(i+1)}^k = \mathcal{F}_i(\hat{\mathbf{p}}_{(i)}^k) = \mathcal{F}_i(\tilde{\psi}_i(\mathbf{z}^k)), \quad i = 0, 1, ... \quad (2)$$

subject to

$$\|\hat{\mathbf{p}}_{(i+1)}^k - \mathbf{p}_{(*)}^k\| < \|\hat{\mathbf{p}}_{(i)}^k - \mathbf{p}_{(*)}^k\| \quad (3)$$

where, $\mathcal{F}_i(\cdot)$ is *an error correction mapping* in the $\mathbf{p}$ space, reducing the errors of the extracted parameter values, $\mathbf{p}_{(*)}^k$ is the parameter vector satisfying $\mathbf{p} = \psi(\mathbf{y})$ (an "ideal" solution). Using this refinement, we actually establish a *recursive parameter correction* procedure as

$$\hat{\mathbf{p}}_{(i)}^k = \tilde{\psi}_i(\mathbf{z}^k) = \mathcal{F}_i(\mathcal{F}_{i-1} \cdots (\mathcal{F}_0(\tilde{\psi}_L(\mathbf{z}^k)))). \quad (4)$$

---
[1]The hat "∧" denotes an estimate of the parameter vector.

As it is seen, the inverse mapping $\mathbf{p} = \psi(\mathbf{y})$ is recursively approximated by

$$\psi_i(\mathbf{y}) \cong \tilde{\psi}_i(\mathbf{y}) = \mathcal{F}_i(\mathcal{F}_{i-1} \cdots (\mathcal{F}_0(\tilde{\psi}_L(\mathbf{y})))). \quad (5)$$

In each step a new error correction mapping $\mathcal{F}_i(\cdot)$ is constructed. Eq. (4) actually corrects the "errors" of the previous estimate of the parameter vector for the kth device to obtain a more accurate parameter estimate and so is referred to as *parameter correction*. Once the correcting functions $\mathcal{F}_i$ are established, no iterative parameter extraction is (e.g., optimization based) needed.

3. SINGULAR VALUE DECOMPOSITION

Singular Value Decomposition(SVD) is applied to the following linear simultaneous equations:

$$\mathbf{A}\mathbf{x} = \mathbf{b} \quad (6)$$

where $\mathbf{A} \in R^{m \times n}, \mathbf{x} \in R^n, \mathbf{b} \in R^m$. Eq.(6) defines $\mathbf{A}$ as a linear mapping from the vector space $\mathbf{x}$ to the vector space $\mathbf{b}$. There are two spaces related to the mapping $\mathbf{A}$: null space and range space. The null space N is a subspace of $\mathbf{x}$ that is mapped to zero: $N = \{\mathbf{y} \in R^n \mid \mathbf{A}\mathbf{y} = \mathbf{0}\}$. The range space R is the subspace of $\mathbf{b}$ which can be "reached" by $\mathbf{A}$: $R = \{\mathbf{c} \in R^m \mid \exists \mathbf{y} s.t., \mathbf{A}\mathbf{y} = \mathbf{c}\}$. The dimension of the range space is called the rank of $\mathbf{A}$; the dimension of the null space is called the nullity of $\mathbf{A}$. There is a related *Fredholm Splitting Theorem*: nullity of $\mathbf{A}$ plus the rank of $\mathbf{A}^T$ is n.

In addition to the above discussion, SVD explicitly constructs orthonormal bases of the null space and range space of $\mathbf{A}$. Suppose $\mathbf{A}$ has the following SVD:

$$\mathbf{A} = \mathbf{U}\mathbf{W}\mathbf{V}^T \quad (7)$$

The columns of $\mathbf{U}$ whose same-numbered diagonal elements w_{jj} in matrix $\mathbf{W}$ are nonzero are an orthonormal set of basis vectors that span the range of $\mathbf{A}$; the columns of $\mathbf{V}$ whose same-numbered diagonal elements w_{jj} are zero are an orthonormal basis for the null space.

Further, SVD of $\mathbf{A}$ can explicitly give the solution for the above linear simultaneous equations. If we denote U_i as the ith column of $\mathbf{U}$, V_i as the ith column of $\mathbf{V}$, r as the number of the first several nonzero singular values of $\mathbf{A}$, then the singular value solution is [2]:

$$\mathbf{x} = \sum_{i=1}^{r} \frac{(U_i, \mathbf{b})}{W_{ii}} V_i \quad (8)$$

If the original equations do have one or more than one solutions, then the above SVD solution gives the smallest length solution; if the original equations do not have any solution, in other words, $\mathbf{b}$ is not in the range of $\mathbf{A}$, then the above SVD solution gives the closest possible solution in the least square sense: the solution $\mathbf{x}$ minimizes the residual: $\| \mathbf{A}\mathbf{x} - \mathbf{b} \|$. The above discussion can be summarized in the following figure

- $\mathbf{A}$ is non-singular

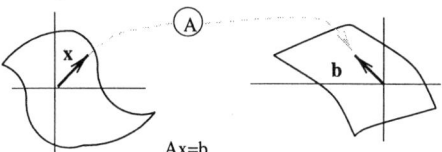

- $\mathbf{A}$ is singular (not full rank)

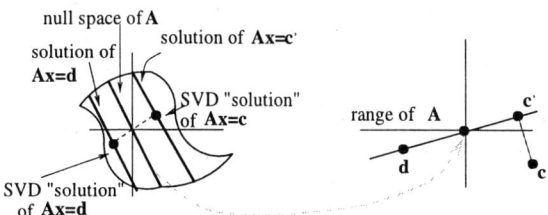

Notes: A singular matrix $\mathbf{A}$ maps a vector space into one of lower dimensionality, here, a plane into a line, called the "range" of $\mathbf{A}$. The "null" space of $\mathbf{A}$ is mapped to zero [2].

4. SINGULAR VALUE DECOMPOSITION TECHNIQUES IN RIA

In this section, a new approach using the Singular Value Decomposition (SVD) techniques is combined with the RIA to eliminate the ill-conditioning problems in RIA and thus creating the Robust Recursive Inverse Approximation (RRIA). The SVD technique is used both during the linear approximation and the error correction. Using linear approximation, the least square problem becomes:

$$\min_{\Delta \mathbf{p}^k \in R^n} \left\{ \varepsilon(\Delta \mathbf{p}^k) = \frac{1}{2} \left[\mathbf{y}_L^k - \mathbf{z}^k\right]^T \left[\mathbf{y}_L^k - \mathbf{z}^k\right] \right\} \quad (9)$$

$$k = 1, 2, \ldots K$$

where, $\mathbf{y}_L^k = \mathbf{J}(\mathbf{p}^0)\Delta\mathbf{p}^k + \mathbf{y}^0$, $\mathbf{y}^0 = \mathbf{y}(\mathbf{p}^0)$, and $\mathbf{p}^0$ is extracted during the nominal parameter extraction (using the existing techniques), and

$$\mathbf{J}(\mathbf{p}^0) = \left[\frac{\partial y_i}{\partial p_j}\right]\bigg|_{\mathbf{p} = \mathbf{p}^0} \quad (10)$$

is the Jacobian matrix evaluated at the nominal point $\mathbf{p}^0$. The SVD of the $\mathbf{J}(\mathbf{p}^0)$ is

$$\mathbf{J}(\mathbf{p}^0) = \mathbf{U}\mathbf{W}\mathbf{V}^T \quad (11)$$

where $\mathbf{U} \in R^{m \times m}, \mathbf{V} \in R^{n \times n}$ are both othonormal matrices, $\mathbf{W} \in R^{m \times n}$ is a diagonal matrix of *singular values* $W_{11} \geq W_{22} \geq \ldots \geq W_{nn} \geq 0$, $\mathbf{V}^T$ means the transpose of $\mathbf{V}$.

Then the least square problem (9) has the solution [2]:

$$\Delta \hat{\mathbf{p}}_{(0)}^k = \sum_{i=1}^{n} \frac{(U_i, \mathbf{z}^k - \mathbf{y}(\mathbf{p}^0))}{W_{ii}} V_i \quad (12)$$

where U_i is the ith column of $\mathbf{U}$, V_i is the ith column of $\mathbf{V}$. In many cases, the normal equations of the least square problem (9) are very close to being singular and the corresponding singular values maybe zero or nearly zero. Once this happens, then the reciprocal $1/W_{ii}$ in equation (12) should be set to zero, not to infinity [2]. As a result, only r terms are left ($r \leq n$) in Eq. (12) and SVD produces a solution which is a subspace projection of the "ideal" solution (Eq. (12) with n replaced by r). Thus, for the measured performance vector $\mathbf{z}^k$ for the kth device, an approximate estimate of the model parameter vector for the kth device is obtained from[2]

$$\hat{\mathbf{p}}_{(0)}^k = \mathbf{p}^0 + \sum_{i=1}^{r} \frac{(U_i, \mathbf{z}^k - \mathbf{y}(\mathbf{p}^0))}{W_{ii}} V_i \quad (13)$$

This linear prediction procedure is denoted by $\hat{\mathbf{p}}_{(0)}^k = \tilde{\psi}_0(\mathbf{z}^k)$. After the linear prediction of the intial parameters, we can then calculate the model performances corresponding to these *estimated initial parameters* (since the mapping $\mathbf{p} \in R^n \mapsto \mathbf{y} \in R^m$ defines the IC model and is known). For the estimated model performances, we *repeat* the linear prediction again and get another set of parameters which are called *simulated parameters*:

$$\tilde{\mathbf{p}}_{(0)}^k = \tilde{\psi}_0(\mathbf{y}(\hat{\mathbf{p}}_{(0)}^k)) \quad (14)$$

The error difference is calculated from the paired parameter data

$$\tilde{\mathbf{d}}_0^k = \hat{\mathbf{p}}_{(0)}^k - \tilde{\mathbf{p}}_{(0)}^k = \hat{\mathbf{p}}_{(0)}^k - \tilde{\psi}_0(\mathbf{y}(\hat{\mathbf{p}}_{(0)}^k)), \quad k = 1, 2, ..., K. \quad (15)$$

The relative errors of the sample means and standard deviations of the paired parameter data are used to check the accuracy of the estimated parameters. When accuracy checking shows that the parameter extraction is not accurate enough, a process, called *Parameter Correction*, is performed to refine the parameter extraction. The data set $\left\{\tilde{\mathbf{d}}_0^k, \triangle\tilde{\mathbf{p}}_{(0)}^k, k = 1, 2, ..., K\right\}$ is used to construct the (scaled) *error correction function* $\mathcal{F}_i(p)$. Here, the quadratic function is postulated to approximate the scaled $\mathcal{F}_i(p)$ (the additional required subscript i is dropped for simplicity)

$$d_t = \sum_{j=1}^{n} \beta_j \triangle p_j + \sum_{j=1}^{n} \sum_{l=j}^{n} \beta_{jl} \triangle p_j \triangle p_l, \quad t = 1, 2, .., n \quad (16)$$

where, $\mathbf{d} = [d_1, d_2, ..., d_n]^T$, $\triangle \mathbf{p} = [\triangle p_1, \triangle p_2, ..., \triangle p_n]^T$, β_j and β_{jl}'s are coefficients[3]. The coefficients can be determined by fitting Eq. (16) to the data set $\left\{\tilde{\mathbf{d}}_0^k, \triangle\tilde{\mathbf{p}}_{(0)}^k, k = 1, 2, ..., K\right\}$ under least-squares criterion. During this process, ill-conditioning problem can again occur, so, the SVD techniques is used to eliminate the problems. The SVD solution formula is similar to the prediction stage.

Once the error correction function $\mathbf{d}_0 = \mathbf{d}_0(\triangle \mathbf{p}_{(0)}^k)$ is established, *parameter correction* on the real measurements is then performed

$$\triangle \hat{\mathbf{p}}_{(1)}^k = \triangle \hat{\mathbf{p}}_{(0)}^k + \mathbf{d}_0(\triangle \hat{\mathbf{p}}_{(0)}^k) \quad (17)$$

$$\hat{\mathbf{p}}_{(1)}^k = \mathbf{p}^0 + \triangle \hat{\mathbf{p}}_{(1)}^k, \quad k = 1, 2, ..., K \quad (18)$$

This procedure is recursively repeated until we arrive at an acceptable result.

---

[2]The subscript "(0)" is used to denote the initial estimate of the parameter vector and hat "∧" is used to denote the estimate.

[3]These coefficients are the ith coefficients of the error correction function $\mathcal{F}_i(.)$. So, if the total of l iterations are performed, there are l sets of these coefficients.

5. EXAMPLE 1

The objective was to extract the 10 parameters of the NMOS transistor HSPICE level 3 model: TOX, XJ, LD, UO, $THETA$, $NSUB$, NFS, $VMAX$, ETA, $KAPPA$, using the I-V data, based on 363 data points on I-V curves for selected biasing voltages V_{DS}, V_{GS} and V_{BS}. The set of these parameters was carefully selected as *independent* model parameters (since several model parameters are internally calculated by HSPICE) to avoid ill-conditioning due to parameter dependencies. Although in practice, these 10 parameters are correlated because they are generated from the more fundamental process parameters, here we assume they are independent so that we can generate these parameter data directly through Monte Carlo (MC) simulations. To check and demonstrate the proposed method, our Monte Carlo (MC) simulation sample size is equal to 100. Using finite-difference approximation, the 363×10 Jacobian matrix was constructed through HSPICE. SVD revealed that the largest/smallest singular ratio was $5.1 \times 10^{-4}/1.0345 \times 10^{-7} = 5.913 \times 10^4$, so, the problem is highly singular and the direct solution using the standard regression will be highly unreliable (or no solution can be found). After SVD-based linear model was constructed, the *accuracy checking* was performed on the estimated parameters obtained from *parameter prediction* (see Table 1). The relative errors of the sample means and standard deviations, δ^m and δ^s, are defined with respect to the previously calculated parameters. It is seen from Table 1, that: (1) all relative errors of sample means, δ^m's, are less than 6.5%; (2) relative errors of standard deviations, δ^s's, are $3 \sim 50$%, i.e., they are quite large.

After two correction steps using (16)-(18), all relative errors have been significantly reduced (see Table 2): (1) all relative errors of sample means, δ^m's, are less than 0.4%, (2) relative errors of standard deviations, δ^s's, are reduced to less than 5%.

A more realistic example should include the process simulation to determine the distributions of model parameters. In other words, more accurate statistical model should be used. In more sophisticated MOSFET model, e.g., BSIM3v3, the relationship between IV curves and model parameters will be complicated. In this situation, one step parameter extraction is impossible and multiple steps are recommended. A multiple step extraction procedure is reported by [6]. And RRIA algorithm can be combined into each extraction step in the above procedure. Also, parameter transformation can be used to avoid unnecessary ill-conditioning. In the case of highly non-linear problem, other nonlinear correction functions (e.g., the reciprocal of linear function) should be considered.

Table 1: Example 1 – Results after *Parameter Prediction*

parameter	sample μ	δ^m	sample σ	δ^s
TOX	2.1471e-08	0.063%	1.2170e-09	3.42%
XJ	1.7840e-07	-6.54%	2.9490e-08	44.6%
LD	1.0792e-07	0.426%	3.6865e-09	49.0%
UO	5.8962e+02	-0.08%	3.6806e+01	6.78%
THETA	1.1718e-01	-1.17%	7.1577e-03	7.38%
NSUB	2.4496e+16	-1.66%	1.5254e+15	12.1%
NFS	5.1164e+12	1.18%	3.4668e+11	28.2%
VMAX	1.9479e+05	-0.73%	1.1452e+04	4.96%
ETA	6.6620e-02	-1.35%	8.2689e-03	46.0%
KAPPA	1.2369e-01	6.05%	2.5363e-02	50.7%

Table 2: Example 1 – Results after two iterations

parameter	sample μ	δ^m	sample σ	δ^s
TOX	2.144e-08	-0.03%	1.101e-09	0.016%
XJ	2.012e-07	0.11%	1.022e-08	0.025%
LD	1.072e-07	0.02%	1.445e-09	4.1%
UO	5.903e+02	-0.02%	3.212e+01	-0.66%
THETA	1.198e-01	0.017%	6.525e-03	0.027%
NSUB	2.530e+16	0.064%	1.375e+15	1.32%
NFS	5.015e+12	-0.019%	2.259e+11	-0.24%
VMAX	1.975e+05	0.006%	1.101e+04	0.15%
ETA	6.892e-02	0.185%	3.274e-03	4.69%
KAPPA	1.084e-01	-0.47%	6.144e-03	4.90%

6. EXAMPLE 2

In this example, the Bandpass filter transfer function is used.

$$H(s) = H_o \left(\frac{s(\omega_o/Q)}{s^2 + s(\omega_o/Q) + \omega_o^2} \right) \left(\frac{s/\omega_z + 1}{s/\omega_p + 1} \right) \quad (19)$$

The objective is to extract parameters $H_o, \omega_o, Q, \omega_p, \omega_z$ in the above equation. 200 points are sampled from its frequency magnitude and phase response and these points are used as one vector in the performance space. 200 BPF devices are generated using Monte Carlo simulation. Similar to the above example, the Jacobian matrix was constructed by finite-difference approximation. The errors after linear prediction and quadratic parameter correction are shown in Table 3 and Table 4. Comparing these two tables, we can see that while some (mostly small) variance errors were increased a little, the big variance errors were decreased, and on average, quadratic parameter correction significantly reduced the prediction errors. (The reasons for the increase of some errors are under investigation.)

Another thing needs to be mentioned is that in the quadratic correction stage, only 4 significant singular values are used instead of 20.

Another observation in this example is that extraction of parameter Q is relatively difficult compared to the other parameters. The reason is obvious: Q is very sensitive to the data collection of performance. Actually, Q is only relevant to the performance around the bandpass region. And only the data from the bandpass region contribute to the extraction of Q. This observation reminds us that statistical model parameter extraction related heavily to the experiment design. Different parameters should be extracted from different data under different conditions.

Table 3: Example 2 – Results after *Parameter Prediction*

parameter	sample μ	δ^m	sample σ	δ^s
H_o	2.9692e+00	-5.47%	3.6557e-01	9.94%
ω_o	3.1551e+04	-0.446%	2.9456e+03	-3.86%
$Q,$	1.8785e+00	-5.95%	2.3882e-01	13.9%
ω_p	2.6523e+05	-1.198%	2.8092e+04	1.63%
ω_z	2.1950e+06	-0.867%	2.3891e+05	-0.257%

7. CONCLUSION AND FUTURE WORK

In all the examples tried, the proposed RRIA method was quite powerful to overcome the ill-conditioning problem and demonstrated good convergence. The success of the SVD technique results from

Table 4: Example 2 – Results after two iterations

parameter	sample μ	δ^m	sample σ	δ^s
H_o	3.1223e+00	4.90%	3.6699e-01	0.385%
ω_o	3.1675e+04	0.391%	3.1341e+03	6.01%
$Q,$	1.9814e+00	5.19%	2.4268e-01	1.59%
ω_p	2.6756e+05	0.87%	2.7928e+04	-0.588%
ω_z	2.2121e+06	0.77%	2.4213e+05	1.33%

the fact that the solution obtained is the subspace projection of the "ideal" solution which is a good approximation to the "real" solution, but it avoids the ill-conditioning problem.

Future research will be concentrated on: investigation of more precise correction functions, proving the method convergence and relevant conditions required, more investigation of other test examples and etc.

8. REFERENCES

[1] Ming Qu and M. A. Styblinski, "Recursive Inverse Approximation - A New Approach to Statistical IC Model Parameter Extraction", *Proc. IEEE/First International Workshop on Statistical Metrology*, Dec, 1995.

[2] William H. Press, Brian P. Flannery, Saul A. Teukolsky and Wialliam T. Vetterling, *Numerical Recipes in C*, Cambridge University Press, 1989.

[3] M. Qu and M. A. Styblinski, "A heuristic global optimization algorithm and its application to CMOS circuit variability minimization," in *Proc. ISCAS*, (Chicago), pp. 1809–1812, May 1993.

[4] M. Qu, *Statistical Modeling of Large Analog Integrated Circuits and Its Applications to Design for Quality and Manufacturability*. PhD Dissert., Texas A&M University, 1995.

[5] M. Qu, "SMIC - Statistical Modeling for Integrated Circuits: User's Guide," Laboratory for Intelligent Design Systems, Texas A&M University, August 1995.

[6] Yuhua Cheng, et al., *BSIM3v3 Manual*, The Regents of the University of California, 1997.

Study of Optimal Importance Sampling in Monte Carlo Estimation of Average Quality Index

Mansour KERAMAT and Richard KIELBASA

Ecole Supérieure d'Electricité (SUPELEC)
Service des Mesures, Plateau de Moulon
F-91192 Gif-sur-Yvette Cédex France
E-mail: mansour.keramat@supelec.fr

ABSTRACT

In this contribution, the optimal *Importance Sampling*, which is a variance reduction technique in Monte Carlo estimation, is theoretically studied in the case of electronic circuits. To the best of our knowledge, this problem has previously not been studied in electronic circuits. In this study, the theoretical basis of the optimal importance sampling is developed in the case where no approximate model of the circuit responses is given. Simulation results show good agreement with the theoretical basis.

I. Introduction

Due to inherent fluctuations in integrated circuits and discrete components manufacturing process, manufactured circuits exhibit a spread in performance values and quality index that results in low Average Quality Index (AQI) [3] or circuit parametric yield. In order to maximize the AQI or the parametric yield, it is imperative to take these manufacturing fluctuations into consideration during circuit design. Since AQI or yield cannot be determined analytically except in trivial cases, we must use some approximation and estimation methods.

The Monte Carlo (MC) method [1] is the most reliable technique for AQI or yield estimation of electronic circuits. However, the MC analysis involves a multitude of circuit simulations and is therefore computationally expensive especially for large circuits. A number of schemes for reducing the required sample size have been reported in the general literature on MC estimation, called *variance reduction techniques* [1], [5], [6].

The *Importance Sampling* is a variance reduction technique [1], [2]. Here, we develop the optimal importance sampling MC estimator in the case where no approximate model of circuit responses is available. By the use of the obtained results the efficiency of *Uniform Warping Method* (UWM) [2] which is a type of importance sampling is studied.

II. Optimal Importance Sampling

The quality of an electronic circuit can be expressed by means of fuzzy sets [3]. Due to the fluctuations in the manufacturing process, the quality of the circuits varies from one to another. Therefore, it is more realistic to consider the Average Quality Index (AQI) [3] as a quality measure. The AQI can be expressed as

$$Q = \int_{R^n} q(\mathbf{x}) f_{\mathbf{x}}(\mathbf{x}) d\mathbf{x} . \quad (1)$$

where $\mathbf{x}$ is an n-vector of circuit parameter's disturbances, $0 \leq q(\mathbf{x}) \leq 1$ denotes the circuit quality index, and $f_{\mathbf{x}}(\mathbf{x})$ is the probability density function (pdf) of the disturbances. The Primitive Monte Carlo (PMC) estimator of AQI is given by

$$\hat{Q} = \frac{1}{N} \sum_{i=1}^{N} q(\mathbf{X}^i), \quad (2)$$

where $\mathbf{X}^i$'s are randomly sampled according to $f_{\mathbf{x}}(\mathbf{x})$.

A. Optimal Sampling PDF

The basic idea of the importance sampling consists of concentrating the distribution of sample points in the parts of the integration domain that are of most "importance" instead of spreading them out evenly. Integral (1) can be rewritten as

$$Q = \int_{R^n} q(\mathbf{x}) \frac{f_{\mathbf{x}}(\mathbf{x})}{g(\mathbf{x})} g(\mathbf{x}) d\mathbf{x}, \quad (3)$$

where $g(\mathbf{x})$ is the new sampling pdf. In order to estimate the integral we take a sample $\mathbf{X}^1, \mathbf{X}^2, ..., \mathbf{X}^N$ from pdf $g(\mathbf{x})$ and substitute its value into the following importance sampling formula

$$\hat{Q}_I = \frac{1}{N} \sum_{i=1}^{N} q(\mathbf{X}^i) \frac{f_{\mathbf{x}}(\mathbf{X}^i)}{g(\mathbf{X}^i)} . \quad (4)$$

The variance of the importance sampling estimator is [1]

$$\mathrm{Var}(\hat{Q}_I) = \frac{1}{N} \int \left[\frac{q(\mathbf{x}) f_{\mathbf{x}}(\mathbf{x})}{g(\mathbf{x})} \right]^2 g(\mathbf{x}) d\mathbf{x} - \frac{Q^2}{N} . \quad (5)$$

It can be shown that the following sample pdf minimizes the variance of the importance sampling estimator [1]

$$g(\mathbf{x}) = \frac{q(\mathbf{x}) f_{\mathbf{x}}(\mathbf{x})}{Q} . \quad (6)$$

It is clear that this sampling pdf cannot be used in practice because it depends on the circuit quality index function.

B. Using the Notion of Circuit Class

In order to obtain an optimal sampling scheme that does not depend on a specific circuit, we introduce the notion of *Circuit Class* by using *Stochastic Process Theory* [8]. In this approach, we can take advantage of general properties of electronic circuit behavior.

The notion of the circuit class is illustrated in Fig. 1(a). Now we consider a class of circuits that is determined by a set of quality index functions. Here, $q(x,\omega)$ is considered as a stationary stochastic process with correlation function $r(.)$. ω is an outcome of the circuit class and $q(x,\omega_i)$ denotes a specific realization for a given circuit. Furthermore, from the fact that $q(.)$ is a quality index, then $0 \le q(x,\omega) \le 1$. Fig. 1(b) shows a sample of the quality index function from the stochastic process.

Definition: *The set of electronic circuits, whose quality indices are special realizations of $q(x,\omega)$, is called a class of circuits.*

In fact, each circuit problem is a special realization of the stochastic process. If we consider the variance of the importance sampling estimator over the circuit class, the following theorem can be stated.

Theorem 1: *If there is no approximate model of the quality function, then the optimal importance sampling estimator is the primitive Monte Carlo estimator.*

The proof of this theorem is given in Appendix. From *Theorem 1*, one can conclude that the disturbance pdf is the best sampling pdf if no approximation of the circuit responses is available.

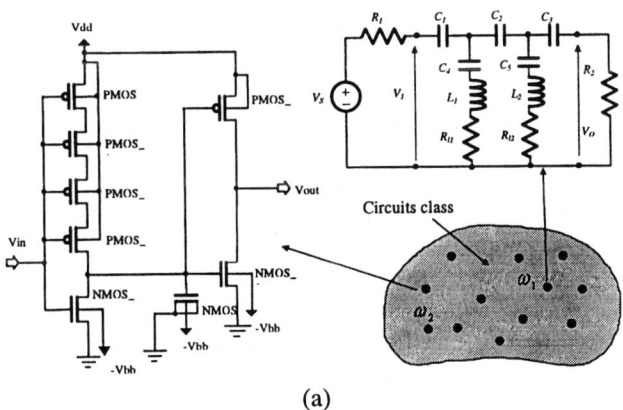

(a)

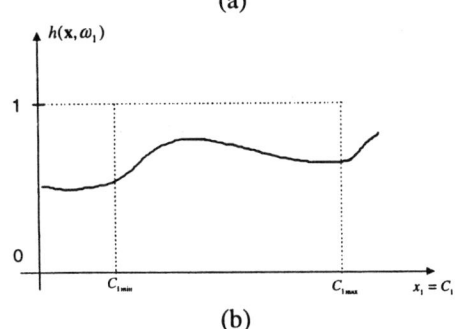

(b)

Fig. 1. Stochastic process. (a) Notion of circuit class. (b) Sample function of the stochastic process.

The Uniform Warping Method (UWM) was classified in the category of importance sampling [2]. In this method, the sampling pdf is uniform. From Theorem 1, one can conclude that the MC estimator with UWM cannot be in general more efficient than the PMC estimator. In addition, the inefficiency of the application of UWM in *Centers of Gravity* (CoG) *Algorithm* (yield optimization method) was reported in [4].

III. Numerical and Circuit Examples

In this section, the applications of the optimal importance sampling scheme to an electronic circuit and a numerical function are presented.

Example 1: Variance Limit of Uniform Warping Yield Estimator [4]

Consider the following importance sampling estimator

$$\hat{Y}_I = \frac{1}{N}\sum_{i=1}^{N} I_P(\mathbf{x}^i)w_i, \qquad (7)$$

where $w = f(\mathbf{x})/g(\mathbf{x})$. In UWM, $g(\mathbf{x})$ is chosen uniform. In order to have an idea about the variance of estimator (7), we suppose that $f(\mathbf{x})$ is a truncated n dimensional Gaussian pdf with independent components and $Y = 1$. By using (5), we have

$$\left(\sigma^2_{\hat{Y}_I}\right)_{Y=1} = \frac{1}{N}\left\{\prod_{j=1}^{n}E\left[w_{i,j}^2\right]-1\right\} = \frac{1}{N}\left\{\left(E\left[w_{i,j}^2\right]\right)^n - 1\right\}, \qquad (8)$$

where

$$w_{i,j} = \frac{f_{x_j}(x_j)}{h_{x_j}(x_j)} = \frac{6}{c\sqrt{2\pi}}e^{-\frac{(x_j-x_j^0)^2}{2\sigma_j^2}}, \qquad (9)$$

and $c = 2\,\text{erf}(3)$. After some manipulations, we obtain

$$E\left[w_{i,j}^2\right] = \frac{6\,\text{erf}(3/\sqrt{2})}{c^2\sqrt{\pi}} = 1.7017. \qquad (10)$$

By substituting (10) into (8), the limit of the variance of the estimator is found to be

$$\left(\sigma^2_{\hat{Y}_I}\right)_{Y=1} = \frac{1}{N}\left[(1.7017)^n - 1\right]. \qquad (11)$$

It is seen that the variance increases exponentially with the dimension of the parameter space. For instance, if $N=100$ and $n=10$, the standard deviation of the yield estimator is 1.42 when $Y = 1$. It is clear that this is not a practical estimator.

Example 2: Pinel's High Pass Filter Circuit

In this example, we consider the fifth-order high pass filter circuit shown in Fig. 2(a), which has served as a test example for many statistical design methods. The nominal circuit response and the performance specifications are shown in Fig. 2(b).

In order to evaluate the efficiency of two types of yield estimators, it is assumed that C_1, C_2, C_3, and C_4 are subjected to a Gaussian statistical variations with a 10% tolerance. The sampling pdf is chosen to be uniform with $N = 100$, i.e., UWM is used. The variances of both yield estimators are shown in Fig. 3. It shows that the standard deviation (SD) of the PMC yield estimator converges to its theoretical value obtained from Eq. (11). The dotted line corresponds to the analytically found value of the SD of the primitive yield estimator (without UWM). The variance of the yield estimator is greater with than without UWM.

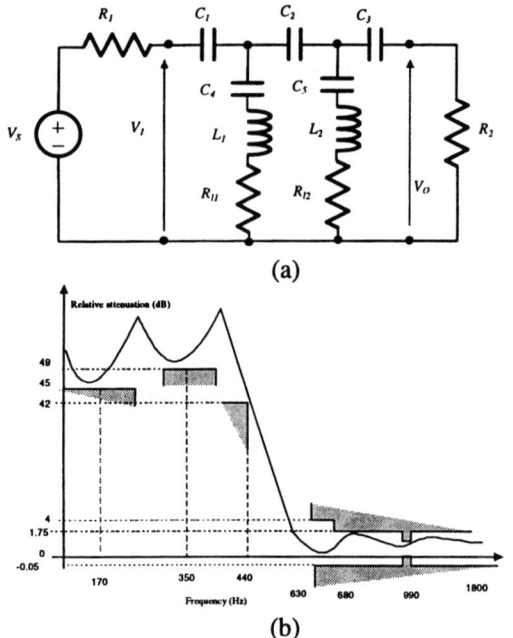

Fig. 2. Pinel's High pass filter circuit. (a) Schematic circuit. (b) Performance specifications.

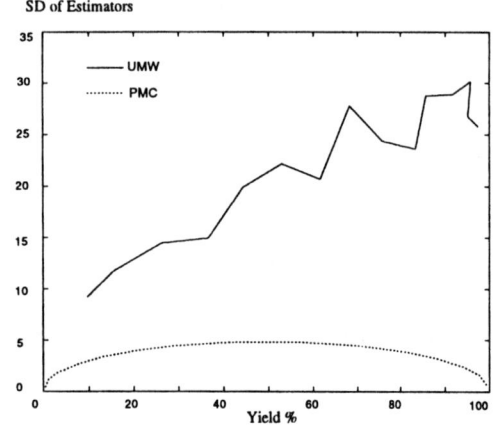

Fig. 3. Standard deviation of the yield estimators by the use of UWM.

In [7], [4], the *Regression Estimator* (RE) and *Weighted Sum* (WS) type estimators have been proposed for center of pass and fail points in the Centers of Gravity (CoG) Algorithm as an efficient estimators. Fig. 4 illustrates the SD of the RE, WS, and PMC estimators with UWM. It is seen that UWM results in a large variance in the estimations.

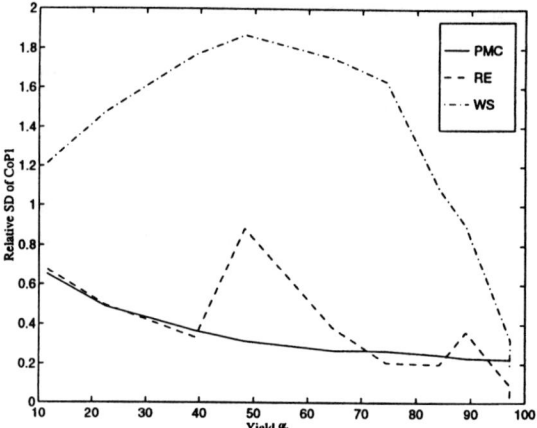

Fig. 4. Pass points CoG estimator standard deviations with UWM.

IV. Conclusions

The importance sampling is a variance reduction technique in Monte Carlo estimation. Here, the problem of finding an optimal importance sampling MC estimator is theoretically studied. This problem is treated in the case where no approximate model of the circuit responses is available, so that simulations are the only way to evaluate the circuit quality function.

In this case, we proved that the best sampling pdf is the parameter's disturbance pdf. This results in the optimality of the PMC estimator in the absence of any approximate model of the circuit response. As an important result, we can conclude that the PMC estimator is generally more efficient than the Uniform Warping Method (UWM) in which the sampling pdf is uniform.

Further research in this area is in progress in order to extend the optimal importance sampling for the case where some approximate model of the circuit responses is available.

Appendix
Optimal Importance Sampling Scheme

Consider that $q(\mathbf{x},\omega)$ is a stationary stochastic process. It is assumed that $0 \leq q(\mathbf{x},\omega) \leq 1$ and that its pdf for a given $\mathbf{x}$ is symmetric with respect to 0.5, i.e., $E_\omega[q(\mathbf{X},\omega)|\mathbf{X}=\mathbf{x}] = 0.5$. The evaluation of the following integral is our aim

$$Q(\omega) = \int_{R^n} q(\mathbf{x},\omega) f_\mathbf{x}(\mathbf{x})d\mathbf{x}, \quad (A.1)$$

where $f_\mathbf{x}(\mathbf{x})$ is a known pdf. The importance sampling estimator is

$$\hat{Q}_I(\omega) = \frac{1}{N}\sum_{i=1}^{N} q(\mathbf{X}^i,\omega)\frac{f_\mathbf{x}(\mathbf{X}^i)}{g(\mathbf{X}^i)}, \quad (A.2)$$

where $g(\mathbf{x})$ is the new sampling pdf. It can be shown that this is an unbiased estimator of $Q(\omega)$. First, we consider the following lemma.

Lemma 1: *Consider the following optimization problem (a variational problem)*

$$\begin{cases} \min_{g(\mathbf{x})} \left\{ \int \dfrac{f_\mathbf{x}^2(\mathbf{x})}{g(\mathbf{x})} d\mathbf{x} \right\}, \\ \text{subject to } \int g(\mathbf{x}) d\mathbf{x} = 1, \end{cases} \quad (A.3)$$

where $f_\mathbf{x}(\mathbf{x})$ *is a given pdf. Then the solution is*

$$g(\mathbf{x}) = f_\mathbf{x}(\mathbf{x}). \quad (A.4)$$

Proof: By using the Cauchy-Schwarz inequality [1], we have

$$1^2 = \left[\int f_\mathbf{x}(\mathbf{x}) d\mathbf{x} \right]^2 = \left[\int \dfrac{f_\mathbf{x}(\mathbf{x})}{[g(\mathbf{x})]^{1/2}} [g(\mathbf{x})]^{1/2} d\mathbf{x} \right]^2 \\ \leq \int \dfrac{f_\mathbf{x}^2(\mathbf{x})}{g(\mathbf{x})} d\mathbf{x} \int g(\mathbf{x}) d\mathbf{x} = \int \dfrac{f_\mathbf{x}^2(\mathbf{x})}{g(\mathbf{x})} d\mathbf{x}. \quad (A.5)$$

By substituting solution (A.4) into (A.3), the integral of the objective function is equal to 1. The result follows immediately. ■

Proof of Theorem 1: From the well-known formula, the variance of the importance sampling estimator can be expanded as

$$\mathrm{Var}(\hat{Q}_l) = \mathrm{Var}_{\tilde{\mathbf{X}}}\left(E_\omega[\hat{Q}_l | \tilde{\mathbf{X}}] \right) + E_{\tilde{\mathbf{X}}}\left[\mathrm{Var}_\omega(\hat{Q}_l | \tilde{\mathbf{X}}) \right], \quad (A.6)$$

where $\tilde{\mathbf{X}} = [\mathbf{X}^1, \mathbf{X}^2, ..., \mathbf{X}^N]$. The first term of the right-hand side can be expanded as follows:

$$\mathrm{Var}_{\tilde{\mathbf{X}}}\left(E_\omega[\hat{Q}_l | \tilde{\mathbf{X}}] \right) = \mathrm{Var}_{\tilde{\mathbf{X}}}\left(\dfrac{1}{N} \sum_{i=1}^{N} 0.5 \dfrac{f_\mathbf{x}(\mathbf{X}^i)}{g(\mathbf{X}^i)} \right) \\ = \dfrac{1}{4N} \mathrm{Var}_\mathbf{x}\left(\dfrac{f_\mathbf{x}(\mathbf{X})}{g(\mathbf{X})} \right) = \dfrac{1}{4N} E_\mathbf{x}\left[\dfrac{f_\mathbf{x}^2(\mathbf{X})}{g^2(\mathbf{X})} \right] - \dfrac{1}{4N}. \quad (A.7)$$

Let σ_q^2 denote the variance of the stationary stochastic process $\mathbf{q}(\mathbf{X}, \omega)$ and $r(.)$ stand for its correlation function. We also have

$$E_{\tilde{\mathbf{X}}}\left[\mathrm{Var}_\omega(\hat{Q}_l | \tilde{\mathbf{X}}) \right] = \dfrac{1}{N^2} E_{\tilde{\mathbf{X}}}\left[\sum_{i=1}^{N} \dfrac{f_\mathbf{x}^2(\mathbf{X}^i)}{g^2(\mathbf{X}^i)} \sigma_h^2 \right. \\ \left. + 2 \sum_{i=1}^{N} \sum_{j>i}^{N} \dfrac{f_\mathbf{x}(\mathbf{X}^i)}{g(\mathbf{X}^i)} \dfrac{f_\mathbf{x}(\mathbf{X}^j)}{g(\mathbf{X}^j)} \mathrm{Cov}_\omega(\hat{Q}_l(\mathbf{X}^i, \omega), \hat{Q}_l(\mathbf{X}^j, \omega)) \right] \\ = \dfrac{\sigma_q^2}{N} E_\mathbf{x}\left[\dfrac{f_\mathbf{x}^2(\mathbf{X})}{g^2(\mathbf{X})} \right] + \dfrac{2}{N^2} \dfrac{N(N-1)}{2} \\ \sigma_q^2 E_{\tilde{\mathbf{X}}}\left[r(\mathbf{X}^1 - \mathbf{X}^2) \dfrac{f_\mathbf{x}(\mathbf{X}^1)}{g(\mathbf{X}^1)} \dfrac{f_\mathbf{x}(\mathbf{X}^2)}{g(\mathbf{X}^2)} \right]. \quad (A.8)$$

The second term of the right hand side of (A.8) is not a function of the sampling pdf $g(\mathbf{x})$, because

$$E_{\tilde{\mathbf{X}}}\left[r(\mathbf{X}^1 - \mathbf{X}^2) \dfrac{f_\mathbf{x}(\mathbf{X}^1)}{g(\mathbf{X}^1)} \dfrac{f_\mathbf{x}(\mathbf{X}^2)}{g(\mathbf{X}^2)} \right] = \\ = \iint r(\mathbf{X}^1 - \mathbf{X}^2) f_\mathbf{x}(\mathbf{X}^1) f_\mathbf{x}(\mathbf{X}^2) d\mathbf{X}^1 d\mathbf{X}^2 \quad (A.9)$$

By substituting (A.7) and (A.8) into (A.6), the problem of finding the best sampling pdf is

$$(\text{P1}) \begin{cases} \min_{g(\mathbf{x})} E_\mathbf{x}\left[\dfrac{f_\mathbf{x}^2(\mathbf{x})}{g^2(\mathbf{x})} \right], \\ \text{subject to } \int g(\mathbf{x}) d\mathbf{x} = 1. \end{cases} \quad (A.10)$$

By the use of *Lemma 1*, the solution of problem (P1) is

$$g(\mathbf{x}) = f_\mathbf{x}(\mathbf{x}). \quad (A.11)$$

Therefore, the best sampling pdf is the primitive parameter's disturbance pdf, and its related estimator is called the Primitive Monte Carlo (PMC) estimator. ■

Acknowledgment

We acknowledge the useful discussions with Dr. M. A. Styblinski. We also wish to thank Dr. G. Fleury for his valuable comments to improve the presentation.

References

[1] R. Y. Rubinstein, *Simulation and the Monte Carlo Method*. John Wiley & Sons, Inc., 1981.

[2] D. E. Hocevar, M. R. Lightner, and T. N. Trick, "A study of variance reduction techniques for estimating circuit yields," *IEEE Trans. Computer-Aided Design*, vol. CAD-2, no. 3, pp. 180-192, July 1983.

[3] J. C. Zhang and M. A. Styblinski, *Yield and Variability Optimization of Integrated Circuits*, Kluwer Academic Publisher, 1995.

[4] M. Keramat, "Study of centers of gravity algorithm in design centering of electronic circuits," Ecole Superieure d'Electricite (SUPELEC), Paris, France, Tech. Rep. No. SUP-1196-16, December 1996.

[5] M. Keramat and R. Kielbasa, "Latin hypercube sampling Monte Carlo estimation of average quality index for integrated circuits," *Analog Integrated Circuits and Signal Processing*, vol. 14, no. 1/2, pp. 131-142, 1997.

[6] M. Keramat and R. Kielbasa, "A study of stratified sampling in variance reduction techniques for parametric yield estimation," in *Proc. IEEE Int. Symp. Circuits Syst.*, Hong Kong, June 1997, pp. 1652-1655.

[7] M. Keramat and R. Kielbasa, "Parametric yield optimization of electronic circuits via improved centers of gravity algorithm," in *Proc. IEEE 40th Midwest Symposium on Circuits and Systems*, Sacramento, CA, August 3-6 1997.

[8] A. Papoulis, *Probability, Random Variables, and Stochastic Process*, 3rd edition. McGraw-Hill, Inc., 1991.

WORST-CASE ANALYSIS OF LINEAR ANALOG CIRCUITS USING SENSITIVITY BANDS

Michael W. Tian and C.-J. Richard Shi

Department of Electrical and Computer Engineering
University of Iowa, Iowa City, IA 52242, USA
{wtian,cjshi}@eng.uiowa.edu

ABSTRACT

A novel approach for frequency-domain worst-case analysis is proposed in this paper. It is based on the fact that if a circuit response is monotonic with respect to the changes in a circuit parameter value, then the extreme values of the response occur at the extreme values of that parameter. The monotonicity is identified by the sensitivity band computation over the parameter space, and is used to reduce the number of uncertain parameters in circuit simulation. These ideas are implemented in a prototype circuit simulator, the experimental results on a state variable filter are described.

1. INTRODUCTION

With the decreasing device size and integration of analog and digital sub-systems, circuit parameter variations caused by manufacture fluctuation and operating condition have significant effects on circuit performance. In the absence of more sophisticated statistical facilities, *worst-case (tolerance) analysis* which calculates the extreme response values, has attracted much attention in recent years [1, 2, 3, 7].

The uncertain circuit parameter p_i and circuit response f_j can be represented as interval quantity p_i^I (with lower bound p_i^L, upper bound p_i^R) and f_j^I (with lower bound f_j^L, upper bound f_j^R) respectively. Given a circuit with m uncertain parameters and n responses to be uncertain, m dimensional interval vector $\mathbf{p}^I$ builds the *parameter space* and n dimensional interval vector $\mathbf{f}^I$ builds the *response space*. Fig. 1(a) shows a simple RLC circuit where $R1$ and $L2$ are uncertain within the interval $[0.9, 1.1]$Ohm and $[0.9, 1.1]$Henry. Fig. 1(b) shows the parameter space built by $R1$ and $L2$, and the values of these two parameters are uncertain within the shaded area. Fig. 1(c) shows the response space built by the real part of $V(2)$ and the imaginary part of $V(2)$ at frequency 1 Hz, and the values of these two responses are uncertain within the shaded area. Worst-case analysis is to calculate the bounds of response space from the parameter space by ignoring the detailed statistical structure.

The main difficulty associated with the worst-case analysis is to identify the *worst-case parameter condition* which leads to the extreme response values. *Traditional vertex analysis* assumes that the extreme response values are resulted from the extreme parameter values [6], *i.e.*, vertices of parameter space. With this assumption, sensitivities computed at the nominal parameter condition are used to find the vertices corresponding to specific extreme values

This research is sponsored by U.S. Defense Advanced Research Projects Agency (DARPA) under grant number F33615-96-1-5601 from the U.S. Air Force, Wright Laboratory, Manufacturing Technology Directorate, and by the National Science Foundation under grant CDA 96-01503.

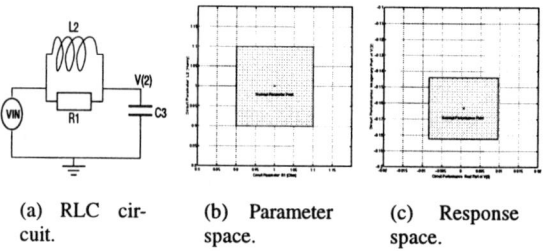

(a) RLC circuit. (b) Parameter space. (c) Response space.

Figure 1: Parameter space vs. response space.

of response [4]. Worst-case analysis also can be performed by *Monte Carlo simulation* [2] or *interval analysis* [7].

It is well known that the traditional vertex analysis [6] is invalid for frequency-domain worst-case analysis, except when the computed response f_j^I is monotonic with respect to the changes in all uncertain parameter values. In this paper, the monotonicity is investigated by *sensitivity bands* over the parameter space. As observed, the common situation is that the computed response f_j^I is monotonic with respect to the changes in the majority uncertain parameter values while non-monotonic to the rest. The monotonicity is used to reduce the number of uncertain parameters in circuit simulation based on the fact if circuit response f_j^I is monotonic with respect to the changes in parameter p_i^I, then the extreme values of f_j^I (f_j^L or f_j^R) occur at the extreme values of p_i^I (p_i^L or p_i^R).

The rest of this paper is organized as follows: Section 2 investigates the validity of traditional vertex analysis by distinguishing two categories of circuits. Section 3 addresses the relationship between monotonicity and worst-case analysis. Section 4 describes the computation of sensitivity bands over the parameter space by solving systems of linear interval equations. Section 5 gives the complete algorithm for worst-case analysis. Section 6 presents the application of the afore mentioned ideas on a state variable filter.

2. THE VALIDITY OF VERTEX ANALYSIS

In this section, the validity of the traditional assumption that the extreme values of response occur at the vertices of parameter space will be investigated by distinguishing two categories of circuits: circuits with and without frequency dependent components.

For a linear circuit which does not contain frequency-dependent components, the circuit equation is a system of *real linear interval equations*, circuit response f_j^I is related to uncertain parameter p_i^I

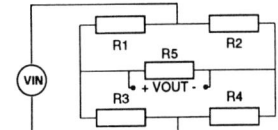

Figure 2: A resistive bridge circuit.

by a *bilinear* function:

$$f_j^I = \frac{n_1 p_i^I + n_2}{d_1 p_i^I + d_2} = \frac{n_1}{d_1} + \frac{n_2 - \frac{n_1 d_2}{d_1}}{d_1 p_i^I + d_2} \qquad (1)$$

At any point in the parameter space, n_1, n_2, d_1 and d_2 are real constants independent to p_i^I. This implies that f_j^I is monotonic with respect to the changes in p_i^I.

Suppose that the lower bound and upper bound of circuit response f_j^I occur at point $\acute{\mathbf{p}}$ and $\grave{\mathbf{p}}$ respectively, then,

$$\begin{array}{l} f_j^L = f_j(\acute{p}_1, \acute{p}_2, \ldots, \acute{p}_i, \ldots, \acute{p}_m) \\ f_j^R = f_j(\grave{p}_1, \grave{p}_2, \ldots, \grave{p}_i, \ldots, \grave{p}_m) \end{array} \qquad (2)$$

where $\acute{p}_i \in p_i^I$ and $\grave{p}_i \in p_i^I$ for $i = 1, 2, \ldots, m$. We have

$$\acute{p}_i \in \{p_i^L, p_i^R\} \text{ and } \grave{p}_i \in \{p_i^L, p_i^R\} \quad i = 1, 2, \ldots, m \qquad (3)$$

Fig. 2 shows a resistive bridge circuit. Suppose that all the five resistors are uncertain within range $[0.9, 1.1]$Ohm and the parameter space is built as $(R_1, R_2, R_3, R_4, R_5)$, it can be easily verified that the lower bound of the voltage gain occurs at vertex $(1.1, 0.9, 0.9, 1.1, 0.9)$ with the value of -0.0529, and the upper bound occurs at vertex $(0.9, 1.1, 1.1, 0.9, 0.9)$ with the value of 0.0529.

For a linear circuit which contains frequency-dependent components, circuit equation is a system of *complex linear interval equations*. The response f_j^I is related to parameter p_i^I by a *bi-quadratic* function:

$$f_j^I = \frac{n_1 (p_i^I)^2 + n_2 p_i^I + n_3}{d_1 (p_i^I)^2 + d_2 p_i^I + d_3} \qquad (4)$$

At any point in the parameter space, n_1, n_2, n_3, d_1, d_2 and d_3 are real constants independent to p_i^I. Due to the quadratic relationship, the monotonicity between circuit response f_j^I and parameter p_i^I is not definitely satisfied. Hence the extreme values of circuit response occur within the parameter space, but not necessarily at the vertices, *i.e.*,

$$\acute{p}_i \in [p_i^L, p_i^R] \text{ and } \grave{p}_i \in [p_i^L, p_i^R] \quad i = 1, 2, \ldots, m \qquad (5)$$

The simple RLC circuit shown in Fig. 1(a) is a circuit which contains frequency-dependent components. Fig. 3(a) shows the comparison of accurate response bounds (approximated by 10^6-trial Monte Carlo simulation) and response bounds computed by traditional vertex analysis. It can be observed that the accurate response differs from the response calculated by vertex analysis in some frequency regions. This is because the worst-case parameter values which result in the extreme values of response do not occur at the vertices of parameter space. In these cases, the response calculated by vertex analysis always under-estimates the accurate response.

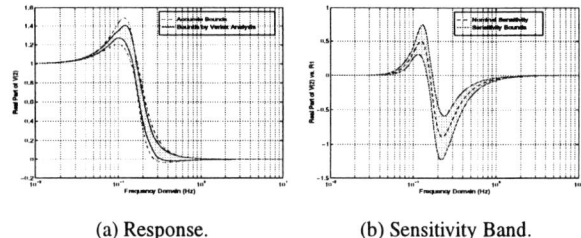

(a) Response. (b) Sensitivity Band.

Figure 3: The real part of $V(2)$.

3. MONOTONICITY

Although the assumption that the extreme values of response occur at the vertices of parameter space is invalid for frequency-domain circuit simulation, it has been observed from practical circuits that this assumption is valid for most parameters at frequency points that are away from poles and zeros. This is explored to reduce the number of uncertain parameters using monotonicity in this section.

For a circuit which contains uncertain parameters, there are three cases of monotonicity between circuit response f_j^I and circuit parameter p_i^I:

- f_j^I is monotonically increasing with respect to the changes in p_i^I over the whole parameter space.
- f_j^I is monotonically decreasing with respect to the changes in p_i^I over the whole parameter space.
- f_j^I is monotonically increasing with respect to the changes in p_i^I for a part of parameter space, while monotonically decreasing with respect to the changes in p_i^I for the rest part of parameter space. This is the *non-monotonic* case.

Further the monotonicity can be related to worst-case analysis according to these three cases.

Proposition 1 *For a circuit, the extreme values of circuit response f_j^I can be calculated according to the monotonicity as follows:*

$$f_j^L(\mathbf{p}) = f_j(\acute{\mathbf{p}}) \qquad f_j^R(\mathbf{p}) = f_j(\grave{\mathbf{p}}) \qquad (6)$$

where

$$\begin{cases} \acute{p}_i = p_i^L, \grave{p}_i = p_i^R & \text{if any } \mathbf{p} \in \mathbf{p}^I, p_i \in p_i^I : \frac{\partial f_j(\mathbf{p})}{\partial p_i} \geq 0 \\ \acute{p}_i = p_i^R, \grave{p}_i = p_i^L & \text{if any } \mathbf{p} \in \mathbf{p}^I, p_i \in p_i^I : \frac{\partial f_j(\mathbf{p})}{\partial p_i} \leq 0 \\ \acute{p}_i = p_i^I, \grave{p}_i = p_i^I & \text{else} \end{cases}$$

$$i = 1, 2, \ldots, m$$

If the monotonicity is satisfied by all uncertain parameters, two nominal circuit simulations with extreme parameter values are needed for worst-cast analysis. However, the most common situations are that the monotonicity is satisfied by the majority of uncertain parameters, while not satisfied by the rest of the uncertain parameters. In these cases, uncertain parameters which satisfy the monotonicity can be substituted by their extreme values. The reduction of uncertain parameters leads to better accuracy and less computational cost of circuit simulation.

These three cases of monotonicity are demonstrated by the RLC circuit shown in Fig. 1(a). Fig. 3(b) plots the sensitivity bands of response $V(2)$ versus parameter $R1$ over the parameter space. From Fig. 3(b), $V(2)$ is monotonically increasing in frequency region $[0, 0.148]$ Hz, monotonically decreasing in $[0.174, \infty]$ Hz, and non-monotonic in $[0.148, 0.174]$ Hz versus the changes in $R1$.

4. COMPUTATION OF SENSITIVITY BANDS

The monotonicity can be identified using the computation of sensitivity bands over the parameter space, which is d by solving systems of linear interval equations.

Over the parameter space, frequency-domain circuit equation are described as systems of complex linear interval equations, which are further converted to systems of real linear interval equations as follows:

$$\begin{bmatrix} \mathcal{G}^I & -\mathcal{C}^I \\ \mathcal{C}^I & \mathcal{G}^I \end{bmatrix} \begin{bmatrix} \mathbf{x}_\mathcal{R}^I \\ \mathbf{x}_\mathcal{I}^I \end{bmatrix} = \begin{bmatrix} \mathbf{w}_\mathcal{R} \\ \mathbf{w}_\mathcal{I} \end{bmatrix} \qquad (7)$$

where subscript $\mathcal{R}$ denotes the real part, and $\mathcal{I}$ the imaginary part, matrix $\mathcal{G}^I$ is contributed by all frequency-independent components, $\mathcal{C}^I$ is contributed by all frequency-dependent components.

In order to evaluate the sensitivities of all elements of vector $\mathbf{x}^I$ versus a single parameter p_i^I, we differentiate Eq.7 with respect to p_i^I to obtain:

$$\begin{bmatrix} \frac{\partial \mathcal{G}^I}{\partial p_i^I} & -\frac{\partial \mathcal{C}^I}{\partial p_i^I} \\ \frac{\partial \mathcal{C}^I}{\partial p_i^I} & \frac{\partial \mathcal{G}^I}{\partial p_i^I} \end{bmatrix} \begin{bmatrix} \mathbf{x}_\mathcal{R}^I \\ \mathbf{x}_\mathcal{I}^I \end{bmatrix} + \begin{bmatrix} \mathcal{G}^I & -\mathcal{C}^I \\ \mathcal{C}^I & \mathcal{G}^I \end{bmatrix} \begin{bmatrix} \frac{\partial \mathbf{x}_\mathcal{R}^I}{\partial p_i^I} \\ \frac{\partial \mathbf{x}_\mathcal{I}^I}{\partial p_i^I} \end{bmatrix}$$
$$= \begin{bmatrix} \frac{\partial \mathbf{w}_\mathcal{R}}{\partial p_i^I} \\ \frac{\partial \mathbf{w}_\mathcal{I}}{\partial p_i^I} \end{bmatrix} \qquad (8)$$

Because vector $\mathbf{w}$ does not depend on p_i^I, the equation set above can be rewritten as:

$$\begin{bmatrix} \mathcal{G}^I & -\mathcal{C}^I \\ \mathcal{C}^I & \mathcal{G}^I \end{bmatrix} \begin{bmatrix} \frac{\partial \mathbf{x}_\mathcal{R}^I}{\partial p_i^I} \\ \frac{\partial \mathbf{x}_\mathcal{I}^I}{\partial p_i^I} \end{bmatrix} = - \begin{bmatrix} \frac{\partial \mathcal{G}^I}{\partial p_i^I} & -\frac{\partial \mathcal{C}^I}{\partial p_i^I} \\ \frac{\partial \mathcal{C}^I}{\partial p_i^I} & \frac{\partial \mathcal{G}^I}{\partial p_i^I} \end{bmatrix} \begin{bmatrix} \mathbf{x}_\mathcal{R}^I \\ \mathbf{x}_\mathcal{I}^I \end{bmatrix} \qquad (9)$$

According to the circuit formulation proposed in [5], since every parameter appears only in a few matrix entries, so the right-hand side of Eq.9 is a sparse vector which can be directly obtained by inspection. The interval analysis algorithm proposed in [5] can be directly used to solve Eq.9. Note also the most time consuming aspect in interval analysis is the manipulation of the left-hand-side (LHS) coefficient matrix of Eq.9, e.g., matrix inversion and multiplication.

Given a circuit with m uncertain parameters, Eq.9 will be performed m times in order to cover all the uncertain parameters. The total computation can be reduced because only the RHS vector changes, and the matrix stays the same.

5. THE COMPLETE ALGORITHM

The worst-case analysis algorithm is described in Fig. 4, and is referred as *enhanced vertex analysis* in this paper. It is based on the sensitivity bands over the parameter space and the relationship between monotonicity and worst-case analysis. Let Ω denotes the frequency point set where the frequency-domain simulation is performed, $\mathbf{f}^I$ the n dimensional response vector, and $\mathbf{p}^I$ the m dimensional parameter vector. For each frequency point ω_k in Ω, the sensitivity bands of response vector $\mathbf{f}^I$ versus all parameters are computed first. Then for each response f_j, the vertices of the parameter space corresponding to the extreme values of circuit response are identified according to the monotonicity. If the extreme values of response occur at the extreme values of all parameters,

ENHANCED_VERTEX_ANALYSIS
1 For each $\omega_k \in \Omega$;
2   for $i = 1, 2, \ldots, m$;
3      $(\frac{\partial \mathbf{f}}{\partial p_i})^I \leftarrow$ computation of the sensitivity bands;
4   for $j = 1, 2, \ldots, n$;
5   for $i = 1, 2, \ldots, m$;
6   if $(\frac{\partial f_j}{\partial p_i})^I \geq 0$;
7          $\acute{p}_i \leftarrow p_i^L$ and $\grave{p}_i \leftarrow p_i^R$;
8   else if $(\frac{\partial f_j}{\partial p_i})^I \leq 0$;
9          $\acute{p}_i \leftarrow p_i^R$ and $\grave{p}_i \leftarrow p_i^L$;
10   else;
11         $\acute{p}_i \leftarrow p_i^I$ and $\grave{p}_i \leftarrow p_i^I$;
12   if $\acute{\mathbf{p}}^L = \acute{\mathbf{p}}^R$ and $\grave{\mathbf{p}}^L = \grave{\mathbf{p}}^R$;
13       $f_j^L \leftarrow f_j(\acute{\mathbf{p}})$ and $f_j^R \leftarrow f_j(\grave{\mathbf{p}})$ by nominal sim.;
14   else;
15       $f_j^L \leftarrow f_j^L(\acute{\mathbf{p}})$ and $f_j^R \leftarrow f_j^R(\grave{\mathbf{p}})$ by MC sim.;
16 End of simulation.

Figure 4: The algorithm of enhanced vertex analysis.

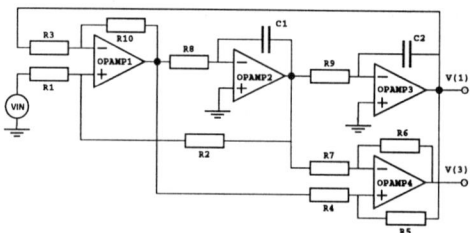

Figure 5: Diagram of state variable filter.

then two nominal circuit simulations with extreme parameter values are performed to obtain the accurate response bounds. If the extreme values of response occur at the extreme values of only a part of the parameters, the total number of uncertain parameters can be reduced according to the monotonicity. In these cases, the resulted circuit simulation is solved by large-trial-number Monte Carlo simulation, due to the reduction of parameter uncertainty.

6. EXPERIMENTAL RESULTS

The proposed ideas have been implemented into a prototype circuit simulator based on SPICE3F5. In this section, a high-sensitive state variable filter is used to demonstrate the simulation results.

Fig. 5 shows the diagram of the state variable filter. The nominal parameter values are given as:

$$r_1 = 10k, r_2 = 20k, r_3 = 10k, r_4 = 10k, r_5 = 100k$$
$$r_6 = 10k, r_7 = 30k, r_8 = 45.5k, r_9 = 2.2k, r_{10} = 10k$$
$$c_1 = 1.6nf, c_2 = 1.6nf$$

All of these parameters are associated with $\pm 1\%$ statistical variations in the following experiments.

For this circuit, the real part of output $V(1)$ is considered as the circuit response. The sensitivity bands over the parameter space are first computed by interval analysis. Fig. 6(a)–Fig. 6(h) plot the real part of sensitivity bands versus uncertain parameter $C1$, $C2$, $R1$, $R2$, $R3$, $R8$, $R9$ and $R10$, respectively. Since the values of $R4$, $R5$, $R6$ and $R7$ do not affect the value of output

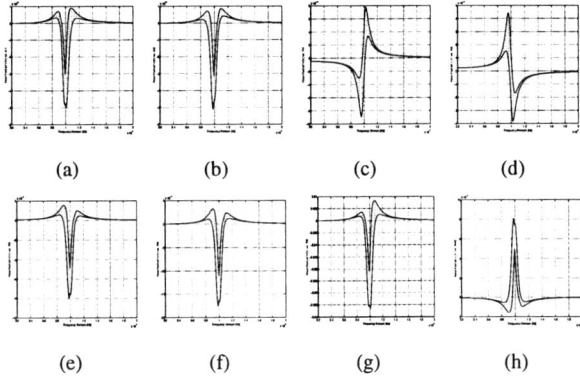

Figure 6: The real part of sensitivity bands.

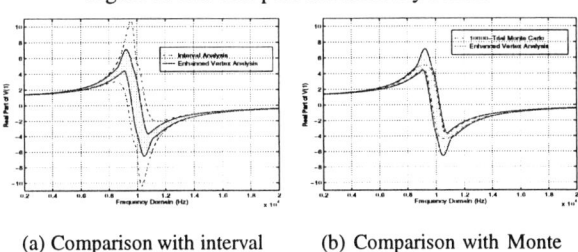

(a) Comparison with interval analysis. (b) Comparison with Monte Carlo.

Figure 7: Simulation results of state variable filter.

$V(1)$, the sensitivities of $V(1)$ versus these four parameters are constant at zero.

The simulation results of the worst-case analysis by the proposed enhanced vertex analysis are compared by the interval analysis and 10^4-trial Monte Carlo simulation as shown in Fig. 7(a) and Fig. 7(b) respectively. From Fig. 7(a) and Fig. 7(b), it can be observed that interval analysis over-estimates the response bounds by slight expansion, and Monte Carlo simulation under-estimates the response bounds when the trial number is not sufficient.

The enhanced vertex analysis is performed two times in order to compute the lower bound of the real part of $V(1)$, and the upper bound of the real part of $V(1)$. For a specific frequency point, if all of the parameters satisfy the monotonicity, the vertex analysis will result in a nominal circuit simulation with parameter values substituted by their extreme values. If only a part of parameters satisfy the monotonicity, the vertex analysis will result in a parameter-uncertainty-reduced circuit simulation, with this part of uncertain parameters substituted by their extreme values. In these cases, circuit simulation is performed by 10^6-trial Monte Carlo simulation to guarantee the accuracy. The number of non-monotonic parameters is circuit-dependent and critical for both speed and accuracy of the enhanced vertex analysis. For the state variable filter, the average number of non-monotonic parameters versus the frequency is shown in Fig. 8. It can be observed that the monotonicity is satisfied by all circuit parameters at most frequency points.

7. CONCLUSIONS

Analog circuits are designed with parameter variations on mind. Worst-case analysis, which calculates the extreme values of response according to the parameter space, is of specific interest to circuit designers. In this paper, an efficient method has been proposed to calculate the sensitivity bands by solving systems of

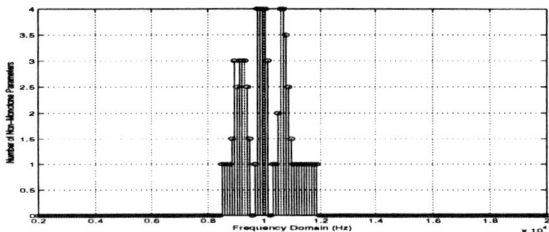

Figure 8: Number of non-monotonic parameters.

linear interval equations. And the computed sensitivity bands are used to identify circuit response monotonicity and to enhance the traditional vertex analysis. Compared to interval analysis and Monte Carlo simulation, the proposed approach produces more accurate results with less computational cost.

Acknowledgment: The authors would like to thank Prof. Xieting Ling of Fudan University, Shanghai, for his constructive suggestions and comments.

8. REFERENCES

[1] K.J. Antreich, H.E. Graeb and C.U. Wieser, "Circuit analysis and optimization driven by worst-case distances," *IEEE Tran. Computed-Aided Design of Integrated Circuits and Systems*, vol. 13, 1994, pp. 57–71.

[2] A. Dharchoudhury and S.M. Kang, "Worst-case analysis and optimization of VLSI circuit performances," *IEEE Tran. Computer-Aided Design of Integrated Circuits and Systems*, vol. 14, 1995, pp. 481–492.

[3] A. Levkorich, E. Zeheb and N. Cohen, "Frequency response envelopes of a family of uncertain continuous-time systems," *IEEE Tran. Circuits and Systems Part I*, vol. 42, 1995, pp. 156–165.

[4] A. Pahwa and R.A. Rohrer, "Band-faults: Efficient approximations to fault bands for the simulation before fault diagnosis of linear circuits," *IEEE Tran. Circuits and Systems*, vol. 29, 1982, pp. 81–88.

[5] C.J.R. Shi and M.W. Tian, "Simulation and sensitivity of linear analog circuits under parameter variations," *ACM Tran. Design Automation of Electronic systems*, to appear.

[6] R. Spence and R.S. Soin, *Tolerance Design of Electronic Circuits*, Addison-Wesley Publishing Company, 1988.

[7] W. Tian, X.T. Ling and R.W. Liu, "Novel methods for circuit worst-case tolerance analysis," *IEEE Tran. Circuits and Systems Part I*, vol. 43, 1996, pp. 272–278.

FAST TIME-DOMAIN NOISE SIMULATION OF SIGMA-DELTA CONVERTERS AND PERIODICALLY SWITCHED LINEAR NETWORKS

Yikui Dong and Ajoy Opal

Department of Electrical and Computer Engineering
University of Waterloo
Waterloo, Ontario N2L 3G1, Canada
ydong@vlsi.uwaterloo.ca   aopal@vlsi.uwaterloo.ca

ABSTRACT

This paper presents fast time domain methods for computer simulation of electrical noise and sampling-clock jitter in sigma-delta data converters and in general periodically switched linear networks, such as, switched capacitor filters at circuit macro-model level. The proposed methods have been implemented in a computer program SDNoise. In this paper, examples of thermal noise simulation in a switched capacitor filter and in a sigma-delta A/D converter are given. Examples of sampling-clock jitter simulation are also given for the sigma-delta A/D converter in the case when it is used to convert baseband signal, and in the case when it is used as IF front-end to convert IF signal by sub-sampling technique. Simulation time in the order of minutes is shown for 74k clock-cycle data of the sigma-delta A/D converter when thermal noise and clock jitter are present.

1. INTRODUCTION

Oversampled sigma-delta modulation ($\Sigma\Delta$ modulation) techniques are very popular in high resolution data converter design. The performance of a sigma-delta modulator is known to be restricted by available signal swing at one end and by circuit noise floor at the other. While power supply, substrate and clock feedthrough noise depend heavily on layout and circuit topology, thermal noise generated by the resistances and the amplifiers is a fundamental constraint. Accurate calculation of electrical noise is important to predict the circuit performance characteristics.

Besides electrical noise, sampling-clock jitter is another major contributor to signal-to-noise-ratio (SNR) degradation of sigma-delta converters, especially in high frequency applications. Sigma-delta data converters are traditionally designed to convert low-frequency, "baseband" analog and digital signal with high resolution. In baseband design, output noise contributed by sampling-clock jitter is most likely masked by electrical noise floor. In high-frequency designs, however, the situation changes. The output noise contributed by sampling clock jitter becomes remarkable.

Recently, with the increasing interest of implementing compact, low cost and low power transceivers for wireless communication systems, various RF (radio-frequency) front-end conversion architectures are proposed as alternatives of the conventional superheterodyne scheme which requires significant amount of discrete components. One of the architecture that receives a lot of attention is the direct conversion scheme which mixes RF signal down to baseband directly by setting sampling clock frequency as incoming RF carrier frequency. Another more promising approach is the sub-sampling technique which down-converts the RF signal by setting the sampling clock frequency as a sub-multiple of the RF carrier frequency. These schemes are under active research in design. Some recent works [1] [2] have used sub-sampling technique in sigma-delta data converter designs to down-convert IF (intermediate frequency) signal for digital signal processing.

RF front-end conversion designs are much sensitive to sampling clock jitter than baseband conversion applications. In sigma-delta modulators that are used for direct RF or IF signal conversion, output noise generated by sampling clock jitter could surpass thermal noise to become the dominant factor that limit conversion accuracy. Precise estimation methods for the effect of sampling clock jitter are much needed to aid RF/IF design.

Noise analysis techniques for linear time-invariant systems in frequency domain are well known. Many studies of noise analysis for periodically switched linear circuits can also be found in literature since the launch of switched-capacitor techniques in 1970s. Noise analysis for nonlinear circuits, however, is always a difficult problem. In recent years, stochastic differential equations are used to analyze noise effects in circuits with modest nonlinearity [3], that is, in which the circuit nonlinear characteristics are differentiable.

When the nonlinearity is harsh, such as the quantizer in the sigma-delta modulator, as we know, there is *no* precise noise estimation method available. In this research, we present a novel time domain noise simulation method for sigma-delta converters. The simulation is at circuit macro-model level, where switches, capacitors, resistors, inductors, comparators, opamps and linear controlled sources are used as building blocks. This technique not only makes circuit noise estimation for sigma-delta converters possible, but also does it in a efficient way. In addition to sigma-delta data converters, this noise simulation technique can also be used on general periodically switched linear networks, such as switched capacitor filters.

The algorithms have been implemented in a computer program SDNoise, to simulate sigma-delta modulators and general periodically switched linear networks when thermal noise and clock jitter present. Designers can input their circuit net-list from terminal or by a input file into SDNoise to quickly explore the effects of different circuit architectures, different switch ON-resistances and different opamp unity gain-bandwidths on the output noise resulted from sampling clock jitter and from thermal noise.

The simulation methods are validated on several linear time invariant or periodically switched linear circuits by comparing simulation results with exact analysis or physical measurement data.

In this paper, thermal noise effect, and clock jitter effect on the conversion accuracy of baseband signals and of sub-sampled IF signals in a sigma-delta A/D converter are studied by SDNoise. Simulation time in the order of minutes is shown.

2. METHODOLOGY

The modified nodal analysis equations for a linear time-invariant circuit are

$$\mathbf{G}\mathbf{v}(t) + \mathbf{C}\frac{d}{dt}\mathbf{v}(t) = \mathbf{d}x(t), \quad \mathbf{v}(0) = \mathbf{v}_0 \quad (1)$$

where $\mathbf{G}$ is the conductance matrix and $\mathbf{C}$ is the capacitance matrix. Vector $\mathbf{d}$ stores connection information from input $x(t)$ to the circuit.

It is shown in sampled data simulation algorithm [4] that the exact solution of (1) at time $t = nT + T$ can be written as

$$\mathbf{v}(nT+T) = \mathbf{M}(T)\mathbf{v}(nT) + \mathbf{P}(T)x(nT) \quad (2)$$

where T is a pre-chosen fixed time step, matrix $\mathbf{M}(T)$ and vector $\mathbf{P}(T)$ are constants. This algorithm is useful in the sense that solution of (1) can be given in an accurate and efficient way at fixed time points $t = nT$, where $n = 1, 2, 3, \cdots$. The computation cost associated with each time point is in the order of $O(m^2)$, where m is the dimension of the system of equations.

2.1. Sampling clock jitter simulation

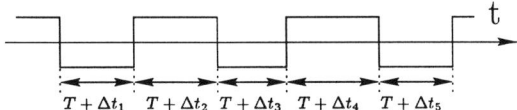

Figure 1: Sampling clock jitter

Sampling clock jitter is the random deviation of the clock-duration from its nominal value, as illustrated in Fig.1. For a sampled data system with clock jitter, accurate solution of (1) is needed at time instants $t = nT + \Delta t$, where T is the fixed time step, and Δt is a small random deviation. One way to compute the system response is to extend the sampled data simulation algorithm as following:

$$\mathbf{v}(t_0 + T + \Delta t) = \mathbf{M}(T + \Delta t)\mathbf{v}(t_0) + \mathbf{P}(T + \Delta t)x(t_0) \quad (3)$$

where matrix $\mathbf{M}(T + \Delta t)$ and vector $\mathbf{P}(T + \Delta t)$ can be approximated by the first k terms of Taylor expansion as:

$$\tilde{\mathbf{M}}(T + \Delta t) = \mathbf{M}(T) + \sum_{i=1}^{k} \frac{(\Delta t)^i}{i!}\frac{d^i}{dt^i}\mathbf{M}(T)$$
$$\tilde{\mathbf{P}}(T + \Delta t) = \mathbf{P}(T) + \sum_{i=1}^{k} \frac{(\Delta t)^i}{i!}\frac{d^i}{dt^i}\mathbf{P}(T) \quad (4)$$

The algorithms for computation of the derivatives of $\mathbf{M}$ matrix and $\mathbf{P}$ vector are developed. These matrixes and vectors are calculated *only once* with $\mathbf{M}(T)$, $\mathbf{P}(T)$ in a preprocessing step. At each simulation point, matrix $\mathbf{M}(T + \Delta t)$ and vector $\mathbf{P}(T + \Delta t)$ are evaluated by (4) and system response $\mathbf{v}(t_0 + T + \Delta t)$ is evaluated by (3). The total cost for each simulation point is still in the order of $O(m^2)$.

2.2. Thermal noise simulation

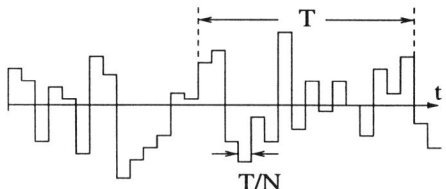

Figure 2: Time-domain noise waveform

Thermal noise arises from the random thermal motion of free electrons in resistive mediums. In practice, a noisy resistor can be modeled as a noiseless resistor R in series with a random voltage source $v_n(t)$ having flat power spectrum $4kTR \; V^2/Hz$, or as its Thévenin equivalent counterpart, a noiseless conductor G in parallel with a random current source $i_n(t)$ having flat power spectrum $4kTG \; A^2/Hz$. In the proposed simulation algorithm, each noise source in the circuit is represented in time domain as a random amplitude pulse waveform as shown in Fig. 2. The pulse width, which is the noise simulation step-size, is inversely related to noise bandwidth, while the pulse amplitude is given by a pseudo-random number generator and is related to circuit noise power. At each simulation step, the circuit response is computed by:

$$\mathbf{v}(nT+T) = \mathbf{M}(T)\mathbf{v}(nT) + \mathbf{P}_s(T)\mathbf{x}(nT) \quad (5)$$
$$+ \sum_i \mathbf{P}_{ni}(T)\phi_i(nT)$$

where $\phi_i(t)$ is random amplitude pulse waveform represents noise source i. The matrix $\mathbf{M}(T)$ and vectors $\mathbf{P}_s(T)$, $\mathbf{P}_{ni}(T)$ are constants that are calculated only once in a preprocessing step.

3. NUMERICAL EXAMPLES

The circuit output spectrums in this section were obtained by 64k data Fast Fourier Transform (FFT), where 1k stands for 1024 data points. All the simulations were performed on a Sun Sparc 20/71 workstation.

3.1. Thermal noise simulation of a periodically switched linear network

Periodically switched linear (PSL) networks, such as switch-capacitor and switch-current circuits are widely used in modern VLSI circuit design. These circuits change their topologies at switching instants only. Within each phase, the network is lumped, linear and time invariant. The proposed time domain noise simulation technique can be applied to PSL circuits. A set of constant matrixes, namely $\mathbf{M}$, $\mathbf{P}_s$ and $\mathbf{P}_n$ are calculated for each phase before simulation starts. The simulation proceeds by calculating circuit response at the end of each phase according to (5). The calculated results are used as initial condition for next phase.

The noise simulation has been speeded up in SDNoise by using different calculation step for noise response and signal response. Within each phase of a PSL network, the circuit is lumped, linear and time-invariant, supper-position rule applies. Noise response is calculated by many small steps, while signal response is calculated only once for each phase. Noise response and signal response are added together at the end of each phase to get the complete response.

The circuit in Fig.3 is a switch-capacitor integrator from [5]. The circuit is clocked at 10 kHz with 4 phases, each with a duration of 0.025 ms. The capacitors in the circuit are 10 pF in value, and the switches have 3.5 $k\Omega$ on-resistances. The opamp has 700 kHz unity gain bandwidth and 1.55 $M\Omega$ equivalent input noise resistance. This circuit is simulated by SDNoise when different noise bandwidths are considered.

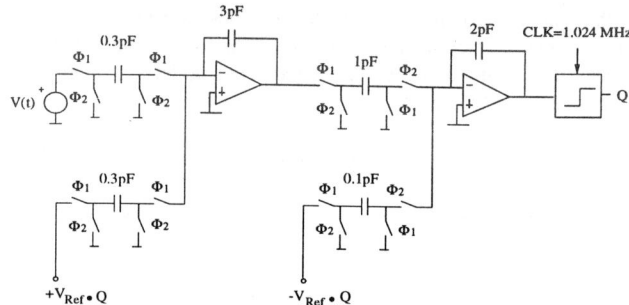

Figure 5: A second order switch-capacitor $\Sigma\Delta$ modulator

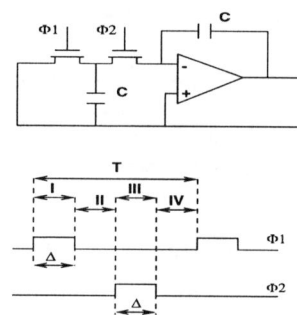

Figure 3: A switch-capacitor integrator

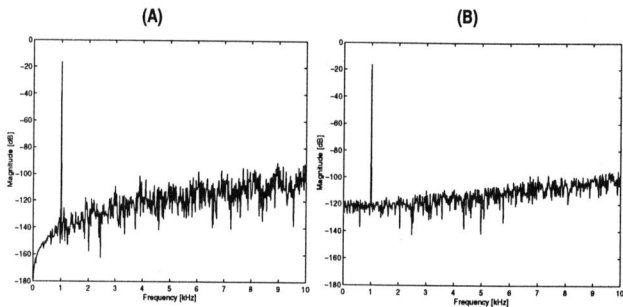

Figure 6: Output spectrum of the SC SDM (A) ideal case (B) when thermal noise present

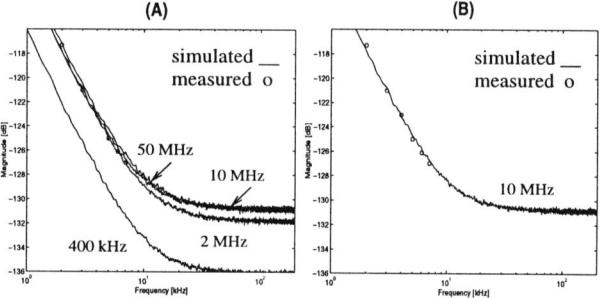

Figure 4: Simulated SC integrator output spectrums (A) with different foldover noise bandwidth (B) matches physical measurement data

In a switch-capacitor circuit, broadband noise will be folded-back to baseband due to under-sampling. As shown in Fig.4(A), the thermal noise power increases as we increase the thermal noise bandwidth. A noise bandwidth of 400 kHz representing foldover of 40 noise bands gives lower noise power in baseband than a noise bandwidth of 10MHz. Increasing the noise bandwidth beyond 10 MHz does not increase the noise power in baseband because of circuit bandwidth limitations. Simulation with a noise bandwidth of 50 MHz gives the same results as the 10 MHz bandwidth. The output noise converged with the increasing of noise bandwidth. As shown in Fig.4(B), the converged noise spectrum matches with physical measurement data [5].

3.2. Thermal noise and clock jitter simulation of a $\Sigma\Delta$ A/D converter

A sigma-delta modulator can be partitioned into two sub-networks, the nonlinear part – quantizers and the linear part – the rest of the circuit. Noise simulation of a SDM proceeds by calculating circuit response for the linear part and updating quantizer state at the end of each clock.

As an example, the circuit in Fig.5 is a second order switch-capacitor sigma-delta modulator with clock frequency of 1.024 MHz. When input signal is 1 kHz sinusoidal of amplitude 0.75V, the circuit is simulated with ideal elements, where the switches are open/short circuits, and the opamps are ideal. It takes SDNoise 18.64 seconds to get 74k clock cycles of the circuit response. This is very fast compared with HSPICE, which may take days to get these results. As shown in Fig.6(A), the output spectrum obtained from FFT follows the expected high-pass quantization noise curve.

The circuit is simulated again by SDNoise when thermal noise is present. In this case, the switch has 3.5 $k\Omega$ ON-resistance and infinite OFF-resistance. The opamp has 7 MHz unity gain-bandwidth and 1.55 $M\Omega$ equivalent input noise resistance. It only takes SDNoise 36.15 minutes to simulate 74k clock cycles of data. Comparable results from SPICE are not available. As shown in Fig.6(B), the output spectrum is flat in low frequency range, showing a thermal noise floor. At higher frequency, quantization noise still dominates. Smoother output spectrum compared with the one in ideal-case is also observed. An explanation is that thermal noise in the circuit acts to some extent as a dithering noise, which breaks limit cycle effect and smoothes output spectrum. The simulation results show that when quantization noise decreases to some extent in signal-band, circuit noise, which is thermal noise generated by MOSFET and opamps in this example, prevails the total output noise power. There is a lower limit on total in-band noise power. On the other hand, the need of low power, low voltage design restrict available signal swing. An upper bound on achievable resolution of sigma-delta modulators is imposed by circuit noise floor at one end and by available signal swing at the other.

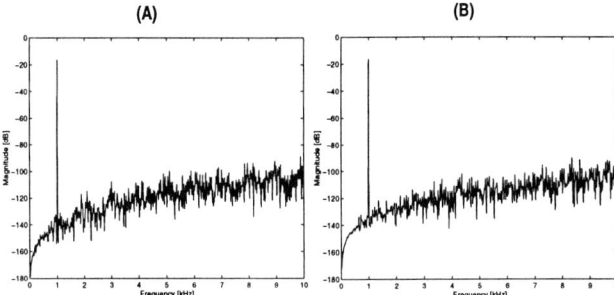

Figure 7: SC SDM output spectrum, (A) with 1 kHz input (B) with 1025 kHz input

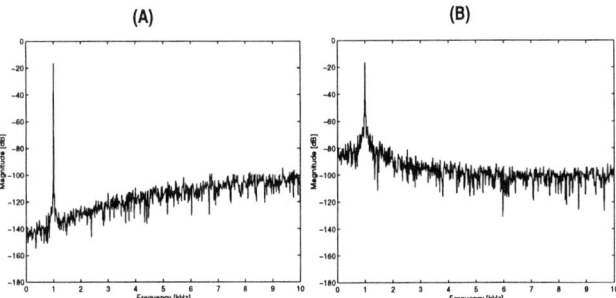

Figure 8: SC SDM output spectrum when 0.01% clock jitter present, (A) with 1 kHz input (B) with 1025 kHz input

We have mentioned in the introduction section that output noise resulted from sampling clock jitter is relatively small in baseband sigma-delta modulator designs. It is most likely masked by electrical noise floor. In RF/IF front-end sigma-delta modulator designs, however, sampling clock jitter noise becomes important. It could be the dominant factor that limits the conversion resolution. The following experiments are used to illustrate above points, and to show our sampling clock jitter simulation for sigma-delta modulators.

The switch-capacitor sigma-delta modulator in Fig.5 is simulated again with 3.5 $k\Omega$ ON-resistance and 7 MHz opamp unity gain-bandwidth. In order to check clock jitter effects only, the thermal noise is not taken into account here. Firstly, a baseband signal, which is a 1 kHz sinusoidal of amplitude 0.75 V is simulated. Secondly, a IF sinusoidal signal of 1.025 MHz with the same amplitude is simulated. The IF signal is expected to be down-converted to 1 kHz in baseband according to sub-sampling theory.

With no clock jitter in the circuit, the IF signal is down-converted to 1 kHz nicely, as shown in Fig.7(B), it has a smoother high-pass quantization noise spectrum than the one shown in Fig.7(A). A possible reason is the high-frequency IF signal keeps the quantizer busy and breaks the limit cycle effects.

When sampling clock jitter is present in the circuit, the output noise spectrum changes. The circuit is simulated again with 0.01% clock jitter, which is a small random variation of 48 ps rms in the sampling clock. The in-band output noise power for the baseband version starts to increase, as shown in Fig.8(A), the output noise power generated by sampling clock jitter appears from 0 to 1 kHz and exhibits a noise "skirt" around signal frequency. This noise, however, is much less than the thermal noise floor which sits at -120 dB as shown in Fig.6(B), and will be buried under the thermal noise floor. In strong contradiction with the baseband version, the in-band output noise power increases dramatically for the IF version. The output noise due to clock-jitter and its "skirt" stay around -80 dB in signal-band. It is 40 dB above the thermal noise floor. The noise generated by clock-jitter will dominate total in-band output noise power and determine the performance measures, such as SNR.

The simulation of clock jitter effect is fast. It only takes SD-Noise 5.15 minutes to simulate circuit response for 74k clock cycles when clock jitter is present.

In summary, the information of output thermal noise power is important in baseband design and the information of output noise power due to clock jitter is important in RF/IF design. These noise estimations are much needed in design-phase. Our method can simulate both effects either separately or simultaneously.

4. CONCLUSION

In this paper, we present thermal noise and sampling clock jitter simulation methods that are developed, *for the first time*, for sigma-delta modulators at circuit macro-model level. The algorithms have been implemented in a computer program, SDNoise, and tested extensively. Simulation time in the order of minutes is shown for 74k clock-cycle data of the sigma-delta A/D converter when thermal noise and clock jitter are present.

Circuit designers can use this program to study the effects of different circuit architectures, different switch ON-resistances and different opamp unity gain-bandwidths on output thermal noise, which is important in baseband sigma-delta modulator design, and on output clock jitter noise, which is important in RF/IF sigma-delta modulator design.

In addition to sigma-delta modulators, these thermal noise and sampling clock jitter simulation techniques can also be used on general periodically switched linear networks, such as switched capacitor filters.

5. REFERENCES

[1] Feng Chen, *Design Techniques for CMOS Low Power Passive Sigma-Delta Modulator*, Ph.D. thesis, University of Waterloo, 1996.

[2] Armond Hairapetian, "A 81-MHz IF receiver in CMOS," *IEEE J. of Solid-State Circuits*, vol. 31, no. 12, pp. 1981–1986, December 1996.

[3] A. Demir, E. Liu and A. L. Sangiovanni-Vincentelli, "Time-domain non-monte carlo noise simulation for nonlinear dynamic cirucits with arbitrary excitations," *IEEE Trans. on CAD of Integrated Circuits and Systems*, vol. 15, no. 5, pp. 493–505, May 1996.

[4] A. Opal, "Sampled data simulation of linear and nonlinear circuits," *IEEE Trans. on CAD of Integrated Circuits and Systems*, vol. 15, no. 3, pp. 295–307, March 1996.

[5] C. A. Gobet and A. Knob, "Noise analysis of switched capacitor networks," *IEEE Trans. Circuits and Systems*, vol. CAS-30, no. 1, pp. 37–43, January 1983.

EFFICIENT UTILIZATION OF ON-CHIP INDUCTORS IN SILICON RF IC DESIGN USING A NOVEL CAD TOOL; THE LNA PARADIGM

Yannis Papananos and Yorgos Koutsoyannopoulos

Microelectronic Circuit Design Group
National Technical University of Athens
9 Iroon Politechniou, GR-157 73, Athens, Greece

ABSTRACT

A CAD tool for modeling planar and multi-layer polygonal integrated inductors on silicon substrates has been developed. The tool can be used in the efficient design of RF ICs containing on-chip inductors. The accuracy and reliability of the software is established through measurement results. The CAD tool is then used in the extraction of design guidelines for the development of inductor structures suitable for a given application. This procedure is demonstrated with the design of an LNA.

1. INTRODUCTION

The recent boom in portable wireless communications applications in combination with the advances in silicon technology make the implementation of silicon RF front-end ICs operating in the low GHz range both technically and economically attractive.

On-chip inductors generally enhance the reliability and efficiency of silicon integrated RF cells; they offer circuit solutions with superior noise performance and contribute to a high level of integration. However, poor integrated inductor modeling has been so far a major obstacle in their extensive utilization. In order to overcome this problem, a complete modeling and CAD tool, called "SISP" (Spiral Inductor Simulation Program), has been developed and presented in [1][2]. Polygonal and multi-layer integrated inductors, as well as transformers on silicon substrates can be accurately modeled up to several gigahertz. As an example, in Fig. 1(b) the accurate prediction of the inductance and the self-resonance frequency of a square two-layer inductor (Fig. 1(a)) is displayed [2]. Another substantial feature of SISP is the precise modeling of the coupling among two or more on-chip inductors [2].

In this paper, the utilization of the above features of the software for the successful exploitation of integrated inductors in silicon RF IC designs will be demonstrated. In section 2 a brief presentation of SISP's features will be given, followed by a comparison between measurement and simulation results. Section 3 contains useful integrated inductor design guidelines in the form of comprehensive nomographs. Section 4 presents a design example employing SISP, and in particular a low noise amplifier operating in the vicinity of 1 GHz. Finally, section 5 contains the conclusions.

2. SISP: A BRIEF PRESENTATION

Spiral Inductor Simulation Program is a PC-based CAD program developed in C++. The core algorithm, which has been analytically presented in [1], extracts a two-port network consisting of eleven lumped elements for every segment of the inductor. The main elements of the two-port are the series inductance, the resistance of the segment and the capacitors formed by the insulating SiO_2 between the inductor and the Si substrate. The algorithm also calculates the coupling capacitances between parallel adjacent segments and the RC networks for the modeling of the substrate layers under the insulator. The mutual inductance between the segments of the spiral is modeled with a transformer. The equivalent circuit of the spiral inductor includes a transformer for every possible couple of segments. The feature that enhances the accuracy of the inductor model, above the first resonance frequency, is the utilization of a full capacitance matrix. The matrix contains a capacitor not only between adjacent segments but also for every possible pair of segments of the inductor. This improvement is currently being added to the core algorithm of SISP employing techniques presented in [3].

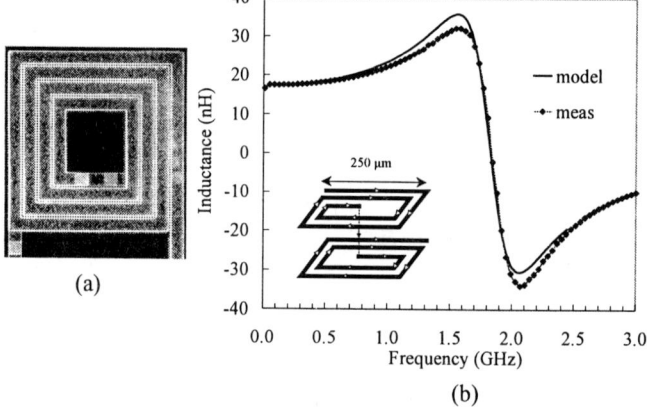

Fig. 1. (a) Two-layer inductor microphotograph (250μm x 250μm)
(b) Modeled and measured inductance [2]

SISP's layout tool can generate spiral inductor structures or alternatively import layout designs in CIF format. The designer should enter the CMOS, BiCMOS or bipolar process parameters before creating a layout. A fast, segment by segment extraction of the equivalent SPICE subcircuit is executed within seconds, while a similar model extraction with any EM software would take days. Three distinct versions of the model are produced to accommodate "typical", "fast" and "slow" technology variations. Passive elements may be frequency dependent to incorporate conductor skin effect, if supported by the SPICE simulator being used. Simulations can be initiated through the user-interface. The software displays simulation results in rectangular, polar or Smith chart plots, against measurement results (from S-parameter sets), if any. As an example, Fig. 2 displays the comparison

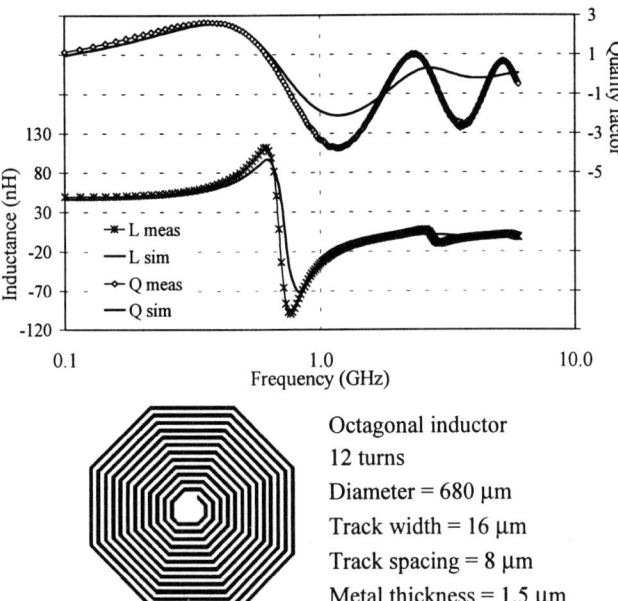

Fig. 2. Simulation vs. measurement of an octagonal inductor

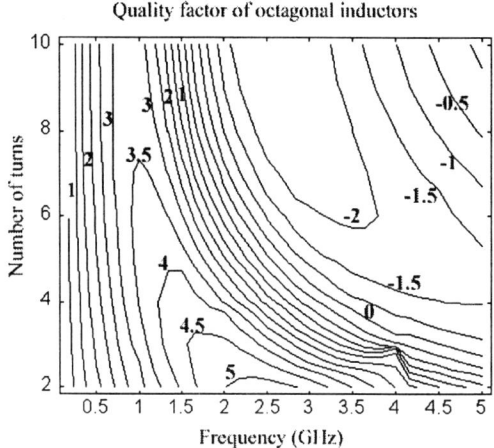

Fig. 3. Quality factor of octagonal spiral inductors with radius of 200μm, track width of 14μm and spacing between tracks of 3μm

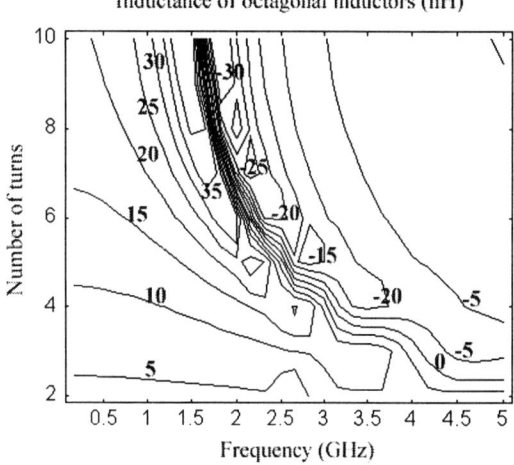

Fig. 4. Inductance of octagonal spiral inductors with radius of 200μm, track width of 14μm and spacing between tracks of 3μm

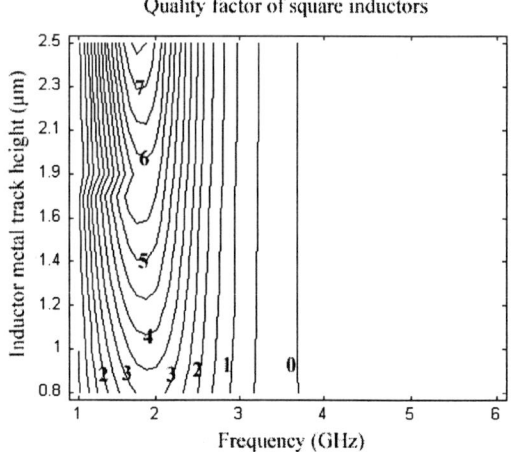

Fig. 5. Quality factor of 5-turn square spiral inductors with outer dimension of 300μm, track width of 14μm and spacing between tracks of 3μm

between measurement and simulation results of a 12-turn octagonal inductor fabricated in a typical silicon bipolar process.

3. INDUCTOR OPTIMIZATION GUIDELINES

Having already established the reliability of the SISP software [1][2], the tool can be used in the performance optimization of on-chip inductors. This can be done in two ways: (a) by optimizing the geometrical characteristics (i.e. number of turns, segment width, segment distance, etc.) of the spiral inductors, and (b) by deriving hints for process modifications, if supported by the foundry. The ease of use and speed of operation of the SISP software allows the generation of the useful nomographs that can prove to be a valuable aid to the RF IC designer in order to optimize his/her circuits. Various RF applications increasingly demand the improvement of the performance of on-chip inductors beyond the current state-of-the-art. Towards this purpose, modifications of certain parameters of a silicon process, such as Al metalization thickness and substrate doping, may be imperative [4].

Octagonal spiral inductors in a three-metal layer digital CMOS process were modeled and simulated to demonstrate how the quality factor (Q) and the inductance (L) depend on the number of turns. Simulation results are summarized in the form of nomographs. Fig. 3 displays the Q and Fig. 4 the L of octagonal inductors with a radius of 200μm, versus frequency and number of turns. From these nomographs, the highest values of L and Q and the respective resonance frequencies are found, while silicon area remains constant and the number of turns varies. Furthermore, Fig. 5 presents the improvement in the quality factor of a square inductor while the thickness of the metal track increases. Important is the fact that the resonance frequency of this family of inductors remains almost constant for the specific range of heights. In similar ways, nomographs can reveal the direction towards which technology parameters should be altered.

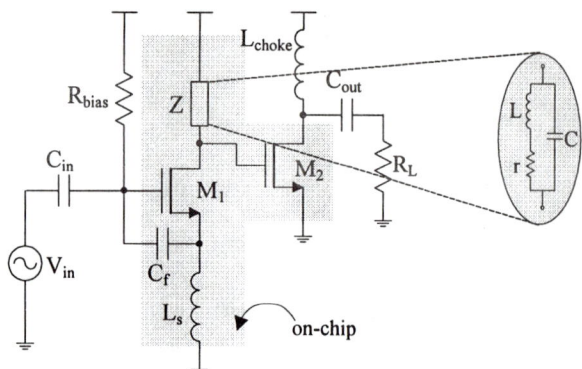

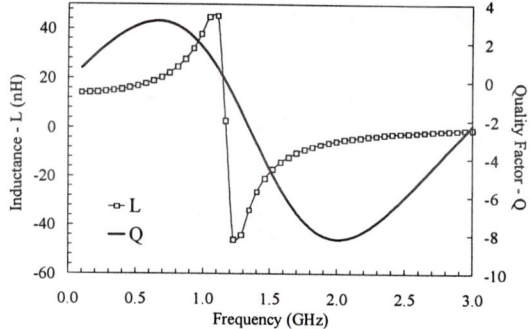

Fig. 6. LNA schematic and simplified model of the LC-tank Z

Fig. 7. Inductance and quality factor of the Z load of the LNA

4. THE LNA PARADIGM

The current trend in 1-2GHz one-chip RF front-ends dictates the existence of more than one inductor on the same chip [5][6]. In this case, the magnetic coupling between individual inductors or in intended transformers can play an important role in the performance of the overall system. Therefore, the layout of the circuit drastically affects its electrical behavior. SISP is capable of predicting related phenomena, leading to reliable and effective SPICE simulations and thus drastically reducing the design effort.

In this paper, the design of a simple tuned low-noise amplifier is presented. The particular design is by no means optimized in terms of noise, gain and silicon area, but it is rather used as a vehicle to demonstrate the software's capabilities. The LNA is designed in a digital sub-micron CMOS process and is operated around 1GHz. The schematic diagram of the LNA is shown in Fig. 6, where the shaded area represents the on-chip components including load Z, that is the LC tank, and L_s, which is impedance matching inductor. The designer can select the inductors' geometry either by using nomographs or by simulation with SISP. Using the simplified model shown in Fig. 6 for the LC tank, the gain of the LNA is given by

$$\text{Gain} = \frac{g_m + jQg_m}{r[(g_d r - \omega^2 LC + 1) + j\omega(rC + g_d L)]} \quad (1)$$

where $Q = \omega L/r$ is the quality factor of the inductor.

It is evident that the quality factor of the inductor (defined mainly by the inductor's series resistance), strongly affects the frequency response of the LNA. Moreover, the gain of the amplifier is optimized if the value of the parallel capacitor C of the load is minimized. For this purpose, operation of the integrated inductor around its self-resonance frequency was chosen in order to eliminate the need for a capacitive element in the tank. This can be done thanks to the ability of SISP to accurately predict the inductor's behavior at high frequencies, even beyond self-resonance frequency, as it has already been shown in Fig. 1 and Fig. 2. Fig. 7 displays the inductance and quality factor of the selected 6-turn square spiral inductor versus frequency. Its outer dimensions are 470µm x 470 µm, the track width is 25µm and the spacing between the tracks is 3µm. The inductor is laid out on the third metal layer of the process.

The on-chip inductor L_s (3-turn square, 470µm x 470 µm) is used in combination with the $C_{gs}//C_f$ capacitor for 50Ω input matching of the LNA. The input impedance of the LNA is calculated by

$$Z_{in} = \frac{1}{sC_{gst}} + sL_s + \frac{g_m L_s}{C_{gst}} \quad (2)$$

where $C_{gst} = C_{gs}//C_f$.

The 50Ω driving capability of the LNA is achieved through an open-drain configuration formed by transistor M_2 as shown in Fig. 6. This open-drain stage lowers the output impedance of the LNA and provides the necessary current to drive a 50Ω instrument port only for measurement purposes. Its gain does not exceed zero dB. For an integrated LNA that is used as a cell in a silicon RF system, the open-drain stage would not be necessary.

The alternative to the above input impedance matching topology is to employ an inductance L_g at the gate of the transistor M_1 as shown in Fig. 8. In this case L_g is used to cancel out the imaginary part of the input impedance caused by the parasitic capacitance C_{gs} of the device M_1. The overall input impedance is given by

$$Z_{in} = s(L_g + L_{bond}) + \frac{1}{sC_{gs}} + \frac{g_m L_{bond}}{C_{gs}} \quad (3)$$

However, this solution has certain drawbacks: (a) the value of the inductor L_g is usually higher than that of L_s in Fig. 6, resulting in larger silicon area and (b) the impedance matching is not so broadband. The value of L_g can be reduced by employing the grounded capacitance C_g connected to the gate of M1, but this decreases the quality factor of the tuned amplifier.

On-chip inductors' coupling effect

The magnetic coupling of the two integrated inductors of the LNA can seriously affect system performance and its improper modeling can lead to a degraded frequency response. Input impedance matching of the amplifier is also affected by the coupling of the two inductors. Fig. 9 depicts the LNA performance variance in terms of gain, under different inductor

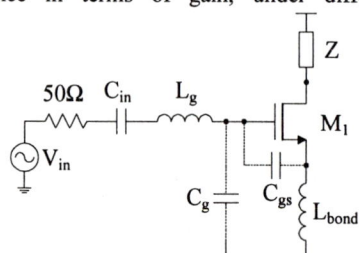

Fig. 8. Alternative input impedance matching topology

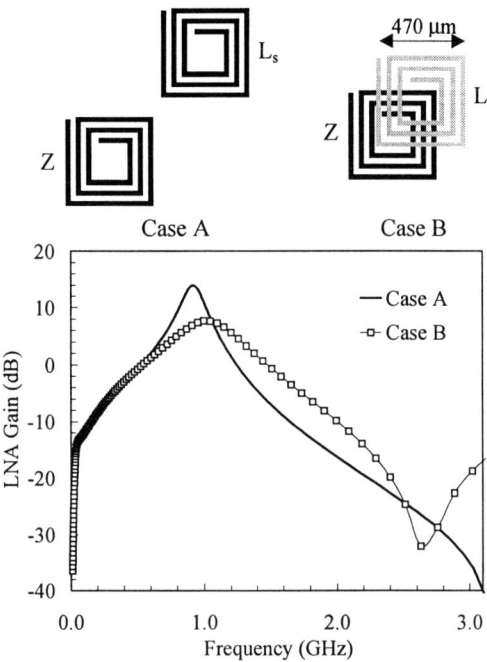

Fig. 9. LNA gain variance for two placement schemes of on-chip inductors

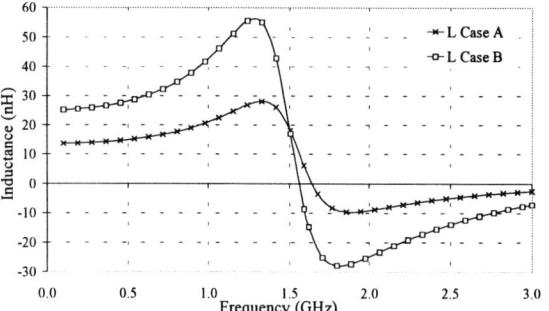

Fig. 10. Inductance of Z that is magnetically coupled to L_s

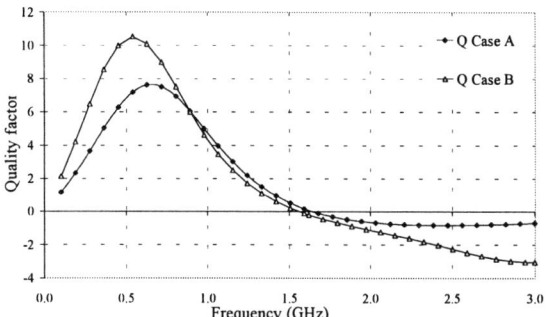

Fig. 11. Quality factor of Z that is magnetically coupled to L_s

placement schemes, as predicted by SISP. In case A the two inductors are laid out on the third metalization layer; in case B L_s is laid out on the second layer and partly under the inductor of load Z. The diagonal placement of the devices provides the necessary magnetic isolation for the optimum operation of the amplifier, at the cost of increased area coverage. However, if the gain degradation is acceptable, case B can lead to a more compact layout.

To give a more illustrative example of how the performance of an inductor is affected by the presence of other inductors in its vicinity, Fig. 10 and Fig. 11 present the L and the Q values of a square inductor (Z) in the two distinct cases that are depicted in Fig. 9. Inductor L_s is driven with a separate current source that has the same phase as the source that drives Z. When L_s is too close to Z, its performance can be relatively changed, due to the magnetic coupling between the two, providing a very simple and valuable tuning tool. In other words, the L and Q values of an inductor can be tuned by varying the current of a second inductor in its vicinity.

5. CONCLUSIONS

An effective method of utilizing on-chip inductors in silicon RF IC design is demonstrated through an LNA paradigm. A novel CAD tool, SISP, is capable of modeling the coupling between adjacent spiral inductors in any metal layer, and predicting the inductor-related performance variance of RF circuits.

Acknowledgment. This work was partially funded by Esprit Project No. 24123 – OCMP

REFERENCES

[1] Y. Koutsoyannopoulos, Y. Papananos, C. Alemanni, S. Bantas, "A Generic CAD Model for Arbitrarily Shaped and Multi-Layer Integrated Inductors on Silicon Substrates", *Proc. 1997 European Solid-State Circuits Conf.*, pp. 320-323, Sep. 1997, Southampton, U.K.

[2] Y. Koutsoyannopoulos, Y. Papananos, "SISP: A CAD Tool for Simulating the Performance of Polygonal and Multi-Layer Integrated Inductors on Silicon Substrates", *Proc. 1997 Int'l Conf. on VLSI and CAD*, pp. 244-246, Oct. 1997, Seoul, Korea.

[3] N. D. Arora, K. V. Raol, R. Schumann, L. M. Richardson, "Modeling and Extraction of Interconnect Capacitances for Multilayer VLSI Circuits", *IEEE Trans. on Computer-Aided Design of Integrated Circuits and Systems*, pp. 58-67, vol. 15, Jan. 1996.

[4] A. C. Reyes, S. M. El-Ghazaly, S. Dorn, M. Dydyk, D. K. Schroder, "Silicon as a Microwave Substrate", *Dig. of 1994 IEEE MTT-S Int'l Microwave Symp.*, pp. 1759-1762, vol. 3, May 1994, San Diego, U.S.

[5] S. Pipilos, Y. Tsividis, J. Fenk, Y. Papananos, "A Si 1.8 GHz RLC Filter with Tunable Center Frequency and Quality Factor", *IEEE Journal Solid-State Circuits*, vol. 31, pp. 1517-1525 Oct. 1996.

[6] K. L. Fong, C. D. Hull, R. G. Meyer, "A Class AB Monolithic Mixer for 900-MHz Applications", *IEEE Journal of Solid-State Circuits*, vol. 32, pp. 1166-1172, Aug. 1997.

OPTOMEGA: An Environment for Analog Circuit Optimization

Mansour KERAMAT and Richard KIELBASA
Ecole Supérieure d'Electricité (SUPELEC)
Service des Mesures, Plateau de Moulon
91192 Gif-sur-Yvette Cédex France
E-mail: mansour.keramat@supelec.fr

ABSTRACT

The development of OPTOMEGA, an environment of analog circuit optimization, is presented. OPTOMEGA communicates with the electric circuit simulator OMEGA and the Matlab engine by the *Interprocess Communications* (IPC) techniques. We have used the Matlab engine as a subenvironment for pre-developing functions and methods and a post-processing tool in order to map the circuit responses on desirable performances. OPTOMEGA consists of the nominal circuit optimization by global optimization algorithms and the parametric yield or *Average Quality Index* (AQI) optimization. The tool includes efficient Monte Carlo analysis algorithms which ensure an extensive verification and characterization of the design. Two circuit optimization examples are given.

I. Introduction

In order to design electronic circuits, particularly integrated circuits, the statistical variations of the circuit parameter values must be considered. These statistical fluctuations due to manufacturing tolerances will involve a variation of the circuit responses. In this case the parametric production yield is a very useful design criterion.

Circuit designers are often given a design specification which consists of ideal circuit responses with lower and upper limits, a set of designable parameters, and a set of random variables characterizing the statistical variations. The goal of circuit design is to determine the values of these designable parameters to achieve better performance and higher yield. For better performance, we want the actual circuit response to match the ideal response. For higher yield, we want the probability that the actual response falls within the given limits to be maximized. Usually, a performance-optimized design is not necessarily a yield-maximized design. A common practice is to find a performance-optimized design first, and then to maximize the parametric yield.

OPTOMEGA is an environment for performance or nominal optimization, and yield or *Average Quality Index* (AQI) [1] optimization of electronic circuits, as well as efficient Monte Carlo (MC) analysis. OPTOMEGA consists of OLGA: an environment of mathematical function optimization [4], OMEGA: an open electric circuit simulator [2], and OMEGAOSCOPE: a tool for visualization and manipulation of the simulator results. OMEGA can be connected to Matlab by *Interprocess Communications* (IPC) technique [3]. In this simulator, we can describe a circuit element model by a set of equations written in a C-based language.

In order to facilitate the development procedure, we have used the Matlab engine in OPTOMEGA as a pre-development environment for the methods and algorithms. After completing a method, we can write a C-based function and integrate it in the library of OPTOMEGA. Therefore, by using this very flexible environment for programming, debugging, and post processing of results, we can reduce the development time of each method.

II. Problem Formulation

In order to have a general notation for discrete and integrated circuits, we use the following notation. $\mathbf{x} = [x_1, x_2, ..., x_n]^T$ denotes the circuit designable parameters (e.g., nominal value of passive RLC elements, and nominal MOS transistor mask dimensions). $\xi = [\xi_1, \xi_2, ..., \xi_l]^T$ is the vector of random variables characterizing the statistical variations, (e.g., statistical variations of RLC elements, and variations of device model parameters), and $f_\xi(\xi)$ stands for their joint probability density function (pdf). The space of these random variables is also called *disturbance space*. $\mathbf{p} = \mathbf{p}(\mathbf{x}, \xi)$ is the vector of circuit parameters.

A. Nominal Optimization

In nominal optimization, the random variables are replaced by their mean values. For a given circuit, we usually define a set of performances and their specifications. For each performance, an error function (e_i) can be defined in such a way that if $e_i \leq 0$, then the corresponding constraint is satisfied. Therefore, we have a multiobjective optimization problem. In order to map these objective functions to a unique cost function, we use the generalized least *p*th type function [5]. The cost function is

$$U(p) = \begin{cases} \left[\sum_{i=1}^{M}(w_i e_i)^p\right]^{1/p} & e_{max} > 0 \\ -\left[\sum_{i=1}^{M}(-w_i e_i)^{-p}\right]^{-1/p} & e_{max} \leq 0 \end{cases}, \quad (1)$$

where w_i's are coefficients of importance for each error function, and $e_{max} = \max_i e_i$. Thus, this is a deterministic cost function and one can use an optimization algorithm in order to minimize it. One of the properties of this cost function is that even after obtaining a solution that satisfies all the constraints, it keeps searching for a *deeper point* in the performance space.

B. Average Quality Index (AQI) Optimization

The quality of a circuit can be defined in various ways. In a fabrication line, the circuit quality changes from one to another. This is why one needs to define an AQI for a circuit production. One of the important quantities of AQI is the manufacturing parametric yield of a circuit.

Assume that we have m circuit performances $\mathbf{y} = [y_1, y_2, ..., y_m]$. For each performance y_i, a *membership function* $\eta_i = \eta_i(y_i)$ can be defined by using the *fuzzy sets*. η_i can be interpreted as a quality index measuring the goodness of performance y_i. A good circuit should have a high value of the quality index η_i for every corresponding performance. The circuit quality index can be defined as

$$\eta(\mathbf{y}) = \varphi[\eta_1(y_1), \eta_2(y_2), ..., \eta_m(y_m)], \quad (2)$$

where $\varphi[.]$ is an appropriate intersection operator [1]. One can formulate AQI in disturbance space as

$$Q = \int_{R^l} \eta(\mathbf{y}(\mathbf{x}, \xi)) f_\xi(\xi) d\xi, \quad (3)$$

where R^l is disturbance space. Furthermore, the parametric yield is a special case of AQI where the quality index only takes on 1 or 0 value (indicator function).

For statistical circuit design we need to calculate AQI. It can be estimated by the efficient Monte Carlo (MC) methods, e.g., *Latin Hypercube Sampling* (LHS) [6], *optimal stratified sampling* [7]. The MC method is the most reliable technique for the statistical analysis of electrical circuits. The unbiased Primitive MC (PMC) based estimator of AQI can be expressed as

$$\hat{Q}_{MC}(\mathbf{x}) = \frac{1}{N}\sum_{i=1}^{N}\eta(\mathbf{y}(\mathbf{x}, \xi^i)), \quad (4)$$

where ξ^i's are independently drawn random samples from $f_\xi(\xi)$, and N is the sample size. In order to optimize AQI, we use the *Generalized Centers of Gravity* (GCoG) algorithm [8], which is applicable to discrete and integrated circuits.

III. OPTOMEGA: Optimization Tool

The conceptual diagram of OPTOMEGA is depicted in Fig. 1. The core of OPTOMEGA is an environment of mathematical function optimization system called OLGA [4]. It communicates with Matlab by IPC technique. Matlab and the circuit simulator OMEGA also communicate with each other by IPC.

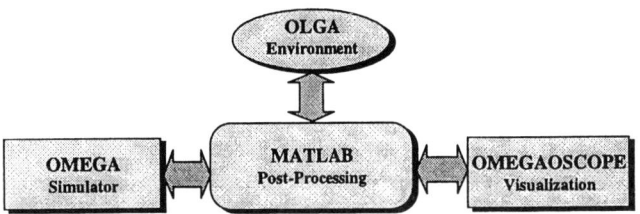

Fig. 1. Conceptual diagram of OPTOMEGA.

A. OLGA: System of Global Optimization

In order to optimize a cost function, we need to have a tool for defining the desired cost function and choosing an appropriate optimizer. OLGA is an environment for optimizing mathematical cost functions. The cost function can be written in C language and it is also possible to use a predefined library which lets us communicate with the Matlab engine.

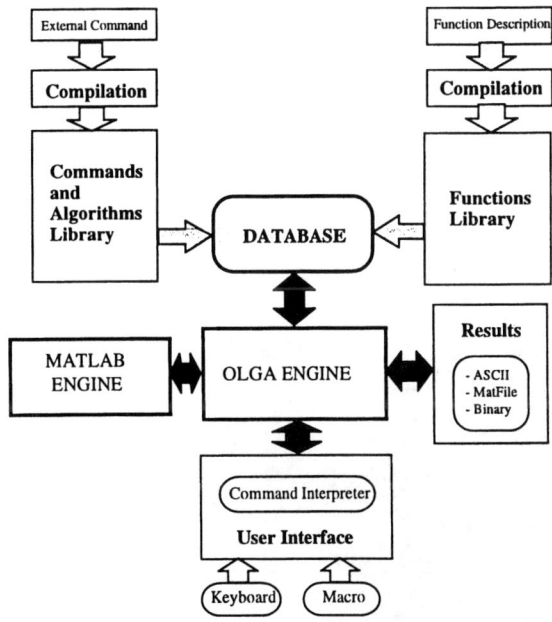

Fig. 2. Overview of OLGA: an environment of mathematical function optimization.

OLGA provides several local and global optimizers which can be selected by the user for an optimization process. The results of an optimization process are displayed on the screen with different display mode options. Moreover, the results can be written in a file with different formats. In addition, this environment provides several procedures for evaluating a stochastic optimizer by using standard test functions. There are two types of instructions: internal and external. The internal instructions are the basic functionality of OLGA. The external instructions are designated to extend the set of instructions and to create personal functions by the user.

Fig. 2 illustrates general aspects of OLGA. The cost function for optimizing can be written in C language and then compiled. In function definition, it is also possible to use the Matlab engine services. The set of compiled functions constitute the library of cost functions which can be selected by the user.

In OLGA, we have a library of commands which can be extended by developing external commands. We can consider a functionality for an external command and then it can be written in C code and compiled in order to be included in the available commands library. In order to minimize a cost function, one can choose one of the following optimizers [9]:

1- Simplex (Nelder and Mead),
2- Simulated Annealing,
3- Enhanced Simulated Annealing,
4- SAPLEX (Simulated Annealing/simPLEX) [5].

The user can communicate with the OLGA engine by keyboard or a *macrocommand file*, which contains a set of OLGA commands. We can also define some input variables for a macrocommand file.

B. OMEGAOSCOPE

OMEGAOSCOPE is a graphical user interface environment for visualization and manipulation of the OMEGA results. The main program of OMEGAOSCOPE was written in C code, and in order to use the Matlab facilities of graphical interface, it uses the Matlab engine. OMEGAOSCOPE has a help menu, file menu, Matlab commands area, visualization area, zoom function, and other facilities (see Fig. 3).

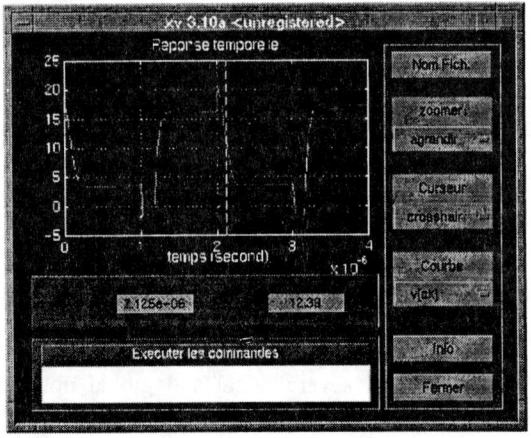

Fig. 3. Snapshot of OMEGAOSCOPE

IV. Circuit Examples

In this section, a nominal circuit design problem and a yield optimization of a CMOS clock driver circuit will be presented.

Example 1: High Pass Filter Circuit [8]

As an example, we have chosen a high pass filter circuit for performance optimization (Fig. 4(a)). Fig. 4(b) shows the desired specifications of voltage transfer function. The parameters to optimization are C_i, $i = 1,...5$, and the optimized values obtained by OPTOMEGA are 20n, 15n, 17n, 40n, and 70n respectively. The values of R_1, R_2, L_1, L_2, R_{l1}, R_{l2} are given 10k, 10k, 3.8, 3.1, 10, and 10 respectively. The optimized circuit meets all the specifications (Fig. 4(c)). Moreover, we have used generalized least *p*th type cost function for centering the circuit in performance space. The cost function is written as a Matlab function for flexibility of defining and testing various criteria.

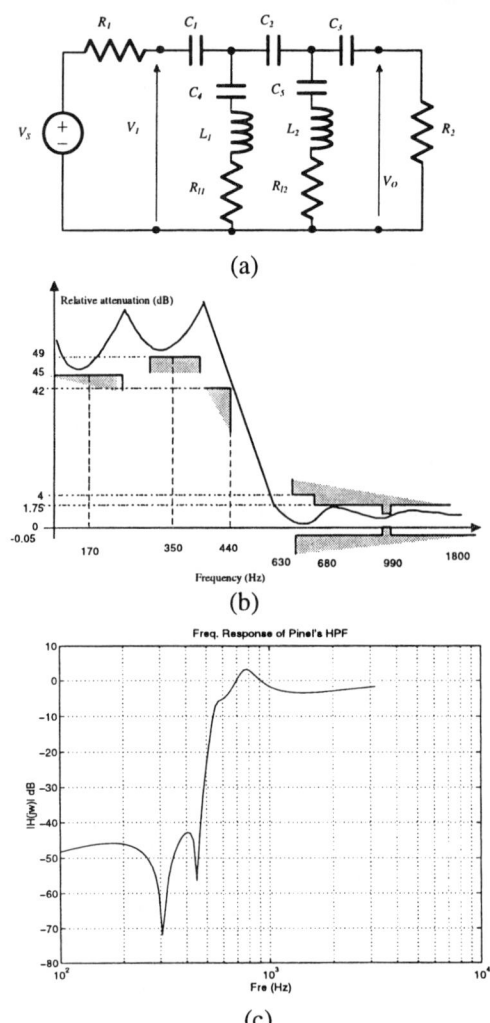

Fig. 4. Circuit and specifications of a high pass filter.

Example 2: CMOS Clock Driver Circuit [1]

A CMOS clock driver is shown in Fig. 5. The clock driver provides two outputs V_{out1} and V_{out2} in opposite phase. The performance of interest is the clock skew. The specifications are that the skew falls in the interval [-1,1] ns. The disturbance space is taken as in [1].

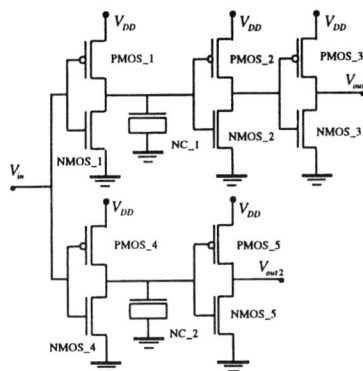

Fig. 5. Schematic diagram of a CMOS clock driver.

In this example, the widths of transistor NMOS_1 and NMOS_4 were selected as designable parameters. The yield optimization results are shown in Fig. 6. In each iteration we used 50 circuit simulations. In Fig. 6(a), it is seen that the optimized circuit has 100% yield (verified by 200 samples). The initial yield was approximately 2%. The evolution of designable parameters during the process of yield optimization is illustrated in Fig. 6(b).

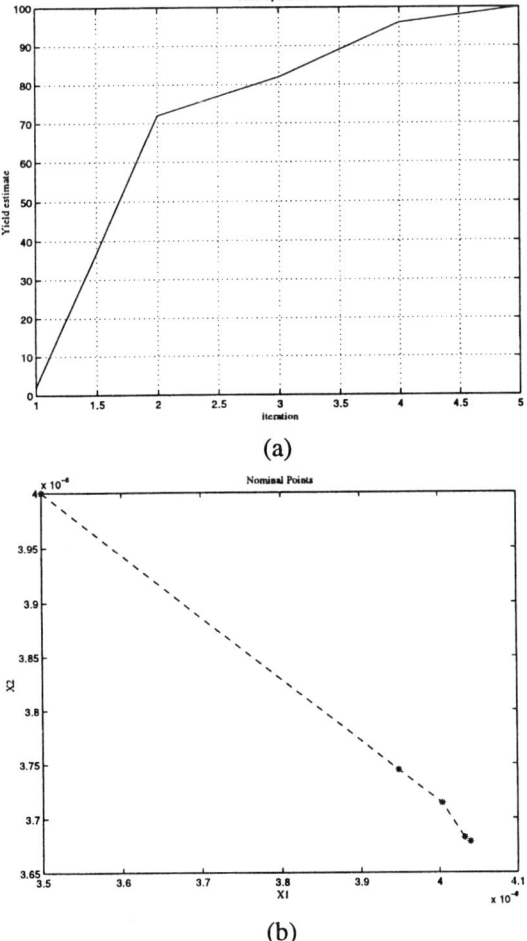

Fig. 6. CMOS clock driver circuit results. (a) Yield estimates vs. iterations. (b) Designable parameters trajectory in GCoG.

V. Conclusions

We have developed a new environment of circuit optimization (OPTOMEGA) which is coupled with the electric simulator OMEGA and Matlab engine by Interprocess Communications (IPC) techniques. OPTOMEGA provides us with the nominal or performance optimization and AQI or yield optimization. The histogram of each performance can be estimated by an efficient Monte Carlo technique (LHS). The Matlab environment provides us with facilities of post-processing, programming and debugging of several consistent methods. The server mode and Mex file are two major facilities in the development of new procedures. For visualizing and manipulating the results of OMEGA, we have developed OMEGAOSCOPE by the use of Matlab server mode, Mex file, and graphical interface instructions.

References

[1] J. C. Zhang and M. A. Styblinski, *Yield and Variability Optimization of Integrated Circuits*, Kluwer Academic Publisher, 1995.

[2] P. Aldebert and R. Klielbasa, "OMEGA: system of electric simulator, user's manual," Ecole Superieure d'Electricite (SUPELEC), Paris, September 1993.

[3] M. Keramat, "Using Matlab in interprocess communications," Ecole Superieure d'Electricite (SUPELEC), Paris, Tech. Rep. No. SUP-0496-7, April 1996.

[4] M. Keramat, "User's guide for OLGA: an environment of optimization," Ecole Superieure d'Electricite (SUPELEC), Paris, Tech. Rep. No. SUP-1295-5, December 1995.

[5] J. W. Bandler and C. Charalambous, "Theory of generalized least pth approximation," *IEEE Trans. Circuit Theory*, vol. CT-19, pp. 287-289, May 1972.

[6] M. Keramat and R. Kielbasa, "Latin hypercube sampling Monte Carlo estimation of average quality index for integrated circuits," *Analog Integrated Circuits and Signal Processing*, vol. 14, no. 1/2, pp. 131-142, 1997.

[7] M. Keramat and R. Kielbasa, "A study of stratified sampling in variance reduction techniques for parametric yield estimation," in *Proc. IEEE Int. Symp. Circuits Syst.*, Hong Kong, June 1997, pp. 1652-1655.

[8] M. Keramat and R. Kielbasa, "Generalized centers of gravity algorithm for yield optimization of integrated circuits," in *Proc. IEEE Int. Symp. Circuits Sys.*, Monterey, CA, May 31-June 3, 1998.

[9] Gérard Berthiau, "Electronic circuit optimization by simulated annealing: adaptation and comparison with other optimization algorithms," Ph.D. dissertation, Ecole Centrale, Paris, France, Dec. 1994 (French).

AC CONSTRAINT TRANSFORMATION FOR TOP-DOWN ANALOG DESIGN

B. G. Arsintescu[†], E. Charbon, E. Malavasi, W. Kao

Cadence Design Systems,
555 River Oaks Pkwy.,
San Jose, CA 95134

[†]Delft University of Technology,
Mekelweg 4, 2628CD Delft,
The Netherlands

ABSTRACT

In mixed designs, analog blocks are often the most difficult parts to realize, due to the large number of high-level constraints and the importance of second order effects. In this paper a method is proposed to automatically transform AC constraints from higher to lower levels of hierarchy within a top-down design methodology. The transformation consists of finding a linear approximation for the analytic expression of the transfer function for flat and hierarchical designs. A state-of-the-art filter design example illustrates the capabilities of the method.

1. INTRODUCTION

Mixed-signal applications are becoming more and more important in today's IC market. High performance specifications and parasitics between the analog and digital parts make mixed-signal designs particularly intricate. Non-idealities introduced by the layout implementation cause second-order effects which become important when tight specifications are required. For this reason, the deviations of performance from specifications must be kept under control. This requires second order effects be transformed into parameters manageable by constraint-driven tools. Moreover, as the complexity of analog circuits grows, hierarchical constraint reuse will reduce the size of the problem, thus effectively decreasing the time to market for mixed-signal ICs.

In order to enforce all specifications and to reduce parasitic effects, constraint-driven tools were developed. These tools require a transformation of high-level constraints into constraints on parameters that a specific DA tool can enforce. Constraint transformation consists of two phases: (a) model generation and (b) constraint generation. During model generation, a model is created, relating performance measures at a given level of hierarchy to lower-level design parameters. During constraint generation, every design parameter is associated with an actual bound. As a simple example, consider a delay constraint $t \leq t_{max}$ on a given net. During model generation we model $t = kRC$, where R and C are the lumped resistance and capacitance of the net. During the constraint generation phase we determine values for the maximum R and C such that $k \cdot R^{max} \cdot C^{max} \leq t_{max}$.

A number of techniques have been proposed in the literature to address the problem of constraint generation. In [1] each high-level performance is associated with the root of a tree, whose nodes represent either lower-level performance variables or independent design parameters. The selection of all performance variables and design parameters is left to the designer, while the reasoning process uses sensitivities to compute the resulting path. The method proposed by [2] consists of obtaining bounds on the performance variables of a given level of hierarchy as a result of a quadratic optimization problem. Constrained optimization alone [3] allows for the removal of all non-critical design parameters and the calculation of parasitic constraints when the bottom level is reached. All mentioned methods rely on user defined constraint models at each level of hierarchy. Constraint transformation becomes instrumental in the solution of complex hierarchical design problems when *constraint-driven tool sets [4]* are available.

In this paper we describe an automated top-down constraint transformation method for a class of linear continuous time circuits. High-level AC specifications are propagated automatically to lower levels of hierarchy. A pole-zero analysis generates the analytic expression for the transfer functions of interest. Sensitivities with respect to design parameters and parasitic effects are then computed analytically at each level of hierarchy, using standard algebraic computation techniques. The generation phase is performed using conventional variable allocation methods for both deterministic and non-deterministic cases [3, 5, 6].

This method can be applied automatically while at the same time control is maintained at every step of the design. As an example, a 4th order OTA-RC filter is presented at the end of this paper to illustrate the advantages of our top-down constraint-driven design methodology, underlining the method strengths with respect to block reuse and mismatch characterization.

The paper is organized as follows. In Section 2 the constraint transformation problem is defined and the main characteristics of the transformation method are outlined. Examples are presented in Section 3.

2. CONSTRAINT TRANSFORMATION MODEL

Following the notations introduced in [4], let us define $[\mathbf{K}]$, the vector of the n performance measures considered in a given design. Moreover, define $[\mathbf{K_0}]$ as the nominal performance vector and $[\Delta\mathbf{K}] = [\mathbf{K}] - [\mathbf{K_0}]$ as the performance degradation of the circuit. Similarly, let us define $[\mathbf{P}]$ as the vector of all m design parameters, $[\mathbf{P_0}]$ the nominal parameter vector and $[\Delta\mathbf{P}] = [\mathbf{P}] - [\mathbf{P_0}]$ the vector of all parameter variations. The problem of finding the mathematical relation between $[\Delta\mathbf{K}]$ and $[\Delta\mathbf{P}]$ corresponds to the model generation problem.

Problem 1 Model Generation Problem. *For a given circuit with performance vector $[\mathbf{K}]$ and parameter vector $[\mathbf{P}]$, find an $n \times m$ matrix $[\mathbf{S}]$ s.t.:*

$$[\Delta\mathbf{K}] = [\mathbf{S}] \cdot [\Delta\mathbf{P}]. \tag{1}$$

$[\mathbf{S}]$ is called the sensitivity matrix of $[\mathbf{K}]$ with respect to $[\mathbf{P}]$. The model in (1) expresses a linear approximation between the degradation vectors $[\Delta\mathbf{K}]$ and $[\Delta\mathbf{P}]$.

Problem 2 Constraint Generation Problem.
For a given circuit with performance vector $[\mathbf{K}]$ and allowed degradation $[\overline{\Delta\mathbf{K}}]$, find a set of bounds $[\Delta\mathbf{P}]^{bound}$ on the components of $[\Delta\mathbf{P}]$ such that:

$$0 \leq [\Delta\mathbf{P}] \leq [\Delta\mathbf{P}]^{bound} \Longrightarrow 0 \leq [\Delta\mathbf{K}] \leq [\overline{\Delta\mathbf{K}}]. \tag{2}$$

In this paper we will focus on Problem 1 for AC performance functions. AC performance functions are specified with respect to one or more transfer functions. The analytic expression of the frequency transfer function $H(\omega)$ of a given circuit is:

$$H(\omega) = \frac{\prod(j\omega - z_i)}{\prod(j\omega - p_i)}, \tag{3}$$

where p_i, z_i are the poles and zeroes of the transfer function, respectively. The constraint transformation method will find the sensitivity matrix for any linear continuous time circuit using the analytic expression of the transfer function, based on the following assumptions: (i) there is a finite number of poles and zeroes in the frequency range of interest, (ii) simulation methodologies to compute the values and multiplicity of the poles and zeroes exist and (iii) a linear approximation of the transfer function around the nominal working point can be done.

The AC constraints are bounds given to the transfer function as shown in Figure 1. Four classes of AC constraints problem are identified in the figure:

1. **Punctual constraints on the transfer function**:
   $\mathcal{C}(H(\omega))$ @ $\omega = \omega_1$,
2. **Range constraints on the transfer function**:
   $\mathcal{C}(H(\omega))$ @ $\omega \in (\omega_1, \omega_2)$,

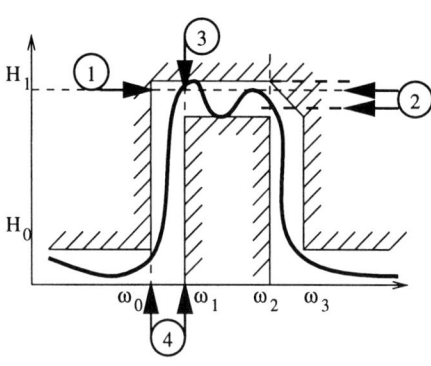

Figure 1: Transfer function constraints

3. **Punctual constraints on frequency**:
   $\mathcal{C}(\omega)$ @ $H(\omega) = H_1$,
4. **Range constraints on frequency**:
   $\mathcal{C}(\omega)$ @ $H(\omega) \in (H_0, H_1)$.

In these definitions, $\mathcal{C}()$ is a constraint operator, which must be continuous and differentiable in the range of interest. For example, the low frequency gain, defined as $\mathcal{C}(H(\omega)) = |H(\omega)| \geq H_0 @ \omega \to 0$, is a punctual constraint on the transfer function. Unity-Gain Bandwidth, defined as $\omega \geq \omega_0 @ |H(\omega)| = 1$, is a punctual constraint on frequency. The maximum ripple constraint, defined as
$\max_{(\omega_0, \omega_1)} H(\omega) - \min_{(\omega_0, \omega_1)} H(\omega) \leq H_\Delta @ \omega \in (\omega_1, \omega_2)$
is a range constraint on the transfer function.

The method consists in finding an analytic expression of the linear approximation around the nominal working point for constraint function $\mathcal{C}$. The linear approximation is computed by derivating the analytic constraint function $\mathcal{C}(\cdot)$ with respect to the circuit parameters $[\mathbf{P}]$:

$$K = \mathcal{C}(\cdot) \Rightarrow \Delta K = \frac{\partial \mathcal{C}}{\partial P} \cdot \Delta P \tag{4}$$

$$\Rightarrow \begin{cases} \Delta K_H = \frac{\partial \mathcal{C}}{\partial H} \frac{\partial H}{\partial P} \cdot \Delta P \\ \Delta K_\omega = \frac{\partial \mathcal{C}}{\partial \omega} \frac{\partial \omega}{\partial P} \cdot \Delta P. \end{cases} \tag{5}$$

In Equation (5), $\frac{\partial H}{\partial P}$ are the circuit sensitivities, hence the constraint transformation problem is solved for all the punctual constraints on the transfer function (class 1):

$$\Delta K_H = \left(\frac{\partial \mathcal{C}}{\partial H}(H(\omega_1))\right) \cdot S_P^H \big|_{\omega=\omega_1} \cdot \Delta P, \tag{6}$$

where $\frac{\partial \mathcal{C}}{\partial H}$ is the analytic expression of the derivative of function $\mathcal{C}$.

For frequency constraints, the frequency ω_0 s.t. $H(\omega_0) = H_1$ is first calculated. The composed derivative of the transfer function is:

$$S_P^H \big|_{\omega=\omega_0} = \frac{\partial H}{\partial P}\bigg|_{\omega=\omega_0} = \left(\frac{\partial H}{\partial \omega} \cdot \frac{\partial \omega}{\partial P}\right)_{\omega=\omega_0}. \tag{7}$$

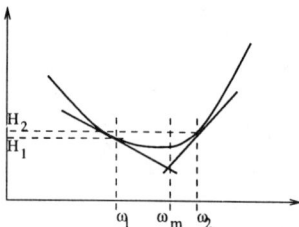

Figure 2: Extreme points of the transfer function.

The analytic expression of the slope of the transfer function can be easily obtained by derivation and the solution for punctual constraints on frequency(class 3) is:

$$\Delta K_\omega = \frac{\partial \mathcal{C}}{\partial \omega}(\omega_0) \frac{1}{\frac{\partial H}{\partial \omega}(\omega_0)} \cdot S_P^H\big|_{\omega=\omega_0} \cdot \Delta P. \quad (8)$$

Equation (8) cannot be calculated for the extremes of the transfer function (i.e. maximum and minimum points) where $\frac{\partial H}{\partial \omega} = 0$. To overcome this problem, the frequency of the maximum ω_m is approximated piecewise linearly as shown in Figure 2, using the neighbor poles and zeroes. Based on the analytic formula obtained for ω_m, the solution for punctual constraint on frequency in the extreme points becomes:

$$\Delta K_\omega = \frac{\partial \mathcal{C}}{\partial \omega}(\omega_m) \cdot \frac{S_P^H\big|_{\omega=\omega_2} - S_P^H\big|_{\omega=\omega_1}}{\frac{\partial H}{\partial \omega}(\omega_1) - \frac{\partial H}{\partial \omega}(\omega_2)} \cdot \Delta P. \quad (9)$$

For the range constraints problems (class 2 and 4), Equations (6, 8, 9) hold over an interval if the constraint function $\mathcal{C}(\cdot)$ is monotonic. Otherwise, the analysis should be done for each monotonic sub-domain of $\mathcal{C}(\cdot)$, resulting in a constraint system. However, the range constraints can be expressed as a norm rather than a function. The constraint transformation method is extended to handle constraints formulated with two norm operators.

1. Unity norm: $\|H(\omega)\|_1 \triangleq \int_{\omega_1}^{\omega_2} H(\omega)\delta\omega$. E.g., the total nonlinearity is calculated as an integral of the transfer function over a given interval.
2. Infinite norm: $\|H(\omega)\|_\infty \triangleq \max_{\omega_1 \leq \omega \leq \omega_2} H(\omega)$. E.g., the ripple constraint is the difference between the transfer function maximum and minimum within the interval.

For the unity norm $\|\cdot\|_1$, let us consider the transfer function is also dependent on circuit parameters $P : H(\omega, P)$. The two linear operators $\partial(), \int()$ can be swapped when ω and $[\mathbf{P}]$ are independent:

$$\begin{aligned}\frac{\partial}{\partial P}\left(\int_{\omega_1}^{\omega_2} H(\omega)\delta\omega\right) = \int_{\omega_1}^{\omega_2} \frac{\partial H}{\partial P}\delta\omega \Rightarrow \\ \frac{\partial}{\partial P}(\|H(\omega)\|_1)\Delta P = \|S_P^H\|_1 \cdot \Delta P.\end{aligned} \quad (10)$$

If ω and $[\mathbf{P}]$ are not independent, the operators cannot be swapped. If the operators are not linear, the transformation is dependent on the sign and monotonicity of both operators, because the triangle formula should be used for linear approximation: $\|a + b\| \leq \|a\| + \|b\|$.

For the infinite norm $\mathcal{C} = \|\cdot\|_\infty$, the general problem is transformed in a system of simultaneous punctual problems. The solution of the problem is the maximum among the *numerical values* of the punctual problems. The expression for the linear approximation around the nominal point is:

$$\frac{\partial}{\partial P}\left(\max_{(\omega_1,\omega_2)} H(\omega)\right) = \left\{\max_{\omega_i}\left(\frac{\partial H}{\partial P}\big|_{\omega=\omega_i}\right)\big|\frac{\partial H(\omega)}{\partial \omega}\big|_{\omega=\omega_i} = 0\right\}, \quad (11)$$

where ω_i are all the frequencies at which $H(\omega)$ has a local maximum. The constraint transformation is solved completely calculating sensitivities in the extreme points:

$$\Delta K_i = \max_{\omega_i}\left\{S_P^H\big|_{\omega=\omega_i} \cdot \Delta P\right\}. \quad (12)$$

Identical equations are obtained for the frequency range constraints. The domain of definition for the constraint function $\mathcal{C}$ is extended from continuous differentiable functions to linear norms independent of the parameters and to infinity norm. No further extension was pursued because the definition domain for $\mathcal{C}(\cdot)$ already covers the need for AC constraint transformations.

The described method is generalized for hierarchical designs. Given a circuit with k modules $\mathcal{M}_j, 1 \leq j \leq k$, each module with a parameter vector $[\mathbf{P_j}]$, a parametric model of the circuit is mapping any performance function K_i into module parameters:

$$K_i = \mathcal{K}_i([\mathbf{P_1}], \ldots, [\mathbf{P_k}]). \quad (13)$$

A linear approximation can be made with respect to the parameter vectors $[\mathbf{P_j}]$:

$$\Delta K_i = \frac{\partial}{\partial P_j}(\mathcal{K}_i([\mathbf{P_1}], \ldots, [\mathbf{P_m}])) \cdot [\Delta \mathbf{P_j}]. \quad (14)$$

This transforms the high-level constraint ΔK_i into cell parameter constraint vectors $[\Delta \mathbf{P_j}]$. In a top-down manner, these become constraints $[\Delta \mathbf{P_j}] = [\Delta \mathbf{K_j}]$ for each cell and can be transformed further using a parametric model of the child cell. The hierarchical transformation can be repeated until device level or it can be used for constraint generation at the library level.

In Figure 2 the hierarchical transformation flow is shown. A pole-zero analysis is required to obtain the analytic expression of the transfer function. The high-level constraints are transformed to a set of punctual transfer function and/or frequency constraints. In case of hierarchical constraints, the parametric cell constraints are transformed to device parameters, thus descending the hierarchy. Sensitivity analysis is performed by the simulator for device parameter transformations.

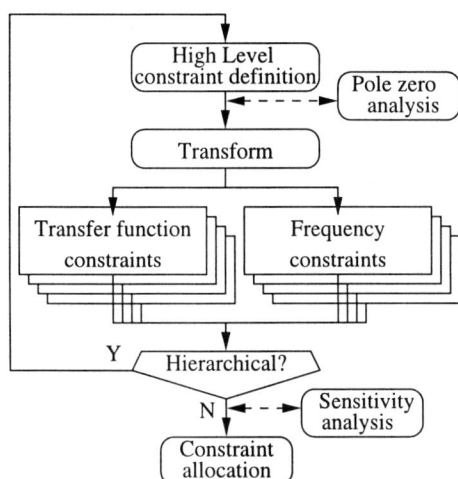

Figure 3: Method flow

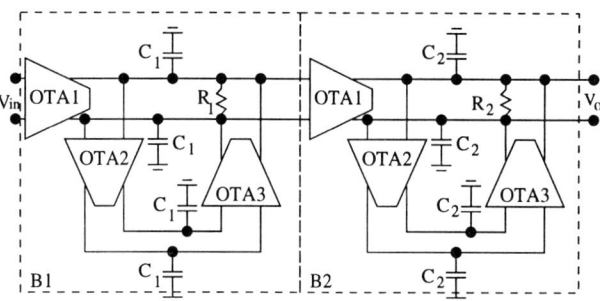

Figure 4: Architecture of the fourth order maximally flat 10.7 MHz bandpass filter

3. EXAMPLES

The top-down constraint transformation method has been applied to a fourth-order filter design [7]. Constraint generation for one of the hierarchical examples shows the suitability of the method for design reuse and mismatch characterization. The architecture of the 4th order filter in Figure 4 is based on two biquadratic sections connected in cascade. The pole frequencies are controlled by OTA2 and the associated capacitors. The quality factor is controlled (ideally) by the resistor and the OTA1 of each biquad. In order to have a similar behavior for both biquads, the AC gain for each filter was designed to be unity, i.e. $|H(\omega_0)| = 1$. In Table 1 a list of high-level constraint transformations to biquad parameters is given. The transfer function H_4 of the 4th order filter is the product of the transfer function of the biquads, H_1, H_2 respectively. The values of the bandwidth and quality factor for the 4_{th} order filter, shown in Table 1, are taken from filter design tables. For this case, the parameter vector is created by the biquad design parameters:

$$[\mathbf{P}]_{biquad} = [\omega_{10}, \omega_{20}, BW_1, BW_2, Q_1, Q_2, |H_1|, |H_2|]^T. \quad (15)$$

Based on Table 1 relations, a transformation/sensitivity matrix $[\mathbf{S}]$ is created to transform the constraint vector $[\mathbf{K}]$ (the first column in Table 1) into parameter constraints. The hierarchical constraint transformation for the 4th biquad is as follows:

$$[\Delta \mathbf{K}]_{4ord} = [\mathbf{S}]_{4ord} \cdot [\Delta \mathbf{P}]_{biquad},$$
$$[\Delta \mathbf{P}]_{biquad} = [\Delta \mathbf{K}]_{biquad} \Rightarrow$$
$$[\Delta \mathbf{K}]_{biquad} = [\mathbf{S}]_{biquad} \cdot [\Delta \mathbf{P}]_{comp},$$
$$[\Delta \mathbf{P}]_{comp} = [\Delta \mathbf{K}]_{comp} \Rightarrow$$
$$[\Delta \mathbf{K}]_{comp} = [\mathbf{S}]_{comp} \cdot [\Delta \mathbf{P}]_{device}. \quad (16)$$

Three levels of hierarchy were identified: the top level, with specifications on the 4th order filter, the second level with biquad specification, and the component level, where design parameters are specification for the OTAs, resistors and capacitors.

Let us consider the block design for the active biquad filter (e.g. B1 in Figure 4). Table 2 shows the expressions of its constraints. The fourth column in Table 2 lists the parametric relations between circuit and modules parameters. $OTA2$ and $OTA3$ are identical (ideally, transconductances $g_{m2} = g_{m3} = g_m$, output conductances $g_{02} = g_{03} = g_0$ and low-frequency gain $A_{V2} = A_{V3} = A_V$). The mismatches between identical devices are denoted by: $g_{m,\Delta} = g_{m2} - g_{m3}$, $C_\Delta = C_1 - C_2, \ldots$. The vector of circuit parameters is the collection of OTA parameters, resistors, capacitors and mis-

K_i	$\mathcal{K}(\cdot)$
ω_0	$\sqrt{\omega_{10}\omega_{20}}$
BW	$f(BW_1, BW_2)$
Q	$f(Q_1, Q_2)$
$|H|@\omega_0$	$|H_1| \cdot |H_2|$

Table 1: 4_{th} order high level constraint mapping to biquad constraints

K_i	ΔK_i	$K(H(\omega))$	$\mathcal{K}(P)$
ω_0	$\omega \in (\omega_{min}, \omega_{max})$	$\omega_0^2 = p_1 \cdot p_2$	$\sqrt{\frac{g_{m2}g_{m3}}{C_1 C_2}} = \frac{g_m}{C}$
BW	$BW \leq BW_{max}$	$BW = p_1 + p_2$	$\frac{g_2}{C}$
Q	$Q \geq Q_{min}$	$Q = \frac{\omega_0}{BW}$	$\frac{\sqrt{g_{m2}g_{m3}}}{g_2}\sqrt{\frac{C_2}{C_1}} = \frac{g_m}{g_2}$
$|H_{V_o}(\omega_0)|$	$|H_{V_o}(\omega_0)| \geq H_{min}$		$\frac{g_{m1}}{g_2}\sqrt{\frac{g_{m2}}{g_{m3}}}\sqrt{\frac{C_2}{C_1}} = \frac{g_{m1}}{g_2}$

Table 2: Second order filter constraints

VI-129

matches:

$$[\mathbf{P}]_{\text{comp}} = [\begin{array}{c} g_{m1}, g_{01}, A_{V1}, g_m, g_0, A_V, \omega_z, \omega_p, \\ g_{m,\Delta}, A_{V,\Delta}, C, C_\Delta, g_2, \ldots \end{array}], \quad (17)$$

where ω_z, ω_p are the dominant zero and pole respectively of the OTA.

For this example, constraint generation was done using PARCAR [3]. The resulting bounds for the OTA parameters and mismatches are shown in the third column of Table 3, together with analytic expressions of the sensitivity matrix. In the table, $A \ldots G$ are constants depending on the parameters of the modules. In case the two biquads are identical, the set of parameters is reduced to the relative nominal parameters (ω_0, BW, etc.) and their mismatches. The full design of the 4th order filter reduces to specifications on two OTA structures, one resistor, one capacitor and matching requirements.

The OTA constraints are further transformed to device parameters for the design of the OTA itself. Based in the analytic expression of the transfer functions and circuit sensitivities, the transformation matrix is completely determined using our automated transformation method.

4. CONCLUSIONS

In this paper a method to automate the constraint model generation for analog AC constraints is presented. Analytic expressions of the transfer function of the circuit are used to obtain the transformation sensitivity matrix, which yields a linearized model of the relation between constraint and design parameters. A top-down hierarchical transformation example shows an application of our method. The method is easily extended to any constraint that can be formulated with an analytic expression.

5. REFERENCES

[1] C. A. Makris and C. Toumazou, "Analog IC Design: Part II - Automated Circuit Correction by Qualitative Reasoning", *IEEE Trans. on Computer Aided Design*, vol. CAD-14, n. 2, pp. 239–254, February 1995.

[2] H. Chang, E. Felt and A. L. Sangiovanni-Vincentelli, "Top-Down, Constraint-Driven Design Methodology Based Generation of a Second Order Σ-Δ A/D Converter", in *Proc. IEEE Custom Integrated Circuit Conference*, pp. 533–536, May 1995.

[3] U. Choudhury and A. L. Sangiovanni-Vincentelli, "Automatic Generation of Parasitic Constraints for Performance-Constrained Physical Design of Analog Circuits", *IEEE Trans. on Computer Aided Design*, vol. CAD-12, n. 2, pp. 208–224, February 1993.

[4] E. Malavasi, E. Charbon, E. Felt and A. L. Sangiovanni-Vincentelli, "Automation of IC Layout with Analog Constraints", *IEEE Trans. on CAD*, vol. 15, n. 8, pp. 923–942, 1996.

[5] G. B. Arsintescu and R. Otten, "Constraints Space Management for the Layout of Analog IC's", in *Design Automation and Test in Europe Conference*, 1998.

[6] E. Charbon, P. Miliozzi, E. Malavasi and A. L. Sangiovanni-Vincentelli, "Generalized Constraint Generation in the Presence of Non-Deterministic Parasitics", in *Proc. IEEE International Conference on Computer Aided Design*, pp. 187–192, November 1996.

[7] J. Silva-Martinez, M. Steyaert and W. Sansen, *High-Performance CMOS Continuous-Time Filters*, Kluwer Academic Publ., Boston, MA, 1993.

[P] \ [K]	[P₀] \ [K₀]	[ΔP] \ [ΔK]	ω_0 10.7MHz 100KHz	BW 500Khz 50Khz	Q 20 4	$|H(\omega_0)|$ 0dB 2dB
g_m	$300\mu A/V$	$[-3.68, 7.4]\mu A/V$	$A\frac{2g_m+g_{m,\Delta}}{g_m}$	0	$D\frac{2g_m+g_{m,\Delta}}{g_m}$	$F\frac{g_{m,\Delta}}{g_m^{3/2}}$
$g_{m,\Delta}$	0%	$[-2.9, 1.96]\%$	$A\frac{g_m}{g_{m,\Delta}}$	0	$D\frac{g_m}{g_{m,\Delta}}$	$-F\frac{\sqrt{g_m}}{g_{m,\Delta}}$
C	4pF	$[-0.94, 0.94]pF$	$-B\frac{2C+C_\Delta}{C}$	$-\frac{g_2}{C(C+C_\Delta)^2}$	$-E\frac{C+C_\Delta}{C}$	$G\frac{C_\Delta}{C^{3/2}}$
C_Δ	0%	$[-1.2, 1.56]\%$	$-B\frac{C}{C_\Delta}$	$-\frac{g_2}{C_\Delta(C+C_\Delta)^2}$	$-E\frac{1}{C_\Delta}$	$-G\frac{\sqrt{C}}{C_\Delta}$
g_{m1}	$15\mu A/V$	$[-1.97, 1.5]\mu A/V$	0	0	0	$\frac{H_0}{g_{m1}^2}$
g_2	$15\mu A/V$	$[-1.74, 1.74]\mu A/V$	0	$\frac{1}{g_2(C+C_\Delta)}$	$-\frac{Q}{g_2}$	$-\frac{H_0}{g_2^2}$

Table 3: Transformation and generation matrix

Maximally Routable Switch Matrices for FPD Design *

Guang-Min Wu and Yao-Wen Chang

Department of Computer and Information Science, National Chiao Tung University, Hsinchu, Taiwan ROC

Abstract

An FPD switch matrix is said to be maximally routable *if it has the maximum routing capacity among all switch matrices of the same size. In this paper, we present two classes of most economical maximally routable switch matrices.*

1 Introduction

Field-Programmable Devices (FPDs) refer to any digital, user-configurable integrated circuits used to implement logic functions. Due to their short production time and low prototyping cost, FPDs have become a very popular alternative to realizing logic designs. Figure 1 shows the architectures of major commercially available FPDs. As illustrated in Figures 1(a) and (b), a *Field-Programmable Gate Array (FPGA)* consists of an array of logic modules that can be connected by general routing resources. The logic modules contain combinational and sequential circuits that implement logic functions. The routing resources consist of horizontal and vertical channels and their intersection areas. An intersection area of a horizontal and a vertical channels is referred to as a *switch module*. A net can change its routing direction via a switch module; this requires using at least one programmable switch inside the switch module. A large circuit that cannot be accommodated into a single FPGA is divided into several parts; each part is realized by an FPGA, and these FPGAs are then interconnected by a *Field-Programmable Interconnect Chip (FPIC)* (see Figure 1(c)). In a *Complex Programmable Logic Device (CPLD)*, logic modules are surrounded by *continuous* horizontal and vertical routing tracks (see Figure 1(d)). Similar to FPGAs, an intersection area of a horizontal and a vertical channels in an FPIC or a CPLD is also referred to as a switch module.

Switch modules are the most important component of the routing resources in FPDs. Studies by [5, 7, 9] have shown that the higher the routability of the switch modules, the smaller the track count is needed to achieve 100% routing completion; that is, increasing the routability of a switch module also improves the area performance of a router. Therefore, it is of significant importance to consider switch-module design. In current technologies, FPD programmable switches usually consume a large amount of area. Due to the area constraint,

---

*This work was partially supported by the National Science Council of ROC under Grant No. NSC-87-2215-E-009-041.

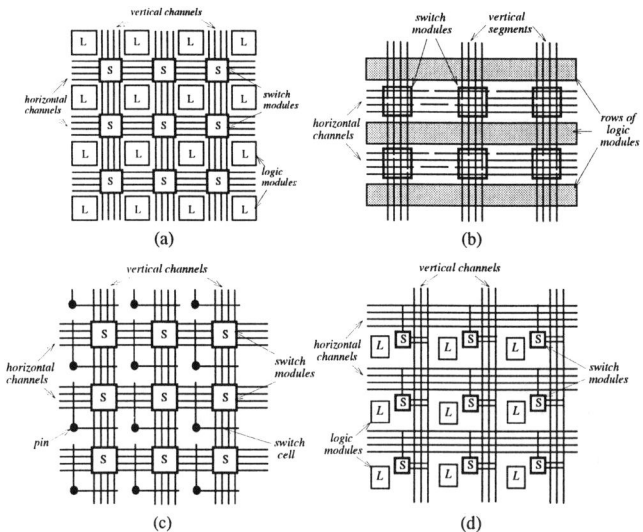

Figure 1: Major FPD architectures. (a) The symmetric-array-based FPGA model. (b) The row-based FPGA model. (c) The FPIC model. (d) The CPLD model.

the number of switches that can be placed in a switch module is usually limited, implying limited routability. Therefore, there is a basic trade-off between routability and area for switch-module architectures.

There are two types of switch modules in commercially available FPDs, *switch matrices* and *switch blocks*. (See Figure 2 for their models.) The effects of switch-module architectures on routing for the symmetric-array-based FPGAs were first studied experimentally by Rose and Brown [9]. A theoretical study of flexibility and routability was later presented based on a stochastic model [5]. The primary conclusion in both of the studies in [5, 9] is that high pin-to-track connectivity combined with relatively low switch-module connectivity is a better solution to the routability and area trade-off. Therefore, the architecture of a switch module is of particular importance, due to a relatively small switch population in a switch module. Chang, Wong, and Wong [7] proposed a class of high-routability switch blocks and analyzed three types of well-known switch blocks; they showed theoretically and experimentally that switch blocks with higher routability usually lead to better area performance, which confirms the findings by [5, 9].

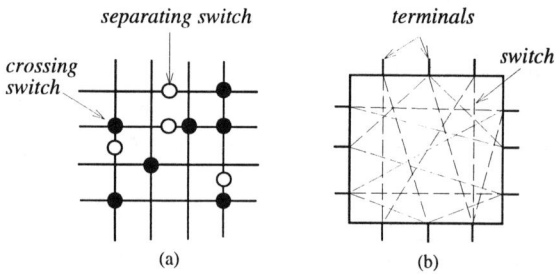

Figure 2: Switch-module models. (a) Switch matrix. (b) Switch block.

In this paper, we focus on switch *matrices*. Not much work has been reported on switch-matrix design. Zhu, Wong, and Chang in [12] first explored the feasibility conditions for switch matrices and presented a design heuristic based on a stochastic approach. Chang, Wong, and Wong in [6] applied a network-flow based heuristic for switch-module design. Sun *et al.* in [10] studied the effects of using the two switch-module architectures—switch matrices and switch blocks—on routing. Based on the study in [10], an FPGA/FPIC with switch matrices in general needs fewer switches but more routing tracks for routing completion than that with switch blocks. This work shows the trade-offs in using the two types of switch modules.

A switch matrix is said to be *maximally routable* (*MR* for short) if it has the maximum routing capacity[1] among all switch matrices of the same size. In this paper, we present most economical MR switch matrices for two classes of switch-matrix models, one with no separating switches and one with separating switches.[2] For the model with no separating switches, we show that *track-covering*[3] switch matrices are MR. It needs only w switches to construct a track-covering switch matrix of size w, compared to a fully populated switch matrix which has w^2 crossing switches. We prove that no switch matrix of size w with less than w switches can be MR. For the model with separating switches, we present *diagonal*[4] switch matrices and show that they are MR. The diagonal switch matrix of size w has only $6w-8$ ($6w-9$) crossing switches and $8w-12$ separating switches if w is even (odd), $w > 1$, compared to a fully populated switch matrix which has w^2 crossing switches and $2w^2-2w$ separating switches. We prove that no switch matrix with less than $6w - 8$ ($6w - 9$ if w is odd) crossing switches and $8w - 12$ separating switches can be MR.

---
[1] A formal definition of routing capacity is given in Section 2.
[2] A formal definition of separating (and crossing) switches is given in Section 2.
[3] A formal definition of track-covering switch matrices is given in Section 3.
[4] A formal definition of diagonal switch matrices is given in Section 4.

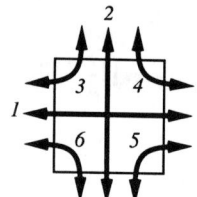

Figure 3: Six types of connections.

2 Problem Formulation

A *switch matrix* is a $w \times w$ square block with w terminals on each side of the block. It consists of a grid of horizontal and vertical tracks. There are two types of switches in a switch matrix, *crossing switches* and *separating switches*. (See Figure 2(a).) If a crossing switch at the intersection of a horizontal and a vertical tracks is 'ON,' the two tracks are connected; if it is 'OFF,' the tracks are not connected and thus are electrically non-interacting. If a separating switch on a track is 'OFF,' the track is split into two electrically non-interacting routing segments so that the terminals on opposite sides can be used independently; if it is 'ON,' the track becomes a single electrical track. In Figure 2(a), the crossing switches are represented by solid circles and the separating switches by hollow circles. Switch matrices are used in various symmetric-array-based FPGAs [4, 11], row-based FPGAs [1, 8], FPICs [3], and CPLDs [2].

A *connection* is an electrical path between two terminals on different sides of a switch module. Connections can be of six types, each of which is characterized by two sides of a module, as shown in Figure 3. For example, type-6 connections connect terminals on the left and the bottom sides of a module. Type-1 and type-2 connections are *straight connections* whereas the others are *bent connections*.

A *routing requirement vector* (RRV) $\vec{n}$ is a six-tuple $(n_1, n_2, \ldots, n_6)$, where n_i is the number of type-i connections required to be routed through a switch module, $0 \leq n_i \leq w$, $i = 1, 2, \ldots, 6$. An RRV $\vec{n}$ is said to be *routable* on a switch module S if there exists a routing for $\vec{n}$ on S. For example, the RRV $(0, 1, 0, 1, 1, 1)$ is routable on the switch matrix shown in Figure 4 by programming the switches 1, 2, 3, and 7 to be ON, and a routing solution is illustrated by the thick lines; on the other hand, the RRV $(2, 2, 1, 0, 1, 0)$ is not routable on the switch matrix.

The *routing capacity* of a switch module S is referred to as the number of distinct routable vectors on S; that is, the routing capacity of S is the cardinality $|\{\vec{n} | \vec{n}$ is routable on $S\}|$. The *maximally routable* (*MR* for short) switch matrix is defined as follows:

Definition 1 *A switch matrix is said to be* maximally routable *(MR) if it has the maximum routing capacity among all switch matrices of the same size.*

Since an MR switch matrix has the maximum routing

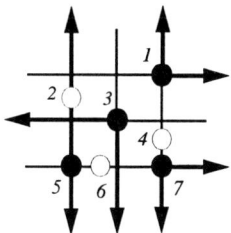

Figure 4: Example of a switch-matrix routing.

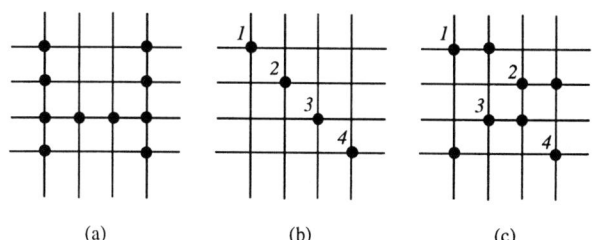

Figure 5: Switch matrices. (a) A simple, non-track-covering switch matrix. (b)(c) Two simple, track-covering switch matrices.

capacity, it is desirable to find such a switch matrix. In the following two sections, we present most economical MR switch matrices for two classes of switch-matrix models, one with no separating switches and one with separating switches.

3 Switch Matrices with No Separating Switches

3.1 Feasibility Condition

To identify an MR switch matrix, we first need to consider the feasibility conditions for switch matrices. We refer to a switch matrix with no separating switches as a *simple* switch matrix. Let Π_o denote the set of simple switch matrices. The following lemma states the feasibility constraint for an RRV to be routable on a simple switch matrix.

Lemma 1 *An RRV $\vec{n} = (n_1, \ldots, n_6)$ is routable on a switch matrix $S \in \Pi_o$ of size w only if $\max\{n_1, n_2\} + \sum_{i=3}^{6} n_i \leq w$.*

Hence, by Definition 1 and Lemma 1, an MR switch matrix in Π_o must be able to accommodate all RRVs satisfying $\max\{n_1, n_2\} + \sum_{i=3}^{6} n_i \leq w$. We proceed to explore such MR simple switch matrices.

3.2 Track-Covering Switch Matrices

A track is *covered* if there is a crossing switch on the track. A switch matrix S of size w is said to be *track-covering* if it consists of a set of w crossing switches which cover all vertical and horizontal tracks of S. The switch matrix depicted in Figure 5(a) is simple, but not track-covering, whereas the two switch matrices shown in Figures 5(b) and (c) are both simple and track-covering. A set of w crossing switches covering all vertical and horizontal tracks in each of Figures 5(b) and (c) is indicated by labels 1, 2, 3, and 4.

We have the following feasibility condition for simple, track-covering switch matrices.

Theorem 1 *An RRV $\vec{n} = (n_1, \ldots, n_6)$ is routable on a simple, track-covering switch matrix S of size w if and only if $\max\{n_1, n_2\} + \sum_{i=3}^{6} n_i \leq w$.*

Corollary 1.1 *A simple switch matrix is MR if and only if it is track-covering.*

For a switch matrix of size w, we label the tracks $1, 2, \ldots, 2w$ from top to bottom for its horizontal tracks and then from left to right for its vertical tracks. Let $\pi_h(i,j)$ ($\pi_v(i,j)$) denote a permutation of two horizontal (vertical) tracks, together with the crossing switches, if any, where $1 \leq i, j \leq w$ ($w+1 \leq i, j \leq 2w$). To identify switch matrices with the same routing capacity, we give the definition of *isomorphic* switch matrices as follows.

Definition 2 *Two switch matrices S_1 and S_2 are isomorphic if S_1 (S_2) can be obtained by performing a sequence of $\pi_h(i,j)$ and/or $\pi_v(p,q)$ operations on S_2 (S_1), where $1 \leq i, j \leq w$ and $w+1 \leq p, q \leq 2w$.*

It is easy to see that performing $\pi_h(i,j)$ and $\pi_v(i,j)$ operations on a simple switch matrix S does not affect the routing capacity of S. Hence, the following theorem holds.

Theorem 2 *Any two simple, isomorphic switch matrices have the same routing capacity.*

Based on Theorem 2, it is easy to identify the whole class of simple, track-covering switch matrices, and thus the whole class of MR switch matrices in Π_o. A simple, track-covering switch matrix of size w needs only w crossing switches on one of its diagonals. In particular, the numbers of switches are also the minimum requirement for a simple switch matrix to be MR.

Theorem 3 *No simple switch matrix of size w with less than w crossing switches can be MR.*

Therefore, the simple, track-covering switch matrix with w crossing switches on one of its diagonals is a "cheapest" MR switch matrix in Π_o. Note that the number of switches required for a simple, track-covering switch matrix is very small, compared to a fully populated switch matrix in Π_o which has w^2 crossing switches.

4 Switch Matrices with Separating Switches

4.1 Feasibility Conditions

In this section, we explore the feasibility condition of switch matrices with separating switches. Let Π_s denote the set of switch matrices with separating switches. The following lemma states the feasibility constraint for an RRV to be routable on a switch matrix in Π_s.

Lemma 2 *An RRV $\vec{n} = (n_1, \ldots, n_6)$ is routable on a switch matrix $S \in \Pi_s$ of size w only if $\vec{n} = (w, w, 0, 0, 0, 0)$ or the following set of inequalities is satisfied:*

$$n_1 + n_3 + n_6 \leq w \quad (1)$$
$$n_2 + n_3 + n_4 \leq w \quad (2)$$
$$n_1 + n_4 + n_5 \leq w \quad (3)$$
$$n_2 + n_5 + n_6 \leq w \quad (4)$$
$$n_1 + n_2 + \max\{n_3 + n_5, n_4 + n_6\} \leq 2w - 1. \quad (5)$$

Hence, by Definition 1 and Lemma 2, an MR switch matrix in Π_s must be able to accommodate all RRVs satisfying Inequalities (1)–(4) and $(w, w, 0, 0, 0, 0)$. We proceed to explore such MR switch matrices.

4.2 Diagonal Switch Matrices

For the purpose of concise description, we refer to the *sub-diagonals* of a switch matrix S as the conceptual slanted lines parallel and adjacent to the two diagonals of S. Therefore there are four sub-diagonals on each switch matrix (see Figure 6). A diagonal switch matrix D is constructed by using the following three rules:

- Rule 1: Place crossing switches on the two diagonals of D;

- Rule 2: Place crossing switches on the four sub-diagonals of D;

- Rule 3: Place separating switches between the diagonals and sub-diagonals of D.

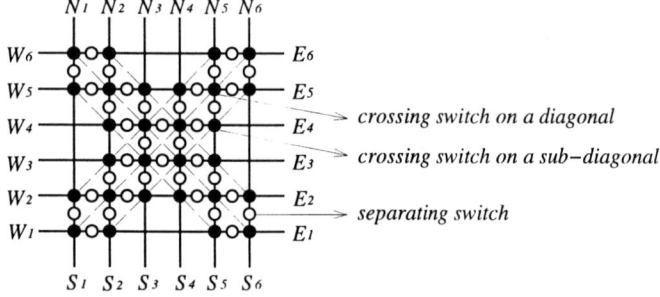

Figure 6: The diagonal switch matrix of size 6.

We have the following feasibility condition for diagonal switch matrices.

Theorem 4 *An RRV $\vec{n} = (n_1, \ldots, n_6)$ is routable on a diagonal switch matrix D of size w if and only if* $\vec{n} = (w, w, 0, 0, 0, 0)$ *or the following set of inequalities is satisfied:*

$$n_1 + n_3 + n_6 \leq w \quad (6)$$
$$n_2 + n_3 + n_4 \leq w \quad (7)$$
$$n_1 + n_4 + n_5 \leq w \quad (8)$$
$$n_2 + n_5 + n_6 \leq w \quad (9)$$
$$n_1 + n_2 + \max\{n_3 + n_5, n_4 + n_6\} \leq 2w - 1. (10)$$

Corollary 4.1 *The diagonal switch matrices are MR.*

It is easy to see that each diagonal switch matrix of size w contains $6w - 8$ ($6w - 9$) crossing switches if w is even (odd), and $8w - 12$ separating switches, $w > 1$. In particular, the numbers of switches are also the minimum requirement for a switch matrix to be MR.

Theorem 5 *No switch matrix in Π_s of size w with less than $6w - 8$ ($6w - 9$ if w is odd) crossing switches and $8w - 12$ separating switches can be MR, $w > 1$.*

Hence, the diagonal switch matrices are "cheapest" MR switch matrices in Π_s. Note that the number of switches required for a diagonal switch matrix is very small, compared to a fully populated switch matrix in Π_s which has w^2 crossing switches and $2w^2 - 2w$ separating switches.

References

[1] Actel Corp., *FPGA Data Book and Design Guide*, 1996.

[2] Altera Corp., *FLEX 10K Handbook*, 1996.

[3] Aptix Inc., *FPIC AX1024D*, Preliminary Data Sheet, Aug., 1992.

[4] AT&T Microelectronics, *AT&T Field-Programmable Gate Arrays Data Book*, Apr. 1995.

[5] S. D. Brown, J. Rose, and Z. G. Vranesic, "A stochastic model to predict the routability of field-programmable gate arrays," *IEEE Trans. Computer-Aided Design*, vol. 12, no. 12, pp. 1827–1838, Dec. 1993.

[6] Y.-W. Chang, D. F. Wong, and C. K. Wong, "Design and analysis of FPGA/FPIC switch modules," in *Proc. IEEE Int. Conf. Computer Design*, pp. 394–401, Austin, TX, Oct. 1995.

[7] Y.-W. Chang, D. F. Wong, and C. K. Wong, "Universal switch modules for FPGA design," *ACM Trans. Design Automation of Electronic Systems*, vol. 1, no. 1, pp. 80–101, Jan. 1996.

[8] A. El Gamal, et al., "An architecture for electrically configurable gate arrays," *IEEE J. Solid-State Circuits*, vol. 24, no. 2, pp. 394–398, Apr. 1989.

[9] J. Rose and S. Brown, "Flexibility of interconnection structures for field-programmable gate arrays," *IEEE J. Solid State Circuits*, vol. 26, no.3, pp. 277–282, Mar. 1991.

[10] Y. Sun, T. -C, Wang, C. K. Wong, and C. L. Liu, "Routing for symmetric FPGAs and FPICs," in *Proc. IEEE/ACM Int. Conf. Computer-Aided Design*, pp. 486–490, Santa Clara, Nov. 1993.

[11] Xilinx Inc., *The Programmable Logic Data Book*, 1996.

[12] K. Zhu, D.F. Wong and Y.-W. Chang, "Switch module design with application to two-dimensional segmentation design," in *Proc. IEEE/ACM Int. Conf. Computer-Aided Design*, pp. 481–486, Santa Clara, Nov. 1993.

FAULT EMULATION WITH OPTIMIZED ASSIGNMENT OF CIRCUIT NODES TO FAULT INJECTORS

Reza Sedaghat-Maman

Institute of Microelectronic Systems
University of Hanover
Callinstr. 34, D-30167 Hanover, Germany
sedaghat@ims.uni-hannover.de

ABSTRACT

Fault injection into an optimized circuit is made possible with the introduction of additional logic called Fault Injectors, which are controlled by a Fault Activator. In order to attain an optimum utilization of FPGA resources a novel technique is presented for the assignment of nodes and corresponding Fault Injectors in the matrix form of the Fault Activator.

1. INTRODUCTION

A new approach to design verification defined as Logic Emulation involves the development of a reprogrammable prototype of a digital circuit. With the use of logic emulators a "real-time" speed can be reached which is 10^3 to 10^6 faster than software simulation [6]. Another advantage of the logic emulator over logic simulators is that it can be connected to the target system as a prototype of the digital circuit. With the application of "real-time" fault injection it is possible to verify the function of the tested system and its reaction to the inserted fault.

A novel approach to fault injection using a commercial emulator was introduced in previous papers [1] [2]. The expansion of a circuit using Fault Injectors and a Fault Activator resulted in an overhead of FPGA resources 3-4 times higher than in the mapped circuit without additional logic. The method here explains how to attain the optimal usage of FPGA-resources for fault emulation in order to reduce this overhead.

Two other methods of fault simulation with logic emulators i.e. Fault Emulation have recently been introduced and are referred to the Fault Grading Method [3] and Serial Fault Emulation SFE [4].

This paper is organized as follows: Section 2 defines the assignment of circuit nodes to Fault Injectors. Section 3 details the experimental results. Section 4 concludes this paper.

2. ASSIGNMENT OF CIRCUIT NODES

Fault Injectors [1] are implemented for modelling Stuck-at-zero and Stuck-at-one faults [5]. The fault injection procedure involves inserting the Fault Injector into the cut nodes. The two signals L_i and C_j control the Fault Injectors. The node in the circuit where the fault is to be inserted is represented by the data in/out variables N_{in} and N_{out}. A further signal EN ($EN = L_i \cdot C_j$) determines the activation/deactivation of the Fault Injectors. A two-dimensional array structure connects the Fault Injectors, reflecting the matrix form of the Fault Activator. The actual control of the Fault Injectors is carried out by the Fault Activator, which in turn is addressed and controlled by x- and y-decoders. The matrix form of the Fault Activator consists of a line decoder and a column decoder. The additional variable MUX controls the Fault and Good Emulation process through the activation/deactivation of these decoders.

High CLB-usage [13] results when circuit nodes are assigned randomly to Fault Injectors. An optimized assignment of nodes to Fault Injectors in the Fault Activator leads to a reduction in CLB-overhead as well as improved FPGA partitioning, mapping, placement and routing.

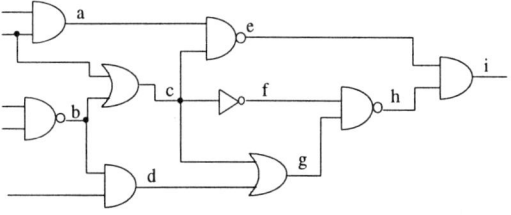

Fig. 1: Circuit without Fault Injectors

The circuit in Fig. 1 is shown in Fig. 2 as an Extra-Node-Graph [7] in which the logic is represented as Instance-Nodes and the nets as Extra-Nodes.

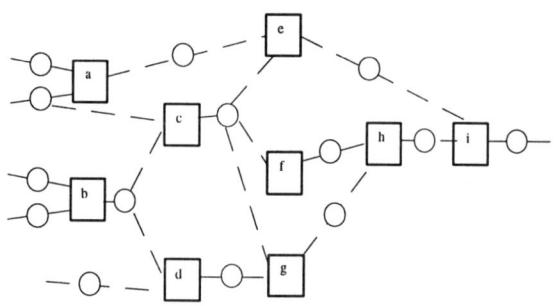

Fig. 2: EXTRA-NODE-GRAPH of Fig. 1

The circuit of Fig. 1 is shown in Fig. 3 with inserted Fault Injectors (S-a-0) not connected to the Fault Activator.

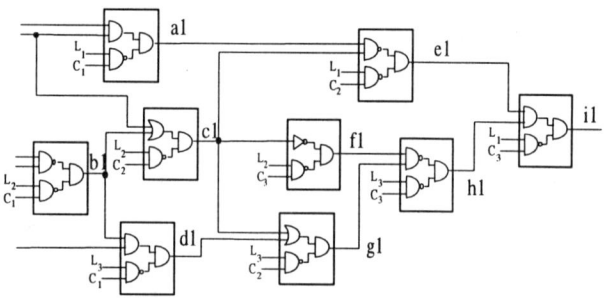

Fig. 3: Circuit with Fault Injectors

Fault Injectors must be connected to eachother in such a way that an optimal partitioning and mapping of the circuit in the logic emulator results. This is an important prerequisite for the later compilation of the circuit in the logic emulator.

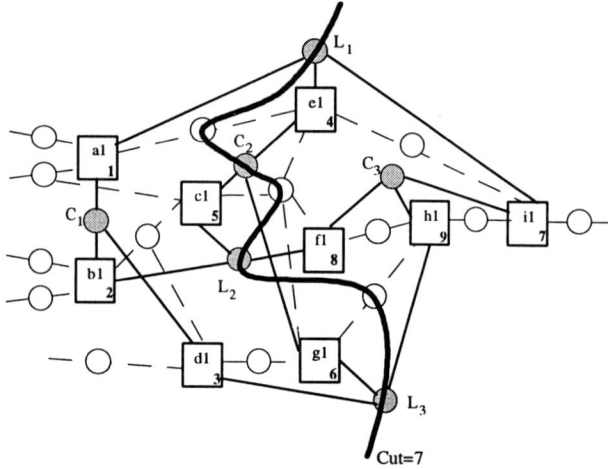

Fig. 4: Partition with optimized assignment and Cut=7

When the circuit with Fault Injectors is represented as a graph (fig. 4) it is evident that an optimal connection of Fault Injectors is reached with an almost equal quantity of instance nodes in each partition. In this case, two partitions with 4 and 5 nodes each resulted in the optimal partitioning of the circuit with 7 cuts. The optimal cut is attained with as few cuts as possible in the Fault Activator net nodes, as shown in Fig.4. Here net C_1 as well as C_3 is in the partition without a cut.

A partitioning of the circuit becomes increasingly difficult in the case of a random, less than optimal connection of Fault Injectors (L_i, C_j). In the case of a random connection of the Fault Injectors a partition with 9 cuts results, as shown in Fig. 5.

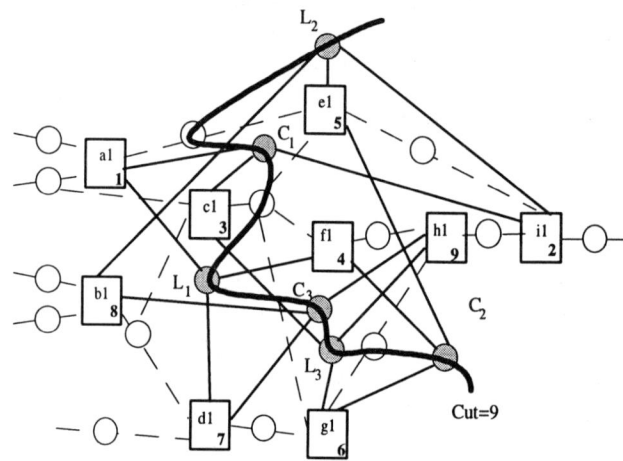

Fig. 5: Partition with random assignment and Cut=9

The following algorithm is suggested for assigning circuit nodes to Fault Injectors:

The Fault Activator with Fault Injectors is modelled as a two-dimensional array (Fig.6,7). The circuit nodes are assigned to all Fault Injectors where faults are to be inserted such that the neighboring nodes are always mapped to the neighboring Fault Injectors in the Fault Activator. The distance between two neighboring Fault Injectors in the Fault Activator is measured as one unit. Assignment is done by minimizing the total wire length λ [10] of the connections (Fig.7). When the nodes (circuit in Fig.1) are assigned randomly to Fault Injectors in the matrix style connections of the Fault Activator a higher wire length λ results (Fig.6) than from an optimized assignment of the nodes to the Fault Injectors (Fig.7).

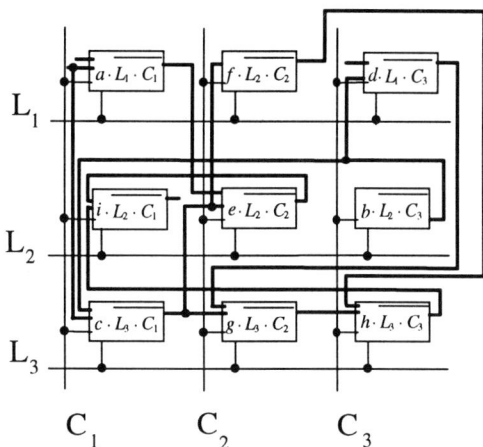

Fig. 6: Random Assignment of Fig. 1 example with Wire Length λ=20

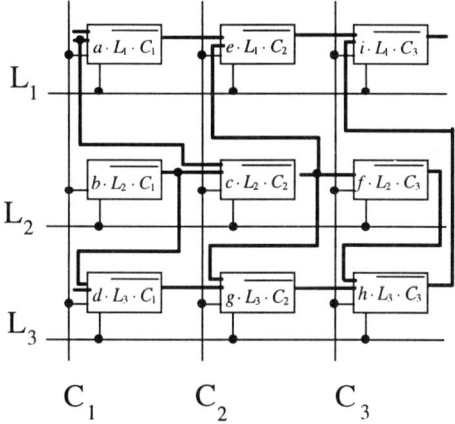

Fig 7: Optimized Assignment of Fig. 1 example with Wire Length λ=14

3. EXPERIMENTAL RESULTS

Fault Emulation results were evaluated using the FPGA (4013 series) from XILINX and Quickturn´s M250 emulation system. Table 1 describes the ISCAS ´85 benchmark circuits and a decoder (32k) for which a test evaluation was performed. When the circuits were emulated with the M250 the maximum emulation speed of the Logic Emulator (11.12 MHz) was reached.

Fault Emulation results are shown in Tables 2 and 3 with Stuck-at-0 (S-a-0) and Stuck-at-1 (S-a-1) Fault Injectors. Using fault collapsing, Fault Injectors were built into the circuits resulting in a reduction in the number of faults [9]. The total CLB-overhead of the expanded circuits is illustrated by the ratio of the number of CLBs(circuit with Fault Injectors) divided by the number of CLBs(circuit without Fault Injectors) and depends on the quantity of Fault Injectors i.e. the number of faults, assignment of circuit nodes and the structure of the circuit.

The min-cut algorithm [10] was utilized for the calculation of the minimal wire length for the assignment of the nodes to Fault Injectors (Fig.6 and 7). Tables 2 and 3 list the number of CLBs for the random(rdm) and optimized(min) assignment methods. Fig. 8 and 9 show the difference between a random assignment and an assignment with the min-cut algorithm. This method of assigning circuit nodes is currently being examined with other algorithms as well as with larger circuits.

Circuit	# Gate	# CLBs	Freq [MHz]
c1908	880	387	11.12
c2670	1192	598	11.12
c3540	1669	732	11.12
c5315	2307	978	11.12
c6288	2416	1047	11.12
c7552	3512	1562	11.12
32k	32754	7423	11.12

Table 1: Data and Logic Emulation of Original Circuit

Circuit	# S-a-0	# CLBs (rdm)	# CLBs (min)	CLB Overhead (rdm)	CLB Overhead (min)	Freq [MHz]
c1908	288	539	525	1.39	1.35	11.12
c2670	705	954	842	1.59	1.4	11.12
c3540	980	3097	2436	4.2	2.49	11.12
c5315	1353	1982	1824	2	1.86	11.12
c6288	5744	4754	4351	4.5	4.1	11.12
c7552	1646	2907	2752	1.86	1.7	11.12
32k	36569	23146	21053	3.6	3.27	11.12
Average				2.71	2.33	

Table 2: Fault Emulation with Stuck-at-0 Fault Injectors

Circuit	# S-a-1	# CLBs (rdm)	# CLBs (min)	CLB Overhead (rdm)	CLB Overhead (min)	Freq [MHz]
c1908	1396	1309	1258	3.38	3.25	11.12
c2670	1781	1667	1471	2.78	2.45	11.12
c3540	2040	3234	2868	4.4	3.9	11.12
c5315	3550	3435	3407	3.51	3.48	11.12
c6288	560	1337	1224	1.27	1.16	11.12
c7552	5149	6649	6294	4.25	4	11.12
32k	36555	22123	20193	3.44	3.1	11.12
Average				3.29	3	

Table 3: Fault Emulation with Stuck-at-1 Fault Injectors

The average overhead for Stuck-at-0 is 2.3 and 3 for Stuck-at-1. Fault Emulation is performed with 20K test patterns. The emulation speed Freq [MHz] can vary depending on the emulation system in use.

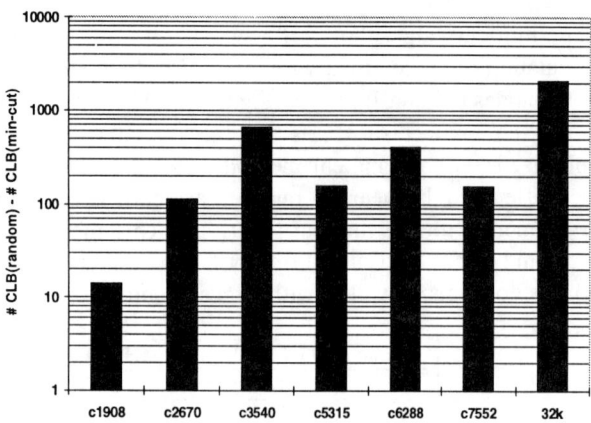

Fig. 8: Improvement in Number of CLBs with Optimized Assignment of Circuit Nodes for S-a-0

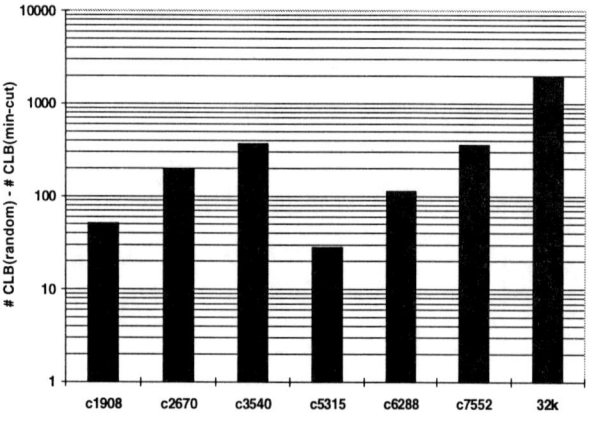

Fig. 9: Improvement in Number of CLBs with Optimized Assignment of Circuit Nodes for S-a-1

4. CONCLUSIONS

Fault injection with additional logic affects CLB-overhead. The optimized assignment of circuit nodes to Fault Injectors leads to improved partitioning of the circuit. As a result CLB-overhead is reduced. Design overhead is not a matter of concern when considering the capacity of present and future logic emulators. As shown, Fault Injectors were easily mapped to FPGAs and primary circuit inputs were minimally expanded by three.

REFERENCES

[1] R. Sedaghat-Maman, E. Barke, "Real Time Fault Injection Using Logic Emulators", ASP-DAC'98 Proc. of Asia and South Pacific Design Automation Conference, 1998

[2] R. Sedaghat-Maman, E. Barke, "A New Approach to Fault Emulation", RSP'97 Proc. of the 8th International Workshop of Rapid System Prototyping, 1997, p. 173-179

[3] K. Cheng, S. Huang, W. Dai, "Fault Emulation: A new Approach to Fault Grading", ICCAD, 1995, p. 681-686

[4] L. Burgun, F. Reblewski, G. Fenelon, J. Barbier, O.Lepapa, "Serial Fault Emulation", Proc. of the 33th Design Automation Conference, 1996, p. 801-806

[5] M. Abramovici, M. A. Breuer, A. D. Friedman "Digital Systems Testing and Testable Design", New York, W.H. Freeman and Company, 1990, p. 131

[6] U. R. Khan, H. L. Owen, J. L. Hughes, "FPGA Architectures for ASIC Hardware Emulator", Proc. 6. IEEE ASIC Conference, 1993, p. 336

[7] XILINX data book, " The Programmable Logic", 1994

[8] C.j. Alpert, A.B. Kahng, "Recent Directions in Netlist Partitioning: A Survey", Integration, The VLSI Journal Bd. 19, 1995, p. 1-93

[9] B.R. Wilkins, "Testing Digital Circuits" London, Chapman & Hall, 1994, p. 54

[10] S. M. Sait, H. Joussef, "VLSI Physical Design Automation", New York, IEEE Press, 1994, p. 142-165

STATE-SPACE TECHNIQUE FOR MINIMAL REALISATION OF ANALOGUE CIRCUITS AND SYSTEMS

H.J. Kadim
University of Hull
Department of Electronic Engineering
Hull HU6 7RX
UK
E-mail: h.j.kadim@e-eng.hull.ac.uk

T. Arslan
Cardiff School of Engineering
Cardiff University of Wales
Cardiff CF2 1XH
UK
E-mail: arslan@cardiff.ac.uk

ABSTRACT

An analogue circuit's behaviour can be represented by a number of natural modes. To ensure the compliance of the circuit to a prescribed specification, it is important to test the circuit against the performance of its natural modes. Unspecified mode's characteristics, which could be due to disturbances in the form of faults (i.e. hard or soft), can indicate abnormality in circuits specified behaviour. The work presented here is twofold: (i) an investigation of systems dynamics to estimate the reliability of the circuit when operating in the presence of faults; (ii) a proposed method for minimal realisation.

1. INTRODUCTION

The functions of an analogue system are complex, not capable of being adequately described in simple terms. The great variety of possible input and output signals, and the number of possible circuit configurations in which an analogue system may be used make it extremely difficult to determine which of the many system parameters are important. Furthermore, accurate analogue simulation has to be performed at the device level and this is very demanding of CPU time and memory. Consequently it can only practically be applied to small analogue designs, otherwise the analysis effort and time requirements would be prohibitive.

There are numerous techniques in the literature for testing analogue/mixed-analogue VLSI circuits, some of these techniques use simulation methods [1] and others use symbolic methods [2], but few attempts have been made to estimate the robustnace of these circuits to faults (for instance they may identify faults which are not important for system behaviour). Techniques which are either aimed at testing [3] or at identifying faults which have no effect on analogue circuits' behaviour [4], or aimed at testability analysis [5] generally have high computational cost even for relatively small circuits because of the number of variables and parameters involved. Faults which have no effect on a circuit's behaviour are an indication of a redundancy in the circuit [6] which is another contributor to the problem suffered by established techniques since the inclusion of a redundant part of a circuit in any analysis process will add to the overall cost in terms of time and effort. The redundancy may be the result of an accidental failure to implement minimal design in which case there is a need to eliminate such redundancy (i.e. to achieve minimal realisation); or it may be included deliberately in order to satisfy some other design criterion, in which case it is important to be identified and excluded during testing or analysis of the design.

Based on the work presented in [7], this paper is to investigate the effect of faults on the dynamic characteristics of VLSI circuits and their tolerance to such faults, and to propose a state-space based method of evaluating a minimal realisation.

2. STATE-SPACE: MATHEMATICAL BACKGROUND

The *state space* representation [6] of a system is given as follows:

$x'(t) = Ax(t) + Bu(t)$ (1)
$y(t) = Cx(t) + Du(t)$ (2)

where,
x(t): state vector (n elements); **y(t)**: output vector (m elements); **u(t)**: input vector (r elements)
A: system-interconnection matrix (n.n); **B**: driving matrix (n.r); **C**: output matrix (n.m);
D: transmission matrix (m.r)

The diagonal matrix of A (i.e. A_d) has its main diagonal representing the distinct poles of the system which can be real or complex (λ_i; i: 1, 2, ..., n).

$A_d = (\lambda_{ij})$, (3)

where $\lambda_{ij} = 0$ when $i \neq j$.

The state space vector representation of A_d is given by

$d' = A_d \, d$ (4)

(d: diagonal-system vector)

With the natural modes being $e^{\lambda_{ij} t}$

The relationship between the actual states of the system and the diagonal states is given by

$x = V \, d$ (5)

(V: a matrix with eigenvectors as columns).

Differentiating both sides

$x' = V \, d'$ (6)

The matrix V can be considered as a transducer matrix transforming the state vector d into the state vector x. The eignvector can be determined using the following equation

$A \, V_w = \lambda_w \, V_w$ (7)

where,
V_w: an eignvector of A
λ_w: the corresponding eignvalue

For undriven system, equation (1) becomes
$$x' = A x \quad (8)$$

From (4), (6) and (8)

$$A = V A_d V^{-1} \quad (9)$$

From equation (9), since V can be chosen arbitrarily, there is an infinite number of systems can be obtained, all having the same eigenvalues.

From the above it is informative that systems with different realisations may have the same natural modes, and hence their characteristics are identical. With reference to a reducible system, faults occurring in such a system may result in reconfiguring the system into its reducible realisation without effecting its natural modes (i.e. there is a nonsingular matrix such that $A = V \hat{A} V^{-1}$; $\hat{A}$: faulty system).

3. FAULTS AND EIGNVALUES

Consider a faulty system such that the fault introduced has no effect on the output.
From equations (1) and (2)

$$x' + \hat{x} = [A + \hat{A}][x' + \hat{x}] + BU$$

$$y = [c_1 \quad c_2][x' + \hat{x}] + DU$$

separating the faulty signals

$$\hat{x} = \hat{A} \hat{x}$$

$$\begin{bmatrix} \hat{x}_1 \\ \hat{x}_2 \end{bmatrix} = \begin{bmatrix} \hat{a}_{11} & \hat{a}_{12} \\ \hat{a}_{21} & \hat{a}_{22} \end{bmatrix} \begin{bmatrix} \hat{x}_1 \\ \hat{x}_2 \end{bmatrix}$$

from equation (7)

$$\begin{bmatrix} \hat{a}_{11} & \hat{a}_{12} \\ \hat{a}_{21} & \hat{a}_{22} \end{bmatrix} \begin{bmatrix} \hat{W}_{11} \\ \hat{W}_{21} \end{bmatrix} = \hat{\lambda}_1 \begin{bmatrix} \hat{W}_{11} \\ \hat{W}_{21} \end{bmatrix} \quad (10)$$

$$\begin{bmatrix} \hat{a}_{11} & \hat{a}_{12} \\ \hat{a}_{21} & \hat{a}_{22} \end{bmatrix} \begin{bmatrix} \hat{W}_{12} \\ \hat{W}_{22} \end{bmatrix} = \hat{\lambda}_2 \begin{bmatrix} \hat{W}_{12} \\ \hat{W}_{22} \end{bmatrix} \quad (11)$$

from (10)

$$\hat{a}_{11} \hat{W}_{11} + \hat{a}_{12} \hat{W}_{21} = \hat{\lambda}_1 \hat{W}_{11} \quad (12a)$$

and

$$\hat{a}_{21} \hat{W}_{11} + \hat{a}_{22} \hat{W}_{21} = \hat{\lambda}_1 \hat{W}_{21} \quad (12b)$$

from (11)

$$\hat{a}_{11} \hat{W}_{12} + \hat{a}_{12} \hat{W}_{22} = \hat{\lambda}_2 \hat{W}_{12} \quad (13a)$$

and

$$\hat{a}_{21} \hat{W}_{12} + \hat{a}_{22} \hat{W}_{22} = \hat{\lambda}_2 \hat{W}_{22} \quad (13b)$$

assume

$$\hat{W}_{11} = \alpha W_{11} \quad \text{and} \quad \hat{W}_{21} = \beta W_{21}$$

where α and β are factors representing the change in the eignvectors W_{11} and W_{21} respectively in the presence of a fault.

Substituting $\hat{W}_{11}$ in equations 12a and 12b, and re-arranging

$$W_{21} / W_{11} = (\alpha/\beta)[(\hat{\lambda}_1 - \hat{a}_{11}) / \hat{a}_{12}] \quad (14a)$$

$$W_{21} / W_{11} = (\alpha/\beta)[\hat{a}_{21} / (\hat{\lambda}_1 - \hat{a}_{22})] \quad (14b)$$

similarly

$$W_{12} / W_{22} = (\gamma/\kappa)[(\hat{\lambda}_2 - \hat{a}_{22}) / \hat{a}_{21}] \quad (15a)$$

$$W_{12} / W_{22} = (\gamma/\kappa)[\hat{a}_{12} / (\hat{\lambda}_2 - \hat{a}_{11})] \quad (15b)$$

where γ and κ are factors representing the change in the eignvectors W_{12} and W_{22} in the presence of fault. Re-arranging equations 14a and 14b, and differentiating with respect to α or β

$$0 = \hat{\lambda}_1 - \hat{a}_{11} \quad \Rightarrow \quad \hat{\lambda}_1 = \hat{a}_{11}, \text{ and } \hat{a}_{12} = 0.$$

Re-arranging equation 15a and 15b, and differentiating with respect to γ or κ

$$0 = \hat{\lambda}_2 - \hat{a}_{22} \quad \Rightarrow \quad \hat{\lambda}_2 = \hat{a}_{22}, \text{ and } \hat{a}_{21} = 0.$$

Hence

$$\begin{bmatrix} \hat{\lambda}_1 & 0 \\ 0 & \hat{\lambda}_2 \end{bmatrix}$$

The fault introduces a change in the elements of the main diagonal of the matrix which represents the eignvalues of the system. For the case of non-irreducible systems, the introduction of a fault may not effect the overall system behaviour [7] [8].

Illustration: The state space representation of a third order system is given as follows:

$$A = \begin{bmatrix} -6 & -11 & -6 \\ 1 & 0 & 0 \\ 0 & 1 & 0 \end{bmatrix} \quad B = \begin{bmatrix} 1 \\ 0 \\ 0 \end{bmatrix}$$

$$C = \begin{bmatrix} 0 & 1 & 2 \end{bmatrix} \quad D = 0$$

The system realisation is shown in figure 1.

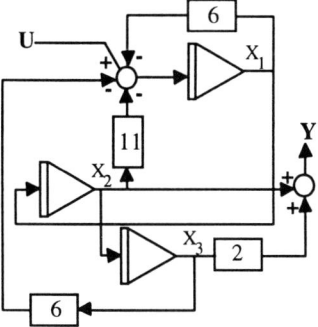

Figure 1 A third order system

$$V = \begin{bmatrix} 0.9435 & 0.8729 & 0.5774 \\ -0.3145 & -0.4364 & -0.5774 \\ 0.1048 & 0.2182 & 0.5774 \end{bmatrix}$$

The system response to a unit step input is shown in figure 2.

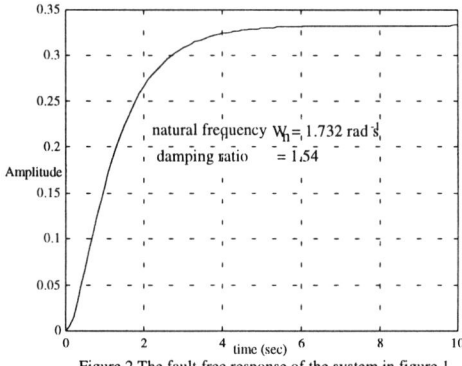

Figure 2 The fault free response of the system in figure 1

A fault is injected into the system such that the output of the integrator of the state variable x_3 is kept at 0 value. Hence, the system is reduced to a second order with only two state variables.

From equations 14a and 15b

$$\frac{W_{11}}{W_{21}} = \frac{a_{12}/\beta}{\lambda_1 - a_{11}/\alpha} \quad (16a)$$

$$\frac{W_{12}}{W_{22}} = \frac{a_{12}/\beta}{\lambda_2 - a_{11}/\alpha} \quad (16b)$$

Manipulating 16a and 16b
$\alpha = 3/2, \quad \beta = 2$
re-arranging equations 13a and 13b in terms of the eignvectors ratios, and matching the coefficients with their correspondent in equations 14a and 15b

$\hat{a}_{11} = -4, \quad \hat{a}_{12} = -3$

Repeating the above procedure for 14b and 15a

$\hat{a}_{22} = -0, \quad \hat{a}_{21} = 1$

The state space representation of the new system (i.e. faulty system) is as follows:

$$\hat{A} = \begin{bmatrix} -4 & -3 \\ 1 & 0 \end{bmatrix} \quad \hat{B} = \begin{bmatrix} 1 \\ 0 \end{bmatrix}$$

$$\hat{C} = \begin{bmatrix} 0 & 1 \end{bmatrix} \quad \hat{D} = 0$$

$$\hat{V} = \begin{bmatrix} -0.9487 & 0.7071 \\ 0.3162 & -0.7071 \end{bmatrix}$$

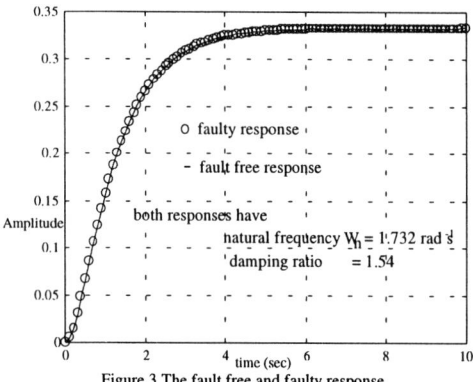

Figure 3 The fault free and faulty response of the system in figure 1

The reduced system realisation (minimal realisation) is shown in figure 4.

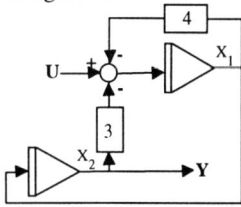

Figure 4 A reduced order system

The eignvector matrix shows that the new system has the same eignvectors ratios and the same eignvalues when compared with the original system.

For the irreducible system in figure 2, a change in any of the eignvalues as a result of a fault(s) would trigger abnormalities in the system's characteristics, as investigated in the following section.

4. FAULT INJECTION

To illustrate the effect of hard and soft faults on the system shown in figure 4, only the integrator representing the state variable x_1 is considered. The analogue representation of the integrator is shown in figure 5.

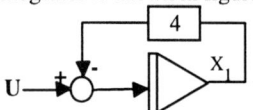

Figure 5 An integrator representing x_1

The analogue representation of the integrator in figure 5 is shown in figure 6

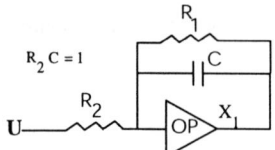

Figure 6 Analogue representation of an integrator

$$\frac{R_1}{R_2} = 4 \qquad (17)$$

(the integer '4' is the value of the feedback coefficient in figure 5)

Any slight or large variations (i.e. soft or hard faults) in either R_1 or R_2 would introduce changes in the value of the feedback coefficient and eventually affect the eignvalue of the state variable x_1. The change in the eignvalue yields a change in the system behaviour as illustrated in figure 7.

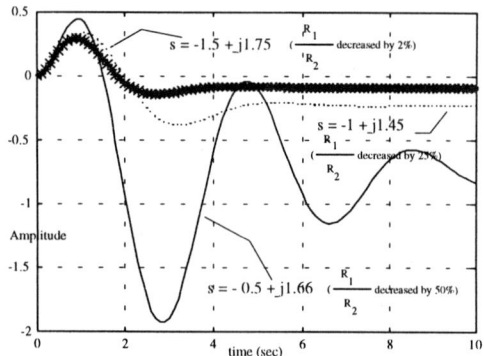

Figure 7 Changes in the behaviour of the system in figure 6 as a result of changes in the coefficient associated with the state variable x_1

With reference to figure 5, It is seen that the characteristics of the poles are severely disturbed by the changes in either R_1 or R_2. With reference to [6] the changes in the system's parameters are shown below.

System parameters	R1/R2 decreased by 2%
natural frequency Wn	1.732 rad s^{-1}
damped frequency Wd	1.75 rad s^{-1}
damping ration ξ	0.65
overshoot Mp	0.064
tp	6.4% overshoot occurs at 1.794 sec
settling time	2.637 sec
No. of oscillations necessary to reach ts	0.558 oscillations

R1/R2 decreased by 25%	R1/R2 decreased by 50%
1.733 rad s^{-1}	1.761 rad s^{-1}
1.65 rad s^{-1}	1.69 rad s^{-1}
0.29	0.28
0.38	0.39
38% overshoot occurs at 1.9 sec	39% overshoot occurs at 1.85 sec
5.97 sec	6.08 sec
1.576 oscillations	1.637 oscillations

The irreducable system obtained above represents an overdamped system, but its behaviour was drifted into oscillation prior to reaching its steady state as a result of the changes in R1 and R2 which in turn affect the state variable x_1 (i.e. the characteristics of a natural mode). If the variations in the characteristics of this mode are within a prescribed limit permitted by the specifications, then the system could be kept in operation without a noticeable compromise in performance.

5. CONCLUSIONS

Systems are vulnerable to faults. Depending on the realisation of the system under investigation, faults occurring in the system may not have an effect on the system's behaviour. Hence, systems under such faults can operate without compromising performance. The method introduced here uses an investigation of systems' dynamic which is represented by a number of natural modes, and is independent of systems' layout. Unexpected variations in a natural mode's characteristics will trigger abnormalities in systems behaviour. Recording such variations and the time span associated with these variations provides necessary information for identifying faults and estimating the reliability of systems when operating in the presence of faults.

In the presence of a redundancy in a circuit, the paper demonstrated a method which makes use of systems' dynamic related information to eliminate hardware redundancy.

6. REFERENCES

[1] M. Zwolinski, "Relaxation Methods for Analogue Fault Simulation", Proc. 20th Int. Conference on Microelectronics, Nis, Yugoslavia, 1995, pp. 467-471.

[2] Z. You, E.S. Sinencio, J.P. Gyvez, "Analogue System-level Fault Diagnosis based on a Symbolic Method in the Frequency Domain", IEEE Trans. Instrum. Meas, vol. 44, No. 1, Feb. 1995, pp. 28-35.

[3] P. Caunegre, C. Abraham, "Fault Simulation for Mixed-Signal Systems", Journal of Electronic Testing, Theory and Applications, v. 8, 1996, pp. 143-152.

[4] C. Chee, I.M. Bell, "Enhancing the Testability of Totally-Self-Checking Analogue Checkers", 3rd IEEE Mixed Signal Testing Workshop, Seattle, 1997, pp.185-192.

[5] M.Salamani, B. Kaminska, "Multifrequency Testability Analysis for Analogue Circuits", Proc. IEEE VLSI Test Symposium, 1994, pp. 54-59.

[6] G.F.Franklin, J. Powell, E. Abbas, "Feedback Control of Dynamic Systems", Addison-Wesley Publishing Company, Inc., USA,1994, , pp. 72-77 pp. 478-483.

[7] H.J. Kadim, B.R. Bannister, "A Dominant Pole Based Approach for Fault Identification in Analogue VLSI", CEEDA'96, Bournemouth,1996, pp. 418-422.

[8] H. Moor, A. Yaqub, "Linear Algebra", HarperCollins Publisher, Inc., USA,1992, pp. 376-103.

PRECISE-MD: A SOFTWARE TOOL FOR RESOURCES CONSTRAINED SCHEDULING OF MULTI-DIMENSIONAL APPLICATIONS*

J. Hua, O. Rashid, N. L. Passos, R. Halverson, R. Simpson

Department of Computer Science
Midwestern State University
Wichita Falls, TX 76308

ABSTRACT

Nested loops, usually found within the multi-dimensional (MD) computation problems, can be modeled as MD data flow graphs (MDFGs). In order to optimize such loops, a scheduling technique able to achieve parallel execution in the loop body is required. This paper presents a software system, Precise-MD, designed to solve such problems. Precise-MD allows the user to input the MD problems represented by MDFGs through a graphical interface and then applies an MD-Scheduling algorithm, OPTIMUS, which is able to obtain the shortest schedule length for a resource constrained system in polynomial time. Experiment results demonstrate the usability and application of this tool.

1. INTRODUCTION

Multi-dimensional (MD) computation, such as image processing and fluid dynamics, usually include codes which contain groups of repetitive operations represented by nested loops. In order to optimize these computations, a scheduling technique is required to change the execution order of some or all of the operations in the loop to allow them to be executed in parallel. This optimization is usually done in a resource constrained environment in which the number of functional units available for the operations is limited. This paper presents a software system, named Precise-MD (Program for REsource ConstraIned SchEduling of MD-applications), which optimizes nested loops by using the push-up scheduling technique [7].

Loop transformations are commonly used to optimize multi-dimensional problems [1, 2, 3, 5, 10, 11]. However, most of these transformations do not produce the schedule and resource allocation data necessary for the circuit design. Some existing techniques for scheduling a MD computation model do not consider the resource constraint condition [9], such as the affine-by-statement technique [3] and the index shift method [6]. The multi-dimensional rotation technique can achieve a short schedule length for each iteration in the loops, however, the optimality of the result depends on the number of algorithm iterations determined from an user input [8]. The push-up scheduling algorithm [7], which is the basis of our software system, is capable of achieving optimal scheduling for a resource constrained MD application in polynomial time. A multi-dimensional computation represented by a nested loop is modeled as a cyclic data flow graph, which is called MD data flow graph (MDFG). In an MDFG, each node represents an operation in the nested loop body. An edge between two nodes represents the dependence existing between the respective operations. Edge labels represent iteration delays. The input to the Precise-MD system is a valid MDFG. After retiming the nodes, i.e. redistributing the MD delays in the MDFG, each operation is reassigned to the earliest control step in which the corresponding functional unit is available. Any delays are updated as necessary. Finally, the resulting MDFG is returned to the user. All these operations are performed in an user friendly way, through a graphical user interface. The flow diagram of the scheduling process is shown in Figure 1.

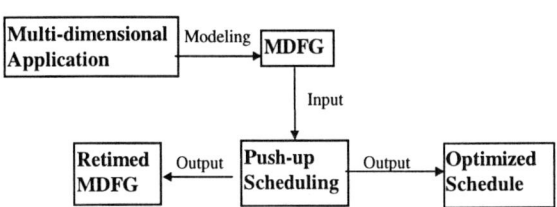

Figure 1. Flow diagram of Precise-MD.

This paper presents the characteristics of Precise-MD beginning with basic concepts in the next section. Section 3 shows the implementation details, followed by an example of the use of this tool. Section 5 summarizes the contents of this paper.

---
*This work was supported in part by the National Science Foundation under Grant NO. MIP 9704276.

2. BACKGROUND

A valid MDFG is a directed graph represented by the tuple (V, E, d, t), where V is the set of operation nodes in the loop body, E is the set of directed edges representing the dependence between two nodes, d is a function that represents the MD-delay between two nodes, and t is the time required for computing a certain node [7]. An example of a valid MDFG and its corresponding loop body is presented in Figure 2.

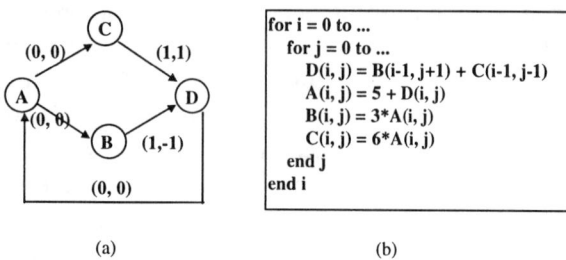

(a)   (b)

Figure 2. (a) A valid MDFG. (b) Its corresponding loop body.

The push-up scheduling technique can be applied to any valid MDFG and achieves the shortest possible scheduling length for a resource constrained MD-application in polynomial time. In this technique, there are three basic functions used in determining the optimal schedule. One is the earliest starting time (control step) for computing a node u, $ES(u)$, which can be obtained by $ES(u) = \max\{1, ES(v_i) + t(v_i)\}$, where v_i is any member of the set of nodes preceding u by an edge e_i and $d(e_i) = (0,0)$. To simplify the example, we assume $t(v_i) = 1$. The second required function is $AVAIL(fu)$, which returns the earliest control step in which the functional unit fu is available.

It has been proven that given an MDFG and an edge $e: u \to v$, such that v can be scheduled to $ES(v)$ and $d(e) = (0, 0)$, a MD retiming of u is required if $ES(v) > AVAIL(fu_v)$. This retiming allows the schedule of v at time $AVAIL(fu_v)$. This implies that a node may be rescheduled several times during the process.

A third function $MC(u)$ gives the number of extra non-zero delays required by u along any zero-delay path to u. This value is then used to calculate the actual retiming function for node u. The push-up scheduling technique generates a retiming vector $r(u)$ for each node in the MDFG in such a way that the retimed delay of each edge is given by
$$d_r(e_j) = d(e_j) + r(u)$$
where e_j is any outgoing edge of u, and
$$d_r(e_i) = d(e_i) - r(u)$$
where e_i is any of incoming edge of u. A new scheduling table is generated by reassigning u to a new control step, i.e. modifying $ES(u)$. The push-up scheduling technique is summarized in the algorithm OPTIMUS [7].

Algorithm OPTIMUS
Input: MDFG
Outputs: Retimed MDFG, Schedule Table
a) Find a schedulable node u in the given MDFG, i.e., a node that satisfies one of the following conditions:
- has no incoming edges
- has all incoming edges with non-zero MD delay
- has all its predecessor nodes with zero-delay edge already scheduled to earlier control steps

b) Schedule node u:
- assign u to the earliest control step when the required functional unit is available
- remove all the outgoing edges of u which have zero-delay
- compute the MC function of u

c) repeat the steps from (a) to (c) until every node has been scheduled.
d) compute the retiming function for each node in the MDFG by
$$\forall u \in V, r(u) = (MC_{maximum} - MC(u))*r,$$
where r is the MD retiming vector.

The essence of the push-up scheduling technique is the chained MD-retiming [9], which pushes the MD-delay from incoming edges of u to its outgoing edges.

3. THE PRECISE-MD SYSTEM

The implementation of the push-up scheduling technique was done in a PC Windows environment. The user can input the MDFG, perform the scheduling task, and output the results in an user-friendly way. The input of Precise-MD is an MDFG drawn by the user. The outputs of Precise-MD include the retiming function for the MDFG, the MC functions for each node, the final schedule table for the problem and the retimed MDFG. Snapshots of the input and output environments are shown in Figures 3 and 4.

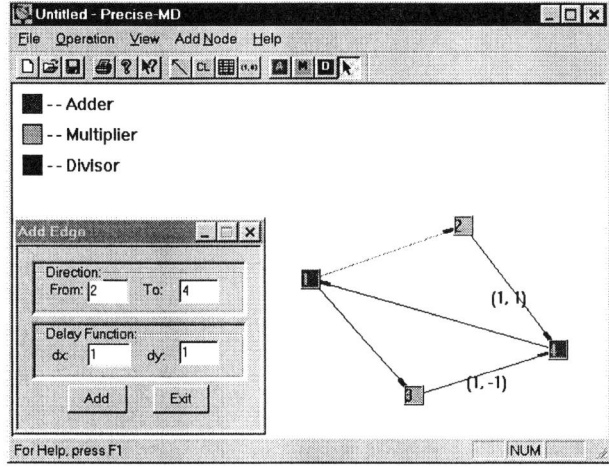

Figure 3. Input of Precise-MD.

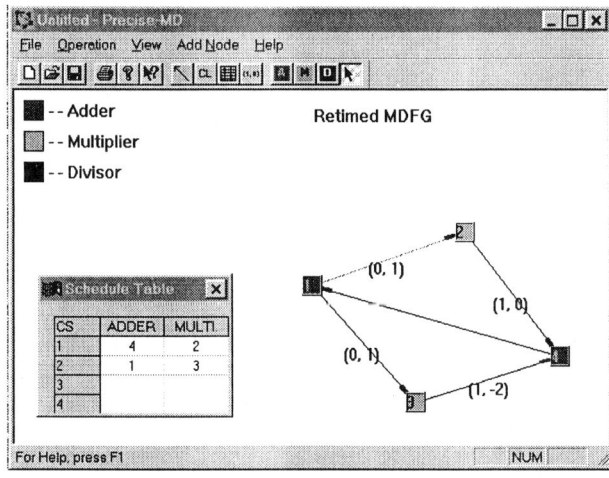

Figure 4. Output of Precise-MD.

In developing the software, the MDFGs were coded as a class in an object oriented implementation. The core of this class is a dynamic array of nodes. Each node contains a set of associated attributes, including the operation code of the node (addition, multiplication, subtraction or division), the function ES, number of incoming edges, the delay counting MC, coordinates that determine the node position in the window and a linked list of outgoing edges. This data structure is shown in Figure 5.

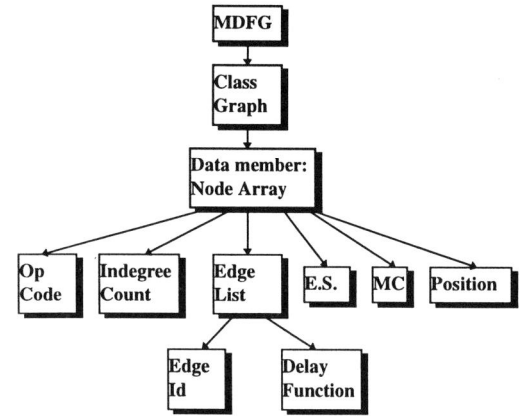

Figure 5. MDFG Data Structure.

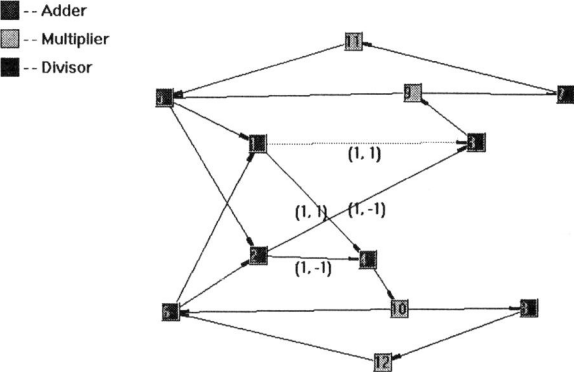

Figure 6. Wave Digital Filter MDFG.

4. EXPERIMENTS

A wave digital filter which was designed to compute the solution for a partial differential equation problem is used as our application example. The MDFG for the digital wave filter is presented in Figure 6 after applying the Fettweis transformations method [4]. In this experiment, we assume that there are two adders and one multiplier available and each of them takes one unit of time to finish one operation.

After inputting the wave digital filter MDFG to Precise-MD, a retiming vector (0, 1) is automatically generated by the system. Using this retiming vector and applying the algorithm OPTIMUS to the original graph, the MC and retiming functions for each node are obtained from the system as shown in Figure 7. Finally, Precise-MD produces a retimed MDFG and a schedule table seen in Figure 8. We notice that the final schedule length of 4 control steps obtained by Precise-MD is the optimal solution for this problem. Comparing such result to 7 control steps acquired by a traditional list scheduling algorithm solution, it is possible to verify that this is a significant reduction.

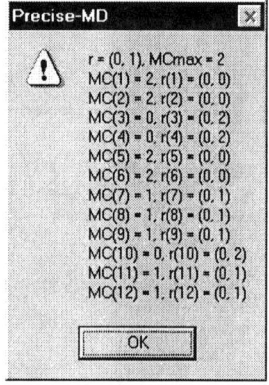

Figure 7. MC functions and retiming functions for Wave Digital Filter problem.

CS	ADDER	ADDER	MULTI.
1	3	4	9
2	7	8	10
3	5	6	11
4	1	2	12

(a)

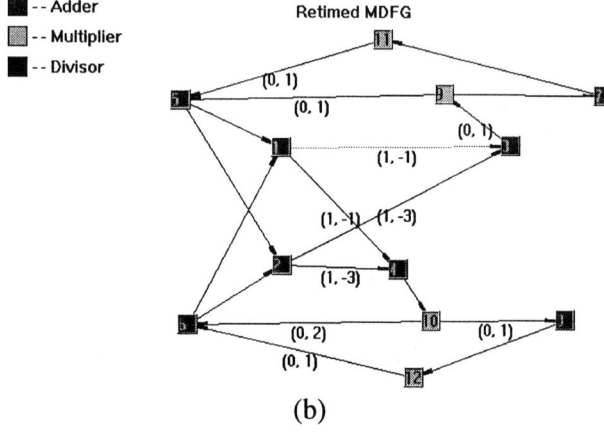

(b)

Figure 8. (a) Schedule table for the Wave Digital Filter problem (b) Retimed MDFG.

5. SUMMARY

In this study, we have developed a software tool with an user friendly interface that executes the push-up scheduling technique to optimize multi-dimensional computation models. The results show that it can achieve the shortest possible scheduling in polynomial time, confirming the theory presented in the literature. This is an efficient software tool for researchers who work in high level synthesis and digital signal processing. Further development of this program will include other scheduling algorithms in order to allow the comparison of results obtained by those different methods.

6. REFERENCES

[1] A. Aiken and A. Nicolau, "Fine-Grain Parallelization and the Wavefront Method", *Lang. and Compilers for Parallel Computing*, Cambridge, MA, MIT Press, 1990, pp. 1-16.

[2] U. Banerjee, "Unicomodular Transformations of Double Loops", *Advances in Languages and Compilers for Parallel Proc.*, Cambridge, MA, MIT Press, 1991, pp. 192-219.

[3] A. Darte and Y. Robert, "Constructive Methods for Scheduling Uniform Loop Nests", *IEEE Transactions on Parallel and Distributed Systems*, Vol. 5, no. 8, 1994, pp. 814-822.

[4] A. Fettweis and G. Nitsche, "Numerical Integration of Partial Differential Equations Using Principles of Multidimensional Wave Digital Filter", *Journal of VLSI Signal Processing*, 3, 1992, pp. 7-24.

[5] G. Goosens, J. Wandewalle, and H. De Man, "Loop Optimization in Register Transfer Scheduling for DSP Systems", *Proceedings of the ACM/IEEE Design Automation Conference*, 1989, pp. 826-831.

[6] L.-S. Liu, C.-W. Ho and J.-P. Sheu, "On the Parallelism of Nested For-Loops Using Index Shift Method", *Proceedings of the International Conference On Parallel Processing*, 1990, Vol. II, pp. 119-123.

[7] N. L. Passos and E. H.-M. Sha, "Push-Up Scheduling: Optimal Polynomial-Time Resource Constrained Scheduling for Multi-Dimensional Applications", *Proceedings of the International Conference On Computer Aided Design*, Nov., 1995, pp. 588-591.

[8] N. L. Passos, E. H.-M. Sha, and S. C. Bass, "Loop Pipelining for Scheduling Multi-Dimensional Systems via Rotation", *Proceedings of the 31st Design Automation Conference*, San Diego, CA, 1994, pp. 485-490.

[9] N. L. Passos and E. H.-M. Sha, "Full Parallelism in Uniform Nested Loops using Multi-Dimensional Retiming", *Proceedings of the International Conf. On Parallel Processing*, Saint Charles, IL, August 1994, Vol. II, pp. 130-133.

[10] M. Rim and R. Jain, "Valid Transformations: a New Class of Loop Transformations", *Proceedings of the International Conference on Parallel Processing*, Saint Charles, IL, August 1994, Vol. II, pp. 20-23.

[11] M. Wolf and M. Lam, "A Loop Transformation Theory and an Algorithm to Maximize Parallelism", *IEEE Transactions on Parallel and Distributed Systems*, Vol. 1, no. 4, 1991, pp. 452-471.

REDESIGNABILITY ANALYSIS OF DIGITAL VLSI CIRCUITS WITH INCOMPLETE IMPLEMENTATION INFORMATION

Chin-Long Wey and Mohammad Athar Khalil

Department of Electrical Engineering, Michigan State University
East Lansing, MI 48824-1226; e-mail: wey@egr.msu.edu

Abstract

This paper describes a new problem of digital circuit design —redesign of digital VLSI circuits with incomplete implementation information., and presents a solution —redesign process. Efficient algorithms are developed to derive the transfer functions of the portion with incomplete implementation information. Thus, the portion can be re-implemented using the derived transfer functions. We do not intend to discover the exact circuit schematic and components that were present in the circuit originally implemented. Rather, the functions originally intended to be present will be identical. A set of simple rules is proposed in this study to quickly analyze the redesignability of a target circuit.

1. Introduction

Of considerable interest in the design automation community is the problem of re-engineering of digital circuits. Re-engineering is the examination and alternation of a system to reconstitute it in a new form, which potentially involves changes at the requirements, design, and implementation level [1]. This paper deals with the problem of recovering the design of digital VLSI circuits with incomplete implementation information. More specifically, given a digital VLSI circuit, the original implementation information is either missing or incomplete. With the partial knowledge in the implementation, an efficient redesign process is developed to recognize the functionality of the missing/incomplete parts and to recover the original design from the existing implementation. In other words, assume that the schematic circuit diagram is given, but some parts are missing.

A circuit is *redesignable* [2] if the transfer function, i.e., inputs/outputs relationship, of each missing part can be derived. Therefore, the missing parts can be re-implemented from the derived transfer functions. Note that we do not intend to discover the exact circuit schematic and components that were present in the circuit originally implemented. Rather, the functions originally intended to be present will be identical.

Basically, a circuit can be described by a system model, *Component Connection Model* (CCM) [3,4], as shown in Figure 1(a), where *a* and *b* are the *component input* and *output variables*, respectively, and *u* and *y* are the primary inputs and outputs, respectively. A digital circuit can also be described by the CCM model, where each component may represent a block of logic gates. For the redesign problem, as illustrated in Figure 1(b), the target circuit is comprised of *Missing Parts*, as indicated by the shaded blocks, and *Known Parts*. Without loss of generality, the blocks can be re-arranged as in Figure 1(c), where B-group contains all unknown blocks, while both A-group and C-group contain the remaining known blocks. For the redesign problem, we make the following assumptions:

1. *The functionality of the target circuit is given; and*
2. *The internal structure of B-group is unknown, but its input/output nodes are given.*

Assumption 1 implies that, for any input vector *u*, the corresponding output vector *y* is attainable. Based on **Assumption 2**, B-group is equivalent to a black-box, where the external nodes are known. One trivial solution is to apply all possible combinations to inputs of B-group and probe the outputs. Here, we assume that the inputs/outputs of B-group may not be all accessible except the primary inputs/outputs.

Based on the assumptions, a target circuit can be partitioned into three groups, as shown in Figure 2, where $p_B=\{p_1,p_2,...,p_w\}$ and $q_B=\{q_1, q_2,...,q_r\}$ are the inputs and outputs of B-group, respectively. Some primary outputs, y_B, of the target circuit may be resulted from B-group. A-group takes the primary inputs, u_{PIA}, and produces the outputs, p_A, and primary outputs, y_A. Note that the inputs p_B are comprised of all elements in p_A, possibly some elements in y_A, and some primary inputs. On the other hand, the inputs of C-group include all elements in q_B, possibly some primary inputs, u_{PIC}, and some primary outputs, y_{IC}, resulted from both A-group and B-group, and the outputs of C-group are the primary outputs, y_C. The primary inputs to these three groups may have some in common. Therefore, the redesign process is to derive the functions F_j's, $j=1,2,..,r$, where $q_j=F_j(p_1, p_2,..,p_w)$, and to re-implement the functions F_j's for the *Missing Parts* in B-group. This paper presents the process for deriving the functions.

In the next section, the process that generates the transfer functions of B-group is described. Section 3 presents circuit partitioning schemes for Redesignability analysis and for reducing computational complexity of the redesign process. Finally, a concluding remark is given in Section 4.

2. Development

The major task of the redesign problem is to find the inputs/outputs relationship of the *Missing Parts* in B-group. For simplicity of presenting the developed redesign process, the example circuit, Z4ml [5], is employed, where the schematic circuit diagram can be found in [2]. The masked portion in Figure 3 represents the missing or incomplete implementation information.

The redesign process first starts with partitioning the target circuit into three groups, where B-group includes all unknown blocks, while A-group and C-group contain the remaining known blocks. In this implementation, A-group and C-group are partitioned in such a way that C-group contains those gates which takes q_i's as their inputs, and their fan-in and fan-out gates, while the remaining known blocks are included in A-group. As shown in Figure 4, B-group has $p_B=\{393,394,449,460,486\}$, $q_B=\{396\}$, and $y_B=\{26\}$. For C-group, the gates associated with q_B, or node 396, are the *NOR2* gate "289" (denoted as the gate with the output node 289), and the *NAND3* gate "476". The gate "289" has a fan-in gate, OR gate "386", and a fan-out gate, *OAI21* gate "25". Therefore, C-group has $u_{PIC}=\{2,5\}$, $y_{IC}=\{24\}$, and $y_C=\{25\}$. Furthermore, A-group includes the remaining gates, i.e., $u_{PIA}=\{1,2,3,4,5,6,7\}$, $p_A=\{393,394,449,460,486\}$, and $y_A=\{24,27\}$.

After partitioning the target circuit into three groups, the next step is to check if the outputs, $q_B=\{q_1,q_2,...,q_r\}$, of B-group are all *observable* from the primary outputs y_C. In Figure 2, the primary output vector y_C is a function u_{PIC}, q_B, and y_{IC}, i.e.,

$$y_C = G_C(u_{PIC}; \{q_1,q_2,...,q_r\}; y_{IC}) \quad (1)$$

For example, in Figure 4,

$$N_{25} = G_C(\{2,5\}; \{396\}; \{24\})$$
$$= \overline{N}_{24}[(u_2+u_5)+N_{396}]+u_2u_5N_{396}. \quad (2)$$

The observability of an input q_i of C-group, i.e., an output of B-group, can be checked if there exists an input vector which sensitizes q_i to any primary outputs in y_C. Otherwise, the outputs are not observable. For example, {396} is observable from the primary output {25}. By (2), one can generate a set of input vectors as follows.

$$N_{25}=N_{396} \text{ if } [(u_2=u_5=0) \text{ \& } N_{24}=0] \text{ or } [(u_2=u_5=1)] \quad (3)$$

Since the gates in A-group are known and thus the outputs, p_A, of A-group can be resulted from the primary inputs in u_{PIA}. Note that the inputs $p_B=\{p_1,p_2,...,p_w\}$ of B-group include all elements in p_A, and may include some elements in y_A and some primary inputs. Therefore, when an input vector is applied to A-group, the corresponding output vector including p_A and y_A results. This implies that an input vector, p_B^a, of B-group results when an input vector u^a is applied to A-group. For simplicity, we refer that p_B^a is *reachable* by u^a. An input vector of B-group is *unreachable* if it cannot be resulted from primary inputs. The following properties conclude.

Property 1.

The outputs, $q_1, q_2, .., q_r$, and y_B, of B-group are "don't cares" if the corresponding input vector is unreachable from primary inputs.

Property 2.

Let $q_B^a=(q_1^a,q_2^a,...,q_r^a)$ denote the corresponding output vector of B-group for a reachable input vector p_B^a. If q_i is not observable, then q_i^a is a "don't care."

Property 3.

Let $q_B^a=(q_1^a,q_2^a,...,q_r^a)$ denote the corresponding output vector of B-group for a reachable input vector p_B^a. If q_i is observable and the sensitized input vector is independent of $q_1, q_2, .., q_{i-1}, q_{i+1}, .., q_r$, then q_i^a can be determined by the derived primary outputs and the known primary inputs.

Property 4.

Let $q_B^a=(q_1^a,q_2^a,...,q_r^a)$ denote the corresponding output vector of B-group for a reachable input vector p_B^a. If q_i is observable and the sensitized input vector depends on $q_1, q_2, .., q_{i-1}, q_{i+1}, .., q_r$, where there exists at least one of q_j's which cannot be pre-determined, then q_i^a is undefined.

Consider an input vector $u^0=(u1,u2,u3,u4,u5,u6,u7)=(0,0,0,0,0,0,0)$. When it is applied to the target circuit, the output $y=(N_{24},N_{25},N_{26},N_{27})=(0,0,0,0)$. Here u^0 satisfies the condition in (3), and thus, by Property 3, $N_{396}=N_{25}=0$. When the input vector is applied to A-group, we obtain $(393,394,449,460,486)=(0,1,1,1,1)$. This implies that, when the inputs $(0,1,1,1,1)$ is applied to B-group, the output $N_{396}=0$ and $N_{26}=0$.

It can be easily verified that $(393,394,449,460,486)=(0,0,0,0,0)$ is unreachable. By Property 1, both N_{396} and N_{26} are don't cares. Table 1 shows all input combinations of $(393,394,449,460,486)$ and their corresponding outputs $(396,26)$, where all possible $2^7=128$ primary input combinations are simulated and the resultant inputs/outputs of B-group are tabulated. Table 1 shows that only 6 combinations are reachable. Therefore, the Boolean expressions of B-group can be derived from Table 1 as follows,

$$N_{396} = \overline{N}_{394}N_{393} + \overline{N}_{460}$$
$$N_{26} = \overline{N}_{393}\overline{N}_{394} + N_{393}N_{394}$$

VI-148

Apparently, the developed B-group may not have exactly the same topological structure as the original one. However, it can be easily verified that the circuit with the developed B-group has the same functionality as the original one. In fact, it is not necessary that the developed B-group has the same functionality as the original B-group as long as the functionality of the circuit with the developed B-group and the original circuit are the same. This leads to a way of identifying redundant nodes and gates. For example, the above expressions require only 3 inputs, i.e., 393,394,460. Thus, nodes 486 and 449 are redundant, so are the corresponding NAND gate and inverter.

3. Redesignability Analysis

The redesign process is developed with two steps: (1) redesignability check, and (2) redesign solution. The former step checks if the circuit is redesignable. If so, the second step provides a redesign solution. On the other hand, if the circuit fails the redesignability check, some other strategies may be needed to improve the redesignability [6].

The following discussions consider a reachable input vector p_B^a of B-group and its corresponding output vector $q_B^a = (q_1^a, q_2^a, .., q_r^a)$. To simplify the redesignability check process, C-group is partitioned into m blocks, as shown in Figure 5. Each block C_i may contain s_i primary inputs, say $u_{Ci} = \{u_{Cij}, j=1,2,..,s_i\} \subseteq u_{PIC}$, where $s_i \geq 0$; z_i primary outputs produced from A-group and B-group, $y_{ICi} = \{y_{ICij}, j=1,2,..,z_i\}$, where $z_i \geq 0$; and t_i outputs of q_B, $Q_{Ci} = \{q_{Cij}, j=1,2,..,t_i\} \subseteq q_B = \{q_1, q_2, ..., q_r\}$, where $t_i \geq 1$ and $t_i \leq r$. $y_{Ci} \subseteq y_C$, where n_{yCi} is the number of primary outputs in y_C and $n_{yCi} \geq 1$.

In this implementation, all blocks C_i's are sorted in an ascending order with t_i. Combining Properties 2 and 3, q_i^a can be determined if q_i is the only unknown parameter in Q_{Ci}. The following lemmas and property result.

Lemma 1.

(a) If q_i is the only unknown parameter in Q_{Ci}, then q_i^a can be determined.
(b) If $Q_{Ci} = \{q_i\}$, i.e., $t_i = 1$, then q_i^a can be determined.

Property 5.

Let $C = \{C_1, C_2, .., C_m\}$ and $q_B = \{q_1, q_2, ..., q_r\}$.
(a) If m = r, the target circuit is redesignable;
(b) If m < r, the target circuit is unredesignable;
   Let t_i^* be the number of undetermined q_i^a's in Q_{Ci} with the updating process
(c) If $t_i^* = 1$ for all i, the target circuit is redesignable;
(d) If there exists at least one $t_i^* > 1$, the target circuit is unredesignable.

To describe the above redesignability analysis process, the target circuit in Figure 3(a) is again considered, but B-group is changed to include the *OAI21* gate with output node 465, the *INV1* gate with output node 449, and the *AND2* gate with the output node 486. Thus, the B-group has the input nodes {10,287,460,393,394} and the output nodes {465,449,486}. As shown in Figure 5, C-group includes the inputs $u_{PIC} = \{u_1, u_2, u_3, u_4, u_5, u_6, u_7\}$, $Q_C = \{465, 449, 486\}$, and $y_{IC} = \{27\}$; and the outputs $y_C = \{24, 25, 26\}$. C-group is partitioned into three sub-groups, C_1, C_2, and C_3, each has $t_i = 1$. Therefore, by Property 5(a), the circuit is redesignable.

Once the circuit is redesignable, the next step is to provide a redesign solution. To derive the redesign solution, we must first check whether $p_B = \{p_1, p_2, .., p_w\}$ is reachable. One simple solution to the reachability check problem is the use of exhaustive simulation. To reduce the computational complexity, a simple circuit partitioning scheme is proposed in this implementation. A-group takes some primary inputs u_{PIA} and produces the outputs $(p_1, p_2, ..., p_w)$ and some primary outputs y_A. The exhaustive simulation approach requires to perform 2^k logic simulations on A-group, where k is the number of primary inputs in u_{PIA}. Suppose that A-group can be partitioned into g blocks, denoted as $A_1, A_2, .., A_g$. Each block A_i, as shown in Figure 6, may take a_i primary inputs, say $u_{Ai} = \{u_{Aij}, j=1,2,..,a_i\} \subseteq u_{PIA}$, where $1 \leq a_i \leq k$, and produce some outputs $p_{Ai} \subseteq p_A$ and $y_{Ai} \subseteq y_A$. If $u_{Ai} \cap u_{Aj} = \emptyset$, for any i and j, then $k = a_1 + a_2 + .. + a_g$. This implies that the number of simulations required is reduced from 2^k simulations on A-group to much smaller $(2^{a_1} + 2^{a_2} + .. + 2^{a_g})$ simulations on smaller sub-circuits of A-group.

As shown in Figure 6, A-group takes the inputs $u_{PIA} = \{u_1, u_3, u_4, u_6, u_7\}$ and produces the outputs $p_A = \{10, 287, 460, 393, 394\}$ and $y_A = \{27\}$. Thus, the exhaustive simulation approach requires $2^5 = 32$ simulations on A-group. Here, A-group is partitioned into two blocks, A_1 and A_2, where, in A_1, $u_{A1} = \{u_3, u_6\}$, $y_{A1} = \emptyset$, and $P_{A1} = \{10, 394, 460\}$; and, in A_2, $u_{A2} = \{u_1, u_4, u_7\}$, $y_{A2} = \{27\}$, and $P_{A1} = \{287, 393\}$. Here, we only need $2^2 = 4$ simulations on A_1-group and $2^3 = 8$ simulations on A_2-group.

4. Conclusion

This paper describes a new problem of redesigning digital VLSI circuits with incomplete implementation information. Given a digital circuit with incomplete implementation information, the proposed redesign process is to recover the original design from the partial known implementation information. In this study, the developed redesign process is comprised of two steps: redesignability check and redesign solution generation. Property 5 provides so simple rules for redesignability/unredesignability check. As mentioned, based on the assumptions, all target circuits are redesignable. However, some redesign solutions may end-up with redesigning the entire circuit which is costly. Therefore, by "redesignable" in this study we mean that the circuit can be redesigned at a reasonably low cost. Many other rules are being developed to comprehensively check the redesignability of given circuits.

References

1. E.J. Chikofsky and J.H. Cross II, "Reverse Engineering and Design Recovery: A Taxonomy," IEEE Software, Vol. 7, pp.13-17, January 1990.
2. C.L. Wey, "Development of Redesign Process for Digital VLSI System," Proc. of 40th Midwest Symp. on Circuits and Systems, August 1997.
3. C. L. Wey, "UUT Modeling for Digital Test - A Self-Test Approach," Proceedings of the IEEE Fourth Annual Phoenix Conference on Computers and Communication, March 1985, pp. 312 - 316.
4. C. L. Wey, "A Searching Approach Self-testing Algorithm for Analog Fault Diagnosis," in *Testing and Diagnosis of Analog Circuits and Systems*, edited by R.-W. Liu, Chapter 6, pp. 147 - 186, North-Holland, 1991.
5. MCNC Library from the 1989 International Workshop on Logic Synthesis, http://www.cbl.ncsu.edu/pub/Benchmark dirs/LGSynth89/ DOCUMENTATION/doc.txt.
6. M.A. Khalil, "Redesign Process of Digital VLSI Circuit with Incomplete Implementation Information," M.S. Thesis, Department of Electrical Engineering, Michigan State University, May 1998.

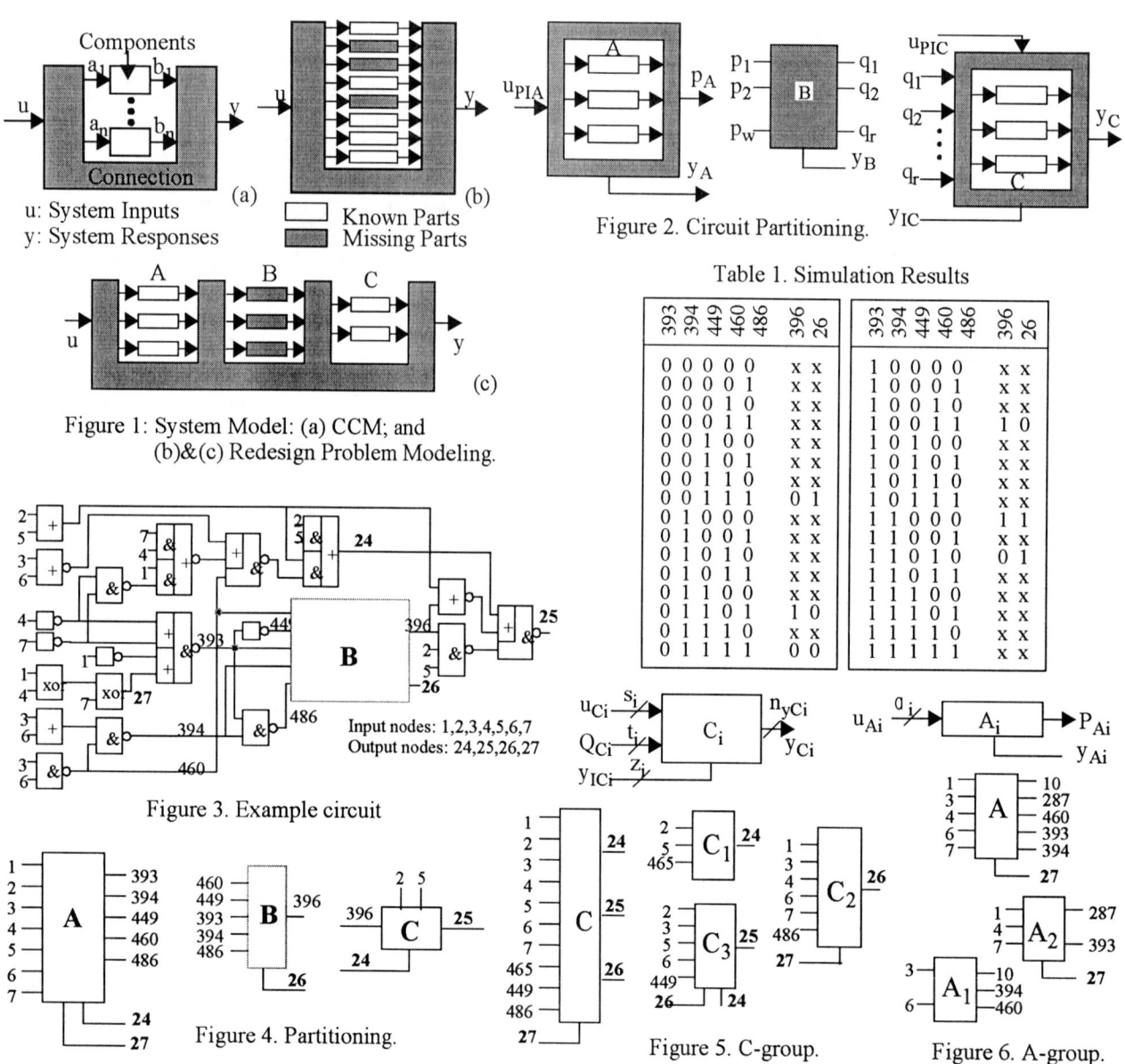

Figure 1: System Model: (a) CCM; and (b)&(c) Redesign Problem Modeling.

Figure 2. Circuit Partitioning.

Figure 3. Example circuit

Figure 4. Partitioning.

Figure 5. C-group.

Figure 6. A-group.

Table 1. Simulation Results

FUZZY MULTIOBJECTIVE DECISION MAKING ON MODELED VLSI ARCHITECTURE CONCEPTS

Hartwig Jeschke

Laboratorium für Informationstechnologie, Universität Hannover
Schneiderberg 32, D 30167 Hannover, Germany

ABSTRACT

This paper discusses a novel approach for the support of decisions on alternative architectural concepts. Multiple objectives for cost and performance are applied. Analytical performance models have been extended by *fuzzy arithmetic*. Known analytical efficiency measures are generalized by a *fuzzy multiobjective decision making* approach. The proposed modeling technique has been implemented by a new tool, the *VSP Decision Program*, which flexibly supports the high-level specification and evaluation using various performance models.

1. INTRODUCTION

The rapid and continuing improvements in semiconductor technology [7] more and more support the monolithic VLSI implementation of complex applications with high computational performance, such as real-time processing of digital video signals [11]. Such VLSI devices are produced at high volumes. Their architectural design, therefore, must aim at the best compromise with respect to multiple objectives, such as low realization costs, high computational performance, low power consumption, or high flexibility for processing of different (or enhanced) applications.

Because of the complexity of the design space, a general system-level synthesis approach does not exist for sophisticated processor architectures, like, e. g., the video signal processor reported in [6]. The functional specification and the high-level simulation of multiple alternative architectural concepts result in a high effort. Therefore in first design steps architectural alternatives are frequently investigated using parametrized and analytical models. These models focus on pipelining, parallelization, multi-level cache concepts [1, 2, 5, 10], or power consumption [12]. Most of these models consider only a single performance or cost criterion, e. g., processing time, throughput rate, speedup, or power consumption.

Other models attempt to incorporate the VLSI costs and the computational performance [3, 8, 11]. In these models an efficiency measure is defined by the ratio of modeled *performance* and *cost*, such as *throughput rate* and *silicon area*. In the case of continuous and unconstrained efficiency functions, their optima can be frequently determined by the related extremum points [5 pp. 70-73, 8]. Even for incompletely specified realization parameters, the resulting parametrized functional descriptions of the optima support parametrized explorations of the design spaces. As a disadvantage, this methodology is only applicable to a restricted class of modeling problems, e. g., data path submodules of processors. Considering complex systems on silicon, the constraints and the non-continuity of the design spaces must not be neglected. Obviously, real-time requirements strictly define lower bounds on the requested computational performance. Upper bounds of the chip sizes are defined by the yields of semiconductor manufacturing processes. Additionally, an appropriate efficiency measure should allow to flexibly incorporate more objectives than only a maximum *throughput rate* per *silicon area*. Thus, an appropriate extension and generalization of the well-established analytical architectural models is of great interest.

This paper introduces a novel and generalized efficiency measure for VLSI architectures. First, analytical architectural models are discussed (Section 2). These models are extended by *fuzzy interval arithmetic* (Section 3). In Section 4, the fulfillment of individual design objectives by their related performance or cost criteria is defined by a new fulfillment measure. A *fuzzy multiobjective decision making* approach supports the specification of a generalized efficiency from the fulfillment measures for multiple objectives. Equivalences and extensions of *fuzzy multiobjective decision making* to known modeling techniques are demonstrated in Section 5.

2. ANALYTICAL ARCHITECTURAL MODELING

Various analytical architectural models have been already defined, such as reported in [10]. In order to discuss efficiency measures for VLSI implementations, some of the most frequently applied analytical modeling functions are shown in *Table* 1, using generic types.

Assume generic functions f_c for costs and f_p for performance, which depend on some realization parameters $c_{k,i}$, and $p_{k,i}$ and on architectural parameters n_i. An efficiency measure η can be defined as

$$\eta = \frac{f_p}{f_c} \qquad (1)$$

In order to focus on the advantages of analytical models, let us consider a simple efficiency function η_1, which depends on the ratio of $f_{c,1}$ (N =1) and $f_{p,1}$ (*Table* 1). The derivation of η_1 by n_1 and

$$\frac{d\eta_1}{dn_1} = 0 \qquad (2)$$

result in the optimum architectural parameter $n_{1,opt}$:

$$n_{1,opt} = \sqrt{\frac{c_{0,1}p_0}{c_{1,1}p_1}} \qquad (3)$$

The resulting graph of *Eq.* 1 - *Eq.* 3 is shown in *Fig.* 1.

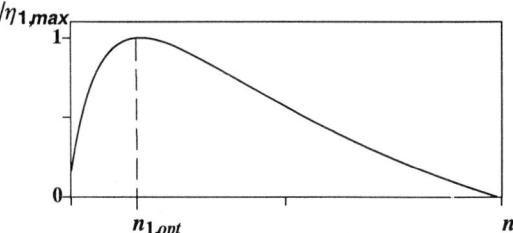

Figure 1: Normalized efficiency as a function of an architectural parameter n_1. $\eta_{1,\max}$: maximum of η_1 for all values of n_1.

In the case of a data path pipelining example [5, pp. 70–73], n_1 denotes the number of pipeline stages. First, for small values of n_1 ($n_1 < n_{1,opt}$), the efficiency increases. On the other hand, pipeline breaks and the resulting latency time more and more reduce the efficiency enhancement. At $n_{1,opt}$ the most efficient parameter value is denoted. Because of pipeline breaks, for all parameters values exceeding $n_{1,opt}$ the efficiency decreases with increasing n_1.

As another example, *Eq. 1– Eq. 3* can be applied to the unconstrained optimization of the number of parallel processor modules. Here the efficiency is defined by the ratio of the *throughput rate* and the *silicon area* [8].

Generic Functional Types	Examples
$f_{c,1} = \sum_{i=1}^{N}(c_{0,i} + c_{1,i} n_i)$	Chip area [5, 8, 11] Pipeline breaks [5 pp. 70–73]
$f_{c,2} = \sum_{i=1}^{N}(c_{0,i} + c_{1,i} n_{1,i} n_{2,i}^2)$	Power consumption [12]
$f_{c,3} = c_0 A\, e^{c_1 A}$	Chip cost per die [5] (A : die area)
$f_{p,1} = 1 / \left(\dfrac{1}{p_0} + \dfrac{1}{p_1 n_1}\right)$	Throughput on : pipelining [5] data parallelization [8] Speedup [1]
$f_{p,2} = \underset{i}{Min}(f_{p,i})$	Throughput on : pipelined submodules functional parallelization
$f_{p,3} = 1 / \sum_{i=1}^{N} 1/f_{p,i}$	Throughput on : sequential submodules

Table 1: A selection on generic functional types for architectural modeling: f_c – cost function, f_p – performance function, i – index on processor submodules, n, A – varied architectural parameters, $p_{k,i}$, $c_{k,i}$ – realization parameters.

With respect to the complexity of the VLSI design processes, at early phases of architectural designs, the full specification of all realization parameters ($p_{k,i}$, $c_{k,i}$) is frequently not possible. Because of the analytical description of the optimum parameter value $n_{1,opt}$ (*Eq. 3*), this uncertainty is no problem. The design space can be sufficiently explored by variations on realization parameter values ($p_{k,i}$, $c_{k,i}$) and by calculations of the functional description of $n_{1,opt}$.

In contrast to the previous examples, complex processor architectures consist of hierarchical compositions of multiple functional modules. An appropriate analytical efficiency measure therefore must incorporate more design goals and additional functions, e. g., $f_{p,2}$ (*Table 1*). Furthermore, constraints must be considered, like a maximum *silicon area* or a minimum *throughput rate*. The resulting efficiency η is non-continuous and constrained. Therefore, a parametrized functional description of the optimum parameter values is no longer possible.

3. EXTENSION OF ANALYTICAL MODELS

The previous section has shown for modeling of complex processor architectures, that in general parametrized functional descriptions of the optimum design parameter values cannot be expected. Thus in early design phases, the handling of uncertainties by parametrized descriptions of design optima cannot be expected, too. Nevertheless the architectural *cost* and *performance criteria* can be composed of basic functions, like the examples in *Table 1*. An appropriate measure therefore is needed, which directly incorporates uncertainties on the parameters of these functional descriptions. In general, uncertainties can be considered by *probability theory*. Because of the rapid changes of semiconductor technology as well as the complexity of VLSI design processes, these uncertainties cannot be sufficiently modeled by statistical processes.

As reported in [15], *fuzzy set theory* supports the specification of uncertainties, too. A modeling parameter x is represented by a *fuzzy set X*. The degree of membership of x in X is specified by a membership function $\mu_X(x)$. In the case of $\mu_X(x)=0$ the parameter values x are certainly not members of X. If $\mu_X(x)=1$, the parameter values x are certainly members of X. For all other parameter values the degree of membership can be continuously specified by real numbers between 0 and 1.

Fuzzy membership functions can be arbitrarily interpreted. Here, a *fuzzy set* first is considered as a measure for possibility [15]. The shape of a *fuzzy possibility measure* is frequently interpreted as an upper bound for a related probability measure [15, pp. 113–114]. Because of this relationship, *fuzzy possibility distributions* may indicate upper bounds of the expected realization characteristics of modeled technical processes.

The shape of *fuzzy membership functions* can be arbitrarily specified. In order to reduce the computational effort, *fuzzy sets* are frequently approximated by trapezoidal shapes (*Fig. 2*). The trapezoidal approximation is used in the following.

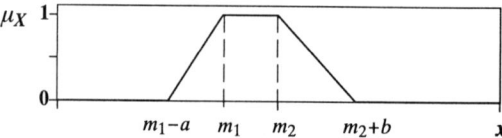

Figure 2: Approximation of fuzzy membership functions by a trapezoidal shape. A fuzzy set X is specified by characteristic parameters: $X = [m_1, m_2, a, b]$.

According to the fuzzy parameter descriptions, algebraic operations on *fuzzy sets* are of interest. Here, the *fuzzy set* theory supports algebraic operations on *fuzzy sets* by the extension principle [15], like, e. g., the *fuzzy addition* :

$$X \oplus Y = [m_{1,X} + m_{1,Y}, m_{2,X} + m_{2,Y}, a_X + a_Y, b_X + b_Y] \quad . \quad (4)$$

Because of their relation to algebraic modeling techniques on real (*crisp*) numbers, the fuzzified parameter sets are frequently considered as *fuzzy numbers* [15]. As a result, within the known analytical architectural models (e.g., those of *Table 1*), all parameters can be replaced by *fuzzy numbers*. The analytical cost and performance functions are analyzed with respect to uncertainty using the extension principle. They result in cost and and performance criteria, which are represented by *fuzzy numbers*, too.

4. GENERALIZATION OF EFFICIENCY BY FUZZY MULTIOBJECTIVE DECISION MAKING

Using the efficiency η (*Eq. 1*), an increasing *silicon area* can be compensated by an increasing *throughput rate*. The justification of a direct extension of η by additional cost factors and performance criteria cannot be expected. Obviously, a direct compensatory and unconstrained ratio of new criteria, such as *flexibility* of an architecture for processing of different applications or *power dissipation*, is not useful.

Yager has proposed a decision making approach for multiple *(M)* objectives *(MODM)*, which is based on *fuzzy sets* [14]:

$$\mu_f = \mu_{f,1}^{w_1} \cdot \mu_{f,2}^{w_2} \cdot \ldots \cdot \mu_{f,M}^{w_M}, \quad 0 \leq \mu_{f,i} \leq 1 \quad . \quad (5)$$

For each criterion i, the degree of fulfillment of its related design objective is measured by a degree of *fuzzy membership* $\mu_{f,i}$. With respect to different degrees of importance, the individual fulfillment measures can be weighted by exponents w_i. The overall degree of fulfillment μ_f is derived from the product of the weighted individual fulfillment measures (*Eq. 5*).

Yager's multiobjective decision making approach has been proposed for investigations in *design for testability* concepts [4]. However, application specific objectives for cost and performance are missing in [4]. The derivation of the fulfillment measure for individual objectives by related cost or performance criteria is not considered in [4, 14], too. Thus, the following is focused on both problems, the specification of objectives and the definition of an appropriate measure for the fulfillment of design objectives.

Considering multiobjective decision making, *fuzzy numbers* can be used for the specification of design objectives (*Fig. 3*). $\mu = 0$ denotes those cost and performance values, which are excluded by constraints. On the other hand, $\mu = 1$ denotes the functional values, which are certainly appropriate for the investigated VLSI design. All other values define the individual design goals by their degree of *fuzzy membership*. Examples for the specification of design objectives by *fuzzy numbers* are shown in *Fig. 3*.

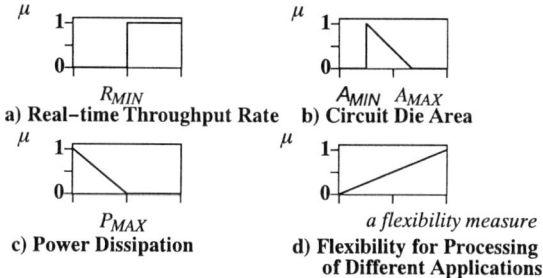

a) Real-time Throughput Rate   b) Circuit Die Area
c) Power Dissipation   d) Flexibility for Processing of Different Applications

Figure 3: Examples for the specification of design objectives (goals and constraints) by *fuzzy numbers*.

In the case of the first objective in *Fig. 3a*, the goal is always to exceed a threshold, the minimum requested throughput rate (R_{MIN}) for real-time processing. In *Fig. 3b*, a design objective for the circuit die area is defined. Because of the yield of the manufacturing process, the area must not exceed A_{MAX}. With respect to the relative costs on chip packages, chips sizes below A_{MIN} are not appropriate, too. For all other values between A_{MIN} and A_{MAX}, a minimum chip size is preferred by the *fuzzy membership* function. *Fig. 3c* considers the demand for low power design. The power dissipation must not exceed P_{MAX}. For values less than P_{MAX}, the goal in power dissipation is characterized by "as small as possible". Assuming that an architectural measure for flexibility can be defined, e. g., by qualitative *fuzzy linguistic variables* [15], a related design objective for flexibility can be specified as "as much as possible" (*Fig. 3d*). According to the specific application, which is to be implemented on silicon, it should be noted, that additional or different fuzzy design goals can be defined, too.

A solution for the specification of a measure for the fulfillment of objectives by cost and performance criteria has been proposed in [13]. First, the intersection of *fuzzy numbers* for an individual objective and its related architectural criterion, such as *chip area*, *throughput rate*, or *power consumption*, is performed by the minimum on the *fuzzy membership* functions. As shown in *Fig. 4a*, one result can be a *fuzzy set* with a trapezoidal shape (the shape of the gray-shaded area). The degree of fulfillment of the objective is specified by the ratio of the (gray-shaded) area of the intersection and the area below the shape of the fuzzy criterion. In the case of no intersection (*Fig. 4b*), the degree of fulfillment is 0. The degree of fulfillment is 1, if a design criterion is fully in accordance with the related objective (*Fig. 4c*). All other combinations of objectives and goals result in a degree of fulfillment between 0 and 1.

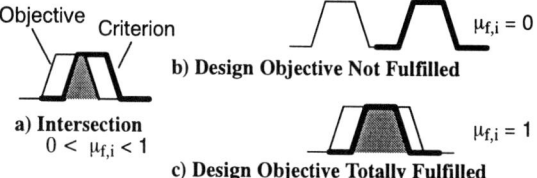

a) Intersection $0 < \mu_{f,i} < 1$
b) Design Objective Not Fulfilled $\mu_{f,i} = 0$
c) Design Objective Totally Fulfilled $\mu_{f,i} = 1$

Figure 4: The fulfillment $\mu_{f,i}$ of a design objective.

5. APPLICATION EXAMPLES

The previous sections have introduced a novel decision making approach for the investigation of VLSI based architectures, which supports multiobjective decision making for analytically modeled architectures (*Table 1*). Because of the computational requirements for the processing of *fuzzy numbers*, a novel software tool has been developed [9].

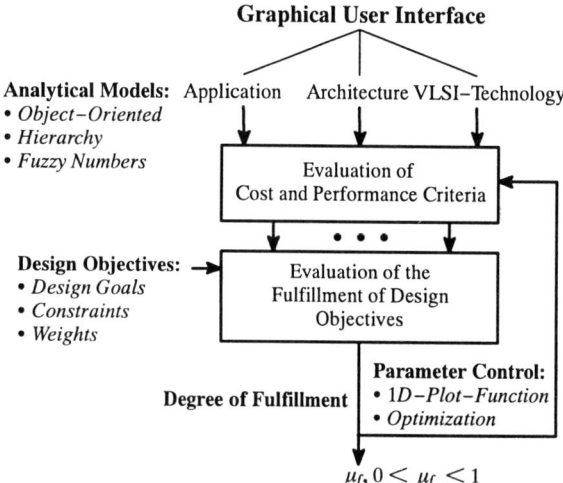

Figure 5: Modeling of architectures: *VSP Decision Program*

Providing a graphical user interface, the *VSP Decision Program* (*Fig. 5*) supports the specification of analytical models. They represent the computational requirements of the target applications, the structure of architectures, to be investigated, and the characteristics of VLSI technologies. Due to a flexible and programmable approach various analytical models can be defined. According to different functional processor modules within the processor architectures, e. g., data paths, controllers, register files, or on-chip memories, individual module types are specified by objects. The objects encapsulate the data structure and the distinct evaluation methods for each module type. A hierarchical decomposition of processor architectures into their submodules is supported, too. The basic data type used for evaluation is the *fuzzy number*. *Fuzzy numbers* are used for the specification of uncertain data as well as design objectives. They are processed according to the extension principle [15]. First, architectural cost and performance criteria are evaluated (Section 2). Then the degree of fulfillment of the design objectives is determined (Section 4).

In order to explore the design space, parameters of an investigated architecture can be varied (*Fig. 5: Parameter Control*). First the degree of fulfillment can be analyzed as a function of a single parameter (1D-Plot-Function). With respect to higher dimensionalities, the degree of fulfillment can be optimized globally for multidimensional parameter spaces by an implemented parameter search algorithm.

In order to demonstrate the capabilities of the proposed approach, in the following, the 1D-Plot-Function is applied to two examples, which are based on the generic functional types of *Table* 1. To reduce the complexity for the presentation, the parameter n_1 is varied, all other parameters remain constant. The first example shows the equivalence of the known analytical modeling techniques and the *fuzzy multiobjective decision making* approach. The second example demonstrates, the first step to consider multiple objectives. The weights w_i are set to 1.

In the case of the first example *(Fig. 6)*, a multiobjective analysis is performed on $f_{c,1}$ and $1/f_{p,2}$ as functions of the parameter n_1. Fig. 6 shows the chosen objectives. The objectives denote, that the goal for both functions $f_{c,1}$ and $1/f_{p,2}$ is to reduce the criteria to the smallest possible value.

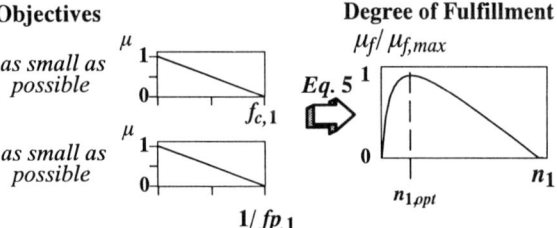

Figure 6: Multiobjective analysis for two criteria : $f_{c,1}$ and $1/f_{p,1}$ *(Table* 1*)*, μ_f – degree of fulfillment of the objectives.

Fig. 6 demonstrates that the maximum degree of fulfillment of the specified objectives results in the same optimum parameter $n_{1,opt}$, as indicated by *Eq.* 3 and *Fig.* 1. Thus, for the selected modeling criteria and objectives *(Fig.* 6*)*, the fuzzy multiobjective decision making approach is equivalent to the known analytical modeling techniques.

The second application example illustrates the extension to multiple cost and performance criteria as well as to multiple objectives *(Fig.* 7*)*. Starting with $n_1 = 1$, the degree of fulfillment of the objectives μ_f is 0. With increasing parameter n_1, μ_f remains at 0, until the minimum threshold is exceeded for all the objectives $(n_{1,a})$. Then, according to $f_{c,1}$ and $f_{c,2}$, μ_f decreases with an increasing n_1, until $n_1 = n_{1,b}$. For $n_1 > n_{1,b}$, the objectives for either $f_{c,1}$ or $f_{c,2}$ result in $\mu_f = 0$.

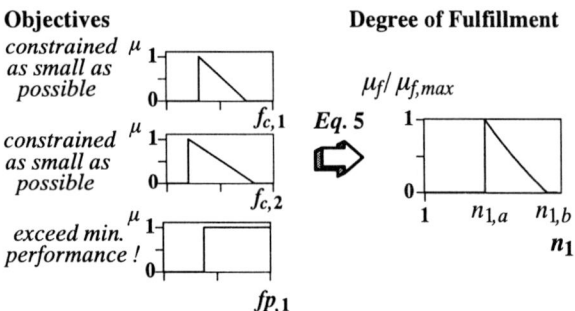

Figure 7: Multiobjective analysis on three criteria : $f_{c,1}, f_{c,2}$ and $f_{p,1}$ *(Table* 1*)*, μ_f degree of fulfillment of the objectives.

The proposed approach has been applied to investigations in the design space of complex video signal processor architectures for multimedia, e. g., the AxPe 640 V, reported in [9]. Currently a modeling library for a set of video signal processor architectures as well as video signal processing applications is under development.

6. CONCLUSION

This paper has introduced a systematic extension of analytical efficiency models on VLSI architectures. The proposed approach is based on *fuzzy set theory*, processing of *fuzzy numbers*, and *fuzzy multiobjective decision making*. The ratio of performance and cost has been extended to a new efficiency measure, which denotes the degree of fulfillment of multiple design objectives. Examples have shown that the proposed method is equivalent to known modeling techniques on efficiency. Furthermore, decision making on multiple design objectives is supported. The proposed approach has been implemented by a software tool, the *VSP Decision Program*.

REFERENCES

[1] Amdahl, G.M., "Validity of the single-processor approach to achieving large scale computing capabilities". In AFIPS Conference Proceedings, Vol. 30 (Atlantic City, N.J., Apr. 18-20). AFIPS Press, Reston, Va., pp. 483-485, 1967.

[2] Benner, R.E., Gustafson, J.L., and Montry, G.R., "Development and analysis of scientific application programs on a 1024-processor hypercube". SAND, 88-0317, Sandia National Laboratories, Feb. 1988.

[3] DFG-Final Report Pi 169/4,"Performancemodellierung von Multiprozessoranordnungen für die Echtzeitvideosignalverarbeitung". Hannover, August 1997.

[4] M. Fares. B. Kaminska, "Exploring Test Spaces with Fuzzy Decision Making". *IEEE Design & Test of Computers*, Vol. 11, No 3, 1994.

[5] M. J. Flynn, "Computer Architecture – Pipelined And Parallel Processor Design". Jones And Bartlett Publishers, 1995.

[6] K. Gaedke, H. Jeschke, P. Pirsch, "A VLSI Based MIMD Architecture of a Multiprocessor System for Real-time Video Processing Applications". *Journal of VLSI Signal Processing*, Kluwer Academic Publishers, Vol. 5, No. 2/3, , pp.159 – 169, April 1993.

[7] L. Geppert, "Solid State". *IEEE Spectrum*, Vol 33, No 1, pp. 51 – 55, 1996.

[8] H. Jeschke, K. Gaedke, P. Pirsch, "Multiprocessor Performance for Real-Time Processing of Video Coding Applications". *IEEE Transactions on Circuits and Systems for Video Technology*, Vol. 2, No. 2, pp. 221 – 230, June 1992.

[9] H. Jeschke, M. Wahle, M. Wienhöfer, "The VSP Decision Program". http://www.mst.uni-hannover.de/Forschung/Projekte/vspdecision, 1997.

[10] C. M. Krishna (Ed.), "Performance Modeling for Computer Architects". *IEEE Computer Society*, Oct. 1995.

[11] P. Pirsch, N. Demassieux, W. Gehrke, "VLSI Architectures for Video Compression – A Survey". Proceedings of the IEEE, Vol. 83, No. 2, pp. 220-246, 1995.

[12] K. Roy, R. Roy, T.-L. Chou, "Design of Low Power Digital Systems". In "Emerging Technologies". Tutorial for ISCAS 1996, Eds. R. Cavin, W. Liu, pp. 137-204, 1996.

[13] M. Wosnitza, "Modellgestützte Architekturbewertung unter Berücksichtigung unscharfer Kosten- und Performance-Parameter". Diploma Thesis, Universität Hannover, 1994.

[14] R. R. Yager, "Fuzzy Decision Making Including Unequal Objectives," Fuzzy Sets and Systems 1, North-Holland, pp. 87 – 95, 1978.

[15] H.-J. Zimmermann, "Fuzzy Set Theory and its Application". 2nd Edition, Kluwer Academic Publishers, Boston, 1991.

PARALLEL COPROCESSOR ARCHITECTURES FOR MOLECULAR DYNAMICS SIMULATION: A CASE STUDY IN DESIGN SPACE EXPLORATION

M. Gerber[1]   *T. Gössi*[2]

Swiss Federal Institute of Technology (ETH), CH-8092 Zürich, Switzerland

[1]Computer Engineering and Networks Lab, gerber@tik.ee.ethz.ch

[2]Electronics Lab, goessi@ife.ee.ethz.ch

ABSTRACT

The purpose of the paper is to describe a new semi-automated design space exploration method based on genetic programming. A new control/dataflow specification method is proposed as well as appropriate models for hardware parts and algorithms. With this method we are able to test many different hardware architectures and algorithms against cost, speed, computation time and other constraints within very short time. The remaining manual work is to exploit the model parameters of the components of the architecture and the algorithm. In contrast to other approaches our method is suited for embedded and distributed systems. The method, models and application are explained in detail by means of a comprehensive case study.

1 INTRODUCTION

Molecular systems are characterized as systems of thousands of particles (molecules or atoms) that interact with each other. Interaction is composed of several physical forces, e.g. the Van der Vaals and Coulombic force. Computer simulations are a common tool to investigate dynamic, thermal and thermodynamic properties in molecular systems. The Molecular Dynamics (MD) simulation method [1] [7] is based on numerical integration of Newton's equation of motion. Within each time step, the interaction forces between the involved particles must be determined. After the integration step the velocities of the particles are known and the new positions can be calculated for the next step. The result is a spatial trajectory for each particle. With these data and by the use of statistical physics laws, all interesting molecular properties can be evaluated. A time step typically represents two picoseconds and during a run about 100 time steps are simulated. For large systems with a lot different interaction types the simulation is very time consuming. A system with 36'000 atoms will take about 40 seconds for one time step (SunUltra1/170). The following discussion serves to explain the algorithm-architecture trade-offs in the design of fast MD simulators.

Three different acceleration techniques are in today's principal focus of research: (1) A lot of researchers all over the world make every effort to find new chemical models providing the same or even a higher numerical accuracy at a lower algorithmic complexity. (2) The MD simulation program can be parallelised for simulations on general purpose multiprocessors [4] [6]. (3) The parallelised program runs on a host machine with an attached dedicated parallel coprocessor. The coprocessor may consist of risc processors, ASIC's, or other processing elements [2] [3] [5].

One example for an algorithmic approach to reduce simulation time is the *pairlist method* [8]: The interaction is not calculated between all possible particle-pairs in the system but only between particles whose distance is smaller than a certain cutoff radius. This can be done because the interaction with a far away particle is very weak and can be neglected. Thus a so-called pairlist is generated containing all currently interacting pairs. Given the fact that we only consider simulations of liquids, the particles do not move far within one time step. Therefore, the pairlist must not be updated every iteration. With the pairlist concept the number of pairs and therefore the simulation time can be reduced significantly. Of course, a small cutoff has an impact on the accuracy of the molecular simulation. The user has to choose a trade-off between accuracy and time consumption of the simulation. A lot of other algorithmic-oriented acceleration techniques are well-known and a good simulation tool offers the user a huge amount of different algorithms.

Special purpose hardware accelerators like the MD-Grape machine [2] suffer from heavily reduced flexibility: In the MD-Grape project an ASIC has been developed to calculate the non-bonded forces. Typically, the non-bonded forces calculation requires about 80-90% of the overall simulation time. Through the use of virtual pipelines the ASIC is able to calculate one pairwise force per clock cycle. The pairlist concept is supported as well as arbitrary coefficients of the polynom for the force calculation. But, in order to keep the glue hardware simple, the symmetry of pairwise forces was neglected leading to the situation that all forces are calculated twice. In addition it is only possible to calculate forces between the same kind of particles on the ASIC. This means that if there are two different types of molecules in the system, the interactions of only one of them may be calculated on the coprocessor. A hardware overcoming these problems would be too complex to fit in one ASIC. Other special purpose hardware projects became outdated [3] as a result of the rapidly increase of computing power of even cheap computers or are very large consisting of thousands of processors supporting only the simplest algorithms. A lot of systems are targeted at general n-body problems (astronomic systems) without the possibility to exploit the advantages of the homogenous particle distribution of liquids.

The goal of our interdisciplinary research project at ETH Zürich is to accelerate MD simulations of liquids by one order of magnitude based on the GROMOS (GROningen MOlecular Simulation package) software [7]. The scope focuses on molecular systems with 10k - 100k particles with typically one or several proteins dissolved in liquid. We want to achieve our goal by using a ~10k$ workstation or PC with dedicated MD coprocessor at an estimated hardware cost of about 10k$.

Typical design trade-offs and problems to find an optimal hardware solution can be summarized as follows:

- The partitioning of the MD algorithm: Which function runs on the host, which one on dedicated hardware? This mapping leads to the major factors determining a solution, e.g. the computation power of the coprocessor, the communication latency and bandwidth, the architecture of the coprocessor, its cost and the host performance. Find the optimal hardware architecture under varying constraints such as cost, maximum computation time, and the possible mappings.

- Choosing the right algorithms. The choice depends on the mapping of a function (host or coprocessor), and their architectures, see the above discussion on algorithm-architecture trade-offs.

We developed a new semi-automated design space exploration method. With this method we are able to test many different hardware architectures and algorithms against cost, speed, computation time and other constraints within very short time. The remaining manual work is to exploit the model parameters of the components of the architecture (e.g. communication bandwidth, computation power of processors) and the tasks of the algorithm (e.g. the number of floating point operations or execution time).

The paper is organized as follows: In section 2, we describe our CAD supported control/dataflow specification method as well as the new models for algorithms, hardware components and communication requirements. In section 3 the design space exploration method using a system synthesis tool based on genetic programming (GP) is explained. In section 4 we present some results with respect to our molecular dynamics application.

2 SPECIFICATION MODELS

2.1 Modelling the Algorithm

The algorithm modelling procedure can be subdivided into four steps:

1. Specification of the algorithm using a combined data/control flow description (CDFG), see fig. 1.
2. Conversion into a simple dataflow graph (DAG) whose operations are executed iteratively corresponding to the iterative nature of the MD simulation (fig. 2, left).
3. Convert the DAG into a problem graph by adding communication nodes (white). (fig. 3, left)
4. Add *function parameters* to function and communication nodes.

For the CDFG specification we use a simplified version of the flow graph model presented by Gupta [12]. A flow graph is a polar acyclic graph consisting of operational nodes and edges for data and control dependencies. A boolean function associated with all edges determines if an edge is enabled or not. Unlike the dataflow graph the flow graph operational nodes are of different types. Our CDFG model uses only a subset of the special flow graph nodes, as listed in table 1.

Operation	Description
nop	No operation
cond	Conditional fork
join	Conditional join
loop	Hierarchical node

Table 1. Operations in the CDFG

An edge may represent a data dependency or a control dependency. Exactly one variable is associated to an edge. In our first approach, a specific variable may not appear in both edge types. A node is activated when the input expression, consisting of either AND or OR operations on the enabling booleans of the input edges, is true. On completion of the operation one or all of the booleans of the output edges are set to true. Cond and join nodes are used if there are more than one predecessor or successor node, else a nop node is used. Loop nodes indicate hierarchy, where the hierarchical element may be called once (procedure call) or multiple times (for-loop). All nodes have a functional implementation depending on the algorithm. The functional implementation is either C Code or VHDL.

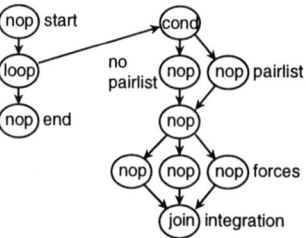

Figure 1. Sample CDFG

The CDFG description has several advantages: (1) Compared to other approaches such as a pure dataflow graph (DAG, not allowing control structures), or a synchronous data flow graph (SDF [9], a control dependent data flow), our model supports the full specification of complex parallel algorithms such as an MD simulation. (2) Our molecular dynamics simulation algorithm as most of control and data flow driven algorithms may be specified in an easy and straightforward way. (3) During design space exploration only models of the tasks are associated to the nodes, the functional implementation is done later (gradual refinement). (4) The CDFG description is independent of the partitioning into hardware and software. (5) Not only embedded systems may be specified in this way, but also distributed parallel systems, e.g. a workstation cluster. (6) The specification is easy to change, e.g. to replace algorithms with more efficient ones, or to introduce more parallelism. (7) The CDFG graphic editor and code generator are embedded in the CodeSign Tool [13].

The next steps include the conversion of the CDFG to be compliant with the design space exploration process. These tools require a problem graph (fig. 3,left) whose operations are executed iteratively corresponding to the iterative nature of the MD simulation. The problem graph is derived from a simple DAG (fig. 2, left) by adding communication nodes instead of edges. The DAG represents one possible dataflow in CDFG graph.

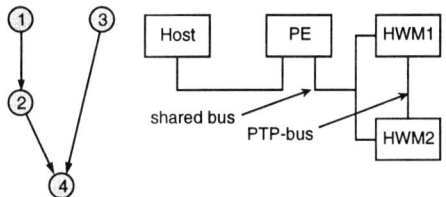

Figure 2. Dataflow graph (left) and architecture

In every step, the *function parameters* are added as follows: Three different models are available to characterize the complexity of a functional node: The number of floating or fixed point operations depends on the *problem parameters* and is a good measure of function complexity. If the function is neither floating nor fixed point dominated, the complexity is measured with a profiling tool resulting in a non-generic model. The amount of input and output data for one function is constant because dynamic data types are not supported. Thus, the communication bandwidth requirement can easily be determined and is associated to the communication nodes in the problem graph.

2.2 Modelling of Architectures

An architecture including functional resources can also be modelled with a directed graph similar to the problem graph. The architecture graph contains physical resources like processors, ASIC's, buses, etc. (fig. 2, right). Models for cost, performance and communication bandwidth are associated with the nodes similar to the problem graph. These *constraint parameters* are derived from measurement, data sheet or user specification.

The architecture graph model (fig. 3, right) is a super-set of all architectures which can be allocated. The knowledge of the designer is essential to provide suitable architectures.

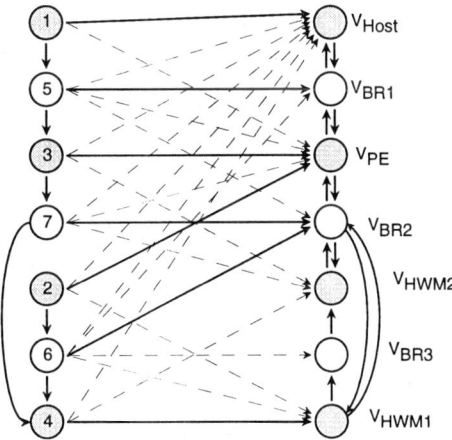

Figure 3. Specification graph

2.3 Modelling of Algorithm-Architecture Relation

This step comprises the combination of the problem graph with the architecture graph. The specification graph in fig. 3 consists of an architecture and problem graph as well as mapping edges which relate the nodes of the two neighbouring graphs. These edges describe all possible bindings of tasks to resources. The insertion of mapping edges is another step requiring the designer's knowledge.

The next step concerns optimization: Find an implementation (a feasible binding plus schedule) with minimal computation time under cost constraints.

3 EXPLORATION

Our goal during design space exploration is to find not only one, but a collection of useful architectures. Generally, the following steps are necessary to obtain one valid implementation: a) select an architecture (allocation), b) map algorithmic tasks onto its components (binding) and c) determine an appropriate schedule of the tasks (scheduling). For step a and b we use a system synthesis tool based on genetic programming (GP) [11], step c is performed with a heuristic algorithm similar to the well-known list-scheduling algorithm.

Our exploration technique takes into account: (1) The communication requirements, (2) finite computation and communication resources and (3) loop pipelining (iterative schedule).

Exploration is an iterative optimization task repeating the steps a, b, c. To start an exploration, one has to specify an optimization goal, e.g. latency minimization (period) under resource constraints (cost). The input for the first iteration is a randomly generated population of feasible bindings. The result of each step is a set of implementations, illustrated as points in fig. 4.

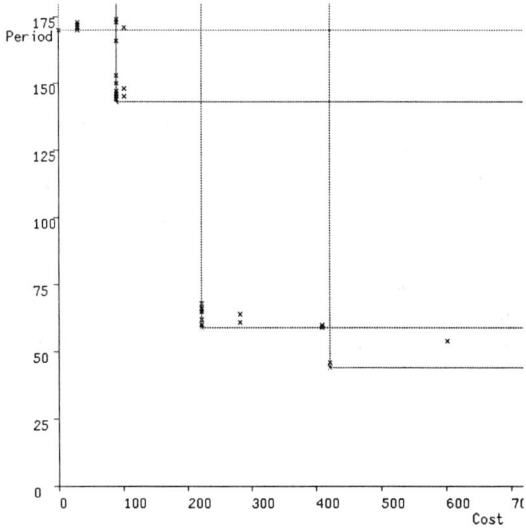

Figure 4. Pareto points

The best solutions are chosen by a *fitness function* (selection). The Pareto points indicated by lines in fig. 4 generally achieve a high score. The initial population size then is restored through recombination (crossover or mutation) in order to exploit new points in the search space. Iteration is aborted when no better implementations are found.

For all solutions in fig. 4 there exist a feasible binding and a schedule such as the one in fig. 5. Depending on cost or time constraints, one or two of the Pareto points serve as final imple-

VI-157

mentation with the appropriate mapping and schedule. Now another exploration with a different specification graph and parameters is performed in order to find more solutions.

4 THE CASE STUDY

In this chapter we present the result for the MD simulation algorithm. As an example we used a problem graph with 19 nodes and an architecture graph with 8 nodes. The problem graph represents one MD time step without the pairlist calculation (the problem graph with pairlist and all other available MD features contains 35 nodes). According to the previous section, the architecture graph is a "super architecture" connected with the host (one processor) via a bidirectional bus. The super architecture is composed as a mesh of processors containing some redundant buses.

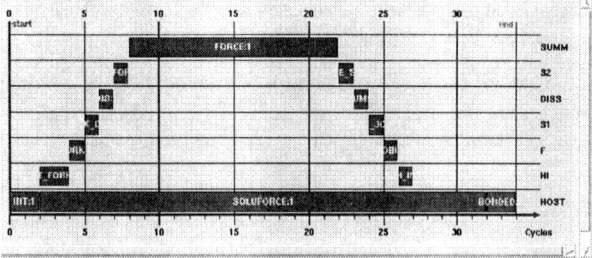

Figure 5. Schedule

The model parameters are for a SUN Ultra 1 host, a 10 MBytes per second host interface and Analog Device's Sharc DSP as coprocessors. The Sharc communication parameters are derived from the data sheets, performance models are deduced from simulations with Analog Device's Sharc simulator. Host parameters are simulation or profiling results. A Sharc processor typically has six on-chip communication channels. Consequently, a hierarchical solution seems obvious for more than five processors. The architecture graph connects a maximum number of 12 coprocessors with the host.

Host performance	Number of coprocessors	Iterations	Cost	Time
Ultra1/170=1	12	30	420	43
1	6	30	220	57
1	2	30	90	143
2	12	30	440	34

Table 2. Comparison

The first three columns in table 2 correspond to the three fastest Pareto points in fig. 4, where all valid implementations after 30 iterations are listed. The fastest implementation is a hierarchy with two Sharc's in the first level connected directly with the host and five Sharc's per first level processor in the lower level. The second column in table 2 is also a two level hierarchy, but with only one processor in the first level and five in the second. The third column is just the first level in the hierarchy of the fastest solution. The last column is derived from another exploration run with all parameters kept the same except the host performance.

The resulting fastest architecture is the same as this in the first column, the corresponding schedule is illustrated in fig. 5, where the step-by-step communication through the hierarchy is apparent. The cycle time is limited by the task *soluforce* which is forced to be executed on the host. To use the performance of 12 coprocessors we need either a faster host or we must allow the evolutionary algorithm to map soluforce to the coprocessors.

After exploring all problem graphs the most promising solutions (hierarchical DSP, ASIC processors, workstation cluster, distributed memory RISC multiprocessor) are further investigated: A better architecture model is combined with a generic model of the MD algorithm (CDFG) and implemented in Mathematica. This task is still manual but necessary if we want to simulate a real MD step. Further work includes the automation of this task and the back annotation of schedules (fig. 5) to the CodeSign tool.

REFERENCES

[1] M.P. Allen, D.J. Tildesley: *Computer Simulation of Liquids*. Oxford University Press, (1987)

[2] T. Fukushige, M. Taiji: *A highly-parallelized special-purpose computer for many-body simulations with an arbitrary ventral force: MD-GRAPE*. The Astrophysical Journal, 468: 51-61, (1996)

[3] A.F. Bakker, C. Bruin: *Design and Implementation of the Delft molecular-dynamics processor*. Special purpose computers, 183-222, Academic Press Inc. (1988)

[4] W. Scott, A. Gunzinger: *Parallel molecular dynamics on a multi signal processor system*. Computer Physics Communication 75, 65-86, (1993)

[5] H. Bekker, H.J.C. Berendsen: *GROMACS: A parallel computer for molecular dynamics simulation*. Physics Computing '92 (Conference proceedings)

[6] W. Smith: *Molecular dynamics on hypercube parallel computers*. Computer Physics Communications 62 (1991)

[7] W.F. van Gunsteren: *Biomolecular Simulation: The GROMOS96 Manual and User Guide*. Hochschulverlag vdf AG an der ETH Zürich, (1996)

[8] W.F. van Gunsteren, H.J.C Berendson: *On searching neighbours in computer simulation of macromolecular systems*. Journal of Computational Chemistry, Vol. 5, No. 3, 272-279, (1983)

[9] E.A. Lee, D.G. Messerschmitt: *Synchronous Dataflow*. Proceedings of the IEEE, 75(9): 1235-1245, (1987)

[10] M. Schöbinger, L. Thiele: *Synthesis of domain specific heterogeneous multiprocessor systems: hybrid video coding schemes*. Proc. IEEE ISCAS Conference, Atlanta, (1996)

[11] T. Blickle, J. Teich, L. Thiele: *System-Level Synthesis using Evolutionary Algorithms*. Journal on Design Automation for Embedded Systems. (1997)

[12] R.K. Gupta: *Co-Synthesis of Hardware and Software for Digital Embedded Systems*. Prentice Hall, (1994)

[13] R. Esser: *An Object Oriented Petri Net Approach to Embedded System Design*. Hochschulverlag vdf AG an der ETH Zürich, 1997.
R. Esser: *CodeSign - Concepts and Tutorial.*
http://www.tik.ee.ethz.ch/~codesign

DUAL EDGE OPERATIONS IN
REDUCED ORDERED BINARY DECISION DIAGRAMS[†]

D. M. Miller
VLSI Design and Test Group
Department of Computer Science
University of Victoria
Victoria, BC, CANADA V8W 3P6
mmiller@csr.uvic.ca

R. Drechsler
Institute of Computer Science
Albert-Ludwigs-University
79110 Freiburg/Breisgau
GERMANY
drechsle@informatik.uni-freiburg.de

ABSTRACT

The use of input and output negation in reducing the size of ROBDDs has been well investigated. Here we consider duality, a fundamental property of Boolean functions, and how it can be used to further reduce ROBDD size. We show how to introduce dual edge markers into a ROBDD package with effectively no storage overhead and a very small increase in per node processing cost.

Our experimental results show that dual markers can reduce the size of the ROBDD, sometimes quite substantially, and even in cases where they offer little reduction, can still lead to increased overall processing speed.

We consider the variable reordering problem and sifting in particular. We show that dual markers can not be used directly in shifting, but show that there is substantial advantage in applying sifting to an ROBDD with output negations, followed by a second phase where dual markers are used to reduce the size of the ROBDD.

1. INTRODUCTION

A Reduced Ordered Binary Decision Diagram (ROBDD), Bryant [3], is a graph structure that has proven very effective for the representation and manipulation of Boolean functions. The size (number of nodes) of the ROBDD is often of critical concern. Since the size of an ROBDD can be quite dependent on the underlying variable ordering, there has been much research on variable reordering methods. In addition, input and output negations can be assigned to certain edges in an ROBDD leading to a reduction in its size.

Duality is an important property of Boolean functions that we here consider in concert with output negation as a means to reducing ROBDD size.[*] Dual markers are added to certain edges indicating the edge points to the dual of the function represented by the subgraph rooted by the destination node rather than the function itself. They are very similar to output negations which indicate the complement of the function. Indeed we show that similar implementation techniques are applicable and the use of dual markers can be very efficient.

---

† This research began while the first author was on sabbatical at TIMA Laboratory, Grenoble, France. Support by way of a Research Grant from the Natural Sciences and Engineering Research Council of Canada is gratefully acknowledged.

* The idea to incorporate dual operators in ROBDDs was initially suggested to the second author by Paul Tafertshofer, Technical University of Munich.

We consider adjacent variable exchange based reordering schemes such as sifting [7] for ROBDDs with dual markers. We show there is an inherent limitation and suggest one approach to dealing with it. The concluding section of the paper suggests areas for further research.

2. BACKGROUND

A *binary decision diagram* (BDD) [1] is a *directed acyclic graph* in which each nonterminal node is labelled by a single variable and has two outgoing edges labelled 0 and 1. Each terminal node is labelled by a constant, either 0 or 1, and has no outgoing edges. At least one nonterminal node is identified as a *top* node and has no incoming edges. For a system of functions, there is one top node per function. We here consider only totally-specified functions.

A BDD is *ordered*, Bryant [1], if each variable appears at most once on any path from a top node to a terminal node. Bryant termed an ordered BDD *reduced*, if (i) no node is redundant *i.e.* no nonterminal has both edges leading to the same function, and (ii) common subgraphs are shared *i.e.* any required subfunction is represented by a unique subgraph. Bryant showed that for a given ordering, the ROBDD for a binary function is unique.

Duality is a well-known property in Boolean functions and we here only review the background necessary for this paper.

The *dual* of a function $f(X)$ is given by

$$f^D(X) = \bar{f}(\bar{X}) \qquad \ldots(1)$$

where the bar over X means the variables are all negated. Note that $(f^D(X))^D = f(X)$. A function is termed *self-dual* (SD) if $f^D(X) = f(X)$.

It can be shown that

$$f^D(X) = \bar{x}_i f_1^D(X_1) + x_i f_0^D(X_1), \qquad \ldots(2)$$

where $X_1 = \{x_1, \ldots x_{i-1}, x_{i+1}, \ldots x_n\}$, and $f_0(X_1)$ and $f_1(X_1)$ are the cofactors of $f(X)$ resulting from setting $x_i = 0$ and $x_i = 1$, respectively. Note the placement of the cofactors in eqn. (2).

3. INVERSION IN ROBDDS

Edge operations [1,2,6] can be used to take advantage of negation in ROBDDs. An *output negation* on an edge indicates that the function represented by the subgraph to which the edge points is to be inverted. The advantage is that a function and its inverse need not both be represented/stored. This leads to several advantages both in terms of size and computation.

To maintain the uniqueness of the ROBDD representation, it is necessary to constrain the use of output negations. The commonly used rules [5,6] are: (i) the single terminal node is labelled 0; (ii) an output negation is not used on a 0-edge. Rule (ii) is possible since an output negation on a 0-edge can be removed by applying negation to the 0-edge, the 1-edge and all edges that point to the node.

Output negations are easily incorporated into an ROBDD package. A single bit is sufficient to indicate whether an edge has an output negation or not. A bit position at the low end of the pointer associated with the edge can be used since pointer addresses are typically multiples of 4 or 8. This bit must be masked out when the pointer value is used. Using this approach, there is no storage and little computation overhead.

Minato et al. [6] suggested a second type of edge attribute called an *input negation* that indicates the edge points to the function found by exchanging the 0 and 1-edges of the node to which the edge points. As with output negations, some constraint is required to preserve uniqueness of the representation.

Minato's [5,6] suggestion is that the address of the node pointed to by the 0-edge be less than or equal to the address of the node pointed to by the 1-edge. Unlike output negations where the normalization is based solely on functional properties, this approach is implementation dependent. We consider this further below and for our approach to employing duality suggest a functional technique that is equally applicable to input negation.

4. DUALITY IN ROBDDS

To exploit duality in an ROBDD, we introduce a *dual marker*, which is an attribute added to an edge to indicate that the edge points to the dual of the function represented by the subgraph rooted by the destination node. Like output negations, the dual marker is a 1-bit flag associated with the edge, hence it too can reside in a low order 'unused' bit of the edge pointer.

Eqn. (2) supplies the basic rule for introducing dual markers in ROBDDs as shown in Fig. 1 (dual markers are shown in black and output negations are shown in white). Adding the dual marker to the ingoing edge requires we add a dual marker to each outgoing edge and *interchange the edges*.

Note that when the initial graph has a dual marker on one edge and not the other, the dual marker remains on that edge after applying the rule illustrated in Fig. 1 due to the edge swap.

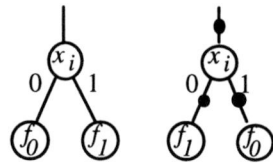

Fig. 1: ROBDD implementation of eqn. (2).

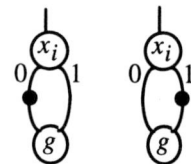

Fig. 2 Representation for self-dual (SD) functions.

Fig. 3: One form of anti-dual (AD) function.

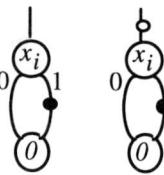

Fig. 4: Representation of a variable and its inverse.

Clearly, after accounting for negation, a self-dual function has one of the two structures shown in Fig. 2. This is helpful as the identification of SD functions is important to ensuring uniqueness of the representation. A variable and the inverse of a variable are SD functions.

A second class of functions where
$$f(X) = \overline{f^D(X)} \qquad \ldots(3)$$
is of equal importance. These are termed *anti-dual* (AD) functions. An AD function has a top node with the 0-edge and 1-edge leading to a common node, there is a dual marker on one of the edges and an output negation on one of the edges – possibly the same edge as shown in Fig. 3. Note that the constant functions 0 and 1 are AD functions.

As is usual, restrictions are required to ensure the uniqueness of the representation. The rules are as follows:
(i) A single terminal node is used and has the value 0. It is labelled an AD function. The constant 1 is an edge to that node with a dual marker.
(ii) A variable is represented by the structure shown in Fig. 4 and is labelled SD. The inverse of the variable is an edge with an output negation.
(iii) $\overline{x_i} f_0(X_1) + x_i f_1(X_1)$ is normalized as follows:
  a) If the weight of f_0 is greater than the weight of f_1, we store
  $$\overline{x_i} f_1^D(X_1) + x_i f_0^D(X_1)$$
  recording that the function being stored is actually the dual of the original. If the two weights are equal, a depth-first search is performed to see if f_0 is *after* f_1 in the sense that $f_1 = 0$ for the first difference encountered in the two functions. If that is the case, the dual is applied and recorded.
  b) If after (a), following the 0-edges (accounting for negations and duals) from the top node of the function to a terminal node would lead to a value of 1, we complement both outgoing edges and record that the function being represented is negated. Note that no search is required for this check as the value reached by following the 0-edge path can be stored in a single bit in the node.
  c) After (b), the node is checked an appropriately tagged if it represents a SD or AD function.
4. Application of a dual marker to an edge leading to a node labelled SD is ignored.

5. Application of an output negation to an edge leading to a node labelled AD is actually entered as a dual marker.

The weight of a function is the number of 1's in its truth table. Function weight is most easily computed as a special case of output probability as described in [4]. For this problem, the output probability for a node is one half the sum of the output probabilities of its children adjusted for output negations and dual markers. This can be computed when the node is created.

The use of the function weight and the associated techniques described above means the normalization is functional and not dependent on node addresses.

The above rules favour a dual marker over an output negation when the choice is available as that facilitates the identification of SD and AD functions. Rule 3b is a generalization of the rule of not putting an output negation on a 0-edge. The above rules are more complex than for output negations alone. However, they can be effectively implemented. The greatest complication is the search required when the output probabilities are equal.

5. VARIABLE REORDERING

The choice of variable ordering can significantly affect the size of an ROBDD. Sifting [7] is a heuristic method for finding a good variable ordering. Sifting is based on adjacent variable interchange. The basic operation is shown in Fig. 5 for the case of no edge negations or dual markers. To interchange variables x and y, the given transformation is applied to every node labelled x. A key issue is to 'reuse' the node originally labelled x in place so that no edges leading into it from higher in the ROBDD need to be modified. This makes the interchange a local operation.

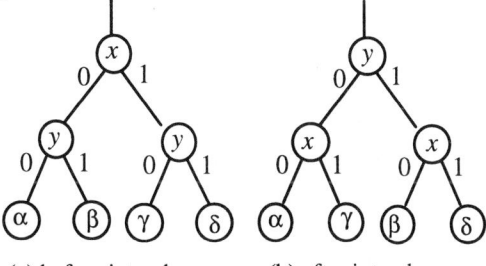

(a) before interchange   (b) after interchange

Fig. 5: Adjacent variable interchange transformation.

When edge negations are used, normalization requires there be no negations on 0-edges. As a result, adjacent variable interchange can be carried out as shown in Fig. 5. with no effect on the edges leading to the structure. However, when dual markers are used this is unfortunately not the case.

When using duals, normalization of the top node after the interchange of the variables can require a dual operation that must be recorded on every edge leading to that node. This destroys the locality of the transformation and applying an approach like sifting to an ROBDD with dual markers becomes much more complex and time consuming. We have investigated solutions to this problem including annotating nodes with output negations and dual markers in addition to the edges, but, while this facilitates a solution, the need to update edges leading to the top node of the structure persists. How to efficiently reorder an ROBDD with dual markers remains a significant open problem.

6. EXPERIMENTAL RESULTS

The techniques described above have been incorporated in an ROBDD package written in C. The usual hashing and computation cache techniques [4] are employed. As noted above, the bottom two bits of each edge pointer store the output negation and dual marker flags. Similarly, two bits are borrowed in the node to record if the function rooted by that node is SD or AD and another is used for the 0-edge path value. Our implementation uses 24 bytes per node including the output probability and the two outgoing edges.

The tables below show ROBDD size with and without sifting and for both cases show the size for ROBDDs with no edge markers, with output negations alone, and with output negations and dual markers.

Note that each ROBDD for the column *dual after sifting* was found by performing a depth first copy of the corresponding ROBDD from *negations after sifting*. Dual markers are introduced as appropriate during the copying process. They were not found by using dual markers during the sifting process due to the normalization problem.

In all examples, iterative sifting is used; *i.e.* the ROBDD is sifted to find a good variable ordering repeatedly until no improvement in ROBDD size is achieved.

			without sifting			with sifting		
	in	out	none	neg	dual	none	neg	dual
con1	7	2	20	18	17	17	15	12
postal	8	1	27	25	23	22	19	16
misex1	8	7	49	41	40	41	35	32
5xp1	7	10	90	74	65	80	42	29
bw	5	28	116	108	98	102	99	85
misex2	25	18	142	136	131	88	80	75
alu2	10	8	151	134	127	94	87	81
sao2	10	4	156	155	155	87	81	69
mdiv7	8	10	240	183	138	171	130	98
clip	9	5	256	226	171	107	87	69
duke2	22	29	978	973	970	359	353	342
sn74181	14	8	997	858	806	732	616	515
vg2	25	8	1061	1044	1023	177	148	142
misex3	14	14	1303	1301	1284	649	521	507
alu4	14	8	1354	1197	1185	747	754	680
Totals			6940	6473	6233	3473	3067	2752

Table I: ROBDD size: non-symmetric functions.

Table I shows the results for a collection of non-symmetric functions. Considering the total number of nodes for these examples without sifting, output negations introduce an improvement of 6.71% while the subsequent introduction of dual markers adds a further improvement of 3.81%. The total overall improvement for the two together is 10.27%. With sifting, output negations introduce an improvement of 11.69% and the subsequent introduction of dual markers adds a further improvement of 10.27%. The total overall improvement for the two together is 20.76%. Sifting alone leads to an improvement of 49.96% while sifting negations and dual markers together yield an overall improvement of 60.35%. These results clearly show that the effectiveness of using dual markers depends on a good variable ordering.

	in	out	none	neg	dual
rd53	5	3	25	17	15
9sym	9	1	35	25	16
rd73	7	3	45	31	23
rd84	8	4	61	42	34
Totals			166	115	88

Table II: ROBDD size: symmetric functions.

Table II shows the results for four symmetric functions. Variable ordering of course has no affect on ROBDD size for symmetric functions. As expected output negations introduce substantial overall improvement, 30.72%. Introducing dual markers adds a further significant improvement of 23.48% for a total overall improvement of 46.98%.

		without sifting			with sifting		
in	out	none	neg	dual	none	neg	dual
4	3	17	13	12	15	11	6
6	4	40	32	27	27	21	12
8	5	87	71	56	42	34	20
10	6	182	150	113	60	50	30
12	7	373	309	226	81	69	42

Table III: ROBDD size: n-bit adders.

Table III shows the result for some *n*-bit adders (2*n* inputs and *n+1* outputs). The advantage of sifting is clear. The adders were specified with the variables from one operand followed by the variables from the second. Sifting identifies an interleaved ordering which is known to be optimal. The advantage of incorporating dual markers into an interleaved ordered ROBDD with edge negations is also clear.

		without sifting			with sifting		
in	out	none	neg	dual	none	neg	dual
4	4	17	15	15	15	12	8
6	6	53	49	50	49	41	30
8	8	154	143	146	141	135	113
10	10	439	350	414	394	347	316
12	12	1247	1189	1158	1084	755	863

Table IV: ROBDD size: n-bit multipliers.

Table IV shows the results for n-bit multipliers ($2n$ inputs and $2n$ outputs) with the same initial variable ordering as the adders. Multipliers are well known to be a difficult case for the ROBDD representation. Bryant [1] in fact showed that the size of the ROBDD is exponential in n for all variable orderings. Not surprisingly, the improvement for multipliers is not as good as it is for adders.

The final entry of this table shows that incorporating dual markers can increase the size of the ROBDD. Multipliers are the only example we have encountered which shows this effect. We have not been able to identify which property of multipliers causes this behaviour.

7. CONCLUDING REMARKS

This paper has introduced the use of dual markers on the edges of a ROBDD. The results show that significant reduction in the size of a ROBDD can be achieved by using dual markers and output negations in concert with a variable reordering method such as sifting, and that the effectiveness of dual markers is dependent on a good variable ordering.

ROBDD size is but one measure of improvement. The time required to manipulate ROBDDs is also important. As an illustration of this, consider the construction of an ROBDD prior to sifting which we do from a cube list by building an ROBDD for each cube and then ORing those together using Bryant's [1] apply operation. Apply is recursive and the number of calls to apply is a first-order measure of the complexity of building an ROBDD from a cube list.

For the functions in Table I, the total number of apply invocations is 360,836 when output negations are used on their own. This reduces to 305,182 when dual markers are added, which is a reduction of 15%, substantially higher than the actual ROBDD size reduction. Somewhat surprisingly, some problems (*e.g.* misex3) show a very small node count improvement (1%) when dual markers are used, but the calls to apply reduce quite significantly (20%). This is because ORing of two ROBDDs spends most computational effort at the bottom of the ROBDDs which is where the dual markers are, in general, most prevalent. Interestingly, the symmetric functions, which show good node count improvement, do not yield much improvement in the number of calls to apply.

Two major research topics are suggested by the work in this paper. Due to the normalization problem noted in section 5, dual markers complicate sifting or any other reordering technique based on adjacent variable interchange. More work is required on how best to address this issue. The advantage that would be gained is indicated by our results on incorporating dual markers after sifting. Also more work is required to understand under which circumstances adding dual markers can increase ROBDD size.

8. REFERENCES

[1] Akers, S.B., "Binary decision diagrams," *IEEE Trans. on Computers*, V. C-27, no. 6, pp. 509-515, 1978.

[2] Brace, K.S., R.L. Rudell and R.E. Bryant, "Efficient implementation of a BDD package," *Proc. ACM/IEEE Design Automation Conference*, pp. 40-45, 1990.

[3] Bryant, R.E., "Graph-based algorithms for Boolean function manipulation," *IEEE Trans. on Computers*, V. C-35, no. 8, pp. 677-691, 1986.

[4] Miller, D.M., "An improved method for computing a generalized spectral coefficient," *IEEE Trans. On Computer-Aided Design*, accepted for publication.

[5] Minato, S., *Binary Decision Diagrams and Applications for VLSI CAD*, Kluwer Academic Publishers, Boston, 1996.

[6] Minato, S., N. Ishiura and S. Yajima, "Shared binary decision diagrams with attributed edges for efficient Boolean function manipulation," *Proc. ACM/IEEE Design Automation Conference*, pp. 52-57, 1990.

[7] Rudell, R.L., "Dynamic variable ordering for ordered binary decision diagrams," *Proc. IEEE/ACM ICCAD*, pp. 43-47, 1993.

ROBDD as a recursively defined periodic bit-string

Seong-Bong Lee, Kwang-Il Yeon, In-Hak Park
Semiconductor Technology Division, ETRI
161 Kajong-dong, Yusong-Gu, Taejeon, 305-350
Tel 042-860-5552, Fax 042-860-6108, Email sblee@etri.re.kr

ABSTRACT

The paper describes bit-string interpretation of ROBDDs. ROBDDs are viewed as recursively defined periodic bit-strings and BDD operations as recursively defined bit-wise operations. Using the periodicity of bit-strings, we prove that the ROBDDs generated by rotating the bit-string of a ROBDD are mutually different. It implies not only the exponential BDD size but also more compact BDD representation. And the interpretation can explain the hidden ideas of variable ordering.

1. INTRODUCTION

Reduced ordered binary decision diagrams (ROBDD)[1] are memory-efficient data structure for Boolean function manipulations, which are widely used in logic synthesis and verification[2]. But the functional interpretation based on the Shannon expansion sheds little insight on the behavior of some Boolean functions, of which BDD size grows exponentially. We describe a ROBDD as a graphical representation of a recursively defined periodic bit-string and show one-to-one correspondence between ROBDDs and special periodic bit-strings. Based on the periodicity of the bit-strings, we prove that all Boolean functions generated by rotating the bit-string of a ROBDD are mutually different, and so their ROBDDs are. That explains the exponential BDD size of some Boolean functions and also leads to more compact BDD representation from the fact that all ROBDD nodes can be partitioned uniquely by the rotation relation. The bit-string interpretation gives simple explanations of BDD operations and the hidden ideas of variable ordering[2,3].

2. PERIODIC BIT-STRING INTERPRETATION

2.1 ROBDD as Periodic bit-string

According to Bryant[1], a ROBDD is defined as follows:

[Def. 1] A ROBDD is a rooted, directed acyclic graph with vertex set V containing two terminal vertexes, denoted as *0* and *1*, and non-terminal vertices. A non-terminal vertex v has as attributes a variable index $index(v) \in \{ 1, ..., n \}$ and two children $low(v)$, $high(v)$ $\in V$ of which indexes are less than $index(v)$. Each non-terminal vertex v with $index(v) = i$ denote a unique i-variable Boolean function f_v defined as $f_v(x_i, ..., x_1) = \overline{x}_i\, f_v(x_i=0, ..., x_1) + x_i f_v(x_i=1, ..., x_1)$.

Since both a ROBDD and a SOM(Sum-Of-Minterms) are canonical forms of a Boolean function, a ROBDD corresponds uniquely to a SOM. Figure 1 shows a ROBDD for the Boolean function f(a,b,c) = a+bc and its corresponding SOM in short notation. Note that the variable order forms a list of the variables and each minterm is expressed in integer in the SOM notation.

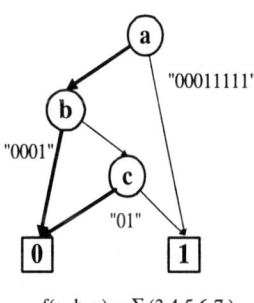

f(a, b, c) = Σ(3,4,5,6,7)

Figure 1. A ROBDD and its SOM

[Def. 2] A bit-string of a n-variable Boolean function f:
  $BS(f) = b_1 b_2 ... b_p$, where $(p = 2^n)$
    and ($b_i = 1$ *if i is a minterm of f*, otherwise $b_i = 0$).
[Def. 3] A *n-fold concatenation* of bit-string $BS(f)$:
  $PBS(1,f) = BS(f)$,
  $PBS(n,f) = PBS(n-1,f)BS(f)$ where $n > 1$, and
  $BPBS(b,f) \equiv PBS(2^b, f)$,
  where n is called as *frequency* and b as *log-frequency*.

Since a SOM is canonical, a bit-string is also canonical

by definition. It means that a ROBDD defines a unique bit-string. And since a subgraph rooted by a node in a ROBDD is also a ROBDD, each node v corresponds to a uniquely determined bit-string denoted as $BS(f_v)$ or simply $BS(v)$. Hereafter we use $PBS(n,v)$ and $BPBS(b,v)$ instead of $PBS(n,f_v)$ and $BPBS(b,f_v)$ respectively.

[Def. 4] For each node v in a ROBDD, an integer $p(v)$, called *log-period*, is defined as $p(v) = n - index(v) + 1$ where n is the number of variables of the ROBDD.

Note that $p(v)$ is the number of variables of f_v since each node is the root node of a subgraph and p(root node) is equal to the number of variables by definition. In other words, $p(v) = log(|BS(v)|)$ where $|x|$ is the string length of x, as the name implies.

Upon closer examination of a ROBDD in Figure 1, we see that $BS(v)$ of each node v can be constructed by concatenating the *properly repeated BS(low(v))* and $BS(high(v))$. For instance, $BS(a) = $ "0001 1111". The first half bit-string "0001" = $BS(b = low(a))$ and the second half "1111" = 4 times repeated $BS(1 = high(a)) = BPBS(2,1)$. Note that the *log-frequency* 2 in $BPBS(2,1)$ is equal to $p(a) - p(1) - 1$, where -1 means the half part of $BS(v)$. The following theorem describes this fact more precisely.

[Theorem 1] For any non-terminal node v of a ROBDD,
$BS(v) = BPBS(p(v)-p(low(v))-1, low(v))$
$BPBS(p(v)-p(high(v))-1, high(v))$

Proof) Let the variable ordering = $(x_n, ..., x_m, ..., x_1)$, $f_v = f(x_n, ..., x_1)$, and $f_{low(v)} = g(x_m, ..., x_1)$ where $n = p(v)$ and $m = p(low(v))$. From the definition of a ROBDD, $f_{low(v)} = f(x_{n=0}, x_{n-1}, ..., x_1) = g(x_m, ..., x_1)$. Since $g(x_m, ..., x_1)$ does not have the variables, $x_n, x_{n-1}, ..., $ and x_{m+1}, $f(x_n=0, x_{n-1}=0, ..., x_{m+1}=0, x_m, ..., x_1) = f(x_n=0, x_{n-1}=0, ..., x_{m+1}=1, x_m, ..., x_1) = ... = f(x_n=0, x_{n-1}=1, ..., x_{m+1}=1, x_m, ..., x_1) = g(x_m, ..., x_1)$. It follows that if I is a minterm of $g(x_m, ..., x_1)$, then $I, I+2^m, I+2 \cdot 2^m, ..., I+(2^{n-m-1}-1)2^m$ are minterms of $f(x_n, ..., x_1)$. In other words, each minterm of f is 2^{n-m-1} times repeated at the interval of $2^m = |BS(low(v))|$. So, the first half of $BS(v) = PBS(2^{n-m-1}, low(v)) = BPBS(n-m-1, low(v)) = BPBS(p(v)-p(low(v))-1, low(v))$. Similarly, the last half of $BS(v) = BPBS(p(v)-p(high(v))-1, high(v))$. ●

From Theorem 1, if the variable index is replaced with the corresponding log-period, a ROBDD represents a recursively-concatenated bit-string, so that the bit-string of a node can be constructed from the two bit-strings of its son nodes. Since ROBDDs have the minimum number of nodes, 2^n-lengthed bit-strings can be represented compactly by an n-variable ROBDD. For example, a maximum-length sequence of a linear feedback shift register can be represented by a ROBDD if a dummy 0 or 1 is inserted in a proper position so that its length = 2^n. The following C-like procedure describes a converting method of a bit-string to a ROBDD.

```
BS2BDD(int n /* log-period */)
{
    if (n == 0)/* terminal node */
        return getNextBit() == 0 ? 0: 1;
    /* next recursive call */
    low = BS2BDD(n-1);
    high = BS2BDD(n-1);
    if (low == high) return high;
    return findOrAddBddNode(n, low, high);
}
```

2.2 Minimal Log-periodic Bit-string

The log-period of a ROBDD node, $p(v)$ is minimal in that there is no other bit-string $BS(w)$ such that $BS(v) = BPBS(n,w)$ with $n > 0$, since $BS(v) = BPBS(n,w) = BPBS(n-1, w)BPBS(n-1,w)$ implies $low(v)=high(v)$ that is impossible in a ROBDD. It follows that the infinitely concatenated periodic bit-string, $PBS(\infty, v)$ is also unique for each node v. Therefore we can view a ROBDD as a periodic bit-string with a minimal log-period, simply called as *minimal log-periodic*. For instance, in Figure 1, $PBS(\infty, a) = $ "0001111100011111 .." with its log-period $p(a)=3$, $PBS(\infty, b) = $ "00010001 .." with $p(b)=2$, and $PBS(\infty, c) = $ "010101 .." with $p(c) = 1$. All nodes have different minimal log-periodic string. We conclude the fact with Theorem 2, of which proof is straightforward, since a ROBDD corresponds to a unique bit-string with minimal log-period, which defines a unique minimal log-periodic bit-string.

[Theorem 2] There is one-to-one correspondence between ROBDDs and minimal log-periodic bit-strings.

We can easily extend $PBS(n,f)$ by introducing '*negative frequency*' to define an infinite periodic bit-string and the infinite log-periodic Boolean function.

[Def. 5] A *negative n-fold concatenation* of bit-string:
$PBS(0,f) = \lambda$ (null string),
$PBS(n,f) = BS(f)PBS(n+1,f)$ if $n < 0$.
Similarly, $BPBS(b,f) = PBS(-2^{-b}, f)$ if $b < 0$.
$BS^\infty(f) \equiv PBS(-\infty, f)PBS(\infty, f)$

[Def. 6] An infinite log-periodic Boolean function:
$F_b(x, BS(f)) = $ x-th symbol of $BS(f)$ if $0 \leq x < 2^b$, and
$F_b(x-2^b, BS(f)) = F_b(x+2^b, BS(f)) = F_b(x, BS(f))$
for integer x and non-negative integer b.

By above definitions, $BS(F_b(x, BS(v))) = BS^\infty(v)$ for a ROBDD node v. Note that $F_b(x+r, BS(v)) = F_b(x, BS(w))$

when $BS(w)$ is the r-bit left-rotated bit-string of $BS(v)$. Then two infinite bit-string $BS^\infty(v)$ and $BS^\infty(w)$ can not be distinguishable when the start position of their period is not given. For instance, in Figure 1, $BS(b)=$ "0001", and $BS^\infty(b)=$ "..01000100..". By 1-bit left-rotation, new bit-string $BS(b_1)=$"0010", and $BS^\infty(b_1)=$ "..01000100..". $BS^\infty(b)$ and $BS^\infty(b_1)$ are same except their start positions of period. This means that $BS^\infty(v)$ can represent $2^{p(v)}$ distinct bit-strings with different numbers of rotation. That is proven by the following two theorems.

[Theorem 3] For a minimal log-periodic bit-string $BS^\infty(v)$, new bit-string $BS^\infty(w)$ which is generated by r-bit left-rotating $BS(v)$ is also log-periodic with the same minimal log-period, where $0 < r < 2^{p(v)}$.
Proof) Let $V(x) = F_{p(v)}(x, BS(f_v))$ and $W(x) = F_{p(w)}(x, BS(f_w))$ in brief. Since $n = p(v)$ is minimal, there is no $m < n$ such that $V(x + 2^m) = V(x)$. Since $W(x) = V(x + r) = V(x + 2^n + r) = W(x + 2^n)$, $BS^\infty(w)$ is also log-periodic, of which log-period is at most n. Suppose that $BS^\infty(w)$ has smaller log-period m than n, i.e., $W(x + 2^m) = W(x)$. Then $V(x + r) = V(x + r + 2^m)$. It follows that $V(x + 2^m) = V(x)$, contradicting no such m. Consequently, $p(w) = p(v)$. •

[Theorem 4] A minimal log-periodic bit-string $BS^\infty(v)$ and all other log-periodic bit-strings generated by r-bit left-rotating $BS(v)$ with $0 < r < 2^{p(v)}$ are mutually different.
Proof) Let $V(x) = F_{p(v)}(x, BS(f_v))$ and $n = p(v)$ in brief. By theorem 2, all $V(x), V(x + 1), ..., V(x + 2^n - 1)$ are log-periodic having the same minimal log-period n. Suppose that $V(x + r) = V(x)$ for $0 < r < 2^n$, then by definition $V(x + k) = V(x + 2^n)$, that means that r must be an proper divider of 2^n which is one of $\{2^0, 2^1, ..., 2^{n-1}\}$. But $V(x + 2^m) \neq V(x)$ for all $m < n$ since n is the minimal log-period. Thus $V(x + k) \neq V(x)$ for all $k < 2^n$. Similarly, suppose that $V(x + k) = V(x + j)$ with $k \neq j$ and $k, j < 2^n$. Let $W(x) = V(x + k)$, $W(x)$ is also an infinite log-periodic function with a minimal log-period by the theorem 3. $V(x + j) = W(x + j - k) = W(x)$, that is impossible for any j-k. Thus all $V(x), V(x + 1), ..., V(x + 2^n - 1)$ are distinct, and so the corresponding bit-strings are. •

From Theorem 2, ROBDDs determine their minimal log-periodic bit-strings uniquely. And all ROBDDs generated by the bit-string rotation, called as *rotated ROBDDs*, are mutually different by Theorem 4. It means that there are $2^{p(v)}-1$ different rotated ROBDDs of a ROBDD. In other words, given a Boolean function with a variable ordering, its bit-string consists of all possible rotated patterns of some bit-string. Then its BDD can grow exponentially. Figure 2 shows 4 rotated ROBDDs. Note that the size of rotated ROBDDs is different as shown in Figure 2. And the bit-string rotation is a restricted form of the cube transformations[4].

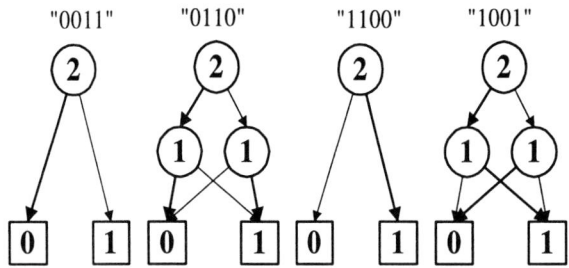

Figure 2. Examples of rotated ROBDDs

From a different viewpoint, Theorem 4 implies that $2^{p(v)}$ rotated nodes can be represented by one ROBDD node with the rotation number. Note that the intersection of any two rotated node sets is empty. It means that all ROBDD nodes can be partitioned uniquely by the bit-string rotation relation. Selecting only one representative ROBDD from each rotated node set, we can construct smaller ROBDD by attaching the rotation number as edge-value. The representative BDD is usually minimum-sized. Figure 3 show an example ROBDD with the rotation number. Note that ROBDD with rotation number is canonical by restricting the representative node set.

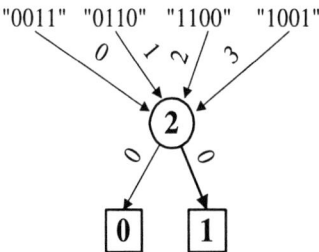

Figure 3. ROBDD with rotation numbers

2.3 Bit-string Operation

BDD operations, such as AND or NOT, can be easily expressed by bit-wise Boolean operations of log-periodic bit-strings. Since the resulting bit-string is also log-periodic, there exist a BDD node v such that $BS^\infty(v) = BS^\infty(x)$ **op** $BS^\infty(y)$ by theorem 2, where **op** denotes a bit-wise Boolean operation. Based on theorem 1, the proof of the followings is so straightforward that can be skipped.

If $p(x) = p(y)$, then
 $BS^\infty(low(v)) = BS^\infty(low(x))$ **op** $BS^\infty(low(y))$ and
 $BS^\infty(high(v)) = BS^\infty(high(x))$ **op** $BS^\infty(high(y))$.
If $p(x) > p(y)$, then
 $BS^\infty(low(v)) = BS^\infty(low(x))$ **op** $BS^\infty(y)$ and
 $BS^\infty(high(v)) = BS^\infty(high(x))$ **op** $BS^\infty(y)$.
If $p(x) < p(y)$, then

$BS^\infty(low(v)) = BS^\infty(x)$ **op** $BS^\infty(low(y))$ and
$BS^\infty(high(v)) = BS^\infty(x)$ **op** $BS^\infty(high(y))$.

These facts lead to the same recursive procedure, *apply*[1] for log-periodic bit-strings manipulations. Note that $p(v)$ is less than or equal to $max(p(x),p(y))$.

Given a variable ordering, each variable is assigned with a specific BDD, called as an *input BDD*, having only one non-terminal node v of which $index(v)$ is the ordering index of the variable, $low(v)= 0$ and $high(v)=1$. Then the BDD of a logic circuit can be constructed from these input BDDs with the *apply* procedure. When $index(v)$ is replaced with its log-period $p(v)$, the input BDD represents a log-periodic bit-string, called as an *input bit-string*, consisting $2^{p(v)-1}$ consecutive 0's and 1's. The bit-string of the circuit can be obtained from these input bit-strings by the *apply* procedure.

3. Variable Ordering

The size of BDDs is highly sensitive to variable ordering. Since smaller BDDs contain more periodic bit-string by theorem 1, a good ordering to minimize BDD size means the one to maximize the periodicity of corresponding bit-string. To maximize the periodicity, it is natural that the input bit-string having most consecutive 1's or 0's is assigned to the variable that affects the largest region of a circuit.

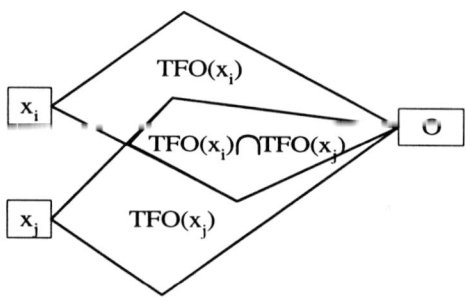

Figure 4. Transitive Fanouts.

When a DAG represents a logic circuit, a *transitive fanout* set of a node v, $TFO(v)$ is defined as a set of nodes that can be reachable from the node v, as shown in Figure 4. The function of a circuit is expected to depends most heavily on the input x_i with maximum $|TFO(x_i)|$. If x_i is assigned with the minimum index, the corresponding input bit-string has the maximum log-period. Since it has the largest consecutive 0's and 1's, the resulting bit-string has large consecutive 0s or 1s, and it also shares large parts of bit-string with other bit-strings. In the same reason, the input bit-string with the next largest log-period is assigned to the input x_j with maximum $|TFO(x_i) \cap TFO(x_j)|$, so that it gives the smallest disturbance to the periodicity of previous chosen input bit-strings.

Note that the input with maximum *TFO* has many reachable fanouts in the circuit, since they have large *TFO*. And the input with maximally intersected *TFO* is usually the neighbor one on the previously selected inputs in depth-first search manner. The ordering to maximize the periodicity is almost same with the fanout-oriented ordering[2]. Note that the computation of $TFO(x)$, especially that of finding a variable with maximally intersected *TFO*, is time-consuming. The depth-first search is a computationally efficient approximation.

4. SUMMARY

In this paper we have described a more informative ROBDD interpretation, in which ROBDDs are viewed as recursively defined periodic bit-strings. We have shown that all ROBDDs generated by the bit-string rotation are mutually different. That explains the exponential BDD size and also leads to more compact BDD representation. The interpretation is also useful to explain the variable ordering heuristics. We are currently working on extending the application of our interpretation and devising efficient ROBDD rotation operations.

5. REFERENCES

[1] R.E. Bryant, "Graph-based Algorithms for Boolean function Manipulation," IEEE Trans. on Computers, 1986, Vol. C-35, No. 8, pp.677-691.

[2] S. Malık, A. R. Wang, R. K. Brayton, and A. Sangiovanni-Vincentelli, "Logic verification using binary decision diagrams in a logic synthesis environments," Proc. ICCAD-88, pp.6-9.

[3] R. Rudell, "Dynamic variable ordering for ordered binary decision diagrams," Proc. ICCAD-93, pp.42-47.

[4] J. Bern, C. Meinel, and A. Slobodava, "Efficient OBDD-Based Boolean Manipulation in CAD Beyond Current Limits," Proc. DAC-95, pp.408-413.

Generation of quasi-optimal FBDDs through Paired Haar spectra

Chip-Hong Chang
Electronics Design Centre, French Singapore Institute,
Nanyang Polytechnic, 180 Ang Mo Kio Ave 8,
Singapore 569830.

Bogdan J. Falkowski
School of Electrical and Electronic Engineering,
Nanyang Technological University,
Nanyang Avenue, Singapore 639798.

Abstract

A polynomial Haar expansion for unnormalized Haar transform of incompletely specified Boolean function has been derived. Based on the Haar expansion, the entropy and equivocation in probability theory have been formulated in terms of some subsets of coefficients from the recently introduced Paired Haar spectrum. A unified and systematic method founded on the concept of entropy has been developed to exploit the don't care sets of incompletely specified Boolean functions for the heuristic minimization of Free Binary Decision Diagrams. The approach is general and can be extended to other combinatorial decision problems.

1. Introduction

Finding the minimal realizations for logic functions is usually associated with the problems of optimizing their reduced representations. For large digital circuits, Free Binary Decision Diagrams (FBDD) [4, 10] is a more succinct representation than the cubical representation of disjunctive sum-of-products expression for a given function in two-levels. Besides being more succinct, reduced FBDD of a fixed complete type is also canonical [10]. Obviously, it is NP-hard to transform the general circuit topology with an NP-complete satisfiability test to an optimal OBDD [2]. This holds also for the computation of the more general form of the minimal FBDD. In this paper, a unified entropy approach operated on Paired Haar spectrum to the heuristic optimizations of FBDD, with effective utilization of the don't care sets for incompletely specified Boolean functions have been developed.

The concept of entropy [11, 14, 16] in probability theory arose from attempt to develop a theoretical model for the transmission of discrete information in noisy channels. Since the introduction of Shannon's theorem on channels with noise in terms of a quantity known as equivocation [11, 16], the exposition of the theory of entropy and equivocation has appeared in various disciplines. In this paper, we exploit the general nature and theoretical significance of this mathematical apparatus to bridge the gap between communication theory and combinatorial decision problems. The concept of entropy and equivocation is applied to the generation of quasi-optimal FBDD of incompletely specified Boolean functions. We show that entropy and equivocation can be elegantly formulated by Paired Haar spectrum. Moreover, the unified and systematic entropy approach that has evolved from the presented theorems to this general decision problem is intuitively appealing.

2. Basic Definitions

A Binary Decision Diagram (BDD) [2, 4, 6, 8-10, 13, 19, 18] is a Rooted Directed Acyclic Graph representation with *Vertex Set V* and *Edge Set E*. The Vertex Set consists of two types of vertices, the *nonterminal* and *terminal vertices*. A nonterminal vertex $v \in V$ has as attributes an *index*, denoted by *index(v)*, to identify an input variable of a function, and two children (or successors), *low(v)* and *high(v)* $\in V$. A terminal vertex (or terminus) $u \in V$ has no child and it has a value, denoted by *value(u)*. *value(u)* = 0, 1 or 0.5 for the functional value of logical zero, one or don't care respectively. The Edge Set consists of two types of edges. A *0-edge* is a link from a node v to its low child *low(v)* and a *1-edge* is one that connects v to *high(v)*. A *root* is the topmost or the first non-terminal vertex in the BDD. A path from a vertex v_1 to a vertex v_2 is a set of vertices and edges traversed from v_1 to v_2. A *Free Binary Decision Diagram* (*FBDD*) [4, 10] is a BDD for which each variable of the function represented by it is encountered at most once along any path from the root to a terminal vertex. An *Ordered Binary Decision Diagram (OBDD)* [2, 6, 8-10, 13, 18] is a special subset of FBDD in which the input variables in all paths appear in a fixed order, and there exists an index function for every nonterminal vertex $v \in V$ such that *index(low(v))* < *index(v)* and *index(high(v))* < *index(v)*.

Property 1 : A path with k vertices represents a $(n-k)$-cube where $k = 1, 2, ..., n$ since an absent vertex corresponds to a vacuous variable in a product term or "–" in a cube notation.

Definition 1 : Let X be a finite space with elementary events X_i and the probability distribution $p(X_i)$ for $1 \le i \le n$ and $\sum_{i=1}^{n} p(X_i) = 1$. The *entropy* $H(X)$ of the finite space X is defined as [16]:

$$H(X) = -\sum_{i=1}^{n} p(X_i)\log_2 p(X_i) \quad (1)$$

where the expression $p(X_i)\log_2 p(X_i)$ is taken to be 0 if $p(X_i) = 0$.

Definition 2 : Let X and Y be two finite spaces with elementary events X_i, Y_j and their probability distributions $p(X_i)$ and $p(Y_j)$, respectively. $1 \le i \le n$, $1 \le j \le m$, $\sum_{i=1}^{n} p(X_i) = 1$ and $\sum_{j=1}^{m} p(Y_j) = 1$. X_i and Y_j may be dependent. The *conditional entropy* $H(Y | X_i)$ of space Y based on the assumption that event X_i has occurred in space X is given by [16] :

$$H(Y|X_i) = -\sum_{j=1}^{m} p(Y_j|X_i)\log_2 p(Y_j|X_i). \quad (2)$$

Since the occurrence of each event X_i results in a specific value of $H(Y|X_i)$, the conditional entropy $H(Y|X_i)$ can be regarded as a random variable defined on the space X. The mathematical expected value of this random variable leads to the definition of *equivocation*.

Definition 3 : The equivocation $H(Y|X)$ [16] is defined as the conditional entropy of the finite space Y averaged over the space X. Mathematically,

$$H(Y|X) = \sum_{i=1}^{n} p(X_i) H(Y|X_i). \quad (3)$$

Definition 4 : The *unnormalized Haar transform* T_N [1, 3, 5-7, 12, 15, 19, 20] of order $N = 2^n$ can be defined recursively as :

$$T_N = \begin{bmatrix} T_{\frac{N}{2}} \otimes \begin{bmatrix} 1 & 1 \end{bmatrix} \\ I_{\frac{N}{2}} \otimes \begin{bmatrix} 1 & -1 \end{bmatrix} \end{bmatrix} \text{ and } T_1 = 1 \quad (4)$$

where $I_{\frac{N}{2}}$ is an identity matrix of order $N/2$ and the symbol '$\otimes$' denotes the right-hand Kronecker product.

Based on the recursive definition of unnormalized Haar transform in (4), a polynomial Haar expansion of an n-variable Boolean function F can be derived.

Theorem 1 :

$$F(X) = \frac{1}{2^n}\left\{ r_{dc} + (-1)^{x_n} r_0^{(0)} + \sum_{l=1}^{n-1} 2^l (-1)^{x_{n-l}} \sum_{k=0}^{2^l-1} r_l^{(k)} \prod_{i=n-l+1}^{n} x_i^{k_{i-n+l}} \right\} \quad (5)$$

where $k_i \in \{0,1\}$ is the i-th bit in the binary l-tuple of the order k; $x_i^j = x_i$ if $j = 1$ and $x_i^j = \bar{x}_i$ if $j = 0$.

For efficient synthesis of incompletely specified Boolean functions, instead of operating on a single spectrum from the R-coded vector, a *Paired Haar transform* has been introduced [5, 7].

Definition 5 : A *Paired Haar transform* (PHT) for an incompletely specified n-variable Boolean function F is a mapping $\chi : (F_{ON}, F_{DC}) \rightarrow (R_{ON}, R_{DC})$, where $R_{ON} = T \times F_{ON}$ and $R_{DC} = T \times F_{DC}$. F_{ON} is obtained by replacing all don't care outputs of F by 0s, and F_{DC} is obtained from F by replacing all true outputs by 0s and don't care outputs by 1s. T is the unnormalized Haar transform defined in (4). The tuple (R_{ON}, R_{DC}) is known as the *Paired Haar spectrum*. Spectral coefficients from spectra R_{ON} and R_{DC} are indicated by lower case letters accordingly.

Definition 6 : Let X_i be an input assignment covered by a cube C whose cardinality is equal to $|C|$, then the output signal probability $p(C)$ under the set of input assignments X_i, $\forall i = 1, 2, \ldots, |C|$ is given by :

$$p(C) = \frac{1}{|C|} \sum_{i=1}^{|C|} F(X_i), \quad (6)$$

where $F(X_i)$ is the R-coded functional value of the input assignment X_i.

The value of $p(C)$ lies between 0 and 1 which indicates the likelihood of the cube C being an implicant of the function F. $p(C) = 1$ if C is an ON cube and $p(C) = 0$ if C is an OFF cube.

When applying the statistical decision theory to logic synthesis problems, we are often interested in comparing the equivocations for different input assignments and select one with the maximum likelihood. Depending on the formulation of the problem, the maximum likelihood decision corresponds to either the maximum or minimum equivocation. Since the conditional entropies $p\log_2 p$ and $p\log_2 p + (1-p)\log_2(1-p)$ encountered in the decision problems are monotonic increasing for $0 \leq p \leq 0.5$ and monotonic decreasing in the range of $0.5 \leq p \leq 1$, an appropriate metric to describe the equivocation for an input assignment C would be a number that is proportional to $|p(C) - 0.5|$ where $|\bullet|$ denotes the absolute value of $\bullet$. We called this number the *likelihood metric*.

Theorem 2 : Let $\rho_i(C)$ be the cube resulting from shifting the cube C by i bits to the right and $\gamma_i(C)$ be the number of '–' in C between bit 1 and bit i inclusive. Then, the likelihood metric, $M(C)$ can be expressed as the summation of selected Paired Haar coefficients:

$$M(C) = \left| M_{ON}(C) + \frac{1}{2} M_{DC}(C) - 2^{n-1}|C| \right| \quad (7)$$

where $M_{ON}(C) = |C|(r_{ON})_{dc} + \sum_{l=0}^{n-1} \delta_{ON}(l)$,

$$\delta_{ON}(l) = \begin{cases} 0 & \text{if } x_{n-l} = \text{'--'} \\ 2^{l+\gamma_{n-l}(C)}(-1)^{x_{n-l}} \sum_{X \in \rho_{n-l}(C)} (r_{ON})_l^{(X)} & \text{if } x_{n-l} \neq \text{'--'} \end{cases}, \quad (8)$$

and $M_{DC}(C) = |C|(r_{DC})_{dc} + \sum_{l=0}^{n-1} \delta_{DC}(l)$,

$$\delta_{DC}(l) = \begin{cases} 0 & \text{if } x_{n-l} = \text{'--'} \\ 2^{l+\gamma_{n-l}(C)}(-1)^{x_{n-l}} \sum_{X \in \rho_{n-l}(C)} (r_{DC})_l^{(X)} & \text{if } x_{n-l} \neq \text{'--'} \end{cases}. \quad (9)$$

In the above equations, $X \in \rho_{n-l}(C)$ denotes the set of input assignments (in decimal number representation) covered by the cube $\rho_{n-l}(C)$.

3. Generation of quasi-optimal FBDD and OBDD through Paired Haar spectrum

Let X and Y denote the random variables associated with the decision variables and the terminal value of a path, respectively. Then, the conditional entropy $H(Y = \varepsilon | X_i)$ is the likelihood or expectancy that the children of the vertex is a terminal vertex with value ε given that a decision variable x_i has been selected. Therefore, a quasi-optimal FBDD can be generated by recursively seeking for a maximum likelihood metric for each path of a FBDD in a depth first traversal. Let $C = \langle c_n c_{n-1} \ldots c_1 \rangle$ be the cube associated with a vertex v of the FBDD, where c_i is the edge value of the decision variable x_i being traversed from the root to v, and the vacuous variables in C, denoted by '–', are all possible decision variables for the vertex v. Then, we have the following propositions:

Proposition 1 : Let C denote the cube associated the path from the root of the FBDD to a vertex v. At the root of the FBDD, C is an n-cube. During depth first traversal of the FBDD, a candidate vacuous variable x_i of C is selected for the vertex v such that $M_{max}(C \cap \dot{x}_i) = \max_{x_s \in \Omega}(C \cap \dot{x}_s)$, where Ω is the set of m vacuous variables of C. The candidate variable x_i is used to decompose C into two $(m-1)$-cubes $C_0 = C \cap \bar{x}_i$ and $C_1 = C \cap x_i$ associated with the children $low(v)$ and $high(v)$, respectively. When $M_{ON}(C) + M_{DC}(C) = 2^n|C|$, the vertex associated with the cube C can be replaced by a 1-terminus. When $M_{ON}(C) = 0$, the vertex associated with the cube C can be replaced by a 0-terminus.

Proposition 2 : To improve the quality of the results obtained from Proposition 1, if there are more than one vacuous variables x_i of C with the maximum likelihood metric $M_{max}(C \cap \dot{x}_i)$, any variable among them with the maximum likelihood metric of $M_{DCmax}(C \cap \dot{x}_i) = \max_{\dot{x}_s \in \Omega'} M_{DC}(C \cap \dot{x}_s)$ is selected as a candidate variable, where Ω' is the set of literals with $M(C \cap \dot{x}_i) = M_{max}(C \cap \dot{x}_i)$.

In Proposition 2, when two variables lead to the same conditional entropy, we select one that leads to the decomposition with more allocable don't care outputs for the children, i.e., one that maximizes the entropy of $H(Y = 0.5|X_i)$. The algorithm for the selection of a good decision variable for a vertex v for an incompletely specified Boolean function from its Paired Haar spectrum, (R_{ON}, R_{DC}) is shown in Fig. 1.

```
Procedure Select_var(R_ON, R_DC, C) {
  for (each x_i ∈ Ω, the set of vacuous variables of C) {
    Calculate M_ON(C∩x̄_i), M_DC(C∩x̄_i), M_ON(C∩x_i) and
      M_DC(C∩x_i) from (R_ON, R_DC);
    M(C∩x̄_i) = |M_ON(C∩x̄_i) + ½M_DC(C∩x̄_i) − 2^(n−1)|C∩x̄_i||;
    M(C∩x_i) = |M_ON(C∩x_i) + ½M_DC(C∩x_i) − 2^(n−1)|C∩x_i||;
    M(C∩ẋ_i) = max(M(C∩x̄_i), M(C∩x_i));
  }
  V = {x_i ∈ Ω | M_max(C∩ẋ_i) = min_{x_s∈Ω} M(C∩ẋ_s) };
  if (|V| ≠ 1)
    Select any variable x_i ∈ V such that
      M_DCmax(C∩ẋ_i) = max_{ẋ_s∈Ω'} M_DC(C∩ẋ_s);
  return index of the selected variable ;
}
```

Fig. 1 Selection of good decision variable for a vertex.

In Procedure **Select_var**, the likelihood metric $M(C \cap \dot{x}_i)$ can be calculated from selected coefficients of Paired Haar spectrum by (7). A computational cache may be used to cache the previously calculated likelihood metrics. Based on Propositions 1 and 2, the algorithm for the generation of a quasi-optimal FBDD for incompletely specified Boolean function is shown in Fig. 2.

```
FBDD_MIN(R_ON, R_DC) {
  Initialize(C, fbdd, unique_table);
  fbdd->root = FBDD_MIN_AUX(R_ON, R_DC, C, unique_table);
  return fbdd;
}

FBDD_MIN_AUX(R_ON, R_DC, C, unique_table) {
  p = probability(R_ON, R_DC, C, &p_ON, &p_DC);
  if (p_ON = 0) return FBDD_ZERO;
  if (p_ON + p_DC = 1) return FBDD_ONE;
  i = Select_var(R_ON, R_DC, C);
  C_0 = C ∩ x̄_i; C_1 = C ∩ x_i;
  low = FBDD_MIN_AUX(R_ON, R_DC, C_0, unique_table);
  high = FBDD_MIN_AUX(R_ON, R_DC, C_1, unique_table);
  if (low = high) return low;
  return unique_table_find(unique_table, x_i, low, high);
}
```

Fig. 2 Generation of quasi-optimal FBDD for an incompletely specified function.

In Fig. 2, the procedure **Initialize** sets up the FBDD structure *fbdd* and a unique node table *unique_table* that keeps only unique vertices generated by the algorithm. The cube C is initialized to be an n-cube where n is the number of input variables. The procedure **FBDD_MIN_AUX** is a recursive routine that generates the vertices of the minimal FBDD by depth first traversal. In **FBDD_MIN_AUX**, *FBDD_ZERO* and *FBDD_ONE* are the 0- and 1- termini, respectively. The procedure **Select_var** in Fig. 1 is used to determine the best top variable x_i for the present vertex. The variables *low* and *high* are the pointer to the low and high children of the present vertex, respectively. The procedure **unique_table_find** searches in the *unique_table* for the vertex with the specified top variable and children. If found, it returns the pointer to the targeted vertex. Otherwise, a new vertex with the specified top variable and children is inserted in *unique_table* and returned.

4. Experimental Results

The algorithm **FBDD_MIN** is implemented in C, and the minimal or near minimal FBDDs are generated for some benchmarks functions from the two-level examples of MCNC benchmark suite. The results are summarized in Table 1. In Table 1, The columns labeled '#inputs' and '#outputs' are the number of input variables and outputs for each system of functions, respectively. The fourth column labeled 'Size' denotes the number of non-terminal vertices of the multi-root FBDD and the fifth column labeled 'Time' is the system execution time in seconds.

5. Conclusion

Paired Haar Transform has been introduced as an extension of unnormalized Haar transform to specially deal with the added complexity in allocating the don't care sets of incompletely specified Boolean functions [5, 7]. In the applications of Paired Haar spectrum to logic minimization, the Free Binary Decision Diagrams have been considered. Since exact minimization of FBDD have been proven to be NP-hard, the algorithms proposed for their optimization are heuristic. By treating them as a general combinatorial decision problem, the concept of entropy and equivocation are adopted and re-formulated in terms of Paired Haar spectrum.

REFERENCES

[1] N. Ahmed and K. R. Rao, *Orthogonal Transforms for Digital Signal Processing*. Berlin: Springer-Verlag, 1975.

[2] B. Bolling and I. Wegener, "Improving the variable ordering of OBDDs is NP-complete," *IEEE Trans. Comput.*, vol. 45, no. 9, pp. 993-1001, Sep. 1996.

[3] A. M. Buron, J. A. Michell and J. M. Solana, "Single chip fast Haar transform at megahertz rates," in *Theory and Applications of Spectral Techniques*, C. Moraga, Ed., University Dortmund Press, pp. 8-17, Oct. 1988.

[4] S. Chakravarty, "A characterization of Binary Decision Diagrams," *IEEE Trans. Comput.*, vol. 42, no. 2, pp. 129-137, Feb. 1993.

[5] B. J. Falkowski and C. H. Chang, "A novel paired Haar based transform: algorithms and interpretations in Boolean domain," in *Proc. 36th Midwest Symp. on Circuits and Systems*, Detroit, Michigan, pp. 1101-1104, Aug. 1993.

[6] B. J. Falkowski and C. H. Chang, "Efficient algorithms for the forward and inverse transformations between Haar spectrum and binary decision diagram," in *Proc. 13th IEEE Int. Phoenix Conf. on Computers and Communications*, Phoenix, Arizona, pp. 497-503, Apr. 1994.

[7] B. J. Falkowski and C. H. Chang, "Properties and applications of Paired Haar transform," in *Proc. 1st IEEE Int. Conf. on Information, Communications and Signal Processing*, Singapore, Sep. 1997.

[8] S. J. Friedman and K. J. Supowit, "Finding the optimal variable ordering for binary decision diagrams," *IEEE Trans. Comput.*, vol. 39, no. 5, pp. 710-713, May 1990.

[9] M. Fujita, Y. Matsunaga and T. Kakuda, "On the variable ordering of binary decision diagrams for the application of multi-level logic synthesis," in *Proc. European Design Automation Conf.*, pp. 50-54, Feb. 1991.

[10] J. Gergov and C. Meinel, "Efficient Boolean manipulation with OBDD's can be extended to FBDD's," *IEEE Trans. Comput.*, vol 43, no. 10, pp. 1197-1209, Oct. 1994.

[11] M. E. Hellman, "An extension of the Shannon theory approach to cryptography," *IEEE Trans. Inf. Theory*, vol 23, pp. 289-294, May 1978.

[12] S. L. Hurst, D. M. Miller and J. C. Muzio, *Spectral Techniques in Digital Logic*. London: Academic Press, 1985.

[13] N. Ishiura, H. Sawada and S. Yajima, "Minimizations of binary decision diagrams based on exchanges of variables," in *Proc. IEEE Int. Conf. on Computer Aided Design*, pp. 472-475, 1991.

[14] A. M. Kabakcioglu, P. K. Varshney and C. R. P. Hartmann, "Application of information theory to switching function minimization," *IEE Proceedings*, vol 137, Pt. E, no. 5, pp. 389-393, Sep. 1990.

[15] M. G. Karpovsky, *Finite Orthogonal Series in the Design of Digital Devices*. New York: John Wiley, 1976.

[16] A. I. Khinchin, *Mathematical Foundations of Information Theory*. New York: Dover Publications, Inc., 1957.

[17] M. R. Mercer, R. Kapur and D. E. Ross, "Functional approaches to generating orderings for efficient symbolic representations," in *Proc. 29th ACM/IEEE Design Automation Conf.*, pp. 624-627, Jun. 1992.

[18] R. Rudell, "Dynamic variable ordering for ordered binary decision diagrams," in *Proc. IEEE Int. Conf. on Computer Aided Design*, pp. 42-47, 1973.

[19] G. Ruiz, J. A. Michell and A. Buron, "Fault detection and diagnosis for MOS circuits from Haar and Walsh spectrum analysis: on the fault coverage of Haar reduced analysis," in *Theory and Applications of Spectral Techniques*, C. Moraga, Ed., University Dortmund Press, pp. 97-106, Oct. 1988.

[20] G. Ruiz, J. A. Michell and A. Buron, "Switch-level fault detection and diagnosis environment for MOS digital circuits using spectral techniques," *IEE Proc.*, Part E, vol. 139, no. 4, pp. 293-307, Jul. 1992.

TABLE 1 Benchmark results for FBDD_MIN.

	#inputs	#outputs	Size	Time (s)
9sym	9	1	33	0.06
5xp1	7	10	104	0.04
sao2	10	4	130	0.11
apex4	9	19	1465	0.54
bw	5	28	139	0.05
clip	9	5	207	0.10
con1	7	2	21	0.03
inc	7	9	89	0.06
misex1	8	7	54	0.08
sqrt8	8	4	41	0.02
ex1010	10	10	1231	0.86
rd84	8	4	59	0.10

Calculation of Paired Haar Spectra for Systems of Incompletely Specified Boolean Functions

Bogdan J. Falkowski
School of Electrical and Electronic Engineering,
Nanyang Technological University,
Nanyang Avenue, Singapore 639798.

Chip-Hong Chang
Electronics Design Centre, French Singapore Institute,
Nanyang Polytechnic, 180 Ang Mo Kio Ave 8,
Singapore 569830.

Abstract

A new algorithm is given that converts a reduced representation of Boolean functions in the form of disjoint cubes to unnormalized Paired Haar spectra for systems of incompletely specified Boolean functions. Since the known algorithms that generate unnormalized Haar spectra always start from the truth table of Boolean functions the method presented computes faster with a smaller computer memory. The method is extremely efficient for such Boolean functions that are described by only few disjoint cubes and it allows the calculation of only selected spectral coefficients, or all the coefficients can be calculated in parallel.

1. Introduction

Haar transform is known to have the smallest computational requirement and has been used mainly for pattern recognition and image processing [1, 12, 14, 15]. Although the properties of Haar spectra have considerable interest and attraction for Boolean functions, the majority of publications to date have employed the Walsh rather than Haar transform in their considerations [10, 15]. It is mainly due to the fact that up to now there is no efficient method of calculating Haar spectra directly from reduced representations of Boolean functions. Recently, efficient symbolic methods based on Binary Decision Diagrams representation for the computation of unnormalized Haar spectra have been developed [4, 9, 13]. These methods can be used efficiently in various CAD systems and the decision diagrams can represent both the original Boolean functions and their spectra. Binary Decision Diagrams [4, 11] have proved to be very convenient data structures for majority of discrete functions representations permitting manipulations and calculation with large discrete functions efficiently in terms of space and time. Therefore they are frequently used to represent data structures in modern CAD VLSI systems. However, some of such systems are based on cubical representation [5, 7, 8, 11] rather than decision diagrams and the current article solves the problem of efficient calculation of Paired Haar spectrum for such CAD systems. The presented algorithm has overcome the inefficiency of the calculation of both spectra directly from the definition of the transforms by matrix multiplication. By allowing to represent the Boolean function in the form of an array of disjoint cubes instead of minterms, the spectral coefficients can be computed more rapidly from such a reduced representation with smaller required memory, while the ability to calculate only partial spectra is still preserved. Hence, the new algorithm has allowed practical applications of Paired Haar transforms for CAD systems using cubical rather than graph based representations of discrete functions. In order to use Boolean functions that are represented as minterms or arrays of non disjoint cubes, the input data are preprocessed by a fast algorithm that generates an array of disjoint ON- cubes (in the case of completely specified Boolean functions) or disjoint ON- and DC- cubes (in the case of incompletely specified functions). The algorithm that generates such an array and its implementation is described in [7]. For each disjoint cube, the appropriate partial spectral coefficients is calculated. The final Paired Haar spectrum is found by adding all the corresponding partial coefficients contributed by the complete array of disjoint cubes.

2. Basic Definitions

A collection of 2^i, $i \in \{0, 1, ..., n\}$ adjacent minterms is called an *i-cube* [11]. A cube can be represented by an *n*-string of symbols 0, 1 and –, where 0 corresponds to the complemented value of the variable, 1 to the affirmative value and – to the missing variable in the cube. The *ON*, *OFF* and *DC cubes* are cubes corresponding to the product terms of ON, OFF and DC minterms, respectively. An *ON array* of cubes of a Boolean function *F*, denoted by *ON(F)*, is defined as a set of cubes for which $F = 1$, an *OFF array*, denoted by *OFF(F)*, is a set of cubes for which $F = 0$, and a *DC array*, denoted by *DC(F)*, is a set of cubes for which $F = -$.

Definition 1 : The *unnormalized Haar transform* T_N of order $N = 2^n$ can be defined recursively as [10, 13, 15] :

$$T_N = \begin{bmatrix} T_{\frac{N}{2}} \otimes \begin{bmatrix} 1 & 1 \end{bmatrix} \\ I_{\frac{N}{2}} \otimes \begin{bmatrix} 1 & -1 \end{bmatrix} \end{bmatrix} \text{ and } T_1 = 1 \quad (1)$$

where $I_{\frac{N}{2}}$ is an identity matrix of order $N/2$ and the symbol '$\otimes$' denotes the right-hand Kronecker product.

For an *n*- variable Boolean function $F(x_1, x_2, ..., x_n)$ Haar spectrum is given by $R = [H_N] F$ where R is Haar spectrum (a column vector of dimension $2^n \times 1$) and F is the *R-coded* truth vector of Boolean function $F(X)$ [8]. In R coding, the false minterms are coded as 0, true minterms as 1 and don't care (DC) minterms as 0.5.

Besides the first two Haar spectral coefficients r_{dc} (so called *dc coefficient* corresponding to *dc function*) and $r_0^{(0)}$, which are globally sensitive to $F(X)$, the remaining 2^n-2 Haar spectral coefficients are only locally sensitive. A spectral coefficient $r_l^{(k)}$ is characterized by its degree l and order k.

Property 1 : For a Haar spectrum of an *n*-variable Boolean function F, there are 2^l spectral coefficients of degree l, each measures a correlation of a different set of 2^{n-l} neighboring minterms where $l = 1, 2, ..., n$. The *dc* coefficient r_{dc} and the zero degree coefficient $r_0^{(0)}$ measure a correlation of 2^n neighboring minterms (the whole Karnaugh map). The value of r_{dc} is equal to the number of minterms of F and the coefficient $r_0^{(0)}$ describes the difference between the number of minterms in the functions $\bar{x}_n$ and x_n.

Definition 2 : A *standard trivial function* (STF), denoted by u_I, $I \in \{0, 1, ..., 2^n-1\}$, associated with each Haar spectral coefficient r_{dc} or $r_l^{(k)}$ describes some set of 2^{n-l} neighboring minterms on a Karnaugh map that has an influence on the value of a spectral coefficient r_{dc} or $r_l^{(k)}$ where $0 \le l \le n-1$ and $0 \le k \le 2^l-1$.

Property 2 : The *degree l* of Haar coefficient indicates the number of literals present in a STF u_I for $I = 1, 2, .., 2^n-1$.

Property 3 : The *order k* of Haar spectral coefficient $r_0^{(0)}$ is the decimal equivalence of the binary *l*-tuple formed by writing a 1 or 0 for each variable in a STF u_I ($I = 2, 3, ..., 2^n-1$) according to whether this literal appears in affirmation or negation. When k is expressed as a binary *l*-tuple, the most significant bit (MSB) corresponds to the literal $\dot{x}_n$ and the least significant bit (LSB) corresponds to the literal $\dot{x}_{n-l+1}$.

For each *index I* of a STF u_I, there exist unique values l and k such that $I = 2^l + k$.

Recently, a *Paired Haar transform* has been introduced [2, 5] to efficiently allocate don't care minterms in logic minimization of incompletely specified Boolean functions.

Definition 3 : A *Paired Haar transform* (PHT) for an incompletely specified *n*-variable Boolean function F is a mapping $\chi : (F_{ON}, F_{DC}) \to (R_{ON}, R_{DC})$, where $R_{ON} = T \times F_{ON}$ and $R_{DC} = T \times F_{DC}$. F_{ON} is obtained by replacing all don't care outputs of F by 0s, and F_{DC} is obtained from F by replacing all true outputs by 0s and don't care outputs by 1s. T is the unnormalized Haar transform. The tuple (R_{ON}, R_{DC}) is known as the *Paired Haar spectrum*. Spectral coefficients from spectra R_{ON} and R_{DC} are indicated by lower case letters accordingly.

Example 1 : For the four-variable incompletely specified Boolean function $F(X) = \Sigma_{ON}(8, 9, 10, 14, 15) + \Sigma_{DC}(1, 4, 5)$, the Paired Haar spectrum $(R_{ON}, R_{DC}) = [((r_{ON})_{dc}, (r_{DC})_{dc})$ $((r_{ON})_0^{(0)}, (r_{DC})_0^{(0)}) ((r_{ON})_1^{(0)}, (r_{DC})_1^{(0)}) ... ((r_{ON})_3^{(7)}, (r_{DC})_3^{(7)})]^T$
$= [(5, 3) (-5, 3) (0, -1) (1, 0) (0, 1) (0, 2) (1, 0) (-2, 0)$
$(0, -1) (0, 0) (0, 0) (0, 0) (0, 0) (1, 0) (0, 0) (0, 0)]^T$.

3. Paired Haar spectrum for system of incompletely specified Boolean functions

In this section, an efficient method for the calculation of Paired Haar spectrum of a system of incompletely specified Boolean functions that can have any number of functions and arbitrary locations of don't care minterms in each of the functions of the system is presented. Let us consider a system of t incompletely specified functions. By ordering the system of t functions to form a binary t-tuple $F_{t-1} F_{t-2} ... F_0$, where F_{t-1} is the MSB, a single multi-valued output function F is obtained. Furthermore, let F_{jON} be the truth vector obtained from F_j by replacing its don't care outputs by 1, and F_{jDC} be the truth vector obtained from F_j by replacing its true outputs by 0s and its don't care outputs by 1s. The functions F_{ON} and F_{DC} can be written as a weighted sum of each individual function F_j as follows :

$$F_{ON} = \sum_{j=0}^{t-1} 2^j F_{jON} \text{ and } F_{DC} = \sum_{j=0}^{t-1} 2^j F_{jDC} \qquad (2)$$

Applying Paired Haar transform to both sides of the above expression, we have

$$R_{ON} = 2^j R_{jON} \text{ and } R_{DC} = 2^j R_{jDC} \qquad (3)$$

where the tuples (R_{ON}, R_{DC}) and (R_{jON}, R_{jDC}) are the Paired Haar spectra of the multiple output function F and its j-th output F_j, respectively. The total spectrum (R_{ON}, R_{DC}) is called the *ordered Paired Haar spectrum* since it is sensitive to the relative position of each output function within the system. Since the weighted sums F_{ON} and F_{DC} are formed from super increasing sequence, it is trivial to show that the ordered Paired Haar spectrum obtained in this way is unique.

Each disjoint cube C consists of an input part ($x_n x_{n-1} ... x_1$) and an output part ($y_{t-1} y_{t-2} ... y_0$), where the input variable $x_i = 0, 1$ or $-$ ($1 \le i \le n$) depending on whether x_i appears as complemented or affirmative form or does not appear in the product term represented by C, and the output variable $y_j = 0$, 1 or $-$ ($0 \leq j \leq t-1$) depending on whether the cube represented by the input part of C is a OFF, ON or DC cube of the function F_j. The procedure **Ordered_Paired_Haar** for the calculation of Paired Haar spectrum for a system of incompletely specified functions computed is given in Fig. 1.

Procedure Ordered_Paired_Haar(Array of disjoint cubes D)
{
   Initialize(PHS);
   foreach (cube $C_j \in D$, $j = 1$ to $npsc$) {
     p = number of '–' in C_j;
     ONweight = DCweight = 0;
     foreach (output variable y_i of C_j, $i = 0$ to $t-1$)
       if ($y_i = 1$) ONweight = ONweight + 2^i;
       else if ($y_i = -$) DCweight = DCweight + 2^i;
     $(r_{ON})_{dc}$ += $2^p \times$ ONweight;
     $(r_{DC})_{dc}$ += $2^p \times$ DCweight;
     for ($l = 0$ to $n-1$) {
       order_list = $\{k \in \mathbf{Z} \mid k \subseteq \rho_{n-l}(C)\}$;
       for (each integer k in order_list) {
         p = number of '–' in C; q = number of '–' in $\rho_{n-l}(C)$;
         if (bit x_{n-l} of C = 0) $v = 2^{p-q}$;
         else if (bit x_{n-l} of C = 1) $v = -2^{p-q}$;
         if (lookup(PHS, l, k, $(r_{ON})_l^{(k)}$, $(r_{DC})_l^{(k)}$) = 0)
           create($(r_{ON})_l^{(k)}$, $(r_{DC})_l^{(k)}$);
         if (C is an ON cube) $(r_{ON})_l^{(k)}$ += ($v*$ONweight);
         else if (C is a DC cube) $(r_{DC})_l^{(k)}$ += ($v*$DCweight);
         if (($r_{ON})_l^{(k)} = 0$ and $(r_{DC})_l^{(k)} = 0$) remove(PHS, l, k);
         else insert(PHS, l, k, $(r_{ON})_l^{(k)}$, $(r_{DC})_l^{(k)}$);
       }
     }
   }
   return PHS;
}

Fig. 1 Algorithm for calculating Paired Haar spectrum.

In Fig. 1, PHS is a link list of non-zero valued Paired Haar spectral coefficients sorted in ascending order of degree l and order k. the routine **Initialize** sets up the link list PHS and initializes the dc coefficient $(r_{ON})_{dc}$ and $(r_{DC})_{dc}$ to 0. The number of partial spectral coefficients $npsc$ is equal to the number of disjoint ON and DC cubes. To conserve disk space, it is sufficient to store only the non vanishing Paired Haar coefficients. The global variable ONweight and DCweight accumulates the weight factors contributed by all ON and DC outputs of each cube C_j. Hence the partial coefficient is calculated once for each cube C_j. The array order_list is an array of integers representing the minterms covered by the cube $\rho_{n-l}(C)$, where $\rho_i(C)$ is the cube obtained by shifting the cube C i bits to the right, and q is the number of '–' in the cube $\rho_i(C)$, i.e., $q = \log_2 |\rho_i(C)|$. The routine **lookup** searches from PHS for any non-zero Paired Haar coefficient of degree l and order k. If found, it returns the coefficient in the tuple $((r_{ON})_l^k, (r_{DC})_l^k)$. Otherwise, the routine **create** is called to allocate new Paired Haar coefficient of degree l and order k. If the computed values of $(r_{ON})_l^k$ and $(r_{DC})_l^k$ are both equal to zero, the routine **remove** is called to remove the Paired Haar coefficient of degree l and order k from PHS. Otherwise, the routine **insert** is called to insert the non-zero coefficient in PHS according to l and k. The partial dc coefficient can be easily computed from the cardinality of the cube C. By summing up the respective partial coefficients contributed by all disjoint cubes, the full Paired Haar spectrum for the n-variable Boolean function F is obtained. The Procedure **Ordered_Paired_Haar** can also be modified to include options to just calculate a selected Paired Haar coefficient or only spectral coefficients for a complete degree. In the former case, the Procedure **partial_coef** can be simplified to accept the desired degree l and order k as arguments. In the latter case, the degree l is supplied as an additional input argument to Procedure **Paired_Haar** and the for loop involving l is omitted.

4. Experimental results

The algorithm **Ordered_Paired_Haar** is implemented in C, and the computation time and space requirement of the Paired Haar spectra for some MCNC benchmark functions are given in Table 1. The MCNC benchmark functions in PLA format are preprocessed by the disjoint cube algorithm [3, 6] before the test. The number of disjoint cubes is given in the fourth column labeled #disjoint in Table 1. The number of input and output variables of each function are also given in the second and third columns, respectively. The column labeled #coefficients is the number of non-vanishing Paired Haar coefficients and the column labeled Time is the system execution time in seconds on a HP Apollo Series 735 workstation.

5. Conclusion

A new algorithm that generates Paired Haar spectrum for system of incompletely specified Boolean functions from the disjoint cube representation has been shown. Since the number of such cubes can be considerably smaller than the number of minterms, the memory requirements can be reduced significantly. The advantages of this kind of representation used frequently in modern CAD VLSI systems, especially the fact that for practical functions the number of disjoint cubes is much smaller than the number of minterms, has been manifested in [2]. The ability to calculate only some spectral coefficients made possible by this research is very important since there are many spectral methods in digital logic design for which the values of only selected spectral coefficients are needed [8, 10].

The fundamental advantage of the presented algorithm is the usage of a reduced representation of Boolean functions in the form of disjoint cubes as the internal data from which the algorithm calculates the spectra. Such an approach gives the presented algorithm the ability to yield solutions to problems of very high dimensions and is applicable to these CAD systems which use cubical representation for discrete functions. The algorithm is very well suited for systolic VLSI realizations, and may be implemented as hardware coprocessor in a manner similar to those used for other binary expansions.

References

[1] N. Ahmed and K. R. Rao, *Orthogonal Transforms for Digital Signal Processing*. Berlin: Springer-Verlag, 1975.

[2] M. J. Ciesielski, S. Yang and M. A. Perkowski, "Minimization of multiple-valued logic based on graph coloring," *Tech. Rep. TR-CSE-90-13*, Dep. Elect. Comp. Eng., Univ. of Massachusetts, Amherst, September 1990. (Earlier version of this paper appeared as : "Multiple-valued minimization based on graph coloring," *Proc. of IEEE Int. Conf. on Computer Design: VLSI in Computers & Processors*, pp. 262-265, October 1989.)

[3] B. J. Falkowski and C. H. Chang, "A novel paired Haar based transform: algorithms and interpretations in Boolean domain," in *Proc. 36th Midwest Symp. on Circuits and Systems*, Detroit, Michigan, pp. 1101-1104, Aug. 1993.

[4] B. J. Falkowski and C. H. Chang, "Efficient algorithms for the forward and inverse transformations between Haar spectrum and binary decision diagram," in *Proc. 13th IEEE Int. Phoenix Conf. on Computers and Communications*, Phoenix, Arizona, pp. 497-503, Apr. 1994.

[5] B. J. Falkowski and C. H. Chang, "Generation of multi-polarity Arithmetic transform from reduced representation of Boolean functions," *Proc. of 28th IEEE Intnl. Symp. on Circuits and Systems*, Seattle, Washington, pp. 2168-2171, May 1995.

[6] B. J. Falkowski and C. H. Chang, "Properties and applications of Paired Haar transform," in *Proc. 1st IEEE Int. Conf. on Information, Communications and Signal Processing*, Singapore, Sep. 1997.

[7] B. J. Falkowski, I. Schaefer and C. H. Chang, "An efficient computer algorithm for the calculation of disjoint cubes representation of Boolean functions," in *Proc. 36th IEEE Midwest Symp. on Circuits and Systems*, Detroit, Michigan, pp. 1308-1311, August 1993.

[8] B. J. Falkowski, I. Schaefer and M. A. Perkowski, "Effective computer methods for the calculation of Rademacher-Walsh spectrum for completely and incompletely specified Boolean functions," *IEEE Trans. Computer-Aided Design*, vol. 11, no. 10, pp. 1207-1226, Oct. 1992.

[9] J. P. Hansen and M. Sekine, "Decision diagram based techniques for the Haar wavelet transform," in *Proc. 1st Int. Conf. on Information, Communications and Signal Processing*, Singapore, Sep. 1997, vol 1, pp. 59-63.

[10] M. G. Karpovsky, *Finite Orthogonal Series in the Design of Digital Devices*. New York: John Wiley, 1976.

[11] T. Sasao and M. Fujita, *Representations of Discrete Functions*. Boston: Kluwer Academic, 1996.

[12] R. S. Stankovic and B. J. Falkowski, "Haar functions and transforms and their generalizations," in *Proc. 1st Int. Conf. on Information, Communications and Signal Processing*, Singapore, Sep. 1997, vol 4, pp. 1-5.

[13] M. Stankovic, D. Jankovic and R. S. Stankovic, "Efficient algorithm for Haar spectrum calculation," in *Proc. 1st Int. Conf. on Information, Communications and Signal Processing*, Singapore, Sep. 1997, vol 4, pp. 15-20.

[14] L. P. Yaroslavsky, *Digital Picture Processing*. Berlin: Springer-Verlag, 1985.

[15] L. A. Zalmanzon, *Fourier, Walsh and Haar Transforms and Their Application in Control, Communication and Other Fields*. Moscow: Nauka, 1989 (in Russian).

TABLE 1. Benchmark Results for **ORDERED_PARIED_HAAR**.

Functions	#inputs	#outputs	#disjoint	#coefficients	Time (s)
9sym	9	1	145	211	0.01
5xp1	7	10	75	128	0.02
alu4	14	8	1043	12008	0.16
sao2	10	4	96	102	0.03
bw	5	28	106	29	0.03
clip	9	5	176	504	0.02
con1	7	2	11	85	0.03
inc	7	9	33	128	0.01
misex1	8	7	32	232	0.03
misex3	14	14	164	3168	0.15
table5	17	15	166	78011	0.32
sqrt8	8	4	40	255	0.03
t481	16	1	887	28231	0.27
b12	15	9	70	28880	0.15
ex1010	10	10	1017	1021	0.05
rd84	8	4	256	256	0.02

Pseudo-Symmetric Functional Decision Diagrams

Malgorzata Chrzanowska-Jeske, Xiang Ying Ma and Wei Wang

Electrical Engineering Department, Portland State University
1800 SW 6th Avenue, Portland, OR 97207 USA

Abstract

A new algorithm for generating a regular logic structure, Pseudo-Symmetric Functional Decision Diagrams (PSFDDs), for completely specified Boolean functions is presented. The diagrams are based on Functional Decision Diagrams (FDDs) and Pseudo-Symmetric Binary Decision Diagrams (PSBDDs). A Davio expansion is used to generate the initial vertex subfunctions which are then modified by a new Join-XOR operation. The operation allows to combine adjacent vertices such that the function is represented as a regular pseudo-symmetric network which can be easily implemented with an array of AND/XOR gates. Due to the regular structure the interconnection length is known from the logic representation so the post-layout delays can be accurately predicted before the layout is completed.[1]

I. INTRODUCTION

In recent years, synthesis methods based on AND/XOR realization have gained more attention. AND/XOR realization proved to be very efficient for large classes of circuits like arithmetic circuits, error correcting circuits and circuits for telecommunication [3]. Functional Decision Diagrams (FDDs), which have been introduced [1, 2] in the last couple of years, represent various multilevel AND/XOR circuit realizations. A vertex in an FDD represents XOR/AND complex gate and, therefore, FDDs can be implemented as an array of such gates. AND/XOR gates are usually available as a single logic blocks in a number of fine-grain FPGA architectures. Unfortunately, the number of vertices in FDDs can increase exponentially with the number of levels, while a number of logic blocks in a two-dimensional array increases linearly in every direction, and therefore mapping is a difficult task. Placement and routing of such non-complete binary trees, FDDs, result in an unpredictable increase in the interconnection length and consequently increased delay. In submicron technologies where a delay of the device is limited by the delay of interconnects all the efforts spend on minimizing a number of gates in a logic representation can be overshadowed by unpredictable routing.

We have implemented an algorithm for generating a regular multi-level, two-dimensional AND/XOR representation for combinational circuits by extending the ideas from Pseudo-Symmetric Binary Decision Diagrams (PSBDDs) [4, 5]. This new data structure, called Pseudo-Symmetric Functional Decision Diagrams (PSFDDs), can be generated for any arbitrary Boolean function. The regular representation on the logic level is directly transferred to a target layout. A layout design, relative positions of gates and wires, is known before the layout is completed. The main advantages of PSFDDs are localized connections, predictable delay, and no placement or routing required. In addition, a known, in a pre-layout phase, interconnection structure can be used to predict power dissipation associated with wires. PSFDDs can be used to implement general functions in technologies where interconnect delay is a limiting factor, as it is a case for submicron technologies. In contrast to FDDs, PSFDDs can also be easily mapped to a number of fine-grain CA-type FPGAs. Some previously mentioned classes of functions should especially benefit from this representation. We generated PSFDDs for a number of MCNC benchmark functions and shown that such representation is feasible.

II. BACKGROUND

A Functional Decision Diagram (FDD), a graph representation of a Boolean function f, is a rooted tree G=(V, E) generated from f by successive application of Davio decompositions. The function of vertex v and its two successors satisfy Davio decompositions.

$f(v) = low(v) \oplus x_i\, high(v)$ Davio I   (positive, pD)
$f(v) = low(v) \oplus x_i'high(v)$ Davio II (negative, nD)

The Functional Decision Diagram for an arbitrary Boolean function generated using only Davio I expansion is shown in Fig.1a. In Fig.1b the FDD for a function with variables a and b being symmetric, $fa'b = fab'$, is presented. It can be noticed that the diagram from Fig.1b is a regular and can be easily map to a two-dimensional array with a root node being placed in the upper left corner of the array, as shown in Fig.2b. In Fig.2a. a mapping of the non-symmetric FDD form Fig.1a is shown and the advantages of the regular logic structure for mapping are obvious by comparing these two figures.

If only Davio I decomposition is used, FDDs are called Positive Polarity FDD. In general, any combination of Davio I and Davio II can be used, and such diagrams are called Fixed Polarity FDDs. To simplify the description we will only consider Positive Polarity FDDs with one expansion variable on a level.

A totally symmetric function can always be represented by an OFDD (Ordered FDD) which can be drawn as a planar regular array. Due to the symmetric properties of all variables in a totally symmetric

---

[1] This work was supported in part by the NSF grant MIP-9629419

function, adjacent vertices in a planar drawing of the OFDD are equal and can be represented as one vertex. Therefore, we can combine such vertices into one and if all adjacent vertices in the tree are combined, a regular two-dimensional structure, shown on Fig.1b, is created. It is, however, not the case for an arbitrary function, so to generate the similar regular representation for non-symmetric function some modifications have to be made. A *Join/XOR* operation which allows to combine adjacent vertices, even if they are not equal, without changing the original function is introduced. New diagrams are called Pseudo-Symmetric Functional Decision Diagrams, PSFDDs, (PSFDDs).

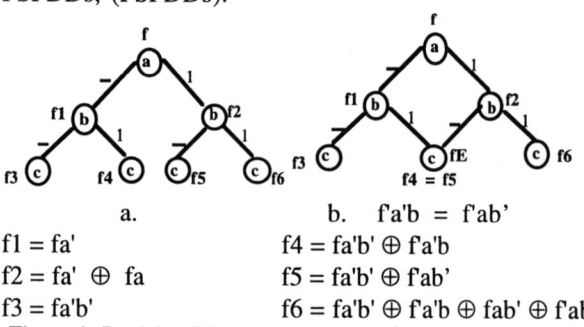

a.
b. $f'a'b = f'ab'$

f1 = fa'
f2 = fa' $\oplus$ fa
f3 = fa'b'

f4 = fa'b' $\oplus$ f'a'b
f5 = fa'b' $\oplus$ f'ab'
f6 = fa'b' $\oplus$ f'a'b $\oplus$ fab' $\oplus$ f'ab

Figure 1. Decision Diagram, a.) *a* and *b* are non-symmetric variables, b.) *a* and *b* are symmetric variables

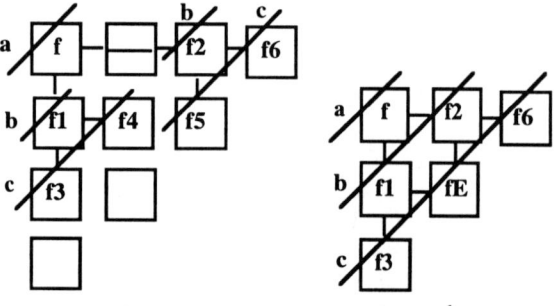

a.
b.

Figure 2. Mapping FDDs to a regular array, a.) diagram from Fig.1a., b.) from Fig. 1b.

III. GENERATING PSFDDs

Using the Join-XOR operation, PSFDDs are created by recursively applying Davio decompositions and subsequently combining adjacent vertices into a single vertex. The idea of the *Join/XOR* operation is based on the *Join/OR* operation in PSBDDs [4, 5] but is a bit more complex. The *Join/XOR* operation for two level expansion is shown in Fig.3. To prove that root functions of the two diagrams in Fig.3, FDD and PSFDD, are the same we show that nodes B and C in the PSFDD are the same as in FDD.

B in PSFDD $= D \oplus x2W$
$= D \oplus x2(x2E \oplus \overline{x2F})$
$= D \oplus x2(x2E \oplus \overline{x2F})$ = **B in FDD**

C in PSFDD $= W \oplus x2Y$

$= x2E \oplus \overline{x2F} \oplus x2(E \oplus F \oplus G)$
$= x2E \oplus \overline{x2F} \oplus x2E \oplus x2F \oplus x2G$
$= \overline{x2F} \oplus x2F \oplus x2G$
$= F \oplus x2G$ = **C in FDD**

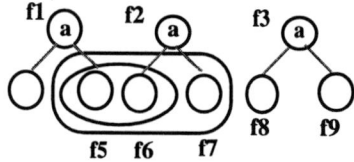

$A = B \oplus x1C$ $W = x2E \oplus \overline{x2F}$
$B = D \oplus x2E$ $Y = E \oplus F \oplus G$
$C = F \oplus x2G$

Figure 3. A two-level expansion for FDD and PSFDD.

As it can be noticed, the *Join-XOR* operation is not limited to two vertices which are joined together but propagates also to the adjacent vertices. In Fig.4, the iterative procedure for performing the *Join-XOR* operation on the entire row of vertices is shown.

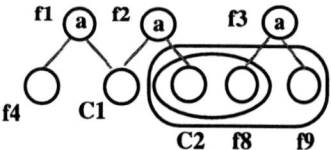

a. generating cofactors

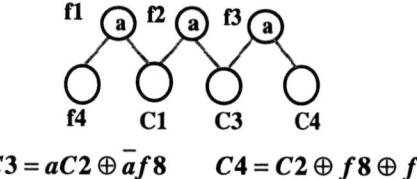

b. $C1 = af5 \oplus \overline{a}f6$ $C2 = f5 \oplus f6 \oplus f7$

c. $C3 = aC2 \oplus \overline{a}f8$ $C4 = C2 \oplus f8 \oplus f9$

Figure 4. Constructing PSFDD - one level

To preserve the structure of a function we perform the *Join-XOR* operation from the most left vertex to the most right vertex, recursively. The left-right direction is used for positive Davio expansion, and for negative Davio expansion the opposite direction from the right to the left has to be performed. The correctness of the both operations can be easily proved. The steps of *the Join-XOR* operation for one level of positive Davio expansion are shown in Fig.4. Since during the *Join-XOR* operation variables are reintroduced back to the function a variable can appear more than once in the same path from the root vertex to the terminal vertex. It can be shown that there exist an upper bound on a number of times each variable needs to be used as a decomposition variable and, therefore, the process of generating such diagrams will always converge. The number of nodes and the number of levels in the PSFDD depends very strongly on the order of variables. Due to the multiple appearance of the same variables the different approaches to ordering variables, similarly as in PSBDDs [5], can be used. PSFDDs are canonical structures in respect to variable order and type of decomposition used.

IV. VARIABLE ORDERING

In addition to the variable ordering another strategies can be used if minimizing the size of the diagram is of the main concern. One such strategy is explained below. In Fig.5 two regular representations of the given function are shown. They differ in the variable orders and in the strategies applied during diagram generations. The PSFDDs, created accordingly to the PSFDD description, with *cabda,* non-restricted, and *caabd,* restricted, variable orders are shown in Fig.5a and 5b, respectively. The non-restricted variable order is the order in which any variable can appear at any position and no variable has to be repeated consecutively until it is removed completely from the function as it is the case for variable *a* in Fig.5b. The only difference between the two variable orders is that variable *a* is repeated consecutively. In this example the number of nodes in two diagrams is the same but it is not the rule.

$$f = a\overline{b}\overline{d} + bc + cd$$

Variable order (c,a,b,d,a)　　　　**(c,a,a,b,d)**

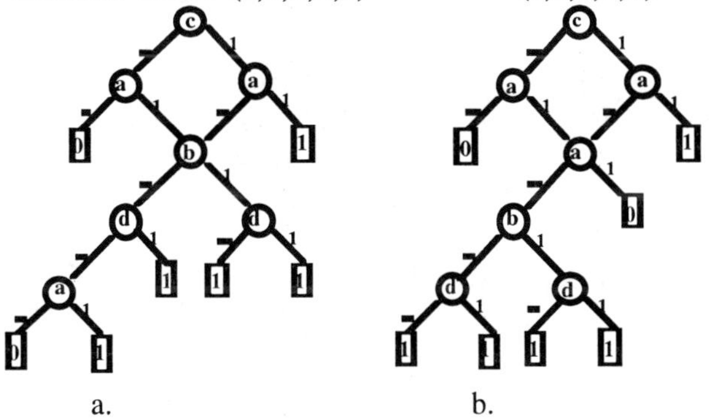

a.　　　　　　　　　　b.

$$f = a \oplus ac \oplus ad \oplus cd \oplus ab \oplus bc \oplus abd \oplus bcd \oplus ac$$
$$f = a \oplus ad \oplus abd \oplus cd \oplus ab \oplus bc \oplus bcd$$

Figure 5. Restricted and non-restricted variable orders

The decision diagram, shown in Fig.6, has been created for the same function as PSFDDs in Fig 5. and using the restricted variable ordering method, as in Fig.5b. Please notice that the variable order is almost the same as in Fig 5a. The only difference is that variable *b* is repeated consecutively in Fig 6. and variable *a* is repeated at the end of the variable order in Fig.5a.

However the generation process has been modified. The modification to the generic PSFDD generation procedure is added by not combining two nodes, indicated with "*" as it should be done according to the PSFDD description. The motivation behind this change is that because there exist only one not constant cofactor for each vertex on the previous level, the *Join-XOR* operation is not needed to maintain the linear growth of the structure. For larger examples this method gives diagrams with smaller number of nodes. However, as could be seen from the comparison between diagrams in Fig.5a and 5b with 7 nodes, and the diagram in Fig.6. with 10 nodes, this method does not always guarantee better results.

Variable order (c, a, b, b, d)

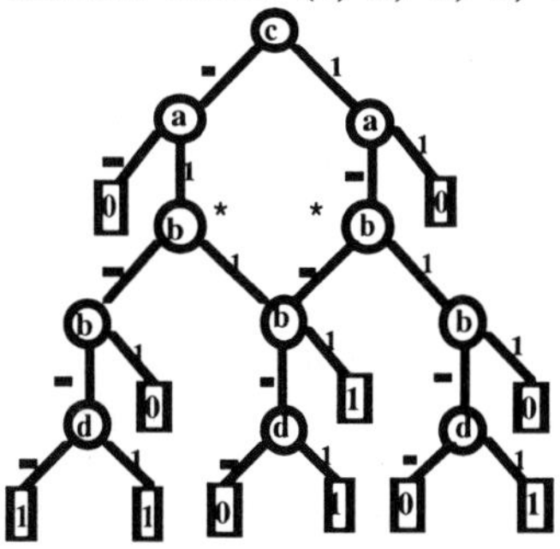

$$f = a \oplus ad \oplus abd \oplus cd \oplus ab \oplus bc \oplus bcd$$

Figure 6. Modified PSFDD

Only positive Davio expansion is used in Fig.5 and Fig.6, therefore, a Reed-Mueller canonical representation of the function can be obtained by flattering the diagrams, and Reed-Muller representations are given at the bottom of each diagram.

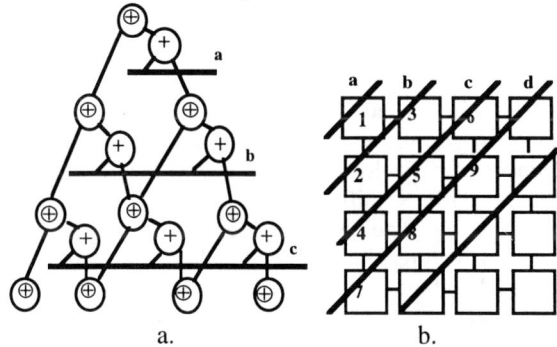

a. 　　　　　　b.

Figure 6. a) PSFDD implementation with XOR/AND gates, b) mapping to an AND/XOR array.

A circuit realization using an array of XOR/AND gates is shown in Fig.6. Only neighbor-to-neighbor connections are used to realized edges of the PSFDD, and expansion variables are assigned to global busses.

V. RESULTS AND CONCLUSIONS

A size of the PSFDD depends very strongly on the variable order of the input variables and on the order of variables reintroduced back to the function. In this approach we first check each function for symmetry sets and then the largest symmetry set is placed at the beginning of the order so we get a regular structure without variable injections, next, some heuristic rules are used to order the remaining variables. Ordering heuristics are not discussed due to the space limitation.

The algorithm, for generating PSFDDs, was coded in the C language and run in the UNIX environment on SPARC workstations. In Table 1 we present the results obtained for the MCNC benchmark functions. Function names are given in the first column, and the name of the specific output is given in the next one. The number of inputs is given in the third column. In *"without ordering"* and *"with ordering"* sections results for algorithms without any ordering heuristic and with simple heuristic rules for non-symmetric variables are given, respectively. The number of nodes and the number of levels are reported in both sections, and *"na"* means that results are not available. Please notice that for a number of functions like for example c8, ttt2 output o0 and sct output b0 the size of PSFDDs was significantly decreased with preordering. For some functions we were able to generate PSFDDs only when using ordering, but in some other cases the ordering heuristic made it worst.

Creating layout with PSFDD is straight forward. As can be seen from the PSFDD implementation in Fig. 6, the area occupied by the PSFDD is proportional to the number of nodes and can be easily estimated by multiplying the number of nodes by the area of a single cell. The final layout created with PSFDDs is very compact. The upper bound on the area occupied by a PSFDD would be 1/2 *(#of levels)•(#of levels)•area*, where *area* is the area of a single cell. The average area could be roughly estimated as *(#of levels/2)•(#of levels/2)•area* which is a half of the upper bound.

The main technical contribution of this paper is the development of the algorithm which allows for generating a new function representation, Pseudo-Symmetric Functional Decision Diagrams, for a completely specified arbitrary Boolean functions with a large number of variables. We also showed that the size of PSBDDs can be significantly decreased by a proper variable ordering. The idea of creating regular two-dimensional structures during logic synthesis has several important advantages. It merges the stages of logic synthesis and layout synthesis into a single stage, making use of the regularity of the structure. No placement or routing is required. This is especially important for submicron technologies where performance is limited by interconnection delay. The length of interconnections is known in the pre-layout stage and therefore the delay and power associated with wires can be very accurately predicted. This information can be used in delay, power and area optimization. We shown that the regular structures for non-symmetric arbitrary Boolean functions can be created using AND/XOR representation, which is superior to AND/OR representation for a number of classes of functions. The ideas presented in this paper for PSFDDs can be easily combined with Pseudo-Symmetric Binary Decision Diagrams presented in [5] to create Pseudo-Symmetric Kronecker Decision Diagrams including Generalized and Free Kronecker Diagrams as presented in [6].

REFERENCES
1. U. Kebschull W. Rosenstiel. "Efficient Graph-based Computation and Manipulation of Functional Decision Diagrams"., *Proc. of EDAC*, pp.278-282, 1993.
2. U. Kebschull, E. Schubert, W. Rosenstiel, "Multilevel logic based on functional decision diagrams", *Proc. of EDAC*, pp. 43-47, 1992.
3. Ch-Ch. Tsai, M. Marek-Sadowska, "Multilevel Logic Synthesis for Arithmetic Functions," *Proc. of DAC*, pp.242-247, 1996.
4. M. Chrzanowska-Jeske, Z Wang, "Mapping of Symmetric and Partially Symmetric Functions to the C A-type FPGAs," *Proc. MSCS,* pp.290-293, 1995.
5. M. Chrzanowska-Jeske, Z. Wang, Y. Xu, "A Regular Representation for Mapping to Fine-Grain, Locally-Connected FPGAs," *Proc. of ISCAS*, vol.4, pp.2749-2752, 1997.
6. M. Perkowski, M. Chrzanowska-Jeske, "Reed-Muller Lattice Diagrams," *Proc. of IFIP W.G. Workshop on Application of RM Expansion in Circuit Design,* , 1997.

Table 1

File name	#out	#ins	with ordering		without ordering	
			#lev	#nods	#lev	#nods
B9	r0	9	16	87	12	41
	b1	9	20	149	15	60
	j1	9	16	115	19	156
c8	u0	11	11	63	26	293
cht	y1	5	9	43	7	27
cm162a	q	10	11	30	12	18
cm163a	t	9	11	37	10	27
count	k0	5	8	37	6	18
	n0	8	13	79	19	159
ex2	k2	6	8	26	na	na
	f3	5	7	19	na	na
mux	v	21	107	5015	na	na
	g3	7	11	51	na	na
pcle	z	10	12	50	12	49
sct	x	7	10	37	9	36
	y	9	11	43	12	54
	z	9	12	49	14	77
	a0	10	13	57	17	111
	b0	11	14	66	20	157
term1	n0	17	79	1544	79	1543
	o0	18	144	6574	144	6573
	p0	19	174	9589	174	9588
	e0	12	15	76	23	207
x2	q	10	13	53	21	184
x4	e5	7	20	192	15	114
ttt2	g0	5	9	39	8	26
	j0	7	8	28	8	23
	o0	14	33	345	110	3494
	r0	7	16	101	22	217

A NEW LOCK BASED STATE CODING METHODOLOGY FOR SIGNAL TRANSITION GRAPHS

Radhakrishna Nagalla
School of Computer Science and Engineering
University of New South Wales
NSW, 2052, Australia

ABSTRACT

In this paper a new approach for enforcing complete state coding (CSC) property in Signal Transition Graph (STG) specifications will be discussed. As a novel contribution we propose a lock based methodology to ascertain whether a given STG has Complete State Coding (CSC) property. Unlike most of the existing methods which operate on a state graph, our method operates on the STG. This approach has the advantage of being either easily automated or easier to visually correlate with the STG specifications. Experimental results with a large number of practical asynchronous bench marks are presented.

1 INTRODUCTION

Signal Transition Graphs (STGs) are a subclass of interpreted Petri nets originally introduced by Chu [1], for the specification of asynchronous control circuits. The problem of synthesizing asynchronous interface circuits from STG specifications has been studied by many researchers [1, 2, 3, 4, 5, 7, 9, 10, 12]. To realize the asynchronous behavior as a logic circuit, the STG specifications must satisfy the complete state coding (CSC) property[5]. Intuitively, the CSC means that the signals specified by the STG completely define the circuit states. If the given STG specification does not satisfy the CSC property, it must be transformed to satisfy the CSC. Based on the main data structure they use for ascertaining the CSC property, all the existing methods can be classified into two distinctive categories.

The first category uses the state graph (SG) derived from the interpreted Petri net as the main data structure. Under this scheme, a number of researchers [2, 3, 10, 12] have proposed a variety of methodologies. Since the size of the SG can be exponential with respect to the number of signal transitions of the circuit, all these methods which use the state graph scheme show an exponential time complexity for their worst case.

In the methods under second category, unique state coding (USC) and CSC properties are checked and enforced at the STG level. Most of the methods [4, 7, 9, 11] specified under this category are run-time efficient and can be applied to satisfy state coding properties in very large and concurrent STG specifications. Majority of the methods under this category can only be applied to marked graph specifications and those which can be applied to free-choice nets [9] are pessimistic in nature.

In this paper, we proposed initially a new lock based methodology on marked graph specifications and later this methodology was extended to free-choice nets and some non-free choice nets. Compared to the existing methods, this approach achieves considerable improvement in terms of the computing time and the implementation area.

2 PRELIMINARIES

A Petri net is a four-tuple $\langle T, P, F, M_0 \rangle$ where T, P and F form a directed bipartite graph and M_0 is the initial *marking*. T is a set of net transitions. P is a set of places which can be used to specify conflict or choice. F gives the flow relation between transitions and places: i.e. $F \subseteq (T \times P) \cup (P \times T)$. A Petri net is a **Marked Graph** (MG) if every *place* in it has exactly one predecessor and one successor. A Petri net is a **State Machine** (SM) if every *transition* in it has exactly one predecessor and one successor. A **Free-Choice net** (FC net) is a Petri net where the input place p must be unique, if two or more transitions share the same p. An FC net is *live* if every transition can be enabled through some sequence of firings from initial marking M_0. An FC net is *safe* if all simple cycles contain at least one token and each token exists in a cycle where it is the only token. A good review of more basic Petri net concepts is given in [6].

An STG is an interpreted free-choice Petri net where the net transitions are interpreted as rising and falling signal transitions in an asynchronous control circuit. Places with only one input and one output are omitted in the STGs. The transitions are described by $t \times \{+,-\}$, where $t+$ represents a 0 to 1 transition of signal t and $t-$ a 1 to 0 transition. An example STG is shown in Figure 1(a)(i). An STG is **live**, if the underlying petri net is live and safe, and the rising and falling transitions of each signal strictly alternate.

To derive a logic circuit, an STG must be converted into a state graph [1]. The synthesis procedure described in [1] uses the signals in a circuit directly as state variables, so that the circuit must be able to tell its global state only from its input and output signals. When two different states are given the same binary representation, the digital circuit cannot distinguish the two states from each other. Thus every state of the SG must be assigned a unique binary vector of the signal values.

A state graph has a **unique state coding** (USC) if there are no two states with the same binary coding. A state graph is said to satisfy the **complete state coding** (CSC) constraint if either it satisfies the USC condition or the transitions of non-input signals enabled in two states with the same binary state assignment are the same. The CSC constraint is the necessary and sufficient constraint to derive the logic circuit functions from the state graph [5]. A CSC violation must be corrected by inserting new signal transitions in the STG, so as to distinguish between the states violating the CSC constraint.

Let S represents the set of all the transitions in a given live STG specifications.

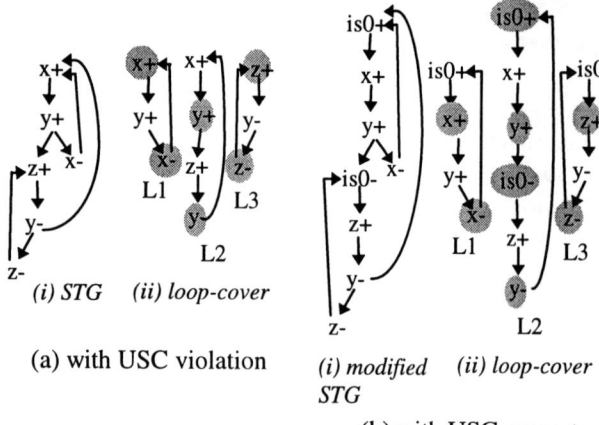

(a) with USC violation (i) modified STG (ii) loop-cover

(b) with USC property

Figure 1. *An example illustrating the significance of the loop components and the loop-cover*

A set of transitions T in a live STG is said to be **feasible** iff there exist a state from which these transitions can fire without firing any of the transitions not belonging to the set.

A set of transitions T_c is said to be **complementary set** iff $T_c \subset S$ (T_c is proper subset of S, i.e. T_c does not contain all the transition of S) and $t^* \in T_c \Rightarrow t^{*\sim} \in T_c$ where $t^{*\sim}$ represents a complementary transition of t^*.

A set of transitions T_c is said to be **feasible complementary set (FCS)** if it is feasible and a complementary set. The set of transitions in S excluding an FCS is called **paired feasible complementary set (PFCS)**. If the given STG is a complete cyclic graph, PFCS will be another FCS.

A set of four transitions t_1^*, t_2^*, t_3^* and t_4^* are **interleaved**, denoted by $I(t_1^*, t_2^*, t_3^*, t_4^*)$ iff the sequence of transitions lies $t_1^* \rightarrow t_3^* \rightarrow t_2^* \rightarrow t_4^*$ in a simple cycle.

The following theorem (proved in [1]) gives the relationship between the USC property and the complementary set T_c.

Theorem 1 : *A live STG has the unique state coding property iff there are no feasible complementary set (FCS) of transitions in S.*

3 STATE CODING IN MARKED GRAPHS

In this section, we first propose a new methodology for enforcing USC/CSC property in marked graph (MG) specifications. We will first derive a set of minimum number of state machine (SM) components that completely cover all the arcs (places) in the given STG. Each SM component will be a simple loop because the original STG is a marked graph. Thus such a set of the SM components or **loop components (LCs)** is called **loop-cover**. The loop-covers for the STGs given in Figures 1(a)(i) and 1(b)(i) are shown in Figures 1(a)(ii) and 1(b)(ii). The SM component in which all transitions of a signal reside, is considered as representing that signal and such transitions are called **members** of the SM component. In Figure 1(a)(i) & 1(b)(ii) the members are highlighted. The loop-cover is derived in a such a way that every signal is member of at least one loop component.

Theorem 2 : *If the loop components in a marked graph STG are not fully locked, there will be always a USC violation [8].*

Figure 1(a) shows an STG with USC violation and its loop-cover. In Figure 1(a)(ii), the member transitions of L1, L2 and L3 are not interleaved with one and another and thus the STG is not fully locked. Figure 1(b) shows a modified STG with USC property and the corresponding LCs which are fully locked.

The above theorem only provides a necessary condition for the USC property. It will only be a sufficient condition if member transitions in each LC are interleaved among each other. Figure 2(a) illustrates an example where LCs are fully locked but the member transitions within the LCs are not interleaved. If the member transitions are not interleaved, they may form FCSs. The FCSs, if exist, will be detected with the help of loop complementary paths in LCs.

In an LC, a path from a member transition t^* to another member transition b^* is defined as a **loop complementary path** ($LCP(t^*, b^*)$) if member transitions between t^* and b^*, including t^* and b^* form complementary set of transitions (and a subset of all transitions in that LC), i.e. if a member transition belong to the path so does the complementary member transition. For example, $LCP(dr-, da+) = \{dr-, da-, lr+, la+, dr+, da+\}$ in the STG shown in Figure 2(a).

An $LCP( )$ is **valid** if it can form a feasible complementary set (FCS) of transitions either by itself or in combination with the LCPs in other LCs, otherwise it is **invalid**. The validity of an $LCP( )$ and the associated FCS can be calculated using the *validate_LCP( )* procedure as shown in Figure 3. The operation of the procedure is illustrated below by the example shown in Figure 2(a). Initially all nodes in the STG are marked as don't-care.

Let us first consider $LCP(zr+, zr-)$ in loop $L1$. While traversing from $zr+$ to $zr-$ the nodes encountered, $\{zr+, za+, zr-\}$ are marked as valid. Because $LCP(dr-, da+) \cup LCP(zr+, zr-)$ completely covers all member nodes in $L1$, the nodes encountered $\{dr-, da-, lr+, la+, dr+, da+\}$, while traversing from $dr-$ to $da+$ are marked as invalid. In the next step, the LC in which a member valid node exists is $L3$ and the LCP in which no invalid node exists is $LCP(zr+, za-)$. Thus nodes$\{zr+, za+, zr-, za-\}$ are marked as valid and the nodes $\{lr+, la+, dr+, lr-\}$ are marked as invalid. Because in the next step there exists no other LC in which a member valid node exist, $LCP(zr+, zr-\}$ and $LCP(zr+, za-\}$ are valid and all the valid nodes form a FCS.

All the nodes in the STG except the valid ones form the corresponding PFCS of an FCS. Find the top and bottom nodes in both FCS and PFCS. The top (bottom) nodes represent those transitions out of which one of them may be enabled first (last) in the sequence transitions. The reason why we are interested in the top nodes in PFCS is that one of these transitions will be enabled immediately after firing all the transitions in the FCS. Thus it can be seen very easily that the top nodes in FCS and PFCS are enabled in different sets of states which has same binary coding. Thus in case of CSC (and not USC) violations, FCS is significant only if non-input transitions in the top nodes of FCS and PFCS are different. Thus the FCS becomes $\{zr+(T), za+, zr-, za-(B) / da+(b), dr-(t), lr-(b) \}$, where $T(t)$ and $B(b)$ indicates top and bottom in FCS (PFCS). We are only interested in the top and bottom nodes of PFCS.

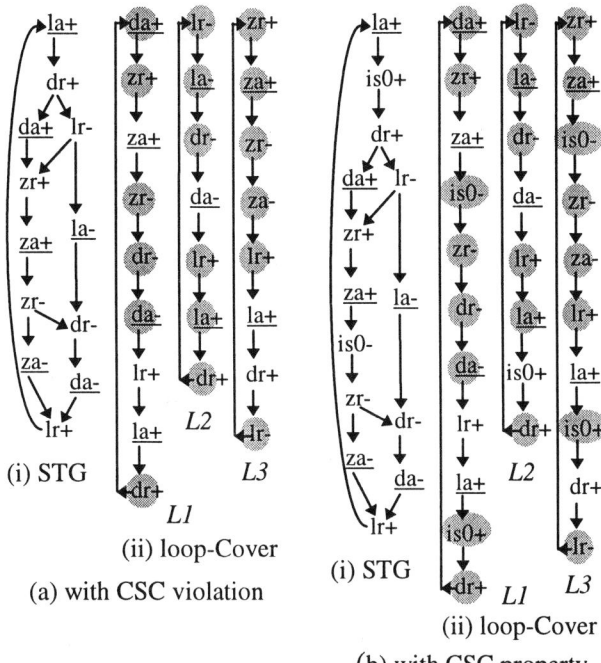

Figure 2. *An example illustrating the significance of the local complementary paths*

Similarly if we start with LCP(*dr-*, *da+*), the LCP(*dr-*, *da+*), the LCP(*lr+*, *lr-*) and the LC, *L2* form another FCS which is equal to {*da+*(B), *dr-*(T), *da-*, *lr+*, *la+*, *dr+*, *lr-*(B), *la-*(X) / *zr+*(t), *za-*(b) } where X represents a don't-care position. A node position can be don't-care if it can be a either top or bottom position. If we start with either LCP(*lr+*, *dr-*) or LCP(*lr-*, *lr+*), we will find that there are no FCSs associated with them, and thus they are invalid LCPs.

The number of potential FCSs in an STG where the LCs are not fully locked can be large. In large concurrent STG specifications, the number of FCSs can be reduced significantly by fully locking all the LCs. Thus the procedure inserts new signal transitions in such a way that the LCs are fully locked and no valid LCPs and in turn no FCSs exist. The FCSs in the example shown in Figure 2(a) are solved by inserting *is0+* and *is0-* as shown in Figure 2(b).

4 STATE CODING IN FREE-CHOICE STGS

Our methodology was extended to free-choice STGs by unfolding all the MG-components and thus creating a new single MG representing all the MG components [8]. Then the techniques described in the previous section were used to enforce USC/CSC property within the MG components and their combinations. Our methodology can also be used to enforce the state coding property in some of the non-free-choice STGs (alex-nonfc[10] example).

5 EXPERIMENTAL RESULTS

Our lock based USC/CSC algorithm for ensuring the correct state encoding in the STG specifications was implemented in the

validate_LCP(LCP(x^*, y^*)) {
    Let LCP(x^*, y^*) be a loop complementary path in LC L_n.
step1: Mark all nodes between x^* and y^* encountered as valid. If there exists an LCP(p^*, q^*) in L_n, such that LCP(p^*, q^*) $\bigcup$ LCP(x^*, y^*) contains all the member nodes of LC L_n, then mark all nodes between p^* and q^* as invalid. Mark the LC L_n as visited.
step2: If there exists an LC L_k in which one of its member nodes a^* is a valid node such that a^* is not a member of already visited LCs **then** {
    If there exists an LCP(u^*, v^*) in L_k such that none of the transitions between u^* and v^* are invalid **then** {
        If validate_LCP(LCP(u^*, v^*)) ==*valid* then
            LCP(u^*, v^*) is *valid*, return *valid*;
        else
            return *invalid*;
    }
    else {
        if none of the transitions in L_k are invalid **then**
            mark transitions in L_k as valid, go to **step 2**.
        else
            return *invalid*;
    }
}
else
    Form an *FCS* with all the valid nodes and form the *PFCS* with the remaining nodes in the STG. Find top and bottom nodes in both FCS and PFCS. return **valid**;
}

Figure 3. *An algorithm for validating local complementary paths and creating any existing FCSs*

C language. We tested our algorithm on a large number of STG benchmarks [3] including both free-choice and non-free choice STGs (highlighted in Table 1). We compared the performance of our algorithm with other algorithms such as e.g. Lavagano et al. [3] and Cortadella et al. [2] algorithms. The results of these experiments are given in Table 1. The results indicate that our algorithm achieves many orders of magnitude of performance improvement in terms of execution time. The reason is that the complexity of our algorithm is a polynomial function of the number of transitions, whereas in case of Lavagno's et al. algorithm and Cortadella's et. al. algorithm the complexity is a function of the number of states. The CSC algorithm also performs on an average better than Lavagno's et al. and is comparable with the Cortadella's et al. in terms of implementation area.

We have used Cortadella's et al. [2] synthesis tool, *petrify,* to obtain the modified STGs with the CSC problem solved using the *petrify -csc* option. In all the three cases, the two level implementation area literals were calculated from the prime-irredundant cover using the **astg_syn -r** option in the U.C. Berkeley logic synthesis tool SIS. Even though SIS is used to calculate the two-level area literals in order to compare all the three methods, the execution times given in Table 1 represent the actual CPU times taken by *SIS, Petrify* and *our method* to satisfy the CSC for all the benchmark examples. In case of *our*

Table 1. *Experimental results with practical benchmarks[3] obtained on a SPARC SLC workstation*

STG Name	Initial specifications			SIS [3]			Petrify [2]			Our method		
	Initial signals	Initial trans.	Initial states	Final signals	2 level Area lit.	CPU time sec.	Final signals	2 level Area lit.	CPU time sec.	Final signals	2 level Area lit.	CPU time sec.
mr0	11	22	302	13	86	1519.7	14	48	356.56	14	47	0.29+32.1
mr1	9	18	190	11	53	323.7	12	43	210.98	12	36	0.13+13.2
mmu1	8	16	82	10	37	58.8	10	32	111.99	10	33	0.15+3.9
sbuf-ram-wrt	10	20	58	12	35	56.1	12	24	192.96	12	27	0.25+4.5
vbe4a	6	12	58	8	41	6.7	8	24	68.46	8	25	0.10+3.0
nak-pa	9	18	56	10	41	25.4	10	21	38.21	10	19	0.19+3.2
pe-rcv-ifc-fc	8	45	46	9	62	13.3	9	53	392.23	10	43	43.89+3.5
ram-read-sbuf	10	20	36	11	23	84.8	11	20	52.20	11	21	0.10+2.7
alex-nonfc	6	21	24	Non free-choice STG			7	31	37.49	7	31	0.16+1.2
sbuf-send-pkt	6	23	21	7	14	8.5	7	18	18.51	7	16	0.46+0.9
duplicator	4	12	20	5	24	1.0	6	22	21.61	6	15	0.10+1.2
sbuf-send-ctl	6	20	20	8	43	3.0	8	29	98.99	8	32	0.60+2.0
atod	6	12	20	7	19	3.5	7	19	10.42	7	18	0.10+1.2
alloc-outbnd	7	20	17	9	23	2.5	9	18	26.52	9	18	0.34+1.2
sbuf-read-ctl	6	12	14	7	15	2.0	7	15	8.43	7	15	0.05+0.07
Total					660	2143.4		527	1853.42		489	47.66+79.2

algorithm, the first component in the execution time indicates the time taken by our algorithm to satisfy USC/CSC and the second component indicates the time taken to generate the logic implementation for the modified STG by SIS.

6 CONCLUSION

An efficient lock based approach for enforcing USC/CSC in the given STG specifications is presented. This approach identifies the unlocked state machine (loop) components and provides a mechanism to lock all the components. Then any existing feasible complementary sets will be detected with the help of local complementary paths and will be removed by appropriately inserting new signal transitions. Unlike most of the existing algorithms which operate on a state graph, this algorithm operates on the STG. This approach has the advantage of being either easily automated or easier to visually correlate with the STG specifications and the calculation can be undertaken with a paper and pencil. Experimental results from a large number of practical asynchronous bench marks indicate that, compared to existing techniques, our visual synthesis method achieves considerable improvement in terms of computing time as well as a reduced implementation area.

REFERENCES

[1] Tam-Anh Chu. Synthesis of Self-Timed VLSI circuits from Graph-theoretic Specifications, PhD thesis, MIT, June 1987.

[2] J. Cortadella, M. Kishinevsky, A. Kondratyev, L. Lavagno and A. Yakovlev. *Methodology and tools for state encoding in asynchronous circuit synthesis*. In Proc. of 33rd DAC, pages 63-66, 1996.

[3] L. Lavagno, C. Moon, R. Brayton, and A. Sangiovanni-Vincentelli. Solving the State Assignment Problem for Signal Transition Graphs. In *Proc. of 29th DAC*, pages 568-572, 1992.

[4] K. J. Lin, and C. S. Lin. Automatic Synthesis of Asynchronous Circuits. In *Proc. of 28th DAC*, pages 296-301, 1991.

[5] C. W. Moon, P. R. Stephan, and R. K. Brayton. Specification, Synthesis and Verification of Hazard-Free Asynchronous Circuits. In *Proc. of ICCAD*, pages 322-325, 1991.

[6] T. Murata. Petri nets: Properties, Analysis and Applications. *Proc. of the IEEE*, 77(4):541-580, April 1989.

[7] R. Nagalla, and G. Hellestrand. Signal Transition Graph Constraints for Synthesis of Hazard-Free Asynchronous Circuits with Unbounded-Gate Delays. In *Formal Methods in System Design*, 5, pages 245-273, 1994.

[8] R. Nagalla. Synthesis of Asynchronous Interface Circuits from Signal Transition Graph Specifications, PhD thesis (submitted), UNSW, Australia, 1997.

[9] E. Pastor, and J. Cortadella. An Efficient Unique State Coding Algorithm for Signal Transition Graphs, In *Proc. of ICCD*, pages 174-177, 1993.

[10] R. Puri, and J. Gu. A Modular Partitioning Approach for Asynchronous Circuit Synthesis. In *Proc. of DAC*, 1994.

[11] P. Vanbekbergen, F. Catthoor, G. Goossens, and H. De Man. Optimized Synthesis of Asynchronous Control Circuits from Graph-theoretic Specifications. In *Proc. of ICCAD*, pp 184-187, 1990.

[12] P. Vanbekbergen, B. Lin, G. Goossens, and H. De Man. A Generalized State Assignment Theory for Transformations on Signal Transition Graphs. In *Proc. of ICCAD*, pages 112-117, 1992.

MULTI–INPUT/MULTI–OUTPUT BLOCK DIAGRAM GRAMMAR

Yoshihiro Adachi, Suguru Kobayashi, and Kensei Tsuchida

Department of Information and Computer Sciences, Toyo University
2100, Kujirai, Kawagoe, Saitama, 350, Japan
TEL:+81-492-39-1442, FAX:+81-492-33-9788, Email:adachi@eng.toyo.ac.jp

ABSTRACT

A multi–input/multi–output block diagram grammar for a block diagram with multiple inputs and/or multiple outputs is formalized in terms of a context–sensitive graph grammar. This grammar is defined by adding to the single-input, single-output block diagram grammar we proposed previously dummy nodes and new productions for dealing with multiple inputs and/or multiple outputs. A parser based on the multi–input/multi–output block diagram grammar is also implemented; it uses a bottom–up parallel algorithm to parse diagrams with multiple inputs and/or multiple outputs. The block diagram grammar defined in this paper makes a fundamental and important formal model for system analysis and design using block diagrams on a computer.

1. INTRODUCTION

The block diagram is widely used as a graphical tool for system analysis and design in control engineering, digital (analogue) filter design, and other fields. We have proposed a block diagram grammar that is formalized for the syntax of a block diagram with a single input and a single output [1, 2].

In modern control engineering, however, engineers must often deal with systems with multiple inputs and/or multiple outputs. Therefore, a formal model for generating and analyzing block diagrams with multiple inputs and/or multiple outputs on a computer must be developed.

In this paper, a multi–input/multi–output block diagram grammar (BDG) for a block diagram with multiple inputs and/or multiple outputs is formalized in terms of a context–sensitive graph grammar. This grammar is defined by adding dummy nodes and new productions for dealing with multiple inputs and/or multiple outputs to the grammar for a block diagram with a single input and a single output. Adding dummy nodes makes it possible to deal analogously with a block diagram with multiple inputs and/or multiple outputs and a block diagram with a single input and a single output. A parser based on the BDG is also implemented; it uses a bottom–up parallel algorithm to parse diagrams with multiple inputs and/or multiple outputs.

2. GRAPH GRAMMAR AND DERIVATION

Referring to work edited by Rozenberg[3], we define a graph grammar using subgraph rewriting based on subgraph isomorphism. This grammar can deal with context–sensitive productions.

Definition 1 (Graph) Let Σ be an alphabet of nodes. A *graph* over Σ is a 3-tuple $D = (V, E, \lambda)$, where

1. V is a finite nonempty *set of nodes*.
2. $E \subseteq \{(v, w) \mid v, w \in V, v \neq w\}$ is a *set of edges*.
3. $\lambda : V \to \Sigma$ is a *node labeling function*.

Let $\mathcal{G}_\Sigma$ be a set of graphs over Σ. A *graph language* $\mathcal{D}$ over Σ is a subset of $\mathcal{G}_\Sigma$. □

Two graphs $H = (V_H, E_H, \lambda_H)$ and $K = (V_K, E_K, \lambda_K)$ are *isomorphic* if there is a bijection $\theta : V_H \to V_K$ such that $E_K = \{(\theta(v), \theta(w)) \mid (v, w) \in E_H\}$ and, for all $v \in V_H$, $\lambda_K(\theta(v)) = \lambda_H(v)$. Then θ is called *isomorphism from H to K*. H and K are *mutually disjoint* if $V_H \cap V_K = \phi$.

Definition 2 (Graph grammar) A *graph grammar* is a 4-tuple $G = (\Sigma_n, \Sigma_t, S, P)$, where

1. Σ_n is a *nonterminal node alphabet*.
2. Σ_t is a *terminal node alphabet*. Σ_n and Σ_t are finite nonempty mutually disjoint sets. $\Sigma = \Sigma_n \cup \Sigma_t$ is a *total node alphabet*.
3. $S \in \Sigma_n$ is a *start label*. The *start graph* G_s is a graph that consists of a single node labeled S and no edge.
4. P is a finite nonempty set of *productions*. Each element in P is a 5-tuple $p = (A, X, B, Y, C)$, where

    (a) $A = (V_A, E_A, \lambda_A)$ is a graph over Σ and $X = (V_X, E_X, \lambda_X)$ is a graph over Σ_n such that X is an induced subgraph of A. $B = (V_B, E_B, \lambda_B)$ is a graph over Σ, where $|V_A| \leq |V_B|$, and $Y = (V_Y, E_Y, \lambda_Y)$ is a graph over Σ such that Y is an induced subgraph of B.

    (b) These four graphs A, X, B, and Y satisfy $A = X$ and $B = Y$, otherwise there exists an induced subgraph K such that $K = A - X = B - Y$. K is called *context subgraph*.

    (c) $C \subseteq V_X \times V_Y \times \{\text{in, out}\}$ is a *connection relation*.

A production $p = (A, X, B, Y, C)$ will also be denoted by $((A, X) ::= (B, Y), C)$. (A, X) is the *left-hand side* of p and (B, Y) is the *right-hand side* of p. We write $lhs(p) = (A, X)$ and $rhs(p) = (B, Y)$. Each element (v_x, v_y, α) of C is a *connection instruction* of p. □

Two production $p_1 = ((A_1, X_1) ::= (B_1, Y_1), C_1)$ and $p_2 = ((A_2, X_2) ::= (B_2, Y_2), C_2)$ are called *isomorphic* if there are isomorphism θ_l from A_1 to A_2 and isomorphism θ_r from B_1 to B_2, and $C_2 = \{(\theta_l(v_x), \theta_r(v_y), d) \mid (v_x, v_y, d) \in C_1\}$. Then (θ_l, θ_r) is called *isomophism from p_1 to p_2*.

We will assume that P does not contain distinct isomorphic productions. By $copy(p)$ we denote the set of all productions that are isomorphic to a production p in P. An element of $copy(p)$ will be called a *production copy* of p. $copy(P) = \cup_{p \in P} copy(p)$.

Definition 3 (Derivation) Let $G = (\Sigma_n, \Sigma_t, S, P)$ be a graph grammar, $H = (V_H, E_H, \lambda_H)$ be graph in $\mathcal{G}_\Sigma$, and M be an induced subgraph of H. Let $p \in P$ be a production of G, $p' = ((A', X') ::= (B', Y'), C')$ be a production copy of p where $X' = (V_{X'}, E_{X'}, \lambda_{X'})$, $Y' = (V_{Y'}, E_{Y'}, \lambda_{Y'})$, and H and Y' are mutually disjoint. We write $H \underset{p}{\Rightarrow} H'$ or just $H \Rightarrow H'$, if there exists $A' = M$ and H' is the graph $(V_{H'}, E_{H'}, \lambda_{H'})$ in $\mathcal{G}_\Sigma$ such that

$$V_{H'} = (V_H - V_{X'}) \cup V_{Y'},$$
$$E_{H'} = \{(v, w) \in E_H \mid v, w \in V_H - V_{X'}\} \cup E_{Y'} \cup \{(v, y) \mid v \in V_H - V_{X'}, (v, x) \in E_H, (x, y, \text{in}) \in C'\} \cup \{(y, v) \mid v \in V_H - V_{X'}, (x, v) \in E_H, (x, y, \text{out}) \in C'\},$$
$$\lambda_{H'}(x) = \begin{cases} \lambda_H(x), & x \in V_H - V'_X \\ \lambda_{Y'}(x), & x \in V_{Y'}. \end{cases}$$

$H \underset{p}{\Rightarrow} H'$ is called *derivation step* and a sequence of derivation steps is called *derivation*.

A derivation $H_0 \underset{p'_1}{\Rightarrow} H_1 \underset{p'_2}{\Rightarrow} \cdots \underset{p'_n}{\Rightarrow} H_n$, $n \geq 0$, is *creative* if the graphs H_0 and Y'_i ($1 \leq i \leq n$), where $rhs(p'_i) = (B'_i, Y'_i)$, are mutually disjoint. We will restrict ourselves to creative derivations.

The *graph language generated by G* is
$$\mathcal{L}(G) = \{H \in \mathcal{G}_{\Sigma_t} \mid G_s \overset{*}{\Rightarrow} H\}$$
where $\overset{*}{\Rightarrow}$ is a transitive closure of $\underset{p}{\Rightarrow}$ ($p \in copy(P)$). □

3. BLOCK DIAGRAM LANGUAGE

We define a BDG in terms of the graph grammar defined in the previous section.

3.1. Multi–input/multi–output block diagram grammar

Definition 4 (BDG)
Let $G_{\text{Block}} = (\Sigma_{n_{\text{Block}}}, \Sigma_{t_{\text{Block}}}, [\text{BD}], P_{\text{Block}})$ be a graph grammar, where the nonterminal node alphabet Σ_n and the terminal node alphabet Σ_t are shown in Fig. 1 and the set P_{Block} of productions is shown in Fig. 2.

The *block diagram language generated by G_{Block}* is the set of terminal graphs derived from the start graph which consists of a single node labeled [BD]. Each element of the block diagram language is a *block diagram*. We call the grammar G_{Block} *block diagram grammar* (BDG). □

Both the nonterminal and terminal node alphabets shown in Fig. 1 consist of dummy node labels (the part above the dotted line) and node labels defined for a single–input, single–output block diagram[1, 2] (the part below the dotted line).

Combinations of nodes labeled 'sum', 'plus', and 'minus' are used to represent adders and subtracters as shown in Fig. 4.

Nonterminal node alphabet $\Sigma_{n_{\text{Block}}}$	Terminal node alphabet $\Sigma_{t_{\text{Block}}}$
[D_In] : nonterminal dummy input	▷ : dummy input
[D_Out] : nonterminal dummy output	▷ : dummy output
[Fork] : nonterminal fork	▷ : fork
[Junc] : nonterminal junction	▷ : junction
[BD] : block diagram	◎ : input
[Elem] : element	● : output
[Branch] : nonterminal branch	block : block
[Sum] : nonterminal sum	● : branch
	Ⓢ : sum
	⊕ : plus
	⊖ : minus

Figure 1: The node alphabet of the block diagram grammar

3.2. Example of block diagram generation

For the block diagram example with two inputs and two outputs shown in Fig. 4, a derivation based on the BDG is illustrated in Fig. 5. When the dummy nodes are ignored, the derived diagram can be regarded as the one shown in Fig. 4.

4. BLOCK DIAGRAM PARSER

We have implemented a block diagram editor and a block diagram parser in Prolog. The former generates an internal symbolic representation of a diagram input by means of a GUI on an X window. The latter parses the internal representation on the basis of the BDG.

Parsing diagrams on the basis of the BDG requires the reverse application of a production copy.

Definition 5 (Reverse application)
Let $G = (\Sigma_n, \Sigma_t, S, P)$ be a graph grammar, $H = (V_H, E_H, \lambda_H)$ be a graph in $\mathcal{G}_\Sigma$, and M be an induced subgraph of H. Let $p \in P$ be a production of G, $p' = ((A', X') ::= (B', Y'), C')$ be a production copy of p where $X' = (V_{X'}, E_{X'}, \lambda_{X'})$, $Y' = (V_{Y'}, E_{Y'}, \lambda_{Y'})$, and H and X' are mutually disjoint. *Reverse application* of the production copy p' to H produces another graph $H' = (V_{H'}, E_{H'}, \lambda_{H'})$ in $\mathcal{G}_\Sigma$ such that

$$V_{H'} = (V_H - V_{Y'}) \cup V_{X'},$$
$$E_{H'} = \{(v, w) \in E_H \mid v, w \in V_H - V_{Y'}\} \cup E_{X'} \cup \{(v, x) \mid v \in V_H - V_{Y'}, (v, y) \in E_H, (x, y, \text{in}) \in C'\} \cup \{(x, v) \mid v \in V_H - V_{Y'}, (y, v) \in E_H, (y, v, \text{out}) \in C'\},$$
$$\lambda_{H'}(x) = \begin{cases} \lambda_H(x), & x \in V_H - V_{Y'} \\ \lambda_{X'}(x), & x \in V_{X'}. \end{cases}$$

□

The parser first finds all inputs and outputs, and automatically adds dummy nodes. Then this preprocessed data

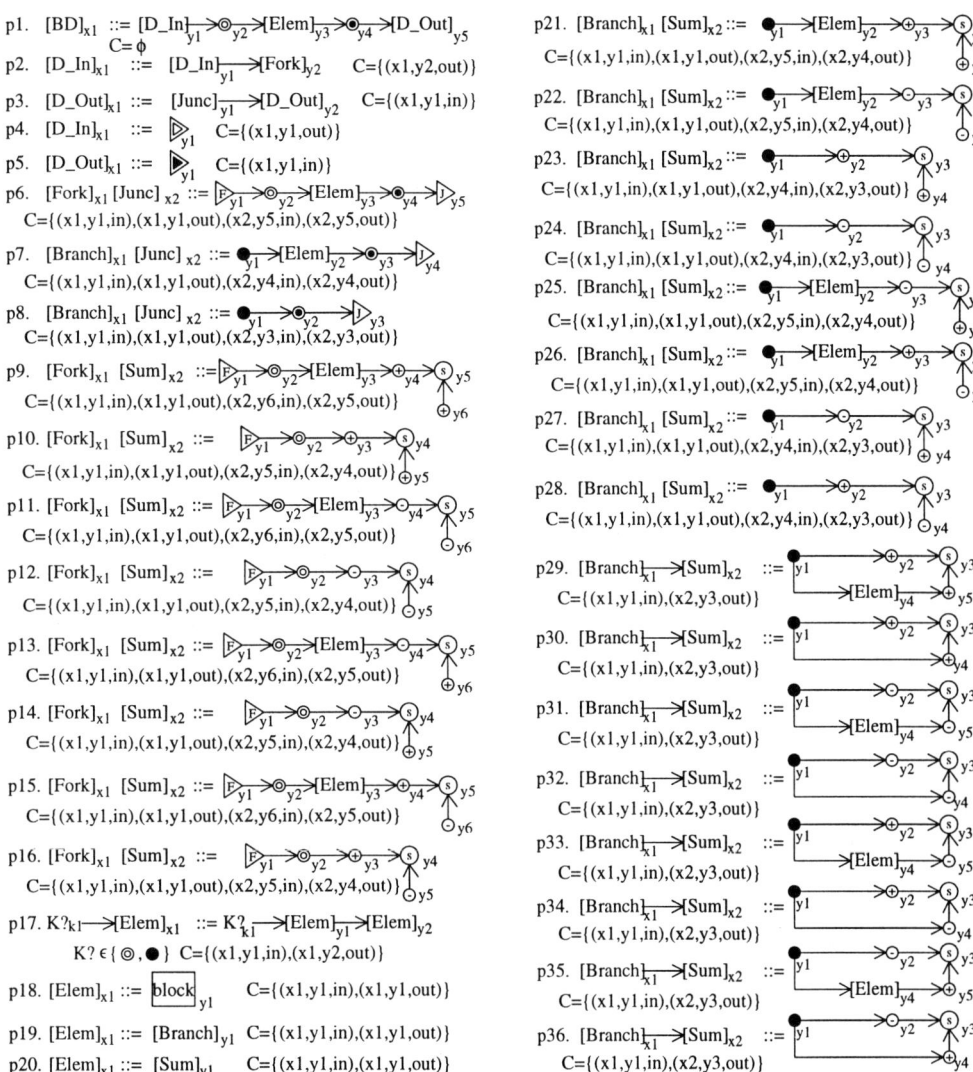

Figure 2: The productions of the block diagram grammar

is parsed based on the BDG using the bottom-up parallel algorithm shown in Fig. 6.

This algorithm has the following important properties: the algorithm terminates for any input diagram (*termination*), for any valid block diagram the algorithm finds a production copy sequence deriving it (*completeness*), and the resultant production copy sequence for a valid diagram is always applicable to the start graph to derive the parsed diagram (*correctness*).

Figure 7 shows the block diagram editor and parser dealing with the diagrams shown in Fig. 4.

5. CONCLUSIONS

We have defined a BDG in terms of a context-sensitive graph grammar. We have also implemented a parser based on the BDG. The parser can parse many diagrams instantly, making it quite practical. The BDG defined in this research makes a fundamental and important formal model for system analysis and design using block diagrams on a computer.

6. REFERENCES

[1] Y. Adachi, S. Kobayashi, K. Anzai, and K. Tsuchida, Block Diagram Grammar and Structure Recognition Based on Graph Rewriting, IFAC/IEEE CACSD'97 pp.257-262(1997).

[2] K. Anzai, Y. Adachi, S. Kobayashi, and K. Tsuchida, Block Diagram Generation and Parsing Based on Graph Grammar, IEEE ISCAS'97 pp.1970-1973(1997).

[3] G. Rozenberg (Eds.), Handbook of Graph Grammars and Computing by Graph Transformations, World Scientific Publishing(1997).

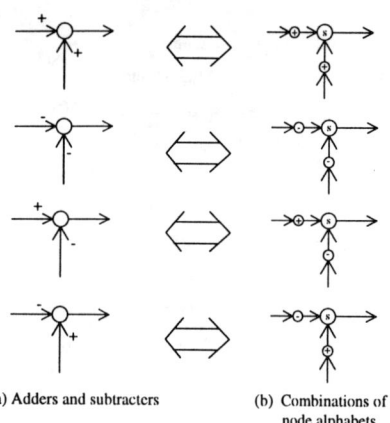

(a) Adders and subtracters   (b) Combinations of node alphabets

Figure 3: Representation of adders and subtracters

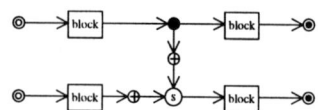

Figure 4: Block diagram example

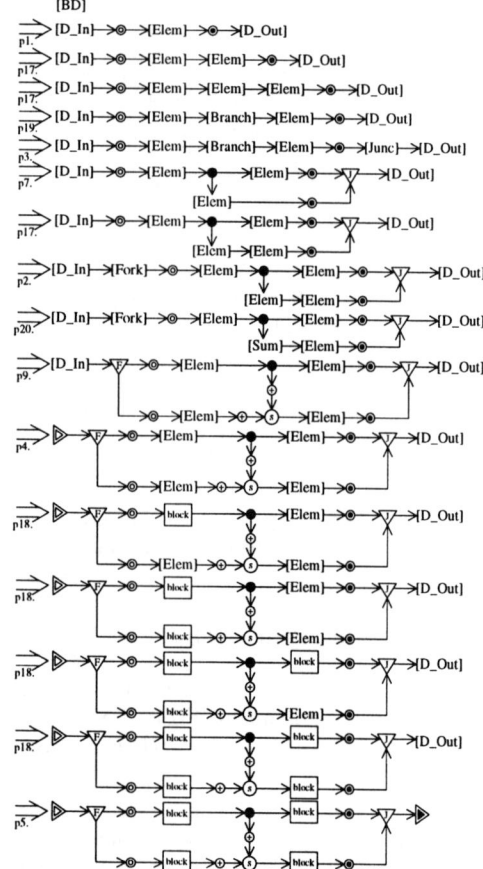

Figure 5: Derivation example

```
parsing{
  input
    G_Block = (Σ_{n_Block}, Σ_{t_Block}, [BD], P_Block) : the BDG
    D = (V_D, E_D, λ_D) : a graph to be parsed
  output
    "valid" or "invalid"
    a production copy sequence
  variable
    S[ ] : an array of sets of graphs
    d : a level of rewriting
    S_map : a set of pair (p, rhs(p')), where p' ∈ copy(p)
  method
    attach the symbol ⊥ to the graph D ;
    d ← 0 ;
    S[d] ← {D} ;
    while {
      if S[d] = φ then
        output("invalid") and exit;
      S[d + 1] ← φ ;
      for every graph G ∈ S[d] {
        S_map ← φ ;
        while ( ∃p ∈ P, p'  = ((A', X') ::=
                (B', Y'), C')  ∈  copy(p), B'
                is an induced subgraph of G,
                (p, rhs(p')) ∉ S_map )  do {
          make the reverse application of p' to G and
          let the resultant graph be G' ;
          G' ⇒_{p'} G'' ;
          if G'' = G then {
            attach the production copy p' to G' ;
            link the production copy attached to G' with the
            production copy (or the symbol ⊥) attached to G ;
            S[d + 1] ← S[d + 1] ∪ {G'} ;
          }
          S_map ← S_map ∪ {(p, rhs(p'))} ;
        }
      }
      d ← d + 1 ;
      if ( there exists a graph X that consists of a
           single node labeled [BD] in S[d] ) then {
        output("valid");
        while tracing the link back to the symbol ⊥,
        output each production copy ;
        exit ;
      }
    }
}
```

Figure 6: Algorithm of the bottom-up parallel parsing

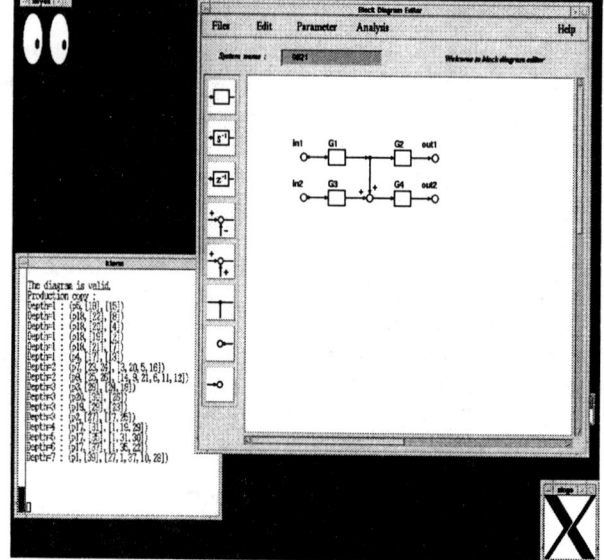

Figure 7: A snapshot of our block diagram editor and parser

IMPROVED MINIMIZATION METHODS OF PSEUDO KRONECKER EXPRESSIONS FOR MULTIPLE OUTPUT FUNCTIONS

Per Lindgren[*]

Division of Computer Engineering
Luleå University of Technology
97187 Luleå, Sweden
pln@sm.luth.se

Rolf Drechsler   *Bernd Becker*

Institute of Computer Science
Albert-Ludwigs-University
79110 Freiburg im Breisgau, Germany
{drechsle/becker}@informatik.uni-freiburg.de

ABSTRACT

Pseudo Kronecker Expressions (PSDKROs) are a class of AND/EXOR expressions. For a Boolean function with a given variable order the minimal PSDKRO can be derived efficiently using Decision Diagram (DD) techniques. The quality, i.e., the number of products in the expression, of the result is known to be dependent on the variable ordering.

This paper proposes several improvements and enhancements to previous minimization methods. A pruning technique that can be tuned to tradeoff quality for computational resources is presented. By applying dynamic ordering methods, significant improvements to many previously reported results are obtained. Furthermore, a new method for the minimization of multiple output functions is outlined. Experiments on a set of MCNC benchmarks confirm the advantages of the presented algorithms.

1. INTRODUCTION

The use of EXOR gates in the synthesis process reduces the hardware costs in many cases. Additionally, EXOR based circuits often have nice testability properties. In contrast to AND/OR minimization - that in the meantime is well understood - in AND/EXOR minimization several restricted classes are considered, like *Fixed Polarity Reed-Muller Expressions* (FPRMs) and *Kronecker Expressions* (KROs). (For an excellent overview see [7].) This subclasses are of interest, since the minimization of general *Exclusive Sum of Product Expressions* (ESOPs) turned out to be computationally very hard.

As one alternative *Pseudo Kronecker Expressions* (PSDKROs) [7] have been proposed, since they are an interesting compromise: the resulting 2-level forms are of moderate size and additionally the minimization process can be handled within reasonable time bounds using Multi-Place DDs or more recently Ordered Binary DDs (OBDDs) [2] as shown in [3]. Furthermore, for minimization of totally symmetric functions, the algorithm is shown to derive the exact minima in polynomial time [3].

[*] The co-operative work was conducted at Albert-Ludwigs-University on a grant from the Swedish Institute.

In this paper we propose a pruning technique for further speeding up the approaches presented in [7, 3]. The method can be tuned to tradeoff quality for CPU and memory resources, producing heuristic results for previously unmanageable problems. The quality of the minimized PSDKRO expression is known to be dependent on the variable order applied during minimization. We employ dynamic variable ordering strategies and give a comparison of quality and efficiency. Furthermore, we have a closer look at the principles how multiple output functions are handled in presented approaches and outline an alternative method relaxing previous decomposition constraints. This method especially becomes interesting if PSDKRO minimization is seen as a preprocessing for general AND/EXOR minimization or multi-level synthesis. Experimental results are reported that demonstrate the efficiency of the proposed algorithms.

2. PSEUDO KRONECKER EXPRESSIONS

In this section we briefly review the essential definitions of *Pseudo Kronecker Expressions* (PSDKROs). (For more details see [7, 3].)

Let f_0 (f_1) denote the *cofactor* of f with respect to $\overline{x}(x)$ and f_2 is defined as $f_2 = f_0 \oplus f_1$, $\oplus$ being the Exclusive OR operation. A Boolean function $B^n \to B$ can then be represented by one of the following formulae:

$$f = \overline{x}f_0 \oplus xf_1 \quad Shannon\ (S) \qquad (1)$$

$$f = f_0 \oplus xf_2 \quad positive\ Davio\ (pD) \qquad (2)$$

$$f = f_1 \oplus \overline{x}f_2 \quad negative\ Davio\ (nD) \qquad (3)$$

If we apply to a function f either S, pD or nD we get two sub-functions. To each sub-function again S, pD or nD can be applied. This is done until constant functions are reached. If we multiply out the resulting expression we get a 2-level AND/EXOR form, called a PSDKRO. In Section 5 we take a closer look at PSDKRO expressions for multiple output functions.

The decompositions are applied with respect to a fixed variable ordering. Note that the choice of the variable ordering in which the decompositions are applied and the choice

```
prod psdkro(node f, int prune) {
1  if (f == 0(1)) return 0(1);
2  if (f.prod defined) return f.prod;
3  f_0 = cofactor_0(f); f_1 = cofactor_1(f);
4  p_0 = psdkro(f_0, prune);
5  (P2) → if (p_0 ≥ prune) return MAXINT;
6  p_1 = psdkro(f_1, prune);
7  (P0,P1,P2) → if (min(p_0, p_1) ≥ prune) return MAXINT;
8  f_2 = EXOR(f_0, f_1);
9  if (memory or time limit reached) return p_0 + p_1;
10 p_2 = psdkro(f_2, pf(prune, p_0, p_1));
11 f.prod = p_0 + p_1 + p_2 - max(p_0, p_1, p_2);
12 return f.prod;
}
```

Figure 1: Sketch of the enhanced algorithm

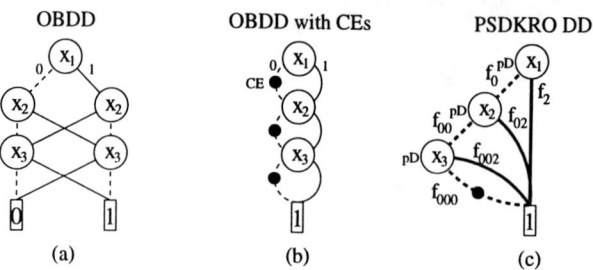

Figure 2: Parity function

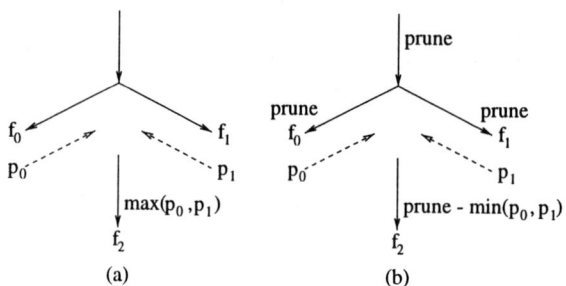

Figure 3: Pruning of search space

of the decomposition per sub-function largely influences the size of the resulting representation, and may vary from linear to exponential.

Example 1 *Consider the parity function $f(x_1, x_2, x_3)$. A minimal (w.r.t. number of products) SOP form is: $f = \overline{x}_1\overline{x}_2 x_3 + \overline{x}_1 x_2 \overline{x}_3 + x_1 \overline{x}_2 \overline{x}_3 + x_1 x_2 x_3$ (4 products, as shown by tracing the 1-paths in Figure 2 (a)), while a minimal PSDKRO form is: $f = x_1 \oplus x_2 \oplus x_3$ (3 products, as shown by tracing the 1-paths in Figure 2 (c)). For an n-variable parity function the minimal SOP form requires 2^{n-1} products, while the minimal PSDKRO requires n products.*

3. PSDKRO MINIMIZATION

For the implementation of the algorithm we used *Ordered Binary Decision Diagrams* (OBDDs) [2]. The starting point of our algorithm is the OBDD representation of the function that is to be minimized. A *1-path* represents a product term with the cost 1.

Starting from the root of the OBDD the graph is recursively traversed towards the terminals {0, 1}. At each node an EXOR operation is carried out, which by the use of OBDDs can be performed in polynomial time [2]. From the fact that for each of the decomposition formulae from Section 2 only two out of the three possible successors f_0, f_1 and f_2 are needed to represent the function, we can choose the two least costly in order to minimize the resulting PSDKRO [1].

For each node f the minimal number of product terms needed for the representation as a PSDKRO is stored in the variable $f.prod$. Thus, each node has to be evaluated only once.

Example 2 *Consider the algorithm in Figure 1 (without the "prune" related code). Assume that $p_0 = 2$, $p_1 = 2$ and*

[1] The basic idea of the algorithm is the same as used in [7] for PSDKRO minimization.

$p_2 = 1$ *(lines 4, 6, and 10). The cost of S, pD and nD decompositions are $p_0 + p_1 = 4$, $p_0 + p_2 = 3$ and $p_1 + p_2 = 3$ accordingly. Line 11 computes the least cost 3. Figure 2 (c) shows a PSDKRO DD for the parity function $f(x_1, x_2, x_3)$ using only pD decompositions. This example relates to the decomposition for x_1.*

3.1. An Enhanced Minimization Algorithm

Previous algorithms exhaustively traverse the search space to find optimal decompositions, [7, 3]. We make the following critical observations: The OBDD representation holds an initial (non-optimal) PSDKRO solution having only Shannon decompositions, $(\overline{x}f_0 \oplus xf_1)$ with the cost $p_0 + p_1$. f_2 is a part of the solution iff $p_2 < \max(p_0, p_1)$, i.e., f_2 must be less costly than at least one other cofactor f_0 or f_1 (see Figure 3 (a)). Given a cost limit "prune", f_2 is a part of the solution iff $p_2 <$ prune $- \min(p_0, p_1)$, since the cost of a Davio decomposition is $\min(p_0, p_1) + p_2$, which in turn must be less than "prune" (see Figure 3 (b)). Our hands on experiences show that a high cost for f_0 or f_1 in many cases also leads to a high cost for f_2.

These observations are exploited in Figure 1, lines 5, 7, 9, 10 and the additional pruning parameter. In the following three pruning methods are considered.

P0 By returning *MAXINT* whenever the cost exceeds the cost limit, further "pointless" traversal is prevented, line 7. The pruning function "pf", line 10, chooses the minimum of (a) and (b) as the cost limit for f_2, i.e., $\min(\max(p_0, p_1), \text{prune} - \min(p_0, p_1))$.

P1 By the choice of a more aggressive pruning function the algorithm can be tuned to tradeoff quality for computational resources. By choosing $\min(p_0, p_1)$ for condition (a), computation of f_2 is aborted as soon as p_2 equals either p_0 or p_1. This realizes the idea to traverse f_2 only where Davio decomposition is clearly favorable.

P2 Intrigued by the last observation, we sought an answer whether it might be sufficient to inspect only one of the cofactors to estimate the cost. This question is manifested in line 5, which aborts traversal whenever p_0 reaches the cost limit. In the following *P2*, is used with the exact pruning function from *P0*.

Furthermore as a last effort to reduce the complexity, line 9 computes the cost of a Shannon decomposition, (which is inherent in the OBDD representation as observed above), whenever computational resources are exhausted.

The motivation to apply pruning and heuristic approaches is to save CPU and memory resources, thus enable us to tackle problems for which previous attempts fail due to time or memory limits.

3.2. Literal Minimization

Simultaneous minimization of literal count is straightforward. Whenever we can choose between decomposition types, (all producing the minimal number of products), we choose the decomposition producing the least number of literals, i.e., $f.lit = \min(l_S, l_{pD}, l_{nD})$ corresponding to the literal count of S, pD and nD decompositions, respectively. Further implementation details are left out because of page limitations.

3.3. Complemented Edges

The use of Complemented Edges (CEs) in OBDD packages [1] has several advantages; firstly the number of nodes needed to represent the function is often reduced due to increased overlapping, and secondly further controlling cases for computation can be utilized. The presented algorithm depends heavily on the efficiency of EXOR operations. By the use of complemented edges the result of $f_0 \oplus f_1$ can be directly derived whenever $f_0 = \overline{f_1}$. In Figure 2 all EXOR computations leading to the minimized PSDKRO (c), can be derived in constant time using CEs (b).

4. INFLUENCE OF VARIABLE ORDERING

In [7] the effect of variable ordering was concluded by enumerating all possible orderings for a given function and minimizing the corresponding PSDKROs. However, no method to approach the ordering problem was reported. In general dynamic variable ordering [6] has proven useful to many DD problems and has been applied in [4] to PSDKRO based Maitra term minimization. In this paper we adopt dynamic variable ordering directly to the problem of PSDKRO minimization and present two different approaches.

R1 Move to top strategy. First we compute the cost using the initial ordering of the input variables, e.g., $(x_1, x_2, x_3, .., x_n)$, where x_1 is the top variable. Then we try all other variables as the top variable and store the best ordering obtained, e.g., $(x_3, x_1, x_2, .., x_n)$. We now repeat the procedure, e.g., with x_3 as the new initial top variable, until no further optimization is obtained or a time limit is reached.

R2 Local variable exchange method. Basically the "sifting" algorithm presented in [6]. Unfortunately the cost cannot be derived from local operations, complete PSDKRO minimization is required for each variable order applied.

5. MULTIPLE OUTPUT FUNCTIONS

In [7] the algorithm for minimizing the PSDKRO for Multiple Output (MO) functions is based on Multi-Place DDs, for which each terminal corresponds to an output vector. However, as used e.g., in [8], the terminals can be encoded by a set of Output Selection (OS) variables, (one variable-node for each output function). This encoding gives us the freedom to change the grouping of output functions by the position of the OS nodes. Moving all OS nodes to the top of the DD allows applying decompositions with respect to each output function separately, while keeping the OS nodes at the bottom of the DD corresponds to the representation used in [7]. In the latter case, decompositions are applied simultaneously to all output functions (analogous to a multi-valued interpretation). Besides these two extremes, OS nodes can be interleaved with the DD nodes. The introduction of output selection variables increases the number of 1-paths (i.e., the cost function), in turn leading to an overestimation of the number of required products. To correctly account for product sharing of MO functions is to the best of our knowledge trivial only when the OS nodes is at the bottom of the diagram.

6. EXPERIMENTAL RESULTS

In this section we present experimental results for a set of MCNC PLA benchmark functions performed on a *Sun Ultra 1* workstation with 256Mb RAM. (All totally symmetric functions are marked by *.) For the implementation we used the CUDD 2.1.2 BDD-package [9]. The peak memory requirement of the performed experiments was measured to 75Mb. Dynamic reordering methods are implemented by synthesis operations, as internal BDD cost functions do not apply.

In a first set of experiments (see Table 1) we show the effect of pruning and heuristic minimization under the variable ordering given by the PLAs. The results obtained by re-implementing the algorithm for MO PSDKRO minimization from [3] are marked *MO*. Columns *P0* and *P1* show the results for the exact and aggressive pruning functions,

name	in/out	Number of Products			CPU Time in Seconds			
		MO/PO	P1	P2	MO	P0	P1	P2
5xp1	7/10	47	51	47	0.01	0.01	0.01	0.01
add6	12/ 7	132	133	132	0.03	0.03	0.03	0.03
bc0	26/11	180	221	180	5.34	2.79	0.81	1.34
co14*	14/ 1	14	14	14	0.01	0.01	0.01	0.01
duke2	22/29	108	157	108	25.49	23.88	9.19	17.09
in2	19/10	117	128	117	2.59	2.49	1.13	2.00
in7	26/10	42	82	42	0.17	0.17	0.16	0.16
inc	7/ 9	31	33	31	0.01	0.01	0.01	0.01
intb	15/ 7	500	1153	500	0.75	0.72	0.59	0.69
misex3	14/14	754	1199	754	5.11	4.29	2.92	3.94
rd53*	5/ 3	20	27	20	0.01	0.01	0.01	0.01
rd73*	7/ 3	63	79	63	0.01	0.01	0.01	0.01
rd84*	8/ 4	107	134	107	0.01	0.01	0.01	0.01
sao2	10/ 4	41	63	41	0.01	0.01	0.01	0.01
t481	16/ 1	13	17	13	0.01	0.01	0.01	0.01
tial	14/ 8	939	1231	939	1.30	1.27	1.21	1.18
vg2	25/ 8	293	401	295	1.54	1.54	1.45	1.38
x6dn	39/ 5	104	161	104	5.90	1.68	0.44	0.38

Table 1: Pruning experiments

name	in/out	Number of Products			CPU Time in Seconds		
		R1	R2	SO	R1	R2	SO
5xp1	7/10	43	42	51	0.05	0.13	0.01
add6	12/ 7	132	132	132	0.32	1.97	0.03
bc0	26/11	175	174	299	43.93	289.58	0.51
co14*	14/ 1	-	-	14	-	-	0.01
duke2	22/29	91	85	153	920.43	3949.75	0.08
in2	19/10	117	117	138	18.83	139.98	0.09
in7	26/10	36	37	51	2.06	34.66	0.02
inc	7/ 9	31	30	47	0.03	0.11	0.00
intb	15/ 7	402	401	390	33.98	62.89	0.21
misex3	14/14	686	638	929	71.89	478.56	0.39
rd53*	5/ 3	-	-	20	-	-	0.01
rd73*	7/ 3	-	-	55	-	-	0.01
rd84*	8/ 4	-	-	90	-	-	0.01
sao2	10/ 4	33	33	48	0.10	0.30	0.01
t481	16/ 1	13	13	13	0.02	0.25	0.01
tial	14/ 8	728	820	752	39.99	104.72	0.36
vg2	25/ 8	250	216	233	75.25	324.79	0.21
x6dn	39/ 5	101	98	118	29.61	196.35	0.14

Table 2: Ordering experiments

respectively. The latter is clearly more efficient while compromising the quality. Under the assumption that inspecting a single cofactor is enough for cost estimation P2 efficiently obtains optimal results for the given set of benchmarks except for the near optimal result for vg2.

In the second set of experiments the results for different variable reordering methods are compared (see Table 2). Columns R1 and R2 show the effect of dynamic reordering methods applied to the MO minimization method. To speed up the cost calculation heuristic P2 is used. In general better results are obtained from the time consuming sifting method R2. The results show significant improvements to many previously reported results MO in Table 1. Columns SO show the results from Single Output (SO) minimization, (corresponding to OS nodes at top, Section 5), using the PLA variable orderings. During the minimization the number of products required is overestimated, thus leading to non-optimal solutions. However, common products are counted only once in the reported results. Compared to MO in Table 1, SO minimization shows quality improvements in many cases. For some benchmarks (e.g., intb, tial and vg2) SO even compares to results obtained by variable reordering. Furthermore, for the benchmark set SO minimization consumes significantly less computational resources than MO minimization.

Based on these promising results it is focus of current work to integrate the PSDKRO minimizer in an EXOR based synthesis tool. Another open problem is the optimization of incompletely specified functions. First results can be found in [5].

Acknowledgment

The authors wish to thank Prof. T. Sasao for inspiring discussions on the PSDKRO minimization problem.

7. REFERENCES

[1] K.S. Brace, R.L. Rudell, and R.E. Bryant. Efficient implementation of a BDD package. In *Design Automation Conf.*, pages 40–45, 1990.

[2] R.E. Bryant. Graph - based algorithms for Boolean function manipulation. *IEEE Trans. on Comp.*, 35(8):677–691, 1986.

[3] R. Drechsler. Pseudo Kronecker expressions for symmetric functions. In *VLSI Design Conf.*, pages 511–513, 1997.

[4] G. Lee. Logic synthesis for cellular architecture FPGAs using BDDs. In *ASP Design Automation Conf.*, pages 253–258, 1997.

[5] G. Lee and R. Drechsler. ETDD-based generation of complex terms for incompletely specified boolean functions. In *ASP Design Automation Conf.*, 1998.

[6] R. Rudell. Dynamic variable ordering for ordered binary decision diagrams. In *Int'l Conf. on CAD*, pages 42–47, 1993.

[7] T. Sasao. AND-EXOR expressions and their optimization. In T. Sasao, editor, *Logic Synthesis and Optimization*, pages 287–312. Kluwer Academic Publisher, 1993.

[8] T. Sasao and J.T. Butler. A method to represent multiple-output switching functions by using multi-valued decision diagrams. In *Int'l Symp. on multi-valued Logic*, pages 248–254, 1996.

[9] F. Somenzi. *CUDD: CU Decision Diagram Package Release 2.1.2*. University of Colorado at Boulder, 1997.

COMPUTATIONAL EXPERIENCE WITH A PRIMAL-DUAL INTERIOR POINT METHOD FOR SMOOTH CONVEX PLACEMENT PROBLEMS

Andrew Kennings[1]

Ryerson Polytechnic University,
Toronto, Ontario, Canada, M5B 2K3

Mark Frazer[2], *Anthony Vannelli*[3]

University of Waterloo,
Waterloo, Ontario, Canada, N2L 3G1

ABSTRACT

We present a primal-dual interior point method (IPM) for solving smooth convex optimization problems which arise during the placement of integrated circuits. The interior point method represents a substantial enhancement in flexibility verses other methods while having similar computational requirements. We illustrate that iterative solvers are efficient for calculation of search directions during optimization. Computational results are presented on a set of benchmark problems for an analysis of the method.

1. INTRODUCTION

Placement is a critical step in the physical implementation of a circuit. Fixed decisions made during architectural design, logic synthesis, and so forth require high quality placements to guarantee success during subsequent stages of the design procedure. In placement, one seeks to position the cells of a circuit while minimizing the total wire length and placement area subject to constraints on valid cell positions. Constraints are a result of the technology and layout style used (e.g., in row-oriented placement, cells must be placed into rows without overlap). Delay constraints may be included [1].

Placement is intractable and numerous heuristics have been proposed [2, 3, 4]. We consider optimization-based methods (i.e., from the family of "quadratic methods") in which placement is accomplished via several iterations of optimization interleaved with partitioning. The optimization determines relative cell positions while ignoring placement restrictions. Since wire length minimization pulls cells together, partitioning is used to "push" cells into less utilized regions of the placement area. Several iterations of optimization and partitioning provides a "good idea" of where cells belongs in the placement. Finally, a legalization heuristic is used to adjust cell positions (i.e., satisfy placement restrictions) – iterative improvement may also be applied. Finding cell positions via optimization represents the computational burden of these methods; their flexibility and efficiency relies on the continual development of more effective solution methodologies.

In this paper, we propose a convex *interior point method* (IPM) for optimization-based placement. The IPM is a flexible and efficient alternative to other methodologies [1, 4, 5, 6] and offers several benefits: (i) it is a proven and efficient state-of-the-art optimizer for large and sparse optimization problems, and (ii) it can handle a variety of constraints including general convex constraints required for future formulations. The IPM relies on the solution of large systems of linear equations and is therefore *computationally equivalent* to other less flexible methodologies. We consider modern *iterative solvers* [6, 7] to enhance computational efficiency. We illustrate that iterative solvers perform well on the optimization problems which arise in cell placement.

In Section 2, formulations for cell placement are presented. Section 3 presents the IPM and illustrates the need to solve a sequence of large and sparse symmetric indefinite systems of linear equations. Iterative solvers are described in Section 4. Numerical results are presented in Section 5 to demonstrate the performance of the IPM for cell placement. We summarize in Section 6.

2. PROBLEM FORMULATION

Optimization-based placement heuristics take as input a circuit netlist and produce a placement of cells as output. Typically, the circuit netlist is transformed into an equivalent weighted graph [4, 6] and optimization is used to keep highly connected cells together. To make the problem tractable, placement restrictions are relaxed (e.g., cells overlap), but constraints are included to encourage an even cell distributions throughout the placement area.

2.1. Previous Formulations

In **GORDIAN** [5], cell positions are computed by solving the quadratic program given by

$$\min_{x} \left\{ \sum_{i,j} a_{ij}(x_i - x_j)^2 : Hx = b \right\} \quad (1)$$

where $a_{ij} \geq 0$ indicates the strength of connection between cells i and j, x denotes the unknown cell positions, H denotes a matrix of first moment constraints and b is a right-hand side vector. This problem is with respect to the x-direction and an identical problem is solved in the y-direction. This problem can be transformed into an equivalent unconstrained problem whose solution is obtained by solving a positive definite (although dense) system of equations using the conjugate gradient method [8].

To obtain a linear estimate of wire length (and a better placement), **GORDIAN-L** [4] has been proposed. The quadratic program in (1) is solved to obtain an initial vector of cell positions x^0. The edge weights are modified and the problem given by

$$\min_{x^k} \left\{ \sum_{i,j} \frac{a_{ij}}{|x_i^{k-1} - x_j^{k-1}|}(x_i^k - x_j^k)^2 : Hx^k = b \right\} \quad (2)$$

---

[1] Partially supported by a Ryerson Polytechnic University Starter Grant 97EE101.
[2] Partially supported by a Natural Sciences and Engineering Research Council of Canada (NSERC) PGS-A graduate scholarship.
[3] Partially supported by a Natural Sciences and Engineering Research Council of Canada (NSERC) operating grant (OGP 0044456) and an Information Technology Research Centre of Ontario (ITRC) operating grant.

is resolved. Here, x^{k-1} and x^k denote the cell positions at the last and current iterations, respectively. The result of this approach is twofold: (i) as the vector x ceases to change from one iteration to another, the solution satisfies $\frac{(x_i^k - x_j^k)^2}{|x_i^{k-1} - x_j^{k-1}|} \approx |x_i - x_j|$, and (ii) the **GORDIAN** solver can be used since the problem is approximated as a quadratic program at each iteration. This approach requires more computational effort due to the additional quadratic programs which must be solved.

Finally, β-regularization has been proposed [6] in which the problem given by

$$\min_x \left\{ \sum_{i,j} a_{ij} \sqrt{(x_i - x_j)^2 + \beta} : Hx = b \right\} \quad (3)$$

is solved with $\beta > 0$. For small values of β, the objective function approximates $|x_i - x_j|$, except around the origin where β serves to avoid the non-differentiability of $|x_i - x_j|$ and the discontinuities in (2). In [6], efficient variations of Newton's method are proposed. A search direction to update the solution is obtained by solving a system of linear equations of the form

$$\begin{bmatrix} Q & H^T \\ H & 0 \end{bmatrix} \begin{bmatrix} \delta x \\ \delta y \end{bmatrix} = \begin{bmatrix} \zeta_1 \\ \zeta_2 \end{bmatrix} \quad (4)$$

where Q is the Hessian of the objective function and y is a vector of dual variables. This system of equations in symmetric and indefinite, often referred to as the *augmented equations*. Matrices of this form appear often in numerical optimization and much attention has been paid to their solution. The augmented equations are sparse when compared to those systems solved by **GORDIAN** or **GORDIAN-L**. Iterative solvers for symmetric indefinite matrices are different from those used for positive definite matrices – one is required to use Krylov subspace solvers such as [9, 10]. In all cases, it is the solution of systems of linear equations which represent the computational burden of all these methods.

2.2. Our Formulation

We consider the formulation given by

$$\begin{aligned} \min \quad & \sum_{ij} a_{ij} \sqrt{(x_i - x_j)^2 + (y_i - y_j)^2 + \beta} \\ \text{s.t.} \quad & H_x x \geq b_x, H_y y \geq b_y \\ & l \leq x, y \leq u, \end{aligned} \quad (5)$$

where x and y denote the x- and y- positions of the cells, respectively. Our formulation is similar to [6], but with several differences: (i) we couple the x- and y- directions in our objective function to capture the true two-dimensional nature of the problem and (ii) we impose general inequality constraints and variable bounds which are more flexible than strict linear equality constraints (although equality constraints are *easily* included) whereas other methodologies rely on the presence of *only* equality constraints [4, 5]. Our formulation also extends to general convex constraints required for certain delay constraints [1]. In this work, the variable bounds restrict cells to fall within the placement area and the inequality constraints include first moment constraints [11].

We select β as in [6]. Let L_x and L_y represent that maximum cell separation in the x- and y- directions, respectively. We let $\beta = \beta_r(L_x^2 + L_y^2)$ which implies the objective function can be written as

$$a_{ij} K \sqrt{\frac{(x_i - x_j)^2 + (y_i - y_j)^2}{L_x^2 + L_y^2} + \beta_r} \quad (6)$$

where $K = \sqrt{L_x^2 + L_y^2}$. In this case, $\frac{(x_i - x_j)^2 + (y_i - y_j)^2}{L_x^2 + L_y^2} \leq 1$ and the problem is normalized over the unit circle. This motivates selecting $\beta_r \in (10^{-1}, \cdots, 10^{-7})$ where smaller values result in a more "linearized" objective function. As β_r is decreased, the objective function loses convexity and the problem becomes ill-conditioned. Selection of β_r represents a tradeoff in the quality of the approximation versus the difficulty of the problem.

3. INTERIOR POINT METHODS

The placement problem may be expressed concisely as

$$\min_x \{ f(x) : Hx \geq b, l \leq x \leq u \} \quad (7)$$

where $f(x)$ denotes our wire length estimate and x now denotes the unknown cell positions in both the x- and y- directions.

We consider a primal-dual IPM based on the extension of an IPM for linear programming [12]. We associate a family of barrier problems with the original problem and, by slowly reducing the barrier parameter to zero, we obtain the solution to the original problem. The barrier problems are given by

$$\begin{aligned} \min \quad & f(\mathbf{x}) - \mu \sum_i^m \log s_i - \mu \sum_i^n \log p_i - \mu \sum_i^n \log v_i \\ \text{s.t.} \quad & Hx - s = b, \\ & x + p = u, \\ & x - v = l, \\ & s, p, v > 0 \end{aligned} \quad (8)$$

where $\mu > 0$ is the barrier parameter and s, p, and v are primal slack variables. The KKT conditions for the barrier problem are given by

$$\begin{aligned} -\nabla_x f(x) + H^T y - q + w &= 0 \\ Hx - s - b &= 0 \\ x + p - u &= 0 \\ x - v - l &= 0 \\ Ys - \mu e &= 0 \\ Tp - \mu e &= 0 \\ Wv - \mu e &= 0 \end{aligned} \quad (9)$$

where y, t and w denote the dual variables. Assuming the KKT system has a unique solution for each $\mu > 0$, the set of solutions defines a smooth curve that converges to the optimal solution of the original problem.

To obtain an approximate solution to the KKT conditions, we use one dampened Newton step. Assuming an initial point with $s, p, v, y, t, w > 0$ is provided, a search direction is obtained by solving the indefinite system of linear equations given by

$$\begin{bmatrix} -\nabla^2 f - TP^{-1} - WV^{-1} & H^T \\ H & Y^{-1}S \end{bmatrix} \begin{bmatrix} dx \\ dy \end{bmatrix} = \begin{bmatrix} \zeta_1 \\ \zeta_2 \end{bmatrix}, \quad (10)$$

where ζ_1 and ζ_2 are vectors computed from the KKT conditions. Directions ds, dp, dt, dv and dw are obtained with matrix-vector operations.

The IPM requires an initial solution with s^0, v^0, p^0, y^0, t^0, $w^0 > 0$. We compute an initial primal feasible solution based on the physical meaning of the primal variables – we select x^0 such that cells are initially at their center of gravity and then select $s^0, v^0, p^0 > 0$ to satisfy the primal constraints. A dual solution is computed as

$$(y^0, t^0, w^0) = (m + 2n)((S^0)^{-1} e, (P^0)^{-1} e, (V^0)^{-1} e) \quad (11)$$

which ensures that $y^0, t^0, w^0 > 0$. This initial solution is centered [12] since the complementary pairs (y_i, s_i), (t_i, p_i) and (w_i, v_i) all equal $(m + 2n)$.

After computation of the search direction, we compute a step size which ensures non-negativity on the appropriate variables. We compute

$$\begin{aligned} \alpha_p &= \tau \max\{\alpha \geq 0 : (s,v,p) + \alpha(ds, dv, dp) \geq 0\} \\ \alpha_d &= \tau \max\{\alpha \geq 0 : (y,t,w) + \alpha(dy, dt, dw) \geq 0\} \end{aligned} \quad (12)$$

where $\tau \in (0,1)$ is a backtrack parameter. The actual step size taken is $\alpha = \min\{1.0, \alpha_p, \alpha_d\}$.

We update μ at each iteration using the rule

$$\mu = \gamma \frac{Ys + Tp + Wv}{m + 2n} \quad (13)$$

where γ is the *centering parameter* fixed at $\gamma = 0.1$.

Finally, IPM iterations terminate once the infinity norm of the KKT conditions and the value of μ drop below a preselected threshold $\varepsilon \in (10^{-7}, 10^{-1})$. Further details of IPM implementations can be found in [12].

4. SEARCH DIRECTIONS

The computational burden of the IPM is the solution of the indefinite system of linear equations in (10). Hence, the IPM is *computationally equivalent* to other methods. Since direct solvers [13] are computationally unattractive due to the large amounts of fill incurred when factoring large matrices, we consider modern *iterative solvers* [10, 9] which require less storage and less computational effort due to the inexact nature of the resulting solutions.

Iterative solvers require *preconditioning matrices* to accelerate their convergence. There is a tradeoff – accurate preconditioners reduce iteration counts, but require more storage, time to compute and work per solver iteration. Less accurate preconditioners are cheaper to compute, require less storage and work per solver iteration, but require more solver iterations (less work per iteration is good, unless the *number* of iterations is large). Empirical testing is required to determine "good combinations" for particular problems. Iterative solvers have been considered previously for cell placement [6, 11, 14].

We consider preconditioners based on incomplete LDL^T *drop tolerance* factorizations which are generated by retaining (or rejecting) entries in L based on their numerical values [7] – our own testing has indicated that drop tolerance is superior to other preconditioning techniques. Moreover, we use *a priori* matrix ordering (e.g., minimum degree ordering) to reduce fill – this is important for IPMs as ordering can be done *once* prior to any IPM iterations, implying a savings in computational effort.

Direct LDL^T factorizations exist using *a priori* orderings, [13] (no zero pivots occur in the matrix D). However, this may not be the case during an incomplete factorization – zero pivots in D indicate failure. Fortunately, zero pivots can be avoided:

Proposition [15]: For a symmetric positive definite matrix, zero and negative pivots are avoided by appropriately modifying diagonal elements in D during an incomplete factorization.

This leads to the following required proposition for symmetric indefinite matrices:

Proposition: For the symmetric indefinite matrix in (10), zero pivots are avoided by appropriately modifying diagonal elements in D during an incomplete LDL^T factorization.

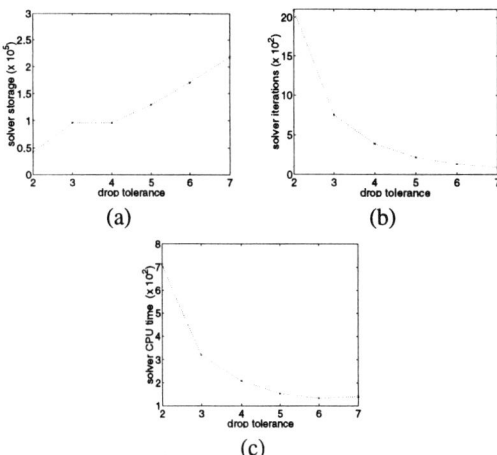

Figure 1: Iterative solver performance for *biomed*: (a) storage, (b) iterations, and (c) time.

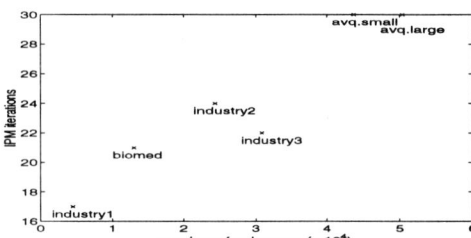

Figure 2: IPM iterations verses problem size.

In addition, the number of positive and negative entries in D (i.e., the inertia of D) is equal to that obtained using a direct LDL^T factorization.

5. NUMERICAL RESULTS

We tested the performance of the iterative solver and the IPM on circuits available from MCNC [16]. Results are included for circuit *biomed* due to space limitations. Similar trends were obtained for all other circuits. Results were produced on a 167 MHz Sun Ultra 1 with 192M memory. Times are reported in CPU seconds. The iterative solver used was **CGS** [9].

Figure 1 illustrates the storage requirements, iteration counts and CPU times for drop tolerance preconditioning with $\beta_r = 10^{-1}$ and $\varepsilon = 10^{-2}$. The drop tolerance was varied within $(10^{-7}, 10^{-2})$ (axes labelled "drop tolerance", "beta setting" and "IPM tolerance" should be interpreted as 10^{-v}, where v is the axis value). Iteration counts and CPU times are cumulative over all IPM iterations. The iterative solver was terminated when the relative residual error dropped below 10^{-12}. When compared to a direct factorization, only $4.1\% - 21.0\%$ of the storage space was required depending on the drop tolerance. Iteration counts and CPU times decrease as the drop tolerance decreases and a more accurate preconditioner is obtained. Savings in time are not significant beyond a drop tolerance of 10^{-5}.

The effect of problem size (i.e., the number of unknown cell positions) on the IPM are presented in Figure 2. The number of IPM iterations grows slowly with problem size and is a desireable property of IPMs.

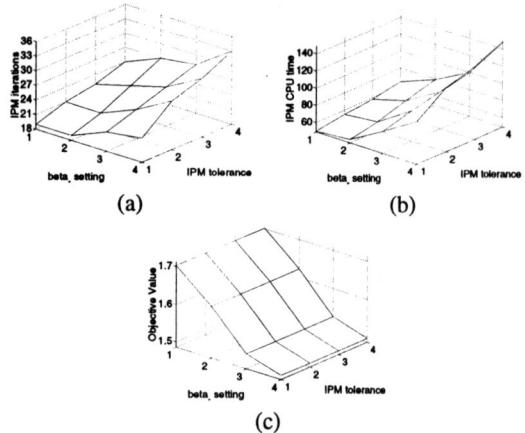

Figure 3: IPM performance versus β_r and ϵ for *biomed*: (a) iterations, (b) time, and (c) objective value.

Figure 3 illustrates the effect of variations in β_r and ϵ in terms of the IPM iterations, CPU time, and final objective value. Recall that β_r affects the convexity and the conditioning of the problem whereas ϵ determines the solution accuracy. Decreases in β_r and ϵ result in more IPM iterations and CPU time. The increase in iterations is reasonable in that the surface is "flat". A more significant observation is that, for a fixed β_r, changes in ϵ do not significantly influence CPU time, whereas for a fixed ϵ, changes in β_r do influence CPU time. Investigations revealed that the additional time is used by the iterative solver to determine search directions.

Finally, for a fixed β_r (which determines the unique optimal solution) the objective value does not vary significantly verses ϵ – accuracy is not a difficulty and early IPM termination is possible which saves time. We confirmed early termination by considering the physical meaning of the solutions verses ϵ (i.e., the cell positions) and did not notice any significant differences.

6. CONCLUSIONS

An IPM for smooth convex placement problems has been described which is capable of handling large placement problems. Computational requirements are similar to other optimization-based methods. The IPM extends optimization-based placement method by facilitating the inclusion of convex equality and inequality constraints (although we have considered only linear constraints) – this leads to the possibility of more complicated formulations.

We have also illustrated the benefits of iterative methods for search direction calculations. Tradeoffs in storage, computational effort and iterative solver iterations have been presented.

7. REFERENCES

[1] T. Hamada, C. K. Cheng, and P. M. Chau. Prime: A timing-driven placement tools using a piecewise linear reisitive network approach. In *30th ACM/IEEE Design Automation Conference*, pages 531–536, 1991.

[2] W. J. Sun and C. Sechen. Efficient and effective placements for very large circuits. *IEEE Transactions on Computer-Aided Design of Integrated Circuits and Systems*, 14(3):349–359, 1995.

[3] D. J. H. Huang and A. B. Kahng. Partitioning-based standard-cell global placement with an exact objective. In *International Symposium on Physical Design*, April 14-17 1997.

[4] G. Sigl, K. Doll, and F. M. Johannes. Analytical placement: A linear of quadratic objective function? In *23rd ACM/IEEE Design Automation Conference*, pages 57–62, 1991.

[5] J. M. Kleinhans, G. Sigl, F. M. Johannes, and K. J. Antreich. GORDIAN: VLSI placement by quadratic programming and slicing optimization. *IEEE Transactions on Computer-Aided Design of Integrated Circuits and Systems*, 10(3):356–365, 1991.

[6] C. J. Alpert, T. F. Chan, D. J. H. Huang, A. B. Kahng, I. L. Markov, P. Mulet, and K. Yan. Faster minimization of linear wirelength for global placement. In *International Symposium on Physical Design*, 1997.

[7] J. K. Dickinson and P. A. Forsyth. Preconditioned conjugate gradient methods for three-dimensional linear elasticity. *International Journal for Numerical Methods in Engineering*, 37:2211–2234, 1994.

[8] W. W. Hager. *Applied Numerical Linear Algebra*. Prentice Hall, Englewood Cliffs, New Jersey, 1988.

[9] P. Sonneveld. CGS, a fast Lanczos-type solver for nonsymmetric linear systems. *SIAM Journal on Scientific and Statistical Computing*, 10:36–52, 1989.

[10] H. A. van der Vorst. BI-CGSTAB: A fast and smoothly converging variant of BI-CG for the solution of nonsymmetric linear systems. *SIAM Journal on Scientific Statistical Computing*, 13(2):631–644, 1992.

[11] A. Kennings and A. Vannelli. An efficient interior point approach for QP and LP models of the relative placement problem. In *Midwest Symposium of Circuits and Systems*, 1996.

[12] J.-P. Vial. Computational experience with a primal-dual interior-point method for smooth convex programming. *Optimization Methods and Software 3*, pages 285–316, 1994.

[13] R. J. Vanderbei and T. J. Carpenter. Symmetric indefinite systems for interior point methods. *Mathematical Programming*, 58:1–32, 1993.

[14] C. J. Alpert, T. F. Chan, D. J. H. Huang, I. L. Markov, and K. Yan. Quadratic placement revisited. In *Design Automation Conference*, 1997.

[15] M. A. Ajiz and A. Jennings. A robust incomplete choleski-conjugate gradient algorithm. *International Journal for Numerical Methods in Engineering*, 20:949–966, 1984.

[16] K. Kozminski. Benchmarks for layout synthesis – evolution and current status. In *28th ACM/IEEE Design Automation Conference*, pages 265–270, 1991.

AN INITIAL PLACEMENT ALGORITHM FOR 3-D VLSI

Michiroh Ohmura

Department of Electrical Engineering, Faculty of Engineering
Hiroshima Institute of Technology
2-1-1 Miyake, Saeki-ku, Hiroshima 731-51 JAPAN
ohmura@cc.it-hiroshima.ac.jp

ABSTRACT

As manufacturing technology has advanced in recent years, a 3-D IC in which circuits are piled on top of each other has been the focus of attention in the device area. However no initial placement technique has been researched in the layout design of 3-D VLSI.

In this paper, we propose a 3-D initial placement algorithm which places strongly connected modules close to each other including adjacent layers by introducing the gains for modules. This algorithm also takes account of layer assignment by multiplying a constant k by the distance in the direction of Z.

1. INTRODUCTION

As manufacturing technology has advanced in recent years, a 3-D IC [3, 9] in which circuits are piled on top of each other has been the focus of attention in the device area. However, only a few detailed routing techniques [1, 5, 8] for 3-D channels have been proposed and no initial placement technique has been researched in the layout design of 3-D VLSI.

One method of designing initial placement of 3-D VLSI by applying conventional 2-D algorithms [4, 6, 7] consists of (1) the layer assignment by the min-cut method [2] and (2) optimization of each layer by the pairwise interchange [10]. This method places strongly connected modules close in each layer, but it does not utilize the adjacent layers in order to decrease the wire length. On the other hand, we cannot completely ignore the layer assignment because the cost of the routing (the resistance of the wires) in the Z direction is higher than that in the X or the Y direction.

In this paper, we propose a 3-D initial placement algorithm which places strongly connected modules close each other including adjacent layers by introducing the gains for modules. This algorithm also takes account of layer assignment by multiplying a constant k by the distance in the direction of Z.

Since the adjacent layers are utilized while optimizing the wire length, our algorithm can determine the 3-D initial placement in which the wire length is minimized. The layer assignment is also taken into account. Experimental results show that our algorithm can produce a good initial placement for 3-D VLSI.

Features of the proposed method are as follows.

(1) 3-D chip region:

A set of modules is placed in the 3-D chip region where several layers are piled in the direction of Z.

(2) Introduction of the gains for modules:

Generally, a net connects several modules. The center of each net is determined by the average coordinates of those modules. A module may belong to several nets. The gain of each module is calculated based on the sum of distances from the centers of those nets. Our algorithm interchanges two modules which give the maximum sum of gains and the total wire length is minimized in the 3-D region.

(3) Consideration of layer assignment:

In 3-D VLSI, the cost of the routing (the resistance of the wires) in the Z direction is higher than that in the X or the Y direction. By multiplying a constant k by the distance in the direction of Z, our algorithm also takes account of layer assignment while minimizing total wire length.

(4) Multi-terminal nets and arbitrary number of modules:

The proposed algorithm can treat multi-terminal nets and arbitrary number of modules.

2. PRELIMINARIES

2.1. 3-D chip region

A circuit C which is achieved on the 3-D chip is defined as a 2-tuple $C = (M, N)$ where $M = \{M_i\}$ is a set of modules and $N = \{n_j\}$ is a netlist. We assume that each module M_i is a cube of the same size and it has a virtual terminal in the center. Each net $n_j \in N$ is denoted by the set of modules instead of the set of virtual terminals of modules.

The placement region in the 3-D VLSI is referred to as the 3-D chip region R. R is divided into slots (see Fig. 1), and each module is placed in the slot in R. The placement

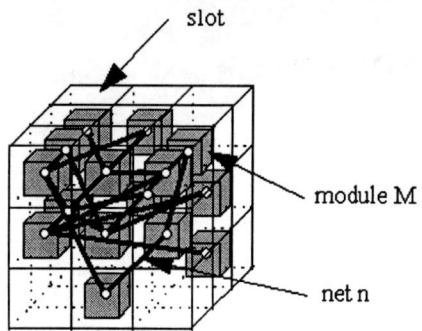

Figure 1: 3-D chip region R.

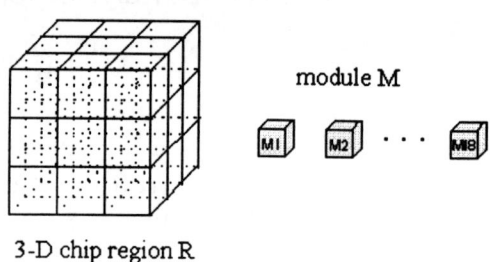

Figure 2: Input of Example 1.

of the module M_i is defined as the coordinates of the center of the corresponding slot and it is denoted by (x_i, y_i, z_i).

2.2. 3-D initial placement problem 3DI

Now, we explain the wire length of each net in R. Let xl_j and xs_j be the maximum and the minimum X coordinates of modules which belong to the net $n_j \in N$. yl_j, ys_j, and zl_j, zs_j are assumed in the same way. The total wire length L is defined as

$$L = \sum_{n_j \in N} ((xl_j - xs_j) + (yl_j - ys_j) + k(zl_j - zs_j))$$

where k is a constant.

The cost of the net (the resistance of the wire) which passes the layers along the Z direction is usually higher than that in the same layer along the X or the Y directions. The cost in the Z direction is taken into account by multiplying a constant k by the wire length along the Z direction.

3-D initial placement problem is formulated as follows.

[Problem 3DI] Given a circuit C and a 3-D chip region R, find the placement which minimizes the objective function L.

[Example 1] An example of a circuit $C = (M, N)$ and a 3-D chip region R ($3 \times 3 \times 3$) is shown in Fig.2, where a set of modules $M = \{ M_1, M_2, ..., M_{18} \}$, a net list $N = \{ n_1, n_2, n_3 \}$, $n1 = \{ M_1, M_2, M_3, M_4, M_5, M_6, M_7 \}$, $n2 = \{ M_8, M_9, M_{10}, M_{11}, M_{12}, M_{13} \}$, $n3 = \{ M_{14}, M_{15}, M_{16}, M_{17}, M_{18} \}$. Figure 3 shows an output of Problem 3DI for this circuit.

3. OUTLINE OF PROPOSED ALGORITHM

3.1. Gain of module

Let us consider the net n_1 shown in Fig. 4. Two modules of this net are contributed to $xl_1, xs_1, yl_1, ys_1, zl_1,$ and zs_1. In this case, it is impossible to reduce the wire length by the pairwise interchange which replace two modules at the same time.

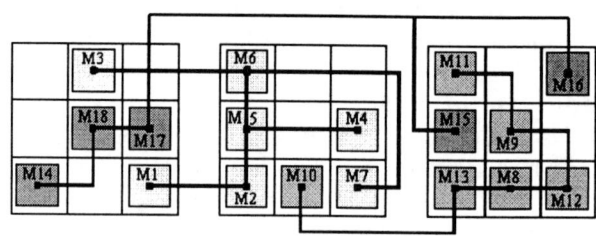

Figure 3: Output of Example 1.

In order to avoid this problem, our method introduces the gain for modules. Generally, a net connects several modules. The center (xg_j, yg_j, zg_j) of each net $n_j \in N$ is defined as the average coordinates of those modules. A module may belong to several nets. For each module M_i, the set of these nets is represented by N_i. The gain of M_i in the X direction is defined as

$$ax_i = \frac{1}{|N_i|} \sum_{n_j \in N_i} (xg_j - x_i)$$

and the gain of M_i in the Y direction is defined in the same

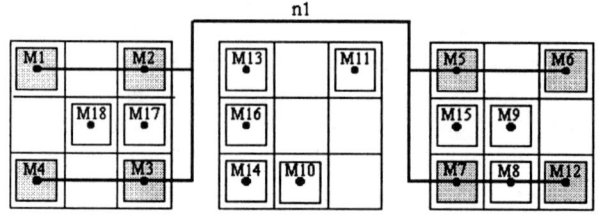

Figure 4: net n_1.

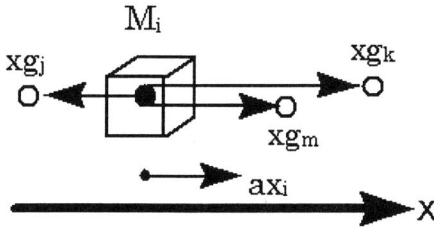

Figure 5: Example 2.

way. The gain of M_i in the Z direction is defined as

$$az_i = \frac{k}{|N_i|} \sum_{n_j \in N_i} (xg_j - x_i)$$

where k is a constant. k corresponds to the constant in the definition of the total wire length L.

[Example 2] Let us assume that the module M_i belong to the nets $N_i = \{ n_j, n_k, n_m \}$. If $xg_j = -2$, $xg_k = 5$, $xg_m = 7$, and $x_i = 1$, the gain of the module M_i in the direction X is

$$ax_i = \frac{1}{3}((-2 - 1) + (5 - 1) + (7 - 1)) = 2.33$$

Next, we define the gain ax_{ij} of adjacent modules M_i, M_j ($x_i < x_j$) in the X direction as follows.

$$ax_{ij} = ax_i - ax_j$$

The gains ay_{ij}, az_{ij} are defined in the same way.

[Example 3] Let us assume that the modules M_i and M_j are adjacent in the 3-D region R. If $ax_i = 5$, $ax_j = -4$, the gain of the adjacent modules M_i and M_j in the direction X is

$$ax_{ij} = 5 - (-4) = 9.$$

3.2. Outline of algorithm

The proposed algorithm applies the procedure 3DIA repeatedly and tries to reduce the total wire length.

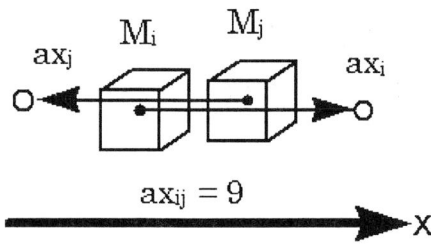

Figure 6: Example 3.

[Procedure 3DIA]

Step 1: Obtain a set MA of all adjacent modules;

Step 2: Compute the center (xg_j, yg_j, zg_j) of each net $n_j \in N$;

Step 3: Compute the gains ax_i, ay_i, az_i for each module $M_i \in M$;

Step 4: Compute the gains $ax_{ij}, ay_{ij}, az_{ij}$ for each adjacent modules M_i and $M_j \in MA$;

Step 5: $max \leftarrow \{$ maximum value among all ax_{ij}, ay_{ij}, and $az_{ij} \}$;

Step 6: if ($MA = \phi$ or $max < 0$) then Step 11;

Step 7: Interchange the adjacent modules which give the max;

Step 8: Recalculate the center (xg_k, yg_k, zg_k) for each net n_k which connects the modules M_i and M_j;

Step 9: Recalculate the gains $ax_{ij}, ay_{ij}, az_{ij}$ for the modules M_i and M_j;

Step 10: $MA = MA - \{(M_i, M_j)\}$, then Step 5;

Step 11: Calculate the total wire length L (Stop).

The time complexity of this procedure is $O(|M||N|)$ where $|M|$ is the number of modules and $|N|$ is the number of nets. The proposed algorithm applies this procedure 3DIA repeatedly until the placement is not improved.

4. EXPERIMENTAL RESULTS

To evaluate the performance of the proposed heuristic technique, the proposed algorithm has been implemented in C on DECstation 5000/120 (21.7 MIPS). Table 1 shows the comparison between the proposed algorithm and the conventional method which consists of the layer assignment by the min-cut [2] and the optimization of wire length on each layer by the pairwise interchange [10](see Fig.7).

If the constant k of the problem formulation is very large, that means each layer is designed separately, the conventional method may produce a good solution. However, k will be small in the future, and all layers will be utilized as a 3-D region. In this case, our proposed method can produce a better solution.

The simulation experiments with $k = 1.5$ is shown in Table 1. The total wire length produced by the proposed method is an average of 17.4% shorter than that obtained by the conventional method. The number of repetition of the procedure 3DIA is also shown in Table 1.

Table 1: Experimental results.

data	3-D chip region R			logic circuit C		wire length L	
	layers	rows	columns	$|M|$	$|N|$	conv.	ours (repetition)
1	2	2	2	8	10	59	22 (3)
2	2	2	3	12	10	34	30 (5)
3	2	4	3	24	15	56	45 (7)
4	4	2	3	24	20	84	64 (4)
5	2	3	5	30	20	76	66 (5)
6	2	4	3	24	25	88	68 (6)
7	2	3	4	24	30	102	89 (5)
8	4	3	3	24	40	175	157 (8)
9	2	5	5	50	50	241	194 (11)
10	4	4	4	64	60	359	317 (11)

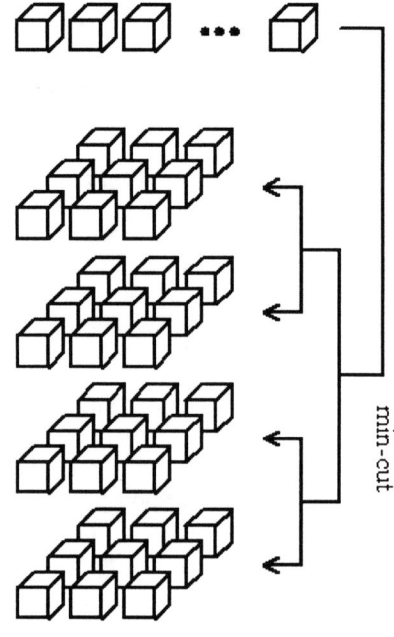

Figure 7: Conventional method.

5. CONCLUSION

In this paper, we propose a 3-D initial placement algorithm which places strongly connected modules close to each other including adjacent layers by introducing the gains for modules. This algorithm also takes account of layer assignment. The effectiveness of the proposed algorithm was shown by the simulation experiments.

6. REFERENCES

[1] R. J. Enbody, G. Lynn, and K. H. Tan, *Routing the 3-D chip*, Proc. 28th DA Conf., pp.132-137, 1991.

[2] C. M. Fiduccia and R. M. Mattheyses, *A linear-time heuristic for improving network partitions*, Proc. 19th DA Conf., pp.175-181, 1982.

[3] T. Kunio, K. Oyama, Y. Hayashi, and M. Morimoto, *Three dimensional ICs, having four stacked active device layers*, Technical Digest, 1989 IEEE IEDM, pp.837-840, 1990.

[4] T. Lengauer, *Combinatorial Algorithms for Integrated Circuit Layout*, John Wiley & Sons, 1990.

[5] M. Ohmura, *3-D router for irregular channels*, Proc. ISCAS'97, pp.1692-1695, 1997.

[6] S. M. Sait and H. Youssef, *VLSI Physical Design Automation - Theory and Practice*, IEEE press, 1995.

[7] N. Sherwani, *Algorithms for VLSI Physical Design Automation - Second Edition*, Kluwer Academic Publishers, 1995.

[8] M. Sode and T. Yoshimura, *Multi-layer channel router*, IEICE Tech. Rep. VLD92-40, 1992 (in Japanese).

[9] A. Wada, K. Morimoto, and Y. Tomita, *4-layer 3-D IC technologies for parallel signal processing*, Technical Digest, 1990 IEEE IEDM, pp.599-602, 1990.

[10] W. Wolf, *Modern VLSI Design A Systems Approach*, Prentice Hall, 1994.

A NOVEL METHODOLOGY FOR POWER CONSUMPTION REDUCTION IN A CLASS OF DSP ALGORITHMS

K. Masselos, P. Merakos, T. Stouraitis, C. E. Goutis

VLSI Design Laboratory
Department of Electrical and Computer Engineering
University of Patras, Rio 26500, Greece
e-mail: masselos@ee.upatras.gr

ABSTRACT

In this paper a novel approach for low power realization of DSP algorithms that are based on inner product computation is proposed. Inner product computation between data and coefficients is a very common computational structure in DSP algorithms. The proposed methodology is based on an architectural transformation that reorders the sequence of evaluation of the partial products forming the inner products. The total hamming distance of the sequence of coefficients, which are known before realization, is used as the cost function driving the reordering. The reordering of computation reduces the switching activity at the inputs of the computational units. Experimental results show that the proposed methodology leads to significant savings in switching activity and thus in power consumption.

1. INTRODUCTION

The recent rapid advances in the areas of wireless communications and multimedia technology made available a large number of portable battery-operated systems such as cellular phones, pagers, wireless modems, portable videophones and hand-held digital video cameras. All these systems make extensive use of DSP. Since power consumption is the overriding issue in the design of portable systems, low power DSP became an increasingly important research area.

Inner product computation between data and coefficients forms an important part of the total power budget of a DSP system [1]. Inner products are usually implemented in hardware using a multiply accumulator based computational unit.

The power consumption in digital CMOS circuits is ought to three sources [2], the dynamic (or switching), the short circuit, and the leakage power dissipation. The switching component of power dissipation can be optimized in the high levels of abstraction where the most significant power savings can be achieved [2]. Power optimization techniques for DSP have been proposed in [3-4].

In this paper a systematic methodology for power consumption reduction in multiply accumulator-based implementations of DSP algorithms is described. The main idea is the application of an architectural transformation, the computation reordering aiming at reducing the switching activity at the inputs of the computational units. Different categories of algorithms are identified.

The rest of the paper is organized as follows: In section 2 two different categories of algorithms based on inner product computation are identified, while in section 3 the mathematical model of the proposed methodology is presented along with the problem formulation for the two categories. In section 4 the target architecture models are described. Experimental results are given in section 5, while conclusions are offered in section 6.

2. CATEGORIES OF ALGORITHMS

a) Category 1

The main type of computation in the algorithms of this category is the convolution between data and coefficient vectors. A typical example belonging to this category is the FIR filtering. An N-tap FIR filter performs the following convolution:

$$Y_n = \sum_{i=0}^{N-1} C_i X_{n-i} \quad (1)$$

where C_i's are the coefficients of the filter (forming an N-point coefficient vector) and X_n, Y_n are the n_{th} terms of the input and output sequences respectively. Thus the evaluation of one point of the output sequence requires computation of an N-point inner product between data and coefficient vectors. The main characteristic of the computation performed by the algorithms of this category is that for the evaluation of the output terms the same coefficient vector and different data vectors are used. The computation required for one output point (as described by eqn. 1) is defined as the basic computation for this category of algorithms. Another algorithm belonging to this category is the Discrete Wavelet Transform.

b) Category 2

The main type of computation in the algorithms of this category is the matrix-vector multiplication between a coefficient matrix and a data vector. Typical examples of algorithms belonging to this category are the common DSP transformations (like DCT and DFT-FFT). An N-point transformation performs the following computation:

$$Y = CX \Leftrightarrow \begin{bmatrix} Y_0 \\ Y_1 \\ \vdots \\ Y_{N-1} \end{bmatrix} = \begin{bmatrix} C_{00}, & C_{01}, & \dots C_{0N-1} \\ C_{10}, & C_{11}, & \dots C_{1N-1} \\ & & \\ C_{(N-1)0}, & C_{(N-1)1}, & \dots C_{(N-1)(N-1)} \end{bmatrix} \times \begin{bmatrix} X_0 \\ X_1 \\ \vdots \\ X_{N-1} \end{bmatrix} \quad (2)$$

where Y is the N-point output data (Y_i's) vector, C is the $N \times N$ coefficient (C_{ij}'s) matrix, and X is the N-point input data (X_i's) vector. Evaluation of each term Y_i of the output column vector requires computation of an N-point inner product between the i_{th} row of the coefficient coefficient matrix and the input data vector. The main characteristic of the basic computational structure of the algorithms of the second category is that the coefficient vector used for the computation of the output points is not always the same. The number of the different coefficient vectors equals to N i.e. the transformation size. On the other hand the same data vector is used for all the N output points of the transformation. The computation required for N output points as described by eqn. 2 is defined as the basic computation for the second category of algorithms.

3. MATHEMATICAL MODELS AND PROBLEM FORMULATION

a) Category 1

The convolution operation, for the evaluation of an output term, is described by the following equation:

$$y_n = \sum_{f(k)=0}^{N-1} C_{f(k)} X_{n-f(k)} \quad (3)$$

where y_n is the n_{th} term of the output sequence, C is the coefficient vector and X is the vector of the input data. The computation is performed according to the ordering function $f(k)=k$, $k=0,...,N-1$. The total hamming distance of the sequence of evaluation of the partial products (forming the inner product) as determined by the ordering function $f(k)$ is given by the following equation:

$$TotalHD(f(k)) = \sum_{f(k)=0}^{N-1} HD(p_{f(k)}, p_{f(k)+1}) + HD(p_0, p_{N-1}) \quad (4)$$

where $p_{f(k)}$ is the partial product $c_{f(k)}x_{n-f(k)}$. The hamming distance for a pair of partial products $HD(p_k,p_l)$ is given as

$$HD(p_k, p_l) = HD(c_k, c_l) \quad (5)$$

In eqn. 4 the term $HD(p_0, p_{N-1})$ denotes the hamming distance between the first (f(k)=0) and the last (f(k)=N-1) partial product. This term is included because after the computation of the last partial product (p_{N-1}) for the output term y_n the first partial product (p_0) for the term y_{n+1} must be computed. The aim of the proposed methodology is to derive a new ordering function $g(k)$, $k=0,...,N-1$, such that the total hamming distance of the convolution computation as determined by g(k), given by eqn. 5 when f(k) is replaced by g(k), is minimum and the inner product value is computed according to the equation

$$y_n = \sum_{g(k)=0}^{N-1} C_{g(k)} X_{n-g(k)} \quad (6)$$

The problem of computation reordering is formulated as a Travelling Salesman Problem (TSP). The graph G (V, E) of the problem consists of the set V of N vertices corresponding to the partial products required for the computation of an inner product, and the set E of edges which model the unconstrained transition from one partial product to another. The problem's graph is complete i.e. each vertex pair is connected by an edge. To each edge of the graph a cost, which is the HD between the two partial products that the edge connects as defined by eqn. 5 for a specific ordering function, is assigned. A closed path over all vertices (partial products) must be found, without passing from a vertex more than once, resulting in a minimum HD cost. The costs of the edges are well-bounded positive integer numbers. Their lower bound is zero and the higher bound equals to the number of bits used for the representation of the coefficients.

Several algorithms have been proposed for the solution of the TSP problem. If the size of the problem is relatively small, an exact solution can be found in short time. For larger problem sizes the NP-complete class of the problem motivated the research for heuristic algorithms. Christofides in [5] proposed a heuristic that requires $O(n^4)$ time to find a nearly optimal solution.

b) Category 2

The computation of an output point of a transformation of length N is described by the following equation

$$y_{f(i)} = \sum_{g(i,j)=0}^{N-1} c_{f(i),g(i,j)} x_{g(i,j)}, \quad for \ f(i)=0,1,........N-1 \quad (7)$$

where y's are the output points, c's are the N coefficients of one row of the coefficient matrix and x's are the N elements of the input data vector. Since the coefficient structure is two-dimensional the computation is performed according to two ordering functions, $f(i)=i$ and $g(i,j)=j$, $i, j=0,1,2,...N-1$. It can be said that the $f(i)$ function determines the order in which the inner products are computed while $g(i,j)$ function determines the order in which the partial products that form an inner product are computed and it is different for each inner product (row of the coefficient matrix). The total hamming distance of the sequence of evaluation of the partial products that constitute the N inner products as determined by the ordering functions $f(i)$, $g(i,j)$, is given by the following equation:

$$Total\ HD(f,g) = \sum_{f(i)=0}^{N-1} \sum_{g(i,j)=0}^{N-1} HD(p_{f(i),g(i,j)}, p_{f(i),g(i,j)+1}) + HD(p_{f(i),N-1}, p_{f(i)+1,0}) \quad (8)$$

where $p_{f(i),g(i,j)}$ is the partial product $c_{f(i)g(i,j)}x_{g(i,j)}$. The hamming distance for a pair of partial products is given as

$$HD(p_{f(i),g(i,j)}, p_{f(k),g(k,l)}) = HD(c_{f(i),g(i,j)}, c_{f(k),g(k,l)}) \quad (9)$$

In eqn. 8 the term $HD(p_{f(i),N-1}, p_{f(i)+1,0})$ denotes the hamming distance between the last partial product of the row f(i) and the first partial product of the row f(i)+1. This term is included because after the computation of the last partial product ($p_{f(i),N-1}$) for the output term $y_{f(i)}$ the first partial product ($p_{f(i)+1,0}$) for the term $y_{f(i)+1}$ must be computed.

The aim of the proposed methodology is to derive new ordering functions $p(i)$, $r(i,j)$, such that the total hamming distance of a matrix-vector product computation as described by eqn. 8 (when f, g are replaced by p, r) is minimized and the inner product value is computed according to the equation

$$y_{p(i)} = \sum_{r(i,j)=0}^{N-1} c_{p(i),r(i,j)} x_{r(i,j)}, \quad for \ p(i)=0,1,........N-1 \quad (10)$$

The formulation of the computation reordering problem is harder for the algorithms of the second category. In a first step the minimum cost ordering of the inner products IP_i must be determined. In the second step the minimum cost ordering of the partial products p_{ij} that constitute each inner product, must be found. The problem is tackled in the following way: For all possible pairs of inner products IP_i, IP_j, the minimum cost (switching activity) connection is determined. The minimum cost connection is defined as the pair of partial products p_{ik}

(belonging to the inner product IP_i, k=0,1,...N-1) and p_{jl} (belonging to the inner product IP_j, l=0,1,...N-1) that minimizes the hamming distance between all possible pairs of partial products. The graph G (V, E) of the problem consists of the set V of N vertices corresponding to the N inner products while E is the set of edges modeling the unconstrained transitions from one inner product to another. To each edge the hamming distance of the minimum cost connection of the inner products connected by the edge is assigned as a cost. The determination of a minimum cost ordering of the inner products is formulated as a Travelling Salesman Problem (TSP) on this graph. The strategies proposed for the algorithms of the first category can be used. As soon as the minimum switching ordering of the inner products is determined, the partial products that constitute each inner product must be ordered to minimize the switching required for its computation. Each vertex (inner product) of the problem's graph is a graph with vertices corresponding to the partial products that constitute the inner product represented by the initial vertex. The hamming distances between the partial products are assigned as costs to the edges of the graphs. Thus the derivation of the minimum cost ordering of the partial products of an inner product can be formulated as a restricted minimal spanning tree problem, since only an open path over the vertices of the graph must be found and the starting and ending vertex of the path are determined by the inner product ordering (first step). An exact solution can be found if the number of the vertices of the graph is relatively small. Two widely known algorithms for the solution of the minimal spanning trees problem are the algorithms of Kruskal [6] and Prim [6]. In all cases the costs of the edges are well-bounded positive integer numbers. Their lower bound is zero and the higher bound equals to the number of bits used for the representation of the coefficients.

4. TARGET ARCHITECTURE MODEL

The structure of the proposed architecture for the algorithms of the first class is shown in figure 1. The data are stored in a background memory. From the definition of the convolution it is obvious that an overlapping by N-1 terms of the data vectors, required for the computation of successive output terms of the convolution, exists. This data reuse [7] may lead to significant power savings if the appropriate memory hierarchy is introduced. To exploit the data reuse a memory hierarchy is created by introducing a set of N foreground registers for storing the data terms required for the evaluation of one output term of the convolution. The use of the foreground registers favors power consumption reduction by replacing the background memory accesses by register accesses. After the n-1_{th} output term of the convolution is computed the data term in each register is shifted to the previous register, the new data term is transferred from the background memory to the first register (register[n]) and the computation of the n_{th} output term of the convolution may start. Since the data are read from the foreground registers according to the derived ordering function no switching penalty in the address lines is introduced.

The coefficients are usually stored in a coefficient memory (usually a ROM). The computation of the output terms of the convolution is performed on a multiply accumulator based computational unit. The number of multiply accumulators used is in general determined by the performance requirements of the application. The proposed architecture model described in this section fits well to both custom hardware and instruction set architectures. The same architecture model can be used for the second category of algorithms. The N input data terms required for the computation of a specific output vector are transferred to these registers.

The main effect of the proposed methodology on the power consumption is the reduction of the switching activity at the inputs of the computational units by deriving an activity optimal schedule for the evaluation of the partial products. The switching activity inside the computational units and thus the power consumption is proportional to the input switching activity [8]. Thus the computation reordering transformation leads to important power reduction in the computational units.

5. EXPERIMENTAL RESULTS

The proposed methodology was applied to some widely used DSP algorithms. The simulation results are included in table 1 (where – denotes reduction of switching activity while + denotes increase). The FIR filters were simulated using random data, while the 2-D DCTs, 1-D DFT, 1-D FFTs, and 2-D Wavelet transforms using real image data and finally the 1-D 16-point DCT using real speech data. For the coefficients, a 16-bit two's complement fixed point representation was assumed. It is assumed that the basic computation (for both categories) is executed on the same piece of hardware. In such a case the number of the available computational hardware resources does not affect the savings achieved by the computation reordering. If the basic computation is assigned to a number of different resources the savings are reduced as the number of resources increases.

The switching activity in the coefficient inputs of the computational units is significantly reduced (19% worst case, 75% best case). The effect of the computation reordering is different in the data inputs of the computational units. In some cases, the computation reordering introduces significant penalty (worst case 14%). This is because the reordering destroys the data correlation, especially when the transform length is relatively large. In the case of the FIR filters no penalty is introduced by the reordering because of the random nature of data. For the fast DCTs [9] and the FFTs [10], the length is reduced to small sizes (4 points) and thus no significant penalty is introduced in data activity, while in some cases small savings are observed. However, even in cases where a penalty in data activity is introduced, this is much smaller in comparison to the savings achieved, making the use of the proposed methodology advantageous.

6. CONCLUSIONS

In this paper a systematic approach for power consumption reduction in hardware realizations of DSP algorithms requiring inner product computation was presented. Two different categories of algorithms were identified.

The proposed methodology derives a low power target architecture for each category of algorithms. An architectural transformation the reordering of computation is proposed to reduce the switching activity at the inputs and inside the computational units. Information related to algorithm's coefficients which are statically determined is used for this reason.

Experimental results prove that the reordering of computation leads to significant activity savings in the computational units (data paths) even when the length of the inner product computation required by the algorithm is relatively small.

7. REFERENCES

[1] M. Tien-Chien Lee, V. Tiwari, S. Malik, and M. Fujita, "Power Analysis and Minimization Techniques for Embedded DSP Software", IEEE Transactions on VLSI Systems, Vol. 5, No. 1, pp. 123-135, March 1997.

[2] J. M. Rabaey, M. Pedram, "Low Power Design Methodologies", Kluwer Academic Publishers 1995.

[3] N. Sankarayya, K. Roy, D. Bhattacharya, "Algorithms for Low Power and High Speed FIR Filter Realization Using Differential Coefficients", IEEE Transactions on Circuits and Systems II, Vol. 44, No. 6, June 1997, pp. 488-497.

[4] A. Chatterjee, R. Roy, "Synthesis of Low Power Linear DSP Circuits using Activity Metrics", proc. of the 7th Intl. Conference on VLSI Design-January 1994, pp. 265-270.

[5] C. H. Papadimitriou, K. Steiglitz, "Combinatorial Optimisation: Algorithms and Complexity", Prentice-Hall, Inc., Englewood Cliffs, 1982.

[6] T. H. Cormen, C. E. Leiserson, R. L. Rivest, "Introduction to Algorithms", Second Edition, McGraw-Hill, 1990.

[7] J. P. Diguet, S. Wuytack, F. Catthoor, H. DeMan, "Hierarchy Explorartion in High Level Memory Management", in proc. of the 1997 International Symposium on Low Power Electronics and Design, Monterey CA, August 18-20.

[8] C. Y. Tsui, K. K. Chan, Q. Wu, C. S. Ding, M. Pedram, "A Power Estimation Framework for Designing Low Power Portable Video Applications", proc. of the 1997 Design Automation Conference (DAC 97), Anaheim, California.

[9] V. Bhaskaran, K. Konstantinides, "Image and Video Compression Standards", Kluwer Academic Publishers, 1994.

[10] W. Smith, J. Smith, "Handbook of Real-Time Fast Fourier Transforms", IEEE Press 1995.

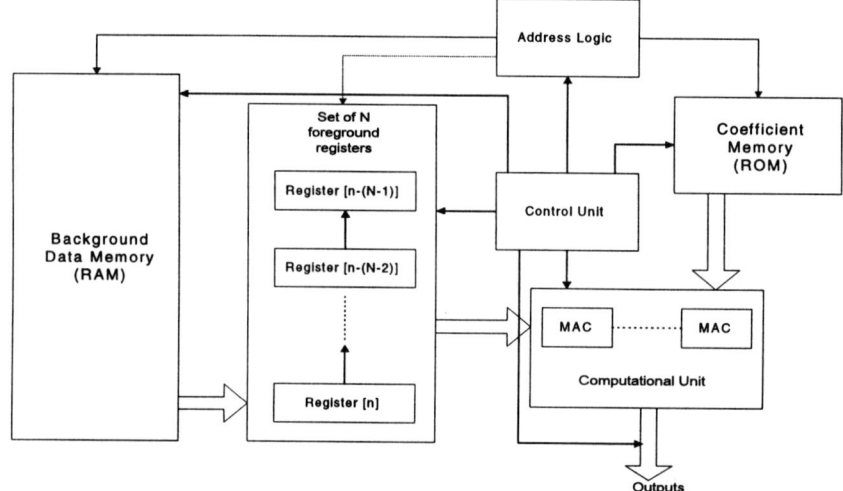

Fig. 1: Target architecture model for the first class of algorithms.

Algorithm	Change in coefficient activity (# transitions)	Change in coefficient activity (%)	Change in data activity (# transitions)	Change in data activity (%)
14-tap FIR filter	-115296	-75	-2517	-1.9
63-tap FIR filter	-180181	-62	-1827	-0.4
2-D 8 point DCT (R-C)	-2720862	-36	+856967	+10
2-D 8 point fast DCT	-1376254	-33	+83789	+2
1-D 16 point fast DCT	-1631	-20	+76	+0.5
1-D 8 point DFT	-5046266	-68	+501453	+17
1-D 9 point PTL FFT	-307155	-19	+30887	+1
1-D 7 point Singleton FFT	-333579	-26	-3554	-0.5
1-D 9 point Swift FFT	-415748	-23	-43428	-1
2-D Wavelet (15-tap filters)	-23224486	-59	+1675437	+14
2-D Wavelet (11-tap filters)	-14128501	-52	+1269121	+10
2-D Wavelet (9-tap filters)	-10886373	-50	+1100986	+9

Table 1: Simulation results

PERFORMANCE MODELING FOR SYSTEM DESIGN: AN MPEG A/V DECODER EXAMPLE

Dale E. Hocevar, Sundararajan Sriram, and Ching-Yu Hung

DSPS R&D Center, Texas Instruments, Inc.
P.O. Box 655303, MS 8368
Dallas, TX 75265
hocevar, sriram, or hung @hc.ti.com

ABSTRACT

This paper describes a system level performance simulation methodology for VLSI system design of complex signal processing devices. This methodology also provides a means for HW/SW co-simulation and co-design. An MPEG audio/video decoder example illustrates this approach. Through this example we demonstrate an extremely fast simulation method that can process multiple frames of MPEG-2 compressed video per minute, and provides the necessary information for developing and evaluating the decoder architecture. This also allows for rapid simulation over numerous test bitstreams. Our methodology allows us to measure many different performance metrics, quickly construct and alter the simulation model, process actual bitstreams, and generate test cases. A pathway for developing detailed simulation models of the lower levels of the design process is also discussed.

1. INTRODUCTION

VLSI Signal processing systems are often complex, high throughput, real time systems requiring considerable time to design. These architectures can consist of a mix of programmable processors, task specific hardware elements, register files, and assorted on-chip and off-chip memories. System design concerns include the division of work between hardware and software, buffer placement and sizing, memory bandwidth allocations, task scheduling, synchronization mechanisms, hardware utilization and clock rate determination, timing analysis, and performance robustness.

Analysis via pencil and paper for these system level issues is usually too simplistic to be reliable. On the other hand, simulation of the full design, even as high as the RT level in an HDL, often proves to be very slow, costly, and it is difficult to verify the system level performance, let alone evaluate architectural tradeoffs. Also, full simulation cannot be performed until late in the design process.

An accurate system level performance simulation model can be of great benefit in designing such architectures and verifying their performance. Such models can execute very quickly and are fairly simple to build.

This paper presents an architecture level performance simulation approach for complex signal processing systems. The fast simulation this approach provides allows for the above system level design issues to be determined, a systematic means for evaluating tradeoffs, and for the performance to be verified. We illustrate our approach with an example: an MPEG-2 audio/video decoder for which this approach has been employed. We also discuss a methodology for nearly seemless continued design and simulation down to synthesized VHDL.

2. SIMULATION APPROACH

Our approach models at the level of data block transfers, synchronization timepoints, task processing times and module computation times. Modeling can be entirely high-level behavioral or it can include some functional computation to model data-dependent behavior. Though time is measured in clock cycles, the simulation is not computed at the cycle level.

This modeling is based on building a *"process oriented"* simulation model specified in C++ using a library of simulation facilities called CSIM [1]. This library allows the various concurrently running hardware modules and software tasks to be specified as processes written in a behavioral fashion in C++. These processes run in parallel within the simulation framework and they synchronize with one another using event and mailbox constructs, e.g. *set()*, *wait()*, *send()* and *receive()* functions. Methods for modeling resources such as busses are also available. If needed, processes can be dynamically created and deleted.

Delay functions allow for direct passage of simulation time within HW/SW processes. Waiting on event or mailbox synchronization and resource reservations provide for indirect passage of simulation time. Hardware module and software task processing times used in the model are based on either the current module/task design, design requirements, or our own estimations.

Because the simulator consists only of C/C++ code linked with the CSIM library, and CSIM entails minimal simulation overhead, execution speed can be extremely fast. Performance metrics such as various buffer usage statistics, memory/bus bandwidths, hardware/software module utilization, buffer overflow/underflow conditions, timing analysis such as presence of sufficient time slacks to ensure robust operation, etc., can easily be measured within the simulation model. Time for constructing the simulation model varies considerably; from man-weeks for very simple behavioral models, to man-months for fairly complex hybrid behavioral/functional models. In addition, this approach is flexible and allows the simulation model to be easily rearranged to evaluate different design ideas.

Our approach facilitates HW/SW co-design and co-simulation. Methods have been developed for modeling the multiple software tasks that execute on embedded processors, including interrupt handling and context switching. Each necessary software section is modeled with a process and interconnected with event constructs, similar to the hardware modules. The software sections typically are interrupt service routines and partitions of the main code into sections such that the switching between these is minimized.

To accomplish this one extra process is created called the *coordinator* which operates as a custom kernel. It starts and stops the execution of all the other software processes, only <u>one</u> of which can be active at a time on a single embedded processor. The *coordinator* is written specifically to control the software processes in a predefined manner, and it handles all CPU interrupts. Usually, upon receiving an interrupt, it stops the current process (via the interconnecting events) and passes control over to the correct interrupt process if necessary. Upon completion of that process, it processes other pending interrupts, and then continues the initial stalled process. More on this in Section 4.

There are other system level simulation tools and methods that one could use to build the types of performance models described here, some with sophisticated graphical user interfaces. Most, however, operate on different simulation mechanisms than CSIM. One well known tool is SES/workbench [3] which is based upon *transaction* or *token processing* concepts. Ptolemy [2], is a well know academic tool that allows multiple computation models (such as dataflow and discrete event) to co-exist. A very new and useful tool is Cosmos [5] which is also token based. And of course VHDL could also be used. For more information on system modeling see [4]. We chose to use CSIM because of its ease of use and its *process-oriented* approach, which allows complex systems to be behaviorally specified. In addition, CSIM can be simply linked with a C++ program, resulting in very efficient simulation.

3. MPEG A/V DECODER EXAMPLE: BACKGROUND

The example we use as a case study is a single chip VLSI implementation of an audio/video (A/V) decoder for MPEG coded data; specifically MPEG-2 for the video. Such a decoder forms a key component for several digital television products, such as DSS & DBS, DVD, digital cable settop box decoders, etc.

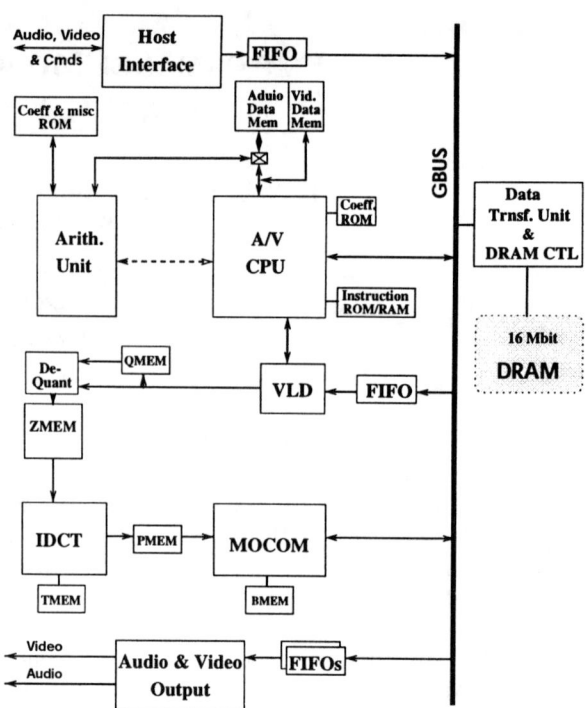

Figure 1. A/V decoder hardware block diagram.

In the design of Fig. 1, the audio and video parsing were implemented within a single programmable processor, called the A/V CPU. The A/V CPU uses a hardware variable length decoder (VLD) unit for video parsing, and an Arithmetic Unit (AU) coprocessor for audio processing. The A/V CPU is responsible for extracting header information from the MPEG-2 bitstream, computing motion vectors, conveying parameter information to various hardware elements (such as the motion compensation module), and several other video functions. In addition, the CPU also handles audio parsing functions, and sends the parsed blocks of audio samples to the AU. The audio and video parsing operations run as multitasked processes on the A/V CPU.

The design parameters in this decoder are FIFO sizes between modules (such as VLD and IDCT), access priority to DRAM, functionality that can be placed in SW in the A/V CPU, amount of memory allocated to frame and bitstream storage, and the speed at which the different modules in the system run. Several parameters in the combined A/V decoder must be measured before we can ascertain that the decoder meets performance requirements. These include bandwidth to DRAM, correct synchronization between modules, timeliness of decoding, adequate buffer sizes, etc.

4. A/V DECODER EXAMPLE: SIMULATION DETAILS

The performance simulator model for this example demonstrates the approach of Section 2. Fig. 2 shows the partitioning chosen, and shows all the hardware or software *'processes'* used in this model; i.e., the rounded corner boxes. Only the critical FIFOs and buffers are shown; the DRAM is modeled but not shown. All event constructs are depicted as lines with arrows; those with nearby small mailboxes are the mailbox constructs which allow message passing with each event and also provide FIFO queuing of events. This construct is used quite often to model the passing of data blocks through a buffer that holds multiple blocks. Messages sent to a mailbox at initialization time sets the number of data blocks that the corresponding buffer can hold. Arrays of events are used to allow a process to wait on several events simultaneously; this is heavily used for CPU interrupts and CPU process context switching.

Notice the A/V CPU contains several software processes including the *coordinator* process. The Field, HV_Sync and portions of the AP (Audio Parsing) process are interrupt service routines; the remaining processes are executed in the prescribed pattern by the *coordinator*, along with enforcing the other synchronizing conditions. The array of events labeled cp_events_in[ ] contains all the events that trigger the *coordinator*, and cp_events_out[ ] the events that the *coordinator* sets. Notice that some of the arrayed events are used only within the CPU (see the named indices in the CPU box); these allow the *coordinator* to turn these processes on or off to simulate context switching. Other events, depicted below the CPU box, are real external interrupt signals that trigger some action by the *coordinator*.

Certain hardware modules consume a fixed amount of time for processing data independent of the actual data values. These modules are assigned fixed delay times based on estimates of their computation times. Examples of these are IDCT, dequantization, and motion compensation. Other modules, such as variable length decoding, exhibit data-dependent behavior. These modules process actual data in the performance simulator implementation. Thus we route data through only those portions of the simulator that need to explicitly process data to maintain correct timing. Bitstream parsing, including run/level data that is passed on to the expansion/DQ unit, is done precisely. This is because we have to extract picture types (I, P, or B), motion vector information, and variable length coded DCT coefficients from the bitstream. This information is crucial because it governs the operation of the decoder. Such a strategy leads to efficient simulation since we are modeling each module only at a requisite level of detail. This results in a simulation speed of 2-5 frames of MPEG-2 video per minute, as opposed to 12-24 hours per frame for full design VHDL (i.e. synthesizable).

5. A/V DECODER EXAMPLE: SIMULATION RESULTS

The first and most important outcome of the performance simulation model is the video decoding synchronization scheme. We used the simulator to develop, fine-tune, and verify the synchronization among the A/V CPU and dedicated datapath modules. The model feeds bitstream

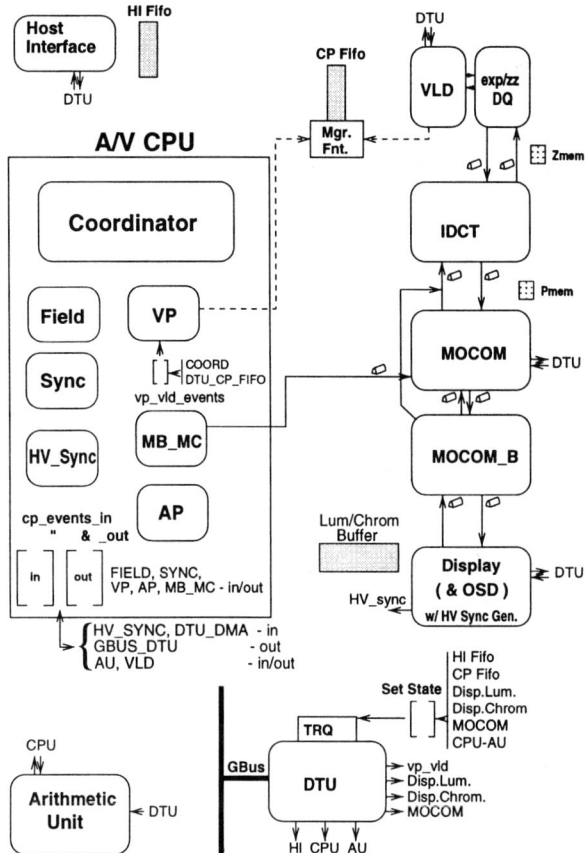

Figure 2. Schematic of the performance simulator; rounded corner boxes are HW or SW processes.

data to the decoder at the bit-rate specified in the bitstream, the modules coordinate and communicate through signals and data buffers, and the display module consumes data according to actual display rate (NTSC or PAL). Thus, when the model processes through an entire bitstream file without causing a buffer underflow or overflow error, or other error conditions, we have verified that the system would successfully decode that particular bitstream in real time.

In addition to this built-in pass/fail indication, we collect timeliness measurements, e.g. the *slice slack time*, the minimal time distance, in a macroblock row, between any data item being produced by the decoder and being consumed by the display. For all the test bitstreams, the slice slack time is at least 5 times the average macroblock decoding time.

We also collected resource utilization measurements during the simulation. In our early architecture design stage, we obtained some rough estimates for the most critical utilization numbers, such as external DRAM bandwidth and processor utilization. Utilization data from the simulation allowed us to check against our earlier estimates, providing sanity-checks against modeling inaccuracies. As the architecture evolves further during design, the utilization measurements can help to make more design tradeoffs.

Fig. 3 shows major module utilization for four large bitstream files, 15-31 frames each.

The simulation model is also very useful to predict what-if scenarios in system design. For instance, one could add modeling for customer programming, OSD (on-screen display), error correction, and possibly some decryption computation. These external activities may interfere with the video decoding operation by taking up processor time, DRAM bandwidth, etc. In Fig. 4, we show the simulated effect of OSD loads on the DRAM bandwidth on one bitstream. A very heavy OSD load of 1/2 screen, 24 bits/pixel is used in this experiment.

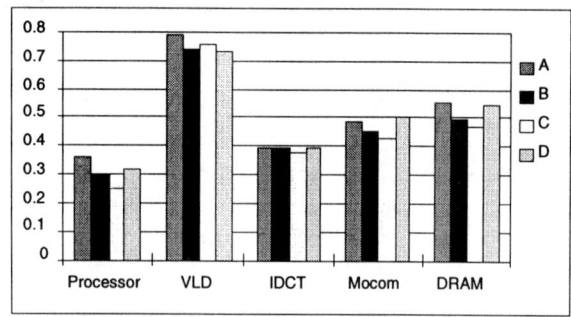

Figure 3. Module utilization for 4 large bitstreams.

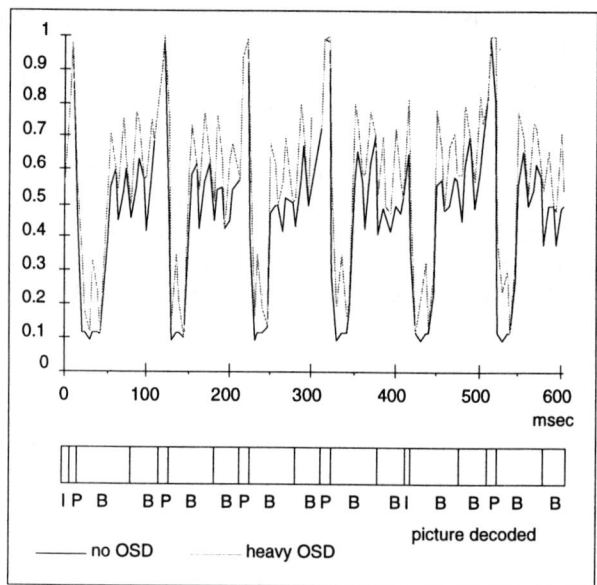

Figure 4. External DRAM utilization versus time with and without OSD.

6. PATH FOR PROGRESSIVE SIMULATION & DESIGN

In addition to performance simulation, it is very useful to have a pathway that progresses through more detailed levels of design and simulation seamlessly, and with minimal tool transitioning time. This also allows the more efficient simulation vehicle to be used <u>maximally</u> at each phase (e.g. Quickturn emulation with full design VHDL), and <u>minimizes</u> the simulation time that may have to be spent in inefficient simulation environments (e.g. full design VHDL simulation on a workstation). Such an approach is discussed in [5]; below is a brief description.

After the performance simulation and architectural design phase, the simulator can be converted to a cycle based (CB) model by replacing all synchronization mechanisms with interconnection signals. Then clock events are added, and fully clock accurate module interface signals are designed and implemented, after which, bit/cycle accurate, functionally correct, internal module codes are developed or altered as necessary. A cycle accurate simulator for a CPU/μP in the design, if available, can be integrated into the CB model; this allows detailed HW/SW co-simulation.

At this point the design has matured considerably, and what remains is to complete the implementation of all modules. RTL C/C++ codes for hardware modules can be developed first, tested within the CB simulator, and later used to develop the synthesizable VHDL. Alternatively, the RTL VHDL can be written first and tested using the input/output signals surrounding that module, captured by the CB simulator. The top level VHDL can then be written (which is mostly just interconnection placements of the modules) and the VHDL module codes are then incorporated into it. It is then possible to move to an extremely fast Quickturn based emulation, after preliminary checking is completed using the full VHDL simulation.

7. CONCLUSIONS

In this paper we discussed a simulation methodology based on utilizing performance modeling to facilitate system level design. An MPEG A/V decoder architecture with its simulation model provided an example of such an approach. Various performance metrics that were obtained were presented, such as module and DRAM utilization. The performance simulator ran at a simulation speed of 2-5 frames per minute, as opposed to 12-24 hours per frame for full design level VHDL simulation. This allowed us to efficiently carry out architecture evaluations and refinements. We also presented a methodology for transitioning from this model to a clock-cycle based model, and finally to a VHDL model.

REFERENCES

[1] H. Schwetman, "Using CSIM to model complex systems," *Proc. of the 1988 Winter Simulation Conference*, 1988, pp. 246-253.
[2] J. Pino, S. Ha, E. A. Lee, and J. T. Buck, "Software Synthesis for DSP Using Ptolemy," *Journal of VLSI Signal Processing*, Jan. 1995.
[3] SES/workbench[TM]: A Multilevel Design Environment for Modeling and Evaluation of Complex Systems," Scientific and Engineering Software, Inc., Austin, TX, May 1989.
[4] R. Goering, "Designers reach for a higher level," and "Emerging tools aid high-level design," (two part series), *Elec. Eng. Times*, August 3&10, 1992.
[5] Cosmos User's Manual, Omniview Design, Inc., Pittsburgh, PA, 1997.
[6] D.E. Hocevar, C-Y Hung, D. Pickens and S. Sriram, "Top-Down Design Using Cycle Based Simulation: an MPEG A/V Decoder Example," *Great Lakes Symposium on VLSI*, Feb. 1998.

GRAPH TRANSFORMATION FOR COMMUNICATION MINIMIZATION USING RETIMING

Michael Sheliga, Zhihong Yu, Fei Chen, Edwin H.-M. Sha

Dept. of Computer Science & Engineering
University of Notre Dame
Notre Dame, IN 46556

ABSTRACT

Nested loops are normally the most time intensive tasks in computer algorithms. These loops often include multiple dependencies between arrays that impose communication constraints when used in multiprocessor systems. These dependencies may be between dependent arrays (loop dependencies), or between independent arrays (data dependencies). In this paper, reducing the communication caused by data and loop dependencies for perfect nested loops is explored. It is shown that for a given partition data dependencies may be treated as a specialized form of loop dependencies. Once this is done, previous results on scalable loop tiling can be used to calculate the final total communication. Next, the effects of changing the partition for both loop and data communication are examined. Using these results, the optimal partition for a number of cases are examined. Results are shown which illustrate the efficiency of the system as well as the savings achieved.

1. INTRODUCTION

Applications such as digital signal processing, image processing, fluid mechanics and real-time adaptive controller require high computer performance. Many of them need the designs of dedicated highly parallel systems. As one of the design-point teams, the *PetaFlop* project studies super parallel architectures in order to achieve 10^{15} floating point operations per second. It is known that in highly parallel architectures, communication is often the bottleneck limiting execution speed. For example, the experimental design we did using the processor-in-memory (PIM) EXECUBE array as a co-multiple processor for video compression shows that communication imposes significant overhead on the overall performance. It is seen that no matter how fast the processing circuits are, the communication part gives the performance bottleneck. Therefore, this paper studies the fundamental theories and techniques which can transform the input program specified by a graph model to reduce the communication while it is executed in a parallel system. The proposed techniques can be used in the first step as the *graph transformation* step either for application specific design or for code generation for highly parallel architectures.

While the problem of calculating and minimizing communication costs due to loop data dependencies has been widely studied, such research has involved changing the way iterations are partitioned, not modifying the graph or loop dependencies themselves. This paper presents algorithms that minimize loop communication

THIS WORK WAS SUPPORTED BY NSF MIP 95-01006 AND NSF/ACS96-12028

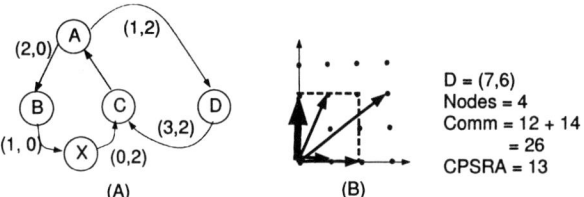

Figure 1: (A) Original data flow graph. (B) Iteration space and 4-node partition for the left graph.

for multi-dimensional graphs (a common graph model for multi-dimensional DSP or nested loops in a program) by modifying the structure of the input graph and changing the distribution of the loop dependencies themselves, using the *multi-dimensional retiming* concept.

This paper calculates both the "shape" and "size" of the iterations to be executed on a particular processor, for a given set of loop dependencies. This paper also minimizes the communication imposed by uniform inter-array loop dependencies by legally transforming the dependencies themselves. Three techniques, *graph simplification, angle modification and delay reduction*, are developed to reduce the communication requirements. First, the structure of the graph is simplified where possible so as to combine dependencies and eliminate nodes. Second, the communication is reduced by using the method of multi-dimensional retiming. Third, the overall magnitude of the dependencies is reduced so as to lessen the communication. To the author's knowledge, this is the first paper to attempt to minimize inter-partition communication by transforming the dependencies of the original graph.

Much of previous results on retiming focus on one-dimensional scheduling problems. Multi-dimensional retiming research has been focused around the improvement of parallelism inherent in multi-dimensional applications [3, 7, 8] not on the improvement of the resulting communication volume. Previous research on communication minimization, on the other hand, has concentrated on decreasing the communication by changing the partition, not by modifying the delays themselves [1, 5].

2. BASIC CONCEPT

We will use Figure 1 to Figure 4 to introduce the concepts of our algorithm. Figure 1(A) presents a simple multi-dimensional data flow graph (DFG). The vectors next to the edges in the graph represent multi-dimensional delays that indicate what iteration the data

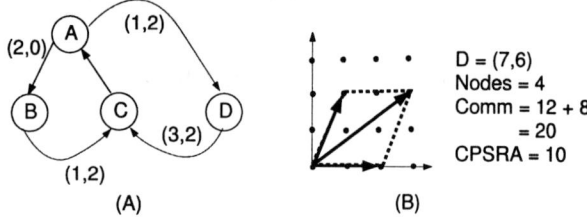

Figure 2: (A) The graph after graph simplification. (B) The iteration space for the left graph.

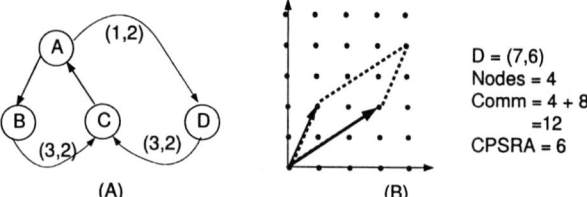

Figure 3: (A) The graph after angle modification. (B) The iteration space for the left graph.

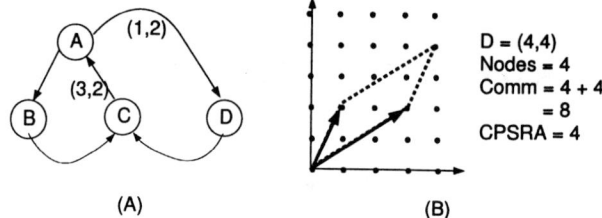

Figure 4: (A) The graph after delay reduction. (B) The iteration space for the left graph.

will be used in. Figure 1(B) shows the associated iteration space where delays are represented by the lines with arrows, while the group of iterations, called a partition, that are to be executed by a single processor is bounded by dashed lines. In this example, 4 nodes are in a partition.

The communication across each edge of the partition for a single delay is well known to be the dot product of the vectors representing the edge of the partition, and the normal to the delay which has the same magnitude as the delay. For example, communication for each node in the partition across the counterclockwise(CCW) boundary (e.g. $(0, 2)$) in figure 1(B) is $(0, 2) \cdot (1, 2) + (0, 2) \cdot (3, 2) + (0, 2) \cdot (2, 0) + (0, 2) \cdot (1, 0) + (0, 2) \cdot (0, 2) = (0, 2) \cdot ((1, 2) + (3, 2) + (2, 0) + (1, 0) + (0, 2)) = (0, 2) \cdot (7, 6) = (0, 2) \cdot D = 12$. Similarly the communication across the clockwise (CW) boundary (e.g. $(2, 0)$) is $(2, 0) \cdot (7, 6) = 14$. Hence the total communication is $12 + 14 = 26$.

The best possible partition should be aligned with the *outermost* vectors that represent delays in the graph [1]. Hence, in figure 1(B), the partition is aligned with the vectors $(2, 0)$ and $(0, 2)$.

For a given two-dimensional graph and the resulting optimal partition angles, the communication per square root of area (CPSRA) is constant, regardless of the area, if the length of the sides of the graph are optimal. For example, if the length of each side is tripled, the tile size will increase ninefold, while the communication will triple. Hence, the communication per square root of area (CPSRA) will remain the same. Therefore, this measure may be used to compare iteration spaces of unequal areas. So, for Figure 1 (B), the effective communication, $CPSRA$, is $\frac{26}{\sqrt{2*2}} = 13$.

Figure 2 demonstrates the concept of *graph simplification*. During this process, certain nodes are eliminated from the graph. For example, in figure 2(A), node X is eliminated and a new edge, $e_{B,C}$, is produced. Note that in the associated iteration space shown in figure 2(B), with the resulting effective communication reduced to 10.

Figure 3 demonstrates the concept of *angle modification*. During this process, *multidimensional retiming* is used to redistribute the delays of the graph so as to reduce the angle of the outermost delay vectors. It can be shown that the angle of the outermost delay vectors is one of the most important factors affecting communication cost. Generally speaking, we would like the angle to be reduced. In figure 3(A) delay $(2, 0)$ is retimed through node B. The effective communication is decreased to 6.

Figure 4 demonstrates the concept of *delay reduction* in which we decrease the sum of all delays in the graph. By pushing the $(3, 2)$ delay through node C, we decrease the sum of all delays in the graph from $(7, 6)$ in figure 3(A) to $(4, 4)$ in figure 4(A). The new effective communication, as shown in figure 4(B), is 4, which is 30.8% of the communication in figure 1(B).

As you can see, communication minimization requires not only an efficient strategy for the partitioning of computational tasks, but also an optimized method of distributing delays in the graph(retiming) so as to reduce the amount of data that needs to be moved between iterations.

3. COMMUNICATION MINIMIZATION USING RETIMING

The overall communication minimization algorithm begins with graph simplification, followed by angle modification and then delay reduction. Since deciding on the exact retiming vectors is a difficult problem, it is worthwhile to reduce the complexity of the graph using *graph simplification*. Once this is done, MD retiming is used to reduce the communication in two ways. First, it is possible to modify the extreme vectors of the graph using MD retiming, called as *angle modification*. This enables the shape of the partition to be changed, and the resulting communication to be reduced. Second, it is possible to reduce the overall magnitude of the delay vectors in the graph, called *delay reduction*, thereby reducing the overall communication.

3.1. Graph Simplification

Linear operations are represented in a DFG by a series of connected nodes with indegree and outdegree (the number of incoming and outgoing edges of a node, respectively) of one. These nodes may be combined to produce a simpler graph, justified by the following property.

Property 3.1 *Given two delays, the angle of the sum of the delays will be between the angles of the two delays.*

Since we may combine delays for a node with indegree one and outdegree one without affecting the rest of the graph, and without increasing the angle of the outermost delay vectors, it is always worthwhile to do so. The class of nodes which may be included in the simplification may be expanded if we consider nodes with indegree one and arbitrary outdegrees.

Figure 5 shows this process in detail. Figure 5 (A) is the original graph, while (B) is the same graph with nodes with indegree and outdegree equal to one eliminated (nodes 1, 4, 9 and 10). In (C) we have eliminated node 5, which has indegree one, and its associated edges, and replaced them with two new edges, $e_{7,2}$ and $e_{7,6}$. The value of $d(e_{7,2})$, for example, is equal to $d(e_{7,5}) + d(e_{5,2}) = (5,5) + (1,1) = (6,6)$. Figure 5(D) shows the final graph after nodes with outdegree one have been replaced.

3.2. Angle Modification

During angle modification the extreme delay vectors of the graph are changed so as to be closer to each other. Two algorithms are presented here. The first algorithm, *extreme vector modification*, simply retimes those nodes that are the current extreme vectors. The second, *loop modification*, takes a more global approach and retimes entire loop at once.

We refer to the sum of the delays in a loop as the loop vector and denote it by d_L for loop L. For any given loop, there must be a delay vector either aligned with or outside the loop vector. Therefore, there is no point in trying to move the extreme delay vectors inside the extreme loop vectors. The loop associated with the extreme loop vector in the clockwise (CW) direction is referred to as the CW critical loop, with the counterclockwise (CCW) critical loop defined similarly. As an example, the extreme loop vector in the CCW direction for figure 5(D) is $(3,7)$, which is the sum of loop $L_{2,3}$.

Extreme vector modification alters a graph by modifying one node at a time. Since there are two extreme vectors, the one that is further from the corresponding extreme loop vector is modified first. We first try the modification that would change the angle of the vector the most.

This algorithm continues until either both extreme vectors have been made equal to the corresponding extreme loop vector, or no more retimings are possible that would reduce the extreme vectors.

As an example of this algorithm, consider figure 5 (D). If we were only interested in the CCW direction, we would first retime edge $e_{2,3} = (0,4)$ to become $(1,4)$ and then $(2,4)$ leaving edge $e_{3,2} = (1,3)$ as the extreme CCW vector. No more local retiming would help reduce the extreme vector. Therefore, at this point we would push the entire extreme vector across node 2, resulting in $e_{2,3} = (3,7)$ and $e_{3,2} = (0,0)$. This step is done only if doing so does not increase the extreme vector.

Loop modification takes a different approach than extreme vector modification. It considers the delay sum of entire loops instead of individual vectors. The intuition behind this approach is to retime the loops that have the worst angles first, since these loops will be the hardest to adjust.

First, the delay sum for each loop, d_L, is calculated. The extreme loops are then retimed beginning with the loop that is furthest from the current total delay sum for the entire graph. For example, if the extreme loop consisted of four edges, and d_L was $(4,8)$, each edge would be retimed so that the resulting delay would be $(1,2)$. After all critical loops are retimed, non-critical loops are retimed.

3.3. Delay Reduction

During delay reduction we push delays through nodes that have more incoming edges than outgoing edges, or more outgoing edges

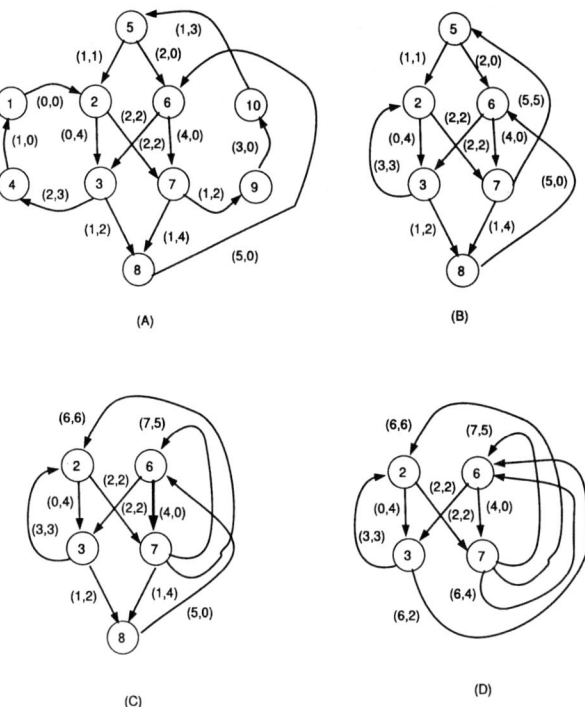

Figure 5: An example of the graph simplification algorithm

than incoming edges. However, this is only done when it does not change the extreme vectors of the graph. Delay reduction begins by undoing some of the effects of graph simplification algorithm. In particular, nodes and the associated edges that had indegree equal to one and outdegree greater than one, or outdegree equal to one and indegree greater than one are reintroduced into the graph.

The value on the reintroduced edges is set to $(0,0)$. We then attempt to push as large of a value of delay through appropriate nodes as possible, without changing the value of the extreme vector (delays are actually changed repeatedly by $(0,1)$ or $(1,0)$). For example, suppose the current extreme vectors were $(2,1)$ and $(0,8)$ and node A had incoming edges with values $(8,5)$ and $(9,9)$, and an outgoing edge with value $(0,0)$. In this case we would push $(2,2)$ across node A resulting in incoming edges of $(6,3)$ and $(7,7)$. Notice that of the resulting incoming edges the angle of $(6,3)$ is equal to the extreme vector $(2,1)$. Also notice that the total delay sum of the entire graph has been reduced by $(2,2)$.

The algorithms presented in this section may be expanded to higher dimensions with minor adjustments to the extreme vector modification and loop modification algorithms. The graph simplification and delay reduction algorithms will largely remain the same.

4. EXPERIMENTAL RESULTS

This section presents experimental results for several two dimensional graphs. The graph from figure 5 is used, as well as different versions of the infinite impulse response filter, and the wave digital filter. These graphs may be found in [6], and filter name contains the page number.

Filter	Inputs				EVM Results					Loop Results				
	CW	CCW	DSum	Com	CW	CCW	DSum	Comm	Im%	CW	CCW	DSum	Com	Im%
Figure 5	(2, 0)	(0,4)	(26,25)	46.9	(5,1)	(1, 3)	(32,26)	44.3	**5.5**	(5,1)	(1, 3)	(32,26)	44.3	**5.5**
IIR4442a	(1,-1)	(-1,2)	(13,25)	43.3	(1,0)	(0, 1)	(3, 7)	9.2	**78.8**	(1,0)	(0, 1)	(3, 7)	9.2	**78.8**
IIR4443b	(8,-15)	(-2,4)	(41,87)	122.3	(1,0)	(0, 1)	(3, 7)	9.2	**92.5**	(1,0)	(0, 1)	(3, 7)	9.2	**92.5**
IIR62	(1,-1)	(-3,4)	(34,52)	59.4	(1,0)	(0, 2)	(3, 7)	9.2	**84.5**	(1,0)	(0, 2)	(3, 7)	9.2	**84.5**
WDF37	(1, 0)	(-6,1)	(38,4)	18.3	(1,1)	(-1,1)	(4, 4)	5.7	**68.9**	(1,1)	(-1,1)	(4, 4)	5.7	**68.9**
WDF63a	(8,-31)	(1,-4)	(66,248)	88.4	(1,-1)	(1, 1)	(4, 4)	5.7	**93.6**	(1,-3)	(1, 1)	(9, 4)	8.4	**90.5**
WDF119	(3, 0)	(-5,1)	(27,4)	15.5	(1, 1)	(-1,1)	(4, 4)	5.7	**63.2**	(1, 1)	(-1,1)	(4, 4)	5.7	**63.2**
WDF121	(2, 0)	(-3,1)	(15,4)	13.3	(1, 1)	(-1,1)	(4, 4)	5.7	**57.1**	(1, 1)	(-1,1)	(4, 4)	5.7	**57.1**

Table 1: Results for several filters using extreme vector modification and loop modification

Inputs			Vector		Loop	
Nodes	Edges	MAG	Spd	Comm	Spd	Comm
10	2	2	2.97	88.4	2.44	79.2
10	2	10	2.88	152.3	2.55	144.3
10	4	2	3.03	164.2	2.13	142.2
10	4	10	2.95	309.7	2.02	271.3
20	2	2	3.04	167.6	2.21	143.4
20	2	10	3.06	352.4	2.18	265.2

Table 2: Comparison of extreme vector modification and loop modification.

Table 1 shows the results for the above filters using the extreme vector and loop modification algorithms. The second and third columns represent the extreme vectors of the input graph in the CW and CCW directions, respectively. The fourth column represents the sum of the absolute values of all input delays. This column gives an idea of how many delays are in the graph. The fifth column represents the total effective communication for the input graph. Columns six through nine and eleven through fourteen are similar to columns two through five, but show the corresponding values for the final graph, while columns ten and fifteen represent the percentage of reduction of the effective communication. The average reduction in communications is **68%**.

It should be noted that for these relatively small graphs, extreme vector modification is very efficient. In Table 1 it is able to find the best delay configuration in all cases. Since extreme vector modification operates locally, and the input graphs are relatively small, this is as expected.

In order to further compare extreme vector modification and loop modification using a variety of larger graphs, a series of random graphs were generated as shown in Table 2. The nodes column in this table is the number of nodes in the graph. The edges column indicates the number of edges per node in the graph. The MAG column represents the average absolute value of the magnitude of the delays in the graph.

These results show that as the graph becomes larger, extreme vector modification becomes less and less effective since it operates locally. On large graphs, it is likely to become stuck on a local minima, and not be able to find the optimal result. Loop modification is able to perform more consistently for larger graphs.

5. CONCLUSION

In highly parallel computers, communication is often the bottleneck limiting execution speed. While the problem of calculating and minimizing communication costs due to loop data dependencies has been previously studied, such research has involved changing the way iterations are partitioned, not modifying the graph or loop data dependencies themselves. This paper presents algorithms that minimize loop data communication for multi-dimensional graphs by modifying the structure of the input graph and changing the distribution of the loop dependencies themselves, using *multi-dimensional retiming*. Results that illustrate the savings in communications are presented for several practical input systems.

6. REFERENCES

[1] P. Boulet, A. Darte, T. Risset and Y. Robert, "(Pen)-ultimate Tiling," *Proc. Scalable High Performance Computing Conference*, pp. 568-576, May 1994.

[2] P.-Y. Calland and T. Risset, "Precise Tiling for Uniform Nested Loops,", pp. 330-337, 1995.

[3] A. Darte and Y. Robert, "Constructive Methods for Scheduling Uniform Loop Nests," *IEEE Transactions on Parallel and Distributed Systems*, 1994, Vol. 5, no. 8, pp. 814-822.

[4] E. Hodzic and W. Shang, "On Supernode Transformation with Minimized Total Running Time,", pp. 402-414.

[5] F. Irigoin and R. Triolet, " Supernode Partitioning," *Proc.15th Annual ACM Symposium on Principles of Programming Languages*, Jan 1988, pp. 319-329.

[6] N. I. Passos "Improving Parallelism on Multi-Dimensional Applications: The Multi-Dimensional Retiming Framework". Ph.D. dissertation, University of Notre Dame, 1996.

[7] N. Passos and E. H.-M. Sha, "Scheduling of Uniform Multi-Dimensional Systems under Resource Constraints," (regular paper) Accepted for publication in *IEEE Transactions on VLSI Systems*.

[8] N. Passos and E. H.-M. Sha, "Synchronous Circuit Optimization via Multi-Dimensional Retiming," (regular paper) in *IEEE Transactions on Circuits and Systems, vol II - Analog and Signal Processing*, Vol. 43, No. 7, July 1996, pp. 507-519.

GATE TO CHANNEL SHORTS IN PMOS DEVICES: EFFECTS ON LOGIC GATE FAILURES

M. Shaheen Sayeed and Samiha Mourad
Department of Electrical Engineering
Santa Clara University
Santa Clara, CA 95053
smourad@scu.edu
408 554-4163, 408 554-5474 (fax)

ABSTRACT

Gate to channel shorts via oxide layer have enormous effect on VLSI circuits and may render devices useless. However, there is no sure way of detecting and locating such faults due to its random nature. In order to test VLSI circuits, a working model is essential in the event a gate oxide short occurs. A model for gate oxide shorts in PMOS transistors is presented in this paper and used to determine the behavior of digital gates in the presence of these shorts. Except when the short resistance is very low, the failure is not detectable by voltage testing and current or delay testing is necessary.

1. INTRODUCTION

The effect of gate oxide breakdown on the operation of logic gates is very important because this failure mechanism is one of the main causes of circuit malfunctioning in manufacturing and in the field. Gate oxide shorts are unintended and undesirable electrical connection between gate and channel, source or drain terminals of an MOSFET through the oxide layer. Fabrication defect and other electrical stresses may result in oxide shorts and poses a serious reliability problem of the device. Gate oxide thickness is one of the limitations to device scaling [3]. In addition, a correct model is essential in testing digital circuit. Finally, gate oxide breakdown has been examined in the case of NMOS device [1] but very little has been reported in the literature about the effect of the same defects in PMOS devices.

In this paper, we first present a model for gate oxide shorts in PMOS devices with p- and n-doped polysilicon gates. Then, we use this model to study the behavior of some CMOS logic gates such as inverter, NAND and transmission gates in the presence of such shorts. The goal of this research work is to analyze the effects of gate oxide shorts in the context of logic gates by using a simple but realistic model, which closely represent the physical reality.
In the case of p-doped polysilicon gate, the model for shorts in PMOS may be extended from that for the NMOS [7]. There is, however, a significant difference between these two models. While in NMOS, the breakdown may be realistically modeled as a purely resistive [4], in PMOS, it is also represented by dominant PN junction device as shown in Fig. 1. It is this device that limits the drain current I_{DD} and makes the failure less dramatic.

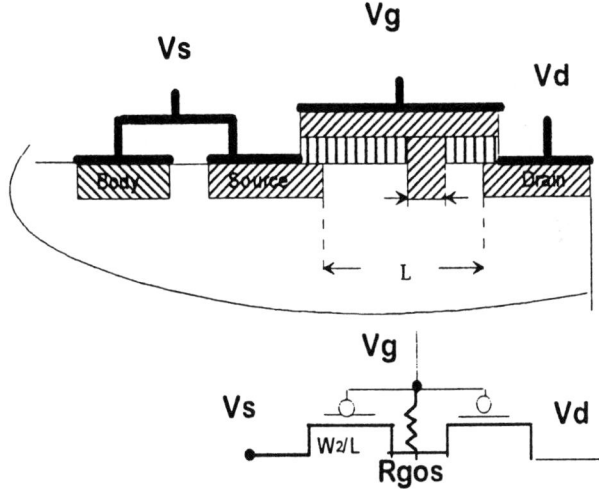

Figure 1. P-doped Polysilicon Gate: GC Model

For the n-doped polysilicon gate, we propose the model as presented in Fig. 2. We validated this model using Technology Modeling Associate's device simulation tool DaVinci. Figure 3 shows a sample of the IV characteristics as obtained from DaVinci and those obtained by spice simulation. We then performed an extensive spice simulation study using this model in several logic gates. In all circuits under test (CUT), we represented the resistive short by Rgos. Also, each circuit under test (CUT) was preceded and followed by a buffer as illustrated in Fig 4. As CUT, we used an inverter, a 2-input NAND gate and a transmission gate.

2. MODELS FOR GATE OXIDE SHORTS

Different models have been reported in literature along with the causes of the gate oxide shorts [2], [6]. Some models based on experimental results also appear in literature [5]. Gate oxide shorts are primarily caused by

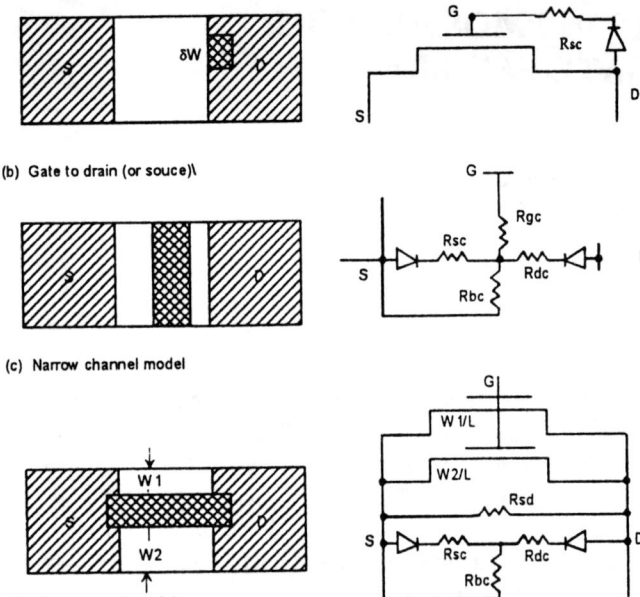

(b) Gate to drain (or source)

(c) Narrow channel model

(d) Short channel model

Figure 2. N-doped Polysilicon Gate: Model for Various Short locations.

irregular and defective growth or breakdown of oxide during processing or by stress due to excessive voltage while the device is in operation. These defects could best be characterized as random and irregular in size, shape and relative location in the channel. Gate oxide is the heart of the MOSFET operation. A breakdown of the oxide in the channel region disrupts the isolation of the channel (inverted or otherwise) from the gate.

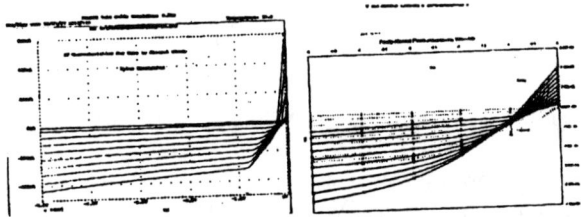

Figure 3. IV characteristics for PMOS gate to channel short: a) Spice results, b) DaVinci's results.

Depending on the severity of the breakdown, this phenomenon can be represented by a variable short resistance accompanying the derived devices due to split of the channel. A bigger size oxide breakdown may cause the device to cease proper operation while a nominal oxide short may be masked. Besides the gate to channel short, there may be breakdowns in the gate overlap resulting a short between gate and source/drain. The later is more easily to model than the former. So, we are seeking a reliable model for PMOS gate to channel oxide short.

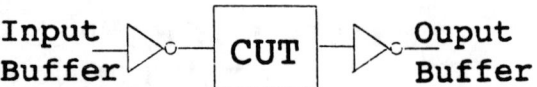

Figure 4. Experimental Setting.

Figs 1 and 2 represent two models for gate to channel shorts in PMOS transistor. The first model is for p-doped poly gate, a simple variable resistance is lumped near the shorted gate region[7]. The second model is for the n-doped poly gate. This model is much complicated by the fact that the inverted channel of PMOS and the gate are appositely doped. Also location and orientation of the defect may render the original channel shortened or narrowed as shown. These possible combinations are modeled in Fig. 2.

3. SIMULATION RESULTS

Two types of simulations are used in this work. Device simulation is done to analyze the behavior of the PMOS transistor having gate to channel oxide short of varying size and shape at different locations of the channel. The IV characteristics of the PMOSFET from the device simulation plays as a reference for the cross validation of our proposed model. The model's IV relation from spice simulation and the device simulation matches very closely as shown in Fig. 4. Our simulation result shows that, for high values of R_{gos}, no deterioration in the output voltage is observable irrespective of the location of the defect within the PMOS gate. However, the switching current is increased. For lower values of R_{gos} the defect can easily be modeled as stuck at fault. The effect of the short resistance is shown in Figs. 5 and 6. In essence, the following important results were observed:

1. The deterioration of the circuit performance is more pronounced when the short is from gate to drain (GD) than it is for gate to the source (GS). In Fig. 5, the output voltage and current for an inverter are shown for both GD and GS shorts for various values of R_{gos}. For GS short, the inverter switched correctly although with noticeable delays. In the case of GD short, as R_{gos} decreases, the voltage level is also decreased and eventually the circuit no longer inverted the input.

2. The location of the defect within the channel influences the voltage and current as shown in Fig. 6. This failure cannot be detected with voltage testing, but it is easily detectable with current testing (I_{DDQ}).

3. The malfunctioning of a simple gate in the presence of gate oxide short is clearly depicted in Fig.7. Here the output of the 2-input NAND gate deviates from the expected voltage level with the variation of short

resistance which models the size of the gate oxide short.

4. For transmission gates, the effect of shorted PMOS transistor is pattern dependent. But, unlike CMOS gates, the defect is easily observable from the output. These are illustrated in Fig. 8, where the output of a transmission gate is shown when the gate to channel short is induced in the PMOS of the pass gate.

4. CONCLUSION

We presented a simple model for gate oxide breakdown in PMOS devices, and showed how its usage is important in digital testing of logic gates that are becoming susceptible for gate oxide breakdown. In particular, traditional stuck at fault models are not sufficient in analyzing the gate oxide shorts. Current or delay testing can be effectively used in the analysis and design of digital circuits under oxide breakdown.

5. ACKNOWLEDGEMENT

We are grateful to T. Bau for the DaVinci simulation of the gate-to-channel model in the PMOS transistor.

6. REFERENCES

[1] J. C. Chen and S. Mourad, "Characterization of gate to channel shorts in BiCMOS logic gate," *VLSI Test Symposium*, pp. 440-445, 1994.

[2] C. F. Hawkins and J. M. Soden, " Electrical properties and detection methods for CMOS IC defects", *Proceedings of the 1st European Test Conference*, pp. 159-167, Apr. 1989.

[3] C. Hu, "Reliability of thin SiO2", *Semiconductor Science and Technology*, No. 9, pp. 969-1004, 1994.

[4] R. Rodriguez-Montanes, "Current vs. logic testing of gate oxide short," *Int'l Test Conference*, pp. 510-519, 1991.

[5] J. Segura, C. De Benito et.al., "A detailed analysis and electrical modeling of gate oxide shorts on MOS transistors," *Journal of Electrical Testing*, pp. 229-239, Aug. 1996.

[6] J. M. Soden and C. F. Hawkins, "Test Considerations for gate oxide shorts in CMOS IC's ," *IEEE Design & Test of Computers*, vol.3, pp. 56-64, Aug. 1986.

[7] M. Syrzycki, "Modeling of spot defects in MOS transistors," *Int'l Test Conference*, pp. 148-157, 1987.

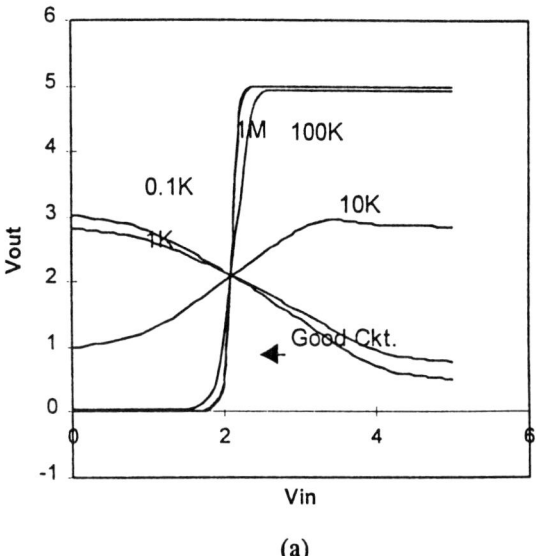

(a)

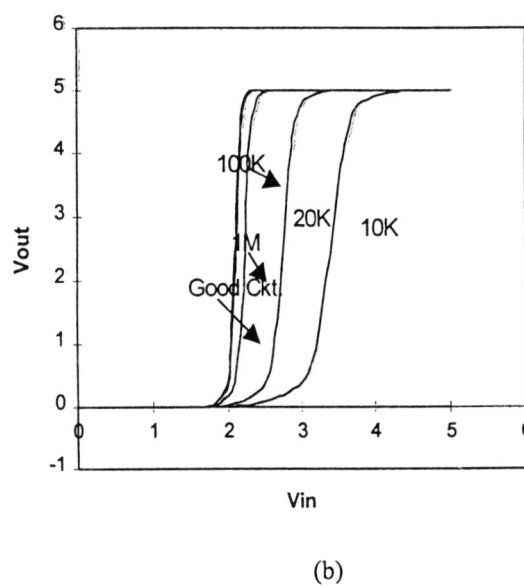

(b)

Figure 5. VTC of an Inverter for Various Rgos: a) Gate to Drain, b) Gate to Source short.

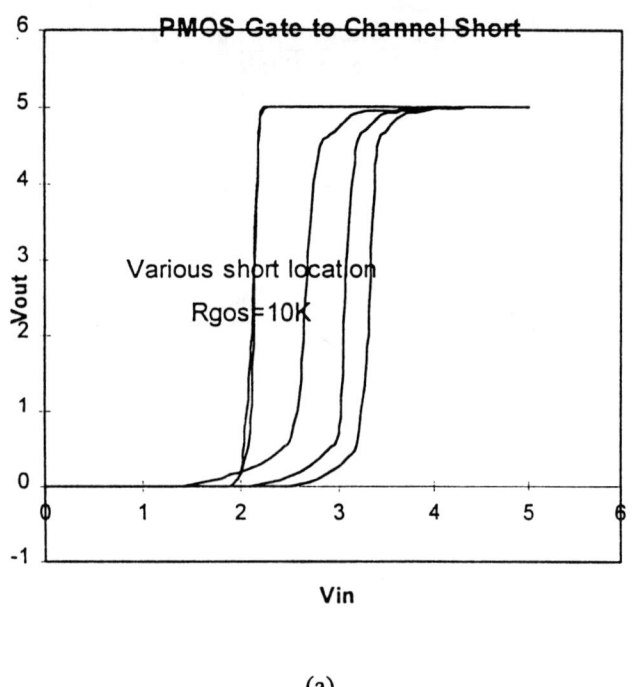

(a)

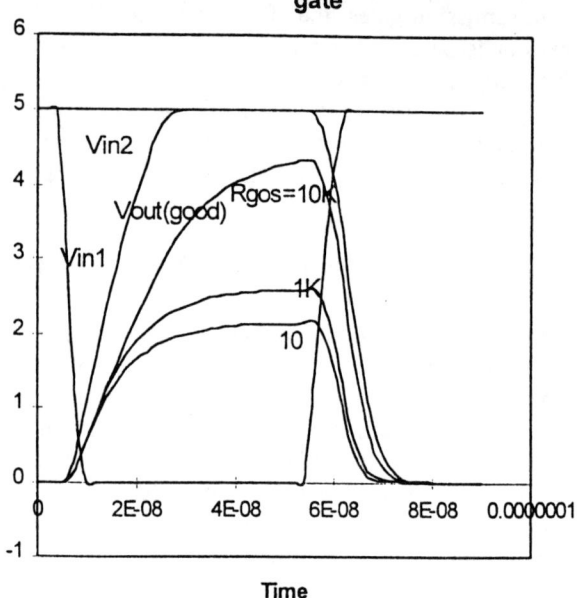

Figure 7. Transient behavior of 2-input NAND Gate with one of the PMOS with gate to channel short.

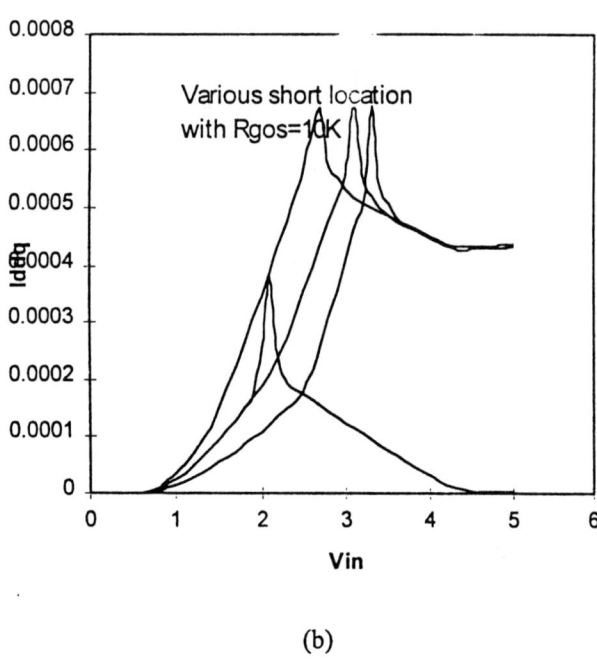

(b)

Figure 6. Transfer curve of an inverter for gate to channel short at different locations: a) Voltage, b) Current.

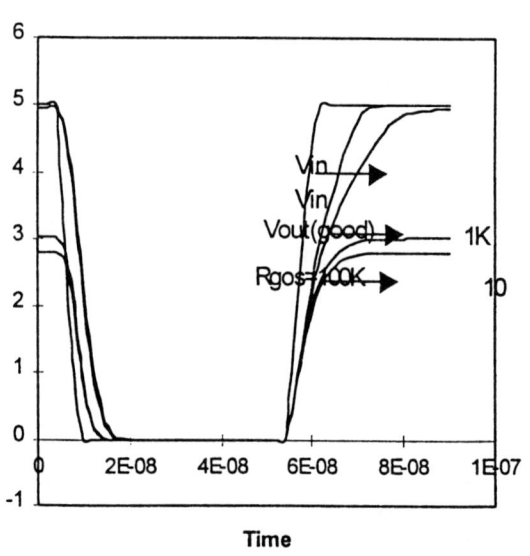

Figure 8. Transient behavior of transmission Gate, the PMOS with gate to channel short.

Realistic Delay Modeling in Satisfiability-Based Timing Analysis

Luís Guerra e Silva, João P. Marques Silva, Luís Miguel Silveira and Karem A. Sakallah[*]

Cadence European Laboratories/INESC
Instituto Superior Técnico
R. Alves Redol, 9
1000 Lisboa, Portugal

[*]Electrical Engineering and Computer Science Dept.
Advanced Computer Architecture Lab.
University of Michigan
Ann Arbor, MI 48109-2122

Abstract

Circuit delay computation taking into account the existence of false paths represents a significant and computationally complex problem. Existing research work has focused mainly on path sensitization models and algorithms, and on gate and interconnect delay models. Nevertheless, work in these two main areas has evolved separately, and so most path sensitization models and algorithms assume very rudimentary gate and interconnect delay models. In this paper we propose a modeling framework for circuit delay computation as a sequence of instances of propositional satisfiability. This framework is used to capture several path sensitization models under the unit delay model. Moreover, several algorithms for propositional satisfiability are evaluated seeking to illustrate the computational challenges posed by the circuit delay computation problem. Finally, realistic delay models taking into account extracted interconnect delays and fanout data are incorporated into the proposed circuit delay computation framework in order to experimentally evaluate its applicability.

1 Introduction

Recent years have seen an ever increasing need for more accurate delay estimation methodologies in digital circuits, in particular due to the decisive role that delay estimation plays in determining limiting operating clock frequencies. A key problem associated with circuit delay estimation is the existence of false paths, which cause straightforward and efficient topological path analysis procedures to yield potentially conservative delay estimates. In contrast with topological delay estimation, solving the false path problem is computationally hard, being an NP-complete problem [15]. Research work on false paths has been extensive and, among others, several promising modeling and algorithmic approaches have been proposed [1, 2, 6, 9, 15, 15, 18, 22]. Despite this research effort, we believe that a comprehensive and unified computational study of different models and algorithms for solving the false path problem is still missing. In this paper we propose to partially solve this problem by studying a set of path sensitization criteria, under the assumption of floating mode circuit operation. This work is undertaken within a unified framework for solving the false path problem, which is based on propositional satisfiability models and algorithms. Furthermore, we explore more realistic delay modeling within the proposed framework, thus evaluating how SAT-based circuit delay computation is dependent upon the delay model considered. The computational study described in this paper is necessarily incomplete, since several relevant models and algorithms are not covered. Nevertheless, this study proposes an experimental procedure which can be generalized for those other models. Furthermore, we note that false paths can exist in both combinational and sequential circuits, even though in this paper we will exclusively consider combinational false paths.

The organization of the paper is as follows. We start with a few brief definitions, and then describe how to capture path sensitization using propositional satisfiability models. This section follows closely the work of [16], but a significantly simpler approach is used to derive the SAT models for the viability path sensitization criterion [15]. In addition, the SAT modeling approach of [16] is shown to be easily extended to other floating mode path sensitization criteria, namely static sensitization [5] and exact floating-mode sensitization [6]. Afterwards, in Section 4, the experimental procedure is described and experimental results are analyzed. Conclusions resulting from the proposed analysis are given in Section 5.

2 Definitions

In the following we shall assume a combinational circuit M, with PI primary inputs, PO primary outputs, composed of simple gates (AND, NAND, OR, NOR, NOT), where for a circuit node f, $c(f)$ denotes the controlling logic value of f and $nc(f)$ denotes the non-controlling logic value of f. For each circuit node f, $FI(f)$ denotes the fanin nodes of f and $FO(f)$ denotes the fanout nodes of f. The delay between the fanin node g of a circuit node f and f is denoted by $d(g,f)$. A *complete path* (or simply a path) in a circuit is a sequence of nodes connecting a primary input to a primary output. A *partial path* denotes a connected sequence of nodes within a path.

The circuit delay computation problem consists of identifying the *largest* path delay value in a given circuit along which a signal transition is able to propagate from the primary input to the primary output of the path, under a chosen propagation model and for some primary input vector.

3 Path Sensitization Conditions

The conditions under which signals propagate from the primary inputs to the primary outputs in a combinational circuit are generally referred to as path sensitization conditions. Path sensitization conditions depend on the model of operation assumed for the circuit, in particular the different forms of stimuli on the primary inputs, and the waveform model assumed at each node in the circuit. Even though detailed and precise models can be considered, we shall restrict ourselves to floating mode operation, under which all nodes are assumed to undergo a single known transition, from an initial unknown value to a final *stable* known value. Most criteria defined under floating mode operation are conservative (e.g. viability [15] and the *exact* criterion under floating mode operation [6]), thus overestimating the circuit delay in some situations. Nevertheless, as shown in [15], viability and floating mode sensitization are *robust*, thus providing upper bounds on the circuit delay under the bounded gate delay model (i.e. assuming that each gate delay is within some interval $[0, d_{max}]$).

A characterization of different sensitization criteria for floating-mode operation for simple gates, under the assumption of single path sensitization, is illustrated in Figure 1, and identifies logical and temporal constraints on the side inputs to each node x in a path. $\tau(x)$ denotes the propagation delay of a signal transition to node x along a given path. The side inputs values can either be *controlling* (c) or *non-controlling* (nc). Symbol C indicates that a given circuit node value is unknown and may experience changes in time. For static sensitization, the side

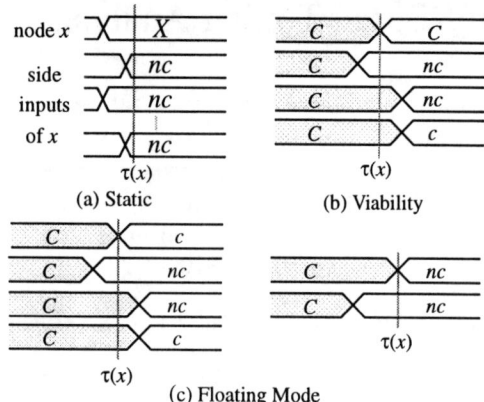

Figure 1: A characterization of path sensitization criteria

inputs are required to assume non-controlling values for propagation of a signal transition to occur. For viability, the side inputs are required to either be non-controlling or stabilize later that the node on the path. Finally, for floating-mode operation, it is assumed that the initial value of each primary input is unknown and changes to a known logic value at the specified arrival time. In the floating-mode sensitization criterion, a node y in the fanout of a node x stabilizes as a direct consequence of node x stabilizing if x is either the *earliest* controlling value to stabilize or all fanin nodes assume non-controlling values and x is the *latest* node to stabilize.

3.1 Satisfiability Models for Path Sensitization

In this section we show how to capture different path sensitization conditions using satisfiability models. Basically, the objective is to define conditions under which a given circuit node can stabilize at a given time instant.

Definition 1. We define the Boolean function $\chi^{f,t}(c)$ such that $\chi^{f,t}(c) = 1$ if and only if circuit node f stabilizes at a time greater than or equal to t when input vector c is applied to the primary inputs.

Clearly the definition of $\chi^{f,t}(c)$ leads naturally to the following observations.

Lemma 1. For a given input vector c and a circuit node f, the following conditions must hold:

1. $(\chi^{f,t}(c) = 1) \Rightarrow (\chi^{f,\tau}(c) = 1)$ for all $\tau \leq t$.
2. $(\chi^{f,t}(c) = 0) \Rightarrow (\chi^{f,\tau}(c) = 0)$ for all $\tau \geq t$.

Moreover, for a given circuit delay Δ, and considering the set of primary outputs PO, we have the condition,

$$\sum_{g \in PO} \chi^{g,\Delta}(c) = 1 \quad (1)$$

for some input vector c. This condition must be satisfiable to ensure that at least one path with delay Δ is sensitizable under the path sensitization model assumed. Furthermore, the definition of function $\chi^{f,t}(c)$ will differ for different sensitization conditions, as we will see in the following sections.

3.2 Viability

Given the interpretation of viability for simple gates in Figure 1-(b) and considering the generalization for multiple paths with the same delay values, we have the following conditions for a given circuit node f to stabilize at a time no earlier than a given delay t for some input vector c:

1. At least one fanin node g of f, with delay $d(g,f)$ between g and f, must stabilize at a time no earlier than $t - d(g,f)$. (This condition permits the existence of multiply sensitized partial paths.)
2. Furthermore, either a fanin node assumes a non-controlling value or it stabilizes at a time no earlier than $t - d(g,f)$, thus being passive regarding propagating a signal transition from g to f. Formally, we have,

$$\chi^{f,t}(c) = \sum_{g \in FI(f)} \chi^{g, t-d(g,f)}(c) \cdot \\ \cdot \prod_{h \in FI(f)} (\chi^{h, t-d(h,f)}(c) + (h = nc(f))) \quad (2)$$

which is basically equivalent to the viability condition proposed in [16]. Furthermore, observe that each function $\chi^{f,t}(c)$ can be viewed as a node in a combinational circuit. Given T. Larrabee's well-known mapping [14] from circuits into CNF formulas and from condition (1) it is straightforward to generate a CNF formula for capturing the sensitization conditions for all paths with delay no smaller than a given threshold delay Δ. It can easily be concluded that the CNF formula size is polynomial in the number of $\chi^{f,t}(c)$ functions considered.

3.3 Static Sensitization

For static path sensitization, using the model illustrated in Figure 1-(a), and again taking into account that multiple signal transitions can propagate from the fanin nodes to a given node f, we get the following definition of $\chi^{f,t}(c)$:

$$\chi^{f,t}(c) = \sum_{g \in FI(f)} \left(\chi^{g, t-d(g,f)}(c) \cdot \prod_{h \in FI(f) - \{g\}} (h = nc(f)) \right) \quad (3)$$

which basically requires that at least one fanin node g of f to stabilize no earlier than $t - d(g,f)$ and such that the remaining fanin nodes assume non-controlling values. Clearly this condition must hold for any of the fanin nodes. Moreover, and as with viability, creating the CNF formula for static sensitization becomes immediate by using conditions (1) and (3).

3.4 Floating Mode Sensitization

In order to capture the exact path sensitization model under the floating mode of operation [6], the following observations are useful:

1. If the fanin node in any path being studied assumes a controlling value, then the floating mode condition is equivalent to viability.
2. Otherwise, all input nodes must be non-controlling. In this situation, propagation from any potential fanin node g only requires that a transition reaches that node, i.e. $\chi^{g, t-d(g,f)}(c) = 1$ and that all other inputs assume non-controlling values.

These observations lead to the following definition of $\chi^{f,t}(c)$:

$$\chi^{f,t}(c) = \sum_{g \in FI(f)} \chi^{g, t-d(g,f)}(c) \cdot \\ \left((g = c(f)) \cdot \prod_{h \in FI(f)} (\chi^{h, t-d(h,f)}(c) + (h = nc(f))) + \\ + \prod_{h \in FI(f)} (h = nc(f)) \right) \quad (4)$$

Observe that since a fanin node is required to satisfy $\chi^{g, t-d(g,f)}(c) = 1$, then at least one of these nodes will guarantee $\chi^{f,t}(c) = 1$ provided all inputs assume non-controlling values.

4 Experimental Results

The circuit delay computation algorithm consists solely of iteratively generating and solving instances of SAT for decreasing circuit delays starting from the largest topological path delay

Circuit [18]	LTP	Δ	Iterations	Viability		Floating-Mode	
				Clauses	Variables	Clauses	Variables
C432	17	17	1	942	330	1335	450
C499	11	11	1	585	226	661	247
C880	24	24	1	627	242	932	341
C1355	24	24	1	2053	704	3029	1025
C1908	40	37	7	5755	1932	8962	2936
C2670	32	30	5	6559	2255	10132	3402
C3540	47	46	3	5732	2027	7318	2532
C5315	49	47	5	5401	1887	8085	2773
C6288	124	123	2	13244	4239	19299	6257
C7552	43	42	3	2180	760	3386	1158
CBP.12.2	40	23	77	2081	707	3757	1261
CBP.16.4	44	27	89	1698	592	2957	1009
CLA.16	34	34	1	479	183	812	294
TAU92EX1	27	24	31	1213	450	2026	711
MULT-CSA	78	78	1	11415	3496	12309	3982

Table 1: Statistics for the benchmark circuits

Circuit	LTP/Δ	grasp	rel_sat	posit	tegus	h2r	csat	dpl
C432	17/17	0.03	0.03	0.02	0.13	0.18	0.10	0.45
C499	11/11	0.02	0.22	0.01	0.11	518.08	68.60	0.17
C880	24/24	0.04	0.02	0.01	0.11	0.08	0.10	0.15
C1355	24/24	0.12	0.09	0.19	0.32	2.27	0.50	*
C1908	40/37	0.26	1.43	0.48	3.29	4.37	5.40	107.37
C2670	32/30	2.83	3.21	*	*	*	*	*
C3540	47/46	0.54	0.29	0.69	4.72	2.81	3.10	519.63
C5315	49/47	1.27	0.54	1.03	*	6.44	5.90	*
C6288	124/123	11.19	42.79	*	86.73	206.50	203.90	*
C7552	43/42	0.17	0.44	0.05	1.02	0.63	0.80	3.27
CBP.12.2	40/23	1.53	5.63	0.73	23.95	17.67	15.40	1228.67
CBP.16.4	44/27	1.03	6.66	0.56	16.40	16.53	17.70	310.65
CLA.16	34/34	0.04	0.02	0.00	0.07	0.05	0.00	0.03
TAU92EX1	27/24	0.63	2.73	0.06	5.73	2.65	1.90	13.27
MULT-CSA	78/78	5.90	13.89	4.43	518.06	88.83	21.80	*

Table 2: CPU times for viability

and until a satisfiable instance of SAT is found, which corresponds to the circuit delay. All the satisfiability models described in the previous sections have been implemented and used for generating a large number of instances of SAT, each of which denotes the sensitization conditions for a given target circuit delay for a chosen circuit. In this section we provide results of a large number of satisfiability algorithms [3, 4, 7, 10-13, 19, 21][1] on these instances of SAT. For the results shown in Table 2 a SUN Sparc 5/85 machine, with 64 MByte of physical memory, was used. Furthermore, we also study the effects on computed circuit delay and algorithm execution time when a more realistic delay model is used. The results of this study are presented in Table 3 and were obtained on a SUN UltraSparc 1 with 384 MByte of physical memory.

4.1 Statistics for the Benchmark Circuits

One potential problem of SAT-based circuit delay computation algorithms is the size of the CNF formulas. In Table 1, we provide statistics for the different benchmark circuits, under the unit gate delay model. LTP denotes the largest topological path delay and Δ denotes the circuit delay under the viability and floating mode path sensitization criteria. For the most significant path sensitization criteria, i.e. viability and floating-mode, Table 1 includes the largest number of clauses for any given iteration of the algorithm, as well as the number of variables in that situation. The number of iterations until a sensitizable path delay is found is also included. As can be concluded, the worst-case number of clauses is reasonable, given the original circuit sizes.

4.2 Results for Viability

The results for viability are shown in Table 2. (A more comprehensive set of results, involving other criteria can be found in [20].) Entries with a '*' indicate that the respective algorithm did not finish in less than 3,000 CPU seconds. It is interesting to observe that the vast majority of the generated instances of SAT are extremely easy to solve with most SAT algorithms. The exceptions to this rule are the less sophisticated SAT algorithms, in particular the Davis-Putnam procedure, which is unable to solve a large number of benchmarks. On the other hand, for a few benchmarks, only a few SAT algorithms are able to compute the circuit delay in a reasonable amount of time. In general, GRASP and rel_sat are by far the most efficient algorithms for solving this class of instances of SAT. Finally, we observe that for a few benchmarks and for viability using TEGUS we have been unable to reproduce the results of [16]. One possible justification is that the CNF formulas used in this paper and in [16] are necessarily different. Furthermore, the version of TEGUS used in [16] may have been optimized for the circuit delay computation problem, whereas the results of TEGUS included in the paper are obtained with the version that is available in SIS [21].

4.3 Realistic Delay Modeling

The previous results were obtained assuming a unit delay model for each gate. However, in general we need to consider more realistic delay models which should take into account the following constraints:

1. Different delay values for different types of gates.
2. Variation of delay with the number of fanouts/fanins.
3. Interconnect delay estimation (for circuits for which layout information is available).

Gate delays and delay variation with the number of fanouts/fanins can be easily modeled using information available from an IC library databook. Interconnect delay, however is hard to estimate for the benchmark circuits available, which are only described at the gate level. To obtain this information the benchmark circuits were mapped using the standard-cell library ECPD07 (ES2/Atmel)[2] [11], and the parasitic capacitances of the interconnect were extracted. For each gate, the (load-dependent) propagation delay (t_p) is given by:

$$t_p = t_{p_i} + dt_p \cdot C_l \quad (5)$$

where t_{p_i} is the intrinsic propagation delay, dt_p is the differential (load-dependent) propagation delay and C_l is the load capacitance at the gate output. Further, this load capacitance is given by:

$$C_l = C_i + \sum C_g \quad (6)$$

where C_i is the lumped interconnect capacitance and $\sum C_g$ is the sum of the input capacitances of all the fanouts. The interconnect resistance for this technology is very small, resulting in a negligible interconnect delay that has been discarded. However the interconnect capacitance is significant and was used to more realistically model the load-dependent propagation delay of each gate. We further note that all delays were computed with two-digit precision. Clearly, this gate delay model leads to a significantly larger number of path delays, which increases the number of iterations of the circuit delay computation algorithm.

---

1. A description of SAT algorithms and an evaluation of their use for solving the circuit delay computation problem is given in [20].

2. Mapping to this library requires that each gate with more than 4 inputs has to be expanded into a sequence of gates each with no more than 4 inputs.

Circuit	Unit Delay				Realistic Delay			
	LTP/Δ	grasp	rel_sat	tegus	LTP/Δ	grasp	rel_sat	tegus
C432	20/20	0.02	0.02	0.04	20.20/19.90	1.26	0.75	25.31
C499	12/12	0.01	0.11	0.03	16.67/16.64	0.01	0.24	0.03
C880	24/24	0.02	0.01	0.02	18.59/18.59	0.01	0.01	0.03
C1355	25/25	0.06	0.04	0.08	22.38/21.97	0.21	3.87	2.33
C1908	42/39	0.17	0.95	0.91	32.44/29.68	75.35	#V	427.57
C2670	34/32	0.63	0.53	8.62	40.31/38.62	65.95	155.06	#B
C3540	47/46	0.33	0.15	6.07	45.19/43.10	1994.69	#V	#B
C5315	49/47	0.57	0.30	7.12	58.57/57.36	4.36	2.34	113.97
C6288	124/123	6.64	28.60	5.40	73.82/73.06	4672.90	#V	#B
C7552	43/42	0.11	0.28	0.26	38.57/36.39	142.44	50.85	285.34
CSA.16.4	41/22	0.08	11.25	2.27	36.00/20.10	0.36	24.92	14.16
CBP.12.2	40/23	0.71	3.29	7.72	22.65/13.94	25.10	67.45	301.74
CBP.16.4	44/27	0.44	15.20	7.05	25.84/16.52	4.31	39.29	70.74
CLA.16	34/34	0.02	0.00	0.01	21.68/21.65	0.08	0.25	0.20
MULT-CSA	78/78	3.35	3.35	4.82	81.10/80.87	3277.26	#V	#B

Table 3: CPU times for unit vs. realistic delay models, using viability

In Table 3 we present the CPU times for checking satisfiability for different SAT algorithms, obtained on the technology mapped circuits assuming both a unit delay model and a realistic delay model. For this experiment the path sensitization criterion used was viability. In the column for rel_sat, entries with "#V" indicate that the maximum number of variables (23,000) was exceeded. In the column for TEGUS, entries with "#B" indicate that the maximum number of backtracks was exceeded. As can be observed, the CPU times increase significantly because the number of iterations of the algorithm also increases accordingly. However, this added complexity signifies that we are now able to obtain much more accurate delay estimates for the circuit delay.

5 Conclusions

In this paper we propose a unified propositional satisfiability modeling and algorithmic framework for studying circuit delay computation methodologies. Different path sensitization models were considered and reasonably efficient results were obtained. Regarding the SAT algorithms used, one class of algorithms provides by far the most efficient and robust results. Both algorithms in this class (GRASP and rel_sat) use a number of search pruning techniques, which are shown to be particularly effective for solving circuit delay computation problems. Moreover, more realistic delay models, which take into account extracted interconnect delays and fanout data, were incorporated into the proposed modeling and algorithmic framework. Preliminary results suggest that the approach is still feasible, though necessarily more inefficient.

Additional work involves concluding the experiments described in this paper, as well as experimenting with a larger number of benchmarks. Another experiment that will be useful in identifying which SAT algorithms can actually be used in practice for circuit delay computation is to study families of circuits, for which we can increase the problem complexity by selecting larger members of that family. A well-known example is the family of carry-skip adders [15], where measures of size/complexity include the number of bits in the adder as well as the number of bits per block. Furthermore, additional experiments ought to include more detailed gate delay models with the goal of evaluating how large the CNF formulas can become, and how the different SAT algorithms handle larger CNF formulas, with more path sensitization options.

References

[1] P. Ashar, S. Malik and S. Rothweiler, "Functional Timing Analysis using ATPG," in *Proceedings of the European Design Automation Conference*, 1993.

[2] R. I. Bahar, E. A. Frohm, C. M. Gaona, G. D. Hachtel, E. Macii, A. Pardo and F. Somenzi, "Algebraic Decision Diagrams and Their Applications," in *Proceedings of the International Conference on Computer-Aided Design*, November 1993.

[3] P. Barth, "A Davis-Putnam Based Enumeration Algorithm for Linear Pseudo-Boolean Optimization," Technical Report MPI-I-95-2-003, Max-Planck-Institut für Informatik, January 1995.

[4] R. Bayardo Jr. and R. Schrag, "Using CSP Look-Back Techniques to Solve Real-World SAT Instances," in *Proceedings of the National Conference on Artificial Intelligence (AAAI-97)*, 1997.

[5] J. Benkoski, E. Vanden Meersch, L. Claesen and H. De Man, "Efficient Algorithms for Solving the False Path Problem in Timing Verification," in *Proceedings of International Conference on Computer-Aided Design*, pp. 44-47, 1987.

[6] H.-C. Chen and D. H. Du, "Path Sensitization in Critical Path Problem," *IEEE Transactions on Computer-Aided Design*, vol. 12, no. 2, pp. 196-207, February 1993.

[7] J. Crawford and L. Auton, "Experimental Results on the Cross-Over Point in Satisfiability Problems," in *Proceedings of the 11th National Conference on Artificial Intelligence (AAAI-93)*, pp. 22-28, 1993.

[8] M. Davis and H. Putnam, "A Computing Procedure for Quantification Theory," *Journal of the Association for Computing Machinery*, vol. 7, pp. 201-215, 1960.

[9] S. Devadas, K. Keutzer and S. Malik, "Computation of Floating-Mode Delay in Combinational Circuits: Practice and Implementation," *IEEE Transactions on Computer-Aided Design*, vol. 12, no. 12, pp. 1924-1936, December 1993.

[10] O. Dubois, P. Andre, Y. Boufkhad and J. Carlier, "SAT versus UNSAT," *Second DIMACS Implementation Challenge*, David S. Johnson and Michael A. Trick (eds.), DIMACS Series in Discrete Mathematics and Theoretical Computer Science, 1993.

[11] ES2 ECPD07 Library Databook, July 1995.

[12] J. W. Freeman, *Improvements to Propositional Satisfiability Search Algorithms*, Ph.D. Dissertation, Department of Computer and Information Science, University of Pennsylvania, May 1995.

[13] D. S. Johnson and M. A. Trick (eds.), *Second DIMACS Implementation Challenge*, DIMACS Series in Discrete Mathematics and Theoretical Computer Science, 1993. DIMACS benchmarks available in ftp://Dimacs.Rutgers.EDU/pub/challenge/sat/benchmarks/cnf.

[14] T. Larrabee, *Efficient Generation of Test Patterns Using Boolean Satisfiability*, Ph.D. Dissertation, Department of Computer Science, Stanford University, STAN-CS-90-1302, February 1990.

[15] P. C. McGeer and R. K. Brayton, *Integrating Functional and Temporal Domains in Logic Design: The False Path Problem and its Implications*, Kluwer Academic Publishers, 1991.

[16] P. McGeer, A. Saldanha, P. R. Stephan, R. K. Brayton and A. L. Sangiovanni-Vincentelli, "Timing Analysis and Delay-Test Generation Using Path Recursive Functions," in *Proceedings of the International Conference on Computer-Aided Design*, November 1991.

[17] P. McGeer, A. Saldanha, R. K. Brayton and A. L. Sangiovanni-Vincentelli, "Delay Models and Exact Timing Analysis," in *Logic Synthesis and Optimization*, T. Sasao (Ed.), 1993.

[18] J. P. M. Silva and K. A. Sakallah, "Efficient and Robust Test-Generation Based Timing Analysis," in *Proceedings of the International Symposium on Circuits and Systems*, pp. 303-306, 1994.

[19] J. P. M. Silva and K. A. Sakallah, "GRASP—A New Search Algorithm for Satisfiability," in *Proceedings of the International Conference on Computer-Aided Design*, November 1996.

[20] L. G. Silva, J. P. M. Silva, L. M. Silveira and K. A. Sakallah, "Satisfiability Models and Algorithms for Circuit Delay Computation," in the ACM Workshop on Timing Issues in the Specification and Synthesis of Digital Systems (TAU), December 1997.

[21] P. R. Stephan, R. K. Brayton and A. L. Sangiovanni-Vincentelli, "Combinational Test Generation Using Satisfiability," Memorandum no. UCB/ERL M92/112, Department of Electrical Engineering and Computer Sciences, University of California at Berkeley, October 1992.

[22] H. Yalcin and J. P. Hayes, "Hierarchical Timing Analysis Using Conditional Delays," in *Proceedings of the International Conference on Computer-Aided Design*, November 1995.

Enhancing Circuit Performance under a Multiple-Phase Clocking Scheme*

Yaun-chung Hsu[†] Shangzhi Sun[‡] David H.C. Du Xuedao Chu[§]
Department of Computer Science
University of Minnesota
Minneapolis, MN 55455

Abstract

For general synchronous circuits, input and output data are stored in latches or flip-flops which are triggered by the clock signal, so the clock period is a measurement of circuit performance. Previous studies on this issue are restricted by the assumption of triggering all inputs at the same clock phase. We propose a new method to reduce the clock period without the above assumption. The proposed method allows the existence of clock skew and produces a better clock period. The improvement in circuit performance is demonstrated by our experimental results.

1. Introduction

For general synchronous circuits, input and output data are stored in latches or flip-flops which are triggered by the clock signal. Therefore, the clock period is a measurement of circuit performance. Based on the knowledge of the role of the clock signal in synchronous circuits, we choose the approach of reducing the clock period to enhance circuit performance.

Generally, the lower bound of a clock period is set as the the longest path delay of the circuit [1, 3, 4]. This clock period setting is safe in the environment of a single-phase clocking scheme. However, this value has been shown to be overestimated [6] in multi-phase clocking scheme environments. If we allow input and output data to be triggered by different clock phases, the lower bound of the clock period could be reduced to the difference between longest path delay and shortest path delay [5, 6, 7].

The basic assumption of these settings [5, 6, 7] is that input data need to be triggered by one clock phase and output signals triggered by another clock phase. However, due to the existence of clock skews, it is necessary to adjust the arriving time of input clock signals. Besides the problem of clock skew, these settings are not correct if inputs come from different blocks.

Therefore, we propose a new method to reduce the clock period without the above assumption. In this approach, input and output signals are latched by flip-flops, with latched signals at the rising edge. The proposed method allows the existence of clock skew and produces better a clock period. We demonstrate this in the following example.

In panel of Figure 1, there is a circuit with inputs A, B to single output O. The longest(shortest) path delay

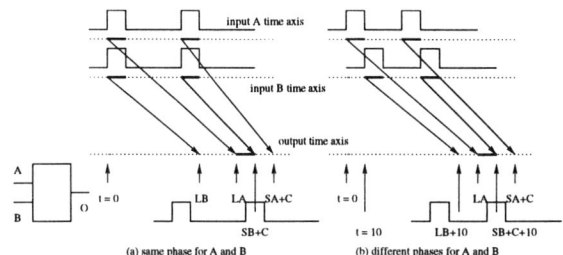

Figure 1: Multi-phase clocking example.

from A to O is denoted as $L_A(S_A)$. We also denote the longest(shortest) path delay from B to O as $L_B(S_B)$. Clock perios will equal to 80ns when $L_A = 100$ns, $L_B = 50$ns, $S_A = 40$ns, and $S_B = 20$ns, and input A and B are triggered by the same clock phase, from the research result in [5]. Now we consider A and B can be triggered by different clock phases. Suppose the rising time of clock phase for A is 0ns and for B is 10ns. The latest possible arriving time of the output signal is 100ns if the data signal starts from input A. The earliest arriving time of the next output signal is (30ns + clock period) if the input signal starts from B. The clock period can be reduced to 70ns in this way.

2. Calculating Clock Period

There are two type of latching units, level-sensitive latch and edge-triggered D-flip-flop(ETDFF) [2], used in synchronous circuits. input and output structures as the latch unit, We adopt the rising ETDFF as the synchronous unit through circuits. The clock used in synchronizing ETDFFs is a sequence of signals with periodic pulse current and the *clock period* is the distance from the current pulse to the next. Before the rising or falling edge, the flip-flop needs time to set up its circuit. This time is defined as *set up time*. After the active range of the clock, the flip-flop still needs time to hold data. This time is defined as *hold time*. The relation of set up time and hold time to the clock period setting is illustrated in Figure 2(a). Depending on the number of phases, the clocking scheme can be categorized as *single − phase clocking* or *multi − phase clocking*.

We notice that different phases may start from different times during a clock period. Take the multi-phase clocking scheme in Figure 2(b) as an example. There are three different clock phases, A, B, and C, which are attached on flip-flops LA, LB, and LC respectively. The

---
*Research supported in part by NSF grant MIP-9007168.
[†]Current address: IBM, Rochester, MN 55901
[‡]Current address: Synopsys, Inc., Mountain View, CA 94043.
[§]Current address: Institute of Automation, Qufu Normal University, Qufu, Shangdong, P.R. China.

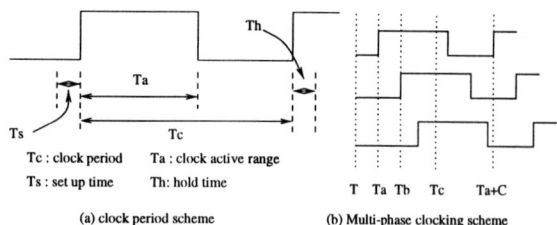

Figure 2: Phase clocking scheme.

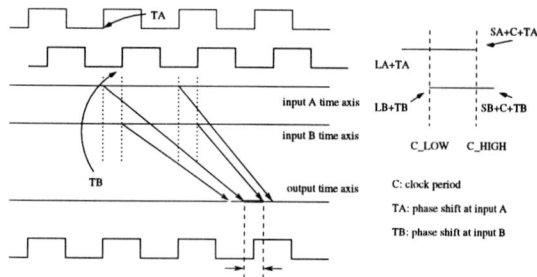

Figure 3: Input pins are triggered by different clock phases.

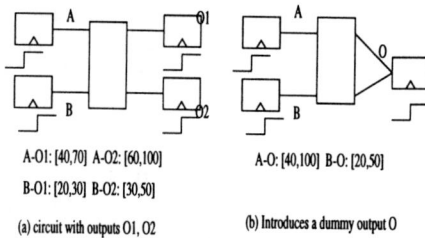

Figure 4: Circuit outputs are triggered by same clock phase.

active range of phase A starts at time T_a if the clock period begins at time T. The shift time of $(T_a - T)$ is defined as the *phase shift* time of phase A on flip-flop LA. Likewise, $(T_b - T)$, and $(T_c - T)$ are defined as the *phase shift* time of B and C respectively.

We use the circuit example in Figure 1(a) to demonstrate our idea. If inputs A and B are triggered by the same clock phase, the clock period of this circuit is calculated as $C=\max\{L_A, L_B\} - \min\{S_A, S_B\}$ [5, 6] which equals 80ns. However, if inputs A and B are triggered by different clock phases, the clock period C can be further reduced. Let the phase shift times of A and B be T_A and T_B respectively(refers to Figure 3). If we set T_A as 0ns, and T_B as 20ns, the clock period can be reduced to 60ns since there is still a stable range on output O.

For the current cycle, the stable ranges of input data on inputs A and B are $[T_A, T_A+C]$ and $[T_B, T_B+C]$ respectively. To ensure there is a intersection of stable rages, the following relations must be kept:
$T_A + L_A \leq T_A + C + S_A$, $T_A + L_A \leq T_B + C + S_A$,
$T_B + L_B \leq T_B + C + S_B$, $T_B + L_B \leq T_A + C + S_B$

After replacing the actual values of the longest and shortest path delays and phase shift time, the feasible range of C is $[60ns, \infty)$. Compared to the clock period in single phase clocking, 80ns, there is a noticeable improvement in performance.

If there are n inputs in this circuit, n output data stable ranges are generated when applying input vectors to the circuit [5]. Let output data stable range SR_i represent the corresponding range from input I_i. SR_i can be formulated as $SR_i = [L_i + t_i, S_i + C + t_i]$ where L_i is the longest path delay, S_i is the shortest path delay, C is the clock period, and t_i is the phase shift time on I_i input flip-flop. Let L_{max} and S_{min} denote the lower and upper bounds of the intersection. They can be calculated as $L_{max} = \{L_i + t_i \mid \forall i = 1, \ldots, n\}$, $S_{min} = \{S_i + C + t_i \mid \forall i = 1, \ldots, n\}$ The output stable range must be large enough to allow the set up and hold operation of flip-flops.

3 Outputs Triggered by One Clock Phase

To generalize the solution of clock period and phase shift time, we formulate the equations in a set of linear equations and solve the equations to obtain the optimal solution with n inputs, $I_1, \ldots, I_n$, and one output O(dummy output) in the circuit. We also use C, t_s, and t_h to represent clock period, set up time and hold time respectively. Meanwhile, L_i, S_i, and t_i represent longest path delay from I_i to O, shortest pat delay from I_i to O, and phase shift time at flip-flop of I_i, $\forall I_i, i = 1, \ldots, n$.

If we only consider input I_i, $(t_i + L_i)$ is the latest arriving time of the current cycle's data to the output flip-flop. We use C_LOW to represent the possible latest data arriving time throughout the circuit.
$$C\_LOW = max\{t_i + L_i \mid \forall i = 1, \ldots, n\}$$
The value of C_LOW will serve as the lower bound of the possible output stable range. On the other hand, $(t_i + C + S_i)$ is the earlist arriving time of the next cycle's data which comes from input I_i. Likewise, we use C_HIGH to denote the earliest possible data arriving time during the next cycle.
$$C\_HIGH = min\{t_i + C + S_i \mid \forall i = 1, \ldots, n\}$$
In other words, C_HIGH is used as the upper bound of the output stable range. Because the minimum time for a latching operation is $t_s + t_h$, the output stable range $[C\_LOW, C\_HIGH]$ must meet this requirement to provide enough time for setting up and holding. Besides this restriction on C_LOW and C_HIGH, the clock period C and phase shift time t_i must be greater than or equal to zero.

$LP1$: (same clock phase)

$$\begin{aligned} Minimize \quad & C \\ subject\ to \quad & t_s + t_h \ < t_i + C + S_i - t_j - L_j, \\ & \forall\, i, j = 1, \ldots, n \\ & t_i, C \ \geq 0, \forall\, i = 1, \ldots, n \end{aligned}$$

Next theorem provides a method to obtain an optimal clock period under the condition that all output pins are triggered by same clock phase.

Theorem 1 *The optimal solution of a clock period which fulfills LP1 is* $C = max\{L_i - S_i \mid \forall\, i = 1, \ldots, n\} + t_s + t_h$.

We use C_{min} to represent the optimal clock period, which generated from this theorem, in following discussion. This optimal solution exists only when we can adjust phase shifts of inputs to meet the requirements of $LP1$. If there is an upper bound on phase shifts, the previous solution is not valid in some cases. We introduce a parameter D to serve as the upper bound of phase shifts. We propose a solution in the next corollary when the phase shifts have an upper bound D.

Corollary 1 *If phase shifts have an upper bound D, the optimal solution of the clock period which fulfills $LP1$ equals $max\{C_{min}, C_{min} + \Delta C\}$. This solution exists when $\Delta C = max\{L_{max} - S_i - C_{min} - D\}$, and $L_{max} = max\{L_i \mid \forall i = 1,..., n\}$.*

If ΔC is greater than 0, C_{min} is not a valid clock period of the circuit since the output flip-flops cannot latch correct data.

4 Outputs Triggered by Different Clock Phases

The previous set of linear equations are designed to solve the problem of minimizing the clock period of circuits under the condition that all output pins are triggered by the same clock phase. If we remove this constraint the clock period may be further reduced. We use the circuit illustrated in Figure 4 as an example to demonstrate this point.

Let the example circuit have inputs A, B and outputs O_1, O_2. If we restrict outputs O_1 and O_2 to be triggered by the same clock phase and adopt the solution from $LP1$, the resulting clock period will be 60ns. Now we consider the case when outputs O_1 and O_2 can be triggered by different clock phases. Let phase shift time of A be t_A and phase shift time of B be t_B. The stable range of O_1 becomes $[max\{70+t_A, 30+t_B\}, min\{40+t_A+C, 20+t_B+C\}]$. At the same time, the stable range of O_2 becomes $[max\{100+t_A, 50+t_B\}, min\{60+t_A+C, 30+t_B+C\}]$. The clock period, C, must be set to prevent the above two stable ranges from becoming empty ranges. Therefore, the minimum clock period calculated from these stable ranges is 40ns when $t_A = 0ns$ and $t_B = 30ns$.

Suppose there are n inputs, $I_1, ..., I_n$, and m outputs, $O_1, ..., O_m$, in the circuit. L_{ij} represents the longest path delay between input i_i and output O_j. S_{ij} is defined as the shortest path delay between input i_i and output O_j. If there is no path between I_i and O_j, L_{ij} is assigned to 0 and S_{ij} is assigned to ∞. We also use C_HIGH_p and C_LOW_q to represent $min\{t_i + C + S_{ip} \mid \forall i = 1, ..., n\}$ and $max\{t_i + L_{iq} \mid \forall i = 1, ..., n\}$.

$LP2$: (*different clock phases*)

$$Minimize \quad C$$
$$subject\ to \quad t_s + t_h < C\_HIGH_p - C\_LOW_q,$$
$$\forall p, q = 1, ..., m$$
$$t_i, C \geq 0, \forall i = 1, ..., n$$

The optimal clock period can be obtained by solving $LP2$ with a linear programming technique. However, there is a simpler way to calculate the optimal solution if the circuit meets certain criteria. From the proof of Theorem 1, we find out that the feasible phase shifts of inputs can be a range of values. The range is called the *feasible range of phase shift* (RPS). Let

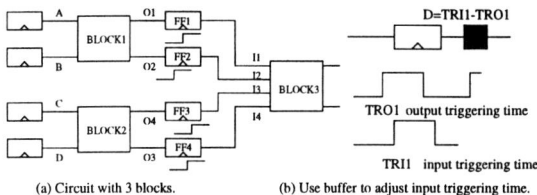

Figure 5: Circuit inputs come from different combinational circuit blocks.

$L_{mj} = max\{L_{ij} \mid \forall i = 1, ..., n\}$ be the longest path which ends at output O_j. We also set $S_{mj} = S_{kj}$ if L_{mj} occurs when $i = k$. Another term that we will use later is the proposed clock period, C_m. We set C_m as the value of $max\{L_{ij} - S_{ij} \mid \forall i = 1, ..., n, j = 1, ..., m\}$. RPS_{ij} represents the feasible range of phase shift from input I_i to output O_j. If we choose a value from RPS_{ij} as the phase sift time on input I_i, the output data stable range intersects with range $[L_{mj}, S_{mj}]$. We use [lower_bound, upper_bound] to represent this feasible range of phase shift in the following discussion. The lower_bound, upper_bound can be calculate from the relationship between $[L_{ij}, S_{ij} + C_m]$ and $[L_{mj}, S_{mj}]$. To ensure RPS_{ij} to be a valid range, we set the bounds as follows:

$$lower\_bound = max\{L_{mj} - S_{ij} - C, 0\}$$
$$upper\_bound = S_{mj} + C - L_{ij}$$

RPS_i represents the feasible range of phase shift on input I_i in the whole circuit.
$$RPS_i = RPS_{i1} \cap RPS_{i2} \cap RPS_{i3} \cap ... \cap RPS_{im}$$

Theorem 2 *The optimal clock period which fulfills $LP2$ is $C = max\{L_{ij} - S_{ij} \mid \forall i = 1, ..., n, j = 1, ..., m\}$ when $RPS_i \neq \emptyset, \forall i = 1, ..., n$.*

We can use this simple formula to calculate the clock period if the longest and shortest path pairs of the circuit meet the criteria described in the theorem. For the circuits which cannot meet the criteria, we use linear programming techniques to solve the problem.

5. Extension to General Synchronous Circuits

The circuit we discuss in the sections above is a combinational circuit with inputs and outputs. However, general synchronous circuits may compose some circuit blocks, which are normally combinational circuits. Latches or flip-flops are used as buffer between circuit blocks. One example of this type of circuit is pipelined circuits. Each combinational circuit block corresponds to a stage and the clock signal can be used to trigger the latches between blocks.

If the circuit contains more than one combinational circuit block, some latches will serve both as output and input triggers. We demonstrate this in Figure 5(a). The flip-flops FF_1 and FF_2 are output latches of $BLOCK_1$, and also input flip-flops to $BLOCK_3$. Let us consider the output and input clock phase on flip-flop FF_1. The output O_1 is triggered by the output clock phase of $BLOCK_1$. The input I_1 is latched by the input clock phase of $BLOCK_3$. If the output and input clock share the same phase, the output data from $BLOCK_1$ can be available correctly for the input of $BLOCK_3$ in the next cycle. Otherwise, the correct output data cannot arrive to $BLOCK_3$ in time. To prevent this situation, we

```
Procedure Buffer_Insertion() {
    /* ΔT_i: buffer size inserted after FF_i. */
    TR = max{TRO_i - TRI_i | ∀ i = 1,...,k};
    if TR < 0 TR = 0;
    for (i = 1; i ≤ k; i + +) {
        if (TRI_i > TRO_i)
            ΔT_i = TRI_i - TRO_i + TR;
        else
            ΔT_i = TR + TRI_i - TRO_i; } }
}
```

Figure 6: Procedure Buffer_Insertion

| Circuit | # of | # of | # of | # of | Longest | Shortest |
Circuit	Gate	Leads	PI	PO	Delay	Delay
c432	250	426	36	7	57.40	4.50
c499	555	928	41	32	53.30	4.00
c880	443	729	60	26	53.00	4.00
c1355	587	1064	41	32	49.90	4.30
c1908	913	1498	33	25	76.60	4.60
c2670	1426	2076	233	140	86.90	0
c3540	1719	2939	50	22	98.70	4.00
c5315	2485	4386	178	123	99.30	1.40
c6288	2440	4800	32	32	319.90	2.50
c7522	3719	6144	207	108	85.30	0

Table 1: Characteristics of ISCAS85 benchmark circuits.

| Circuit | $LP1$ | $LP3$ | | | |
| Name | C | C | C | C | C |
		$D=5$	$D=10$	$D=20$	$D=40$
c432	51.50	51.50	51.50	51.50	51.50
c499	49.30	49.30	49.30	49.30	49.30
c880	48.30	48.30	48.30	48.30	48.30
c1355	45.60	45.60	45.60	45.60	45.60
c1908	72.00	72.00	72.00	72.00	72.00
c2670	74.90	80.50	75.50	74.90	74.90
c3540	90.40	90.40	90.40	90.40	90.40
c5315	90.60	92.90	90.60	90.60	90.60
c6288	317.40	317.40	317.40	317.40	317.40
c7522	69.50	78.90	73.90	69.50	69.50

Table 2: Experiment results of $LP1$ and $LP3$

| Circuit | $LP2$ | $LP4$ | | | |
| Name | C | C | C | C | C |
		$D=5$	$D=10$	$D=20$	$D=40$
c432	50.80	52.90	52.90	50.80	50.80
c499	40.40	49.10	49.00	44.20	40.40
c880	35.40	49.00	41.30	40.50	35.40
c1355	36.70	45.60	45.20	36.70	36.70
c1908	59.70	71.00	62.50	62.50	59.70
c2670	52.50	81.40	70.30	62.60	53.50
c3540	63.00	94.00	79.90	76.40	63.70
c5315	49.80	97.90	90.60	68.30	51.90
c6288	310.00	316.90	315.40	315.40	310.80
c7522	53.30	85.10	80.80	67.20	53.30

Table 3: Experiment results of $LP2$ and $LP4$

add buffers after the latches to adjust for the difference of input and output triggering time. To generalize the buffer insertion procedure, we assume there are k flip-flops between the current stage and the next stage. Let the k flip-flops be $FF_1, ..., FF_k$. For FF_i, TRO_i represents the output triggering time of the current stage and TRI_i represents the input triggering time of the next stage. We use the procedure illustrated in Figure 6 to obtain the buffer size which is inserted into the circuit.

After the buffers are inserted into the circuit, the data arriving time to the next stage is adjusted to meet the input triggering time of the next combinational circuit block. Therefore, the proposed clock period setting can be applied to the whole circuit via applying this buffer insertion procedure to all blocks.

6. Experiments

The solutions derived from the linear systems above are implemented by C language and *Mathematica* on Sun Spark Machine. The circuits used to demonstrate the experimental results are the benchmark set IS-CAS85. Table 1 shows the characteristics of each circuit of the benchmark set. In this implementation, the set up time and hold time are assumed to be zero and are added to the clock period later.

We have discussed different situations for a multiple phase clocking scheme in Section 4. The original equations in $LP1$ and $LP2$ have no upper limit restriction on the phase shift time. It is practical to put a range restriction on these parameters. Therefore, we add the equation, $t_i \leq D, \forall i = 1,...,n$, to $LP1$ and $LP2$. We use $LP3$ and $LP4$ to represent these modified linear systems.

The test results of $LP1$ and $LP3$ are listed in Table 2. The first column shows the resulting clock period after solving $LP1$. Columns 2 through column 5 show the results of $LP3$. $LP1$ is a special case of $LP3$ with an unlimited upper bound. There was no improvement in the clock period for circuits c499, c1355, c1908, and c6288. We found out that some stable ranges from inputs of these circuits had the same range length as the output stable range. This is because the longest path and shortest path both started from the same input pins.

The experimental results of the clock period after solving $LP2$ and $LP4$ are listed in Table 3. We first check the criteria described in Theorem 2. If the test circuit meets the criteria, the clock period can be calculated by the formula provided in Theorem 2. Otherwise, we still to need the linear programming method to compute the optimal clock period. If we set the value of D larger, the resulting clock period is near the one with unlimited phase shift time. When comparing the results in Table 2 with those of Table 3, we find that $LP2$ produces a shorter clock period than $LP1$. The same situation is also found between $LP4$ and $LP3$.

References

[1] M. Flynn and S. Waser. Introduction to Arithmetic for Digital Systems Designers. CBS College Publishing, 1982.

[2] S. Unger and C.-J. Tan. Clocking Schemes for High-Speed Digital Systems. In *IEEE Trans. on Computers*, October, pages 880-895, 1986.

[3] B. Ekroot. Optimization of pipelined processors by insertion of combinational logic delay. Ph.D. thesis, Department of Electrical Engineering, Stanford University, September 1987.

[4] J. Fishburn. Clock skew optimization. In *IEEE Transactions on Computers*, 39(7):945-951, 1990.

[5] L. Liu, H. Chen, and David Du. The calculation of signal stable ranges in combinational circuits. In *IEEE International Conference on Computer-Aided Design*, 1991.

[6] S. Cheng, H. Chen, D. Du, and A. Lim. The role of long and short paths in circuit performance. In *29th Design Automation Conference*, pages 543-548, 1992.

[7] S. Sun, D. Du, and Y. Hsu. On Valid Clocking for Combinational Circuits. In *International Conference on Computer Design*, 1994.

APPLICATION OF THE VARIABLE DIMENSION NEWTON METHOD TO LARGE SCALE CIRCUITS

S.W.Ng
The Hong Kong Polytechnic University
Hung Hom, Kowloon
swng@encserver.en.polyu.edu.hk

ABSTRACT

The Variable Dimension Newton Method (VDNR) is used to solve the DC solutions of various kinds of circuits. The problems encountered and the corresponding solutions are discussed. The simulation results show that the Variable Dimension Newton Method is much robust than the conventional Newton method and those homotopy alternatives used in SPICE3f3.

1. INTRODUCTION

We consider the problem of finding a solution or all the solutions for the set of nonlinear equations $F(x) = (f_1..f_n)' = 0$. Throughout this paper, we assume that the problem $F(x)=0$ has at least one solution. The classical algorithm for this problem is the Newton method. The classical Newton Raphson method consists of the iteration equation

$$x^{i+1} = x^i - h \frac{\partial F}{\partial x}\bigg|_{x=x^i}^{-1} F(x^i)$$

where h is the step size. The Newton Raphson method is well known for its good convergent rate but its convergence is only local and may require a very good initial guess of the solution. Various variants that can improve the convergence property have been proposed[1-7]. The Variable Dimension Newton method (VDNR) was firstly proposed in [8]. It is based on the Generalized Newton method of Ben-Israel in [9-10] and then independently rediscovered by the author in [8]. The advantage of the Generalized Newton method over the classical method is its enlarged convergence zone. In the following the Variable Dimension Newton method reported in [8] will be briefly described. Then applications of this method to large scale circuits will be discussed and is followed by the benchmark results.

2. THE VARIABLE DIMENSION NEWTON METHOD

The basic approach of the Variable Dimension Newton method is to solve a succession of systems $F_1(x) = 0,..., F_m(x) = 0,...$ until $F_n(x) = F(x) = 0$, where $F_m(x) = (f_1..f_m)'$. The dimension of the problem will be increased or decreased accordingly to track the solution. When we have $m \le n$, $F_m(x)$ has one or more than one solutions. The solution set forms an $n-m$ dimensional hypersurface. The basic iteration step of the Generalized Newton method [8] is

$$x^{i+1} = x^i - J_i^+ F_m(x^i)$$

On the assumption that J_i has rank m, J_i^+ is the pseudo inverse of J_i and is defined to be

$$J_i^+ = J_i'(J_i J_i')^{-1}$$

where $J_i = \frac{\partial F_m(x^i)}{\partial x}$ and rank of $J = m$.

To find the solution of the complete problem $F_n(x) = F(x) = 0$, we must track the solution of $F_m(x) = 0$ in the correct direction until we meet the solution of $F_{m+1}(x) = 0$. The basic tracking equation is

$$x^{i+1} = x^i + (I - J_i^+ J_i)\xi$$

where ξ is the tracking direction.

When $(I - J_i^+ J_i)\xi = 0$, the tracking step fails and the algorithm should be restarted with a lower dimension.

3. THE SIMULATION OF LARGE SCALE CIRCUITS

When the VDNR is applied on large scale circuits, quite a few practical problems were discovered. They are :

The tracking step $(I - J_i^+ J_i)\xi$ cannot be computed efficiently. On the assumption that $F_m(x^i) = 0$, the tracking equation can be simplified to $x^{i+1} = x^i + \xi$. Then the testing of tracking step failure should be modified to $\underset{i=m+1..n}{Min} \frac{(x_i^{k+1} - x_i^k)}{\xi_i} < \varepsilon_t$ accordingly (ε_t is a user defined tolerance).

When the algorithm fails to track the solution along a path and results in the removal of a function from tracking, the new tracking step may detect a zero crossover on the function just removed as shown in Fig.1. Therefore, this crossover should be ignored.

When the tracking step fails, it is necessary to know which of the functions under tracking should be removed. By making use of the knowledge that at least one of the function to be removed can be computed by $\gamma' = \xi' J^+$. Then the corresponding function of each $|\gamma_k| > 0.9\|\gamma\|_\infty$ should be removed.

The biggest problem encountered is the large number of nonlinear equations to be solved. In order to reduce the number to equations to be solved, the circuits are pre-processed. When transistor models with non-zero contact resistances are used, many internal device nodes will be created. These nodes have node voltages close to the external terminal node voltages. Therefore, the number of nodes can be reduced by setting the contact resistances to zero. This pre-processing step reduces remarkably the potential number of nodes to be traced. When the biasing point is found on the reduced circuit, the solution can be used as an initial condition for another run of simulation on the full circuit by the classical Newton method.

Since in most of the circuits, the presence of a power supply voltage is essential to ensure a meaningful biasing point or circuit state, the VDNR start by tracking those independent voltage source nodes. Effectively, the VDNR is doing a source stepping. However, it is different from the conventional source stepping because the step size is different for each voltage source. This will be explained in detail later.

4. THE ALGORITHM

The VDNR is implemented on a SPICE like simulator. The Modified Nodal Matrix is used to model a circuit. The VDNR algorithm is :

1. Start VDNR using $x^0 = 0$ and $h=0.1$;
2. Compute $F(x^k)$ and set $\alpha = \{i : f_i = 0\}$.
3. Set $\xi_i = 0$ for $i \in \alpha$ otherwise
   $$\xi_i = \begin{cases} h & \text{if } f_i(x) < 0 \\ -h & \text{if } f_i(x) \geq 0 \end{cases}$$
4. $x^{c,0} = x^k + \xi$; $j=0$; Repeat {
   $y = (F_\alpha(x^j), 0)'$; $\Delta x^j = H^{-1} y$;
   $x^{j+1} = x^j - \Delta x^j$; $j=j+1$; } until
   $\underset{i=1..m}{Max} |f_i(x^{c,j})| < \varepsilon_f$. If the iteration limit is exceeded, goto step (12) ;
5. If step (4) converge, continue from here.
6. $x^{k+1} = x^{c,j}$;
7. for $j=1..n$, If $(f_j(x^{k+1}) f_j(x^k)) < 0$ and $j \notin \alpha \cup \beta$), { increase the dimension by $\alpha = \alpha \cup \{j\}$ and goto step (9) ; }
8. If $\underset{i=m+1..n}{Min} \frac{(x_i^{k+1} - x_i^k)}{\xi_i} \leq \varepsilon_t$, goto step (12) ; else $k=k+1$ and goto step (4) ;
9. $x^{c,0} = x^{k+1}$; $j=0$; Repeat { $y = (F_\alpha(x^j), 0)'$;
   $\Delta x^j = H^{-1} y$; $x^{j+1} = x^j - \Delta x^j$; $j=j+1$; }
   until $\underset{i=1..m}{Max} |f_i(x^{c,j})| < \varepsilon_f$. If the iteration limit is exceeded, goto step (12) ;
10. If step (9) converge, continue from here.
11. Set $\beta = \phi$ and $k=k+1$; go to step (2) ;
12. $\xi' H^{-1} = (\gamma', \theta)$; $\beta = \{k \| \gamma_k | > \mu \|\gamma\|_\infty\}$;
    $\alpha = \alpha \setminus \beta$; Goto step (2) ;

Steps (3)-(6) above are the tracking steps. Thus, the effective source stepping size is $x^{k+1} - x^k$. Obviously, the step size of each independent source will not identical and this differentiates the VDNR from the traditional source stepping method. Step (7) detects a sign reversal in any unsolved function. Step (9) ensures the new set of functions have $F_\alpha(x) \approx 0$. Step (12) handles any failure in the algorithm.

In step (4) above, failure in convergence will result in a lowering of dimension. This is a simplified version of the actual coding used in the simulator. The actual coding allows the algorithm to restart step (3) with a smaller prediction step size h until h is smaller than a user defined tolerance. The discussion of optimum step size control strategy is beyond the scope of this paper and so its actual implementation in the program will not be explained in detail here.

5. BENCHMARK RESULTS

On easy circuits, the source stepping approach is sufficient to find the biasing point. On stiff circuits, the source stepping approach will fail and step (12) of the algorithm above will come into effect. Table 1 below is a summary of the simulation results of 14 benchmark circuits from the WWW site zodiac.cbl.ncsu.edu. The simulation costs of VDNR are listed as '1st run+2nd run'. The '1st run' is the simulation cost of the reduced circuits. The '2nd run' is the simulation cost of the full circuit using the results of the reduced circuit as initial guesses. The BJT and MOSFE models used in here are identical to that of SPICE3f3. This allows a genuine comparison between the algorithm in SPICE3f3 and the VDNR.

Because of the node by node tracking approach, the VDNR is inherently less efficient than other methods by nature. In comparison with the SPICE3f3 or other homotopy based methods, the simulation cost is relatively higher but acceptable. On stiff circuits, the robustness of VDNR is obviously superior over SPICE3f3. The SPICE3f3 either converged to an unstable biasing point or failed to converge even after the GMIN stepping and source stepping alternatives were used. Both alternatives are homotopy based methods. Therefore, the VDNR is a good alternative to SPICE or other homotopy methods.

Because of the exponential relationship between the voltage and current, BJT circuits are more difficult to track than MOSFET circuits. All the benchmark circuits here were run using the same step size control strategy on h to maintain consistence. The strategy used here is optimized for the BJT circuits. The step size control strategy tends to use smaller step size. This is less efficient on MOSFET circuits.

A higher order prediction algorithm may reduce the computational cost of the VDNR. However, the development of a better tracking algorithm is beyond the scope of this paper.

Although there is no guarantee that the VDNR will not converge to an unstable solution, experimental results show that the chance is small. There are 4 bistable circuits in Table 1. They are 'astabl', 'latch', 'gm17' and 'rich3'. All of them converge to the stable solutions. It is because the behavior of a circuit around the unstable biasing point violates the passivity assumption of the VDNR, ie $\Delta f_k(x) \Delta x_k < 0$ for some k. Therefore, the VDNR tends to move away from the unstable biasing point.

6. CONCLUSION

This paper demonstrates the feasibility of applying VDNR on large scale circuits. It is shown that this algorithm has definite advantages over the classical Newton method and other algorithms, eg. the homotopy methods.

7. REFERENCES

[1] I.W.Sandberg. An algorithm for solving the equations of monotone nonlinear resistive networks. Int. J. of Circuit Theory and Applications, vol. 22, 1994, pp357-361.

[2] D.Wolf, S.Sander. Multi-parameter homotopy methods for finding DC operating points of nonlinear circuits. IEEE Int. Symp. on Circuits and Systems, Chicago, May 1993, pp2478-2481.

[3] R.C.Melville, L.Trajkovic, S.C.Fang, L.T.Watson, Artificial parameter homotopy methods for the DC operating point problem, IEEE T.CAD, vol. 12, no6, pp861-877, June 1993.

[4] L.V.Kolev, M.Mladenov. An interval method for finding all operating points of nonlinear resistive circuits. Int. J. of Circuit Theory and Applications, vol. 18, 1990, pp257-267.

[5] L.V.Kolev. Finding all solutions of nonlinear resistive circuit equations via interval analysis. Int. J. of Circuit Theory and Applications, vol. 12, 1984, pp175-178.

[6] M.J.Chien, E.S.Kuh. Solving piecewise linear equations for resistive networks. Int. J. of Circuit Theory and Applications, vol. 4, 1976, pp3-24.

[7] L.Vandenberghe, J.Vandewalle. Variable dimension algorithms for solving resistive circuits. Int. J. of Circuit Theory and Applications, vol. 18, 1990, pp443-474.

[8] S.W.Ng. A variable dimension Newton method. IEEE Int. Symp. on Circuits and Systems, London, vol. 6, pp129-132, May 1994.

[9] A.Ben-Israel. A Newton-Raphson method for the solution of systems of equations, Journal of Mathematical Analysis and Applications, vol. 15, pp243-252, 1966.

[10] A.Ben-Israel, in "Generalized Inverses and Applications", edited by M. Zuhair Nashed, pp245-302, 1976, Academic Press.

Table 1 Simulation Results on Benchmark Circuits

Circuit	# of mosfets	# of bjts	# of user nodes	# of device nodes	VDNR # of model evaluations (1st run +2nd run)	SPICE3f3 # of model evaluations	Remark
gm1	46		31	92	420+1	16	
astabl		2	6	4	185+3	13	unstable solution
gm2	7		5	14	362+2	9	
bjtinv		12	26	12	269+3	25	
latch		14	23	42	228+3	12	unstable solution
todd3	13		13	26	212+3	22	
mike2	12		11	24	861+2	13	
gm3	30		17	60	340+2	9	
rca		11	18	11	397+2	6	
gm17	56		31	112	492+12	42	unstable solution
bias		12	12	39	601+3	163	
nagle		23	25	23	2222+6	23	
rich3	106		51	212	1237+5	N/A	failed
vreg		20	19	0	2558	N/A	failed

Note : Total # of nodes = # of user nodes + # of device nodes

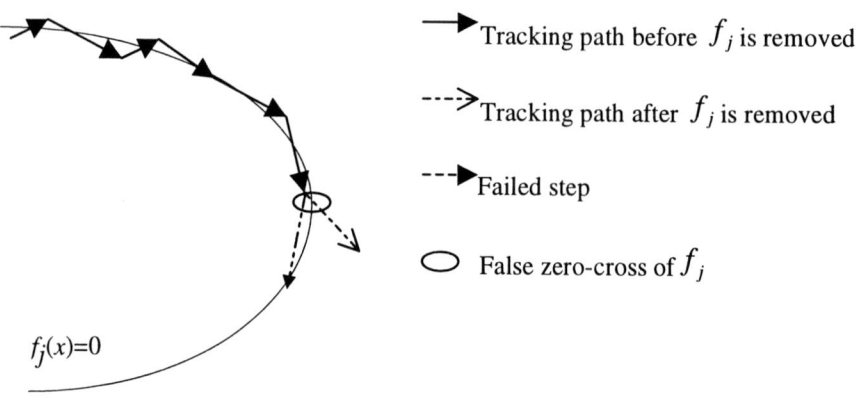

Figure 1. Crossing of $f_j(x) = 0$ when f_j is removed from tracking.

HOMSPICE: SIMULATOR WITH HOMOTOPY ALGORITHMS FOR FINDING DC AND STEADY-STATE SOLUTIONS OF NONLINEAR CIRCUITS

Ljiljana Trajković *

School of Engineering Science
Simon Fraser University
Burnaby, British Columbia
Canada V5A 1S6
ljilja@cs.sfu.ca

Eula Fung [†] *and Seth Sanders*

EECS Department
University of California
Berkeley, CA 94720-1770
eula@sequence.stanford.edu
sanders@eecs.berkeley.edu

ABSTRACT

We describe the use of homotopy (also called parameter embedding and continuation) methods for finding dc operating points and steady-state solutions of BJT and MOS transistor circuits. Past implementation of homotopy algorithms in proprietary industrial circuit simulators proved that they were viable options to resolving convergence difficulties for finding circuits' dc operating points. In this paper we describe a software implementation of publicly available homotopy algorithms (from the software package called HOMPACK) in the UCB SPICE circuit simulator. The new simulator, called **HomSPICE**, provides options for finding a circuit's dc operating points and steady-state solutions via three homotopy algorithms. We illustrate the performance of **HomSPICE** on several simulation examples.

1. INTRODUCTION

Parameter embedding methods are robust and accurate numerical techniques for solving nonlinear algebraic equations [1], [2]. They can be used to find multiple solutions of equations that possess more than one solution. A class of embedding algorithms called probability-one homotopy algorithms that promise global convergence [3] have been implemented in a publicly available software package HOMPACK [4]. Past research and implementations of homotopy algorithms for finding circuit dc operating points indicated promising results [5], [6], [7]. These algorithms have been used to find solutions to highly nonlinear circuits that could not be simulated using conventional numerical methods. They are also useful in finding dc operating points of multistable circuits. The main drawback of homotopy methods is their computational intensity. Therefore, they are most suitable for solving difficult nonlinear problems where initial solutions are hard to estimate, or when multiple solutions are desired. For circuits that fall in this category, homotopy algorithms offer a very attractive alternative.

We have implemented several homotopy algorithms in SPICE 3F5 [8], the latest version of "Berkeley SPICE" [9], [10]. The new simulator called **HomSPICE**, uses homotopy algorithms implemented in software package HOMPACK [4]. The three algorithms, FIXPDF, FIXPNF, and FIXPQF, are based on the ordinary differential equation, the normal flow, and the augmented Jacobian matrix methods, respectively. The advantage of using SPICE and HOMPACK is their publicly available source code. Further benefit of using SPICE [8] is that most commercial circuit simulators, being derived from SPICE, employ similar numerical algorithms for solving circuit equations.

Homotopy algorithms were added as options for two types of SPICE analyses: calculating a circuit's dc operating points and calculating a circuit's periodic steady-state response. Finding the dc operating point is essential for circuit simulation: both steady-state and transient analyses require a priori knowledge of a circuit's dc operating point. Finding a circuit's steady-state response is essential when designing power-conversion circuits, oscillators, and RF modulators. important to circuit designers. Many circuit performance measures, such as distortion and power dissipation, can only be measured accurately when the circuit has reached steady-state. Nevertheless, steady-state calculation is often not directly addressed by circuit simulators. In SPICE and most SPICE-like simulators, the steady-state response can only be found via a transient analysis. The steady-state is reached when transient effects become negligible. Unfortunately, this process may take an intolerably long time for lightly damped circuits, circuits with time constants much larger than their switching periods, and for high Q circuits. For such circuits, a direct steady-state solver is a better option. Two steady-state algorithms were implemented in SPICE 3C1 [11] and these routines have been included in **HomSPICE**.

This work was supported by the *NSF VPW Grant GER-9550153 and the [†]NSF Graduate Fellowship.

2. HOMOTOPY METHODS: BACKGROUND

Homotopy methods are used to solve systems of nonlinear algebraic equations and can be applied to a large variety of problems. We are most interested in solving the zero finding problem

$$\mathcal{F}(\mathbf{x}) = \mathbf{0}, \qquad (1)$$

where $\mathbf{x} \in \mathcal{R}^n$, $\mathcal{F} : \mathcal{R}^n \to \mathcal{R}^n$. (Note that the fixed point problem can be easily reformulated as a zero finding problem.)

We create the homotopy function $\mathcal{H}(\mathbf{x}, \lambda)$ by embedding $\mathcal{F}(\mathbf{x})$ into an equation of higher dimension

$$\mathcal{H}(\mathbf{x}, \lambda) = \mathbf{0}, \qquad (2)$$

where $\lambda \in \mathcal{R}$, $\mathcal{H} : \mathcal{R}^n \times \mathcal{R} \to \mathcal{R}^n$. At $\lambda = 0$,

$$\mathcal{H}(\mathbf{x}, 0) = \mathbf{0} \qquad (3)$$

is an easy equation to solve, and at $\lambda = 1$,

$$\mathcal{H}(\mathbf{x}, 1) = \mathbf{0} \qquad (4)$$

we recover the original problem (1). The parameter λ is called the continuation or homotopy parameter.

A simple example of a homotopy is

$$\mathcal{H}(\mathbf{x}, \lambda) = (1 - \lambda)(\mathbf{x} - \mathbf{a}) + \lambda \mathcal{F}(\mathbf{x}), \qquad (5)$$

where $\mathbf{a} \in \mathcal{R}^n$ is a known constant vector. Hence, $\mathcal{H}(\mathbf{x}, 0) := \mathbf{x} - \mathbf{a} = \mathbf{0}$ has an easy solution $\mathbf{x} = \mathbf{a}$, while $\mathcal{H}(\mathbf{x}, 1) := \mathcal{F}(\mathbf{x}) = \mathbf{0}$ is our original problem. By following solutions of $\mathcal{H}(\mathbf{x}, \lambda) = \mathbf{0}$ as λ varies from 0 to 1, we reach the solution to $\mathcal{F}(\mathbf{x}) = \mathbf{0}$.

The solutions trace a path known as the zero curve. Various numerical problems may occur depending on the behavior of this curve. One problem occurs if the curve folds back. At the turning point the values of λ decrease as the path progresses. Increasing λ from 0 to 1 results in "losing" the curve. The difficulty is resolved by making λ a function of a new parameter: the arc length s. This method is known as the arc length continuation [4], [2].

A variation [5] of the standard homotopy (5)

$$\mathcal{H}(\mathbf{x}, \lambda) = (1 - \lambda)\mathbf{G}(\mathbf{x} - \mathbf{a}) + \mathcal{F}(\lambda \mathbf{x}) \qquad (6)$$

was implemented in the **HomSPICE** dc operating point analysis. Diagonal matrix $\mathbf{G}$ allows scaling of the linear term $(1 - \lambda)(\mathbf{x} - \mathbf{a})$ in (6). In **HomSPICE**, the elements of $\mathbf{G}$ were chosen as the coefficients of the corresponding x_i terms in $\mathcal{F}(\mathbf{x})$ with the idea that scaling the linear term might help convergence. For example, for the equation

$$h_i(\mathbf{x}, \lambda) = (1 - \lambda)g_{ii}(x_i - a_i) + f_i(\mathbf{x}), \qquad (7)$$

g_{ii} is chosen to be equivalent to the coefficient of x_i in $f_i(\mathbf{x})$. In the case when the coefficient is too small ($< 10^{-8}$), causing the Jacobian matrix to lose rank, the g_{ii} value is set to 1.

3. DC OPERATING POINT ANALYSIS

Three steps were essential for implementing homotopy algorithms for finding a circuit's dc operating points:

- extract the nonlinear function $F(x)$ and its Jacobian from the data structure of the circuit simulator
- pass the information to the curve tracking algorithms, and
- create well behaved homotopies.

The first step was easily solved because SPICE [8] employs the Newton-Raphson method that requires knowing $\mathcal{F}(\mathbf{x})$ and its Jacobian. In the SPICE data structure, at every iteration (i), the Jacobian is explicitly saved in a matrix form, called here $\mathbf{A}$, along with a right hand side vector $\mathbf{b}$ that satisfies

$$\mathbf{A}|_{\mathbf{x}^{(i)}}\mathbf{x}^{(i)} - \mathcal{F}(\mathbf{x}^{(i)}) = \mathbf{b}|_{\mathbf{x}^{(i)}}. \qquad (8)$$

Hence, at every iteration,

$$\mathcal{F}(\mathbf{x}) = \mathbf{A}\mathbf{x} - \mathbf{b}, \qquad (9)$$

where $\mathcal{F}(\mathbf{x})$ are the Modified Nodal Analysis circuit equations [12].

Creating homotopies is rather straightforward, except for choosing which x_j in the linear term $(1-\lambda)g_{ii}(x_j - a_i)$ needs to be added to each equation $f_i(\mathbf{x})$ when creating the homotopy (6). The proper choice is crucial or the Jacobian matrix of the homotopy may lose rank during curve tracking. While we theoretically know to which equation each variable should be added, we do not explicitly know the order of equations and variables in the simulator's data structure. In SPICE, the nodal current equations are written first, followed by the voltage equations. The variables are ordered with the voltage variables first followed by the current variables. However, the ordering of the current variables relative to the voltage variables is not transparent.

A simple, although not necessarily rigorous, algorithm for choosing the correct variable to add to each equation is to add x_j to the equation f_i where x_j has the largest coefficient among all the variables in $f_i(\mathbf{x})$. In case of Nodal Analysis equations written for node i, the variable x_i will have the only positive coefficient. In contrast, the Modified Nodal Analysis voltage equations contain only one nonzero coefficient, which corresponds to the variable we want to add when creating the homotopy function. In the Modified Nodal Analysis current equations, the largest coefficient usually corresponds to the current variable that we want to add. The algorithm implemented in **HomSPICE** uses a swapping technique to ensure that the same variable is never added twice.

The performance of **HomSPICE** is illustrated by simulating the triple Schmitt trigger cascade circuit [13]. The results are shown in Table 1.

4. STEADY-STATE ANALYSIS

Three standard techniques for computing the steady-state response directly are: finite difference, shooting, and harmonic balance [14]. The shooting method is the most attractive for implementation in SPICE. The finite difference method creates large systems of equations with many unknowns that can easily consume too much memory, and harmonic balance is a frequency domain technique, and, thus, does not take advantage of many of the features available in SPICE.

For a non-autonomous T-periodic system of the form

$$\dot{\mathbf{x}} = f(\mathbf{x}, t), \tag{10}$$

where $f(\mathbf{x}, t) = f(\mathbf{x}, t+T)$ for all t, a T-periodic steady-state solution $\tilde{\mathbf{x}}$ satisfies the two-point boundary condition

$$\tilde{\mathbf{x}}(0) = \tilde{\mathbf{x}}(T). \tag{11}$$

The solution to equation (10), subject to an initial condition constraint $\mathbf{x}(0) = \mathbf{x}_0$, can be expressed with the state transition function as $\mathbf{x}(t) = \phi(\mathbf{x}_0, 0, t)$, which is a function of time, initial state, and initial time. Any solution to the T-periodic steady-state problem will satisfy

$$\phi(\tilde{\mathbf{x}}(0), 0, T) = \tilde{\mathbf{x}}(0). \tag{12}$$

The shooting method focuses on solving equation (12), which can be viewed as searching for a fixed point of the map $\phi(\cdot, 0, T)$. The function $\phi(\cdot, 0, T)$ maps the state of the circuit starting from some initial state $\mathbf{x}_0$ at time 0 to its final state after the circuit variables have evolved for T seconds. The state variables are the currents through the inductors and voltages across the capacitors. The map $\phi(\cdot, 0, T)$ is called the *first return map* or *Poincaré map*.

The shooting problem can be solved by various methods, including Newton-Raphson [15] and extrapolation [16]. Both methods are locally convergent. Using homotopy methods promises global convergence, though the computation time will take longer. In both Newton-Raphson and globally convergent homotopies, the state transition function and its Jacobian matrix need to be computed for various values of the state variables. The state transition function at a given state x_0 is easy to compute using SPICE because it is equivalent to running a transient analysis on the circuit over one period using x_0 as the initial condition. The Jacobian matrix of $\phi(\cdot, 0, T)$ can be computed using sensitivity circuits. It is most efficiently computed during the transient analysis that already generates many required values. The details of calculating the Jacobian, otherwise known as the sensitivity matrix, can be found in [11] and [14].

To apply globally convergent homotopies to the shooting problem requires that the Poincaré map be:

- C^2 and

- $\phi(\cdot, 0, T)$ maps a convex, compact set into itself.

However, we cannot guarantee that any of these conditions are met. For power electronic circuits which contain state-controlled switches, the Poincaré map may be everywhere differentiable, continuous and piecewise differentiable, or even discontinuous. In general, qualitative knowledge of the circuit's behavior should be sufficient to determine if the circuit will be an appropriate candidate for homotopy steady-state analysis.

The steady-state **HomSPICE** solutions for a dc power supply circuit [16] are shown in Table 2.

The two cycles in stead-state regime of each waveform are shown in Fig. 1 along with the solutions found by the homotopy analyses

The solution traces of the FIXPNF algorithm are shown in Fig. 2. The traces generated by the other algorithms are similar. The traces indicate the efficiently of the implemented homotopy for the given problem. Indeed, most of the curves are almost linearly approaching the solution, implying a good choice of homotopy.

5. CONCLUDING REMARKS

We have combined homotopy algorithms from the HOMPACK suite of globally convergent homotopy algorithms with the SPICE 3F5 circuit simulator, and have applied them successfully to finding dc operating points and steady-state solutions of transistor circuits. We used simple standard homotopy function that exploited the information readily available from the data structure of a circuit simulator. Our implementation, which employs public domain packages, illustrates that simple homotopies prove adequate for solving some difficult benchmark circuits. More sophisticated homotopies, such as "variable gain" (for BJT circuits) and "variable threshold" (for MOS circuits) promise even better performance.

6. REFERENCES

[1] E. L. Allgower and K. Georg, *Numerical Continuation Methods: An Introduction.* New York: Springer-Verlag Series in Computational Mathematics, 1990, pp. 1–15.

[2] L. T. Watson, "Globally convergent homotopy algorithms for nonlinear systems of equations," *Nonlinear Dynamics,* vol. 1, pp. 143–191, Feb. 1990.

[3] S. Chow, J. Mallet-Paret, and J. A. Yorke, "Finding zeroes of maps: homotopy methods that are constructive with probability one," *Mathematics of Computation,* vol. 32, no. 143, pp. 887–899, July 1978.

[4] L. T. Watson, S. C. Billups and A. P. Morgan, "Algorithm 652: HOMPACK: A suite of codes for globally

Solution	FIXPDF		FIXPNF	
	No. of Jacobian eval.	λ	No. of Jacobian eval.	λ
1.	918	1.000004	250	1.000000
2.	1867	1.000000	450	0.999983
3.	2121	1.000001	508	1.000000

Table 1: The number of Jacobian evaluations required for finding each operating point by using FIXPDF and FIXPNF for the triple Schmitt trigger cascade circuit.

State variable	Transient	FIXPDF	FIXPNF	FIXPQF
c1	-1.759777e+01	-1.779588e+01	-1.778526e+01	-1.778065e+01
c2	1.775704e+01	1.776682e+01	1.775577e+01	1.775117e+01
c3	1.787620e+01	1.788487e+01	1.787421e+01	1.787068e+01
l1	1.767396e-02	1.773145e-02	1.769987e-02	1.772588e-02
No. of Jacobian eval.		32	13	10
λ		1.000124	1.000094	0.999803

Table 2: The result of the transient analysis at time 1.98 seconds, and the steady-state initial conditions found by **HomSPICE** for the dc power supply circuit [16].

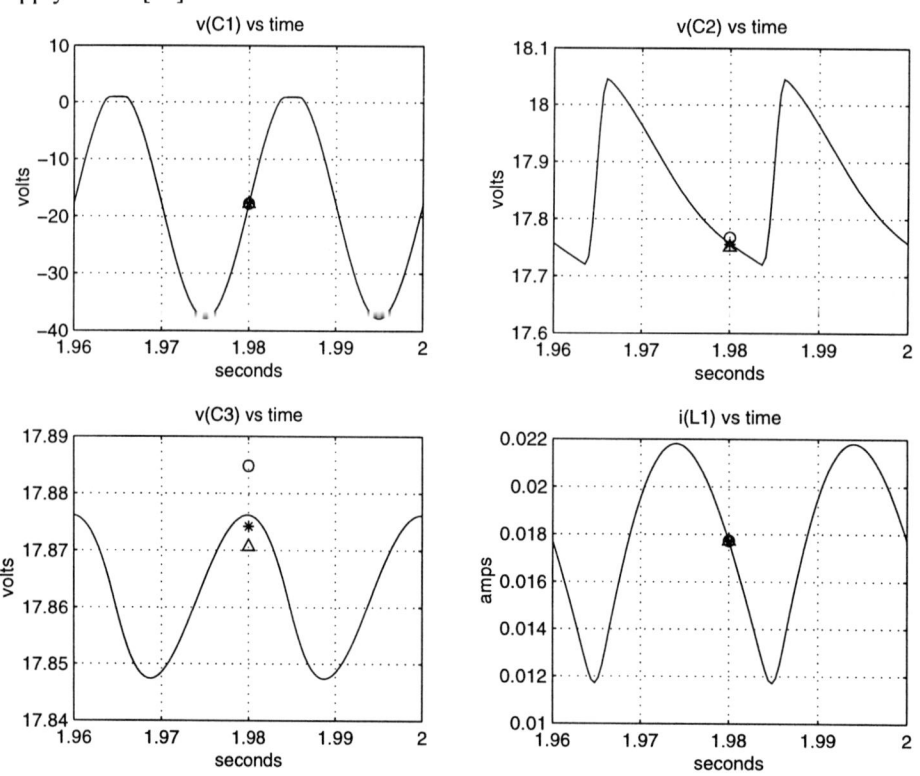

Figure 1: The steady-state initial conditions for the dc power supply [16] found by the homotopy methods plotted against the circuit's steady-state waveforms: ○ are the FIXPDF solutions, ∗ are the FIXPNF solutions, and △ are the FIXPQF solutions.

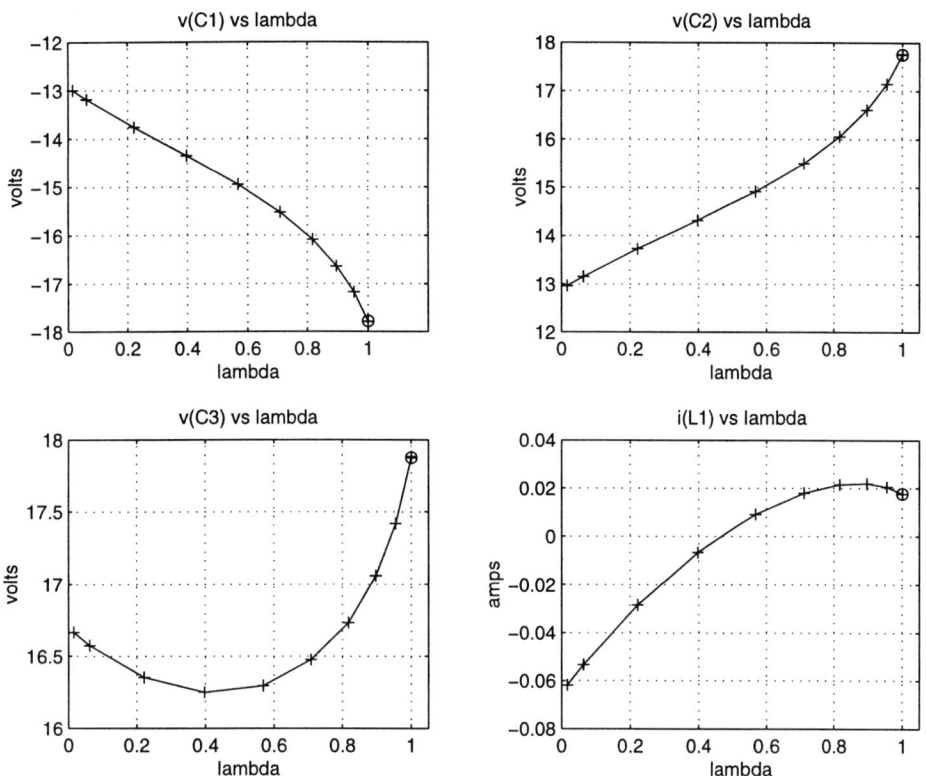

Figure 2: The solution curves for the dc power supply [16] generated by the FIXPNF algorithm.

convergent homotopy algorithms," *ACM Trans. Mathematical Software*, vol. 13, no. 3, pp. 281–310, Sept. 1987.

[5] R. C. Melville, Lj. Trajković, S. C. Fang and L. T. Watson, "Artificial parameter homotopy methods for the dc operating point problem," *IEEE Trans. on CAD*, vol. 12, no. 6, pp. 861–877, June 1993.

[6] Lj. Trajković and W. Mathis, "Parameter embedding methods for finding dc operating points: Formulation and implementation," *Proc. NOLTA '95*, Las Vegas, NV, Dec. 1995, pp. 1159–1164.

[7] D. Wolf and S. Sanders, "Multiparameter homotopy methods for finding dc operating points of nonlinear circuits," *IEEE Trans. Circuits Syst.*, vol. 43, pp. 824–838, Oct. 1996.

[8] T. L. Quarles, A. R. Newton, D. O. Pederson, and A. Sangiovanni-Vincentelli, "SPICE 3 Version 3F5 User's Manual," Department of EECS, University of California, Berkeley, March 1994.

[9] T. L. Quarles, "The SPICE3 Implementation Guide," *Memorandum No. UCB/ERL M89/44*, Department of EECS, University of California, Berkeley, April 24, 1989.

[10] A. Vladimirescu, *The SPICE Book*. New York: John Wiley & Sons, Inc., 1994.

[11] P. N. Ashar, "Implementation of algorithms for the periodic steady-state analysis of nonlinear circuits," *M. S. Thesis*, UC Berkeley, March 1989.

[12] C. W. Ho, A. E. Ruehli, and P. A. Brennan, "The modified nodal approach to network analysis," *IEEE Trans. Circuits Syst.*, vol. CAS-22, pp. 504–509, Jan. 1975.

[13] P. Horowitz and W. Hill, *The Art of Electronics*. Cambridge, MA: Cambridge University Press, 1983.

[14] K. S. Kundert, J. K. White, and A. Sangiovanni-Vincentelli, *Steady-State Methods for Simulating Analog and Microwave Circuits*. Boston, MA: Kluwer Academic Publishers, 1990.

[15] T. J. Aprille and T. N. Trick, "Steady-state analysis of nonlinear circuits with periodic inputs," *Proc. IEEE*, vol. 60, pp. 108–14, January 1972.

[16] S. Skelboe, "Computation of the periodic steady-state response of nonlinear networks by extrapolation methods," *IEEE Trans. Circuits Syst.*, vol. CAS-27, pp. 161–175, March 1980.

CONVERGENCE CONDITIONS OF WAVEFORM RELAXATION METHODS FOR CIRCUIT SIMULATION

Yao-Lin Jiang[1]   Omar Wing[2]

[1]*Institute of Information and System Sciences, School of Science, Xi'an Jiaotong University,*
Xi'an, P. R. China. E-mail: yljiang@xjtu.edu.cn
[2]*Department of Information Engineering, the Chineses University of Hong Kong,*
Hong Kong, P. R. China. E-mail: owing@ie.cuhk.edu.hk

Abstract

For two general classes of circuits which are described by nonlinear differential-algebraic equations and linear differential-algebraic equations respectively, we present convergence conditions of the waveform relaxation methods, in which the proofs are based on the operator spectral theory and are identical. These convergence conditions reveal the types of splittings of the equations for which the waveform relaxation methods will converge.

1 Introduction

The waveform relaxation (WR) methods for solving circuit equations are an attractive alternative to the direct methods [1]. These methods, which are sometimes called dynamic iteration methods [2], are based on the extension of the classical relaxation methods to function spaces. The advantages of WR come from potential exploitation of multirate behavior and latency [3] and its parallel implementation in a distributed environment [4].

The "classical" *waveform* relaxation methods have been successfully applied to MOS circuits simulation, in which the circuits are weakly and unidirectionally coupled. However, for tightly coupled systems such as bipolar transistor circuits, the convergence of these methods can be extremely slow. For some circuits, they may not converge at all. Recently the waveform Krylov subspace methods and the SOR waveform relaxation method with convolution have been shown to be more robust in that circuits can be found on which the latter two methods will converge but the classical methods do not [5 - 7].

It follows that it is necessary to have a deeper understanding of the convergence properties of WR before it can be accepted as a reliable circuit simulation technique.

It is well-known that splitting of the circuit equations is crucial to the convergence of the waveform relaxation methods. One always wants to know in theory what splitting makes a waveform relaxation method converge and what splitting does not, and which variables in the split iterative system will contribute to the convergence and which ones do not.

At the present time, the convergence theory on WR is mainly aimed at circuits describable by ordinary differential equations (ODE) [2, 8 - 11], though the original WR method was intended for nonlinear differential-algebraic equations (DAE) [1]. Further, the known convergence conditions for nonlinear ODE or nonlinear DAE systems are almost all based on the construction of a weighted norm under which the corresponding waveform relaxation operators are strictly contractive. Thus the well-known contraction mapping principle in some function space equipped with this new norm can be used to prove the convergence of the iterative waveform sequence produced by the WR methods [1, 8 - 9, 12 - 13].

In this paper, we present the convergence conditions of the waveform relaxation methods for two general classes of circuits: those which are described by nonlinear DAE and those by linear DAE. The proofs of these conditions for the nonlinear and linear cases are identical and direct. They are based on the spectral theory of linear operators in function spaces and we do not need to construct a new norm when the nonlinear case is treated. The convergence conditions reported here are new and less restrictive than those previously published.

The organization of this paper is as follows. In Section 2, convergence conditons of the waveform relaxation methods for a system described by nonlinar differential-algebraic equations are provided. Conditions for the linear case are presented in Section 3. Finally, conclusion is given in Section 4.

2 Convergence conditions of waveform relaxation methods for a nonlinear differential-algebraic equation

In this section, we consider a system of nonlinear differential-algebraic equations given as follows:

$$\begin{cases} \dot{x}(t) = \tilde{f}(x,\dot{x},y,e_1,t)(t), & x(0) = x_0, \\ y(t) = \tilde{g}(x,\dot{x},y,e_2,t)(t), & t \in [0,T] \end{cases} \quad (1)$$

where t is a time variable, $x_0 \in \mathbf{R}^{n_1}$ is an initial value, $[0,T]$ is a given finite time interval, $x(t) \in \mathbf{R}^{n_1}$ and $y(t) \in \mathbf{R}^{n_2}$ are to be computed, $e_1(t) \in \mathbf{R}^{m_1}$ and $e_2(t) \in \mathbf{R}^{m_2}$ are known input functions. In electronic engineering, many nonlinear circuits can be described by System (1). In fact, System (1) is quite general since it may include delay differential equations and integro-differential equations as special cases.

The general form of the waveform relaxation methods for System (1) is given as

$$\begin{cases} \dot{x}^{(k+1)}(t) = f(x^{(k+1)}, x^{(k)}, \dot{x}^{(k+1)}, \dot{x}^{(k)}, y^{(k+1)}, \\ \qquad\qquad\qquad y^{(k)}, e_1, t)(t), \\ y^{(k+1)}(t) = g(x^{(k+1)}, x^{(k)}, \dot{x}^{(k+1)}, \dot{x}^{(k)}, y^{(k+1)}, \\ \qquad\qquad\qquad y^{(k)}, e_2, t)(t), \\ x^{(k+1)}(0) = x_0, \quad t \in [0,T], \quad k = 0, 1, \cdots \end{cases} \quad (2)$$

where $[x^{(0)}(\cdot), y^{(0)}(\cdot)]^t$ is a given initial iteration, and the nonlinear splitting functions $f : (\mathbf{R}^{n_1})^4 \times (\mathbf{R}^{n_2})^2 \times \mathbf{R}^{m_1} \times [0,T] \mapsto \mathbf{R}^{n_1}$ and $g : (\mathbf{R}^{n_1})^4 \times (\mathbf{R}^{n_2})^2 \times \mathbf{R}^{m_2} \times [0,T] \mapsto \mathbf{R}^{n_2}$ satisfy

$$f(x,x,x^\wedge,x^\wedge,y,y,e_1,t)(t) = \tilde{f}(x,x^\wedge,y,e_1,t)(t)$$

and

$$g(x,x,x^\wedge,x^\wedge,y,y,e_2,t)(t) = \tilde{g}(x,x^\wedge,y,e_2,t)(t)$$

where $x, x^\wedge \in \mathbf{R}^{n_1}$, $y \in \mathbf{R}^{n_2}$, $e_1 \in \mathbf{R}^{m_1}$, $e_2 \in \mathbf{R}^{m_2}$ and $t \in [0,T]$. In circuit simulation, usually one adopts some typical splittings such as Jacobi splitting and Gauss-Seidel splitting etc. [1, 8] similar to those in the classical relaxation method.

Now, we assume that the splitting functions f and g satisfy the following Lipschitz conditon as in [13]:

Condition L. For any given input functions $e_1 : [0,T] \mapsto \mathbf{R}^{m_1}$ and $e_2 : [0,T] \mapsto \mathbf{R}^{m_2}$, there are constants a_l and $b_l (l = 1, 2, \cdots, 6)$ such that

$$\|f(u_1,u_2,\cdots,u_6,e_1,t)(t) - f(v_1,v_2,\cdots,v_6,e_1,t)(t)\| \le \sum_{l=1}^6 a_l \|u_l - v_l\|$$

and

$$\|g(u_1,u_2,\cdots,u_6,e_2,t)(t) - g(v_1,v_2,\cdots,v_6,e_2,t)(t)\| \le \sum_{l=1}^6 b_l \|u_l - v_l\|$$

where $t \in [0,T]$, $u_l, v_l \in \mathbf{R}^{n_1} (l = 1, 2, \cdots, 4)$ and $u_l, v_l \in \mathbf{R}^{n_2} (l = 5, 6)$.

Let A_1 and A_2 be two simple 2×2 matrices as follows,

$$A_1 = \begin{bmatrix} a_3 & a_5 \\ b_3 & b_5 \end{bmatrix}, \qquad A_2 = \begin{bmatrix} a_4 & a_6 \\ b_4 & b_6 \end{bmatrix}$$

where $a_l, b_l (l = 3, 4, \cdots, 6)$ are Lipschitz constants appearing in Condition L. We denote $A_0 = (I - A_1)^{-1} A_2$.

Theorem 1 Let (i) Condition L be satisfied, (ii) the matrix $(I - A_1)$ have nonnegative inverse (i.e., $(I - A_1)^{-1} \ge 0$), and (iii) the spectral radius $\rho(A_0) < 1$. Then, the sequence $\{[x^{(k)}, y^{(k)}]^t\}$ produced by the algorithm (2) will converge to the solution $[x^{(*)}, y^{(*)}]^t$ of System (1), with a linear convergence rate $\rho(A_0)$.

The key step of the approach is to prove that the following inequality for the algorithm (2) is valid for $t \in [0,T]$,

$$\begin{bmatrix} \|\dot{x}^{(k+1)}(t) - \dot{x}^{(*)}(t)\| \\ \|y^{(k+1)}(t) - y^{(*)}(t)\| \end{bmatrix} \le \mathcal{L} \begin{bmatrix} \|\dot{x}^{(k)} - \dot{x}^{(*)}\| \\ \|y^{(k)} - y^{(*)}\| \end{bmatrix}(t)$$

where $\mathcal{L} : C([0,T]; \mathbf{R}^2) \mapsto C([0,T]; \mathbf{R}^2)$ is a bounded linear operator such that $\rho(\mathcal{L}) < 1$.

The proof appears in [14]. The approach adopted there is direct and is completely different from that in [12 - 13]. It is less restrictive than any of the conditions previously published and is also more general than one of the original waveform relaxation methods [1]. See [14] for details and further discussions.

3 Convergence conditions of waveform relaxation methods for a linear differential-algebraic equation

In this section, we consider a system of linear differential-algebraic equations in semi-explicit form:

$$\begin{cases} M\dot{x}(t) + Ax(t) + By(t) = f_1(t), & x(0) = x_0, \\ Cx(t) + Ny(t) = f_2(t), & t \in [0,T] \end{cases} \quad (3)$$

where $x(t) \in \mathbf{R}^{n_1}$ and $y(t) \in \mathbf{R}^{n_2}$, M and N are $n_1 \times n_1$ nonsingular and $n_2 \times n_2$ nonsingular matrices respectively, A is an $n_1 \times n_1$ matrix, B is an $n_1 \times n_2$ matrix, C is an $n_2 \times n_1$ matrix, $f_1(t) \in \mathbf{R}^{n_1}$ and $f_2(t) \in \mathbf{R}^{n_2}$ are two input functions, and $x_0 \in \mathbf{R}^{n_1}$ is an initial value.

Much of the results of this section are found in [7]. They are included here for comparison with the nonlinear case.

In order to compute numerically the solution of System (3) by the waveform relaxation methods, we use the following iterative process :

$$\begin{cases} M_1\dot{x}^{(k+1)}(t) + A_1 x^{(k+1)}(t) + B_1 y^{(k+1)}(t) \\ \quad = M_2 \dot{x}^{(k)}(t) + A_2 x^{(k)}(t) \\ \qquad + B_2 y^{(k)}(t) + f_1(t), \\ C_1 x^{(k+1)}(t) + N_1 y^{(k+1)}(t) \\ \quad = C_2 x^{(k)}(t) + N_2 y^{(k)}(t) + f_2(t), \\ x^{(k+1)}(0) = x_0, \quad t \in [0,T], \quad k = 0,1,\cdots \end{cases} \quad (4)$$

where $M = M_1 - M_2$, $A = A_1 - A_2$, $B = B_1 - B_2$, $C = C_1 - C_2$, $N = N_1 - N_2$, and $[x^{(0)}(\cdot), y^{(0)}(\cdot)]^t$ is a given initial iteration. Further, M_1 and N_1 are nonsingular matrices.

Denoting $D_1 = A_1 - B_1 N_1^{-1} C_1$. Let $(\mathcal{R}_1)_c : C([0,T]; \mathbf{R}^{n_1}) \mapsto C([0,T]; \mathbf{R}^{n_1})$ and $(\mathcal{R}_2)_c : C([0,T]; \mathbf{R}^{n_2}) \mapsto C([0,T]; \mathbf{R}^{n_1})$ be Volterra integral operators as follows,

$$((\mathcal{R}_1)_c u)(t) = \int_0^t e^{-M_1^{-1} D_1 (t-s)} M_1^{-1} (A_2 \\ - B_1 N_1^{-1} C_2 - D_1 M_1^{-1} M_2) u(s) ds$$

and

$$((\mathcal{R}_2)_c v)(t) = \int_0^t e^{-M_1^{-1} D_1 (t-s)} M_1^{-1} (B_2 \\ - B_1 N_1^{-1} N_2) v(s) ds$$

where $u \in C([0,T]; \mathbf{R}^{n_1})$ and $v \in C([0,T]; \mathbf{R}^{n_2})$. Further, we also denote that

$$\varphi_1(t) = e^{-M_1^{-1} D_1 t}(I - M_1^{-1} M_2) x_0 \\ + \int_0^t e^{-M_1^{-1} D_1 (t-s)} M_1^{-1} [f_1(s) - B_1 N_1^{-1} f_2(s)] ds$$

and $\varphi_2(t) = -N_1^{-1} C_1 \varphi_1(t) + N_1^{-1} f_2(t)$.

Thus, for any fixed k we can solve Eq. (4) in function space $C([0,T]; \mathbf{R}^{n_1+n_2})$, namely

$$\begin{bmatrix} x^{(k+1)} \\ y^{(k+1)} \end{bmatrix} = \begin{bmatrix} M_1^{-1} M_2 & 0 \\ D_2 & N_1^{-1} N_2 \end{bmatrix} \begin{bmatrix} x^{(k)} \\ y^{(k)} \end{bmatrix} \\ + \begin{bmatrix} (\mathcal{R}_1)_c & (\mathcal{R}_2)_c \\ -N_1^{-1} C_1 (\mathcal{R}_1)_c & -N_1^{-1} C_1 (\mathcal{R}_2)_c \end{bmatrix} \begin{bmatrix} x^{(k)} \\ y^{(k)} \end{bmatrix} \\ + \begin{bmatrix} \varphi_1 \\ \varphi_2 \end{bmatrix} \quad (5)$$

where $D_2 = N_1^{-1} C_2 - N_1^{-1} C_1 M_1^{-1} M_2$. Then, in this function space, we can write tightly Eq. (5) as

$$z^{(k+1)} = \mathcal{R} z^{(k)} + \varphi, \quad k = 0, 1, \cdots$$

where $z^{(l)} = [x^{(l)}, y^{(l)}]^t (l = k, k+1)$, $\varphi = [\varphi_1, \varphi_2]^t$ and $\mathcal{R} : C([0,T]; \mathbf{R}^n) \mapsto C([0,T]; \mathbf{R}^n)$ is a bounded linear operator. Based on this iterative operator equation, we have

Theorem 2 The waveform relaxation solution of System (3) according to the splitting of (4) will converge if

$$\rho(M_1^{-1} M_2) < 1 \quad \text{and} \quad \rho(N_1^{-1} N_2) < 1$$

In practice, most circuit systems and some resulting systems of demiconductor devices (e.g., MOSFET etc. [5]) after spatially discretizing are nonlinear equations. Applying Newton's method in function spaces [15] we approximate these nonlinear equations by the following linear system with time-varying coefficients

$$\begin{cases} M(t)\dot{x}(t) + A(t)x(t) + B(t)y(t) = f_1(t), \\ C(t)x(t) + N(t)y(t) = f_2(t), \\ x(0) = x_0, \quad t \in [0,T] \end{cases} \quad (6)$$

where $M(t)$ and $N(t)$ are nonsingular matrices for all $t \in [0,T]$. We assume that the coefficient matrix-value functions in System (6) are continuously differentiable.

To compute the solution of System (6) by the waveform relaxation methods, we obtain an iterative system as the iterative process (4) except the coefficients are time-varying. We use a splitting such that $M_1(t)$ and $N_1(t)$ in the decoupled system are nonsingular matrices for $t \in [0,T]$, and in this iterative system all coefficient matrix-value functions are continuously differentiable. Similarly, we have that if

$$\max_{t \in [0,T]} \rho(M_1^{-1}(t) M_2(t)) < 1$$

and

$$\max_{t \in [0,T]} \rho(N_1^{-1}(t) N_2(t)) < 1$$

are satisfied, then the corresponding waveform relaxation method converges.

Further, to speed up the convergence of the iterative process (4), one should adopt those splittings of M and N such that $\rho(M_1^{-1} M_2) < 1$ and $\rho(N_1^{-1} N_2) < 1$ be as small as possible. See [16].

4 Conclusion

We have presented convergence conditions of the waveform relaxation methods for classes of circuits described by either nonlinear differential-algebraic equations or linear differential-algebraic equations. In both case, the approach to obtaining these conditions are to rewrite the circuit equations as operator equations in function spaces and then to apply the spectral theory of linear operator to the split equations.

From the convergence conditions, we gain a better understanding on how the equations should be split in order to guarantee convergence when waveform relaxation is applied to circuit simulation.

Acknowledgement

This research was supported by the Chinese University of Hong Kong under a special postdoctoral fellowship scheme and by the Hong Kong Research Grants Council grants CUHK 253/94E and CUHK 4147/97E.

References

[1] E. Lelarasmee, A. Ruehli, and A. L. Sangiovanni-Vincentelli, "The waveform relaxation method for time-domain analysis of large scale integrated circuits," *IEEE Trans. on CAD of IC and Systems*, vol. 1, no. 3, pp. 131–145, July 1982.

[2] U. Miekkala and O. Nevanlinna, "Convergence of dynamic iteration methods for initial value problem," *SIAM J. Sci. Stat. Comput.*, vol.8, no. 4, pp. 459–482, July 1987.

[3] R. A. Saleh and A. R. Newton, "The exploitation of latency and multirate behavior using nonlinear relaxation for circuit simulation," *IEEE Trans. on CAD of IC and Sys.*, vol. 8, no. 12, pp. 1286–1298, December 1989.

[4] A. Lumsdaine, J. M. Squyres, and M. W. Reichelt, "Waveform iterative methods for parallel solution of initial value problems," in *Scalable Parallel Libraries Conference*, Mississippi State, MS, October 1994.

[5] A. Lumsdaine, M. W. Reichelt, J. M. Squyres, and J. K. White, "Accelerated Waveform Methods for Parallel Transient Simulation of Semiconductor Devices," *IEEE Trans. on CAD of IC and Sys.*, vol. 15, no. 7, pp. 716–726, July 1996.

[6] W.-S. Luk and O. Wing, "Waveform Krylov subspace methods for tightly coupled systems," in *Proc. IEEE Int'l Symp. on Circuits and Systems*, May 1996, vol. IV, pp. 536–539.

[7] Y.-L. Jiang, W.-S. Luk, and O. Wing, "Convergence-theoretics of classical and Krylov waveform relaxation methods for differential-algebraic equations," *IEICE Transactions on Fundamentals of Electronics, Communications and Computer Sciences*, vol. E80-A, no. 10, pp. 1961–1972, October 1997.

[8] J. K. White and A. Sangiovanni-Vincentelli, *Relaxation Techniques for the simulation of VLSI circuits*, Kluwer Academic Publishers, Boston 1987.

[9] P. Debefve, F. Odeh, and A. E. Ruehli, "Waveform techniques," in *Circuit Analysis, Simulation and Design, Part 2* (A. E. Ruehli Ed.), Elsevier Science Publishers B. V. (North-Holland), 1987.

[10] R. Wang and O. Wing, "Transient analysis of dispersive VLSI interconnects terminated in nonlinear loads," *IEEE Trans. on CAD of IC and Sys.*, vol. 11, no. 10, pp. 1258–1277, October 1992.

[11] J. Janssen and S. Vandewalle, "Multigrid waveform relaxation on spatial finite element meshes: The continuous-time case," *SIAM J. Numer. Anal.*, vol. 33, no. 2, pp. 456–474, April 1996.

[12] K. R. Schneider, "A remark on the wave-form relaxation method," *Internat. J. Circuit Theory Appl.*, vol. 19, pp. 101–104, 1991.

[13] Z. Jackiewicz and M. Kwapisz, "Convergence of waveform relaxation methods for differential-algebraic equations," *SIAM J. Numer. Anal.*, vol. 33, no. 6, pp. 2303–2317, December 1996.

[14] Y.-L. Jiang and O. Wing, "On the convergence of waveform relaxation algorithm for nonlinear differential-algebraic equations," in preparation, 1997.

[15] D. J. Erdman and D. J. Rose, "Newton waveform relaxation techniques for tightly coupled systems," *IEEE Trans. on CAD of IC and Sys.*, vol. 11, no. 5, pp. 598–606, May 1992.

[16] Y.-L. Jiang and O. Wing, "Splitting techniques to speed up the convergence of waveform relaxation methods for tightly coupled circuit systems," Proceedings of 1997 European Conference on Circuit Theory and Design, pp. 1054–1058, Budapest, September 1997.

A time-frequency algorithm for the simulation of the initial transient response of oscillators

H. G. Brachtendorf
Bell Laboratories, Murray Hill, New Jersey, USA
brachtd@research.bell-labs.com
G. Welsch, R. Laur
University of Bremen, Bremen, Germany
{welsch, laur}@item.uni-bremen.de

Abstract

The paper presented here deals with a novel algorithm for calculating the initial transient response of an oscillator circuit. The method is based on reformulating the original system of ordinary differential-algebraic equations (DAEs) by a system similar to partial differential equations (PDEs). The time-scales of the solution of the PDEs are unlike the original DAEs not widely seperated. The PDE approach is therefore much better suited for CAD. Unlike existing methods the novel algorithm has a sound mathematical basis.

1 Introduction

The simulation of analog circuits which exhibit stiffness is one of the most demanding challenges for circuit simulation tool. The reason is that the time-step is fixed by the highest frequency of interest, whereas the time interval for which a transient solution has to be calculated is fixed by the smallest time constant of the system. Traditional integration formulas use expansions in form of Taylor series or polynomial approximations. Because these approximations are not well-suited for oscillatory solutions the time-step must be much smaller than the period of oscillation. Cases for which the settling time of a circuit is decades larger than the period of the asymptotic steady state solution include high Q-filters and oscillators. Figure (4) shows as a typical example the initial transient response of a van der Pol oscillator. The oscillation frequency is much higher than the time constant of the envelope which leads to extraordinary run-times. This is always true for quasi-sinusoidal oscillators such as quartz or cavity oscillators. The lower the distortion the longer the settling time. Hence, traditional time-domain methods are prohibitive.

Several attempts have been made for simulating ordinary differential equations which exhibit highly oscillatory solutions by special integration formulas, e. g.[1, 2], or extrapolation formulas [3] in the time domain. These techniques are often referred to as two time-scale harmonic balance [4], envelope following techniques [13], the multirevolution method [5] or the method of multiple scales [6, 7, 8, 9, 10] which have its roots in perturbation theory.

In the papers of Roychowdhury [11, 12] and Brachtendorf et al [14, 15, 16] a novel algorithm has been derived which is based on reformulating the original set of ordinary differential-algebraic equations (DAEs) which results from modified nodal analysis (MNA) by a set of partial differential-algebraic equations (PDEs). This technique which has been originally developed for the simulation of quasiperiodic steady states [14] has been used in [11, 12] for the simulation of initial value problems where the period of oscillation is either known or estimated a priori.

In the paper presented here an algorithm is developed for the calculation of the transient response of autonomous circuits. The novel technique might be considered as a special envelope following or multiple time-scales algorithm. Unlike existing method this algorithm has a sound mathematical basis. The technique is not restricted only to autonomous or mildly nonlinear circuits. Instead, the modification for evaluating the transient response of a non-autonomous circuit is straightforward. The novel algorithm reduces to the method presented in [11, 12] when the frequency is fixed and hence no secular term occurs. For ease of presentation we derive the method only for circuits which exhibit only one oscillating frequency, e. g. one single pair of complex conjugate eigenvalues in the complex plane. The extension to systems with several pairs of conjugate eigenvalues is straightforward. A modification of the method [17] is also suited for studying forced oscillators, e. g. mode locking effects. Results will be presented in another paper.

In the sequel lower case letters represent time domain und upper case letters time-frequency domain waveforms.

2 Derivation of the method

The modified nodal analysis (MNA) leads to a system of linear or nonlinear ordinary DAE's of dimension N

$$f(\dot{v}(t),v(t),t) := i(v(t)) + \frac{d}{dt}q(v(t)) + i_s(t) = 0 \quad (1)$$

where $v : \mathbf{R} \rightarrow \mathbf{R}^N$ is the vector of node voltages and some branch currents. $q : \mathbf{R}^N \to \mathbf{R}^N$ is the vector of charges and fluxes and $i : \mathbf{R}^N \to \mathbf{R}^N$ the vector of sums of currents entering each node and branch voltages, both depending on $v(t)$. We assume that $i(v)$ and $q(v)$ are differentiable functions. Furthermore, $i_s : \mathbf{R} \to \mathbf{R}^N$ is the vector of input sources, both voltages and currents. We assume that (1) has a unique solution for any initial condition. In the following the vector of unknowns $v(t)$ is referred to as the vector of node voltages. Because we assume that the circuit is autonomous the stimulus vector is constant in time:

$$i_s(t) = const. \quad (2)$$

Note that the fundamental frequency of the solution is unknown a priori. We assume here for ease of presentation that (1) exhibits only one fundamental oscillating frequency, e. g. one pair of complex conjugate eigenvalues In this case the DAEs (1) are reformulated by a set of PDEs of dimension two which is shown below. If there are n pairs of complex eigenvalues of the (linearized) system the DAE is reformulated by an $(n+1)$-dimensional PDE. This extension is straightforward. Sometimes the eigenvalues are clustered. Then the dimension of the PDE can be reduced to the number of clusters plus one.

Instead of solving the set of ordinary DAEs (1) which exhibit the multirate behavior we solve for the PDE

$$f(D_1 v(t_1,t_2), D_2 v(t_1,t_2), v(t_1,t_2), t_1, t_2) := \quad (3)$$
$$i(v(t_1,t_2)) + D_1 q(v(t_1,t_2)) + D_2 q(v(t_1,t_2)) + i_s = 0$$

with identical stimulus i_s of the original DAE and identical current-voltage and charge-voltage relations. $v(t_1,t_2)$ is assumed being periodic in t_1 with unknown period $T(t_2)$ (s. below). The definitions of D_1 and D_2 are given below. It can be shown that (3) has only a unique solution if and only if (1) has a unique solution [17]. The proof is here omitted for brevity.

The quasiperiodic waveforms $x(t_1,t_2)$ of (3) are expanded for a fixed t_2 by a Fourier polynomial [1]:

$$x(t_1,t_2) = \sum_{k=-\infty}^{\infty} X(k,t_2) e^{jk\omega(t_2)t_1} \quad (4)$$

with unknown fundamental angular frequency $\omega(t_2)$ which depends in general on t_2. The operators D_1 and D_2 applied to $x(t_1,t_2)$ are defined as follows:

$$D_1 x(t_1,t_2) := \sum_{k=-\infty}^{\infty} jk\omega(t_2) X(k,t_2) e^{jk\omega(t_2)t_1}$$

$$D_2 x(t_1,t_2) := \sum_{k=-\infty}^{\infty} \frac{\partial X(k,t_2)}{\partial t_2} e^{jk\omega(t_2)t_1} +$$

$$\sum_{k=-\infty}^{\infty} jk \frac{\partial \omega(t_2)}{\partial t_2} t_2 X(k,t_2) e^{jk\omega(t_2)t_1} \quad (5)$$

Please note the non-standard formulation of the secular term [2] $jk \frac{\partial \omega(t_2)}{\partial t_2} t_2$. The waveforms (4) and (5) are by definition periodic in t_1 with an a priori unknown time-varying frequency. Hence, the solution must not be calculated in the complete (t_1, t_2) plane but rather along a strip $[0, T(t_2)[\times[0, \infty[$ which makes the algorithm presented here so efficient. Also note that the Fourier coefficients of the negative frequencies are complex conjugate to the corresponding positive ones. For a numerical treatment we have to approximate the solution (4) by a subspace which is spanned by a finite number of Fourier polynomials:

$$x(t_1,t_2) = \sum_{k=-K}^{K} X(k,t_2) e^{jk\omega(t_2)t_1} \quad (6)$$

The nonlinear device constitutive equations are evaluated for fixed t_2 by a collocation method on a predefined grid, e. g. the time-frequency approach (6) is backtransformed into the time domain by employing the inverse Discrete Fourier Transform (IDFT). Then the device equations are evaluated for distinct time-points on the mesh and finally backtransformed into the time-frequency domain by the DFT. This is common practice for the well-known harmonic balance technique and is therefore not considered here in detail.

Using the fact that the partial derivative with regard to t_1 for the Fourier polyomial (6) is simply the multiplication with $k\omega$ for the k-th harmonic, results into the system of ordinary differential equations in t_2 (for ease of presentation t_2 has been replaced by t):

$$F(V(t),t) = I(V(t)) + j\Omega\, Q(V(t)) +$$
$$jt \frac{d}{dt}\Omega\, Q(V(t)) + \frac{d}{dt}Q(V(t)) + I_s = 0 \quad (7)$$

where ($j = \sqrt{-1}$ and $\omega(t) = 2\pi/T(t)$)

$$\Omega := [\Omega_{nm}]$$
$$\Omega_{nm} := \begin{cases} \Omega_{nn} & \text{if } n=m \\ 0 & \text{if } n \neq m \end{cases}$$
$$\Omega_{nn} := \text{diag}\{-K\omega,\ldots,-\omega,0,\omega,\ldots,K\omega\} \quad (8)$$

Note that the fundamental frequency ω is unknown a priori. This is the only significant difference for calculating the initial transient response of autonomous and non-autonomous systems. Equation (7) is solved for initial conditions $V(t_0) = V_0$ and $\omega_0 = \omega(0)$ using any integration formula. Because the oscillation is approximated by a Fourier polynomial, (7) does not exhibit any multirate behavior (presupposed that there is only one significant oscillation). Furthermore, because ω is an unknown, we have exactly one additional unknown than equations in (8). For getting an isolated solution of the oscillator waveforms it is common practice to set the imaginary

---

[1] An expansion in trigonometric basis functions is done only for ease of presentation. Any other solution technique can be employed as well

[2] The name secular term originates from perturbation techniques applied to differential equations in astronomy

part of a fundamental frequency of one node voltage equal to zero.

3 Results

We consider firstly the symmetric oscillator given by the system below:

$$\frac{d}{dt}\begin{bmatrix} x_1 \\ x_2 \end{bmatrix} - \omega_0 \begin{bmatrix} c_0(1-x_1^2) & -1 \\ 1 & c_0(1-x_2^2) \end{bmatrix} \begin{bmatrix} x_1 \\ x_2 \end{bmatrix} = 0 \quad (9)$$

If $|c_0| \ll 1$ the time constant of the envelope is decades larger than the period of the oscillation. Hence, simple time domain methods are prohibitive. Fig. (1) shows the time domain solution of the motions x_1 and x_2 for a relatively large $c_0 = -0.1$. Fig. (2) shows the simulation curve of the real and imaginary parts of the fundamental Fourier coefficient of the unknown x_1 using the novel algorithm outlined in the section before. Note that the envelope is smooth and therefore traditional integration formulas for the novel system of ordinary DAEs (7) are very efficient. Time steps can be used which are orders of magnitude larger than the period of oscillation. By a proper choice of the time step with regard to efficiency and accuracy the simulation time is nearly independent of the damping factor c_0. Fig. (3) shows simulation results for an initial damping factor $c_0 = -0.01$.

As a second example we consider the van der Pol equations:

$$\frac{d}{dt}\begin{bmatrix} x_1 \\ x_2 \end{bmatrix} - \omega_0 \begin{bmatrix} 0 & -1 \\ 1 & c_0(1-x_1^2) \end{bmatrix} \begin{bmatrix} x_1 \\ x_2 \end{bmatrix} = 0 \quad (10)$$

Because the van der Pol coefficient matrix is unsymmetric the frequency depends slightly on the actual state of the oscillator. The time domain solution is depicted in figure (4) for a relatively large damping constant $c_0 = -0.1$. The real and imaginary parts of the fundamentals of the waveform x_1 are shown in figure (5). As before the simulation time using the novel technique is nearly independent of the damping constant. Figure (6) shows the calculated fundamental frequency depending on t_2 normalized to the systems parameter ω_0. As expected there is a slight dependency of ω with respect to t_2 and ω differs slightly from ω_0 at the beginning.

Conclusions

In this paper a novel algorithm has been derived for calculating initial transient responses of systems of ordinary DAEs which exhibit high frequency oscillations. The method is based on reformulating the ordinary DAEs by partial DAEs with a suitable incorporation of the secular terms. Unlike existing techniques the algorithm presented here has a rigorous mathematical basis. The dimension of the partial DAEs depends on the number of significant fundamental eigenfrequencies of the system. The resulting set of equations does not exhibit high-frequency waveforms and can therefore be efficiently solved by conventional integration formulas. Furthermore, the algorithm is not limited to mildly nonlinear circuits. Especially for the simulation of oscillators exhibiting all these problems mentioned above this algorithm has been applied successfully which has been demonstrated by typical examples. This algorithm is not restricted to simulating electronic circuits. Instead, similar problems arise when calculating e. g. orbits of artifical satellites.

Acknowledgement

The authors would like to thank Mrs. A. Bunse-Gerstner and Mr. R. Stöver of the University of Bremen Mr. P. Feldmann, Mr. B. Melville and Mr. J. Roychowdhury from Bell Laboratories for fruitfull discussions.

References

[1] L. Petzold: "An Efficient Numerical Method for highly Oscillatory Ordinary Differential Equations." SIAM. J. Numer. Anal. Vol. 18, No. 3, pp. 455-479, June 1981.

[2] G. Denk: "A new efficient numerical integration scheme for highly oscillatory electric circuits." Int. Series of Numerical Math., Vol. 117, pp. 1-15, Birkhäuser Verlag, Basel, 1994

[3] Q. Zheng: "The Transient Behavor of an Oscillator." Int. Series of Numerical Math., Vol. 117, pp. 143-164, Birkhäuser Verlag, Basel, 1994

[4] J. L. Summers, M. D. Savage, "Two timescale harmonic balance. I. Application to autonomous one-dimensional nonlinear oscillators." Phil. Trans. R. Soc. Lond. A 340, 1992, pp. 473-501.

[5] B. Melendo, M. Palacios, "A new approach to the construction of multirevolution methods and their implementation." Appl. Num. Math., Elsevier, vol. 23, pp. 259-274, 1997.

[6] A. H. Nayfeh, Introduction to Perturbation Techniques. John Wiley, New York, 1981.

[7] A. H. Nayfeh, D. T. Mook, "Non-Linear Oscillations. John Wiley, New York, 1979.

[8] R. Grimshaw, Nonlinear Ordinary Differential Equations. Blackwell Scientific Publications, Oxford, 1990.

[9] P. A. Lagerstrom, Matched Asymptotic Expansion. Springer, Berlin, 1988.

[10] R. E. Mickens, Oscillations in planar dynamic systems. World Scientific, Singapore, 1996.

[11] Roychowdhury, J.: "Efficient Methods for Simulating Highly Nonlinear Multi-Rate Circuits." Design Automation Conference, Anaheim, CA, June 1997.

[12] Roychowdhury, J.: "A Unified Method for Analyzing Multi-Rate Circuits." Bell Laboratories, Internal Technical Memorandum, ITD-97-31503R, February 1997.

[13] J. Roychowdhury, P. Feldmann, "Computation of circuit waveform envelopes using an efficient, matrix-decomposed harmonic balance algorithm." Proc. Custom Integrated Circuits Conf., 1996.

[14] Brachtendorf, H. G.: Simulation des eingeschwungenen Verhaltens elektronischer Schaltungen. Aachen: Shaker 1994.

[15] Brachtendorf, H.G.; Welsch, G.; Laur, R.; Bunse-Gerstner, A.: "Numerical steady state analysis of electronic circuits driven by multi-tone signals." Electronic Engineering, Vol. 79, pp. 103-112, 1996.

[16] H. G. Brachtendorf, G. Welsch, R. Laur, "A novel time-frequency method for the simulation of the steady state of circuits driven by multi-tone signals." IEEE Proc. Int. Symp. on CAS, Vol. 3, pp. 1508-1511, Hongkong, June 9-12.

[17] H. G. Brachtendorf "On the relation of certain classes of ordinary differential algebraic equations with partial differential equations." Bell Laboratories, Internal Technical Memorandum, 1131G0-971114-19TM, November, 1997.

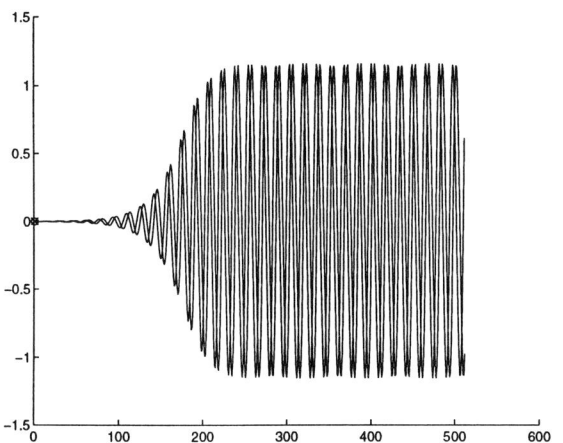

Figure 1: Time domain solution of the differential equation (9)

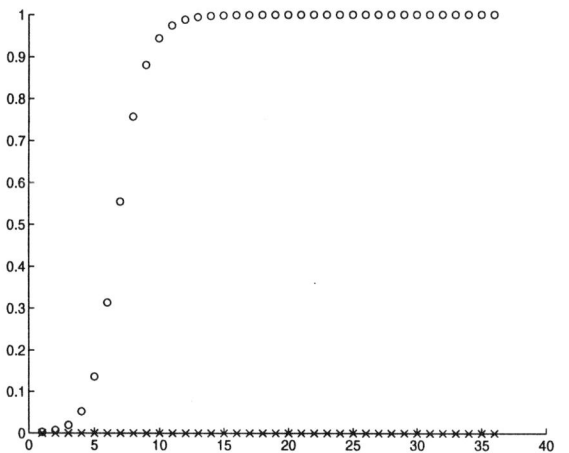

Figure 2: Real ('x') and Imaginary part ('o') of the fundamental frequency solution of x_1 for $c_0 = -0.1$

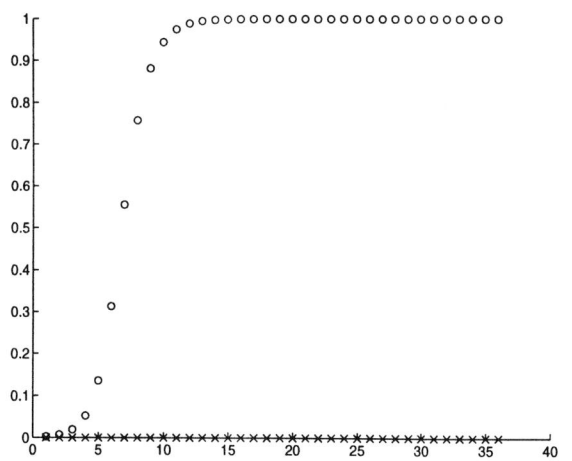

Figure 3: Real ('x') and Imaginary part ('o') of the fundamental frequency solution of x_1 for $c_0 = -0.01$

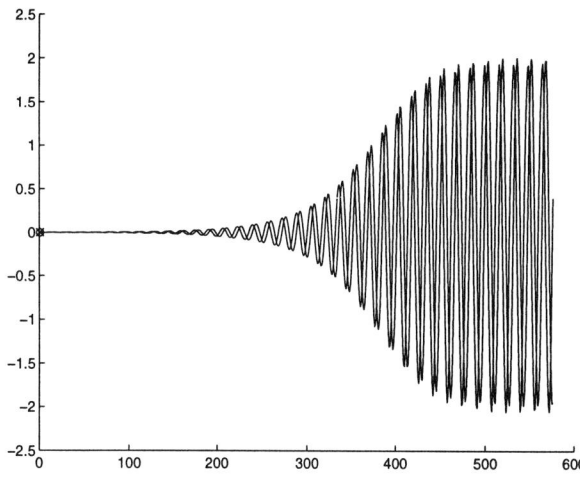

Figure 4: Time domain solution of the van der Pol equations (10)

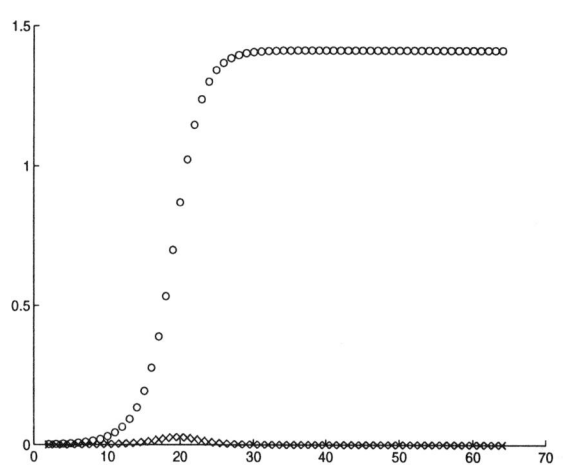

Figure 5: Real ('x') and Imaginary part ('o') of the fundamental frequency solution of x_1 for $c_0 = -0.1$ of the van der Pol equations (10)

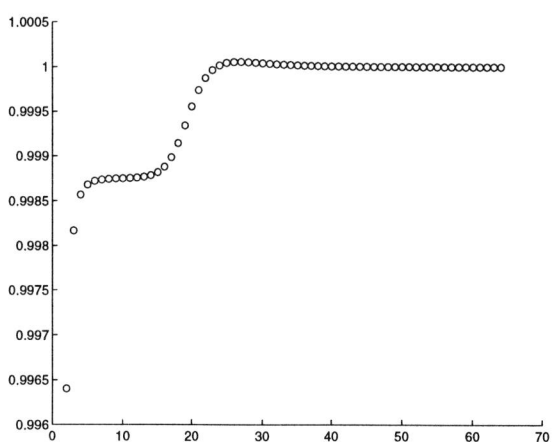

Figure 6: Shift of the fundamental frequency with time

ON ORIENTATION METRIC AND EUCLIDEAN STEINER TREE CONSTRUCTIONS

Y.Y. Li   K.S. Leung   C.K. Wong [†]

Department of Computer Science and Engineering
The Chinese University of Hong Kong
Shatin, N.T., Hong Kong
email: {yyli, ksleung, wongck}@cse.cuhk.edu.hk

ABSTRACT

We consider Steiner minimal trees (SMT) in the plane, where only orientations with angle $\frac{i\pi}{\sigma}$, $0 \leq i \leq \sigma - 1$ and σ an integer, are allowed. The orientations define a metric, called the orientation metric, λ_σ, in a natural way. In particular, λ_2 metric is the rectilinear metric and the Euclidean metric can be regarded as λ_∞ metric. In this paper, we provide a method to find an optimal λ_σ SMT for 3 or 4 points by analyzing the topology of λ_σ SMT's in great details. Utilizing these results and based on the idea of loop detection first proposed in [8], we further develop an $O(n^2)$ time heuristic for the general λ_σ SMT problem, including the Euclidean metric. Experiments performed on publicly available benchmark data for 12 different metrics, plus the Euclidean metric, demonstrate the efficiency of our algorithms and the quality of our results.

1. INTRODUCTION

Given a set P of n points in the plane, a tree interconnecting points of P and a set of arbitrary points Q is called *a Steiner tree* (ST). Points of P are called *demand points* and points of Q are called *Steiner points*. A Steiner tree with minimum length is called *a Steiner minimal tree* (SMT). The *Steiner tree problem* is to find a Steiner minimal tree for the n given demand points, which will be referred to as S_n. In the past, Steiner problems in the Euclidean and rectilinear metrics have attracted much attention due to their applications in industrial engineering problems and VLSI wiring problems, where wires in general are required to run horizontally and vertically [7-8], [10-12]. Now it is quite common for VLSI artwork to contain 45° and 135° line segments. The generalization from traditional problems to the problems involving objects with several fixed orientations and the metrics naturally induced by these fixed orientations can be found in [1-6].

Given a set Θ of at least two orientations (angles) in $\Re^2$, pairwise different, we represent the orientations by the angles with the x-axis of the corresponding straight lines. Sort the orientations by the angles in a counterclockwise direction and call them $\beta_1, \beta_2, \ldots, \beta_k$ with β_1 cyclically following β_k. See Fig. 1(a). The orientation distance between any two points p_1, p_2 in $\Re^2$ can be defined as follows [2]:

$\forall p_1, p_2 \in \Re^2$,

$$d_A(p_1, p_2) = \begin{cases} d_E(p_1, p_2) & \text{if } p_1 \text{ and } p_2 \text{ lie on a line,} \\ & \text{whose orientation is in } \Theta \\ \min_{p_3 \in \Re^2}\{d_A(p_1, p_3) + d_A(p_3, p_2)\} & \text{otherwise} \end{cases}$$

where d_E denotes the Euclidean distance. From [2] we know that d_A induces a metric, denoted by A_Θ in $\Re^2$ for any given Θ. Furthermore, $d_A(p_1, p_2)$ can be directly calculated as follows: Draw k lines in the orientation of $\beta_1, \beta_2, \ldots, \beta_k$ at p_1 and p_2 respectively. Let β_i and β_j be the orientation lines closest to the line p_1p_2. See Fig. 1(b). Then

$$d_A(p_1, p_2) = d_E(p_1, p_3) + d_E(p_3, p_2),$$

where p_3 is the intersection of the two orientation lines β_i and β_j. The parallelogram shown in Fig. 1(b) will be referred to as the *parallelogram of* $\{p_1, p_2\}$. In the special case when the σ orientations form angles $\frac{i\pi}{\sigma}$ with the x-axis, $0 \leq i \leq \sigma - 1$ and $\sigma \geq 2$ an integer, we call the metric λ_σ *metric*. In particular, $\sigma = 2, \infty$ correspond to the well-known rectilinear and Euclidean metrics respectively.

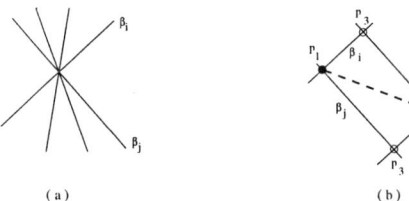

Figure 1: A_Θ metric

In this paper we consider the Steiner tree problem for the whole family of λ_σ metric, namely, for $2 \leq \sigma \leq \infty$. After establishing some fundamental properties for the orientation metric, we propose a heuristic to find near-optimal λ_σ solutions for S_n efficiently.

2. DEFINITIONS AND PRELIMINARY RESULTS

Given G, a solution of S_n, denote the Steiner point set of G by Q_G, and the degree of point q by $deg(q)$. For $p_1, p_2 \in \Re^2$, denote their λ_σ distance by $|p_1p_2|$. Define $\theta = \frac{\pi}{\sigma}$.

Lemma 1 *Given a λ_σ MST for P, for any $p_i \in P$ and any two neighbors p_j, p_k of p_i, we have, $\angle p_jp_ip_k \geq 2\arcsin(\frac{1}{2}\cos(\frac{\theta}{2}))$, where $1 \leq i, j, k \leq n$.*

---

[*] Research partially supported by the Strategic Research Program at the CUHK under Grant No. SPR 9505 and by two HK Government RGC Earmarked Grants, Ref. No. CUHK 333/96E and CUHK 352/96E.

[†] On leave from IBM T.J. Watson Research Center, Yorktown Heights, NY 10598, U.S.A.

Using Lemma 1, it is easy to obtain

Lemma 2 *Given a λ_σ minimum spanning tree (MST) for P, $1 \leq deg(p_i) \leq 8$ for any $p_i \in P$.*

If an ST T can be transformed to another ST T', so that they have the same total length, then they are called *equivalent ST's* and those transformations are called *equivalent transformations*. An equivalent transformation in λ_3 metric and one in λ_6 metric are shown in Fig. 2(a), (b) respectively.

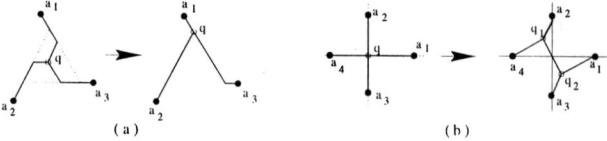

Figure 2: Equivalent transformations in λ_3 metric and λ_6 metric

Let p_1, p_2 be the two vertices of an edge. If point p_1 is on an orientation line of point p_2, we say $p_1 p_2$ *is a straight edge*. Otherwise, $p_1 p_2$ is called *a non-straight edge*. Define $A_q = \{a_1, a_2, \ldots, a_{deg(q)}\}$ as the set of points adjacent to q and $i(q)$ as the size of set $\{\hat{a} \in A_q, \hat{a}q \text{ is non-straight }\}$. For any $a_1, a_2 \in A_q$, let $\angle_{\lambda_\sigma} a_1 q a_2$ be the angle which is in the interval $[0, \pi)$ and is formed by the two nearest edges of the two parallelograms of $\{q, a_1\}$ and $\{q, a_2\}$ respectively. See Fig. 3(a). Define $\Gamma_q = \{\gamma | \gamma = \angle_{\lambda_\sigma} a_j q a_k, a_j, a_k \in A_q, a_j q \text{ is adjacent to } a_k q\}$. Note that $\sum_{\gamma \in \Gamma_q} \gamma \leq 2\pi$.

For S_n, we want to know how q is connected to its neighbors for any $q \in Q_G$. In particular, we hope to obtain the angle $\gamma_\sigma = \inf\{\angle_{\lambda_\sigma} a_1 q a_2\}$ for any $a_1, a_2 \in A_q$. In general, q may be connected to its neighbors with non-straight edges as shown in Fig. 3(a). To estimate γ_σ, we first consider a special case as described in Lemma 3 and shown in Fig. 3(b), where all points are connected by straight edges, i.e. m_1, m_2, n_1 and n_2 are integers.

Lemma 3 *Given a point p_1, if there exist points p_2, p_3, q and a structure as shown in Fig. 3(b) such that all of qp_1, p_1p_2, p_1p_3 are straight edges and $\angle p_2 p_1 p_3 = m\theta$, where m is an integer, then $|qp_1| + |qp_2| + |qp_3| < |p_1p_2| + |p_1p_3|$, iff $m \leq M(\sigma)$, where*

$$M(\sigma) = \begin{cases} 2t - 2, & \text{when } \sigma = 3t, \\ 2t - 1, & \text{when } \sigma = 3t + 1, \\ 2t, & \text{when } \sigma = 3t + 2. \end{cases}$$

and t is an integer.

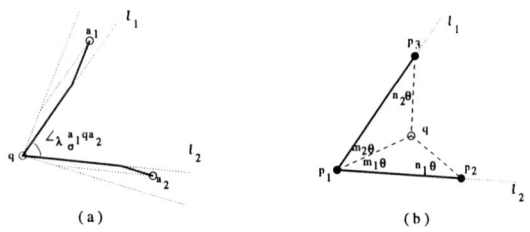

Figure 3:

From Lemma 3, we have,

Corollary 1 *In the cases when $\sigma = 3t + 1, 3t + 2$, if $\angle p_2 p_1 p_3 \geq (M(\sigma) + 1)\theta$ or in the case $\sigma = 3t$, if $\angle p_2 p_1 p_3 \geq (M(\sigma) + 2)\theta$, then $|qp_1| + |qp_2| + |qp_3| > |p_1 p_2| + |p_1 p_3|$. When $\sigma = 3t$, if $\angle p_2 p_1 p_3 = (M(\sigma) + 1)\theta$, then we have $|qp_1| + |qp_2| + |qp_3| = |p_1 p_2| + |p_1 p_3|$.*

Here we can prove the following more general result:

Theorem 1 *Given G a solution of S_n in λ_σ metric, for any $q \in Q_G$ and any $a_1, a_2 \in A_q$, we have $\angle_{\lambda_\sigma} a_1 q a_2 \geq (M(\sigma) + 1)\theta$, where*

$$M(\sigma) = \begin{cases} 2t - 2, & \text{when } \sigma = 3t, \\ 2t - 1, & \text{when } \sigma = 3t + 1, \\ 2t, & \text{when } \sigma = 3t + 2. \end{cases}$$

And we can prove $\gamma_\sigma = (M(\sigma) + 1)\theta$.

Corollary 2 *Given a λ_σ solution G, for any $q \in Q_G$, $3 \leq deg(q) \leq 4$.*

Corollary 3 *Given a λ_σ solution G ($\sigma \neq 3$), for any $q \in Q_G$ with $deg(q) = 4$, A_q forms a cross and $i(q) = 0$.*

Corollary 4 *There exists G, a λ_σ solution of S_n, for $\sigma \geq 5$, such that for any $q \in Q_G$, $deg(q) = 3$.*

3. λ_σ SMT FOR S_3 AND S_4

In this section we consider the construction of SMT's for S_3 and S_4, which will form the key parts of our overall algorithm. A problem must be solved here to obtain solutions for S_3 is: Given three demand points, $P = \{p_1, p_2, p_3\}$, and suppose an MST for P is formed by $p_1 p_2$ and $p_1 p_3$, what should the critical value of $\angle_{\lambda_\sigma} p_2 p_1 p_3$ be to ensure the existence of a Steiner point? Employing similar proof techniques as in Theorem 1, and directly utilizing some results there, we can prove:

Theorem 2 *If $\{p_1 p_2, p_1 p_3\}$ form an MST for P, then there exists a Steiner point q for P iff $\angle_{\lambda_\sigma} p_2 p_1 p_3 \leq M(\sigma)\theta$.*

We know that the lines passing through the n demand points along the σ orientations may produce new intersection points, which are called *the first generation points*. As we continue adding missing orientation lines through the first generation points, some new intersection points called *the second generation points* are obtained. We can obtain the ith generation points by repeating this procedure. Designate O_i as the set of the ith generation points. Using Theorem 1 and Corollary 1, we have the following result for S_n:

Theorem 3 *Given a λ_σ solution G for S_n, there exists an equivalent solution G', such that for any $q \in Q_{G'}$, there exist two points $a_1, a_2 \in A_q$, such that q must be an intersection point of two orientation lines of a_1 and a_2 respectively.*

Corollary 5 *For any solution G of $P = \{p_1, p_2, p_3\}$, there must be an equivalent solution G', such that either $Q_{G'} = \emptyset$, or for $q \in Q_{G'}$, $q \in O_1$.*

Corollary 6 *For any solution G of $P = \{p_1, p_2, p_3, p_4\}$, there must be an equivalent solution G', such that,*
  1. $Q_{G'} = \emptyset$, *or*
  2. $Q_{G'} = \{q\}$, *then $q \in O_1$, or*
  3. $Q_{G'} = \{q_1, q_2\}$, *then $q_1 \in O_1$ and $q_2 \in O_1 \bigcup O_2$.*

Based on Theorem 2, Corollary 5 and Corollary 6 we can construct SMT's for S_3 and S_4. And the complexity of the calculation does not increase as σ increases. Details are omitted due to page limitation.

4. ALGORITHM AND IMPLEMENTATION

1. Outline of the algorithm We propose a very efficient algorithm to generate ST's for the whole metric family, namely, from λ_2 to λ_∞. It is based on the loop detection heuristic introduced in [8]. The overall idea is: Starting from a λ_σ MST, we introduce good Steiner points into the demand point set incrementally and generate an MST for the enlarged demand point set.

We use an example in λ_∞ metric to illustrate the concept. For the MST shown in Fig. 4(a), we consider edges e_1 and e_4. The *Candidate Steiner points* (CSP's) q_1 and q_2 are found in Fig. 4(e) by adding edges $q_1p_1, q_1p_2, q_1q_2, q_2p_4, q_2p_5$ and three loops are formed after adding these edges. They are $q_1p_1p_2q_1, q_2p_4p_5q_2$ and $q_1p_2p_3p_4q_2q_1$. The λ_σ MST will be improved if we can remove one edge (not necessarily the longest one) in each loop and the total cost of removed edges is larger than the total length of the added edges. These removed edges are called *dependent edges* of the corresponding CSP's.

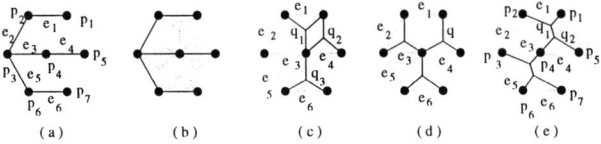

Figure 4: An Example in the Euclidean Metric

Basically we need to consider each pair of edges in the MST and find all possible refinements. There are two cases.

Case 1. Two edges, say, e_2, e_3, share a vertex in the MST as shown in Fig. 4(a). If there exists a Steiner point q for the three demand points, then e_2, e_3 are dependent edges of CSP q. The Steiner point q will be referred to as *a local CSP*.

Case 2. Edges e_1, e_4 do not share any vertex as shown in Fig. 4(a). If there exist Steiner point(s) for the four vertices, then at most three loops will be formed after adding the SMT of the four vertices to the MST. In one case, e_1, e_4 and one edge from the other loop, say, e_2, can be the dependent edges of the CSP's q_1, q_2 in Fig. 4(e) while in the other case only one of e_1, e_4 together with one edge from the other loop may be the dependent edges of the CSP, e.g., e_4 and e_2 can be the dependent edges of q in Fig. 4(d). The Steiner points q_1, q_2, q will be referred to as *global CSP's*.

A pseudo-code description of our algorithm is given here.

procedure λ_σ-ST-Refinement
1 *initialization*;
2 *generate MST for P*;
3 **while** (*found CSP(s) with positive gain*
    and *iteration* $\leq C_1$)
4   *collect all CSPs with positive gain into a set S*;
5   *select a maximal independent legal set S_c of S*;
6    $P = P \bigcup S_c$;
7   *generate MST for P*;
8   *delete unnecessary CSPs*;
9   *adjust each Steiner point by the Adjust operation*
       (*to be defined later*);

2. Implementation The algorithm begins with generating an MST for P using Prim's algorithm [13]. The algorithm repeats C_1 times to find a set of good Steiner points, where C_1 is a constant independent of σ. The iteration size can be fixed since most of the improvement over the MST is achieved within the first few iterations. In each iteration, after putting the CSP's with positive gain into a set S, *a maximal independent legal subset* is chosen as follows. After sorting the CSP's by gain and removing redundant CSP's (those having the same coordinates and dependent edges), we generate C_2 subsets of CSP's such that the dependent edges in each subset are different from each other. Here C_2 is also a constant. We consider each CSP in decreasing order of gain. It can be put into an existing subset, when the dependent edges of the CSP are different from those of any other CSP's in the subset. When a CSP can not be put into any existing subset, if there are less than C_2 subsets, we create a new subset for the CSP; otherwise the CSP is discarded. After this, we choose the subset with the largest total gain to test its legality. The CSP subset is *legal* if adding these CSP's and removing their dependent edges will still produce a tree. Fig. 4 illustrates the legality of a subset of CSP's by an example in the Euclidean metric, which is easier to understand than other metrics. An MST is given in (a). (b) shows all CSP's with positive gain. (c) is an independent subset of CSP's with maximal gain. The relations of CSP's with their dependent edges in (c) are $(q_1, (e_1, e_2)), (q_2, (e_4, e_3)), (q_3, (e_6, e_5))$. But the subset $\{q_1, q_2, q_3\}$ is not legal because it does not produce any tree. (d) shows another subset with the same total gain as that in (c) but is legal. (e) is produced by the maximal independent subset procedure proposed in [8] and is obviously not an optimal ST. After we generate an MST for the enlarged demand point set, those Steiner points with degree one or two are useless and removed.

It is easy to see that the first iteration of our algorithm only produces first and second generation Steiner points. For each Steiner point q with degree three we replace q with the solution of A_q. This process is called the Adjust operation and is proposed since most Steiner points have degree three in λ_σ metric, for $3 \leq \sigma \leq \infty$, according to Corollary 4. In our implementation, the Adjust operation in λ_∞ metric is replaced by the more efficient Association-Reconstruction procedure [12, 14], which is not directly applicable to λ_σ metric. Experimental results show that the Adjust operation and the Association-Reconstruction procedure are quite effective in accelerating convergency.

3. Efficiency of our algorithm We propose a heuristic, which will examine only a portion of all these refinements in order to speed up the algorithm, without sacrificing the quality of the solutions significantly.

Given two edges p_1p_2, p_3p_4, define the distance from the two edges to their CSP's as $x - |p_1p_2| - |p_3p_4|$, where x is the cost of an SMT for $\{p_1, p_2, p_3, p_4\}$. It is obvious that the CSP's should not have any positive gain if their distance from the two edges is larger than the cost of the longest edge in the MST. This is called the *trim condition*.

For any edge of the MST, e_i, we consider the CSP's of e_i with any other edge e_j as follows, where $1 \leq i, j \leq n - 1$ and $i \neq j$. The local CSP's of e_i and e_j are put in set S directly. For global CSP's satisfying the trim condition, we always keep the twelve nearest CSP's after computing the distances to e_i and e_j. Then the twelve CSP's are tested by loop detection and those CSP's with positive gain are put in set S. The time complexity is $O(n)$ for keeping twelve nearest CSP's as well as for loop detection. Therefore, given e_i, the number of chosen CSP's is $O(1)$ because the

COMPUTATIONAL COMPLEXITY ANALYSIS OF SET-BIN-PACKING PROBLEM

Tomonori Izumi, Toshihiko Yokomaru, Atsushi Takahashi, and Yoji Kajitani

Department of Electrical and Electronic Engineering
Tokyo Institute of Technology
Ookayama, Meguro, Tokyo, 152–8552 Japan
E-mail : chiron@ss.titech.ac.jp

ABSTRACT

Given a set of items and a set of bins of the same capacity, the *Set-Bin-Packing Problem (SBP)* is to pack all the items into the bins where each item is associated with a set and a bin can contain items as long as the number of distinct elements in the union of the sets does not exceed the capacity. One of applications is in FPGA technology mapping, which is our initial motivation. In FPGA terminology, an item, an element in an item, a bin, capacity correspond to a gate, an input signal of a gate, an LUT, and the number of input terminals of an LUT. In this paper, the computational complexity of SBP is studied with respect to three parameters; the number of input terminals of an LUT, the upper bound of the number of input signals of a gate, and the upper bound of the number of fanout gates of a signal, respectively. Our result reveals that SBP remains NP-hard for small values of these parameters. The results are summarized on a 3D map of computational complexities with respect to these three parameters.

1. INTRODUCTION

The *Packing problem* is to pack given items into given containers as efficiently as possible under various constraints. Since it is fundamental and significant with variations and applications, there have been many researches including computational complexity analysis and development of exact, approximate, and/or heuristic algorithms[8]. The

---

This work has been a part of a project in the Research Body of CAD21 at Tokyo Institute of Technology.

Bin-Packing Problem is one of such packing problems: Pack given a set of items each of which has its own size into as few bins which have the same capacity.

One of applications of Bin-Packing is found in VLSI circuit clustering (or partitioning, technology mapping) where an item and a bin correspond to a gate (or a cell, a module) and a cluster (or a block), respectively. Since the main concern has been in the area traditionally in the field of VLSI layout design, the size of an item and the capacity of bins correspond to the area of the gate and the area of clusters, respectively.

Recent increase of the density of circuit devices causes "pin-crisis", that is, the constraint of the number of terminals becomes more and more significant. It is because the possible number of terminals placed at the periphery of the layout area is approximately proportional to the square-root of the density. Moreover, the constraint is the most critical in a specific programmable device such as *Field Programmable Gate Allays (FPGAs)*.

Since one input signal which is common to more than one gates in a cluster occupies only one terminal of the cluster, the terminals needed may be less than the sum of the numbers of input signals of the gates. From this set-theoretic property of the terminals contrast to the algebraic one of the areas, the *Set-Bin-Packing (SBP)* is abstracted: Every item is associated with a set of elements and a bin can contain items as long as the number of distinct elements in the union of the sets does not exceed the capacity.

An straightforward application of SBP to the

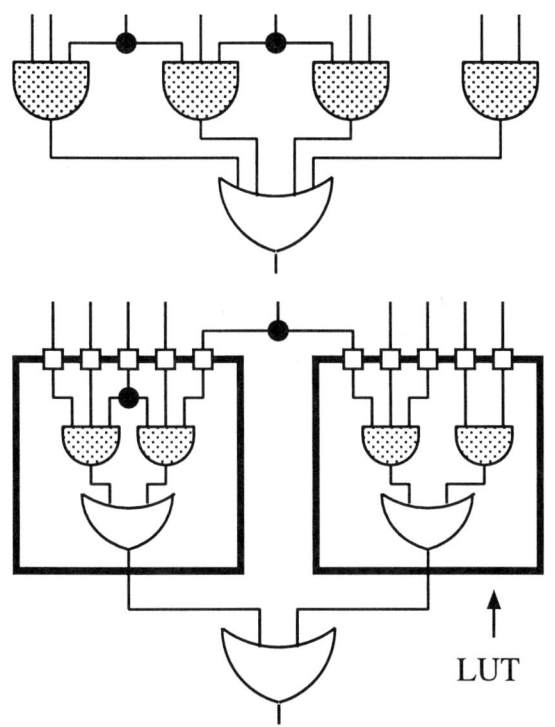

Figure 1: An example of mapping: Four gates are mapped into two 5-input LUTs.

practical problems is seen in the technology mapping of gates into *Look-Up Tables (LUTs)* of an FPGA [9][2]. An LUT has the fixed number α of input terminals and one output terminal, called the α-input LUT, and any logic circuit with α or less input signals and with an output signal is able to be implemented in an α-input LUT. For example, four AND-gates in Fig 1 (above) are packed into two 5-input LUTs as shown in Fig 1 (below). In the following, we discuss on SBP using the terms in FPGA technology mapping, such as 'gate', 'signal', and 'LUT', for the sake of practical image.

If there are no common signals, or if the advantage of common signals are ignored, SBP is reduced to *Integer-Bin-Packing (IBP)* [1, 3, 4, 5] where sizes of items are integers and they follow conventional algebra. Although IBP has been known to be NP-hard, it is not only known to be polynomial time solvable when the capacity of bins is fixed [6] but also was proved that a very simple algorithm *First Fit Decreasing (FFD)* outputs an exact solution if the capacity α is 6 or less[10], which is large enough in FPGA technology mapping.

Motivated by these circumstances, this paper analyzes the problem SBP from computational complexity. Let γ be the upper bound of the number of input signals of a gate and and δ be the upper bound of the number of fanout gates of a signal (gates which have the input signal). Although SBP is NP-hard in general, it is expected that SBP has polynomial time algorithms as well when the parameters α, γ and δ are small values.

In this paper, the computational complexity of SBP with respect to these parameters is discussed and we determine for almost all the cases if SBP is NP-hard or polynomial time solvable. As opposed to our expectation, SBP remains mostly hard even for the small values of the parameters.

The rest of this paper is organized as follows. After preliminaries in Section 2, our results are summarized on the 3-dimensional map of the computational complexities of SBP with the parameters α, γ, and δ in Section 3. Section 4 concludes the work.

2. PRELIMINARIES

Let $S = \{s_1, s_2, \ldots, s_{N_s}\}$ be a set of signals and $G = \{g_1, g_2, \ldots, g_{N_g}\}$ be a set of logic gates. A set of *input signals* of a gate g is denoted by input(g). The *size* of a gate g is defined as $|\text{input}(g)|$ and denoted simply by $|g|$. The set of gates which has an input signal s is referred to as the *fanout gates* of s and denoted by fanout(s). The *fanout* of a signal s is defined as $|\text{fanout}(s)|$ and denoted simply by $|s|$. Let $\Pi = \{\pi_1, \pi_2, \ldots, \pi_\beta\}$ be a partition of G into clusters π_i's, that is, $\pi_i \subseteq G$ for $1 \leq i \leq \beta$, $\pi_i \cap \pi_j = \emptyset$ for $i \neq j$, and $\bigcup_{1 \leq i \leq \beta} \pi_i = G$. The set of *input signals* of a cluster π is defined as input$(\pi) = \cup_{g \in \pi}$input(g). The *size* of a cluster π is defined as $|\text{input}(\pi)|$. An *i-cluster* is a cluster whose size is i or less. The number α of input terminals of LUTs is referred to as the *capacity* of LUTs. A cluster π must be an α-cluster to be mapped into an α-input LUT. Set-Bin-Packing is defined as follows.

Set-Bin-Packing (SBP)

Instance: A set S of signals, a set G of gates, the capacity α of LUTs, and the number β of LUTs.

Question: Is there any partition Π of G into β or less α-clusters?

SBP is known to be NP-complete in general[9]. We consider SBP under limitations as follows.

1. The capacity of LUTs is α.
2. The size of every gates is at most γ.
3. The fanout of every signal is at most δ.

These parameters are constants (not input values). SBP with respect to parameters α, γ, and δ is denoted by $SBP(\alpha, \gamma, \delta)$.

If there exists a pair of gates g and g' such that $\text{input}(g') \subseteq \text{input}(g)$, they can be mapped in the same LUT without increasing the number of LUTs. Therefore, we assume that there is no such pair of gates. If there exists a gate g such that $|g| > \alpha$, there is no partition into α-clusters. Therefore, we assume that there is no such gate, that is, $\gamma \le \alpha$.

3. 3D MAP OF COMPUTATIONAL COMPLEXITIES

We present the 3D map of computational complexities with respect to α, γ, and δ as shown in Fig. 2. The map consists of three planes corresponding to $\delta = 1$ (top), $\delta = 2$ (middle), and $\delta \ge 3$ (bottom). Each area specified by (α, γ, δ) is labeled 'P', 'NP-c', or '?' indicating that the computational complexity of $SBP(\alpha, \gamma, \delta)$ is polynomial time solvable, NP-complete, or unknown, respectively.

$SBP(\alpha, 1, \delta)$ is trivially solved for any α and δ. Thus, the areas $(\alpha, 1, \delta)$ is labeled 'P'.

The label of an area $SBP(\alpha, \gamma, \delta)$ such that $\gamma = \alpha$ is same as the label of the area $SBP(\alpha, \gamma - 1, \delta)$. This is by the following property: Let $I_{SBP} = (S, G, \beta)$ be an instance of $SBP(\alpha, \gamma, \delta)$ where $\gamma = \alpha$ and G' be a set of gates of size α in G; The answers for I_{SBP} and for $(S, G \setminus G', \beta - |G'|)$ are the same.

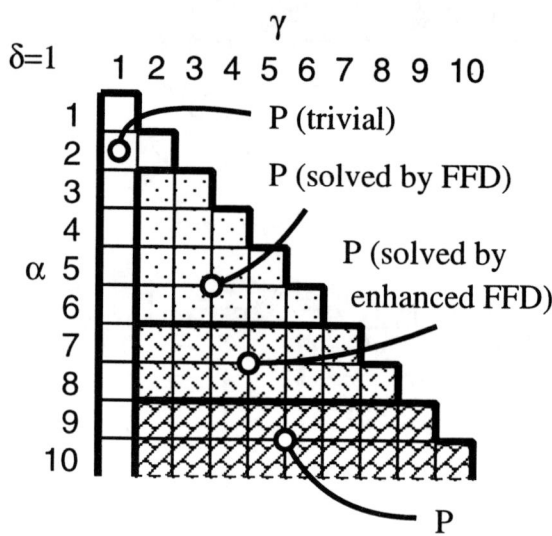

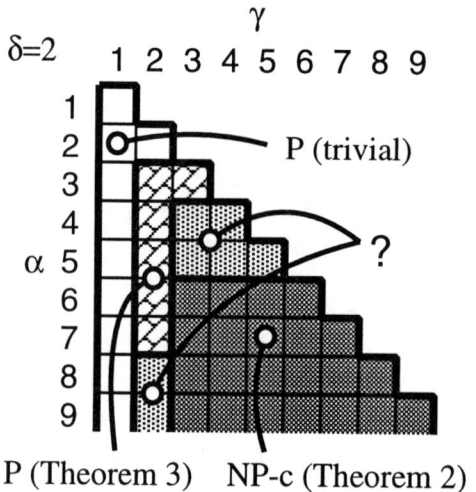

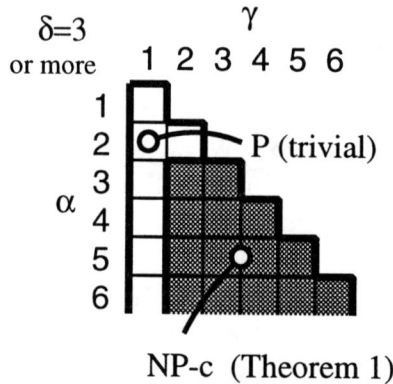

Figure 2: 3D map of the computational complexities of $SBP(\alpha, \gamma, \delta)$.

The case of $\delta = 1$ which is equivalent to IBP is well studied as follows.

Theorem [10] The FFD algorithm solves IBP in $O(|G|\log|G|)$ time for $\alpha \leq 6$.

Theorem [10] The enhanced FFD algorithm in [10] solves IBP in $O(|G|\log|G|)$ time for $\alpha \leq 8$.

Theorem [6] IBP is polynomial time solvable for any constant α.

By these theorems, the areas on the plane of $\delta = 1$ are labeled 'P'.

Our main results are the following theorems for the case of $\delta \geq 2$. The proofs are omitted here for the space(See [7]).

Theorem 1 $SBP(\alpha, \gamma, \delta)$ *is NP-complete for* $\alpha \geq 3$, $\gamma \geq 2$, *and* $\delta \geq 3$.

Theorem 2 $SBP(\alpha, \gamma, \delta)$ *is NP-complete for* $\alpha \geq 6$, $\gamma \geq 3$, *and* $\delta \geq 2$.

Theorem 3 $SBP(\alpha, \gamma, \delta)$ *is polynomial time solvable for* $\alpha \leq 7$, $\gamma \leq 2$ *and* $\delta \leq 2$.

4. CONCLUSION

We analyzed the computational complexity of the Set-Bin-Packing problem under limitations by capacity α of LUTs, upper bound γ of a gate size, and upper bound δ of a fanout of signal. Our main results are Theorems 1, 2, and 3 whose contributions are to fill almost the area for $\delta \geq 2$. However, the 3D map of computational complexities has not been completed remaining some areas still open. Among them, $SBP(4,3,3)$, $SBP(5,3,3)$, $SBP(5,4,3)$ and $SBP(\alpha,2,2)$ for $\alpha \geq 8$ are essential since $SBP(4,4,3)$ and $SBP(5,5,3)$ are reduced to $SBP(4,3,3)$ and $SBP(5,4,3)$, respectively. As opposed to our initial expectation that SBP is solvable for small parameters, it was revealed that SBP is mostly hard. The non-trivial solvable cases are only $SBP(\alpha,2,2)$ which may cover few practical cases.

5. REFERENCES

[1] S. D. Brown, R. J. Francis, J. Rose, and Z. Vranesic. *Field-programmable gate arrays*. Kluwer Academic Publishers, 1992.

[2] J. Cong and Y.-Y. Hwang. Structural gate decomposition for depth-optimal technology mapping in LUT-based FPGA design. In *Proc. 33rd Design Automation Conf.*, pages 726–729, 1996.

[3] R. J. Francis, J. Rose, and K. Chung. Chortle: A technology mapping program for lookup table-based field programmable gate arrays. In *Proc. 27th Design Automation Conf.*, pages 613–619, 1990.

[4] R. J. Francis, J. Rose, and Z. Vranesic. Chortle-crf: Fast technology mapping for lookup table-based FPGAs. In *Proc. 28th Design Automation Conf.*, pages 227–233, 1991.

[5] R. J. Francis, J. Rose, and Z. Vranesic. Technology mapping of lookup table-based FPGAs for performance. In *Proc. International Conf. on CAD*, pages 568–571, 1991.

[6] M. R. Garey and D. S. Johnson. *Computers and intractability : A guide to the theory of NP-completeness*. W. H. Freeman and Company, 1979.

[7] T. Izumi, T. Yokomaru, A. Takahashi, and Y. Kajitani. Computational complexity analysis of set-bin-packing problem. *IEICE Trans. on Fundamentals of Electronics, Communications and Computer Science*, May 1998. to appear.

[8] S. Martello and P. Toth. *Knapsack problems*. John Wiley & Sons Ltd., 1990.

[9] R. Murgai, R. K. Brayton, and A. S.-Vincentelli. Cube-packing and two-level minimization. In *Proc. International Conf. on CAD*, pages 115–122, 1993.

[10] T. Yokomaru, T. Izumi, A. Takahashi, and Y. Kajitani. Solution of integer bin packing problem with fixed capacity by FFD. Technical Report 95-DA-76-1, IPSJ, 1995. in Japanese.

A NEW APPROACH TO OVER-THE-CELL CHANNEL ROUTING*

Ting-Chi Wang[1], *Shui-An Wen*[2+], *D. F. Wong*[3], *and C. K. Wong*[4]

[1] Department of Information and Computer Engineering, Chung Yuan Christian University, Chungli, Taiwan, R.O.C.
[2] Computer & Communications Research Laboratories, Industrial Technology Research Institute, Hsinchu, Taiwan, R.O.C.
[3] Department of Computer Sciences, University of Texas at Austin, Austin, TX 78712, U.S.A.
[4] Department of Computer Science and Engineering, Chinese University of Hong Kong, Hong Kong

ABSTRACT

In this paper, we study the over-the-cell channel routing problem for the cell model in which all the pins are positioned along a horizontal line inside the corresponding cell. We present an efficient approach to the problem under the assumption that two metal layers are available for routing in each of the two over-the-cell regions and three metal layers are available for routing in the channel. The idea of our approach is to treat the three routing regions as a two-layer expanded channel, and to generate a two-layer channel routing solution first. The solution is then transformed into the final over-the-cell channel routing solution with the objective of minimizing the resulting width of the original channel. The transformation problem is reduced to the constrained two-processor scheduling problem for which we develop a polynomial time optimal algorithm. Our approach has been implemented in C language, and experimental results are also provided to support it.

1 INTRODUCTION

Channel routing is an importing step in VLSI physical design, and has been extensively studied in the past [1, 6, 17, 23-25]. Due to recent advances in fabrication technology, over-the-cell routing has become feasible for further reduction of routing area. A major consideration in the design of over-the-cell channel routing algorithms is the cell model. In the traditional type of cell models, all the pins are along the top and bottom cell boundaries. Algorithms in [3-4, 7-8, 11-13, 15-16, 18] were designed for this type of cell models. Recently, a new type of cell models where all the pins are inside the cell has been introduced. Algorithms for this type of cell models can be found in [9,14, 19-22].

In this paper, we consider the over-the-cell channel routing problem for the cell model in which all the pins are assumed to be along a horizontal line inside their corresponding cell. This cell model was first mentioned in [20]. Let T and B denote the top and bottom cell rows, respectively, and let C denote the horizontal channel between T and B. Let L_T and L_B denote the horizontal lines inside T and B, respectively. Let R_T denote the over-the-cell region between L_T and the bottom boundary of T, and let R_B denote the over-the-cell region between L_B and the top boundary of B. Only the regions inside R_T and R_B are considered for over-the-cell routing with respect to channel C. Without loss of generality, let $\{t_1, t_2, ..., t_m\}$ and $\{b_1, b_2, ..., b_n\}$ denote the sets of pins that are on L_T and L_B, respectively, and are to be connected

---

*This work was partially supported by the National Science Council of R.O.C. under grant NSC-84-2215-E-033-004.

+The author was previously with the Department of Information and Computer Engineering, Chung Yuan Christian University, Taiwan, R.O.C..

in R_T, R_B, and C. The above terms are illustrated in Figure 1, and will be used throughout the rest of the paper.

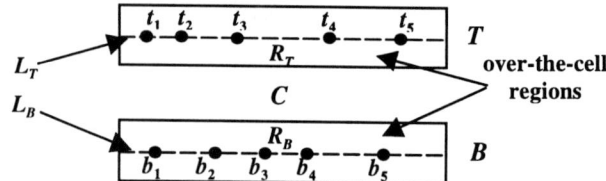

Figure 1: An illustration of the terms.

The over-the-cell channel routing problem for this cell model is to route the two sets of pins, i.e., $\{t_1, t_2, ..., t_m\}$ and $\{b_1, b_2, ..., b_n\}$, using the variable area to be generated in C, and the fixed area given in R_T and R_B in such a way that the resulting width in C is as small as possible. The authors in [20] considered the problem for the case where only one metal layer is available for routing in each of R_T and R_B, and gave a heuristic algorithm without details. Later, the authors in [21] presented an optimal algorithm that generates a planar routing in each of R_T and R_B while minimizing the resulting channel density. The authors in [20] also considered another case where two metal layers are available for routing in each of R_T and R_B but without the presence of vias between the two layers, and presented a heuristic method.

We study in this paper the over-the-cell channel routing problem for the case in which there are three metal layers available for routing in C, and there are two metal layers (i.e., the second and third layers) available for routing in each of R_T and R_B. Among the three metal layers, the first and the third layers, called *horizontal* layers, are reserved only for horizontal wire segments, and the second layer, called *vertical* layer, is reserved only for vertical wire segments. We also make the following four assumptions for the problem. Firstly, all the pins are located on the second layer. Secondly, a via can be used to connect two wire segments of the same net on two adjacent layers. (This means that vias are allowed in R_T, R_B, and C.) Thirdly, the two cell rows are of equal length. (If the two cell rows are of unequal length, we can properly extend their vertical boundaries outwardly to make them equal.) Finally, if the channel has any exit pin, the exit pin is allowed to be positioned along a vertical boundary of R_T, or R_B, or C. Under the above assumptions, we present an over-the-cell channel routing approach that consists of three steps. The first step is to treat the three routing regions (i.e., the channel and the two over-the-cell regions) as a two-layer expanded channel, and to apply any well-designed existing two-layer channel router to generate a routing solution that consists of a set of horizontal tracks. The second step is to solve *the track assignment problem* that assigns each track to a routing region as well as a corresponding horizontal layer such that without violating any constraint, the resulting width of C is as small is possible. The track assignment problem is reduced to *the constrained two-processor scheduling problem* for which we develop a polynomial time optimal algorithm. The last step of our approach is to complete the overall over-the-cell channel routing solution by connecting corresponding vertical wire segments on the second layer to the tracks with vias. Our approach has been implemented in C language, and experimental results are also provided to support it.

The rest of this paper is organized as follows. In Section 2, we define the constrained two-processor scheduling problem, and present a polynomial time algorithm to solve it optimally. In Section 3, our approach to the over-the-cell channel routing problem is described in detail. Finally, in Section 4, we present the experimental results and conclude this paper.

2 CONSTRAINED TWO-PROCESSOR SCHEDULING PROBLEM

In this section, we define the constrained two-processor scheduling problem, and present a polynomial time algorithm that solves it optimally. As shall be seen in Section 3, the track assignment problem considered in the second step of our over-the-cell channel routing approach can be reduced to the scheduling problem.

2.1 Problem Formulation

An instance of the constrained two-processor scheduling problem is specified by a tuple $<G, k_1, k_2>$, where $G=(V,E)$ is a DAG, called the *job precedence graph*, and k_1 and k_2 are two non-negative integers with $k_1 + k_2 \leq |V|$. Each vertex v_i in V represents a job that requires one unit of execution time on any processor. Each edge (v_i, v_j) in E specifies a *precedence constraint* that means v_i is a *predecessor* of v_j, v_j is a *successor* of v_i, and v_i must be completed before v_j starts. For each edge (v_i, v_j) in E, if there exists no vertex v_k such that both (v_i, v_k) and (v_k, v_j) are in E, then we say v_j is an *immediate successor* of v_i. For each vertex v_i in V, we use $Succ(v_i)$ to denote the set of all immediate successors of v_i. The graph G is said to be *transitively reduced* if for each edge (v_i, v_j) in E, v_j is an immediate successor of v_i.

We assume that once a processor begins to execute a job, the execution will be continued until the job is completed. When a processor completes a job, it becomes ready right away. A *schedule* for $<G, k_1, k_2>$ is defined as follows: For each job v_i, determine a pair $(ST(v_i), PI(v_i))$ which assigns job v_i to be executed at time $ST(v_i)$ on processor $PI(v_i)$. Suppose the execution starts at time 0. Then, the total execution time of a schedule is defined to be $total\_time = \max_{v_i \in V}\{ST(v_i)+1\}$. A schedule will partition the set V into three disjoint sets S_1, S_3 and S_2, where $S_1 = \{v_i \mid ST(v_i) \leq k_1 - 1, \text{ and } v_i \in V\}$, $S_3 = \{v_i \mid k_1 \leq ST(v_i) \leq total\_time - k_2 - 1, \text{ and } v_i \in V\}$, and $S_2 = \{v_i \mid total\_time - k_2 \leq ST(v_i) \leq total\_time - 1, \text{ and } v_i \in V\}$.

A schedule must also satisfy the following three conditions:

(1) There are exactly k_1 jobs in S_1, there are exactly k_2 jobs in S_2, and these jobs are executed under the single-processor environment. (We assume processor p_1 is the only processor assigned to execute these jobs, and hence $PI(v_i) = p_1$ for all $v_i \in S_1 \cup S_2$.)

(2) There are exactly $|V| - k_1 - k_2$ jobs in S_3, and they are executed under the two-processor environment. (That is, both processors p_1 and p_2 are available for executing these jobs.)

(3) None of the precedence constraints imposed on the $|V|$ jobs is violated in the schedule. That is, for each edge (v_i, v_j) in E, we have $ST(v_i) < ST(v_j)$.

Given an instance of the constrained two-processor scheduling problem, the objective is to find a schedule whose total execution time is minimum. Such a schedule is called an *optimal schedule*. When $k_1 = k_2 = 0$, the problem becomes the *traditional two-processor scheduling problem* [2, 10].

2.2 Review of Algorithm A [2]

Before presenting our algorithm for the constrained two-processor scheduling problem, we need to briefly review a polynomial time algorithm, called Algorithm A, which was proposed by Coffman and Graham, and solves the traditional two-processor scheduling problem optimally [2]. Algorithm A requires the precedence graph G to be transitively reduced beforehand. Once G is transitively reduced, Algorithm A first computes an integer label $\alpha(v_i)$ for each vertex v_i, where $1 \leq \alpha(v_i) \leq |V|$. The label for each vertex is defined as follows.

(a) An arbitrary vertex v_i with $Succ(v_i) = \phi$ is chosen, and $\alpha(v_i)$ is defined to be 1.

(b) Suppose for some $r \leq |V|$, the integers $1, 2, \ldots, r-1$ have been assigned to be the labels of $r-1$ vertices. For each vertex v_i whose immediate successors all have been assigned labels, let $N(v_i)$ denote the decreasing sequence of integers from the set $\{\alpha(v_j) \mid v_j \in Succ(v_i)\}$. (Note that $N(v_i)$ could be an empty set.) Among all such v_i's, choose the one whose $N(v_i)$ is the smallest with respect to the lexicographic numbering order, and define $\alpha(v_i)$ to be r.

(c) Repeat (b) until each vertex gets assigned a label.

See Figure 2(a) for an illustration of the above definition. Based on the labels, the set V is re-arranged as an ordered set $V' = \{v'_{|V|}, v'_{|V|-1}, \ldots, v'_1\}$, where V' is a permutation of V, and $\alpha(v'_i) = i, 1 \leq i \leq |V|$. Then, Algorithm A begins to schedule G using the following rule: At any time a processor is idle, it scans the set V' and begins to execute the job that has the largest label among all *ready* jobs. (A job is said to be ready at time t if all of its predecessors have been completed by time t.) If no ready job exists, the processor keeps idle for another unit of time. Algorithm A also makes the convention that if both processors p_1 and p_2 simultaneously attempt to execute the same job, then the job is always executed by p_1. Figure 2(b) shows the optimal schedule generated by Algorithm A for the DAG of Figure 2(a).

Based on the schedule generated by Algorithm A, a few important notations can be defined as follows.

Definition 1: The jobs U_i and W_i are recursively defined as follows:

(1) U_0 is defined to be the job executed by p_1 with $ST(U_0) = total\_time - 1$. W_0 is defined to be the (possibly empty) job executed by p_2 with $ST(W_0) = total\_time - 1$.

(2) For $i \geq 1$, W_i is defined to be the (possibly empty) job v for which $\alpha(v) < \alpha(U_{i-1}), ST(v) < ST(U_{i-1})$ and $ST(v)$ is maximal. For $i \geq 1$, U_i is defined to be the job executed by p_1 with $ST(U_i) = ST(W_i)$.

Definition 2: Suppose W_i can be defined for $0 \leq i \leq m$. Then, for $0 \leq i \leq m$, block $B_i \subseteq V$ is defined to be the set $\{v \mid ST(U_{i+1}) < ST(v) \leq ST(U_i), v \neq W_i, \text{ and } v \in V\}$ that is arranged as an ordered set according to the increasing order of each vertex's label.

See Figure 2(c) for an illustration of the above definitions. It is clear that the number of jobs in each block B_i is odd. According to the above definitions, the following property can be proven: For each i, $0 \leq i \leq m-1$, if $v \in B_i$ and $v' \in B_{i+1}$, then v' is the predecessor of v. This property is used to prove the optimality of Algorithm A. The time complexity of Algorithm A is $O(|V|^2)$, assuming the precedence graph G is transitively reduced. Please refer to [2] for the details of Algorithm A.

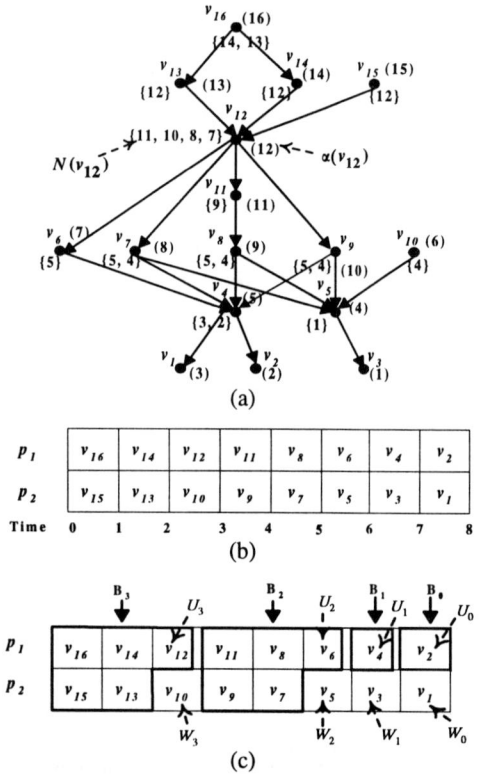

Figure 2: (a) A transitively reduced DAG and the associated labels. (b) The optimal schedule generated by Algorithm A. (c) Illustration of Definitions 1 and 2.

2.3 Algorithm CTPS

Now, we are ready to describe our algorithm, called the *Constrained Two-Processor Scheduling* (abbreviated as *CTPS*) algorithm, that optimally solves the constrained two-processor scheduling problem in polynomial time. To simplify the presentation, the input to Algorithm *CTPS* is assumed to be a tuple $<G, k_1, k_2>$, and the output is also a tuple $<S_1, S_3, S_2>$. The $S_1(S_2)$ denotes the set consisting of the first k_1 (the last k_2) jobs in the schedule that are executed under the single-processor environment (i.e., by p_1), and is arranged as an ordered set according to the increasing order of the time at which each job begins to be executed. The S_3 denotes the set consisting of the remaining $|V| - k_1 - k_2$ jobs that are executed by p_1 and p_2, and is arranged as an ordered set of pairs (v_i, v_j)'s according to the increasing order of the time at which each job begins to be executed. (Any two jobs v_i and v_j in S_3 that are executed at the same time forms a pair (v_i, v_j).) Without loss of generality, for each pair (v_i, v_j) in S_3, we assume jobs v_i and v_j are executed by p_1 and p_2, respectively. (Note that due to the precedence constraints, there may exist no v_j that is executed by p_2, and hence v_j is the empty job in this case.) Since S_1, S_3 and S_2 are all ordered sets, it is easy to determine the schedule based on them.

Algorithm *CTPS* is a greedy algorithm and its main idea is to gradually add jobs to S_1 and S_2 (without violating any precedence constraint) until S_1 has k_1 jobs and S_2 has k_2 jobs. The jobs to be added to S_1 and S_2 are obtained by repeatedly applying Algorithm A. Algorithm *CTPS* is outlined as follows. (Note that $INV(G)$ will be used throughout the rest of the paper to denote the graph obtained by reversing the direction of each edge in a given DAG G.)

Step 0: Initialize S_1, S_2 and S_3 to be empty sets.
Step 1: If G is not transitively reduced, then reduce it.
Step 2: While $k_1 > 0$, do the following.
    2.1 Run Algorithm A on G and get the blocks $B_m, B_{m-1}, ..., B_0$.
    2.2 Let $k = \min\{k_1, |B_m|\}$.
    2.3 Set L to be the set consisting of the first k jobs in B_m.
    2.4 Append L to S_1 from the end.
    2.5 Update G by removing all the vertices in L and their incident edges from G.
    2.6 Set $k_1 = k_1 - k$.
End while.
Step 3: While $k_2 > 0$, do the following.
    3.1 Run Algorithm A on $INV(G)$ and get the blocks $B_n, B_{n-1}, ..., B_0$.
    3.2 Let $k = \min\{k_2, |B_n|\}$.
    3.3 Set L to be the set consisting of the first k jobs in B_n.
    3.4 Append L to S_2 from the end.
    3.5 Update G by removing all the vertices in L and their incident edges from G.
    3.6 Set $k_2 = k_2 - k$.
End while.
Step 4: Run Algorithm A on G, and set S_3 to be the generated schedule.
Step 5: Reverse the jobs in S_2, and then return $<S_1, S_3, S_2>$.

Figure 3 illustrates how algorithm *CTPS* works when G is the DAG as shown in Figure 2(a), and $k_1 = k_2 = 2$.

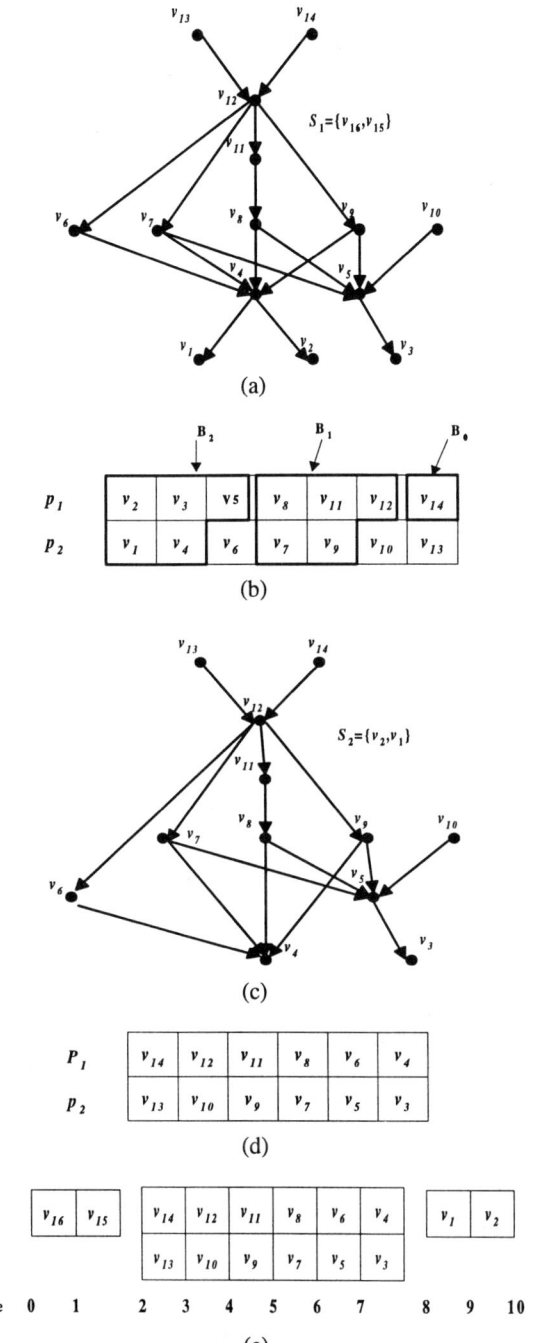

Figure 3: (a) The resulting S_1 and G after Step 2. (b) The optimal schedule of $INV(G)$ generated after Step 3.1. (c) The resulting S_2 and G after Step 3. (d) The resulting S_3 after Step 4. (e) The overall schedule.

2.3.1 Time Complexity of Algorithm *CTPS*

The time complexity of Algorithm *CTPS* can be analyzed as follows. Step 1 can be implemented in $O(|V|^3)$ time, where $|V|$ denotes the number of vertices in the job precedence graph [26,27]. (Note that the bound is not tight and can be improved.) Steps 2 and 3 each iterate at most k_1 and k_2 times, respectively, each iteration takes $O(|V|^2)$ time [2], and hence Steps 2 and 3 totally take $O((k_1 + k_2)|V|^2)$ which can written as $O(|V|^3)$ since $k_1 + k_2 \leq |V|$. Step 4 takes $O(|V|^2)$ time, and Step 5 takes $O(k_2)$ time. Therefore, the total time complexity of Algorithm *CTPS* is $O(|V|^3)$. In fact, Steps 2 and 3 can be modified such that they both can be implemented in $O(|V|^2)$ time. However, the overall time complexity of Algorithm *CTPS* is still dominated by Step 1.

2.3.2 Optimality of Algorithm *CTPS*

Since Algorithm *CTPS* is a greedy algorithm, if we can prove that it has the greedy choice (as stated in Lemmas 1 and 2) and the optimal substructure (as stated in Lemmas 3 and 4) properties, then we can easily prove its optimality [27]. Due to space limitation, all the proofs are omitted. In the following lemmas, we use $A(G)$ to denote the set of blocks with respect to the schedule generated by Algorithm A for a given DAG G.

Lemma 1: Given $<G, k_1, k_2>$, let $A(G)$ be $\{B_m, B_{m-1}, ..., B_0\}$, k be $\min\{k_1, |B_m|\}$, and L be the set consisting of the first k jobs in B_m. Then, there exists an optimal schedule $<S_1, S_3, S_2>$ such that if $v \in L$, then $v \in S_1$.

Lemma 2: Given $<G, k_1, k_2>$, let $A(INV(G))$ be $\{B_m, B_{m-1}, ..., B_0\}$, k be $\min\{k_2, |B_m|\}$, and L be the set consisting of the first k jobs in B_m. Then, there exists an optimal schedule $<S_1, S_3, S_2>$ such that if $v \in L$, then $v \in S_2$.

Lemma 3: Given $<G, k_1, k_2>$, let $A(G)$ be $\{B_m, B_{m-1}, ..., B_0\}$, k be $\min\{k_1, |B_m|\}$, and L be the set consisting of the first k jobs in B_m. Suppose $<S_1, S_3, S_2>$ is an optimal schedule for $<G, k_1, k_2>$ such that the set of the first k jobs in S_1 equals L. Then $<S_1 - L, S_3, S_2>$ is an optimal schedule for $<G', k_1 - k, k_2>$ where G' is the graph obtained by removing all the vertices in L and their incident edges from G.

Lemma 4: Given $<G, k_1, k_2>$, let $A(INV(G))$ be $\{B_m, B_{m-1}, ..., B_0\}$, k be $\min\{k_2, |B_m|\}$, and L be the set consisting of the first k jobs of B_m in the reverse order. Suppose $<S_1, S_3, S_2>$ is an optimal schedule for $<G, k_1, k_2>$ such that the set of the last k jobs in S_2 equals L. Then $<S_1, S_3, S_2 - L>$ is an optimal schedule for $<G', k_1, k_2 - k>$ where G' is the graph obtained by removing all the vertices in L and their incident edges from G.

Based on Lemmas 1-4, we can prove the following theorem.

Theorem 1: *Algorithm CTPS solves the constrained two-processor scheduling problem optimally.*

3 THE OVER-THE-CELL CHANNEL ROUTING APPROACH

In this section, we present our three-step approach, called the *Over-The-Cell Channel Routing* (abbreviated as *OTCCR*) algorithm, to the over-the-cell channel routing problem. The first step of Algorithm *OTCCR* is to treat the three routing regions (i.e.,

C, R_T and R_B) as a two-layer expanded channel, and to use any well-designed existing two-layer channel router to generate a two-layer channel routing solution. Let the set of tracks in the routing solution be $S = \{s_1, s_2, ..., s_w\}$, where w is the number of tracks, and track s_i is placed above track s_j if $i < j$. See Figure 4(a) for an example. As pointed out in [5], a DAG $G_S = (V_S, E_S)$, called the *track ordering graph* of S, can be constructed as follows. Each track s_i corresponds to a vertex in V_S, and there is a directed edge (s_i, s_j) in E_s if track s_i and track s_j each have a via at the same column and track s_i is placed above track s_j. Each edge (s_i, s_j) in E_s specifies a *via constraint* between tracks s_i and s_j. Let $S' = \{s'_1, s'_2, ..., s'_w\}$ be a permutation of S where track s'_i is placed above track s'_j if $i < j$. Then S' is said to be a *valid* permutation if there is no edge (s'_i, s'_j) in E_S for $i > j$. (That is, no via constraint is violated between any two tracks in S').

Given any two-layer channel routing solution denoted by a set of tracks, it can be transformed into a three-layer channel routing solution using the *track permutation* technique as proposed in [5]. This technique is to find a valid permutation of the tracks, say $S' = \{s'_1, s'_2, ..., s'_w\}$, and to put every two consecutive tracks (not in any pair yet) into a pair (if possible) starting from the first track such that the number of pairs obtained is minimum. For every two consecutive tracks, say s'_i and s'_{i+1}, to be paired, if they do not have vias at the same column, then they become a pair; otherwise, track s'_i is paired with an empty track (denoted by ϕ), and track s'_{i+1} and track s'_{i+2} become the next candidate to be paired. After the pairing is done, the tracks in each pair, called *a folded pair of tracks*, are assigned to the same track on the two horizontal layers, and the transformation is done. It has been proven that this transformation problem can be reduced to the traditional two-processor scheduling problem using the track ordering graph as the job precedence graph [5]. We will generalize this idea in the second step of Algorithm *OTCCR*.

Suppose the number of available tracks on the horizontal layer of R_T is k_1, the number of available tracks on the horizontal layer of R_B is k_2, and the number of tracks generated in the first step is w. Without loss of generality, we may assume $k_1 + k_2 \leq w$. The second step of Algorithm *OTCCR* is to solve the track assignment problem defined as follows. Given the set of tracks generated in the first step, the objective of the problem is to assign k_1 tracks to R_T, k_2 tracks to R_B, and the remaining ($w - k_1 - k_2$) tracks to C such that without violating any constraint, the resulting width of C obtained by the track permutation technique is as small is possible. The width of C is defined to be the number of folded pairs of tracks in C. An instance of the track assignment problem is denoted by a tuple $<S, k_1, k_2>$, and its output is denoted by another tuple $<T_1, T_3, T_2>$, where T_1, T_3 and T_2 denote the ordered sets of tracks (from top to bottom) assigned to R_T, C and R_B, respectively. Note that each element in T_1 or T_2 is a track, and each element in T_3 is a folded pair of tracks. Figure 4 (excluding the vertical wire segments) shows an instance of the track assignment problem, and two different solutions, assuming $k_1 = k_2 = 1$. Clearly, Figure 4(c) is an optimal solution.

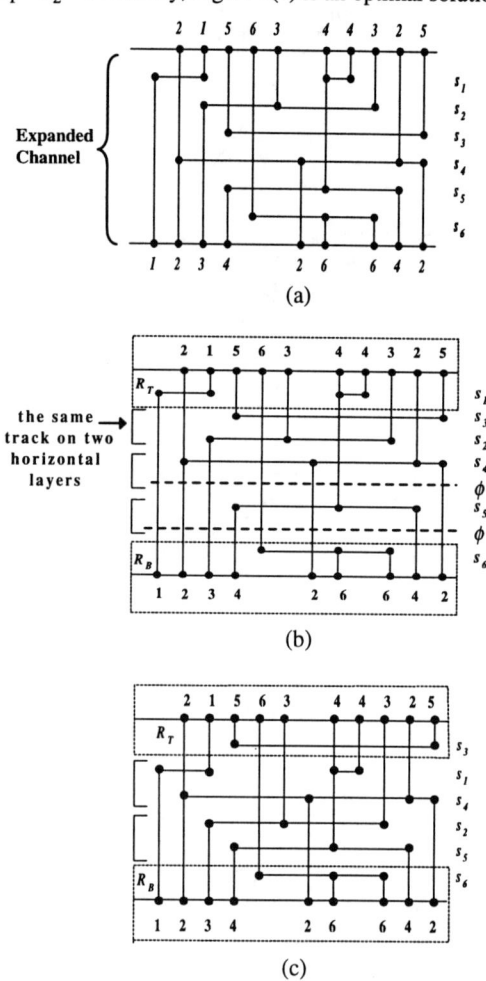

Figure 4: (a) A two-layer channel routing solution with 6 tracks. (b) An over-the-cell channel routing solution with 3 tracks in the channel. (c) An over-the-cell channel routing solution with 2 tracks in the channel.

The following lemma states that the track assignment problem can be reduced to the constrained two-processor scheduling problem.

Lemma 5: *Let $<S, k_1, k_2>$ be an instance of the track assignment problem, and G_S be the track ordering graph of S. Then $<T_1, T_3, T_2>$ is an optimal solution to $<S, k_1, k_2>$ if and only if $<T_1, T_3, T_2>$ is an optimal schedule for the instance $<G_S, k_1, k_2>$ of the constrained two-processor scheduling problem.*

Based on lemma 5, the track assignment problem is solved by constructing G_S first, and then running Algorithm *CTPS* for $<G_S, k_1, k_2>$. Therefore, we have the following theorem.

Theorem 2: *The track assignment problem can be optimally solved in $O(w^3)$ time, assuming the track ordering graph is given, and w is the number of tracks.*

After the second step of Algorithm *OTCCR* is done, the last step is to complete the overall over-the-cell channel routing solution by connecting corresponding vertical wire segment on the second layer to the tracks with vias. (See Figure 4 (c) for an example.)

Finally, we analyze the time complexity of Algorithm *OTCCR*. Let w be the number of tracks generated in the first step, and L be the number of columns in the channel. The time complexity of the first step depends on which two-layer channel router is used. The second step takes $O(w^2 L)$ time to construct the track ordering graph, and takes $O(w^3)$ time to solve the constrained two-processor scheduling problem. The last step takes $O(wL)$ time.

4 EXPERIMENTAL RESULTS AND CONCLUDING REMARKS

We have implemented Algorithm *OTCCR* in C language on a SPARC 10 workstation running Unix operating system. We directly started from the second step of Algorithm *OTCCR* by using several published two-layer channel routing solutions [23-25] as the inputs. For each test example, we assumed that each pin is located at a line inside its corresponding cell row. Several different values of k_1 and k_2 were tried for each test example, and the results are reported in Table 1. To the best of our knowledge, no previous work has been done for exactly the same problem that we have considered in this paper, and hence no comparisons can be made.

Table 1: Experimental results.

Test Example	# of tracks in two-layer channel	# of tracks in three-layer channel				
		$k_1=2$ $k_2=2$	$k_1=3$ $k_2=3$	$k_1=4$ $k_2=4$	$k_1=5$ $k_2=5$	$k_1=6$ $k_2=6$
yk3a[23]	15	6	5	4	3	2
yk3b[23]	17	9	7	6	5	4
yk3c[23]	18	8	7	6	5	4
deutsch1[24]	19	12	10	9	7	6
deutsch2[23]	28	22	20	19	17	15
deutsch3[23]	20	11	9	9	7	6
deutsch4[25]	19	12	10	9	8	6

It is known that the traditional two-processor scheduling problem can be optimally solved in linear time [10,28], and hence our future work is to investigate whether the constrained two-processor scheduling problem can be also optimally solved in linear time.

REFERENCES

[1] M. Burstein and R. Pelavin, "Hierarchical Channel Router", *Integration, the VLSI Journal*, vol. 1, pp. 21-38, 1983.

[2] E. G. Coffman and R. L. Graham, "Optimal Scheduling for Two-Processor Systems", *Acta Information*, pp. 200-213, 1972.

[3] J. Cong and C. L. Liu, "Over-the-Cell Channel Router", *IEEE Trans. on Computer-Aided Design of Integrated Circuits and Systems*, vol. 9, no. 4, pp. 408-418, 1990.

[4] J. Cong, B. Preas and C. L. Liu, "General Models and Algorithms for Over-the-Cell Routing in Standard Cell Design", in *Proc. Design Automation Conf.*, pp.709-715, 1990.

[5] J. Cong, D. F. Wong and C. L. Liu, "New Approach to Three- or Four-Layer Channel Routing", *IEEE Trans. on Computer-Aided Design of Integrated Circuits and Systems*, pp. 1094-1104, 1988.

[6] D. N. Deutsch, "A Dogleg Channel Router", in *Proc. Design Automation Conf.*, pp. 425-433, 1976.

[7] D. N. Deutsch, and P. Glick, "An Over-the-Cell Router", in *Proc. Design Automation Conf.*, pp.32-39, 1980.

[8] S. Danda, X. Liu, S. Madhwapathy, A. Panyam, N. Sherwani and I. G. Tollis, "Optimal Algorithms for Planar Over-the-Cell Routing Problems", *IEEE Trans. on Computer-Aided Design of Integrated Circuits and Systems*, vol. 15, no. 11, pp. 1365-1378, 1996.

[9] T. Fujii, Y. Mima, T. Matsuda and T. Yoshimura, "A Multi-Layer Channel Router with New Style of Over-the-Cell Routing", in *Proc. Design Automation Conf.*, pp.585-588, 1992.

[10] H. N. Gabow, "An Almost-Linear Algorithm for Two-Processor Scheduling", *Journal of ACM*, vol. 29, no. 3, pp. 766-780, July 1982.

[11] N. D. Holmes, N. A. Sherwani and M. Sarrafzadeh, "New Algorithms for Over-the-Cell Channel Routing Using Vacant Terminals", in *Proc. Design Automation Conf.*, pp. 126-131, 1991.

[12] N. D. Holmes, N. A. Sherwani and M. Sarrafzadeh, "Algorithms for Three-Layer Over-the-Cell Channel Routing", in *Proc. International Conf. on Computer-Aided Design*, pp. 428-431, 1991.

[13] H. E. Krohn, "An Over-the-Cell Gate Array Channel Router", in *Proc. Design Automation Conf.*, pp. 665-670, 1983.

[14] J. Kim and S.-M. Kang, "A New Triple-Layer OTC Channel Router", *IEEE Trans. on Computer-Aided Design of Integrated Circuits and Systems*, vol. 15, no. 9, pp. 1059-1070, 1996.

[15] M. S. Lin, H. W. Perng, C. Y. Hwang and Y. L. Lin, "Channel Density Reduction by Routing over the Cells", in *Proc. Design Automation Conf.*, pp. 120-125, 1991.

[16] S. Natarajan, N. A. Sherwani, N. D. Holmes and M. Sarrafzadeh, "Over-the-Cell Channel Routing for High Performance Circuits", in *Proc. Design Automation Conf.*, pp.600-603, 1992.

[17] R. L. Rivest and C. M. Fiduccia, "A 'Greedy' Channel Router", in *Proc. Design Automation Conf.*, pp.418-424, 1982.

[18] Y. Shiraishi and Y. Sakemi, "A Permeation Router", *IEEE Trans. on Computer-Aided Design of Integrated Circuits and Systems*, vol. CAD-6, pp. 462-471, 1987.

[19] M. Terai, K. Takahashi, K. Nakajima and K. Sato, "A New Model for Over-the-Cell Channel Routing with Three Layers", in *Proc. International Conf. on Computer-Aided Design*, pp. 432-435, 1991.

[20] B. Wu, N. A. Sherwan, N. D. Holmes and M. Sarrafzadeh, "Over-the-Cell Routers for New Cell Model", in *Proc. Design Automation Conf.*, pp. 604-607, 1992.

[21] T.-C. Wang, D. F. Wong, Y. Sun and C. K. Wong, "On Over-the-Cell Channel Routing", in *Proc. of EURO-DAC*, pp. 110-115, 1993.

[22] T.-C. Wang, D. F. Wong and C. K. Wong, "A New Channel Pin Assignment Algorithm and Its Application to Over-the-Cell Routing," in *Proc. IEEE International Symposium on Circuits and Systems*, pp. 1560-1563, 1997.

[23] T. Yoshimura and E. S. Kuh, "Efficient Algorithms for Channel Routing", *IEEE Trans. on Computer-Aided Design of Integrated Circuits and Systems*, vol. CAD-1, pp. 25-35, 1982.

[24] M. Burstin and R. Pelavin, "Hierarchical Wire Routing". *IEEE Trans. on Computer-Aided Design of Integrated Circuits and Systems*, vol. CAD-2, no. 4, pp. 223-234, Oct. 1983.

[25] D. N. Deutsch, "Compacted Channel Routing", in *Proc. International Conf. on Computer-Aided Design*, pp. 223-225, 1985.

[26] A. V. Aho, M.R. Garey and J. D. Ullman, "The Transitive Reduction of a Directed Graph," *SIAM J. Comput.*, pp. 131-137, 1972.

[27] T. H. Cormen, C. E. Leiserson and R. L. Rivest, *Introduction to Algorithms*, MIT press, 1990.

[28] H. N. Gabow and R. E. Tarjan, "A Linear Time Algorithm for a Special Case of Disjoint Set Union," *J. Comput. Syst. Sci.*, vol. 30, pp. 209-221, 1985.

A HEURISTIC ALGORITHM TO SOLVE CONSTRAINED VIA MINIMIZATION FOR THREE–LAYER ROUTING PROBLEMS

Kazuhiro Takahashi and Toshimasa Watanabe

Department of Circuits and Systems, Faculty of Engineering, Hiroshima University
4-1, Kagamiyama 1-chome, Higashi-Hiroshima, 739 Japan
E-mail: watanabe@infonets.hiroshima-u.ac.jp

ABSTRACT

The constrained via minimization problem is the problem of minimizing the number of vias by changing the layer assignment of nets whose routing are given. It has already been known that the problem is NP-complete even for three layer routing. The subject of the paper is to propose a heuristic algorithm VMBF, based on the breadth-first search, to solve the constrained via minimization for the three layer routing problem. Experimental results are provided to show capability of the proposed algorithm.

1. INTRODUCTION AND MOTIVATION

Multi-layer routing is inevitable in the current stage of VLSI and PCB design. It requires minimizing the number of vias in order to avoid higher production cost, too large routing area and deterioration of reliability of resulting boards. The via minimization problem is a problem of constructing layouts in which the number of vias is minimized. It has two kinds of subproblems: the unconstrained via minimization problem (UVM) [10] and the constrained via minimization one (CVM) [3, 4, 5, 6, 7]. The former incorporates via minimization into routing, while the latter tries to minimize the number of vias by changing the layer assignment of nets under the assumption that routing of these nets is given. Here we consider only routing in horizontal and/or vertical directions. Under this restriction every connecting path (or a wire) between any pair of terminals of each net is partitioned into one or more horizontal/vertical straight portions (each called a *horizontal/vertical segment*). Each point at which a horizontal segment and a vertical one are in contact with each other is called a *junction*. The *junction degree* (of a junction) is the number of segments containing this junction.

Algorithms of [4, 5, 6] accept any layout as their input. However they have a critical defect such that their assignment may leave some wire segments passing through vias in the resulting layout. Although this should obviously be undesirable, none of them have any mechanism to avoid it.

In this paper, we propose a new heuristic algorithm VMBF(Via Minimization by Breadth First search) for 3CVM: it is based on the breadth-first search, does not have any restriction on layouts given as input, and prevents any wire segment from passing through vias. In addition to decreasing the number of vias, also required is reducing crosstalk in order to prevent the deterioration of the reliability and in order to enhance the performance of the resulting boards. The term "crosstalk" means amplification of signal delay and noise caused by increase in coupling capacitance among wiring resources. VMBF can reduce not only the number of vias as mentioned above but also crosstalk. Experimental results provided in the paper show that it has high capability enough to be used in practical systems for printed wiring board design.

In [4], heuristic algorithms for a CVM are classified into two categories: global sense algorithms and local sense ones. VMBF also consists of these two kinds of algorithms. All heuristic algorithms proposed in [4, 5, 6, 7] restrict vias to be placed at junctions. [7] proved that 3CVM is NP-hard even under this restriction. We consider the same constraint: there is an exceptional case, which will be mentioned later.

2. GRAPH MODELS

Each of heuristic algorithms in [3, 4, 5, 6] constructs a graph model from a given layout(Figure 1), and tries to minimize the number of vias by utilizing this graph. The following two kinds of graphs are used.

1. *Via − crossing graph* (VCG) [3].

2. *Segment − crossing graph* (SCG)[4](Figure 2).

    This is constructed as follows. First, by considering the given layout to be layerless, each segment in a layout is represented as an individual vertex. Two vertices are connected by an edge, called a *cross edge*, if and only if the corresponding two segments are crossing; two vertices are connected by another edge, called a *via edge*, if and only if the corresponding two segments share a via (that is, every junction is assumed to have a via). Layers assigned to any pair of vertices connected by a cross edges should be *different*, and it is desirable to assign every pair of vertices connected by a via edge to the *same* layer. Figure 2 shows the SCG constructed from the layout of Figure 1. In this paper, we slightly modify SCG so that it may represent the constraints mentioned above.

3. CONSTRAINED VIA MINIMIZATION FOR THREE-LAYER ROUTING BASED ON BREADTH-FIRST SEARCH

VMBF has two main characteristics. The one is to consider an adjacent via, and the other is to utilize breadth-first search in order to reduce the number of vias.

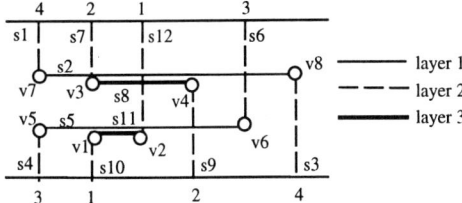

Figure 1: An example of a layout

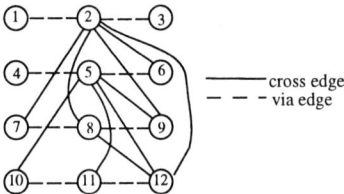

Figure 2: The SCG [4] constructed from Figure 1.

3.1. Adjacent Vias and a Modified SCG

Figure 3 shows examples of situations in layouts of three-layer routing problems. In Figure 3 (a), the segment 3 of the net b passes through a via that has already been used by another net a. Suppose that the segments 1, 2 and 3 are assigned to the layer 1, to the layer 3 and to the layer 2, respectively. Then the segment 3 is in contact with this via. We call such a via an *adjacent via*. Obviously existence of any adjacent via has to be avoided. However no method proposed in [4, 5, 6] has any mechanism to prevent appearance of adjacent vias. Our method restricts any pair of segments, sharing an adjacent via, to be assigned to either the same layer or two successive layers. For example, in Figure 3 (a), the segment 1 cannot be assigned to the layer 3 if the via is an adjacent one and the segment 2 is assigned to the layer 1.

Although vias are supposed to lie at junctions in this paper, there is an exceptional case: in Figure 3 (a), it may happen that we can move the via from the junction as in Figure 3 (b). We do so as long as it is possible. But this is not always the case: in Figure 3 (c), we can not move the via from the junction because the segment 1 of the net a is crossing over the segment 4 of the net c. So we consider a via as an adjacent one only if it cannot be moved elsewhere as in Figure 3 (c).

We impose the following restriction on terminals and segments, similarly to adjacent vias: if some segment x of a net a passes through any terminal t to which a segment y of another net b is incident as shown in Figure 4, then t and b have to be assigned to either the same layer or the two successive one.

We slightly modify an SCG so that it may represent adjacent vias. We simply add edges, called *adjacent via edges*, to an original SCG: two vertices are connected by an adjacent via edge if and only if both of the corresponding segments are incident to an adjacent via. The resulting SCG is called a *modified SCG*(MSCG: Figure 5).

3.2. Breadth-First Search

We utilize the breadth-first search (BFS), which is applied to a subgraph of a modified SCG.

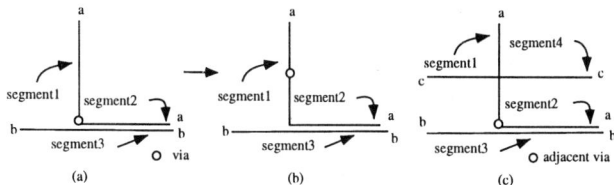

Figure 3: An example requiring an adjacent via.

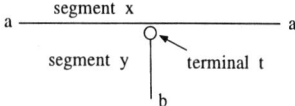

Figure 4: A terminal t passed through by a segment.

In this paper, we modify the way of visiting vertices of each level as follows: all vertices visited from *one* vertex of the current level are of the next level, as shown in Figure 6 (b). Clearly, labeling depends upon the choice of vertices of the current level. We call the vertex from which all vertices of the next level are visited the *source* of the next level. For example, the vertex b is the source of the level 2 in Figure 6 (b).

3.3. The Algorithm VMBF

3.3.1. The outline

We only state the outline of the proposed algorithm VMBF. A *trivial wire segment*, introduced in [7], is a segment which can always be assigned to a layer that is different from those to which segments crossing it are assigned. (Hence no via is created.)

Algorithm VMBF;

step 1: Pre-processing.

step 2: Construct a modified SCG and then delete all trivial wire segments.

step 3: Repeat from step 4 to step 6 until we get a layout satisfying the condition for termination.

step 4: Execute the proposed global sense algorithm.

step 5: Post-processing (assignment of layers to trivial wire segments).

step 6: Execute the local sense algorithm proposed by [6].

We briefly outline only step 4.

3.3.2. The condition for termination
(Omitted)

3.3.3. The proposed method

This is a global sense algorithm and is composed of three procedures, *Select*, *Coloring* and *Backtrack*. *Select* chooses a vertex from the nontrivial MSCG as a candidate for next assignment of a layer. *Coloring* assigns a layer to the vertex selected by *Select*. If *Coloring* fails to assign it a feasible layer, we try to find a feasible assignment by means of *Backtrack*. These operations will be repeated until every vertex of the nontrivial MSCG is assigned a feasible layer. We briefly explain each of the three procedures.

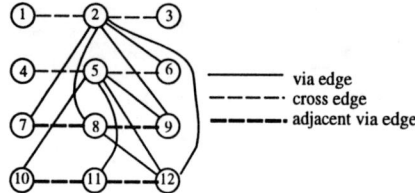

Figure 5: The modified SCG constructed from Figure 2.

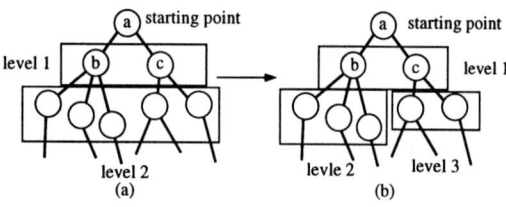

Figure 6: Levels by BFS.

1. **Procedure** *Select*. The procedure repeats selecting one vertex for assignment of a layer from the set of vertices of the same level given by BFS, according to the priorities from the highest (a) to the lowest (f). The details are omitted. If all vertices of the current level are assigned to layers, then vertices of the next level will be the candidates.

2. **Procedure** *Coloring*. The procedure assigns a layer to the vertex u selected by *Select*. We denote the number of layers that cannot be assigned to the vertex u by $x(u)$. The operation depends on the value $x(u)$. Here, we use the value $|r_\tau(u)|$ defined in [4] and the value $|V_\tau(u)|$ introduced in this paper. We have

   $|r_\tau(u)|$: the number of vias that can be removed by assigning the layer τ to u.

   Let P be a path of length 2 consisting of two edges $e_1 = (u,v)$ and $e_2 = (v,w)$. We call P a *via path* from u if P satisfies the following two conditions:

   (a) the inner vertex v is not yet assigned to any layer;

   (b) one of $\{e_1, e_2\}$ is a cross edge and the other is a via edge (or an adjacent via edge).

   Figure 7 shows a situation in the coloring step, where characters besides vertices are the segment numbers and the numbers written inside vertices are the layer numbers assigned to these vertices. Let

   $V_\tau(u) = \{w|$ there is a via path from u to w w and is assigned to the layer τ $\}$.

   Then we have

   $|V_\tau(u)|$: the number of vias whose removal will become impossible if the layer τ is assigned to u (Figure 7).

   For example, the $|V_2(u)| = 2$ in Figure 7. (Coloring operation)

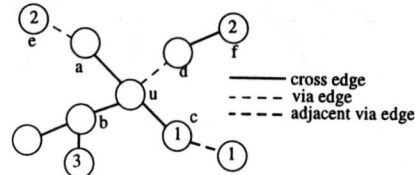

Figure 7: A situation in *Coloring*. The characters show the segment numbers; the figures do the assigned layer number.

   (a) $x(u) \in \{0, 1\}$: The vertex u is to be assigned to a layer τ having the maximum value of $|r_\tau(u)|$. If there exist at least two such layers then we select a layer with the minimum value of $|V_\tau(u)|$. If we have at least two such layers then, even at this stage, we choose layer 2 if it is available; otherwise we choose layer 1 if it is available; otherwise layer 3 will be selected.

   (b) $x(u) = 2$: There is only one layer left to be assigned to u, and this assignment will be done.

   (c) $x(u) = 3$: *Coloring* fails to assign any layer to u and, therefore, we proceed to *Backtrack*, in which we try to assign another layer to u.

3. **Procedure** *Backtrack*. This operation tries to assign a layer to any vertex u to which *Coloring* failed to get feasible assignment, by changing assignment that has already been obtained so far, where we have $x(u) = 3$. The method proposed in [4] is improved and is utilized in this paper: all vertices that failed to get feasible assignment are considered as candidates for reassignment in this paper. The details are omitted.

3.3.4. Post-processing

(Omitted)

3.3.5. Improvement through a local sense algorithm

(Omitted)

4. MINIMIZING CROSSTALK

Crosstalk depends on signal transmission time as well as coupling capacitance. Here, by the term "reducing crosstalk", we mean minimizing only coupling capacitance: the definition of [8] is adopted. We construct a graph, called the *crosstalk graph* (CTG), from a given layout, and we utilize it during execution of *Select*, *Coloring* and *Backtrack* of the previous section. Because of space limitation, we omit the details.

5. EXPERIMENTAL RESULTS

The four methods mentioned in the following have been implemented on a personal computer GATEWAY2000 (CPU: Pentium/120MHz, OS: FreeBSD 2.1) with the C programming code. We generated randomly 100 input data with the number of nets 50, 100, 200, 300, 400 and 500, respectively, for three layer routing problems. The Bruell's method [11] is used to obtain initial layouts for each of the proposed methods. We apply the following four methods to these layouts and compare the results.

procedure ab: the proposed method based on breadth-first search considering adjacent vias;

procedure abc_total: combination of procedure ab and Method 1 (reducing the total CT value);

procedure abc_net: combination of procedure ab and Method 2 (reducing the maximum CT value among those of nets);

procedure HVH: the Ahn's method [7].

The first three methods are proposed in this paper. Although procedure HVH is exclusive to HVH_CVM, we compare results by our methods with those by procedure HVH, because all input data in our experiment are HVH instances. Unfortunately we were unable to compare the results of this paper with those by the heuristic algorithms of [4, 5, 6], because almost all of layouts given by them have adjacent vias. The four values used in comparison are explained.

Via rate: the percentage given by

$$(first\_via - output\_via)/(first\_via) \times 100,$$

where *first_via* is the number of vias existing in a given initial layout and *output_via* is the number of vias appeared in a layout output by each of these methods. The greater percentage, the better.

Total CT: the total CT value (in the number of grids) appeared in a layout output by each method. The smaller value, the better.

Net CT: the maximum CT value (in the number of grids) among those of all nets in a layout output by each method. The smaller value, the better.

CPU: the computation time in second.

We attach "♯" at the beginning of each value, such as ♯ Via rate, to denote the total number of data for which each method produced the best value among the four methods. Hence the greater value, the better. We have already obtained the results for 100 input data, each of which has 50, 100, 200, 300, 400 and 500 nets, respectively. Hence we only show results for 500 nets: other cases have the similar tendency. Table 1 shows comparison of the number of cases with the best output, while Table 2 does the average over 100 input data. In every table, "*" denotes the best among the four candidates. As for *Via rate*, it is observed that **procedure ab** gives the best results and that **procedure abc_net** shows almost the same results as **procedure ab**. As for *Net CT* and *Total CT*, **procedure abc_net** produces better results than **procedure ab**, even though these results are much worse than those by **procedure HVH**. These results shows that **procedure abc_net** has the best capability among the four methods in the following points.

- Both the number of vias and crosstalk appeared are (almost) smallest.
- Leaving no adjacent vias in all layouts produced in this experiment.

6. REFERENCES

[1] R.Y.Pinter, "Optimal Layer Assignment for Interconnect", *Journal of VLSI and Computer Systems*, Vol. 1, No.2, pp. 123-137, 1984.

Table 1: Comparison of the number of cases with the best output when the number of nets is 500

	ab	abc_total	abc_net	HVH
♯ Via rate	*55	12	45	4
♯ Net CT	0	0	0	*100
♯ Total CT	2	*56	10	32
♯ CPU	0	0	0	*100

(The greater the number in this table, the better.)

Table 2: Comparison of the average over 100 input data when the number of nets is 500

	ab	abc_total	abc_net	HVH
Via rate(%)	*12.1	11.8	*12.1	11.4
Net CT (grid)	2884	2859	2898	*2305
Total CT (grid)	215003	*213461	214425	213857
CPU(second)	629	1582	1679	19

(The greater Via rate and the smaller Net CT and Total CT, the better.)

[2] X.M.Xiog and Ernest S.Kuh, "The Constrained Via Minimization Problem for PCB and VLSI Design", *Proceedings of 25th ACM/IEEE Design Automation Conference*, pp. 573-578, 1988.

[3] K.C.Chang and H.C.Du, "Layer Assignment Problem for Three-Layer Routing", *IEEE Transactions on Computers*, Vol. 37, No.5, pp. 625-632, 1988.

[4] K.E.Chang, H.F.Jyu and W.S.Feng, "Constrained via minimization for three-layer routing", *Computer-Aided Design*, Vol. 21, No.6, pp. 346-354, July/August 1989.

[5] S.C.Fang, K.E.Chang and W.S.Feng, "Via Minimization with Associated Constraints in Three-Layer Routing Problem", *IEEE Transactions on Computer-Aided Design of Integrated Circuits and Systems*, pp. 1632-1635, 1990.

[6] S.C.Fang, K.E.Chang, W.Shiung and S.J.Chan, "Constrained Via Minimization with Practical Considerations for Multi-Layer VLSI/PCB Routing Problems", *Proceedings of 28th ACM/IEEE Design Automation Conference*, pp. 60-65, 1991.

[7] K.Ahn and S.Sahni, "Constrained Via Minimization", *IEEE Transactions on Computer-Aided Design of Integrated Circuits and Systems*, Vol. 12, No.2, pp. 273-282, 1993.

[8] S.Thakur, K.Y.Chao and D.F.Wong, "An Optimal Layer Assignment Algorithm for Minimizing Crosstalk for Three Layer VHV Channel Routing", *Proceedings of 1995 IEEE International Symposium on Circuits and Systems*, pp. 207-210, 1995.

[9] T.Miyoshi, S.Wakabayashi, T.Koide and N.yoshida, "An MCM Routing Algorithm Considering Crosstalk", *Proceedings of 1995 IEEE International Symposium on Circuits and Systems*, pp. 211-214, 1995.

[10] J.Cong and C.L.Liu, "On the k-Layer Planar Subset and Topological Via Minimization Problems", *IEEE Transactions on Computer-Aided Design of Integrated Circuits and Systems*, Vol. 10, No.8, pp. 972-981, 1991.

[11] P.Bruell and P.Sun, "A "Greedy" Three Layer Chanel Router", *Proceedings of IEEE International Conference on Computer-Aided Design*, pp. 298-300, 1985.

UTILITY FUNCTION BASED HYBRID ALGORITHM FOR CHANNEL ROUTING

Hussein A. Etawil and Anthony Vannelli

Department of Electrical and Computer Engineering
University of Waterloo
Waterloo, Ontario
Canada N2L 3G1

ABSTRACT

This paper presents a two layer channel router with no doglegs, based on a hybridization of Stochastic Evolution and Tabu search methods. The problem-domain knowledge expressed in the form of utility functions is used to guide the exploration of the search space. Unlike previous search heuristic based routers, the use of utility functions in our router provides a powerful tool to determine the best moves that guarantee convergence in shorter times. The algorithm begins with an initial placement of nets, generated such that the nets are relatively in conform with the vertical constraint graph. Vertical and horizontal constraints are observed during the search process. The feasibility of the ideas is demonstrated using *five* benchmark problems. Optimal solutions are found in each case.

1. INTRODUCTION

[1] In circuit layout, a channel is a routing region bounded by two parallel rows of terminals [13]. The top and bottom rows are called *top boundary* and *bottom boundary* respectively. Each terminal is assigned a number between 0 and N. Terminals with the same number i ($1 \leq i \leq N$) must be connected by net i, while those with number 0 designate unconnected terminals. The horizontal and vertical dimensions of a channel are called *channel length* and *channel height* respectively. The horizontal line along which a net is placed is called *track*. Figure (1) illustrates an example of a netlist [13].

A solution to channel routing problem (CRP) is a set of horizontal and vertical segments for each net. The solution specifies the channel height in terms of the total number of *tracks* required for routing. Thus, the main objective is to minimize the channel height which implies minimizing the number of tracks and consequently the channel area.

There are two key constraints which must be satisfied while assigning the horizontal and vertical segments, namely, horizontal and vertical constraints. These constraints are represented by a directed graph (Vertical Constraint Graph, VCG) and indirected graph (Horizontal Constraint Graph, HCG). Figures 2(a) and 2(b) illustrate the VCG and HCG for the netlist in Figure (1) [13]. Channel routing has been extensively investigated and many channel routers have been developed. The algorithms developed so far can be broadly classified as either *greedy* in nature [6, 12] or based

[1] The research of the second author is partially supported by the Natural Sciences and Engineering Research Council of Canada (NSERC) operating grant number OGP 004456 and an operating grant from Information Technology Research Centre of Ontario.

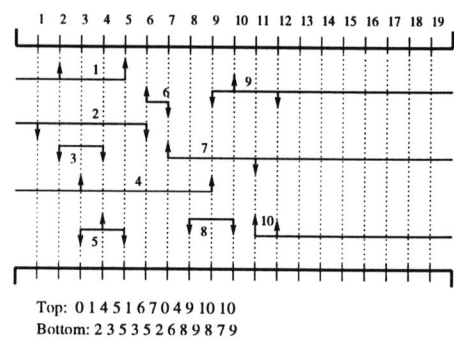

Figure 1: Netlist representation for routing requirements.

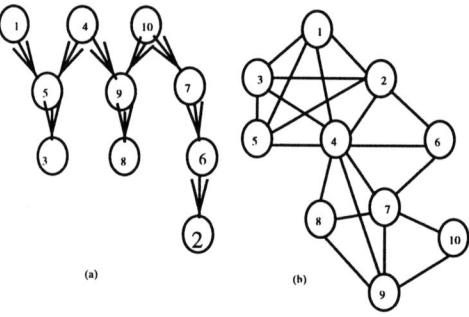

Figure 2: (a) VCG (b) HCG for the netlist in Fig 1. In HCG, maximal cliques are (1,2,3,4,5), (2,4,6), (4,6,7), (4,7,8,9) and (7,9,10).

on search heuristics[9, 2]. Greedy algorithms route the channel in a constructive manner. They are efficient, but an optimal solution is not necessarily guaranteed. Search heuristic techniques employ an iterative improvement approach. They offer a way of alleviating local minima problem by allowing some uphill moves in a controlled manner. This paper presents a channel router based on a search heuristic approach. Specifically, it is based on a hybridization of Stochastic Evolution (SE) and Tabu Search (TS) methods. SE is a stochastic method used for combinatorial optimization[14]. The SE algorithm is an instance of a more general class of *adaptive heuristics* [14]. It is an extension of the evolution concept suggested by Kling *et al.* in[8]. TS is a metaheuristic designed for solving combinatorial optimization problems [5]. The underlying idea is to forbid some search directions (moves) at a present iteration in order to avoid cycling, but to be able to escape from a local optimal point. Tabu Search is capable of exploiting the solution space effectively and it has given consistently better solutions across a wide range of problems compared to other methods previously applied to these problems[7, 1].

Based on the above observations, a hybrid algorithm based on combining SE and TS is chosen as a search engine in our channel router. In this approach, SE is guided by TS such that duplication of solutions is prohibited. That is, during the generation of neighborhood solutions, if a move is chosen and its status is Tabu, another move will be chosen unless the aspiration criteria determines otherwise. *Aspiration* forces SE to backtrack to previous solutions to refine the search in those regions. Problem-domain information expressed in the form of utility functions, are also employed to determine the best moves during the search process.

2. UTILITY FUNCTION BASED MODEL

In decision-making theory, multi-attribute utility functions order preferences of different decision outcomes [4, 10]. The ordering of the classification preferences is prescribed by the estimates and assumptions in the decision model. In the context of channel routing, the decisions involve assigning nets to tracks such that no horizontal and vertical constraints are violated and the number of tracks is minimum. In the previously developed search techniques based routers, the neighborhood solutions are generated by randomly selecting nets and tracks [9, 3, 11, 2]. In this case, the algorithm is highly likely to get trapped in a local minima. In this work, we propose combining a variety of problem-domain information using utility functions such that (during the generation of neighborhood solutions), nets and tracks are selected based on their respective utilities. The net utility function expresses information about the goodness of assigning a net to a track. The track utility function combines information about the goodness of cluster of nets assigned to a track and about the sparsity of the track.

The vertical constraint graph (VCG) provides an insight about the ideal position of a net (ideal track) such that no vertical constraint violations are caused. To be more specific, in a typical netlist, nets that come before net n_i in $p(n_i)$ [$p(n_i)$ is the longest path passing through the vertex corresponding to net n_i in VCG] must be assigned to tracks above n_i track, and nets that follow net n_i in $p(n_i)$ must be assigned to tracks below n_i track.

The certainty of assigning a net n_i to track r is given by the following utility function

$$\pi_i^r = \exp(-0.5 z^2)$$

where

$$z = \frac{r - d(n_i)}{p(n_i)}, r = 1, \cdots, k$$

If we add two fictitious nodes to the VCG; i.e, a *starting node s* and a *termination node t*, $d(n_i)$ is the longest path from s to the node corresponding to net n_i in VCG and k is the maximum number of tracks. The certainty of a cluster of nets assigned to track r is determined as follows

$$\mathcal{U}^r = \psi^r \gamma^r$$

$$\psi^r = \Pi_{i=1}^{q} \pi_i^r$$

$$\gamma^r = \exp(-0.5 \delta^2)$$

$$\delta = 1 - \frac{\sum_{j=1}^{q} l_j}{\mathcal{L}}$$

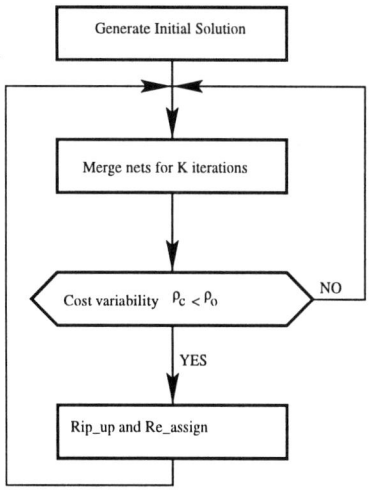

Figure 3: Outline of the SETS-CR.

Here q is the number of nets placed in track r, δ is a measure of the sparsity of a track, l_j is the span of net n_j and $\mathcal{L}$ is the channel length. Each of the single-attribute functions $\{\pi_i^r, \gamma^r\}$ is restricted to the interval $[0, 1]$. For a net n_i, π_i^r reflects the certainty or the goodness of assigning n_i to track r. π_i^r equals 1 correspond to an ideal situation; i.e. track r is the ideal track for net n_i. For track r, γ^r is a certainty measure of how good and effective track r is utilized. Again, γ^r equals 1, indicates that track r is fully utilized. By multiplication of the values of single-attribute utility functions in the interval $[0, 1]$, the resultant multi-attribute utility functions will also be restricted to the same interval.

3. ALGORITHM DESCRIPTION

Figure 3 illustrates an outline of the hybrid channel router (SETS-CR). The following is a detailed description of the different phases of SETS-CR.

3.1. Initial Solution

Given a set of nets $\mathcal{S} = \{l_1, l_2, \cdots, l_n\}$ where l_j is the horizontal span of net j, $\mathcal{S}$ is partitioned into three subsets $\mathcal{S}_t$, $\mathcal{S}_m$ and $\mathcal{S}_b$ such that $\mathcal{S}_t$ includes the nets connected to the top boundary of the channel, $\mathcal{S}_m$ includes the nets connected to the top and the bottom boundaries of the channel and $\mathcal{S}_t$ includes the nets connected to the bottom boundary of the channel. The initial solution is generated by assigning the nets in $\mathcal{S}_t$ to the channel starting from the top track, then nets in $\mathcal{S}_m$ followed by nets in $\mathcal{S}_b$. Using this strategy to assign the nets guarantees distributing the nets such that they are relatively in conform with the VCG and their vertical segments are relatively short.

3.2. Merging Phase

Following the generation of the initial solution, the merging phase is executed. In the merging phase, nets are selected for merging as follows. The certainty, π_i^r, of assigning net n_i to each track r is determined. The tracks are then sorted in a descending order based on their respective values of π_i^r. Net n_i is attempted for merging with other clusters of nets assigned to other tracks that exhibit higher certainty compared to the current track (in which n_i is already placed in). The move is accepted if the difference in cost ΔC between the newly generated solution S_{new} and the present solution S_{pre} is positive, otherwise the move is only accepted if $\Delta C > a$ where a is a random number $\in [0, -p]$ [14]. Once a net is moved to a new track, it will not be allowed to move to any other track for a number of iterations equals the *Tabu List* length $\mathcal{T}$ unless the move satisfies the required *Aspiration Level*. The Aspiration Level used is the cost of the *best solution* obtained so far. The process of merging the nets continues for K iterations (typically $5 \leq K \leq 10$). At the end of the Kth iteration, the cost variability ρ_c is computed for the K values of the cost function; i.e:

$$\rho_c = \frac{\sigma_c}{\mu_c}$$

where μ_c and σ_c are the average and standard deviation of the K values of the cost. If the value of ρ_c is less than a threshold ρ_o (typically $\rho_o < 0.03$), it is evident that the algorithm might have been trapped in a local minima. In this case, the *Ripup and Reassign* phase is executed to help the algorithm escape the local minima.

The cost function used is given as follows:

$$C = \alpha_1 w + \alpha_2 n_s$$

where w is the number of tracks and n_s is the number of sparse tracks. α_1 and α_2 are positive weights to control the importance of w and n_s respectively (typically $\alpha_1 = 10, \alpha_2 = 5$). A track is considered sparse if the sparsity of the track δ is less than a threshold δ_0 (typically $\delta_0 = 0.1$).

3.3. Ripup and Reassign Phase

In this phase, low certainty tracks are identified and clusters of nets assigned to these tracks are selected for ripping up and reassign. A certainty $\mathcal{U}^r$ of track r is considered low if it is less than a specified threshold $\mathcal{U}_o$ (typically $\mathcal{U}_o = 0.2$). Tracks that exhibit low certainty values are selected for ripping up their nets for reassignment to other higher certainty track. A net n_i, placed in track r, is ripped up and attempted for reassignment (provided that the net Tabu List is 0, or the move satisfies the Aspiration Level) if its assignment certainty π_i^r is less than specified threshold π_o (typically $\pi_o = 0.2$). Moving net n_i to, say track t, is accepted as a valid move if the new computed track certainty, $\mathcal{U}^t$, is higher than the previous one.

4. IMPLEMENTATION AND RESULTS

The algorithm was implemented in C++ on a SUN SparcStation 2. A set of benchmark problems taken from an existing paper [13] are used to evaluate the performance of the algorithm. The statistics for these benchmarks are shown in Table (1). Two scenarios are conducted using these benchmarks. In the first scenario, the performance of the algorithm is evaluated based on the effectiveness of combining TS and SE, and in the second scenario, it is based on the effectiveness of using utility functions to select the best moves. To ensure the validity of the approach, the algorithm is executed for 20 trials in each scenario [2].

In the first scenario, two experiments are conducted. In the first experiment, the algorithm employs only SE as a search engine,

---

[2] In [2], EPCHR was executed for 10 trials to ensure the effectiveness of the approach.

Benchmark	No. of nets	Global optimum
ex1	21	12
ex3a	45	15
ex3b	47	17
ex4c	54	18
Deutsch ex.	72	28

Table 1: Statistics for the different benchmarks.

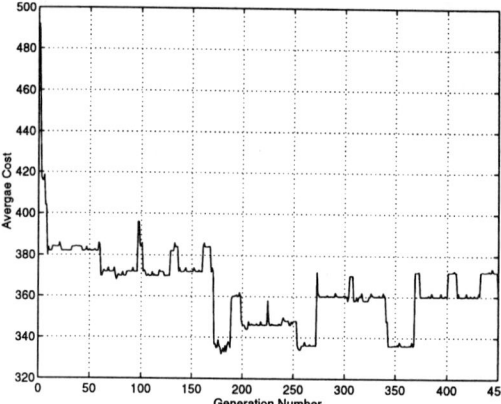

Figure 4: The variation of the average cost value as the algorithm progresses over generations for difficult Deutsch problem.

and in the second experiment the SETS hybrid is employed as a search engine. Nets and destination tracks are randomly selected. The algorithm converges to the global optimal routing width in both experiments, except for *ex3a* benchmark for which the optimal answer obtained is one track beyond the global optimum. For all the benchmarks, the number of generations, *AVG-NUM*, needed to converge to the optimal solution, *OPT-SOL*, in the second experiment is less compared to the first experiment, see Table (2). This observation demonstrates the significance of SETS hybrid as a search engine.

Benchmark	AVG-GEN			OPT-SOL		
	SE	SETS	STUF	SE	SETS	STUF
ex1	4	4	4	12	12	12
ex3a	335	160	310	16	16	15
ex3b	297	227	206	17	17	17
ex4c	67	56	45	18	18	18
Deutsch ex.	341	217	172	28	28	28

Table 2: For all the benchmarks, average number of generations (AVG-GEN) required to converge to an optimal solution (OPT-SOL) when the search engine is (i)only SE; (ii)SETS hybrid; (iii)SETS hybrid and utility functions (STUF).

In the second scenario, the effectiveness of using utility functions in selecting the best moves is investigated. The algorithm is executed with utility functions used to choose candidate nets and candidate tracks for a move. The results obtained are again reported in Table (2). The global optimum for each benchmark, including *ex3a* benchmark, is found. Also, the AVG-NUM required to obtain the global optimum is smaller compared to the previous scenario. This indicates that using utility functions to determine best moves is a quite effective approach.

Figure (4) illustrates the variation of the *average cost* versus the *generation number* for Deutsch difficult example. By observing the average cost contour, one can easily see that the algorithm climbs hills and descends valleys of the solution space. This suggests that the algorithm explores the solution space effectively and convergence to the global optimum is very likely.

The major limitation of the SETS-CR channel router is in the choice of the control parameters; i.e, Tabu Length $\mathcal{T}$, control parameter p_o and the threshold criteria ρ_o, δ_o, π_o and $\mathcal{U}_o$. Our experience shows that: $\mathcal{T} \in \{4, 5, \cdots, 7\}$; $p_o \in \{1, 2\}$; $\rho_o = 0.03$; $\delta_o = 0.1$; $\pi_o = 0.2$ and $\mathcal{U}_o = 0.2$ yields sufficiently good results. Regarding the Aspiration Criterion used, again, our experience shows that such a criterion is quite effective in forcing the SE to backtrack previous solutions and refine those search regions.

The implemented version of the SETS based router may be improved in many ways; i.e: (1)Developing a learning mechanism that allows for data-driven updating of the parameters; (2)allow intermediate solutions (infeasible solutions) may help in exploring the search space more effectively. In this case, including an effective mechanism to resolve the violations is also worth investigation.

5. CONCLUSIONS

A hybrid channel routing algorithm based on SE and TS is presented. Problem-domain information expressed in the form of utility functions is used to guide the search engine to explore the search space in an effective way compared to previous strategies. The effectiveness of the approach was demonstrated by experimenting with several benchmark problems. Global optimal solutions are obtained for all the benchmarks. Unlike Simulated Annealing, Stochastic Evolution does not climb so many hills, and at the same time it proves to be very effective in finding global optimal solutions. Combining TS with SE is very simple and it does not need any extra effort. The use of utility functions to guide the exploration of a problem search space is quite general. To be more specific, utility functions can be used with heuristic techniques (SA, GAs, EP) to solve any combinatorial problem.

6. REFERENCES

[1] S. M. Areibi. *Towards Optimal Circuit Layout Using Advanced Search Techniques*. PhD thesis, University of Waterloo, Ont. Canada, 1995.

[2] L. M. Patnaik B. B. Prahlada and R. C. Hansdah. Epchr: An extended evolutionary programming algorithm for vlsi channel routing. In *Proc. of the 4th Annual Conf. on Evolutionary Programming*, pages 521–544, 1995.

[3] R. J. Brouwer and P. Banerjee. A parallel simulated annealing algorithm for channel routing on a hypercube multiprocessor. In *IEEE Conf. on Comp. Desig.*, pages 4–7, 1988.

[4] D.W. Bunn. *Applied Decision Analysis*. McGraw-Hill, 1984.

[5] F. Glover. Future paths for integer programming. *Computers and Operatios Research, 1:3*, pages 533–549, 1986.

[6] A. Hashimoto and S. Stevens. Wire routing by optimizing channel assignment within large apertures. In *In proc. 8th, Design Automation Workshop*, pages 155–169, 1971.

[7] A. Hertz. Finding a feasible course schedule using tabu search. *Discrete Applied Mathematics 35*, pages 255–270, 1992.

[8] R. Kilng and P. Banerjee. Esp: Placement by simulated evolution. *IEEE Trans. on Computer Aided Design, 8(3)*, pages 245–256, 1989.

[9] D. F. Wong H. W. Leong and C. L. Liu. *Simulated Annealing for VLSI Design*. Kluwer Academic publishers, 1988.

[10] G. M. Paoli. Estimating certainty in classification of motor unit action potentials. Master's thesis, University of Waterloo, 1993.

[11] P. Rao and R. C. Hansdah. Extended distributed genetic algorithms for channel routing problem. In *Proc. of the 5th IEEE symp. on Parallel and Distributed Processing*, pages 726–733, 1993.

[12] R. L. Rivest and C. M. Fiduccia. A greedy channel router. In *Proc. 19th Design Automation Conf.*, pages 418–424, 1982.

[13] T. Yoshimura and E. Kuh. Efficient algorithms for channel routing. *IEEE Trans. on Comp. Aided Design, CAD-1*, pages 25–35, January 1982.

[14] S. G. Youssef and V. B. Rao. Combinatorial optimization by stochastic evolution. *IEEE Trans. on Computer Aided Design, 10(4)*, pages 525–535, April 1991.

AN AGE-CONTROLLED EVOLUTIONARY ALGORITHM FOR OPTIMIZATION PROBLEMS IN PHYSICAL LAYOUT

Andreas Huber and Dieter A. Mlynski

Institut für Theoretische Elektrotechnik und Messtechnik
Universität Karlsruhe, Kaiserstr. 12, D-76128 Karlsruhe, Germany
e-mail: huber@tem.etec.uni-karlsruhe.de

ABSTRACT

We present the first evolutionary approach using an age-controlled population model for solving layout problems. In this approach the age is counted individually. Age-intrinsic information and possible conclusions are discussed elaborately. The age-intrinsic information is exploited to improve and accelerate the optimization. Selection and mutation benefit from this approach. Due to industrial needs the optimization characteristic is controlled by a single strategy variable σ. This strategy is best-suited for discrete and combinatorial optimization problems. It facilitates efficient controlling of genetic operators. The presented strategy has been implemented and successfully applied to the assignment problem on Analog Transistor Arrays. This paper focuses on the general problem-independent aspects.

1. INTRODUCTION

Evolutionary algorithms (EA) are well-suited for multi-parameter optimization. They have been successfully applied to uncounted problems, conclusively in physical layout. The layout problem on Analog Transistor Arrays (ATA) is mapped into a high-dimensional non-continuous optimization problem. The complexity is np-complete [1]. An EA has been applied in order to solve this intractable problem. As a novelty in physical layout we count the individuals age and use its information to improve computation.

1.1. Transistor Array Design Style

An ATA is described by a set $\mathbf{D}$ of $\mu(\mathbf{D})$ devices. All component parts of ATA are called devices. Generally d different type of devices are defined. Subsets $\mathbf{D_i}$, $i = 1..d$ are formed by $\mu(\mathbf{D_i})$ devices of the same type.

$$\sum_{i=1}^{d} \mu(\mathbf{D_i}) = \mu(\mathbf{D})$$

The layouts and the locations of the $\mu(\mathbf{D})$ devices on an ATA are fixed. Only the metal layer for signal interconnection needs to be individually designed for different circuits.

1.2. Layout Strategy for ATA

The ATA design style is characterized by the fact that the layout of the master chip is invariable except for the surfacial metal layer. This layer is patterned individually to personalize different circuits. Requested circuits are personalized by applying the metal layer. Only few different resistors are available, other values must be composed. Consequently the design process of the connection layer is divided. First unavailable resistors must be composed by existing values. This yields a modified equivalent circuit. Second the component parts of this equivalent circuit are assigned to device locations on the ATA and third the connections are routed secludingly.

2. EVOLUTIONARY ALGORITHMS FOR OPTIMIZATION

Evolutionary Algorithms (EA) or Evolutionary Programs (EP) [2], namely Genetic Algorithms (GA) [3] [4] and Evolution Strategies (ES) [5] [6], are algorithms which imitate the principles of natural evolution and heritage to solve optimization problems. EAs maintain a population of individuals for iteration, they have selection processes based on the fitness of individuals, and recombination operators to generate new individuals. According to the natural paragon of evolution EAs maximize a fitness function. A fitness function is just simply a reciprocal cost function. Usually there are two genetic operators called mutation and crossover. Each individual represents a potential solution to the problem. It is evaluated to give a measure of its fitness. Then, based on the individual fitness, a new generation is formed by selection. Maintaining multiple solutions at each time-step decreases the risk of being captured by a local minimum.

2.1. Age Intrinsic Information

In this paper the age is counted individually. The age intrinsic information is exploited for the sake of faster computation and better results.

Some aspects of the age intrinsic information are discussed now. First the relation between individual age and optimization progress afterwards possible conclusions are discussed. The population model supposed for this considerations works as follows:
• The population size is constant for all generations.
• New individuals are created by mutation.
• New individuals are compared with their ancestors immediately. The better one survives, the less fit is deleted.

Let p_{IMP} be the probability to improve a solution by random mutation. It is assumed to be constant. Consequently the probability to find an improvement in the n-th attempt is $p(n)$:

$$p(n) = (1 - p_{IMP})^{n-1} \cdot p_{IMP}$$

The average number of attempts n_{AVR} that are necessary to improve an individual yields $\frac{1}{p_{IMP}}$.

$$n_{AVR} = \sum_{n=1}^{\infty} n \cdot p(n)$$
$$= p_{IMP} \cdot \sum_{n=1}^{\infty} n \cdot (1 - p_{IMP})^{n-1}$$
$$= \frac{1}{p_{IMP}}$$

In a population model that creates in every generation a variation from all individuals the average individual's maximum age is $n_{AVR} = \frac{1}{p_{IMP}}$. In other words: the more difficult it is to find better solutions, the higher becomes the individual's maximum age. To perform a qualitative investigation we use the 1D-cost-function $c_{1D}(x)$. A possible behavior of $c_{1D}(x)$ is shown in figure 1.

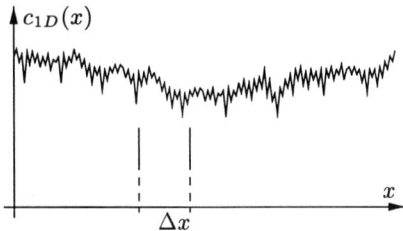

Figure 1: 1d cost function

Δx is the mutation radius. This is the range of variation that is possible with one mutation. Δx is assumed to be fix. If a local domain isn't searched for local minimum yet the probability to find better solutions p_{INP} is high. Individual are quickly extruded by better individuals. They don't become old. The better an area is scanned, the more decreases p_{INP} and thus individuals live longer till they are extruded by better offsprings. They become older and older. Consequently to scan an local area elaborately demands high individual maximum age. In other words the individual's maximum age indicates how good a certain area is scanned.

The interpretation of this must be carefully done. In particular the individual's current age mustn't be mixed up with the individual's maximum age. Every individual starts with the age of zero. In the beginning of it's lifetime no statement about the maximum age is possible.

The main problem in optimization is that the algorithm sticks in local minima. The individuals become older and older without finding an escape. Therefore a major task is to guide the algorithm out from the dominion of local minima.

This can easily be done limiting the individual age. Worsen is possible if individuals are deleted after a limited number of attempts, even though no better individual have been found meanwhile. This facilitates hill climbing and escaping.

The effect of small individual's maximum age is a rough but fast exploration of a broad region. This is required in the beginning of the optimization run where the location of the deepest local minimum within the search space is searched. High individual's maximum age provides detailed but time consuming exploration of local regions. This is required towards the end of the algorithm and applies to the residual local minima.

Obviously $maxage$ corresponds to the control parameter temperature in the simulated annealing optimization strategy.

From this considerations the strategy for our population model is derived:

In the beginning just small individual's maximum age is allowed. During the optimization, the allowed $maxage$ increases steady as well as individual fitness is expected to do too.

3. ALGORITHM IMPLEMENTATION

An algorithm has been developed and implemented in C-language which uses the ideas presented as yet. It is briefly outlined in the following. A more detailed description is given in [7] [8] [9]. Given a basic circuit description (net-list and component-list) and the ATA description. First the initial population is determined. Therefore every individual gets random assignment and suboptimal decomposition of resistors. Then the evolution optimization strategy, alternately using recombination and selection, starts. After some number of generations the program converges.

The modified net-list, the modified component-list and the assignment list is given as result, determined by the best generated individual.

3.1. Population Model

An evolution algorithm's behavior essentially depends on the chosen population model. Thus we offer an extension to the overlapping population model [2]. In analogy to the natural paragon the age of every individual is counted separately.

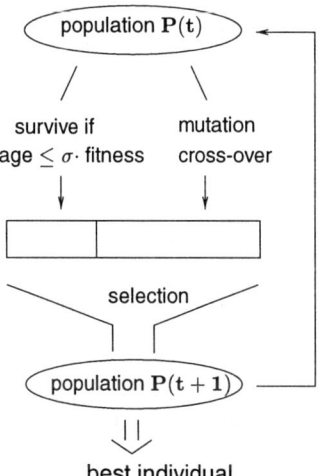

Figure 2: age-controlled population model

The developed age-controlled population model (figure 2) takes advantage of the age-intrinsic information as described in 2.1 . Starting with the population $\mathbf{P(t)}$ new individuals are created by existing individuals undergoing the recombination operators. The population $\mathbf{P(t+1)}$ is formed by a selection step. For this selection step unchanged individuals are allowed if their age is less than a maximum value denoted with $maxage$.

$$age \leq maxage$$

In 2.1 a time variant $maxage$ is motivated. $maxage$ and $fitness$ should behave similar. In order to reach this $maxage$ is chosen linearly dependent from the current fitness value $fitness$.

$$maxage = \sigma * fitness \quad (fitness \geq 1)$$

It is sensible to use $fitness$ in the sense of relative fitness, instead of absolute values. That means $fitness$ is normalized with respect to the initial fitness value. This strategy reduces the influence of different circuits. Making $maxage$ linearly dependent from $fitness$ causes useful features:

1. The characteristic of search changes during optimization as claimed in 2.1. In the beginning $fitness$ is small and hill-climbing is intensified. Towards the end $fitness$ rises and hill-climbing regresses. Thus the algorithm sticks in a residual minimum which is searched through for this reason.
2. $fitness$ carries direct information about progress. It's therefore the best variable to control $maxage$. Moreover no separate variable is necessary.
3. The algorithm has a single strategy variable σ to adjust optimization characteristic. σ can even be changed during the optimization. This is a powerful control mechanism. The intended industrial use demands effective control of the optimization behavior. The σ-mechanism provides this demanded also.

4. INDIVIDUAL

Individuals are possible solutions to the problem. Thus they must represent an entire description of a solution. In our case these are a decomposition of unavailable resistors and an assignment list. Each individual contains:

- assignment list
- decomposition information
- age

The assignment list contains entries for every device on the macrocell. Each entry contains:

- component part: the specified component part is assigned to the device location
- decomposition information: pointer to the next resistor in a decomposed resistor
  kind of decomposition (serial / parallel)

Each individual has its specific decomposition information. Consequently the algorithm can optimize the decomposition of resistors and the assignment simultaneously.

4.1. Fitness Function

In this approach the fitness function consists of three cost terms k_{cut}, k_{len} and k_{sym}. With the overall cost K

$$K = \alpha_1 * k_{cut} + \alpha_2 * k_{len} + \alpha_3 * k_{sym}$$

and the worst individual's cost K_0 the individual fitness is:

$$fitness = \frac{K_0}{K} \quad (\geq 1)$$

Due to normalization the initial value of fitness is one.
Since the problem is not continuous the fitness-function is it neither. The main problem that arises from the non-continuous fitness function are difficulties to control the mutation step.

4.2. Mutation / cross-over

New individuals are created from existing individuals undergoing genetic operators. Due to the chosen data representation and the discrete problem, mutation is the main operator.
In [5] Rechenberg introduced an Evolutionary Algorithm called ES. ES are tagged by the fact that mutation is the main operator. In fact, the first ES used the mutation mechanism exclusively. This is opposite to the first 'Genetic Algorithms' whose main operator was cross-over.
For theoretical investigation of mutation-dominated optimization algorithms Rechenberg introduced the so-called Evolution Window (German: Evolutionsfenster). It is outlined in figure 3

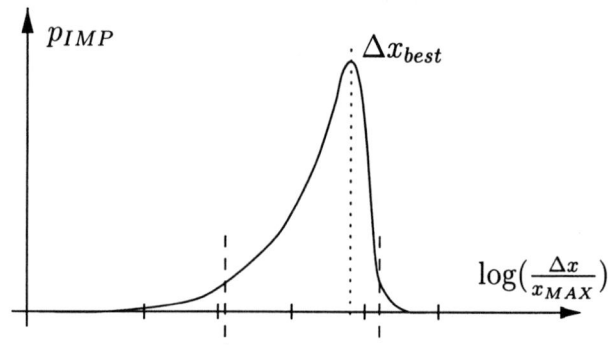

Figure 3: evolution window [5]

The probability of improvement p_{IMP} is given with respect to the variation $\frac{\Delta x}{x_{MAX}}$. He found that there is a small band of variation with a remarkable probability p_{IMP} to find better solutions. To the right of this band p_{IMP} disappears since the variation is too big. Small variations are represented to the left of the band. This means conservation and stagnation. p_{IMP} therefore decreases. As a rule of thump $\frac{1}{10} - \frac{1}{5}$ are good values for p_{IMP}.
Figure 3 reveals an interesting fact: The progress is very sensitive to large variation. For variations bigger than Δx_{best} we have a steep decline in progress. For smaller Δx we have a broad range (2-3 decades) of acceptable progress.
Rechenberg recognized mutation and step-width are crucial for successful optimization. It's therefore important to control the step width of mutation [6]. To adapt Rechenberg's idea to our problem we have to mind that we deal with a discrete problem. This has two consequences: First continuous variation Δx is impossible. Second distinct types of variation have different repercussion. For example: displacement of a resistor usually has much more effect than a simple rotation.
From this considerations the mutation operator is derived. In a mutation a certain number of so-called basic mutation take place. basic mutations are:
• rotation (resistors only)
• displacement
• decomposition / uniting (resistors only)
A basic mutation is the smallest possible mutation. At least one basic mutation is necessary. Investigations did show that numbers less than 10 are reasonable.
Controlling the step width of mutation is feasible by controlling the number of basic mutations. This can easily be done using the individual's age. Depending on the individual age age a certain

number n_{MUT} of basic mutation take place.

$$n_{MUT} = irandom(1 + ld(age))$$

$irandom(x)$ is defined for $x \geq 1$. It provides a random integer value between 1 and x.

At the beginning of the lifetime of the individual n_{MUT} equals 1, and the mutation radius is the smallest possible. Together with the age the mutation radius increases. Due to the fact that progress is less sensitive to smaller variation each individual tries to approach it's best mutation step width cautiously.

5. RESULTS

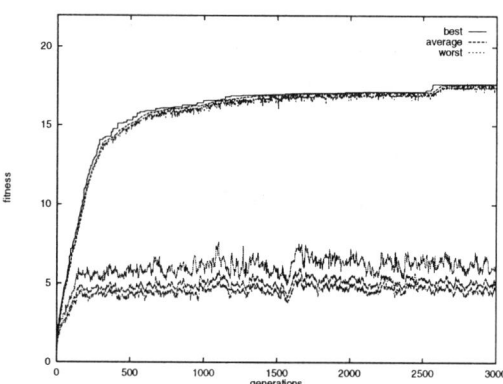

Figure 4: fitness using $\sigma = 10$ and $\sigma = 0.1$

Figure 4 shows the best, the average and the worst fitness during optimization using $\sigma = 10$ and $\sigma = 0.1$. Using $\sigma = 0.1$ leads to a rough shape. This indicates an aimless search but not an optimization. Thus $\sigma = 0.1$ is inapplicable. Using $\sigma = 10$ causes an almost disappearing gap between best and worst individual. The gap is characteristic for evolutionary algorithms and should never disappear. Lately it's an measure for the hill climbing capability. $\sigma = 10$ is chosen too big for our problem.

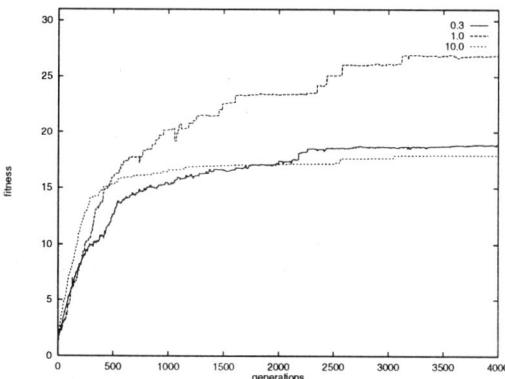

Figure 5: influence ot strategy variable σ

Figure 5 compares the best fitness for $\sigma = 0.3$, $\sigma = 1$ and $\sigma = 10$. The influence of different values of σ is obvious. Best results are achieved using $\sigma = 1$. A value of about $\sigma = 1$ proved to be best-suited for our problem.

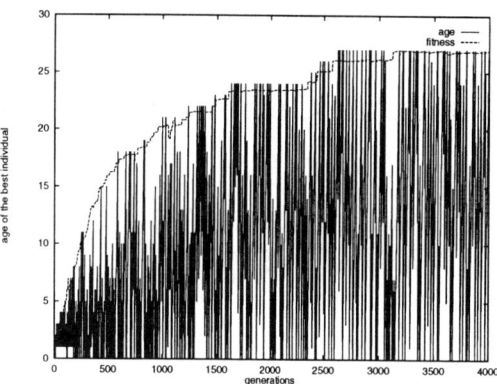

Figure 6: age of the best individual

In figure 6 age and fitness of the actually best individual is given. The effect of the presented age-controlled mechanisms are clearly visible. σ has been set to 1 as age never exceeds the integer next to the current value of the fitness.

At about generation 1400 for example no progress in fitness is recognizable due to a dominating local minimum. In this situation both mechanisms become obvious:

First the best individual is eliminated several times. This is due to the strategy variable σ. Second the average age increases and the mutation radius does respectively.

After some generations an escape is found. Thus fitness increases and the average age of the best individual decreases till the algorithm sticks for the next time.

6. REFERENCES

[1] Garey, Michael R. and Johnson, David S. *Computers and Intractability. A guide to the theorie of NP-completeness; Second Printing 1980*. W. H. FREEMAN AND COMPANY, New York, 1979.

[2] Michalewicz Zbigniew. *Genetic Algorithms + Data Structures = Evolution Programs*. Springer Verlag, Berlin, 1992.

[3] D. E. Goldberg. *Genetic Algorithms in Search, Optimization and Machine Learning*. Addison Wesley, 1989.

[4] J. H. Holland. *Adaption in Natural and Artificial Systems; Second Printing 1992*. MIT Press, Cambridge, 1992.

[5] I. Rechenberg. *Evolutionsstrategie; Neuauflage 1994*. Friedrich Frommann Verlag, Stuttgart, 1975.

[6] H.-P. Schwefel. *Numerische Optimierung von Computer-Modellen mittels der Evolutionsstrategie*. Birkhäuser, Basel Stuttgart, 1977.

[7] Andreas Huber and Dieter A. Mlynski. An efficient age-controlled evolution approach solving the assignment problem on analog transistor arrays. *MWSCAS August '97*, 1996.

[8] Andreas Huber, Hans G. Wolf, and Dieter A. Mlynski. Evas: A new evolution algorithm solving the assignment problem in analog layout with automatic consideration of matching constraints. *ICECS October '96*, 1996.

[9] Hans G. Wolf and Dieter A. Mlynski. A new genetic single-layer routing algorithm for analog transistor arrays. *ISCAS May '96*, 1996.

Timing Optimization of Mixed Static and Domino Logic

Min Zhao Sachin S. Sapatnekar

Department of Electrical and Computer Engineering
University of Minnesota, 200 Union Street SE, Minneapolis 55455, USA.
contact: sachin@ece.umn.edu

Abstract

A timing optimization algorithm dealing with circuits containing mixed domino and static logic is described. Transistor-level node timing constraints of domino logic is described. The optimization procedure preserves the requirements of maintaining adequate noise margins by constraining the sizing procedure. After sizing, charge-sharing problems are identified with a new method and rectified.

1 Introduction

Domino logic is one of the most effective circuit configurations for implementing high speed logic designs. Domino logic has the advantage of small area, fast operation and low power. However, it has drawbacks which include an inherently non-inverting nature, strict timing constraints, charge sharing and noise susceptibility. All of these factors have restricted applications of domino logic to the timing-critical regions of high-performance designs. However, recently there has been a vast amount of interest in using this logic style. The goal of the paper is to present a tool that performs optimal sizing for mixed domino and static logic circuits.

In this paper, we provide a treatment that considers the domino circuit as combinational logic embedded in a sequential circuit. The analysis technique developed here shows that techniques similar to the static combinational circuit analysis can be applied to domino circuits. Given a circuit consisting of flip-flops, domino gates and static gates, the circuit segment between flip-flops is considered here, and is optimized subject to timing constraints. This circuit segment may consist of static, domino or mixed logic.

Although several sizing algorithms have been published in the past (a survey is provided in [1]), most of them have not considered domino logic. The work by Chen and Kang in [2] and by Wurtz [3] perform sizing for domino circuits, treating them as combina-

---
[1]This work was supported in part by the National Science Foundation under award MIP-9502556 and a gift from Intel Corp.

tional circuits. However, both techniques perform local optimizations, optimizing only one domino block at a time. This work extends the timing analysis technique used in Venkat et al. [4] and van Campenhout et al. [5] to identify timing problems, and develops a sizing technique to rectify them.

2 Domino Logic Timing Constraints

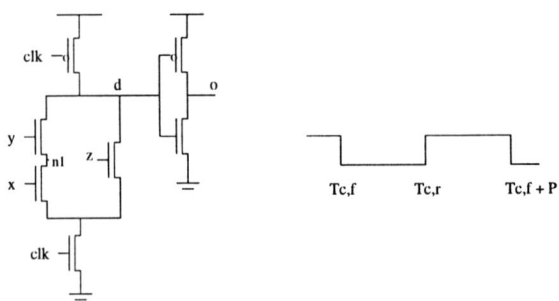

Figure 1: A Typical Domino Circuit

A representative domino gate configuration is shown in Figure 1. When the clock input is low, the gate precharges, charging the dynamic node d to logic 1. In the next half-cycle of the clock when it goes high, the domino gate evaluates, i.e., the dynamic node either discharges or retains the precharged state, depending on the values of the input signals. The two-step mode of operation with a precharge and an evaluate phase causes the timing relationships in domino logic to be more complex than those for static logic.

We list the node timing constraints for domino logic as follows in terms of the signal arrival times and the clock arrival time. In case of multiple clocks for the domino logic, the clock signal c should be set to be the clock signal that feeds the gate that is currently under consideration.

(i) Any falling event at a data input should meet the setup-time requirement to the rising edge of the evaluate clock. If $T_f(in)$ refers to the falling event time of the input node, then we require that

$$T_f(in) \leq T_{c,r} - T_{setup} \qquad (1)$$

where the setup time T_{setup} is a constant that acts as a safety margin.

(ii) The rising event of the output node of the domino gate must be completed before the falling edge of evaluate clock. If $T_r(out)$ refers to the rising event time at the output node, then the circuit operates correctly only if

$$T_r(out) \leq T_{c,f} + P \quad (2)$$

In other words, before the beginning of the precharge for next cycle, the correct evaluation result must have traveled to the output node.

For example, in Figure 1, the rising event at the output node o of the domino gate must satisfy (2). Since we can write

$$T_r(o) = \max((T_r(x) + D_f(x,d), T_r(y) + D_f(y,d),$$
$$T_r(z) + D_f(z,d), T_{c,r} + D_f(c,d))) + D_r(d,o) \quad (3)$$

where $T_r(x), T_r(y), T_r(z)$ are the rising event times at inputs x, y and z, respectively, $D_f(i,d)$ represents the delay of a falling transition at the dynamic node d due to a rising transition at input $i \in \{x,y,z\}$, and $D_r(d,o)$ represents the rise delay of the inverter feeding the gate output node o. Therefore for $i \in \{x,y,z\}$, we get

$$D_f(i,d) + D_r(d,o) - P \leq T_{c,f} - T_r(i) \quad (4)$$
$$D_f(i,d) + D_r(d,o) - P \leq T_{c,f} - T_{c,r} \quad (5)$$

(iii) The rising event d of the domino gate must be completed before the rising edge of the evaluation clock, i.e.,

$$T_r(d) \leq T_{c,r} \quad (6)$$

If we denote the rise time of the dynamic node through the p-transistor fed by the clock as $D_r(c,d)$, then the rising event time can be expressed as:

$$T_r(d) = T_{c,f} + D_r(c,d) \quad (7)$$

This leads us to the constraint given by

$$D_r(c,d) \leq T_{c,r} - T_{c,f} \quad (8)$$

This constraint implies that the pulse width of precharge clock must be capable of pulling up the output node.

3 Sizing Algorithm
3.1 Overview

The problem is solved in two phases. In the first phase, the sizing problem is solved subject to timing and noise margin constraints. In the second phase, any charge-sharing problems that were created as a result of the sizing procedure are resolved. The sizing problem is formally stated as follows:

$$\text{minimize} \quad Area \quad (9)$$
subject to
$$\max(T_r(o), T_f(o)) \leq T_{spec} \quad \forall o \in PO$$
$$T_f(in) \leq T_{c,r} - T_{setup} \quad \forall in \in I_{domino}$$
$$T_r(out) \leq T_{c,f} + P \quad \forall out \in O_{domino}$$
$$T_r(d) \leq T_{c,r} \quad \forall d \in D_{domino}$$
$$K_1 \leq \frac{W_p}{W_n} \leq K_2 \quad \forall \text{ gates in the circuit.}$$

where $Area$ is the area of the circuit and, as in other work on transistor sizing [6], is approximated as a sum of transistor sizes, PO is the set of primary outputs, I_{domino}, O_{domino} and D_{domino} are, respectively, the set of inputs, outputs dynamic nodes of all of the domino gates in the circuit.

3.2 Timing Analysis

The timing analysis procedure described here is based on the PERT procedure and uses an table-lookup delay model for delay calculation. The rising and falling event arrival times for each node v are calculated as follows:

$$T_r(v) = \max(T_f(u) + D_r(u,v)) \quad (10)$$
$$T_f(v) = \max(T_r(u) + D_f(u,v)) \quad (11)$$

where $T_r(u)$ and $T_f(u)$ are, respectively, the rising and falling event times for nodes u and v, and $D_f(u,v)$, $D_r(u,v)$ are, respectively, the worst fall delay and rise delay from input u to output v.

The domino clock input node is treated in the same way as any primary input node, and the rising or the falling edge of the clock provide the corresponding event times for the clock node. The rising and falling event arrival times at the output node of a domino gate can be obtained similarly to the static gate arrival time computations, using (10) and (11). The only difference is that the rising event at the dynamic node is related only to the falling edge of the domino clock and is independent of the other input nodes. The constraint graph is modified to capture the fact that (unlike static gates) the falling transition at an input node cannot influence the rising transition at the dynamic node by setting the value of D_r from each input node of the domino gate to the output node as $-\infty$.

3.3 Sizing Algorithm

The sizing algorithm used here is an adaptation of the TILOS algorithm [6]. Beginning with a circuit where all transistors are minimum-sized, each iteration selects one transistor and increases its size by a constant factor.

In each iteration, a timing analysis is performed to identify the constraint $g(\mathbf{w}) \leq 0$ with the largest violation, where $g(\mathbf{w})$ denotes the fact that the constraint g is a function of the vector $\mathbf{w}$ of transistor widths. The traceback procedure described above is used to determine the critical path of the circuit, which corresponds to that constraint. The sensitivity of the constraint function g to each transistor width is computed, and the width of the transistor with the most negative sensitivity is increased. The iterations continue until the timing specifications are all met, or until no further improvement is possible.

3.4 Noise Margins

Noise margin constraints are applicable to both static and dynamic gates. In [2], Chen and Kang describe a technique for deriving bounds K_1 and K_2 on W_p/W_n that will ensure that noise margin constraints are satisfied:

$$K_1 \leq Ratio = W_p/W_n \leq K_2 \quad (12)$$

For an inverter, it is a simple matter to verify whether $Ratio$ satisfies the specified bounds or not. For complex gates, each domino gate is reduced to an equivalent inverter corresponding to the largest and smallest value of $Ratio$. During the sizing process, these are compared with K_1 and K_2, respectively, to ensure that during the sizing process, these bounds are not violated.

In other words, the constraint above corresponds to the following two constraints that are always maintained during sizing:

$$Ratio_{min} = W_{p(min)}/W_{n(max)} \geq K_1$$
$$Ratio_{max} = W_{p(max)}/W_{n(min)} \leq K_2$$

The value $W_{p(max)}(W_{n(max)})$ corresponds to the equivalent inverter width when all pmos (nmos) transistors in the complex gate are on, and $W_{p(min)}$ ($W_{n(min)}$) is the equivalent inverter width when only the largest resistive path [1] of the complex gate is on.

4 Charge Sharing Algorithm

4.1 Estimation of the Worst-Case

Charge-sharing noise is produced by charge redistribution between a dynamic evaluation node and internal nodes within the gate. The usual way [2, 7, 8] of estimating worst case charge sharing is as follows. During the precharge stage, the uppermost device of every n-stack is assumed to be off, so that only the capacitance at the dynamic node, C_d is precharged. In the evaluate stage, the bottommost devices in the n-stack are configured to be off, and all devices above these in the n-stack are assumed to be on. The total capacitance that now shares charge with the dynamic output node is $C_d + C$, where C is the sum of all internal node capacitors.

However, this may be too pessimistic. If the worst-case arrival time for each input is known, and if we can identify a node n such that there is a path from the dynamic node d to n on which the rise transition on all transistors is guaranteed to arrive sufficiently before time $T_{c,r}$, then node n will be precharged and will not trigger charge-sharing. If C_{pre} is the total capacitance of all such nodes n, then we can arrive at a less conservative estimate of charge sharing that states that

$$V_{worst} = V_{dd} \cdot \frac{C_d + C_{pre}}{C_d + C} \quad (13)$$

The calculation of C_{pre} is illustrated by the example of Figure 2, which is taken from a fast adder [9]. The value of C is $C_1 + C_2 + C_3 + C_4 + C_5 + C_6 + C_7 + C_8$. If we know that signal a4 arrives before $T_{c,r}$ and that the arrival of a3, b3 is later than $T_{c,r}$, then we know C_1, C_3 should be precharged and C_2 may not be precharged. Therefore, the value of C_{pre} in (13) is $C_1 + C_3$. If instead, b3 were to arrive before $T_{c,r}$, then C_2 can also be added to C_{pre} and this would correspond to a smaller value of V_{worst}.

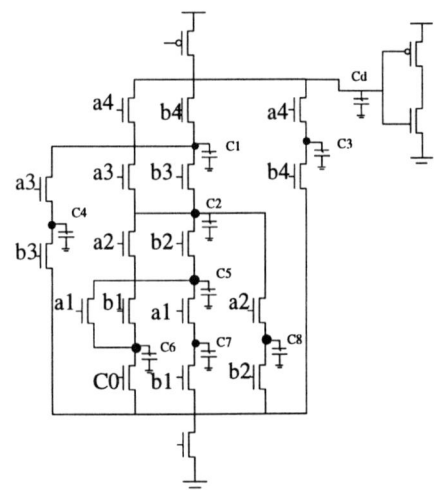

Figure 2: An Example for Charge Sharing

4.2 Reducing Charge Sharing

We note that if two nodes are connected by a transistor whose input arrives before evaluation, then precharging one node will also precharge the other node; otherwise, then precharging one node would not precharge the other. We refer to any set of such nodes as a *channel-connected precharge set*. Our algorithm finds the channel connected precharge set with the largest total capacitance and connects a pmos transistor to a node in that set. This procedure requires one traversal of the graph representing the channel-connected component [1]. This total capacitance is then added to C_{pre} and the worst case voltage due to charge sharing is calculated.

5 Experimental Results

The CAD tool is implemented in C++, and takes an input in the form of a transistor netlist. The constraints applied on the circuits include specifications on the clocks, output timing specifications, technology parameters, upper and lower bounds on the size of each transistor and constraints on the worst-case voltage V_{worst} due to charge sharing.

A summary of the results on several sample circuits is shown in Table 1. For each circuit, the number of transistors $|T|$ is listed. For various specifications on the output arrival time, T_{spec} (in ns) and for various domino clock specifications listed in the "Clk" column, the results of sizing are listed. The clock specification is in the format $(T_{c,r}, T_{c,f}, T_{c,r} + P)$, as in Figure 1, where all numbers are specified in ns. The area is reported as "-" if the specifications are too tight to be satisfied.

Table 1: Results of Transistor Sizing

Circuit/ Unsized Area	T_{spec} (ns)	Clk (ns)	Optimized Area	CPU Time
test2/ 28	0.39	(0,0.15,0.39)	47	0.09s
	0.36	(0,0.15,0.36)	71	0.10s
	0.35	(0,0.11,0.35)	52	0.09s
	0.32	(0,0.11,0.32)	93	0.11s
adder2/ 160	0.55	(-0.2,0,0.55)	177	0.18s
	0.45	(-0.2,0,0.45)	222	0.55s
	0.40	(-0.2,0,0.45)	317	0.81s
	0.38	(-0.2,0,0.38)	-	0.77s

The results of the application of the charge sharing algorithm on the example in Figure 2 is illustrated in Table 2. The set of signals that arrive before time T_{cr} are listed as the early signals. The second column shows the parameter $\frac{V_{worst}}{V_{dd}}$, which is defined in (13). The node at which a pmos transistor is added is defined in column Pn, and the updated value of $\frac{V_{worst}}{V_{dd}}$ is listed after the addition of the first pmos and the second pmos, respectively.

Table 2: Charge Sharing Algorithm Result

# N($< T_{cr}$)	Ratio	1st time		2nd time	
		Pn	Ratio	Pn	Ratio
NULL	0.210	c5	0.361	c2	0.514
a1,b1,a3,b3,a4,b4	0.635	c5	0.939	c8	1
a1,b1,a2,b2,a3,b3	0.210	c1	0.939	c3	1
a1,b2,a3,a4,b4	0.939	c8	1	-	-
a1,a2,a3,a4	1	-	-	-	-

References

[1] S. S. Sapatnekar and S. M. Kang, *Design automation for timing-driven layout synthesis.* Boston, MA: Kluwer Academic Publishers, 1993.

[2] H. Y. Chen and S. M. Kang, "A new circuit optimization technique for high performance CMOS circuits," *IEEE Transactions on Computer-Aided Design*, vol. 10, pp. 670–676, May 1991.

[3] L. T. Wurtz, "An efficient scaling procedure for domino CMOS logic," *IEEE Journal of Solid-State Circuits*, vol. 28, pp. 979–982, Sept. 1993.

[4] K. Venkat, L. Chen, I. Lin, P. Mistry, and P. Madhani, "Timing verification of dynamic circuits," *IEEE Journal of Solid-State Circuits*, vol. 31, pp. 452–455, Mar. 1996.

[5] D. van Campenhout, T. Mudge, and K. A. Sakallah, "Timing verification of sequential domino circuits," in *Proceedings of the IEEE/ACM International Conference on Computer-Aided Design*, pp. 127–132, 1996.

[6] J. P. Fishburn and A. E. Dunlop, "TILOS: A posynomial programming approach to transistor sizing," in *Proceedings of the IEEE/ACM International Conference on Computer-Aided Design*, pp. 326–328, 1985.

[7] K. Venkat, L. Chen, I. Lin, P. Mistry, P. Madhani, and K. Sato, "Timing verification of dynamic circuits," in *Proceedings of the IEEE Custom Integrated Circuits Conference*, pp. 271–274, 1995.

[8] K. L. Shepard and V. Narayanan, "Noise in deep submicron digital design," in *Proceedings of the IEEE/ACM International Conference on Computer-Aided Design*, pp. 524–531, 1996.

[9] Z. Wang, G. A. Jullien, W. C. Miller, J. Wang, and S. S. Bizzan, "Fast adders using enhanced multiple-output domino logic," *IEEE Journal of Solid-State Circuits*, vol. 32, pp. 206–213, Feb. 1997.

OPTIMIZING CIRCUITS WITH CONFIDENCE PROBABILITY USING PROBABILISTIC RETIMING

S. Tongsima, C. Chantrapornchai, E. H.-M. Sha

Dept. of Computer Science and Engineering
University of Notre Dame
Notre Dame, IN 46556, USA

Nelson L. Passos

Dept. of Computer Science
Midwestern State University
Wichita Falls, TX 76308, USA

ABSTRACT

VLSI circuit manufacturing results in theoretically identical components that actually have varying propagation delays. A "worst-case" or even "average-case" estimation of such delays during the design procedure may be overly pessimistic and will lead to costly and unnecessary redesign cycles. This paper presents a new optimization methodology, called probabilistic retiming, which transforms a circuit based on statistical timing data gathered either from component production histories or from a simulation of the fabrication process. Such circuits are modeled as graphs where each vertex represents a combinational element that has a probabilistic timing characteristic. A polynomial-time algorithm, applicable to such a graph, is developed which retimes a circuit in order to produce a design operating in a specified cycle time within a given confidence level. Experiments show that probabilistic retiming consistently produces faster circuits for a given confidence level, as compared with the traditional retiming algorithm.

1. INTRODUCTION

In VLSI design, engineers are normally facing the problem of designing a circuit able to achieve a given time constraint. Nevertheless, the information about the propagation delay of each circuit component is uncertain due to the VLSI fabrication process fluctuations or component input variation [2, 11]. Those constraint variations prevent the designers from accurately estimating the timing value. Under current design environment, the "worst-case" or "average-case" assumption are normally used to approximate the timing of the logical circuit elements. However, the use of such values of the propagation delay in the synthesis process, to guide the design, may not always be correct and, therefore, may result in unnecessary and costly redesign cycles. With this motivation, a good synthesis tool should consider the different possibilities of the final system during the initial design phase, providing the engineer with an architecture able to achieve an expected performance within a given, qualitatively provable, confidence level. Such a tool would provide a more realistic initial design that could dramatically reduce the cost and time spent in development process.

This paper presents a polynomial-time circuit transformation algorithm, called *probabilistic retiming*, which considers the probabilistic nature of the timing characteristic of basic circuit elements. It produces a solution that satisfies the timing constraints within a given confidence level. Furthermore, by considering the uncertain component timing characteristics at the beginning of the design cycle, time consuming redesign iterations can be minimized. As traditional retiming has had a great impact on the optimization problem of various applications, this new technique will have a similar impact on various applications with uncertain timing characteristics.

Many researchers implicitly adopted the worst case timing information as one of their design assumptions. In particular, Ishii modified the retiming technique in such a way that it can also handle precharged structures and gated clock signal circuitry [4]. Dey, Potkonjak and Rothweiler proposed a method to enhance retiming by transforming sequential circuits [1]. A large number of applications of retiming were explored, always making the use of the same assumption, such as the work of Shenoy and Brayton [13], Lockyear and Ebeling [10], and Liu et al. [9]. Lower bounds were computed, such as the work by Papaefthymiou, based on the minimum clock period presented in terms of the delay-to-register ratio, which can be found by retiming a circuit without considering its probabilistic behavior [12].

More comprehensive delay models were proposed by Soyata, Friedman and Mulligan [14, 15]. These models do not, however, consider the variance of circuit delays due to the manufacturing process. Recently, Karkowski and Otten introduced a model to handle the imprecise propagation delay of events [5]. In their approach, fuzzy set theory [16] was employed to model imprecise delays, but with only three possible values. Multiple objective linear programming [7] was chosen to reduce the uncertainty of such delays. Nonetheless, this model is restricted to a simple triangular fuzzy distribution and does not consider probability values which are usually easy to obtain from the quality control of the fabrication process.

In order to generalize the idea of having imprecise delays, in this paper, propagation delays are assumed to be random variables that are associated with probability distributions. The retiming technique is then extended to optimize circuits under probabilistic environments. In order to formalize this methodology, a circuit is represented as a graph where the vertex set V is a collection of combinational logic components associated with random variables representing propagation delays. Edges in the graph represent the interconnections between components which may route through zero or more registers. The probability distribution of the propagation delay being the value x is denoted by $\mathcal{P}(X = x)$, where X represents the propagation delay for a component and x is a particular value values [6].

A simple example is presented to show the usefulness of probabilistic retiming. Figure 1(b) illustrates a graph G in which the set of vertices (or nodes) is $\{A, B, C, D, \text{ and } E\}$. Assume the nodes have the timing characteristics as shown in Figure 1(a). Figure 1(c) shows a graph which considers the execution time of each vertex

---

PARTIALLY SUPPORTED BY THE NSF (MIP 95-01006) AND THE ROYAL THAI GOVERNMENT SCHOLARSHIP.

based on the worst case analysis of the propagation delays. The maximum propagation delay of all paths that contains no registers for graph G, called the clock period or cycle period, and denoted by $\Phi(G)$, is 9. Based on worst case assumptions, traditional retiming gives the "best" circuit shown in Figure 1(d). But due to the variances in execution time, the best retiming obtained using worst-case assumptions may produce a non-optimal circuit, if not all circuits produced need to meet the cycle period constraint. One might wish to retime G in order to obtain $\Phi(G) = 3$. However, such a desired clock period cannot be achieved from the worst case analysis. However, since the propagation delay associated with each component is a random variable, a designer might wish to produce a circuit in such a way that, with at least 90% confidence, the final produced systems could operate at a clock period less than or equal to 3.

$$\begin{array}{c|c|c}
A : \mathcal{P}(T_A = x) & 0.1 \text{ if } x = 1 & 0.9 \text{ if } x = 3 \\
B : \mathcal{P}(T_B = x) & 0.3 \text{ if } x = 1 & 0.7 \text{ if } x = 2 \\
C : \mathcal{P}(T_C = x) & 0.4 \text{ if } x = 1 & 0.6 \text{ if } x = 2 \\
D : \mathcal{P}(T_D = x) & 0.3 \text{ if } x = 1 & 0.7 \text{ if } x = 2 \\
E : \mathcal{P}(T_E = x) & 0.9 \text{ if } x = 1 & 0.1 \text{ if } x = 4
\end{array}$$

(a) Timing characteristic

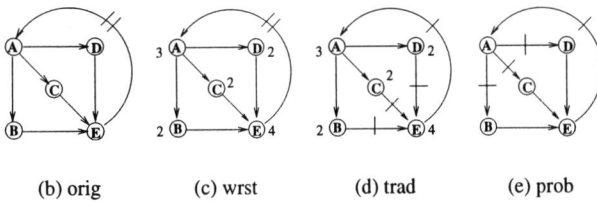

(b) orig   (c) wrst   (d) trad   (e) prob

node graph

Y be the random variable representing the *maximum* of the cumulative propagation delays along any path that has no registers on it. Then, in Figure 1(e), $\mathcal{P}(Y = 2) = 0.00324$, $\mathcal{P}(Y = 3) = 0.89676$, and $\mathcal{P}(Y \geq 4) = 0.10$. Thus, the retimed graph in Figure 1(e) has satisfied the designer requirements that the probability of the final circuit's clock period being less than or equal to 3 is equal to 90%. On the other hand, after substituting back those probability distributions to the vertices of the graph in Figure 1(d), the probability of the clock period being less than or equal to 3 for such a graph is only 9%.

2. PRELIMINARIES

A circuit that contains functional elements with associated propagation delay probability distributions can be modeled as a *probabilistic graph* (PG). A probabilistic graph (PG) is a vertex-weighted, edge-weighted, directed graph $G = \langle V, E, d, T \rangle$, where V is the set of vertices representing circuit elements, E is the set of edges representing the data dependencies between vertices, d is a function from E to $\mathbb{Z}^+$, the set of positive integers, representing the number of registers on an edge, and T_v is a random variable representing the propagation delay of a node $v \in V$. Each vertex $v \in V$ is weighted with a *probability distribution function* (pdf) of

the propagation delay, given by T_v, where T_v is a discrete random variable associated with the set of possible propagation delays of the vertex v such that $\sum_{\forall x} \mathcal{P}(T_v = x) = 1$. A shorthand $u \xrightarrow{e} v$ represents edge $e \in E$ from u to v, $u, v \in V$, and a path p starting from u and ending at v is denoted by the notation $u \xrightarrow{p} v$. Also, the number of registers of a path p ($d(p)$) where $p = v_0 \xrightarrow{e_0} v_1 \xrightarrow{e_1} \cdots \xrightarrow{e_{k-1}} v_k$ is computed by $d(p) = \sum_{i=0}^{k-1} d(e_i)$.

Retiming operations rearrange registers in a circuit so that the behavior of the circuit is preserved while achieving a faster circuit [8]. The optimization goal is to reduce the *clock period* $\Phi(G)$, representing the execution time of the longest path, in terms of propagation delay, (referred to as the critical path) that has all non-zero register edges, i.e., defined by the equations $\Phi(G) = \max\{t(p) : d(p) = 0\}$ where $p = v_0 \xrightarrow{e_0} v_1 \xrightarrow{e_1} \cdots \xrightarrow{e_{k-1}} v_k$, $t(p) = \sum_{i=0}^{k} t(v_i)$, and $d(p) = \sum_{i=0}^{k} d(e_i)$. The retiming of a graph $G = \langle V, E, d, t \rangle$ is a transformation function from vertices to the set of integers, $r : V \mapsto \mathbb{Z}$. The retiming function describes the movement of registers with respect to the vertices so as to transform G into a new graph $G_r = \langle V, E, d_r, t \rangle$ where d_r represents the number of registers on the edges of G_r. The positive (or negative) value of the retiming function determines the movement of the registers. During retiming the same number of registers are pushed from all incoming (outgoing) edges of a node to all outgoing (incoming) edges. For instance, $r(u) = 1$ implies that one register is pushed from all incoming edges of node $u \in V$, to all outgoing edges of node u. If a register is pushed from all outgoing to all incoming edges of u, then $r(u) = -1$. The following summarizes some essential properties of the retiming transformation.

Property 2.1
1. r is a legal retiming if $d_r(e) \geq 0, \forall e \in E$.
2. $\forall \ u \xrightarrow{e} v$, where $u, v \in V$, $d_r(e) = d(e) + r(u) - r(v)$.
3. $\forall \ u \xrightarrow{p} v$, where $u, v \in V$, $d_r(p) = d(p) + r(u) - r(v)$.
4. In any directed cycle (l) of G and G_r, $d_r(l) = d(l) > 0$.

3. PROBABILISTIC RETIMING

Recall the definition of a probabilistic graph (PG). Since the propagation delay of each vertex of a PG is a random variable, the traditional notion of a fixed global clock period, $\Phi(G)$, for PG G is no longer valid. Therefore, the random variable $\mathsf{mrt}(G)$, called the maximum reaching time of graph G, which represents the probabilistic clock period for graph G is introduced. Similarly, $\mathsf{mrt}(u, v)$ represents the probabilistic clock period for the portion of the graph between nodes u and v. The requirements for a probabilistic graph are usually described by an acceptable propagation delay for the final circuit, denoted by c, and a confidence level $\theta = 1 - \delta$, where δ is the acceptable probability of not achieving the required performance. In this paper, the requirement is expressed as $\mathcal{P}(\mathsf{mrt}(G) \leq c) \geq \theta$, or $\mathcal{P}(\mathsf{mrt}(G) > c) > \delta$. The goal of probabilistic retiming is to transform a PG such that the requirement can be satisfied.

Since it is complex to efficiently compute a function of dependent random variables, Algorithm 3.1 which computes the $\mathsf{mrt}(G)$ assumes the random variables are independent. Fortunately, by using this approach, the algorithm can give an upper bound of the exact solution of the mrt when considering the operation of dependent random variables. In the algorithm, two dummy vertices with zero propagation delays, v_s and v_d, are added to the graph. A set of zero-register edges is used to connect vertex v_s to all root-nodes, and to connect all leaf-nodes to vertex v_d. Thus, the $\text{mrt}(v_s, v_d)$ gives the overall maximum reaching time of the graph $\text{mrt}(G)$. In order to compute the mrt, only the graph portion that has zero register edges, i.e., a directed acyclic graph (DAG), is considered.

Algorithm 3.1 (Maximum reaching time)

Input : PG $G = \langle V, E, d, T \rangle$
Output: $\text{mrt}(G) = \text{tmp}_{\text{mrt}}(v_s, v_d)$

1 $G_0 = \langle V_0, E_0, d, T \rangle$ such that $V_0 = V + \{v_s, v_d\}$,
2 $E_0 = E - \{e \in E | d(e) \neq 0\} + \{v_s \xrightarrow{e} v \in V_r, u \in V_l \xrightarrow{e} v_d\}$
3 $\forall u \in V_0, \text{tmp}_{\text{mrt}}(v_s, u) = 0, T_{v_s} = T_{v_d} = 0, Queue = v_s$
4 while $Queue \neq \emptyset$ do
5      $get(u, Queue)$
6      $\text{tmp}_{\text{mrt}}(v_s, u) = \text{tmp}_{\text{mrt}}(v_s, u) + T_u$
7   foreach $u \xrightarrow{e} v$ do
8          $indegree(v) = indegree(v) - 1$
9          $\text{tmp}_{\text{mrt}}(v_s, v) = \max(\text{tmp}_{\text{mrt}}(v_s, u), \text{tmp}_{\text{mrt}}(v_s, v))$
10   if $indegree(v) = 0$ then $put(v, Queue)$
11   end_do
12 end_do

Algorithm 3.1 traverses the DAG portion (G_0) of the graph in topological order and compute the mrt of each node v with respect to v_s. At Line 9, the mrt of all parents of a node are maximized. Line 6 will then add T_v to the tmp_{mrt} of node v producing the final mrt with respect to all paths reaching node v. Finally, after node v_d is visited, the $\text{mrt}(G)$ contains the final mrt of the graph.

Algorithm 3.2 presents the probabilistic retiming algorithm. The algorithm retimes vertices whose probability of propagation delays being greater than c is larger than the acceptable probability value. Lines 7–18 traverse the DAG in a breath-first search manner and update the tmp_{mrt} for each node as in Algorithm 3.1. After updating a vertex, the resulting tmp_{mrt} is tested to see if the requirement, $\mathcal{P}(\text{tmp}_{\text{mrt}}(G) > c) \leq \delta$, is met. Line 15, then decreases the retiming value of any vertex v that violates the requirement unless the vertex has previously been retimed in a current iteration. The algorithm then repeats the above process using the retimed graph obtained from the previous iteration.

Algorithm 3.2 (Probabilistic retiming)

Input: PG $G = \langle V, E, d, T \rangle, c, \delta$.
Output: Retiming function r.

1 $\forall$ vertex $v \in V$, set $r(v) = 0$
2 for $i = 1$ to $|V|$ do
3      $G_r = Retime(G, r); mod_flag = $ false
4      $G_0 = \langle V_0, E_0, d, T \rangle$ where $V_0 = V + \{v_s, v_d\}$
5      $E_0 = E - \{e \in E | d(e) \neq 0\} + \{v_s \xrightarrow{e_1} v \in V_r, u \in V_l \xrightarrow{e_2} v_d\}$
6      $\forall u \in V_0, \text{tmp}_{\text{mrt}}(v_s, u) = 0, Queue = v_s, T_{v_s} = T_{v_d} = 0$
7   while $Queue \neq \emptyset$ do
8          $get(u, Queue)$
9          $\text{tmp}_{\text{mrt}}(v_s, u) = \text{tmp}_{\text{mrt}}(v_s, u) + T_u$
10   foreach $u \xrightarrow{e} v$ do
11             $indegree(v) = indegree(v) - 1$
12             $\text{tmp}_{\text{mrt}}(v_s, v) = \max(\text{tmp}_{\text{mrt}}(v_s, u), \text{tmp}_{\text{mrt}}(v_s, v))$
13   if $\mathcal{P}(\text{tmp}_{\text{mrt}}(v_s, v) > c) > \delta$
14   then if u has not been retimed at current iteration
15   then $r(u) = r(u) - 1; mod_flag = $ true;
16   if $indegree(v) = 0$ then $put(v, Queue)$
17   end_do
18   end_do
19   if $mod_flag = $ false then Report r, break
20 end_do

4. EXPERIMENTAL RESULTS

In each experiment, for a given confidence level $\theta = 1 - \delta$, Algorithm 3.2 is repeatedly applied to search for the best clock period. Table 1 shows the results for traditional retiming using worst-case propagation delay assumptions (column worst) and the probabilistic model with varying confidence levels. The first column in Table 1 lists the tested benchmarks. Column worst in the table presents the optimized clock period obtained from applying traditional retiming using the worst-case propagation delay of each adder (24ns) and each multiplier (30ns). In these experiments, the distributions of the propagation delays for the basic components (adder and multiplier) in these circuits were obtained from [3], and were assumed to fit normal distributions [2]. Columns 4 through 8 show the best clock period c for the confidence levels 0.9 down to 0.5. Notice that for all benchmarks the clock periods with $\theta = 0.9$ are still smaller than the propagation delays in Column 3. The "%" columns list the feasible clock period reduction percentages with respect to the worst-case method. With $\theta = 0.9$ all benchmarks have more than an 18% improvement, while with $\theta = 0.5$ all benchmarks have more than a 26% reduction.

Table 2 compares the probabilistic retiming algorithm to the traditional retiming algorithm with average-values used for individual components. In each experiment, the expected value is 16ns for an adder and 20ns for a multiplier [3]. Traditional retiming is applied to these graph, assuming the expected values (G_{avg}). The clock period of $G_{\text{avg}}\Phi(G_{\text{avg}})$ is listed in column 3 of Table 2. $E(\text{mrt}(G_{\text{avg}}))$ in column 4 is an expected value of the maximum reaching time of G_{avg}, if the pdfs for each component are taken into account. Since the probabilistic nature of each individual component must be considered to calculate the actual propagation delay for the entire graph (column 4), using the average timing value for each component may produce misleading results. This scenario illustrates that using traditional retiming with average-case values produces poor results that are significantly larger than $\Phi(G_{\text{avg}})$, when the probabilistic nature of each component is included.

In order to make a comparison to probabilistic retiming, results for the benchmarks are also computed using Algorithm 3.2 with confidence levels of 0.9 to 0.5. The expected (average) clock periods for these situations are shown in columns 5 to 9. These expected clock periods are consistently smaller than $E(\text{mrt}(G_{\text{avg}}))$. In addition, these values are normally much closer to $\Phi(G_{\text{avg}})$ than to $E(\text{mrt}(G_{\text{avg}}))$, and are actually less than $\Phi(G_{\text{avg}})$ in some cases. Hence, the approach of using the expected values for each component is neither a good heuristic in the initial design phase nor does it give any quantitative confidence to the resulting circuits.

5. CONCLUSION

VLSI circuit manufacturing results in devices with varying propagation delays. The estimation of such delays during the design procedure may not be totally accurate, leading to extra design cycles. We present a new transformation technique, called probabilistic retiming, which considers varying timing delays, and takes $\mathcal{O}((cn)^2|V|^2|E|)$ time. From experiments, considering realistic cases, this algorithm can effectively optimize circuits while utilizing manufacturing probability information and the designer requirements of a desired clock period c and a confidence level θ. This methodology is expected to have an impact in many areas,

Benchmark	num. nodes	c worst	$\mathcal{P}(\mathrm{mrt}(v_s, v_d) \leq c) \geq 1 - \delta = \theta$									
			$\theta = 0.9$		$\theta = 0.8$		$\theta = 0.7$		$\theta = 0.6$		$\theta = 0.5$	
			c	%	c	%	c	%	c	%	c	%
Biquad IIR	8	78	60	23	57	26	56	28	52	33	51	35
Diff. Equation	11	118	81	31	77	35	76	36	73	38	72	39
3-stage direct IIR	12	54	44	19	41	24	40	26	39	28	36	33
All-pole Lattice	15	157	120	24	117	25	115	27	113	28	112	29
4th order WDF	17	156	116	26	112	28	109	30	108	31	106	32
Volterra	27	276	216	22	212	23	208	25	205	26	202	27
5th Elliptic	34	330	240	28	236	29	233	30	230	31	228	31
All-pole Lattice (uf=2)	45	468	350	25	346	25	343	27	340	27	338	28
All-pole Lattice (uf=6)	105	1092	811	26	806	26	802	27	799	27	796	27
5th Elliptic (uf=4)	170	1633	1185	27	1174	28	1169	29	1164	29	1160	29

Table 1: Probabilistic retiming versus worst case traditional retiming

Benchmark	num.	$\Phi(G_{\text{avg}})$	$E(\mathrm{mrt}(G_{\text{avg}}))$	Expected clock period (Algorithm 3.2)				
				$\theta = 0.9$	$\theta = 0.8$	$\theta = 0.7$	$\theta = 0.6$	$\theta = 0.5$
Biquad IIR	8	52	70.40	52.64	52.30	53.18	51.02	51.64
Diff. Equation	11	72	76.05	73.07	72.50	72.32	71.29	70.94
3-stage direct IIR	12	36	41.90	37.70	38.36	38.97	38.40	36.27
All-pole Lattice	15	104	114.45	111.77	111.40	111.00	110.41	110.04
4th order WDF	17	104	106.73	106.44	105.98	105.28	104.99	104.17
Volterra	27	200	204.00	202.44	202.00	201.22	200.31	198.06
5th Elliptic	34	220	233.30	228.41	227.59	226.95	226.01	225.19
All-pole Lattice (uf=2)	45	312	342.17	338.11	337.62	337.00	336.08	335.27
All-pole Lattice (uf=6)	105	728	800.51	794.02	793.39	792.53	791.57	790.31
5th Elliptic (uf=4)	170	1100	1204.50	1160.80	1159.30	1158.17	1156.58	1154.90

Table 2: Probabilistic retiming versus average case analysis

e.g., high-level synthesis, loop scheduling, testing, etc., similar to traditional retiming.

6. REFERENCES

[1] S. Dey, M. Potkonjak, and S. G. Rothweiler. Performance optimization of sequential circuits by eliminating retiming bottlenecks. In *Proceedings of the 1992 IEEE/ACM International Conference on Computer Aided Design*, pages 504–509, 1992.

[2] S. Director, W. Maly, and A. Strojwas. *VLSI design for manufacturing: yield enhancement*. Kluwer Academic Publishers, 1990.

[3] Texas Instruments. *The TTL data book*, volume 2. Texas Instruments Incorporation, 1985.

[4] A. T. Ishii. Retiming gated-clocks and precharged circuit structures. In *Proceedings of the 1993 IEEE/ACM International Conference on Computer Aided Design*, pages 300–307, 1993.

[5] I. Karkowski and R. H. J. M. Otten. Retiming synchronous circuitry with imprecise delays. In *Proceedings of the 32nd Design Automation Conference*, pages 322–326, San Francisco, CA, 1995.

[6] E. Kreyszig. *Advanced Engineering Mathematics*, chapter 23. John Wiley and Sons, New York, 6th edition, 1988.

[7] Y.-J. Lai and C.-L. Hwang. A new approach to some possibilistic linear programming problems. *Fuzzy Sets and Systems*, 49:121–133, 1992.

[8] C. E. Leiserson and J. B. Saxe. Retiming synchronous circuitry. *Algorithmica*, 6:5–35, 1991.

[9] L.-T. Liu et al. Performance-driven partitioning using retiming and replication. In *Proceedings of the 1993 IEEE/ACM International Conference on Computer Aided Design*, pages 296–299, 1993.

[10] B. Lockyear and C. Ebeling. The practical application of retiming to the design of high-performance systems. In *Proceedings of the 1993 IEEE/ACM International Conference on Computer Aided Design*, pages 288–295, 1993.

[11] P. Mozumder and A. Strojwas. Statistical control for VLSI fabrication processes. In W. Moore, W. Maly, and A. Strojwas, editors, *Yield Modeling and Defect Tolerance in VLSI*. Adam Hilger, Bristol and Philadelphia, 1987.

[12] M. C. Papaefthymiou. Understanding retiming through maximum average-delay cycles. *Mathematical Systems Theory*, 27:65–84, 1994.

[13] N. Shenoy and R. K. Brayton. Retiming of circuits with single phase transparent latches. In *Proceedings of the 1991 International Conference on Computer Design*, pages 86–89, 1991.

[14] T. Soyata and E. Friedman. Retiming with non-zero clock skew, variable register and interconnect delay. In *Proceedings of the 1994 IEEE/ACM International Conference on Computer Aided Design*, November 1994.

[15] T. Soyata, E. Friedman, and J. Mulligan. Integration of clock skew and register delays into a retiming algorithm. In *Proceedings of the International Symposium on Circuits and Systems*, pages 1483–1486, May 1993.

[16] L. A. Zadeh. Fuzzy sets as a basis for a theory of possibility. *Fuzzy Sets and Systems*, 1:3–28, 1978.

A Matlab based tool for bandpass continuous-time sigma-delta modulators design

Philippe BENABES, Patrick ALDEBERT, Richard KIELBASA

SUPELEC, Service des Mesures, Plateau de Moulon
F91192 GIF/YVETTE France
Email : philippe.benabes@supelec.fr , patrick.aldebert@supelec.fr

Abstract: A methodology for synthesis and analysis of bandpass sigma-delta (ΣΔ) converters has been developed and integrated in a Matlab toolbox. It allows to synthesize ΣΔ modulators with continuous time filters from discrete time topologies. The analysis method is based on discretization of continuous-time models. It uses a discrete time simulator, more efficient than an analog simulator. All tools are included in a fully interactive, graphic and open framework in which user-developed modules can be added up.

I. Introduction

Sigma-delta (ΣΔ) analog-to-digital converters are very attractive because they achieve high accuracy with few critical analog components. Nowadays, ΣΔ are often built from discrete-time switched capacitor integrators which are well suited for VLSI integration. Unfortunately, when using a standard technology, the sampling frequency of the modulator is limited by 10 to 50 MHz which results in a signal bandwidth lying from 50 kHz up to a few MHz.

An alternative to discrete-time filters are continuous-time filters. Although continuous-time modulators are not easy to integrate, they exhibit a key advantage over their discrete-time counterparts : the sampling operation is inherently done inside of the modulator loop so that the restriction of the mentioned maximum Nyquist frequency is removed. On the other hand, continuous-time circuits are more difficult to design and simulate than discrete-time circuits.

Recently published continuous-time modulators operates between tens of MHz up to a few GHz [1]-[3]. In [1] and [2], analog simulations exhibits a large computation time. A synthesis method of continuous-time modulators based on the discrete-time ones was described by Schreier in [3], but the effect of non ideal functionality of the Digital to Analog Converter (DAC) feedback is not studied. Gosslau studies this effect in [4] but doesn't give any way to compensate it.

A method for designing continuous-time lowpass sigma-delta modulators with non ideal DAC was proposed [5]. It has been extended to bandpass topologies and integrated in a Matlab toolbox. The delay introduced by the modulator feedback has been taken into account since its synthesis.

Analysis of the final modulator can be accomplished using a discrete-time simulator more efficient than an analog simulator. For the while all non-idealities have not been implemented yet (especially non linear ones), and analog simulations are still necessary, but our tool can easily eliminate some bad candidates.

II. Equivalency between Continuous-time and Discrete-time Filters

The behavior of ΣΔ modulators employing discrete-time filters has been widely studied [6]. In this synthesis method, we suppose that a ΣΔ modulator with discrete-time filter has been designed and should be transformed into a continuous-time one.

A classical topology of continuous-time ΣΔ modulator is shown in Fig. 1.

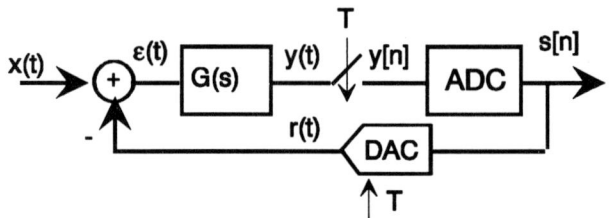

Fig. 1. Continuous time ΣΔ modulator with sampled input signal.

Let's suppose that the input signal equals to 0. The continuous-time filter $G(s)$ excited by the signal $r(t)$ obtained from a sample and hold (DAC output) with sampling frequency $f_s = 1/T$, where T denotes the sampling period (Fig. 2). The output signal $y(t)$ of that filter is sampled at the same frequency and without any delay, (resulting $y[n] = y(nT)$).

The system can be presented as an equivalent discrete-time system shown in Fig. 3, where $F(z)$ denotes the discrete-time transfer function. The relationship between $G(s)$ and $F(z)$ can be expressed by the well-known formula [4]

$$F(z) = \left(1 - z^{-1}\right) Z_T \left\{ L^{-1}\left[\frac{G(s)}{s}\right] \right\}, \quad (1)$$

where L^{-1} stands for the inverse Laplace transform, Z for the z-transform, and $Z_T\{y(t)\} = Z\{y[n]\}$.

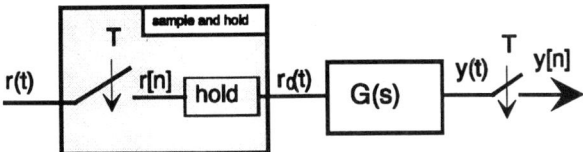

Fig. 2. Block diagram of a continuous filter with sampled and held input signal

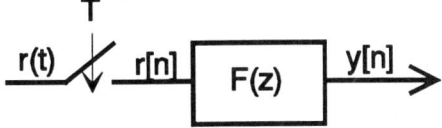

Fig. 3. Block diagram of equivalent discrete-time model.

The behavior of a non ideal DAC is illustrated in (Fig. 4). If it is assumed that non idealities of DAC can be modeled by a delay and a first order transfer function (exponential shape), the equivalent discrete-time model of $G(s)$ with non ideal DAC can be expressed as follows:

$$F^*(z) = (1-z^{-1})Z_T\left\{L^{-1}\left[\frac{B(s).G(s)}{s}\right]\right\}, \quad (2)$$

where $B(s)$ contains the non ideal part of DAC functionality. If the output stage of DAC circuit is supposed to behave in a linear way, the step response of non ideal part of DAC can be approximated by (3).

$$r_u(t) = u(t-d)\left(1 - e^{\frac{t-d}{\tau}}\right), \quad (3)$$

where $r_u(t)$ is the output of non ideal DAC. Delay d can be seen as a simple model of the transmit time in the middle stages of DAC, and the first order transfer function as a model of the output stage response. Its impulse response given by (4).

$$B(s) = \frac{e^{-ds}}{(1+\tau s)}. \quad (4)$$

If the output stage op-amp of the DAC is not sufficiently fast, the slew rate limitation cannot be ignored. A typical step response ($s[n] = u[n]$) of DAC is shown in Fig. 4 (linear shape). For the sake of simplicity, the op-amp step response is approximated by pure slewing rate behavior (the feedback signal is approximated by a piecewise linear function).

The delay υ in the step response depends on the absolute value of input signal. (it is a nonlinear system). Fortunately, the input of the DAC of a one bit $\Sigma\Delta$ modulator can only be 1 or -1. Thus the absolute value of input signal is constant and it implies that the delay υ is constant. Consequently, if $d+\upsilon$ is less than sampling period then the non ideal part of DAC can be described as a linear system with the following step response

$$b(t) = u(t-d)\left(\frac{t-d}{\upsilon}\right) - u(t-d-\upsilon)\left(\frac{t-d-\upsilon}{\upsilon}\right). \quad (5)$$

From (5), the impulse response is obtained and we have

$$B(s) = \frac{e^{-ds} - e^{-(d+\upsilon)s}}{\upsilon s}. \quad (6)$$

So that the equivalent discrete-time filter will have the following z-transfer function

$$F_k^*(z) = (1-z^{-1})Z_T\left\{L^{-1}\left[\frac{G_k(s)(e^{-ds} - e^{-(d+\upsilon)s})}{\upsilon s^2}\right]\right\}. \quad (7)$$

These equations can be easily extended even though $d+\upsilon$ is higher than one sampling period.

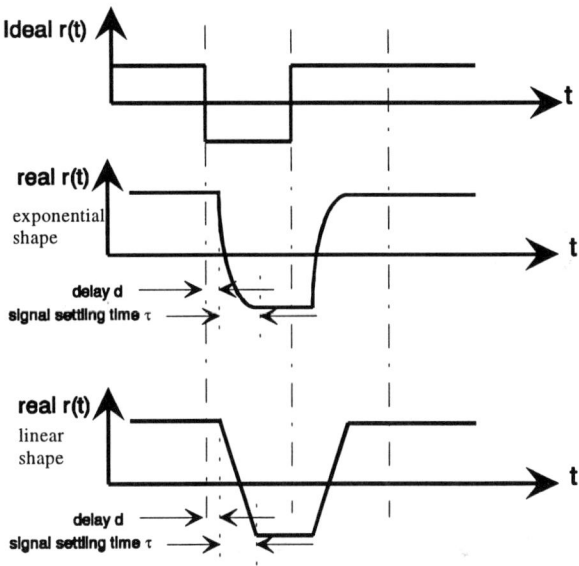

Fig. 4. Functionality of a non-ideal DAC.

III. MATLAB Toolbox

A toolbox was developed under MATLAB graphical environment.

The main window (Fig. 5) contains menus and boxes to capture interactively the specifications of the initial discrete-time modulator: topology (one is based upon Schreier's toolbox [7], the other is based upon MSCL topology [8] & [9]), topology-dependant parameters, bandpass or lowpass modulator, central frequency, order of modulator, oversampling ratio, DAC and ADC number of bits, Q factors of resonators, and input signal used for simulations. Some new combinations can be computed such as lowpass-bandpass or multiple-band topologies.

At this level, several characteristics such as output signal and noise spectrum, poles and zeros are automatically computed.

Some other can be computed on request : SNR function of input signal power, impulse response of open-loop modulator filter, impulse response of modulator (Fig. 6), poles locus.

Some other parameters are used for the synthesis of the continuous-time modulator : position of resonators, delay introduced in the feedback loop, feedback signal shape (linear or exponential), topology of the continuous-time modulator.

When the discrete-time topology fills the requirements, the continuous time modulator can be synthesized. This process is fully automatic and leads to a graphical block-diagram of the resulting continuous-time modulator (Fig. 7).

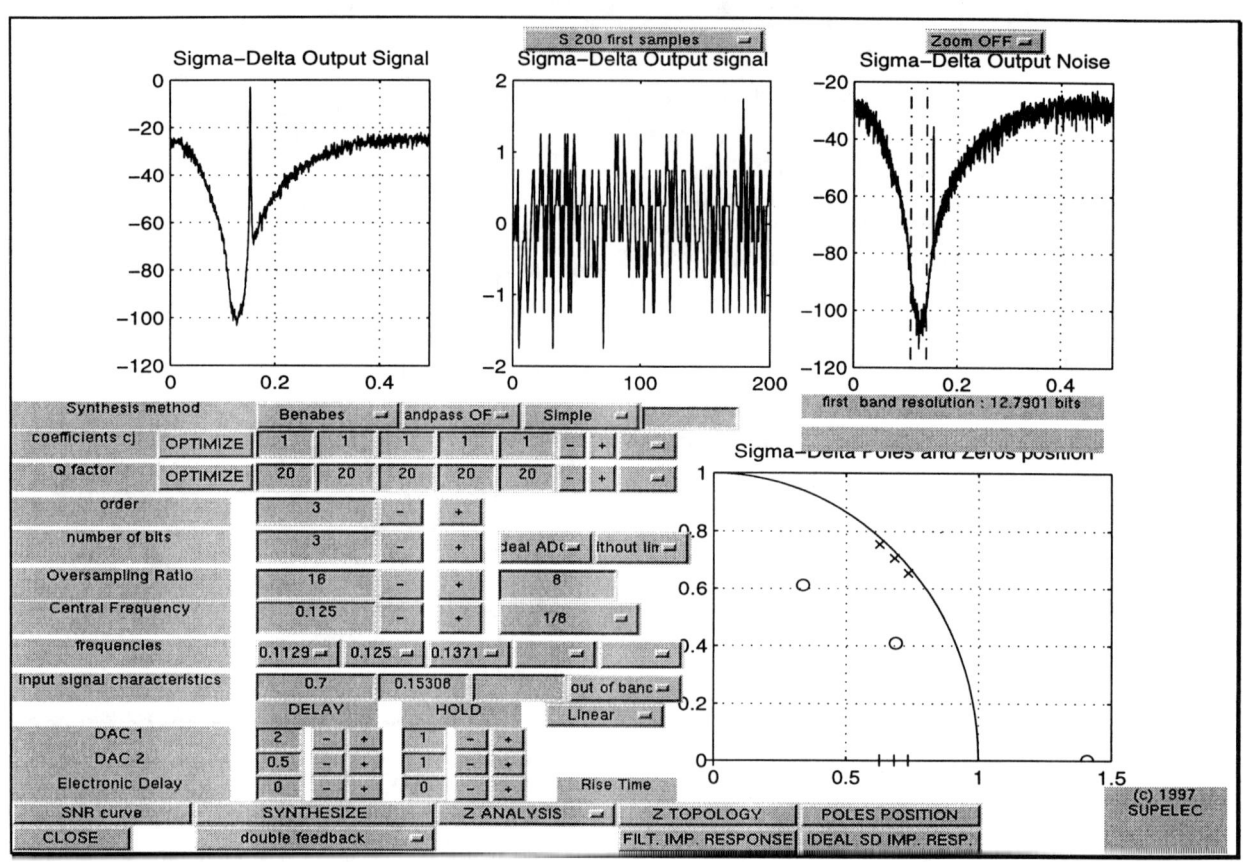

Fig. 5 : The toolbox main window

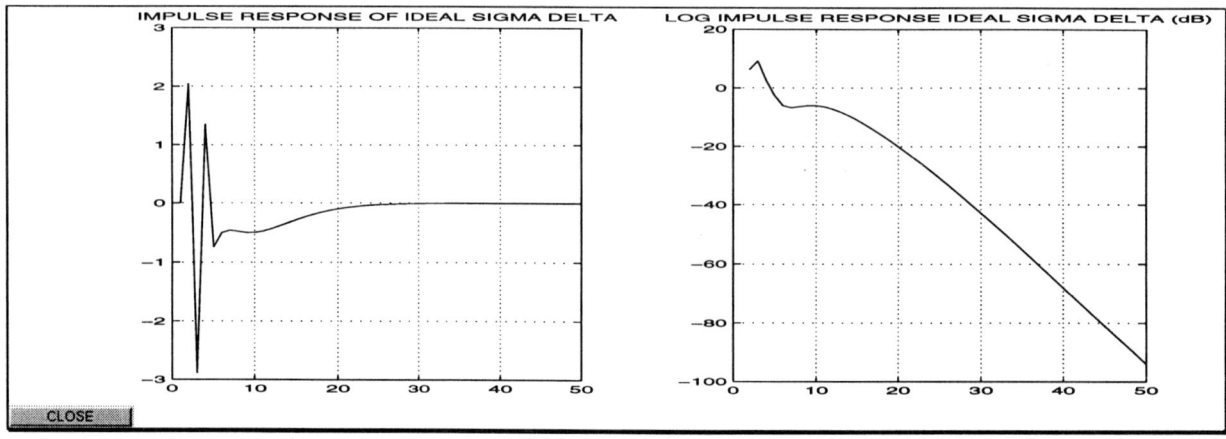

Fig. 6 : Impulse response of the modulator

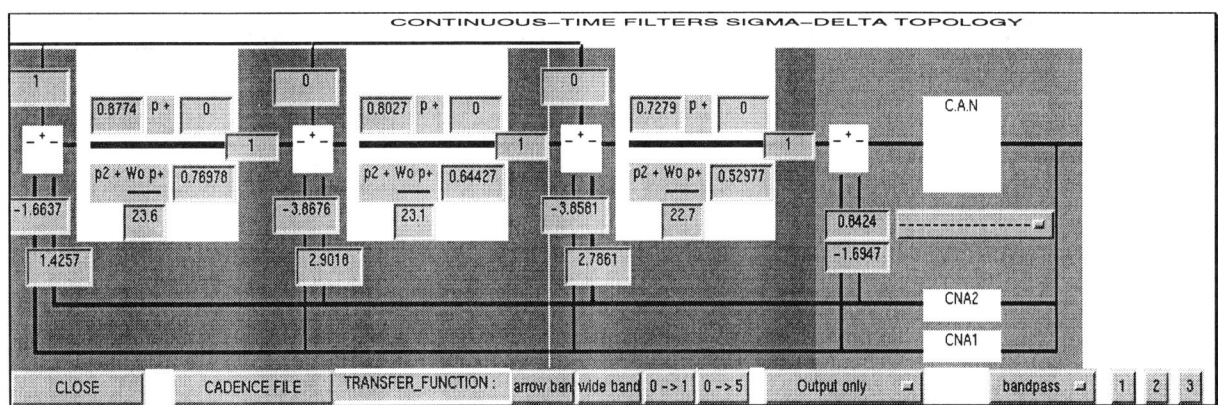

Fig. 7 : continuous-time block diagram

During the synthesis phasis, no computation is performed about signal transfer function. In the continuous-time topology window, signal coefficients can be introduced manually, or computed automatically by optimizing the signal transfer function.

The resulting signal transfer function can be displayed graphically, as also intermediary resonator outputs (Fig. 8)

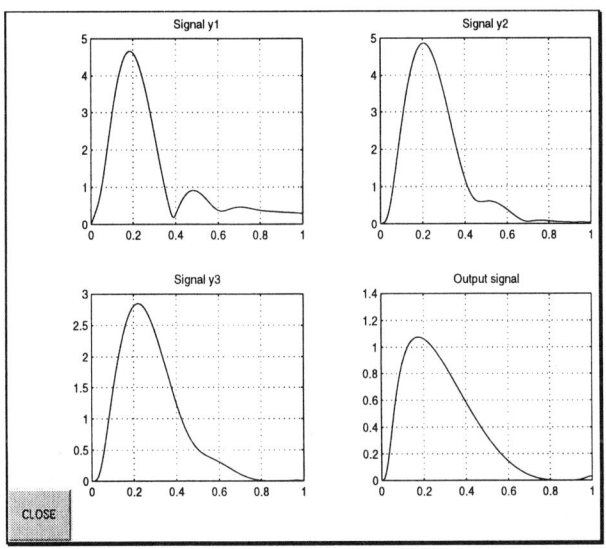

Fig. 8 : signal transfer functions

Some simulations of the continuous time modulator can be performed. One method uses a discrete-time equivalent model of the synthesized modulator by the inverse transformation process of (1). With this method, one point per output sample is calculated for all intermediary signals. The major advantage of this method is its high speed Unfortunately, non-linearities can hardly be taken into account.

A second method based upon bilinear transformation of continuous-time filters can be used for more precise simulations. Some amplifiers non-linear characteristics can be introduced. This method requires more memory and computation power as more than 1 point per output sample is calculated (typically 8 or 16).

IV. Conclusion

A new tool for analysis and synthesis of $\Sigma\Delta$ modulators with continuous-time filters was described in this paper. Some non-idealities such as feedback DAC delay and rise/fall time can be modeled by an discrete-time equivalent model. This model allows to take into account these parameters within synthesis process of continuous-time modulators and to perform faster simulations. All tools are integrated in an user-friendly environment.

References

[1] R. Koch, B. Heise, F. Eckbauer, E. Engelhardt, J. A. Fisher, and F. Parzefall, "A 12 bit $\Sigma\Delta$ analog-to-digital converter with a 15 MHz clock rate," *IEEE Journal of Solid-State Circuits*, vol. 21, pp. 1003-1010, December 1986.

[2] J. F. Jensen, A. E. Cosand, and R. H. Walden "A 3.2-GHz second-order delta-sigma modulator implemented in InP HBT Technology," *IEEE Journal of Solid State Circuits*, vol. 30, pp 103-106, October 1995.

[3] R. Schreier and B. Zhang, "Delta-Sigma modulators employing continuous-time circuitry," *IEEE Trans. Circuit & Systems-I: Fundamental Theory and Applications*, vol. 43, pp. 324-332, April 1996.

[4] A. Gosslau, A. Gottwald, "*Optimisation of a $\Sigma\Delta$ modulator by the use of a slow adc*", IEEE ISCAS'88, pp. 2317-2320, 06/88

[5] P. Benabes, M. Keramat and R. Kielbasa, "*A Methodology for designing continuous-time sigma-delta modulators,*" European Design and Test Conference (ED&TC 97), pp46-50, Paris, 1997

[6] J.C. Candy and G. C. Temes, Oversampling Delta-Sigma Data Converters. IEEE Press, New York, 1991.

[7] R. Schreier and B. Zhang, "Delta-Sigma modulators employing continuous-time circuitry," *IEEE Trans. Circuit & Systems-I: Fundamental Theory and Applications*, vol. 43, pp. 324-332, April 1996.

[8] P. Benabes, *New Bandpass Sigma-Delta Modulators*, Ph.D. dissertation, Service des Mesures, SUPELEC, Paris, France, September 1994 (in French).

[9] P. Benabes, A. Gauthier and R. Kielbasa, "A Multistage Closed-Loop Sigma-Delta Modulator (MSCL)", *Analog Integrated Circuits and Signal Processing*, vol. 11, N° 3, pp.95-204, November 1996

BICMOS VERSUS CMOS TECHNOLOGY IN FULLY DIFFERENTIAL OTA DESIGNS

H. Recoules, R. Bouchakour, P. Loumeau

Ecole Nationale Supérieure des Télécommunications, Département électronique
46 Rue Barrault, 75634 Paris, Cedex 13, France
Tel : 33 (1) 45 81 72 11, Fax : 33 (1) 45 80 40 36

ABSTRACT

A design program evaluates CMOS and BICMOS fully differential OTA's performance independently of the application. As a validation an amplifier has been realized. Measurements and simulations are shown and compared. Finally a comparison is presented between CMOS and BICMOS OTA when long or short transistor lengths are used.

1. INTRODUCTION

The present trend toward high speed - high density analog circuit, tends to make technology choice a primary concern. Nowadays, the BICMOS technology advantage on CMOS is not really determined. It has been shown in specific applications how BICMOS circuits can have greater performance compared to CMOS ones [1] [2]. However it seems interesting to compare these technologies whatever the application. In this paper, frequency performance of folded cascode OTA (figure 1) are discussed in CMOS and BICMOS technologies independently of the application. A program has been developed to easily obtain the frequency behavior as a function of the current consumption (which determine the application domain). The first part of this paper briefly introduces the program and presents the results obtained with long dimension devices (L=1μm). In section 2 design results have been exploited and a BICMOS OTA has been realized in order to validate the program and the first part results. Measurements are shown and compared to the simulations. In the last part, the program is used to evaluate the frequency performance of OTA when minimal transistor lengths are used and finally the application domain of each technology is discussed.

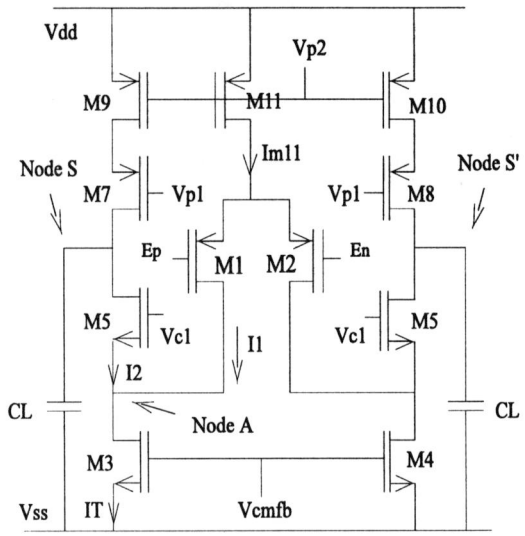

Figure 1: Transconductance amplifier

2. PERFORMANCE COMPARISON FOR L=1μM

The current consumption of the circuit is related to the IT variable (DC bias current, figure 1) by the following expression :

$$Current\ consumption = 2 * IT$$

The program developed in Matlab computes the performance parameters versus the DC bias current [3]. This procedure allows us to generate graphics illustrating various combinations of performance parameters, as the gain-bandwidth product or the phase margin versus the DC bias current of the circuit, and so on. This kind of procedure is very practical to observe the parameter evolutions and the design limits. With the program we can also impose the current IT and make it deduce the performance corresponding to this particular design.

The difficulty to compare two technologies resides in

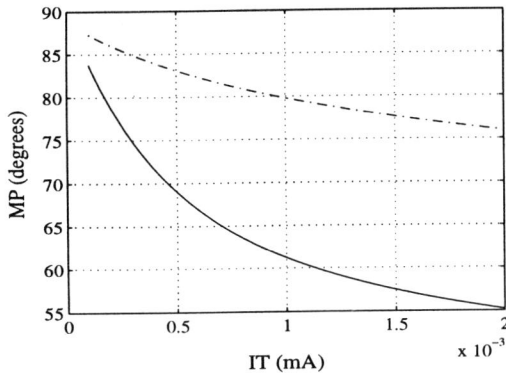

Figure 2: Phase margin versus DC bias current CMOS $1\mu m$ (-) and BICMOS $1\mu m$ (- .)

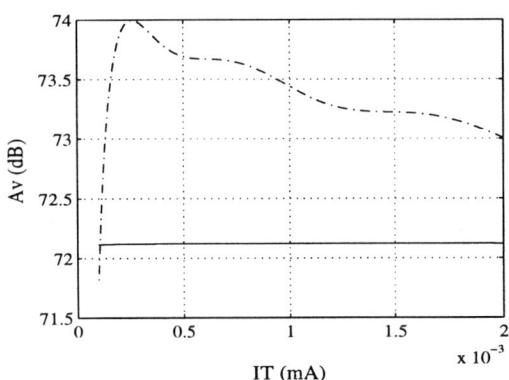

Figure 4: DC gain versus DC bias current, CMOS $1\mu m$ (-) and BICMOS $1\mu m$ (- .)

evaluating relevant parameter variations. We choose to illustrate the gain-bandwidth, phase margin and gain parameters as a function of the DC bias current IT. This is a manner to characterize the frequency performance of the folded cascode OTA in multiple applications. The CMOS and BICMOS circuits differ from transistors 5 and 6 (figure 1) which have been replaced by bipolar devices with two base electrode access. In the following discussion we first assume that

- currents are equally distributed between the output and input branches ($I1 = I2 = \frac{IT}{2}$)
- the length of the devices is set to $1\mu m$

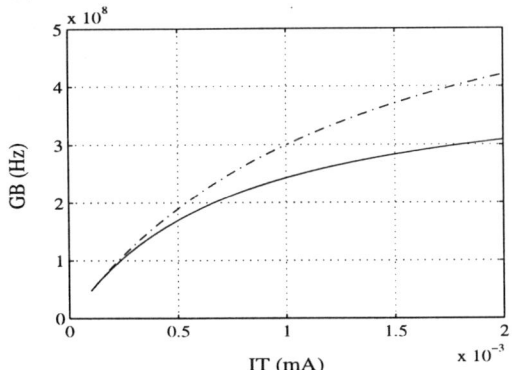

Figure 3: gain-bandwidth versus DC bias current, CMOS $1\mu m$ (-) and BICMOS $1\mu m$ (- .)

We can see from figure 2 how the phase margin is improved when a BICMOS circuit is used. The difference with CMOS increases with IT because the parasitic capacitors at node A become higher (due to the important MOS transistor dimensions) leading to a degradation of the second pole in the transfer function. It demonstrates the well-known high driving capabilities of bipolar devices compared to CMOS ones. Nevertheless the advantage of the BICMOS technology is not obvious if we consider the gain-bandwidth product in figure 3. The differences are significant just for high current consumptions. For example, to increase GB from 580Mhz to 760MHz, IT must be fixed to 2mA which just characterizes high consumption applications. In the same way, the improvement of the static gain doesn't justify the use of bipolar devices if low phase margins are needed (figure 4). In conclusion, except if the load capacitors can be changed in the specifications, the use of bipolar devices is not very interesting if the application doesn't impose the OTA to have a high phase margin.

3. PROGRAM VALIDATION FOR L=1μM

In order to validate the program results, performance specifications have been defined (table 1) and a realization has been carried out. This circuit is supposed to be inserted in a $\Sigma\Delta$ modulator soon. The amplifier has been realized in a $0.5\mu m$ BICMOS technology which is really adapted to perform these goal specifications. It is clear this is a restricted validation if we consider the low gain-bandwidth product required, but the circuit environment is not adapted to high frequency measurements (pad capacitor, differential probe ...).

GB	Ao	MP	Dvout
70Mhz	75dB	$> 80°$	Max
CMR	SR	PSRR	CL
Max	$> 150\mu V/s$	High	1pf

Table 1: Performances required for the OTA

Thanks to the program the design parameters have been found easily (the performance results gave 75Mhz for GB, 73dB for Ao and 86° for MP whereas the simulations respectively gave 72Mhz, 75 dB and 85°). All $Vgs - Vt$ of the devices are equal to 0.2V in order to maximize the output swing and the input common mode range. The static polarizations such as Vp1 and Vc1, which also influence Dvout, are respectively set to the highest and lowest value they can reasonably have. The approximated limits are conditioned by the saturation of M10 and M4. Concerning the current polarization in the OTA branches, an equal distribution of I1 and I2 is adopted, IT is set to $170\mu A$ in order to achieve the desired slew-rate value. Finally a switch capacitor network is connected between S, S' and Vcmfb to control the common mode voltage of this fully differential circuit. This leads to an increase of the current consumption and of the output capacitor loads (1.3 pf instead of 1pf). Note that the program already takes into account these output parasitic capacitors when the circuit corresponding to the desired performances is computed. Table 2 gives all the necessary design material to be inserted in a netlist file (note that the power is ±1.65V).

M1	M4	M8	M10	M11
115:1	57:1	110:1	188:1.5	375:1.5
Vp2	**Vp1**	**Vc1**	**Vcmfb**	
0.8V	0.36V	-0.52V	-0.77V	

Table 2: Design materials

For the measurement, the amplifier were placed into a 24-pin DIL package. The circuit has been configured as a derivator with a very low dominant pole. This allows to observe very similar characteristics than the open-loop gain frequency response (the resistors slightly decrease the static gain and the gain-bandwidth product, however thanks to simulations the real performance values can be deduced). The outputs S and S' are connected to the HP1141A differential probe. Its outputs present a capacitance of 2pf and a resistance of $10M\Omega$ between the two connectors. In conjunction with the probe, an HP41504A gain and phase analyzer is used to directly visualize the Bode diagram measurement. The results are shown in figure 5. Two simulations are also presented in order to compare the estimated results with the measurements. They correspond to the amplifier response in the test conditions with the use of two models : MASTAR [4] (dashed line) which is close to the model used in the program, and mm9 from Philips (circles). As we can see a static gain of 70dB is achieved instead of 65 dB given by the two simulations. The gain-bandwidth product is 20% percent less than the expected value (this is due to the layout mismatches

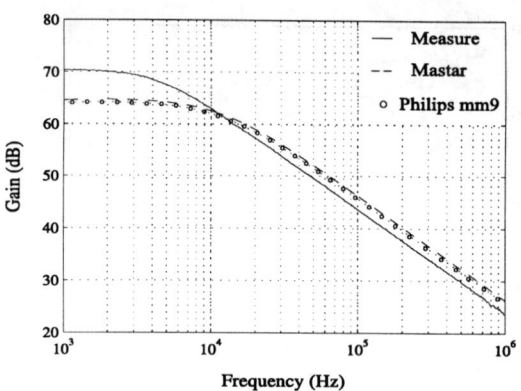

Figure 5: Bode diagram measurement (-) and simulation with Mastar (- -) and mm9 (o)

and the model accuracy). From here we can deduce the real parameter GB corresponding to the amplifier alone, we have found around 50Mhz which is a pretty good result when compared to the program and the simulation estimations. In conclusion, we can get reasonable calculations with the program , the DC gain is well estimated and a fluctuation of 20% maximum can be considered on the two curves in figure 3.

4. PERFORMANCE COMPARISON FOR L=0.5μM

In order to compare the limits of the amplifiers in term of performance, we present here the same graphics than in part 2 but the length of the devices has been fixed to $0.5\mu m$. We also have reported the results with L equal to $1\mu m$ in the same figures to see if the BICMOS technology is more suited for long or short devices. As expected figure 6 shows how the phase

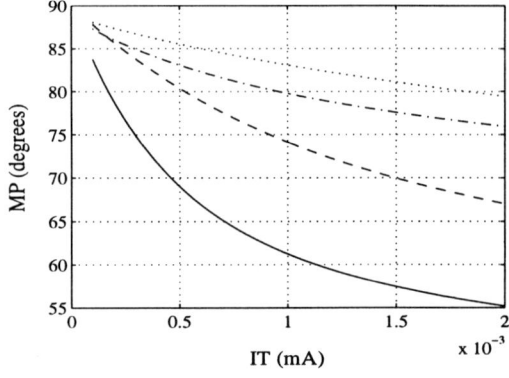

Figure 6: Phase margin versus DC bias current, CMOS: 0.5 μm (- -), 1 μm (-) and BICMOS: 0.5 μm (...), 1 μm (- .)

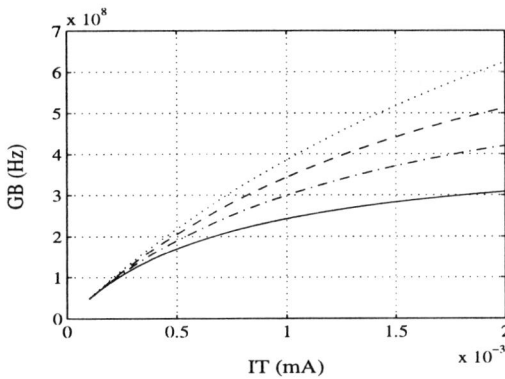

Figure 7: gain-bandwidth product versus DC bias current, CMOS: 0.5 μm (- -), 1 μm (-) and BICMOS: 0.5 μm (...), 1 μm (- .)

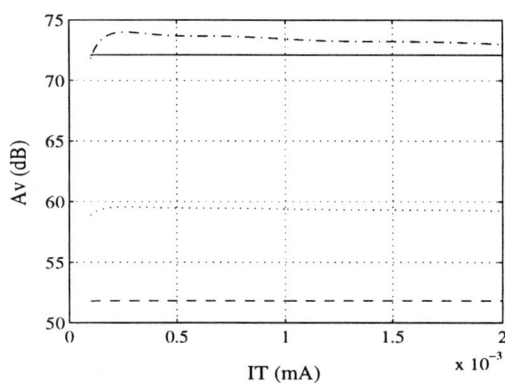

Figure 8: DC gain versus DC bias current, CMOS: 0.5 μm (- -), 1 μm (-) and BICMOS: 0.5 μm (...), 1 μm (- .)

margin decreases with the length. We can also observe that the use of short transistor imposes a high phase margin need. Concerning the gain-bandwidth product (figure 7), we can clearly see the augmentation of the parameter when minimal lengths are used. An excess of 100Mhz can be achieved if IT equals 2mA. On the contrary, if we look at figure 8 the DC gain difference between the two technologies decreases when the transistors are longer. This means in this case, that the use of bipolar transistors is only interesting for L close to $0.5\mu m$. Consequently the choice has to be made in function of the three following key parameters: GB, MP and the power consumption (proportional to IT). In conclusion, assuming that the load capacitor CL is imposed, the use of BICMOS amplifiers is well suited for :

- very high phase margin and low current consumption applications. The choice of L will depend on the gain-bandwidth product and the static gain in the specifications.

- most of high current consumption applications.

Finally the application domain where the CMOS technology still competes, concerns the low current consumption circuits with poor constraints on the phase margin.

5. CONCLUSION

An evaluation of the frequency performances of folded cascode OTA implemented in CMOS and BICMOS technologies has been presented thanks to a design procedure. An experimental realization of a BICMOS OTA gave a validation of the program. Comparisons of the key performance parameters allowed to define the application domains related to both technologies. In general, the use of BICMOS OTA suits well to improve the frequency performances (GB, Av, ...) especially if the load capacitors can be changed in the specifications [5] [6] [2]. If not, the advantages to its CMOS counterpart essentially concern the increase of the phase margin, the static gain (only when short devices are used) and all the performance parameters at very high currents.

6. REFERENCES

[1] F. Larsen and M. Ismail. "the design of high performance low cost bicmos op-amps in a predominantly cmos technology". *IEEE*, 1995.

[2] M. Ismail and T. Fiez. *Analog vlsi signal and information processing*. Mc Graw-Hill, 1995.

[3] H. Recoules, B. Bouchakour, and P. Loumeau. "Optimization of Bicmos fully differential OTA's gain-bandwidth and comparison with CMOS technology". *MIDWEST*, 1997.

[4] T. Skotnicki, C. Denat, P. Senn, G. Merckel, and B. Hennion. "A new analog/digital CAD model for sub-halfmicron MOSFETs". *IEDM94*.

[5] M. Nayebi and B. A. Wooley. "A 10-bit video BICMOS track-and-hold amplifier". *IEEE J.SSC*, 24(6):1507–1516, Dec. 1989.

[6] G. Nebel, U. Klein, and H. J. Pfleiderer. "Large bandwidth BICMOS operational amplifiers for SC-Video-applications". *IEEE Proc.ISCAS94*, pages 85–88, 1994.

ANALYSIS OF MODULATOR CIRCUITS BASED ON MULTI-DIMENSIONAL FOURIER TRANSFORMATION

Akio USHIDA, Yoshihiro YAMAGAMI, Yoshifumi NISHIO
Department of Electrical and Electronic Engineering,
Tokushima University, Tokushima, 770 JAPAN

ABSTRACT

There are many communication circuits driven by multi-tone signals such as modulator and mixer. If the input frequency components are largely different in each other, the brute force numerical method will take an enormous computation time to get the steady-state responses. In this paper, we show a SPICE oriented algorithm based on multi-dimensional Fourier transformation, where all of the circuit analyses such as dc- and ac-analysis in the algorithm are carried out with SPICE. On the other hand, a very simple sensitivity analysis and 2-dimensional FFT are carried out by a Fortran program (or C program). We found that the convergence ratio of our algorithm is sufficiently large, and can be applied to wide class of communication circuits.

1. INTRODUCTION

Many communication circuits, such as modulators, mixers and frequency converters, are driven by multi-tone signals. There are 2 basic approaches for the computation of the steady-state responses: (1) frequency-domain approach [1-2] and (2) time-domain approach [3-4]. The former can be applied only to weakly nonlinear circuits, because the scale of determining equations becomes very large for strongly nonlinear circuits. The latter is based on numerical integration techniques, that can be efficiently applied to circuits having a few number of the state variables.

Generally, modulators and mixers are driven by two input signals, namely, the high frequency carrier and the low frequency modulating signal. In this case, if we use a *brute force method* (transient analysis) for getting the steady-state response, it will take an enormous computational time, because the step size must be chosen sufficiently small depending on the carrier signal. Consider an example such that a ratio of the two input frequencies is $f_2/f_1 = 1000$. If the step size is chosen $h = T_2/100$ for $T_2 = 1/f_2$, it will take 1000×100 numerical integrations for only one period ($T_1 = 1/f_1$).

A new SPICE oriented method is presented in this paper which is based on both *2-dimensional Fourier transformation* and *frequency-domain relaxation* methods. Assume that a given circuit is composed of nonlinear resistive sub-networks and reactive elements such as capacitors and inductors. At first, using the *substitution sources*, the circuit is partitioned into two groups, namely, the nonlinear resistive subnetworks and the reactive elements. The substitution waveforms are described by 2-dimensional Fourier expansions, and the coefficients are calculated by the relaxation method.

We have developed a very simple simulator consisting of SPICE and a Fortran program, where all of the circuit analysis are implemented by SPICE. On the other hand, very simple sensitivity analysis and 2-dimensional FFT are carried out by the Fortran program.

2. BASIC APPROACH

To focus on the main idea of our relaxation method, consider a circuit as shown in Fig.1(a)[1]. Now, assume the two inputs $e(t)$ and $j(t)$ contain two independent frequency components ω_1 and ω_2.

Then, the substitution sources at the partitioning point will be generally assumed of the form

$$v_C(t) = V_{C,0} + \sum_{k=1}^{M} \{V_{C,2k-1} \cos \nu_k t + V_{C,2k} \sin \nu_k t\} \quad (1.1)$$

$$i_L(t) = I_{L,0} + \sum_{k=1}^{M} \{I_{L,2k-1} \cos \nu_k t + I_{L,2k} \sin \nu_k t\} \quad (1.2)$$

$$\nu \equiv m_{1k}\omega_1 + m_{2k}\omega_2 \quad (1.3)$$

---
[1]Generally, integrated circuits are composed of capacitors and resistive elements such as transistors and diodes. If in this case, it contains large capacitances, the transient response will continue for a large period, and it take long computation time to get the steady-state by the brute force method. Therefore, we partitioned the circuit into two groups containing nonlinear resistive circuit and reactive elements respectively, as shown in Fig.1(b).

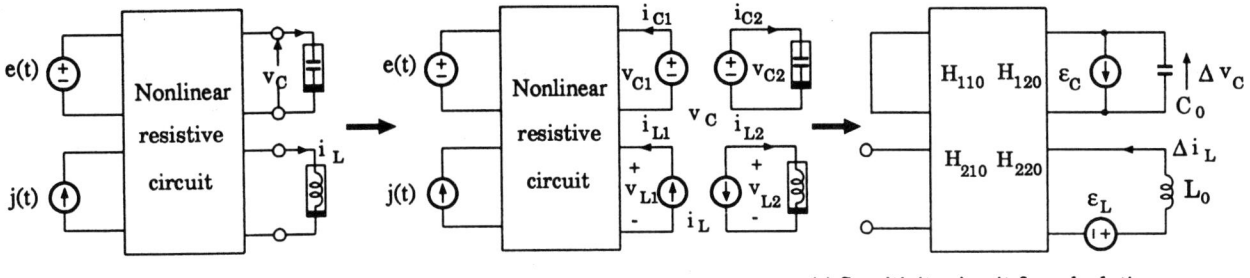

(a) Nonlinear circuit (b) Circuit partition (c) Sensitivity circuit for calculation of the variational value

Fig.1 Schematic diagram of our relaxation method

where m_{1k}, m_{2k} are integers satisfying

$$|m_{1k}| \leq B, \quad |m_{2k}| \leq B \tag{1.4}$$

for some sufficiently large B.

Assuming that the original circuit in Fig.1(a) has a unique steady-state solution described by (1), then $v_C(t)$ and $i_L(t)$ satisfies the following *determining equation*:

$$F_1(v_C, i_L) \equiv i_{C1}(t) + i_{C2}(t) = 0 \tag{2.1}$$

$$F_2(v_C, i_L) \equiv v_{L1}(t) - v_{L2}(t) = 0 \tag{2.2}$$

Now, assume the nonlinear capacitor and inductor are described by

$$q_{C2} = \hat{q}_{C2}(v_{C2}), \quad \phi_{L2} = \hat{\phi}_{L2}(i_{L2}) \tag{3}$$

Then, we have

$$i_{C2} = \frac{\partial \hat{q}_{C2}}{\partial v_{C2}} \frac{dv_{C2}}{dt}, \quad v_{L2} = \frac{\partial \hat{\phi}_{L2}}{\partial i_{L2}} \frac{di_{L2}}{dt} \tag{4}$$

Let us calculate the steady-state response using an iteration technique. Assume the solution at the jth iteration is given by

$$v_C^j(t) = V_{C,0}^j + \sum_{k=1}^{M} \left\{ V_{C,2k-1}^j \cos \nu_k t + V_{C,2k}^j \sin \nu_k t \right\} \tag{5.1}$$

$$i_L^j(t) = I_{L,0}^j + \sum_{k=1}^{M} \left\{ I_{L,2k-1}^j \cos \nu_k t + I_{L,2k}^j \sin \nu_k t \right\} \tag{5.2}$$

To evaluate the solution at the $(j+1)$th iteration, put

$$v_C^{j+1}(t) = v_C^j(t) + \Delta v_C(t), \quad i_L^{j+1}(t) = i_L^j(t) + \Delta i_L(t) \tag{6}$$

where the variations $\Delta v_C(t)$ and $\Delta i_L(t)$ are described by

$$\Delta v_C(t) = \Delta V_{C,0} + \sum_{k=1}^{M} \left\{ \Delta V_{C,2k-1} \cos \nu_k t + \Delta V_{C,2k} \sin \nu_k t \right\} \tag{7.1}$$

$$\Delta i_L(t) = \Delta I_{L,0} + \sum_{k=1}^{M} \left\{ \Delta I_{L,2k-1} \cos \nu_k t + \Delta I_{L,2k} \sin \nu_k t \right\} \tag{7.2}$$

Substituting $v_C^{j+1}(t)$, $i_L^{j+1}(t)$ from (6) into (2), and neglecting the higher-order terms of $\Delta v_C(t)$ and $\Delta i_L(t)$ in the Taylor expansion of nonlinear terms, we obtain

$$F_1(v_C^j + \Delta v_C, i_L^j + \Delta i_L) = i_{C1}(v_C^{j+1}, i_L^{j+1}) + i_{C2}(v_C^{j+1})$$

$$\approx i_{C1}(v_C^j, i_L^j) + \frac{\partial i_{C1}(v_C, i_L)}{\partial v_C} \Delta v_C + \frac{\partial i_{C1}(v_C, i_L)}{\partial i_L} \Delta i_L$$

$$+ i_{C2}(v_C^j) + \frac{\partial \hat{q}_{C2}(v_C)}{\partial v_C} \frac{d\Delta v_C}{dt} = 0 \tag{8.1}$$

$$F_2(v_C^j + \Delta v_C, i_L^j + \Delta i_L) = v_{L1}(v_C^{j+1}, i_L^{j+1}) - v_{L2}(i_L^{j+1})$$

$$\approx v_{L1}(v_C^j, i_L^j) + \frac{\partial v_{L1}(v_C, i_L)}{\partial v_C} \Delta v_C + \frac{\partial v_{L1}(v_C, i_L)}{\partial i_L} \Delta i_L$$

$$- v_{L2}(i_L^j) - \frac{\partial \hat{\phi}_{L1}(i_L)}{\partial i_L} \frac{d\Delta i_L}{dt} = 0 \tag{8.2}$$

Now, define the *residual sources* as follows:

$$\varepsilon_C^j(t) \equiv i_{C1}(v_C^j, i_L^j) + i_{C2}(v_C^j) \tag{9.1}$$

$$\varepsilon_L^j(t) \equiv v_{L1}(v_C^j, i_L^j) - v_{L2}(i_L^j) \tag{9.2}$$

Since the relation (8) for calculating Δv_C and Δi_L is a linear time-varying system, it is not easy to solve even if it is linear. Therefore, we approximate the equations by the *time-invariant systems* as follows:

$$\begin{pmatrix} H_{110} & H_{120} \\ H_{210} & H_{220} \end{pmatrix} = \begin{pmatrix} \frac{\partial i_{C1}(v_C, i_L)}{\partial v_C} & \frac{\partial i_{C1}(v_C, i_L)}{\partial i_L} \\ \frac{\partial v_{L1}(v_C, i_L)}{\partial v_C} & \frac{\partial v_{L1}(v_C, i_L)}{\partial i_L} \end{pmatrix} \tag{10.1}$$

for $v_C = v_{C0}$, $i_L = i_{L0}$, and

$$C_0 = \left. \frac{\partial \hat{q}_{C2}}{\partial v_{C2}} \right|_{v_{C0}}, \quad L_0 = \left. \frac{\partial \hat{\phi}_{L2}}{\partial i_{L2}} \right|_{i_{L0}} \tag{10.2}$$

where v_{C0} and i_{L0} are dc solutions at the operating points. Thus, the relation (8) can be described as follows:

$$\begin{pmatrix} H_{110} & H_{120} \\ H_{210} & H_{220} \end{pmatrix} \begin{pmatrix} \Delta v_C \\ \Delta i_L \end{pmatrix} + \begin{pmatrix} C_0 & 0 \\ 0 & -L_0 \end{pmatrix} \begin{pmatrix} \Delta \dot{v}_C \\ \Delta \dot{i}_L \end{pmatrix}$$
$$= -\begin{pmatrix} \varepsilon_C^j(t) \\ \varepsilon_L^j(t) \end{pmatrix} \quad (11)$$

Thus, we have the equivalent sensitivity circuit shown by Fig.1(c). It can be easily solved by the phasor technique. Observe that, although the convergence ratio may be decreased for the strongly nonlinear circuits, the algorithm is very simple and produces the exact solution after convergence.

The iterations will continue until the variational values $\Delta v_C(t)$ and $\Delta i_L(t)$ satisfy the following *stopping condition*:

$$\| \Delta V_C \| + \| \Delta I_L \| < \delta \quad (12)$$

$\Delta V_C \equiv [\Delta V_{C,0}, \ldots, \Delta V_{C,2M}], \Delta I_L \equiv [\Delta I_{L,0}, \ldots, \Delta I_{L,2M}]$ for a sufficiently small δ. Furthermore, if the residual current does not satisfy the following condition:

$$\| \varepsilon^j(t) \| \equiv \sqrt{\frac{1}{T} \int_0^T (\varepsilon_C^j(t))^2 + (\varepsilon_L^j(t))^2 dt} < \epsilon \quad (13)$$

for a large T and a small ϵ, then we need to increase the Fourier terms M in (1).

Note that the nonlinear resistive network in Fig.1(a) may have small parasitic capacitors. If they cannot be neglected at the high frequency, we need to take account of the nonlinear capacitors in Fig. 1(a). Thus, the computer efficiency of the algorithm will be decreased according to the number of nonlinear capacitors. In our many examples, we recommend to partition a circuit into subcircuits at only coupling capacitors, whose capacitors voltages are considered as substition voltage sources in Fig.1(b).

3. SPICE IMPLEMENTATION

Nowaday, SPICE is widely used for many circuit simulation purposes such as dc-analysis, ac-analysis, transient analysis and so on. Our simulator is implemented by ac- and dc-analysis of SPICE, and a very simple Fortran program (or C-program).

Implementation algorithm

0. A given circuit is partitioned into nonlinear resistive circuits and reactive elements with substitution sources. Considering the amplitudes of the signal and carrier inputs [7], set the highest harmonic M in (1).

Next, choose sufficiently small stopping conditions δ and ϵ in (12) and (13), respectively. At first, we draw the dc circuit diagram, and solve it by SPICE [2]. Thus, each capacitor is replaced by a substituting voltage source with a dc-voltage source V_{C0}^0, and each inductor by a substituting current source with I_{L0}^0.

1. Set $v_C^0(t) = V_{C0}^0$ and $i_L^0(t) = I_{L0}^0$. Applying ac-sweep of SPICE, determine $H_{110}, \ldots, H_{220}, C_0$ and L_0 at zero frequency. Set $j = 0$

2. Solve the nonlinear resistive circuit with $(e(t), j(t), v_C^j(t), i_L^j(t))$ by dc-analysis of SPICE. In this case, 2-dimensional FFT can be carried out by the application of the one-dimensional FFT to the ω_1-components and ω_2-components, separately [7]. Thus, we have 2-dimensional Fourier expansions of $i_{C1}^j(t)$ and $v_{L1}^j(t)$.

3. Calculate the responses of nonlinear reactive elements, and describe them by the 2-dimensional Fourier expansions of $i_{C2}^j(t)$ and $v_{L2}^j(t)$. Note that if the reactive elements are linear, we need not apply 2-dimensional FFT.

4. Estimate $\varepsilon_C^j(t)$ and $\varepsilon_L^j(t)$ given by (9). Thus, the relation (11) can be easily solved by the phasor technique, and get $\Delta v_C(t)$ and $\Delta i_L(t)$. If $\| \Delta V_C^j \| + \| \Delta I_L^j \| < \delta$, go to 5.
Otherwise, set $v_C^{j+1}(t) = v_C^j(t) + \Delta v_C$ and $i_L^{j+1}(t) = i_L^j(t) + \Delta i_L$, and $j = j + 1$. Go to Step 1.

5. Estimate $\| \varepsilon^j(t) \|$ in (13). If $\| \varepsilon^j(t) \| < \epsilon$, **stop**. Otherwise, increase B in (1.4) and go to Step 0.

We have carried out the algorithm with IBM PC loading PSPICE of MicroSim co.

4. AN ILLUSTRATIVE EXAMPLE

Consider a mixer circuit shown by Fig.2(a). It has two inputs of

$$e_1(t) = 0.01 \sin 2\pi \times 50 \times 10^6 t [V]$$

$$e_2(t) = 0.01 \sin 2\pi \times 51 \times 10^6 t [V]$$

We partition the circuit at the two capacitors C_1 and C_2, because they are considered as sufficiently large compared

---
[2]Introduce the compensational resistors R_c and $-R_c$ if the nonlinearity is strong [8].

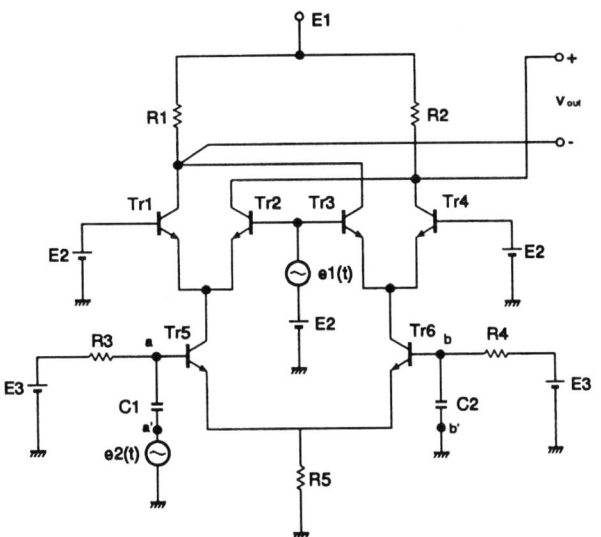

Fig.2(a) Mixer circuit
$R_1 = R_2 = 100\Omega$, $R_3 = R_4 = 10k\Omega$, $C_1 = C_2 = 0.01\mu F$
$R_5 = 200\Omega$, $E_1 = 5V$, $E_2 = 2.5V$, $E_3 = 12V$

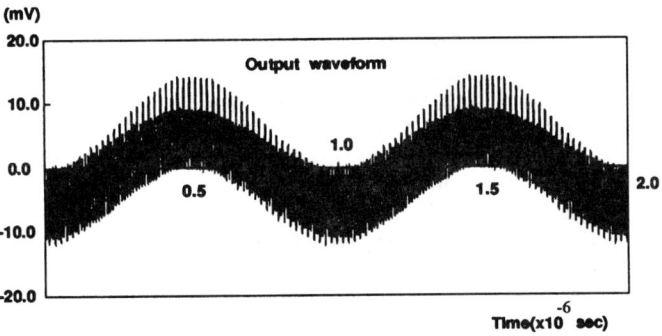

Fig.2(b) Steady-state output waveform

to the parasitic capacitances in the transistors. We assume the waveforms as follows:

$$v_i(t) = V_{i,0} + \sum_{k=1}^{7} \sum_{n=1}^{7} \{V_{i,kn,c1} \cos(k\omega_1 + n\omega_2)t$$
$$+ V_{i,kn,c2} \cos(k\omega_1 - n\omega_2)t + V_{i,kn,1s} \sin(k\omega_1 + in\omega_2)t$$
$$+ V_{i,kn,2s} \sin(k\omega_1 - n\omega_2)t\}, \quad i = 1, 2$$

We found that $V_{1C0}^0 = V_{2C0}^0 = 2.388[V]$ by the dc-analysis of SPICE. The steady-state response can be obtained in 3 iterations of our algorithm. Note that if we apply a *brute force method*, it will take an enormous computation time because the relative frequency difference of two inputs is very small, and the response contains a very low frequency of $1MHz$.

5. Conclusions and Remarks

In this paper, we have presented that 2-dimensional Fourier transformation can be efficiently applied to calculate the steady-state response driven by 2-frequency input signals such as modulators and mixers. The efficiency does not depend on the frequency values. Furthermore, it can be easily modified to the analysis of multiple-frequency inputs greater than two, and to the noise analysis.

We have developed a very simple simulator cosisting of SPICE and a Fortran program, where all of the circuit analyses are implemented by SPICE. Another simple sensitivity analysis and 2-dimensional FFT are carried out by the Fortran program.

Note that, for very high frequency, we need to take into account parasitic capacitors of transistors. If the number is increased, the computer efficiecy will be decreased the computer efficiency. This is a future research problem.

REFERENCES

[1] K.S.Kundert, G.B.Sorkin and A.Sangjovanni-Vincentelli, "Applying harmonic balance to almost-periodic circuit," *IEEE Trans. Microwave Theory Tech.*, vol.MTT-36, pp.366-378, 1988.

[2] A.Ushida and L.O.Chua, "Frequency-domain analysis of nonlinear circuits driven by multi-tone signals," *IEEE Trans. Circuits Syst.*, vol.CAS-31, pp.766-779, 1984.

[3] M.Okumura, T.Sugawara and H.Tanimoto, "An efficient small signal frequency analysis method of nonlinear circuits with two frequency excitations," *IEEE Trans. Computer-aided Design*, vol.CAD-9, no.3, pp.225-235, 1990.

[4] M.Okumura, H.Tanimoto, T.Itakura and T.Sugawara, "Numerical noise analysis for nonlinear circuits with a periodic large signal excitation including cyclostationary noise sources," *IEEE Trans. Circuits Syst.-I: Fundamental Theory and Applications*, vol.CAS-40, no.9, pp.581-590, 1993.

[5] A.Ushida, T.Adachi and L.O.Chua, "Steady-state analysis of nonlinear circuits based on hybrid method," *IEEE Trans. Circuits Syst.-I: Fundamental Theory and Applications*, vol.CAS-39, no.8, pp.649-661, 1992.

[6] R. Telichevesky and K.Kundert, *SpectreRF Primer*, Cadence Design System, san Jose, California, July, 1996.

[7] A.Ushida, L.O.Chua and T.Sugawara "A substitution algorithm for solving non-linear circuits with multi-frequency components," *Int. Jour of Circuit Theory and Applications*, vol.15, pp.327-355, 1987.

[8] A.Ushida and L.O.Chua "Steady-state response of nonlinear circuits: A frequency-domain relaxation method," *Int. Jour of Circuit Theory and Applications*, vol.17, pp.249-269, 1989.

JOINT OPTIMIZATION OF MULTIPLE BEHAVIORAL AND IMPLEMENTATION PROPERTIES OF ANALOG FILTER DESIGNS

Niranjan Damera-Venkata and Brian L. Evans[*]

Dept. of Electrical and Computer Eng.
Engineering Science Building
The University of Texas at Austin
Austin, TX 78712-1084 USA
damera-v@ece.utexas.edu
bevans@ece.utexas.edu

Miroslav D. Lutovac and Dejan V. Tošić

Faculty of Electrical Engineering
Bulevar Revolucije 73
University of Belgrade
11000 Belgrade, Yugoslavia
lutovac@iritel.bg.ac.yu
tosic@telekom.etf.bg.ac.yu

ABSTRACT

This paper presents an extensible framework for optimizing analog filter designs for multiple behavioral and implementation properties. We demonstrate the framework using the behavioral properties of magnitude response, phase response, and peak overshoot, and the implementation property of quality factors. We represent the analog filter in terms of its poles and zeroes. We match the constrained non-linear optimization problem to a sequential quadratic programming (SQP) problem, and develop symbolic mathematical software to translate the SQP formulation into working MATLAB programs to optimize analog filter designs. The automated approach avoids errors in algebraic calculations and errors in transcribing the mathematical equations in software. The packages are freely distributable.

1. INTRODUCTION

Classical analog filter design techniques optimize one filter property subject to constraints on the magnitude response. In designing and implementing analog filters, several behavioral properties (e.g. magnitude response, phase response and peak overshoot) and implementation properties (e.g. quality factors) may be important. For example, anti-aliasing filters require a near linear phase response while meeting a set of magnitude specifications [1].

This paper presents a formal extensible framework for optimizing analog filter designs for multiple behavioral and implementation properties. We demonstrate the framework using the behavioral properties of magnitude response, phase response, and peak overshoot, and the implementation property of quality factors. The framework takes an existing analog filter design, e.g. one designed using a classical numeric approach or a modern symbolic approach [2], and jointly optimizes any combination of these four properties subject to constraints on these four properties. We match the constrained non-linear optimization problem to a sequential quadratic programming (SQP) problem and develop symbolic mathematical software to translate the SQP formulation into working MATLAB programs that can optimize analog filter designs. SQP requires that the objective function [3] and the constraints [4] be real-valued and twice continuously differentiable with respect to the free parameters. The free parameters are the pole and zero locations, which we use to represent the analog filter. SQP relies on the gradients of the objective function and constraints. SQP methods have been previously applied to optimizing loss and delay in digital filter designs [5] and optimizing even-order all-pole filter designs [6]. In this paper, we reformulate our results in [6] to include an even number of zeros in the analog filter to be optimized.

Using the symbolic mathematics environment Mathematica, we program the objective function and constraints, compute their gradients symbolically, and generate MATLAB code for the objective function and constraints as well as their gradients. The generated MATLAB code calls the SQP procedure `constr` in the Optimization Toolbox to solve the constrained non-linear optimization problem. In Mathematica, a designer can add, delete, and change cost measures and constraints for a given property, and our symbolic software will then regenerate the MATLAB numerical optimization code. We have bridged the gap between the symbolic work designers do on paper and the working computer implementation, thereby eliminating algebraic errors in hand calculations and bugs in coding the software implementation. Our software is available at http://www.ece.utexas.edu/~bevans/projects/syn_filter_software.html.

Section 2 reviews notation. Section 3 derives a family of weighted, differentiable objective functions to measure the deviation in magnitude response, deviation in linear phase response, quality factors, and peak overshoot of the step response, of an analog filter. In the derivation, we find a new analytic approximation for the peak overshoot. Section 4 converts filter specifications into differentiable constraints. Section 5 gives an example of an optimized filter design.

2. NOTATION

We represent an analog filter by its n complex conjugate pole pairs $p_k = a_k \pm jb_k$ where $a_k < 0$ and its r complex conjugate zero pairs $z_l = c_l \pm jd_l$ where $c_l < 0$, such that

---

[*] This research was supported by an NSF CAREER Award under Grant MIP-9702707.

$r \leq n$. The magnitude and unwrapped phase responses of an all-pole filter, expressed as real-valued differentiable functions, are

$$|G(j\omega)| = \prod_{k=1}^{n} \frac{a_k^2 + b_k^2}{\sqrt{a_k^2 + (\omega+b_k)^2}\sqrt{a_k^2 + (\omega-b_k)^2}}$$
$$= \prod_{k=1}^{n} \frac{a_k^2 + b_k^2}{\sqrt{(\omega^2 + 2(a_k^2 - b_k^2))\omega^2 + (a_k^2 + b_k^2)^2}} \quad (1)$$

$$\angle G(j\omega) = \sum_{k=1}^{n} \arctan\left(\frac{\omega - b_k}{a_k}\right) + \arctan\left(\frac{\omega + b_k}{a_k}\right) \quad (2)$$

We factor the polynomial under the square root in (1) into Horner's form because it has better numerical properties. Together with the zero pairs, the magnitude and unwrapped phase responses are

$$|H(j\omega)| = |G(j\omega)| \times \prod_{l=1}^{r} \frac{\sqrt{(\omega^2 + 2(c_l^2 - d_l^2))\omega^2 + (c_l^2 + d_l^2)^2}}{c_l^2 + d_l^2} \quad (3)$$

$$\angle H(j\omega) = \angle G(j\omega) - \sum_{l=1}^{r} \arctan\left(\frac{\omega - d_l}{c_l}\right) + \arctan\left(\frac{\omega + d_l}{c_l}\right) \quad (4)$$

In this paper, Q represents quality factors, ϵ represents a small positive number, σ denotes deviation, m represents slope of a line, and t is time.

3. OBJECTIVE FUNCTIONS

In this section, we derive measures of closeness to an ideal magnitude and phase response, quality factors, and peak overshoot. The objective function is a non-negative function that it is weighted combination of these measures.

3.1. Deviation in the Magnitude Response

Based on the notation in Figure 1, the five components of the objective function relating to the deviation from an ideal magnitude response in the least squares sense are:

$$\sigma_{sb1} = \int_{0}^{\omega_{s1}} F_{s1}(\omega) |H(j\omega)|^2 \, d\omega \quad (5)$$

$$\sigma_{tb1} = \int_{\omega_{s1}}^{\omega_{p1}} F_{t1}(\omega) \left(|H(j\omega)| - (m_1 \omega - m_1 \omega_{s1})\right)^2 d\omega \quad (6)$$

$$\sigma_{pb} = \int_{\omega_{p1}}^{\omega_{p2}} F_p(\omega) \left(|H(j\omega)| - 1\right)^2 d\omega \quad (7)$$

$$\sigma_{tb2} = \int_{\omega_{p2}}^{\omega_{s2}} F_{t2}(\omega) \left(|H(j\omega)| - (m_2 \omega - m_2 \omega_{s2})\right)^2 d\omega \quad (8)$$

$$\sigma_{sb2} = \int_{\omega_{s2}}^{\infty} F_{s2}(\omega) |H(j\omega)|^2 \, d\omega \quad (9)$$

where $F_p(\omega)$, $F_{t1}(\omega)$, $F_{t2}(\omega)$, and $F_s(\omega)$ are integrable weighting functions, and m_1 and m_2 are the slopes of the ideal response in the transition regions defined as $m_1 = 1/(\omega_{p1} - \omega_{s1})$ and $m_2 = 1/(\omega_{p2} - \omega_{s2})$.

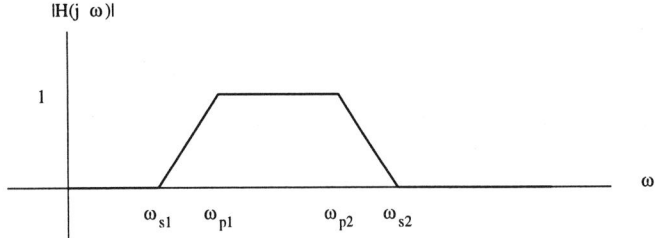

Figure 1: The ideal magnitude response

3.2. Deviation in the Phase Response

For the passband response, the objective function measures the deviation from linear phase over some range of frequencies (usually over the passband):

$$\sigma_{phase} = \int_{\omega_1}^{\omega_2} (\angle H(j\omega) - m_{lp}\omega)^2 \, d\omega \quad (10)$$

where m_{lp} is the ideal slope of the linear phase response. Unfortunately, one does not know the value of m_{lp} à priori. We can compute it as the slope of the line in ω that minimizes (10):

$$\min_{m_{lp}} \int_{\omega_1}^{\omega_2} (\angle H(j\omega) - m_{lp}\omega)^2 \, d\omega \quad (11)$$

In (11), the $H(j\omega)$ term does not depend on m_{lp}, so the integrand is quadratic in m_{lp}. To find the minimum, we take the derivative with respect to m_{lp}, set it to zero, and solve for m_{lp}:

$$m_{lp} = \frac{\int_{\omega_1}^{\omega_2} \angle H(j\omega) \, \omega \, d\omega}{\int_{\omega_1}^{\omega_2} \omega^2 \, d\omega} \quad (12)$$

After evaluating the integrals,

$$m_{lp} = \frac{3}{2(\omega_2^3 - \omega_1^3)} \times \left[\sum_{k=1}^{n} [f_{lp1}(\omega_2) - f_{lp1}(\omega_1)] - \sum_{l=1}^{r} [f_{lp2}(\omega_2) - f_{lp2}(\omega_1)] \right] \quad (13)$$

where $f_{lp1}(\omega)$ is

$$f_{lp1}(\omega) = 2\omega a_k + (b_k^2 - a_k^2 - \omega^2) \times \left(\arctan\left(\frac{\omega - b_k}{a_k}\right) + \arctan\left(\frac{\omega + b_k}{a_k}\right) \right) + a_k b_k \left(\log\left(1 + \frac{(\omega - b_k)^2}{a_k^2}\right) - \log\left(1 + \frac{(\omega + b_k)^2}{a_k^2}\right) \right)$$

and $f_{lp2}(\omega)$ is

$$f_{lp2}(\omega) = 2\omega c_k + (d_k^2 - c_k^2 - \omega^2) \times \left(\arctan\left(\frac{\omega - d_k}{c_k}\right) + \arctan\left(\frac{\omega + d_k}{c_k}\right) \right) + c_k d_k \left(\log\left(1 + \frac{(\omega - d_k)^2}{c_k^2}\right) - \log\left(1 + \frac{(\omega + d_k)^2}{c_k^2}\right) \right)$$

Using Mathematica, we computed the definite integrals in (12) and verified the answers. Now that we have a closed-form solution for m_{lp}, we can substitute (13) into (10) to obtain a rather complicated but differentiable expression for the deviation from linear phase.

3.3. Filter Quality

The quality factor measures the relative distance of a filter pole from the imaginary frequency axis. The lower the quality factor, the less likely that the pole will cause oscillations in the output. The quality factor Q_k for the kth second-order section with conjugate poles $a_k \pm jb_k$ (with $a_k < 0$) and the effective overall quality factor Q_{eff} are

$$Q_k = \frac{\sqrt{a_k^2 + b_k^2}}{-2a_k} \qquad Q_{\text{eff}} = \left(\prod_{k=1}^{n} Q_i \right)^{\frac{1}{n}} \qquad (14)$$

where $Q_k, Q_{\text{eff}} \geq 0.5$. $Q_k = 0.5$ corresponds to a double real-valued pole ($b_k = 0$), and $Q_k = \infty$ corresponds to an ideal oscillator ($a_k = 0$). We define Q_{eff} as the geometric mean of the quality factors, and other measures could be used. We use $Q_{\text{eff}} - 0.5$ to measure the filter quality.

3.4. Peak Overshoot in the Step Response

From the step response, we can numerically compute the peak overshoot and the time t_{peak} at which it occurs. In order to make the peak overshoot calculation differentiable, this section derives an analytic expression that approximates t_{peak} in terms of the pole-zero locations. The derivation assumes that there are no multiple poles.

The Laplace transform of the step response is

$$\frac{H(s)}{s} = \frac{1}{s} \left[\prod_{k=1}^{n} \frac{a_k^2 + b_k^2}{s^2 - 2a_k s + a_k^2 + b_k^2} \right] \times \left[\prod_{k=1}^{n} \frac{s^2 - 2c_k s + c_k^2 + d_k^2}{c_k^2 + d_k^2} \right] \qquad (15)$$

Assuming no duplicate poles, partial fractions yields

$$\frac{H(s)}{s} = \left[\frac{A}{s} + \sum_{k=1}^{n} \frac{C_k s + D_k}{s^2 - 2a_k s + a_k^2 + b_k^2} \right] \qquad (16)$$

$$\begin{aligned} C_k &= 2|B_k|\cos(\angle B_k) \\ D_k &= -2|B_k|(a_k \cos(\angle B_k) + b_k \sin(\angle B_k)) \\ B_k &= [H(s)(s - p_k)]_{s=p_k} = |B_k| e^{j\angle B_k} \\ A &= [H(s) \times s]_{s=0} = 1 \end{aligned}$$

$|B_k|$ and $\angle B_k$ can be expressed as real-valued differentiable functions of the pole and zero locations.

After inverse transforming (16), the step response is

$$h_{\text{step}}(t) = 1 + \sum_{k=1}^{n} e^{a_k t} \left[C_k \cos(b_k t) + \left(\frac{D_k + C_k a_k}{b_k} \right) \sin(b_k t) \right] \qquad (17)$$

By analyzing the kth term in the summation in (16), the kth peak overshoot occurs at time

$$t_{\text{peak}}^k = -\frac{1}{b_k} \left[\arctan\left(\frac{(D_k + 2C_k a_k) b_k}{C_k(a_k^2 - b_k^2) + D_k a_k} \right) + \pi \right] \qquad (18)$$

We construct the following function to approximate t_{peak} for the purposes of computing derivatives:

$$t_{\text{peak}} \approx \frac{1}{n} \sum_{k=1}^{n} t_{\text{peak}}^k \Rightarrow t_{\text{peak}} = \beta \frac{1}{n} \sum_{k=1}^{n} t_{\text{peak}}^k \qquad (19)$$

Here, β is set to the true value of t_{peak} (found numerically) divided by the approximation $\frac{1}{n} \sum_{k=1}^{n} t_{\text{peak}}^k$. We validated (19) using the SQP routine on several designs. We measure the peak overshoot cost by using $(h_{\text{step}}(t_{\text{peak}}) - 1)^2$.

4. CONSTRAINTS

This section discusses two sets of constraints. The first specifies the magnitude response, quality, and peak overshoot, and the second prevents numerical instabilities in the computations. We sample the magnitude response at a set of passband frequencies $\{\omega_i\}$ and stopband frequencies $\{\omega_l\}$:

$$1 - \delta_p \leq |H(j\omega_i)| \leq 1, \forall i \quad \text{and} \quad |H(j\omega_l)| \leq \delta_s, \forall l \quad (20)$$

We compute the maximum overshoot by finding the maximum value of step response in (17) by searching over $t \in [\min_k t_{\text{peak}}^k, \max_k t_{\text{peak}}^k]$. Before finding the gradient of this constraint, we substitute the analytic approximation for t_{peak}, given by (19), into (17).

When the analog filter is implemented, the second-order sections will typically be cascaded in order of ascending quality factors. The earlier sections will attenuate input signals so as to minimize the oscillatory behavior of the final sections. The implementation technology imposes an upper limit on the quality factors, Q_{max}. For macro components, we set Q_{max} to 10 for $\omega_{p2} < 2\pi(10)$ kHz, and 25 otherwise:

$$\frac{\sqrt{a_k^2 + b_k^2}}{-2a_k} < Q_{max} \quad \text{for} \quad k = 1 \ldots n \qquad (21)$$

Since a_k and c_k appear in the denominator in (2), (4), and (13), and b_k appears in the denominator in (17), we constrain these negative-valued parameters to be a neighborhood away from zero:

$$\begin{aligned} a_k &< -\epsilon_{div} < 0 \quad \text{for} \quad k = 1 \ldots n \\ b_k &< -\epsilon_{div} < 0 \quad \text{for} \quad k = 1 \ldots n \\ c_l &< -\epsilon_{div} < 0 \quad \text{for} \quad l = 1 \ldots r \end{aligned}$$

where ϵ_{div} is 2.2204×10^{-14} for MATLAB. To ensure the numerical stability of the denominators of $|B_k|$ and $\angle B_k$ in (16),

$$\sqrt{a_k - a_m} > \epsilon_{div} \quad \text{for} \quad k = 1 \ldots n \text{ and } m = k+1 \ldots n$$

These constraints are analogous to preventing duplicate poles and poles spaced too closely to one another.

5. AN EXAMPLE FILTER DESIGN

We will minimize the peak overshoot and deviation from linear phase of a lowpass filter. The specifications on the magnitude response are $\omega_p = 20$ rad/sec with $\delta_p = 0.21$ and $\omega_s = 30$ rad/sec with $\delta_s = 0.31$. In the objective function, we weight the linear phase cost by 0.1 and overshoot cost by 1. The optimization took 13 seconds to run using MATLAB 5 on a 167 MHz Ultrasparc workstation. The non-negative objective function is reduced from an initial value of 2.87 to 4.33×10^{-5}. Table 1 and 2 list the initial and final poles and zeros, respectively. Figure 2 plots the frequency and step responses for the initial and final filters. Figure 2 illustrates that the optimization procedure effectively trades off transition bandwidth in the magnitude response for more linear phase in the passband and a lower overshoot. The peak overshoot is reduced from 25% to 10%.

Q	Poles	Zeros
1.7	$-5.3553 \pm j16.9547$	$\pm j20.2479$
61	$-.1636 \pm j19.9899$	$\pm j28.0184$

Table 1: Pole-zero locations for the initial filter

Q	Poles	Zeros
0.68	$-11.4343 \pm j10.5092$	$-3.4232 \pm j28.6856$
10	$-1.0926 \pm j21.8241$	$-1.2725 \pm j35.5476$

Table 2: Pole-zero locations for the optimized filter

6. CONCLUSION

We have developed a formal, extensible framework for optimizing multiple behavioral and implementation properties of analog filter designs. We have implemented the framework as a set of Mathematica programs that generate MATLAB programs to perform the optimization. Both the algebraic derivations and programming tasks would be nearly impossible for a human to carry out correctly. By performing both processes together, we can validate that the assumptions in the algebraic derivations are legitimate and that the source code is generated properly. Furthermore, the algebraic abstraction empowers the researcher to create new filter design programs by simply redefining the cost function— our software will take care of recomputing the derivatives and regenerating the source code.

7. REFERENCES

[1] T. Saramaki and K.-P. Estola, "Design of linear-phase partly digital anti-aliasing filters," in *Proc. IEEE Int. Conf. Acoust., Speech, and Signal Processing*, (Tampa, FL), Mar. 1985.

[2] M. Lutovac, D. V. Tosic, and B. L. Evans, "Algorithm for symbolic design of elliptic filters," in *Int. Workshop on Symbolic Methods and Applications to Circuit Design*, (Leuven, Belgium), pp. 248–251, Oct. 1996.

[3] S. Wright, "Convergence of SQP-like methods for constrained optimization," *SIAM Journal on Control and Optimization*, vol. 27, pp. 13–26, Jan. 1989.

[4] K. Schittkowski, "NLPQL: A Fortran subroutine solving constrained nonlinear programming problems," *Annals of Operations Research*, vol. 5, no. 1-4, pp. 485–500, 1986.

[5] S. Lawson and T. Wicks, "Improved design of digital filters satisfying a combined loss and delay specification," *IEE Proceedings G: Circuits, Devices and Systems*, vol. 140, pp. 223–229, June 1993.

[6] B. L. Evans, D. R. Firth, K. D. White, and E. A. Lee, "Automatic generation of programs that jointly optimize characteristics of analog filter designs," in *Proc. of European Conference on Circuit Theory and Design*, (Istanbul, Turkey), pp. 1047–1050, Aug. 1995.

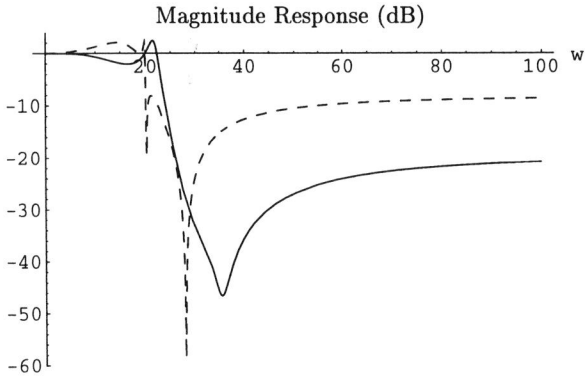

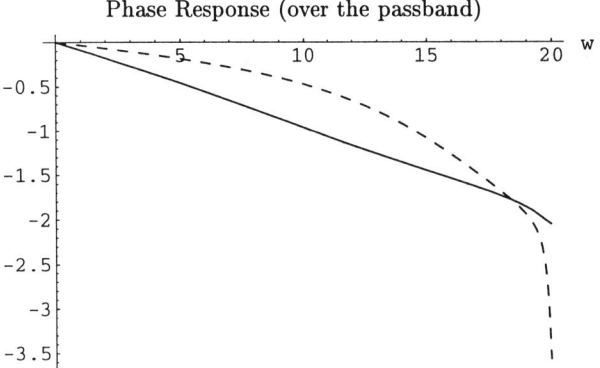

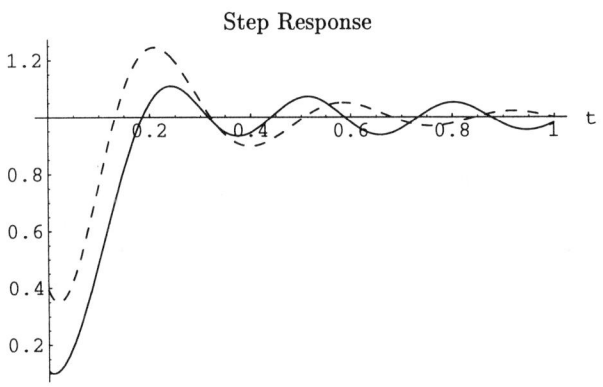

- - - initial filter ——— optimized filter

Figure 2: Two fourth-order lowpass filters to meet the magnitude specifications $\omega_p = 20$ rad/s, $\delta_p = 0.21$, $\omega_s = 30$ rad/s, and $\delta_s = 0.31$. The initial filter is an elliptic filter, and the final filter is optimized for phase and step response. We are trading linear phase response over the passband and peak overshoot in the step response for magnitude response, while keeping the magnitude response within specification. For the optimization, we set the maximum quality factor $Q_{\max}$ to be 10. Even though the initial guess is infeasible because its maximum Q value is 61, the SQP procedure in Matlab adjusted the initial guess to be a feasible solution.

New Description Language and Graphical User Interface for Module Generation in Analog Layouts

M. Wolf, U. Kleine, J. Schulze

Otto-von-Guericke-University of Magdeburg
Institute for Measurement Technologies and Electronics (IPE)
PO Box 4120, D-39016 Magdeburg, Germany

ABSTRACT

This paper presents a new description language and a graphical user interface for a module generator environment. The description language MOGLAN is adapted to the problem of writing analog module generators and provides an easy-to-read, short source code. The graphical user interface supports the writing, translating, executing, and debugging of modules. With these tools, analog designers are able to write module generators and to bring in their analog specific knowledge. This increases the quality of automatic layout solutions and decreases the time consuming process of manual layout generation.

1. Introduction

Today the layout of an analog integrated circuit is often still manually designed by experts. Although in recent years some tools for the automated layout generation have been presented [1-4] they are rarely accepted because they often bypass the designer. These tools automatically generate analog layouts, but if the designer wants to change details in the automatic solution, usually much effort is necessary. Due to the various personal layout styles, which result from personal experiences, and due to the different requirements for analog circuits, different subjective optimal layouts exist for an analog topology. According to experience, designers want to be able to control each geometry of the automatic layout but do not want to become programmers. For optimal solutions it is necessary to include the knowledge of the designers into the layout generation. To overcome this problem, a new description language adapted for analog layout generation and a graphical user interface including a debugger have been developed. With these tools the analog designer himself is able to use the powerful functions for layout generation presented in [5, 6] and to write module generators which automatically create the layout of clustered circuit elements.

2. Procedural Description Language MOGLAN

In this section the new procedural description language MOGLAN (**MO**dule **G**enerator **LAN**guage) is described in more detail. The goal of MOGLAN is to provide a natural language for the description of generators for analog integrated circuits.

2.1 Concept of MOGLAN

MOGLAN allows the copying of the constructive layout style of designers by generating simple geometries and putting them together in complex layouts. In contrast to existing layout description languages, which are C [7-9] or Pascal [10-11] based, MOGLAN is more natural, adapted for layout description and therefore easy to learn. A natural layout language is also presented in [12], but this approach takes so called blocks from a library of optimized built-in-blocks and places and routes these blocks to an entire circuits. In contrast, MOGLAN is a description language for these blocks, in order to be able to write optimal modules for different circuit applications.

While the designer must regard the design rules in manual layout generation himself, using MOGLAN this is automatically done. Only a relative placement of objects is performed, and the minimal distances and exact coordinates of objects are automatically calculated by using a special compactor [5]. Even complex rule checks as the latch-up-rule, which defines the placement of substrate contacts, can easily be performed in MOGLAN. As a general rule, minimal widths or dimensions according to the design rules are selected, if no dimension is specified. This feature minimizes the length of the source code.

Several layout optimization steps are possible in MOGLAN: The edges of geometrical objects without defined dimensions are variable, if the designer did not fix them. Variable edges can be moved in subsequent generation steps in order to optimize the layout. If a variable edge defines the minimal distance between two objects in a compaction step, the compactor automatically moves this edge away until it is no longer relevant. The result is that the objects can be more densely placed.

The main algorithms for module generation [5] have been implemented in the language C++. This object oriented implementation makes it possible to maintain such a program package. The drawback of using C++ for the description of module generators is that the analog designer must still use a "conventional" programming language. To avoid this, a new description language has been implemented. This new language is not only a simple translation of command names into C++ but has its own grammar which is adapted to the problem of module generation. The main issues of MOGLAN are:

- Variables can be used without declaration.
- Parameters can be used and omitted in an arbitrary order.
- No marker for end of command is required.
- No main-program with complex declarations is required.
- The modular description supports the hierarchical and successive design style of analog designers.
- Control structures and loops are supported.

A compiler has been implemented using the UNIX tools lex and yacc [13] to translate the MOGLAN source code into C++ code, which is then compiled to an executable generator.

2.2 Description structures in MOGLAN

A module consists of several hierarchical built entities. These entities are defined by the following syntax:

```
ENT entityName([parameter1 [= default1]]
            [, parameterN [= defaultN]])
...
END ENT
```

An entity is defined between the keywords ENT and END ENT. After the name of the entity (defined by *entityName*) an optional parameter list for this entity can be defined in parentheses. After each parameter a default value can be specified which is chosen if this parameter has been omitted in an entity call. The number of parameters is arbitrary. An entity is called for instantiation as follows:

```
variable = entityName([parameter1 = value1]
              [, parameterN = valueN])
```

An entity is instantiated with its name and a list of parameters. The number of the specified parameters and the order is arbitrary. The specified parameters in an entity call are identified by the parameter name (*parameter1*) in front of the value (*value1*). The created entity can be stored in *variable* for subsequent processing. After the hierarchical definition of entities one entity, whose geometries are provided by a generator call, must be selected. This is done by the following MAIN statement in which one entity is called in the same manner as for instantiation.

```
MAIN
    entityName([parameter1=value1] [, parameterN=valueN])
END
```

Within the definition of entities various functions for creating, placing, shifting, and accessing objects can be used. These functions can be divided into the following groups:

- **Geometric primitives**: The functions for creating geometric primitives [5] generate geometrical objects in an entity, for example rectangles around or inside a structure, arrays of rectangles, two overlapping rectangles, and a ring around a structure. The creation of these geometries can be controlled by several function parameters. The use is eased because parameters can arbitrarily be set or omitted. If geometrical dimensions are omitted, the minimal possible value defined in the design rules will be taken to minimize the layout area.

- **Compaction**: An instantiated entity can be compacted into another entity for abutting objects. The compaction step is performed in one dimension and only two objects are involved, hence the result is predictable. This is an important property for analog designers. The relevant rules, which define the location of the compacted object, will automatically be found and read from the design rules.

- **Shift and rotation**: Before compacting objects, it may be necessary to shift them orthogonal towards the compaction direction or to rotate them. An entity can be placed with several shift or offset functions for example at the same maximal north edge of another entity or horizontally centered at a reference entity. Furthermore, it is possible to rotate and to mirror objects.

- **Identification of objects**: Each entity and even each rectangle of an object can be identified by a name. This name is uniquely given by the environment or can be defined by the designer. Each entity or rectangle can be addressed by this name in the hierarchical object structure.

- **Database access**: Properties of all generated entities and rectangles (e. g. width, area or maximal north edge) can be accessed by MOGLAN functions. It is also possible to read technology data from the design rules, although this is not necessary for normal applications.

- **Constraints**: The designer can define special constraints, which must be fulfilled as well as the design rules. With the help of these constraints, the designer can control the layout generation.

- **Message functions**: The module writer can use different functions to write messages into the command interpreter window of Cadence Design Framework II (DFII). It is also possible to cancel the generator execution.

Furthermore, the programming of for- and while-loops, if-then-else conditions and switch statements is allowed in MOGLAN.

3. Graphical User Interface

A graphical user interface has been implemented to facilitate the writing of parameterizable and technology independent modules. The user interface consists of a text editor for the module description, some features for the automatic generation of module source code and control functions for translating, executing and debugging the modules. It communicates directly with DFII in order to monitor the results of the module generation in the adequate editor.

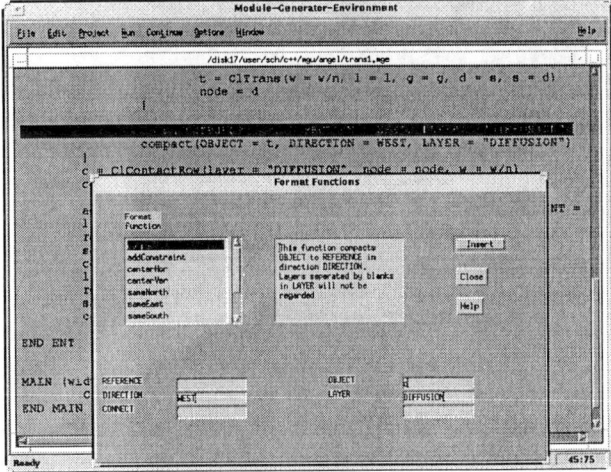

Fig. 1: Automatic generation of source code

In Fig. 1 the automatic source code generation for a module description is depicted. For each function the corresponding parameter names are displayed in a dynamic parameter field. After entering the values, the functions are syntax correctly inserted. The layout results will be displayed in the layout editor of DFII. Furthermore, it is possible to execute the generator "step by step". In this mode each generation step is displayed in the layout editor. Instead of implementing an interpreter for the language as in [7, 9] for C, an own debugging capability has been implemented in the user interface with interprocess communication in order to work directly on the new language without debugging the C++ code. Each generation step is displayed in the layout editor of DFII. The corresponding line of the source code is highlighted in the text editor and the values of variables are displayed in an output window. With the help of this debugger the user can pursue the module generation easily and find mistakes, if the result is not as expected.

4. Results

In this section, some layout examples generated with MOGLAN will be presented. Firstly, a complex module with a source code extract will be explained and secondly, a layout of an entire operational amplifier will be shown.

4.1 Cascode Current Mirror

The use of the new procedural description language drastically eases and shortens the module description. Therefore it is possible to write and maintain even complex modules. Fig. 2 shows one example for a complex module and the corresponding schematic is depicted in Fig. 3. The source code for this parameterizable and technology independent cascode current mirror consists of approximately 150 lines. Former approaches needed more lines for even a module of a simple folded transistor.

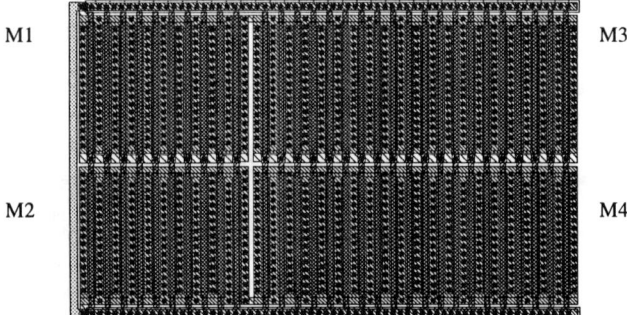

Fig. 2: Layout of a cascode current mirror

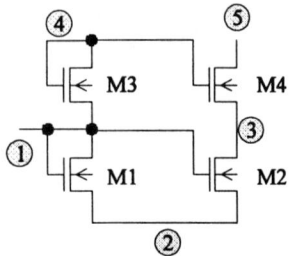

Fig. 3: Schematic of a cascode current mirror

Fig. 4 shows an extract of the source code of the layout depicted in Fig. 2. Some entities, which are used in this example, are omitted in this extract but have been written in the same manner.

The line numbers do not belong to the source code. They have been added in order to explain the commands. The definition of the entity Rect in lines 1-3 is displayed in this extract to explain the hierarchical use of entities. The main entity cascode is defined between line 5 (keyword ENT) and line 27 (keyword END ENT). The parameters for this entity are the width w1 of the lower transistors (M1, M2), the width w2 of the upper transistors (M3, M4), the length l of all transistors, the number of foldings n of the lower transistors and the doping type of the module defined by the variable pmos.

```
1  ENT Rect(layer, w, l, node)
2    inRect(LAYER=layer, WIDTH=w, LENGTH=l, NODE=node)
3  END ENT
4  ...
5  ENT cascode(w1, w2, l, n, pmos)
6    t1 = TwoTrans(w=w1, l=l, n=n, g1=1, d1=1, s1=2,
                   g2=1, d2=3, s2=2, sw1=1, sw2=1)
7    t2 = TwoTrans(w=w2, l=l, n=n*w2/w1, g1=4, d1=4,
                   s1=1, g2=4, d2=5, s2=3)
8    compact(OBJECT=t1, DIRECTION=WEST)
9    compact(OBJECT=t2, DIRECTION=WEST, LAYER="METAL1")
10   if (pmos==1)
11   {  outRect(LAYER="PPLUS")
12   }
13   else
14   {  outRect(LAYER="NPLUS")
15   }
16   sub1 = ContactRow(layer="DIFFUSION", l=getLength(),
                       node=2, sub=!pmos)
17   centerHor(OBJECT=sub1)
18   compact(OBJECT=sub1,DIRECTION=SOUTH,LAYER="METAL1")
19   sub1 = clone(OBJECT=sub1)
20   compact(OBJECT=sub1,DIRECTION=NORTH,LAYER="METAL1")
21   r = Rect(layer="METAL2", w=getWidth(), node=2)
22   centerVer(OBJECT=r)
23   compact(OBJECT=r, DIRECTION=EAST, LAYER="METAL2")
24   if (pmos==1)
25   {  outRect(LAYER="NTUB")
26   }
27 END ENT
```

Fig. 4: Source code extract for cascode current mirror

In line 6, one object (t1) of the entity TwoTrans, which generates two folded transistors with facing gate contacts, is instantiated for the transistors M1 and M2. The first three parameters define the geometrical dimensions, the subsequent six parameters specify the electrical potential (indicated in Fig. 3) of the transistors and the last two parameters (sw1, sw2) determine that the source terminals of both transistors are not external connections of this module. Due to the definition of the electrical potentials the gates of the two transistors are automatically connected and M1 becomes a diode connected MOS-transistor (g1=d1=g2=1). There is no special evaluation of the potentials in the description for the transistors, but the connections automatically occur due to the relative placement. The entity TwoTrans can be reused in other modules because of the general potential declaration.

One further object (t2) of this entity is instantiated in line 7. t1 and t2 are relatively placed by two compact commands in lines 8 and 9. The parameter LAYER in the second compact command specifies that rectangles on the layer METAL1, which have the same electrical potential, are automatically connected in this compaction step. This builds the interconnections between M1/M2 and M3/M4. The next six lines (10-15) build a condition, which like loops is supported in the language, and create either a PPLUS or an NPLUS rectangle around the entire structure depending on the variable pmos. Actually, the marking of n-doped diffusion areas is not necessary in the used technology, but this technology independent description is more general. In a mapping file for the used technology it is defined, that the layer NPLUS is not generated.

A substrate contact row on the same electrical potential as the source terminals of M1 and M2 is created in line 16, centered horizontally on the entire structure in line 17 and compacted into direction SOUTH to the structure in line 18. A copy of this contact row is compacted into direction NORTH in lines 19 and 20. For the connection of the substrate contact rows a rectangle on METAL2 is created, centered vertically and compacted in lines 21-23. A rectangle for the well in the case of a pmos module is placed around the structure in lines 24-26. As can be seen from this example, neither design rules nor exact coordinates of geometries appear in the source code. The design rules are automatically evaluated by the commands of the description language.

4.2 Rail-to-Rail Operational Amplifier

With the help of MOGLAN, module generators have been written to create the layout of a rail-to-rail operational amplifier, whose schematic is depicted in Fig. 5. The partitioning is indicated by the rectangles. Modules for a differential pair, a 2, 3 and 5 fold current mirror, a double 2 fold current mirror and single transistors have been created. The layout for this circuit is presented in Fig. 6. A post-layout simulation shows that this result meets the specifications of the circuit.

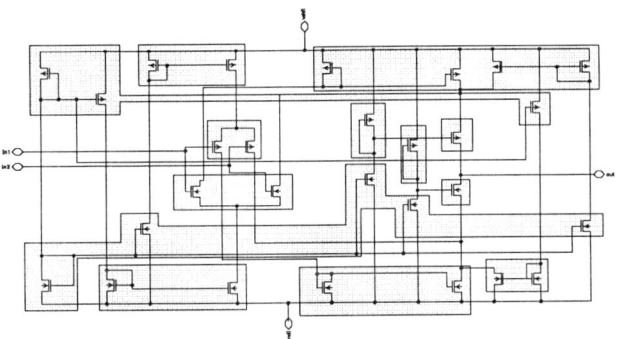

Fig. 5: Partitioning of the operational amplifier

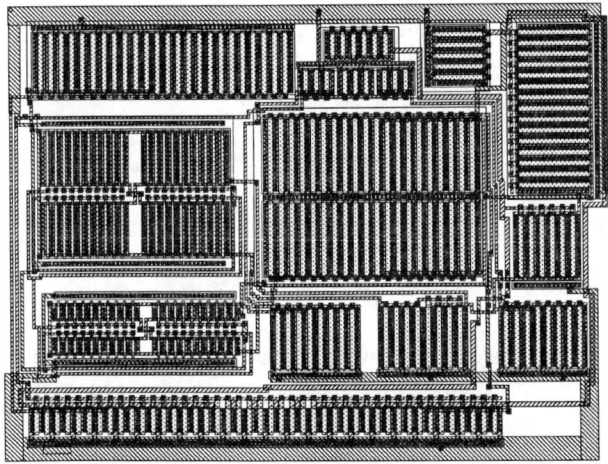

Fig. 6: Layout of the operational amplifier

Acknowledgment: The authors acknowledge Th. Pasch for providing the circuit of the operational amplifier and R. Zinke for evaluating the tools and generating the layout.

5. Conclusion

In this paper the new procedural description language MOGLAN developed for the automatic layout generation for analog circuits has been presented. Module generators written in this language have a short and easy-to-read source code. This eases the problem of maintaining a technology independent and parameterizable module generator library. A graphical user interface supports the writing, translating, executing and debugging of module generators. With these tools, the analog designer himself is able to automate the time consuming process of manual layout generation.

6. References

[1] J. Rijmenants, et al. "ILAC: An Automated Layout Tool for Analog CMOS Circuits". *IEEE J. Solid-State Circuits*, Vol. 24, No. 2, pp. 417-425, April 1989.

[2] H. Y. Koh, et al. "OPASYN: A Compiler for CMOS Operational Amplifiers". *IEEE Trans. Computer-Aided Design*, Vol. 9, No. 2, pp. 113-125, Feb. 1990.

[3] J. M. Cohn, et al. "KOAN/ANAGRAM II: New Tools for Device-Level Analog Placement and Routing". *IEEE J. Solid-State Circuits*, Vol. 26, No. 3, pp. 330-342, March 1991.

[4] V. Meyer zu Bexten, et al. "ALSYN: Flexible Rule-Based Layout Synthesis for Analog IC's". *IEEE J. Solid-State Circuits*, Vol. 28, No. 3, pp. 261-268, March 1993.

[5] M. Wolf, et al. "A Novel Analog Module Generator Environment". *Proc. The European Design & Test Conference*, pp. 388-392, March 1996.

[6] M. Wolf, et al. "Application Independent Module Generation in Analog Layouts". *Proc. The European Design & Test Conference*, p. 624, March 1997.

[7] A. Greiner, F. Pétrot. "Using C to Write Portable CMOS VLSI Module Generators". *EURO-DAC 1994*, pp. 676-681, 1994.

[8] J. M. Mata. "ALLENDE: A Procedural Language for the Hierarchical Specification of VLSI Layouts". 22nd *Design Automation Conference*, pp. 183-189, 1985.

[9] P. A. D. Powell, et al. "The Icewater Language and Interpreter". 21st *Design Automation Conference*, pp. 98-102, 1984.

[10] W. E. Cory. Layla. "A VLSI Layout Language". 22nd *Design Automation Conference*, pp. 245-251, 1985.

[11] R. J. Lipton, et al. "ALI: a Procedural Language to Describe VLSI Layouts". 19th *Design Automation Conference*, pp. 467-474, 1982.

[12] B. R. Owen, et al. "BALLISTIC: An Analog Layout Language". *IEEE Custom Integrated Circuits Conference*, pp. 3.5.1-3.5.4, 1995.

[13] B. W. Kernighan, et al. "The UNIX Programming Environment". 1st Edition 1984, Prentice-Hall.

ANALYSIS AND COMPENSATION OF OTA NON-IDEAL EFFECTS IN VIDEO-FREQUENCY CMOS SINC(X) EQUALIZERS

F. Dudek, B.M. Al-Hashimi and M. Moniri

School of Engineering and Advanced Technology
Staffordshire University
Beaconside, Stafford ST18 0AD, UK

ABSTRACT

This paper presents a detailed analysis and minimization of OTA non-ideal effects in the performance of sinc(x) equalizers operating at video-frequencies. To compensate these effects, a set of equalizer design equations, expressed in terms of the active devices input capacitance, output resistance and the polynomial coefficients used to correct the sinc(x) distortion is derived, facilitating the synthesis process. Simulation based on CMOS OTAs demonstrates the effectiveness of the design equations in compensating accurately the error introduced by the OTAs in the ideal equalizer. This is demonstrated with reference to a D/A converter with sampling rate of 27MHz over 10MHz bandwidth.

1. INTRODUCTION

D/A converters introduce sinc(x) distortion into the signal being converted. This distortion is especially significant for low sampling-to-signal frequency ratios. For example, a 2.1dB loss is introduced in the filter passband of a 10MHz video signal with the standard sampling rate $F_s = 27$MHz. Video filters have stringent passband specification, typically ≤ 0.1dB ripple, and hence require sinc(x) correction. Effective methods to correct this distortion include cascading an amplitude equalizer with the filter in order to produce gain boost in the filter passband of opposite shape to the sinc(x) distortion. Most equalizer circuits are based on op-amps, for example [1]. The transconductance-capacitor (g_m-C) approach is now preferred in the design of continuous-time circuits operating in the MHz region. Some OTA based equalizers have been reported in literature [2-3]. In [2], a 2nd-order canonical amplitude equalizer with 4 OTAs and 2 floating capacitors was described. For IC implementation, grounded capacitors are not only easier to integrate and less affected by parasitic errors than floating capacitors, but they are also advantageous in mixed-signal designs since cost-effective single-poly CMOS processes are normally employed. In [3], a 2nd-order amplitude equalizer having 5 OTAs and 2 grounded capacitors was proposed. However, its correction accuracy is limited since it only realizes real transmission zeros. Recent work [4] has described a more efficient g_m-C amplitude equalizer and its CMOS implementation. The equalizer uses the same number of components as the one reported in [3] but achieves increased flexibility in the shaping of the voltage transfer function and hence better correction accuracy due to a pair of complex transmission zeros. Although the work in [4] presented an efficient equalizer design, however it does not take into account the non-ideal effects of the active devices and their influence on the sinc(x) correction accuracy. This paper analyzes and incorporates the OTA non-ideal parameters in the equalizer transfer function and minimizes their effect by deriving a set of design equations suitable for video frequency applications.

2. EQUALIZER DESIGN CONSIDERATIONS

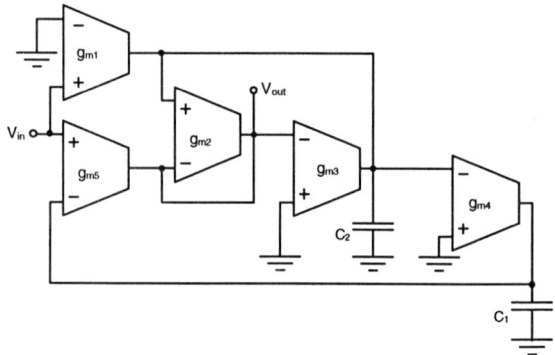

Figure 1: New amplitude equalizer

Assuming $g_{m1} = g_{m2} = g_{m3} = g_{m4} = g_m$, the ideal transfer function of the amplitude equalizer [4] shown in Fig. 1 is:

$$H(s) = \frac{V_{out}}{V_{in}} = \frac{\dfrac{g_{m5}}{g_m}s^2 + \dfrac{g_m}{C_2}s + \dfrac{g_m g_{m5}}{C_1 C_2}}{s^2 + \dfrac{g_m}{C_2}s + \dfrac{g_m g_{m5}}{C_1 C_2}} = \frac{\alpha_2 s^2 + \alpha_1 s + \alpha_0}{s^2 + \alpha_1 s + \alpha_0} \quad (1)$$

where

$$g_m = \frac{\alpha_0}{\alpha_1 \alpha_2}C_1 \qquad g_{m5} = \frac{\alpha_0}{\alpha_1}C_1 \qquad C_2 = \frac{\alpha_0}{\alpha_1^2 \alpha_2}C_1 \quad (2)$$

The design process of a sinc(x) equalizer for correcting a particular distortion involves ensuring that the equalizer transfer function is the reciprocal of the sinc(x) transfer function. To achieve high correction accuracy, the design problem is formulated by applying curve matching optimization technique [5] based on a minimax error function. As an example, consider the design of a sinc(x) equalizer when the ratio of the D/A converter sampling rate, ω_s, to the filter bandwidth, ω_p, is 2.7 : 1. Table.1 gives the polynomial coefficients α_i of Eqn. 1 for this example.

	α_2	α_1	α_0
ideal equalizer	0.3368	1.0565	3.0000

Table 1: ideal equalizer polynomial coefficients for $\omega_s : \omega_p = 2.7 : 1$

At video frequencies, the performance of the equalizer is limited by the non-ideal characteristics of the active devices. Previous work has shown that input capacitance and output resistance of an OTA form the significant parasitics in g_m-C circuits [6]. A video-frequency OTA model is shown in Fig. 2, where C_{in} is the input capacitance and g_{out} is the output conductance.

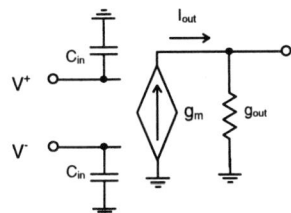

Figure 2: video frequency OTA model

Using this model for all the equalizer OTAs and assuming that $g_{m1} = g_{m2} = g_{m3} = g_{m4} = g_m$, circuit analysis yields the following transfer function:

$$H(s) = \frac{V_{out}}{V_{in}} = \frac{\beta_6 \cdot s^2 + \beta_5 \cdot s + \beta_4}{\beta_3 \cdot s^3 + \beta_2 \cdot s^2 + \beta_1 \cdot s + \beta_0} \quad (3)$$

where

$$\beta_6 = g_{m5}(C_{in}(2C_{in} + 2C_1 + C_2) + C_1 C_2)$$
$$\beta_5 = g_{out}g_{m5}(4C_{in} + 2C_1 + C_2) + g_m^2(C_{in} + C_1)$$
$$\beta_4 = g_m^2(g_{out} + g_{m5}) + 2g_{m5}g_{out}^2 \quad (4)$$
$$\beta_3 = 2C_{in}(C_1 C_2 + C_{in}(2C_{in} + 2C_1 + C_2))$$
$$\beta_2 = C_1 C_2(2g_{out} + g_m) + C_{in}(2C_1 + C_2)(4g_{out} + g_m) + 2C_{in}^2(6g_{out} + g_m)$$
$$\beta_1 = (2g_{out}^2 + g_{out}g_m)(2C_1 + C_2 + 4C_{in}) + g_m^2(C_1 + C_{in}) + 4g_{out}^2 C_{in}$$
$$\beta_0 = 4g_{out}^3 + 2g_m g_{out}^2 + g_m^2(g_{out} + g_{m5})$$

This shows, that the effect of the OTA non-ideal parameters on the ideal equalizer not only modifies the complex pair of pole-zero positions but also introduces an extra real pole as shown in Table.3. The polynomial coefficients of the non-ideal equalizer are given in Table.2, when the ratio of the D/A sampling rate to the filter bandwidth is 2.7 : 1.

	non-ideal equalizer
β_6	0.1511
β_5	0.4439
β_4	0.9769
β_3	0.0977
β_2	0.4587
β_1	0.6177
β_0	1.0000

Table 2: non-ideal equalizer polynomial coefficients for $\omega_s : \omega_p = 2.7 : 1$

equalizer	Poles	Zeros
ideal	-0.5284 ± 1.6495	-1.5689 ± 2.5390
non-ideal	-0.4795 ± 1.5841 -3.7365	-1.4692 ± 2.0755

Table 3: ideal and non-ideal equalizer pole-zero locations for $\omega_s : \omega_p = 2.7 : 1$

Note, in analyzing the non-ideal equalizer performance, it has been assumed that all the equalizer OTAs have the same C_{in} and g_{out}. This assumption is valid, since 4 of the 5 OTAs in the circuit have identical transconductance values (g_m). Although the remaining OTA has a different g_m, simulation has shown that the OTA input capacitance and output conductance values do not vary significantly, provided the OTA operates in the linear region.

In order to absorb the OTA parasitics into the equalizer components and hence minimize their effects, the equalizer components should be expressed in terms of the polynomial coefficients used to correct the sinc(x) distortion, as well as the input capacitance and output conductance of the OTA devices. Manipulating Eqn. 4 results in the following equalizer design equations, taking into account the OTA non-ideal parameters:

$$g_{m5} = \frac{2C_{in}\beta_6}{\beta_3} \quad (5)$$

$$g_m = \frac{2g_{out}^2(g_{m5} - 2g_{out}) + \beta_0 - \beta_4}{2g_{out}^2} \quad (6)$$

$$C_1 = \frac{\beta_5(g_m + 2g_{out}) + g_m^2 C_{in}(g_{m5} - 2g_{out} - g_m) + g_{m1}(4g_{out}^2 C_{in} - \beta_1)}{g_m^2(2g_{out} + g_m - g_{m5})} \quad (7)$$

$$C_2 = \frac{\beta_2 g_{m5} - \beta_6(g_m + 2g_{out}) - 4g_{out}g_{m5}C_{in}(2C_{in} + C_1)}{2g_{out}g_{m5}C_{in}} \quad (8)$$

The polynomial coefficients ($\beta_1, \beta_2, \ldots, \beta_6$) are found using curve matching optimization by ensuring that the equalizer transfer function (Eqn. 3) is the reciprocal of the sinc(x) distortion curve. Furthermore, the parameters C_{in} and g_{out} are usually known depending on the OTA chosen for implementation.

3. DESIGN EXAMPLE AND SIMULATION

To confirm the theoretical analysis of the non-ideal equalizer, consider correcting the sinc(x) distortion of a D/A converter with a sampling rate of $F_s = 27$MHz over 10MHz bandwidth. The sinc(x) distortion is shown in Fig. 4. Using Table.1, Eqn. 2 and impedance and frequency scaling, the ideal equalizer values are given in Table.4. Based on these values and ideal OTAs (i.e. G component), the equalizer SPICE simulation is shown in Fig. 4. The correction accuracy of the ideal equalizer when combined with the sinc(x) distortion is < 0.01dB.

equalizer	ideal	compensated
C_1 (pF)	2.000	0.500
C_2 (pF)	15.960	7.000
g_m (μS)	1059.380	656.700
g_{m5} (μS)	356.840	301.200
C_{in} (pF)	0.000	1.550
g_{out} (μS)	0.000	62.500

Table 4: ideal and compensated equalizer component values for $\omega_s : \omega_p = 2.7 : 1$

equalizer	g_m		g_{m5}	
	W_N	W_P	W_N	W_P
ideal	81.6μm	235.6μm	16.8μm	48.3μm
compensated	50.6μm	146.1μm	14.1μm	40.7μm

Table 5: ideal and compensated equalizer W/L ratios for $\omega_s : \omega_p = 2.7 : 1$ ($L_N = L_P = 1.2$μm)

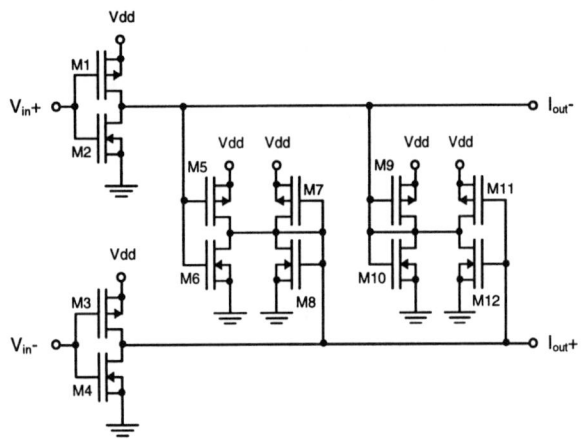

Figure 3: Fully-balanced CMOS OTA

It is well known that monolithic g_m-C circuits require fully balanced structures. A fully balanced high-frequency CMOS OTA [7] is shown in Fig. 3. Based on the CMOS OTA of Fig. 3 and using typical values of the AMS 0.8μm double-metal double-poly CMOS process and SPICE level 6 transistor models, Fig. 4 shows the uncompensated equalizer response. As can be seen, it is significantly different from the ideal response, now resulting in an equalizer sinc(x) correction accuracy of < 0.8dB, which is clearly not acceptable in video applications. The transistor W/L dimensions for this example are given in Table.5. Extensive simulation of the CMOS OTA with different values of g_ms have shown that the OTA has $C_{in} = 1.55$pF and $g_{out} = 62.5$μS. Using these values, Table.2 and Eqns. 5-8, the compensated equalizer values are given in Table.4. The corresponding transistor W/L dimensions are given in Table.5. Fig.5 shows the simulated response of the compensated equalizer where the correction accuracy is < 0.05dB, which compares favorably with the correction achieved by the ideal equalizer. This confirms the effectiveness of the proposed equalizer design equations.

4. CONCLUDING REMARKS

This paper has presented a detailed analysis and minimization of OTA non-ideal effects in video frequency sinc(x) equalizers. The compensation has been achieved by deriving a set of design equations incorporating the OTA input capacitance and output resistance and the polynomial coefficients used to correct the sinc(x) distortion.

5. REFERENCES

[1] Sharma, S., Taylor, J.T. and Haigh, D.G.
"Stray-free second-order circuit for correction of sample-and-hold amplitude distortion in switched-capacitor filters"
Electronics Letters, 1988, Vol. 24, No.16, pp. 1007-1008

[2] Malvar, H.S.
"Electronically Controlled Active-C Filters and Equalizers with Operational Transconductance Amplifiers"
IEEE Transactions on Circuits and Systems, Vol. CAS-31, No. 7, July 1984, pp. 645-649.

[3] Wyszynski, A. and Schaumann, R.
"A current-mode biquadratic amplitude equalizer",
Analog integrated circuits and signal processing (AICSP), 1993, No. 4, pp. 161-166.

[4] Dudek, F., Al-Hashimi, B.M. and Moniri, M.
"CMOS equalizer for compensating sinc(x) distortion of video D/A converters"
Electronics Letters, 1997, Vol. 33, No.19, pp. 1618-1619

[5] MATLAB™: *Optimization Tool Box*, The Math Works, Inc.

[6] Alarcon, E., Podeva, A. & Vidal, E.
"A Complete OTA Frequency Model"
IEEE Proceedings of the 39th Midwest Symposium on Circuits and Systems, 1997, pp. 455-458

[7] Nauta, B. and Seevinck, E.
"Linear CMOS Transconductance Element for VHF Filters"
Electronics Letters, 1989, Vol. 25, No. 7, pp. 448-450

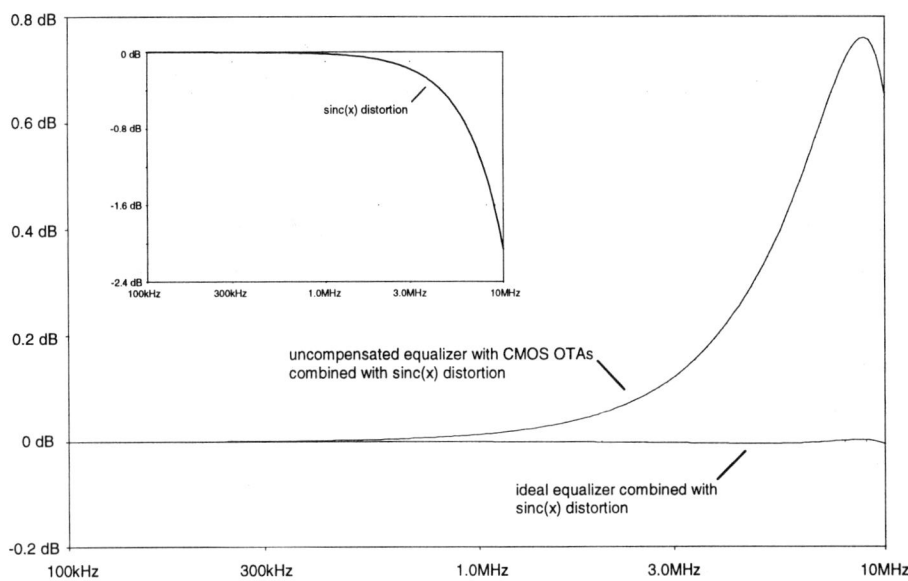

Figure 4: Simulated frequency response of ideal equalizer and uncompensated CMOS equalizer when combined with 27MHz D/A converter sinc(x) distortion

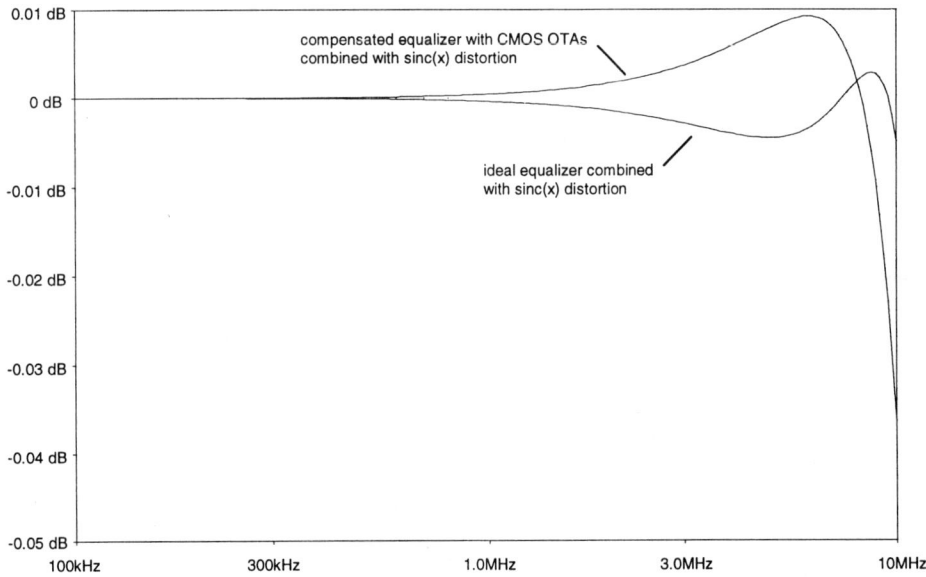

Figure 5: Simulated frequency response of ideal equalizer and compensated CMOS equalizer when combined with 27MHz sinc(x) distortion using the proposed equalizer design equations

ADJOINT NETWORK OF PERIODICALLY SWITCHED LINEAR CIRCUITS

Fei Yuan and Ajoy Opal

VLSI Research Group, Department of Electrical and Computer Engineering
University of Waterloo, Waterloo, Ontario, Canada, N2L 3G1

ABSTRACT

This paper presents a general theory of the adjoint network of multiphase periodically switched linear (PSL) circuits in frequency domain. The phasor representation of Tellegen's theorem [1] for PSL circuits in steady-state is introduced and the adjoint networks of elements typically encountered in PSL circuits are developed. It is shown that the transfer functions from multiple input sources to one output of a PSL circuit can be computed efficiently by solving the adjoint network of the circuit only once. More importantly, it is demonstrated that the aliasing transfer functions from the input sources in the side bands to the output in the base band can be obtained efficiently by computing the corresponding high-order frequency components of the response of the adjoint network of the circuit.

1. INTRODUCTION

Periodically switched linear (PSL) circuits, such as non-ideal switched-capacitor (SC) filters, switched-current (SI) filters, modulators, mixers, etc. are widely used in telecommunications. Noise analysis of these circuits is critical, especially for low-power and wireless applications. In analyzing the output noise power of these circuits, an excessive amount of computation is required due to a large number of broad-band noise sources in the circuits and the fold-over effect caused by the under-sampling of these noise signals. Adjoint network is well known as one of the most efficient techniques in computing the transfer functions from multiple input sources to single output of linear time-invariant (LTI) circuits [2, 3, 4]. Its advantages in noise analysis of LTI and ideal sampled-and-held (S/H) SC circuits were also demonstrated [5, 6, 7]. This paper presents a general theory of the adjoint network of PSL circuits.

2. TELLEGEN'S THEOREM FOR PSL CIRCUITS IN PHASOR DOMAIN

The only time-varying elements in PSL circuits are MOSFET switches. In practice, they are often modeled as an ideal switch in series with a LTI resistor whose resistance is the typical channel resistance of the MOSFET in the triode region. The ideal switch is further modeled as a short/open-circuited branch such that the topology of the circuit remains unchanged during switching. As a result, it

---

This work was financially supported by the Natural Sciences and Engineering Research Council of Canada.

---

can be shown that two PSL circuits N and $\hat{N}$ having the same graph satisfy

$$\mathbf{v}_b^T(t)\hat{\mathbf{i}}_b(\tau) = 0 \quad (1)$$

where t and τ are the time variables of N and $\hat{N}$, respectively. $\mathbf{v}_b(t)$ and $\hat{\mathbf{i}}_b(\tau)$ are the respective branch voltage and current vectors of N and $\hat{N}$. It follows from Eq.(1) that

$$\mathbf{v}_b^T(t)\hat{\mathbf{i}}_b(\tau) - \mathbf{i}_b^T(t)\hat{\mathbf{v}}_b(\tau) = 0 \quad (2)$$

Eqs.(1) and (2) are called the *strong* and *weak* forms of Tellegen's theorem for PSL circuits in the time domain, respectively. The weak form of Tellegen's theorem incorporates the branch voltages and currents of N and $\hat{N}$. It is of particular usefulness in characterizing the relationship between the network variables of the two circuits.

As is well known, LTI circuits in *steady-state* can be analyzed conveniently in the phasor domain. Consequently, Tellegen's theorem for LTI circuits can also be written in the phasor domain [4]. Unlike LTI circuits, due to periodic switching, the network variables of PSL circuits contain an infinite number of frequency components even though the input of the circuits is a single-tone [8]. It can be shown that in steady-state the network variables of a PSL circuit with input $e^{j\omega_o t}$ can be represented by the Fourier series

$$v(t) = \sum_{n=-\infty}^{\infty} V_n e^{j(\omega_o + n\omega_s)t} \quad (3)$$

where V_n is the phasor of $v(t)$ of the n-th order, ω_o and ω_s are the input and clock frequencies of the circuit, respectively.

Consider two different PSL circuits N and $\hat{N}$ having the same graph. Let the clock and input frequencies of N and $\hat{N}$ be identical. Using Eq.(3), we can write Eq.(2) as follows

$$\sum_{n=-\infty}^{\infty}\sum_{m=-\infty}^{\infty} (\mathbf{V}_n^T\hat{\mathbf{I}}_m - \mathbf{I}_n^T\hat{\mathbf{V}}_m)e^{j\omega_o(t+\tau)}e^{j\omega_s(nt+m\tau)} = 0 \quad (4)$$

where $\mathbf{V}_n$, $\mathbf{I}_n$, $\hat{\mathbf{V}}_m$ and $\hat{\mathbf{I}}_m$ are the phasors of the branch voltages and currents of N and $\hat{N}$, respectively. To extract the relationship between the phasors in Eq.(4), let us set

$$t + \tau = 0 \quad (5)$$

and integrate Eq.(4) with respect to t from 0 to $T_s = 2\pi/\omega_s$. Because

$$\int_0^{T_s} e^{j(n-m)\omega_s t}dt = T_s\delta_{mn} \quad (6)$$

where δ_{mn} is the Kronecker delta function, i.e. $\delta_{mn} = 1$ if $m = n$ and 0, otherwise, we therefore obtain

$$\sum_{n=-\infty}^{\infty} (\mathbf{V}_n^T \hat{\mathbf{I}}_n - \mathbf{I}_n^T \hat{\mathbf{V}}_n) = 0 \qquad (7)$$

Eq.(7) characterizes the relationship of the phasors of the network variables of N and $\hat{N}$ with reversed time variables. It is important to note that $t + \tau = 0$ implies that the switching clock sequence of $\hat{N}$ is reversed, as compared with that of N.

Theorem 1 *Given two PSL circuits having the same graph and reversed time variables, the weak form of Tellegen's theorem for PSL circuits in the phasor domain is given by (7).*

3. ADJOINT NETWORK

For a given PSL circuit N, an apparent approach to derive its adjoint network $\hat{N}$ is that for each element in N, construct the interreciprocal counterpart in $\hat{N}$, which is connected to the same nodes to which the element in N is connected, such that

$$\sum_{n=-\infty}^{\infty} (V_n \hat{I}_n - I_n \hat{V}_n) = 0 \qquad (8)$$

is satisfied. The adjoint network constructed in this way manifests its advantages in computing transfer functions and aliasing transfer functions of PSL circuits, as will be shown shortly. In this section, we will derive the adjoint networks of elements typically encountered in PSL circuits.

Ideal Switches To derive the adjoint network of an ideal switch, let us make use of

$$v(t)\hat{i}(\tau) - i(t)\hat{v}(\tau) = 0 \qquad (9)$$

where $v(t)$ and $i(t)$, $\hat{v}(\tau)$ and $\hat{i}(\tau)$ are the voltages and currents of the switch and its adjoint network, respectively. When the switch is OPEN, the current through the switch is zero, i.e. $i(t) = 0$. To validate Eq.(9) for arbitrary $v(t)$, we set $\hat{i}(\tau) = 0$. So the adjoint network of an OPEN switch is also an OPEN switch. In a similar fashion, one can also show that the adjoint network of a CLOSED switch is a CLOSED switch. Having derived the adjoint network of ideal switches, we will further show that an ideal switch and its adjoint network satisfy Eq.(8). Representing the voltages and currents of the switch and its adjoint in Fourier series using Eq.(3) and substituting them into Eq.(9) yield

$$\sum_{n=-\infty}^{\infty} \sum_{m=-\infty}^{\infty} (V_n \hat{I}_m - I_n \hat{V}_m) e^{j\omega_s(n-m)t} = 0 \qquad (10)$$

Note that Eq.(5) was utilized in deriving Eq.(10). Following the same approach as we did before, one can show that

$$\sum_{n=-\infty}^{\infty} (V_n \hat{I}_n - I_n \hat{V}_n) = 0 \qquad (11)$$

LTI Capacitors and Inductors The constitutive equation of a LTI capacitor in PSL circuits in the phasor domain is given by

$$I_{c,n} = j(\omega_o + n\omega_s)CV_{c,n} \qquad (12)$$

for $n = 0, \pm 1, \ldots$ where C is the capacitance of the capacitor. $I_{c,n}$ and $V_{c,n}$ are the phasors of the current and voltage of the capacitor, respectively. Substituting the voltages and currents of the capacitor and its adjoint into Eq.(8) for a specific n, we obtain

$$V_{c,n}[\hat{I}_{c,n} - j(\omega_o + n\omega_s)C\hat{V}_{c,n}] = 0 \qquad (13)$$

To validate Eq.(13) for arbitrary $V_{c,n}$, a natural choice is therefore

$$\hat{I}_{c,n} - j(\omega_o + n\omega_s)C\hat{V}_{c,n} = 0 \qquad (14)$$

We therefore conclude that the adjoint network of a LTI capacitor in PSL circuits is also a LTI capacitor of the same capacitance. In a very like manner, it can be shown that the adjoint network of a LTI inductor in PSL circuits is also a LTI inductor of the same inductance.

Controlled Sources and opamps The adjoint networks of four types of controlled sources (VCVS, VCCS, CCCS, CCVS) can be derived conveniently using Eq.(8). The results are the same as those in [6]. Readers are referred to the cited reference for details. Opamps are usually analyzed using the so-called macro-models. These macro-models are essentially LTI circuits consisting of resistors, capacitors and controlled sources. The adjoint network can thus be constructed conveniently using the methods detailed here.

4. TRANSFER FUNCTION THEOREM

Given a PSL circuit N, let us construct the adjoint network $\hat{N}$ on the basis of interreciprocity such that Eq.(8) is satisfied for every interreciprocal element in the circuits. Because independent sources, open-circuited branches and short-circuited branches are not interreciprocal, to simplify analysis, let us separate the interreciprocal elements from the rest. The summation in Eq.(7) over the interreciprocal elements vanishes. The only terms left over are those that are associated with the non-interreciprocal elements. This leads to

$$\sum_{n=-\infty}^{\infty} (\mathbf{V}_{p,n}^T \hat{\mathbf{I}}_{p,n} - \mathbf{I}_{p,n}^T \hat{\mathbf{V}}_{p,n}) = 0 \qquad (15)$$

where the subscript p specifies network variables of the non-interreciprocal elements. Without losing generality, let N have only one current input $i_s(t) = e^{j\omega_o t}$ and one voltage output $v_o(t)$. Let the input of $\hat{N}$ be a current source of unity strength $\hat{i}_o(\tau) = e^{j\omega_o \tau}$, which is applied to the port corresponding to the output of N and the output be the voltage across the branch corresponding to the input of N (the current source is removed). Substituting these quantities into Eq.(15) and noting that (i) the inputs of both N and $\hat{N}$ are single-tones at ω_o, thus, $I_{s,n} = 1$ if $n = 0$; and 0 otherwise; $\hat{I}_{o,n} = 1$ if $n = 0$; and 0 otherwise. (ii) $I_{o,n} = 0$ and $\hat{I}_{s,n} = 0$, for $n = 0, \pm 1, \ldots$, we thus obtain

$$V_{o,0} = \hat{V}_{s,0} \qquad (16)$$

So the desired transfer function from the input current source to the output of N at ω_o is given by the output of $\hat{N}$ at ω_o explicitly. The transfer functions for other input/output configurations can also be derived in a similar manner and the results are summarized in Table 1. These results can be extended to circuits having multiple inputs and single output conveniently.

Theorem 2 (transfer function theorem) *The transfer functions from the multiple input current (voltage) sources to the single voltage (current) output of a PSL circuit N at frequency ω_o are given explicitly by the voltages across (currents through) the branches in the adjoint network $\hat{N}$, corresponding to the input branches in N, provided that the input of the adjoint network is a current (voltage) source of unity strength at the same frequency.*

With the transfer function theorem, we need to solve the adjoint network at the frequency at which the transfer functions are to be evaluated *only once* to obtain all the transfer functions of the original circuit.

5. FREQUENCY REVERSAL THEOREM

Noise sources encountered in PSL circuits, such as thermal and shot noise, are white in nature. Due to the undersampling of these broad-band noise signals, the amount of computation required is excessive. In this section, we will introduce an elegant theorem that enables us to compute the aliasing transfer functions in a very efficient manner.

Consider a PSL circuit N having only one input current source $i_s(t) = e^{j(\omega_o + m\omega_s)t}$ and one voltage output. In practice, because one is only interested in the output noise power in the base band, the aliasing transfer functions characterizing the relationship between the input noise signals in the side bands and the output in the base band are needed. Let us construct the adjoint network $\hat{N}$ as before and further let the input of $\hat{N}$ be $\hat{i}(\tau) = e^{j\omega_o \tau}$. Substituting them into Eq.(15) and noting that (i) the summation over the interreciprocal elements vanishes, (ii) the output branches of both N and $\hat{N}$ are open-circuited, thus, $I_{o,n} = 0$ and $\hat{I}_{s,n} = 0$ for $n = 0, \pm 1, \ldots$, we therefore have

$$\sum_{n=-\infty}^{\infty} (V_{o,n}\hat{I}_{o,n} - I_{s,n}\hat{V}_{s,n}) = 0 \quad (17)$$

Because the inputs of N and $\hat{N}$ are single-tones at $\omega_o + m\omega_s$ and ω_o, respectively, thus, $I_{s,n} = 1$ if $n = m$, and 0 otherwise; $\hat{I}_{o,n} = 1$ if $n = 0$, and 0 otherwise. Incorporating these conditions, Eq.(17) is simplified to

$$V_{o,0} = \hat{V}_{s,m} \quad (18)$$

Eq.(18) reveals that the desired aliasing transfer function from the input current source in the m-th side band to the output in the base band of N is given explicitly by the m-th order frequency component of the response of $\hat{N}$. The aliasing transfer functions for other input/output configurations can also be derived in a similar fashion and the results are given in Table 2. These results can be extended conveniently to circuits having multiple inputs and single output.

Theorem 3 (frequency reversal theorem) *Given a PSL circuit N and its adjoint $\hat{N}$, the aliasing transfer function from the input current (voltage) sources at frequency $\omega_o + m\omega_s$ to the voltage (current) output at frequency ω_o of N is given explicitly by the m-th order frequency component of the voltage across (current through) the branches in $\hat{N}$, corresponding to the input branches in N, provided that the input of the adjoint network is a current (voltage) source of unity strength at ω_o.*

With the frequency reversal theorem, the aliasing transfer functions from all relevant side bands to the base band can be obtained by performing *only one* frequency analysis on the adjoint network at the frequency in the base band and finding the corresponding high-order frequency components of the response of $\hat{N}$. Because the computation of high-order frequency components of the output of PSL circuits is much less expensive as compared with solving the circuit at multiple frequencies [8], the computational gain is clearly substantial.

6. NUMERICAL EXAMPLE

Consider a bi-phase parasitic-insensitive SC integrator and its adjoint network shown in Fig.1 [10]. All MOSFET switches are modeled with an ideal switch in series with a noisy LTI resistor of $3.5k\Omega$. The duty cycles of phases 1 and 2 are 0.488 and 0.512, respectively. The clock frequencies of N and $\hat{N}$ are 100 kHz. The circuits were solved using WATSNAP [8] and the results are presented in Tables 3 and 4. As can be seen that (i) the transfer function from $I_s(\omega_o)$ to $V_o(\omega_o)$ of N is the same as that from $\hat{I}_o(\omega_o)$ to $\hat{V}_s(\omega_o)$ of $\hat{N}$. (ii) The aliasing transfer functions of N match the corresponding frequency components of the output of $\hat{N}$ exactly.

7. SUMMARY

A general theory on the adjoint network of multiphase PSL circuits in the frequency domain is developed. The theory provides a powerful means of reducing the computational cost associated with the noise analysis of PSL circuits in two aspects (i) it derives the transfer functions from the multiple input sources to the single output of the circuits at a frequency by solving its adjoint network at the same frequency *only once*, (ii) it computes the aliasing transfer functions by evaluating the corresponding high-order frequency components of the response of the adjoint network.

Table 1: Transfer function theorem

Original		Adjoint		Transfer function
Input	Output	Input	Output	
$i_s(t)$	$v_o(t)$	$\hat{i}_o(\tau)$	$\hat{v}_s(\tau)$	$V_{o,0} = \hat{V}_{s,0} I_{s,0}$
$v_s(t)$	$v_o(t)$	$\hat{i}_o(\tau)$	$\hat{i}_s(\tau)$	$V_{o,0} = -\hat{I}_{s,0} V_{s,0}$
$i_s(t)$	$i_o(t)$	$\hat{v}_o(\tau)$	$\hat{v}_s(\tau)$	$I_{o,0} = -\hat{V}_{s,0} I_{s,0}$
$v_s(t)$	$i_o(t)$	$\hat{v}_o(\tau)$	$\hat{i}_s(\tau)$	$I_{o,0} = \hat{I}_{s,0} V_{s,0}$

Table 2: Frequency reversal theorem

Original		Adjoint		Aliasing
Input	Output	Input	Output	transfer function
$i_s(t)$	$v_o(t)$	$\hat{i}_o(\tau)$	$\hat{v}_s(\tau)$	$V_{o,0} = \hat{V}_{s,m} I_{s,m}$
$v_s(t)$	$v_o(t)$	$\hat{i}_o(\tau)$	$\hat{i}_s(\tau)$	$V_{o,0} = -\hat{I}_{s,m} V_{s,m}$
$i_s(t)$	$i_o(t)$	$\hat{v}_o(\tau)$	$\hat{v}_s(\tau)$	$I_{o,0} = -\hat{V}_{s,m} I_{s,m}$
$v_s(t)$	$i_o(t)$	$\hat{v}_o(\tau)$	$\hat{i}_s(\tau)$	$I_{o,0} = \hat{I}_{s,m} V_{s,m}$

Table 3: Transfer function and aliasing transfer functions of parasitic-insensitive SC integrator

Input freq. (Hz)	Output freq. (Hz)	Output	
		Real part	Imag. part
889	889	2.68218E-02	-8.02467E-02
100889	889	5.29485E-02	1.6045E-02
200889	889	-9.60776E-04	2.04816E-03
300889	889	1.7885E-02	3.83242E-03
400889	889	-5.77075E-04	2.07626E-03
500889	889	1.07444E-02	1.43950E-03
600889	889	-3.42938E-04	2.08695E-03
700889	889	7.59345E-03	4.25669E-04
800889	889	-1.47032E-04	2.08153E-03
900889	889	5.77376E-03	-1.21182E-04

Table 4: Frequency response of the adjoint network of parasitic-insensitive SC integrator

Input freq. (Hz)	Output freq. (Hz)	Output	
		Real part	Imag. part
889	889	2.68218E-02	-8.02467E-02
889	100889	5.29485E-02	1.6045E-02
889	200889	-9.60776E-04	2.04816E-03
889	300889	1.7885E-02	3.83242E-03
889	400889	-5.77075E-04	2.07626E-03
889	500889	1.07444E-02	1.43950E-03
889	600889	-3.42938E-04	2.08695E-03
889	700889	7.59345E-03	4.25669E-04
889	800889	-1.47032E-04	2.08153E-03
889	900889	5.77376E-03	-1.21182E-04

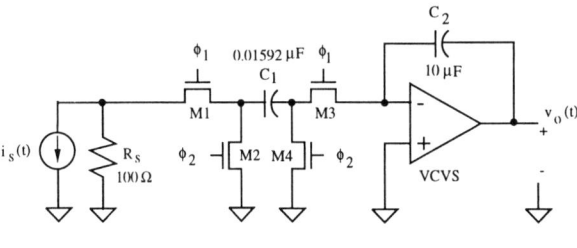

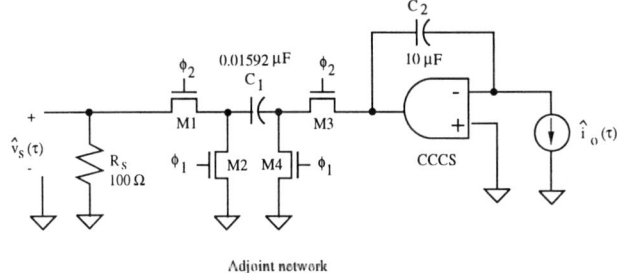

Figure 1: The schematics of parasitic-insensitive SC integrator and its adjoint network

8. REFERENCES

[1] B. D. H. Tellegen, "A general network theorem and applications", *Philips Research Report*, 7, pp. 259-269, 1952.

[2] S. W. Director and R. A. Rohrer, "The generalized adjoint network and network sensitivity", *IEEE Trans. Circuit Theory*, Vol. CT-16, No. 3, pp. 318-323, 1969.

[3] A. K. Seth, "Comments on time-domain network sensitivity using the adjoint network concept", *IEEE Trans. Circuit Theory*, July, pp. 367-370, 1972.

[4] T. N. Trick, *Introduction to circuit analysis*, John Wiley and Sons, New York, 1979.

[5] R. Rohrer, L. Nagel, R. Meyer and L. Weber, "Computationally efficient electronic-circuit noise calculations", *IEEE Journal of Solid-State Circuits*, Vol. SC-6, No. 4, pp. 204-213, 1971.

[6] R. D. Davis, "A derivation of the switched-capacitor adjoint network based on a modified Tellegen's theorem", *IEEE Trans. Circuits and Systems*, Vol. CAS-29, No. 4, pp. 215-220, 1982

[7] J. Vandewalle, H. J. De Man and J. Rabaey, "The adjoint switched capacitor network and its application to frequency, noise and sensitivity analysis", *International Journal of Circuit Theory and Applications*, Vol. 9, pp. 77-88, 1981.

[8] A. Opal and J. Vlach, "Analysis and sensitivity of periodically switched linear networks", *IEEE Trans. Circuits and Systems*, Vol. CAS-36, No. 4, pp. 522-532, 1989.

[9] J. L. Bordewijk, "Interreciprocity applied to electrical networks", *Applied Scientific Research*, B6, pp. 1-74, 1956.

[10] R. Unbehauen and A. Crchocki, *MOS switched-capacitor and continuous-time integrated circuits and systems*, Springer-Verlag, Germany, 1989.

LAYOUT DRIVEN MACROMODEL OF AN OPERATIONAL AMPLIFIER

Chung-Yuk Or

The Chinese University of Hong Kong
cyor@ee.cuhk.edu.hk

Jose E. Franca

Instituto Superior Técnico, Portugal
franca@ecsm4.ist.utl.pt

ABSTRACT

This paper describes the generation of the macromodel of an operational amplifier from the layout database. Unlike traditional schematic based macromodels, this makes it possible to capture all the complex parasitic effects in the layout and hence providing a more accurate representation of the database submitted to fabrication. Results show that such macromodel provides a much faster simulation than the simulation of the netlist extracted from the layout and yet achieves close accuracy with the latter. Therefore, the proposed layout-driven macromodel allows the practical simulation of large layout databases of mixed-signal systems containing in a more time-efficient and accurate manner than can be presently achieved.

1. INTRODUCTION

In the typical analog integrated circuit design cycle [1], schematic level simulations are carried out during the design phase until the specifications are met and then a layout implementation of the circuit is obtained using some popular CAD tools. However, since such layout generates a large number of parasitic elements that are not truly modelled at the schematic level, it is mandatory to perform a simulation of the extracted layout netlist in order to obtain the final verification of the circuit. When the analog circuit is a subsystem of a large analog-digital chip, possibly containing DSP, microprocessor and memory cores as well as analog-to-digital and digital-to-analog converters, among many other blocks, the post layout simulation of the whole chip is a rather time-consuming and sometimes impossible task. Although the digital cores can be efficiently modelled and simulated taking into account realistic layout information, the analog subsystems can still be a bottleneck for the post-layout simulation since they are represented at the transistor level. This situation could be significantly improved if the analog circuit actually laid-out could be represented by a macromodel and then embedded in the higher system level simulation. This is the concept of *layout driven* macromodel generation proposed and illustrated in this paper considering the example of a two-stage internally-compensated CMOS operational amplifier [2]. The proposed modelling methodology is implemented in a program (called LDOM) which accepts the layout netlist and generates the HSPICE subcircuit file of the macromodel. Simulation results are presented to give a performance comparison of the macromodel and the corresponding netlist in terms of simulation time and accuracy.

2. METHODOLOGY

The major difference between a schematics and a layout netlist is that the latter contains the parasitic capacitances that are inherently present in the actually fabricated circuit. The layout netlist is, thus, a more *realistic* representation of the actual fabricated circuit. In order to take into account such information in the macromodel of an analog cell, we have developed a program (LDOM) that automatically performs the procedures represented in figure 1. First, from the node table of the netlist, the program identifies each of the corresponding transistors in the cell and hence the circuit topology. This step is important for locating the critical transistors and nodes for use in later steps. Then, with the knowledge of the physical geometries (W/L) and technology defined process parameters, the bias current, transconductances, resistances and capacitances of the macromodel are calculated. Next, the extraction of the parasitic capacitances is carried out. Since not all the parasitic capacitances are equally relevant, the program identifies the parasitics associated with the critical nodes to be accounted for in the model. Finally, with all the parameters in the macromodel, a HSPICE subcircuit file is generated automatically. This subcircuit can be used to replace the original netlist in the higher level system simulation.

3. MACROMODELLING THE OPERATIONAL AMPLIFIER

In order to illustrate the proposed methodology we consider here the example of generating a layout-driven macromodel for the internally compensated two-stage CMOS operational amplifier shown in figure 2. The macromodel structure is the two poles small signal model [3] with the modifications that the controlled current sources have limited supply current to account for the slew rate limitation and the output voltage is limited to the supply voltage. The schematic diagram of the macromodel circuit is shown in figure 3, where the lumped capacitors C_1, C_2 and C_F represent the combined presence of all the parasitic capacitances associated with the nodes. The transconductances g_{m1} and g_{m2} represent the small signal transconductances of the transistors M_1 and M_5 respectively, and can be obtained from [4]

$$g_{m1} = \alpha\sqrt{2k_n\frac{W_1}{L_1}I_{D1}} \qquad (1)$$

$$g_{m2} = \sqrt{2k_p\frac{W_5}{L_5}I_{D5}} \qquad (2)$$

where k_n and k_p are determined by the technology defined process parameters [5] and α is a *fitting* parameter to account for the highly nonlinear behavior of the transistor. It is found from simulations

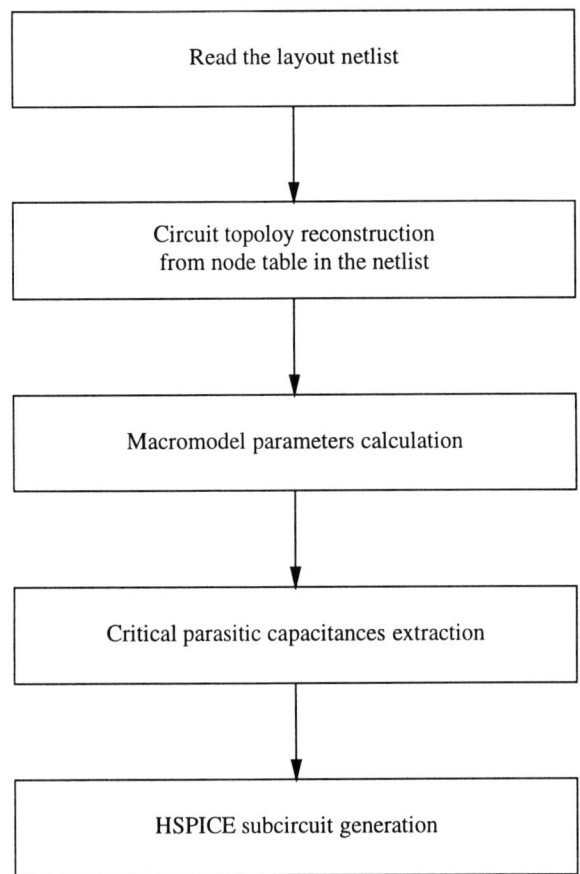

Figure 1: Sequence of procedures of the methodology for the generation of layout driven macromodels of analog cells.

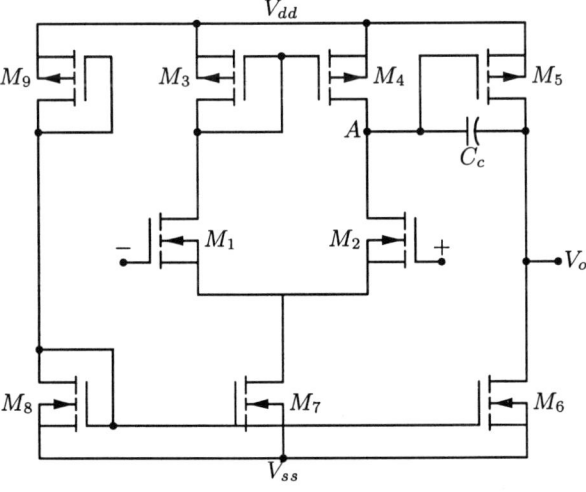

Figure 2: Schematic of the two-stage CMOS operational amplifier.

that this fitting parameter is about 0.8. The maximum currents that can be supplied by the two voltage-controlled current sources are the maximum currents that can flow through M_1 and M_5.

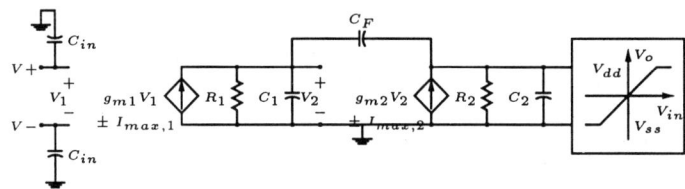

Figure 3: The macromodel for the two-stage operational amplifier.

Therefore,

$$I_{max,1} = I_{D7} \quad (3)$$
$$I_{max,2} = I_{D6} \quad (4)$$

The resistors R_1 and R_2 are equivalent resistances seen at node A and output node of the operational amplifier. They are given by

$$R_1 = r_{ds2} \parallel r_{ds4} \quad (5)$$
$$R_2 = r_{ds5} \parallel r_{ds6} \quad (6)$$

As mentioned before, the capacitors C_1 and C_2 include the gate-source, gate-drain capacitances and the parasitic capacitances of the two most critical amplifier nodes and are determined [2] by

$$C_1 = C_{gd2} + C_{gd4} + C_{gs5} + C_{1p} \quad (7)$$
$$C_2 = C_{gd6} + C_{2p} \quad (8)$$

where C_{1p} and C_{2p} are parasitic capacitances extracted from the layout netlist. The capacitor C_F is the summation of the compensating capacitor and the parasitic capacitance across the gate-drain of the transistor M_5. The two C_{in}'s are the gate capacitances of the input differential pair M_1 and M_2 which account for the input capacitances of operational amplifier.

4. SIMULATION RESULTS

The operational amplifier connected as an inverting amplifier driving a capacitive load is used as the test circuit for the simulations of the layout netlist, the macromodel and the schematics without parasitics for comparison. The circuit is shown in figure 4.

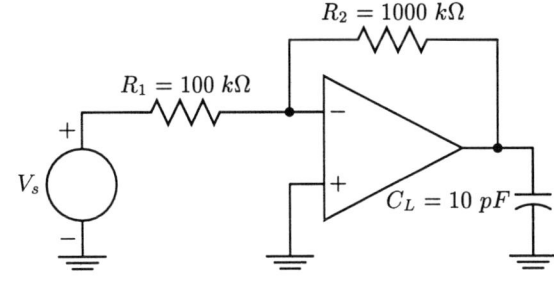

Figure 4: An inverting amplifier circuit for performance simulation.

A transient simulation of the circuit with the netlist, the macromodel and the schematics as the operational amplifier is shown in figure 5. The curves for the macromodel are with name 'laymodel' whereas the curves for the netlist are with name 'layout'. The curves for the schematics are with name 'twostage'. It can be seen that there is virtually no difference between the two curves corresponding to the netlist and macromodel representations. On the other hand, the simulation of the schematics done with no parasitic effects shows large derivation from the other two curves.

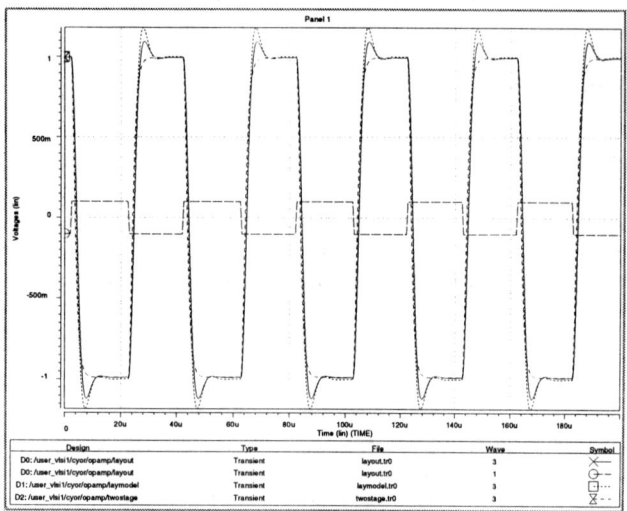

Figure 5: A transient simulation of the test circuit.

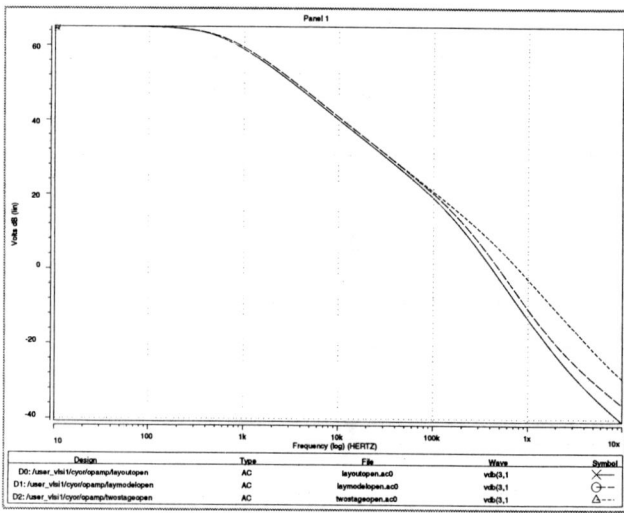

Figure 6: A magnitude response of the test circuit.

The open loop magnitude and phase responses of the opamp shown in figure 6 and 7, respectively, also indicate that a close matching between the netlist and the macromodel is attained. As in the case of transient simulation, the schematic representation of the operational amplifier behaves very differently in the magnitude and phase responses of the test circuit. Discrepancies between the netlist and macromodel simulations appear only outside the frequency band of interest of the operational amplifier which is about 1 MHz. This is due to the effect of parasitic capacitances at the non-critical nodes and thus the non-dominant poles get more and more important as frequency goes beyond the gain-bandwidth product of the operational amplifier. In the simulations of the HSPICE Transient and AC analyses running in a ULTRA Sparc workstation, the macromodel is 39.4% faster than the netlist simulation without loss of accuracy. Such saving in simulation time can be rather significant when the circuit becomes much more complicated.

5. CONCLUSIONS

In conclusion, this paper proposes a methodology to generate macromodels of analog circuits with the input of their layout netlists. The procedures are carried-out automatically by the LDOM program. A two-stage internally compensated CMOS operational amplifier is used as the example to illustrate the key steps to obtain the macromodel. Simulation results show excellent matching of the macromodel and the netlist behaviors and yet a faster simulation is achieved using the macromodel. Future works should be done to develop marcomodels of large analog block such as analog-to-digital converter (ADC).

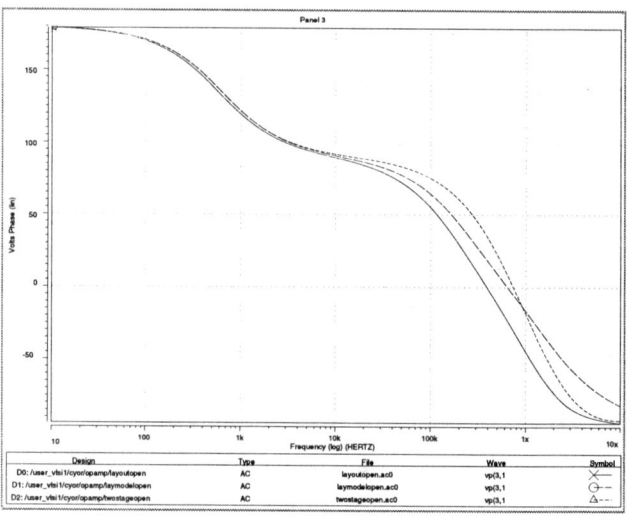

Figure 7: A phase response of the test circuit.

6. REFERENCES

[1] Henry Chang et al., "A Top-Down, Constraint-Driven Design Methodology for Analog Integrated Circuits", *In Proc. Custom Integrated Circuit Conference*, pages 841-846, Boston, May 1992.

[2] Randall L. Geiger, Phillip E. Allen and Noel R. Strader, *VLSI Design Techniques for Analog and Digital Circuits.* McGraw-Hill, 1990.

[3] P. R. Gray and R. G. Meyer, "MOS operational amplifier design – A tutorial overview," *IEEE J. Solid-State Circuits*, vol. SC-17, pp. 969-982, Dec. 1982.

[4] P. R. Gray and R. G. Meyer, *Analysis and Design of Analog Integrated Circuits.* New York:Wiley, 1977.

[5] *HSPICE level 6 MOSFET model*, ATMEL-ES2, Oct 1993.

ON THE ALGEBRAIC REUSE OF HARDWARE DESIGN

Ana C. V. de Melo

Department of Computer Science – IME
State University of Sao Paulo (USP)
Rua do Matao, 1010 – Cidade Universitária – 05508-900
Sao Paulo – SP – Brazil
e–mail: acvm@ime.usp.br
fax: +55 11 818 6134

ABSTRACT

The widespread use of computers in a diversity of activities today demands complex computational systems to be produced efficiently. This factor has led to a requirement for new methods and techniques to enhance controllability, quality and productivity of systems. Reusability is recognized as a basic principle for enhancing productivity and quality of engineering products. Additionally, formal development of software/hardware has emerged as an approach to ensure quality and help handle the complexity of description of such systems. So, for reasons of economy, productivity, quality and time to market, it is highly desirable to formally reuse hardware/software components.

This paper presents a foundation for formal reuse of synchronous processes using a process algebra (EPA [2]). Assuming the existence of a library of formally verified components, we propose to make effective reuse of these existing elements when creating new systems. The strategy used here is to formally create an *interface element* with which a library process is composed in order to implement the desired component. In doing so, the verification task of the whole system is reduced to verifying the *interface element*.

Keywords: Reusability, High–Level Synthesis, Process Algebras, Bisimulation, Interface Equation.

1. MOTIVATION

To make engineering products competitive to the market, techniques must be provided to allow them to be produced at a low cost, in a short period of time and with a desirable quality. These prerequisites for engineering products

http://www.ime.usp.br/

can be achieved by enhancing their productivity and quality. The application of the reuse principles drives to a better productivity. To enhance the quality of such products however, a formal development is required to assure their "correctness". So, it is highly desirable to formally reuse hardware/software components.

Formal methods have been employed to provide verification of hardware such as HOL Proof System [9] and the process algebra Circal [11]. In order to formally reuse hardware, a process algebra based on SCCS (Synchronous CCS [12]), EPA [1, 2], is used to specifying and formally reasoning about synchronous hardware.

Some efforts have been employed to accomplish formal reuse. A wide form of reuse hardware components has been addressed by techniques to synthesize hardware design, from system down to logic level. Also, some attention has been paid to the formalization of this task as noted by Busch, Nusser and Rössel in [4]. The system-level synthesis is very concerned with *partitioning* of systems by adding structure to behavioural objects (in general, the systems are partitioned into standard components). By contrast, reusable components are not required to fit a natural partitioning of the desired components; a library component could, for example, embed some functions not required by the specification element and still be useful for reuse. System–level synthesis is thus an instance of reusability in the sense that the existing component could eventually be a subsystem of the desired one. Apart from this restricted form of reuse addressed by synthesis, not much attention has been paid to reuse complex hardware components.

In the software fields, the efforts employed to reuse complex systems either address methods roughly informal or concentrate on the development of specification languages to allow reuse. The methods of reuse not based on formal means suffer from not ensuring the quality of products.

Despite being essential for the formal reuse, most developments on specification languages do not pay attention to mechanisms for *partial identification* of processes, which is also important to make reuse effective.

Considering the deficiencies of the current reuse methods, this work presents a foundation for formal reuse of synchronous hardware. A process algebra is taken to model hardware, and we concentrate on the mechanisms for abstractly identifying processes. The main goal is to show an algebraic treatment of reusable processes.

2. THE PROBLEM STATEMENT

The reuse of software/hardware design involves managerial as much as descriptive information [7]. Due to the formal approach being taken for reuse in the present work, we confine ourselves to reusing the descriptive information of hardware components.

Different approaches can be adopted for reusing software and hardware [3, 17]. We propose reuse of synchronous processes by "transforming" an existing process into the desired behaviour. Basically, two approaches can be adopted to transforming processes: either modifying the design of an existing component, or composing such component with an interface to achieve the desired behaviour. The task involved in the first approach is *redesigning* an existing system into a new one. In the second approach, the *Interface Approach*, the existing process is reused without any change in design, but a new process (the *interface process*) must be developed. Hence, the existing process has its behaviour "transformed" by composition with the new process.

From a practical point of view, the interface approach is attractive for reuse because processes are reused without being re–verified. Another aspect which makes this approach attractive is the possibility of having automatic procedures to find the interface process.

Taking the interface approach, the interface process must be composed ($|$) with an existing component (C) to provide the behaviour of the desired process (*Spec*). Considering that strong bisimulation ($\sim$) [12] is adopted as the behavioural equivalence, the resulting composition must be strong bisimilar to the desired process. The *Interface Equation* for this particular reuse problem is defined as follows:

$$(C \mid X) \restriction_c \mathsf{Chs}(Spec) \sim Spec \quad (1)$$

Spec is the desired process while C is the existing component, and X is the interface process to be found. Note that the observable channels of the compound process are restricted to the set of channels of the desired process ($\restriction_c$ Chs(*Spec*)); all the other communication ports are internalized.

3. THE TECHNIQUES INVOLVED

To give an algebraic treatment to the reuse problem, a combinator (*decomposition operator*) must be created to build the interface process. This operator is aimed at playing the same role as the "division" operator for numbers. For example, consider having to find a solution for equation

$$x \star 2 = 10. \quad (2)$$

How can this equation be solved? One approach is to use the algebra defined for the division and multiplication operators, for $m, n, t \in \mathbb{Z}$ and $n \neq 0$:

$$m \star n = t \quad \text{if and only if} \quad m = t \div n,$$

and thus find the following solution:

$$x = 10 \div 2.$$

To obtain a similar algebraic treatment of processes, a *decomposition operator* (/) is defined.

The creation of interface processes through formal means involves: a model for representing concurrent processes, a way to decide similarities between the desired and the existing processes, and finally, the decomposition operator. As mentioned above, EPA is used to model concurrent processes, and a notion of similarities between processes useful for reuse is created. Then, based on these similarities, a *decomposition operator* is defined to build generic interface processes.

3.1. Similarities Between Processes

Consider having the trivial hardware component *Nor–gate* in the library:

process $Nor_p(c_1, c_2: bool/c_3: bool) \triangleq$
$((c_1(\mathtt{t}), c_2(\mathtt{t})/c_3(\mathtt{f}))+$
$(c_1(\mathtt{t}), c_2(\mathtt{f})/c_3(\mathtt{f}))+$
$(c_1(\mathtt{f}), c_2(\mathtt{t})/c_3(\mathtt{f}))+$
$(c_1(\mathtt{f}), c_2(\mathtt{f})/c_3(\mathtt{t}))) :: Nor_p(c_1, c_2/c_3).$

Process Nor_p above communicates with the environment using the input channels c_1, c_2 of type *bool* and the output channel c_3. It moves for one of the four initial actions (as with the choice operator ($+$), only one action can be performed at a time) and then evolves into $Nor_p(c_1, c_2/c_3)$.

Suppose a nor gate that delays its output for one step, the *Nor–gate–delay*, is now required:

process $NorDel_f(c_1, c_2: bool/c_4: bool) \triangleq$
$((c_1(\mathtt{t}), c_2(\mathtt{t})/c_4(\mathtt{f}))+$
$(c_1(\mathtt{t}), c_2(\mathtt{f})/c_4(\mathtt{f}))+$
$(c_1(\mathtt{f}), c_2(\mathtt{t})/c_4(\mathtt{f}))) :: NorDel_f(c_1, c_2/c_4)+$
$(c_1(\mathtt{f}), c_2(\mathtt{f})/c_4(\mathtt{f})) :: NorDel_t(c_1, c_2/c_4)$

process $NorDel_t(c_1, c_2: bool/c_4: bool) \triangleq$
$((c_1(\mathtt{t}), c_2(\mathtt{t})/c_4(\mathtt{t}))+$
$(c_1(\mathtt{t}), c_2(\mathtt{f})/c_4(\mathtt{t}))+$
$(c_1(\mathtt{f}), c_2(\mathtt{t})/c_4(\mathtt{t}))) :: NorDel_f(c_1, c_2/c_4)+$
$(c_1(\mathtt{f}), c_2(\mathtt{f})/c_4(\mathtt{t})) :: NorDel_t(c_1, c_2/c_4)$

Roughly speaking, these components have some similarities (they are nor gates) but are clearly not equivalent. Despite being defined over different sets of output channels, the similarities between these processes can still be formally checked. In fact, Nor_p can simulate actions of $NorDel_f$ if both processes are restricted to channels $\{c_1, c_2\}$. Such Channel Restricted Simulation (cr–simulation) is denoted as $\sqsubseteq^c$ (its formal definition is found in [5]). Then, we may realize that $NorDel_f \sqsubseteq^c Nor_p$ and $NorDel_t \sqsubseteq^c Nor_p$.

3.2. A Decomposition Operator

In the previous section (Section 3.1), the component Nor_p was checked a cr–simulation of $NorDel_f$. Also, Nor_p is deterministic for all actions that simulate $NorDel_f$. So, an interface process can be found to simulate $NorDel_f$ when composed with Nor_p.

Action $Nor_p \xrightarrow{c_1(\mathtt{f}), c_2(\mathtt{f})/c_3(\mathtt{t})} Nor_p$, for example, simulates $NorDel_f \xrightarrow{c_1(\mathtt{f}), c_2(\mathtt{f})/c_4(\mathtt{f})} NorDel_t$. Also, $NorDel_f \sqsubseteq^c Nor_p$ and $NorDel_t \sqsubseteq^c Nor_p$. Thus, the "division" process $NorDel_f / Nor_p$ comprises

$(NorDel_f / Nor_p) \xrightarrow{(c_1(\mathtt{f}), c_2(\mathtt{f})/c_4(\mathtt{f})) \div (c_1(\mathtt{f}), c_2(\mathtt{f})/c_3(\mathtt{t}))} (NorDel_t / Nor_p),$

where

$(c_1(\mathtt{f}), c_2(\mathtt{f})/c_4(\mathtt{f})) \div (c_1(\mathtt{f}), c_2(\mathtt{f})/c_3(\mathtt{t})) =$
$c_1(\mathtt{f}), c_2(\mathtt{f}), c_3(\mathtt{t})/c_4(\mathtt{f}).$

The application of the decomposition operator builds the entire interface process such that

$(Nor_p \mid (NorDel_f / Nor_p)) \lceil_c \mathsf{Chs}(NorDel_f) \sim NorDel_f.$

The interface process given by $(NorDel_f / Nor_p)$ can be minimized [6] to another trivial hardware component: a delay process.

4. RESULTS AND DISCUSSION

The main technical result here is the definition of a particular decomposition operator which has been proved to be a solution for the interface equation when applied to the desired and existing processes.

The reuse of hardware components through formal means is still in its infancy. Most works in the hardware area which address a kind of reuse are synthesis–based, or based on an informal approach, such as [13]. Even in the software domain, formal reuse has only been attempted by a few works that concentrate on partial definition of processes [8, 10, 19, 15, 20]. In fact, abstract definition of processes is essential to obtain effective reuse. In the present work we have assumed an existing model for representing processes to concentrate on mechanisms to identify them by making certain abstractions. This work has indicated that besides abstraction on representation of processes, mechanisms based on abstractions to identifying processes are also useful for reuse.

Reuse undertaking the interface approach is very much concerned with the idea of decomposing the desired process into submodules, so that one of those submodules matches the existing component. This decomposition problem has already been solved, by Shields [18] and his peers [16, 14], for deterministic processes represented in CCS and taking weak bisimulation as the equational equality. Here, a solution for the decomposition problem considering *nondeterministic* processes represented in EPA and taking *strong bisimulation* as the equivalence relation is presented.

A formal reuse of existing components leads to a decompositional verification of processes. As with the reuse of components formally verified, only the new elements being constructed need to be verified. Since composition of the interface with the existing process is proved bisimilar to the desired specification, verification of the interface element ensures verification of the whole system. In fact, the topmost definition of the interface process is correct by construction, and only its refinements must be verified.

5. REFERENCES

[1] H. Barringer, G. Gough, B. Monahan, and A. Williams. An algebraic framework for action and process. D2.3c, Formal Verification Support for ELLA, IED project 4/1/1357, University of Manchester, May 1993.

[2] H. Barringer, G. Gough, B. Monahan, and A. Williams. A design and verification environment for ELLA. In *Proceedings of ASP-DAC/CHDL/VLSI'95*, pages 685–691, Chiba, Japan, Aug 1995.

[3] T. J. Biggerstaff and A. J. Perlis. *Software Reusability – Concepts and Models*, volume 1. Addison–Wesley Publishing Company, first edition, 1989.

[4] H. Busch, H. Nusser, and T. Rössel. Formal methods for synthesis. In P. Michael, U. Lauther, and P. Duzy, editors, *The Synthesis Approach to Digital System Design*. Kluwer Academic Publishers, 1993.

[5] A. C. V. de Melo and H. Barringer. A foundation for formal reuse of hardware. In P. Camurati and H. Eveking, editors, *Correct Hardware Design and Methods 95*, volume 987 of *LNCS*. Springer Verlag, 1995.

[6] A. C. V. de Melo and H. Barringer. Minimization of concurrent systems preserving bisimulation. In *SBC-CI'97*, 1997.

[7] P. Freeman. A perspective on reusability. In P. Freeman, editor, *Tutorial: Software Reusability*. IEEE – The Computer Society Press, 1987.

[8] J. A. Goguen. Principles of parameterized programming. In T. J. Biggerstaff and A. J. Perlis, editors, *Software Reusability – Concepts and Models*, volume 1. Addison–Wesley Publishing Company, 1989.

[9] M. J. Gordon. Why higher–order logic is a good formalism for specifying and verifying hardware. In G. J. Milne and P. A. Subrahmanyam, editors, *Formal Aspects of VLSI Design*. North–Holland, 1986.

[10] S. Katz, C. A. Richer, and Khe-Sing The. PARIS: A system for reusing partially interpreted schemas. In T. J. Biggerstaff and A. J. Perlis, editors, *Software Reusability – Concepts and Models*, volume 1. Addison–Wesley Publishing Company, 1989.

[11] G. Milne. *Formal Specification and Verification of Digital Systems*. McGraw–Hill, first edition, 1994.

[12] R. Milner. *Communication and Concurrency*. Prentice–Hall, first edition, 1989.

[13] N. S. Nagaraj. OPSYN – OASYS based pseudosynthesis tool. In *VLSI Design 1992 – The fifth International Conference on VLSI Design*. IEEE – The Computer Society Press, 1992.

[14] J. Parrow. Submodule construction as equation solving in CCS. *Theoretical Computer Science*, 68:175–202, 1989.

[15] N. S. Prywes and E. D. Lock. Use of the model equational language and program generator by management professionals. In T. J. Biggerstaff and A. J. Perlis, editors, *Software Reusability – Applications and Experience*, volume 2. Addison–Wesley Publishing Company, 1989.

[16] H. Qin and P. Lewis. Factorization of finite state machines under observational equivalence. In J. C. M. Baeten and J. F. Groote, editors, *CONCUR'91*, volume 527 of *LNCS*. Springer–Verlag, 1991.

[17] W. Schäfer, R. Prieto-D´az, and M. Matsumoto. *Software Reusability*. Ellis Horwood, 1994.

[18] M. W. Shields. Implicit system specification and the interface equation. *The Computer Journal*, 32(5), 1989.

[19] D. M. Volpano and R. B. Kieburtz. The templates approach to software reuse. In T. J. Biggerstaff and A. J. Perlis, editors, *Software Reusability – Concepts and Models*, volume 1. Addison–Wesley Publishing Company, 1989.

[20] M. Wirsing, R. Hennicker, and R. Stahl. Menu – an example for the systematic reuse of specifications. In *2nd European Software Engineering Conference*, volume 387 of *LNCS*. Springer Verlag, 1989.

ASSESSING THE UNIQUENESS OF THE DC SOLUTIONS BY TEARING OF CACTUS GRAPHS VIA DETECTION OF HINGING STRUCTURES

Arturo Sarmiento Reyes, Rafael Vargas Bernal

Instituto Nacional de Astrofísica Óptica y Electrónica,
Electronics Department, CAD Group
P.O. Box 51, 72000 Puebla, Pue., Mexico.

ABSTRACT

Cactus graphs are used to determine whether or not a circuit may posses multiple dc solutions. This work introduces a tearing method for cactus graphs that can be used to decompose the initial cactus into smaller cactus subgraphs which are hinging altogether. The method is focussed on circuits containing independent current and voltage sources, positive linear resistors, and BJTs, but it can be easily modified in order to cope with devices that can be modelled by equivalent circuits composed by coupled branches. Besides, the concept of path-matrices is introduced in order to determine if the whole graph of the circuit can be split into subgraphs. In a further step, the subgraphs can be analysed separately in order to detect multiple dc operating points.

1. INTRODUCTION

Cactus graphs have been used in the past [1–3] to cope with the problem of assessing the uniqueness of the dc operating points. The concept of cactus graph [3] can be explained by using the general graph of the circuit. In figure 1, the graph of a circuit shows an internal subgraph G_L related to the interconnection pattern of the non-coupled part of the circuit. Each coupled branch — transactor — is represented by a pair of edges, namely, the controlling edge $\langle 1 \rangle$ and the controlled edge $\langle 1' \rangle$.

After achieving some graph manipulations to the complete graph of the circuit, it is possible to reduce the starting graph into a final graph composed by the pair edges concerning all transactors. The resulting graph is thus formed by sets of branches resembling the leaves of a cactus.

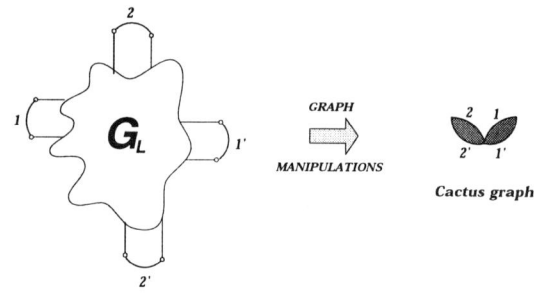

Figure 1: A cactus graph.

R. Vargas Bernal is holder of a scholarship from CONACyT/Mexico under contract 71960/124548.

If every BJT is modelled by the well-known Ebers and Mol schema [4], then it contributes with a pair of cactus leaves, i.e. the initial cactus graph of a general BJT network (see figure 2) is given by:

$$G = G_L \cup G_q \qquad (1)$$

where G_L and G_q are the graph of the linear circuit and the the graph of transistors respectively. Besides:

$$G_L = G_L(V_L, E_L) \qquad G_q = G_q(V_q, E_q)$$

where V_L and E_L are the set of vertices and edges of the graph of the linear part of the circuit, and V_q and E_q are the set of transistor terminals and the edges of the transistor cactuses.

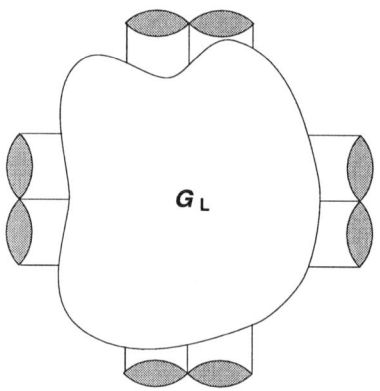

Figure 2: Initial BJT cactus graph.

However, the main disadvantage of the use of cactus graphs resides in the fact that in order to obtain the final cactus graph(s) a total of 2^r graph combinations must be tested, where r is the number of total resistors.

1.1. Assessing the uniqueness.

The uniqueness of the dc solution is assessed by applying the method of Chua [3] to every reduced cactus subgraph in a slightly modified form. This step consists in finding an embedded cactus graph associated to a flip-flop alike structure as shown in the figure 3, by applying some graph operations to the branches of G_L, namely open-circuit conversion — $\mathcal{O}(\bullet)$ — and short-circuit conversion — $\mathcal{S}(\bullet)$.

At the end, if such a graph can be found, then the circuit may posses multiple dc solutions.

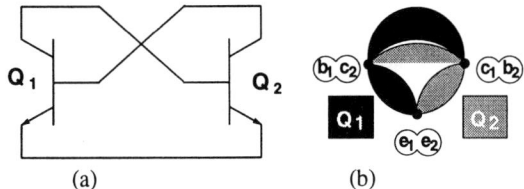

Figure 3: Flip-flop alike structure and its cactus.

2. TEARING PROCEDURE

Because the figure 2^r may become a number impossible to handle, even for small transistor circuits, then a graph partitioning procedure will be carried out in order to reduce the number of graph manipulations.

The tearing procedure is based on the concept of hinging cactus subgraph. These graphs are subgraphs that are hinging *"naturally"* in the original circuit graph. In fact they are embedded in the whole graph, and it is the aim of the tearing procedure to detect them.

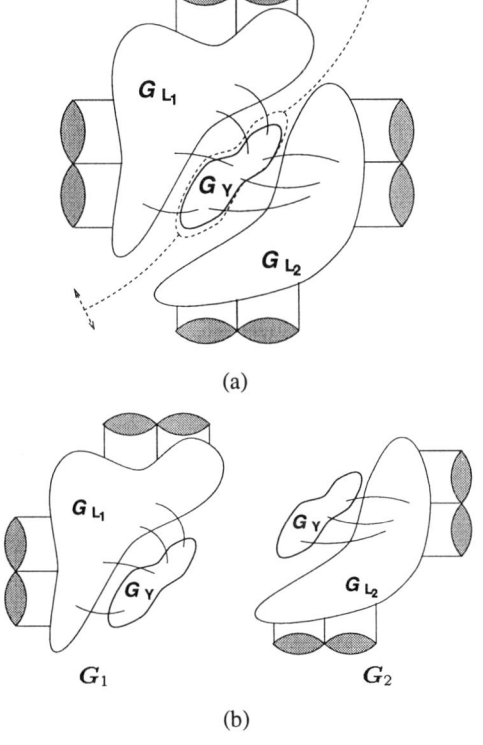

Figure 4: Tearing of cactus graphs.

The figure 4-(a) shows schematically this concept: the hinge formed by the subgraph G_Y acts in fact as the link that joins the subgraphs G_{L_1} and G_{L_2}. It clearly results (see the figure 4-(b)) that the whole graph can be separated in two subgraphs:

$$G_1 = G_{L_1}(V_1, E_1) \cup G_{q_1} \cup G_Y$$
$$G_2 = G_{L_2}(V_2, E_2) \cup G_{q_2} \cup G_Y$$

(2)

where V_i and E_i are the set of vertices and edges respectively of the i-th subgraph. Furthermore G_{q_1} and G_{q_2} are the set of cactus leaves for each partitioning.

Some special cases arise:

1. One-vertex hinging subgraphs.

   Given a graph $G(V, E)$, where V is the set of vertices, and E is the set of edges, then it is said that the subgraphs $G_1(V_1, E_1)$ and $G_2(V_2, E_2)$ are hinging at one vertex, if $E_1 \cup E_2 = E$, and $V_1 \cap V_2$ contains one and only one vertex.

   The tearing procedure is applied to such a graph by finding the common vertex of the subgraphs G_{L_1} and G_{L_2}.

2. One-edge hinging subgraphs.

   Given a graph $G(V, E)$, where V is the set of vertices, and E is the set of edges, then the subgraphs $G_1(V_1, E_1)$ and $G_2(V_2, E_2)$ are said to be one-edge hinging subgraphs, if $E_1 \cap E_2$ contains one and only one edge, and $V_1 \cap V_2$ contains only the pair of vertices being incident to the common edge.

   Similarly, the tearing procedure is applied to such a graph by finding the edge common to the subgraphs G_{L_1} and G_{L_2}.

2.1. Path matrices

Now, we have to determine a method in order to find the the subgraph G_Y that allows us to carry out the tearing procedure. This problem can be easily solved by resorting to the paths[1] connecting the terminals between transistor pairs, and detecting either a minimum common graph, i.e. a vertex or an edge, or a complete subgraph.

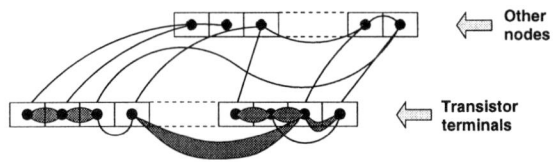

Figure 5: Paths in the graph.

The path matrices contain the information of the set of vertices or set of edges that allows us to connect the terminals of one transistor in the circuit with the terminals of another transistor or with the nodes of the non-coupled part of the circuit. As a result, it can be stated that the paths will contain only edges belonging to the positive linear resistors which are crossing from the hyper-plane of terminals to the hyper-plane of the rest of the nodes, as shown in the figure 5. Two path matrices arise: the matrix with the transistor terminals (see figure 6-(a)) and the matrix with the rest of nodes (see

---

[1] A path is subgraph in which all vertices have degree 2, except the first and the last vertices, that have degree 1.

figure 6-(b)). The path matrices are given as:

$$M_q = [m_{ij}] \qquad M_l = [\widetilde{m}_{ij}]$$

with the following characteristics:

$$m_{ij} = \mathcal{L}_{ij} \qquad \widetilde{m}_{ij} = \widetilde{\mathcal{L}}_{ij} \qquad (3)$$

where $\mathcal{L}_{ij}$ is the set of paths between transistor terminal i and transistor terminal j, and $\widetilde{\mathcal{L}}_{ij}$ is the set of paths between vertex i and vertex j. It clearly results that $m_{ij} = m_{ji}, \widetilde{m}_{ij} = \widetilde{m}_{ji}, m_{ii} = \{\phi\}$, and $\widetilde{m}_{ii} = \{\phi\}$.

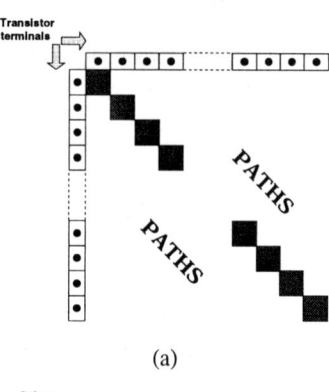

(a)

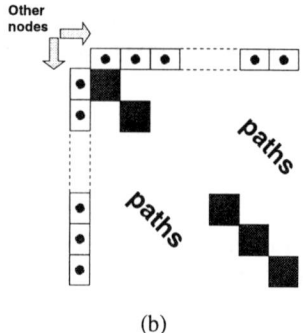

(b)

Figure 6: Path matrices.

The path matrix of the transistors have a block structure as follows:

1. Blocks of paths associated to a single transistor.

$$M_{Q_i} = \begin{bmatrix} \mathcal{L}_{e_i b_i} & \mathcal{L}_{e_i c_i} \\ & \mathcal{L}_{b_i c_i} \end{bmatrix}$$

2. Blocks of paths between two transistors.

$$M_{Q_i,Q_j} = \begin{bmatrix} \mathcal{L}_{e_i e_j} & \mathcal{L}_{e_i b_j} & \mathcal{L}_{e_i c_j} \\ \mathcal{L}_{b_i e_j} & \mathcal{L}_{b_i b_j} & \mathcal{L}_{b_i c_j} \\ \mathcal{L}_{c_i e_j} & \mathcal{L}_{c_i b_j} & \mathcal{L}_{c_i c_j} \end{bmatrix}$$

Therefore $M_{Q_i,Q_j} = M_{Q_j,Q_i}^T$.

3. IMPLEMENTING THE ALGORITHM

The algorithm can be recast in the next steps:

1. Form the graph

2. Separate the vertices: transistor terminals, and other vertices.

3. Form the path matrices.

4. Start a search procedure in order to select sets of paths between two group of transistors.

5. Determine as well the set of other nodes.

6. The intersection of both sets is the hinging graph G_Y.

7. Two graphs are formed:

$$G_1 = G_{L_1} + G_Y$$
$$G_2 = G_{L_2} + G_Y$$

8. Convert G_Y into a super-node, by applying $\mathcal{S}(\bullet)$ to its edges.

9. Repeat the procedure (if possible) to both G_{L_1} and G_{L_2} from step 2.

10. If not possible, then apply $\mathcal{S}(\bullet)$ and/or $\mathcal{O}(\bullet)$ to the edges of both G_{L_1} and G_{L_2} in order to detect if a flip-flop alike cactus such as shown in figure 3-(b) is present.

4. EXAMPLE

The circuit [5] shown in the figure 7 will be used to illustrate the method. The circuit is actually composed by two Schmitt-triggers connected *back-to-back*. However, from its cactus graph (figure 8), it is quite difficult to notice the existence of a hinging subgraph.

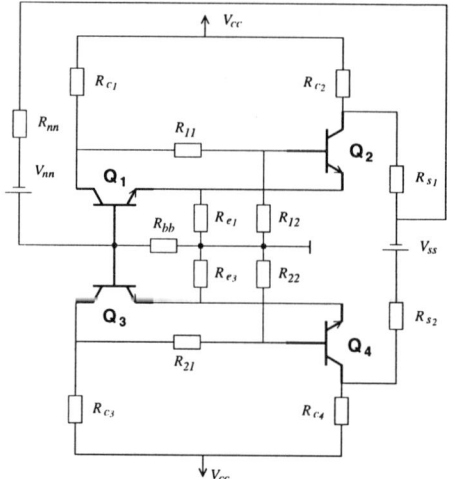

Figure 7: Circuit example

The application of the method yields a hinge formed by the subgraph shown in the figure 9. Consequently, the separated subgraphs are those shown in the figure 10.

After converting the hinge into a super-node, i.e. all branches being considered as short circuits, the flip-flop alike structure can be obtained by applying:

$$\mathcal{S}(R_{11}) \quad \text{and} \quad \mathcal{O}(R_{c_1}, R_{e_1}, R_{12})$$

or

$$\mathcal{S}(R_{21}) \quad \text{and} \quad \mathcal{O}(R_{c_3}, R_{e_2}, R_{22})$$

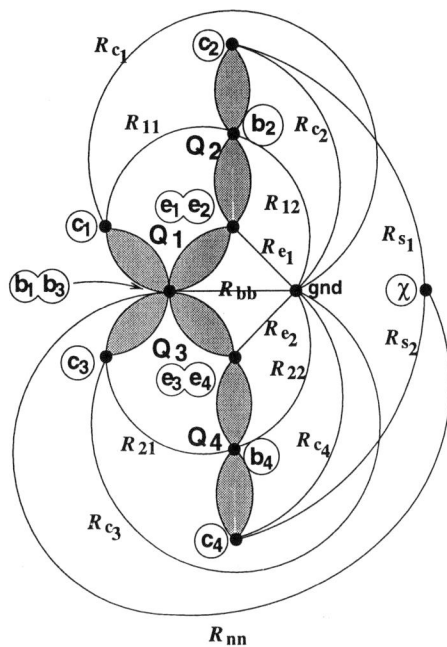

Figure 8: Cactus graph of the dead network.

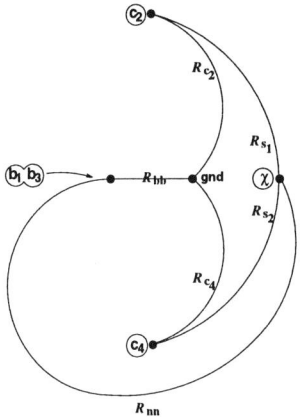

Figure 9: The hinging subgraph.

5. CONCLUSIONS

The present work allows us to find the subgraphs involved in the partition of the cactus graph of the circuit. It resorts to concepts of graph theory in order to form a pair of matrices containing the paths within the circuit. The application of this algorithm with cactus graph concepts allows us to asses the uniqueness of the dc solutions for circuits with more than two transistors.

6. REFERENCES

[1] Tetsuo Nishi and Leon O. Chua. Topological proof of the Nielsen-Willson theorem. *IEEE Transactions on Circuits and Systems*, CAS-33(4):398–405, April 1986.

[2] Tetsuo Nishi. On the number of solutions of a class of nonlinear resistive circuit. *Proceedings of the IEEE International Symposium on Circuits and Systems, Singapore*, pages 766–769, 1991.

[3] Tetsuo Nishi and Leon O. Chua. Uniqueness of solution for nonlinear resisitive circuits containing CCCS's or VCVS's whose controlling coefficients are finite. *IEEE Transactions on Circuits and Systems*, CAS-33(4):381–397, April 1986.

[4] Ian E. Getreu. *Modelling the Bipolar Transistor*. Computer-Aided Design of Electronic Circuits. Elsevier Scientific Publishing Company, 1971.

[5] L. O. Chua and A. Ushida. A switching-parameter algorithm for finding multiple solutions of nonlinear resistive circuits. *International Journal of Circuit Theory and Applications*, 4(3):215–239, July 1976.

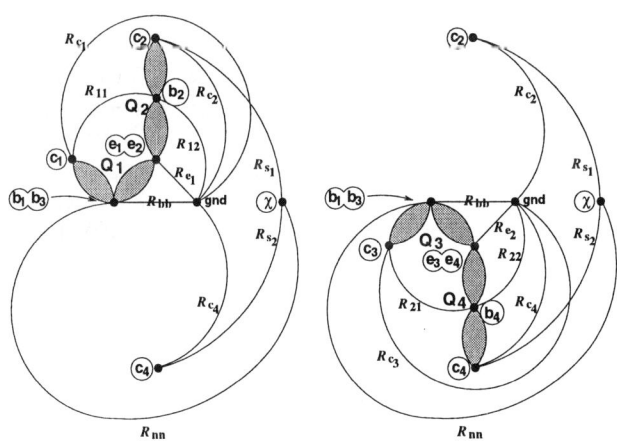

Figure 10: The separated graphs.

On the High Level Canonical Representation of Piecewise Linear Functions

Pedro Julián, *Student Member, IEEE*, Alfredo Desages and Osvaldo Agamennoni

Abstract— In this work, we propose a *representation basis* for the set of PWL functions $f : \mathbf{D} \mapsto \mathbf{R}^1$ defined over a simplicial partition when **D** is a rectangular compact set in $\mathbf{R}^n$. The functions of the basis exhibit several types of *nested* absolute value functions. In addition, an efficient numerical method is given for the resolution of the parameters of the High Level Canonical representation.

I. Introduction

Continuous piecewise linear (PWL) functions have been widely used for analysis and modelling of nonlinear systems. The Canonical Piecewise Linear (CPWL) expression $F : \mathbf{R}^n \mapsto \mathbf{R}^m$

$$F(\mathbf{x}) = \mathbf{a} + \mathbf{B}\mathbf{x} + \sum_{i=1}^{\sigma} \mathbf{c}_i \left| \alpha_i' \mathbf{x} - \beta_i \right|, \qquad (1)$$

introduced by Chua and Kang [1], [2] produced a significant progress in PWL representation, due to the fact that (1) is able to represent any PWL from $\mathbf{R}^n$ to $\mathbf{R}^m$, possessing the *consistent variation property* as defined in [3], with a minimal number of parameters. One limitation of (1) is that it is only able to represent *any arbitrary* PWL function $F : \mathbf{R}^1 \mapsto \mathbf{R}^m$. In general, the representation in high order dimension domains requires the use of multiple *nestings* of absolute value functions [4].

The availability of such *High Level (HL) Canonical PWL representation* was proposed in [5] and rigorously proved in [4]. However, these results are mainly theoreticall, and do not provide neither the explicit expression of the required functions, nor a constructive methodology. In this work, we propose a *representation basis*, consisting of *HL CPWL* functions, for the set of PWL functions $f : \mathbf{D} \mapsto \mathbf{R}^1$ defined over a simplicial partition (see [6]) when **D** is a rectangular compact set in $\mathbf{R}^n$. In addition, we present a systematic and efficient numerical approach to obtain the parameters of a HL CPWL function that represents a given PWL function.

A. Basic Definitions and Notation

Indexes are assumed to be integers, and the notation $i \in \{1, n\} \uparrow$, indicates that the index i takes the values $\{1, ..., n\}$ increasingly from 1 to n. The symbol "′" is used to denote transpose. Several times in the paper, it will be necessary to order a set of elements with some specific structure, according to one of the following two possibilities.

Definition 1 (Ordering A) A set of k indexes $r_i \in \{1, ..., n\}$, $i \in \{1, ..., k\}$, where $r_1 \in \{1, ..., n-k+1\}$ and $r_j \in \{r_{j-1}+1, ..., n-k+j\}$, are A-ordered if they are generated as follows: $r_1 \in \{1, n-k+1\} \uparrow$; for every fixed r_1, $r_2 \in \{r_1+1, n-k+2\} \uparrow$; for every fixed r_2, $r_3 \in \{r_2+1, n-k+3\} \uparrow$; ...; for every fixed r_{j-1}, $r_j \in \{r_{j-1}+1, n-k+j\} \uparrow$.

In this way, if $n = 4$ and $k = 2$, the set of indexes $\{r_1, r_2\}$, with $r_1 \in \{1, ..., 3\}$, $r_2 \in \{r_1+1, ..., 4\}$ can be A-ordered as follows: $\{1, 2\}, \{1, 3\}, \{1, 4\}, \{2, 3\}, \{2, 4\}, \{3, 4\}$.

Definition 2 (Ordering B) A set of n indexes $s_1 \in \{a_1, ..., b_1\}, ..., s_n \in \{a_n, ..., b_n\}$, where a_i, b_i are integers satisfying $a_i < b_i \; \forall i$, are B-ordered if they are generated as follows: $s_1 \in \{a_1, b_1\} \uparrow$; for every s_1 fixed $s_2 \in \{a_2, b_2\} \uparrow$; ...; for every s_{n-1} fixed $s_n \in \{a_n, b_n\} \uparrow$.

In this way, the set of indexes $\{s_1, s_2\}$, with $s_1 \in \{0, 1\}$, $s_2 \in \{3, ..., 5\}$ can be B-ordered as follows: $\{0, 3\}, \{0, 4\}, \{0, 5\}, \{1, 3\}, \{1, 4\}, \{1, 5\}$.

Let a domain **D** be partitioned into a set of convex polyhedrons called *regions* $R^{(i)}$, $i \in \{1, \kappa_1\}$ so that $\mathbf{D} = \bigcup_{i=1}^{\kappa_1} \bar{R}^{(i)}$, by a set

$$H := \{H_i \subset \mathbf{D}, i \in \{1, ..., h\}\} \qquad (2)$$

of a finite number of $n-1$ dimensional hyperplanes, also called *boundaries*

$$H_i = \{\mathbf{x} : \pi_i(\mathbf{x}) := \alpha_i' \mathbf{x} - \beta_i = 0\}, \qquad (3)$$

where $\alpha_i \in \mathbf{R}^n$ and $\beta_i \in \mathbf{R}^1$. Then, a PWL function f is defined by the local (linear) functions $f^{(i)}(\mathbf{x}) = J^{(i)}\mathbf{x} + w^{(i)}$, where $J^{(i)} \in \mathbf{R}^{1 \times n}$, $w^{(i)} \in \mathbf{R}^1$, and $f(\mathbf{x}) = f^{(i)}(\mathbf{x})$ for any $\mathbf{x} \in \bar{R}^{(i)}$. In this paper, only continuous PWL functions will be considered.

Let $\mathbf{x}_0, \mathbf{x}_1, ..., \mathbf{x}_n$ be $n+1$ points in the n-dimensional space. A *simplex* $\Delta_{\bar{\mathbf{x}}}(\mathbf{x}_0, ..., \mathbf{x}_n)$ is defined by the set $\{\mathbf{x} \in \mathbf{R}^n : \mathbf{x} = \bar{\mathbf{x}} + \sum_{i=0}^{n} \mu_i \mathbf{x}_i\}$, where $0 \leq \mu_i \leq 1$, $\forall i \in \{1, ..., n\}$ and $\sum_{i=0}^{n} \mu_i = 1$. A simplex is said to be proper if and only if it can not be contained in an $(n-1)$ dimensional hyperplane.

II. Problem Statement

Consider a rectangular domain **D** partitioned by a set of boundaries H, and let $PWL_H[\mathbf{D}]$ be the set of all continuous PWL mappings $f : \mathbf{D} \mapsto \mathbf{R}^1$, with the boundary configuration H. It is easy to see that if addition and multiplication by a scalar $r \in \mathbf{R}^1$ are defined as

a) $(f+g)(\mathbf{z}) = f(\mathbf{z}) + g(\mathbf{z}), \forall \mathbf{z} \in \mathbf{D}$
b) $(r \cdot f)(\mathbf{z}) = r \cdot f(\mathbf{z}), \forall \mathbf{z} \in \mathbf{D}$

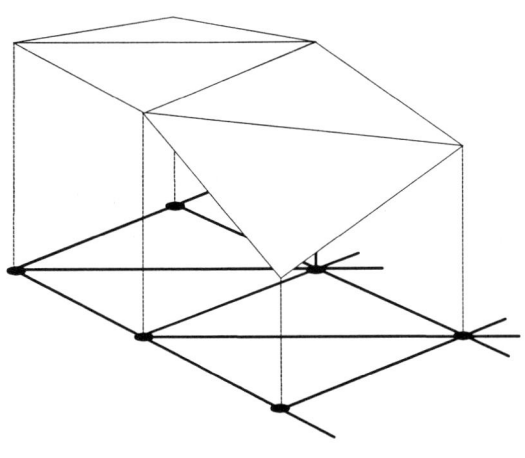

Fig. 1. Constructive approach in $\mathbf{R}^2$

then the set $PWL_H[\mathbf{D}]$ is a *linear vector space (LVS)*. Consequently, the objective of the paper is to find a *basis* for the LVS $PWL_H[\mathbf{D}]$ when H is a simplicial partition. We propose a constructive approach which is motivated by the following procedure.

A. Construction Strategy

Let us suppose, for simplicity, that the domain $\mathbf{D}$ belongs to $\mathbf{R}^2$. If a boundary configuration is defined so that the entire domain is partitioned into proper simplices and a function value is associated to each intersection (given by a vertex), as illustrated by fig. 1, it is possible to define a PWL function with the following characteristics:

a) Considering the function values assigned to each vertex, a *unique* linear (local) function is defined for each simplex.

b) The different linear (local) functions, that are continuous on the boundaries of the partition, define a continuous PWL function.

The extension of this idea to a n-dimensional domain leads us to define proper simplices of $n+1$ vertices. Then, we associate one function value to each vertex. After that, it is possible to uniquely determine a linear (local) function for each one of the simplices, in such a way that the collection of all the local functions form a continuous PWL function. From this procedure, it is clear that any arbitrary PWL function $f:\mathbf{D}\mapsto\mathbf{R}^1$, defined over the simplicial boundary configuration introduced, is uniquely determined by its values on the vertices.

III. Characterization of the Domain

We consider a rectangular compact domain of the form

$$\mathbf{S}:=\{(x_1,...,x_n):0\le x_i\le m_i\delta, i\in\{1,...,n\}\},\quad (4)$$

where δ is the grid step and $m_i\in Z$, $\forall i\in\{1,...,n\}$. First, $\mathbf{S}$ is subdivided into $\prod_{i=1}^n m_i$ hypercubes of the form $[0,\delta]^n$. After that, each hypercube is subdivided into simplices.

Consider first a unitary hipercube $[0,1]^n$. A simplex in $[0,1]^n$ can be uniquely defined by its $n+1$ vertices. Then, by choosing one common vertex as the origin and the remaining n as the summation of unitary vectors, every simplex

$$\Delta\left(\mathbf{v}_{r_0},\mathbf{v}_{r_1},...,\mathbf{v}_{r_n}\right)\quad (5)$$

in $[0,1]^n$ is formally defined by selecting

$$\mathbf{v}_{r_k}=\sum_{i=0}^k\mathbf{e}_{r_i},\ \forall k\in\{0,n\},\quad (6)$$

where $\mathbf{e}_{r_i}$ is the r_i-th unit vector, $\mathbf{v}_{r_0}=\mathbf{e}_{r_0}=\mathbf{0}$, $r_i\in\{1,...,n\}$, ($r_i\ne r_j$ if $i\ne j$). The uniqueness of this subdivision is a consequence of the following lemma, which can be found in [6].

Lemma 1: Every $\mathbf{z}\in[0,1]^n$ has a unique representation $\mathbf{z}=\sum_{i=0}^m\lambda_i\mathbf{v}_{r_i}$ where $\lambda_i>0$, $\forall j=0,1,...,m(\le n)$, and $\sum_{i=0}^m\lambda_i=1$.

Next, $\mathbf{S}$ is subdivided into the simplices

$$\delta\cdot\Delta_\mathbf{p}\left(\mathbf{v}_{r_0},\mathbf{v}_{r_1},...,\mathbf{v}_{r_n}\right)\quad (7)$$

where $\mathbf{p}=(p_1,...,p_n)$, with $p_j\in\{0,...,m_j-1\}$, $\forall j\in\{1,...,n\}$. The following lemma presents the boundary configuration which produces this subdivision in simplices.

Lemma 2: The simplicial boundary configuration H defined by the set of hyperplanes

$$\begin{aligned}&\text{a) }\left\{\mathbf{x}:\pi_q^{(k_q\delta)}(\mathbf{x})=x_q-k_q\delta=0\right\},\\&\text{b) }\left\{\mathbf{x}:\pi_{i,j}^{(0,k_j\delta)}(\mathbf{x})=x_i-(x_j-k_j\delta)=0\right\},\quad(8)\\&\text{c) }\left\{\mathbf{x}:\pi_{i,j}^{(k_i\delta,0)}(\mathbf{x})=(x_i-k_i\delta)-x_j=0\right\}\end{aligned}$$

and the boundary of $\mathbf{S}$, namely $\partial\mathbf{S}$, where $q\in\{1,...,n\}$, $k_q\in\{0,...,m_q-1\}$, $k_i\in\{0,...,m_i-1\}$, $k_j\in\{0,...,m_j-1\}$, and $j\in\{i+1,...,n\}$, $\forall i\in\{1,...,n-1\}$, subdivide $\mathbf{S}$ in the simplices defined in (7).

Proof: See [7].

When no confusion arise, the dependence on the evaluation point will be dropped to simplify the notation, i.e., $\pi_q^{(k)}(\mathbf{x})=\pi_q^{(k)}$. Finally, note that this subdivision determines a set P of $q_N=\prod_{i=1}^n(m_i+1)$ points, defined as $P:=\{(x_1,...,x_n):x_i=k_i\delta,k_i\in\{0,...,m_i\},i\in\{1,...,n\}\}$

A. Points Ordering

To simplify the proof of the main result (lemma 5), the points of P are ordered in classes as follows:

a) Class V^0: The origin $(x_1,...,x_n):x_i=0$, $\forall i\in\{1,...,n\}$.

b) Class V^r: The points

$$\left\{\mathbf{x}\in\mathbf{R}^n:\mathbf{x}=\sum_{p=1}^r q_{k_p}\delta\cdot\mathbf{e}_{k_p}\right\},\quad(9)$$

where $k_1,k_2,...,k_r$, with $k_1\in\{1,...,n-r+1\}$, $k_j\in\{k_{j-1}+1,...,n-r+j\}$, $\forall j\in\{1,...,r\}$, are A-ordered. In addition, for every fixed set of indexes $\{k_1,...,k_r\}$, the indexes $q_{k_1},q_{k_2},...,q_{k_r}$, where $q_{k_i}\in\{1,...,m_{k_i}\}$ are B-ordered.

In this way an ordered set of points $V=\{V^0,...,V^n\}$ results, containing the same points than P, but in an appropriate order.

IV. Nested Absolute Value Functions

In this section, we introduce the functions that are necessary to construct the basis of the LVS $PWL_H[\mathbf{D}]$. First of all, consider the linear functions $\pi_i^{(j_i\delta)}$, $i \in \{1,...,n\}$, $j_i \in \{0,...,m_{j_i}-1\}$, and a generating function

$$\gamma(\pi_i^{(j_i\delta)}, \pi_k^{(j_k\delta)}) =$$
$$\tfrac{1}{4}\left\{\left||-\pi_i^{(j_i\delta)}|+\pi_k^{(j_k\delta)}|-|-\pi_i^{(j_i\delta)}+|\pi_k^{(j_k\delta)}|\right|+\right. \quad (10)$$
$$\left.+\left|-\pi_i^{(j_i\delta)}|+|\pi_k^{(j_k\delta)}|-|-\pi_i^{(j_i\delta)}+\pi_k^{(j_k\delta)}|\right|\right\}$$

which satisfies

$$\gamma\left(\pi_i^{(j_i\delta)},\pi_k^{(j_k\delta)}\right) = \begin{cases} \pi_i^{(j_i\delta)}, \text{ if } 0 \leq \pi_i^{(j_i\delta)} \leq \pi_k^{(j_k\delta)}, \\ \pi_k^{(j_k\delta)}, \text{ if } 0 \leq \pi_k^{(j_k\delta)} \leq \pi_i^{(j_i\delta)} \\ 0, \text{ if } \pi_i^{(j_i\delta)} \leq 0 \text{ or } \pi_k^{(j_k\delta)} \leq 0. \end{cases} \quad (11)$$

The following functions derived from the γ function are proposed: $\gamma^0 = 1$, $\gamma^1\left(\pi_i^{(j_i\delta)}\right) = \gamma\left(\pi_i^{(j_i\delta)},\pi_i^{(j_i\delta)}\right)$, $\gamma^2\left(\pi_{i_1}^{(j_{i_1}\delta)},\pi_{i_2}^{(j_{i_2}\delta)}\right) = \gamma\left(\pi_{i_1}^{(j_{i_1}\delta)},\pi_{i_2}^{(j_{i_2}\delta)}\right)$ and in general

$$\gamma^k(\pi_{i_1}^{(j_{i_1}\delta)},...,\pi_{i_k}^{(j_{i_k}\delta)}) = \gamma(\pi_{i_1}^{(j_{i_1}\delta)},\gamma^{k-1}(\pi_{i_2}^{(j_{i_2}\delta)},...,\pi_{i_k}^{(j_{i_k}\delta)})) \quad (12)$$

The following properties of the γ^k functions can be directly inferred from (11) and (12).

Property 1: The function γ^k verifies

$$\gamma^k(\pi_{i_1}^{(j_{i_1}\delta)},...,\pi_{i_k}^{(j_{i_k}\delta)}) =$$
$$\begin{cases} \text{a) } \pi_{i_1}^{(j_{i_1}\delta)}, \text{ if } 0 \leq \pi_{i_1}^{(j_{i_1}\delta)} \leq \gamma^{k-1}(\pi_{i_2}^{(j_{i_2}\delta)},...,\pi_{i_k}^{(j_{i_k}\delta)}), \\ \text{b) } \gamma^{k-1}(\pi_{i_2}^{(j_{i_2}\delta)},...,\pi_{i_k}^{(j_{i_k}\delta)}), \text{ if } \\ \qquad 0 \leq \gamma^{k-1}(\pi_{i_2}^{(j_{i_2}\delta)},...,\pi_{i_k}^{(j_{i_k}\delta)}) \leq \pi_{i_1}^{(j_{i_1}\delta)}, \\ \text{c) } 0, \text{ if } \pi_{i_p}^{(j_{i_p}\delta)} \leq 0 \text{ for some } i_p \in \{1,...,k\}. \end{cases}$$
$$(13)$$

Property 2: Consider the function γ^k evaluated on the arguments ζ_i, $i \in \{1,...,k\}$. If $\zeta_i = \zeta_j = \zeta$, holds $\forall i,j$ then $\gamma^k(\zeta,...,\zeta) = \zeta$.

Property 3: Consider the function γ^k evaluated on the arguments ζ_i, $i \in \{1,...,k\}$. If there exists at least one $i \in \{1,...,k\}$ such that $\zeta_i \leq 0$, then $\gamma^k(\zeta_1,...,\zeta_k) = 0$.

A. The Nesting Degree

An important characteristic of these functions is the number of *nestings* of absolute value functions. In the sequel, this property will be referred to as the *nesting degree* (*n.d.*). By simple inspection, it can be seen that γ^0 has $n.d. = 0$, γ^1 has $n.d. = 1$ and, in general, γ^k has $n.d. = k$.

Analyzing the $n.d. = 0$ function $\gamma^0 = 1$, it is easy to see that it is constant for every $\mathbf{x} \in \mathbf{D}$. In contrast, any $n.d. = 1$ function $\gamma^1(\pi_r)(\mathbf{x})$ is equal to $\pi_r(\mathbf{x})$ in the region $\{\mathbf{x}: \pi_r(\mathbf{x}) \geq 0\}$ but zero everywhere else. A $n.d. = 2$ function $\gamma^2(\pi_{r_1}, \pi_{r_2})(\mathbf{x})$ is equal to $\pi_{r_1}(\mathbf{x})$ or $\pi_{r_2}(\mathbf{x})$ in the region $\{\mathbf{x}: \pi_{r_1}(\mathbf{x}) \geq 0, \pi_{r_2}(\mathbf{x}) \geq 0\}$ but zero everywhere else. By extending this reasoning, we can see that a $n.d. = k$ function $\gamma^k(\pi_{r_1},...,\pi_{r_k})(\mathbf{x})$ is equal to $\pi_{r_1}(\mathbf{x})$, or $\pi_{r_2}(\mathbf{x})$, ..., or $\pi_{r_k}(\mathbf{x})$, in the region $\{\mathbf{x} \in \mathbf{R}^n : \pi_{r_1}(\mathbf{x}) \geq 0, ..., \pi_{r_k}(\mathbf{x}) \geq 0\}$ but zero everywhere else. From the inspection of the above mentioned properties, observe that depending on the *n.d.* each function "acts" on a particular region of the domain. Moreover, different *n.d.* functions act in regions of completely different nature. This characteristic is of fundamental importance in understanding the construction principle of the basis for PWL functions and gives some intuitive idea of independence between the basis elements.

B. The Functions of the Basis

We are now ready to introduce the set of functions which generate the basis of $PWL_H[\mathbf{S}]$. We group them in a vector $\Lambda = \begin{bmatrix} \Lambda^{0^T} & ... & \Lambda^{n^T} \end{bmatrix}^T$, ordered according to its *n.d.*. The construction of Λ is as follows:

a) Λ^0: ($n.d. = 0$) The function $\gamma^0 = 1$.

b) Λ^r: ($n.d. = r$, $r \in \{1,...,n\}$) The functions

$$\gamma^r(\pi_{i_1}^{(j_{i_1}\delta)},...,\pi_{i_r}^{(j_{i_r}\delta)}), \quad (14)$$

where $i_1, i_2, ..., i_r$, satisfying $i_1 \in \{1,...,n-r+1\}$ and $i_j \in \{i_{j-1}+1,...,n-r+j\}$, $\forall j \in \{2,r\}$, are A-ordered. In addition, for every fixed set of indexes $i_1, i_2, ..., i_r$, the indexes $j_{i_1}, j_{i_2}, ..., j_{i_r}$ where $j_{i_p} \in \{0,...,m_{i_p}-1\}$ are B-ordered.

V. Main Result

Before introducing the main result, some previous results are necessary. The proofs are given in [7].

Lemma 3: The number of points of V coincides with the number of component functions in Λ.

Lemma 4: The PWL function components of Λ are defined over the boundary configuration H in $\mathbf{S}$, and no additional boundaries are generated.

Lemma 5: The matrix A whose rows are obtained by the application of function Λ over the points of V, is lower triangular and non singular.

Now, we show that any PWL function $f \in PWL_H[\mathbf{S}]$ can be uniquely represented as a linear combination of the elements of Λ. Associated to every vector function $\Lambda^i : \mathbf{S} \mapsto \mathbf{R}^{q_i}$, let us define a parameter vector $C_i \in \mathbf{R}^{q_i \times 1}$. After that, stacking each one of the C_i vectors we form a parameter vector $C \in \mathbf{R}^{q_N \times 1}$, $C := \begin{bmatrix} C_0^T & C_1^T & ... & C_n^T \end{bmatrix}^T$. In this way, a function which is a linear combination of the elements of Λ can be written as $\alpha(\mathbf{x}) = C^T \Lambda(\mathbf{x})$.

Theorem 1: The PWL component functions of Λ are a basis of the linear vector space $PWL_H[\mathbf{S}]$.

Proof: Any PWL function $f : \mathbf{S} \mapsto \mathbf{R}^1$ over the boundary H is uniquely defined by its values on the points V. In consequence, it is possible to formulate the following equation system which results of setting $f = C^T \Lambda$ over the

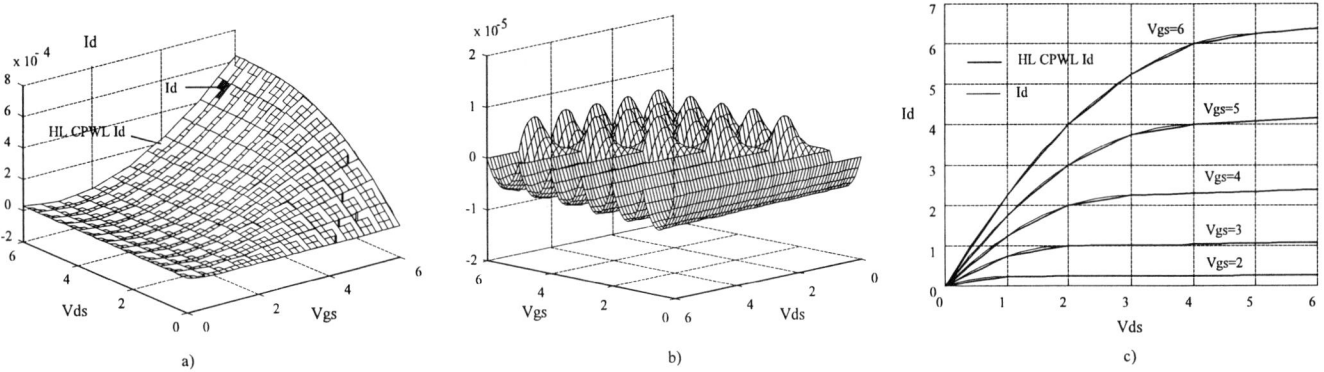

Fig. 2. a) Plot of the exact current I_d and its HL CPWL approximation. b) Approximation error $\left(\|e\|_\infty = 7.2e^{-6}\right)$. c) Comparison of both families of currents

points V

$$\begin{bmatrix} f(V^0) \\ f(V^1) \\ \vdots \\ f(V^n) \end{bmatrix} = \begin{bmatrix} \Lambda^{0^T}(V^0) & \cdots & \Lambda^{n^T}(V^0) \\ \vdots & \ddots & \vdots \\ \Lambda^{0^T}(V^n) & \cdots & \Lambda^{n^T}(V^n) \end{bmatrix} \begin{bmatrix} C_0 \\ C_1 \\ \vdots \\ C_n \end{bmatrix}, \quad (15)$$

where $f(V^i)$ ($\Lambda^j(V^i)$ resp.) is an abbreviated notation for the vector that results of evaluating function f (Λ^j resp.) on the points V^i.

Note that the resulting matrix in (15), is the matrix A introduced in Lemma 5.

Finally, considering that: a) by Lemma 3, matrix A is square; b) by Lemma 4, function Λ posses the same boundary configuration that the PWL function f and c) by Lemma 5 matrix A^{-1} exists, so that it is possible to obtain the parameter vector as $C = A^{-1}B$, where $B = \left[f(V^0)^T \ldots f(V^n)^T\right]^T$; we can conclude that $C^T \Lambda(\mathbf{x}) \equiv f(\mathbf{x})$, $\forall \mathbf{x} \in \mathbf{S}$, and the Theorem is proved.

Remark: An important property of the method proposed follows from the fact that the matrix A in Lemma 5, is lower triangular. This let us to solve the parameter vector C, without the inversion of matrix $A \in R^{q_N \times q_N}$, by substitution as

$$c_i = a_{ii}^{-1}\left(b_i - \sum_{p=1}^{i-1} a_{ip}c_p\right), \quad i \in \{1, q_N\} \quad (16)$$

where a_{ij}, b_i, c_i are the components of matrices A, B and C respectively.

VI. EXAMPLE

Consider the drain current expression of a MOSFET transistor, on the domain $\mathbf{S} = \{V_{gs}, V_{gd} : 0 \leq V_{gs} \leq 6, 0 \leq V_{gs} \leq 6\}$, according to the Shichman-Hodges model:

$$I_d = k\left((V_{gs} - V_t)V_{ds} - 0.5V_{ds}^2\right), \text{ if } V_{gs} - V_t \geq V_{ds}$$
$$I_d = 0.5k(V_{gs} - V_t)^2(1 + \lambda(V_{ds} - V_{gs} + V_t)),$$

if $V_{gs} - V_t < V_{ds}$, where $k = 50\mu A/V^2$, $V_t = 1V$, $\lambda = 0.02V^{-1}$. A HL CPWL function was designed to approximate I_d using a simplicial subdivision of $\mathbf{S}$, with $\delta = 1$. The results are shown in fig. 2.

VII. CONCLUDING REMARKS

Even though the necessity of HL CPWL functions were already emphasized in [5] and [4], numerical representation schemes have not been reported in the literature for the case of a general domain dimension. In this paper we presented a representation basis for the set of arbitrary PWL functions defined over a simplicial subdivision of a compact domain $\mathbf{D} \subset \mathbf{R}^n$. In addition, the method proposed exhibit efficient numerical properties, provided that the parameters associated to a HL CPWL function are obtained from the resolution of a lower triangular linear equation system. The use of a simplicial partition is motivated by the systematic formulation of hyperplanes and functions. However, further research is necessary to extend the results presented here to the case of an arbitrary partition of the domain.

REFERENCES

[1] L. O. Chua and S. Kang, "Section-wise piecewise-linear functions: canonical representation, properties and applications," *Proc. IEEE*, vol. 65, pp. 915–929, 1977.

[2] S. Kang and L. O. Chua, "A global representation of multidimensional piecewise-linear functions with linear partitions," *IEEE Trans. Circuits Syst.*, vol. 25, pp. 938–940, 1978.

[3] L. O. Chua and A. Deng, "Canonical piecewise-linear representation," *IEEE Trans. Circuits Syst.*, vol. 35, pp. 511–525, 1988.

[4] J. N. Lin, H. Xu, and R. Unbehahuen, "A generalization of canonical piecewise-linear functions," *IEEE Trans. Circuits Syst.*, vol. 41, pp. 345–347, 1994.

[5] C. Kahlert and L. O. Chua, "The complete canonical piecewise-linear representation-part i: the geometry of the domain space," *IEEE Trans. Circuits Syst.*, vol. 39, pp. 222–236, 1992.

[6] M. Chien and E. Kuh, "Solving nonlinear resistive networks using piecewise-linear analysis and simplicial subdivision," *IEEE Trans. Circuits Syst.*, vol. CAS-24, pp. 305–317, 1977.

[7] P. Julián, A. Desages, and O. Agamennoni, "High level canonical piecewise linear representation using a simplicial partition." Submitted to IEEE Trans. on Circ. and Syst., 1997.

HIERARCHICAL SYMBOLIC ANALYSIS OF LARGE ANALOG CIRCUITS WITH DETERMINANT DECISION DIAGRAMS

Xiangdong Tan and C.-J. Richard Shi

Department of Electrical and Computer Engineering
University of Iowa, Iowa City, Iowa 52242, U.S.A.
Email: {xtan,cjshi}@eng.uiowa.edu

ABSTRACT

A new hierarchical approach is proposed to symbolic analysis of large analog circuits. The key idea is to use a graph-based representation, called Determinant Decision Diagram (DDD), to represent the symbolic determinant and cofactors of the MNA matrix for each subcircuit block. By exploiting the inherent sharing and sparsity of symbolic expressions, DDD is capable of representing a huge number of symbolic product terms in a canonical and highly-compact manner. Further, it enables cofactoring and sensitivity computation to be performed with time linear in the size of DDD. Experimental results have demonstrated that our method outperforms the best-known hierarchical symbolic analyzer SCAPP and even numerical simulator SPICE for small-signal AC analysis.

1. INTRODUCTION

Symbolic analysis is to calculate the behavior or the characteristic of a circuit in terms of symbolic parameters. It is important for many applications such as optimum topology selection, design space exploration, behavioral model generation, and fault detection [2]. However, symbolic analysis has not been widely used by analog designers. The root of the difficulty is apparently: the number of product terms in a symbolic expression may increase exponentially with the size of a circuit. Any manipulation and evaluation of symbolic expressions will require CPU time at best linear in the number of terms, and therefore have both the time and space complexities exponential in the size of a circuit.

One way to cope with the circuit-size limitation problem of symbolic analysis is by means of hierarchical decomposition. Hierarchical decomposition is to generate symbolic expressions in a nested form [3, 7]. There are two methods known as topological analysis method [7] and network approach [3]. Both are based on the *sequence-of-expressions* concept to obtain the transfer functions. Unfortunately, the number of expressions can grow rapidly so that even the compilation of the generated expressions could take enormous time. Manipulation (other than evaluation) of the resulting sequences of expressions is known to be complicated and often requires dedicated efforts, e.g., sensitivity calculation in [4] and lazy approximation in [6].

In this paper, we present a new hierarchical approach to exact symbolic analysis. It takes advantage of both hierarchical decomposition and a recently introduced graph-based representation for

---

This work is sponsored by U.S. Defense Advanced Research Projects Agency (DARPA) under grant number F33615-96-1-5601 from the United States Air Force, Wright Laboratory, Manufacturing Technology Directorate.

symbolic determinants called Determinant Decision Diagram [5]. By exploiting the *sparsity* and *sharing* among expressions, DDD can store a symbolic determinant compactly. For example, 1.01×10^8 symbolic product terms can be represented by a diagram with only 767 vertices. More importantlly, manipulations such as cofactoring and sensitivity can be performed in linear time in the size of DDD.

The rest of the paper is organized as follows: Section 2 provides an overview of the hierarchical symbolic analysis procedure. Sections 3 and 4 describe the DDD-based approach. Section 5 presents experimental results. Section 6 concludes the paper.

2. OVERVIEW OF HIERARCHICAL ANALYSIS

For a linear(ized) time-invariant analog circuit, its system equation can be formulated by, for example, the modified nodal analysis (MNA) approach in the following general form [9]:

$$\begin{bmatrix} A & C \\ C & D \end{bmatrix} \cdot \begin{bmatrix} v \\ i \end{bmatrix} = \begin{bmatrix} j \\ e \end{bmatrix}, \qquad (1)$$

where **v** is the vector of node voltage variables, **i** is the vector of the branch current variables, **A** is the modified nodal admittance matrix, **B**, **C**, **D** are contributions of the branch relationship, **j** represents the external current sources and **e** represents the independent voltage sources. We assume the presence of the predefined subcir-

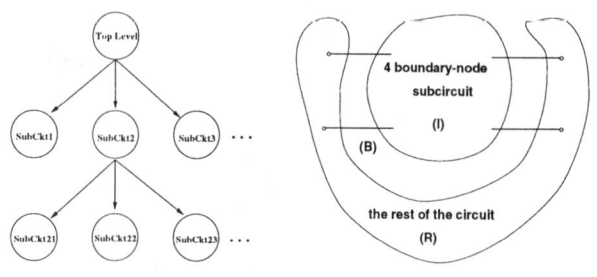

(a) Circuit hierarchy.   (b) Partition of a subcircuit.

Figure 1: Model of hierarchical analysis.

cuits in the circuit hierarchy. The circuit hierarchy can be viewed as a rooted tree shown in Fig. 1(a). A circuit may have one or more subcircuits at each hierarchical level. Consider a subcircuit with some internal structure and terminals, as illustrated in Fig. 1(b). The circuit unknowns—the node-voltage variables **v** and branch-current variables **i** — can be partitioned into three disjoint groups

$\mathbf{x}^I$, $\mathbf{x}^B$, and $\mathbf{x}^R$, where the sup-scripts *I, B, R* stand for, respectively, internal variables, boundary variables and the *rest* of variables. *Internal* variables are those local to the subcircuit, *boundary* variables (also called *tearing variables*) are those related to both the subcircuit and the rest of the circuit. Note that boundary variables also include those variables required as outputs. Then, the system-equation set (1) can be rewritten in the following form:

$$\begin{bmatrix} \mathbf{A}^{II} & \mathbf{A}^{IB} & \\ \mathbf{A}^{BI} & \mathbf{A}^{BB} & \mathbf{A}^{BR} \\ & \mathbf{A}^{RB} & \mathbf{A}^{RR} \end{bmatrix} \begin{bmatrix} \mathbf{x}^I \\ \mathbf{x}^B \\ \mathbf{x}^R \end{bmatrix} = \begin{bmatrix} \mathbf{b}^I \\ \mathbf{b}^B \\ \mathbf{b}^R \end{bmatrix}. \quad (2)$$

The basic idea that underlines all hierarchical analysis methods is to eliminate the number of equations and the number of variables from the equation-set above until a set of equations involving only the desired variables remains. The physical meaning of such elimination is to eliminate subcircuits internal to all input/output nodes in a circuit. So we call this *subcircuit elimination*. The resulting set of equations can be written as follows:

$$\begin{bmatrix} \mathbf{A}^{BB*} & \mathbf{A}^{BR} \\ \mathbf{A}^{RB} & \mathbf{A}^{RR} \end{bmatrix} \begin{bmatrix} \mathbf{x}^B \\ \mathbf{x}^R \end{bmatrix} = \begin{bmatrix} \mathbf{b}^{B*} \\ \mathbf{b}^R \end{bmatrix}, \quad (3)$$

where

$$\mathbf{A}^{BB*} = \mathbf{A}^{BB} - \mathbf{A}^{BI}(\mathbf{A}^{II})^{-1}\mathbf{A}^{IB}, \quad (4)$$

and

$$\mathbf{b}^{B*} = \mathbf{b}^B - \mathbf{A}^{BI}(\mathbf{A}^{II})^{-1}\mathbf{b}^I. \quad (5)$$

Subcircuit elimination can be performed for all the subcircuits by visiting the circuit hierarchy in a bottom-up fashion. Hierarchical *numerical* analysis performs subcircuit elimination numerically by partial LU decomposition [8]. Hierarchical *symbolic* analysis is to introduce intermediate variables to represent subcircuit elimination (4) and (5) using a *sequence of expressions* [3, 7]. However, expressions resulting for $\mathbf{A}^{BI}(\mathbf{A}^{II})^{-1}\mathbf{A}^{IB}$ and $\mathbf{A}^{BI}(\mathbf{A}^{II})^{-1}\mathbf{b}^I$ are usually very complicated, and the size of such expressions may grow rapidly even the compilation of the generated expressions could take a very long time.

3. BASIC IDEA

Let $\mathbf{A}$ be an $n \times n$ matrix. It may be denoted as $[a_{u,v}], u,v = 1,...,n$. The *determinant* of matrix $\mathbf{A}$ is denoted by $det(\mathbf{A})$. According to linear algebra, the inverse of matrix $\mathbf{A}$ can be written as

$$\mathbf{A}^{-1} = \frac{1}{det(\mathbf{A})}[\Delta_{u,v}], \quad (6)$$

where

$$\Delta_{u,v} = (-1)^{u+v} det(\mathbf{A}_{a_{u,v}}).$$

and $[\Delta_{u,v}]$ is called the *adjoint* matrix of $\mathbf{A}$. $\Delta_{u,v}$ is the first-order *cofactor* of $det(\mathbf{A})$ with respect to $a_{u,v}$, $\mathbf{A}_{a_{u,v}}$ is the $(n-1) \times (n-1)$-matrix obtained from the matrix $\mathbf{A}$ by deleting row u and column v. Note that each entry in the adjoint matrix is a cofactor of the original matrix, and thus the adjoint matrix is a *full* (dense) matrix.

Suppose that the number of internal variables is m, and the number of boundary variables is n. For practical circuits, n usually is small given a good partitioning. More importantly, $\mathbf{b}^I$ is a zero vector, $\mathbf{A}^{BI}$ and $\mathbf{A}^{IB}$ are very *sparse* submatrices. This implys that only a few of the first-order cofactors of A^{II} are needed. Applying (6) to (4), we have the expanded form of (4):

$$a_{u,v}^{BB*} = a_{u,v}^{BB} - \frac{1}{det(\mathbf{A}^{II})} \sum_{k_1,k_2=1}^{m} a_{u,k_1}^{BI} \Delta_{k_2,k_1}^{II} a_{k_2,v}^{IB}, u,v=1,...,n. \quad (7)$$

Note that we need first-order cofactors Δ_{k_2,k_1}^{II} only when a_{u,k_1} and $a_{k_2,v}$ are both non-zeros, and at same time a_{u,k_1} and $a_{k_2,v}$ are zero for most time due to sparsity of A^{BI} and A^{IB}. So the key problem of hierarchical symbolic analysis is how to represent symbolically the determinant $det(\mathbf{A}^{II})$ and *a few* of its first-order cofactors in a compact and efficient manner. We will show in the next section that this problem can be solved by using a recently-introduced graph representation called Determinant Decision Diagrams.

4. DETERMINANT DECISION DIAGRAMS FOR HIERARCHICAL SYMBOLIC ANALYSIS

Determinant Decision Diagrams is a canonical and compact graph-based representation of symbolic-matrix determinants [5]. It has enabled the exact symbolic analysis of such circuits like μA-741 operational amplifiers for the first time [5].

Formally, a determinant decision diagram *DDD* is a signed rooted directed acyclic graph with two terminal vertices, namely the 0-terminal vertex and the 1-terminal vertex. It has two outgoing edges, called 1-edge and 0-edge, pointing, respectively, to D_{a_i} and $D_{\overline{a}_i}$. A determinant graph having root vertex a_i denotes a matrix determinant D defined recursively as follows:

1. if a_i is the 1-terminal vertex, then $D = 1$,
2. if a_i is the 0-terminal vertex, then $D = 0$,
3. Otherwise(non-terminal vertex), $D = a_i \cdot s(a_i) \cdot D_{a_i} + D_{\overline{a}_i}$,

where $s(a_i)$ is a sign $\{+, -\}$ associated with each non-terminal vertex a_i. It can be determined recursively as follows.

1. Let $P(v)$ be the set of DDD vertices that originate the 1-edges in any path rooted at v to the 1-terminal. Then

$$s(v) = \prod_{x \in P(v)} sign(r(x) - r(v)) \, sign(c(x) - c(v)), \quad (8)$$

where $r(x)$ and $c(x)$ refer to the row and column indices of vertex x in the matrix, and $sign(u) = 1$ if $u > 0$, and $sign(u) = -1$ if $u < 0$.

2. If v has an edge pointing to the 1-terminal vertex, then $s(v) = +1$.

We have shown that DDD is a graph representation of the expansion of matrix determinant $det(\mathbf{A})$ in a way similar to binary decision diagrams (BDDs) for Shannon expansion of Boolean functions. Each vertex in a DDD represents a matrix expansion:

$$det(\mathbf{A}) = a_{r,c}(-1)^{r+c} det(\mathbf{A}_{a_{r,c}}) + det(\mathbf{A}_{\overline{a}_{r,c}}). \quad (9)$$

where $det(\mathbf{A})$ is the determinant represented by this vertex, the vertex pointed by its 1-edge represents its cofactor $(-1)^{r+c} det(\mathbf{A}_{a_{r,c}})$ with respect to $a_{r,c}$, and the vertex pointed by its 0-edge represents $det(\mathbf{A}_{\overline{a}_{r,c}})$, which is the *reminder* of $det(\mathbf{A})$ with respect to $a_{r,c}$.

For example, consider the following determinant

$$det(\mathbf{M}) = \begin{vmatrix} a & b & 0 & 0 \\ c & d & e & 0 \\ 0 & f & g & h \\ 0 & 0 & i & j \end{vmatrix} = adgj - adhi - aefj - bcgj + bchi.$$

(10)

Figure 2 illustrates the corresponding DDD representation under the expansion order: $a, c, b, d, f, e, g, i, h$ and j. Symbolic expressions represented by each vertex are also given near the vertices in the figure. In DDDs, each path from root vertex (a in our case) to 1-terminal is called *1-path*. Each 1-path uniquely define a product term which includes all vertices (symbols) from which the 1-edges in the 1-path originate. We note that subterms ad, gj, and hi appear in several product terms of the matrix determinant, and they are shared in the DDD representation.

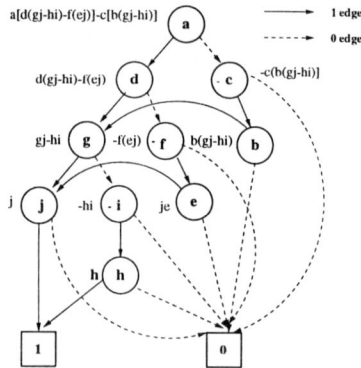

Figure 2: A determinant decision diagram for matrix **M**.

A key issue is that how to find a suitable expansion order such that the DDD has as few vertices as possible. This is called DDD vertex ordering. An efficient heuristic has been proposed in [5], which gives the optimal order for ladder-structured circuits and usually good order for practical analog circuits.

5. IMPLEMENTATION AND EXPERIMENTAL RESULTS

The proposed method has been implemented in a symbolic analyzer that can read the circuit description in the SPICE format. The analysis is performed by depth-first traversal of the circuit hierarchical tree shown in Fig. 1(a). At each node, circuit equation set is built and its subcircuits are suppressed using equation (7). Then the analysis is moved upwardly. At the top level of the circuit hierarchy, symbolic transfer function is derived using Cramer's rule and all the symbolic determinants and cofactors generated are represented by DDDs.

The results from two examples are presented. The first example is an active low-pass filter circuit shown in Fig. 3. It has four identical subcircuits, named $X1$ to $X4$. Figure 4 shows the detailed structure of a subcircuit. Each subcircuit contains two Opamps. We have tested our program on two different implementation of Opamps: a linear model of 741 Opamp circuit shown in Fig. 5(a) and a miller-compensated two-stage Opamp circuit shown in Fig. 5(b). For the miller-compensated Opamp circuit, all the MOS transistors are replaced by their corresponding small-signal models at the DC operating point computed by SPICE. The

AC analysis is performed by depth-first traversals of all the DDD vertices used to represent all symbolic expressions at each frequency point. The numerical value of the determinant is obtained when its root is reached. Since each DDD vertex only needs one visit for each frequency point, the time complexity of the DDD-based numerical evaluation is directly proportional to the DDD size.

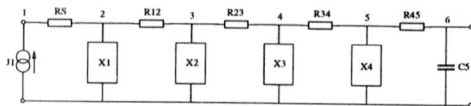

Figure 3: An active low-pass filter.

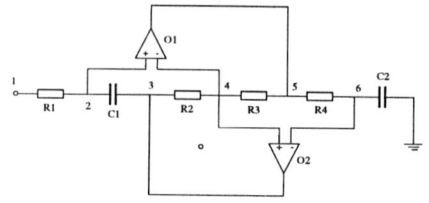

Figure 4: An FDNR subcircuit.

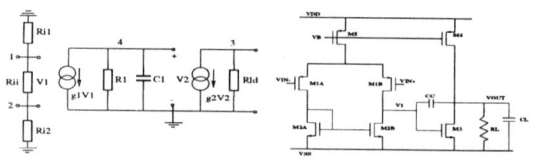

(a) Linear model of 741 Opamp. (b) Miller-compensated two-stage Opamp.

Figure 5: Two implementations of an Opamp.

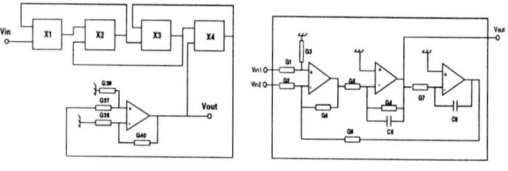

(a) The partitioned RC band-pass filter. (b) The subcircuit in band-pass filter.

Figure 6: An active RC band-pass filter

The second example is a band-pass filter circuit (Fig. 6(a)) which was widely used for hierarchical analysis [3, 7]. It consists of four topologically identical subcircuits $X1$ to $X4$ shown in Fig. 6(b). For each Opamp in the subcircuits, we also test two implementations in Fig. 5(a) and Fig. 5(b), respectively. We have conducted two sets of experiments on a SUN SPARCstation 5 with 32M memory. We first compare our program with SPICE on repetitive numerical evaluation. For each circuit, 1000 frequency points are simulated. The results are summarized in Table 1, where *#term* is the actual number of distinct product terms generated, $|DDD|$ is the number of DDD vertices used to represent all the symbolic expressions. Columns 4, 5, and 6 list, respectively, the CPU time in seconds used by the proposed DDD-based method, SPICE, and the speedup of the proposed method over SPICE for each test circuit. From Table 1, we can see that the proposed DDD-based method

Table 1: Comparison against SPICE in numerical evaluation.

circuit	#term	|DDD|	DDD	Spice	Speedup
LP(linear)	23	43	1.07	6.58	6.15
LP(miller)	67	58	1.56	14.82	9.50
BP(linear)	562	228	3.07	9.70	3.16
BP(miller)	622	243	3.72	17.76	4.77

LP is for low-pass filter and BP is for band-pass filter.
linear or *miller* denotes the corresponding Opamp model used.

outperforms SPICE for all the test cases. Further, the speedup increases with the size of the circuit.

We compare our method with SCAPP—a best-known hierarchical symbolic analyzer [3]. We construct the test circuits by cascading, respectively, the first 1, 2, 3 and 4 subcircuit blocks(Fig. 6(a)). The Opamp subcircuit is implemented by the miller-compensated Opamp circuit. SCAPP uses the flatten circuit description, and exploits an automatic optimized partitioning strategy[1].

The results are summarized in Table 2. Columns 1 and 2 list, respectively, the number of subcircuits cascaded for each test case, and the total number of node voltage and branch current variables in the flatten circuits. Columns 3 and 4 describe the number of DDD vertices, and the number of product terms represented. Columns 5 and 6 give the numbers of additions and multiplications used in the expressions generated by SCAPP. Note that the number of additions and multiplications used by the proposed DDD-based method is exactly the number of DDD vertices. From Table 2,

Table 2: Comparison against SCAPP.

#subckt	#var	DDD-based		SCAPP	
		#vertices	#terms	#add	#mul
1	23	146	500	556	274
2	39	369	2.60×10^4	1339	854
3	55	568	1.60×10^6	1772	1074
4	71	767	1.01×10^8	2479	1639

we can observe that the DDD based representation is much more compact than the sequence-of-expressions representation used in SCAPP. Note that the number of DDD vertices is about 2-4 times less than the number of expressions in SCAPP, and the storage of each expression takes much more space than that for one DDD vertex. We also see that the DDD size grows almost linearly in the circuit size, despite that the number of product terms grow exponentially.

Table 3 shows the statistics of using the DDD-based symbolic method and SCAPP for repetitive numerical evaluation. For SCAPP, we compile the generated symbolic expressions and then execute the code for simulation. For each test case, we report in Columns 2 and 3 the CPU time required to construct the DDD and then the CPU time taken for simulation from the constructed DDD. For SCAPP, we report respectively in Columns 4, 5, and 6 the CPU time for SCAPP to generate the sequence-of-expressions (*analy*), the compilation time (*comp*), and the actual simulation time (*sim*). The last two columns give the matrix setup time and simulation time used by SPICE. We can see from Table 3 that the proposed DDD-based method is more efficient than both SCAPP and SPICE. Although it may take less time for SCAPP to obtain the sequence of expressions, it takes a much longer time to compile

---
[1] We have not exploited optimal partitioning yet.

Table 3: Comparison against SCAPP and SPICE in CPU time.

#subckt	DDD-based		SCAPP			SPICE	
	const.	sim.	analy.	comp.	sim.	setup	sim.
1	0.37	2.09	0.81	13.1	2.60	1.10	5.34
2	1.01	4.75	2.09	33.3	7.49	2.70	8.98
3	2.42	6.91	3.69	44.2	10.37	3.12	15.58
4	12.75	9.19	5.54	64.7	12.06	3.42	22.10

the generated expressions. We also note that our program is interpreted, while SCAPP is compiled. We have not exploited optimal partitioning, while SCAPP already did. Therefore, we expect that the compiled and optimized version of our DDD-based method could offer even more improvement over SCAPP.

6. CONCLUSIONS

A new hierarchical method for symbolic analysis of large analog circuit has been presented and implemented. Experimental results have shown that the proposed method compares very favorably with the best-known symbolic analyzer SCAPP and numerical simulator SPICE for small-signal AC analysis.

Acknowledgment: The authors wish to thank Prof. G. Gielen of KUL for several helpful discussions on symbolic analysis and Prof. M. Hassoun of Iowa State University for making SCAPP code available to us.

7. REFERENCES

[1] F. V. Fernández and A. Rodríguez-Vázquez "Symbolic analysis tools—the state of the art", pp. 798–801 in *Proc. IEEE Int. Symp. Circuits and System*, 1996.

[2] G. Gielen, P. Wambacq and W. Sansen, *Symbolic analysis methods and applications for analog circuits: A tutorial overview*, Proc. IEEE, vol. 82, no. 2, pp. 287–304, Feb. 1994.

[3] M. M. Hassoun and P. M. Lin, "A hierarchical network approach to symbolic analysis of large scale networks", *IEEE Trans. Circuits and Systems*, vol. 42, no. 4, pp. 201–211, April 1995.

[4] P. M. Lin, "Sensitivity analysis of large linear networks using symbolic program", pp. 1145–1148 in *Proc. IEEE Int. Symp. Circuits and Systems*, 1992.

[5] C. J. Richard Shi and X. Tan, "Symbolic analysis of large analog circuits with determinant decision diagrams", in *Proc. IEEE Int. Conf. Computer Aided Design (ICCAD)*, Nov., 1997.

[6] S. J. Seda, M. G. R. Degrauwe and W. Fichtner, "Lazy-expansion symbolic expression approximation in SYNAP", pp. 310–317 in *Proc. IEEE Int. Conf. Computer Aided Design (ICCAD)*, 1992.

[7] J. A. Starzky and A. Konczykowska, "Flowgraph analysis of large electronic networks", *IEEE Trans. Circuits and Systems*, vol. 33, no. 3, pp. 302–315, March 1986.

[8] M. Vlach, "LU decomposition algorithms for parallel and vector computation", pp. 37–64 in *Analog Methods for Computer-Aided Circuit Analysis and Diagnosis*, T. Ozawa (ed.), Marcel Dekker, New York, 1988.

[9] J. Vlach and K. Singhal, *Computer Methods for Circuit Analysis and Design*, Van Nostrand Reinhold, New York, 1994.

REDUCING OPERATION COMPLEXITY IN SYMBOLIC TECHNIQUES THROUGH PARTITIONING

Gianpiero Cabodi   Paolo Camurati   Stefano Quer

Politecnico di Torino
Dip. di Automatica e Informatica
Turin, ITALY

ABSTRACT

Binary Decision Diagrams (BDDs) are the state-of-the-art core technique for the symbolic representation and manipulation of Boolean functions, relations and finite sets. Many applications resort to them in the field of CAD, but size and time complexity are a strong limitation to a wider applicability.

In this paper we primarily address the problem of memory limits. In particular, we first include an experimental observation of memory usage and running time for some basic operators used in reachability analysis of Finite State Machines. Then we describe how disjunctive partitioning allows us to decompose large problems into sub-problems. Finally, we show the benefits in terms of memory requirements, CPU time, and overall performance.

1 INTRODUCTION

The manipulation of Boolean functions, relations and sets is one of the most important operations in several areas of CAD such as logic synthesis, verification, testing, etc. Its manipulation efficiency depends on the data structure used for representing such entities.

In the last decade Reduced Ordered Binary Decision Diagrams (ROBDDs or simply BDDs) have been intensively used for their efficiency, and their application covers several fields of symbolic computation.

Although quite successful, symbolic methods reach their limits as BDDs require too much memory and their manipulation is computationally expensive. As far as memory is the issue, virtual memory is not a good solution. BDDs are acyclic graphs, and conventional BDD manipulation packages traverse them on a "depth-first" basis: This results in random access to memory, with a very large number of page faults when the available memory is smaller than the working set.

In the literature, one can find several approaches to deal with this disadvantage: Approximate solutions, alternative representation techniques (other than BDDs), modified memory access methods, etc. They are usually quite expensive, because of the need to rewrite large parts of the code, and none is general enough.

In the approach presented in [1] we improve on standard techniques by a "divide-and-conquer" methodology. When monolithic BDDs become too large or when computations are too expensive, we decompose BDDs. Expensive operations are then carried out on the decomposed form. This allows us to deal with just one sub-problem at a time. Here, we further analyze this methodology. First of all, we motivate the approach with an experimental observation of memory usage and running time for some basic operators used in reachability analysis of Finite State Machines (FSMs). Secondly, we present the technique, focussing on "single" BDD operators. Finally, we show that the overall method is particularly effective on large problems with a detailed analysis of memory and time performance.

2 MOTIVATION AND OVERVIEW

We suppose a basic knowledge of BDDs, supports, sets, characteristic functions, FSMs, and reachability analysis (i.e., breadth-first traversal and iterative squaring). The reader should refer to the literature for an introduction to these arguments [1, 2, 3].

Let us start with an analysis of the costs of a few complex BDD operators. We report data for two different application cases (image computation and iterative squaring). Problem costs (memory, time, and number of page faults) are expressed as a function of the size of the operands. Size of the hash/cache table managed by the BDD package is used as a parameter.

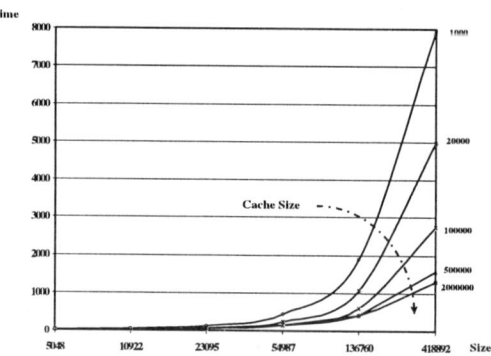

Figure 1: Image with (conjunctively) clustered transition relation: CPU time.

Figure 1 shows the time cost of the image computation pro-

AND–ABSTRACT operations, each producing an intermediate result by conjoining a cluster and "early" quantifying some variables. We use circuit s1423 and we represent data from traversal level 7 to 12. Time complexity is almost linear only with a perfect cache size (of about 2,000,000 entries) whereas it is exponentially worse with smaller cache sizes.

Figure 2 reports the average number of page faults (for ten different runs), and the global amount of main memory used with the same experiment but for all first 12 levels of traversal. In this case the cache table has a "standard" size of about 100,000 entries. To use the same scale, on the vertical axis, we express memory in Kbytes.

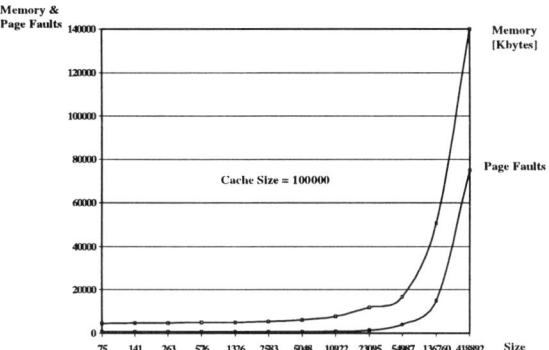

Figure 2: Image with (conjunctively) clustered transition relation: Memory usage and Number of Page Faults.

Figure 3 reports a few squaring iterations on circuit ocl_fix [3]. T^{n+1} is computed as $\exists_z(T^n(s,z) \cdot T^n(z,y))$, for n equal to 1, 4, 8, 12 and 16. Time complexity is higher than in the previous case. Squaring steps for increasing transition relations are in fact characterized by growing size of both operands of AND–ABSTRACT operations. So time complexity shows a quadratic cost with a large cache, and a rapidly decreasing performance as cache size is reduced. Memory is less critical in this case, with a rather linear profile.

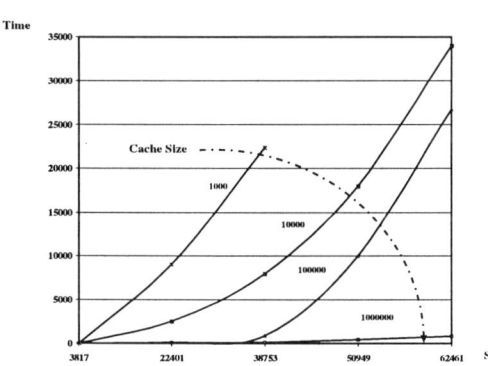

Figure 3: Single steps of iterative squaring: CPU time.

All the diagrams show a threshold after which efficiency decreases enormously.

Now considering state-of-the-art performance improvements tion can be found in literature (and in practice), besides increasing cache table size. But this implies using more memory, that comes out to be the truly critical resource in BDD based applications. Moreover, paging techniques are almost useless in this case, due to the "random" distribution in memory of the BDD nodes accessed by all recursive operators. Our target is to reduce operation complexity working below the thresholds, i.e., in the linear part of each graph.

In [1] we propose *"partitioning"* (problem decomposition) as a divide–and–conquer approach to tackle memory limitations, with benefits also in terms of CPU time, coming for a more efficient usage of the available memory.

As the problems we are concerned with support *disjunctive* decomposition, for a generic binary operator f op g, we write:

$$f \text{ op } g = (f_1 \text{ op } g_1) + (f_2 \text{ op } g_2)$$

with $f = f_1 + f_2$, and $g = g_1 + g_2$. In particular, we propose partitioning based on (recursive) splitting variable selection:

$$f \text{ op } g = (\overline{v} \cdot f \text{ op } \overline{v} \cdot g) + (v \cdot f \text{ op } v \cdot g) \quad (1)$$

(where v is a variable in the true support of f and/or g). In the more general case v is any partitioning function (called "window" function in [4]). We restrict to v variables as a compromise between the need to produce "good" partitions and "easy–to–find" splitting functions. Being our goal to obtain small sub-problems, a "good" decomposition means a minimal number of "small" partitions. Recursive selection of splitting variable(s) is a natural choice, that follows the common trend of recursive BDD operations.

Before proceeding, let us point out the two main aspects that differentiate our method from a combined variable reordering and standard variable splitting of recursive BDD operators:

- We allow different splitting variables when recurring in different partitions. This offers a wide range of solutions for partitioning, with a boundary case represented by the *Free–BDD* scheme [2].

- Subproblems are completely solved in different partitions, before combining their results. This has low impact on APPLY–like operators (based on a single BDD traversal), but we reap huge benefits when large intermediate results are produced before obtaining the final one. We only deal with subsets of intermediate results, and we exploit dynamic reordering to optimize single subproblems.

3 PARTITIONING BDD OPERATORS

We focus here on *single* operators. Results on full traversals, exploiting partitioning within inner steps and various optimizations, are presented in [1].

Following Equation (1), a partitioned AND–ABSTRACT operation is expressed as:

$$\exists_x(f \cdot g) = \exists_x(\overline{v} \cdot f \cdot g) + \exists_x(v \cdot f \cdot g)$$

load, because problem locality is increased. A further optimization is possible when the splitting variable is one of the quantifying ones ($v = x_i \in x$):

$$\exists_x(f \cdot g) = \exists_x(f_{\overline{x_i}} \cdot g_{\overline{x_i}}) + \exists_x(f_{x_i} \cdot g_{x_i})$$

where we exploit the cofactor property $v \cdot f = v \cdot f_v$ and we quantify the splitting variable.

A similar decomposition applies to image computation. Let us indicate with $C(s)$ a set of states, with $TR(s,x,y)$ the transition relation, and with $\text{IMG}(TR, C)$ the computation of the image of set C according to TR. In the case of domain variables, since these are quantifying variables, we apply the optimization previously described for AND–ABSTRACT.

$$\begin{aligned}\text{IMG}(TR, C) &= \exists_{s,x}(TR \cdot C) \\ &= \exists_{s,x}(TR_{\overline{s_i}} \cdot C_{\overline{s_i}}) + \exists_{s,x}(TR_{s_i} \cdot C_{s_i}) \\ &= \text{IMG}(TR_{\overline{s_i}}, C_{\overline{s_i}}) + \text{IMG}(TR_{s_i}, C_{s_i})\end{aligned}$$

The above formula expresses transition relation based image computation as a single AND–ABSTRACT operation with a *monolithic* transition relation. State-of-the-art reachability analysis uses conjunctively partitioned transition relations, i.e., an image operation is performed as a sequence of AND–ABSTRACT steps. This is a reason for expecting memory gains by partitioning. Moreover, in this case single internal steps, i.e., each single AND–ABSTRACT operation, can be optimized through partitioning.

A further application for partitioning is an entire traversal. One of the intermediate reachable state sets computed by the traversal TRAV might be partitioned, and several disjoined traversals activated on subsets

$$\text{TRAV}(\delta, C) = \text{TRAV}(\delta, C_1) + \text{TRAV}(\delta, C_2)$$

with $C = C_1 + C_2$. In the most general case image computations on partitioned sets produce overlapping results. So a common variable splitting cannot be applied to C and TR, and the same (full) TR is required by all image computations. Reachable state set may overlap, and, as a consequence it may happen that reachable state (and computational work) overlapping dominates partitioning, and that a very low gain is attained. Strategies to partition transition relations, and to deal with partitioned traversals are described in [3, 4], where not purely breadth-first traversals are also discussed. Here, following [1], we do not partition full traversals. We only partition images and we recombine (or re-split) state sets after each image operation. The technique is complementary to alternative state space traversal policies, because it is aimed at optimizing the cost of image computation.

As far as iterative squaring is concerned, we want to optimize step $\exists_z(T^n(s,z) \cdot T^n(z,y))$, and we follow two approaches:

- Splitting separately $T^n(s,z)$ and $T^n(z,y)$.
- Splitting $T^n(s,z)$ and $T^n(z,y)$ conjointly on a common variable of the set of variables z.

The first choice produces better splitting results whereas the second one allows us to better optimize conjunctions[1].

[1] In the product $f \cdot g = (f_1 + f_2) \cdot (g_1 + g_2)$, if the same splitting variable is used on both the terms only two terms, the diagonal products, have to be considered.

Our program is written on top of the Colorado University Decision Diagram (CUDD) package. All the experiments refer to a 200 MHz DEC Alpha with a 256 Mbyte main memory, with a 228 Mbyte working memory limit.

Level 3			
$|From| = 13414, |To| = 78914$			
# Part.	1	2	4
# Garbage	6	8	6
Peak BDD size	1032621	800134	428213
Mem. [Mbyte]	59.2	46.0	42.0
Time [s]	545	536	510

Table 1: Partitioned Operators Results on circuit s1269.

Tables 1, 2 and 3 show the relation among partitioning, memory and time during image computation.
Part. indicates the number of partitions: When # Part. equals 1 we have the original monolithic representation. # Garbage shows the number of activations of the garbage collection procedure. Peak BDD Size is the size of the biggest BDD we produce during image computation. Mem. reports the amount of memory needed, in Mbytes. Threshold indicates the number of nodes above which we partition sets. # Page Faults is the number of pages faults[2]. Time is reported in seconds or hours.

In all these experiments we leave the cache size free to assume the value established by the CUDD package. The number of activations of the garbage collection procedure gives us an idea about the complexity of the problem and about the memory optimization that the BDD package can guarantee.

Level 11				
$|From| = 141255, |To| = 425061$				
# Part.	1	2	4	8
# Garbage	8	11	13	18
Peak BDD size	704359	366369	200758	110568
Mem. [Mbyte]	42.3	35.6	27.1	23.7
Time [s]	533	459	387	403

Level 12				
$|From| = 425061, |To| = 1361263$				
# Part.	1	4	16	32
# Garbage	9	16	26	34
Peak BDD size	1956036	531808	165386	113435
Mem. [Mbyte]	108.9	67.6	65.64	54.0
Time [s]	2442	1551	1232	1293

Level 13				
$|From| = 1361263, |To| = 4226259$				
# Part.	2	8	14	17
Threshold	1000000	200000	150000	125000
Peak BDD size	ovf	918679	689007	500894
Mem. [Mbyte]	ovf	142	110	97
Time [h]	>10	2.2	1.9	1.8
# Page Fault	$> 5 \cdot 10^6$	132445	2467	1245
Elapsed Time [h]	>50	3.3	1.9	1.8

Table 2: Partitioned Operators Results on circuit s1423. ovf means overflow on BDD nodes and data not available.

For example for circuit s1269, in Table 1, with garbage collection off and one partition 121.8 Mbytes of main memory

[2] Major page faults in a Unix System.

procedure on only 59.2 Mbytes are needed.

Level 9			
$|From| = 763729, |To| = 1319332$			
# Part.	1	10	21
Threshold	1000000	100000	50000
Peak BDD size	3291342	399706	262510
Mem. [Mbyte]	214	41	37
Time [h]	8.3	7.4	7.7
# Page Fault	618658	890	280
Elapsed Time [h]	11.9	7.4	7.7

Table 3: Partitioned Operators Results on circuit s3271.

In general increasing the number of partitions increases the number of activations of the garbage collection procedure but at the same time the maximum amount of memory necessary is reduced. In parallel also the BDD peak size is smaller and this gives an idea of how much the overall complexity is reduced.

The advantage in terms of time tends to increase with the size of the problem as Table 2 shows. For circuit s1423 after level 12 only the partitioned approach is possible and the standard approach could lead us to far more than 1,000,000 pages faults for a single run. This implies tens of hours just dealing with paging. A higher number of partitions may decrease the number of page faults by more than 3 orders of magnitude. The time reported is the CPU time, i.e., user time and system time together, and it approximately coincides with user time. The time spent in page faulting is not kept into account and just the run or elapsed time is reported (column Elapsed Time). All the experiments run using approximately 100% of the CPU time and data shows that every page fault is dealt with in approximately 20 milliseconds.

Table 3 reports similar data on circuit s3271.

	Time [s]			
$|T^2| = 12933, |T^3| = 34701$				
# Part.	1	2	4	8
Cache Size				
1000003	22	21	18	23
500009	31	21	18	23
300007	45	21	18	23
200003	458	22	18	23
150001	12318	44	18	23
125003	28711	71	19	23
100003	ovf	4273	20	23
75011	ovf	79611	25	24
50021	ovf	ovf	80	25
25013	ovf	ovf	2305	32
15013	ovf	ovf	14950	550
10007	ovf	ovf	ovf	12700

Table 4: Squaring on circuit s1512 from T^2 to T^4. ovf means overflow in time, i.e., a CPU time larger than 24 hours.

Table 4 reports data on iterative squaring on circuit s1512 with different cache sizes and numbers of partitions. We compute T^3 starting from T^2. Performance is very dependent on cache size and partitioning can be of great benefit in this case when memory requirements are strict. To have good performance with just one partition we must use

shows how partitioning greatly reduces this number. Decomposition is also useful in other operations.

Order	# Part. = 1		# Part. = 4	
	# Nodes	Time [s]	# Nodes	Time [s]
Initial	1361263		345522 (1361263)	
Final	801724	2075	194109 (810107)	503

Table 5: Re-ordering on a large set of states.

For example Table 5 reports an experiment on re-ordering the reachable state set of circuit s1423 at level 12 (1361263 nodes). Directly re-ordering the set (column # Part. = 1) requires 2075 seconds to produce a final set of 801724 nodes (41% improvement). On the other hand we can decompose the set, let say in 4 subsets (column # Part. = 4), and re-order just one of the subsets. Reordering the subset of 345522 nodes requires just 244 seconds and produces on that subset a BDD of 194109 nodes. Using the re-shuffling operation on the initial set produces a set of 810107 nodes (40% improvement) in 259 seconds. Practically we produce the same result in $244 + 259 = 503$ seconds instead of 2075. The operation is actually faster also considering the splitting time, 620 seconds.

5 CONCLUSIONS

The current limit of BDDs and symbolic techniques resides in memory requirements, i.e., in the inability to compute and represent very large functions, relations or sets.

In this paper we analyze memory performance of a decomposition strategy recently presented. Experimental results show that this approach is applicable to standard operations usually found in reachability analysis. Moreover it is easy to apply, it works well with large sets, it is more effective than virtual memory, and more general than existing approaches.

References

[1] G. Cabodi, P. Camurati, and S. Quer. Improved Reachability Analysis of Large Finite State Machine. In *Proc. IEEE/ACM ICCAD'96*, pages 354–360, San Jose, California, November 1996.

[2] R. E. Bryant. Binary Decision Diagrams and Beyond: Enabling Technologies for Formal Verification. In *Proc. IEEE/ACM ICCAD'95*, pages 236–243, San Jose, California, November 1995.

[3] G. Cabodi, P. Camurati, L. Lavagno, and S. Quer. Disjunctive Partitioning and Partial Iterative Squaring: an effective approach for symbolic traversal of large circuits. In *Proc. EDA/SIGDA/ACM/IEEE DAC'97*, pages 728–733, Anaheim, California, June 1997.

[4] A. Narayan, A. J. Isle, J. Jain, R. K. Brayton, and A. Sangiovanni-Vincentelli. Reachability Analysis Using Partitioned-ROBDDs. In *Proc. IEEE/ACM ICCAD'97*, San Jose, California, November 1997.

WAVEFORM APPROXIMATION TECHNIQUE IN THE SWITCH-LEVEL TIMING SIMULATOR BTS

Molin Chang, Wang-Jin Chen, Jyh-Herng Wang* and Wu-Shiung Feng

Department of Electrical Engineering
National Taiwan University
Taipei, Taiwan, R.O.C.

*National Center for High-Performance Computer
7, R&D Rd. VI, Science-Based Industry Park
HsinChu, Taiwan, R.O.C.

ABSTRACT

In this paper an accurate and efficient switch-level timing simulator is described. The high accuracy is attributed to a new waveform approximation technique, which includes delay estimation and slope estimation. Efficient delay and slope calculations are accomplished through a switch-level simulation instead of using a transistor-level simulation. A new approach for delay estimation is presented, and it models the delay behavior of an RC tree by two equations: a dominant delay equation and an offset delay equation. Both are derived by a special process to fit the surface built by experimental data measured from the actual delay behavior of a CMOS gate. The results show good agreement with SPICE.

1. INTRODUCTION

BTS (Binary-tree Timing Simulator), which is two to three orders faster than SPICE, is an event-driven switch-level timing simulator and performs more accurate waveform approximation during the transient state.

Most switch-level algorithms emphasized how to calculate the time constant of charging/discharging the load capacitance more accurately. There are many researches on this topic [1-3]. However, all of above can not offer us more accurate waveform information in transient state; we want to know not only whether the logic gate changes state or not, but also when the output voltage begins to change and how fast it will change. Therefore, the waveform approximation technique is divided into two parts that are the delay estimation and the slope estimation. The delay estimation tells us when the output begins to change and the slope estimation tells us how fast the output will change. An uncertain amount of overshoot, chiefly due to parasitic capacitors, will almost always be produced at the output node while an event is happening at the input. The width of overshoot is the keypoint; if it can be predicted well, and then the delay will be estimated accurately. Then, the slope relates closely to the RC time constant of the discharging/charging path. Lin and Mead [4] proposed an efficient method that can be implemented in a recursive way. Furthermore, another important feature of BTS is that the delay and slope calculations are considered with internal charges and charge sharing effects [5,6]. The Internal charges stored in the internal nodes of a MOS circuit will increase the delay time about 20% when the tested circuit is a five-input NAND gate with four fully charged internal nodes [7]. Therefore, the effect of internal charges should also be considered when the delay time is estimated.

The remainder of this paper is organized as follows. First, we describe the MOS model used in BTS in Section 2. Next, the method of waveform approximation is presented in Section 3. Then, the delay and the slope estimations are discussed in Sections 4 and 5, respectively. Finally, the simulation results are given in Section 6 and summary in Section 7.

2. MOS MODEL

The MOS model in BTS is composed of voltage-controlled switch, effective resistance R_{eff} and equivalent grounded capacitances. The transistor is *on* (the switch conducts) if and only if the gate voltage of the NMOS transistor is higher than its threshold voltage V_t. The turn-on effective resistor is distinguished by two cases: R_{on} (in steady state) and R_t (in transient state), because the MOS transistor (denoted by MOSt) has different response under different gate state. Therefore, the value of R_{eff} may be one of the three cases:
- *Infinity*: if $V_{gs} < V_t$
- R_{on}: if V_{gs} is high (in steady state)
- R_t: if V_{gs} changes from L to H (in transient state)

The values of R_{on} and R_t depend on the physical parameters and the load capacitance, and R_t depends also on the slope of the signal at the gate.

3. WAVEFORM APPROXIMATION

The approximation work can be simplified if we cut off the overshoot and use a linear segment followed by an exponential tail to approach the falling (or rising) signal [8]. We use two equations as follows to plot the transient waveform.

- for a rising signal

$$f = \begin{cases} 0.2t/T & \text{for } t<3T \\ 1-0.4\exp(-(t-3T)/2T) & \text{for } t\geq 3T \end{cases} \quad (1)$$

- for a falling signal

$$f = \begin{cases} 1-(0.2t/T) & \text{for } t<3T \\ 0.4\exp(-(t-3T)/2T) & \text{for } t\geq 3T \end{cases} \quad (2)$$

where T is half of the time spent by the signal between 90% (for a falling signal) or 10% (for a rising signal) and 50% of the steady state. If the value of T can be obtained, the transient waveform will then be easily plotted.

The difference between the time when the output signal begins to change and the time when the input signal begins to change is defined as delay, which is denoted by D. The changing rate after the output begins to change is defined as slope, which is denoted by S.

4. DELAY ESTIMATION

4.1 Overshoot

Owing to the electrical characteristics of a MOS transistor, there are many parasitic capacitors existing inside a CMOS gate, e.g., Cgs, Cgd and so on. So the waveform of the drain of a MOSt depends not only the turn-on mechanism of MOSt but also the path formed by Cgd. The overshoot of output waveform, which can be treated as the excessive charge stored in the output node, is caused by the differential gate capacitor current. Observe that the amount of overshoot is determined by four factors as follows: (1) the slope S_i of input signal, (2) the size of C_{gd}, (3) the load capacitance Cl of output, and (4) the resistance R_p of discharging path in the N tree (or charging path in the P tree).

By analyzing some sample circuits using SPICE and varying the values of factors as mentioned above, we measure the data of delay time and then we can model the delay behaviors of CMOS gates by two equations.

(1) *Dominant delay equation:*
Cl is fixed, so this equation describes the relationship among delay, S_i, and R_p. It is easy to change S_i but R_p is not. Therefore, an alternative method is used. We increase the number of MOSt's in N tree circuit in order to change R_p discontinuously, and then R_p is replaced with N_p. In other words, we use the circuits such as inverter, two-input NAND gate, three-input NAND gate, and so on, as the primitive cases. The effect of internal charges that we probably meet in the actual circuits are extracted as an independent problem (see subsection 4.2).

For each primitive case, changing the input slope will produce a set of discrete two dimension curve, called NANDx-curve (x is the number of input). By collecting all the sets of data, we can plot a three-dimensional surface as shown in Fig. 1 and can use a hyperbolic surface (Eq. 3) to fit it.

$$D_D = (0.0292 N_p + 0.369)(S_i + 0.3) + 0.12 \qquad (3)$$

The deriving procedure is described as below:
Step 1: Use a straight line to fit a *NANDx-curve* in D_D-S_i plane, called curve α.
Step 2: Use a straight line to fit the curve, called *SLOPEy-curve* (y is the value of input slope), in D_D-N_p plane, and then normalize this curve, called curve β, which is used to modulate the curve α in the direction of N_p-axis.
Step 3: D_D=(curve α)(curve β)+offset.

(2) *Offset delay equation:*
N_p is fixed, so this equation describes the relationship among offset delay (an offset value with respect to D_D), S_i, and Cl. This equation is used for compensating the value of delay time calculated by the dominant delay equation, which does not consider the effect of the changing factor Cl. If N_p is adjusted, we obtain a set of surfaces. It means that we can obtain a discrete three-dimensional surface for each primitive case. The method for constructing this surface is the same as mentioned above. Similarly, we can also use a set of hyperbolic surfaces

$$D_O = f(N_p)(0.293 S_i C_l + 0.023) \qquad (4)$$

to fit them, where $f(N_p)$ represents the coefficients that are the function of N_p. The surfaces when N_p=1 are shown in Fig. 2, which include the surfaces built by experimental data and derived approximate surface.

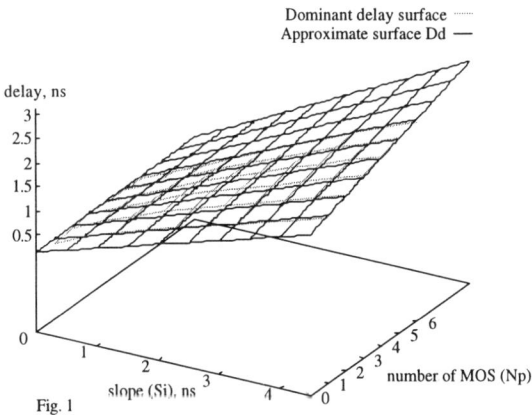

Fig. 1

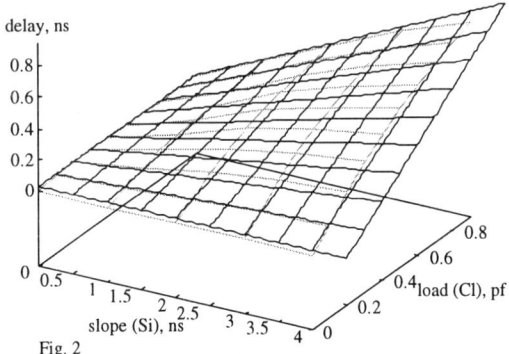

Fig. 2

4.2 Internal Nodes

The delay due to the internal charges can be calculated approximately as

$$dt = Q/I_{av} = (Q/V_s) 2R \qquad (5)$$

where I_{av} is the average current, V_s is the voltage swing, R is the effective resistance of the conducting path and Q is the charge stored in the internal nodes. More than one internal nodes may be going to charge or discharge in the series-parallel tree, and these nodes must be taken into account when calculating the switching delay. Thus, Eq. 5 is rewritten as

$$D_I = \Sigma \, 2R_i (Q_i/V_s) \qquad (6)$$

where R_i is the effective resistance of internal node v_i with respect to ground, Q_i is the charge stored in the internal node v_i.

After all, the total delay is summed up by the delay times caused by the effect of overshoot and the internal nodes, including the charge sharing effect [5,6].

$$D_{total} = D_D + D_O + D_I \qquad (7)$$

5. SLOPE ESTIMATION

If the output waveform can be treated as a simple RC waveform, then the parameter T in Eqs. 1 and 2 can be calculated by the equation: $T = (t_{50\%} - t_{10\%})/2 = 0.294 RC$. In BTS we defined slope as the time spent by the signal voltage dropping one volt, i.e. in units of time/volt, and then $T = S$ when $Vdd = 5V$. Therefore, $T = S = 0.294 RC$.

The equivalent RC time constant of active tree can be computed by the equations as described in reference 4 and implemented by a recursive algorithm while traversing the whole RC tree. Furthermore, two effects, called *non-active tree effect* and *bottle-neck effect*, should also be considered together.

5.1 Non-active tree effect

Because the output slope of falling signal is affected by not only the N tree but also some internal nodes connected to output node in the P tree, the algorithm for computing slope should consider both and then estimate the total effect. When estimating the slope, the internal nodes in the *non-active tree* should be considered, and its influence should be added into the component obtained from calculating the RC time constant of the active tree. For example, if the output state is changed from HIGH to LOW in the circuit as shown in Fig. 3(a), there is at least one discharging path existing in the *active tree* (the N tree in this case) and no charging path in the *non-active tree* (the P tree). In this case, the effect of the nodes v_5^* and v_6^* should be added when estimating the slope, and both internal nodes can be viewed as a charge supplier that can supplement the charge loss at the load capacitor.

5.2 Bottle-neck effect

A bottle-neck always exists in the discharging path and will form the highest barrier to prevent the discharging current flowing on it. Therefore, the discharging rate of the output node is dominated by the bottle-neck. However, it is a complicated work to find the bottle-neck of the charging/discharging (C/D) path because the C/D path is changed dynamically depending upon the input patterns. Furthermore, parallel connections in the C/D path will increase the difficulty of this problem. Finding the location of bottle-neck becomes a major work when estimate the bottle-neck effect. In general, we can not expect that the bottle-neck is a MOSt at most of time except that the C/D path is a pure series connection.

The transistors in the bottle-neck should be replaced with Rt and all others should be replaced with Ron while calculating the slope value. From simulation results by SPICE, we find that all MOSt's in transient state not located in the bottle-neck can not affect the slope value of output waveform.

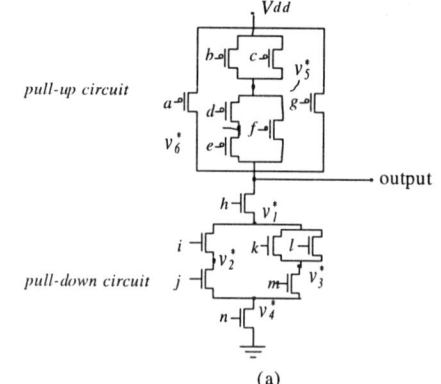

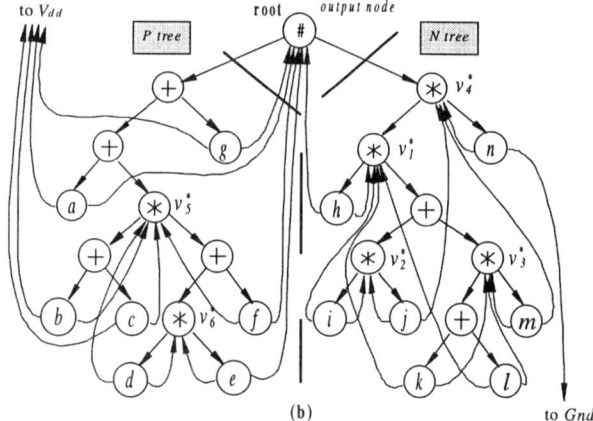

Fig. 3: A series-parallel circuit and its corresponding MTB tree.

6. RESULTS

This method has been tested extensively for basic modules such as counters, decoders, adders, and ALU's. The CPU time comparisons are summarized in Table 1.

An one-cluster circuit, also using the circuit as shown in Fig. 3(a) as an example, is simulated by using SPICE and our timing simulator BTS. The results are compared as shown in Fig. 4. The bold solid lines are the results obtained from our simulator. Five input patterns are applied to this circuit, which are case 1: (/,0,0,0,1,1,1), case 2: (1,0,0,0,/,1,1), case 3: (1,0,0,0,1,/,1), case 4: (1,1,0,0,1,1,/), and case 5: (1,1,/,0,1,/,1), and the associated output waveforms are labeled A, B, C, D and E, respectively. Note that '/' represents the ramp input signal with the rising time 1ns from LOW to HIGH. The waveform E is extracted independently for its different slope, and compared with the waveform C to distinguish its difference. There are small errors presented in the simulation results because the target circuit is simply an one-stage circuit.

7. SUMMARY

An accurate waveform approximation technique is proposed, which is achieved by the new approach of delay estimation and the modified slope estimation. The new approach of delay estimation improves the previous version of BTS that converted the overshoot effect to the turn-on-time of MOSt (the value of V_T was shifted to 3.1V), and can offer a better adaptability for a wide range of circuit and input specification. For each different fabrication process, the equations for delay estimation are derived only once. The deriving procedure is a simple and quick work because it can be achieved only by a few samples. Of course, this procedure can also be aided by a program on computer. The modified slope estimation also increases the accuracy of the transient waveform prediction, including under some special circumstances.

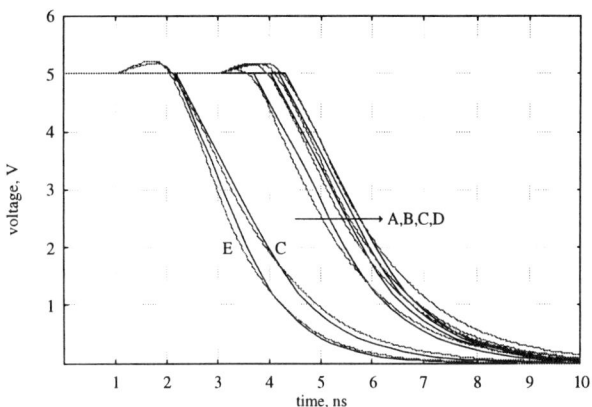

Fig. 4: The simulated waveforms of the circuit as shown in Fig. 3(a). Bold line: BTS. Light line: SPICE

8. REFERENCES

[1] R. E. Bryant, "A Switch level model and simulator for MOS digital systems," IEEE Computers, vol. C-33, pp. 160-177, 1984.

[2] C. J. Terman, "RSIM - A Logic-Level Timing Simulator," Proceedings of the IEEE International Conference on Computer Design, New York, pp. 437-440, November, 1983.

[3] J. Rubinstein, P. Penfield, and M. A. Horowitz, "Signal delay in RC tree networks," IEEE Trans. on Computer-Aided Design, vol. CAD-2, NO. 3, pp.202-211, 1983.

[4] T. M. Lin, and C. A. Mead, "Signal delay in general RC networks," IEEE Trans. on Computer-Aided Design, vol. CAD-3, No.4, pp.331-349, 1984.

[5] Molin Chang, S,-J Yih and Wu-Shiung Feng, "Algorithm based on modified threaded binary tree for estimating delay affected by internal charges in CMOS gates", Electronics Letters, Vol. 32, No. 20, pp. 1877-1879, 26th September 1996.

[6] Molin Chang, S,-J Yih and Wu-Shiung Feng, "Recursive algorithm for calculating effective resistances in RC tree", Electronics Letters, Vol. 33, No. 2, pp. 131-133, 16th January 1997.

[7] J. H. Wang, Molin Chang, and W. S. Feng, "Binary-tree timing simulation with consideration of internal charges". IEE Proceedings-E, vol. 140, No.4, pp. 211-219, July 1993.

[8] F.C.Chang, C.F.Chen, and P.Subramaniam, "An accurate and efficient gate level delay calculator for MOS circuits," Proceedings of 25th ACM/IEEE conference on Design automation, Anaheim, CA, USA, pp.282-287, 1988.

Circuit	MOS no.	CPU time on PC (DX4-100), secs BTS	Pspice	Speed ratio	Primary input event no.
complex gate (Fig. 1(a))	12	0.11	2.53	0.043	2
inverter chain (100 stages)	200	0.28	241.18	0.0012	1
74138	88	0.33	102.22	0.0032	11
7483	258	0.99	774.64	0.0013	13
74381	584	1.10	1670.98	0.00066	14

Table 1: Comparisons between BTS and Spice

New and Efficient Method for the Multitone Steady-State Circuit Simulation

R. Larchevêque*, P. Bolcato* and E. Ngoya**

* ANACAD/MGC 11A, Chemin de la Dhuy 38240 Meylan (France)

** IRCOM, Université de Limoges 123 Avenue Albert Thomas 87060 Limoges (France)

Abstract

A new method termed Compressed Time is presented to compute the steady-state response of nonlinear circuits with periodic or quasi-periodic excitation. It is different from standard or enhanced Shooting method or Harmonic Balance (HB) and allows the simulation of large systems such as large circuits and large number of harmonics. It is based on a time domain formulation of HB equations written in a preconditioned form. The generated system is then solved with a nonlinear block relaxation algorithm. Performances and simulation results are shown for a typical RF circuit.

1. Introduction

Multitone steady-state analysis of large NonLinear (NL) circuits has lately become a new challenge in the simulation area, due to the rapid growth of the RF IC market. To compute the steady-state of electronic circuits, two different algorithms were developed in the 70's.

The first one, named *Shooting method* [1], operates in the time domain. It is based on a normal transient analysis (Time Domain Integration or TDI) with a technique that tries to kill or shoot the transient phase by finding the appropriate initial conditions putting TDI directly in the steady-state. *Shooting method* is definitely a single tone algorithm. This means that it cannot handle circuits driven by multiple nonharmonically related periodic input sources. When Newton method is used to solve the Shooting system, a full Jacobian matrix needs to be constructed and factored, which is a very expensive operation not suitable for large circuits.

On the other hand, the second algorithm, called *Harmonic Balance* (HB) [2] operates in the frequency domain. It computes the steady-state response as the solution of a NL algebraic equation where the signal is represented by its truncated Fourier Series. The size of the solved system is n.Nf (where n is the circuit dimension and Nf the number of Fourier coefficients). HB is a natural multitone algorithm because of the use of multi-dimension Fourier Transform but Newton method inside HB has to factor a big and rather dense Jacobian matrix; which is then not suitable to handle large circuits.

Recently, Krylov subspace iterative solvers have been used to overcome the bottleneck (circuit size) of HB [3-4] as well as Shooting [5]. However Krylov based Shooting is still a one tone algorithm and HB requires a good preconditioner to achieve convergence with NL circuits.

We present in this paper a new algorithm to compute the circuit steady-state. It is based on a time domain formulation of HB equations written in a preconditioned form. The generated system is then solved with a NL block relaxation algorithm. This is a multitone algorithm and the convergence capabilities are insensitive to the level of NL and it is also able to handle large circuits.

2. New algorithm description

This method has been entitled *Compressed Time* (CT) because during the resolution it seems that the steady-state has been reached after a compressed (i.e. short) transient phase like in a compressed time domain integration. This method is not based on TDI but rather on convolution as will be described later.

2.1 CT formulation

The starting point is a piecewise formulation of Harmonic Balance:

$$F(\omega_k) + Y(\omega_k)X(\omega_k) = G(\omega_k) \qquad (1)$$

Where X is the vector of circuit variables (Fourier coefficients of voltages, currents, charges and fluxes), F is the contribution of circuit NL components (Fourier coefficients of voltages, currents, charges and fluxes), G is the vector of stimuli and Y is the nodal admittance matrix representing the linear part of the circuit.

The system of equations (1) is equivalent to the following system (obtained after a multiplication by Z):

$$X(\omega_k) + Z(\omega_k)F(\omega_k) = Z(\omega_k)G(\omega_k) \qquad (2)$$

Where $\omega_k = k\omega_0 \quad k = 0,\dots,\pm N$, N is the number of harmonics, ω_0 is the fundamental pulsation and Z is the nodal impedance matrix representing the linear part of the circuit ($ZY = I$, I is the identity matrix).

Equation (2) can be written in the time domain [6] (after inverse Fourier Transform (FT)):

$$x(t) + z(t) * f(t) = z(t) * g(t) \quad (3)$$

Where * denotes convolution:
$c(t) = h(t) * s(t) = \int_0^{\tau_{max}} h(\tau)s(t-\tau)d\tau$, τ_{max} is the time duration of $h(t)$.

One important point to notice from equation (2) is that the steady-state response of the circuit depends solely on the value taken by $Z(\omega)$ at $\omega = \omega_k$. The philosophy of CT is to find an auxiliary impedance matrix $\hat{Z}(\omega)$ satisfying (2) and exhibiting *well behaved* impulse responses (IR) $\hat{z}(t)$. *Well behaved* means short and well confined to the time origin ($\hat{z}(0) >> \hat{z}(t)$ for $t \neq 0$). These customized IR $\hat{z}(t)$ then generate a well conditioned system producing a diagonally dominant Jacobian matrix (as explained later) suited for solution by an iterative linear solver. Next section (2.2) explains how IR $\hat{z}(t)$ are computed and section 2.3 describes the iterative method used to efficiently solve (3) and find the steady-state response of the circuit. These two next sections consider the one tone case for notational simplicity but the extension to multitone is straightforward and is briefly explained in section 2.4.

2.2 Fitting confined IR

After testing many causal functions, we have found that the most simple and efficient ones to describe confined IR have the general form below:

$$h(t) = p(t) \sum_{k=-N}^{N} B_k e^{jk\omega_0 t} \quad (4)$$

$$\text{Where} \quad p(t) = \frac{1}{T_0} e^{-\alpha \frac{t}{T_0}} \text{rect}\left(\frac{t - \frac{T_0}{2}}{T_0}\right) \quad (5)$$

The IR duration is T_0 ($T_0 = \frac{2\pi}{\omega_0}$) imposed by the rectangle window. $p(t)$ is an exponential decay confining function where α is a positive argument used to control the decay of $h(t)$ (decay causes IR confinement). The B_k are 2N+1 complex coefficients ensuring that $H(\omega)$ (Fourier transform of $h(t)$) matches $Z(\omega)$ at all the different frequencies ω_k of the steady-state solution spectrum.

The B_k coefficients are computed by taking the Fourier transform of $h(t)$ and writing $H(\omega_k) = Z(\omega_k)$ ($k = 0,...,\pm N$). B_k are then the solution of the following linear system: $[P]\vec{B} = \vec{Z}$, where $\vec{B}$ and $\vec{Z}$ are vectors of B_k and $Z(\omega_k)$ values respectively.
$[P]$ is the de-embedding matrix, $[P]_{ij} = P((i-j)\omega_0)$, where $P(\omega)$ is the Fourier transform of $p(t)$.

Illustration of confined IR may be found in [7]

As $h(t)$ is a continuous time function, $c(t) = h(t) * s(t) = \int_0^{T_0} h(\tau)s(t-\tau)d\tau$. $s(t)$ is an arbitrary signal (representing for instance $f(t)$ in equation (3)). The above convolution integral is calculated with a polynomial approximation of $s(t)$, $c(t)$ can be described in a discretized form:

$$c(n) = \sum_{m=0}^{2N} \hat{h}(m)s(n-m) \quad (6)$$

Where $\hat{h}(n)$ is obtained from $h(t)$ (details of the computation can be found in [8]).

Introducing (6) into (3) and considering a steady-state period lead to the following system:

$$\begin{cases} x(n) + \sum_{m=0}^{2N} \hat{z}(n-m)f(m) = \overline{g}(n) \\ n = 0,...,2N \end{cases} \quad (7)$$

Where $\overline{g}(n) = \sum_{m=0}^{2N} \hat{z}(n-m)g(m)$

and the NRM Jacobian matrix:

$$J = I + \begin{bmatrix} \hat{z}(0) & \hat{z}(2N) & \cdots & \hat{z}(1) \\ \hat{z}(1) & \hat{z}(0) & & \vdots \\ \vdots & & \ddots & \vdots \\ \hat{z}(2N) & \cdots & & \hat{z}(0) \end{bmatrix} \begin{bmatrix} f'(0) & & & 0 \\ & f'(1) & & \\ & & \ddots & \\ 0 & & & f'(2N) \end{bmatrix} \quad (8)$$

Where $f'(n) = \frac{\partial f}{\partial x}(n)$.

The effects of IR confining ($\hat{h}(0) >> \hat{h}(n)$ for $n \neq 0$) clearly produce a block-diagonally dominant J matrix. Next section shows how system (7) is efficiently solved using iterative methods.

2.3 Solving the steady-state equation

The most common way to solve NL systems such as equation (7) is to use NRM, where the Jacobian matrix -(prediagonalized by IR confinement)- may be successfully treated with Gauss-Seidel (GS) relaxation method. A more efficient alternative is to exchange NRM and GS loops [9]. This implementation, termed as CT algorithm, to solve the NL equation (7) is described further on figure 0.

Solving (7) with GS-NRM as explained above leads to compute the solution time step after time step. At each time step the NL system (9) is solved with NRM. One GS iteration corresponds to the computation of 2N+1 time points (equivalent to one period of the input signal). This algorithm seems to be very similar to TDI where the number of periods to reach steady-state has been dramatically reduced by compressing IR. This is the reason why this algorithm has been called *Compressed Time*, but it should be noticed that it is not TDI: it is based on convolution and it can handle the multitone case.

CT algorithm:

Guess a solution $x^0 = (x^0(0) \cdots x^0(2N))^T$.
For p=1 to Gauss-Seidel limit {
  For n=0 to 2N {
    Solve equation (9) (see annex below) for $x^p(n)$ with NRM.
  }
  if $\|x^p - x^{p-1}\| < tolerance_{steady-state}$ return.
}

Annex:

$x^p(n) + \hat{z}(0) f^p(n) = \overline{g}(n) - g_{last}(n) - g_{next}(n)$
where:
$g_{last}(n) = \sum_{m=0}^{n-1} \hat{z}(n-m) f^p(m)$
$g_{next}(n) = \sum_{m=n+1}^{2N} \hat{z}(n-m) f^{p-1}(m)$

Figure 0: CT Algorithm description

2.4 The multitone case

For the multitone case equation (3) is obtained from equation (2) via multi-dimension FT. Convolution in equation (3) becomes multi-dimensional convolution :

$$c(\vec{t}) = \int_{\tau_1} \cdots \int_{\tau_Q} h(\vec{\tau}) s(\vec{t} - \vec{\tau}) d\vec{\tau} \quad \text{and}$$

$$c(\vec{n}) = \sum_{m_1=0}^{2N_1} \cdots \sum_{m_Q=0}^{2N_Q} \hat{h}(\vec{m}) s(\vec{n} - \vec{m})$$

Q is the number of fundamental frequencies, $\vec{t} = (t_1 \cdots t_Q)^T$, t_i is the time instant along the ith direction and N_i is the number of considered harmonics of ith fundamental frequency.
Confined IR are obtained by multi-dimensional fitting; the general form of the Q-dimension function used is:

$h(\vec{t}) = p(\vec{t}) \sum_{\vec{k}} B_{\vec{k}} e^{j\vec{t}^T [\Omega] \vec{k}}$ where

$$p(\vec{t}) = \prod_{i=1}^{Q} \frac{1}{T_i} e^{-\alpha_i \frac{t_i}{T_i}} rect\left(\frac{t_i - \frac{T_i}{2}}{T_i}\right)$$

T_i is the period of ith fundamental frequency,
$\vec{k} = (k_1 \cdots k_Q)^T / |k_i| \leq N_i$, $[\Omega] = \begin{bmatrix} \omega_1 & & 0 \\ & \ddots & \\ 0 & & \omega_Q \end{bmatrix}$ and ω_i is the ith fundamental pulsation.

The CT algorithm, described for single tone in section II.2, can be used as well for multitone simulation. The only difference is the "for n=1 to N" loop replaced by multiple nested loops.

3. Example and simulation results

The above described CT algorithm has been implemented in the Eldo electrical simulator [10] and validated with several RF circuits. The use of the CT method is illustrated in this paper with a typical RF circuit, a 2GHz Image Reject Mixer. Figure 1 shows the schematic diagram of the circuit and more details may be found in [11]. It consists of a Low Noise Amplifier (LNA), two mixers, three phase shifters, a combiner and a bias cell. It has 250 nodes, 60 bipolar and MOS transistors and about 300 linear elements. A two tone simulation has been performed with a LO frequency of 1.8GHz (100mV amplitude) and a sweep of the input signal from 1.2GHz to 2.2GHz (10mV amplitude), considering 10 harmonics for each fundamental frequency.

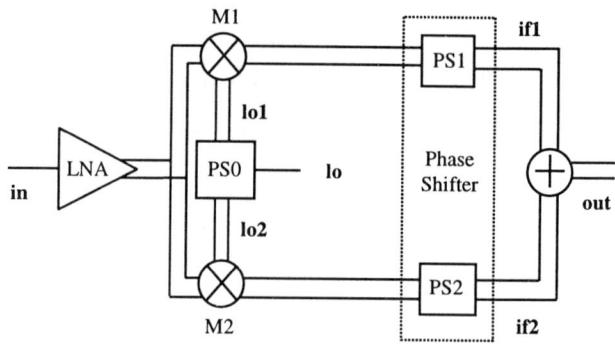

Figure 1: Schematic diagram of the circuit

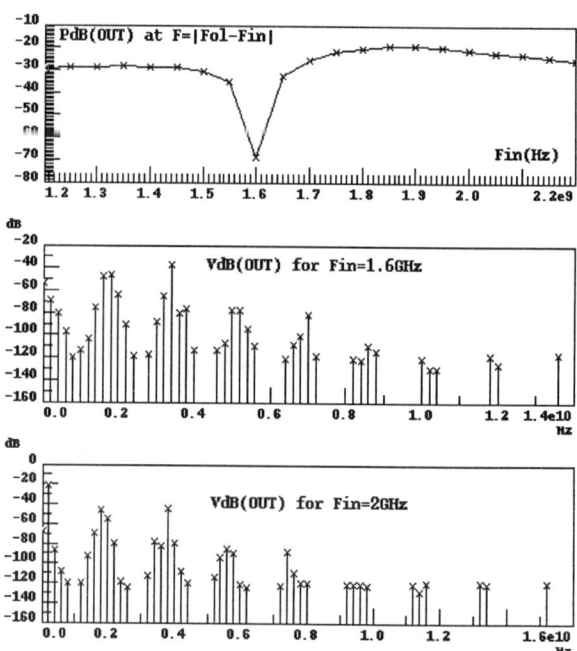

Figure 2: Steady-state simulation results of the Image Reject Mixer circuit

The simulation results are displayed on figure 2. The first curve shows the output power at frequency $|F_{LO} - F_{IN}|$ versus input frequency. We can clearly see the Image Frequency (1.6GHz) rejection of 35dB conforms to experimental measurements [11]. The second and third curves show the output spectrum for two different values of the input frequency (respectively 1.6GHz and 2GHz). The input level used in simulation (10mV) corresponds to a relatively high level compared to the normal use of the circuit to show the ability of CT algorithm to handle nonlinearity. One steady-state simulation of the presented circuit (a single value of input frequency) requires 6 GS iterations and takes 1min20s. CPU on Sun Ultra1 Sparc Work-Station. The complete sweep simulation (20 input frequency points) takes 12min.

4. Conclusion

This paper describes a new algorithm (termed *Compressed Time* or *CT*) to efficiently simulate steady-state of nonlinear circuits submitted to multitone excitations. It is based on a preconditioned time domain formulation of HB suited to be solved with a NL block relaxation method (GS-NRM). It can therefore handle large systems (large circuits as well as large number of harmonics) and convergence is not affected by strong nonlinearity. CT has been implemented in an industrial electrical simulator and validated with several RF circuits. Simulation results and performances are shown for a typical RF application.

5. Acknowledgements

The authors would like to thank Joël Besnard for the implementation into Eldo and Denis Pache for providing lots of test circuits. We would also like to thank Driss Yachou for the PC and Jean Rousset for advice on text policy.

6. References

[1] T. J. Aprille and T. Trick, "Steady State Analysis of Nonlinear Circuits with Periodic Inputs", Proceedings of the IEEE, Vol. 60, n°1, pp. 108-114, Jan. 1972.

[2] M. S. Nakhla and J. Vlach, "A Piece-Wise Harmonic Balance Technique for Determination of Periodic Response of Nonlinear Systems", IEEE Trans. Circuits and Systems, CAS-23, pp. 85-91, Feb. 1976.

[3] H. G. Brachtendorf and al, "Fast Simulation of the Steady State of Circuits by the Harmonic Balance Technique", Proceedings of the ISCAS 95, pp. 1388-1390, June 1995.

[4] D. Long and al, "Full-chip Harmonic Balance", Proceedings of the CICC 97, pp. 379-382, May 1997.

[5] R. Telichevesky and al, "Efficient Steady State Analysis based on Matrix-Free Krylov-Subspace Methods", Proceedings of the 32nd DAC, pp. 480-484, June 1995.

[6] A. Buonomo, "Time Domain Analysis of Nonlinear Circuits with Periodic Excitation", Electronic Letters, Vol. 27, n°1, pp.65-66, Jan. 1991.

[7] R. Larchevêque and E. Ngoya, "Compressed Transient Analysis Speeds Up the Periodic Steady-state Analysis of Nonlinear Microwave Circuits", Proceedings of IEEE MTT-S, THB4, June 1996.

[8] J. S. Roychowdhury and al, "Algorithms for the Transient Simulation of Lossy Interconnect", IEEE Trans. On CAD of Integrated Circuits and Systems, Vol. 13, n°1, pp.96-104, Jan. 1994.

[9] M. P. Desai and I. N. Hajj, "On the Convergence of Block Relaxation Methods for Circuit Simulation", IEEE Trans. Circuits and Systems, Vol. CAS-36, n 7, pp. 948-958, July 1989.

[10] Eldo v4.6 User's Manual, ANACAD EES / MGC, copyright 1997.

[11] D. Pache and al, "An Improved 3V 2GHz BICMOS Image Reject Mixer IC", Proceedings of the CICC 95, pp 95-98, May 1995.

Generalized Centers of Gravity Algorithm for Yield Optimization of Integrated Circuits

Mansour KERAMAT and Richard KIELBASA
Ecole Supérieure d'Electricité (SUPELEC)
Service des Mesures, Plateau de Moulon
F-91192 Gif-sur-Yvette Cédex France
E-mail: mansour.keramat@supelec.fr

ABSTRACT

The aim of Design Centering is to minimize or even eliminate the undesirable effect of manufacturing process variations by changing the designable parameter values. The Centers of Gravity (CoG) algorithm is a yield optimization algorithm that was originally developed on a heuristic basis for discrete component circuits. In this contribution, the generalized CoG algorithm which can be applied to integrated circuits is presented.

I. Introduction

In contrast to discrete component circuits, where the performance is directly influenced by discrete electrical components, the factors influencing integrated circuit performance are the *fabrication process parameters* and the *device geometry*. These parameters are subjected to manufacturing process variations. In circuit level design, the nominal value of device geometry can be determined. For a given technology, the nominal value of process parameters which correspond to the many model parameters (e.g., oxide thickness, flat band voltage, sheet resistance, and lateral diffusion) are fixed.

For discrete circuits, the fact that designable parameters and random variables (noise factors) are in the same space has led to several direct large-sample yield optimization methods. The CoG algorithm was originally developed on a heuristic basis for discrete circuits [1]. However, its implementation is very simple and it gives satisfactory results in practice. That is why it was implemented in several commercial packages, e.g., [6]. In our recent work [3], [4], the theoretical basis and some optimality aspects of this algorithm, and the improved CoG algorithm was described. In [5], some extension of CoG to integrated circuits as a yield enhancement algorithm was reported. However, it can be shown that the extended algorithm converges to a local optimum in the sense of the CoG design centering. Here, the general extension of this algorithm, called the generalized CoG algorithm, is presented.

II. Principle of Centers-of-Gravity Algorithm

In order to have a general notation for discrete and integrated circuits, we use the following notation. $\mathbf{x} = [x_1, x_2, \ldots, x_n]^T$ denote the circuit designable parameters (e.g., nominal value of passive RLC elements, nominal MOS transistor mask dimensions). $\xi = [\xi_1, \xi_2, \ldots, \xi_m]^T$ is the vector of random variables characterizing the statistical variations, (e.g., statistical variations of RLC elements, variations of device model parameters) and $f_\xi(\xi)$ stands for their joint probability density function (pdf). The space of these random variables is also called *disturbance space*. $\mathbf{p} = \mathbf{p}(\mathbf{x}, \xi)$ is the vector of circuit parameters. The general notion of acceptability region and disturbance space is illustrated in Fig. 1.

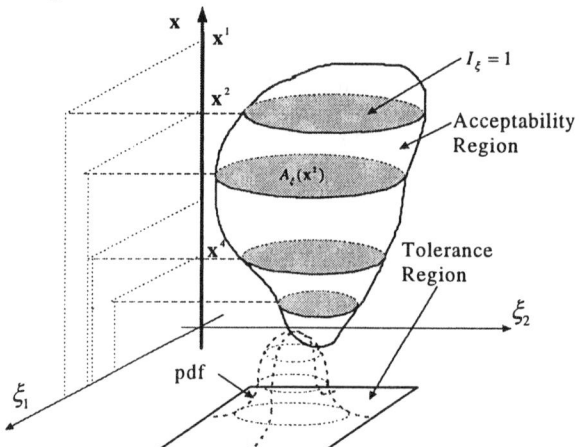

Fig. 1. Notion of acceptability region in integrated circuits.

Design Centering aims at increasing the manufacturing parametric yield by varying the designable circuit parameters while keeping $f_\xi(\xi)$ fixed. Mathematically it can be described as

$$\underset{\mathbf{x}}{Max}\left\{Y(\mathbf{x}) = \int I_p(\mathbf{p}) f_p(\mathbf{p}) d\mathbf{p}\right\}, \quad (1)$$

where $f_p(\mathbf{p})$ is the joint pdf of circuit parameters and $I_p(\mathbf{p})$ takes on 1 when all of the performance specifications are met, and 0 otherwise. In disturbance space, the yield can be expressed as

$$Y(\mathbf{x}) = \int_{R^m} I_p(\mathbf{x}, \xi) f_\xi(\xi) d\xi. \quad (2)$$

In the case of discrete circuits, we have $\mathbf{p} = \mathbf{x} + \xi$. A relatively efficient method for optimizing circuit yield is the *Centers of Gravity* (CoG) algorithm [1], which makes use of spatial distribution of pass and fail circuits in parameter space after a Monte Carlo analysis. It is an iterative process which by nature requires the prediction of direction of optimization. In the CoG algorithm, it is given by

$$\mathbf{x}_{M+1} = \mathbf{x}_M + \kappa(\hat{G}_A - \hat{G}_F), \quad (3)$$

where κ is a coefficient of step length which can be chosen $(1 - \hat{Y})$ [1], $\hat{G}_A$ and $\hat{G}_F$ are respectively the estimate of pass and fail points CoG. The weighted CoG of pass points is given by

$$G_A = \frac{1}{Y}\int_{R_T} \mathbf{p} I_P(\mathbf{p}) f_\mathbf{p}(\mathbf{p}) d\mathbf{p}, \quad (4)$$

where R_T is the tolerance region of parameter values. It can be shown that [1]

$$(1-Y)G_F + YG_A = \mathbf{x}. \quad (5)$$

An unbiased estimator of the pass center of gravity (4) can be expressed as

$$\hat{G}_A = \frac{1}{N}\sum_{i=1}^{N}\mathbf{p}^i I_P(\mathbf{p}^i), \quad (6)$$

where $\mathbf{p}^i$'s are drawn randomly according to $f_\mathbf{p}(\mathbf{p})$. In the same way, $\hat{G}_F$ can be obtained.

III. Extension to Integrated Circuits

In this section, the extension of CoG for integrated circuits is discussed in detail.

A. Using Projection Methods

The CoG algorithm can be extended to the problems in which the parameters with constant nominal value and a given distribution are allowed. These parameters are usually concerned with the fabrication process parameters, such as *oxide thickness, flat band voltage, length reduction, and width reduction*.

One of the most common techniques for handling constraints in an optimization problem is to use a *descent method* in which the direction of descent is chosen to decrease the cost function and to remain within the constraint region. To this end, the movement direction is projected onto the constraint region. This technique is called *Projection Method* [2]. Fig. 2 illustrates the application of the projection method in CoG algorithm in order to handle some parameters with fixed nominal value and given distribution. In practice, this scheme can easily be implemented. This scheme has implicitly been used by some authors (e.g., [5]) without mentioning the projection methods.

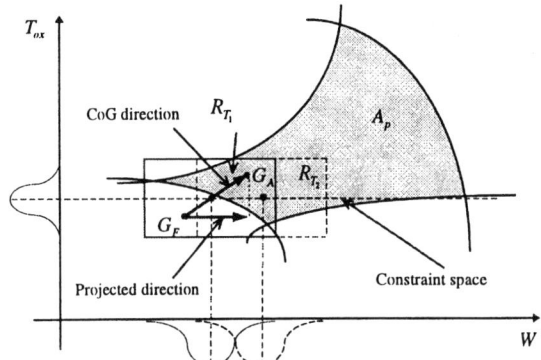

Fig. 2. Projection method in the CoG algorithm.

B. Using Random Perturbation Method

In integrated circuits, some of the parameters are strongly correlated. For example, the width of the transistors are random variables, but they have a correlation coefficient near unity (without consideration of mismatch effects). The strong correlation coefficient results in a restriction in the exploration space of parameters. Thus the solution of optimization is suboptimal. This situation is shown in Fig. 3. In this example, it is assumed that we have two designable parameters and there is just one statistical variation (width reduction). It is seen that the 100% parametric yield is not obtainable by the conventional CoG algorithm. The drawback of the yield enhancement algorithm reported in [5] is the same as mentioned above.

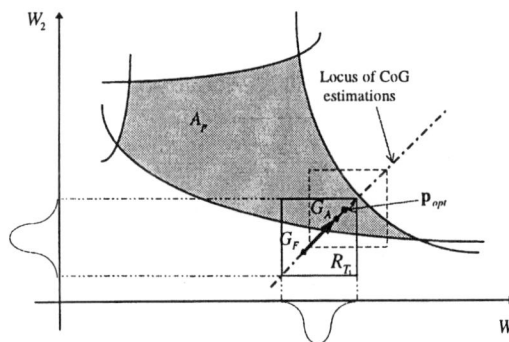

Fig. 3. Correlated parameters problem in integrated circuits.

In order to overcome this problem, we can add an independent random variable to each designable parameters. This method is known as *Random Perturbation Method Technique* [7], [8]. These additive random variables allow us to obtain some information about the yield gradient with respect to the designable parameters.

We perturb the designable parameters $\mathbf{x}$ with some additional random spread vector η of known pdf $h(\eta, \beta)$, with $E[\eta] = 0$ and η is independent of ξ. β is a parameter for controlling the standard deviation of η. The perturbed yield can be defined as [7]

$$\tilde{Y}(\mathbf{x},\beta) = \int_{R^n}\left[\int_{R^m} I_p(\mathbf{x}-\eta,\xi)f_\xi(\xi)d\xi\right]h(\eta,\beta)d\eta$$
$$= \int_{R^n} Y(\mathbf{x}-\eta)h(\eta,\beta)d\eta. \quad (7)$$

In fact, this is the convolution of Y with the kernel function $h(\eta,\beta)$. It is also called a *smoothed functional* of yield [8]. One of the possible choice of kernel function is the multigaussian pdf

$$h(\eta,\beta) = \frac{1}{(2\pi)^{n/2}\beta^n \prod_{i=1}^{n}\sigma_i}\exp\left[-\frac{1}{2}\sum_{i=1}^{n}\left(\frac{\eta_i}{\beta\sigma_i}\right)^2\right], \quad (8)$$

where σ_i's are the standard deviation of η_i's. It is clear that when $\beta \to 0$ then the kernel function approaches the Dirac delta function. Consequently,

$$\lim_{\beta\to 0}\tilde{Y}(\mathbf{x},\beta) = Y(\mathbf{x}). \quad (9)$$

In practice, $Y(.)$ is a continuous function with some degree of smoothness. Thus for some sufficiently small value of β, we have $\tilde{Y}(\mathbf{x},\beta) \approx Y(\mathbf{x})$.

In yield gradient calculation, we first use the projection methods. Therefore, the yield gradient can only have non zero components over the designable parameters. The gradient of yield with respect to designable parameters is obtained by

$$\nabla_\mathbf{x}\tilde{Y}(\mathbf{x},\beta) = \int_{R^n}\nabla_\mathbf{x}Y(\mathbf{x}-\eta)h(\eta,\beta)d\eta$$
$$= \int_{R^n} Y(\mathbf{x}-\eta)\nabla_\eta h(\eta,\beta)d\eta \quad (10)$$
$$= E_{\eta,\xi}\left[I_p(\mathbf{x}-\eta,\xi)\mathbf{C}^{-1}\eta\right],$$

where $\mathbf{C}$ is a positive definite diagonal covariance matrix with $c_{ii} = \beta^2\sigma_i^2$. By using (4) and (5), Eq. (10) can be rewritten as

$$\nabla_\mathbf{x}\tilde{Y}(\mathbf{x},\beta) = \mathbf{C}^{-1}\frac{\tilde{Y}}{\tilde{Y}}\left\{E_{\eta,\xi}\left[I_p(\mathbf{x}-\eta,\xi)(\mathbf{x}+\eta)\right]\right.$$
$$\left. - E_{\eta,\xi}\left[I_p(\mathbf{x}-\eta,\xi)\mathbf{x}\right]\right\} = \tilde{Y}\mathbf{C}^{-1}(G_{A\mathbf{x}}-\mathbf{x}) \quad , \quad (11)$$
$$= \tilde{Y}(1-\tilde{Y})\mathbf{C}^{-1}(G_{A\mathbf{x}}-G_{F\mathbf{x}}),$$

where $G_{A\mathbf{x}}$ and $G_{F\mathbf{x}}$ are projections of pass and fail points CoG, respectively. The yield maximum occurs at the point $\mathbf{x}_{opt}$ for which the condition $\nabla Y(\mathbf{x}_{opt}) = 0$ holds. After (11) this is only possible if $G_{A\mathbf{x}} = G_{F\mathbf{x}} = \mathbf{x}$. This property justifies the way of the optimization procedure of the CoG algorithm.

The primitive Monte Carlo (PMC) estimator of $G_{A\mathbf{x}}$ is

$$\hat{G}_{A\mathbf{x}} = \frac{1}{N\hat{\tilde{Y}}}\sum_{i=1}^{N}(\mathbf{x}+\eta^i)I_p(\mathbf{x}+\eta^i,\xi^i), \quad (12)$$

where (η^i,ξ^i) are randomly drawn according to the joint pdf $f_\eta(\eta)f_\xi(\xi)$, and

$$\hat{\tilde{Y}} = \frac{1}{N}\sum_{i=1}^{N}I_P(\mathbf{x}+\eta^i,\xi^i). \quad (13)$$

The PMC estimator of $G_{F\mathbf{x}}$ can be obtained in the same way.

Thus, estimating the pass and fail points CoG allows us to use the CoG optimization procedure. It should be noted that the value of β is decreased by successive iterations. In what follows, we call this algorithm the Generalized CoG (GCoG).

IV. Simulation Results

In this section, the application of the GCoG algorithm to a numerical function and a CMOS clock driver circuit is presented.

Example 1: *3 dimensional Acceptability Region* [7]

Consider the following acceptability region:

$$A_p = \left\{\mathbf{x},\xi \mid 100\left[x_2-(x_1+\xi)^2\right]^2+\left[1-(x_1+\xi)\right]^2 \leq 8\right\}, (14)$$

where $\mathbf{x} = [x_1,x_2]^T$ are designable parameters and $\xi \in R^1$ is the random variable uniformly distributed in the interval [-1,1], with $E[\xi] = 0$. The initial designable parameter values are $\mathbf{x}^0 = [-0.95,0.4]^T$ with $Y(\mathbf{x}^0) = 19.74\%$. The solution to this problem is $x_{1opt} \in [-0.26,0.3]$, $x_{2opt} \approx 0.26$ with $Y(\mathbf{x}_{opt}) \approx 72\%$ [7]. It can be shown that the yield derivative at optimal point is not continuous. A two dimensional cross section of the acceptability region is shown in Fig. 4.

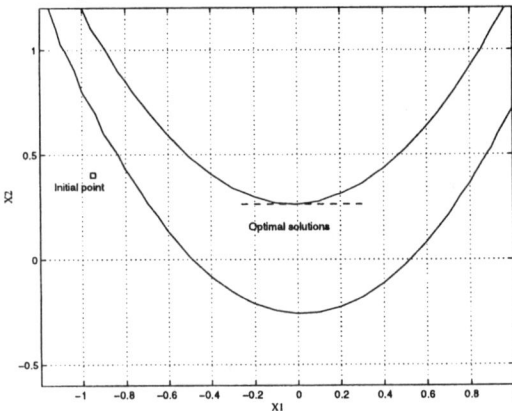

Fig. 4. The cross section of the acceptability region of Example 1 for $\xi = 0$.

The results of yield optimization by GCoG are shown in Fig. 5. The sample size of yield and CoG estimations was chosen to be 50. The optimal solution is $\mathbf{x}_{opt} = [-0.18,0.23]^T$ with $Y(\mathbf{x}_{opt}) = 70\%$ (by 1000 Monte Carlo samples), which is very close to the analytical optimal solution. The standard deviation of additive random variables are $\sigma_1 = \sigma_2 = 0.7$ with Gaussian pdf. The values of β were [1, 0.8, 0.7, 0.6, 0.4, 0.3, 0.2, 0.1, 0.05, 0.01]. It is seen that the total number of simulations is equal to 500. However, in [7], the total number of simulations for this problem with *Stochastic Approximation Algorithm* was 1310.

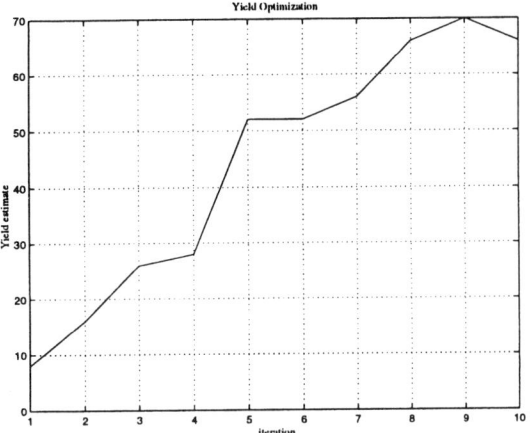

Fig. 5. Yield optimization of Example 1.

Example 2: *CMOS Clock Driver Circuit [8]*

A CMOS clock driver is shown in Fig. 6. The clock driver provides two outputs V_{out1} and V_{out2} in opposite phase. The performance of interest is the clock skew between two outputs. The specifications are that the skew falls in the interval [-1,1] ns. The model parameters used to characterize CMOS manufacturing process disturbances are described in [8]. These variables are considered independent and of Gaussian probability distribution.

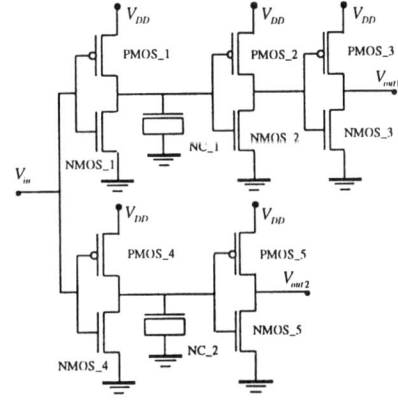

Fig. 6. Schematic circuit of a CMOS clock driver.

In this example, the width of transistor NMOS_1 and NMOS_4 were selected as designable parameters. The optimization results are shown in Fig. 7. In each iteration we used 50 circuit simulations. In Fig. 7, it is seen that the optimized circuit has 100% yield (verified by 200 samples and β=0). The initial yield was approximately 2%. It should be noted that the estimate presented in Fig. 7 is an estimation of $\tilde{Y}(\mathbf{x},\beta)$. The value of β was decreased in the same way as in Example 1.

V. Conclusions

The Centers of Gravity (CoG) algorithm was heuristically developed for design centering of discrete

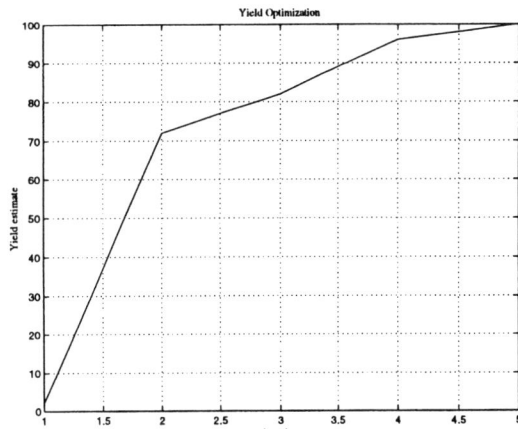

Fig. 7. Yield estimates vs. Iterations of the CMOS clock driver circuit.

circuits. This algorithm can easily be implemented and gives satisfactory results in practice. Here, the limitations of CoG for integrated circuit applications were studied and some theoretical solutions have been proposed by using the *Projection Methods* and the *Random Perturbation Method*. The extension of CoG to integrated circuits is called Generalized CoG (GCoG) algorithm. The successful application of GCoG to a 3-dimensional numerical example and a CMOS clock driver circuit was described.

Future research will be focused on an automatic choice of β and different smoothing functionals in the perturbation scheme.

References

[1] R. Spence and R. S. Soin, *Tolerance Design of Electronic Circuits*. Addison-Wesley Publishers Ltd., 1988.

[2] D. G. Luenberger, *Optimization by Vector Space Methods*. John Wiley & Sons, Inc, 1967.

[3] M. Keramat and R. Kielbasa, "Optimality aspects of centers of gravity algorithm for statistical circuit design," in *Proc. IEEE 40th Midwest Symp. Circuits Syst.*, Sacramento, CA, August 3-6 1997.

[4] M. Keramat and R. Kielbasa, "Parametric yield optimization of electronic circuits via improved centers of gravity algorithm," in *Proc. IEEE 40th Midwest Symp. Circuits Syst.*, CA, August 3-6 1997.

[5] M. Singha and R. Spence, "The parametric yield enhancement of integrated circuits," *Int. J. Circuit Theory and Applications*, vol. 19, pp. 565-578, 1991.

[6] ANACAD, *ASPIRE User's Manual*, November 1996.

[7] M. A. Styblinski and L. J. Opalski, "A random perturbation method for IC yield optimization with deterministic process parameters," in *Proc. IEEE Int. Symp. Circuits Syst.*, Canada, May, 1984, pp. 977-980.

[8] J. C. Zhang and M. A. Styblinski, *Yield and Variability Optimization of Integrated Circuits*, Kluwer Academic Publisher, 1995.

STATISTICAL DESIGN OF INTEGRATED CIRCUITS USING MAXIMUM LIKELIHOOD ESTIMATION OF THE COVARIANCE MATRIX

A. Seifi, J. Vlach and K. Ponnambalam

Department of Electrical and Computer Engineering
University of Waterloo, Waterloo, Ontario, Canada, N2L 3G1
Email: aseifi@vlsi.uwaterloo.ca

ABSTRACT

A new formulation is proposed for statistical design of integrated circuits with correlated input parameters.
The method uses a polyhedral approximation of the feasible region and finds the maximum volume ellipsoid contained in that polyhedron. The orientation of the ellipsoid is fixed by a maximum likelihood estimate (MLE) of the correlation matrix. The ellipsoid center is a nominal design with the maximum yield. The covariance estimation is formulated as a semidefinite program which uses the sampling observations as input data. The design centering problem is presented as a second-order cone programming and solved by a special interior-point optimization algorithm. The optimal design of a switched-capacitor filter is presented.

1. INTRODUCTION

The statistical design optimization concentrates on finding the best nominal values and tolerances for the input parameters and operating conditions which result in an improvement of the yield. One possible approach is to construct an approximation of the feasible region. Most *design centering* [1] and *statistical sampling* methods are based on this general idea [15]. The assumptions of bounded-ness and convexity of the feasible region are essential for this approach. The fundamental questions are: (1) which method provides a better approximation for the feasible region? (2) how efficiently can the approximating problem be solved by the solution methodology? and (3) where does the correlation matrix come into play?

Polyhedral (linear) approximation is widely used because of its simplicity. Quadratic approximation may be more suitable but is certainly more expensive [15]. An ellipsoidal approximation is proposed in [1] but the problem has an order of n^2 variables. In this paper, the first-order second-moment (FOSM) reliability method is used to construct a polyhedral approximation of the feasible region [8].

This research is a continuation of the work in [11]. It is related to the ellipsoid methods in [1, 10, 15] and to maximizing the norm body introduced in [2, 6]. The technique proposed in [1] generates a sequence of ellipsoids containing the feasible region until it converges to the minimum volume one. The method in [15] finds some points on the boundary of the region which define an ellipsoidal approximation of the region. A common shortcoming of these two methods is that some part of the approximating ellipsoid falls outside the region and as a result, the yield may be overestimated. We propose a method which finds the maximum volume ellipsoid contained in the approximated region. The orientation of the ellipsoid is fixed by the estimated correlation matrix. It can be shown that the ellipsoid center is a design with maximum entropy, assuming Gaussian distribution for the input variables with a given covariance matrix [5]. Another related idea is to find the *largest Hessian ellipsoid* inscribed in the polytope [10]. The Hessian ellipsoid is characterized by the Hessian of a log-barrier function and has no connection to the covariance structure of the problem.

The proposed method may be considered as a new formulation to *design centering* with correlated variables. It is motivated by the recent developments in interior-point methods for semidefinite and second-order cone programming problems. The main advantages of the new formulation are: (1) it connects the yield maximization problem to the underlying correlation structure of the design variables; (2) it allows optimization to be performed in the space of design variables rather than the space of the performance functions which leads to major savings in circuit simulation; and (3) it can be solved by an efficient and robust interior-point algorithm.

2. PROBLEM FORMULATION

Suppose that $h(x)$ indicates the performance function of a set of correlated random variables x and the requirement is $h(x) \geq 0$. The analytic form of $h(x)$ is usually unknown, but the sensitivity information ($g_j = \frac{\partial h(x)}{\partial x_j}$; $j = 1, 2, \cdots, n$) can be obtained. Let there be m design requirements $h_i(x) \geq 0, i = 1, 2, \cdots, m$, to be satisfied by the nominal design μ. Replacing $h_i(x)$ by its linear approximation at x_i^* leads to a polyhedral region defined by:

$$\mathcal{P} = \{x \in \Re^n | g_i^T(x - x_i^*) \geq 0; \ i = 1, 2, \cdots, m\}. \quad (1)$$

It is required that $\mathcal{P}$ be convex and bounded. In practice, one can add simple lower and upper bounds on x to make it bounded. The reference point x_i^* is found so that it is on the surface of $h_i(x) = 0$, and has the minimal distance β_i,

$$\beta_i = \frac{g_i^T(\mu - x_i^*)}{(g_i^T C g_i)^{1/2}}, \quad (2)$$

from the nominal point μ. Note that β_i is also a measure of the yield as explained in [8]. Finding x_i^* is a least-squares subproblem which can be solved using an iterative algorithm explained in [11]. Let us now consider a multivariate Gaussian distribution function with the mean μ and covariance matrix C for the design variables x:

$$f(x) = ((2\pi)^n \det C)^{-\frac{1}{2}} \exp\left(-\frac{1}{2}(x-\mu)^T C^{-1}(x-\mu)\right).$$

The level sets (equidensity contours) of $f(x)$ are concentric ellipsoids defined by

$$\mathcal{E}(\mu, C, \gamma) = \{x \in \Re^n | \ (x-\mu)^T C^{-1}(x-\mu) \leq \gamma^2\}, \quad (3)$$

where γ identifies a particular ellipsoid of this type. Note that γ must be less than or equal to all β_i's in order for $\mathcal{E}$ to be contained in $\mathcal{P}$. The covariance matrix C can be viewed as $C = \Sigma R \Sigma$, where Σ is a diagonal matrix whose entries are the standard deviations σ and R is the matrix of correlation coefficients. The orientation of the ellipsoid $\mathcal{E}$ is determined by R, and its volume, V, is proportional to:

$$V = \alpha_n (\det C)^{1/2} = (\prod_{j=1}^{n} \lambda_j(C)^{1/2}), \quad (4)$$

where $\lambda_j(C)$ denotes the j-th eigenvalue of C and α_n is a constant. Maximizing the ellipsoid volume is therefore equivalent to minimizing the determinant of C^{-1} or minimizing the logarithmic barrier function

$$\phi(C) = \log \det C^{-1}. \quad (5)$$

The function ϕ is strictly convex over the set of positive definite matrices and has many other nice properties (see [13] for more details).

In practice, it suffices to have $\beta_i \geq 3$ which translates to the yield of 99.86%. It is shown below that the yield constraints ($\beta_i \geq 3$; $i = 1, 2, \cdots, m$) are equivalent to the ellipsoid $\mathcal{E}$ being contained in the polyhedron $\mathcal{P}$ with $\gamma = 3$. Using an affine transformation $x = By + \mu$, where B is a symmetric positive definite matrix ($C = B^2$), the ellipsoid $\mathcal{E}$ can be expressed as:

$$\mathcal{E} = \{By + \mu \mid \|y\| \leq \gamma\}, \quad (6)$$

where $\|.\|$ denotes the standard Eucleadian norm. Then, the contained-ness condition can be written as:

$$\mathcal{E} \subseteq \mathcal{P} \iff \forall y \in \mathcal{E}, \|y\| \leq \gamma \Rightarrow g_i^T(By + \mu) - g_i^T x_i^* \geq 0$$

$$\iff \sup_{\|y\| \leq \gamma} g_i^T By \leq g_i^T \mu - g_i^T x_i^* \quad (7)$$

$$\iff \gamma \|B g_i\| \leq g_i^T(\mu - x_i^*) \quad (8)$$

$$\iff \gamma \leq \frac{g_i^T(\mu - x_i^*)}{(g_i^T B^2 g_i)^{1/2}}. \quad (9)$$

It can be seen from (2) and (9) that all the yield constraints will be satisfied if we set $\gamma = 3$.

The statistical yield optimization is concerned with finding a nominal design μ such that, under random variations characterized by the covariance matrix C, the overall yield is maximized. Having replaced the feasible region by its polyhedral approximation $\mathcal{P}$, the problem becomes equivalent to finding the largest volume ellipsoid, centered at μ and inscribed in the polytope $\mathcal{P}$. Such an ellipsoid can be uniquely found if $\mathcal{P}$ is convex and bounded.

The central argument here is that the matrix R is inherent to the physical nature of the problem and cannot be optimized as desired. Therefore, we first find a maximum likelihood estimate of R using sampling observations and will then proceed to optimization with a fixed correlation matrix. The assumption of Gaussian probability distribution is essential for the suggested approach. If such an assumption cannot be validated or no sampling is possible, one has to resort to other robust optimization or non-probabilistic reliability methods [3, 4].

2.1. MLE OF THE COVARIANCE MATRIX

Suppose that it is possible to get a sample of N observations, $x^{(1)}, x^{(2)}, \cdots, x^{(N)}$, from the process with a known initial nominal design μ_0. Then, a sample covariance matrix S can be calculated by

$$S = \frac{1}{N} \sum_{k=1}^{N} (x^{(k)} - \mu_0)(x^{(k)} - \mu_0)^T. \quad (10)$$

The MLE of the covariance matrix C, denoted by $\hat{C}$, can be found by maximizing the log-likelihood function $\log \prod_{k=1}^{N} f(x^{(k)})$ which is equivalent to solving:

$$\max_C \quad \log \det C^{-1} - \frac{1}{N} \sum_{k=1}^{N} (x^{(k)} - \mu_0)^T C^{-1} (x^{(k)} - \mu_0)$$

$$\text{s.t.:} \quad C \succ 0.$$

The constraint $C \succ 0$ requires that the matrix C be positive definite. This Semi-Definite Programming (SDP) problem is suggested for covariance estimation in [13]. It has an analytical solution $C = S$ if S is nonsingular. If S is singular, when the sample size is too small for example, the problem becomes unbounded and one has to add extra constraints such as lower and upper bounds on the diagonal elements of C or its inverse. It is also possible to set some entities of C to fixed values using prior information. The resulting SDP problem can be readily solved using MAXDET software provided in [13]. After $\hat{C}$ has been estimated, the correlation matrix $\hat{R}$ can simply be extracted from it as follows:

$$\hat{\Sigma} = \text{Diag}\{\hat{C}_{ii}\}^{1/2}, \quad \hat{R} = \hat{\Sigma}^{-1}\hat{C}\hat{\Sigma}^{-1}. \quad (11)$$

2.2. DESIGN CENTERING WITH FIXED CORRELATIONS

The optimization problem here is to find the largest inscribed ellipsoid with a fixed orientation, characterized by the estimated matrix $\hat{R}$. The complexity of this problem is much less than the general maximum volume ellipsoid problem since the number of optimization variables is reduced to two n-vectors of μ and σ. Therefore, we do not have to resort to an SDP formulation which involves an order of n^2 number of variables. This design centering problem is formulated here as a Second-Order Cone Programming (SOCP) which accepts constraints in the form of (8) (see [9] for an introduction). Having fixed the matrix R, maximizing the ellipsoid volume leads to minimizing the log-barrier function $(-\log(\sigma - \sigma_0))$. However, a linear cost function of σ is considered here as required by the SOCP formulation. The function may reflect the relative cost of tolerances in various components. The SOCP model is then given by:

$$\min_{\mu,\sigma} \quad c^T\sigma$$
$$\text{s.t.} \quad 3\|Bg_i\| \le g_i^T(\mu - x_i^*); \ i = 1, 2, \cdots, m, \quad (12)$$
$$\sigma > \sigma_0,$$

in which $B = (\Sigma R \Sigma)^{1/2}$, $\Sigma = \text{Diag}\{\sigma_j\}$ and σ_0 is a lower bound on σ. The reference points x_i^*'s and gradient vectors g_i's are found in the linearization step and kept fixed here. This problem can be solved directly by a special interior-point algorithm given in [7]. A similar model was used in [11] with a different objective function and solved by a sequential quadratic programming (SQP) algorithm. The results of these two algorithms are compared in the next section.

Once the optimal solution to the SOCP model is found, one can update the approximating polyhedron $\mathcal{P}$ and repeat the process until μ converges within a desired accuracy.

3. TEST PROBLEM RESULT

The example shown here is a 4-th order switched capacitor low pass filter. The edge of the pass band is at 1 kHz and the switching frequency is 8 kHz. The filter was taken from [12] and for our purposes was modified to the form given in Figure 1. Element values for the initial equiripple pass-band response μ_0 were obtained by WATSCAD [14].

The operational amplifiers are ideal and the elements are assumed to have normal distributions with some correlation matrix. For statistical design, the pass band response is required to be between the boundary shown by the solid lines in Figure 2. The suppression at frequency 2 kHz should be at least 0.019 (-34.4 db). The problem has $n = 14$ variable capacitors and we set $m = 7$ output requirements at 7 frequency points.

The initial values for the standard deviations σ were set to 1% of the initial equiripple design μ_0 (The initial values of tolerances were $\frac{3\sigma_0}{\mu_0} = 3\%$). Since no real sampling data was available for correlation estimation, a correlation matrix was set up artificially as follows. We assumed that correlation would exist only between capacitors in each of the five vertical sections in Figure 1, and not between these sections. The correlation coefficients were generated as the ratios of capacitor initial values (C_i) in each section, $R_{ij} = \min\{C_i, C_j\}/\max\{C_i, C_j\}$, based on the initial equiripple design μ_0 (see [11] for more details).

Each major iteration included a linearization step and an optimization step. The linearization step was done using a Matlab implementation of the iterative algorithm explained in [11]. The optimization model in (12) was solved with the given matrix R using the SOCP software provided in [7]. This software was run by a Matlab (version 4.0) calling interface on a Sun Sparc 20/71 Workstation. Each optimization step took about 29 iterations and less than one second while each linearization step took about 4 seconds. The linear approximations were then updated using the optimal solution found in the optimization step. It took the total time of about 35 seconds and 7 major iterations to converge to the final optimal design point.

This is a significant improvement over 443 seconds taken for producing the same results using SQP method, previously reported in [11]. The main reason for such a speedup is solving the SOCP model by a special interior-point algorithm provided in [7]. The initial equiripple (curve a) and the optimal response (curve b) are depicted in Figure 2.

4. CONCLUSION

This paper describes a method for statistical design optimization with correlated variables. The main contribution is in the development of a new formulation of the problem which can be solved efficiently by interior-point methods. Connections to statistical methods for yield estimation in the presence of correlation have been made transparent. The computational results show significant improvements over previously reported results.

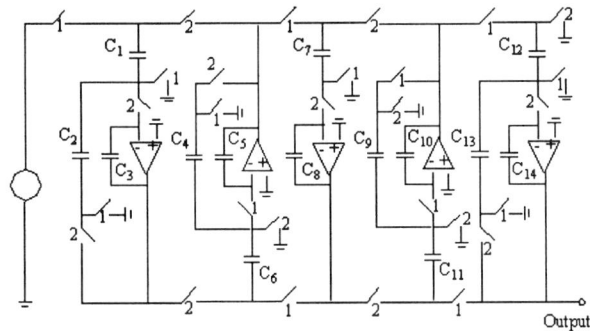

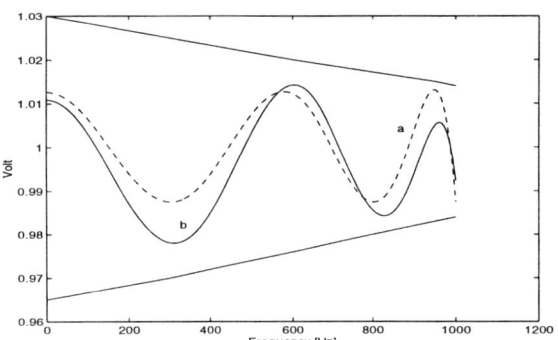

Figure 2: (a) Initial and (b) Optimal responses of the switched capacitor filter.

Figure 1: The switched capacitor filter.

5. REFERENCES

[1] H. L. Abdel-Malek and A.-K. S. O. Hassan, "The ellipsoidal technique for design centering and region approximation," IEEE Transactions on Circuits and Systems, vol. 10, pp. 1006-1014, 1991.

[2] K. J. Antreich, H. E. Graeb and C. U. Wieser, "Circuit analysis and optimization driven by worst-case distances," IEEE Transactions on Circuits and Systems, vol. 13, no. 1, pp. 57-71, 1994.

[3] Y. Ben-Haim, "Robust reliability in the mechanical sciences," Springer-Verlag, 1997.

[4] A. Ben-Tal and A. Nemirovski, "Robust convex programming," Working Paper #5/95, Faculty of Industrial Engineering and Management, Technion Institute of Technology, Israel, October 1995.

[5] A. P. Dempster, "Covariance selection," Biometrics 28, pp. 157-175, 1972.

[6] S. W. Director, W. Maly and A. Strojwas, "VLSI design for manufacturing: yield enhancement," Kluwer Academic, Boston, MA, 1990.

[7] M. S. Lobo, L. Vandenberghe and S. Boyd, "Second-order cone programming," Information Systems Laboratory, Electrical Engineering Department, Stanford University, Stanford CA94305, May 1997 (Submitted to Linear Algebra and Applications).

[8] H. O. Madsen, S. Krenk and N. C. Lind, "Methods of structural safety," Prentice-Hall, New Jersy, 1986.

[9] Y. Nesterov and A. Nemirovskii, "Interior-point polynomial algorithms in convex programming," SIAM Studies in Applied Mathematics, Volume 13, Philadelphia, PA, 1994.

[10] S. S. Sapatnekar, P. M. Vaidya and S. M. Kang, "Convexity-based algorithms for design centering," Technical Report ISU-VLSI-93-SS01, Iowa State University, 1993.

[11] A. Seifi, K. Ponnambalam and J. Vlach, "Probabilistic Design of Integrated Circuits with Correlated Input Parameters," Proceedings of European conference on Circuit Theory and Design, Budapest, pp. 996-1001, 1997.

[12] J. Taylor and J. Mavor,"Exact design of stray-insensitive switched-capacitor LDI ladder filters from unit element prototypes", IEEE Transactions on Circuits and Systems, vol. CAS-33, no. 6, pp. 613-622, June 1986.

[13] L. Vandenberghe, S. Boyd and S-P Wu, "Determinant maximization with linear matrix inequality constraints," Information Systems Laboratory, Electrical Engineering Department, Stanford University, Stanford CA94305, April 1996 (To appear in SIAM Journal on Matrix Analysis and Applications, April 1998).

[14] J. Vlach, "WATSCAD - A program for analysis of switched-capacitor networks," Manual, University of Waterloo, Department of Electrical and Computer Engineering, Waterloo, Ontario, Canada N2L 3G1.

[15] J. M. Wojciechowski and J. Vlach, "Ellipsoidal method for design centering and yield estimation," IEEE Transactions on Circuits and Systems, vol. 12, no. 10, pp. 1570-1579, 1993.

MODELING THE DYNAMIC BEHAVIOR OF SERIES–CONNECTED MOSFETS FOR DELAY ANALYSIS OF MULTIPLE–INPUT CMOS GATES

L. Bisdounis and O. Koufopavlou

VLSI Design Laboratory, Department of Electrical & Computer
Engineering, University of Patras, GR-26500 Patras, Greece.
e-mail: bisdouni@ee.upatras.gr

ABSTRACT

In this paper the dynamic behavior of series-connected MOSFETs is studied, in order to compute the propagation delay of multiple-input static CMOS gates. A method for the reduction of series-connected MOSFETs to a simple MOSFET with the same behavior is proposed. The effective width of the equivalent transistor is not constant as in some previous works. So all cases of input slopes, the load capacitance, the number and the position of the switching inputs, and the body effect, are considered in order to determine the equivalent transistor's width. Along with the reduction process, an accurate analytical inverter timing model is used to compute the propagation delay of multiple-input static gates. The produced results are in very good agreement with SPICE simulations.

1. INTRODUCTION

Many different techniques for timing modeling of VLSI circuits have been proposed in order to improve the speed of circuit simulators. The reliability of these approaches depends on the accuracy with which the transient response and consequently the propagation delay of basic circuits can be evaluated.

In [1] we introduced an accurate analytical timing model for the CMOS inverter. The analytical nature of this model results in high computational speed, provided that an accurate and fast method to analyze multiple-input gates is to reduce them to equivalent inverters. Many reduction techniques [2]–[8] have been proposed in the literature. Traditionally, the equivalent width of series-connected transistors is approximated by W/N (W: channel width of one transistor, N: number of series-connected transistors) [2],[3], without taking into account the effects of the load and the input transition time. This is the main reason of inaccuracy, because it is valid only for step input waveforms, or when all transistors operate in the linear region. Another source of delay errors in some existing methods [3],[4] is the use of simple quadratic equations for the transistor currents, which are not valid for the recent sub-micron technologies.

Recently, a reduction technique in order to determine the effective width of serial-connected transistors has been proposed in [5], where several empirical parameters are used. However, the conventional model ($W_{eff} = W/N$) is used when the inputs are fast or all inputs are switched together. In [6], an approach to generalize an inverter-based timing model to multiple-input gates is proposed, where the serial transistors are handled by repeated dc analyses using SPICE every time a transition sets a new path. However, the modeling of transient phenomena by dc analysis results in inaccuracies. A number of dc analyses is also required in the reduction technique presented by Kong et al. [7], in order to determine the effective transconductance of series-connected transistors. Moreover, in [7] the current of the short-circuiting block is neglected in the delay calculation. Both techniques [6],[7] are limited to single input switching. The reduction technique suggested in [8], is demonstrated only for step inputs, and the calculation of the transistors' effective resistance is not discussed.

In this paper, a new reduction technique based on the dynamic behavior of series-connected MOSFETs, is presented. In the determination of the equivalent transistor's width, the influences of the output load, the input transition time, the number and the position of the switching inputs, and the body effect, are taken into account. Also, the additional delay due to the internal and coupling capacitances is included. Along with the reduction process, an accurate inverter timing model [1] which includes most of the factors that influence the inverter operation such as the currents through both transistors, the load capacitance, the input slope, and the input-to-output coupling capacitance, is used to compute the propagation delay of multiple-input CMOS gates. For the transistor currents, the α-power MOS model [9] is used, in order to include the velocity saturation effect of recent short-channel devices.

2. REDUCTION TECHNIQUE

Consider the multiple-input NAND gate shown in Fig.1. In the following we describe the reduction technique for series-connected nMOS transistors. The treatment of pMOS transistors is analogous. For the transistor drain currents, the following expressions of the α-power law MOSFET model [9] are used.

$$I_D = \begin{cases} P_C (W/L)(V_{GS} - V_{TH})^{\alpha}, & V_{DS} \geq V'_{DO}, \text{ Saturation} \\ P_L (W/L)(V_{GS} - V_{TH})^{\alpha/2}, & V_{DS} < V'_{DO}, \text{ Linear} \end{cases},$$

where $V'_{DO} = P_V(V_{GS} - V_{TH})^{\alpha/2}$ is the drain saturation voltage, and $P_L = P_C/P_V$. P_C, P_V and α (velocity saturation index) are extracted from the I-V static characteristics of the device. For the determination of the device threshold voltage (V_{TH}) a linear approximation of the body effect is used [6], which results to the following simple formula

$$V_{TH} = V_{TO} + \gamma_1 V_{SB},$$

where V_{TO} is the zero-bias threshold voltage, and γ_1 is the body effect coefficient.

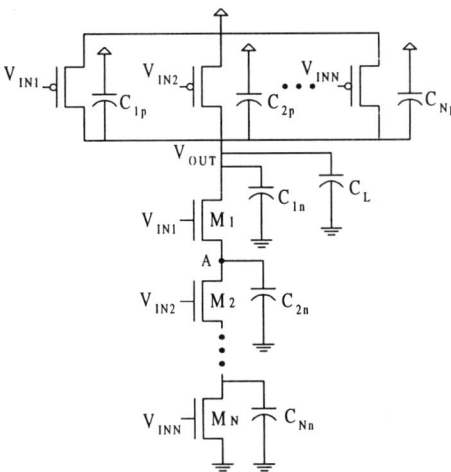

Fig.1: Multiple-input NAND gate

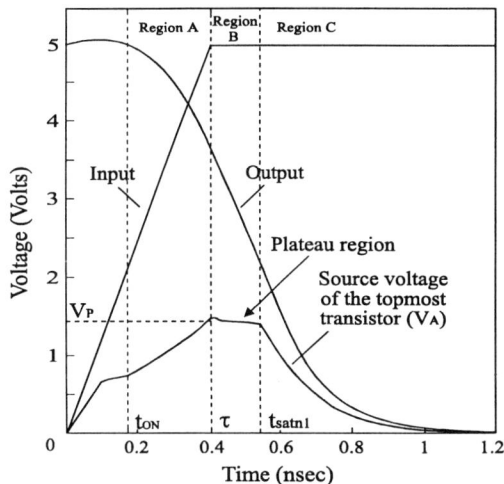

Fig.2: Voltage waveforms for fast inputs

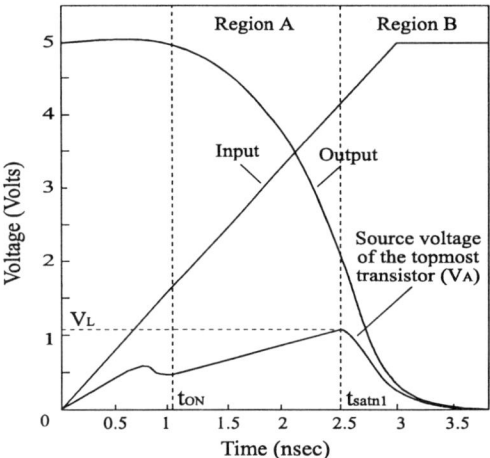

Fig.3: Voltage waveforms for slow inputs

First, the case when all the transistors are switching together, which is the worst case scenario, is analyzed. The input voltages are assumed to be ramps, with input rise time τ. When the serial array performs the discharge operation, initially the topmost transistor operates in the saturation region, while the rest operate in the linear region. In the case when the topmost transistor is still saturated after the end of the input transition (fast input transition compared with the output one), its source node (A) is charged up to a plateau voltage [4] and maintains that voltage level until all transistors enter in the linear region (Fig.2). Then it follows the output voltage of the gate to ground. Since, the transistors M_2 to M_N operate in the linear region they approximated by an equivalent transistor M_K with channel width equal to W_K

$$\frac{1}{W_K} = \frac{1}{W_2} + \frac{1}{W_3} + \cdots + \frac{1}{W_N}.$$

When the voltage of node A is in the plateau region, no current flows in or out the internal capacitance of the node. Thus, the drain currents of the transistors M_1 and M_K are equal,

$$I_{D1} = I_{DK},$$

$$P_C(W_1/L)\left[V_{DD} - V_{TO} - (1+\gamma_1)V_P\right]^{\alpha_1} = P_L(W_K/L)\left(V_{DD} - \overline{V}_{TK}\right)^{\alpha_k/2} V_P, \quad (1)$$

where V_P is the plateau voltage. $\overline{V}_{TK}$ is the average value of the threshold voltages for the transistors M_2 to M_N, an approximated value of which is given by

$$\overline{V}_{TK} = \left(\sum_{i=1}^{N-1} V_{THi}\right) / (N-1), \quad \text{where} \quad V_{THi} = V_{TO} + \gamma_1 V_{Si}, \text{ and}$$

$$V_{Si} = \frac{(N-i)(V_{DD} - V_{TO})}{2N}.$$

Note, that the parameter α_k is extracted using the channel width W_K because it depends on the transistor width [9]. In order to solve (1), a Taylor series expansion of its left part around the point $V_P = [(N-1) V_{DD} / 2N]$, up to the second order coefficient is used. After that, V_P becomes the root of a simple quadratic equation.

In the following, the voltage at node A is considered linear for the interval between the time t_{ON} and τ. t_{ON} is the time where the chain of the serial transistors starts conducting. It is calculated by analyzing the influence of the gate-drain and gate-source capacitances on the chain operation [10]. After the calculation of the time t_{ON} and the value of V_A at this time, the slope of the voltage waveform at node A can be determined.

In the case when the topmost transistor enters in the linear region before the end of the input transition (slow input transitions), its source node voltage exhibits a peak value (V_L, see Fig.3) lower than the plateau one, before the input reaches its final value. This peak value occurs when the topmost transistor is entering the linear region. SPICE simulations indicate that the slope of V_A in this case is approximately the same with that calculated assuming the existence of plateau region. For slow inputs, V_A is considered linear between the time t_{ON} and t_{satn1}. t_{satn1} is the time when the topmost transistor is entering the linear region (Fig.3).

The discharge current through the serial array when the topmost transistor operates in the saturation region, is given by

$$I_D = P_C(W_1/L)\left[V_{DD} - V_{TO} - (1+\gamma_1)V_A\right]^{\alpha_1},$$

$$I_D = P_C(W_1/L)\left[1 - \frac{(1+\gamma_1)V_A}{V_{IN} - V_{TO}}\right]^{\alpha_1}(V_{IN} - V_{TO})^{\alpha_1}.$$

The above equation has the same form as the current equation in the saturation region of a single transistor, the equivalent width of which is given by

$$W_{eq} = W_1\left[1 - \frac{(1+\gamma_1)V_A}{V_{IN} - V_{TO}}\right]^{\alpha_1} \quad (2)$$

Similarly, when the topmost transistor operates in the linear region the equivalent transistor's width becomes

$$W_{eq} = \frac{W_1}{N}\left[1 - \frac{(1+\gamma_1)(N-1)V_{OUT}}{N(V_{IN} - V_{TO})}\right]^{\alpha_1/2} \quad (3)$$

Since all transistors operate in the linear mode, the array is considered as a voltage divider ($V_A = (N-1)V_{OUT}/N$), in the above equation. The last step is to determine the value of W_{eq} in each region of the chain operation.

In the case where the plateau region exists (Fig.2), three regions are studied.

Region A, $t_{ON} \leq t \leq \tau$: The equivalent width is calculated by equation (2) at the time $t = (t_{ON} + \tau)/2$.

Region B, $\tau < t \leq t_{satn1}$: The equivalent width is calculated by equation (2) for $V_{IN} = V_{DD}$, and $V_A = V_P$. t_{satn1} is calculated by equating the output voltage expression of the inverter model [1] in region 5A with the drain saturation voltage of the topmost transistor.

Region C, $t > t_{satn1}$: The equivalent width is the average between that calculated by equation (3) for $t = t_{satn1}$ and that calculated for $V_{OUT} = 0$.

In the case when the topmost transistor is entering the linear region before the end of the input transition (Fig.3), two regions are studied.

Region A, $t_{ON} \leq t \leq t_{satn1}$: W_{eq} is calculated by equation (2) at the time $t = (t_{ON} + t_s)/2$. t_s is an approximation of t_{satn1}, and is calculated by equating the drain-source voltage ($V_{OUT} - V_A$) with the drain saturation voltage of the topmost transistor. The output voltage expression is determined by solving the differential equation resulting from the application of the Kirchoff's current law at the output node, with the assumption of negligible pMOS current. The use of t_s instead of t_{satn1} results in an error lower than 2% in the calculation of W_{eq}.

Region B, $t > t_{satn1}$: W_{eq} is calculated by equation (3), as in the region C of the previous case. t_{satn1} is calculated by equating the output voltage expression of the inverter model [1] in region 4 (or in region 3 for slower inputs) with the drain saturation voltage of the topmost transistor.

The output response of a multiple-input gate is a function of the number and the position of the switching transistors in the serial chain. When the input transition is sufficiently faster than the output, the topmost terminal switching shows faster discharge operation. This is because the lower transistors must discharge the upper transistors' internal capacitances. As the input transition becomes slower the lower terminal switching shows faster operation. This is because the transistor nearest to the ground has a smaller threshold voltage, while the magnitude of its gate-source voltage is greater than the other transistors in the chain. Hence, it will have a higher channel conductance than the other switching transistors, and the discharge operation will become faster.

The output waveforms for different combinations of switching inputs are translational [4], i.e. the shape of the curve is preserved except that its transition edge is shifted to the right or to the left depending on the combination of input signals. On the basis of this observation the equivalent width for each combination of switching inputs can be determined by multiplying the one of the worst case (all inputs switching together) with a single empirical factor m, during the input transition. m depends on the position and the number of the switching transistors, and on the relation between the input and the output waveforms. As a good metric of this relation, the single lumped parameter $G = (I_{DO}\tau)/(V_{DD}C_L)$ is used. I_{DO} is the drain current at $V_{GS} = V_{DS} = V_{DD}$ of a device with channel width equal to W/N (W: channel width of one transistor, N: number of series-connected transistors). Simulation results show that m changes exponentially with respect to G. This enables us to use the following equation for the determination of the coefficient m

$$m = m_{vf} + (m_{vs} - m_{vf})[1 - e^{-d(G-0.2)}], \quad (4)$$

where m_{vf} is the weight coefficient for very fast inputs, m_{vs} is the weight coefficient for very slow inputs, and d is a constant. m_{vf} and m_{vs} are given by a look-up table (part of which is given in Table I). The values of the look-up table are extracted from SPICE simulations by adjusting the transistor size of an inverter to give the same output with the multi-input gate in the vicinity of $V_{DD}/2$, for all the combinations of switching inputs. The values of Table I was obtained using a 0.8-micron technology, for G = 0.2 (very fast inputs) and G = 10 (very slow inputs). The constant d is equal to 0.42 for the used technology process. This value is almost independent of N, at least for the range $2 \leq N \leq 5$.

In the case of overlapping inputs, the existing waveform representation techniques [2],[5] can be used, which reduce the overlapping input signals of a multiple-input gate to a single effective signal. The equivalent channel width of parallel-connected transistors can be extracted by adding the widths of the switching transistors, and in the case of overlapping inputs can be found as in [5].

In the series-connected transistors, since the charge variation due to each of $C_{1n},...,C_{Nn}$ capacitances (see Fig.1) occurs through a different number of channels, their contribution to the output capacitance depend on their relative position in the chain. In the parallel transistors the capacitances $C_{1p},...,C_{Np}$ are added to the output load because they tied directly to the output node.

3. RESULTS AND CONCLUSIONS

In this section we illustrate the accuracy of the proposed approach for the evaluation of the transient response and the propagation delay of multiple-input CMOS static gates. In

Table I: Coefficients m_{vf} and m_{vs}

Switching Inputs	m_{vf}		m_{vs}	
	Number of serial transistors			
	3	2	3	2
1	1.12	1.07	1.41	1.29
2	1.06	1.02	1.65	1.60
1, 2	1.03	1	1.12	1
3	1.03		2.07	
1, 3	1.09		1.15	
2, 3	1.02		1.29	
1, 2, 3	1		1	

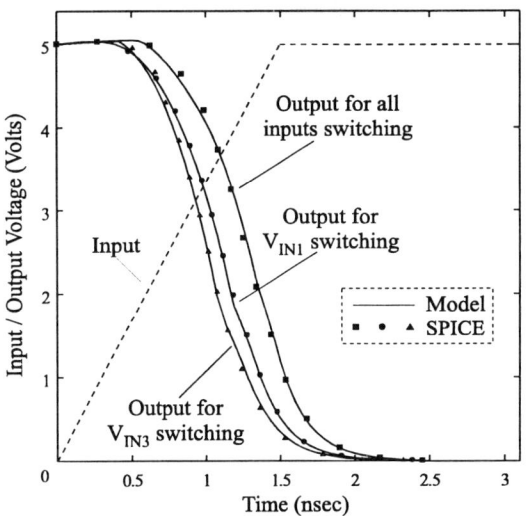

Fig.4: Output waveforms of a 3-input NAND gate

Fig.4 the output voltage waveforms of a static 3-input NAND gate for various input terminals switching with $C_L = 0.2$ pF, $\tau = 1.5$ ns and $V_{DD} = 5$ V, are shown. A 0.8-micron technology has been used, with transistor widths $W_n = 12$ μm and $W_p = 3$ μm. The output waveforms produced by SPICE simulations are added for comparison. It can be observed that the analytical waveforms are very close to those produced by SPICE simulations. This occurs because our model for the reduction of series-connected MOSFETs includes the influences of the output load, the input transition time, the number and the position of the switching inputs, and the body effect. In Fig.4 the input slope is smaller than that of the output waveforms. Thus, as mentioned in section 2, the topmost terminal (V_{IN1}) exhibits slower operation than the last one (V_{IN3}). In Fig.5, the propagation delay at the 50% voltage level of a 4-input NAND gate with the same characteristics, is plotted as a function of the input rise time. Results derived using the conventional model ($W_{eq} = W/N$) are also given. It is shown that, the conventional model gives inaccurate results, especially in the cases when only one terminal is switching.

4. REFERENCES

[1] L. Bisdounis, S. Nikolaidis, O. Koufopavlou, "Analytical transient response and propagation delay evaluation of the CMOS inverter for short-channel devices", *IEEE J. Solid-State Circuits*, vol.33, pp. 302-306, February 1998.

[2] Y.H. Jun, K. Jun, S.B. Park, "An accurate and efficient delay time modeling for MOS logic circuits using polynomial approximation", *IEEE Trans. CAD*, vol.8, pp. 1027-1032, September. 1989

[3] Y.H. Shih, Y. Leblebici, S.M. Kang, "ILLIADS: A fast timing and reliability simulator for digital MOS circuits", *IEEE Trans. CAD*, vol.12, pp.1387-1402, September. 1993.

[4] S.M. Kang, H.Y. Chen, "A global delay model for domino CMOS circuits with application to transistor sizing", *International Journal Circuit Theory & Applications*, vol.18, pp.289-306, May 1990.

[5] A. Nabavi-Lishi, N.C. Rumin, "Inverter models of CMOS gates for supply current and delay evaluation", *IEEE Trans. CAD*, vol.13, pp.1271-1279, October 1994.

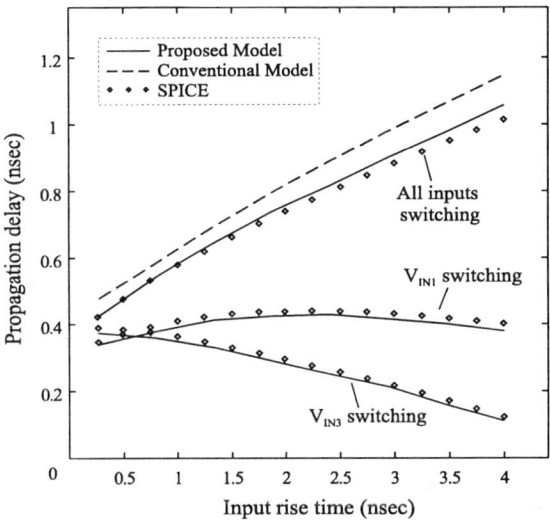

Fig.5: Propagation delay of a 4-input NAND gate

[6] T. Sakurai, A.R. Newton, "Delay analysis of series-connected MOSFET circuits", *IEEE J. Solid-State Circuits*, vol.26, pp.122-131, February 1991.

[7] J-T. Kong, D. Overhauser, "Methods to improve digital MOS macromodel accuracy", *IEEE Trans. CAD*, vol.14, pp.868-881, July 1995.

[8] D. Deschacht, M. Robert, D. Auvergne, "Synchronous-mode evaluation of delays in CMOS structures", *IEEE J. Solid-State Circuits*, vol.26, pp.789-795, May 1991.

[9] T. Sakurai, A.R. Newton, "Alpha-power law MOSFET model and its applications to CMOS inverter delay and other formulas", *IEEE J. Solid-State Circuits*, vol.25, pp. 584-594, April 1990.

[10] A. Chatzigeorgiou, S. Nikolaidis, "Collapsing the transistor chain to an effective single equivalent transistor", in Proc. Design, Automation and Test in Europe, February 1998.

A NEW CURVE FITTING TECHNIQUE FOR ANALYSIS OF FREQUENCY-DEPENDENT LOSSY TRANSMISSION LINES

Yuichi TANJI, Yoshifumi NISHIO and Akio USHIDA

Dept. of Electrical and Electronic Engineering, Tokushima University
2-1 Minami-Josanjima, Tokushima 770, JAPAN
E-mail: tanji@ee.tokushima-u.ac.jp

ABSTRACT

Analysis of frequency-dependent lossy transmission lines is very important for designing the high-speed VLSI, MCM and PCB. The frequency-dependent parameters are always obtained as tabulated data. In this paper, a new curve fitting technique of the tabulated data for the moment matching technique in interconnect analysis are presented. This method based on Chebyshev interpolation enhances the efficiency of the moment matching technique.

1. INTRODUCTION

The high-speed performance of microwave or digital circuit systems is limited by the interconnect effects rather than the switching speed of semiconductor devices. When the operating frequency is increase, the current density of conductor tends to be great around the surface of the conductor. Due to the high packing density, the interconnects such as VLSI, MCM and PCB are closely placed on each other, and the current density is also great at the near side between conductors. They are known as the skin effect and proximity effect [1], respectively, thus the interconnects of high-speed integrated circuits have frequency-dependent characteristics. The frequency-dependent parapeters are always obtained by any numerical procedure and as tabulated data in real frequency. Therefore, the analysis of frequency-dependent lossy transmission lines with tabulated data is very important for accurate analysis VLSI's circuits, MCM and PCB.

For such analysis, FFT based algorithm is very accurate. However, this method is not useful from computational point of view, because the system to be analyzed contain very large number of transmissions lines, and FFT based algorithm needs large number of data points. The moment matching technique [3], [4] are efficient and accurate for the interconnect analysis. Recently, these methods are extended to the frequency-dependent case [5], [6]. Since the moment matching techniques are essentially Padé approximation of any Laplace functions, if any transfer function are described in power series of complex s, the moment matching technique can be applied to the analysis. Thus the key technique in reference [5], [6] is how the tabulated data in real frequency is described in power series of complex s, and the piecewise polynomial approximation in [6] and the least square approximation in [5] are used.

In this paper, we provide a new curve fitting technique for the moment matching scheme in interconnect analysis. The proposed method is based on Chebyshev interpolation technique. Chebyshev polynomial is considered as an almost minimax approximate polynomial. Hence the proposed method based on Chebyshev interpolation gives a good approximation than one by means of the least square fitting [5]. Moreover, since the proposed method does not require for any matrix operation, this method does not suffer from the singularity problem in the least square fitting [7]. The polynomial must be constructed as having real coefficients due to realistic impedance or admittance functions. The discrete orthogonal property of Chebyshev polynomial allows us to construct the continuous polynomial with real coefficients, different from the piecewise one in [6].

In the numerical examples, the proposed method gives reliable results to the tabulated data in real frequency.

2. FREQUENCY-DEPENDENT LOSSY TRANSMISSION LINES

The frequency-dependent transmission lines are described by the Telegrapher's equation in the Laplace-domain:

$$\frac{d}{dx} \begin{bmatrix} \mathbf{V}(s,x) \\ \mathbf{I}(s,x) \end{bmatrix} = \mathbf{D}(s) \begin{bmatrix} \mathbf{V}(s,x) \\ \mathbf{I}(s,x) \end{bmatrix} \quad (1)$$

where

$$\mathbf{D}(s) = \begin{bmatrix} 0 & -\mathbf{Z}(s) \\ -\mathbf{Y}(s) & 0 \end{bmatrix}$$

$$\mathbf{Z}(s) = \mathbf{R}(s) + s\mathbf{L}(s), \quad \mathbf{Y}(s) = \mathbf{G}(s) + s\mathbf{C}(s).$$

The parameters $\mathbf{R}(s)$, $\mathbf{L}(s)$, $\mathbf{C}(s)$, $\mathbf{G}(s)$ are per unit length registance, inductance, capacitance, conductance matrices, respectively, and these matrices are arbitrary functions of complex s. Actually, these matrices are not given as a funtion of complex s, but also tabulated data to some points, $j\omega_i$'s on the imaginary axis.

In this paper, our aim is how to apply the moment matching scheme [3], [4] for solving (1). If any transfer functions are described in power series of complex s, we can apply the moment matching scheme to the analysis, because the moment matching scheme is essentially Padé approximation. Hence, the input($x = 0$)-output($x = l$) relations of transmission lines are described in power series of complex s is the key technique. Assuming that the parameter matrices $\mathbf{R}(s)$, $\mathbf{L}(s)$, $\mathbf{C}(s)$, $\mathbf{G}(s)$ are power series of complex s, the matrix exponential method [5] [6] is very powerful technique to describe the input-output relation. Here we briefly modify the matrix exponential method to increase the efficiency.

Applying the matrix exponential method, the input output relation of Eq. (1) is given by

$$\begin{aligned}\left[\begin{array}{c}\mathbf{V}(s,l)\\ \mathbf{I}(s,l)\end{array}\right] &= \exp(\mathbf{F}(s)l)\left[\begin{array}{c}\mathbf{V}(s,0)\\ \mathbf{I}(s,0)\end{array}\right]\\ &= \sum_{n=0}^{\infty}\frac{1}{n!}(\mathbf{F}(s)l)^n\left[\begin{array}{c}\mathbf{V}(s,0)\\ \mathbf{I}(s,0)\end{array}\right]\\ &= \sum_{n=0}^{\infty}\mathbf{T}_ns^n\left[\begin{array}{c}\mathbf{V}(s,0)\\ \mathbf{I}(s,0)\end{array}\right].\end{aligned} \quad (2)$$

In reference [3], the convergence of infinite series (2) is illustrated grately depending on the length l of transmission lines. If the convergence is smooth, the transmission lines must be divided in some regin. (This implies that $\exp(\mathbf{F}(s)l_1)$ rapidly converges than $\exp(\mathbf{F}(s)l_2)$, if $l_1 < l_2$.) In this case, the following relation is very useful in order to get the whole characteristics of transmission lines:

$$\exp(\mathbf{F}(s)l) = \exp(\mathbf{F}(s)\frac{l}{2})\cdot\exp(\mathbf{F}(s)\frac{l}{2}). \quad (3)$$

However, dividing the transmission lines requires for more computational cost, because this means that some equations are added to the circuit equation. Alternatively, $\exp(-\mathbf{F}(s)\frac{l}{2})$ is multiplied from left side of Eq. (2) instead of dividing, and we can get the following relations:

$$\exp(-\mathbf{F}(s)\frac{l}{2})\left[\begin{array}{c}\mathbf{V}(s,l)\\ \mathbf{I}(s,l)\end{array}\right] = \exp(\mathbf{F}(s)\frac{l}{2})\left[\begin{array}{c}\mathbf{V}(s,0)\\ \mathbf{I}(s,0)\end{array}\right]. \quad (4)$$

The relation (4) represents continuity of the voltages and currents at the center point of the transmission lines, whereas the relation (2) gives a relation of the output variables to the input. This means that the relation (4) is more effective than (2), because the complexity depends on the length of the transmission lines.

Assuming $\mathbf{F}(s)$ as M degree matrix polynomial of complex s, the coefficients of the matrix exponential (2) is obtained in recursive mannar [5], [6]:

$$\mathbf{T}_n = \begin{cases}\sum_{j=-1}^{\infty}\mathbf{T}_{0,j} & (n=0)\\ \sum_{j=\mathrm{int}(\frac{i-1}{M})}^{\infty}\mathbf{T}_{i,j} & (n\neq 0)\end{cases} \quad (5)$$

where

$$\mathbf{T}_{i,j} = \frac{l}{j+1}\sum_{k=0}^{\min(i,M)}\mathbf{F}_k\mathbf{T}_{i-k,j-1}$$
$$(i=0,\ldots,jM, j\neq 0)$$

$$\mathbf{T}_{i,j} = \frac{l}{j+1}\sum_{k=i-jM}^{M}\mathbf{F}_k\mathbf{T}_{i-k,j-1}$$
$$(i=jM+1,\ldots,(j+1)M, j\neq 0)$$

$$\mathbf{T}_{i,0} = \mathbf{F}_il \quad (i=0,\ldots,M)$$
$$\mathbf{T}_{i,0} = \mathbf{I}.$$

$\exp(-\mathbf{F}(s)l)$ can be calculated by multiplying $\mathbf{T}_{i,j}$ by $(-1)^{j+1}$. Although the matrix exponential (2) congerves after 40-50 terms, $\exp(\mathbf{F}(s)l/2)$ rapidly converges than $\exp(\mathbf{F}(s)l)$, because the length of transmission lines is half.

Interchanging the elements of (4), we can get the ports relation of the transmission lines as follows:

$$\sum_{n=0}^{\infty}\mathbf{P}_ns^n\left[\begin{array}{c}\mathbf{V}(s,0)\\ \mathbf{V}(s,l)\end{array}\right] + \sum_{n=0}^{\infty}\mathbf{Q}_ns^n\left[\begin{array}{c}\mathbf{I}(s,0)\\ \mathbf{I}(s,l)\end{array}\right] = 0. \quad (6)$$

3. CHEBYSHEV INTERPOLATION SCHEME OF FREQUENCY-DEPENDENT PARAMETERS

In the previous section, the matrix exponential method is applied to the moment generation of lossy transmission lines with frequency-dependent parameters. Here, it is assumed that the parameters given as tabulated data to some point on imaginary axis are able to write in power series of complex s. So, the procedure for making the power series from the tabulated data is provided in this section.

3.1. Curve Fitting Algorithm

Let $r(s)$, $l(s)$, $c(s)$, $g(s)$ be (i,j) element of $\mathbf{R}(s)$, $\mathbf{L}(s)$, $\mathbf{C}(s)$, $\mathbf{G}(s)$, respectively, where these values are given as tabulated data to some points, $j\omega_i$'s on the imaginary axis. The (i,j) element $z(s) = r(s) + jsl(s)$ of the series impedance matrix and $y(s) = g(s) + jsc(s)$ of the parallel admittance matrix are determined so that they satisfies

$$z(j\omega_i) \approx \sum_{k=0}^{N}z_k(j\omega_i)^i, \quad y(j\omega_i) \approx \sum_{k=0}^{N}y_k(j\omega_i)^i \quad (7)$$
$$(i = 0, 1, \ldots, N)$$

where N is the number of the data. Moreover the coefficients z_k and y_k are assumed as real numbers, which is a reasonable assumption due to realistic impedance or admittance functions.

Let us consider $z(s)$ only, and $y(s)$ can be obtained by the same procedure. In reference [5], eliminating the lossless part of l(s) is introduced in order to approximate accurately, namely, the lossless part $l(\infty)$ is separated from $l(j\omega)$ such as

$$l'(j\omega) = l(j\omega) - l(\infty). \quad (8)$$

Then, $z'(j\omega_i) = r(j\omega_i) + j\omega_il'(j\omega_i)$ is intepolated by the Chebyshev series. A transform $x = \omega/\omega_m$ is used to convert $\omega \in [0,\omega_m]$ into $x \in [0,1]$. Assuming $z'(-j\omega)$ is complex conjugate to $z'(j\omega)$, the interpolated polynomial is obtained by

$$z'(j\omega_m x) = \sum_{k=0}^{N-1}{}'a_kT_k(x) \quad (9)$$

where the symbol $\sum'$ denotes the summation with the first component divided by 2 and $T_k(x) = \cos k\theta$. From the discrete orthogonal property of Chebyshev polynomial, a_k ($k = 0, 1 \ldots, N-1$) are given as follows:

if N is odd,

$$a_k = \begin{cases} \dfrac{2}{N} \sum_{i=0}^{\frac{N-3}{2}} \{2r(j\omega_m \cos\theta_i)\cos k\theta_i \\ \qquad\qquad + r(0)\cos\frac{k\pi}{2}\} & \text{(k: even)} \\ j\dfrac{2}{N} \sum_{i=0}^{\frac{N-3}{2}} 2\omega_m \cos\theta_i l(j\omega_m \cos\theta_i)\cos k\theta_i \\ & \text{(k: odd)} \end{cases}$$
(10.a)

if N is even,

$$a_k = \begin{cases} \dfrac{2}{N} \sum_{i=0}^{\frac{N}{2}-1} 2r(j\omega_m \cos\theta_i)\cos k\theta_i & \text{(k: even)} \\ j\dfrac{2}{N} \sum_{i=0}^{\frac{N}{2}-1} 2\omega_m \cos\theta_i l(j\omega_m \cos\theta_i)\cos k\theta_i \\ & \text{(k: odd)} \end{cases}$$
(10.b)

where $\theta_i = \frac{2i+1}{N+2}\pi$ $(i=0,1,\ldots,N-1)$ are the Chebyshev points. Note that if N is even number, the information at $s=0$ does not reflect the fitting curve. Hence, N is prefer to be odd number.

The coefficients of a_k are real part or imaginary part only, thus we can derive the power series of $j\omega_m x$ having real coefficients. First, $z'(j\omega_m x)$ of (9) is converted into a power series with respect to x. Using the recurrence formula of Chebyshev polynomial,

$$\begin{cases} T_0(x) = 1, \quad T_1(x) = 1, \\ T_{k+1}(x) = 2xT_k(x) - T_{k-1}(x), \end{cases} \quad (11)$$

Chebyshev polynomial $T_k (i=2,3,\ldots)$ is obtained by

$$\begin{aligned} T_2(x) &= 2x^2 - 1, \\ T_3(x) &= 4x^3 - 3x, \\ T_4(x) &= 8x^4 - 8x^2 + 1, \\ T_5(x) &= 16x^5 - 20x^3 + 5x, \\ T_6(x) &= 32x^6 - 48x^4 + 18x^2 - 1, \\ &\vdots \end{aligned} \quad (12)$$

As a result, the finite Chebyshev series (9) is converted into a power series with x:

$$z'(j\omega_m x) = \sum_{k=0}^{N-1} b_k x^k. \quad (13)$$

From (12), $T_{2m}(x)$ and $T_{2m+1}(x)$ are even and odd functions, respectively. Thus, b_{2m} and b_{2m+1} in (13) are respectively real and imaginary part only as a_{2m} and a_{2m+1} in (9). Consequently, $z'(j\omega_m x)$ is expressed in power series of $j\omega_m x$ with real coefficients:

$$z'(j\omega_m x) = \sum_{k=0}^{N-1} z'_k (j\omega_m x)^k, \quad (14)$$

Table 1. Coefficients of power series given by the proposed method.

	value		value
c_0	3.448	c_6	-3.292×10^{-8}
c_1	4.670	c_7	1.956×10^{-10}
c_2	-1.684×10^{-2}	c_8	-1.601×10^{-11}
c_3	5.196×10^{-4}	c_9	3.295×10^{-14}
c_4	-3.196×10^{-7}	c_{10}	-2.917×10^{-15}
c_5	4.447×10^{-7}		

where

$$z'_k = \begin{cases} (-1)^{\frac{k}{2}} \dfrac{b_k}{\omega_m^k} & \text{(k: even)} \\ (-1)^{\frac{k+1}{2}} \dfrac{jb_k}{\omega_m^k} & \text{(k: odd)} \end{cases}$$

From (8), (14), a element of the series impedance matrix of transmission lines is described by

$$z(s) = z_0 + s(z_1 + l(\infty)) + \sum_{k=2}^{N-1} z'_k s^k. \quad (15)$$

where all coefficients of s^k are real numbers.

3.2. Shifted Coefficients of Power Series

When the multi-point Padé approximation [3], [4] is used to getting dominant poles, the shifted moments, the coefficients of Taylor expansion at an arbitrary point s_k is needed. Thus, in section II $F(s)$ in (2) and the matrix exponential must be a matrix polynomial of complex $\sigma = s - s_k$.

Let be $\mathbf{F}(s)$ M degree matrix polynomial as (2), then $\mathbf{F}(s)$ is convered into a matrix polynomail of complex $\sigma = s - s_k$:

$$\mathbf{F}(s) \equiv F(\sigma) = \sum_{m=0}^{M} \sum_{i=0}^{m} \binom{m}{i} \mathbf{F}_m s_k^{m-i} \sigma^i. \quad (16)$$

4. NUMERICAL EXAMPLES

To show the efficiency of our method, the 3-conductors transmission lines provided by M. Celik are considered. The frequency-dependent parameters are listed as the tables II, III in reference [5]. Using the proposed method in Sect. 3, (2,2) element of the series impedance matrix is given by the 10-degree power series of complex s:

$$z_{22}(s) = \sum_{i=0}^{10} c_i s^i \quad (17)$$

where each coefficient is listed in Table 1.

For comparison, the values of z_{22} in $s = j\omega$ and the tabulated data are shown in Fig. 1. In this figure, the frequencies range is from 0 to 7 GHz. The proposed method gives a reliable result. The time- and frequency-domain responses in the example [5] can be calculated by the matrix exponential method in Sect. 2 and complex frequency hopping [3]. The circuit includes two identical 3-conductor transmission lines, 4 resisters, 7 capacitors and 1 inductors. Transient responses to a pulse input (0.8 [ns] pulse width, 0.1 [ns] rise

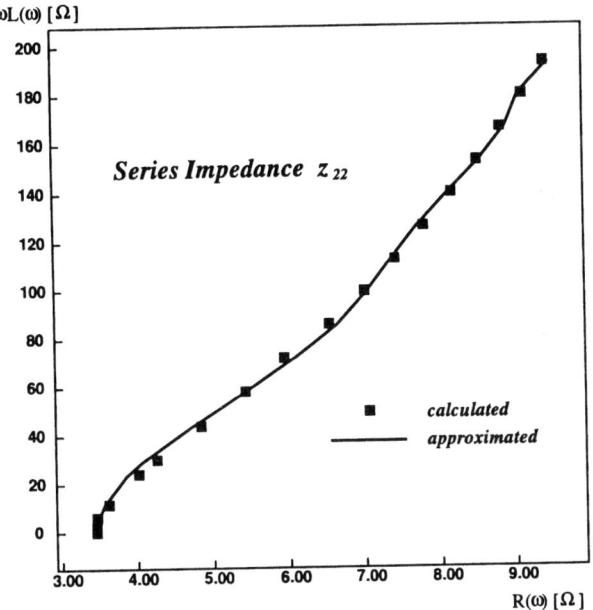

Figure 1. (2,2) element of the series impedance matrix of the transmission lines provided by M. Celik [3].

and fall time) and frequency response to a impulse input are shown in Fig. 1, 2, respectively. These result are compared with the result by the frequency-domain method [8] and single Padé approximation [2]. Here, in complex frequency hopping, the maximum frequency is selected by 5 [GHz] and 9 expansion points is considered.

5. CONCLUSIONS

A new curve fitting technique for analysis of frequency-dependent lossy transmission lines have been presented. This method is efficiently incorporated with the moment matching technique [3], [4]. Although the object of this paper is turned to the moment mathing technique, this method is easyly applied to the method of characteristics by means of a technique in reference [9]. This is our future work.

REFERENCES

[1] L. T. Hwang and I. Turlik, " A review of the skin effect as applied to thin film interconnects," *IEEE Trans. Comp. Hybrids Manuf. Technol.*, vol. 15, vol. 15, no. 1, pp. 43-54, Feb. 1992.

[2] L. T. Pillage and R. A. Rohrer, "Asymptotic waveform evaluation for timing analysis," *IEEE Trans. Computer-Aided Design*, vol. 9, no. 4, April 1990.

[3] E. Chiprout and M. S. Nakhla, "Analysis of interconnect networks using complex frequency hopping (cfh)," *IEEE Trans. Computer-Aided Design*, vol. 14, no. 2, Feb. 1995.

[4] M. Celik, O. Ocali, M. A. Tan, and A. Atalar, "Improving awe accuracy using multipoint padé, approximation" *Proc. ISCAS'94*, vol. 1, pp. 379-382, 1994.

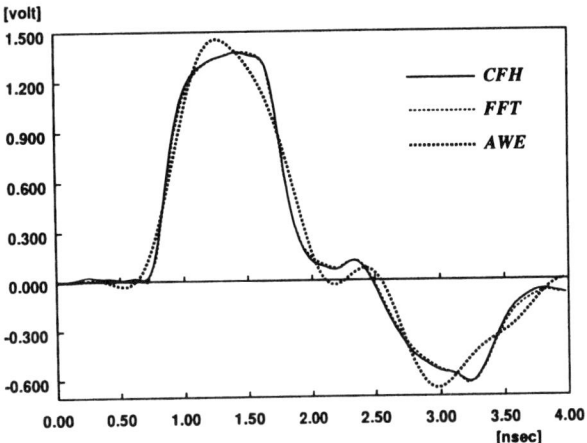

Figure 2. Transient response to a pulse input.

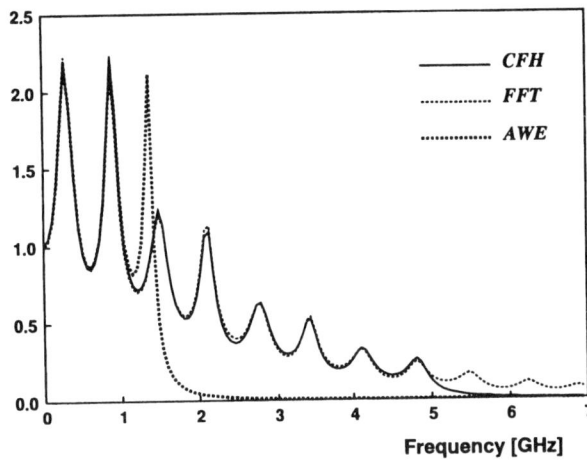

Figure 3. Frequency response to a impulse input.

[5] M. Celik and A. C. Cangellaris "Efficient transient simulation of lossy packaging interconnects using moment-matching techniques," *IEEE Trans. Comps., Pack., & Manuf. Technol., Part-B*, vol. 19, no. 1, pp. 64-73, Feb. 1996.

[6] R. Khazaka, J. Poltz, M. Nakhla, Q. J. Zhang, "A fast method for the simulation of lossy interconnects with frequency dependent parameters," *Proc. IEEE Multi-Chip Module Conf.*, pp. 95-98, Feb. 1996.

[7] G. H. Golub and C. F. Van Loan, *Matrix Computation*, the Johns Hopkins University Press, 1983.

[8] Y. Tanji, L. Jiang and A. Ushida, "Analysis of pulse responses of multi-conductor transmission lines by a partitioning technique," *IEICE Trans. on Fundamentals*, vol. E77-A, No. 12, pp. 2017-2027, Dec. 1994.

[9] T. Watanabe, A. Kamo and H. Asai, "Time-domain simulation of lossy coupled transmission lines based on delay evaluation technique," *Proc. Euro. Conf. on Circuits Theory and Design*, vol. 2, pp. 517-520, Sept. 1997.

TWO-POLE APPROXIMATION FOR HIGH SPEED INTERCONNECT DESIGN

Jianhua Shao and Richard M M Chen

Department of Electronic Engineering
City University of Hong Kong
83 Tat Chee Avenue, Kowloon, Hong Kong
Email: jhshao@ee.cityu.edu.hk

ABSTRACT

In this paper, an approach to find two approximate poles for high speed interconnect design is presented. The approach overcomes the instability problem associated with Padé approximation. These two poles are used for rapid evaluation of line parameter changes. Since transmission line system is a stiff system with some poles of small magnitude and some poles of large magnitude, in general, it is impossible to match the simulated waveform with the calculated waveform using small number of poles. The objective here is to evaluate efficiently the effect of parameter changes in our optimization process. Numerical examples show that the proposed approach can meet our objective.

1 INTRODUCTION

Signal delay is largely affected by interconnects than by gates in high performance systems and inductance effect becomes significant and important. Interconnects are modeled as lossy transmission lines to reflect the effect of inductance.

Due to the distributed nature of interconnects, poles of such systems are transcendental and infinite in number. Since the second order system is the simplest one that can reflect the effect of inductance, many researchers have attempted to capture the system behavior using two-pole approximation [1-6]. Padé approximation is one popular approach utilized, which converts a power series into a rational function. The other approach uses lumped elements to approximate the interconnects, which is computationally expensive.

There are two problems associated with Padé approximation. One is numerical instability and the other is inherent instability. Both generate positive poles for passive systems. Although the first problem can be solved to some extent by utilizing extended precision, the second problem is more difficult to handle. A method to enhance the stability by employing higher order moments has been proposed in [7].

In this paper, an approach to find two approximate poles for interconnect design is presented. The proposed approach overcomes the instability problem associated with Padé approximation.

2 TWO-POLE APPROXIMATION

Moments of a time-domain waveform, $v(t)$, are defined via the Laplace transformation of the waveform as follows:

$$V(s) = \int_0^\infty e^{-st} v(t)\, dt = m_0' + m_1' s + m_2' s^2 + \cdots \quad (1)$$

where m_k', $k = 0, 1, 2, \cdots$, are the Maclaurin series coefficients of $V(s)$, and the k-th moment is

$$m_k' = \frac{(-1)^k}{k!} \int_0^\infty t^k v(t)\, dt \quad (2)$$

We calculate the moments utilizing the method proposed by Yu and Kuh [8].

In a high performance system, input signals undergo a pure delay $\tau = \sum_{i=1}^n d_i \sqrt{L_i C_i}$ to arrive at the output end, where n is the number of lines on the signal path, L_i and C_i are the inductance and capacitance of the i-th line per unit length respectively, and d_i is the length of the corresponding line.

When a rational function is used to approximate the delay in frequency domain, 20 or more poles may be needed [9]. In order to approximate the system behavior better, the pure delay must be excluded from the moments. The new moments have the following relation with original moments:

$$(m_0 + m_1 s + \cdots)e^{-s\tau} = m_0' + m_1' s + \cdots \quad (3)$$

which results:

$$\begin{aligned}
m_0 &= m_0' \\
m_1 &= m_1' + m_0' \tau \\
m_2 &= m_2' + m_1' \tau + \frac{1}{2} m_0' \tau^2 \\
&\cdots
\end{aligned} \quad (4)$$

For a system with N poles, the k-th moment has the following relation with poles and residues:

$$m_k = \sum_{i=1}^N \frac{k_i}{p_i^{k+1}} \quad (5)$$

where p_i and k_i are the i-th pole and its corresponding residue respectively. Using Padé approximation, $2N$ moments are required to find the approximate poles and residues.

By (5), we can see that poles of large magnitude contribute less to high order moments. So moment-matching using higher order moments tends to find the actual poles of smaller magnitude. Furthermore, the pole of the smallest magnitude can be found as:

$$p_1 = \lim_{k \to \infty} \frac{m_k}{m_{k+1}} \qquad (6)$$

where $|p_1| < |p_2| < \cdots < |p_N|$ [7].

For RC circuits, all poles are real, and the ratios m_k/m_{k+1} monotonically approache the value of p_1 [10]. However, for RLC circuits, the ratios of successive moments do not have the similar properties when the inductance effect is significant.

Considering the circuit shown in Fig. 1, parameters for Line 1 are $d = 6cm$, $R = 3.06\Omega/cm$, $L = 1.30nH/cm$, $C = 2.98\ pF/cm$, and for Line 2 are $d = 5cm$, $R = 15.00\Omega/cm$, $L = 3.33nH/cm$, $C = 1.15pF/cm$. Fig. 2 shows the simulated step response using Spice3f5 together with the response of a 10-order system obtained using Padé approximation. 20 moments from m_0 to m_{19} are used for the approximation. We can see that the calculated response of the 10-order system cannot match with the simulated waveform exactly.

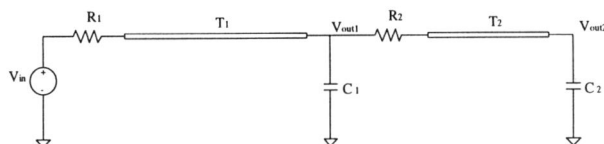

Fig. 1 Example Circuit
$R_1 = 1.6\Omega, R_2 = 8.9\Omega, C_1 = 10pF, C_2 = 5pF$.

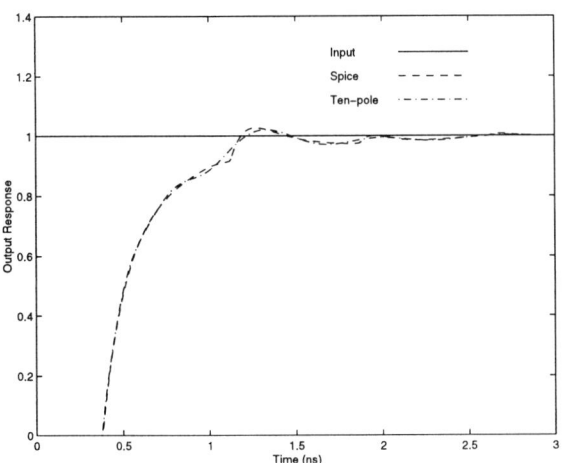

Fig. 2 Output Responses at Vout1.

Table 1 lists the 10 approximate poles, residues, and the first 11 moments and their successive ratios. The line parameters L and C are scaled by 10^9 to overcome the precision problem. Because the condition number of the moment matrix is very large, the result is not exact.

Transmission line system is a stiff system with some poles of small magnitude and some poles of large magnitude. In general, it is impossible to match the simulated waveform with the calculated waveform using small number of poles. Usually, the response is dominated by poles of small magnitude. However, residues also play an important role in affecting the response. Poles with small residues cannot dominate the response as we can see from p_{10} in Table 1.

Table 1 Poles, Residues, Moments and Ratios

i (k)	p_i	k_i	m_k	m_k/m_{k+1}
0	nil	nil	1	-4.28
1	-55.3	3.16+4.83e-41i	-2.33e-1	-3.60
2	-4.94-17.6i	0.610+0.575i	6.48e-2	-3.64
3	-4.94+17.6i	0.610-0.575i	-1.78e-2	-9.36
4	-1.96-3.17i	0.426+0.465i	1.90e-3	0.625
5	-1.96+3.17i	0.426-0.465i	3.04e-3	-0.895
6	-1.90-0.830i	0.372+0.254i	-3.40e-3	-1.46
7	-1.90+0.830i	0.372-0.254i	2.33e-3	-1.85
8	-1.50-9.38i	0.270+0.467i	-1.26e-3	-2.22
9	-1.50+9.38i	0.270-0.467i	5.66e-4	-2.73
10	-1.60e-12	6.62e-54-3.85e-94i	-2.07e-4	nil

When two-pole approximation is used, we can write the transfer function H(s) as:

$$H(s) = \frac{\omega_n^2}{s^2 + 2\zeta\omega_n s + \omega_n^2} \qquad (7)$$

where ω_n is the natural undamped frequency and ζ is the damping ratio [11]:

$$\begin{aligned} \omega_n &= \sqrt{p_1 p_2} \\ \zeta &= -\frac{1}{2}\frac{p_1 + p_2}{\sqrt{p_1 p_2}} \end{aligned} \qquad (8)$$

The damping condition is controlled by ζ. $0 < \zeta < 1$, $\zeta = 1$, and $\zeta > 1$ correspond to underdamped, critically damped, and overdamped responses respectively. For underdamped case, the percent overshoot is

$$PO = e^{-\frac{\zeta\pi}{\sqrt{1-\zeta^2}}} \qquad (9)$$

An optimization scheme based on the two parameters ζ and ω_n has been proposed in [12]. A small $\zeta (< 0.5)$ means large overshoot and a large $\zeta (> 1.5)$ represents slow response. A larger ω_n leads to a shorter delay time when ζ is specified. Consequently, in order to have fast response with acceptable amount of overshoot, we want ω_n to be large and ζ to be close to 1.

Generally, the line parameters, resistance, inductance, and capacitance, are functions of the line width. The design process is to change the line widths to maximize ω_n while ζ remaining in a specified range. The optimization process needs an efficient way to evaluate the effect of line parameter changing.

The objective here is to find two approximate poles that can reflect the properties of the response, giving a smaller ζ when the overshoot is larger and a larger ζ when the response is slower. We do not attempt to match the simulated waveform with the waveform of a two-pole approximation since the matching generally will fail.

When the inductance effect is significant, the ratios of successive moments do not converge. Since complex poles always exist in a transmission line system, we assume that the two approximate poles are complex conjugate pair.

Using (5), the ratio of two successive moments is

$$\frac{m_k}{m_{k+1}} = p_1 \frac{1 + \frac{k_2}{k_1}\left(\frac{p_1}{p_2}\right)^{k+1} + \cdots + \frac{k_N}{k_1}\left(\frac{p_1}{p_N}\right)^{k+1}}{1 + \frac{k_2}{k_1}\left(\frac{p_1}{p_2}\right)^{k+2} + \cdots + \frac{k_N}{k_1}\left(\frac{p_1}{p_N}\right)^{k+2}} \quad (10)$$

Let p_1 and p_2 be the two approximate poles that we want to find and the corresponding k_1 and k_2 are much larger than other residues so that the two poles are dominant. For a second order system, we have $k_1 = -k_2$ because $h|_{t=0} = 0$. We also assume that the other poles are of much larger magnitude. Representing the poles using polar coordinates, we have $p_1 = r\,e^{i\theta}$ and $p_2 = r\,e^{-i\theta}$. As a result, (10) becomes:

$$\frac{m_k}{m_{k+1}} = p_1 \frac{1 - e^{i\theta(2k+2)}}{1 - e^{i\theta(2k+4)}} \quad (11)$$

Because using high order moments tends to find the actual poles of small magnitude, we only use the low order moments, which can be calculated accurately and efficiently.

Eliminating p_1 using m_k/m_{k+1} and m_{k+1}/m_{k+2}, we have:

$$\frac{m_k\,m_{k+2}}{m_{k+1}\,m_{k+1}} = \frac{(1 - e^{i\theta(2k+2)})(1 - e^{i\theta(2k+6)})}{(1 - e^{i\theta(2k+4)})(1 - e^{i\theta(2k+4)})} \quad (12)$$

Take $k = 4$, the right side of (12), represented as $f(\theta)$, becomes:

$$f(\theta) = 1 - \frac{1}{4\alpha^2(256\alpha^8 - 512\alpha^6 + 352\alpha^4 - 96\alpha^2 + 9)} \quad (13)$$

where $\alpha = cos(\theta)$. Because the poles are in the left plane, we can first restrict θ in the range of $(\pi/2, \pi)$. For $2.619 < \theta < 3.141$, $f(\theta)$ monotonically increases from -6831.4 to 0.97222. Thus, we can search for θ in this range according to the ratio of m_4/m_5 and m_5/m_6. The magnitude r is obtained using (11). If m_4/m_5 gives negative magnitude, m_5/m_6 should be used. Then, we have both p_1 and p_2.

Because of the searching range we used, ζ is very close to 1. Therefore, a correction term is needed. The current setting is $-400000\,e^{(10-30\zeta^2)}$ for $\zeta \leq 0.93$.

It is also possible to use a larger k value. Since a larger k value results in a complex conjugate pair closer to the real axis, more correction to ζ is needed.

When the line parameters, resistance and capacitance, are dominant, the ratios of successive moments will converge, the dominant poles are real and the above approach will not work. For this case, we can use the method proposed by Tutuianu instead [10].

3 NUMERICAL EXAMPLES

Next, we demonstrate the effectiveness of our method by changing the line parameters of the circuit in Fig. 1. We consider three cases and the corresponding output responses are shown in Fig. 3, Fig. 4, and Fig. 5, respectively. The results are summarized in Table 2 and compared with the results of Padé approximation. We use moments m_4, m_5, m_6, and m_7 for Padé approximation.

In Table 2, R, L, and C are line parameters. $p_{1,2}$ represent the two approximate poles. The natural undamped frequency and the damping ratio are obtained using (8). Since in our method, a correction to ζ is needed, ζ' gives the corrected ζ for our method and the calculated ζ for Padé approximation. PO' lists the percent overshoot obtained according to (9) using ζ', and PO gives the percent overshoot obtained from the simulated results.

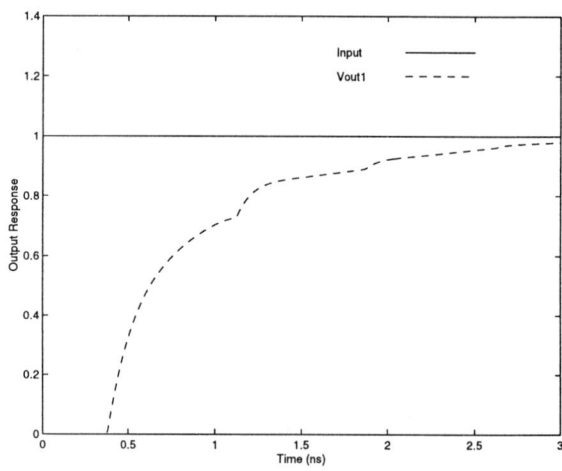

Fig. 3 i) Slow Response.

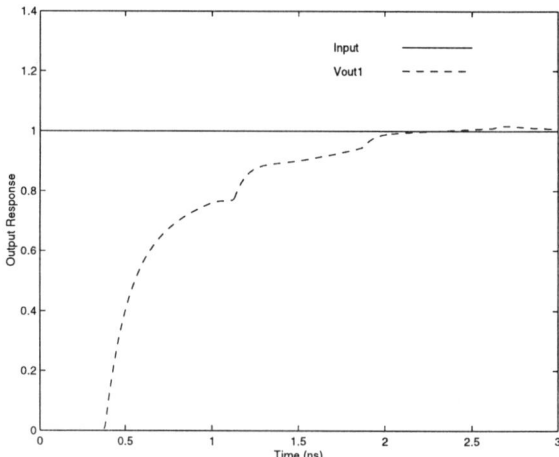

Fig. 4 ii) Response with Small Amount of Overshoot.

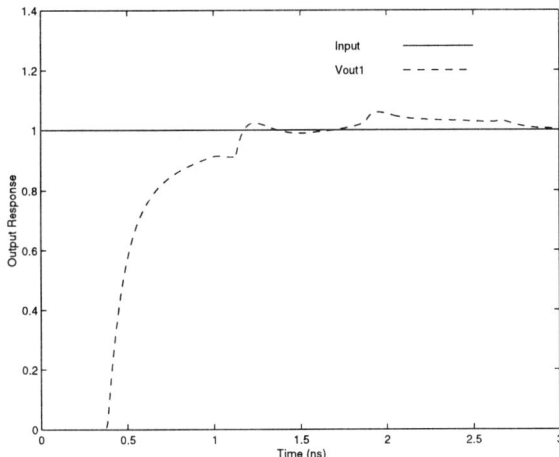

Fig. 5 iii) Response with Larger Amount of Overshoot.

4 CONCLUSIONS

From Table 2, we can see that Padé approximation is more accurate for slow response and less accurate for fast response while our method can evaluate all three cases appropriately. Since slightly underdamped response is good to system performance, we intend to reach such a situation, which will help us to choose a complex conjugate pair as the two approximate poles. For Padé approximation, there is no definite choice of moments, it is a process of trial-and-error. For our method, the choice of moments is determined. The proposed method has been used in our interconnect optimization scheme for rapid evaluation of design parameters.

ACKNOWLEDGMENT

The authors would like to thank City University of Hong Kong for the support to this research work.

REFERENCES

[1] Y. Sugiuchi, B. Katz, and R. A. Rohrer, "Interconnect Optimization Using Asymptotic Waveform Evaluation (AWE)," in Proc. MCMC 94, pp.120-125, 1991.

[2] D. Zhou, S. Su, F. Tsui, D. S. Gao, and J. S. Cong, "A Simplified Synthesis of Transmission Lines with a Tree Structure," Int. J. Analog Integrated Circuits Signal Process., pp.19-30, Jan., 1994.

[3] J. Wang and W Dai, "Optimal Design of Self-Damped Lossy Transmission Lines for Multichip Modules," in Proc. ICCD, pp.594-598, Oct., 1994.

[4] A. B. Kahng and S. Muddu, "Optimal Equivalent Circuits for Interconnect Delay Calculations Using Moments," in Proc. European DAC, pp. 164-169, 1994.

[5] A. B. Kahng and S. Muddu, "Two-Pole Analysis of Interconnect Tree," in Proc. MCMC 94, pp. 105-110, 1994.

[6] T. Xue, E. S. Kuh, and Q. J. Yu, "A Sensitivity-Based Wiresizing Approach to Interconnect Optimization of Lossy Transmission Line Topologies," in Proc. MCMC 96, pp.117-122, 1996.

[7] D. F. Anastasakis, N. Gopal, S. Y. Kim, and L. T. Pillage, "Enhancing the Stability of Asymptotic Waveform Evaluation for Digital Interconnect Circuit Applications," IEEE Trans. CAD, vol. 13, no.6, pp.729-735, June, 1994.

[8] Q. J. Yu and E. S. Kuh, "Moment Models of General Transmission Lines with Application to MCM Interconnect Analysis," in Proc. MCMC 95, pp.158-163, 1995.

[9] J. E. Bracken, "Interconnect Simulation with Asymptotic Waveform Evaluation," PhD dissertation, Carnegie Mellon University, 1994.

[10] B. Tutuianu, F. Dartu, and L. Pileggi, "An Explicit RC-Circuit Delay Approximation Based on the First Three Moments of the Impulse Response," in Proc. DAC, pp. 611-616, 1996.

[11] B. C. Kuo, *Automatic Control Systems*, 7th Edition, Prentice-Hall, 1995.

[12] J. H. Shao and R. M. M. Chen, "MCM Interconnect Design Using Two-Pole Approximation," in Proc. Design, Automation and Test in Europe, 1998.

Table 2 Results and Comparison

case	line	R (Ω/cm)	L (nH/cm)	C (pF/cm)	method	$p_{1,2}$	ζ	ω_n	ζ'	PO' (%)	PO (%)
i	1	5.55	1.98	1.95	Ours	-1.689±0.445i	0.967	1.747	0.967	≈ 0	≈ 0
	2	7.50	2.38	1.62	Padé	-1.614±0.420i	nil	1.667	0.968	≈ 0	
ii	1	3.75	1.52	2.56	Ours	-0.464±0.207i	0.913	0.508	0.792	1.70	1.66
	2	7.89	2.45	1.57	Padé	-1.534±1.010i	nil	1.837	0.835	0.85	
iii	1	2.05	0.95	4.11	Ours	-1.507±0.725i	0.901	1.672	0.668	5.96	5.88
	2	6.00	2.08	1.85	Padé	-1.511±1.481i	nil	2.116	0.714	4.06	

ANALYSIS OF INTERCONNECTED LUMPED DISTRIBUTED MULTI-BRANCH MULTI-STAGE NETWORKS

Iwata Sakagami

Muroran Institute of Technology
Mizumoto-Cho, Muroran-shi, 050 Japan

ABSTRACT

Lumped distributed multi-branch networks have been used to model interconnections of logical circuits and computer-networks for the suppression of multiple reflection waves which cause distorted signals.

In this paper, a method of deriving network functions(NFs) is discussed by considering the interconnection of three multi-branch networks, because fast output calculations are expected from the analytical NFs. The NFs are confirmed by calculating the frequency characteristics of a broad-band improved hybrid-ring with a multi-stage structure, and also by calculating the step responses of a three-stage network which includes an inhomogeneous coupled section.

1. INTRODUCTION

Lumped distributed networks with branches have been used to model interconnections of logical circuits [1-3] and computer-networks[4] as a method of analyzing multiple reflection waves which cause distorted signals, increases in effective rise time, or cross talks. It has been reported that multiple reflectionsin waves can be suppressed in properly terminated transmission line and resistor networks [5,6].

Recently, network functions (NFs) such as network voltage reflection coefficients (NVRCs) and network voltage transmission coefficients (NVTCs) have been presented analytically in a vector form by formulating the flow of the multiple reflections in a N non-commensurate lines junction network with arbitrary terminal impedances (N-port) [6,7]. The NFs of a two-satge multi-port network consisting of a N-port and a M-port have been discussed [8], however, the details are still insufficient.

In this paper, a method of deriving NFs of lumped distributed multi-branch multi-stage networks is presented by considering the interconnection of three multi-branch networks. In the first example, a broad-band improved hybrid-ring of which an even (or odd) mode equivalent circuit is divided into three 2-ports and two 3-ports will be taken up [9]. The NFs are confirmed by calculating the frequency characteristics. In the second example, the NFs of a three-stage network which includes an inhomogeneous coupled section [3] are confirmed in the time domain by calculating the step responses.

2. NETWORK FUNCTIONS OF AN INTERCONNECTED NETWORK

In Fig.1, three multi-branch networks of N-, M-, and L-ports are cascaded. The boundaries p-q and q-r indicate the connected interfaces. T_{pN}, T_{qM} and T_{rL} represent the voltage scattering matrices at the junction s.

2.1 NFs of a N-Port in the First Stage Network

Let us briefly describe the N-port shown in Fig.1, because the NFs are already stated in [7] and [8]. The voltage scattering matrix of the junction network of ports 1_{pc}, 2_{pc}, ... and N_{pc} are given by

$$\mathbf{T}_{pN} = \begin{bmatrix} \Gamma_{11} & T_{21} & \cdots & T_{N1} \\ T_{12} & \Gamma_{22} & \cdots & T_{N2} \\ \vdots & \vdots & \cdots & \vdots \\ T_{1N} & T_{2N} & \cdots & \Gamma_{NN} \end{bmatrix} \quad (1)$$

When a unit impulse $\delta(t)$ is applied to port 1_p as a forward traveling voltage signal, the reflection at port 1_p, G_{11}, and the transmission from port 1_p to port i_p (i=2,3,...,N), G_{1i}, are given by

$$\mathbf{G}_N = [G_{11} \quad G_{12} \quad \cdots \quad G_{1N}]^t$$
$$= \mathbf{L}_N (\mathbf{E}_N - \Gamma_N)^{-1} \mathbf{W}_{1,N} \quad (2)$$

where

$\mathbf{L}_N = \mathrm{diag}(\lambda_{kp}), \quad \lambda_{kp} = (1 + \Gamma_{kpt})\xi_{kp}$

$\Gamma_N = \mathbf{T}_{pN}\mathbf{D}_N$

$\mathbf{W}_{1,N} = \xi_{1p}[\Gamma_{11} \quad T_{12} \quad \cdots \quad T_{1N}]^t$

$\mathbf{D}_N = \mathrm{diag}[\Gamma_{kp}\xi_{kp}^2]$

$\xi_{kp} = \exp(-s\tau_{kp})$

VI-354

$s = j\omega$

τ_{kp} : propagation delay time of the k-th conductor.

$\Gamma_{kpt}, \Gamma_{kp}$: Reflection coefficients looking into the load Z_{kpl} from port k_p of the N-port (k=1, 2,.., i-1, i+1, .., N).

$\Gamma_{ipt}, \Gamma_{ip}$: Reflection coefficient looking into the interface p-q from the left side (k=i).

Although both of Γ_{kpt} and Γ_{kp} are defined in the same way, they are used in a different way on deriving NFs of interconnected networks.

The reflection coefficient Γ_{ip} looking into the interface p-q can be given by

$$\Gamma_{ip} = G_{1q1q} \{\Gamma_{1q}=\Gamma_{1qt}= 0,$$
$$\Gamma_{jq} = G_{1r1r}(\Gamma_{1r}=\Gamma_{1rt}=0)\}, \quad (3)$$

where G_{1q1q} and G_{1r1r} are (1,1) elements of vectors $\mathbf{G}_M$ and $\mathbf{G}_L$ of M- and L-ports, respectively.

2.2 NFs from N-Port to M-Port

The NVTC G_{1pmq} from port 1_p of N-port to port m_q (m=2,3,...,M) of M-port is given by

$$G_{1pmq} = G_{1pip}(\Gamma_{ipt}=0)G_{1qmq}(\Gamma_{1q}=\Gamma_{1qt}=0), \quad (4)$$

where G_{1pip} and G_{1qmq} indicate a (i,1) element of $\mathbf{G}_N$ and a (m,1) element of $\mathbf{G}_M$, respectively.

G_{1pip} is the NVTC from port 1_p to port i_p at the interface p-q. $G_{1pip}(\Gamma_{ipt}=0)$ means that Γ_{ipt} which belongs to the matrix $\mathbf{L}_N$ is zero but Γ_{ip} determined by (3) is used as it is in the matrix $\mathbf{D}_N$. Since Γ_{ipt} in $\mathbf{L}_N$ is to give true voltages at port i_p, $G_{1pip}(\Gamma_{ipt}=0)$ represents all the forward impulses passing through the interface p-q from the left side to the right side. In this paper, a voltage induced by both of an incident voltage wave and its reflection is called true voltage [10]. $G_{1qmq}(\Gamma_{1q}=\Gamma_{1qt}=0)$ represents true voltages that a unit forward impulse traveling from the interface p-q generates at port m_q of the M-port. Therefore, (4) is obtained by the product of $G_{1pip}(\Gamma_{ipt}=0)$ and $G_{1qmq}(\Gamma_{1q}=\Gamma_{1qt}=0)$.

2.3 NFs from N-Port to L-Port

The NVTC G_{1plr} from port 1_p to port l_r (l=2,3,...,L) of L-port is given by

$$G_{1plr} = G_{1pjq}(\Gamma_{jqt}=0)G_{1rlr}(\Gamma_{1r}=\Gamma_{1rt}=0). \quad (5)$$

3. EXAMPLES

3.1 Case of a Broad-Band Hybrid-Ring

In Fig. 2a, the 'open' boundary shows the even mode equivalent circuit of the broad-band hybrid-ring [9], and the 'short' the odd mode.

The equivalent circuit is divided into five sub-circuits, <a>, <b>,...,<e>. Circuits <a>, <d> and <e> stand for 2-ports, and circuits <b> and <c> stand for 3-ports. The numbers 1_a and 2_a represent the port number of the circuit <a>.

The curves in Fig.2b are obtained from the method stated above, and show the same frequency characteristics as in [9].

3.2 Case of a Network with an Inhomogeneous Coupled Section

The network in Fig.3a is presented in [3]. The conductors #4 and #5 represent a symmetric inhomogeneous coupled section. The network can be regarded as an interconnected network consisting of 3-, 4- and 2- ports as shown in Fig.3b, where the port numbers of Fig.3b are rewritten according to

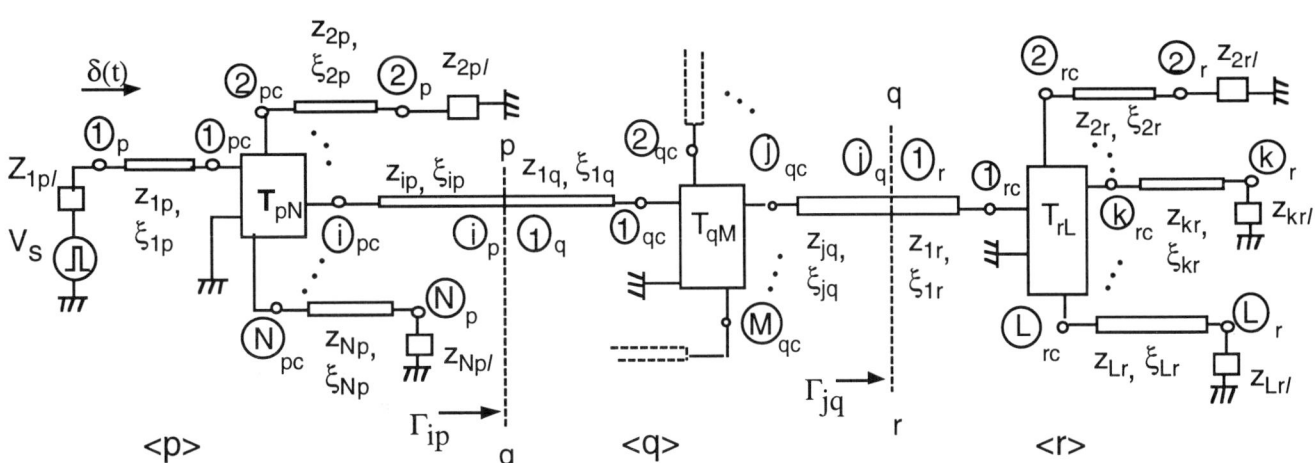

Fig.1 Interconnection of three multi-branch networks.

the style of Fig.1 and four shaded fictitious transmission lines of characteristic impedance Z_{fi} are newly inserted.

With respect to the coupled section, the even and odd mode characteristic impedances Zev and Zod and the even and odd mode propagation delay time Tev and Tod can be derived from the parameters, k_L, k_C, ρ, and τ as follows:

Tev = 30.7nsec, Tod = 24.2nsec,
Zev = 146.4Ω, and Z_{od} = 62.0Ω.

Referring to Fig.1 and Fig.3b, it is understood that the coupled section corresponds to a junction network of the 4-port with port numbers 1q, 2q, 3q, and 4q. The voltage scattering matrix T_4 of the coupled section can be determined from [11]. Therefore, NFs of Fig.3b can be derived using (2) repeatedly as described in Chap.2. Next, NFs of Fig.3a can be determined from the NFs of Fig.3b by replacing the propagation delay time of the fictitious transmission line τ_{fi} to be zero.

The step responses in Fig.3c were calculated from the NFs from ports 1 to 4 and from ports 1 to 6 of Fig.3a using a fast Laplace transform (FLT) routine. Analytically, the responses $V_4(t)$ and $V_6(t)$ have to converge to 0.833 and zero according to the DC analysis in the stationary state. The waveforms in Fig.3c satisfy this condition, and show good agreement with [3].

4. CONCLUSION

The analytical process of deriving network functions has been demonstrated as to rather complicated lumped distributed

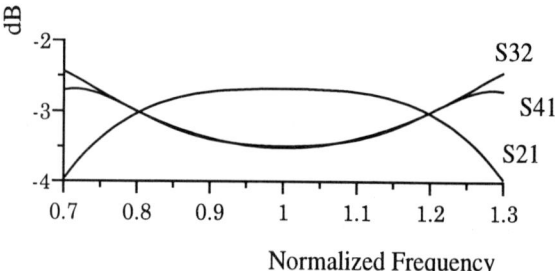

Fig.2b Frequency characteristics

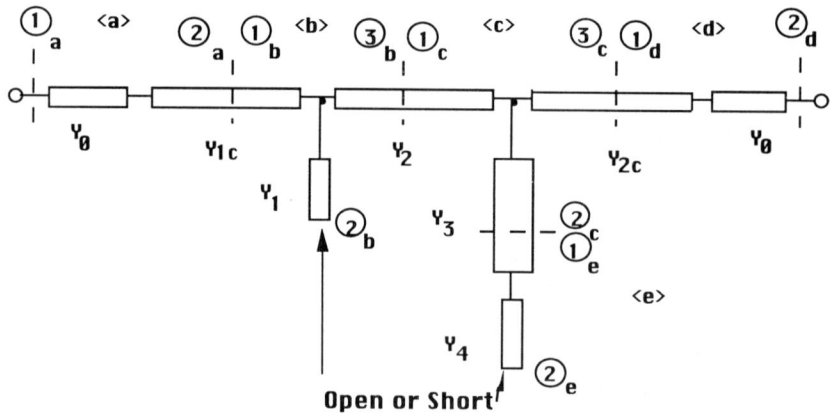

Fig.2a Even (or odd mode) equivalent circuit of the broad-band hybrid-ring [9]

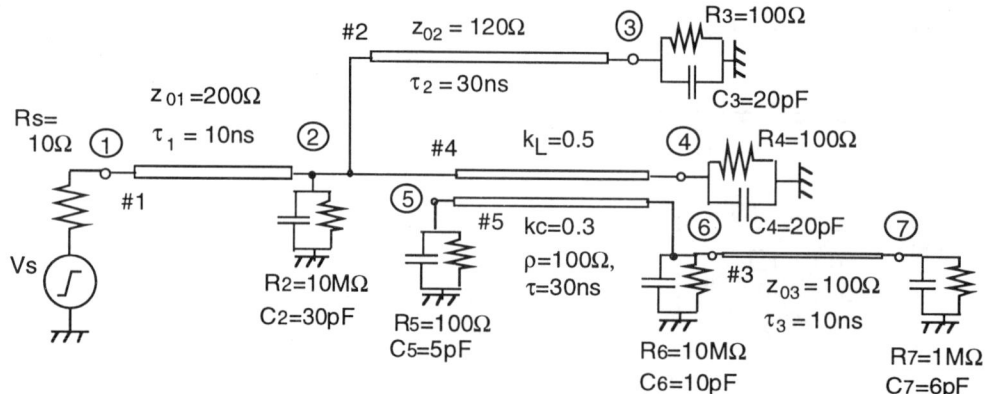

Fig.3a A multi-stage multi-branch network with an inhomogeneous coupled section [3]

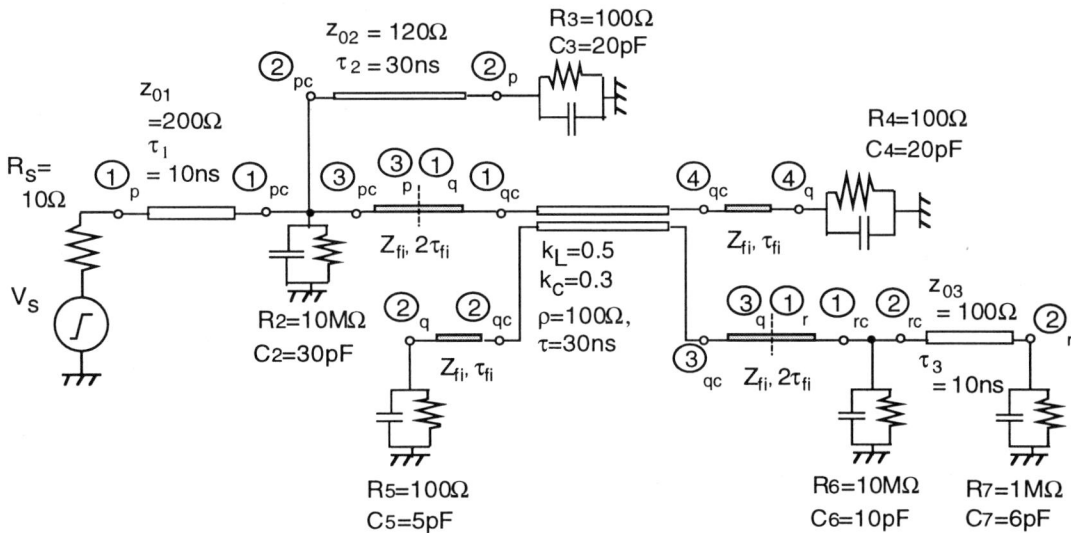

Fig.3b Interconnection of 3-, 4-, and 2- ports using ficfitious transmission lines

cascaded multi-branch coupled-line networks. The calculated responses have shown good agreement with other methods in the frequency domain and also in the time domain. This paper is considered to be useful to analyze the cross-talk, or to suppress the cause of waveform distortions in high speed digital systems.

5. REFERENCES

[1] Fairchild, <u>ECL DATA BOOK</u>, Chaps. 4, 5, 6 (1977).
[2] M.Kato and H.Inose, "Analysis of multiple reflections in branched transmission line by means of z-transform," Trans. IPS Japan, Vol.13,5, pp.287-293 (May 1972).
[3] V.Dvorak, "Computer-aided analysis of pulse signal propagation in digital systems," Int. J. Electronics, Vol.45, No.6, pp.657-665, 1978.
[4] C.W.Trueman, "An electromagnetic course with EMC applications for computer engineering students," IEEE Trans., Educ., vol.E-33, pp.119-128, Feb.1990.
[5] HITACHI, <u>SEMICONDUCTOR DATA BOOK ECL</u>, HITACHI, Sept.'83(1983).
[6] I.Sakagami and A.Kaji, "On the realization of resistively matched three-ports and the ramp-waveform responses of resistive, signal-split three-port transmission-line networks," IEEE Trans., Microwave Theory Tech., vol.41, No.2, pp.234-243, Feb.1993.
[7] I.Sakagami, A.Kaji and T.Usami, "Analysis of multiple reflections by transfer functions of transmission line networks with branches and its application," IEICE Trans. Commun., vol.E75-B, No.3, pp.157-164, March 1992.
[8] I.SAKAGAMI, "Network Reflection and Transmission Coefficients for the Interconnection of Multi-Port Multi-Line Junction Networks", Special Selection on Karuizawa Workshop, IEICE Trans., on Fundamentals, Vol.E79-A, No.3, March 1996.
[9] D.I.Kim and Y.Naito, "Broad-band design of improved hybrid-ring 3-dB directional couplers," IEEE Trans., Microwave Theory and Tech., vol.30, No.11, pp.2040-2046, Nov. 1982.
[10] H.J.Carlin and Giordano, <u>NETWORK THEORY -An introduction to reciprocal and nonreciprocal circuit-</u>, Prentice-Hall, Sec.4.2, 1964.
[11] G.I.Zysman and A.K.Johnson,"Coupled transmission line networks in an inhomogeneous dielectric medium," IEEE Trans., Microwave Thoery and Tech., vol.MTT-17, No.10, pp.753-759, Oct. 1969.

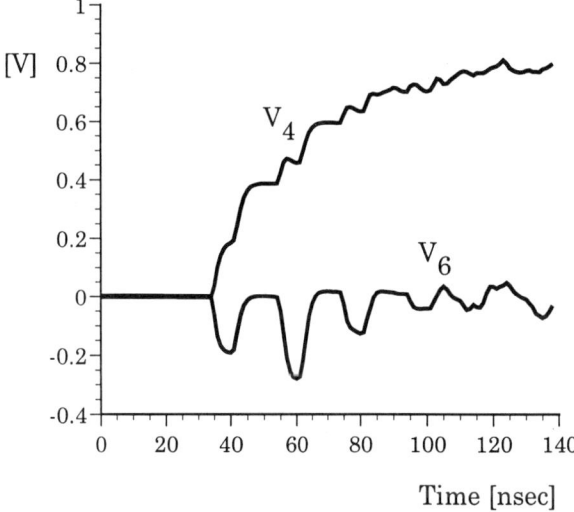

Fig.3c Step responses at ports 4 and 6.

PRECORRECTED-DCT TECHNIQUES FOR MODELING AND SIMULATION OF SUBSTRATE COUPLING IN MIXED-SIGNAL IC'S

João Paulo Costa [*]   *Mike Chou* [†]   *L. Miguel Silveira* [*]

[*] INESC / Cadence European Laboratories
Dept. of Electrical and Computer Engineering
Instituto Superior Técnico
Lisboa, 1000 Portugal

[†] Research Laboratory of Electronics
Dept. of Electrical Engineering and Computer Science
Massachusetts Institute of Technology
Cambridge, MA, 02139, U.S.A.

ABSTRACT

Industry trends aimed at integrating higher levels of circuit functionality have triggered a proliferation of mixed analog-digital systems. Magnified noise coupling through the common chip substrate has made the design and verification of such systems an increasingly difficult task. In this paper we present a new method based on a precorrected-DCT algorithm that extends an eigendecomposition-based technique and can be used to accelerate operator application in BEM methods. This method is shown to avoid storage of a dense matrix, as is typical in BEM methods, while at the same time taking all of the substrate boundary effects into account explicitly. This technique can be used for accurate and efficient modeling of substrate coupling effects in mixed-signal integrated circuits.

1. INTRODUCTION

An emphasis on compactness in consumer electronic products and a widespread growth and interest in wireless communications, have triggered a proliferation of mixed analog-digital systems. Single chip mixed-signal designs combining digital and analog blocks built over a common substrate are advantageous in terms of power dissipation, package count, etc. However, coupling problems resulting from the combined requirements for high-speed digital and high-precision analog components make the design of such systems an increasingly difficult task. Noise coupling through the common chip substrate, has been identified as a significant difficulty in mixed-signal design [1, 2, 3, 4].

Methodologies based on costly trial and error techniques are not adequate for modeling such coupling effects, as they require the ability to perform multiple fabrication runs and are dependent upon the expertise and experience of the designer. Boundary-Element methods (BEM) are very appealing for the solution of this type of problems because the size of the matrix to be solved is dramatically reduced since only the relevant boundary features are discretized. Recently an eigendecomposition-based method was presented that eliminates the need for dense-matrix storage and can be used to accurately compute substrate models [5]. Nevertheless its computation time for large, dense circuits is still high.

In this paper we present a novel precorrected-DCT technique that extends the eigendecomposition-based technique and that when used in a Krylov subspace solver, speeds up operator-vector application significantly. This method, when compared to the eigendecomposition method, trades off a slight accuracy decrease for a substantial speedup in terms of computation time. Thus, it allows for the fast but still accurate extraction of substrate models in problems containing several hundred surface unknowns.

In the next few sections we present some background into the problem of modeling substrate coupling and specifically on BEM methods. We briefly review the functional eigendecomposition technique and then describe our algorithm based on a precorrected-DCT technique. We show how it can be used to dramatically speed up the extraction process producing a model that is easily incorporated into standard circuit simulators such as SPICE to perform coupled circuit-substrate simulation. In this abstract we show a simple example that illustrates the efficiency and accuracy of the new technique presented.

2. BACKGROUND

For typical mixed-signal circuits operating at frequencies below a few gigahertz, the substrate behaves resistively [3, 6]. Assuming this electrostatic approximation, the substrate is therefore usually modeled as a stratified medium composed of several homogeneous layers characterized by their conductivity, as shown in Figure 1. On the top of this stack of layers a number of ports or contacts, assumed planar, are defined, which correspond to the areas where the designed circuit interacts with the substrate.

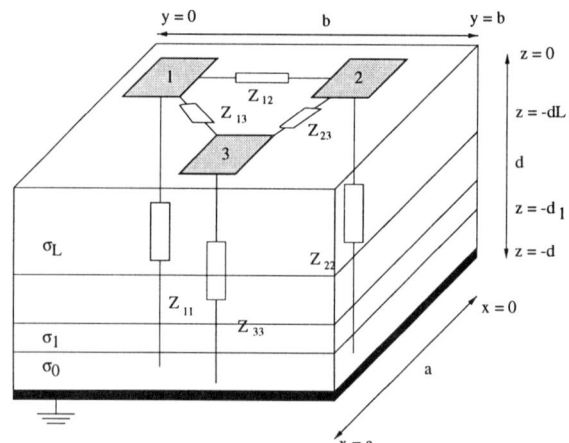

Figure 1: Cross-section of substrate showing a 3D model as an homogeneous multilayered system with contacts on the top surface.

The bottom of the substrate is either attached through some large contact to some fixed voltage (usually ground) or left floating. Figure 1 shows a cross-section of the substrate as a multilayered system with contacts defined on the top surface and an example of the resistive coupling model.

In the electrostatic case the governing equations reduce to the well known Laplace equation

$$\nabla \cdot (\sigma \nabla \Phi) = 0 \qquad (1)$$

inside the substrate volume, where Φ is the electrostatic potential and σ is the substrate conductivity. Application of Green's theorem, assuming a modified Green's function G which accounts for the problem's boundary conditions, gives the potential at some observation point r due to a unit current injected at some source point r'. Usage of the medium's Green's function greatly simplifies the problem by implicitly taking into account the boundary conditions, making it unnecessary to discretize the boundaries. The substrate Green's function for substrates with a grounded backplane has been previously computed in analytical form [3] and shown to be

$$G = \sum_{m,n=0}^{\infty} f_{mn} \cos(\alpha x) \cos(\alpha x') \cos(\xi y) \cos(\xi y') \qquad (2)$$

where $\alpha = m\pi/a$, $\xi = n\pi/b$, a and b are the substrate lateral dimensions and the f_{mn} can be computed with the help of recursion formulas (see [3] for details).

Given a set of m contacts, we seek a model that relates the currents on those contacts, I_c to their voltage distribution V_c. In practice, for reasons of accuracy it is necessary to discretize each of the contacts into a collection of panels. A set of equations relating the currents and potentials on all panels in the system is then formulated

$$\Phi_p = Z_p I_p \qquad (3)$$

Obtaining each entry in this impedance matrix $Z_p \in \mathbb{R}^{n \times n}$ requires computing an integral involving the Green's function over the appropriate panel surfaces. In [7] it was shown that computation of the Green's function can be performed in a very efficient way by truncating the series and rewriting the resulting equation such that each of the matrix elements can be obtained from careful combinations of appropriate terms of a two-dimensional DCT. Since a DCT can be efficiently computed with an FFT, this technique leads to a significant speedup in computation time.

For most problems, the dominant factor in the computational cost will be that of solving the system in Eqn. (3) m times to find the appropriate contact-to-contact resistances. If Gaussian elimination (i.e. LU-factorization) is used then this cost will be $\mathcal{O}(n^3)$ which is overwhelming. Iterative algorithms and namely Krylov-subspace algorithms can be used to speedup the computation of (3). An example of such methods is the Generalized Minimum Residual algorithm, GMRES [8]. GMRES solves the linear system by minimizing the norm of the residual $r^k = \Phi_p - Z_p I_p^k$ at each step k, of the iterative process. The major cost of these algorithms is the computation of a matrix-vector product which is required at each iteration.

3. SPARSIFICATION VIA EIGENDECOMPOSITION

In the functional eigendecomposition method recently proposed [5], direct computation of the matrix-vector product, $Z_p I_p^k$ required at each GMRES iteration is avoided. This operation corresponds in essence to computing a set of average panel potentials given a substrate injected current distribution. This can be done by means of an eigenfunction decomposition of the linear operator that relates injected currents to panel potentials. In this method the current distribution in the N by N panels [1] resulting from the discretization of the top of the substrate can be represented by

$$q(x,y) = \sum_{m=0}^{N-1} \sum_{n=0}^{N-1} q_{mn} \vartheta(x - \alpha', y - \xi') \qquad (4)$$

where q_{mn} is the total current at panel (m, n), $\vartheta(x,y)$ is a square-bump function that serves as an averaging function, defined as

$$\vartheta(x,y) = \begin{cases} \frac{N^2}{ab} & \Leftarrow -\frac{a}{2N} \leq x \leq \frac{a}{2N}, -\frac{b}{2N} \leq y \leq \frac{b}{2N} \\ 0 & \Leftarrow elsewhere \end{cases}$$

and $\alpha' = (m + 1/2)a/N$, $\xi' = (n + 1/2)b/N$.

It is also assumed that the same current distribution function $q(x, y)$ can be decomposed into a sum of functions of the form

$$q(x,y) = \sum_{i=0}^{\infty} \sum_{j=0}^{\infty} a_{ij} \, \varphi_{ij}(x,y) \qquad (5)$$

where the a_{ij} are the coefficients of the decomposition. By choosing $\varphi_{ij}(x, y)$ as the eigenfunctions of the linear operator $\mathcal{L}$ which takes us from currents to potentials, then by definition, the potential can be written as

$$\Phi(x,y) = \sum_{i=0}^{\infty} \sum_{j=0}^{\infty} \lambda_{ij} \, a_{ij} \, \varphi_{ij}(x,y) \qquad (6)$$

where λ_{ij} are the eigenvalues of $\mathcal{L}$. Therefore, if the eigenpairs (eigenfunctions and eigenvalues) of the operator $\mathcal{L}$ implied by Poisson's equation are known, and an eigendecomposition of the injected substrate currents can be obtained such as (5), then the potentials are trivially obtained from (6).

For the substrate problem, the eigenfunctions and the eigenvalues of the impedance operator $\mathcal{L}$ can be derived from Poisson's equation and the knowledge of the boundary conditions. It can easily be shown [5], taking into account the boundary condition on the substrate, that the particular form of the functions is

$$\varphi_{ij}(x,y) = \cos\left(\frac{i\pi x}{a}\right) \cos\left(\frac{j\pi y}{b}\right) \qquad (7)$$

with the eigenvalues given by (for $m, n \neq 0$),

$$\lambda_{mn} = \frac{\beta_L \sinh(\gamma_{mn} d) + \Gamma_L \cosh(\gamma_{mn} d)}{\sigma_L \gamma_{mn} \left(\beta_L \cosh(\gamma_{mn} d) + \Gamma_L \sinh(\gamma_{mn} d)\right)} \qquad (8)$$

where $L + 1$ is the number of layers in the substrate profile with conductivities $\sigma_i, i = 0, \cdots, L$, and d its thickness. The values of Γ_L and β_L can be computed in a recursive manner as in [7].

The current distribution given by Eqn 4 is transformed into a representation such as Eqn. (5) by eigenfunction decomposition, and the a_{ij} coefficients are determined in the usual way, multiplying both sides of (5) by $\varphi_{kl}(x, y)$ followed by integration in x between 0 and a and in y between 0 and b, yielding, after some algebra,

$$a_{ij} = A'_{ij} T_{ij} = A'_{ij} \sum_{m=0}^{N-1} \sum_{n=0}^{N-1} q_{mn}$$
$$\cos\left(\frac{(m+1/2)\pi i}{N}\right) \cos\left(\frac{(n+1/2)\pi j}{N}\right) \qquad (9)$$

---

[1] Without loss of generality we will consider square layouts.

where the A'_{ij} were derived in [5].

Inspection of Eqn. (9) reveals that, for $0 \le i, j \le N-1$, the coefficients a_{ij} are the result of a 2D type-2 DCT on the set q_{mn}. Such an operation can be efficiently performed by means of an FFT. For a_{ij}, $i, j \ge N$ it has been shown, by using the symmetry properties of the DCT, that all coefficients can be related to the first $N \times N$ cosine mode coefficients, $a_{ij}, 0 \le i, j \le N-1$. This process is termed unfolding. Thus by simple computation of the DCT implied in (9), it is possible to obtain an arbitrary number of cosine mode coefficients without incurring in any substantial extra cost. The average potential in each panel can then easily be computed by taking the inner product between Eqn. (6) and $\vartheta(x, y)$ over the given panel, limiting the summation indices to the available a_{ij}. Refolding of these coefficients can be performed to obtain the average panels potentials. Specifically,

$$\overline{\Phi}_{pq} = \sum_{i=0}^{N-1} \sum_{j=0}^{N-1} T_{ij} K_{ij} \cos\left(\frac{(p+1/2)\pi i}{M}\right) \cos\left(\frac{(q+1/2)\pi j}{N}\right) \quad (10)$$

where T_{ij} is the 2D DCT of q_{mn} as seen in (9). For details, including the form of K_{ij} see [5].

4. ACCELERATING POTENTIAL COMPUTATION VIA PRECORRECTED-DCT

Even though the eigendecomposition algorithm reviewed is extremely accurate and more efficient than the Green's function method, for large circuits this methodology can become very costly in terms of computation time. It is possible to significantly speedup the model computation process at the cost of a small accuracy decrease by means of the precorrected-DCT (PcDCT) algorithm.

The main idea behind the PcDCT algorithm is to realize that the effect of an injected current in a panel on the potential of another far away panel can be considered the same for small variations in the distance between panels. Thus the PcDCT algorithm is similar to other precorrected algorithms previously published [9].

Consider a group of panels at some location (call it group g_i) and another group, g_j, far away from g_i. If a current is injected in some panel in g_i, this current will have a similar effect on every panel of g_j, as long as the distance between groups is *large*, where *large* depends of the substrate's profile. Similarly, if the same current is injected in another panel in g_i the effect will be the same.

With the above considerations we can construct an algorithm to approximately compute a potential distribution due to an injected current distribution. This algorithm is composed of four main steps:

1. Constructing a coarse representation of the detailed panel current distribution.
2. Computing a coarse potential distribution resulting from the coarse current representation.
3. Interpolating the detailed panel potentials from the coarse potential distribution.
4. Calculate the nearby panel interactions exactly, and make the appropriate corrections.

The first two steps in the PcDCT algorithm are accomplished with an eigendecomposition of an approximated current distribution. The substrate top surface is divided into a coarse grid of cells unrelated to the underlying panel discretization. Note that this panel discretization is **not** required to be uniform. The panel currents are then projected onto those cells. For two distant cells, as stated above, the current injected in any of the panels from the first cell, i, will have nearly the same effect in all panels of the second cell, j, which allows us to say that the influence of i's panel currents on j's panel potentials is the same as if the total sum of i panel currents was uniformly distributed in i. Thus we can compute a grid cells' current vector I_g from a panels' current vector I_p as

$$I_g = W I_p \quad (11)$$

where $W \in \mathbb{R}^{s^2 \times n}$, s^2 is the number of grid cells and n is the number of panels. W is thus a simple incidence matrix indicating whether a panel belongs to a given cell or not.

Once a vector of cell currents is obtained, the corresponding vector of cell potentials can be computed. For this operation, the eigendecomposition technique previously reviewed is used. This operation is represented as

$$\Phi_g = \mathcal{L} I_g \quad (12)$$

where $\mathcal{L}$ is again the linear operator that represents Poisson's equation, now applied to the coarse grid of cells.

From the vector of cell potentials, Φ_g, the corresponding panel potentials can be obtained. If all cells are distant from cell i we can say that all panels in i are equally influenced by the injected currents so the potentials on all panels in cell i are equal to the cell potential. That is

$$\Phi_p = W^T \Phi_g \quad (13)$$

At this point one has a vector of approximated panel potentials due to a vector of panel currents at the substrate top surface. These panel potentials were obtained under the assumption that all cells were distant enough from each other. However, this is clearly not true everywhere, since for instance cells are not far away from themselves nor from their nearest neighbor cells. For neighbor cells the result of the current to potential operator must be computed directly. But since an approximate effect has already been computed under the "far away" assumption, it is necessary to remove from every cell potential the influence of the neighbor cell's currents. Thus it is necessary to compute the contribution of the current on a specific cell (m, n) to the average potential of a specific cell pq, ψ_{pq}^{mn} (now using two letters to denote a cell in 2D space). This can be shown to be

$$\psi_{pq}^{mn} = \sum_{i=0}^{N-1} \sum_{j=0}^{N-1} K_{ij} \, q_{mn} \quad (14)$$
$$\cos(\theta(m)i) \cos(\theta(n)j) \cos(\theta(p)i) \cos(\theta(q)j)$$

where $\theta(w) = (w + 1/2)\pi/N$ and K_{ij} comes from (10). Eqn. (15) can be rewritten as

$$\psi_{pq}^{mn} = q_{mn} \, h_{pq}^{mn} \quad (15)$$

with

$$D_{pq} = \sum_{i=0}^{N-1} \sum_{j=0}^{N-1} K_{ij} \cos\left(\frac{p\pi i}{N}\right) \cos\left(\frac{q\pi j}{N}\right) \quad (16)$$

$$h_{pq}^{mn} = \frac{1}{4} \left(D_{(m-p)(n-q)} + D_{(m-p)(n+q+1)} + D_{(m+p+1)(n-q)} + D_{(m+p+1)(n+q+1)} \right) \quad (17)$$

We should emphasize that we need to compute and store the h_{pq}^{mn} coefficients only for self and nearest-neighbor coarse-cell interactions. D_{pq} is a type-1 DCT which needs to be computed only once on K_{ij} and discarded afterwards.

Eqn. (15) can be decomposed into a vector of panel potentials by noting that q_{mn} is the sum of all currents in cell (m,n) so the individual contributions may be written as a vector and

$$\Psi_{pq}^{mn} = h_{pq}^{mn} I_{p_{mn}} \qquad (18)$$

where $I_{p_{mn}}$ is the vector of panel currents in (m,n) and Ψ_{pq}^{mn} is the vector of panel potentials in cell (p,q) due to cell (m,n) panel currents.

For every cell k (we return to one letter naming) we have 4, 6 or 9 neighbor cells, so if we call the total contribution to k's panel potentials from neighbor cells Ψ_k we have

$$\Psi_k = \sum_{i \in \mathcal{N}} h_k^i I_{p_i} \qquad (19)$$

where $\mathcal{N}$ represents the set of neighbor cells.

All that is left to do now is to compute the exact near panels interactions for cell k, which we will call V_k. This computation can be made in several ways. We use the Green's function to determine the individual relations between panel currents and potentials, represented as a set of small matrices Z_k^i were each coefficient Z_{pq} relates panel p's potential in cell k to panel q's current in cell i. So V_k can be written as

$$V_k = \sum_{i \in \mathcal{N}} Z_k^i I_{p_i} \qquad (20)$$

and the corrected potential of cell k's panels, $\tilde{\Phi}_{p_k}$, as

$$\tilde{\Phi}_{p_k} = \Phi_{p_k} - \Psi_k + V_k. \qquad (21)$$

Finally the relation between cell k's panel potentials and the set of panel currents is obtained from

$$\tilde{\Phi}_{p_k} = W_k^T \mathcal{L}_k W I_p - \sum_{i \in \mathcal{N}} h_k^i I_{p_i} + \sum_{i \in \mathcal{N}} Z_k^i I_{p_i} \qquad (22)$$

5. MEMORY AND COST COMPARISON

In this section we use m as the number of contacts, n as the number of panels, P as the number of cells per side in the eigendecomposition method and s as the number of PcDCT cells in each dimension.

For large circuits with several hundreds of contacts, the dominant cost of using the Green's function method is $\mathcal{O}(m\, K_G\, n^2)$ where K_G is the average number of GMRES iterations. Similarly, the cost of using the eigendecomposition method is $\mathcal{O}(4\, m\, K_E\, P^2 \log(P))$ where K_E is the average number of GMRES iterations. Since n is typically on the order of P^2, the eigendecomposition method is clearly more efficient than the Green's function method.

For the PcDCT method the cost is mainly dependent on the number of cells s^2 and the number of neighbor panels. Assuming an uniform distribution of panels on the surface, the average number of panels per cell is n/s^2. The dominant cost factor in the computation of a GMRES iteration is $\mathcal{O}(4s^2 \log s + 5n^2/s^2)$. The total cost is then $\mathcal{O}\left(mK_P(4s^2 \log s + 5n^2/s^2)\right)$. For small s the second term is clearly dominant and the cost will decrease with increasing s. For increasing s the first term becomes dominant and the cost gradually becomes independent of n. For some intermediate values of s both terms are comparable and the total cost is seen to be much smaller than that of the eigendecomposition method (since $s \ll P$).

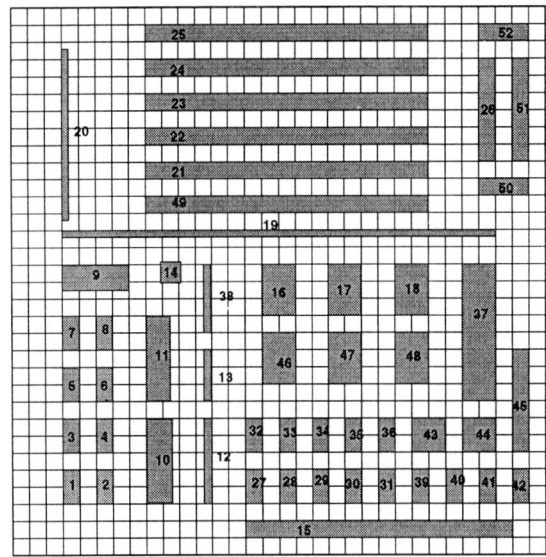

Figure 2: Symbolic layout for an example problem showing a significant number of substrate contacts.

The storage requirements for all three methods are readily obtained. The Green's function method requires $\mathcal{O}(n^2)$ (mostly for storage of the dense Z_p), while the eigendecomposition method requires $\mathcal{O}\left((u+1)P^2\right)$ with u being the the number of cosine modes per cell. Storage requirements for the PcDCT method is $\mathcal{O}(s^2 + 5n^2/s^2)$ where the first term corresponds to the DCT matrix, and the second term to the precorrection matrices. Dominance of each term as a function of s follows the behavior outlined for the computational cost and thus for appropriate values of s the PcDCT method is also more memory efficient.

It is easy to see that for layouts where the contacts are grouped into several separated clusters, an appropriate choice can be made such that the number of direct interactions is reduced and the method's performance will be superior. This coupled with the fact that for the precorrected-DCT method non-uniform panel discretizations are allowed, makes the method extremely efficient and allows for the extraction of larger, more complex circuits.

6. EXPERIMENTAL RESULTS

In this section we present an example that shows the accuracy and efficiency of the substrate coupling extraction algorithm presented in this paper. Figure 2 shows the layout for an example problem where the substrate contacts are marked and numbered. The substrate profile used in this example was taken from [7] and is described in Figure 3. This substrate is used in various BiCMOS processes. For the example shown, extraction was performed and a resistive model was obtained in the form of a matrix relating the resistance from every contact to every other contact.

As can be seen from Figure 2 the contacts for this problem are of varying dimensions, which is typical of mixed-signal designs. Accuracy constraints will limit the discretization employed for producing the set of panels that describe the problem. In this example, usage of a uniform discretization for the Green's function method would produce a problem with too many panels and the computation time and memory requirements would be overwhelm-

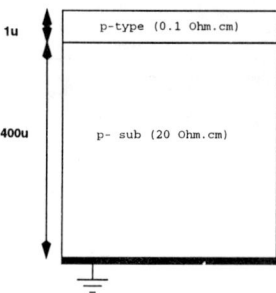

Figure 3: Substrate cross-section for example problem.

Name	Contacts	Green's	EigenD.	PcDCT[1]	PcDCT[2]
R0	1 - BP	45971.7	56270.7	54146.8	56163.6
R1	1 - 2	396.842	416.286	423.396	440.932
R637	15 - BP	22875.2	25068.6	24409.9	24584.4
R678	16 - 19	328.087	475.285	406.608	425.244
R712	17 - BP	27654.6	73693.3	78045.3	76783.4
R784	19 - 20	445.761	465.918	503.91	460.498
R839	20 - 42	2.03e+6	1.5e+6	1.50e+6	1.57e+6
R878	21 - 49	57.7918	67.713	72.8311	69.7378
R1242	37 - BP	24042.2	29267.4	30971.2	29902.8
R1250	37 - 45	251.934	239.167	276.581	247.965
R1255	37 - 50	2085.73	3063.67	3301.29	3422.54

Table 1: Selected set of extracted resistances for the example layout. Node numbers reflect contact numbers and node BP refers to the grounded backplane contact.

ing. Therefore an efficient non-uniform discretization algorithm was devised and employed in this problem. Table 1 shows a selected set of relevant resistances computed, using the method based on direct application of the substrate Green's function, the eigendecomposition based method and our PcDCT method with two DCT sizes for a non uniform discretization. The eigendecomposition method has the highest accuracy because a minimal uniform discretization was used. For the PcDCT method the choice of s has, theoretically, a negligible effect on accuracy. The error in that method mainly depends on the *order* of the approximation scheme (related to the definition of neighborhood of a cell).

Table 2 summarizes some relevant parameters from the extraction, such as the number of panels, the memory used, the CPU time necessary, etc. The results in Table 2 indicate a speedup of around 180 times the Green's function method with better accuracy and 12 times the eigendecomposition method. The memory requirements for the PcDCT algorithm are seen to be considerably smaller than those of the Green's function method, as was expected. In this example some savings in memory were also obtained with respect to the eigendecomposition algorithm, a situation that is dependent upon the choice of s and the structure of the layout.

7. CONCLUSIONS

In this paper we presented a new substrate modeling technique based on a precorrected-DCT algorithm that extends an eigendecomposition-based technique and can be used to accelerate operator application in BEM methods. The method was shown to be both computationally and memory efficient. Speedups of up to two orders for magnitude together with some memory savings

Method	Green's	EigenD.	PcDCT[1]	PcDCT[2]
Discretization	non-unif.	uniform	non-uniform	
# contacts	52	52	52	52
# panels	2647	17764	2647	2647
Avg # panels/contact	51	341	51	51
Size of DCT	512×512	256×256	32×32	64×64
Memory usage	144.6MB	25.5MB	7.1MB	7.4MB
# GMRES iterations	8030	2930	2239	2399
Avg per contact	154	56	43	46
Computation Times (seconds on an Ultra Sparc 1)				
discretization	0.06	0.54	0.06	0.06
Green's function DCT	10.33	N/A	8.93	8.91
Total setup time	14241.1	9.92	20.5	34.9
Solve cost (GMRES)	107278.0	8405.7	493.0	598.5
Total extr. time	123630	8405.6	687.4	830.5

Table 2: Summary of relevant extraction parameters obtained for the given substrate profile.

were obtained on the example presented, while acceptable accuracy is maintained. The technique presented can therefore be used for accurate and efficient modeling of substrate coupling effects in large portions of mixed-signal integrated circuits.

This work was partially supported by the Portuguese JNICT programs PRAXIS XXI and FEDER under contract 2/2.1/T.I.T/1661/95 and grant BM-6853/95

8. REFERENCES

[1] David K. Su, Marc J. Loinaz, Shoichi Masui, and Bruce A. Wooley. Experimental results and modeling techniques for substrate noise in mixed-signal integrated circuits. *IEEE Journal of Solid-State Circuits*, 28(4):420–430, April 1993.

[2] T. A. Johnson, R.W. Knepper, V. Marcellu, and W. Wang. Chip substrate resistance modeling technique for integrated circuit design. *IEEE Transactions on Computer-Aided Design of Integrated Circuits*, CAD-3(2):126–134, 1984.

[3] Ranjit Gharpurey. *Modeling and Analysis of Substrate Coupling in Integrated Circuits*. PhD thesis, Department of Electrical Engineering and Computer Science, University of California at Berkeley, Berkeley, CA, June 1995.

[4] Bram Nauta and Gian Hoogzaad. How to deal with substrate noise in analog cmos circuits. In *European Conference on Circuit Theory and Design*, pages Late 12:1–6, Budapest, Hungary, September 1997.

[5] João Paulo Costa, Mike Chou, and L. Miguel Silveira. Efficient techniques for accurate modeling and simulation of substrate coupling in mixed-signal IC's. In *Proceedings of DATE'98 - Design, Automation and Test in Europe, Exhibition and Conference*, Paris, February 1998.

[6] Nishath K. Verghese, David J. Allstot, and Mark A. Wolfe. Verification techniques for substrate coupling and their application to mixed-signal ic design. *IEEE Journal of Solid-State Circuits*, 31(3):354–365, March 1996.

[7] Ranjit Gharpurey and Robert G. Meyer. Modeling and analysis of substrate coupling in integrated circuits. *IEEE Journal of Solid-State Circuits*, 31(3):344–353, March 1996.

[8] Y. Saad and M. H. Schultz. GMRES: A generalized minimal residual algorithm for solving nonsymmetric linear systems. *SIAM Journal on Scientific and Statistical Computing*, 7:856–869, July 1986.

[9] J. Phillips and J. White. A precorrected-FFT method for capacitance extraction of complicated 3-D structures. In *Proceedings of the Int. Conf. on Computer-Aided Design*, November 1994.

ANALYSIS OF THE TRANSISTOR CHAIN OPERATION IN CMOS GATES FOR SHORT CHANNEL DEVICES

A. Chatzigeorgiou and S. Nikolaidis [1]

Computer Science Department, [1]Department of Physics
Aristotle University of Thessaloniki
54006 Thessaloniki, Greece

ABSTRACT

A detailed analysis of the transistor chain operation in CMOS gates is presented. The chain is diminished to a transistor pair taking into account the actual operating conditions of the structure. The output waveform is obtained analytically, without linear approximations of the output voltage and for ramp inputs. The a-power transistor current model which takes into account second order effects of submicron devices is used, while previous inconsistencies in the chain currents are eliminated by introducing a drain-to-source voltage modulation factor. The exact time when the chain starts conducting is efficiently calculated removing a major source of errors. The calculated output waveform results according to the proposed model are in excellent agreement with SPICE simulations.

1. INTRODUCTION

Since the need for analytical methods which can accurately perform timing simulations of digital integrated circuits is growing as the minimum feature sizes decrease and the number of transistors per chip increases, modeling of CMOS gates is becoming important. It has been extensively pointed out, that simulators such as SPICE which are based on numerical methods, are excessively slow for large designs. Motivated by the previous observations, much research effort has been devoted to the investigation of the behavior of the CMOS inverter and well defined expressions for its output response have been obtained [1], [2], [3].

However little has been done on more complicated gates such as NAND/NOR gates because of their multinodal circuitry and multiple inputs. Modeling of these gates is intricated mainly by the operation of the transistor chain through which the output load is discharged (NAND) or charged (NOR). Since the timing behavior of such a chain cannot be obtained by solving a differential equation at each node of the structure, the inherent properties and operating conditions of the chain have to be exploited. All previous attempts to model the transistor chain can be categorized in two main groups :

The most usual one is the replacement of the complete chain by a single equivalent transistor. As a rule of thumb, the width of the equivalent transistor is calculated by a single m-times transconductance reduction, where m is the number of the devices in the chain. Although attempts have been made in order to improve the efficiency of this model incorporating parasitic capacitances [4], the single equivalent transistor replacement generally fails to reproduce the output waveform of the chain, since it does not take into account the actual operating conditions of the structure.

The next step that has been taken in search for a better modeling technique was to replace a part of the transistor chain, namely those devices which operate always in the linear region, by an equivalent resistor. Such models have been presented by [5], [6]. However these techniques are based on simplified approximations and lead to prohibitively inaccurate results.

It should be mentioned that all previously reported methods ignore second order effects that are present in submicron devices, assume only step inputs and present inconsistency in the chain currents, which is the main error in existing modeling techniques [7].

In this paper, a different approach is followed, overcoming the inaccuracies of all previous works. Nonsaturated devices are replaced by an equivalent transistor whose width is calculated efficiently without leading to inconsistent currents. The method is presented for non-zero transition time inputs, short channel transistor current models and the exact time point when the chain starts conducting is calculated, eliminating another main source of errors.

2. TRANSISTOR CHAIN OPERATION

In order to study the operation of the transistor chain in CMOS gates, let us consider the circuit of Fig. 1a where the discharging of a load capacitance (C_L) through the NMOS transistor chain is examined. Charging through a PMOS chain is symmetrical. The parasitic capacitances formed by the drain/source diffusion areas are also shown. A common ramp input is applied to the gates of all transistors in the chain :

$$V_{in} = \begin{cases} 0 & t < 0 \\ \dfrac{V_{DD}}{\tau} \cdot t & 0 \leq t \leq \tau \\ V_{DD} & t \geq \tau \end{cases} \qquad (1)$$

where τ is the input transition time. All internal nodes are considered to be initially discharged. In case the nodes are charged at $t=0$, the output waveform can be obtained by shifting it in time according to the charge that was initially stored in all nodes [5] and will not be discussed here.

In order to take into account second order effects of submicron devices, the a-power model [2] has been used for the transistor currents :

$$I_D = \begin{cases} 0 & V_{GS} \leq V_{TN}, \quad cutoff \\ k_l(V_{GS} - V_{TN})^{\alpha/2} V_{DS}, & V_{DS} < V_{D-SAT}, \quad linear \\ k_s(V_{GS} - V_{TN})^{\alpha} & V_{DS} \geq V_{D-SAT}, \quad saturat. \end{cases} \qquad (2)$$

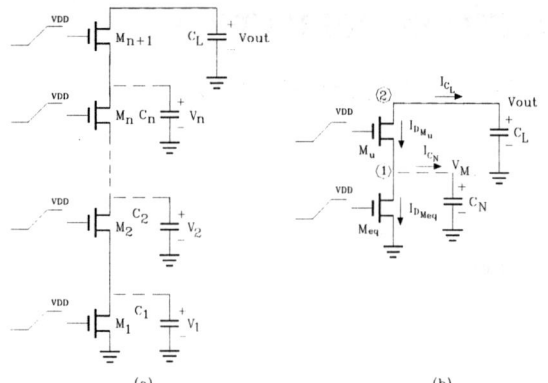

Fig. 1: (a) Complete transistor chain and (b) two-transistor equivalent chain

where V_{D-SAT} is the drain saturation voltage, k_l, k_s are the transconductance parameters which depend on the width to length ratio of a transistor, a is the carrier velocity saturation index and V_{TN} is the threshold voltage which is approximated by its first order Taylor series approximation around $V_{SB}=1$V, $\widetilde{V}_{TN} = \theta + \delta \cdot V_{SB}$.

The topmost transistor in the chain (M_{n+1}) operates initially in saturation since its drain-to-source voltage (V_{DS}) is higher than the drain-to-source saturation voltage (V_{D-SAT}). As the output load capacitance discharges and the internal node voltages rise, transistor M_{n+1} will enter the linear mode of operation when $V_{DS}=V_{D-SAT}$. All other transistors of the chain operate always in linear mode, since after time t_1 when the chain starts conducting their V_{DS} never exceeds the drain saturation voltage [6].

From the time point τ when the input reaches its final value and until the time point t_2 when the topmost transistor exits saturation (in case $t_2 > \tau$), all node voltages remain constant. That is because if the node voltages were decreasing, the saturation current of the topmost transistor would increase, thus increasing the node voltages. On the other hand, if the node voltages were increasing the current of the topmost transistor would decrease thus decreasing the node voltages. Consequently, all node voltages remain at their initial potential at time τ, and this state which is known as the "plateau" state [5] is apparent for fast inputs or large output loads (Fig. 2a). During the plateau state all parasitic currents at the internal nodes are eliminated since the voltages remain constant. In this way the currents of all transistors in the chain are equal.

In order to calculate the plateau voltage of the chain, let us consider the circuit of Fig. 1a and assume that the same ramp input is applied to all transistors. Although the analysis here refers to fast input ramps where the plateau state appears, the derived results are also valid for slow inputs. A first approximation is used for the width W'_{eq} of the equivalent transistor M_{eq} in Fig. 1b, which replaces all the nonsaturated transistors and is given by :

$$\frac{1}{W'_{eq}} = \frac{1}{W_1} + \frac{1}{W_2} + \ldots + \frac{1}{W_n} \quad (3)$$

The plateau voltage at the source of the top transistor, V_p, occurs at the end of the input ramp ($V_{in}=V_{DD}$) where the current of the top transistor ceases to increase. Thus, V_p can be calculated by setting the saturation current of the top transistor (M_u) equal to the current of the bottom transistor (M_{eq}) which operates in linear mode :

$$k_s\left(V_{DD} - \theta - (1+\delta)V_p\right)^a = k_{l_{eq}}\left(V_{DD} - V_{TO}\right)^{a/2} V_p \quad (4)$$

The above equation can be solved with very good accuracy using a second order Taylor series approximation around $V_p=1$ V.

The approach of previous works is based on the assumption that there is a uniform distribution of the source voltage of the top transistor among the drain/source nodes of the rest transistors in the chain operating in linear mode. However, this is not a valid assumption as the gate-to-source voltage and the threshold voltage of each transistor in the chain are different and consequently they would not be able to drive the same current if they had equal drain-to-source voltages. For example, equating the currents through the two closer to ground transistors (for the same transistor width) for $V_{in}=V_{DD}$ and setting the same V_{DS} for each transistor gives :

$$I_1 = I_2 \Rightarrow$$
$$k_l(V_{DD} - \theta)^{a/2} V_{DS} = k_l(V_{DD} - \theta - (1+\delta)V_1)^{a/2} V_{DS} \quad (5)$$

which results in $(1+\delta)V_1 = 0$ where V_1 is the drain voltage of the bottom transistor. This is an invalid expression, because always $\delta > 0$. Trying to keep the current of each transistor in the chain constant, the reduction in V_{GS} and the increase in V_{TN} of a transistor closer to the output is compensated by an increase in its V_{DS}. Considering a gradual increment of V_{DS} by a constant factor v ($v>1$), called *drain-to-source voltage modulation factor*, as we are moving closer to the output, results in very good agreement with SPICE simulations. This means that for two adjacent transistors it is $V_{DS_{(j+1)}} = v \cdot V_{DS_{(j)}}$, where the index shows the position of the transistor in the chain (Fig. 1a). In this way, equation (5) can be rewritten as :

$$k_l(V_{DD} - \theta)^{a/2} V_{DS_1} = k_l(V_{DD} - \theta - (1+\delta)V_{DS_1})^{a/2} v V_{DS_1} \quad (6)$$

In order to solve the above equation, a first order approximation of the V_{DS_1} term inside the parenthesis in the right hand side of eq. (6) is used. Considering the part of the transistor chain which contains the nonsaturated devices as a voltage divider, that term V_{DS_1} can be set equal to V_p/n (for the case that all transistors have the same width) and eq. (6) can be solved for v resulting in :

$$v = \left[\frac{V_{DD} - \theta}{V_{DD} - \theta - (1+\delta)(V_p/n)}\right]^{a/2} \quad (7)$$

Consequently, the plateau voltage of the chain is : $V_p = (1+v+\ldots+v^{n-1}) \cdot V_{DS_1}$. Equating the current that flows through the equivalent transistor (M_{eq} in Fig. 1b) with the current through the closest to the ground transistor of the chain (M_1 in Fig. 1a), the final width of the equivalent transistor is obtained:

$$W_{eq} = \frac{W_1}{1+v+\ldots+v^{n-1}} \quad (8)$$

which is used in the mathematical analysis.

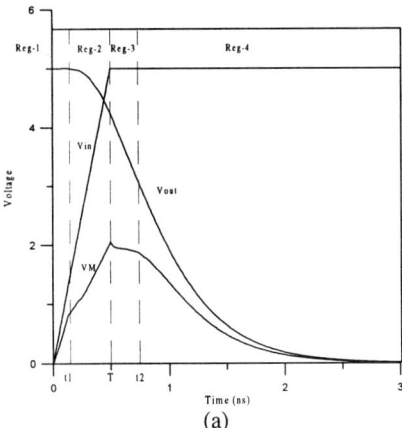 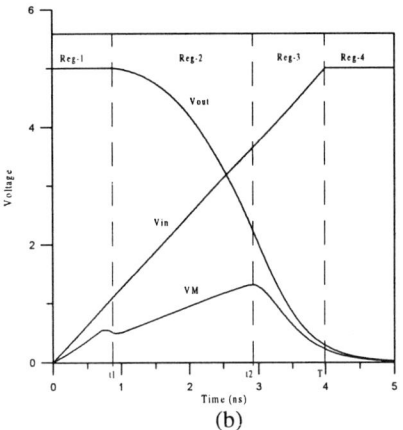

Fig. 2: Regions of operation for (a) fast and (b) slow input ramps

The accuracy of the proposed width for the equivalent transistor is validated by comparison between the output responses of the complete chain and the two transistor chain model, as shown in Fig. 3 for an HP 0.5 μm technology. Also, a comparison with the output response, when the equivalent transistor width is calculated in the conventional way described by eq. 3 and when the nonsaturated devices are replaced by a resistor [5] is also presented in Fig. 4. The superiority of the proposed method is obvious. Consequently, the multinodal analysis problem is now diminished to a two node-analysis which decreases the complexity of the solution significantly.

3. OUTPUT WAVEFORM ANALYSIS

Because of coupling capacitance (C_M) between transistor gates and the drain/source nodes, drain voltages tend to follow the input ramp until all lower transistors start conducting. Until the time point where the transistor below a node starts conducting, the voltage waveform of that node, as it is isolated between two cut-off transistors, is derived by equating the current due to the coupling capacitance of the node, $I_{C_{M_i}}$, to the charging current of the parasitic node capacitance I_{C_i}:

$$I_{C_{M_i}} = I_{C_i} \Rightarrow$$
$$C_{M_i}\frac{dV_{in} - dV_i}{dt} = C_i \frac{dV_i}{dt} \Rightarrow V_i[t] = \frac{C_{M_i}}{C_{M_i} + C_i} V_{in}[t] \quad (9)$$

After the time at which all transistors below the i-th node start to conduct (t_{s_i}) and until the time at which the complete chain starts to conduct (t_1), this node is subject to two opposite trends. One tends to pull the voltage of the node high and is due to the coupling capacitance between input and the node and is intense for fast inputs and high coupling to node capacitance ratio. The other tends to pull its voltage down because of the discharging currents through all lower transistors and is more intense for nodes closer to the ground. For simplicity, here, the two trends are considered to be counterweighted which gives good results in most practical cases. Therefore, the voltage of each node after the time where all the lower transistors start conducting and until time t_1, is considered to be constant and equal to the node voltage at the beginning of this time interval.

By solving $V_{GS_i} - V_{TN_i} = 0$ for each transistor in the chain, the time at which the i-th transistor starts conducting (t_{s_i}) is given by the recursive expression:

$$t_{s_i} = \tau \cdot \frac{\theta + (1+\delta)\dfrac{C_{M_{i-1}}}{C_{M_{i-1}} + C_{i-1}} \dfrac{V_{DD}}{\tau} t_{s_{i-1}}}{V_{DD}} \quad (10)$$

where the index i corresponds to the position of the transistor in the chain and starts counting ($i=1$) from the bottom transistor. ($t_{s_0} = 0$). From the above expression, the time at which the chain starts conducting $t_{s_{n+1}} = t_1$, can be easily obtained.

It has been observed by SPICE simulations that the voltage (V_M) at the source of the top transistor is almost linear between time t_1 and time τ. According to the above, V_M will have a value V_s at time t_1 and V_p at time τ. Thus, V_M for the time interval t_1-τ can be expressed as: $V_M[t] = V_a + m \cdot t$,

where $V_a = V_s - \dfrac{V_p - V_s}{\tau - t_1} t_1$ and $m = \dfrac{V_p - V_s}{\tau - t_1}$.

Although the slope of V_M was calculated for fast inputs, it can be found exactly in the same way for slower inputs [8].

The differential equations that describe the operation of the circuit in Fig. 1b are derived by applying Kirchhoff's current law at nodes 2 and 1:

$$I_{C_L} = -I_{D_{M_u}} \Rightarrow C_L \frac{dV_{out}}{dt} = -I_{D_{M_u}} \quad (11)$$

$$I_{D_{M_u}} = I_{D_{M_{eq}}} + I_{C_N} \Rightarrow -C_L \frac{dV_{out}}{dt} = I_{D_{M_{eq}}} + C_N \frac{dV_M}{dt} \quad (12)$$

where V_M is the voltage at the intermediate node and C_N is the lumped capacitance of all diffusion capacitances of the internal nodes in the chain. Each node capacitance, C_{node}, is calculated as a function of "base" area and "sidewall" periphery [9].

The above differential equations are solved resulting in the expressions for the output voltage waveform for each operating region of the transistors in the chain.

Two cases, fast and slow input ramps are considered. For the fast (slow) case, the intermediate node voltage V_M attains its maximum value when (before) the input ramp reaches V_{DD}.

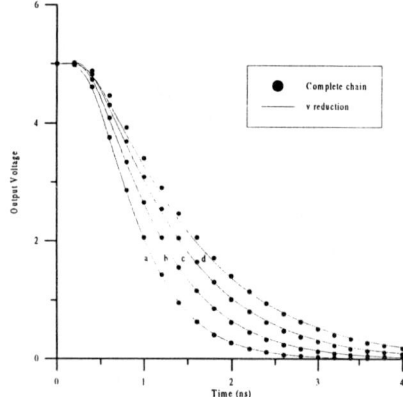

Fig. 3: Output waveform comparison between complete chain and two transistor chain, for a=3, b=4, c=5, d=6 transistors in the chain

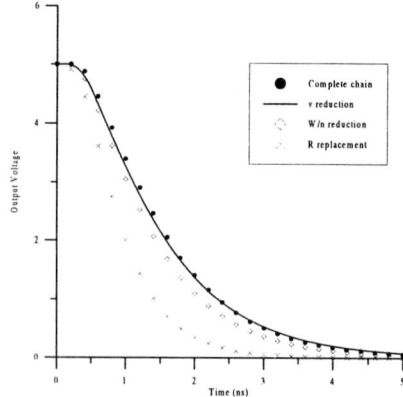

Fig. 4: Comparison between the output waveform of the complete chain and the two transistor chain model using the v factor, the n-times transconductance reduction and replacement by a resistor, for a 6 transistor chain

A. Fast input ramps

Region 1. The top transistor M_u is cut off. This region extends from time $t=0$ until $t=t_1$ when transistor M_u starts conducting and enters saturation. The output voltage remains at V_{DD} (Fig. 2a). This is also validated by SPICE simulations: no overshoot is observed because of the very small gate-to-drain coupling capacitance of a transistor in cut-off.

Region 2. The upper transistor is saturated and the bottom operates in linear mode. This region extends from time t_1 until $t=\tau$ when the input reaches its final value. Since the system of differential equations that describes the operation of the circuit cannot be solved analytically, V_M is considered to be linear.

Substituting $V_M[t] = V_a + m \cdot t$ into eq. (11) and solving the resulting equation gives :

$$V_{out} = c_1 + (q_1 \cdot t - q_2)^a \frac{k_s}{C_L(1+a)} \left[\frac{q_2}{q_1} - t \right] \quad (13)$$

where $q_1 = (V_{DD}/\tau) - (1+\delta)m$, $q_2 = \theta + (1+\delta)V_a$ and $c_1 \approx V_{DD}$.

Region 3. The input ramp has reached V_{DD}, the top transistor is in saturation and the bottom in the linear mode of operation. The limit of this region is time t_2 when the top transistor exits saturation and until that time, the intermediate node remains at the plateau voltage. Since $V_M = V_p$, differential eq. (11) gives :

$$V_{out} = c_2 - \frac{k_s}{C_L} \left[V_{DD} - \theta - (1+\delta) V_p \right]^a \cdot t \quad (14)$$

where $c_2 = V_{out}\big|_{t=\tau} + \frac{k_s}{C_L} \left[V_{DD} - \theta - (1+\delta) V_p \right]^a \cdot \tau$.

The limit of this region is computed by solving $V_{D-SATN}[t_2] = V_{out}[t_2] - V_p$ for the upper transistor, where

$$V_{D-SATN}[t] = \frac{k_s}{k_l} (V_{GS} - V_{TN})^{a/2}$$ according to [2].

Region 4. Both transistors operate in linear mode. The system of differential equations becomes :

$$C_L \frac{dV_{out}}{dt} = -k_{lu} (V_{DD} - \theta - (1+\delta) V_M)^{a/2} (V_{out} - V_M) \quad (15)$$

$$-C_L \frac{dV_{out}}{dt} = k_{lb} (V_{DD} - V_{TO})^{a/2} V_M + C_N \frac{dV_M}{dt} \quad (16)$$

where k_{lu}, k_{lb} specify the linear region transconductances for the upper and bottom transistors respectively. Since the above system cannot be solved analytically, V_M in eq. (15), in the term that is powered to $a/2$, is replaced by its average value $V_p/2$. Solving eq. (15) for V_M, substituting the resulting expression in eq. (16), and setting $g_1 = k_{lu} \left(V_{DD} - \theta - (1+\delta) \frac{V_p}{2} \right)^{a/2}$ and

$g_2 = k_{lb}(V_{DD} - V_{TO})^{a/2}$ results in a second order differential equation which has the solution :

$$V_{out} \cong c_3 \cdot e^{\frac{-p_2 + \sqrt{p_2^2 - 4p_1 p_3}}{2p_1} t} \quad (17)$$

where $p_1 = \frac{C_N \cdot C_L}{g_1}$, $p_2 = \frac{C_L \cdot g_2}{g_1} + C_L + C_N$, $p_3 = g_2$

and c_3 is calculated by equating the above equation for $t=t_2$ with $V_{out}[t_2]$ which is obtained from the previous region.

B. Slow input ramps

For slow input ramps the analysis can be performed in the same way, except for region 3 ($t_2 < t < \tau$) since the top transistor exits saturation before the input reaches V_{DD} (Fig. 2b). For this time interval the input has to be approximated by its average value and the analysis can proceed as in region 4 for fast inputs.

Whether an input ramp is slow or fast can be determined by solving $V_{D-SATN}[t_2] = V_{out}[t_2] - V_M[t_2]$ in the second region. If the top transistor exits saturation before the input reaches its final value ($t_2 < \tau$), the input is slow, otherwise it should be considered fast.

The previous analysis was based on the assumption that normalized inputs, i.e. inputs which have the same starting point and transition time are applied to the transistors of the chain. In case non-normalized inputs are applied, an input mapping algorithm [8] can be employed in order to map every possible input pattern to a set of normalized inputs.

4. RESULTS AND DELAY CALCULATION

The calculated output waveforms of the two transistor equivalent chain, match very well the SPICE simulation results

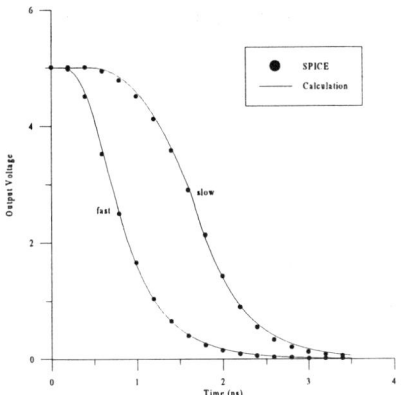

Fig. 5: Output waveform comparison between simulated and calculated values for fast and slow input ramps and for a 0.5μm HP technology

of the complete chain, as shown in Fig. 5. A comparison of the chain output response calculated according to the proposed method to that produced by the approach of [4], where the chain is replaced by a single transistor with its transconductance reduced by the number of the transistors in the chain is also included. In Table I, approximation errors in the calculation of the output waveforms for the two approaches at half-V_{DD} point when the same ramp input is applied to all transistors are presented. Moreover, a comparison for the case of tapered chains is also given. From this comparison it is obvious that the proposed two-transistor equivalent chain models the behavior of the complete chain with excellent accuracy and is much more reliable than the replacement by a single transistor : not only the average error of the proposed approach (4.1 %) is much smaller than the average error in the simple *n*-times transconductance reduction (15.5 %), but furthermore the latter method presents a higher error variance.

Since the output waveform expression for each of the regions of operation is known, propagation delay for the discharging case (t_{PHL}) can be calculated as the time from the half-V_{DD} point of the input to the half-V_{DD} point of the output. The region in which $V_{DD}/2$ of the output occurs, can be found by comparing it with $V_{out}[t_2]$ and $V_{out}[\tau]$. Using this definition, delay results for several input waveforms and transistor chains have been obtained and compared with simulation results. It was observed that in all cases the propagation delay computed using the analytical expressions is within 4 % of that computed by SPICE when the same ramp input was applied to all transistors.

5. CONCLUSION

A detailed analysis for the operation of the transistor chain in CMOS gates was introduced. All nonsaturated devices in the chain are replaced by an equivalent transistor whose width is efficiently calculated taking into account the operating conditions of the structure. The exact time when the transistor chain starts conducting is obtained and analytical expressions for the output response to non-zero transition time inputs are extracted using short channel transistor current models which take into account second order effects of submicron devices. The calculated output waveform and delay results present very small errors compared to SPICE simulation values.

6. REFERENCES

[1] L. Bisdounis, S. Nikolaidis and O. Koufopavlou, "Analytical Transient Response and Propagation Delay Evaluation of the CMOS Inverter for Short-Channel Devices", *IEEE J. Solid-State Circuits*, vol. 33, no. 2, pp. 302-306, February 1998.

[2] T. Sakurai and A. R. Newton, "Alpha-Power Law MOSFET Model and its Applications to CMOS Inverter Delay and Other Formulas", *IEEE J. Solid-State Circuits*, vol. 25, no. 2, pp. 584-594, April 1990.

[3] N. Hedenstierna and K. O. Jeppson, "CMOS Circuit Speed and Buffer Optimization", *IEEE Trans. Computer-Aided Design*, vol. CAD-6, no. 2, March 1987.

[4] A. Nabavi-Lishi and N. C. Rumin, "Inverter Models of CMOS Gates for Supply Current and Delay Evaluation", *IEEE Trans. Computer-Aided Design of Integrated Circuits and Systems*, vol. 13, no. 10, pp. 1271-1279, October 1994.

[5] S. M. Kang and H. Y. Chen, "A Global Delay Model for Domino CMOS Circuits with Application to Transistor Sizing", *Int. J. Circuit Theory and Applicat.*, vol. 18, pp. 289-306, 1990.

[6] B. S. Cherkauer and E. G. Friedman, "Channel Width Tapering of Serially Connected MOSFET's with Emphasis on Power Dissipation", *IEEE Trans. Very Large Scale of Integration (VLSI) Systems*, vol. 2, no. 1, pp. 100-114, March 1994.

[7] J.-T. Kong and D. Overhauser, "Methods to improve digital MOS macromodel accuracy", *IEEE Trans. Computer-Aided Design of Integrated Circuits and Systems*, vol. 14, pp. 868-881, July 1995.

[8] A. Chatzigeorgiou and S. Nikolaidis, "Collapsing the Transistor Chain to an Effective Single Equivalent Transistor", *Proc. Design Automation and Test in Europe Conference (DATE)*, Paris, France, February 1998.

[9] J. M. Rabaey, *"Digital Integrated Circuits : A Design Perspective"*, Upper Saddle River, NJ : Prentice Hall, 1996.

Table I: Approximation error (%) in calculation of a 4-transistor chain output response for the two-transistor and single-transistor equivalent approaches, at $V_{DD}/2$. L and W are given in μm.

L	W	τ=0.5ns		τ=1ns		τ=2ns	
		Prop.	Conv.	Prop.	Conv.	Prop.	Conv.
0.5	4.5	4.751	7.852	5.769	5.897	7.168	1.477
	9	4.200	18.202	5.794	16.887	7.655	21.204
1	12	0.979	20.533	5.534	21.637	6.502	19.316
	18	1.771	42.511	3.059	39.580	4.446	36.564
0.5, a=0.7	W_b=9	1.996	6.347	1.169	4.344	3.089	0.938
1, a=0.7	W_b=18	2.072	3.780	4.032	6.652	5.000	5.980

DELAY AND POWER ESTIMATION FOR A CMOS INVERTER DRIVING RC INTERCONNECT LOADS

S. Nikolaidis, A. Chatzigeorgiou [1] and *E.D. Kyriakis-Bitzaros* [2]

Department of Physics, [1]Computer Science Department

Aristotle University of Thessaloniki

54006 Thessaloniki, Greece

[2]Institute of Microelectronics, NCSR "Demokritos",

15310 Agia Paraskevi, Greece

ABSTRACT

The resistive-capacitive behavior of long interconnects which are driven by CMOS gates is analyzed in this paper. The analysis is based on the π-model of an RC load and is developed for submicron devices. Accurate and analytical expressions for the output voltage waveform, the propagation delay and the short circuit power dissipation are derived by solving the system of differential equations which describe the behavior of the circuit. The effect of the coupling capacitance between input and output and that of short circuit current are also incorporated in the proposed model. The calculated propagation delay and short circuit power dissipation are in very good agreement with SPICE simulations.

1. INTRODUCTION

As the minimum feature sizes for integrated circuits scale downwards, the resistive component of the interconnect loads becomes comparable to the gate output impedance and a single capacitor is no longer a valid gate load model. More accurate load models have to be used for taking into account the increased role of the resistance in the determination of the load behavior and consequently the propagation delay of the driving CMOS gates.

Much research effort has been devoted and very powerful methods have been proposed during the last years for modeling CMOS gates driving simple capacitive loads [1], [2]. Expressions for the propagation delay of CMOS gates driving RC loads have also been derived [3], [4], [5] but they present significantly lower accuracy mainly because they are based on simplified assumptions for the transistor operation and use simple models for the representation of the interconnect loads. Recently, in [6] an analytical method with emphasis on the short-circuit power dissipation has been presented for an inverter driving an RC π load.

In order to find analytical expressions for the propagation delay and the output waveform shape, an interconnect load may be modeled in different ways [7]. Such an expression for the propagation delay of a load modeled simply by a resistor in series with a capacitor, was derived in [3]. However, the driving transistor was considered to operate always in linear mode, the influence of the short circuit current was ignored and the simplified case of step input was examined, thus resulting in limited accuracy.

In [4] an effective capacitance in order to replace the RC output load was calculated by an iteration procedure based on simplified assumptions for the shape of the output response. The real output waveform was approximated by the charging/discharging of the effective capacitance until the time point where the output voltage becomes equal to $V_{DD}/2$. Capturing of the remaining portion of the output response is achieved by a simple resistive model.

A time varying Thevenin equivalent model was proposed in [5] for the estimation of the gate delays. The gate was replaced by an equivalent circuit model composed of a linear voltage source and a linear resistor where their values were determined by using empirical factors thus reducing the accuracy, especially for submicron technologies.

A good approximation for an interconnect load is obtained with the π model, achieving an accuracy better than 3% in delay calculations [7]. In this way analytical expressions for the propagation delay and the output waveform can be found if the load is replaced by its π equivalent and the corresponding system equations are solved. This is the method followed in this paper in order to capture with higher accuracy the performance of CMOS gates driving RC interconnect loads.

2. TRANSIENT RESPONSE ANALYSIS

A circuit composed of an inverter driving an equivalent π-model is considered, where the gate-to-drain coupling capacitance, C_m, is taken into account (Fig. 1). The α-power law model [1], which considers the velocity saturation effect of short channel devices, is used for the transistor current representation:

$$I_D = \begin{cases} 0 & V_{GS} \leq V_{TO} : \textit{cutoff region} \\ k_l(V_{GS} - V_{TO})^{a/2} V_{DS} & V_{DS} < V_{D-SAT} : \textit{linear region} \\ k_s(V_{GS} - V_{TO})^{a} & V_{DS} \geq V_{D-SAT} : \textit{saturation region} \end{cases} \quad (1)$$

where V_{D-SAT} is the drain saturation voltage [1], k_l, k_s are the transconductance parameters, a is the velocity saturation index and V_{TO} is the zero bias threshold voltage.

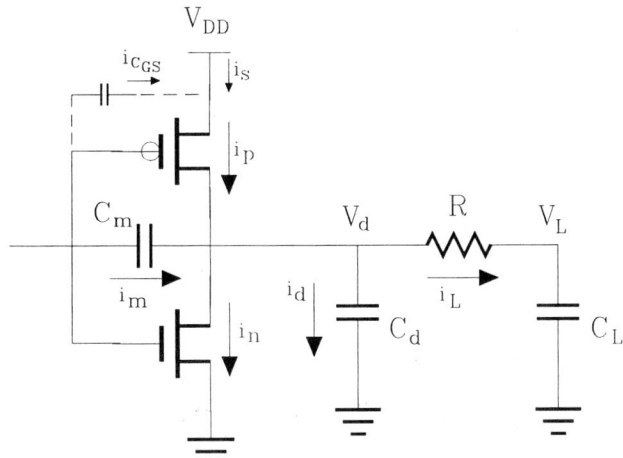

Fig. 1 Inverter driving the π-model of an RC load

A rising ramp input with transition time τ is applied to the transistor gates. The case for a falling ramp is symmetrical. The differential equations that describe the operation of the circuit in Fig. 1 are obtained by applying the Kirchhoff's voltage law in the loop of the π subcircuit:

$$V_d = V_L + V_R \Rightarrow V_d = V_L + RC_L \frac{dV_L}{dt} \quad (2)$$

and Kirchhoff's current law in the transistors drain node using eq. (2):

$$i_n + i_d + i_L - i_m - i_p = 0 \Rightarrow$$

$$\frac{dV_L}{dt} + C_2 \frac{d^2V_L}{dt^2} - C_3 \frac{dV_{in}}{dt} + \frac{i_n}{C_1} - \frac{i_p}{C_1} = 0 \quad (3)$$

where:

$$C_1 = C_L + C_d + C_m, \quad C_2 = \frac{RC_L(C_d + C_m)}{C_1}, \quad C_3 = \frac{C_m}{C_1}$$

In order for the above differential equation to be solved analytically, the parasitic current through the pMOS transistor is initially considered negligible. At the end of the analysis its influence on the output response will be determined.

Two main cases for input ramps are considered: for fast (slow) inputs the nMOS device is in saturation (in the linear region) when the input voltage reaches its final value. In order to obtain the output voltage expression analytically, four regions of operation are considered.

Fast input ramps

Region 1 ($0<t<t_1$). The nMOS transistor is cut-off and differential equation (3) becomes:

$$\frac{dV_L}{dt} + C_2 \frac{d^2V_L}{dt^2} - C_4 = 0 \quad (4)$$

with initial conditions $V_L(0) = V_{DD}, \frac{dV_L}{dt}(0) = 0$ and $C_4 = C_3 V_{DD}/\tau$. The output waveform expression is given by:

$$V_L(t) = V_{DD} + C_4 t - C_2 C_4 \left(1 - e^{-\frac{t}{C_2}}\right) \quad (5)$$

This expression describes the small overshoot of the output waveform due to the coupling capacitance C_m. Generally for the case of driving long interconnection lines, since $C_m \ll C_d$ the overshoot value is almost negligible and V_{out} can be considered equal to V_{DD} without significant error in this region. This region extends until time $t_1 = \frac{V_{TO}\tau}{V_{DD}}$ where $V_{in}=V_{TO}$.

Region 2 ($t_1<t<\tau$). The nMOS device operates in saturation and the input signal is still in transition. Equation (3) becomes:

$$\frac{dV_L}{dt} + C_2 \frac{d^2V_L}{dt^2} - C_3 \frac{dV_{in}}{dt} + \frac{k_s}{C_1}\left(\frac{V_{DD}}{\tau}t - V_{TO}\right)^a = 0 \quad (6)$$

which can not be solved analytically. In order to obtain an analytical expression for the output waveform in this region, the current term is approximated by a second order Taylor series at $t=\tau/2$ where $V_{in}=V_{DD}/2$ with excellent accuracy (error < 1.5%) as $\frac{i_n}{C_1} = A_0 + A_1 t + A_2 t^2$. The differential equation is solved resulting in the following expression for the output waveform:

$$V_L(t) = C[1] + C_5 t + C_6 t^2 + C_7 t^3 + C[2] e^{-\frac{t}{C_2}} \quad (7)$$

where: $C_5 = C_4 - A_0 + 2C_2\left(\frac{A_1}{2} - C_2 A_2\right)$,

$$C_6 = C_2 A_2 - \frac{A_1}{2}, \quad C_7 = -\frac{A_2}{3}$$

and $C[1]$, $C[2]$ are the integration constants.

Region 3 ($\tau<t<t_2$). The input has reached its final value and the nMOS transistor is still in saturation. Equation (3) becomes:

$$\frac{dV_L}{dt} + C_2 \frac{d^2V_L}{dt^2} + \frac{k_s}{C_1}(V_{DD} - V_{TO})^a = 0 \quad (8)$$

resulting in:

$$V_L(t) = C[3] - K_1 t + C[4] e^{-\frac{t}{C_2}} \quad (9)$$

where $K_1 = \frac{k_s}{C_1}(V_{DD} - V_{TO})^a$ and $C[3]$, $C[4]$ are the integration constants.

This region extends until time t_2 when the nMOS transistor exits saturation. The time point t_2 is calculated by the equation:

$$V_d(t_2) = V_L(t_2) + C_L R \frac{dV_L}{dt}(t_2) = V_{D-SATN}(t_2) \quad (10)$$

which can be solved without any approximation. V_{D-SATN} is the drain saturation voltage of the nMOS device.

Region 4 ($t>t_2$). The nMOS transistor operates in linear mode and the solution of equation (3) becomes:

$$V_L(t) = C[5] e^{-\frac{1+\sqrt{1-4C_8 K_4}}{2C_8}t} + C[6] e^{-\frac{1-\sqrt{1-4C_8 K_4}}{2C_8}t} \quad (11)$$

where $K_2 = \frac{k_l}{C_1}(V_{DD} - V_{TO})^{\frac{\alpha}{2}}$, $K_3 = 1 + K_2 C_L R$, $C_8 = \frac{C_2}{K_3}$, $K_4 = K_2/K_3$

Slow input ramps

The operating conditions of the structure in regions 1 and 2 are the same as for fast inputs, however region 2 extends from time t_1 to time t_2, where $t_2<\tau$.

Region 3 ($t_2<t<\tau$). The nMOS transistor operates in linear mode while the input is still a ramp. The differential equation describing the output evolution in this region is given by:

$$\frac{dV_L}{dt}+C_2\frac{d^2V_L}{dt^2}-C_4+\frac{k_l}{C_1}(V_{in}-V_{TO})^{\frac{a}{2}}\left(V_L+RC_L\frac{dV_L}{dt}\right)=0 \quad (12)$$

which can not be solved analytically. For this reason, V_{in} is replaced by its average value $\widetilde{V}_{in}=\frac{V_{in}(t_2)+V_{DD}}{2}$. This is a valid approximation since for most of the practical cases the duration of this region is very small and thus V_{in} takes values very close to that average value.

According to this, the solution of eq. (3) is:

$$V_L(t)=-\frac{C_{10}}{K_6}+C[7]e^{-\frac{1+\sqrt{1-4C_9K_6}}{2C_9}t}+C[8]e^{-\frac{1-\sqrt{1-4C_9K_6}}{2C_9}t} \quad (13)$$

where $K_5=\frac{k_l}{C_1}(\widetilde{V}_{in}-V_{TO})^{a/2}$, $K_6=\frac{K_5}{1+K_5RC_L}$,

$C_9=\frac{C_2}{1+K_5RC_L}$, $C_{10}=-\frac{C_4}{1+K_5RC_L}$

Region 4 is solved exacty as for fast inputs.

A comparison of the output response, V_L, calculated by the proposed method with that derived by SPICE simulations for fast and slow inputs is given in Fig. 2 for an HP 0.5 µm technology, W_n=30 µm and W_p=50 µm. The accuracy of the proposed analysis is obvious.

3. THE EFFECT OF THE SHORT-CIRCUIT CURRENT ON PROPAGATION DELAY

In the above analysis, the current through the pMOS transistor, called short circuit current, was considered negligible. Generally, this is a valid assumption because the capacitive load in long interconnection lines is large enough so that the output voltage doesn't change significantly until the time the pMOS transistor becomes off. This means that the drain-to-source voltage of the pMOS transistor remains small and its current also takes small values. Consequently, ignoring the short-circuit current, in order to simplify the mathematical analysis, does not have any significant effect on the accuracy of the presented analysis. However, a method for taking into account its influence in the estimation of the propagation delay of gates driving long interconnections is presented.

The short-circuit current through the pMOS transistor exists in the interval $[t_{ov}, t_p]$ where t_{ov} is the time where the voltage overshoot at the output of the inverter ceases. That is because during the voltage overshoot, the pMOS current is flowing towards V_{DD} and thus no current path exists between V_{DD} and ground. Time t_{ov} can be calculated by setting the voltage expression for the inverter output, V_d, in region 2 equal to V_{DD}. t_p is the time when the pMOS transistor turns off (when

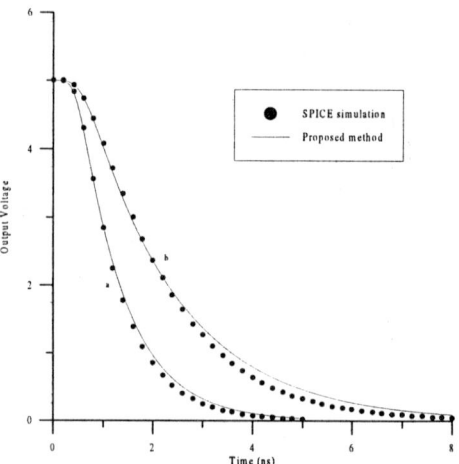

Fig. 2 Output waveform comparison between simulated and calculated values for (a) slow (τ=0.5 ns, R=400 Ω, C_d=C_L=1.5pF) and (b) fast (τ=0.5 ns, R=100 Ω, C_d=C_L=5 pF) cases.

$V_{in}=V_{DD}-|V_{TP}|$). The existence of the pMOS current results in a decrease of the discharging current and thus in an increase of the propagation delay. It acts like an amount of charge Q_e initially stored in the output node and which has to be removed through the nMOS transistor. Consequently, the equivalent charge can be calculated by integrating the current of the pMOS device from time t_{ov} to time t_p. Considering that the pMOS transistor operates for half of the interval $[t_{ov}, t_p]$ in linear mode and that the current waveform is symmetrical around the middle of this interval [6], Q_e can be calculated as :

$$Q_e=\int_{t_{ov}}^{t_p}i_p dt=2\int_{t_{ov}}^{\frac{t_{ov}+t_p}{2}}k_l\left(\frac{V_{DD}}{\tau}t-V_{TP}\right)^{\frac{a}{2}}|V_d-V_{DD}|dt \quad (14)$$

The pMOS drain-to-source voltage (V_d-V_{DD}) derived in the previous section is used in this integral.

In this way, the increase in the propagation delay is found as the time needed to remove the equivalent charge Q_e. An average value for the discharging current, I_{dis}, should be used. However, it can be approximated by the nMOS transistor current at time $t_p/2$, $I_{dis}=i_n[t_p/2]$, which is known from the previous analysis. Thus, the time needed to discharge this extra charge which causes the additional propagation delay can be calculated as $t_{ad}=\frac{Q_e}{I_{dis}}$.

4. ESTIMATION OF SHORT-CIRCUIT POWER DISSIPATION

The short-circuit power which is dissipated during the output switching is due to the current i_s (Fig. 1), which is drawn from V_{DD} towards the source of the pMOS transistor. Current i_s can be found by applying Kirchhoff's current law at the source of the pMOS transistor :

$$i_s=i_p-i_{C_{GS}} \quad (15)$$

where $i_{C_{GS}}=C_{GS}\frac{dV_{in}}{dt}$ is the current through the gate-to-source coupling capacitance.

Energy starts being dissipated at time t_s when i_s starts flowing towards the source of the pMOS transistor. Time t_s can be calculated by setting $i_s = 0$ using the linear region expression for the pMOS current. The pMOS transistor starts its operation in linear mode and then enters saturation at approximately $t_{sat} = \dfrac{t_{ov} + t_p}{2}$ [6], where i_p and consequently i_s reach their maximum value. Assuming that the pMOS current and consequently i_s is symmetrical around t_{sat}, the dissipated energy due to the short-circuit current is given by :

$$E_{SC} = 2 \cdot V_{DD} \int_{t_s}^{t_{sat}} i_s \, dt \qquad (16)$$

Consequently, the short circuit power dissipation for a symmetrical driver and for a system clock frequency f, is :

$$P_{sc} = 2 \alpha f E_{SC} \qquad (17)$$

where α is the switching activity of the output node.

The logic stages following a large RC load will dissipate significant amounts of short-circuit power due to the degraded waveform which they receive as input. Connecting the 20% and 80% point of the output waveform, an effective ramp input for the following stages is obtained which can be used in the corresponding formulas [1] for the calculation of the short circuit power dissipation in these stages.

5. RESULTS AND CONCLUSIONS

Since the output waveform expression for each of the regions of operation is known, propagation delay can be calculated as the time from the half-V_{DD} point of the input to the half-V_{DD} point of the output. Using this definition, the propagation delay has been calculated for several output loads (Fig. 3). It has been observed that in all cases the calculated values are in very good agreement with the delay derived by SPICE simulations. The error was less than 3.5 % while in [3] the error in propagation delay that has been mentioned, for many cases exceeded 40 %. It has also been found that the effect of the input slope on the propagation delay is significant, proving that delay models which consider step input are inadequate.

A comparison between the short-circuit energy dissipation per output transition calculated using the proposed method and the energy which is measured by SPICE is given in Fig. 4. The calculated energy dissipation is very close to that computed by SPICE and it should be mentioned that it is always overestimated which is required since design specifications should always be met.

6. REFERENCES

[1] T. Sakurai and A. R. Newton, "Alpha-Power Law MOSFET Model and its Applications to CMOS Inverter Delay and Other Formulas", *IEEE J. Solid-State Circuits*, vol. 25, no. 2, pp. 584-594, April 1990.

[2] L. Bisdounis, S. Nikolaidis and O. Koufopavlou, "Analytical Transient Response and Propagation Delay Evaluation of the CMOS Inverter for Short-Channel Devices", *IEEE J. Solid-State Circuits*, vol. 33, no. 2, pp. 302-306, February 1998.

[3] V. Adler and E.G. Friedman, "Delay and Power Expressions for a CMOS Inverter Driving a Resistive-Capacitive Load", *Proc. of IEEE Int. Symp. on Circuits and Systems (ISCAS)*, pp.101-104, 1996.

[4] J. Qian, S. Pullela and L. Pillage, "Modeling the ''Effective Capacitance'' for the RC Interconnect of CMOS Gates", *IEEE Trans. Computer-Aided Design of Integrated Circuits and Systems*, vol. 13, No. 12, pp. 1526-1535, Dec. 1994.

[5] F. Dartu, N. Menezes and L. T. Pileggi, "Performance Computation for Precharacterized CMOS Gates with RC Loads", *IEEE Trans. Computer-Aided Design of Integrated Circuits and Systems*, vol. 15, no. 5, pp. 544-553, May 1996.

[6] A. Hirata, H. Onodera and K. Tamaru, "Estimation of Short-Circuit Power Dissipation for Static CMOS Gates Driving a CRC π Load", *Proc. International Workshop Power and Timing Modeling, Optimization and Simulation (PATMOS)*, pp. 279-290, 1997.

[7] T. Sakurai, "Approximation of Wiring Delay in MOSFET LSI", *IEEE J. Solid-State Circuits*, vol. SC-18, pp. 418-426, Aug. 1983.

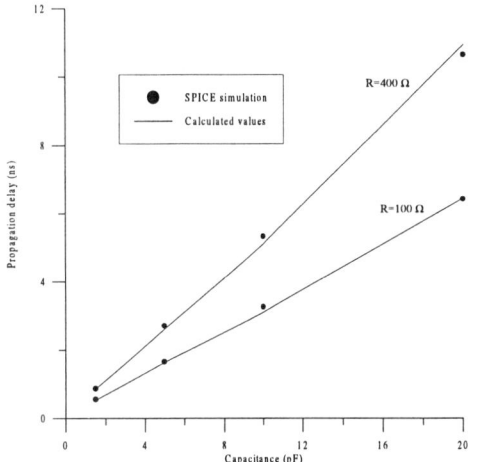

Fig. 3 Comparison between propagation delays measured with SPICE and calculated values using the proposed method for several output capacitances and two different resistance values. The input transition time is 0.5 ns

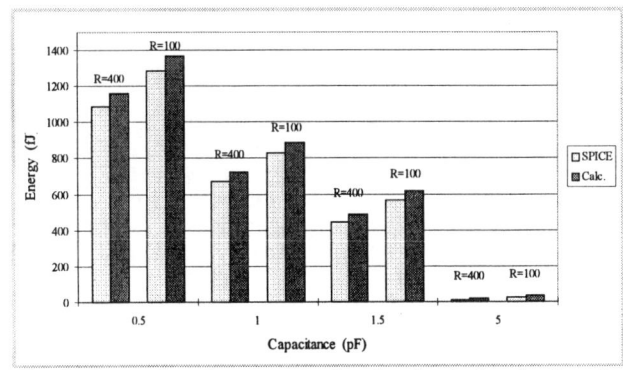

Fig. 4 Comparison between simulated and calculated values for short-circuit energy dissipation, for several capacitances and resistance values (Ω). The input transition time is 0.5 ns.

PARALLEL ALGORITHMS FOR SIMULTANEOUS SCHEDULING, BINDING AND FLOORPLANNING IN HIGH-LEVEL SYNTHESIS

Pradeep Prabhakaran

Center for Reliable and
High-Performance Computing
Coordinated Science Laboratory
University of Illinois
1308 West Main St.
Urbana, Illinois 61801
(pradeep@crhc.uiuc.edu)

Prithviraj Banerjee

Center for Parallel and
Distributed Computing
Department of Electrical and Computer Engineering
Northwestern University
2145 Sheridan Road
Evanston, Illinois 60208
(banerjee@ece.nwu.edu)

ABSTRACT

With small device features in sub-micron technologies, interconnection delays play a dominant part in cycle time. Hence, it is important to consider the impact of physical design during high level synthesis. In comparison to a traditional approach which separates high-level synthesis from physical design, an algorithm which is able to make these stages interact very closely, would result in solutions with lower latency and area. However, such an approach could result in increased runtimes. Parallel processing is an attractive way of reducing the runtimes. In this paper, two parallel algorithms for simultaneous scheduling, binding and floorplanning algorithm are presented. A detailed hardware model is considered, taking into account multiplexor and register areas and delays. Experimental results are reported on an IBM SP-2 multicomputer, with close to linear speedups for a set of benchmark circuits.

KEYWORDS: Parallel algorithms, High-level synthesis, timing driven synthesis, floorplanning.

1. INTRODUCTION

With the rapid improvement in VLSI technology, circuit design is becoming extremely complex and is placing increasing demands on CAD tools. Parallel processing is fast becoming an attractive solution to reduce the inordinate amount of time spent in VLSI circuit design [4].

Scheduling and module binding are the major steps in high level synthesis. Scheduling assigns the nodes in the control data flow graph (CDFG) to specific time steps, and binding assigns the nodes to specific functional units. Floorplanning determines the actual positions of modules in a physical design. The latency of a schedule depends on the cycle time and the number of cycles in the schedule. The cycle time is determined by the longest path in a schedule step, which includes the functional unit delays, multiplexor delays,

register delays and interconnect delays. The number of cycles is determined by the schedule. A smaller number of cycles might require some operations to be chained together in a cycle, for example, but it could increase the cycle time. The area of the floorplan includes the areas of functional units, registers, multiplexors, control and wiring.

With sub-micron technologies, interconnect delay between modules is increasingly becoming a major part of the cycle time. Interconnect delays strongly depend on the number and sizes of modules used and their relative positions. The area of the generated floorplan is determined by the number of modules used, and their positioning, which is partly decided by how small the overall latency should be. Thus, to minimize the execution time, it is important to consider the scheduling and binding information during floorplanning. Similarly, during scheduling and binding, the effect on floorplanning should be considered.

Figure 1 shows an example of how scheduling and binding without considering floorplanning can increase the latency of a schedule. It can be seen in Figure 1(a) that binding operation 10 to $M1$ and 11 to $M3$ results in a cycle delay of 26, since modules (M_1, M_3) and (M_3, M_4) are farther away in the floorplan, whereas binding 10 to $M1$ and 11 to $M2$ as in Figure 1(b) results in a cycle time of 18 since modules (M_1, M_2) and (M_2, M_4) are closer in the floorplan.

A simulated annealing approach which combines scheduling, binding and floorplanning has been shown to be effective in reducing the latency and area in comparison to a two-step approach which separates the problems of scheduling and binding, and floorplanning [5]. However, such an approach leads to increased runtime. Parallel processing has been shown to be very effective in finding solutions fast, with comparable quality to that a sequential algorithm for many CAD problems [4]. In this work, we present two parallel algorithms which performs combined scheduling, binding and floorplanning in the high-level synthesis process. We report experimental results on an IBM SP-2 multicomputer.

In Section 2 a brief survey of existing work is presented. In Section 3, a summary of the combined scheduling, binding and floorplanning algorithm is presented. In Section 4, two parallel algo-

---

This research was supported in part by the National Science Foundation under grant MIP-9320854, the Defense Advanced Research Projects Agency under contracts DAA-H04-94-G-0273, and DABT-63-97-0035 administered by the Army Research Office.

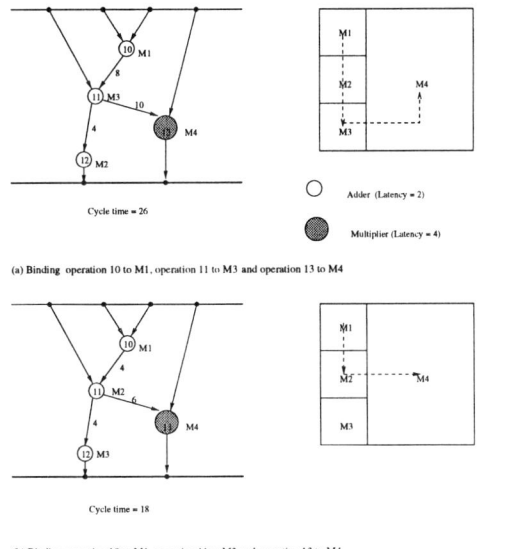

Figure 1: Effect of considering floorplanning during scheduling and binding

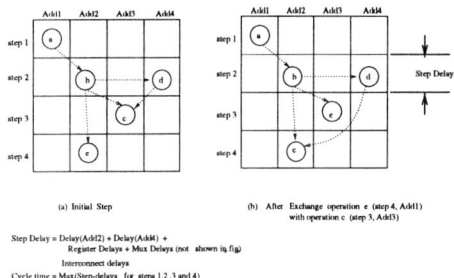

Figure 2: Representation of scheduling and binding problem as a placement problem in a two-dimensional table

rithms for combined scheduling, binding and floorplanning are presented. In Section 5, the results are presented for a set of high-level synthesis benchmarks.

2. RELATED WORK

Previous researchers have addressed the problem of incorporating physical design information in high level synthesis. Fang and Wong [7] describes a integrated binding and floorplanning algorithm which performs a constructive binding for each move of a simulated annealing based floorplanning algorithm.

Weng and Parker [9] describe an algorithm for simultaneous scheduling, binding and floorplanning. They perform the floorplanning of modules in critical paths, followed by an iterative improvement phase which performs rebinding to reduce the latency.

In the past, various tools have been developed to solve the problems at lower levels in the VLSI CAD hierarchy using parallel algorithms, which includes placement, logic synthesis, test generation, fault simulation [4]. However, in the field of high-level synthesis, such effort had been rather limited. J. Roy et .al [19] have proposed a distributed version of force directed list scheduling (FDLS). An algorithm for performing force directed scheduling in parallel appears in [20].

3. THE BASIC COMBINED SCHEDULING, BINDING AND FLOORPLANNING ALGORITHM

The combined scheduling,binding and floorplanning is a simulated annealing based algorithm. It transforms the combined scheduling and binding into a placement problem in a two-dimensional table [6], [5]. The rows represent the steps and columns represent functional units as shown in Figure 2. There are two types of moves:

- Move an operation from one position in the table to another

- Interchange two operations

Each move is performed only if the precedence constraints are satisfied, and at most one operation is assigned to a functional unit. Two operations are chained together in a step if there is a precedence constraint from one of them to the other and the latter operation is an immediate successor of the former. Each move is evaluated based on a cost function which is a linear function of area and the latency of the schedule.

$$C = \alpha L + \beta A$$

where area A is given by the layout generated by the constructive timing-driven floorplanning algorithm described briefly in the following section, and the latency L is the product of the number of cycles and the delay of the longest cycle, where the delay is the longest path in a step including the latencies of functional units, mux delays, register delays and the interconnect delays. α and β are user defined constants which reflects the relative importance of area and latency.

3.1. The Floorplanning Algorithm

The constructive timing driven floorplanning algorithm, used for estimating the area and interconnect delays, considers criticality of interconnections between modules in addition to the overall area of the floorplan.

The algorithm has three main phases. In the first phase, the slacks of connections between modules are calculated from the schedule. This is done by first constructing an extended signal flow graph [7], [5] for each step whose nodes correspond to the functional units and the edges correspond to the interconnections. Let T be the longest delay of step s, assuming zero interconnect delays. The slacks of each edge are calculated as the difference between the earliest time the source of the edge can put the data on the interconnection and the latest time the sink can consume it, such that the maximum delay for the step remains equal to T. The slacks are now converted to costs that can be assigned to individual edges using a method similar to the *zero slack algorithm* [16] for standard cell placement). These weights are a measure of urgency of the interconnect corresponding to the edge. The higher the weight, the closer the modules should be in the floorplan to minimize the latency.

After the weights of nodes in the extended $SDFG$ are calculated for each step, the overall weight for the interconnection between each pair of modules is calculated by combining the weights for that edge across all steps.

In the second phase, modules are clustered using a greedy clustering algorithm in a bottom-up manner, based on the costs calculated in the previous phase. This is done using a heuristic aimed at minimizing the latency and area [5]. Along with the clustering, the augmented bounding curve [5] of each cluster is determined in a bottom up manner. Each point in the augmented bounding curve is a minimal point in the design space which corresponds to a particular slicing structure and orientation of the modules in the cluster. The minimum cost point of the augmented bounding curve determines the slicing structure of the floorplan generated by the algorithm.

In the final phase, the slicing structure is constructed in a top down manner based on the bounding curves computed in the previous phase. These modules are then reoriented in a bottom-up manner inorder to further reduce the interconnect delay, employing a technique similar to one proposed in [12].

The interconnect delays between two modules for the constructed floorplan are calculated based on the Manhattan distance between the centers of the modules as described in [7], [5]. Further details of the floorplanning algorithm can be found in [5].

4. PARALLEL ALGORITHM FOR SIMULATED ANNEALING USING MULTIPLE MARKOV CHAINS

Multiple Markov chains [2], [1], [3] have been shown to be an effective way to perform simulated annealing in parallel. In this section we briefly describe the multiple Markov chain approach.

Simulated annealing can be thought of as a search path whose moves are accepted or rejected depending on particular cost evaluations. This search path can be considered as a Markov chain. Given p processors, a straightforward way of running simulated annealing in parallel is to initiate simulated annealing with a different seed on different processors, let each processor perform moves independently and then finally select the best solution from those computed by all the processors.

Although this approach could potentially result in better solution, it may not result in any speedup. We use an approach similar to the multiple Markov chain approach proposed by Lee and Lee [1] for graph partitioning and Chandy and Banerjee [3] for VLSI cell placement. Let N be the number of moves performed by the simulated annealing algorithm. The overall idea is to make each processor perform an independent simulated annealing with a different seed, but make it perform only N/p moves. Periodically, the processors exchange solutions and each processor will either continue with its own solution or with a different processor's solution. The number of moves performed between these exchanges should be such that each processor should get sufficient time to advance its own annealing process by a considerable amount. Based on the way this exchange is performed, the multiple Markov chain approach can be synchronous or asynchronous [1], [3].

In the synchronous approach, all processors synchronize at regular intervals and determine the best solution found so far. Each processor then updates its local database with the best solution and continue. A disadvantage of this approach is that it could get trapped in a local minimum. In the asynchronous approach, a processor is designated as the master processor which stores the globally best solution. The other processors periodically checks the global best

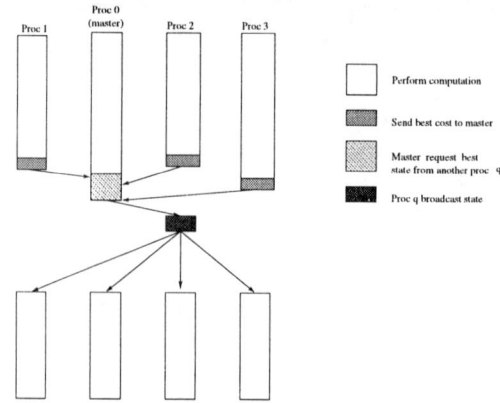

Figure 3: Synchronous Multiple Markov chain

solution and performs an update of its local solution if necessary. In this case, the processors do not have to synchronize and could potentially escape local minima.

5. PARALLEL ALGORITHMS FOR COMBINED SCHEDULING, BINDING AND FLOORPLANNING

In this section, we describe two parallel algorithms for simultaneous scheduling, binding and floorplanning all of which are based on the simulated annealing algorithm described in section 2. The first scheme utilizes synchronous multiple Markov chains and the second schemes uses a variation of the asynchronous multiple Markov chain approach. Let P denote the number of processors.

5.1. Synchronous Multiple Markov Chains

In this algorithm, each processor performs N/P moves per temperature where N is the number of moves performed by the sequential algorithm per temperature. Periodically, each processor sends a cost measure of its current solution to processor 0. Processor 0 selects the best cost of the ones received from all other processors and compares it with the cost of the current global best solution. If it received a better cost than the global best, it sends a message to the processor which has the current best global solution. When that processor receives that message, it broadcasts the solution to all other processors. All the processors update their local solution with that received from the broadcast, and continues annealing from that solution. A barrier synchronization is implicitly achieved as a result of the above step. Figure 3 shows a typical scenario during the synchronous multiple Markov chain based approach.

5.2. Asynchronous Multiple Markov Chains

In this section we describe the asynchronous multiple Markov chain approach. Processor 0 is designated as the master processor which stores the global best solution. Each processor is assigned to perform N/P moves per temperature. At fixed intervals the slave processors sends the cost of its current solution to the master processor. The master processor periodically probes for any messages from other processors. If there are no messages, the master processor continues the annealing process. Otherwise, it receives the

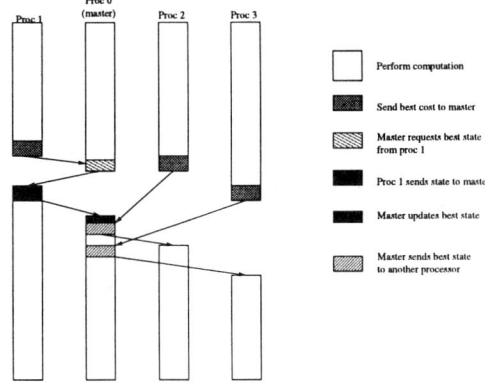

Figure 4: Asynchronous Multiple Markov chain

message and compares the cost with the cost of the global best solution and performs the following action. If the slave processor's solution is better than the current global best, the master processor sends a message to that processor requesting it to send its solution. The slave processor sends its solution to the master and the master updates the global best solution. The slave processor continues annealing from its current solution. If the cost of the current global best solution is better that the one received from the slave processor, the master sends the global best solution to the slave processor, which updates its current state with that received from the master and then continues from that point. Figure 4 shows the flow of messages for asynchronous multiple Markov chain based approach.

6. EXPERIMENTAL RESULTS

Experiments were performed for three high-level synthesis benchmarks using two different design libraries given in table 1. The parallel algorithms were run on 2,4 and 8 processors on an IBM SP2 message-passing multiprocessor. The cost function used for comparison of quality of the solutions is given by $\alpha L + \beta A$, where L is the latency of the schedule, considering the interconnect delays, and A is the area of the floorplan and α and β are user-defined positive constants. Thus, a lower value for the cost function indicates that the solution is of better quality. The quality of the solutions and runtimes for the benchmark ciruits for the sequential algorithm are shown in table 2. The quality of the solutions generated by 2,4 and 8 processors(Q) and the speedups(S) are shown in tables 3 and 4 for the synchronous and asynchronous parallel algorithms. In the tables, Q is given by $Cost_{seq}/Cost_{par}$ where $Cost_{seq}$ is the cost of the solution generated by the sequential algorithm and $Cost_{par}$ is that of the solution generated by the parallel algorithm. The speedups reported are relative to runtimes of the sequential algorithm on IBM SP-2, shown in table 2. It can be seen that both synchronous and asynchronous schemes give almost the same quality of solution as the sequential approach. In some cases, they were able to arrive at a better quality solution than the sequential algorithm. Asynchronous scheme was able to give better solution than the synchronous scheme in many cases. This is due to the fact that it has higher probability of avoiding being trapped in local minima when compared to the synchronous algorithm. In most cases, the speedups are close to linear. In some cases, we get super-linear speedup. This is due to the fact that simulated annealing can be considered to be a search process. When multiple threads of search are initiated in different parts of the solution space in parallel, the expected time for one of them to converge to the global minimum is smaller than that for a single processor executing those threads one after the other.

7. CONCLUSION

With small device features in sub-micron technologies, since the interconnect delays play a dominant role in the cycle time, it is important to consider the effect of physical design on high-level synthesis. We have presented parallel algorithms which combines physical design and high level synthesis using simulated annealing. Experimental results show that the approach is quite effective in obtaining solutions of almost the same or better quality as the sequential algorithm, while giving close to linear speedups.

8. REFERENCES

[1] S. Y. Lee and K. G. Lee, "Asynchronous communication of multiple Markov chains in parallel simulated annealing," in *Proceedings of the International Conference on Parallel Processing*, Aug.1992.

[2] E. H. L. Aarts, F. M. J.de Bont, E. H. A. Habers, and P. J. M. van Laarhoven, "Parallel implementations of the statistical cooling algorithm," *Integration, the VLSI journal*, vol. 4 Sept. 1986.

[3] J. A. Chandy and P. Banerjee, "Parallel Simulated Annealing Strategies for VLSI Cell Pacement," *Proceedings of the 9th International Conference on VLSI Design*, Jan. 1996

[4] P. Banerjee, "Parallel Algorithms for VLSI Computer-aided Design Applications," *Englewood-Cliffs, NJ:Prentice Hall*, 1994.

[5] P. Prabhakaran and P. Banerjee, "Simultaneous Scheduling, Binding and Floorplanning in High-level Synthesis," *Tech Report CRHC-97-16*, Sept 1997

[6] S. Devadas and A. R. Newton, "Algorithms for Hardware Allocation in DataPath Synthesis," *IEEE trans. on CAD*, vol. 8, No. 7, 1989

[7] Y. Fang and D. F. Wong, "Simultaneous Functional Unit Binding and Floorplanning," *Proc. of ICCAD*, pp. 317-321, 1994.

[8] V. Moshnyaga, H. Mori, H. Onodera, K. Tamary, "Layout-Driven Module Selection for Register-Transfer Synthesis od Sub-micron ASICs," *Proc. of ICCAD*, pp. 100-103, 1993.

[9] J. Weng and A. C. Parker, "3D Scheduling : High-Level Synthesis with Floorplanning," *Proc. of 28th DAC*, pp. 668-673, 1991.

[10] D. W. Knapp, "Fasolt: A Program for Feedback-Driven Data-Path Optimization," *IEEE Trans. on CAD*, vol. 11, No. 6, pp. 677-695, 1992.

[11] W. Dai and E. S. Kuh, "Simultaneous Floor Planning and Global Routing for Hierarchical Building Block Layout," *IEEE trans on CAD*, 1987

[12] D. P. La Potin, and S. W. Director, "Mason: A Global Floorplanning Approach for VLSI Design," *IEEE trans on CAD*, 1986

[13] L. StockMeyer, "Optimal Orientations of Cells in Slicing Floorplan Designs", *Information and Control, Vol 59,(1983)*, 510-522.

[14] R. H. J. M. Otten, "Efficient Floorplan Optimization,", *Proc. ICCD 83*, pp 96-98.

[15] T. Yamanouchi, K. Tamakashi and T. Kambe, "Hybrid Floorplanning Based on Partial Clustering and Module Restructuring", *International Conference on Computer Aided Design*, 1996.

[16] R. Nair, C. L. Berman, P. S. Hauge and E. J. Yoffa, "Generation of Performance Constraints for Layout," *IEEE Trans. Computer-Aided Design*, 1989.

[17] H. Jang and B. Pangrle, "A Grid-Based Approach for Connectivity Binding with Geometric Costs," *Proc. of ICCAD*, pp. 94-99, 1993.

[18] D. Zhou, F. P. Preparata and S. M. Kang, "Interconnection Delay in Very High-Speed VLSI," *IEEE Trans. on Circuits and Systems*, Vol 38, No. 7, pp.779-790, 1991.

[19] J. Roy, N. Kumar, R. Dutta, R. Vemuri,"DSS: A distributed high-level synthesis system", in *IEEE design and test of computers*, June 1992.

[20] P. Prabhakaran and P. Banerjee, "Parallel Algorithms for Force Directed Scheduling of Flattened and Hierarchical Signal Flow Graphs," *Proceedings of the International Conference on Computer Design*, 1996.

Table 1: The library set

Fabrication	16 bit adder		16 bit multiplier		16 bit register	
Technology	Delay	Area	Delay	Area	Delay	Area
1.6 μm	18	746875	200	8711250	3	597500
1.2 μm	13	420000	150	4900000	3	336000

Table 2: Quality measures and runtimes for the sequential algorithm

Circuit	Fabrication Technology	Latency (ns)	Area (sqmm)	Cost	Runtime (seconds)
FIR	1.6(16 bit)	574.395	57.19	947.03	1.01×10^4
2-step	1.2(16 bit)	430.688	31.75	542.95	1.01×10^4
FIR	1.2(8 bit)	753.700	46.87	669.97	6.625×10^3
3-step	1.0(8bit)	568.600	25.86	391.47	7.112×10^3
Elliptic	1.2(8 bit)	371.189	15.60	232.54	3.449×10^3
8-step	1.0(8bit)	284.899	8.91	141.65	3.557×10^3

Table 3: Quality and speedup comparisons for Synchronous Markov-chain approach

Circuit	Fabrication Technology	2 proc		4 proc		8 proc	
		Q	S	Q	S	Q	S
FIR	1.6(16 bit)	1.000	2.06	1.000	3.90	1.00	7.82
2-step	1.2(16 bit)	1.000	1.61	1.000	3.82	1.000	7.26
FIR	1.6(16 bit)	1.000	1.97	1.000	4.24	1.000	8.17
3-step	1.2(16 bit)	1.000	2.34	1.000	3.82	1.000	8.36
Elliptic	1.6(16 bit)	0.978	1.78	0.979	4.32	0.979	7.40
8-step	1.2(16 bit)	0.982	1.77	0.971	4.04	0.965	10.0

Table 4: Quality and speedup comparisons for Asynchronous Markov-chain approach

Circuit	Fabrication Technology	2 proc		4 proc		8 proc	
		Q	S	Q	S	Q	S
FIR	1.6(16 bit)	1.000	2.30	1.000	3.25	1.000	7.95
2-step	1.2(16 bit)	1.000	1.81	1.000	3.04	1.000	6.44
FIR	1.6(16 bit)	1.000	2.05	1.000	4.00	1.000	8.04
3-step	1.2(16 bit)	1.000	1.83	1.000	3.79	1.000	7.85
Elliptic	1.6(16 bit)	1.000	2.06	0.990	3.38	0.988	7.93
8-step	1.2(16 bit)	1.007	1.95	1.000	4.63	0.975	10.37

A SIMPLE ALTERNATIVE FOR STORAGE ALLOCATION IN HIGH-LEVEL SYNTHESIS

Mohammed Aloqeely

Computer Engineering Dept., King Saud University
P.O. Box 51178, Riyadh, 11543, Saudi Arabia
aloqeely@{cat.syr.edu, ccis.ksu.edu.sa}

ABSTRACT

A recent trend in high-level synthesis is to introduce special structure memory elements as an alternative to RAMs which suffer from address generation and decoding overhead. In this paper, an alternative, called Sequential FIFO Memory (SFM) is investigated. The problem of allocating variables to SFMs is studied thoroughly including the theoretical background, mapping algorithms and experiments. Moreover, the combinatorial optimization problems related to SFMs were found to be tractable in contrast to those of previous non-RAM alternatives.

1. INTRODUCTION

Memory allocation (or assignment) is an important step in high-level synthesis. The memory allocation problem maps intermediate results (variables) and constants onto a set of memory elements in the targeted architectural model [1,2]. Traditionally, the used memory elements were registers or register files. RAM based register files and memories are very flexible and easy to allocate in the synthesis process but they suffer from the penalty of address generation and decoding especially if their size is large. The need and the feasibility of introducing new non-RAM storage structures in high-level synthesis was first elucidated by Aloqeely and Chen in [3,4]. The idea was initiated by the regularity in access patterns observed in many applications which could be supported by special non-RAM memories.

The allocation of sequencers, e.g., queues, stacks and bi-directional queues was presented in [3-5]. The algorithms were mainly targeting applications with a high degree of computational regularity and were not applied to irregular applications. Subsequent work by other researchers however, was even applied on applications which are less regular. Another type of sequencers, the circular queue was studied in [6] by Bennour and Aboulhameed. Sequential read/write memories (SRWMs) were proposed as an alternative to register files in [2] by Gerez and Woutersen. SRWMs, as reported in [2] and [7] occupy smaller area and require less power than queue structure proposed earlier, but nonetheless, the allocation process is no less complex. Finally, Ahmed et al. have proposed the use of dual-stacks in [8]. This paper is proposing yet another alternative, namely Sequential FIFO Memory (SFM) which is a modified version of the SRWM proposed in [2].

If there is one subtle common feature of all new memory alternative, it would be the computational complexity involved in mapping variables to them. In the case of SRWMs it was reported in [2] that the realizability problem (just checking if a set of variables can be mapped to the same SRWM) is NP-Complete. The minimal grouping problem for SRWM and for stacks was also reported to be NP-Complete in [2] and [4]. In some other cases (e.g., queues and circular queues), although not mathematically proven, the researchers have a great suspicion about the NP-completeness of the realizability problem. The computational complexity of the allocation process is induced by the restricted nature of operation

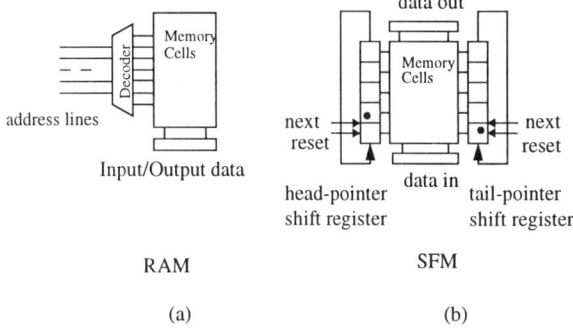

Figure 1: SFM vs. conventional RAM. organization.

of these devices. Unfortunately, computational complexity limits the chances of arriving at good final solutions for large-size designs.

The alternative used in this paper, although is structurally close to the SRWM studied in [2], has less constraints on the allocation process. As explained later, this fact simplifies the mapping process drastically.

The remainder of this paper is as follows. The next section introduces SFM. Section 3 investigates the problem of allocating SFMs in data path synthesis. Allocation algorithms and experimental results are given in Sections 4 and 5. Limitations and future directions are given in Section 6. Finally the paper is concluded with comments and conclusions in Section 7.

2. SFM

To understand how a SFM works, recall the general structure of a RAM, shown in Figure 1 (a). An array of storage locations is arranged so that each word is accessed by activating an enable line which is directly obtained from an address decoder. On the other hand, a SFM works on the same principle of a RAM but the address decoder is replaced by two single-bit shift registers as depicted in Figure 1 (b). Only one bit in each shift register is allowed to have a "1" value and hence it can in effect select one word of the array. The location of this "1" word pointer is controlled by two signals *next* and *reset*. *Reset*, puts the "1" bit at the lowest memory location whereas *next* advances the "1" one position and hence makes the pointer point to the next higher memory location by triggering a shift operation. When the pointer reaches the highest position, it goes back to the starting point in a cyclic fashion. Notice that the shift register need not be physically cyclic to do this since this operation could be achieved by using the reset signal.

One of the two shift registers points at the word to be read while the other one points at the word to be written, in contrast to a

SRWM which has only a single shift register for both write and read. As a result, a SFM can function as a perfect wrap-around FIFO memory. The "pointer-based" structure of FSMs (similar to SRWMs) makes them better in power consumption than shift-register chains (the data is not pumped up between registers in every clock cycle). See [2] and [7] for details.

3. PROBLEM FORMULATION

The objective of memory allocation is to find a mapping from a set of variables to a set of SFMs that optimizes some cost function. The following two problems have been defined in [2] and are restated in the context of SFM:

Realizability: Given a set of values S_k, find out if all values in S_k can be assigned to the same SFM.

Minimal Grouping: Given a set of variables S, where $S=\{v_1..v_n\}$ partition S into a minimal number of disjoint subsets $S_1, S_2,...,S_k$ such that each subset is realizable in a single SFM (i.e., all the variables in a subset can be assigned to the same SFM).

It has to be pointed out that in this paper, we are focusing solely on operating a SFM as a perfect FIFO memory although its structure permits using it in slightly different fashions. Further, the formulations and algorithms apply to any other structure that acts as a perfect FIFO memory.

The following are some basic definitions and assumptions followed in the formulation.

Given a set $S=\{v_1..v_n\}$ of variables define:

Definition 1: The write time of variable v, denoted as $W(v)$, is the time at which v is first defined (i.e., is stored in a memory element); and the read time of variable v, denoted as $R(v)$, is the time at which v is used (i.e., is fetched from a memory element).

In case a variable is written or read more than once variable splitting [5] may be used to make sure that all variables are in the single-write single-read category. We will see how to extend the formulation to variables with a single write and multiple reads at Section 6.

Definition 2: Variable v_1 *precedes* v_2 if $W(v_1) < W(v_2)$ and $R(v_1) < R(v_2)$

The *precede* relation is both transitive and anti-symmetric. Therefore it constructs a partial ordering relation on V.

3.1 Realizability

As is the case in all non-RAM memory elements, two types of conflicts can prevent two variables from being assigned to the same SFM: access conflicts and control (or sequence) conflicts:

Access conflict: The two variables are read/written at the same time.

Sequence conflict: The two variables are violating the FIFO access pattern. What FIFO means is first written first read, otherwise we have a conflict.

Fortunately and unlike the case with queues, circular queues and SRWMs, since a SFM behaves like a perfect FIFO memory, (similar to *p-queues* in [11]) we don't have to worry about the relative locations of variables within the SFM since it will automatically be valid as long as the condition in the following Lemma is satisfied.

Lemma 1: A set of variables can be grouped in the same SFM if for any two variables v_1 and v_2 in S: Either v_1 *precedes* v_2 or v_2 *precedes* v_1.

It is obvious that the realizability problem is tractable. In fact it can be achieved by merely sorting variables according to write time and checking if they are sorted according to read time. This is remarkable, recalling that the realizability problem for SRWM is NP-complete [2].

3.2 Minimal Grouping

In this paper we present two approaches to the minimal grouping problem. The first approach is based on a formal mathematical model while the second one is a graph theoretic algorithm. The two approaches are given next in Section 4.

4. MINIMAL GROUPING APPROACHES

4.1 A Formal mathematical approach

The minimal grouping problem can be solved using integer linear programming (ILP). It is based on converting a set of observations into a set of mathematical equations and inequalities as follows:

1. If none of the variables v_i and v_j *precedes* the other then they cannot share the same SFM.
2. Every variable v_i has to be assigned to a single SFM.
3. The number of variables assigned to a SFM should not exceed a maximum allowed limit.

Consider n variables that are to be allocated to m SFMs. The maximum allowed size of a SFM is M. The variables used in the formulation are the following:

y_j is a 0-1 integer variable associated with SFM_j such that $y_j=1$ if SFM_j is required, otherwise $y_j=0$; $(1 \leq j \leq m)$.

$x_{i,j}$ is a 0-1 integer variable associated with variable v_i and SFM_j. $x_{i,j}=1$ if v_i is mapped to SFM_j; otherwise, $x_{i,j}=0$; $(1 \leq i \leq n, 1 \leq j \leq m)$.

The problem can be formulated as:

$$\text{Minimize} \quad \sum_{j=1}^{m} y_j \quad \text{subject to:}$$

$$x_{i,j} + x_{k,j} \leq 1 \quad (1 \leq j \leq m) \quad \forall i,k$$

s.t. neither v_i nor v_k precedes the other (1)

$$\sum_{j=1}^{m} x_{i,j} = 1 \quad \forall i \quad (2)$$

$$\left(\sum_{i=1}^{n} x_{i,j}\right) - y_j \cdot M \leq 0 \quad \forall j \quad (3)$$

Constraint 1 insures that no incompatible variables will be mapped to the same SFM. Constraint 2 insures that every variable will be mapped to a SFM and constraint 3 states that the number of variables mapped to a SFM should be less than or equal to the maximum allowed limit for SFMs.

The above formulation for minimizing the total number of SFMs is pretty simple yet it does not include any other factor in the cost (e.g., total number of registers). It is possible however with more variables and constraints to include some other factors as well, similar to the ILP formulation for p-queues (or perfect FIFOs) proposed in [11] which I came to know recently. Nevertheless, this comes at a price of longer execution time and less suitability for larger size problems.

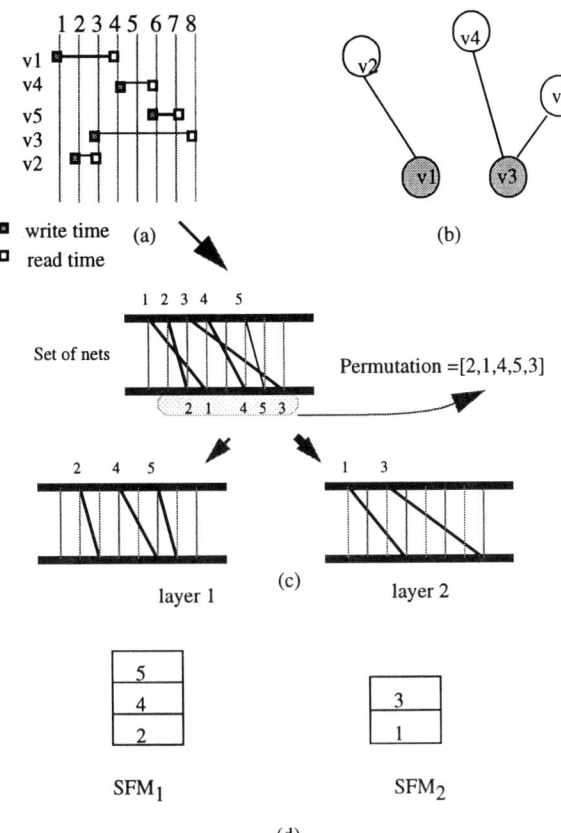

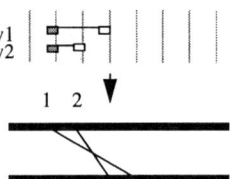

Figure 3: Handling variables with equal write/read time.

sultant set of nets could be represented by a permutation $P=\{p_1,p_2,...,p_n\}$ of the numbers 1, 2,...,n and hence comes the name "permutation" graph. Simply, number the nets according to their upper terminals and obtain the permutation at the lower side as shown in Figure 2 (c). The minimal grouping problem is then transformed to routing the nets to a minimal number of layers such that each layer is used by a group of non-intersecting nets as illustrated in Figure 2 (c).

If more than one variable is accessed at the same time step then we have to create superficial points to make sure that no two nets share the same upper or lower terminal. In creating superficial points, care is taken to make sure that incompatible variables will result in incompatible (*crossing*) nets as shown in Figure 3. Finally we end up with a permutation of the numbers 1,2,...n which corresponds to the set of variables. Now sorting the permutation using minimal number of queues (i.e., coloring the permutation graph) solves the minimal grouping problem.

In summary, the set of input variables is transformed into a set of nets which can be represented by a permutation and then the permutation is fed to the following canonical coloring algorithm for permutation graphs obtained directly from [9].

Algorithm: Allocate to SFMs

$k=0$;
  for $j:=1$ to n do begin
    $i:=$ index of first allowable SFM /* i is the smallest value s.t
                $p_j >$ Last(i) which is the last entry of SFM$_i$ */
    Color (P_j) := i;   /* assign v_j to SFM$_i$ */
    Last (i) := P_j;   /* p_j becomes the last entry in SFM$_i$ */
    $k:=\max(k,i)$;   / * k = number of SFMs used so far   */
  end
return k

It is fairly easy to see that applying the above algorithm on the example in Figure 2 assigns v_1 and v_3 to a SFM and v_2, v_4 and v_5 to another one as illustrated in the Figure 2 (d).

5. EXPERIMENTAL RESULTS

The ILP formulation is solved using the *lp-solve* package [12] and the algorithm based on node coloring has been implemented on a Pentium II machine running Windows95. A number of examples from the literature have been tried to illustrate the proposed approach and to compare the use of SFMs against previous work. High-level synthesis benchmarks were not used since they are specified for a complete synthesis process so the lifetimes of variables obtained are scheduler dependent which prevents accurate comparison between allocation procedures as pointed out in [2]. Therefore we will only consider already scheduled test cases.

The first test case is a simple scheduled example used in [8]. The results based on both ILP and graph based algorithm are summa-

Figure 2: Graph based model: (a) Input variables (b) Conflict graph
        (c) Corresponding set of nets routed using two layers
        (d) The allocation of variables to SFMs

4.2 Minimal grouping based on node coloring

The minimal grouping problem can best be treated using a graph theoretic model by constructing a resource conflict graph $G=[V,E]$ where V is a set of vertices corresponding to the set of variables S, and E is the set of edges defined as follows:

(i,j) is in E iff neither variable v_i precedes variable v_j nor does v_j precede v_i (i.e., v_i and v_j cannot be grouped in the same SFM). Now the problem of minimal grouping is solved by solving the coloring problem for the conflict graph as illustrated in Figure 2 (a) and (b). Graph coloring is NP-complete for general graphs but fortunately, this particular graph falls in a special category of graphs called *permutation graphs* or *queue sorting graphs* [9] for which many combinatorial optimization problems are easy. In fact the minimal grouping problem at hand is strikingly similar to the queue sorting problem [9]. Allocating variables to minimal number of SFMs is achieved by coloring a permutation graph which can be done in $O(n.\log n)$ using the algorithm described in [9].

To make things easier to understand, we will transform the problem to a special version of the single-layer river routing problem in which we have a number of nets to be routed such that each net corresponds to a variable. Each net has a single terminal at the upper side and a single terminal at the lower side. The upper and lower terminals correspond to the write time and read time of the variable respectively. A straight line is drawn between the upper terminal and the lower terminal to form a net. Notice that the re-

rized in Tables 1 along with a comparison with stack based approaches. The second test case is a set of matrix transposition benchmarks suggested by Parhi [13]. Table 2. summarizes the results obtained by the proposed algorithms compared to those obtained by SRWMs [2] assuming single phase clocking. Notice that different ILP results (only in number of registers) were obtained by reordering data. Several other examples have been tried as well. In general, the use of SFMs yielded better or equivalent results than other non-RAM structures in most test cases.

Table 1: Comparison of results for single stacks, dual stacks and SFMs

Case	Mapping to Single Stacks	Mapping to dual stacks	Mapping to SFMs
# Modules	5	3	3
Total # Registers[a]	6	5	5

a. assuming a 2-phase clock.

Table 2: Results of the matrix transposition example

Case	Using SRWM [2]		SFM using ILP		SFM by node coloring	
	# Modules	# locations	# Modules	# locations	# Modules	# locations
3x3	2	7..8	2	5-6	2	5
4x4	4	13..14	3	10-12	3	11
5x5	4..5	22..24	4	17-20	4	18
6x6	5..6	33..35	NA	NA	5	28
7x7	6	45..49	NA	NA	6	39
8x8	7	61..64	NA	NA	7	53

6. FINAL REMARKS

Thus far in this paper, only single-write single-read variables were considered. The formulation can be extended to handle single-write multiple read variables without the need for variable splitting. Basically *Definition* 2 is modified as follows:

v_1 *precedes* v_2 if $W(v_1) < W(v_2)$ and $R_{last}(v_1) < R_{first}(v_2)$

In other words we have to make sure if v_1 is written before v_2 then the latest read of v_1 must occur before the earliest read of v_2. In this way if a variable is accessed more than once then the read pointer is kept pointing at its location during the period from its *first* read until its *last* read. The resulting compatibility graph is transitively orientable (comparability graph [9,10]) and its clique problems are easy. The minimal grouping problem can now be solved using the method presented in [1] and [9] which transforms the clique partitioning problem of comparability graphs into the well known network flow problem. There is a pitfall however! If a large number of variables have multiple reads that occur at "distant" time steps then the grouping might not be efficient because of the high potential for conflict. Therefore, it might be sometimes desirable to perform variable splitting and as such, a good preprocessing step might be to have an intelligent variable splitting phase which decides if and where to perform variable splitting. Such a step could be an interesting future extension.

The second point is regarding applications characterized with infinite iterations (mainly DSP) in which operations are repeated every T steps where T is the iteration period. The allocation algorithms has to be modified a little bit to accommodate this application domain. Basically we have to make sure that no conflict exists between variables from different iterations.

Third, minimizing the total number of registers was not an explicit objective in the minimal grouping approaches. Nevertheless, it can be done as a post-processing step by performing pair-wise exchanges between variables mapped to different modules. Simulated annealing is one possible candidate for this task. The same thing could be done to reduce interconnection cost.

Finally, it should be emphasized that non-RAM memories, including SFMs, are not intended to replace RAMs all the time but rather as alternatives that could be used when the application best suites them.

7. CONCLUSIONS

The problem of allocating variables to SFMs is addressed in this paper. First a theoretical investigation of the problem uncovers important results about the computational complexity associated with the allocation problem and then two approaches are proposed. The first approach is a mathematical ILP formulation and the second one is a classical algorithm that exploits the specific structure of the conflict graph associated with SFMs mapping.The analytical reasoning and experimental results indicate that SFMs are more flexible than previous special structure memories, and can be effective in eliminating address generation and decoding and their associated overhead.

8. REFERENCES

[1] L. Stock, "Data Path Synthesis," *INTEGRATION, the VLSI Journal*, Elsevier, 18 (1994) pp. 1-71

[2] S.H. Gerez and E.G. Woutersen, "Assignment of Storage Values to Sequential Read-Write Memories," *In Proc. of the European Design Automation Conference*, 1996.

[3] M. Aloqeely and C.Y.R. Chen, "Sequencer-based data-path synthesis of regular iterative algorithms," *In ACM/IEEE 31st Design Automation Conf.*, pages 394-399, 1994.

[4] M. Aloqeely and C.Y. R. Chen, "A new approach for exploiting regularity in data-path synthesis," *In the European Design Automation Conference*, Sept. 1994.

[5] M. Aloqeely, Sequencers, a new alternative for data-path synthesis. Ph.D. dissertation, Dept. of Electrical and Computer Engineering, Syracuse University, June 1995.

[6] I.E. Bennour and E.M. Aboulhamid, "Register allocation using circular FIFOs," *In IEEE International Symposium on Circuits and Systems*, pages 560-563, 1996.

[7] A. Heubi, S. Grassi, M. Ansorge, and Pelandini. A low power VLSI architecture with an application to adaptive algorithms for digital hearing aids. In M.J. Holt et al, editors, *Signal Processing VII: Theories and Applications (Proc. of EUSIPCO-94 7th Euro. Sig. Proc. Conf.)*, pages 1975-1878, 1994.

[8] I. Ahmed, M. Dhodhi, and F. Ali, "Allocation of dual-stacks in data path synthesis," *In Proc. of IEEE International Conference on Computer Design*, October 1996.

[9] M.C. Golumbic, *Algorithmic Graph Theory and Perfect Graphs*. Academic Press 1980.

[10] N. Sherwani, *Algorithms for VLSI design automation*. Kluwer Academic Publishers, 1992.

[11] H. Khanna and M. Balakrishnan, "Allocation of FIFO Structures in RTL Data Paths," *In Proc. of the 10th Int. Conf. on VLSI Design*, Jan. 1997, pp. 130-133.

[12] *lp-solve* package. Endhoven Univ. of Tech., Design Automation Section, ftp cite: ftp.es.ele.tue.nl

[13] K. Parhi, "Systematic synthesis of DSP data format converters," *IEEE Trans. on Circuits and Systems-II*, 39(7):423-440, July 1992.

A New Partitioning Framework for Uniform Clock Distribution During High-Level Synthesis

H. Krishnamurthy[1], M. B. Maaz, and M. A. Bayoumi
The Center for Advanced Computer Studies
The University of Southwestern Louisiana, Lafayette, Louisiana 70504

Abstract — In this paper, we present a new partitioning framework that significantly provides uniform clock distribution during high-level sysnthesis. This weighted cluster partitioning technique is aided by an innovative resource allocation technique which provides a uniform resource utilization. This results in a uniform power requirement distribution across the synthesized chip which is practically beneficial as well. The resource cluster information obtained from this new partitioning synthesis framework could be used to aid the hierarchical placement and routing tools as well as the clock-router to design and produce a close to optimal uniform clock routing. This new framework applies the idea of local (intra-cluster) communication, while minimizing global communication busses. This resulted in a low-power design as well. The partitioner cost function considers three factors: the number of clock lines per unit area, the number of cuts between clusters, and the size of partitions. Experimental results show that this partitioning framework provides uniform clock distribution among clusters as well as fairly uniform sized partitions.

I. Introduction

Traditionally, clock routing is done after the placement and routing of individual blocks/clusters that are identified by the placement tool or very rarely, by the logic synthesis tools. While a great amount of current research work is being focussed on developing clock routing tools to obtain a close to optimal uniform/balanced clock distribution with minimal skew, little or no effort has been invested on exploring the HLS (High-Level Synthesis) domain to guide a clock distribution/clustering driven synthesis. The primary focus of our proposed high-level synthesis framework is to target uniform clock distribution/requirement among the clusters obtained after resource allocation. This would result in hardware clusters with a fairly uniform clock requirement among them, in turn aiding the clock router to design a close to optimal clock routing for the entire chip.

The minimization of global busses and implementing local intra-cluster communication is also considered. This would minimize power consumption due to switching on long global busses (inter-cluster communication) [1] [2]. Also, while allocating resources to operations in the input scheduled graph, uniform resource utilization is considered as well. A modified clique partitioning technique based on the algorithm in [1] has been used to perform resource allocation. A weighted cluster partitioning technique based on the traditional ratio-cut partitioning approach [3] with a new cost function has been implemented for partitioning the netlist to individual clusters/blocks.

The following section lists of the assumptions being considered, and the input to the framework. In section three, the functional flow of the partitioner is presented. The resource allocation technique being employed in this work is discussed in section four. Section five discusses the weighted cluster partitioning technique with the cost function. In section six, the experimental results are presented, and in section seven possible avenues for future research and conclusions regarding this framework are discussed.

II. Assumptions and Input to the Framework

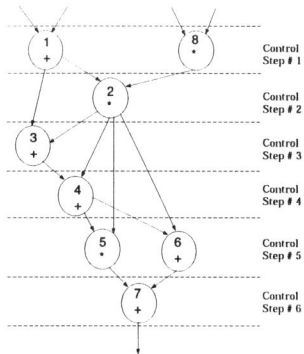

Fig. 1 A scheduled input data flow graph as an example

During the design of this new partitioning framework, certain assumptions are made, primarily for ease of implementation and to demonstrate the feasibility of this approach :

1. The input is an already scheduled data flow graph
2. Only addition and multiplication operations in the input graph
3. The area and clock requirement/s of hardware functional units (for adders and multipliers in this case) are provided up front
4. No feedback loops in the input graph
5. All operations/nodes in the input graph have two inputs and possible multiple fanouts

Scheduling a given data flow graph has been considered only as an elementary phase in the overall resource allocation context in several contemporary work in high-level synthesis. Though scheduling would influence the result of the resource allocation phase, for this work scheduled data flow graphs with some degree of freedom for the nodes is assumed and in fact, taken advantage of to generate multiple resource allocation solutions.

The information about area and clock requirement/s of individual functional units is vital to this project since the ratio-cut partitioning [3] tool needs to evaluate the merit of partitions/clusters that it generates. The assumption about each node having two inputs and multiple fanouts is practical in almost all the functional units that are designed in current systems.

---
[1] Currently, he is with Intel., Oregon.

The input to the framework is a scheduled data flow graph represented in ASCII format. The nodes in the graph represent operations (not functional units). Figure 1. shows an input scheduled DFG which will be considered as an example through out the paper.

III. Functional Flow of the Framework

The functional flow of the framework is shown in figure 2. In this new framework, resource allocation to nodes (representing operations) in the input graph is being performed before partitioning in order to specifically target uniform clock requirement among the hardware clusters. Some contemporary work in the area of partitioning graphs with min-cut objectives [4], have targeted partitioning on the input data flow graph while performing resource allocation later on [5]. This would not give an accurate picture of the clock requirement among the obtained clusters since the clusters at this stage (after partitioning) still contain the nodes/operations which could map to different hardware resources. Therefore, it is imperative to perform resource allocation before partitioning the obtained netlist in order to target a uniform clock distribution/requirement among clusters [6] [7]. In order to explore the available design space in multiple ways, the resource allocation phase has to generate different possible solutions to the allocation problem.

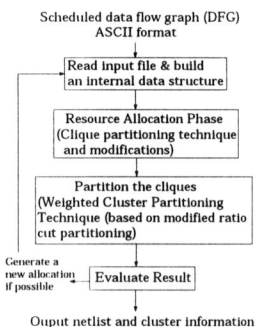

Fig. 2 Functional flow of the HLS partitioning framework

IV. The Resource Allocation Phase

The basic clique partitioning algorithm [2] is based on a neighborhood heuristic and is one of the most popular algorithms for resource allocation apart from the move based approaches like simulated annealing and stochastic evolution. The primary objectives of the resource allocation phase are:
1. Mapping operations (nodes in the input graph) to as minimum resources (functional units) as possible
2. Capable of generating multiple allocation solutions for the same given input graph to better explore the design space
3. Balancing resource utilization (the number of operations mapped to each functional unit)
4. Encouraging local communications to be mapped to the same resource/functional unit
5. Capable of utilizing the degree of freedom (ASAP — ALAP values of the nodes in the scheduled input graph) [2] to generate different allocation solutions
6. Capable of handling multiple types of operations (additions and multiplications for instance) in the input graph and mapping them to respective types of resources (adders and multipliers respectively in this case)

The basic steps of the conventional clique partitioning algorithm [2] are as follows:
1. Compute the number of common neighbors for all edges in the graph and also identify the number of edges that have to be deleted in case the nodes of each edge are merged to one cluster. When merging two nodes, the following edges are deleted from the graph:
   a. edge between the two nodes to be merged
   b. one of the edges to their common neighbors
   c. if any node in the graph is a neighbor of only one of the two nodes to be merged, then the edge is deleted as well
2. Choose the edge which has the maximum number of common neighbors for merging (based on the neighborhood heuristic, thereby encouraging local communication as well). If there are more than one edges with the same number of common neighbors, then choose the one which has the least number of edge deletions in case the two nodes are merged together.
3. After forming a cluster, update the graph and repeat the process until all edges are eliminated and all nodes clustered.

The basic clique partitioning algorithm [2] has some drawback, and a new modified resource allocation could be developed to address the issues and also help explore a wider design space during allocation. Consider the following example for instance:

If the original input graph is composed of nodes, say addition operations and all of them are scheduled in different control steps [10], applying the original clique partitioning algorithm would have resulted in the allocation of two different adders while one would have sufficed. In the modified algorithm, an edge to one of the nodes being merged (if it's not a common neighbor) is not deleted and retained with additional control step violation checks performed before clustering nodes to cliques. Some of the salient modifications/enhancements proposed in this work are:
1. Retain edges to single neighbors of edges merged
2. While forming cliques, check for control step violation on the resource
3. When single floating nodes are formed due to edge deletions, try clustering them with the already formed cliques
4. Exploit the nodes' degree of freedom in the original schedule to better the solution if possible
5. For each clique formed, check for the number of operations mapped to that resource so that a more uniform resource utilization could be obtained
6. Pick edges for clustering at random if the other factors are common (edges 2–3 and 3–4 in the first example above). This would result in a different allocation and perhaps, a wider design space

Figure 3 shows the result of applying the modified clique partitioning algorithm. The cliques shown in the graph were the output of the resource allocation phase. As noticed in the figure, some cliques contain only addition or multiplication operations, whereas one of the cliques (clique 2) contains both addition and multiplication operations (if cluster contains both adder and multiplier modules). This implies that an adder and a multiplier are mapped to the same clique.

V. Weighted Cluster Partitioning Algorithm

The output of the resource allocation phase is the netlist where cliques represent hardware units instead of operations. This output

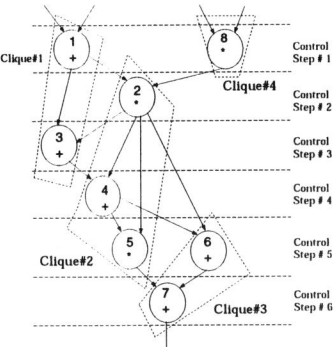

Fig. 3 Applying clique partitioning algorithm to a DFG

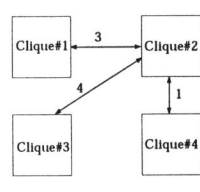

Clique#	Type	Operations mapped to clique
1	+	1,3
2	+,*	2,4,5
3	+	6,7
4	*	8

Fig. 4 The input representation to the partitioning phase

is then fed into the partitioning phase of the framework. The output of the resource allocation phase represented in Fig.3 is shown in figure4. Cliques are joined together by edges which represent the data dependencies between different nodes of the original graph mapped on to the respective cliques. The directions of these dependencies would not make any difference because the goal is to minimize the number of cuts between partitions [6].

In order to target uniform clock distribution as well as min-cut between clusters[6] [8], and uniform area among the clusters obtained, weighted cluster partitioning technique is used. This algorithm is based on ratio-cut partitioning algorithm [3] with modified cost function. The cost function is given in equation 1. as follow:

$$CostFunction: Cf\alpha(W1Cl/Ai, W2Ci, W3Ai) \qquad eq. 1$$

where Cl/Ai represents the number of clock signals per area, Ci represents the number of cuts between clusters, Ai is the area of each cluster. W1, W2, W3 are the weights of the cost function. $0<=W1,W2,W3<=1$. These weight values varies depending on which one of the above three factors need to be targeted most. Because our main focus is to obtain uniform clock distribution, W1 will have the highest weight value, followed by W2, and then W3. This weighted cluster technique is based on the ratio cut algorithm which is divided into two phases: initialization, and iterative shifting.

V.1. Initialization

This phase of the ratio-cut algorithm [3] is designed to determine two cliques as references which are called (*ref_clique*) to apply the ratio-cut algorithm on them. The first reference (*ref_clique_s*) is assumed to be the first clique in the graph, or it can be randomly picked. The *Breadth-First* algorithm [10] is applied to traverse the graph in order to find the farthest clique from *ref_clique_s*, and let this clique be (*ref_clique_t*). The second step of this phase is to determine two initial clusters, each containing either (*ref_clique_s*) or (*ref_clique_t*) but not both. This is done to assure the proper partitioning sequence. Starting with (*ref_clique_s*) add one clique at a time with the *ref_clique_s* and compute the ratio cut obtained by this move, until all the cliques beside the other (*ref_clique_t*) are moved to the partition containing the first ref_clique. Even if a move results in an increase in the ratio-cut value, more cliques are moved to the first seed. The combination of the cliques with *ref_clique_s* that result in the smallest ratio cut value is then chosen. The same step is then repeated for *ref_clique_t*, and another minimal ratio-cut partition is formed. Both results are compared and the smallest ratio-cut value is chosen, and two partitions are formed. For our example, cluster#1 includes clique#1 which is found to be the (*ref_clique_s*), and cliques 2, 4. cluster#2 includes only clique#3 which is found to be the (*ref_clique_t*). Because cluster#1 has more cliques mapped to it than cluster#2, then the sequence s->t is used, which refers to the shifting direction to be considered in phase II of the algorithm. If more cliques are mapped to cluster#2 than cluster#1, then t->s shifting direction will be considered in phase II of the algorithm.

V.2. Iterative Shifting

This phase basically determines as close to optimal partitioning solution as possible. The initial clusters achieved from the initialization phase are used here. If the initialization part, the sequence s->t is chosen then right shift is applied, otherwise if t->s sequence is chosen then left shift is applied. The reason for this sequencing approach is to limit the number of shifting to be performed as well as eliminating redundant shifting that may worsen the partitioning result. Shifting means to move one clique at a time from one cluster to another. If left shifting is first being utilized, then we keep on monitoring the value of the ratio-cut after each shift. Whenever a new minimum ratio_cut value is obtained, the sequence of the shifting will be changed to right shifting. This is done to obtain a feasible answer. The sequences required for iterative shifting is shown in figure 5. Figure 6 shows the result of applying partitioning steps to our illustrative example.

VI. Output Results of The Partitioner

By analyzing the above obtained results, we notice that placing cliques 1,2,and4 in cluster#1, and clique 3 in cluster#2, or placing cliques 1, 4 in cluster#1, and cliques 2,3 in cluster#2 result in the same value of cuts between cluster which found to be 4 (data dependencies). However, the second solution which was the result of applying the iterative shifting phase of the partitioner gives more uniform clock distribution, as well as close to uniform areas of clusters 1,and 2, while maintaining the same number of dependencies between clusters/partitions. For the above example, two partitions are found to be sufficient, but in most cases, we may need to divide certain partitions into smaller clusters/partitions by applying the same partitioning algorithm. This is required in order to obtain uniform clock distribution across the chip.

The output of the partitioner could be used to aid hierarchical placement and routing tools, as well as clock routing tools. It is important to place all modules that share a lot of dependencies close to one another, in order to minimize routing. This require knowledge of determining dependencies, as well as the total sequence of partitioning. Our framework assigns a sequence of bits to represent each partition/cluster. This placement and routing idea is illustrated in figure 7. The first partition is assigned a value of 1 (this partition contains all the modules). After applying the partitioning algorithm, we obtain two new clusters, both are the children of the first cluster. They are assigned values 11, and 10 to represent them. The placement of "0" and "1" depends on the partitioning. The value "0" refers to right child of the parent partition, where value "1" refers to left child of the parent partition. Partitions that share the same common parent are placed near each other because they usually share a lot of dependencies.

VII. Conclusion

In this paper, we have presented a new framework that provides uniform clock distribution during high-level synthesis. This weighted cluster partitioning technique[3] was aided by an innovative resource allocation phase based on modified clique

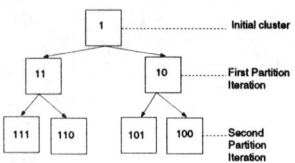

Fig. 7 Output cluster Information from the Partitioner

partitioning [2]. The output of the framework would also produce a uniform power requirement distribution across the synthesized chip. The resource cluster information obtained from this framework could be used to aid hierarchial placement and routing tool, as well as the clock router to design, to produce close to optimal uniform clock routing. The framework encourages local (intra-cluster) communication, while minimizing global communication busses which directly result in lower power [9].

Acknowledgment

The authors acknowledge the support of U.S Department of Energy (DoE), EETAPP program.

References

1. R.Mehra, L.M.Guerra and J.M.Rabaey, "A Partitioning Scheme for Optimizing Interconnect Power," in *IEEE Journal of Solid-State Circuits*, Vol.32, No.3, March 1997, pp. 433–443
2. C.J.Tseng and D.P.Siewiorek, "Automated Synthesis of Data Paths in Digital Systems," in *IEEE Trans. on Computer-Aided Design*, Vol. CAD-5, No.3, July 1986, pp. 379–395
3. Y.C.Wei and C.K.Cheng, "Ratio Cut Partitioning for Hierarchical Designs," in *IEEE Trans. on Computer-Aided Design*, Vol.10, No.7, July 1991, pp. 911–920
4. G.Vijayan, "Partitioning Logic on Graph Structures to Minimize Routing Cost," in *IEEE Trans. on Computer-Aided Design*, Vol.9, No.12, December 1990, pp. 1326–1334
5. Y.G.Saab and V.B.Rao, "Fast Effective Heuristics for the Graph Bisectioning Problem," *IEEE Trans. on Computer-Aided Design*, Vol. 9, No. 1, January 1990, pp.91–98
6. B.Krishnamurthy, "An Improved Min-Cut Algorithm For Partitioning VLSI Networks," in *IEEE Trans. on Computer-Aided Design*, Vol. C-33, No.5, May 1984, pp.438–446
7. S. Devadas and A.R.Newton, "Algorithms for Hardware Allocation in Data Path Synthesis," in *IEEE Trans. on Computer-Aided Design*, Vol. 8, No.7, July 1989, pp. 768–781
8. H.H.Yang and D.F.Wong, "Efficient Network Flow Based Min-Cut Balanced Partitioning," in *IEEE Trans. on Computer-Aided Design of Integrated Circuits and Systems*, Vol.15, No. 12, December 1996, pp.1533–1540
9. R.V.Cherabuddi, M.A.Bayoumi and H.Krishnamurthy, "A Low Power Based System Partitioning and Binding Technique for Multi-Chip Modules," in *Proc. of the Seventh Great Lakes Symposium on VLSI, Urbana-Champaign, IL*, March 1997, pp. 156–162
10. Y.Alavi *et al*, "Graph Theory, Combinatorics and Applications — volume 2," in the*Proc. of the Sixth Quadrennial International Conference on the Theory and Applications of Graphs*, John Wiley & Sons, Inc. 1991

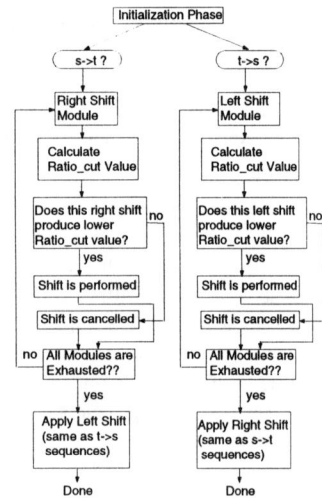

Fig. 5 Flow chart for the Iterative Shifting phase of the partitioning algorithm

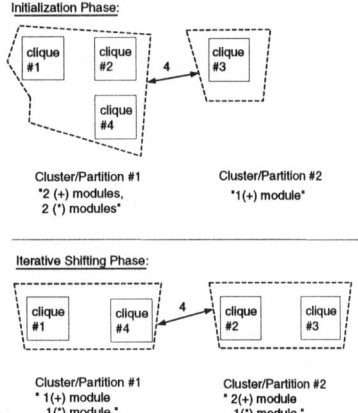

Fig. 6 The result of applying partitioning steps to the illustrative example in figure 4

A BINDING ALGORITHM FOR RETARGETABLE COMPILATION TO NON-ORTHOGONAL DATAPATH ARCHITECTURES

Masayuki YAMAGUCHI[†,††], Nagisa ISHIURA[††], and Takashi KAMBE[†]

[†]Precision Technology Development Center, SHARP Corporation
2613-1 Ichinomoto-cho, Tenri, Nara, 632-8567 Japan
Phone: +81-743-65-2531, Fax: +81-743-65-4968
E-mail: {masa,kambe}@edag.ptdg.sharp.co.jp

[††]Dept. Information Systems Eng., Osaka University
2-1 Yamada-Oka, Suita, Osaka, 565-0871 Japan
Phone: +81-6-879-7807, Fax: +81-6-875-5902
E-mail: {yamaguti,ishiura}@ise.eng.osaka-u.ac.jp

ABSTRACT

This paper presents a new binding algorithm for a retargetable compiler which can deal with diverse architectures of application specific embedded processors. The architectural diversity includes a "non-orthogonal" datapath configuration where all the registers are not equally accessible by all the functional units. Under this assumption, binding becomes a hard task because inadvertent assignment of an operation to a functional unit may rule out possible assignment of other operations due to unreachability among datapath resources. We propose a new BDD-based binding algorithm to solve this problem. In the experiments, a feasible binding which satisfies the reachability is found or the deficiency of datapath is detected within a few seconds.

1. INTRODUCTION

A retargetable compiler is a useful tool in the design of systems using application specific instruction processors, for it enables (1) efficient software development using programming language, (2) software development in concurrence with hardware design, and (3) easier software transportation from an architecture to newly designed or modified ones. Motivated by these merits, a lot of researches have been carried out [1-8] to enhance retargetability and to improve code quality of such compilers.

Embedded processors used in commercial audio and video products often employ so radical architectures that conventional retargetable compilers aiming at simple extension of general purpose processors can not deal with them. One of the characteristics that makes the retargeting difficult is a "non-orthogonal" datapath configuration. Figure 1 shows an example of such a datapath. It contains so called "heterogeneous registers" which are not equally accessible by all the functional units but work for some special purposes associated with specific functional units.

CHESS system [7] has presented a graph-based approach to deal with such architectures. It tries to map DFGs representing a source program onto a datapath graph by binding and scheduling methods similar to those used in high-level synthesis. Binding of the DFG onto a non-orthogonal datapath is reduced to a path search problem.

However, we found that it leaves an important problem unresolved: a problem of "resource unreachability." Let us explain this on the datapath of Figure 1. Suppose we have two instructions ALU_ADD and ADRS_ADD which execute addition using ALU and ADD, respectively, and we want to read RAM1 using the address obtained by addition of two values stored in registers ACC0 and ACC1. The desirable solution is to use ADRS_ADD because only ADRS_ADD can store the result to register A of RAM1. However, a compiler may assign the addition to the ALU since it takes a lot of clock cycles to transfer the data in ACC0 and ACC1 to ADD. In this case, the result

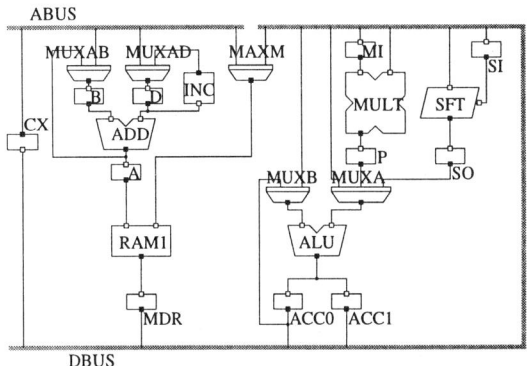

Figure 1: Processor with non-orthogonal architecture

of the addition is unreachable to the address port of RAM1, and thus compilation fails.

Although problems in this particular example may be resolved by a simple look-ahead technique, reconvergences in a DFG and asymmetry in a datapath could make the problem very tricky. Thus, binding becomes one of the problems of the highest priority in developing compilers retargeting non-orthogonal datapath architectures.

In this paper, we propose a binding algorithm incorporating resource unreachability problem in non-orthogonal datapaths. We solve the binding problem in two steps. In the first phase, each operation node in the DFG is assigned to a functional unit in the datapath, so that the existence of the paths between functional units corresponding to the edges in the DFG are guaranteed (if they exist at all). We use BDD-based algorithm to solve this search problem and enumerate the feasible combination of operation assignment. In the second binding phase, assignment of data paths to the edges in DFG is attempted for one of the operation assignments obtained in the first phase.

We first show the model of target architectures in Section 2. The binding problem and our algorithm are described in Section 3. Section 4 shows experimental results.

2. MODELING OF TARGET ARCHITECTURES

A target architecture is represented by the combination of a datapath graph (DPG) and parallel constraints (PC).

A datapath graph DPG S consists of set R of resources and set C of interconnections. The DPG is a graph representation of a datapath where nodes and edges represent resources and connections, respectively. R is divided into two disjoint subsets R_F and R_R. R_F is a set of functional units like ALUs and multipliers. For $r \in R_F$ a set of the operations which r can execute is defined. ROMs, RAMs, and selectors (buses) are also classified into this category. R_R is a set of register files which operate in synchronous to the clock. For $r \in R_R$ defined is the capacity of r, denoted by capacity(r), indicating the number of data r can hold. Each r in R has an output and inputs. The output and the k-th input of r are denoted by $r.out$ and $r.in_k$, respectively. We say there is a path from output $r_o.out$ to input $r_m.in_k$ if $r_m.in_k$ is reachable from $r_o.out$ by going through registers functional units which provide "through operation." The through operation, for example, refers to data transfer from one of the inputs to the output of a multiplier by setting 1 to the other input.

Parallel constraints PC are constraints posed on the datapath, which limit the simultaneous activities of the resources and interconnections. Let x_r ($r \in R$) and y_c ($c \in C$) be Boolean variables where $x_r = 1$ ($y_c = 1$) iff r (c) is activated. Then the parallel constraints are expressed in the form of a logical expression PC consisting from x_r and y_c. It declares that activities which makes $PC = 1$ is inhibited. For example, $x_{\text{ADD}} \wedge x_{\text{MULT}}$ means that ADD and MULT can not operate at the same control step.

Target architectures are specified either in terms of

(1) the instruction set,

(2) the datapath configuration, or

(3) the combination of (1) and (2).

In either case, the target architecture is modeled by the combination of DPG and PC. If we have an instruction set, we can construct a DPG by listing all the primitive operations and all the combinations of source and destination registers for each operation from the instruction set. Parallel constraints are used to specify the effect of instruction formats [9]: If a combination of operations or register accesses is not included in the instruction set because of some limitations such as sharing of instruction fields or lack of definitions, PC is added to the model. On the other hand, if we have a datapath architecture, we can construct a DPG in a straightforward manner by tracing the datapath structure.

3. BINDING PROBLEM

3.1. Approach

As we have shown in Introduction, we must assign an addition to the adder attached to an RAM in order to use the result of the addition as an address to the RAM. The difficulty is not limited to the binding of operations to functional units. Binding of data transfers to paths in DP is also a hard task. For example, since ALU in Figure 1 is connected to a single bus and does not have registers to latch input data, it is required that either of the two data be supplied from a path not containing DBUS. The parallel constraints must be also considered at the same time.

Since we think the two binding problems above are also too hard to be solved in a single step, we decided to employ a two-phase binding strategy. In the first phase, we search possible solutions for operation binding, and in the second phase, we try path binding to consummate the binding task.

3.2. Binding of Operations

A problem of operation binding may be well explained using a *binding space graph (BSG)* for a given DFG and a DP. The Figure 2 shows an example of DFG and its BSG. The nodes of the BSG are grouped into subsets $A_1, \cdots, A_7$ where A_i consists of nodes representing possible assignments of operation o_i. The edges in BSG indicate the existence of paths between bound operation resources. Our goal is to find a combination of assignments which provides all the necessary paths to the data dependency edges. A subgraph depicted by bold lines indicates a solution.

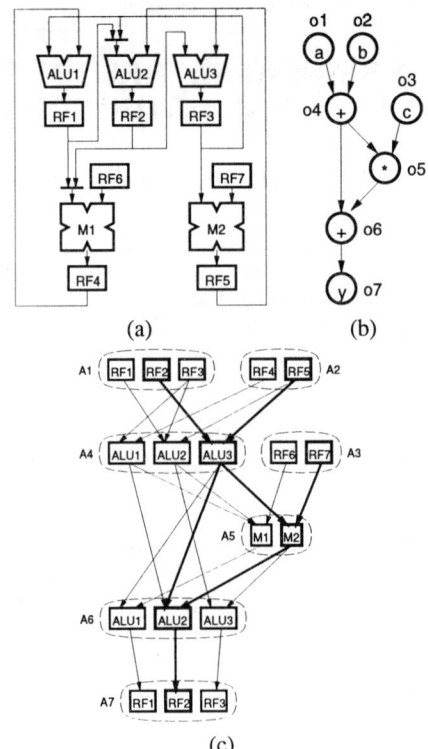

Figure 2: DFG and its binding space graph BSG.

Formally, BSG for a DFG (O, E) and a DP (R, C) is a directed acyclic graph (A, T). The node set A comprises of $n = |O|$ disjoint subsets $A = A_1 \cup A_2 \cup \cdots \cup A_n$ where A_i is associated with operation $o_i \in O$. If $\{r_{f_1}, r_{f_2}, \cdots, r_{f_{m_i}}\}$ is a set of functional resources which can execute operation o_i, then $A_i = \{a_{i,f_1}, a_{i,f_2}, \cdots, a_{i,f_{m_i}}\}$ where a_{i,f_j} represents that o_i is bound to r_{f_j}. Edge $(a_{p,f}, a_{i,f_j})$ in T exists iff $(o_p, o_i) \in E$ is the k-th incoming edge of o_i and there is a path from $r_f.at$ to $r_{f_j}.in_k$.

The operation binding problem is formulated as to find an embedding of a DFG (O, E) into the BSG (A, T). Namely, operation binding is a pair of mappings $\alpha = (\alpha_o, \alpha_e)$ (where $\alpha_o : O \to A$ and $\alpha_e : E \to T$) which satisfies:

1. $\alpha_o(o_i) \in A_i$, and
2. $\alpha_e((o_i, o_j)) = (\alpha_o(o_i), \alpha_o(o_j))$.

We have developed a BDD-based algorithm for solving this problem. We introduce Boolean variable x_{i,f_j} for each node a_{i,k_j}. Variable x_{i,f_j} becomes 1 iff $\alpha_o(o_i) = a_{i,f_j}$. Then we can set up the following two conditions:

1. Since only one element of $A_i = \{a_{i,f_1}, a_{i,f_2}, \cdots, a_{i,f_{m_i}}\}$ is chosen as $\alpha_o(o_i)$, only one of $x_{i,f_1}, x_{i,f_2}, \cdots, x_{i,f_{m_i}}$ becomes 1. Thus we have,

$$\text{ord}_1(A_i) = (\sum_{j=1}^{m_i} x_{i,f_j} = 1). \quad (1)$$

$$\text{ord}_1 = \bigwedge_{A_i \in A} \text{ord}_1(A_i). \quad (2)$$

Here $\sum$ stands for arithmetic summation.

2. In order that a_{i,f_j} be chosen as $\alpha_o(o_i)$, all the incoming edges to o_i must be mapped onto some edge $(a, a_{i,f_j}) \in T$ where a

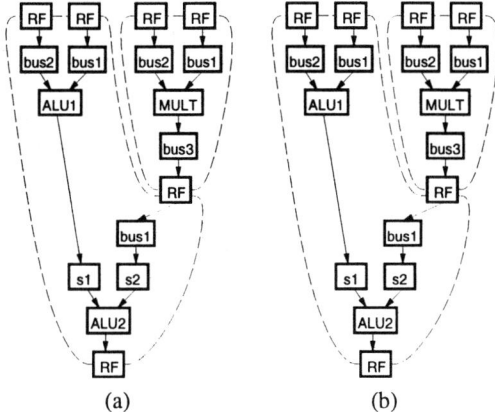

Figure 3: BDFG and simultaneous components.

is also chosen as $\alpha_o(o)$ for some $o \in O$. Let K_i be the number of incoming data dependency edges to o_i and $T_{i,f_j,k}$ be the set of the incoming edges to a_{i,f_j} that are associated with the k-th incoming edges to o_i. Let $x_{p(e)}$ be the Boolean variable of the source node of edge $e \in T$. Then the condition is formulated as:

$$cond_2(a_{i,f_j}) = x_{i,f_j} \rightarrow (\bigwedge_{k=1}^{K_i} \bigvee_{e \in T_{i,f_j,k}} x_{p(e)}). \quad (3)$$

$$cond_2 = \bigwedge_{a_{i,f_j} \in A} cond_2(a_{i,f_j}). \quad (4)$$

The operator "$\rightarrow$" means implication and $x \rightarrow y$ is equivalent to $\overline{x} \vee y$.

The necessary and sufficient condition for the existence of the solution is written as

$$cond = cond_1 \wedge cond_2. \quad (5)$$

The satisfiability of this condition is computed by constructing a BDD for $cond$. The resulting BDD represents the set of all feasible operation bindings.

3.3. Binding of Data Transfer Paths

In this phase, we assign for each edge in the DFG to a path in the DP. We must consider feasibility of solution with respect to path assignment.

The feasibility is easily checked by constructing a *bound DFG* (*BDFG*), decompose it into "simultaneous components," and check resource constraints, parallel constraints, and register constraints for each component.

BDFG is an augmentation of a DFG D whose differences are:

1. Each operation node is labeled by the hardware resource in R to which the the node is mapped.
2. It contains nodes corresponding to registers and selector resources that forms the path to which the data dependency edge is mapped.

Figure 3 shows examples of BDFG and their simultaneous components.

Simultaneous components are connected components in BDFG that are separated by the nodes labeled by register resources. For notational convenience, a register labeled node belongs to every components it disconnects. Simultaneous components are indicated by broken lines in Figure 3.

We only have to check the following constraints for every simultaneous component to see if an binding of functional units and paths are feasible or not:

resource constraints: If a simultaneous component contains the same functional resources or selector resources, the binding is not feasible.

register constraints: If the number of writes to any register resource r exceeds the capacity of r, the binding is not feasible. Moreover, if the number of writes (reads) to any register resource r exceeds the limitation of simultaneous writes (reads) of r, the binding is not feasible.

parallel constraints: If parallel constraints are violated in a simultaneous component, the binding is not feasible.

Now our goal is to find a mapping of data dependency edges to paths which results in a feasible binding. We employ the following straightforward procedure for this search.

1. Choose one operation binding α_o from the set of solutions obtained in the previous phase.
2. Based on α_o, assign one of the shortest paths to each edge in E.
3. Check the constraints for each simultaneous component. If OK then the feasible binding is found.
4. If NG, then mark the resources and the edges that are associated with the violation. For each marked edge, search a path in the DP, which does not go through the marked resources, and replace the original path by it.
5. If possible replacements are exhausted, then try another operation binding.

4. EXPERIMENTAL RESULTS

We implemented the proposed binding algorithm and evaluate its effectiveness. We tried to map some programs onto the datapath structure shown in Figure 1.

The first example "SUM" (Figure 4 (a)) is a program which summates array data by incremental accessing of data memory. We used a loop expansion option to generate the DFG shown in Figure 4 (b). The node labels show the names of the bound resources. Additions for the address increments and displacements are bound to INC and ADD which are indirectly connected to the address port of RAM1. On the other hand, additions for the data summation are bound to ALU. The procedure of this example took 0.7 seconds on Sparc Station 20.

The second example is a program FFT (Figure 5), a fraction of an FFT program, to which no feasible binding exists. Since there is no path from MULT to SHIFT in the datapath, there is no way of mapping the sequence of a multiplication ($*$) and a shift ($>>$) operation onto the datapath. The compiler proved the deficiency of the datapath in 0.5 seconds. We tried recompilation after adding a connection from MUXA to DBUS. This time, there was a solution for binding and scheduling and compilation finished in 0.5 seconds.

Table 1 is a summary of experiments on other programs such as ELLIP (elliptical filter), EDGE (edge detection of image data), and so on. We used the modified datapath structure in the previous experiment. "#lines," shows the line count of each program. "#DFG" and "#BSG" shows the total number of DFG nodes and BSG nodes, respectively. In each case, a feasible binding was found. "CPU" show the CPU time (on Sparc Station 20) for getting a binding result or for proving the deficiency of the datapath.

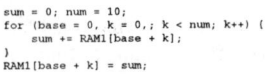

```
sum = 0; num = 10;
for (base = 0, k = 0,; k < num; k++) {
    sum += RAM1[base + k];
}
RAM1[base + k] = sum;
```

(a) C program of SUM.

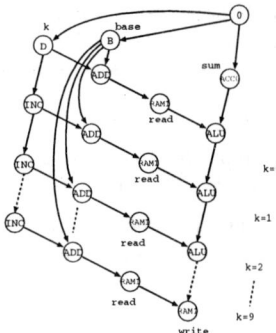

(b) DFG and its binding result.

Figure 4: Experimental results of SUM.

```
fft(n, k)
int    n, k;
{
    int    a, b, c, d, e, f;
    int    coefs, coefc;

    coefs = 512;
    coefc = n + 512;
    sin = RAM1[coefs + k];
    cos = RAM1[coefc + k];
    a = (cos * RAM1[1]) >> 4;
    b = (sin * RAM1[3]) >> 4;
    c = (cos * RAM1[3]) >> 4;
    d = (sin * RAM1[1]) >> 4;
    e = a - b;
    f = c - d;
    RAM1[0] = RAM1[0] + e;
    RAM1[1] = e - RAM1[1];
    RAM1[2] = RAM1[2] + f;
    RAM1[3] = f - RAM1[3];
}
```

Figure 5: C program of FFT'.

Table 1: Results of binding.

Program	#Lines	#DFG	#BSG	CPU (s)
SUM	22	43	74	0.7
FFT	21	38	58	0.5
ELLIP	45	110	134	0.5
WMAHA	58	111	962	0.8
EDGE	84	107	995	1.1

5. CONCLUSION

We have presented a problem of resource unreachability in binding for highly retargetable compilation, and have proposed a new BDD-based binding algorithm to solve it.

Development of a retargetable compiler for non-orthogonal datapath is under way, where we are facing a lot of challenges. Scheduling should take care of the capacity constraints and spill code insertion for heterogeneous registers. Handling of function calls also involves difficult problems to solve.

6. ACKNOWLEDGEMENT

The authors would like to thank Prof. Isao Shirakawa of Osaka University for his advice on this research. We would like to thank Mr. Tetsusaburo Yamamoto, Mr. Yasushi Hattori of Osaka University for their discussion.

7. REFERENCES

[1] G. Araujo and S. Malik: "Optimal Code Generation for Embedded Memory Non-Homogeneous Register Architecture," *Proc. Int. Symp. on System-Level Synthesis (ISSS)*, pp. 36–41 (Sept. 1995).

[2] B. Wess: "Automatic Instruction Code Generation based on Trellis Diagrams," *Proc. Int. Symp. on Circuit and Systems (IS-CAS)*, pp. 645–648 (1992).

[3] H. Akaboshi and H. Yasuura, "COACH: A Computer Aided Design Tool for Computer Architects," *IEICE Trans. Fundamentals, Japan*, vol. E76-A, no. 10, pp. 1760–1769 (Oct. 1993).

[4] C. Lien, P. Paulin, M. Cornero, and A. Jerraya: "Industrial Experience Using Rule-Driven Retargetable Code Generation for Multimedia Applications," *Proc. Int. Symp. on System-Level Synthesis (ISSS)*, pp. 60–65 (Sept. 1995).

[5] P. Marwedel and G. Goossens: *Code Generation for Embedded Processors*, Kluwer Academic Publishers, 1995.

[6] R. Leupers and P. Marwedel: "Retargetable Generation of Code Selectors from HDL Processor Models," *Proc. European Design & Test Conf. (ED&TC) '97*, pp. 140–144 (March 1997).

[7] J. Van Praet, D. Lannwwe, G. Goossens, W. Geurts, and H. De Man: "A Graph Based Processor Model for Retargetable Code Generation," *Proc. European Design & Test Conf. (ED&TC) '96*, pp. 102–107 (March 1996).

[8] J. Sato, Y. Honma, T. Nakata, A. Shiomi, N. Hikichi and M. Imai: "PEAS-I: A hardware/Software Codesign System for ASIP Development," *IEICE Trans. Fundamentals, Japan*, vol. E77-A, no. 3, pp. 483–491 (March 1994).

[9] M. Yamaguchi, A. Yamada, T. Nakaoka, and T. Kambe: "Architecture Evaluation Based on the Datapath Structure and Parallel Constraint," in *IEICE Trans. Fundamentals, Japan*, vol. E80-A, pp. 1853–1860 (Oct. 1997).

Modeling, Extraction and Simulation of CMOS I/O Circuits under ESD Stress *

Tong Li, Ching-Han Tsai, Elyse Rosenbaum and Sung-Mo Kang

Department of Electrical and Computer Engineering
Coordinated Science Laboratory
University of Illinois at Urbana-Champaign
1308 W. Main St., Urbana, IL 61801

Abstract— A CAD tool set for VLSI CMOS I/O circuit design is developed. It includes a circuit simulator, a layout extractor and a substrate resistance solver. This paper presents a new layout extractor for CMOS I/O circuits and a new method for modeling the substrate resistance. With these tools, for the first time, full I/O circuits can be simulated accurately at the circuit-level with the substrate-coupling effects taken into consideration. The CAD tools are demonstratively applied to an industrial circuit.

I. INTRODUCTION

Input/output (I/O) circuits critically affect VLSI reliability and signal integrity. The I/O circuits provide protections against electrostatic discharge (ESD) [1]. Traditionally, the design of ESD protection devices and circuits is approached empirically. The chip ESD reliability is measured with a tester. When a chip fails in the ESD test and has to be redesigned and fabricated again, the cost is high in both time and dollars. Therefore, an ESD reliability CAD tool is critically needed. We have developed a circuit-level simulator iETSIM for ESD protection circuit design verification [2]. In this work, we present a new layout extractor which generates the input deck for iETSIM simulation from the circuit layout. Furthermore, we present a new substrate resistance model and an extraction method which uses a 3D finite difference solver. The resistance extraction procedure is automated and incorporated into the layout extractor.

This paper is organized as follows. In the next section, the layout extractor is described. In section 3, we present the method for substrate resistance extraction. In section 4, an industrial I/O protection circuit is used to illustrate the extraction procedure, and simulation results are presented.

II. I/O LAYOUT EXTRACTION

Commercial layout extractors are only capable of performing limited extraction of the I/O layout. Their limitation stems from the following. First, they only extract the circuit schematic under the normal operating condition, i.e., when the chip is powered up. However, the circuit is typically not powered during ESD events. An ESD event can generate stress current in excess of 1 A to an I/O pin. Conventional device models are not applicable because the devices operate in the high current regime and may be biased differently from the normal operating modes. Therefore, the circuit schematic for I/O reliability simulation must be determined based on the ESD stress condition. Second, they can not extract parasitic devices such as parasitic BJTs which are formed due to side effects of

---

*This research was supported in part by Semiconductor Research Corp. (SRC97-DP-109), JSEP (N00014-97-J1270) and Rome Laboratory (F30602-97-1-0006).

the layout, but may play an important role during ESD stress events.

We have developed a systematic layout extraction procedure. Its block diagram is shown in Fig. 1. In this section, the parts within the dashed box are described.

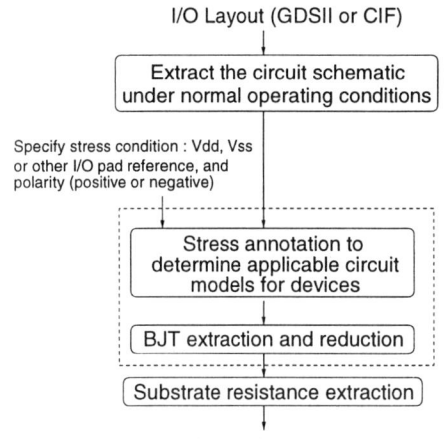

Figure 1: The I/O circuits extraction flow diagram.

A. Stress Annotation

According to the Mil Standard 883C method 3015.7 [3], the Human Body Model (HBM) test should zap all possible pin combinations of a chip in both positive and negative polarities, and the chip must pass 2 kV HBM-ESD level for all stresses. When the ESD zapping is performed between two pads, all other pads are kept floating. Each I/O pad must pass the required protection level with respect to V_{dd}, V_{ss} and any other pad for both polarities.

We propose a static analysis technique, called *stress annotation*, to determine the applicable circuit model for each device during an ESD event. Fig. 2 illustrates the stress annotation performed on an I/O circuit for a specified stress condition, i.e., positive stress on the pad with respect to V_{ss}. First, the circuit schematic for this layout under normal operating conditions must be extracted. Next, the stress annotation is conducted using a breadth-first search to propagate the stress current from the stressed pad. The stress current passes through forward-biased pn junctions and semiconductor resistors. It stops when a reverse-biased pn junction is reached. Each interconnect net in the current path is annotated with its *stress strength* (SS). The relative voltage levels between two interconnect nets can be compared by checking their stress strengths. In addition, the stress annotation will help identify parasitic BJT devices,

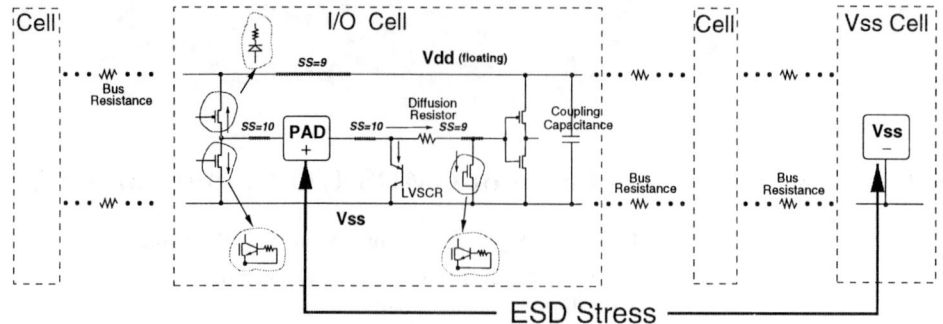

Figure 2: The stress annotation to identify each device's bias condition, and determine its circuit model.

as will be discussed later. Starting with an initial value, such as ten used in Fig. 2, the stress strength is reduced by one whenever the stress current passes a resistive device.

Next, with the bias condition known from the stress annotation, a proper circuit model will be determined for each device. When a device is under ESD current stress, a high current model must be used, such as an NMOS model which covers the snapback regime and a resistor model which considers the velocity saturation effects. Transistor junctions may be forward biased during an ESD event. Consider the driver PMOS transistor shown in Fig. 2. The junction formed by the drain and the n-well is forward biased. As a result, the n-well is charged up and the V_{dd} power line may be pulled high. Therefore, the serial combination of a forward-biased diode and a semiconductor well resistor is substituted for the PMOS transistor.

B. BJT Extraction

Next, extraction of parasitic BJTs is described. There are many such devices in a typical CMOS layout. We propose a systematic approach to identify only those parasitic BJTs which are possibly in the on-state.

Lateral BJT Extraction

Ideally, the ESD current should be conducted through intentionally designed protection devices such as thick field devices (TFDs), low voltage triggering SCRs (LVSCRs) or NMOS transistors. The turn-on of unintentional lateral BJTs is a common cause of chip ESD failure [1][4][5]. An example of parasitic BJT identification is shown in Fig. 3. Once the stress annotation has

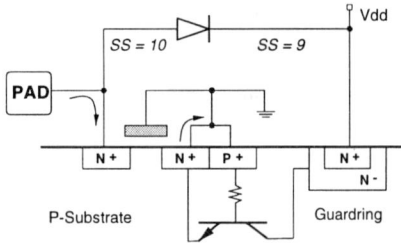

Figure 3: Extraction of lateral parasitic BJTs. The source of NMOS transistor (emitter) and guard ring (collector) form a parasitic BJT.

been performed, the forward biased pn junctions are identified as possible emitters of BJTs. Those diffusions or wells which are stressed with high voltages are identified as possible npn collectors (low voltage nodes are examined for the extraction of lateral PNPs). If there are p emitters and q collectors, the number of possible BJTs is pq. However, since not all of them will

conduct significant current, further reduction is needed. The gain β of the parasitic BJT is used as the criterion for reduction. β is inversely proportional to the square of the base width W. The set of reduction rules is listed below. The referenced configurations are illustrated in Fig. 4.

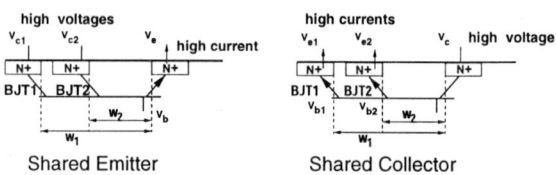

Figure 4: Configurations of parasitic BJTs for reduction.

1. Shared emitter rule:

    (a) If $V_{c1} \leq V_{c2}$, and $\beta_1 < \beta_2$, remove BJT1.

    (b) If $V_{c1} < V_{c2}$, and $\beta_1 \leq \beta_2$, remove BJT1.

    (c) Otherwise, no reduction.

2. Shared collector rule:

    (a) If $V_{e1} \geq V_{e2}$, and $\beta_1 < \beta_2$, remove BJT1.

    (b) If $V_{e1} > V_{e2}$, and $\beta_1 \leq \beta_2$, remove BJT1.

    (c) Otherwise, no reduction.

3. Minimum β rule (technology dependent):

    - If $\beta < \beta_{threshold}$ ($W > W_{threshold}$), no BJT.

Note that the reduction rules consider V_{cb}. To the first order, collector current is independent of the value V_c. In the shared emitter configuration, there exist two current paths, from C2 to C1 then to the shared emitter, or from C2 to the emitter. $V_{c1} < V_{c2}$ and $\beta_1 \leq \beta_2$ indicates that the path from C2 to the emitter is less resistive, so that the current is shunted through C_2. The minimum β rule is technology dependent and the value can be specified by users.

Vertical BJT Extraction

Vertical BJTs can also impact ESD circuit performance. Currents injected by vertical BJTs into the substrate will raise the local substrate potential and may change the substrate bias conditions of other devices. When a diffusion in a well is forward-biased, a vertical BJT will be extracted under either of the two situations, i.e., the diffusion is at high potential while the substrate is grounded, or the diffusion is grounded while the substrate is under the negative stress. The diode in Fig. 2 is replaced by the vertical BJT Q1 as shown in Fig. 5. The

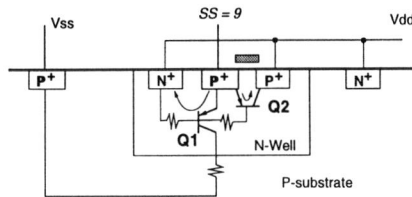

Figure 5: BJTs formed by the PMOS transistor in Fig. 2.

lateral BJT Q2 formed by the drain and source of the PMOS transistor is extracted according to the lateral BJT extraction method. There are two current paths from PMOS drain to V_{dd}, i.e., through Q1 and Q2. If the well resistance is high, Q2 may conduct a significant amount of current [6]. If the size of PMOS transistor is small and can not sustain the high current, circuit failure will occur.

III. SUBSTRATE RESISTANCE EXTRACTION

Substrate resistances impact the substrate potential distribution and the on/off state of various devices. It is a critically important parameter for protection circuit simulation. In this section, we present a substrate resistance model. In addition, we describe a parameter extraction method using a 3D finite difference solver.

A. Substrate Resistance Model

Current flow in one portion of the substrate can cause potential variations in other portions of the substrate. *Transfer resistance* has been proposed to predict the voltage drops in the substrate [8]. It is defined as the surface potential at a point outside the injector divided by the injector current.

Using the concept of transfer resistance, we propose the substrate resistance model shown in Fig. 6. Assume that three BJT

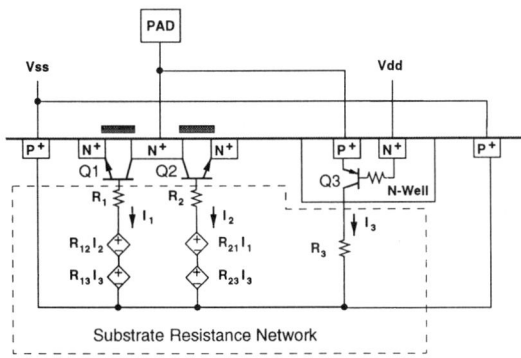

Figure 6: The general substrate resistance model for ESD simulation.

devices Q1, Q2 and Q3 are identified from the layout extraction described in the previous section. There exist three substrate current sources, namely impact ionization currents from the collector-base junctions of Q1 and Q2 and the collector current of Q3. Now let's focus on BJT Q1. The base resistance model of Q1 consists of three components. $R1$ is the resistance from the base of Q1 to the substrate contacts. The other two components are voltage sources controlled by currents I_2 and I_3. R_{ij} is the transfer resistance which can be obtained by dividing the base-emitter voltage of BJT Q_i by the current I_j (other current sources are set to 0). The effect of currents I_1 and I_2 on the collector voltage of Q3 is neglected in this model, because the collector current of Q3 is primarily controlled by its base current.

B. Resistance Extraction

We employ the finite-difference method to extract the transfer resistance. Given a layout structure, we first partition the substrate body in the x, y and z directions, forming a network of grids as shown in Fig. 7. The 3D resistive network is formed

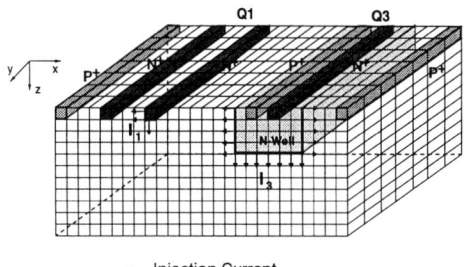

Figure 7: Network of grids for the circuit in Fig. 6 (Q2 is not shown here, and the substrate contact is a ring structure surrounding the layout).

by the grid system. The resistance for each grid can be determined by the doping profile of the substrate. The resistances for the grids inside diffusions or wells are set to infinity (open circuit). To correctly model the boundary condition, the simulated substrate size should be much larger than the I/O circuit region.

We assume that the current is uniformly distributed on the surface of the injection source (the distribution may be adjusted by users), as shown in Fig. 7, and then calculate the voltage distribution across the substrate. Note that the transfer resistance is calculated as the highest local substrate voltage near the emitter junction divided by the total injection current. In total, we need to solve the grid system using the finite difference method n times, where n is the number of current sources.

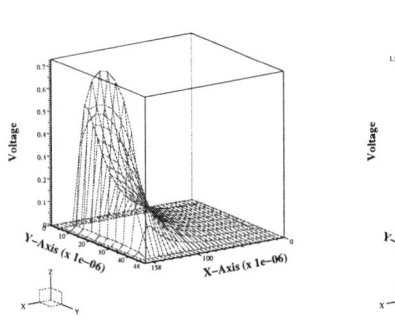 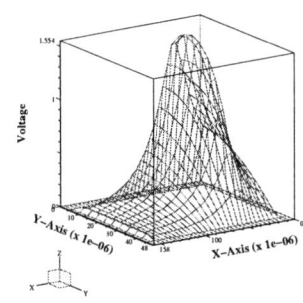

Figure 8: (a) The surface potential distribution under 1mA current injection from NMOS transistor Q1. (b) The surface potential distribution under 1mA current injection from the n-well of the lateral diode Q3.

To calculate the transfer resistances in the substrate network, a 3D finite difference solver is called to solve the resistive grid system in Fig. 7. Sample surface potential distributions are shown in Fig. 8. It takes about 200 seconds to obtain one voltage profile on a SUN SPARCstation 5.

IV. EXPERIMENTAL RESULTS

In this section, an industrial circuit from the literature is used to illustrate the proposed extraction methodology. Also, we will show that the extracted circuit and its simulated results agree with the results of experiments conducted on fabricated test structures and products.

As in most input protection circuits, the circuit shown in Fig. 9 contains a primary protection and a secondary protection device. The secondary protection is provided by gate-grounded NMOS (GGNMOS) transistors to protect the gate oxide of the input transistor. The circuit uses a TFD as the primary protection element and a diffusion resistor as the isolation resistor [4][7].

We apply the proposed methodology to analyze the layout under the positive stress to pad w.r.t. V_{ss}. As illustrated in Fig. 9, diffusions 2 and 5 are forward-biased. They become the sources for minority carrier injection into the substrate. Since diffusions 1, 3 and 4 are stressed with high voltages, there are six possible lateral BJTs, namely (1, 5) (4, 5), (3, 5), (1, 2), (4, 2) and (3, 2). The BJTs are reduced as shown in the figure. Because the TFD and NMOS high current models include BJTs (1, 5) and (3, 2), **an unintended lateral parasitic BJT (4, 2) is detected.**

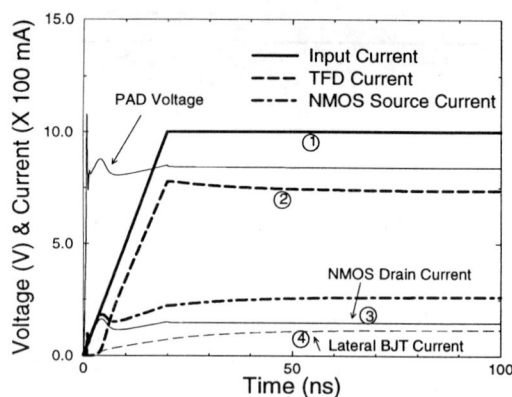

Figure 11: Simulation results under a ramped current input. Most of the stress current flows through the TFD. The NMOS source current is greater than its drain current because part of the source current is collected at the junction connected to pad through the lateral BJT (4, 2).

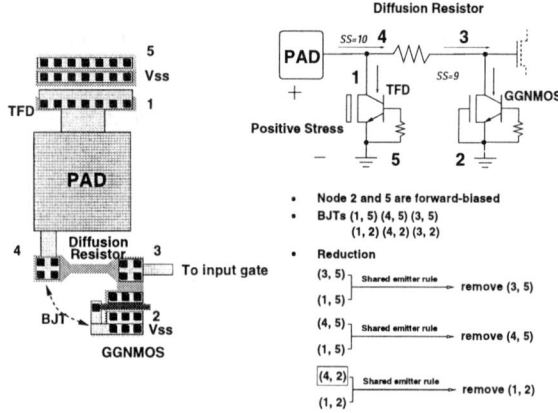

Figure 9: Lateral BJT (4, 2) is detected under the positive stress to pad w.r.t. V_{ss}.

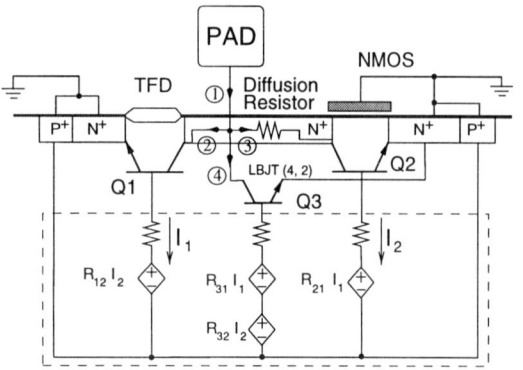

Figure 10: The circuit schematic with the substrate resistance network under positive stress. The coupling effects from Q3 to Q1 and Q2 are neglected.

The circuit schematic is drawn in Fig. 10 along with the cross-section of silicon substrate. The model parameters for devices such as NMOS and BJT are extracted from the measurements of test structures. The substrate resistance is estimated by the 3D finite difference solver. Circuit simulation results in Fig. 11 indicate that there is indeed a significant amount of stress current conducting through the lateral parasitic BJT (4, 2). This confirms the experimental and failure analysis results reported in [4][7].

V. SUMMARY

In this paper, we have presented new CAD tools for CMOS I/O circuit design. To our knowledge, this is the first layout extractor developed for CMOS I/O circuit reliability. Unlike normal extractors, the circuit schematic is extracted based on the specified ESD conditions. In addition, parasitic BJTs are extracted for circuit simulation. We also presented a method for modeling and extracting the substrate resistance which is needed for accurate I/O circuit simulation under ESD.

REFERENCES

[1] C. Duvvury and A. Amerasekera, "ESD: A Pervasive Reliability Concern for IC Technologies," *Proc. of the IEEE,* vol. 81, no. 5, pp. 690–702, May 1993.

[2] C. Diaz, S. M. Kang and C. Duvvury, "Circuit-level Electrothermal Simulation of Electrical Overstress Failures in Advanced MOS I/O Protection Devices," *IEEE Trans. on CAD,* vol. 13, no. 4, pp. 482–493, 1994.

[3] MIL-STD-883C, Electrostatic Discharge Sensitivity Classification, Technical Report, Notice 8, DoD, March 1989.

[4] C. Duvvury and R. Rountree, "A Synthesis of ESD Input Protection Scheme," *EOS/ESD Symposium,* pp. 88–97, 1991.

[5] J. P. LeBlanc and M. D. Chaine, "Proximity Effects of Unused Output Buffers on ESD Performance," *IEEE International Reliability Symposium,* pp. 327–330, 1991.

[6] C. Duvvury and R. Rountree, "Output ESD Protection Techniques for Advanced CMOS Processes," *EOS/ESD Symposium,* pp. 206–211, 1988.

[7] Y. Fong and C. Hu, "Internal ESD Transients in Input Protection Circuits," *IEEE International Reliability Symposium,* pp. 77–81, 1989.

[8] R. R. Troutman, *"Latchup in CMOS Technology : The Problem and Its Cure,"* Kluwer Academic Publishers, 1986.

SINGLE-EVENT EFFECTS IN MICROMACHINED PMOSFETS

Ashraf A. Osman, Mohammad Mojarradi and Kartikeya Mayaram

School of Electrical Engineering and Computer Science
Washington State University, Pullman, WA 99164-2752

ABSTRACT

Single-event effects in micromachined PMOSFETs in a $2\mu m$ standard CMOS process are examined using device simulation. A comparison with the bulk and SOI PMOSFETs with comparable structures is also provided. The effects of N-well depth, angle of incidence, and N-well contact have been investigated in micromachined transistors. The substrate current and collected charge in micromachined PMOSFETs with thin N-wells are significantly lower compared to bulk PMOS devices.

1. INTRODUCTION

Micromachined CMOS devices are attractive for high frequency circuits due to reduced substrate parasitic effects, reduced leakage current, and latch-up immunity. Another advantage of these devices is that they can be readily fabricated without extra process complexity by post-processing steps in a standard CMOS process [1]. The micromachined transistors have a structure similar to Silicon-on-Insulator (SOI) structures with a thick backside insulator (the air region in this case) as shown in Figure 1. Thus, one can expect micromachined MOSFETs to have similar radiation hardened properties as SOI devices when compared to the bulk MOSFETs. This is due to the small sensitive volume of silicon in micromachined and SOI MOSFETS where a large volume of silicon is replaced by the backside insulator.

This paper presents the first investigation of single-event upset (SEU) effects in micromachined PMOS devices. Three-dimensional device simulations have been used to understand SEU effects in these devices. A comparison is also provided between SOI [2] and bulk transistors in the same technology with comparable geometries.

2. THE MICROMACHINING PROCESS

A new two step micro-machining process is used to suspend PMOS transistors in a standard CMOS process and reduce N-well thickness. This process utilizes Tetramethyl Ammonium Hydroxide (TMAH), an electrochemical silicon etchant that selectively removes P-type material while

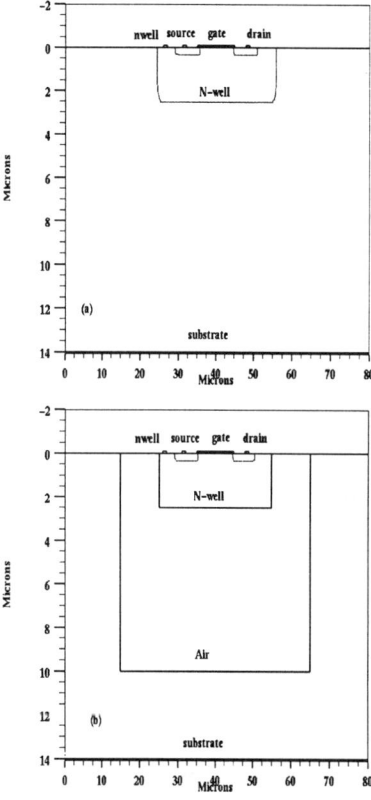

Figure 1: Test device structures for (a) bulk and (b) micromachined PMOSFETs used in the device simulations.

avoiding electrically passivated N-type junctions [3]. The N-type junctions are passivated by the application of a reverse bias voltage which creates a depletion region around the junction and alters the concentration in this vicinity. The resulting change in concentration produces a barrier around the N-type junction that prevents the etchant from attacking this area. The suspended N-well structures are electrically and thermally isolated, and have no parasitic coupling to the substrate (Figure 1 (b)). A second unbiased etch step is used to thin down the N-well. The second etch operation is monitored by on-chip etch monitor structures and terminates when all, or a majority, of the undesired N-type material is removed. Thus the depth of the N-well can be controlled as desired.

This procedure can be applied to individual devices or circuits. As an example, consider a micromachined ring oscillator as shown in the die photo in Figure 2. The ring oscillator is built using PMOS devices in an N-well with windows for post-processing. A TMAH etch is used to remove the substrate under the N-well regions, whereby the structure is suspended over a pit in the silicon substrate.

Figure 2: Die photograph of a micromachined ring oscillator. The oscillator is designed with PMOS devices in an N-well in a standard CMOS process.

3. SIMULATION DETAILS

Three-dimensional (3D) device simulations have been carried out using TMA's Davinci simulator [4]. Standard models were used for carrier mobilities and lifetimes [5]. Simulations were performed to identify the effect of N-well depth, angle of incidence, and N-well contact on the collected drain current and charge. An understanding of these effects provides insight into improved device design.

In these simulations, the single-event effect was simulated with an α-particle track that is incident at the drain contact with a length of 5μm inside the device and a charge density of 1×10^{18} cm^{-3}. The test devices were based on the available 2μm N-well CMOS process doping profiles. The channel length of the simulated MOSFET was 5μm which corresponds to the test devices that were fabricated and are currently being tested. The devices were biased with -5V on the drain and substrate terminals while all the other terminals were grounded.

4. RESULTS

The simulation data and an interpretation of the results are presented in this section for micromachined, SOI, and bulk PMOSFETs.

4.1. N-Well Depth

The transient drain current and the collected charge waveforms for the three different transistors and normal incidence of the track are shown in Figure 3 for an N-well depth of 2.5μm. The depth of the N-well is from the actual process information and is a parameter that can be controlled only in the micromachined and SOI MOSFETs.

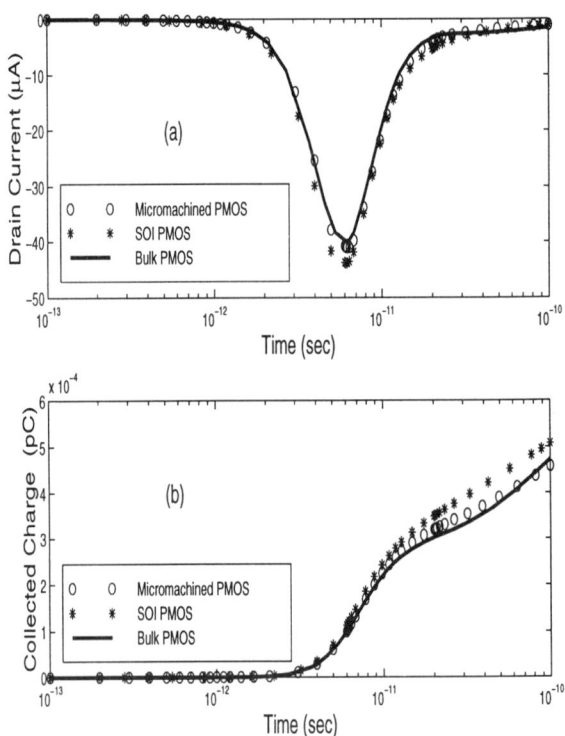

Figure 3: Simulated transient (a) drain current and (b) collected charge at normal incidence for the micromachined, SOI, and bulk PMOS devices for an N-well depth of 2.5μm. All devices show equivalent amounts of collected current and charge.

From the data in Figure 3 it is clear that all of these devices behave similarly under a single-event upset (SEU). The charge collection is due to the funneling effect described in [6]. For each device, the funneling length is identical resulting in the same amount of collected charge. Thus, there is no advantage in terms of SEU immunity in micromachined or SOI PMOSFETs with deep N-well regions. The substrate current waveforms for the three devices are shown in Figure 4. A large substrate current is observed for the bulk MOSFET due to a funneling effect at the N-well substrate junction [7].

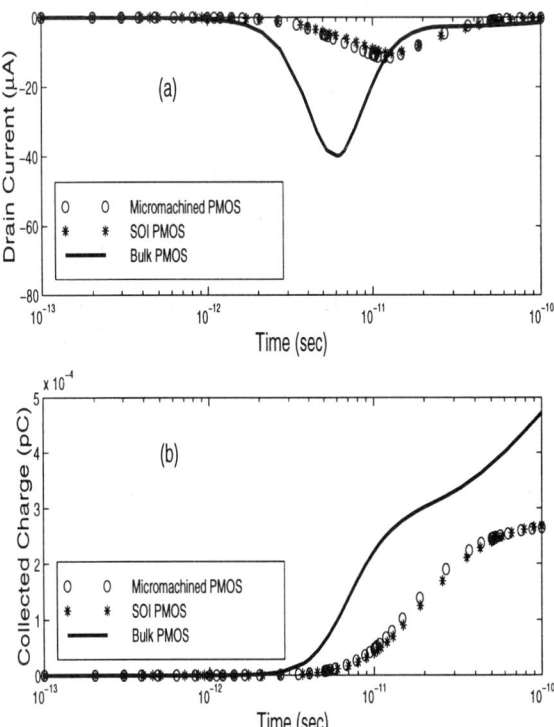

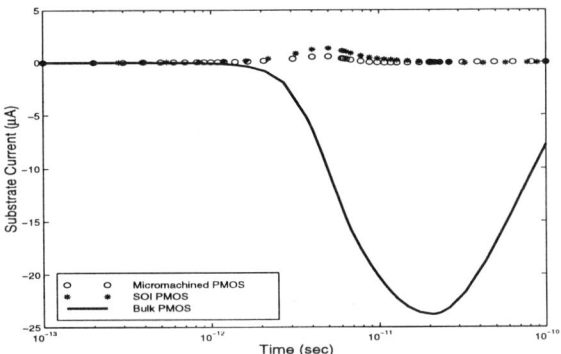

Figure 4: Simulated transient substrate current in the three devices. The bulk PMOSFET shows a large substrate current compared to the micromachined and SOI PMOSFETs due to the funneling effect at the N-well to substrate junction.

Since the N-well depth can be controlled in the micromachined and SOI devices, we have also considered N-well depths of $1.0\mu m$ and $0.5\mu m$, respectively. For an N-well depth of $1.0\mu m$ there is no advantage in terms of SEU immunity. However, with an N-well that is $0.5\mu m$ thick a significant improvement in performance is observed. The simulated collected current and charge waveforms are shown in Figure 5. It is seen that the collected current is smaller in magnitude resulting in a smaller collected charge. For this reason, devices that have N-well depths that are less than $0.5\mu m$ in this technology will offer improved immunity to single-event upsets.

A smaller N-well depth results in a shorter collection length since the funneling effect that this is responsible for collecting charge is limited by a smaller volume of silicon. The hole concentration contours are shown in Figure 6 and clearly demonstrate the limited funneling in micromachined devices compared to the bulk MOSFET.

4.2. Angle of Incidence

The effect of the angle of incidence is investigated by letting the charge beam hit the drain region at various angles.

Figure 5: Simulated transient (a) drain current and (b) collected charge at normal incidence for the micromachined, SOI, and bulk PMOS devices for an N-well depth of $0.5\mu m$. The drain current pulse in the case of the micromachined and SOI devices is considerably lower than that of the bulk MOSFET. The final amount of collected charge is 2×10^{-4} pC for both micromachined and SOI devices while it is 5×10^{-4} pC for the bulk PMOSFET

Three cases have been considered: normal incidence, $30°$ from the normal, and $60°$ from the normal. The collected current and charge waveforms are shown in Figure 7 for these angles of incidence for a device with an N-well depth of $1.0\mu m$. As the angle of incidence is increased, the current pulse and the collected charge increase. This is because of the longer track length within the silicon N-well. The depth of the N-well does not affect this dependence on the angle of incidence.

4.3. N-Well Contact

To evaluate the effect of the N-well contact on the charge collection process the micromachined PMOS structure was simulated without the N-well contact. The collected current and charge waveforms for devices with and without the N-well contact are the same in our simulations. This is because the PMOSFET is a long channel device in which no bipolar amplification takes place even in the absence of the N-well

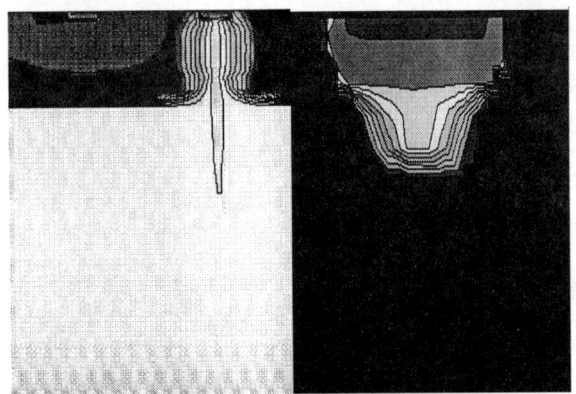

Figure 6: The hole distribution for normal incidence 6ps after the α-particle strikes the drain region in both the bulk and micromachined devices. The contours for the micromachined device are expanded around the drain region.

contact. The excess holes in the N-well region are removed by recombination in both cases.

5. CONCLUSIONS

Micromachined PMOSFETs have been characterized in terms of their single-event upset behavior and compared with the SOI and bulk PMOSFETs. Three-dimensional device simulations have been used for these investigations. The effect of the N-well depth, angle of incidence, and N-well contact on the collected drain current and charge have been studied. It has been shown that the N-well depth has to be sufficiently small for the micromachined PMOSFETs to have better SEU immunity compared to the bulk PMOSFET. In this respect, the micromachined transistors are comparable to the fully depleted SOI MOSFETs. Future work will focus on mixed-mode circuit and device simulations for evaluation of single-event upset in SRAM cells with micromachined transistors.

6. ACKNOWLEDGMENTS

The authors thank TMA for providing the device simulation tools and Randal Thornley and Dorin Patru for the design and processing of the micromachined ring oscillator circuit. This work is supported in part by NSF grant No. EEC9708324.

7. REFERENCES

[1] R. Reay, E. Klassen, and G. Kovacs, "A micromachined low-power temperature-regulated bandgap voltage reference," *IEEE J. Solid-State Circuits*, vol. 30, no. 12, pp. 1371-1381, December 1995.

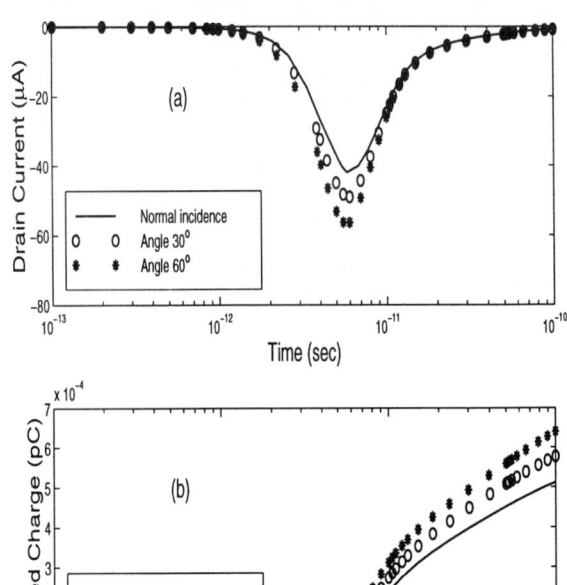

Figure 7: Simulated transient (a) drain current and (b) collected charge for various angles of incidence for the micromachined PMOSFET with an N-well depth of 1.0μm. The current pulse and the collected charge increase with the angle of incidence.

[2] O. Musseau, "Single event effects in SOI technologies and devices," *IEEE Trans. on Nuclear Science*, vol. 43, no. 2, pp. 603-613, April 1996.

[3] R. Reay, E. Klaassen, and G. Kovacs, "Thermally and electrically isolated single crystal silicon structures in CMOS technology," *IEEE Electron Device Letters*, pp. 399-401, Oct. 1994.

[4] Technology Modeling Associates, *Davinci*, February 1997.

[5] P. E. Dodd, "Device simulation of charge collection and single-event upset," *IEEE Trans. on Nuclear Science*, vol. 43, no. 2, pp. 561-575, April 1996.

[6] C. Hu, "Alpha-particle-induced field and enhanced collection of carriers," *IEEE Electron Device Letters*, vol. EDL-3, no. 2, pp. 31-34, February 1982.

[7] J. H. Chern, J. A. Seitchik, and P. Yang, "Single event charge collection modeling in CMOS multi-junctions structure," *IEDM-86 Digest of Tech. Papers*, pp. 538-541, December 1986.

A GENERALIZED HSPICE[1] MACRO-MODEL FOR PSEUDO-SPIN-VALVE GMR MEMORY BITS

Bodhisattva Das and William C. Black, Jr.

Department of Electrical and Computer Engineering,
Iowa State University, Ames, Iowa 50011-3060.[2]

ABSTRACT

Nonvolatile semiconductor storage using Giant-Magneto-Resistance (GMR) memory bits has the potential for revolutionizing both high density and high speed memory applications with devices exhibiting unlimited write endurance and very low required write energy. This work presents the first generalized circuit macro-model for a *pseudo-spin-valve* GMR memory bit. The macro-model is realized as a four terminal sub-circuit which emulates GMR bit behavior over a wide range of sense and word line currents. The non-volatile and nonlinear nature of GMR memory bits are accurately represented by this model and simulations of non-volatile GMR latch structures with HSPICE show expected outcomes. The model is flexible and relatively simple: ranges of the write /read currents and bit resistance values are incorporated as parameterized variables and no semiconductor devices are used within the model.

1. INTRODUCTION

Since the first observation of the GMR effect in magnetic multilayers in 1988 [1], considerable research has been directed towards materials and structures that are capable of showing this phenomenon. GMR materials are now being used as highly sensitive magnetic sensors [2] and are responsible for a continued revolution in disk drive performance and storage density [3]. They also show promise of economical high density data storage in Magnetic Random Access Memory (MRAM) devices [4].

GMR devices employ two ferromagnetic layers separated by a very thin non-magnetic spacer layer. If properly fabricated the resistance of this composite structure is a significant function of the difference between the magnetic moments of the two layers. These layer magnetizations are themselves at least somewhat dependent upon the local magnetic field and typically have two or more stable magnetic states.

An interesting type of GMR device is popularly known as a *spin-valve* [5]. In this structure the magnetization of one ferromagnetic layer is pinned in one direction along the longitudinal direction of the stripe with a layer of anti-ferromagnetic material (such as MnO or MnFe). The magnetization of the other layer is free to rotate but as the bits become very narrow, it tends towards either a parallel or anti-parallel alignment relative to the pinned layer in reproducible and stable states [6]. These orientations correspond to the '0' or '1' states of the magnetic memory bit.

In a *pseudo-spin-valve* structure [7][8][9], neither magnetic layer is pinned but one layer (the "harder" one) has a higher switching field than the other. So, the application of a comparatively weak magnetic field can only alter the magnetic orientation of the "softer" layer, whereas a strong magnetic field can switch both layers.

For both spin-valve and pseudo-spin-valve structures when the pair of ferromagnetic layers are magnetized in the same (parallel) direction, the resistance of the stripe is lower than when they are magnetized in opposite (anti-parallel) directions. In this work, we model a pseudo-spin-valve GMR memory bit with a circuit macro-model that can easily be incorporated into HSPICE simulations. This work is an extension of the modeling of spin-valve bits done by us before [10], but is much more involved due to the complicated nature of pseudo-spin-valve bits. It is interesting to note that though there has been plethora of research on GMR materials or application circuits, we are not aware of any publication on simple circuit-models for pseudo-spin-valve GMR structures. *Our proposed model is thus apparently the first one for pseudo-spin-valve GMR bits.*

2. BEHAVIOR OF A GMR BIT

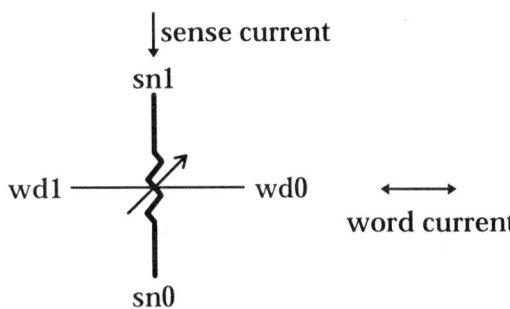

Figure 1. GMR bit with word line.

A GMR element can be considered as a four terminal device (Fig. 1). Two of the terminals connect directly to the GMR resistor (sense lines) while the other two provide a magnetic bias field for reading and writing of the bit (word lines). The word line and sense line are *not* connected to each other. A current through the word line induces a magnetic field on the GMR resistor which can change its resistance value. The effective value of the resistance is sensed between the two sense line terminals.

2.1 Pseudo-Spin-Valve Behavior

The resistance R vs. word current I_w graph of a typical pseudo-spin-valve GMR bit is shown in Fig. 2 [7][8][9]. The hysteretic nature of the graph can be easily observed: it is a combination of system N and system P. System N in turn can be divided into two

---

[1] HSPICE is an authorized trademark of Avant! Corporation.
[2] This work was sponsored in part by Solid-State Electronics Center of Honeywell Inc., Plymouth, MN; and DARPA.

graphs: Na and Nb (Fig. 3); similarly for P: Pa and Pb (Fig. 4). Ni's are points on system N and Pi's are points on system P. The R vs. I_w characteristic is symmetrical about a I_w=constant line through point C. For a conventional pseudo-spin-valve GMR, C is the zero-word-current point (I_{wc}=0); however, our macro model is flexible enough to handle a non-zero center point ($I_{wc} \neq 0$).

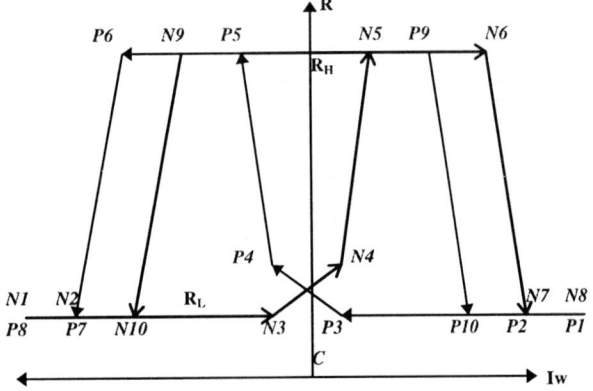

Figure 2. Pseudo-spin-valve R vs. I_w characteristic.

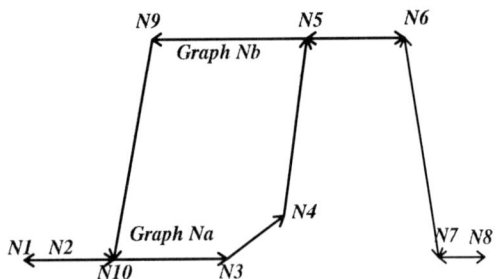

Figure 3. The System N characteristics: a combination of graphs Na and Nb.

If I_w is increased beyond N7 (which is the same as P2), the R vs. I_w characteristic gets onto the system P. Any subsequent magnetization caused by all word current values between points P2 (N7) and P7 (N2) will keep the R-I_w curve on system P only. If the word current value goes beyond point P7 (N2), the R-I_w curve changes its track and gets onto system N. Now, it will stay on system N for any values of word current between points N2 (P7) and N7 (P2). When we are on system N between points N5 and N6 (Figs. 2 & 3), and if now the word current is reduced before reaching N6, the characteristic does not retrace back to N3 via N5 and N4. Rather, the resistance stays high till a point N9 (somewhere between points P5 and P6) where it starts diminishing and finally reaches N10 and then goes back to N2 if the current is decreased (increased in the negative direction) further. Similarly, if we are on system P between points P5 and P6 (Figs. 2 & 4), and the current is increased (decreased in the negative direction) before reaching P6, the resistance stays high till a point P9 between N5 and N6. Then it starts falling down to point P10. So eventually each of the systems N and P can be conceived as a combination of two graphs. For system N, graph Na is called the *major loop* and graph Nb the *minor loop* (Fig. 3). Similarly for system P.

This hysteretic nature of the R vs. I_w characteristic makes the pseudo-spin-valve GMR useful as a memory element. We can arbitrarily assign state 0 to system N and state 1 to system P (or vice versa). When $I_w > I_{P2}$, we say that the GMR is "written" into state 1; and when $I_w < I_{N2}$, it is "written" into state 0. The GMR can be "read" at any word current between I_{N2} and I_{P2} ($I_{N2} \leq I_w \leq I_{P2}$) and the resistance sensed will depend on which state the GMR had been written into before.

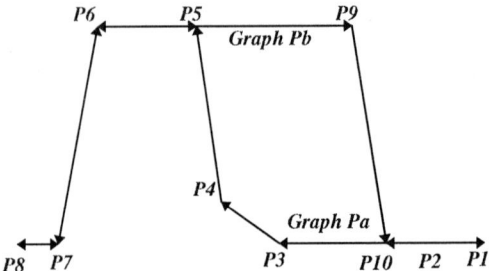

Figure 4. The System P characteristics: a combination of graphs Pa and Pb.

3. THE MACRO-MODEL

The proposed macro-model is conceived as a 4-terminal sub-circuit in HSPICE. Two of the four terminals (wd1 and wd0) are for the word line and the other two (sn1 and sn0) for the sense line (Fig. 1). The circuit can be divided into four simple parts:
1. input (word) circuit
2. bistable multivibrator or Schmitt Trigger
3. decision circuit
4. output (sense) circuit.

In Fig. 5, we annotate a part of the circuit: essentially the part that models system N.

3.1 Input (Word) Circuit

It comprises only one constant-value resistor R_w. The input word current I_w flows through R_w and creates a voltage proportional to I_w between nodes wd1 and wd0.

$$V(wd1, wd0) = R_w \cdot I_w.$$

So, V(wd1, wd0) can be viewed as a scaled version of the word current itself.

3.2 Bistable Multivibrator or Schmitt Trigger

This circuit is realized (Fig. 5) as a very-high-gain (>>1) operational amplifier in a positive feedback configuration with the non-inverting output (node 4) fed back to the non-inverting input (node 5) through resistor R_2. The inverting input and output are shorted to wd0 point, and node wd1 is connected to the non-inverting input (node 5) via a resistor R_1. So, essentially, V(wd1, wd0) is fed to the non-inverting input (node 5) through R_1. The voltage difference between the differential outputs of the op-amp has upper saturation level V_{max} (L_+) and lower saturation level V_{min} (L_-). This circuit behaves as a bistable multivibrator depending on the value of V(wd1, wd0) [11]. In HSPICE this very-high-gain operational amplifier is realized by a Voltage Controlled Voltage Source (VCVS) named *Eopamp*. This bistable circuit has two stable voltage levels for node 4: L_+ and L_- [11]. As V(wd1,wd0) goes higher than $-L_-(R1/R2)$, the circuit regenerates and V(4) reaches upper stable state L_+. Similarly as $V(wd1,wd0) < -L_+(R1/R2)$, V(4) reaches lower stable state L_-.

In the entire macro-model, there are 3 bistable multivibrators: one for system N (to select between graphs Na and Nb), one for system P (to select between graphs Pa and Pb), and the third one for selecting between systems N and P.

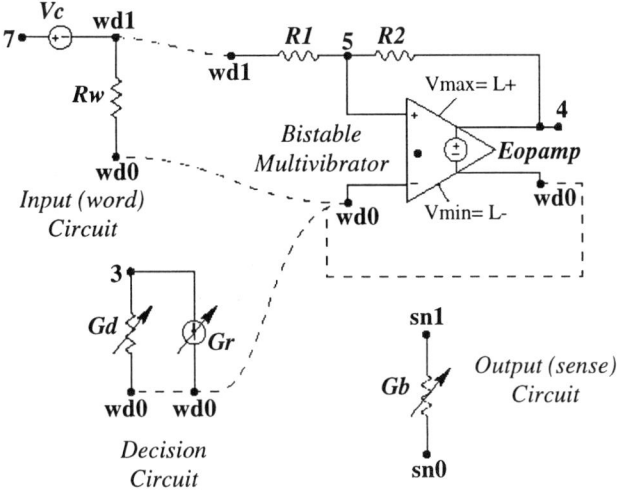

Figure 5. Part of the macro-model.

3.3 Decision Circuit

This circuit comprises G_d, a Voltage Controlled Resistor (VCR) and G_r, a Voltage Controlled Current Source (VCCS).

G_d is a VCR whose resistance is a one-to-one function of the controlling voltage V(wd1,wd0). More specifically, $R(G_d) = 1*V(7,wd0)$ where $V(7,wd0) = V(wd1,wd0) + V_c$; V_c being a constant voltage. So, V(7,wd0) is an origin-shifted version of V(wd1,wd0). *Ideally, we should have made* $G_d = V(wd1,wd0)$, but we cannot, because, HSPICE cannot handle any negative resistance. V_c is chosen to be a value such that V(7,wd0) never goes below zero and hence clipping of the 1:1 VCR characteristic of G_d at the negative region of the controlling voltage is avoided.

The VCCS G_r works as a unity magnitude current source, which has a value +1 or −1 depending on whether the Schmitt Trigger is at the upper stable level (L_+) or at the lower one (L_-). This current essentially creates a positive or negative voltage across G_d (between nodes 3 and wd0), based on the bistable state; i.e.,

$V(3,wd0) = V(7,wd0)$ if $V(4,wd0) = L_+$; and,
$V(3,wd0) = -V(7,wd0)$ if $V(4,wd0) = L_-$.

Once the characteristic is on system N, the current $I(G_r)$ governs which graph (Na or Nb) the GMR characteristic should take depending on the particular bistable state.

There is another similar group of V_c, G_d and G_r for system P. There is a switching circuit to select between systems N and P. The switching circuit is very simple (not shown in Fig. 5): two voltage controlled voltage sources (VCVS) having voltage values in 1:1 relation with the two V(3,wd0)'s; and two voltage controlled resistor (VCR) switches to select between them. The switches are complement to each other: they are ON/OFF depending on whether the third bistable multivibrator (the selector of the two systems) is at the high level or low level; the ON value of the VCRs being unity. The switched (selected) current is then passed through a unity resistor to generate the same voltage as V(3,wd0).

3.4 Output (Sense) Circuit

The sense circuit comprises a Voltage Controlled Resistor (VCR) G_b connected between sense line terminals (points sn1 and sn0). The resistance of G_b varies as a Piece-Wise Linear (PWL) function with the controlling voltage V(3,wd0). G_b is essentially the GMR bit; any external circuitry connected to sn1 and sn0 sees the resistance of G_b as the effective GMR resistance across the sense line terminals (between sn1 and sn0).

The R vs. I_w characteristic of the GMR is mapped into the PWL characteristic of G_b. We essentially have four different graphs (two N's and two P's) all varying between same boundary points and all of them have to be mapped into the same PWL of G_b. So we have to make sure that these four graphs are mapped into mutually-exclusive (non-coinciding) regions of the PWL function. Interestingly, the only modification that is needed to make the model suitable for different pseudo-spin-valve GMR structures is the PWL function of G_b: the entries for the controlling voltage and the resistance values have to be changed accordingly.

4. THE SPECIFIC CASE OF SYMMETRIC MAJOR AND MINOR LOOP

If we consider a simpler assumption that for system P (or N) the upward gradient of the major loop is symmetrical to the downward gradient of the minor loop, we have a characteristic which looks like Fig. 6. Here we can observe the points P9, P10, N9, N10 have different positions from the more generalized case of Fig. 2. Actually Fig. 6 is one particular case of Fig. 2. In this case, we can divide the characteristic into two curves: one for the actual top (*softer*) layer of GMR and the other for the actual bottom (*harder*) layer (Figs. 7 & 8). Then we can just combine the two curves (take absolute difference of the two and add to the nominal low value (R_L) of the GMR resistor) to obtain the Fig. 6 characteristic. This buys us a major advantage in terms of modeling: we can model the characteristic of Fig. 6 with only two bistable multivibrators, instead of the three we otherwise had for the more general case of Fig. 2.

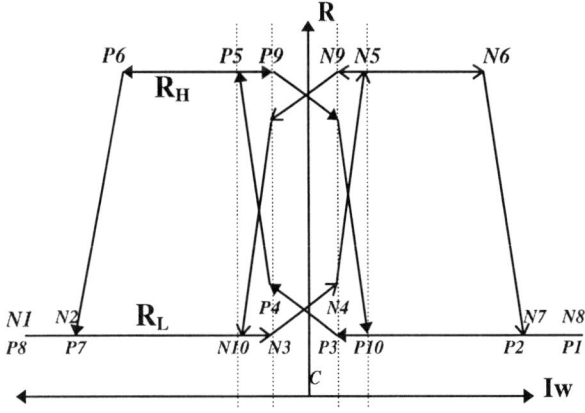

Figure 6. Symmetric major and minor loop.

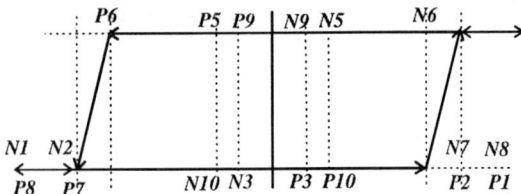

Figure 7. Bottom (*harder*) layer magnetization curve.

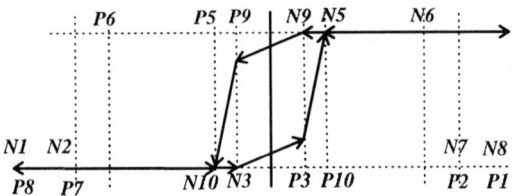

Figure 8. Top (*softer*) layer magnetization curve.

5. THE HSPICE NETLIST FILE

The simplicity of the macro-model and its compatibility with HSPICE format makes the task of writing a netlist file of the circuit very easy in HSPICE [12]. All the variables in the circuit can be written as .PARAM statements; hence they are all parameterized. The sub-circuit is not comprised of any component that needs a power supply, so its behavior is not influenced by any simulated power up or down. *This gives the model the non-volatile nature.* Moreover, there is no semiconductor device in the macro-model, hence possible complications due to different models for MOS or BJT devices is avoided. The entire sub-circuit is written as an HSPICE include file (*.inc).

6. SIMULATION

The GMR sub-circuit was simulated with HSPICE [12]. First a dc analysis was done for the GMR model only to verify that the model works fine for the whole range of word currents. Next, a transient analysis was done to verify that the model by itself has a proper transient response. One small limitation was detected. The node 5 in the bistable multivibrator sub-sub-circuit (Fig. 5) has to be initialized (once for all) to a small positive or negative voltage so that the bistable multivibrator can attain its positive or negative stable level. Otherwise, HSPICE doesn't understand what to do with the initial voltage at the node 5 and hence, keeps it at zero, eternally. This bottleneck was avoided by forcing a small voltage (positive or negative) as an initial condition (.IC statement) to node 5 in the macro-model itself. After this, the "acid test" of the macro model was done by replacing simple resistors with our GMR sub-circuit in a cross-coupled dynamic latch/sense-amplifier structure and simulating (transient) with HSPICE. After initial reset the latched states always showed the expected result. Also, the sub-circuit was put into novel pseudo-spin-valve GMR memory structures and simulated: the GMR bits showed proper state changes at write and read steps. The ΔR for the GMR bit was chosen to be 5% with $R_L=100\Omega$ and $R_H=105\Omega$. The non-volatile nature of the GMR bit was also accurately demonstrated by the model.

7. CONCLUSION

To the authors' knowledge, this is the first published HSPICE model of a GMR bit for pseudo-spin-valve structures. This macro-model will be extremely useful for designers and researchers working on MRAM and in related fields. The flexibility of the model can be easily exploited by the users as they can customize this model by simply altering sub-circuit parameters. The simplicity of this model is a virtue, it doesn't use any of the published rigorously mathematical models of hysteresis [13][14].

8. REFERENCES

[1] M. N. Baibich, J. M. Broto, A. Fert, F. Nguyen Van Dau, F. Pettroff, P. Etienne, G. Creuzer, A. Friederich, and J. Chazelas, "Giantmagneto-resistance of (001)Fe/(001)Cr magnetic superlattice," *Phys. Rev. Lett.*, vol. 61, pp. 2472, 1988.

[2] J. Daughton, J. Brown, E. Chen, A. Pohm, and A. Kude, "Magnetic field sensors using GMR multilayers," *IEEE Trans. Magn.*, vol. 30, pp. 4608-10, 1994.

[3] E. Grochowski, and D. A. Thomson, "Outlook for maintaining areal density growth in magnetic recording," *IEEE Trans. Magn.*, vol. 30, no. 6, pp. 3797-3800, 1994.

[4] A. V. Pohm, J. M. Daughton, J. Brown, and R. Beech, "The architecture of a high performance mass store with GMR memory cells," *IEEE Trans. Magn.*, vol. 31, no. 6, pp. 3200, November 1995.

[5] B. Dieny, V. S. Speriosu, S. Metin, S. S. P. Parkin, B. A. Gurney, P. Baumgart, and D. R. Wilhoit, "Magnetotransport properties of magnetically soft Spin-Valve structures," *J. Appl. Phys.* vol. 69, 47744 (1991).

[6] D. D. Tang, P. K. Wang, V. S. Speriosu, S. Le, and K. K. Kung, "Spin-Valve RAM cell," *IEEE Trans. Magn.*, vol. 31, no. 6, pp. 3206, November 1995.

[7] B. A. Everitt, A. V. Pohm, and J. M. Daughton, "Size dependence of MRAM switching thresholds," *J. Appl. Phys.* vol. 81, 4020 (1997).

[8] A. V. Pohm, B. A. Everitt, R. S. Beech, and J. M. Daughton, "Bias field and end effects on the switching thresholds of Pseudo Spin Valve memory cells," *IEEE Trans. Magn.*, vol. 33, no. 5, September 1997, pp. 3280-3282.

[9] B. A. Everitt, and A. V. Pohm, "Single domain model for Pseudo-Spin-Valve MRAM cells," presented at the *1997 Intermag Conference, IEEE Trans. Magn.*, vol. 33, no. 5, September 1997, pp. 3289-3291.

[10] B. Das, and W. C. Black, Jr., "A generalized HSPICE macro-model for spin-valve GMR memory bits," presented at the *1997 Midwest Symposium on Circuits and Systems*, to be published in the *Trans. MWSCAS* 1997, in press.

[11] A. S. Sedra and K. C. Smith, *Microelectronic Circuits (3rd edition)*, Oxford University Press.

[12] *HSPICE User's Manual (HSPICE version H92)*, vols. 1, 2 and 3, Avant! Corporation.

[13] H. G. Brachtendorf, C. Eck and R. Laur, "Macromodeling of hysteresis phenomena with SPICE", *IEEE Trans. Circuits and Systems - II: Analog and Dig. Sig. Processing*, vol. 44, no. 5, pp. 378, May 1997.

[14] M. Parodi, and M. Storace, "A PWL ladder circuit which exhibits hysteresis", *Intl. J. Circuit Theory and Applications*, vol. 22, pp. 513-526 (1994).

Compact SPICE Modeling and Design Optimization of Low Leakage a-Si:H TFTs for Large-Area Imaging Systems

R.V.R. Murthy, B. Park, D. Pereira, K. Benaissa*, A. Nathan and S.G. Chamberlain**

Department of Electrical and Computer Engineering, University of Waterloo
Waterloo, Ontario N2L 3G1, Canada
* Texas Instruments Inc., Dallas, Texas 75243, USA
** DALSA Inc., 605 McMurray Rd., Waterloo, Ontario N2V 2E9, Canada

ABSTRACT

We present a SPICE model that takes into account the different mechanisms underlying the reverse leakage current in hydrogenated amorphous silicon (a-Si:H) thin film transistors (TFTs). The main source of leakage current in these devices appears to be the parasitic reverse-biased p-i-n diode at the vicinity of the drain. At low gate voltages, the diode's reverse current can be attributed to thermal generation of electrons from the valence to conduction bands through mid-gap states in the a-Si:H. At high gate voltages, the reverse current is due to trap-assisted tunneling, whereby electrons tunnel to the conduction band through mid-gap states. This bias dependent behavior has been modeled and implemented in SPICE using simple circuit elements based on voltage controlled current sources. Simulated and measured reverse leakage current characteristics are in reasonable agreement.

1. INTRODUCTION

The a-Si:H TFT is used extensively as a switching element in active matrix liquid crystal displays (AMLCDs) and large area matrix addressed sensor arrays for imaging of optical and X-ray signals. In these arrays, the leakage current of the TFT should be as small as possible in order to retain the charge that is collected on the sensor (see Fig. 1), which takes place during the OFF-state of the TFT. Thus a study of the mechanisms underlying leakage (or OFF) current in TFTs and modeling of its behavior for SPICE simulations is crucial to the design of large area high performance imaging systems.

The reverse leakage current in a-Si:H TFTs has been studied by several groups for different fabrication [1-3] and operating bias conditions [4], including dependence on temperature [5]. However, an understanding of possible mechanisms underlying leakage, and subsequent modeling for SPICE simulation is somewhat limited, if not inconsistent. The leakage current has been identified to stem from various sources. In inverted staggered TFT structures (see Fig. 2), a (parasitic) back n-channel forms with $V_G \leq 0$, at the top a-SiN$_x$:H/a-Si:H interface. This can give rise to current flow when $V_{DS} \geq 0$. Also, under these bias conditions, because of hole accumulation at the gate a-SiN$_x$:H/a-Si:H interface, there is formation of a parasitic reverse-biased p-i-n diode between the top and bottom interfaces. Here, the i region denotes the intrinsic a-Si:H layer. The reverse conduction in the p-i-n diode is strongly related to the density of states in the energy gap of the a-Si:H layer as well as at both the a-Si:H/a-SiN$_x$:H interfaces. In particular, the conduction through the intrinsic region can arise from different mechanisms, most notably thermal generation and trap-assisted tunneling. The relative dominance of the different mechanisms under different bias conditions along with model equations and the circuit model for SPICE simulations is discussed in the sections that follow.

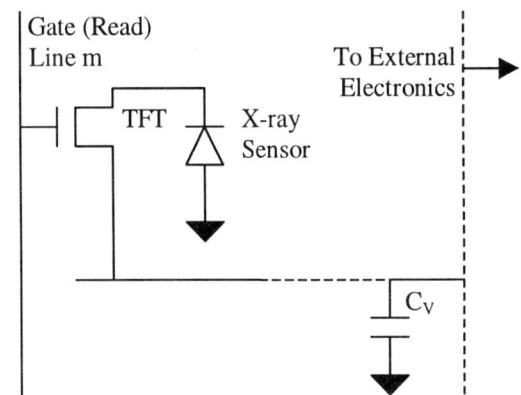

Fig.1 Pixel read-out in one row of an x-ray imaging array.

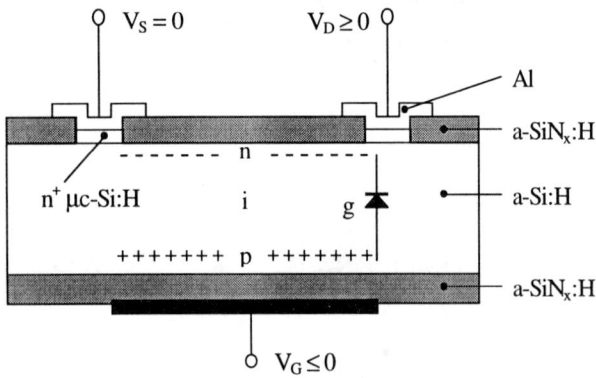

Fig. 2 Cross section of an inverted staggered a-Si:H TFT.

2. EXPERIMENTAL

The TFTs used in this work are based on the inverted staggered structure which is widely used in a-Si:H large area displays and imaging systems (see, e.g., [6]). A fully wet etch process has been used in fabrication. Here, Corning 7059 glass wafers are used as the substrate material, molybdenum (Mo) for the gate metal, and a-SiN$_x$:H as the gate insulator and passivation material. The gate nitride, the active a-Si:H, and top passivation a-SiN$_x$:H films are deposited within one vacuum-pump-down cycle to minimize the density of defect states at interfaces. The thicknesses of the a-Si:H and a-SiN$_x$:H layers are 500Å and 2500Å, respectively. The associated deposition temperatures are 260°C and 320°C, respectively. At source and drain regions, a highly doped micro-crystalline (n$^+$ μc-Si:H) layer (resistivity, 0.2 ohm-cm) is employed to reduce contact resistance. The deposition temperature of the n$^+$ μc-Si:H layer is lower than that of a-Si:H to preserve the integrity of the latter, and its deposition parameters were determined to be near-optimal in terms of material conductivity [7]. For the source and drain contacts, we employ a 1 μm Al film, which also serves as the interconnect and pad metallization. In our process, we employ the passivation layers to serve as etch-masks and etch-stops in patterning the a-Si:H and n$^+$ μc-Si:H layers, using the selective and controllable KOH-based solution. The process is designed to be flexible and general enough to permit integration of other circuit components such as capacitors, resistors, photo-TFTs, and other TFT based structures on the same glass wafer.

A variety of TFTs were fabricated in-house and characterized for their current-voltage and transfer characteristics. The TFT samples were annealed at 170°C prior to DC characterization. All measurements were performed using the DC parametric test system comprising the Keithley 236 source measure units. The variation in device characteristics, for a large variety of samples of different aspect ratios, W/L, was less than 5%. The gate current of the TFTs, for various gate voltages, was below the limit of current sensitivity (~ 10 fA) of the measurement system indicating a high gate nitride quality. All measurements were performed with the TFT source grounded.

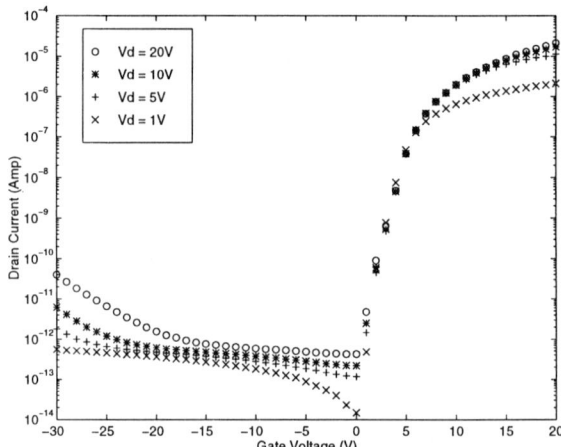

Fig.3 Drain current characteristics of a-Si:H TFTs as function of gate voltage for different drain voltages.

Figure 3 shows the transfer characteristics (I_{DS} vs V_{GS}) at different V_{DS}. We observe a rapid transition from the OFF to the ON state at a gate voltage of less than 5 V. The OFF current is less than 0.1 pA and the ON/OFF ratio is better than 10^7. The reduced series resistance observed stems from the low resistivity n$^+$ μc-Si:H layer at the source and drain regions. Although, not shown, the device exhibits a reasonable square law characteristic ($I_{DS}^{1/2}$ varies almost linearly with V_{GS}) for gate voltages larger than the threshold voltage. The extracted values of threshold voltage and device field effect mobility, based on a fit to measured data, are approximately 2 V and 1 cm^2/Vs, respectively. These values, including those for the OFF current and ON/OFF ratio, are comparable to those reported for TFTs fabricated using state-of-the-art fabrication technologies [8].

In the negative gate voltage region of the transfer characteristics, we can identify two regions of behavior for the V_G-dependence of leakage current (I_{DS}). This can be attributed to different conduction mechanisms in the reverse p-i-n diode at the drain junction. At low V_G and low V_D, the conduction is due to thermal generation. Here, due to the small barrier height at the drain junction, electrons are thermally excited to the conduction band via mid gap states in the a-Si:H layer. This mechanism dominates when the barrier width is large (i.e. small V_{DG}) or for very large thicknesses of the intrinsic a-Si:H layer. At higher gate and/or drain voltages, conduction is due to trap-assisted tunneling, whereby because of the small barrier width at the drain junction, electrons at mid-gap states (traps) in the a-Si:H layer tunnel into the conduction band. This mechanism dominates when the barrier height is large (i.e. large V_{DG}) or for very small thicknesses of the intrinsic a-Si:H layer. The contribution to TFT leakage from induced charge (back channel at the a-Si:H/a-SiN$_x$:H interface) is insignificant in our a-Si:H TFT samples. Device characterization performed under illumination conditions show an independence of the leakage current (I_{DS}) on the gate voltage due to pinning of Fermi level by interface states [9].

3. SPICE MODEL

Following the above observations, it appears that the leakage current is largely due to the presence of the parasitic p-i-n diode at the drain end, whose conduction is due to, and limited by, the injection of holes from the reverse bias drain junction. The injection is determined by the voltage drop (V_o) across the drain junction. Following the gradual channel approximation, the expression for the leakage current [5] can be adapted to read as

$$(L/W)\, I_{Leakage} = \int_{V_G - V_D + V_o}^{V_G} g(\Psi)\, d\Psi \quad (1)$$

where L denotes the channel length, g is a conductance, and Ψ is an integration variable. As seen in Fig. 3 for the negative gate voltage region, the conductance is a function of both gate and drain voltages, whether the dominant mechanism is thermal excitation or tunneling. In both cases, the associated conductances are an exponential function (see [5]) of the voltages and can be stated as

$$g_{th} = g_{tho} \exp(-a\Psi) \quad (2)$$

$$g_{tn} = g_{tno} \exp(-b\Psi). \quad (3)$$

Here, g_{th} and g_{tn} are conductances due to thermal generation and trap-assisted tunneling, respectively, and g_{tho}, g_{tno}, a, and b are coefficients to be extracted from a fit to the measured data. Equations (1) - (3) lead to the following expression for the reverse leakage current:

$$(L/W)\, I_{Leakage} = I_{th} + I_{tn} \quad (4)$$

where

$$I_{th} = (g_{tho}/a)\{\exp[-a(V_G - V_D + V_o)] - \exp[-aV_G]\} \quad (5)$$

$$I_{tn} = (g_{tno}/b)\{\exp[-b(V_G - V_D + V_o)] - \exp[-bV_G]\} \quad (6)$$

and

$$V_o = (0.9 - 0.22V_D)V_G - 0.9V_D - 0.5. \quad (7)$$

At low gate voltages, the barrier height at the drain junction is small, giving rise to thermal excitation of electrons from the valence to conduction bands through mid-gap states in the intrinsic a-Si:H layer. This is described by equation (5). At high gate voltages, the barrier width of the intrinsic a-Si:H region is decreased, and the leakage current arises from tunneling of electrons to the conduction band through mid-gap states in the intrinsic a-Si:H layer. Here, the leakage current is described by equation (6).

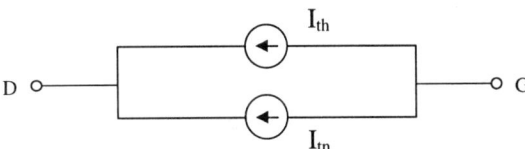

Fig.4 The circuit model for simulation of the reverse leakage current in a-Si:H TFTs due to the parasitic p-i-n diode at the drain junction.

The leakage current as described by the model equations (5) and (6) can be simulated using two voltage controlled current sources in parallel as depicted by the circuit model in Fig. 4. A comparison of simulated and measured characteristics (see Fig. 5) for various drain voltages yields reasonably good agreement. The values of extracted parameters, following a best fit, are: $g_{tho} = 9 \times 10^{-15}$ A/V, $g_{tno} = 5 \times 10^{-21}$ A/V, $a = 0.03$ V^{-1}, and $b = 0.3$ V^{-1}.

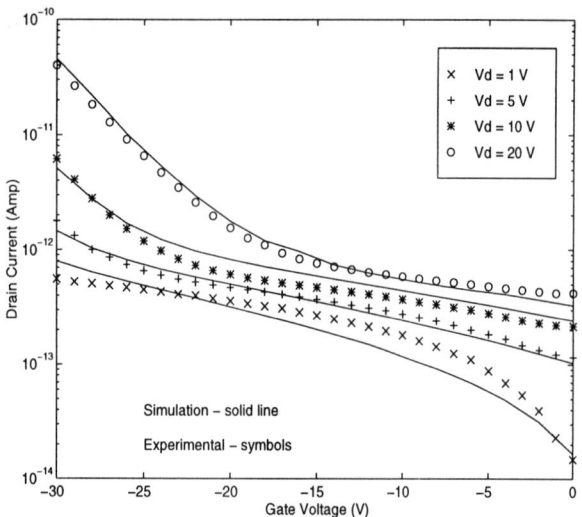

Fig.5 Comparison of simulated and measured leakage current characteristics of a-Si:H TFTs.

Following systematic characterization of TFTs with different a-Si:H layer thicknesses, we found that the optimal thickness for low leakage is around 50 nm. Here, we observe that $I_{Leakage}$, for fixed V_{DG}, increases when the intrinsic layer thickness is either larger or smaller than the optimal value. With the former, due to the decreased E-field, and hence barrier height, in the i-region, there is an enhancement in thermal excitation. In the case of the latter, there is enhancement of trap-assisted tunneling. Both cases lead to an increase in leakage current [9].

4. CONCLUSIONS

The main source of leakage current in a-Si:H TFTs appears to be from the parasitic reverse-biased p-i-n diode between the drain and gate terminals. At low V_G and low V_D, the reduced barrier height at the drain junction gives rise to thermal excitation of electrons to the conduction band via mid-gap states in the intrinsic a-Si:H layer. At high V_G and/or V_D, because of the high, but narrow, energy barrier at the drain junction, electrons tunnel into the conduction band through traps. The optimal thickness of the intrinsic layer in our fabricated samples was found to be around 50 nm. Values larger or smaller than this optimum value lead to an increase in leakage current due to enhanced thermal excitation and tunneling, respectively. Based on the analysis of the measurement data, a SPICE model was developed using voltage controlled current sources. The simulation results closely follow the measured characteristics of the leakage current.

5. REFERENCES

[1] K. Kobayashi, H. Murai, M. Hayama and T. Yamazaki, "The application of hydrogenation to amorphous silicon thin film transistors for the decrease of the OFF current," Mat. Res. Soc. Symp. Proc., vol. 219, pp. 321-326, 1991.

[2] K. S. Lee, J. H. Choi, S. K. Kim, H. B. Jeon, and J. Jang, "Low OFF-state leakage current thin film transistor using *Cl* incorporated hydrogenated amorphous silicon," Appl. Phys. Lett., vol. 69, pp. 2403-2405, 1996.

[3] J. H. Kim, W. S. Choi, C. H. Hong and H. S. Soh, "Investigation of the OFF-current in amorphous silicon thin film transistors for SiO_2 and SiN_x gate insulators," Mat. Res. Soc. Symp. Proc., vol. 424, pp. 85-90, 1997.

[4] M. Hack, H. Steemers and R. Weisfield, "Transient leakage currents in amorphous silicon thin-film transistors," Mat. Res. Soc. Symp. Proc., vol. 258, pp. 949-954, 1992.

[5] G. E. Possin, "High temperature OFF current in a-Si TFTs - Effect of process and structure," Mat. Res. Soc. Symp. Proc., vol. 219, pp. 327-332, 1991.

[6] M.J. Powel, "The physics of amorphous-silicon thin film transistors," IEEE Trans. Electron Devices, vol. 36, pp. 2753 - 2763, 1989.

[7] B. Park, R.V.R. Murthy, K. Benaissa, K. Aflatooni, A. Nathan, R.I. Hornsey and S. G. Chamberlain, "Effect of deposition temperature on the structural properties of n^+ µc-Si:H films," J. Vac. Sci. Technol. A, 16(2), Mar/Apr, 1998, in press.

[8] C.Y. Chen and J. Kanicki, "High field-effect-mobility a-Si:H TFT based on high deposition rate PECVD material," IEEE Trans. Electron Devices vol. 17, pp. 437-439, 1996.

[9] R.V.R Murthy, B. Park, A. Nathan and S. G. Chamberlain, "Compact SPICE modeling and design optimization of low leakage a-Si:H TFTs for large-area imaging systems," to be presented at Mat. Res. Soc. Symp., April 1998.

SPICE MODEL FOR MECHANICALLY STRESSED DEVICE/CIRCUIT SIMULATION

C. Maier, R. Steiner, M. Mayer, R. Vogt, and H. Baltes
Physical Electronics Laboratory, ETH Zürich
Hönggerberg HPT–H6, CH–8093 Zürich, Switzerland
Phone: +41 1 6332089; Fax: +41 1 6331054; Email: maier@iqe.phys.ethz.ch

ABSTRACT

We present a SPICE–compatible circuit model to predict the effects of mechanical stress on the electrical characteristics of devices and integrated circuits (ICs). The model is assembled from unit cells, which consist of resistors and voltage controlled current sources. We verified the circuit model with measurements on an n-well magnetic sensor structure on (100) silicon. The model predicts the influence of stress to less than 10% discrepancy from measurement.

INTRODUCTION

Mechanical stress in ICs stems from packaging (die bonding, molding) of the silicon die. In addition, fabrication process conditions associated with growth and deposition of the thin films overlying the electrical active region contribute to stress. Mechanical stress can seriously degrade the performance of microsensors and ICs. For example, in magnetic Hall microsensors, the stress gives rise to device offset, which deteriorates the detection ability of static and low frequency magnetic fields. In ICs, stress–induced changes in resistance and injection conditions of transistors undermine the highly needed device matching in circuit design. In all of these cases, the effects of stress must be considered an integral part of the design process. Here, the critical design issues are related to the optimization of shape, size, placement, and orientation of microsensor or device in the circuit. These design issues can only be addressed with simulation, particularly since we are dealing with relatively complex device geometries, stress–induced anisotropy in resistance, and stress distributions that are arbitrary [1].

The circuit model presented here has two major advantages. First, a resistive device under stress is described by a SPICE netlist. Thus, it can be treated as a sub–circuit, along with the readout circuit, and device–circuit interactions can be analyzed in a SPICE environment. Second, layout extraction tools can be extended to include model description files of the device under stress, whose mechanical properties are made available in the technology file. This will allow iterative layout optimization of device geometry, placement, and orientation to limit the effects of mechanical stress.

MODEL DERIVATION

The electrical field E and the current density J are related by Ohm's law $J = \sigma \cdot E$, with the electrical conductivity σ. In the presence of stress, the conductivity is no longer isotropic, but becomes directionally dependent and is described by a symmetric 3×3 matrix. For sufficiently low stress, the conductivity can be assumed to be linearly dependent on the mechanical stress T, with the piezoresistance coefficients π as proportionality constants. In terms of reduced matrix notation (see [2]), σ can be written as

$$\sigma_i = \sigma_{Vol} \cdot \left(1 - \sum_{j=1}^{6} \pi_{ij} \cdot T_j \right) \quad (1)$$

with $i, j = 1\ldots6$, where σ_{Vol} denotes the volume conductivity of the unstressed material.

In view of the planar nature of integrated circuits, the generalized Ohm's law in monocrystalline silicon can be reduced to two dimensions:

$$\begin{pmatrix} J_x \\ J_y \end{pmatrix} = \begin{pmatrix} \sigma_1 & \sigma_6 \\ \sigma_6 & \sigma_2 \end{pmatrix} \cdot \begin{pmatrix} E_x \\ E_y \end{pmatrix} \quad (2)$$

with

$$\sigma_1 = \sigma_0(1 - \pi_{11} \cdot T_1 - \pi_{12} \cdot (T_2 + T_3)) \;, \quad (3)$$

$$\sigma_2 = \sigma_0(1 - \pi_{11} \cdot T_2 - \pi_{12} \cdot (T_1 + T_3)) \;, \quad (4)$$

$$\sigma_6 = -\sigma_0 \cdot \pi_{44} \cdot T_6 \;, \quad (5)$$

assuming that the coordinate axes are aligned with the (100) crystal axes. The subscripts for the conductivity are according to Eqn. (2), and σ_0 and J denote the *sheet* conductivity and current density, respectively. Note that the off–diagonal terms σ_6 result in a current flow perpendicular to the applied electric field.

CIRCUIT IMPLEMENTATION

Using equations (2) through (5), and a discretization scheme originally proposed for the modeling of Hall devices [1], passive circuit elements of arbitrary geometry can be represented by a two–dimensional array of unit cells. The discretization scheme is based on charge conservation under steady–state conditions and in the absence of current sources, $\nabla \cdot J = 0$. In electrical network analysis this relation translates into Kirchhoff's current law for a center node C (see Fig. 1), which is solved by SPICE.

In order to represent the relation (2) as a finite set of circuit elements, the electric field strength and sheet current density within a rectangular plate of width $l_{LC} + l_{RC}$ and height $l_{UC} + l_{DC}$ (see Fig. 1) are approximated as:

$$V_{UC} = E_y \cdot l_{UC} \quad (6)$$
$$V_{CL} = E_x \cdot l_{LC} \quad (7)$$
$$V_{CD} = E_y \cdot l_{DC} \quad (8)$$
$$V_{RC} = E_x \cdot l_{RC} \quad (9)$$

and

$$I_{UC} = J_y \cdot (l_{LC} + l_{RC}) \quad (10)$$
$$I_{CL} = J_x \cdot (l_{UC} + l_{DC}) \quad (11)$$
$$I_{CD} = J_y \cdot (l_{LC} + l_{RC}) \quad (12)$$
$$I_{RC} = J_x \cdot (l_{UC} + l_{DC}) \quad (13)$$

The equivalent circuit, the designation of nodes, and the conductivity equations for arbitrary angles of coordinate orientations with respect to the crystal axes on (100) silicon wafers are shown in Fig. 1. The cell consists of resistances and controlled sources, the latter stemming from the presence of off–diagonal terms in the conductivity matrix (2).

The number of such unit cells needed to represent the device to be modeled is determined by device structure, stress distribution, and spatial variations in resistance.

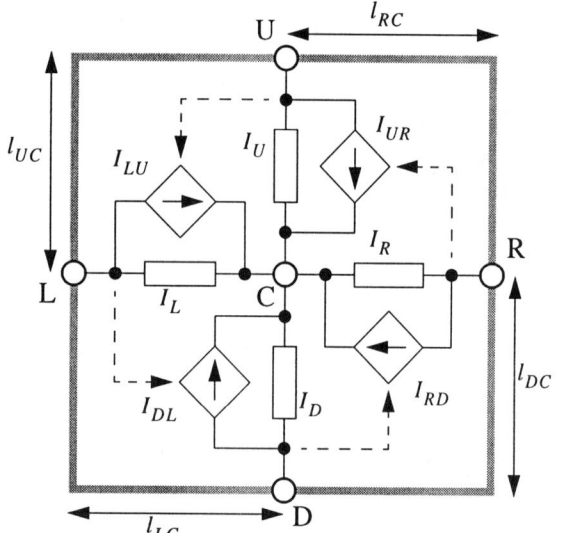

$$I_U = G_{vert} \cdot V_{UC} \cdot \frac{l_{LC} + l_{RC}}{l_{UC}} \qquad I_{UR} = G_{diag} \cdot V_{RC} \cdot \frac{l_{LC} + l_{RC}}{l_{RC}}$$

$$I_L = G_{horiz} \cdot V_{CL} \cdot \frac{l_{UC} + l_{DC}}{l_{LC}} \qquad I_{LU} = G_{diag} \cdot V_{UC} \cdot \frac{l_{UC} + l_{DC}}{l_{UC}}$$

$$I_D = G_{vert} \cdot V_{CD} \cdot \frac{l_{LC} + l_{RC}}{l_{DC}} \qquad I_{DL} = G_{diag} \cdot V_{CL} \cdot \frac{l_{LC} + l_{RC}}{l_{LC}}$$

$$I_R = G_{horiz} \cdot V_{RC} \cdot \frac{l_{UC} + l_{DC}}{l_{RC}} \qquad I_{RD} = G_{diag} \cdot V_{CD} \cdot \frac{l_{UC} + l_{DC}}{l_{DC}}$$

$$G_{vert} = 2\sigma_0(1 - \pi_{11} \cdot T_2 - \pi_{12} \cdot (T_1 + T_3) + C \cdot P \cdot T_d + S \cdot P \cdot T_o)$$
$$G_{horiz} = 2\sigma_0(1 - \pi_{11} \cdot T_1 - \pi_{12} \cdot (T_2 + T_3) - C \cdot P \cdot T_d - S \cdot P \cdot T_o)$$
$$G_{diag} = 2\sigma_0((-\pi_{44}/2 + C \cdot P)T_o - S \cdot P \cdot T_d) \;,$$
$$C = 1 - \cos 4\phi \;, S = \sin 4\phi \;, T_d = T_1 - T_2 \;, T_o = 2T_6 \;,$$
$$P = \frac{1}{4} \cdot (\pi_{44} + \pi_{12} - \pi_{11}) \;.$$

V_{AB}: voltage *from* node A *to* node B

ϕ: angle between stress and crystal coordinate systems

Figure 1: Equivalent circuit model to account for mechanical stress (left) with relations for the stress dependent currents (right).

VERIFICATION OF THE MODEL

As an example, the conductivity of a cross shaped Hall sensor, implemented as buried n–well resistor on a (100) CMOS wafer has been measured under different stress conditions. The sensor has been modeled with 5 square unit cells to account for the device geometry in a simple way (see Fig. 2). The device was subjected to a well–defined stress distribution obtained with the four–point bending bridge method [3]. The dominant component of stress was tensile stress T_1 parallel to the surface of the die. In order to verify the dependence of piezoresistivity on the angle between current, stress, and crystal orientation, the resistance has been measured and simulated with a continuous spinning current scheme [4]. To this end, sine and cosine currents are applied to the contact pairs of the device under test,

$$I_{LR} = I_0 \cdot \sin\psi \text{ and } I_{UD} = I_0 \cdot \cos\psi, \quad (14)$$

and the resulting terminal voltages are translated into components parallel and orthogonal to the current flow,

$$V_{parallel} = V_{LR} \cdot \sin\psi + V_{UD} \cdot \cos\psi, \quad (15)$$

and

$$V_{orthogonal} = V_{UD} \cdot \sin\psi - V_{LR} \cdot \cos\psi. \quad (16)$$

Fig. 3 shows measured and simulated voltages $V_{parallel}$ and $V_{orthogonal}$ as a function of the angle of the current vector ψ for different stress distributions.

With increasing tensile stress parallel to the die surface, the average conductivity increases. The component of conductivity proportional to the *second* harmonic of the current rotation frequency increases as well. These effects are reproduced by the circuit model.

In addition to the stress–induced conductivity modulation, there is also a component proportional to the *fourth* harmonic of the current rotation frequency, caused by the cross shape of the device. This is accounted for by a circuit model which consists of 5 unit cells with a sheet conductivity σ_0 depending on the depletion layer width as

$$\sigma_0 = \frac{\sigma_{Vol}}{t_{nwell}} \cdot \sqrt{\frac{2 \cdot \varepsilon_{Si}}{q \cdot N_D} \cdot (\Phi_{BI} + V_{CS})} \quad (17)$$

with n–well volume conductivity σ_{Vol}, doping level N_D, metallurgical thickness t_{nwell}, permeability of silicon ε_{Si}, and built–in junction potential Φ_{BI}. The voltage V_{CS} between the well and the substrate is measured from the center node of each unit cell.

Even though only the two most dominant effects of change in resistivity are modeled, the change in resistance due to stress is modeled with less than 10% error. Other effects affecting the resistance of the sensor [5] can be modeled by adding resistances and voltage controlled current sources depending on other physical quantities in parallel to the circuit elements shown in Fig. 1.

CONCLUSIONS

The influence of stress on the resistance of a device is modeled by the structure of a unit cell. Effects due to the device geometry which could not be modeled by a

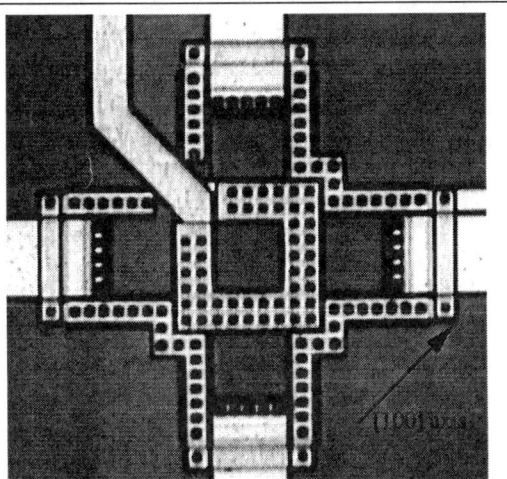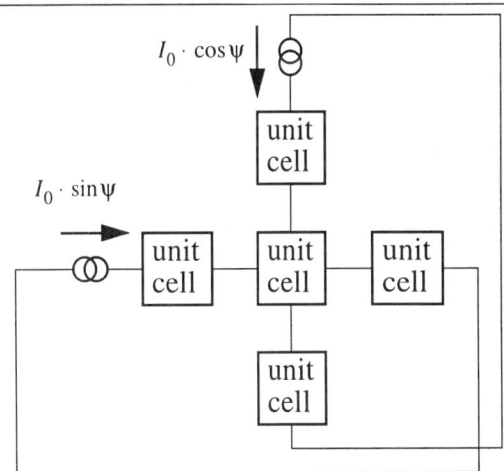

Figure 2: Photograph of the cross–shaped Hall sensor (left) and circuit model of the Hall sensor (right), consisting of 5 unit cells as shown in Figure 1.

single unit cell have been taken into account by an appropriate arrangement of unit cells. With this circuit model, the discrepancy between the model and measurement is less than 10% of the change in resistance. In a similar way, other interactions, e. g. galvanomagnetic effects, can be incorporated as needed [1] into this versatile, yet simple circuit model.

ACKNOWLEDGMENTS

We would like to thank Professor Arokia Nathan of the University of Waterloo, Canada, and Sandra Bellekom, Delft University of Technology, The Netherlands, for helpful discussions. This work has been funded by the European ESPRIT project "MagIC" through Grant no. 20360 (BBW contract no. 95.0109.1).

REFERENCES

[1] A. Nathan and H. Baltes, *Microtransducer CAD, Physical and Computational Aspects*, Chapter 9; Springer Verlag, in print.

[2] D. A. Bittle, J. C. Suhling, R. E. Beaty, R. C. Jaeger, R. W. Johnson, "Piezoresistive Stress Sensors for Structural Analysis of Electronic Packages", *Journal of Electronic Packaging*, 113:203–215, September 1991

[3] M. Mayer, O. Paul, and H. Baltes, "Complete set of piezoresistive coefficients of CMOS n+–diffusion". *Eighth Micromechanics Europe Workshop (MME'97)*, pp. 203–206

[4] R. Steiner, A. Häberli, F.–P. Steiner, and H. Baltes, "Offset Reduction in Hall Devices by Spinning Current Method". *1997 International Conference on Solid–State Sensors and Actuators (Transducers '97)*, pp. 381–384

[5] P. J. A. Munter, "Spinning–current method for offset reduction in silicon Hall plates", *Ph.D. thesis*, Delft University of Technology, 1992

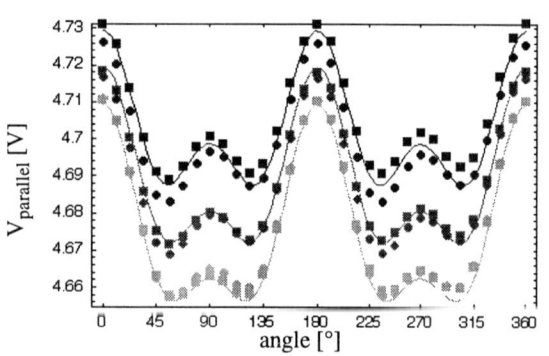
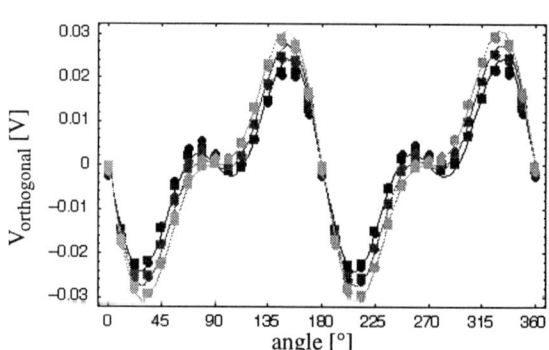

Figure 3: Comparison of simulation (solid lines) and measured data of two sensor ICs (circles and squares) for $V_{parallel}$ and $V_{orthogonal}$ in accordance to equations (15) and (16), at stress distributions dominated by the tensile stress component T_1 of 90MPa (black), 114MPa (dark grey), and 138MPa (light grey).

RAPID EXTRACTION OF CAPACITANCE IN a-Si IMAGING ARRAYS

Hoan H. Pham and Arokia Nathan

Department of Electrical and Computer Engineering
University of Waterloo, Waterloo, Ontario
Canada N2L 3G1

ABSTRACT

We present a new technique for computation of charge density for a multi-conductor system embedded in homogeneous or multiple dielectric media. The charge density distribution determines the parasitic coupling capacitance in large-area imaging arrays or ULSI interconnects, as well as the electrostatic interaction in MEMS. The proposed scheme employs the exponential-expansion-based method for efficient evaluation of the three-dimensional potential and electric field. Here, the memory requirement is independent of the desired degree of accuracy, which is an important feature for large-scale simulation involving panel numbers in the range of a few hundred thousand or several million.

1. INTRODUCTION

With the rapid increase in both component density on chips and operating frequency, the parasitic coupling capacitance associated with interconnects poses serious design issues in high density and high performance integrated circuits. These issues are also common to large-area high resolution amorphous silicon (a-Si) arrays for applications in X-ray imaging (see Fig. 1) or active-matrix liquid crystal displays. Here, the added difficulty stems from the presence of the floating potential of the glass substrate (see Fig. 2). To gain insight into the effect of parasitic coupling capacitance on the overall array performance, one needs to be able to extract the capacitance with a high degree of accuracy and in an efficient manner. In addition, this aids in further development of equivalent circuit models for effective SPICE-like simulations for sensitivity analysis and design optimization at the system level.

The capacitance is calculated from the distribution of charge density on the surfaces of the conductors. In general, given a conductor system $S_1, S_2, \ldots, S_n$ with the applied voltages $V_1, V_2, \ldots, V_n$, we need to determine the charge density σ on these conductors' surfaces. Here, we assume that the conductors are embedded in homogeneous or multiple dielectric media. The charge density problem can be for-

---

This work is supported by the DALSA/NSERC Industrial Research Chair Program, the Information Technology Research Center (ITRC), and the Natural Sciences and Engineering Research Council of Canada.

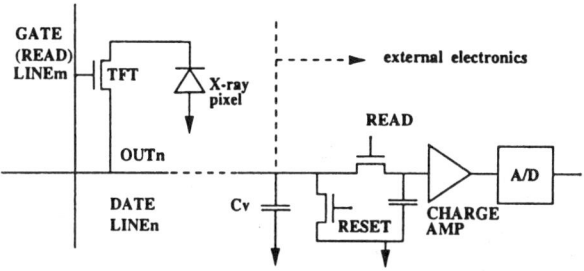

Figure 1: Pixel readout circuit with thin film transistor (TFT) in one row of a matrix imaging area.

mulated in terms of a differential equation, which can then be solved using finite difference or finite element methods; the resultant matrix is sparse but large due to discretization of the whole volume. Alternatively, an integral equation formulation can be employed, leading to a denser but smaller matrix due to discretization of the surfaces only. In this paper, we take the latter approach. In a homogeneous medium, the single-layer potential $\phi(x)$ due to a charge density σ on a surface S is given by:

$$\phi(x) = \int_S G(x,y)\sigma(y)dy, \quad (1)$$

where G(x,y) is the associated Green's function:

$$G(x,y) = \frac{1}{4\pi\|x-y\|}. \quad (2)$$

Using description (1), we can obtain the charge density by solving the equation:

$$\int_S G(x,y)\sigma(y)dy = V_i, \quad x \in S_i. \quad (3)$$

The surface $S = S_c = \bigcup_{i=1}^{n} S_i$ covers all conductor regions. In the case of a multiple dielectric medium, we also need to include the charge density at the interface S_d between different dielectric layers, and the associated equation of flux continuity reads [1]:

$$\epsilon^+ \frac{d\phi(x)}{dn} = \epsilon^- \frac{d\phi(x)}{dn}. \quad (4)$$

Here, ϵ^+ and ϵ^- denote dielectric constants of the respective dielectric layers. Following standard convention, we assume $\epsilon^+ < \epsilon^-$ and the normal vector $\mathbf{n}$ at the interface points toward the dielectric layer of smaller dielectric constant (i.e., ϵ^+). The above equation can be cast into an integral equation of the form:

$$\sigma(x) + \frac{1}{2}\lambda \int_S \frac{\partial G(x,y)}{\partial n_x}\sigma(y)dy = 0, \quad x \in S_d. \quad (5)$$

The surface $S = S_c \cup S_d$ covers both conductor and interface regions; $\frac{\partial G(x,y)}{\partial n_x}$ is the normal derivative at a point x on the interface S_d; and λ, which is in the range $[0, 1]$, is defined as:

$$\lambda = \frac{\epsilon^- - \epsilon^+}{\epsilon^- + \epsilon^+}. \quad (6)$$

The governing equation for the multiple dielectric medium is described by a system comprising (3) and (5):

$$\begin{cases} \int_S G(x,y)\sigma(y)dy = V_i, \quad x \in S_i \\ \sigma(x) + 2\lambda \int_S \frac{\partial G(x,y)}{\partial n_x}\sigma(y)dy = 0, \quad x \in S_d. \end{cases} \quad (7)$$

In the panel method [2], the surfaces (including interfaces) are meshed into N panels. The charge density on each panel is assumed to be constant and takes the value at the centroid. Equations (3) or (7) leads to a system of linear equations:

$$A\sigma = b. \quad (8)$$

The $N \times N$ matrix A is dense and we are interested in dealing with large N (e.g., in the range of a few hundred thousand or several million). In this range of N, it becomes prohibitively expensive to store all entries of the matrix A and to solve eq. (8) using a direct method such as Gauss elimination. For the former, the memory requirement is of $O(N^2)$ and for the latter, the computational time is of $O(N^3)$. For this reason, it is more efficient to use an iterative method, whose key operation is matrix-vector multiplication. Using iterative methods, we need to take into account:

- The efficiency and accuracy of matrix-vector multiplication in terms of computational time and memory requirement;
- The convergence rate.

In the context of our problem, the matrix-vector multiplication reduces simply to finding the potential and electric field given a charge distribution. The evaluation of the potential and electric field can be performed, with some loss of accuracy, using multipole-expansion (MP) [3] or exponential-expansion (EE) [4] based methods. The convergence rate can be improved by (i) applying preconditioning [5, 6], and/or (ii) increasing the accuracy of the approximation of matrix-vector multiplication. Alternatively, we can employ Fredholm integral equations of the second kind for both homogeneous and multiple dielectric media that yields a less ill-conditioned matrix [7].

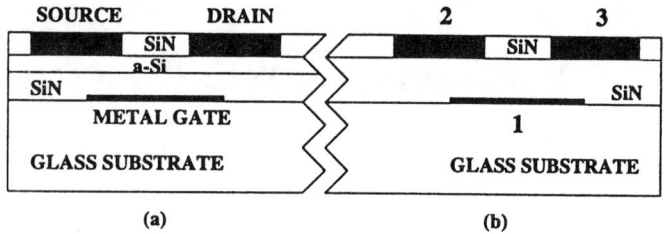

Figure 2: Schematic cross-sections illustrating (a) TFT and (b) interconnect.

2. EVALUATION OF POTENTIAL AND ELECTRIC FIELD

The state-of-the-art algorithms for evaluation of the potential share, more or less, the same common structure:

- Approximation of the Green's function $\frac{1}{r}$, which varies from algorithm to algorithm;
- Creating a tree-like hierarchy to easily identify clusters of particles or centers;
- Performing translation operations to replace a cluster of particles or centers with a new one.

2.1. Multipole-expansion-based method

This method is based on the approximation of Green's function $\frac{1}{r}$ in terms of spherical harmonics. Among the popular algorithms is the Greengard and Rokhlin fast multipole algorithm (FMA) [3]. The first implementation of FMA for capacitance calculation was reported in [8]. With this method, to evaluate the potential with accuracy of $O(\frac{1}{2^p})$, where p is the order of the multipole expansion, the computational time is of $O(p^3 N)$ and memory requirement is of $O(p^2 N)$.

For large N, this memory requirement of $O(p^2 N)$ is problematic if a high degree of accuracy is expected on the potential and the electric field. In fact, this is required in many cases, including:

- At interfaces of different dielectric media, where we need to compute the electric field, it is crucial that the potential be approximated with higher accuracy than that required for the electric field. It is observed that the inaccuracy in the computed electric field is drastically reduced as the potential is computed to higher degrees of accuracy. For example, if the accuracy in the computed potential is of the order of 10^{-3}, then the corresponding accuracy in the electric field is barely 10^{-1}. However, if the potential is computed to an accuracy of 10^{-8}, then the corresponding accuracy of the electric field is 10^{-7}.

- As the number of panels N increases, the matrix A can become ill-conditioned. Consequently, the matrix-vector multiplication Ax is very sensitive to small changes in x. In the MP or EE method, the error in matrix-vector multiplication can easily be amplified by A, leading to deterioration of convergence rate or even to unreliable results. The use of a preconditioner can reduce the number of iterations; however, the associated memory requirement may turn out to be expensive. A more accurate approximation of the potential (i.e. expansion of higher order) may be used to reduce the effect of ill-conditioning. Alternatively, the convergence rate may be improved by use of a different formulation for the charge density [7] that yields an integral equation of the second kind for both homogeneous and multiple dielectric media, thus rendering itself more suitable for iterative methods. The current approaches yield either an integral equation of the first kind (eq. (3)) or a combination of first kind and second kind (eq. (7)), resulting in a more ill-conditioned matrix.

- As a consistency check, one should verify that the approximation employed for the computation of the potential is indeed sufficient. Here, we recompute the potential at higher accuracy and compare the resultant charge density distributions. If the change is within acceptable limits, then we have confidence that the approximation on potential is adequate; therefore, any discrepancy in the computed charge density may be attributed to discretization and related errors.

2.2. Exponential-expansion-based method

The mathematical background of this method is based on an integral representation of Green's function $\frac{1}{r}$:

$$\frac{1}{r} = \frac{1}{\sqrt{x^2+y^2+z^2}}$$
$$= \int_0^\infty e^{-\lambda z} J_0(\sqrt{x^2+y^2}\lambda) d\lambda, (z > 0) \quad (9)$$

which is the Laplace transform of Bessel function J_0 of the first kind with order 0 [9]. An approximation of (9) using Gauss quadratures yields:

$$\frac{1}{r} \approx \sum_{l=1}^{S_{app}} \beta_l E_l(x,y,z), \forall (x,y,z) \in D, \quad (10)$$

where β_i's are constants, S_{app} denotes the size of the approximation, D is a domain of interest, and $E_l(x,y,z)$'s are functions on D that are independent of each other and have the following property:

$$E_l(x_1+x_2, y_1+y_2, z_1+z_2) =$$
$$E_l(x_1,y_1,z_1) E_l(x_2,y_2,z_2). \quad (11)$$

The EE method was first briefly presented in [4] and its use for the charge density problem was detailed in [10]. The computational time and memory requirement for evaluating the potential and electric field is of $O(S_{app}N)$ and $O(N)$, respectively. An important feature of the EE method is that the memory requirement is independent of the desired degree of accuracy. This can be explained by observing that the functions $E_l(x,y,z)$ are independent of each other; therefore, they can be evaluated in parallel or in sequence. In the latter, one memory unit can be used for all $E_l(x,y,z)$. With this property, the concern about accuracy and memory requirement can be addressed more efficiently, thus permitting large-scale simulation involving panel numbers in the range of hundred thousand to several million, independent of the desired degree of accuracy on the approximation of potential. The accuracy can range from a few digits to machine precision.

3. ILLUSTRATIVE EXAMPLES

As a demonstration of the use of the EE method, we present two examples which involve calculation of the coupling capacitance in an a-Si thin film transistor (TFT) and an interconnect. Here, a key issue is the high aspect ratio of the thin film layer thickness relative to other physical dimensions. As a result, a large number of panels are needed just to barely resolve the surfaces. Also a large portion of the total panel count is dedicated to meshing the glass substrate. Thus, if a Green's function for the (infinite) glass substrate is incorporated, the total panel count can be reduced or suitably distributed to reduce the overall discretization error.

The first example is a simple three-dimensional a-Si TFT (Fig. 2a). The length, width, and thickness of the structure considered in simulations is $100\mu m \times 100\mu m \times 11.3\mu m$. Here, only $10\mu m$ of the glass substrate ($\epsilon_r = 3.9$) is considered to reduce panel count. In going from bottom to top, the dielectric layers are glass substrate, gate SiN, a-Si, and top SiN; the respective layers thickness and dielectric constant are listed in table 1. The length of the source and drain regions is $25\mu m$, while that of the gate is $20\mu m$. The overlap between gate and the source or drain is $3\mu m$. The structure considered in the second example (Fig. 2b) is virtually identical, in terms of geometrical dimensions, except that the thin a-Si layer is now replaced by the gate SiN layer. In both examples, we are interested in computing the coupling capacitances among the three conductors: gate (1), source (2), and drain (3).

The average relative error in the approximation of the potential and electric field is 10^{-7} and 10^{-5}, respectively; this corresponds to approximation size $S_{app} = 116$. With an initial coarse mesh and an initial guess of zero charge den-

	gate SiN	a-Si	top SiN	source & drain	gate
thickness	0.25	0.05	1.0	1.0	0.12
ϵ_r	4.50	11.9	3.1		

Table 1: Thickness (μm) of various layers (see Fig. 2) and their dielectric constant.

	gate	source	drain
gate	1.98×10^2	-9.76×10^1	-9.76×10^1
source	-9.77×10^1	1.05×10^2	-9.03×10^{-1}
drain	-9.77×10^1	-9.03×10^{-1}	1.05×10^2

Table 2: Coupling capacitances (fF) in the TFT example.

	(1)	(2)	(3)
(1)	1.631×10^2	-8.121×10^1	-8.121×10^1
(2)	-8.121×10^1	8.358×10^1	-6.986×10^{-1}
(3)	-8.121×10^1	-6.986×10^{-1}	8.358×10^1

Table 3: Coupling capacitances (fF) in the interconnect example.

sity, the charge density is obtained after several runs. Each run consists of the following steps: (i) solution of eq. (8) for charge density; (ii) mesh refinement with each panel subdivided by four using a tree hierarchy; (iii) interpolation of the computed charge density for generation of new initial guess. The final number of panels used in the two examples are 386,444 and 319,372, respectively.

The computed values of the capacitance are listed in tables 2 and 3. The relative change in the values in the last two runs is observed to be less than 5×10^{-3}. Therefore it is expected that the relative error in capacitance is around 5×10^{-3}. For the level of precision considered, we see that the values of mutual capacitance are symmetric for the interconnects (table 3), but not so for the TFT (table 2). Here, the entries (C_{12}, C_{21}) and (C_{13}, C_{31}), which represent the gate-source and gate-drain coupling capacitances, differ by one unit in the third digit. This can be attributed to the charge density on the panel, which is scaled by the respective dielectric constants of the surrounding media. This differs in the case of the gate as compared to the source and drain regions. The discrepancy can be reduced by use of a finer mesh.

4. CONCLUSION

We have presented a new scheme for computing the charge density and hence the coupling capacitance associated with a multi-conductor system embedded in homogeneous or multiple dielectric media. The scheme is based on the EE method, which allows evaluation of the three-dimensional potential and electric field distribution for a wide range of accuracy. Furthermore, the method accommodates use of a large number of panels without exorbitant requirements in memory. The discretization error can be improved by adaptive grid refinement and by use of a higher order approximation of the charge density on the panel.

5. REFERENCES

[1] S. Rao, T. Sarkar, and R. Harrington, "The electrostatic field of conducting bodies in multiple dielectric media," *IEEE Trans. on Microwave Theory and Techniques*, vol. 32, pp. 1441–1448, 1984.

[2] R. F. Harrington, *Field Computation by Moment Methods*. IEEE Press, New Jersey, 1993.

[3] L. Greengard, *The Rapid Evaluation of Potential Fields in Particle Systems*. MIT Press, 1988.

[4] H. Pham and A. Nathan, "Rapid evaluation of the potential fields in three dimensions using exponential expansion," *Canadian Journal of Physics*, vol. 75, pp. 689–693, 1997.

[5] S. Vavasis, "Preconditioning for boundary integral equations," *SIAM J. Matrix Analysis and Applications*, vol. 13, no. 3, pp. 905–925, 1992.

[6] F. L. K. Nabors, F. Korsemeyer and J. White, "Preconditioned, adaptive, multipole-accelerated iterative methods for three-dimensional potential integral equation of the first kind," *SIAM J. Sci. Stat. Comput.*, vol. 15, no. 3, pp. 713–735, 1994.

[7] H. Pham and A. Nathan, "An integral equation of the second kind for the charge density problem in homogeneous and multiple dielectric media," Tech. Rep. UW E&CE 98-03, Department of Electrical and Computer Engineering, University of Waterloo, January 1998.

[8] K. Nabors and J. White, "Fastcap: A multipole-accelerated 3-D capacitance extraction program," *IEEE Trans. on Computer-Aided Design of Integrated Circuits and Systems*, vol. 10, pp. 1447–1459, 1991.

[9] G. Watson, *A Treatise of the Theory of Bessel functions*. Cambridge University Press, 2nd ed., 1966.

[10] H. Pham and A. Nathan, "Solving the charge density problem using exponential expansion," Tech. Rep. UW E&CE 98-02, Department of Electrical and Computer Engineering, University of Waterloo, January 1998.

AN EFFICIENT MOS TRANSISTOR CHARGE/CAPACITANCE MODEL WITH CONTINUOUS EXPRESSIONS FOR VLSI

Steve H. Jen, Bing J. Sheu, Alex Y. Park
Department of Electrical Engineering
and Integrated Media Systems Center
University of Southern California, Los Angeles, CA 90089-0271, USA.

Abstract

A unified modeling approach for the submicron MOS transistor charge/capacitance characteristics in all operation regions is presented. The development of MOS charge model is based on the charge density approximation to reduce the complexity of the expression. The unified charge densities in gate, channel, and bulk are obtained with assistance of the sigmoid, hyperbola, and exponential interpolation techniques. By carrying out the integration of the charge densities along the channel area, the terminal charges associated with gate and bulk can be obtained. The non-reciprocal capacitance behavior is well realized in this model. Good agreement between the measurement data and simulation results is obtained.

1 Introduction

With rapid advances in submicron silicon fabrication technologies, CMOS circuits are widely used in communication, control, and signal processing applications. As shown in Fig. 1, the feature size of the CMOS technology has been improved to the quarter-micron range. New generation of microprocessors [1] has achieved the 600 MHz clock operation rate with 2 V power supply voltage in a 0.35 μm CMOS technology. In wireless communication applications, the operation frequency of CMOS analog circuits is increasing to 1 GHz range [2]. On the other hand, as the power supply voltage of mixed-signal circuits decreases, many transistors tend to be biased around the on-set of strong inversion from the weak inversion region in order to reduce power dissipation and to increase voltage gain. Therefore, accurate modeling of MOS charge-capacitance characteristics in all operation regions is very important for small-signal AC analysis as well as large-signal transient analysis.

In this paper, a unified MOS transistor charge model valid in all operation regions is described. Instead of calculating the surface potential by solving the Poisson's equation directly [3], charge density approximation is used to obtain the gate, channel, and bulk charge densities to reduce the complexity of the expressions. The sigmoid function and hyperbola techniques are used in this model to unify the charge density expressions, and provide the smooth transition between different regions. The terminal charges are calculated by integrating the charge densities across the channel area. Good agreement between the measured data and simulation results is obtained.

2 The Charge/Capacitance Model

Derivation of MOS charge model is based on the quasi-static approximation [4]. The charge of an MOS transistor is made up of three fundamental components: the charge residing on the gate electrode, q_g, the fixed charge residing in the bulk depletion layer, q_b, and the mobile channel charge residing in the channel region, q_c. The charge densities are derived by using gradual-channel approximation and depletion approximation [5]. In order to ensure the continuity of the charge and high accuracy of modeling the transitions betweens different operation regions, instead of separate charge expressions in each operation region, the unified charge density expressions are obtained by applying the hyperbola [6, 7] and sigmoid [7] techniques,

$$\begin{aligned} q_g &= C_{OX}[(V_{GSth} + f_C(V_{th} - V_{FB} - \phi_S) - f_C \cdot V_{ch}) \\ &+ \frac{\gamma_1^2}{2}\left(-1 + \sqrt{1 + \frac{4(V_{GFh} - V_{GSth} - V_{FS})}{\gamma_1^2}}\right) \\ &+ (V_{GS} - V_{GFh})] \end{aligned} \quad (1)$$

$$q_c = -C_{OX}(V_{GSth} - \alpha_x \cdot f_C \cdot V_{ch}) \quad (2)$$

$$q_b = -(q_g + q_c) \qquad (3)$$

where

$$f_C = \frac{1}{2}\left(1 + \frac{V_{GS} - V_{th}}{\sqrt{(V_{GS} - V_{th})^2 + K_c}}\right). \qquad (4)$$

Here, V_{FB} is the flat-band voltage, $V_{FS} = V_{FB} + V_{BS}$ is defined as flat-band voltage referring to source terminal potential, and f_C is a sigmoid function which has the value of 1 in the strong-inversion region and smoothly transits to 0 in the weak-inversion region. K_c is defined as a model parameter to accurately predict the behavior at the transition between the weak- and strong-inversion regions.

The conductance-degradation coefficient, α_x, is obtained by taking inverse value of the differential term of saturation voltage, V_{DSAT}, with respect to the gate voltage, and the expression is

$$\alpha_x = \left(\frac{\partial V_{DSAT}}{\partial V_{GS}}\right)^{-1}$$

$$= \left(1 - \frac{\gamma_1^2}{2\sqrt{V_{GSth} + V_{th} - V_{FS} + \frac{\gamma_1^2}{4}}}\right)^{-1} \qquad (5)$$

The total charge stored in each of the gate, channel, and bulk regions is obtained by integrating the distributed charge densities, q_g, q_c, and q_b, over the channel area,

$$Q_G = W_{eff}L_{eff}C_{OX}\left[\left(V_{GSth} + f_C(V_{th} - V_{FB} - \phi_S) - f_C \cdot V_{DSATh}\frac{4T_C - 1}{6T_C}\right)\right.$$
$$\left. + \frac{\gamma_1^2}{2}\left(-1 + \sqrt{1 + \frac{4(V_{GFh} - V_{GSth} - V_{FS})}{\gamma_1^2}}\right) + (V_{GS} - V_{GFh})\right], \qquad (6)$$

$$Q_C = -W_{eff}L_{eff}C_{OX}$$
$$\cdot \left(V_{GSth} - \alpha_x \cdot f_C \cdot V_{DSATh}\frac{4T_C - 1}{6T_C}\right) \qquad (7)$$

$$Q_B = -(Q_G + Q_C), \qquad (8)$$

where

$$T_C = 1 - \frac{\alpha_x}{2}\frac{V_{DSATh}}{V_{GSth}}. \qquad (9)$$

To complete the charge model, expressions for the drain and source terminal charges should also be included. This can be done by channel charge partitioning. A physically meaningful 40/60 partitioning scheme as for drain and source terminal charges, developed by Ward [8], is used. By carrying out the integration, the charges associated with the drain and source terminals are obtained,

$$Q_D = -W_{eff}L_{eff}C_{OX}V_{GSth}\left(\frac{-1}{2} + T_C\right.$$
$$\left. + \frac{(1 - T_C)(1 + 3 \cdot T_C + 6 \cdot T_C^2)}{30 \cdot T_C^2}\right), \qquad (10)$$

$$Q_S = -W_{eff}L_{eff}C_{OX}V_{GSth}\left(\frac{1}{2} + \frac{(1 - T_C)^2}{3 \cdot T_C}\right.$$
$$\left. - \frac{(1 - T_C)(1 + 3 \cdot T_C + 6 \cdot T_C^2)}{30 \cdot T_C^2}\right). \qquad (11)$$

The inter-nodal capacitances are represented as derivatives of the terminal charges, Q_G, Q_B, Q_D, and Q_S, with respect to the terminal voltages, i.e.,

$$C_{ij} = x_{ij}\frac{\partial Q_i}{\partial V_j}. \qquad (12)$$

where the indices i and j represent any of the four terminals, gate, bulk, drain, or source. $x_{ij} = -1$ for $i \neq j$, and $x_{ij} = 1$ for $i = j$. The differentiation of these charge expressions is also unified through all operation regions.

3 Experimental Results and Discussion

The unified MOS transistor charge/capacitance model has been developed and compared to the measured data. Figure 2 shows the comparison of simulated results and measured data of conductance-degradation coefficient, α_x, which decreases as increasing the gate voltage. Figure 3 shows the normalized plots of the four terminal charges of a short-channel MOS transistor with the 40/60 channel-charge partitioning method. The normalization factor is $(W_{eff}L_{eff}C_{OX})$. Notice that the unified charge expressions are continuous over all operation regions. The sigmoid function, f_c, facilitates the smooth transition between the weak- and strong-inversion regions. Here, fitting parameter, K_c, helps to model the curvature of capacitances at the transition portion, as shown in Fig. 4.

In the time-domain large-signal analysis and frequency-domain small-signal analysis for the charge-based approach, 16 charge derivatives are needed in the assembly of the nodal admittance matrix. These derivatives may be regarded as capacitances. To form the matrix, nine of the 16 capacitances are independent [8]. Figure 5(a) and (b) show the normalized plots of nine capacitances against the gate voltage and drain voltage, respectively. Notice that the capacitance curves are smooth throughout all operation regions. The nonreciprocal capacitance property is clearly shown. Comparison of measured and calculated results of the capacitances associated with gate and bulk terminals for an $L_{eff} = 0.5~\mu m$ transistor is shown in Fig. 6. Good agreement between the measured data and calculated results has been obtained.

4 Summary

The unified MOS transistor charge/capacitance model has been developed to provide continuous characteristics in all operation regions. Since the transition between the weak- and strong-inversion regions is carefully considered and accurately modeled in the unified expressions, this model is very useful for low-voltage and low-power IC design.

Acknowledgement

This work was partially supported by NSF under Grant ECS-9322279.

References

[1] B. A. Gieseke, et al., "A 600 MHz superscalar RISC microprocessor with out-of-order execution," Tech. Digest of IEEE Int'l Solid-State Circuits Conference, vol. 40, pp. 176-177, Feb. 1997.

[2] J. C. Rudell, et al., "A 1.9 GHz wide-band IF double conversion CMOS integrated receiver for cordless telephone applications," Tech. Digest of IEEE Int'l Solid-State Circuits Conference, vol. 40, pp. 304-305, Feb. 1997.

[3] H. J. Park, P. K. Ko, C. Hu, "A charge sheet capacitance model of short channel MOSFET's for SPICE," IEEE Trans. on Computer-Aided Design, vol. 10, no. 3, pp. 376-389, Mar. 1991.

[4] Y. P. Tsividis, Operation and Modeling of the MOS Transistor, McGraw-Hill: New York, NY, 1987.

[5] B. J. Sheu, W.-J. Hsu, P. K. Ko, "A MOS transistor charge model for VLSI design," IEEE Trans. on Computer-Aided Design, vol. 7, no. 4, pp. 520-527, Apr. 1988.

[6] A. Chatterjee, C. F. Machala, P. Yang, "A Submicron DC MOSFET Model for Simulation of Analog Circuits," IEEE Trans. Computer-Aided Design of Integrated Circuits and Systems, vol. 14, no. 10, pp. 1193-1207, Oct. 1995.

[7] S. H. Jen, B. J. Sheu, Y. Oshima, "A unified approach to submicron DC MOS transistor modeling for low-voltage ICs," J. Analog Integrated Circuits and Signal Processing, Kluwer Academic Publishers, vol. 7, no. 2, pp. 107-118, Feb. 1997.

[8] D. E. Ward, "Charge-based modeling of capacitance in MOS transistor," Ph.D. dissertation, Tech. Rep. G 201-11, Integrated Circuit Laboratory, Stanford University, June 1981.

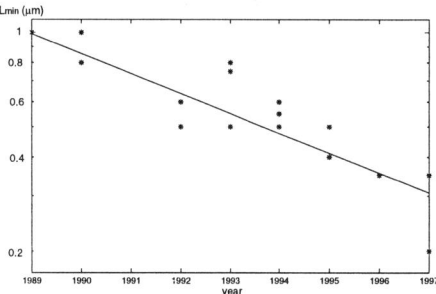

Figure 1: Plot of the minimun channel length of CMOS technology versus year.

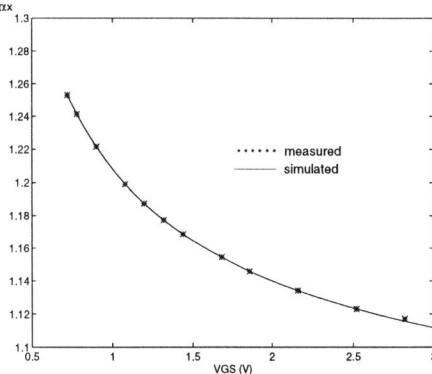

Figure 2: Plots of conductance-degradation coefficient, α_x, vs. V_{GS}.

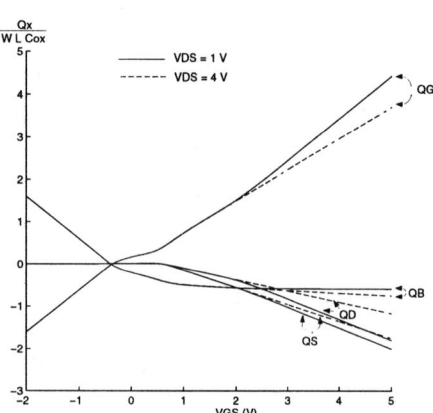

Figure 3: Normalized terminal charges versus gate voltage for two drain voltages with $V_{BS} = 0\ V$.

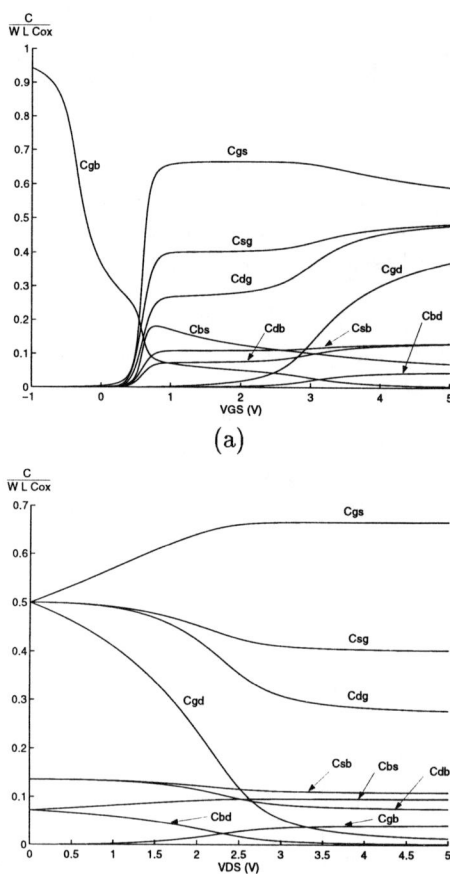

Figure 5: Plots of nine normalized capacitances. (A) Capacitances versus V_{GS} with $V_{DS} = 2\ V$ and $V_{BS} = 0\ V$. (B) Capacitances versus V_{DS} with $V_{GS} = 3.5\ V$ and $V_{BS} = 0\ V$.

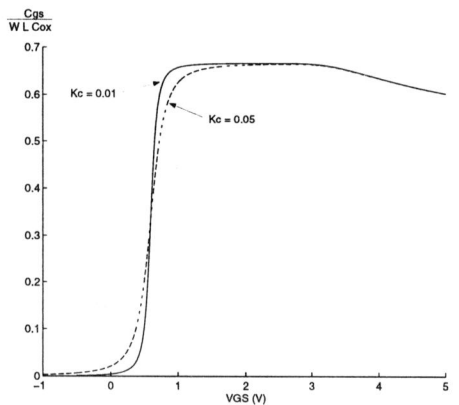

Figure 4: Plots of normalized C_{gs} versus V_{GS} for two K_c values.

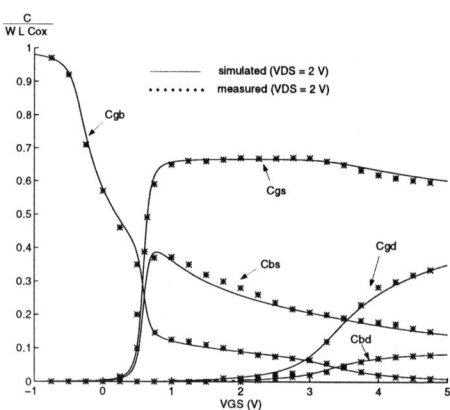

Figure 6: Plots of normalized capacitances versus V_{GS} of an NMOS transistor of W/L = 25 μm/0.5 μm for two V_{DS} values.

WAVELET-BASED GALERKIN METHOD FOR SEMICONDUCTOR DEVICES SIMULATION

Fung-Yuel Chang, Fellow, IEEE [1] *and Chung-Kei Thomas Chan* [2]

The Chinese University of Hong Kong, Hong Kong
[1]fychang@ee.cuhk.edu.hk, [2]ckchan1@ee.cuhk.edu.hk

ABSTRACT

Using wavelet methods, local high order schemes can be constructed near the singularities. Moreover, the stiffness matrix is sparse and can readily be inverted due to the compact support property of wavelets. An adaptive Galerkin-wavelet method for semiconductor devices simulation is presented. A set of wavelet bases can be chosen adaptively for each iteration according to the error levels. So, computational time is saved and hence a more accurate result can be obtained by including higher order terms. Also, an elegant way to handle the boundary conditions is provided. A simulation of an abrupt P-N junction is used to demonstrate this effectiveness in this paper.

1. INTRODUCTION

In semiconductor simulation, singularities occur at the interfaces of materials with different doping concentrations and material properties. The numerical solution of the governing equations is popularly solved by finite difference methods. In order to treat those singularities properly, the step size of the finite difference methods has to be reduced and thus in turn slowing down the computation. One of the solutions to treat those singularities is to use a smaller step size near the singularities and a larger step size where the solution is smooth. In this paper, another alternative to solve this singularity problem is provided using wavelets.

Wavelet bases are suitable candidates for treating the singularity problems because their biorthogonality property allows multi-resolution approximation of the solution. Therefore, local high order schemes can be constructed near the singularities and local low order schemes can be constructed in smooth regions. A self adaptive algorithm is feasible to choose a set of wavelet bases in different positions and resolutions for each iteration. So, computational time is saved and hence a more accurate result can be obtained by including higher order terms. Moreover, because of their properties of localizations in time and space, wavelet methods are also suitable for multi-resolution transient simulation of semiconductor devices.

During this decade, wavelet methods are applied to solve differential equations. The methods can be classified into two classes: collocation method and Galerkin method [1]. The DBWM for semiconductor devices simulation was proposed in [2] using collocation method. So, in this paper, the application of Galerkin-wavelet methods for semiconductor devices simulation is studied. Wavelets functions are used directly as basis functions to approximate solutions [4]. The resulting stiffness matrix is a sparse matrix with diagonal preconditioner, thus allowing $\mathcal{O}(N)$ algorithms for solving the corresponding linear systems of equations.

The paper is organized as follows. A brief review of the governing equations for semiconductor devices simulation and their boundary conditions is given in Section 2. In Section 3, a modified Gummel's iteration scheme is suggested. A brief summary of the essentials of multi-resolution decomposition is described in Section 4. In Section 5, Galerkin method is applied on the modified Gummel's scheme using wavelet approximation of the solution. An adaptive scheme for selecting a set of wavelet bases is presented in Section 6. Numerical results of an example are presented and discussed in Section 7. A brief conclusion is drawn in Section 8.

2. BASIC EQUATIONS

In this section, a brief review of the governing equations for semiconductor devices simulation is presented. There are basically five equations which describe the carrier, current, and field distributions in semiconductor devices [6], [7].

The current densities of holes and electrons are given by the superposition of a diffusion term and a drift term:

$$J_n = -q\mu_n n \frac{\partial \psi}{\partial x} + qD_n \frac{\partial n}{\partial x} \quad (1)$$

$$J_p = -q\mu_p p \frac{\partial \psi}{\partial x} - qD_p \frac{\partial p}{\partial x} \quad (2)$$

For the one-dimension case, the continuity equations of hole and electrons are given by:

$$\frac{\partial n}{\partial t} = -\mathcal{U} + \frac{1}{q}\frac{\partial J_n}{\partial x} \quad (3)$$

$$\frac{\partial p}{\partial t} = -\mathcal{U} - \frac{1}{q}\frac{\partial J_p}{\partial x} \quad (4)$$

Where $\mathcal{U}$ is the recombination rate.

The relationship between the charge distribution and the electric potential ψ is given by the Poisson's equation:

$$\frac{\partial^2 \psi}{\partial x^2} = \frac{q}{\varepsilon_s \varepsilon_o}(n - p + N_a - N_d) \quad (5)$$

The carrier concentrations and the potentials are governed by Maxwell-Boltzmann statistics approximation for non-degenerate semiconductor devices. i.e. hole concentration, p, electron concentration, n, the electric potential, ψ, hole quasi-Fermi potential, φ_p and electron quasi-Fermi potential, φ_n are related by:

$$n = n_i e^{(\psi - \varphi_n)/\phi_T} \quad (6)$$

$$p = n_i e^{(\varphi_p - \psi)/\phi_T} \quad (7)$$

Assume there is a set of a single level Shockley-Hall recombination centers in the middle of the bandgap. Then the recombination rate $\mathcal{U}$ is given by [8]:

$$\mathcal{U} = \frac{pn \Leftrightarrow n_i^2}{\tau_p(n+n_i) + \tau_n(p+n_i)} \qquad (8)$$

For steady state simulation, equations (1), (2), (6), (7) and (8) are substituted into equations (3), (4) and (5). After putting $\frac{\partial p}{\partial t} = 0$ and $\frac{\partial n}{\partial t} = 0$; and using the Einstein relation, a reduced set of equations for semiconductor simulation with variables ψ, φ_p and φ_n is obtained:

$$\psi'' = \frac{qn_i}{\varepsilon_s \varepsilon_o}(e^{(\psi-\varphi_n)/\phi_T} \Leftrightarrow e^{(\varphi_p-\psi)/\phi_T} \Leftrightarrow N) \qquad (9)$$

$$D_n(e^{\psi/\phi_T})'(e^{-\varphi_n/\phi_T})' + D_n(e^{\psi/\phi_T})(e^{-\varphi_n/\phi_T})''$$
$$= \frac{e^{(\varphi_p-\varphi_n)/\phi_T} \Leftrightarrow 1}{\tau_p(e^{(\psi-\varphi_n)/\phi_T}+1) + \tau_n(e^{(\varphi_p-\psi)/\phi_T}+1)} \qquad (10)$$

$$D_p(e^{-\psi/\phi_T})'(e^{\varphi_p/\phi_T})' + D_p(e^{-\psi/\phi_T})(e^{\varphi_p/\phi_T})''$$
$$= \frac{e^{(\varphi_p-\varphi_n)/\phi_T} \Leftrightarrow 1}{\tau_p(e^{(\psi-\varphi_n)/\phi_T}+1) + \tau_n(e^{(\varphi_p-\psi)/\phi_T}+1)} \qquad (11)$$

Where $N(x) = (N_d(x) \Leftrightarrow N_a(x))/n_i$.

In the later example, an abrupt P-N junction is simulated using the above equations. There are totally three variables and two boundaries, and therefore six boundary conditions have to be satisfied [7]:

$$\varphi_{n0} = V_0 \qquad (12)$$
$$\varphi_{p0} = V_0 \qquad (13)$$
$$\varphi_{n1} = V_1 \qquad (14)$$
$$\varphi_{p1} = V_1 \qquad (15)$$

$$\psi_0 = V_0 \Leftrightarrow \phi_T \ln(\sqrt{(N(0)/2)^2+1} \Leftrightarrow N(0)/2) \qquad (16)$$

$$\psi_1 = V_1 \Leftrightarrow \phi_T \ln(\sqrt{(N(1)/2)^2+1} \Leftrightarrow N(1)/2) \qquad (17)$$

Where V_0 and V_1 are terminal voltages at two ends of the junction.

3. ITERATION SCHEME

Gummel's iteration scheme [7] is modified in this section. In his work, the variable set is $\{\psi, \varphi_p, \varphi_n\}$. Variables are updated one by one. In slightly coupled system, Gummel's iteration scheme is computationally faster than Newton's.

In Gummel's paper, the Poisson's equation is expanded in terms of the difference between the available trial solution ψ and the exact solution. Neglecting terms of higher orders, the Poisson's equation is considered as a linear differential equations. We would like to do the same procedure to the other two non-linear equations. A special variable set is used to simplify the expression. The variable set in this paper is $\{\psi, e_p, e_n\}$, where $e_p = e^{\varphi_p/\phi_T}$ and $e_n = e^{-\varphi_n/\phi_T}$.

Three linearized equations for simulation are obtained after simple algebraic derivation and approximation:

$$\delta_s'' \Leftrightarrow H_1 \delta_s = H_2 \qquad (18)$$

$$\Leftrightarrow F_5 \delta_n + F_2 F_6 \delta_n' + F_3 F_6 \delta_n'' = F_4 \Leftrightarrow F_1 F_6 \qquad (19)$$

$$\Leftrightarrow G_5 \delta_p + G_2 G_6 \delta_p' + G_3 G_6 \delta_p'' = G_4 \Leftrightarrow G_1 G_6 \qquad (20)$$

Where,

$$H_1 = \frac{qn_i}{\varepsilon_s \varepsilon_o \phi_T}(e^{\psi/\phi_T} e_n + e_p e^{-\psi/\phi_T}) \qquad (21)$$

$$H_2 = (\Leftrightarrow \psi'' + e^{\psi/\phi_T} e_n \Leftrightarrow e_p e^{-\psi/\phi_T} \Leftrightarrow N) \qquad (22)$$

$$F_1 = D_n(e^{\psi/\phi_T})' e_n' + D_n(e^{\psi/\phi_T}) e_n'' \qquad (23)$$

$$F_2 = D_n(e^{\psi/\phi_T})' \qquad (24)$$

$$F_3 = D_n(e^{\psi/\phi_T}) \qquad (25)$$

$$F_4 = e_p e_n \Leftrightarrow 1 \qquad (26)$$

$$F_5 = e_p \qquad (27)$$

$$F_6 = \tau_p(e^{\psi/\phi_T} e_n + 1) + \tau_n(e_p e^{-\psi/\phi_T} + 1) \qquad (28)$$

The expressions for G's are similar to those for F's, so they are not repeated here.

The overall scheme is summarized below:

Step 1: A guess at the trial solution is made.

Step 2: Use equation (18) to update ψ.

Step 3: Use equation (19) to update e_n.

Step 4: Use equation (20) to update e_p.

Step 5: Repeat Steps 2, 3 & 4 until the solution converges.

4. MULTI-RESOLUTION DECOMPOSITION

A brief summary of the essentials of the wavelet expansion and multi-resolution decomposition is described [10]. Multi-resolution decomposition starts with a basic function called the scaling function ϕ which possesses translation and dilation properties:

$$\phi_{j,k} = 2^{j/2} \phi(2^j x \Leftrightarrow k) \qquad (29)$$

A set of subspaces, $\{V_j = \text{span}\{\phi_{j,k} \mid k \in \mathbf{Z}\}\}_{j \in \mathbf{Z}}$ is called a multi-resolution approximation of $L^2(\mathbf{R})$ if it possesses the following properties [4]:

$$V_j \subset V_{j+1}, \quad \forall j \in \mathbf{Z} \qquad (30)$$

$$\bigcup_{j \in \mathbf{Z}} V_j \text{ is dense in } L^2(\mathbf{R}), \quad \bigcap_{j \in \mathbf{Z}} \mathbf{V_j} = \emptyset \qquad (31)$$

$$f(x) \in V_j \quad f(2x) \in V_{j+1}, \quad \forall j \in \mathbf{Z} \qquad (32)$$

$$f(x) \in V_j \quad f(x \Leftrightarrow 2^{-j} k) \in V_j, \quad \forall j, k \in \mathbf{Z} \qquad (33)$$

Any square integrable function can be approximated as closely as desired by a function in space V_j. An approximation of a function $f \in L^2(\mathbf{R})$ at resolution of 2^{-j} can be represented by the projection:

$$f_j(x) = P_j f(x) = \sum_k \phi_{j,k}(x) <\phi_{j,k}, f> \qquad (34)$$

The resolution of the approximation improves as j increases. At the extreme case, when j tends to infinity, the projected function is the same as the original one. The construction of wavelets starts by considering the orthogonal complement, W_j, of V_j in V_{j+1}. The resolution of the jth approximation of a function improves one level by including the jth details of the function. i.e. the orthogonal projection, Q_j, of the function to V_j in V_{j+1}.

$$f_{j+1}(x) = P_{j+1} f(x) = P_j f(x) + Q_j f(x) \qquad (35)$$

Wavelet, $\psi_{j,k}$, is an orthogonal basis of subspace W_j. It also possesses translation and dilation properties:

$$\psi_{j,k} = 2^{j/2}\psi(2^j x \Leftrightarrow k) \quad (36)$$

Therefore, $f \in L^2(\mathbf{R})$ can be approximated by its approximation and its details.

$$f_m(x) = \sum_k \alpha_k \phi_{j,k}(x) + \sum_{i=j}^{m-1} \sum_k \beta_{i,k} \psi_{i,k}(x) \quad (37)$$

Where $\alpha_k = <\phi_{j,k}, f>$ and $\beta_{i,k} = <\psi_{i,k}, f>$.

5. GALERKIN-WAVELET METHOD

In a boundary value problem, a finite number of expansions is sufficient to approximate $f \in L^2([0,1])$ with appropriate edge functions [11]. In [11], there are edge functions for the approximation spaces $V_j[0,1]$ and the complement spaces $W_j[0,1]$. To reduce the computation time, the edge functions for the complement spaces $W_j[0,1]$ can be omitted. This simplification can be justified if the solution near the boundaries is smooth. Therefore, $f \in L^2([0,1])$ can be approximated by "interior" scaling functions and "interior" wavelets with "edge" functions at the coarsest scale:

$$f_m = \sum_{n=1}^{N} \gamma_n \phi_{j,n}^l + \sum_{k=0}^{M_j} \alpha_k \phi_{j,k} + \sum_{i=j}^{m-1} \sum_{k=0}^{M_i} \beta_{i,k} \psi_{i,k} + \sum_{n=1}^{N} \lambda_n \phi_{j,n}^r \quad (38)$$

Where $M_j = 2^j \Leftrightarrow 2N \Leftrightarrow 1$ and N is the no. of vanishing moments.

The modified semiconductor equations (18), (19) and (20) can be represented by:

$$L(f) = g \quad (39)$$

By applying the Galerkin method with the expansion (38), the problem then reduces to a linear system of the form:

$$M\underline{u} = \underline{b} \quad (40)$$

The first equation and the last equation in the system (40) are replaced by the following boundary equations to implement the boundary conditions:

$$\sum_{n=1}^{N} \delta_n^l \phi_{j,n}^l(0)$$
$$= \begin{cases} \psi_0 \Leftrightarrow \sum_{n=1}^{N} \gamma_n \phi_{j,n}^l(0) & \text{for EP} \\ e^{-\varphi_{n0}/\phi_T} \Leftrightarrow \sum_{n=1}^{N} \gamma_n \phi_{j,n}^l(0) & \text{for EQFP} \\ e^{\varphi_{p0}/\phi_T} \Leftrightarrow \sum_{n=1}^{N} \gamma_n \phi_{j,n}^l(0) & \text{for HQFP} \end{cases} \quad (41)$$

$$\sum_{n=1}^{N} \delta_n^r \phi_{j,n}^r(1)$$
$$= \begin{cases} \psi_1 \Leftrightarrow \sum_{n=1}^{N} \gamma_n \phi_{j,n}^r(1) & \text{for EP} \\ e^{-\varphi_{n1}/\phi_T} \Leftrightarrow \sum_{n=1}^{N} \gamma_n \phi_{j,n}^r(1) & \text{for EQFP} \\ e^{\varphi_{p1}/\phi_T} \Leftrightarrow \sum_{n=1}^{N} \gamma_n \phi_{j,n}^r(1) & \text{for HQFP} \end{cases} \quad (42)$$

6. ADAPTIVE SCHEME

A set of wavelet bases with different positions and resolutions can be adaptively chosen for each iteration.

The scheme starts with the observation that if the original device equations (9), (10) and (11) are satisfied, the right hand sides of equations (18), (19) and (20) are equal to zero. So, g(x), in equation (39), indicates the error between the trial solution and the exact solution. And g(x) is projected to $\underline{b}$, in equation (40), by a fast wavelet transform.

Since $\underline{b}$ indicates the error between the trial solution and the exact solution for each wavelet coefficients, a set of wavelet bases with higher errors can be adaptively chosen for the construction of the stiffness matrix at each iteration with reference to the vector $\underline{b}$.

The scheme for each iteration is summarized below:

Step 1: Calculate g(x) using previous trial solution.

Step 2: Project g(x) into $\underline{b}$ using FWT.

Step 3: Select a set of wavelet bases with greater errors.

Step 4: Solve the equations using reduced set of bases.

7. RESULTS AND DISCUSSION

An abrupt P-N junction is simulated by the Galerkin-Wavelet method. The results of the simulation are presented and discussed in this section. A P-N junction with abrupt doping profile represents the worst case in terms of truncation errors.

The Si step-junction diode is doped with $N_a = 2.5 \times 10^{15} \text{cm}^{-3}$ on the p-side and $N_d = 2.5 \times 10^{17} \text{cm}^{-3}$ on the n-side. The length of p-side is 9.8304×10^{-4} cm and that of n-side is 3.2768×10^{-4} cm. The steady state solution of this diode at 0.6V forward bias in room temperature is simulated using Galerkin-Wavelet method with Daubechies 5 wavelets.

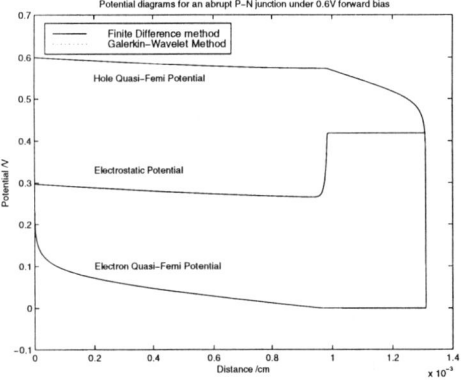

Figure 1: Potential diagrams for an abrupt P-N junction under 0.6V forward bias

Fig. 1, 2 and 3 show the results of the simulation. We can hardly observe any difference between two methods in the potential diagrams (fig. 1) and in the carrier densities (fig. 3). Therefore, the solution of the Galerkin-Wavelet method can closely approximate the solution for the variables ψ, φ_n and φ_p. However, Daubechies wavelets are functions generated by numerical method and therefore there are numeric errors in the calculation of their derivatives. Hence, there is an acceptable error in the calculation of current densities (fig. 2).

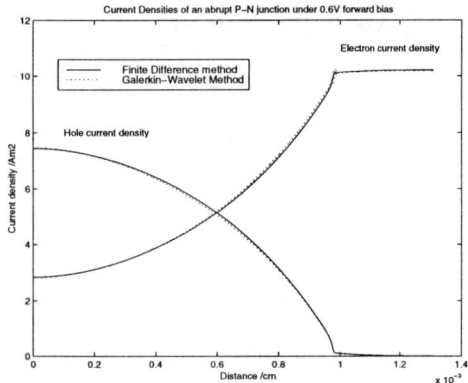

Figure 2: Current densities for an abrupt P-N junction under 0.6V forward bias

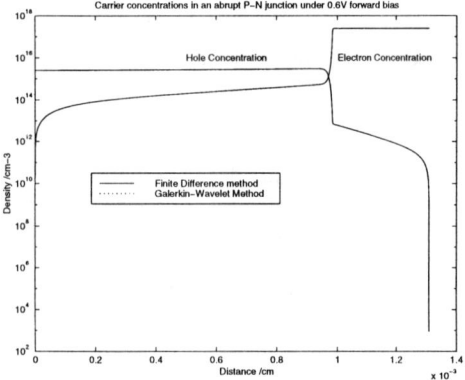

Figure 3: Carrier densities for an abrupt P-N junction under 0.6V forward bias

In order to show the adaptive feature of the Galerkin-Wavelet method, a very small step size of 1×10^{-8}cm is used in this example. 131073 of grid points are required by the finite difference method at this resolution. 16336 wavelet bases are used to approximate the solution in this example. There is a 7/8 reduction in the number of variables if the Galerkin-Wavelet method is used. Table 1 shows the number of wavelet bases adaptive selected for computation for each iteration. With the adaptive feature of the Galerkin-Wavelet method, more than 7/8 reduction in the number of variables can be achieved.

Iteration	ψ	φ_n	φ_p
1	179	190	161
2	495	514	630
3	244	540	337
4	401	347	585
5	362	378	268
6	478	354	501

Table 1: Number of bases for each iteration

In fact, the modified Gummel's iterative scheme presented in the section 3 allows fast algorithm for the finite difference method. After the application of the finite difference, all of the three decoupled equations (18), (19) and (20) become systems of equations in the form:

$$T\delta = B \qquad (43)$$

Where matrix T is a tridiagonal matrix.

With the modified Gummel's iterative scheme, the calculation of the solution with the finite difference method is very fast because a fast iterative algorithm is available for the inversion of a tridiagonal matrix. With a large reduction in the number of variables in Galerkin-Wavelet method, we can achieve further 10% reduction in CPU time.

8. CONCLUSION

Gummel's iterative scheme is modified in this paper. An adaptive Galerkin-Wavelet method is applied on this modified scheme to solve the Dirichlet boundary value problem. A simulation of an abrupt junction diode is used to demonstrate the effectiveness of this method. Other wavelet basis can be used to improve the accuracy in the future.

9. REFERENCES

[1] B. Silvia, N. Giovanni, C. R. Jean, "Wavelet methods for the Numerical Solution of Boundary Value Problems on the Interval". *Wavelets: Theory, Algorithms, and Applications*, pp. 425-448, Academic Press, 1994.

[2] F. Y. Chang, K. P. Pun, "Discrete B-Spline Wavelet Method for Semiconductor Devices Simulation", *Proceedings of the International Symposium on Circuit and Systems, Hong Kong, 1997*, Vol. 1, pp 193-196, 1997.

[3] J. C. Xu, W. C. Shann, "Galerkin-Wavelet Methods for Two-Point Boundary Value Problems", *Numerische Mathematik*, Vol. 63, 1992.

[4] B. Z. Steinberg, Y. Leviatan, "On the Use of Wavelet Expansions in the Method of Moments", *IEEE Trans. on Antennas and Propagation*, Vol. 41, pp. 610-619, May 1993.

[5] R. F. Harrington, *Field Computation by Moment Methods*, New York, IEEE Press, 1993.

[6] E. S. Yang, *Microelectronic Devices*, McGRAW-Hill, 1988.

[7] H. K. Gummel, "A Self-Consistent Iterative Scheme for One-Dimensional Steady State Transistor Calculations". *IEEE Trans, Electron Devices*, Vol. ED-11, pp. 445-465, Oct. 1964.

[8] C. T. Sah, R. N. Noyce, W. Shockley, "Carrier Generation and Recombination in P-N Junctions and P-N Junction Characteristics", *Proc. IRE*, Vol. 45, pp. 1228-1243, September 1957.

[9] I. Daubechies, "Orthonormal Bases of Compactly Supported Wavelets". *Comm. Prue Appl. Math.*, Vol. 41, pp. 909-996, 1988.

[10] I. Daubechies, *Ten Lectures on Wavelets*, SIAM Publications, Philadelphia, 1992.

[11] A. Cohen, I. Daubechies, P. Vial, "Wavelets on the interval and Fast Wavelet Transforms", *Appied and computational harmonic analysis*, Vol 1, pp. 54-81, 1993.

USING PLAS TO DESIGN UNIVERSAL LOGIC MODULES IN FPGAS *

K.K. Lee   D.F. Wong

Department of Computer Sciences
The University of Texas at Austin

ABSTRACT

We consider implementing Universal Logic Modules in Field Programmable Gate Arrays (FPGAs) with Programmable Logic Arrays (PLAs) using a reduced number of programmable switches. These have the advantages of a regular structure and are very easy to program. By suitably reducing the number of switches in a PLA, the total number of switches is reduced tremendously, while the PLA still remains functionally complete. Since switch features take up more space than other logic elements, the savings can translate to a reduction in area. The reduction in programmable switches implies that a smaller number of programming bits is required for each ULM. We obtained 3-input and 4-input ULMs using 5 and 13 programmable switches respectively, matching previous results [3] but in a smaller area. Technology mapping is also very simple. We also obtain *approximate* ULMs with very high coverage. We obtained an approximate ULM using 11 programming switches which covers 99% of all 4-input functions, using a much smaller area than [1].

1 INTRODUCTION

FPGA architecture consists of logic modules and routing resources, replicated in large numbers. These are used to implement logic functions by suitably configuring them during the technology mapping phase. It is desirable to have a small function capable of implementing a large set of functions to reduce the number of logic modules required for a circuit. On one end of the spectrum, look-up tables (LUTs) are functionally complete and can implement any function of a fixed number of inputs, m, but its size grows exponentially with m. At the other end, Actel's multiplexor-based macro-cells are not functionally complete; the ACT2 modules can implement 85% of all 3-input functions and only 36% of all 4-input functions [1].

A number of researchers considered using ULMs as logic blocks in FPGAs [1, 2, 3]. The *approximate* ULM in [1] has 8 inputs and a high 99% coverage for 4-input functions. The main objective in [3] was to reduce the number of programming bits while remaining functionally complete, differentiating configuration inputs from function inputs. By combining BDDs into super-BDDs and deciding which switches can be removed or combined, they constructed ULM.3 and ULM.4 using 5 and 13 programming bits respectively. No technology mapping algorithm was given, however.

PLAs are useful for implementing combinational logic due to its versatility, simplicity, regularity and predictable delay. A typical PLA has many inputs and outputs but usually only a small number of minterms. In order to compare with other ULMs, only m-input, 1-output PLAs, called Universal PLAs (UPLAs) are considered. UPLAs are easy to configure and have both hardwired and programmable switches (called *h-switches* and *p-switches* respectively).

---
*This work was partially supported by the Texas Advanced Research Program under Grant No. 003658288.

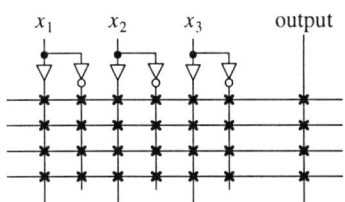

Figure 1: A schematic diagram of a 3-input PLA

This paper describes the construction of UPLA.3s and UPLA.4s with a 29% and 33% reduction in the number of switches compared to a PLA.3 and PLA.4 respectively. The number of p-switches was reduced by 67% and 75% respectively. Since switch features take up more space than other elements in a PLA, this reduction can lead to a reduction in area, and a smaller configuring circuitry. The mapping algorithm is very simple, a benefit not easily derivable for other ULMs other than LUTs. We also show how to further reduce the number of p-switches to match the result of [3], but in a smaller area. The method in [3] has also not been generalized to higher dimensions, which appears to be quite difficult. The result in [1] corresponds to an *approximate* UPLA.4 with 12 rows and 12 p-switches. We designed an approximate ULM.4 that uses only 8 rows and achieves a higher coverage than their ULM, but in a much smaller area.

This paper is organized as follows : Section 2 gives some definitions, Section 3 discusses architectural considerations, properties and design of UPLAs. A simple mapping algorithm is also given. Section 4 discusses how to further reduce the number of switches under NPN-equivalence and give a simple mapping algorithm under NPN-equivalence. Section 5 considers approximate ULMs with a smaller number of switches.

2 PROBLEM DEFINITIONS AND PRELIMINARIES

Let $X_m = \{x_1, \ldots x_m\}$ be a set of m variables, F_m the set of functions of X_m, and M_m the set of minterms of X_m. Following [1], we use ULM.m to denote a ULM of m-input variables. This notation is also extended to UPLAs and SUPLAs shown later.

Definition: A sequence of permutations and complementations is called a *transformation*. When applied to the inputs of a PLA, it is called a *restricted specialization*. If instantiating inputs to a constant (**0** or **1**) and bridging of inputs are allowed, it is called a *general specialization*. $\begin{pmatrix} 1 & 2 & \cdots & m \\ \pi(1) & \pi(2) & \pi(m) \end{pmatrix}$ denotes a transformation where x_i is assigned $x_{\pi(i)}$. Complementation is denoted by $\overline{\pi(i)}$. Thus, $\begin{pmatrix} 1 & 2 & 3 & 4 \\ \overline{1} & 4 & 2 & 3 \end{pmatrix}$ means x_1 is assigned $\bar{x}_1$, x_2 assigned x_4, x_3

assigned x_2 and x_4 assigned $\bar{x}_3$. For convenience, we also define the equivalence $\pi(x_i) = x_{\pi(i)}$.

Definition: The *parity* of a minterm is even if the number of negative (complemented) literals in the minterm is even and odd otherwise. The sets $M_m^o \subset M_m$ and $M_m^e \subset M_m$ denote the set of odd and even minterms of X_m respectively. A transformation has the same parity as the number of negative literals in the transformation.

Definition: Two minterms m_i and m_j are *adjacent* if they differ in polarity in only one literal. Thus, $x_1\bar{x}_2x_3$ and $x_1\bar{x}_2\bar{x}_3$ are adjacent, but $x_1\bar{x}_2\bar{x}_3$ and $x_1x_2x_3$ are not.

Definition: A PLA g is *restrictedly PLA-specializable* to a function f if g can be transformed to f by performing restricted specialization and independently setting a given set of p-switches on g. They are *generally PLA-specializable* if we use general specialization instead. If g is PLA-specializable to f, then g can implement f. Two functions f and f' are *NPN-equivalent* if we can transform f to f' by negating and/or permuting the inputs and output of f. If g can implement f, then g can also implement f' using NPN-equivalence. In PLA terminology, a row in the PLA is also referred to as a *pterm* or *product term*.

Definition: The set of minterms realizable in a row of a PLA is the *minterm cover* of that row. The set of functions realizable in a PLA g is the *function cover* of g, which is just the OR of its constituent minterm covers. A function is *functionally complete* or *universal* for m inputs if it can implement all m-input functions.

Given a UPLA.m, g, we define two problems related to UPLAs :

UPLA Optimization Problem : Given g, what is the minimum number of switches required for g to be functionally complete?

UPLA Cover Problem : Given g and a fixed number of switches, how many functions can be covered by g?

This paper deals mainly with the first problem, with some interesting results for the second in the Section 5.

3 UPLA DESIGN

3.1 Architecture of UPLA

We now consider the capabilities of PLAs used. Each row in the AND-plane implements a minterm with switches (Fig. 1). If the switch is on, the variable is fed into the minterm of that row. There is no logical difference between the NOR-NOR and the AND-OR implementation except in the switch positions, although physical properties differ and are not dealt with here. The appropriate choice should be made according to physical design criteria. We will use the straightforward AND-OR implementation to illustrate our ideas.

The switches in a PLA are labeled s_{ix_j} where i is the row number and x_j the variable name of that column, provided a switch exist at that position. For convenience, a switch is also interchangeably called by the variable name (and the row) it is driven by. A traditional PLA has a fully populated array of switches at each intersection of a row (pterm) and a column (variable or its complement). Since we assume output complementation is available for ULMs, for fairness, we also equip PLAs with such capability. Thus a universal PLA will have 2^{m-1} rows (pterms) and each row will have $2m$ switches in the AND-plane. For a one output PLA, it will also have 2^{m-1} switches in the OR-plane. The minterm cover of each row in a PLA is the same and is able to realize any minterm of m variables.

UPLAs are similar except that these have a reduced number of switches. We also assume that output complementation is available, and that the functions **0** and **1** are excluded, since these can be obtained more efficiently and directly from other sources.

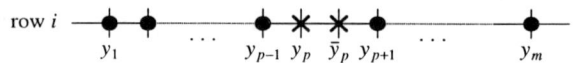

s_{iy_p}	$s_{i\bar{y}_p}$	minterm(s) implemented
on	on	0
on	off	$y_1 \ldots y_{p-1}y_p y_{p+1} \ldots y_m$
off	on	$y_1 \ldots y_{p-1}\bar{y}_p y_{p+1} \ldots y_m$
off	off	$y_1 \ldots y_{p-1} y_{p+1} \ldots y_m$

Figure 2: Configuration of UPLA-adjacent cells

3.2 Properties of PLA based ULMs (UPLAs)

We now give a few straightforward lemmas that influence our design of UPLAs. Details of the proofs can be found in [4].

Lemma 1 *A PLA requires at least 2^{m-1} rows to be universal.*

Lemma 2 *Let m_i be a minterm and π a transformation. If m_i and π have the same parities, then $\pi(m_i)$ is even and odd otherwise.*

Corollary 3 *Let M be a set of minterms of the same parity. For any transformation, M will be mapped to another set of minterms M' such that all its minterms have the same parity.*

Lemma 4 *Any UPLA.m must implement, in different rows, all the minterms of either M_m^o or M_m^e.*

Lemma 5 *A PLA that has the minterms of either M_m^o or M_m^e only in its function cover is insufficient to be a UPLA.m.*

3.3 Straightforward reduction of rows and switches

From the above lemmas, the UPLA.m uses the minimum 2^{m-1} rows. We now consider reducing the number of switches. Consider two adjacent minterms $m_i = y_1 \ldots y_{p-1}y_p y_{p+1} \ldots y_m$ and $m_j = y_1 \ldots y_{p-1}\bar{y}_p y_{p+1} \ldots y_m$, where y_k is either x_k or $\bar{x}_k$. To combine these, we do the following : if each of the switches at $y_k, k \neq p$, are hardwired, and we want to have all four combinations of the values of these two minterms, we can equip p-switches at y_p and $\bar{y}_p$ (see Fig. 2: dots denote h-switches and crosses denote p-switches). The table gives the minterms obtained depending on the configuration of the switches s_{iy_p} and $s_{i\bar{y}_p}$. The notation $f_r = y_1 \ldots y_{p-1}\{0|1|y_p|\bar{y}_p\}y_{p+1} \ldots y_m$ denotes the minterm cover of row r. The '|' denotes *exclusive* selection of the items within the braces. The minterm cover of row 1 in Fig. 3(a) is $\{0|1|x_1|\bar{x}_1\}x_2x_3$.

Consider the graph $G = (M_m, E)$ where $m_i, m_j \in E$ iff m_i and m_j are adjacent. Combining adjacent minterms as described above, a row in the UPLA corresponds to a matched edge in G. Every perfect matching then defines a UPLA. In Fig. 3, two different UPLA.3s and UPLA.4s are given based on different perfect matchings. Since there are many perfect matchings (hence UPLAs), which one to use may depend on other factors like layout considerations. The UPLA obtained has 2^{m-1} rows (Lemma 1). Thus, the number of switches in the AND-plane is reduced by $(m-1)2^{m-1}$. The function cover of the UPLAs in Fig. 3(c) & (d) are thus :

$$\begin{aligned}F_4 =\ & \{0|1|x_1|\bar{x}_1\}\bar{x}_2\bar{x}_3\bar{x}_4 + \{0|1|x_1|\bar{x}_1\}\bar{x}_2\bar{x}_3x_4 +\\ & \{0|1|x_1|\bar{x}_1\}\bar{x}_2x_3\bar{x}_4 + \{0|1|x_1|\bar{x}_1\}\bar{x}_2x_3x_4 +\\ & \{0|1|x_1|\bar{x}_1\}x_2\bar{x}_3\bar{x}_4 + \{0|1|x_1|\bar{x}_1\}x_2\bar{x}_3x_4 +\\ & \{0|1|x_1|\bar{x}_1\}x_2x_3\bar{x}_4 + \{0|1|x_1|\bar{x}_1\}x_2x_3\bar{x}_4\end{aligned}$$

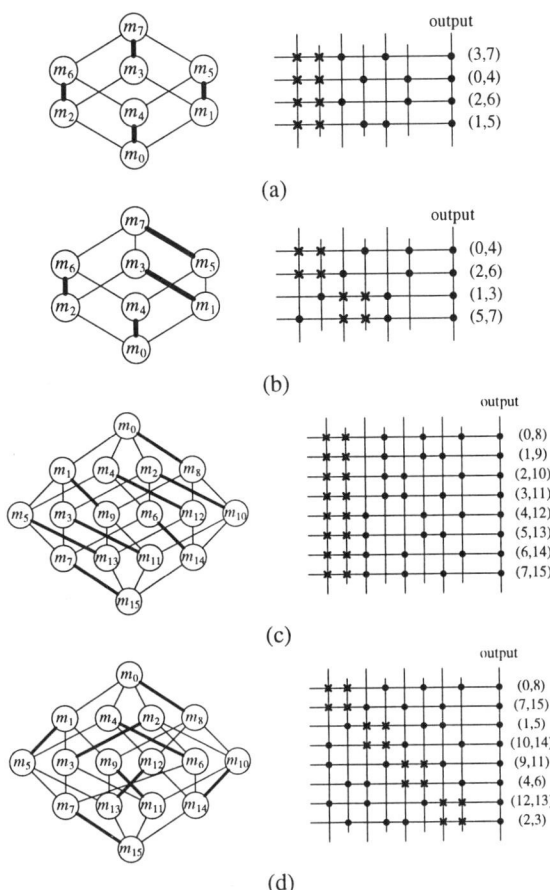

Figure 3: Using graph matching to determine position of p-switches

$$F_4' = \{0|1|x_1|\bar{x}_1\}\bar{x}_2\bar{x}_3\bar{x}_4 + \{0|1|x_1|\bar{x}_1\}x_2x_3x_4 +$$
$$\bar{x}_1\{0|1|x_2|\bar{x}_2\}\bar{x}_3x_4 + x_1\{0|1|x_2|\bar{x}_2\}x_3\bar{x}_4 +$$
$$x_1\bar{x}_2\{0|1|x_3|\bar{x}_3\}x_4 + \bar{x}_1x_2\{0|1|x_3|\bar{x}_3\}\bar{x}_4 +$$
$$\bar{x}_1\bar{x}_2x_3\{0|1|x_4|\bar{x}_4\} + x_1x_2\bar{x}_3\{0|1|x_4|\bar{x}_4\}$$

The algorithm for specializing a UPLA.m to an m-input function f is thus very straightforward. We need only determine the configuration of each pair of UPLA-adjacent minterms in f according to the table in Fig. 2 and set the switches appropriately :

Algorithm Map (function f)
1. **for** each pair of UPLA-equivalent minterms in f **do**
2.   determine configuration according to Fig. 2.
3.   set the switches appropriately.

4 SUPLA DESIGN

With NPN-equivalence, further reduction of the number of p-switches is possible. The resulting UPLAs are called *smaller* UPLAs (SU-PLAs). Given a SUPLA g and a function f, we want to find a specialization of g to implement f.

4.1 Technology Mapping Algorithm

We first present a simple scheme to implement a given function on a given SUPLA, then show that if only two minterms are not directly covered, only three transformations need to be checked for function

equivalence instead of $m!2^m$ different combinations. Thus, together with its complement, we can map a function in just 6 comparisons. First, given a function, f, we convert it to a 2^m bit binary number (denoted $bin(f)$). For $m \leq 5$, one computer word suffices on most systems. Let $a_i \in \{0, 1\}$ be the value of m_i in f. For example, $bin(XOR4)$ is equal to

a_{15}	a_{14}	a_{13}	a_{12}	a_{11}	a_{10}	a_9	a_8	a_7	a_6	a_5	a_4	a_3	a_2	a_1	a_0
0	1	1	0	1	0	0	1	1	0	0	1	0	1	1	0

Now suppose certain minterms are *uncovered* in a PLA g, i.e., not in the function cover. To implement f, we use transformations so that each of the transformed minterms of f can be implemented by a separate row in g. In practice, we transform g instead, since given a PLA g, we can precompute all the different uncovered minterms in g for each transformation before the mapping phase and remove equivalent configurations, something hard to do with f. For example, consider a 4-input function with uncovered minterms m_9 (i.e., $x_1\bar{x}_2\bar{x}_3x_4$) and m_{10} (i.e., $x_1\bar{x}_2x_3\bar{x}_4$). The UPLA-adjacent minterms are m_1 and m_2 respectively, and are *critical*. A possible function cover of g is :

$$g = \{0|1|x_1|\bar{x}_1\}\bar{x}_2\bar{x}_3\bar{x}_4 + \{1|\bar{x}_1\}\bar{x}_2x_3\bar{x}_4 + \{0|1|x_1|\bar{x}_1\}\bar{x}_2x_3x_4 +$$
$$\{1|\bar{x}_1\}\bar{x}_2\bar{x}_3x_4 + \{0|1|x_1|\bar{x}_1\}x_2\bar{x}_3\bar{x}_4 + \{0|1|x_1|\bar{x}_1\}x_2\bar{x}_3x_4 +$$
$$\{0|1|x_1|\bar{x}_1\}x_2x_3x_4 + \{0|1|x_1|\bar{x}_1\}x_2x_3\bar{x}_4$$

When switch $s_{2\bar{x}_1}$ is off, row 2 covers m_2 and m_{10} (i.e., $a_2 = a_{10} = 1$), and when on covers only m_2 (i.e., $a_2 = 1, a_{10} = 0$). Similarly, when $s_{4\bar{x}_1}$ is off, row 4 covers m_1 and m_9, and when on covers only m_1. If $a_1 = a_2 = 1$, rows 2 and 4 can implement them. If both are zero, we complement the output to implement it. The values of other minterms are immaterial. Thus g fails to cover a function f directly iff $a_1 \neq a_2$ in $bin(f)$. If we cannot implement f directly, we use transformations on g to change the critical minterms so that they all have the same value. For example, consider $\pi = \begin{pmatrix} 1 & 2 & 3 & 4 \\ 2 & 4 & 1 & 3 \end{pmatrix}$. Then $\pi(m_1) = m_{14}$ and $\pi(m_2) = m_4$, and the critical minterms are now m_4 and m_{14}. Thus, if $a_4 = a_{14}$, we have found a way to specialize g to f. Therefore, we can try all possible transformations to see if g is PLA-specializable to f. In other words, we want to find a π such that all the $\pi(m_j)$ have the same value, where m_j is a critical minterm.

For efficiency, we associate with each transformation a mask of 2^m bits where bits corresponding to critical minterms have value 0's while all others have value 1's. Thus 1011111111101111 is the mask for π which has a 0 in bit positions 4 and 14, and 1 otherwise. With this mask, we do a bitwise OR with $bin(f)$. If the result is all 1's, we have found a specialization for f. What remains is to set the switches as earlier outlined. Therefore, by precomputing this mask for each of the transformation, it takes at most $m!2^m$ comparisons to find a transformation if one exist. For $m = 4$, this is 384 transformations. We will show later that, if the number of critical minterms is 2, a dramatic reduction to 3 transformations and their complements are sufficient, regardless of the value of m.

Lemma 6 *All m input functions can be implemented on a SUPLA.m that has 2 critical minterms.*

Proof : Given a transformation and a function, consider how to determine if there is a match. WLOG, assume that m_1 and m_2 are critical. Then $m_1 = \bar{x}_1 \ldots \bar{x}_{m-3}\bar{x}_{m-2}\bar{x}_{m-1}x_m$ and $m_2 = \bar{x}_1 \ldots \bar{x}_{m-3}\bar{x}_{m-2}x_{m-1}\bar{x}_m$. A match occurs if $a_1 = a_2$. Thus g

Algorithm Map-NPN (function f)
1. $p := bin(f)$;
2. $matched := \text{false}$;
3. $i := 1$;
   /* N is the number of precomputed masks */
4. **while** ($matched = \text{false}$) **and** ($i \leq N$) **do**
5.   **if** (($mask[i] \vee p$) = $\bar{0}$) **then**
6. $matched := \text{uncomplemented}$;
7.   **else if** (($mask[i] \vee \bar{p}$) = $\bar{0}$) **then**
8. $matched := \text{complemented}$;
9. $i := i + 1$;
10. **return** SwitchPattern($p, mask[i], matched$);

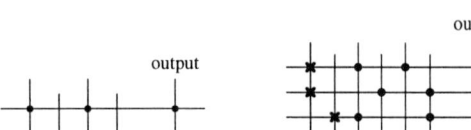

Figure 4: SUPLA.2   Figure 5: SUPLA.3

under $\pi_0 = \begin{pmatrix} 1 & \cdots & m-3 & m-2 & m-1 & m \\ 1 & & m-3 & m-2 & m-1 & m \end{pmatrix}$ the identity transformation, can implement f if ($a_1 = a_2$). Now consider the transformations $\pi_1 = \begin{pmatrix} 1 & \cdots & m-3 & m-2 & m-1 & m \\ 1 & & m-3 & m-1 & m & m-2 \end{pmatrix}$ and $\pi_2 = \begin{pmatrix} 1 & \cdots & m-3 & m-2 & m-1 & m \\ 1 & & m-3 & m & m-1 & m-2 \end{pmatrix}$. The critical minterms under π_1 are m_4 and m_1 and under π_2, m_2 and m_4. Therefore, under the transformations π_0, π_1 and π_2, PLA g can specialize to function f if ($a_1 = a_2$) + ($a_1 = a_4$) + ($a_2 = a_4$). Note that this condition is always true. Hence, the SUPLA.m with two critical minterms can specialize to any m-input function f. □

Furthermore, once we know which of the condition ($a_i = a_j$) for $i, j \in \{1, 2, 4\}$ is true, we also know how to implement the function f. This means that we can specialize a PLA with 2 critical minterms to implement any function f in just 6 comparisons when its complements are also considered.

Corollary 7 *A SUPLA.m requires at most $2^m - 2$ p-switches and $m2^{m-1}$ h-switches.*

4.1.1 SUPLA.2

Lemma 8 $g = x_1 x_2 + \{0|\bar{x}_1 \bar{x}_2\}$ *is a ULM.2 under restricted specialization if the equivalence class $\{x_1\}$ are excluded. It is ULM.2 under general specialization.*

Proof : g is the SUPLA.2 shown in Fig. 4. The two remaining equivalence classes $\{x_1 x_2 + \bar{x}_1 \bar{x}_2\}$ and $\{x_1 x_2\}$ can be implemented by turning the only p-switch on for the first class and off for the second class. Hence, it is a ULM.2.

To implement x_1 using general specialization, simply bridge both inputs to x_1 and turn off the only p-switch in the SUPLA.2. The function becomes $x_1 x_1 + 0 \cdot \bar{x}_2 = x_1$. □

Corollary 9 *SUPLA.2 requires at most 5 h-switches and 1 p-switch.*

4.1.2 SUPLA.3

Theorem 10 *A SUPLA.3 requires at most 12 h-switches and 5 p-switches under restricted specialization if the NPN-equivalent class $\{x_1\}$ is excluded. It can implement functions from the class $\{x_1\}$ under general specialization.*

Proof : Consider the case where g is a function cover with three critical minterms with the same parity, $\{m_{i_1}, m_{i_2}, m_{i_3}\}$. When can a given function f be successfully mapped into that configuration? From Cor. 3, for any transformation π, the mapped critical minterms must all have the same parity. Thus the match is successful if the values of $\pi(m_{i_j})$ are the same. In other words, the minterms form at least two groups, namely $M_m^e = \{0, 3, 5, 6\}$ and $M_m^o = \{1, 2, 4, 7\}$ under all transformations. For the case of $m = 3$, these two groups suffices for our proof here (in fact, these are the only two groups for $m = 3$). Using the same technique as in Lemma 6, we can find transformations such that given a function $f = a_7 a_6 \ldots a_0$, it can be mapped to g if the condition, S, is satisfied :

$$\begin{aligned} S = &\ (a_0 = a_3 = a_5) + (a_0 = a_3 = a_6) + (a_0 = a_5 = a_6) + \\ &\ (a_3 = a_5 = a_6) + (a_1 = a_2 = a_4) + (a_1 = a_2 = a_7) + \\ &\ (a_1 = a_4 = a_7) + (a_2 = a_4 = a_7) \end{aligned}$$

The first four terms corresponds to the values of even minterms and the rest, the odd minterms. Now consider the function such that $a_0 = a_1 = a_2 = a_3 = 0$, $a_4 = a_5 = a_6 = a_7 = 1$. Notice that these values cause condition S to evaluate to false, and so this particular function cannot be covered by g.

It is not difficult to see that the condition S fails only when $a_{i_1} = a_{i_2} \neq a_{i_3} = a_{i_4}$ and $a_{j_1} = a_{j_2} \neq a_{j_3} = a_{j_4}$ where $a_{i_k} \in \{a_0, a_3, a_5, a_6\}$ and $a_{j_l} \in \{a_1, a_2, a_4, a_7\}$ and each a_{i_k} and a_{j_l} are distinct. These functions are from the equivalence class of $\{x_1\}$. In other words, functions from the equivalence class of $\{x_1\}$ are the only functions that cause S to fail when there are 3 critical minterms. Hence a SUPLA.3 uses at most $8 - 3 = 5$ programming bits if functions from the NPN-equivalent class of $\{x_1\}$ are ignored.

Now, one can easily verify that the SUPLA.3 in Fig. 5 implements all functions under restricted specialization other than those from the NPN-equivalent class of $\{x_1\}$. To implement x_1, simply bridge all inputs to x_1 and turn off the switch on row 2. The function implemented is thus $f = x_1 x_1 + x_1 \bar{x}_1 \bar{x}_1 + x_1 \bar{x}_1 + \bar{x}_1 x_1 = x_1$. Thus at most 12 h-switches and 5 p-switches are required under general specialization. □

4.1.3 SUPLA.4

Further improvements require a slightly different technique, already seen in SUPLA.2. Instead of putting all switches of critical minterms so that the minterm cover is of the form $y_1 \ldots y_{i-1} \{1|y_i\} y_{i+1} \ldots y_m$, we can also put a p-switch in the OR-plane to cancel the presence of a hardwired minterm. In other words, the minterm cover is of the form $\{0|y_1 \ldots y_m\}$. For example, in row 2 of the SUPLA.2 in Fig 4, if the switch is on, we get the minterm $\bar{x}_1 \bar{x}_2$, otherwise we get **0**.

Using a mixture of these two techniques, we designed a SUPLA.4 that uses 32 h-switches. It also uses 13 p-switches, the same number as used in [3]. Consider the situation where a row has a p-switch in the OR-plane and some other p-switch(es) in the AND-plane. When the switch in the OR-plane is turned off, the minterm is switched off, regardless of the configuration of the switches in the AND-plane. On the other hand, when the switch is on, the values in the AND-plane are then fed to the output. If there are n p-switches in that row, the

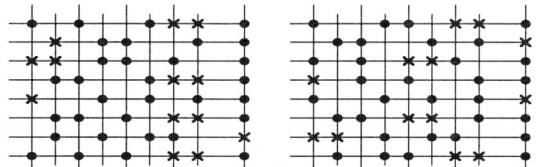

Figure 6: Two different SUPLA.4s

m	$\lvert M_m \rvert$	#fns	ZV [3]	SUPLAs
2	4	16	–	1
3	8	256	5	5
4	16	65536	13	13
5	32	4.3×10^9	–	30
6	64	1.8×10^{19}	–	62

Figure 7: Results on #programming bits

	PLA	UPLA			SUPLA		
m	total	#h	#p	total	#h	#p	total
2	10	4	4	8	5	1	6
3	28	12	8	20	12	5	17
4	72	32	16	48	32	13	45
5	176	80	32	112	80	30	110
6	426	192	64	256	192	62	254

Figure 8: Number of switches required. #h = number of h-switches, #p = number of p-switches.

number of possible configurations have been reduced to $2^{n-1} + 1$ instead of 2^n, where the 1 comes from **0** that results when the switch in the *OR*-plane is off. This implies that in order to maximize the coverage of minterms, if there is a p-switch in the *OR*-plane of a particular row, there should not be any p-switch in the *AND*-plane of that row. Bearing this observation in mind, we then take a UPLA from the previous section and search for a row with two p-switches in the *AND*-plane, convert one of them to a h-switch and then change the h-switch in the *OR*-plane to a programmable one. We observe its effect on coverage as this happens. We also tried to put more than one p-switch in the *OR*-plane (on different rows) using the same technique above. Fig. 6 shows just two of the many different SUPLA.4s that we have found.

5 APPROXIMATE ULM.4

Once the number of p-switches is less than 13, the SUPLA is no longer universal. To determine the coverage of each configuration, we wrote a C/C++ program that performs technology mapping on all 4-input functions. The number of rows is fixed at 2^{m-1}. Each row originally implements one of the odd minterms. We observe the coverage as we vary the number of p-switches. Before testing each configuration, we keep only those transformations that give us unique critical minterms so as to reduce the number of transformations needed to be checked when mapping each function. For each number of p-switches, we try all possible placements of the switches to determine the minimum and maximum number of functions covered by some configuration of the setting under restricted specialization. The results are summarized in Fig. 9.

With 12 switches and using only restricted specialization, we can cover 65490 functions, representing a 99.9% coverage of the 65536 functions of 4 inputs. Thus, the SUPLA.4 is almost universal. We

	smallest coverage		largest coverage	
#p	#fns †	% coverage	#fns †	% covered
13	65278	99.6%	65534	100.0%
12	49150	75.0%	65490	99.9%
11	36506	55.7%	65050	99.3%
10	12635	19.3%	61210	93.4%
9	8891	13.5%	41354	63.1%
8	1019	1.6%	21226	32.4%

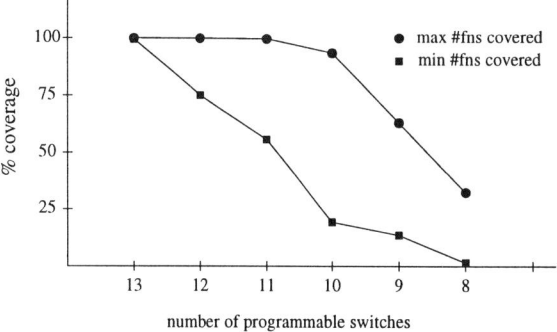

Figure 9: Number of p-switches vs coverage of 4-input functions under restricted specialization. The number of minterms is fixed at 8. †functions **0** and **1** are excluded.

note that this is a 1% increase from the ULM in [1]. In fact, the function $fn294$ that implements their ULM has 12 minterms, while ours only had 8. The coverage is still more than 99% when the number of p-switches was reduced to 11. A simple estimate of the ULM sizes based on the number of transistors required show that our implementation has a smaller area than that of [1] and [3].

6 CONCLUDING REMARKS

We have presented a new design of logic modules for FPGA architecture using a restricted version of PLAs that have a much smaller number of switches. PLAs have the desirable property of regularity, predictable delays and easy mapping. The UPLAs and SUPLAs presented in this paper uses smaller area than traditional PLAs and compares favorably with LUTs and other ULM designs. Also, the simplicity of the mapping algorithm is maintained. Approximate ULM.4s of very high coverage are given for varying number of programmable switches.

7 REFERENCES

[1] Thakur, S. and Wong, D. F., *On Designing ULM-Based FPGA Logic Modules*, In Proc. 3rd Internatinal Symposium on FPGAs, Feb 1995, pp. 3-9.

[2] Lin, C.C., Marek-Sadowska, M. and Gatlin D., *Universal Logic Gate for FPGA Design*, In Proceedings of ICCAD, 1995, pp. 164-168.

[3] Zilic, Z. and Vranesic, Z. G., *Using BDDs to Design ULMs for FPGAs*, In Proc. 4th Internatinal Symposium on FPGAs, 1996, pp. 24 to 30.

[4] Lee, K. K. and Wong, D. F., *On the Use of PLAs in Designing Universal Logic Modules in FPGAS*, Technical Report, Department of Computer Sciences, University of Texas, Austin, 1998, to be published.

FPGA MAPPING OF SEQUENTIAL CIRCUITS WITH RETIMING

Jun-yong Lee

Dept. of Computer Engineering
Hongik University
Seoul, Korea 121-791

Eugene Shragowitz

Dept. of Computer Science
University of Minnesota
Minneapolis, MN 55455

ABSTRACT

Constructive and iterative steps of FPGA mapping algorithms for sequential circuits are enhanced by retiming technique and fuzzy logic. Multiple criteria measured for design data are connected by a hierarchical structure of fuzzy logic rules for decision making. The discussed mapper outperforms commercial tools in both area and timing for substantial set of MCNC benchmarks.

1. INTRODUCTION

While many technology mapping algorithms have been proposed recently for combinational circuits, only a few algorithms have been introduced for sequential circuits. The number of technology mapping algorithms applicable to mapping of sequential circuits to FPGAs is even smaller.

Among them are Xilinx's XNFMAP and PPR, AT&T's ATOM. Xilinx's XNFMAP and PPR perform mapping for XC3000 and XC4000 FPGA architectures for both combinational and sequential circuits. Also, according to [12], AT&T's system called ATOM is capable to handle sequential circuits for XC3000 and AT&T's FPGAs. These technology mapping algorithms are mainly focused on reducing the number of utilized CLBs. The ATOM system considers flipflops as one of the nodes types and maps the sequential circuits in the way similar to one applied to combinational circuits.

Miyazaki et. al. [10] proposed a general performance improvement technique for mapping sequential circuits. To improve clock speed and data throughput, registers were inserted in the given sequential circuit without changing the structure of a mapped graph. Two operations are performed: "loop shrinking" and "register insertion." "Loop shrinking" is applied to a given input graph G as a preprocess for mapping. In this step, the register insertion points are determined by solving a graph-theoretical problem.

Then, procedure "register insertion" inserts as many registers as possible into every path to reduce the clock cycle. This technique can improve the throughput of the circuit. But it increases the number of clock cycles or latency. This technique often fails to insert registers into the graph because it cannot handle constraints imposed by FPGA architecture.

The majority of technology mapping methods for FPGAs are designed to map a Boolean network into the XILINX XC3000 FPGA circuit structure, where each CLB has 2^5 bits of SRAM LUT capable of implementing Boolean equations with up to 5 inputs. As the FPGAs have become more popular, the FPGA architecture has evolved. In the new SRAM-based FPGA architectures,

---

This work was supported by MIC under the grant AB-97-G-0524

such as XILINX XC4000 [13] or AT&T's ORCA [3], a configuration of a single CLB becomes more complex. This Xilinx XC4000 series, which is one of the most popular types of FPGAs today, is selected as targets for mapping. In Xilinx XC4000, the CLB consists of 3 LUTs, which are used for up to 9 inputs.

In this paper the FMS (Fuzzy Logic Mapper for Sequential Circuits) for FPGAs with complex CLB structures is introduced.

The FMS technology mapping algorithm applies fuzzy logic for mapping of gates to CLBs. The fuzzy logic approach was successfully used in CAD earlier for several layout applications [7],[4], and for mapping combinational circuits into FPGAs in our earlier works [6].

2. PROBLEM DEFINITION

Input to the technology mapping system is presented in the form of directed graph. Nodes of the graph model are simple combinational gates and memory elements. The memory elements included in the sequential circuits are usually D-type flipflops.

The technology mapping system described here produces FPGA circuits that implements the input graph. The system assigns the nodes of a given graph to the CLBs of the target FPGA, in such a way that the resulting FPGA circuit satisfies all the constraints introduced by the target FPGA architecture.

The directed edges E of the graph model interconnections between functional elements. Edges connect outputs of elements to inputs of other functional elements and with external world. Edges are weighted with the number of registers along the connection.

Technology mapping is driven by several objectives. The most critical of objectives is improvement in timing without increase in number of CLBs. Timing is a difficult problem for look-up-table-based FPGAs because of substantial fixed delay associated with each Look-up tables used in the solution. Improvement in timing can be achieved, if the maximal number of CLBs on the critical path is reduced.

Retiming is a procedure of a circuit optimization by relocating the registers with the purpose of minimizing clock period while preserving logic functions. In a ground breaking work on this subject, Leiserson et. al. [8] studied various aspects of this problem. They proposed a polynomial time algorithm for determining a retiming that minimizes clock period and an algorithm for minimizing of total number of registers.

This algorithm can be effectively used for optimization of general sequential circuits with multi-level logic. But applicability of this algorithm to technology mapping of FPGAs is limited. In this work, we describe a system that applies combination of numeric methods with fuzzy logic capable of representation of heuristic knowledge in the well-organized and structured form of fuzzy logic rules.

3. OVERVIEW OF GRAPH MAPPING ALGORITHM

The algorithm described in this paper consists of two major steps: graph mapping and retiming. The graph mapping algorithm constructs the solution by mapping given graph of combinational and memory elements into the FPGA structure while optimizing simultaneously utilization and timing.

Our technology mapping algorithm for mapping of combinational circuits was extended to handle sequential circuits. The basic difference is that registers are also mapped into the CLBs, and that the timing delays are computed between flipflops rather than between PI's and PO's.

The retiming algorithm uses results of the initial step to optimize timing further by logically equivalent transformations of graph. This algorithm trades improvement in timing for marginal degradation of the utilization.

There are several objectives in FPGA mapping: area, delay, routability and others. Our previously developed mapper for combinational circuits [6,7] maps a given DAG (Directed Acyclic Graph) of Boolean equations to an FPGA format in a way that satisfies multiple objectives, with preferences given to some of them. The initial mapping algorithm belongs to the class of constructive heuristics, i.e., it builds a solution by successive augmentations of new elements to the partial solution.

An approach selected for solving this combinatorial problem is a decomposition into successive mapping of individual CLBs, while optimizing area, timing and routability.

Selection of this approach is justified by the fact that constraints for individual CLBs are difficult to satisfy during any iterative procedure. More traditional approach to decomposition of DAG into subtrees [2] individually mapped into CLBs neglects either packing or timing aspect of a problem.

Simultaneous optimization of area (F_1) and timing (F_2) by a greedy algorithm that adds nodes to a selected CLB is complicated by the fact that F_1 and F_2 are not monotone functions, i.e., addition of the next node to CLB may not change values of the objective functions for partial solutions, and, therefore, can not be used to select a next node to map.

In the same time, domain specialists are well aware about dependencies between optimized objectives and properties of underlying Boolean networks and are capable of formulating these dependencies in the linguistic terms.

These linguistic dependencies, representing a domain knowledge are quantitatively evaluated according to rules of fuzzy logic and are used to approximate selection of nodes for mapping according to values of functions F_1 and F_2.

From a given graph, an algorithm selects the best node to start mapping to a CLB. In this selection process, a set of fuzzy logic criteria is used for choosing the best candidate.

Starting from a seed node, mapping is continued to the nodes connected to the already mapped nodes in the CLB until no more node can be mapped into the CLB. For each candidate node from graph, constraints are checked to keep a solution feasible. Among the three look-up table of CLB, H-LUT is packed first with the seed node, and then F-LUT and G-LUT are packed.

In the mapping process, if an input to the already mapped node comes from the output of a flipflop, then this flipflop together with

```
Initialize membership functions for Fuzzy Rules and Preference Rules
ending_condition = 0
Call CLB_map ( ) ;
    /* procedure CLB_map maps a DAG to a network of CLBs */
Analyse the result and compare it with stated goals
If ( the result satisfy goals  )
    then ending_condition = 1 ;

While ( ending_condition = 0 ) {
    Call Retiming( ) ;
    Call CLB_map ( ) ;
    Analyse the result and compare it with stated goals
    If ( the result satisfy goals  or  no more improvement is expected  )
        then ending_condition = 1 ;
}
```

Figure 1: The pseudo-code of mapping algorithm for sequential circuits with retiming

its input node becomes a seed for a new CLB. Only combinational nodes can be added to the CLB with the selected seed node.

Fuzzy logic technique is used in selecting nodes to be mapped into CLBs as in the combinational circuit mapping. The fuzzy logic rules and criteria used in the initial mapping of sequential circuit mapping are the same as those in the combinational circuit mapping [6].

After mapping is completed for one CLB, a new seed node is chosen for a new CLB, and the above sequence is repeated until all the nodes of the given DAG are mapped into a CLB network.

The technology mapper for combinational circuits has look-ahead features to eliminate inefficiencies produced by greedy approach. In some cases mapping does not produce efficient utilization of the whole CLB capacity. For instance, when a subnet for the chosen node is small enough to fit into a single F-LUT, it is more efficient to map this subnet into a single F-LUT rather than to use the entire CLB of three look-up tables.

The process of initial mapping consists of a selection of a new CLB with the following mapping of graph nodes into it. A node to be mapped into a new CLB is determined according to a maximal value of a decision function $D(x)$, constructed by fuzzy logic rules. These rules connect measurable criteria defined for graph with membership in fuzzy sets of solutions with good timing and small area.

Fuzzy logic rules formalize a knowledge about dependencies between optimized functions and underlaying data. The more detailed description of the mapping algorithm for combinational circuits can be found in [6].

4. OUTLINE OF THE RETIMING STEP

Retiming module relocates flipflops so that the modified circuit produces smaller maximal timing delay between consecutive flipflops. As it was mentioned earlier, the existing algorithms for delay optimization are not applicable to look-up table based FPGAs, where the delay of a look-up table is constant regardless of the number of gates mapped into it. Here we define logically equivalent local circuit transformations which may improve timing in mapping procedures.

In this model, flipflops are added as a special type of nodes to graph. These nodes are moved backward or forward along the edges to reduce the timing delay. Figure 2 illustrates the basic idea of our retiming algorithm.

4.1. Basic Operations

Four basic retiming operations were defined for local transformations, which relocate flipflops across the combinational gates or fanout points. It is assumed that all other constraints on LUTs and CLBs are satisfied when an operation is accepted.

1. Move-backward a register (or Move-forward a gate)

   This operation moves a register from the output of the gate to the inputs of the gate as depicted in Figure 3. Using the general retiming notation as in [8], a value of r for gate g is

   $$r(g) = 1$$

   Therefore, for out-edge e_{out} and each in-edge e_{in}, the edge weights become

   $$w_r(e_{out}) = w(e_{out}) - 1$$
   $$w_r(e_{in}) = w(e_{in}) + 1$$

   Presence of the register on the out-edge indicated that initially $w_r(e_{out}) \geq 1$. After the register was moved backward, it still satisfies the condition, $w(e_{out}) > 0$, for legal retiming. This operation can be viewed as moving of a gate forward over the register.

2. Move-forward a register (or Move-backward a gate)

   This operation is the inverse of the "Move-backward a register" operation. All the registers on the input side are moved forward across the gate and merged into a single register. The retiming for a gate g is

   $$r(g) = -1$$

   which results in

   $$w_r(e_{out}) = w(e_{out}) + 1$$
   $$w_r(e_{in}) = w(e_{in}) - 1$$

   In order to apply this operation all the inputs of the gate should come from outputs of registers. In this case, $w(e_{in}) > 0$ for all input edges of the gate g, which satisfies the condition of legal retiming. It also can be interpreted as moving a gate backward while merging the input registers into one.

3. Split a register

   When a circuit contains a fanout from the output of the register, the register can be split into the fanout branches as in Figure 4. This operation itself is not a retiming operation, but it is often necessary to satisfy precondition for "Move-forward register" operation. This operation increases the number of utilized registers to the number of fanouts. This operation satisfies a condition $w_r(e_{out}) \geq 0$.

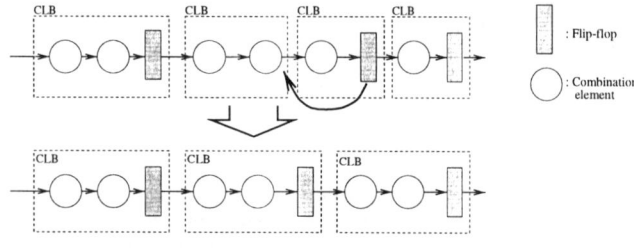

Figure 2: Retiming in FPGA mapping

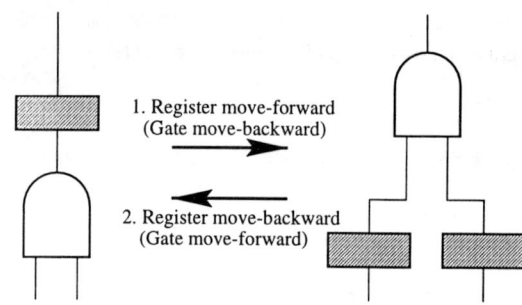

Figure 3: Basic operations for retiming

4. Merge registers

   This is the inverse operation of "Split a register." This operation is used to merge redundant registers or to enable The "move-backward" operation by removing a fanout point between registers and a gate. This transformation retains the same number of registers on the path from the output of a one gate to the input of another gate.

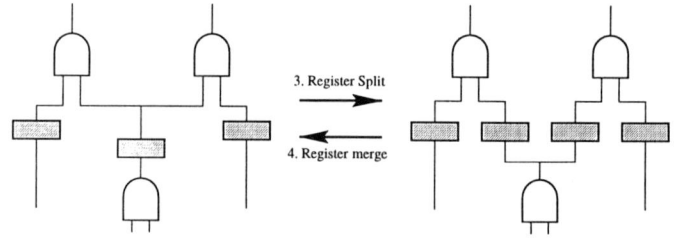

Figure 4: Basic operations for retiming (Cont.)

5. DESCRIPTION OF RETIMING ALGORITHM

For the input circuit, the initial mapping of the combinational and sequential part into CLBs is performed first. In the process of mapping, selected criteria can be emphasized by preference rules.

After the initial mapping, retiming procedure is applied, which tries to relocate flipflops or nodes to reduce the timing delay. The CLB-map procedure and the retiming procedure are repeated until the retiming fails to produce an improvement with respect to the timing delay.

The retiming procedure modifies the graph to improve timing after mapping a graph into an FPGA format. From the result of mapping, the number of maximal levels of CLBs is computed. The goal is to modify the graph so that the number of maximal levels is reduced in the next run of mapping.

Once the maximal number of CLB levels is computed after initial mapping, the CLBs which are on the path with the maximal level are found. Then nodes which were mapped to the CLBs on the longest path are identified.

This set of nodes forms a subgraph of combinational gates with the outputs of gates connected to the inputs of registers or POs. The inputs in the subgraph come from the registers or PIs.

From this subgraph, the candidate nodes for retiming are selected by fuzzy logic rules according to the timing criticality of each node. These nodes are supposed to be relocated to reduce the size of a subgraph. In the proposed system, we consider relocating nodes rather than relocating flipflops, which are equivalent operations as it was shown in the previous section.

To relocate the nodes, they should be moved forward or backward from their locations. If the node is a root node in the subgraph, then the output of the node is connected to a register or to a PO. If a node is directed to a register, then the node can be moved forward over the register, reducing the size of the subgraph.

If the candidate node is a leaf node of the subgraph, then the inputs of the node come from registers or PI's. If all the inputs of the node are from registers, then the node can be moved backward over the register, reducing the size of the subgraph. But relocation of the candidate nodes is not always possible because of constraints imposed by the FPGA architecture.

To perform the "move-forward" operation on a node, it is required that the output of the node is connected directly to the register. When the output of the node is connected to more than one register, the registers should be merged into a single register by the "merge-register" operation.

To execute the "move-backward" operation on a node, it is required that each input of the node comes from a register without any fanout. When the register has fanout to several nodes, it should be split into several registers-one for each fanout by "split-a register" operation. If the retiming procedure for the selected candidate node fails, the next candidate node is investigated. The result of an attempt is accepted if new timing is not worse than the previous one.

6. EXPERIMENTAL RESULTS

The input test cases for proposed FMS system were described in BLIF (Berkeley Logic Interchange Format). In order to demonstrate performance of the mapper, 16 benchmark circuits were randomly selected from the set of MCNC test cases for sequential circuits. Using these test cases, FMS system performed technology mapping and retiming. The final technology mapping results of FMS system were transformed into Xilinx's XNF (Xilinx's Netlist File) format. These results were completed by Xilinx's placement and routing system (PPR 5.2).

On the other hand, selected MCNC test cases in BLIF format were directly transformed to Xilinx's XNF format and used for Xilinx's technology mapping, placement and routing system.

The results of experiments are shown in Table 1. The first three columns report the numbers of used CLBs, maximal CLB levels and the delay of the critical path performed by the proposed mapper. The second three columns present the results of a final mapping after iterations of mapping and retiming. The last three columns present results achieved by the Xilinx system (mapping+physical design) for the same test cases.

The overall results show that the proposed system outperforms Xilinx's technology mapping system in both area and timing. The FMS system reduced the number of the maximal CLB levels by 18.4% which produce 15.6% in reduction of actual delay after layout. Our system also reduced the number of used CLBs by 6.2%.

7. REFERENCES

[1] R. K. Brayton, R. Rudell, A. Sangiovanni-Vincentelli, and A. R. Wang, "MIS, a multiple-level logic optimization system," *IEEE Transactions on Computer-Aided Design*, vol. CAD-6, pp.1062-1081, Nov. 1987.

[2] R. J. Francis, J. Rose, and Z. Vranesic, "Chortle-crf: Fast Technology Mapping for Lookup Table-Based FPGAs," *Proc. 28th ACM/IEEE Design Automation Conference*, pp. 227-233, 1991.

Circuit Name	FMS			Xilinx's PPR 5.2		
	No. of CLBs	Max. levels	Delay (ns)	No. of CLBs	Max. levels	Delay (ns)
s1488	155	4	51.3	168	6	82.5
s1494	157	4	52.5	167	6	80.6
s208	17	4	46.7	19	4	46.1
s298	26	4	42.9	20	4	51.2
s344	24	5	55.1	27	6	61.3
s349	24	5	56.4	28	6	56.9
s382	28	4	51.1	29	4	50.9
s386	38	3	43.5	39	6	55.3
s400	28	4	50.6	30	4	56.6
s420	33	5	57.1	46	6	69.5
s444	32	6	64.3	36	6	64.6
s510	55	4	53.7	64	7	78.8
s526	50	4	50.2	46	4	57.2
s820	88	3	45.8	90	4	58.5
s832	92	3	44.9	90	5	61.3
s838	76	8	82.9	98	10	100.7
ratio	0.938	0.816	0.844	1	1	1

Table 1: Experimental Results

[3] D. Hill, B. Britton, B. Oswald, N.-S. Woo, S. Singh, T. Poon, B. Krambeck, "A new Architecture for High-Performance FPGAs," *Int. Workshop on FPGA*, Sep. 1992.

[4] E. Kang, R. Lin and E. Shragowitz, "Fuzzy Logic Approach to VLSI Placement," *IEEE Transactions on VLSI Systems*, vol. 2, pp. 489-501, Dec. 1994.

[5] J.-Y. Lee and E. Shragowitz, "Performance Driven Technology Mapper for FPGAs with Complex Logic Block Structures," *Proceedings of IEEE International Symposium on Circuits and Systems (ISCAS)*, vol. 2, pp. 1219-1222, Seattle, WA, April 1995.

[6] J.-Y. Lee and E. Shragowitz, "Technology Mapping for FPGAs with Complex Block Architectures by Fuzzy Logic Technique," *Proceedings of the Asia South-Pasific Design Automation Conference (ASP-DAC)*, pp. 295-300, Makuhari, Japan, August 1995.

[7] R. Lin and E. Shragowitz, "Fuzzy logic approach to placement problem," *Proc. of the ACM/IEEE 29th Design Automation Conference*, pp. 153-158, 1992.

[8] C. E. Leiserson and J. B. Saxe, "Retiming Synchronous Circuitry," *Algorithmica*, vol. 6, no. 1, pp. 5-35, 1991.

[9] S. Malik et. al., "Retiming and Resynthesis: Optimizing Sequential Networks with Combinational Techniques," *IEEE Transaction on CAD*, Vol. 10, No. 1 January 1991.

[10] T. Miyazaki et. al., "Performance Improvement Technique for Synchronous Circuits Realized as LUT-Based FPGAs," *IEEE Transaction on VLSI Systems*, Vol. 3, No. 3, September 1995.

[11] E. Shragowitz, J.-Y. Lee and E. Kang, "Application of Fuzzy Logic to Computer-Aided Design of Electronic Systems," (Invited Paper) *Proceedings SPIE's 1996 International Symposium on Aerospace/Defence Sensing and Control*, Orlando, FL, April, 1996.

[12] N.-S. Woo, "ATOM: Technology Mapping of Sequential Circuits for Lookup Table-based FPGAs," Technical Report, AT&T Bell Laboratories, November 1991.

[13] *Xilinx Programmable Logic Data Book*, Xilinx Inc., San Jose 1994.

[14] R. Yager, "Multiple objective decision-making using fuzzy sets," *Int. Journal of Man-Machine Studies*, vol. 9 pp. 375-382, 1977.

[15] L. A. Zadeh, "Fuzzy sets," *Information and Control 8*, pp. 338-353, 1965.

RAISE: A DETAILED ROUTING ALGORITHM FOR SRAM BASED FIELD-PROGRAMMABLE GATE ARRAYS USING MULTIPLEXED SWITCHES

V. Baena-Lecuyer, M. A. Aguirre, A. Torralba, L. G. Franquelo and J. Faura

Dpto. de Ingeniería Electrónica
Escuela Superior de Ingenieros,
Camino de los descubrimientos s/n, Sevilla–41092 (SPAIN)
e–mail: baena@gte.esi.us.es

ABSTRACT

This paper describes a new detailed routing algorithm, specially designed for those architectures that are found in most recent generations of Field-Programmable Gate Arrays (FPGAs). The algorithm also brings a solution for those architectures where multiplexed switches are used in order to decrease the chip area like the recently proposed FIPSOC FPGA [1]. The algorithm, called RAISE, can be applied to a broad range of optimizations problems and has been used for detailed routing of symmetrical FPGAs, whose routing architecture consists of rows and columns of logic cells interconnected by routing channels, with or without the use of multiplexed switches.

RAISE (Router using AdaptIve Simulated Evolution) searches not just for a possible solution, but tries to find the one with minimum delay. Excellent routing results have been obtained over a set of several benchmark circuits getting solutions close to the minimum number of tracks.

1. INTRODUCTION

In the last years, Field-Programmable Gate Arrays (FPGAs) have been widely accepted as an attractive means of implementing digital circuits. There is a wide range of commercial FPGAs, but one of the most important types is the symmetrical FPGA, which consists of rows and columns of logic blocks with horizontal routing channels between rows and vertical routing channels between columns. This type of FPGAs was first introduced by Xilinx in 1986, but currently it can be found, for instance, in some of the Altera and Quicklogic families.

Symmetrical FPGAs reach very high logic capacities; a key problem in the design of this kind of FPGAs is the structure of their routing channels. Short segments use less chip area (less segment length is wasted using short segments), but to provide long connections, a succession of short segments linked via programmable routing switches is required, reducing speed. On the other hand, long segments waste chip area but improve speed (less segments are required to make long connections passing through only a few switches).

A second problem is the design of the switch matrixes, using many switches increases the routability but also increases the chip area due to the amount of SRAM required to store the switches states. On the other hand, using just a few switches reduces chip area but also reduces routability. In FPGAs where area is critical, a possible solution is the use of multiplexed switches; for example a group of eight switches can be controlled by only three SRAM cells using a decoder; however with this solution only one out of eight switches can be closed. This solution requires a more complex routing software.

This paper presents RAISE, a new detailed router adapted for generic symmetrical FPGAs with or without multiplexed switches.

2. RAISE: ROUTER USING ADAPTIVE SIMULATED EVOLUTION

RAISE is based on SILK [4], a simulated evolution program for channel routing. Before running RAISE, for each point to point net, a set of possible paths is generated (for example, using the technique called Coarse Graph Expansion (CGE) [2] [3]). If this set is too large, a pruning process is initiated to eliminate paths of those nets that are supposed to be more routable (a net with a set of 100 paths is supposed to be more routable than one with 50 path). The criteria used in the pruning process is to minimize the number of common segments between paths of the same nets, making possible the use of different segments if one of them is used by other multipoint nets.

RAISE then searches for a path subset that makes possible the routing of all the nets, while minimizing delays.

Three steps are carried out by RAISE:

1. Initial Routing.
2. Rip-Up and Rerouting.
3. Post-optimization.

2.1. Initial Routing

The algorithm, of statistical nature, uses an initial solution to start an iterative process. This solution, does not need to be feasible, i.e., conflicts, if any, will be solved in the following steps. For the initial solution, RAISE selects for each point to point net the path with minimum delay. The delay is calculated with the RC-Tree algorithm of [5].

2.2. Rip-Up and Rerouting

The rip-up and rerouting phase solves the conflicts generated in the initial routing phase. To this purpose, RAISE uses the Simulated Evolution technique [4].

A cost is generated, for each point to point net using a cost function, which accounts for the path delays and the conflicts with other nets; then this cost is scaled in the range [0.1, 0.9]. For each

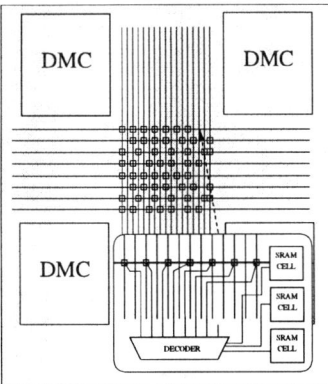

Figure 1: Multiplexed switches in S-Blocks

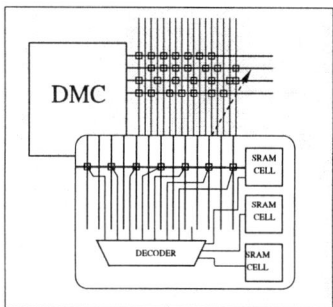

Figure 2: Multiplexed switches at Logic-Blocks outputs

point to point net, a random number between 0 and 1 is generated, if this number is less than the scaled cost of the routed path, the path is removed. After end of this process, there will be a set of routed point to point nets and another set of non-routed point to point nets. Next, for each multipoint net, in a random order, all the non-routed point to point nets are routed, choosing the path with minimum cost. This process is repeated until a solution with no conflicts is obtained or until a maximum number of iterations is reached.

Using a random number generator to select the non-routed nets, allows the algorithm to avoid local minima. Note that in the selection process, the nets with higher costs have a higher probability of being removed. However randomly removing some good nets also helps to avoid getting stacked at a local minima.

A key point in such an algorithm is the cost function. This function should contain at least a delay and a conflict term. But other terms can be added to improve the convergence:

From the problem definition, we know that point to points nets from the same multipoint net can share segments. To improve chip area, the number of shared segments in a point to point net should be maximized.

Besides, it would be desirable to exploit some advantages from the experience gained in previous iterations.

All this terms are in the following heuristic function cost:

$$\begin{aligned} cost = &\alpha \cdot (num\_nets\_using\_wires) \\ &+ \alpha \cdot (switch\_conflict) \\ &+ \beta \cdot (history\_cost) \\ &+ \gamma \cdot (\frac{path\_delay}{min\_path\_delay}) \\ &+ \delta \cdot (num\_segments\_used) \end{aligned}$$

In that follows SEGA [2] - [3] terminology is used when possible.

Each term is explained as follows:

- *num_nets_using_wires*: number of multipoint nets that share segments with this point to point net.
- *switch_conflict*: multiplexer conflicts, (this term is used only in the case of multiplexed switches).
- *history_cost*: it accounts for the demand of each segment of this point to point net in previous iterations.

- *path_delay*: self explanatory.
- *min_path_delay*: minimum path delay of the set of possible paths for this multipoint net.
- *num_segment_used*: number of segments used in this path.

The *num_nets_using_wires* term is calculated as follows:

$$num\_nets\_using\_wires = \sum_{i=0}^{n} NumMPNetUsingWire(W_i, k)$$

Where $NumMPNetUsingWire$ is the number of multipoint nets that use segment W_i, k is the number of present iteration and n is the number of wire segments that use the path.

The switch conflict term accounts for the use of several switches of a single multiplexer (this term is only used when the FPGA uses multiplexed switches). In order to help the algorithm to exit from local minima, this term increases with each iteration when a switch conflict occurs. In FIPSOC architecture, the multiplexed switches are located in logic blocks outputs and inputs, and in rows of S-Blocks (figures 1 and 2) meaning that an output/input must be connected to only one segment of the routing channel and that horizontal segments can be connected to only one vertical segment in an S-Block (note that vertical segments can be connected to more than one horizontal segment in an S-Block). In the case of the inputs, conflicts cannot occur, logic block input pins must be connected to only one net. In the case of the outputs, conflicts can only occur within the point to point nets of a multipoint net (obviously two different multipoint nets cannot have a switch conflict at an output of a Logic Block). In S-Blocks, conflicts can occur between nets of different multipoint nets; nevertheless, such conflicts imply shared segments between multipoint nets; as another term exist to account for shared segments in the function cost yet, we will only consider the switch conflicts that occur between nets of the same multipoint net. Instead of using a conflict value for each switch in the FPGA, a value is assigned for each pair of point to point nets of the same multipoint net, i.e., the $switch\_conflict(i,j)$ value is incremented if paths i and j of the same multipoint net causes a conflict. If we are calculating the cost of path i and this path has a switch conflict with paths j and k, then the $switch\_conflict$ term of the function cost is::

$$switch\_conflict(i,j) + switch\_conflict(i,k)$$

Circuit	CLBs	Ch. Occup. (%)	P. P. Nets	M. P. Nets	Iterations	Av. Delay	Max. Delay
4-ALU	38	87	253	118	2012	4.90	36.16
SPI	12	94	176	74	1439	3.40	25.30

Table 1: Results for FIPSOC circuits

Circuits	Channel Density	RAISE	SEGA Area	SEGA Speed	Sega Anneal
9symml	9	9	10	13	11
term1	10	10	11	12	11
C499	10	12	13	15	12
C1355	11	12	13	17	13
vda	14	14	14	19	15

Table 2: Minimum number of tracks per channel required for a successfully routing

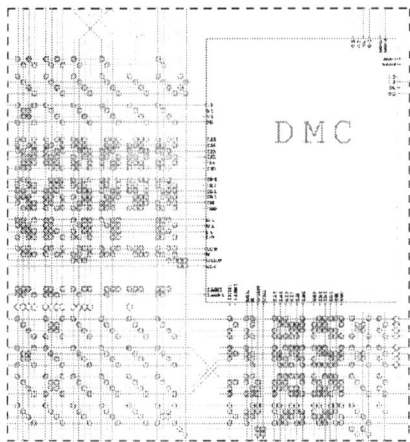

Figure 3: FIPSOC Architecture

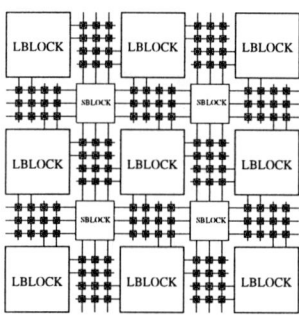

Figure 4: FPGA structure

The *history_cost* term can be easily calculated if we remember which segments were shared in previous iterations. In our case, it is calculated as follows:

$$ShareHist(W_i, K) = 0.5 \cdot ShareHist(W_i, K-1) + NetsUsingW(W_i, K)$$

$$history\_cost = num\_nets\_using\_wires \cdot \sum_{i=0}^{n} ShareHist(W_i, k)$$

where $NetsUsingW$ is the number of multipoint net that use wire W_i, K is the number of the present iteration and n is the number of wire segments used in this path.

The α, β, γ and δ parameters have to be tuned to reduce the number of iterations and to get a fast convergence, however the results are good in a wide range of their values.

2.3. Post-optimization

This phase is reached when a feasible solution has been obtained. Then, for each point to point net in a random order, the paths with the least delay from those that do not conflict with present solution, are selected for routing, reducing the net delay. This phase is repeated until no a change is accepted in an iteration.

3. RESULTS

To test the performances of RAISE, different routing solutions have been obtained with a set of benchmark circuits. The results have been classified in two parts, one with architectures that use multiplexed switches, and the other without them. For the first case, FIPSOC architecture is used (figure 3), by now there is no other FPGA with multiplexed switches and benchmarks for this new architecture don't exist at the moment; however two circuits have been used to test RAISE with multiplexed switches, a 4 bits ALU and an SPI (Serial Peripheral Interface) channel. The results are shown in table 1, where *Num. P. P. Nets* is the number of point to point nets of the circuits, *Num. M. P. Nets* is the number of multipoint nets of the circuits, *Av. Net Delay* is the average net delay and *Max. Net Delay* is the maximum net delay of the circuit, both of them measured in nanosecond. *Sega speed* [2] - [3] algorithm has been modified to accept multiplexed switches but get no solution for these circuits.

For the second case, i.e., architecture without multiplexed switches, the FPGA structure we used can be seen in figure 4, the C blocks have a switch for each segment, i.e. in SEGA terminology, fc=W; the routing structure of an S-Block is shown in figure 5: the connections between top and left segments are shown in (a), in (b) the connections between top and left segments, and in (c) the connections between top and bottom segments. The S-Block is symmetrical with $f_s = 5$. For simplicity, the vertical

W	RAISE			SEGA Area		
	Av. Delay	Max. Delay	Exec. time (s)	Av. Delay	Max. Delay	Exec. time (s)
9	3.538	28.651	499	-	-	-
10	3.885	38.318	4.13	4.018	42.637	0.60
11	3.969	44.066	2.17	4.416	49.714	0.67
12	4.080	34.872	1.70	4.169	41.914	0.69
13	4.405	40.501	1.80	4.535	44.271	0.92
14	4.393	49.396	1.85	4.840	52.316	1.08
15	4.308	33.394	2.02	5.124	44.850	1.93

Table 3: Average and maximum delays generated by RAISE and SEGA Area for 9symml and different channel density

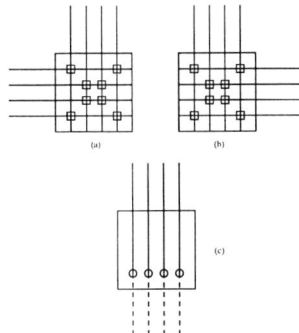

Figure 5: S block routing structure

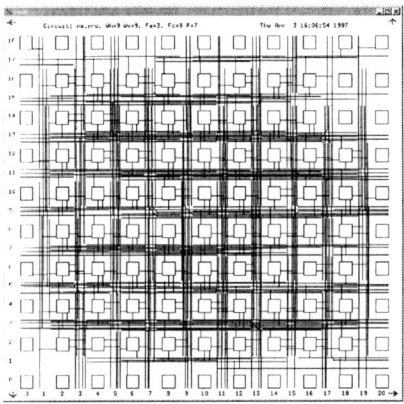

Figure 6: 9symml RAISE routing solution with nine track per channel

and horizontal routing channels have only one track group with W segments, offset 1, and length 3.

α parameter is set to 2.0, β parameter to 0.5, γ to 1.0 and δ to 1.0.

We can see the results in table 2 for a set of benchmark circuits. For this FPGA architecture, the number of wire-segments in each routing channel, required to route the circuits is very close to the minimum number predicted by the global router. Note that RAISE reaches solutions that other routers can't find. In figure 6 we show a RAISE solution for the 9symml circuit with nine tracks per channel. From table 3, we see the maximum and average path delay for different numbers of wire-segments per channel, also for the 9symml circuit. We can see that RAISE normally obtains better solutions than SEGA Area router and can be used to find solutions in difficult circuits with hugely saturated channels. The price to be paid for this better performances is compute-time cost. Like other statistical based optimization programs, RAISE takes a time searching for new solutions, as can be seen in the execution time column of table 3.

4. CONCLUSIONS

This paper has presented RAISE, a simulated evolution router for FPGAs, who gives solutions for symmetrical FPGAs with or without multiplexed routing switches. RAISE uses a statistical technique to explore the solutions space. It has been shown that RAISE normally obtains better solutions than different versions of SEGA. Furthermore it finds solutions that other routers can't find.

5. ACKNOWLEDGMENTS

The authors would like to acknowledge financial support by the European Union through the ESPRIT project FIPSOC and by CI-CYT through the TIC86-0860 project.

6. REFERENCES

[1] Julio Faura, Chris Horton, Phuoc Van Duong, Jordi Madrenas, Miguel Angel Aguirre, and Josep Maria Insenser, "A Novel Mixed Signal Programmable Device With On-Chip Microprocessor", *Proceedings of the IEEE 1997 Custom Integrated Circuits Conference CICC'97*, pp 103-106.

[2] Stephen Dean Brown, "Routing Algorithms and Architectures for Field-Programmable Gate Arrays", *Thesis, Department of Electrical Engineering*, University of Toronto, Canada. January 1992.

[3] G. Lemieux and S. Brown, "A Detailed Router for Allocating Wire Segments in FPGAs", *ACM Physical Design Workshop*, Lake Arrowhead, California, pp. 215-226. April 1993.

[4] Youn-Long Lin, Yu-Chin Hsu, and Fur-Shing Tsai, "SILK: A Simulated Evolution Router", *IEEE Transactions on Computer-Aided Design*, Vol. 8. NO. 10. October 1989.

[5] M. Khellah, S. Brown, and Z. Vranesic, "Modelling Routing Delays in SRAM-Based FPGAs", *Proc. 1993 CCVLSI*, Banff, Canada, pp. 6B.13-6B.18, Nov.1993.

RAPID PROTOTYPE OF A FAST DATA ENCRYPTION STANDARD WITH INTEGRITY PROCESSING FOR CRYPTOGRAPHIC APPLICATIONS

Hassina GUENDOUZ & Samir BOUAZIZ

Abstract

In this paper we present the design of a chip for real-time cryptographic processing in industrial applications. The chip acts as a co-processor of a system for the automatic ciphering and data integrity processing : it implements the DES/MAC algorithm (Data Encryption Standard/Message Authentication Code) in the same silicon area with high speed performance. These research come within a process aiming at reaching an implementation onto specialised pipeline and parallel architecture from algorithm specification.

The design has been performed in rapid prototype ACTEL Programmable Gate Array Device using an approach based on high level transformations on the VHDL specifications.

This paper is in part of an European Program which aims to design a macro cell library for cryptographic applications (*)

Index Terms : DES, MAC, Pipeline, FPGA, cipher, cryptography.

INTRODUCTION

Systems for encryption or key transfer, such as DES [1] require fast processing. So, software solution is not interesting for these systems. The implementation onto specialised hardware chip speeds up cryptographic calculation and grows the security level. To this purpose, custom commercial chips are available, [2] [3]; nevertheless, no one fits our computational and constraints needs.

Usually, in cryptographic systems, the user cipher his message, store the results on disk and then call the integrity process to calculates the MAC on the stored message. On this manner, the message is loaded twice which decrease speed performance for a long message. In this paper, we propose an original implementation of the DES algorithm which allows us to process in parallel the cipher and MAC phases.

This paper is organised as follows. After a brief description of DES algorithm in the next section, the second section presents the architectural mechanisms required for an efficient and original implementation. The third section presents the proposed architecture.

---

* *This work was supported by the Joint European Submicron Semiconductor Initiative (JESSI) working CRITT-CCST group*

2. DATA ENCRYPTION ALGORITHM :

The DES algorithm is designed to encipher and decipher blocks of data consisting of 64 bits under control of a 64-bit key. This algorithm is based on two main operations :

1) The cipher-core of the DES
2) The Key schedule calculation

The core of the DES realise the encipher/decipher in such a way

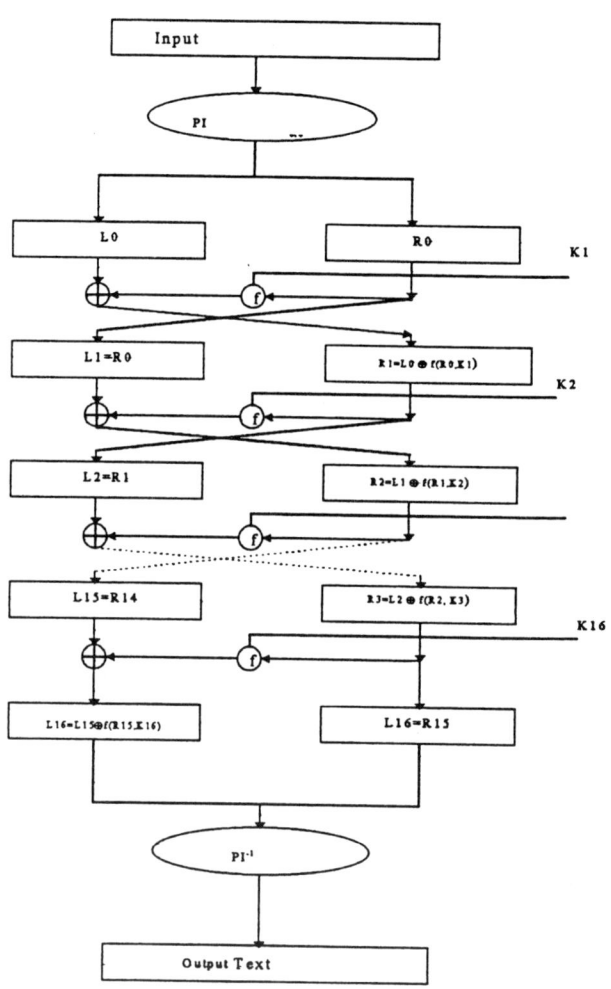

Fig.1. Data Encryption Standard Algorithm

A block to be enciphered is subjected to an initial permutation IP, then to a complex key-dependent computation and finally to a permutation IP^{-1}

key-dependent computation can be simply defined in terms of a function f, called the cipher function, and a function Ks, called the key schedule. The definition of the cipher function f is given in terms of primitive functions which are called the selection functions Si.

user to change frequently his secret key. This implementation grows the security level. In the next section, we describe architectural mechanisms required for an efficient implementation of this algorithm..

Fig. 2 calculation of f(R,K)

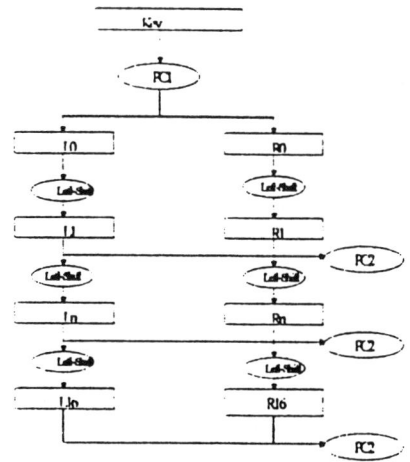

Fig 3: Key Schedule Calculation

The sketch of the calculation of $f(R,K)$ used in Fig.1 is explained in Fig.2. This process is performed in 16 iterations which correspond to key-cycle generation.

In Fig.2, E denotes a function which takes a block of 32 bits as input and yields a block of 48 bits as output. Each of the unique selection functions $S1$, $S2$, ...$S8$, takes a 6-bit block as input and yields a 4 bit block as output.

The key schedule calculation is represented by the sequence $K1$, $K2$, ... $K16$ which correspond to the generated 48 bits keys from the secret key Ks. This is described in Fig3. $PC1$ and $PC2$ correspond to Permuted Choice. This key is held by each member of a group of authorised users. The cryptographic security of data depends on the security provided for the key used to encipher and decipher data.

Three things must be undertaken carefully to provide high level security :

1) Key generation ;
2) Key transfer between each member ;
3) The possibility for the user to change easily and frequently his secret Key.

In our implementation, the key schedule process, shown on Fig.3, operates in parallel with the DES cipher_core. These two parts are implemented in the same module which represent the DES core. Thus, the secret key calculation is always in

3. HARDWARE IMPLEMENTATION

The circuit may be partitioned on three hierarchical

parts :

1) the first one implements the DES core which comprise cipher_core and key schedule process;
2) the second one implements the cipher_processor interface which control the several DES mode operations;
3) the last one represents the I/O interface.

3.1. The DES Core :

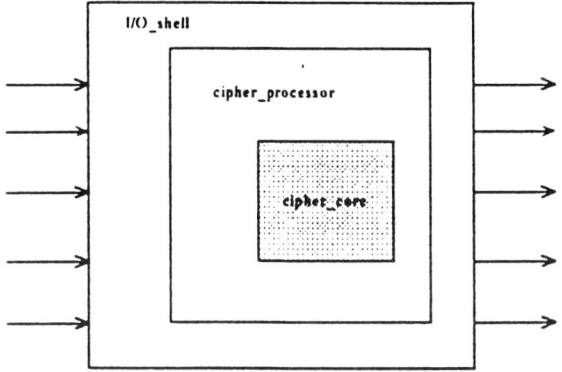

Fig. 4. Common three levels hierarchical architecture

The DES core is partitioned on two complex parts as described in the second section: the cipher_core and the key calculation. Each part described above is suited to a pipeline implementation. This implementation style divide complex low processes into fast elementary one. The time computation (in term of clock cycle) is the same but the frequency grows which speed up the data processing.

The influence of the pipeline stages, on architecture speed efficiency, is measured through numerous simulations. To provide a best compromise between speed and silicon area, we pipeline the main loop of the DES core on 2 stages.

On the other parts, this choice make easy the MAC implementation for integrity process. In fact, the first stage is used for ciphering process and the second one for integrity code calculation. Usually, the user cipher his message, store the result on disk and then call the integrity process to calculates the MAC on the stored message. On this manner, the message is loaded twice which decrease speed performance for a long message. Our implementation allows us to compute in parallel the two processes which minimise access disk. At the end of each 16 iterations, we stored cipher text and his MAC.

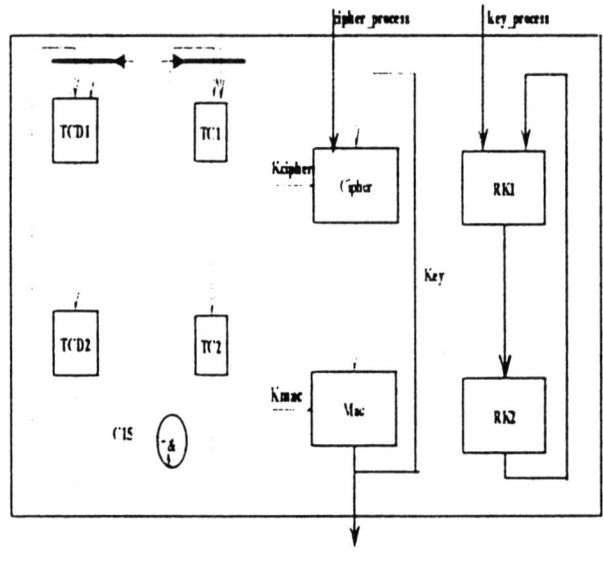

Fig. 5. DES_CORE

In order to take all the benefit of the pipeline, we use a Tag Control «TC» associated to counter «C» which brings us, at each time, the number of iteration done by the cipher_process. Three cases have to be undertaken:

1) C=15 and TC =1 : Pipeline stage and input register are free
2) C<>15 and TC=1 : Pipeline busy and input register free
3) C<>15 and TC<>1 : Pipeline and input register are busy

Another Control Tag named TCD, for Code/Decode, inform us for enciphering or deciphering process to apply at the pointed stage.

3.2. Cipher_Processor :

The Data Encryption Standard may be implemented in different modes. The usually ones are :

1) Electronic Code Book mode (ECB), which corresponds to the elementary one ;
2) Cipher Block Chaining mode (CBC) which uses the first one for ciphering the next plain text ;
3) Cipher Feedback mode (CFB) ;
4) Output Feedback mode (OFB) ;
5) EDE DES mode (Encrypt-Decrypt-Encrypt) ;

Each one of these modes are based on a basic bloc : the cipher-core described above. The proposed architecture for this cipher_core may be extended to treat each mode by using the cipher_processor. This can be made by finite state machine implementation (control path) which select a data path in the DES core for each selected mode.

3.3. INPUT/OUTPUT (I/O) INTERFACE :

The DES algorithm is designed to encipher and decipher blocks of data consisting of 64 bits under control of a 64-bit key. To minimise I/O pins, each block is splited on 16 bits and data acquisition is done through FIFO (First In, First Out) storage. To avoid speed degradation, data storage and data ciphering processes operate on parallel.

The use of two 8 bits coded-registers facilitates the interface with another components like microprocessor which send an order to the DES chip to realise an operation like :

1) storage key
2) cipher or decipher text.....

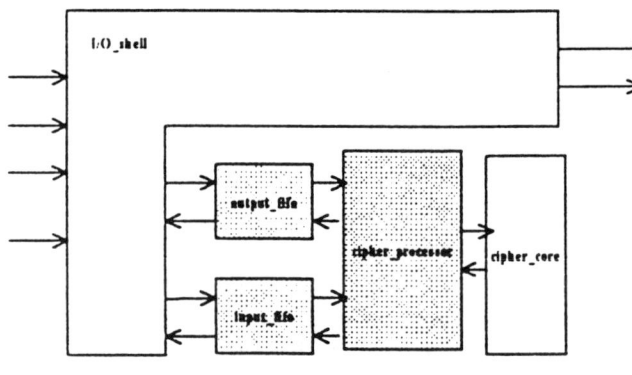

Fig 6. Data Exchange between several blocks

4. DESIGN METHODOLOGY

The design methodology for developing the rapid prototype is shown in Fig. 7. The chip architecture was described at RT-level using VHDL language (Model Technology). We modelled in a synthetizable way all logic blocks and, using a logic synthesis tools (ASYL+), we mapped the architecture onto a real circuit in the target ACTEL (ACT3 family) FPGA technology for rapid hardware prototyping. The ACTEL family of FPGA was chosen based on efficiency of implementation and speed.

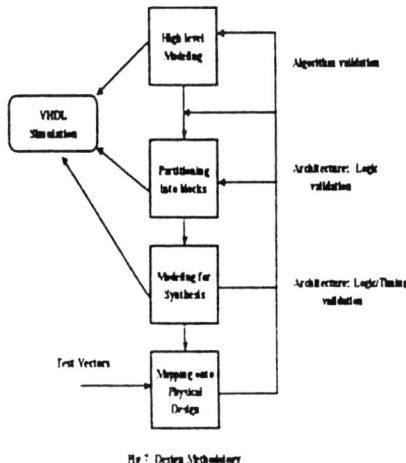

Fig 7: Design Methodology

The validation of the resulting chip was performed using specific test pattern files. This validation is not able to gives us the timing performance. It's only useful to verify that the prototype was functionally correct. On the place/route phase we performed the critical timing analysis and all verification of circuits operating. The critical path obtained is about 70 ns.

The chip complexity is of about 12,000 gates and throughput rate of 86 Mbits/s is evaluated for ECB or CBC mode with MAC calculation.

This evaluation results from the following formula, where « n » represent the number of blocks to cipher. Each block size is 64 bits.

Mode	formula	Throughput
ECB/CBC/MAC	343 n/[11 + (n-2)*4]	86 Mbit/s
EDE	228 n/[27 + (n-2)*4]	57 Mbit/s

The first block is ciphered with MAC calculation in about 40 cycles (2*16 + 4 + 4), where the number « 16*2 » corresponds to the DES algorithm iteration in two pipeline stage. The number « 4 » corresponds to the FIFO input and output stages. The next block is obtained in about 44 cycles. For each other block, we need only to add 32 cycles which leads to [44+ 32(n-2)/2]

For ASIC implementation the throughput is approximately three or four times greater than results obtained onto FPGA prototype which leads to 250 Mbits/s for ECB/CBC with MAC calculation. The critical path may also decrease to 25 ns which corresponds to 40Mhz.

5. CONCLUSION

The rapid prototype would provide full functionality, allowing any design errors.

Our proposed architecture can throughput ciphering message with MAC calculation in about 250Mbits/s for ASIC implementation.

The results obtained are very encouraging when compared with existing chips [3] [2] , and we project to develop an ASIC with an industrial partner.

REFERENCES

[1] Specifications for the *Data Encryption Standard*, Federal Information Processing Standards PUB 46, January 1977.

[2] Pijnenburg MicroElectronics & Software. *PCC100 Data Encryption Standard Device*, Datasheet Product. Feb 1992.

[3] SuperCrypt (CE99C003A). *99C003A Data Encryption Standard Device*, Datasheet Product. Feb. 1992.

[4] Bruce Shneier. *Cryptographie appliquée*, International Thomson Publishing. Nov. 1994.

Design and Analysis of Quasi-Switched-Capacitor Step-up DC/DC Converters

Henry Chung, *Member, IEEE*

Department of Electronic Engineering
City University of Hong Kong
Tat Chee Avenue, Kowloon Tong
Kowloon, Hong Kong

Abstract - A new switched-capacitor (SC)-based step-up DC/DC converter is proposed. It has the prominent features of continuous input current waveform and having better regulation capability than the traditional SC converters. The problem of conducted electromagnetic interference (EMI) with the supply network is minimized. Concept of energy transfer is achieved by using dual basic SC step-up converter cells operating in anti-phase. The voltage conversion ratio is controlled by a current control scheme in order to adjust the charging profile of the capacitors. A generalized *n*-stage converter is presented and is analyzed by a simplified third order state-space equation set. The static and dynamic behaviors and the design constraints of the proposed converter are derived. A prototype of 30W, 5V/12V, 2-stage converter has been built, giving an overall efficiency of 78% with power density of 15W/in^3.

I. INTRODUCTION

In recent years, a new class of SC-based DC/DC converters for low-power applications were proposed in [1, 2]. The major characteristic of these converters is that they do not require inductive element in the power conversion process and are amenable to monolithic integration. Although they exhibit many advantages, they have the following common intrinsic and extrinsic drawbacks : 1) the input current is pulsating, 2) the regulation capability is weak since the output voltage will vary with the input voltage, and 3) the dc voltage conversion ratio is primarily determined by the physical structure of the converter.

Some of the above aspects have been improved in [3]-[5], which present converters with adjustable voltage conversion ratio. They can give a wide range of desired output voltage values from a given input voltage. The basic idea is to control the capacitor charging duration through a PWM duty-cycle control scheme. Although all these converters exhibit many good performance behaviors, the input current waveform still contains considerable current peaks. The semiconductor switches are under high current stress of short duration during the charging period. This will particularly cause conducted EMI problem to the supply network [6]. Though the problem can be minimized by using an input reservoir capacitor as an auxiliary supply source in charging the capacitors, the resulting physical size will then be increased.

This paper presents a generalized *n*-stage step-up DC/DC converter topology. The converter solves the problem of pulsating input current and remains the prominent characteristics of the previous SC-based boost converters in [4, 5]. The principle of operation will be described in Section II. A third order state-space equations are derived to represent the system. By applying the state-space averaging technique, the steady-state behavior and the small-signal dynamic analysis are presented in Section III. Section IV will give the experimental performances of a 5V/12V 2-stage step-up converter prototype. The conclusion follows in the last section.

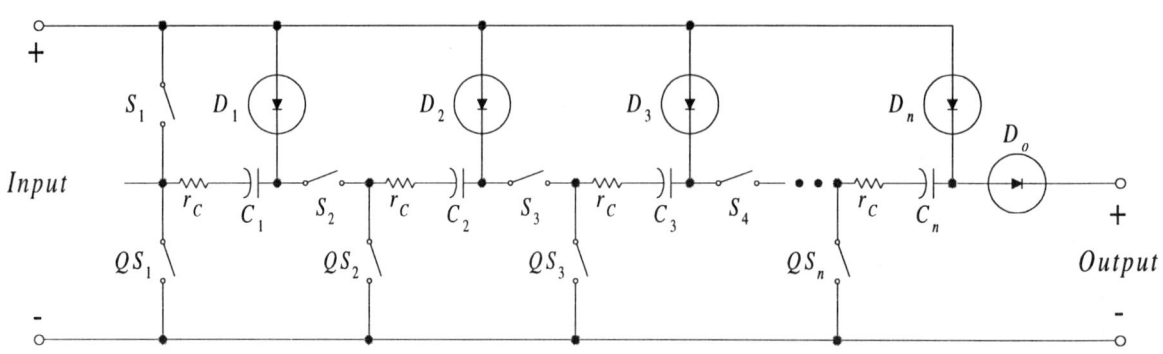

Fig. 1 Basic structure of a *n*-stage step-up DC/DC converter cell.

The research work was supported by RGC/UGC Grant 9040207.

II. SC-BASED STEP-UP DC/DC CONVERTER

A. Basic n-stage SC step-up converter cell

Fig. 1 shows a *n*-stage step-up DC/DC converter cell. All capacitors are similar with same value C. Each capacitor has an equivalent series resistance (ESR) r_C. QS_1 to QS_n, are operated in active region, acting as a voltage-controlled current source. The current magnitude is determined by the gate-source voltage. S_1 to S_n are operated as static switches.

The basic operating sequence of the two types of switches (i.e. QS_1 to QS_n and S_1 to S_n) are in complementary triggering. Within one switching period T_S, both of them are operated for same duration of $T_S/2$. The operation is composed of two topologies. In the first one, QS_1 to QS_n are in operation and S_1 to S_n are open. C_1 to C_n will be linearly charged by constant current, determined by QS_1 to QS_n, from the supply voltage through D_1 to D_n. In the second one, QS_1 to QS_n are open and S_1 to S_n are closed. All the capacitors are connected in series with the input voltage to supply the output load.

B. Realization of n-stage step-up DC/DC converter

A general *n*-stage step-up converter, as shown in Fig. 2, is realized by connecting two cells in parallel. The two cells are supplying from the same input source and to the same output load. Each cell is operated for $T_S/2$ and in anti-phase in one cycle. Since the two cells will be charged from the input source alternately for same duration, giving an overall continuous input current. In Topology 1, all cell 1 capacitors (i.e. C_{11} to C_{1n}) will be linearly charged by a constant current I_{ch} for $T_S/2$. At the end of this topology, all cell 1 capacitors will become at a voltage slightly higher than

$$\frac{v_{out} - v_{in}}{n} \quad (1)$$

for compensating the parasitic losses in Topology 2. v_{in} and v_{out} are the input voltage and output voltage respectively. The parasitic losses include the voltage drops across the on-resistance of the switches (r_{on}), ESR of the capacitors (r_C), and the diode voltage on D_{1o} (v_D). Within Topology 1, all cell 2 capacitors (i.e. C_{21} to C_{2n}) are connected in series with the input voltage through S_{21} to S_{2n} to supply electric energy to the output load R_L. In Topology 2, all cell 1 capacitors will stop charging and are connected in series with the input voltage through S_{11} to S_{1n} to supply electric energy to the output load. All cell 2 capacitors are charged linearly to a voltage slightly higher than the value in (1) in order to compensate the parasitic losses in Topology 1 in next cycle.

Since the capacitors within their respective cell undergoes similar charging and discharging processes, the capacitor voltage waveform within the same cell is similar to each other. The capacitor voltages are linearly increasing in the charging process and are in exponential decay trajectory in the discharging process. v_{out} is composed of the discharging phases of both cell 1 and cell 2. The output capacitor C_o is used to reduce the output voltage ripple. The converter input current i_{in} will then include all the charging currents of the capacitors in one cell and the output current. That is,

$$i_{in} = n\, I_{ch} + I_{out} \quad (2)$$

Since I_{ch} and I_{out} exhibit slightly constant magnitude throughout the operation, the input current is continuous.

C. Comparisons with previous SC-based converters

The following aspects can be distinguished from previous converters:

1) The input current waveform of the traditional ones is pulsating while the waveform of this converter is constant. The conducted EMI problem is subsequently suppressed.

2) The duty-time of all switches is kept at $T_S/2$ while the on-time of the charging switch in the previous ones is controlled by PWM modulator to adjust the output voltage. Their charging time is short at light load or at low output voltage condition, which is practically hard to be implemented. Thus, the regulation capability is improved.

3) The conversion efficiency of the proposed circuit can be proved to be same as the one in [5]. This gives the same performance index as the previous converter but presents better behaviors that have been discussed in (1) and (2).

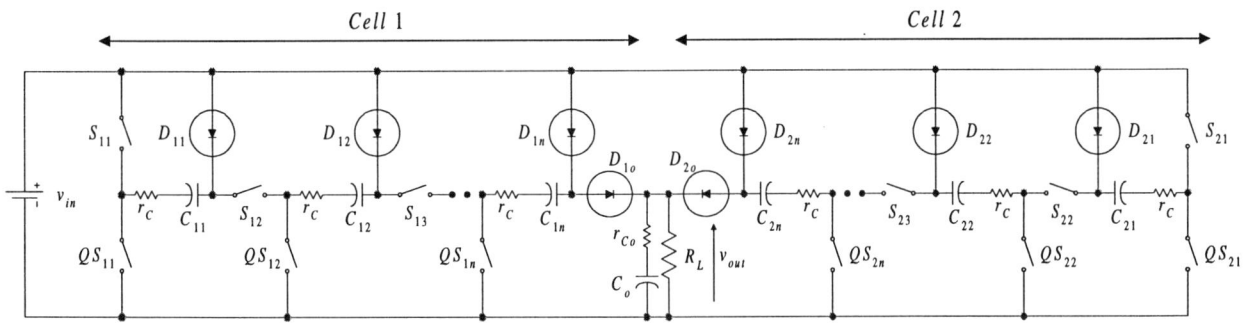

Fig. 2 Complete realization of a *n*-stage SC-based step-up DC/DC converter.

III. ANALYSIS OF THE SC STEP-UP DC/DC CONVERTER

The converter is analyzed by applying the state-space averaging technique in [7]. As all capacitors within the same cell undergo same charging and discharging processes, their voltage trajectories will behave identically. If each capacitor is treated by an individual state variable, the size of the state matrix will be $(2n+1) \times (2n+1)$. The analysis is complicated and time-consuming. As shown below, a set of third-order equations for this n-stage converter is derived. Each cell is represented by a single state variable for capacitor voltage in that cell. If v_{C_1} represents cell 1 capacitor voltage and v_{C_2} represents cell 2 capacitor voltage,

$$\dot{x} = A_{av} x + B_{av} u$$
$$v_{out} = C_{av} x + D_{av} u \qquad (3)$$

where $x = [v_{C_1} \ v_{C_2} \ v_{C_o}]^T$, $u = [I_{ch} \ v_{in} \ v_D]^T$,

$$A_{av} = \begin{bmatrix} -\frac{n(R_L + r_{C_o})}{2C\beta} & 0 & \frac{R_L}{2C\beta} \\ 0 & -\frac{n(R_L + r_{C_o})}{2C\beta} & \frac{R_L}{2C\beta} \\ \frac{nR_L}{2C_o\beta} & \frac{nR_L}{2C_o\beta} & -\frac{[R_L + n(r_C + r_{on})]}{C_o\beta} \end{bmatrix} ; B_{av} = \begin{bmatrix} \frac{1}{2C} & -\frac{R_L + r_{C_o}}{2C\beta} & \frac{R_L + r_{C_o}}{2C\beta} \\ \frac{1}{2C} & -\frac{R_L + r_{C_o}}{2C\beta} & \frac{R_L + r_{C_o}}{2C\beta} \\ 0 & \frac{R_L}{C_o\beta} & -\frac{R_L}{C_o\beta} \end{bmatrix} ;$$

$$C_{av} = \begin{bmatrix} \frac{nR_L r_{C_o}}{2\beta} & \frac{nR_L r_{C_o}}{\beta} & \frac{2R_L(r_C + r_{on})}{\beta} \end{bmatrix} ; \text{ and } D_{av} = \begin{bmatrix} 0 & \frac{R_L r_{C_o}}{\beta} & -\frac{R_L r_{C_o}}{\beta} \end{bmatrix}.$$

where $\beta = \{(R_L + r_{C_o})[R_L + n(r_C + r_{on})] - R_L^2\}$.

A. Formulation of the steady-state characteristics

The steady-state output voltage is obtained by substituting $\dot{x} = 0$ into (3). Thus,

$$v_{out} = \{C_{av}(-A_{av}^{-1} B_{av}) + D_{av}\} u$$
$$= \frac{a_0}{b_0} I_{ch} \qquad (4)$$

where $a_0 = nR_L[n \ r_{C_o}^2 + r_{C_o} R_L + n(r_C R_L + r_{C_o} r_{on} + R_L r_{on})]$
and $b_0 = n[n \ r_{C_o}^2 + r_{C_o} R_L + n(r_C R_L + r_{C_o} r_{on} + R_L r_{on})]$.

If all QS switches are in active mode, equation (4) shows that v_{out} is independent of the input voltage and forward voltage drop of the diodes and is merely determined by the drain current of the QSs and the parasitic resistances of the components. Moreover, the disturbance in the output load can be compensated by controlling I_{ch}, which is determined by the gate-source voltage applied to the QSs, for keeping a constant output voltage.

B. Determination of the conversion efficiency

If C and C_o are assumed to be similar, it gives $r_C = r_{C_o}$ and $a_0 = b_0 R_L$ in (4). Therefore, v_{out} can be modified into

$$v_{out} = I_{ch} R_L = i_{out} R_L$$
$$\Rightarrow i_{out} = I_{ch} \qquad (5)$$

Thus, the conversion efficiency η of the converter becomes

$$\eta = \frac{v_{out} \ i_{out}}{v_{in} \ i_{in}} = \frac{v_{out} \ I_{ch}}{v_{in} \ (nI_{ch} + I_{ch})} = \frac{v_{out}}{(n+1) v_{in}} \qquad (6)$$

It gives consistent results as that derived in [5]. The conversion efficiency of the proposed converter is same as the previous SC step-up converters but the proposed converter shows better performance characteristics.

C. Formulation of the dynamic characteristics

A small-signal variation of the charging current $\hat{i}_{ch}$ on its steady-state value I_{ch}, i.e., $i_{ch} = I_{ch} + \hat{i}_{ch}$, and of the input voltage $\hat{v}_{in}$ on its steady-state value V_{in}, i.e. $v_{in} = V_{in} + \hat{v}_{in}$ are imposed, then a small-signal converter output variation $\hat{v}_{out}$ will be superimposed on its steady-state output V_{out}, i.e. $v_{out} = V_{out} + \hat{v}_{out}$. By using (4), the input-to-output transfer function [i.e. $G_{og} = \hat{v}_{out}(s)/\hat{v}_{in}(s)$] and control-to-output transfer functions [i.e. $G_{oc} = \hat{v}_{out}(s)/\hat{i}_{ch}(s)$] are obtained. The procedures are similar to that performed in [6], where

$$G_{og}(s) = \frac{\hat{v}_{out}(s)}{\hat{v}_{in}(s)} = \begin{bmatrix} 0 & 1 & 0 \end{bmatrix} \begin{bmatrix} C_{av}(sI - A_{av})^{-1} B_{av} + D_{av} \end{bmatrix}$$
$$= \frac{s(\alpha_1 + \alpha_2 s)}{b_0 + b_1 s + b_2 s^2} \qquad (7)$$

and

$$G_{oc}(s)=\frac{\hat{v}_{out}(s)}{\hat{i}_{ch}(s)}=\begin{bmatrix}1 & 0 & 0\end{bmatrix}\begin{bmatrix}C_{av}(sI-A_{av})^{-1}B_{av}+D_{av}\end{bmatrix}$$
$$=\frac{a_0+a_1 s}{b_0+b_1 s+b_2 s^2} \quad (8)$$

where

$$\alpha_1=2CR_L(nr_{C_o}^2+r_{C_o}R_L+nr_C R_L+nr_{C_o}r_{on}+nR_L r_{on});$$
$$\alpha_2=2C_o C r_{C_o} R_L \gamma; \quad a_1=nC_o r_{C_o} R_L \gamma$$
$$b_1=(nC_o r_{C_o}+2nCr_{C_o}+2CR_L+nC_o R_L+2nCr_{on})\gamma;$$
$$b_2=2C_o C\gamma^2;$$
$$\gamma=(nr_{C_o}r_C+r_{C_o}R_L+nr_C R_L+nr_{C_o}r_{on}+nR_L r_{on});$$

I is the identity matrix. It can be seen from (7) that the converter output voltage is with less influence by the low frequency variation in the supply voltage, featuring low audio-susceptibility to the supply voltage.

IV. Experimental Prototype

A 5V/12V, 30W, 2-stage step-up regulator with power density of 15W/in^3 has been built in the laboratory. It is operating at 220kHz with overall efficiency of 78% at rated load (including the required power for the driving circuit). The values of the components are all 220μF and their ESRs are 0.01Ω. Fig. 3 shows the experimental output voltage and input current at rated load operation. It can be seen that the input current is continuous. The theoretical small-signal input-to-output and control-to-output responses are shown in Fig. 4, together with the experimental measurements.

V. Conclusion

A generalized *n*-stage SC-based step-up DC/DC converter, which uses no magnetic element, was designed and analyzed. The converter presents all the positive characteristics of previous SC converters. It also provides adjustable voltage conversion ratio, which is independent of the circuit structure, and gives better input current waveform and regulation capability than previous SC converters. A 2-stage 5V/12V step-up converter prototype has been built at a nominal output power of 30W, demonstrating the superior features of this converter.

References

[1] F. Ueno, T. Inoue, T. Umeno, and I. Oota, "Analysis and application of switched-capacitor transformers by formulation," *Electron. Commun. Japan*, pt. 2, vol. 73, pp. 91-103, 1990.

[2] F. Ueno, T. Inoue, I. Oota, and I. Harada, "Power supply for electroluminescene aiming integrated circuits," in *Proc. IEEE Int. Symp. Circuit Systs.*, May 1992, pp. 1903-1906.

[3] S.V. Cheong, S.H. Chung, and A. Ioinovici, "Inductorless dc-to-dc converter with high power density," *IEEE Trans. Ind. Electron.*, vol. 41, no. 2, pp. 208-215, 1994.

[4] O.C. Mak, Y.C. Wong, and A. Ioinovici, "Step-up dc power supply based on a switched-capacitor circuit," *IEEE Trans. Ind. Electron.*, vol. 42, no. 1, pp. 90-97, 1995.

[5] G. Zhu and A. Ioinovici, "Switched-capacitor power supplies: dc voltage ratio, efficiency, ripple, regulation," in *Proc. IEEE Int. Symp. Circuits Systs.*, May 1996, pp. 553-556.

[6] H. Ott, *Noise Reduction Techniques in Electronic Systems*, John Wiley and Sons, 1989.

[7] R. D. Middlebrook and S. Cuk, "A general unified approach to modelling switching-converter power stages," in *Proc. IEEE Power Electron. Spec. Conf. Rec.*, Jun. 1976, pp. 18-34.

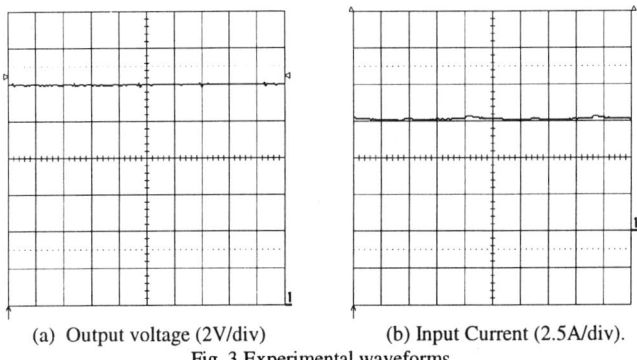

(a) Output voltage (2V/div)   (b) Input Current (2.5A/div).

Fig. 3 Experimental waveforms

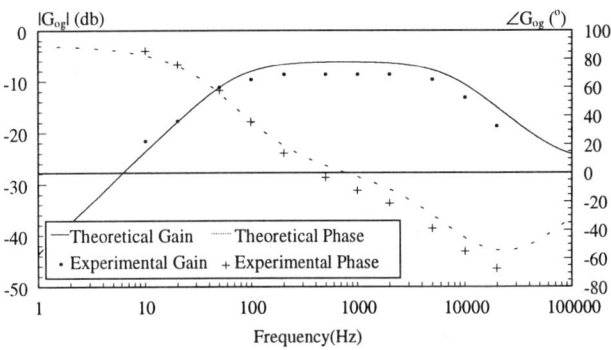

(a) Input-to-output characteristics

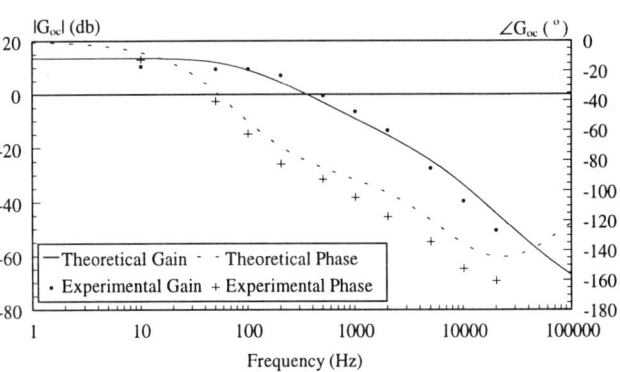

(b) Control-to-output characteristics.

Fig. 4 Small-signal frequency responses.

HIGH EFFICIENT PWM ZERO-VOLTAGE-TRANSITION DC-DC BOOST CONVERTER

J. Berkovici and A. Ioinovici

Electrical and Electronics Engineering Dept., Center for Technological Education Holon
52 Golomb St., Holon 58102, Israel

ABSTRACT

A soft-switching boost converter with duty-cycle control is proposed. Compared to its hard-switching PWM counterpart, the new switch-mode converter contains four additional elements: a resonant inductor, an active switch and two diodes. The parasitic capacitances of the main and auxiliary active switch serve as resonant capacitors. Multi-resonance is created in the circuit, thus allowing for zero-voltage-switching of the active switches and rectifier diode. The commutation of all switches takes place with zero capacitive turn-on losses; consequently, the switching losses are small. The current through the resonant inductor is not allowed to increase over the load current; the resonant stage is delayed until the inductor has been discharged. As a result, the voltage and current stresses on the devices are at the same level as on their counterparts in conventional converters, implying normal-rated switches, and normal conduction losses. The overall efficiency is very high, as proved by the measurements on a prototype realized in the laboratory.

1. PREVIOUS WORK AND GOALS STATEMENT

The soft-switching converters offered the possibility of a high-frequency operation and, thus, of minimization of size and weight. But the reduction of switching losses was mitigated by a difficult frequency-control. By adding an active-switch in the topology of a quasi-resonant converter (QRC), duty-cycle controlled soft-switching techniques have been obtained. Unfortunately, both the active and passive switches were subject to voltage and current stresses significantly higher than those in their hard-switching PWM counterparts. Zero-voltage-transition (ZVT) PWM converters have been proposed in [1]-[3]. In these circuits, both the main switch and the rectifier diode were commutating at zero-voltage switching (ZVS). But turn-on capacitive loss was accompanying the commutation of the auxiliary switch. In [1], the voltages across the additional switches were reaching two times the nominal value. In [2], the resonant current through the auxiliary switch was reaching a peak higher than the input current. In [3], two auxiliary active switches were introduced. These increased voltage/current stresses were leading to an increase in conduction losses. In [4], a new zero-voltage-transition duty-cycle controlled buck converter was developed, in which a) the main switch, the auxiliary active switch and the rectifier diode are commutated at ZVS with no capacitive turn-on losses; b) the stresses on the switches are similar to those on their counterparts in hard-switching converters.

The aim of this paper is to develop a similar boost converter. The proposed converter and its cyclical switching behavior are described in Section 2. The analysis results are given in Section 3, followed by the SPICE simulated waveforms and the experimental results in Section 4. The conclusions are the subject of the last section.

2. ZVT-PWM BOOST CONVERTER

The proposed soft-switching converter operating at constant switching frequency is shown in Fig.1. Compared to a conventional hard-switching PWM converter, it contains four additional elements: a resonant inductor, L, an active switch, S_2, and two diodes, D_3 and D_4. D_1 and D_2, respectively, are the body diodes of the active switches, S_1 and S_2. The parasitic capacitances of the main switch S_1 and auxiliary switch S_2, C_1 and C_2, respectively, are used as resonant capacitors. The three reactive elements, L, C_1 and C_2 allow for a multi-resonance process in the circuit, which creates the

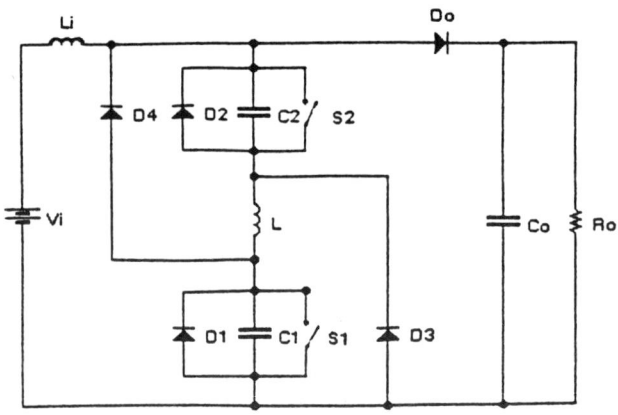

Fig.1 - Diagram of Proposed PWM ZVT Converter

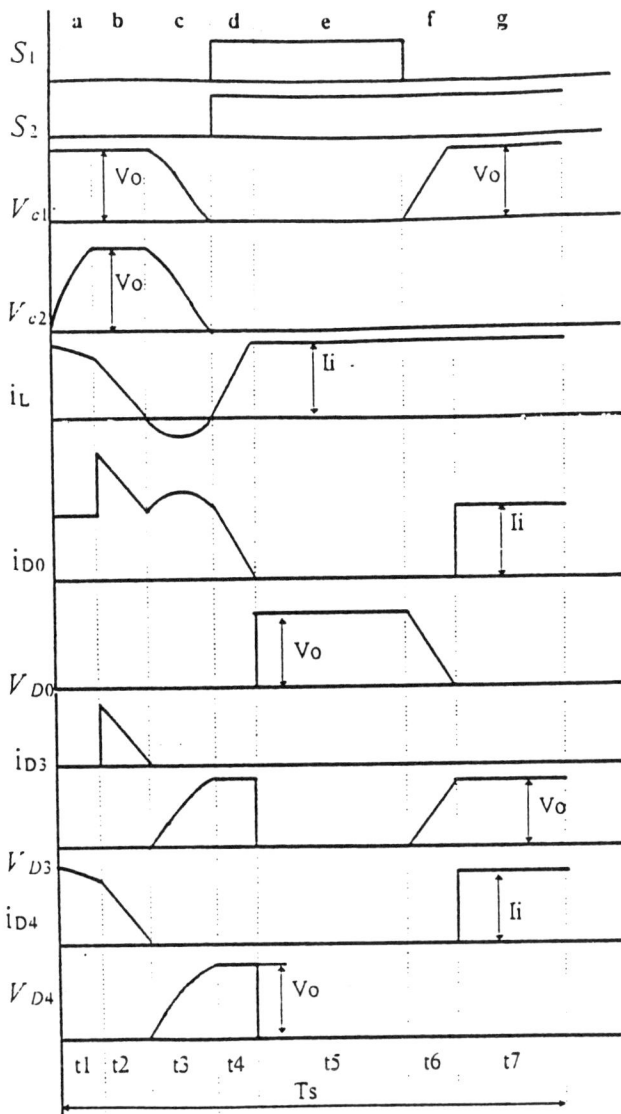

Fig 2. Key Waveforms of Proposed Converter

necessary condition for soft-switching of the switches.
The main problem in QRCs was the resonant inductor current, which was reaching a resonant peak much higher than the load current. The role of D_3 and D_4 in the present circuit is to prevent such current peaks: the inductor current is not allowed to increase over the load current; the resonant stage is delayed until the resonant inductor has been discharged through a path created by D_3 and D_4.

The timing diagram of the externally controlled switches S_1 and S_2 is given in Fig.2. In the same figure, the theoretical waveforms of the resonant inductor current (i_L), resonant capacitor voltages $(V_{C1}$ and $V_{C2})$, the currents through the diodes D_3, D_4 and rectifier diode D_0 and the voltages across the same diodes (V_{D3}, V_{D4}, V_{D0})

are shown for a steady-state cycle. T_s denotes the switching period ($T_s = 1/f_s$, f_s being the switching frequency). The converter goes through seven switching topologies in a cycle. The seven equivalent circuits are shown in Fig.3, where I_i denotes the input current and V_0 the output voltage. L_0 is large enough to allow the replacement of V_i, L_0 by a current source I_i.

A cycle begins by switching off S_2.

Stage 1: C_2- capacitor charging. In this stage $i_L = I_i \cos\omega_1 t$, $V_{C2} = I_i \sqrt{L/C} \sin\omega_1 t$, where $\omega_1 = 1/\sqrt{LC}$. When V_{C2} reaches V_0, V_{D3} drops to zero and D_3 starts conducting at ZVS. The stage duration is

$$t_1 = \frac{1}{\omega_1} \sin^{-1} \frac{V_0}{I_i \sqrt{L/C}}.$$

Stage 2: Inductor discharging. The inductor is discharging through D_3, D_4 and load, so that a future resonant current peak is avoided. D_0 carries the current I_i plus the decreasing inductor current, given by $i_L = i_L(t_1) - \frac{V_0}{L}t$. $V_{C2} = V_{C1} = V_0$. When i_L drops to zero, D_3 and D_4 stop conducting with both ZCS and ZVS. The stage duration is $t_2 = \frac{i_L(t_1)L}{V_0}$.

Stage 3: Three-element resonance. The multi-resonance process takes place in the C_1-L-C_2 circuit, i_L changes its orientation through L. The waveforms are given by the equations:

$$V_{C2} = V_{C1} = \frac{1}{2}V_0(1+\cos\omega_3 t), \quad i_L = V_0\sqrt{\frac{C}{2L}}\sin\omega_3 t,$$

where $\omega_3 = \frac{1}{\sqrt{LC/2}}$. V_{C2}, V_{C1} and i_L reach zero at the same time, giving the cycle duration t_3 of π/ω_3. At the same time, V_{D3} and V_{D4} reach V_0. At this moment, the feedback circuit dictates the turn-on of S_1 and S_2, at ZVS, with zero capacitive turn-on losses, because C_1 and C_2 have been discharged completely.

Stage 4: Inductor charging, $i_L = \frac{V_0 t}{L}$. The stage ends when i_L reaches I_i, giving the stage duration $t_4 = I_i L/V_0$. The current through D_0 drops to zero, thus D_0 turns off at ZCS.

Stage 5: Controlled-duration stage. In this stage $V_{C1} = V_{C2} = 0$, $i_L = I_i$. This is a typical boost converter topology, in which C_0 maintains the output voltage. The end of the stage, of duration t_5, is dictated by the feedback circuit through a PWM control: depending on the values of the line and load, the control circuit dictates the moment to switch-off S_1, with ZVT.

Stage 6: C_1-capacitor charging. In this stage $V_{C1} = \frac{I_i t}{C}$, $V_{C2} = 0$, $i_L = I_i$. When V_{C1} reaches the value V_0, V_{D0} drops to zero, D_0 and D_4 start conducting at ZVS.

Stage 7: Free-wheeling stage. i_L flows through D_4 and S_2 in a closed circuit and a constant flux of energy is transferred from line to load.

3. ANALYSIS OF THE CONVERTER

By using standard derivation based on equalizing the input and the output energies, it was found the input-to-output DC conversion ratio M,

$$M = \frac{1}{\frac{1}{2}(1-D)a_C + \sqrt{\frac{1}{4}(1-D)^2 a_C^2 + a_C a_L}},$$

where $a_C = \frac{1}{1 + RC/2T_s}$, $a_L = \frac{L}{RT_s}$,

where D is the duty ratio $(t_4 + t_5)/T_s$.

The only additional current stress in this circuit appears in Stage 2, when

$$i_{D0} = I_i + i_L = I_i + I_i \cos \sin^{-1} \frac{V_0}{I_i \sqrt{L/C}} - \frac{V_0}{L} t.$$ At

the beginning of the stage the maximum stress is

$$I_{D0.\,max} = I_i \left(1 + \sqrt{1 - \frac{V_0^2 C}{I_i^2 L}}\right).$$

4. SIMULATION AND EXPERIMENTAL RESULTS

The designed circuit was simulated by using PSPICE. The simulated characteristics of the voltages across S_1 and S_2 and of the inductor current are shown in Fig. 4.

The proposed converter was implemented in the laboratory by using MOSFETs IRF540 for S_1 and S_2 ($C_1 = C_2 = 3000$pF), SBL1030 for diodes D_3, D_4, D_0 and an inductor L = 8µH ($L_i = 100$µH, $C_0 = 50$µF). The converter was operated at $f_s = 100$kHz. The converter was designed for the nominal values $V_i = 15$V, $V_0 = 32$V, $P_0 = 50$W. The experiments waveforms of V_{C1}, V_{C2}, I_L, V_{D0}, and V_0 are given in Fig. 5.

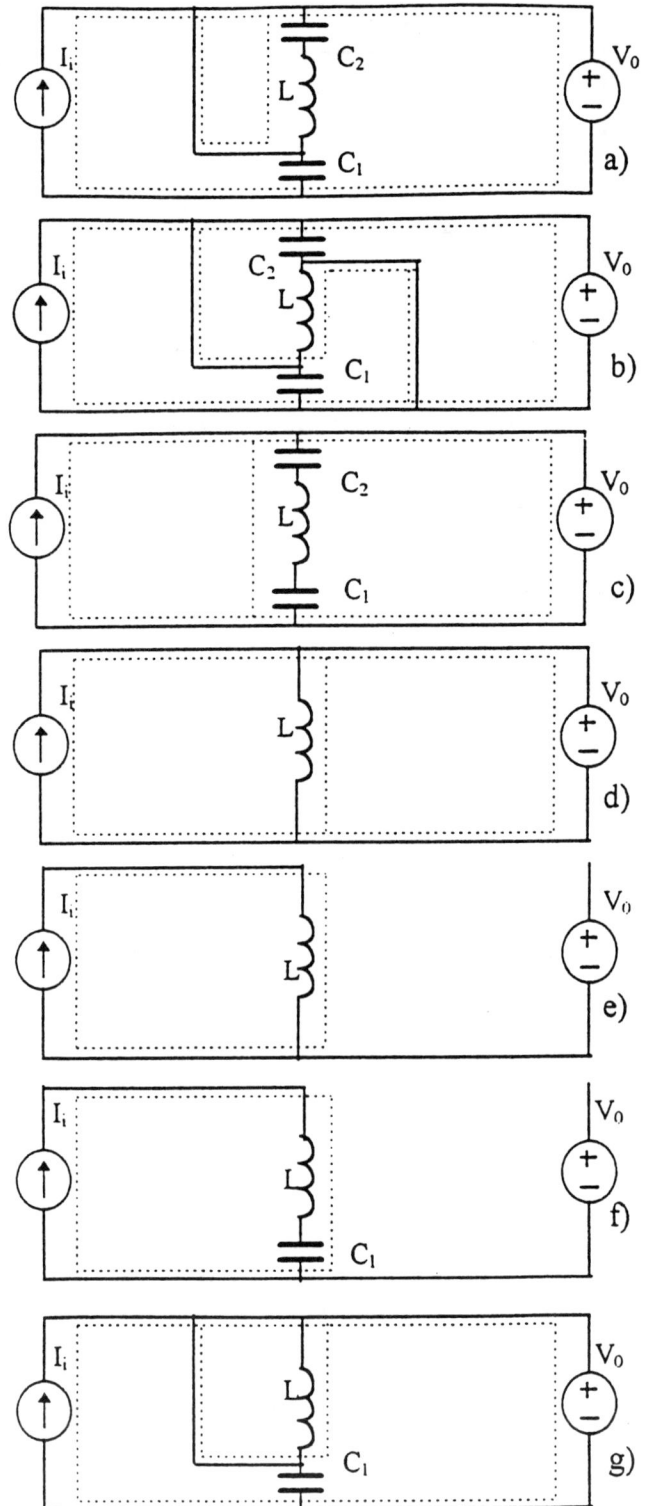

Fig 3. Equivalent Switching Networks. (a) C_2-Capacitor Charging. (b) Resonant Inductor Discharging. (c) Three-Element Resonance. (d) Resonant Inductor Charging. (e) Controlled Stage. (f) C_1-Capacitor Charging. (g) Free-wheeling Stage.

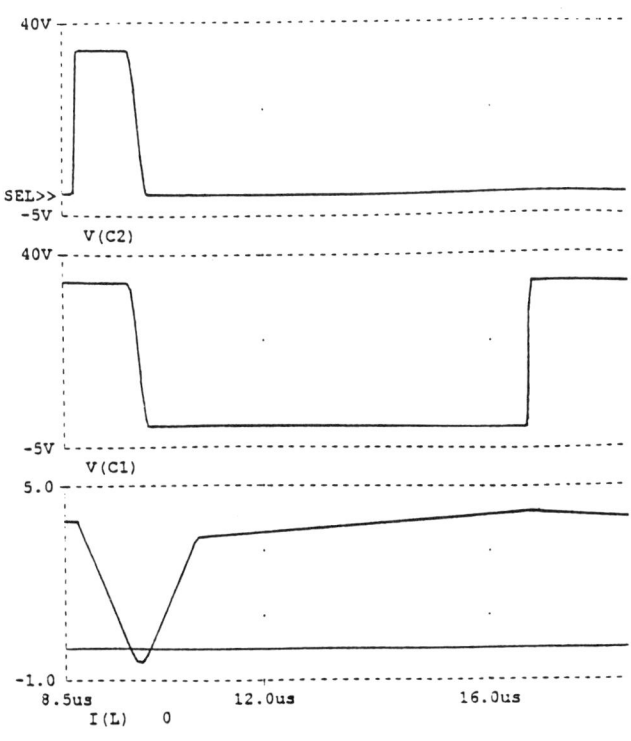

Fig 4. PSPICE Simulation of Voltages V_{C1}, V_{C2} and Inductor Current I_L in the Circuit of Fig. 1. for Vin = =15V, L_i= 100μH, C_0=50μF, R_0= 20ohm, L = 8μH, C_1 = =C_2= 3000pF

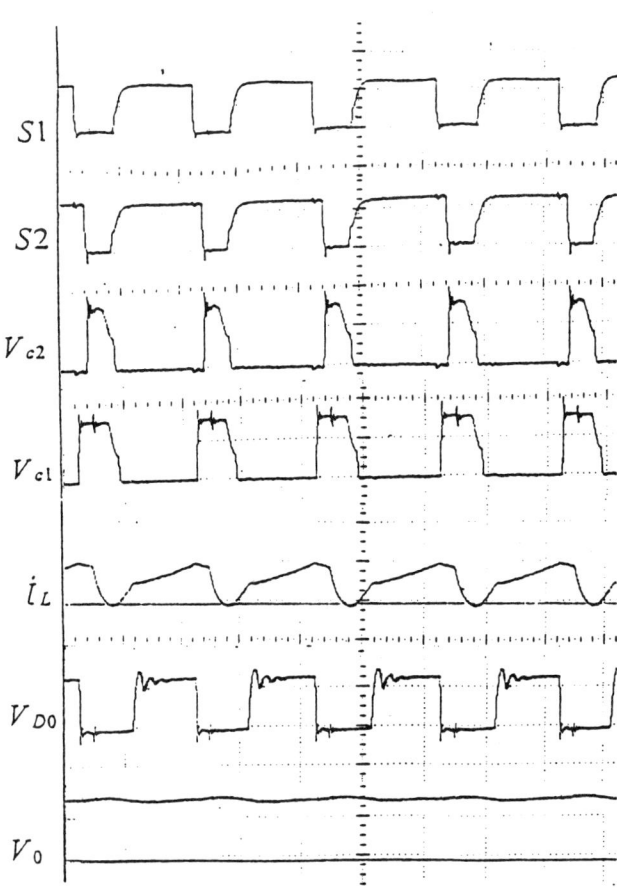

Fig 5. Oscillograms of main Waveforms of converter. Voltages for S_1, S_2 - 10V/div, for V_{C1}, V_{C2}, V_{DO}, V_0 - 20V/div; current : i_L - 5A/div

5. CONCLUSIONS

The proposed converter achieves:

a) ZVS at turn-on of the main and auxiliary active switches, as well as of the rectifier diodes and auxiliary diodes, with no turn-on capacitive losses.
b) all the switches are subjected to voltage and current stresses at the same level as those on their counterparts in a conventional PWM converter, allowing for normal conduction losses.
c) good regulation and high efficiency, as rendered evident by experiments.
d) an operation at constant switching frequency.

ACKNOWLEDGMENT

Berkovici thanks the Israeli Ministry of Absorption for a fellowship which allowed him to work on this research project.

REFERENCES

[1] L. Yang and C.Q. Lee, "Analysis and design of boost zero-voltage-transition PWM converter," in *IEEE 8th Applied Power Electronics Conf. Proc.*, 1993, pp.707-713.

[2] G. Hua, C.S. Leu, Y. Jiang, and F.C.Y. Lee, "Novel zero-voltage-transition PWM converter," *IEEE Trans. Power Electron.*, Vol.9, pp.213-219, March 1994.

[3] H.Wei and A. Ioinovici, "DC-DC zero-voltage-transition converter with PWM control and low stresses on switches," in *IEEE 10th Applied Power Electronics Conf. Proc.*, 1995, pp.523-529.

[4] B.P. Divakar and A. Ioinovici, "Zero-voltage transition converter with low conduction losses operating at constant switching frequency", in *IEEE Power Electronics Specialists Conf. Proc.*, 1966, pp.1885-1891

BIDIRECTIONAL BUCK-BOOST CONVERTER WITH VARIABLE OUTPUT VOLTAGE

Bhaskar Krishnamachari

Department of Electrical Engineering
The Cooper Union
New York, NY 10003
krishn2@cooper.edu

Dariusz Czarkowski

Department of Electrical Engineering
Polytechnic University
Brooklyn, NY 11201
dcz@pl.poly.edu

ABSTRACT

An increasing number of manufacturing processes rely on ultra-high speed and accuracy machines. Piezoceramic actuators are being utilized as parts of such machines. So far, only linear and switched-capacitor power supplies have been used for driving piezoceramic actuators in such applications. This paper proposes a switch-mode power supply to reduce cost and increase system efficiency. In the proposed design, the traditional PWM buck-boost topology is modified to accommodate bidirectional operation, and dynamic compensation is applied between the reference and the output to ensure good frequency response and low steady-state error of the variable output voltage. The converter operation is verified by Saber simulations.

1. INTRODUCTION

Piezoceramic actuators are used in ultra-high speed and accuracy machinery. A power supply with a variable output voltage capability is required for driving these actuators. Although switch-mode power supplies (SMPS) are lightweight and efficient, due to design difficulties, they have not been used for such applications so far.

An equivalent model of a piezoelectric actuator can be represented at low frequencies (below 1 kHz) as a capacitance of the dielectric [1]. The value of this capacitance for actuators investigated in this study is in the order of $C = 10$ μF.

The power supply system should be able to convert a 120 V_{rms} ac line voltage to a stabilized and regulated dc output voltage V_{out}. The dc range of V_{out} is 0–250 V. In a stand-by mode, the voltage across the actuator is kept in the middle of the voltage range, that is, at about 125 V. For an actuator maximum operating frequency $f = 500$ Hz, the maximum output current of the power supply can be calculated as

$$I_{Omax} = \pi f C V_{out,max} = 3.93 \text{ A}. \quad (1)$$

The ac line voltage is rectified in a peak rectifier consisting of a diode bridge and a large capacitor. The peak rectifier provides an unstabilized dc voltage of about $V_{in} = 170$ V to the input of a dc-dc converter. The task of the dc-dc converter is to supply a stabilized dc voltage to the actuator and to regulate the output voltage according to a motion-control goal. The required magnitude of the output voltage is indicated by a reference voltage input to the power supply. Most SMPSs are built to provide a stabilized dc output voltage of a constant value. There are no reports on SMPSs with output voltage tracking capabilities over the large range required for such capacitive loads as piezoelectric actuators. For this application, it is also desirable to have high speed operation so that the output voltage follows the reference even at frequencies up to 500 Hz. Large changes in the output voltage and fast dynamic response make topology selection and the design of the converter very challenging.

2. TOPOLOGY SELECTION

There are several methods of controlling SMPSs, e.g., pulse-width modulation (PWM), frequency control, phase control, and cycle-by-cycle control. A PWM dc-dc converter is proposed for the investigated application because of its simple structure, well-known dynamic behavior, and possibility of a pulse-by-pulse current limiting and instantaneous shutdown. The output voltage in PWM converters is controlled against line and load variations by adjusting the duty ratio D of the switches

$$D = \frac{t_{on}}{t_{on} + t_{off}} = \frac{t_{on}}{T}. \quad (2)$$

The buck-boost topology is selected as a basic power-conversion cell. This topology has the ability to provide an output voltage higher or lower than the input voltage, can be easily implemented using a few circuit elements, and is well-researched and established in its conventional unidirectional form. The simplified transfer function for the buck-boost converter is given by

$$\frac{V_{out}}{V_{in}} = \frac{-D}{1-D}. \quad (3)$$

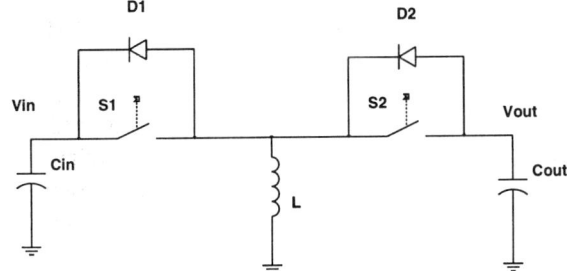

Figure 1: Bidirectional buck-boost topology.

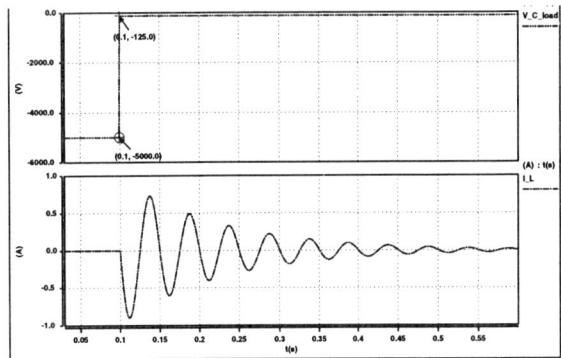

Figure 2: Bidirectional current flow in inductor L due to load disturbance (averaged model).

Usually, it is considered a drawback of the buck-boost topology that it provides a negative output voltage. In this application, however, the actuator can be appropriately biased by simply reversing the terminals because it does not require referencing to ground.

Although the topology of the conventional buck-boost converter is successfully used for constant-output-voltage power supplies, it is not suitable for use with piezoelectric actuators. In the conventional circuit, the inductor current can only charge the output capacitor C_{out}. The discharge of the capacitor is due to the load current. Such a slow and uncontrollable discharge dynamics is not acceptable for an actuator. Moreover, the actuator may be charged from the load side during mechanical oscillations. Hence, a controllable, bidirectional power flow to and from the output capacitor is needed.

3. DESIGN FOR BIDIRECTIONAL OPERATION

To achieve bidirectional operation [2], the conventional buck-boost topology needs to be augmented by addition of an anti-parallel diode to the input switch and a controllable switch to the output diode as seen in Fig. 1. The two switches, which can be implemented using MOSFETs, are operated in a complementary fashion, i.e., when switch S1 is on, S2 is off and vice-versa. With this modification, a negative current through the inductor L is now possible which enables the recovery of mechanical energy from the load, its conversion to electrical energy and subsequent storage in the input filter capacitor C_{in}. The bidirectional arrangement of switches results in a synchronous rectifier topology which also increases the efficiency of the converter, especially for low output voltages. It requires, however, a more complicated control circuit.

Challenging design issues arise in compensating the closed loop for control of the device. The k-factor method described by Venable [3] is applied to obtain the resistance and capacitance values in the compensating error amplifier used in voltage control mode. Computational software tools, such as Saber and Matlab, are used at each stage for design, testing, and analysis. The design process is carried out in incremental stages. The initial circuit designs utilize averaged models of PWM switches [4] that allow for fast simulations; the next step involves using ideal switches with a separate PWM controller; in the final simulations, these ideal switches are replaced with models of MOSFETs.

A damped-resonant R_m-L_m-C_m circuit with a resonant frequency of 20 Hz and 0.03% damping represents the mechanical load. The values of the load model components are $R_m = 2.5$ Ω, $L_m = 63.3$ H, $C_m = 1.0$ μF.

Fig. 2 demonstrates the bidirectional operation of this power supply. To simulate external disturbance, the capacitor C_m is charged to an initial voltage of -5000 V. At this moment, C_m is isolated from the output capacitor C_{out} which represents the piezoelectric actuator. Then, the load is connected to the output at the time instant $t = 0.1$ s. After the disturbance, the output current shows an exponentially decaying sinusoidal pattern with frequency equal to the characteristic frequency of the load, and amplitude and time constant of the decay determined by the settings of the converter control circuitry. The simulation results show that the inductor current reverses direction, transferring energy back and forth between the input and output of the converter.

The presented simulation example together with analytical considerations show that the designed converter is stable. However, the requirement of reference-to-output tracking at high speeds requires further attention.

4. REFERENCE-TO-OUTPUT COMPENSATION

Fig. 3 shows a block diagram of the closed-loop bidirectional buck-boost converter where T_1 is the control to output transfer function of the power stage, and Z_1 and Z_2 represent values of impedances chosen for the compensating Op-Amp. $T_c = -Z_2/Z_1$ represents the compensation obtained earlier for the closed loop under the assumption that the non-inverting terminal is held constant. The following simple manipulations yield a transfer function between the

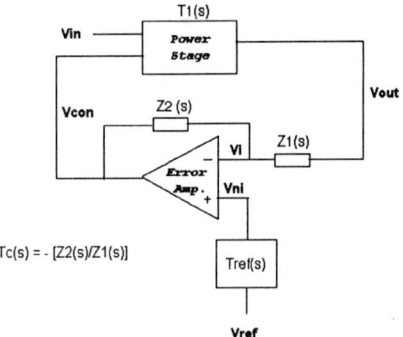

Figure 3: Block diagram of bidirectional power supply.

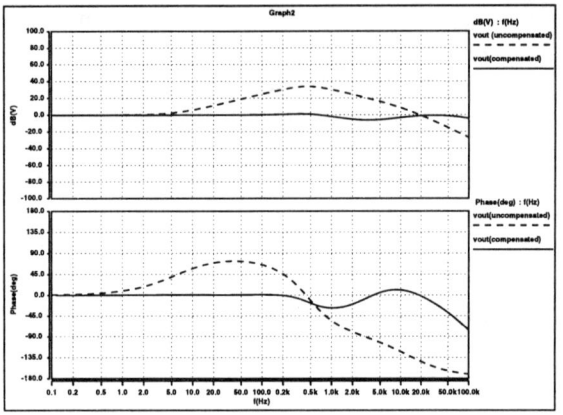

Figure 4: Reference-to-output frequency compensation.

non-inverting terminal and the output node:

$$V_{ni} \approx V_i$$

$$\frac{V_{out} - V_i}{Z_1} = \frac{V_i - V_{con}}{Z_2} \quad (4)$$

$$V_{con} = \frac{V_{out}}{T_1}.$$

Hence,

$$\frac{V_{out} - V_{ni}}{Z_1} = \frac{V_{ni} - V_{out}/T_1}{Z_2}$$

$$-T_c T_1 (V_{out} - V_{ni}) = T_1 V_{ni} - V_{out} \quad (5)$$

$$V_{out}(1 - T_c T_1) = V_{ni}(T_1 - T_c T_1)$$

and, finally,

$$T_{ni} \equiv \frac{V_{out}}{V_{ni}} = \frac{T_1 - T_c T_1}{1 - T_c T_1}. \quad (6)$$

The dashed lines in Fig. 4 present the amplitude (top) and the phase (bottom) characteristics of the transfer function T_{ni}. It can be seen that T_{ni} has a zero at about 5 Hz and ence signal components in the frequency range from 5 Hz to 20 kHz are amplified in comparison to the dc component.

The considered application demands a flat reference-to-output gain for frequencies up to 500 Hz. Hence, an additional compensation T_{ref} is required in the reference signal path as shown in Fig. 4. Solid lines in Fig. 4 present the amplitude and phase characteristics of the compensated transfer function $V_{out}/V_{ref} = T_{ref} T_{ni}$. It can be noticed that the desired flat amplitude response is achieved up to 50 kHz. T_{ref} has been implemented with two simple second-order compensators in series. The complete circuit diagram of the proposed bidirectional buck-boost converter is shown in Fig. 5.

5. DYNAMIC BEHAVIOR SIMULATION

To check the output tracking capabilities of the converter, a 500 Hz sinusoidal reference signal has been applied. The parameters of the reference signal have been selected in such a way that the desired output is a sinusoid with -125 V average value and amplitude of 100 V. Fig. 6 compares the desired (dashed) and actual (solid) output voltages. It can be observed that the amplitude tracking error is less than 5% with a phase shift of about 20°.

Fig. 7 shows the converter response to a step change in the reference voltage which decreases the desired (dashed line) output voltage level by 5 V. Solid lines in Fig. 7 represent the actual output voltage (top) and the inductor current (bottom).

The presented simulations have been obtained with Saber hybrid simulator version 4.2. The power circuit components used in the simulation were $V_{ac} = 170 \sin(2\pi \times 60t)$ V, $C_{in} = 100\,\mu\text{F}$, $L = 100\,\mu\text{H}$, $C_{out} = 10\,\mu\text{F}$, and APT40M50JN MOSFET models. The switching frequency was selected to be 100 kHz.

6. CONCLUSION

The selection of an appropriate topology of a SMPS for a piezoelectric actuator application is a complicated process involving both rigorous and heuristic approaches. Often contradictory requirements for static and dynamic performance of the power supply as well as for the interaction with the electromechanical environment demand great skills from the designer.

A bidirectional PWM buck-boost converter is proposed to serve as a power supply for piezoelectric actuators. Design and simulation results in both frequency and time domain show that the proposed converter is able to provide the required dynamic performance. Experimental verification of the concepts presented here is planned as a next step of this research. For practical reasons, it may be easier to implement the proposed bidirectional converter using a trans-

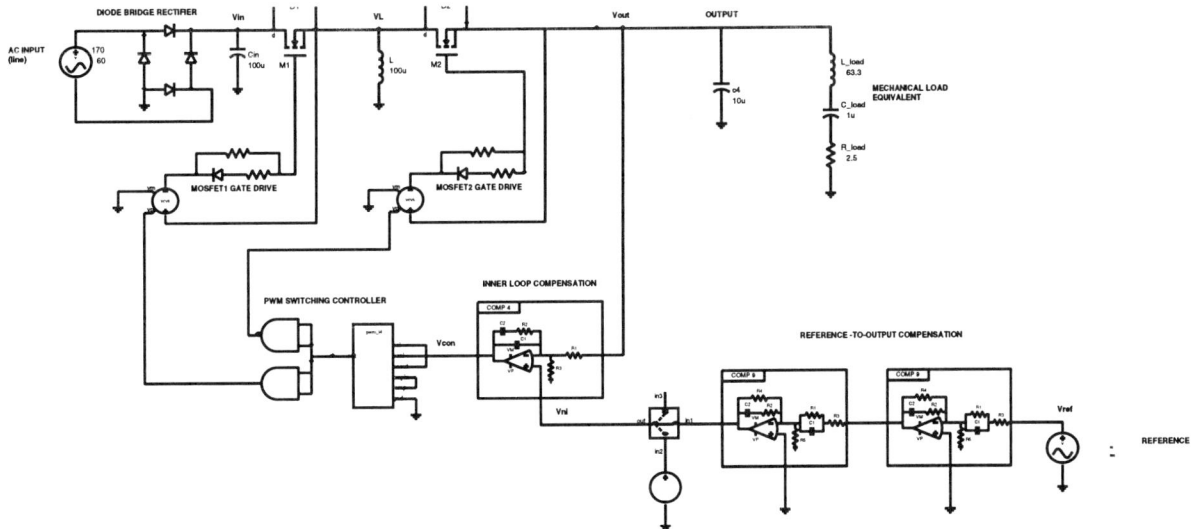

Figure 5: Circuit diagram of the bidirectional buck-boost converter with compensation in the reference signal path.

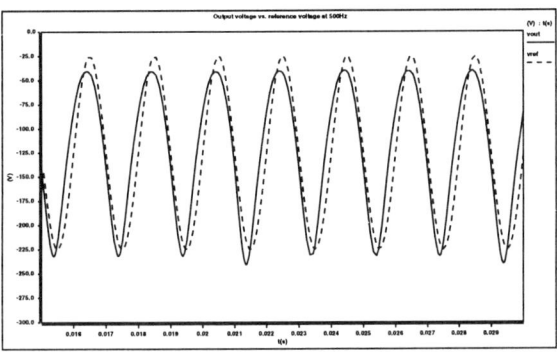

Figure 6: Desired (dashed) and actual (solid) output voltage for a 500 Hz reference signal (averaged model).

former version of the buck-boost topology, namely, the flyback converter.

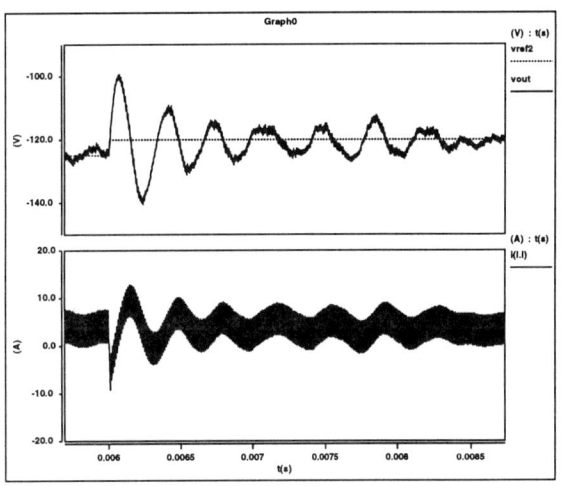

Figure 7: Effect of step change in reference voltage (model with MOSFETs).

7. ACKNOWLEDGMENTS

Bhaskar Krishnamachari's work was supported by the National Science Foundation under the Research Experience for Undergraduates Grant EEC-9619749.

8. REFERENCES

[1] C. Kasuga, T. Nishimura, F. Harashima, and H. Ezuhara, "Characteristics analysis method of multilayer piezoelectric actuator," *Proc. of the IEEE International Conf. in Industrial Electronics, Control, Instrumentation, and Automation (IECON'92)*, San Diego, CA, November 9-13, 1992, vol. I, pp. 336-339.

[2] D. Czarkowski and M. K. Kazimierczuk, "Application of state feedback with integral control to pulse-width modulated push-pull DC-DC convertor," *IEE Proc., Pt. C, Control Theory Appl.*, vol. 141, No. 2, pp. 99-103, March 1994.

[3] Venable, D. H., "The k-factor: a new mathematical tool for stability analysis and synthesis,"*Proceedings of Powercon 10*, San Diego, CA, March 22-24, 1983.

[4] Vorperian, V., "Simplified analysis of PWM converters using averaged model of PWM switch – part I: continuous conduction mode," *IEEE Trans. on Aerospace and Electronic Systems*, vol. 26, no. 3, pp. 490-496, May 1990.

SLIDING MODE CONTROL OF A BUCK CONVERTER FOR AC SIGNAL GENERATION

*Domingo Biel, **Enric Fossas, ***Francesc Guinjoan and *Rafael Ramos

*Dpt. d'Enginyeria Electrónica. E.U.P.V.G. UPC. C/ Víctor Balaguer s/n. 08800- Vilanova i la Geltrú (Barcelona).
**Dpt. de Matemàtica Aplicada i Telemàtica. UPC.Mòdul C3.Campus Nord. C/Gran Capitán s/n. 08034-Barcelona
***Dpt. d'Enginyeria Electrónica. UPC.Mòdul C4.Campus Nord. C/Gran Capitán s/n. 08034-Barcelona

ABSTRACT

This work is devoted to the design of a sliding feedback control scheme for a Buck converter in AC signal generation task where amplitude, frequency and offset can be externally adjusted. A sliding control law over an autonomous switching surface is proposed. As a result, the control scheme is found to be robust with respect to parameters variations and external disturbances. Simulations validating the design and some considerations on the control law implementation are also presented.

1. INTRODUCTION

Sliding-mode control techniques have been proposed as an alternative to PWM control strategies in DC-DC switching regulators since they make these systems very robust to perturbations, namely variations of the input voltage and/or in the load [3],[5]. These techniques have also been applied to the design of high-efficiency inverters, where a switching DC-DC converter is forced to track, by means of an appropriate sliding-mode control action, an external sinusoidal reference [1].

The work here reported proposes an autonomous (time independent) switching surface and a sliding control law for AC signal generation in a Buck converter, where no external reference is needed. The paper is organized as follows: in the first section a sliding surface for the obtaining of an AC output voltage in a Buck converter is proposed; then the sliding domain in the phase plane is deduced in terms of both the converter and the AC signal parameters; subsequently a switching control law is proposed and its robustness analyzed. Finally, simulation results and a EPROM implementation of the control law are also presented.

2. SLIDING SURFACE AND SLIDING DOMAIN

Consider the Buck converter depicted in Figure 1, where an input bridge has been added to ensure the bipolarity of the AC output.

The system can be represented by the following set of differential equations:

$$\frac{d}{dt}\begin{bmatrix}i\\v\end{bmatrix}=\begin{bmatrix}0 & -\frac{1}{L}\\ \frac{1}{C} & -\frac{1}{RC}\end{bmatrix}\cdot\begin{bmatrix}i\\v\end{bmatrix}+\begin{bmatrix}E/L\\0\end{bmatrix}\cdot u \quad (1)$$

where u acts as the control input, taking values in the discrete set $u \in \{-1,1\}$ depending on which switches are active, namely

$u=-1$ means input filter voltage equal to -E

$u=+1$ means input filter voltage equal to +E

As it's well known, the first step to perform in the sliding-mode control technique is the choice of a sliding surface $\sigma(i,v)$ which provides the desired asymptotic behavior when the converter is forced to evolve over it. In this work, the desired output is an AC signal defined by:

$$v(t) = A \cdot sin(\omega t) + B \quad (A>0) \quad (2)$$

where the state variable v satisfies the following differential equation

$$\dot{v}^2 + \omega^2 \cdot (v-B)^2 = \omega^2 \cdot A^2 \quad (3)$$

Therefore, by choosing the sliding surface $\sigma(i,v)$ as:

$$\sigma := \dot{v}^2 + \omega^2 \cdot (v-B)^2 - \omega^2 \cdot A^2 = 0 \quad (4)$$

the desired AC output will be obtained when $\sigma(i,v)=0$. This is the outline of the control policy.

On the other hand, by imposing the invariance condition,

$$\left(\frac{d\sigma}{dt}\right)_{u=u_{eq}} = 0$$

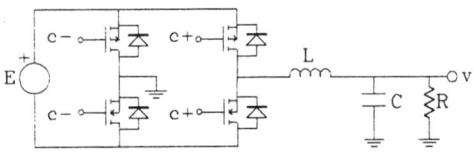

Figure 1. Buck converter with input bridge

$$\frac{d\sigma}{dt} = 2\dot{v}\ddot{v} + 2\omega^2\dot{v}(v-B) = 0 \Rightarrow \begin{cases} \dot{v} = 0 \\ \text{ó} \\ \ddot{v} + \omega^2(v-B) = 0 \end{cases} \quad (5)$$

the equivalent control u_{eq} describing the dynamical behavior of the converter over the sliding surface can be obtained, as it can be seen in the next equation where u_{eq} is expressed as a function of v and $\dot{v}$:

$$E \cdot u_{eq} = \frac{L}{R}\dot{v} + (1 - LC\omega^2) \cdot (v - B) + B \quad (6)$$

Finally, the sliding domain into which the sliding motion is ensured will be given by

$$\min\{u^-, u^+\} < u_{eq} < \max\{u^-, u^+\} \quad (7)$$

In our case:

$$\begin{cases} \dot{v}^2 + \omega^2 \cdot (v-B)^2 = \omega^2 \cdot A^2 \\ -E < \frac{L}{R}\dot{v} + (1 - LC\omega^2)\cdot(v-B) + B < E \end{cases} \quad (8)$$

It can be noted that there exist a sliding regime over any switching surface obtained by modifying the parameters A, B and ω, if this surface is located into the region bounded by the two straighlines defined in (8). Figure 2. shows these boundaries for $\omega = 1/\sqrt{LC}$.

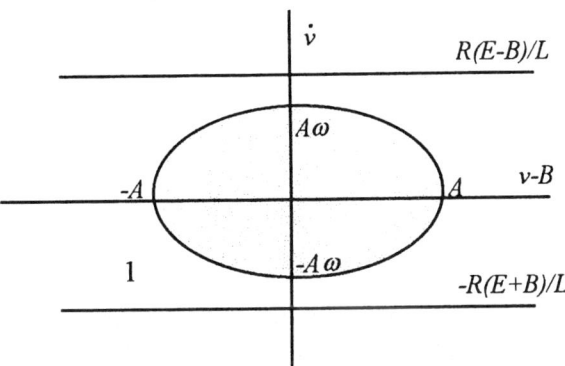

Figure 2. Sliding surface and sliding domain in the phase plane for $\omega = 1/\sqrt{LC}$.

In terms of the output signal parameters (A, B, ω) the previous inequality can be rewritten as:

$$-E < \frac{L}{R}\cdot\omega\cdot A\cos(\omega t) + (1 - LC\omega^2)\cdot A\sin(\omega t) + B < E \quad (9)$$

or, equivalently

$$A < \frac{E - |B|}{LC} \cdot \frac{1}{\sqrt{\frac{\omega^2}{(RC)^2} + \left(\omega^2 - \frac{1}{LC}\right)^2}} \quad (10)$$

which constitutes a design restriction.

If an output signal without offset ($B=0$) is desired, the previous restriction can be reduced to:

$$A < E \cdot \gamma(\omega)$$

where

$$\gamma(\omega) = \frac{1}{LC} \cdot \frac{1}{\sqrt{\frac{\omega^2}{(RC)^2} + \left(\omega^2 - \frac{1}{LC}\right)^2}}$$

is the frequency response of the converter output filter.

Figure 3 shows the plot of A/E versus the output signal frequency, taking the load value R as a parameter. Given a value of R, the sliding regime is ensured for the values of A/E which fall below the corresponding curve. It can also observed from this diagram that a sinusoidal output signal with an amplitude A higher than the input voltage of the converter E, can be obtained.

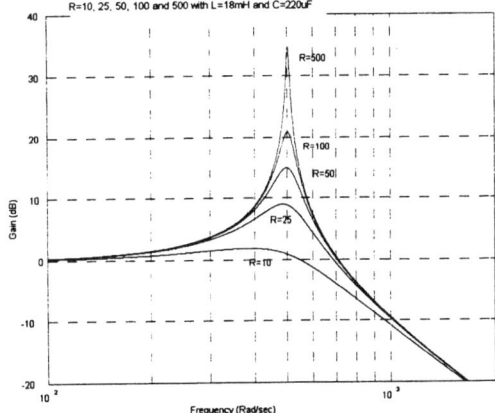

Figure 3. Plot of A/E versus output signal frequency.

When the load has a reactive component it can be proved that the equivalent control is given by the next expression:

$$u_{eq}(s) = \frac{V(s)}{E} \cdot \frac{L\cdot s + (1 - \omega^2 LC)\cdot Z_i(s)}{Z_i(s)}; \quad B = 0 \quad (11)$$

where $Z_i(s)$ is the load impedance, and the sliding domain will be done by $-1 < u_{eq} < 1$.

3. SWITCHING CONTROL LAW

From the sliding-mode control technique, the power converter can reach the sliding surface σ if

$$\frac{d\sigma^2}{dt} < 0$$

Therefore, from (5)

$$\frac{d\sigma^2}{dt} = 2\sigma\frac{d\sigma}{dt} = 2\sigma 2\dot{v}\left(\ddot{v} + \omega^2(v-B)\right) \quad (12)$$

thus,

$$\frac{d\sigma^2}{dt} = 4\sigma\dot{v}\left(\frac{E}{LC}u - \left(\frac{1}{RC}\dot{v} + \left(\frac{1}{LC} - \omega^2\right)v + \omega^2 B\right)\right) \quad (13)$$

Taking into account the expression of $u=u_{eq}$ given by (6) and recalling that $-1<u_{eq}<1$, the switching control law can be defined as:

$$u = \begin{cases} +1 & \text{si } \sigma\cdot\dot{v} < 0 \\ -1 & \text{si } \sigma\cdot\dot{v} > 0 \end{cases} \quad (14)$$

It can be pointed out that both, the control law and the sliding surface are robust to the variations of the converter parameters, thus leading to a robust scheme.

On the other hand, an undesired phenomena can arise near the intersection between the sliding surface and the curve of the equilibrium points of the system, defined by $\dot{v}=0$, when $\sigma>0$: if the system trajectory reach the curve of equilibrium points before than the sliding surface, the system will remain on the classic regulation operation mode ($\dot{v}=0$).

This phenomena can be avoided by modifying the control action in a region of a small width k near the equilibrium curve when $\sigma>0$, as shown in Figure 4. It can be demonstrated that, over this new curve ($\dot{v}=k$) the system slides towards the desired sliding surface σ.

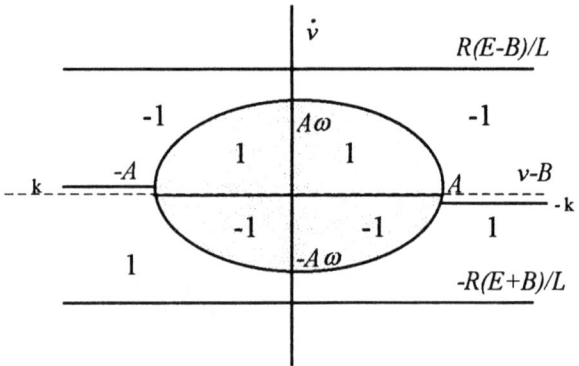

Figure 4. Value of the control variable u in the phase plane.

4. CONTROL LOOP IMPLEMENTATION

Due to the nonlinearity of the sliding surface and the

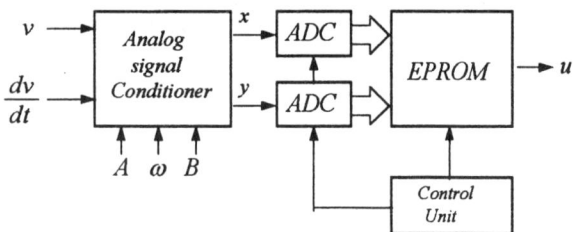

Figure 5. Control loop scheme.

phase plane partition leading to the modified control law described in the former section, the control loop has been performed through a look-up table recorded on a EPROM device.

The control loop scheme is depicted in Figure 5.

First of all, the variable v and $\dot{v}$ of the Buck converter are scaled according to the following expressions

$$y = \frac{1}{A\omega}\cdot\frac{dv}{dt}; \quad x = \frac{(v-B)}{A} \quad (15)$$

Thus enabling a simple design of an externally programmable signal conditioner by means of standard analog circuitry.

$$\sigma(x,y) = x^2 + y^2 - 1 = 0 \quad (16)$$

Through these changes of variable the new phase plane portrait of the control law is modified as shown in Figure 6.

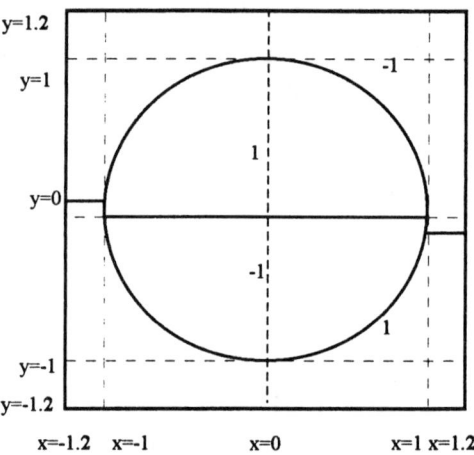

Figure 6. Phase Plane XY.

From the digitized pairs (x,y) the EPROM output gives the value of the control variable u according to Figure 6. It should be noted, that, with this procedure the EPROM needs to be programmed only once, independently of the output parameter values A, B and ω.

5. EXAMPLES

The previous control law has been simulated in a Buck converter with the following parameters: $L= 18$mH, $C= 220$ µF and $E=28$ volt. Figures 7 and 8 show the closed loop system response and the phase plane, respectively for different values of the output signal parameters A, B and ω. The amplitude has been changed from 10 to 30 volts, the offset from -5 to 5 volts and the frequency from 300 to 600 rad/s, simultaneously a pulsating load varying between 40 and 100 Ω at a rate of 150 rad/s is applied. As it can seen these results validate the proposed design. Figures 9 and 10 show the closed loop system response and the phase plane XY when the load has a reactive component L=100 mH.

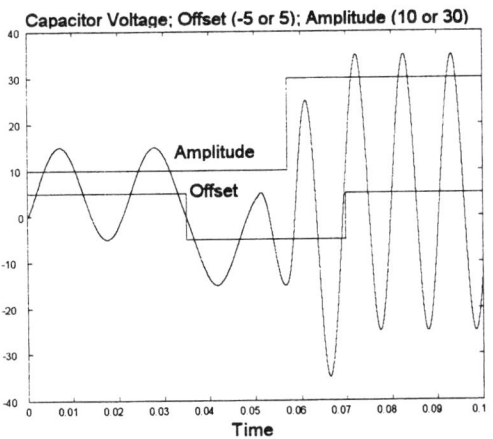

Figure 7. Output signal with different values of R, A, B y ω.

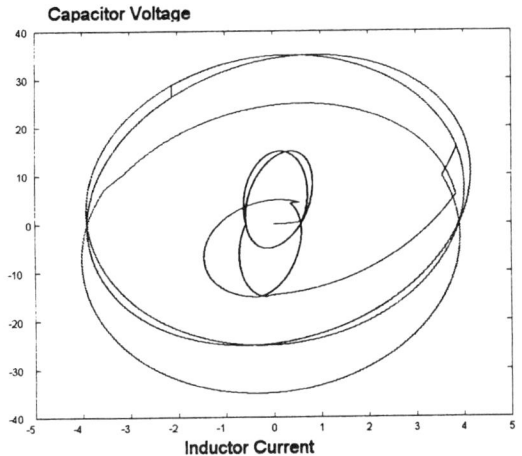

Figure 8. Phase plane with different values of R, A, B y ω.

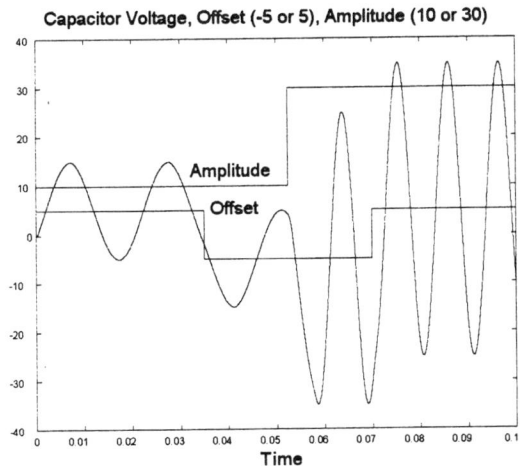

Figure 9. Output signal with different values of R, A, B y ω, when the load has an inductor componet L=100 mH.

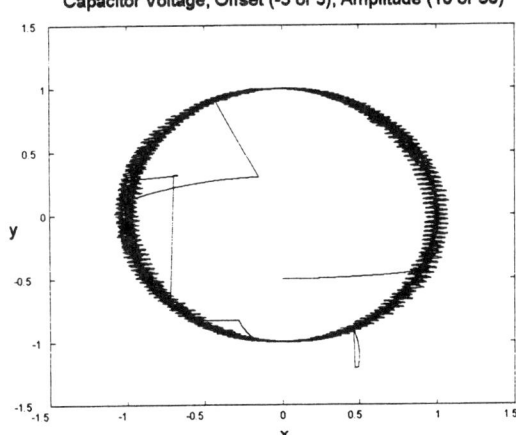

Figure 10. Phase Plane XY, when the load has an inductor component L=100 mH.

6. CONCLUSION

In this paper, a robust sliding-mode control scheme for programmable AC signal generation in a Buck converter has been proposed. The design procedure presented in this work, suggest both a sliding surface and a switching control law leading to the obtaining of a programmable AC output. The analysis has been validated by simulations results.

ACKNOWLEDGMENT

This work has been partially sponsored by the Comisión Interministerial de Ciencia y Tecnología CICYT TAP97-0969-CO3-03,01. The authors would like to thank A. Jiménez Marquez and J. Gavaldá Ferré their suggestions to this work.

7. REFERENCES

[1] Fossas E. and Olm J.M. "Generation of Signals in a Buck Converter with Sliding Mode Control". Proceedings of ISCAS'94. London 1994. pp. 157-160.

[2] Fossas, E and Biel, D. "A Sliding mode approach to robust generation on dc-to-dc converters". Conf. on Dec. and Control. Kobe, Japan. 1996. pp. 4010-4012.

[3] Sira-Ramirez, H. "Sliding motions in bilinear switched networks". IEEE Trans. on Circuits and Systems. V. cas 34 N. August 1987. pp. 919-933.

[4] Utkin, V.I. "Sliding mode and their applications in variable structure systems". Mir. Moscow, 1978.

[5] Venkataramanan R, Sabanovic A. and Cuk S. "Sliding mode control of DC-to-DC converters". Proceedings IECON 1985. pp. 251-258.

NOVEL PWM CONTROL METHOD OF SWITCHED CAPACITOR DC-DC CONVERTER

Tadashi Suetsugu

Department of Electronics, Fukuoka University
8-19-1,Nanakuma, Johnan, Fukuoka 814-0180, Japan.
TEL:+81-92-871-6631 (ext. 6394)
FAX:+81-92-865-6031
e-mail: sue@suetsugu.tl.fukuoka-u.ac.jp

ABSTRACT

Switched capacitor converter becomes important as an inductor-less power supply circuit. The big input current ripples and small control width are the problems of switched capacitor converter. PWM method of switched capacitor converter was introduced already, but the control width was not big. In conventional method, the output voltage adjusted by the ratio of discharging and charging time value of the condenser. In this proposed method, the output voltage is controlled by regulating overall length of the dead time. The simulation result shows that the control width of output voltage increased about 1.6 times compared with the conventional method. Furthermore, the ripple of input current also decreased.

1. INTRODUCTION

Switched capacitor converter (SC converter) is paid its attention to as a powerful candidate of integration and already has been made as hybrid IC [1]. Moreover only switch and the drive circuit part practically used as IC and power supply of digital circuit of low electric power capacity. Recent SC converter study is made about a circuit which establishes voltage conversion rate by series and parallel connection of a capacitor, and it begins from the switched-capacitor transformer application for a DC-DC converter by Oota, Inoue, Ueno in 1983 [2]. Circuit which improved input current ripple [3], AC-DC converter circuit [4], circuit which used an inductor for noise cleaning partly [5] and circuit which can control voltage conversion rate by PWM [6] [?] were proposed afterwards.

In SC converter, it becomes a pending problem to make the control width of voltage conversion rate large and the rejection of a ripple in input current and rejection of noise of output voltage. The ripple in input current was improved large by [3]. Input current ripple can be made very small if switching duty ratio is the neighborhood of 0.5. However, a ripple by input current becomes large when the switching duty ratio changes large from 0.5. The PWM control of SC converter was done by controlling ratio of discharging and charging time of the capacitor. However, duty ratio must have been changed large in order to change voltage conversion rate when connected load is light. As the result, the maximum control width of output voltage is determined by minimum conduction time of switch device. Also, there are also the problems that input current ripple increases because duty ratio slips off large from 0.5.

In this paper, a PWM control method is proposed to get small input current ripple and larger voltage conversion rate by improving the timing of switching used in [1]. This method is the control method to control output voltage by controlling width of the *"dead time"* when all switches become off. The maximum control width of output voltage with this method is determined by the minimum conduction time of switch device which is same as the conventional method. However, voltage variation for equal conduction time is larger than the conventional method by this proposed method. The voltage variations increased in about 1.6 times larger than the conventional method. Further, it will be shown that maximum of magnitude of input current ripple of this proposed method is smaller than that of conventional method.

2. SC CONVERTER CIRCUIT

The well known circuit configuration of SC converter in [1] is shown in Fig. 1. Figure 2 shows switches act by a timing in the conventional PWM method. Switch $S1$ and switch $S4$ are on, and switch $S2$ and $S3$ are off in

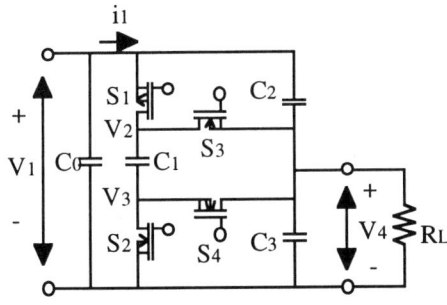

Figure 1: Circuit diagram of a switched-capacitor converter

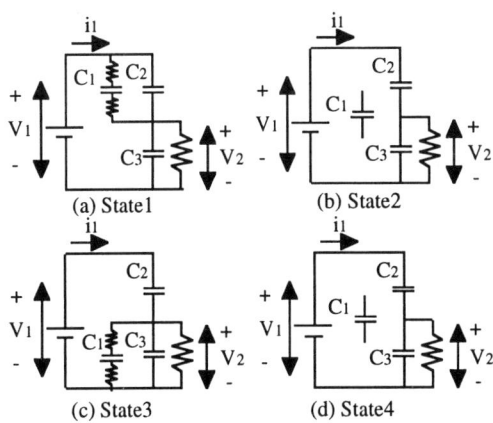

Figure 3: Equivalent circuits for four switching states

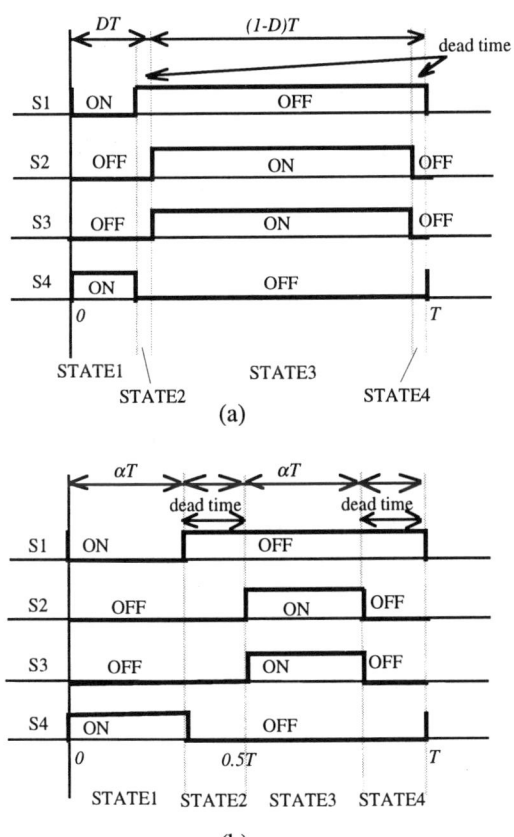

Figure 2: Switch operations (a)Conventional PWM method (b)Proposed PWM method

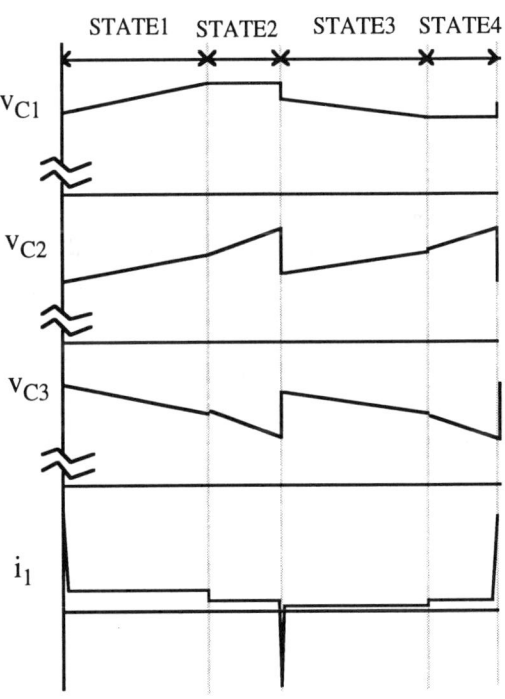

Figure 4: Waveforms of proposed switching pattern

state 1. Equivalent circuit of state 1 becomes Fig. 3 (a). Voltage of capacitor C_3 decreases so that direct current flows to load resistance and then the voltage of C_1 and C_2 rises. State 2 is the dead time when all switches are off. Equivalent circuit in this state becomes Fig. 3 (b). In this state, voltage of capacitor C_3 decreases and C_2 increases. C_1 does not change. In state 3, switch $S2$ and $S3$ are on and switch $S1$ and $S4$ are off. Equivalent circuit of this time becomes Fig. 3 (c). Voltage of C_1 is higher than that of C_3 in the first of state 3. Therefore charge moves suddenly to capacitor C_3 from C_1, and voltage of C_3 rises first and decreases slowly after. Voltage of C_2 suddenly decreases at the start and increases slowly. State 4 is the dead time when all switches are off. Voltage of C_2 increases same as state 2, and voltage of C_3 decreases. Next, in the first of state 1, voltage of C_2 is higher than that of C_1 again. Voltage of capacitor C_2 suddenly decreases and then increases slowly as charge moves. Voltage of C_3 suddenly increases at first and then decreases slowly.

3. NEW PWM CONTROL METHOD

The dead time is very small in the conventional PWM method as shown in Fig. 2 (a). In the proposed PWM method, the output is controlled by the ratio of sum of state 1 and state 3 and sum of state 2 and state 4.

In other words, the output is controlled by the ratio of the conducted and not conducted time of capacitor. Figures 2 (a) and 4 show the switching pattern and waveforms of the proposed PWM method respectively. Period of state 1 and state 3 is defined as αT (where $0 < \alpha < 0.5$), and the circuit is controlled by increasing and decreasing with α. In the conventional methods, output voltage was determined by ratio of charging and discharging time of the capacitor. In reference [6], period of dead time was not constant. The period of state 1 and 2 were varied but the period of state 3 was fixed to 0.5. When the connected load is light, the capacitor has charged fully because there is little discharge current. Therefore, ratio between time of charging and discharge has no effect.

In the proposed PWM method, output voltage is determined by conduction time instead of the ratio of charging and discharging time. Output voltage can be reduced by decreasing discharging time even though there is little discharge current. Therefore larger voltage control width can be kept in light-load.

4. SIMULATION RESULT

A comparison of the conventional, proposed method and the method shown in [6] by a numerical simulation

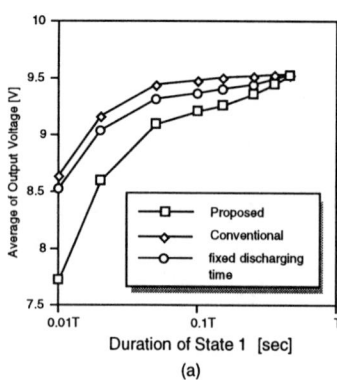

Figure 5: Output voltage corresponding to α when $R_L = 200\Omega$

by shooting method in reference [8] about the circuit is described below. Exactly, the method in [6] uses different circuit topology; an equivalent of a symmetrical pair of circuit in Fig.1. In this paper, the comparison was done with the circuit in Fig.1. Used circuit parameters are $V_1 = 20V$, $C_1 = 20.8\mu F$, $C_2 = 8.8\mu F$, $C_3 = 31.0\mu F$ and switching frequency is $1kHz$. Switch on and off resistance was 0.5Ω and $1M\Omega$ for all switches respectively. Output voltage of the conventional and proposed method corresponding to duration of state 1 calculated at a circuit of 200Ω load resistance are shown in Fig. 5. The the duration of state 1 for the conventional and proposed method corresponds to DT and αT respectively. The purpose of this comparison is finding maximum limit of voltage variations which is determined by minimum conduction time of switch device. In this circuit, output voltage is $10V$ when the dead time is zero. Output voltage variations are $2.28V$ for the proposed and $1.37V$ for the conventional method when $\alpha = 0.02$. This means that there is about 1.6 times larger voltage variation for the proposed method than the conventional. It is also larger than the method in [6] (noted as 'fixed discharging time' in Fig.5-8). It has similar rate for other α values. The maximum value of input current corresponding to the duration of state 1 is shown in Fig. 6. Current shown here is an absolute value of i_1 in Fig. 1. The current in the proposed method decreases compared to other methods. Figure 7 shows the output voltage corresponding to the duration of state 1 when load is 40Ω. Maximum of input current for this case is shown in Fig. 8. Voltage variation and current ripples are also improved to other methods in this case.

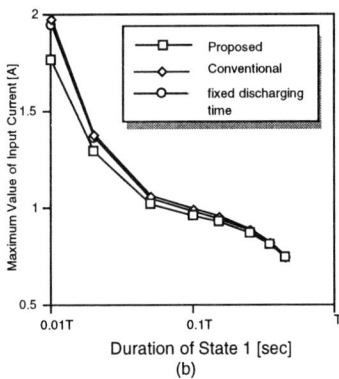

Figure 6: Input current ripple corresponding to α when $R_L = 200\Omega$

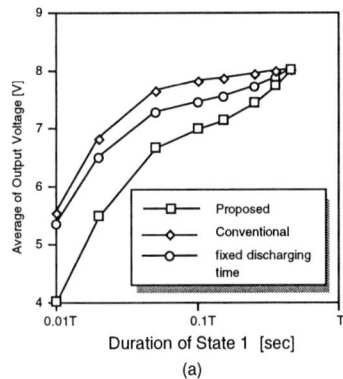

Figure 7: Output voltage corresponding to α when $R_L = 40\Omega$

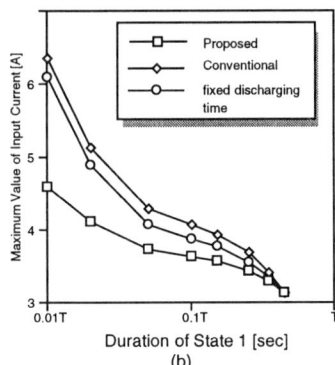

Figure 8: Input current ripple corresponding to α when $R_L = 40\Omega$

5. CONCLUSION

The control width of output voltage of SC converter was improved. It is because time constant for the capacitor is small while the dead time and this PWM method uses long dead time. The input current ripple was also improved. It is because both charging and discharging time are short in this method. Therefore, voltage variations of the capacitor is smaller than that of other methods.

6. REFERENCES

[1] F. Ueno, T. Inoue, I. Oota and T. Umeno, "Analysis of switched-capacitor DC-DC converter and its hybrids –Small volume and high power–," Tech. Report IEICE, PE98-52, 1989.(Japanese),pp.43-49.

[2] I Oota, T. Inoue and F. Ueno, "A realization of low-power supplies using switched-capacitor transformers and its analysis," Trans. IEICE Vol.J 66-C, No.8, Aug. 1983, pp.576-583.

[3] T. Umeno, T. Takahashi, I. Oota, F. Ueno and T. Inoue, "New switched capacitor DC-DC converter with low input current ripple and its hybridization," Proc. 33rd IEEE Midwest Symp. Circuit, Syst. Aug.1990, pp.1091-1094.

[4] F. Ueno, T. Inoue, and I. Oota, "Realization of a switched-capacitor AC-DC converter using a new phase controller," Proc. IEEE Int. Symp. Circuits, Syst., May 1992, pp.1903-1906.

[5] K. Kuwabara, E. Miyachika and M. Kohsaka, "New switching regulators derived from switched-capacitor DC-DC converters," Tech. Report IEICE, PE88-5, 1988, pp.31-37.

[6] S. Cheong, H. Chung and A. Ioinovici, "Inductorless DC-to-DC converter with high power density," IEEE Trans. Ind. Elec., Vol.41, No.2 Apr. 1994, pp.208-215.

[7] K.D.T. Ngo and R. Webster, "Steady-state analysis & design of a switched-capacitor DC-DC converter," IEEE PESC'92 Record, pp.378-385, 1992.

[8] K. Kundert, J. White and A. Sangiovanni-Vincentelli, "Steady-State Methods for Simulating Analog and Microwave Circuits," Kluwer Academic Publishers, 1990.

TRUE-WORST-CASE EVALUATION IN CIRCUIT TOLERANCE & SENSITIVITY ANALYSIS USING GENETIC ALGORITHMS AND AFFINE MATHEMATICS

L.Egiziano, N.Femia, G.Spagnuolo and G.Vocca

Dipartimento di Ingegneria dell'Informazione ed Ingegneria Elettrica - D.I.I.I.E.
Università di Salerno - I 84084 Fisciano (SA) - ITALY
Tel. (089) 964279 - Fax. (089) 964218 - e-mail: femia@diiie.unisa.it

ABSTRACT

New methods for circuits Tolerance and Sensitivity Analysis (TSA) are presented in this paper. Genetic Algorithms (GA) and Affine Arithmetic (AA) techniques have been adopted to find respectively the inner and the outer solution in True Worst-Case (TWC) problems for circuits where the strong uncertainty of parameters yields large changes of performances. It is shown that GA and AA allow a sharp determination of the TWC in TSA problems, even of great complexity. Some circuit examples are presented to highlight the accuracy and the efficiency of the new methods, whose application is best suited for the CAD of electronic circuits submitted to performances and regulations constraints to be fulfilled in presence of large parameters uncertainties.

1. INTRODUCTION

The increasing demand of high-performances low-pollution electronic circuits imposes the adoption of new CAD tools featuring efficient parameters uncertainties handling capabilities. Indeed, high performances involve tight constraints on circuit parameters, especially those ones connected to operating frequency and parasitic phenomena. Besides, there are intrinsic large uncertainties in the models of electric and electronic circuits deriving from the critical conditions connected to the miniaturization and to high frequency operation. For example, self and mutual parasitic inductances and capacitances in PCB-SMPSs are strongly dependent on the hardware layout arrangement [1] while in microelectronics there are even parasitic parameters located potentially everywhere. *Tolerance* and *Sensitivity Analysis* (TSA) are topics deeply discussed in circuit analysis related literature [2]. A great variety of problems can be framed into the TSA context. In general TSA means to solve an *evaluation problem*, that can be framed as follows. Let a model of a system be given:

$$(1) \quad \begin{aligned} f(\partial \underline{x}, \underline{x}, \underline{c}, \xi) &= 0 \\ \underline{x}(\xi_o) &= \underline{x}_o \end{aligned}$$

where ξ is the independent variable vector (e.g. time t, space position vector r), $\underline{x}$ is the vector of the unknown variables, $\underline{c}$ is a vector of coefficients, each being in general a function $c_k = c_k(p_1,..,p_N)$ of circuit's parameters $p_1...p_N$, and ∂ indicates differential operations. Let $\underline{x}(\xi)$ be the solution of equations (1). The characteristic of the circuit we are interested in can in general be expressed by means of a quantity related to $\underline{x}(\xi)$: for example the value $\underline{x}(\xi)$ at a given point ξ, the whole solution $\underline{x}(\xi)$ or the maximum value $\underline{x}_{max}(\xi)$ or the point $\xi^* \ni \underline{x}(\xi^*)=0$ over a given interval $[\xi_L, \xi_U]$. All these quantities depend on the values of circuit's parameters $p_1...p_N$, which can be affected by uncertainty. We can define then as *evaluation function (e.f.)* the function expressing the relation between the quantity we are interested in and the system's parameters. The objective of TSA lies in studying the range of the e.f. over given intervals of circuit's parameters. Basically, we can distinguish *Small-Change TSA* (SC-TSA) from *Large-Change TSA* (LC-TSA) problems, and *Non-Iterative TSA* (NI-TSA) from *Iterative TSA* (I-TSA) problems.

The first classification concerns the entity of parameters uncertainty. SC-TSA is a classic topic covered in literature [2]: in direct tolerance analysis a nominal value is known for circuit parameters with a relatively small tolerance and the effect on the e.f. is sought; in inverse tolerance analysis for a given e.f. tolerance the corresponding tolerance on the parameters is sought; in sensitivity analysis the variational effects of parameters on the e.f. are sought. LC-TSA is only marginally treated [3][4] in literaure. It concerns all those situations where parameters can exhibit large range variations (up to 100%): electro-thermal stress, failure and EMC related analysis problems can fall in this context.

The second classification is focused on the type of computation involved by the evaluation function. In some cases the e.f. is available in explicit form and the dependence of the coefficients c_k on the circuit's parameters is expressed through simple analytical functions which involve affordable computations: for example, let the e.f. be the value of the capacitor voltage during the step transient of a series RC linear circuit at any given instant. In this case the evaluation of the e.f. with respect to the parameters R and C is *non iterative* as the effect of parameters uncertainties can be detected at any instant independently from the others. In case of more complex linear circuits, or non linear circuits, the step transient solution can, or must, be put in sample-data discrete-time domain form, which involves *iterative* computations, so that the value at any instant depends on previous values.

In general, even if the e.f. is available in explicit form the coefficients can be involved non linear functions of the circuit's parameters either because of the system non-linearity or because of the system complexity, so that they can be evaluated in numerical way only. The classical methods for TSA, like the basic [5] or improved [6] Monte-Carlo ones, or the incremental-, adjoint- and variational-equations [2][3] ones, suffer from severe limitations in SC-TSA and NI-TSA problems, since they allow to study the dependence upon one parameter at time only, or the small-changes condition is required, or they may involve huge computation in presence of numerous uncertain parameters. In [4] it is shown that the large-changes analysis too is possible for linear circuits using ad adjoint net. The aforementioned methods are inadequate, for the safety of results and for computation times, in LC-TSA and I-TSA problems. Then, new numerical methods are necessary to carry out efficient and reliable TSA in presence of large and diffused parameters uncertainties and strong non linear evaluation functions, thus making possible the optimization of circuits under design through the forecasting of performance fluctuations in the most critical conditions. In the next sections it is shown how Affine Arithmetic and Genetic Algorithms can be fruitfully employed in this context.

2. AFFINE ARITHMETIC & GENETIC ALGORITHMS.

In TWC-TSA problems we are interested in the evaluation of the true range of an e.f. over given intervals of variations of its parameters. Excepted few lucky easy situations, these problems must be solved numerically. MC methods allow the computation of an underestimation of the true range (inner solution) only, whereas Interval Mathematics (IM) [7]-[10] provides always an overestimation (outer solution). The reliability of such methods is good in SC-TSA and NI-TSA problems, while their performances strongly decline in TWC LC-TSA and I-TSA problems. In particular, besides the mentioned limits of MC methods, the main open question for IM methods lies in the excessive overestimation of the true range. This is due to the so-called *width-dependency* pathology affecting long chains IM computations [8], mostly encountered when iterative models are adopted, as in *transient tolerance analysis* [7][11][12]. Inner and outer solutions in TWC problems must be as close as possible in order to get a good estimation of the true range. In this paper new methods are proposed to overcome the limits of MC and IM based techniques in hard TWC TSA problems. Affine Arithmetic (AA) and Genetic Algorithms (GA) have been applied to this purpose. AA is a variant of Interval Arithmetic. It is best suited for the calculation of the outer solution, especially in LC-TSA and I-TSA problems. Besides, GA [15] enable efficient calculation of the inner solution. The main result obtained with the joined AA-GA approach, highlighted in this paper, lies in the possibility of minimizing the distance between the inner and the outer solutions in hard TWC TSA problems. A brief description of the main elements characterizing the AA and the GA methods is given next.

Affine Arithmetic. Affine Arithmetic has been introduced in the recent years [13] with the aim of making numerical computations involving operations on intervals more reliable with respect to Interval Arithmetic. This result has been attained using for each quantity affected by uncertainty the following polynomial first degree form (affine form):

$$(2) \quad x = x_o + \sum_{k=1}^{n} x_k \varepsilon_k$$

where $\varepsilon_k = [-1,1]$ $\forall\ k=1,..,n$ are dummy interval variables acting as uncertainty, or noise, sources and x_k are the related weights. Representing the uncertain coefficients of the circuit model in affine form enables to account for correlations among them using the same noise variables. This greatly reduces the width of the results of interval computations as the same noise sources can appear in several, or even all, the uncertain coefficients, being all related to the same uncertain circuit parameters. This makes computations much more reliable than those carried out by IM in cases of long computation chains as the propagation of the overestimation error is strongly reduced. The main drawback of AA is the greater amount of computations required with respect to IM. This makes mandatory a clever setting of some degrees of freedom characterizing the AA-based simulations. This aspect is put in evidence in section 3.

Genetic Algorithms. A Genetic Algorithm (GA) is a general search, optimization and learning method based on the Darwinian principles of biological evolution. An *evolutionary pressure* is driven by ad hoc "genetic operators" acting over a population of individuals each one representing a possible solution for the given problem. The main advantage of GA lies in the intrinsic good balance between the need of an exhaustive visit of the possible solutions space and the will of following an eventually found path of possible good solutions. A typical GA search starts usually with a randomly generated population of "individuals" or "chromosomes" typically containing a coded set of parameters. For each one of them a performance index, the *fitness*, is computed based on an *objective function* that we want to optimize. The higher the fitness the higher the individual surviving probability. Evolutionary pressure is given by a *crossover* operator, which combines the parents genes to produce offspring while a *mutation* operator ensures a wide visit of the solutions space. In TSA problems an individual is simply a set of values of parameters, each one taken within its own uncertainty interval. The fitness is calculated considering the number of points of the solution corresponding to that individual where the result (inner solution) is better than the envelope of the solutions obtained up the current generation.

3. EXAMPLES.

Two examples of TWC TSA problems are presented in this section. The first example concerns the step transient time domain response of the RLC circuit of fig.1 in two different situations: (a): ±10% uncertainty on all R, L and C parameters; (b): ±50% uncertainty on the capacitance C and ±5% uncertainty for the remaining parameters. Let the e.f. be the capacitor voltage over a 2ms time interval. The solution of the state-space model is:

$$(3) \quad x(t) = e^{At} x_o + A^{-1}(e^{At} - I) b\, u(t)$$

$$A = \begin{bmatrix} -(R_L + R \cdot R_C /(R + R_C))/L & -R/L(R+R_C) \\ R/C(R+R_C) & -1/C(R+R_C) \end{bmatrix} \quad b = \begin{bmatrix} 1/L \\ 0 \end{bmatrix}$$

The form (3), however, requires high order truncated series to get good approximations for matrix exponential functions. Consequently it is more convenient to adopt the sample-data discrete-time domain form:

$$(4) \quad x(k+1) = e^{A\Delta t} x(k) + A^{-1}(e^{A\Delta t} - I) b\, u(k)$$

Fig.1. $R_L = 0.5\Omega$, $L = 1mH$, $R_C = 0.5\Omega$, $C = 10\mu F$, $R = 20\Omega$, $u(t) = 10V\ 1(t)$.

From the expression (4) it is clear the strong non linear dependence of the e.f. with respect to the circuit parameters. This, in addition to the iterative kind of computation, makes the IM approach not reliable to get the outer solution. Resorting to MC-based computation for the inner solution would not necessarily give "good" results even with numerous trials. Approaching the problem with the joined AA-GA method gives the results shown in fig.2. The continuous line curves give the true range of the e.f. for case (a): the inner (GA) and the outer (AA) solutions are coincident. The inner solution has been obtained after 60 generations with GA. The GA adopted involves non-standard operators for both mutation and crossover. Due to the nature of the problem, a real-chromosomes representation of each individual has been adopted. This, compared to the bit-string representation, ensures higher precision and a lighter structure of the algorithm. The BLX-α crossover operator [14] has been adopted: this ensures an effective recombination of the information stored in the parents preserving the genomes structure. At the beginning of the evolution, privilege is given to the visit of the space of the solutions. Afterwards, when individuals become more similar, everyone driving to the optimal solution, the BLX-α allows to preserve the good schemata in each of them. To emphasise this useful behaviour, a *non-uniform* mutation operator [15] is chosen: it allows wider changes on the non-specialised "young" individuals while producing small variations on chromosomes present in aged generations. This behaviour allows also to produce a kind of "fine-tuning" around a good even non-optimal solution.

The outer solution has been obtained with AA using 150 time samples on the 2ms time interval. The sampling rate is dictated by the specific characteristics of the transient considered. In fig.3 it is shown how the results given by the AA for the outer solution are strongly influenced by the sampling rate and by the partitioning of parameters uncertainty intervals. This last operation consists in dividing the uncertainty interval of each parameter in a number of subintervals and then making the union of the range of the e.f. computed over the whole possible combinations of subintervals. Partitioning is mandatory when the e.f. is strongly non linear and is a valuable tool for reducing the overestimation. However it may cause a significant increase of computations.

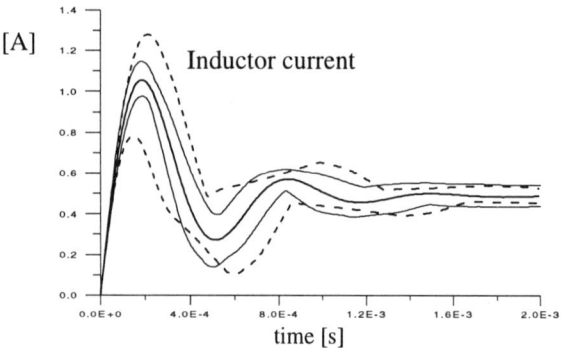

Fig.2. — = AA&GA-computed TWC with ±10% uncertainty upon all parameters; --- = AA&GA-computed TWC with ±50% uncertainty upon the capacitance C and ±5% upon all the other parameters; middle line = nominal solution.

In some cases, like the one we are examining, choosing a suitable sampling rate can be more convenient, as indicated by the plots of fig.3. In fact using 150 samples without partitioning (curves 2) gives a much sharper estimation than using 65 samples with uncertainty intervals partitioned in 3 segments (curves 3).

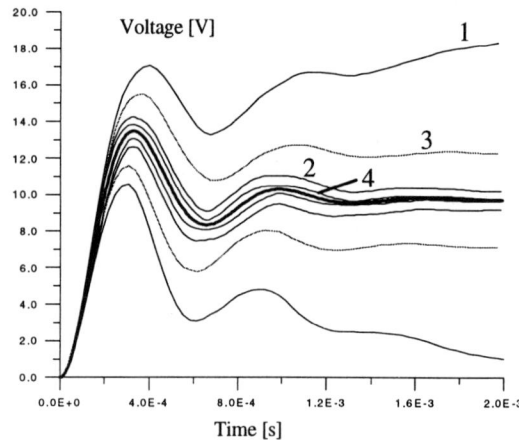

Fig.3. Capacitor voltage calculated by AA (IP = interval partitioning; SR = sampling rate): (1) no IP + low SR; (2) no IP + high SR; (3) IP + low SR; (4) IP + high SR. Nominal response plotted in the middle.

The plot concerning the case (b) is also included in fig.2. The dashed lines give also in this case the TWC solution as the inner GA solution and the outer AA solution are coincident.The settings are the same as for the case (a). It is worth noting that, unlike case (a), in case (b) the variation of the capacitance is so wide that the natural frequency of the circuit is sensibly modified. This explains why the upper and the lower bound of the solution do not have the same shape. Previous time-domain simulations of RLC networks with uncertain parameters found in literature are limited to very soft problems only [7][8]. In order to highlight the potentialities of the AA-GA approach in sensitivity problems, the peak transient value of the inductor current has been evaluated for different levels of parameters uncertainties around nominal values. The results are shown in fig.4.

The second example concerns the buck converter of fig.5 which, for some values of parameters exhibits sub-harmonics and chaos. Fig.6 shows the time-domain solution of the circuit in presence of ±5 parameters uncertainty around nominal values.

Among the uncertain parameters the voltage loop gain K is included. The plots indicates that the periodicity of the solution is lost. This is even more clearly highlighted in fig.7, where the samples values only of the inductor current at the instants multiple of the switching period are reported. In this example the impact of computation chains is even heavier due to the switched nature of the circuit and to the presence of a feedback loop.

The major benefits offered by the new approaches can be exploited when an objective function is known in explicit analytical form, but it is strongly non linear with respect to numerous and large-uncertain parameters, or when the objective function is not available at all so that a system of non linear equations must be solved by iterations.

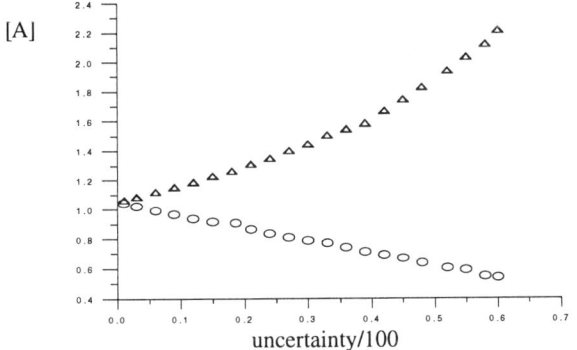

Fig.4. Sensitivity of step-transient peak inductor current for the circuit of fig.1 with respect to parameters variations: Δ = upper bound, o = lower bound.

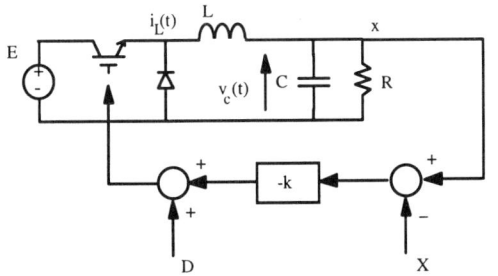

Fig.5. Voltage-loop compensated buck converter: E=33V, T=333.33μs, X=25V, L=208μH, C=222μF, R=12.5Ω.

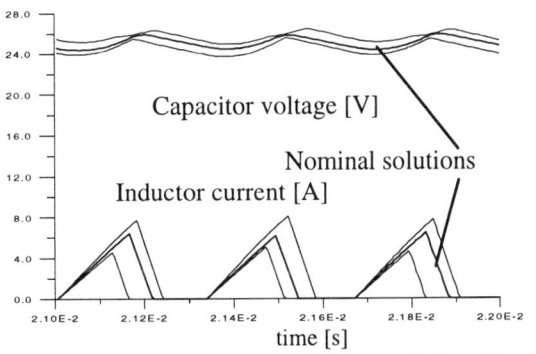

Fig.6. Variation of the steady-state solution for the buck converter of fig.5 in presence of parameters uncertainties.

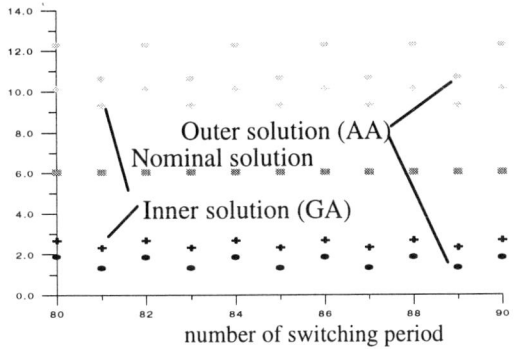

Fig.7. Variation of the inductor current samples due to a ±15% variation of parameters.

CONCLUSIONS

In this paper the usefulness of Affine Arithmetic and Genetic Algorithms in hard circuit TWC-TSA problems has been proved. The examples proposed show that, both in apparently simple and evidently complex cases, the innovative approach ensures safe TWC computations, even in presence of strongly non linear evaluation functions depending on numerous and largely uncertain parameters.

ACKNOWLEDGEMENTS

This work has been supported by M.U.R.S.T. 40% - 60% funds.

REFERENCES

[1] N.Dai, F.C.Y.Lee:" Characterization and Analysis of Parasitic Parameters and Their Effects in Power Electronics Circuit", Proc. of 1996 IEEE PESC, Baveno, Vol.II, pp.1370-1375.
[2] L.O.Chua, P.Lin, Computer-Aided Analysis of Electronic Circuits, Prentice-Hall, 1975.
[3] J.Ogrodzki, Circuit Simulation Methods and Algorithms, CRC Press, Boca Raton, 1994.
[4] D.Aeyels, J.L.Willems:" The Large-Change Sensitivity Network ", Int. Journal of Circuit Theory and Applications, Vol.18, 1-9 (1990).
[5] I.M.Sobol, A Promer for the Monte Carlo Method, CRC Press, Boca Raton, 1994.
[6] T.Koskinen, P.Y.K.Cheung:" Hierarchical Tolerance Analysis Using Statistical Behavioral Models, IEEE Trans on CAD of Integr. Circ. and Syst., Vol.15, No.5, 1996, pp.506-516.
[7] L.V.Kolev, Interval Methods for Circuits Analysis, World Scientific Publishing, Singapore, 1993.
[8] A.N.Michel, H.F.Sun:" Analysis of Systems Subject to Parameters Uncertainties: Application of Interval Analysis", Circuits Systems Signal Process, Vol.9, No.3, 1990, pp.319-341.
[9] A.Cirillo, N.Femia, G.Spagnuolo:" An Interval Mathematics Approach to Tolerance Analysis of Switching Converters", Proc. of IEEE PESC'96, Vol.II, pp.1349-1355.
[10] N.Femia, G.Spagnuolo:" Identification of DC-DC Switching Converters Characteristics for Control Systems Design Using Interval Mathematics", Proc. of 1996 IEEE Workshop on Computers in Power Electronics.
[11] C.L.Harkness, D.P.Lopresti: "Interval Methods for Modeling Uncertainty in RC Timing Analysis", IEEE Trans. on C.A.D., Vol.11, No.11, pp.1388-1401, 1992.
[12] W.Tian, X.-T.Ling, R.-W.Liu:" Novel Methods for Circuit Worst-Case Tolerance Analysis", IEEE Trans on Circuits and Systems-I Fund. Theory and Appl., Vol.43, No.4, April 1996, pp.272-278.
[13] J.L.D.Comba, J.Stolfi:" Affine Arithmetic and its Applications to Computer Graphics", Proc. of VI SIBGRAPI, 1993, pp.9-18.
[14] Eshelmann, L.J., Schaffer, J.D.: *Real-coded genetic algorithms and interval-schemata*, 2nd Workshop on the Foundations of Genetic Algorithms and Classifier Systems, Morgan Kauffman Publ., San Mateo, CA, 1993
[15] Davis, L. (Editor): *Handbook of Genetic Algorithms*, Van Nostrand Reinhold, New York, 1991

ANALYTICAL SOLUTION TO HARMONIC CHARACTERISTICS OF PWM H-BRIDGE CONVERTERS WITH DEAD TIME

C.M. Wu, W.H. Lau and H. Chung
Department of Electronic Engineering
City University of Hong Kong
Tat Chee Ave., Kowloon, Hong Kong

ABSTRACT

This paper presents an analytical method to calculate the output harmonic characteristics of PWM H-bridge converters with dead time. A three-dimensional (3-D) model derived for generating the pulse-width-modulated (PWM) pulse train is used to consider the effects of dead time. The harmonic characteristics are obtained by applying the double Fourier series to the output PWM waveform. The results obtained by this new technique are verified with simulations using PSpice.

1. Introduction

Inserting a dead time [1] to the PWM H-bridge converters is a common practice to avoid cross-conduction current through the leg of the converter during the crossover between the upper and lower switches. The dead time causes the output waveform to deviate from an ideal PWM waveform and that produces undesirable spectral distortions in the output [2,3]. Many methods have been developed to evaluate these distortions. Some approaches [4,5] use sophisticated numerical techniques to find the PWM pulse position and width. The dead time pulse train can then be computed and its harmonic spectrum is obtained by applying the Fourier analysis. However, these methods cannot show the detailed composition of the harmonic structure.

Based on a 3-D model developed in [8], the analysis of the harmonic characteristics of an ideal H-bridge converter has been given in [6]-[7]. A general expression for the harmonic components of the PWM pulse train has been derived and the information of the position and width of each PWM pulse are implicitly described by the Bessel functions. This paper presents an analytical method for calculating the harmonic characteristics of the output voltage of an H-bridge converter with the consideration of dead time. The 3-D model described in [8] is modified to incorporate the dead time for generating the PWM signal. Double Fourier analysis is then applied to the 3-D model to obtain an elegant function to describe the harmonic characteristics. The detailed derivation is illustrated with the single-sided naturally-sampled PWM (SSNS-PWM) waveform.

2. 3-D Model of the PWM pulse Train

The basic structure of a PWM H-bridge converter is shown in Figure 1. Without considering the dead time, the generation of the trailing-edge SSNS-PWM waveform for a modulation signal $-\sin \omega_v t$ is shown in Figure 2. The time axis can also be represented by two different angular abscissas, $\theta = \omega_c t$ and $\phi = \omega_v t$, where ω_c and ω_v are the angular frequencies of the carrier and modulation signal respectively. The ratio between ω_c and ω_v is denoted by M_R, i.e., $M_R = \omega_c / \omega_v = \theta / \phi$. The PWM pulse width $\Omega(\phi)$ projected on the θ-axis for a particular cycle of the carrier signal can be expressed as

$$\Omega(\phi) = \pi(1 - M \sin \phi) \quad (1)$$

where M is the modulation index. Within the region $0 \leq \theta \leq 2\pi$ and $0 \leq \phi \leq 2\pi$, a 3-D function $F(\theta,\phi)$ can be defined as

$$F(\theta,\phi) = \begin{cases} V_d & 0 \leq \theta \leq \Omega(\phi) \\ 0 & \Omega(\phi) < \theta \leq 2\pi \end{cases} \quad (2)$$

The PWM waveform function $g(t)$ representing the output voltage v_a of one leg is defined by $F(\theta,\phi)$ and a plane P governed by M_R, i.e.

$$v_a = g(t) = F(\theta,\phi) = F(\omega_c t, \omega_v t) \quad (3)$$

Figure 3(a) shows the 3-D model and how the PWM signal is generated. The projection of the intersections between the 3-D function $F(\theta,\phi)$ and plane P on the $F(\theta,\phi)$-θ plane forms the PWM signal. The generation process can also be illustrated in a 2-D plane as shown in Figure 3(b). The curves represented by (1) in each carrier cycle and the line with slope $1/M_R$ can be plotted on the $\theta - \phi$ plane as shown in Figure 3(b). The projection of the intersection points between these curves on the θ axis clearly represents the PWM pulse train. In fact, this 2-D representation is equivalent to present Figure 2 in angular domains θ and ϕ. For simplicity reason, this 2-D representation will be used to illustrate the derivation of the harmonic characteristics of the output PWM signal.

For the following analysis, we assume that (1) the switches are ideal; (2) the load is inductive and i_a lags V_a with φ ($0 < \varphi < \pi/2$); and (3) the dead time is inserted prior to the rising edges. With the dead time, the output waveform will deviate from the normal one. For simplicity reason, we only consider one leg of the PWM H-bridge converter. During the dead time, both switches in a leg are off. For inductive load, the output voltage of the H-bridge depends on the polarity (or direction) of the output current which causes either the upper or lower diodes to conduct as shown in Figure 1. When ϕ is in the range from φ to $\pi+\varphi$ and $i_a < 0$, a gain in voltage occurs at the pulse falling edge. The region corresponding to the falling edges is modified as shown in Figure 4. Curve AB is replaced by curve $AA'B'B$ and the slope of the line AA' is $1/M_R$. Curve $A'B'$ is defined as

$$\Omega'(\phi) = \Omega(\phi - \Delta_\phi) + \Delta_\theta \quad (4)$$

where Δ_θ denotes the dead time and $\Delta_\phi = \Delta_\theta / M_R$ (Δ_θ and Δ_ϕ are dead time in angular domain in relation to ω_c and ω_v respectively). The projection from the intersection points, which are formed when plane P cuts the curves AB and $A'B'$, on the axis θ equals to the dead time Δ_θ. This indicates that an additional voltage is gained for negative output current. Furthermore, the signal frequency is usually much less than the carrier frequency and it can reasonably be assumed that $M_R \gg 1$. Thus the line AA' can be considered in parallel with the axis θ and $F(\theta,\phi)$ is then given in (5) within the region $\varphi \leq \phi < \pi + \varphi$.

$$F(\theta,\phi) = F_-(\theta,\phi) = \begin{cases} V_d & 0 \leq \theta \leq \Omega'(\phi) \\ 0 & \Omega'(\phi) < \theta \leq 2\pi \end{cases} \quad (5)$$

When ϕ is in the range from $\pi+\varphi$ to $2\pi+\varphi$ and $i_a > 0$, a loss of voltage occurs at the pulse rising edges. The region corresponding to the rising edges is modified as shown in Figure 4. The line CD is replaced by the line $CC'D'D$ and the slope of the line CC' is $1/M_R$. With the same assumption that $M_R \gg 1$, $F(\theta,\phi)$ is given in (6) within the region $\pi + \varphi \leq \phi \leq 2\pi + \varphi$.

$$F(\theta,\phi) = F_+(\theta,\phi) = \begin{cases} 0 & 0 \leq \theta \leq \Delta_\theta \\ V_d & \Delta_\theta < \theta \leq \Omega(\phi) \\ 0 & \Omega(\phi) < \theta \leq 2\pi \end{cases} \quad (6)$$

The effects caused to the SSNS-PWM pulse train due to the dead time is shown in Figure 4.

3. The harmonic characteristics

To calculate the harmonic characteristics of the distorted PWM waveform, double Fourier series [7] approach is used. $F(\theta,\phi)$ can be expressed as:

$$F(\theta,\phi) = \frac{1}{2}A_{00} + \sum_{n=1}^{\infty}[A_{0n}\cos(n\phi) + B_{0n}\sin(n\phi)]$$

$$+ \sum_{m=1}^{\infty}[A_{m0}\cos(m\theta) + B_{m0}\sin(m\theta)] \quad (7)$$

$$+ \sum_{m=1}^{\infty}\sum_{n=\pm 1}^{\pm\infty}[A_{mn}\cos(m\theta + n\phi) + B_{mn}\sin(m\theta + n\phi)]$$

where

$$A_{mn} = \frac{1}{2\pi^2}\int_0^{2\pi}\int_0^{2\pi} F(\theta,\phi)\cos(m\theta + n\phi)d\theta d\phi$$

$$B_{mn} = \frac{1}{2\pi^2}\int_0^{2\pi}\int_0^{2\pi} F(\theta,\phi)\sin(m\theta + n\phi)d\theta d\phi \quad (8)$$

The time function of PWM pulse train can be obtained by replacing θ and ϕ in (7) with $\omega_c t$ and $\omega_v t$ respectively. Substituting (1)-(6) to (8) and after simplification, the signal, carrier and cross-modulated harmonics described by A_{mn} and B_{mn} can be written in terms of an ideal part plus a correction part, i.e., $A = A_{ideal} + A_{cor}$ and $B = B_{ideal} + B_{cor}$, and they are given as follows:

Ideal parts:

$$A_{00} = V_d, \quad A_{0n} = A_{m0} = A_{mn} = 0,$$
$$B_{01} = -MV_d/2, \quad B = 0(n \neq 1),$$
$$B_{m0} = \frac{(-1)^{m+1}V_d}{\pi m}J_0(\pi mM) + \frac{V_d}{m\pi}, \quad (9)$$
$$B_{mn} = \frac{(-1)^{m+1}V_d}{\pi m}J_n(\pi mM)$$

where $J_n(x)$ is the n order Bessel functions of the first kind.

Correction parts:

$$A_{00(cor)} = \frac{MV_d}{\pi}(1 - \cos\Delta_\phi)\cos\varphi - \frac{MV_d}{\pi}\sin\Delta_\phi \sin\varphi$$

$$A_{0n(cor)} = \frac{V_d\Delta_\theta}{\pi^2 n}\left[(-1)^n - 1\right]\sin n\varphi$$
$$+ \frac{MV_d}{2\pi}\int_\varphi^{\pi+\varphi}(\sin\phi - \sin(\phi - \Delta_\phi))\cos n\phi d\phi$$

$$A_{m0(cor)} = -\frac{V_d}{2\pi m}\sin(m\Delta_\theta)$$
$$+ \frac{(-1)^{m+1}V_d}{2\pi^2 m}\int_\varphi^{\pi+\varphi}\left[\sin(\pi mM\sin(\phi - \Delta_\phi)\right.$$
$$\left. - m\Delta_\theta) - \sin(\pi mM\sin\phi)\right]d\phi$$

$$A_{mn(cor)} = \frac{V_d}{2\pi^2 mn}\left[(-1)^n - 1\right]\left[\cos(n\varphi) - \cos(m\Delta_\theta + n\varphi)\right]$$

$$+ \frac{V_d(-1)^{m+1}}{2\pi^2 m}\int_\varphi^{\pi+\varphi}\left[\sin(\pi mM \sin(\phi - \Delta_\phi)\right.$$

$$\left. - m\Delta_\theta - n\varphi) - \sin(\pi mM \sin\phi - n\varphi)\right]d\phi$$

$$B_{0n(cor)} = -\frac{V_d \Delta_\theta}{\pi^2 n}\left[(-1)^n - 1\right]\cos n\varphi$$

$$+ \frac{MV_d}{2\pi}\int_\varphi^{\pi+\varphi}\left[\sin\phi - \sin(\phi - \Delta_\phi)\right]\sin(n\phi)d\phi$$

$$B_{m0(cor)} = \frac{V_d}{2\pi m}\left[\cos(m\Delta_\theta) - 1\right]$$

$$+ \frac{V_d(-1)^{m+1}}{2\pi^2 m}\int_\varphi^{\pi+\varphi}\left[\cos(\pi mM \sin(\phi - \Delta_\phi)\right.$$

$$\left. - m\Delta_\theta) - \cos(\pi mM \sin(\phi))\right]d\phi$$

$$B_{mn(cor)} = \frac{V_d}{2\pi^2 mn}\left[(-1)^n - 1\right]\left[\sin(n\varphi) - \sin(m\Delta_\theta + n\varphi)\right]$$

$$+ \frac{V_d(-1)^{m+1}}{2\pi^2 m}\int_\varphi^{\pi+\varphi}\left[\cos(\pi mM \sin(\phi - \Delta_\phi)\right.$$

$$\left. - m\Delta_\theta - n\phi) - \cos(\pi mM \sin\phi - n\phi)\right]d\phi$$

(10)

If the dead time is zero, it is obvious that the correction part given in (10) will be zero and $g(t)$ becomes the solution for an ideal PWM situation. With different m and n, equations (9) and (10) give a complete analytical description of the harmonic characteristics of the output voltage of a PWM converter with and without the dead time.

4. Verification of result

In order to verify the analytical solution, simulations with ideal devices using Pspice have been carried out. The results are compared with those calculated using formula (9)-(10). The detailed conditions for simulation are V_d=10V, f_c=4 kHz, f_v=50 Hz, $\Delta_\theta = 0.16\pi$ (=20 μs), M=0.4, φ=30°. The first 5 main harmonics of V_a are given in Table 1. It can be seen that these results are very close and the validity of the analytical solution is confirmed.

5. Conclusion

An analytical solution to give a precise description of the output harmonic characteristics of PWM H-bridge converters with dead time is derived and its validity is confirmed by simulations. This solution is an important tool to provide an in-depth insight into the effects of the dead time to harmonics and help develop correction circuit to suppress their undesirable effects.

5. References

1. N. Mohan, T. M. Undeland and W. P. Robbins, "Power electronics," *John Wiley & Sons,* 1995
2. Y. Murai, T. Watanabe, and H. Iwasaki, "Waveform distortion and correction circuit for PWM inverter with switching lag-time," *IEEE Trans. Ind. Applicat.*, Vol. 23, No. 5, pp. 881-886, Sep./Oct. 1987
3. D. Leggate, and R. J. Kerkman, "Pulse-based dead-time compensator for PWM voltage inverters," *IEEE Trans. Ind. Applicat.*, Vol. 44, No. 2, pp. 191-197, Apr. 1997
4. P. D. Evans, P. R. Close, "Harmonic distortion in PWM inverter output waveforms," *IEE Proc.-Electr. Power Appli.*, Vol. 134, Pt. B, No 4, July 1987
5. J. A. Taufiq, and J. Xiaoping, "Fast accurate computation of the DC-sided harmonics in a traction VSI drive," *IEE Proc.*, Pt. B, Vol. 136, No. 4, July 1989
6. H. Mellor, S. P. Leigh, and B. M. G. Cheetham, "Reduction of spectral distortion in class D amplifiers by an enhanced pulse width modulation sampling process," *IEE Proceedings-G*, Vol. 138, No. 4, pp. 441-448, Aug. 1991
7. J. Shen, J. A. Taufiq, and A. D. Mansell, "Analytical solution to harmonic characteristics of traction PWM converters," *IEE Proc.-Electr. Power Appli.*, Vol. 144, No. 2, pp. 158-168, March 1997
8. W.R. Bennett. "New results in the calculation of modulation products," *Bell Syst. Tech. J.*, 1933, 12, pp. 228-243

	$A_{0n} + jB_{0n}$		
	Ideal case	With dead time	
n	Analytical	Analytical	Simulation
1	2.00 ∠ -90°	1.2260 ∠ -65.8°	1.2250 ∠ -65.6°
3	0	0.3393 ∠ 0.0°	0.3415 ∠ -0.112°
5	0	0.2037 ∠ -60.1°	0.2037 ∠ -60.1°
7	0	0.1455 ∠ -120.0°	0.1431 ∠ -123.3°
9	0	0.1132 ∠ -180.0°	0.1125 ∠ -179.2°

Table 1: The first 5 main harmonics obtained from analytical solutions and simulations

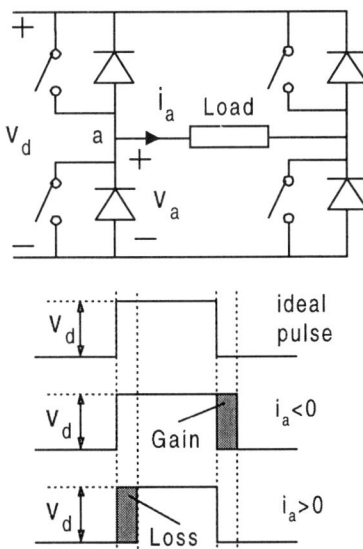

Figure 1: The basic configuration of an H-bridge converter

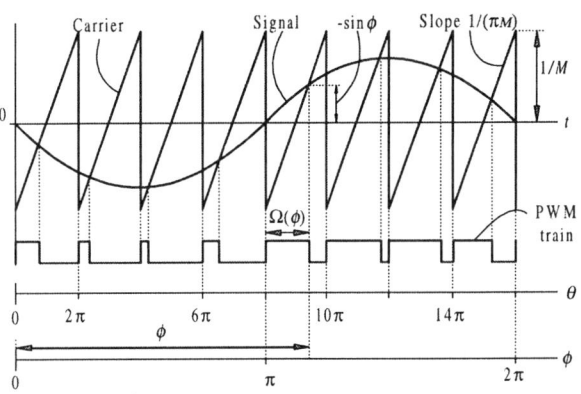

Figure 2: The conventional representation of PWM signal

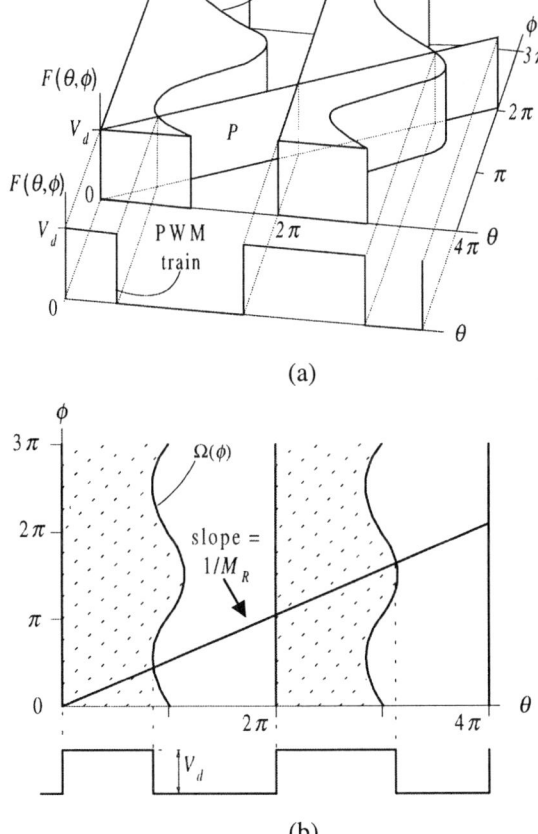

Figure 3: (a) 3-D Model of the SSNS PWM, and (b) the 2-D representation of the 3-D model

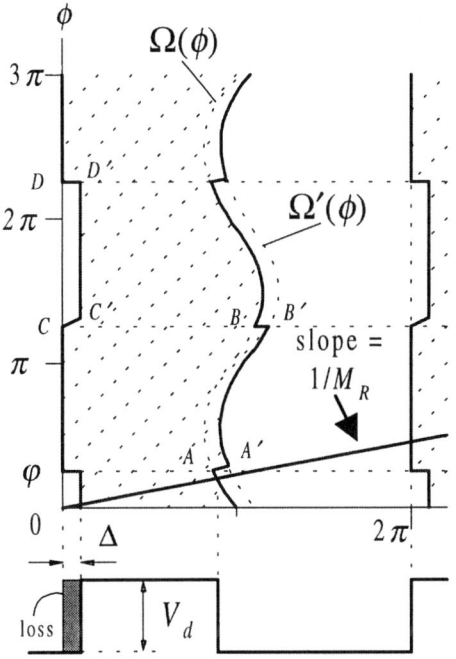

Figure 4: The 2-D representations of the 3-D model for SSNS-PWM with dead time

VI-465

GENERAL PURPOSE SLIDING-MODE CONTROLLER FOR BIDIRECTIONAL SWITCHING CONVERTERS

Alfonso Romero*, Luis Martínez-Salamero*, Hugo Valderrama*, Oscar Pallás** and Eduardo Alarcón**

* Departament d'Enginyeria Electrònica, Elèctrica i Automàtica.
Escola Tècnica Superior d'Enginyeria.
Universitat Rovira i Virgili
Carretera de Salou s/n
43006 Tarragona, Spain

** Departament d'Enginyeria Electrònica.
Escola Tècnica Superior d'Enginyers de Telecomunicació.
Campus Nord - Mòdul C-4.
C/. Gran Capità s/n
08034 Barcelona, Spain.

ABSTRACT

A standardized sliding- mode contoller for bidirectional switching converters is described. The proposed circuit is suitable for large-signal applications and can be used in a large class of dc-to-dc switching converters. As a result, robustness and fast dynamic response against supply, load and parameter variations is obtained. The controller implementation has been performed by means of standard operational amplifiers (OA's) and it has been tested experimentally in elementary and complex switching power converters. An integrated realization of the controller has been simulated at transistor level by means of HSPICE exhibiting similar characteristics to those presented by the OA's-based implementation.

1. INTRODUCTION

The solution of the control problems in dc-to-dc switching converters has been traditionally undertaken by means of conventional techniques based on the linearization of the converter bilinear model. These techniques are based on a linear low frequency characterization of the switching converter so that frequency-domain design methods can be applied to solve the control problem. The controller design following that approach can be specially difficult if the converter has several poles and one or more right half plane zeroes. In addition, the low frequency small-signal model cannot predict the large-signal behavior or the stability of the system.

On the other hand, there are large-signal control techniques [1] which have provided excellent performances in bidirectional converters, their application having led to the generation of different types of power waveforms in the converter output and opened the way to new designs of power amplifiers and sinusoidal inverters. These techniques are based on a nonlinear discrete-time recurrence obtained through the integration of the converter state equations along the switching period and the subsequent introduction of the control constraints.

An alternative approach based on the application of variable structure theory and the associated sliding regimes has been developed in the last years [3] - [4]. The sliding-mode control is a time-domain technique suitable for large-signal applications exhibiting low sensitivity to external perturbations and parameter variations.

We describe in this paper the main characteristics of a general purpose sliding-mode controller which can be used to regulate a large class of switching converters. Namely, buck, boost, buck-boost, Cuk, buck with input filter, boost with output filter and SEPIC converter.

2. O.A.'S - BASED STANDARIZED MODULE

It has been shown in the last years [2] - [5] that linear combinations of state variables used as sliding surfaces provides stable equilibrium points in second a fourth order converters. The simplicity of the control law for a large family of converters is exploited in this paper to implement a general-purpose controller of the type shown in Fig. 1

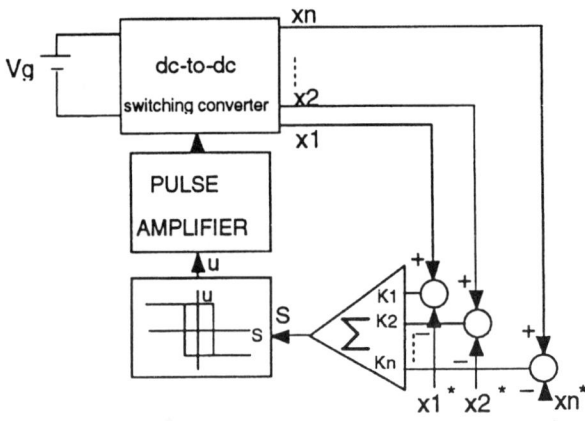

Fig.1 Block diagram of the generalized sliding-mode controller

The practical implementation of the controller requires three different stages. Namely, input stage for measuring converter floating signals or referred to ground, stage for generation of the sliding surface and output stage based on a hysteretic comparator with buffer circuits.

The input stage consist of several differential structures of the type shown in Fig.2. The resistances of the circuit are choosen to implement the coefficients of the sliding surface. Voltage V_{ref} is required to generate the error signal associated to the state variable measured by the first differential amplifier. The differential structure of Fig.2 is repeated for each converter variable needed in the generation of the sliding surface.

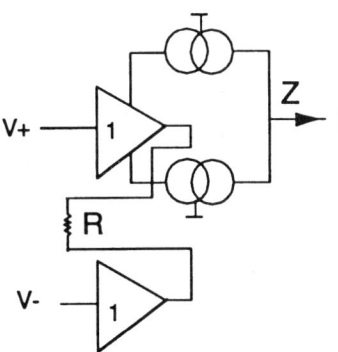

Fig 3. CCII with high impedances.

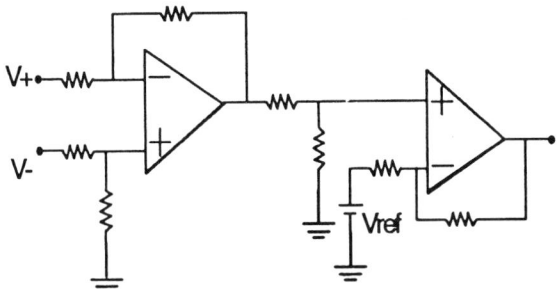

Fig 2. Basic differential structure in the input stage.

The second stage implements the sliding surface that will eventually be used in the converter control. It uses the outputs of the first stage to make linear combinations of them by means of operational amplifiers. Therefore, the expression of the voltage at the output of the second stage will be of the form: $S(x)=K_1(x_1-x_1^*)+...+K_n(x_n-x_n^*)$, where $K_1,..., K_n$ are the sliding surface coefficients, $x_1,..., x_n$ are the converter variables and $(x_1-x_1^*),...,(x_n-x_n^*)$ are the state variable errors. The surface uses up to four state variables and admits the inclusion of a compensating network [8].

The third stage is a hysteretic comparator implemented by means of an operational amplifier with positive feedback. The hysteresis width can be externally controlled and the corresponding output signal is a square wave. To obtain a binary signal that activates/ deactivates the converter switch, two buffer circuits are added to the comparator output in order to adjust the power level of the comparator output to the required level in the converter input.

3. INTEGRATED APPROACH.

The integrated version of the controller has been designed at transistor level with 0.8 µm CMOS technology and has been performed at simulation level by means of HSPICE. The design is based on current-mode due to the inherent benefits of this technology. First, slew rate problems are considerable reduced in comparison to voltage mode. Secondly, linear combinations of signals are significantly simplified. Finally, currents are mor immune to noise and interferences than voltages. This last property is particularly important in the context of switching converters in which fast switchings can create important interferences.

The final architecture of the integrated approach uses the same number of stages than the OA's-based module. The acquisition stage is implemented by means of Current Conveyor type (CCII) structures in which one of the inputs is modified to have high impedance as depicted in Fig.3.

As shown in Fig. 3, the structure inputs are voltage buffers and the difference between both input voltages is provided by the output current at node Z. This current can be weighted by means of resistor R. There are different alternatives to implement the voltage buffers. We have obtained good results with OA's of large bandwith in voltage follower configuration. However, for an integrated design the whole operational amplifier is not required to implement the buffer. Class AB buffers of 4 transistors as shown in Fig. 4 have been employed exhibiting excellent performances, specially as regards to slew rate behavior.

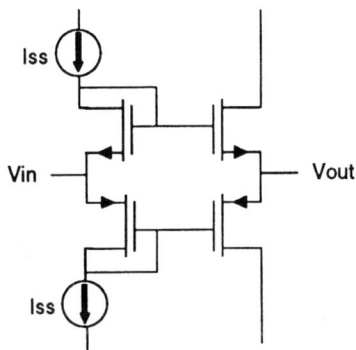

Fig. 4. Class AB buffer

Concerning current mirrors, the best behavior in the simulations has been exhibited by those of regulated cascode type. These current mirrors operate with low distorsion up to 1 MHz. They also have an output impedance of several MΩ which guarantees a very good approximation of voltage-controlled current source. This basic structure has been used to obtain the information of each of the variables that are combined to create the sliding surface. The corresponding scheme is represented in Fig. 5, i.e. two class AB buffers and two regulated cascode mirrors.

The second stage implements the sliding surface of the controller. The use of current-mode technology facilitates the generation of linear combinations since a single node represents an algebraic addition of currents. In case that substraction is needed, a simple change of polarity in one of the currents is sufficient.

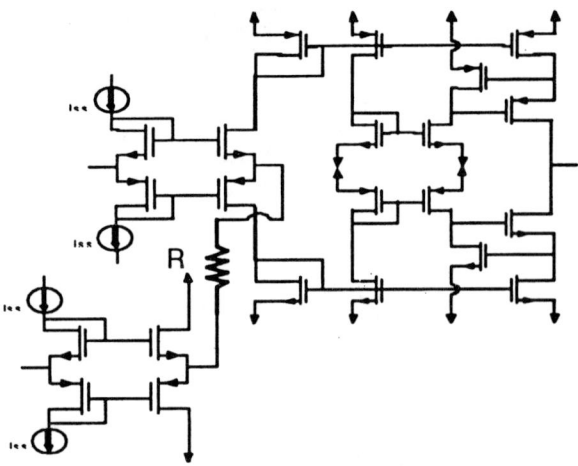

Fig.5 Stage for signal acquisition.

The third stage is a hysteretic comparator with input current and output voltage and externally controlled hysteresis width. Also, a very small input impedance is required since the stage input signal is a current. Taking into account these considerations, we have selected the structure depicted in Fig. 6. The design of this structure is based on a voltage inverter configuration using a two transistor feedback to obtain low input impedance. The hysteresis is achieved by means of nonlinear feedback through the differential pair.

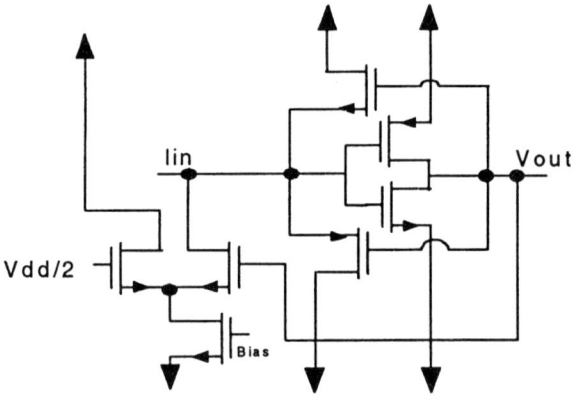

Fig. 6. Hysteresis comparator

4. RESULTS

The OA's-based standardized module has been simulated by means of PSPICE and then experimentally verified for different sliding surfaces in the control of a buck converter operating in continuous conduction mode. The experimental results are comparable to those provided by commercial controllers based on PWM operation.

The integrated design has been simulated on a SUN workstation of the type ULTRASPARC 167. The models used for transistors have been of level 47 obtained from manufacture's data for AMS technology of 0.8 μm. DC supply is ±5 V and the bias currents are 50 μA. The simulation results are in good agreement with the experimental waveforms obtained using the OA's-based controller. As an example, Fig. 7 shows the start-up of the output voltage in a buck converter which uses the sliding surface $S=0.67 (i_L-1) + 0.33 (v_C-3)$ where i_L is the inductor current and v_C is the converter output voltage. Fig. 8 shows the output of the hysteresis comparator for the same surface. Finally, Fig. 9 shows the corresponding converter behavior in the phase-plane.

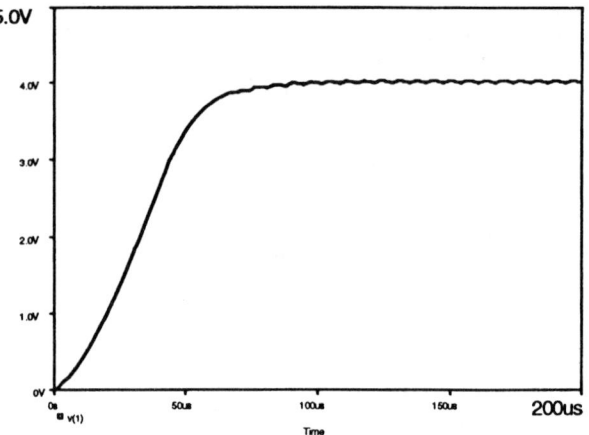

Fig. 7. Output voltage response during start-up.

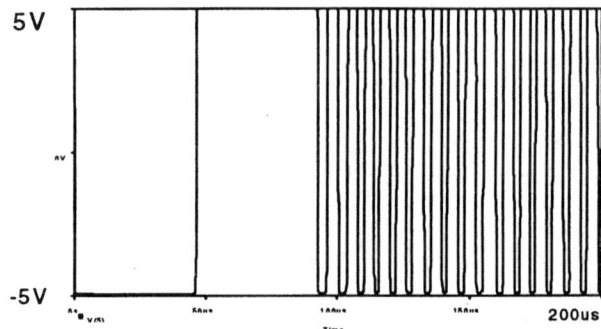

Fig. 8. Hysteresis comparator output

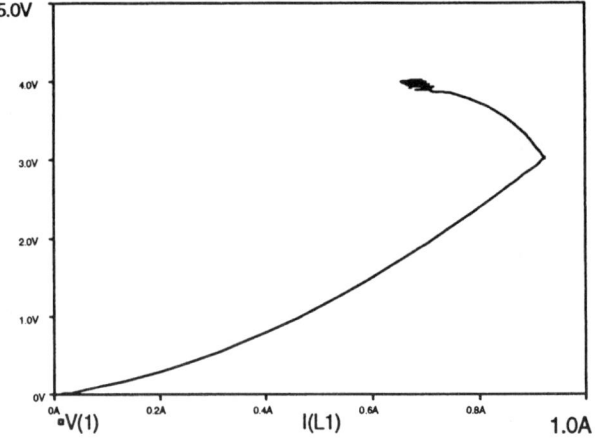

Fig.9. Phase-plane representation

5. CONCLUSIONS

Two different configurations for the implementation of sliding-mode controllers have been studied. The first structure is based on OA's and operates in voltage-mode. The second approach is based on a transistor level design and operates in current-mode. Both architectures provide satisfactory results and show the feasibility of the proposed controllers. The second architecture has some important advantages over the first one due to the current-mode technology employed in the design that ensures a larger bandwidth. Work in progress contemplates the integration of the second structure.

6. ACKNOWLEDGMENT

The authors want to express their acknowledgment to Dr. Sonia Porta for her helpful suggestions in the controller simulation.

7. REFERENCES

[1] J. Majó, L. Martínez, A. Poveda, L. García de Vicuña, F. Guinjoan, A.F. Sánchez, J.C. Marpinard and M. Valentin. "Large-Signal Feedback Control of a Bidirectional Coupled-Inductor Cuk Converter". IEEE Transactions on Industrial Electronics. October 1992, Vol 39, No 5 pp 429-436.

[2] L. Martínez, A. Poveda, J. Majó, L.García de Vicuña, F. Guinjoan, J.C. Marpinard and M. Valentin. "Lie Algebras Modeling of Bidirectional Switching Converters" Proceedings of ECCTD'93 pp 1425-1429.

[3] F. Domínguez, E. Fossas, R. Giral and L. Martínez. "Boost converter with output filter . A sliding approach" Proceedings of 37th Midwest Symposium on Circuits and Systems. Lafayete, Louisiana, August 3-5, 1994. pp 1265-1268.

[4] F. Domínguez, E. Fossas and L. Martínez. "Stability analysis of a buck converter with input filter via sliding-mode approach" Proceedings of IECON'94. pp 1438-1442.

[5] J. Hernanz, L. Martínez, A. Poveda and E. Fossas. "Analysis of a sliding-mode controlled SEPIC converter". Transactions of the Institute of Electronic Enginers of Japan. Vol 116-D, No 11, November 1996 pp 1140-1144.

[6] V.I. Utkin, Sliding modes and their application in variable structure systems, MIR Publishers, Moscow 1978.

[7] H. Sira-Ramírez, "Sliding motions in bilinear switched networks" IEEE Transactions on Circuits and Systems, 1987, CAS-34, No-8. pp 919-933.

[8] R. Giral, L. Martínez, J, Hernanz, J. Calvente, F. Guinjoan, A. Poveda and R. Leyva. "Compensating Networks for Sliding-Mode Control". Proceedings of ISCAS'95, IEEE International Symposiums on Circuits and Systems, Seattle, Wa, April 29 - May 3, 1995. pp 2055 -2058.

A 1,5 kW TWO TRANSISTORS FORWARD CONVERTER USING A NON-DISSIPATIVE SNUBBER

C.H.G. Treviso.

A. A. Pereira.

V.J. Farias.

J.B.Vieira Jr. (IEEE Member).

L.C. de Freitas (IEEE Member) (*).

(*) Corresponding Author.

Universidade Federal de Uberlândia
Centro de Ciências Exatas e Tecnologia
Departamento de Engenharia Elétrica
Campus Santa Mônica - Bloco 3N
38400-902 - Uberlândia - MG - Brazil
Fone: (034)236-5099   Fax: (034)236-5099
Email: freitas@ufu.br

Abstract - **This paper will present a 1,5 kW operation with 90% efficiency at full load of a Two Transistors Forward Converter with non-dissipative snubbers.**

The proposed approach allows building a converter with high frequency operation. A detailed analysis of the operation will be presented. The output voltage is controlled by PWM with a constant frequency.

The complete operating principles, theoretical analysis, relevant equations, simulation and experimental results will be presented in this paper.

I - INTRODUCTION

The power supply unit is the main circuit for almost all electronic equipment, because it provides the necessaries voltages for correct works of these electronic equipments, for it, in the last years, size and weight reduction in the Switching Mode Power Supply (SMPS) design, has been the workhorse of the industry and academic researches. For achieving these features in SMPS it is necessary increasing operating frequency.

The achievement of high-frequency in SMPS requires reduction of switching losses. Conventional Resonant and Quasi-Resonant Converters [2] provide ZCS (Zero-Current Switching) and/or ZVS (Zero-Voltage Switching) and the converters can operate with high-frequency, but these techniques have limitation of load, because the current and/or voltage peaks over the switches and range of frequency control, making difficult the filter components design.

Thus one good way to reach high-frequency and high-power operation is to use the non-dissipative structure showed in the Fig. 1. This circuit provide:

Soft Switching for full load range;

Conduction losses are almost the same as those observed in the hard PWM converter.

II - THE PROPOSED FORWARD CONVERTER

Fig. 1 shows a simplified schematic circuit of the a two transistors Forward converter with non-dissipative snubber that operates without commutation losses.

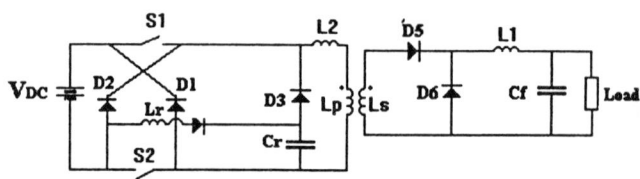

Fig. 1 - Simplified schematic circuit of the a two transistor Forward converter with non-dissipative snubber.

III - PRINCIPLE OF OPERATION

Following the theoretical analysis for the a two transistor Forward converter with snubber non-dissipative showed in Fig. 1 will be presented.

This converter has seven operating stages in a switching cycle, these stages are as follows:

FIRST STAGE (t_0, t_1) - It begins when the switches S_1 and S_2 are turned on, in the ZCS form by L2. During this stage the C_R voltage rises from $-V_i$ to V_{CR1} in a resonant way with L_R. In this stage i_{LR} rises from zero to i_{LR1} in a resonant way too, finishing when the diode D6 is blockaded. It's a very quickly stage.

SECOND STAGE (t_1, t_2) - In this stage, the current in L_R rises of i_{LR1} until maximum and decrease until zero. During this stage the power is transferred to transformer secondary. The voltage on the resonant capacitor rises, in a resonant way, from V_{CR1} to V_{DC}.

THIRD STAGE (t_2, t_3) - This stage begins when the voltage on the resonant capacitor reach V_{DC} and finish when S_1 and S_2 are turned off in the ZVS form, due the C capacitor clamped in V_{dc}.

FOURTH STAGE (t₃, t₄) - This stage begins when S1 and S2 are turned off in the ZVS form, due the C_R capacitor clamped in V_{DC}. In this stage the voltage on the resonant capacitor decrease linearly, with constant load current, until to reach zero, when finish this stage.

FIFTH STAGE (t₄, t₅) - This stage begins when voltage v_{CR} reach 0 (zero). During this stage C_R oscillates with L_m (magnetization inductance) and ends when $v_{CR}= -V_{DC}$, polarizing D1 and D2. There is not power transferred to transformer secondary due diode D5 is inverse polarity.

SIXTH STAGE (t₅, t₆) - In this stage, the voltage v_{CR} is $-V_{dc}$. The magnetizing current is reset across D1, D2 and V_{DC}. This stage finish when the magnetizing current reach zero.

SEVENTH STAGE (t6, t7) - This stage begins when the magnetizing current is reset and finishes when the switches S_1 and S_2 are turned on beginning a new switching cycle. During this stage $V_{CR}= -V_{dc}$.

Fig. 2 shows the theoretical waveforms of one switching cycle.

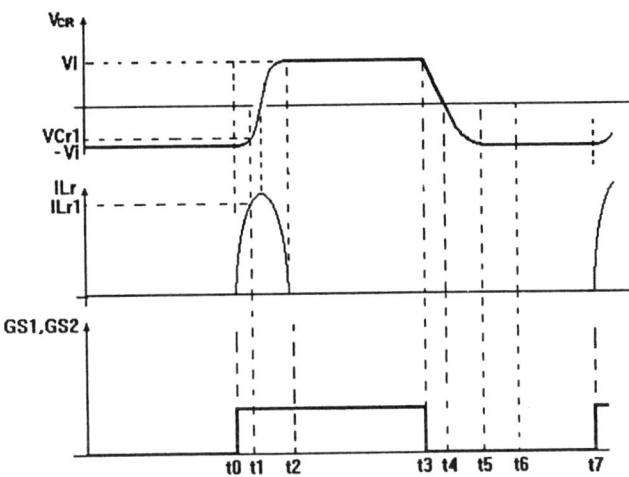

Fig. 2 - Principal waveforms for the two transistors Forward converter with non-dissipative snubber.

The output voltage V_0, can be obtained by the analytical study of the operating stages and with the following assumptions:
- All components and switches are ideal;
- The input voltage V_{DC} is ripple-free;
- The magnetization inductance L_m is very large;
- The transformer leakage inductance is negligible.
- A unity turns ratio transformer is assumed.

$$\frac{Vo}{Vdc} = d - Fs \cdot \frac{\alpha}{\omega o} + \frac{Fs}{2 \cdot \omega o \cdot \alpha} \quad (1)$$

where:

Fs = switching frequency; Fo = resonant frequency;

$$\omega o = 2\pi Fo = \frac{1}{\sqrt{L_R C_R}} \quad (2)$$

$$\alpha = \frac{(I_o)}{Vdc}\sqrt{\frac{L_R}{C_R}} \quad (3)$$

Fig. 3 show the conversion range g for equations above.

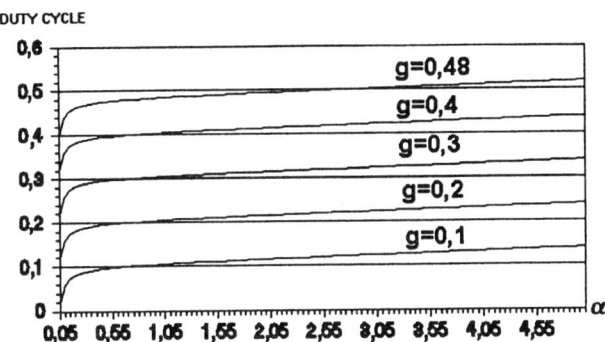

Fig. 3 - Conversion range of the converter.

How was showed in the Fig. 3, for α until 0.55, the converter has the conduct like resonant and after 0,55 like a PWM.

The fig. 4 shows the equivalents circuits for each stage.

IV - THE CONTROL STRATEGY

Fig. 5 shows the block diagram of control strategy fo laboratory implementation.

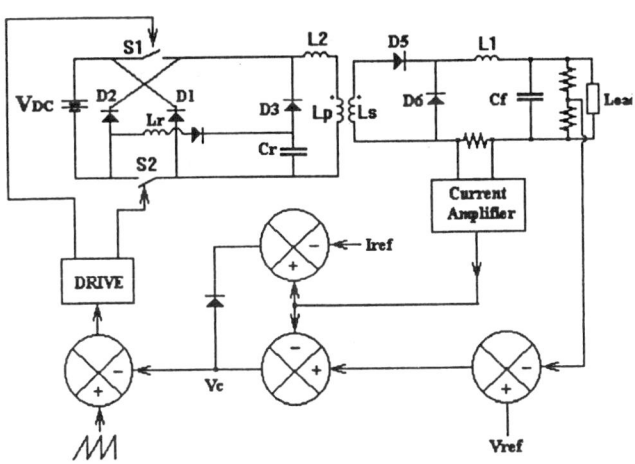

Fig. 5 - The Block Diagram of Control Strategy.

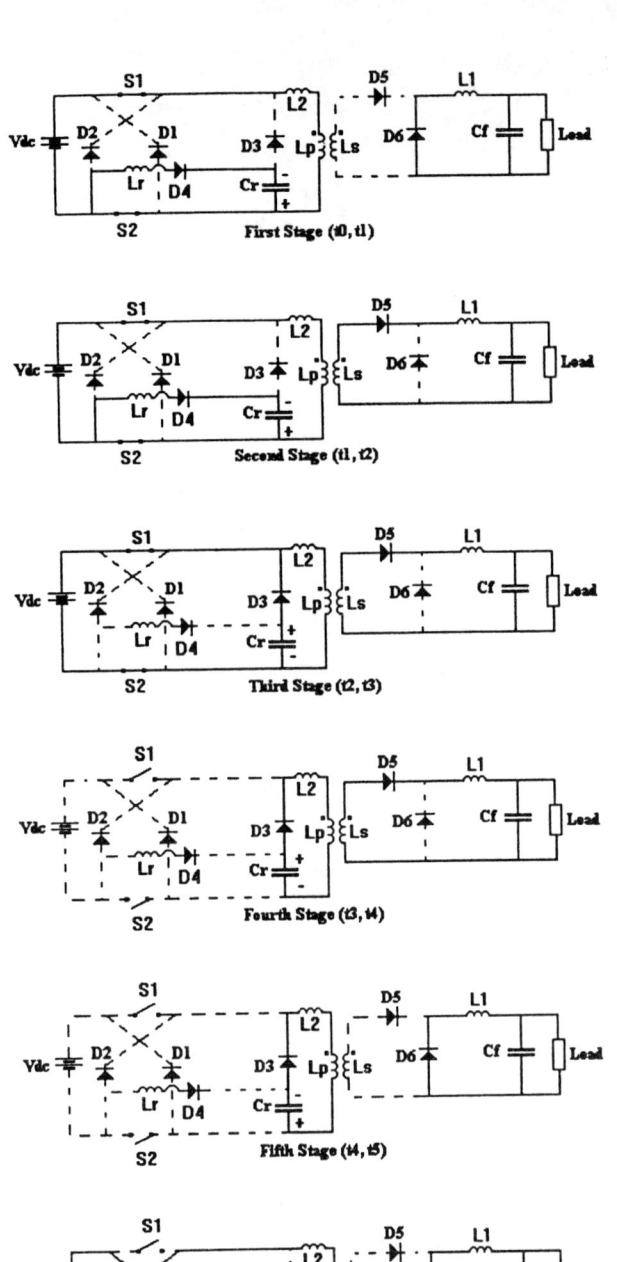

Fig. 4 - Equivalents circuits for each stage.

V - SIMULATION AND EXPERIMENTAL RESULTS

The proposed Forward converter of Fig. 1, was studied by simulation and a prototype was breadboarded. At simulation and a prototype has been used the following parameters set:

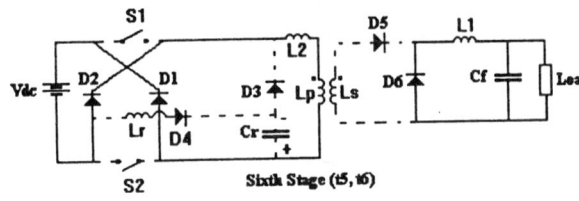

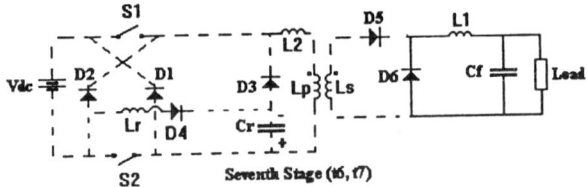

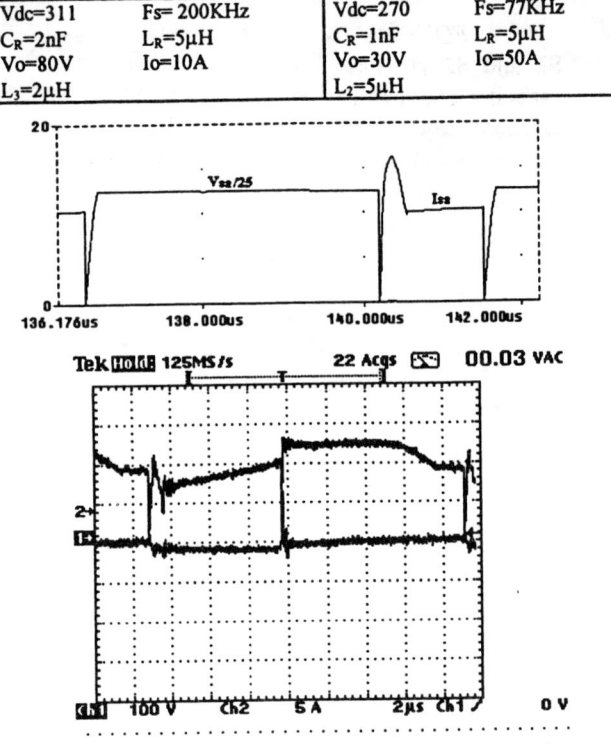

Fig. 6 - Simulation Results (top trace), and Experimental Results (bottom trace).

The time interval t_2-t_0 (resonant interval) can be made as short as desired, depending of the L_R and C_R values with respect to the switching cycle. So, the converter can operate as a conventional hard PWM converter during most part of the cycle, but without commutation losses.

The proposed converter shown in Fig. 1, was breadboarded with the following parameters set:

MOSFET - IRF740 (4 in each switch)	
DIODE D1,D2,D3,D4 - UF5404	
DIODE D5,D6 - MUR3020	
INDUCTOR L2 - 5uH	RESONANT INDUCTOR - 5uH
Fs - 77kHz	RESONANT CAPACITOR - 1nF
OUTPUT POWER - 1500W	OUTPUT VOLTAGE - 30V
INDUCTOR L1 - CORE EE65/33/26 - THORNTON - 45uH - 9 turns	
TRANSFORMER: CORE EE-65/33/26 THORNTON -PRIMARY 14 turns -SECONDARY 4 turns	

Fig. 6 shows the efficiency curve of converter.

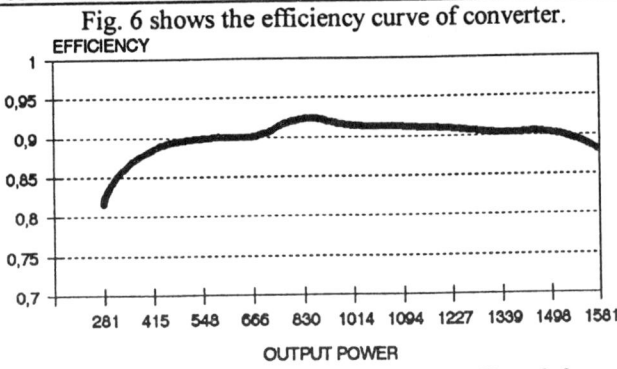

Fig. 7 - Efficiency for the structures breadboarded.

Fig. 8a and 8b show the dynamics responses for load changing from 50% to 100% in the frequency of 100 Hz and 1kHz.

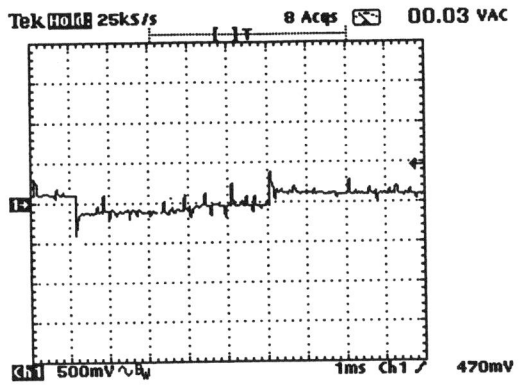

Fig. 8a - Response Dynamic for 100 Hz with load change of 50% to 100% (750W to 1500W).

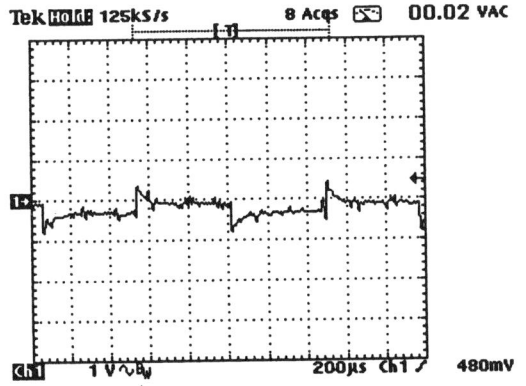

Fig. 8b - Response Dynamic for 1 kHz with load change of 50% to 100% (750W to 1500W).

VI - THE LAYOUT AND PROTOTIPE

Fig. 9a and 9b show the layout and prototype photographs of the studied converter.

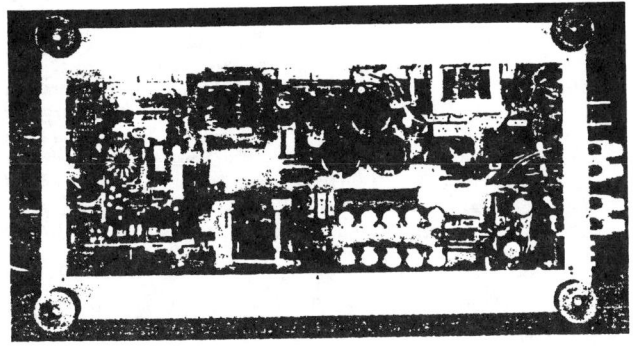

Fig. 9a - Photograph of the layout in the final version.

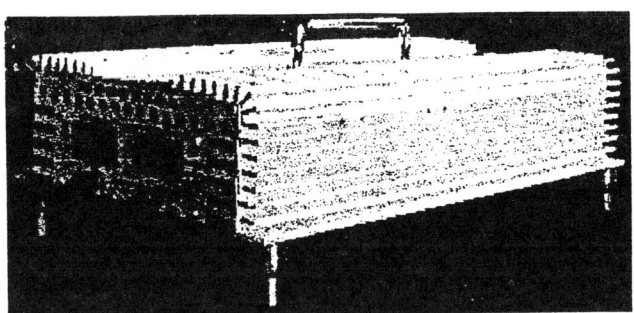

Fig. 9b - Photograph of the Prototype in the final version

VII - CONCLUSIONS

This paper presented a non-dissipative Forward converter working in 1.5 kW load sitguation. As result the switching losses are reduced. The proposed approach allows to obtain better operating performances than those already known for the hard- switching PWM operating in higher switching frequencies and with wide range of power.

The switches are not submitted to over voltages and the peak of current across the switch S_1 and S_2, can be adjusted by the suitable choose of the resonant capacitor and inductor (C_R, L_R).

VIII - BIBLIOGRAPHY

[1] - Pressman, A. I.; "Switching Power Supply Design" McGraw Hill International Editions, Engineering Series, 1992, Singapore.

[2] - Lee, F. C.; "High-Frequency Quasi-Resonant Converte Technologies", Proceeding on the IEEE, vol. 76, nº 4 April 1988.

[3] - de Freitas, L.C.; Farias, V. J.; Vieira Jr., J. B.; Hey, F L. and Cruz, D. F.; "An Optimum ZVS-PWM DC-to DC Converter Family: Analysis, Simulation an Experimental Results", IEEE Power Electronic Specialists Conference - PESC'92 Record, pp. 229-235

[4] - de Freitas, L. C.; Farias, V. J. and Pereira Filho Nicolau. "A Novel Family of DC-DC PWM Converter Using the Self Resonant Principle." PESC'94 - Record pp. 1385-1391.

[5] - de Freitas, L. C.; Pereira, A. A.; Andres, J. L. "A Hig Operating Self-Resonant - PWM Forward Converter APEC'94.

[6] -Pereira, A. A., Farias, L. C. And Vieira Júnior, J. B., " New soft commutated Forward Converter." Midwest'9 - Rio de Janeiro.

IX - ACKNOWLEDGEMENT

The authors acknowledge THORNTON INPEC f the ferrite cores support and SIEMENS for the capacito support.

A LOW-POWER VLSI DESIGN METHODOLOGY FOR HIGH BIT-RATE DATA COMMUNICATIONS OVER UTP CHANNEL

Manish Goel and Naresh R. Shanbhag

Coordinated Science Lab./ECE Department,
Univ. of Illinois at Urbana-Champaign,
1308 W. Main Street, Urbana IL-61801.
E-mail : [mgoel,shanbhag]@uivlsi.csl.uiuc.edu

ABSTRACT

Presented in this paper is a systematic methodology to design low-power integrated transceivers for broadband data communications over unshielded twisted-pair (UTP) channels. The design methodology is based upon two algorithmic low-power techniques referred to as *Hilbert transformation* and *strength reduction* and a high-speed pipelining technique referred to as *relaxed look-ahead transformation*. Finite-precision requirements and power savings are presented. The application of these techniques to design low-power and high-speed 155.52 Mb/s ATM-LAN and 51.84 Mb/s VDSL transceivers is illustrated.

1. INTRODUCTION

Numerous high-bit rate digital communication technologies are currently being proposed that employ unshielded twisted-pair (UTP) wiring. These include asymmetric digital subscriber loop (ADSL), high-speed digital subscriber loop (HDSL), very high-speed digital subscriber loop (VDSL), asynchronous transfer mode (ATM) LAN [2] and broadband access [3]. The above mentioned transmission technologies are especially challenging from both algorithmic and VLSI viewpoints. This is due to the fact that high data rates (51.84 Mb/s to 155.52 Mb/s) need to be achieved over severely bandlimited (less than 30 MHz) UTP channels which necessitates the use of highly complex digital communications algorithms. Low-cost solutions require an integrated approach whereby algorithmic concerns such as signal-to-noise ratio (SNR) and bit-error rate (BER) along with VLSI constraints such as power dissipation, area, and speed, are addressed in a joint manner.

In this paper, we present a systematic design methodology which incorporates algorithmic and VLSI architectural issues in the design of low-power transceivers for broadband data communications over UTP channels. The application of these techniques to the design of 155.52 Mb/s ATM-LAN and 51.84 Mb/s VDSL is demonstrated.

2. THE CHANNEL

In this section, we will describe the UTP-based channel for ATM-LAN and VDSL.

2.1. The Channel Response

In the LAN environment, the two major causes of performance degradation for transceivers operating over UTP wiring are *propagation loss* and *crosstalk* generated between adjacent wire pairs. The worst-case propagation loss is given in the TIA/EIA-568 draft standard for category 3 cable [1] as

$$L_P(f) = 2.320\sqrt{f} + 0.238f, \qquad (1)$$

where the propagation loss $L_P(f)$ is expressed in dB per 100 meters and the frequency f is expressed in MHz.

Similarly, the worst-case NEXT loss model for a single interferer is also given in the TIA/EIA draft standard [1]. This loss can be expressed as:

$$L_N(f) = 43 - 15\log f, \qquad (2)$$

where the frequency f is in megahertz.

In case of VDSL, the performance degradation is due to propagation loss and far-end crosstalk (FEXT). The propagation loss characteristics of a BKMA cable, which is typically employed in this environment, are similar to that of a category 5 cable specified in the TIA/EIA-568A Standard [1], as follows:

$$L_P(f) = 3.597\sqrt{f} + 0.043f + 0.0914/\sqrt{f}, \qquad (3)$$

where the propagation loss $L_P(f)$ is expressed in dB and the frequency f is expressed in MHz.

The equal-level FEXT ($EL-FEXT$) loss in a FEXT dominated environment can be written as:

$$EL - FEXT == \frac{K}{\Psi f^2 d} \qquad K = (49/N)^{0.6}, \qquad (4)$$

where Ψ is the coupling constant which equals 10^{-10} for 1% equal level 49 interferers, d is the distance in kilofeet, f is the frequency in kilohertz and N is the number of interferers.

2.2. The Carrierless Amplitude/Phase (CAP) Modulation Scheme

In the following, we describe a bandwidth-efficient two-dimensional passband transmission scheme referred to as carrierless amplitude/phase modulation (CAP). Note 64-CAP modulation scheme is the standard for ATM-LAN over UTP-3 at 155.52 Mb/s and 16-CAP is the standard for 51.84 VDSL [3].

The block diagrams of a digital CAP transmitter and receiver are shown in Fig. 1. In the transmitter (see Fig. 1(a)), the bit stream to be transmitted is first passed through a scrambler in order to randomize the data. The scrambled bits are then fed into a CAP encoder, which maps blocks of m bits into one of $k = 2^m$ different complex symbols $a(n) = a_r(n) + ja_i(n)$. After the encoder, the symbols $a_r(n)$ and $a_i(n)$ are fed to digital shaping filters. The outputs of the filters are subtracted and the result is passed through a digital-to-analog converter (DAC), which is followed by an interpolating low-pass filter (LPF).

A typical CAP receiver (see Fig. 1(b)) consists of an analog-to-digital converter (ADC) followed by a parallel arrangement of two adaptive digital filters. The adaptive filters in Fig. 1 are referred to as a T/M fractionally spaced linear equalizers (FSLEs). In addition to the FSLEs, a CAP receiver can have a NEXT canceller and a decision feedback

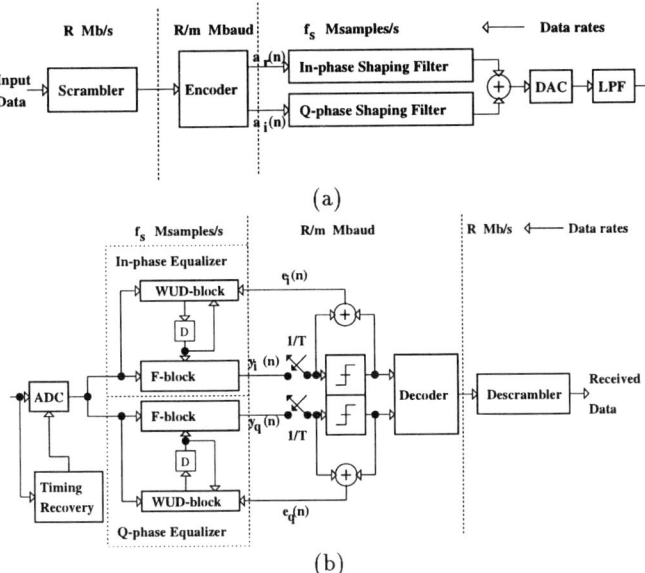

Figure 1: Carrierless Amplitude Phase (CAP) transceiver (a) transmitter and (b) receiver.

equalizer (DFE). The decision to incorporate a NEXT canceller and/or a DFE depends upon the channel impairments and the capabilities of an FSLE. For example, a NEXT canceller is required for 155.52 Mb/s ATM-LAN to cancel NEXT. Similarly, in case of 51.84 Mb/s VDSL, the presence of radio frequency interference (RFI) necessitates the use of a DFE.

3. LOW-POWER AND HIGH-SPEED EQUALIZER ARCHITECTURES

For high bit-rate applications, the equalizers in Fig. 1(b) need to be implemented as high sample-rate adaptive filters, which are typically power-hungry. In this section, we show how the techniques of Hilbert transformation [4] and strength reduction [5] can be employed to design low-power receivers. Also discussed is the technique of *relaxed look-ahead transformation* [7] pipelining the strength-reduced equalizer.

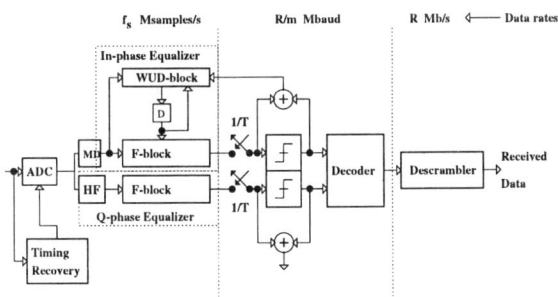

Figure 2: Hilbert-based CAP equalizer.

3.1. Low-Power FSLE Architecture via Hilbert Transformation

It can be shown that the in-phase and the quadrature-phase equalizers of the CAP receiver in Fig. 1 are Hilbert transforms of each other. In this subsection, we exploit this relationship to obtain a low-power structure.

If the in-phase and the quadrature-phase equalizer filter impulse responses are denoted by $f(n)$ and $\tilde{f}(n)$, respectively, then
$$\tilde{f}(n) = h_I(n) * f(n), \quad (5)$$
where the symbol "$*$" denotes convolution. Let $y_i(n)$ and $y_q(n)$ denote the in-phase and the quadrature-phase components of the receive filter output, respectively, and $x(n)$ denote the input. Employing (5), the equalizer outputs can be expressed as
$$y_i(n) = f(n) * x(n),$$
$$y_q(n) = \tilde{f}(n) * x(n) = f(n) * [h_I(n) * x(n)]. \quad (6)$$
From (6), we see that $y_q(n)$ can be computed as the output of a filter, which has the same coefficients as that of the in-phase filter with the Hilbert transform of $x(n)$ as the input. Hence, the CAP receiver in Fig. 1(b) can be modified into the form [4] as shown in Fig. 2, where **HF** is the Hilbert filter. The ideal Hilbert filter is approximated by an M-tap windowed version. From a power dissipation perspective, the Hilbert filter length should be as small as possible. The power dissipation versus performance trade-off has been explored in [4] where it was shown that for 51.84 Mb/s ATM-LAN, assuming $N = 32$, we obtain SNR_o values of more than 25 dB, when the Hilbert transformer length M is more than 33. This results in a 21% power savings.

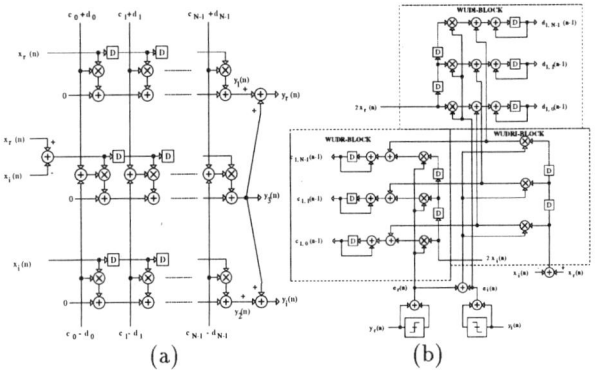

Figure 3: Strength-reduced equalizer (a) **F**-block and (b) **WUD**-block.

3.2. Low-power Complex Adaptive Filter via Strength Reduction

The product of two complex number $(a+jb)$ and $(c+jd)$ is given by $(a+jb)(c+jd) = (ac-bd) + j(ad+bc)$. A direct-mapped architectural implementation of this would require a total of four real multiplications and two real additions to compute the complex product. Application of strength reduction involves reformulating the above multiplication as follows

$$(a-b)d + a(c-d) = ac - bd, \quad (a-b)d + b(c+d) = ad + bc, \quad (7)$$

where we see that strength reduction reduces the number of multipliers by one at the expense of three additional adders. Typically, multiplications are more expensive than additions and hence we achieve an overall savings in hardware. The strength reduced **SR** architecture [5] is obtained by applying strength reduction transformation at the algorithmic level.

Consider an N-tap serial complex LMS filter described by the following equations

$$e(n) = y_d(n) - \mathbf{W}^H(n-1)\mathbf{X}(n),$$
$$\mathbf{W}(n) = \mathbf{W}(n-1) + \mu e^*(n)\mathbf{X}(n), \quad (8)$$

where $\mathbf{W}(n) = [w_0(n), w_1(n), \ldots, w_{N-1}(n)]^T$ is the weight vector with $\mathbf{W}^H(n)$ being the Hermitian (transpose and complex conjugate), $\mathbf{X}(n) = [x(n), x(n-1), \ldots, x(n-N+1)]^T$ is the input vector, $e^*(n)$ is the complex conjugate of the adaptation error $e(n)$, μ is the step-size, and $y_d(n)$ is the desired signal.

Traditionally, the complex LMS algorithm is implemented via the cross-coupled **CC** architecture, where four real **F**-blocks and four real **WUD**-blocks are employed requiring a total of $8N$ multipliers and $8N$ adders. Applying strength reduction to (8) results in the following equations, which describe the **F**-block computations of the **SR** architecture [5]:

$$y_1(n) = \mathbf{c}_1^T(n-1)\mathbf{X}_r(n), \; y_2(n) = \mathbf{d}_1^T(n-1)\mathbf{X}_i(n)$$
$$y_3(n) = -\mathbf{d}^T(n-1)\mathbf{X}_1(n),$$
$$y_r(n) = y_1(n) + y_3(n), \; y_i(n) = y_2(n) + y_3(n), \quad (9)$$

where $\mathbf{X}_1(n) = \mathbf{X}_r(n) - \mathbf{X}_i(n)$, $\mathbf{c}_1(n) = \mathbf{c}(n) + \mathbf{d}(n)$, and $\mathbf{d}_1(n) = \mathbf{c}(n) - \mathbf{d}(n)$. Similarly, the **WUD** computation is described by,

$$\mathbf{c}_1(n) = \mathbf{c}_1(n-1) + \mu[\mathbf{eX}_1(n) + \mathbf{eX}_3(n)]$$
$$\mathbf{d}_1(n) = \mathbf{d}_1(n-1) + \mu[\mathbf{eX}_2(n) + \mathbf{eX}_3(n)], \quad (10)$$

where $\mathbf{eX}_1(n) = 2e_r(n)\mathbf{X}_i(n)$, $\mathbf{eX}_2(n) = 2e_i(n)\mathbf{X}_r(n)$, $\mathbf{eX}_3(n) = e_1(n)\mathbf{X}_1(n)$, $e_1(n) = e_r(n) - e_i(n)$, $\mathbf{X}_1(n) = \mathbf{X}_r(n) - \mathbf{X}_i(n)$. It is easy to show that the **SR** architecture (see Fig. 3) requires only $6N$ multipliers and $8N+3$ adders. This is the reason why the **SR** architecture results in $21-25\%$ power savings [5] over the **CC** architecture.

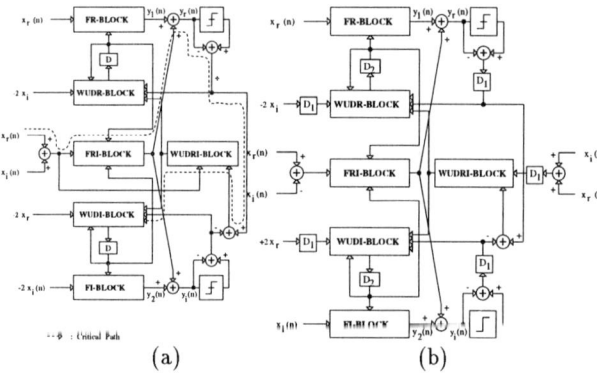

Figure 4: Strength-reduced equalizer (a) Serial and (b) Pipelined.

3.3. Pipelined Strength-reduced (PIPSR) Architecture

The block diagram of the **SR** architecture is shown in Fig. 4(a). The dotted line in Fig. 4(a) indicates the critical path of the **SR** architecture. As explained in [5], both the **SR** as well as **CC** architectures are bounded by a maximum possible clock rate due the computations in this critical path. This throughput limitation is eliminated via the application of the *relaxed look-ahead transformation* [7] to the **SR** architecture (see (9-10)). Application of relaxed look-ahead to the **SR** architecture in (9-10) results in the following equations that describe the **F**-block computations in the **PIPSR** architecture:

$$y_1(n) = \mathbf{c}_1^T(n-D_2)\mathbf{X}_r(n), \; y_2(n) = \mathbf{d}_1^T(n-D_2)\mathbf{X}_i(n)$$
$$y_3(n) = -\mathbf{d}^T(n-D_2)\mathbf{X}_1(n)$$
$$y_r(n) = y_1(n) + y_3(n), \; y_i(n) = y_2(n) + y_3(n), \quad (11)$$

where D_2 is the number of delays introduced before feeding the filter coefficients into the **F**-block. Similarly, the computation of the **WUD** block of the **PIPSR** architecture are given by

$$\mathbf{c}_1(n) = \mathbf{c}_1(n-D_2) + \mu \sum_{i=0}^{LA-1} \left[\mathbf{eX}_1(n-D_1-i)\right.$$
$$\left. + \mathbf{eX}_3(n-D_1-i)\right] \quad (12)$$

$$\mathbf{d}_1(n) = \mathbf{d}_1(n-D_2) + \mu \sum_{i=0}^{LA-1} \left[\mathbf{eX}_2(n-D_1-i)\right.$$
$$\left. + \mathbf{eX}_3(n-D_1-i)\right], \quad (13)$$

where $\mathbf{eX}_1(n)$, $\mathbf{eX}_2(n)$ and $\mathbf{eX}_3(n)$ are defined in the previous subsection, $D_1 \geq 0$ are the delays introduced into the error feedback loop and $0 < LA \leq D_2$ indicates the number of terms considered in the sum-relaxation. A block level implementation of the **PIPSR** architecture is shown in Fig. 4(b) where D_1 and D_2 delays will be employed to pipeline the various operators such as adders and multipliers at a fine-grain level. The high-throughput of the **PIPSR** architecture can also be traded-off with supply voltage reduction resulting in additional power savings [5] of $40-69\%$. Therefore, the **PIPSR** architecture results in $60-90\%$ power savings as compared to the serial **CC** architecture. In a similar manner, the relaxed look-ahead transformation can be employed to pipeline the real adaptive filters.

4. SYSTEMATIC DESIGN METHODOLOGY AND EXAMPLES

In this section, we present our design methodology for the high-speed communication system design. We determine the encoder and shaping filters in the transmitter and adaptive filter lengths, precisions and pipelining levels in the receiver.

4.1. Design Methodology and Applications

The design methodology consists of the following steps:

Step 1: Determine the specifications for the signal constellation and the spectrum based on the channel constraints.

Step 2: Determine adaptive filters lengths via the floating-point simulations. Apply low-power transformations given in section 3. Let N be the tap-length and $SNR_{o,fl}$ be the output SNR in dB of the floating point algorithm after this step.

Step 3: Determine the precision for the adaptive filters such that the fixed-point SNR is close to the floating-point SNR employing the following equations [6] for an **SR** architecture,

$$B_{F,SR} > \frac{1}{2}\log_2\left(\frac{N\sigma_x^2}{4\beta\sigma_d^2}\right) + \frac{SNR_{o,fl}(dB)}{6} \quad (14)$$

$$B_{WUD,SR} \geq \frac{1}{2}\log_2\left(\frac{1}{\mu^2\sigma_x^2\sigma_d^2}\right) + \frac{SNR_o(dB)}{6}, \quad (15)$$

where σ_x^2 is the input power to the **F**-block, σ_d^2 is the power of symbol constellation (or the desired signal) and $\beta \ll 1$ is the ratio between the quantization noise at the output and floating point MSE. Also, SNR_o is the desired SNR at the output. Typically, we choose $SNR_o = SNR_{o,fl}$ because we assume that the quantization error at the output is small. Similar expressions can be obtained for the real adaptive filters also.

Step 4: Calculate critical path delay for the fixed-point architectures, and pipeline (if needed) via relaxed look-ahead transformation (see section 3.3) by choosing appropriate values of D_1, D_2 and LA.

4.2. 155.52Mb/s ATM-LAN

In this subsection, we will employ the proposed methodology to design 155.52 Mb/s ATM-LAN transceiver for UTP-3 wiring. Given the data rate of $R = 155.52$ Mb/s and the FCC imposed limit of 30 MHz on the transmit spectrum, we choose a value of $m = 6$ to get the symbol rate $1/T$ to be 25.92 Mbaud. Choosing $0 - 30$ MHz spectrum, we obtain a value of $\alpha = 0.15$. Clearly, the sample rate f_s needs to be greater than 60 MHz. We choose the next nearest multiple of the symbol rate or 77.76 MHz as the value of f_s. As the excess bandwidth is less than 100%, a NEXT canceller is required to suppress NEXT as shown in Fig. 5.

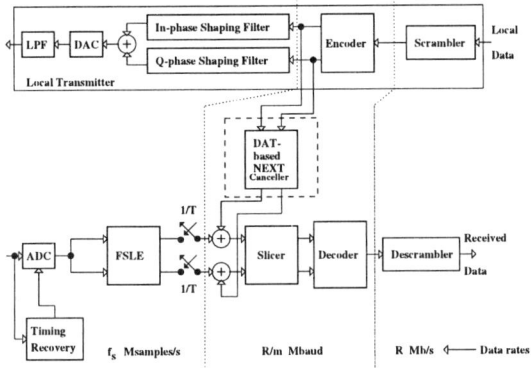

Figure 5: 155.52 Mb/s CAP transceiver.

Next, we decide the lengths of the adaptive filters according to **Step 2**. It has been observed in [2] that the NEXT canceller span is weakly dependent upon the equalizer and that it should be about $1\mu s$ or greater. Given the symbol rate of 25.92 Mbaud, we find that the number of NEXT canceller taps $N_{nextc} \geq 26$. We chose a value of $N_{nextc} = 32$ for our simulations. The equalizer length is determined via the simulations, which show that with symbol span equal to 40, $SNR_o = 36.1$ dB is obtained. Note that this is also the value of $SNR_{o,fl}$ employed in **Step 3** to determine the precisions. Employing (14) and (15), we choose the **F**-block precision in the NEXT canceller $B_{F,nextc} = 10$ bits and the **WUD**-block precision in the NEXT canceller $B_{WUD,nextc} = 14$ bits. A similar analysis can be applied to the equalizer. In doing so, we obtained the **F**-block precision $B_{F,eq} = 12$ bits and $B_{WUD,eq} = 18$ bits for the equalizer.

Given the tap lengths, the precisions (derived above) and the assumption of a 1 bit adder delay of 1 ns, it can be shown that pipelined architectures for the NEXT canceller and the equalizer becomes necessary. Hence, we employ the pipelined strength reduced adaptive filter architecture for the NEXT canceller with $D_1 = 16$, $D_2 = 4$, and $LA = 1$ and the pipelined FSLE architecture with $D_1 = 84$, $D_2 = 4$, and $LA = 1$. The simulation results for the finite-precision, pipelined 155.52 Mb/s ATM-LAN receiver is shown in Fig. 6(a), where we see that the final $SNR_o = 35.3$ dB provides a noise margin of 5.8 dB, which is quite sufficient for practical purposes.

4.3. 51.84 Mb/s VDSL

For VDSL, we need a DFE with feedforward equalizer (FFE) span of 16 symbol periods and a feedback equalizer (FBE)

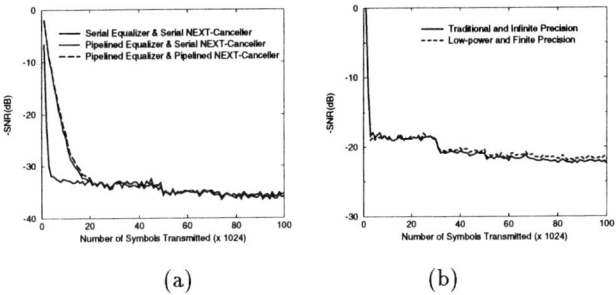

Figure 6: Simulation results for (a) 155.52 Mb/s ATM-LAN and (b) 51.84 Mb/s VDSL.

with 8 taps. Clearly, other combinations of FFE and FBE spans are also possible. The precisions of the equalizer were: $B_{F,ffe} = 10$ bits and $B_{WUD,ffe} = 16$ bits. The low-power strength reduced DFE had precisions: $B_{F,fbe} = 9$ bits and $B_{WUD,fbe} = 13$ bits. A Hilbert filter with length $M = 65$ and coefficient precision of 9 bits was chosen. With these parameters, simulation results (see Fig. 6(b)) indicate that the finite-precision VDSL receiver can achieve an $SNR_o = 21.8$ dB. This is 0.3 dB lower than the SNR_o of the floating point receiver model. Note that with $SNR_{o,ref} = 21.5$ dB (corresponding to a $BER = 10^{-7}$ for a 16-CAP transceiver), we have a noise margin of 0.3 dB. Clearly, it is harder to achieve the desired data rates with comfortable noise margins in a VDSL environment as compared to the ATM-LAN environment.

5. ACKNOWLEDGMENTS

This work was supported via National Science Foundation CAREER Award MIP-9623737.

6. REFERENCES

[1] *Commercial Building Telecommunications Cabling Standard*, TIA/EIA-568-A Standard, 1994.

[2] G. H. Im and J. J. Werner, "Bandwidth-efficient digital transmission up to 155 Mb/s over unshielded twisted-pair wiring," *IEEE J. Select. Areas Comm.*, vol. 13, no. 9, pp. 1643-1655, Dec. 1995.

[3] D. D. Harman, G. Huang, G. H. Im, M.-H. Nguyen, J.-J. Werner, and M. K. Wong, "Local distribution for interactive multimedia TV," *IEEE Multimedia Magazine*, pp. 14-23, Fall, 1995.

[4] R. Hegde and N. R. Shanbhag, "A low-power phase-splitting adaptive equalizer for high bit-rate communications systems," *Proc. of IEEE Workshop on Signal Processing Systems*, Nov. 1997, Leicester, U.K..

[5] N. R. Shanbhag and M. Goel, "Low-power adaptive filter architectures and their application to 51.84 Mb/s ATM-LAN," *IEEE Trans. on Signal Processing*, vol. 45, no. 5, pp. 1276-1290, May 1997.

[6] M. Goel and N. R. Shanbhag, "A pipelined strength reduced adaptive filter: Finite precision analysis and application to 155.52 Mb/s ATM-LAN," *1997 Proc. Midwest Symposium on Circuits and Systems*, Sacramento, CA, August 1997.

[7] N. R. Shanbhag and K. K. Parhi, *Pipelined Adaptive Digital Filters*. Kluwer Academic Publishers, 1994.

[8] B. R. Petersen and D. D. Falconer, "Minimum mean square equalization in cyclostationary and stationary interference: Analysis and subscriber line calculations", *IEEE Journal on Selected Areas in Communications*, vol. 9, no. 6, pp. 931-940, Aug. 1991.

VLSI DESIGN OF AN ATM SWITCH WITH AUTOMATIC FAULT DETECTION

Louis Chung-Yin Kwan, Chi-Ying Tsui, Chin-Tau Lea
Department of Electrical and Electronic Engineering
Hong Kong University of Science and Technology
Clear Water Bay, Hong Kong

ABSTRACT

This paper describes a VLSI implementation of a multistage self-routing ATM switch fabric. The size of the switch prototype is 16x16 and is designed to handle the OC-12 (622 Mbps) link rate. Based on a bit-slice architecture, the entire 16x16 switch is implemented using four identical chips. The switch has multiple paths, created by a randomizer in front of the routing stages, between each input-output pair. The switch uses an input/output-buffering scheme and contains no buffers inside the fabric. To facilitate fault detection and isolation, we add automatic fault detection schemes at the node, chip, and system levels of the design.

1. INTRODUCTION

Shared-buffer architecture has dominated the field of commercial ATM switches over the years for its simplicity. In this type of ATM switches, input cells from different ports are time multiplexed into a centralized memory and read out in the same fashion. A central control unit is used to maintain the queues and control the flow. The advantage of shared buffer is that smaller physical memory is needed. The disadvantage is that the speed of the memory has to be scaled up with the number of ports. This limit is also encountered in the shared medium approach. As the demand of capacity soars, multistage architectures have become an inevitable trend in ATM switch design. Of particular interest to us is the Banyan-type network [1]. The control of the switching is distributed over each switching element. So, multiple input/output transmission happens in the same cycles. As a result, the switch is more scalable with respect to the size of the switch. Buffers can be placed at the input, output and inside the switch [2,3]. The performance, in term of throughput and cell loss rate, is different in each configuration. Also, the way to handle multicast traffic in multistage switch is difficult and not fixed yet. So, the optimal architecture of an ATM switch is still an open issue. However, the common criteria of an ATM switch are low cell loss rate, high throughput, high scalability and cost effective.

In this paper, we describe a VLSI implementation of a multistage self-routing ATM 16x16 switch fabric which is designed for mixed ATM and IP traffic [4,5]. It is designed to handle the line rate of OC-12 (622 Mbps). The switch is featured by simple routing strategy, internal non-buffering, multiple in/out paths, and automatic fault detection. Multiple paths are created by a randomizer in front of the routing stages. The switch uses in/out-buffering scheme, and has no buffer inside the switch fabric. As a result, the order of cell transmission is always maintained. Although randomization is an efficient scheme to beat the

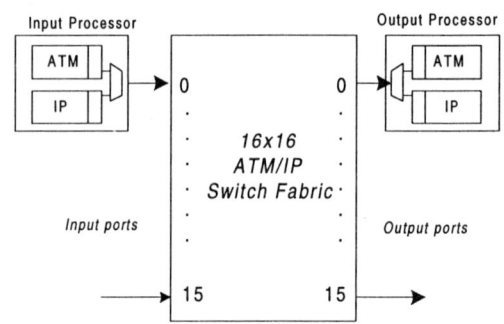

Figure 1. 16x16 ATM/IP Switch

pathetic traffic patterns from degrading the performance of the switch, it makes fault detection difficult since cells can be sent to the output port through different paths. Thus special attention has to be given to fault detection in the switch.

2. A 16x16 ATM/IP SWITCH

The 16x16 ATM/IP switching system is shown in Fig.1. The switch is designed to handle ATM cells and IP packets simultaneously. The I/O processors contain ATM and IP queue multiplexed into the switch fabric. Although an IP packet is fragmented into ATM cells, inside the network no IP packet interleaving occurs, i.e., consecutive IP cells belong to the same IP packet. This greatly simplifies the IP processing. But this also presents a major challenge for the switch design. It is accomplished with a combination of priority, unbuffered switch fabric and correct buffer management in the port processor. It is achieved by assigning the lowest priority to the first cell of an IP packet. It guarantees that once the first cell of an IP packet reaches a destination port, the remaining cells will be transmitted with a higher priority than other IP packet cells destined for the same output port. The other IP packets can only be received until the current IP packet transmission is over [4,5].

Latency is an important design issue for unbuffered switches. Cells which are blocked will re-try in the next time slot. As a result, the sum of the latency of the switch fabric and the time for the acknowledgement signal to transmit back has to be less than the whole cell transmission time. The proposed switch fabric is a multistage interconnection network(MIN) as shown in Fig.2. Four 4x4 switching elements(SE) are used in each stage instead of the usual 2x2 SEs. For the same input/output port number, more stages are needed for a smaller SE design. So, both the latency and internal blocking probability are higher in a switch with 2x2 than the one with 4x4 SEs. In the proposed switch, two extra stages are placed in front of the 2 routing stages. The path

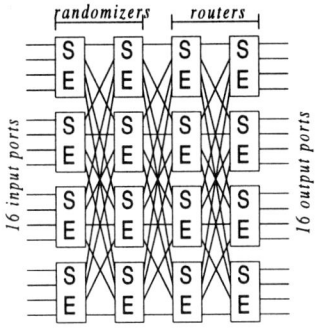

Figure 2. Multistage Self-Routing Switch.

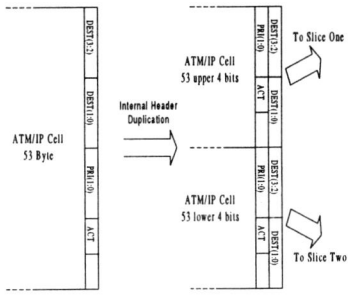

Figure 3. Internal Header Duplication and field information

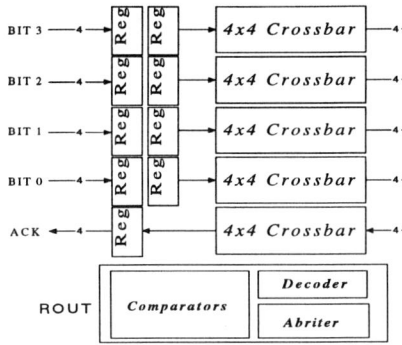

Figure 4. 4x4 Router Structure

setup in these 2 extra stages is randomized. For a particular virtual circuit connection, the path the cells go through inside the switch is different for different time slot. It is shown that randomization is necessary for internal blocking switch to provide immunity to congestion[5].

The operation of the switching system is as follows, the Input Processor processes the input cells. It captures the ATM/IP cell header and then look up information in the translation table. An Internal tag will be attached in the head of cell. Firstly, the cells go through 2 stages of randomizer. The randomizer randomly selects one non-blocking in/out combination and set-up the paths in the crossbar accordingly. The two randomizing stages provide paths from each input port to any one of the 16 entrances of the routing stages. In the routing stage, the control determines the routing from the tag information which contains destination, priority and activity. Since multiple paths are created by the randomizer, for each virtual circuit connection, the path it goes through is different every cell time slot.

3. SWITCH ELEMENT DESIGN

The data transmission is designed to be either 8-bit or 16-bit in parallel. The switch design uses a bit-slice approach. It is formed by multiple 4-bit 16x16 switch chips to minimize the number of I/O on each chip. For a 8-bit system, as shown in Fig.3, two identical slices are used and they operates independently. Consistency in the internal path setup is ensured by duplicating the cell tag to each slice and making the pseudo-random sequences in the randomizer of each slice to be the same. The basic component of the switch is the 4x4 switch element which can be used for both the router and the randomizer. The SE is a 4-bit 4x4 crossbar switch with an extra one bit crossbar for the acknowledgement signal to be sent back from the output port.

3.1 Router Structure and Control

Fig.4 shows the structure of a 4x4 router. There are five one bit datapath which contains pipeline registers and a crossbar. Two stages of registers are required as the control unit needs the tag, which is available in two consecutive clock cycles, to setup the paths in the crossbar. The control unit(ROUT) compares the destination fields from the 4 inputs. The results showing any output port conflict between 2 inputs are restored in a register. In

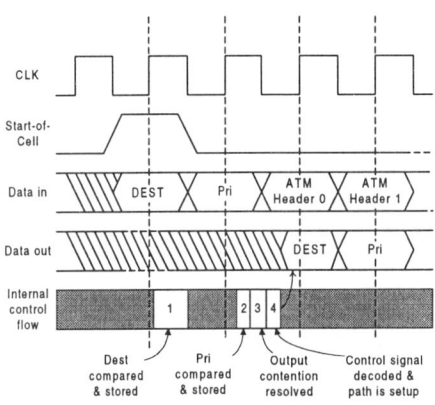

Figure 5. Router Control Timing Diagram

the next cycle, the comparator compares the priority fields to determine the relative priority between any 2 inputs. The arbiter will then resolve the destination and priority conflict and the decoder decodes the destination and the masked activity field to generate 16 path control signals for the crossbar setup. Fig.5 shows the timing diagram of the control in ROUT.

3.2 Randomizer Structure and Control

For the 4x4 randomizer SE, the structure is basically the same as the router except that only one stage of register is needed as the header is not required for the path setup. A counter and ROM table are used to generate the pseudo-random sequence. The ROM table stores the predefined path set-up control signals. For non-blocking output path set-up, the number of combination is equal to $_4P_4(24)$. As a result, the ROM table has only 24 pre-determined entries.

A counter is used to generate the address for the ROM. It counts from 0 to 23 and counts whenever a new cell transmission cycle starts. The counter receives a start-of-cell signal from the source and a new destination pattern is then loaded. It is desirable to provide a different pseudo-random destination sequence in different randomizers in a slice. In this way, the correlation of path setup in different randomizers is reduced. In our design, two bits from the tag are arbitrarily chosen as an offset to the counter whenever a new cell transmission cycle starts. The offset is the same for the corresponding randomizers in all bit slices. The modified count value is then applied to the ROM address input. The advantage of this method is two-fold. Firstly, the destination pattern is different in every randomizer in a chip. Secondly, a fixed sequence no longer exists and the quality of randomness is assured.

4. FAULT DETECTION, LOCATION AND RECOVERY

4.1 Overview of Fault-Tolerant design in ATM switch

For an ATM switching system, high-reliability and high-availability is preferred because the cost is great if the network is down. As a result, fault-tolerance is as important as performance in the switch system. Fault tolerance is achieved by fault detection, location and then recovery. In the proposed switch, multiple identical chips are used in parallel because of the bit-slice approach. It provides a convenient way for fault detection by duplication and comparison scheme. A precise fault location scheme is also proposed which can locate faults in the crossbar in a very short time. Redundancy is provided in each crossbar and automatic fault recovery can be carried out in the SE level. However, for control logic fault, recovery needs to be done in higher level because the cost for control redundancy is too high.

4.2 Fault Models

The proposed ATM switch is a modular one of which the building component is a SE. Inside each SE, there are a control unit and a 4-bit slice crossbar. In the crossbar, the structure is highly regular and thus fault behavior can be analyzed practically. Only permanent faults are considered. Common fault models, short(SS), open(OS) and 1/0-stuck(S1/S0), are used. Under SS, a faulty connection is set-up between an input and output pair. Bit is misrouted to that output and it may crash with another input which is destined to that column. Under OS, path from the input to the output is always open. Bit which are supposed to pass through this path are considered blocked. Bit in that column will enter high impedance state. Under S1/S0, an input or output channel is shorted to low or high state. In this case, data passing through the faulty link is always corrupted. In any case, a bit error may turn out to be a global routing error as the tag is corrupted or the cell is misrouted. Not only the faulty SE is affected, but also the succeeding stages as the routing is dependent on the previous stage output. It leads to serious error multiplication in the worst case.

For the router control unit, the same fault models can still be applied but the functional error at the control output cannot be generalized as it is built by logic gates without a regular structure. A fault will usually result in incorrect path setup. Fault in the randomizer control like improper ROM address generation will not affect the information integrity and routing correctness in latter stage. However, consistency in paths setup is not guaranteed in multiple chip slices. If a bit corruption occurs in the ROM table which stores the path setup signals patterns, there is a high probability that two input will be switched to the same output and result in a data crash.

4.3 Fault Detection

The switch inherently contains hardware duplication. Multiple identical slices are used in parallel. As shown in Fig. 3, the same tag will be applied to both slices and the output tags should also be consistent. Once any inconsistency is detected, it indicates that a fault occurs in one of the slice. This is a system level fault detection(SLFD). It can only indicate one of the chips is faulty but cannot exactly determine the faulty one. In additional to SLFD, inside each router SE, there is an auxiliary fault detection unit(AFDU). This AFDU is used to check whether the destination bits of an router SE output tag matches with the port ID. In case of a mismatch, it suggests that SE is faulty. Using the AFDU can reduce the fault detection and location time because it can provide a fault signal earlier than the system level checking and can indicate explicitly which SE is faulty. However, the fault coverage of AFDU is lower since only error at 2 of the 4 bits can be detected since each router stage uses only 2 tag bits in the destination field to route. If fault occurs in the other 2 bits or in the randomizer stages, it cannot be detected by the AFDU. In these cases, fault location procedure has to be invoked to determine the faulty chip if SLFD indicates a fault.

4.4 Fault Location

When an AFDU detects a fault, fault location procedure only runs on the crossbar of that SE. For a fault detected at the system level. The faulty SE and also the faulty chip is not known. Fault location procedure checks every SE in each chip until the fault is detected. Each fault model is tested sequentially. A fault location unit supplies input test vectors to the crossbar and then analyzes the output responds to find out the faulty node. The simplified configuration is shown in Fig.6. When the crossbar enters the test-mode, the PMOS will switch on and pull up the output column. Then, the testing for different fault models begins.

- *Short-stuck* : Every node in the crossbar is switched off. Then a '0' will be applied to the row under test(row X) while the other rows have '1' at the input. If there is SS in (X,Y), The Yth column output will be '0'. The test continues until each row is tested.

- *Stuck-1/0 & open stuck* : In both cases, the similar detection procedure used for SS fault can be used. Every node in the row under test(X) is switched on. A '0' is applied to the input. Normally, every input to the fault detection unit will change to '0'. In case of a stuck-1 and open stuck fault in (X,Y), column Y remains '1' and the fault is detected. In case of the input is stuck at '1' or open, all the output columns will remain '1' and the fault is detected. The input signal then goes back to '1'. Stuck-0 in (X,Y) can be detected if the Y column does not change back to '1'. In case of the input is stuck at '0', all the output columns will remain '0' and the fault is detected.

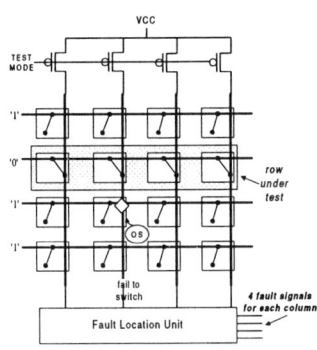

Figure 6. Fault Location Circuit

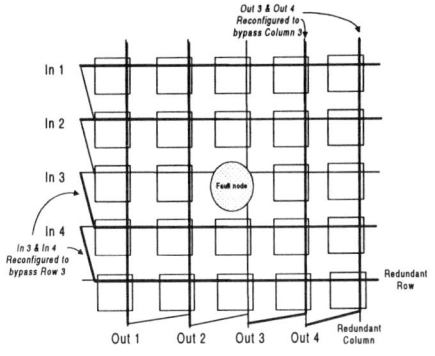

Figure 7. Crossbar with redundancy

Operating Voltage	3.3V
Transistors Count	39,500
Chip Area	4x3 mm^2
I/O number	166
Maximum Frequency	100MHz

Table 1. Chip Design Summary

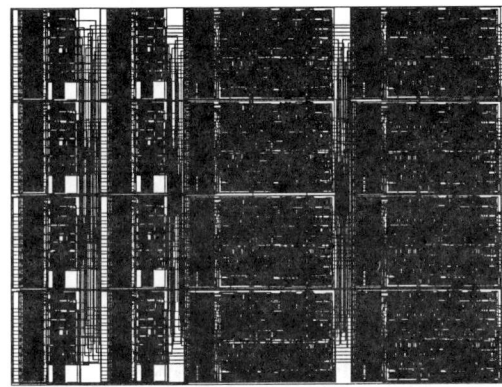

Figure 8. VLSI Layout of the 16x16 ATM/IP Switch

To complete a testing sequence, *4(for SS)+8(for S-I/O & OS)= 12* clock cycles are needed for each crossbar. If all the crossbars in each switch element are free from fault. It suggests fault occurs in the control logic and fault recovery has to be done in the system level.

4.5 Fault Recovery

Fault recovery can be done after fault is located. A crossbar with a redundant row and column can be used for fault recovery [7]. It is shown in Fig.7. Normally, the redundant row and column are not used. When the node(X,Y) is faulty, it can be located by the testing sequence discussed in section 4.4. Then, the connections to the inputs and outputs are reconfigured to bypass the faulty row and column. Fig.7 shows an example. By doing so, faults in a single row and column are recovered. Fault tolerance is achieved and performance is not degraded.

5. RESULT

The 16x16 switch has been implemented using 0.8um CMOS technology. Fig.8 shows the layout of the design. The chip is simulated to run at 100MHz. The maximum throughput is 1.6 Gbps per channel for a 4 chip, 16 bit switch. Because of blocking [8], the effective throughput is less, but is still more than adequate to handle the line rate of 622Mbps. Table.1 summarized the other parameters of the chip.

6. CONCLUSION

In this paper, we proposed and designed a multistage, self-routing 16x16 ATM switch which can handle the line rate of OC-12. The switch is built by cascading columns of 4x4 Switch Elements. Bit-slice approach is used to reduce the number of I/O and provide a simple fault detection mechanism. Identical chips can be used in parallel to increase the effective bandwidth. Internal blocking is minimized by using a randomization stage to provide multipath for any input-output pairs. Fault detection, location and recovery schemes which can improve the robustness and fault tolerant capability of the switch were also discussed.

7. REFERENCES

[1] C-L Wu and T-Y Feng, "On a class of multistage interconnection networks," IEEE Trans. On Computers, vol-29, no. 8, pp. 694-702, Aug. 1980.

[2] Ra'ed Y. Awdeh and H.T. Mouttah, "Survey of ATM switch architectures", Computer Networks and ISDN Systems, vol. 27, pp. 1567-1613, 1995.

[3] I. Lliadis and W.E. Densel, "Performance of Packet Switches with Input and Output Queueing", Int. Conf. Communication, Vol. 2, pp. 747-753, April 1990.

[4] C-T Lea, B. Li, and C-Y Tsui, "A/I Net: A network that integrates ATM and IP", submitted to IEEE Networks.

[5] Chin-Tau Lea, "A Multicast Broadband Packet Switch", IEEE Transactions on Comm., pp.621-630, April 1993.

[6] A. Youssef, "Randomized Routing Algorithms for Clos Networks", Computers & Electrical Engineering, vol. 19,No.6,pp.419-429,1993.

[7] Dhiraj K. Pradhan, "Fault-Tolerant Computer System Design", Prentice-Hall,1996.

[8] J. H, Patel, "Performance of processor to memory interconnections for multiprocessors", IEEE Trans. Computer, Vol. C-30, pp. 771-780, Oct. 1981.

A SIGNALING PROTOCOL ARCHITECTURE FOR AN ATM MOBILE SIMUALATOR

Jeong-Ju Yoo*, Jea-Hoon Nah*, Jea-Hoon Yoo*, Yoon-Ju Lee*, David Hutchison**

* Electronics and Telecommunications Research Institute(ETRI)
Mobile Switching Section, Switching Technology Division, ETRI
161 Kajong-Dong, Yusong-Gu, Taejon, 305-350, KOREA

** Lancaster University
Computing Department, Lancaster University
Lancaster LA1 4YR, United Kingdom

ABSTARCT

Our ATM mobile simulator to be presented is a device to include the mobility functions to be interacted with an ATM switching controller. In this paper, we summarized how to easily implement mobile signaling protocol of the ATM mobile simulator to verify mobility functions to be located within an ATM mobile switching controller. A basic functional model is proposed in association with mapping the model onto our mobile ATM simulator. This paper also presents the signaling protocol stack and signaling protocol procedures for our ATM mobile simulator to support mobile terminal access control, bearer control and call control.

1. INTRODUCTION

IMT-2000(International Mobile Telecommunications-2000) is intended to provide both the worldwide roaming and offer services such as multiple rate, multimedia services at a maximum of 2Mbps depending on user requirements. In addition, it should be compatible with various fixed network services as the same quality and security of the fixed network. Adaptation of ATM into IMT-2000 will be of interest because ATM has greatly advanced technologies capabilities in the current fixed network. There were attempts to integrate mobile network architectures and signaling procedures into B-ISDN[1][2].

In ITU(International Telecommunication Unit), meanwhile, IMT-2000 standardization of radio part has been developed in ITU-R TG6/1 and network part in ITU-T SG11/WP3. Currently, work on IMT-2000 network standardization is focused on stabilizing the basic functional modeling and the network reference model[3]. As there are still many problems to be solved, the completion of a standard seems a long way off. One of reasons is that the basic functional modeling and the network reference model are often changed at meetings. The information flow among the functional entities within these models is yet to be resolved, as it depends on the basic functional model. Therefore, it is desirable to move forward in both development and standardization.

This paper shows how to implement easily a current IMT-2000 functional model onto our mobile ATM simulator. In addition, the system configuration of our ATM mobile simulator is introduced in the chapter 2. The chapter 3 presents the signaling protocol stack and the design framework for our simulator. In the chapter 4, the handling of each protocol entity is described using the message sequence chart.

Finally, we like to conclude with some discussions about our further work so as to make progress towards a complete ATM mobile network in future.

2. PROTOCOL ARCHIETCTURE AND SYSTEM CONFIGURATION

Figure 1 shows a functional model to support the mobility of IMT-2000 at the upper side and a protocol architecture to be implemented within physical entities of a mobile station, a basic station and a mobile switching section at the lower side.

ITU-T IMT-2000 reference model consists of two planes - RACP(Radio Access Control Plane) and CCP(Communication Control Plane). The CCP is responsible for the overall communication control and the RACP is in charge of assigning and supervisory radio resource. The CCP seems to be so ambiguous, meaningless and fat that most functional entities are concentrated on this plane. It is difficult to instantly understand the overall network concept and it is not easy to inter-work with other networks such as B-ISDN and AIN(Advanced Intelligent Network). Neither any guidance for implementation nor layered structure is available in the IMT-2000 model. It is difficult to match between the functional model and the protocol architecture. On the other hand, the separation of the connection control from the call control function is characterized by locating independent functional entities, CnCAF, CnCF, CCAF and CCF in this model. The separation of call and connection control in UMTS was discussed [4]. The separation was addressed in IN(intelligent Network) call modeling for UMTS, and required the changes on the current IN model and proposed new architecture of SSF(Service Switching Function) for the handover. Heeralall[5] proposed the need of call and connection separation for multimedia services and handover in IMT-2000. IMT-2000 should support various services such as multimedia services in the near future. Such multimedia services require to set up and release connections during a call. In addition, call continuity has to be guaranteed during handover by changing of connections. Therefore multimedia services and the handover function in IMT-2000 require the separation of call and connection.

Each plane communicates with each other over plane interaction when they need it. The CCP includes eleven functional entities at a network side and four functional entities at a mobile side. The RACP has six functional entities, two at a mobile side and four at a network side. Call control function(CCF) is located at the network side and call control agent function(CCAF) at the mobile side, and both provide with call control processing. Connection control agent function(CnCAF) at a mobile side and connection control function(CnCF) at a network side function connection control. Therefore, it is possible that both the call and connection functions are controlled independently. Work describing the other functional entities can be found in [6].

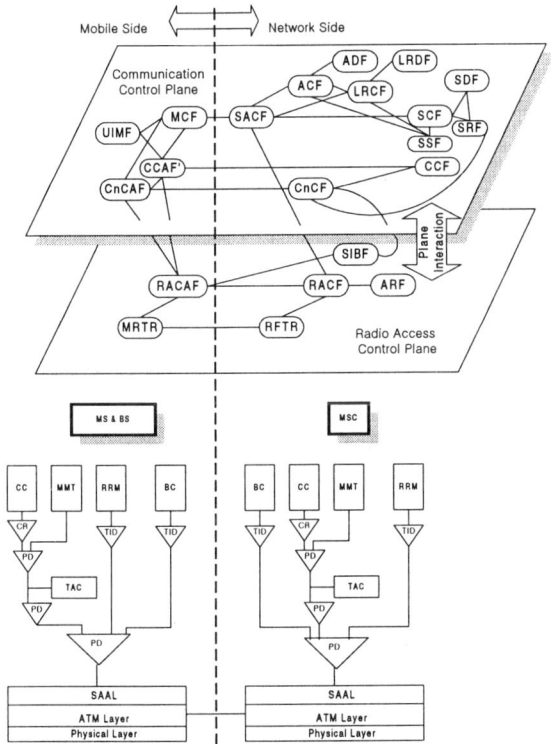

Figure 1. Functional Model & Protocol Architecture

Functional entities described above have to map onto protocol architecture at the lower side of Figure 1 so as to provide with ATM mobile signaling. Terminal identity(TID), call reference(CR) and protocol discriminator(PD) are used for discriminating among protocol entities to be described in the following chapter. Each protocol entity is mapped onto each protocol entity, for example CnCAF and CnCF to BC, and CCAF and CCF to CC. This protocol architecture is used in our project. An exiting ATM signaling has the well-defined interface such as DSS2(Digital Signaling Set 2) and ISUP(ISDN User Part) of No.7 signaling system and it also has call control and bearer control.

Figure 2 shows the system configuration to include the simulator. A signaling virtual channel(SVC) is used for signaling between the simulator and an ATM IMT-2000 exchange(A-IMX). On the other hand, permanent virtual channel(PVC) is used for the communication path between the A-IMX and a web server. In addition, IP over ATM is supposed to be used between mobile stations and the simulator. Mobile communications in this system configuration could take place in two different forms : fixed web server-to-mobile station and mobile station-to-mobile station. These communications are supposed to be demonstrated services in our project.

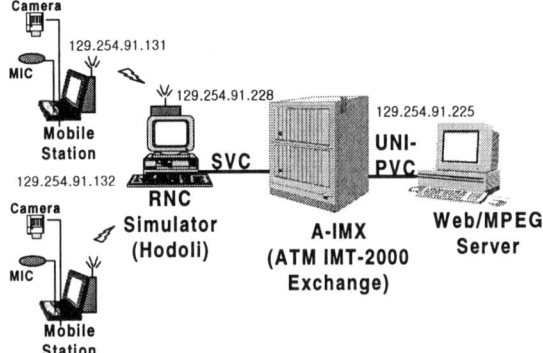

Figure 2. System Configuration of a Simulator

3. SIGNALING PROTOCOL FOR AN ATM MOBILE SIMULATOR

The following figure 3 shows the signaling protocol stack of an ATM mobile simulator and a mobile switching controller.

Establishing ATM wireless development environment has many difficulties at initial implementation phase. This section describes our project named an ATM mobile simulator. This simulator has mobile protocol entities so as to test mobile functions located in the mobile access local switch of ATM mobile switching controller(MSC).

Current B-ISDN UNI protocol stack is a single protocol over a simple point-to-point or point-to-multipoint fixed interface. However in the wireless mobile network, the connection of a mobile terminal must be changed from one access network to another access network through a handover process. The handover functionality assumes that the radio access network has the capability to dynamically setup and release bearer connections during the call. These features imply the need for call and bearer separation at the UNI. Moreover, call control association should uniquely identify each active mobile terminal. In Figure 3, signaling protocol stack has the modulated structure that allows the separation among these functions - call, connection, terminal access, location registration, and mobility management. Therefore, it is easy to implement each function. It is also possible for each protocol entities to move from this simulator to other physical entity. Protocol entities communicate with each other using the predefined internal messages in order to support mobile services.

Blocks indicated as a rectangular box in Figure 3, a implementation unit, are described as follows:

- Processing GUI(Graphic User Interface)

GUI is responsible for the Man Machine Interface to process all interfaces between a operator and the simulator. GUI to be implemented by JAVA should be able to show the predefined graphic motions, based on the processing output from application parts(APs) like CC, BC, and so on. Each mobility function within the MSC can be checked its operations.

- Interface between GUI and AP

This provides with the interface between GUI and protocol entities. It analyzes the message from GUI and then decides the designated protocol entity to be received the message among protocol entities. It also makes a role of a mediator to and from GUI and AP after changing into the suitable message format.

- BC(Bearer Control)

BC controls the bearer between MSC and the simulator and it also maintains radio and network resources. It also support the function of a hand-over.

- CC(Call Control)

CC handles both mobile incoming and outgoing calls. CC is implemented using the mobility signaling protocol based on Q.2931.

- MMT(Mobility Management- Terminal)

MMT controls terminal mobility to support terminal authentication, and terminal registration and update.

- TAC(Terminal Access Control)

TAC handles the association of the mobile terminal to the network such as the terminal and network relation setup function and the paging function.

- RNC(Radio Access Network)

RNC functions the interface between AP(Application Part) and ATM protocol. It analyzes a message from AP or Signaling ATM Adaptive Layer using a protocol discriminator and then it decides the appropriate block to be received a message. It makes a role of a mediator between the BS and MSC, so it gives the interface between the BS and MSC.

- SAAL, ATM and PL

These blocks are implemented with B-ISDN signaling protocol- AAL, ATM and Physical Layer. It is installed using the commercial ATM adaptive card. Inter-working is performed between SAAL to AC by exchanging the predefined API(Application Programming Interface).

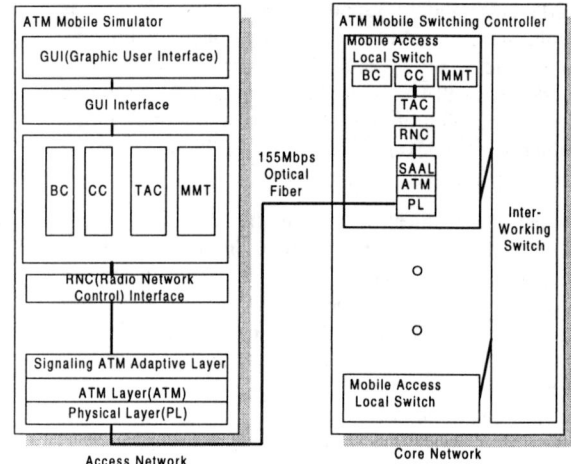

Figure 3. Signaling Protocol Stack for a Simulator

The MSC will have the symmetrical protocol stack with an ATM mobile simulator. For signaling protocol between an ATM mobile simulator and ATM mobile switching controller, the advanced B-ISDN signaling protocol is supported. The optical fiber at 155 Mbps between the MSC and the simulator is the physical interface.

4. SIGNALING PROTOCOL PROCEDURE

A signaling protocol framework for our simulator is designed on SDL(specification and description language) editor using ObjectGEODE[7]. The framework contains six major blocks which are graphic user interface(GUI), GUI protocol entity(GUI_PE), call control(CC), bearer control(BC), mobility management terminal and terminal access control(MMT-TAC), and radio access control(RNC). The blocks are connected with each other by channel instance that is a means of conveying signals. Two kinds of interfaces are categorized as static and dynamic interfaces as follws:

- Static interface

Signals are used to exchange information with other blocks, call control, bearer control, mobility management terminal and terminal access control, and radio access control interface blocks. Signal "AAL_RNC_BS" has a byte string format and it is used to convey signaling messages between the block RNC and Q.SAAL. Signals, "RNC_CC_BS", "RNC_BC_BS" and "RNC_MM_BS", have the same signal structure of the signal "AAL_RNC_BS". They support to transfer the signaling messages between the RNC and the other blocks such as CC, BC, and MMT_TAC.

- Dynamic interface

Each block performs the dynamic behavior to connect the

ATM mobile simulator to the MSC. For example, the block BC of the simulator communicates with the BC located within the MSC to establish and release connections

Figure 4 shows the signaling procedure among blocks, GUI, GUI_PE and MMT_TAC for the location registration, the mobile-to-fixed terminal communications, and the mobile-to-mobile communications. The signaling protocol procedure below is focused on blocks, MMT and TAC. Interface "A" is the reference point of the interaction between GUI and GUI Interface, and reference point "C" indicates the interaction between GUI Interface and TAC/MMT.

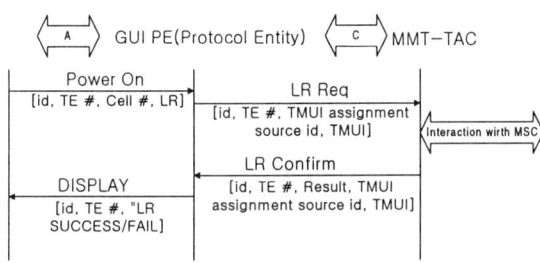

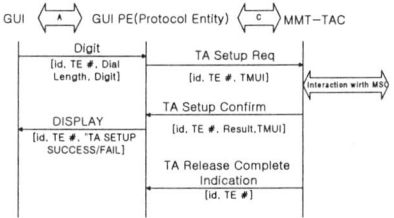

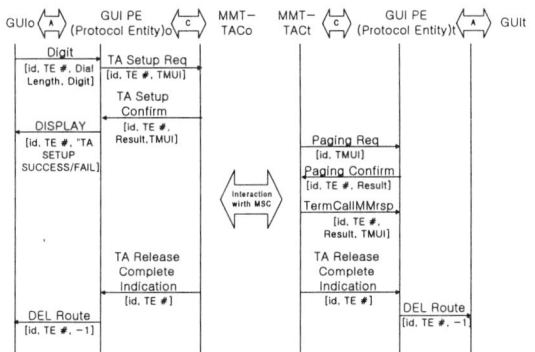

Figure 4. Message Sequence Chart for MMT and TAC

In case of "B" and "C", signaling procedures include both the setup establishment and release processing. All interworking signals among blocks, excepting the signal "Paging Req", include signaling identity(ID) and terminal equipment number(TE) as the common information element. IDs are identified as "Power On", "DISPLAY", "DIGIT", "TA Setup Req", "TA Setup Confirm", "TA Release Complete Indication" and "DEL Route". Blocks, CC and BC, have such a similar signaling procedure as one shown in figure 4. Therefore, the total signaling procedure for mobile communications will be completed through the interaction among these five blocks.

5. CONCLUSION

One of issues is to integrate a fixed network and a wireless network. The ATM gives a good solution to achieve an integrated network even though some requirements such as mobility function are required. There is advancement of the IMT-2000 standardization by ITU but the current status is not stable. We can't expect when the complete specification will come out soon. It takes a long time and it seems to have some problems in the basic model.

Our signaling protocol architecture show how to implement the IMT-2000 functions within the ATM mobile simulator and how to map the functional model onto the protocol architecture. We presented the signaling protocol architecture of a mobile ATM simulator with modular protocol entities. This signaling protocol was described using signaling protocol procedures. The practical implementation of the mobile ATM simulator is our ongoing project.

Our basic objective of this project is to demonstrate the feasibility of a radio interface independent ATM mobile switching controller through the prototype implementation of a mobile ATM simulator.

We are also interested in developing the wireless and wired integrated signaling standardization, adapting AIN to a mobile ATM network and providing compatibility to a mobile network in order to interact with existing other networks.

6. REFERENCES

[1] B. Marchent, 1995, "Performance Evaluation of Network Architectures and Signaling Procedures for Integration of UMTS into B-ISDN/IN", Mobile Telecommunication-1995, 377-381.
[2] A. Bora and D. Cox, 1996, "Handling Mobility in a Wireless ATM Network", INFOCOM'96, 1405-1413.
[3] Paj Pandya, et al 5, 1997, "IMT-2000 Standards : Network Aspects", IEEE Personal Communications. August 1997, 20-29
[4] W. van der Brock and P. Massafra, 1995, "UMTS requirements on IN call modeling", Mobile Telecommunications 1995 Summit, 334-349
[5] S. Heeralall, 1997, "A mobile network view on call/connection separation", ITU-T, TD3/11 BATH 052
[6] E. Chien, 1997, "Version 8.0 of draft new recommendation Q.FNA, Network functional model for IMT-2000", ITU-T, TD3/11 BATH 172.
[7] Cellware, 1997, "CELL-EXPRESS", Technical Manual, September 1997.

COMPILE-TIME PRIORITY ASSIGNMENT AND RE-ROUTING FOR COMMUNICATION MINIMIZATION IN PARALLEL SYSTEMS

David R. Surma, Edwin H.-M. Sha, and Peter M. Kogge

Department of Computer Science and Engineering
University of Notre Dame
Notre Dame, IN 46556

ABSTRACT

The performance gains of massively parallel systems can be significantly diminished by the inherent communication overhead. This overhead is caused by the required message passing resulting from the task allocation scheme. To minimize this overhead, a *hybrid* static-dynamic scheduling technique is presented. The static phase makes use of *a priori* information at compile-time to assign priorities to each message transmission. The priorities are determined using the recently developed *Collision Graph* model and are utilized at run-time to arbitrate the message transmissions. Determining an optimal priority scheme is an *NP-Complete* problem. Therefore the developed techniques employ heuristics and a flexible routing scheme to deal with a general case model of message traffic. Experiments performed show a significant improvement over baseline approaches.

1. INTRODUCTION

With the advent of massively parallel machines there have been considerable gains made in reducing task processing times. However, these gains are significantly diminished by the inherent communication overhead. As one of the point design teams to develop *Petaflop* super-computers [1], our research group encountered such a problem while implementing a parallel solution for simulating partial differential equations, representing fluid dynamics problems. With the platform being a *tightly-coupled* architecture such as the processor-in-memory *EXECUBE* [2], we realized that the communication overhead impeded our efforts to obtain an optimized execution time. To reduce this overhead, we present a study of the communication incurred when nodes transfer information. Our novel technique involves both compile-time analysis and run-time scheduling. Experiments show significant improvement compared to baseline approaches.

The creation of a new scheduling technique was required since most existing scheduling methods do not consider the communication characteristics of the problem [3, 4] and are unable to achieve an optimal schedule. Furthermore, most techniques developed for parallel compilers do not consider this overhead [3, 5]. This research assumes that a suitable task allocation scheme has been used and deals specifically with the ordering and routing of the message transmissions. Therefore, the new scheduling technique is much different from *traditional* multi-processor scheduling [4] because it schedules at a lower level. Static techniques, while being able to achieve an optimal or near-optimal solution, require

THIS WORK WAS SUPPORTED IN PART BY NSF MIP 9501006, NSF ACS 96-12028, AND JPL 961097.

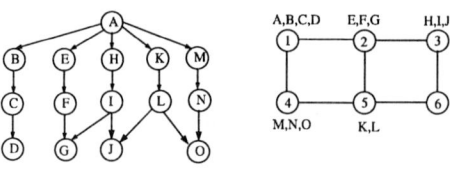

Figure 1: (a) Task Flow DAG. (b) Tasks assigned to processing nodes

known information about the message traffic. Unfortunately, this *a priori* information may be unavailable or inaccurate. Dynamic scheduling techniques suffer from being unable to utilize information that might be known about the processing environment. Thus, this research presents a *hybrid* technique utilizing the appealing components of both approaches.

To exemplify this type of scheduling, consider the task *directed acyclical graph*, or *DAG*, of Figure 1. Figure 1(b) shows one possible assignment of this graph to a two-dimensional mesh network of six processors. While tasks assigned to the same processor require no internode communication, this assignment scheme indicates that messages must be exchanged. For example, node 1 sends messages to nodes 2, 3, 4, and 5 corresponding to edges $A \rightarrow E, A \rightarrow H, A \rightarrow M$, and $A \rightarrow K$ of the *DAG*. Since there is only a single bidirectional link between each node, network *collisions* occur. By *collisions* we mean that messages will compete for at least one physical link in the network. Table 1 gives possible orderings of the resulting message traffic when *XY-routing* is used. Messages on the same line may be sent in parallel without collisions. In *worm-hole* routed networks, the time to transmit a message is relatively distance insensitive [6] so we can assume that equal length messages will take the same amount of time, t, to traverse the network. Thus, schedule 1 gives an ordering which completes at time $4t$ while schedule 2 completes at time $3t$, a savings of 25% based on the communication schedule. An even greater amount of improvement can be obtained if message $(A \rightarrow K)$ is re-routed to traverse in a YX direction. This new schedule is also shown in Table 1 with the re-routed message denoted as $A \rightarrow K'$. The completion time of this new ordering is $2t$. Thus, this work addresses the ordering or *scheduling* of the messages as well as the re-routing of some of them to reduce the overall completion time.

The term used for this research is *communication scheduling*. It not only encompass routing aspects and path selection issues as discussed in [6,7], it also determines the order that the messages in the system should be sent. There have been several studies related to this problem. One effort develops a `traffic scheduling' algorithm for multi-processor networks to balance the network links

Schedule 1	Schedule 2
$A \to E; A \to M$	$A \to K'; A \to M$
$A \to H$	$A \to H; L \to J; L \to O$
$A \to K\ I \to G$	$A \to E; I \to G$
$L \to O; L \to J$	
Re-routed Schedule	
$A \to K'; A \to H$	
$A \to E; A \to M; I \to G; L \to O$	

Table 1: Example Communication Schedules

Message ID	Est. Departure Time	Source	Destination
1	1.0	(3,1)	(7,7)
2	2.0	(2,5)	(5,7)
3	3.0	(3,4)	(5,6)
4	4.0	(2,2)	(7,8)
5	1.0	(3,1)	(6,4)
6	6.0	(2,1)	(6,3)
7	7.0	(2,1)	(5,4)

Table 2: Example message list

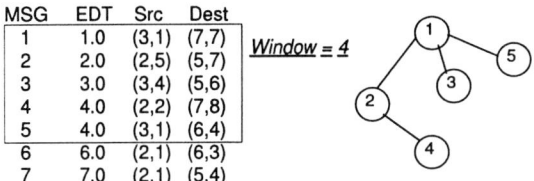

Figure 2: Collision Graph for S with window = 4.

Definition 1 *A message is defined to be* $M = (m_{edt}, m_S, m_D)$ *where m_{edt} is the estimated departure time of the message, m_S is the source node of the message, and m_D is the destination node of the message.*

The first step in the communication minimization process is to determine priorities for each message. The algorithm to do this is called the *Priority Mapping and Re-routing, or PRIMAR* algorithm and it begins by transforming the problem into a graph model, called a *collision graph* or *CG*.

Definition 2 *A CG is defined as* $G = (V, E)$ *where V is the set of nodes $v_1, v_2, ...v_N$ representing messages $M_1, M_2, ...M_N$; and $E = \{(v_i, v_j)|$ the paths of M_i and M_j intersect.$\}$.*

Since the estimated departure times vary throughout the message list, it is possible that two messages can traverse the same paths without colliding if these times are sufficiently far apart. Consequently, a CG is not constructed for the entire message list. Rather, the message list is first sorted by estimated departure time and then processed in sections determined by a user input parameter called a *window*. This window is used as the range for the message traffic departure times to be operated on as a set, S. Figure 2 shows a CG constructed for the nodes in S from Table 2 when the window parameter is 4. To get the ordering from the undirected CG, arrows indicating message precedence must be added to the graph. An edge directed from $v_1 \to v_2$ denotes that the message corresponding to v_1 is to be scheduled before the message corresponding to v_2. If no edge exists between any two nodes they may be scheduled in parallel.

Once an edge orientation has been established, the actual priorities are determined by first finding the node(s) without any incoming edges and assigning them the highest priority. Next, these nodes and their edges are removed from the graph, and the process repeats assigning the next highest priority and so on for all messages. Thus, the major problem is determining the edge orientation for the CG that yields a priority scheme which produces the best performance. Central to getting the best performance is finding the *maximum* number of messages that can be transmitted in parallel at any one time. This correlates to finding the *maximum independent set* from the CG. Since finding a maximum independent set is an *NP-Complete* problem, our problem is also NP-Complete, and heuristics are needed to arrive at a solution.

Consider again the CG of Figure 2. The maximum independent set is 3 comprised of nodes 2, 3, 5. Those messages will be assigned priority 0 (highest) and are said to be in S'. The other nodes in $S - S'$ have collisions with the nodes in S'. Therefore, to enlarge S' *re-routing* of the messages in $S - S'$ is considered. Re-routing in a process where the message routing path is changed from XY to YX. However, since deadlocks are a concern in wormhole routed networks, some restrictions are required. 8 turns

based on the fact that a large number of messages must eventually be delivered [8]. Their work, however, uses a *First-Come First-Served, FCFS,* approach and does not perform any *scheduling* of the individual message transmissions. Lee and Kim perform path selection in a wormhole routed network but they search for *unique* paths for pairs of communicating nodes [7]. Kandlur and Shin [9] present a work similar to [7] in that dedicated paths are found. The problem with these techniques is that the dedicated paths can cause other messages to follow longer paths even though the dedicated links are unused. Additionally, no scheduling is done which can improve the overall performance. Recent work by Eberhart and Li [10] does perform a type of dynamic communication scheduling. However, their work is restricted to analyzing communication patterns that are commonly used in data parallel applications. The work presented here can apply to any type of message-passing activity.

This paper presents a hybrid technique which uses known information about the required message traffic to *statically* determine priorities for the individual messages. Then, at run time when a node has several messages to transmit along the *same* physical link, *preference* is given to the message with the highest priority. The basis for the priority determination is the recently developed *collision graph* model [11]. The communication scheduling problem has been addressed previously in a purely static manner using fixed routing and a specific message traffic model [12]. This research greatly improves this effort by presenting a technique for a general model of message traffic which allows re-routing of messages and operates dynamically.

2. COLLISION GRAPH AND PRIORITY ASSIGNMENT

This starting point is a list of N messages to be transmitted by the network nodes. The goal is to find an optimal communication schedule which reduces the overall processing time. Table 2 shows a sample message list to be executed on a $10X10$ two-dimensional mesh processor network. This work considers single packet messages composed of an arbitrary number of flits. Nodes of the multiprocessor system are attached to all-port routers and the routing scheme is XY as the default or a re-routed scheme which will be discussed shortly.

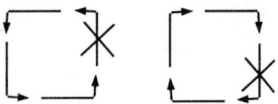

Figure 3: Illustration of allowable routing turns

are possible in two-dimensional mesh networks and XY routing is deadlock free by prohibiting 4 turns. We only restrict the 2 turns shown in Figure 3. Thus, our term for this type of routing is XY and *restricted YX* routing.

In the example, nodes 1 and 4 are eligible to be re-routed since they do not violate the turn restrictions. Node 1 is arbitrarily selected first for re-routing and it can be routed in a YX direction without colliding with any message in S'. Thus, it will be assigned priority 0, added to S', and its *routing flag* set to YX. This flag is part of the flit header and each router must be able to interpret it for proper routing. Next node 4 is considered. Since if it is re-routed it will collide with a member of S' (re-routed message 1), it cannot be re-routed. After the nodes with top priority have been determined, they will be eliminated from the graph and the nodes in $S - S'$ will be *aged*. (Node 4 is this example.) Aging is a process where messages have their departure times updated to a later time. The value used for aging is determined by the length of the standard message. Next, the entire list of remaining messages are resorted and the process repeats assigning priority 1. This is done until all messages have an assigned priority. The algorithm is executed with several window sizes, a metric produced and the best priority scheme used. The *PRIMAR* algorithm is now formally presented.

Algorithm 1 *PRIMAR*

Input: $G=(V,E)$, and M
Output: $M(v)_{pri} \; \forall v \in V$

begin
    $pri = 0$;
    Input window from user;
    $I = \emptyset$;
    repeat until $V = 0$;
        sort V by estimated departure time, edt;
        limit1 = earliest estimated departure time of a node $v \in V$;
        limit2 = limit1 + window;
        Build $G_t = (V_t, E_t)$
            such that $V_t = \{v | \; limit1 \leq M(v)_{edt} \leq limit2\}$
            and $E_t = \{e | \; u \xrightarrow{e} v \; and \; u,v \in V_t\}$;
        Determine the maximum independent set, $I \subset G_t$;
        $\forall v \in I, M(v)_{pri} = pri$;
        $\forall v \in G_t \notin I$, Explore re-routing for each v
        If re-routing can be done, $M(v)_{pri} = pri$,
            path direction = YX, and add $M(v)$ to I;
        $\forall v \in (V_t - I) \; M(v)_{edt} = M(v)_{edt} + age$;
        $pri = pri + 1$;
        $V = V - I$;
    end loop;
end algorithm PRIMAR

3. HYCORE TECHNIQUE AND RESULTS

The *Hybrid Communication Scheduling with Re-routing, or HYCORE,* technique utilizes the results of the *PRIMAR* algorithm. At run-time each node selects a message to transmit based on several factors. If a node has only one message ready to transmit, it checks the routing flag and if the appropriate link is available the message is transmitted. However, if the node has several messages that are ready to be transmitted, the priority is used as the *arbiter*. A simulation program was developed to determine the time a message reaches its destination and a performance metric established. The timing information was derived from using 1.0 unit for a single flit [6] to transfer one hop in the network. The actual time corresponding to this *unit* value is dependent on the size of the flits and the transmission speeds in the interconnect. Uniform message lengths composed of 10 flits are assumed. The performance metric is the average completion time, or *ACT*, for all messages transmitted. The *ACT* is used because our focus is on the *individual* message transfers. While we are interested in having the shortest final completion time we also want to have as many messages transmit as soon as possible. Thus, by using the *ACT* we can distinguish between two schedules which have equivalent final schedule completion times.

In the example message list of Table 2, the *ACT* for a statically determined schedule is 25.0. A *FCFS* approach has a time of 28.86 while our hybrid approach without re-routing yields a value of 26.14. Utilizing re-routing the static approach value decreases to 21.57 while the *HYCORE* technique is 22.14. Thus, the improvement gained by the *HYCORE* technique over a *FCFS* approach is a significant 23.28%. The statically determined algorithm being the best makes sense because if exact information is known *a priori* about the message traffic a schedule can be optimized. However, obtaining this information with much accuracy is difficult. Consequently, in experiments a *variance* is introduced which accounts for network uncertainties, congestion, and other performance fluctuations. This variance is distributed uniformly over the estimated departure times of all messages.

Two models of message traffic were considered in our experiments. First, LU factorization, matrix multiplication, and bitonic sorting were analyzed to determine the message passing that occurs when they are mapped to a two-dimensional mesh architecture. *ACT* values are given in Table 3 for the results of the *SCORE* static scheduling algorithm utilizing re-routing [13], a *FCFS* approach both with and without re-routing, and the *HYCORE* techniques. In this table the variance is 0 so the static approach again performed the best. Further note that the *HYCORE* technique outperforms the *FCFS* approach by approximately 20%. Table 3 also shows the results when the variance is 4. Static scheduling no longer works best as it must compensate for worst case times, and *HYCORE* still works better than *FCFS* although the percentage is not as great. This is due to the deteriorating accuracy of the information used to determine the priorities. It is still better indicating that having some knowledge, albeit not totally accurate, improves the performance.

Table 4 shows results obtained when applying the scheduling techniques to randomly generated traffic patterns consisting of 30 messages. A *hotspot* index was used to vary the amount of collisions by causing the message destinations to be in a certain area with a given percent. The results are averages of 100 trials for each case. Note that the differences in the amount of improvement that can be obtained depends on the nature of the message traffic. The *HYCORE* technique works best on traffic where there is a moderate amount of collisions. At low collisions (10% hotspot index in the table), there is not much parallelism to exploit and consequently the improvement that can be obtained, while still significant, is comparably low. At high amounts of collisions, the *CG* resembles a *clique* where the *FCFS* approach will begin to work as well as other approaches. Since the comparison is with this *FCFS* approach, as the amount of collisions increases, the amount

Operation	Msgs Sent	SCORE	FCFS	Re-routed FCFS	HYCORE	% HYCORE Improvement	SCORE	FCFS	Re-routed FCFS	HYCORE	% HYCORE Improvement
LU Factorization	42	20.31	28.48	24.64	21.98	22.82	29.24	28.11	25.63	24.02	14.55
Matrix Multiply	96	20.97	26.09	22.13	21.45	17.78	32.71	27.84	27.44	23.97	13.90
Bitonic Sorting	200	60.71	80.28	71.38	67.31	16.15	89.43	81.34	73.56	70.79	12.97
		Variance = 0					Variance = 4				

Table 3: Comparison of scheduling techniques

Hotspot Index	SCORE	FCFS	Resched FCFS	HYCORE	Percent Imprved
10%	40.71	48.31	44.88	42.78	11.45
25%	44.28	55.17	47.71	45.63	17.29
50%	63.90	86.91	73.90	69.41	20.14
75%	100.33	134.23	109.76	105.98	21.05
90%	128.56	175.69	150.12	141.34	19.55

Table 4: Experiments with 30 messages and variance = 0

Msgs sent	SCORE	FCFS	HYCORE	Percent Imprved
10	43.55	37.95	34.63	8.75
20	48.89	45.24	38.94	13.93
30	56.46	54.21	44.18	18.50
40	61.91	58.30	46.56	20.14
50	64.87	62.78	50.89	18.94

Table 5: Experiments with 40% hotspot index and variance = 4

of improvement that can be obtained decreases. In the table note the falloff in improvement when the hotspot index exceeds 75%. In between these extremes, however, the improvement obtained by the *HYCORE* technique steadily increases to a maximum of 21%.

Two parameters are changed to study the effects of additional messages transmissions and also the introduction of a variance. Table 5 shows results for experiments using a 40% hotspot index and varying the *amount* of messages transmitted when the variance is 4. From this table it can be seen that the static *SCORE* technique performs poorly while the *HYCORE* technique is again better than the *FCFS* approach. Note that the amount of improvement begins to diminish when the number of messages is greater than 40. This is the case because more messages results in more collisions for a fixed hotspot index. As shown in the previous analysis, once the number of collisions becomes great, the performance diminishes.

4. CONCLUSION

This paper presents a new framework for studying communication scheduling. The *HYCORE* technique combines static and run-time elements along with re-routing to reduce the communication overhead by over 20% for both application-specific message traffic and for randomly generated message traffic. This technique will almost always perform better than a *FCFS* approach due to its using re-routing and since it acts first to schedule its messages on a *FCFS* basis. In the presence of variances, this technique will outperform baseline static scheduling techniques as well.

5. REFERENCES

[1] P. M. Kogge, S. C. Bass, J. B. Brockman, D. Z. Chen, and E. Sha, "Pursuing a Petaflop: Point design for 100 TF computers using PIM technologies," in *Proceedings of Frontiers of Massively Parallel Computation*, October 1996.

[2] P. M. Kogge, "EXECUBE- A New Architecture for Scalable MPPs," in *1994 International Conference on Parallel Processing*, vol. I, pp. 77–84, August 1994.

[3] H. Kasahara and S. Narita, "Practical multiprocessor scheduling algorithms for efficient parallel processing," *IEEE Transactions on Computers*, vol. c-33, pp. 1023–1029, November 1984.

[4] H. El-Rewini, T. G. Lewis, and H. H. Ali, *Task Scheduling in Parallel and Distributed Systems*. Englewood Cliffs, NJ: Prentice Hall, 1994.

[5] S. Shukla, B. Little, and A. Zaky, "A compile-time technique for controlling real-time execution of task-level dataflow graphs.," in *1992 International Conference on Parallel Processing*, vol. II, pp. 49–56, 1992.

[6] L. M. Ni and P. McKinley, "A survey of wormhole routing techniques in direct networks," *IEEE Computer*, vol. 26, pp. 62–76, February 1993.

[7] S. Lee and J. Kim, "Path selection for communicating tasks in a wormhole-routed multicomputer," in *1994 International Conference on Parallel Processing*, vol. 3, pp. 172–175, 1994.

[8] R. P. Bianchini and J. P. Shen, "Interprocessor traffic scheduling algorithm for multiple-processor networks," *IEEE Transactions on Computers*, vol. C-36, pp. 396–409, April 1987.

[9] D. D. Kandlur and K. G. Shin, "Traffic routing for multicomputer networks with virtual cut-through capability," *IEEE Transactions on Computers*, vol. c-41, pp. 1257–1270, October 1992.

[10] A. Eberhart and J. Li, "Contention-free communication scheduling on 2d meshes," in *1996 International Conference on Parallel Processing*, pp. 44–51, August 1996.

[11] D. R. Surma and E. Sha, "Collision graph based communication scheduling for parallel systems," *International Journal of Computers and their Applications*, March 1998.

[12] D. R. Surma and E. Sha, "Efficient communication scheduling with re-routing based on collision graphs," in *International Symposium on High Performance Computing Systems*, July 1997.

[13] D. R. Surma and E. Sha, "SCORE: An efficient technique to reduce congestion in parallel systems," in *To be presented at the Tenth International Conference on Parallel and Distributed Computing Systems*, September 1997.

A VLSI DESIGN OF DUAL-LOOP AUTOMATIC GAIN CONTROL FOR DUAL-MODE QAM/VSB CATV MODEM[1]

*Muh-Tian Shiue, Kuang-Hu Huang, Cheng-Chang Lu, Chorng-Kuang Wang, and *Winston I. Way*

Department of Electrical Engineering
National Central University, Chung-li, Taiwan, ROC

*Department of Communication Engineering
National Chiao Tung University, Hsin-chu, Taiwan, ROC

ABSTRACT

A digitized automatic gain control (DAGC) whose loop bandwidth can be automatically regulated by a digital quantizer is presented in this paper. The designed quantizer that only costs tens of gates provides the DAGC both wide loop bandwidth for fast acquisition and narrow loop bandwidth for low AGC gain jitter in stable steady-state. The receive bandpass filter, variable gain amplifier (VGA), and digital control circuits have been implemented in VLSI using $0.8\mu m$ CMOS technology. For both 64-QAM and 8-VSB signals, the closed-loop experimental results show that the designed DAGC has input dynamic range from $22mV_{pp}$ to $456mV_{pp}$, transient mode bandwidth 1kHz, steady-state bandwidth 90Hz, settling time of step response less than 2ms using 10MHz clock for digital control chip.

1. INTRODUCTION

Traditionally, CATV networks provide low-cost unidirectional broadcast distribution of 6MHz NTSC analog TV signal to the homes over coaxial cable. Future residential cable data services are expected to deliver broadband services and Internet access. CATV networks of the next-generation is HFC networks to provide additional emulated point-to-point services of telephone companies. In order to provide the broadband services, many facilities must be evolved to have broadband characteristics. All of those, the high speed digital CATV modem is one of the key facilities that successfully provide the broadband services. Quadrature Amplitude Modulation (QAM) and Vestigial SideBand (VSB) Modulation are two competing candidates for downstream broadcast of digital TV on HFC networks. For seamless evolution, the transmission spectrum of ATSC in U.S. is standardized to use the conventional NTSC 6MHz bandwidth. Furthermore, the Grand Alliance (GA) has standardized the VSB transmission scheme proposed by Zenith for the terrestrial broadcasting of HDTV [1] programs due to its better test results than QAM. Although DAVIC has standardized QAM scheme for digital video transmission over the last-mile of HFC network and QAM system is possible to be the standard of IEEE 802.14 Working Group, there is a debate about whether the VSB or QAM technology will be adopted for the digital CATV modem.

Since the transmission distance is different, the received signal amplitude will vary over a wide dynamic range. Besides, the phase-locked loop performances usually depend on the received signal level since the gains of carrier and timing phase-error detectors are usually proportional to the signal amplitude. To simplify the digital circuits and to facilitate the carrier and timing recovery circuits properly operate in the vicinity of their designed point, it is necessary to provide a signal amplitude controller to automatically adjust the received signal to a constant level.

Basically, limiting amplitude is not suitable for multiple level systems such as the 64-QAM and the 8-VSB signals for digital CATV modems [2, 3, 4]. AGC employs certain type of signal variation detector to generate the proper gain control voltage of VGA to achieve high performance signal amplitude control. Therefore, the feedback AGC is chosen.

In the rest sections of this paper, the proposed dual-loop DAGC (Digital AGC) VLSI architecture for dual-mode QAM/VSB digital CATV modems and its linear model analysis will be described first. Section III presents the closed-loop experimental results using the fabricated analog VGA /BPF /DAC chip, digital control chip, and commercial ADC chip. Finally, the summary is given in Section IV.

2. DUAL LOOP DIGITAL AGC

2.1. VLSI AGC Architecture

As mentioned above, there are two competing systems, which are QAM (Quadrature Amplitude Modulation) and VSB (Vestigial SideBand modulation), to provide high speed digital transmission for the digital CATV broadcast [1, 5]. Both QAM and VSB systems are bandwidth efficient, and use the conventional 6MHz bandwidth of NTSC (National TV System Committee) system. QAM system is originally proposed by GI/MIT, and VSB system is proposed by Zenith.

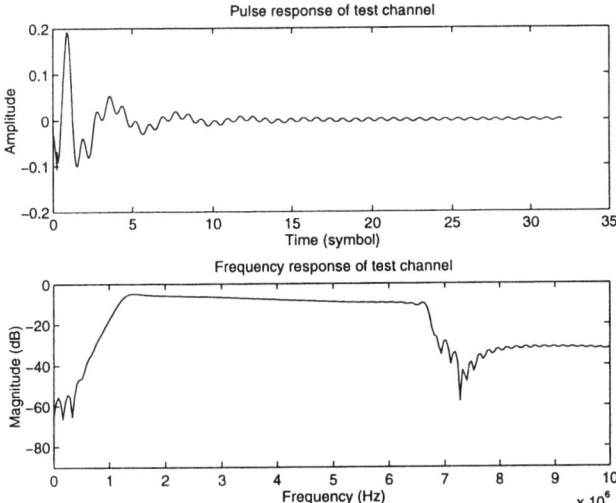

Figure 1: Characteristics of the test channel for CATV modems.

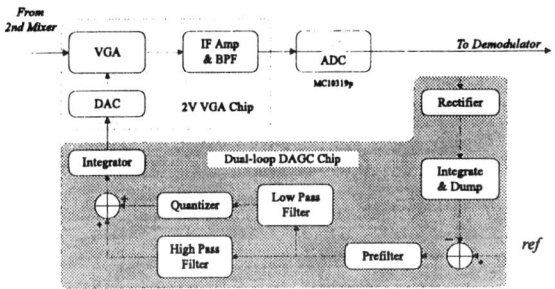

Figure 2: Block diagram of the proposed DAGC

In general, two classes of signal attenuation detectors, coherent and noncoherent detectors are often used. When the carrier phase estimation is available, the coherent detector is generally superior in performance. Fig. 1 shows a test channel response for digital CATV modems. The carrier of QAM system is allocated at the center of the pass band. On the other hand, the carrier of VSB system is positioned at the corner depending on the upper or lower sideband is used. Even there is a coherent carrier phase reference, the coherent detector is not suitable for both the QAM-based and VSB-based digital CATV modems since the magnitude of center and corner components are dramatically different. For such a case or a phase coherent reference not available, an envelope detector may be used to generate the gain control voltage of VGA.

A traditional analog gain control structure for AGC is less flexible than a digital one. Fig. 2 shows the proposed dual-loop DAGC VLSI architecture where the combination of digital gain control loop and high frequency analog signal path make the control loop more flexible by using DSP techniques. Basically, the proposed architecture employs the bandwidth regulation scheme [6] to achieve fast convergent acquisition and precise steady state response. The proposed DAGC architecture consists of analog circuits in the forward path as a conventional analog AGC and digital circuits in the feedback control path. With the tendency of mixed analog-digital circuit integration of IF front-end,
digitizing the received signal in IF band rather than in baseband is preferable in order to decrease the delay of AGC forward path. The output signal of ADC is rectified first, then integrated and dumped to estimate the power level of the amplified signal. The reference input that is used to set the desired power level of ADC output signal is subtracted from the dumped value to extract gain error information for automatically adjusting the VGA gain.

The loop filter is a key component to determine the AGC performance. In this proposed architecture, a dual-loop digital filter and a quantizer are employed to provide a wide loop bandwidth at the initial acquisition state while a narrow loop bandwidth at the steady state. In the beginning, the amplified signal power level is far from the reference input. The path of low-pass filter through quantizer provides a wider loop bandwidth that can converge fast but it induces more in-band noise. However, the low-pass filter in our design to provide wider bandwidth will be shut down automatically when the residual error signal is smaller than the quantization threshold after the initial acquisition procedure. In the steady state, only the conventional narrow bandwidth loop filter works alone. Notice that a first-order Butterworth low-pass prefilter is introduced to filter out the data pattern dependent noise, and to minimize the steady state fluctuation of quantizer output. This behavior forces the proposed AGC architecture to be very stable and only need short acquisition time.

2.2. First-Order Approximation Analysis

The output signal level of VGA is a function of the control voltage V_c. The exponential model for VGA is adopted in this design since the small signal VGA gain k_{vga} is proportional to the control voltage V_c only and independent on the input signal V_i. Fig. 3 shows the first-order approximation model of proposed dual-loop DAGC. Here, the s-domain model is used to illustrate the bandwidth regulated dual-loop DAGC for simplicity. The closed-loop transfer function $H(s)$ of the AGC output respected to the reference

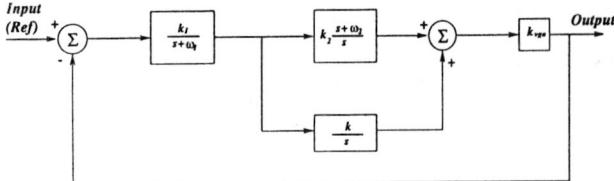

Figure 3: Linear model of bandwidth regulated dual-loop AGC

voltage V_{ref} is

$$H(s) = \frac{k_1 k_2 k_{vga} s + k_1 k_{vga}(k_2 w_2 + k)}{s^2 + (w_1 + k_1 k_2 k_{vga})s + k_1 k_{vga}(k_2 w_2 + k)}$$
$$= \frac{k_1 k_2 k_{vga} s + k_1 k_{vga}(k_2 w_2 + k)}{s^2 + 2\zeta w_n s + w_n^2} \quad (1)$$

where the natural frequency

$$w_n = \sqrt{k_1 k_{vga}(k_2 w_2 + k)} \quad (2)$$

and the damping factor

$$\zeta = \frac{w_1 + k_1 k_2 k_{vga}}{2\sqrt{k_1 k_{vga}(k_2 w_2 + k)}}. \quad (3)$$

Therefore, if $k = 0$ while the error signal is below the quantizer threshold, the natural frequency w_n will become

$$w_n = \sqrt{k_1 k_2 k_{vga} w_2} \quad (4)$$

and the damping factor ζ will become

$$\zeta = \frac{w_1 + k_1 k_2 k_{vga}}{2\sqrt{k_1 k_2 k_{vga} w_2}}. \quad (5)$$

The reduction of the closed-loop bandwidth and increase of the damping factor improve the stability in the steady state. Notice that when $k \neq 0$ the proposed architecture provides a large loop bandwidth and a proper low damping factor, which help speeding up the initial acquisition.

3. CMOS CIRCUITS DESIGN AND EXPERIMENTAL RESULTS

System behavior simulations were done in SPW, while VerilogXL simulations are for gate-level simulation of digital control circuit. The finite word length and ADC latency delay have been considered in behavior simulations. For saving adders, the absoluter, integrator-and-dumpor, and subtractor are assembled together. In real implementation, the absolute input value is subtracted from a residual value, after repeating 64 sample-times the result is dumped and then the reference is preset as the initial value of residual-value registers. All the digital filter coefficients are power of two, so the designed digital control circuit can avoid the multiplier via shifting and adding the input data. The dual-loop digital control circuits are implemented using standard cells of CMOS 0.8 μm standard cells. Cell Ensemble CADENCE is employed for auto-place and auto-routing. The total chip area is $4.7 mm^2$. For open-loop testing of digital control chip, the ADC simulation data are transferred into the designed chip by a pattern generator. The measured outputs are identical to the simulation results. The average power dissipation is 14mW under 21.52MHz clock rate using 5V power supply.

Considering the dynamic range of input signal from 40 mV_{pp} to 400 mV_{pp}, the DAGC has to provide at least 20dB passband gain control capability to maintain the output signal at $200 mV_{pp}$ for the following ADC. The analog chip consists of a twelfth order 4.035 MHz low-IF bandpass filter, at least 20dB dynamic range VGA, and 7-bit control DAC. The analog chip is implemented in 2V 0.8μm double poly double metal CMOS technology. The receive BPF is partitioned to several sub-filters. The VGA is accomplished by the variable transconductor in front of each of sub-filters [7, 8, 9]. The analog VGA /BPF /DAC chip is also tested individually using 2V power supply. The measured results show that this analog chip has an active area $2.1 mm^2$, passband from 1.2MHz to 7MHz, 27dB gain-control range from $22 mV_{pp}$ to $456 mV_{pp}$, THD 48dB, and dissipates 30mW using 2V power supply.

In order to demonstrate the attractive characteristics, the proposed DAGC architecture is realized in hardware using our digital control chip, analog circuits VGA /BPF /DAC chip, and commercial discrete components such as ADC and general purpose OP-amp to provide 20dB constant gain for the adopted commercial ADC input. For measuring the input signal dynamic range, the sinusoidal input is generated by a function generator. Both 64-QAM and 8-VSB signals were used to test the closed-loop dynamic characteristics of step response such as settling time, bandwidth in transient state, bandwidth in steady-state, and so on. The closed-loop measurements are shown in Figure 5. The behavior of hardware is almost the same as that of simulations to prove the architecture of proposed dual-loop DAGC and our VLSI chips. Using off-shelf ADC clocked at 20MHz, the experimental results show that the realized DAGC has an input dynamic range from 22 mV_{pp} to 456 mV_{pp}, transient mode bandwidth 1 kHz, steady-state bandwidth 90Hz, settling time of full dynamic range step response less than 2ms.

4. CONCLUSIONS

In this paper, a dual-loop digital AGC architecture suitable for VLSI implementation has been described. The proposed

DAGC subsystem can provide distinct loop bandwidth to achieve fast convergence and stable performances at the same time. The loop bandwidth is controlled by a quantizer that only costs tens of logic gates overhead. This digital control circuit and key analog component VGA/BPF/DAC have been implemented and tested individually. The closed-loop experiments have been conducted to measure the characteristics of proposed DAGC architecture. Using the off-shelf ADC clocked at 10MHz, the experimental results show that the realized DAGC has an input dynamic range from $22mV_{pp}$ to $456mV_{pp}$, transient mode bandwidth 1kHz, steady-state bandwidth 90Hz, settling time of step response less than 2ms. If the clock is 21.52MHz that is the original specification, it can be expected that this realization of proposed DAGC has transient mode bandwidth larger than 2kHz, steady-state bandwidth 180Hz, settle time of step response less than 1ms.

5. REFERENCES

[1] Zenith Electronics Corporation, Grand Alliance, "VSB Transmission System", Technical Details, Feb. 18, 1994.

[2] H. Meyr and G. Ascheid, "Synchronization in digital communication", vol. 1, Wiley.

[3] C. K. Wang, P. C. Huang, and C. Y. Huang, "A BiCMOS limiting amplifier for SONET OC-3", *IEEE J. Solid-State Circuits,* vol. 31, pp. 1197-1200, Aug. 1996.

[4] M. T. Shiue and C. K. Wang, "A VLSI architecture design of dual-mode QAM/VSB transceiver for high speed digital CATV modem", *to be submitted to IEEE Trans. Circuits and Systems.*

[5] K. Laudel, et al., "Performance of a 256-QAM demodulator equalizer in a cable environment", *NCSA Technical Papers,* pp. 283-304, 1994.

[6] M. T. Shiue and C. K. Wang, "Discrete-continuous bandwidth regulator for the timing recovery and the AGC system", R.O.C. Patent applying.

[7] B. Nauta, E. Klumperink, and W. Kruiskamp, "A CMOS triode transconductor", *Proceedings ISCAS'91, Singapore,* pp. 2232-2235, June 1991.

[8] B. Nauta, "A CMOS transconductance-C filter technique for very high frequencies", *IEEE J. Solid-State Circuits, Vol. 27, No. 2,* pp. 142-153, Feb. 1992.

[9] F. Yang, P. Loumeau, and P. Senn, "Novel output stage for DC gain enhancement of OPAMP and OTP", *Electron Lett., Vol. 29, No. 11,* pp. 958-959, May 1993.

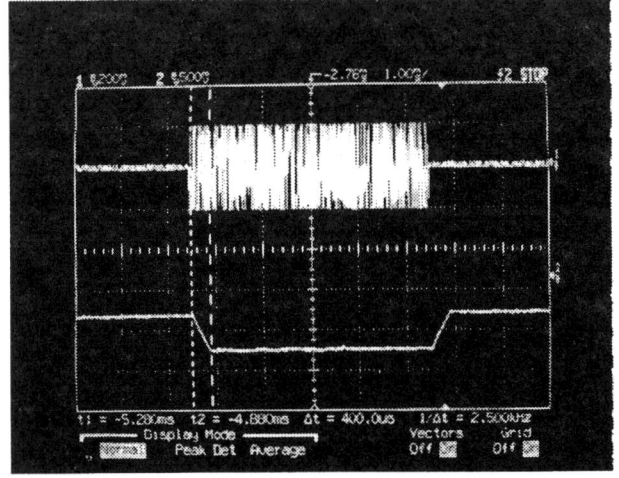

Figure 4: Voltage of control input of the 2V VGA chip.

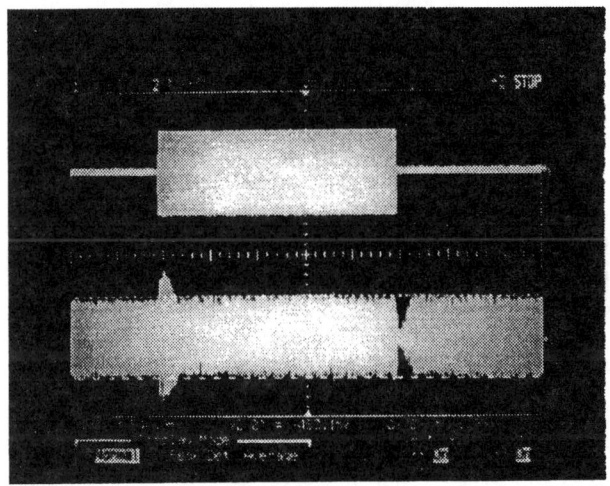

Figure 5: Step response of the 2V VGA chip output for the proposed dual-loop DAGC.

ROUTING MULTIPOINT CONNECTIONS IN COMPUTER NETWORKS

Sun Wensheng, Liu Zemin

P.O.Box 57, Beijing University of Posts and Telecommunications,
Beijing, 100876, P.R.China

ABSTRACT

This paper studies the problem of constructing minimum-cost multicast trees with end-to-end delay and delay variation constraints to meet the quality of service (QoS) requirements of real-time interactive applications operated in high-speed packet-switched environments. The paper first discusses the routing problem with bounded delay along the paths from the source to each destination and bounded variation among the delays along these paths, and then presents a new algorithm based on Hopfield neural networks to optimize the multicast tree with delay and delay variation constraints. The simulations show that the proposed algorithm achieves its best performance in constructing constrained multicast tree in computer networks.

1. INTRODUCTION

In high-speed computer networks, multicast is becoming an important requirement to support multimedia services. Multicast applications, such as teleconference, education, entertainment distribution, distributed data processing, etc., usually involve a single source node and a group of destination nodes with a point-to-multipoint connection established between the source and the destinations. These applications require low end-to-end delays or high transmission reliability.

Multicast routing algorithms can be classified into two categories [1]. The first category is the shortest path algorithms which minimize the cost of each path from the source node to a destination node. Bellman-Ford's algorithm and Dijkstra's algorithm are two well known shortest path algorithms. The other category is the *minimum Steiner tree* (MST) algorithms. Their objective is to minimize the total cost of the multicast tree, which is known to be *NP-complete*. Hwang [2] provides a survey of both exact and heuristic minimum Steiner tree algorithms. Gelenbe [3] presents an improved neural heuristic based on MSTH algorithm developed by Kou [4] or ADH algorithm developed by Rayward-Smith, which uses a random neural network to find potential Steiner vertices that are not ready in the solution returned by the MSTH or ADH. The average execution time of this algorithm is relatively large. If the destination set of a minimum Steiner tree includes all nodes in the network, the problem is reduced to the minimum spanning tree problem which can be solved in polynomial time.

In order to support real-time applications, network protocols must be able to provide QoS guarantees. For example, a guaranteed upper bound on end-to-end delay must be provided to certain distributed multimedia applications. These are called delay-constrained algorithms to distinguish them from other algorithms. Several delay-constrained heuristics have been proposed in [5][6].

There are several situations in which the need for bounded variation among the path delays arises [7]. During a teleconference, it is important that the speaker be heard by all participants at the same time, otherwise, the communication may lack the feeling of an interactive face-to-face discussion. When multicast messages are used to update multiple copies of a replicated data item in a distributed database system, minimizing the delay variation would minimize the length of time during which the database is in an inconsistent state. Rouskas [7] presented a heuristic algorithm which is used to construct multicast trees that guarantee certain bounds on the end-to-end delays from the source to the destination nodes, as well as on the variation among these delays. But they do not attempt to optimize the multicast tree in terms of cost. This paper develops their heuristic to minimize the multicast tree cost by using Hopfield neural networks.

This paper is organized as follows. In Section 2 we present a model for multicast communication in packet-switched networks and propose an algorithm of constructing minimum multicast tree with delay constraints. In Section 3 we present an approach to construct constrained Steiner tree. We present the performance of the proposed algorithm in Section 4, and conclude the paper in Section 5.

2. CONSTRAINED MINIMUM STEINER TREE

2.1 Network Model for Multicasting

The network is modeled as a weighted graph $G = (V, E)$, where V denotes the set of nodes and E denotes the set of edges corresponding to the set of communication links connecting the nodes. Let $L = |E|$ denote its size. Without loss of generality, we only consider graphs in the case that there is at most one edge between each node pair. We define a *link-delay function* $F: E \to \Re^+$ and a *cost-function* $C: E \to \Re^+$, respectively. The value F_l associated with link $l \in E$ is a measure of the delay that packets experience on that link. The value C_l denotes the cost of link l. Multicast packets are routed from the source s to every destinations in D ($D \subseteq V - \{s\}$) which is called the *destination set* or *multicast group*, via links of a *multicast tree* $T = (V_T, E_T)$, which is a subgraph of G spanning s and the nodes in D. Let $P_T(s, v)$ denote the path from source s to destination $v \in D$ in the tree, then multicast packets from s to v experience a total delay of $\sum_{l \in P_T(s,v)} F_l$.

Two parameters are defined to characterize the quality of the tree as perceived by the application performing the multicast. These parameters relate the end-to-end delays along individual source-destination paths to the desired level of quality of service.

- *Source-destination delay tolerance*, Δ: Parameter Δ represents an upper bound on the acceptable end-to-end delay along any path from the source to a destination node. This parameter reflects the fact that the information carried by multicast packets becomes at most stale Δ time units after its transmission at the source.
- *Interdestination delay variation tolerance*, ξ: Parameter ξ is the maximum difference between the end-to-end delays along the path from the source to any two destination nodes that can be tolerated by the application. In essence, this parameter defines a synchronization requirements for the various receivers.

2.2 Constrained Minimum Steiner Tree

Let Δ and ξ be the delay and delay variation tolerances, respectively, specified by a higher level application. Our objective is to construct a *minimum* Steiner tree in which all source-destination paths are within the two tolerances. This optimum delay and delay variation bounded multicast tree problem is called *constrained minimum Steiner tree* (CMST) problem and is formally described as follows

CMST Problem: Given a network topology $G=(V,E)$ with a source node $s \in V$, a multicast group D, a link-delay function $F: E \to \Re^+$, a link-cost function $C: E \to \Re^+$, a delay tolerance Δ, and a delay variation tolerance ξ, then construct a constrained minimum Steiner tree $T = (V_T, E_T)$ spanning $D \cup \{s\}$, such that

$$\sum_{l \in P_T(s,v)} F_l \leq \Delta \qquad \forall v \in D \qquad (1)$$

$$\left| \sum_{l \in P_T(s,v)} F_l - \sum_{l \in P_T(s,u)} F_l \right| \leq \xi \qquad \forall u,v \in D \qquad (2)$$

$$\text{Minimize } Cost(T) \qquad (3)$$

where $Cost(T)$ denotes the total cost of the tree T.

As in [7], we will refer to (1) as the *source-destination delay constraint*, while (2) as the *interdestination delay variation constraint*. Formula (3) is the new constraint which minimizes the mutilcast tree cost, which is called *cost constraint*. Constraints (1) and (2) represent two conflicting objectives. Indeed, the delay constraint (1) dictates that short paths should be used. But choosing the shortest paths may lead to a violation of the delay variation constraint among nodes that are close to the source and nodes that are far away from it. Consequently, it may be necessary to select longer paths for some nodes in order

to satisfy (2). The problem of finding a feasible tree is one of selecting paths in a way that strikes a balance between the two objectives. Rouskas has proved that the problem is *NP-complete*.

In this paper, we combines constraint (3) to constraints (1) and (2) to minimize the multicast tree. We now present an algorithm to construct a tree satisfying the above three constraints for the given values of the path delay and the interdestination delay variation tolerances. We assume that the source node has complete information regarding the network topology to construct a multicast tree. This information may be collected and updated using an existing topology-broadcast algorithm or other techniques.

The proposed algorithm consists of two major steps. Step 1 selects feasible paths for each source-destination pair, which satisfy constraints (1) and (2). The step involves:

1) Calculate *k*-shortest delay paths from s to each node $d_i \in D$ (Let $M = |D|$) using Dijkstra's algorithm, which all satisfy delay constraint (1). Assume Ω_{d_i} is the calculated set of paths from s to d_i, which is sorted by delay in increasing order.

2) If $\exists \Omega_{d_i} = \emptyset$, the delay tolerance Δ is too tight. In this case, negotiation should be made between source and destinations to determine a looser value for Δ, then go to 1).

3) Select all paths which satisfy constraint (2), refine Ω_{d_i} for each $d_i \in D$ by deleting no feasible paths. Let $N_i = |\Omega_{d_i}|$, where N_i is the path number in set Ω_{d_i}. If one set Ω_{d_i} is empty, negotiation should be also made about the delay variation violation, then go to 1).

After Step 1, all paths in each Ω_{d_i} satisfy constraints (1) and (2). Step 2 constructs minimum multicast tree by selecting one path in each Ω_{d_i}. In the paper, if *j*th path in Ω_{d_i} is selected, $V_{ij} = 1$; otherwise, $V_{ij} = 0$. In Step 2, using the constraint (3) as an objective function, the minimum multicast tree problem is formulated as the following constrained optimization problem

Given   network topology $G=(V,E)$, source node s, multicast group D, link-cost function $C: E \to \Re^+$, and Ω_{d_i} for each $d_i \in D$

Minimize $Cost(T)$

over $V = (V_{00}\ V_{01} \cdots V_{0N_1} \cdots V_{MN_M})$

Subject to $V_{ij} \in \{0,1\} \qquad \forall 1 \leq i \leq M,\ 1 \leq j \leq N_i$

where V is a vector of V_{ij}. In the following, we propose a Hopfield type neural network to solve the optimum problem. For the sake of its parallel processing mechanics, call setup may be finished in a very short time.

3. OPTIMUM STEINER TREE BASED ON NEURAL NETWORKS

The use of neural networks to solve constrained optimization problems was initiated by Hopfield and Tank. They proposed a neural network model to get good solutions to discrete combinatorial optimization problems, and demonstrate the computational power of their network by applying their model to TSP and other types of combinatorial optimization problems.

```
(1,1)   ○   ○   ...   ○  (1,N₁)
        ...             ...
(i,1)   ○   ○   ...   ○  (i,Nᵢ)
        ...             ...
(M,1)   ○   ○   ...   ○  (M,N_M)
```

Figure 1. Hopfield neural model

As shown in Figure 1, the neural network proposed in this paper consists of neuron matrix. Each neuron is modeled as a nonlinear device with a sigmoid monotonic increasing function relating the output V_{ij} of the neuron to its input U_{ij}. The output V_{ij} is allowed to take on any value between 0 and 1. The sigmoid function is given by

$$V_{ij} = g_{ij}(U_{ij}) = \frac{1}{1+e^{-\lambda_{ij}U_{ij}}} \quad \forall i \in M, j \in N_i \quad (4)$$

In order to solve the optimization problem by using Hopfield model, we define the following energy function whose minimization process drives the neural network into its lowest energy state:

$$E = \frac{\mu_1}{2}\sum_{i=1}^{M}\sum_{j=1}^{N_i}\sum_{\substack{k=1\\k\neq j}}^{N_i}V_{ij}V_{ik} + \frac{\mu_2}{2}\sum_{i=1}^{M}(\sum_{j=1}^{N_i}V_{ij}-1)^2$$

$$+\frac{\mu_3}{2}\sum_{i=1}^{M}\sum_{j=1}^{N_i}\sum_{\substack{p=1\\p\neq i}}^{M}\sum_{q=1}^{N_i}V_{ij}V_{pq}\sum_{l=1}^{L}C_l(\rho_{ij,l}\oplus\rho_{pq,l}) \quad (5)$$

where if link l is in the jth path of the set Ω_{d_i}, $\rho_{ij,l}=1$; otherwise, $\rho_{ij,l}=0$.

In the above, the μ_1 term is minimized if each row contains at most a single 1, which corresponds to at most one path selected in each set Ω_{d_i}. The μ_2 term ensures that there will be exactly M 1's in the final solution, which means there is only one path selected in each set Ω_{d_i}. When combined together, the μ_1 and μ_2 terms ensure that each row will have exactly a single 1. The μ_3 term minimizes the cost of the multicast tree, in which $\oplus$ is logical "OR" operator.

To simulate the evolution of the neural network, we must construct the equations of motion. As mentioned before, the neural network converges to a minimum with the monotonically decrease of the energy function. In terms of the energy function E, the dynamics of the neuron are described by

$$\frac{dU_{ij}}{dt} = -\frac{U_{ij}}{\tau} - \frac{\partial E}{\partial V_{ij}} \quad \forall i \in M, j \in N_i \quad (6)$$

where τ is the time constant connected to each neuron.

By substituting, the equation of motion of the neural network is readily obtained

$$\frac{dU_{ij}}{dt} = -\frac{U_{ij}}{\tau} - \frac{\mu_1}{2}\sum_{\substack{k=1\\k\neq j}}^{N_i}V_{ik} - \frac{\mu_2}{2}(\sum_{k=1}^{N_i}V_{ik}-1)$$

$$-\frac{\mu_3}{2}\sum_{\substack{p=1\\p\neq i}}^{M}\sum_{q=1}^{N_i}V_{pq}\sum_{l=1}^{L}C_l(\rho_{ij,l}\oplus\rho_{pq,l}) \quad (7)$$

Now, rewriting energy function in such a way as to take into account the representation of the neurons with double indices, we get

$$E = -\frac{1}{2}\sum_{i=1}^{M}\sum_{j=1}^{N_i}\sum_{p=1}^{M}\sum_{q=1}^{N_i}W_{ij,pq}V_{ij}V_{pq} - \sum_{i=1}^{M}\sum_{j=1}^{N_i}V_{ij}I_{ij} \quad (8)$$

By comparing the corresponding coefficients in (5) and (8), the connection strengths and the biases are derived

$$W_{ij,pq} = -\mu_1\delta_{ip}(1-\delta_{jq}) - \mu_2\delta_{ip}$$

$$-\mu_3(1-\delta_{ip})\sum_{l=1}^{L}C_l(\rho_{ij,l}\oplus\rho_{pq,l})$$

$$I_{ij} = \mu_2$$

where $\delta_{ip} = \begin{cases} 1 & i=p \\ 0 & i \neq p \end{cases}$

One major issue related to the efficiency of the Hopfield model in solving combinatorial optimization problem is the lack of rigorous guidelines in selecting appropriate value of the energy function coefficients. By experiment we have chosen the weighting coefficients as follows

$$\mu_1 = 200, \quad \mu_2 = 3100, \quad \mu_3 = 1050$$

4. SIMULATION RESULTS

The performance of the proposed neural algorithm is simulated via a 6-node, 10-edge network shown in Figure 2. Given the initial neurons' input voltages U_{ij}'s at time $t=0$, the time evolution of the state of the neural network is simulated by numerically solving (7). This corresponds to solving a system of nonlinear differential equations, where the variables are the neuron's output voltages V_{ij}.

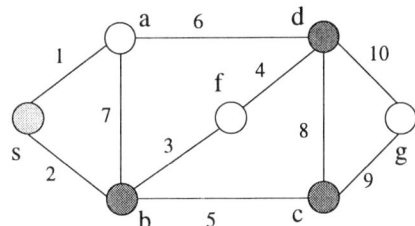

Figure 2. Network model

Accordingly, the simulation consists of observing and updating the neuron output voltages at incremental time steps δt. In addition, the time constant τ of each neuron is set to 1 without any loss of generality and for simplicity it is assumed that $\lambda_{ij} = \lambda$, $g_{ij} = g$ all independent of the subscript (i,j). The simulation has shown that a good value for δt is 10^{-4}. Reducing this tolerably small value increases the simulation time without improving the results. Experimentally we found that a large transfer parameter λ gives rise to a fast neural response for which the solution is not always a global minimum, a small λ yields a slower response which can guarantee an optimum solution. In fact, a good solution is enough for most needs. In this paper, in order to allow the neurons' dynamics to wander freely in their state space, in the search for global minimum, we have chosen $\lambda=2$.

Another important parameter in simulation is the neuron initial input voltages U_{ij}'s, which should be set to zero. However, some initial random noise $\delta U_{ij} \in [-0.0002,+0.0002]$ will help to break the symmetry caused by symmetric network topologies. The algorithm is stopped when the system reaches a stable final state. This is assumed to occur when all neuron output voltages do not change by more than a threshold value from one update to the next.

Assume there is one multicast request with a source node s and a multicast group $D = \{b, c, d\}$. Table 1 shows the paths available for each source-destination. Without lossing generality, we assume that all paths satisfy constraints (1) and (2). The algorithm is executed 1000 times in different link cost, the simulation results are shown in Table 2. 68.7% of the results constructed best multicast trees, 28.5% of that constructed good trees, only 2.8% failed in finding multicast trees. The global optimum is obtained in all runs within 300 iterations.

5. CONCLUSION

An improved multicast routing algorithm based Rouskas's heuristic has been proposed by using a Hopfield type neural network. The general principles involved in the design of the proposed neural network have been discussed. The proposed algorithm can construct a minimum Steiner tree in which all source-destination paths are within the delay and delay variation tolerances. The algorithm combines many features, such as a very good convergence, a relatively low time complexity and an ability to operate in real time. From the simulation results, obtained under different network topologies, it can be concluded that the proposed algorithm is both efficient and effective in constructing multicast tree.

Table 1 Available paths for each source-destination

multicast requests (source, destination)	available paths (sequence of edge)
(s, b)	{2}, {1,7}, {1,6,4,3} {1,6,8,5}, {1,6,10,9,5}
(s, c)	{1,6,8}, {1,7,5}, {1,7,3,4,8} {2,5}, {1,6,10,9}, {1,6,4,3,5} {1,7,3,4,10,9}
(s, d)	{1,6}, {2,3,4}, {1,7,3,4} {2,7,6}, {2,5,9,10}, {2,7,6} {2,5,8}, {1,7,5,8}

Table 2 Simulation results

best trees	good trees	failure
68.7%	28.5%	2.8%

6. REFERENCES

[1] H.F. Salama, D.S. Reevws, and Y. Viniotis, "Evaluation of multicast routing algorithm for read-time communication on high-speed networks," *IEEE JSAC*, vol.15, no.3, pp.332~345, April, 1997.

[2] F.K. Hwang, and D.S. Richards, "Steiner tree problems," *Networks*, vol.22, no.1, pp.55~89, Jan., 1992.

[3] E. Gelenbe, A.Ghanwani, V.Srinivason, "Impoved neural heuristics for multicast routing," *IEEE JSAC*, vol.15, no.2, pp.147~155, Feb., 1997.

[4] L. Kou, G. Markowsky, and L. Berman, "A fast algorithm for Steiner trees," *Acta Informat.*, vol.15, pp.141~145, 1981.

[5] Q. Zhu, M. Parsa, and J.J. Garcia-Luna-Aceves, "A source-based algorithm for delay-constrained minimum-cost multicasting," in *Proc. IEEE INFOCOM'95*, pp.377~385, 1995.

[6] V.P. Kompella, J.C. Pasquale, and G.C. Polyzos, "Multicasting for multimedia applications," in *Proc. IEEE INFOCOM'92*, pp.2078~2085, 1992.

[7] G.N. Rouskas, and I. Baldine, "Multicast routing with end-to-end delay and delay variation constraints,", *IEEE JSAC*, vol.15, no.3, pp.346~356, April., 1997.

DYNAMIC ROUTING ALGORITHMS IN ATM NETWORKS

Gang Feng, Zemin Liu

Box 57, Beijing University of Posts and Telecommunications
Beijing 100876, P.R. China

ABSTRACT

In our previous work, we elaborated the multistage virtual path(VP) control strategy and the VP topology optimization problem in which we assume a set of alternative routes exist between each source-destination (SD) pair. In the first part of this paper, a supplementary algorithm, which can yield all possible routes between two nodes and can be used for general routing problems, is provided. Based this algorithm, a dynamic VC routing policy is studied. The proposed policy distinguishes from other strategies in that it is considered in a generic networking environment. In the last part of this paper, we present a dynamic VP routing algorithm which is an important component of the dynamic VC routing policy. Detailed analyses and experimental results demonstrate these algorithms are correct, practical and useful.

1. INTRODUCTION

In asynchronous transfer mode(ATM) networks, the virtual path(VP) concept has been proposed for the purpose of managing network resources more efficiently[1-3]. As we know, ATM cells are transmitted through two types of connection, virtual path connection (VPC) and virtual channel connection (VCC). A VPC is defined as a logical link between two nodes which are termed VP terminators. Generally, a VPC consists of a group of VCCs, while a VCC is a concatenation of several VPCs. A collection of VPs is called VP subnetwork. For a physical network, some researchers envision there are a number of VP subnetworks which are classified according to the characteristics of the traffic transported in them[4], but others think all VPs makes up a single VP subnetwork[5]. In the later case, VP subnetwork is equivalent to VP topology.

VPC and VCC are identified by virtual path identifier(VPI) and virtual channel identifier (VCI), respectively. Since VP allows VCCs to be handled in bundles and only at VP terminators will the VCIs be translated, processing costs at transit nodes are greatly reduced. On the other hand, the decrease of the network throughput also ensues from the introduction of the VP mechanism since it makes the statistical multiplexing capability be deteriorated. Therefor, an efficient VP control strategy should be developed to mediate the discrepancy and exploit the best utilization of the VP concept.

In [6], we discussed the multistage VP control method in which the adjustment of VP is executed in three stages with specific tasks for each stage. In brief, the objective in the first stage is to initiate the VP subnetwork based on known traffic requirements. In the second stage, the VP topology is locally modified to track the instantaneous change of traffic state. After a period of local operations, the efficiency of VP topology may be aggravated. Thus, the aim in the third stage is to ensure the network meets the current traffic requirements by globally reconfiguring the VP topology.

The feasibility and effectiveness of the multistage VP control method has been investigated and identified. Besides, we also provided an efficient algorithm for the tasks in the first and third stages. The algorithm, however, is based on an assumption that there exist a set of alternative routes between each source-destination (SD) pair. (Please refer to [6] for more details.) In Section II of this paper, we will present a supplementary algorithm which can yield all possible routes between two nodes. Furthermore, the dynamic routing problem, the task in the second stage of the multistage VP control, will be discussed. An efficient VC routing policy is presented in Section III. In Section IV, we focus on the problem how to route the VPs dynamically so that the desired network performance can be maintained. Our problem formulation is based on a generic network model. The simple dynamic VP routing algorithm is based on a cost function in which the influences of the load balance and offered capacity are considered. In Section V, some experimental results, which demonstrate our algorithms are correct and effective, will be provided.

2. ALGORITHM I: SEARCHING ALL POSSIBLE ROUTES BETWEEN TWO NODES

Given the topology of an arbitrary network model, our algorithm for searching all possible routes between two nodes can be described as the following recursive function:

Function f(Parameter: node p)
{ IF p=destination node THEN
    {write S; RETURN; }
IF p doesn't belong to S THEN
    {add p to S;

*FOR each neighboring node **q** of node **p** DO f(**q**);*
*delete **p** from S;*
  }
}

In the above algorithm, S is a set storing the nodes included in a route. It is empty at the initial state. When we execute this algorithm, one of the two nodes is taken to be the input parameter of function $f(\cdot)$, while the other is defined as the destination node. Analyzing the algorithm, we can find out that all the nodes included in set S forms a route once the condition <p=destination node> is satisfied. Since this is a recursive procedure, all possible routes can be found when it is finished.

When implementing this algorithm in C++ language, we use a number to represent a node and a two-dimensional array to characterize the network topology. The effectiveness of the algorithm has been verified by applying it to a variety of network models which are automatically generated. The size of network model varies from several nodes to more than 100 nodes. Hundreds of experiments have proved the algorithm not only can correctly find all routes between two nodes, but also can respond with high speed for networks of moderate size. For instance, when it is run on a pentium-100 computer and is applied to a 60-node network with the maximal nodal degree of 6, 4413 routes between two randomly selected nodes are found within 2.19 seconds. Among these routes, the longest includes 31 nodes, while the shortest passes only 9 nodes. With the increase of the network size, the number of routes and the time to be consumed for searching will increase rapidly. When it is used to test a 70-node network, for example, more than 200,000 routes are found between two nodes and the time consumed is 4.5 minutes.

By means of Algorithm I, many traditional routing algorithms, such as the shortest path algorithm and the least loaded routing(LLR) algorithm can be easily realized. Furthermore, as applied to the VP topology optimization problem in our previous work[6], this algorithm can be used for a general routing problem to generate alternative routes which are necessary in successive algorithms. In this case, there is no time constraint. The dynamic VP routing algorithm which will be presented in Section IV is also based on Algorithm I.

3. THE VC ROUTING POLICY BASED ON A GENERIC NETWORK MODEL

Almost all of the previous work regarding the dynamic VC routing problem is based on an assumption that the network model is a fully VP-connected network, i.e. there is at least one direct VP between any two nodes. This requirement may be met for a network with moderate size. However, it is impossible for a large network since the VPIs may be fast exhausted[7]. Furthermore, too many VPs are disadvantageous for improving the capacity utilization efficiency. Therefor, for a generic network model, a reasonable VP topology is generally not fully interconnected. Under this circumstance, the dynamic VC routing problem become more difficult. As introduced in [4,8], for a fully VP-connected network, the routing policy can be designed to route a VCC on a route consisting of at most 2 VPs, and the admission decision can be made at local nodes. This method not only guarantees the network's high responsiveness, but also guarantees a satisfactory VCC admission rate. In the case of a generic network model, however, calls may occur between a SD pair where no direct VP or two-VP routes exist. We shouldn't hope to route a VCC on multiple VPCs in a distributed manner since the network may be overwhelmed by the large amount of information interchanged between local nodes. Routing a VCC on multiple VPs at the central node isn't practical either, because the state of the utilization of the VP capacity is unknown there. Hence, other actions should be taken to ensure the quality of service (QoS) requirements to be satisfied.

Our method to solve the above problem is illustrated in Figure 1. When a VCC request can't be met at a local node, it is sent to the central node where the dynamic VP routing algorithm is employed to attempt to establish a new VPC or a complete physical route. In this way, the call admission rate is expected to be improved substantially. Furthermore, since new VPCs meeting the up-to-date traffic requirements can be set up (of course, the obsolete VPCs may be torn down by another mechanism), the VP topology is able to adapt to traffic changes in time.

Some issues arising in this routing strategy need further explanations. First, a connection decision to be made at the central node may consume much time and need much information to be interchanged between the local node and central node. Therefor, the highest priority for establishing direct VPCs should be given to the SD pairs very sensitive to admission time and those between which calls occur most frequently. Second, if there exists a direct VP or two-VP route, but the VP bandwidth is insufficient, a request for increasing the VP bandwidth should be sent to the central node first. The method of borrowing and returning capacity proposed in[9] is also applicable to our policy. Third, routing a VCC on a complete physical route is based on the consideration that there are no adequate VPIs while a physically connected route, which has enough free capacity and satisfies the time delay requirements, is available.

4. THE DYNAMIC VP ROUTING ALGORITHM

When VCC requests can not be granted at local nodes, when new services are added to the network, or when link/node failures take place, new VPCs are needed to be dynamically established. The dynamic VP routing problem considered in this paper is based on the assumption the required VP bandwidth has been determined. Therefor, it can be described as follows: *For an arbitrary network, given the capacity of each physical link, and the bandwidth and physical route of each existing VP, route a new VP with specified bandwidth between two nodes with an objective that the future network throughput can be maximized.*

The analyses in our prior work[6] demonstrate the optimal physical route of the new VP should satisfy two underlying requirements in order to maximize the network throughput: (1) the capacity utilization rates of physical links be optimally equilibrated, and (2) the physical route length of the new VP, or the capacity offered to it be as small as possible.

The first requirement can be converted to that the variance of the capacity utilization rates is minimal. In this way, the two requirements can be combined into a single cost function, while our final goal is to find a route from a set of alternative routes which leads to the minimal value of the cost function. As mentioned in Section II, the alternative routes may be obtained by means of Algorithm I. However, from all the possible routes generated by Algorithm I, only those having enough free capacity and satisfying the time delay requirements are taken to be effective alternative routes.

For a K-link network, suppose there are N alternative routes between the two target nodes. On condition that the ith route is chosen to be the physical route of the new VP, we assume the capacity utilization rate (defined as the ratio of the offered capacity to the total capacity of the link) of the kth physical link is U_{ik}, and thus the variance of the capacity utilization rates can be expressed as

$$\sum_{k=1}^{K}(U_{ik}-m_i)^2$$

where $m_i = \frac{1}{K}\sum_{k=1}^{K} U_{ik}$ is the average of the capacity utilization rates. If we use L_i to denote the length of the ith alternative route (generally defined as the number of physical links making up the ith route), the cost function can be given by

$$f_i = \alpha \cdot \sum_{k=1}^{K}(U_{ik}-m_i)^2 + \beta \cdot L_i$$

where α and β are cost factors. Our objective is to find a route with the minimal value of the above function, i.e.

$$\operatorname*{Min}_{i=1}^{N} f_i.$$

5. EXPERIMENTAL RESULTS

As illustrated in Figure 2, the experimental network employed to test our algorithms consists of 8 nodes and 11 links. The capacity of each link is supposed to be 100 units and the offered capacity of each link is given in Table I. We hope to route a new VP between nodes A and H with a required bandwidth of 30 units.

By means of Algorithm I, we can obtain all the 8 routes between A and H as enumerated in Table II. In this experiment, we assume the traffic that will be transmitted in the new VP requires the physical route consist of at most four links in order to satisfy the time delay requirement. Therefor, the first three routes are not effective. The forth one is not effective either, since the first physical link included in this route has no adequate free capacity.

To select one best route from the rest effective routes, we assume the two cost factors are $\alpha=100$ and $\beta=10$, and thus we can calculate the cost of each effective route as given in Table II. Finally, the sixth route is selected to accommodate the new VP.

Table I Offered Capacity of Each Link

Link No.	1	2	3	4	5	6	7	8	9	10	11
Offered Capacity	80	50	40	65	30	20	60	40	50	60	55

Table II Alternative Routes and Costs

Route No.	Nodes included	Effectiveness	Cost
1	ABCGFEDH		
2	ABCGFH		
3	ABFEDH		
4	ABFH		
5	ADEFH	Yes	88.409091
6	ADH	Yes	63.227273
7	AEDH	Yes	106.409091
8	AEFH	Yes	84.136364

6. CONCLUSIONS

The future ATM based broadband integrated service digital network (B-ISDN) is expected to support a variety of traffics with diverse statistical characteristics. Traffic patterns are likely to change due to many unpredictable reasons. Therefor, dynamically routing VC and VP is very critical for maintaining the desired network performance. In this paper, a dynamic VC routing policy and a dynamic VP routing method are discussed in detail. Some experiments are also conducted to demonstrate the practicability, feasibility and effectiveness of the algorithms. Furthermore, a supplementary algorithm for searching all possible routes between two nodes is also provided. By means of these algorithms, an efficient dynamic mechanism can be established.

REFERENCES

[1] J. Burgin, D. Dorman, "Broadband ISDN resource management: the role of virtual paths," *IEEE Communication Magazine*, Sep. 1991

[2] K. Sato, S. OHTA, I. TOKIZAWA, "Broad-band ATM network architecture based on virtual paths," *IEEE Trans. Commun.*, vol.38, no. 8, pp. 1212-1221, August 1990

[3] V.J. Friesen, J.J. Harms, J.W. Wang, "Resource management with virtual paths in ATM networks," *IEEE Network*, Sep./Oct., 1996

[4] H.W. Chu, D.H.K. Tsang, "Dynamic routing algorithms in VP-based ATM networks", *ICC'95*, 1995, pp.1364-1368

[5] C.J. Hou, "Routing virtual circuits with timing requirements in virtual path based ATM networks," *Proc. INFOCOM'96*, 1996, pp.320-328

[6] G.Feng, Z.M.Liu, "Virtual path topology optimization using a neural network approach in multistage VP control," to be published on *IEICE Trans Commun.*

[7] E. Gelenbe, A. Ghanwani, V. Srinivasan, "Improved neural heuristics for multicast routing," *IEEE J. Select. Areas Commun.*, vol.15, no.2, Feb. 1997

[8] S. Gupta, K.W. Ross, "Routing virtual path based ATM networks," *Proc. GLOBECOM'93*, 1992, pp571-575

[9] R.H. Hwang, "LLR routing in homogeneous VP-based ATM networks," *Proc. INFOCOM'95*, 1995, pp.587-593

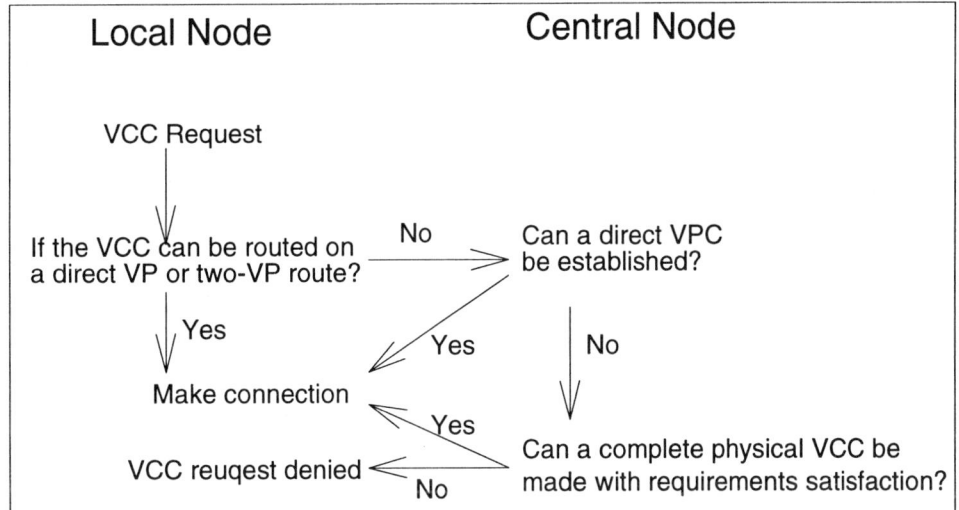

Figure 1. VC Routing Policy

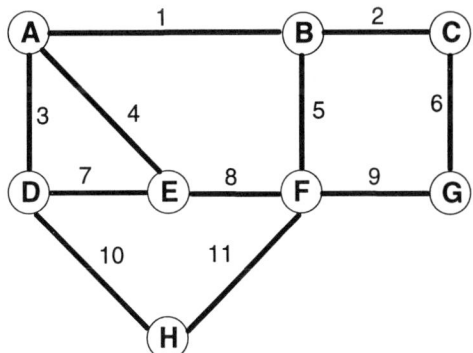

Figure 2. Physical Topology of the Experimental Network

A NEW MEMORY CONTROLLER FOR THE SHARED MULTIBUFFER ATM SWITCH WITH MULTICAST FUNCTIONS

Robert C. Chang and Chih-Yuan Hsieh

Dept. of Electrical Engineering,
National Chung-Hsing University,
Taichung, Taiwan, ROC

ABSTRACT

A novel ATM switch memory controller incorporated the shared multibuffer architecture is proposed. By applying the cyclic N method at the address queues, the blocking effect is eliminated and no memory and cross-point switch speed up is required. Multicast functions are efficiently carried out via a multicast queue. Each multicast packet only occupies one space in buffer memory and no additional copy circuit or counter is needed. Thus, the buffer utilization is improved and the hardware complexity is reduced. With the aid of the input traffic adaptive controller, multicast packets are dynamically sent to the output ports according to the input traffic pattern so that the unfairness problem due to employing the multicast queue is alleviated. By adopting the new memory controller, the throughput can be elevated to about 99.2%.

1. INTRODUCTION

Asynchronous transfer mode (ATM) has been standardized as the transporting and switching technique used for future broadband integrated services digital networks (B-ISDN). One of the key technologies used for an ATM network is the ATM switch. Numerous ATM switch architecture has been proposed [1-12]. They can be classified into two major approaches, namely space division type and time division type switches. Space division type switches transfer packets in parallel. Two buffering strategies, input buffering and output buffering, may be adopted. However, input buffering suffers from HOL blocking effect while output buffering encounters the output contention problem. Furthermore, buffer utilization is poor in either approaches. On the other hand, time division type switch architectures are known to have the best buffer utilization and lowest cell-loss probability [5]. However, the memory access time becomes a bottleneck since the memory bandwidth is proportional to the I/O port number. Thus the switch size is limited. A shared multibuffer switch architecture that combined the space division and time division type switch architecture has been proposed [8]. In the shared multibuffer switch, all the buffer modules are logically considered as a large united buffer memory so that buffer sharing is still provided. Fig. 1. Shared multibuffer architecture. Cross-point switches are used instead of multiplexer and demultiplexer so that the memory access time constraints can be relaxed. However, due to the blocking effect at the read out stage, the throughput is degraded to about 60%. Three read and one write operations within one cell slot is needed to achieve a throughput of 97% [8]. This paper presents a new memory controller for the shared multibuffer architecture to eliminate the blocking effect and further elevate the throughput while only one read and one write operation is performed within each cell slot.

Multicast functions are essential for an ATM switch in order to support various kinds of applications like video on demand and video conference. Several multicast solutions can be found in the literature [7-12]. Among these solutions, the multicast queue method [9,10,11] is the most hardware-efficient one. But it is unfair to the unicast packets since the multicast packets are always sent in prior [12]. We describe a dynamic multicast scheme that takes the input traffic patterns into consideration, and thus the unfairness problem is improved.

This paper is organized as follows. In section 2, an overview of the shared multibuffer ATM switch system is presented. Section 3 gives the details of the proposed cyclic address queue method and the dynamic multicast scheme. Discussions and comparisons are given in section 4. Section 5 gives some conclusion remarks.

2. OVERVIEW OF THE ATM SWITCH SYSTEM

Among the various ATM switch architectures, the shared multibuffer ATM switch is notable for its successfully combined the space and time division type switch architectures as shown in Fig. 1. By assigning the input packets into emptier buffer memories in prior, the physically separated buffers are logically considered as a

---

* This work is supported by the National Science Council of Taiwan, ROC under grant NSC 87-2218-E-005-021.

united large buffer memory, and thus buffer sharing is achieved. By using cross-point switches instead of multiplexer and demultiplexer, parallel transmission can be accomplished. Hence, it gains the advantages of both the space division and time division type switches.

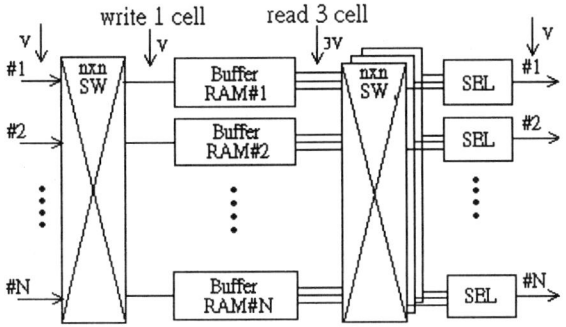

Fig.1. Shared multibuffer architecture.

Our proposed cyclic shared multibuffer switch architecture is shown in Fig. 2. I/O circuits are used to receive/send signals from/to communication channels. Header Conversion (HD_CNV) decode the VPI/VCI of each input ATM cell, and find the corresponding new VPI/VCI or multicast group channel and other routing information for that cell. At the read out stage, HD_CNV transforms the internal cell format to the ATM cell format. Also, HD_CNV governs the external Multicast Table (MT) through its local controller. According to the occupancy of the buffers and routing information found by HD_CNV, the memory controller routes the input cells to the individual buffer memories, and appends each cell to the corresponding output queue. Multicast cells are translated by a table lookup operation at the output stage. According to the content at the MT, nulticast cells can be routed to their destination outputs alone with their new VPI/VCI.

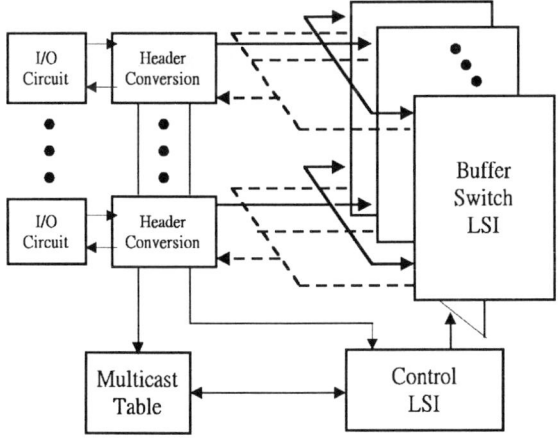

Fig. 2. The proposed cyclic shared multibuffer architecture.

3. ARCHITECTURE OF THE MEMORY CONTROLLER

In order to prevent the blocking effect which causes the both degradation of throughput and increase of hardware complexity, the situation that two or more cells read out from the same buffer memory to different output ports should be avoided. The basic idea of our proposed memory controller is to adopt the cyclic N method at the address queues to conquer the above mentioned problems. Besides, to support various multicast applications, a multicast queue (MA) is employed to handle multicast cells. By using an Input Traffic Adaptive Controller (ITAC), multicast cells are dynamically sent to the output ports according to the input traffic pattern. Thus, adopting the multicast queue won't increase the delay and cell-loss probability of the unicast cells, and the unfairness problem is improved. The block diagram of the proposed memory controller is shown in Fig. 3.

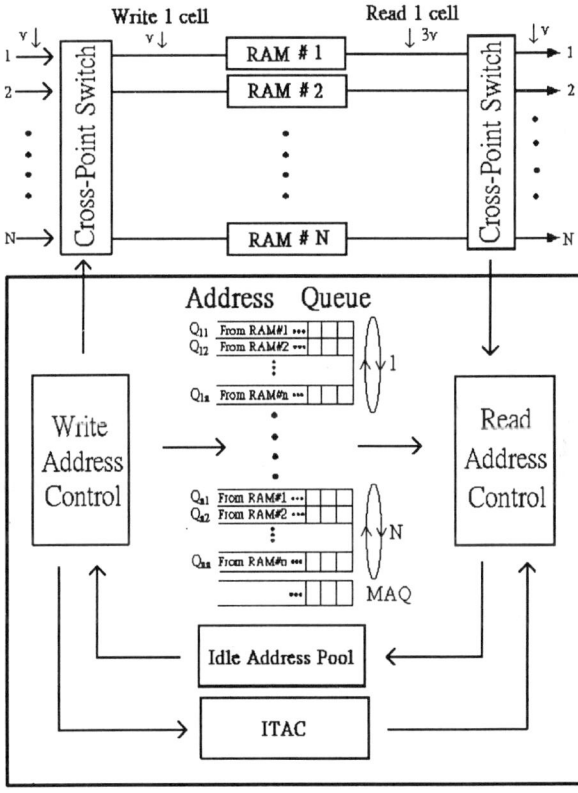

Fig. 3. Block diagram of the proposed memory controller.

3.1 Cyclic Address Queues

The basic concept of the cyclic address queues is to

decompose each output queue into N sub-queues for an NxN switch. Q_{11} stores the addresses of the packets that are stored in RAM#1 and destined to output port 1. Q_{12} stores the addresses of the packets that are stored in RAM#2 and destined to output port 1. In general, Q_{xy} stores the addresses of packets that are stored in RAM#y and destined to output port x. Besides, a multicast queue (MQ) is used to store the addresses of the multicast packets.

At every cell slot, each output reads one of the N sub-queues according to the cyclic N algorithm. With this algorithm, packets will be read out from different RAMs to different outputs at every cell slot. Therefore, one read operation per cell slot is enough since there won't be two packets read out to different output ports from the same buffer RAM. Detailed switch operations are described in the following.

(1) N cell slots are considered as a cyclic unit. N is the switch size.
(2) In the 1st cell slot, output port i reads out the packet according to Q_{ii}.
(3) In the 2nd cell slot, output port i reads out the packet according to $Q_{i[(i+1) \bmod N]}$
(4) In the jth cell slot (1<j<N), output port i reads out the packet according to $Q_{i[(i+j-1) \bmod N]}$.
(5) In the kth cell slot,
  (i) The addresses stored in MQ are sent out to the read address control.
  (i) Multicast packet is sent to the output ports according to the MT. It is a bit map table, every bit represents one output port. A selector is introduced to select between unicast and multicast packets. Multicast packets are selected prior because it has higher priority than unicast packets. Those output ports that multicast packets are not destined to then select unicast packets. Block diagram of this selection is shown in Fig. 4.
  (ii) If the multicast packet is stored in RAM#m, then according to the cyclic N algorithm, the output port x (x=[(m-k+1) mod N]) either accepts the multicast packet if the multicast packet is also destined to the output port x, or accepts no packet if the multicast packet is not destined to the output port x. The unicast packet whose address stores in Q_{xm} is not allowed to transfer to the output port x in order to avoid the need of memory speed up.
(6). Repeat steps (2)~(5).

3.2 Dynamic Multicast Scheme

In order to prevent the unfairness problem due to MQ as previously mentioned, the multicast packets read out cycle should be adjustable depending on the input traffic patterns. Thus, it is more fair for unicast packets as far as delay and cell-loss probability is concerned. The value of the multicast packet read out cycle k is determined by the input traffic adaptive controller (ITAC) block as shown in Fig. 5. Two counters are used separately to count the number of unicast and multicast packets that enter the switch. The ratio R of unicast and multicast packets are calculated by the divider. Assume that a multicast packet is destined to N/2 output ports in average, then (k-1) x N and N/2 unicast packets are read out during the first (k - 1) cell slots and the kth cell slot respectively. The unicast and multicast read out packets' ratio during the k cell slots should be equal to R, (i.e. [(k - 1) x N + N/2] / [N/2] = R). Thus, the value of k should be dynamically adjusted to (R + 1) / 2 to best suit the input traffic pattern. With this method, the input traffic pattern are adaptively employed by the output scheduling, and hence the unfairness problem mentioned in the previous section is improved.

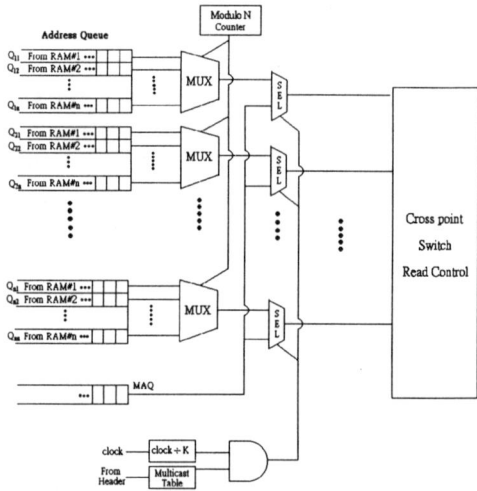

Fig. 4. Block diagram of the selection between unicast and multicast packets.

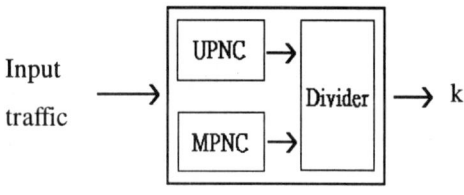

UPNC: Unicast packets number counter.
MPNC: Multicast packets number counter.
Fig. 5. Block diagram of input traffic adaptive controller.

3.3 Queue Implementation

Though the number of queues of our proposed memory controller increase from N to N^2 for an NxN switch. The storage size won't increase by implementing the queue in a linked list fashion [13]. So the increasing of the queue numbers won't become a problem.

4. DISCUSSION AND COMPARISON

It can be seen clearly from the above operations, only the output port x $\{ x = [(m - k + 1) \mod N] \}$ at the kth cell slot is possible to be blocked. Assume that there are always packets in each queue, a simple calculation shows that when $N = k = 8$, a throughput of $(8 \times 7 + 7 + 0.5) / 8 \times 8 = 99.2\%$ can be achieved while memory and cross-point switch speed up is not needed.

From the above discussions, we can see the proposed memory controller incorporating the shared multibuffer switch architecture provides a more cost-effective and higher throughput solution. Since the non-blocking property induced by the cyclic N method is size independent, the buffer access time constraints are removed, and the required buffer memory bandwidth now only depends on the line speed, and the throughput won't degrade as the switch size grows. As a result, a bigger switch element containing larger buffer memories with higher line speed is expected to be carried out as the process technology advances.

5. CONCLUSION

A new memory controller incorporated the shared multibuffer ATM switch architecture is presented in this paper. By applying the cyclic address queue method, the blocking effect is eliminated. The throughput is elevated and the hardware complexity is reduced. Adopting the dynamic multicast scheme, the unfairness problem is improved with only a little extra hardware. Furthermore, no additional counters, address copy circuits or copy networks which are required in other multicast switches are needed. In addition, because the buffer memory are shared by all output ports and multicast packets occupies only one space in buffer, it results an excellent utilization of the buffer. Since all the above mentioned properties are size independent, it is potential that the proposed cyclic shared multibuffer switch architecture can be expandable.

6. REFERENCE

[1] F. A. Tobagi, "Fast Packet Switch Architectures for Broadband Integrated Services Digital Networks," *Proceedings of the IEEE*, vol. 78, no. 1, pp. 132-167, Jan. 1990.

[2] E. W. Zegura, "Architectures for ATM Switching Systems," *IEEE Comm. Magazine*, pp. 28-37, Feb. 1993.

[3] J. Garcia-Haro and A. Jajszczyk, "ATM Shared-Memory Switching Architectures," *IEEE Network*, pp. 18-26, Jul/Aug. 1994.

[4] A. Pattavina, "Nonblocking Architectures for ATM Switching," *IEEE Commu. Magazine*, pp. 38-48, Feb. 1993.

[5] M. Hluchyj and M. Carol, "Queueing in High-Performance Packet Switch," *IEEE J. Select. Area Comm.*, vol. 6, no. 9, pp. 1587-1597, Dec. 1988.

[6] Y. Shobatake, et al., "A One-Chip Scalable 8*8 ATM Switch LSI Employing Shared Buffer Architecture," *IEEE J. Select. Area Comm.*, vol. 9, pp. 1248-1254, Oct. 1991.

[7] J. S. Turner, "Design of a Broadcast Packet Switching Network," *IEEE Trans. Comm.*, vol. 36, no. 6, pp. 734-743, Jun. 1988.

[8] H. Kondoh, et al., "A 622-Mb/s 8x8 ATM Switch Chip Set with Shared Multibuffer Architecture," *IEEE J. Solid-State Circuits*, vol. 28, no. 7, pp. 808-815, Jul. 1993.

[9] Takahiko Kozaki, et al., "32x32 shared buffer type ATM switch VLSI's for B-ISDN's," *IEEE J. Select. Area Comm.*, vol. 9, no. 8, pp. 1239-1247, Oct. 1991.

[10] S. Kumar and D. P. Agrawal, "On Multicast Support for Shared-Memory-Based ATM Switch Architecture," *IEEE Network*, pp. 34-39, Jan/Feb. 1996.

[11] Jin Li and Chuan-Lin Wu, "Design and implementation of a multicast-buffer ATM switch," *Proceedings Int's Conf. on Network Protocols*, pp. 84-91, 1995.

[12] T.-H. Lee and S.-J. Liu, "Multicast in a shared buffer memory switch," *IEEE Region 10 Conf. on Commu.*, vol. 41, no. 1, Jan. 1993.

[13] Y-S Lin and C. B. Shung, "Queue management for shared buffer and shared multi-buffer ATM switches," *Proc IEEE INFOCOM '96*, pp. 688-95, Mar. 1996.

SCHEDULING ALGORITHM FOR REAL-TIME BURST TRAFFIC USING DYNAMIC WEIGHTED ROUND ROBIN

Taeck-Geun Kwon, Sook-Hyang Lee, and June-Kyung Rho

R&D Lab., LG Information & Communications, LTD.
533 Hogye-dong, Anyang
Kyungki-do, 430-080, Korea

ABSTRACT

In this paper, we propose a new scheduling algorithm, called the Dynamic Weighted Round Robin (DWRR) scheduling discipline, which is suitable for real-time variable bit rate (rtVBR) service as well as other services such as constant bit rate (CBR), non real-time VBR (nrtVBR), available bit rate (ABR) services in a high-speed network. We employ dynamic weight to serve a queue associated with active and busy connection with strict delay and bandwidth requirements. This adaptive scheduling algorithm enables a network to provide multimedia service characterized by integrated and burst traffic. We finally compare our new algorithm with existing one in terms of fairness and Quality of Service (QoS) guarantee by simulation study.

1. INTRODUCTION

Many multimedia applications such as video-on-demand and video conferencing require the network to provide various QoS guarantees in terms of bandwidth, delay, and delay jitter. This is because there exists heterogeneity in multimedia source traffic characteristics. Asynchronous Transfer Mode (ATM) might be a solution for delivery of the multimedia contents consisting of different types of media such as constant bit rate (CBR) audio, variable bit rate (VBR) video, and best effort traditional data.

The only a few traffic parameters (i.e., peak cell rate and sustainable cell rate) has been intended to be used in the early ATM network, furthermore, policing and conformity check should be enforced at the network entry points [2, 10]. Recently, new service disciplines that aim to provide different QoS for each connection have been proposed [1, 9, 11]. Previous solutions are based on either a time-framing strategy or a sorted priority queue mechanism. Though sorted priority queues provide flexibility, they are too complex to implement in ATM [7, 9]. In addition, most previous schemes are considered for traditional packet switch and/or router but not fixed small-size packet, called *cell*, in ATM switch.

There has been a growing concern about the potential impact of rtVBR traffic on ATM traffic engineering. In this paper, we present a new service discipline called *Dynamic Weighted Round Robin (DWRR) scheduling algorithm* that provides implementation simplicity and flexibility in both the dynamic allocation of bandwidth for bursty traffic and the management of non FIFO-based delay sensitive cell scheduling. Basically, DWRR is based on a weighted round robin scheduling algorithm with introducing adaptive quantum of service varying traffic conditions. While cells are arrived at peak rate in a queue belonging to a connection, the queue is allow to send more cells in order to avoid cell overflow and reduce cell transfer delay.

The rest of the paper is organized as follows. In Section 2 we compare our idea with previous works, and we describe our scheduling algorithm to apply rtVBR traffic in Section 3. In Section 4 we present simulation results to justify the significance of our scheme. Finally, we state our conclusions and future works in Section 5.

2. BACKGROUND

In this section, we briefly outline the characteristics of the rtVBR service. Hence, we formalize a traffic model to describe the service, then we describe a service model to provide QoS guaranteed service in our system.

2.1. Real-time Variable Bit Rate Traffic

Unlike both conventional telecommunication network and legacy data network, an emerging network should allow to transfer real-time multimedia data with satisfying various QoS. These requirements enforce to change traffic management policy of those traditional networks. Since delay sensitive data should be scheduled to depart the system in time, this requires non FIFO scheduling in network elements. In addition, versatile resource allocation is needed for bandwidth guaranteed service. Consider the compressed video traversing a network, both intelligent queuing and sophisticated bandwidth reservation scheme are required for QoS guarantee. Furthermore, only a simple but powerful scheduling mechanism is applicable for ATM network in order to implement the algorithm in hardware [3]. Note that one cell processing time is about 2.8 μs in the primary rate of ATM, i.e., 155 Mbit/s. To provide performance guarantee, the network needs to allocate resources on a per-connection basis. Such per-connection queuing can isolate the effects of illegal behavior of traffic source.

In this paper, we use the (ρ, $\mathcal{B}$) traffic parameter to represent burst traffic, where ρ is the average load of the system and $\mathcal{B}$ denotes the burst length. For simplicity, we assume that back-to-back cells arrive at the system during burst period in each connection.

2.2. Delay and Bandwidth Guaranteed Service

Since ATM networks allow to transfer various types of data, the ATM switching system should distinguish different QoS and control individual flow to obey the traffic contract which is negotiated on setting up the connection. Intuitively, the system must handle deadline of individual cell, thus several previous researches focuses on sorting deadlines in per-connection queues [3, 6, 7]. As pointed out the significant drawback of these sorted priority queue mechanisms such as Delay Earliest-Due-Data (Delay-EDD), Virtual Clock, Fair Queuing, etc. [4, 5, 6], the algorithmic complexity of these scheme is $O(\log N)$ in general; where N is denoted as the number of virtual connections in ATM (see [11]). Although some simple version of scheduling algorithms have been proposed[7, 8], it is very difficult to apply burst traffic with varying bandwidth during a session.

3. DYNAMIC WEIGHTED ROUND ROBIN

3.1. Algorithm

Now, we concentrate on cell scheduling algorithm for ATM switches to provide rtVBR traffic effectively. Assume that per-connection queue (Q_i) stores cells from the same source in a flow, and a threshold value (TH_i) is defined corresponding with transfer delay in each queue. The scheduler gives more chance to be served in order to reduce waiting time, and the threshold is used to detect a "busy" flow in the system.

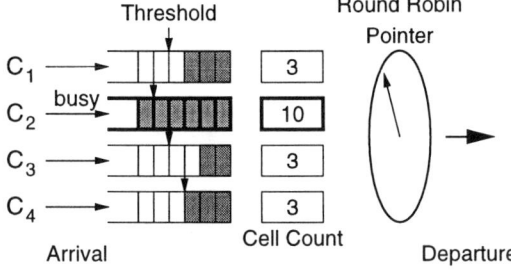

Figure 1: Dynamic Weighted Round Robin

Definition 1 *A flow i is busy if an arriving cell is stored into the queue belonging to the flow exceeding threshold TH_i. That is, the size of Q_i exceeds the threshold TH_i in a busy flow.*

Figure 1 illustrates a conceptual architecture of traffic management to implement the proposed algorith. Each queue has own threshold and cell count, and it regulates ingress cells as described in the following algorithm.

Algorithm for Dynamic Weighted Round Robin
Constants:
    N: number of connections;
    Q_i: per-connection queue ($i = 0, 1, \cdots, N-1$);
    TH_i: threshold of Q_i;
    r_i: minimum cell counter to be served in a time slot;
    R_i: maximum cell counter to be served in a time slot;
    δ_i: decreasing factor;
Variables:
    W: available weight;
    w_i: current weight of Q_i (cell counter);
Initialization:
    for $\forall i$ **do**
        $w_i \leftarrow r_i; Q_i \leftarrow NULL$;
    od
En-queuing: [on arrival of cell $c(i)$ belonging to flow i]
    ENQUEUE ($Q_i, c(i)$);
    if SIZE (Q_i) $> TH_i$ **then** (* busy *)
        $quatum \leftarrow \min(R_i, w_i + W) - w_i$;
    else (* non-busy *)
        $quatum \leftarrow \max(r_i, w_i - \delta_i) - w_i$;
    fi
    $w_i \leftarrow w_i + quantum; W \leftarrow W - quantum$;

Dequeuing:
    $i \leftarrow 0$;
    while FOREVER **do**
        for $k \leftarrow 1$ **to** w_i **do**
            SEND (DEQUEUE (Q_i));
            if IS_EMPTY (Q_i) **then**
                $w_i \leftarrow r_i$;
                $i \leftarrow (i+1)$ modulo N; (* round-robin *)
                break;
            fi
        od
    od

When a cell belonging to flow i arrives at queue Q_i, the scheduler determines whether the flow is busy or not. If the flow is busy, the scheduler increases the value of cell count which represents the weight of service. Otherwise, the scheduler makes the cell count to the initial state. In order to prevent frequent state change, the value of cell count

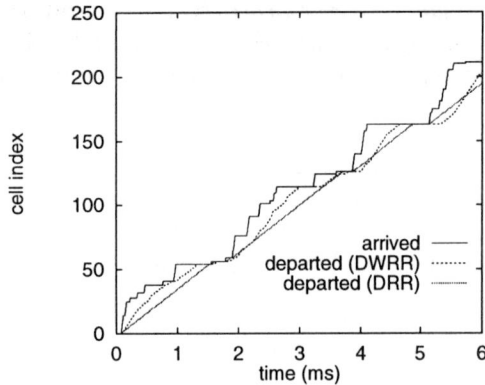

Figure 2: Comparison of DRR and DWRR

Table 1: Experiments

Exp.	connections	ρ	$\mathcal{B}$	TH_i	δ_i
I	$C_1 \sim C_3$	0.3	30	90	9
II	$C_1 \sim C_{10}$	0.045	5	20	2
	$C_{11} \sim C_{20}$	0.045	2	10	1
III	C_1	0.6	50	30	9
	C_2	0.25	10	10	9
	C_3	0.05	5	5	9

can be decreased by pre-defined parameter, called decreasing factor δ_i.

3.2. Implementation Issues

While servicing a queue, at most *cell count* w_i cells can be departed from the queue; this mechanism avoids starvation of cells in other queues. In our scheme, approximated delay control can be supported with the threshold of queue. Assuming that there exists a connection with tight delay requirement, the threshold should be fixed at the short length of a queue. Therefore, the scheduler must determine the threshold with consideration of limited transfer delay of cells, scheduling granularity, and traffic characteristics. On servicing, the computational complexity is $O(1)$ as shown in the DWRR algorithm.

4. NUMERICAL RESULTS

Figure 2 shows the differences between static scheduling and dynamic scheduling scheme, i.e., *Deficit Round Robin* (DRR) with a possible ATM extension and DWRR. In DRR, a quantum of service assigned to each connection queue, called *deficit counter* is fixed regardless of the number of cells backlogged. Our DWRR, however, the quantum is adaptive to the size of connection queue and the amount of cells departed will be increased for the busy connection.

In this section, we present the simulation results for the performance of DWRR in terms of fairness, delay, and throughput. Since our scheme based on a time-framing strategy but not sorted priority queue mechanism, the simulated results of DWRR is compared with that of DRR with an extension for fixed-size packet switched network. To determine the characteristics of a suitable simulation result, we considered bursty traffic sources in the alteration of a periodic discrete-time cell emission process with the following parameters:

- average cell rate (ρ): the load of a system for a flow ($0 \leq \rho \leq 1$)

- burst size ($\mathcal{B}$): the average number of cells transmitted at peak rate

Figure 2 shows a typical example of burst cell arrivals with $\mathcal{B} = 10$ and $\rho = 0.1$. Either back-to-back cells or no cell arrive at the scheduler. In DRR, a constant amount of cells departs from the queue; in DWRR, however, the number of cells departed is increased or decreased corresponding to the queue length. From Figure 2, we see that the horizontal distance between arrived and departed curves denotes the number of cells waiting in the queue. In addition, the vertical distance between two curves represents queuing delay in the scheduler. Clearly, the objective of scheduler should reduce the difference, which can achieve both guaranteeing real-time service and reducing the buffer occupancy.

4.1. Simulation Model

In this section we classify 3 examples for simulation: (1) high-speed bursty traffic, (2) low-speed bursty and non-bursty traffic, and (3) mixed traffic. Table 1 presents the actual simulation parameters for each traffic pattern. Note that the aggregated traffic loads are all 0.9 although the source generates either Poisson or constant traffic.

4.2. Results

We apply the worst-case fairness index WFI_i that has been proposed in the previous works to determine fairness over all connections [8]; WFI_i denotes the ratio of maximum and ideal share to be obtained by flow i. Formally,

$$WFI_i = \frac{\hat{S}_i}{S_i}$$

where $\hat{S}_i$ is the worst-case ratio of the number of cells sent by flow i to the total number of cells by all flows, and S_i is the ideal quotient for flow i.

Consequently, the fairness of our scheme is the same as that of DRR; that is, $WFI_i = 1$. This is because both DRR and DWRR can be work-conserving.

Although the size of the various *Quantum* variables in DRR determines the load of flow, cells can not have higher

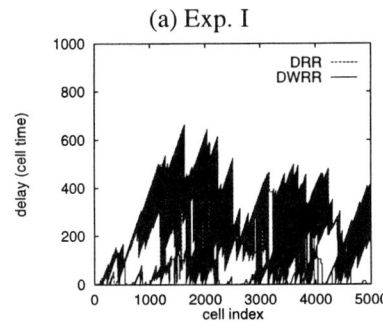 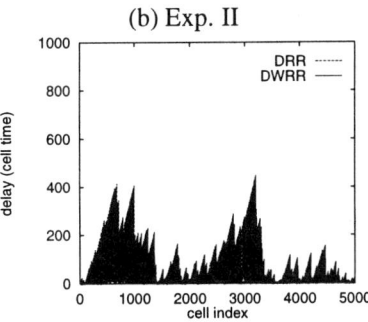 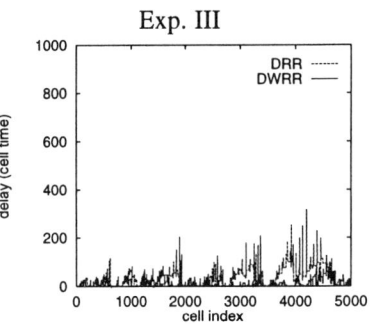

Figure 3: Delay

priority to be scheduled during congestion of flow. Unlike DRR, *Quantum* variables in DWRR can vary with the status of flow during lifetime of connections. Figure 3 shows the difference of departure cell delay. Under heavy load of certain flow, DWRR provides more opportunity for waiting cells in the busy flow. In addition, we have observed that throughput and long-term fairness of DWRR are the same as those of DRR.

5. CONCLUSIONS

In this paper, we have proposed a simple and effective scheduling algorithm, called the Dynamic Weighted Round Robin (DWRR) algorithm, for guaranteeing QoS with respect to delay and bandwidth for real-time burst traffic. We believe this topic is one of the most important issues for transmission of compressed video in ATM network. The key idea is to detect busy flow and increase the value of cell count which is the maximum number of cells to be served in a scheduling time unit. To justify the proposed scheduling algorithm, we have simulated 3 categories of mixed traffic.

There are still important but orthogonal issues left to be investigated. When the scheduler chooses a number of cells in a queue, reshaping mechanism is required before sending cells through links [2]. In addition, several system parameters such as threshold and increasing factor must be examined carefully corresponding to the characteristics of actual traffic.

6. REFERENCES

[1] F. Bonomi and K.W. Fendick, "The Rate-Based Flow Control Framework for the Available Bit Rate ATM Service," *IEEE Network Mag.*, pp. 25 – 39, Mar./Apr. 1995.

[2] P.E. Boyer, F.M. Guillemin, M.J. Servel, and J.-P. Coudreuse, "Spacing Cells Protects and Enhances Utilization of ATM Network Links," *IEEE Network Mag.*, pp. 38 – 49, Sept. 1992.

[3] H.J. Chao, H. Cheng, Y.-R. Jenq, and D. Jeong, "Design of a Generalized Priority Queue Manager for ATM Switches," *IEEE JSAC'97*, 15(5), pp. 867 – 880, 1997.

[4] P. Goyal and H.M. Vin, "Generalized Guaranteed Rate Scheduling Algorithms: A Framework," *Technical Report CS-TR-95-30*, Dept. of Computer Science, University of Texas, Austin, 1995.

[5] P. Goyal, H.M. Vin, and H. Cheng, "Start-time Fair Queuing: A Scheduling Algorithm for Integrated Services Packet Switching Network," *Technical Report TR-96-02*, Dept. of Computer Science, Univ. of Texas at Austin, 1996.

[6] A.K. Parekh and R.G. Gallager, "A Generalized Processor Sharing Approach to Flow Control in Integrated Services Networks: The Single-Node Case," *IEEE/ACM Trans. on Networking*, 1(3), pp. 344 – 357, 1993.

[7] J. Rexford, F. Bonomi, A. Greenberg, and A. Wong, "Scalable Architectures for Integrated Traffic Shapping and Link Scheduling in High-Speed ATM Switches," *IEEE JSAC'97*, 15(5), pp. 938 – 950, 1997.

[8] M. Shreedhar and G. Varghese, "Efficient Fair Queuing using Deficit Round Robin," *Proc. of ACM SIGCOM'95*, pp. 231 – 242, 1995.

[9] A. Stamoulis and J. Liebeher, "S2GPS: Slow-Start Generalized Processor Sharing," *Technical Report CS-97-03*, Dept. of Computer Science, University of Virginia, 1997.

[10] J. Turner, "New Directions in Communications, or Which way to the Information Age?" *IEEE Commun. Mag.*, vol. 24, pp. 8 – 15, 1986.

[11] H. Zhang and D. Ferrari, "Rate-Controlled Service Disciplines," *J. High Speed Networks*, 3(4), pp. 389 – 412, 1994.

Design and Implementation of an ATM Segmentation Engine with PCI Interface

Chan Kim, Jong-Arm Jun, Kyou-Ho Lee, Hyup-Jong Kim

Electronics and Telecommunications Research Institute
161 Kajeong-dong, Yuseong-goo, Taejon, KOREA

ABSTRACT

This paper describes the design and implementation of an ATM Segmentation Engine including PCI interface. This engine, which was designed for an ASIC called ASAH-NIC, contains DMA read machine and segmentation machine. It uses local memory where control information is stored in schedule table, VC table, buffer descriptor and status queue. It has several spepcial features like processing split packet buffers and automatic alignment and packing of transmit data.

1. INTRODUCTION

This paper describes the implementation of an ATM segmentation engine with PCI bus which is part of an ASIC called ASAH-NIC which performs segmentation for any arbitrary number of connections at the same time with 155Mbps performance(with reassembly engine in one chip). This engine can also tightly control the CBR, VBR burst cell generation using scheduler, and also supports UBR traffic with timer control. This paper discusses the segmentation part with its DMA master read function. Figure 1 shows the block diagram of the ASAH-NIC chip. The DMA slave and local memory interface and re-assembly function will not be discussed.

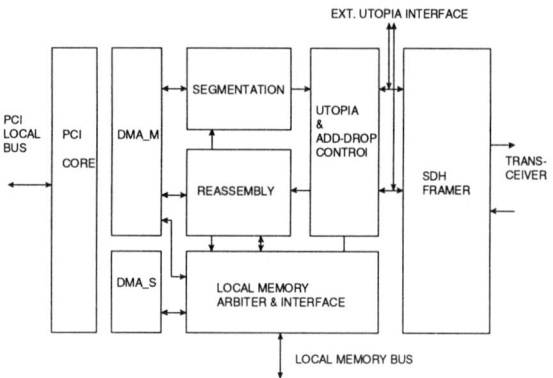

Figure 1. ASAH-NIC ASIC's Block Diagram

2. BASIC OPERATION

The basic operation relies on some data structure maintained in the local memory. They are VC table, buffer descriptor table, buffer descriptor queue, status queue, and schedule table.

VC Table contains the control information for each connection such as the AAL type, whether it's buffers contain complete cells, and whether it's UBR or not. It also keeps the CURR_DSCR and LAST_DSCR which point to the current and last element of the buffer descriptor linked list connected to the VC table. It also keeps the BUFF_RD_CNT which shows how many bytes have been read from the current buffer. It is to be noted that only the VC table is updated and written back to memory after a connection's service thus reducing local memory usage. The VC table also has UBR next pointer to link the VC table belonging to UBR into a cyclic linked list and timer control values. Partial CRC32 value is also kept in the VC table.

When there is data to send, the host CPU writes buffer descriptor for that data to the descriptor queue in the local memory while incrementing the write pointer. Buffer descriptor contains the starting address and the size of the buffer in byte. The buffer descriptor should also contain packet trailer and EOP(end of packet) indication. It also has the VCC index which shows what connection the data belongs to. Since VCC index is the order of the VC table in the pre-assigned SRAM address space, we can get the start address of the corresponding VC table with the VCC index. The buffer descriptors sent to the descriptor queue is taken out and linked to the corresponding VC table. When linked, the VCC index field of the buffer descriptor is used as pointer to form linked list.

The order of service for each connection is determined by the schedule table. The schedule table's entry contains the start address of the VC table to be serviced. Only the CBR or VBR connection is registered in this schedule table.

The segmentation engine also writes status information into status queue and the host CPU reads from the status queue. The status information includes the finished buffer address and VCC index, and other status information.

Figure 2 shows the free buffer list and method of linking buffer descriptors to the VC table.

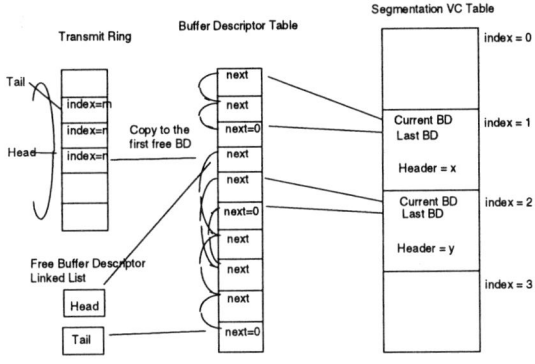

Figure 2 method of buffer descriptor linking.

3. HARDWARE ARCHITECTURE

3.1 Segmentation Processor

This section describes each processing block's action.

3.1.1 Buffer Descriptor Link Manager(BLM)

The descriptor queue level is increased when the host CPU writes one or more BD(s) to the BDQ and decremented when a descriptor is taken out and linked to a VC table. The BLM continues linking the BD in the BDQ while the BDQ level is not zero.

The BLM reads one BD from the BDQ and copies the data into a new free BD's address which was fetched from FBM. This new address is linked to the VC table corresponding to the VCC index using the current and last descriptor pointer of the VC table and next pointer of the BD. Figure 3 shows a typical linking action when there was two BDs already linked.

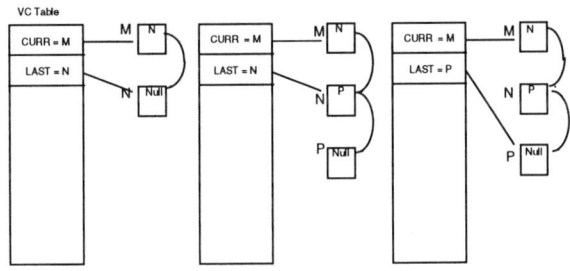

Figure 3. Example of a Buffer Linking Procedure

3.1.2. Free Buffer Descriptor Manager(FBM)

The free buffer descriptor manager manages the linked list of free buffer descriptor. When requested from the BLM, it gives the head pointer of the free list and updates the head pointer to the next free one. The next address is always pre-fetched to get the next pointer before it is overwritten with null by the BLM. In addition, when requested by the schedule and segmentation circuit, the FBM appends the address of the used descriptor to the tail of the free linked list.

3.1.3 Scheduling and Segmentation Manager(SSM)

SSM comprises of VC table and buffer descriptor register files, main controller, and segmentation cell buffer. The segmentation cell buffer has 10 cell buffers and writes DMA read cell data or unassigned cells into the FIFO as requested and sends the cell to the UTOPIA interface when there is at least one cell in the FIFO. Because the write controller and the main controller monitors the FIFO state before any (DMA) cell writing, FIFO overflow is prevented.

VC table and buffer descriptor register file keeps the VC table and buffer descriptor read from the memory and supplies various signals for the controller and requests DMA with retrieved parameters when commanded by the controller. The VC table is updated at the end of the DMA.

The number of remaining bytes is obtained by subtracting BUFF_RD_CNT from the buffer size. For AAL 5, if the remaining bytes are more than 48, 48 bytes are requested. If the EOP bit of the buffer descriptor is set and the number of remaining bytes is less than 40, DMA is requested for these remaining bytes with EOP cell indication set. (Figure 4 (a)) But if the remaining bytes are more than 40 and less than or equal to 48, the remaining bytes are requested with EOP cell indication deasserted. In this case, the number of bytes requested for the next DMA will be zero to indicate all padding cell.(Figure 4. (b))

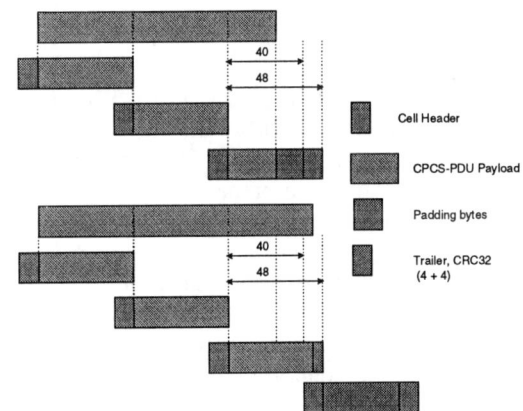

Figure 4. Typical AAL 5 segmentation
(a) above : remaining <= 40
(b) below : 40 < remaining <= 48

As a special case, if EOP is not set for the buffer and the number of remaining bytes is less than 48 but there is a following buffer already linked to the VC table, it assumes that the following buffer has the needed data and two DMA action is performed for one cell. We call this 'split DMA' in this engine. This can be called a data gathering function for segmentation.

Figure 5 shows the simplified state diagram of the main segmentation controller. It should be understood that VC table information for the decision is updated at the end of DMA completion. And data availability includes the split DMA case.

Figure 5. State Diagram of the Main Controller

The controller starts reading schedule table processing "insert unassigned cell", and "jump" and "pass" operations. If the OP code is "immediate transmit", the controller reads the VC table pointed to by the schedule table. If the CURRENT value is not zero, (meaning at least one BD is linked to the VC table), it loads the buffer descriptor but if it is not, the controller jumps to UBR pointer fetch state. If the UBR pointer is available, it switches to UBR service but if it is not, it commands insertion of 1 unassigned cell and proceeds to next entry of the schedule table.

After loading the VC table and BD, if it is decoded that there's enough data for DMA, the controller commands DMA request. At the end of DMA completion, the VC table is updated with prepared values and new information is available. In all cases when a VC table is newly loaded, the burst count is preset to the defined burst value and the count is decremented at every cell transferring DMA completion and checked for burst continuation. And if the current buffer data is running out by the DMA, the controller frees the current descriptor.

After every DMA completion, the controller checks if another DMA should be performed for the connection. The first case is for split DMA and the second case is when the burst counter is not zero. For each case the controller jumps to appropriate states of reading BD or writing VC table or checking size. When the burst ended due to the lack of data, it returns to UBR service after writing back the VC table.

During the UBR service, timer value is also checked before reading buffer descriptor and the controller jumps directly to unassigned cell insertion if data is not available for DMA. And after a DMA completion, it goes to VC table write back state without looking at burst counter.

The minimum cell generation time distance control for UBR traffic is done using a timer and two fields of the VC table. The timer is incremented every given period of, for instance, 1 usec. Whenever a cell is transmitted, the current timer value plus time distance value for that connection is added and written into the VC table. To generate a cell for any UBR traffic, the timer should be equal to or greater than this time set at its last transmission time.

3.1.4. UBR Pointer Manager(UPM)

The UBR pointer manager searches the UBR linked list VC table and get ready for the UBR pointer having buffer. When UBR pointer is fetched by the SSB it starts searching again. The UPM's pointer searching is enabled only when there is at least one buffer descriptor linked to UBR traffic.

3. 1. 5. Status Report Manager(SRM)

When the status report request bit of the buffer is set, the main control requests status report to the status report manager providing status information. The status report manager latches and writes the status information into the status queue and generates interrupt.

3. 2 DMA Master READ Processor

Basically, the DMA read block receives DMA request with DMA parameters and after reading requested data from the PCI memory of the PCI core, returns the ATM cell with enable signal.

DMA parameter includes cell_request signal, EOP cell indication, and other control signals as well as the start address and size of the DMA read, cell header, AAL type, trailer, CRC32, FCT indication. For AAL5, partial CRC32 value is also returned to the segmentation circuit at the end of DMA.

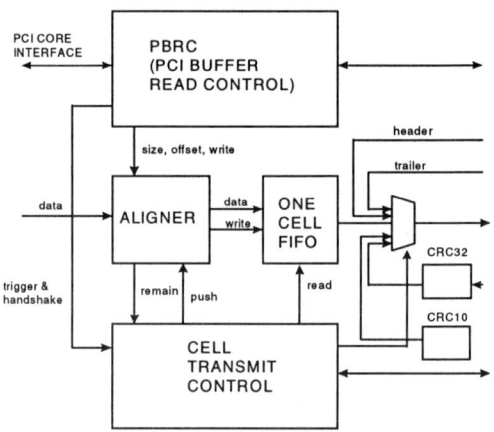

Figure 6. DMA Master Read Circuit

Figure. 6 is the diagram of DMA READ processing block. **PCI read controller** transfers the DMA parameter of start address and size to the PCI core and requests memory read. When passing the parameter to PCI core, the PCI read controller changes the start byte address and byte size to word equivalent values so that all the requested bytes can be contained in the words requested to the PCI core.

Whenever the FIFO is not empty it repeats reading the PCI FIFO until all the requested data is written to the alignment circuit. When sending the unaligned words to the alignment circuit, the PCI read controller calculates the start location(offset) and number(size) of the bytes to be taken from the word and supplies this information with enable signal. Figure 7 shows the offset and size values for a split DMA case where 17 bytes from PCI address BA0045 and 16 bytes from PCI address 79205C is read to form 33 bytes for the cell payload(remaining bytes are filled with padding bytes).

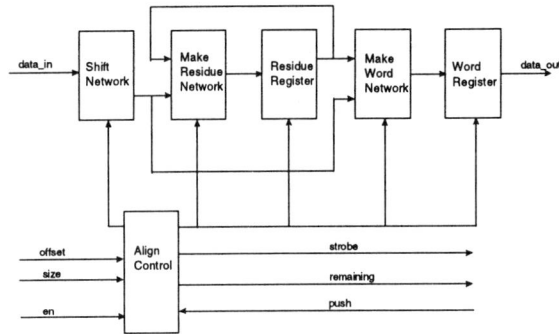

Figure. 8 Alignment Circuit

After the data has been moved to the FIFO, the cell transmit controller reads the data and makes ATM cell by multiplexing ATM cell header, FIFO output data, CRC10, CRC32, and trailer. Before starting cell forming routine, it pushes the occasional remaining word in the alignment circuit out to the FIFO. The padding is done automatically by reading the FIFO 10 or 12 times regardless of the amount of data in the FIFO which is designed to output padding pattern when empty.

4. CONCLUSION

The segmentation engine designed for an ASIC called ASAH-NIC, performs all the protocol processing using local SRAM and reads directly from host or PCI memory to make ATM cells thus relieving the host CPU from processing burdens. Since there is no limitation on the start address and the size, there is no need to move the data to a separate location after processing a higher layer application processing. This engine can tightly control the CBR, VBR bursts and UBR cell generation using schedule and timer.

The ASAH-NIC ASIC has PCI function, SAR and ATM functions for various AAL types, and 155Mbps physical layer function with external UTOPIA interface and provides one-chip solution for ATM adapter.

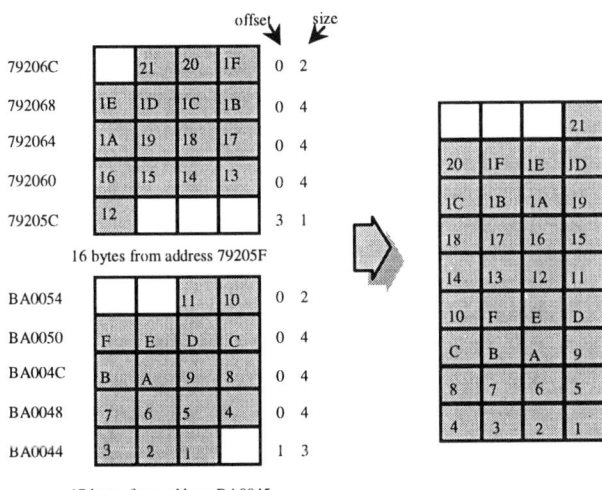

Figure 7. Word Alignment Circuit's Input and Output

Figure 8 shows the block diagram of alignment circuit. The residue register keeps the bytes which are not yet assembled into an aligned word. When a new word has been assembled, the word register latches the aligned word and the strobe signal is pulsed. The controller controls the amount of shift and select and latch actions according to the number of residual bytes and the offset, and size of the input word.

Acknowledgments

We would like to thank Dr. Jae-geun Kim, Dr. Munkee Choi and Yeong-won Hwang, for their valuable advices and encouragement

References

[1] ITU-T Rec. I.432, I.361, I.363, I.371
[2] PCI Local Bus Specification, Rev 2.1. June. 1995

A NOVEL NEURAL ESTIMATOR FOR CALL ADMISSION CONTROL AND BUFFER DESIGN IN ATM NETWORK

Liang Zhang, Zemin Liu, Senior Member, IEEE
57#, Beijing Uni. of Posts & Telecommunications, Beijing, 100088, P.R.China

ABSTRACT

In call admission control of ATM network, it is difficult for the conventional method to judge the accepting boundary accurately and dynamically, for the imprecision description of the traffic parameters and the different requirement of the allowed QoS. In this paper we propose a novel neural network structure as an intelligent control scheme to perform ATM admission control. The neural estimator can learn the probability distribution of the CLR, thus can control the ATM traffic very accurately and dynamically. The disperse structure of the neural estimator make it easy to learn and operate. The trained neural network can also be used as a buffer estimator in the reference design of ATM system. The simulation result show the advantage of this neural estimator.

1. INTRODUCTION

The broadband integrated services digital network (B-ISDN) is expected to provide the infrastructure for a new generation of digital communications services. The network will carry a wide ranges of traffic types, including voice, data, video, etc. The asynchronous transfer mode (ATM) has been recommended by ITU-T as the main transfer method for future B-ISDN, for it can offer the flexibility that is needed to handle traffic with widely varying characteristics and service requirements.

However, due to the uncertainties of broadband traffic patterns, unpredictable statistical fluctuations of traffic flows can cause congestion in the ATM network. In ATM network, call admission control (CAC) decides whether to accept or reject the call based upon availability of capacity required to support the quality-of-service (QoS) according to the parameters declared by the terminal. There are two main schemes available for call admission control in an ATM network. The conventional methods applies a parameter model of the traffic being offered, either by requiring each call to provide an accurate description of its traffic parameters, or by measuring the observed traffic and fitting it to a model, and then infers the cell loss rate (CLR) from the model. However, there are tremendous difficulties in implementing CAC by conventional methods due to these reasons: 1) it is difficult to characterize and model the high variability of the multimedia traffic by simple parameters; 2) it is difficult to check the status of current network to get a reliable availability of capacity.

The second type of call admission scheme is executed by the artificial neural networks (ANN). The most prominent advantage of applying neural networks to adaptive control mechanisms in ATM network is that neural networks are capable of learning the nonlinear relationships between the inputs and the outputs. A.Hiramatsu proposed a ANN model to solve the problem of CAC and link capacity control[1][2]. J.E.Neves proposed a fully connected feedforward neural network to predict the ATM traffic parameters in the switching nodes and in the transmission links[3]. E.Nordstrom applied the neural networks in a subroutine called link admission control (LAC)[4].

In this paper, we mainly discuss about the usage of Neural Network in ATM network call admission control, and propose a novel neural estimator based on the discussion. With the disperse structure combined by a certain amount of sub neural networks, the neural estimator can not only find out a precise accepting boundary dynamically, but also estimate the buffer length in the design of a ATM switcher. The organization of this paper is as follows: in Section 2, we first discuss the general purpose of a neural network in admission control of ATM network, and propose a novel structure which can judge the accepting boundary dynamically in Section 3. In Section 4 the neural estimator is used in the reference design of ATM switcher as a buffer estimator. The simulation results are shown in Section 5, and the conclusion is shown in Section 6.

2. THE GENERAL USE OF NN IN CAC

2.1 Neural network

A neural network is generally a multiple-input multiple-output nonlinear mapping circuit, which can learn an unknown non-linear input-output relation from a set of examples. It consists of many neurons connected to each other; each neuron is a multiple-input single-output nonlinear circuit. The connection strengths are called weights, and the neural network input-output relation can be modified by changing the set of weight values. In this paper, we propose a general feedforward neural network. The three-layered neural network consists of an input layer, a hidden layer, and an output layer. Each layer is a group of many neurons, and the output of a neuron in one layer is input to all the neurons in the next layer. In the working phase, the neural network produces the output values according to the input values and the weight vector. While in the training phase, the input examples and the desired output values are simultaneously set and the neural network adjust the weight values according to certain learning algorithms, among which the most common used is back-propagation (BP) algorithm[5].

2.2 Neural CAC

The neural networks used in admission control of an ATM network are mainly feedforward neural networks, where the inputs can be the traffic parameters together with net condition, and the outputs can be the decision, which divide the inputs space into two sides, one belonging to the accepting type, and the other rejected. The main principle of the neural network is to act as a multiple-input multiple-output nonlinear mapping circuit, which can learn an unknown non-linear input-output relation from a set of examples. Whether the neural controller can work or not depend mostly upon the precision of the division, which separate the high-loss types from low-loss types. So the selection of the inputs and outputs of the neural network become a key problem in the application of neural network in CAC.

3. A NOVEL NEURAL CAC ESTIMATOR

The neural estimator based on a three-layer feedforward network can approximate any continuous function theoretically. However, the more parameters exist in the input and output, the more complicated the neural network will be. And, if the input parameter can't represent the real input pattern accurately, then it is hard to describe the nonlinear function between the input pattern and the output pattern.

As the outputs of the neural network in CAC, most previous work select cell loss rate (CLR) for the CLR is the most important parameters in estimating the quality of the services. However, the neural network learn only the average of observed CLR data, while the observed CLR data are expected to disperse because a wide variety of cell generation patterns exits even in a transmission rate class. This means that average learning may not guarantee the allowed CLR in CAC. Moreover, it is hard to estimate exactly the CLR since the value of the CLR is usually very small and a small error can make great influence to the whole estimator. In addition, if the requirement of the allowed CLR changes, then all the sampling data will have to discard and the neural network have to be trained again.

For the previous reasons, we choose the number of the different kind of connections as the inputs of the neural estimator. As to the service type, we divide it into three types: audio, data and video, for the services in each type have the similar traffic parameters. We didn't take traffic parameters in for whether the traffic parameters can represent the service accurately is a considerable problem. With the number of different type of connections as the inputs of the neural network, the only thing the neural network should do is to find out which kind of combination could be an accepting one.

As to the output of the neural estimator, we still choose the CLR as the target for it is the most important parameter in the QoS. But we didn't count the CLR directly for the previous reason. We choose the probability distribution of the cell queue length as the outputs of the neural estimator, for the probability distribution of the cell queue length can easily change to the probability distribution of the CLR. First we define a virtue buffer which is long enough to hold all the possible cell queues. With the buffer length given, the summary of the probability distribution of the cell queue length longer than the given buffer length in the virtue buffer will be counted as the probability

distribution of the CLR. Thus the probability distribution of the cell queue length changes to the probability distribution of the CLR. Suppose that $p(i)$ represent the Probabilities of the cell queue length which is longer than the real buffer length, then $\sum_i p(i)$ will equal to the CLR.

After the inputs and outputs of the neural network have been decided, we give the structure of the neural estimator as follow:

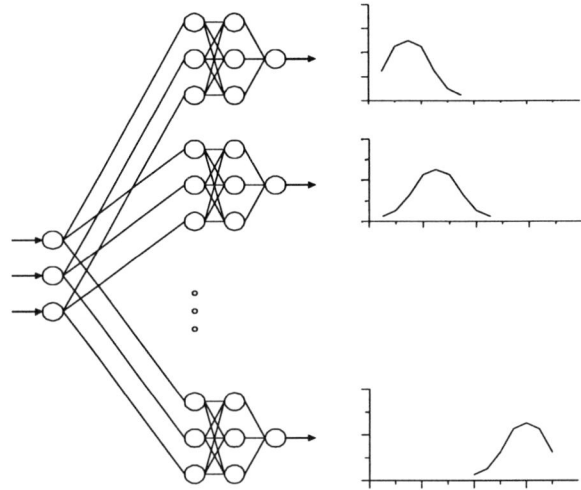

Fig. 1 The disperse structure of the neural estimator and the output of each sub neural network

Here we use a certain number of sub neural network instead of a large one, for each sub neural network is a simple feedforward neural network, easy to learn and operate. The total number of the sub neural network depends on how many intervals we use to describe the probability distribution of the cell queue length in the virtue buffer. usually we can divide the virtue buffer into 10 or 20 average intervals, each interval is used to count the probability of the cell queue length falling into it.

The connections are defined as audio, data and video respectively, so the input pattern can be described by a matrix $P(n1, n2, n3)$, n representing the number of each type of connections. The pattern P is input to all the sub neural networks, while the output of each sub neural network will be the probability distribution of the cell queue length.

The input-output relation of each sub neural network are described as the right part of the Fig. 1. The X axis represents the equivalent capacity of the input pattern, while the Y axis represents the probability of the according cell queue length falling into that certain interval. We can see that a certain intervals will contain the probability of a certain pattern more than that of the other patterns. As we can expect, when the equivalent capacity of the input pattern increases, the probability of the according cell queue length falling into the longer interval also increases.

While to each input pattern P, the outputs of all sub neural networks are as follows:

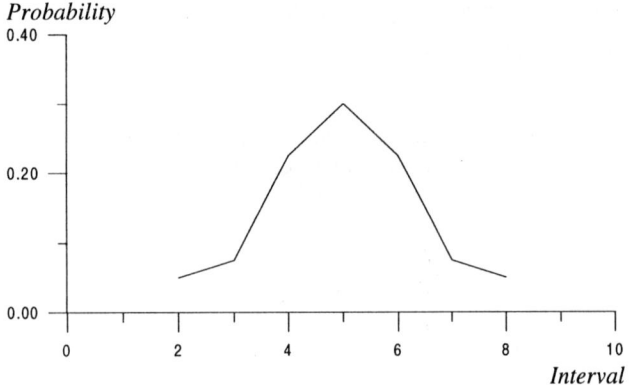

Fig. 2 The outputs of the neural estimator with a certain input pattern

From the relation curve in Fig. 2 we can see that such kind of relations in Fig. 1 will result in a curve that represent the probability distribution of cell queue length of a certain pattern. So, with the real buffer length given, according to the different demand of the allowed CLR, the neural estimator can judge whether to let a certain input pattern be an accepting one, thus determine the accepting boundary dynamically.

4 THE USAGE OF NN IN REFERENCE DESIGN OF ATM SYSTEM

The neural estimator described in Section 3 can be in turn used as a buffer estimator in the reference design of an ATM network. Here we mainly consider the cell loss rate (CLR), and take the cell delay and jitter for granted. The main purpose of the NN in the reference design of ATM network is to estimate the length of the queuing buffer, so that the ATM system can satisfy the QoS regarding to the CLR and the channel utilization.

When the cell arrives, if there are queuing cells still waiting for transferring, then the coming cell has to stay in the buffer and wait for the former cells to be dealed first. Here we don't consider the priority of the cell, such that the queuing buffer is merely a FIFO system. In the estimation of the buffer length using neural estimator, we regard the max payload of the ATM system as link capacity, so that the purpose of the buffer estimator is simplified to find out the length of the buffer, which can accept as much connections as demand, while satisfying the QoS of the connections under different conditions.

As we have described in Section 3, the probability distribution of the cell queue length has a steady relation with the input pattern. With a certain link capacity, and the real buffer length of the ATM system given, we can deduce the CLR from the probability distribution of the cell queue length of that input pattern. Suppose that $p(i)$ represent the probabilities of the cell queue length which is longer than the real buffer length, then CLR equals to $\sum_i p(i)$.

Now let's use the same neural network which has been trained as in Section 3. If we can find such an minimum bounds in the virtue buffer that the summon of the probabilities of the cell queue length which is longer than that bounds, described as $\sum P_i \leq \text{CLR}$, then the bounds will certainly be the length of the real buffer. So we count the $p(i)$ from the bottom of the virtue buffer, plus each of the probabilities of the cell queue length which equals to each intervals, until $\sum_i p(i) > \text{CLR}$. Then the position of interval before that one will be the real length of the buffer.

This kind of measuring is made for one special input pattern. While to different kind of input patterns, there will be different length through counting. Among these values, we pick up the maximum value which satisfying the channel utilization of the system as the buffer length.

5 SIMULATION AND NUMERICAL RESULTS

5.1 Neural estimator in CAC

First, extensive simulations were performed to obtain the data set for the training phases. The data sources were treated as average services, the audio sources were simulated using ON/OFF model, and the video sources were simulated using AR model. The total link capacity is 155.52 Mbps.

We use 10 sub neural networks in the structure of the neural estimator. For each input pattern P, we sample the cell queue length in the virtue buffer for 1000 times. The virtue buffer has no length limited, thus all of the possible cell queue length will be counted. From the sampling data we choose a max buffer length, which is longer than the max possible cell queue length, and also satisfying the following condition: that after we divide the max buffer length into 10 average intervals, the real buffer length we will use is N times of the length of the interval, where N is an integer. This choice of the position of the real buffer in the virtue buffer is to make it easy for the neural estimator to count the CLR, since the data outside the real buffer will be counted as loss cells. In our numerical simulations, we choose the max buffer length of 100 cells, while the real buffer length of 60 cells. The data sampled were divided into 10 types, each belonging to 1 interval. The numbers of the data falling into the 10 intervals were normalized by 1000, and the 10 normalized data will be the output of the 10 sub neural networks.

With enough sampling data, the 10 sub neural networks were trained respectively. Because each sub neural network is a simple forward neural network, it is easy for them to convergence.

In the operation phase, when a call request arrives, the connection combination plus the incoming one are input to the neural estimator, and the outputs of each sub neural network are input to a controller, see Fig. 3, where the probability distribution of the cell queue length is changed to CLR. Compared with the allowed CLR, the controller will decide to accept or reject the input connection combination, which means to accept or reject the incoming call request.

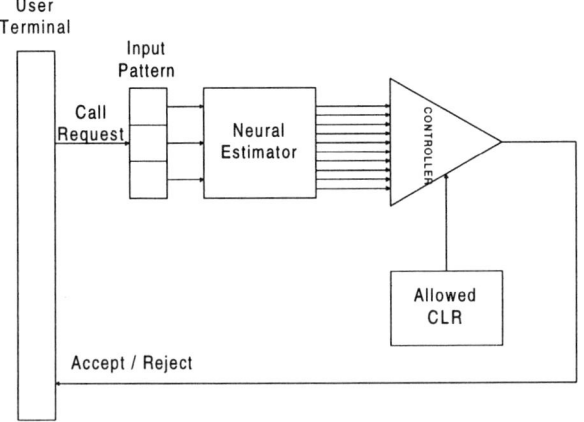

Fig. 3 The structure of the neural estimator in ATM network

Fig. 4 is the channel utilization of the Neural Estimator scheme (NE) compared with the Peak Rate scheme (PR) and Equivalent Width scheme (EW)[6]. At the same requirement of the QoS, the channel utilization of the NE is better than the others.

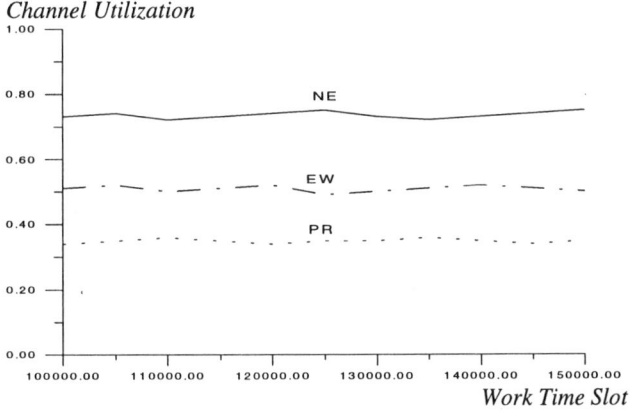

Fig. 4 The channel utilization of the three schemes when $CLR < 10^{-9}$

5.2 Buffer estimator in ATM reference design

In the reference design of ATM network, we still use the neural network as described in the previous Section. There are some little difference, since we don't know the real buffer length beforehand. The length of the virtue buffer must be considered carefully, and the more intervals the better.

In order to secure the precision, we chose the virtue buffer length at an enough value of 200 cells, and divide it into 20 intervals. Assume that the switch system has the link capacity of 60%, which means that the system has a steady output rate of 155.52Mbps × 80% = 124.416Mbps. And if we want to utilize the channel to 70%, while keep the QoS of the connections below 10^{-9}, then we first train the neural network at the link capacity of 124.416Mbps just as the previous Section. In the work time, we select a certain number of input patterns, whose sum of demanding capacity is equal to the channel utilization, 155.52Mbps × 70% = 108.864Mbps here. From those input patterns, we got a group of lengths, from which we select the max value as the real buffer length. In our reference design, the value of 80 cells was enough.

6. CONCLUSION

In call admission control of ATM network, it is difficult for the conventional method to judge the accepting boundary accurately and dynamically, for the imprecision description of the traffic parameters and the different requirement of the allowed QoS. As an alternative approach, in this paper we propose a novel neural estimator as an intelligent control mechanism in order to perform ATM call admission control. The neural network in this model can learn the nonlinear relationships between the number of different type of connections and the probability distribution of the CLR, and thus can control the ATM traffic very accurately and dynamically even if the allowed CLR changes. The simulation results show the effectiveness of the neural estimator in ATM network as an intelligent and adaptive control scheme. The same neural estimator can also be used in the reference design of ATM system as a buffer estimator, and can estimate the buffer length according to different type of ATM systems and variant demanding. We can expect that this control and estimate scheme can be applied to the more realistic ATM networks.

7. REFERENCES

[1] A. Hiramatsu, "ATM Communications Network Control by Neural Networks," IEEE Transaction in Neural Networks, vol. 1, 1990.

[2] A. Hiramatsu, "Integration of ATM Call Admission Control and Link Capacity Control by Distributed Neural Networks," IEEE J. Select Areas Commun., vol. 9, no. 7, Sep. 1991

[3] J. E. Neves, L. B. Almeida, and M. J. Leitco, "B-ISDN Connection Admission Control and Routing Strategy with Traffic Prediction by Neural Networks," in Proc. SUPERCOM ICC 94, New Orleans, Louisiana, May 1994.

[4] E. Nordstrom and J. Carlstrom, "A Reinforcement learning Scheme for Adaptive Link Allocation in ATM Networks", in Proc. of the Intel Workshop on Applications of Neural Networks to Telecommunications 2 (IWANNT-95), Stockholm, Sweden, 1995.

[5] D.E. Rumelhart, G. E. Hinto and R. J. Williams, "Learning Internal Representations by Error Propagation," Parallel Distributed Processing, vol. 1, 1986

[6] J. J. Base, "Survey of traffic control schemes and protocols in ATM networks," Proc. of IEEE, vol. 79, No. 2, 1990.

DCT BASED ERROR CONCEALMENT FOR RTSP VIDEO OVER A MODEM INTERNET CONNECTION

Yon Jun Chung, Jong Won Kim and C.-C. Jay Kuo

Integrated Media Systems Center
Department of Electrical Engineering-Systems
University of Southern California
Los Angeles, California 90089-2564

ABSTRACT

Data lost in the form of dropped packets is a major obstacle to actualizing real time Internet video via modem connection with frame rates as high as 30 frames per second. Error concealment (EC) techniques must restore the lost data. Due to rigid system specifications of this application, only the computationally simplest error concealment methods are plausible. Replication, which is the simplest temporal error concealment (TEC) method, fits well in this situation. There are however circumstances when replication is inadequate and linear interpolation, the simplest spatial EC, is used in its place. In this work, we consider an alternative spatial EC to linear interpolation, i.e. the DCT-coefficient based EC algorithm, with comparable computational cost and consistently better image quality in real-time Internet video applications.

1. INTRODUCTION

Recent introduction of time-based Internet protocols such as RTSP (real time streaming protocol) [1] and low bitrate video codecs such as H.263 [2] has enabled researchers to entertain the possibility of full duplex video transmission via the Internet modem connection. These advances forecast the day when videophone-like applications with high framerates can transmit over the plain old telephone system (POTS) using the Internet as the main means of transport. Currently, the achievable frame rate for commercially available Internet videophones over modem connection is still too low to be satisfactory. Efforts are underway to increase the achievable frame rate, but the task of increasing the frame rate needs supporting mechanisms. One such mechanism is error concealment. This paper describes an error concealment (EC) algorithm suitable for this purpose.

If one were to set out to create a modem-capable real time video application with high frame rates, the necessary ingredients would be a video codec that compresses sequences at a high frame rate and outputs at a low bit rate, a time-based Internet protocol that is compatible with existing infrastructure, and a modem that has a sufficient bandwidth. Surprisingly, all three components already exist. For sequences which contain relatively little motion content such as the test sequence akiyo, H.263 video codec is already capable of compressing a QCIF (176x144 pixels) sized sequence down to as low as 23 Kbps average bit rate at 30 frames per second (fps) with acceptable spatial and temporal visual quality. Unlike resource allocating Internet protocols such as RSVP (resource reservation protocol) [3], RTSP already transports time-based medium throughout the current Internet infrastructure without any router specification. Besides, 56 Kbps modems have become common these days.

The absence of high frame rate modem-capable real time video applications in the presence of necessary ingredients points to obstacles in system integration. The methods described in [4] are efforts to overcome such obstacles in system integration. One of the largest obstacles comes in the form of lost packets. Depending on Internet traffic conditions, packets are indiscriminately dropped or lost as a means of congestion alleviation. The data contained within these dropped packet are irrecoverably lost, and methods like error correction coding for wireless networks are not applicable. The prescribed environment of Internet full duplex video communication at a high frame rate over a modem connection further exacerbates the lost packet problem by markedly constricting operating margins. A block diagram of the video system environment is shown in Fig. 1. First of all, full duplex communication dictates that overall delay be preferably less than one second. This means that the elapsed time, from the instant of video capture at the server to display at the client has to take less than one second. Such a strict time margin makes timely detection and retransmission of lost packets practically impossible.

Since retransmission of lost packets is impractical, lost data in dropped packets must be reconstructed at the client side with EC techniques. Unfortunately, rigid system specifications restrict the selection of possible EC techniques. The desired high frame rate limits the amount of time allowed for EC techniques. Let n fps be the target frame rate. Then, it leaves $1/n$ second for the client to uncompress, detect, localize and conceal the error. This implies that, for a 30 fps frame rate, the EC technique can consume only a portion of about 33 ms for all intermediate steps to display a single frame. Any EC method chosen to meet such a time restriction must balance between its image restorative capability and its computational complexity.

Regardless of the restorative capability, if an EC algorithm cannot be executed within the duration intermittent to two successive frames, it cannot be taken into consideration. The simplicity of the replication method was the

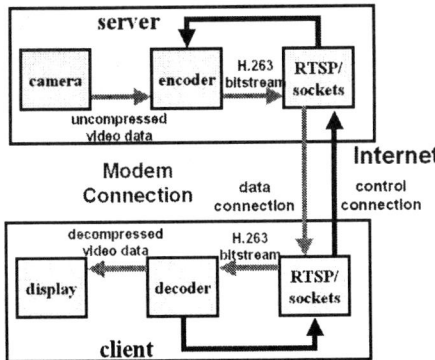

Figure 1: Blockdiagram of real-time Internet video applications.

primary reason for its selection in [4]. However, blindly replicating all image pieces due to lost packets was less than an acceptable solution. How to overcome its shortcomings serve as the major motivation for this work.

This paper is organized as follows. In Section 2, localization of lost data corresponding to dropped packets will first be described. Afterwards, the spatial error concealment algorithm based on DCT coefficient interpolation will be presented in detail as well as employment scenarios. Section 3 will present the experimental results of the proposed EC scheme. Concluding remarks will be given in Section 4.

2. ERROR CONCEALMENT BASED ON DCT COEFFICIENT INTERPOLATION

In keeping with H.263 terminology, pictures and frames will be used interchangeably. Before any EC technique can be employed, the picture location corresponding to the data lost due to dropped packets must be determined. There are two synchronizing structures in H.263: picture start code (PSC) and group of blocks start code (GBSC). They signal the start of a picture and a horizontal strip group of blocks (GOB), respectively. For QCIF sized sequences, each picture consists of nine GOBs, each of which consists of eleven 16x16 macroblocks (MB).

The actual occurrence of lost data is first detected by tracking the packet sequence number. If successively arriving packets do not have consecutive packet sequence numbers, then one packet or packets have been lost. Packets that arrive out of sequence are useless because of the real time nature of this application, and are thus treated as lost packets as well. By comparing the last correctly transmitted GBSC with the next available GBSC, the number of missing MBs can be inferred. Since the finest synchronization within one picture is the GBSC in H.263, lost packets manifest themselves in missing horizontal strips. The challenge of the EC method is to restore these missing strips to its best ability within a very limited amount of time.

Most EC methods fall into one of two categories: temporal or spatial error concealment. In fact, the replication method mentioned earlier [4, 6] is the simplest form of temporal EC. While well suited for most error concealment situations, it cannot be universally applied for all types of data loss. Examples of such loss include missing GOBs which takes place in the middle of scene changes and MBs with motion vectors (MV) greater than 8 pixels. For these loss types, an alternative EC to replication must be employed, possibly an EC in the spatial domain. One of the computationally simplest spatial EC is known as linear interpolation error concealment (LEC). However, the end result often has the look of blurred or runny luminescence which is not desirable. In this work, we propose a DCT coefficient based spatial EC (DEC) algorithm with comparable complexity to LEC but providing better output image quality.

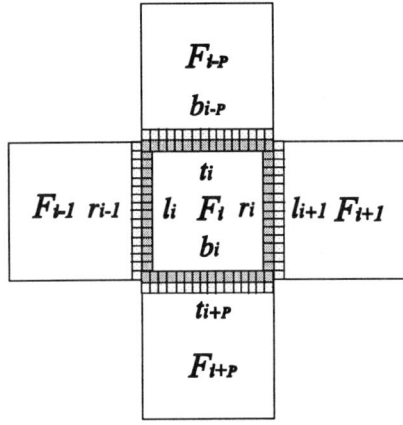

Figure 2: Illustration of missing MB (F_i) with its four boundary vectors (t_i, b_i, l_i, and r_i) and their neighboring boundary vectors($b_{i-P}, t_{i+P}, r_{i-1}$, and l_{i+1}) located in the four adjacent MBs ($F_{i-P}, F_{i+P}, F_{i-1}$ and F_{i+1}), respectively.

Only the basic idea of the DEC algorithm is presented here. We refer to [5] for greater detail and more regirous mathematical developement. Consider the missing MB represented by the center square F_i in Fig. 2. For QCIF image sequences, there are 9x11 MBs per picture so $P = 11$, and each MB has 16×16 pixels so that F_i is NxN with $N = 16$. It is reasonable to assume that the boundary vectors $[t_i, b_i, l_i, r_i]$ of the missing MB are smoothly connected to their respective adjacent vectors

$$b_{i-P}, t_{i+P}, r_{i-1}, l_{i+1}.$$

Therefore, a suitable cost function can be expressed as:

$$\begin{aligned}\psi &= \parallel t_i - b_{i-P} \parallel^2 + \parallel b_i - t_{i+P} \parallel^2 \\ &+ \parallel l_i - r_{i-1} \parallel^2 + \parallel r_i - l_{i+1} \parallel^2.\end{aligned} \quad (1)$$

As mentioned earlier, lost packets in H.263 result in missing horizontal strips. Thus the horizontal adjacent vectors r_{i-1} and l_{i+1} are not available. They need to first be filled by linearly interpolating pixel pairs

$$(b_{i-(P+1)}(N-1), t_{i+P-1}(N-1)) \text{ and } (b_{i-(P-1)}(0), t_{i+P+1}(0)),$$

respectively. By expressing the cost function in (1) in the frequency domain, and setting its first derivative to zero, we obtain the optimal solution X_i which satisfies

$$\Psi(X_i)' = 0, QX_i = C_i, \quad (2)$$

where Ψ is the DCT of ψ, Q is an $N^2 x N^2$ matrix containing DCT kernel functions. After some manipulation, we can derive the following:

$$X_i = [a_{i,0,0} \cdots a_{i,0,N-1} a_{i,1,0} \cdots a_{i,1,N-1} a_{i,2,0} \cdots a_{i,N-1,N-1}]^T$$

and

$$\begin{aligned} C_i &= \phi(0) \otimes \mathbf{b}_{i-P} + \phi(N-1) \otimes \mathbf{t}_{i+P} \\ &+ \mathbf{r}_{i-1} \otimes \phi(0) + \mathbf{l}_{i+1} \otimes \phi(N-1), \end{aligned}$$

where $\phi(n)$ are the DCT kernel functions and vectors

$$\mathbf{b}, \mathbf{t}, \mathbf{r}, \mathbf{l}$$

are the 1-DCT coefficient vectors of their spatial counterparts, $\otimes$ is the Kronecker-delta product. Simply speaking, X_i is an $N^2 x 1$ vector of the DCT coefficients of F_i and $a_{i,n,m}$ the 2D DCT coefficients of the missing MB F_i while C_i is an $N^2 x 1$ vector computed from DCT kernels and DCT of adjacent vectors. Eq. (2) can be expanded as

$$\begin{aligned} QX_{i,n,m} &= 2(\phi_n^2(0) + \phi_m^2(0))a_{i,n,m} \\ &+ \sum_{l=0, l \neq \frac{n}{2}}^{\frac{N}{2}-1} 2\phi_n(0)\phi_{2l}(0)a_{i,2l,m} \\ &+ \sum_{k=0, k \neq \frac{m}{2}}^{\frac{N}{2}-1} 2\phi_m(0)\phi_{2k}(0)a_{i,n,2k} \\ &= C_{i,n,m} \quad (3) \end{aligned}$$

Let us examine these equations. There are N^2 unknown 2D DCT coefficients while only $4N$ known border adjacent 1D DCT coefficients. This leads to a set of severely underdetermined linear equations. The connection necessary for reducing this set of equations turns out to be the even and odd symmetries of DCT kernels, i.e.

$$\begin{aligned} \phi_{even}(N-1) &= \phi_{even}(0) \\ \phi_{odd}(N-1) &= -\phi_{odd}(0). \quad (4) \end{aligned}$$

By separating the set of equations in Eq. (2) into four groups according to even-odd pairings, i.e. Q_{ee}, Q_{eo}, Q_{oe} and Q_{oo}, as shown in Fig. 3(a), we can exploit the properties in (4) to down-select subsets of $N-1$ linearly independent equations within each group. In this manner, the reduced matrices Q_{xx} become invertible so that $4(N-1)$ unknowns can be solved.

It is computationally expensive to solve for all $4(N-1)$ unknown DCT coefficients. As explained in [5], it is reasonable to solve 20 out of the 60 for the case with $N = 16$. All other DCT coefficients are set to zero. In Fig. 3(b), we show the selected DCT coefficients in our implementation, where the index i has been dropped for brevity. The selection of $a_{n,m}$ is equivalent to choosing rows and columns of $\mathbf{A_i} = [a_{i,n,m}]$, the matrix to which IDCT must be performed to obtain the missing block F_i. Ever greater details can be added by selecting additional $a_{n,m}$ terms, but only from the same row or column so linear independency is maintained.

The fact that the inversion of the four Q_{xx} matrices can be performed off-line is the main contribution to the

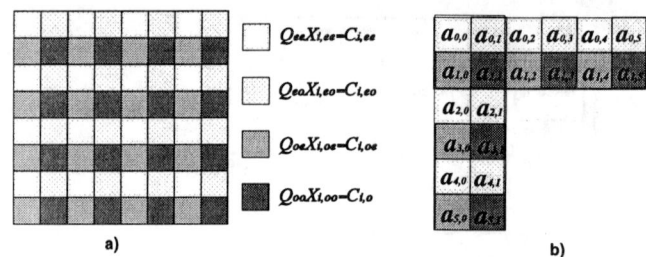

Figure 3: (a) The DCT coefficient location of the four groups and (b) the selected 20 DCT coefficients in the implementation.

computational simplicity of DEC. The only other major remaining operation is the 2-D IDCT operation of matrix $\mathbf{A_i}$. However, since many of the terms (236 out of 256) in $\mathbf{A_i}$ are set to zero the 2-D IDCT operation is reduced to $20N^2$ multiplications plus some additions.

Using the number of multiplications as a measure of complexity, LEC requires $2N^2$ multiplications to restore an $N \times N$ block for the horizontal types of losses common to H.263. On the other hand, the proposed DEC uses $20N^2$ multiplications to restore a similar sized loss. At first, this may seem like a substantial complexity increase, but actually it is consistent with other computations taking place side by side. For example, an IDCT for the same sized $N \times N$ block (in H.263 format, an $N \times N$ macroblock is represented by 6 $\frac{N}{2} \times \frac{N}{2}$ blocks) requires $12N^2$ multiplications. By comparison, MV based TECs are much more computationally expensive. They use a comparison metric such as mean absolute difference (MAD) that require substantially more time to restore a similarly sized $N \times N$ block.

It is worthwhile to remember that the proposed DEC is specifically targeted for frames with very large MVs and intra-coded frames. In H.263, an intra-coded frame usually occurs during a scene change, with no temporal correlation with its previous adjacent frame. In this scenario the applicability of temporal EC methods is limited, although efforts such as [7] are underway to circumvent this limitation. In addition to lacking temporal information, intra-coded frames are more prone to loss packets by virtue of their larger size. They are often more than ten times the size of inter-coded frames, and would suffer a proportionate amount of packet loss in environments with relatively constant packet loss rate.

3. EXPERIMENTAL RESULTS

In the experiment, we examine the Foreman sequence with packet losses and three error concealment methods. They are illustrated in Fig. 4 (a)-(d), where (a) is the Foreman sequence with loss while (b)-(d) are restored with replication error concealment, LEC, and the proposed DEC, respectively. The proposed DEC can capture frequency characteristics such as edges because it interpolates from DCT coefficient. This can be seen by comparing Fig. 4(c) with Fig. 4(d) with special attention to the vertical beam on the crane. The DEC restored vertical beam of Fig. 4(d) has this edge partially reconstructed whereas the LEC restored

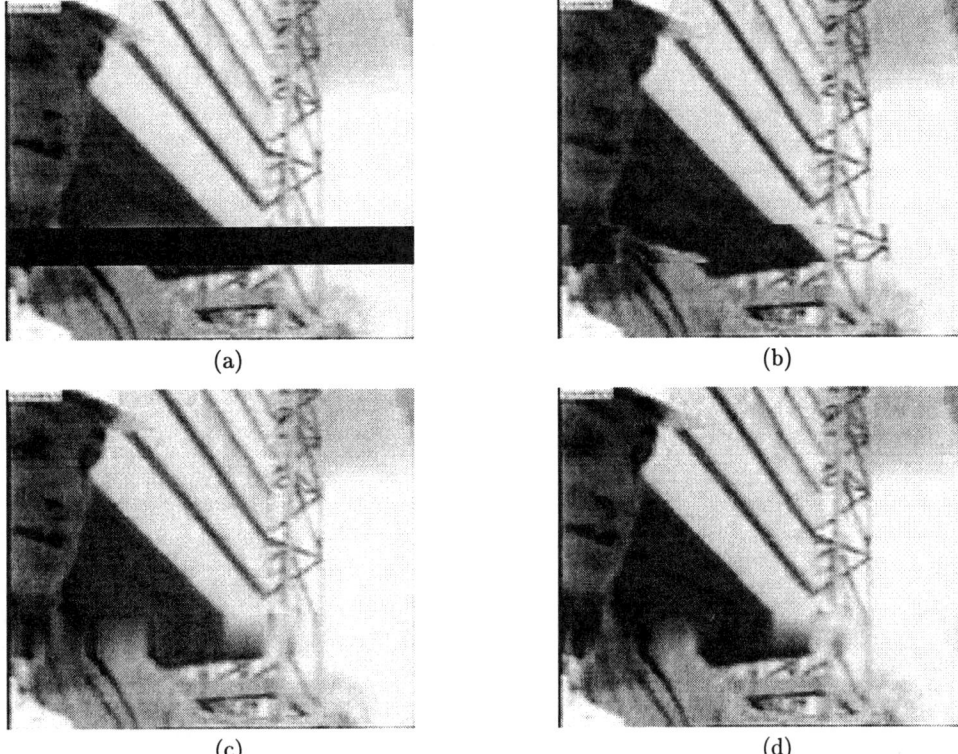

Figure 4: Foreman sequence with packet losses and three error concealment methods. (a) Foreman with loss. (b) Foreman restored with replication error concealment. (c) Foreman restored with LEC. (d) Foreman restored with the proposed DEC.

beam of Fig. 4(c) just looks smudgy. Similar observations can be made around the lapel and shoulder area. The proposed DEC method strives to restore line orientation using the DCT information of the border pixels. For sequences with missing MBs whose MV magnitudes are larger than 8, DEC performs better than replication EC, as is evident in Fig. 4 and Tab. 1.

Foreman	error-free	replication	LEC	DEC
MSE	310.68	371.19	336.66	335.27
PSNR	23.207	22.434	22.858	22.877

Table 1: MSE and PSNR comparison table amongst the error-free transmission, replication EC, LEC and our proposed DEC

4. CONCLUSION

In this paper, we introduced a DCT coefficient based error concealment method as an alternative spatial error concealment to linear interpolation for real time Internet video applications. Due to its low computational cost, the proposed scheme is well suited for the time critical environment such as duplex RTSP video via modem connection at the expense of image quality. The proposed method has computational complexity compatible to linear interpolation method while providing better image results consistently.

5. REFERENCES

[1] H. Schulzrinne, A. Rao and R. Lanphier, *Internet Draft IETF, Real Time Streaming Protocol (RTSP), draft-ietf-mmusic-rtsp-04.txt*, September 17, 1997.

[2] *DRAFT ITU-T Recommendation H.263: Video coding for low bit rate communication*, July, 1997.

[3] R. Braden, L. Zhang, S. Berson, S. Herzog and S. Jamin, *RSVP: Resource ReSerVation Protocol, RFC 2205*, September 1997.

[4] Y. Chung and J. Kuo, "Non-Disruptive RTSP Video Over the Internet using a Modem Connection", in *SPIE Visual Communication and Image Processing'98*, vol. 3309, pp. 517–524, San Jose, Jan. 1998.

[5] J. Park, J. Kim and S. Lee, "DCT Coefficient Recovery Based Error Concealment Technique and Application to the MPEG-2 Bit Stream Error", *IEEE Cir & Sys for Video Technology*, vol. 7, pp. 845–854, Dec. 1997.

[6] Q. Zhu, Y. Wang and L. Shaw, "Coding and Cell-loss Recovery in DCT-based Packet Video" *IEEE Trans. Circuits and Systems for Video Tech.*, vol. 3, pp. 248–258, June 1993.

[7] S. Aign, "A Temporal Error Concealment Technique for I-Pictures in an MPEG-2 Video-Decoder", in *SPIE Visual Communication and Image Processing'98*, vol. 3309, pp. 405–416, San Jose, Jan. 1998.

A TRANSMITTING, AND RECEIVING METHOD FOR CDMA COMMUNICATIONS OVER INDOOR ELECTRICAL POWER LINES

Hideaki Okazaki*, Mitsusato Kawashima**

*Gifu National College of Technology, Motosu-gun, 501-0495 Japan, okazaki@gifu-nct.ac.jp
**Ibiden Industries Co., Ltd., Ogaki, 503-0936 Japan, mk302251@marinet.or.jp

ABSTRACT

Transmitting and receiving method for CDMA communications through 2-wire or 3-wire power line cord over indoor power lines are proposed. The key idea for transmitting method is to use a fundamental model of a coupling transformer-less transmitting circuit including the influence of the distribution transformer and to find out the combinations of model parameters giving the acceptable distortion of the transmitted waveforms for CDMA communications. That for receiving method is to use a filtering method that separates arbitrary waveforms of transmitted data signal from the superposition waveforms of the sinusoidal high power signal and transmitted data signal without distorting arbitrary waveforms of transmitting signal at all. These methods are experimentally confirmed to be effective.

1. INTRODUCTION

The demand for localized communication networks for office automation, security monitoring, environmental management in buildings, computer communications, and other applications continues to increase. Networks utilize physical links which often involve substantial installation costs, appearance compromises to building interiors, inconvenience, or limitations on equipment locations. Alternatives to physical links include radio or infrared channels; however, radio implies licensing and interference, while infrared requires line of sight transmission. Electrical power distribution circuits provide reasonably universal channels with a simple and standard interface in the form of a wall-socket plug. Disadvantages include limited bandwidth in comparison with cable or fiber-optic links, high noise levels, and uncertain and varying levels of impedance, attenuation, and noise. In power line communications, conventional and inexpensive modulation schemes for data transmission such as amplitude shift keying (ASK) or frequency shift keying (FSK) are well known to be ruled out. Spread spectrum techniques have been confirmed to be effective instruments to provide reliable communications over (especially indoor) electrical power lines[1]-[3]. Recently some modems for indoor power line communications are realized and sold. However, by using spread spectrum techniques, code division multiple access (CDMA) communications over indoor electrical power lines have not been discussed and realized. The purpose of this paper is to propose a transmitting, and receiving method for CDMA communications over indoor electrical power lines and to report experimental results of the proposed method. In indoor electrical power line communications proposed and realized until present, the data is usually transmitted to (or received from) the power line through the line coupling networks that consist of an audio (or pulse) transformer for isolation and impedance matching, as well as a high-pass filter blocking the 50 or 60 Hz power signal. Besides, indoor electrical powers in family residences and buildings are normally sent to through distribution transformers from outdoor high voltage power lines supplied by electric power corporations (shown in Fig.1). Both of these coupling and distributing transformers cause much distortion to the effective waveforms for realizing CDMA communications in using DS spread spectrum techniques ,e.g. waveforms like multi-steps. Therefore, in order not to distort the effective waveforms for CDMA communications possibly, we consider the influences of the distribution transformers on the transmitting waveforms in power line communications and alternative line coupling transmitting and receiving circuits without any coupling transformers.

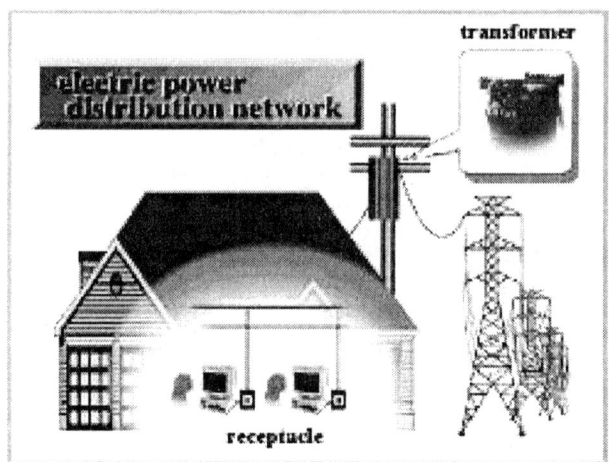

Figure 1 Electric power distribution network

2. MODEL OF A COUPLING TRANSFORMER-LESS TRANSMITTING CIRCUIT

2.1 A Fundamental Model

Consider the fundamental model for realizing CDMA communications over indoor power lines (shown in Fig.2). In this model, e(t) and Z_3 represent voltage signal source and signal source impedance, respectively. C_1 and C_2 are coupling capacitors. Z_1 and Z_2 are impedances for adjusting transmitted waveforms over indoor power line. L_1, L_2, and M represent the self and mutual inductances of the distribution transformer. Z_0 represents outdoor power line impedance. In order to investigate the combinations of model parameters giving the acceptable distortion of the transmitted waveforms for CDMA communications, we assume the conditions for constructing simpler essential model as follows:

- the dimensions of the indoor power distribution circuit are small compared with the wavelength associated with the highest frequency of the signal source;
- the influence of indoor power (2-wire or 3-wire) line cord length is modeled by a capacitance in parallel with line cord;
- the indoor power line load model is given by the load resistance R_L;
- Z_1 and Z_2 are regarded as only lines (i.e. zero-impedances); and
- Z_0 and Z_3 are regarded as the resistances R_0 and r

Under the above assumptions, the following simple model equations are derived:

$$\frac{d e_R}{dt} = \frac{-2}{C \cdot R_L} e_R - \frac{1}{C \cdot r} v_{c1} - \frac{2}{C} i_1 + \frac{1}{C \cdot r} e(t),$$

$$\frac{d v_{c1}}{dt} = \frac{-1}{C \cdot R_L} e_R - \frac{1}{C \cdot r} v_{c1} - \frac{1}{C} i_1 + \frac{1}{C \cdot r} e(t),$$

$$\begin{pmatrix} \frac{d i_0}{dt} \\ \frac{d i_1}{dt} \end{pmatrix} = \frac{1}{L_1 \cdot L_2 - M^2} \begin{pmatrix} -M & -L_2 \cdot R_0 \\ L_1 & M \cdot R_0 \end{pmatrix} \cdot \begin{pmatrix} e_R \\ i_0 \end{pmatrix}$$

$$+ \frac{A \sin 2\pi ft}{L_1 \cdot L_2 - M^2} \begin{pmatrix} L_2 \\ -M \end{pmatrix}. \quad (1)$$

Here v_{c1}, e_R, $e(t)$, i_0, and i_1 are the voltage across C_1, the voltage source signal, current through L_1 and current through L_2. $A\sin(2\pi ft)$ represents the outdoor AC supply model.

2.2 Typical Simulation Results and Possibility of CDMA communications

In case that $e(t)$ is given by a square wave signal source, typical examples of the relation between waveforms of transmitted signal over power line and combinations of the model parameters are illustrated in Fig.3-6, by using fourth order Runge-Kutta numerical integration method. These simulation results are obtained under the following model parameters based on the experimental electric power line in the section 3: $L_1 = 143.5$[mH], $L_2 = 38.7$[mH], $M = 74.15$[mH], $C = C_1 = C_2 = 0.01$[μF], $R_L = 10$ [kΩ], $r = 10$[Ω], $r_0 = 0.001$[Ω], $f = 60$[Hz], $A = 280$[v], calculation time step : 0.1[μs], V_A : amplitude of $e(t)$, fs : frequency of $e(t)$.

- In Fig.4(b) and 6(b), the amplitude of square wave signals and the voltage level of the transmitted signals over power line are recognized to be correlated.

- Further, in Fig.4(a) and 6(b), both waveforms of the square wave signals and the voltage across C_1 are correlated.

(1) **2-wire line cord CDMA communications**: the first result indicates the possibility that under appropriate conditions, CDMA communications through 2-wire line cord are realized by using the correlation of some voltage levels of the source signal and the transmitted signal over power line.

(2) **3-wire line cord CDMA communications**: the second results indicates the large possibility that under appropriate conditions, CDMA communications through 3-wire line cord with neutral terminal (green) are realized in case that the neutral terminal is connected to the signal ground in Fig.2 which is determined by the voltage divider consisting of C_1 and C_2 in parallel with power line.

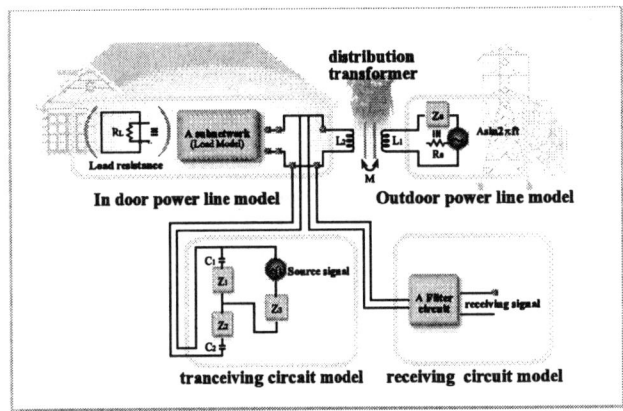

Figure 2. A fundamental model of a coupling transformer-less transmitting circuit and receiving circuit model.

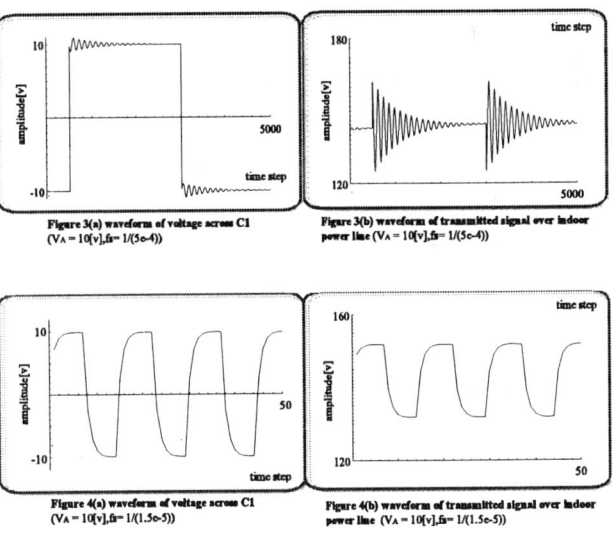

Figure 3(a) waveform of voltage across C1 ($V_A = 10$[v], fs= 1/(5e-4))

Figure 3(b) waveform of transmitted signal over indoor power line ($V_A = 10$[v], fs= 1/(5e-4))

Figure 4(a) waveform of voltage across C1 ($V_A = 10$[v], fs= 1/(1.5e-5))

Figure 4(b) waveform of transmitted signal over indoor power line ($V_A = 10$[v], fs= 1/(1.5e-5))

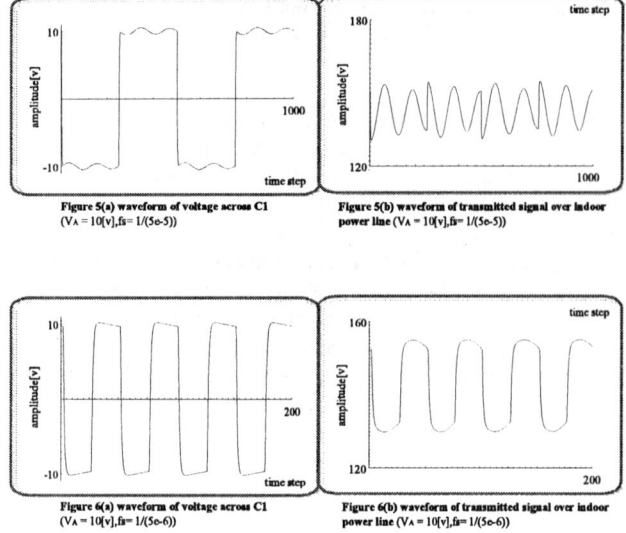

Figure 5(a) waveform of voltage across C1
($V_A = 10[v], fs= 1/(5e-5)$)

Figure 5(b) waveform of transmitted signal over indoor power line ($V_A = 10[v], fs= 1/(5e-5)$)

Figure 6(a) waveform of voltage across C1
($V_A = 10[v], fs= 1/(5e-6)$)

Figure 6(b) waveform of transmitted signal over indoor power line ($V_A = 10[v], fs= 1/(5e-6)$)

capacitance of indoor power (2-wire or 3-wire) line cord length, we set $C_1 = C_2 = 0.01[\mu F]$ and $C_3 = 10000[pF]$ in the circuit in Fig.7.

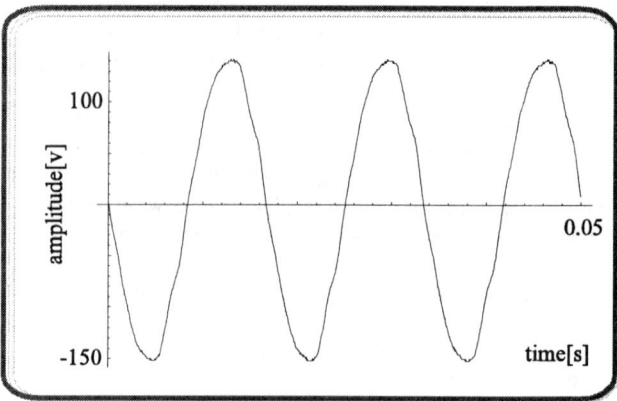

Figure 8 waveform of power supply AC

3.2 Experimental Results Related to CDMA Communications

The experimental results in this section and the following section 4.2 are measured and converted to digitized data by Tektronix digital real-time oscilloscope TDS340P (500MS/s), and are illustrated by using Mathematica, and FreeHand softwares. Figure 8 shows the waveform of indoor power-supplying AC in our college building. This waveform is distorted a little. Using the proposed transmitting circuit in Fig. 7, we obtain some experimental results related to both of (1) 2-wire and (2) 3-wire line cord communications mentioned in the preceding section 2.2 (see Fig. 9 and 10).

(i) Figure 9(d) indicates that the amplitude of square wave signals and the voltage level of the transmitted signals over power line are correlated actually under appropriate conditions. Then the experimental results are illustrated in Fig. 10, by using the correlation of some voltage levels of the source signal and the transmitted signal over power line, and converting the multiple signals used by a kind of spread code into multi-step waves in proportion to the multiple signal pattern voltage level as shown in the left in Fig.10 "0"-"9". In Fig. 10, The top left, the bottom left, and the right represent the examples of four channel multiple signal patterns by using a kind of spread code, the converted multi-step waves, and transmitted signals over power line of which cord length is 30[m], respectively, when the converted multi-step waves (the min peak to max peak voltages are $-10[v]$ and $10[v]$. the frequency span of the minimal square unit composing the multi-step waves is $240[kHz]$) are signal sources. Figure 10 indicates that the waveforms of the converted multi-step waves and those of the transmitted signals over power line are correlated under appropriate conditions. The positive possibility is experimentally shown that CDMA communications through 2-wire line cord are realized

(ii) Figure 11 also indicates that both waveforms of pattern "0" (the multiple signal pattern that all the data is low level in case of four channel CDMA communication) more complex than the square wave signals and the voltage across C_1 are

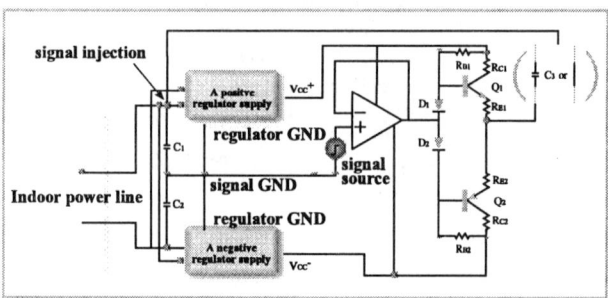

Figure 7. A realization of a coupling transformer-less transmitting circuit.

3. EXPERIMENTAL RESULTS OF THE TRANSMITTING CIRCUIT FOR CDMA COMMUNICATIONS

3.1 Experimental Electric power line and Transmitting Circuit

In western honshu island of Japan, the electric power in a single family residence is usually 100v - 60Hz and is normally sent to through a distribution transformer from 200v – 60Hz outdoor power line supplied by a electric power corporation. Thus we construct the simple experimental power line as shown in Fig.2 that consists of 2-wire or 3-wire with neutral line cord regarded as 100v indoor power line model, 100v – 200v distribution transformer for 1.5kVA ($L_1 = 143.5[mH]$, $L_2 = 38.7[mH]$, $M = 74.5[mH]$), and 200v – 60Hz power-supplying AC in our college building. Next, we propose the actual transmitting circuit as shown in Fig. 7 that consists of voltage follower and push-pull transistor circuit. In the transmitting circuit, the capacitor C_3 between injection point and output of the push-pull circuit, is for protection from power line voltage. Considering about the

correlated actually under appropriate conditions. The more positive possibility is experimentally shown that CDMA communications through 3-wire line cord with neutral terminal (green) are realized in case that the neutral terminal is connected to the signal ground in Fig.7.

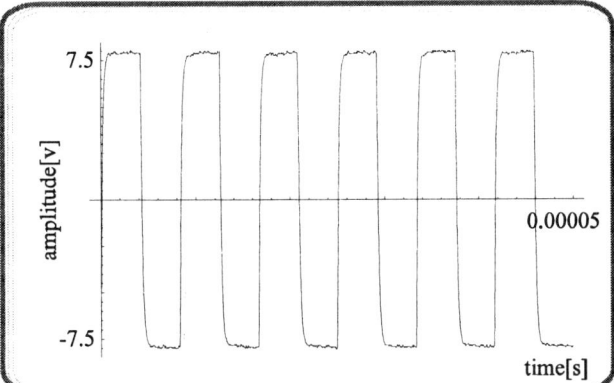

Figure 9(a) waveform of signal (after push-pull circuit)
(square:V_A=8[v], fs=120[kHz])

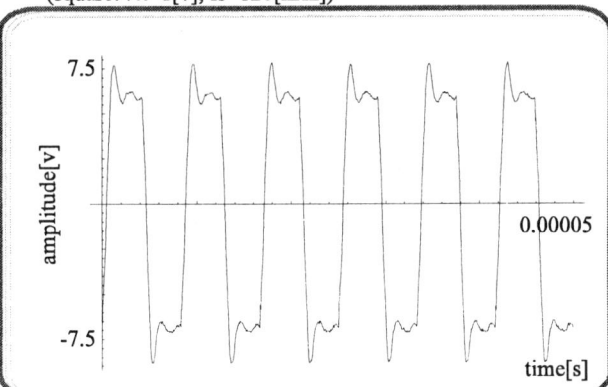

Figure 9(b) waveform of voltage across C1
(square:V_A=8[v], fs=120[kHz])

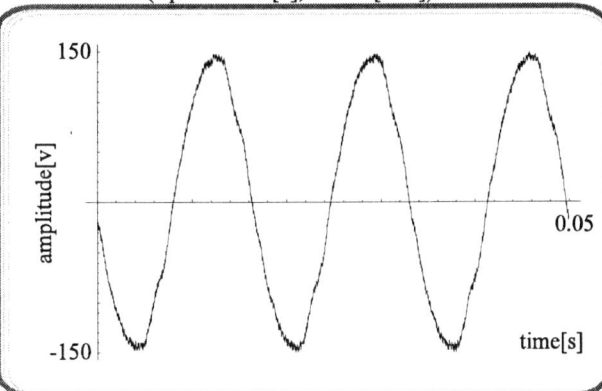

Figure 9(c) waveform of voltatge across C2

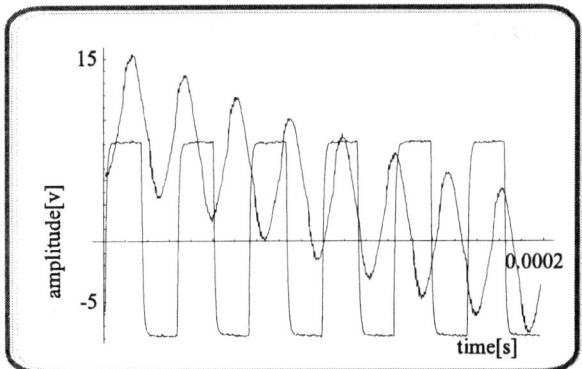

Figure 9(d) waveform of transmitted signal over indoor power
(A:signal soure after push-pull circuit(V_A=8[v], fs=120[kHz]))
(B:transmitted signal over indoor power line of which cord length is 30[m])

4. A FILTERING METHOD FOR SEPARATING ARBITRARY WAVEFORMS OF TRANSMITTED DATA SIGNAL

4.1 A Realization of Receiving Circuit

We propose the filtering method shown in Fig.12 that separates arbitrary waveforms of transmitted data signal from the superposition waveforms of the sinusoidal high power signal and transmitted data signal without distorting arbitrary waveforms of transmitted signal at all.

The algorithm of the filter, especially the phase shifter in the block diagram in Fig.12 is described below.

Under the condition that f_h is sufficiently different from f,
$F_1(A \sin(2\pi ft) + H(t)) = b_1 A \sin(2\pi ft+\phi)$, and
$F_2(A \sin(2\pi ft) + H(t)) = b_2 A \sin(2\pi ft+\varphi)$ hold.
Then, from rotating transformation matrix :

$$\begin{pmatrix} \cos(2\pi ft + \theta) \\ \sin(2\pi ft + \theta) \end{pmatrix} = \begin{pmatrix} \cos(\theta) & -\sin(\theta) \\ \sin(\theta) & \cos(\theta) \end{pmatrix} \bullet \begin{pmatrix} \cos(2\pi ft) \\ \sin(2\pi ft) \end{pmatrix},$$

$\sin(2\pi ft+\phi) = \sin(\phi)\bullet\cos(2\pi ft) + \cos(\phi)\bullet\sin(2\pi ft)$
$\sin(2\pi ft+\varphi) = \sin(\varphi)\bullet\cos(2\pi ft) + \cos(\varphi)\bullet\sin(2\pi ft)$ hold.
We have

$$\begin{pmatrix} \cos(2\pi ft) \\ \sin(2\pi ft) \end{pmatrix} = \begin{pmatrix} \sin(\phi) & \cos(\phi) \\ \sin(\varphi) & \cos(\varphi) \end{pmatrix}^{-1} \bullet \begin{pmatrix} \sin(2\pi ft + \phi) \\ \sin(2\pi ft + \varphi) \end{pmatrix}$$

$$= \frac{1}{\sin(\phi - \varphi)} \begin{pmatrix} \cos(\varphi) & -\cos(\phi) \\ -\sin(\varphi) & \sin(\phi) \end{pmatrix} \bullet \begin{pmatrix} \sin(2\pi ft + \phi) \\ \sin(2\pi ft + \varphi) \end{pmatrix}. \quad (2)$$

Therefore, there exist $e = \sin(\phi)/\sin(\phi-\varphi)$ and $d = -\sin(\varphi)/\sin(\phi-\varphi)$ such that $\sin(2\pi ft) = e\bullet\sin(2\pi ft+\varphi) + d\bullet\sin(2\pi ft+\phi)$.

The notation is given below.

t: time; phase shift of first linear filter:ϕ; phase shift of second linear filter:φ; Amplitude rate of first linear filter: b_1; Amplitude rate of second linear filter: b_2; First and second linear filter operators: $F_1(x)$, and $F_2(x)$; Amplitude operator:$(1/b) \equiv x/b$; Sinusoidal signal: $A\sin(2\pi ft)$, where A, and f are amplitude, and frequency respectively; Arbitrary waveform signal with

Figure 10 four channel multiple signal patterns by using a kind of spread code and their transmitted patterns over indoor power line of which cord length is 30[m]

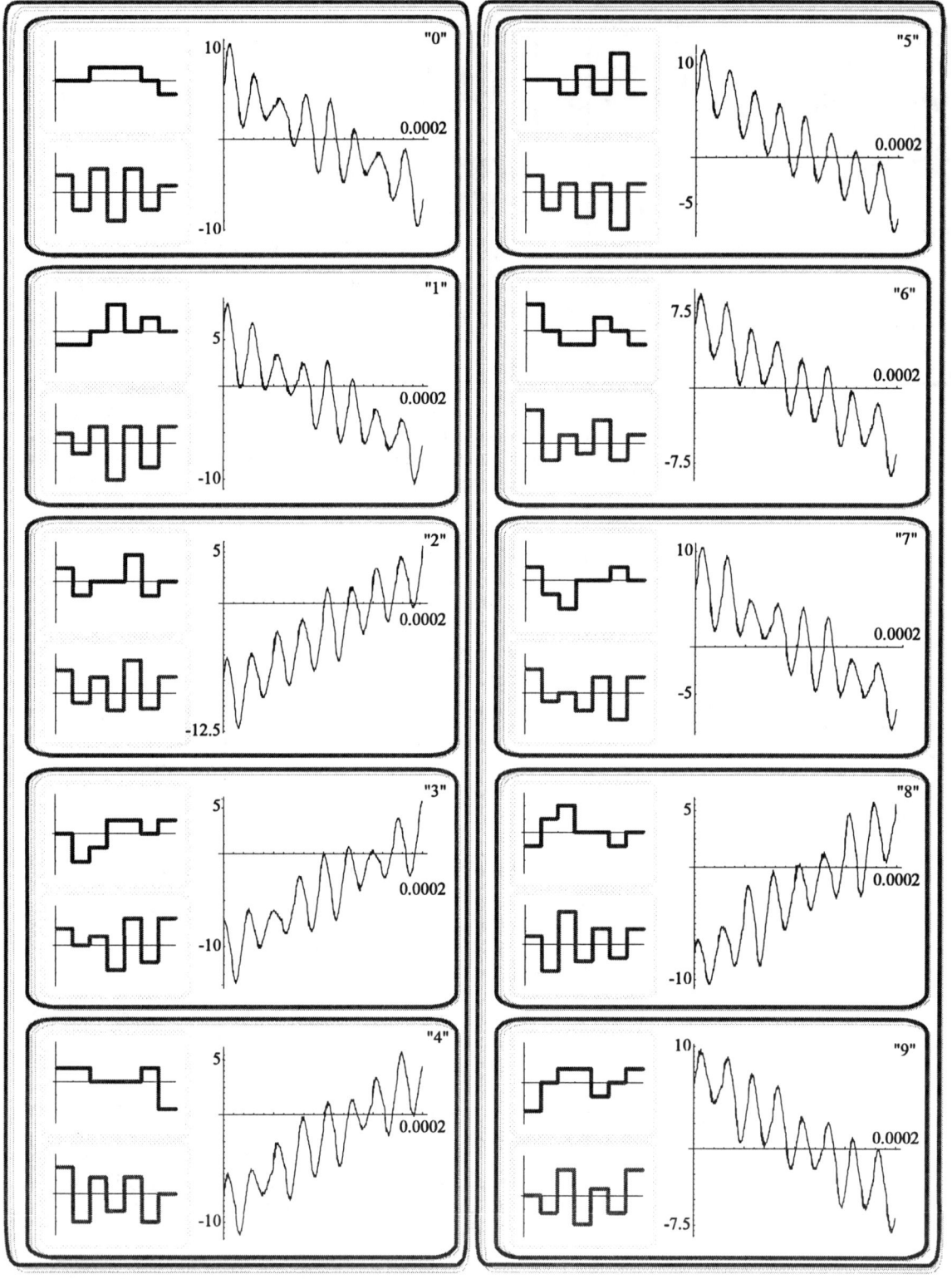

the fundamental frequency fh : H(t); Superposition waveforms of the sinusoidal high power signal and transmitting data signal: $A\sin(2\pi ft) + H(t)$.

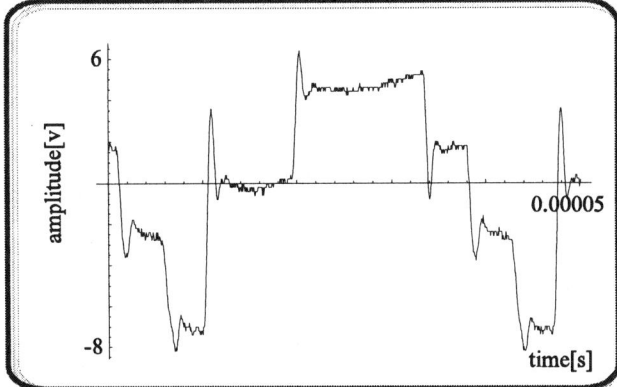

Figure 11(a) waveform of transmitted signal over power line across C1 (30[m] cord length) by using CDMA pattern "0"

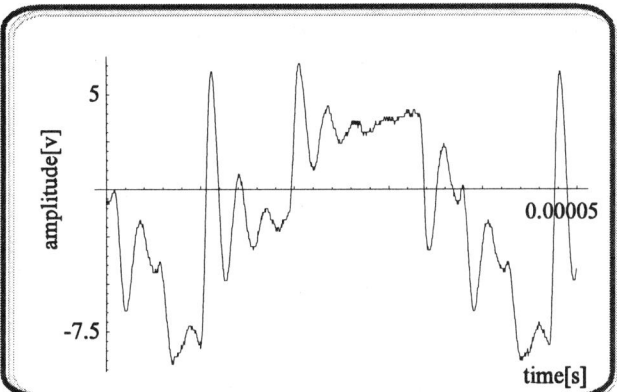

Figure 11(b) waveform of transmitted signal over power line across C1 (90[m] cord length) by using CDMA pattern "0"

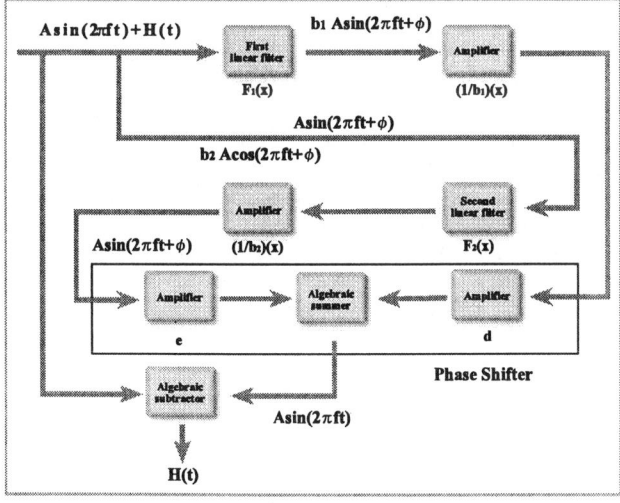

Figure 12. Block diagram of the proposed filtering method.

4.2 Experimental result related to the effectiveness

Figure 13(a) ,Fig.13(b) , and Fig.13(c) show the typical example of source signal , the transmitted signal over power line and the received signal used by the proposed filter circuit mentioned in the preceding section, respectively. In this experiment, the filter circuit is constructed by using normal Op amp techniques. The first and second linear filters are given by RC low-pass filter and Infinite integrator, respectively. Figure 13 is obtained by using the voltage divider consisting of the resistances $10[k\Omega]$ and $1[M\Omega]$ in parallel with power line. Since the waveform of the voltage across C_1 well conforms to that of the received signal by the filter circuit, it is experimentally confirmed that in spite of the distorted waveform of indoor power-supplying AC (shown in Fig.8), the waveform of the transmitted signal over power line conforms to that of the received signal. Therefore, the proposed filter circuit is considered to be effective for separating waveforms of transmitted signal from the superposition waveforms of the sinusoidal high power signal and transmitted signal.

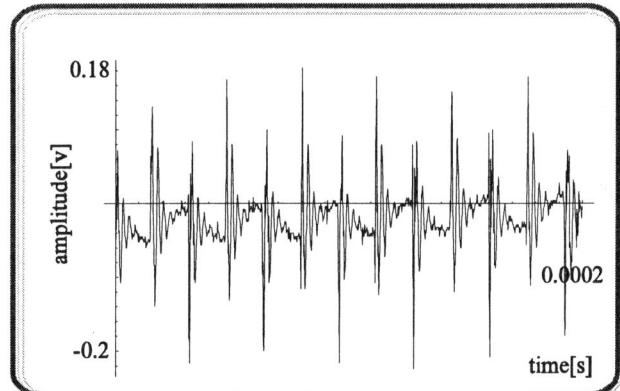

Figure 13(a) signal source waveform of voltage across C1 (V_A=10[v], fs=50kHz)

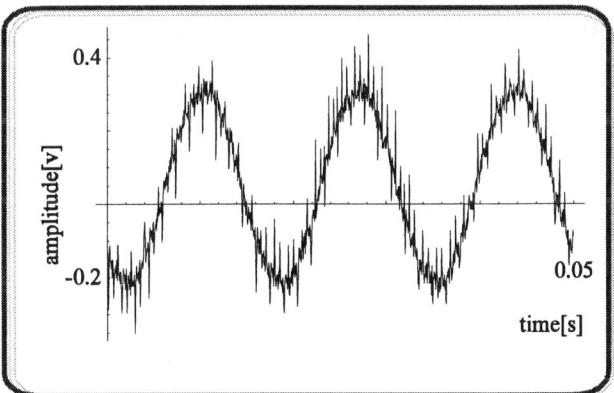

Figure 13(b) waveform of transmitted signal over indoor power line across C2 (V_A=10[v], fs=50kHz)

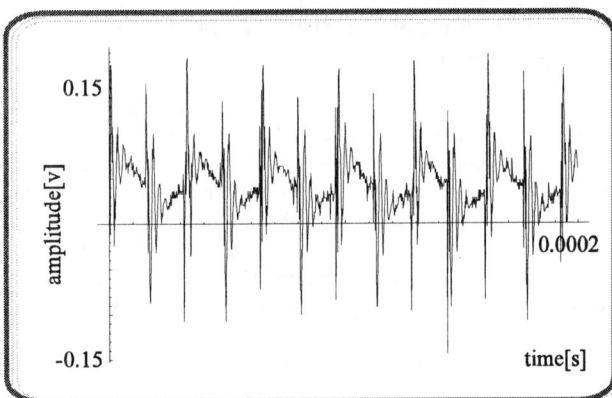

Figure 13(c) waveform of received signal by using the proposed filter (V_A =10[v], fs=50kHz)

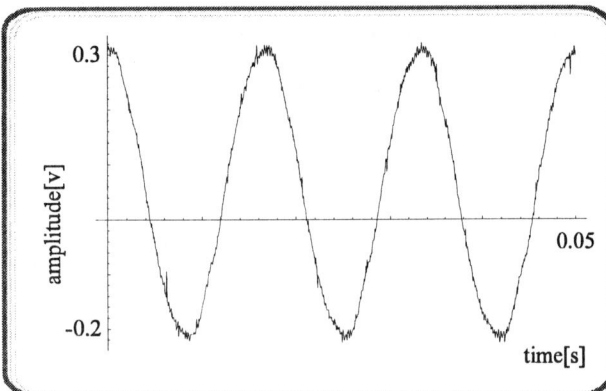

Figure 13(d) waveform of indoor power line across C_2 without transmitted signals
(V_A=10[v], fs=50kHz)

5. CONCLUSIONS

Using the results in the preceding sections, we construct an actual communication system for CDMA communications through 3-wire line cord that consists of the coupling transformer-less transmitter and receiver, and DS spread spectrum modulator / demodulator of which spread sequences are a kind of M sequences (8 bits) for realizing 7 channels CDMA communications at least. Although we use the general purpose type and on board type switching regulator supplies for the transmitting circuit, four channels CDMA communications cord over indoor electrical power lines are confirmed to be realized experimentally in not only the author's experimental power line but also the power distribution networks of other buildings.

The following results are obtained:

(1) The simulation results of the fundamental model of the transmitter give the useful circuit parameter information for constructing actual transmitter.
(2) Under appropriate conditions, It is experimentally confirmed that the waveforms of the transmitted signals over power line and those of the converted multi-step wave source signals to which are converted the multiple signals used by a kind of spread code into, in proportion to the multiple signal pattern voltage level, are correlated.
(3) Under the conditions such that both waveforms of the square wave signals and the voltage across C_1 are correlated, it is experimentally confirmed that both waveforms of the voltage across C_1 and the multiple signal patterns used by a kind of spread code such as pattern "0" (the multiple signal pattern that all the data is low level in case of four channel CDMA communication) more complex than the square wave signals are correlated.
(4) The proposed transmitting circuit is confirmed to be effective for realizing CDMA communications through 2-wire line cord and more effective for realizing CDMA communications through 3-wire line cord.
(5) The proposed filtering method is confirmed to be effective for separating waveforms of transmitted signal from the superposition waveforms of the sinusoidal high power signal and transmitted signal.
(6) Four channels CDMA communications through 3-wire line cord over indoor electrical power lines are confirmed to be realized experimentally.

ACKNOWLEDGEMENT

The authors would like to thank General Manager Kazumasa Adachi (Tech. & Develop. Div. Ibiden Co., LTD.) for bringing them this interesting research theme and thank Director Katsumi Kuwabara (Marketing & Planning Div. Ibiden Industries Co., LTD) for his help and promotion in carrying forward their research. They are also grateful to Mr. Shouichi Yamanaka (Elec. Tech. & Develop. Div. Ibiden Industries Co., LTD) for his help in the experiments.

REFERENCES

[1] P.K.Van der Gracht, and R. W. Donaldson," Communication Using Pseudonoise Modulation on Electric Power Distribution Circuits", IEEE Trans. on Comm., vol. 33 no. 9, pp. 964- 974, 1985.
[2] K. Dostert, " Frequency hopping spread spectrum modulation for digital communications over electrical power lines", IEEE Journal on Selected Areas in Comm., vol. 8 no. 4, pp. 700- 710, 1990.
[3] K. Dostert, "A Novel Frequency Hopping Spread Spectrum Scheme for Reliable Power Line Communications", IEEE Second International Symposium on Spread Spectrum Techniques and Applications (ISSSTA 92), Yokohama, Japan, 9-3, pp. 183-186, 1992.
[4] H. Okazaki," A Filtering Method and System for Separating Arbitrary Waveform of Transmitted Signal from the Superposition Waveforms of a Sinusoidal Signal and Transmitted Data Signal ", JAPAN PATENT APPLICATION NO. HEI9(1997)-125030
[5] H. Okazaki, and M. Kawashima," A Method of Transmitting Signals for Communications Over Indoor Alternative Current Power Lines "JAPAN PATENT APPLICATION NO. HEI9(1997)-267760

Network Design and Control for Multipont-to-Multipoint Communications

Kazuhiko Kinoshita[1], Junichiro Soeda[2], Nariyoshi Yamai[3],
Tetsuya Takine[1] and Koso Murakami[4]

1 Department of Information Systems Engineering, Graduate School of Engineering, Osaka University
2-1 Yamadaoka, Suita, Osaka 565, Japan

2 System Engine Group, Multimedia Development Center, Matsushita Electric Industrial Co., LTD
1006 Kadoma, Kadoma, Osaka 571, Japan

3 Computer Center, Okayama University
3-1-1 Tsushima-naka, Okayama, Okayama 700, Japan

4 Computation Center, Osaka University
5-1 Mihogaoka, Ibaraki, Osaka 567, Japan

ABSTRACT

This paper considers the network design and control for multi-point-to-multipoint communications. We propose a new routing control method which includes the member connection method and the route setup algorithm. The former implies that each member has its own route to send information to other members. Thus it can avoid traffic concentration in a particular route. Further, in the route setup algorithm, all routes are generated in parallel, by adding a link, which have lower-cost and lower-load, one by one, based on grouping by the number of destination nodes on the route. An interesting feature of the proposed algorithm is that every route can be established with a unified algorithm, even when newcomers join ongoing communications. In addition to the above method, we also propose a network design algorithm. It can make the proposed routing control method more effective. Simulation experiments show that the proposed framework is suitable for multipoint-to-multipoint communications.

1. INTRODUCTION

In recent years, while networks have high speed transmission capability by virtue of optical communication technology, multimedia applications among many people such as TV conferences become interesting. Since the present networks are designed for supporting only point-to-point communications, point-to-multipoint and multipoint-to-multipoint communications for these multimedia applications consist of many point-to-point communications. Note that routes for these point-to-point communications form a tree, and that each branch node is in charge of duplicating the received messages in order to reduce traffic on the common links on the tree [4,5].

Multipoint-to-multipoint communications can be performed by using a route generated for point-to-multipoint communications. In this case, only one tree is used for all communications, and therefore traffic from all members flows on each link on the tree. As a result, only the links that have enough capacity to support simultaneous communications are available. On the other hand, when each tree is used for the communication from each member, not only the links that have enough capacity, but also links of less available capacity can be utilized. In this method, however, when some new members join an ongoing multipoint-to-multipoint communication, some procedure is required to solve the problem for connecting the new member to the present route. This paper proposes a new route setup algorithm applicable to any cases.

The rest of this paper is divided into three sections. In section 2, a new routing control algorithm is proposed. Section 3 discusses the network design suitable for the proposed routing control algorithm. Finally, in section 4, simulation results are provided to confirm the efficiency of the proposed network design and route control algorithm.

2. ROUTING CONTROL

2.1. Member Connection Method

We first propose the member connection method to establish connections among members. As seen in Figure 1, in this method, members send messages to other members on their own route. Note that, by transmitting messages on individual routes, traffic can be distributed over many routes, which leads to an efficient use of network resources.

For the member connection method proposed in this paper, each node needs to be able to duplicate the transferred messages. As for point-to-multipoint communications with a message duplication function, Steiner tree problems with several restrictions are frequently adopted [6, 7]. Steiner tree problem is, when an undirected graph G and a node set

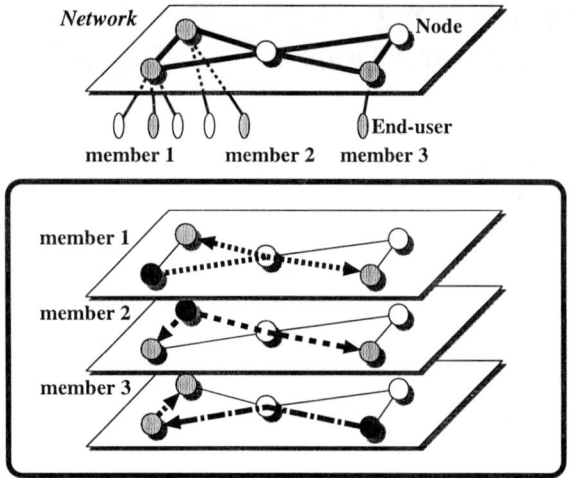

Figure 1: The member-connection method.

D are given, to find a tree of G with minimal cost spanning all nodes in D [2, 3, 4, 5]. This problem is known as NP-complete, and a heuristic algorithm is used to solve it in practice [6, 7, 8].

We define the connection-oriented Steiner tree problem as a multicast routing problem and formulate it as follows. We define cost $C_S(T)$ as the number of total links in a tree T, $L_M(T)$ as the maximal link load in a tree T, and $L_A(T)$ as the average of link loads in a tree T. Given a connected graph $G = (V, E)$ with a set of destination nodes D, we define S_T as a set of trees, where each tree T in S_T spans all nodes in D and has the minimum $C_S(T)$. Let $S_{T'}$ denote a subset of S_T, where each tree T' in $S_{T'}$ has the minimum $L_M(T')$. We define the connection-oriented Steiner tree as a tree T'' belonging to $S_{T'}$ and having the minimum $L_A(T'')$. The connection-oriented Steiner tree problem is to find a tree T''. Cost minimization leads to supporting more communications and load minimization leads to distributing the load more effectively.

In multipoint-to-multipoint communications, the connection-oriented Steiner trees should be established for all members who join communications. In the elementary method the connection-oriented Steiner tree problems are solved one by one. When some newcomers join ongoing communications, however, we have to resolve two problems. Old members should re-establish the connection-oriented Steiner trees to reach the newcomers. On the other hand, the newcomers have to construct the connection-oriented Steiner trees spanning all members. Thus, in general, it needs two different algorithms, which is not fit for practical use. Therefore, we propose the routing algorithm which can not only distribute the traffic load effectively but also establish each route in the same manner even when some newcomers join ongoing communications.

2.2. Proposed Routing Algorithm

A connection-oriented cheapest path between v and w is the least cost path. If there are several paths with the same cost, we choose the one whose maximal link load is the minimal among the paths. If there are still several paths, the path with the least average link load path is chosen. Given a graph $G = (V, E)$, the number n of members who join the communication, a set D of nodes connected with members who join the communication, and each member's initial tree T_i ($i = 1, 2, \ldots, n$), the details of the proposed algorithm are as the followings, where we call a node in D a *member node*.

1. Classify all initial trees by the number of member nodes in the tree. And let G_i denote a set of trees with i member nodes.

2. Choose a tree T_k from the group G_m with minimal m, in the decreasing order of the link costs one by one and perform:

   (a) Generate the connection-oriented cheapest paths between each node in T_k and each node in D but not in T_k.

   (b) Choose the least cost path from the paths generated in (a). If there are several paths with the same cost, the tie-breaking choice is performed based on the path loads as mentioned above.

   (c) Connect T_k and the member node with the path chosen in (b).

3. Merge G_m and G_{m+1} into G_{m+1}.

4. If not all trees belong to $G_{|D|}$, go to 2.

Figure 2 shows an example of the process to construct all routes when three members begin to communicate.

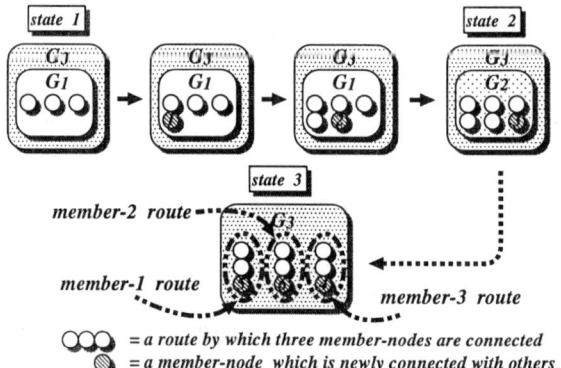

Figure 2: An example of the process to construct each route.

The proposed algorithm has three merits: (i) each route is constructed in the same manner even when newcomers join ongoing communication, (ii) the loads are well distributed over all routes due to constructing each route gradually, (iii) point-to-point communications and point-to-multipoint communications can be treated as special cases of multipoint-to-multipoint communications. It is because the routing algorithm for point-to-point communications corresponds to the case $n = |D| = 2$ in the proposed algorithm and the routing algorithm for point-to-multipoint communications corresponds to focus on a certain tree.

3. NETWORK DESIGN

Next, we investigate the network design suitable for the proposed route control algorithm. In the following discussion, for simplicity, we suppose the capacities of all links are identical. To support multipoint-to-multipoint communications efficiently, essential requirement on establishing a connection is availability of alternative routes for the connection. In particular, as for multiple nodes with high communication demand, supplying several routes among these nodes is desirable for the proposed routing algorithm. Because the routing algorithm selects the route with the least load when there are several routes with the same cost, it is expected to avoid congestion.

Moreover, if an indispensable link exists between a certain pair of nodes, the load on such a link cannot be shared. Therefore, this link becomes a bottleneck. Thus, it makes sense that nodes with a little communication demand play a role of relay nodes for other communications with high demand.

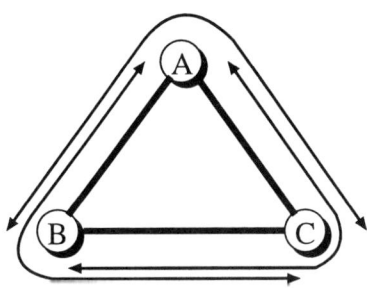

Figure 3: Structure that connects three nodes.

The connecting pattern of three nodes is a minimum unit that satisfies our design policy described above. For example, Figure 3 shows that the demand of links between A and B is high, both a direct path between them and an alternative path ($A - C - B$) are available. In this case, the routing algorithm can select the path with the least load. Therefore, the proposed design algorithm adopts the connecting patterns of three nodes as a unit.

The input, output and algorithm in this network design algorithm are as the followings.

- Input: occurrence probability table for each pair of nodes, the average communication time, the maximum number of links allowed to use in a network L.

- Output: a network G suitable for the proposed routing algorithm.

[Create the connected graph]

1. Compute total traffic for each pair of nodes by multiplication occurrence probability by average communication time.

2. Let $N = L$ and $r = 1$.

3. Compute the sequence of sets $M_1, M_2, \cdots, M_n$ of three nodes in the decreasing order in terms of total traffic in all pairs of three nodes. Consider each node to be a connected graph consisting of only the node itself.

4. If all nodes in M_r belong to the same connected graph, increase r by one and repeat this step. Otherwise go to 5.

5. Add a link between every pair of nodes in M_r, if the link does not exist yet. Subtract the number of the added links from N.

6. If all nodes are connected, go to 7. Otherwise increase r by one and go to 4.

[Invest the remaining resources]

7. Compute the sequence $M'_1, M'_2, \cdots, M'_n$ of sets of three nodes in the decreasing order in terms of total traffic in all node pairs without one or more direct links between them. Let $r = 1$.

8. If $N = 0$ or $r = n$, stop. Otherwise go to 9.

9. Let n_r be the number of node pairs without direct links in M'_r. If $n_r \leq N$, go to 10. Otherwise, go to 11.

10. Add a link between every pair of nodes without a direct link in M'_r, subtract n_r from N and go to 7.

11. Increase r by one, and go to 8.

At the beginning, by the steps 3 through 6, it prepares paths for all pairs of nodes based on the given table. See Figure 4. In state 1, three nodes 1, 2 and 5 are connected. Next, in the state 2, nodes 2, 3 and 4 are connected. In the state 3, M_3 and M_4 are skipped because all nodes in M_3 and M_4 have already been connected, and then nodes 1, 2 and 6 of M_5 are connected.

Thus, the network topology is constructed where there exists at least one path between any pair of nodes. Next, in the steps from 7 to 11, some additional links are supplied based on the amount of total traffic in all node pairs in a set M'_k without one or more direct links up to a given number L of links. In this example, nodes 1, 2 and 3 in M'_1 are connected.

4. PERFORMANCE EVALUATION

In this section, we evaluate the performance of the proposed network by simulation experiments. Fixing the maximum number of links, we design a network topology by the proposed network design algorithm. Then routes are generated

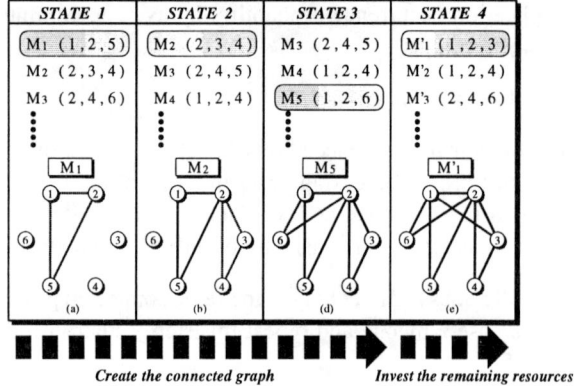

Figure 4: States in designing network topology.

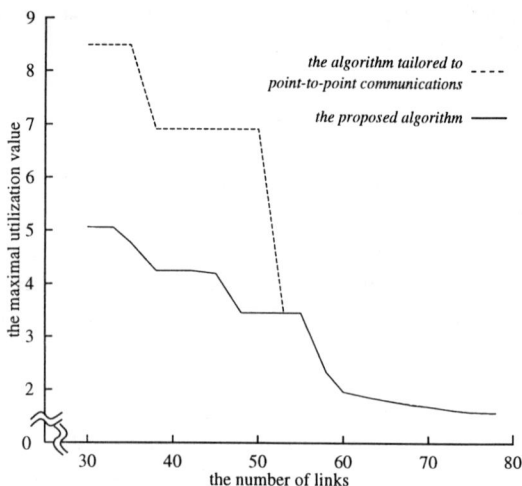

Figure 5: Simulation result.

by the routing algorithm proposed in section 2 for all multipoint-to-multipoint communications. Let N_U denote the utilization value which indicates the amount of traffic flow on each link, namely $N_U = P \times T$ where P and T denote occurrence probability and the average communication time, respectively. Thus the link with large utilization value is likely to be congested. We call the largest utilization value of those among all links *congestion value* which is the performance metric hereafter. Note that a small congestion value implies that traffic is well distributed over the network. Therefore, we evaluate whether this design algorithm reduces congestion value effectively when the maximum number of links is increased. In Figure 5, we compare the proposed design algorithm (based on three-node connection) with the algorithm modified by two-node connection. We claim that the latter satisfies only design policy for point-to-point communications: "it is usually economical that the design prepares a direct path between two nodes with high communication demand and provides a path through some nodes between two nodes without enough communication demand." The horizontal axis represents the maximum number of links, and the vertical axis congestion value.

From Figure 5, the proposed algorithm reduces congestion value more effectively than the algorithm based on two-node connection. In particular, in this simulation, if the number of links is less than 50, congestion value by the proposed algorithm can be half compared with the other. Also, if the number of links is larger than 50, the gaps between those two algorithm shrink, because the network structure approaches to a complete mesh and both structures become almost the same. Thus we conclude that the proposed algorithm reduces congestion value more effectively than the algorithm tailored to point-to-point communications.

5. CONCLUSIONS

In this paper, we proposed a new routing control method for multipoint-to-multipoint communications. It can distribute the link loads over all routes by constructing each route gradually. Moreover, every route can be established with a unified algorithm, even when newcomers join ongoing communications.

In addition, we also proposed a network design algorithm suitable for the proposed routing control method. Simulation experiments show that our framework is better than that tailored for point-to-point communications.

As further studies, we start to make a practical system and expand the proposed network design algorithm to be adapted for non-identical link capacities.

6. REFERENCES

[1] H. Tode, Y. Sakai, M. Yamamoto, H. Okada, Y. Tezuka, "Multicast Routing Algorithm for Nodal Load Balancing," *Proc. IEEE INFOCOM '92*, pp.2086–2095, New York, 1992.

[2] P. Winter, "Steiner Problem in Networks: A Survey," *Networks*, vol.17, pp.129–167, 1987.

[3] S. Dreyfus, R. Wagner, "The Steiner problem in graphs," *Networks*, vol.1, pp.195–207, 1971.

[4] S. Hakimi, "Steiner's problem in graphs and its implications," *Networks*, vol.1, pp.113–133, 1971.

[5] B. M. Waxman, "Routing of Multipoint Connections," *IEEE J. Select. Areas Commun.*, vol.6, pp.1617–1622, 1988.

[6] L. Kou, G. Markowsky, L. Berman, "A fast algorithm for Steiner trees," *Acta Informatica*, vol.15, pp.141–145, 1981.

[7] V. J. Rayward-Smith, A. Clare, "On Finding Steiner Vertices," *Networks*, vol.16, pp.283–294, 1986.

[8] S. Voss, "Steiner's problem in graphs: heuristic methods," *Discrete Applied Mathematics*, vol.40, pp.45–72, 1992.

A MULTI-RATE CHANNELIZED WIRELESS LAN SYSTEM WITH FIXED CHANNEL ASSIGNMENT

Chi-Wai Lam and Tsz-Mei Ko

Hong Kong University of Science and Technology
Department of Electrical and Electronic Engineering
Clear Water Bay, Hong Kong

ABSTRACT

We consider a multi-rate channelized wireless LAN system for a typical office building. The problem is formulated as a fixed channel assignment problem with multi-rate traffic. We generalized several known fixed channel assignment algorithms which were originally designed for single rate cellular networks. A new lower bound called the generalized clique bound is obtained. The generalized algorithms are compared via simulations for several typical office environments. It is found that these algorithms perform almost optimally, i.e., use the least bandwidth to satisfy the call requirements, when compared with the generalized clique bound.

1. INTRODUCTION

Wireless LANs are gaining its popularity in office buildings. Figure 1 shows a typical in-building LAN that is made up of several wireless LANs. Each wireless LAN serves a room or partition and is connected to a wired network through a base station called a mobile support station (MSS). A mobile unit can communicate with a host on the wired network or another mobile unit in a different cell via the local MSS. The MSS handles the necessary resource management and channel allocation. The traffic between the MSS and the mobile units are often asymmetric in nature due to the fact that users often send short messages to request large files and/or images. The uplink communication (from the mobile units to the MSS) which consists of mostly short messages is best handled by a contention based MAC protocol. On the other hand, the downlink communication (from the MSS to the mobile units) is best handled by a multiple access scheme such as TDMA, FDMA or CDMA in order to satisfy some quality of service (QoS) requirements for real-time data [1].

Frequency division multiple access (FDMA) is a mature technology and can easily be implemented [2]. We consider the channel allocation problem for the downlink communication using FDMA. The in-building LAN (Fig. 1) can be viewed as a cellular network containing N cells in which a cell is a wireless LAN. Since different applications may require different amount of bandwidths, we consider a multimedia environment with K classes of traffic in which a class k call ($1 \le k \le K$) requires b_k units of bandwidth. We assume that the traffic pattern is known and quasi-static. Thus channels can be allocated to each cell in advance. When a class k call arrives (i.e., a request to set up a downlink of b_k bandwidth units from the MSS to a mobile unit) in a cell, the MSS will check and allocate a class k channel for this downlink connection if such a channel is available.

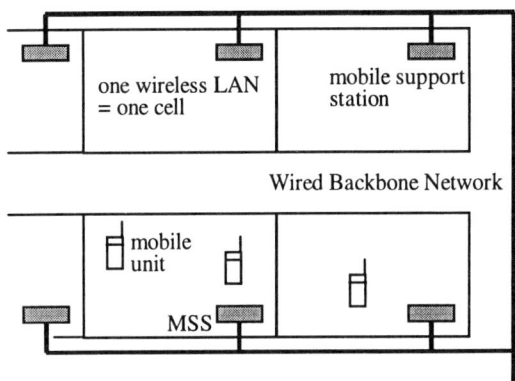

Figure 1. An In-building LAN that consists of several wireless LANs.

Communication in a wireless LAN is interfered by other calls. Thus the allocated spectrum must also satisfy co-site constraints due to interferences within the cell and adjacent/nearby constraints due to interferences from adjacent/nearby cells. These interference constraints can be specified by a separation matrix [3].

2. THE MULTI-RATE FCA PROBLEM

The multi-rate FCA problem considered in this paper is summarized as follows. As usual, frequencies are represented by positive integers 1, 2, 3, ...

Given:

N: the number of cells in the system.
K: the number of traffic classes.
b_k, $1 \le k \le K$: the bandwidth requirement for a class k call.
m_{ik}, $1 \le i \le N$, $1 \le k \le K$: the number of class k channels required in cell i.
c_{ij}, $1 \le i, j \le N$: the minimum separation required between a frequency interval used in cell i and a frequency interval used in cell j.

Find:

$f_{ikm} = \left[f_{ikm}^l, f_{ikm}^u \right]$, $1 \le i \le N$, $1 \le k \le K$, $1 \le m \le m_{ij}$:
the frequency intervals assigned to the mth requirement for class k channels in cell i such that the frequency span

$$W = \max_{i,k,m} f^u_{ikm}$$

is minimized subject to the separation or compatibility constraints

$$d(f_{ikm}, f_{jln}) \geq c_{ij}$$

for all i, j, k, l, m, n except $i = j$, $k = l$ and $m = n$. In the above inequality, $d(f_{ikm}, f_{jln})$ denotes the separation distance between the two assigned frequency intervals $f_{ikm} = \left[f^l_{ikm}, f^u_{ikm}\right]$ and $f_{jln} = \left[f^l_{jln}, f^u_{jln}\right]$. That is,

$$d(f_{ikm}, f_{jln}) = \begin{cases} f^l_{jln} - f^u_{ikm} & \text{if } f^l_{jln} > f^u_{ikm} \\ f^l_{ikm} - f^u_{jln} & \text{if } f^l_{ikm} > f^u_{jln} \\ 0 & \text{otherwise} \end{cases}$$

Example 1. A system has $N = 3$ cells with separation matrix

$$C = \begin{pmatrix} c_{1,1} & c_{1,2} & c_{1,3} \\ c_{2,1} & c_{2,2} & c_{2,3} \\ c_{3,1} & c_{3,2} & c_{3,3} \end{pmatrix} = \begin{pmatrix} 5 & 2 & 0 \\ 2 & 5 & 2 \\ 0 & 2 & 5 \end{pmatrix}.$$

There are two traffic classes with bandwidth requirements $B = [b_1 \ b_2] = [1 \ 5]$ bandwidth units. Suppose the channel requirements are

$$M = \begin{pmatrix} m_{1,1} & m_{1,2} \\ m_{2,1} & m_{2,2} \\ m_{3,1} & m_{3,2} \end{pmatrix} = \begin{pmatrix} 1 & 1 \\ 0 & 3 \\ 1 & 3 \end{pmatrix},$$

i.e., we need to assign $m_{1,1} = 1$ class 1 channel in cell 1, $m_{1,2} = 1$ class 2 channel in cell 1, etc. One possible assignment (such that the separation constraints are satisfied) is

$$f_{1,1,1} = 1; f_{1,2,1} = [6, 10];$$
$$f_{2,2,1} = [12, 16], f_{2,2,2} = [21, 25], f_{2,2,3} = [30, 34];$$
$$f_{3,1,1} = 1; f_{3,2,1} = [6, 10], f_{3,2,2} = [36, 40], f_{3,2,3} = [45, 49].$$

This assignment has a frequency span $W = 49$.

3. LOWER BOUND

In this section, we describe and prove a lower bound for the frequency span W for the multi-rate FCA problem. In [3,6], the single-rate FCA problem is shown to be equivalent to a graph number-coloring problem. We can similarly show that the multi-rate FCA problem stated in Section 2 is equivalent to a graph interval-coloring problem. Each cell can be viewed as a vertex such that vertex i is labeled with the co-site separation constraint c_{ii}. For any two cells, say cell i and cell j, with separation requirement $c_{ij} \neq 0$, we join the two corresponding vertices by an edge with label c_{ij}. Then we assign b_k consecutive positive integers (i.e., length b_k intervals) for each required class k channel to the appropriate vertices subject to the separation constraints defined on the vertices and edges. That is, the multi-rate FCA problem becomes the generalized graph interval-coloring problem that we need to assign m_{ik} length b_k intervals to vertex i, $1 \leq i \leq N$ such that

(i) for any two intervals assigned to two adjacent vertices, the intervals must be separated by at least the label on the edge joining the two vertices.

(ii) for any two intervals assigned to the same vertex, the intervals must be separated by a distance not less than the label on the vertex.

(iii) the chromatic number, i.e., the largest number assigned to any vertex, is minimized.

Let $\chi(G, M)$ denote the chromatic number for a graph G with requirement M (where M is the requirement matrix with elements m_{ik}, $1 \leq i \leq N$, $1 \leq k \leq K$). Note that the frequency span W in the multi-rate FCA problem is equal to the chromatic number $\chi(G, M)$. The generalized clique bound can be stated as follows.

Theorem 1 (The Generalized Clique Bound):

$$\chi(G, M) \geq \max_{\text{all cliques } V'} \left\{ \sum_{i:v_i \in V'} \sum_{k=1}^K m_{ik} b_k + \left(\sum_{i:v_i \in V'} \sum_{k=1}^K m_{ik} - 1 \right)(c_{\min} - 1) \right\}$$

where a *clique* V' is a complete subgraph of G and $c_{\min}$ is defined as

$$c_{\min} = \min_{i,j: v_i, v_j \in V'} c_{ij}.$$

Proof: Let V' be a clique of G. Since V' is fully connected, all the assigned intervals must be disjoint, i.e., the intervals would totally consume

$$W_{V', \text{intervals}} = \sum_{i:v_i \in V'} \sum_{k=1}^K m_{ik} b_k$$

frequency numbers. Furthermore, any two assigned intervals must be separated by at least $c_{\min}$, i.e., each separation consumes at least $c_{\min} - 1$ frequency numbers. So the total number of bandwidth units used for all the separations must be at least

$$W_{V', \text{space}} \geq \left(\sum_{i:v_i \in V'} \sum_{k=1}^K m_{ik} - 1 \right)(c_{\min} - 1).$$

The total frequency span $W_{V'}$ for a clique V' must be at least

$$W_{V'} = W_{V', \text{interval}} + W_{V', \text{space}}$$
$$\geq \sum_{i:v_i \in V'} \sum_{k=1}^K m_{ik} b_k + \left(\sum_{i:v_i \in V'} \sum_{k=1}^K m_{ik} - 1 \right)(c_{\min} - 1).$$

Since this inequality must be satisfied by all cliques V',

$$\chi(G, M) \geq \max_{\text{all cliques } V'} W_{V'}.$$

and the theorem follows.

Note that for the special case $K = 1$, $b_1 = 1$ and $m_{i,1} = m_i$, the generalized clique bound is reduced to the clique bound

$$\chi(G, M) \geq \max_{\text{all cliques } V'} \left\{ c_{\min} \left(\sum_{i:v_i \in V'} m_i - 1 \right) + 1 \right\}$$

for a single rate FCA problem. Thus we call Theorem 1 the *generalized clique bound* for fixed channel assignment in a wireless system with multi-rate traffic. We will use it to benchmark the performances for the channel assignment algorithms discussed in Section 4.

Example 2. The system in Example 1 can be represented by the following graph G_1.

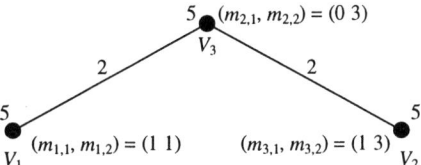

For this system, there are 5 cliques $\{v_1\}$, $\{v_2\}$, $\{v_3\}$, $\{v_1, v_2\}$ and $\{v_2, v_3\}$ with lower bounds, respectively,

$$W_{\{v_1\}} \geq 1 \cdot 1 + 1 \cdot 5 + (2 - 1) \cdot 4 = 10,$$

$$W_{\{v_2\}} \geq 0 \cdot 1 + 3 \cdot 5 + (3 - 1) \cdot 4 = 23,$$

$$W_{\{v_3\}} \geq 1 \cdot 1 + 3 \cdot 5 + (4 - 1) \cdot 4 = 28,$$

$$W_{\{v_1,v_2\}} \geq 1 \cdot 1 + 4 \cdot 5 + (5 - 1) \cdot 1 = 25,$$

$$W_{\{v_2,v_3\}} \geq 1 \cdot 1 + 6 \cdot 5 + (7 - 1) \cdot 1 = 37.$$

Therefore a lower bound for this system is

$$\chi(G_1, M) \geq \max\{10, 23, 28, 25, 37\} = 37.$$

4. MULTI-RATE FCA ALGORITHMS

The algorithms below are generalizations of the single rate FCA algorithms considered in [8]. For a multi-rate system, the cells are first ordered using a generalized degree of difficulties. Then we assign frequency intervals class by class, from those require more bandwidth to less bandwidth. For example, if there are traffic classes 1, 2, 3 requiring bandwidths 1, 5, 9 respectively, we will assign channels to class 3 traffic first, then class 2 and lastly class 1. We briefly summarize the different ordering methods and assignment strategies [6,8] in the following. We first re-define the degree of cell i as

$$d_i = \sum_{j=1}^{N} \sum_{k=1}^{K} m_{jk} c_{ij} - c_{ii}.$$

Then we order the cells using either node-degree or node-color orderings [8] described as follows.

(i) *Node-degree ordering*: cells are ordered in decreasing order of their degrees.

(ii) *Node-color ordering*: Of the N cells, the cell with the least degree is placed at the Nth position and then eliminated from the system. Of the remaining $N - 1$ cells, the degrees are re-calculated. The cell with the least degree is placed at the $(N - 1)$th position and eliminated from the system. The process is continued until all the cells are ordered.

With the cell orderings, the channels can be ordered with or without interleaving as follows. Create K matrices A_k ($1 \leq k \leq K$), one for each traffic class, such that matrix A_k has dimension $N \times \max_{1 \leq i \leq K} \{m_{ik}\}$. In matrix A_k, fill in class k call requirements for the highest degree cell in the first row starting at the first column. Then fill in class k call requirements for the second highest degree cell in the second row starting at column $m_k' + 1$, where m_k' denotes the class k call requirement for the highest degree cell. The remaining calls are similarly filled in cyclically until the last row. Now we can obtain call orderings in two ways, *column-wise* or *row-wise* from the matrices A_k.

After ordering the calls, we are now ready to assign frequency intervals to the calls using one of the following two strategies.

(i) *Frequency Exhaustive strategy*: starting at the top of the list, assign to each call the least possible frequency interval, without violating the separation constraints until all the requirements are satisfied.

(ii) *Requirement Exhaustive strategy*: starting with frequency 1, exhaust the list from the top to the bottom and assign frequency interval (with the appropriate length) that contains frequency 1 to those compatible calls. After that, use frequency 2 and exhaust the list again, and so on until all the requirements are satisfied.

Example 3: We again consider Example 1 to illustrate two FCA algorithms:

(i) Node-**D**egree, **R**ow-wise, **R**equirement exhaustive (*DRR*);

(ii) Node-**C**olor, **C**olumn-wise, **F**requency exhaustive (*CCF*).

DRR: We first compute the degrees $(d_1, d_2, d_3) = (11, 22, 21)$. Node-degree gives the cell ordering $\{v_2, v_3, v_1\}$. So

$$A_1 = \begin{pmatrix} \\ f_{3,1,1} \\ f_{1,1,1} \end{pmatrix} \quad \text{and} \quad A_2 = \begin{pmatrix} f_{2,2,1} & f_{2,2,2} & f_{2,2,3} \\ f_{3,2,1} & f_{3,2,2} & f_{3,2,3} \\ f_{1,2,1} & & \end{pmatrix}$$

where f_{ikm} is the mth requirement for class k calls in cell i. Then we list the calls in a row-wise manner, class by class, to obtain the call ordering

$$f_{2,2,1}, f_{2,2,2}, f_{2,2,3}, f_{3,2,1}, f_{3,2,2}, f_{3,2,3}, f_{1,2,1}, f_{3,1,1}, f_{1,1,1}.$$

The requirement exhaustive strategy gives the assignment $f_{2,2,1} = [1,5]$, $f_{3,2,1} = [7,11]$, $f_{1,2,1} = [7,11]$, $f_{2,2,2} = [13,17]$, $f_{3,2,2} = [19, 23]$, $f_{1,1,1} = [19,19]$, $f_{2,2,3} = [25,29]$, $f_{3,2,3} = [31,35]$, $f_{3,1,1} = [40,40]$ with frequency span $W = 40$.

CCF: With degrees $(d_1, d_2, d_3) = (11, 22, 21)$, v_1 is first placed at the bottom of the list and eliminated. The degrees for the remaining cells are re-calculated as $(--, 18, 21)$. So the node-color cell ordering is (v_3, v_2, v_1). Consequently,

$$A_1 = \begin{pmatrix} f_{3,1,1} \\ \\ f_{1,1,1} \end{pmatrix} \quad \text{and} \quad A_2 = \begin{pmatrix} f_{3,2,1} & f_{3,2,2} & f_{3,2,3} \\ f_{2,2,1} & f_{2,2,2} & f_{2,2,3} \\ f_{1,2,1} & & \end{pmatrix}$$

with column-wise ordering

$$f_{3,2,1}, f_{2,2,1}, f_{1,2,1}, f_{3,2,2}, f_{2,2,2}, f_{3,2,3}, f_{2,2,3}, f_{3,1,1}, f_{1,1,1}.$$

The frequency exhaustive strategy gives the assignment $f_{3,2,1} = [1,5]$, $f_{2,2,1} = [7,11]$, $f_{1,2,1} = [1,5]$, $f_{3,2,2} = [13, 17]$, $f_{2,2,2} = [19,23]$, $f_{3,2,3} = [25,29]$, $f_{2,2,3} = [31,35]$, $f_{3,1,1} = [37,37]$, $f_{1,1,1} = [13,13]$ with frequency span $W = 37$.

The CCF algorithm yields optimal result for Example 3 since the largest assigned integer 37 is equal to the *clique bound* obtained in Example 2. Simulation results for some reasonably sized systems are given in the next section.

5. RESULTS AND CONCLUSIONS

We model a 3-floor, 27-room typical office environment (Figure 2) to obtain a 27 cell system with two separation constraints C_1 and C_2. Constraint C_1 has co-site constraints $c_{ii} = 5$ ($1 \leq i \leq 27$) and adjacent constraints $c_{ij} = 2$ (for cell i adjacent to cell j, e.g., $c_{1,2} = c_{1,9} = c_{1,10} = 2$). Constraint C_2 has co-site constraints 2 and adjacent constraints 1. Constraint C_1 models systems with poor receivers and/or transmissions using high carrier frequencies so that the Doppler shift becomes significant and thus necessary for having large separation constraints. Constraint C_2 models systems with reasonably good receivers.

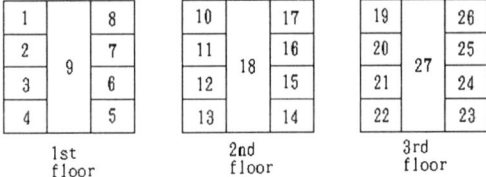

Figure 2. Floor plan for a typical office environment with 3 floors.

We assume there are three traffic classes requiring bandwidths $B = [1\ 5\ 9]$. Three requirement patterns are simulated. Requirement M_1 (Fig. 3) models a typical office environment with more low-rate data such as emails than high-rate graphics data. Requirement M_2 (with $m_{i,k} = 7$ for $1 \leq i \leq 27$, $1 \leq k \leq 3$) models an office environment with PCs sharing graphical applications and wireless printers so that high-rate data is as much as low-rate data. Requirement M_3 (Fig. 4) models an office building with inhomogeneous traffic distribution among cells.

Cell	1	2	3	4	5	6	7	8	9	10	11	12	13	14
class 1	5	5	5	5	5	5	5	5	5	5	5	5	5	5
class 2	3	3	3	3	3	3	3	3	3	3	3	3	3	3
class 3	2	2	2	2	2	2	2	2	2	2	2	2	2	2

Cell	15	16	17	18	19	20	21	22	23	24	25	26	27
class 1	5	5	5	5	5	5	5	5	5	5	5	5	5
class 2	3	3	3	3	3	3	3	3	3	3	3	3	3
class 3	2	2	2	2	2	2	2	2	2	2	2	2	2

Figure 3. Requirement M_1: Office environment with mainly low rate data (homogeneous traffic among cells).

Cell	1	2	3	4	5	6	7	8	9	10	11	12	13	14
class 1	7	7	7	14	7	7	7	5	2	7	7	7	14	7
class 2	7	7	7	14	7	7	7	5	2	7	7	7	14	7
class 3	7	7	7	14	7	7	7	5	2	7	7	7	14	7

Cell	15	16	17	18	19	20	21	22	23	24	25	26	27
class 1	7	7	5	2	7	7	7	14	7	7	7	5	2
class 2	7	7	5	2	7	7	7	14	7	7	7	5	2
class 3	7	7	5	2	7	7	7	14	7	7	7	5	2

Figure 4. Requirement M_3: Inhomogeneous traffic among cells.

The frequency spans for the various FCA algorithms are given in Figure 5. We follow the conventions used in [8] to use the first letter *D* or *C* to represent Node-*D*egree or Node-*C*olor ordering. The second letter represents *C*olumn-wise or *R*ow-wise ordering. The third letter represents *R*equirement or *F*requency Exhaustive strategy. As an example, DCR indicates Node-*D*egree, *C*olumn-wise, *R*equirement Exhaustive strategy. The *clique bounds* (CB) are also obtained for comparison.

Case	CB	DCR	DCF	DRR	DRF	CCR	CCF	CCR	CRF
M_1, C_1	377	382	**381**	436	483	382	**381**	436	483
M_1, C_2	315	**316**	316	349	362	**316**	316	349	362
M_2, C_1	143	148	**147**	156	168	148	**147**	156	168
M_2, C_2	114	**115**	115	120	122	**115**	115	120	122
M_3, C_1	503	549	537	530	603	536	**517**	529	603
M_3, C_2	420	430	438	427	447	427	**420**	430	444

Figure 5. Frequency spans for FCA algorithms.

We can choose the best assignment obtained from the eight algorithms for each case. From the simulation results, the best assignment among the eight is almost optimal as noted by the fact that the frequency spans are very closed to the generalized clique bound for all the six cases. Also the algorithms considered have complexity of only $O(n^2)$ and thus are very efficient. Since the algorithms are fast and perform almost optimally, we conclude that the algorithms are efficient and effective for multi-rate fixed channel assignment.

6. ACKNOWLEDGMENTS

This work is supported by a grant from the University Grant Council, Hong Kong.

7. REFERENCES

[1] N. Abramson, " Multiple Access Techniques for Wireless Networks," *Proceedings of IEEE*, vol. 82, no. 9, 1994.

[2] D.F. Bantz and F.J. Bauchot, " Wireless LAN Design Alternatives," *IEEE Network*, vol. 8, no. 2, pp. 43-53, 1994.

[3] F. Box, " A Heuristic Technique for Assigning Frequencies to Mobile Radio Nets," *IEEE Transactions on Vehicular Technology*, vol. 27, pp. 57-64, May, 1977.

[4] M. Duque-Anton, D. Kunz and B. Ruber, " Channel assignment for cellular radio using simulated annealing, *IEEE Transactions on Vehicular Technology*, vol. 42, no. 1, pp. 14-21, Feb, 1993.

[5] A. Gamst, " Some Lower Bounds for a Class of Frequency Assignment Problems," *IEEE Transactions on Vehicular Technology*, vol. 35, pp. 8-14, Feb, 1986.

[6] A. Gamst and W. Rave, " On Frequency Assignment in Mobile Automatic Telephone Systems," *Proceedings of Globecom '82*, pp. 309-315, 1982.

[7] M.R. Garey and D.S. Johnson, *Computers and Intractability: A Guide to the Theory of NP-Completeness*, W. H. Freeman and Co., 1979.

[8] K.N. Sivarajan, R.J. McEliece, and J.W. Ketchum, " Channel Assignment in Cellular Radio," *39th IEEE Vehicular Technology Conference*, pp. 846-850, May, 1989.

[9] C. W. Sung and W. S. Wong, " A graph theoretic approach to the channel assignment problem in cellular systems," *IEEE Veh. Tech. Conf. '95*, Chicago, July, 1995.

AN INTRODUCTION TO MULTI-SENSOR DATA FUSION

James Llinas
Research Professor
Center for Multisource Information Fusion
State University of New York at Buffalo
Buffalo, NY 14260

and

David L. Hall
Senior Research Associate
Applied Research Laboratory
The Pennsylvania State University
State College, PA 16802

ABSTRACT

Multi-sensor data fusion is an emerging technology applied to Department of Defense (DoD) areas such as automated target recognition, battlefield surveillance, and guidance and control of autonomous vehicles, and to non-DoD applications such as monitoring of complex machinery, medical diagnosis, and smart buildings. Techniques for multi-sensor data fusion are drawn from a wide range of areas including artificial intelligence, pattern recognition, statistical estimation, and other areas. This paper provides a tutorial on data fusion, introducing data fusion applications, process models, and identification of applicable techniques. Comments are made on the state-of-the-art in data fusion.

1. INTRODUCTION

In recent years, multi-sensor data fusion has received significant attention for both military and non-military applications. Data fusion techniques combine data from multiple sensors, and related information from associated databases, to achieve improved accuracies and more specific inferences than could be achieved by the use of a single sensor alone [1, 2, 3, 4]. The concept of multi-sensor data fusion is hardly new. Humans and animals have evolved the capability to use multiple senses to improve their ability to survive. For example, it may not be possible to assess the quality of an edible substance based solely on the sense of vision or touch, but evaluation of edibility may be achieved using a combination of sight, touch, smell, and taste. Similarly, while one is unable to see around comers or through vegetation, the sense of hearing can provide advanced warning of impending dangers. Thus, multi-sensory data fusion is naturally performed by animals and humans to achieve more accurate assessment of the surrounding environment and identification of threats, thereby improving their chances of survival.

While the concept of data fusion is not new, the emergence of new sensors, advanced processing techniques, and improved processing hardware make real-time fusion of data increasingly possible [5, 6]. Just as the advent of symbolic processing computers in the early 1970's provided an impetus to artificial intelligence [7], recent advances in computing and sensing have provided the ability to emulate, in hardware and software, the natural data fusion capabilities of humans and animals. Currently, data fusion systems are used extensively for target tracking, automated identification of targets, and limited automated reasoning applications. Spurred by significant expenditures by DoD, data fusion technology has rapidly advanced from a loose collection of related techniques, to an emerging true engineering discipline with standardized terminology (see Fig. 1), collections of robust mathematical techniques [2, 3, 4], and established system design principles. There is even beginning to be commercial software available for data fusion applications [8].

Applications for multi-sensor data fusion are widespread. Military applications include: automated target recognition (e.g., for smart weapons), guidance for autonomous vehicles, remote sensing, battlefield surveillance, and automated threat recognition systems, such as identification-friend-foe-neutral (IFFN) systems [9]. Non-military applications include monitoring of manufacturing processes, condition-based maintenance of complex machinery, robotics [10], and medical applications. Techniques to combine or fuse data are drawn from a diverse set of more traditional disciplines including: digital signal processing, statistical estimation, control theory, artificial intelligence, and classic numerical methods [11, 12, 14]. Historically, data fusion methods were developed primarily for military applications. However, in recent years these methods have been applied to civilian applications.

In principle, fusion of multi-sensor data provides significant advantages over single source data. In addition to the statistical advantage gained by combining same-source data (e.g., obtaining an improved estimate of a physical phenomena via redundant observations), the use of multiple types of sensors may increase the accuracy with which a quantity can be observed and characterized. For example, a radar provides the ability to accurately determine the aircraft's range, but has a limited ability to determine the angular direction of the aircraft. By contrast, an infrared imaging sensor can accurately determine the aircraft's angular direction, but is unable to measure range. If these two observations are correctly associated, then the combination of the two sensors data provides an improved determination of location than could be obtained by either of the two independent sensors. This results in a reduced error region as shown in the fused or combined location estimate. A similar effect may be obtained in determining the identity of an object based on observations of an object's attributes. For example, there is evidence that bats identify their prey by a combination of factors that include size, texture (based on acoustic signature), and kinematic behavior.

2. A DATA FUSION PROCESS MODEL

One of the historical barriers to technology transfer in data fusion has been the lack of a unifying terminology, which crosses application-specific boundaries. Even within military applications, related but different applications such as identification-friend-foe (IFF) systems, battlefield surveillance, and automatic target recognition, have used different definitions for fundamental terms such as correlation and data fusion. In

Fusion	The integration of information from multiple sources to produce the most specific and comprehensive unified data about an entity.
Alignment (Level 1)	Processing of sensor measurements to achieve a common time base and a common spatial reference.
Association (Level 1)	A process by which the closeness of sensor measurements is completed.
Correlation (Level 1)	A decision-making process which employs an association technique as a basis for allocating sensor measurements to the fixed or tracked location of an entity.
Correlator-Tracker (Level 1)	A process which generally employs both correlation and fusion component processes to transform sensor measurements into updated states and covariance for entity tracks.
Classification (Level 1)	A process by which some level of identity of an entity is established, either as a member of a class, a type within a class, or a specific unit within a type.
Situation Assessment (Level 2)	A process by which the distributions of fixed and tracked entities are associated with environmental, doctrinal, and performance data.
Threat Assessment (Level 3)	A structured multi-perspective assessment of the distributions of fixed and tracked entities which result in estimates of (e.g.): • expected courses of action,• unit compositions and deployment, • environmental effects, • enemy lethality, and • functional networks (e.g., supply, comms)

Figure 1. Table of terminology and definitions.

order to improve communications among military researchers and system developers, the Joint Directors of Laboratories (JDL) Data Fusion Working Group, established in 1986, began an effort to codify the terminology related to data fusion. The result of that effort was the creation of a process model for data fusion, and a Data Fusion Lexicon [11, 12]. The top level of the JDL data fusion process model is shown in Fig. 2. The JDL process model is a functionally-oriented model of data fusion and is intended to be very general and useful across multiple application areas. While the boundaries of the data fusion process are fuzzy and case-by-case dependent, generally speaking the input boundary is usually at the post-detection, extracted parameter level of signal processing. The output of the data fusion process is (ideally) a minimally ambiguous identification and characterization (viz., location and attributes) of individual entities, as well as a higher level interpretation of those entities in the context of the application environment.

The JDL Data Fusion Process model is a conceptual model which identifies the processes, functions, categories of techniques, and specific techniques applicable to data fusion. The model is a two-layer hierarchy. At the top level, shown in Fig. 2, the data fusion process is conceptualized by: sources of information, human computer interaction, source pre-processing, Level 1 processing, Level 2 processing, Level 3 processing, and Level 4 processing. Each of these is summarized below.

•*Sources of Information*. The left side of Fig. 2 indicates that a number of sources of information may be available as input including: (1) local sensors associated with a data fusion system (e.g., sensors physically associated with the data fusion system or organic sensors physically integrated with a data fusion system platform); (2) distributed sensors linked electronically to a fusion system; (3) national data collection systems; and (4) other data such as reference information, geographical information, etc.

•*Human Computer Interaction (HCI)*. The right side of Fig. 2 shows the human computer interaction (HCI) function for fusion systems. HCI allows human input such as commands, information requests, human assessments of inferences, reports from human operators, etc. In addition, HCI is the mechanism by which a fusion system communicates results via alerts, displays, and dynamic overlays of positional and identity information on geographical displays. In general, HCI incorporates not only multi-media methods for human interaction (graphics, sound, tactile interface, etc.), but also methods to assist humans in direction of attention, and overcoming human cognitive limitations (e.g., difficulty in processing negative information).

•*Source Preprocessing (Process Assignment)*. An initial process allocates data to appropriate processes and performs data pre-screening. Source preprocessing reduces the data fusion system load by allocating data to appropriate processes (e.g., locational and attribute data to Level 1 object refinement, alerts to Level 3 processing, etc.). Source preprocessing also forces the data fusion process to concentrate on the data most pertinent to the current situation. Extensive signal processing and detection theory may be required [13, 14]. A special case of source preprocessing is the synthesis of multiple component sensory array data to estimate the location and velocity of a target. A summary of these techniques is shown in Fig. 3.

•*Level 1 Processing (Object Refinement)*. This process combines locational, parametric, and identity information to achieve refined representations of individual objects (e.g., emitters, platforms, weapons, or geographically constrained military units). Level 1 processing performs four key functions: (1) transforms sensor data into a consistent set of units and coordinates; (2) refines and extends in time estimates of an object's position, kinematics, or attributes; (3) assigns data to objects to allow the application of statistical estimation techniques; and (4) refines the estimation of an object's identity or classification.

•*Level 2 Processing (Situation Refinement)*. Level 2 processing develops a description of current relationships among objects and events in the context of their environment. Distributions of individual objects (defined by Level 1 processing) are examined to aggregate them into operationally-meaningful combat units and weapon systems. In addition, situation assessment focuses on relational information (i.e., physical proximity, communications, causal, temporal, and other relations) to determine the meaning of a collection of entities. This assessment is performed in the context of environmental information about terrain, surrounding media, hydrology, weather, and other factors. Situation assessment appears to be vaguely defined because it addresses the interpretation of data, analogous to how a human might interpret the meaning of sensor data. Situation assessment uses both formal and heuristic techniques to examine, in a conditional sense, the meaning of Level 1 processing results.

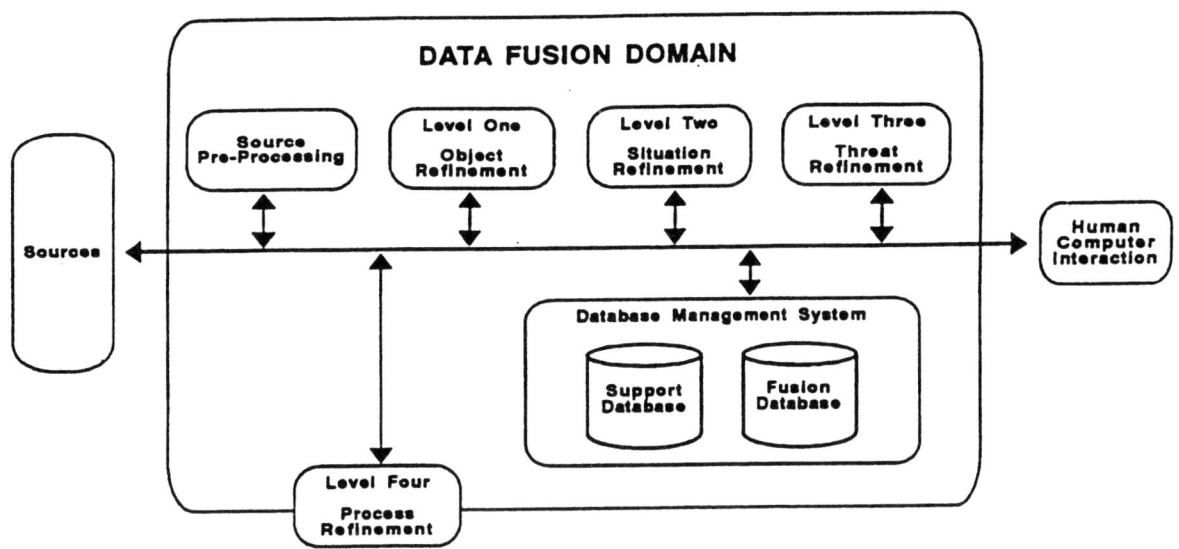

Figure 2. Top level data fusion process model.

Sources	The sources provide information at a variety of levels ranging from sensor data to *a priori* information from databases to human input.
Process Assignment	Source preprocessing enables the data fusion process to concentrate on the data most pertinent to the current situation as well as reducing the data fusion processing load. This is accomplished via data pre-screening and allocating data to appropriate processes.
Object Refinement (Level 1)	Level 1 processing combines locational, parametric, and identity information to achieve representatives of individual objects. Four key functions are: • transform data to a consistent reference frame and units; • estimate or predict object position, kinematics, or attributes; • assign data to objects to permit statistical estimation; and • refine estimates of the objects identity or classification.
Situation Refinement (Level 2)	Level 2 processing attempts to develop a contextual description of the relationship between objects and observed events. This processing determines the meaning of a collection of entities and incorporates environmental information, *a priori* knowledge, and observations.
Threat Refinement (Level 3)	Level 3 processing projects the current situation into the future to draw inferences about enemy threats, friendly and enemy vulnerabilities, and opportunities for operations. Threat assessment is especially difficult because it deals not only with computing possible engagement outcomes, but also assessing an enemy's intent based on knowledge about enemy doctrine, level of training, political environment, and the current situation.
Process Refinement (Level 4)	Level 4 processing is a *meta-process*, i.e., a process concerned about other processes. The three key Level 4 functions are: • monitor the real-time and long-term data fusion performance; • identify information required to improve the multi-level data fusion product; and • allocate and direct sensor and sources to achieve mission goals.
Database Management System	Database management is the most extensive ancillary function required to support data fusion due to the variety and amount of managed data, as well as the need for data retrieval, storage, archiving, compression, relational queries, and data protection.
Human-Computer Interaction	In addition to providing a mechanism for human input and communication of data fusion results to operators and users, the human-computer interfaction (HCI) includes methods of directing human attention as well as augmenting cognition, e.g., overcoming the human difficulty in processing negative information.

Figure 3. JDL process model components.

• *Level 3 Processing (Threat Refinement).* Level 3 processing projects the current *situation* into the future to draw inferences about enemy threats, friendly and enemy vulnerabilities, and opportunities for operations. Threat assessment is especially difficult because it deals not only with computing possible engagement outcomes, but also assessing an enemy's intent based on knowledge about enemy doctrine, level of training, political environment, and the current situation. The overall focus is on intent, lethality, and opportunity. Level 3 processing develops alternate hypotheses about an enemy's strategies and the effect of uncertain knowledge about enemy units, tactics, and the environment. Game theoretic techniques are applicable for Level 3 processing.

• *Level 4 Processing (Process Refinement).* Level 4 processing may be considered a *meta-process*, i.e., a process concerned about other processes. Level 4 processing performs four key functions: (1) monitors the data fusion process performance to provide information about real-time control and long-term performance; (2) identifies what information is needed to improve the multi-level fusion product (inferences, positions, identities, etc.); (3) determines the source specific data requirements to collect required information (i.e., which sensor type, which specific sensor, which database); and (4) allocates and directs the sources to achieve mission goals. This latter function may be outside the domain of specific data fusion functions. Hence, Level 4 processing is shown as partially inside and partially outside the data fusion process.

• *Data Management.* The most extensive support function required to support data fusion processing is database management. This collection of functions provides access to, and management of data fusion databases, including data retrieval, storage, archiving, compression, relational queries, and data protection. Database management for data fusion systems is particularly difficult because of the large and varied data managed (i.e., images, signal data, vectors, textural data) and the data rates both for ingestion of incoming sensor data, as well as the need for rapid retrieval.

A summary of the JDL data fusion process components are shown in Fig. 3. Each of these components can be hierarchically broken down into subprocesses. For example, Level 1 processing can be subdivided into four types of functions: data alignment, data/object correlation, object positional, kinematic, and attribute estimation, and finally, object identity estimation. The object positional, kinematic, and attribute estimation function is further subdivided into system models, defined optimization criteria, optimization approaches, and basic processing approach. At this lowest level in the hierarchy, specific methods such as Kalman filters, alpha-beta filters, covariance error estimation, etc. are identified to perform each function.

3. SUMMARY

The data fusion community is rapidly evolving. Significant investments in DoD applications, rapid evolution of microprocessors, advanced sensors, and new techniques have led to new capabilities to combine data from multiple sensors for improved inferences. Applications of data fusion range from DoD applications such as battlefield surveillance and automatic target recognition for smart weapons to non-DoD applications such as condition-based maintenance and improved medical diagnosis. Implementation of such systems requires an understanding of basic terminology, data fusion processing models, and architectures. This paper is intended to provide an introduction to these areas as a basis for further study and research.

4. REFERENCES

[1] E. Waltz, "Data Fusion for C3I: A Tutorial," *Command, Control, Communications Intelligence (C3I) Handbook*, EW Communications, Inc., Palo Alto, CA, pp. 217-226, 1986.

[2] J. Llinas, E. Waltz, *Multisensor Data Fusion*, Artech House, Inc., 1990.

[3] D. Hall, *Mathematical Techniques in Multisensor Data Fusion*, Artech House, Inc., 1992.

[4] L.A. Klein, *Sensor and Data Fusion Concepts and Applications*, SPIE Optical Engineering Press, Tutorial Texts, Vol. 14, 1993.

[5] D.L. Hall, J. Llinas, "A Challenge for the Data Fusion Community I: Research Imperatives for Improved Processing," *Proceedings of the Seventh National Symposium on Sensor Fusion*, Albuquerque, NM, March 1994.

[6] J. Llinas, D. L. Hall, "A Challenge for the Data Fusion Community II: Infrastructure Imperatives," *Proceedings of the Seventh National Symposium on Sensor Fusion*, Albuquerque, NM, March 1994.

[7] J. Gelfaud, *Selective Guide to Literature on Artificial Intelligence and Expert Systems*, American Society for Engineering Education, 1992.

[8] D.L. Hall, R.J. Linn, "A Taxonomy of Algorithms for Multisensor Data Fusion," *Proceedings of the 1990 Tri-Service Data Fusion Symposium*, pp. 13-29, April 1991.

[9] D.L. Hall, R.J. Linn, J. Llinas, "A Survey of Data Fusion Systems," *Proceedings of the SPIE Conference on Data Structure and Target Classification*, Vol. 1470, pp. 13-36, Orlando, FL, April 1991.

[10] M.A. Abidi, R.C. Gonzalez, *Data Fusion in Robotics and Machine Intelligence*, Academic Press, Boston, MA, 1992.

[11] Kessler et al., *Functional Description of the Data Fusion Process*, report prepared for the Office of Naval Technology, published by the Naval Air Development Center, Warminster, PA, January 1992.

[12] Data Fusion Lexicon, published by the Data Fusion Subpanel of the Joint Directors of Laboratories Technical Panel for C3 (F. E. White, Code 4202, NOSC, San Diego, CA), 1991.

[13] H.L. Van Trees, *Detection, Estimation, and Modulation Theory, Part I: Detection, Estimation, and Linear Modulation Theory*, John Wiley & Sons, 1968.

[14] H.L. Van Trees, *Detection, Estimation, and Modulation Theory, Part II: Radar-Sonar Signal Processing and Gaussian Signals in Noise*, Krieger, 1992.

FROM GI JOE TO STARSHIP TROOPER: THE EVOLUTION OF INFORMATION SUPPORT FOR INDIVIDUAL SOLDIERS

David L. Hall, Ph.D.
Applied Research Laboratory
The Pennsylvania State University
University Park, PA

and

James Llinas, Ph.D.
The State University of New York at Buffalo
Buffalo, NY

ABSTRACT

As we approach the millennium, numerous programs and research plans have focused on the use of advanced technology to improve the effectiveness of military forces. Examples include: the *Army XXI* concepts and the *Army After Next* (http://cacfs.army.mil/aan.html) concepts for military technology beyond the year 2020. Technologies to improve the effectiveness of soldiers include advanced tactical communications, the Global Positioning Satellite (GPS) system for precise geolocation, fire and forget *smart* weapons, improved sensors, and multi-sensor data fusion systems. The interest in such technology is exemplified by the recent movie, *Starship Troopers*, based on a science fiction book by Heinlein [9]. This paper summarizes some of the innovations provided to soldiers to transform them from the era of GI Joe to an anticipated era of the *starship trooper*. In particular, we analyze information technologies such as advanced sensors, multi-sensor data fusion, and automated reasoning.

1. INTRODUCTION

In recent years, there have been numerous changes in the mission, focus, and deployment of military units. Force structures have changed from a focus on massive tactical and theater *many-on-many* engagements which attempt to overcome adversaries by superior force or weapons, to an environment in which small units are rapidly deployed to perform functions such as intervention in local disputes, protection of civilians and facilities, and crisis intervention of terrorist activities. These changes include an increased tempo of tactical conflicts, *anywhere/anytime* potential conflicts, non-traditional and unpredictable enemy doctrine, and a political demand for zero casualties. Tactical environments have become very complex, with worldwide visibility of tactical engagements available over news sources such as CNN. In Bosnia, for example, U. S. Marines found themselves in an environment in which they were concurrently filmed by CNN and assaulted by snipers amidst civilian and onlookers.

Advances in computers, communications, materials, information science, and multi-sensor data fusion have improved the ability to rapidly, precisely, and effectively deploy small units for limited objectives with minimum or zero casualties. These advances have sought to improve all aspects of an individual soldier's (and small unit) performance including improved sensing (to augment and extend human senses), improved inter-soldier communications and location, improved speed and maneuverability, improved weapon systems, and increased access to data and improved modeling and reasoning. These are described briefly below.

The development of sensors such as acoustic, seismic, radar, optical, and magnetic sensors has extended the bandwidth and range with which a soldier can observe the battlefield environment. While the use of remote sensing is not new (e.g., use of observers in balloons with binoculars during the U.S. Civil War), new nanofabrication technology, advanced computing capabilities, and improved tactical communications allows the creation and deployment of semi-autonomous information collection systems using unmanned autonomous vehicles. These multi-sensor systems provide the ability to collect enormous amounts of information in a *vacuum cleaner* approach to provide information on the environment (e.g., terrain, weather conditions, etc.), the location and identification of enemy units and sensors, and ancillary information concerning supporting units, civilian locations, etc. The new technologies have extended the human senses in distance, spectral frequency coverage, acuity, resolution, and dynamic range. The individual soldier has the potential to *hear all* and *see all*.

Improvements in tactical communications via anti-jam, spread spectrum systems and worldwide communications networks (Feher [3]) allow individual soldiers to communicate among themselves, with military superiors, and with supporting resources throughout the world. These communications systems provide increasingly wide bandwidths to deliver unprecedented amounts of information including textual information, sensor data, video, and reference data. Moreover, the use of the GPS system for accurate self-location allows individuals to know where they are (and where their compatriots are) within a few meters accuracy. In principle, small military units have the capability to work in concert over an extended area as if they were within direct visual or aural communication distances.

Another area which has affected modern warfare is the increased speed and maneuverability of ground-based troops. General Norman Schwarzkopf [14] described how the rapid deployment of U.S. troops significantly affected the outcome of the Desert Storm conflict in 1991 and presented a new environment in which speed, maneuverability, and logistics became a primary tactical advantage.

The combination of new sensors, computing capability, and materials have also improved the ability of a single individual to control a weapon system to destroy enemy forces and facilities. Dupuy [1] has described the evolution of weaponry from muscle-based weapons such as bows and arrows to nuclear weapons of mass destruction. Recently, the use of so-called *smart* weapons provides the capability for a soldier to precisely direct force against a target many miles distance using a *fire and forget* approach. Other remotely guided weapon systems such as laser guided weapons allow an individual to control the application of force to an individual enemy system (e.g., tank) or even a single room within a remote building.

Finally, new technology has provided the basis for augmentation of human memory and cognitive abilities. Access to a tactical Internet, rapidly increasing computer storage capability and rapidly increasing computational capability augment human memory and cognitive processing. Huge amounts of data can be provided to individuals at a local level, with rapid access to worldwide databases. Information on military history, tactics, schematics on enemy entities, terrain and weather data, pictures of equipment, etc. can be provided to individual soldiers. In addition, computing capability is available to model target dynamics, engagements, and complex signal propagation phenomena to augment a soldier's memory and "thinking" ability.

With these improvements in speed, power, memory, cognition, geolocation, and communications, one might be led to believe that the individual soldier and team are only as good as their utilization of advanced technology — hence, the assertion that wars and conflicts are won in the laboratory rather than on the battlefield. Yet, as stressed by Major General Robert H. Scales, Jr., Commandant of the Army War College (cited by H. Summers [15]), real engagements are characterized by volatility, uncertainty, complexity and ambiguity. The effectiveness of individual soldiers is predicated upon their training and ability to think and act in real environments. Interestingly enough, Summers [15] points out that this argument concerning the extent to which modern science affects the results of war is hardly new. William Tecumseh Sherman [Civil War General], in an address to the West Point graduating class of 1869, said, "who honestly believes that one may, by the aid of modern science, sit in comfort and ease in his office chair, and with little blocks of wood to represent men, or even figures or algebraic symbols, master the great game of war". "I think this is an insidious and most dangerous mistake." General Sherman's admonition is as true today as it was in 1869.

2. THE ROLE OF INFORMATION FUSION

In the modern *information age* of military operations, a key element is the ability to fuse or combine multiple sources of information. Information fusion provides the means to aggregate, interpret, and filter vast amounts of potential information about an evolving situation. The goal of information fusion is to use sensors and computers to mimic (and surpass) the ability of humans and animals to combine information from multiple senses to improve the ability to detect and identify the presence of threats, i.e., to process multiple sensor data to understand the dynamic context of an evolving situation. This capability has evolved in humans and animals to improve their ability to survive. A general overview of multi-sensor data fusion is provided by Hall and Llinas [7] and a detailed discussion of information fusion (including mathematical techniques) is provided by Hall [4] and Waltz and Llinas [17].

The concept of information fusion for small units or individual soldiers involves the collection and processing of information from multiple sources. For individual soldiers and small units, data may be collected by deployed unattended sensor systems, by autonomous or human controlled drones, by remote collection systems (such as aircraft or other systems), by local sensors (perhaps even organic to the individual soldier), and processed data provided over a tactical Internet. The information and data must be processed in a hierarchical sense. At the lowest level, signal processing or image processing techniques may be used to detect the existence of an enemy unit or target. Subsequently, statistically-based methods may be used to determine target position of velocity. At a still higher level of abstraction, the characteristics and identity of the target may be obtained via pattern recognition techniques, followed by an analysis of the target behavior (both as an individual target and in the context of other entities, the environment, etc.). Finally, the overall situation and threat may be analyzed to determine potential enemy actions and hypothesized future events. In order to span this hierarchy and transform sensor data into information (and ultimately knowledge), a whole set of methods are required ranging from signal processing methods (e.g., time domain and frequency domain processing) to statistical estimation, to pattern recognition and to automated reasoning methods such as rule-based expert systems. The selection, utilization, and interpretation of these methods is a major challenge for multi-sensor data fusion. To assist the understanding and application of data fusion processing techniques, the Department of Defense Joint Director's of Laboratories (JDL) group has developed a taxonomy of algorithms, a data fusion process model, and has established a data fusion lexicon (for example, see Kessler et al. [10]).

During the past 30 years, information fusion research has proceeded rapidly, funded by large-scale DoD programs such as the U.S. Army's All Source Analysis System (ASAS) program. Numerous prototype data fusion systems have been implemented for applications such as target tracking, automatic target recognition, Identification-Friend-Foe-Neutral (IFFN) systems, Electronic Support Measures (ESM), and automated situation assessment (see Hall et al. [6]). Such systems have shown considerable success in the ability to perform functions such as target detection and tracking and the ability to perform automated recognition of complex targets. Context-based reasoning for automated threat assessment and situation assessment has been much less successful.

3. CHALLENGES FOR HUMAN REASONING

While prototype expert systems have been developed to support military analysts and units (e.g., situation assessment programs, threat assessment programs), humans continue to greatly exceed the capability of these prototype automated reasoning systems. Despite several decades of research in artificial intelligence and enormous increases in computing capability, human reasoning continues to exceed the ability of even large-scale expert systems involving many thousands of rules. It is beyond the scope of this paper to address reasons of these deficiencies (e.g., involving problems with knowledge representation, inference techniques, representation of so-called real-world knowledge, challenges with knowledge engineering, etc.). A partial review of these issues is provided by Hall, Hansen, and Lang [5].

Unfortunately, while automated reasoning systems do not compete effectively with human reasoning, human reasoning itself is demonstrably suspect. There are many problems and biases associated with human reasoning. Examples of biases include: (1) humans tend to "leap to conclusions" (e.g., quickly choose a plausible hypothesis to explain observed information); (2) confirmation bias in which a hypothesis is selected and subsequent contradictory evidence is ignored; (3) consistent overconfidence of adopted hypothesis; (4) changing estimates of the likelihood of a hypothesis based on how the hypothesis is "framed"; (5) misplaced causality; (6) conjunction effects (in which more complex hypotheses are deemed more likely than subsets of those hypotheses); (7) *Modus Tollens* inference biases (in which reasoning can be performed with positive information, but not with negative information); and (8) a bias towards false pattern recognition in which patterns are "discerned" when no patterns in fact exist. A discussion of these types of biases are provided by J. St. B. T. Evans et al. [2] and Piattelli-Palmarini [12].

Problems in reasoning are exacerbated by stressful situations such as those found in combat situations. Under stress, humans become increasingly prone to errors in judgment. Research has demonstrated (for example, see T. L. Ruble [13] and O. Svenson and A. J. Maule [16]) that under stress, people resort to sub-optimal coping mechanisms. For example, they may either ignore information which would assist in decision making, or frantically search for more information while avoiding making a decision. Unfortunately, the very advisory systems designed to assist tactical decision makers may in fact induce the very sub-optimal decision styles that they are trying to avoid. In particular, the use of data displays to show extensive amounts of information, alarm bells or lights, or complex human computer interfaces may cause operators "freeze up" or induce them to ignore the advisory system. An extensive assessment of the issues involved with "human-in-the-loop" decision making has been performed by Llinas et al. [11].

In addition to these issues, one must consider the effects of varying decision styles and preferential modes of accessing information. Studies have shown that various styles of decision making affect the efficacy of decision making. Key aspects include the techniques with which a decision maker accesses information and how they relate to other decision makers. Finally, M. J. Hall [8] has discussed the issue of information access modes. This effect involves the extent to which individuals have preferred modes of information access (e.g., visual, aural, or kinesthetic). Presentation of material to an individual in a non-optimal mode reduces the efficacy of communication and can induce stress.

4. RESEARCH IMPLICATIONS

It is clear that extensive research will be required before information systems will truly enhance the capabilities of individual soldiers. Current attempts to broadcast information to individual soldiers ignore the fact that the human is ultimately "a low pass filter" and that human decision making is fraught with biases. The net result of such advisory systems is to make soldiers "smart but dead". Application oriented research should be performed to develop "man-in-the-loop" advisory systems to assist in the interpretation of data, development, and evaluation of alternate hypotheses and guided decision making to account for human biases. Components of such research would include the following topics: basic theory of uncertainty; basic architectures for approximate reasoning; new humanistic-based Human Computer Interface (HCI); support systems for dealing with cognitive illusions and weaknesses; new cognitive models and approaches for automated reasoning; techniques to address the human trust issues; and new understanding of cognitive limitations on information overload, stress, decision styles, cooperative decision making and biases, etc. Performance of such research should pave the way for increasingly *smart* soldiers, rather than *smart* weapons.

5. REFERENCES

[1] Dupuy, T. N. *Understanding Defeat*. Paragon House Publishers, 1990.

[2] Evans, J. St. B. T., Newstead, S. E. and Byrnes, R. M. J. *Human Reasoning: The Psychology of Deduction*. Lawrence Erlbaum Associates, Hillsdale, 1993.

[3] Feher, K. *Digital Communications: Satellite/Earth Station Engineering*. Prentice-Hall, Inc., Englewood Cliffs, NJ, 1983.

[4] Hall, D. L. *Mathematical Techniques in Multi-Sensor Data Fusion*. Artech House, Inc., Boston, MA, 1992.

[5] Hall, D. L., Hansen, R. J. and Lang, D. C. "The negative information problem in mechanical diagnostics". *Transactions of the ASME*, 119:370-377, April 1997.

[6] Hall, D. L., Linn, R. J. and Llinas, J. "A survey for data fusion systems". *Proceedings of SPIE Conference on Data Structure and Target Classificaiton*, 1470:13-36, Orlando, FL, April 1991.

[7] Hall, D. L. and Llinas, J. "An introduction to multi-sensor data fusion". *Proceedings of IEEE*, 85(1):6-23, January 1997.

[8] Hall, M. J. "A multi-mode informational approach to improved computer-based training". *Intelligent Ships Symposium II*, ASNE, pages 123-150, Philadelphia, PA, November 1996.

[9] Heinlein, R. A. *Starship Troopers*. G. Putnam, USA, 1959.

[10] Kessler, O. et al. *Functional Description of the Data Fusion Process*. Technical Report, Office of Naval Technology, Naval Air Development Center, Warminster, PA, January 1992.

[11] Llinas, J., Drury, C., Bialas, W. and Chen, A. Studies and Analyses of Vulnerabilities in Aided Adversarial Decision Making. Technical Report, SUNY Buffalo, Department of Industrial Engineering, February 1997.

[12] Piatteli-Palmarini, M. *Inevitable Illusions*. John Wiley and Sons, Inc., New York, NY, 1994.

[13] Ruble, T. L. "Effects of cognitive styles and decision setting on performance." *Organizational Behavior and Human Decision Processes*, 46(2):283, August 1990.

[14] Schwarzkopf, H. M. and Petre, P. *It Doesn't Take a Hero*. Bantam Books, New York, NY, 1992.

[15] Summers, H. "Modern science can't make a soldier". *Washington Times*, page 35, 8 February 1998.

[16] Svenson, O. and Maule, A. J. *Time Pressure and Stress in Human Judgment and Decision Making*. Plenum Press, New York, NY, page 335, 1993.

[17] Waltz, E. and Llinas, J. *Multi-Sensor Data Fusion*. Artech House, Inc., Boston, MA, 1990.

A MULTI SENSOR DATA FUSION ALGORITHM FOR THE USCG'S VESSEL TRAFFIC SERVICES SYSTEM

Sean A. Midwood and Ian N. Glenn
National Defense Headquarters
101 Colonel By Drive
Ottawa, Ontario K1A 0K2

Murali Tummala
Department of Electrical & Computer Engineering
Naval Postgraduate School
Monterey, California

ABSTRACT

This paper describes the development of an algorithm to fuse redundant observations due to multiple sensor (type and location) coverage in order to provide a significant reduction in duplicate track information provided to Vessel Traffic Services (VTS) operator displays. The design of the algorithm allows acceptance of inputs from any type of sensor (radar, acoustic, GPS, system generated and manual tracks) as long as the basic decision criteria elements are provided. [1] The result of this effort is a computationally efficient and cost effective software solution to a significant system deficiency that impacts greatly on overall waterway safety. The algorithm is tested with real data collected from the VTS system at Puget Sound in September 1996. [2]

1. INTRODUCTION

The United States Coast Guard uses the US Navy's Joint Maritime Command Information System (JMCIS) software as the core software in their Vessel Traffic Services (VTS) system. This software allows numerous sensors of various types, primarily radar, to make reports to the central supervisory and controlling site, the Vessel Traffic Center (VTC). At the VTC, the sensor information is plotted as tracks on the displays of operators who are tasked with monitoring vessel traffic and providing advisories to vessels in transit or anchoring in key waterways. Current VTS software lacks a mechanism to correlate duplicate sensor tracks which would reduce the amount of superfluous information presented to each operator. This paper proposes a fuzzy association approach to the fusion of this multisensor data.

2. APPROACH

The algorithm performs central level fusion on data from various sensor sources providing vessel tracks for display and archival purposes. The algorithm is a refinement of a previously proposed algorithm [3] to fuse the outputs of sensors providing overlapping coverage. The algorithm has been generalized to accept and fuse an arbitrary number of tracks from any available sensor that can provide any of the following feature information: latitude, longitude, course, speed, and optionally any other feature such as size. The data collected are fused to create a single unified track table for display to the VTS operators and for maintenance of an historical record. The fusion process consists of several levels in order to achieve an integrated data set. The relevant features are extracted and the most recent sensor observations isolated. The sensor tracks are now ready to be correlated and fused where necessary. Let us first present an overview of fuzzy association as it applies to fusion and then detail its application to VTS.

3. FUZZY ASSOCIATION FOR FUSION

The goal of the fusion algorithm is to combine or fuse tracks of the same vessel observed and reported to the system by different input devices whether from radar processors or some other sensor. These fused tracks can then be associated with a unique platform identifier represented in the system by a unique platform number and a unique platform icon. The fuzzy membership is used to achieve this fusion. The membership function from fuzzy set theory provides a mechanism to measure correlation between observation or track pairs.

Data fusion is a process dealing with association, correlation and combination of data from multiple sources to achieve a refined position and identity estimation [4]. The aim of the data fusion is to derive more information in the final result than is present in only a single source of information. The combination of multiple sensors has the added benefit of redundancy of reporting. The failure of a single sensor then becomes non critical for coverage of an area. In addition, multiple sensors provide improved spatial coverage of an area with improved resolution over that offered by a single sensor.

Data fusion is usually classified into three types: positional fusion, identity fusion and threat assessment [5]. Positional fusion endeavors to determine an improved position estimate of a target by combining parametric data, such as azimuth, range, and range rate. Identity fusion uses known characteristics to determine the identity of a target. Threat assessment is the highest level of data fusion and is used for military or intelligence fusion systems to determine the meaning of the fused data from an adversarial point of view. The application of data fusion to JMCIS and VTS requires only positional fusion, and the method by which this is achieved will now be discussed.

4. POSITIONAL FUSION

Initial positional fusion is accomplished by an adaptive Kalman filter tracker operating at each remote radar site. This is sensor level fusion. The proposed algorithm assumes that the sensor level fusion is being performed correctly and that valid tracks are being generated and sent to the central site for further processing. Central level positional fusion is performed at the central site with the aim of eliminating the redundancies in observations or tracks being generated by each of the sensor level fusion algorithms. These redundancies occur when there is overlapping coverage provided by sensors (e.g., two radars that cover the same waterway). Each radar gets returns on the target, starts a track and forwards the track information to the central site for display and historical record keeping.

Additional redundant observations can result from the input of tracks from the Automated Dependent Surveillance (ADS) system [6] or Estimated Positions (EPs) for vessels based on Standard

Routes (SRs) generated by the Predictive Decision Support Aids (PDSA) [7]. Each vessel observation appears in the Track Database Manager (Tdbm) database [8] along with a date/time stamp. Each source of track information includes sufficient information to generate the following attributes: position (latitude and longitude), course, and speed.

The fuzzy association system takes these attributes and determines membership or similarity by correlation. This is accomplished as follows. Fuzzy set theory considers the partial membership of an object in a set. A membership function is used to grade the elements of a set in the range [0,1]. The grade of membership is a measure of the correlation of an object to a defined set. The closer the object is graded to one, the higher the membership of the object is in the set and the more compatible with the set being considered.

Design of a fuzzy association system involves the following four steps: determining the universe of discourse of inputs and outputs; designing membership functions; choosing fuzzy rules to relate the inputs and outputs; and determining a defuzzifying technique.

When comparing the latitudes of two separate radar tracks to see if they are similar a geometric membership can be constructed that takes into account the errors present in the system inherent to each remote site generating a track. A triangular shaped membership function is a good choice for a positional comparison because of the accuracy of the radars in reporting the target position. The latitude given in one track is subtracted from the latitude given in another track held as the reference. The difference in latitude is used to determine the membership value. Figure 1 shows the membership functions used in the algorithm.

In general, the design of membership functions is based on the attributes inherent to those aspects being compared. Since both radar and ADS positions reported to the system are relatively accurate, the triangular membership function is appropriate. For other attributes where there is less accuracy such as speed, broadening the roof of the membership function to include a greater range of values is valuable. It is also useful to truncate the membership function at a given value as in the case of the course membership function. It utilizes a trapezoidal shape to allow a generous association within a reasonable range of values but not outside of a fixed range.

Next, in order to evaluate each of the membership values returned, a threshold needs to be established that reflects the physical limitations. In the case of radar returns, a variable threshold is set that takes into account accuracy limitations of the radar dependent on the range of the target.

Once all of the attributes for the track pair being assessed have been assigned membership values, they can be checked to see that they exceed the designated threshold. Each value is checked sequentially starting with latitude to ensure that it exceeds the threshold. If it does not, no further checks are made and association fails. This method has the advantage of computational efficiency. If all values exceed the assigned threshold, association is made as indicated by a binary output of '1' from the defuzzifier. Figure 2 schematically shows the action of the fuzzy associative system. If the membership values, θ_i, all exceed the single threshold, ϕ, the two tracks would be associated. If the ϕ exceeds

any of the θ_i, no association would be made. The result is a single unified set of tracks representing a unique set of vessels present in the system in that time window. In the fused tracks, the original reporting sensor and its assigned track number are maintained for archival purposes as well as to assist in maintaining a unique platform number.

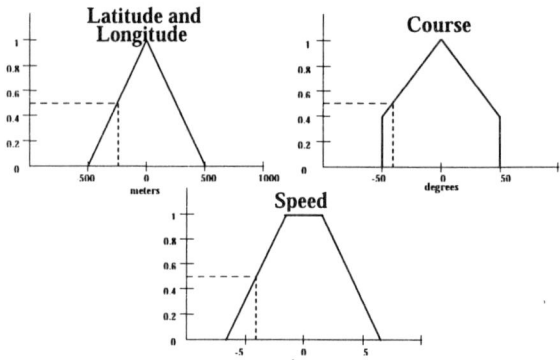

Figure 1 Membership Functions Used

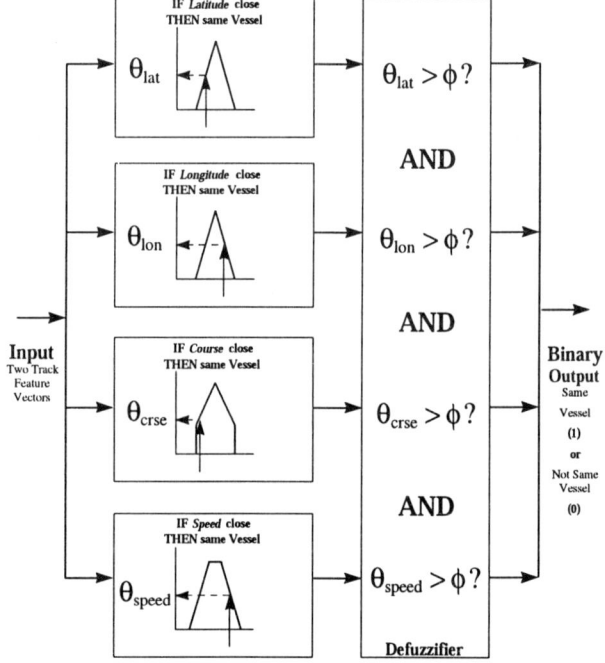

Figure 2 Fuzzy Associative System

5. DATABASE FUSION

The data set is now ready to be used to update the Tdbm. The track number is used to determine if this track being added is new to the system. If the search of the track number in the Tdbm is successful, the associated platform number is appended to the track in question. If the search fails, a new platform track number is generated and the operator can be alerted to the new "unknown" track. At this point the multilevel sensor fusion cycle is complete. The output of the various sensors have been related to each other, and the unified set has been related to the previous sets (the

Tdbm). The data window can now be moved forward in time to gather in the next batch of sensor tracks and the process repeated.

6. RESULTS

Data from an operational VTS system was collected at Puget Sound in September 1996. The availability of "real world" information negated the requirement to simulate data as had been the case in the previous work. [1] This data allowed for thorough testing of the algorithm for a variety of real life scenarios which were chosen purposely by USCG and ADS contractor personnel. These scenarios realistically depict the redundancy issues faced by system operators with overlapping information from multiple radar and ADS tracks.

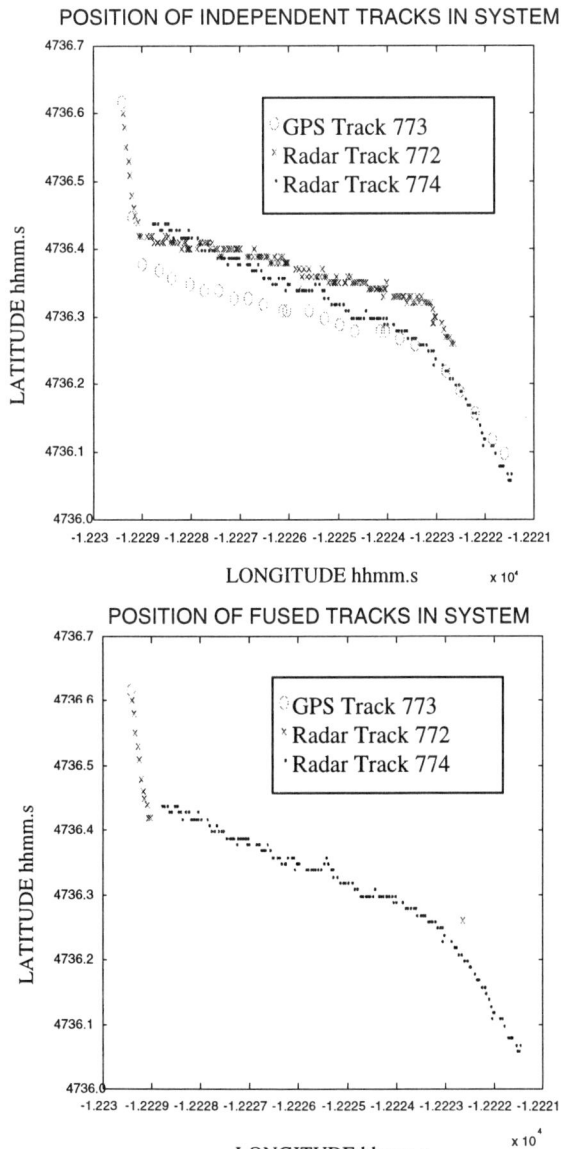

Figure 3 Overlapping Radar and GPS Scenario

The algorithm performed correctly under all test scenarios. The redundant tracks would stay fused as long as each track pair being assessed had a data point within the observation window. There were no problems associated with vessels that were turning and the algorithm always selected the superior sensor. The algorithm had no trouble dealing with a large amount of tracks and or interruptions in data streams. The following is representative of the type of situation the USCG would like resolved within the VTS:

> Overlapping radar coverage (772 and 774) and ADS coverage (Track 773) on a single vessel. Track 773 is the first to acquire the vessel but hands it off to track 772, once 772 acquires the track due to its superior status. Track 774 then acquires track and takes a hand off from 772 due to 774's superior status. See Figure 3.

Many other scenarios were examined and the algorithm performed well in all circumstances. In summary: the algorithm fused all tracks that were in the overlap region that met the fusion criteria; would change reporting responsibility for a track to the next inferior sensor if the superior sensor ceased reporting; would change reporting responsibility for a track to a more superior sensor if that sensor started to report on a vessel which was currently assigned to a less capable sensor; and had no trouble with crossing or passing situations. Marginal situations were easy to discriminate as the algorithm would defuse immediately upon failure of the fusion criteria.

7. OBSERVATIONS

The key observations to be made are the affects that the individual membership functions had on the results. If the membership function was not sufficiently broad enough the decision to fuse two tracks was not made. This is particularly true for the course membership function. Vessels that are going extremely slow and/or turning tend to have widely varying headings from the radar reports. The addition of fusion parameters, such as size and track quality, would certainly provide a greater degree of confidence in situations where position, course and speed are very close. While the data collected did not contain this type of situation, it is reasonable to assume that this scenario is common in the busy harbors and waterways under the USCG management. These findings are consistent with the simulated overlapping radar results reported in [1].

The algorithm's current performance is limited by the number of attributes that could be used to determine association. Only *Latitude*, *Longitude*, *Course* and *Speed* were adopted to determine a level of "sameness" between vessels. The membership functions for *Latitude*, *Longitude* and *Speed* were triangular in nature as these attributes were considered to be reported accurately by both radar and ADS. The *Course* attribute is not reported with reasonable accuracy by radar when vessels are turning at reasonable speeds. The tracking algorithms in the radar processors are the primary cause of this problem. Due to this problem, the membership function for *Course* is trapezoidal in shape allowing for a more generous association within a reasonable range but not outside a fixed value. ADS, on the other hand, reports *Course* very accurately. The algorithm can easily be modified to accept information from any sensor type as long as the specified attribute is available. The evaluation for a specified attribute can be turned off should it not be present in the data from a given sensor. Most importantly, the algorithm can be modified to accept additional attributes which would further refine and improve the fusion decision making process.

8. CONCLUSION

This paper described an algorithm to fuse redundant observations due to multiple sensor (type and location) coverage in order to provide a significant reduction in duplicate track information provided to the Vessel Traffic Services (VTS) operator displays. This **proof-of-concept** algorithm is a continuation of the work reported in [1] and [3]. The results presented are ready for final verification and validation by the USCG. The design of the algorithm allows acceptance of inputs from any type of sensor (radar, acoustic, GPS, system generated and manual tracks) as long as the basic decision criteria elements are provided. The result of this effort is a computationally efficient and cost effective software solution to a significant system deficiency that impacts greatly on overall waterway safety. The algorithm was tested with real data collected from the VTS system at Puget Sound in September 1996. The testing showed that the algorithm correctly fuses redundant sensor observations on the same vessel resulting in a significant reduction in the amount of unnecessary information presented to the VTS operator.

The fusion algorithm performed as expected. The performance of the algorithm can be enhanced by adding other attributes from which measures of similarity could be determined. *Size* and *Track Quality* would appear to be likely candidates as this data could easily be extracted and/or reported from both radar and ADS sensors. These additional measures would allow for greater flexibility in applying the fusion process to a given set of track reports. With six measures to choose from, a weighting scheme relative to the importance of each membership function could be implemented. This is not feasible with the current four attributes as once you determine the relative "sameness" of position to each other via *Longitude* and *Latitude* you are left with just *Speed* and *Course*. *Speed* is a reasonably stable and accurate measure but the *Course* attribute has far too much variation for it to be dependable. Additional features would mitigate this problem somewhat. One other way to improve the algorithm without adding features is to improve the tracking capabilities of each radar's RSP to improve the accuracy and reliability of the *Course* data.

9. IMPLEMENTATION UPDATE: FEBRUARY 1998

Concurrent with with the testing presented here, a commercial system was developed based on the results of the simulation [1]. The US Naval Air Warfare Center, Aircraft Division *SureTrak*™ Vessel Traffic Information System was designed with this fusion approach as its core algorithm and is fusing the inputs from four radars and two GPS equipped range safety boats in the Chesapeake Test Range area. Similar implementations are under consideration for Vanderberg Air Force Base and the Naval Underwater Warfare Center, Key Port, WA. The US Coast Guard are continuing their efforts to integrate the algorithm into JMCIS for the Vessel Traffic Services System.

10. REFERENCES

[1] Glenn, Ian N. "Multilevel Data Association for the Vessel Traffic Services System and the Joint Maritime Command Information System," Masters Thesis, Naval Postgraduate School, Monterey, CA, December 1995.

[2] Midwood, Sean A., "A Computationally Efficient and Cost Effective Multisensor Data Fusion Algorithm for the United States Coast Guard's Vessel Traffic Services System," Master's Thesis, Naval Postgraduate School, Monterey, CA, September 1997.

[3] Ruthenberg, Thomas M., "Data Fusion Algorithm for the Vessel Traffic Services System: A Fuzzy Associative System Approach," Master's Thesis, Naval Postgraduate School, Monterey, CA, March 1995.

[4] Waltz, E. and Llinas, J., Multi-Sensor Data Fusion, Artech House Inc., Boston, MA, 1990.

[5] Hall, David L., Mathematical Techniques in Multisensor Data Fusion, Artech House Inc., Norwood, MA, 1992.

[6] Inter-National Research Institute (INRI), Functional Description Document for Automated Dependent Surveillance (ADS), July 25, 1995.

[7] Range Directorate Chesapeake Test Range, "System Design Document For The Coast Guard Vessel Traffic Service System," MIPR Number DTCG23-92-F-TAC111, Naval Air Warfare Center, MD, March, 1994.

[8] Inter-National Research Institute (INRI), Tdbm Service Application Programmer's Interface (API) For the Unified Build (UB) Software Development Environment (SDE), SPAWARSYSCOM SDE-API-TDBM-2.0.11.5, March 31, 1994.

MICROSIMULATION AS A TOOL FOR TARGET TRACKING AND STATE ESTIMATION

Donald E. Brown and C. Louis Pittard

Department of Systems Engineering
University of Virginia
Charlottesville, VA 22903 USA

ABSTRACT

Data fusion systems process large amounts of data into information for decision support. One of the fundamental components of data fusion is state estimation which provides estimates of current and future environmental states. Traditionally the state estimation procedures in data fusion have been accomplished by Kalman filtering techniques. However, these methods have difficulties with domains such as ground operations where past behavior does not correlate as highly with future behavior. This paper provides an overview to the use of microsimulations for state estimation in this domain.

1. INTRODUCTION

Traditional state estimation procedures in data fusion rely heavily on past observations of objects but do not account for local information. These approaches work well when the objects of interest are non-maneuvering aircraft or projectiles. However, they fail to account for the behavior of intelligent objects in terrain. In this paper, we present an algorithm for state estimation of objects moving over terrain using both behavioral and geographic information system (GIS) information. The proposed model specifically accounts for the influence of the environment (rivers, roads, vegetation, elevation, etc.) on the propagation of the probability densities. We make a number of assumptions about the motion of targets, but do not necessarily assume knowledge of goal locations.

The traditional technique used for state estimation in data fusion come from the Kalman filter (KF) family of algorithms based on the equations of motion of targets. These equations allow the recursive computation of the state of a system based solely on past observations [1][2]. Historically, the first state estimation and tracking techniques have relied almost exclusively on target locations provided by sensors such as radars [3][4]. Later, more advanced systems have incorporated information pertaining to the orientation, velocity, and acceleration of the target [5][6][7]. This evolution suggests that increasing the amount of information included in the state-space can improve the quality of the tracking process. In applications involving ground targets, a map of terrain features affecting target behavior (e.g., roads, rivers, obstacles) is usually available, and heuristic rules relating these features to target motion can often be expressed in a motion model. However, such models typically involve non-linear state equations which in turn induce highly non-gaussian probability densities. These densities are difficult to describe and represent, and do not lend themselves well to integral calculations for estimation and association computations. They also rule out the use of a Kalman filter.

Additionally much of the evidence in modern data fusion systems derives from human-in-the-loop systems. While the original data may and normally do come from sensors, somewhere in the processing before they reach the data fusion systems a human analyst or operator has frequently processed these data and produced a report. Hence, the data fusion systems must be specifically designed to convert human generated reports into probabilistic evidence for use in automated reasoning. Because reporting may vary considerably with intelligence discipline, the data fusion system needs to employ tailored evidence conditioning and mediation for the specific class of messages and discipline: SIGINT, IMINT, HUMINT, and MASINT. However, there are also core elements of this task that remain constant across disciplines. This paper will consider this core element, and not the details needed within the individual disciplines.

The remainder of the paper is organized as follows. The next section describes evidence conditioning and the role of state estimation. Section 3 then gives an overview to the location processing needed for state estimation of intelligent objects in terrain. Section 4 provides a short discussion of behavioral estimation and Section 5 contains our conclusions. As an overview paper, there are few details on actual processing steps provided here, but we do give references for the interested reader.

2. EVIDENCE CONDITIONING

State estimation is sometimes viewed as a task that follows correlation between messages or sensor data. However, before we can begin the task of data fusion we must convert data in messages into a useful representation and perform state estimation using the existing evidence. Hence, we view the state estimation tasks we describe in this paper as appropriate both before and after correlation. For example, incoming messages arrive at different times and in order for fusion to occur we need to synchronize the estimates derived from the messages. The techniques in this paper are particularly relevant to this problem. However, these same state estimation approaches we describe here can also be done with the results of the correlation process to give updated estimates and predictions of the situation.

In order to further understand the role of state estimation before correlation, we need brief description of these pre-correlation steps. We use the term evidence conditioning for this pre-correlation process. Evidence conditioning converts message data into well-conditioned probabilistic statements.

We divide the evidence conditioning task into three subtasks: interpretation, conditioning, and mediation. Interpretation provides access to the results of message parsing (including free text parsing) and the resulting data structures. Conditioning adds or infers the probabilistic representations and relationships needed for the state estimation tasks. Finally mediation provides the interfaces to later processing steps. For example, the conditioning step may produce a raster-based distribution for a convoy location. The mediator might then convert this representation into a single bivariate gaussian approximation to enable later processing elements to work with the results. Thus, the mediator has knowledge of the requirements of later processing and can convert the conditioning results appropriately.

Within this three part design for evidence conditioning state estimation falls into the second component: state estimation. This step produces the conditional probabilities for the current and future states. As noted the techniques we described in the next two sections can be employed both within this evidence conditioning process as well as with the results of the correlation processes.

3. LOCATION ESTIMATION

The problem is to determine the best estimate of the probability density function of target location for successive time increments 1, 2,...., T by combining the greatest amount of information available on terrain characteristics and target behavior. The system is assumed to have the Markovian property, i.e., the density functions at time $n+1$ only depend on the density functions at time n. Further, we define a state space approach to the problem, which we summarize here in the pictorial description given in figure 1.

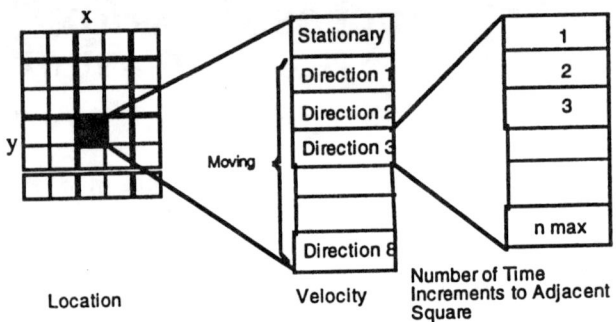

Fig. 1. State space representation

A number of assumptions is made concerning the comportment of the vehicles being tracked:

- There are no *a priori* known goal states for targets
- The targets have a tendency to move in a constant direction.
- If roads are present in the environment and the target is moving on a road, then there is a high likelihood that it will remain on this road. Also, if the target is in a non-road state but at least one of the eight adjacent cells is a road, then there is a high likelihood that the target will move onto this road.
- A vehicle is either moving or stationary (Fig. 1). It has a known probability - which may depend upon its location - of switching states at every time increment.
- It follows from the choice of the state-space that the probability of finding an object at time k in cell (x, y) as a sum over the conditional probabilities of remaining stationary and moving from one of the adjacent locations.

Details on the exact equations for this process, the methods for track initiation, as well as performance metrics and be found in [8].

While the above assumptions effectively incorporate local terrain information into our state estimation process they do not account for the actual behavior of intelligent objects moving in terrain. This is because most targets we are interested in tracking have more than local information. To account for this we need a much richer model formulation.

We accomplish this in our approach using potential surfaces. These potential surfaces provide a global description of the information about target motion. Building a potential surface requires: (i) the efficient computation of the distance from any pixel to any significant category of pixels (road, water, city, marsh, etc.) and (ii) the functional form of the attractive and repulsive potential fields. Terrain features may have different distances of influence, which the designer must determine. These distances will depend on the domain and are a function of the availability and use of maps.

As a preprocessing step, we calculate the potential surface for the region of interest as the sum of the effects of each obstacle and attractor on every pixel (where a pixel is the lowest level of resolution of interest in the problem) in the region of interest. In [8] we show how this calculation is performed and argue that it does not create spurious local minima or maxima since our calculation does not allow the effects of an attractor, say, to override the surface created by an obstacle.

Once we have created the potential surface, we then initiate a microsimulation involving all objects within the region to determine their estimated locations over the desired time interval. Notice that this process produces probability surfaces conditional on the time increment. Figure 2 shows an example for a single object which begins in an off road state and quickly converges to an on road state. Notice that for an object without initial momentum the probability surface would look quite different(see figure 3). The use of our potential surface approach has produced a more realistic estimate for the behavior of an intelligent object with more global terrain information

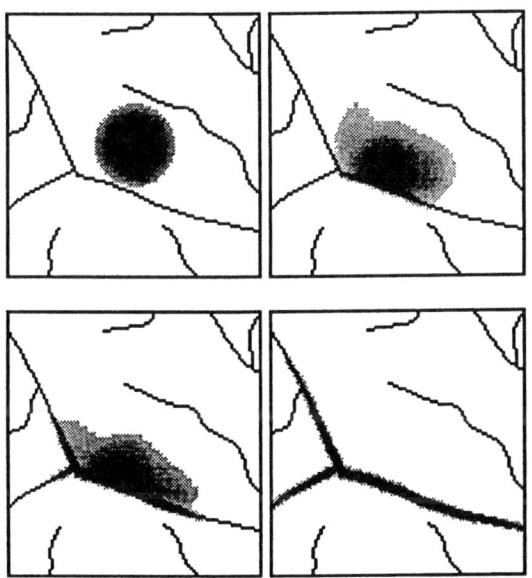

Fig.2. Propagation over terrain using a potential function.

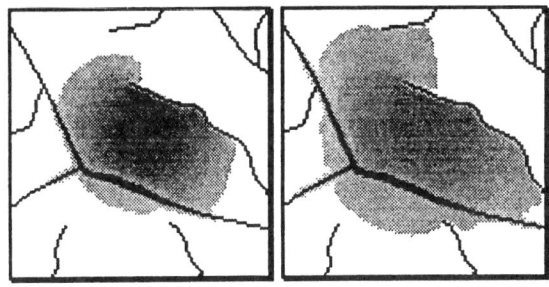

Fig. 3. Propagation with using a potential function

4. BEHAVIOR ESTIMATION

Behavior estimation requires the use of network models to simulate the activity of targets. These network models show the behavioral patterns of intelligent objects and how these behaviors are scripted into situations.

Figure 4 shows an example network model for a target. Each arc in this representation contains the likelihood of transitioning from one state to the next.

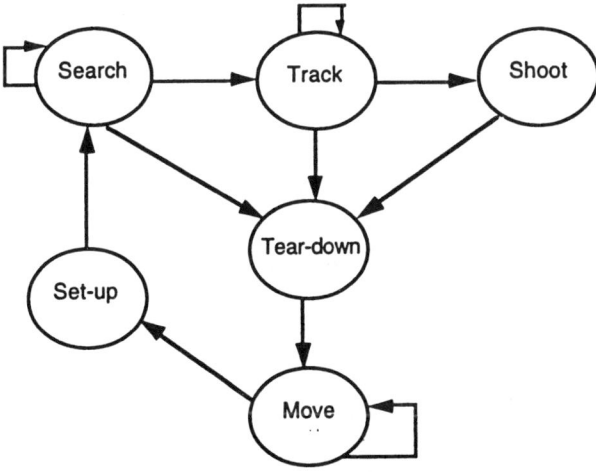

Figure 4. Example State Transition Network

As with location modeling we compute the probabilities based on a series of assumptions:

- There are no *a priori* known goal states. As with location estimation, this means that we do not begin the process with knowledge of the ultimate behavior of the target. However, if this information is available here, we can exploit it.
- The targets behavior is described by a known set of discrete states. Further each behavior state in this set is reachable from the other behavior states.
- Our knowledge of transitions between states is probabilistic.

With these assumptions (and a few other technical assumptions) we can easily model the process using

Markov models. With single objects, we can calculate the need probabilities directly from state transition matrices. However, the interactions between objects require us to run simulations. As with location estimation, we run a simulation of each object in the environment and derive the probability distributions for activity based on the results of the simulations over the time increment desired.

5. CONCLUSIONS

This paper has provided an overview to the use of a microsimulation approach to state estimation. The fundamental idea is to use combine all available evidence into the state estimation process through the use of a simulation. This contrasts with traditional approaches which have relied only on past locations, for example, to predict future locations.

In addition to location estimates the approach is also useful for behavioral estimates. This extension is possible by using state transition diagrams to show how behaviors relate for each target. Again we can run simulations using these state transition diagrams to give us estimates for target behavior.

We have run tests using the location estimation procedure on real data collected on the motion of M1 tanks during training. These results show significant improvement in estimation performance(as measured by root mean square error) for this procedure versus traditional Kalman filtering techniques. We have not yet tested the behavioral estimation procedure, but hope to do so over the next year.

6. REFERENCES

[1] A. Gelb, *Applied Optimal Estimation,* Cambridge, MA, MIT Press, 1984.

[2] C. B. Chang and J. A. Tabaczynski, "Application of State Estimation to Target Tracking," *IEEE Transactions on Automatic Control,* AC-29, No. .2, 1984.

[3] R. A. Singer, "Estimating optimal tracking filter performance for manned maneuvering targets," *IEEE Transactions on Aerospace and Electronic Systems,* vol. AES-6, July 1970.

[4] R. L. Moose, H. F. Vanlandingham, and D. H. McCabe, "Modeling and estimation for tracking maneuvering targets," *IEEE Transactions on Aerospace and Electronic Systems,* vol. AES-15, 1979.

[5] Barker, A., D.E. Brown, and W.N. Martin, "Bayesian Estimation and the Kalman Filter," *Computers and Mathematics with Applications,* Vol. 30, No. 10, 1995, 55-77.

[6] R. J. Fitzgerald, "Simple Tracking Filters: Position and Velocity Measurements," *IEEE Transactions on Aerospace and Electronic Systems,* vol. AES-18, 1982.

[7] D. D. Sworder and R. G. Hutchings, "Maneuver estimation using measurements of orientation," *IEEE Transactions on Aerospace and Electronic Systems,* vol. AES-9, pp. 625-637, July 1990.

[8] Nougues, P. O. and D.E. Brown, "We Know Where You Are Going: Tracking Objects in Terrain," *IMA Journal of Mathematics Applied in Business and Industry,* Vol 8 (1997), pp. 39-58.

INFORMATION UNDERSTANDING:
INTEGRATING DATA FUSION AND DATA MINING PROCESSES

Edward L. Waltz
ERIM International
1975 Green Road
Ann Arbor, MI 48105, USA
waltz@erim-int.com   phone 313-994-1200 ext. 2618

ABSTRACT

Data fusion and data mining are complementary processes that contribute abductive-inductive (learning and discovery) and deductive (detection) capabilities. Not to be confused with each other, the processes offer distinct and reciprocal capabilities that, once integrated, provide powerful tools for improving the automatic recognition of targets by complex, non-literal signatures from multiple sources. ERIM International is developing and applying this integrated technology to improve the performance of traditional data fusion processes.

1. INTRODUCTION

Data fusion and data mining are two knowledge creation processes that contribute automated process technologies to a variety of application domains, including: intelligence, military surveillance and reconnaissance, commercial marketing analysis, business process automation, weather analysis, environmental monitoring and commerce.

The defense community has developed data fusion processes, seeking improvements over single-sensor intelligence assessment and automatic target recognition (ATR) processes by using multiple sensors and sources of information. The U.S. Joint Directors of Laboratory (JDL) Data Fusion Subpanel has for over a decade, developed functional architecture models and coordinated technology exchange in this area. (See [1, 2] for a systems-level overview of the technology) .

On the other hand, the more recent data mining research has been widely applied by the business community to learn (or discover) business-related knowledge in vast data bases of process and sales data [3]. The emphasis of data mining is on the discovery of complex relationships in large data sets.

For national security applications, the two processes offer the promise of increasing the performance of systems for discovering target models, and then detecting targets with those models. While data fusion focuses on automatic target recognition (using target models), data mining focuses on automatic model discovery.

The current state of the art in multisensor ATR and data fusion depends upon relatively intuitive or literal target signatures, used as the models for detection and classification. Distinct signatures of target shape, emissions, or kinematic behavior, developed by manual analysis forms the basis for detection and classification. The fusion process associates observations from multiple sources (correlated with a common physical target object) and combines the data.

As military targets have moved to become less observable, and the domains of phenomena for sensing have expanded, the *joint* signature in time, space and spectrum has become more complex, and not at all intuitive. This is where data mining processes can contribute to the discovery and development of target models to support data fusion.

2. MINING AND FUSION

Data mining and fusion are processes that implement fundamental reasoning processes that are mutually supportive in modeling targets, and detecting instances of those models in observed data.

2.1 Reasoning Processes

The fundamental reasoning processes (Table 1) include two basic stages:
- *Induction-Abduction-* these *discovery* processes detect the presence of "interesting" patterns or relationships in data sets that *may* be general models or signatures that characterize events or objects of interest. The model hypotheses are validated by human analysis, and, once validated for generality, applied to data fusion [4].

- *Deduction* - this *detection* process applies learned models as templates to infer the existence of an instance of a modeled event or object in a set of data.

Table 1- Fundamental Reasoning Processes

Reasoning Processes	General Properties	Implementation
Abduction	Create model hypotheses for *specific* sets of data to explain that *specific* set	Mining (Discovery of models)
Induction	Extend model hypotheses for representative sets of data to make a *general* assertion or explanation	
Deduction	Apply models to create hypotheses to detect and classify (explain) the existence of target	Fusion (Detection using models)

For literal target signatures (e.g. discrete target objects in electro-optical imagery) the mining process has been performed by manual analysis (e.g. statistical clustering, heuristic rule generation, etc.) using human analysis and detailed modeling of the targets.

Non-literal signatures are not at all intuitive or apparent, however, *across* multiple data domain. In these cases, automated discovery applying abduction-induction must be performed across the disparate data domains to discover subtle relationships that may become detection models for fusion. Consider, for example, the discovery of a simple relationship across data sources relating events prior to an illicit narcotics transaction. In this the "target" of surveillance is the transaction events:

- Bank transfers Account X to Y (t-3)
- Telephone traffic J to Q or L (t-2∓1)
- Vehicle types p1 or p2 in surveillance video at locations 34, 34a or 56 (t-0.5 ∓ 0.25)
- Vehicle type T4 depart facility k21 (t+1∓1)

The non-literal pattern model in this case may have a literal and logical explanation, *once found*, but the issue is discovery of reliable patterns.

2.2 Data Mining

Data mining techniques automate the processes to search large volumes of raw data to locate potential relationships between data elements based upon very general quantitative definitions of a "relationship" [5, 6]. Statistical clustering, neural network, n-dimensional visualization and other tools are applied to hypothesize (*abduct*) the presence of relationships that may be sufficiently general - and important - to provide a model for detection. Discovered relationship hypotheses are presented to human experts to assess their general validity and to hypothesize (*induce*) their broad applicability as a signature model for fusion. A variety of commercial mining tools have been developed to support business applications where the multisource data sets generally include numerical (statistical) and textual data. Intelligence and military applications expand the envelope of mining requirements to include data across multimedia (i.e static imagery, video, text messages, time sequence events, and signals intelligence data).

2.3 Data Fusion

Data fusion processes include the alignment of data sets, correlation of sets that can be associated from common sources (objects or events) and combination of the sets to make detection or classification decisions. The combination may occur in at least two ways:

- Level-1 fusion - a decision process (Boolean, Bayesian, etc.) creates a composite decision from the set of discrete detection and classification decisions from sensors.
- Level- 0 fusion - extracted (segmented) data sets from potential target objects are combined to create a single detection or classification decision. The data sets may be raw data (e.g. pixels or pulses) or processed data (e.g. thresholded pulses, or computed signature vector from a segmented region of pixels).

The performance of fusion (measured in terms of P_{DET}, false alarm rate, classification accuracy) is, of course dependent, on the validity of the target models (measured in terms of coverage, completeness, necessity and sufficiency of pattern conditions) delivered by the mining process.

Table 2 contrasts the inputs, functions and processes of mining and fusion.

Table 2 - Comparison of Data Fusion and Data Mining Processes

Process:	DATA MINING	DATA FUSION
Knowledge Created	**Discovery** of the *existence* of previously unrecognized patterns associated with entities or events in time or space	**Detection** of the *presence* of known entity or event types in time or space applying previously learned models
Input Data	**Pre-Operational**- Statistically large collection of representative data containing observations of known targets with ground truth	**Operational** - Data collected by multiple sensors and sources
Reasoning Process	**Abduction-Induction**: Discovery of sufficient correlated relationships in data to infer a general *description* (or rule set) that may be always or generally (to some quantified degree) true	**Deduction**: Detection of previously known patterns in data to infer the *presence* and *identity* of the entity or event represented by that pattern
Knowledge Patterns (Models) Used	**Unknown**: (general) model of "interesting" data properties is used as template to detect qualifying candidates for "new" models of targets of interest	**Known**: (specific) models are used as templates to detect similar patterns in data.
Object of Detection/ Discovery Process and Knowledge Gained	• **Discovery** of "interesting" general relationships and patterns of behavior, which may be, validated as general models of relationships or behavior. • **Discovery** of new types of entities or events, by previously unidentified and unknown patterns, in large volumes of data	• **Detection** of individual and related sets of entities and events. • **Detection** of the presence, type and location of known types of entities or events in large volumes of data
Output Knowledge	**Models** of relationships or behaviors to describe target entities or events	**Detected targets** based on matching sensor data with known models of entities or events

3. INFORMATION UNDERSTANDING

ERIM International is conducting research, exploring the integration of mining and fusion techniques to expand the ability to detect and classify non-literal target signatures, hidden in disparate data sets, including:

- Imagery data (Static panchromatic, IR, multispectral, hyperspectral)
- Spatial data (maps, charts, terrain models)
- Video imagery
- Statistical data sets
- Textual data sets containing key words, phrases or concepts

This *information understanding* process tightly couples the discovery and detection processes using all available source data to provide cues and clues to intelligence and business analysts tasked with challenging investigative problems. The integrated process (Figure 1) couples the discovery and detection processes at corresponding levels of information abstraction:

- Level - 0 models provide relationships at the signal level ("pixels", "phrases" and "pulses") that provide evidence about components of objects. These models include image and spatial patterns that occur at the pixel and segmented feature levels [7].
- Level -1 models provide relationships that provide evidence about complete objects.
- Level- 2 models provide evidence about situations that are explained by temporal, behavioral or structural relationships between objects.

The understanding process is tailored to unique applications and the mining-fusion processes are mutually supportive. The performance of the mining processes (and validity of abducted models) is evaluated by the effectiveness of fusion in cueing analysts to valid targets or situations in new data.

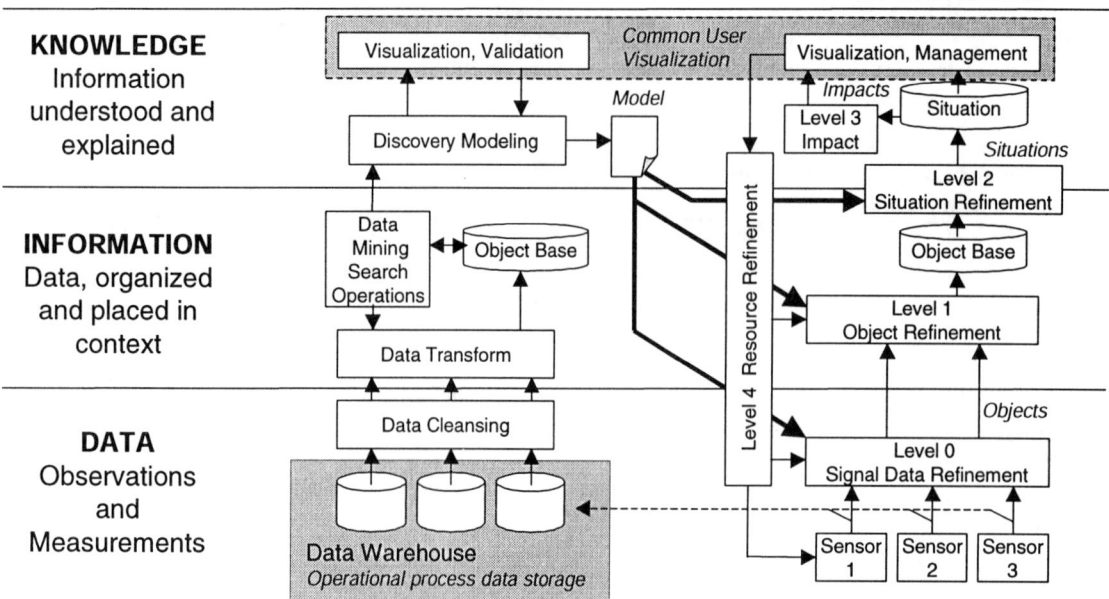

Figure 1 - Integrated Data Mining and Fusion permits discovery and detection of non-literal targets in disparate data

The effectiveness of fusion is evaluated by the detection performance when tested against the data set used for discovery.

4. SUMMARY

The driving factors of decreasing target observability and complexity of data sets required to detect those subtle targets have fostered the complementary technologies of data mining and data fusion. ERIM International's *Information Understanding* research is developing the technology to integrate these powerful tools to efficiently exploit disparate sensor data to discover subtle multi-phenomenology target signature models and then to apply those models to the detection of critical targets.

5. REFERENCES

1. Waltz, Edward, L. and Llinas, James, *Multisensor Data Fusion*, Norwood MA, Artech, 1990.
2. Hall, David L. and Llinas, James, "An Introduction"
3. The term "data warehousing" is applied to large data bases of operational data, organized and integrated with decision support tools to permit manual analysis, on-line analytic processing (OLAP) and automated mining.
4. The distinction between abduction and induction has to do with the applicability of the model - to the "best explanation" of a limited data set, or to all possible instantiations of the object modeled, respectively.) C. S. Peirce defined abduction as a separate stage of critical thinking (although not an element of formal symbolic logic) which hypothesizes a model to explain only the training set. Induction extends the hypothesis beyond the training set to general application.
5. Piatetsky-Shapiro, Gregory, and Frawley, William J.(eds), *Knowledge Discovery in Databases*, Menlo Park CA, AAAI Press/MIT Press, 1991.
6. Fayyad, Usama M. and Piatetsky-Shapiro, Gregory (eds) and padhr Smyth, *Advances in Knowledge Discovery and Data Mining* Cambridge, MA, MIT Press, 1996.
7. Waltz, Edward, "The Principles and Practice of Image and Spatial Data Fusion", *Proc. Of 8th National Data Fusion Conf. Dallas TX, 15-17 March 1995.*

STRATEGICALLY-CONTROLLED INFORMATION FUSION

Dr. Steven M. Flank

Information Systems Office (ISO), Defense Advanced Research Projects Agency (DARPA)
3701 North Fairfax Drive
Arlington, Virginia 22203-1714, USA

ABSTRACT

Core Approach for Strategically Controlled Information Fusion

Define performance metrics that quantitatively characterize fusion engine (FE) output. Define context metrics that quantitatively characterize FE performance as a function of input data. Quantitatively map FEs' performance space (how predicted performance depends on context). Encapsulate FEs so that they share data representations, services, and queries. Control this confederation of encapsulated FEs with an intelligent fusion strategist that maximizes overall performance as a function of context by: a) Selecting the FE with the best-predicted performance for a given real-time context; b) Sequencing FEs to maximize performance by receiving data sets outputs from other FEs, thus improving the region of the performance space in which they operate; c) Adjusting FE control and modeling parameters to achieve peak performance over a broader range of contexts; d) Reacting to FE outputs by comparing to predicted performance and adjusting parameters, sequences, and selections to maintain overall performance.

I. INTRODUCTION: PROGRAM GOALS

The joint warfighter is over-loaded by high volumes of data from broadcast intelligence, multiple sensors, and multiple exploitation sites, compounding the "fog of war" with uncorrelated, redundant, incomplete and ambiguous information. Advanced technologies can produce clear and actionable pictures of the battlespace using existing and emerging information sources tailored to each warfighter's needs. The Dynamic Multi-user Information Fusion (DMIF) program will design, build, and demonstrate a battlespace awareness information architecture that (1) utilizes all available sources of battlefield information at the entity level and above, (2) strategically controls distributed fusion assets to produce, update and synchronize situation hypotheses, and (3) provides tailorable mission-focused situation information products to support a wide variety of user missions and tasks. These automated capabilities support joint intelligence operations and can be integrated into stand-alone systems or inserted in existing ops/intel automated support systems.

Operational Challenges

- Reducing Information Overload -- Users are drowning in data, but starving for information. Large quantities of surveillance data (anecdotally, up to 80%) are never processed or are routed automatically to stand-alone databases without being incorporated into current situation estimates. Large quantities of the information that does make it into situation awareness tools (anecdotally, up to 80%) are redundant or out-of-date.
- Overcoming Barriers to Interoperability -- Today's 400+ tactical data processing applications, including the dozens of information-level fusion applications in daily use, provide only fragmentary views into the battlefield situation. They are designed to accept only selected data types, and their performance adds value only in narrow areas. Nor can they currently share their data or take advantage of each other's specialized capabilities. The only method for obtaining a general situation estimate today is integration by video screen -- that is, assembling a large number of applications on separate workstations, and building up one's conclusions by walking around the room and examining each computer screen.

- Improving Speed, Cost, and Reusability of Development -- Each fusion development effort today must start from scratch. Replicating the databases, sys admin functions, interfaces, and process controls that are required for each new fusion engine leaves only 10-25% of the effort to be devoted to algorithm development. Stand-alone designs also cannot rely on existing FEs to provide a base capability, but have to build yet-another-Kalman-filter. A common fusion infrastructure can mean 75% lower development costs with those costs now focused on new, advanced algorithms.

II. OPERATIONAL GOALS

Reduce information overload

- Reduce the total number of hypotheses presented to the user by eliminating redundant observations, combining reports from multiple sensors and sources, and by raising the level of aggregation of the output information as appropriate. Goal: x3 reduction.
- Reduce the number of incorrect hypotheses presented to the user by avoiding regions of low performance for individual FEs, by passing data through more processing stages (each stage adding more inference capabilities), and by explicitly accounting for uncertainty and multiple hypotheses. Goal: 50% reduction.
- Reduce the number of out-of-date hypotheses presented to the user by linking observations from different times into explicit tracks and by using multiple hypothesis techniques to prune out hypotheses that are not supported by subsequent data. Goal: 30% reduction.
- Respond only to user or mission needs specified at either build or run time by setting explicit information goals for the fusion strategist to meet, by filtering information according to space, time, and force type, and by applying additional inference in Product Finishers needed to draw the conclusions specific to that user or mission. Goal: x2 reduction.

Overcome barriers to interoperability

- Exchange information among fusion engines or situation awareness applications across multiple organizations by encapsulating the majority of major FEs in use today. Goal: all three services plus national agencies (DISA, NSA).
- Enable preservation of service- or mission-unique functionality while still providing full DII COE compatibility and utilization of DII COE Correlation Services by encapsulating GCCS and by inserting the Fusion Strategist into the DII COE. Goal: No reduction in existing performance of stand-alone applications, plus performance benefits of FE confederation as stated above.
- Feed sensor and intelligence data (after processing) directly to ops applications by translating ops users' information requirements and real-time priorities into a form that the Fusion Strategist can use to drive FE processing and by filtering database contents through the Product Finishers. Goal: Targeting and force deployment applications (e.g., AFATDS, JFACC), meeting 80% of user queries, with latencies of a few minutes.

Improve speed, cost, and reusability of development

- Asymptotically approach low cost and time to encapsulate the nth Fusion Engine, by using CORBA and common schema methods to ensure common access and data exchange. Goal: Less than $150k, under 8 weeks per FE.
- Increase the proportion of development costs devoted to improved algorithms instead of system infrastructure by publishing APIs, data standards, and engineering guidelines for new FEs. Goal: 90% of development cost focused on algorithms (resulting in approximately 75% lower overall cost to develop).
- Apply existing FEs to new domains (new theaters, new target sets, new mission objectives) without the two-year re-development cycle now required. Goal: Extend generic models and translate for each FE within a few weeks.
- Improve ops-intel integration by linking existing ops application directly to situation information. Goal: Build new Product Finishers in under two months, at less than one-quarter the cost of a new dedicated situation awareness application of equivalent performance.

III. TECHNICAL GOALS

Context and performance characterization

Challenge: Systematize a complex information space and a complex set of operations on data.

Develop and validate new metrics for characterizing context in real-time. Select and refine existing performance metrics for characterizing entity-level and aggregated entity-level fusion.

Tasks:
- Establish a set of context metrics that meet the following criteria for necessity and sufficiency: metrics span the data space; are maximally orthogonal; can be computed in real-time without reference to ground truth; are a small enough set to be computed efficiently; are a small enough set to avoid combinatorial explosions in the fusion strategist's optimization routines.
- Establish a set of performance metrics that meet the following criteria: metrics span the performance space for the evolving confederation of fusion engines; are sufficiently orthogonal to differentiate FE strengths; are operationally meaningful; can be computed efficiently given current test data generation capabilities.
- Construct methods using the following sequence: Association and estimation metrics first, then prediction metrics; JDL Level 1 metrics first (i.e., vehicle level), then JDL Levels 2 and 3 (i.e., unit and threat level); correlators first, trackers second, aggregators third, then assessors.

Success criteria:
- Prove equivalence classes (FE performance stable for multiple data sets with equivalent context metrics);
- Prove FE differentiability (distribution of FE performance over the performance space);
- Demonstrate implementability (real-time context measurements and affordable FE characterization).

Intelligent fusion strategist

Challenge: Apply existing control theory and optimization techniques to the new arena of fusion strategies, building on the context and performance characterization. Use these techniques to map problem space (user information requirements, current observations and hypotheses) to solution space (inference capabilities of available fusion engines), given constraints on available algorithms, time, bandwidth, and processing power. Initial implementations focus on classical optimization techniques (e.g., selection via modus ponens). Later implementations include fuzzy logic approaches to optimization under uncertainty, and generative planning approaches for creating new control logic in response to external stimuli (such as new information requirements arising from users or new FEs becoming available).

Tasks:
- Implement a fusion strategist that can achieve maximum overall performance by selecting, sequencing, adjusting control parameters, and reacting to real-time performance of multiple encapsulated fusion engines.
- Construct strategist capabilities by implementing classical selection algorithms first, then adding parameter optimization methods, then extending the selection algorithms with fuzzy logic methods, then improving the optimization methods by adding resource constraints, and finally adding methods to create new FE sequences on the fly (dynamic plan generation).

Success criteria:
- Prove that a confederation of FEs performs better than any single FE;
- Demonstrate extended performance envelopes for even single FEs, using control parameter adjustments and FE sequencing;
- Demonstrate robustness of performance as incoming data, information goals, and available resources change;
- Demonstrate ease of insertion of new FEs and resulting improvement in performance for the enhanced confederation of FEs.

Fusion engine encapsulation

Challenge: Use emerging object-modeling and wrapping techniques, such as UML and CORBA, to create interoperability among legacy FEs. Focus on information representation and data methods that can translate between multiple legacy representations and accommodate constant evolution without requiring labor-intensive revision or re-engineering. For example, implement an attribute-value pair schema tied to a formal ontology so that data structures remain the same even as semantic representations change.

Tasks:
- Wrapping: Develop engineering methods for wrapping migration FEs for common access, control, and communication. Publish these methods as APIs, data standards, and

implementation guidelines for use by other FE developers. Implement for a set of at least 12 migration FEs.
- Performance Modeling: Apply context and performance metrics to model the performance of at least 12 migration FEs using the highest-fidelity test data available.

Success criteria (for wrapping):
- Data translation methods with zero information loss (using round-trip translations);
- Model translation and data access methods accurately regression tested against the original FEs;
- Learning curve statistics with an asymptotic encapsulation cost of under $150k in less than 8 weeks.

Success criteria (for performance modeling):
- Consistency of performance over multiple test sets;
- Enough performance points to interpolate performance prediction curves that the fusion strategist can use;
- Confirm FE differentiability (distribution of FE performance over performance space).

Agile FE model parameters

Challenge: Develop new methods for generically representing the knowledge about the battlefield that fusion engines need to perform their automated inference. For example, use the newly developed approach for combining Bayesian networks and frames (first-order predicate logic representations). Develop new methods for evolving this knowledge as the outside world, or our state of knowledge about it, evolves. For example, use the newly developed algorithms for automatically combining Bayes net fragments (adding, splitting, splicing, and inverting). Engineer a set of methods for translating this generic representation into the form needed by each encapsulated FE (such as .h parameter files).

Tasks:
- Construct generic models (knowledge about the outside world used for automated inference) for Missile Order of Battle first, Ground Order of Battle second, emphasizing heavy forces initially, then adding light and irregular forces.
- Construct translation methods to transform these models into the parameters used by the encapsulated FEs. Use real-time update methods (such as re-loading .h parameter files) where possible.
- Characterize the impact of modeling parameters on FE performance, for use by the strategist in real-time optimization.
- Provide model-editing tools (such as map-based visualization for expressing spatial constraints, or anomaly and model mismatch detection), so that end-users can update the generic models in battle-relevant time frames.

Success criteria:
- The generic models include all parameters used by the encapsulated fusion engines;
- Revisions to the generic model are correctly and automatically propagated to all FEs consuming that model;
- FE performance curves are extended 5-50% by the real-time adjustment of most modeling parameters;
- Experiments with representative users show realistic time-lines for editing and constructing generic models (updates in less than one day for moderately trained users).

Transition

DMIF provides an enabling architecture for improved performance and interoperability of existing and emerging situation awareness applications. Therefore, the most important transitions are to include that architecture in the infrastructure of the large service and agency C3I systems, specifically the Army ABCS (including ASAS), the Air Force TBMCS (including the Situation Assessment and Awareness segments), and the DISA DII COE (including the correlation services segment).

THE CANADA-NETHERLANDS COLLABORATION ON MULTISENSOR DATA FUSION AND OTHER CANADA-NATO MSDF ACTIVITIES

Éloi Bossé and Jean Roy

Defence Research Establishment Valcartier
2459 Blvd. Pie XI North
Val-Bélair, QC G3J 1X5 CANADA
Email: eloi.bosse@drev.dnd.ca

ABSTRACT

This paper describes a collaborative effort between Canada and The Netherlands in analyzing multi-sensor data fusion systems. In view of the overlapping interest in studying and comparing applicability and performance of advanced state-of-the-art Multi-Sensor Data Fusion (MSDF) techniques, the two research establishments involved (DREV and TNO/FEL) have decided to join their efforts in the development of MSDF testbeds. This resulted in the Joint-FACET (Fusion Algorithms & Concepts Exploration Testbed), a highly modular and flexible series of applications that is capable of processing both real and synthetic input data. In addition to the bilateral collaboration with The Netherlands, this paper presents a survey of the other Canadian MSDF activities conducted under NATO Research Study Groups.

1. INTRODUCTION

Major ongoing activities undertaken by the Decision Technology Section at Defence Research Establishment Valcartier (DREV) and the Observation Systems Division at TNO Physics and Electronics Laboratory (TNO-FEL) are the investigation of sensor management, integration, and Multi-Sensor Data Fusion (MSDF) techniques that could apply to the current Canadian and Dutch Frigates Above Water Warfare (AWW) sensor suite, as well as their possible future upgrades, in order to improve their performance against the predicted future threats. Fundamental issues in developing an MSDF system are the selection of an appropriate architecture and the choice of efficient and dedicated fusion algorithms to fulfill its role in the ship combat direction system. By joining their efforts, Canada and the Netherlands are mutually increasing their capability to explore a wider range of design philosophies, as well as the opportunity to benefit from the participants' previous experiences and lessons learnt.

The approach retained for this joint effort is to employ sufficiently representative sensor and phenomenological simulations in the development of a MSDF testbed. A highly modular and flexible testbed is being developed as the result of the evolution of two existing testbeds: the DREV CASE_ATTI (Concept Analysis and Simulation Environment for Automatic Target Tracking and Identification) and the TNO-FEL MT3 (Multiple Target Tracking Testbed). The resulting higher capacity testbed is hereafter referenced as Joint-FACET (Fusion Algorithms & Concepts Exploration Testbed). Joint-FACET will be used by Canada and the Netherlands in analyzing the various performance trade-offs required in the selection of the best multi-sensor data fusion architecture and algorithms applicable to their respective frigates.

This paper describes Joint-FACET and presents a brief survey of the other collaborative activities ongoing under NATO Research Study Groups.

2. A GENERIC MSDF SYSTEM

Figure 1 shows a generic MSDF system where the key functions are identified. A detailed description of all these functions is provided in Ref. 1. The processing can be divided into blocks such as: 1) data alignment (spatial, time); 2) data association; 3) target kinematics data fusion; 4) target identity data fusion; 5) track management process; 6) cluster management process; 7) input data preparation; 8) track database; 9) fusion configuration monitoring and control.

In any MSDF system, a sensor data alignment in time and space must take place before any fusion can be performed. Navigation data are used to estimate and remove the effects of sensor motion from the received data. The functions of data association (labeling measurements from different origins and/or sensors, at different times, that correspond to the same object or feature) and data fusion (combining measurements from different times and/or different sensors) are also required in one form or another in essentially all multiple sensor fusion applications: data association determines what information should be fused, the fusion function performs the fusion.

In addition to the MSDF system which detects, localizes, and identifies targets, the local sensor management, on the basis of an evolving picture, and under the command of the overall Command and Control (C2) resource management, manages the information that the MSDF might receive by pointing, focusing, maneuvering, and adaptively selecting the modalities of its sensors and sensor platforms. The overall C2 resource management has the responsibility of first examining and prioritizing what is unknown in the context of the situation and threat and then developing options for collecting this information by sending priorities to the MSDF local sensor management function. The interaction of the local sensor management and the MSDF function is done through the fusion configuration, monitoring and control which is responsible for the initialization of the MSDF system, the setup and the adjustment of the various MSDF algorithm parameters to control the quality of the MSDF product.

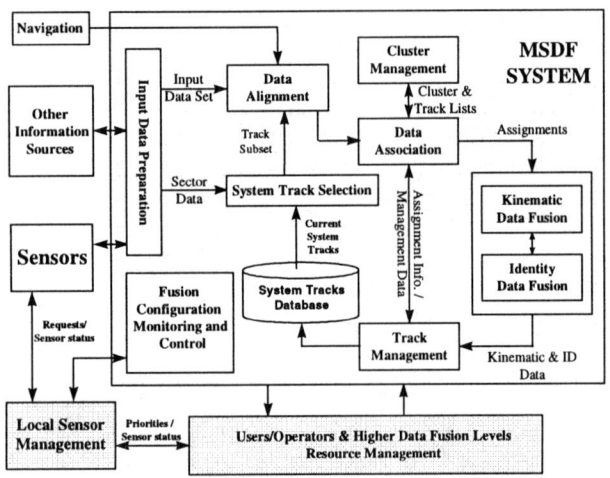

FIGURE 1 - A generic MSDF system

The management of sensors may require that different sensors cooperate to acquire measurements on a common target. The two primary cooperative functions are cueing and hand-off. *Cueing* is the process of using the contacts (i.e., contact-level cueing) or tracks (i.e., track-level cueing) from one sensor (A) to point another sensor (B) towards the same target or event. *Hand-off* occurs when sensor A has cued sensor B for transferring surveillance or fire control responsibility from A to B. Two processes must occur for cueing or hand-off: (1) the cueing sensor must provide the cued sensor data that contains sufficient information to point to the target and identify it as the specific target being cued, (2) the cued sensor must search for the target of interest and verify that it has been acquired.

One of the key issues in developing an MSDF system is the question of where in the data flow to actually combine or fuse the data. The MSDF architecture is an important issue since the benefits are different depending on the way the sensor or other source data are combined. There are four broad alternatives to fusing positional information: at the signal level, at the contact (plot) level, at the track level and finally a hybrid approach which allows fusion of either contact or track data. For identity fusion, there are several types of architectures which can be used: 1) data level fusion, 2) feature level fusion, and 3) decision level fusion.

There is no universal architecture which is applicable to all situations or applications. The selection of the MSDF architecture type should be aimed at optimizing the target detection, tracking and identification performance required for a specific platform given its missions. However, the selection is also constrained by the technological capabilities (both hardware and software). It depends on the quality of the sensors being fused, the availability of computer processing power, the bandwidth of the available data transmission paths, and the degree to which operator intervention is required or desired.

3. THE CANADIAN CASE_ATTI TESTBED

The CASE_ATTI system (Ref. 2) is a highly modular, structured and flexible simulation environment providing the algorithm-level test and replacement capability required to study and compare the technical feasibility, applicability and performance of advanced, state-of-the-art MSDF techniques. The global structure of the CASE_ATTI (Fig. 2) testbed comprises a sensor module which is responsible for providing the realistic measurement data to the tracking algorithms. Given a user-defined scenario, it generates true target positions and measured target positions, which are subsequently made available to the tracking module. Currently, the module supports surveillance radar, IFF, ESM and IR sensor simulations. The current tracking module supports a wide variety of tracker architecture types, varying from a simple single sensor tracker to an arbitrary complex hierarchical multiple sensor topology. The sensor-level trackers currently implemented include:

- Multiple-Hypothesis Tracker (MHT)
- Track-Split filter
- Nearest-Neighbor type trackers (both Munkres-based and optimal)
- (Joint) Probabilistic Data Association filter (JPDA/PDA).

The MHT implementation is capable of handling multiple simultaneous reports from different sensors; the other trackers are capable of handling single sensor reports at a time. A global track fuser using a version of the MHT tracker for assignment has also been provided to fuse the sensor-level tracks to form global tracks. In addition, feedback of these global tracks to the local sensor-level trackers is allowed. Current efforts include provisions for advanced cluster management schemes.

tracking module widened. At this instant, the tracking module is also able to process data that originate from tracking radars (i.e., systems that are capable to aim their antenna at the target of interest) and it can be used to track features in a sequence of optical images.

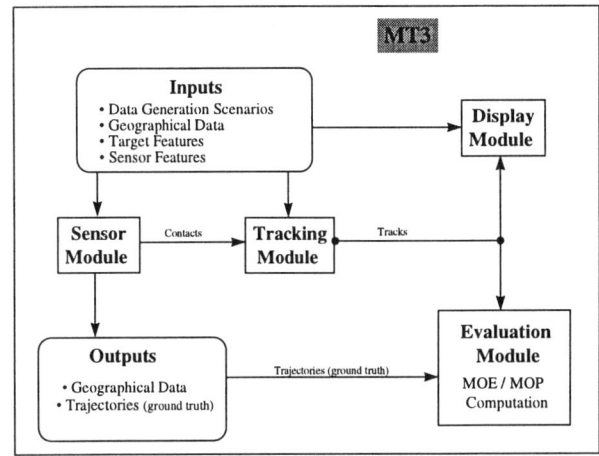

FIGURE 3 - The Dutch MT3.

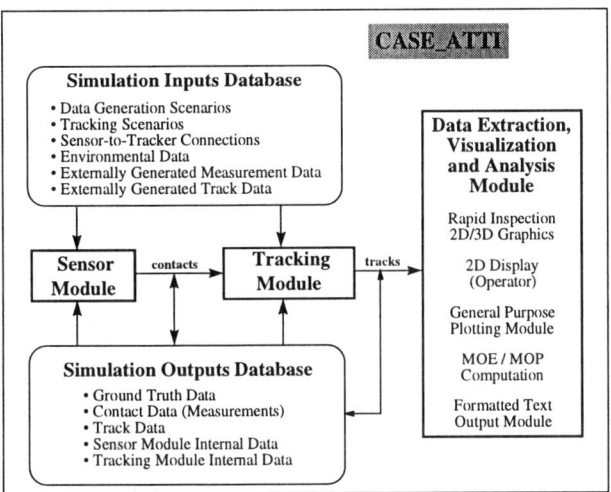

FIGURE 2 - The global structure of CASE_ATTI

4. THE DUTCH MT3 TESTBED

At TNO-FEL the Multi-Target Tracking Testbed (MT3) (Ref. 3) has been developed since early 1996 to support research in the field of tracking and sensor fusion. Currently, the phrase MT3 refers to a series of computer applications that are relevant to assess performance of tracking algorithms utilizing both recorded ('live') data and synthetically generated data. The components of MT3 are: a sensor module, a Tracking Module (TM), a Display Module (DM), an Evaluation Module (EM). The relationship between these modules is shown in Fig. 3. The tracking module and the display module run concurrently and utilize socket communication for the transfer of display data (i.e., track and plot messages). Originally, the tracking module main requirement was to perform multi-target tracking on plot data that were gathered by surveillance radars: track-while-scan (TWS). These systems employ a regular scanning pattern and the tracking module supports systems with rotating antennas. Furthermore, a highly modular design was requested so that a variety of tracking algorithms could be accommodated. As more research groups at TNO-FEL became interested in the MT3 suite of applications, the scope of the

5. THE JOINT-FACET PROJECT

CASE_ATTI and MT3 are multiple target tracking and identification testbeds. The primary objective of these testbeds is to measure the ability of various MSDF systems to generate the estimated tactical picture that accurately reproduces the ground truth tactical picture. The Joint-FACET analysis process shall scan the MSDF solution space to identify the most promising MSDF algorithms, architectures, concepts or knowledge that are worth to further development. The high-level requirements of Joint-FACET are: - the development shall be pursued independently at physically remote locations without necessarily sharing the same code and platforms; - each version (Canadian and Dutch) shall have the capability to run the same scenarios and process the same test cases; - each version shall have some common results visualization tools; - each version shall have the capability to compare the performance of an algorithm running at one site with respect to a different algorithm running independently at the other site working on the same set of data.

Figure 4 shows the global structure of Joint-FACET. It consists of the co-operation of CASE_ATTI and MT3 with two different **MSDF** modules and where commonality is achieved at critical points in the testbed:

in the Scenario and Data Generation (SDG) module and in the Performance Evaluation (PE) module.

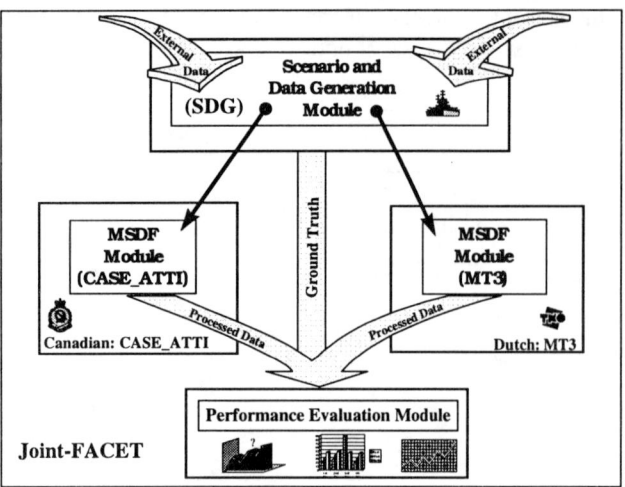

FIGURE 4 - Global structure of the Joint-FACET system

6. OTHER CA-NATO PROJECTS

Canada is involved in several research groups under NATO Defense Research Group (DRG) organization that are looking at the MSDF problem. The NATO AC/243 (Panel 10) on long-term research related to air defense set up an Exploratory Group to investigate the need for cooperative research in the area of sensor fusion for short range air defense (SHORAD). It was revealed that there was a significant gap between the traditional C^4I and the detailed study of data fusion algorithms. The DRG approved the establishment of a new Panel 10/RSG.18 on sensor fusion. The RSG-18 has during its first phase investigated qualitative aspects of sensor fusion for SHORAD weapon systems. The second phase will establish quantitative techniques to assess the performance of the fused sensors and the application of these techniques to SHORAD weapon systems.

As complementary to RSG-18 which was directed to short range air defense, a new RSG on multisensor fusion for maritime wide area surveillance under AC/243 (Panel 3) has been proposed. The main emphasis for this group is on the maritime environment, surface and air, for surveillance and ship survivability. The efforts encompasses all relevant sensors with an initial emphasis on radar and FLIR. The specific objectives are:

1. evaluate quantitatively the benefits for sensor fusion at the system level;
2. improve understanding of cost/performance benefits and trade-off of various sensor combinations.

Over recent years, there has been an increasing trend in military systems for the future battlefield to make use of data from all possible sources. Improved stealth technologies, coupled with the significant reduction target signatures and emissions will require the use of more sophisticated suite of sensors able to operate in different spectral bands and capable of detecting different types of electro-magnetic emissions. It has become clear that the improvements in single sensor performance are reaching the technical limits of both the sensor and the algorithms, and further improvements will tend to be marginal and at a considerable computational costs. A new NATO AC/243 (Panel 3/RSG.22) on multisensor image exploitation has been proposed to address some of the problems outlined above. In particular, the RSG will investigate ways of performing multisensor segmentation and how the results of such techniques can be evaluated. The purpose is to see where and how image analysis can provide the most military benefit in future conflicts.

7. CONCLUSION

This paper described a collaborative effort between Canada and the Netherlands in analyzing MSDF systems, for potential applications to their respective frigates. In view of the overlapping interest in studying and comparing the applicability and performance of advanced state-of-the-art MSDF techniques, the two research establishments involved have decided to join their efforts in the development of MSDF testbeds. This resulted in the Joint-FACET, a highly modular and flexible series of applications that is capable of processing both real and synthetic input data. By joining their efforts, Canada and the Netherlands are mutually increasing their potential for exploring a wider range of design philosophies, as well as the opportunity to benefit from the participants' previous experiences and lessons learnt.

8. REFERENCES

1. Roy, J., Bossé, É., "A Generic Multi-Source Data Fusion System", DREV-R-9719, March 1998.

2. Roy, J., Bossé, É., Dion, D., "CASE_ATTI: An Algorithm-Level Testbed for Multi-Sensor Data Fusion ", DREV-R-9411, May 1995.

3. Theil, A., Huizing, A.G., "The Multi-Target Tracking Testbed: Functionality of the Tracking Module Version 1.0 ", TNO report FEL-97-A232, November 1997.

RECRUITMENT COMPENSATION AS A HEARING AID SIGNAL PROCESSING STRATEGY

Jont B. Allen (www.research.att.com/info/jba)

AT&T Labs–Research, 180 Park Ave
Florham Park, New Jersey 07932

ABSTRACT

Although Fowler is commonly given credit for discovering recruitment [14], Steinberg and Gardner at Bell Labs [25] were the first to understand its true significance. Today recruitment is poorly understood, and is generally *misdefined* as the *abnormally rapid growth of loudness*. It is not well known that loudness in sones does *not* grow more rapidly in the recruiting ear; rather it is the intensity of an equally loud tone (i.e., the loudness–level in phons) that grows more rapidly. Regardless of the definition, recruitment is the most basic manifestation of sensory–neural hearing loss. Recruitment is due to the loss of outer hair cell (OHC) function. The sound–detecting inner hair cells (IHC) within the cochlea have a limited dynamic range of less than 60 dB. The OHCs nonlinearly compress the dynamic range of the signal excitation to the IHC, extending its dynamic range. Thus the normal function of OHCs plays a critical role in loudness and speech coding. Distortion product otoacoustic emissions (OAE) are an objective measure of OHC nonlinear compression, while loudness growth is a subjective measure. This report discusses the application of multi-band compression (MBC) to the compensation of loudness recruitment. This technology was "reinvented" at Bell Labs between 1983 and 1987 and is now sold by the ReSound Corporation. Our basic strategy with multi-band compression is to restore the normal dynamic range which is lost due to OHC misfunction. I describe why MBC works and what the hair cells do. In the oral presentation I shall describe how the OHCs might act as compressors (see my web site). Finally I review the history of the MBC hearing–aid development at Bell Labs.

1. LOUDNESS GROWTH

Acoustical signal *intensity* is defined as the flow of acoustic energy. *Loudness*, in *sones* or *loudness units* (LU),[1] is the name given to the perceptual attribute corresponding mainly to acoustic signal intensity. Loudness depends in a complex manner on a number of acoustical variables, such as intensity, frequency, spectral bandwidth, and on the temporal properties of the stimulus, as well as on the mode of listening (e.g., in quiet or in noise, binaural or monaural stimulation). Iso–loudness contours, which define the *loudness–level*, or *phon* scale, were first determined in 1927 by Kingsbury [15, 9, (p. 227)]. These curves describe the relation between equally–loud tones or narrow bands of noise at different frequencies.

In 1924 Fletcher and Steinberg published an important paper on the measurement of the loudness of speech signals [13]. In this paper, when describing the growth of loudness, the authors state

the use of the above formula involved a *summation of the cube–root of the energy rather than the energy.*

This cube–root dependence had first been described by Fletcher the year before [8].

Today, any power–law relation between the intensity of the physical stimulus and the psychophysical response is referred to as *Stevens' law* [23, 4]. Fletcher's 1923 loudness growth equation established the important special case of loudness for Stevens' approximate, but more general, psychological "law." Weber's "law" states that $\Delta I / I$ is constant, where I is the intensity, and ΔI is the just noticeable difference (JND) in the intensity. Weber's law is known to be only approximately correct for pure tones [22]. Fechner's "law," that loudness is proportional to the log of the intensity, is based on the idea that the JND may be integrated to obtain a psychophysical scale, which in this case is loudness [6]. By 1961, Stevens was arguing that Fechner's law was fundamentally incorrect [26]. As described previously, by 1924 Fletcher had shown that Fechner's law did not hold for speech signals (they discusses this point in the 1924 paper) and for tones in 1933. The relation between the loudness growth law and the JND has a long history [24, 4].

2. COCHLEAR NONLINEARITY: HOW?

What is the source of Fletcher's cube–root loudness growth (i.e., Stevens' Law)? Today we know that the basilar membrane motion is nonlinear, and that cochlear outer hair cells (OHC) are the source of the basilar membrane nonlinearity, resulting in the cube–root loudness growth observed by Fletcher.

From noise trauma experiments on animals and humans, we may conclude that recruitment (abnormal loudness growth) occurs in the cochlea [7]. In 1937, Lorente de No theorized that recruitment is due to hair cell damage [19]. Animal experiments have confirmed this prediction and have emphasized the importance of outer hair cell (OHC) loss [18, 17]. This loss of OHCs causes a loss of the basilar membrane compression as first described by Rhode in 1971 [2, 21, (p. 291)]. It follows that the cube–root loudness growth starts with the nonlinear compression of basilar membrane motion due to stimulus dependent voltage changes within the OHC.

We still do not know precisely what controls the basilar membrane nonlinearity, although we know that it is related to outer–hair–cell length changes, which are controlled by the OHC membrane voltage. This voltage is determined by shearing displacement of the hair cell cilia by the tectorial membrane. Based on ear canal impedance measurements, the most likely cause of nonlinear basilar membrane mechanics is due to changes in the microme-

---
[1] Sones and LU are related by a scale factor. One sone is 975 LU.

chanical impedances within the organ of Corti, which result from OHC length changes. This conclusion logically follows from ear canal impedance measurements expressed as the nonlinear power reflectance, defined as the retrograde to incident power ratio [5], which shows that the relative local basilar membrane impedance is stimulus–level dependent.

3. COCHLEAR NONLINEARITY: WHY?

This leaves unanswered *why* the OHCs compress the signal on the basilar membrane. The answer to this question has to do with the large dynamic range of the ear.

IHC dynamic range. Based on the Johnson (thermal) noise within an inner hair cell, it is possible to accurately estimate a lower bound on the RMS voltage within the inner hair cell. From the voltage drop across the cilia we may estimate the upper dynamic range of the cell. The total dynamic range of the IHC must be less than this ratio, namely less than 65 dB. The dynamic range of hearing, on the other hand, is frequently stated to be 120 dB. Thus, *the inner hair cell does not have a large enough dynamic range to code the dynamic range of the input signal.* Spread of excitation models and neuron threshold distribution of neural rate do not address this fundamental problem. Nature's solution to this problem is the OHC–controlled basilar membrane compression.

The formula for the Johnson RMS thermal electrical noise voltage $|V_c|$ due to cell membrane leakage currents is given by

$$|V_c|^2 = \int_{-\infty}^{\infty} \frac{2kTRdf}{1+(2\pi fRC)^2}, \quad (1)$$

where R is the cell membrane leakage resistance and C is the membrane capacitance C. It follows that

$$|V_c| = \sqrt{\frac{4kT}{C}}. \quad (2)$$

The cell capacitance has been determined to be about 9.6 pF for the IHC [16]. Thus from Eq. 2 we find that $V_c = 21\ \mu V$.

Although the maximum DC voltage across the cilia is 120 mV, the maximum RMS change in cell voltage that has been observed is about 30 mV (I. J. Russell, personal communication). The ratio of 30 mV to the noise floor voltage (21 μV), expressed in dB, is 63 dB. Thus it is impossible for the IHC to code the 120 dB dynamic range of the acoustic signal. Because it is experimentally observed that, taken as a group, IHCs *do* code a wide dynamic range, the nonlinear motion of the basilar membrane must be providing compression within the mechanics of the cochlea prior to IHC detection.

4. LOUDNESS ADDITIVITY

In 1933, Fletcher and Munson published their seminal paper on loudness. It details *1)* the relation of iso–loudness across frequency (loudness–level or phons), *2)* their loudness growth argument (in loudness–units, or sones), *3)* a model showing the relation of masking to loudness, *4)* and the basic idea behind the critical band (critical ratio) [12]. The arguments they used were elegant, and the results were important.

Rather than thinking directly in terms of loudness growth, they tried to find a formula describing how the loudnesses of several stimuli combine. From loudness experiments with low– and high–pass speech and complex tones [13, 9], and other unpublished experiments over the previous 10 years, they found that loudness adds.

Today this model concept is called *loudness additivity*. Their hypothesis was that when two equally loud tones that do not mask each other are presented together, the result is "twice as loud." This method is sometimes referred to as the *indirect method*. Fletcher verified his additivity hypothesis by use of the *direct method* in which subjects are asked to turn up the sound until it is "twice as loud." A further verification of this assumption lies in the predictive ability of this additivity assumption. For example, they showed that 10 tones that are all equally loud, when played together, are 10 times louder, as long as they do not mask each other. Fletcher and Munson found that loudness additivity held for signals "between the two ears" as well as for signals "in the same ear." When the tones masked each other (namely, when their masking patterns overlapped), additivity still held, but over an attenuated set of patterns [12]. Their 1933 model is fundamental to our present understanding of auditory sound processing.

Fletcher's working hypothesis was that each signal is *nonlinearly compressed* by the cochlea, neurally coded, and the resulting cochlear nerve neural rates are added. The 1933 experiment clearly showed how loudness (i.e., the neural rate, according to Fletcher's model) adds. Fletcher and Munson also determined the *compression function* $G(p)$. Their experiment did not prove that $G(p)$ must result from the nonlinear action of the cochlea but it was consistent with it.

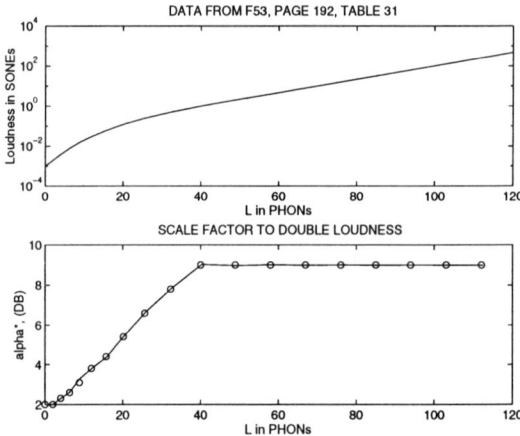

Figure 1: This figure shows the loudness growth and α^* from p. 192, Table 31 [11] as a function of the loudness level, in phons. When α^* is 9 dB, loudness grows as the cube–root of intensity. When α^* is 3 dB, loudness is proportional to intensity.

From their formulation Fletcher and Munson found that at 1 kHz, and above 40 dB SPL, the pure tone loudness G is proportional to the cube–root of the signal intensity ($G(p) = (p/p_{\text{ref}})^{2/3}$, because two equally loud simultaneous tones, are equally loud to a single tone 9 dB greater in intensity. In other words, if the pressure p is increased by 9 dB, the loudness is doubled. Below 40 dB SPL, loudness was frequently assumed to be proportional to the intensity ($G(p) = (p/p_{\text{ref}})^2$, $\alpha^* = 2^{1/2}$, or 3 dB).[2]

---

[2]The parameter α^* is the gain applied to the pressure p to double the loudness.

Figure 1 shows the loudness growth curve and α^* given in [11, Table 31, page 192]. As may be seen from the figure, the measured value of α^* at low levels was not 3 dB, but was closer to 2 dB. Fletcher's statement that loudness is proportional to intensity (α^* of 3 dB) was an idealization that was appealing, but not supported by actual results. The basic idea, and the cube–root dependence on intensity above 40 dB SPL, was first published in Fletcher (1923).

5. RECRUITMENT AND THE RATE OF LOUDNESS GROWTH

In Fig. 2 we show a normal loudness growth function along with a simulated recruiting loudness growth function. It is necessary to plot these functions on a log-log (log–loudness versus dB SPL) scale because of the dynamic ranges of loudness and intensity. The use of dB and log–loudness have resulted in a misinterpretation of recruitment. In the figure we see that for a 5 dB change in intensity at 60 dB SPL, the loudness changes by 2.22 sones in the normal ear and 0.585 sones in the recruiting ear. If we define $\beta = 10\log(I/I_{\text{ref}})$ as the relative intensity in dB SPL, then $d\beta = 10 dI/I$. Thus at a given intensity, the ratio of two slopes, defined in terms of dB SPL ($dL/d\beta$), will be the same as those defined in terms of intensity (dL/dI).

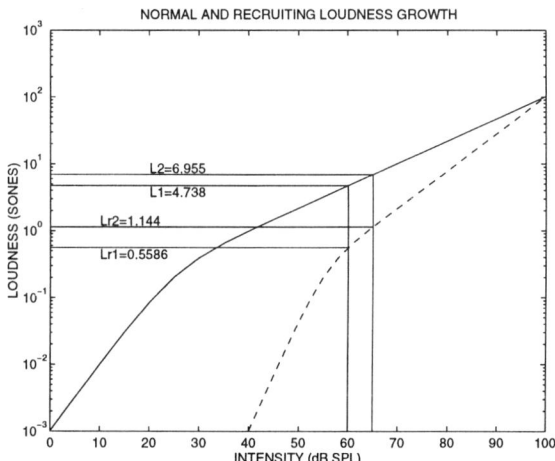

Figure 2: We show here a recruitment type loss corresponding to a variable loss of gain on the dB scale. The upper curve corresponds to the normal loudness curve whereas the lower curve corresponds to a simulated recruiting hearing loss. For an intensity level change between 60 and 65 dB, the loudness change is smaller for the recruiting ear. The belief that the loudness slope in the damaged ear is greater led to the belief that the JND in the damaged ear should be smaller (e.g., this was the rationale behind the SISI test) [20, page 160]. Both conclusions are false.

The relative loudness in the recruiting and normal ear changes by a factor of $0.585/2.22 = 0.26$ in this example. Although the visual slope of the log–loudness growth function is steeper for the recruiting ear, the computed slope in sones per watt (dL/dI) is smaller. When the loudness growth is plotted as loudness–level (i.e., in terms of phons) the computed slope is greater. Thus it is proper to define recruitment as "the abnormally rapid growth of loudness–level." In fact this is frequently how recruitment is described, in terms of loudness–level. However, the formal definition is given in terms of loudness. Thus, the misunderstanding results from confusing the phon and sone scales.

Steinberg and Gardner clearly understood what they were dealing with, as is indicated in the following quote [25, page 20]

> Owing to the expanding action of this type of loss it would be necessary to introduce a corresponding compression in the amplifier in order to produce the same amplification at all levels.

This model of hearing and hearing loss, along with the loudness models of Fletcher [12], are basic to an eventual quantitative understanding of cochlear signal processing and the cochlea's role in detection, masking, and loudness in normal and impaired ears. The work by [10] and [25], and work on modeling hearing loss and recruitment [1], support this view.

6. MULTI–BAND COMPRESSION

Ed Villchur recognized the importance of Steinberg and Gardner's observations, and vigorously promoted the idea of compression amplification. Supporting the cost of the research from his very successful loudspeaker business, he contracted David Blackmer of **dbx** to produce a multi-band compression hearing aid for experimental purposes. Using the experimental multi-band compression hearing aid, Villchur experimented on hearing impaired individuals, and found that Steinberg and Gardner's predictions were correct [27].

Fred Waldhauer, an analogue circuit designer of some considerable ability, heard Villchur speak about his experiments in multi-band compression. After the breakup of the Bell System in 1983, Waldhauer proposed to AT&T management that Bell Labs design and build a multi-band compression hearing aid as an internally funded venture. I soon joined Waldhauer in this proposed venture, which was internally funded some months later.

While Waldhauer and his team looked into new analogue circuit designs, my group did the algorithm design and hardware and software simulations of several signal processing architectures. Eventually we built a digital wearable hearing aid prototype based on a subset of these early designs. It quickly became apparent that the best processing strategy compromise was a two–band compression design that was generically similar to the Villchur scheme. With lots of help from my colleague Joe Hall of AT&T, and Patricia Jeng, Harry Levitt, and many others from CUNY, we designed a fitting procedure, and ran several field trials.

AT&T licensed its hearing aid technology to ReSound on February 27, 1987. Four AT&T people went to ReSound to continue the development. I continued to interact with them as a member of the ReSound Scientific Advisory Board. Eventually, after solving many difficult practical problems, Fred's newly formed ReSound team was successful in finalizing the analogue compression processing chip that is now the heart of the ReSound hearing aid. The chip design and the loudness–based LGOB (Loudness Growth in Octave Bands) fitting procedure [3] turned out to be the most challenging aspects of the final hearing aid.

It was, and remains, difficult to show quantitatively the nature of the improvement of MBC in a way that can convince the world. It is probably fair to say that only with the success of ReSound's MBC hearing aid in the marketplace has the world come to accept Villchur's claims. Why is this? I think the problem has to do with the two views of what MBC is and how it works. These views strongly influence how people think about compression. They are the *articulation index (AI) view* and the *loudness view*.

The AI–view is based on the observation that speech has a dynamic range of about 30 dB in 1/3 octave frequency bands. The assumption is that the speech AI will increase in a recruiting ear, as the compression is increased, if the speech is held at a fixed loudness. This view has led to unending comparisons between the optimum linear hearing aid and the optimum compression hearing aid.

The loudness–view is based on restoring the natural dynamic range of all sounds to the listener to provide the impaired listener with all the speech cues in a more natural way. Soft sounds for normals should be soft for the impaired ear, and loud sounds should be loud. According to this view, loudness is used as an index of audibility, and complex arguments about JNDs, speech discrimination, and modulation transfer functions just confound the issue. This view is supported by the theory that OHCs compress the IHC signals.

Neither of these arguments deals with important and complex issues such as the changing of the critical band with hearing loss, or the dynamics of the compression system. The analysis of these important details are interesting only *after* the signals are placed in the audible range. Imagine for example a listening situation with a soft speaker and a loud speaker, having a conversation. In this situation, the compressor gain must operate at syllabic rates to be effective. The use of multiple bands ensures that a signal in one frequency band does not control the gain in any other band.

When properly designed and fitted, MBC signal processing has proven to be *the* most effective speech processing strategy we can presently provide. The reason it works is because it supplements the OHC compressors provided by Mother Nature, which are damaged in sensory–neural hearing loss.

7. REFERENCES

[1] J. B. Allen. Modeling the noise damaged cochlea. In P. Dallos, C. D. Geisler, J. W. Matthews, M. A. Ruggero, and C. R. Steele, editors, *The Mechanics and Biophysics of Hearing*, pages 324–332, New York, 1991. Springer-Verlag.

[2] J. B. Allen and P. F. Fahey. Using acoustic distortion products to measure the cochlear amplifier gain on the basilar membrane. *Journal of the Acoustical Society of America*, pages 178–188, July 1992.

[3] J. B. Allen, J. L. Hall, and P. S. Jeng. Loudness growth in 1/2-octave bands (LGOB)—A procedure for the assessment of loudness. *J. Acoust. Soc. Am*, 88(2):745–753, August 1990.

[4] J. B. Allen and S.T. Neely. Modeling the relation between the intensity JND and loudness for pure tones and wideband noise. *Journal of the Acoustical Society of America*, 102(6):3628–3646, December 1997.

[5] J. B. Allen, G. Shaw, and B. P. Kimberley. Characterization of the nonlinear ear canal impedance at low sound levels. *ARO*, 18(757):190, February 1995.

[6] E.G. Boring. *History of Psychophysics*. Appleton–Century, 1929.

[7] William F. Carver. Loudness balance procedures. In Jack Katz, editor, *Handbook of clinical audiology, 2^d edition*, chapter 15, pages 164–178. Williams and Wilkins, Baltimore MD, 1978.

[8] Harvey Fletcher. Physical measurements of audition and their bearing on the theory of hearing. *Journal of the Franklin Institute*, 196(3):289–326, September 1923.

[9] Harvey Fletcher. *Speech and Hearing*. D. Van Nostrand Company, Inc., New York, 1929.

[10] Harvey Fletcher. A method of calculating hearing loss for speech from an audiogram. *Journal of the Acoustical Society of America*, 22:1–5, January 1950.

[11] Harvey Fletcher. Speech and Hearing in Communication. In Jont B. Allen, editor, *The ASA edition of Speech and Hearing in Communication*. Acoustical Society of America, New York, 1995.

[12] Harvey Fletcher and W.A. Munson. Loudness, its definition, measurement, and calculation. *Journal of the Acoustical Society of America*, 5:82–108, 1933.

[13] Harvey Fletcher and J.C. Steinberg. The dependence of the loudness of a complex sound upon the energy in the various frequency regions of the sound. *Physical Review*, 24(3):306–317, September 1924.

[14] E.P Fowler. A method for the early detection of otosclerosis. *Archives of otolaryngology*, 24(6):731–741, 1936.

[15] B.A. Kingsbury. A direct comparison of the loudness of pure tones. *Physical Review*, 29:588–600, April 1927.

[16] C.J. Kros and A.C. Crawford. Potassium currents in inner hair cells isolated from the guinea–pig cochlea. *Journal of Physiology*, 421:263–291, 1990.

[17] M.C. Liberman. Single-neuron labeling and chronic cochlear pathology III. Stereocilia damage and alterations of tuning curve thresholds. *Hearing Research*, 16:55–74, 1984.

[18] M.C. Liberman and N.Y.S. Kiang. Acoustic trauma in cats. *Acta Otolaryngologica*, Suppl. 358:1–63, 1978.

[19] R. Lorente de No. The diagnosis of diseases of the neural mechanism of hearing by the aid of sounds well above threshold. *Transactions of the American Otological Society*, 27:219–220, 1937. Discussion of E. P. Fowler's paper on recruitment.

[20] Frederick N. Martin. *Introduction to audiology*. Prentice–Hall, Englewood Cliffs, NJ 07632, 3 edition, 1986.

[21] J.O. Pickles. *An introduction to the physiology of hearing*. Academic Press Inc., London, England, 1982.

[22] R.R. Riesz. Differential intensity sensitivity of the ear for pure tones. *Physical Review*, 31(2):867–875, 1928.

[23] Walter Rosenblith. Sensory preformance of organisms. *Reviews of Modern Physics*, 31:485–491, 1959.

[24] Bertram Scharf. Loudness. In E.C. Carterette and M.P. Friedman, editors, *Handbook of Perception, Vol. IV*, chapter 4, pages 187–234. Academic Press, Inc., New York, 1978.

[25] J.C. Steinberg and M.B. Gardner. Dependence of hearing impairment on sound intensity. *Journal of the Acoustical Society of America*, 9:11–23, July 1937.

[26] S.S. Stevens. To honor Fechner and repeal his law. *Science*, 1961.

[27] Edgar Villchur. Signal processing to improve speech intelligibility in perceptive deafness. *Journal of the Acoustical Society of America*, 53(6):1646–1657, 1973.

A FLEXIBLE FILTERBANK STRUCTURE FOR EXTENSIVE SIGNAL MANIPULATIONS IN DIGITAL HEARING AIDS

Robert Brennan   *Todd Schneider*

dspFactory, Waterloo
Ontario, Canada, N2J 4V1

ABSTRACT

Filterbanks for digital hearing aids must use significantly different criteria than those designed for coding applications. For digital hearing aids, the filterbank channel gains must be adjustable over a large dynamic range to compensate for the hearing loss. This adjustability violates the alias cancellation properties of critically sampled filterbanks designed for coding. This paper describes a filterbank designed exclusively for hearing aid applications. Consideration will be given to the extremely limited memory, low delay and low power requirements that must be met in a typical hearing aid application.

1. INTRODUCTION

It has long been known that hearing loss is a function of both frequency and input level. A filterbank (Fig. 1) provides a natural decomposition of the input signal into frequency bands, which may be processed independently to best compensate for the hearing loss and meet prescriptive targets. Although filterbanks have traditionally been constructed using analog techniques, digital filterbanks have a great number of advantages including precise control over the phase response, which greatly facilitates signal reconstruction.

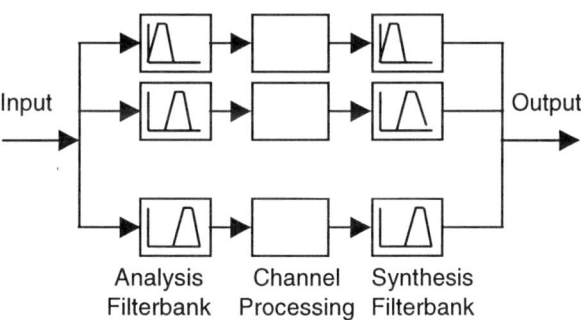

Figure 1. Filterbank signal processing

In this paper we describe an extremely flexible framework for separating the input signal into frequency bands which forms the basis for a multichannel compression hearing aid developed by the authors [1]. Filterbank design for hearing aids must address the extremely limited memory available, low delay requirement and the flexibility for accurate fitting. An oversampled, weighted overlap-add filterbank, designed to meet these requirements, will be presented in the paper.

Although non-uniform auditory critical bands are a better fit to hearing physiology, fast modulation techniques are not directly applicable. Much greater computational efficiency is obtained when the filter response shapes are realized as a series of modulations of a low-pass prototype filter covering the entire frequency range. This modulation produces identical filter shapes resulting in a uniformly spaced filterbank. Although a greater number of bands are needed to achieve sufficient resolution at low-frequencies, this additional computational expense is more than compensated for by the use of a fast modulation technique.

To achieve a further improvement in low-frequency fitting, an *even/odd* stacking modification was implemented which allows a selectable shift of the filterbank center-frequencies by one-half band (Fig. 2). This doubles the number of potential band edges compared to the number of bands. This filterbank has been implemented and operates on a Motorola DSP56301-based portable platform. It is suitable for low-power, real-time operation.

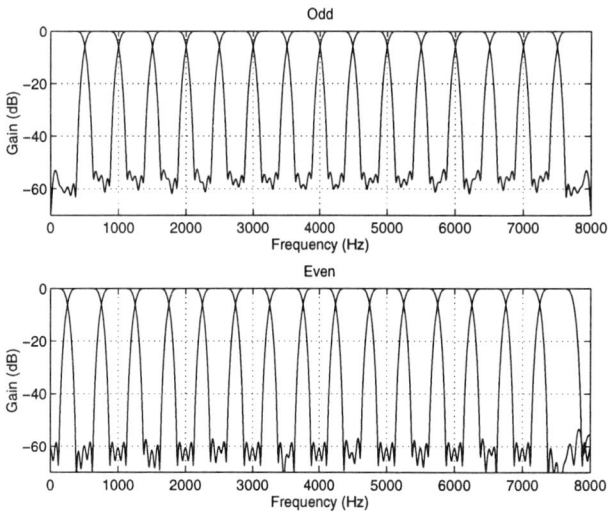

Figure 2. Frequency response of filter bank channels for odd and even channel stacking arrangements

2. CRITERIA

2.1 Coding Background

A filterbank decomposes the input signal into a series of separate frequency bands. By using designs that minimize the overlap between adjacent bands, the resulting representation is approximately orthogonal. It is natural that filterbanks have found extensive use in the low bit-rate coding of speech and image signals and have been heavily optimized for this case. The MPEG coder is an example used for high fidelity speech coding [2, 3].

For sub-band coding purposes, reliance is placed on the fact that typical spectra are not flat. This enables the coder to allocate more bits to the more perceptually important high-energy regions and fewer bits to the less perceptually important low energy regions of the spectrum.

At first glance, an M-band filterbank would appear to increase the data rate because the single input data stream has been split into M separate bands generating M times more data. It is possible, however, to *decimate* the data streams because of their reduced bandwidth. In the important case of *critical* sampling, only 1 in M samples is used (for an M-band uniform filterbank) in each band. The total data rate is thus unchanged through the filterbank, which is intuitively satisfying since no new information was added. The problem is that overlap between adjacent bands results in the generation of *aliasing* distortion, since any residual response in the adjacent bands is folded back into the original band by the decimation procedure.

In general, the most concise alias-free representation is only obtained if the filter bands directly abut each other with complete frequency coverage and no overlap. Although this is impossible, developments have led to filterbanks with slightly overlapping bands designed in such a way that aliasing distortion generated in the analysis stage is exactly canceled by imaging distortion in the synthesis stage [4, 5].

In the absence of coding (quantization of the bands), these filterbanks produce transparent results. Under increasingly coarse quantization, the effective gain changes (from the quantization step) result in the imaging distortion not completely canceling the aliasing distortion. In practice, this does not degrade the perceptual audio quality because the coarsely quantized bands are lower in energy and the noise floor masks the uncanceled aliasing distortion.

2.2 Hearing Aid Application

For hearing aid use, the frequency splitting is performed for the purpose of modifying the spectral shape of the input signal. Hearing aid fitting typically requires a wide gain adjustment range. In a compression system, the input signal level, which can be measured as the overall level, channel level or a combination, controls these gains [1].

Given the requirement for wide gain adjustment, the alias cancellation theory is invalid and critical sampling is insufficient. This problem necessitated the development of an oversampled filterbank. Although oversampling increases the data rate, it is the price that must be paid for gain adjustability without aliasing.

In a compression system, gain changes are dynamic. This may cause anomalies in the overall frequency response if phase differences exist between adjacent bands. To avoid these undesirable frequency response notches or peaks at the band edges (which frequently occur in analog systems), it is necessary to constrain the filter channel impulse responses to be linear phase and of equal delay.

Thus, an ideal filterbank for a hearing aid application would: (1) allow precise fitting of prescriptive targets, (2) have short delay (3) be computationally efficient and (4) use a minimal amount of memory.

3. OVERSAMPLED WOLA FILTERBANK

The criteria mentioned in the latter part of the last section are contradictory. For example, better filter responses are obtained only at the expense of delay. Both cannot be simultaneously satisfied. Thus, the filterbank is an engineering compromise best arrived at through careful consideration of the performance parameters.

Initial experiments were conducted with critically sampled filterbanks. As mentioned before these filterbanks achieve good performance only in cases with mild channel gain changes. To achieve high-quality reproduction under widely varying gain changes, it was necessary to oversample by at least a factor of two to reduce the level of uncanceled aliasing generated when band gains differ greatly.

The selected design uses an oversampled, weighted overlap-add (WOLA) DFT filterbank [6, 7, 8] to split the input signal into 16 frequency bands. This filterbank uses modulation via the DFT to replicate a single prototype filter into 32 complex filter bands. This modulation produces identical filter shapes and results in a uniform filterbank. At a sampling frequency of 16 kHz, the resulting bands are 500 Hz wide. Total computational complexity is 46 multiply-accumulates per output point. Group delay is 12.5 ms.

To achieve sufficient frequency resolution at low frequencies, a uniform filterbank requires a larger number of bands than would be required by a non-uniform filterbank. Fortunately, the DFT method described above generates a large number of bands at low computational expense. This large number of bands is required to achieve a good fit to audiometric data at low frequencies (which is normally given on a log-frequency scale). As a result of the linear frequency spacing and large number of channels, the high frequency band spacing may be greater than necessary. Thus, it may be advantageous to group bands at high frequencies for gain adjustment purposes.

It is important to realize that better frequency resolution is available only at the expense of greater signal delay for *any* filterbank method. Long delays are not desirable–delays of 6-8 ms are reported to be just noticeable. Delays longer than 20 ms may cause interference between speech and visual integration [9]. Clearly, less delay is better.

For audiometric fitting, a useful compromise is available in the form of the *even* or *odd* stacking of the filterbank (Fig. 2). Effectively, this procedure doubles the effective length of the FFT to 2*N* points but selects only half the number of bins.

Mathematically, the forward and inverse even and odd DFT transform pairs (Equations 1 and 2) are given by:

1) $$Y_k = \sum_{n=0}^{N-1} y(n) W_N^{-(k+v)n}, k = 0,1,\ldots,N-1$$

2) $$y(n) = \frac{1}{N}\sum_{k=0}^{N-1} Y_k W_N^{(k+v)n}, n = 0,1,\ldots N-1$$

Where

$$W_N = e^{-k(2\pi/N)}$$

FFT algorithms were developed for both cases. For even stacking, ν is zero and the conventional *N*-point FFT may be used. For odd stacking, ν is 0.5 and the conventional FFT must be extended to be an odd FFT [10]. This additional choice for ν doubles the number of bin locations without requiring a longer analysis and allows band edges to be placed within one-half the width of the filterbank channel.

4. FILTERBANK OPERATION

4.1 Overview

The frequency response of the analysis window (i.e., the prototype low-pass filter) is modulated by the odd (even) FFT to produce channel responses as illustrated in Fig 2. The individual channel signals are decimated by *N/OS* where *N* is the FFT size and *OS* is the oversampling factor. It is critical that the analysis lowpass filter be sufficiently sharp to minimize the aliasing distortion generated by the decimation step.

The spectral shape of the input signal is modified at this point by applying suitable gains to the frequency channel signals. This is followed by the corresponding inverse odd (even) FFT, interpolation, synthesis window weighting and the overlap-add procedure. This window (a low-pass filter which is the counterpart to the analysis window), minimizes the spectral imaging distortion created during the interpolation step.

To conserve memory, it is highly desirable that the synthesis window be derived from the analysis window. Fortunately, the oversampling of the channel signals relaxes the requirements on the synthesis window. The channel images are spaced at intervals of *OS* rather than 1 as in critical sampling. Since the synthesis window low-pass function need only reject images *OS*-channels away it can be set as a decimated version of the analysis window. The decimated window also has the advantage that synthesis delay (half the synthesis window length) is significantly reduced.

4.2 Analysis/Synthesis Window Design

Both time and frequency domain constraints must be placed on the analysis and synthesis windows. For a WOLA DFT filterbank with *N/2* frequency bands (i.e., one that uses an N-point FFT) and band outputs that are decimated by *N/OS* (*OS*=2 was used above), *time domain constraints* must be placed on the analysis window coefficients such that a zero appears every M_1 samples (where M_1=*N*) [6]. The synthesis window must have zeros every M_2 samples where (M_2=*N/OS*). For *OS*=2, this constraint (i.e., a synthesis window with zeros at half the spacing of the analysis window) can be met by decimating the analysis window coefficients by a factor of two.

Frequency domain constraints must be placed on the combined analysis/synthesis window frequency response such that both windows are "good" *M*-band filters. The analysis window must have a cutoff frequency of π/M_1 and the synthesis window must cutoff at $\pi/M_2 = 2\pi/M_1$. This constraint can be met by designing an analysis window, that does not "droop" at half it's cutoff frequency (i.e., $\pi/2M_1$), which becomes the cutoff frequency of the (decimated) synthesis filter.

A recently developed method of designing *M*-band filters that allows simultaneous time and frequency constraints to be placed on a design is the eigenfilter approach [11]. This approach was used to design the combined analysis/synthesis window.

4.3 Analysis

An illustration of the analysis portion of the filterbank is shown in Fig. 3. The main sequence of events is

- Read *R* input block samples
- Read sign from analysis sign table at the sign table pointer
- Apply sign to samples
- Circularly increment input sign table pointer
- shift input FIFO and add *R* new samples
- Apply window and time-fold to *N* samples
- Apply circular shift of (*n* mod *N*)
- Take *N*-point FFT (even or odd)
- Apply channel gains to (complex) frequency data

4.4 Synthesis

The schematic of the synthesis operations is shown in Fig. 4. The operations are

- Take inverse FFT of (complex) input (even or odd)
- Apply circular shift of (*n* mod *N*)
- Periodically extend to L/DF samples
- Apply synthesis window
- Accumulate into output FIFO, shift out *R* samples
- Read sign from synthesis sign table at the sign table pointer

- Apply sign to the *R* shifted out samples
- Circularly increment sign table pointer to next sign value
- Circularly increment *n* (mod *N*) by *R/OS*

Where:

- *R* is the input block size
- *L* is the input window size
- *N* is the FFT size
- *DF* is the decimation factor
- *OS* is the oversampling factor

The sign table contains the sign factors in a repeating pattern (+1, +1, -1, -1, +1, +1, -1, -1) used to modulate the input sequence and the output sequence in the case of odd stacking. For even stacking, the sign factors are all +1.

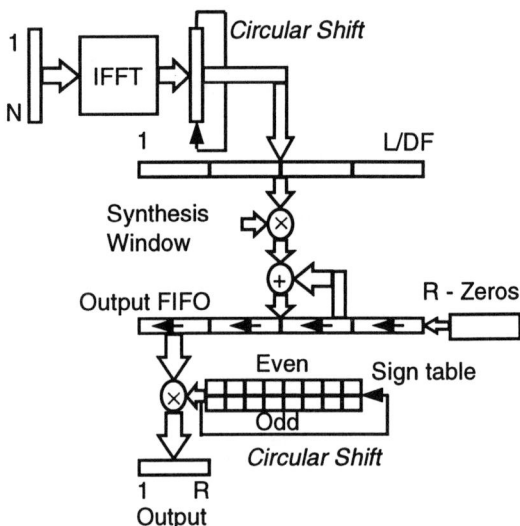

Figure 4. Schematic of filterbank synthesis

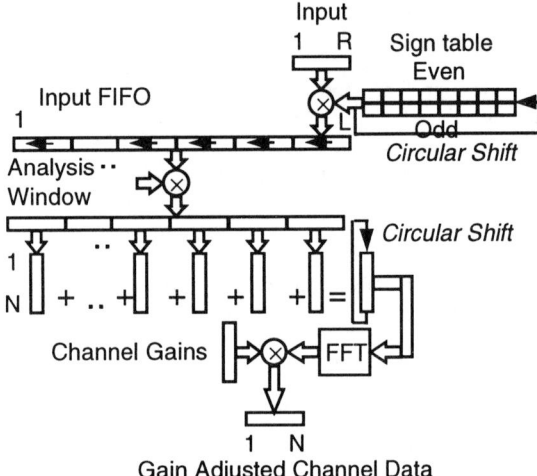

Figure 3. Schematic of filterbank analysis

5. CONCLUSIONS

A highly flexible filterbank structure has been described in this paper. Several tradeoffs have been made to make this a practical filterbank that can meet the requirements of a hearing aid application.

The filterbank structure is *M*-band with uniform bands. This structure provides many advantages in a hearing aid context. To conserve memory, only a single analysis window is stored. A DFT is used to modulate and replicate this lowpass prototype. To further reduce memory, the synthesis window is created by decimating the analysis window, subject to time and frequency domain constraints that can be satisfied by using an eigenfilter design method. This filterbank was implemented and runs in real-time on a Motorola DSP56301-based portable system. It has been used successfully for a real-time compression system implementation [1].

6. REFERENCES

[1] Schneider, T., Brennan R.L., "A Multichannel Compression Strategy for a Digital Hearing Aid," *Proc. ICASSP*-97, Munich, Germany, pp. 411-415

[2] Pan, D.Y., "A Tutorial on MPEG/Audio Compression", *IEEE Multimedia Magazine*, Summer 1995, pp. 60-74.

[3] Pan, D.Y., "Digital Audio Compression," *Digital Technical Journal*, Vol. 5., No. 2, Spring 1993, pp. 28-40.

[4] Chu, P.L., "Quadrature Mirror Filter Design for an Arbitrary Number of Equal Bandwidth Channels." *IEEE Trans. on ASSP*, Vol ASSP-33, No. 1, 1985, pp.203-218..

[5] Rothweiler, J.H., "Polyphase Quadrature Filters – A New Subband Coding Technique," *Proc ICASSP-83*, Boston, MA, pp. 1280-1283.

[6] Crochiere, R.E. and Rabiner, L.R., *Multirate Digital Signal Processing*. Prentice-Hall Inc., 1983.

[7] Vaidyanathan, P.P., "Multirate Digital Filters, Filter Banks, Polyphase Networks, and Applications: A Tutorial," *Proc. IEEE*, Vol. 78, No. 1, pp. 56-93, January 1990.

[8] Vaidyanathan, P.P., *Multirate Systems and Filter Banks*. Prentice-Hall Inc., 1993.

[9] Agnew, J. "An Overview of Digital Signal Processing in Hearing Instruments," *The Hearing Review,* July 1997.

[10] Bellanger, M., *Digital Processing of Signals*. John Wiley and Sons, 1984, pp. 82-89.

[11] Vaidyanathan, P.P. and Nguyen, T.Q., "Eigenfilters: A new Approach to Least-squares FIR Filter Design and Applications including Nyquist Filters," *IEEE Trans. on Circuits and Systems*, Vol. CAS-34, No. 1, January 1987, pp. 11-23.

MULTIBAND COMPRESSION HEARING AIDS: DEVELOPING A PERFORMANCE METRIC

Jon C. Schmidt* and Janet C. Rutledge**

*Starkey Laboratories, Inc., 6600 Washington Ave. So., Eden Prairie, MN, 55344
**Division of Otolaryngology-HNS, Univ. of Maryland Medical School, 16 S. Eutaw St., Baltimore, MD, 21201

ABSTRACT

When attempting to compare different compression algorithms used in multiband compression hearing aids, or different parameter settings of a given compression algorithm, there is a need to compare equivalent amounts of compression. Since there are several parameters that influence the actual amount of compression that occurs, it is desirable to know which set of parameters will create the least perceptual difference from the original signal for a given amount of compression. Unfortunately, there is not a well defined approach for determining the amount of compression that has been imparted to a signal. Two degree of compression metrics are proposed that can be used to determine the amount of compression by direct analysis of the audio signal before and after compression. These metrics give excellent results in predicting the ratings of subject-based testing on the audio quality of several different audio segments across several parameter variations.

1. INTRODUCTION

The dynamic range compression of audio signals is characterized by a standard set of compression parameters: an I/O curve consisting of one or more thresholds and compression ratios, and attack and release time constants. While these parameters are a good first order approximation of the signal processing that is involved, they are by no means a complete characterization. Interactions between these parameters and the exact mechanisms by which the temporal parameters are implemented have a significant effect on the actual compression behavior applied to real-world, non-sine tone, signals [1]. Moving to a multiband compression algorithm adds even more implementation-specific influences. When attempting to compare the performance of competing compression algorithms, one cannot simply set a few key parameters to similar settings and assume that similar compression will occur. To pursue these important comparisons, a metric must be used that quantifies the dynamic variation of an audio signal.

2. PEAK/RMS RATIO

The peak/RMS ration is frequently used to describe the variation in the dynamics of a signal. A lower ratio corresponds to less dynamic variation. This can be a misleading figure if it is applied to perceptual loudness variations. Table 1 shows the peak/RMS ratios of a source signal, a single band compressor output, and a multiband compressor output [2]. Notice that while the overall peak/RMS ratio (top line labeled "Sum") is considerably lower for the single band compressor, for every frequency band the multiband approach yields a lower peak/RMS ratio. As shown in Figure 1 the multiband compressor produces an output signal that perceptually has less dynamic variation, but in fact looks and sounds more similar to the original signal.

Table 1. Peak/RMS ratios for a source signal and two compressor outputs: single band and multiband.

Band	Source	Single	Multi
Sum	10.23	4.76	7.83
2	8.74	6.90	4.94
4	9.33	5.33	4.97
6	6.80	5.87	4.37
8	5.46	5.26	3.97
10	7.31	6.59	4.74
12	7.18	5.90	4.74
14	10.01	6.91	5.76
16	8.34	6.59	5.49
20	14.71	13.07	8.00
24	12.75	16.05	7.86
28	17.84	19.74	10.30

3. DEFINING A DEGREE OF COMPRESSION METRIC

For applications where perceptual concerns are important, a degree of compression metric should incorporate frequency bands, possibly as narrow as critical bands. The metric should also consider the signal in a time-segmented manner since the perceived loudness of a sound depends not only on the absolute level of the signal but on how long that level is maintained. Numerically this involves calculating the RMS level of the signal over some short window of time. The calculations are performed across time then frequency, or across frequency then time. With either approach, the key issue is the numeric expression of the dynamic variation of the signal. Ultimately, the goal is to characterize the entirety of the dynamic variation of a signal with a single number.

When working along the time divisions, two different calculations are used to characterize the dynamic variation. The first calculation is to divide the maximum level by the mean level. While the emphasis of a single peak in the waveform has been removed by using the RMS level of a short time window, this dynamic variation calculation may still be heavily influenced by a single time segment having an unusually large level. The second calculation is the standard deviation of the set of levels. A histogram would truly characterize the dynamic behavior of a signal, and the standard deviation is the best single value to represent the distribution.

Each metric also has several parameter variations. The length of the time window over which the RMS level is calculated has four possible lengths: 10, 20, 35 and 60 msec. The level calculations include the optional use of logarithms, square roots, or cube roots. When summing values across the frequency bands, six different weightings are used. These weightings are: 1) Unweighted: each band is given equal importance; 2) Loudness Contours: the weights are based on the ear's sensitivity at 65 phons; 3) Maximum Level: the maximum of the windowed RMS levels for each band of the original uncompressed audio signal; 4) Log Maximum Level: log of weighting #3; 5) Mean Level: the mean of the windowed RMS levels of the original uncompressed audio signal; and, 6) Log Mean Level: log of weighting #5. Several additional metrics using other dynamic variation approaches but having similar parameter variations were also developed and evaluated.

The final result of each degree of compression metric is a single number showing the relative closeness of a compressed signal to the original uncompressed audio signal. The compressed signals with lower metric values should have undergone less compression and should sound more similar to the original signal. The compressed signals with higher metric values should have undergone greater compression and should sound less like the original signal.

4. EVALUATING THE METRICS

A multiband compression scheme with 28 bands was used to generate a set of 12 compressed stimuli [2]. There were two variations in the level estimate/time constant and six variations in the I/O curve. The I/O curve had three segments, a linear portion, a compressive portion, and a limiting portion. The limiting portion was set to two different levels, and the threshold/compression ratio was varied as a pair for three different configurations. Four different audio segments were used: jazz, vocal/guitar

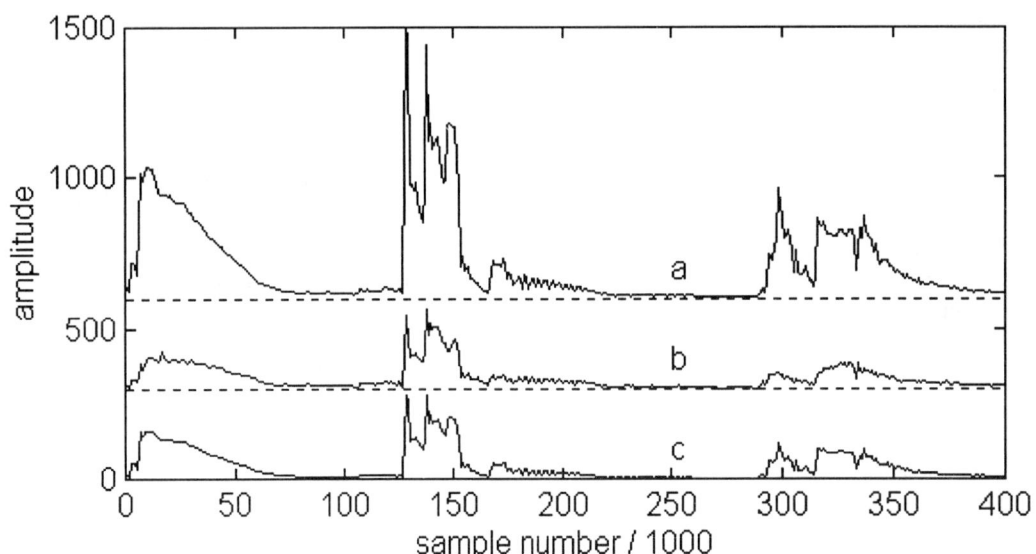

Figure 1. Acoustic guitar amplitude envelopes for frequency band from 790-935 Hz: a) original signal, b) single band compression, c) multiband compression. The amplitudes are offset for visual clarity of the waveforms.

(James Taylor), classical (Bartok), and simultaneous female and male speakers. The 20 subjects were not known to have any hearing impairments and were from one of three groups: professional recording engineers, students in music related fields, or students in other fields. The tests consisted of paired comparisons where the subjects were asked to rate the similarity of each pair of stimuli on a scale from 50 to 100, with 100 being most similar.

The subject testing data was used to place the stimuli on a quality scale with the distances between the stimuli corresponding to the subject's scoring. A quality scale was also generated for each metric. An average score across all subjects was generated for each audio segment and was used as the control score. Each individual subject quality scale was rated against the control score, and outlying subjects were removed.

Figure 2 shows the control score for the jazz segment and the quality scale for one variation of metric 21. The uncompressed stimulus is seen on the left labeled as 0. The compressed stimuli are noted by the values 1-12. Note that there is a strong similarity between the human subject judgements and the objective metric results. The primary differences are the position of stimulus 1 and the order of the 8,9,10 grouping.

Each metric, with all parameter variations, was rated against the control score for each audio segment. The performance of some of the metrics is shown in Table 2. The letter grades used for the metrics are: A: < 15, B: 15 - 20, C: 20-25, D: 25-35, F: > 35, where lower scores are better. For comparison, the average rating of the human subjects against the control score for each audio segment was 20.4 for jazz, 21.8 for James Taylor (JT), 22.0 for Bartok, and 15.0 for speech.

There are several instances where a metric performs quite well on one audio segment but quite poorly on others. This demonstrates the importance of using multiple well chosen audio segments. A useful metric will need to perform well across a range of audio segments. As shown in Table 2 this is the case with certain variations of metrics 21 and 13, which are clearly the best metrics. Metric 21 uses the standard deviation across the time windows after summing across the frequency bands, and Metric 13 uses the maximum RMS level divided by the mean RMS level within each band prior to summing across frequency bands.

While some weights are obviously better for a specific metric, different weights work well for other metrics. The data is quite scattered for the metrics using the logarithm of the levels. Some metrics achieve better results using the logarithm, other metrics are strongly the opposite, and for others this factor is not critical. The influence of window length also varies from metric to metric. For many metrics window length is not a factor.

Metric 21 works best when the logarithm of the levels is not used. There is a significant degradation to performance, especially on the JT segment, when the logarithm of the levels is used. Using the cube root or square root of the RMS level reduces the effectiveness of the metric. It performs best when using the shorter time windows, and it achieves nearly identical ratings using weightings 1, 4 and 6. It is the only metric to achieve a superb score across all

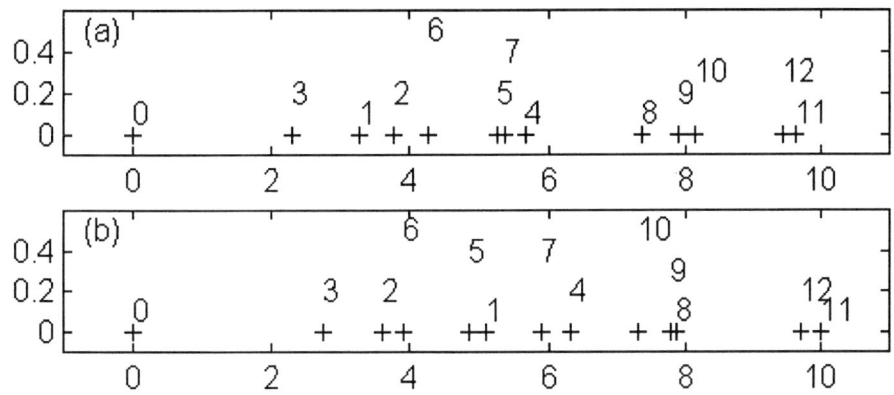

Figure 2. Quality scales for the jazz segment. (a) The control score derived from the average human subject quality scale. (b) The quality scale for Metric 21, window = 20 msec, no logarithm, weighting = 1.

four segments, and the nineteen best variations of this metric have an average rating that is better than 15.

Metric 13 performs best when using the logarithm of the levels, longer windows, and weightings 3 and 5. The performance drops fairly quickly outside of these parameter variations. This metric scored superbly on three of the segments, with the score on the Bartok segment just barely out of the superb range for a few of the variations.

5. CONCLUSIONS

On the strength of average rating, and consistency across all four audio segments, metric 21 would earn the ranking of best metric. The consistency across all four audio stimuli cannot be over emphasized. A valid, useful metric must perform well independent of the audio signal that is being compressed.

Based on several criteria, the four audio segments used in this research effort achieved the desired property of being different types of audio stimuli. First, the subject testing revealed a different stimuli distribution for each of the four segments. Second, the metrics also produced different distributions for each of the segments. Finally, many of the metric variations performed well on some audio segments while performing quite poorly on other audio segments. There was no common pattern in this behavior.

The differences between a metric and the subject rating are equally important. Assuming that a metric has validity, a stimulus that is found to be more compressed than another stimulus, yet ranks higher on the criteria, gives us direction on how to best set compression parameters. The goal of compression is to provide the least perceptual difference from the original for a specified amount of compression. If the compression parameters can be set so that more compression is achieved, yet the output sounds more similar to the original signal, a strong step towards this goal has been achieved.

A valid degree of compression metric, such as either of the two presented here, can be used to generate meaningful, quantitative comparisons of the compression imparted to an audio signal by any signal processing algorithm. These metrics can be useful tools for research comparing compression algorithms and determining optimal parameter sets for compression algorithms

6. REFERENCES

[1] Kates J.M. "Optimal Estimation of Hearing-Aid Compression Parameters". *Journal of the Acoustical Society of America*, 94:1-12, 1993.

[2] Schmidt J. C. and Rutledge J. C. "Multichannel dynamic range compression for music signals". *Proceedings of the IEEE International Conference on Acoustics, Speech and Signal Processing*, 2:1013-1016, 1996.

Table 2. Metric Performance. Window, Log, and Weight are parameter variations within the metric. Jazz, James Taylor (JT), Bartok, and Speech are the four audio stimuli. The more similar a metric's quality scale is to the control score, the lower the score and the higher the grade generated by the rating procedure. "Average" is the average score across the four stimuli.

Metric	Window	Log	Weight	Jazz	JT	Bartok	Speech	Average
13	longer	Yes	5,3	A	A	A-	A	13-17
13	longer	Yes	6,4	A	A-	C	A	14-21
13	longer	No	5,3	B-	B	B-	C-	19-30
13	longer	Yes	1,2	B-	D-	F	A-	23-25
13	longer	No	6,1,2,4	D-	F	F	B-	30-45
21	shorter	No	1,4,6	A	A	A	A	12-13
21	shorter	No	2>5,3	A-	B-	B	A	14-19
21	longer	Yes	4,1,6,2	C	F	A-	B	23-25
21	longer	Yes	3,5	D-	F	A	B	24-27
12	shorter	No	6,4>5	A	A	B	D	16-17
23	NA	No	3,5	C	F	A	A	22-23
26	mid	No	3	B	A	F	A	22

MULTICHANNEL COMPRESSION IN THE NORMAL EAR AND AS A SIGNAL PROCESSING ALGORITHM FOR THE HEARING IMPAIRED

E. William Yund

Sensorineural Hearing Loss Research Laboratory
Veterans Affairs Northern California Health Care System
150 Muir Road
Martinez, CA 94553-4612, USA

ABSTRACT

The nature of sensorineural hearing loss (SNHL) indicates that full-range multichannel compression (FRMCC) signal processing will help the hearing-impaired. Our recent studies of speech perception in hearing impaired subjects support the value of FRMCC with at least 8 channels, especially when the signal-to-noise ratio (S/N) is low. Results from other laboratories, however, have been less favorable. The particular acoustic conditions used in these experiments, plus the restricted time subjects have to acclimatize to each signal processing algorithm, seem to account for the differences among studies, but field testing of FRMCC is needed. Developments in digital signal processing (DSP) have made it possible to plan extensive field tests of binaural 8-channel FRMCC. Hearing-aid users will be able to evaluate the FRMCC in all the acoustic environments they normally encounter, both during and after full acclimatization to the signal processing.

1. INTRODUCTION

The purpose of this report is to provide a simple model for the common SNHL that comes with age and/or noise exposure and to discuss the signal processing algorithm that is essentially the inverse of this model of hearing loss, FRMCC. There is a triple purpose in discussing SNHL in this way: (1) to provide a good first approximation description of the hearing loss, (2) to describe FRMCC and the logic behind its development, and (3) to discuss the problems of testing a signal processing algorithm for application to hearing aids. A critical question in this context is why experimental tests of FRMCC have not more clearly demonstrated its value for improving speech perception in hearing-impaired subjects. Although we should not expect FRMCC to convert the hearing-impaired person into a normal-hearing person, we would expect FRMCC to be better than other amplification schemes whose design seems to ignore the nature of the hearing impairment. This mismatch between expectation and observation suggests that there are problems either in the understanding of SNHL, in the experimental methods, and/or in our interpretation of the results. Resolving problems in any of these areas is important for applying FRMCC, as well as other signal processing algorithms, for the benefit of hearing-impaired patients. A brief discussion of the nature of speech-perception problems in SNHL is also included to facilitate the application of signal-processing solutions to these problems of speech perception.

2. SNHL and FRMCC

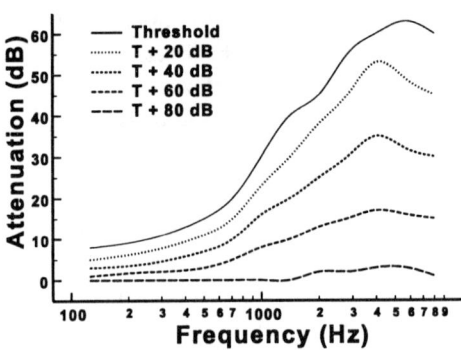

Figure 1: Variation of hearing impairment with frequency at different response levels.

2.1 SNHL Model

Figure 1 illustrates a simple model for SNHL that can be useful in understanding the complexity of designing a hearing aid to compensate for the loss [1][2][3]. This particular example is a very common type of loss among hearing-aid users, a sloping high-frequency hearing loss. The impairment is modeled as attenuation located prior to the transduction of sound energy to neural response, with the attenuation at threshold response given by the top, solid curve in the figure. The curve shows the attenuation required at each frequency to give a normal-hearing subject this threshold impairment, ranging from less than 10 dB at 125 Hz to over 60 dB at 4-8 kHz. The values plotted in this threshold curve are the same as those of the standard audiogram, but the shape of the curve would be inverted because higher threshold values are usually lower on the ordinate in the audiogram. None of the other curves correspond to measurements made in the typical audiological examination and the methods by which they are measured will not be discussed in detail here.

As the level of the response increases (successively lower curves in the Fig. 1), the required attenuation to yield a normal response not only decreases, but decreases differentially at different frequencies. Thus, although the sensitivity of the system to different frequencies varies by over 50 dB at the lowest response level (top curve), the sensitivity is near normal and varies across frequency by less than

5 dB at the highest response level (bottom curve). In effect, there is a hearing loss *only at high frequencies and low intensities*; at high intensities and/or at low frequencies, the response magnitudes are essentially normal (0-20 on this scale corresponds to the range of thresholds for normal hearing). In a different type of patient who had a greater threshold elevation at low frequencies, there would be a low-intensity hearing loss at *all* frequencies but the high-intensity responses would remain essentially normal at all frequencies.

A further complexity not illustrated in the figure is the frequency specificity of the attenuation: At frequency differences around 10%, attenuation is essentially independent. Thus, the normal response to low-frequency components of a complex sound, and the severely attenuated response to high-frequency components of the sound, can occur at the same time.

Of course, this model should not be taken completely literally because the hearing loss is not the installation of a complex attenuator, but rather frequency-specific damage to a bank of narrow-band compressive amplifiers—the Outer Hair Cells (OHCs) of the cochlea (see [4] for review). Although we did not know about OHC function when Villchur first worked on this model of hearing impairment [1], we now know that the model would be more literally correct if the label on the ordinate of Fig. 1 were changed from "Attenuation" to "Loss of Amplification". But changing the label on the ordinate, would not change the problem of hearing-aid design. To compensate for this hearing loss, the amplification of each component of sound must depend on frequency and on the intensity at that frequency, but not on the intensities at other frequencies.

One final consideration in this model for SNHL is the time constant of the attenuation variation or, perhaps more correctly, of the compression of the normal OHCs. The time constant is of interest here, primarily to help determine the parameters of the FRMCC or other signal processing that might be able to compensate for the reduction or loss of OHC function. In this context, the OHC-compression time constant is very short, allowing virtually instantaneous adjustment, but apparently without the distortions a compressive amplifier with such short time constants would produce in the auditory frequency range. The position of the OHC within narrow-band subsections of the sound-energy-to-neural-response transduction process may permit these seemingly contradictory properties. In an external signal processing algorithm, however, time constants less than a few ms should produce too much distortion. From the speech-signal perspective, time constants shorter than a few ms also would not be required: Time constants of a few ms will be short enough to permit amplification adjustment of the various frequency bands of single phonemes (vowels or consonants) as they occur in the flow of speech—presumably giving the hearing-impaired listener as good a chance as possible to identify each phoneme.

2.2 FRMCC

The model of hearing loss summarized in Fig. 1 has motivated research on FRMCC, primarily because FRMCC is the signal-processing inverse of this SNHL model. In FRMCC, the input signal is divided into a number of frequency bands and then the signal within each band is amplified with maximum amplification at the lowest intensity and gradually decreasing amplification as the intensity increases. In a true full-range algorithm, the amplification at normal-hearing threshold in each frequency band is enough to bring the intensity of that frequency band up to the impaired threshold and then the amplification is reduced by a constant factor (the compression ratio) until there is no amplification at some high level, e.g., the maximum comfortable intensity in that band. The effect in each frequency band is to map the range of intensity available in normal hearing into the range of intensity remaining in the hearing impairment.

Applying compression throughout the intensity range, as in full-range compression, has two advantages: (1) It corresponds to our understanding of the deficit, as illustrated in Fig. 1. And (2), it permits the use of smaller compression ratios, so that the compression itself will not interfere with speech perception [5]. In contrast to full-range compression, compression limiting uses high levels of amplification with no compression at low intensities and large compression ratios at high intensities, turning our understanding of the hearing impairment upside down and limiting the range of potentially useful intensity information that can be transmitted to the hearing-impaired listener. The rationale for compression limiting has been that *any* compression would be devastating to speech perception, and thus that its use should be restricted as much as possible. This rationale is weakened by failed attempts to find evidence that mild compression interferes with speech perception (see [5] for further discussion).

In his pioneering studies of 2-channel FRMCC, Villchur [6] did not attempt to measure attenuation functions, as in Fig. 1, for individual hearing-impaired subjects and later experiments confirm the lack of need to do so. Barfod [7] measured equal-loudness contours across frequencies for his subjects and concluded that growth-of-loudness functions at each frequency could be adequately estimated as straight lines from measured thresholds. Intensity-response functions measured with an entirely different method (calculated from dichotic pure-tone pitch interactions [8]) were equally consistent with straight-line functions (see also [9] for a brief description of a comparison of FRMCC based on measured-curve and straight-line intensity-response functions). In sum, FRMCC should be beneficial for hearing-impaired patients, even without extensive, precise measurements of suprathreshold auditory sensitivity.

3. EXPERIMENTAL DIFFICULTIES

Although the logical argument in favor of the use of FRMCC in hearing aids is very strong, it has been quite difficult to obtain supporting experimental evidence for the efficacy FRMCC. This experimental difficulty is of general interest in the context of signal processing in hearing aids because most of the specific problems are not unique to testing FRMCC signal processing. One set of problems stems from the scientific preference and/or the technological necessity to test complex signal-processing algorithms under a limited range of well-controlled laboratory conditions. Such limited-range testing may be particularly ineffective for an algorithm like FRMCC because its principal logical advantage is that it will maintain near-optimal performance in the broadest range of rapidly changing acoustic conditions. Furthermore, any chosen set of test conditions may have an unintended bias for or against a signal-processing algorithm. Indeed, such an unintended bias may have had a major effect on research and development of FRMCC with more than a few channels.

3.1 FRMCC, Number of Channels, and S/N

The first major study of FRMCC that included a large number of channels [10] compared the speech recognition of hearing-impaired subjects with 16-channel compression and frequency-shaped linear amplification. The speech was presented at high signal-to-noise ratios (S/Ns) and at high intensities (so that frequency components of the speech remained above threshold as much as possible for the linear amplification). Under those conditions, speech recognition with the 16-channel algorithm was a little worse than that with linear amplification. In contrast, major studies of 2-channel compression [11][12][13] have used different methods, where key measurements are made at low S/N, and the results show 2-channel compression to be somewhat better than linear amplification. Together, these result had been interpreted as evidence against the efficacy of FRMCC with so many channels [4][14][15].

Our more recent results [16] provide evidence for an alternate explanation. The same 8-channel FRMCC that is only almost as good as frequency-shaped linear amplification at high S/N is far superior to that linear amplification at low S/N, suggesting that the S/N difference between the 16- and 2-channel experiments may account for *all* of the performance difference seen across the experiments. Indeed, other results suggest that the 16-channel algorithm would be superior to 2-channel algorithms at low S/N. Direct comparisons of FRMCC with 4, 6, 8, 12, and 16 channels demonstrate improved performance up to 8 channels and the same performance for FRMCC with 8 and 16 channels [9]. Of course, problems can arise in comparing algorithms across studies based on a single parameter of the processing. For example, there are a number of studies of multichannel compression (MCC) algorithms that are not FRMCC algorithms. The critical difference between 6-channel compression-limiting MCC and 8-channel FRMCC is probably not in the number of channels.

3.2 Frequency-Shaped Linear Amplification

Another problematic aspect of limited-range testing concerns the advantage it gives to frequency-shaped linear amplification—a common control-amplification against which a new algorithm is evaluated. Accurate prior knowledge of the stimuli to be presented permits the experimenter to optimize the level of amplification at each frequency *for these particular stimuli*. Indeed, if the frequency spectrum and intensity of stimuli are very limited (e.g., one voice at one intensity, as in [7]), then the simple linear amplifier gains an almost-FRMCC-like ability to adjust to the different stimuli across experiments. Of course, the FRMCC algorithm itself gains nothing from limited stimulus variation, because it adjusts its amplification continuously, independent of long-term stimulus similarity.

The experimenter's adjustment of the linear control amplification offers an explanation for its excellent performance at high S/N, where nothing but an appropriate frequency response prevents the linear amplifier from transmitting most of the speech information, at intensities that are above the listener's elevated thresholds. The fact that this frequency response and level of amplification would be inappropriate for other stimuli, is not a problem here. In the case of FRMCC at high S/N, however, there are three possible sources of less than optimal speech recognition: (1) minor distortion introduced by the signal processing, (2) greater amplification of the lower-level noise, and (3) speech information enhanced by the processing, that the impaired listener is not accustomed to hearing. The situation changes at low S/N where,(1) minimal distortion is masked by the noise, (2) noise will not be enhanced, and (3) any bit of additional information may be the difference between identifying and failing to identify the speech.

3.3 Speech and Acclimatization

One final source of problems in testing hearing-aid signal-processing algorithms should be considered. The vast majority of hearing-impaired patients have no complaint except that they cannot understand speech as well as they could, especially under noisy conditions. As a result, speech is the only stimulus set that can be used in hearing aid evaluations, but speech is a complex, highly-redundant information code. The redundancy of the speech code can be a problem because different individuals can use different cues to identify the same speech sound and hearing loss causes people to shift their cue utilization [17]. In this sense, the hearing-impaired listener is not just a normal-hearing listener with reduced information, but a listener with reduced information who uses that remaining information in a different way. If you could give this hearing-impaired listener all of the information available to the normal-hearing listener, you could not expect all of that information to be used "normally" until the hearing-impaired listener had time to adjust to the change.

Readjustment of cue utilization may be part of the explanation for the months an individual needs to acclimatize to a new hearing aid [18]. Whether or not shifting cue utilization and acclimatization are related, however, both indicate the need for extended experience with a new algorithm before we can measure the maximum benefit a patient would receive. In addition, acclimatization to complex signal processing algorithms may take longer than to the conventional hearing aids for which it has been studied [4]. Highly significant improvements in speech perception have been found for 16-channel FRMCC after extended listening experience in the laboratory [19], but such complex algorithms have not been studied outside of the laboratory. Our field studies of binaural 8-channel FRMCC [20] currently being planned include measurements of acclimatization, and also of specific consonant confusions to indicate changes in cue utilization.

4. FUTURE RESEARCH

Further work on FRMCC, as well as research on other signal processing algorithms for hearing-aids, will have to include field testing as the major evaluation method. Now that portable, programable DSP units are available, there is no reason to accept the difficulties imposed by restricting hearing-aid evaluation to the laboratory. Of course, some laboratory testing will be necessary to establish that a new algorithm can achieve its goal under the conditions for which it was designed, but then the primary evaluation should be shifted to the field. Field evaluation not only avoids the types of problems discussed earlier from the history of FRMCC research, but also tests the idea of applying the algorithm to the real problems of hearing impairment. Some signal processing algorithms may solve problems, but at the same time create others in situations commonly encountered by the hearing impaired. It is clearly possible, for example, that FRMCC might create variations in the signals that are too complex for stable acclimatization to occur and thus that it would yield significant recurring phoneme confusions (although previous results [19] suggest otherwise). Similarly, a noise-reduction algorithm that facilitates speech perception in frequency-shaped white noise may interfere when the noise consists of a small number of other voices. For any particular algorithm, the specific advantages and disadvantages will be different. Only with extended field testing can we be confident that the vast majority of such problems will be revealed.

In addition to initial evaluation of algorithm design goals, laboratory testing might be valuable in understanding or verifying problems or successes found in the field. For example, if subjects had reported from field experience that FRMCC was especially beneficial at low S/N prior to [9], such a follow-up laboratory study would have been useful because the result was unexpected in the context of previous FRMCC theory. Overall, however, the primary focus of future hearing-aid research must be shifted from laboratory study to field testing.

ACKNOWLEDGMENTS

This work was supported by VA Rehabilitation Research and Development Service and the VA Medical Research Service.

REFERENCES

[1] Villchur, E. "Simulation of the effect of recruitment on loudness relationships in speech". *Journal of the Acoustical Society of America*, 56:1601-1611, 1974.

[2] Duchnowski, P. and Zurek, P.M. "Villchur revisited: Another look at AGC simulation of recruiting hearing loss". *Journal of the Acoustical Society of America*, 98:3170-3181, 1995.

[3] Yund, E.W., and Crain, T.R. "Voiced stop consonant discrimination with multichannel expansion hearing loss simulations". in *Modeling Sensorineural Hearing Loss*, edited by W. Jesteadt, Lawrence Erlbaum Associates, Mahwah, NJ, 1997, pages 149-167.

[4] Van Tasell, D.J. "Hearing loss, speech, and hearing aids," *Journal of Speech Hearing Research*, 36:228-244, 1993.

[5] Crain, T.R. and Yund, E.W. "The effect of multichannel compression on vowel and stop-consonant discrimination in normal-hearing and hearing-impaired subjects". *Ear and Hearing*, 16:529-543, 1995.

[6] Villchur, E. "Signal processing to improve speech intelligibility in perceptive deafness,". *Journal of the Acoustical Society of America*, 53:1646-1657, 1973.

[7] Barfod, J. "Multichannel compression hearing aids: Experiments and considerations on clinical applicability". in *Sensorineural Hearing Impairment and Hearing Aids*, edited by C. Ludvigsen and J. Barfod, Scandinavian Audiology Supplement 6, 1979, pages 315-340.

[8] Yund, E.W., Simon, H.J., and Efron, R. "Speech discrimination with an 8-channel compression hearing aid and conventional aids in speech-band noise". *Journal of Rehabilitation Research and Development*, 24(4):161-180, 1987.

[9] Yund, E.W., and Buckles, K.M. "Multichannel compression hearing aids: Effect of number of channels on speech discrimination in noise". *Journal of the Acoustical Society of America*, 97:1206-1223, 1995.

[10] Lippmann, R., Braida, L., and Durlach, N. "Study of multichannel amplitude compression and linear amplification for persons with sensorineural hearing loss". *Journal of the Acoustical Society of America*, 97:1206-1223, 1995.

[11] Laurence, R.F., Moore, B.C.J., and Glasberg, B.R. "A comparison of behind-the-ear high fidelity hearing aids and two-channel compression aids, in the laboratory and in everyday life". *British Journal of Audiology*, 17:31-48, 1983.

[12] Moore, B.C.J., Laurence, R.F., and Wright, D. "Improvements in speech intelligibility in quiet and in noise produced by two-channel compression hearing aids". *British Journal of Audiology*, 19:175-187, 1985.

[13] Moore, B.C.J., and Glasberg, B.R. " A comparison of four methods of implementing automatic gain control (AGC) in hearing aids". *British Journal of Audiology*, 22:93-104, 1988.

[14] CHABA Working Group on Communication Aids for the Hearing-Impaired. "Speech-perception aids for hearing-impaired people: Current status and needed research". *Journal of the Acoustical Society of America*, 90:637-684, 1991.

[15] Moore, B.C.J. "Characterization and simulation of impaired hearing: Implications for hearing aid design". *Ear and Hearing*, 12:154S-161S, 1991.

[16] Yund, E.W., and Buckles, K.M. "Enhanced speech perception at low signal-to-noise ratios with multichannel compression hearing aids". *Journal of the Acoustical Society of America*, 97:1224-1240, 1995.

[17] Hedrick, M. "Effects of acoustic cues on labeling fricatives and affricates". *Journal of Speech, Language, and Hearing Research*, 40:925-938, 1997.

[18] Gatehouse, S. "The time course and magnitude of perceptual acclimatization to frequency responses: Evidence from monaural fitting of hearing aids". *Journal of the Acoustical Society of America*, 92:1258-1268, 1992.

[19] Yund, E.W., and Buckles, K.M. "Intelligibility of multichannel-compressed speech in noise: Long term learning in hearing-impaired subjects". *Ear and Hearing*, 16:417-427, 1995.

[20] Magotra, N., Sirivara, S. and Yund, E.W. "Real-time digital look-ahead multichannel compression (MCC) hearing aid". *Hearing Aid Research and Development Conference*. Bethesda, Sep 1997.

MULTICHANNEL ADAPTIVE NOISE REDUCTION IN DIGITAL HEARING AIDS

N. Magotra[†], P. Kasthuri[†], Y. Yang[†], R. Whitman[†], F. Livingston[‡]

Department of Electrical and Computer Engineering [†]
University of New Mexico
Albuquerque, New Mexico 87131
email: magotra@houdini.eece.unm.edu
Ph:505-277-0808; Fax:505-277-1439

Texas Instruments Incorporated [‡]
12203 Southwest Freeway
Stafford, TX 77477

Abstract

We have developed a digital binaural hearing aid using Texas Instrument's floating point TMS320C3X digital signal processing (DSP) chip. The device is referred to as the Digital Programmable Hearing Aid (DIPHA). It has the capability of being programmed to perform a variety of speech processing tasks and is in essence a *digital hearing lens*. It permits matching the processing strategy to an individuals specific hearing loss. This paper deals with the adaptive noise reduction algorithm employed by DIPHA. Typically the audiologist/therapist selects a number of bandpass filters to design (in effect) a custom binaural equalizer for the hearing impaired subject and cascades the noise reduction algorithm with the equalizer. This paper explores the concept of performing the adaptive noise reduction in individual bands of the equalizer, a multichannel adaptive noise reduction strategy for the hearing impaired.

1. INTRODUCTION

We have developed a digital binaural hearing aid using Texas Instrument's floating point TMS320C3X digital signal processing (DSP) chip. The device is referred to as the Digital Programmable Hearing Aid (DIPHA). DIPHA uses a wide bandwidth (0-16 KHz) and incorporates speech enhancement (noise reduction) as an integral part of its operation. It is capable of sampling up to two input speech channels at a variable sampling rate. We are currently using a sampling rate of 20 KHz for each channel. It is a real-time processing system, implying that there is no delay between input and output and it updates itself with *every new data sample* obtained.

It has the capability of being programmed to perform a variety of speech processing tasks. It permits matching the processing strategy to an individuals specific hearing loss. This paper presents an extension of the adaptive noise reduction algorithm employed by DIPHA to the multichannel case.

Typically the audiologist/therapist interactively programs the binaural equalizer (spectral shaping) algorithm used to compensate for the subjects hearing loss. The equalizer is implemented as two banks of bandpass filters, one for each channel (ear). The audiologist can interactively (in real time) choose the number of filters in each bank and select their frequency characteristics to tune them for each individual patient. Further details of this aspect may be found in [1].

The adaptive signal enhancement algorithm used in DIPHA has been designed to work with just one input data channel (single microphone), so as not to encumber the patient with multiple microphones associated with a beamforming approach. Hence we restricted our attention to working with dual (right ear and left ear) single-input single output processing blocks. As shown in Figure 1, the input signal can be conditioned by a highpass filter to compensate for the low frequency spectral tilt in speech signals [2]. The HPF is a simple first order infinite impulse response (IIR) filter with tunable cutoff frequency. Note that the placement of the HPF may be varied.

The following section describes the adaptive speech enhancement algorithm, while section three presents some results and discussion and finally section four presents the conclusions and future work being pursued.

2. SPEECH ENHANCEMENT

The core of the adaptive speech enhancement algorithm is the Real-time Adaptive Correlation Enhancer (RACE) algorithm. RACE is essentially an adaptive finite impulse response (FIR) filter [3]. As shown in Figure 1, the speech input (x(m)) is used to update the RACE coefficients. These coefficients consist of the estimated autocorrelation coefficients ($\hat{R}_{xx}(m,l)$) of the input channel. The autocorrelation coefficients are updated using a recursive estimator as given by the following equation

$$\hat{R}_{xx}(m,l) = \beta \hat{R}_{xx}(m-1,l) + (1-\beta)x(m)x(m+l) \quad (1)$$

Equation (1) represents a recursive estimator which corresponds to sliding an exponential window over the data with a time constant (τ in seconds) given by $\tau = 1/((1-\beta)f_s)$ where f_s represents the sampling frequency (sps) and

m: time index
l: lag index $|l| \leq L$
L: maximum lag value
β : smoothing constant ($0 << \beta < 1$)

The Z-transform of the adaptive filter can then be expressed as

$$H(z) = a_o(m) + a_1(m)z^{-1} \cdots + a_{2L}(m)z^{-2L} \quad (2a)$$

where

$$a_i(m) = \hat{R}_{xx}(m, L-i) \quad i = 0, 1, ..2L \quad (2b)$$

The input channel is then filtered using H(z) to obtain the enhanced output as shown in Figure 1. We have shown that for a narrowband signal the amplitude gain and signal-to-noise ratio (SNR) gain are both equal to approximately half the filter length or L. In terms of convergence considerations we have shown that RACE is able to converge rapidly enough so that the short term stationarity of the speech signal [2] does not cause any problems for the algorithm.

Since RACE is inherently an open loop adaptive system we include a gain (g(m)) control block that permits the following choices for g(m):
1) g(m)=1
2) g(m)=$1/(L\,\sigma_x^2(m))$
3) g(m)= $\sqrt{\sigma_{x_h}^2(m)/\sigma_{x_e}^2(m)}$

The variances defined above are also estimated via the recursive equation

$$\hat{\sigma}_z^2(m) = \beta \hat{\sigma}_z^2(m-1) + (1-\beta)z^2(m) \quad (3)$$

where z(m) is set appropriately to x(m), $x_h(m)$ (the input to the adaptive filter) or $x_e(m)$ (the output of the adaptive filter). The program implementing the algorithm also applies some control logic that sets g(m) as per the choice made.

To implement the multichannel version of RACE (MRACE), as shown in Figure 2, the bandpass filtered output of each filter is either processed through the RACE stage or delayed by the appropriate lag value. This scheme allows the user, flexibility to decide as to which frequency bands need to be enhanced. The frequency bands themselves are set by the therapist working with the test subject. The decision as to which frequency bands need enhancement is based on the noise spectrum. Only those filtered outputs where the noise level is predominant need to be run through RACE. The levels which do not need noise reduction are simply delayed by the maximum lag value used in RACE before the outputs of all the filters are combined.

The following section describes the implementation aspects of MRACE and a discussion of some test results obtained from a particular implementation of the algorithm.

3. RESULTS

For the results presented in this section the test input consisted of the spoken sentence "Do you think that she should stay out so late" corrupted with additive noise consisting of recorded cafeteria noise. Both were sampled at 20 KHz and the noise file was scaled to simulate a 0 dB signal-to-noise ratio (SNR). The binaural equalizer consisted of a bank of seven bandpass filters with the following cut-off frequencies (all in Hz.): 70-250, 250-400, 400-1000, 1000-2000, 2000-3000, 3000-5000, 5000-9000. These cut-off frequencies were selected from the data file of one of our test subjects. For the purposes of this paper, each filter gain was set at unity. Each filter was a 50 tap linear phase FIR with 80 dB band rejection. Spectral analysis of the cafeteria noise data indicated that its power level was dominant in the low frequency range from 0-5 kHz beyond which it shows an attenuation close to -40 dB. This indicated that only those filters which lie within the 0-5 kHz range needed to have their outputs enhanced. AS an experiment we ran two trials - one in which we simply enhanced the outputs of the first five filters and a second trial in which the output of every single filter was enhanced. In each case the RACE parameters (L=6, $\beta = 0.987$, HPF cut-off=700Hz, no scaling) were kept constant.

Figures 3 to 6 show some data plots illustrating the results obtained as described above. Figure 3 shows the time trace over the phrase "Do you" in the clean recording, Figure 4 shows the same data after corruption with cafeteria noise, Figure 5 presents the corresponding output when all filter outputs were enhanced, while Figure 6 shows a plot corresponding to the output when only the first five filter outputs were enhanced. To get a feel for the relative amounts of improvement provided by the two different trials, we computed the SNR improvement in both cases. When all seven filter outputs were enhanced the SNR improvement was approximately 4 dB while enhancing only the first five filter outputs yielded a SNR improvement of approximately 14 dB.

4. CONCLUSIONS AND FUTURE WORK

Test results using DIPHA without the multichannel option on hearing impaired test subjects indicate that the algorithm improves their speech discrimination ability. As indicated by the results presented in section three, different MRACE configurations can yield different results, as measured through defined quantitative measures. However, the perception of speech is highly subjective in nature and in the final analysis it is the patient's response that determines the success or failure of the signal processing used. Hence, as part of our on-going research we intend to evaluate MRACE on a large number of test subjects.

Currently our real-time hearing aid is based on a 5 volt TMS320C3X DSP chip (C30 on the PC patient testing platform and C31 on the wearable (walkman sized) unit). An important goal of our future work is to try and reduce the size/power of the portable DIPHA-XP2 unit. Hence we are planning to port our algorithms to Texas Instruments latest DSP chip (TMS320C54X). The porting process will include a detailed quantization analysis.

The TMS320C54X [4] DSP chip is a 16-bit, fixed-point DSP and a member of the TMS320 family of DSPs. It features a hardware-intensive CPU with a high degree of parallelism and a powerful, flexible instruction set. The 54X also features on-chip memory and peripherals for increased operational speed, reduced system power requirements, and increased system integration. Finally, the 54X is fabricated with advanced IC processing technology for enhanced performance and low power consumption - the latest versions can operate with a 1 volt power supply. Also, the architecture and instruction set of the 54X lend themselves to the algorithms present in DIPHA. The multi-band equalization portion of the DIPHA, for example, could be effectively handled using the 54X's FIRS instruction requiring only one cycle per filter tap. The on-chip memory and peripherals of the 54X would be useful in minimizing the physical size of DIPHA. Adding more on-chip peripherals to the 54X core using TI's customizable DSP (cDSP) technology would reduce the physical size and power requirements of DIPHA even further. Finally, the low power consumption provided by the 54X would increase the battery life of DIPHA.

References

[1] N. Magotra, J. Stewart, K. Bricker, R. Weaver , "Digital Auditory Device using the TMS320C30," International Conference on Signal Processing Applications and Technology, Cambridge MA, Nov. 1992, pp. 15-19.

[2] L. R. Rabiner, R. W. Schafer, "Digital Processing of Speech Signals," Prentice-Hall, 1978.

[3] N. Magotra, "Seismic Event Detection and Location using Single Station (three-component) Data", PhD dissertation, Department of Electrical and Computer Engineering University of New Mexico, 1986.

[4] TMS320C54X DSP: CPU and Peripherals: Reference Set, Volume #1, Texas Instruments, 1996, (SPRU131C).

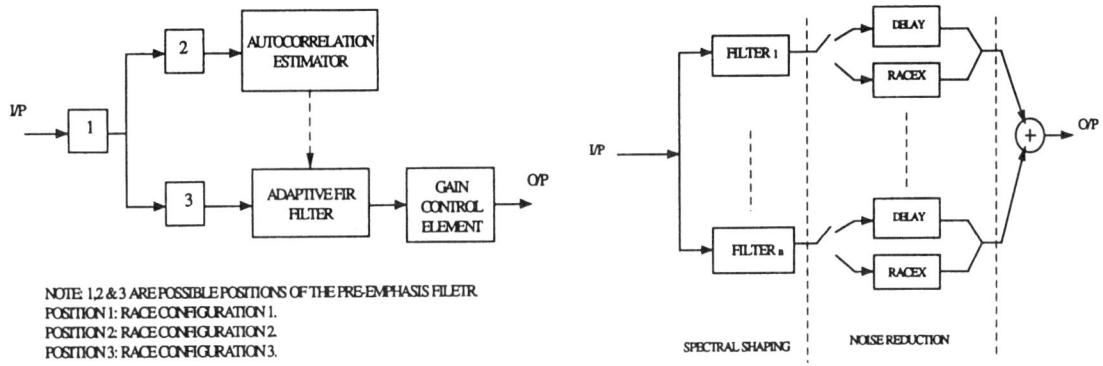

NOTE: 1,2 & 3 ARE POSSIBLE POSITIONS OF THE PRE-EMPHASIS FILTER.
POSITION 1: RACE CONFIGURATION 1.
POSITION 2: RACE CONFIGURATION 2.
POSITION 3: RACE CONFIGURATION 3.

Figure 1: Different RACE Configurations.

Figure 2: Overall Block Diagram

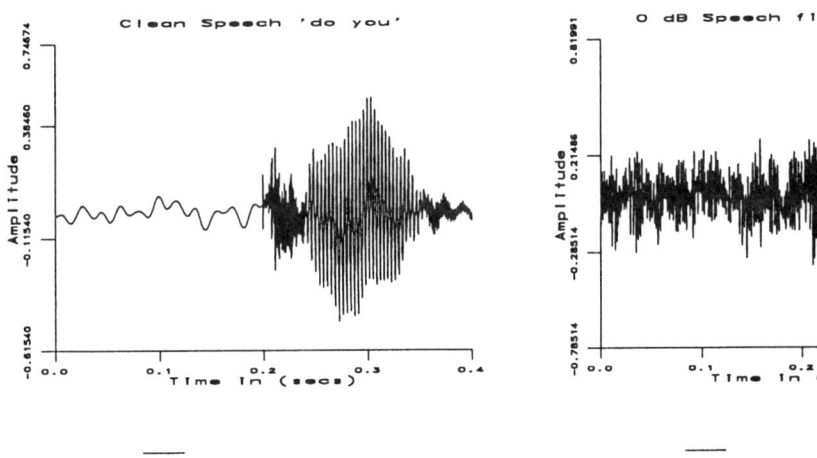

Figure 3

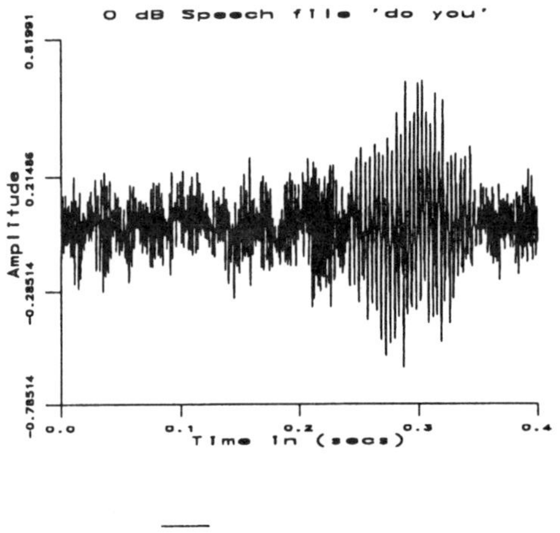

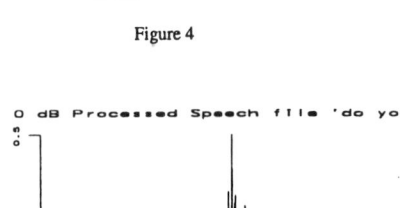

Figure 4

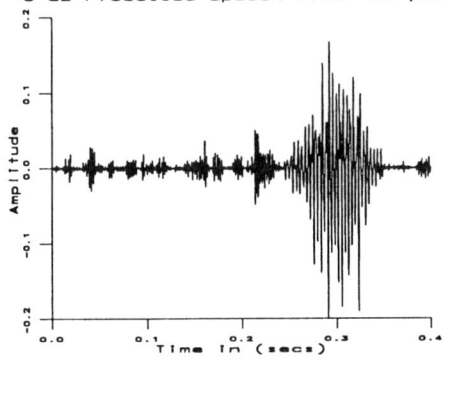

— Output of all filters run through RACEX

Figure 5

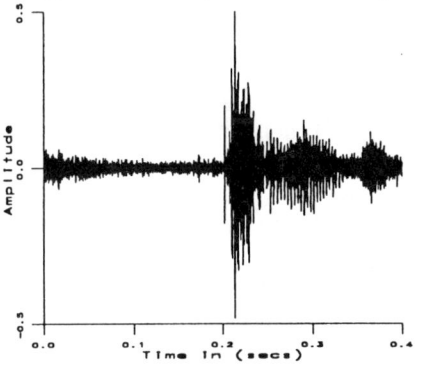

— Output of first 5 filters run thru RACEX

Figure 6

SIGNAL PROCESSING TECHNIQUES FOR A DSP HEARING AID

Brent W. Edwards

ReSound Corporation
220 Saginaw Drive
Redwood City, CA 94063

ABSTRACT

The recent development of commercial hearing aids with digital signal processing (DSP) capabilities opens the door for an explosive growth in hearing aid sophistication. Current limitations on signal processing design due to hearing aid constraints, and the specialized needs of listeners with hearing loss, result in certain design specifications that any DSP hearing aid must meet. This paper will review the latest state-of-the-art with respect to current DSP hearing aids on the market, with an emphasis on the ReSound programmable DSP hearing aid, developed in a collaboration with GN Danavox and AudioLogic, Inc. Multiband compression, single-microphone processing, multi-microphone processing, and feedback reduction will be discussed.

1. INTRODUCTION

Until the past decade and a half, commercial hearing aid technology had developed little beyond simple linear amplification with peak clipping. Low power and small size requirements, combined with unclear scientific evidence for how to properly compensate for sensorineural hearing loss, kept the sophistication of hearing aid signal processing to a minimum. Today, the most sophisticated aids available cover a variety of nonlinear processing architectures for hearing loss compensation and some include simple noise reduction processing. They are small enough to fit completely inside the auditory canal, run on a 1.3V battery and draw less than 2mA for a battery life of over 100 hours.

While the introduction of hearing aids with DSP chips allow the application of sophisticated signal processing techniques to these devices, their capabilities are still limited by a processing power of a few MIPS due to the inherent constraints of a hearing aid platform. The following will attempt to detail the current state of the art in the hearing aid industry, with an emphasis on the technology being introduced with a programmable DSP system developed by a consortium formed by ReSound Corp., AudioLogic, Inc. and GN Danavox (the device hereafter being referred to as the RAD aid).

2. MULTIBAND COMPRESSION

Sensorineural hearing loss is primarily characterized by a loss of sensitivity to sounds that is more severe for low-level than high-level signals. In order for a hearing aid to restore loudness to normal levels, the gain supplied must vary with the level of the signal [10]. The gain must also adjust fast enough such that different amounts of gain is provided to successive phonemes of differing level. If the gain is reduced for a high-level vowel, for example, the gain must increase quickly so that a following consonant is audible to the hearing aid wearer and not forward-masked by the vowel. This level-dependent gain is referred to as fast-acting or syllabic compression and can be characterized by the compression ratio, the kneepoint level at which the gain transitions from linear to compression (usually set at a low level such that speech falls in the compression region), and the gain at the kneepoint.

Figure 1 shows the loudness curve for a hearing-impaired subject from which the gain necessary for normal loudness restoration can be calculated. The data was gathered using a loudness scaling procedure with half-octave bands of noise [1]. As indicated by the arrows, gains of 25-dB and 8-dB should be applied to stimulus levels of 50-dB SPL and 80-dB SPL, respectively.

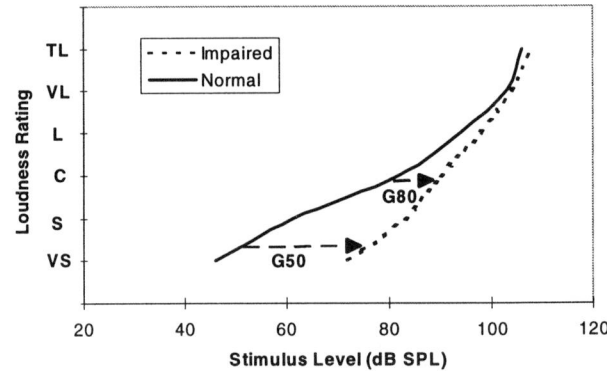

Figure 1. Typical loudness curves for both normal-hearing and hearing-impaired subjects. Loudness categories correspond to: Very Soft, Soft, Comfortable, Loud, Very Loud, Too Loud.

Because hearing loss varies with frequency, the gain and compression ratios must also vary with frequency. This is typically implemented by filtering the signal into different frequency bands and applying separate gain and compression to the signal in each band. One drawback to this system is illustrated in Figure 2, which shows the gain response of a three-band processor intended to provide equal gain at all frequencies, measured with both an 80-dB SPL tone-sweep and a broadband noise with 80-dB SPL in each band [2]. Identical I/O functions are used for each band so that equal gain should be applied across frequency for these signals. Because of the skirts of the filters used, the nonlinear processing near the crossover frequency between two bands provides more gain for narrowband signals than for broadband signals and violates the design requirement of a flat gain function. This effect is made worse as the number of bands increases since the number of crossover regions increases as well. Problems may also occur as the bandwidths of the filters approach the frequency spacing of a harmonic signal since the harmonics may fall near the transition regions between bands and the gain applied to each harmonic will depend on how many harmonics fall within any given band.

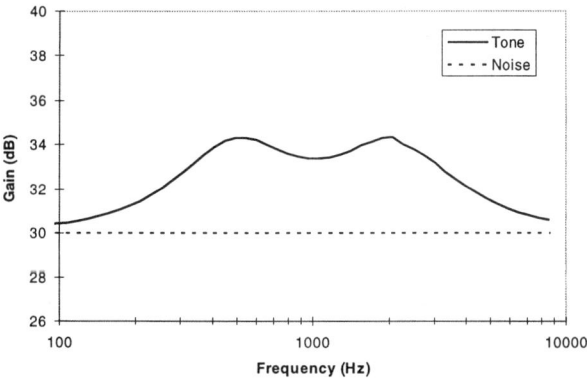

Figure 2. Gain response of a 3-band compressor, with the gain and compression ratio set at 30 dB and 3:1, respectively, for the stimulus level used. The response was measured with both a tone sweep and broadband noise.

This problem is solved by the sufficient overlapping of bands, such that the gain at any given frequency is determined by multiple bands [5]. The bandpass filtering can be treated as a sampling of the power spectrum and thus a minimum sampling rate is required to eliminate aliasing in the autocorrelation domain. By overlapping the bands, the crossover effect is not only reduced but problems resulting from, for example, the gain varying with the number of harmonics within a band is also reduced.

In the RAD aid, the multiband compressor that compensates for the listener's hearing loss is composed of 14 significantly-overlapping bands that minimize the problems discussed above. The gain and power calculations are performed in the frequency domain by using shift-and-sum FFT processing. This technique allows other FFT-based algorithms to be performed in parallel with the compressor and saves processing cycles. It is important that the overall delay of the hearing aid is on the order of milliseconds in order to preserve synchrony with visual cues and to keep the perception of the listener's own voice in synchrony with their speech production. This necessitates short frame lengths in the FFT structure.

3. SINGLE-MICROPHONE PROCESSING

Hearing-impaired listeners have a more difficult time understanding speech in noise than normal-hearing listeners and may require a speech-to-noise ratio (SNR) up to 10 dB higher than normal for the same recognition performance[6]. Many techniques exist for improving the SNR in a single-microphone system, but none have been shown to improve speech intelligibility for human listeners [4]. In fact, some algorithms increase the SNR while reducing intelligibility. As such, single-microphone hearing aid techniques for improving speech in noise are designed to improve subjective measures of benefit, referred to in the field as sound comfort.

Slow-acting automatic gain control (AGC) systems have been implemented in commercial hearing aids which reduce the level of sounds in frequency regions with high levels of noise and little to no levels of speech. This processing is performed both in regions with and without hearing loss. While many have suggested that this technique reduces upward spread of masking in the presence of high-level low-frequency noise, making high-frequency speech cues more audible, it has not been shown to have any benefit towards improved intelligibility [9]. Thus, improved comfort is the goal of this technique.

The signal envelopes in different frequency regions provide a useful statistic for estimating the presence of speech and the level of noise in each band. If the statistics of the envelope in a band is similar to that which is expected for speech in quiet, then no attenuation in that band is applied. As the envelope statistics approach that of noise, whether mechanical noise or speech babble, the gain in that band is reduced. This technique is embodied in the RAD aid using the envelopes in each of the 14 bands of the compressor. Using a large number of bands with this technique maximizes the ability to take advantage of spectral mismatches between the noise and speech. The different time constants of the syllabic compressor and noise reduction system minimize the interaction of the two AGCs such that they do not counteract each other. This noise reduction technique also addresses the common hearing aid problem of audible microphone noise in quiet environments when the compressor gain is maximum.

Speech understanding in noise is exacerbated by auditory systems with damaged by outer hair cells since the resulting loss of lateral suppression and broader auditory filters reduces perceived spectral contrast (e.g., [8]). Place cues that rely on spectral envelope shape become less distinguishable because of the smoothed spectrum in the perceptual domain. Attempts to sharpen the spectrum with single-microphone processing techniques, however, have had little success [7].

4. MULTI-MICROPHONE PROCESSING

Array processing can be used with hearing aids in order to provide direction-dependent gain. The design typically assumes that the target signal is in front of the hearing aid wearer, and the directional system can be designed to maximize the attenuation from all other directions. Because of cosmetic constraints, arrays with several microphones that span a useful distance and are separate from the body of the hearing aid have not been feasible as widely-acceptable commercial products. Two omni-directional microphones mounted on a hearing aid, however, can provide a significant amount of directionality that results in improved speech understanding in noisy environments.

. The size of behind-the-ear cases on which the microphones must be mounted limits the maximum separation between microphones to approximately 1.5 cm; in-the-ear cases limit microphone separation to approximately 1 cm. The small distance between microphones prevents the application of standard beamforming--passing the signal unchanged from the direction of the beam by adding the microphone signals together--since useful directionality only occurs above 6 kHz at these separation distances. Directionality is achieved by subtracting the back microphone from the front one, steering the direction of the response null by adding a delay to the back microphone. With this processing, the on-axis (frontal) frequency response has a highpass characteristic, rolling off at 6-dB/octave, which provides a tinny characteristic to the sound and may make low-frequency signals inaudible. The RAD aid, which has two microphones, achieves the back-microphone delay with an all-pass filter and uses digital filtering to compensate for the reduced low-frequencies.

Because of the location of the microphones, a significant amount of directionality is provided simply from the headshadow effect. Figure 3 shows the polar patterns of a two-microphone directional system and a single omni-directional microphone on a behind-the-ear aid worn on a mannequin. The improvement in speech intelligibility is approximately 5 dB for a speech signal in front of the listener and diffuse background noise, meaning that signal-to-noise ratio is 5-dB less than that needed for the same intelligibility with an omni-directional microphone. To take into account the importance of different frequency regions to speech intelligibility, the articulation index can be incorporated with a directivity index to obtain an overall measure of the estimated benefit from the directionality of an aid with respect to understanding speech in noise [3].

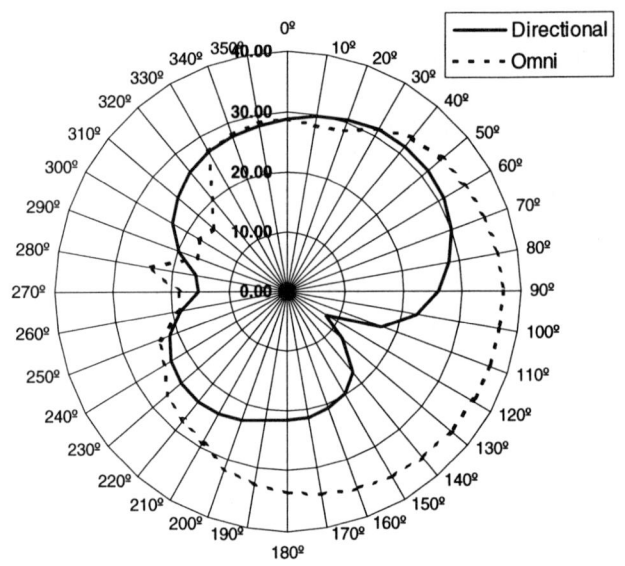

Figure 3. Polar plots for a behind-the-ear hearing aid with an omni-microphone and a two-microphone directional array, measured with the aid on a mannequin.

Directionality becomes less effective when there are phase and amplitude mismatches between the two microphones on a directional aid, and effect of the mismatch is more significant as the distance between microphones decreases. Thus, the responses of the two microphones in a directional aid must be as closely matched as possible. The RAD aid has two A/D converters such that the output of each microphone is sampled and digitized, the rear microphone signal is delayed, and the rear signal is subtracted from the front signal. By subtracting the two microphones digitally instead of electrically before the A/Ds, digital filters can be used to match the two microphones such that their amplitude and phase responses are as close as possible, as is done in the RAD aid. This also allows techniques for tracking any changes in their responses over time and adaptively compensating for these changes.

Noise cancellation can also be attempted by substituting the rear omni-directional microphone with a directional microphone that is directed behind the listener. An adaptive filter is applied to the signal from the directional microphone to minimize the error signal obtained when subtracting the filtered rear signal from the front signal. Laboratory experiments have shown that maximum benefit is achieved when the adaptive filter is only updated when no speech (target) signal is detected from the on-axis direction [11]. This ensures that that the directional microphone is only measuring noise and the target signal is not canceled.

5. FEEDBACK

Feedback is a significant problem for many hearing aid wearers. Leaks around an ear mold and through the vent in the ear mold can cause feedback, which may be made worse by the presence of an object brought near the aid such as a phone receiver, hat or hand. Because vents are often used to prevent an occlusion effect of the listener's own speech sounding boomy, the amount of gain that an aid provides may be limited to below that necessary for proper hearing loss compensation due to the feedback path through the vent or leaky earmold.

One solution that many manufacturers have used is to simply reduce the hearing aid gain in a broad frequency band that includes the feedback signal. Notch filters have also been used to solve this problem. Both of these solutions have the drawback that they reduced the gain in the problem frequency region, making any speech cues that may exist in that reduced frequency region inaudible to the hearing aid wearer. The RAD aid uses digital processing for adaptive feedback cancellation whereby the feedback path between the receiver and microphone is estimated and a filtered version of the signal to the receiver is subtracted from the output of the microphone. A brief, internally-generated noise burst is used to obtain an initial estimate of the feedback path and the parameters of an adaptive filter used as a part of the feedback cancellation is calculated using correlation techniques. During the normal functioning of the aid, the receiver input is passed through the digital filter and then subtracted from the microphone output to cancel the feedback signal. The filter is also adapted during the normal use of the aid, without any internally-generated noise probe, in order to track both sudden and slowly-varying changes to the feedback path such as the placement of a phone near the ear or a loosening ear mold, respectively.

6. CONCLUSIONS

DSP chips allow for significant improvements to hearing aid processing, but care must be taken that the specialized needs of the hearing-impaired users are accounted for. Specifically, any processing such as noise reduction or directionality must not counter the effect of the hearing-loss compensation processing. While improved speech intelligibility is one of the primary goals of a hearing aid, algorithms which improve subjective characteristics such as sound quality are also goals for signal processing.

7. REFERENCES

1. Allen J.B., Hall J.L. and Jeng P.S. *"Loudness growth in 1/2-octave bands--A procedure for the assessment of loudness"*. Journal of the Acoustical Society of America 88: 745-753, 1990.

2. Edwards B.W. and Struck C.J. *"Device characterization techniques for digital hearing aids"*. Journal of the Acoustical Society of America 100: 2741, 1996.

3. Greenberg J. and Zurek P. *"Evaluation of an adaptive beamforming method for hearing aids"*. Journal of the Acoustical Society of America 91: 1662-1676, 1992.

4. Lim J.S. *Speech Enhancement*. Prentice Hall, Englewood Cliffs, NJ, 1983.

5. Lindemann E. *"The Continuous Frequency Dynamic Range Compressor"*. IEEE Workshop on Applications of Signal Processing to Audio and Acoustics, New Paltz, New York, 1997.

6. Plomp R. *"Auditory handicap of hearing impairment and the limited benefit of hearing aids"*. Journal of the Acoustical Society of America 63: 533-549, 1978.

7. Stone M.A. and Moore B.C.J. *"Spectral feature enhancement for people with sensorineural hearingimpairment: Effects on speech intelligibility and quality"*. Journal of Rehabilitation Research and Development 29: 39-56, 1992.

8. Summers V. and Leek M.R. *"The internal representation of spectral contrast in hearing-impaired listeners"*. Journal of the Acoustical Society of America 95: 3518-3528, 1994.

9. Van Tasell D.J., Larsen S.Y. and Fabry D.A. *"Effects of an adaptive filter hearing aid on speech recognition in noise by hearing-impaired subjects"*. Ear and Hearing 9: 15-21, 1988.

10. Villchur E. *"Signal processing to improve speech intelligibility in perceptive deafness"*. Journal of the Acoustical Society of America." 53: 1646-1657, 1973.

11. Weiss M. *"Use of an adaptive noise canceler as an input preprocessor for a hearing aid"*. Journal of Rehabilitation Research and Development 24: 93-102, 1987.

WIRELESS POWER TRANSFER FOR A MICRO REMOTELY PILOTED VEHICLE

David C. Jenn, Robert L. Vitale

Naval Postgraduate School
Department of Electrical and Computer Engineering
833 Dyer Rd, Code EC/Jn
Monterey, CA 93943, USA

ABSTRACT

A prototype rectifying antenna (rectenna) to provide wireless power transfer (WPT) to a micro-remotely piloted vehicle (MRPV) is developed. Microwave radiation at 1.3 GHz is converted into DC to drive a small motor and spin a mockup helicopter rotor blade. The rectenna serves a dual purpose as the antenna and outer body of the proposed vehicle and allows efficient reception of power over 360 degrees around the vehicle.

Wireless power transfer to the mockup MRPV has been demonstrated with less than 1 watt of transmitted power at near field ranges. Rectification efficiencies up to 30% were measured for two rectifier circuit configurations using 1.3 GHz continuous wave (CW) and pulse modulated transmitted signals.

1. INTRODUCTION

One important application of a micro-remotely piloted vehicle (MRPV) is the video surveillance of an area where people are not allowed or physically cannot go. Examples include the interior of buildings that may be controlled by unfriendly personnel, areas where environmental dangers are present or perhaps where it is more economical to use robotic surveillance.

A MRPV with the smallest possible size and greatest endurance is desired to meet the challenge of remote surveillance inside buildings. The chief aspect that prevents size reduction of a MRPV is the volume and weight devoted to carrying onboard fuel or batteries. Mission endurance requirements set onboard fuel/battery requirements, which in turn increase engine/motor and support structures requirements, all adding, size and weight. A remotely powered (via WPT) MRPV can reduce or eliminate onboard fuel or battery requirements and provide large-scale decreases in the size of the MRPV while permitting endurance to be independent of fuel storage requirements.

A proposed deployment of a MRPV system using beam tagging [1] to synchronize the signals from multiple

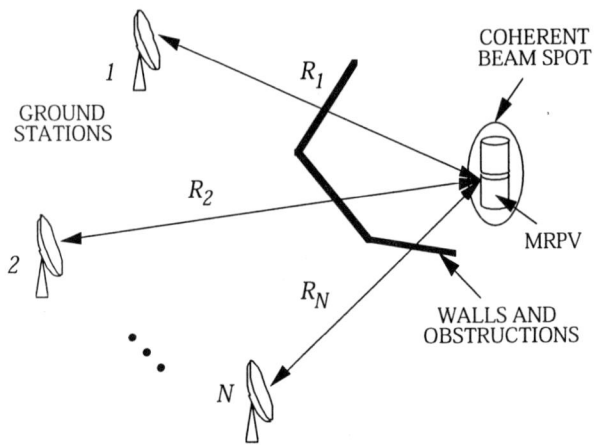

Figure 1: Proposed deployment of wireless powered micro-RPV

ground stations is shown in Figure 1. Multiple transmitting sites provide signal diversity and maximize power density at the MRPV while reducing individual ground station power requirements and antenna size.

Previous WPT rectenna designs have concentrated on using either flat array antennas or flat rectangular or circular patch antennas [2-3]. Many of these designs have achieved a high RF-DC conversion efficiency (85%), however they are not suited to the current application because the antenna structures lack a uniform radiation pattern in azimuth and therefore restrict the orientation and maneuvers of the MRPV. For a wireless powered MRPV for use in the interior of a building, it is desired to eliminate these restrictions and therefore the development of a rectenna with uniform azimuth radiation pattern was undertaken. A rectenna of this type is shown in Figure 2. Previous research into the tradeoffs involved in frequency selection has led to the choice of 1-2 GHz region [4]. A frequency of 1.3 GHz was chosen for the demonstration model because it corresponds to the frequency of an exiting high power radar located at the Naval Postgraduate School.

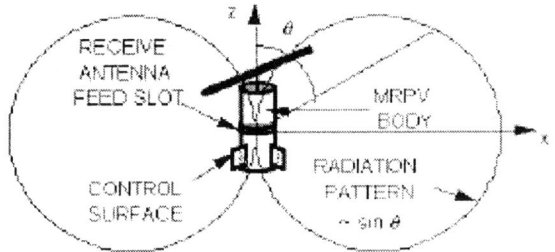

Figure 2: MRPV rectenna radiation pattern

2. MRPV RECTENNA DESIGN

The MPRV rectenna development consists of two parts: the antenna design and the development of the microwave rectifying circuitry.

The antenna structure was chosen to be a fat dipole with an offset slot feed. The radius of the dipole replicates a full scale MRPV prototype under development by Lutroics Corporation. Design parameters investigated include the overall dipole length, feedpoint, dipole arm separation and geometry of the region separating the arms. In an effort to reduce cost, commercially available off the shelf materials were used for the antenna. Since the prototype is non-flying, weight was not an issue and therefore copper plumbing pipe and endcaps were used. The copper pipe was easily cut to size on a lathe and thus reduced MRPV model development time. The antenna structures for functional flying MRPVs will use metal films or coatings layered over a lightweight composite structure.

Electromagnetic computer models of the MRPV rectenna were generated to determine an antenna structure with a driving point impedance close to that of the rectifier transmission line (50 ohms). Electromagnetic modeling was accomplished using the GNEC version of NECWINPRO. GNEC uses the Numerical Electromagnetic Code Version 4.1 engine (NEC 4.1) developed by Lawrence Livermore National Laboratory.

The full-scale body dimension of the Lutronics vehicle determined the length of the MRPV antenna. GNEC models that varied the feed position and separation of the dipole arms were generated and the impedance results analyzed until a VSWR of 1.55 was achieved at 1.3 GHz.

The rectifying circuit is a microstrip structure designed around a commercially available surface mount GaAs Schottky barrier diode manufactured by Hewlett Packard. The package is configured as an unconnected pair in a standard SOT-143 low profile package. Microstrip circuits using both one and two diodes connected in parallel were designed. A knowledge of the diode package scattering parameters (S-Parameters) at the operating frequency was required to design impedance matching circuitry for maximum power transfer. Low power (15 dBm) impedance characteristics of the HP GaAs Schottky diode package were determined from HP8510C vector network analyzer (VNA) measurements of the diode package surface mounted onto a 50 ohm microstrip test fixture fabricated from glass-epoxy (FR4) PC board material (Figure 3).

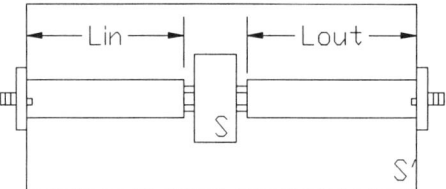

Figure 3: Microstrip test fixture for SOT-143 SMT Schottky barrier diode package

The diode scattering parameters were determined by shifting the reference planes of the microstrip test fixture VNA measurements using

$$[S] = \begin{bmatrix} e^{j2\theta_1} S'_{11} & e^{j(\theta_1+\theta_2)} S'_{12} \\ e^{j(\theta_1+\theta_2)} S'_{21} & e^{j2\theta_2} S'_{22} \end{bmatrix} \quad (1)$$

$$\theta_{1\,or\,2} = L_{in\,or\,out} \frac{2\pi}{\lambda_{guide}} \qquad \lambda_{guide} = \frac{\lambda_{free\,space}}{\sqrt{\varepsilon_{effective}}}$$

The effective dielectric constant of the microstrip substrate ($\varepsilon_{effective}$) is related to the relative dielectric constant (ε_r) of the dielectric substrate and the geometry of the microstrip transmission line by

$$\varepsilon_{effective} = \frac{\varepsilon_r + 1}{2} + \frac{\varepsilon_r - 1}{2} \left(\frac{1}{\sqrt{1 + 12\frac{d}{w}}} \right) \quad (2)$$

The relative permittivity of the substrate material and thickness (d) to width (w) ratio of the microstrip line are physical properties of the microstrip circuit. A microstrip thru line fixture was simulated using EEsof Touchstone. The relative permittivity (ε_r) was varied until the return loss (S_{11}) matched that measured on the VNA. Touchstone simulations provided the good agreement to VNA measurements when ε_r was 2.5.

A microstrip rectification circuit with a shorted microstrip stub for input impedance matching was designed using the scattering parameters determined from the VNA measurements. Two principal microstrip circuits were fabricated using this approach. One used two diodes of the HSMS-8525 diode package and the second only one of the diodes. Additional variations of these circuits were fabricated using a chip resistor (37 ohms) and a miniature DC motor as loads. The microstrip rectifier circuits with chip resistor loading were used to measure microwave rectification efficiency. The circuits with the DC motors were used to demonstrate MRPV operation. A sample microstrip rectifier circuit under test on the VNA is shown in Figure 4.

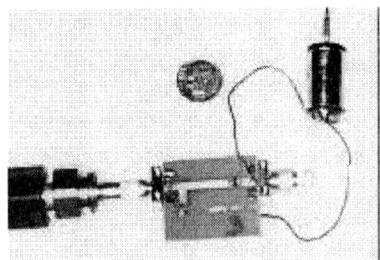

Figure 4: MRPV microstrip rectifier with DC motor

3. TESTING

3.1 Summary of Testing

MRPV testing consisted of rectifier efficiency measurements and of wireless powered motor operational testing. An objective is to provide transmitting power by the use of an available high power radar transmitter. The AN/SPS-58 air search radar transmitter was chosen as the transmitter due to its frequency (1.3 GHz) and high peak power (12 kW). Disadvantages of this transmitter include low pulse repetition frequency (PRF) of 3 kHz and low duty cycle (2.4%). However, due to the high peak power, the average power (290 watts) appeared to be great enough to encourage free space testing. Prior work [1-3] used CW signals, therefore measurements using pulse modulated microwave energy were required observe the effects of a pulsed radar signal on the rectifier efficiency. Efficiency measurements of simulated radar signals with characteristics of 7 µs pulse width at PRFs of 36 kHz, 13 kHz and 3 kHz. Signals were coupled directly into the MRPV rectifier circuit (no antenna attached). Average input power, average reflected power and DC voltage across the chip resistor load were measured. A block diagram of the test setup is shown in Figure 5.

3.2 Rectifier Efficiency Measurements

Measurements of DC power supplied to the chip resistor load as a function of absorbed power (input power less reflected power) were made and converted to rectification efficiency. Measurements were conducted for microstrip circuits containing a single diode and both diodes in the HP-8525 GaAs Schottky Barrier diode package. Efficiency measurements using 1.3 GHz CW as the input

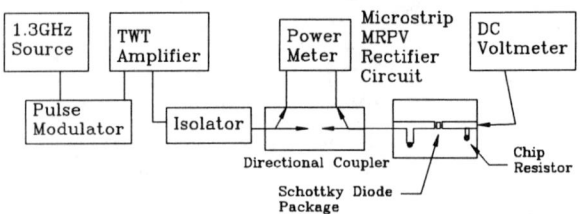

Figure 5: Block diagram of MRPV rectifier efficiency measurement using simulated radar signals

signal were also made for reference. Efficiency measurements were also made for pulse modulated (PM) signals to represent the effects of a high-powered search radar. The PM signals had a pulse width of 7 µs and pulse repetition rates of 36, 13 and 3 kHz. The 3 kHz PRF corresponded to the AN/SPS-58 Radar transmitter signal. Experimentation using various loading capacitances was conducted to determine if efficiency could be improved by storing energy for use while the pulse was off or during the negative half cycle of the received signal. Efficiency plots are shown in Figure 7 (single diode) and Figure 8 (dual diodes) for the various input signals. Solid lines indicate a 37-ohm chip resistor load and dashed lines the chip resistor with 11 µF capacitor in parallel.

3.3 DC Motor Operation using Microwave Signals

A second set of rectifier circuits terminated with a miniature DC motor (replacing the chip resistor) were tested using the setup of Figure 5. DC voltage measurements were made for the same input signals as previously described and are shown in Figure 9 for the one diode configuration. Motor armature voltage (V_a) is a function of supplied current (I_a), motor rotor speed (ω_r) and permanent magnet DC machine flux constant (K_v) and is given by

$$V_a = I_a \cdot R_{dc} + K_v \cdot \omega_r \qquad (3)$$

K_v can be determined by open circuit voltage measurement when the machine is run as a generator at a known speed.

3.4 Wireless Powered Operation of Mockup MRPV

Wireless powered operation of the mockup MRPV was demonstrated using 1.3 GHz CW and 36 kHz PRF signals feeding a 16 dB horn antenna. Motor operation was demonstrated using WPT at a distance of 33 inches for CW and 3 inches for 36 kHz PRF. Average transmitting antenna power was 1.8 watts for CW and 0.44 watts for 36 kHz PRF. Both signals had field strength of approximately 0.4 mW/cm^2 at the MRPV.

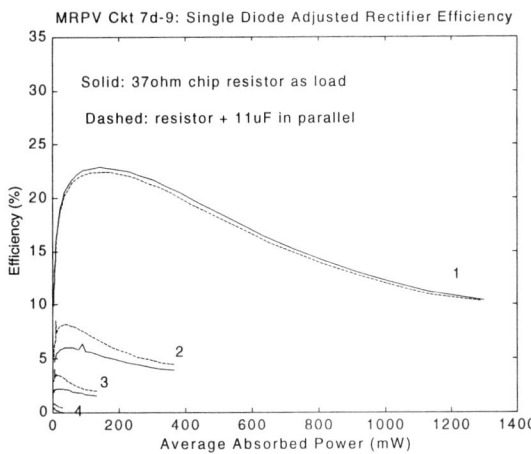

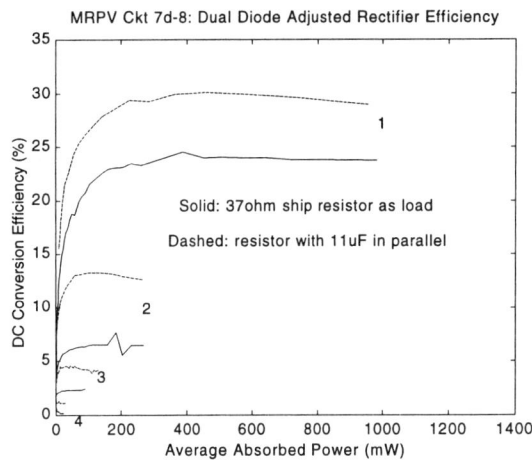

Figure 7: Microwave rectification efficiency for a single SMT RF Schottky barrier diode circuit. (1)-CW; (2)-prf=36 kHz; (3)-prf=13 kHz; (4)-prf=3 kHz

Figure 8: Microwave rectification efficiency for dual SMT RF Schottky barrier diode circuit. (1)-CW; (2)-prf=36 kHz; (3)-prf=13 kHz; (4)-prf=3 kHz

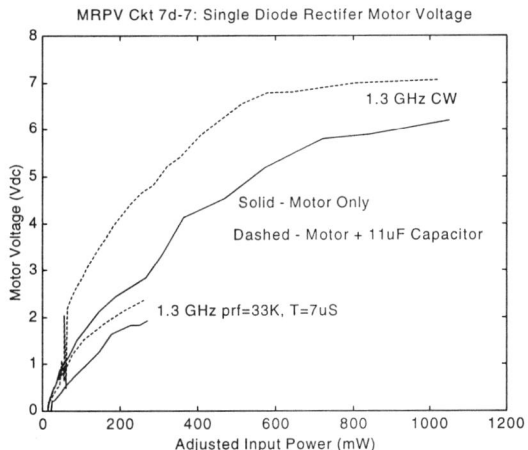

Figure 9: Motor Voltage using MRPV rectifier and microwave input signals, single diode configuration

Rectification efficiency tended to decrease for pulse modulated (radar) signals and can be attributed to the low duty cycle of the signal. However as additional diodes are introduced along with some capacitance, efficiencies may approach that obtained using CW signals as indicated by the low power region of Figure 8.

Distances further away from the transmitting antenna require much greater transmitted power and/or multiple sources as previously discussed in [1]. Greater transmit power testing at longer ranges is planned using an air search radar. However, rectifier efficiency for the low PRF pulse modulated signals need to be improved.

4. CONCLUSIONS

Remote powering or fueling of a MRPV via wireless power transmission (WPT) has been demonstrated for a mockup MRPV. Transmitted microwave energy needed to be at least 0.4 mW/cm^2 at the MRPV location for operation.

Microwave rectifier efficiency tended to improve as incident power on the diode increased until the diode saturated. After saturation, efficiency decreased unless additional diodes were introduced in parallel to divide the current. As expected significant increases in rectifier efficiency were obtained when capacitance was placed in parallel with the load.

REFERENCES

[1] Jenn, D.C., "RPV", *IEEE Potentials*, vol 16, No. 5, pages 20-22, Dec. 1997.

[2] Bharj, SS, Carnisa, R., Grober, S., Wozniak, F. Pendleton, E., "High Efficiency C-Band 1000 Element Rectenna Array for Microwave Powered Applications", *IEEE Antennas and Propagation Society International Symposium*, vol. 1, pages 123-125, July 1992.

[3] McSpadden, J.O., Yoo, T., Chang, K., "Theoretical and Experimental Investigation of a Rectenna Element for Microwave Power Transmission" *IEEE Transactions on Microwave Theory and Techniques*, vol. 40 No. 12, pages 2359-2366 Dec 1992.

[4] Gibson, T.B., "Propagation Loss Study and Antenna Design for the Micro-Remotely Powered Vehicle (MRPV)", Master's Thesis, Naval Postgraduate School, Monterey, CA, September 1995.

Control of Elastic Multi-Link Manipulators Based on the Dynamic Compensation Method [1]

Victor A. Utkin
Institute of Control Sciences, Profsoyuznaya 65,
117806, Moscow, Russia, E-mail: vicutkin@ipu.rssi.ru

Abstract

The paper is concerned with the development of the sliding mode control for a multi-link flexible manipulator based on the pole assigment approach. The problem of chattering taking place in the system with switching control is solved by the introduction of components of the dynamic compensator state vector into the sliding surfaces. The motion separation method is used for the reconstruction of state vector by the methods of asymptotical observers and dynamic compensation.

1. Introduction

The paper presents a new approach to the design of rigid-flexible structure systems, based on a sliding mode technique. It is known, that robustness and order reduction are the most important virtue of variable structure control (VSC) with sliding mode [1]. The structure of rigid-flexible system can be presented as two mutually connected subsystems: the first subsystem describes the behavior of rigid coordinates, the second subsystem describes the vibration of flexible coordinates. The common approach to the control problem of flexible system, both for a continuous control algorithms [2], and for VSC' algorithms [3] consists of solution of two problems: the problem of stabilization of rigid subsystem and the restriction of flexible variables excitation. In this paper we suggest the control strategy, which gives a possibility to solve the stabilisation problem of flexible subsystem directly. In the base of developing approach lies a method of motion decomposition [4], allowing split a decision of problems of large dimension on independently decided subproblems of smaller dimension. Use VSC' algorithms in problems of robot control is presented natural on the strength of that, that, as a rule, control influences of executive devices have a lock-and-key nature. A known drawback of system with discontinuous control is "chattering" phenomena. For eliminating this drawback exist two possibility: i) increasing a frequency of switching the control in sliding mode motion and ii) control amplitude reduction. As shown in [1], increasing a frequency of control switching possible to obtain to the account of using the asymptotical observers-in this case sliding motion appear in the chain of feedback through the observers and, thereby, are exclude dynamic nonidealites of control object. On the other hand, in [5] is offered method of invariant system syntheses based on the VSC philosophy, allowing reduce discontinuous control amplitude (Section 3.2). Main idea of given work is conclude in use of both specified approaches. As follows, for the evaluation flexible variable is use theory of asymptotic observers (Section 3.3), but in the control problem (Section 4) this observer is consider as a dynamic compensator.

2. Modeling of manipulators by rigid and flexible coordinates subsystems

The dynamics equations of manipulators with distributed mass are described by the rigid-body and elastic coordinates, which reflect flexible nature of slim arms:

$$H(x)\ddot{x} = h(x,\dot{x}) + B\tau, \qquad (1)$$

where $x = (\theta^t, q^t)^t$ is the vector of the generalised coordinates $\theta \in R^n$ and the flexure (elastic) coordinates $q \in R^k$, $H(x)$ is the positive definite mass matrix, τ is the vector of the joint torques, $h(x,\dot{x})$ is the vector of generalised forces arising from centrifugal, coriolis, gravity effects and it associated with the elastic coordinates, B is the control distribution matrix. With the assumption of small elastic deflection, the dynamic model (1) can be described as

$$\begin{bmatrix} H_{11} & H_{12} \\ H_{21} & H_{22} \end{bmatrix} \begin{bmatrix} \ddot{\theta} \\ \ddot{q} \end{bmatrix} = \begin{bmatrix} h_1(x,\dot{x}) \\ h_2(x,\dot{x}) \end{bmatrix} + \begin{bmatrix} \tau \\ Wq \end{bmatrix}, \quad (2)$$

where the matrix W corresponds to the natural vibration modes, provided that all motors are locked ($\tau = 0$). The system (2) can be divided into two subsystems describing the behavior of rigid-body (eq.(3))

---

[1] This work has been supported by the INTAS under Grant No. 94-0965

and elastic coordinates ((eq.(4))

$$\bar{\theta} = H_1^{-1}[H_{12}H_{22}^{-1}(h_2 - Wq) - (h_1 - \tau)], \quad (3)$$

$$\bar{q} = H_2^{-1}[H_{21}H_{11}^{-1}(h_1 - \tau) - (h_2 - Wq)], \quad (4)$$

where $H_1 = H_{11} - H_{12}H_{22}^{-1}H_{21}$ and $H_2 = H_{22} - H_{21}H_{11}^{-1}H_{12}$. The model (3),(4) is used for control design. In the next section, a state vector evolution problem is considered. In Section 4, the variable structure technique based on the dynamic compensation method is applied to the control of a flexible structure.

3. State estimation

3.1. Nonspillover observation

The motion equation for a flexible structure can be written [6] as follows:

$$\dot{\theta} = A\theta + Bu, \quad (5)$$

where $\theta \in R^n$ is a controlled state vector and $u \in R^m$ is an input vector.

$$\dot{q} = Wq + B_1 u, \quad (6)$$

where $q \in R^k$ is a residual state vector. The output equation is:

$$y = D_1\theta + D_2 q, y \in R^p \quad (7)$$

Suppose that the linear system (5) is controllable and overall system (5), (6) is state-observable with respect to output variables y. As known, if observer is constructed on the basis of equation (5) only, the occurrence of residual modes leads to observation and control spillover phenomena, which can result in the instability of the controlled system. We below consider the nonspillover effect procedure of observer synthesis, using a full object model and sliding modes control strategy. Let us constructed the observer as follows:

$$\dot{\bar{\theta}} = A\bar{\theta} + Bu + L_1(y - D_1\bar{\theta} - D_2\bar{q}), \quad (8)$$

$$\dot{\bar{q}} = W\bar{q} + B_1 u + L_2(y - D_1\bar{\theta} - D_2\bar{q}), \quad (9)$$

Assuming that the sliding mode occurs along the switching surface $s = C\bar{\theta}$ we solve the control problem under assumption $\bar{\theta} = \theta$. Let us write the closed system of equations for variables $\bar{\theta}$, $\Delta\theta = \theta - \bar{\theta}$ and $\Delta q = q - \bar{q}$:

$$\begin{vmatrix} \dot{z} \\ \Delta\dot{\theta} \\ \Delta\dot{q} \end{vmatrix} = G \begin{vmatrix} z \\ \Delta\theta \\ \Delta q \end{vmatrix}, \quad (10)$$

where

$$G = \begin{bmatrix} A - C^*A & C^*(A - L_1 D_1) & CL_1 D_2 \\ 0 & A + L_1 D_1 & L_1 D_2 \\ 0 & L_2 D_1 & W + L_2 D_2 \end{bmatrix},$$

where $C^* = B(CB)^{-1}C$.

Note that the equivalent control $u_{eq} = -(CB)^{-1}C[Az - (L_1 D_1 - A)\Delta z + L_1 D_2 \Delta q]$ derived from the equation $\dot{s} = 0$ [6] is substituted in the equation (5). By virtue of suitable choices of the matrix C and pair (L_1, L_2) independently, the eigenvalues of the system (10) may be allocated as desired.

In offered above algorithm of syntheses of observer essential difficulties can cause high (n+k)-dimensional problems of choice of matrixes L_1, L_2. Offered in [7] method of motion separation in problems of observing allows split its decision on independently decided subproblems with dimension not above dimension of vector output variable (p).

3.2. The dynamic compensation method in the observation problem

Consider a linear dynamic system

$$\dot{\theta} = A\theta + B(u + \Lambda q), \dot{q} = Wq + B_1 u, \quad (11)$$

where variables are considered as follows: vector θ is the state vector, u is the control vector and q is the vector of disturbances. It is required to ensure invariance of the state vector x (under assumption that the vector x is measurable) with respect to external disturbances q. Let us express the control u in the form [5]:

$$u = -\Lambda z + v, \dot{z} = Wz + B_1 u + Lv, \quad (12)$$

where $z \in R^k$ is the state vector of the compensator and v is the vector of new control. By combining (11) and (12), the motion equations for closed system are given by

$$\dot{\theta} = A\theta + B\Lambda\varepsilon + Bv, \dot{\varepsilon} = W\varepsilon + Lv, \quad (13)$$

where $\varepsilon = z - q$. The control v is chosen in a class of systems with discontinuous control or high gains

$$v = -Msign(s) \text{ or } v = -ks, s = C\theta, \quad (14)$$

where $s \in R^m$, C is $(m \times n)$ constant matrix, M and k are positive scalars. Under sufficiently large M or k, it is possible to divide the overall motion in system (13) into two phases: at the first stage the state vector reaches the sliding surface (or fast motion), whereas at the second stage the sliding (or slow) mode occurs along this surface. According to the method of equivalent control, the equations of sliding modes (or slow motion) can be written as follows

$$\dot{\theta} = [A - B(CB)^{-1}CA]\theta, s = C\theta = 0,$$
$$\dot{\varepsilon} = -L(CB)^{-1}CA\theta + (W - L\Lambda)\varepsilon. \quad (15)$$

The characteristic polynomial of the diagonal cell matrix (15) coincides with the product of the characteristic polynomials of the diagonal elements. We assume that the pair (A, B) is controllable and, without loss of generality, the pair (Λ, W) is observable.

Hence, by appropriate choice of the matrices C and L it is possible to arbitrarily assign n-m nonzero eigenvalues of the matrix $A - B(CB)^{-1}CA$ and k eigenvalues of the matrix $W - L\Lambda$. Note that the system (15) is homogeneous and, hence, the stability in the system (15) can be assured with the aid of a linear control with finite gains.

Also, note that the amplitudes of discontinuities of controls that ensure the fulfillment of the conditions of occurrence of motions on the sliding manifold will be nonzero. After the end of the transient process, the undesirable self-oscillations (named as the chattering phenomena) will occur in the system. On the other hand let us note that by virtue of the system stability the equivalent control will tend to zero. In this case the amplitude of the control discontinuities can be also diminished to zero. Thus, it is possible to eliminate the chattering phenomena.

3.3. The observation problem for flexible structures

Now consider the observer design problem for estimation of residual coordinates q in the system (3), (4) under assumption that the variables θ and $\dot{\theta}$ are measurable. Also, we use a simplified model, in which the nonlinear terms with residual coordinates and their time derivatives and the terms in the inertia matrix with residual coordinates are dropped [2]. Using nonsingular change of the residual variables

$$\bar{q}_1 = q + L_1(\theta)\dot{\theta}, \quad \bar{q}_2 = \dot{q} + L_2(\theta)\dot{\theta}$$

the system (4) can be written about a new coordinate as follows

$$\dot{\bar{q}}_1 = \dot{L}_1(\theta,\dot{\theta})\dot{\theta} + \bar{q}_2 - L_2(\theta)\dot{\theta} + L_1(\theta)\ddot{\theta}, \quad (16)$$
$$\dot{\bar{q}}_2 = \dot{L}_2(\theta,\dot{\theta})\dot{\theta} + q + L_2(\theta)\ddot{\theta},$$

or

$$\dot{\bar{q}}_1 = \bar{q}_2 - L_1 \bar{H}_1 W \bar{q}_1 + \bar{h}_1, \quad (17)$$
$$\dot{\bar{q}}_2 = (\bar{H}_2 - L_2 \bar{H}_1)W\bar{q}_1 + \bar{h}_2,$$

where
$\bar{H}_1 = H_1^{-1} H_{12} H_{22}^{-1}, \bar{H}_2 = H_2^{-1} H_{21} H_{11}^{-1}$
$\bar{h}_1 = L_1 \bar{H}_1 (L_1 \dot{\theta} + h_2) + L_1 H_1^{-1}(h_1 + \tau) + \dot{L}_1(\theta,\dot{\theta})\dot{\theta} - L_2(\theta)\dot{\theta},$
$\bar{h}_2 = \bar{H}_2(h_1 - \tau) - H_2^{-1} h_2 + L_2 \bar{H}_1 h_2 + L_2 H_1^{-1}(\tau - h_1) + \dot{L}_2(\theta,\dot{\theta})\dot{\theta},$ Describe the observer dynamics by equation

$$\dot{z}_1 = \bar{z}_2 - L_1 \bar{H}_1 W z_1 + \bar{h}_1, \quad (18)$$
$$\dot{z}_2 = (\bar{H}_2 - L_2 \bar{H}_1)W z_1 + \bar{h}_2,$$

and write equations (17) and (18) in the variables $\varepsilon_i = \bar{q}_i - \bar{z}_i, i = 1,2$.

$$\dot{\varepsilon}_1 = \varepsilon_2 - L_1(\theta)D(\theta)\varepsilon_1, \quad \dot{\varepsilon}_2 = [\bar{H}_2 W - L_2(\theta)D(\theta)]\varepsilon_1. \quad (19)$$

where $D(\theta) = \bar{H}_1 W$. Consider the stabilization problem for the system (19), using a suitable choice of the function matrices $L_1(\theta)$ and $L_2(\theta)$ on the basis of the pole assignment approach under the assumption that the system (19) for $L_1 = L_2 = 0$ is observable with respect to the output variables $y = D\bar{\varepsilon}_1$ (this is equivalent to the fact that the pair $(\bar{H}_2 W, D)$ is observable), and $rank(\bar{H}_2 W) \le rank(\bar{H}_1 W) = rank(D) = n$ for any θ. Let us assume that, instead of the system (19), the problem of observer design has been decided for the observable system with constant coefficients described by equations

$$\dot{\varepsilon}_1 = \varepsilon_2 - L_{o1}D_o \varepsilon_1, \quad \dot{\varepsilon}_2 = (D_{o1} - L_{o2}D_o)\varepsilon_1. \quad (20)$$

where $rank(D_o) = n, rank(D_{o1}) = rank(\bar{H}_2 W) \le rank(D)$. Then, it is readily verified that, by the suitable choice of the matrices $L_1(\theta)$ and $L_2(\theta)$ in the system (19) according to the relationships $L_1 = L_o D_o D^t (DD^t)^{-1}$ and $L_2 = (L_{o2}D_o - D_{o1} + \bar{H}_2 W)D^t(DD^t)^{-1}$, the system (19) can be reduced to the system (20). Thus, the variables $\bar{z}_i, i = 1,2$ may be given as estimates of the variables $\bar{q}_1$ and $\bar{q}_2$ and, consequently, we have solved the problem of reconstruction of the variables q and $\dot{q}$.

4. Control design

In the Sections 3.1 and 3.2, the observation problem has been solved using control resources that restrict the solution of the stability problem of the overall system. In this section, we assume that the estimate of the state vector is obtained on the basis of the theory of asymptotical observers (as in Section 3.3). The procedure of design of invariant systems described in Section 3.2, directly shown a possibility of correcting the dynamics of the disturbances model. Indeed, after the substitution of equivalent control (13)

$$v_{eq} = -B^{-1}A\theta - \Lambda\varepsilon$$

into the second equation (11), we get

$$\dot{q} = (W - B_1\Lambda)q - B^{-1}A\theta$$

and, since the variables θ are asymptotically stable (eq.(15)), the behavior of the disturbances model is corrected by the component $B_1 \Lambda q$. This result can practically be used, if there is the possible to change parameters of matrices B_1 and W.

Consider the problem of stabilization of the system (3),(4) on the basis of the separation motion method [3] under assumption that this system is controllable and $rank B = n$. The idea of synthesis procedure will be demonstrated for linear case, then the mathematics model is presented as follows.

$$\dot{\theta}_1 = \theta_2, \dot{\theta}_2 = A_{21}\theta_1 + A_{21}\theta_2 + Bu, \dot{q} = Wq + B_o u, \quad (21)$$

where $Q_1, Q_2 \in R^n$ - vectors of rigid-body coordinates, $q \in R^k$ -vector of flexible coordinates. The following procedure decomposes the control problem into three independent subproblems of smaller dimension.

On the first step by nonsingular change of the variables $\bar{q} = q + M_1\theta_1 + M_2\theta_2$, where $M_2 = -B_0 B^{-1}, M_1 = WM_2 - M_2 A_{22}$, the three subsystem (21) is written as follows

$$\dot{\bar{q}} = W\bar{q} + M\theta_1, \qquad (22)$$

where $M = M_2 A_{21} - WM_1$. Since the system (22), where θ_1 is regarded as a control, is controllable, there exists a feedback matrix F such that matrix $W + MF$ has any desired eigenvalue distribution. Thereby, by means of choice matrixes F in correlation

$$\theta_1 = F\bar{q} \qquad (23)$$

can be solved a problem to stabilization of system (22).

For ensuring an equality (23), on the second step of the procedure necessary to ensure equality a zero variables $\bar{\theta}_1 = \theta_1 - F\bar{q}$, describe on the strength of (21) and (22) by equations

$$\dot{\bar{\theta}}_1 = (FW + MF)\bar{q} + M\bar{\theta}_1 + \theta_2. \qquad (24)$$

Shall consider variable θ_2 in equation (24) as fictituous control and will place their equal

$$\theta_2 = -(MF + FM)\bar{q} - M\bar{\theta}_1 + K\bar{\theta}_1 \qquad (25)$$

Then, after substitution of this expression in (24) by choice matrixes K is ensured stabilizations of variable $\bar{\theta}_1$

$$\dot{\bar{\theta}}_1 = K\bar{\theta}_1.$$

Finally, on the third step of procedure necessary to ensure an equality (25) or convergence to zero variable

$$\bar{\theta}_2 = \theta_2 + (MF + FM)\bar{q} + M\bar{\theta}_1 - K\bar{\theta}_1$$

which will comply with equation of type

$$\dot{\bar{\theta}}_2 = \bar{A}_{12}\bar{\theta}_1 + \bar{A}_{22}\bar{\theta}_2 + \bar{W}\bar{q} + B_2 u. \qquad (26)$$

Problem to stabilisations a system (26) can be solved (similarly second step) by the choice of control u in the class of continuous functions or in the class of discontinuous functions by means of introduction of sliding mode on the surface $s = \bar{\theta}_2 = 0$. Note that in practice realize moments (variable u), created by executive devices in the class of discontinuous functions is not presented possible. Model of control object must be complement by equations own dynamics of executive devices. In this case, intimate above hierarchical procedure can be continue up to real discontinuous controls [8].

Note, that using the method of theory of sliding mode ensures invariance of stabilisation problem to uncertainties of description of control object and to external disturbances. In particular, this fact allows synthesise a stabilisation problem for nonlinear systems in the suggestion, that are known only bounds of nonlinear members.

5. Conclusions

In given work are present results on the syntheses of problems of flexible structure control on the base of method of motion separation [3], allowing divide problems of large dimension on independently decided subproblems of smaller dimension. In spite of that greater volume of the work is refer to linear systems, procedure of syntheses of observer for the nonlinear case (section 3.3) and remark at the end of section 4 for introduction the sliding modes in nonlinear systems, show a possibility of using the brought results for the nonlinear system syntheses.

6. References

[1] Utkin V.I. (1992) Sliding modes in control and optimization. Sprigner Verlag.

[2] Dong Li (1994) Nonlinear control design for tip position tracking of a flexible manipulator arm. Int. J. Control, vol. 60,no.6, pp. 1065-1082.

[3] Young K. and Ozguner U.(1990) Frequency shaped variable structure control. Proc. VSS'90, pp. 181-185, Sarajevo, Yugoslavia, 19-20 March.

[4] Drakunov S.V., Isosimov D.B., Luk'yanov A.G., Utkin V.A. and Utkin V.I. (1990) The block control principle, I. Automation and Remote Control, Vol.51, No.5, Part 1, pp. 601-609.

[5] Utkin V.A. and Utkin V.I. (1983) Design of invariant systems by the method of separation of motions. Automation and Remote Control, Vol. 44, No.12, Part 1, pp. 1559 - 1566.

[6] Balas M.J.(1978) Feedback Control of Flexible systems, IEEE Transactions on Automatic Control, vol. AC-23,no. 4, pp.673-679.

[7] Utkin V. A. (1990) Method of separation of motions in observation problems. Automation and Remote Control, Vol. 44, No.12, Part 1, pp.300-308.

[8] Utkin V.A. (1994) Constrained robot control based on the method of movements separation. Tampere Int. Conf. on Machine Automation Mechatronics Spell profitability, Tampere, Finland, Febr. 15-18, pp.86-97.

NEW BROADBAND 100-MBPS SWITCH SYSTEM USING BROADBAND PIN-BOARD SWITCH AND HIGH-PRECISION PIN-HANDLING MECHANISM

Shuichiro Inagaki, Takashi Yoshizawa, and Keiichi Kobayashi

NTT Opto-electronics Laboratories

3-9-11, Midori-cho, Musashino, Tokyo 180-0012, Japan

Abstract

A broadband pin-board switch system using high-precision pin-handling technology was studied. It can crossconnect 320 twisted pair cables with 240 twisted pair cables. Experiments showed that the error rates of the system applied to a 100-Mbps Fast Ethernet LAN and a 150-Mbps ATM LAN meet the criteria of the LAN links.

1. Introduction

There are several switch systems, such as a pin-board switch system [1], switching IC system, and relay system, that can be used to automate jumpering in distribution frames and as a switch for backup systems, etc. A broadband switch system is becoming necessary as higher-speed communications are being provided. However, the switching IC has a low breakdown voltage and requires power to hold its switching state and the relay cannot transmit high-speed signals. On the other hand, the pin-board switch, which consists of a matrix board with crosspoint holes (about 1 millimeter in diameter) and small connecting pins and can make a signal path between an input port pair and an output port pair by inserting a pin into a designated hole of the matrix board, is expandable in bandwidth by optimizing the wiring pattern of the matrix board. Therefore, we studied a broadband pin-board switch system using high-precision pin-handling technology.

2. System configuration

Figure 1 shows the configuration of this system. The system consists of several switching units and a controller. Each unit consists of the broadband pin-board switch and a pin-handling mechanism composed of a hand assembly, a positioning sensor, and a positioner. Figure 2 shows the broadband pin-board switch and the pin-handling mechanism of the prototype system. The prototype system can crossconnect 320 twisted pair cables with 240 twisted

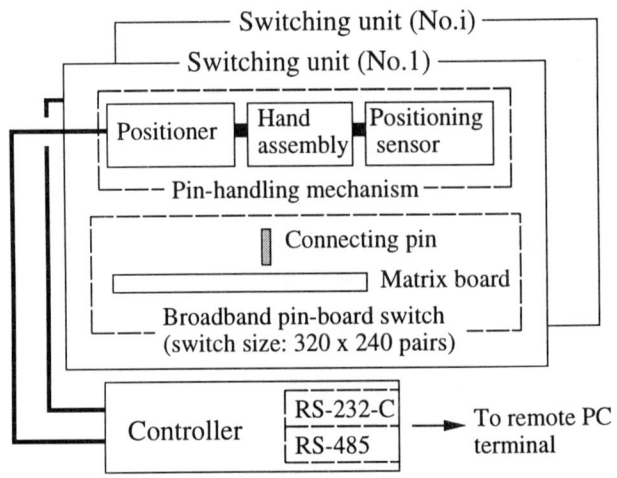

Figure 1 System configuration.

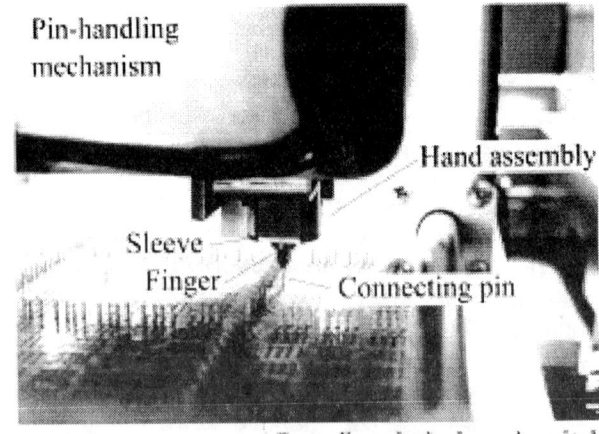

Figure 2 Prototype broadband pin-board switch and high-precision pin-handling mechanism.

3. Broadband pin-board switch

The main objective of the system investigation was to make a broadband pin-board switch. The wiring pattern in the matrix board was optimized. Furthermore, so that the transmission characteristics of the broadband pin-board switch under service conditions may satisfy the requirements of ISO-defined category 5 LAN [2], a twisted pattern structure and a stub pattern shortening structure were devised. Figure 3 shows perspective views of the structures. The twisted pattern structure can cancel crosstalk noise in switch areas 1 and 2. Extracting a pin from the hole of the stub pattern shortening structure separates switch area 4 from switch area 3. The stub pattern shortening structure enables an open stub pattern that causes reflection and corresponds to switch area 4 being separated from a signal path set in the broadband pin-board switch. Therefore, these two structures considerably reduce the near-end crosstalk and the return loss of the broadband pin-board switch. However, the twisted pattern structures may cause polarity inversion between the input and output of the broadband pin-board switch. Therefore, a polarity compensation structure, which is also shown in figure 3, was devised. By inserting a pin into either hole 1 or 2, the polarity compensation structure keeps the broadband pin-board switch polarity constant. These structures for improving the data transmission characteristics and the normal switching structure can be fabricated in the same manufacturing process and all these structures work with the pin-handling mechanism. Hence, this is very effective for reducing the system cost.

Figure 4 shows the transmission characteristics of typical paths set in the prototype broadband pin-board switch with 5 meter length input/output cables. Furthermore, experiments showed that the error rates for both a 100-Mbps Fast Ethernet LAN and a 150-Mbps ATM LAN meet the criteria of the LAN links, and the floor cables of the LANs can be extended over 100 meters. Certainly, the prototype broadband pin-board switch can withstand 200 volts. Therefore, the

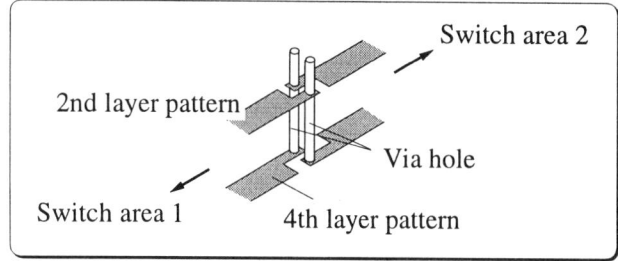

(1) Twited pattern structure

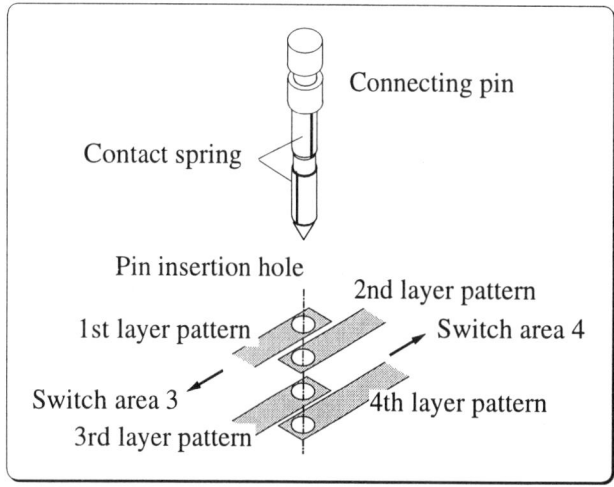

(2) Stub pattern shortning structure

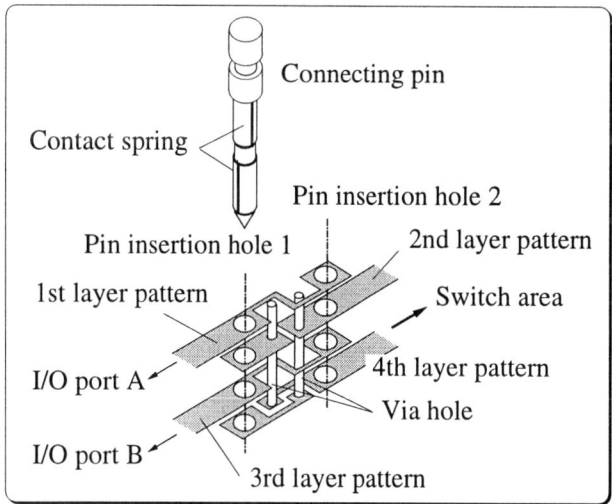

(3) Polarity compensator structure

Figure 3 Twisted pattern structure, stub pattern shortening structure, and polarity compensator structure.

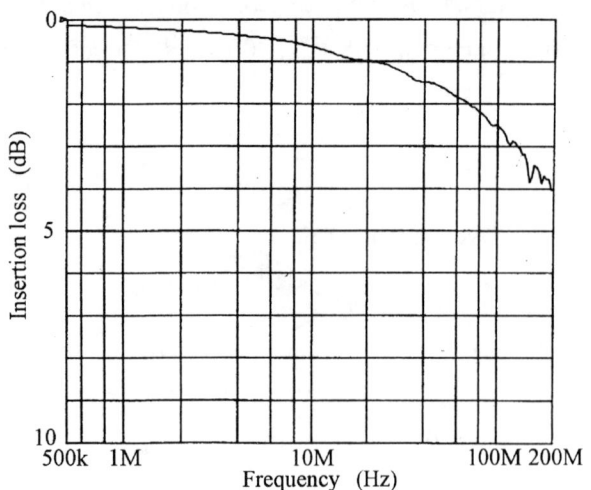

(1) Insertion loss

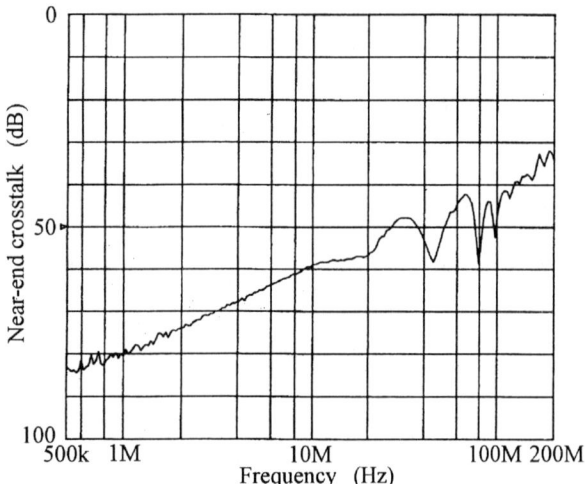

(2) Near-end crosstalk

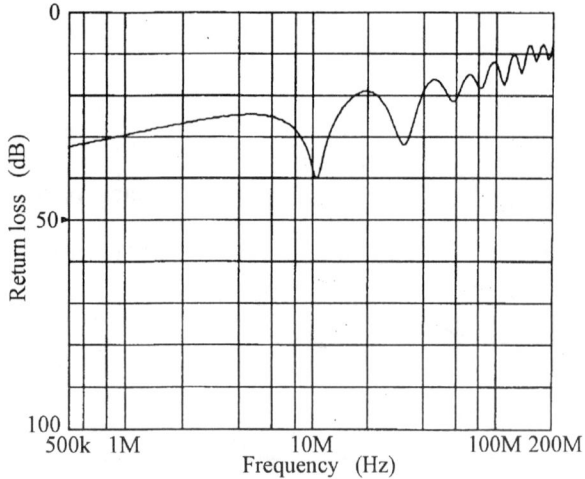

(3) Return loss

Figure 4 Typical transmission characteristics of broadband pin-board switch with 5 meter length input/output cables.

prototype broadband pin-board switch can transmit signals ranging from analog telephone signals to 100-Mbps digital signals.

4. High-precision pin-handling mechanism

The pin-handling mechanism manipulates the pin, whose body is plastic and rather fragile [3] and which must be inserted into (or extracted from) holes located every 1.5 millimeters in the matrix board. To make pin insertion and extraction reliable, the pin-handling mechanism requires four thin high precision fingers for grasping the pin.

In order to make a high-precision hand mechanism with four fingers and to reduce its assembly time, the one-piece structure shown in figure 5 was used and manufactured by electric discharge machining. Since each finger has a spring part, the fingers are closed when the tapered sleeve which contacts the outer regions of the fingers is pushed down. In this way, the hand can grasp a pin. To prevent damage to the pin, the hand assembly has a compliance mechanism which limits the pin insertion force. By optimizing the finger tip shape and making the finger actions symmetrical, we improved the positioning margin of the hand assembly for pin insertion and extraction to approximately plus or minus 0.35 millimeters.

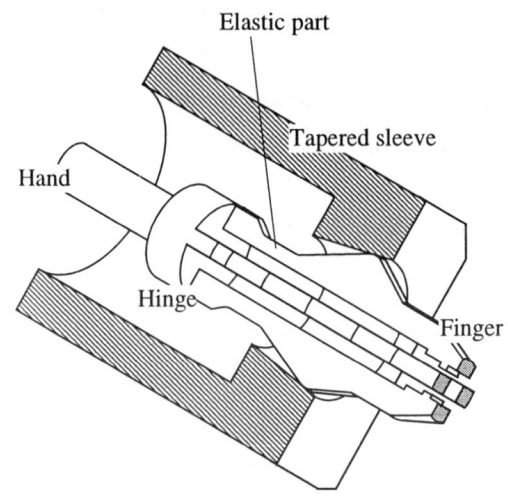

Figure 5 One-piece hand structure and sleeve.

Both accuracy and low cost are required for the positioner. Considering the hole position deviation (about 0.1 millimeters) and assembly tolerances, the accuracy of the hand assembly positioning must be kept within 0.2 millimeters over the whole travel range of the positioner (about 500 millimeters). In order to reduce the system cost, stepping motors and timing belts are used for the positioner. However, it is difficult for the belts to keep the positioning accuracy because of their creep and thermal expansion characteristics. In order to compensate for the positioning error caused by the belts, the safeguards mentioned below are used to maintain reliable positioning. First, a reflection-type fiber sensor using an LED is used as the positioning sensor. The sensor beam is focused on the matrix board surface through a convex lens. This sensor configuration can detect the hole and the pin head. Positioning failure can be prevented by using the sensor to count the number of holes along the path being traveled. Second, by using the sensor output for gating, the detected number of pulses of the stepping motor photo-encoder is utilized to check the traveling length and positioning of the pin-handling mechanism at the center of a hole. Furthermore, the positioner action is compensated for by information about the matrix board tilt and the belt expansion obtained from the sensor at system initialization. Third, in order to increase the pin manipulation reliability, the existence of the pin at the designated hole is checked by the sensor before and after pin manipulation.

5. Path management

Path information, i.e. path connection status, is needed for cable management. Considering the total cost of the system, we chose a remote PC terminal to manage this information. The controller translates commands sent from the terminal - such as 'set path' and 'clear path' - into a simple series of commands - such as 'move', 'open hand', and 'close hand' - and drives the pin-handling mechanism. The prototype controller is a custom-designed circuit board on which a single-chip CPU is mounted. For remote and multiple control by the terminal, the controller has an RS-232-C interface and an RS-485 interface, as shown in figure 1. Maintenance software tools, such as a pin location checker for the indicated path and path initialization for recovery from abnormal pin insertion status due to power failure, are provided. An error retry function is used to recover from a temporary error.

6. Conclusion

We studied a broadband pin-board switch system using high-precision pin-handling technology. New structures were devised to improve the data transmission characteristics. A one-piece structure was used to make a high-precision hand mechanism with four fingers. Furthermore, a reflection-type fiber sensor using an LED was used to compensate for the positioning error caused by the belts. The prototype broadband pin-board switch system can crossconnect 320 twisted pair cables with 240 twisted pair cables and can transmit signals ranging from analog telephone signals to 100-Mbps digital signals.

References

[1] S.Hosokawa, et al.:"Robot automation of MDF-jumpering operation," Trans. of IEICE, Vol.J73-B-I, no.12, pp.922-931, Dec. 1990.

[2] "Information technology - Generic cabling for customer premises," ISO/IEC 11801:1995(E), July 1995.

[3] S.Umemura, et al.:"Design of high density pin board matrix switches for automated main distribution frame systems," IEEE Trans. CHMT, vol.CHMT-15, no.2, pp.266-277, April 1992.

AN AGENT-BASED STRUCTURE FOR MOBILE ROBOTS USING VISION AND ULTRASONIC SENSORS

T. F. Bastos-Filho, R. A. C. Freitas, M. Sarcinelli-Filho, H. A. Schneebeli

Department of Electrical Engineering, Federal University of Espirito Santo
Av. Fernando Ferrari, s/n
Vitoria, Espirito Santo 29060-900, BRAZIL

ABSTRACT

Object recognition is an important task associated to mobile robots navigation. Upon detecting any obstacle, the recognition system must be able to say which obstacle is in the robot trajectory, so that the control system is able to plan a new trajectory for the robot, deviating from the detected obstacle or following it, depending on which is the obstacle. It is normally necessary to recognize a few obstacles that are commonly present in the robot operating environment. In this paper, a system is proposed to perform the task of recognizing the objects present in the trajectory of a mobile robot. This system is based on information coming from ultrasonic sensors and a digital monochromatic camera. The operation of this system is addressed, as well as an example of object detection and recognition is presented.

1. INTRODUCTION

A differential-drive robot is being developed at the Department of Electrical Engineering (DEL) of the Federal University of Espirito Santo (UFES), Brazil. The robot is a round platform with four wheels: two of them are mounted on a common axis and are independently driven by two DC motors, and the two other are free wheels. Figure 1 shows that mobile robot, called Brutus.

This robot uses an Motorola MC-68332 microcontroller to control the several processes involved in its movement, like the internal sensorial perception and the trajectory planning. The robot is designed to avoid collisions when moving itself in a semi-structured environment. This is an environment whose contours are not previously known, but is normally closed and has a regular surface. This kind of environment normally contains a small number of known objects (like walls, edges, corners, chairs and tables), which must be avoided during the robot movement, but unknown objects can also appear in the trajectory of the robot, and must also be detected.

The robot is also equipped with a ring of sixteen ultrasonic sensors distributed along the outer circle of the robot platform (see Figure 1), which are responsible for detecting that an obstacle is close to Brutus (only obstacles closer than 1 m are considered). On detecting the presence of an obstacle closer than 1 m, this ultrasonic sensing system informs to the control unit onboard of the robot the distance and the direction of the detected obstacle. The direction is defined by the angle between the axis of movement and the acoustic transmission axis of the sensor that detected the obstacle (the angular displacement between adjacent sensors is 22.5 degrees). To implement the ultrasonic sensing system, a 68HC11 microcontroller is used to control the 16 ultrasonic sensors. Then, after receiving the information that an obstacle was detected, a recognition system is activated, in order to recognize the obstacle (if it is one among the small number of objects aforementioned). This recognition system will be installed in another MC-68332 microcontroller board, whose main task will be to process an image acquired by a monochromatic camera (Sony model XC-77). This camera is located at the top of the robot platform, and is used to obtain more information about the environment of the robot, and to recognize the objects present in its trajectory. All this mixed sensor system is designed as an agent-based system [1]. The necessity of recognizing the obstacles in the trajectory of the robot is that the trajectory should be changed whenever an obstacle is detected, and this change depends on the object (e. g., it is possible to pass under a table, but not through a wall).

In this work, an agent-based control structure is implemented in order to process the information provided by the artificial vision system and the ultrasonic sensors, thus defining any trajectory change to be accomplished. The paper is organized as follows: this introductory section; Section 2, describing the agent-based control system used to control the mobile robot; Section 3, describing the object recognition system; and, finally, Section 4, containing the final remarks.

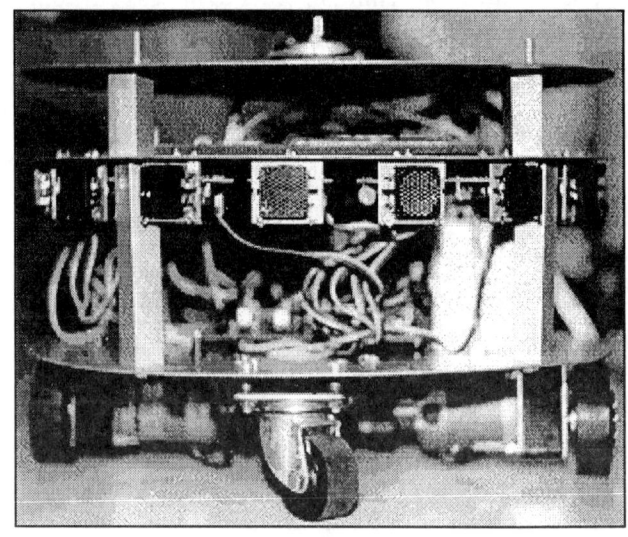

Figure 1: The differential-drive mobile robot Brutus.

2. THE AGENT-BASED CONTROL SYSTEM

An agent-based architecture is an alternative approach to the traditional mobile robot control problem. In [2] it is presented an abstraction for agent-based controllers, in which the agents are represented by concurrent and interconnected modules, distributed in three categories, as shown in Figure 2: sensor agents (primitive sensor agents and virtual sensor agents), behavior agents and actuator agents (virtual actuator agents and primitive actuator agents). In this approach, like in Brooks' approach [3], a group of modules is responsible for an activity (behavior) to be executed by the robot. Each activity represents a system goal. Mechanisms of interaction and mediation between modules are responsible for defining the main goal.

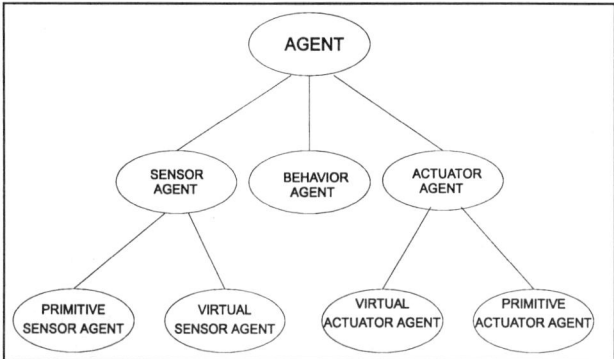

Figure 2: The categories of agents.

This module categorization makes the purpose of each module closer to its representation, giving to each agent category some particular characteristics, related to status and communication, associated to their specific role. For example: the actuator agents have the characteristic of being commanded by just one behavior agent at any time. The sensor drivers are represented by primitive sensor agents while tasks like *build maps* are represented by virtual sensor agents. The modules that define actions or behaviors are represented by behavior agents and tasks like *move forward* are represented by virtual actuator agents. The actuators output drivers are represented by primitive actuator agents [4].

The structure for building agent-based control systems developed in [4] implements the abstractions in [2] using the C++ programming language and concurrent libraries. So, a concurrent object-oriented structure is built. This structure is very suitable for behavior-based systems modeling because it joins characteristics such as modularity, encapsulation, inheritance, concurrency and the C++ efficiency.

Once the mobile robot here mentioned uses the behavior-based control architecture proposed in [4], the sensing system is also designed as an agent-based system [1]. In this system, four kinds of agents are implemented: a *Local Scheduler* sensor agent, an *Arbitrator* sensor agent, 16 *S* sensor agents, related to each one of the existing ultrasonic sensors, and a *Discriminator* sensor agent, related to the camera used, which can be dynamically created, whenever it is necessary to recognize an object [1]. Figure 3 shows the accomplished implementation.

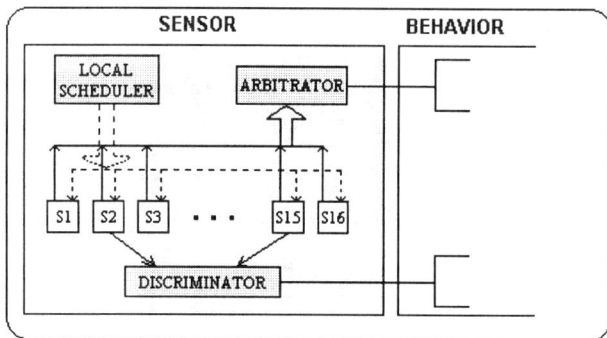

Figure 3: Agents of the sensor system.

3. THE STRATEGY FOR OBJECT RECOGNITION

The object recognition task is a basic one to a mobile robot. This task helps navigation (obstacle avoidance), and allows trajectory planning or behavior activation whenever an obstacle is detected. As examples of behaviors to activate, we can cite to follow a wall, to cross a door, to contour an unknown obstacle, etc.

A cheap solution for the recognition problem of a limited set of objects (what is typical in a semi-structured environment) can be implemented using only an ultrasonic system [5-6]. The main problem is that ultrasonic sensors have limited capacity of recognition due to the signal wavelength as well as to their sensitivity to several of environmental factors [6-8]. For this reason, recognition systems based on artificial vision have been normally used [9-11]. However, there is a problem associated to the use of artificial vision for recognizing objects also: the task of image processing is too demanding to be carried out by the onboard processing unit, thus making the navigation slow [6].

The recognition system here presented is a mixed one, partially based on ultrasonic sensors and partially based on artificial vision. The 16 ultrasonic sensors installed in the mobile robot allow to cover all the environment around the robot. However, their frequency of activation (to be activated here means to irradiate the ultra-sonic signal and to capture the echo reflected from an object), depends on a priority associated to each one: when the robot is moving ahead, the priority associated to the sensor that is in its front is the biggest one, and so on [1]. Thus, when an object is detected, more than one sensor may receive its echo signal, but only that one that received the biggest peak value is arbitrated as the sensor that detected the object. It measures, then, the distance to the object, based on the time of flight of the signal. The direction in which the object is apparently located is given by the angular displacement of that sensor related to the frontal sensor. Then, if the distance measured is lower than 1 m, the camera mounted in a stepping motor driven system located at the top of the robot platform (see Figure 4) is rotated towards the sensor that detected the object. Then, a frontal image of the detected object is captured. In the language of the agent-based control system, this means that the *S* sensor agent that detected the obstacle requests the creation of a copy of the *Discriminator* sensor agent [1], which rotates the camera in the right direction. Then, the system starts carrying out the recognition of the detected object, by processing the acquired image.

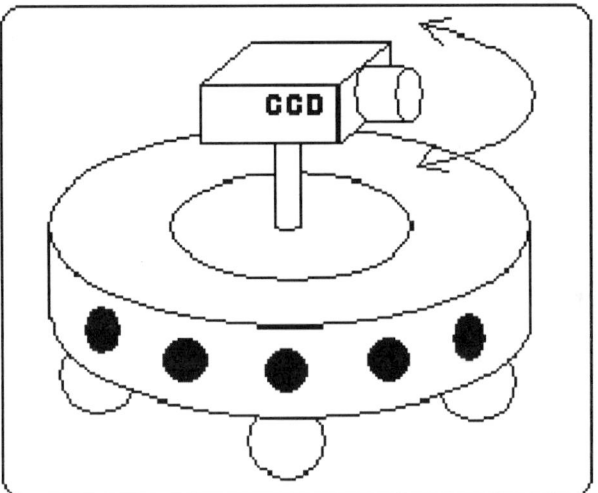

Figure 4: Onboard camera location.

Thus, notice that only upon the detection of any obstacle the camera is activated in order to acquire an image of that obstacle (remember that only objects closer than 1m to the robot external circle is considered as a detected obstacle). This implies in a low rate of image acquiring and processing.

The acquired image of a detected obstacle is then filtered using a mask to enhance the vertical contours, which are associated to the forms of corners, table feet, chair feet, etc. The resulting image is then submitted to a threshold mask used to transform the image in a binary one (black and white levels only). Then, an algorithm is employed to extract the more important quasi-vertical contours present in the image, what is done by eliminating those that are not meaningful, in terms of its length in the scene (remember that the final interest is to detect contours that correspond to the set of objects to be recognized, and that the acquired images are always frontal images).

The task of object recognition is then performed, based only on the segmented image containing the main vertical contours of the detected object. It is important to mention that the computation of this segmented image is a very simple one, which does not overload the onboard processing unit. The recognition is performed in the following way: if no quasi-vertical contour is present in the segmented image, the object detected is an infinite plane (a wall); if more than two quasi-vertical contours are detected, the object is an unknown one (for the set of objects aforementioned, at most two quasi-vertical contours are possible); if only one quasi-vertical contour is detected, the object is a corner, an edge or a corner with a door; and if two quasi-vertical contours are detected the object is a table foot or a chair foot, depending on how wide is the region inside the two contours. Figures 5a to 5d show an example of this image treatment, when applied to the captured image of an edge.

Figure 5a: Original image corresponding to an edge.

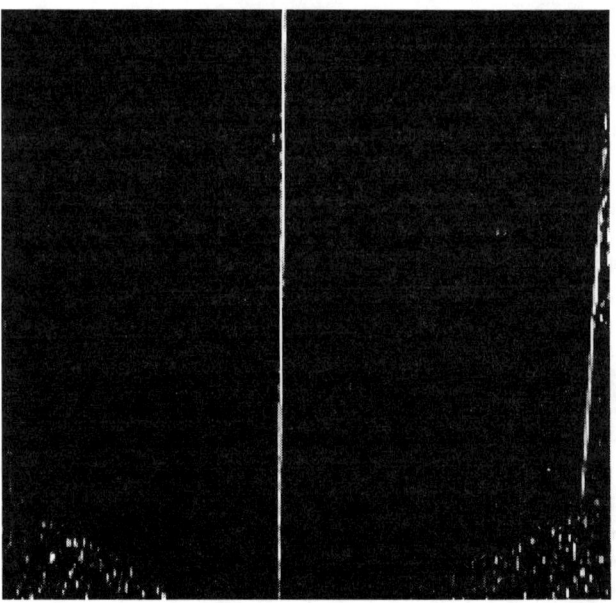

Figure 5b: The image after enhancing the vertical edges.

As it can be seen after comparing the Figures 5c and 5d, the algorithm used to detect the principal vertical edges ignores discontinuity in a line, thus interpreting many segments in the same vertical line as one line, as well as the vertical lines that are not meaningful in extension (see the second line at the rightmost side of Figure 5c, which disappeared in Figure 5d). In this example, a line is not considered a meaningful one if it does not cover seventy percent of the image height.

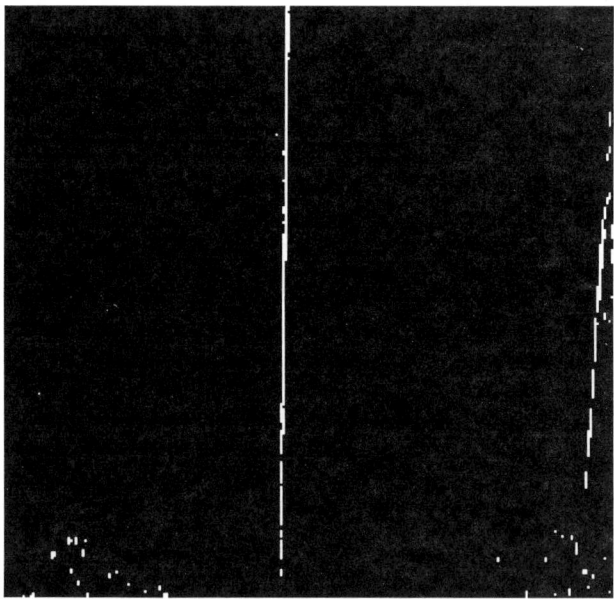

Figure 5c: The image after using the threshold mask.

Figure 5d: The final processed image.

4. FINAL REMARKS

The ability of recognizing some obstacles in the robot operating environment gives more flexibility to its control system, and allows the design of a great number of behaviors to be programmed in the robot, thus allowing the robot to perform more complex tasks.

This work presented an agent-based structure for object recognition, in order to help the navigation of a mobile robot. The system implemented is based on ultrasonic sensors combined with artificial vision. The recognition task is a very simple one, once image acquiring and processing is performed only when an object is detected closer than 1m to the robot. Besides this, the used algorithms for processing the images themselves are very simple, when the set of recognizable objects contains is a small one. Other features accomplished by the proposed recognition system are robustness, compactness and efficiency, what make it a very interesting one, mainly if considering the limited computational capacity and energy availability of Brutus.

The next step on developing this recognition system is to install it onboard of the robot and to expand it to recognize other objects.

5. ACKNOWLEDGEMENT

The authors would like to thank Professors Benjamin Kuchen and Ricardo Carelli, of the Institute of Automatic, National University of San Juan, Argentina, for the cession of the software DECVISION, used for processing the images here presented.

6. REFERENCES

[1] Freire, E. O. and Bastos-Filho, T. F. and Xavier, J. E. M. and Schneebeli H. A., "Agent-Based Ultrasonic Sensing System for Mobile Robots". *7th Latin-American Congress of Automatic Control*. Buenos Aires, Argentina, pages 910-915. September, 1996 (in Portuguese).

[2] Schneebeli, H. A., "The Control of a Multi-fingered Hand". *Ph.D. thesis*, Universität Karlsruhe, Germany, 1992 (in German).

[3] Brooks, R. *Achieving Artificial Intelligence Trough Building Robots*, AI Memo 899, MIT Press, 1986.

[4] Xavier, J. E. M. and Schneebeli, H. A. "A Structure for Constructing Agent-Based Control Systems for Mobile Robots". *2nd Brazilian Symposium on Intelligent Automation*, Curitiba, Brazil, pages 97-102. September, 1995 (in Portuguese).

[5] Indelicato, A. and Bastos-Filho, T. F. "A Structure for the Recognition of Geometrical References for Mobile Robots Using Fuzzy Logic and Ultrasonic Sensors". *7th International Fuzzy Systems Association World Congress*. Prague, Czech Republic, pages 361-364. June 1997.

[6] Anaya, J. J. and Fritsch, C. and Ruiz, A. and Ullate, L. G. "A High Resolution Object Recognition Ultrasonic System". *Sensors and Actuators A*, 37-38(4):644-650, 1993.

[7] Bastos-Filho, T. F. "Seam Tracking and Analysis of Arc Welding Environments Automated Trough Ultrasonic Sensors". *Ph.D. thesis*, Universidad Complutense de Madrid, Spain, 1994 (in Spanish).

[8] Bastos-Filho, T. F. and Calderón, L. and Ceres, R. and Martin, J. M. "Ultrasonic Signal Variation Caused by Thermal Disturbances". *Sensors and Actuators A*, 44:131-135, 1994.

[9] Brooks, R. A. "Visual Map for a Mobile Robot". *IEEE International Conference on Robotics and Automation*, pages 824-829. 1985.

[10] Horn, B. K. P. *Robot Vision*, McGraw-Hill, New York, USA, 1989.

[11] Kriegman, E. T. and Binford, T. O. "Stereo Vision and Navigation in Buildings for Mobile Robots". *IEEE Transactions on Robotics and Automation*, 5(6):792-803, 1989.

CMOS 1D and 2D N-well Tetra-lateral Position Sensitive Detectors

M. F. Chowdhury+, P. A. Davies* and P. Lee*

+ LSI Logic Europe Ltd., Greenwood House,
London Road, Bracknell RG12 2UB, UK.
* Electronic Engineering Laboratory, University of Kent at Canterbury,
Canterbury, Kent CT2 7NT, UK.

ABSTRACT

In this paper characteristics of one dimensional (1D) and two dimensional (2D) N-well tetra-lateral Position Sensitive Detectors (PSD) structures are reported. The PSDs have been fabricated using standard CMOS process. Measurements of the device parameters (dark current, resistance, and junction capacitance), responsivity, position resolution, position linearity and variations due to optical spot size shows that such PSD structures exhibits unique characteristics that are worth considering when designing the biasing and linearization circuits. Overall observation shows that both PSD structures can be successfully fabricated using standard CMOS technology. These measurements were primarily taken to determine the range of operating conditions for each device. This information is intended for chip designers to deduce specifications for designing on-chip signal conditioning and signal processing circuits to improve the performance of the device. Applications of PSDs include optical lateral, rotational or angular position sensing, surface smoothness measurements, range-finding instrumentation, 3D imaging and micro-mechanics.

1. INTRODUCTION

An optical position sensitive detector is a device used to measure the position of an optical beam incident to the surface of a planar p-n junction referred here as the active area. When the optical beam falls on the active area of the PSD a photo-potential is induced and this causes photo-current to flow through the circuit connected to electrodes X1 and X2 when operated in reverse bias condition as shown in Figure 1.

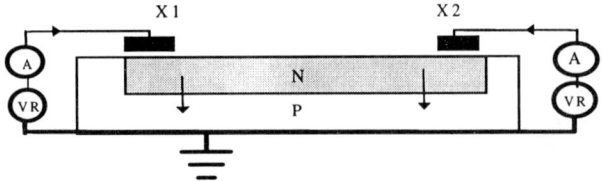

Figure 1: Current flow through N-well PSD structure

Current flowing through X1 and X2 is proportional to the position of the incident light beam. The magnitude of this current is determined by the intensity of the optical source and physical properties of the PSD structures[2]. Difference to sum ratio of the two currents is usually taken [5] to produce a signal which should be a linear function of the position of the light beam and independent to the intensity of the optical source:

$$I = \frac{(Ix1 - Ix2)}{(Ix1 + Ix2)}$$

In practice, however this signal is found to be non-linear and further processing is required to linearise the device.

The lateral photoeffect was first observed by Shottky in 1930 [1] and later by Wallmark in 1957 [2]. The analysis and improvements of the effects have been reported by a number of authors [3, 4, 5] for different geometry and biasing conditions. Measurements of the PSD properties have shown that there are considerable distortion present in the device due to variations in device structures and physical properties of the silicon used for the sensing area. This distortion is found to get progressively worse towards the extreme edges of the active area. In this paper we report that there are other factors which influence this distortion. These are: responsivity, position resolution, and variations due to optical spot size.

These measurements can be particularly useful for the circuit designers involved in designing on-chip front-end biasing and signal conditioning circuits using CMOS/BiCMOS technologies. All measurements were taken using a purpose-built IC chip containing an number of experimental PSD structures. The chip were fabricated using AMS 1.2µm CMOS technology with p-substrate always connected to ground.

The chip contained four different photo-diodes and six different PSD structures. In this paper only the measurements taken for a selected samples of 1D and 2D rectangular N-well PSD structures are reported.

2 PSD structures

Figure 2 shows the cross-sectional views of the two types of PSD structures considered in this paper.

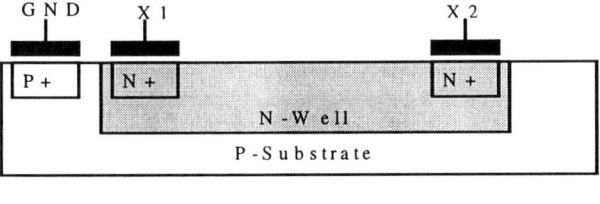

(a) N-well PSD

Figure 2: PSD structure cross-sectional views

As can be seen from Figure 2 the active (or the position sensing) area of the PSD is the N-well. Dimensions of 1D structure is 200µm X 1600 µm with two electrode connected opposite longer ends. Dimensions of 2D structure is 800µm X 800µm with electrodes connected to all four edges.

	Resistance X1 to X2 (Ω)	Res. per Sq Expected
N-WELL 2D	0.631K	2.2K
N-WELL 1D	15.1K	2.2K

Comparing resistance per square for 2D PSD structure showed that it is significantly lower than the initial expected value. According to the AMS specification the resistance is expected to be 2.2KΩ per square. The reason for this reduction can be explained from the simplified model of the 2D PSD structure shown in Figure 3. For 1D PSD, the measured resistance per square is 1.9KΩ.

From Figure 3 the measured resistance between adjacent electrodes was found to be approximately 650Ω. Thus, the resistance measured between X1 and X2 is the parallel combination of all the resistance and approximately agrees with the measured values. For the 1D PSD this parasitic resistance do not appear since the PSD contains only two electrodes placed on the opposite end of the active area.

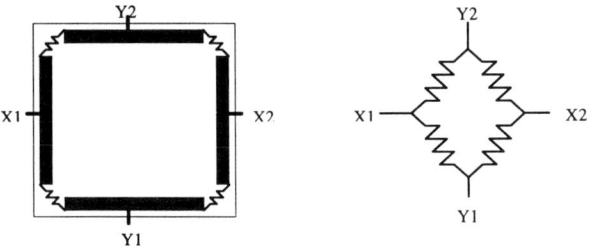

Figure 3: Simplified 2D PSD model

Measurements showed that 2D CMOS PSD has low active area resistance and this makes the device extremely sensitive to biasing voltage [6]. Any small offset in bias voltage will cause a DC offset current to flow through the PSD and this can significantly reduce the sensitivity of the device and can offset the actual centre position of the PSD. Design of the PSD biasing circuits including some practical considerations are reported by [4,5,6].

3 Measurements

3.1 Responsivity Measurements

When designing a circuit using optical sensors, it is necessary to know the range of current the sensor can produce with respect to the intensity of the optical source.

Figure 4 shows a plot of Power (P) verses Current (I) taken for each device with the optical beam positioned at the centre of the PSD. The responsivity of each PSD can be estimated by taking the gradient of each line.

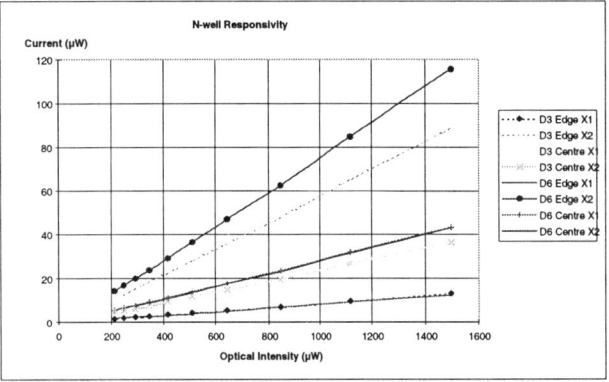

Figure 4: P-I characteristics of 1D and 2D PSD structures

It is interesting to note that for a PSD, value of A/W would change depending on the position where the incident optical beam is illuminated on the active area. This observation has been confirmed from the experimental measurements taken as shown in Figure 5.

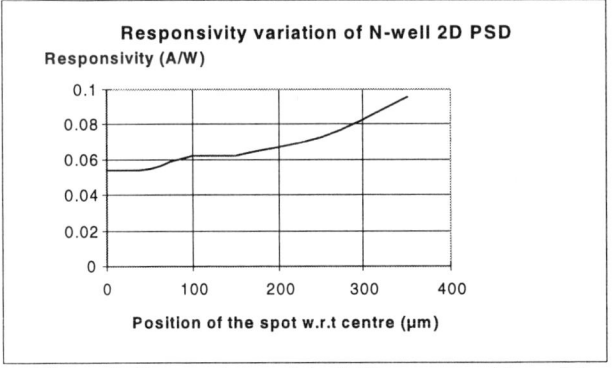

Figure 5: Responsivity of the PSDs w.r.t position of the optical beam

3.2 Variations due to spot-size

The purpose of this measurement is to show the effect on the PSD sensitivity when the spot size is varied. The sensitivity of the device was investigated by observing the variations in the difference between the currents flowing through opposite electrodes. This difference gives a measure of the dynamic range of the device from which the overall sensitivity of the device can be determined.

The measurements were taken by positioning the spot at X distance from the edge of one of the electrode (X is the maximum radius of the spot to be measured) and recording current through opposite electrodes as the spot size is varied. The difference between the two currents was taken and the relative percentage deviation from the smallest spot size was calculated:

$$\% \operatorname{Re} duction = \left(\frac{(Ix2 - Ix1)_o - (Ix2 - Ix1)_i}{(Ix2 - Ix1)_o} \right) \times 100$$

Where: $Ix1$ and $Ix2$ are the currents through opposite electrodes, subscript (o) denotes the first smallest (or the reference) spot size and i represents the ith spot size. The plot of spot size variation is as shown in Fig. 6.

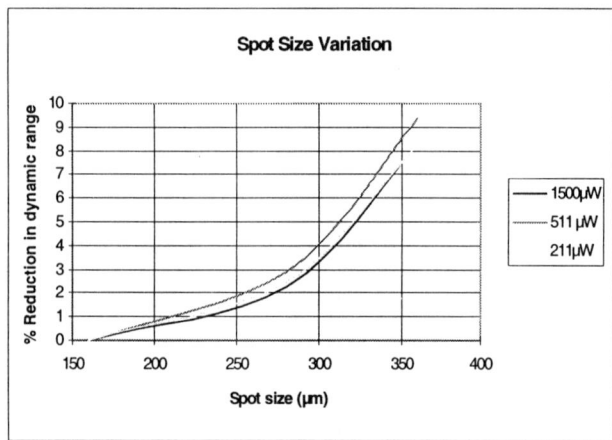

Figure 6: Variation of PSD currents at different spot sizes

Results illustrates that as the spot size is increased, the dynamic range of the PSD is reduced. Results show that as much as 9% reduction in device sensitivity can be expected if the spot is increased from 160 μm to 360 μm.

3.3 Linearity measurements

The chip containing the PSDs was mounted on Newport X-Y translation stages and each PSD was scanned in turn and measurements of current through opposite electrodes was recorded automatically using a desk-top PC. A semiconductor laser TOLD 9521 with a collimator was used as a light source. The diameter of the spot size was approximately 40μm. The wavelength of the emitted light was 670nm. Linearity measurements were taken at minimum, medium and maximum optical intensity levels. The PSDs were biased with VR = 5, with TTi 1906 DMM connected in series with the supply and the electrodes to measure the currents (see Figure 1). Figure 7 shows a typical linearity measurement example.

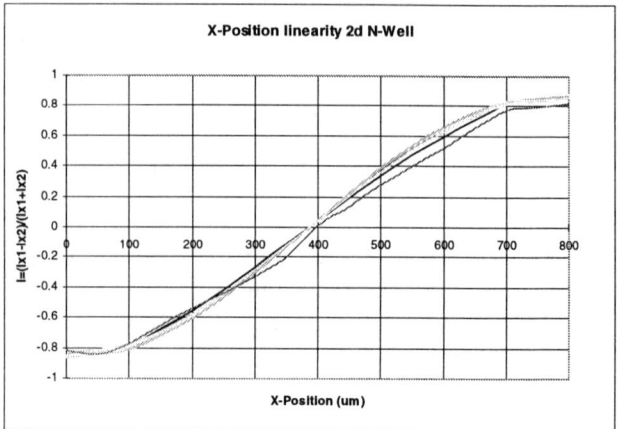

Figure 7: Example of position linearity of 2D N-Well PSD

In order to compute the PSD linearity independent to the intensity of the optical beam, difference to sum ratio of the current was computed using:

Measured position = ((X1 - X2)/(X1 + X2)) * (L/2) .... (1)

Where X1 and X2 are the currents or voltages measured at the end terminals of the PSD. L is the length of the PSD active area.

The measurements shows that when tested at three different power levels, as expected the device is fairly linear near the centre and the position accuracy is worst near the edges or at the extreme ends of the active regions.

Ideally, difference to sum ratio of the PSD currents should give values between +/- 1. However, due to inherent biasing offsets and device imperfections, in practice, the values were found to lie between +/- (0.95 to 0.85). This could lead to an initial 5 - 15% error. To remove this error, equation (1) was normalised by a factor (1/A). Where A is the measured maximum (or minimum) value expected for the selected PSD. Measured position is thus, given by:

New Measured position (M) = ((X1 - X2)/(X1 + X2)) * (L/2) * (1/A) .... (2)

The position error was thus calculated by taking the difference between the normalised position computed using equation (2) and the actual position of the translation stages.

The measurements shows that the devices tested has fairly linear characteristics near the centre region. Initial estimation of the results shows that near the centre approximately +/- 1% position error can be expected, whilst near the edges error could increase by more than +/- 12%. Furthermore the results shows that the linearity of the Pinch well device is better than N-well device. For the pinch-well devices improvements of 2% to 6% was observed.

3.4 Position resolution

When using the PSD for practical instrumentation application it is necessary to know the resolution at which the device will be able to sense the changes in position of the optical beam. The position resolution measurements were taken by scanning (spot size 5μm) the most-linear centre zone of the PSDs with the step size of the Newport Translation stages set to 0.2μm. The translation stages were carefully calibrated for each measurements in order to ensure maximum accuracy. To measure with such accuracy, it was also necessary to use Keithly pico-ammeter. As before, difference-to-sum ratio of the currents were computed and position resolution was estimated at different power levels. The results showed that minimum position resolution of 0.5μm can be achieved at 150μW optical power. The sensitivity of the PSD decreases with increasing optical power.

4 Conclusions

In this paper it has been shown that both 1-D and 2-D PSD structures can be successfully fabricated on standard CMOS process. As expected, the performance of 1-D PSD was found to be much better than the 2-D PSD structures and this due to resistance per square for 1-D structure was significantly higher than 2-D structures. Lower than expected resistance per square for 2-D structures were due to effective bridge formation of parasitic resistance of the adjacent electrodes.

The measurements for spot size variations showed that the sensitivity of the device reduces with increasing spot size. Linearity measurements showed that most linear region of the PSD is near the centre of the device. There is very little difference in linearity between 1D and 2D N-well structures. However, results clearly shows that linearity of the 2-D PSD structures were significantly worse than 1-D structures. In particular, all devices exhibited highly non-linear characteristics near the extreme edges of the active area.

Position resolution measurements showed that 1-D PSD structure can achieve 0.5μm resolution with 5μm spot size at 150μW optical intensity. For the same spot size and optical intensity position resolution of 0.8μm can be achieved for 2-D PSD structures. In general position resolution for 1D PSD structure was found to be marginally better than N-well structures.

Acknowledgements

Authors would like to acknowledge EPSRC and AMS for their support to conduct this research work. Dr. Chowdhury would also like thank LSI Logic. for providing the facilities to writing and submitting this paper.

References

[1] "Uebber den Entsstehungsort der Photoelectronen in Kupfer-Kupferoxyddulphotosellen" W. Schottky. Phys. Z Vol. 31, pp 913-925, 1930.

[2] "A new semiconductor photocell using lateral photoeffect", J. T. Wallmark, Proc. IRE, Vol. 45, pp 474-483, 1957.

[3] "Photoeffects in Nonuniformly Irradiated p-n Junctions", G. Lucovsky, Journal of Applied Physics, Vol. 31, No. 6, June 1960.

[4] "Single- and Dual-Axis Lateral Photodetectors of Rectangular Shape", H. J. Woltring, IEEE Trans. Electronic Devices, Vol. ED-22, pp. 581-590, 1975.

[5] "A Method for Measurement of Multiple Light Spot Positions on One Position-Sensitive Detector PSD)", D. Qian, W. Wang and B. Buckman, IEEE Trans. On Instrumentation and Measurement, Vol. 42, No. 1, pp.14-18, 1993.

[6] "An Integrated Position Sensor using JFETs as a Buffer for PSD Output Signal", H. Muro and P. J. French, Sensors and Actuators, pp.544-522, 1990.

HIGH PERFORMANCE CMOS POSITION-SENSITIVE PHOTODETECTORS (PSDs)

A. Mäkynen, T. Ruotsalainen, T. Rahkonen, J. Kostamovaara

University of Oulu, Department of Electrical Engineering and Infotech Oulu,
FIN-90570 Oulu, FINLAND

ABSTRACT

Five different constructions of optical position-sensitive detectors (PSDs) implemented using standard CMOS technology are reported. The results show that CMOS technology provides means for implementing high performance PSDs.

1. INTRODUCTION

Combining signal processing and photosensing on the same chip provides a possibility to implement truly small-sized optical sensors. The most cost-effective means of implementing signal processing for such a sensor system is to use standard CMOS technology. Therefore, various kinds of high-performance CMOS-compatible photodetectors should be available. Here five different constructions of optical position-sensitive detectors (PSDs) are reported. The PSDs are implemented using 1.2 µm double metal, single poly n-well CMOS technology. The applications of these miniature optical position sensors include shooting simulators in military marksmanship training and triangulation based distance sensors used in industrial and consumer products such as compact cameras, for example.

2. CMOS LEPs

Optical position sensing using the operating principle of the lateral effect photodiode (LEP) provides an accurate and simple means for detecting light spot position. The position response of a LEP structure is inherently linear and its accuracy is not affected by spot truncation or nonoptimal sampling, as may be the case with sensors composed of discrete photodetectors [1,2]. A single-axis LEP is a large-area continuous photodiode with two extended electrodes at opposite edges of a resistive layer acting as the anode or cathode of the detector. A tetralateral 2-axis LEP has correspondingly four extended electrodes at the edges of a single resistive sheet. A duolateral 2-axis detector has a resistive sheet on both sides of the diode, and correspondingly one pair of electrodes on both sheets. The current carriers generated in the illuminated region are divided between the electrodes in proportion to the conductance (distance) of the current paths between the illuminated region and the electrodes. The deflection of the spot from the centre of the detector is simply

$$\Delta x = \frac{L}{2}\frac{i_b - i_d}{i_b + i_d} \quad \text{and} \quad \Delta y = \frac{L}{2}\frac{i_a - i_c}{i_a + i_c}, \quad (1)$$

where i_a, i_b, i_c and i_d are the average signal currents of the electrodes a, b, c and d, and L the length of the side of the active area of the LEP (Fig. 1). The standard deviation of position sensing results is

$$\sigma \approx \frac{L\sqrt{4kTB/R}}{P_S S} \quad (2)$$

where P_S is the RMS-value of signal power, S the responsivity and R the interelectrode resistance. The position sensing resolution improves as R is increased. The maximum value is limited by the requirement according to which the sensor bandwidth $B=\pi/2RC$ must be higher than the signal bandwidth in order not to deteriorate linearity [3]. Assuming that the noise bandwidth is $\pi/2$ times B, the best achievable resolution in the case of wide-band operation (pulse-modulated signal) can be approximated using

$$\sigma_{min} = \frac{LB\sqrt{4kTC}}{P_S S}. \quad (3)$$

According to (2) and (3) the best resolution is achieved by using the highest resistance that ensures the sensor bandwidth to be larger than signal bandwidth, the smallest active area suitable for the particular application (also C∝area) and a photodetector with high responsivity and low capacitance.

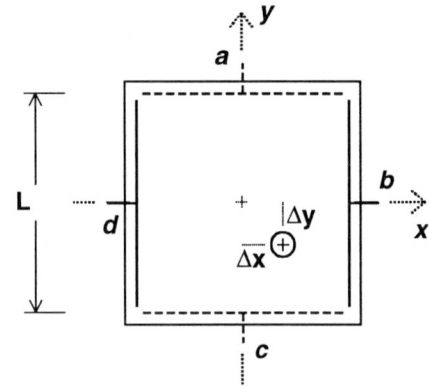

Figure 1. 2-axis conventional LEP

CMOS photodetectors, such as substrate-well photodiodes, are suitable for optical sensing applications where a responsivity of about 0.2 to 0.4 A/W is sufficient, where relatively large (~5% standard deviation) spatial and spectral nonuniformity of the responsivity can be tolerated, and where high speed is not of importance [4-6]. One could come to a conclusion that the fairly large spatial responsivity variation, which directly alters the "electrical" centroid position, might be detrimental to PSD linearity. Experimental studies have shown, however, that CMOS PSDs utilizing the LEP structure offer the same linearity as the best quality conventional LEPs (0.1 to 0.3 % standard deviation with respect to full scale) [7,8]. A serious drawback, however, is the high junction capacitance (typically 10 to 100 aF/μm^2) of the photodiode which decreases the best achievable resolution at least tenfold as compared to conventional PSDs.

The conventional LEP structure in CMOS is noisy since the sheet resistance values of applicable layers (well/diffusion) are relatively low as compared to conventional devices (1 - 5 kΩ/□ vs. 10 - 50 kΩ/□), which together with low responsivity leads to a high NEP figure. It is also difficult to implement 2-axis detectors with good linearity using standard CMOS. In the basic tetralateral structure a large amount of nonlinearity is caused by the current division between four contacts instead of two, and the linear structures, such as duolateral and pincushion, are practically impossible to implement in standard CMOS.

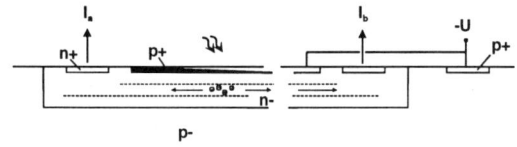

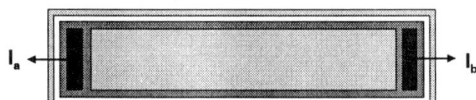

Figure 2. Single-axis CMOS LEP.

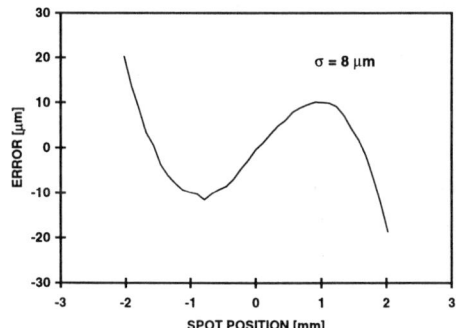

Figure 3. Nonlinearity of the single-axis LEP.

To demonstrate the performance achievable with a CMOS LEP, the implementation and test results of a single-axis and a 2-axis device are presented. The active area of the single-axis LEP measures 5.0 x 0.2 mm^2. It uses a p-substrate-n-well junction as a photodetector and a pinched well as the current dividing layer (Fig. 2). The responsivity of the well-substrate photodiode was 0.42 A/W (850 nm), and the measured interelectrode resistance 152 kΩ at 5 V bias corresponding to the typical sheet resistance of a pinched well resistor (~6 kΩ/□). The NEP (1.6 pW/$\sqrt{Hz}$) represents a power value by which SNR~1 is achievable simultaneously at each output electrode. It is calculated on the basis of the thermal noise and responsivity alone since the noise due to dark current (~20 pA) is negligible. The nonlinearity depicted in Fig. 3 is about 8 μm (0.2 %) when 80 % of the active dimension was used for measurements. The test results summarized in Table 1 show that the achievable linearity is about the same but the NEP and the characteristic time constant of the CMOS LEP are larger than those of a conventional LEP [9,10].

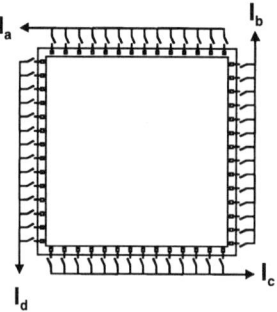

Figure 4. Simplified outline of the 2-axis LEP.

The 2-axis LEP was implemented using tetralateral geometry. The lateral response was linearized by changing it effectively to a single-axis detector and by performing 2-axis measurements using two successive single-axis measurements. The idea was originally presented and demonstrated by Morikawa et al. [11]. Here we present a modified CMOS implementation of it. The basic construction is similar to a single-axis CMOS LEP, but instead of continuous, extended contacts each electrode is composed of a row of 100 minimum-sized n+ diffusion contacts positioned about 25 μm from each other. Each of the contact spots is connected to a common signal rail by a simple MOS switch. The operation is set up by alternately disconnecting one pair of the electrodes and measuring one dimension at a time. A simplified outline of the PSD measuring 2.5 x 2.5 mm^2 is depicted in Fig 4.

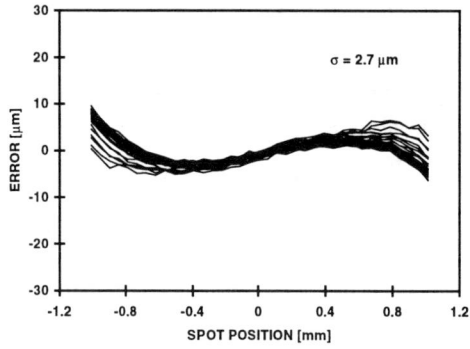

Figure 5. Nonlinearity of the 2-axis LEP in the alternate mode.

The alternate single-axis operating principle and the usage of "discrete" electrodes provides lower noise than the conventional tetralateral structure. The resistance seen by each preamplifier increases when the virtual ground from the obsolete electrode pair is disconnected, and when the low-resistive path formed by the conventional strip-like electrodes breaks off.

Nonlinearities in the alternate (2.7 µm, 0.14 %, Fig. 5) and tetralateral (68 µm, 7.2 %) modes correspond to those of a single-axis CMOS LEP and a conventional tetralateral LEP as expected. The tests also showed that the bare existence of the additional pair of electrodes, even if the MOS switches were conducting, did not cause any nonlinearity as long as no current was drawn from them. An interelectrode resistance and a NEP of 4.5 kΩ and 10 pW/√Hz were achieved in the alternate measurement mode. In the tetralateral mode the values were 0.85 kΩ and 47 pW/√Hz, respectively. The linearity in the alternate mode is about the same, whereas the NEP and the characteristic time constants are worse than those of a conventional 2-axis LEP (Table 1).

3. AREA ARRAY PSDs

To demonstrate the possibility of implementing more sensitive PSDs in standard CMOS, we present two simple detectors both of which are composed of an area array of photodetectors and of current dividing resistance residing outside the array active area. Using this structure it was possible to take advantage of resistors and photodetectors other than those suitable for the continuous LEP structure, and therefore to provide means for improving detector sensitivity either by increasing interelectrode resistance or by increasing the effective responsivity of the photodetectors.

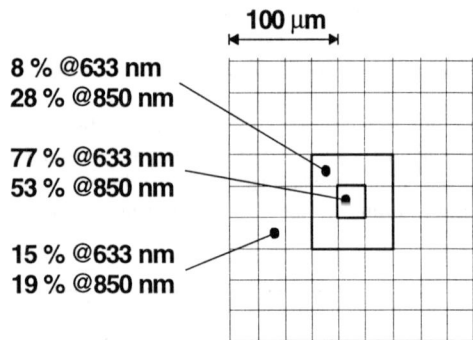

Figure 6. Measured photocurrent distribution in a well-substrate photodiode array.

An array of discrete photodetectors instead of a continuous one can cause additional error in the centroid estimation. Alexander et al. have shown that the centroid of a spatially digitized light spot is free of systematic error only if the maximum spatial frequency of that spot is less than the sampling frequency of the array [1]. Thus problems usually arise when the spot size approaches that of a single photodiode. Note, however, that the relative sampling frequency of the well-substrate photodiode array is increased by cross-talk between adjacent photodiodes due to free carrier diffusion in the substrate. This means that the light spot size can be close to that of the photodiode without introducing any major position sensing error. The effect of cross-talk is depicted in Fig. 6 that shows the photocurrent distribution in a well-substrate photodiode array (pitch 25 µm, fill factor 68 %) when only the photodiode at the centre is illuminated. Note the wavelength dependence and the large amount of cross-talk when the 850 nm illumination is used.

3.1 Phototransistor PSD [12]

According to (2), the resolution of a PSD can most effectively be improved by increasing responsivity. Here phototransistors are used as photodetectors, and the resolution for a single-axis device is

$$\sigma \approx \frac{L\sqrt{4kTB/R + q(P_S + P_B)S\beta^2 B}}{P_S S\beta}, \quad (4)$$

where β is the small signal current gain, S the primary responsivity, P_B the background illumination power and R the resistance used for current division. It can be concluded that even a moderate value of β clearly improves resolution as compared to the conventional LEP, provided that background illumination is low.

The construction of the 2-axis PSD is presented in Fig. 7. It is composed of 100 x 100 CMOS-compatible vertical phototransistors occupying a total area of 2.5 x 2.5 mm^2 and the array pitch and fill factor of it are 25 µm and 68 %. Of the phototransistors in each row, the emitters are alternately connected to the row and the column current lines. The row and column lines, again, are connected to two arrays of resistors composed of one hundred 30 Ω polysilicon resistors each. The output currents of the phototransistor PSD are processed as if they were those of a conventional LEP.

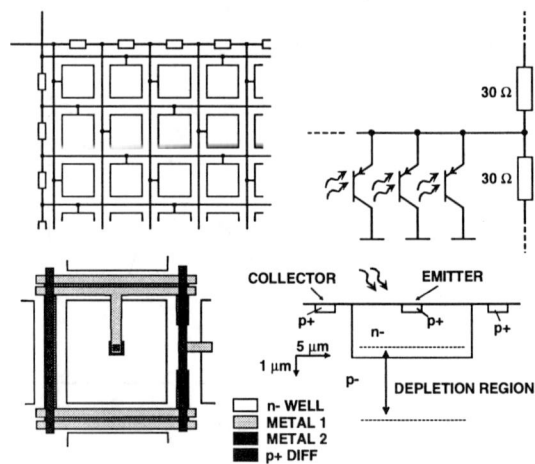

Figure 7. Phototransistor PSD.

The tests were performed using low background illumination, square wave modulated carrier of 1.2 kHz, signal bandwidth of 30 Hz and wavelength of 850 nm. When compared experimentally to a conventional 2-axis LEP (R~10 kΩ), it was found that the responsivity of the phototransistor PSD is high enough to provide an approximately seven times better resolution (P_S~10 nW) at best, and that equal resolution

($\sigma/L \sim 1‰$) can be achieved using ten times less signal power at best (Fig. 8). Responsivity can still be increased at low signal levels using background or a signal to optically bias the phototransistors. Due to a smaller illuminated area the latter results in a lower total bias current and thus in lower noise. By using 100 pW signal bias, better than 10 A/W responsivity was achieved even at very low (<10 pW) signal levels. The NEP without bias current was about 0.5 pW/$\sqrt{Hz}$.

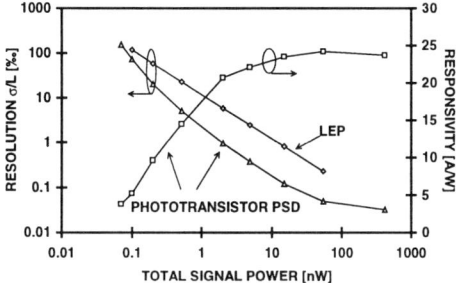

Figure 8. Resolution of the phototransistor PSD vs. conventional LEP.

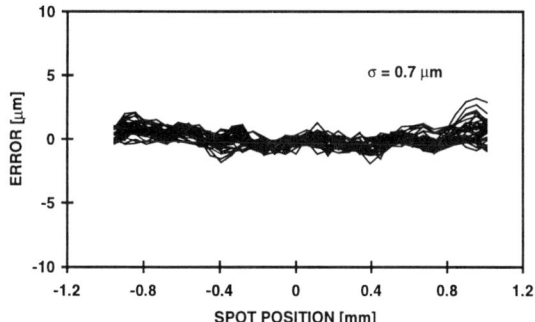

Figure 9. Nonlinearity using 300 µm spot.

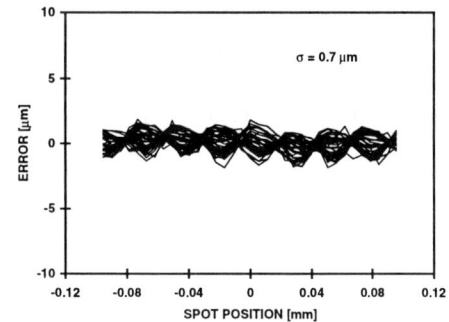

Figure 10. Ripple caused by small (50 µm) spot size.

The measured standard deviation of position sensing error is 0.7 µm (0.04 %, Fig. 9) and 1.1 µm (0.06 %) when a light spot of 300 µm and 50 µm were used, respectively. In case of the 50 µm spot the position sensing error is composed of periodic ripple on top of a slightly curved trend. The ripple is caused by nonoptimal spatial sampling of the array (~50 µm spot sampled with 25 µm pitch), and the curvature most probably by the nonuniformity of the resistors because the general shape of the error characteristics measured from different positions of the array resemble one another. The correspondence between the ripple frequency and the array pitch is shown in Fig. 10. The measurements cover 8 % of the array active dimension and show 0.7 µm standard deviation (1/28 of the pitch). According to simulations, the lowest possible centroiding error in the case of the 50 µm spot should have been worse than 1/10 of the array pitch [13]. The low value of additional ripple is probably a result of cross-talk between phototransistors. In typical position sensing applications where spot sizes vary from 0.1 to 1 mm, discrete sampling has essentially no effect on PSD accuracy.

3.2 Tracking PSD

The main idea of the tracking PSD is to combine the best properties of the conventional PSDs, the high electrical resolution of a four quadrant (4Q) PSD and the linear measurement field and low sensitivity to atmospheric turbulence of a LEP [14]. A conventional 4Q PSD consists of four photodiodes (quadrants) positioned symmetrically around the centre of the detector and separated by a narrow gap. The position information is derived from the optical signal powers received by the quadrants whose electrical contribution then serves to define the relative position of the light spot with respect to the gap position. The operation of the proposed tracking PSD is similar to a conventional 4Q PSD with the exception that the gap position electrically tracks the light spot within the active area of the PSD (Fig.11).

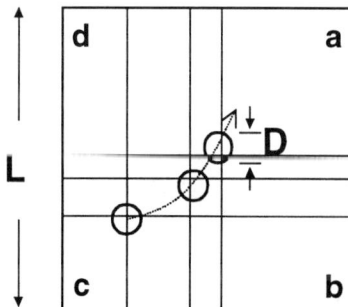

Figure 11. Operating principle of the tracking PSD.

The high accuracy (<<µm) of the photodetector array geometry is used to establish good integral linearity while high resolution is based on the "subspot" interpolation performed using inherently nonlinear but highly sensitive 4Q function. The approximations of the displacements with respect to the gap position are

$$\Delta x \approx \frac{\pi D}{8} \frac{(i_a + i_b) - (i_c + i_d)}{i_a + i_b + i_c + i_d} \quad \Delta y \approx \frac{\pi D}{8} \frac{(i_a + i_d) - (i_c + i_b)}{i_a + i_b + i_c + i_d} \quad (5)$$

where i_a, i_b, i_c and i_d are the average signal currents of the quadrants a, b, c and d, and D the diameter of a uniform spot. The corresponding standard deviation of the measured centroid positions is

$$\sigma \approx \frac{\pi D i_n}{4 P_s S}, \qquad (6)$$

where i_n is the dominating noise typically caused by the feedback resistance of the transimpedance preamplifier. Note that in a LEP the scale factor is solely determined by the extent of the detector active dimensions (L), whereas in the tracking PSD the scale factor of the interpolator is determined by the size of the light spot (D). Thus the resolution ratio between a conventional LEP and the tracking PSD assuming equal responsivity and infinitely narrow gap is

$$\frac{\sigma_{TPSD}}{\sigma_{LEP}} \approx \frac{D}{L} \frac{\pi}{4} \frac{i_{nTPSD}}{i_{nLEP}}, \qquad (7)$$

where i_{nTPSD} and i_{nLEP} are the noises related to one quadrant and edge electrode, respectively. Since D/L and i_{nTPSD}/i_{nLEP} both are typically about 0.1, the tracking PSD resolution should be about 100 times better than with a conventional LEP.

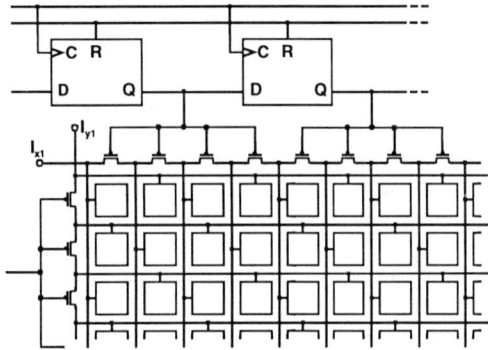

Figure 12. Construction of the tracking PSD.

The construction of the tracking PSD depicted in Fig. 12 is quite similar to that of the phototransistor PSD but is composed of well-substrate photodiodes instead of phototransistors, and of PMOS -transistors instead of polysilicon resistors. The total area, pitch and fill factor are the same as in a phototransistor PSD. The transistors are used to divide the array in to four quadrants. In our prototype the transistor gates are driven by the logic signals of a shift register. Each shift register output drives four PMOS transistors, and so the smallest possible gap width/step is 100 µm. The switching transistor on-resistance is not critical because the series resistance of the array does not generate any noise. But since we wanted to use the array also as a LEP in which MOS -transistors are used for current division, the aspect ratio and the size of the transistors (W/L = 100/20) were designed to provide a uniform interelectrode resistance of about 100 kΩ at 5V bias voltage.

The lateral response in the tracking mode shows excellent integral linearity as expected (Fig. 13). The inherent nonlinearity of the interpolation function causes ripple the standard deviation of which is 3 µm (0.15 %). Here a spot size of 300 µm was used to properly cover the minimum gap step (100 µm). If a smaller gap width were used, the error would probably be smaller. The measured NEP was 1.3 pW/√Hz and the position sensing resolution about 40 times better when compared to a conventional PSD (R~10 kΩ) equal in size. The difference is smaller than predicted by (7) due to lower effective responsivity of the tracking PSD. Note that the resolution of the tracking PSD can not be directly deduced from its NEP since the resolution is also dependent on the D/L-ratio. In the LEP operating mode the measured interelectrode resistance was about 300 kΩ at 5 V bias and the corresponding NEP about 3.1 pW/√Hz. The standard deviation of position sensing error was 2.8 µm (0.14 %, Fig. 14).

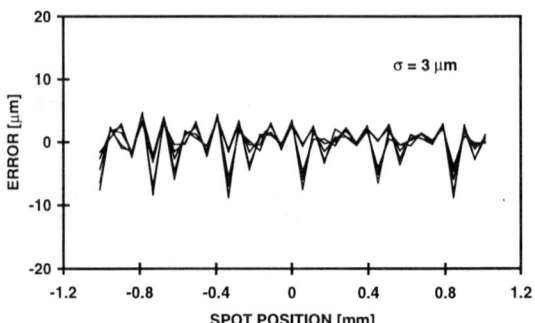

Figure 13. Nonlinearity of the tracking PSD.

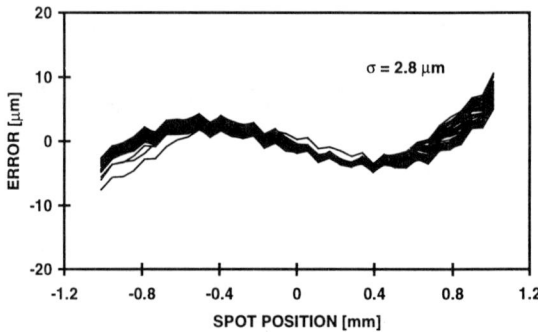

Figure 14. Nonlinearity of the tracking PSD when used in LEP mode.

4. BINARY PSD

As demonstrated above, highly accurate PSDs can be implemented using standard CMOS technology. Problems may arise, however, when signal processing electronics should be implemented. High demands are set for the sensitivity and linearity of the electronics especially if the dynamic range of the input signal is large, as is the case in long-range outdoor applications, for example. One solution to simplify the implementation of signal processing could be to realize a fully digital PSD in which an essential part of the signal processing electronics is integrated within the pixel itself and only simple combinatorial logic is needed to provide a fully digital connection to the signal processor. A binary PSD array having 50 µm pitch and 30 % fill factor has been implemented by the authors using 1.2 µm CMOS technology [15]. Subpixel position sensing accuracy was achieved by spreading the spot over several pixels and by calculating the centroid position. The standard deviation of the calculated centroid position is

$$\varepsilon \approx 0.21 \frac{p}{\sqrt{q}}, \qquad (8)$$

where the spot diameter is $D=qp$ and p the pixel pitch [16]. The method is quite effective up to the accuracy of about 1/10 of a pixel, where after the total signal power needed to improve accuracy raises rapidly. The binary PSD provided the accuracy of 4.3 μm when a spot of 280 μm in diameter, pulse width of 8 ms and signal power of 8 nW were used. The performance corresponds roughly to that of a conventional PSD.

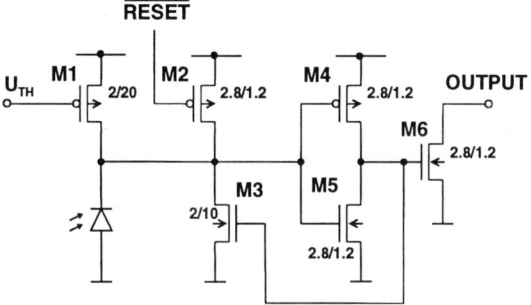

Figure 15. Original binary pixel.

The pixels of the original PSD were operated in the continuous current mode in which larger amount of signal power is used to activate a pixel than what is effectively needed. Here the usage of charge integration with a modified pixel circuit is proposed to increase the sensitivity and uniformity of the binary PSD.

The original pixel circuit is depicted in Fig. 15. It comprises a photodiode, current source for setting the threshold current, RS latch and the output stage. In the continuous current mode the latch is set when the photocurrent exceeds the threshold current set by M1. In the charge integrating mode the signal current is used just to discharge the capacitance present at the input of the latch, and the threshold current is used only to prevent integration of the background photocurrent during the time between reset and readout. Besides better sensitivity the charge integrating mode is expected to provide better uniformity since it does not suffer as much from the large spatial variation of the threshold currents as the continuous current mode. In order to have good efficiency, short (μs) but powerful light pulses, the photocurrent of which clearly exceeds the threshold current level, should be used as the signal. When a well-substrate photodiode measuring 38 x 38 μm^2 is used, approximately half of the capacitance originates from the photodiode and the other half from the circuit. According to simulations the latch sets when its input voltage is reduced to about 2.1 V (U_{DD}=5 V). To improve the sensitivity of the original circuit, two transistors and one bias voltage were added (Fig. 16). M8 isolates the photodiode capacitance from the latch input and M7 increases the switching voltage to about 3.8 V (U_B = 1 V) thus decreasing the energy needed for activating the pixel.

According to the test results the energy needed for activating the original pixel is 1 pJ, and 0.23 pJ for the improved pixel (U_{DD}=5V, U_B=1V). The same sensitivity was achievable essentially irrespective of the pulse width and pulse repetition frequency (PRF). The shortest light pulse and the highest PRF used were 200 ns and 1 MHz, respectively (Fig. 17). Lowering U_B to 0.6 V, introducing background bias current of 1 nA or using high duty factor (pulse width multiplied by PRF) improved the sensitivity by about 20 %. For high speed operation, however, U_B had to be at least 0.9 V.

According to the measurements, a binary PSD composed of integrating pixels should be as fast and accurate as a conventional PSD but at the same time ten times more sensitive. Note that the binary PSD also provides an opportunity to detect centroids of multiple spots as well as possibility to perform spatial filtering for distorted spot images in order to improve accuracy. Such properties are particularly useful in various 3D motion and shape measurements and in outdoor measurements where atmospheric turbulence restricts the measurement resolution [14].

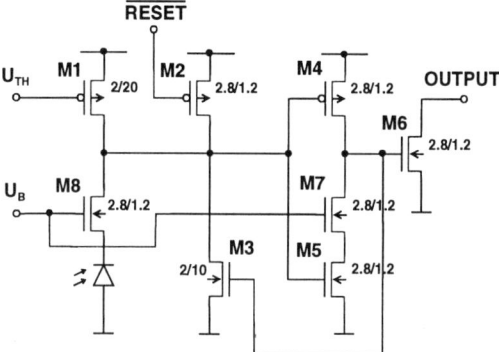

Figure 16. Modified binary pixel.

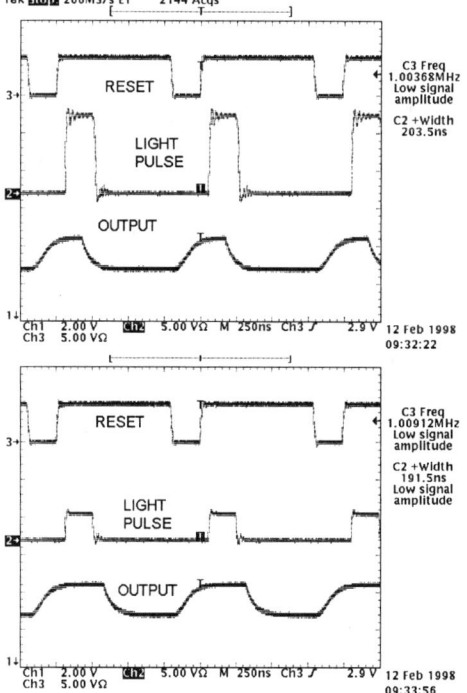

Figure 17. Pulse shapes of the modified binary pixel.

5. CONCLUSIONS

CMOS PSDs utilizing continuous and discrete structures were implemented and tested. The results show that CMOS technology provides viable means for implementing optical PSDs and that the achievable performance is comparable or even better than that of PSDs implemented with dedicated semiconductor technologies.

6. REFERENCES

[1] B. F. Alexander, K. C. Ng,"Elimination of systematic error in subpixel accuracy centroid estimation", *Optical Eng.*, vol. 30, pp. 1320-1331, Sept. 1991.

[2] D. J. W. Noorlag, S. Middelhoek, "Two-dimensional position-sensitive photodetector with high linearity made with standard IC-technology", *IEE Journal on Solid State and Electron Devices*, vol. 3, pp. 75 – 82, May 1979.

[3] C. A. Klein, R. W. Bierig, "Pulse-response characteristics of position-sensitive Photodetectors", *IEEE Trans. Electron Devices*, vol. 21, pp. 532-537, Aug. 1974.

[4] P. Aubert, H. Oguey, R. Vuilleumier, "Monolithic optical position encoder with on-chip photodiodes", *IEEE J. Solid-State Circuits*, vol. 23, pp. 465-473, Apr. 1988.

[5] G. Soncini, M. Zen, M. Rudan, G. Verzellesi,"On the electro-optical characteristics of CMOS compatible photodiodes", in *Conf. Rec. Melecon '91*, Ljubjana, Yugoslavia, pp. 111-113, May 1991.

[6] P. Palojärvi, T. Ruotsalainen, G. Simin, J. Kostamovaara, "Photodiodes for high frequency applications implemented in CMOS and BiCMOS processes", in *Proc. SPIE*, vol. 3100, Munchen, pp. 119-126, June 1997.

[7] A. J. Mäkynen, J. T. Kostamovaara and T. E. Rahkonen, "CMOS photodetectors for industrial position sensing", *IEEE Trans. Instrum. Meas.*, vol. 43, pp. 489-492, June 1994.

[8] J.Kramer, P Seitz, H. Baltes, "Industrial CMOS technology for the integration of optical metrology systems (photo-ASICs)", *Sensors and Actuators*, vol. A 34, pp. 21-30, July 1992.

[9] SiTek Electro Optics, "SiTek PSD position sensing detectors," *Product Catalogue*, 1993.

[10] Hamamatsu Photonics, Technical data sheets of position-sensitive detectors, 1988 - 1995.

[11] Y. Morikawa, K. Kawamura, "A small-distortion two-dimensional position-sensitive detector (PSD) with on-chip MOSFET switches", *Sensors and Actuators*, vol. A 34, pp. 123-129, July 1992.

[12] A. Mäkynen, T. Ruotsalainen, J. Kostamovaara, "High accuracy CMOS position-sensitive photodetector (PSD)," *Electronics Letters*, vol.33, pp.128-130, Jan. 1996.

[13] J. A. Cox," Point source location sensitivity analysis", in *Proc. SPIE*, vol. 686, pp. 130-137, 1986.

[14] A. Mäkynen, J. Kostamovaara,"Accuracy of lateral displacement sensing in atmospheric turbulence using a retroreflector and a position-sensitive detector", *Optical Eng.*, vol. 36, pp. 3119-3126, Nov. 1997.

[15] A. Mäkynen, T. Rahkonen, J. Kostamovaara,"A binary photodetector array for position sensing", *Sensors and Actuators*, vol. A 65, pp. 45-53, Feb.1998.

[16] C. Bose, I. Amir,"Design of fiducials for accurate registration using machine vision", *IEEE Trans. Pattern Analysis Machine Intelligence*, vol.12, pp. 1196-1200, Dec. 1990.

Table 1. Comparison between a conventional LEP and the implemented CMOS PSDs.

Property		Conventional LEP 1-axis 5 x 0.2 mm² 2-axis 2.5 x 2.5 mm²	1-axis CMOS LEP (U_b=5V)	2-axis CMOS LEP (U_b=2V)	Photo-transistor PSD	Tracking PSD LEP mode	Tracking PSD 4Q mode
Responsivity S (A/W)		0.6	0.42	0.38	>20[*]	0.4	
Sheet resistance (kΩ/□)		10 ... 50	6	4.5	-	-	
Resistance R (kΩ)	1-axis	250 ... 1250	152	4.5	-	-	-
	2-axis	10 ... 50	-	0.85[**]	3	300	-
NEP (pW/√Hz)	1-axis	0.4 ... 1	1.6	10	-	-	-
	2-axis	2 ... 5	-	47[**]	0.5	3.1	1.3
Capacitance C (pF)	1-axis	1 ... 10	160	-	-	-	-
	2-axis	6 ... 60	-	1300	-	~400	-
Time constant (ns)	1-axis	25 ... 1300	2500	-	-	-	-
RC/π^2	2-axis	6 .. 300	-	610	~100 k	13 k	-
Nonlinearity (%)	1-axis	~0.1	0.2	0.14	-	-	-
$FS = 0.8 \times L$	2-axis	~0.3	-	7.2[**]	0.04...0.06[*]	0.14	0.15

[*] P_S >1nW, spot diameter 50μm, [**] tetralateral mode

MEASURING TEMPERATURE CALIBRATION FREE WITH BIPOLAR TRANSISTORS

O. Kanoun

Institut für Meß- und Automatisierungstechnik,
Universität der Bundeswehr München
85577 Neubiberg, Germany
tel.: +89-6004-3752, fax: -2557,
email: Olfa.Kanoun@UniBw-Muenchen.de

ABSTRACT

The use of pn-junctions for temperature measurement has gone through several stages from the simple temperature sensor to the integrated sensor. In this paper we introduce the use of bipolar transistors for a calibration free temperature measurement. The elimination of the need to calibrate enables to reduce the production and maintenance costs of sensors.
Not all known temperature measurement methods using bipolar transistors are able to realize a calibration free temperature measurement. Even methods realizing a calibration free temperature measurement like the well-known method by Verster have been used in practice with calibration in order to realize an acceptable accuracy. We propose to use an accurate i-u characteristic model and to take more measured i-u points into account. This improves the accuracy of measurement without affecting the calibration free behaviour.

1. INTRODUCTION

The pn-junction have a pronounced sensitivity towards temperature which allows to realize temperature sensors. The favourable properties of transistors and diodes for this application are due to the highly predictable and time independent way in which the pn-junction voltage is related to temperature.

Temperature sensors based on pn-junctions exist since the 60^{ties}. A lot of investigations have been made with regard to different aspects. At first problems related with technology and modelling of the devices were in the focus of interests.
Later on the easy integration of pn-junctions leaded to a lot of integrated temperature sensors based on pn-junctions [12].

In this paper the temperature measurement based on transistors is considered under the aspect of the calibration. Calibration is a process during which the sensor output signal is detected by well-known input values. The necessity of calibration increases the production and maintenance costs. Therefore both manufacturers and appliers prefer to use sensors, which dispose of calibration.

The method by Verster [1] offers the possibility of a calibration free temperature measurement. However, this fact couldn't be used in practice because of the low accuracy reached. Calibration free temperature measurement must guarantee a certain accuracy class to be profitable for some applications with corresponding requirements.

2. MEASUREMENT PRINCIPLES, PREVIOUS METHODS

We can classify the temperature measurement methods using bipolar transistors in two classes (figure 1).
In the first class (section 2.1), temperature is considered as the main input of the device. In this case the bias current through the basis-emitter pn-junction is used to adjust the behaviour of the $U_{be}(T)$ characteristic.

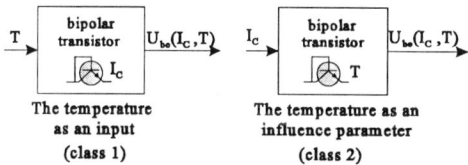

Figure 1. Classification of the temperature measurement methods based on bipolar transistors

In the temperature measurement methods dealt with in section 2.2, temperature is calculated from the i-u characteristic of the basis-emitter pn-junction. Thereby temperature is considered as a parameter influencing the behaviour of the i-u characteristic.

2.1 Methods based on the $U_{be}(T)$ characteristic

The direct way to investigate the behaviour of a temperature sensor is to consider the dependance of its output signal on temperature. In the case of bipolar transistors, the output signal is the basis-emitter voltage. Because it depends also on the collector current, the temperature measurement is generally carried out by defined conditions of the bias current I_c.

The $U_{be}(T)$ characteristic of a transistor can be approximated by line with negative slope beginning at a constant value at 0 K (figure 2) [2]:

$$U_{be}(T) \approx 1{,}27\ V - CT$$

To realize a good accuracy the behaviour of the $U_{be}(T)$ characteristic has been investigated and modelled mathematically [2, 3, 4].

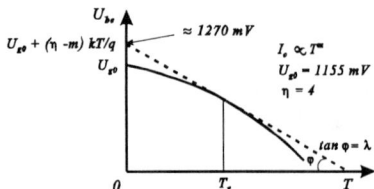

Figure 2. The base-emitter voltage U_{be} vs. temperature T [2]

An example of an accurate $U_{be}(T)$ characteristic model is given in (1) [2]. There by the bias current through the pn-junction has been chosen proportional to T^m.

$$U_{be}(T) = U_{g0}\left(1-\frac{T}{T_r}\right) + \left(\frac{T}{T_r}\right) - (\eta-m)\left(\frac{kT}{q}\right)\ln\left(\frac{T}{T_r}\right) \quad (1)$$

with T_r = calibration temperature, U_{g0} = extrapolated band-gap voltage at 0 K and η, m constants.

Because of the pure mathematical modelling of the $U_{be}(T)$ characteristic, the developed models contain a set of parameters assumed to be determined from calibration data.

2.2 Methods based on the i-u characteristic

The model (2) developed by Shockley [5] is the fundamental model of the i-u characteristic of a pn-junction:

$$I(U,T) = I_S(T)\cdot\left[\exp\left(\frac{eU}{kT}\right) - 1\right] \quad (2)$$

With k = Boltzmann's constant, e = the electron charge, T = temperature in Kelvin and I_s = saturation current of the pn-junction.

In this model the saturation current I_s is dependent on the temperature. The direct use of the Shockley model to temperature measurement requires the determination of the saturation current I_s using the calibration data [6].

Verster [1] eliminated the saturation current I_s in the shockley model with using the difference ΔU between two voltages measured with a pn-junction operated at two different currents I_1 and I_2:

$$\Delta U = U_2 - U_1 = \frac{kT}{e}\cdot\ln\left(\frac{I_2}{I_1}\right)$$

With U_1, U_2 the voltages measured by I_1, I_2.

The typical uncertainty of measurement in using the method by Verster without calibration is about 3 K. This is due to the deviation between the Shockley model (2) and the real i-u characteristic of a pn-junction (figure 3).

To increase the measurement accuracy, Goloub [7] introduced the resistance effects into the Shockley model to describe the base-emitter-junction (3) in a bipolar transistor with a base-collector short-circuit:

$$U(I,T) = \frac{kT}{e}\cdot\ln\left(\frac{I}{I_S(T)}\right) + RI \quad (3)$$

$$R = r_e + \frac{r_b}{\beta}$$

With r_e the emitter resistance, r_b the base resistance and β the forward gain.

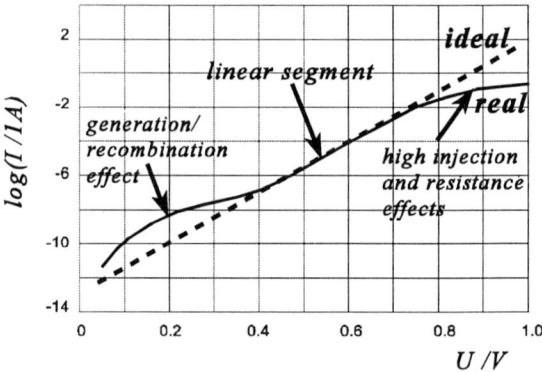

Figure 3. The real i-u characteristic of the pn-junction (exaggerated difference to the ideal i-u characteristic by Shockley)

In this case the basic signal for temperature measurement is the difference between the potential differences (4) of three voltages of the pn-junction operated by the currents I_0, $(I_0+\Delta I)$, and $(I_0-\Delta I)$ (4). Through the twofold difference, both parameters representing the resistance effects and the saturation current are eliminated.

$$\Delta(\Delta U) = (U_1 - U_2) - (U_2 - U_3) \quad (4)$$

$$= \frac{kT}{e}\cdot\ln\left(\frac{(I_0-\Delta I_0)(I_0+\Delta I_0)}{I_0^2}\right)$$

The method by Goloub enables us to gain a measurement failure of about 1.7 K without calibration (figure 4).

To correct the remaining systematical temperature error, the models were generally multiplied by an empirical factor m (5). This parameter shall compensate the remaining failures between the measured voltages and the calculated voltages:

$$U_{corrected}(T) = m\cdot U_{modeled}(T) \quad (5)$$

The parameter m needs to be determined through at least a one-point calibration.

3. CALIBRATION

In order to investigate the behaviour of a sensor, a calibration process is necessary to detect the sensor output signal by well-known input values. The calibration is primarily a mean to examine the real behaviour of the sensor. The received data during the calibration process are also useful for adjusting some individual sensor attributes during the production process. These data are also used for calculating some individual sensor parameters which are to be taken into account during use.

The calibration of sensors is frequently carried out in the industrial production process and during maintenance of active systems. This is why calibration is generally regarded as a matter of course, although it is an arduous process.
During the calibration process, the sensor has to be taken out from the application in which it is in use and then inserted into precise calibration equipments. This can be complicated in some cases because of difficulties in stopping the applications or in the accessibility of the sensors. Therefore calibration is in most cases an undesirable expensive process.

Temperature sensors are very often used so that a reduction of production and maintenance costs through elimination of the calibration costs will be very advantageous. Particularly because of the high costs of an equipment holding constant and well-defined temperatures.

The very accurate calibration free temperature measurement methods like gas thermometers or acoustic thermometers [8] are complicated an unpractical to be used for normal applications.
For other kinds of sensors like resistance temperature sensors, important sensor attributes are defined through one or more geometrical measures. Because of the uncertainty of production processes, the sensors must be calibrated and adjusted during the production process until it reaches a specified degree of accuracy.

In the next section we will introduce how pn-junctions could realize a calibration free temperature measurement with a quite good accuracy.

4. A NEW CONCEPT FOR AN ACCURATE CALIBRATION FREE TEMPERATURE MEASUREMENT

To measure a quantity without needing calibration, the used basic model must satisfy some conditions with respect to its structure and to the nature of its parameters. A model contains in general nature constants, directly measurable parameters, and unknown parameters. For a calibration free measurement the quantity being measured must be calculable without needing to predetermine any of the other unknown parameters in the model.

Not all temperature measurement methods based on pn-junctions are suitable for a calibration free measurement. With methods based on the $U_{be}(T)$ characteristic it will be difficult to realize a calibration free measurement because of the structure of the models. The $U_{be}(T)$ characteristic was modelled mathematically, so that the model equation will contain a range of unknown parameters that are assumed to be calculated from calibration data. The value of the basis-emitter voltage by 0 K is known and physically well-defined. To realize an accurate temperature measurement we will need at least a 1-point calibration to define the slope of the $U_{be}(T)$ characteristic.

In the case of methods based on the i-u characteristic, we know that a calibration free measurement is possible with the methods by Verster or by Goloub but it leads to a limited accuracy level. This is due is the insufficient agreement between the used models and the real behaviour of the i-u characteristic involving secondary effects like generation/recombination effects, high injection effects and resistance effects of the semiconductor device (figure 3). An easy correcting of the used models through multiplication with an empirical parameter (4) affects the calibration free behaviour of the measurement method.

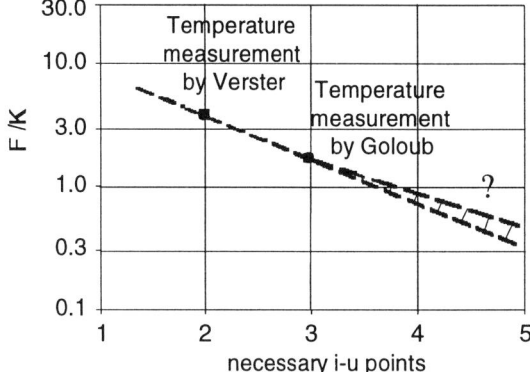

Figure 4. The average of the measurement failure F vs. the number of the necessary i-u-points

The extrapolation of the results of the experimental measurements carried out for about 80 transistors and shown in figure 4 predict, that the measurement accuracy could be increased through the consideration of more i-u points from the pn-junction characteristic and the use of a more precise model describing the real behaviour of the i-u characteristic.

The proposed new method in this paper consists of using an accurate i-u characteristic model. It is to be expected that this model has more unknown parameters. Therefore it will be possible to calculate temperature only with enough measurement data and an adequate signal processing.
The Gummel-Poon model [9] is a candidate for an accurate model describing the pn-junction i-u characteristic. A reduced version of the Gummel-Poon model (figure 5) for this case has exactly six unknown parameters [10] including temperature.

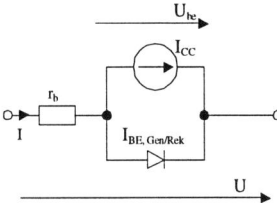

Figure 5. Reduced Gummel-Poon model [10]

The investigation of the reduced Gummel-Poon model for the application in temperature measurement demonstrates, that the

Gummel-Poon model is able to model the secondary effects by small currents like generation/recombination effects, and the secondary effects by middle and high currents like high injection and resistance effects [11].

To fulfill the conditions of a calibration free measurement with an accurate model containing a range of parameters, the calculation of all unknown parameters together with temperature will be carried out *online*. We propose to integrate the i-u-points measurement and the signal processing extracting the model parameters into a smart temperature sensor (figure 6). This will not increase the costs a lot due to today's availability of cheap micro controllers.

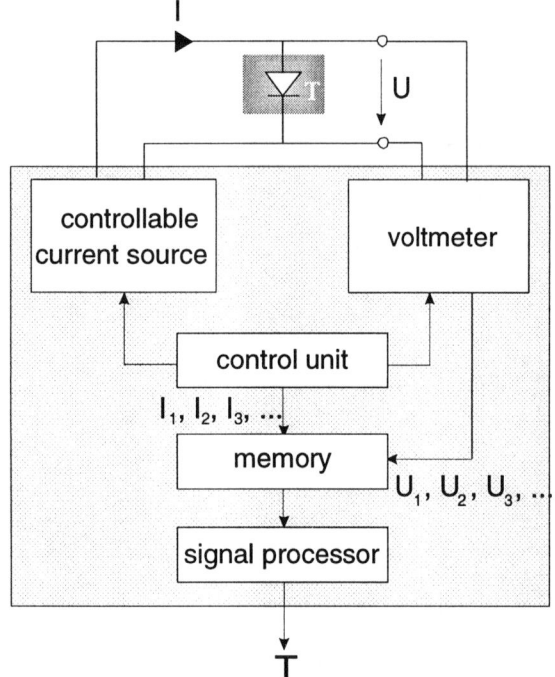

Figure 6. Principle of the proposed smart sensor for calibration free temperature measurement with bipolar transistors

The diagram in figure 6 is to be seen as the principle structure of the proposed calibration free smart temperature sensor. In dependance of the application in which the calibration free sensor will be used, it will be necessary to adapt this concept. For example, if the total measurement time is critical, the i-u-points measurement could be parallelized through use of more than one transistor of the same typ. This idea is implemented in the realization of some integrated temperature sensors [12]. In other cases with a slow process dynamic it will be possible to carry out the i-u point measurement in a sliding time window.

5. SUMMARY

Not all well-known temperature measurement methods based on pn-junctions are adequate, when we have to realize a calibration free temperature measurement method.

To reach a good accuracy, pure mathematical developed models like models of the $U_{be}(T)$ characteristic must contain a lot of parameters to be adjusted through calibration.

Previous methods based on the i-u characteristic like the method by Verster [1] and the method by Goloub [7] could realize a calibration free temperature measurement but they didn't reach an acceptable accuracy level due to the deviations between the used models and the real behaviour of the i-u characteristic. This fact prevented the calibration free use of both methods.

We propose to use an accurate i-u characteristic model taking into account secondary effects affecting the i-u characteristic of pn-junctions like generation/recombination effects, high injection effects and resistance effects like the reduced Gummel-Poon model [10].

The i-u-points measurement together with the signal processing will be integrated within the sensor so that no parameter are to be predetermined through calibration.

6. ACKNOWLEDGMENTS

I wishes to express my gratitude towards Prof. Hans-Rolf Tränkler and Dr. Michael Horn for useful discussions.
This work was supported by the "Deutsche Forschungsgemeinschaft" under the code TR 232 /7-1.

7. REFERENCES

[1] T. C. Verster, *pn-junction as an ultralinear calculable thermometer,* Electronic Letters, May 1968, Vol. 4, No. 9, pp. 175-176

[2] G. C. M. Meijer, *Thermal sensors based on Transistors,* Sensors and actuators, 10 (1986), pp 103-125,

[3] Palett, J. E., *An Electrical Thermometer,* Electronic Engineering, May 1963, 35, pp. 313-315

[4] A. Ohte, M. Yamagata, *A Precision Silicon Transistor Thermometer,* IEEE Transactions on Instrumentation and Measurement, vol. IM-26, December 1977, pp. 335-341

[5] W. Shockley, *The Theory of P-N-Junctions in Semiconductors and P-N-Junction Transistors,* Bell Systems Technical Journal, 28, 1949, pp. 435-489

[6] P. Ipsen, *Thermometrie mit Silizium-Dioden,* Bosch-Technische Berichte 2, Heft 3, Nov. 1967, pp. 137-144

[7] B. Goloub, O. Goloub, A. Baran, *Genauigkeitserhöhung für Transistor-Temperatursensoren,* Proceedings of Sensor 97, Nürnberg, 13.-15. 5. 1997, Vol. III, pp. 183-188, ACS Org. GmbH, Wunstorf

[8] T. J. Quinn, *Temperature,* Academic Press, 1983, S. 61-72

[9] H. K. Gummel, *A Charge Control Relation for Bipolar Transistor,* Bell Systems Technical Journal, Jan. 1970, pp. 115-120

[10] R. Holmer, H.-R. Tränkler, *Temperaturmessung mit Bipolar-Transistoren,* Sensoren und Meßsysteme, Bad Nauheim, 11.-13. 3. 1996, pp. 107- 112, VDI Berichte Nr. 1255, VDI-Verlag-Düsseldorf

[11] O. Kanoun, H.-R. Tränkler, *Kalibrationsfreie Temperaturmessung auf der Basis von pn-Übergängen,* Temperatur '98, 16./17. 02. 1998, Berlin

[12] *AD 590,* Analog devices data sheet Rev. B, 1997

Two Temperature Sensors Realized in BiCMOS Technology

I. M. Filanovsky
University of Alberta
Edmonton, Alberta, Canada, T6G 2E1

Abstract -- Two temperature sensors realized in BiCMOS technology are described. The first sensor is using, as a temperature sensitive element, a circuit of the threshold extractor. The output of this circuit provides the p-channel transistor threshold voltage that linearly changes with temperature. The second sensor is using a bridge that includes polysilicon and base-diffused resistors. A one-stage operational amplifier amplifies the output voltage of the sensors. The temperature simulation of the sensor circuits shows that the circuit with extractor has a more linear temperature characteristics than the circuit with the bridge.

I. INTRODUCTION

Thermal sensors are being integrated along with other IC sensors for on the spot temperature measurement. Electronic temperature sensing is a mature art [1], yet each new technology brings new possibility for innovation. Two circuits considered in this paper were designed for realization in BiCMOS 0.8 μm technology.

The first sensor is using a linear dependence of the threshold voltage V_{TP} on temperature [2]. To realize this sensor a simple circuit of V_{TP} extractor [3] operating in a wide range of bias currents was created. The output of this circuit, if not loaded, may be used directly as a temperature sensor. To allow loading a version with a simple one-stage amplifier attached to the threshold extractor was also designed.

The second sensor is using opposite temperature coefficients of polysilicon and base diffusion layers. A Wheatstone bridge with a one-stage amplifier attached to the bridge output was designed.

The sensor including the threshold extractor as a temperature sensitive element may be realized in a CMOS technology as well.

II. TEMPERATURE SENSOR CIRCUITS

A. Threshold extractor circuit

It is assumed here that the p-channel devices have separate wells, and the n-channel devices have a common substrate.
Let us consider a series connection of two transistors shown in Fig. 1, a. In this configuration T_1 operates in saturation and T_2 is in linear (triode) region of operation. In common with [2,3], assume square law transistor behavior. Then, if T_1 and T_2 are matched the circuit is described by the following equations

$$|V_{GS1}| = |V_{TP}| + \sqrt{\frac{2I}{K_1}} \quad (1)$$

$$(1+a)I = K_2 |V|(|V_{GS2}| - |V_{TP}|) - \frac{K_2 V^2}{2} \quad (2)$$

and

$$|V_{GS2}| = |V_{GS1}| + |V| \quad (3)$$

In (1)-(3), $K_i = \mu_p C_{ox}(W/L)_i$ as usual, V_{TP} is the threshold voltage for p-channel transistors, $i = 1,2$. The body effect is, obviously, eliminated. Substituting (1) and (3) into (2) and solving the obtained square equation one can find that

$$|V| = \sqrt{2I}\left(\sqrt{\frac{1}{K_1} + \frac{1+a}{K_2}} - \sqrt{\frac{1}{K_1}}\right) \quad (4)$$

Now, assume that in the circuit of Fig. 1, b transistor T_3 is in saturation. Then

$$|V_{GS3}| = |V_{TP}| + \sqrt{\frac{2I_3}{K_3}} \quad (5)$$

If the battery V in the drain circuit has the voltage

$$V = \sqrt{\frac{2I_3}{K_3}} \quad (6)$$

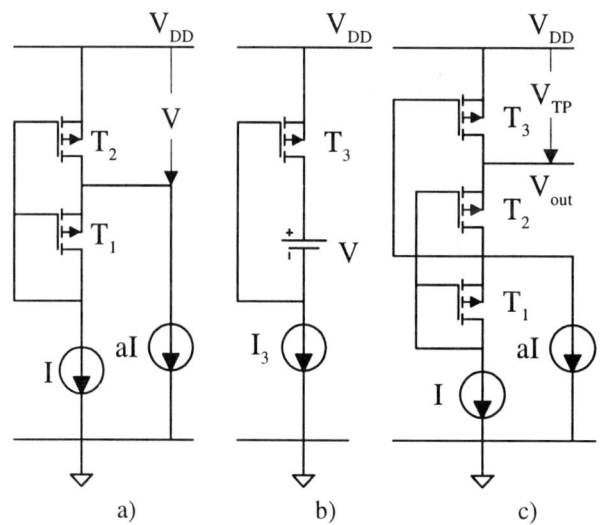

Fig. 1. V_{TP} extractor and its subcircuits
a) series connection of two transistors
b) transistor with reduced drain-source voltage
c) three-terminal V_{TP} extractor circuit

then the difference between the drain voltage of the transistor T_3 and V_{DD} line will have exactly the value of V_{TP}. One has to check, of course, that the condition of saturation

$$|V_{DS3}|=|V_{GS3}|-|V|\geq|V_{GS3}|-|V_{TP}| \qquad (7)$$

is satisfied. It will be satisfied if $|V|\leq|V_{TP}|$, i.e. if

$$\sqrt{\frac{2I_3}{K_3}}\leq|V_{TP}| \qquad (8)$$

Finally, let us consider the circuit of Fig. 1, c, where the battery in the drain of T_3 is substituted by the top transistor of Fig. 1, a. In the circuit of Fig. 1, c the current $I_3=(1+a)I$. Let this value and the voltage V given by (4) be substituted in (6). Then one obtains that if the condition

$$\sqrt{\frac{1}{K_1}+\frac{1+a}{K_2}}-\sqrt{\frac{1}{K_1}}=\sqrt{\frac{1+a}{K_3}} \qquad (9)$$

is satisfied, then the output voltage V_{out} of the circuit of Fig. 1, c referenced to V_{DD} line will be exactly equal to V_{TP}. The bias current I should satisfy the condition

$$\sqrt{\frac{2(1+a)I}{K_3}}\leq|V_{TP}| \qquad (10)$$

The equation (9) has many solutions with different spread of transistor aspect ratios. A convenient solution (which was used for simulation and realization) is to take $a=2$, $K_2=K_1$. Then one obtains $K_3=K_1$.

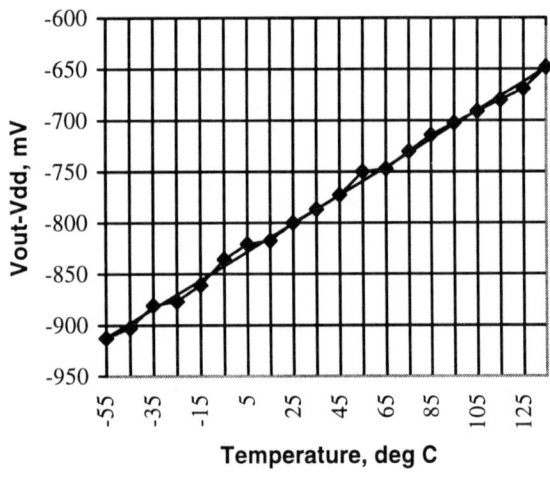

Fig. 2. V_{TP} temperature dependence

The preliminary simulation of this extractor [4] allowed to establish that the optimal transistor current I is about 10 μA and the V_{TP} temperature coefficient is around 1 mV/°C. The extractor was fabricated and tested [6]. The temperature dependence of the V_{TP} voltage is shown in Fig. 2. The trendline shows that, indeed, the dependence of the V_{TP} voltage on temperature is rather linear, and the temperature coefficient is about 1.4 mV/°C.

Other electrical characteristics of this circuit and some statistics can be found in [6]. An attempt to built a V_{TN} extractor and to use it as a temperature sensor was unsuccessful [5]. The circuit of that extractor was more complicated, more sensitive to voltage supply variations, and did not show a linear dependence of the circuit output voltage (also measured with respect to the V_{DD} line) versus temperature.

B. *Wheatston bridge circuit*

The BiCMOS technology allows to create a temperature sensitive Wheatstone bridge (Fig. 3).

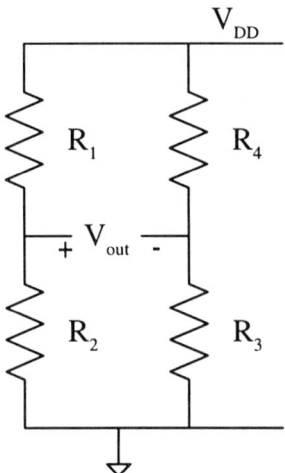

Fig. 3. Wheatstone bridge

Indeed, if this bridge is fabricated using polysilicon resistor layer (RPOLY) for realization of resistors R_1 and R_3, and base diffusion layer (RDIFF) for realization of R_2 and R_4 then the output voltage of the bridge will be temperature dependent. In accordance with the resistor model parameters of BiCMOS 0.8 μm technology [7] the base diffusion resistors have a positive temperature coefficient of $\alpha_1=1.34*10^{-3}$ 1/grad, the polysilicon resistors have a negative temperature coefficient of $\alpha_2=-3.51*10^{-4}$ 1/grad. The output voltage of the bridge is

$$V_{out}=\left(\frac{R_2}{R_1+R_2}-\frac{R_3}{R_3+R_4}\right)V_{DD} \qquad (11)$$

Assuming that in this bridge $R_1=R_3=R_0[1+\alpha_1(T-T_0)]$ and $R_2=R_4=R_0[1+\alpha_2(T-T_0)]$ one obtains that

$$V_{out}=\frac{V_{DD}(\alpha_1-\alpha_2)(T-T_0)}{2+(\alpha_1-\alpha_2)(T-T_0)} \qquad (12)$$

The output characteristic of this bridge calculated using (12) for V_{DD} = 5 V in the temperature range of –55 to 145 °C is shown in Fig. 4, a.

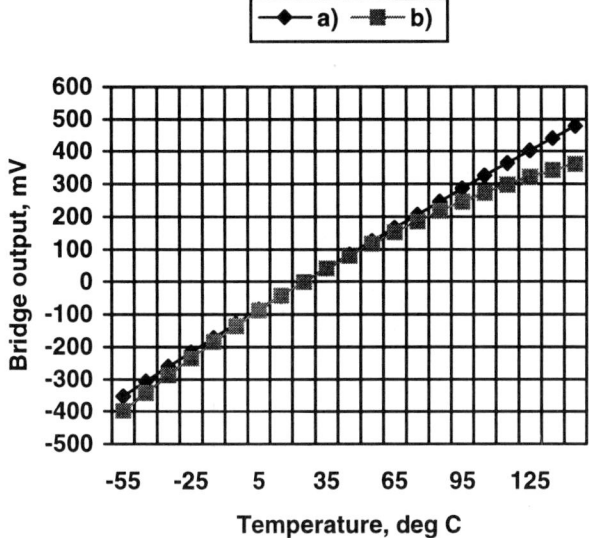

Fig. 4. Bridge temperature characteristics

The temperature dependencies of the resistors unfortunately have a quadratic component as well, i.e.

$$R_i = R_0[1+\alpha_i(T-T_0)+\alpha_{iq}(T-T_0)^2] \quad (13)$$

Here i=1,2. Again, in accordance with the process model data $\alpha_{1q} = 3.95*10^{-6}$ 1/grad² for polysilicon resistors and $\alpha_{2q} = 8.38*10^{-7}$ 1/grad² for base diffusion resistors. The dependence of the bridge output voltage, when the quadratic components are considered as well, is also shown in Fig. 4, b.

III. REALIZATION WITH OUTPUT AMPLIFIERS

Both sensors were also realized with simple signal conditioning amplifiers.

The circuit with temperature-sensitive extractor is shown in Fig. 5. It includes the bias divider (transistors T_4, T_5, and T_6) which sets the current I=12.5 µA in the transistor T_8 at the room temperature (27 °C). The temperature simulation shows that the divider current changes from 15.2 to 10.5 µA when the temperature changes from –50 to 150 °C. This variation of current does not introduce any errors in the V_{TP} threshold voltage [5] which appears at the output of the extractor circuit (transistors T_1, T_2, and T_3). This voltage is applied to the level

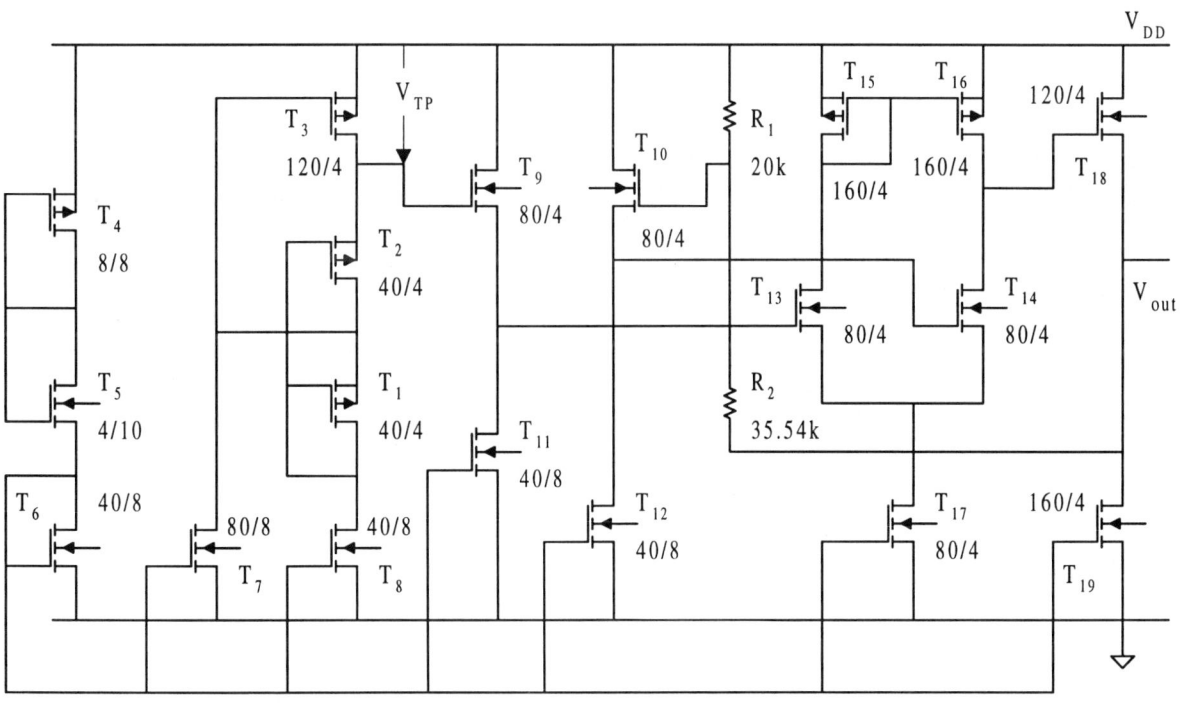

Fig. 5. Extractor sensor with amplifier

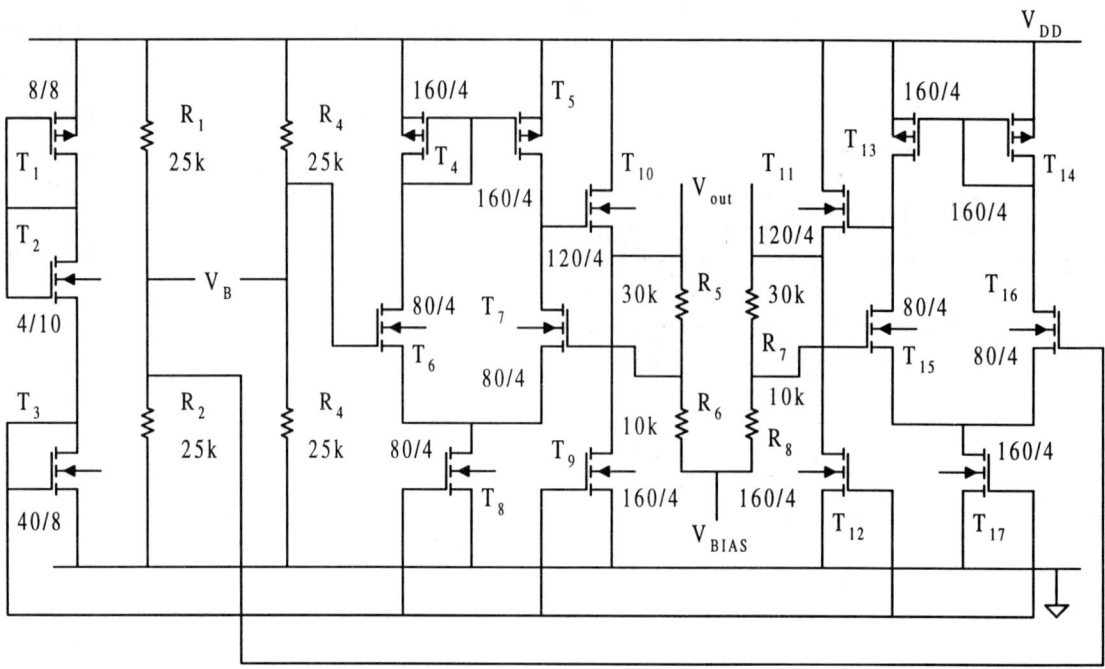

Fig. 6. Resistive bridge and amplifier

shifter (transistors T_9 and T_{11}) and amplified by one-stage operational amplifier (transistors T_{13} to T_{19}). The feedback voltage is applied via a symmetric level shifter (transistors T_{10} and T_{12}). The feedback voltage divider includes two polysilicon resistors R_1 and R_2.

The circuit with the temperature-sensitive resistive bridge is shown in Fig. 6. It includes the bias divider (transistors T_1, T_2, and T_3) which sets the current $I=50$ μA in the transistors T_8 and T_{17} of two symmetric one-stage operational amplifiers (transistors T_4 to T_8 and T_{13} to T_{17}). The feedback circuits in both amplifiers are closed via polysilicon resistive dividers.

The physical design of both circuits was done as well. At the present time the circuits are accepted for fabrication.

Acknowledgements

The author thanks Mr. N. Jantz and Mr. M. Margala for their help with Cadence design software, and Dr. W. Tinga for the possibility to use the Missimers temperature cabinet. Manufacturing of the chips was arranged by the Canadian Microelectronic Corporation (CMC).

References

[1] W. Rasmussen, "Thermal Sensors", *Sensor technology and devices*, Chapter 8, (L. Ristic, Editor), Artech House, Boston, 1994.

[2] S.M. Sze, *Physics of Semiconductor Devices*, J. Wiley, New York, 1981.

[3] Y.P. Tsividis, R.W. Ulmer, "Threshold voltage generation and supply-independent biasing in c.m.o.c. integrated circuits", *Electronic Circuits and Systems*, vol. 3, No 1, pp. 1-4, 1979.

[4] R. Alini, A. Baschirotto, R. Castello, F. Montecchi, "Accurate MOS threshold voltage detector for bias circuitry", *Proc. IEEE Int. Symp. on Circuits and Systems*, San Diego, pp. 1280-1283, May, 1992.

[5] I. M. Filanovsky, An input-free V_T extractor circuit using a series connection of three transistors, *Int. J. Electronics*, vol. 82, No 5, pp. 527-532.

[6] I. M. Filanovsky, An input-free V_{TP} and $-V_{TN}$ extractor circuits realized on the same chip, *Proc. 40th Midwest Symposium on Circuits and Systems*, Sacramento, CA, Aug. 3-6, 1997 (in print).

[7] *CMC BiCMOS Design Kit V0.3 for Cadence Analog Artist*, Canadian Microelectronic Corporation, Ottawa, 1995.

A CMOS INTEGRATED INFRARED RADIATION DETECTOR FOR FLAME MONITORING

P. Bendiscioli[1], F. Francesconi[2], P. Malcovati[1], F. Maloberti[1], M. Poletti[1] and R. Valacca[1]

[1] Integrated Microsystem Laboratory, University of Pavia, Via Ferrata 1, 27100, Pavia, Italy
Tel. +39 382 505 205, Fax. +39 382 505 677
[2] Micronova Sistemi S.r.l., Piazza G. Marconi 4, 27020 Trivolzio (Pavia), Italy
Tel. +39 382 930 701, Fax. +39 382 930 701

ABSTRACT

This paper presents an integrated infrared radiation detector for flame monitoring applications, fabricated in CMOS technology. The system discriminates the radiation of the flickering flame in an oil burner from the steady background radiation generated by the furnace by considering only the harmonic components of the infrared signal in the band from 50 Hz to 250 Hz. In order to maximize the flexibility and the robustness of the system, most of the signal processing is performed in the digital domain. Experimental results on an integrated prototype, confirming the validity of the proposed approach, are reported.

1. INTRODUCTION

Nowadays the regulations and the requirements concerning security in domestic and industrial apparates are becoming more stringent and severe. In particular gas and oil burners require necessarily flame monitoring systems, in order to detect the presence of the flame and avoid dangerous leakages of gas or oil. Very simple and cheap ionization probes are typically used in gas burners, in view of the relatively clean environment. These sensors, however, degrade very quickly in oil burners because of the dirt combustion residues. Other available flame detectors placed inside the furnace of oil burners [1], including thermocouples, microphones and video cameras, provide a low signal-to-noise ratio and easily degrade in the hostile furnace ambient. By contrast, flame monitors based on radiation detection [2, 3, 4] can be placed outside the furnace, in a less harsh and polluted environment, thus increasing significantly long-term stability and reliability of the system.

In particular, silicon infrared (IR) sensors (photodiodes) can be fabricated using the standard CMOS layers, thus allowing the integration of complex signal processing functions on the sensor chip at very low cost. Of course, in the IR region of the spectrum (from 800 nm to 1000 nm), the emission intensity of the flame is at least one order of magnitude lower than the emission intensity of the furnace. The IR radiation power produced by the furnace, however, is concentrated at DC (or at least at very low frequency), while the unavoidable flickering of the flame spreads the IR radiation power of the burning oil in a frequency band ranging from 50 Hz to 250 Hz. It is, therefore, possible to detect the presence of the flame with a large signal-to-noise ratio by monitoring the IR radiation in this frequency band, while suppressing the large DC component. Thanks to the CMOS compatibility of IR photodiodes, the additional hardware required to perform the filtering can be integrated on the sensor chip, thus making the resulting microsystem very convenient in terms of cost and performance.

2. INTEGRATED PHOTODIODE

The cross-section of a CMOS compatible IR photodiode is shown in Fig. 1. The active junction of the diode (*n*-well/*p*-substrate) is sufficiently deep to allow the generation of electron/hole pairs due to IR radiation [5, 6].

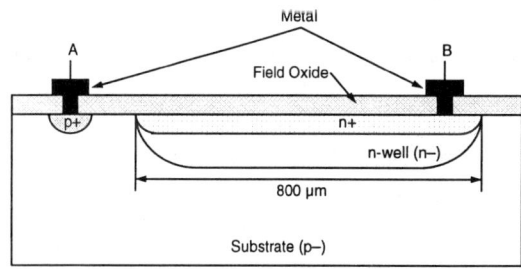

Figure 1. Cross-section of the CMOS integrated photodiode

The responsivity of a photodiode at wavelength λ is defined as

$$s(\lambda) = \frac{I_{ph}(\lambda)}{\Phi(\lambda)}, \quad (1)$$

where $I_{ph}(\lambda)$ denotes the generated photocurrent and $\Phi(\lambda)$ the corresponding incident radiation power. Therefore, the generated photocurrent can be expressed as

$$I_{ph} = \int_{\lambda_1}^{\lambda_2} \Phi(\lambda) s(\lambda) d\lambda. \quad (2)$$

In order to determine the expected photocurrent in flame monitoring applications, the responsivity of the photodiode used in the proposed system has been measured (Fig. 2). From the measurements we obtained an effective photocurrent I_{ph} of the order of few tens of µA. It is worth to note from Fig. 2 that the integrated photodiode shows a large responsivity also in the visible region of the spectrum, as expected. This additional information may be useful in flame monitoring applications, however, if necessary, it can be removed by placing an optical filter on top of the sensor.

Parameter	Value
Technology	CMOS
Power Supply	5 V
Frequency Band	50 Hz ÷ 250 Hz
Radiation Wavelength	800 nm ÷ 1000 nm
Output Signals	Digital
Flame Presence	1 Bit
Flame Intensity	8 Bits
DC Component Suppression	–40 dB
High Frequency Rejection	–30 dB
Flame Detection Threshold	Programmable

Table 1. Specifications of the IR detector for flame monitoring

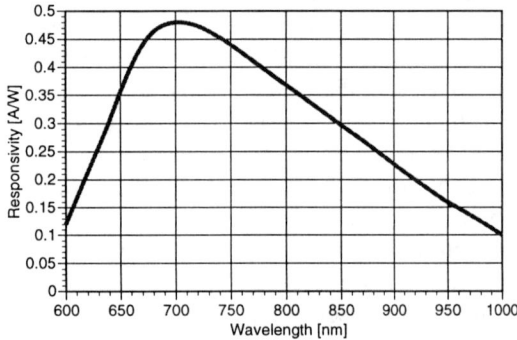

Figure 2. Measured spectral responsivity of the integrated photodiode

3. SYSTEM DESCRIPTION

The most important specifications of an IR radiation detector for flame monitoring are summarized in Tab. 1. In view of the low frequency of the considered signals, a digital implementation of the filters, which does not require external components to realize large time constants, is preferable with respect to a classical analog solution and, in addition, it is more flexible and easily programmable.

The block diagram of the proposed IR detector is shown in Fig. 3. The system consists of an integrated inversely biased IR photodiode, a readout circuit, an analog-to-digital converter (ADC), the digital filters, a digital signal processing circuit and a feedback loop with digital-to-analog converter (DAC) for suppressing the DC component of the IR signal.

The readout circuit for the integrated photodiode has two main functions: provide a suitable biasing voltage for the diode and transform the diode current into a voltage. Both of these tasks are accomplished by an operational amplifier with resistive feedback. The used operational amplifier is based on a standard two-stage class AB structure. The feedback resistor R and the biasing voltage V_B are external, in

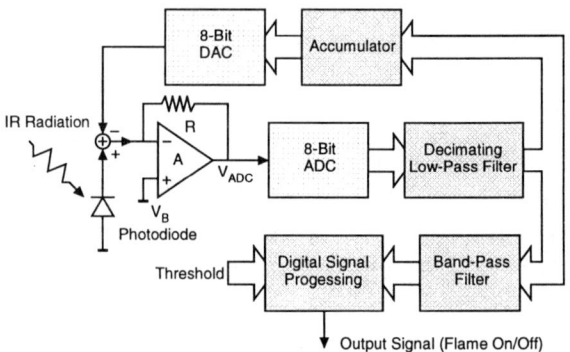

Figure 3. Block diagram of the infrared radiation detector

order to allow sensitivity and operating point adjustments.

The current generated by the photodiode, transformed into a voltage and amplified, is converted into the digital domain by an 8-bit resistor-string-based successive approximation ADC with sampling frequency $F_{SH} = 32$ kHz. The resulting digital word is then down-sampled (decimated) with $F_{SL} = 1$ kHz, in order to reduce the complexity of the subsequent digital filtering section. The output word of the decimator is delivered either to the feedback loop and to a digital band-pass filter. The feedback loop extracts the DC component of the signal with a digital accumulator and subtracts it directly from the sensor signal by means of an 8-bit current mode DAC (binary weighted current sources). At the same time, the band-pass filtered signal is delivered to the digital signal processor, whose block diagram is shown in Fig. 4, where it is rectified, accumulated and compared with a programmable threshold, in order to produce an output bit representing the status of the flame.

The block diagram of the decimating filter is shown in Fig. 5. The filter consists of two accumulators sampled at $F_{SH} = 32$ kHz followed by two differentiators sampled at

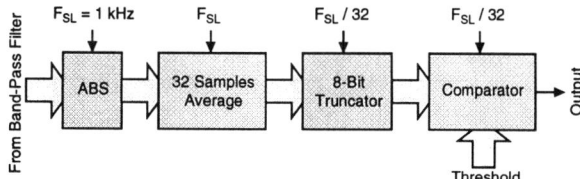

Figure 4. Block diagram of the digital signal processing circuit

$F_{SL} = 1$ kHz [7, 8] and implements the z-domain transfer function

$$H_D(z) = \left(\frac{1 - z^{-D}}{D(1 - z^{-1})}\right)^2, \quad (3)$$

where D = 32 is the decimating factor.

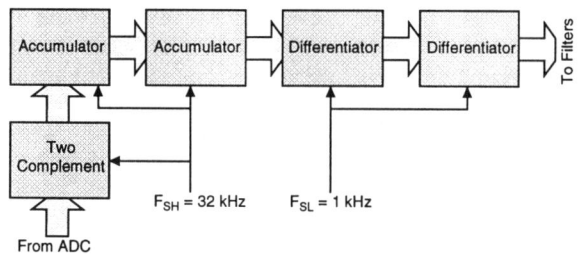

Figure 5. Block diagram of the decimating low-pass filter

The digital band-pass filter, whose block diagram is shown in Fig. 6, is obtained by cascading a high-pass filter with z-domain transfer function

$$H_{LP}(z) = 1 - z^{-3} \quad (4)$$

and a five-tap FIR low-pass filter. The coefficients A, B and C have been approximated with powers of 2 ($A = -2^{-9}$, $B = -2^{-6} - 2^{-7} - 2^{-9}$ and $C = 2^{-1} + 2^{-5}$), in order to replace the expensive multipliers with simple shift registers and adders [9]. The resulting transfer function

$$H_{LP}(z) = Az^{-5} + Bz^{-4} + Cz^{-3} + Cz^{-2} + Bz^{-1} + A, \quad (5)$$

is graphically shown in Fig. 7.

4. EXPERIMENTAL RESULTS

The proposed integrated IR detector has been integrated in a standard 0.8 μm double-poly, double-metal CMOS process. The microphotograph of the chip is shown in Fig. 8. The total die area, including pads is 3.7 mm × 2.5 mm, while the IR photodiode are is 800 μm × 800 μm.

The measured operational amplifier output voltage (V_{ADC}) as a function of the diode photocurrent (I_{ph}) with and with-

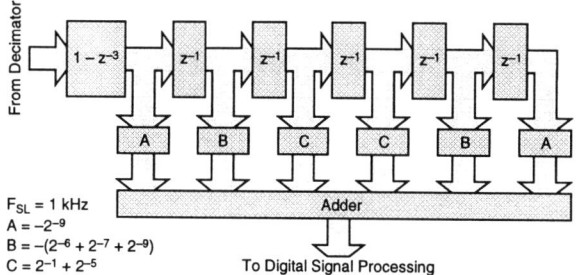

Figure 6. Block diagram of the band-pass filter

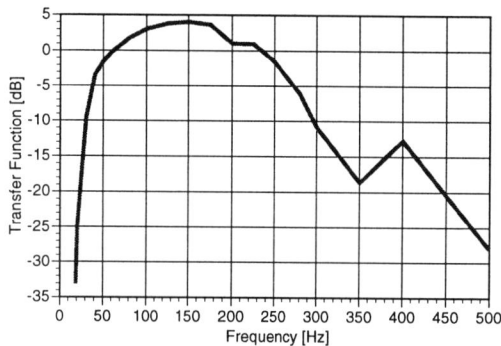

Figure 7. Transfer function of the band-pass filter

Figure 8. Microphotograph of the infrared detector chip, including the photodiode, the analog readout circuit and a digital section

out the DC cancellation feedback loop is shown in Fig. 9. It can be observed that the feedback loop attenuates the DC component more than 30 dB, thus avoiding the saturation of the operational amplifier. The residual DC component is then removed by the digital band-pass filter.

Fig. 10 and Fig. 11 show the measured transient behavior of the system digital output (*Output*) and of the operational amplifier output voltage (V_{ADC}) when the flame is turned off and when the flame is turned on, respectively. To perform these measurements, the flame has been emulated with an IR diode operated at 50 Hz.

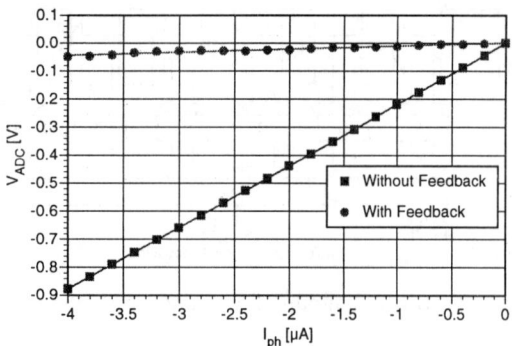

Figure 9. Measured output voltage of the operational amplifier (V_{ADC}) with and without the DC cancellation feedback loop

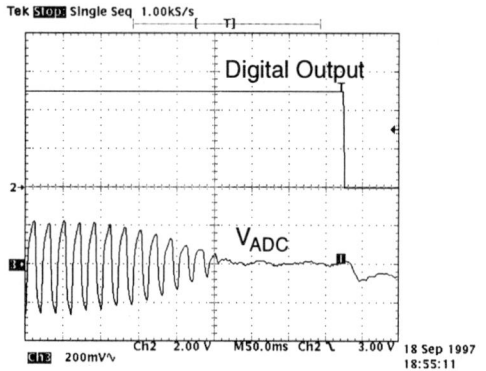

Figure 10. Measured digital output of the system (*Output*) and output voltage of the operational amplifier (V_{ADC}) when the flame (emulated with a diode) is turned off

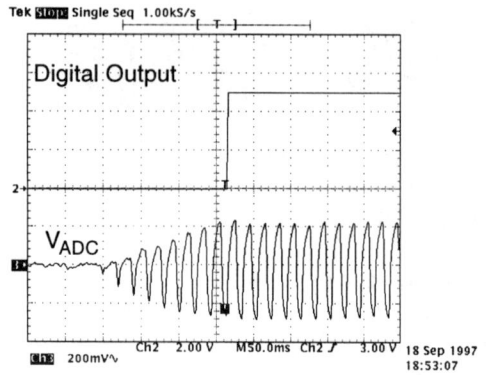

Figure 11. Measured digital output of the system (*Output*) and output voltage of the operational amplifier (V_{ADC}) when the flame (emulated with a diode) is turned on

5. CONCLUSIONS

This paper presented a fully integrated infrared radiation detector for flame monitoring applications. The system discriminates the flickering flame radiation from the steady background radiation generated by the furnace by considering only the harmonic components of the infrared signal in the band from 50 Hz to 250 Hz. This solution allows a large signal-to-noise ratio to be obtained although the flame radiation is lower than the background radiation. In order to maximize the flexibility and the robustness of the system, most of the signal processing is performed in the digital domain. Experimental results obtained from an integrated prototype demonstrate the validity of the proposed approach.

ACKNOWLEDGMENTS

The authors wish to thank Dr. Simona Brigati for the useful discussions as well as Dr. Valerio Annovazzi Lodi and Dr. Sabina Merlo for the help in characterizing the photodiodes.

REFERENCES

[1] A. Jones, "Flame Failure Detection and Modern Boilers", *J. Phys. E: Sci. Instrum.*, vol. 21, pp. 921-928, 1988.

[2] M. Cresser, *Flame Spectrometry in Environmental Chemical Analysis*, The Royal Society of Chemistry, Cambridge, UK, 1994.

[3] J. Chase and G. Lovejoy, "New Scanner Uses Visible Light to Detect Flame," *Power Enginering*, vol. 86, pp. 68-71, 1982.

[4] D. Bolliger, P. Malcovati, A. Häberli, P. Sarro, F. Maloberti and H. Baltes, "Integrated Ultraviolet Sensor System with On-Chip 1 GΩ Transimpedence Amplifier", *ISSCC '96 Digest of Technical Papers*, San Francisco, CA, pp. 328-329, 1996.

[5] S. M. Sze, *Semiconductor Devices: Physics and Technology*, John Wiley & Sons, New York, 1981.

[6] A. Ambroziak, *Semiconductor Photoelectric Devices: An Introduction to Design*, London Iliffe Books LTD, London, 1968.

[7] S. R. Norsworthy, R. Schreier and G. C. Temes, *Delta-Sigma Data Converters: Theory, Design and Simulation*, IEEE Press, Piscatawy, NJ, 1997.

[8] J. C. Candy, "Decimation for Sigma-Delta Modulation", *IEEE Trans. Communications*, vol. Com-34, 1986.

[9] Q. Zaho and Y. Tadakoro, "A Simple Design of FIR Filters with Powers-of-Two Coefficients", *IEEE Trans. Circuits and Systems*, vol. 35, pp. 556-570, 1988.

A Multi-mode X-ray Imager for Medical and Industrial Applications

R.E. Colbeth, M.J. Allen, D.J. Day, D.L. Gilblom, R. Harris, I.D. Job, M.E. Klausmeier-Brown,
J. Pavkovich, E.J. Seppi, E.G. Shapiro, M.D. Wright, J.M.Yu

Varian Associates, Inc.
3075 Hansen Way
Palo Alto CA, 94304 USA

ABSTRACT

This paper describes a multi-mode, digital imager for real-time x-ray applications. The imager has three modes of operation: low dose fluoroscopy, zoom fluoroscopy, and high resolution radiography. These modes trade-off resolution or field-of-view for frame rate and additionally optimize the sensitivity of the imager to match the x-ray dose used in each mode. This large area sensing technology has a form factor similar to that of an x-ray film cassette, no geometric image distortion, no sensitivity to magnetic fields, a very large dynamic range which eliminates repeat shots due to over or under exposure, 12 bit digital output and the ability to switch between operating modes in real-time. Medical applications include fluoroscopy and radiography, while non-destructive test (NDT) applications might include solder joint verification on printed circuit boards and parts inspection in the automotive and aerospace industries.

1.0 INTRODUCTION

Over the last four decades there has been an increasing shift toward digital techniques in radiology. CT, MRI, SPECT, PET and ultrasound are all digital imaging modalities in common use today. The development of these systems was motivated by the additional diagnostic information provided. Today, as office automation and networking have become pervasive, the fact that the information is in digital form offers other potential benefits such as reduced cost of storage, ease of image distribution throughout the hospital, automated image analysis, as is under development for the identification of microcalcifications in mammograms, and teleradiology, where diagnosis can be handled by specialists from remote locations.

Despite the benefits of digital acquisition, radiographic x-ray technology, which accounts for roughly 80% of the imaging done in a radiology department, is still based on the chemical processing of film, while video x-ray technology currently uses a vacuum tube based image intensifier coupled to a video camera. Although in each case, it is possible to digitize the information after it is acquired, an additional step in the imaging process is required which adds cost, time and image degradation.

The imager described below is a solid-state digital acquisition technology capable of both 30 frame-per-second video with resolution of 2 lp/mm (line pair per mm) and high resolution radiographic imaging at 4 lp/mm and repeat rates up to 7.5 shots per second. In addition to combining multiple modalities into one package, large area sensing technology has a form factor similar to that of a film cassette. In the case of radiography, this implies a technology that will be easily integrated into existing equipment. In fluoroscopy (video) applications, the form factor advantage is much more pronounced. In a typical system, shown in Fig. 1a, there is an image intensifier tube (IIT) which converts the x-rays to visible photons, a video camera for recording the output of the IIT, a Spot film changer, which allows the insertion of an x-ray film cassette to acquire high resolution snap shots, a 35 mm Cine camera, which is used to record sequences of images with high resolution, and a box containing the mirror which flips the output of the IIT between these different acquisition devices. Depending on the size of the image intensifier tube, the entire assembly could occupy in the neighborhood of 10 cubic feet. This space is of particular importance in medical applications, since it restricts access to the patient. In contrast, a sensor panel based system, depending on the size of the panel, requires approximately 0.5 cubic feet. Since the panel can be configured as a primary barrier to x-rays, the physician can safely stand behind the sensor, gaining unobstructed access to the patient. Fig. 1b is an artist's rendition of how a sensor panel might replace an IIT based imaging chain on a C-Arm. Beyond the reduction in space required for image acquisition, the sensor panel represents a significant reduction in weight, which facilitates lighter and less costly support structures that have greater ease of movement. The weight reduction advantage is most apparent in comparison to large diameter IIT systems whose total weight can be in the hundreds of pounds, versus less than 50 pounds for the largest panels.

Other advantages of large area sensing technology include:
- no geometric distortion in the image
- no sensitivity to magnetic fields
- intrinsic dynamic range of 10,000:1
- 12 bit digitization at the detector
- real-time switching between operating modes
- superior small area contrast (low crosstalk)
- higher MTF (i.e. greater resolution)
- no blooming from regions of high illumination
- portability

Fig. 2 demonstrates the multi-mode capability of large sensing technology. Fig. 2a shows a radiograph of a hand phantom acquired at a frame rate of 2 fps at the full 4 lp/mm resolution of the imager. This image was taken with an entrance exposure at the receptor of 0.5 mR. The x-ray tube voltage was 50kVp and the tube current was 0.5 mA. The source-to-detector distance (SID) was 100cm and the phantom was directly in contact with the front surface of the detector. Fig. 2b shows an image taken of the same phantom; however, the imager is operating in fluoroscopy mode at 30 fps and a spatial resolution of 2 lp/mm. In this case, the entrance exposure to the imager was 5.6 µR/frame. The image shown is a single frame taken with a technique of 50kVp, 0.5mA and 0.016 inches of Cu filtration in the x-ray beam. For the image in Fig. 2b a post acquisition 3x3 triangular filter has been applied for noise reduction. In addition to showing the multi-mode capability of the imager, theses images demonstrate the large dynamic range inherent in the device. The radiograph used a dose per frame nearly 100 times larger than that of the low dose fluoroscopic image. The anti-blooming feature can also be seen in Fig 2a in the skin area of the image. Despite being adjacent to the unattenuated x-ray beam, there is good image fidelity with a clearly defined edge at the outer surface of the hand phantom.

2.0 Imager Configuration

Fig. 3 shows the configuration of the imager in the context of the overall imaging system. The imager consists of three modules: the Command Processor, the Power Supply and the Receptor. The receptor can be located up to 30 meters from the Command Processor. The connecting cable is 0.5 inches in diameter and carries all power and mode control signals to the receptor. This same cable contains the 60MB/sec serial data link carrying the raw 16 bit digital video back to the Command Processor. The receptor, which is 12.5" x 10.5" x 1.7" including the mounting flange, has been designed for the minimum possible border on the active imaging area, maximizing access to the patient.

The Command Processor is the interface between the imager and the imaging system. The imager operation is controlled via ethernet or one of two serial ports, using software commands. The set of possible imager control operations is supplied to systems integrators in a C library of callable functions. The communications interface is at the level of IP sockets, and hence control of the imager is platform independent. In addition to the software control capability, the Command Processor has a set of uncommitted I/O signals that can be used for hardware handshaking and synchronization of critical tasks. The primary output of the Command Processor is corrected 16 bit digital video. The raw data from the receptor must be corrected for offset and gain variations on a pixel-by-pixel basis. Furthermore, the amorphous silicon panel has a finite number of dead pixels and lines which must be

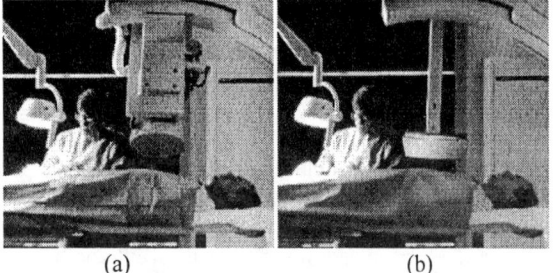

Fig. 1   A comparison of an IIT imager and a sensor panel in a C-arm application. The image with the sensor panel was created using digital editing.

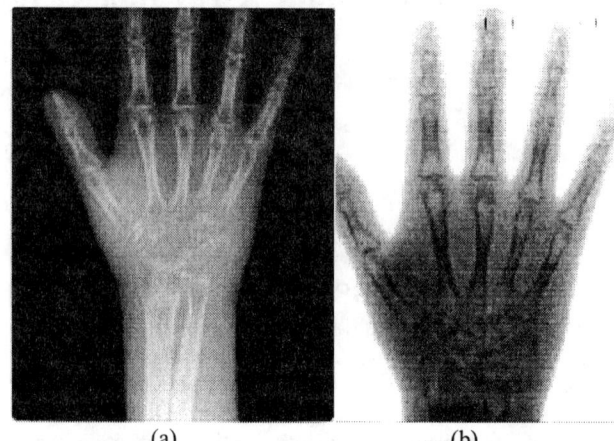

Fig. 2   (a) Radiograph of a hand phantom, 0.5 mR exposure to the detector. (b) Single frame from a 30fps video sequence at 5.6 µR/frame exposure to the detector.

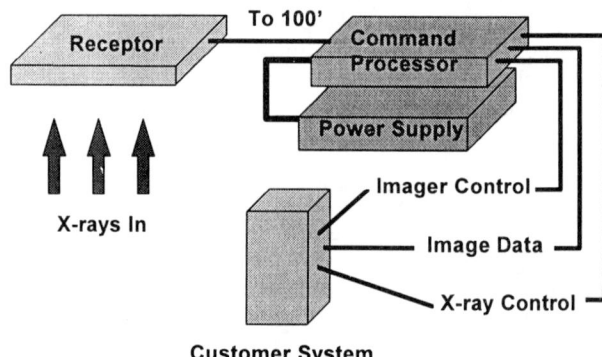

Fig. 3   Imager configuration and its relationship to the complete imaging chain.

corrected through interpolation of the nearest neighbors. The Command Processor also provides a recursive filter for smoothing frame-to-frame noise, or alternately in radiography mode, frame averaging is available. In addition to the 16 bit digital data, there are analog and digital 8 bit video outputs, which conform to the AIA standards for cameras. Brightness and exposure signals are provided to facilitate automatic control of the x-ray generator.

Fig. 4 is a block diagram of the internal configuration of the Command Processor. The core task of the Command Processor is to receive commands over the serial or ethernet ports, then configure the imager for the appropriate operating mode of the receptor, image processing algorithm, and response to hardware I/O signals. The heart of the Command Processor is a MIPS 4300 RISC processor, which is running the VxWorks real-time operating system (RTOS). Using an RTOS allows the imager to respond to events on a frame basis. For example, when an operating mode change is requested, the switch over to the new mode can be guaranteed to occur within two frames.

The Command Processor contains three banks of memory, which are each expandable. There is DRAM associated with the MIPS CPU, high speed Synchronous DRAM (SDRAM) used in the image processing section, and non-volatile flash which stores the run-time application as well as the offset and gain correction values used on system start-up. By using flash memory in this way, the imager is ready to acquire images within 30 seconds of power on.

As discussed above, the image processing section (IPS) corrects for the offset and gain variation and defective pixels on a pixel-by-pixel basis. The source of background and gain variations stems from non-uniformity in the dark current on the array and also the differences between the 1920 readout amplifiers. The defective pixels originate on the array. In order to correct for these phenomenon, the IPS has a pipelined architecture operating at up to 40 MHz pixel rate, which uses calibration data previously stored in SDRAM. The offset and gain correction algorithm can be reduced to the following:

Image Data = $\dfrac{\text{(raw Image data)} - \text{(Offset data)}}{\text{(raw Norm data)} - \text{(Offset data)}}$

The Offset data is simply an image taken with no illumination, i.e. a dark file. The Norm data is an image file taken with uniform illumination to every pixel. To minimize the error introduced by the correction data, the Offset and Norm data is the average of at least 64 frames. The calibration protocol is very straightforward and relatively fast. Essentially two images must be collected, one in the dark, the other with all objects removed from in-front of the Receptor. This type of correction also has the added advantage of removing the spatial non-uniformity in the x-ray beam profile.

The recursive filtering capability provided by the IPS uses SDRAM to store a weighted average of the prior image frames. The recursive filter algorithm is:

Output Frame = α **(New Frame)** + $(1-\alpha)$ **(Old Data)**

where α can take on values between 0 and 1. Recursive filtering introduces a controlled amount of lag into the video output for the purpose of noise reduction. This is a common technique in fluoroscopy, where the signal levels are very low and there is significant noise introduced by the statistics of the x-ray beam itself.

The Power Supply provides all the dc power necessary for both the Command Processor and the Receptor. It is design to conform to the UL 2601 regulation for medical devices. The Power Supply connects to any standard wall outlet and is connected to the Command Processor through a 50 pin sub-miniature D connector. The Command Processor and Power Supply each have a foot print of 10.125" x 11" and when stacked for rack mounting occupy a height of 5.25".

3.0 Receptor Technology

The detector is based on amorphous silicon technology which is very similar to that used in flat panel displays. Because x-rays cannot be easily focused, it is a requirement that the detector be as large as the area to be imaged. In CCD based systems, one way around this requirement is to use a large x-ray conversion screen emitting visible radiation, which is coupled

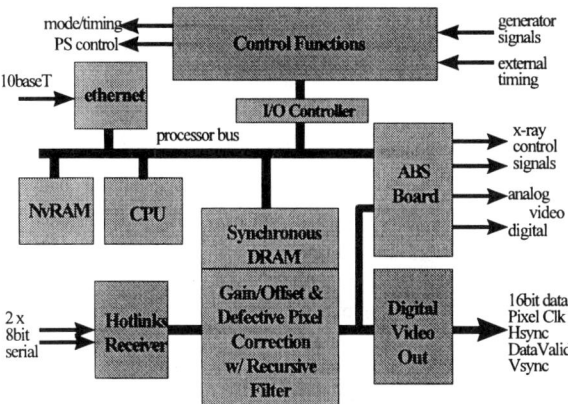

Fig. 4   Block diagram of the Command Processor.

through a mirror and lens to the camera. The drawback to such a system is the substantial loss in signal due to the relatively small solid angle in which light is collected. The larger the area to be imaged, the greater the loss. In addition, the system is necessarily bulky, particularly if the CCD camera needs to be out of the direct path of the radiation. By using large area sensing technology based on amorphous silicon panels, an optimal coupling between the x-ray conversion screen and the photo-detector is achieved.

Fig. 5 shows the internal configuration of the Receptor. The core of the detector is an array of amorphous silicon pin photo-diodes and thin film transistors (TFTs), which are arranged as shown in Fig. 6. On top of the amorphous silicon (a-Si) array is an x-ray scintillator, which converts the x-ray photons to visible radiation. This scintillator is typically a removable Gd_2O_2S screen or thallium-doped CsI which is deposited directly on the a-Si array. These x-ray conversion screens emit near 550 nm, which corresponds to the peak quantum efficiency of the photo-diodes. The a-Si devices are fabricated on a glass substrate in the same manner as the TFTs in active-matrix flat panel displays. The glass substrate is mounted in a base plate, which also holds the readout and drive electronics for the panel.

Amorphous silicon devices have a number of features that make it well suited to an x-ray sensor application. The foremost advantage is that the a-Si can be deposited over a very large area onto glass substrates. It is possible with current technology to conceive of a 17 x 17 inch sensor panel, which is likely to be the largest size required in radiology. Another advantage is the low dark current of the pin diodes and the low leakage current of a-Si TFTs. A typical dark current for the photodiodes is on the order of $2pA/mm^2$. The OFF current of a TFT is less than 0.1pA for temperatures below 50 °C. Such low dark currents make the devices well suited to a charge integrating detector, allowing frame rates on the order of many seconds per frame. The large intrinsic dynamic range of the amorphous silicon pixel comes from the large charge capacity of the photodiode. For a 127 μm pixel the charge capacity is 0.7pF, which gives over 20 million electrons for a 5V reverse bias. And finally, amorphous silicon devices have been shown to be extremely radiation hard, capable of tolerating in excess of 10,000Gy (1 Mrad) total dose [1]. This feature makes a-Si technology attractive for both diagnostic energy imaging (40 to 150kVp) as well as mega-voltage energy imaging such as might be done in radiation therapy and high energy NDT applications, where the dose rate from the source typically exceeds 100 cGy/min.

The same lack of long range order in the amorphous silicon material that makes it radiation hard, also means that charge does not move easily in amorphous silicon. Since the electron mobility in a-Si is less than 1 cm^2/V-sec, it is not practical to build extensive readout electronics into the a-Si plate. As a consequence, every row and column connection must brought out to the edge of the array where it is connected to conventional integrated circuits. This poses a number of unique challenges when compared to CCD or CMOS active pixel sensors where much of the sensitive readout electronics can be built into the detector. The first challenge is simply the mechanical connection to the thousands of signals which come out to the edge of the array. The second and more difficult challenge is that the architecture forces the readout circuit to operate in the presence of the large parasitic resistance and capacitance of the a-Si array. As discussed in reference [2], the dominant noise sources in the imager arise from these parasitics.

In the same way that the a-Si sensor technology is leveraging the huge investment that has been made in displays, the interconnect technology used in flat panel displays is also available for sensors. Typically the row selection and readout electronics are put in TAB (tape automated bonding) packages. These flexible packages are heat sealed to a conductive landing pattern on the glass using a z-axis conducting epoxy. The chip is mounted somewhere in the middle of the package, then at the other end of the package the signal traces are either soldered or heat sealed to a printed circuit board. The advantage of a TAB package over a simple flexible circuit interconnect is the dramatic reduction in the number of pins required at the PCB end of the package. For example, the readout chips are heat sealed at the glass to 128 columns, where the landing pattern has a pitch of 100 µm. The readout chip multiplexes these 128 inputs down to 1 output, so on the PCB end of the package only 30 pins at a pitch of 20 thousands of an inch are necessary to carry the power, control signals and output signal. Another advantage of the TAB package is that the electronics can be wrapped around to the side or the backside of the array, as shown in Fig. 5 for one of the gate driver chips.

The a-Si array is readout a single row at time and can be scanned progressively or interlaced. Referring to Fig. 6, the gate driver chips select which row in the image is accessed by applying a positive voltage to a line of TFT gates. With the TFT's ON, charge collected on individual photo-diodes in the selected row is then discharged onto the corresponding dataline. Each dataline is held at a constant potential by a corresponding charge integrating amplifier. In the imager under discussion, there are 1536 rows and 1920 columns at a pixel pitch of 127 µm. Therefore 1920 charge integrating amplifiers on an effective pitch less than 127 µm are required

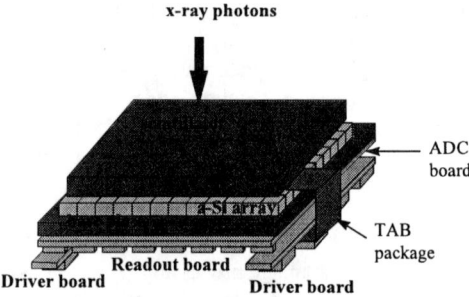

Fig. 5   Internal configuration of the Receptor.

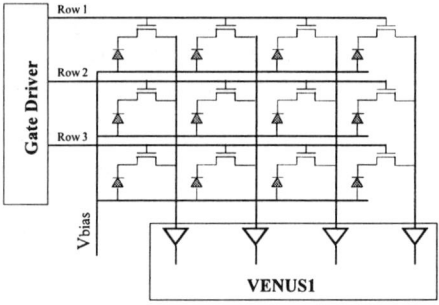

Fig. 6   a-Si TFT/Photodiode array architecture.

to readout the imager. A custom 128 channel readout chip, the VENUS1, was designed in a BiCMOS process for this purpose. The VENUS chips capture an entire row of signals simultaneously, after which the data is multiplexed out to 12 bit analog-to-digital converters (ADC's).

To summarize the signal conversion chain:
1. X-rays are converted to visible photons by the scintillator.
2. The visible photons are absorbed by the pin photo-diodes and converted to electron-hole pairs, which collect on the capacitance of the photo-diodes.
3. The pin photo-diodes are discharged when the pixel's TFT is turned ON. The charge is collected by an integrating amplifier and converted to a voltage.
4. The signal voltage has a programmable gain applied to it which depends on the operating conditions.
5. The output voltage is converted to bits by an ADC.

4.0 Future Work

On-going work will center on larger active area and performance optimization for specific applications. In particular, enhancements to the a-Si array and external electronics, offer the potential to reduce the minimum required entrance dose significantly.

5.0 References

1. Boudry, Antonuk, "Radiation damage of amorphous silicon, thin-film, field-effect transistors," Med Phys. **23**, pg 743-754 (1996).
2. Colbeth et al, "Flat panel imaging system for fluoroscopy applications, " Proc. SPIE Medical Imaging 1998.

NOVEL LOW POWER CLASS-B OUTPUT BUFFER

Pang-Cheng Yu, Jiin-Chuan Wu

Department of Electronics Engineering
National Chiao-Tung University
Engineering Building 4th, 1001 Ta-Hsueh Road
Hsinchu, Taiwan 30050, Republic of China
TEL: +886-3-5731862 FAX: +886-3-5715412
E-mail: pcyu@alab.ee.nctu.edu.tw

ABSTRACT

This paper describes the design of a Class-B output buffer for driving large capacitance load in flat panel display. Due to the large number of output buffers on a column driver chip, the quiescent current of the output buffer must be reduced. A comparator which produces full-swing output is used in the negative feedback path to eliminate quiescent current in the last stage. The proposed circuit was implemented in a $0.8\mu m$ CMOS process. The measured maximum static current is $54\mu A$. With 5V supply voltage and 600pF load, the tracking error voltage is less than $\pm 8mV$ and the output swing is from 0.5V to 5V. The settling time for 4V swing to 0.2% is $8\mu s$, which is more than adequate for driving 1204*1280 pixels LCD panel with 86Hz frame rate.

1. INTRODUCTION

In the matrix type flat panel displays, e.g. liquid crystal display (LCD) and field emission display (FED), the signal required on the row lines is a bi-level signal, but an analog signal is required on the column lines. Thus, analog output buffers must be used in the column driver chip (typically more than 120 buffers in each chip), shown in Fig. 1. The output load of the analog buffer generally consists of a large capacitor, in series with a resistor so that a low output resistance buffer with large output current is needed.

This paper presents a low power Class-B CMOS output buffer. This output buffer demonstrates low static power dissipation, low distortion, and large output swing. The following will describe the design strategy to eliminate the quiescent current of the last output stage. Finally, the simulated and experimental results are shown.

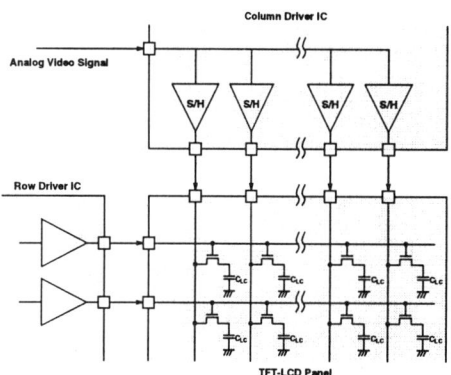

Figure 1: Driving scheme of a TFT-LCD panel.

2. CIRCUIT DESIGN

Figure 2 shows the block diagram of the proposed Class-B output buffer. The two comparators, Cmp1 and Cmp2, are identical which produce a binary output representing the sign of the differential input. The comparator, Cmp1, and output transistor, Mp, form the buffer amplifier for the positive slope of the output voltage swing. Similarly, the comparator, Cmp2, logic inverter, INV2, and output transistor, Mn, form the buffer amplifier for negative slope of the output voltage swing. Since the output transistors are complementary common source type circuit, it is possible to avoid crossover distortion. The operation of the negative slope circuit is the inverted mirror image of the position slope circuit, so the output buffer is operating in Class-B push-pull mode. In addition, when input voltage is equal to output voltage, the output of Cmp1 and Cmp2 are both at the supply voltage and node N is at 0V. That is, output transistors Mp and Mn are both completely turned off, there is no quiescent current and output voltage is held by the load capacitance. Whenever input and output signal voltages are not equal, one of the output transistors is turned on to force the output to follow the input.

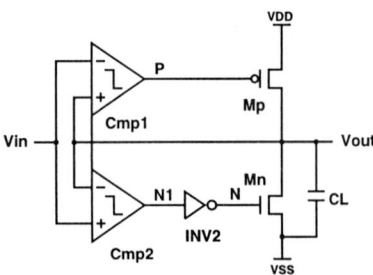

Figure 2: Circuit diagram of Class-B output buffer.

From above discussion, the characteristic of comparator can be described as

$$V_{out} = \begin{cases} V_{DD} & , when\ V_+ \geq V_- \\ 0 & , when\ V_+ < V_- \end{cases} \quad (1)$$

where V_+ is non-inverting input and V_- is the inverting input. Fig. 3 shows a complete schematic of buffer amplifier circuit. The comparator, used for Cmp1 and Cmp2, is a simple n-channel input differential amplifier with push-pull CMOS inverter amplifier. To meet the characteristic of (1), the logic threshold voltage of inverter must be higher than the common mode output voltage of the n-channel input differential amplifier. The common mode output voltage of the differential amplifier due to process variations are shown in Fig. 4.

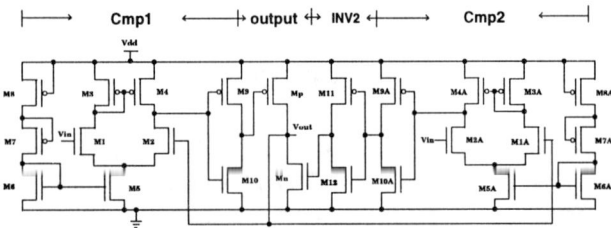

Figure 3: Circuit of the complete Class-B output buffer.

Considering the typical-N typical-P case, when the inputs are between 0.8V and 5V, the output voltage changes from 3.65V to 3.71V, due to the channel length modulation of the current source. When inputs are less than 0.8V, the output voltage increases rapidly from 3.7V to 4.1V, because the input transistors are in sub-threshold region. The voltage transfer curves of the push-pull CMOS inverter for different process variations are shown in Fig. 5. Since the output high voltage of the comparator must be able to turn off the output PMOS, the logic threshold of the inverter is taken as the input voltage when output is 4.2V. With this definition, and consider only the input voltage range between 0.8V and 5V, the common mode output of the differential pair is about 0.35V lower than the logic threshold of the inverter. Since

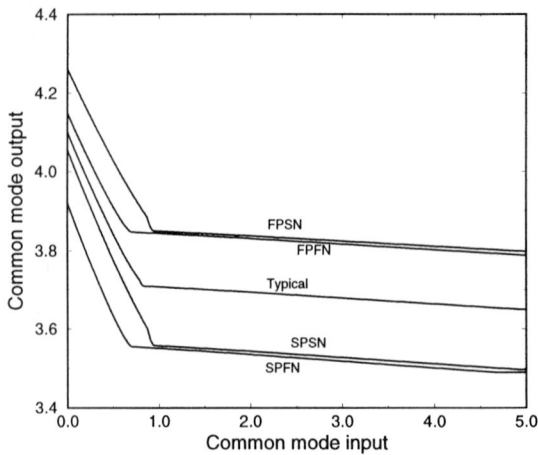

Figure 4: The common mode output voltages of the differential amplifier for 5 different process variations.

the voltage gain of the differential pair is 34, the input resolution of the output buffer is ±10mV, i.e., the difference between the input and output must be larger than ±10mV to turn on one of the output transistors. Since the maximum gray level requirement in the flat panel display is 256 levels, for an output swing of 4V, one LSB is 16mV. Thus, a resolution of 10mV is about 0.6 LSB. When process varies, the output common mode voltage of the differential pair and the inverter's logic threshold vary in the same direction. Thus, the input resolution of the output buffer remains at about ±10mV.

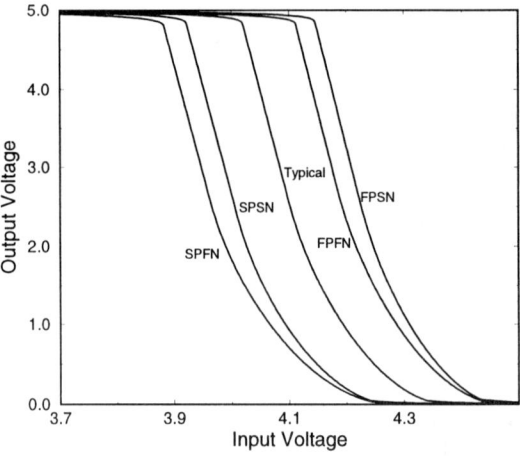

Figure 5: The voltage transfer curves of the inverter INV2 for 5 different process variations.

The simulated transient response of the output buffer with 600pF load is shown in Fig. 6. The settling time for rising and falling are 6.8μs and 4.2μs, respectively.

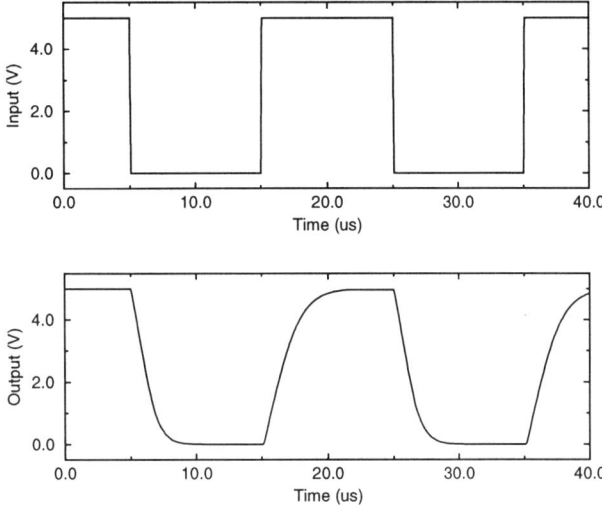

Figure 6: The simulated transient response of output buffer.

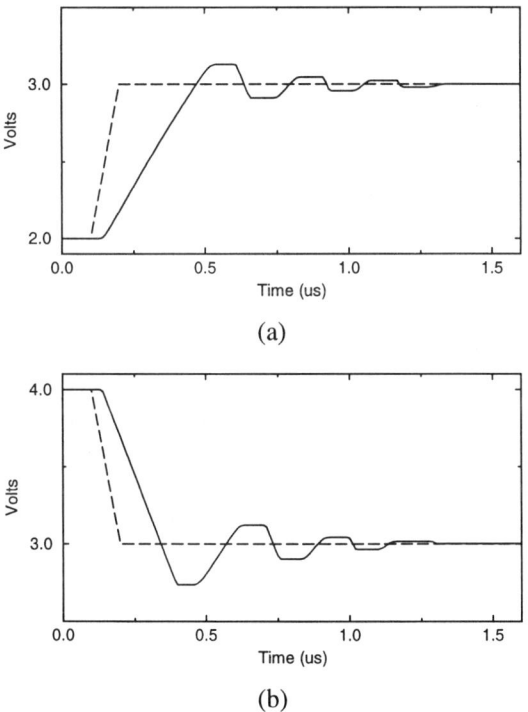

Figure 7: The simulated result of output buffer with 600pF.

The simulated transient response of the output buffer with input changing from 2V to 3V and from 4V to 3V is shown in Fig. 7. Overshoot, undershoot, and ringing are seen. For rising and falling inputs, the offset voltages are 1mV and 8mV, respectively. From Fig. 7, the output buffer is obviously under-damped.

3. EXPERIMENTAL RESULTS

The output buffer was fabricated using a 0.8μm CMOS technology. The die photograph is shown in Fig. 8. The active area is $230*140\mu m^2$ excluding pads. The measurement results show that the maximum static current is 54μA.

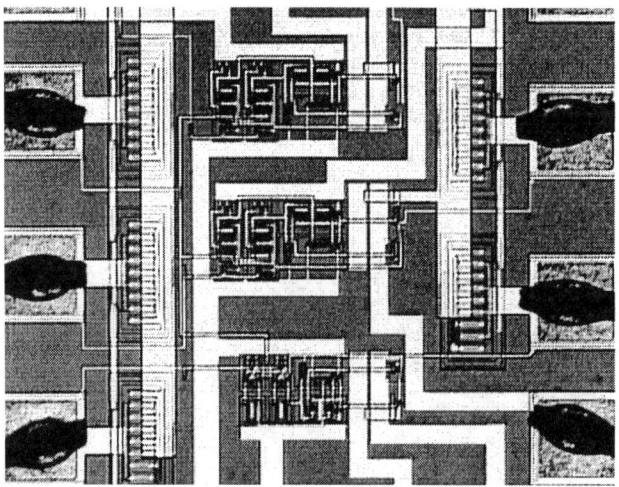

Figure 8: The microphotograph of the die.

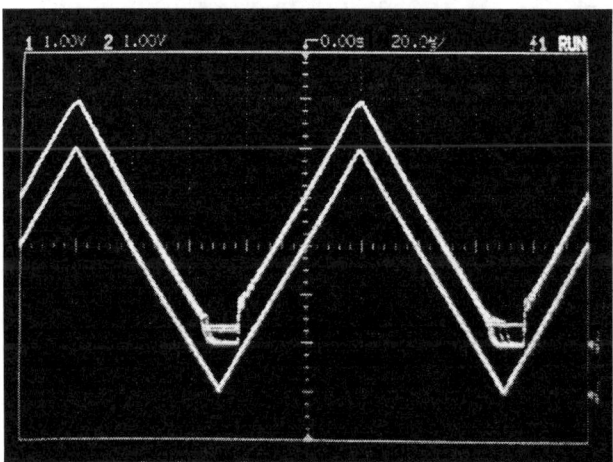

Figure 9: The 10kHz triangle wave of input vs. the output voltage of output buffer.

Fig. 9 shows the measured result of the output buffer with 10KHz triangular input and 600pF load capacitance. The lower trace is the input and the upper trace is the output. The output is basically the same as the input except for input voltage lower than 0.5V, the output is distorted, because it is outside the input common mode range of the comparator.

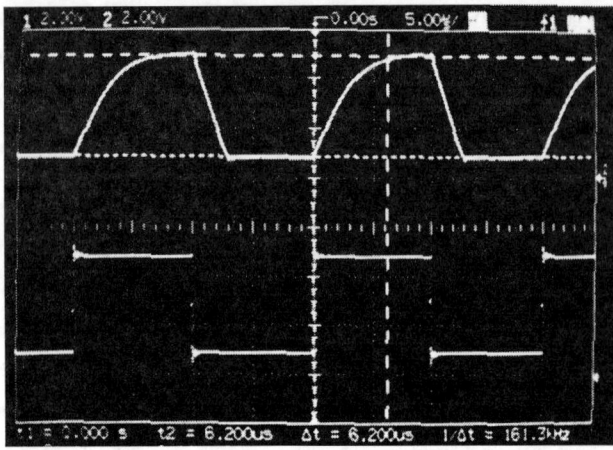

Figure 10: The step response of Class-B output buffer for 600pF capacitance.

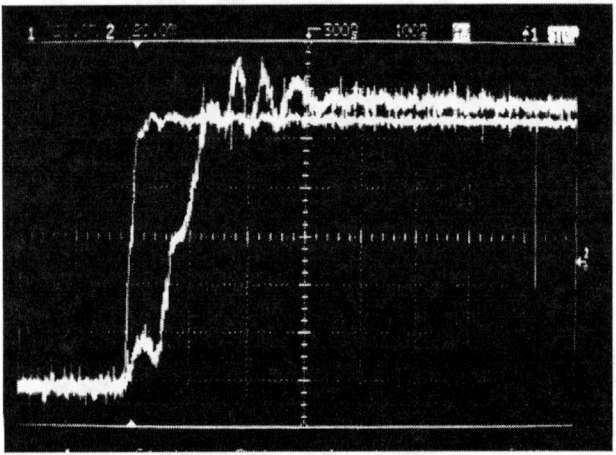

Figure 11: The step response for small input signal.

The step response with 50KHz square wave input and 600pF load capacitance is shown in Fig. 10. The lower trace is the input whose voltage changes between 1V and 5V. The upper trace is the output. The settling times to within 0.2%(0.5LSB) are $8\mu s$ and $3\mu s$ for the rising and falling edges, respectively. The steady state error of the output buffer is shown in Fig. 11. In order to show the small error voltage, the signals are AC coupled. The input is a 100mV swing square wave. The error for the rising edge is about 8mV, and the error for the falling edge is 0. The tracking error of the output buffer for input voltages ranging from 1V to 5V is shown in Fig. 12. The input voltage is changed in 0.1V steps quasi-statically. The errors are all within $\pm 8mV$. For inputs between 1.2V and 5V, the errors are smaller (within $\pm 4mV$). Note that as discussed before, the error is dependent on the rate and magnitude of the input change, thus the results shown in Fig. 12 do not cover all possible cases.

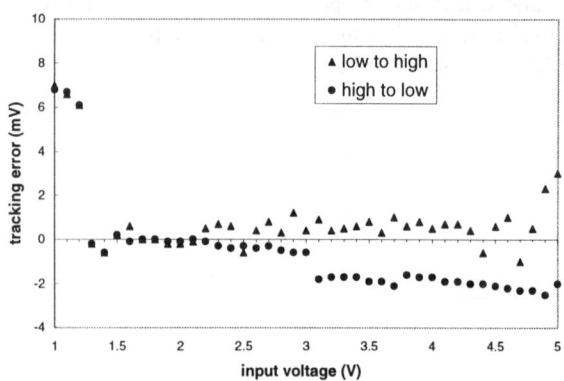

Figure 12: The tracking error versus the input voltage for output buffer.

4. CONCLUSION

A Class-B output buffer amplifier for driving large capacitance load in flat panel display is presented. Due to the large number of output buffers on a column driver chip for flat panel, the quiescent current of the output buffer must be reduced. A comparator which produces full-swing output is used, in stead of an error amplifier using conventional output buffer, in the negative feedback path to eliminate quiescent current in the last output stage. The measured static current is $54\mu A$. With 5V supply voltage and 600pF load capacitance, the maximum tracking error voltage is less $\pm 8mV$, the output voltage swing is from 0.5V to 5V. The settling time for 4V swing to 0.2% is $8\mu s$, which is more than adequate for driving 1204*1280pixels LCD panel with 86Hz frame rate.

Acknowledgments

The authors would like to thank the Chip Implementation Center of National Science Council for their support in chip fabrication. This work was supported by the National Science Council, Republic of China, under Grant NSC87-2215-E-009-027.

A NOVEL IMAGE SENSOR WITH FLEXIBLE SAMPLING CONTROL

Yasuhiro Ohtsuka, Takayuki Hamamoto, Kiyoharu Aizawa, and Mitsutoshi Hatori

Department of Electrical Engineering,
The University of Tokyo, Hongo, Bunkyo-ku, Tokyo 113, Japan

ABSTRACT

We propose a new sampling control system on image sensor. Contrary to the random access pixels, the proposed sensor is able to read out spatially variant pixels at high speed, without inputting pixel address for each access. The sampling positions can be changed dynamically by rewriting the sampling position memory. Since the proposed sensor has sampling position memory that stores the sampling control order, it is able to easily control sampling position. We can achieve any spatially varying sampling patterns.

1. INTRODUCTION

In biological vision, the retina is equivalent to the imaging sensor. It is organized into a space-variant sampling structure including a high-resolution, small central fovea, and a periphery whose resolution linearly decreases in steps. By this characteristic, it able to centralize and distribute the processing loads in the earliest stage of vision.

Therefore it is essential for smart sensing to integrate functions of spatially variant flexible sampling control system onto a computational sensor. The integration on the sensor focal plane results in enhancement of performance of an image sensor.

In this paper, we propose a new sampling control system on image sensor for space-variant sensing. We will present the principles, circuit designs and a prototype of the sensor.

2. COMPUTATIONAL SENSOR

In a traditional image processing system, image acquisition elements are independent from processing elements. A great deal of effort is put into each elements to speed it up. But, since the signal from the sensor is a time-sequence signal of high band width made by the scanning the whole 2D image, communication bottle neck between sensor and peripherals is one of the most serious barrier for high pixel rate imaging such as high frame rate or high resolution imaging.

A novel approach is to integrate processing and sensing to solve such problems. This is called computational sensor or smart sensor. The integration is able to make use of the parallel nature of the image signal on the image sensor plane so that the processing gets remarkably faster compared to conventional image processing systems which follow sense-read-and-process paradigm. The integration also enables sensors to enhance image signals so that the performance of sensing is improved.

3. FLEXIBLE SAMPLING CONTROL ON SENSOR

One of the previous approaches to space variant sampling is random access pixels. For example, there is an investigation into random access PD-Array sensor[1]. In this sensor, feeding of the pixel coordinates is always needed. After decoding of each input coordinate by a Row Decoder or Column Decoder, corresponding pixel value is read out. Only one-pixel can be selected at each access and address of each pixel must be given to read it out. For high speed access, the above become a bottleneck.

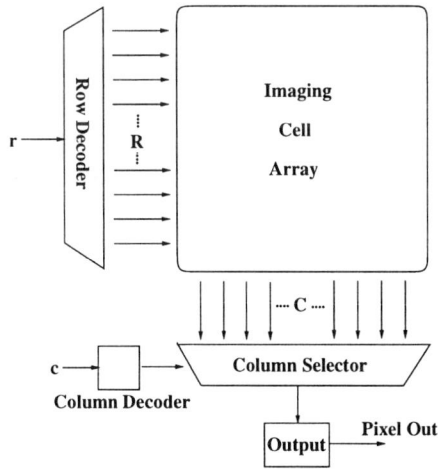

Figure 1: Random Access PD-array

There is also an investigation into controlling the access by row or column unit[2]. In this case, though we can read out pixel values in blocks or in regular sub-sampling pattern, arbitrary control of space-variant control is not possible.

There is a type of polar coordinates style sensor which can provide image of foveated pattern[3]. It consists of a central sensor(fovea) and peripheral sensor independently. Here the central sensor is at a fixed location. In order to get a high resolution output from a different part of the image, the sensor needs to be moved physically.

In this paper a new space variant sampling control which uses sampling position memory is proposed. A memory element of sampling position memory corresponds to a pixel and it contains a binary data to determine the pixel is read out.

A smart scanning shift register (Fig.2) [4] is used to read out pixel values, and the data of the memory is used as a control signal to the register.

In this system, pixel value can be read out at high speed, without inputting pixel address for each access. By rewriting the sampling position memory, the sampling position can be dynamically changed. Thus the position of the fovea can be freely moved on the sensor. Therefore if we use the sensor in active vision systems, the fast feature extraction is possible. Fig.3 illustrates a diagram of the sensor.

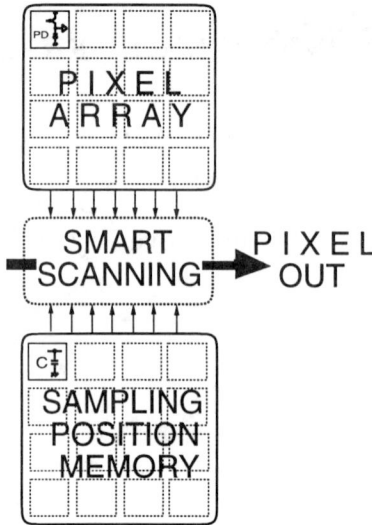

Figure 3: Diagram of sampling control system

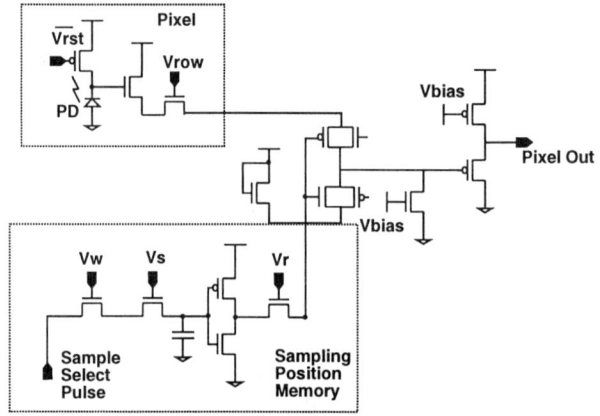

Figure 4: A design of pixel circuit and sampling position memory

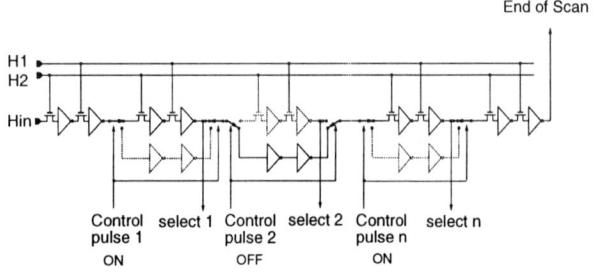

Figure 2: Smart scanning shift register

3.1. Circuit Design of A Prototype

Fig.5 shows the block diagram of the sampling control sensor. This sensor mainly consists of two parts. They are pixel array and sampling position memory. Fig.4 shows an analog circuit of the prototype.

Pixel array consists of pixel circuits. A pixel circuit consists of a few transistors, so that we can get a practical fill factor. Pixel values are transmitted to the horizontal shift register, and selected based on memory values of sampling position memory. And values of only the selected pixels are read out.

The proposed sensor has two different type of horizontal shift registers: a normal and a smart scanning shift registers. One of the two is selected by the mode selection signal. In the case of the smart scanning shift register, only the selected pixels are read out and non-selected pixels are skipped without reading. In order to reconstruct the output image, address data is required, wherein known sampling pattern may be used. In the prototype, the sampling position data as flag signals are also output from sensor by using the bottom horizontal shift register at a rate higher than the output rate of the pixel values.

Sampling position memory mainly consists of a capacitance and switches and can be dynamically rewritten.

Both pixel array and sampling position memory have vertical shift registers which are driven by the same signals to select corresponding pair of rows. The sensor has another horizontal shift register at the bottom so that it can rewrite independently sampling position memory within a horizontal scanning period. With this bottom horizontal shift register, control bits can be read out and rewritten at the same time.

In order to write control bits in sampling position memory, the output signal from the horizontal shift register is controlled by sample selection signal ($Smode$). If $Smode$ is "1", the output signal from the bottom shift register is transmitted to the memory and its value is set to "1". If $Smode$ is "0", the memory is reset. Using $wrmode$ signal, the signal from shift register is input only at the time of writing.

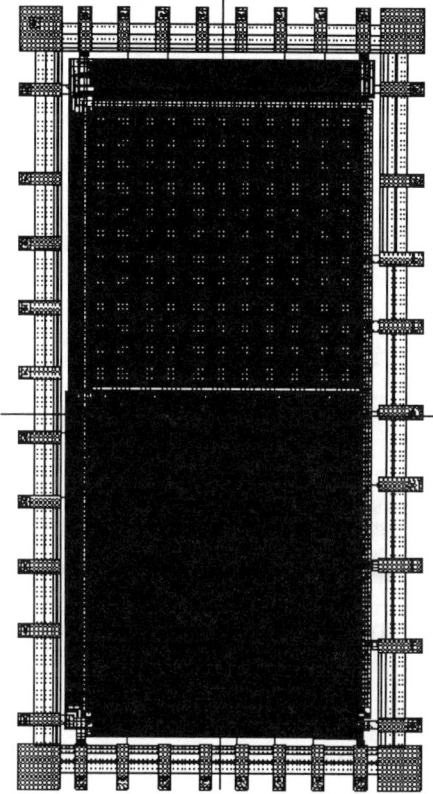

Figure 6: chip layout of a prototype

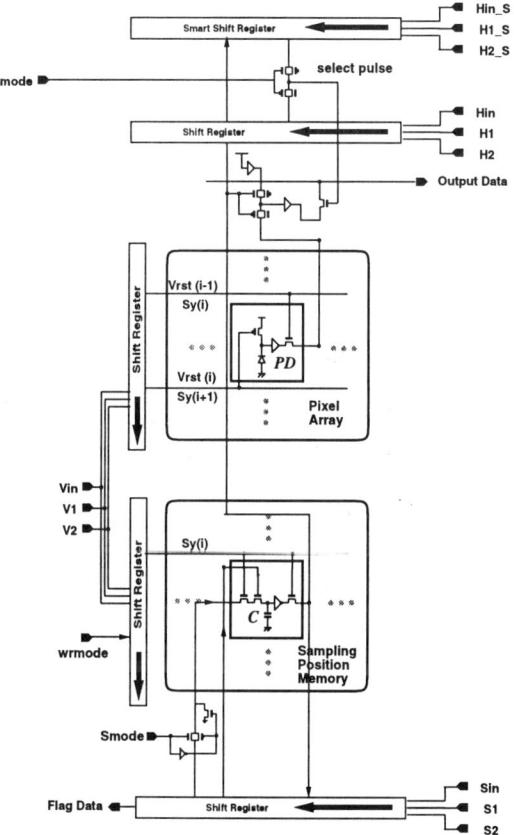

Figure 5: Block diagram of sampling control image sensor

Figure 7: A prototype chip.

3.2. A Prototype Chip

Fig.6 shows a chip layout of a prototype, and Fig.7 shows a picture of the prototype chip. Table 1 shows the characteristics of it. It is designed under 1-poly 2-metal CMOS $0.7\mu m$ rule. Table 1 shows the characteristics of the prototype. The number of the transistors

Figure 8: an image obtained by the prototype

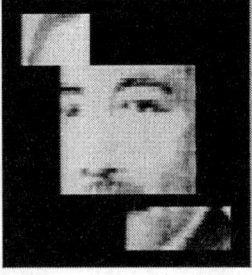

Figure 9: output images selectively sampled by the prototype

in the pixel array element is three, which is equal to the conventional CMOS sensor [5]. Number of pixels is 64 × 64.

Fig.8 shows an output image obtained by the prototype. Fig.9 shows images obtained by the prototype in skip or block access mode. The left is the image when every other column is sampled and read out by the smart horizontal shift register. The right is the one when every other column and every other row is sampled. The center is the one when specific block areas are selectively sampled. The skipped pixels are shown black.

The following advantages is obtained from the sensor.

- It can flexibly set sampling positions.
- Sampling at high resolution at center and low resolution at peripheral pixels, can get image similar to foveated vision.
- It can also output in blocks or by regular subsampling patterns.

4. SUMMARY

Proposed is a new image sensor with space-variant flexible sampling control integrated on a sensor focal plane. The principles of processing, the designs of their circuits based on column parallel architecture, and the

Table 1: Characteristics of the prototype

♯ of pixels	64 × 64 pixels
♯ of transistors	pixel : 3 trs. / pixel
	memory : 9 trs. / pixel
fill factor	25%
power dissipation	5mW / chip

prototype chip have been presented. The prototype is now under further experiments in a system that enables it to take spatially variant images. The detail of obtained results will be presented in the conference.

5. REFERENCES

[1] Orly Yadid-Pecht, Ran Ginosar, "A Random Access Photodiode Array for Intelligent Image Captur", *IEEE Trans Electron Devices*, pp. 1772–1780, Aug. 1991.

[2] T.Nakamura, K.Saitoh, "Recent Progress Of CMD Imaging", *IEEE Proc. Workshop on CCD and Advanced Image Sensors*, R14, 1997.

[3] C.Colombo, M.Rucci, and P.Dario: "Integrating Selective Attention and Space-Variant Sensing in Machine Vision", Image Technology J.L.C.Sanz Ed, Springer, 1996.

[4] K.Aizawa, Y.Egi, T.Hamamoto, M.Hatori, M.Abe H.Maruyama and H.Otake, "Computational Image Sensor For On Sensor Compression", *IEEE Trans.on Electron Devices Special Issue "Solid State Image Sensors"*, Vol.44, No.10, pp. 1724–1730, 1997.

[5] F.Ando, K.Taketoshi, K.Nakamura and M.Imai, "AMI; A New Amplifying Solid State Imager", *ITEJ*, Vol.41, No.11, pp. 1075–1082, 1987.

A 128X128 IMAGING ARRAY USING LATERAL BIPOLAR PHOTOTRANSISTORS IN A STANDARD CMOS PROCESS

Robert W. Sandage and J. Alvin Connelly

School of Electrical and Computer Engineering
Georgia Institute of Technology
Atlanta, GA 30332-0250 USA

ABSTRACT

A 128x128 element photodetector array has been fabricated using lateral bipolar phototransistors in a standard 1.2 μm digital CMOS process. One-bit image processing for the array is accomplished through the use of a current mode comparator. Images focused on to the photodetector array have been successfully captured and are presented in this paper.

1. INTRODUCTION

This paper discusses the design of an imaging array which is part of a fingerprint detection system shown in Fig. 1. In this paper the authors report many improvements over previous work [1]. Particular attention was paid to the choice of photodetective device and to the characterization of the responsivity of the device. Current-mode processing of the array signals and imaging results are also discussed.

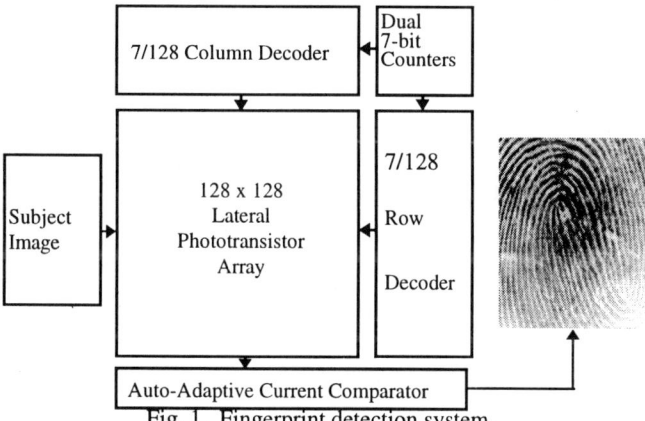

Fig. 1. Fingerprint detection system.

Historically, imaging systems have used photodiodes or charge-coupled devices (CCDs) to convert optical power incident on an array into an image. Arrays using these devices have achieved tremendous packing densities. [2] However, both devices have certain drawbacks depending upon the application. The *pn*-junction photodiode is the simplest photodetecting device and is easily integrated into a standard digital CMOS process. However, it has low responsivity. The charge-coupled device relies on multiple FET devices per cell with floating gates, which, when properly biased, allow charge transfer to be achieved. The requirement for the floating gate necessitates the use of additional processing steps in the fabrication the device. Also, large gate voltages (5 - 15 volts) are necessary to achieve high charge transfer efficiency [2]. Low power design criteria often eliminate CCDs from further consideration.

The lateral bipolar phototransistor (LPT) and the vertical bipolar phototransistor (VPT) are two alternative devices easily implemented in a standard digital CMOS process. The VPT is easily fabricated by adding an additional diffusion to the *pn*-junction diode. However, this device normally has the disadvantage that the substrate becomes the collector, limiting its use to that of the common-collector configuration [3]. The LPT is also easily implemented in a standard CMOS process [4], and the collector is not necessarily common to the substrate. A typical lateral bipolar device suffers from lower β and collector efficiency than the vertical bipolar. Modifications to the typical bipolar transistor have resulted in a device which is much more responsive. Thus, the lateral bipolar phototransistor was selected as the photodetector element. Because *n*-well CMOS technology was required for the specific application and because it is more prevalent, a *pnp* photodetector was implemented.

2. LATERAL PHOTOTRANSISTOR CELL

2.A. Operation

The lateral *pnp* phototransistor is shown in Fig. 2. A rectangular PFET structure is formed with a *p*-diffusion emitter surrounded by a *p*-diffusion collector situated in an *n*-well base [5]. The rectangular structure increases the cross-sectional area of the emitter. Because the base region is more lightly doped than the collector, the collector-base depletion layer extends almost completely into the base, and care must be taken that this depletion layer does not reach the emitter. A wide base width is then necessary, and β is reduced since it is inversely proportional to base width. Equation 1 predicts β [6], where W_B is the base width, W_E is the emitter width, L_p is the diffusion length for holes in the base, and x_c is the depth of the collector p-diffusion.

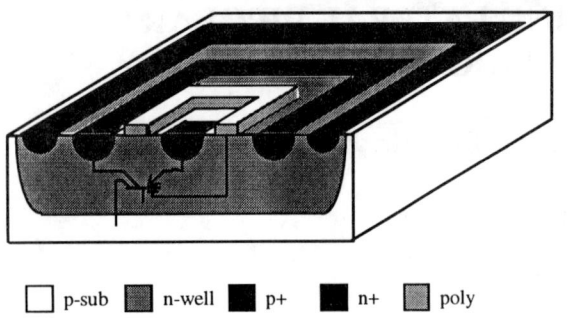

Fig. 2. Lateral PNP transistor.

$$\beta \approx \frac{2 * L_p * x_c}{W_E * W_B} \quad (1)$$

A reasonable estimate of β using typical parameters from the MOSIS HP 1.2 μm process where W_B = 1.2 μm, W_E = 2.4 μm, x_c = 2 μm, and L_p = 100 μm gives an approximate β of 139. The addition of the minimum width polysilicon gate controls and minimizes the base width, thus dramatically increasing the current gain versus a standard lateral bipolar *pnp* transistor. Connecting the gate to the most positive potential in the circuit causes majority carriers in the base to accumulate directly under the gate, limits the widening of the collector-base depletion region, and sets the base width to that of the polysilicon gate.

The lateral phototransistor cell of Fig. 2 also contains a parasitic vertical *pnp* substrate transistor. The effect is a reduction in the collector current efficiency of about 30% to 40% [4], [5], i.e., only about 2/3 of the emitter current reaches the collector, and 1/3 of the emitter current is lost to the substrate. In a standard bipolar fabrication process, this additional parasitic is inhibited through the use of an *n+* buried layer. However, this option usually is not available in most standard CMOS processes. For the application at hand, this is not a problem because the collector is common to the substrate, and photodetector current is drawn from the emitter.

The base of the LPT is left floating and base bias current is provided when incident photons of sufficient energy create electron-hole pairs. The absorption process is very dependent on the wavelength of the incident light, and on the depletion region depth [7], [8]. The photodetector current taken from the emitter is (β+1) times the base current. Thus, photocurrent is increased by (β+1) compared to a similar sized photodiode for a given light intensity. Because the base is floating, photovoltaic operation of the base-collector pn-junction diode occurs [9], and the emitter-base voltage is given by

$$V_{EB} = \frac{kT}{q} \ln\left(\frac{I_{SC}}{I_O} + 1\right) \quad (2)$$

where I_{SC} is the photo-induced current which would result from shorting the emitter-base junction, and I_O is the reverse saturation current of the emitter-base junction.

Mincai, et.al [10], have modeled an npn phototransistor as a standard npn transistor with a photodiode existing between base and collector. Expanding on this work, an equivalent circuit model for the Georgia Tech LPT and its select switch is shown in Figure 3.

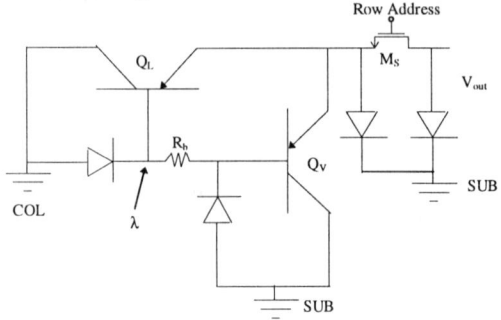

Fig. 3. LPT pixel equivalent circuit model.

The model includes the lateral phototransistor, Q_L, the parasitic vertical phototransistor, Q_V, and the pixel select NFET, M_S. The spectral response of all junction photodiodes and the characteristic curves of the transistors were used to produce a complete wavelength dependent model.

2.B. LPT Experimental Results

The primary parameters of interest for the LPT are responsivity and spectral response. The system for testing the LPT spectral response used a white light source modulated by a rotating wheel with a hole as shown in Fig. 4.

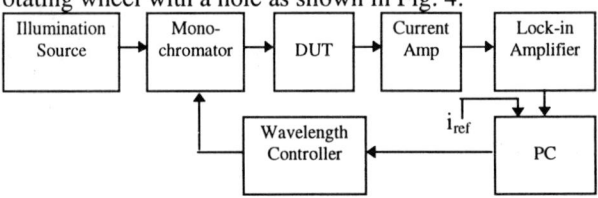

Fig. 4. Experimental setup for measuring LPT spectral response.

A 5 nm bandwidth of this light was then selected using a monochromator. The resulting modulated light was then projected onto the LPT. The detector produced a small AC current which was amplified using a lock-in amplifier. Comparison was made to a precision current reference and recorded using a PC. Typical results for the lateral phototransistor spectral response are shown in Figure 5. The peak wavelength for the response is at 730 nm, which is typical for silicon devices. The response was normalized by using the area of the LPT such that the units are given in Amps/Watt. A typical photodiode has a peak spectral response of about 0.3 to 0.5 A/W, thus the peak response for the LPT is about β times higher. The lateral pnp output characteristics of Figure 6 were determined for four illumination levels using a parameter analyzer, a HeNe laser, a variable neutral density filter, and an optical power meter. The characteristics in Figure 6 show the collector current versus emitter-collector voltage for increasing levels of illumination power. The results of Figures 5 and 6, along with results for the parasitic vertical pnp phototransistor and the select FET are used in the model which was discussed earlier. The model is then suitable for circuit simulation with wavelength and illumination intensity parameters input by the user.

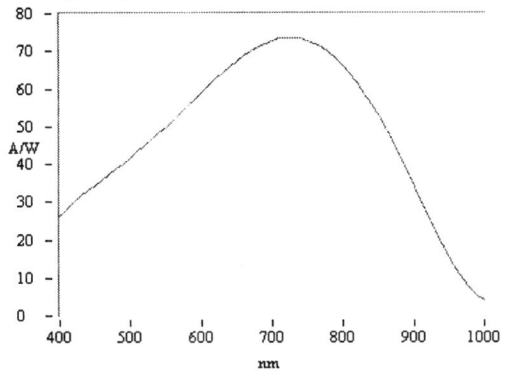

Fig. 5. LPT spectral response.

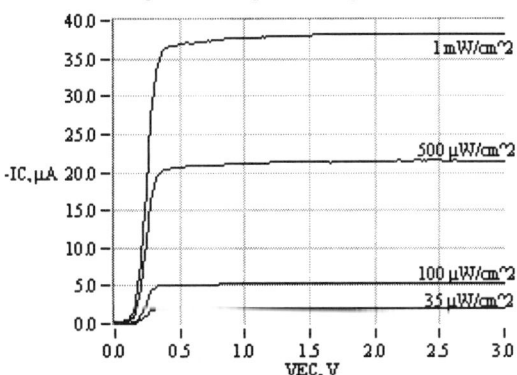

Fig. 6. LPT output characteristics.

3. CURRENT-MODE ARRAY PROCESSING

3.A. Overview

A threshold must be established so that pixel current values may be compared in order to give a black/white image. Three popular methods of thresholding are global, local, and moving window. Global thresholding was implemented by placing the threshold at the average of all (n) pixel currents. The average value is calculated after all pixel signals are read. Methods are currently being researched to determine the global average pixel current in real-time fashion.

A one-bit (black/white) determination of pixel signal level is required for some fingerprint detection applications. A current comparator was used to threshold each pixel current versus the global average of all array pixel currents. This current comparator is shown in Figure 7. Transistors M1-M4 form the basic current comparator subcircuit. If the LPT pixel current is greater than the average current, M4 can not sink the amount of current which M2 is trying to source, and M2 operates in the ohmic region. The voltage at the drains of M2 and M4 is thus high. Likewise, if the pixel current is less than the average current, M4 is operating in the ohmic region, and the voltage at the drains of M2 and M4 is low. Due to the wide range of expected illumination intensities (and thus photocurrents), the current comparator output voltage swing was originally too small for low input currents. Thus, additional gain was achieved by adding two inverter stages, M5-M8.

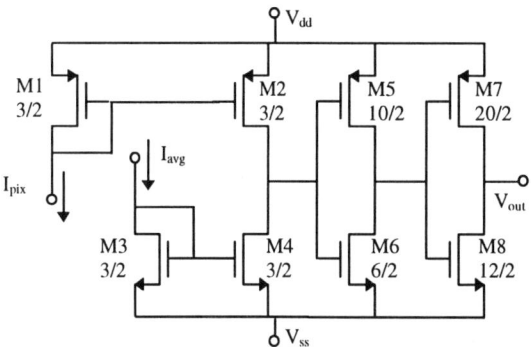

Fig. 7. Current comparator subcircuit.

3.B. Simulation Results

The current comparator was simulated in PSPICE using 2 µm ORBIT model parameters and produced the performance parameters shown in Table I. Simulations were performed at typical low and high values of photocurrent for the LPT. Sensitivity is defined as the input current difference needed to cause the output voltage to undergo a 10% to 90% transition. Propagation delay is defined as the time difference between the 50% output voltage value and 50% input voltage value.

TABLE I
PSPICE SIMULATED CURRENT COMPARATOR PARAMETERS

Reference Current (µA)	Sensitivity (nA)	Propagation Delay (ns)
0.1	4.3	359
10.0	53	51

Two previous designs of the current comparator have been tested. The first design was essentially transistors M1-M4 of Figure 7, and suffered from too small of an output swing for low input current levels. The second design suffered from large offset currents due to an asymmetrical threshold of the inverter stages. Experimental results confirmed this, and new circuits are in fabrication to correct this.

4. IMAGING ARRAY EXPERIMENTAL RESULTS

The 128x128 cell LPT imaging array was designed and fabricated in an HP 1.2 µm n-well process. The chip photo is shown in Figure 8, with each cell being 36 µm x 36 µm and each pad 100 µm on a side. Decoding logic and counters are located on the perimeter of the chip.

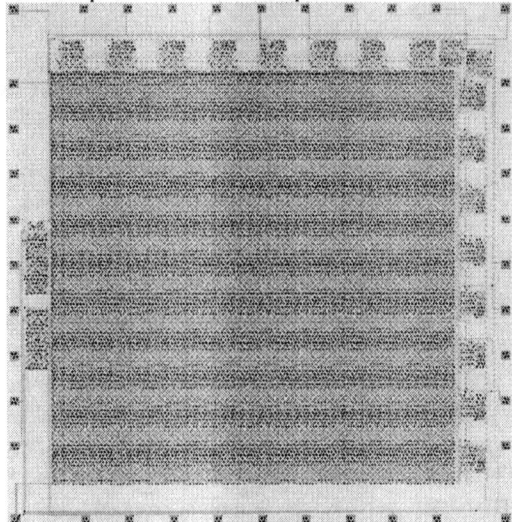

Fig. 8. 128x128 LPT imaging array.

Fig. 9. Image acquired from 128x128 LPT array.

By using a camera lens in front of the array chip and focusing an image onto the array, the 12-level gray-scale image in Figure 9 was acquired and displayed using LabVIEW® software. The range of output currents is about 0.5 µA.. Clock and clear inputs are necessary to scan the array, and the current comparator output represents valid CMOS logic levels.

The array scan clock is provided by a data acquisition board whose maximum clock rate is about 50 kHz, the limiting factor to the speed at which the array can be scanned. At a 50 kHz rate, the time required is about 300 ms.

5. CONCLUSIONS

A lateral phototransistor detection cell has been designed and fabricated in a standard digital CMOS process. Replications of this cell into arrays of dimensions 16 x 16, 64 x 64, and 128 x 128 have successfully imaged test patterns on microscopic masks and using camera lenses. Data has been collected and analyzed via external software. Future work will focus on techniques for establishing the threshold with minimal additional circuitry or processing time.

In summary, a photodetection imaging system has been developed in a standard digital CMOS process. This system is fully testable using digital hardware, thus, further reducing cost through the use of standard digital testing stations.

6. REFERENCES

[1] R. W. Sandage and J. A. Connelly, "A Fingerprint Opto-Detector Using Lateral Bipolar Phototransistors in a Standard CMOS Process," *IEEE IEDM Technical Digest*, 1995, pp. 171-174

[2] E. R. Fossum, "Assessment of Image Sensor Technology for Future NASA Missions," in *Proc. of the SPIE*, vol. 2172, pp. 1-16, 1994.

[3] P. R. Gray and R. G. Meyer, Analysis and Design of Analog Integrated Circuits, New York, NY: Wiley, 1984.

[4] E. A. Vittoz, "MOS Transistors Operated in the Lateral Bipolar Mode and Their Application in CMOS Technology," *IEEE J. Solid-State Circuits*, vol. SC-18, pp. 273-279, June 1983.

[5] W. T. Holman and J. A. Connelly, "A Compact Low Noise Operational Amplifier for a 1.2 µm Digital CMOS Technology," *IEEE J. Solid-State Circuits*, vol. SC-30, pp. 710-714, June 1995.

[6] E. Yang, *Microelectronic Devices*, New York, NY: McGraw-Hill, 1988.

[7] A. H. Sayles and J. P. Uyemura, "An Optoelectronic CMOS Memory Circuit for Parallel Detection and Storage of Optical Data," *IEEE J. Solid-State Circuits*, vol. SC-26, pp. 1110-1115, August 1991.

[8] J. P. Lavine, et. al., "Steady-State Photocarrier Collection in Silicon Imaging Devices," *IEEE Transactions on Electron Devices*, vol. ED-30, September 1983, pp. 1123-1134.

[9] J. I. Pankove, *Optical Processes in Semiconductors*, Englewood Cliffs, NJ: Prentice-Hall, 1971.

[10] H. Mincai, L. Li, H. Qijun, C. Binruo, and C. Changsheng, "Injection Phototransistors," *Sensors and Actuators A*, vol. 40, pp. 167-172, 1994.

SINGLE CHIP CMOS IMAGE SENSORS FOR A RETINA IMPLANT SYSTEM

M. Schwarz, R. Hauschild, B.J. Hosticka, J. Huppertz, T. Kneip, S. Kolnsberg, W. Mokwa, and H.K. Trieu

Fraunhofer Institute of Microelectronic Circuits and Systems,
Finkenstr. 61, Duisburg, Germany

ABSTRACT

This work describes the architecture and realization of microelectronic components for a retina implant system that will provide visual sensations to patients with photoreceptor degeneration by applying electrostimulation of the intact retinal ganglion cell layer. Special circuitry has been developed for a fast single-chip CMOS image sensor system which provides high dynamic range of more than seven decades (without mechanical shutter) corresponding to the performance of the human eye. This image sensor system is directly attachable to a digital filter and a signal processor that compute the so-called receptive-field function for generation of the stimulation data. These external components are wireless linked to an implanted flexible silicon multielectrode stimulator which generates electrical signals for electrostimulation of the intact ganglion cells. All components, including additional hardware for digital signal processing and wireless data and power transmission have been developed for fabrication using our in-house standard CMOS-technology.

1. INTRODUCTION

After it has been found that electrical stimulation of ganglion cells at the inner surface of the human retina yields visual sensations [1, 2] development of a microelectronic prosthesis for blind patients suffering from retinitis pigmentosa seems to be feasible based on the advances in technology and medicine [3, 4]. For this reason we have started developing a system for electro-stimulation of the human retina [4, 5]. Our system consists of five major microelectronic components: three for the external retina encoder and two for the implanted retina stimulator as shown in Figure 1. In addition to the implanted stimulator which is built in CMOS technology including the programmable stimulation pulse generator, the external retina encoder will be described more in detail in this contribution. The encoder includes a power and data telemetry unit, a signal processor for computing the so called *receptive field function* (RF-function) which replaces the basic functionality of the retinal layers, e.g. photoreceptor layer, horizontal, bipolar, and amacrine cell layers and, of course, the photodetector array which includes all components for image acquisition and readout of arbitrary regions of interest required for the RF-computation using an on-chip signal processor interface [6].

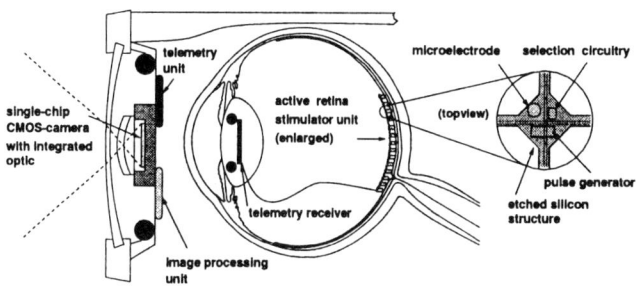

Figure 1: Architecture of the retina implant system for epiretinal ganglion cell electrostimulation

2. RETINA ENCODER CMOS IMAGE SENSOR

The external part of the retina implant system is the so called retina encoder. It provides image acquisition, computation of the RF-function of the retinal cell layers, which can be understood as spatio-temporal filtering [7], and the wireless power and data telemetry unit including channel coding of the stimulus pattern sequences (Fig. 1).

For image acquisition a dedicated CMOS image sensor chip with a high dynamic range with respect to illumination intensity has been developed. Besides the photodetector matrix this chip includes all components necessary for readout of an full image frame, readout of regions of interest, and random memory-mapped pixel access. The photodetector matrix consist of CMOS-compatible photodiodes (formed between drain diffusion an p-well) with associated readout and sensor selection circuits. These picture elements yield a sensitivity range covering more than seven decades (≥ 140 dB) of illumination with a signal-to-noise ratio of 56 dB without any global electronic or mechanical shutter. Due to the logarithmic pixel characteristic the full dynamic range can be used within a single image frame without any distortions like blooming, smearing, or time lag. The spectral sensitivity of the photodiode has been measured to be better than 60% quantum efficiency at 650 nm and better than 40% between 500 and 850 nm. Comparison of measurement and simulation has shown that the photocurrent is primarily generated near the devices surface. Finally a 3 dB pixel cut-off frequency was measured at 1 MHz (0 dB at 3 MHz), while a minimum detectable irradiance of less than $10^{-6} W/m^2$ has been found. Circuit schematic and measurements of sensitivity of the pixel element are shown in Fig. 2 and 3.

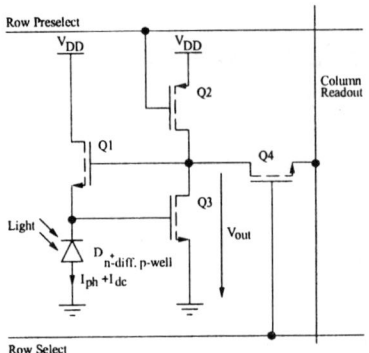

Figure 2: Schematic of the active pixel element

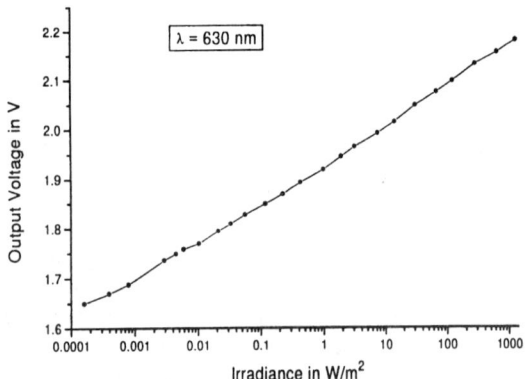

Figure 3: Measurement of pixel light sensitivity

Two CMOS image sensors have been developed using this pixel principle. In order to realize a circularly symmetric receptive field spatial filter kernel (RF-spatial filter kernels can be described by a difference of 2D-gaussian functions [5]) the pixel elements of the first 128×128 pixel CMOS sensor have been arranged on a hexagonal grid structure which yields a reduction of required samples of 13.5% in comparison to rectangular sampling used for the second 400×300 pixel sensor [8]. This is a benefit for lower resolution images, but on the other hand it requires about 5% more pixel area for metal wiring.

Compilation of blocks for row selection (including additional preselection circuits for each row), first stage readout amplifiers, column select multiplexers, and final driver stages yields a similar chip layout for both the hexagonal 128×128 and orthogonal 400×300 pixel CMOS sensor chips as shown in the principal layout in Figure 2 and block diagram shown in Figure 4.

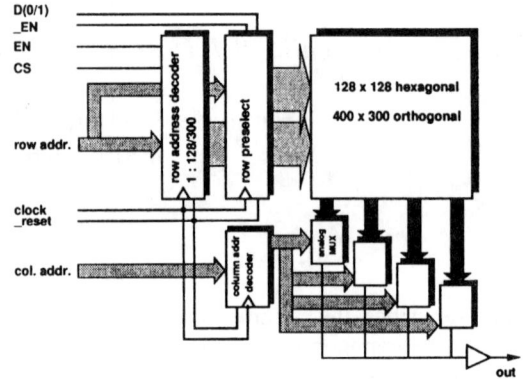

Figure 4: Block diagram of the 128×128 and 400×300 pixel CMOS sensors

Both the hexagonal 128×128 and the orthogonal 400×300 pixel CMOS image sensors have been successfully tested at video frame rates and also operated at frame rates up to 150 Hz with enhanced illumination. Test images captured with both devices are shown in Figure 5.

Figure 5: Test images captured with the integrated 128×128 and 400×300 pixel CMOS single chip sensor (with contrast and contour enhancement)

Both chips operate at a single 5 V power supply and exhibit a sensor gain of $86 mV/decade$. The chip photo of the 128×128 pixel CMOS sensor is shown in Figure 6.

The on-chip standard signal processor interface eliminates the need for an additional frame buffer as required for conventional CCD devices. Thus the use of random and subregion addressing for readout of the pixels from distant regions of the array, e.g. necessary for computation of the receptive fields (RF) that are individually related to the sparse distributed stimulation electrodes of the stimulator positioned on the retina surface (Fig. 1 and 9), increases the total readout rate by a factor of 10 for the Retina-Implant application when compared to full frame readout. This is necessary since not only spatial, but also local temporal filtering operations (within milliseconds) have to be implemented for realization of RF functions.

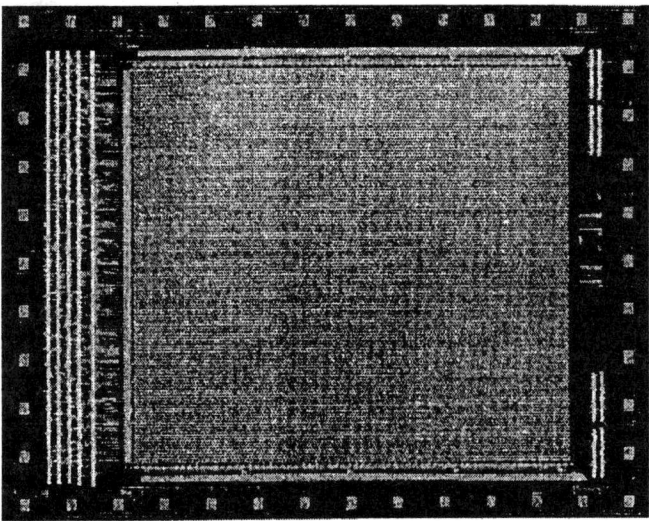

Figure 6: Chip microphotograph of the 128×128 pixel image sensor fabricated in 1μm CMOS technology

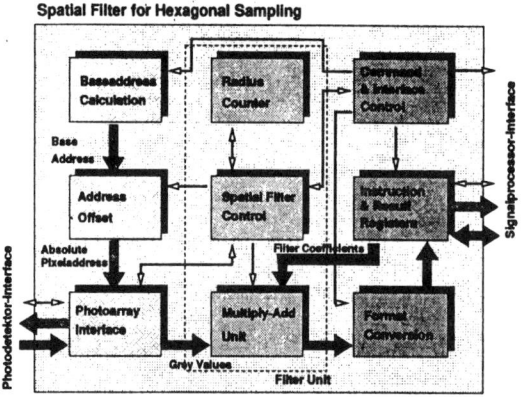

Figure 7: Block diagram of the spatial filter used for implementation of the on- off-center RF-spatial filter (realized as CMOS standard cell design)

Preselection of arbitrary photodetector lines makes the device well suitable for portable applications such as the retina implant system and does not conflict with full frame sensor mode, since power-on for each row (6 $mW/preselected row$) must preceed the current scan-line only by a few rows.

3. RETINA ENCODER SYSTEM

Housing the sensor chip in a package with integrated short focus optics as cover will result in a lightweight miniaturized sensor system which is well applicable for our retina implant system where the camera chip is mounted on a spectacle frame that can easily be placed and removed by the patient (Fig. 1).

A further device developed for the retina implant is the spatial filter necessary for hardware implementation of the RF-functions spatial filter. Figure 7 shows the block diagram of the spatial filter used for realization of the on- and off-center RF-function. It can be inserted between the sensor chip and the signal processor since it uses the digital memory-like interface of the CMOS image sensor for random pixel addressing. It carries out hardware computation of pixel coordinates according to an RF-center-point, simultaneous fetching of associated filter coefficients, and calculation of the fixed-point convolution using its hardware multiply-add unit. The hexagonal filter hardware has been designed using Verilog-HDL and synthesized for the realization in the same standard CMOS technology as used for the sensor chip.

Besides of the spatial filter function, a biological RF-function also performs temporal filtering and spike train generation by retinal ganglion cells, which actually have been implemented off-chip using a digital signal processor (DSP). This DSP also controls the telemetry unit required for wireless transmission of stimulus data (i.e. encoded spike duration, polarity, and electrode address information) and power for the flexible silicon multielectrode electrostimulator implant.

4. RETINA STIMULATOR SYSTEM

The second major subsystem of the retina implant system is the implantable flexible silicon multielectrode electrostimulator (Fig. 1, 8, and 9).

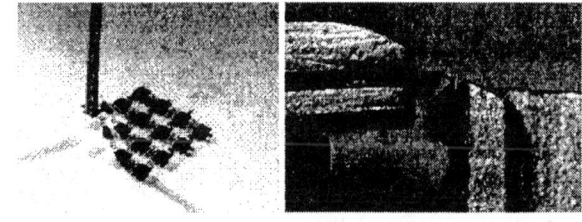

Figure 8: Highly flexible silicon test structures with small regions of thick silicon for circuitry and electrodes fabricated using a modified backside etching process compatible with standard CMOS technology (left: flexible silicon grid structure lifted with vacuum handler, right: etched silicon chip bended at 90°)

The fabrication of this stimulator is also based on our standard CMOS-technology here used for implementation of circuitry for separation of power and stimulation data, error correction as well as for realization of programmable current sources and electrode selectors. The mechanical flexibility of the electrostimulator silicon grid structure shown in Figure 8 has been achieved by applying backside etching of the silicon wafer to thin the silicon substrate at selected sites. The electrodes and electronics are located at thick "islands" connected by thin crosspieces that ensure the flexural response of the stimulator (see physical arrangement and block-diagram in Fig. 9).

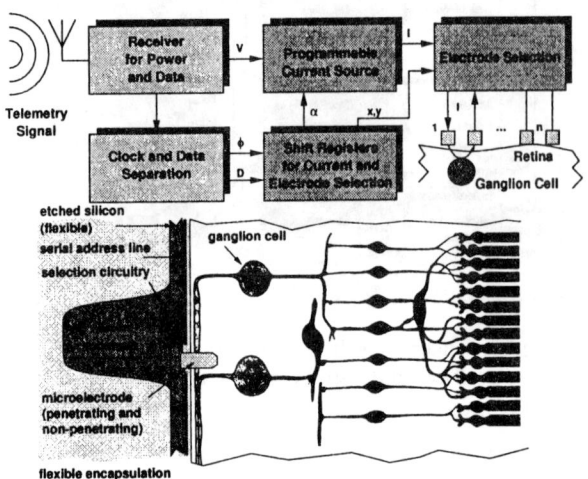

Figure 9: Block diagram and physical arrangement (sectional view) of the flexible silicon multielectrode electrostimulator with circuitry for separation of power and stimulation data, error correction as well as for generation of programmable current pulses and electrode selection

The circuitry required for electrostimulation as shown in Figure 9 has been designed to generate pulses with a programmable pulse width ($10 - 300\mu s$), pulse polarity (including bipolar pulses) pulse current ($10 - 100\mu A$), and pulse rate ($\leq 500Hz$).

5. SUMMARY AND OUTLOOK

The hardware architecture of the retina implant system described in this contribution greatly exceeds the complexity of current bio-electronic systems, e.g. of cochlea implants [4], because both the photo-receptor cells and all of the neural functionality of the retina will be replaced by microelectronic, micro-optical, and micro-mechanical hardware.

The CMOS image sensor chip presented here represents one of five components that have to fulfil special requirements. Its high dynamic range of more than seven decades of illumination has been realized by employing a dedicated pixel with logarithmic characteristic. Other improvements include memory-mapped random and subregion addressing for efficient on- and off-center receptive field computation, and low power dissipation by preselection of only a small number of scan lines during full operation. Besides of the CMOS sensor chip also dedicated hard- and software for realtime emulation of biological receptive field functions and wireless data and power transmission for electrostimulation of ganglion cells have been implemented and tested. Finally, a highly flexible silicon multielectrode stimulator structure has been developed by applying backside etching to the silicon wafer. It should be also mentioned here, that strong work on biocompatible encapsulation of the flexible silicon multielectrode structure and on the optimization of the implantation ophthalmological surgery is carried out by our biomedical and medical partners.

Our future research will concentrate on single-chip CMOS-cameras with higher resolution as well as on on-chip implementation of global and local brightness and contrast adaptation, aperture correction, color processing, and application specific interfaces. All these features are not available in CCD devices which always require external analog and digital circuits for amplification, filtering, and timing generation for pixel shifting. We surmize that our miniaturized CMOS image sensor systems will also be well applicable for automotive systems like rear-mirror mounted cameras for obstacle and pre-crash detection as well as low-cost guidance systems for autonomous vehicles since all these applications can benefit from the high dynamic range of more than seven decades.

6. ACKNOWLEDGEMENTS

The authors greatfully acknowledge useful discussions with the other members of the Retina Implant Team that is developing the retina implant device under contract from the German Federal Ministry of Education, Science, Research and Technology (BMBF), Bonn. This work is based on the results of the feasibility study "*Neurotechnology Report II*" [4].

7. REFERENCES

[1] J.L. Wyatt, J.L. Rizzo, A. Grumet. et. al.: Development of a silicon retinal implant: Epiretinal stimulation of retinal ganglion cells in the rabbit. Invest. Opthal. & Vis. Sci. 35 (Suppl.), 1380, 1994.

[2] A. Benjamin, M.S. Humayun, S. Hickingbotham, E. De Juan et. al.: Characterization of retinal responses to electrical stimulation of retinal surface of rana catesbeiana, Invest. Opthal. & Vis. Sci. 35 (Suppl.), 1832, 1994.

[3] M. S. Humayun, R. Probst, E. De Juan, et. al.: Local electrical stimulation of the human retina: is an intraocular visual prosthesis feasible, Science, 1994.

[4] R. Eckmiller et. al.: Neurotechnologie-Report - Machbarkeitsstudie - Leitprojekt-Vorschlag II, BMBF, Bonn, 1995.

[5] M. Schwarz, B.J. Hosticka, R. Hauschild, W. Mokwa, M. Scholles, and H.K. Trieu: Hardware Architecture of a Neural Net Based Retina Implant for Patients Suffering from Retinitis Pigmentosa, In Proc. IEEE International Conference on Neural Networks, pp. 653–658, Washington, D.C., June 1996.

[6] J. Huppertz, R. Hauschild, B.J. Hosticka, T. Kneip, S. Müller, and M. Schwarz: Fast CMOS Imaging with High Dynamic Range, In Proc. of IEEE Workshop on Charge Coupled Devices & Advanced Image Sensors '97, pp. R7-1–R7-4, Bruges (Belgium), June 1997.

[7] H. Kolb, The Architecture of Functional Neural Circuits in the Vertebrate Retina, Invest. Opthal. & Vis. Sci. 35, pp. 2385–2404, 1994.

[8] R. Mersereau, The Processing of Hexagonally Sampled Two-Dimensional Signals, In: Proc. of the IEEE 67, pp. 930–949, June 1979.

AN ANALOG VLSI VELOCITY SENSOR USING THE GRADIENT METHOD

Rainer A. Deutschmann and Christof Koch

139-74 California Institute of Technology
Pasadena, CA 91125, USA
rainer@klab.caltech.edu

ABSTRACT

Smart vision sensors that unify imaging and computation on one single chip offer great advantage over conventional sensor systems, where the computational part is usually performed separately and serially on a digital computer. Using parallel analog VLSI design principles, compact, low power, inexpensive and real time sensors can be built even in standard CMOS processes.

In this paper we present the first working analog VLSI implementation of a 1-D velocity sensor that uses the gradient method for spatially resolved velocity computation. We use a novel floating gate wide linear range amplifier and a floating gate division circuit. The division by zero problem of the gradient method has been solved. The sensor velocity output linearly codes the stimulus velocity over a wide range, is independent of contrast down to 20% contrast and indicates the correct direction-of-motion down to 4% contrast. In a circular pixel arrangement the sensor reports the rotational velocity up to 350 rpm.

1. INTRODUCTION

Moving scenes are a rich source of information. From the optical flow important parameters such as ego-motion, time-to-contact and the focus-of-expansion can be inferred. Tracking systems can use velocity information, and object segmentation and figure-ground segmentation based on edge detection can be improved by using discontinuities of the optical flow field.

Image velocity estimation has successfully been performed in software, but high frame rates require powerful computers. More recently VLSI systems have been developed where imaging and velocity computation are integrated on one single chip. There are two types of algorithms for determining the velocity field. In the *gradient method* the local velocity is obtained by dividing temporal and spatial derivatives of the local light intensity distribution, whereas in *correlation based methods* image features, such as intensity patterns or edges, are extracted and correlated.

R. D. is now at the Walter Schottky Institut, Technische Universität München, Am Coulombwall, 85748 Garching, GERMANY

Up to now no successful implementation of the gradient method has been reported. One of the earliest analog VLSI velocity chips [1] was supposed to solve the optical flow equation (see section 2), but due to the implementation of the algorithm only one global velocity vector could possibly be obtained.

Subsequently research has concentrated on correlation based methods for velocity estimation [2] [3] [4]. A successful class of sensors measures time of travel of a token [5] [6] [7].

In this paper we present the first analog VLSI sensor that computes the spatially resolved image velocity by a real time division of the temporal and spatial derivatives of the local light intensity. The algorithm is derived in the following section. In section 3 the implementation is described, and in section 4 experimental results are presented.

2. ALGORITHM

An equation which relates the change in image brightness at a point in a plane to the motion of the brightness pattern can be derived from the assumption, that the brightness of a particular point in the pattern is constant:

$$\frac{d}{dt}I(x,y,t) = 0 \quad \Rightarrow \quad \frac{\partial I}{\partial x}v_x + \frac{\partial I}{\partial y}v_y + \frac{\partial I}{\partial t} = 0, \quad (1)$$

where $I(x,y,t)$ is the image intensity on the focal plane and (v_x, v_y) the velocity vector.

From this equation one readily obtains by rearranging for one spatial dimension

$$v(x,t) = -\frac{\partial I/\partial t}{\partial I/\partial x}, \quad (2)$$

i.e. the local velocity can be computed from the division of the temporal and spatial derivatives. It can be seen that velocity information is obtained continously in time as long as temporal and spatial changes are present. Token based methods, though, can only respond to specific features, and after detection the velocity information has to be stored.

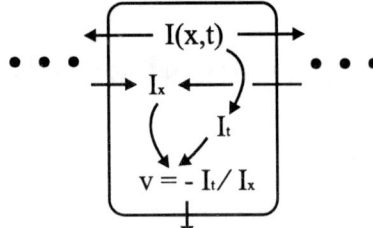

Figure 1: Blockdiagram of one pixel.

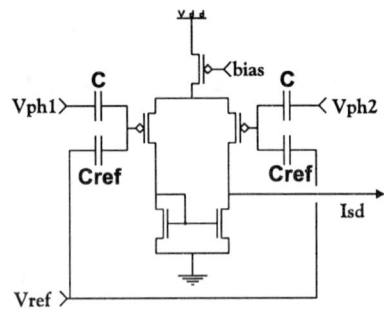

Figure 3: Wide linear range amplifier circuit.

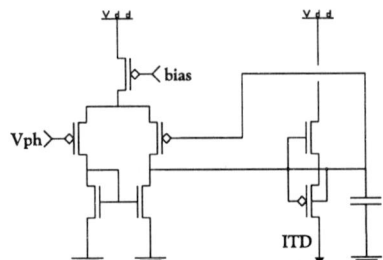

Figure 2: Temporal derivative circuit.

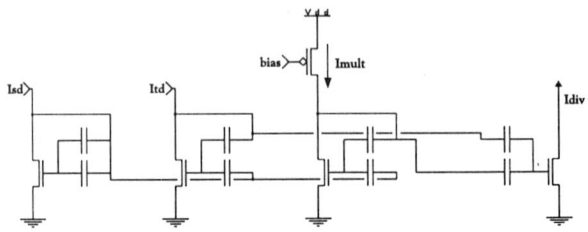

Figure 4: Division circuit.

3. IMPLEMENTATION

Our implementation of equation 2 in analog VLSI hardware consists of four major building blocks, as shown in Figure 1: An adaptive photoreceptor [8], temporal and spatial derivative circuits, and a division circuit. The local light intensity $I(x,t)$ is used for the temporal derivative I_t and is fed to the neighbouring pixels for computing the spatial derivative I_x. The velocity v is obtained from a division of temporal and spatial derivatives.

The temporal derivative circuit (Figure 2) follows [9] and provides a current I_{td} proportional to the rate of change of the photoreceptor voltage V_{ph} for decreasing light intensities.

For the spatial derivative a new wide linear range amplifier was developed (Figure 3). The inputs are coupled to a differential pair through floating gate capacitive dividers. Neglecting parasitic capacitances, the linear range is increased by a factor $1+C_{ref}/C$. This factor was chosen such that the current I_{sd} was linear in the voltage $V_{ph2} - V_{ph2}$ for all natural photoreceptor voltages.

The division circuit (Figure 4) was developed following translinear design principles [10]. The circuit involves only 4 transistors and one current source. Because the input currents can only be unidirectional, the spatial derivative current I_{sd} is rectified before the division (circuit not shown). The output current I_{div} is computed from I_{td}, I_{sd} and a reference current I_{mult} such that

$$I_{div} = I_{mult} \times \frac{I_{td}}{I_{sd}}. \quad (3)$$

as long as all currents stay subthreshold, i.e. $<$100nA. The sign of the velocity output current I_{div} is then corrected in a simple switching circuit (not shown).

We used a commercially available 2.0μm CMOS process. The pixel size was 147μm$\times$270μm but can be optimised. On a 2.2mm $\times$ 2.2mm chip we circularly arranged 20 pixels, each one of which could be accessed through an on-chip scanner. Besides a 5V power supply and a few bias voltages only a lens is required for operation.

4. RESULTS

For debugging reasons not only the computed velocity signal but also the photoreceptor voltage and the temporal and spatial derivatives can be accessed from each pixel. In Figure 5 these signals were recorded over time from one pixel while a moving sinusoidal gray value pattern was presented to the sensor. The top traces show the photoreceptor voltage together with the velocity output. The photoreceptor voltage reflects the sinusoidal stimulus. It can be seen that the sensor reports a constant stimulus velocity during the time when the temporal derivative is computed (falling light intensities), otherwise zero velocity is reported. The height of the plateau indicates the stimulus velocity. The fact that the sensor reports a constant velocity is remarkable considering that both temporal and spatial derivatives are varying according to the stimulus shape, as shown in the two lower traces. The spatial derivative matches the theoretical expectation (phase shifted rectified sinusoid), and the temporal

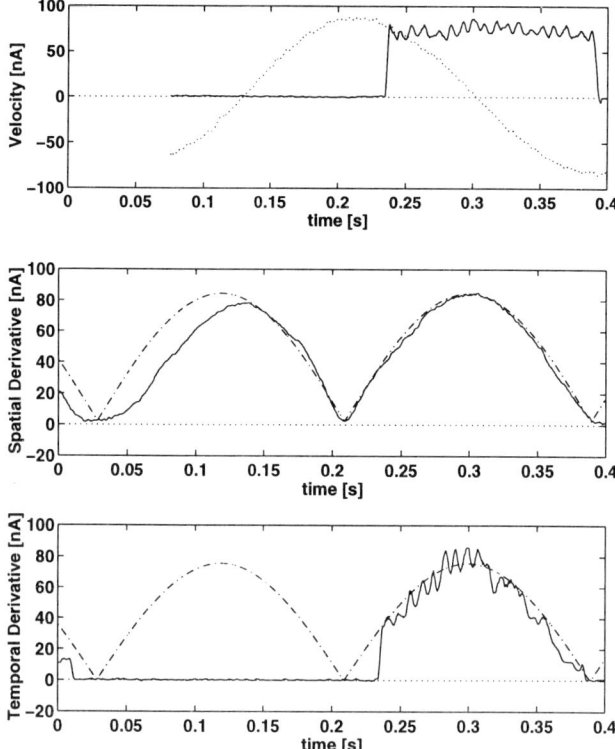

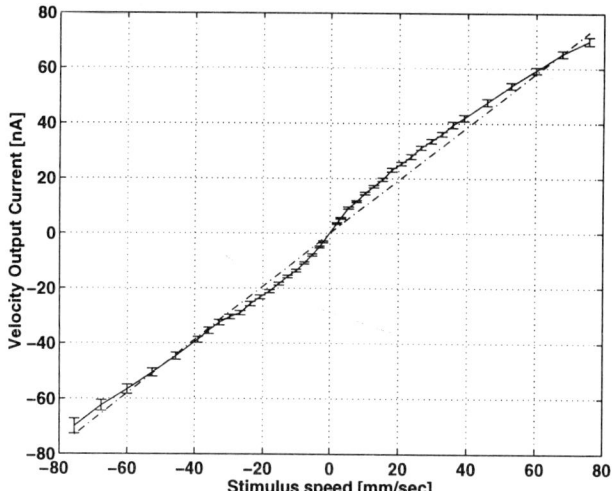

Figure 6: Velocity dependence.

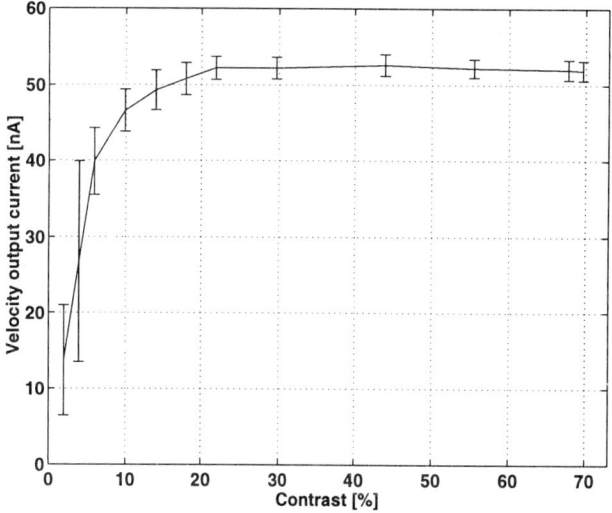

Figure 7: Contrast dependence.

Figure 5: Velocity output and photoreceptor output (dotted) from one pixel for sinusoidal stimulus (top traces). Rectified spatial derivative (middle trace) and temporal derivative (bottom trace) with fit (dashed lines).

derivative is computed correctly as half wave sinusoid during falling intensities.

In Figure 6 the height of the plateau is plotted against the stimulus velocity from 2 mm/sec up to 76 mm/sec. A sine wave stimulus of 72.5% contrast and spatial frequency 0.05 cycles/deg was used. The reported velocity is almost linear with the stimulus velocity over the whole range. The errorbars represent one standard deviation on each side and were obtained over multiple presentations of the same stimulus. With one bias voltage set slightly different an almost linear relationship between stimulus velocity and reported velocity is also obtained for velocities as low as 0.07 mm/sec (almost invisible) up to 2 mm/sec. The same experiments have been performed with saw tooth stimuli and yielded the same linearity.

In Figure 7 the velocity output is plotted for a sinusoid stimulus of constant velocity but different contrasts. The velocity output is independent of the stimulus contrast over a wide range down to 20% contrast. For very small contrasts in order to avoid the division by zero problem in equation 2 a small current is added to the spatial derivative and a different current is subtracted from the temporal derivative. Therefore the velocity output approaches zero for small contrasts. The correct direction-of-motion is reported down to contrast of 4%.

In the previous experiments only one pixel was read out at a time. In a different scanning mode the velocity outputs of all pixels can be summed together. No clock is then needed. If the pixels are arranged in a line, the global translatory velocity can robustly be detected. On one sensor chip we circularly arranged 25 pixels (see Figure 8), which allows for a robust measurement of the rotational velocity. Figure 9 shows the sensor output for a rotating stimulus: Up to 2120 deg/sec (353 rpm) the stimulus velocity is reported almost linearly.

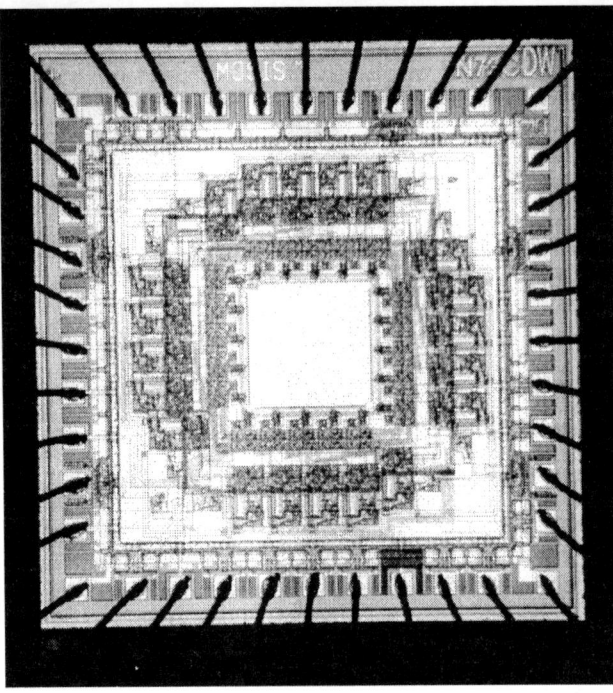

Figure 8: Micro photograph of the sensor chip.

5. DISCUSSION

The first analog CMOS sensor is presented that implements the gradient method for spatially resolved velocity computation. The sensor features a wide dynamic range, good linearity and robust operation. In a linear or circular pixel arrangement the sensor can additionally compute the global translatory or rotatory velocity, respectively. As predicted by the theory for a discrete implementation of the gradient method, the velocity output increases for spatial frequencies close to the Nyquist frequency. It is therefore desirable to further optimise the pixel size. Based on the circuits presented here the gradient method can be implemented for a 2-D velocity sensor.

6. ACKNOWLEDGEMENTS

This work was supported by the Center for Neuromorphic Systems Engineering as a part of the National Science Foundation's Engineering Research Center program.

7. REFERENCES

[1] J. Tanner and C. Mead. "An integrated optical motion sensor". *VLSI Signal Processing II, IEEE Press*, pages 59–76, 1986.

[2] A.G. Andreou, K. Strohbehn, and R.E. Jenkins. "Silicon retina for motion computation". *Proc. IEEE Int. Symp. on Circuits and Systems*, 3:1373–1376, 1991.

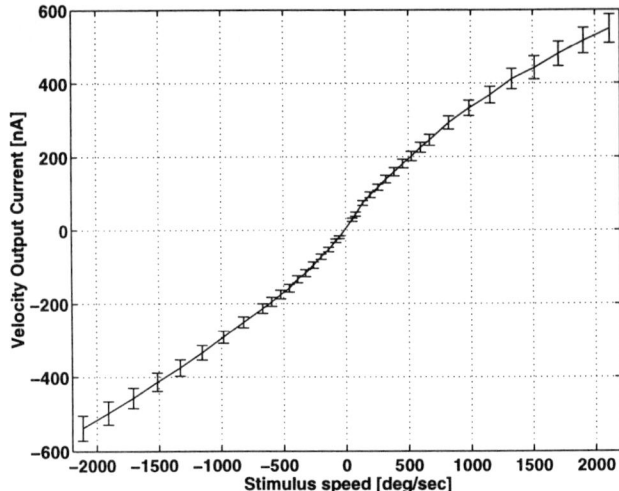

Figure 9: Measuring rotational velocity.

[3] T. Delbrück. "Silicon retina with correlation-based, velocity-tuned pixels". *IEEE Trans. on Neural Networks*, 4:529–541, 1993.

[4] T.K. Horiuchi, W. Bair, B. Bishofberger, A. Moore, and C. Koch. "Computing motion using analog VLSI vision chips: an experimental comparison among different approaches". *Intern. Journal of Computer Vision*, 8:203–216, 1992.

[5] R. Etienne-Cummings, J. Van der Spiegel, and P. Mueller. "A focal plane visual motion measurement sensor". *IEEE Trans. Circuits and Systems 1*, 44:55–66, 1997.

[6] J. Kramer. "Compact integrated motion sensor with three-pixel interaction". *IEEE Trans. Pattern Anal. Machine Intell.*, 18:455–560, 1996.

[7] J. Kramer, R. Sarpeshkar, and C. Koch. "Pulse-based analog VLSI velocity sensors". *IEEE Trans. Circuits and Systems II*, 44:86–101, 1997.

[8] T. Delbrück and C. Mead. "Analog VLSI Phototransduction". *CNS Memo No.30, Caltech*, pages 139–161, 1994.

[9] T. Horiuchi. "Analog VLSI-based, neuromorphic sensorimotor systems: modeling the primate oculomotor system". *California Institute of Technology, PhD Thesis*, 1997

[10] B. A. Minch, C. Diorio, P. Hasler and C. Mead. "Translinear Circuits using subthreshold floating-gate MOS-transistors". *Analog Integrated Circuits and Signal Processing*, 9:167–179, 1996

A SINGLE FOURIER SERIES TECHNIQUE FOR THE SIMULATION AND ANALYSIS OF ASYNCHRONOUS PULSE WIDTH MODULATION IN MOTOR DRIVE SYSTEMS

R.A. Guinee

Cork Institute Of Technology
Cork, IRELAND

C. Lyden

NMRC
University College
Cork, IRELAND

ABSTRACT

A generalized single Fourier series description is presented for the simulation and analysis of asynchronous pulsewidth modulation (PWM) in brushless motor drive (BLMD) systems. This simple expedient eliminates the need for very small time steps in resolving the PWM waveform edge transition times, within a switching interval, without loss of simulation accuracy. Graphed simulation data indistinguishable from that for analog comparator circuit implementation is presented for three forms of PWM based on the generalized Fourier series representation. Substantial savings of up to fiftyfold in motor drive simulation time are achieved without impairment to simulation accuracy.

1. INTRODUCTION

Pulsewidth modulation (PWM) is widely used for efficient power transfer in switch mode power supplies and in industrial machine drives where high torque and precision control are required. In the latter case [1] motor control is achieved through amplitude and frequency regulation of the stator winding voltages by means of a sinusoidal reference PWM inverter power stage using asynchronous natural sampling. This mode of carrier modulation is preferred because of its inherent simplicity and ease of implementation using analog circuitry.

The method of spectral analysis hitherto employed in the study of pulse duration modulation (PDM) in these systems is mathematically complex [2] and based on the well understood double Fourier series 'wall' model [3] first introduced by Bennett [4]. The Fast Fourier Transform (FFT) simulation algorithm has been used to avoid this analytical difficulty in determining the PWM spectral content. The FFT is not however suited to two level pulse sequences as it requires a large number of sample points to accurately resolve the variable PWM edge transitions resulting in increased computation time. Simplification of the PWM process using a single Fourier series for spectral evaluation purposes only has been proposed by [5] for periodic reference waveforms by simulating the action of a uniform sampled digital PWM circuit. This method however does not account for natural sampling and switch transition accuracy is limited by the degree of carrier quantization used. There is a further requirement, in addition to spectral analysis, for a simple and accurate model of the PWM inverter stage for efficient BLMD model simulation. This is necessary in system identification for optimal controller design, in wide bandwidth servodrive applications, in which there is substantial dead time for safe inverter operation. Simplification of the PWM model would considerably reduce the computation (CPU) overhead and eliminate the need for accurate inverter switching resolution.

This paper presents a compact and accurate analytical model for PWM using a generalized single Fourier series description of the natural sampling process. This can then be effectively deployed for software modelling in power electronic applications and in simulation of complex three phase BLMD systems [6] without the need for costly CPU resources. Three asynchronous simulation methods are presented which include leading edge, trailing edge and double edge PWM. In addition to periodic references bandlimited arbitrary modulating signals can also be incorporated in the Fourier model description for PWM. The single Fourier series method eliminates the need for an artificial three dimensional 'wall' model which results in a considerable simplification of the PWM process that is more tractable for analytical purposes. Waveform generation by the proposed generalized Fourier method for each of the three cases of PWM is used to establish model veracity and accuracy when compared with that from a simulated analog comparator circuit. The simulation of a frequency modulated BLMD step response feedback current transient and its comparison with measured data is used to establish confidence in and validation of the proposed PWM inverter model.

2. SINGLE FOURIER SERIES METHOD

A simple and accurate PWM simulation model can obtained by deploying a single Fourier series description of the comparator operation in the width modulation process. The analog comparison of a modulating signal with an asynchronous symmetrical triangular carrier in the natural sampling mode, with switching period T_s and amplitude A_{tri}, results in a symmetrical double edge PWM waveform as shown in Fig. 2. The comparator output (o/p) for the unmodulated case is a rectangular pulse train of amplitude U_d with fixed aperture time $\tau = T_s/2$ which fulfills the rôle of an unmodulated carrier. The extent of the difference in the slope magnitudes of the negative and positive going ramps, constituting one period of the triangular waveform, determine the degree of movement in the leading and trailing edges during modulation. If the leading and

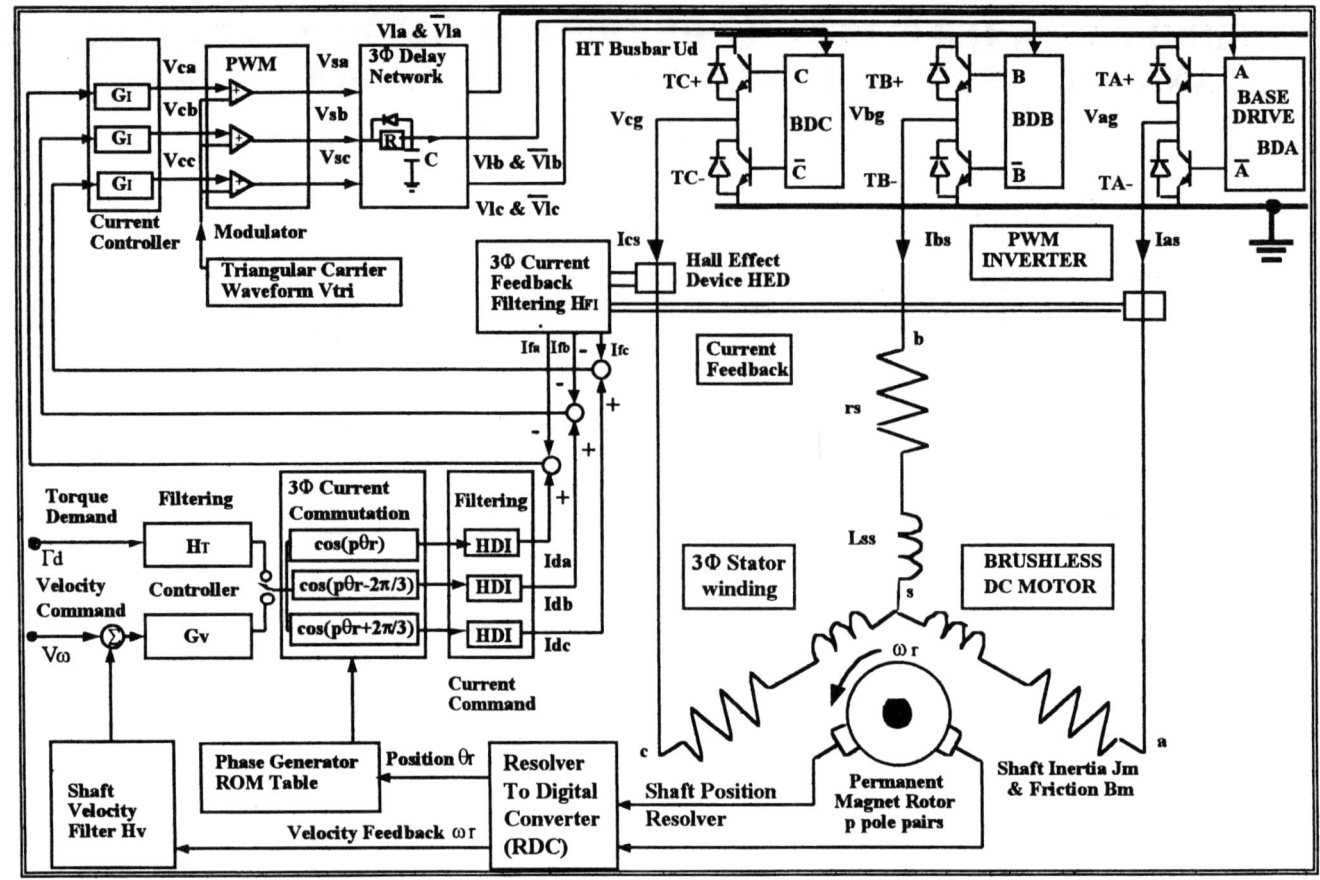

Figure 1 Network structure of a typical brushless motor drive system

trailing edge displacements of the unmodulated pulse sequence corresponding to the negative and positive going ramps, with respect to the triangular axis of symmetry as shown in Fig. 2, are τ_1 and τ_2 then the fixed aperture time is given by

$$\tau = \tau_1 + \tau_2 = T_s/2 \qquad (1)$$

The modulation effect is articulated in its most general form by the Fourier series expansion of the unmodulated asymmetrical pulse sequence given by, with $\tau_c = (\tau_2 - \tau_1)/2$ and $\tau_2 \neq \tau_1$,

$$p(t) = \frac{U_d}{T_s}\tau + \frac{2U_d}{\pi}\sum_{n\geq 1}\left(\frac{1}{n}\right)\sin(n\pi f_s \tau)\cos(n\omega_s(t-\tau_c)) \qquad (2)$$

The pulse duration τ is then altered in accordance with the instantaneous amplitude of the modulating signal $v_m(t)$. An overall modulated aperture time is obtained linearly varying with an arbitrary normalized modulating control signal $\bar{v}_m(t)$ between 0 and T_s as:

$$\tau_m = \tau\left[1 + m_f \bar{v}_m(t)\right] \qquad (3)$$

where $v_m(t) = m_f A_{tri} \bar{v}_m(t)$ with modulation index m_f.
Pulse time modulation results in individual edge translations of

$$\tau_{1m} = \tau_1[1 + m_f \bar{v}_m(t)] \text{ and } \tau_{2m} = \tau_2[1 + m_f \bar{v}_m(t)]$$

with respect to the axis of symmetry corresponding to the points of intersection of the signal with the dual ramp carrier. The resultant PWM inverter phase a o/p waveform from (2) is

$$v_s(t) = \frac{U_d \tau}{T_s}\left[1 + m_f \cos(p\omega_r t)\right]$$
$$+ \frac{2U_d}{\pi}\sum_{n\geq 1}(\frac{1}{n})\sin\left(n\pi f_s \tau\left[1 + m_f \cos(p\omega_r t)\right]\right)\cos\left(n\omega_s(t - \tau_{cm})\right)$$
$$(4)$$

with $\tau_{cm} = \tau_c\left[1 + \cos(p\omega_r t)\right]$ and periodic reference given by

$$v_m(t) = m_f A_d \cos(p\omega_r t) \qquad (5)$$

for a typical BLMD system. Three important cases of pulse edge modulation result from (4) based on the relative magnitudes of τ_1 and τ_2 and include:

- Asymmetrical double edge PWM with $\tau_1 \neq \tau_2$ for which
$\tau = \tau_1 + \tau_2 = T_s/2$ and $\tau_c = (\tau_2 - \tau_1)/2$

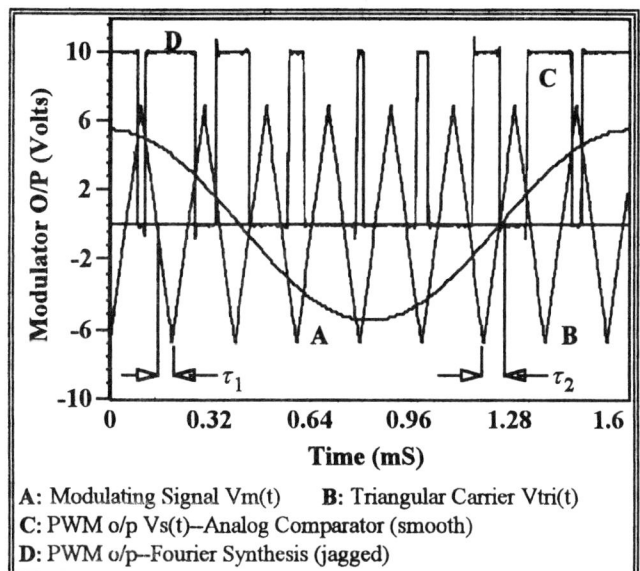

Figure 2 Symmetrical PWM waveform generation for phase a

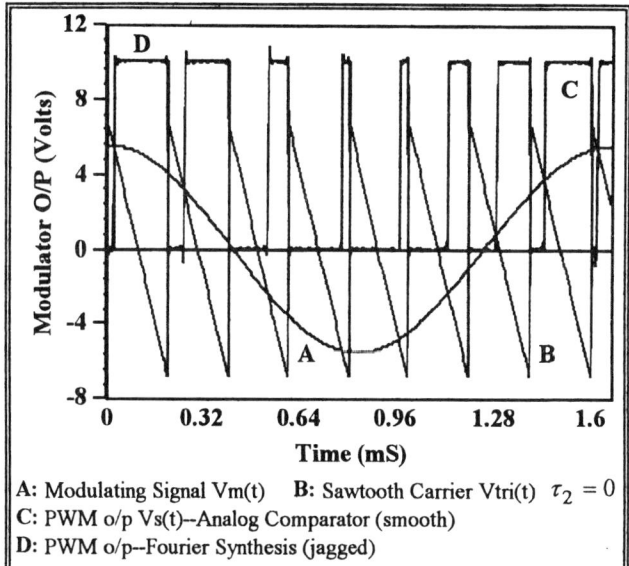

Figure 3 Leading edge PWM generation

Symmetrical double edge PWM with $\tau_1 = \tau_2$ for which $\tau = 2\tau_1 = 2\tau_2 = T_s/2$ and $\tau_c = 0$ as shown in Fig. 2.

- Leading edge PWM with $\tau_2 = 0$ for which
  $\tau = \tau_1 = T_s/2$ and $\tau_c = -\tau_1/2$ as shown in Fig. 3.

- Trailing edge PWM with $\tau_1 = 0$ for which
  $\tau = \tau_2 = T_s/2$ and $\tau_c = \tau_2/2$ as shown in Fig. 4.

In the last two cases the pulse centerline offset τ_c from the triangular axis of symmetry plays an important rôle in anchoring the static reference edge during the modulation phase. When the pulse duration is altered in this instance, with respect to the signal amplitude, its centerline is translated accordingly so that the relevant pulse edge reference is immobilized. This attribute is encoded into the carrier harmonics as modulated time/phase shift in (4) and will result in significant additional harmonic distortion, when expanded into Bessel function components, by comparison with the symmetrical double edge method. Consequently with $\tau_c = 0$, leading to reduced distortion, and due to the ease of circuit implementation the symmetrical double edge method of PWM generation is the preferred option in practical systems. In this case the comparator circuit generation of a symmetrical PWM o/p, for a periodic reference input, can be represented by (4) with $\tau_c = 0$ as

$$v_s(t) = \frac{U_d \tau}{T_s}\left[1 + m_f \cos(p\omega_r t)\right] + \frac{2U_d}{\pi}\sum_{n\geq 1}(\frac{1}{n})\sin\left(n\pi f_s \tau\left[1 + m_f \cos(p\omega_r t)\right]\right)\cos(n\omega_s t) \quad (6)$$

The first term in (6) consists of the dc level shifted PWM encoded signal $v_m(t)$ with an amplitude that is proportional to the modulation depth (MI). This provides a useful means of voltage control in variable speed BLMD applications. The second term in (6) is identical to that given by the double Fourier method in [3] and consists of the distortion components

$$\frac{2U_d\tau}{T_s}\sum_{n=1}^{\infty}J_0(\frac{m_f n\omega_s \tau}{2})\mathrm{sinc}(nf_s\tau)\cos(n\omega_s t)$$
$$+\frac{2U_d\tau}{T_s}\sum_{n\geq 1}\left[2J_1(\frac{m_f n\omega_s\tau}{2})\frac{\cos(n\pi f_s\tau)}{(n\pi f_s\tau)}\cos(n\omega_s t)\cos(p\omega_r t)\right.$$
$$\left.-2J_2(\frac{m_f n\omega_s\tau}{2})\frac{\sin(n\pi f_s\tau)}{(n\pi f_s\tau)}\cos(n\omega_s t)\cos(2p\omega_r t)+\cdots\right] \quad (7)$$

These components include the carrier and its odd harmonics at nf_s, with amplitudes that decrease with increasing m_f, about which multiple (m) side frequency pairs $(nf_s \pm mpf_r)$ exist for $\tau = T_s/2$. The side frequency distribution consists of odd order pairs for even harmonics and even order pairs for odd carrier harmonics.

3. BLMD INVERTER MODELLING

The PWM inverter switching frequency f_s is chosen in BLMD applications such that the carrier and its harmonics with attendant side frequencies are heavily suppressed by the stator winding inductances. In the absence of inverter dead time the voltage supplied to the stator winding phase a input and ground can be approximated by the first term in (6) for a periodic current reference (5) as $v_{ag} \approx U_d \tau_m/T_s$. The effect of inverter switch delay is easily modelled as a blanking period δ at the leading edge of each modulated pulse if winding current flow is positive. This inverter o/p in this instance is given by

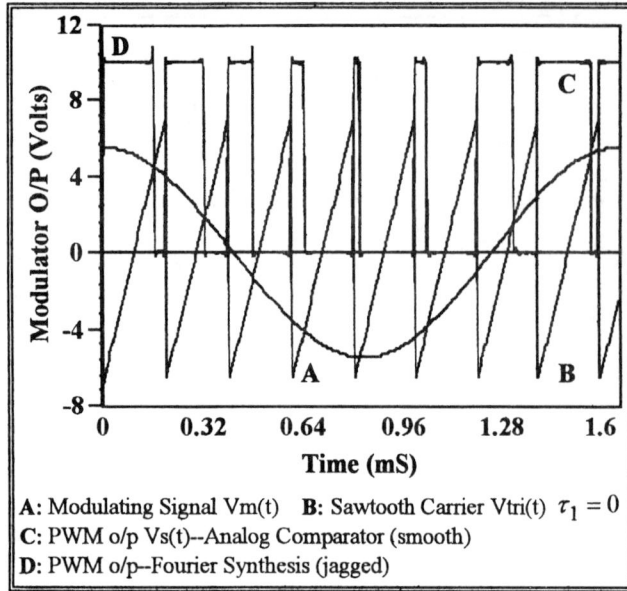

A: Modulating Signal Vm(t)   B: Sawtooth Carrier Vtri(t) $\tau_1 = 0$
C: PWM o/p Vs(t)--Analog Comparator (smooth)
D: PWM o/p--Fourier Synthesis (jagged)

Figure 4 Trailing edge PWM simulation

$$v_{ag} \approx (U_d/T_s)(\tau_m - \delta) \text{ for } i_{as} \geq 0 \qquad (8)$$

Conversely if the phase current flow is negative during the inverter turn-on delay then a pulse of duration δ must be included to model winding connection to the dc busbar with

$$v_{ag} \approx \frac{U_d}{T_s}(\tau_m + \delta) \text{ for } i_{as} < 0 \qquad (9)$$

The above generalized PWM Fourier method along with the equations necessary [1] for the motor drive dynamics constitute a compact BLMD model description for simulation purposes.

4. SIMULATION RESULTS

The generalized single Fourier series expression, with arbitrary $v_m(t)$, forms a very compact model description for PWM applications without the necessity for an accurate resolution of modulated pulse edge transitions. The simulation accuracy is easily controlled to a greater or lesser extent by adjusting the modulated carrier harmonic content through manipulation of the number of components included in second term of (4) as in the case of a periodic reference. Comparison of the three types of PWM waveform generation using the generalized Fourier method with that from a simulated analog comparator circuit in Fig. 2 to Fig. 4 reveal that the o/p simulation traces are indistinguishable from each other at 1μS intervals. These simulation records, which attests to the veracity of the generalized technique, are based on a harmonic index n of 80 using a single cycle 600 Hz signal with MI of 0.8 and a 5kHz carrier. The integrity of the method is apparent from the above simulated PWM waveforms with pulse shape retention identical to that for the analog comparator o/p in all cases. A typical BLMD model step response feedback current step simulation Ifa

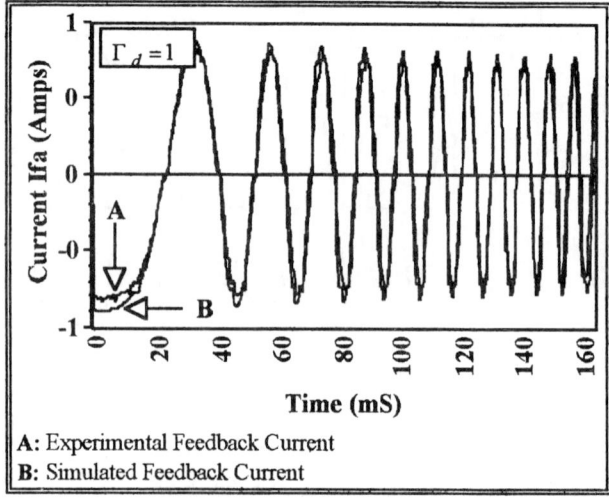

A: Experimental Feedback Current
B: Simulated Feedback Current

Figure 5 Motor step response current feedback

relying on the proposed symmetrical PWM model provides an excellent fit in terms of frequency and phase coherence to actual measurement data shown in Fig.5. A compression factor of fifty for BLMD model simulation time is readily achieved, with a fixed inverter turnon delay of 20μS taken into consideration, using the PWM Fourier method. The simulation is based on a large step size of $23\% T_s$ without impairment to accuracy by comparison with a smaller reference interval of $0.5\% T_s$.

5. CONCLUSIONS

A simplified model for three cases of natural sampled PWM based on a generalized single Fourier series expansion of a pulse train, with identical periodicity to that of the triangular/sawtooth carrier waveform, has been presented. PWM simulation waveforms, for a periodic reference, establish the reliability of the proposed Fourier method. Simulated step responses, with fiftyfold reduction achieved in computation time, of a typical BLMD system attest to the PWM model accuracy when compared with measured data.

6. REFERENCES

[1] Leu, M.C., Liu, S., and Zhang, H.: "Modelling, Analysis and Simulation of Brushless DC Drive System", Winter Meeting of ASME, 89-WA/DSC-1, Dec 1989.
[2] Bowes, S.R. and Mount, M.J.:"Microprocessor control of PWM inverters", IEE Proc., Vol. 128, Pt. B, No. 6, 1981.
[3] Black, H.S.: "Modulation Theory", Van Nostrand, 1953.
[4] Bennett, W.R., "New Results in the Calculation of Modulation Products", The Bell Sys Tech J, Vol. 12, 1933.
[5] Mirkazemi-Moud, M., Williams, B.W., and Green, T.C.: "A Novel Simulation Technique for the Analysis of Digital Asynchronous Pulse Width Modulation", IEEE Trans. Industry Applicat., Vol. 30, No. 5, Sep/Oct 1994.
[6] "MOOG Brushless Technology User Manual: D31X-XXX, T158-01X Controllers, T157-001 Power Supply", MOOG Ltd, 1989

AUTHOR INDEX

A

Aas, Einar Johan..........MPA14-3.........II-101
Abel, Andreas.............MAA7-5..........III-279
Abel, Andreas.............TAA5-2..........IV-465
Aberbour, Mourad..........WPA14-8.........III-199
Abou-Allam, Eyad..........MAA13-10........IV-373
Abshire, Pamela...........WPA14-22........III-251
Abur, Ali.................MPA12-8.........III-448
Acar, Emrah...............TPA1-3..........V-265
Achar, R..................MPA10-1.........VI-66
Achar, R..................MPA10-2.........VI-70
Acharya, Tinku............WAB6-4..........II-272
Adachi, Yoshihiro.........TAA15-14........VI-183
Adams, Michael David......MPA13-1.........IV-93
Agathoklis, Pan...........MPA1-4..........V-78
Aguirre, Miguel Angel.....WPB4-3..........VI-430
Ahamdi, Majid.............MPA15-3.........I-197
Ahmad, Ishfaq.............WAA12-6.........III-603
Ahmad, M. O...............TAB13-3.........V-217
Ahmad, M. Omair...........MAA2-5..........V-49
Ahmad, M. Omair...........WAA4-2..........IV-277
Ahmadi, Majid.............MPA3-5..........III-41
Ahmadi, Majid.............MPA15-4.........I-201
Ahmadi, Majid.............MPA15-3.........I-197
Ahn, Seung Han............MPA14-5.........II-109
Ahn, Su Jin...............WPA15-1.........II-328
Ahn, Su Jin...............WPA12-5.........III-659
Ahn, Su Jin...............MAA15-22........I-112
Ahn, Youngho..............WAA14-15........II-449
Ahola, Rami...............MPA9-1..........I-155
Aizawa, Kiyoharu..........TPA15-8.........VI-637
Akansu, Ali N.............MPA2-6..........V-118
Akers, Lex................WPA14-9.........III-203
Akers, Lex A..............MPA3-2..........III-33
Akin, I.A.................TPA14-7.........I-397
Akkarakaran, Sony.........MAA2-6..........V-53
Aksin, Devrim Yilmaz......MAA15-15........I-84
Al-Hashimi, Bashir........MAA15-23........I-115
Al-Hashimi, Bashir M......WAA13-24........V-445
Al-Hashimi, Bashir M......WAA15-6.........VI-294
AL-Jumah, Abdullah........WPA6-8..........II-304
Alattar, Adnan Mohammed...TPA13-5.........IV-249
Albicki, Alexander........MAA14-12........III-338
Alippi, Cesare............TPA3-4..........III-103
Alku, Paavo...............TAA4-2..........IV-194
Allen, Jont B.............TAA11-1.........VI-565
Allen, Phillip E..........MAA13-22........IV-421
Aloqeely, Mohammed A......WPA4-2..........VI-377
Alquie, G.................MAA9-4..........VI-13
Alvandpour, Atila.........WAA6-3..........II-252
Álvarez-Marquina, Agustín.TAA13-5.........V-178
Alves, Vladimir Castro....MAA12-4.........II-45
Amirtharajah,, Rajeevan...WPA5-6..........IV-604
Amourah, Mezyad M.........MPA9-5..........I-171
Andreou, Andreas..........MAA15-4.........I-551
Andreou, Andreas..........MPA13-1.........IV-93
Antoniou, Andreas.........MAA2-3..........V-41
Antoniou, Andreas.........MPA2-2..........V-102
Apsel, Alyssa.............WAA3-1..........III-107
Aravena, Jorge L..........WPA1-1..........V-449
Arena, Paolo..............WPA3-7..........III-163
Arik, Sabri...............WPA14-5.........III-187
Arikan, Orhan.............TPA1-3..........V-265
Arnaud, Alfredo...........MPA15-21........I-269
Arokia, Nathan............WPA11-4.........VI-401
Aronhime, Peter...........MAA15-4.........I-41
Arrigo, Jeanette F........MAA13-21........IV-417
Arsintescu, Bogdan G......TAA12-8.........VI-126
Arslan, T.................WPA6-8..........II-304
Arslan, Tughrul...........TAA15-3.........VI-139

Arslan, Tughrul...........TAB13-8.........V-237
Ashton, R.W...............MAA14-24........III-387
Ashton, R.W...............TAB14-2.........III-489
Ashton, R.W...............MAA14-13........III-342
Ashton, R.W...............MAA14-22........III-379
Ashton, R.W...............TAB14-11........III-526
Asmanis, Georgios.........WAA8-3..........I-444
Au, Oscar C...............WPA13-9.........V-550
Au, Oscar C...............WAA12-9.........III-615
Au, Oscar C...............WAA12-9.........VI-470
Au, Oscar C...............WAA4-3..........IV-281
Au, Oscar C...............MPA13-8.........IV-122
Auvergne, Daniel..........MPA10-7.........VI-90
Azemard-Crestani, Nadine..MPA10-7.........VI-90

B

Baccarani, Giorgio........WAA3-6..........III-127
Baccarani, Giorgio........MPA15-15........I-245
Baena, Vicente............WPB4-3..........VI-430
Báez-López, David J.M.....MAA15-12........I-72
Bakken, Tim W.............WPA12-1.........III-643
Baltes, Henry.............WPA11-5.........VI-405
Banerjee, Prithviraj......WPA4-1..........VI-372
Banerjee, Prithviraj......MAA10-7.........VI-57
Barbaro, Massimo..........WPA14-14........III-219
Barker, Clashow M.........MPA12-2.........III-427
Baru, Marcelo Daniel......MPA15-21........I-269
Baschirotto, Andrea.......MPA6-2..........II-69
Baschirotto, Andrea.......TAA14-10........II-220
Bastos-Filho, Teodiano Freire..TPA11-4..VI-602
Basu, Sankar..............MPA1-3..........V-74
BatteryWala, Shabbir Hussain...MPA10-5..VI-82
Baturone, Iluminada.......WPA8-3..........I-520
Bauer, Andreas............MAA7-4..........III-275
Bax, Walt T...............TPA5-2..........IV-498
Bayoumi, Magdy............WPA4-3..........VI-381
Bayoumi, Magdy A..........MAA11-8.........IV-57
Bayoumi, Magdy A..........MPA14-16........II-153
Becker, Bernd.............TAA15-15........VI-187
Bego, Lauro Jardim........MAA12-7.........II-57
Begovich, O...............TAB14-7.........III-510
Benabes, Philippe.........WAA15-1.........VI-274
Benboudjema, Kamel........MAA9-4..........VI-13
Bendiscioli, Paolo........TPA15-5.........VI-625
Benini, Luca..............MAA6-2..........II-5
Beraldin, J-Angelo........WPA2-2..........V-488
Berg, Yngvar..............MAA12-2.........II-37
Berns, Daniel W...........WPA7-1..........III-619
Bertazzoni, Stefano.......TAB13-1.........V-210
Besl, Paul J..............WPA2-1..........V-484
Bisdounis, Labros.........WAA15-18........VI-342
Bistritz, Yuval...........MPA1-8..........V-94
Bitter, D.................MPA1-2..........V-70
Black, William C..........WPA11-3.........VI-397
Black, William C..........MPA9-4..........I-167
Bo, G.M...................MPA3-6..........III-46
Bobba, Sudhakar...........MAA10-1.........VI-33
Boche, Holger.............MAA14-9.........III-326
Boemo, Eduardo............TAB6-4..........II-240
Bohanon, Jon..............WPA5-2..........IV-585
Bonet-Dalmau, Jordi.......TAA7-3..........III-460
Bosse, Eloi...............MPA11-9.........VI-561
Bouaziz, Samir............WPB4-4..........VI-434
Bouguet, Jean-Yves........WPA2-2..........V-494
Boukadoum, M..............MAA9-4..........VI-13
Bouridane, Ahmed..........TAB13-4.........V-221
Brambilla, Angelo.........WAA14-1.........II-394
Branciforte, Marco........WPA3-7..........III-163
Brennan, Robert...........TAA11-2.........VI-569
Bresch, Helmut............MPA15-5.........I-205
Bright, Marc Stephen......MAA10-4.........VI-45

Briozzo, Luciano..........TPA3-4..........III-103
Brodersen, Robert.........WPA5-4..........IV-593
Brown, Donald E...........MPA1-5..........VI-549
Bruun, Erik...............WPA9-8..........I-567
Buhmann, Sitta............WAA2-6..........V-345
Bull, David R.............WAA13-19........V-429

C

Cai, Jianfei..............MPA13-5.........IV-110
Campolucci, Paolo.........TAA3-6..........III-82
Cantin, Marc-Andre........WAA14-17........II-458
Cardarilli, G.C...........WAA8-5..........I-452
Cardarilli, Gian Carlo....WAA14-6.........II-414
Cardarilli, Gian Carlo....TAB13-1.........V-210
Carlosena, Alfonso........TAA14-7.........II-208
Carmeli, S................TAB14-8.........III-514
Carneiro, Noel Carlos F...WAA14-14........II-445
Carreto-Castro, M.F.......MPA6-5..........II-81
Carro, Luigi..............MAA12-7.........II-57
Carroll, Thomas L.........WAA10-2.........IV-558
Carroll, Thomas L.........WAA10-3.........IV-564
Carvalho, Delmar B........WPA15-14........II-355
Castleman, Kenneth R......WAA2-1..........V-325
Castleman, Kenneth R......WAA2-2..........V-329
Cattet, Stephane..........MAA15-11........I-69
Cauwenberghs, Gert........MAA3-6..........III-17
Cauwenberghs, Gert........MPA3-9..........III-58
Cauwenberghs, Gert........WPA14-22........III-251
Cauwenberghs, Gert........WPA3-3..........III-147
Cauwenberghs, Gert........WAB7-3..........I-508
Caverly, Robert H.........MAA13-1.........IV-337
Caviglia, D.D.............MPA3-6..........III-46
Cetin, A. Enis............WPA1-8..........V-480
Cha, Jin-Jong.............TPA1-1..........V-257
Chae, Soo-Ik..............TPA14-2.........I-377
Chaiken, Seth.............MPA7-2..........III-395
Chakrabarti, Chaitali.....WAA14-12........II-437
Chan, Cheong-Fat..........WPA15-19........II-371
Chan, Chi-Kwong...........WPA14-4.........III-183
Chan, Chung-Kei Thomas....WPA11-8.........VI-417
Chan, Philip C.H..........MAA12-1.........II-33
Chan, Shing Chow..........WPA13-3.........V-546
Chan, Shueng-Han Gary.....MPA4-1..........IV-61
Chan, Yuk-Hee.............MAA4-5..........IV-17
Chandra, Charu............TAA4-5..........IV-206
Chandrakasan, Anantha.....WPA5-6..........IV-604
Chandramouli, Rajarathnam.MAA5-4..........IV-317
Chandramouli, Rajarathnam..MPA13-17.......IV-158
Chang, Chip-Hong..........TAA15-11........VI-171
Chang, Chip-Hong..........TAA15-10........VI-167
Chang, Fung-Yuel..........WPA11-8.........VI-417
Chang, Hao-Chieh..........MPA13-11........IV-134
Chang, Hao-Chieh..........MPA4-2..........IV-65
Chang, Hao-Chieh..........WPA13-4.........V-530
Chang, Hao-Chieh..........MAA11-4.........IV-41
Chang, Hao-Chieh..........MPA13-21........IV-174
Chang, Hun-Hsien..........TAA14-9.........II-216
Chang, Hyun Man...........WAA14-10........II-429
Chang, Mao-Lin (Molin)....WAA15-14........VI-326
Chang, Robert Chen-hao....WPB13-5.........VI-502
Chang, Shue-Lee...........TAA13-12........V-206
Chang, Tian-Sheuan.......WAA14-2..........II-398
Chang, Tsun-Chen..........WAA14-3.........II-402
Chang, Yao-Wen............TAA15-1.........VI-131
Chantaporncrhai, Chantana.WAA11-8.........VI-270
Chao, K.S.................WPA10-4.........I-583
Charbon, Edoardo..........TAA12-8.........VI-126
Charry, Edgar.............WPA15-2.........II-312
Chatzigeorgiou, Alexander.WAA15-24........VI-368
Chatzigeorgiou, Alexander.WAA15-23........VI-363
Chau, Lap-Pui.............TAB13-5.........V-225

AUTHOR INDEX

Chau, Paul M.............MAA13-21.......IV-417
Chen, Chang Wen.........MPA13-5.......IV-110
Chen, Chang Wen.........MAA4-3.......IV-9
Chen, Chien-In Henry......MPA6-4.......II-77
Chen, Chien-Yau..........WPA14-17.......III-231
Chen, Chuen-Yau..........WPA14-20.......III-243
Chen, Dong...............WPA7-6.......III-635
Chen, Fei................TAA15-21.......VI-207
Chen, Feng...............WAB7-1.......I-500
Chen, Gang...............TAB13-2.......V-213
Chen, Guanrong...........MAA14-6.......III-318
Chen, Guanrong...........WPA7-7.......III-639
Chen, Guanrong...........WPA7-1.......III-619
Chen, Hsu-Tung...........MPA13-21.......IV-174
Chen, Jian-Song..........MPA15-14.......I-241
Chen, Jie................MPA13-10.......IV-130
Chen, Jiunn-Tsair........MAA5-5.......IV-321
Chen, Kuang-Yuan.........TPA12-2.......III-554
Chen, Kun-Nern...........WAA14-3.......II-402
Chen, Li.................MPA13-14.......IV-146
Chen, Liang-Gee..........WPA13-4.......V-530
Chen, Liang-Gee..........MPA4-2.......IV-65
Chen, Liang-Gee..........MAA11-4.......IV-41
Chen, Liang-Gee..........MPA13-21.......IV-174
Chen, Oscal T.-C.........MAA13-24.......IV-429
Chen, Oscal T.-C.........WAA13-16.......V-417
Chen, Oscal T.-C.........TAA4-7.......IV-213
Chen, Oscal T.-C. Chen...TAA4-8.......IV-217
Chen, Pei-Yin............MPA13-13.......IV-142
Chen, Richard M M........WAA15-20.......VI-350
Chen, Sau-Gee............TAA1-4.......V-142
Chen, Sau-Gee............TAB13-6.......V-229
Chen, Sau-Gee............TPA4-3.......IV-225
Chen, Shung-Chih.........MAA10-6.......VI-53
Chen, Tsuhan.............WAA12-7.......III-607
Chen, Tsuhan.............MAA11-6.......IV-49
Chen, Tung-Yang..........TAA14-8.......II-212
Chen, Xiang..............WPA7-2.......III-623
Chen, Yiqin..............MPA9-6.......I-176
Cherry, James A..........WPA10-7.......I-596
Cherry, James A..........WPA10-5.......I-587
Cheung, Chok-Kwan........WAA4-7.......IV-297
Cheung, Chun-Ho..........MPA13-4.......IV-106
Cheung, Kwok-Wai.........MPA13-4.......IV-106
Cheung, Paul Y.S.........WPA6-4.......II-288
Chiang, David, H.........MPA8-1.......I-123
Chiang, Jen-Shiun........TPA12-2.......III-554
Chiang, Jen-Shiun........MPA14-20.......II-169
Chiang, Jen-Shiun........MPA14-21.......II-173
Chibl`e, H...............MPA3-6.......III-46
Chickamenahalli, Shamala A..MPA12-2..III-427
Chilakapati, Uma.........WAA9-7.......I-492
Chin, Wai................WPA7-6.......III-635
Ching, Pak-Chung.........MAA1-3.......V-9
Chiou, Li-Yu.............TAA8-2.......I-285
Chiricescu, Silviu.......TAB6-1.......II-232
Cho, Kwang-Bo............MPA4-4.......IV-73
Choi, Jinho..............MAA1-5.......V-17
Choma, John..............WPA10-3.......I-579
Choma, Jr., John.........MPA15-12.......I-233
Choma, Jr., John.........WPA12-1.......III-643
Chong, Jong-Hwa..........WPA15-3.......II-316
Chou, Mike...............WAA15-22.......VI-358
Chowdhury, Mohamed Foysol..TPA15-1.VI-606
Christensen, Kåre Tais...MAA13-7.......IV-360
Chrzanowska-Jeske,M.....TAA15-12.......VI-175
Chu, Xuedao..............TAA15-24.......VI-219
Chuang, Justin C-I.......TPA5-3.......IV-502
Chun, Byungjin...........TAA13-10.......V-198
Chung, Henry S H.........WAA7-3.......III-574

Chung, Henry S H.........TAA2-1.......VI-438
Chung, Henry S H.........WAA7-1.......III-566
Chung, Yon Jun...........WPB13-9.......VI-518
Chung-Yuk, Or............WAA15-8.......VI-302
Ciezki, John G...........MAA14-13.......III-342
Ciezki, John G...........TAB14-2.......III-489
Ciezki, John G...........MAA14-22.......III-379
Ciezki, John G...........MAA14-24.......III-387
Ciezki, John G...........TAB14-11.......III-526
Cijvat, Ellie............MPA6-1.......II-65
Cilingiroglu, Ugur.......MAA15-15.......I-84
Ciminiera, Luigi.........MAA13-16.......IV-397
Ciocoiu, Iulian B........TAA3-8.......III-86
Cioffi, John M...........WAA12-3.......III-590
Clements, Mark...........WAA14-20.......II-470
Coffman, James W.........MAA11-1.......IV-33
Cohen, Marc..............MAA3-6.......III-17
Cohen, Marc H............WPA14-22.......III-251
Colbeth, Richard E.......TPA15-6.......VI-629
Colodro, Francisco.......WPA14-11.......III-211
Connelly, J. Alvin.......TAA14-3.......II-197
Conti, Massimo...........WPA9-2.......I-543
Copeland, Miles A........TPA5-2.......IV-498
Cornish, Jack............MPA15-12.......I-233
Corron, Ned J............WAA10-6.......IV-576
Cortelazzo, G.M..........WPA2-9.......V-518
Costa, Joao Paulo........WAA15-22.......VI-358
Cousseau, Juan E.........WAA1-7.......V-317
Cremoux, Severine........MPA10-7.......VI-90
Crippa, Paolo............WPA9-2.......I-543
Criscione, M.............WAA3-7.......III-131
Czarkowski, Dariusz......TAA2-3.......VI-446

D

Dachselt, Frank..........TPA9-3.......IV-518
Dai, Liang...............WAA9-3.......I-476
Damera-Venkata, Niranjan....WAA15-4.......VI-286
Damper, Robert Ivan......WAA3-3.......III-115
Darley, Merrick H........WAA6-4.......II-256
Darley, Merrick H........WAA6-2.......II-248
Das, Bodhisattva.........WPA11-3.......VI-397
Davies, Anthony C........MAA14-1.......III-298
Davies, Anthony Christopher....TPA7-2...III-538
Davis, Alan J............WPA10-8.......I-600
Davis, Dennis............WAA1-2.......V-297
de Figueiredo, Rui J.P...MAA14-19.......III-366
de Figueiredo, Rui J.P...MAA3-1.......III-1
De Lima, Jader Alves.....WPA15-20.......II-374
de Medeiros, Manoel Firmino..TAB14-6..III-505
de Queiroz, Antonio C. M....WPA15-21...II-378
Debevec, Paul Ernest.....WPA2-8.......V-514
Debyser, Geert...........MAA9-6.......VI-21
Dec, Aleksander M........MAA13-9.......IV-369
Dec, Aleksander M........MPA15-10.......I-225
Dedieu, Herve............MAA14-4.......III-310
Dedieu, Herve............TPA9-4.......IV-522
Degrugillier, Dominique..MAA15-9.......I-61
Delgado-Restituto, Manuel....TAA5-4.......IV-477
Delgado-Restituto, Manuel....TAA5-4.......IV-473
Demosthenous, Andreas....MPA15-8.......I-217
Demosthenous, Andreas....MPA15-6.......I-209
Deng, Jie................MAA15-4.......I-41
Deng, Yining.............WAA12-4.......III-595
Deng, Yining.............WAA12-4.......VI-450
Deprettere, Ed F.........TAB13-10.......V-245
Deshpande, Sachin G......WAA12-8.......VI-466
Deshpande, Sachin G......WAA12-8.......III-611
Deshpande, Sachin G......MPA13-6.......IV-114
Deutschmann, R. A........TPA15-11.......VI-649
DeWeerth, Stephen P......WAA3-2.......III-111
DeWeerth, Stephen P......WPA14-16.......III-227

Di Grazia, Pietro........WPA3-7.......III-163
Diepenhorst, Marco.......WAA14-18.......II-462
Diniz, Paulo S...........WAA1-7.......V-317
Diniz, Paulo Sergio Ramirez...WAA13-3...V-363
Diniz, Paulo Sergio Ramirez....TAA13-1..V-162
Diorio, Chris............MPA3-1.......III-29
Diorio, Chris............TPA3-1.......III-90
Diorio, Chris............MAA15-16.......I-88
Diorio, Chris............MPA8-3.......I-131
Diorio, Chris............WPA8-5.......I-527
Diorio, Chris............WPA15-13.......II-351
Djahanshahi, Hormoz......MPA3-5.......III-41
Dmitriev, Alexander S....TAA5-6.......IV-481
Doblinger, Gerhard.......WAA1-4.......V-305
Dobrovolny, Petr.........MAA9-8.......VI-29
Donate, Pedro D..........WAA1-7.......V-317
Doretto, G...............WPA2-9.......V-518
Drakakis, E.M............TAA9-3.......I-317
Draper, Jeffrey..........WPA10-3.......I-579
Drechsler, Rolf..........TAA15-15.......VI-187
Du, David H. C...........TAA15-24.......VI-219
Du, Min..................MAA14-17.......III-358
Dudek, Frank.............WAA15-6.......VI-294
Dufort, Benoit...........WPA6-3.......II-284
Dujardin, Eric...........WAA13-20.......V-433
Dunlap, Steven K.........MPA15-1.......I-189
Dyer, Kenneth............MAA8-5.......I-13

E

Ebrahimi, Touradj........WPA2-4.......V-498
Edwards, Brent W.........TAA11-7.......VI-586
Edwards, Robert Timothy...WPA12-3...III-651
El-Masry, Ezz I..........TAA9-2.......I-313
El-Shafei, A. A.-R. H....WPA3-8.......III-167
Elmasry, Mohamed I.......MAA6-3.......II-9
Elwan, Hassan O..........WPA15-8.......II-335
Embabi, S.H.K............WPA8-2.......I-518
Endo, Tetsuro............TAA7-8.......III-481
Endo, Tetsuro............MAA14-11.......III-334
Endo, Tetsuro............MAA14-3.......III-306
Enz, Christian C.........TAA9-6.......I-329
Erdogan, Ahmet Teyfik....WAA13-23.......V-441
Erdogan, Ahmet Teyfik....WAA13-18.......V-425
Eriksson, Patrik.........MPA6-1.......II-65
Eshraghi, Aria...........WPA15-23.......III-386
Eskikurt, Halil Ibrahim..TAB13-8.......V-237
Etawil, Hussein A........WAA11-5.......VI-258
Evans, Brian L...........TPA2-2.......V-277
Evans, Brian L...........WAA15-4.......VI-286

F

Falkowski, Bogdan J......TAA15-11.......VI-171
Falkowski, Bogdan J......TAA1-3.......V-138
Falkowski, Bogdan J......TAA15-10.......VI-167
Fang, L..................WPA10-4.......I-583
Fant, Karl...............MAA12-8.......II-61
Farooqui, Aamir Alam.....WAB6-1.......II-260
Faura, Julio.............WPB4-3.......VI-430
Favalli, Lorenzo.........TPA13-8.......IV-261
Fedi, Giulio.............MAA9-3.......VI-9
Feely, Orla..............MAA14-15.......III-350
Femia, Nicola............TAA2-6.......VI-452
Feng, Gang...............MAA1-7.......V-25
Feng, Gang...............WPB13-4.......VI-498
Fernandez, Francisco V...MAA9-5.......VI-17
Ferrer, Enrique..........WPA12-6.......III-663
Ferri, G.................WAA8-5.......I-452
Ferri, Giuseppe..........WAA8-1.......I-436
Fettweis, Gerhard........MAA5-2.......IV-309
Fidler, J. Kel...........MAA15-18.......I-96
Fiez, Terri..............WAA9-7.......I-492

AUTHOR INDEX

Fiez, Terri................MPA15-1......I-189
Fiez, Terri................WPA10-6......I-591
Fiez, Terri................TAA10-6......I-360
Filanovsky, Igor M.......TPA15-4......VI-621
Filiol, Norman M.........TPA14-10.....I-408
Fiori, Simone............MPA5-4.......IV-441
Fiori, Simone............TAA3-2.......III-66
Fischer, Godi............WPA10-8......I-600
Fischer, Wolf-Joachim....WPA6-1.......II-276
Fitch, Osa...............WPA7-5.......III-631
Flank, Steven M..........MPA11-8......VI-557
Fong, Wai Ching..........TPA13-2......IV-237
Fortuna, Luigi...........WPA3-7.......III-163
Fox, Robert M............TPA12-1......III-550
Fox, Robert M............WPA12-6......III-663
Franca, Felipe Maia Galvao....MAA12-4...II-45
Franca, Jose E...........MAA9-2.......VI-5
Franca, Jose E...........WAA9-2.......I-472
Franca, Jose E...........WAA15-8......VI-302
Franca, José E...........TAA14-12.....II-228
Franca, Jose Epifaneo da......TAA9-5...I-325
Francesconi, Fabrizio....TPA15-5......VI-625
Franchi, Eleonora........WAA3-6.......III-127
Franzon, Paul............WAA14-20.....II-470
Frazer, Mark J...........TAA15-16.....VI-191
Freitas, Roger Alex de Castro...TPA11-4..VI-602
Freking, Robert A........WAA14-16.....II-453
Frey, Douglas............TAA9-4.......I-321
Friedman, Eby G..........WAA6-1.......II-244
Fu, Jyun-Horng (Alex)....WPA7-4.......III-627
Fujii, Nobuo.............WAA13-11.....V-395
Fukui, Yutaka............TAA13-3......V-170
Fukui, Yutaka............MAA13-23.....IV-425
Fukui, Yutaka............MAA13-15.....IV-393
Furukawa, Toshihiro......WPA13-2......V-526

G
Galias, Zbigniew.........MAA14-20.....III-370
Galias, Zbigniew.........WAA10-4......IV-568
Gallivan, Kyle A.........WAA1-5.......V-309
Galton, Ian..............WAB7-2.......I-504
Galton, Ian..............TAA10-7......I-365
Galup-Montoro, Carlos....MAA15-2......I-33
Galvez-Durand, Federico..MAA15-17.....I-92
Gandhi, Rajeev...........WAA13-17.....V-421
Garcia Franquelo, Leopoldo...MPA12-5..III-435
Garrido, N. M. de F......TAA9-5.......I-325
Gatti, Umberto...........MAA8-6.......I-17
Gay-Bellile, Olivier.....WAA13-20.....V-433
Ge, Yongmin..............MAA13-4......IV-349
Geiger, Randall..........WPA9-5.......I-555
Geiger, Randall..........MPA9-6.......I-176
Gerber, Martin...........TAA15-7......VI-155
Gerek, Omer Nezih........WPA1-8.......IV-480
Giakoumis, Ioannis.......TPA13-10.....IV-269
Gielen, Georges..........MAA9-6.......VI-21
Gierkink, Sander L.J.....MPA9-8.......I-185
Gilli, Marco.............WPA12-8......III-671
Gingras, Donald F........WAA5-1.......IV-526
Giomi, Riccardo..........MAA9-3.......VI-9
Girard, Patrick..........WPA6-6.......II-296
Giustolisi, Gianluca.....MPA15-20.....I-265
Glenn, Ian N.............MPA11-3......VI-545
Gnudi, Antonio...........MPA15-15.....I-245
Goel, Manish.............TPA6-1.......VI-474
Goessi, Thomas...........TAA15-7......VI-155
Goetz, Marco.............TAA5-2.......IV-465
Goetz, Marco.............MAA7-5.......III-279
Goh, Chee Kiang..........TAA1-8.......V-158
Goldgeisser, Leonid B....TPA12-3......III-558
Gomez-Castaneda, Felipe..WPA14-1......III-171
Gómez-Vilda, Pedro.......TAA13-5......V-178
Gondim, P. R. de L.......WAA5-4.......IV-538
Gonzalez-Altamirano, G...WPA15-22.....II-382
Goodman, James...........WPA5-6.......IV-604
Gothenberg, Andreas......TPA14-8......I-401
Granja, Edson do Prado...MAA12-4......II-45
Grassi, Giuseppe.........MAA7-6.......III-283
Grayver, Eugene..........MAA13-14.....IV-389
Grayver, Eugene..........TAB13-9......V-241
Green, Michael M.........TPA12-3......III-558
Green, Michael M.........MAA15-5......I-45
Grogan, Paul.............WAA14-5......II-410
Guaitini, Giovanni.......WPA9-2.......I-543
Guendouz, Hassina........WPB4-4.......VI-434
Gui, Xiang...............WAA5-2.......IV-530
Guinee, Richard Anthony..TPA15-12.....VI-653
Guo, Jyh-Huei............WAA14-22.....II-478
Guo, Jyh-Huei............MPA14-22.....II-177
Gurkaynak, Frank Kagan...MPA14-4......II-105
Gustavsson, Mikael.......MAA8-7.......I-21

H
Haenggi, Martin..........WPA3-5.......III-155
Hafed, Mohamed Mohamed...MPA10-6......VI-86
Hahs, Daniel W...........WAA10-6......IV-576
Hajimiri, Ali............MPA14-15.....II-149
Hajj, Ibrahim N..........MAA10-1......VI-33
Hajj, Ibrahim N..........MAA6-1.......II-1
Hajjar, Ara..............WPA6-2.......II-280
Hakkinen, Juha Tapio.....MAA13-20.....IV-413
Hall, David L............MPA11-2......VI-541
Hall, David L............MPA11-1......VI-537
Halonen, Kari............WPA8-1.......I-512
Halonen, Kari............WPA14-23.....III-255
Halonen, Kari............MPA9-2.......I-159
Halonen, Kari............MPA9-7.......I-181
Halonen, Kari............MPA15-17.....I-253
Halonen, Kari............MAA13-19.....IV-409
Halonen, Kari............MPA9-1.......I-155
Halonen, Kari A..........MPA6-3.......II-73
Halonen, Kari A..........MPA15-18.....I-257
Halverson, Ranette H.....TAA15-4......VI-143
Hamamoto, Takayuki.......TPA15-8......VI-637
Hamilton, Alister........MPA15-13.....I-237
Hamilton, Samuel Norman..TAA14-1......II-189
Han, kyungtae............MAA13-13.....IV-385
Hang, Hsueh-Ming.........MPA13-3......IV-102
Hang, Hsueh-Ming.........MPA4-6.......IV-81
Hanna, Magdy Tawfik......TAA1-1.......V-130
Haridasan, Radhakrishnan.WAA4-5.......IV-289
Harjani, Ramesh..........WAA9-3.......I-476
Harnefors, Lennart.......WAA13-8......V-383
Harris, John G...........MPA15-23.....I-277
Harris, John G...........MPA15-2......I-193
Harris, John G...........TAB14-3......III-493
Harrison, Reid R.........TAA14-6......II-204
Hartimo, I.O.............TAB13-11.....V-249
Hasan, Moh'd Abdel Majid.TPA13-4......IV-245
Hasan, Moh'd Abdel Majid.MPA13-18.....IV-162
Hasegawa, Akio...........MAA14-3......III-306
Haseyama, Miki...........WPA14-24.....III-259
Haseyama, Miki...........TAB13-12.....V-253
Hasler, Martin...........MAA7-3.......III-271
Hasler, Paul.............TPA3-1.......III-90
Hasler, Paul.............MPA8-3.......I-131
Hasler, Paul.............WPA8-5.......I-527
Hasler, Paul.............WPA15-13.....II-351
Hasler, Paul.............MAA15-16.....I-88
Hasler, Paul.............MPA3-1.......III-29
Hasler, Paul.............WAA3-1.......III-107
Hasler, Paul.............TAA14-6......II-204
Hassoun, Marwan M........MAA9-1.......VI-1
Hatori, Mitsutoshi.......TPA15-8......VI-637
Hatzopoulos, Alkiviades A.WPA15-5......II-324
Hauck, Oliver Friedrich..WPA13-7......V-542
Hauschild, R.............TPA15-10.....VI-645
Hayashi, Takanori........MPA12-4......III-431
Hayatleh, K..............WPA8-8.......I-539
He, Yong.................WAA12-6......III-603
Hegazi, Emad Mahmoud.....WPA12-2......III-647
Helfenstein, Markus......MAA15-1......I-29
Hematy, Arman............TAA9-1.......I-309
Henkelmann, Heiko........MPA14-12.....II-137
Hentschel, Tim...........MAA5-2.......IV-309
Herpers, Rainer..........WAA2-7.......V-349
Herrera, Ruben D.........TPA7-4.......III-546
Hill, Anthony............WAA6-2.......II-248
Hinamoto, Takao..........MPA1-7.......V-90
Hinamoto, Takao..........WPA14-19.....III-239
Hinamoto, Takao..........MPA1-5.......V-82
Hiraiwa, Atsunobu........MPA13-2......IV-97
Hiskens, Ian A...........MPA12-6......III-439
Ho, Chun ying, Murphy....TAA3-4.......III-74
Hocevar, Dale E..........TAA15-19.....VI-203
Hölling, Matthias........TAA13-4......V-174
Hong, Chang-Yu...........TPA4-2.......IV-221
Horta, Nuno Cavaco Gomes.MAA9-2.......VI-5
Hosticka, B.J............TPA15-10.....VI-645
Hsieh, Hong-Yean.........WAA14-20.....II-470
Hsieh, Jeff Yeu-Farn.....WAA4-8.......V-301
Hsu, Jah-Ming............MPA5-5.......IV-445
Hsu, Yaun-chung..........TAA15-24.....VI-219
Hua, Jia.................TAA15-4......VI-143
Huang, A.................TAB13-11.....V-249
Huang, Chung-Lin.........MAA3-5.......III-13
Huang, Jiwu..............MPA13-14.....IV-146
Huang, Kuang-Hu..........WPB13-2......VI-490
Huang, Po-Chiun..........TAA8-2.......I-285
Huang, Sheng-Chieh.......MPA13-21.....IV-174
Huang, Sheng-Chieh.......WPA13-4......V-530
Huang, Sheng-Chieh.......MPA4-2.......IV-65
Huang, Yih-Fang..........MAA4-2.......IV-5
Huang, Yuejin............WAA5-5.......IV-542
Huber, Andreas...........WAA11-6......VI-262
Hudson, Forrest..........TAA10-4......I-352
Hui, Ronny C. C..........MAA8-4.......I-9
Hung, Ching-Yu...........TAA15-19.....VI-203
Huppertz, J..............TPA15-10.....VI-645
Huss, Sorin Alexander....WPA13-7......V-542
Hutchison, David.........TPA6-4.......VI-482
Hwang, Inchul............TPA13-1......IV-233
Hwang, Jenq-Neng.........WAA12-8......III-611
Hwang, Jenq-Neng.........WAA12-6......III-466
Hwang, Jenq-Neng.........MPA13-6......IV-114
Hwang, Kuo-Fuo...........WAA13-6......V-375
Hwang, Rain-Ted..........WPA6-5.......II-292

I
Iannuccelli, Manuele.....TAB13-1......V-210
Ichige, Koichi...........WPA1-2.......V-453
Igarashi, Ryo............MAA14-3......III-306
Iizuka, Fumitaka.........WAA7-2.......III-570
Ikeda, Hiroaki...........WAA7-4.......III-578
Ikehara, Masaaki.........WAA13-5......V-371
Ikehara, Masaaki.........MPA2-8.......V-126
Ikehara, Masaaki.........TAA1-2.......V-134
Inagaki, Shuichiro.......TPA11-3......VI-598
Ishii, Junya.............TAB14-9......III-518
Ishii, Rokuya............WPA1-2.......V-453
Ishikawa, Jun-ich........WAA7-4.......III-578
Ismail, Mohammed.........WPA15-7......II-331
Ismail, Mohammed.........WAA8-4.......I-448

AUTHOR INDEX

Ismail, Mohammed............WPA15-8.......II-335
Ismail, Yehea....................WAA6-1.......II-244
Isshiki, Tsuyoshi............WAA14-21.......II-474
Itoh, Yoshio..................MAA13-15.......IV-393
Itoh, Yoshio....................TAA13-3.......V-170
Iwata, Atsushi...................TPA14-5.......I-389
Izquierdo, Ebroul.............WPA2-5.......V-502
Izumi, Tomonorim..........WAA11-2.......VI-244

J

Jain, V.K............................WAB6-2.......II-264
Jain, Vijay K....................WPA1-7.......V-473
Jakimoski, Goce..............TPA9-2.......IV-514
Jako, Zoltan.......................TAA5-1.......IV-461
Jamali, M.M.......................MPA1-2.......V-70
James, S..........................WPA14-9.......III-203
Jannesari, S...................MAA14-21.......III-374
Jen, Chein-Wei...............WAA14-2.......II-398
Jenkins, W. Kenneth........WAA1-5.......V-309
Jenkins, W. Kenneth........WAA1-3.......V-301
Jenn, David........................TPA11-1.......VI-590
Jensen, Henrik..................WAB7-2.......I-504
Jeong, Gab Joong...........MPA14-5.......II-109
Jeschke, Hartwig.............TAA15-6.......VI-151
Jian, Ming......................WPA13-10.......V-554
Jian, Ming.........................WAA1-1.......V-293
Jiang, Hai-Yun..................TPA1-4.......V-269
Jiang, Hsin-Chin..............MPA4-5.......IV-77
Jiang, Li........................WAA14-21.......II-474
Jiang, Xicheng....................MAA8-3.......I-5
Jiao, Licheng.....................WAA3-8.......III-135
Jiao, Licheng.....................WAA3-4.......III-119
Jin, Liang..........................MAA5-8.......IV-333
Jin, Liang..........................MAA5-7.......IV-329
Jinno, Kenya......................TAA7-7.......III-477
Jinno, Kenya......................TAA7-6.......III-473
Jiu, Juing-Ying..................MAA11-4.......IV-41
Jo, Han-Cheol.................WPA15-3.......II-316
Joergensen, Allan............MAA13-7.......IV-360
Johansson, Hakan..........WAA13-22.......V-437
Johansson, Henrik O......WPA15-1.......II-308
Johns, David A..................TAA8-8.......I-305
Johns, David A..................MAA5-3.......IV-316
Johnson, Andrew Edie....WPA2-6.......V-506
Johnson, Gregg A...........WAA10-2.......IV-558
Joho, Marcel.....................TAA13-2.......V-166
Johsnsson, Haken............MAA2-4.......V-45
Jones, Bill......................WPA15-1.......II-308
Jonsson, Bengt E..............WPA8-4.......I-524
Jordan, Fred D.................WPA2-4.......V-498
Jou, Jer Min...................MPA13-13.......IV-142
Jou, Jer Min....................MAA10-6.......VI-53
Ju, Wann-Shyang.............MPA5-6.......IV-449
Julián, Pedro.................WAA15-11.......VI-314
Jullien, Graham A.............MPA3-5.......III-41
Jullien, Graham A...........MPA15-3.......I-197
Jun, Jong Arm..................WPB13-7.......VI-510
Jun, Sibum.....................WPA12-5.......III-659
Jun, Sibum.....................MAA15-22.......I-112
Jun, Sibum......................WPA15-6.......II-328
Junji, Kawata..................MAA14-4.......III-310

K

Kageyuki, Kiyose..............WAA13-5.......V-371
Kaiser, Andreas..............MAA15-14.......I-80
Kajitani, Yoji...................WAA11-2.......VI-244
Kamada, Masaru..............WPA1-2.......V-453
Kan, Kai-chiu.....................TAA3-3.......III-70
Kanan, Riad.....................TAA14-2.......II-193
Kananen, Asko Tapio....WPA14-23.......III-255
Kanata, Yakichi.................TPA3-2.......III-94

Kandlur, Dilip D...............WAA12-1.......III-582
Kang, In.........................WPA15-3.......II-316
Kang, Wei.......................WPA7-5.......III-631
Kanoun, Olfa....................TPA15-3.......VI-617
Kansara, M......................TAB14-10.......III-522
Kao, Chi-Chou.................WAA14-3.......II-402
Kao, Chia-Hsiung.............WPA6-5.......II-292
Kao, Hong-Sing...............MAA13-11.......IV-378
Kao, Min-Chi.....................TAA1-4.......V-142
Kao, William....................TAA12-8.......VI-126
Kapoor, Bhanu................MPA13-20.......IV-170
Kapoor, Bhanu.................MAA10-3.......VI-41
Karlsson, Magnus............MPA14-7.......II-117
Karsilayan, Aydin Ilker....MAA15-10.......I-65
Kasthuri, P........................TAA11-5.......VI-582
Kawakami, Hiroshi............TAA7-4.......III-465
Kawase, Takehiko.............MAA7-8.......III-291
Kawashima, Mitsusato.....WPB13-10.......VI-522
Kawata, Junji....................MAA7-2.......III-267
Keady, Aidan G...............WPA10-1.......I-571
Kennedy, Michael Peter....WPA12-8.......III-671
Kennedy, Michael Peter....WAA14-5.......II-410
Kennedy, Michael Peter....TAA5-1.......IV-461
Kennings, Andrew A........TAA15-16.......VI-191
Ker, Ming-Dou..................TAA14-9.......II-216
Ker, Ming-Dou..................TAA14-8.......II-212
Keramat, Mansour..........WAA15-16.......VI-334
Keramat, Mansour............TAA12-7.......VI-122
Keramat, Mansour............TAA12-3.......VI-106
Khalil, Mohammad Athar....TAA15-5.......VI-147
Khellah, Muhammad M......MAA6-3.......II-9
Khoo, I-Hung....................MPA1-1.......V-66
Kielbasa, Richard.............TAA12-3.......VI-106
Kielbasa, Richard...........WAA15-16.......VI-334
Kielbasa, Richard.............TAA12-7.......VI-122
Kim, Beomsup....................MPA9-3.......I-163
Kim, Beomsup..................TAA13-10.......V-198
Kim, Chan......................WPB13-7.......VI-510
Kim, Hong-sun..................MPA6-8.......II-89
Kim, Hyup Jong...............WPB13-7.......VI-510
Kim, Jae-Gon....................TPA13-6.......IV-253
Kim, Jae-Wan...................TAA10-2.......I-344
Kim, Jong-il......................TPA2-2.......V-277
Kim, Kyung-Hoon..............TPA4-2.......IV-221
Kim, Kyung-Soo.................TPA1-1.......V-257
Kim, Kyung-Soo..............WPA15-3.......II-316
Kim, Lee-Sup..................WAA14-7.......II-417
Kim, Lee-Sup...................TPA4-2.......IV-221
Kim, Soo-Won..................TAA10-2.......I-344
Kim, Soo-Won...................TPA13-1.......IV-233
Kimijima, Tadaaki.............TAA13-8.......V-190
Kinoshita, Kazuhiko........WPB13-11.......VI-529
Kirac, Ahmet.....................TAA1-7.......V-154
Kis, Gabor.........................TAA5-1.......IV-461
Kitajima, Hideo...............TAB13-12.......V-253
Kitaoka, Yoshihiro...........WPA13-2.......V-526
Kiya, Hitoshi.....................TAA13-8.......V-190
Kiya, Hitoshi.....................MPA1-6.......V-86
Kiyose, Kageyuki.............WAA13-5.......V-371
Kleine, Ulrich...................WAA15-5.......VI-290
Klumperink, Eric A. M........TAA8-3.......I-289
Kneip, T..........................TPA15-10.......VI-645
Ko, Tsz-Mei...................WPB13-12.......VI-533
Kobayashi, Masaki...........TAA13-3.......V-170
Kobayashi, Suguru..........TAA15-14.......VI-183
Kocal, Osman Hilmi............TPA1-2.......V-261
Kocarev, Ljupco................TPA9-2.......IV-514
Koch, Christof................TPA15-11.......VI-649
Kogge, Peter M...............WPB13-1.......VI-486
Koli, Kimmo......................MPA9-2.......I-159

Koli, Kimmo......................MPA9-7.......I-181
Koli, Kimmo.....................WPA8-1.......I-512
Koli, Kimmo....................MAA13-19.......IV-409
Kolnsberh, S..................TPA15-10.......VI-645
Kolumban, Geza................TAA5-1.......IV-461
Komuro, Motomasa...........MAA14-3.......III-
Koneru, Satyaki..............MPA15-7.......I-213
Kornegay, Kevin T..........MPA15-14.......I-241
Kosunen, Marko.............MAA13-19.......IV-409
Kosunen, Marko..............WPA8-1.......I-512
Kotropoulos, Constantine.....MAA4-6.......IV-21
Koufopavlou, Odysseas.........MAA6-4.......II-13
Koufopavlou, Odysseas.....WAA15-18.......VI-342
Koufopavlou, Odysseas G.....MAA10-8.......VI-62
Kousaka, Takuji.................TAA7-4.......III-465
Kousuke, Katayama............TPA3-3.......III-99
Koutsoyannopoulos, Yorgos....TAA12-6.......VI-118
Kranz, Ernst-Georg..........WPA6-1.......II-276
Krishnamachari, Bhaskar....TAA2-3.......VI-446
Krishnamurthy, Harsha......WPA4-3.......VI-381
Krishnapura, Nagendra......WAA9-4.......I-480
Krishnapura, Nagendra......MPA8-6.......I-143
Krishnapura, Nagendra....MPA15-16.......I-249
Ku, Chung-Wei.................MAA11-4.......IV-41
Kuh, Anthony....................TAA3-1.......III-62
Kuh, Ernest S..................MPA10-3.......VI-74
Kukk, Vello....................WPA12-7.......III-667
Kunieda, Hiroaki.............WAA14-21.......II-474
Kunieda, Nobuyuki...........TPA13-7.......IV-257
Kunt, Murat......................WPA2-4.......V-498
Kuntman, H. Hakan.........WPA15-7.......II-331
Kuo, Kuo-Tang...................TPA4-3.......IV-225
Kuo, Tzu-Chieh................MPA14-2.......II-97
Kuroda, Ichiro.................WAA14-4.......II-406
Kurokawa, Hiroaki............TAA3-4.......III-74
Kwon, Taeck-Geun..........WPB13-6.......VI-506
Kyriakis-Bitzaros, E. D....WAA15-24.......VI-368

L

Lahti, Jukka A.................MAA13-18.......IV-405
Lai, Hon Seng.................MPA13-16.......IV-154
Lai, Hon Seng...................TPA13-3.......IV-241
Lai, Yen-Tai....................WAA14-3.......II-402
Lai, Yung-Kai....................MAA4-8.......IV-29
Lam, Chi-Wai.................WPB13-12.......VI-533
Lam, Kin-Man...................MPA4-7.......IV-85
Lampinen, Harri.............MPA14-24.......II-185
Lan, Mao-Feng..................WPA9-5.......I-555
Lancaster, Jason David....MAA15-23.......I-115
Lande, Tor Sverre............MAA12-2.......II-37
Lang, Mathias C...............WAA13-4.......V-367
Lapic, Stephan K...............WAA5-1.......IV-526
Lapinoja, Mikko J..............WPA9-6.......I-559
Larcheveque, Remi.........WAA15-15.......VI-330
Larson, Lawrence E..........WPA5-8.......IV-608
Lau, W.H..........................TAA2-7.......VI-462
Laur, Rainer A..................TPA10-4.......VI-236
Lazzaroni, M....................TAB14-8.......III-514
Le-Ngoc, Tho...................TPA13-9.......IV-265
Leblebici, Yusuf..............WPA14-6.......III-191
Leblebici, Yusuf................MPA14-4.......II-105
Lee, Chang-Hyeon...........MPA15-12.......I-233
Lee, Chen-Yi..................WAA14-24.......II-486
Lee, Chew Peng..............WPA15-24.......II-390
Lee, Eel-Wan...................TPA14-2.......I-377
Lee, Haeng-Woo................TPA1-1.......V-257
Lee, Jin-Aeon..................WAA14-7.......II-417
Lee, Jong-Yeol................MAA10-2.......VI-37
Lee, Jun-yong..................WPB4-2.......VI-426
Lee, Junsoo....................MAA10-5.......VI-49
Lee, Kok Kiong................WPB4-1.......VI-421

AUTHOR INDEX

Lee, Kyou Ho..................WPB13-7........VI-510
Lee, Mankoo....................WAA6-4........II-256
Lee, Mankoo....................WAA6-2........II-248
Lee, Moon Key.................MPA14-5........II-109
Lee, Seokjun....................TAA4-6........IV-209
Lee, Seong-Bong..............TAA15-9........VI-163
Lee, Sook-Hyang..............WPB13-6........VI-506
Lee, Yew-San..................WAA14-24........II-486
Lee, Yong Hoon..............TAA13-10........V-198
Lee, Yoon Ju....................TPA6-4........VI-482
Lee, Yung-Pin..................MAA11-4........IV-41
Leelavattananon, Kritsapon........WPA8-6........I-531
Leenaerts, Domine...........MAA14-8........III-322
Leong, Choon Haw...........MPA8-2........I-127
LeRiguer, E.....................MPA14-13........II-141
Letaief, K.B......................TPA5-3........IV-502
Leung, Bosco..................MAA13-17........IV-401
Leung, K.S......................WAA11-1........VI-240
Leung, Lap Chi................MAA13-2........IV-341
Li, Chung-Sheng..............WAA12-5........III-599
Li, Dandan......................MPA8-4........I-135
Li, Dongju......................WAA14-21........II-474
Li, Dongju......................MAA15-6........I-49
Li, Hongzhi....................MAA4-3........IV-9
Li, Jiang.........................MAA5-8........IV-333
Li, Shenghong................WPA14-15........III-223
Li, Shipeng.....................TPA2-3........V-281
Li, Simon Cimon.............WPA15-12........II-347
Li, Tong.........................MPA14-6........II-113
Li, Tong.........................WPA11-1........VI-389
Li, Wei..........................MAA1-4........V-13
Li, Weiping....................TPA2-3........V-281
Li, Weiping....................MAA4-4........IV-13
Li, Wenzhe....................MPA13-22........IV-178
Li, Xiaowei.....................WPA6-4........II-288
Li, Y. Y..........................WAA11-1........VI-240
Liao, Jun-Yao.................MPA14-20........II-169
Liao, Jun-Yao.................MPA14-21........II-173
Lidgey, John...................WPA8-8........I-539
Lidgey, John...................WAA8-7........I-460
Lim, Kyoohyun...............MPA9-3........I-163
Lim, Shao-Jen................MPA15-23........I-277
Lim, Yong Ching..............TAA1-8........V-158
Lim, Yong-Ching..............MAA2-7........V-57
Lim, Young-Kwon...........MPA13-23........IV-182
Lím, Drahoslav...............WPA3-1........III-139
Lin, Chi-Hung.................WAA8-4........I-448
Lin, Chun-Fu..................TAB13-6........V-229
Lin, David W..................MPA13-7........IV-118
Lin, Horng-Dar...............WAA12-3........III-586
Lin, Horng-Dar...............WAA12-2........VI-442
Lin, Hung-Jen.................WPA6-7........II-300
Lin, L.............................WAB6-2........II-264
Linares-Barranco, Bernabe......WPA9-4........I-551
Linares-Barranco, Bernabe....WAA14-19........II-466
Lindfors, Saska...............WPA14-23........III-255
Lindfors, Saska...............MPA9-1........I-155
Lindfors, Saska J.............MPA15-18........I-257
Lindfors, Saska J.............MPA6-3........II-73
Lindgren, Per..................TAA15-15........VI-187
Ling, Fan.......................MAA4-4........IV-13
Lionetto, A.....................WAA3-7........III-131
Liou, Ming L...................TPA5-3........IV-502
Liou, Ming L...................WAA12-6........III-603
Litmanen, Petteri Matti....MAA13-6........IV-357
Liu, Bin-Da....................WPA14-17........III-231
Liu, Bin-Da....................WPA14-20........III-243
Liu, Chi-Min..................MPA13-19........IV-166
Liu, Derong...................WPA14-12........III-215
Liu, Sheng-Wei...............WPA1-3........V-457

Liu, Wentai....................WAA14-20........II-470
Liu, Zemin.....................WPB13-8........VI-514
Livingston, F..................TAA11-5........VI-582
Llinas, James..................MPA11-2........VI-541
Llinas, James..................MPA11-1........VI-537
Lo, Chun-Keung.............MAA12-1........II-33
Lo, Chun-Keung.............MAA15-6........I-49
Lockwood, John W.........MAA11-7........IV-53
Long, Stephen I..............MPA6-3........II-73
Louis, Loai.....................TAA10-8........I-369
Loumeau, Patrick............MPA15-22........I-273
Low, Seo-How...............MAA2-7........V-57
Lu, Cheng-Chang............WPB13-2........VI-490
Lu, Jialiang....................WPA7-7........III-639
Lu, Jianhua....................TPA5-3........IV-502
Lu, Wu-Sheng................MAA14-14........III-346
Lu, Wu-Sheng................MPA2-2........V-102
Lu, Wu-Sheng................WAA5-3........IV-534
Lu, Wu-Sheng................MAA2-3........V-41
Lu, Zanjun.....................WPA14-12........III-215
Lubkin, Jeremy...............MPA3-9........III-58
Lucchese, L....................WPA2-9........V-518
Lucke, Lori E..................WPA15-4........II-320
Luh, Louis.....................WPA10-3........I-579
Lun, Pak-Kong..............TAB13-5........V-225
Luong, Cam...................MAA13-2........IV-341
Luong, Howard Cam........MAA8-4........I-9
Lutovac, Miroslav D........WAA15-4........VI-286
Lyden, C........................TPA15-12........VI-653
Lyden, Colin..................WPA10-1........I-571
Lynch, William E............TPA13-9........IV-265

M

Ma, Chor Tin..................MPA5-1........IV-433
Ma, Jun.........................TAB13-10........V-245
Ma, Stanley Jeh-Chun......TAA10-3........I-348
Maaz, Mohamad.............WPA4-3........VI-381
MacEachern, Leonard A...TPA4-4........IV-229
Maeda, Yutaka................TPA3-2........III-94
Maggio, Gian Mario........WPA12-8........III-671
Magotra, N....................TAA11-5........VI-582
Magrath, A.J...................TAA4-1........IV-190
Magrath, A.J...................TPA14-4........I-385
Maier, Christoph.............WPA11-5........VI-405
Maio, Ivano Adolfo.........MAA14-23........III-383
Maio, Ivano Adolfo.........MPA7-6........III-411
Mak, C..........................TAB14-2........III-489
Makkey, Mostafa............WAA9-5........I-485
Mäkynen, Anssi Jaakko...TPA15-2........VI-610
Malavasi, Enrico.............TAA12-8........VI-126
Malcovati, Piero.............TPA15-5........VI-625
Maloberti, Franco...........TPA15-5........VI-625
Maloberti, Franco...........TAB13-2........V-213
Malvar, Henrique S........MPA2-1........V-98
Manaresi, Nicolo'............WAA3-6........III-127
Mancini, Paolo................WAA14-1........II-394
Manduchi, Roberto..........WAA2-3........V-333
Manetti, Stefano.............MAA9-3........VI-9
Manganaro, Gabriele........MAA15-3........I-37
Manjunath, Bangalore S...WAA12-4........III-595
Manjunath, Bangalore S...WAA12-4........VI-450
Manku, Tajinder..............MAA13-10........IV-373
Manku, Tajinder..............TPA4-4........IV-229
Mansoori, Sana Ahmed....TAA1-1........V-130
Mariscotti, A..................MAA14-16........III-354
Marjanovic, Slavoljub......MPA7-7........III-415
Marschner, Uwe..............WPA6-1........II-276
Martinez, Jose Silva........MPA6-5........II-81
Martínez-Olalla, Rafael....TAA13-5........IV-178
Martins, Jorge Manuel.....MAA15-19........I-108
Martins, Jorge Manuel.....TPA14-1........I-373

Martins, Jorge Manuel.....MAA15-19........I-100
Marvasti, Farokh Alim.....TAA4-3........IV-198
Marvasti, Farokh I...........TPA13-4........IV-245
Marvasti, Farokh I...........MPA13-18........IV-162
Masayuki, Yamaguchi......WPA4-4........VI-385
Mascolo, Saverio............MAA7-6........III-283
Masselos, Konstantinos...TAA15-18........VI-199
Mathiazhagan, Chakravarthy...MPA15-16........I-249
Mathis, Wolfgang............MPA15-5........I-205
Matsumoto, Hiroki..........WPA13-2........V-526
Maundy, Brent................MAA15-4........I-41
Mayaram, Kartikeya........MAA13-4........IV-349
Mayaram, Kartikeya........WPA11-2........VI-393
Mayer, Michael...............WPA11-5........VI-405
Maziarz, Bogdan M.........WPA1-7........V-473
Mazzini, Gianluca............TAA5-8........IV-485
McCanny, J....................MPA14-13........II-141
McCarthy, Oliver.............TPA12-4........III-562
McClellan, Kelly..............MPA15-12........I-233
McEachen, John C..........MAA11-1........IV-33
Melnikov, Gerry..............TPA2-5........V-289
Meng, Teresa H..............WAA4-8........IV-301
Meng, Teresa H..............WPA5-5........IV-600
Messina, A.Roman..........TAB14-7........III-510
Miao, Guoqing................TPA14-11........I-412
Michaelis, Markus............WAA2-7........V-349
Michiroh, Ohmura...........TAA15-17........VI-195
Mikhael, Wasfy B...........WAA1-2........V-297
Mikkelsen, Sindre............MAA12-2........II-37
Milanovic, Veljko............MAA13-5........IV-353
Miller, D. Michael............TAA15-8........VI-159
Miller, Neil Linton............MPA14-18........II-161
Miller, William C.............MPA15-3........I-197
Millerioux, Gilles.............TPA9-1........IV-510
Min, Byung-Moo.............TAA10-2........I-344
Minch, Bradley A.............WPA8-5........I-527
Minch, Bradley A.............MPA8-3........I-131
Minch, Bradley A.............MAA15-16........I-88
Minch, Bradley A.............TPA3-1........III-90
Minch, Bradley A.............WPA15-13........II-351
Minch, Bradley A.............MPA3-1........III-29
Minch, Bradley A.............TAA14-6........II-204
Minot, Sophie.................MAA15-9........I-61
Mira, Christian................TPA9-1........IV-510
Miró-Sans, Joan Maria....TAA7-3........III-460
Mirzai, Bahram...............WPA3-6........III-159
Mitra, Sanjit...................WAA13-17........V-421
Mitra, Sanjit K................MAA4-7........IV-25
Mitra, Sanjit K................TAA4-5........IV-206
Miyanaga, Yoshikazu......WAA14-13........II-441
Mizutani, Yoko................WAA7-4........III-578
Mlynek, Daniel................WPA14-6........III-191
Mlynek, Daniel................MPA14-4........II-105
Mo, Y.-S........................MAA2-3........V-41
Mohan, Rakesh...............WAA12-5........III-599
Moiola, Jorge L...............WPA7-1........III-619
Mojarradi, Mohammad....WPA11-2........VI-393
Mok, Wai Hung...............MPA13-12........IV-138
Mokhtari, Mehran............MAA13-12........IV-382
Mokhtari, Mehran............MAA13-7........IV-345
Mokunaka, Naoki............MPA12-1........III-423
Mokwa, W.....................TPA15-10........VI-645
Moniri, Mansour.............TPA14-3........I-381
Moniri, Mansour.............MAA15-23........I-115
Monteiro, Fabrice............TPA5-1........IV-494
Monteiro, Jose C............MAA6-5........II-17
Monti, Antonello.............TAB14-8........III-514
Moon, Gyu.....................MPA6-8........II-89
Moore, Michael S...........TAA4-5........IV-206
Moreira, Jose' Pedro......MPA8-7........I-147

AUTHOR INDEX

Moreno, W...............................MAA11-3........IV-37
Mori, Hiroyuki....................MPA12-7........III-444
Mori, Hiroyuki....................MPA12-4........III-431
Mori, Hiroyuki....................WAA7-2........III-570
Mori, Shinsaku....................MAA14-5........III-314
Morie, Takashi......................TPA7-1........III-534
Moro, Seiichiro....................MAA14-5........III-314
Morris, Tonia G..................WPA14-16........III-227
Moschytz, George S............TAA13-2........V-166
Moschytz, George S............MAA15-8........I-57
Moschytz, George S............WPA3-1........III-139
Moschytz, George S............WPA3-6........III-159
Moschytz, George S............MAA15-1........I-29
Moschytz, George S............WPA3-5........III-155
Moshnyaga, Vasily.............MPA4-8........IV-89
Mota, Antonio S..................MAA6-5........II-17
Mu, Fenghao......................WAA6-3........II-252
Mu, Zhen...........................MPA10-4........VI-78
Mukherjee, Debargha..........MAA4-7........IV-25
Mulder, Jan........................TAA9-8........I-337
Muller, Karsten...................WPA2-5........V-502
Muneyasu, Mitsuji...............MPA1-7........V-90
Muneyasu, Mitusji...............WPA14-19........III-239
Murakami, Kazuhito............TAB14-9........III-518
Murakami, Koso..................WPB13-11........VI-529
Muramatsu, Shogo..............MPA1-6........V-86
Musil, Vladislav...................MAA15-13........I-76

N
Nagai, Takayuki..................TAA1-2........V-134
Nagalla, Radhakrishna........TAA15-13........VI-179
Nagaraj, Krishnaswamy......WAA9-4........I-480
Nagata, Makoto..................TPA14-5........I-389
Nagisa, Ishiura....................WPA4-4........VI-385
Nah, Jae Hoon...................TPA6-4........VI-482
Nahm, Seunghyeon............WAA14-15........II-449
Nahm, Seunghyeon............MAA13-13........IV-385
Naiknaware, Ravindranath...WPA10-6........I-591
Naiknaware, Ravindranath...TAA10-6........I-360
Nakaguchi, Toshiya.............TAA7-7........III-477
Nakanishi, Isao...................TAA13-3........V-170
Nakano, Hideo....................MAA7-8........III-291
Nakhai, Mohammad Reza...TAA4-3........IV-198
Narayanan, H.....................MPA10-5........VI-82
Nathan, Arokia....................WPA11-6........VI-409
Neag, Marius......................TPA12-4........III-562
Neff, Joseph Daniel.............TPA7-3........III-542
Neinhaus, H........................MAA11-3........IV-37
Netto, Sergio Lima..............WAA13-3........V-363
Neves, Jose L....................WAA6-1........II-244
Neves, Rui Ferreira.............WAA9-2........I-472
Newcomb, Robert W..........MAA3-4........III-21
Newcomb, Robert W..........WPA14-2........III-175
Ng, Andrew........................MAA15-6........I-49
Ng, Andrew........................MAA15-24........I-119
Ng, C.K..............................MPA14-23........II-181
Ng, Shek-Wai.....................TPA10-1........VI-223
Ng, Tung Sang...................WAA5-2........IV-530
Ng, Tung Sang...................WAA5-5........IV-542
Nguyen, Truong..................MAA2-2........V-37
Nguyen, Truong..................MAA4-1........IV-1
Niamat, Mohammed Y........MPA1-2........V-70
Nicoletti, Guy M..................WPA3-4........III-151
Niemistö, Matti...................MAA13-18........IV-405
Nieto-Lluís, Victor................TAA13-5........V-178
Nijhuis, J.A.G.....................WAA14-18........II-462
Nikolaidis, Spyridon............WAA15-24........VI-368
Nikolaidis, Spyridon............WAA15-3........VI-363
Nikolic, Borivoje..................MAA12-6........II-53
Nikolic, Borivoje..................MPA7-7........III-415
Nishihara, Akinori................WAA13-11........V-395

Nishikawa, Kiyoshi..............TAA13-8........V-190
Nishimura, Shotaro.............TPA1-4........V-269
Nishio, Yoshifumi................MAA14-4........III-310
Nishio, Yoshifumi................WAA15-19........VI-346
Nishio, Yoshifumi................MAA7-2........III-267
Nishio, Yoshifumi................WAA15-3........VI-282
Nishitani, Takao..................WAA14-4........II-406
Njoelstad, Tormod..............MPA14-3........II-101
Noceti Filho, Sidnei.............WPA15-14........II-355
Nosratinia, Aria..................MPA2-4........V-110
Nossek, Josef A.................WPA13-6........V-538
Nowotny, Ulrich..................TAA14-5........II-201
Nowrouzian, Behrooz.........WAA13-10........V-391
Nowrouzian, Behrooz.........WAA13-9........V-387
Nunnari, G.........................WAA3-7........III-131

O
O'Donnell, John..................WAA14-5........II-410
O'Dwyer, Tom....................WAA14-5........II-410
Obote, Shigeki...................MAA13-23........IV-425
Obote, Shigeki...................MAA13-15........IV-393
Occhipinti, L......................WAA3-7........III-131
Ogorzalek, Maciej J............TPA9-4........IV-522
Ogunfunmi, Tokunbo..........TAA13-12........V-206
Ohkubo, Jun'ya..................WAA14-13........II-441
Ohm, Jens-Rainer..............WPA2-5........V-502
Ohmacht, Martin................MAA11-5........IV-45
Ohno, Wataru....................MAA14-11........III-334
Ohtsuka, Yasuhiro..............TPA15-8........VI-637
Okazaki, Hideaki................MAA7-8........III-291
Okazaki, Hideaki................WPB13-10........VI-522
Okello, James....................TAA13-3........V-170
Oklobdzija, Vojin G............MAA12-6........II-53
Oklobdzija, Vojin G............WAB6-1........II-260
Okuda, Masahiro...............WAA13-5........V-371
Oliaei, Omid......................MPA15-22........I-273
Oliveira, Arlindo L..............MAA6-5........II-17
Olmos, Alfredo...................WPA10-2........I-575
Ono, Toshio.......................WPA15-18........II-367
Opal, Ajoy..........................WAA15-7........VI-298
Opal, Ajoy..........................TAA12-5........VI-114
Orailoglu, Alex...................TAA14-1........II-189
Oraintara, Soontorn............MAA2-2........V-37
Orcioni, Simone.................WPA9-2........I-543
Osa, Juan I.......................TAA14-7........II-208
Osman, Ashraf A...............WPA11-2........VI-393
Ostermann, Joern..............TPA2-1........V-273
Ozgur, Mehmet..................MAA13-5........IV-353

P
Paasio, Ari Juhani..............WPA14-23........III-255
Pace, P.E...........................TAA14-11........II-224
Pace, P.E...........................TPA14-7........I-397
Paek, Seung-Kwon.............WPA1-6........V-469
Page, Kevin J....................MAA13-21........IV-417
Pai, M.A............................MPA12-6........III-439
Palà-Schönwälder, Pere....TAA7-3........III-460
Palmisano, Giuseppe.........WPA15-17........II-363
Pammu, Sridhar.................WPA14-21........III-247
Panagiotaras, George, S....MAA10-8........VI-62
Papadakis, Vasilios............TPA13-9........IV-265
Papananos, Yannis............TAA12-6........VI-118
Papathanasiou, Konstandinos..MPA15-13........I-237
Parhi, Keshab K................MPA5-7........IV-453
Parhi, Keshab K................WAA14-16........II-453
Parhi, Keshab K................WAA13-1........V-354
Parhi, Keshab K................TAB13-10........V-245
Parhi, Keshab K................WAA14-4........II-406
Parisi, Raffaele..................WPA14-7........III-195
Park, Byeong-Ha................MAA13-22........IV-421

Park, Chan-Hong...............MPA9-3........I-163
Park, Joonbae...................MPA14-10........II-129
Park, Sangbeom................MAA8-1........I-1
Park, Sung Min..................TAA8-4........I-293
Parlitz, Ulrich.....................TPA9-2........IV-514
Parssinen, Aarno T............MPA6-3........II-73
Parssinen, Aarno T............MPA15-18........I-257
Passos, Nelson L...............WAA11-8........VI-270
Passos, Nelson L...............TAA15-4........VI-143
Pastore, Stefano................TAA7-1........III-452
Pavan, Shanthi..................MPA15-16........I-249
Pavan, Shanthi..................WPA15-9........II-339
Payne, A.J........................TAA9-3........I-317
Pearlman, William A..........MPA2-5........V-114
Pecora, Louis M................WAA10-3........IV-564
Pei, Soo-Chang.................TPA5-4........IV-506
Pellegrini, Aurelio...............MPA15-15........I-245
Pennala, Riku-Matti............MPA15-11........I-229
Pennisi, Salvatore..............WAA8-6........I-456
Pennisi, Salvatore..............WPA15-11........II-343
Pérez-Castellanos, M.-M...TAA13-5........V-178
Perona, Pietro...................WPA2-3........V-494
Petraglia, Antonio..............WAA9-1........I-468
Petraglia, Antonio..............WAA9-8........I-496
Petrie, Craig Steven...........TAA14-3........II-197
Pham, Hoan H..................WPA11-6........VI-409
Phang, Khoman.................TAA8-8........I-305
Piazza, Francesco.............MPA5-4........IV-441
Piazza, Francesco.............TAA3-2........III-66
Piccirilli, Maria Cristina.......MAA9-3........VI-9
Picun, Gonzalo..................MPA15-21........I-269
Pihl, Johnny......................MAA12-3........II-41
Pimentel, Max Chianca......TAB14-6........III-505
Pineda de Gyvez, Jose......MAA13-8........IV-365
Pineda de Gyvez, Jose......MAA15-3........I-37
Pinto, Rodrigo Luiz de Oliveira...TAA8-7........I-301
Pirsch, Peter.....................MAA11-5........IV-45
Pitas, Ioannis....................TPA13-10........IV-269
Pitas, Ioannis....................MAA4-6........IV-21
Pittard, C. Louis.................MPA11-5........VI-549
Plotkin, Eugene I................MAA1-2........V-1
Po, Lai-Man.......................MPA13-4........IV-106
Po, Lai-Man.......................WAA4-7........IV-297
Poletti, Matteo...................TPA15-5........VI-625
Poor, H. Vincent................WAA5-6........IV-546
Poor, H. Vincent................WPA5-3........IV-589
Porra, Veikko.....................TAA5-4........IV-473
Porra, Veikko.....................TAA5-4........IV-477
Porta, S............................WPA8-8........I-539
Porta, Sonia......................TAA14-7........II-208
Prabhakaran, Pradeep Kumar...WPA4-1........VI-372
Premoli, Amedeo...............TAA7-1........III-452
Pu, Chiang-Jung................MPA15-2........I-193
Punzenberger, Manfred......TAA9-6........I-329
Python, Dominique............MPA8-5........I-139

Q
Qin, Huashu......................MAA14-6........III-318
Quan, X............................WPA8-2........I-518
Quddus, Azhar..................MPA13-15........IV-150
Quer, Stefano....................WAA15-13........VI-322
Quero Reboul, Jose Manuel...MPA12-5........III-435
Quigley, Steven Francis......WPA14-21........III-247
Quigley, Steven Francis......MPA14-18........II-161

R
Raahemifar, Kaamran.........MPA15-4........I-201
Rabiner, Wendi..................WPA5-6........IV-604
Raffo, Luigi.......................WPA14-14........III-219
Ragaie, Hani Fikry.............WPA12-2........III-647
Rahkonen, Timo E.............WPA9-6........I-559

AUTHOR INDEX

Rajan, P. K. .... MPA1-1 .... V-66
Ramamurthi, Bhaskar .... MPA15-16 .... I-249
Ramírez-Angulo, Jaime ...... WPA15-22 .... II-382
Ramírez-Angulo, Jaime ...... WAA14-14 .... II-445
Ramkumar, Mahalingam .... MPA2-6 .... V-118
Ramprasad, Sumant .... MAA6-1 .... II-1
Ramstad, Tor A .... WPA1-4 .... V-461
Ranganathan, N .... MAA11-3 .... IV-37
Rapakko, Harri Antero .... TAA10-1 .... I-341
Rashid, Obaidur .... TAA15-4 .... VI-143
Ratakonda, Krishna .... MPA4-3 .... IV-69
Re, M .... WAA8-5 .... I-452
Recoules, Hector .... WAA15-2 .... VI-278
Reddy, Hari C .... MPA1-1 .... V-66
Redmill, David Wallace .... WAA13-19 .... V-429
Reina, Rodrigo .... WPA15-2 .... II-312
Reißig, Gunther .... MPA7-8 .... III-419
Reißig, Gunther .... MAA14-9 .... III-326
Rhee, Woogeun .... MPA6-6 .... II-85
Rho, June-Kyung .... WPB13-6 .... VI-506
Ridge, D .... MPA14-13 .... II-141
Rieder, Peter .... WPA13-6 .... V-538
Ringer, W.P .... TAA14-11 .... II-224
Rjoub, Abdoul .... MAA6-4 .... II-13
Roberts, Gordon W .... MAA15-20 .... I-104
Roberts, Gordon W .... TAA9-1 .... I-309
Roberts, Gordon W .... WPA6-2 .... II-280
Roberts, Gordon W .... TAA10-8 .... I-369
Roberts, Gordon W .... MPA8-2 .... I-127
Roberts, Gordon W .... WPA6-3 .... II-284
Roche, Christian .... TAA1-6 .... V-150
Rodellar-Biarge, M.-V .... TAA13-5 .... V-178
Rodriguez-Vazquez, Angel ... TAA5-4 .... IV-473
Rodriguez-Vazquez, Angel ... MAA9-5 .... VI-17
Rodriguez-Vazquez, Angel ... TAA5-4 .... IV-477
Roman, Jaime R .... WAA1-2 .... V-297
Roos, Janne Wilhelm .... MPA7-4 .... III-403
Routama, Jarkko .... MPA9-1 .... I-155
Routama, Jarkko Antero .... MPA9-2 .... I-159
Routama, Jarkko Antero .... MPA9-7 .... I-181
Rovatti, Riccardo .... TAA5-8 .... IV-485
Rovatti, Riccardo .... WAA3-6 .... III-127
Roy, Jean .... MPA11-9 .... VI-561
Roy, Sumit .... MAA10-7 .... VI-57
Rumin, Nicholas .... MPA10-6 .... VI-86
Rutledge, Janet C .... TAA11-3 .... VI-573
Ryu, Chul .... MPA13-24 .... IV-186
Ryynanen, Jussi H .... MPA15-18 .... I-257
Ryynanen, Jussi H .... MPA6-3 .... II-73

S

Saed, Aryan .... MPA15-3 .... I-197
Said, Amir .... MPA2-5 .... V-114
Saini, J.P .... MPA5-8 .... IV-457
Saito, Toshimichi .... MPA12-1 .... III-423
Sakagami, Iwata .... WAA15-21 .... VI-354
Sakallah, Karem A .... TAA15-23 .... VI-215
Sakimura, Noboru .... TPA14-5 .... I-389
Salam, Fathi M .... MPA3-8 .... III-54
Salam, Fathi M .... MAA3-2 .... III-5
Salama, C. Andre T .... MAA6-7 .... II-25
Salama, C. Andre T .... TAA10-3 .... I-348
Salcedo-Suner, Jorge .... TAA8-5 .... I-297
Salles, Ronaldo Moreira .... WAA5-4 .... IV-538
Salmeri, Marcello .... TAB13-1 .... V-210
Salsano, Adelio .... TAB13-1 .... V-210
Sanchez, Edgar N .... TAB14-7 .... III-510
Sanchez-Sinencio, Edgar ...... MAA13-4 .... III-365
Sanchez-Sinencio, Edgar ...... WPA8-2 .... I-518
Sanchez-Sinencio, Edgar ...... MAA6-8 .... II-29
Sandage, Robert, W .... TPA15-9 .... VI-641
Sandberg, Irwin W .... MAA14-18 .... III-362
Sandler, Mark B .... TAA4-1 .... IV-190
Sandler, Mark B .... TPA14-4 .... I-385
Santos, Paulo Jorge .... TAA14-12 .... II-228
Sanubari, Junibakti .... TAA13-11 .... V-202
Sapatnekar, Sachin .... WAA11-7 .... VI-266
Saramaki, Tapio Antero .... WAA13-12 .... V-399
Saramaki, Tapio Antero .... WAA13-13 .... V-404
Sarcinelli-Filho, Mario .... TPA11-4 .... VI-602
Sargeni, Fausto .... WPA3-2 .... III-143
Sarmiento-Reyes, L. A .... WAA15-10 .... VI-310
Sasase, Iwao .... MAA14-5 .... III-314
Satakopan, S .... WPA14-9 .... III-203
Sauerwein, Helmut .... WPA13-7 .... V-542
Schaumann, Rolf .... TAA9-7 .... I-333
Schimpfle, Christian V .... WPA13-6 .... V-538
Schmid, Alexandre .... WPA14-6 .... III-191
Schmid, Hanspeter .... MAA15-8 .... I-57
Schmidt, Jon C .... TAA11-3 .... VI-573
Schmitz, Christopher Dale ... WAA1-3 .... V-301
Schneebeli, Hansjorg Andreas .. TPA11-4 .... VI-602
Schneider, Márcio Cherem .... TAA8-7 .... I-301
Schneider, Márcio Cherem .... MAA15-2 .... I-33
Schneider, Todd .... TAA11-2 .... VI-569
Schreier, Richard .... TAA10-4 .... I-352
Schulze, Jens .... WAA15-5 .... VI-290
Schuppener, Gerd .... MPA14-11 .... II-133
Schwarz, M .... TPA15-10 .... VI-645
Schwarz, Wolfgang .... TAA5-2 .... IV-465
Schweizer, Joerg .... TAA5-3 .... IV-469
Sciuto, Donatella .... WAA14-8 .... II-425
Sciuto, Donatella .... WAA14-8 .... II-421
Seara, Rui .... WPA15-14 .... II-355
Sedaghat-Maman, Reza .... TAA15-2 .... V-135
Seifi, Abbas .... WAA15-17 .... VI-338
Sekiya, Hiroo .... MAA14-5 .... III-314
Sellami, Louiza .... MAA3-4 .... III-21
Serrano-Gotarredona, T .... WAA14-19 .... II-466
Serrano-Gotarredona, T .... WPA9-4 .... I-551
Setti, Gianluca .... TAA5-8 .... IV-485
Setty, Suma .... WAA8-8 .... I-464
Sewell, John Isaac .... MAA15-6 .... I-49
Sewell, John Isaac .... MAA15-24 .... I-119
Sha, Edwin .... WPB13-1 .... VI-486
Sha, Edwin .... TAA15-21 .... VI-207
Sha, Edwin .... WAA11-8 .... VI-270
Shaaban, Khaled M .... TAA3-5 .... III-78
Shalash, Ahmed F .... MPA5-7 .... IV-453
Shalfeev, Vladimir D .... WAA10-7 .... IV-579
Shams, Ahmed M .... MPA14-16 .... II-153
Shana'a, Osama .... TAA9-7 .... I-333
Shanbhag, Naresh R .... MAA6-1 .... II-1
Shanbhag, Naresh R .... TPA6-1 .... VI-474
Shanbhag, Naresh R .... MAA10-1 .... VI-33
Shao, Jianhua .... WAA15-20 .... VI-350
Sharaf, Atif Ibrahim .... TPA13-4 .... IV-245
Sharif-Bakhtiar, M .... MPA7-3 .... III-399
Sheliga, Michael .... TAA15-21 .... VI-207
Sheu, Bing .... WPA14-10 .... III-207
Sheu, Bing J .... MAA3-3 .... III-9
Sheu, Bing J .... MPA3-4 .... III-37
Sheu, Bing J .... WAA12-2 .... VI-442
Sheu, Bing J .... WAA12-2 .... III-586
Sheu, Bing J .... MPA4-4 .... IV-73
Sheu, Bing J .... WPA11-7 .... VI-413
Sheu, Jia-Lin .... MPA14-8 .... II-121
Sheu, Ming-Hwa .... WPA1-3 .... V-457
Sheu, Ming-Hwa .... MPA5-6 .... IV-449
Sheu, Ming-Hwa .... MPA5-8 .... II-121
Shi, C.-J. Richard .... WAA15-12 .... VI-318
Shi, Yun Q .... MPA13-14 .... IV-146
Shieh, Bai-Jue .... WAA14-24 .... II-486
Shieh, Ming-Der .... WPA1-3 .... V-457
Shieh, Ming-Der .... MPA5-6 .... IV-449
Shieh, Ming-Der .... MPA14-8 .... II-121
Shim, Jae Wook .... MPA14-5 .... II-109
Shimamura, Tetsuya .... TPA13-7 .... IV-257
Shin, Hyun Chul .... WAA14-7 .... II-417
Shin, Sung-Hyuk .... WAA5-8 .... IV-550
Shin, Taehwan .... TPA13-6 .... IV-253
Shingo, Kawahara .... TPA3-3 .... III-99
Shinohara, Shigenobu .... WAA7-4 .... III-578
Shinomiya, Norihiko .... WPA12-4 .... III-655
Shiraishi, Shin-ichi .... TAB13-12 .... V-253
Shiue, Muh-Tian .... WPB13-2 .... VI-490
Shiue, Wen-Tsong .... WAA14-12 .... II-437
Shojaei, M .... MPA7-3 .... III-399
Shui, PengLang .... WPA1-5 .... V-465
Shui, Tao .... TAA10-4 .... I-352
Silva, Christopher Patrick .... TAA5-9 .... IV-489
Silva, Joao Marques .... TAA15-23 .... VI-215
Silva, Luis Guerra e .... TAA15-23 .... VI-215
Silva-Martinez, Jose .... TAA8-5 .... I-297
Silva-Martínez, José .... MPA8-8 .... I-151
Silveira, Fernando .... MPA15-21 .... I-269
Silveira, Luis Miguel .... TAA15-23 .... VI-215
Silveira, Luis Miguel .... WAA15-22 .... VI-358
Simek, Petr .... MAA15-13 .... I-76
Simonelli, Osvaldo .... TAB13-1 .... V-210
Simoni, Mario .... WAA3-2 .... III-111
Simpson, Richard P .... TAA15-4 .... VI-143
Singh, P .... MAA11-3 .... IV-37
Siohan, Pierre .... TAA1-6 .... V-150
Siu, Wan-Chi .... TAB13-5 .... V-225
Siu, Wan-Chi .... MAA4-5 .... IV-17
Smith, John R .... WAA12-5 .... III-599
Snelgrove, W. Martin .... WPA10-5 .... I-587
Snelgrove, W. Martin .... WPA10-7 .... I-596
Sobelman, Gerald E .... MAA12-8 .... II-61
Sobhy, Mohamed Ibrahim .... WAA9-5 .... I-485
Sobhy, Mohamed Ibrahim .... WPA3-8 .... III-167
Soeda, Junichiro .... WPB13-11 .... V-529
Sommer, Gerald .... WAA2-7 .... V-349
Sommer, Ralf .... MAA9-7 .... VI-25
Song, C T Peter .... WPA14-21 .... III-247
Song, Hwangjun .... WAA4-6 .... IV-293
Song, Leilei .... WAA14-4 .... II-406
Song, Xudong .... MPA13-9 .... IV-126
Soni, Robert A .... WAA1-5 .... V-309
Soudris, Dimitrios J .... MPA14-9 .... II-125
Spaanenburg, L .... WAA14-18 .... II-462
Sridhar, Ramalingam .... MPA14-19 .... II-165
Sriram, Sundararajan .... TAA15-19 .... VI-203
Srivastava, M.C .... MPA5-8 .... IV-457
Stanford, Theron .... WAA3-1 .... III-107
Stanford, Theron .... WPA15-13 .... II-351
Starkov, Sergei Olegovich .... TAA5-6 .... IV-481
Steensgaard, Jesper .... MAA9-6 .... I-488
Steiner, Ralph .... WPA11-5 .... II-405
Stevens, Kenneth S .... MAA6-6 .... II-21
Stojanovski, Toni .... TPA9-2 .... IV-514
Stoyanov, Georgi Kostov ..... WAA13-7 .... V-379
Streitenberger, Martin .... MPA15-5 .... I-205
Strintzis, Michael G .... WPA2-7 .... V-510
Stubberud, Allen R .... MPA1-1 .... V-66
Styer, D .... TPA14-7 .... II-397
Styer, D .... TAA14-11 .... II-224
Stylianou, Yannis .... MAA1-1 .... V-5
Su, Kuan-Cheng .... TAA14-8 .... II-212
Su, Wenjun .... WAA8-7 .... I-460

AUTHOR INDEX

Su, Yih-Ming..................MAA3-7......III-25
Suetsugu, Tadashi............TAA2-5......VI-448
Sultan, Labib.....................WAA3-5......III-123
Sumanen, Lauri..................MAA13-19......IV-409
Sumi, Yasuaki....................MAA13-15......IV-393
Sumi, Yasuaki....................MAA13-23......IV-425
Summerfield, Stephen.......WAA13-15......V-413
Sun, Hongqiao...................TPA2-3......V-281
Sun, Huifang.....................TPA2-4......V-285
Sun, Ming-Ting..................WAA12-8......III-611
Sun, Ming-Ting..................WAA12-8......VI-466
Sun, Ming-Ting..................WAA4-4......IV-285
Sun, S.-W...........................TAA14-8......II-212
Sun, Shangzhi....................TAA15-24......VI-219
Sun, Tao.............................TPA14-9......I-405
Sun, Zhaohui.....................MPA13-5......IV-110
Sundararajan, Vijay...........WAA14-11......II-433
Sundsbo, Ingil....................WPA1-4......V-461
Sung, Wonyong.................MAA13-13......IV-385
Sung, Wonyong.................WAA14-15......II-449
Sung, Wonyong.................TAA4-6......IV-209
Sunwoo, Myung H............WAA14-9......II-425
Sunwoo, Myung H............WAA14-10......II-429
Surma, David R.................WPB13-1......VI-486
Suszynski, Robert..............TPA14-6......I-393
Suter, Bruce.......................MAA6-6......II-21
Suvakovic, Dusan...............MAA6-7......II-25
Suyama, Ken......................WAA9-4......I-480
Suykens, Johan A.K...........WAA10-5......IV-572
Suzuki, Jouji......................TPA13-7......IV-257
Suzuki, Kazuhiro................MPA4-8......IV-89
Suzuki, Tai.........................WAA6-3......II-252
Swamy, M.N.S....................TAB13-3......V-217
Swamy, M.N.S....................WAA4-2......IV-277
Swamy, M.N.S....................MAA2-5......V-49
Swamy, M.N.S....................MAA1-2......V-1
Swidzinski, Jan Feliks........TAA12-1......VI-98
Syoubu, Kouichi................MAA13-23......IV-425
Syoubu, Kouichi................MAA13-15......IV-393
Szabo, Adrian....................TAB14-10......III-522

T

Takahashi, Atsushi............WAA11-2......VI-244
Takahashi, Shin-ichi..........WAA13-5......V-371
Takanori, Matsushita.........MAA7-1......III-263
Takashi, Kambe.................WPA4-4......VI-385
Takatama, Hirokazu..........MPA7-1......III-391
Takeuchi, Tomoaki............TAA1-2......V-134
Takine, Tetsuya.................WPB13-11......VI-529
Tamaru, Keikichi...............MPA4-8......IV-89
Tan, Meng Tong.................MPA15-19......I-261
Tan, Meng Tong.................MPA15-24......I-281
Tan, Nianxiong..................MAA8-8......I-25
Tan, Nianxiong..................MPA6-1......II-65
Tan, Nianxiong..................MAA8-7......I-21
Tan, Xiangdong..................WAA15-12......VI-318
Tanaka, Mamoru................TAA7-7......III-477
Tanaka, Mamoru................TAA7-6......III-473
Tang, Howard....................TAA14-8......II-212
Tang, Pushan.....................TPA14-11......I-412
Tang, Pushan.....................WPA15-16......II-359
Tanji, Yuichi......................WAA15-19......VI-346
Tanskanen, J.M.A...............TAB13-11......V-249
Tao, Yufei...........................MAA15-18......I-96
Tarim, Tuna B....................WPA15-7......II-331
Tavares, Maria Cristina......TAB14-4......III-497
Tavares, Maria Cristina......TAB14-5......III-501
Tavsanoglu, Vedat..............WAA2-6......V-345
Tay, David Ban Hock..........TAA1-5......V-146
Teh, K. C............................MAA5-1......IV-305
Tenhunen, Hannu..............MPA6-1......II-65

Tenhunen, Hannu..............TPA14-8......I-401
Teo, Patrick C....................WAA2-4......V-337
TerHaseborg, Henrickus....WAA14-18......II-462
Terry, John D.....................WAA1-6......V-313
Teuscher, Craig.................WPA5-4......IV-593
Thanachayanont, Apinunt.....WAA8-2......I-440
Tian, Michael W.................TAA12-4......VI-110
To, Cheuk-Him..................WPA15-19......II-371
Tobagi, Fouad....................MPA4-1......IV-61
Tochinai, Koji....................WAA14-13......II-441
Tokuda, Keiichi.................TAA13-11......V-202
Tongsima, Sissades...........WAA11-8......VI-270
Toral Marin, Sergio L........MPA12-5......III-435
Torikai, Hiroyuki...............WAA10-1......IV-554
Torralba, Antonio..............WPB4-3......VI-430
Toshimichi, Saito...............TPA3-3......III-99
Tosic, Dejan V...................WAA15-4......VI-286
Totaro, S............................WPA2-9......V-518
Toth, Laszlo.......................MPA8-6......I-143
Toumazou, C......................TAA9-5......I-317
Toumazou, Chris...............TAB14-12......III-530
Trajkovic, Ljiljana..............TPA10-2......VI-227
Treichler, John..................WPA5-2......IV-585
Trieu, H.K..........................TPA15-10......VI-645
Truong, Nguyen.................MPA2-8......V-126
Tsai, Kun-Chu....................TAB6-3......II-236
Tsai, Richard H..................MPA3-4......III-37
Tsai, Tsung-Han.................WPA13-4......V-530
Tsai, Tsung-Han.................MPA4-2......IV-65
Tsai, Tsung-Han.................MPA13-21......IV-174
Tsao, Ju-Ying.....................WPA14-20......III-243
Tsao, Y.-F...........................TAA14-9......II-216
Tse, C. K............................WPA15-16......II-359
Tsekeridou, Sofia...............MAA4-6......IV-21
Tseng, Yuh-Kuang.............MAA12-5......II-49
Tsividis, Yannis.................WAA9-4......I-480
Tsividis, Yannis.................MPA8-6......I-143
Tsividis, Yannis.................WPA15-9......II-339
Tsubone, Tsdashi...............TAA7-5......III-469
Tsuchida, Kensei...............TAA15-14......VI-183
Tsui, Chi-Ying....................TPA6-2......VI-478
Tsuji, Kohkichi...................MAA14-10......III-330
Tu, Steve Hung-Lung.........TAB14-12......III-530
Tufan, Emir.......................WAA2-5......V-341
Tuqan, Jamal.....................MAA2-1......V-33
Tuqan, Jamal.....................MPA2-7......V-122
Turchetti, Claudio..............WPA9-2......I-543

U

Ueta, Tetsushi....................TAA7-4......III-465
Unbehauen, Rolf................WAA13-14......V-409
Ushida, Akido....................WAA15-3......VI-282
Ushida, Akio......................MAA7-2......III-267
Ushida, Akio......................WAA15-19......VI-346
Ushida, Akio......................MAA14-4......III-310
Utkin, Victor A...................TPA11-2......VI-594

V

Vai, M. Michael..................TAB6-1......II-232
Vaidyanathan, P.P..............MPA2-7......V-122
Vaidyanathan, P.P..............MAA2-6......V-53
Vaidyanathan, P.P..............TAA1-7......V-154
Vaidyanathan, P.P..............MAA2-1......V-33
Vainio, Olli........................MPA14-24......II-185
Valacca, Roberto...............TPA15-5......VI-625
Valle, M.............................MPA3-6......III-46
Valtonen, Martti Erik........MPA7-4......III-403
Vankka, Jouko...................MAA13-19......IV-409
Vannelli, Anthony..............TAA15-16......VI-191
Vannelli, Anthony..............WAA11-5......VI-258
Varho, Susanna..................TAA4-2......IV-194

Vasilescu, G.......................MAA9-4......VI-13
Vasseaux, Tony..................MPA15-22......I-273
Vázquez-González, Alejandro....MPA8-8...I-151
Veillette, Benoit R..............MAA15-20......I-104
Venkatachalam, Vidya.......WPA1-1......V-449
Verhoeven, Chris J. M.......MPA8-7......I-147
Vesma, Jussi......................MAA2-8......V-61
Vesterbacka, Mark.............MPA14-7......II-117
Vetro, Anthony..................TPA2-4......V-285
Vidal-Verdú, Fernando......MPA3-7......III-50
Vieira de Melo, Ana Cristina..WAA15-9..VI-306
Vlcek, Miroslav..................WAA13-14......V-409
Vogt, Rolf...........................WPA11-5......VI-405
Voltz, Peter J.....................WAA5-8......IV-550
Vouilloz, Alexandre............MAA5-6......IV-325
Vucic, Mladen....................TAB14-1......III-485

W

Wada, Masahiro.................MAA7-2......III-267
Wada, Yuji.........................WPA14-19......III-239
Wakabayashi, Shin'ichi.....WAB6-3......II-268
Walker, Paul D..................MAA15-5......I-45
Waltari, Mikko..................MAA13-19......IV-409
Waltari, Mikko Eljas..........MPA15-17......I-253
Waltz, Edward...................MPA11-6......VI-553
Wan, Yi..............................TAB13-7......V-233
Wang, Bo...........................WPA9-3......I-550
Wang, Chen-Chia...............TAA14-9......II-216
Wang, Chih-Liang..............MAA10-6......VI-53
Wang, Chin-Liang..............MPA5-5......IV-445
Wang, Chin-Liang..............MPA14-22......II-177
Wang, Chin-Liang..............WAA14-22......II-478
Wang, Chorng-Kuang........TAA8-2......I-285
Wang, Chorng-kuang.........WPB13-2......VI-490
Wang, Chua-Chin...............WPA6-5......II-292
Wang, Chung-Neng............MPA13-19......IV-166
Wang, Hua O.....................WPA7-6......III-255
Wang, Jhing-Fa..................MAA3-7......III-25
Wang, Jin-Sheng................TPA14-12......I-416
Wang, Jin-Sheng................MAA15-7......I-53
Wang, Jinn Shyan..............MPA14-1......II-93
Wang, Michelle Y..............MAA3-3......III-9
Wang, Michelle Y..............WAA12-2......III-586
Wang, Michelle Y..............WAA12-2......VI-442
Wang, Ting-Chi..................WAA11-3......VI-248
Wang, Xiao-Feng................WAA5-3......IV-534
Wang, Yao.........................TPA2-4......V-285
Wang, Yuhe.......................MAA13-21......IV-417
Wang, Yuke.......................TAB13-3......V-217
Wang, Yunti......................MAA8-3......I-5
Wanhammar, Lars.............WAA13-22......V-437
Wanhammar, Lars.............MPA14-7......II-117
WanHammer, Lars............MAA2-4......V-45
Wantanabe, Hitoshi...........WPA12-4......III-655
Ward, E.S..........................TAB14-10......III-522
Watanabe, Hitoshi.............MPA7-1......III-391
Watanabe, Toshimasa.......WAA11-4......VI-254
Wawryn, Krzysztof.............TPA14-6......I-393
Way, Winston I..................WPB13-2......VI-490
Weeks, Michael Clark........MAA11-8......IV-57
Wei, Che-Ho.......................MPA5-3......IV-437
Wei, Shyue-Win..................WAA14-23......II-482
Weiss, Laurens...................MPA7-5......III-407
Welch, Ryan Joseph...........MPA10-8......VI-94
Wen, Shui-An.....................WAA11-3......VI-248
Wensheng, Sun..................WPB13-3......VI-494
Wey, Chin-Long..................TAA15-5......VI-147
Wey, Chin-Long..................TPA14-12......I-416
Wey, Chin-Long..................MAA15-7......I-53
Wey, Chin-Long..................TAB13-7......V-233
Whitman, R.......................TAA11-5......VI-582

AUTHOR INDEX

Wikner, J. Jacob...................MAA8-8.......I-25
Williams, Douglas B.............WAA1-6.......V-313
Willson, Alan N....................MPA14-2.......II-97
Willson, Jr., Alan N................MAA8-3.......I-5
Wilson, Charles S............WPA14-16.......III-227
Wing, Omar.......................TPA10-3........VI-232
Wisland, Dag T....................MAA12-2.......II-37
Wittenburg, Jens-Peter.........MAA11-5.......IV-45
Wolf, Markus....................WAA15-5.......VI-290
Wolf, Tod D......................WPA13-1.......V-525
Won, Jae-Hee....................MPA14-17.......II-157
Wong, C. K......................WAA11-1.......VI-240
Wong, C. K......................WAA11-3.......VI-248
Wong, D. F......................WAA11-3.......VI-248
Wong, Justy W.C................WPA13-9.......V-550
Wong, Kwok-wo....................TAA3-3.......III-70
Wong, Martin D. F................WPB4-1.......VI-421
Wong, Peter H.W.................WPA13-9.......V-550
Woods, R........................MPA14-13.......II-141
Wooten, E. Curran K...........WPA14-2.......III-175
Worapishet, Apisak................WPA8-7.......I-535
Wornell, Gregory W.............WPA5-1.......IV-583
Wrixon, Adrian..................WAA14-5.......II-410
Wu, An-Yeu......................WAA13-6.......V-375
Wu, An-Yeu......................TAA13-9.......V-194
Wu, Angus.......................MPA14-23.......II-181
Wu, Chai Wah....................MAA7-7.......III-287
Wu, Chai Wah...................MAA14-2.......III-302
Wu, Cheng-Shing................TAA13-9.......V-194
Wu, Chi-Feng....................WPA6-5.......II-292
Wu, Chien-Hsing................MPA14-8.......II-121
Wu, Chien-Ming.................MPA5-6.......IV-449
Wu, Chung-Yu..................MAA13-11.......IV-378
Wu, Chung-Yu....................MPA4-5.......IV-77
Wu, Chung-Yu...................MAA12-5.......II-49
Wu, Chung-Yu...................TAA14-8.......II-212
Wu, Guang-Min..................TAA15-1.......VI-131

Wu, J............................TAA9-2.......I-313
Wu, Jiin-Chuan..................TPA15-7.......VI-633
Wu, Lin..........................MPA9-4.......I-167
Wu, Qiang.......................WAA2-1.......V-325
Wu, Zhixiong....................TPA2-3.......V-281
X
Xia, Cheng-Quan................MAA14-17.......III-358
Xiong, Kaiqi....................TAA7-2.......III-456
Xiong, Zixiang..................WAA4-1.......IV-273
Xu, Gonggui.....................TAA12-1.......VI-98
Xu, Gonggui.....................TAA12-2.......VI-102
Y
Yagyu, Mitsuhiko...............WAA13-11.......V-395
Yamada, Akihiko..................MPA1-6.......V-86
Yamagami, Yoshihiro..........WAA15-3.......VI-282
Yamai, Nariyoshi...............WPB13-11.......VI-529
Yang, Howard C.................TPA14-11.......I-412
Yang, Rui.......................WAA13-2.......V-359
Yang, Xuguang...................MPA2-3.......V-106
Yang, Y.........................TAA11-5.......VI-582
Yao, Minli......................MAA5-7.......IV-329
Yap, Keng C.....................WPA7-7.......III-639
Yasukawa, Hiroshi...............TAA4-4.......IV-202
Yau, Sze Fong...................MPA5-1.......IV-433
Yau, Sze Fong...................TAA13-7.......V-186
Ye, Hua.........................WAA1-8.......V-321
Yee, Dennis.....................WPA5-4.......IV-593
Yemetz, Sergei..................TAA5-6.......IV-481
Yeng, Horng-Ru..................TAA14-9.......II-216
Yeon, Kwang-Il..................WPA15-3.......II-316
Yeung, Tak Keung................TAA13-7.......V-186
Yin, Qinye......................MAA5-7.......IV-329
Yin, Qinye......................MAA5-8.......IV-333
Yokomaru, Toshihiko............WAA11-2.......VI-244
Yoo, Jang-Sik...................TAA10-2.......I-344
Yoo, Jea Hoon...................TPA6-4.......VI-482
Yoo, Jeong Ju...................TPA6-4.......VI-482

Yoshida, Hirofumi...............WAA7-4.......III-578
Yoshizawa, Hiroyasu...........MPA14-14.......II-145
Young, Albert M.................TAA5-9.......IV-489
Yu, Baiying......................WPA9-7.......I-563
Yu, Gwo-Jeng...................WPA14-17.......III-231
Yu, Hongyi......................MAA1-8.......V-29
Yu, Li..........................TAA10-5.......I-356
Yu, Pang-Cheng.................TPA15-7.......VI-633
Yu, Wei........................MAA13-17.......IV-401
Yu, Zhihong....................TAA15-21.......VI-207
Yuihara, Atsushi................MPA12-7.......III-444
Yund, E. William................TAA11-4.......VI-578
Yung, H. C. Nelson.............MPA13-16.......IV-154
Yung, Hon Ching, Nelson....MPA13-12.....IV-138
Yung, Nelson....................TPA13-3.......IV-241
Z
Zaghloul, Mona E................MAA3-4.......III-21
Zaghloul, Mona E................MAA13-5.......IV-353
Zahradnik, Pavel...............WAA13-14.......V-409
Zan, Jinwen.....................WAA4-2.......IV-277
Zeng, X........................WPA15-16.......II-106
Zerzghi, Amanuel................MPA1-3.......V-74
Zhang, Huaizhou................MAA14-6.......III-318
Zhang, Liang...................WPB13-8.......VI-514
Zhang, Ning.....................WPA5-4.......IV-593
Zhang, Qingwen.................WAA1-2.......V-297
Zhang, Yanning.................WPA14-3.......III-179
Zhang, Ying.....................MPA13-4.......IV-106
Zhao, Qifang...................WPA13-5.......V-534
Zheng, Wei Xing.................TAA13-6.......V-182
Zheng, Wei Xing.................MAA1-6.......V-21
Zhixiong, Wu....................TPA-2-3.....V-281
Zhu, Wei-Ping...................WAA4-2.......IV-277
Zhu, Wei-Ping...................MAA2-5.......V-49
Zhuo, Wei......................MAA13-8.......IV-365
Zincke, Christian...............MAA13-5.......IV-353